UCSMP
SCOTT, FORESMAN

The University of Chicago School Mathematics Project

Precalculus and Discrete Mathematics

Teacher's Edition

Authors

Anthony L. Peressini
Susanna S. Epp
Kathleen A. Hollowell
Susan Brown
Wade Ellis, Jr.
John W. McConnell
Jack Sorteberg
Denisse R. Thompson
Dora Aksoy
Geoffrey D. Birky
Greg McRill
Zalman Usiskin

Technology requirements:

Students should have access to graphics calculators or computers equipped with graphing software at almost all times, including homework and tests.

Teachers should have access to comparable technology for in-class demonstrations. (See pages T40-T42.)

About the Cover

The motion of a robot arm can be described by three-dimensional vectors. Such robots today are typically computer controlled. Thus this application involves both an understanding of the mathematics of motion for which calculus is so important and the mathematics of computers for which discrete mathematics is a foundation.

Scott, Foresman

Editorial Offices: Glenview, Illinois Regional Offices: Sunnyvale, California • Atlanta, Georgia • Glenview, Illinois • Oakland, New Jersey • Dallas, Texas

Acknowledgments

Authors

Anthony L. Peressini
Professor of Mathematics, University of Illinois, Urbana-Champaign

Susanna S. Epp
Associate Professor of Mathematics, DePaul University, Chicago

Kathleen A. Hollowell
Coordinator of Secondary Mathematics Inservice Programs, University of Delaware, Newark

Susan Brown
Mathematics Teacher, York H.S., Elmhurst, IL

Wade Ellis, Jr.
Mathematics Instructor, West Valley College, Saratoga, CA

John W. McConnell
Instructional Supervisor of Mathematics, Glenbrook South H.S., Glenview, IL

Jack Sorteberg
Mathematics Teacher, Burnsville High School, Burnsville, MN

Denisse R. Thompson
Assistant Professor of Mathematics Education, University of South Florida, Tampa

Dora Aksoy
UCSMP

Geoffrey D. Birky
UCSMP

Greg McRill
UCSMP

Zalman Usiskin
Professor of Education, The University of Chicago

UCSMP Production and Evaluation

Series Editors: Zalman Usiskin, Sharon L. Senk

Managing Editor: Daniel Hirschhorn

Technical Coordinator: Susan Chang

Director of Evaluation: Denisse R. Thompson

We wish to acknowledge the generous support of the **Amoco Foundation**, the **General Electric Foundation**, and the **Carnegie Corporation of New York** in helping to make it possible for these materials to be developed and tested.

ISBN: 0–673–33367–1

Printing Key 3456789–RRW–989796959493

Contents • Teacher's Edition

The complete Contents for the Student Edition begins on page *vi*.

Note:
The **Professional Sourcebook** is located at the back of the Teacher's Edition.

UCSMP Helps You Update Your Curriculum and Better Prepare Your Students!

As reports from national commissions have shown, students currently are not learning enough mathematics, and the curriculum has not kept pace with changes in mathematics and its applications.

In response to these problems, UCSMP has developed a complete program for grades 7-12 that upgrades the school mathematics experience for the average student. The usual four-year high-school mathematics content—and much more—is spread out over six years. The result is that students learn more mathematics and they are better prepared for the variety of mathematics they will encounter in their future mathematics courses and in life.

In addition, UCSMP helps students view their study of mathematics as worthwhile, as full of interesting information, as related to almost every endeavor. With applications as a hallmark of all UCSMP materials, students no longer ask, "How does this topic apply to the world I know?"

For a complete description of the series, see pages **T19-T52** at the back of the **Teacher's Edition.**

In short, UCSMP...

- Offers a wide variety of content and applications.
- Prepares students to use mathematics effectively in today's world.
- Develops independent thinking and learning.
- Helps students improve their performance.
- Provides the practical support you need.

"For my students, ***Precalculus and Discrete Mathematics*** *is the course that has unified the mathematics they have learned before. This course puts everything together."*

Judith Kemler, Teacher, J. Frank Dobie High School, Houston, TX

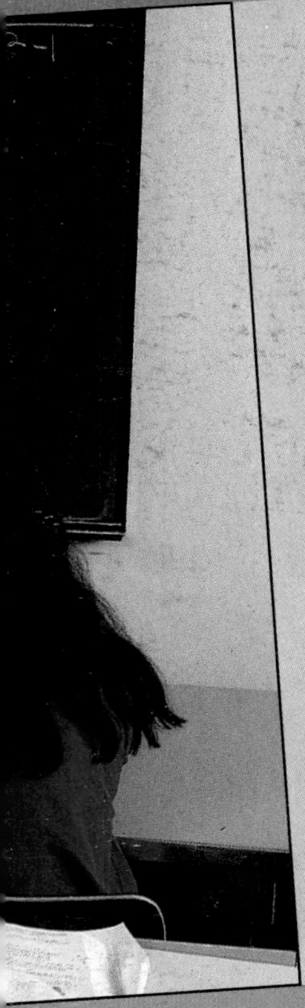

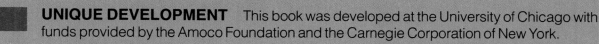

Imagine using a text that has been developed as part of a coherent 7-12 curriculum design, one that has been tested on a large scale *before* publication, and most important, a text that has bolstered students' mathematical abilities, which is reflected in test scores. Read on to find out how UCSMP has done that, and more.

For a detailed discussion of the development and testing of ***Precalculus and Discrete Mathematics,*** see pages ***T45-T49.***

Years of field-testing and perfecting have brought impressive results!

UNIQUE DEVELOPMENT This book was developed at the University of Chicago with funds provided by the Amoco Foundation and the Carnegie Corporation of New York.

PLANNING Initial planning was done with input from professors, classroom teachers, school administrators, and district and state supervisors of mathematics, with attention to the recommendations by national commissions and international studies. The UCSMP secondary curriculum is the first full mathematics curriculum to implement the recommendations of the NCTM Standards committees.

AUTHORSHIP Authors were chosen for expertise in the relevant areas of school mathematics and for classroom experience. All of the authors of ***Precalculus and Discrete Mathematics*** are experienced teachers and are active in the field of mathematics education.

FIELD-TESTING AND EVALUATION Pilot testing began with teaching by the initial team of authors and revising based on their firsthand experiences. Then, there was evaluation based on local studies, and the materials were revised again. Further evaluation and revisions were based on national studies. And finally, there were the **Scott, Foresman** enhancements, such as color photography and about 600 blackline masters, to better meet the needs of teachers and students.

Offers a Wide Variety of Content and Applications

"When I am studying mathematics, I like to see why it works and why it relates to the rest of the world."

Precalculus and Discrete Mathematics student,
Breck School, Minneapolis, MN

Wider Scope

Precalculus and Discrete Mathematics integrates the conceptual underpinnings of calculus with the topics of discrete mathematics. In this course students find the themes that unify their understandings and prepare for the courses they will meet in college.

Precalculus provides the opportunity for students to informally investigate the traditional concepts of calculus, such as maxima, minima, infinite sequences, limits, derivatives, and integrals. In addition, students work with the algebraic manipulation they will need in future courses. All this is continually applied to and illustrated by real-world applications of the topics.

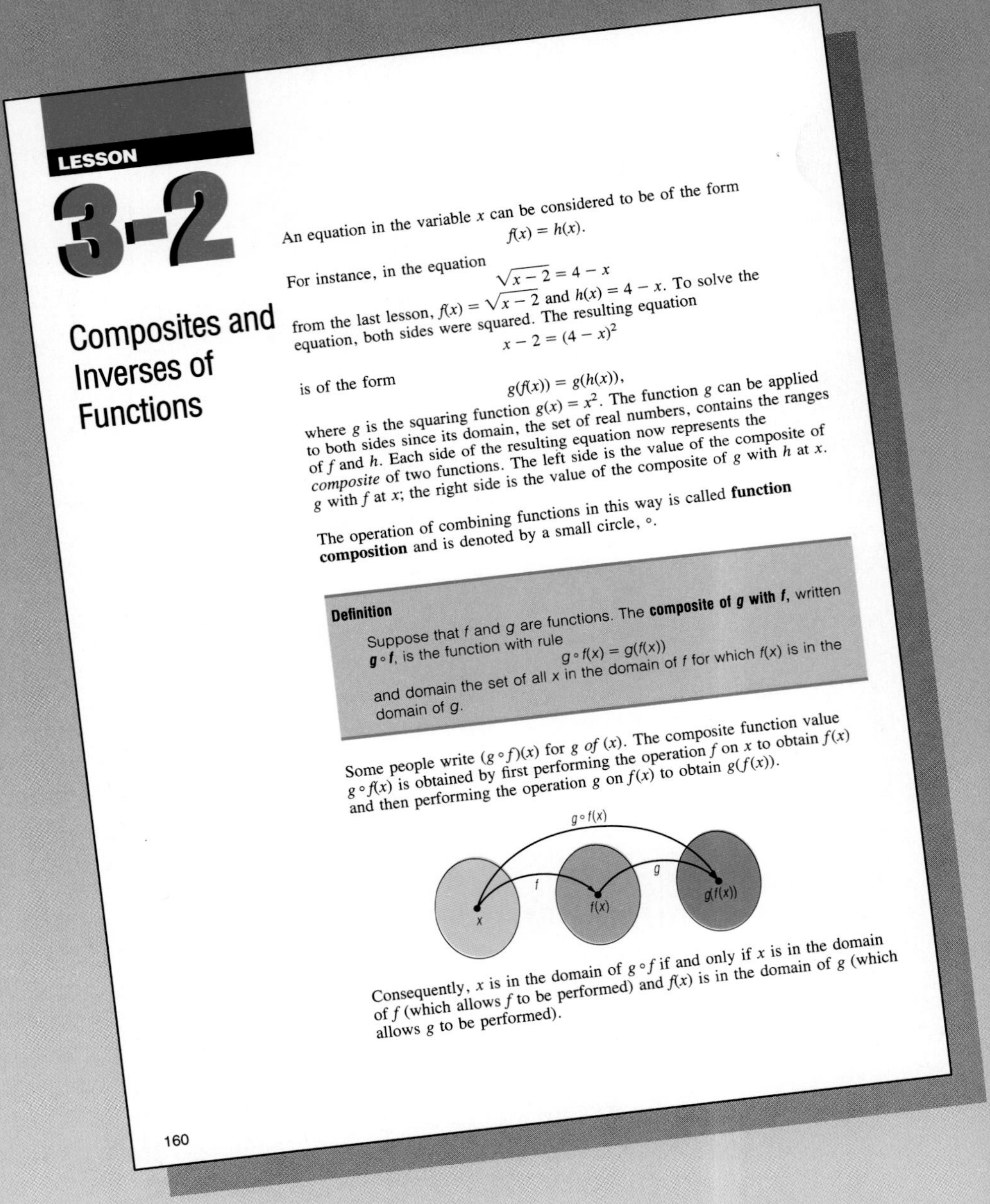

LESSON 3-2

Composites and Inverses of Functions

An equation in the variable x can be considered to be of the form
$$f(x) = h(x).$$
For instance, in the equation
$$\sqrt{x-2} = 4 - x$$
from the last lesson, $f(x) = \sqrt{x-2}$ and $h(x) = 4 - x$. To solve the equation, both sides were squared. The resulting equation
$$x - 2 = (4 - x)^2$$
is of the form
$$g(f(x)) = g(h(x)),$$
where g is the squaring function $g(x) = x^2$. The function g can be applied to both sides since its domain, the set of real numbers, contains the ranges of f and h. Each side of the resulting equation now represents the *composite* of two functions. The left side is the value of the composite of g with h at x. g with f at x; the right side is the value of the composite of g with h at x.

The operation of combining functions in this way is called **function composition** and is denoted by a small circle, $\circ$.

Definition
Suppose that f and g are functions. The **composite of g with f,** written **$g \circ f$**, is the function with rule
$$g \circ f(x) = g(f(x))$$
and domain the set of all x in the domain of f for which $f(x)$ is in the domain of g.

Some people write $(g \circ f)(x)$ for g *of* (x). The composite function value $g \circ f(x)$ is obtained by first performing the operation f on x to obtain $f(x)$ and then performing the operation g on $f(x)$ to obtain $g(f(x))$.

Consequently, x is in the domain of $g \circ f$ if and only if x is in the domain of f (which allows f to be performed) and $f(x)$ is in the domain of g (which allows g to be performed).

160

Discrete Mathematics is recognized as vital for students in a world evermore dependent on the use of computers. As finite, discrete machines, computers (and computer technology) call for additional emphasis on discrete mathematics topics, such as formal logic, recursion, mathematical induction, combinatorics, and graph theory. In this course, students study these and other topics in applications that they understand and see as valuable.

LESSON

11-1

Modeling with Graphs

To solve the Königsberg bridge problem of the preceding page, Euler observed that, for this problem, each land mass could be represented by a point since it is possible to walk from any part of a land mass to any other part without crossing a bridge. The bridges could be thought of as arcs joining pairs of points. Thus the situation of Königsberg could be represented by the following geometric model consisting of four points and seven arcs.

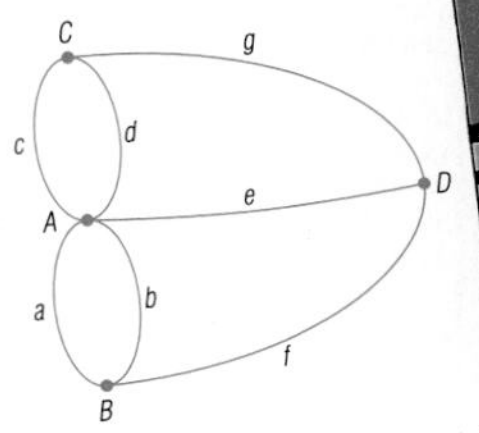

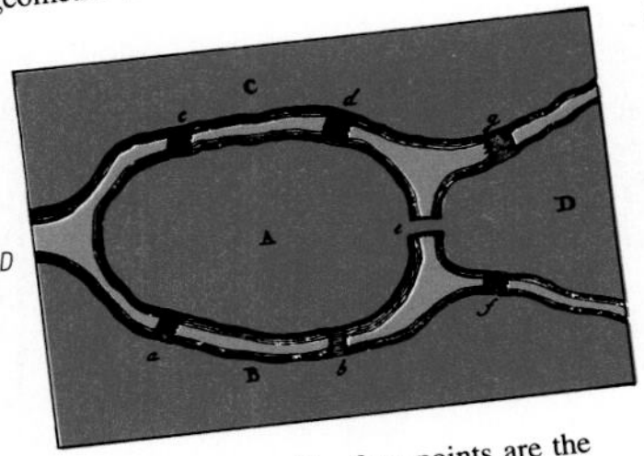

This type of geometric model is called a graph. The four points are the *vertices* and the seven arcs are the *edges* of the graph. In terms of this graph, the Königsberg bridge problem can be stated:

> Is it possible to trace this graph with a pencil, traveling each edge exactly once, starting and ending at the same vertex, without picking up the pencil?

Euler's solution to the Königsberg bridge problem is given in Lesson 11-4.

Another famous puzzle for which a graph is a helpful model was invented in 1859 by the Irish mathematician Sir William Rowan Hamilton (1805-1865). The puzzle consisted of a wooden block in the shape of a regular dodecahedron, as shown here.

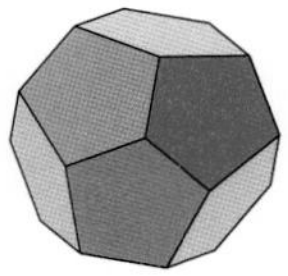

This polyhedron has 12 regular pentagons as its faces, 30 edges, and 20 vertices. Hamilton marked each vertex of the block with the name of a city, and the object of the puzzle was to find a travel route along the edges of the block that would visit each of the cities once and only once. A small pin protruded from each vertex so that the player could mark a route by wrapping string around each pin in order as its city was visited.

By thinking of the polyhedron as transparent, as shown below at the left, you can count to determine that Hamilton's problem involves a graph with 20 vertices and 30 edges in which 3 edges meet at each vertex. The graph below at the right shows the same relationships of vertices and edges in a 2-dimensional diagram; the two graphs are **equivalent**.

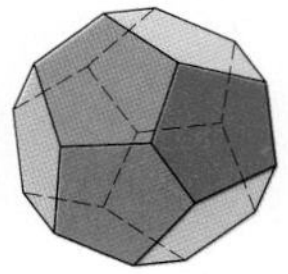

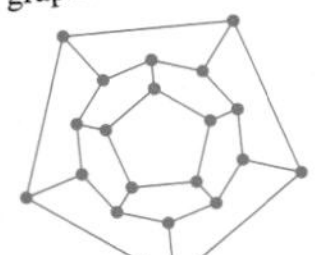

Hamilton's puzzle can be stated in terms of either graph as the following problem:

> Is it possible to trace this graph with a pencil, traveling through each *vertex* exactly once, without picking up the pencil?

In fact, Hamilton also sold the 2-dimensional version of his puzzle since it was easier to work with. Notice the similarity between the problems of Euler and Hamilton; in both the goal is to traverse all objects of one kind in the graph exactly once; in Euler's problem the objects are the edges; in Hamilton's the objects are the vertices.

Several practical situations give rise to problems like Hamilton's.

Example 1 One mailbox is located at each intersection in a city and a postal truck is required to collect the mail from all the boxes in the 42-block region in the map at the right.

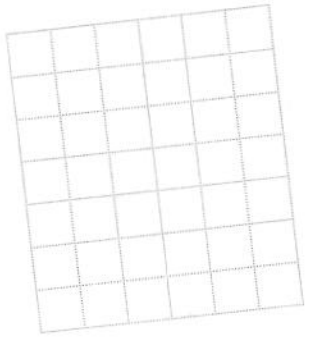

Can the driver plan a pick-up route that begins and ends at the same place and allows collection of the mail from all of the boxes without passing one that has already been collected?

Solution The problem can be modeled by a graph, with the mailboxes being the vertices and the streets being the edges. Experimentation gives several suitable routes, such as the one pictured here.

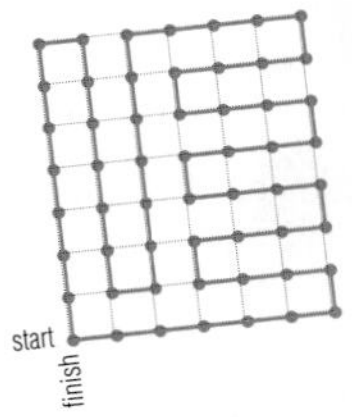

All of the examples you have seen so far result from situations that themselves are geometric. The next example is quite different; it shows the use of graphs in scheduling a complex task. First, some background information is necessary.

Prepares Students to Use Mathematics Effectively in Today's World

"Doing the projects, I learn how math can relate to everyday problems. It's helpful to know I'll use what I'm learning instead of just learning and never knowing how I'll use it."

Precalculus and Discrete Mathematics student, Churchill High School, Eugene, OR

Technology

The availability of scientific calculators, graphics calculators, and computers makes it possible for students to work with functions and statistics at a level never before possible.

With technology, a broader range of both pure and applied problems is available to students. It is also possible to create a classroom environment that encourages exploration and experimentation.

Evidence shows that appropriate use of technology enhances student understanding and problem-solving skills.

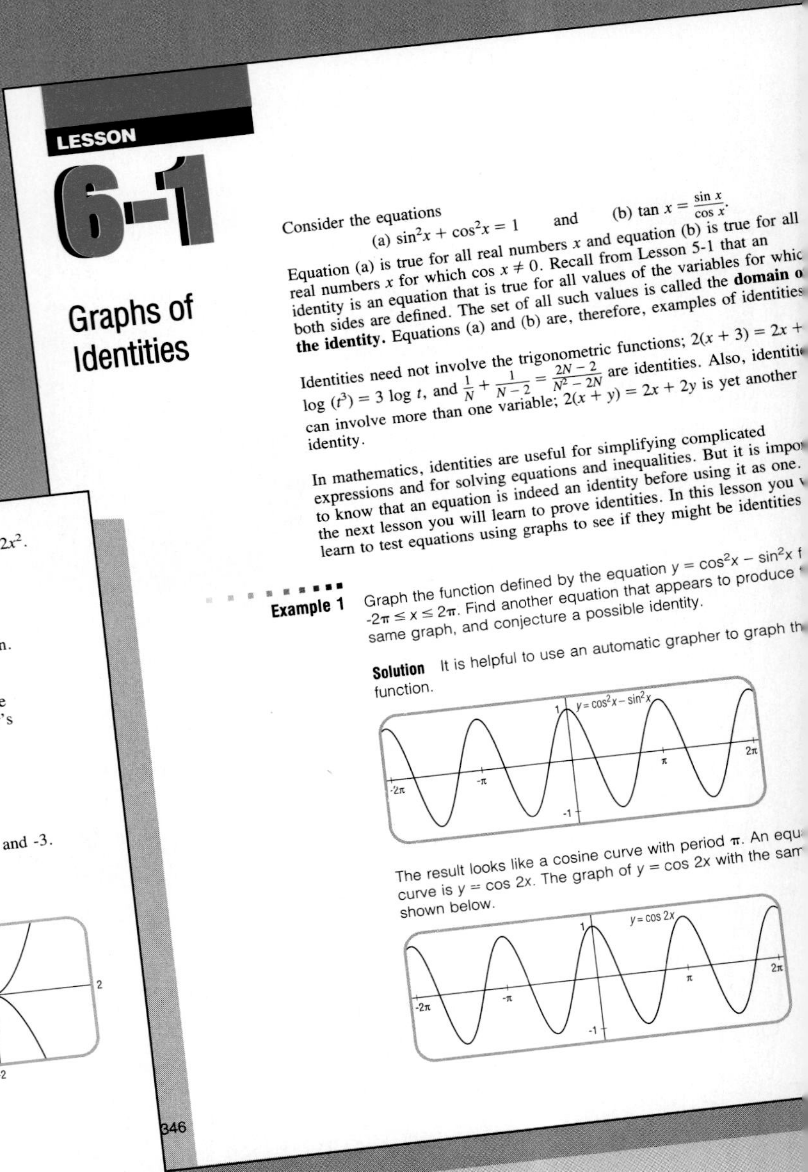

LESSON 6-1

Graphs of Identities

Consider the equations

(a) $\sin^2 x + \cos^2 x = 1$ and (b) $\tan x = \frac{\sin x}{\cos x}$.

Equation (a) is true for all real numbers x and equation (b) is true for all real numbers x for which $\cos x \neq 0$. Recall from Lesson 5-1 that an identity is an equation that is true for all values of the variables for whic[h] both sides are defined. The set of all such values is called the **domain o[f] the identity.** Equations (a) and (b) are, therefore, examples of identities

Identities need not involve the trigonometric functions; $2(x + 3) = 2x +$ ... $\log (t^3) = 3 \log t$, and $\frac{1}{N} + \frac{1}{N-2} = \frac{2N-2}{N^2-2N}$ are identities. Also, identiti[es] can involve more than one variable; $2(x + y) = 2x + 2y$ is yet another identity.

In mathematics, identities are useful for simplifying complicated expressions and for solving equations and inequalities. But it is impo[rtant] to know that an equation is indeed an identity before using it as one. In the next lesson you will learn to prove identities. In this lesson you [will] learn to test equations using graphs to see if they might be identities

Example 1 Graph the function defined by the equation $y = \cos^2 x - \sin^2 x$ f[or] $-2\pi \le x \le 2\pi$. Find another equation that appears to produce [the] same graph, and conjecture a possible identity.

Solution It is helpful to use an automatic grapher to graph th[e] function.

The result looks like a cosine curve with period π. An equ[ation for this] curve is $y = \cos 2x$. The graph of $y = \cos 2x$ with the sam[e window is] shown below.

346

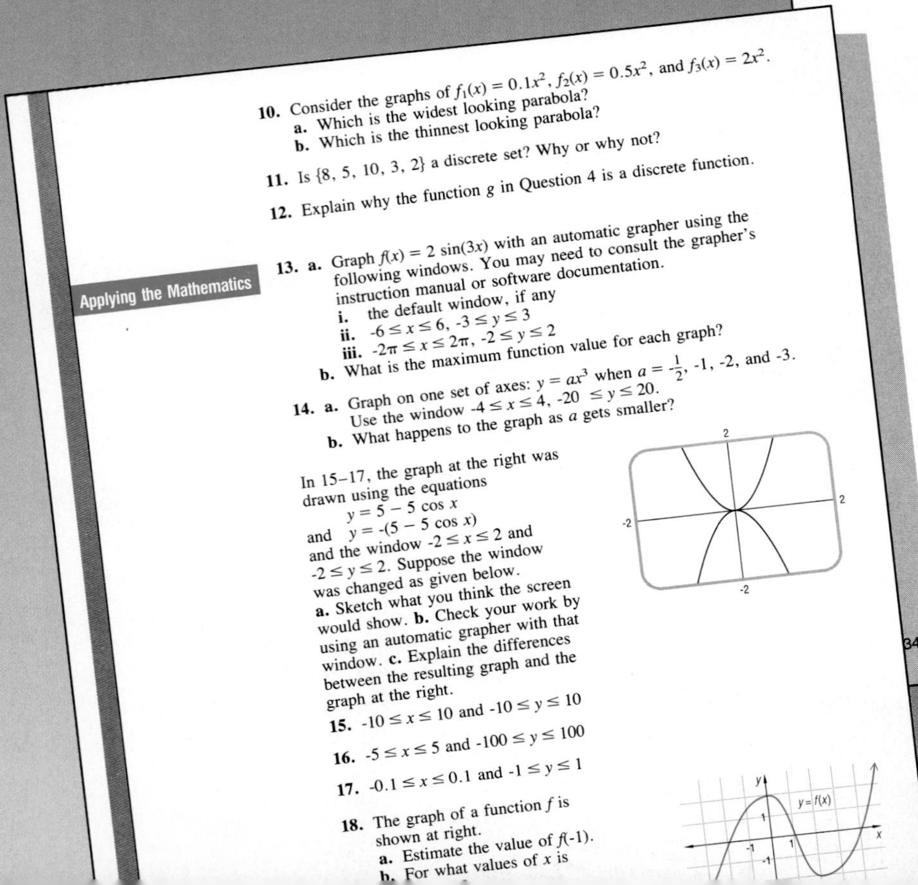

10. Consider the graphs of $f_1(x) = 0.1x^2$, $f_2(x) = 0.5x^2$, and $f_3(x) = 2x^2$.
 a. Which is the widest looking parabola?
 b. Which is the thinnest looking parabola?

11. Is {8, 5, 10, 3, 2} a discrete set? Why or why not?

12. Explain why the function g in Question 4 is a discrete function.

Applying the Mathematics

13. a. Graph $f(x) = 2 \sin(3x)$ with an automatic grapher using the following windows. You may need to consult the grapher's instruction manual or software documentation.
 i. the default window, if any
 ii. $-6 \le x \le 6$, $-3 \le y \le 3$
 iii. $-2\pi \le x \le 2\pi$, $-2 \le y \le 2$
 b. What is the maximum function value for each graph?

14. a. Graph on one set of axes: $y = ax^3$ when $a = -\frac{1}{2}$, -1, -2, and -3. Use the window $-4 \le x \le 4$, $-20 \le y \le 20$.
 b. What happens to the graph as a gets smaller?

In 15–17, the graph at the right was drawn using the equations
$y = 5 - 5 \cos x$
and $y = -(5 - 5 \cos x)$
and the window $-2 \le x \le 2$ and $-2 \le y \le 2$. Suppose the window was changed as given below. **a.** Sketch what you think the screen would show. **b.** Check your work by using an automatic grapher with that window. **c.** Explain the differences between the resulting graph and the graph at the right.

15. $-10 \le x \le 10$ and $-10 \le y \le 10$

16. $-5 \le x \le 5$ and $-100 \le y \le 100$

17. $-0.1 \le x \le 0.1$ and $-1 \le y \le 1$

18. The graph of a function f is shown at right.
 a. Estimate the value of $f(-1)$.
 b. For what values of x is

Reading Lessons

These lessons provide opportunities for students to learn mathematics without the pressure of having to master the content. These special lessons, appearing at the end of each chapter and indicated with a blue band at the top of the page, include a broad variety of topics, such as an examination of infinity and an exploration of Markov chains.

Projects

Projects allow students to become engaged in extended tasks. Students may collect and analyze data or research mathematical ideas. Students are responsible for reporting the results of their work. One set of projects is included at the end of each chapter. These projects offer a wonderful opportunity for students to work cooperatively in small groups.

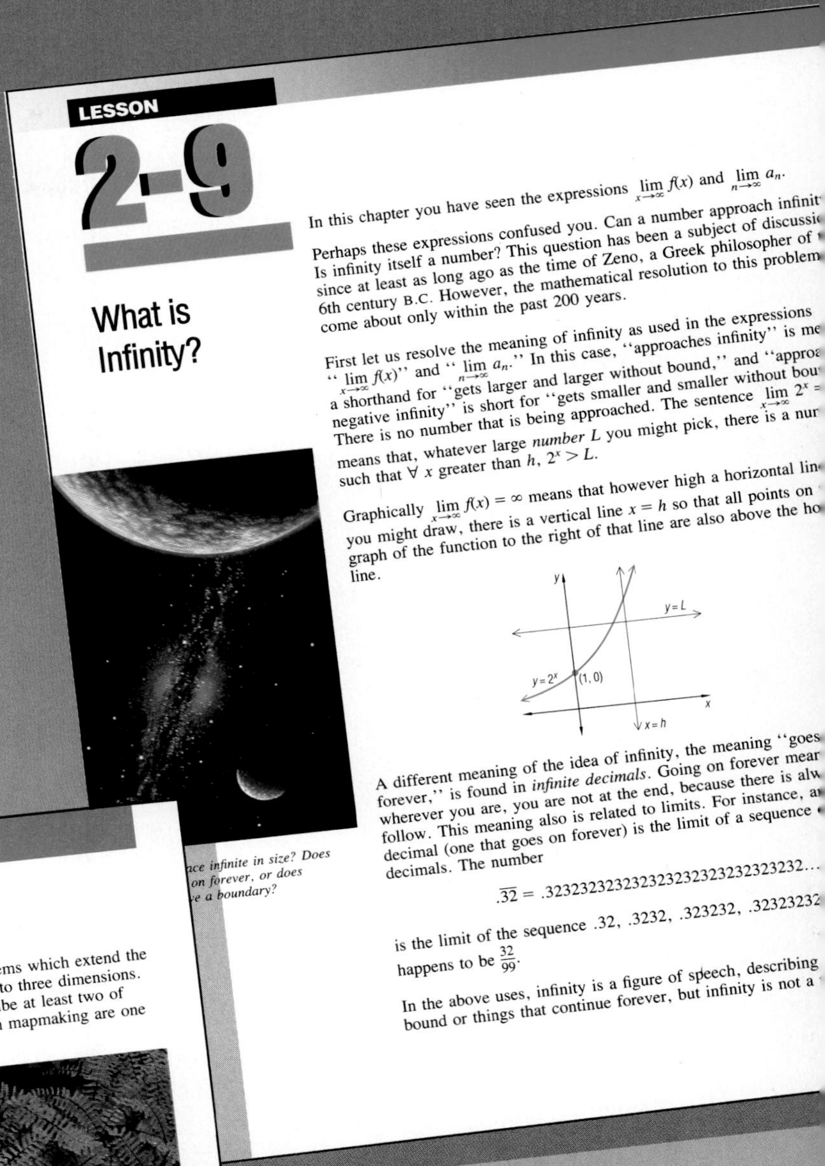

LESSON 2-9

What is Infinity?

In this chapter you have seen the expressions $\lim_{x\to\infty} f(x)$ and $\lim_{n\to\infty} a_n$.

Perhaps these expressions confused you. Can a number approach infinit… Is infinity itself a number? This question has been a subject of discussi… since at least as long ago as the time of Zeno, a Greek philosopher of … 6th century B.C. However, the mathematical resolution to this problem … come about only within the past 200 years.

First let us resolve the meaning of infinity as used in the expressions "$\lim_{x\to\infty} f(x)$" and "$\lim_{n\to\infty} a_n$." In this case, "approaches infinity" is me… a shorthand for "gets larger and larger without bound," and "approa… negative infinity" is short for "gets smaller and smaller without bou… There is no number that is being approached. The sentence $\lim_{x\to\infty} 2^x =$ … means that, whatever large *number* L you might pick, there is a nur… such that $\forall$ x greater than h, $2^x > L$.

Graphically $\lim_{x\to\infty} f(x) = \infty$ means that however high a horizontal lin… you might draw, there is a vertical line $x = h$ so that all points on … graph of the function to the right of that line are also above the ho… line.

A different meaning of the idea of infinity, the meaning "goes forever," is found in *infinite decimals*. Going on forever mear… wherever you are, you are not at the end, because there is alw… follow. This meaning also is related to limits. For instance, a… decimal (one that goes on forever) is the limit of a sequence … decimals. The number

$$.\overline{32} = .3232323232323232323232323232…$$

is the limit of the sequence .32, .3232, .323232, .32323232… happens to be $\frac{32}{99}$.

In the above uses, infinity is a figure of speech, describing … bound or things that continue forever, but infinity is not a …

…ace infinite in size? Does …on forever, or does …e a boundary?

CHAPTER 8

Projects

1. Four representations of complex numbers are in this chapter: $a + bi$, rectangular, polar, and trigonometric. Show that a fifth representation, the matrix $\begin{bmatrix} a & -b \\ b & a \end{bmatrix}$ for $a + bi$, satisfies the following properties of the complex numbers.
   a. The rules for complex addition and multiplication hold.
   b. The modulus of $a + bi$ is related to the determinant of its matrix.
   c. The additive and multiplicative inverses of a complex number are related to the additive and multiplicative inverses of this matrix.
   d. The matrix form has special matrices which correspond to 1 (the multiplicative identity) and 0 (the additive identity).
   e. The argument of a complex number is related to its matrix representation.
2. Report on one of the following applications of complex numbers.
   a. acoustics
   b. electricity
   c. higher mathematics, such as differential equations
4. There are coordinate systems which extend the idea of polar coordinates to three dimensions. Find out about and describe at least two of these systems. (Books on mapmaking are one source.)

Northern maidenhair fern, a spiral in nature

5. Spirals occur frequently in nature. From a library, copy pictures of chambered nautiluses, sunflowers, and other natural objects to show your class, and explain their mathematical properties.
6. The Irish mathematician William Rowan Hamilton attempted to reconcile the diverse phenomena of eighteenth century physics with a new mathematical system he called *quaternions*. This system extended the ordered pair concept to ordered 4-tuples. Find out how he defined addition and multiplication, and what the relationships among his "quaternionic units" are.
7. Write a report on chaotic phenomena found in fields outside mathematics. Describe what … phenomena chaotic, and when, … nature was

Develops Independent Thinking and Learning

"This book promotes independent thinking on the part of the student to an extent that I have not seen in any other textbook."

William Kring, Teacher, Davis High School, Yakima, WA

Reading

Because explanations are clear and well written, students move away from a passive learning stance. By reading the text and using it as a reference tool, students develop the independence they will need in future mathematics classes and in their lives.

Because students can read and understand the text, teachers have the freedom to teach in a variety of ways, to use the class time more creatively, to spend more time on open-ended exploration, and to discuss alternate strategies for solving problems. When students read, the learning environment is enriched.

Problem Solving

In every lesson, students use a variety of problem-solving strategies. The combination of real-world problems, discrete topics, and calculator and computer technology, all combine to create a setting that models future problem-solving that students will encounter in and after mathematics classrooms.

LESSON
7-1

Two Famous Problems

In this lesson you will read about two famous problems whose solutions involve recursive thinking. The first was invented by the French mathematician Edouard Lucas in 1883, and it has come to be known as the **Tower of Hanoi.** (Hanoi is now the capital of Vietnam.) Here is the story made up by Lucas.

According to legend, at the time of Creation, 64 golden disks, each with a small central hole, were placed on one of the three golden needles in a temple in Hanoi. No two of the disks were the same size, and they were placed on the needle in such a way that no larger disk was on top of a smaller disk.

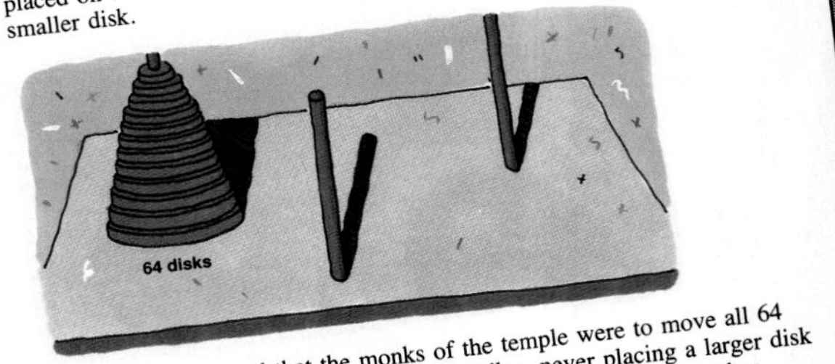

The Creator ordained that the monks of the temple were to move all 64 disks, one by one, to one of the other needles, never placing a larger disk on top of a smaller disk. When all the disks are stacked on the other needle, the world will end, the faithful will be rewarded, and the unfaithful will be punished. If the monks work very rapidly, moving one disk every second, how long will it be until the end of the world?

To solve this problem, imagine that you know the solutions to smaller problems of the same type, and then figure out a way to use those solutions to solve the given problem. The aim is to reduce the given problem repeatedly to obtain problems that are so small that their solutions are obvious. In this case, for example, you could start by supposing you know the length of time needed to transfer a tower of 63 disks. Then you can figure out the time needed to transfer 64 disks. But in order to figure out the time needed to transfer 63 disks, you would need to know the time it takes to transfer 62 disks. This reasoning would proceed until you have reduced the problem to finding the time it takes to transfer one disk. To be specific, you could reason as follows.

To transfer all 64 disks to a second needle, the monks first must have moved the top 63 disks, one by one, to the third needle (Step 1). Then they can move the bottom disk to the second needle (Step 2). Finally they can transfer the 63 disks on top of the bottom disk on the second needle (Step 3). The diagram on page 401 illustrates these steps.

400

Four Kinds of Questions

in each lesson engage students in a variety of activities designed to enhance learning and performance.

COVERING THE READING questions examine what students have read in the lesson.

APPLYING THE MATHEMATICS questions offer a broad range of applications and extensions of the lesson concepts.

REVIEW questions, keyed to past lessons, help students maintain and improve performance on important skills and concepts, and preview ideas to prepare students for topics that will be studied later.

EXPLORATION questions further extend the ideas of the lesson by including applications, generalizations, open-ended experiments, or research.

Questions

Covering the Reading

In 1–3, consider the congruence classes $R0$, $R1$, and $R2$ modulo 3.

1. In which class is the number 4321?

2. In which class is -42?

3. An integer of the form $3k + 1$ is in which of these classes?

4. a. How is the sentence $x \equiv y$ *(mod 4)* read?
b. Describe the congruence classes modulo 4.

5. If your birthday falls on a Thursday this year, on what day will it fall next year, 365 days later? (This assumes there is no leap year day in between.)

6. *True* or *false?*
a. $139 \equiv 59$ (mod 10)
b. $139 \equiv 59$ (mod 9)
c. $139 \equiv 59$ (mod 4)

In 7–10, give the smallest positive value that makes the congruence true.

7. $x \equiv 97$ (mod 5) **8.** $y \equiv 46$ (mod 30)

9. $n \equiv 736$ (mod 360) **10.** $z \equiv -1$ (mod 10)

11. Describe a situation about time that is related to the fact that $9 + 5 \equiv 2$ (mod 12).

12. The ISBN code for a test version of this book was 0-936745-33- __?__, where the blank is the check digit. Fill in the blank.

13. a. Give the first three digits of 18^{10}.
b. Give the last three digits of 18^{10}.

14. Rewrite in the language of modular arithmetic: *x is an even integer*.

Applying the Mathematics

15. Find the last two digits of 4^{2001}.

16. In geometry, rotations differing by 360° are considered equal. How can this be described in the language of modular arithmetic?

17. Prove the Subtraction Property of Congruence.

18. Let $M_3(n)$ be the smallest nonnegative integer congruent to n (mod 3).
a. Graph the function M_3 for integers n between -10 and 10.
b. What is the period of this function?

19. Use the properties of congruence to prove that the square of any odd number is 1 more than a multiple of 8. (Hint: Consider congruence classes mod 8.)

240

20. Consider the real numbers modulo 1.
a. Which numbers are in the same congruence class as π?
b. What does the result in part **a** tell you about the congruence classes of the real numbers modulo 1?

Review

21. *True* or *false?* Every integer n can be written in one and only one of the following forms: $n = 5q + 0$, $n = 5q + 1$, $n = 5q + 2$, $n = 5q + 3$, or $n = 5q + 4$ for some integer q. *(Lesson 4-2)*

22. Suppose $x^9 - 4x^8 + 3x^6 - 48x^4 + x - 4 = (x - 4) \cdot p(x)$ for some polynomial $p(x)$. What is the degree of $p(x)$? *(Lesson 4-1)*

23. Simplify. *(Previous course)*
a. $\frac{6x^5}{2x^3}$ **b.** $\frac{8y^8}{4y^4}$ **c.** $\frac{10z^3}{z}$

24. Consider the polynomial $p(a) = (2a + 5)^3 - (2a)^3 - 125$.
a. Write in expanded form.
b. Write in factored form.
c. Solve $p(a) = 0$. *(Lesson 4-1)*

25. Solve: $2x^2 + 5x \geq 3$
a. graphically **b.** algebraically. *(Lessons 3-6, 3-7)*

26. Prove the following conjecture or disprove it by finding a counterexample. *(Lesson 4-1)*
∀ positive integers a, b, and c, if a divides b and a divides b + c then a divides c.

27. Determine whether the following argument is valid or invalid. Justify your answer.
If a ship hits an iceberg, then it sinks.
The Titanic hit an iceberg.
∴ The Titanic sank.
(Lessons 1-6, 1-8)

Exploration

28. If you command a calculator to display a large power (such as 17^9 this lesson), it may put the answer in scientific notation. How can determine if the calculator is storing more digits than it displays?

29. The calculator which gave the value $1.185878765 \cdot 10^{11}$ for 17^9 s a different value: $1.1858787647 \cdot 10^{11}$. This value is incorrect in rightmost place—the actual value is $1.1858787650 \cdot 10^{11}$ (rounde the nearest) or $1.1858787649 \cdot 10^{11}$ (truncated). What accounts f calculator's error?

LESSON 4-3 Modular Arithme

Helps Students Improve Their Performance

"One of the strengths of this book is that it keeps going back, looking at ideas and problems from different points of view, and connecting one idea with another idea."

Mary Hahn, Teacher, Woodward High School, Cincinnati, OH

Progress Self-Test

This provides the opportunity for feedback and correction — before students are tested formally. The Student Edition contains full solutions to these questions to allow for accurate self-evaluation.

Chapter Review

The main objectives of the chapter are organized into sections corresponding to the four main types of understanding this book promotes: Skills, Properties, Uses, and Representations. Thus, the Chapter Review extends a multi-dimensional approach to understanding, offering a broader perspective that helps students put everything in place.

SKILLS include simple and complicated procedures for getting answers. The emphasis is on how to carry out algorithms.

PROPERTIES cover the mathematical *justifications for* procedures and other theory. To fully understand ideas, students must answer the common question, "But why does it work that way?"

USES include real-world applications of the mathematics. To effectively apply what they learn, students must know when different models or techniques are relevant.

REPRESENTATIONS cover ways to picture what is being studied. Visual images range from graphing functions to choosing the most effective way to display statistical data.

Quizzes and Test Masters

Quizzes and Test Masters in the Teacher's Resource File offer further help to assess mastery. You choose the test format that best suits your needs.

Quiz and Test Writer

The Quiz and Test Writer provides computer-generated quizzes and tests for additional assessment flexibility.

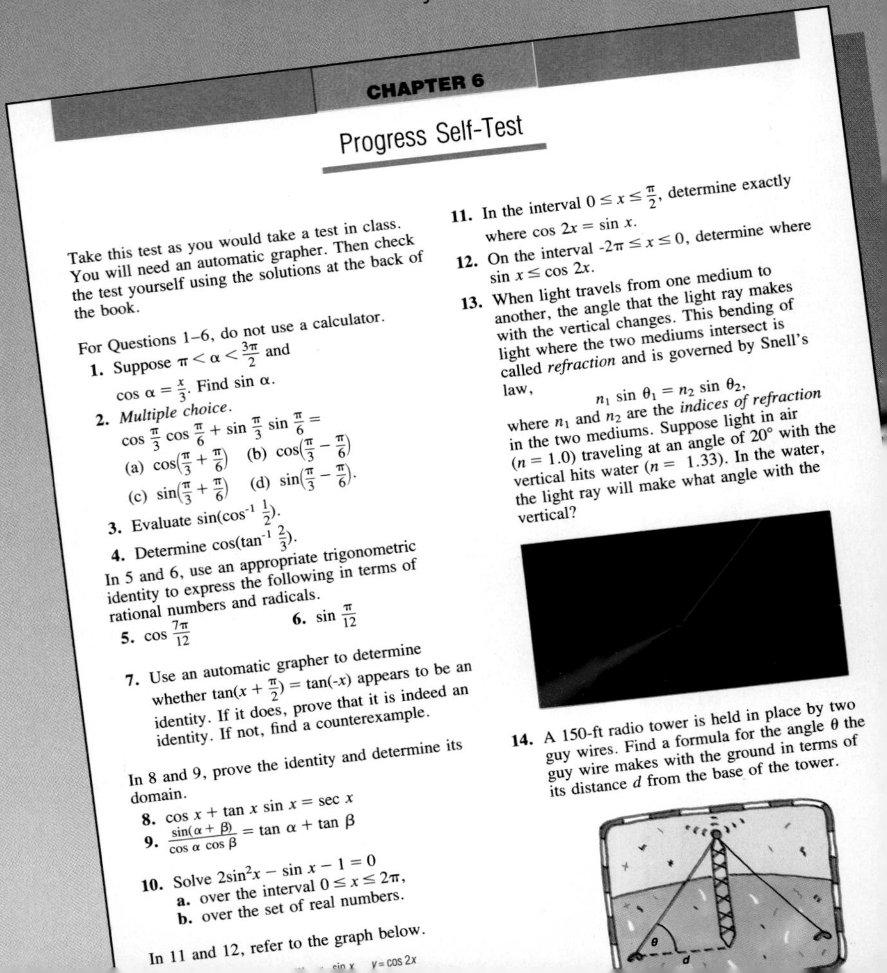

CHAPTER 6

Progress Self-Test

Take this test as you would take a test in class. You will need an automatic grapher. Then check the test yourself using the solutions at the back of the book.

For Questions 1–6, do not use a calculator.

1. Suppose $\pi < \alpha < \frac{3\pi}{2}$ and $\cos \alpha = \frac{x}{3}$. Find $\sin \alpha$.
2. *Multiple choice.* $\cos \frac{\pi}{3} \cos \frac{\pi}{6} + \sin \frac{\pi}{3} \sin \frac{\pi}{6} =$
   (a) $\cos(\frac{\pi}{3} + \frac{\pi}{6})$ (b) $\cos(\frac{\pi}{3} - \frac{\pi}{6})$
   (c) $\sin(\frac{\pi}{3} + \frac{\pi}{6})$ (d) $\sin(\frac{\pi}{3} - \frac{\pi}{6})$.
3. Evaluate $\sin(\cos^{-1} \frac{1}{2})$.
4. Determine $\cos(\tan^{-1} \frac{2}{3})$.

In 5 and 6, use an appropriate trigonometric identity to express the following in terms of rational numbers and radicals.

5. $\cos \frac{7\pi}{12}$
6. $\sin \frac{\pi}{12}$

7. Use an automatic grapher to determine whether $\tan(x + \frac{\pi}{2}) = \tan(-x)$ appears to be an identity. If it does, prove that it is indeed an identity. If not, find a counterexample.

In 8 and 9, prove the identity and determine its domain.

8. $\cos x + \tan x \sin x = \sec x$
9. $\frac{\sin(\alpha + \beta)}{\cos \alpha \cos \beta} = \tan \alpha + \tan \beta$

10. Solve $2\sin^2 x - \sin x - 1 = 0$
    a. over the interval $0 \le x \le 2\pi$,
    b. over the set of real numbers.

In 11 and 12, refer to the graph below.

11. In the interval $0 \le x \le \frac{\pi}{2}$, determine exactly where $\cos 2x = \sin x$.
12. On the interval $-2\pi \le x \le 0$, determine where $\sin x \le \cos 2x$.
13. When light travels from one medium to another, the angle that the light ray makes with the vertical changes. This bending of light where the two mediums intersect is called *refraction* and is governed by Snell's law,
    $$n_1 \sin \theta_1 = n_2 \sin \theta_2,$$
    where n_1 and n_2 are the *indices of refraction* in the two mediums. Suppose light in air ($n = 1.0$) traveling at an angle of 20° with the vertical hits water ($n = 1.33$). In the water, the light ray will make what angle with the vertical?
14. A 150-ft radio tower is held in place by two guy wires. Find a formula for the angle θ the guy wire makes with the ground in terms of its distance d from the base of the tower.

Chapter Review

Questions on SPUR Objectives

SPUR stands for **S**kills, **P**roperties, **U**ses, and **R**epresentations.
The Chapter Review questions are grouped according to the SPUR Objectives for this chapter.

SKILLS deal with the procedures used to get answers.

Objective A: *Without a calculator, use trigonometric identities to express values of trigonometric functions in terms of rational numbers and radicals. (Lessons 6-2, 6-3, 6-4, 6-5)*

In 1–3, suppose x is in the interval $\pi < x < \frac{3\pi}{2}$ and $\sin x = \frac{-3}{8}$. Use trigonometric identities to find each value.

1. $\cos x$ **2.** $\tan x$ **3.** $\csc x$

In 4–6, use appropriate identities to express the following in terms of rational numbers and radicals.

4. $\sin \frac{3\pi}{8}$ **5.** $\cos\frac{\pi}{8}$ **6.** $\tan \frac{\pi}{12}$

7. *Multiple choice.* $\cos \frac{\pi}{4} \cos \frac{\pi}{6} - \sin \frac{\pi}{4} \sin \frac{\pi}{6} =$

(a) $\sin\left(\frac{\pi}{4} + \frac{\pi}{6}\right)$ (b) $\cos\left(\frac{\pi}{4} + \frac{\pi}{6}\right)$

(c) $\sin\left(\frac{\pi}{4} - \frac{\pi}{6}\right)$ (d) $\cos\left(\frac{\pi}{4} - \frac{\pi}{6}\right)$.

In 8–11, suppose x is in the interval $\frac{\pi}{2} < x < \pi$ with $\cos x = \frac{-1}{3}$ and y is in the interval $0 < y < \frac{\pi}{2}$ with $\sin y = \frac{2}{5}$. Use this informatio[n] ... value.

8. $\cos(x + y)$ **9.** $\sin(x$ ...

10. $\sin 2x$ **11.** $\sin \frac{x}{2}$ ...

12. a. Use the identity for $\sin(x$ ... $\sin \frac{5\pi}{12}$.

b. Use the identity $\cos 2x = 1 - 2 \sin^2 x$ to fi[nd] ...

c. Show that your answers to ... are equal.

Objective B: *Evaluate inverse trigonometric functions with or without a calculator. (Lesson 6-6)*

In 13–16, find the exact value without using a calculator.

13. $\sin^{-1}\left(\frac{-\sqrt{2}}{2}\right)$ **14.** $\cos^{-1} 1$

15. $\sin(\tan^{-1} 1)$ **16.** $\sin\left(\sin^{-1}\left(\frac{-1}{2}\right)\right)$

17. Draw an appropriate triangle to determine $\sin\left(\cos^{-1} \frac{2}{5}\right)$.

Objective C: *Solve trigonometric equations and inequalities algebraically. (Lesson 6-7)*

In 18–20, solve over the interval $0 \le x \le 2\pi$.

18. $\cos x = \frac{1}{2}$ **19.** $\tan x = -1$

20. $\sin x = \frac{2}{-\sqrt{3}}$

21. Solve over the set of real numbers: $(\sin x + 1)(\tan x - 1) = 0$.

22. Solve $2\sin^2 x - \cos x - 1 = 0$

a. over the interval $0 \le x \le 2\pi$,

b. over the set of real numbers.

PROPERTIES deal with the principles behind the mathematics.

Objective D: *Prove trigonometric identities and identify their domains. (Lessons 6-2, 6-3, 6-4, 6-5)*

In 26–30, complete each blank so that the resulting equation is an identity.

26. $\cos^2 x + \sin^2 x =$ ___?___

27. $\cos x \sin y - \sin x \cos y =$ ___?___

28. $\sin 2x =$ ___?___

29. $\cos^2 x - 1 =$ ___?___

30. $\sin(x + y) =$ ___?___

In 31–38, prove the identity and identify its domain.

31. $\cos\left(\frac{3\pi}{2} + x\right) = \sin x$

32. $\sec x \cot x = \csc x$

33. $\sin\left(\frac{\pi}{2} + x\right) = \cos x$

34. $\frac{1}{1 + \cos \alpha} + \frac{1}{1 - \cos \alpha} = 2 \csc^2 \alpha$

35. $\cos(\alpha - \beta) - \cos(\alpha + \beta) = 2\sin \alpha \sin \beta$

36. $\sec x + \cot x \csc x = \sec x \csc^2 x$

37. $\cos 4x = \cos^4 x - 6\cos^2 x \sin^2 x + \sin^4 x$

38. $\tan^2 x = \frac{1 - \cos 2x}{1 + \cos 2x}$

USES deal with applications of mathematics in real situations.

Objective E: *Solve problems using inverse trigonometric functions. (Lesson 6-6)*

39. A child 3 ft tall flies a kite on a 200-foot straight string. Find a formula for the angle θ that the string makes with the horizontal in terms of the height of the kite above the ground.

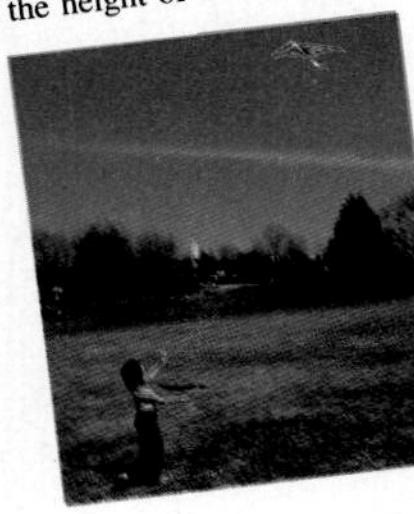

... a spring so ... es from the ... d upward so ... inches. When ... p and down ... es from its ... for one ... hen the ... e distance, in ... uilibrium ... t.

41. A ship travels on a bearing of θ degrees, where θ is measured clockwise from due north. When the ship has traveled 100 miles north of its original position, describe its bearing in terms of its distance x east of its original position.

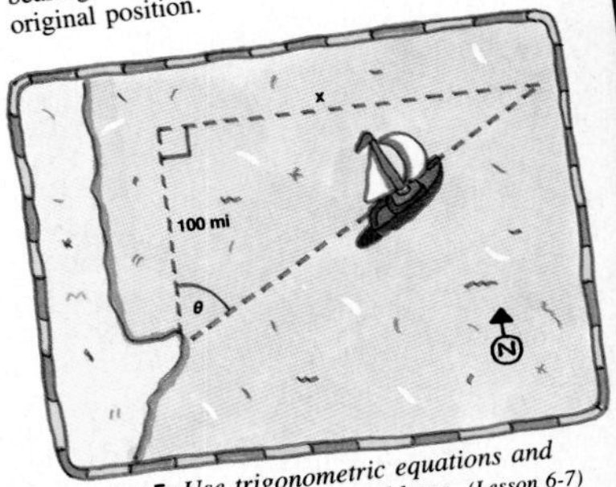

Objective F: *Use trigonometric equations and inequalities to solve applied problems. (Lesson 6-7)*

42. In many situations, the value 9.8 m/sec² is used for acceleration due to gravity. Actually, the equation $g = 9.78049(1 + 0.005288 \sin^2 \theta - 0.000006 \sin^2 2\theta)$ estimates the acceleration g due to gravity (in m/sec²) at sea level as a function of the latitude θ in degrees.

a. Use an automatic grapher to estimate the latitude at which g is 9.8 m/sec².

b. For what latitudes is the acceleration due to gravity greater than 9.81 m/sec²?

In 43 and 44, a quarterback throws a football with an initial velocity of 64 ft/sec. The range is approximated by the equation from Lesson 6-5:

$$R = \frac{v_0^2}{32} \sin 2\theta.$$

43. If the quarterback wants to make a 40 yard pass (120 ft), at approximately what angle should the football be thrown?

44. For what angle values is the range more than 30 yards (90 ft)?

REPRESENTATIONS deal with pictures, graphs, or objects that illustrate concepts.

Objective G: *Use an automatic grapher to test proposed trigonometric identities. (Lessons 6-1, 6-2)*

In 45–47, use an automatic grapher to determine whether the proposed identity appears to be an identity. If it does, prove it algebraically. If not, give a counterexample.

45. $1 + \cot^2 x = \csc^2 x$

46. $\cos 2x = 2\cos x$

47. $\tan(\pi + \gamma) = \tan \gamma$

48. Use an automatic grapher to determine over what domain $\cos x$, with x in radians, can be approximated by $f(x) = 1 - \frac{x^2}{2!} + \frac{x^4}{4!} - \frac{x^6}{6!}$ to within .01.

49. How can you use an automatic grapher to determine whether the proposed identity $\sin(\alpha + \beta) = \sin \alpha \cos \beta - \cos \alpha \sin \beta$ is true? (Do not actually use the automatic grapher, just describe the procedure.)

50. a. Graph the function $f(x) = \sin x \sec x$.

b. What single trigonometric function has a similar graph?

c. What identity is suggested?

Objective H: *Use graphs to solve trigonometric equations and inequalities. (Lesson 6-7)*

51. a. Use an automatic grapher to find all solutions to $\sin x - \cos x = \frac{1}{2}$ (to the nearest tenth) over the interval $0 \le x \le 2\pi$.

b. Use your answer to part **a** to solve $\sin x - \cos x < \frac{1}{2}$.

52. Refer to the graph below of $y = 1.5$ and $y = \cos^2 x + 1$.

a. Solve $\cos^2 x + 1 = 1.5$ over the interval from -2π to 2π.

b. Solve the inequality $\cos^2 x + 1 < 1.5$ over the interval $0 < x < \pi$.

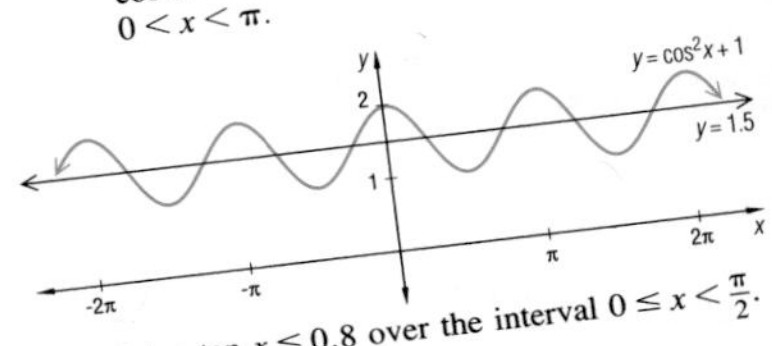

53. Solve $\tan x \le 0.8$ over the interval $0 \le x < \frac{\pi}{2}$.

Provides the Practical Support You Need

"It's been a pleasure to teach PDM. It's been interesting to use the textbook and have a new, fresh approach. Some of the problems are more interesting than any problems I've seen before."

John Adkinson, Teacher, Thornton Fractional South High School, Lansing, IL

Continual involvement of teachers and instructional supervisors—in planning, writing, rewriting, and evaluating—has made this program convenient and adaptable to your needs.

Before each chapter you'll find the following:

- **OBJECTIVES** are letter-coded and keyed to Progress Self-Test, Chapter Review, Lesson Masters, and Chapter Tests, Forms A and B—showing a direct correspondence between what is taught and what is tested.
- The **OVERVIEW** anticipates and addresses your needs for the upcoming chapter.
- The **PERSPECTIVES** are a unique feature that provides the rationale for the inclusion of topics or approaches, provides mathematical background, and makes connections within UCSMP materials. This is interesting information you'll really use.

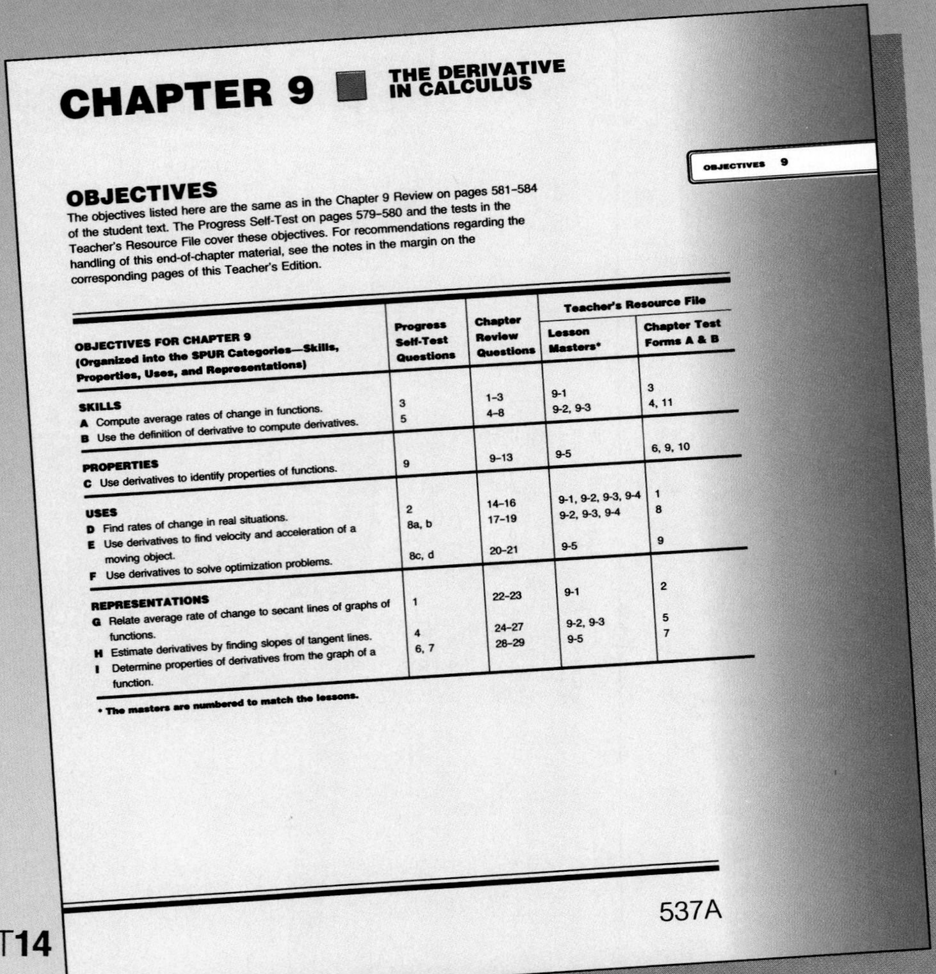

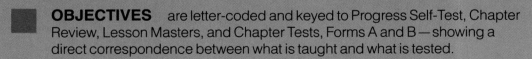

CHAPTER 9 ■ THE DERIVATIVE IN CALCULUS

OBJECTIVES 9

OBJECTIVES

The objectives listed here are the same as in the Chapter 9 Review on pages 581–584 of the student text. The Progress Self-Test on pages 579–580 and the tests in the Teacher's Resource File cover these objectives. For recommendations regarding the handling of this end-of-chapter material, see the notes in the margin on the corresponding pages of this Teacher's Edition.

OBJECTIVES FOR CHAPTER 9 (Organized into the SPUR Categories—Skills, Properties, Uses, and Representations)	Progress Self-Test Questions	Chapter Review Questions	Teacher's Resource File: Lesson Masters*	Teacher's Resource File: Chapter Test Forms A & B
SKILLS				
A Compute average rates of change in functions.	3	1–3	9-1	3
B Use the definition of derivative to compute derivatives.	5	4–8	9-2, 9-3	4, 11
PROPERTIES				
C Use derivatives to identify properties of functions.	9	9–13	9-5	6, 9, 10
USES				
D Find rates of change in real situations.	2	14–16	9-1, 9-2, 9-3, 9-4	1
E Use derivatives to find velocity and acceleration of a moving object.	8a, b	17–19	9-2, 9-3, 9-4	8
F Use derivatives to solve optimization problems.	8c, d	20–21	9-5	9
REPRESENTATIONS				
G Relate average rate of change to secant lines of graphs of functions.	1	22–23	9-1	2
H Estimate derivatives by finding slopes of tangent lines.	4	24–27	9-2, 9-3	5
I Determine properties of derivatives from the graph of a function.	6, 7	28–29	9-5	7

* The masters are numbered to match the lessons.

537A

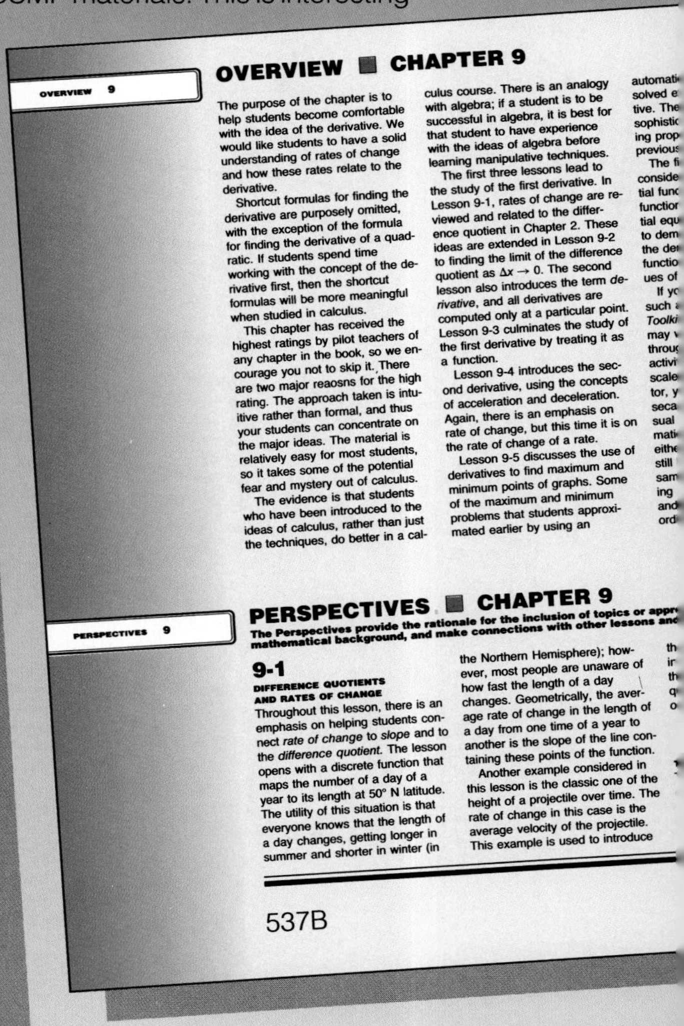

OVERVIEW 9

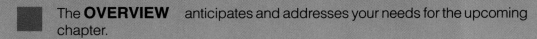

OVERVIEW ■ CHAPTER 9

The purpose of the chapter is to help students become comfortable with the idea of the derivative. We would like students to have a solid understanding of rates of change and how these rates relate to the derivative.

Shortcut formulas for finding the derivative are purposely omitted, with the exception of the formula for finding the derivative of a quadratic. If students spend time working with the concept of the derivative first, then the shortcut formulas will be more meaningful when studied in calculus.

This chapter has received the highest ratings by pilot teachers of any chapter in the book, so we encourage you not to skip it. There are two major reaosns for the high rating. The approach taken is intuitive rather than formal, and thus your students can concentrate on the major ideas. The material is relatively easy for most students, so it takes some of the potential fear and mystery out of calculus.

The evidence is that students who have been introduced to the ideas of calculus, rather than just the techniques, do better in a calculus course. There is an analogy with algebra; if a student is to be successful in algebra, it is best for that student to have experience with the ideas of algebra before learning manipulative techniques.

The first three lessons lead to the study of the first derivative. In Lesson 9-1, rates of change are reviewed and related to the difference quotient in Chapter 2. These ideas are extended in Lesson 9-2 to finding the limit of the difference quotient as $\Delta x \to 0$. The second lesson also introduces the term *derivative*, and all derivatives are computed only at a particular point. Lesson 9-3 culminates the study of the first derivative by treating it as a function.

Lesson 9-4 introduces the second derivative, using the concepts of acceleration and deceleration. Again, there is an emphasis on rate of change, but this time it is on the rate of change of a rate.

Lesson 9-5 discusses the use of derivatives to find maximum and minimum points of graphs. Some of the maximum and minimum problems that students approximated earlier by using an

PERSPECTIVES 9

PERSPECTIVES ■ CHAPTER 9

The Perspectives provide the rationale for the inclusion of topics or appr… mathematical background, and make connections with other lessons and…

9-1

DIFFERENCE QUOTIENTS AND RATES OF CHANGE

Throughout this lesson, there is an emphasis on helping students connect *rate of change* to *slope* and to the *difference quotient*. The lesson opens with a discrete function that maps the number of a day of a year to its length at 50° N latitude. The utility of this situation is that everyone knows that the length of a day changes, getting longer in summer and shorter in winter (in the Northern Hemisphere); however, most people are unaware of how fast the length of a day changes. Geometrically, the average rate of change in the length of a day from one time of a year to another is the slope of the line containing these points of the function.

Another example considered in this lesson is the classic one of the height of a projectile over time. The rate of change in this case is the average velocity of the projectile. This example is used to introduce

537B

The **DAILY PACING CHART** shows you at a glance two alternate ways to pace the chapter.

TESTING OPTIONS list the chapter quizzes and tests for ease of planning.

Professional Sourcebook for UCSMP

Also at your fingertips is a wealth of information that includes valuable material on topics ranging from research to review. See pages **T19-T52** at the back of the **Teacher's Edition.**

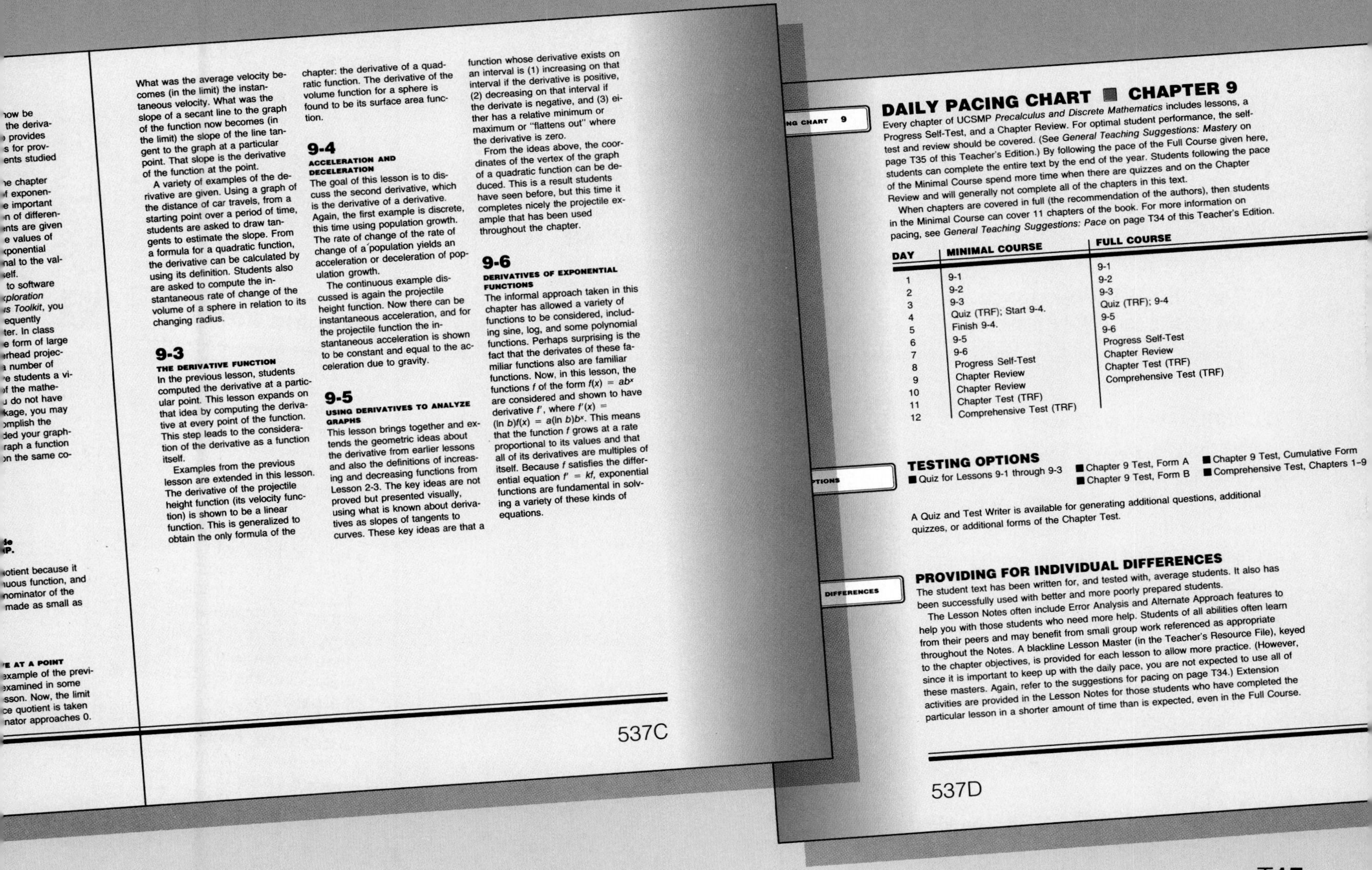
What was the average velocity becomes (in the limit) the instantaneous velocity. What was the slope of a secant line to the graph of the function now becomes (in the limit) the slope of the line tangent to the graph at a particular point. That slope is the derivative of the function at the point.

A variety of examples of the derivative are given. Using a graph of the distance of car travels, from a starting point over a period of time, students are asked to draw tangents to estimate the slope. From a formula for a quadratic function, the derivative can be calculated by using its definition. Students also are asked to compute the instantaneous rate of change of the volume of a sphere in relation to its changing radius.

9-3
THE DERIVATIVE FUNCTION
In the previous lesson, students computed the derivative at a particular point. This lesson expands on that idea by computing the derivative at every point of the function. This step leads to the consideration of the derivative as a function itself.

Examples from the previous lesson are extended in this lesson. The derivative of the projectile height function (its velocity function) is shown to be a linear function. This is generalized to obtain the only formula of the chapter: the derivative of a quadratic function. The derivative of the volume function for a sphere is found to be its surface area function.

9-4
ACCELERATION AND DECELERATION
The goal of this lesson is to discuss the second derivative, which is the derivative of a derivative. Again, the first example is discrete, this time using population growth. The rate of change of the rate of change of a population yields an acceleration or deceleration of population growth.

The continuous example discussed is again the projectile height function. Now there can be instantaneous acceleration, and for the projectile function the instantaneous acceleration is shown to be constant and equal to the acceleration due to gravity.

9-5
USING DERIVATIVES TO ANALYZE GRAPHS
This lesson brings together and extends the geometric ideas about the derivative from earlier lessons and also the definitions of increasing and decreasing functions from Lesson 2-3. The key ideas are not proved but presented visually, using what is known about derivatives as slopes of tangents to curves. These key ideas are that a function whose derivative exists on an interval is (1) increasing on that interval if the derivative is positive, (2) decreasing on that interval if the derivate is negative, and (3) either has a relative minimum or maximum or "flattens out" where the derivative is zero.

From the ideas above, the coordinates of the vertex of the graph of a quadratic function can be deduced. This is a result students have seen before, but this time it completes nicely the projectile example that has been used throughout the chapter.

9-6
DERIVATIVES OF EXPONENTIAL FUNCTIONS
The informal approach taken in this chapter has allowed a variety of functions to be considered, including sine, log, and some polynomial functions. Perhaps surprising is the fact that the derivates of these familiar functions also are familiar functions. Now, in this lesson, the functions f of the form $f(x) = ab^x$ are considered and shown to have derivative f', where $f'(x) = (\ln b)f(x) = a(\ln b)b^x$. This means that the function f grows at a rate proportional to its values and that all of its derivatives are multiples of itself. Because f satisfies the differential equation $f' = kf$, exponential functions are fundamental in solving a variety of these kinds of equations.

537C

DAILY PACING CHART ■ CHAPTER 9

Every chapter of UCSMP *Precalculus and Discrete Mathematics* includes lessons, a Progress Self-Test, and a Chapter Review. For optimal student performance, the self-test and review should be covered. (See *General Teaching Suggestions: Mastery* on page T35 of this Teacher's Edition.) By following the pace of the Full Course given here, students can complete the entire text by the end of the year. Students following the pace of the Minimal Course spend more time when there are quizzes and on the Chapter Review and will generally not complete all of the chapters in this text.

When chapters are covered in full (the recommendation of the authors), then students in the Minimal Course can cover 11 chapters of the book. For more information on pacing, see *General Teaching Suggestions: Pace* on page T34 of this Teacher's Edition.

DAY	MINIMAL COURSE	FULL COURSE
1	9-1	9-1
2	9-2	9-2
3	9-3	9-3
4	Quiz (TRF); Start 9-4.	Quiz (TRF); 9-4
5	Finish 9-4.	9-5
6	9-5	9-6
7	9-6	Progress Self-Test
8	Progress Self-Test	Chapter Review
9	Chapter Review	Chapter Test (TRF)
10	Chapter Review	Comprehensive Test (TRF)
11	Chapter Test (TRF)	
12	Comprehensive Test (TRF)	

TESTING OPTIONS

- Quiz for Lessons 9-1 through 9-3
- Chapter 9 Test, Form A
- Chapter 9 Test, Form B
- Chapter 9 Test, Cumulative Form
- Comprehensive Test, Chapters 1-9

A Quiz and Test Writer is available for generating additional questions, additional quizzes, or additional forms of the Chapter Test.

PROVIDING FOR INDIVIDUAL DIFFERENCES

The student text has been written for, and tested with, average students. It also has been successfully used with better and more poorly prepared students.

The Lesson Notes often include Error Analysis and Alternate Approach features to help you with those students who need more help. Students of all abilities often learn from their peers and may benefit from small group work referenced as appropriate throughout the Notes. A blackline Lesson Master (in the Teacher's Resource File), keyed to the chapter objectives, is provided for each lesson to allow more practice. (However, since it is important to keep up with the daily pace, you are not expected to use all of these masters. Again, refer to the suggestions for pacing on page T34.) Extension activities are provided in the Lesson Notes for those students who have completed the particular lesson in a shorter amount of time than is expected, even in the Full Course.

537D

Super Teaching Support

"It was a good feeling to see how physics and math go together, and how the same problem can be figured out different ways."

Precalculus and Discrete Mathematics student, Mainland Regional High School, Linwood, NJ

1 **RESOURCES** save you time by coordinating all of the ancillaries with the lesson.

2 **OBJECTIVES** are letter-coded for easy reference.

3 **TEACHING NOTES** provide everything you need for lesson planning. These include suggestions for the examples, dialogue to stimulate higher-order thinking, and more.

4 **ALTERNATE APPROACH** offers a different strategy for presenting the lesson.

5 **ERROR ANALYSIS** pinpoints typical student errors and provides remediation techniques.

6 **MAKING CONNECTIONS** helps you connect present content and ideas to material covered in an earlier or later lesson, chapter, or text.

7 **TECHNOLOGY** suggests ways to involve students with computers and graphics calculators. Combine Computer Masters and the new Scott, Foresman software for valuable and interesting activities.

8 **ADDITIONAL EXAMPLES** provide examples parallel to those in the text for added flexibility.

9 **NOTES ON QUESTIONS** highlight important aspects of questions and provide helpful suggestions to enhance learning.

10 **ADDITIONAL ANSWERS** locate answers to questions in the lesson. Shorter answers are overprinted in red near the questions.

11 **MORE PRACTICE** lists the Lesson Master which can be used for additional practice.

12 **EXTENSION** offers high-interest activities for all students, as well as enrichment activities for students needing additional challenge. These well-liked activities provide ideas for technology, careers, connections with other fields of study, and additional applications.

13 **SMALL GROUP WORK** includes activities and projects suitable for groups of students.

14 **EVALUATION** tells what you need to know about the quizzes and tests. It also provides **Alternative Assessment** suggestions to encourage different formats, such as oral presentation and cooperative learning.

15 **LESSON MASTERS** are pictured where you need them for your convenience.

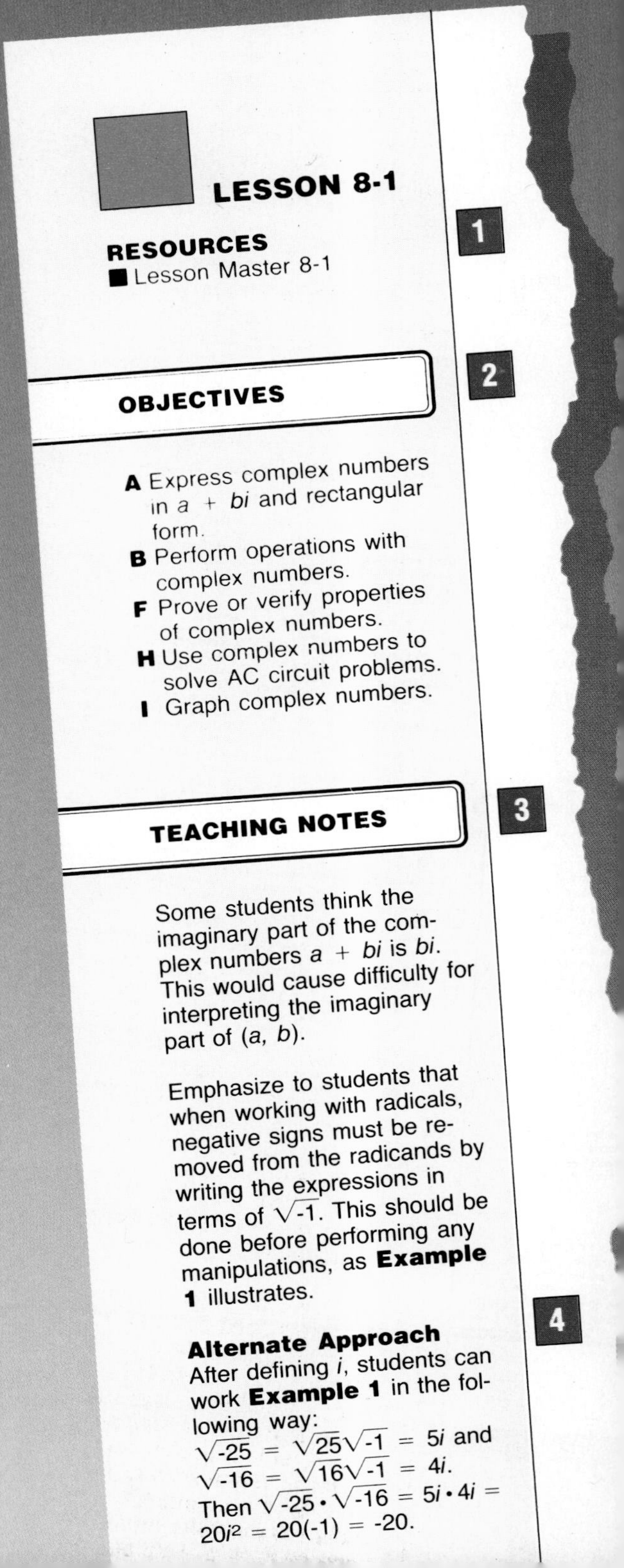

LESSON 8-1

RESOURCES 1

■ Lesson Master 8-1

OBJECTIVES 2

A Express complex numbers in $a + bi$ and rectangular form.
B Perform operations with complex numbers.
F Prove or verify properties of complex numbers.
H Use complex numbers to solve AC circuit problems.
I Graph complex numbers.

TEACHING NOTES 3

Some students think the imaginary part of the complex numbers $a + bi$ is bi. This would cause difficulty for interpreting the imaginary part of (a, b).

Emphasize to students that when working with radicals, negative signs must be removed from the radicands by writing the expressions in terms of $\sqrt{-1}$. This should be done before performing any manipulations, as **Example 1** illustrates.

Alternate Approach 4
After defining i, students can work **Example 1** in the following way:
$\sqrt{-25} = \sqrt{25}\sqrt{-1} = 5i$ and $\sqrt{-16} = \sqrt{16}\sqrt{-1} = 4i$.
Then $\sqrt{-25} \cdot \sqrt{-16} = 5i \cdot 4i = 20i^2 = 20(-1) = -20$.

5 **Error Analysis** If students try to apply the Square Root of a Product Theorem in **Example 1**, they will obtain $\sqrt{-25} \cdot \sqrt{-16} = \sqrt{400} = 20$, which is incorrect. Stress that they must simplify the radicals first. Then they can apply the familiar rules to the square roots of nonnegative real numbers.

6 **Making Connections**
In Lesson 5-2, students learned that to rationalize a denominator of the form $a + \sqrt{b}$ (b is positive), one could multiply both numerator and denominator by $a - \sqrt{b}$. The process of rewriting the quotient in **Example 2** extends this procedure to instances where b is negative. Specifically, the given fraction equals $\dfrac{6 + 7\sqrt{-1}}{8 + 5\sqrt{-1}}$ or $\dfrac{6 + \sqrt{-49}}{8 + \sqrt{-25}}$, and both numerator and denominator are multiplied by $8 - 5\sqrt{-1}$ or $8 - \sqrt{-25}$.

7 **Technology** Some calculators have operations on complex numbers built in. Regarding **Example 3**, some symbol manipulation software, *Derive* among them, ask the user to select the domain for factoring polynomials. Allowable choices include factoring over polynomials with rational, radical, and complex number coefficients. Thus, depending upon the domain of coefficients, $x^2 + 5$ could be prime or could be factored as $(x + i\sqrt{5})(x - i\sqrt{5})$. This idea will be discussed again in Lesson 8-9.

To assist students in proving theorems such as those implied by **Example 3** and **Questions 16 and 20**, suggest that they begin as follows. If the theorem involves only one complex number, represent it as $z = a + bi$. If two complex numbers are involved, represent [illegible] + bi and the

8 **ADDITIONAL EXAMPLES**

1. Multiply $\sqrt{-36}\sqrt{-9}$.
-18

2. Express the quotient $\dfrac{7 + 2i}{3 - 5i}$ in $a + bi$ form.
$\mathbf{\frac{11}{34} + \frac{41}{34}i}$

3. Prove: For all complex numbers z, $z + \bar{z}$ is a real number equal to twice the real part of z.
Let $z = a + bi$. Then $\bar{z} = a - bi$, and so $z + \bar{z} = a + bi + a - bi = 2a$.

9 **NOTES ON QUESTIONS**
Question 6: Some mathematicians do not consider the set of real numbers to be subset of the set of complex numbers, but a set with operations isomorphic to a subset of the set of complex numbers. The question is philosophical: whether isomorphism means that two sets are identical but just represented differently; or whether it means they are different sets.

Question 9: In this question, students are applying the theorem $\overline{z \cdot w} = \bar{z} \cdot \bar{w}$. Ask if the same holds true for an irrational number such as $5 + 3\sqrt{2}$ and its conjugate. (Yes)

10 **ADDITIONAL ANSWERS**

1. $z = 7 + 3i = (7, 3) = Z$; $o = 0 + 0i = (0, 0) = O$, $w = 4 - 9i = (4, -9) = W$; $z + w = 11 - 6i = (11, -6) = P$
The slope of $\overline{OZ} = \frac{3 - 0}{7 - 0} = \frac{3}{7}$; the slope of $\overline{WP} = \frac{-6 - (-9)}{11 - 4} = \frac{3}{7}$; and so $\overline{OZ} \parallel \overline{WP}$.
The slope of $\overline{ZP} = \frac{-6 - 3}{11 - 7} = -\frac{9}{4}$; the slope of $\overline{OW} = \frac{-9 - 0}{4 - 0} = -\frac{9}{4}$; and so $\overline{ZP} \parallel \overline{OW}$.
Therefore, the figure is a parallelogram.

2. $z = a + bi = (a, b) = Z$; [illegible]

FOLLOW-UP

11 **MORE PRACTICE**
For more questions on SPUR Objectives, use *Lesson Master 8-1*, shown on page 471.

12 13 **EXTENSION**
Small groups can examine the counterparts of **Question 16** for the operations with imaginary numbers. Then ask the same questions for nonimaginary, nonreal complex numbers. (Imaginary numbers are not closed under any of the operations; nor are the nonimaginary, nonreal complex numbers.)

PROJECTS
The projects for Chapter 8 are described on pages 529 and 530. **Projects 1 and 6** are related to the content of this lesson.

14 **EVALUATION**
A quiz covering Lessons 8-1 through 8-3 is provided in the Teacher's Resource File.

15

NAME ______

LESSON **MASTER 8–1**
QUESTIONS ON **SPUR** OBJECTIVES

■ **SKILLS** *Objective A* *(See pages 534–536 for objectives.)*
In 1 and 2, rewrite the complex number as an ordered pair.

1. $3 + 5i$ **(3, 5)**
2. $-i - 2$ **(-2, -1)**

In 3 and 4, rewrite the complex number in $a + bi$ form.

3. $(-5, 0)$ **$-5 + 0i$**
4. $(\frac{1}{4}, -1)$ **$\frac{1}{4} - i$**
5. If the real part and the imaginary part of a complex number are 7 and 15, respectively, write the number in $a + bi$ form. **$7 + 15i$**

■ **SKILLS** *Objective B*
In 6–10, perform the indicated operation. Write the result in $a + bi$ form.

6. $\sqrt{-4} \cdot \sqrt{-81}$ **-18**
7. $\frac{6 - 2\sqrt{9}}{3}$ **0**
8. $(12 - 3i) - i(5 + 7i)$ **$19 - 8i$**
9. $\frac{5i}{1 - i} + 3 - 4i$ **$\frac{1}{2} - \frac{3}{2}i$**
10. $(5 - 8i)(2 + 7i)$ **$66 + 19i$**

In 11 and 12, express the solutions in $a + bi$ form.

11. $z^2 = -45$ **$\pm 3\sqrt{5}i$**
12. $-3 + 2i + w = 8 - 9i$ **$11 - 11i$**

■ **PROPERTIES** *Objective F*
13. Let $z = 6 - 2i$. Verify that $z \cdot \bar{z}$ is a real number.
$(6 - 2i)(6 + 2i) = 36 + 4 = 40$

14. Let $z = a + bi$. Prove that $\bar{\bar{z}} = z$.
$\bar{z} = a - bi$, so $\bar{\bar{z}} = a - (-bi) = a - bi = z$

Continued 87

Components Designed for Ease of Teaching

"Teaching PDM has exceeded everything that I expected!"

Mercedes McGowen, Teacher, Elgin High School, Elgin, IL

Student Edition
full color

Teacher's Edition
annotated with margin notes

Teacher's Resource File
About 600 blackline masters
cover your every classroom need!

Quizzes and Test Masters
Quizzes (at least one per chapter)
Chapter Tests, Forms A and B (parallel forms)
Chapter Tests, Cumulative Form
Comprehensive Tests (four per text, including
Final Exam, primarily multiple choice)

Lesson Masters
(one or two pages per lesson)

Computer Masters

Answer Masters
(provide answers for questions in student text;
oversized type to enable display in class,
allowing students to grade their own work)

Teaching Aid Masters
(masters for overhead transparencies; forms,
charts, and graphs from the P.E.; coordinate
grids; and more)

Additional Ancillaries

Solution Manual
Computer Software (for Macintosh and IBM)
GraphExplorer
Quiz and Test Writer

UCSMP
SCOTT, FORESMAN

The University of Chicago School Mathematics Project

Precalculus and Discrete Mathematics

Authors

Anthony L. Peressini
Susanna S. Epp
Kathleen A. Hollowell
Susan Brown
Wade Ellis, Jr.
John W. McConnell
Jack Sorteberg
Denisse R. Thompson
Dora Aksoy
Geoffrey D. Birky
Greg McRill
Zalman Usiskin

Technology requirements:
Students should have access to graphics calculators or computers equipped with graphing software at almost all times, including homework and tests.

About the Cover
The motion of a robot arm can be described by three-dimensional vectors. Such robots today are typically computer controlled. Thus this application involves both an understanding of the mathematics of motion for which calculus is so important and the mathematics of computers for which discrete mathematics is a foundation.

Scott, Foresman

Editorial Offices: Glenview, Illinois Regional Offices: Sunnyvale, California • Atlanta, Georgia • Glenview, Illinois • Oakland, New Jersey • Dallas, Texas

Acknowledgments

Authors

Anthony L. Peressini
Professor of Mathematics, University of Illinois, Urbana-Champaign

Susanna S. Epp
Associate Professor of Mathematics, DePaul University, Chicago

Kathleen A. Hollowell
Coordinator of Secondary Mathematics Inservice Programs, University of Delaware, Newark

Susan Brown
Mathematics Teacher, York H.S., Elmhurst, IL

Wade Ellis, Jr.
Mathematics Instructor, West Valley College, Saratoga, CA

John W. McConnell
Instructional Supervisor of Mathematics, Glenbrook South H.S., Glenview, IL

Jack Sorteberg
Mathematics Teacher, Burnsville High School, Burnsville, MN

Denisse R. Thompson
Assistant Professor of Mathematics Education, University of South Florida, Tampa

Dora Aksoy
UCSMP

Geoffrey D. Birky
UCSMP

Greg McRill
UCSMP

Zalman Usiskin
Professor of Education, The University of Chicago

UCSMP Production and Evaluation

Series Editors: Zalman Usiskin, Sharon L. Senk

Managing Editor: Daniel Hirschhorn

Technical Coordinator: Susan Chang

Director of Evaluation: Denisse R. Thompson

We wish to acknowledge the generous support of the **Amoco Foundation**, the **General Electric Foundation**, and the **Carnegie Corporation of New York** in helping to make it possible for these materials to be developed and tested.

ISBN: 0–673–33366–3

Printing Key 23456789–RRW–989796959493

It takes many people to put together a project of this kind and we cannot thank them all by name. We wish particularly to acknowledge Carol Siegel, who coordinated the printing of pilot and formative editions of these materials and their use in schools; Peter Bryant, Maura Byrne, Dan Caplinger, Janine Crawley, Lewis Garvin, Kurt Hackemer, Maryann Kannappan, Mary Lappan, Theresa Manst, Yuri Mishina, Lee Resta, Vicki Ritter, Lorena Shih, and David Wrisley of our technical staff; and editorial assistants Matt Ashley, Laura Gerbec, Jon Golub, John Jasek, Eric Kolaczyk, Ben Krug, John McNamara, Teri Proske, Robert Schade, and Matt Solit.

We wish to acknowledge and give thanks to the following teachers who taught preliminary versions of this text, participated in the field testing or formative evaluations, and contributed ideas to help improve this text.

John Adkinson
Thornton Fractional High School South
Lansing, IL

Michael Bowers
Churchill High School
Eugene, OR

Cynthia Crenshaw
University School of Nashville
Nashville, TN

Joyce Evans
Breck School
Minneapolis, MN

Ronald R. Godar
San Marcos High School
Santa Barbara, CA

Mary Hahn
Woodward High School
Cincinnati, OH

Deborah Klipp
Mainland Regional High School
Linwood, NJ

Mercedes McGowen
Elgin High School
Elgin, IL

Rheta N. Rubenstein
Renaissance High School
Detroit, MI

Martin Sanford
Renaissance High School
Detroit, MI

David Tyson
Mercersburg Academy
Mercersburg, PA

We also wish to express our thanks and appreciation to the many other schools and students who have used earlier versions of these materials.

UCSMP

Precalculus and Discrete Mathematics

The University of Chicago School Mathematics Project (UCSMP) is a long-term project designed to improve school mathematics in grades K-12. UCSMP began in 1983 with a six-year grant from the Amoco Foundation, whose support continued in 1989 with a grant through 1994. Additional funding has come from the Ford Motor Company, the Carnegie Corporation of New York, the National Science Foundation, the General Electric Foundation, GTE, Citibank/Citicorp, and the Exxon Education Foundation.

The project is centered in the Departments of Education and Mathematics of the University of Chicago, and has the following components and directors:

Resources	Izaak Wirszup, Professor Emeritus of Mathematics
Primary Materials	Max Bell, Professor of Education
Elementary Teacher Development	Sheila Sconiers, Research Associate in Education
Secondary	Sharon L. Senk, Associate Professor of Mathematics, Michigan State University Zalman Usiskin, Professor of Education
Evaluation	Larry Hedges, Professor of Education

From 1983-1987, the director of UCSMP was Paul Sally, Professor of Mathematics. Since 1987, the director has been Zalman Usiskin.

The text *Precalculus and Discrete Mathematics* was developed by the Secondary Component (grades 7-12) of the project, and constitutes the final year in a six-year mathematics curriculum devised by that component. As texts in this curriculum completed their multi-stage testing cycle, they were published by Scott, Foresman. A list of the six texts follows.

Transition Mathematics
Algebra
Geometry
Advanced Algebra
Functions, Statistics, and Trigonometry
Precalculus and Discrete Mathematics

A first draft of this course, then titled *Pre-College Mathematics*, was begun in the summer of 1987, and completed, edited, and tested in three schools during the 1987-88 school year. A second pilot edition, a major revision based on the first pilot, was written and tested in 1988-89. The present title appeared in that edition. Further changes were made and a field-trial edition was given a formal test in nine schools. In this and all other studies, some students had previous UCSMP courses, some had not. This Scott, Foresman edition is based on improvements suggested by that testing, by the authors and editors, and by some of the many teacher and student users of earlier editions.

Comments about these materials are welcomed. Address queries to Secondary Mathematics Product Manager, Scott, Foresman, 1900 East Lake Avenue, Glenview, Illinois 60025, or to UCSMP, The University of Chicago, 5835 S. Kimbark, Chicago, IL 60637.

To the Teacher

This book differs from other books at this level in six major ways. First, it has **wider scope** in that it deals with discrete mathematics in addition to and interwoven with the precalculus topics normally taught at this level. Discrete mathematics includes topics such as logic, properties of integers, induction, recursion, combinatorics, and graphs and networks — all topics important in understanding current mathematics and its uses, particularly relative to computers. This book gives strong attention to the reasoning processes employed by mathematicians and by those who apply mathematics. Complex algebraic manipulations are necessary in virtually all college mathematics courses, and it makes sense to continue to work on these skills this year to keep them strong for college. An overall goal of this course is to foster an interest in the study of mathematics, and, more generally, an appreciation for the axiomatic and deductive approaches used in many fields of study.

Second, this book requires **technology**. Automatic graphers (computers or graphics calculators) are used as instructional tools throughout the book to serve a number of purposes: to promote a student's ability to visualize functions; to explore relations between equations and their graphs; and to develop the concept of limit. It is expected that a demonstration computer is available in class whenever it is needed. Students, too, are expected to have access to this technology – a graphics calculator or a computer with function graphing software is necessary for many of the questions.

Third, **reading and problem solving** are emphasized throughout. Students can and should be expected to read this book. The explanations were written for students and tested with them. The first set of questions in each lesson is called "Covering the Reading." These questions guide students through the reading and check their coverage of critical words, rules, explanations, and examples. The second set of questions is called "Applying the Mathematics." These questions extend students' understanding of the principles and applications of the lesson. Like skills, problem solving must be practiced; when practiced it becomes far less difficult. To further widen the students' horizons, "Exploration" questions are provided in every lesson. Projects, detailed at the end of each chapter, allow students to tackle larger-scale problems, and thus develop persistence along with group problem-solving skills.

Fourth, there is a **reality orientation** towards both the selection of content and the methods taught students for working out problems. Virtually all of the content of this course is studied in detail for its applications to real-world problems. The variety of content of this book permits many of the questions to be embedded in application settings. The reality orientation extends also to the methods assumed in solving the problems: students are assumed to have scientific calculators at all times.

Fifth, **four dimensions of understanding** are emphasized: skill in analyzing and carrying out various algorithms; developing and using mathematical properties and relationships; applying mathematics in realistic situations; and representing or picturing mathematical concepts. We call this the SPUR approach: **S**kills, **P**roperties, **U**ses, **R**epresentations. With the SPUR approach, concepts are discussed in a rich environment that enables more students to be reached.

Sixth, the **instructional format** is designed to maximize the acquisition of skills and concepts. The book is organized around lessons usually meant to be covered in one day. The lessons have been sequenced into carefully constructed chapters which combine gradual practice with techniques to achieve mastery and retention. Concepts introduced in a lesson are reinforced through "Review" questions in the immediately succeeding lessons. This gives students several nights to learn and practice important concepts. At the end of each chapter, a modified mastery learning scheme is used to solidify acquisition of concepts from the chapter so that they may be applied later with confidence. It is critical that the end-of-chapter content be covered. To maintain skills, important ideas are reviewed again in later chapters.

CONTENTS

Chapter 9 The Derivative in Calculus 537

Chapter 10 Combinatorics 585

Chapter 11 Graphs and Circuits 639

Chapter 12 Vectors 699

Chapter 13 The Integral in Calculus 759

To the Student

Purposes of this book

One purpose of this book is given by its name: to provide you with a knowledge of precalculus and discrete mathematics. Precalculus mathematics deals mainly with infinite and continuous processes; its subject matter includes functions of many kinds, coordinate systems, trigonometry, vectors, limits, and an introduction to the basic ideas of calculus: derivatives and integrals. Discrete mathematics deals with finite and iterative processes; its subject matter includes certain functions, logic, properties of integers, sequences, algorithms, recursion and induction, combinatorics, and graphs and networks.

Even if you never take a course in mathematics itself in college, you are likely to encounter many of the ideas in this book. If you study further in any area in which mathematics is encountered, it is likely that you will need the mathematics presented in this book. Discrete mathematics is fundamental in computer science, and calculus is fundamental to engineering and the physical sciences. Both areas are used in business and in the social sciences.

Another purpose of this book is to review and bring together, in a cohesive way, what you have learned in previous courses. Both precalculus and discrete mathematics cover a wide range of topics. You may find topics close together in this book that you encountered in different chapters of books in previous years or that you even studied in different years. Mathematics is a unified discipline in the sense that what is learned in one area can be applied in many other areas, and we wish you to have that spirit of mathematics.

Materials needed

The discrete mathematics theme of this book is a direct result of the computer revolution. The precalculus you will study here has also been affected by computers. Today, people who need accurate graphs of functions do not do them by hand; they use an *automatic grapher* — either a computer with a program that graphs functions or a graphics calculator. We assume you have access to an automatic grapher and a scientific calculator. Many assignments will take you quite a bit longer to complete and some may even be inaccessible if you do not have such time-saving and accurate technology.

Of course, you need pencils and erasers at all times, and it is best to have some lined paper, some unlined paper, and some graph paper.

Organization of this book

We want you to read this book. The reading in each lesson is designed to introduce, explain, and relate the important ideas of the lesson. Read slowly and carefully; if there are calculations or graphs, verify them as you read. Consequently, it is best to have an automatic grapher and scientific calculator nearby. Use this book's glossary and index if you encounter terms with which you are not familiar. Keeping a dictionary handy is also a good idea.

The Questions following each lesson are designed to get you to think about its ideas and are of four types. Questions "Covering the Reading" follow the vocabulary and examples of the lesson. Questions "Applying the Mathematics" may require you to extend the content of the lesson or deal with the content in an unfamiliar way. Do not expect to be able to do all the questions right away. If your first attack on a problem is unsuccessful, try a different approach. If you can, go away from the problem and come back to it a little later. "Review" questions deal with ideas from previous lessons, chapters, and sometimes even previous courses. "Exploration" questions extend the content of the lesson and deal with situations that usually have many possible ways of being approached. Answers to most of the odd-numbered questions are found in the Selected Answers section in the back of the book.

People who work with mathematics for their living routinely deal with problem situations that take more than a single evening to study and resolve. At the end of each chapter, there is a page or two of project ideas for extended work. Some projects ask you to gather data and look closely at everyday situations; others involve exploring certain mathematical ideas in more depth than usual; still others require that you use a computer to solve mathematical problems that would otherwise be inaccessible. Some projects may result in work suitable for exhibits or science fairs.

Also at the end of each chapter are a variety of features designed to help you master the material of that chapter. A chapter Summary reviews the major ideas in the chapter. It is followed by a list of the vocabulary and symbols you are expected to know. Then there is a Progress Self-Test for the chapter. We strongly recommend that you take this test and check your work with the solutions given in the back of the book. Finally, there is a Chapter Review, a set of review questions organized by the objectives of the chapter. You should do these also and check your answers with the answers for the odd-numbered questions found in the back of the book.

The authors have tried to make this book easy to understand and interesting to read. But we know that learning new content can at times be difficult. You can make your job easier if you ask friends or your teacher for help when ideas are not clear. We hope you enjoy this book and wish you much success.

CHAPTER 1 MATHEMATICAL ARGUMENTS

OBJECTIVES

The objectives listed here are the same as in the Chapter 1 Review on pages 74–78 of the student text. The Progress Self-Test on pages 72–73 and the tests in the Teacher's Resources File cover these objectives. For recommendations regarding the handling of this end-of-chapter material, see the notes in the margin on the corresponding pages of this Teacher's Edition.

OBJECTIVES FOR CHAPTER 1 (Organized into the SPUR Categories—Skills, Properties, Uses, and Representations)	Progress Self-Test Questions	Chapter Review Questions	Teacher's Resource File	
			Lesson Masters*	Chapter Test Forms A & B
SKILLS				
A Identify forms of logical statements.	1, 2, 13, 14	1–4	1-1, 1-5	1, 2, 10
B Write logically equivalent forms of statements.	4, 8, 11, 15	5–14	1-1, 1-3, 1-5	5, 8
C Write the negation of a logical statement.	6, 9	15–20	1-2, 1-3, 1-5	6, 7, 15
D Determine the truth value of a statement.	5, 23	21–25	1-1, 1-2, 1-3, 1-5	4, 13, 14, 16
PROPERTIES				
E Identify properties of logical statements.	12	26–32	1-1, 1-2, 1-3, 1-5	9
F Use substitution to verify specific statements.	7	33–34	1-1	3
G Determine whether arguments are valid or invalid.	16, 17	35–39	1-6, 1-8	11, 12, 17
H Use logic to prove or disprove statements.	19	40–49	1-1, 1-3, 1-6, 1-7	13, 14
USES				
I Determine the truth of quantified statements outside of mathematics.	3	50–56	1-1, 1-2, 1-3, 1-5	9
J Determine whether or not a logical argument outside of mathematics is valid.	18	57–60	1-6, 1-8	12
K Read and interpret computer programs using if-then statements.	10	61–62	1-3, 1-5	18
REPRESENTATIONS				
L Translate logic networks into logical expressions and input-output tables and determine output signals.	20, 21	63–65	1-4	19
M Write truth tables for logical expressions.	22	66–70	1-3, 1-5	20, 21

*** The masters are numbered to match the lessons.**

OVERVIEW 1

OVERVIEW ■ CHAPTER 1

One of the themes of this course is *mathematical thinking.* This chapter deals with one aspect of that thinking, formal logic, and applies that logic to computer logic networks, to the analysis and writing of proofs about integers, and to the logic of everyday thinking.

The content of Chapter 1 is a blend of new and familiar material. The formalization of logic probably will be new for most students. In geometry, students dealt with *if-then* statements and the contrapositive, inverse, and converse of such statements. However, they probably have not dealt with universal or existential statements and the relationship of these statements to conditionals. Nor have they applied these ideas to the examination of computer logic networks. Throughout the chapter, students will need to pay attention to detail; such attention is critical to success in computer science and in calculus.

Both English sentences and mathematical sentences are used to illustrate the logic. The mathematical sentences are taken from geometry and from standard precalculus topics such as properties of exponents, logarithms, and trigonometric functions. As such, they begin an implicit process of review that will be made explicit in Chapter 2. Do not take time to discuss exponential, logarithmic, and trigonometric functions at this point. Simply assign the questions containing these ideas and go over them.

Although in many schools the first week contains many shortened periods, it is important to try to adhere to a lesson-a-day pace. The beginning of the year establishes the rountines that will exist throughout the year. Going too slowly now will probably cause problems later in the year. Rather than spending more than one day on a lesson in this chapter, move at the suggested pace and then spend two or three days at the end of the chapter on the Progress Self-Test and the Chapter Review. At that point, students will have an overall picture of the goals of the chapter and will be in a better position to understand and relate the topics covered in the chapter.

PERSPECTIVES 1

PERSPECTIVES ■ CHAPTER 1

The Perspectives provide the rationale for the inclusion of topics or approaches, provide mathematical background, and make connections with other lessons and within UCSMP.

1-1

STATEMENTS

This lesson covers the distinction between a *statement* and a *sentence.* It introduces the terminology and symbolism of *universal statement* and *existential statement.*

The lesson's pedagogical objectives are: (1) to ease students into the book with content that is new yet somewhat familiar; (2) to encourage students to read the lessons; and (3) to introduce some rigor by means of questions that require attention to detail.

1-2

NEGATIONS

In this lesson, negations of simple, universal, and existential statements are considered. Students should be familiar with the first of these from previous courses, but they may not have studied the others. The ideas in this lesson are critical in determining the truth or falsity of statements and for an understanding of indirect reasoning.

1-3

AND AND *OR* AND DE MORGAN'S LAWS

De Morgan's Laws state that for all statements p and q:
$\sim(p \text{ and } q) \equiv (\sim p)$ or $(\sim q)$; and
$\sim(p \text{ or } q) \equiv (\sim p)$ and $(\sim q)$.
These laws are applied in the next lessons and have surprising application to a variety of topics in the book. The laws are deduced from definitions of *and, or,* and *not* by using truth tables.

1-4

COMPUTER LOGIC NETWORKS

This lesson applies *and, or,* and *not* statements to the study of computer logic networks. Students can see that the input-output tables for the AND, OR, and NOT logic gates are identical to the truth tables for *and, or,* and *not,* except that T has been replaced by 1 and F by 0. The truth tables for studying logical expressions are equivalent to input-output tables for studying networks of gates. In this lesson, students are expected to construct an input-output table for a given network and to determine the logical expression corresponding to a particular network. However, students are not expected to construct a logical network for a particular logical expression.

The kind of analysis introduced in this lesson is used in the construction of electrical circuits of all kinds, from those in large installations to those on computer chips.

1-5

IF-THEN STATEMENTS

Students probably have studied *if-then* statements before, particularly in geometry. In the conditional $p \Rightarrow q$, many texts call p the *hypothesis* and q the *conclusion.* We usually refer to p as the *antecedent* and to q as the *consequent,* reserving the word conclusion for the result of a logical argument.

The treatment of the truth values of a universal conditional and of its negation applies many of the ideas of the previous lessons.

1-6

VALID ARGUMENTS

Three valid argument forms are introduced in this lesson: Modus Ponens, or Law of Detachment; the Law of Transitivity; and Modus Tollens, or Law of Indirect Reasoning. (The principles behind the forms are more important than the names of the forms.) These three forms are the basis for most of the proofs students will encounter in this book. They are proved valid by using truth tables.

1-7

PROOFS ABOUT INTEGERS

One of the major objectives of this book is to have students become proficient in writing simple proofs. The many tryouts of these materials have shown that proofs about even and odd integers are not only accessible to students but are also good confidence-builders for future work. Proofs are written in paragraph form.

1-8

INVALID ARGUMENTS

Some ideas are difficult for students to understand until they have seen both instances and noninstances of them. Valid arguments generally are better understood when invalid arguments are discussed as well.

Three kinds of invalid arguments are considered: inverse error, converse error, and improper induction.

1-9

DIFFERENT KINDS OF REASONING

The last lesson in each chapter is a reading lesson. This lesson is intended to be covered, but its contents are not intended to be mastered nor are the ideas tested in this book. The reading lesson essentially provides an opportunity for students to take a nonthreatening look at extensions of the content of the chapter.

The reasoning used outside of mathematics differs in certain ways from that used within mathematics. Sometimes, thinking that is similar to invalid mathematical reasoning can produce useful conclusions in other areas, particularly areas where deduction is not possible. Therefore induction, diagnostic reasoning, and probabilistic reasoning are discussed in this lesson.

DAILY PACING CHART 1

DAILY PACING CHART ■ CHAPTER 1

Every chapter of UCSMP *Precalculus and Discrete Mathematics* includes lessons, a Progress Self-Test, and a Chapter Review. For optimal student performance, the self-test and review should be covered. (See *General Teaching Suggestions: Mastery* on page T35 of this Teacher's Edition.) By following the pace of the Full Course given here, students can complete the entire text by the end of the year. Students following the pace of the Minimal Course spend more time when there are quizzes and on the Chapter Review and will generally not complete all of the chapters in this text.

When chapters are covered in full (the recommendation of the authors), then students in the Minimal Course can cover 11 chapters of the book. For more information on pacing, see *General Teaching Suggestions: Pace* on page T34 of this Teacher's Edition.

DAY	MINIMAL COURSE	FULL COURSE
1	1-1	1-1
2	1-2	1-2
3	1-3	1-3
4	Quiz (TRF); Start 1-4.	Quiz (TRF); 1-4
5	Finish 1-4.	1-5
6	1-5	1-6
7	Quiz (TRF); Start 1-6.	Quiz (TRF); 1-7
8	Finish 1-6.	1-8
9	1-7	1-9
10	1-8	Progress Self-Test
11	1-9	Chapter Review
12	Progress Self-Test	Chapter Test (TRF)
13	Chapter Review	
14	Chapter Review	
15	Chapter Test (TRF)	

TESTING OPTIONS

TESTING OPTIONS

- ■ Quiz for Lessons 1-1 through 1-3
- ■ Quiz for Lessons 1-4 through 1-6
- ■ Chapter 1 Test, Form A
- ■ Chapter 1 Test, Form B

A Quiz and Test Writer is available for generating additional questions, additional quizzes, or additional forms of the Chapter Test.

INDIVIDUAL DIFFERENCES

PROVIDING FOR INDIVIDUAL DIFFERENCES

The student text is written for, and tested with, average students. It also has been successfully used with better and more poorly prepared students.

The Lesson Notes often include Error Analysis and Alternative Approach features to help you with those students who need more help. Students of all abilities often learn from their peers and may benefit from small group work referenced as appropriate throughout the Notes. A blackline Lesson Master (in the Teacher's Resource File), keyed to the chapter objectives, is provided for each lesson to allow more practice. (However, since it is important to keep up with the daily pace, you are not expected to use all of these masters. Again, refer to the suggestions for pacing on page T34.) Extension activities are provided in the Lesson Notes for those students who have completed the particular lesson in a shorter amount of time than is expected, even in the Full Course.

Logic

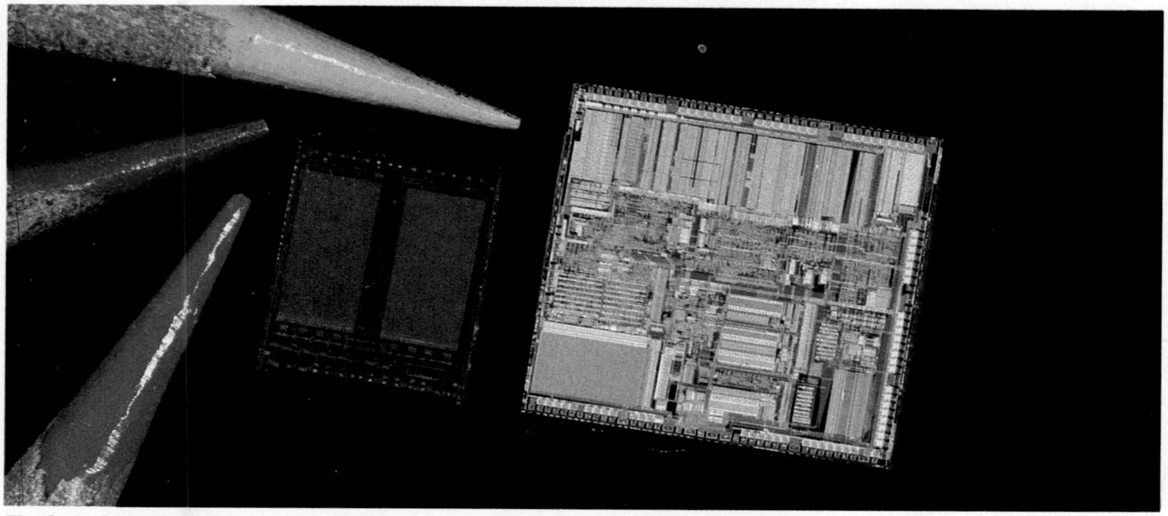

The flash chip (left) and the microprocessor chip are similar to those used in computer logic networks.

Mathematics is a beautiful subject whose results provide a sure, precise help in clear thinking. Yet sometimes something occurs that seems to violate these views. Consider the following "proof" that $1 + 1 = 1$.

Let $x = 1$ and $y = 1$. Then

$$y = x$$
$$-y^2 = -xy$$
$$x^2 - y^2 = x^2 - xy$$
$$(x + y)(x - y) = x(x - y)$$
$$x + y = x$$

Substitute back: $1 + 1 = 1.$

Either something is wrong or mathematics has failed! Finding the cause of the problem requires the examination of mathematical arguments and logic.

Distinguishing valid from invalid conclusions requires careful logical analysis. From the statement

There is a risk of an accident if nuclear power is used.

it is correct to conclude

To have no risk, you cannot use nuclear power.

but incorrect to conclude

There is no risk if you use another power source.

In this chapter on logic, you will touch upon many components of mathematical thinking: the use of careful language, the meaning of a generalization, and the criteria for a valid proof. You will also see how this thinking is employed in computer logic networks and everyday reasoning.

CHAPTER 1

We recommend 12–15 days be spent on this chapter: 9 to 11 days on the lessons and quizzes; 1 day for the Progress Self-Test; 1 or 2 days for the Chapter Review; and 1 day for a Chapter Test. (See the Daily Pacing Chart on page 3D.)

At the beginning of the school year, it is not uncommon to have some teaching periods cut short. Remember that students are expected to read, and do not delay moving on to a new lesson or eliminate certain questions simply because the material was not covered in class. The shortened period is a wonderful reason to expect students to read on their own.

USING PAGE 3

From geometry, students should know that a mathematical system consists of defined and undefined terms, postulates, and a system of logic that allows for the development and proof of theorems. The logic of mathematics is precise and demanding. The error in the proof is subtle; it is identified and explained in Lesson 1-1. Ask students to find the error before they read that lesson.

Have students read the "To the Student" pages and know what supplies they need for this class.

If students have not taken a previous UCSMP course, point out the four types of questions in each lesson, describe the end-of-chapter materials, and show them the location of the glossary and index.

Assign the reading and all questions of Lesson 1-1 for homework.

LESSON 1-1

RESOURCES
- Lesson Master 1-1

OBJECTIVES

Letter codes refer to the SPUR Objectives on page 3A.

A Identify forms of logical statements.
B Write logically equivalent forms of statements.
D Determine the truth value of a statement.
E Identify properties of logical statements.
F Use substitution to verify specific statements.
H Use logic to prove or disprove statements.
I Determine the truth of quantified statements outside of mathematics.

TEACHING NOTES

Students should be aware that they are expected to read the text. The Covering the Reading questions take the students through the lesson in order; if they have difficulty with any of these questions they should return to the lesson to find help. The Applying the Mathematics questions require students to extend the content of the lesson or to put ideas together in new ways.

It is important that students understand the function of the Review questions. These questions provide a continual review of material so that students continue to work on ideas after they have been presented. Continued work on the ideas of Chapter 1 occurs through the later chapters.

LESSON

1-1

Statements

natural numbers – 1,2,3,4,5

whole numbers – 0,1,2,3,4,5

integers – -3,-2,-1,0,1,2,3

rational numbers – 0,1,-7, $\frac{2}{3}$, $1\frac{9}{11}$, $-\frac{34}{10}$, $\sqrt{16}$, 9.618, 0.0004 — numbers represented as ratios of the form $\frac{a}{b}$, where a and b are integers and $b \neq 0$

real numbers – 0,1,-7, π, $\sqrt{5}$, 2.34 numbers that can be represented by decimals

The logic of *statements* helps to find and explain the mathematical problem on the previous page. In logic and mathematics, a **statement** is a sentence that is true or false but not both.

The sentence *1 + 1 = 2* is a statement because it is true.
The sentence *1 + 1 = 1* is a statement because it is false.
The sentence *x + 1 = 5x* is not a statement because it is true for some values of x, false for others.

In this book when a sentence is being discussed for its logical properties, it is written in *italics*. Any sentence may be represented by a single letter such as *p*. If the sentence contains a variable such as *x*, then it may be represented by a symbol such as *p(x)*. For instance, let *p(x)* be the sentence *x + 1 is an integer*. Notice that *p(x)* is not a statement because

p(3): 3 + 1 is an integer

is true while

p(0.5): 0.5 + 1 is an integer

is false.

If the words *for all integers x* are put at the beginning of the sentence *p(x)*, the result

For all integers x, x + 1 is an integer.

is a statement because it has a truth value (true). The phrase "for all" is called a **quantifier** and is represented in logic by the symbol $\forall$. With this symbol, the statement can be written

$\forall$ *integers x, x + 1 is an integer.*

Letting *I* stand for the set of integers, the statement can be rewritten even more briefly.

$\forall$ *x in I, x + 1 is an integer.*

This statement is true.

Statements asserting that a certain property holds for all elements in some set—such as the set of all integers or the set of all triangles—are called *universal statements*.

Definitions

Let S be a set and $p(x)$ a property that may or may not hold for any element x of S. A **universal statement** is a statement of the form

For all x in S, p(x),

or, symbolically,

$\forall$ *x in S, p(x).*

A universal statement is true if and only if $p(x)$ is true for every element x in S; otherwise it is **false.**

Universal statements are powerful because they assert that a certain property holds for *every* element in a set. Thus if you are given any particular element of the set, you can deduce that the property holds for that element. In formal logic, this is known as the **Law of Substitution**. For instance, since the sum of the measures of the angles of *any* triangle equals 180°, then the sum of the measures of the angles of the particular triangle ABC shown below is 180°.

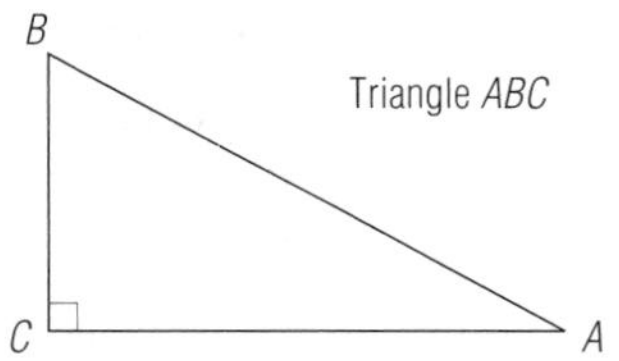

With these ideas, the mistake in the "proof" that $1 + 1 = 1$ can be explained. The argument began with statements assumed to be true: $x = 1$ and $y = 1$. From these statements, the following four statements can be deduced.

$$y = x$$
$$-y^2 = -xy$$
$$x^2 - y^2 = x^2 - xy$$
$$(x + y)(x - y) = x(x - y)$$

The next step is where the mistake arises. It assumes the justification: $\forall$ *real numbers d, if both sides of an equation are divided by d, then the quotients are equal.* But this universal statement is not true for all real numbers because it is not true when $d = 0$. When both sides are divided by $x - y$, it may not look like division by 0, but $x - y = 1 - 1 = 0$. The result

$$x + y = x$$

is seen to be false when the number 1 is resubstituted for x and y.

Answers to all odd-numbered questions are found in the back of the student text, except for (1) Covering the Reading questions (whose answers are literally contained in the lesson); (2) questions that ask for an opinion (and thus have no correct answer); and (3) the Exploration questions.

In the "proof" that $1 + 1 = 1$, you may wish to go through the justification for each step. First, both sides of the original equation are multiplied by $-y$ (Multiplication Property of Equality). Then x^2 is added to both sides (Addition Property of Equality). Each side is factored (Distributive Property). Next, both sides are divided by $x - y$ $\left(\text{or multiplied by } \frac{1}{x - y}\right)$. Then the Law of Substitution is used.

In mathematics, we make a distinction between a statement, which must be true or false, and other types of sentences. For instance, $3 + 2 = 5$ is a (true) statement, and $6 + 2 = 5$ is a (false) statement; $x + 2 = 5$ is not a statement. However, $x + 2 = 5$ can be made into a statement by specifying the value of x or by affixing appropriate quantification. *For all integers x, x + 2 = 5* is a (false) statement. *There exists an integer x such that x + 2 = 5* is a (true) statement.

Students may wonder why we use the notation $p(x)$ to represent a property that may or may not be true for any element x in some set. The letter p is used often because $p(x)$ is a "proposition in x." You may compare the $p(x)$ notation with the $f(x)$ notation common to functions. The symbol $p(x)$ indicates that for each particular x, $p(x)$ is a statement. Strictly speaking, p is a statement-valued function. The use of variables in logic parallels the use of variables in algebra. For instance, just as replacing x with 7 in the function $f(x) = x^2 - 3x + 7$ gives $f(7) = 35$, replacing x with 7 in the sentence $p(x)$: $x^2 > x + 1$ yields the statement $p(7)$: $7^2 > 7 + 1$.

Emphasize that it takes only one example to show that an existential statement is true. A universal statement can be shown to be true by example if the set S over which the statement is defined is finite, and the statement is verified for all elements of S. The proof is then called a *proof by exhaustion.*

The set S is critical to determining the truth or falsity of a universal statement. Consider the inequality $x^2 \geq x$. The universal statement

$\forall$ *natural numbers* x, $x^2 \geq x$

is true, whereas

$\forall$ *reals* x, $x^2 \geq x$

is false, with infinitely many counterexamples (for any real x such that $0 \leq x \leq 1$).

Small Group Work Have students work together to come up with as many different ways as possible to state the property "The square of a real number is nonnegative."

Making Connections As a preview of Lesson 1-2, students might try to add a row to the table at the top of page 8 with the heading "false: if there is a single x in S for which $p(x)$ is false; if there is no x in S for which $p(x)$ is true."

A universal statement may involve more than one variable. For instance, let $p(a, b)$ be the sentence $(a - b)(a + b) = a^2 - b^2$ and let R be the set of real numbers. Then the universal statement

For all real numbers a and b, $(a - b)(a + b) = a^2 - b^2$.

can be written symbolically as

$$\forall a \text{ and } b \text{ in } R,\ p(a, b).$$

Using the Law of Substitution, you can assert $p(100, 3)$, which is

$$(100 - 3)(100 + 3) = 100^2 - 3^2.$$

You can also assert $p(x, y)$, which is

$$\forall x \text{ and } y \text{ in } R,\ (x - y)(x + y) = x^2 - y^2.$$

You can also deduce other statements.

Example 1 Suppose $k \geq 0$. Use substitution to show that $(\sqrt{k+1} - \sqrt{k})(\sqrt{k+1} + \sqrt{k}) = 1$.

Solution The sentence

$$(a - b)(a + b) = a^2 - b^2$$

holds for all real numbers a and b. Hence, in particular, it holds when $a = \sqrt{k+1}$ and $b = \sqrt{k}$ because when $k \geq 0$, both $\sqrt{k}$ and $\sqrt{k+1}$ are real numbers. Substituting for a and b yields

$$(a - b)(a + b) = a^2 - b^2.$$
$$(\sqrt{k+1} - \sqrt{k})(\sqrt{k+1} + \sqrt{k}) = (\sqrt{k+1})^2 - (\sqrt{k})^2$$
$$= (k + 1) - k$$

Thus,

$$(\sqrt{k+1} - \sqrt{k})(\sqrt{k+1} + \sqrt{k}) = 1.$$

There is an English saying that "the exception proves the rule." This may be true for some rules, but it is certainly not true in mathematics. One exception, or *counterexample*, to a universal statement, however, proves that the universal statement is false.

Definition

Given a universal statement

$\forall x$ *in S, p(x).*

A value of x in S for which $p(x)$ is false is called a **counterexample** to the statement.

Example 2 Is the following statement true or false?
For all real numbers x, $x^4 \geq 1$.

Solution Here $p(x)$ is the sentence $x^4 \geq 1$ and S is the set of real numbers.

As a counterexample, take $x = \frac{1}{2}$.

$p(\frac{1}{2})$ is the statement $(\frac{1}{2})^4 \geq 1$. Because $\frac{1}{16} < 1$, $p(\frac{1}{2})$ is false and so the given statement is false.

Example 2 illustrates that: saying that the universal statement

$\forall$ *real numbers x,* $x^4 \geq 1$.

is false is equivalent to saying that

There is at least one real number x such that $x^4 < 1$.

is true. The phrase "there is" or "there exists" is another quantifier and is represented by the symbol $\exists$, read "there exists." The last statement above can be written

$\exists$ *a real number x such that* $x^4 < 1$.

Statements asserting that there exists at least one element of a set for which a certain property holds are called *existential statements*.

Definitions

Suppose that S is a set and that $p(x)$ is a property that may or may not hold for elements x of S. An **existential statement** is a statement of the form

There exists x in S such that p(x).

or, symbolically,

$\exists$ *x in S such that p(x).*

An existential statement is true if and only if $p(x)$ is true for at least one element x in S; otherwise it is **false.**

Example 3 Is this statement true or false?
$\exists$ *an integer n such that* $n^2 = 2$.

Solution The only numbers whose square is 2 are $\sqrt{2}$ and $-\sqrt{2}$. These are not integers. So there is no integer n such that $n^2 = 2$. Thus the statement is false.

To prove the truth of a universal statement, you must show that the statement is true for all members of the appropriate set S. However, to prove the truth of an existential statement, you only need to find one member of the appropriate set S for which the statement is true. For instance, the statement $\exists$ *a real number n such that* $n^2 = 2$ is true because $\sqrt{2}$ is a real number. (Of course, so is $-\sqrt{2}$.)

ADDITIONAL EXAMPLES
You may want to have students work in **small groups** for many of the additional examples given throughout this text.

1. Use substitution to show that for any real numbers x and y, $(x^2 - y^2)(x^2 + y^2) = x^4 - y^4$.
The sentence $(a - b) \cdot (a + b) = a^2 - b^2$ is true for all real numbers a and b. Hence, in particular, it is true when $a = x^2$ and $b = y^2$ for any real numbers x and y, since if x and y are real numbers, so are x^2 and y^2. Substituting for a and b yields

$$(a - b)(a + b) = a^2 - b^2$$
$$(x^2 - y^2)(x^2 + y^2) = (x^2)^2 - (y^2)^2 = x^4 - y^4.$$

Thus, $(x^2 - y^2)(x^2 + y^2) = x^4 - y^4$.

2. Is the following statement true or false?
For all real numbers x, $x^2 > x$.
$p(x)$ is the sentence $x^2 > x$, and S is the set of real numbers. As a counterexample, let $x = \frac{1}{2}$. $p(\frac{1}{2})$ is the statement $(\frac{1}{2})^2 > \frac{1}{2}$. Because $\frac{1}{4} < \frac{1}{2}$, this statement is false.

3. Is this statement true or false?
$\exists$ *a real number x such that* $x^2 > x$.
$p(x)$ is the sentence $x^2 > x$ and S is the set of real numbers. Because we have an existential statement, one example suffices to prove it. Consider $p(2)$: $2^2 > 2$. Since $4 > 2$ is true, the original statement is true.

Here is a summary of the properties of universal and existential statements.

statement	universal	existential
form	*For all x in S, p(x).*	*There exists an x in S such that p(x).*
quantifier	for all	there exists
symbol	$\forall$	$\exists$
true	if true for all values of x in S	if true for at least one value of x in S

In English it is possible to say one thing in several different ways. If you let S be the set of all triangles and $p(x)$ be the sentence *the sum of the measures of the angles of x is 180°*, then the universal statement

$$\forall x \text{ in } S,\ p(x)$$

can be phrased in many ways, including the following:

For all triangles x, the sum of the measures of the angles of x is 180°.
For every triangle x, the sum of the measures of the angles of x is 180°.
The sum of the measures of the angles of any triangle is 180°.
If x, y, and z are the measures of the angles of a triangle, then $x + y + z = 180°$.

Often, when a sentence of the form *If p(x) then q(x)* contains a variable x, the sentence is understood to be a claim about all values of x in some set S (which is determined by the context of the situation). For example, a sentence such as

$$\textit{If } \frac{x}{3} > 5, \textit{ then } x > 15.$$

would usually be intended to mean

$$\textit{For all real numbers } x, \textit{ if } \frac{x}{3} > 5, \textit{ then } x > 15.$$

Existential statements may also be written in a variety of different ways. Let S be the set of all real numbers and $p(x)$ be the sentence $x^2 = x$. The existential statement $\exists$ *x in S such that p(x)* can be written in the following equivalent ways.

There exists a real number x such that $x^2 = x$.
There is at least one real number x for which $x^2 = x$.
For some real number x, $x^2 = x$.
For some real numbers x, $x^2 = x$.
There is a real number which is equal to its square.

Some statements contain both "for all" and "there exists." For instance, the Additive Identity Property of Zero can be written

$\exists$ *a real number n such that* $\forall$ *real numbers x,* $x + n = x$.

8

The number n is, of course, 0. The existence of additive inverses uses the quantifiers in reverse order.

$\forall$ *real numbers x,* $\exists$ *a real number y such that* $x + y = 0$.

You know that $y = -x$.

Questions

Covering the Reading

These questions check your understanding of the reading. If you cannot answer a question you should go back to the reading to help you find an answer.

In 1–3, tell whether the sentence is a statement.

1. *There is exactly one solution to the equation* $2x = 5$. statement

2. *The equation* $x^2 = 1$ *has three solutions.* statement

3. $6x^2 = 7x - 1$. not a statement

4. *Multiple choice.* Which is the true statement?

(a) $\forall$ *real numbers x,* $x^9 = x$.

(b) $\exists$ *a real number x such that* $x^9 = x$. (b)

5. Let $p(x)$ be $x^3 - 1 = (x - 1)(x^2 + x + 1)$.

a. State $p(4)$ and tell whether $p(4)$ is true.

b. Write $p(\sqrt{a})$. $(\sqrt{a})^3 - 1 = (\sqrt{a} - 1)(a + \sqrt{a} + 1)$

c. Consider the sentence $\forall$ *real numbers x,* $x^3 - 1 = (x - 1)(x^2 + x + 1)$. Is this sentence a statement? Yes

a) $4^3 - 1 = (4 - 1)(4^2 + 4 + 1)$; True

In 6–8, identify the sentence as a universal statement, existential statement, or neither.

6. *There exists a bird b such that b cannot fly.* existential statement

7. *For all rivers r in North America, r flows into the Mississippi.* universal statement

8. *Chicago is the capital of Illinois.* neither

9. Name the two quantifiers introduced in this lesson. for all ($\forall$) and there exists ($\exists$)

10. The following statement is true:

$\forall$ *real numbers a and b,* $(a + b)^2 = a^2 + 2ab + b^2$.

Use substitution letting $a = \sqrt{75}$ and $b = \sqrt{12}$ to deduce that $(\sqrt{75} + \sqrt{12})^2 = 147$. See margin.

NOTES ON QUESTIONS

Question 10: Students should use **Example 1** as a model for answering this question.

Question 11: If $x \neq 0$, dividing both sides by x^2 yields $x > 1$, which is false for all real numbers $x \leq 1$. Encourage students to describe, when possible, the set of all counterexamples to a false universal statement.

Making Connections for Question 15: A table of values may be helpful. You may also suggest graphing $y = 2^x$ and $y = x^2$.

Questions 24-26: These questions help students prepare for analyzing functions in Chapter 2.

Question 27: The Exploration question may lead to lively class discussion. Here is a similar paradox:

The male barber in a town shaves all the men and only those men in the town who do not shave themselves. Is the sentence "The barber shaves himself." true or false? (If the barber shaves himself, then he does not belong to the group of men he shaves, and so he does not shave himself. If he does not shave himself, then he belongs to the group of men he shaves, and so he shaves himself. The contradictions obtained in each case show that this sentence is neither true nor false and hence is not a statement.)

ADDITIONAL ANSWERS

10. $(a + b)^2 = a^2 + 2ab + b^2$ holds for all real numbers a and b. Because $\sqrt{75}$ and $\sqrt{12}$ are real, the relation holds when $a = \sqrt{75}$ and $b = \sqrt{12}$. Hence, $(\sqrt{75} + \sqrt{12})^2 = (\sqrt{75})^2 + 2\sqrt{75} \cdot \sqrt{12} + (\sqrt{12})^2 = 75 + 60 + 12 = 147$.

FOLLOW-UP

MORE PRACTICE
For more questions on SPUR Objectives, use *Lesson Master 1-1,* shown on page 11.

PROJECTS
The projects for Chapter 1 are described on page 70. **Project 6** is related to **Question 27** of this lesson.

ADDITIONAL ANSWERS
15.c. for 0, 1, and all integers greater than or equal to 5

18. False; for example, if $a = b = 0$, then there is more than one solution for x, or if $a = 0$, $b = 1$, then there are no solutions for x.

19. False; for example, $\log\left(\frac{1}{10}\right) = -1$ and $-1 < 0$.

24.b.

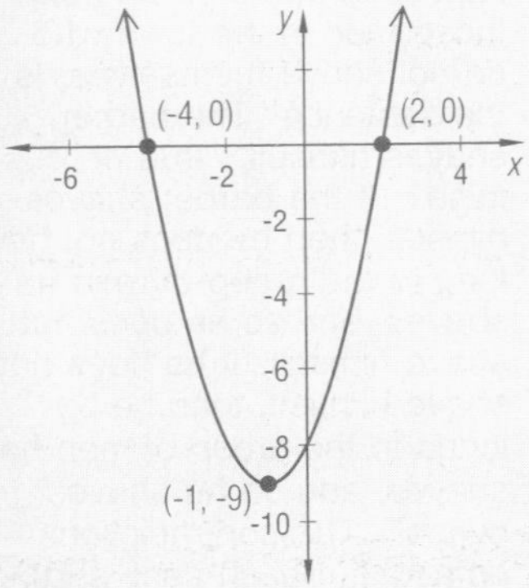

11. Find a counterexample to show that the following statement is false:
$\forall$ *real numbers* x, $x^3 > x^2$. **Sample: $x = 1$.**

12. Consider the statement
For all squares x, x is a rectangle.
Rewrite this statement in each form by filling in the blanks below.
a. $\forall$ _?_ x, _?_. **squares; x is a rectangle**
b. *All* _?_ *are* _?_. **squares; rectangles**
c. *Each* _?_ *is a* _?_. **square; rectangle**
d. *If* _?_ *then* _?_ **x is a square; x is a rectangle**

13. Consider the statement
There exists an even integer that is prime.
Write the statement in each form.
a. $\exists$ _?_ *x such that* _?_. **an even integer; x is prime**
b. *Some* _?_ *is* _?_. **even integer; prime**
c. *At least one* _?_ *is* _?_. **even integer; prime**

14. The Multiplicative Identity Property of one is: any real number times 1 equals that number. Write this property using the symbols $\forall$ and $\exists$.
$\exists$ a real number 1 such that $\forall$ real numbers x, $x \cdot 1 = x$.

Applying the Mathematics

These questions extend the content of the lesson. You should take your time, study the examples and explanations, try a variety of methods, and check your answers with the ones in the back of the book.

15. Let $p(x)$ be $2^x > x^2$.
a. Is $p(x)$ true when $x = 5$? **Yes**
b. Is $p(-1)$ true? **No**
c. For what integer values of x is $p(x)$ true? (You do not have to prove your answer is correct.) **See margin.**

16. Identify a property that is true for all students in your math class. Write this as a universal statement.
Sample: *All students in my math class are teenagers.*

17. Identify a property that is true for some, but not all, students in your math class. Write this property as an existential statement.
Sample: *There exists a student in my math class who owns a car.*

18. *True* or *false*? $\forall$ *real numbers a and b, the equation* $ax = b$ *has exactly one solution.* If false, find a counterexample. **See margin.**

19. *True* or *false*? Justify your answer.
$\forall$ *positive real numbers x,* $\log x > 0$. **See margin.**

20. Consider the statement
Everybody can fool somebody.
Rewrite this statement by filling in the blank below.
$\forall$ *people x,* $\exists$ *a person y such that* _?_. **x can fool y**

21. Find a counterexample to show that the following statement is false:
For all real numbers x there exists a real number y such that $x \cdot y = 1$. **Let $x = 0$; $\forall$ y, $0 \cdot y = 0 \neq 1$.**

Example 3 Negate the statement

Everyone trusts someone.

Solution 1

Here is a solution in words.

p: *Everyone trusts someone.*

~p: *There is someone who does not trust anyone.*

Solution 2

Here is a more formal analysis.

p: ∀ *people x*, ∃ *a person y such that x trusts y.*

~p: ∃ *a person x such that* ∀ *people y, x does not trust y.*

Observe that, in general, the negation of a statement can be generated by reading the statement from left to right and changing ∀ to ∃, changing ∃ to ∀, and changing *p(x, y)* to *not p(x, y)*. The words *such that* are deleted when ∃ is changed to ∀ and are added when ∀ is changed to ∃.

Questions

Covering the Mathematics

1. If a given statement is true, then its negation is _?_. False

2. *True* or *false*? The negation of a universal statement is another universal statement. False

3. *True* or *false*? The negation of

   For all states s, s requires a driver to have a valid driver's license.

   is

   There exists a state s such that s requires a driver to have a valid driver's license. False

In 4–6, write a negation for the statement.

4. *Every person can drive a car.*
   ∃ *a person who cannot drive a car.*

5. *All fractions are rational numbers.*
   ∃ *a fraction which is not a rational number.*

6. ∃ *a real number x such that* $\sin x = \cos x$.
   ∀ *real numbers x,* $\sin x \neq \cos x$.

7. For any statement *p*, *not (not p)* is equivalent to _?_. *p*

8. *Multiple choice.* The negation of
   Some quadratic equations have three solutions
   is
   (a) *Some quadratic equations do not have three solutions.*
   (b) *All quadratic equations have three solutions.*
   (c) *No quadratic equations have three solutions.*
   (d) *There exists a quadratic equation with three solutions.* (c)

2. Write the negation of the statement *Some rectangles are squares.*
p: Some rectangles are squares* is equivalent to *p:* ∃ *a rectangle x, such that x is a square.* Thus, the negation is *~p:* ∀ *rectangles x, x is not a square,* which means that there is no rectangle that is also a square. Thus *~p: No rectangle is a square.

3. Negate the statement *Everybody loves something.*
Solution 1: In the vernacular: *p: Everybody loves something. ~p: There is someone who does not love anything.*
Solution 2: More formally: *p:* ∀ *people x,* ∃ *a thing y such that x loves y.* *~p:* ∃ *a person x such that* ∀ *things y, x does not love y.*

NOTES ON QUESTIONS
Questions 4-6: You may wish to ask students whether the statement or its negation is true. This emphasizes that exactly one of them is true.

NOTES ON QUESTIONS

Question 15: Some students may say that the negation of $n < 11$ is $n > 11$, ignoring $n = 11$. Write on the chalkboard $n < 11$, or $n = 11$, or $n > 11$ by the *Law of Trichotomy* and say that to find the negation of $n < 11$, cover $n < 11$. The negation is what remains: $n = 11$ or $n > 11$, which may be written as $n \geq 11$.

Question 20: Even though choice (b) is true for an infinite number of values for *t*, it is not always true.

ADDITIONAL ANSWERS

10. ***∀ functions f, ∃ real numbers a and b such that $f(a + b) \neq f(a) + f(b)$.***

11.a. people x; a person y; x loves y
b. ∃ a person x such that ∀ people y, x does not love y.

12.a. ∃ a real number x such that $2x + 4 \leq 0$.
b. the negation

13.a. ∃ a man who is not mortal.
b. the given statement

15.a. ∃ n in S such that $n \geq 11$.
b. the negation; 11 is in S and 11 = 11.

16.a. ∀ even integers m, m is not in S.
b. the negation; S contains no even integers.

17.a. ∃ a real number x such that ∀ real numbers y, $\tan x \neq y$.
b. the negation; for $x = \frac{\pi}{2}$, $\tan \frac{\pi}{2}$ is undefined so that ∀ real numbers y, $\tan \frac{\pi}{2} \neq y$.

9. To negate the existential statement ∃ *x in S and y in T such that p(x, y)*, change ∃ to _?_, delete the words *such that*, and change *p(x, y)* to _?_. **∀, not p(x, y)**

10. Write the negation of the following statement:

∃ *a function f such that* ∀ *real numbers a and b*, $f(a + b) = f(a) + f(b)$. **See margin.**

11. Consider the statement
Everybody loves somebody.
a. Write this statement in the form ∀ _?_, ∃ _?_ *such that* _?_.
b. Write the negation of this statement.
See margin.

Applying the Mathematics

In 12 and 13, a statement is given.
a. Write the negation of the statement.
b. Which is true: the given statement or its negation?

12. *p: For all real numbers x,* $2x + 4 > 0$. **See margin.**

13. *q: All men are mortal.* (Socrates) **See margin.**

14. *Multiple choice.* Give the negation of
No one under 21 can legally buy alcohol in Illinois.
(a) *All people under 21 cannot legally buy alcohol in Illinois.*
(b) *All people under 21 can legally buy alcohol in Illinois.*
(c) *Some people under 21 cannot legally buy alcohol in Illinois.*
(d) *Some people under 21 can legally buy alcohol in Illinois.*
(e) *No one under 21 cannot legally buy alcohol in Illinois.* **(d)**

In 15 and 16, $S = \{1, 3, 5, 7, 9, 11\}$. A statement is given.
a. Write the negation of the statement.
b. Which is true: the given statement or its negation? Justify your answer.

15. *r: For all n in S,* $n < 11$. **See margin.**

16. *q: There exists an even integer m such that m is in S.* **See margin.**

17. Consider the statement
∀ *real numbers x*, ∃ *a real number y such that* $\tan x = y$.
a. Write the negation of this statement.
b. Which is true: the statement or its negation? Justify your answer.
See margin.

tan 90 = ∞ so y does not exist

18. *Multiple choice.* Consider the statement
△ABC is isosceles.
Which of the following is the negation of this statement?
(a) *△ABC is a right triangle.*
(b) *△ABC is a scalene triangle.*
(c) *△ABC is an equilateral triangle.*
(d) *△ABC is an obtuse triangle.* **(b)**

Review

19. Find the flaw in the following argument.
Let $x = 2$ and $y = 2$.
Then $x^2 + y = y^2 + x$.
So $x^2 - y^2 = x - y$.
So $(x + y)(x - y) = x - y$.
So $x + y = 1$.
But because $x = y = 2$, $x + y = 4$. *(Lesson 1-1)* See margin.

20. *Multiple choice.* If t is a real number, which of the following statements is true? *(Lesson 1-1)*
(a) $\forall$ *t, cos t = 0.*
(b) $\exists$ *t such that cos t = 0.* (b)

21. Refer to the information in the table below.
At King High School extracurricular activities are offered in four categories: sports teams, foreign language clubs, fine arts clubs, and academic clubs. The activities of a sample of four students are described below.

	sports teams						foreign language clubs				fine arts clubs			academic clubs		
	football	basketball	baseball	soccer	track	swimming	French	German	Russian	Spanish	drama	chorus	band	math	debate	science
Aiko						√		√						√		
Dave	√			√							√			√	√	
Karen					√		√						√	√		
Tomas			√						√					√	√	√

Determine whether each of the following statements is true or false for the students in this sample. Justify your answer. *(Lesson 1-1)*

a. $\forall$ *students s in the sample, s participates on some sports team.*
b. *Some foreign language club has no members from the sample.*
c. $\exists$ *an academic club c such that* $\forall$ *students s in the sample, s is a member of c.*
d. *For every category there is some club which has no members in the sample.*
e. *Every student in the sample is a member of some club in each category.*
See margin.

22. *Multiple choice.* You learned in geometry that
The measure of an angle inscribed in a semicircle is 90°.
Which of the following statements is not equivalent to the one above?
(a) *All angles which are inscribed in a semicircle have a measure of 90°.*
(b) *If* $\angle A$ *is an angle inscribed in a semicircle then* $m\angle A = 90°$.
(c) *For all* $\angle A$, *if* $\angle A$ *is inscribed in a semicircle, then* $m\angle A = 90°$.
(d) *Every angle which is inscribed in a semicircle has a measure of 90°.*
(e) All of the above are equivalent. *(Lesson 1-1)* (e)

19. The flaw is in going from the fourth to the fifth line. Since x − y = 0, you cannot divide both sides of the equation by x − y.

21.a. True, every student in the sample is on at least one sports team.
b. True, no student from the sample is in the Spanish club.
c. True, every student in the sample is in the math club.
d. False, each academic club has at least one member in the sample.
e. False, for example, Dave is not in a foreign language club.

NAME ______

LESSON **MASTER 1-2**
QUESTIONS ON **SPUR** OBJECTIVES

■SKILLS *Objective C (See pages 74–78 for objectives.)*
In 1–4, write the negation of the statement.

1. *All diamonds are made of carbon.*
There exists a diamond that is not made of carbon.

2. $\forall$ *real numbers x, -3 < x.*
$\exists$ x such that -3 ≥ x.

3. *Some equations have no solutions.*
All equations have at least one solution.

4. $\exists$ *a real number x such that* $\forall$ *real numbers y, x is a factor of y.*
$\forall$ real numbers x, $\exists$ a real number y such that x is not a factor of y.

■SKILLS *Objective D*
5. Let *p: No parallelograms have more than two lines of symmetry.*
a. Write *p.* There is a parallelogram with more than two lines of symmetry.
b. Which is true, *p* or ~*p*? ~p

■PROPERTIES *Objective E*
6. *True or false?* The negation of a statement *For all x in S, p(x)* is the statement *For no x in S, p(x).* false
7. Let *p:* $\exists$ *y in R such that q(y).* Write ~*p.* $\forall$y in R, ~q(y)
8. If a statement is false, what can be said about its negation? The negation is true.
9. The negation of an existential statement is an universal statement.

■USES *Objective I*
In 10 and 11, tell which is true, the statement or its negation.
10. $\forall$ *months m, the name for m begins with a consonant.* negation
11. $\forall$ *states s,* $\exists$ *a city c such that c is the capital of s.* statement

 3

NOTES ON QUESTIONS

Question 25: You may need to remind students of the basic definition for converting from logarithmic form to exponential form: $\log_b x = y \Leftrightarrow x = b^y$. Note that implicitly this statement is universal. It is a shorthand form of *∀ positive real numbers x and ∀ real numbers y,* $\log_b x = y \Leftrightarrow x = b^y$.

Question 26: Remind students that while it takes only one example to show Mr. Mailer's claim is false, examples themselves are not sufficient to show it is true.

FOLLOW UP

MORE PRACTICE

For more questions on SPUR Objectives, use *Lesson Master 1-2,* shown on page 17.

23. Consider the true statement

$\forall$ *nonnegative real numbers x,* $\sqrt{x^2} = x$.

Find a counterexample to show that the given statement is false if the condition that x is nonnegative is removed. *(Lesson 1-1, Previous course)*
Sample: $\sqrt{(-1)^2} = 1 \neq -1$

24. Consider the expression $(2x + 3)^2$.

a. What universal statement could you use to expand it?

b. Expand the expression. *(Lesson 1-1, Previous course)*

See below.

25. Evaluate without using a calculator. *(Previous course)*

a. $\log_3 81$ 4 **b.** $\log_5 \frac{1}{125}$ -3 **c.** $\log 1$ 0

Exploration

26. Mr. Mailer claims he can put postage on any letter requiring 8 cents or more postage using a combination of 3-cent and 5-cent stamps.

a. Write his claim in the form

$\forall$ _?_, $\exists$ _?_ *such that* _?_.

b. Check that Mr. Mailer's claim is correct for mail requiring 8¢, 9¢, 10¢, and 11¢.

c. Explore the truth or falsity of Mr. Mailer's claim by checking it for other postal charges. Write a convincing argument to show that his claim is either true or false.

See below.

24. a. *∀ real numbers a and b,* $(a + b)^2 = a^2 + 2ab + b^2$.

b. $(2x + 3)^2 = 4x^2 + 12x + 9$

26. a. *∀ postal charges P ≥ 8 cents, ∃ n and m nonnegative integers such that 3n + 5m = P.*

b. $8 = 5 + 3$; $9 = 3 \cdot 3$; $10 = 5 \cdot 2$; $11 = 3 \cdot 2 + 5$

c. It is true. You can get 8, 9, and 10 cents. By adding a 3 cent stamp to 8, 9, and 10 cents, you can get 11, 12, and 13 cents. By adding more 3 cents stamps you can get all charges over 8 cents.

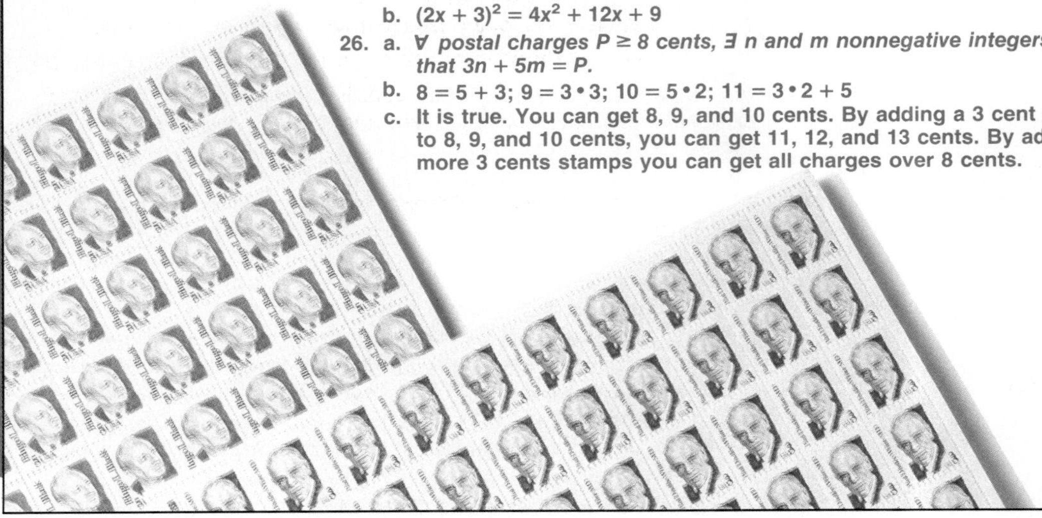

LESSON

1-3

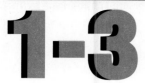

And and *Or* and De Morgan's Laws

In mathematics and in ordinary language, statements are often combined using the words *and, or, not, either . . . or, neither . . . nor*. The following excerpt from the instructions for Schedule SE (Form 1040) of the 1989 Federal Income Tax form illustrates such combinations:

''Generally, you may use this part only if:

A Your **gross** farm income was not more than \$2,400, **or**

B Your **gross** farm income was more than \$2,400 and your **net** farm profits were **less** than \$1,600, **or**

C Your **net** nonfarm profits were less than \$1,600 and also **less** than two-thirds (2/3) of your **gross** farm income.''

Suppose that we use the following notation for the simple statements in these instructions:

f: *your gross farm income was more than \$2400*
p: *your net farm profits were less than \$1600*
n: *your net nonfarm profits were less than \$1600*
g: *your net nonfarm profits were less than two-thirds of your gross farm income*

In this notation, the income tax instructions can be expressed as

You may use this part only if ((not f) or (f and p) or (n and g)).

No wonder so many people are confused by tax forms! Fortunately, not all sentences containing *and*, *or*, and *not* are as complicated as this one.

Example 1 Express the following inequalities by writing out each implied *and*, *or*, and *not*.

a. $x \leq 5$ **b.** $y \not\leq -1$ **c.** $-3 < z < 4$

Solution

a. $x < 5$ or $x = 5$.

b. *not* $(y \leq -1)$, which is equivalent to *not* $(y < -1$ *or* $y = -1)$. As you know, another expression for this is $y > -1$.

c. $-3 < z$ and $z < 4$. The double inequality means that both inequalities are satisfied.

The truth values of sentences that combine two statements with *and* or *or* are as follows.

Definitions

The sentence

p ***and*** q

is true when, and only when, both p and q are true. The sentence

p ***or*** q

is true in all cases except when both p and q are false.

LESSON 1-3

RESOURCES

- Lesson Master 1-3
- Quiz for Lessons 1-1 through 1-3
- Teaching Aid 1 can be used with **Example 2**.
- Computer Master 1

OBJECTIVES

Letter codes refer to the SPUR Objectives on page 3A.

B Write logically equivalent forms of statements.
C Write the negation of a logical statement.
D Determine the truth value of a statement.
E Identify properties of logical statements.
H Use logic to prove or disprove statements.
I Determine the truth of quantified statements outside of mathematics.
K Read and interpret computer programs using if-then statements.
M Write truth tables for logical expressions.

TEACHING NOTES

So that students can see the use of *and* and *or* in an everyday situation, this lesson begins with an excerpt from a Schedule SE1040 tax form. Ask students to examine the form to answer the following questions.

(1) Farmer Jones has a gross farm income of \$2000. May he use this part of the form? Why or why not? (Yes, his gross farm income is less than \$2400.)

(2) Farmer McDonald was correct to use this part. What can you conclude about Farmer McDonald's gross farm income? (any of A, B, or C)

The distinction between the *inclusive or* and the *exclusive or* may motivate an interesting discussion. In ordinary language, the context assists in deciding which use of *or* is intended. In mathematics, ambiguity needs to be avoided and we must state which use of *or* is meant. Some calculators have both an *or* and an *xor* key.

The result of **Example 2** provides the basis for negating statements containing *and* or *or.* Each such statement is negated by changing *and* to *or* and vice versa, and negating the components of the statement. Here is another example.

Negate the following statement: *There are 18,000 movie theaters in the U.S., and there are 151,280 movie theaters in the Soviet Union.* (negation: *There are not 18,000 movie theaters in the U.S., or there are not 151,280 movie theaters in the Soviet Union.)*

Just as two algebraic expressions are equivalent if they yield the same values for all values of the variables, logical expressions are equivalent if and only if they have the same truth values for all values of their statement variables. No matter what specific statements are substituted for the statement variables in equivalent expressions, the resulting statements will either both be true or they will both be false. The recognition of equivalent expressions is not only useful in making some complicated expressions more understandable, it is also vital in the design of computer logic networks (see Lesson 1-4).

These truth values in the definitions of *and* and *or* are summarized by the following truth tables.

Truth Table for *and*

p	*q*	*p and q*
T	T	T
T	F	F
F	T	F
F	F	F

Truth Table for *or*

p	*q*	*p or q*
T	T	T
T	F	T
F	T	T
F	F	F

In ordinary language, we sometimes use the **exclusive or** (one or the other but not both) and sometimes the **inclusive or** (one or the other or both). For example, if a restaurant menu states that "Coffee or tea is free with any sandwich order" you should probably interpret that to mean that you can have either coffee or tea free with your sandwich but that you would pay extra if you wanted both. That is the *exclusive or*. In contrast, if your waiter asked "Cream or sugar?", you would normally take that to mean that he or she is offering cream or sugar or both. That is the *inclusive or*. In mathematics, *or* is always the *inclusive or*.

A **logical expression** is a formula in which variables representing statements are combined in an unambiguous way with *and*, *or*, *not*, or *if-then*. For example

p or (q and r)   *(p or q) and r*

are logical expressions. On the other hand, the formula

p or q and r

is not a logical expression because it is not clear whether it means

p or (q and r), or *(p or q) and r*.

If two logical expressions have the same truth values for all substitutions of statements for their statement variables, we say that the two expressions are **logically equivalent**. The symbol $\equiv$ is sometimes used to denote logical equivalence. For example, $\sim(\sim p) \equiv p$.

Example 2 Use a truth table to prove that *not (p and q)* $\equiv$ *(not p) or (not q).*

Solution Set up a truth table in which the first two columns list the truth values for *p* and *q* and the remaining columns give the truth values for *p and q, not(p and q), not p, not q,* and *(not p) or (not q)*. We show how to fill in the columns in several steps.

p	*q*	*p and q*	*not (p and q)*	*not p*	*not q*	*(not p) or (not q)*
T	T					
T	F					
F	T					
F	F					

LESSON

1-4

Computer Logic Networks

Think of the light in the ceiling of a 2-door car. Unless you have done something with an inside switch, that light will be on when either door is open and off when both are closed. The following table tells whether the light is on or off by the positions of the doors.

Door 1	Door 2	Light
open	open	on
open	closed	on
closed	open	on
closed	closed	off

Does this table look familiar? It may, for it has the same structure as the truth table for *or*, with the words *open* and *on* replacing T and the words *closed* and *off* replacing F. This suggests that logic and electronics are intimately related, and they are. Every time that you press a button on a calculator, or type in a command to a computer, or flip a light switch, you are activating the first *logic gate* in an electronic system. Microprocessors may contain a million logic gates. These gates are interconnected so that they transmit electrical current to produce outputs such as the displays you see after inputting various keystrokes on a calculator.

Quiz 1.1-1.4 Friday

As pieces of computer hardware, logic gates can take a variety of forms. It is not necessary for you to know how these gates are physically constructed in order to understand how they function. Instead, you can think of logic gates as electrical devices with input and output wires. A model of a logic gate is illustrated below.

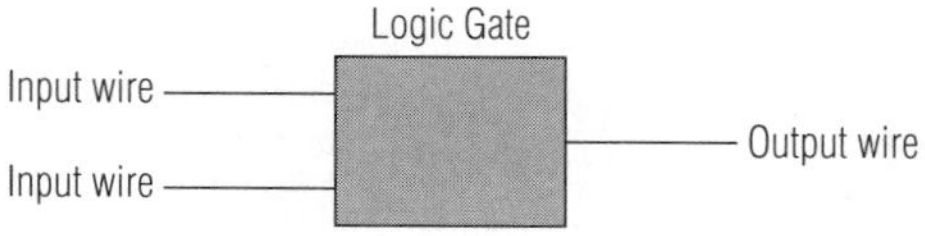

The input and output wires carry electrical signals that are in one of two mutually exclusive states. You can think of these two states as *current ON* or *current OFF* or as *high voltage* or *low voltage*. For convenience, we will refer to these signal states as ***1*** or ***0***. They correspond to *True* and *False*, respectively, in logic. True is 1 is ON; False is 0 is OFF.

A logic gate acts on the input signals that it receives to produce an output signal (1 or 0). Consequently, you will know exactly how a logic gate functions once you know the output signal state that is produced for every possible combination of input signal states. This information can be listed conveniently in an **input-output table** for the logic gate. At the right is such a table for a different logic gate than that for the car door situation.

Input		Output
p	q	
1	1	0
1	0	1
0	1	1
0	0	1

LESSON 1-4

RESOURCES

- Lesson Master 1-4
- Teaching Aid 2 summarizes the operation of logic gates and their inputs and outputs.
- Teaching Aid 3 can be used with **Questions 4-6**.
- Computer Master 2

OBJECTIVE

Letter code refers to the SPUR Objectives on page 3A.

L Translate logic networks into logical expressions and input-output tables and determine output signals.

TEACHING NOTES

The symbolic shapes used for the AND, OR, and NOT logic gates are standard in computer science, although in practice, the words inside the shapes are omitted.

Students may wonder why there is concern with determining whether or not two networks are functionally equivalent. Explain that cost concerns usually require that a network be built for as low a cost as possible while still performing a specific function. For example, one part of the absorption law states that *p or* (*p and q*) $\equiv$ *p*. This law permits the replacement of a part of the network represented by *p or* (*p and q*) by the single gate *p*. Such a replacement decreases both network cost and size. The process of reducing networks is studied extensively by computer design engineers.

Depending on the interest and ability of your students, you may want to discuss the axioms for a Boolean algebra with the entire class. In such an algebra defined on a set S, there are three operations, denoted by $\wedge$, $\vee$, and $'$ that satisfy the following:
For all a, b, and c in S:

Closure
$a \wedge b$, $a \vee b$, and a' are in S.

Associativity
$a \wedge (b \wedge c) = (a \wedge b) \wedge c$
$a \vee (b \vee c) = (a \vee b) \vee c$

Commutativity
$a \wedge b = b \wedge a \quad a \vee b = b \vee a$

Distributivity
$a \wedge (b \vee c) = (a \wedge b) \vee (a \wedge c)$
$a \vee (b \wedge c) = (a \vee b) \wedge (a \vee c)$

Identities There are distinct and unique elements 0 and 1 in S such that
$a \wedge 1 = a \quad a \vee 0 = a$

Complements For each a in S there exists one and only one a' in S such that
$a \wedge a' = 0 \quad a \vee a' = 1$

Notice that the following correspondences can be made between a general Boolean algebra and the specific Boolean algebra of statements.

general	specific
elements in S	statements
$\wedge$	*and*
$\vee$	*or*
$'$	*not*

One consequence of the above axioms is the following theorem.

Idempotent Theorem:
For all elements a in a Boolean algebra, $a \wedge a = a$.
Here is the proof.

$a = a \wedge 1$	Identity
$= a \wedge (a \vee a')$	Comp. Prop.
$= (a \wedge a) \vee (a \wedge a')$	Distrib. Prop.
$= (a \wedge a) \vee 0$	Comp. Prop.
$= a \wedge a$	Identity

You might ask students to supply the reasons.

The table has two input columns labeled p and q, so you can picture the logic gate as

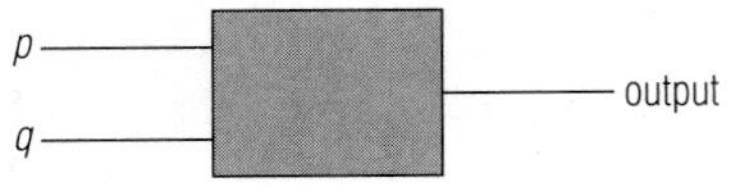

The table tells you that the logic gate will produce an output signal of 0 when the input wire p carries a signal of 1 and the input wire q carries a signal of 1. For any of the other three possible combinations of input signal states, the table tells you that the logic gate will produce an output signal of 1. Thus the input-output table tells you exactly what the logic gate will do with any possible combination of input signals.

The following three logic gates are so basic that they are given special standard symbols.

Gate Name — **Gate Symbol** — **Input-Output Table**

NOT

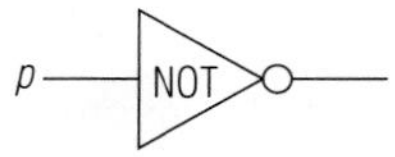

p	Output
1	0
0	1

AND

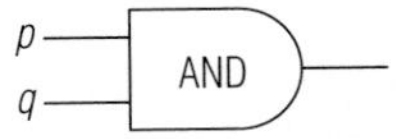

p	q	Output
1	1	1
1	0	0
0	1	0
0	0	0

OR

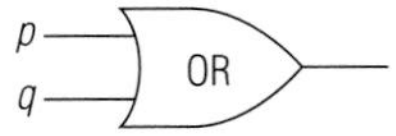

p	q	Output
1	1	1
1	0	1
0	1	1
0	0	0

Notice that
a. the NOT gate output signal is 0 if the input signal is 1, and that the output signal is 1 if the input signal is 0;
b. the AND gate output signal is 1 if the input wire p and the input wire q carry a signal of 1. Otherwise, the output signal is 0;
c. the OR gate output signal is 1 if the input wire p or the input wire q (or both) carries a signal of 1. Otherwise, the output signal is 0.

If you have not noticed this already, you should see that the input-output tables for the NOT, AND, and OR logic gates are essentially the same as the truth tables for *not p*, *p and q*, and *p or q*. The only differences are notational: 1 and 0 are used in place of T and F, respectively, and the last column is labeled with the word *output* instead of with the appropriate logical expression.

NOT, AND, and OR gates are usually connected in such a way that the output signals from some of the gates become input signals for other gates. This is called a *network* of logic gates. The relationship between the input-output tables for NOT, AND, and OR gates and the truth tables for *not p, p and q,* and *p or q* means that for each network of logic gates, you can construct a corresponding logical expression. Also, you can use logic to determine the action of any network.

Example 1 Construct an input-output table that corresponds to the following network:

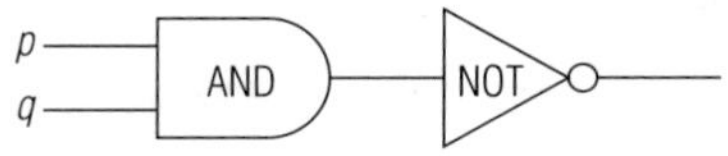

Solution Because the network has two input wires labeled *p, q*, the input-output table should list all possible combinations of signal states for *p, q*.

p	*q*	
1	1	
1	0	
0	1	
0	0	

Input signals first go to an AND gate whose output goes to a NOT gate. Tracking each pair of input signals through the network allows you to determine the appropriate output signal. If *p* is 1 and *q* is 1, then the output from the AND gate is also 1. The NOT gate reverses this value and gives a final output of 0. The other rows of the table are completed in a similar manner.

p	*q*	*p AND q*	output of network *NOT (p AND q)*
1	1	1	0
1	0	0	1
0	1	0	1
0	0	0	1

Check The circuit should represent the truth table for *not (p and q)*.

p	*q*	*p and q*	*not (p and q)*
T	T	T	F
T	F	F	T
F	T	F	T
F	F	F	T

ADDITIONAL EXAMPLES

1. Construct an input-output table that corresponds to the following network.

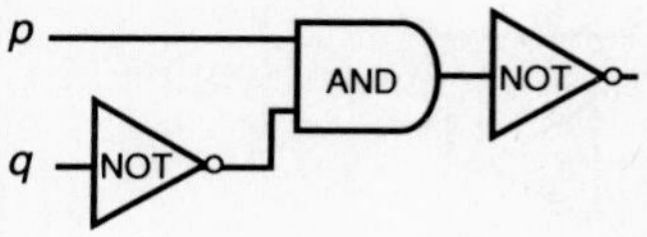

Because the network has two input wires labeled *p* and *q*, the input-output table lists all possible combinations of signal states for *p* and *q*.

p	*q*	*not q*	*p and (not q)*	*not (p and (not q))*
1	1	0	0	1
1	0	1	1	0
0	1	0	0	1
0	0	1	0	1

ADDITIONAL EXAMPLES
2. Verify the network version of Lesson 1-3, Additional Example 2.

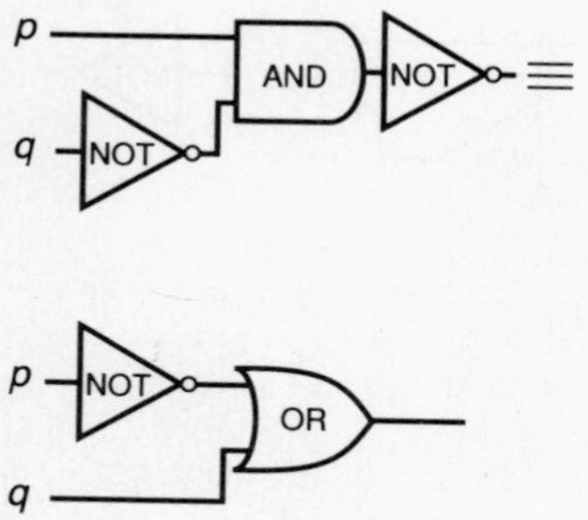

In Additional Example 1 on page 27, we did the input-output table for the network at the top. The table for the equivalent network appears below.

p	*q*	*not p*	*(not p) or q*
1	1	0	1
1	0	0	0
0	1	1	1
0	0	1	1

Because both networks have the same outputs for the given inputs (last columns are identical), the networks are equivalent.

3. Find the logical expression that corresponds to the following network.

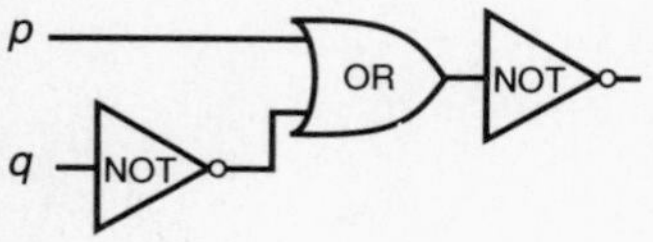

not (p or (not q))

Recall that two logical expressions are logically equivalent if their truth values are always the same. Similarly, if the output columns of the input-output tables for two networks are identical, then the networks produce the same output for each combination of input signals, and so the networks are **functionally equivalent**. The symbol ≡ can be used between functionally equivalent networks just as it can be used between logically equivalent expressions.

Example 2 illustrates the use of functionally equivalent networks to represent one of DeMorgan's Laws in network terms:

$$not\ (p\ and\ q) \equiv (not\ p)\ or\ (not\ q).$$

Example 2 Verify the network version of the given Law of De Morgan:

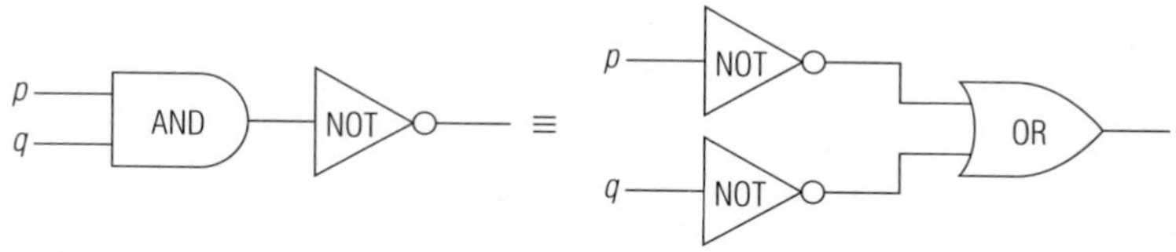

Solution Build the input-output table for each side of the equivalence to show that each possible combination of input signals gives the same output. Here is the table for the right side.

p	*q*	*NOT P*	*NOT q*	*(NOT p) OR (NOT q)*
1	1	0	0	0
1	0	0	1	1
0	1	1	0	1
0	0	1	1	1

The values in the rightmost column are identical to those computed in Example 1 for the network on the left.

In Examples 1 and 2, we started with a network and constructed the corresponding input-output table. It is also possible to start with a network and find the corresponding logical expression. Some people do this by tracing the network from output back to input rather than working from input to output.

Example 3 Find the logical expression that corresponds to the following network.

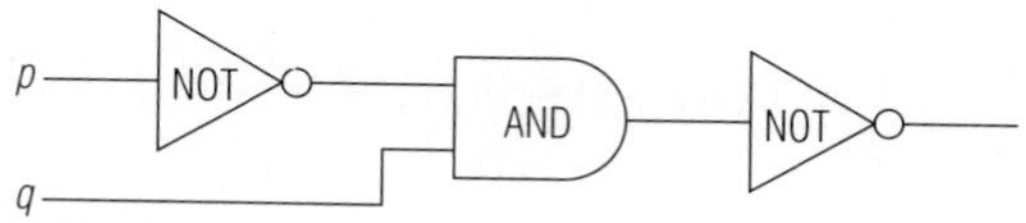

Solution 1 Read the network from right to left and build the expression as you go. The last NOT gate reverses the output of the part prior to it. So,

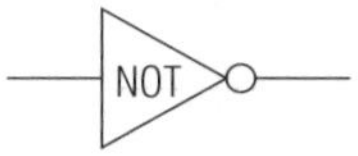

corresponds to *not( ).*

Also

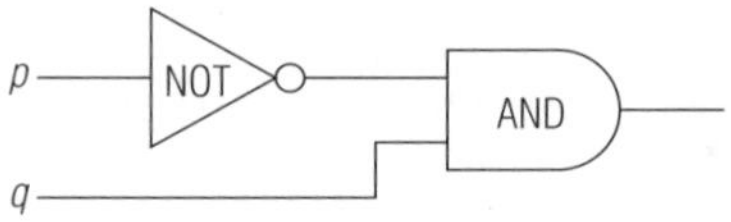

corresponds to *(not p) and q*

Thus the entire network corresponds to *not ((not p) and q)*.

Solution 2 Read the network from left to right, putting each previous step in parentheses.

The first component is *not p*
The second is *(not p) and q*
The last is *not ((not p) and q)*

George Boole

The algebraic properties of *and*, *or*, and *not* were first stated and explored by George Boole (1815-1864) in his book *An Investigation of the Laws of Thought,* which was published in 1853. Although Boole came from a poor family and had only three years of formal education, he went on to become a brilliant scholar who not only contributed new knowledge to several fields in mathematics but also taught Latin and Greek.

Boole discovered that the logical operations of *and, or,* and *not* can form an algebraic system. This discovery has been applied to other situations involving two values like ON-OFF, YES-NO, 1-0. If the values can be combined using operations similar to *and, or,* and *not,* then the system is called a **Boolean algebra.**

Today the Boolean algebra of electronics is an important application of mathematics. It is used to design systems of microprocessors. The first applications of Boolean algebra to the analysis of networks were by A. Nakashima in 1937 and Claude Shannon in 1938. This field continues to be the focus of very active research by engineers, computer scientists, and mathematicians.

Questions

Covering the Reading

1. a. How long ago were the algebraic properties of *and*, *or*, and *not* first studied? **In 1991, it was 138 years ago.**

b. How long ago were these properties first applied to electronic networks? **In 1991, it was 54 years ago.**

NOTES ON QUESTIONS

Questions 2 and 6: Students will have to create input/output (or truth tables) for these questions. A brief discussion on column headings would be helpful for those still not proficient with these tables.

Questions 3-5: Although these questions require students to write a logical expression that corresponds to a network, they are not asked to do the reverse. You may wish to challenge them with a logical expression that they are to convert to the corresponding network. Students should work in **small groups** on the challenge problem.

Question 8: Emphasize that the input/output tables for gates A and B define these gates by giving an output for each possible input. The "*u*" input in the table for Gate B is, of course, the output from Gate A.

FOLLOW-UP

MORE PRACTICE
For more questions on SPUR Objectives, use *Lesson Master 1-4,* shown on page 31.

EXTENSION
Some additional proofs for which you might ask students to supply the reasons are the following:

(1) **Theorem of 0:** For all elements a of a Boolean algebra, $a \wedge 0 = 0$.
Proof: Suppose a is an element of a Boolean algebra.

$a \wedge 0 = a \wedge (a \wedge a')$	Comp. Prop.
$= (a \wedge a) \wedge a'$	Assoc. Prop.
$= a \wedge a'$	Idemp. Thm.
$= 0$	Comp. Prop.

(2) **Theorem of 1:** For all elements a of a Boolean algebra, $a \vee 1 = 1$.
Proof: Suppose a is an element of a Boolean algebra.

$a \vee 1 = a \vee (a \vee a')$	Comp. Prop.
$= (a \vee a) \vee a'$	Assoc. Prop.
$= a \vee a'$	Idemp. Thm.
$= 1$	Comp. Prop.

(3) **Absorption Law:** For all elements a and b of a Boolean algebra,
$a \vee (a \wedge b) = a$.
$a \wedge (a \vee b) = a$.
Proof: (Only the first identity is proved.) Suppose a and b are elements of a Boolean algebra.

$a \vee (a \wedge b)$	
$= (a \wedge 1) \vee (a \wedge b)$	Identity
$= a \wedge (1 \vee b)$	Distrib. Prop.
$= a \wedge 1$	Thm. of 1
$= a$	Identity

PROJECTS
The projects for Chapter 1 are described on page 70. **Projects 5** are related to this lesson.

2. Use the network version below of De Morgan's Law
not (p or q) ≡ *(not p) and (not q)*.

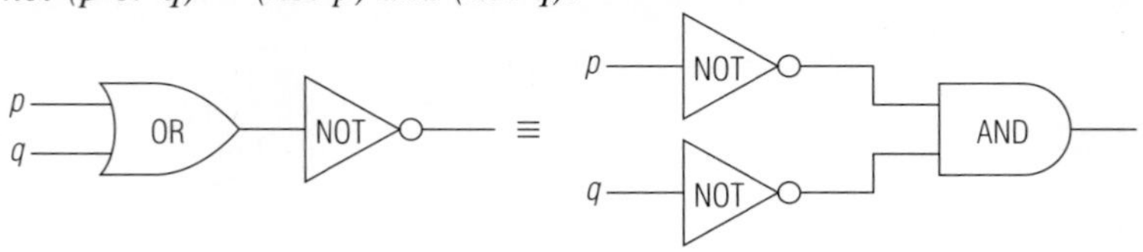

a. Write the input-output table for the left side of the ≡ sign.
b. Write the input-output table for the right side of the ≡ sign.
c. Why do your answers in parts **a** and **b** establish that the two networks are functionally equivalent?
See margin.

3. Write the logical expression that corresponds to the network below.
(p or (not q)) and r

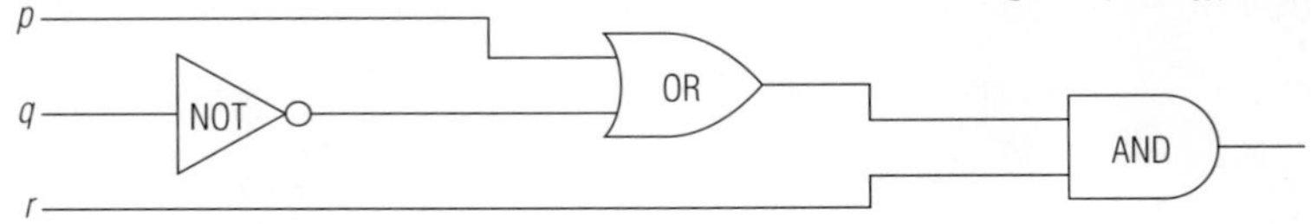

Applying the Mathematics

In 4 and 5, write a logical expression to describe each network.

4.

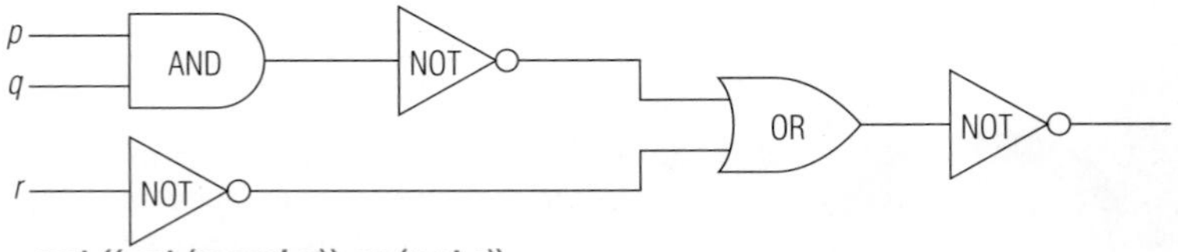

not ((not (p and q)) or (not r))

5.

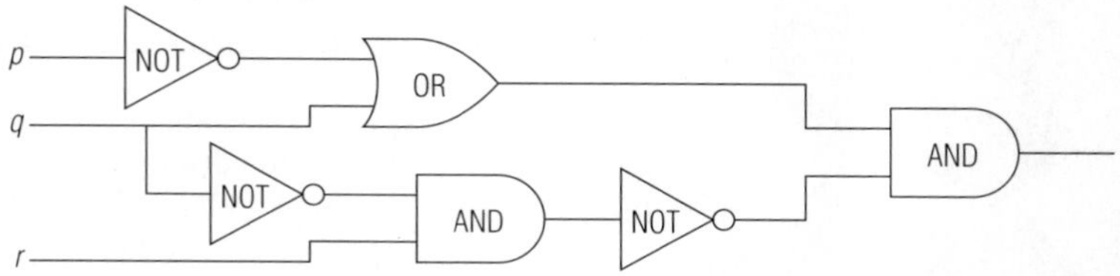

((not p) or q) and (not ((not q) and r))

6. Show that the following network is functionally equivalent to the network of Question 5. **See margin.**

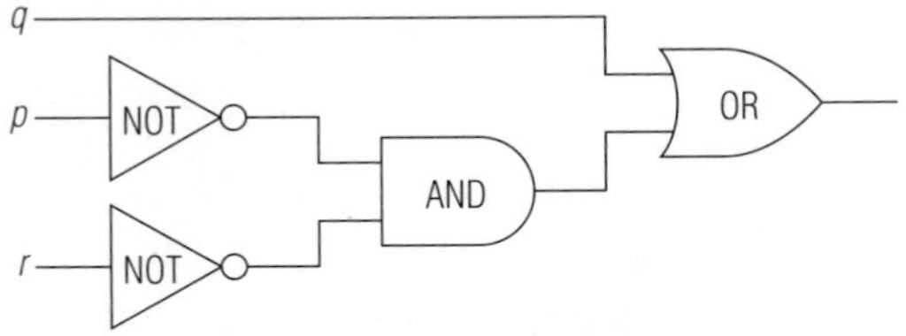

7. Suppose that an AND gate costs 3¢, an OR gate 2¢, and a NOT gate 1¢.
a. What is the cost of the network in Question 5? **11 cents**
b. What is the cost of the network in Question 6? **7 cents**
c. Which is the cheaper network? **the network of Question 6**

8. Two logic gates, Gate A and Gate B, have the following input-output tables.

Gate A

input r	input s	output u
1	1	0
1	0	1
0	1	0
0	0	0

Gate B

input t	input u	output
1	1	1
0	1	1
1	0	0
0	0	0

These gates are wired into the following network.

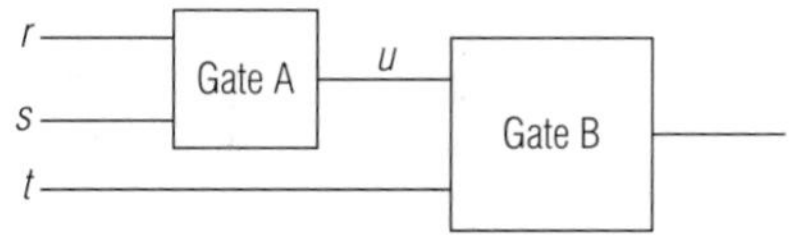

Complete the input-output table at the right for this network.

r	s	t	u	output of network
1	1	1	0	0
1	1	0	0	0
1	0	1	1	1
1	0	0	1	1
0	1	1	0	0
0	1	0	0	0
0	0	1	0	0
0	0	0	0	0

Review

9. Construct a truth table to show all possible truth values of the expression p *and* $\sim q$. Label the columns p, q, $\sim q$, and (p *and* $\sim q$). *(Lesson 1-3)* See margin.

10. Write as an inequality: x is greater than -2 and less than or equal to 4. *(Lesson 1-3)* $-2 < x \leq 4$

In 11 and 12, write the negation. *(Lesson 1-2)*

11. *No symphony orchestra contains a full-time banjo player.*
There is a symphony orchestra with a full-time banjo player.

12. $\forall$ *real numbers x and y*, $x^2 + y^2 > 0$.
$\exists$ real numbers x and y such that $x^2 + y^2 \leq 0$.

13. Is this statement *true* or *false*? Explain your answer. *(Lesson 1-1)*

$$\forall \text{ integers } n\ \exists \text{ an integer } m \text{ such that } n = 2m.$$

False; sample counterexample: let n be 5.

14. Consider the true statement $\forall$ *negative real numbers* y, $|\,y\,| = -y$.

a. Use the law of substitution to show that the statement is true for $y = -5$. $|-5| = -(-5) = 5$

b. Show that the statement is false if the restriction on y is removed. *(Lesson 1-1, Previous course)* Sample: Let $y = 5$. $|5| = 5$, not -5.

Exploration

15. Claude Shannon was a student at the Massachusetts Institute of Technology when he did his pioneering work with the algebra of circuits. Find out more about this great applied mathematician. See margin.

16. Write a paragraph relating the content of this lesson to the ideas of circuits in parallel and circuits in series. See margin.

ADDITIONAL ANSWERS
2., 6., 9., 15., 16. See Additional Answers in the back of this book.

NAME ______

LESSON **MASTER 1-4**
QUESTIONS ON **SPUR** OBJECTIVES

■ **REPRESENTATIONS** *Objective L (See pages 74-78 for objectives.)*

1. Consider the following network.

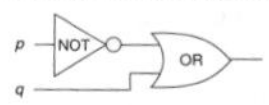

a. Write the logical expression that corresponds to the network. *(not p) or q*

b. Write an input-output table for the network.

p	q	not p	Output (not p) or q
1	1	0	1
1	0	0	0
0	1	1	1
0	0	1	1

2. Fill in the input-output table below for this network.

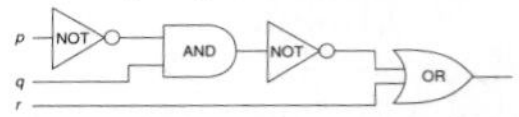

p	q	r	$\sim p$	$(\sim p)$ and q	$\sim((\sim p)$ and $q)$	$\sim((\sim p$ and $q)$ or r
1	1	1	0	0	1	1
1	1	0	0	0	1	1
1	0	1	0	0	1	1
1	0	0	0	0	1	1
0	1	1	1	1	0	1
0	1	0	1	1	0	0
0	0	1	1	0	1	1
0	0	0	1	0	1	1

NAME ______
Lesson MASTER 1-4 (page 2)

3. Suppose an AND gate costs 3¢, an OR gate 2¢, and a NOT gate 1¢.

a. What is the cost of the network in Question 2? 7¢

b. Draw a network that has the same output signals as those in Question 2, but has a lower cost.

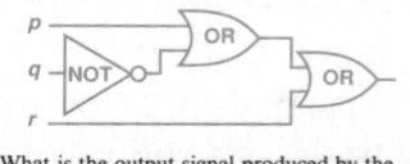

4. What is the output signal produced by the network below if p carries a signal of 1, q a signal of 0, and r a signal of 0? 1

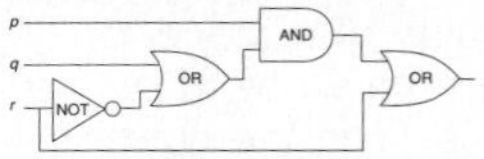

5. Use input-output tables to show that the two networks are functionally equivalent.

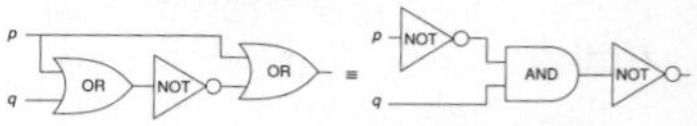

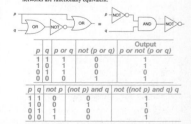

p	q	p or q	not (p or q)	Output p or not (p or q)
1	1	1	0	1
1	0	1	0	1
0	1	1	0	0
0	0	0	1	1

p	q	not p	(not p) and q	not ((not p) and q) q
1	1	0	0	1
1	0	0	1	1
0	1	1	0	0
0	0	1	0	1

LESSON 1-5

RESOURCES
■ Lesson Master 1-5

OBJECTIVES

Letter codes refer to the SPUR Objectives on page 3A.

A Identify forms of logical statements.
B Write logically equivalent forms of statements.
C Write the negation of a logical statement.
D Determine the truth value of a statement.
E Identify properties of logical statements.
I Determine the truth of quantified statements outside of mathematics.
K Read and interpret computer programs using if-then statements.
M Write truth tables for logical expressions.

TEACHING NOTES

Students often think (wrongly) that (1) the truth value of $p \Rightarrow q$ depends upon the truth value of p alone, (2) the truth value of $p \Rightarrow q$ depends upon the truth value of q alone, or (3) the truth value of $p \Rightarrow q$ depends upon the contexts of the statements that p and q represent (not merely their truth values). Each of these beliefs has some justification in the way *if-then* is used formally in mathematics and logic. The discussion of the conditional *If* $x \geq 8$, *then* $x^2 \geq 64$ preceding **Example 1** and the example itself are meant to focus on this point.

LESSON

1-5 *If-then* Statements

(a) *If you knew Peggy Sue, then you'd know why I feel blue.* ©
(b) *If I loved you, time and again I would try to say all I'd want you to know.* ©
(c) *If I had a hammer, I'd hammer in the morning.* ©
(d) *If you want to make the world a better place, take a look at yourself and make that change.* ©

If-then statements are found everywhere. (Can you tell the origins of the four statements written above?) Both inside and outside mathematics, *if-then* statements are present whenever one statement is supposed to follow from another. Within mathematics they form the basis of the language of deduction and proof. In this lesson, we review the language of *if-then* statements that you have studied in previous years and apply the formal logic of the preceding lessons to these statements.

A statement such as any one of the four above, of the form

If p, then q,

is called a **conditional statement,** denoted $\mathbf{p \Rightarrow q}$, and read "*p implies q.*"

Statement p is called the **hypothesis** or **antecedent** and statement q is called the **conclusion** or **consequent**, as in this example.

If a quadrilateral is a rectangle, then its diagonals bisect each other.

hypothesis or antecedent	conclusion or consequent
If a quadrilateral is a rectangle,	*then its diagonals bisect each other.*

How is the truth value of $p \Rightarrow q$ determined by the truth values of p and q? The following example will help to answer this question.

Suppose $p(x) \Rightarrow q(x)$ is the conditional

$$\textit{If } x \geq 8, \textit{ then } x^2 \geq 64.$$

Here $p(x)$: $x \geq 8$ and $q(x)$: $x^2 \geq 64$. For all real numbers x, this conditional is a true statement. Now let us see what truth values are possible for $p(x)$ and $q(x)$.

If $x \geq 8$, then both antecedent and consequent are true. For instance, when $x = 9$, $p(x)$ is $9 \geq 8$ and $q(x)$ is $9^2 \geq 64$.

If $x \leq -8$, then the antecedent is false and the consequent true. For instance, when $x = -10$, $p(x)$ is $-10 \geq 8$ and $q(x)$ is $(-10)^2 \geq 64$.

If $-8 < x < 8$, then both antecedent and consequent are false. For instance, when $x = 7$, then $p(x)$ is $7 \geq 8$ and $q(x)$ is $7^2 \geq 64$.

Thus in a true conditional it is possible to have the following truth values for the antecedent and consequent:

antecedent	consequent
T	T
F	T
F	F

Note that a true conditional can have an antecedent which is false.

This reasoning shows that the only combination of truth values that a true conditional cannot have is a true antecedent and a false consequent.

Now consider the conditional

$$\textit{If } x \geq 8, \textit{ then } x^2 \geq 100.$$

Are there any values of x for which this conditional is a false statement? Of course, the answer is Yes. For instance, if $x = 9$, the antecedent is $9 \geq 8$, which is true, and the consequent is $9^2 \geq 100$, which is false.

The result of this analysis is the following definition.

Definition

Suppose p and q represent statements. The conditional statement $p \Rightarrow q$ is
false whenever p is true and q is false
true in all other cases.

Just as with *not*, *and*, and *or*, this definition can be summarized in a truth table showing the truth values for *if p then q* that correspond to all possible assignments of truth values to p and q.

Truth Table for Conditional

p	q	*if p then q* $p \Rightarrow q$
T	T	T
T	F	F
F	T	T
F	F	T

Example 1 Your teacher makes you the following promise at the beginning of a course. "If your test scores total above 500, then you will get an A for this course." At the end of the course, your test scores total 485 and your teacher gives you an A. Has your teacher kept the promise?

Solution The promise is a conditional statement. The antecedent turned out to be false (485 is not above 500) and the consequent turned out to be true (you got an A). With this combination, the conditional is true. Thus we would say that the teacher did not break the promise and, accordingly, we could say that the teacher kept the promise.

Students tend to have difficulty accepting the truth of a conditional statement with a false antecedent. This may be because in informal speech there are not very many instances of conditional statements with false antecedents. (Here is one: If the answer to this problem is 6, then I'm a monkey's uncle.)

In the standard logic used to do mathematics, a statement must be either true or false. The only way for a sentence of the form *if p, then q* to be false is for *p* to be true and *q* to be false. It follows that for such a sentence to be a statement, it must be true in all other cases—including those for which *p* is false. Some people say that when *p* is false, the conditional *if p, then q* is "true by default."

It is possible to write the Negation of a Universal Conditional more symbolically: $\sim(\forall x, p(x) \Rightarrow q(x)) \equiv \exists x$ such that $p(x)$ and $\sim q(x)$. An example is: *If a country has sent a person to the moon, then that country is the United States.* (negation: *There is a country that has sent a person to the moon and is not the United States*.)

We assume that students are familiar with the meaning of the contrapositive, converse, and inverse of a statement. You may want to illustrate the relationship of the conditional to these other statements by means of a Venn diagram. For instance, consider the conditional statement, *If a car is a Mustang, then it is a Ford.* The truth of this statement can be illustrated by the Venn diagram at the top of the next page, where car 1 is a Mustang. Its contrapositive, *If a car is not a Ford, then it is not a Mustang,* can also be shown to be true via the same diagram by referring to car 2.

Venn diagrams also can be used to illustrate the converse and inverse.

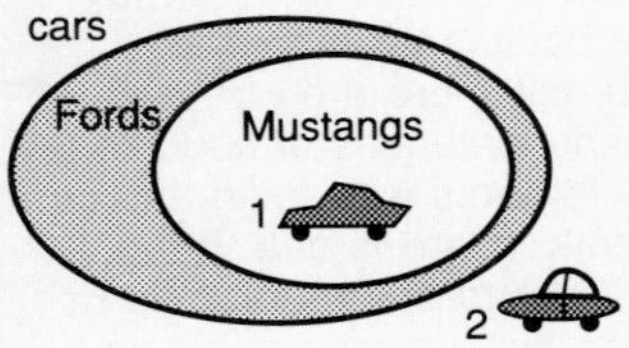

Consider the true statement *p: If a car is a Mustang, then it is a Ford.* The converse of this statement is: *If a car is a Ford, then it is a Mustang.* The diagram below helps illustrate the fact that a statement and its converse are not equivalent. For car 1, both statement *p* and its converse are true. For car 2, statement *p* is true and its converse is false. Thus, a statement may have a different truth value from its converse.

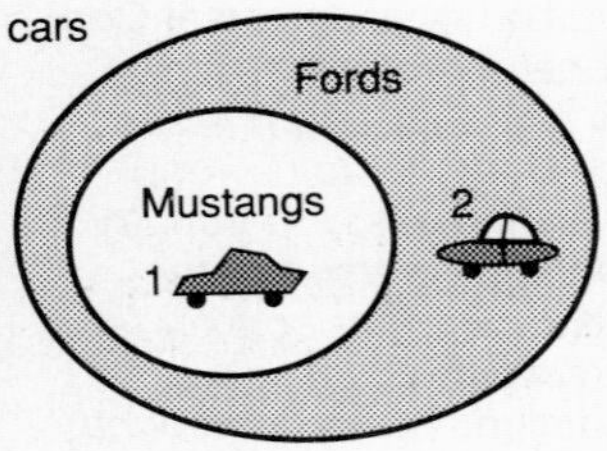

P.S. Did you know that people who write inverse are poets? Or that poetry written by prisoners is called converse?

Stress the following in the discussion of *if and only if* on page 37: (1) *p if and only if q* is often expressed as *p is a necessary and sufficient condition for q.* (2) Two *if-then* statements are contained within this one compound statement.

The negation of a conditional should be true exactly when the conditional is false. Only in the second row of the truth table is $p \Rightarrow q$ false. This occurs when p is true and q is false. However, q is false when *not* q is true. Hence, we have the following theorem.

Theorem (Negation of a Simple Conditional):

The negation of the conditional statement
if p then q
is
p and (not q).

Expressed symbolically:

$$\sim (p \Rightarrow q) \equiv p \text{ and } (\sim q).$$

Caution! The negation of a conditional statement is *not* another conditional statement. Rather, it is an *and*-statement.

Example 2 Write the negation of the conditional statement
If Tom lives in Springfield, then Tom lives in Massachusetts.

Solution Let *p: Tom lives in Springfield* and *q: Tom lives in Massachusetts.* The given statement is a conditional of the form *if p then q*. Therefore its negation has the form *p and (not q)*, or
Tom lives in Springfield and Tom does not live in Massachusetts.

One of the most important types of statements in mathematics is both conditional and universal. It has the form

$$\forall x \text{ in } S, \text{ if } p(x), \text{ then } q(x).$$

For instance, the following universal conditional statement is true.

$$\forall \text{ positive real numbers } x, \text{ if } x^2 > 9, \text{ then } x > 3.$$

But enlarge the domain of x to be the set of all real numbers, and the statement becomes false.

$$\forall \text{ real numbers } x, \text{ if } x^2 > 9, \text{ then } x > 3.$$

The reason that this second statement is false is that there are values of x (for example $x = -4$) for which the antecedent is true ($(-4)^2 > 9$ is true) and the consequent false ($-4 > 3$ is false). Because of the definition of truth and falsity of a conditional, the conditional is false for these values, and so the conditional is not true $\forall$ real numbers x. As with simpler universal statements, -4 is called a counterexample.

The theorem below states this idea symbolically: a universal conditional is false if and only if a counterexample exists.

Theorem (Negation of a Universal Conditional):

Let S be a set and let $p(x)$ and $q(x)$ be statements that may or may not hold for elements x in S. The negation of

$$\forall\ x \text{ in } S, \text{ if } p(x) \text{ then } q(x)$$

is

$$\exists x \text{ in } S \text{ such that } p(x) \text{ and not } q(x).$$

Example 3 Consider the following statement

$\forall$ real numbers a and b, if $a < b$ then $\cos a < \cos b$.

a. Write the negation of this statement.
b. Is the given statement true or false? If false, give a counterexample.

Solution

a. The negation of the given statement is
$\exists$ real numbers a and b such that $a < b$ and $\cos a \not< \cos b$.
b. The given statement is false. As a counterexample, let $a = 0$ and $b = \frac{\pi}{2}$. Then $a < b$ because $0 < \frac{\pi}{2}$, but $\cos a \not< \cos b$ because $\cos a = \cos 0 = 1$, $\cos b = \cos \frac{\pi}{2} = 0$, and $1 \not< 0$.

A conditional is sometimes proved by establishing its *contrapositive*.

Definition

The **contrapositive** of $p \Rightarrow q$ is $\sim q \Rightarrow \sim p$.
The **contrapositive** of *$\forall$ x in S, if p(x) then q(x)* is
$\forall$ x in S, if $\sim q(x)$ then $\sim p(x)$.

The table below shows that the truth values of the conditional statement $p \Rightarrow q$ and its contrapositive $(\sim q) \Rightarrow (\sim p)$ are the same.

p	q	$p \Rightarrow q$	$\sim q$	$\sim p$	$(\sim q) \Rightarrow (\sim p)$
T	T	T	F	F	T
T	F	F	T	F	F
F	T	T	F	T	T
F	F	T	T	T	T

same truth values

p implies q

not p implies not q

Making Connections As a brief review of the previous lesson, have the class construct a logic network for *if p, then q,* based on its truth table.

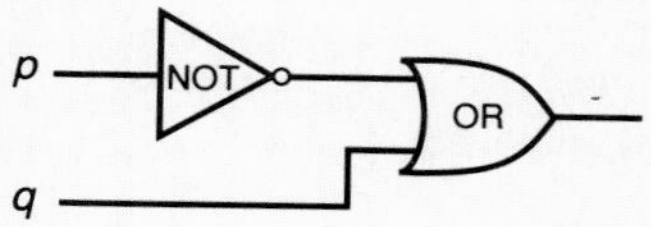

Since the implication $p \Rightarrow q$ is logically equivalent to its contrapositive $\sim q \Rightarrow \sim p$, the proof of the latter implies the former is also true. This is the basis of most indirect proofs.

Error Analysis The negation of a conditional is often confusing to students. It is important that they realize that the negation is not another conditional but is an *and* statement. Have students give the negations of several conditionals aloud, emphasizing the word *and.* Here are some additional examples:
(1) *If Sara is taking this course, then Sara is a senior.* (negation: *Sara is taking this course and is not a senior.*)
(2) *If George Bush is President of the U.S., then George Bush lives in New York City.* (negation: *George Bush is President of the U.S. and George Bush does not live in New York City.*)

ADDITIONAL EXAMPLES

1. Write the negation of the conditional statement: *If Tom is a Packers fan, then he loves the Bears.*
Tom is a Packers fan and he does not love the Bears.

2. Consider the following statement: *$\forall$ real numbers x and y, if $x < y$, then $|x| < |y|$.*
a. Write the negation of this statement.
$\exists$ real numbers x and y such that $x < y$ and $|x| \geq |y|$.

ADDITIONAL EXAMPLES

b. Is the given statement true or false? If false, give a counterexample.
The given statement is false. As a counterexample, let $x = -10$ and $y = 1$. Then $x < y$ because $-10 < 1$, but $|x| \geq |y|$ because $|-10| \geq |1|$.

3. State the contrapositive of the following statement and determine whether the contrapositive is true or false.
a. *∀ real numbers a, if $a > 1$, then $a^2 > a$.*
The contrapositive is ∀ real numbers a, if $a^2 \ngtr a$, then $a \ngtr 1$. Because the original statement is true, the contrapositive is also true.

b. *∀ real numbers b, if $\sqrt{b} = 3$, then $b^2 = 81$.*
The contrapositive is ∀ real numbers b, if $b^2 \neq 81$, then $\sqrt{b}$ is $\neq 3$. Because the original statement is true, the contrapositive is also true.

4. Give the converse and inverse of the universal conditional *∀ functions f and real numbers x, if $f(2x) = 2f(x)$, then $f(4x) = 4f(x)$.*
Its converse is ∀ functions f and real numbers x, if $f(4x) = 4f(x)$, then $f(2x) = 2f(x)$. Its inverse is ∀ functions f and real numbers x, if $f(2x) \neq 2f(x)$, then $f(4x) \neq 4f(x)$.

5. Break the biconditional of the given definition into its two conditionals.
a. *∀ functions f and real numbers a, $f(a) = 0$ if and only if a is a real zero of f.*
∀ functions f and real numbers a, if $f(a) = 0$, then a is a real zero of f. ∀ functions f and real numbers a, if a is a real zero of f, then $f(a) = 0$.

The truth table on page 35 proves the following theorem.

Contrapositive Theorem

A conditional and its contrapositive are logically equivalent. That is, they always have the same truth values.

Example 4 State the contrapositive of the following statement and determine whether the contrapositive is true or false.

∀ real numbers a, if $a^2 = 10$, then $a^6 = 1000$.

Solution The contrapositive is

∀ real numbers a, if $a^6 \neq 1000$, then $a^2 \neq 10$.

Because the original statement is true, the contrapositive is also true.

By either negating or switching the antecedent and consequent of a conditional, but not doing both, two other conditionals may be formed.

Definitions

The **converse** of $p \Rightarrow q$ is $q \Rightarrow p$.
The **converse** of *∀ x in S, if p(x) then q(x)* is
∀ x in S, if q(x) then p(x).

The **inverse** of $p \Rightarrow q$ is $\sim p \Rightarrow \sim q$.
The **inverse** of *∀ x in S, if p(x) then q(x)* is
∀ x in S, if ~p(x) then ~q(x).

Converses and inverses may seem to be similar to contrapositives. But, unlike the contrapositive, neither the converse nor the inverse of a conditional needs to have the same truth value as the original conditional.

Example 5 Give the converse and inverse of the universal conditional

∀ functions f, if f is the cosine function, then $f(0) = 1$.

Solution Its converse is

∀ functions f, if $f(0) = 1$, then f is the cosine function.

Its inverse is

∀ functions f, if f is not the cosine function, then $f(0) \neq 1$.

The original conditional is true because $\cos 0 = 1$. But the converse is not true. There are many functions f with $f(0) = 1$ that are not the cosine function. One is the function f defined by $f(x) = 3x + 1$ for all real numbers x. This same function is a counterexample that shows the inverse is not true either.

36

Given statements p and q,

p if and only if q

means

(*if p then q*) and (*if q then p*)

or, symbolically,

$$p \Rightarrow q \text{ and } q \Rightarrow p.$$

This is naturally written

$$p \Leftrightarrow q$$

and called a **biconditional**. All definitions are biconditionals.

Example 6 Here is the definition of logarithm with base 2. Break this biconditional into its two conditionals.

∀ positive real numbers x, $\log_2 x = y$ *if and only if* $2^y = x$.

Solution

∀ positive real numbers x, if $\log_2 x = y$ *then* $2^y = x$

∀ positive real numbers x, if $2^y = x$ *then* $\log_2 x = y$.

Some additional language associated with conditionals and biconditionals is found in Questions 19-21.

Questions

Covering the Reading

1. Let x be a real number. Consider the statement:

If $x > 1$ *then* $2x^2 + 3x^3 > 1$.

a. Identify the antecedent, the conclusion, the consequent, and the hypothesis. See margin.

b. Is the conditional true or false? True

2. *Multiple choice*. Consider the statement

If Sandra is on the swim team, then she swims every day.

Which would tell you that the statement is false?

(a) *Sandra is not on the swim team, and she swims every day.*

(b) *Sandra is on the swim team, and she does not swim every day.*

(c) *Sandra is not on the swim team, and she does not swim every day.*

(b)

3. *True* or *false*? The negation of

∀ real numbers x, if $2x - 1 > 5$ *then* $x > 2$

is

∃ a real number x such that $2x - 1 > 5$ *and* $x > 2$. False

In 4 and 5, determine whether the conditional is true or false. If false, give a counterexample.

4. *∀ real numbers x, if* $\frac{\pi}{2} < x$ *then cos x is negative.* False
Counterexample: let $x = 2\pi$; $\cos x = 1$, which is not negative.

5. *If a quadrilateral is a square, then it is a parallelogram.* True

b. *∀ angles A and B, A and B are complementary angles if and only if the sum of their measures is 90°.*
∀ angles A and B, if A and B are complementary angles, then the sum of their measures is 90°. ∀ angles A and B, if the sum of their measures is 90°, then A and B are complementary angles.

NOTES ON QUESTIONS

Question 1a: This question asks for the same information twice, in order to familiarize students with all four terms.

Question 9: Students may enjoy discussing the truth values of the converse and inverse of the statement *If it rains, then I carry an umbrella.*

Question 12: Some students will find that examining the truth or falsity of the negation of each statement is easier than looking at the original statement. Finding a counterexample to a false statement is often easier than proving a true statement.

Question 19: Use of the contrapositive may assist students in understanding why *p only if q* is equivalent to *if p, then q.* It is usually not difficult to convince students that *p only if q* is equivalent to *if (not q), then (not p).* For instance, *Hal can make the team only if he gains ten pounds* is equivalent to *If Hal does not gain ten pounds, then he cannot make the team. If (not q), then (not p)* is equivalent to its contrapositive *if p, then q.* Thus, *p only if q* is equivalent to *if p, then q.*

ADDITIONAL ANSWERS

1.a. antecedent, hypothesis: $x > 1$; conclusion, consequent: $2x^2 + 3x^3 > 1$

NOTES ON QUESTIONS
Questions 20 and 21: Point out that to say that *p is a necessary and sufficient condition for q* is equivalent to saying *p if and only if q*.

Question 21: Students might compare this condition with those in your school.

Technology for Question 22: Many versions of BASIC allow an IF . . . THEN . . . ELSE statement that would allow lines 20 and 30 to be combined into the one statement:

```
20 IF N > 0 THEN PRINT N,
    LOG(N) ELSE PRINT N,
    "THE LOG IS
    UNDEFINED"
```

ADDITIONAL ANSWERS
7.a. *If m = 0, then the graph of y = mx + b is not an oblique line.*
b. True

8.a. *If a quadrilateral does not have two angles of equal measure, then the quadrilateral does not have two sides of equal length.*
b. False

9. Converse: *If it will rain tomorrow, then it will rain today.*
Inverse: *If it does not rain today, then it will not rain tomorrow.*

10. *If two supplementary angles are congruent, then they are right angles. If two supplementary angles are right angles, then they are congruent.*

14. *If someone has been convicted of a felony, then that person is not allowed to vote.*

16.a. *If Jon was not at the scene of the crime, then Jon did not commit the crime.*
b. *If one has a true allibi, one is innocent.*

6. Choose the correct word. The (contrapositive, inverse, converse) of *If $e^x = e^y$ then $x = y$* is *If $e^x \neq e^y$ then $x \neq y$.* inverse

In 7 and 8, **a.** write the contrapositive of the given statement, and **b.** determine whether or not the contrapositive is true.

7. *If the graph of $y = mx + b$ is an oblique line, then $m \neq 0$.* See margin.

8. *If a quadrilateral has two sides of equal length, then the quadrilateral has two angles of equal measure.* See margin.

9. Write the converse and inverse of
If it rains today, then it will rain tomorrow. See margin.

10. Given the statement
Two supplementary angles are congruent if and only if they are right angles.
Write two *if-then* conditionals contained in this statement. See margin.

11. Use substitution on the definition of logarithm to complete this sentence.
If $2^5 = 32$, then __?__. $\log_2 32 = 5$

Applying the Mathematics

12. Suppose that $p(n)$ and $q(n)$ are the sentences
$p(n)$: *n is a prime number.*
$q(n)$: *n is an odd number.*
a. Determine the truth or falsity of
$\forall$ *positive integers n,* $p(n) \Rightarrow q(n)$. False
b. Determine the truth or falsity of
$\forall$ *positive integers n,* $q(n) \Rightarrow p(n)$. False

13. Let $p(x)$ be *If $|x| > 6$, then $x > 5$.*
a. Is $p(7)$ true? Yes
b. Is $p(-7)$ true? No
c. Is $p(2)$ true? Yes
d. Is $p(x)$ true for all real numbers x? No

Could be 0, -7

In 14 and 15, rewrite each statement in *if-then* form.

14. *No one who has been convicted of a felony is allowed to vote.* See margin.

15. *Those who can, do.* If one can, then one does.

16. Consider the statement
If Jon committed the crime, then Jon was at the scene of the crime.
a. Write the contrapositive.
b. What is the legal significance of the contrapositive? See margin.

17. Show by a truth table that $p \Rightarrow q$ is not logically equivalent to $q \Rightarrow p$. See margin.

18. Show by a truth table that the converse of $p \Rightarrow q$ is logically equivalent to the inverse of $p \Rightarrow q$. See margin.

19. The statement ***p* only if *q*** is logically equivalent to *if p then q*. Rewrite the following statement in *if-then* form:

A satellite is in orbit only if it is at a height of at least 200 miles above the earth. **If a satellite is in orbit, then it is at a height of at least 200 miles above the earth.**

20. "*p is a* ***sufficient condition*** *for q*" is another way of saying that $p \Rightarrow q$. Write the following statement in *if-then* form:

Having the form 2k for some integer k is a sufficient condition for an integer to be even. **If an integer is in the form 2k for some integer k, then it is even.**

21. "*p is a* ***necessary condition*** *for q*" is another way of saying that $q \Rightarrow p$. Given the following:

Having a GPA of at least 3.5 is a necessary condition for being elected to the honor society.

Write this statement in *if-then* form. **If one is elected to the honor society, then one must have a GPA of at least 3.5.**

22. IF-THEN statements in computer programs have the form IF *condition* THEN *action*. If the condition is satisfied, then the action is performed. If the condition is not satisfied, then the execution of the program moves to the next statement following the IF-THEN statement. Consider the program below.

```
10  INPUT N
20  IF N > 0 THEN PRINT N, LOG(N)/LOG(10)
30  IF N < = 0 THEN PRINT N, "THE LOG IS UNDEFINED"
40  END
```

Determine the output for each input value of N.

a. N = 1000 **1000, 3** **b.** N = -50 **-50, THE LOG IS UNDEFINED**

c. N = 0.1 **0.1, -1**

23. Write the following statement in the form ∀ _?_ L, *if* _?_ *then* _?_.

If a line is vertical, then its slope is undefined. (Lesson 1-2)

∀ lines L, if L is vertical, then its slope is undefined.

Review

24. Complete the input-output table for the logic network below. *(Lesson 1-4)*

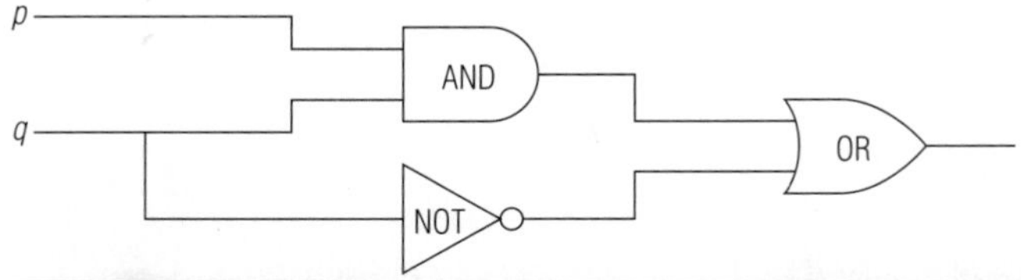

p	q	p AND q	NOT q	(NOT q) OR (p AND q)
1	1	1	0	1
1	0	0	1	1
0	1	0	0	0
0	0	0	1	1

17. See the margin on page 40.

18. See Additional Answers in the back of this book.

NAME ________

LESSON **MASTER 1–5**

QUESTIONS ON **SPUR** OBJECTIVES

■ SKILLS *Objective A (See pages 74–78 for objectives.)*

1. Consider these four statements.

(1) *If the air conditioning is on, then Ed is cold.*
(2) *If Ed is cold, then the air conditioning is on.*
(3) *If the air conditioning is not on, then Ed is not cold.*
(4) *If Ed is not cold, then the air conditioning is not on.*

a. Name a pair of statements that are contrapositives of each other. **[(1) and (4)] or [(2) and (3)]**

b. Which statement is the converse of (3)? **(4)**

■ SKILLS *Objective B*

2. The union of which two statements from Question 1 is equivalent to the following statement?

Ed is cold if and only if the air conditioning is on. **(1) and (2)**

3. Write the inverse of the following statement.

If a parabola has equation $y - k = a(x - h)^2$, then its vertex is (h, k).

If a parabola does not have equation $y - k = a(x - h)^2$, then its vertex is not (h, k).

4. Write the following statement in if-then form, using the symbol ∀.

All numbers divisible by 12 are even.

∀ numbers n, if n is divisible by 12, then n is even.

■ SKILLS *Objective C*

5. Write the negation of the following statement.

If an object is on the moon, then its weight is about $\frac{1}{6}$ its weight on Earth. **There is an object on the moon whose weight is not $\frac{1}{6}$ its weight on earth.**

6. Let *p*: ∀ *real numbers x, if $x^2 = 20$, then $x \neq \sqrt{20}$.*

a. Write ~*p*. **∃ a real number x such that if $x^2 = 20$, then $x \neq \sqrt{20}$.**

b. Which is true, *p* or ~*p*? Justify your answer. **~p is true, because if $x = -\sqrt{20}$ then $x^2 = 20$ but $x \neq \sqrt{20}$.**

NAME ________

■ SKILLS *Objective D*

7. Let *p(x)*: $|x + 2| < 5$ and *q(x)*: $x + 2 < 5$. Is the statement ∀ *x, if p(x), then q(x)* true? **yes**

■ PROPERTIES *Objective E*

8. Under what conditions is the following statement false?

If the team plays at home, then it wins.

when the team plays at home and does not win

9. *Multiple choice.* A conditional statement is logically equivalent to its

(a) converse. (b) inverse. (c) contrapositive. **(c)**

■ USES *Objective I*

10. Find a counterexample to the following statement.

If a food is nutritious, then it tastes good.

sample: Liver is nutritious, and some people think it doesn't taste good.

■ USES *Objective K*

11. Consider the computer program below.

```
10  INPUT A
20  IF (A > -10) AND (A < -5) THEN PRINT "TRUE"
30  END
```

a. What is the output if the input is $A = -2$? **no output**

b. Write line 25 so that the program will print FALSE for all cases in which *A* is not in the interval $-10 < A < -5$.

25 IF (A<=-10) OR (A>=-5) THEN PRINT "FALSE"

■ REPRESENTATIONS *Objective M*

12. Write a truth table for the expression $p \Rightarrow \sim q$.

p	q	~q	p ⇒ ~q
T	T	F	F
T	F	T	T
F	T	F	T
F	F	T	T

NOTES ON QUESTIONS
Question 26: Ask students to graph the original compound inequality and compare the graph obtained with the graph of part **c**.

FOLLOW-UP

MORE PRACTICE
For more questions on SPUR Objectives, use *Lesson Master 1-5,* shown on page 39.

PROJECTS
The projects for Chapter 1 are described on page 70. **Project 4,** dealing with logical semantics in computer languages, is related to the content of this lesson.

ADDITIONAL ANSWERS
17.

p	q	$p \Rightarrow q$	$q \Rightarrow p$
T	T	T	T
T	F	F	T
F	T	T	F
F	F	T	T

↑ ↑
not equivalent

25. Compute the output signal for the logic network below for each of the following input signals. *(Lesson 1-4)*

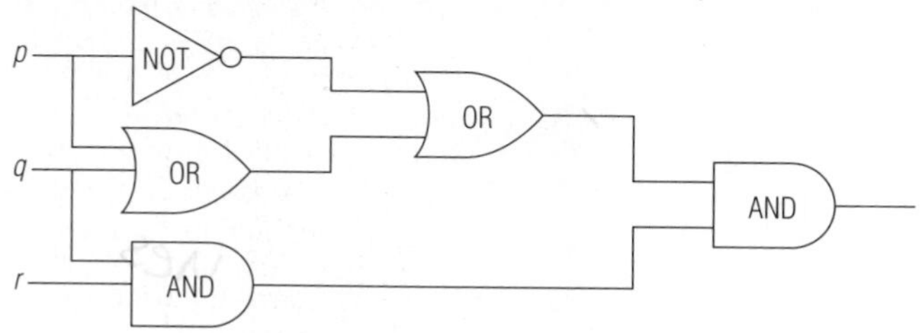

a. $p = 1$, $q = 1$, $r = 1$ **1**
b. $p = 0$, $q = 1$, $r = 1$ **1**
c. $p = 1$, $q = 0$, $r = 1$ **0**

26. Let x be a real number. Given the double inequality
$-5 < x \le 2.$
a. Fill in the blank with *and*, *or*, or *not* so as to make the sentence equivalent to the given inequality.
$-5 < x$ _?_ $x \le 2.$ **and**
b. Write the negation of the statement in part **a.** ***$-5 \ge x$ or $x > 2$.***
c. Graph on the number line the values of x which make the statement in part **b** true. *(Lesson 1-3, Previous course)* **See below.**

Exploration

27. Find a source for each of the four conditionals which begin this lesson.
a) Buddy Holly b) *Carousel* (Rodgers and Hammerstein) c) Peter, Paul and Mary d) Michael Jackson

26. c)

40

LESSON

1-6

Valid Arguments

Lewis Carroll and characters from Alice in Wonderland

A major reason to study logic is to help you make correct inferences or deductions and to help you determine when others have made correct or incorrect deductions. Lewis Carroll (the pseudonym of Charles Lutwidge Dodgson, 1832–1898, an English mathematician, logician, and minister), who is best known as the author of *Alice in Wonderland*, also published two books of logical puzzles. The following problem is adapted from one of those books.

Consider the sentences:

(1) *When I work a logic example without grumbling, you may be sure it is one I can understand.*

(2) *This example is not arranged in regular order like the ones I am used to.*

(3) *No easy examples give me a headache.*

(4) *I can't understand an example that is not arranged in regular order like the ones I am used to.*

(5) *I grumble when I work an example only if I get a headache.*

Suppose each of sentences (1)–(5) is true. Must the following conclusion also be true?

(6) *This example is not easy.*

We will give the answer to this puzzle at the end of this lesson. But first we develop general methods to help solve it.

In logic and mathematics, an argument is not a dispute. An **argument** is a sequence of statements in which all but the final statement are called **premises**, and the final statement is called the **conclusion**. Generally the word *therefore*, or some synonym, or the shorthand symbol ∴ (read "therefore"), is written just before the conclusion.

Consider the following two arguments:

(a) *If Jane solved the problem correctly, then Jane got the answer 10.*
Jane solved the problem correctly.
∴ *Jane got the answer 10.*

(b) *For all real numbers x, if* $x > 3$, *then* $2x^2 - x - 15 > 0$.
$\pi > 3$
$\therefore 2\pi^2 - \pi - 15 > 0$

LESSON 1-6

RESOURCES

- Lesson Master 1-6
- Quiz for Lessons 1-4 through 1-6
- Teaching Aid 1 can be used with **Question 8.**
- Computer Master 3

OBJECTIVES

Letter codes refer to the SPUR Objectives on page 3A.

G Determine whether arguments are valid or invalid.

H Use logic to prove or disprove statements.

J Determine whether or not a logical argument outside of mathematics is valid.

TEACHING NOTES

Students often equate valid arguments with true conclusions. Emphasize that the structure of an argument determines its validity. A valid argument with a false premise can have a false conclusion.

You can use the truth table proof of Modus Ponens (the Law of Detachment) to emphasize that an argument is valid if whenever the premises are true then the conclusion is also true. If the truth table for an argument contains a row of true premises followed by a false conclusion, then the argument is invalid.

To help make Modus Tollens (the Law of Indirect Reasoning) reasonable to students who accept Modus Ponens, you may wish to make a connection between these forms of argument and the fact that *if p, then q* is equivalent to *if*

not q, then not p. They already know that a conditional and its contrapositive are equivalent. The validity of the Modus Tollens argument form can be deduced from Modus Ponens and the equivalence of a statement and its contrapositive, as follows:
Given $p \Rightarrow q$
$\sim q$
we can deduce
$\sim q \Rightarrow \sim p$
$\sim p$
by the equivalence between a conditional and its contrapositive. $\sim p$ follows by Modus Ponens.

The solution to the Lewis Carroll puzzle in the text can be accomplished in two ways. (1) Collapse the conditionals into one conditional by using the Law of Transitivity and then applying the Law of Detachment; or (2) Repeatedly apply the Law of Detachment. Students need to realize that either way is acceptable; there is no one correct way to complete the problem. Here is another Lewis Carroll puzzle that you might want to do with your class:

Babies are illogical.
Nobody is despised who can manage a crocodile.
Illogical persons are despised.
∴ *All babies are incapable of managing crocodiles.*

An alternative conclusion is *A person able to manage a crocodile is no baby.*

In the third paragraph of the lesson, we begin by saying "In logic and mathematics." Logic is often considered to be a branch of mathematics although it is also claimed as a branch of philosophy. Some of your students might be interested in investigating how different a logic course in philosophy might be from a logic course in mathematics.

Alternative Approach In the lesson, we have chosen to

Although the subject matters of arguments (a) and (b) are very different, the forms of the arguments are very similar. The premises are a conditional statement and its antecedent. The conclusion is the consequent. Here are the two versions.

Version (a) Simple Form	Version (b) Universal Form
If p then q	$\forall$ *x, if p(x), then q(x)*
p	*p(c), for a particular c*
∴ *q.*	∴ *q(c).*

The simple form has a very important property: no matter what statements are substituted in place of *p* and *q* in the premises, the truth value of the form is true. The universal form has a similar property: no matter what conditions are substituted in place of *p(x)* and *q(x)* in the premises, the truth value of the form is true. Any form of argument having such a property is called **valid**. The fact that versions (a) and (b) are valid is called the Law of Detachment (because the antecedent is detached from the conditional) or *modus ponens* (which is Latin for "method of affirming").

We prove validity for the simple form of the Law of Detachment below; proving the validity of the universal form requires a technique beyond the scope of this book.

Theorem (Modus Ponens or Law of Detachment)

The following are valid forms of argument:

If p then q	$\forall$ *x, if p(x) then q(x)*
p	*p(c), for a particular c*
∴ *q.*	∴ *q(c).*

Proof The premises are $(p \Rightarrow q)$ and p. To prove the Law of Detachment we must show that the conditional

$$((p \Rightarrow q) \text{ and } p) \Rightarrow q$$

is always true. So we construct a truth table showing all possible truth values for p and q. Then we give truth values for the premises, the conclusion, and the argument form. Because all the rows in the form column are true, the argument is valid.

p	q	$p \Rightarrow q$ (premises)	p (premises)	$(p \Rightarrow q)$ and p	q (conclusion)	$((p \Rightarrow q)$ and $p) \Rightarrow q$ (form)
T	T	T	T	T	T	T
T	F	F	T	F	F	T
F	T	T	F	F	T	T
F	F	T	F	F	F	T

Notice that an entry in the form column would be false only if all premises were true and the conclusion were false. Thus you can also think of a valid argument form in the following way: An argument is valid if and only if when all its premises are true, then its conclusion is true.

The conclusion of a valid argument is called a **valid conclusion**. In a valid argument, the truth of the premises guarantees the truth of the conclusion. However, if one of the premises is false, then the conclusion, while valid, may be false. Thus a valid conclusion is not necessarily a true conclusion.

Consider the following argument:

If a country has over 200 million people, then it imports more than it exports.
Japan has over 200 million people.
∴ *Japan imports more than it exports.*

The argument is valid (by the Law of Detachment), but the conclusion is false. In this case, neither premise is true. In general, even clear thinking from false premises is risky. Do not trust conclusions unless you are certain of the premises from which they are made.

The Law of Detachment enables you to make a single deduction. A second form of valid argument allows you to build chains of deductions. From the premises

If a figure is a square, then it is a parallelogram.
If a figure is a parallelogram, then its diagonals bisect each other.

you can deduce

If a figure is a square, then its diagonals bisect each other.

This fact exemplifies the *Law of Transitivity*. The Law of Transitivity allows you to deduce an *if-then* statement.

Theorem (The Law of Transitivity)

The following are valid forms of argument:

Simple form	Universal form
If p then q	∀ *x, if p(x), then q(x)*
If q then r	∀ *x, if q(x), then r(x)*
∴ *If p then r.*	∴ ∀ *x, if p(x), then r(x).*

The proof of this theorem is outlined below for the simple form.

Proof First write the argument form as a conditional.

$$((p \Rightarrow q) \text{ and } (q \Rightarrow r)) \Rightarrow (p \Rightarrow r)$$

Now construct a truth table and show that this conditional is always true. Because there are three statements *p*, *q*, and *r*, the table has 8 rows. We leave it to you to complete this table and finish the proof.

p	q	r	$p \Rightarrow q$	$q \Rightarrow r$	$(p \Rightarrow q)$ and $(q \Rightarrow r)$	$p \Rightarrow r$	$((p \Rightarrow q)$ and $(q \Rightarrow r)) \Rightarrow (p \Rightarrow r)$
T	T	T					
T	T	F					
T	F	T	F	T	F	T	T
T	F	F					
F	T	T					
F	T	F					
F	F	T					
F	F	F					

demonstrate the validity of an argument form via truth tables. You may also want to use Venn diagrams to illustrate the validity of argument forms. For instance, consider the following argument:

If a pie contains apples, then it is a fruit pie.
This pie contains apples.
∴ *This pie is a fruit pie.*

Analysis of the Venn diagram below shows that if both premises are true, the conclusion also must be true.

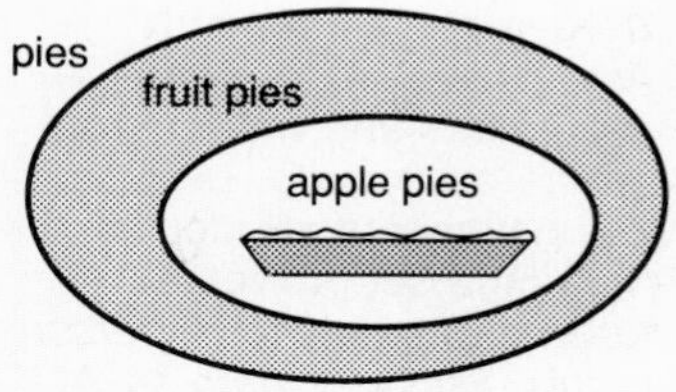

Now consider the following argument:

If a pie is a fruit pie, then it contains apples.
A pie is a fruit pie.
∴ *The pie contains apples.*

This argument has the same form as the first one and therefore is valid also. However, while the conclusion of this argument is valid, it need not be true. There are many examples of fruit pies that do not contain apples. Of course, the reason that a false conclusion can be deduced from this valid argument is that the first premise of the argument is false. Consideration of these two arguments may help students distinguish between valid conclusions and true conclusions.

Making Connections
You may wish to relate the Law of Transitivity to the transitive property of equality: *If* $a = b$ *and* $b = c$, *then* $a = c$, or to the transitive property of inequality: *If* $a < b$ *and* $b < c$, *then* $a < c$. Mathematics contains many examples of transitive properties.

Students who have studied from UCSMP *Geometry* or some other books will have seen the simple forms of the Law of Detachment and the Law of Transitivity.

Error Analysis Some students have difficulty labeling a statement of fact with p or q. Stress the relationship of a statement such as *Jane solved the problem correctly* to the property part in the definition of a universal statement (page 5).

Students also want to call $p(x)$ a function of x, which it is not. Stress that it refers to a *statement* involving a quantity that can only take on values from a given set.

Alternate Approach Ask students to make up statements and write them on the chalkboard. Discuss if the statements are dependent on a set (universal form). For example: *A person in this class can earn an A.* This can be written as: *For all people in this class, Mary is a person in this class, then Mary can earn an A.*

Technology Use a graphics calculator to demonstrate examples such as: *If a point (x, y) is in the third quadrant, then* $3x + 2y < 5$. *(-2, -5) is in quadrant III. Hence,* $3(-2) + 2(-5) < 5$.

Recognizing the form of an argument is an important step in determining whether an argument is valid or not.

Example 1 Write the form of the following argument:

∀ polygons x, if x is a hexagon, then the sum of the measures of the interior angles of x is 720°.
A particular polygon c has an angle sum of 540°.
∴ c is not a hexagon.

Solution Let $p(x)$ and $q(x)$ represent the following statements:

p(x): x is a hexagon.
q(x): the sum of the measures of the interior angles of x is 720°.

Then the argument has the following form.

∀ x, if p(x) then q(x)
not q(c) for a particular c
∴ not p(c).

The law stating that the form of the argument in Example 1 is valid is called *modus tollens* (Latin for "method of denial") or the *Law of Indirect Reasoning*.

Theorem (Modus Tollens or Law of Indirect Reasoning)

The following is a valid form of argument:

Simple form	Universal form
If p then q	*∀ x, if p(x) then q(x)*
not q	*not q(c) for a particular c.*
∴ not p.	*∴ not p(c) for that c.*

The proof of this theorem is left to the Questions.

Example 2 Assume that premises (1) and (2) are both true:

(1) If Mary is sick, then she has a fever.
(2) Mary does not have a fever.

What true conclusion can be deduced?

Solution The premises fit the form of the premises of *modus tollens*. A true conclusion is

Mary is not sick.

We are now ready to return to the Lewis Carroll puzzle. We want to decide if the conclusion *This example is not easy* follows from the premises. In order to apply the theorems in this lesson, we have rewritten the premises in *if-then* form.

(1) *If I work a logic example without grumbling, then you may be sure it is one I can understand.*
(3) *If an example is easy, then it does not give me a headache.*

LESSON

1-7

Proofs about Integers

Each August a twins' convention meets in Twinsburg, Ohio.

In this lesson, the logical principles of the previous lessons are applied to simple proofs. We purposely have picked content you are familiar with so that you can concentrate on the logic. We assume you know the postulates about addition and multiplication of integers: closure, associativity, commutativity, and distributivity, and the addition and multiplication properties of equality.

A **conjecture** is a statement that we believe may be true but have not proved. Consider the following conjecture:

If m and n are any even integers, then $m + n$ is an even integer.

You certainly know that there exist even integers m and n such that $m + n$ is even. But what about all sums of even integers? Are you sure the sum is always even?

To prove the above conjecture, it is essential to have a precise meaning for all the terms used in it. What is the meaning of the term *even integer*? Is 0 even? Is -554 even? Is $\frac{8}{6}$ even? Some people might say that 6, 8, -2, and -10 are even integers but 1, 5, -1, -17 are not. This happens to be true; but it is not a definition of even integer because it does not provide a precise criterion to decide if a number other than those listed is even. The following statement is the usual definition. It says that an even integer is one that is twice some integer.

Definition

An integer n is **even** if and only if $n = 2k$ for some integer k.

Similarly, it is possible to define the notion of odd integer. An odd integer is 1 more than an even integer.

Definition

An integer n is **odd** if and only if $n = 2k + 1$ for some integer k.

LESSON 1-7

RESOURCES
- Lesson Master 1-7

OBJECTIVE

Letter code refers to the SPUR Objectives on page 3A.

H Use logic to prove or disprove statements.

TEACHING NOTES

In this book, students are given experiences with a variety of types of proofs, starting with proofs of properties of even and odd integers in this lesson, and including divisibility proofs in Chapter 4, indirect proofs with rational and irrational numbers in Chapter 5, proofs of trigonometric identities in Chapter 6, and proofs utilizing mathematical induction in Chapter 7. For each of these types of proofs, much attention is given to the structure and words used in the proofs, because we feel that the *writing* of proofs is as difficult as the *conceptualization.* The words we expect students to write are put in a **boldface type**.

We try to keep the presentation of proof as simple as possible. For this reason, direct proofs often start with a sentence beginning "Suppose" and stating the full hypothesis of the theorem to be proved. Occasionally, this sentence is followed by one beginning "We must show that" and stating explicitly what must be shown to complete the proof of the theorem. Mathematicians often omit these sentences, but they are always implicitly contained in proofs.

The field testing of these materials has shown that these proofs about integers are accessible to students of all abilities. Even students who found geometry proofs to be difficult were able to do the proofs. Regardless of the performance level of the class, these proofs should not be skipped.

After defining even integer, some would define an odd integer to be one that is not even. The fact that every integer is even or odd is actually derived in Chapter 4 as a consequence of the Quotient-Remainder Theorem for integers. The definition of odd integer given in this lesson is certainly more convenient for writing proofs. Point out that the definition of *even* and *odd* is frequently used in both directions in such proofs. Also, students need to be aware of the need to use different letters *r* and *s* to refer to two different integers. This issue is addressed in the discussion preceding **Example 2** and also in **Question 14** in which students are asked to find the error in a "proof."

Making Connections
The results in this lesson are simple and are applied in Lesson 7-7 to deduce properties of certain recursively defined sequences. They also are applied to deduce a result in graph theory in Chapter 11.

In college mathematics courses, students are often expected to write proofs. The proofs in college tend to be algebraic, written in paragraphs, and do not have a formal "given" and "to prove" commonly found in geometry.

Alternate Approach Depending on the interest of your students, you may want to relate the kind of reasoning discussed in this lesson to two kinds used in the study of artificial intelligence: forward chaining and backward chaining. To derive the conclusion of a theorem from its

Consider the following examples.

0 is an even integer because $0 = 2 \cdot 0$. Here $n = 0$ and $k = 0$.

-554 is an even integer because $-554 = 2 \cdot (-277)$ and -277 is an integer. Here $n = -554$ and $k = -277$.

-177 is an odd integer because $-177 = 2 \cdot (-89) + 1$ and -89 is an integer. Here $n = -177$ and $k = -89$.

$\frac{8}{6}$ is neither even nor odd because $\frac{8}{6} = \frac{4}{3} = 1\frac{1}{3}$ and so is not an integer.

The same ideas can be used with algebraic expressions.

Example 1 Assume *a*, *b*, *x*, and *y* are integers. Show that:
a. $6x^2y$ is even.
b. $14a + 4b + 3$ is odd.

Solution
a. $6x^2y = 2 \cdot (3x^2y)$ and $3x^2y$ is an integer because the set of integers is closed under multiplication. So, by the definition of *even*, $6x^2y$ is even.
b. $14a + 4b + 3 = 2(7a + 2b + 1) + 1$. Now, $7a + 2b + 1$ is an integer because the set of integers is closed under addition and multiplication. So, by the definition of *odd*, $14a + 4b + 3$ is odd.

Here is a proof of the conjecture that the sum of two even integers is even. Following the proof is a logical analysis of it.

Proof Suppose that *m* and *n* are any even integers. According to the definition of *even*, there are integers *r* and *s* such that $m = 2r$ and $n = 2s$.
By substitution,

$$m + n = 2 \cdot r + 2 \cdot s$$
$$= 2 \cdot (r + s)$$

by the Distributive Property. Because $r + s$ is an integer, it follows by the definition of *even* that $m + n$ is even.

The proof establishes the conjecture as a theorem.

Theorem (Sum of Two Even Integers)
If *m* and *n* are even integers, then $m + n$ is an even integer.

When we put a proof in boldface, we expect you to be able to write one like it. Here is how you could have thought of this proof yourself.

Observe that the theorem can be stated in the form of a universal conditional statement.

∀ integers m and n, if m and n are even, then m + n is even.

The proof begins by supposing that m and n are any even integers. The word *any* is used because the theorem must be proved for *every* pair of even integers, not just for a few examples.

In the proof, both directions of the *if-and-only-if* form of the definition of even are used. At the beginning, the fact that

If t is even then t = 2k for some integer k

was used to deduce that $m = 2 \cdot r$ and $n = 2 \cdot s$ for some integers r and s. We add $2r$ and $2s$ because the consequent deals with $m + n$. The fact that

If t = 2k for some integer k, then t is even

was used to deduce that $m + n$ is even because it has the form of an even integer. And the proof was done.

Observe also that two *different* letters r and s were used in expressing m and n. The reason is that m and n were assumed to be *any* even integers. While it is conceivable that m and n might be equal, the likelihood is that they are not. If the same letter, say r, had been used to represent both m and n, then we would have necessarily had

$$m = 2 \cdot r \text{ and } n = 2 \cdot r \text{ and thus } m = n,$$

which may not be the case.

This method of proof is known as **direct proof** because you proceed directly from the hypothesis to the conclusion. Most proofs in mathematics are direct proofs. Because of its importance, this method is summarized below.

Method of Direct Proof of a Universal Conditional

1. Express the statement to be proved in the form
   ∀ x in S, if p(x) then q(x).
   (This step is often performed mentally.)
2. Start the proof by supposing that x is any element of S for which the antecedent $p(x)$ is true.
3. Use the definitions of the terms that occur in $p(x)$ and $q(x)$ and other known properties to make a chain of deductions ending with $q(x)$.

hypothesis, the conclusion is viewed as a goal to be reached, starting from a certain initial position, the hypothesis. Analysis of this goal leads to the realization that if a certain job is accomplished, then the goal will be reached. Call this job subgoal 1 or SG1 for short. For instance, in the Sum of Two Even Integers Theorem, the goal is: Show that $m + n$ is even. So, SG1: Show that $m + n = 2r$ for some integer r. In more complicated theorems, a longer backward chain may be generated by analyzing each successive subgoal to generate a new subgoal.

At a certain point, backward chaining may become difficult, but analysis of the current subgoal may suggest that it is reachable by a direct line of argument, forward chaining, beginning at the starting point. Using the information in the starting point, another piece of information, N1, is deduced; from that another piece of information, N2, is deduced, and so forth until finally one of the subgoals is reached. For instance, in the Sum of Two Even Integers Theorem, the starting point is: Suppose that m and n are arbitrarily chosen even integers.
N1: $m = 2k$ for some integer k; $n = 2l$ for some integer l.
N2: $m + n = 2k + 2l$, where k and l are integers.
N3: $m + n = 2(k + l)$, where k and l are integers.
When the chain is complete, the theorem is proved. A completed chain is:
Starting point → N1 → N2 → N3 → SG1 → goal.

Small Group Work Either assign or have students draw pieces of paper from a container having one of a variety of numbers written on each piece. Include integers, and positive and negative fractions. Each student could then prove or disprove a number was even or odd. Have students share answers and come up with a strategy of proof.

The following activity also can be done with **small groups.** Make a conjecture that all numbers of the form $m + 3$ are odd numbers. Have students draw an integer from a container and try it. Have them come up with a condition for this conjecture (M must be even) or a different conjecture ($2m + 3$ is an odd number).

ADDITIONAL EXAMPLES

1. Assume that x and y are integers. Show that:

a. $18x^5y^2$ is even.
$18x^5y^2 = 2(9x^5y^2)$. Because the set of integers is closed under multiplication, $9x^5y^2$ is an integer. So, by the definition of *even*, $18x^5y^2$ is even.

b. $12x + 8y + 17$ is odd.
$12x + 8y + 17 = 2(6x + 4y + 8) + 1$. Because the set of integers is closed under addition and multiplication, $6x + 4y + 8$ is an integer. Thus, by the definition of *odd*, $12x + 8y + 17$ is odd.

2. Prove the following theorem:
If m is odd, then $m^2 + 1$ is even.
Suppose m is odd. According to the definition of an odd integer, $m = 2k + 1$ for some integer k. Then,

$$\begin{aligned} m^2 + 1 &= (2k + 1)^2 + 1 \\ &= 4k^2 + 4k + 1 + 1 \\ &= 2(2k^2 + 2k + 1) \end{aligned}$$

Because $2k^2 + 2k + 1$ is an integer, $2(2k^2 + 2k + 1)$ is an even integer. Therefore, $m^2 + 1$ is even.

NOTES ON QUESTIONS

Questions 1-12: It is advisable to go through each of these questions carefully. By the time you have reached the proof in Question 12, all of its parts will have been discussed.

Question 13: A disproof of a universal statement requires only one example, but a proof requires a general argument.

Now we examine the logic of the proof. Here are the steps, written in *if-then* form.

Conclusions	*Justifications*
1. m and n are any even integers $\Rightarrow m = 2r$ and $n = 2s$, where r and s are integers.	Definition of even
2. $m = 2r$ and $n = 2s \Rightarrow m + n = 2r + 2s$	Addition property of equality
3. $m + n = 2r + 2s$ $\Rightarrow m + n = 2(r + s)$ and $r + s$ is an integer	Distributive property; closure of integers under addition
4. $m + n = 2(r + s)$ and $r + s$ is an integer $\Rightarrow m + n$ is even	Definition of even

Each step is an instance of the universal statement that is named at the right. The conclusion *m and n are any even integers* $\Rightarrow$ *m + n is even* follows by the Law of Transitivity.

Suppose $m = 2{,}624{,}316$ and $n = 111{,}778$. Now you know 2,624,316 is even and 111,778 is even and you have the theorem $\forall$*m and n, m and n are even* $\Rightarrow$ *m + n is even*. So by substitution, you have *2,624,316 and 111,778 are even* $\Rightarrow$ *2,624,316 + 111,778 is even*. You may conclude *2,624,316 + 111,778 is even* by the Law of Detachment.

Here is another example of a quite similar direct proof.

Example 2 Prove the following theorem:
If m is even and n is odd, then m + n is odd.

Solution First express the theorem as
$\forall$ *integers m and n, if m is even and n is odd, then m + n is odd.*
Start the proof by supposing that m is even and n is odd. Use the definitions of even and odd integers to go from the start of the proof to the conclusion that $m + n$ is odd.

Suppose m is any even integer and n is any odd integer. According to the definition of an even integer, $m = 2r$, for some integer r. According to the definition of odd integer, $n = 2s + 1$ for some integer s. Then

$$\begin{aligned} m + n &= 2r + (2s + 1) \\ &= 2r + 2s + 1 \\ &= 2(r + s) + 1 \end{aligned}$$

Because $r + s$ is an integer, $m + n$ is an odd integer.

Questions

Covering the Reading

In 1–3, show that the given integer is even or that it is odd.

1. 270 even; $270 = 2(135)$ **2.** 4875 odd; $4875 = 2(2437) + 1$ **3.** -59 odd; $-59 = 2(-30) + 1$

In 4 and 5, show that the integer t is even by expressing it in the form $2 \bullet k$ for some integer k.

4. $t = 6a + 8b$ for some integers a and b. $6a + 8b = 2 \bullet (3a + 4b)$

5. $t = a \bullet b$ where a and b are integers and b is even. See margin.

In 6 and 7, suppose that r and s are integers. Show that the defined integer is odd.

6. $10rs + 7$ See margin. **7.** $6r + 4s^2 + 3$ See margin.

8. Write the definition of odd integer using two *if-then* statements. See margin.

9. Supply the justifications for these conclusions in the proof of Example 2.

a. $m + n = 2r + (2s + 1)$ Addition Property of Equality

b. $2r + 2s + 1 = 2(r + s) + 1$ Distributive Property

10. Supply the missing steps in the proof of the following theorem:

If m and n are any odd integers, then m + n is an even integer.

Suppose ___a.___. Thus there exist integers r and s such that $m = 2r + 1$ and $n =$ ___b.___ according to ___c.___. Then $m + n =$ ___d.___ $= 2 \bullet$ ___e.___. Because ___f.___ is an integer, $m + n$ is an even integer. See margin.

11. Consider the following conjecture:

If c and d are any even integers, then c − d is an even integer.

a. If you were to prove this conjecture with a direct proof, what would be an appropriate sentence with which to begin the proof?

b. What statement must you then show true? $c - d$ is an even integer.

a) Sample: Suppose c and d are any even integers.

Applying the Mathematics

12. Prove the conjecture of Question 11. That is, establish that the conjecture is a theorem. See margin.

13. Find a counterexample to disprove the following:

If r • s is an even integer, then both r and s are even integers.

See margin.

14. a. Find the error in this "proof" of the following conjecture.

If m and n are any even integers, then m + n = 4k for some integer k.

Proof Suppose m and n are any even integers. Because m is even there is an integer k such that $m = 2k$. Also, because n is even, $n = 2k$. It follows that $m + n = 2k + 2k = 4k$, as was to be shown.

b. Find a counterexample to show that the conjecture in part **a** is false. See margin.

274=6 (m+n)
274=2k+2k
6=4k
6/4=k
3/2=k
thus k isn't an integer
has to be integer

Counterexample = to make false

ADDTIONAL ANSWERS

5. Since b is even, let $b = 2m$. $t = a \cdot 2m = 2 \cdot (am)$.

6. $10rs + 7 = 2(5rs + 3) + 1$; $5rs + 3$ is an integer by closure properties. Hence, $10rs + 7$ is odd by definition.

7. $6r + 4s^2 + 3 = 2(3r + 2s^2 + 1) + 1$; $3r + 2s^2 + 1$ is an integer by closure properties. Hence, $6r + 4s^2 + 3$ is odd by definition.

8. *If an integer n is odd, then n = 2k + 1 for some integer k. If for some integer k, n = 2k + 1, then n is an odd integer.*

10.a. m and n are any odd integers
b. $2s + 1$
c. the definition of an odd integer
d. $2r + 1 + 2s + 1$
e. $(r + s + 1)$
f. $r + s + 1$

12. Suppose c and d are any even integers. Thus, there exists integers r and s such that $c = 2r$ and $d = 2s$ according to the definition of an even integer. Then $c - d = 2r - 2s = 2(r - s)$. Because $(r - s)$ is an integer, $c - d$ is an even integer by definition.

13. Counterexample: Let $r = 4$ and $s = 5$. Then $r \cdot s = 4 \cdot 5 = 20$ is an even integer. But s is not an even integer.

14.a. m and n should be any even integers and not necessarily equal. By assigning $m = 2k$ and $n = 2k$, m and n are given the same value.
b. Counterexample: Let $m = 2$ and $n = 4$. $m + n = 2 + 4 = 6 = 4\left(\frac{3}{2}\right)$ But $\frac{3}{2}$ is not an integer.

NOTES ON QUESTIONS
Making Connections for Questions 15 and 16: These are properties that students have known and used for quite some time. The purpose here is to give practice in proof-writing on ideas that students know. Also, the simple ideas used have powerful applications in later lessons.

Question 20: This question could be approached by labeling statements.
p: smoking in the school building
q: assigned detention
The given statement is:
$p \Rightarrow q$ or *if p, then q.*
The contrapositive is
$\sim q \Rightarrow \sim p$.
If you are not assigned to detention, then you were not smoking in the building.

Question 22: For many years, the mission specialists on various space shuttle flights also have been civilians.

ADDITIONAL ANSWERS
15. Suppose m and n are any odd integers. There exists integers r and s such that $m = 2r + 1$ and $n = 2s + 1$ according to the definition of an odd integer. Then $m \cdot n = (2r + 1)(2s + 1) = 4rs + 2r + 2s + 1 = 2(2rs + r + s) + 1$. Because $(2rs + r + s)$ is an integer by closure properties, $m \cdot n$ is an odd integer by definition.

16. Counterexample:
Let $u = 7$ and $v = 3$.
$u - v = 7 - 3 = 4$
But 4 is not an odd integer.

17.a.
Let *p*: *Devin is a boy.*
Let *q*: *Devin plays baseball.*
Let *r*: *Devin is a pitcher.*
$p \Rightarrow q$
$q \Rightarrow r$
$\sim r$
$\therefore \sim p$
b. The argument correctly uses the Law of Indirect Reasoning and the Law of Transitivity.

In 15 and 16, either prove the given statement using a direct proof or disprove it by giving a counterexample.

15. *If m and n are any odd integers, then m • n is an odd integer.* See margin.

16. *If u and v are any odd integers, then u − v is an odd integer.* See margin.

Review

17. Consider the following argument.
If Devin is a boy, then Devin plays baseball.
If Devin plays baseball, then Devin is a pitcher.
Devin is not a pitcher.
∴ Devin is not a boy.
a. Express the argument symbolically.
b. Show that the argument is valid. *(Lesson 1-6)*
See margin.

18. Use the Law of Transitivity to explain why the following Law of Transitivity of Biconditionals is a valid reasoning pattern.
$p \Leftrightarrow q$
$q \Leftrightarrow r$
$\therefore p \Leftrightarrow r$ *(Lessons 1-5, 1-6)* See margin.

19. *Multiple choice.* Find the statement which is logically equivalent to
If figure ABC is a right triangle, then it has exactly two acute angles.
(a) *If ABC has exactly two acute angles, then it is a right triangle.*
(b) *If ABC does not have exactly two acute angles, then it is not a right triangle.*
(c) *If ABC is not a right triangle, then it does not have exactly two acute angles.*
(d) None of the above is logically equivalent to the given statement.
(Lesson 1-5) (b)

20. Given the statement:
Smoking in the school building is a sufficient condition for being assigned to detention.
a. Put this statement in *if-then* form.
b. Write a second *if-then* statement that is logically equivalent to the statement in part **a**. *(Lesson 1-5)* See margin.

21. Consider the universal statement
$\forall$ *real numbers x*, $f(x) = 3x^2 - 5x$.
Use substitution to find
a. $f(-2)$ **b.** $f(y + 2)$ **c.** $f(m + n)$
(Lesson 1-1, Previous course) See margin.

22. Is this universal statement true or false? Explain.
$\forall$ *U.S. astronauts t, t is a member of the military.*
(Lesson 1-1) See margin.

23. Write the statement
If n is an even integer, then $(-1)^n = 1$.
in the form
$\forall$ _?_, *if* _?_ *then* _?_. *(Lesson 1-1)*
$\forall$ integers n, if n is even, then $(-1)^n = 1$.

54

24. Consider the expression below.

$$\frac{y}{y-2} - \frac{y}{y+1}.$$

a. What universal statement could you use to complete the subtraction?

b. Complete the subtraction. *(Lesson 1-1, Previous course)* See below.

25. Evaluate 10!. *(Previous course)* 3,628,800

26. According to this graph of the function f, how many real number solutions are there to the equation $f(x) = 0$? *(Previous course)* 3

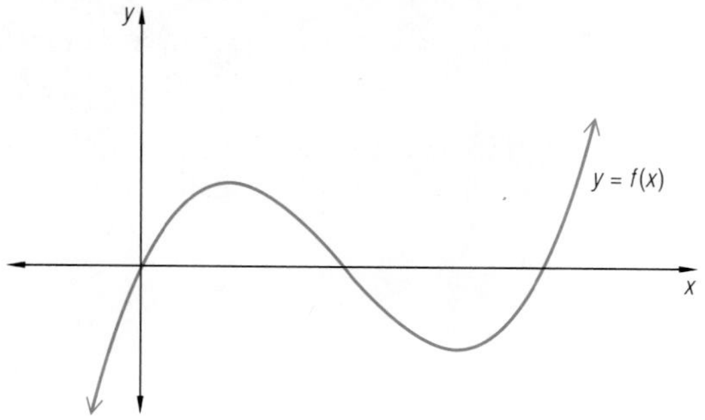

Exploration

27. Explore the even and odd powers of even and odd positive integers. That is, if m and n are positive integers, when is m^n even? When is m^n odd?
$\forall$ positive integers n, m^n is even when m is an even integer.
$\forall$ positive integers n, m^n is odd when m is an odd integer.

24. a) Sample: $\forall$ integers a, b, c, and d with $b \neq 0$ and $d \neq 0$, $\frac{a}{b} - \frac{c}{d} = \frac{ad - bc}{bd}$

b) $\frac{y(y+1) - y(y-2)}{(y-2)(y+1)} = \frac{3y}{y^2 - y - 2}$

FOLLOW-UP

MORE PRACTICE

For more questions on SPUR Objectives, use *Lesson Master 1-7,* shown below.

ADDITIONAL ANSWERS

18. $p \Rightarrow q$ $r \Rightarrow q$
$q \Rightarrow r$ $q \Rightarrow p$
$\therefore p \Rightarrow r$ $\therefore r \Rightarrow p$
$\therefore p \Leftrightarrow r$ by Law of Transitivity

20.a. *If you smoke in the school building, then you will be assigned to detention.*
b. *If you were not assigned to detention, then you were not smoking in the school building.*

21.a. $f(-2) = 3(-2)^2 - 5(-2) = 12 + 10 = 22$
b. $f(y + 2) = 3(y + 2)^2 - 5(y + 2) = 3y^2 + 12y + 12 - 5y - 10 = 3y^2 + 7y + 2$
c. $f(m + n) = 3(m + n)^2 - 5(m + n) = 3m^2 + 6mn + 3n^2 - 5m - 5n$

22. See Additional Answers in the back of this book.

NAME ____________

LESSON MASTER 1–7
QUESTIONS ON **SPUR** OBJECTIVES

■ **PROPERTIES** *Objective H (See pages 74–78 for objectives.)*

1. Show that $14x + 6y$ is even if x and y are integers.
$14x + 6y = 2(7x + 3y)$. Since x and y are integers, $7x + 3y$ is an integer because the integers are closed under addition and multiplication. Therefore, $14x + 6y$ is even.

2. Show that $\forall$ integers a, $20a + 7$ is odd.
$20a + 7 = 20a + 6 + 1 = 2(10a + 3) + 1$. Since a is an integer, $10a + 3$ is an integer, so $2(10a + 3) + 1$ is odd.

3. Supply the missing steps in the proof of the following theorem.

If m is an even integer and n is any integer that is a multiple of 3, then $m \cdot n$ is a multiple of 6.

Suppose m is an even integer and n is an integer that is a multiple of 3.
Then there exists an integer r such that $m =$ $2r$
according to the definition of even number. According to the definition of multiple, there exists an integer s such that $n = 3s$.
By substituting, $m \cdot n =$ $(2r) \cdot (3s)$ $= 6 \cdot$ rs.
Because rs is an integer, $m \cdot n$ is a multiple of 6.

In 4 and 5, give a direct proof of the statement.

4. If m and n are odd integers, then $m - n$ is even.
Suppose m and n are odd integers. Then $\exists$ integers r and s such that $m = 2r + 1$ and $n = 2s + 1$. By substituting, $m - n = (2r + 1) - (2s + 1) = 2r - 2s = 2(r - s)$. Since $r - s$ is an integer, $m - n$ is even.

5. If m is even, m^3 is a multiple of 8.
Suppose m is an even integer. Then $\exists$ an integer r such that $m = 2r$. By substituting, $m^3 = (2r)^3 = 8r^3$. Since r^3 is an integer, m^3 is a multiple of 8.

ADDITIONAL EXAMPLE
Show that the following form of argument is invalid.

If $\sim p$, *then* $\sim q$.
p
$\therefore q$

Construct a truth table which has as its final column heading the form $((\sim p \Rightarrow \sim q)$ ***and*** $p) \Rightarrow q$**. In the second row of the table, both premises are true, but the conclusion is false. Thus, the argument form is invalid. (This is a variant of Inverse Error.)**

Premise ↓   ***Premise*** ↓

p	q	$\sim p$	$\sim q$	$\sim p \Rightarrow \sim q$
T	T	F	F	T
T	F	F	T	T
F	T	T	F	F
F	F	T	T	T

Conclusion ↓

$(\sim p \Rightarrow \sim q)$ *and* p	q
T	T
T	F
F	T
F	F

Form ↓

$((\sim p \Rightarrow \sim q)$ *and* $p) \Rightarrow q$
T
F
T
T

Recall that the inverse of a conditional statement *If p then q* is *If not p then not q*.

The following argument illustrates a second type of invalid argument: the **inverse error**.

> *If a person is a member of the Spanish club, then the person speaks Spanish.*
> *William is not a member of the Spanish club.*
> $\therefore$ *William does not speak Spanish.*

This argument is of the form

> *For all x, if p(x) then q(x).*
> *not p(c), for a particular c.*
> $\therefore$ *not q(c).*

To see why this argument is invalid, notice that it is possible for the two premises to be true while the conclusion is false. The diagram at right shows a situation in which all Spanish club members speak Spanish, and William is not a member of the Spanish club, yet, William speaks Spanish.

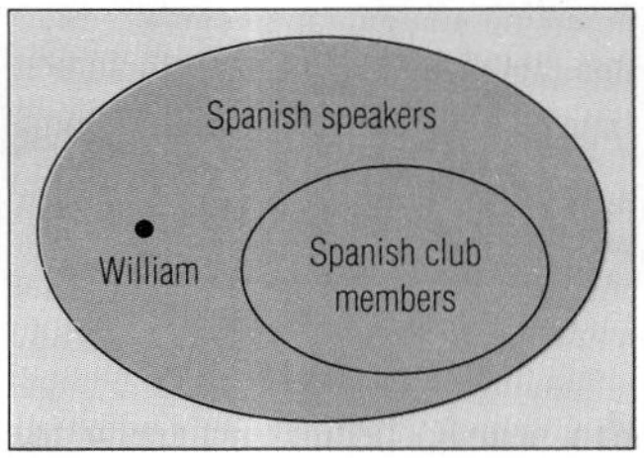

A third type of invalid argument occurs when a generalization is made prematurely. For example, consider the argument

> *For* $f(x) = x^3 - 6x^2 + 12x - 6$,
> $f(1) = 1$
> $f(2) = 2$
> $f(3) = 3$
> $\therefore f(n) = n$ *for all positive integers n.*

First, compute $f(4)$ and note that $f(4) \neq 4$. Therefore the conclusion is false. Because the premises are true but the conclusion is false, this argument is invalid. The error is called **improper induction**. In this type of invalid argument, the premises show that a property is true for some, but not all, elements in a set, and the conclusion asserts that the property is true for all elements in a set. Although improper induction is not a valid method of proof, it is often used by mathematicians to make conjectures which they then try to prove by other means.

For instance, here are premises and a conjecture you might make from them.

Premises: $2^3 = 8$
$8^7 = 2097152$
$(-4)^5 = -1024$

Conjecture: *A positive odd power of any even integer is even.*

It is important to note that an invalid argument may have a true conclusion. (The above conjecture is true.) But invalid arguments may lead to false conclusions, even from true premises. Only in a valid argument with true premises is the conclusion guaranteed to be true.

58

Questions

Covering the Reading

In 1–3, indicate whether the statement is true or false.

1. An argument form is invalid if there exists one argument of that form in which the premises are all true but the conclusion is false. True

2. If an argument is invalid, then it must have a false conclusion. False

3. If a given conditional statement is true, then its inverse must be false. False

4. Given the statement:
If an animal is a whale, then it is a mammal.
a. Draw a diagram to illustrate the relationship between whales and mammals.
b. Write its converse and determine whether it is true or false.
c. Write its inverse and determine whether it is true or false.
See below.

In 5–9, **a.** identify the type of the argument and **b.** tell whether or not it is valid.

5. $\sin 30° = \frac{\sqrt{1}}{2}$

$\sin 45° = \frac{\sqrt{2}}{2}$

$\sin 60° = \frac{\sqrt{3}}{2}$

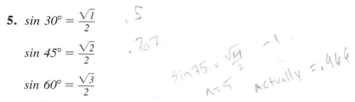

$\therefore \sin (15n)° = \frac{\sqrt{n-1}}{2}$ for all integers $n \geq 2$.
a) improper induction; b) invalid

6. *If you buy a ticket to the concert, you will go.*
You don't buy a ticket to the concert.
∴ *You don't go.*
a) inverse error; b) invalid

7. *If a person does not register to vote, then the person cannot vote.*
John can vote.
∴ *John has registered to vote.*
a) Law of Indirect Reasoning; b) valid

8. *If this is a presidential race, education is a top issue.*
Education is a top issue.
∴ *This is a presidential race.*
a) converse error; b) invalid

9. $\forall$ *real numbers* x, *if* $x > 2$, *then* $x^2 > 4$.
$\forall$ *real numbers* x, *if* $x > 3$, *then* $x^2 > 9$.
$\therefore \forall$ *real numbers* x, *if* $x > n$, *then* $x^2 > n^2$.
a) improper induction; b) invalid

4. a)

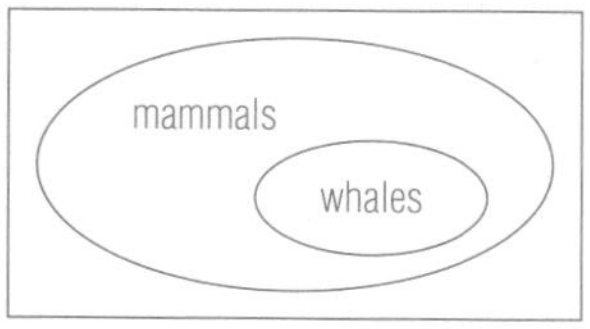

b) *If an animal is a mammal, then it is a whale.* False

c) *If an animal is not a whale, then it is not a mammal.* False

NOTES ON QUESTIONS
Question 5: The pattern shown may help students remember the sines of 30°, 45°, and 60°, but it does not extend to other numbers.

Small Group Work for Question 21: This question could be expanded and done as a **small group.** Mathematical or non-mathematical statements could be used. Results could be presented to the entire class as part of an assessment.

Applying the Mathematics

10. Given the conditional statement $p \Rightarrow q$.

a. Write the converse. $q \Rightarrow p$

b. Write the contrapositive of the converse. $\sim p \Rightarrow \sim q$

c. Identify your answer to part **b** as either the inverse, converse, or contrapositive of the original statement. inverse

11. Assume that Peter has an answering machine attached to his telephone. He turns it on whenever he is leaving home. When Sara called, she got the message on the answering machine. Sara concluded that Peter was not home.

a. Write the form of argument Sara used to draw her conclusion.

b. Is it valid or invalid? Explain.
See margin.

12. Given the following argument.

$\forall\ x,\ x = 3 \Rightarrow x^2 = 9.$
$x \neq 3.$
$\therefore x^2 \neq 9.$

a. Write the form of the argument. See margin.

b. Is the argument valid or invalid? Justify your answer.
invalid; inverse error

13. Given the following argument:

For all persons p, if p is President of the U.S., then p is at least 35 years old.
Queen Elizabeth II is at least 35 years old.
∴ Queen Elizabeth II is President of the U.S.

a. Write the form of this argument. See margin.

b. Are the premises true? Is the conclusion true? Yes, No

c. Is the argument valid or invalid? Justify your answer.
invalid; converse error

In 14–16, write the form of the argument and determine whether the argument is valid or invalid. If the argument is invalid, identify the type of error made in the argument.

14. *If the land is covered with ice, then the land is Antarctica. If the land is Antarctica, then there are research stations there. If the land has research stations, then scientific study is being conducted. Scientific study is being conducted on the land. Therefore, the land is covered with ice.* See margin.

15. *If x is a real number, then* $x^2 \geq 0$. *If x is an imaginary number, then* $x^2 < 0$. *2i is an imaginary number. Therefore 2i is not a real number.* See margin.

16. *If you send a minimum order of \$10 to a mail-order house, then your name is put on a mailing list. If your name is on the mailing list, then you will receive many catalogs in the mail. You receive many catalogs in the mail. Therefore, you must have sent a minimum order of \$10 to the mail-order house.* See margin.

17. By considering all possible truth values that p and q may have:

a. Show that $((p \Rightarrow q)$ *and* $\sim p) \Rightarrow \sim q$ is not always true. See margin.

b. What type of error does this form exemplify? inverse error

ADDITIONAL ANSWERS

11.a. Let *p*: *not home*, and *q*: *answering machine is on.*
$p \Rightarrow q$
q
$\therefore p$
b. invalid; converse error

12.a. Let *p*: $x = 3$, and *q*: $x^2 = 9$.
$p \Rightarrow q$
$\sim p$
$\therefore \sim q$

13.a. Let *q(p)*: *p is President of the U.S.* Let *r(p)*: *p is at least 35 years old.* Let Q be Queen Elizabeth.
$\forall\ p,\ q(p) \Rightarrow r(p)$
$r(Q)$
$\therefore q(Q)$

14. Let *p*: *the land is covered with ice.* Let *q*: the land is Antartica. Let *r*: *there are research stations there.* Let *s*: *scientific study is being conducted.*
$p \Rightarrow q$
$q \Rightarrow r$
$r \Rightarrow s$
s
$\therefore p$
invalid; converse error

15. Let *p(x)*: *x is a real number.* Let *q(x)*: $x^2 \geq 0$. Let *r(x)*: *x is an imaginary number.*
Let *c*: $x = 2i$.
$\forall\ x,\ p(x) \Rightarrow q(x)$
$\forall\ x,\ r(x) \Rightarrow \sim q(x)$
$r(c)$
$\therefore \sim p(c)$
valid; ($r(c) \Rightarrow \sim q(c)$ by the Law of Detachment, and $\sim q(c) \Rightarrow \sim p(c)$ by the Law of Indirect Reasoning)

16. Let *p*: *you send a minimum order of \$10 to a mail-order house.* Let *q*: *your name is put on a mailing list.* Let *r*: *you receive many catalogs in the mail.*
$p \Rightarrow q$
$q \Rightarrow r$
r
$\therefore p$
invalid; converse error

17.a. See Additional Answers in the back of this book.

18. Refer to the two arguments below. (We define a trapezoid as a quadrilateral with at least one pair of parallel sides.)

I. *All squares are rectangles.*
All squares are trapezoids.
$\therefore$ *All rectangles are trapezoids.*

II. *All squares are rectangles.*
All squares are rhombuses.
$\therefore$ *All rectangles are rhombuses.*

a. For argument I, are the premises true? Is the conclusion true?
b. For argument II, are the premises true? Is the conclusion true?
c. Write the form of these arguments. See margin.
d. Use your results from parts **a** and **b** to determine whether your argument form for part **c** is valid or invalid. invalid

a) Yes, Yes; b) Yes, No

Review

19. Prove the following theorem:
If m is any even integer and n is any odd integer, then $m \cdot n$ is even. *(Lesson 1-7)* See margin.

20. Given premises (1)-(5), deduce a valid conclusion and justify your reasoning.
(1) p
(2) $q \Rightarrow r$
(3) $not\ t \Rightarrow not\ s$
(4) $p \Rightarrow q$
(5) $not\ s \Rightarrow not\ r$ *(Lesson 1-6)* See margin.

21. Give an example of two statements p and q such that p *and* q and $p \Rightarrow q$ have different truth values. *(Lesson 1-3, 1-5)*
See margin.

22. Write the following statement as an *if-then* statement.
$\forall$ *integers a and b*, $\frac{a}{b} = \sqrt{2}$ *only if* $\frac{a^2}{b^2} = 2$. *(Lesson 1-5)*
See margin.

23. Write the negation of the statement
If Vanna White is the hostess, then the show is Wheel of Fortune.
(Lessons 1-3, 1-5)
Vanna White is the hostess, and the show is not Wheel of Fortune.

24. Match each property with its name. *(1-1, Previous course)*

a. $\forall x$ *and* y, $xy = yx$. (iii)
b. $\forall x$ *and* y, *if* x *and* y *are integers, then* $x + y$ *is an integer.* (v)
c. $\forall$ x, y *and* z, $z(x + y) = zx + zy$. (iv)
d. $\exists y$ *such that* $\forall x$, $x + y = x$. (ii)
e. $\forall x$, $\exists y$ *such that* $x + y = 0$. (i)

(i) Additive Inverse Property
(ii) Additive Identity Property
(iii) Commutativity of Multiplication
(iv) Distributivity of Multiplication over Addition
(v) Closure of Addition
(vi) Associativity of Addition

18.c. Arguments I and II both have the form below.
$p \Rightarrow q$
$p \Rightarrow r$
$\therefore q \Rightarrow r$

19., 20., 21., 22. See Additional Answers in the back of this book.

NAME ____________

LESSON **MASTER 1–8**
QUESTIONS ON **SPUR** OBJECTIVES

■ **PROPERTIES** *Objective G (See pages 74–78 for objectives.)*
In 1 and 2, a. is the argument valid? b. is the conclusion true?

1. *If $a < 5$, then $a^2 < 25$.*
$-8 < 5$
$\therefore 64 < 25$
a. yes **b.** no

2. *If the freezing point of water is 212°F, then the moon is made of green cheese.*
The moon is not made of green cheese.
$\therefore$ The freezing point of water is 212°F.
a. no **b.** no

3. Under what conditions is an argument guaranteed to produce a true conclusion?
when the premises are true and the argument is valid

4. Joe noticed the following facts:
3, 4, and 5 form a Pythagorean triple, and $3 \cdot 4 \cdot 5$ is divisible by 60.
5, 12, and 13 form a Pythagorean triple, and $5 \cdot 12 \cdot 13$ is divisible by 60.
7, 24, and 25 form a Pythagorean triple, and $7 \cdot 24 \cdot 25$ is divisible by 60.
Joe concluded that if a, b, and c form a Pythagorean triple, then abc is divisible by 60. Is Joe's conclusion valid? Why or why not?
No, it is an example of Improper Induction.

5. Give an example of a valid argument that has a false conclusion.
sample: *If an American has "Ronald" as his first name, then he was once president. Mr. McDonald is an American whose first name is Ronald. $\therefore$ Mr. McDonald was once president.*

6. Give an example of an invalid argument that has a true conclusion.
sample: *If an American has "Ronald" as his first name, then he was once president. Mr. Reagan was once president. $\therefore$ Mr. Reagan's first name is Ronald.*

NAME ____________

■ **USES** *Objective J*

7. a. Write a symbolic form for the argument below.
If a year is divisible by 10, then the U.S. census is taken that year.
The census was taken last year.
$\therefore$ Last year was divisible by 10.

$p \Rightarrow q$
q
$\therefore p$

b. Is the argument valid? no

8. Consider the following statement.
All math teachers can balance a checkbook.

a. Illustrate this situation with a diagram.

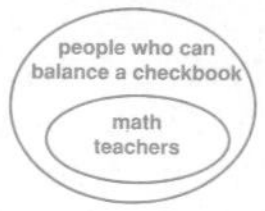

b. Write the converse of the given statement.
Anyone who can balance a checkbook is a math teacher.

c. Write the inverse of the given statement.
Anyone who is not a math teacher cannot balance a checkbook.

d. Suppose Mr. Smith is not a math teacher. Name the type of argument that would lead to the conclusion that Mr. Smith cannot balance a checkbook. inverse error

e. Is the type of argument named in part **d** valid? no

FOLLOW-UP

MORE PRACTICE
For more questions on SPUR Objectives, use *Lesson Master 1-8,* shown on page 61.

EXTENSION
A well-known improper induction example comes from geometry. Begin with a circle, mark n points on it, connect the points with chords, and count the number of regions created.

Circle	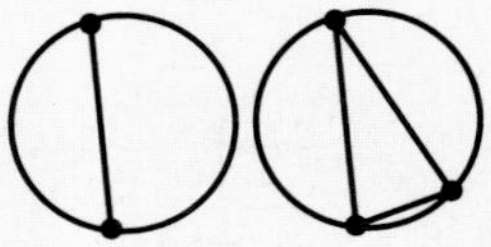	
Number of points	2	3
Number of regions	2	4
Circle	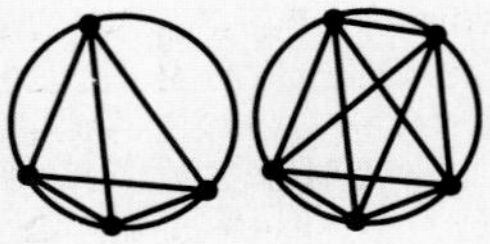	
Number of points	4	5
Number of regions	8	16

Students will conjecture that n points will result in 2^{n-1} regions. In fact, for $n = 6$, there are only 31 regions!

PROJECTS
The projects for Chapter 1 are described on page 70. **Project 5**, dealing with valid and invalid arguments, is related to the content of this lesson.

ADDITIONAL ANSWERS
26.a. See Additional Answers in the back of this book.

25. a. Write a universal statement of the form ∀ *x and a*, ... that could be used to factor the expression $x^2 - a^2$.
b. Use substitution with the statement from part **a** to factor
i. $x^2 - 16$ **ii.** $9y^4 - z^2$
c. Use substitution with the statement from part **a** to quickly calculate the value of $48^2 - 52^2$. *(Lesson 1-1, Previous course)*
See below.

Exploration

26. In this lesson a circle diagram was used to represent premises in an argument. These diagrams are often called **Euler circles** or **Venn diagrams.** Here is how premises are represented by Euler circles.

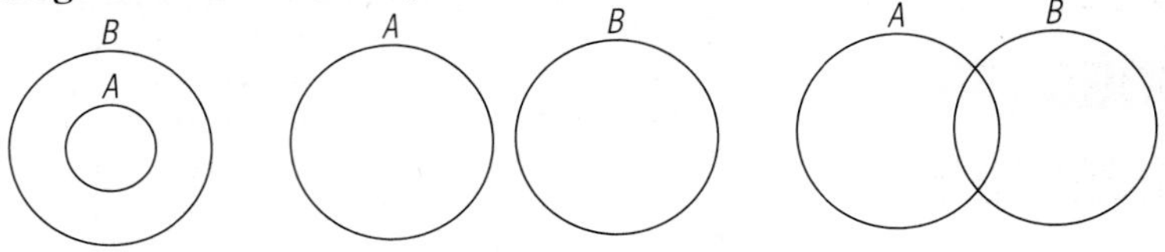

All A are B. *No A are B.* *Some A are B.*

a. Represent each argument below using Euler circles and determine whether the argument is valid or invalid. See margin.
i. *If an animal is a reptile, then it is cold-blooded.*
All dinosaurs are reptiles.
∴ All dinosaurs are cold-blooded.
ii. *Some artists paint in oil.*
Some artists paint with watercolors.
∴ Some oil painters also paint with watercolors.
iii. *All isosceles triangles have reflection symmetry.*
No triangle with reflection symmetry is scalene.
∴ No isosceles triangle is scalene.
b. Compose three arguments and represent their premises by Euler circles. Indicate how you can decide whether each argument is valid or invalid, using the method of Euler circles.
Answers will vary.

25. a) ∀ *real numbers* x *and* a, $x^2 - a^2 = (x - a)(x + a)$
b) i. Let $a = 4$. $x^2 - 16 = (x - 4)(x + 4)$
ii. Let $x = 3y^2$ and $a = z$. $9y^4 - z^2 = (3y^2 - z)(3y^2 + z)$
c) Let $x = 48$ and $a = 52$. $48^2 - 52^2 = (48 - 52)(48 + 52) = -4 \cdot 100 = -400$

LESSON 1-9

Different Kinds of Reasoning

The previous lessons of this chapter have discussed how reasoning is used in mathematics, and many of the examples have shown how mathematical reasoning and logic can also be of use in the world outside mathematics.

The reasoning used in mathematical proof is called **deductive reasoning**. It follows stricter standards of validity than the standards applied in daily pursuits or in science or medicine or other fields such as psychology. The logically valid nature of deductive reasoning is a major reason why people attempt to use mathematics in other endeavors. However, other kinds of reasoning are employed informally in mathematics. For instance, inductive reasoning is often used for coming up with conjectures. In order for a conjecture to become a theorem, however, it must be proved using valid forms of argument.

Reasoning that would be invalid if used in a mathematical proof may be used quite successfully outside of mathematics. You use **inductive reasoning** whenever you make a generalization based on the evidence of many examples. For instance, by the third month of the school year you may have concluded that every time Mr. Smith says "Put away all your books and papers," he gives a pop quiz. This reasoning is invalid mathematically (it is an example of improper induction, for Mr. Smith could change his mind), but it is very common and is often quite reliable.

Scientists use inductive reasoning when they make hypotheses based on data gathered from their observations on experiments. The Italian scientist Galileo Galilei (1564 – 1642) was one of the first to use this method systematically, in his study of the motion of falling bodies. He dropped objects from a number of places, including the Leaning Tower of Pisa, and measured the time it took these objects to fall certain distances. By observing the pattern of the numbers he obtained, he formulated a law of falling bodies that we write today as

$$d = \frac{1}{2}gt^2,$$

where d is the distance an object falls in t seconds under the influence of the acceleration due to gravity g.

Leaning Tower of Pisa

LESSON 1-9

OBJECTIVES

There are no SPUR objectives for any reading lesson.

TEACHING NOTES

The last lesson of each chapter is a "reading lesson," devoted to content that is as important as other content of the chapter but not as amenable to being mastered or tested. Typically, these lessons provide an opportunity to learn about interesting applications of the ideas of the chapter (as in this lesson) or to treat an interesting mathematical idea in further depth. (See page T37 for more detail. Students who have studied from *Functions, Statistics, and Trigonometry* will be familiar with this feature.

Students should read this lesson and do the questions just as with other lessons, and a day should be devoted to discussing it in class. However, no questions from this lesson are on the Progress Self-Test, the Chapter Review, or other tests published separately for this book.

The reading lessons serve a dual purpose: they provide an opportunity to discuss mathematics in a less threatening atmosphere ("it will not be on the test") and they provide a day to do review questions from the rest of the chapter in preparation for the Progress

Self-Test. To allow more time to review, there are no questions in the category "Applying the Mathematics."

You might begin the lesson by asking students for differences between deductive, inductive, and probabilistic reasoning. The first two of these are the subject of **Question 1**; the latter is required when an element of chance is present in the situation. Then ask whether this view of inductive reasoning agrees with the scientific method that they have discussed in their science classes.

Actually, both diagnostic reasoning and inductive reasoning have probabilistic components. In diagnostic reasoning, a person might conclude that a patient has a particular illness or a problem is due to such-and-such, but seldom are definitive statements made. In inductive reasoning, as the number and variety of true instances increase, faith in the conclusion continues to grow.

To aid in a discussion of diagnostic reasoning, you might ask students to bring in copies of owner manuals for appliances that they have at home to determine which ones have trouble-shooting sections. Typical items whose manuals include charts for trouble-shooting are cameras, answering machines, cassette recorders, and televisions. (See **Question 20**.)

You might have students consider what type of reasoning is used by the National Transportation Safety Board as it investigates air or rail disasters to determine the cause of an accident. Data and observations from many sources are collected in order to draw a conclusion using both diagnostic and inductive reasoning.

Scientists are usually careful to say that no individual experiment or series of experiments can *prove* beyond any doubt that a certain law or scientific theory holds in general. Albert Einstein once said: "No amount of experimentation can ever prove me right; a single experiment can prove me wrong." However, a large number of experiments testing many different consequences of the law or theory make it *more likely* that the theory is correct. A theory such as Galileo's law of falling bodies has now been tested in many thousands of individual instances. Although we may not be 100% sure it is correct, we consider it to be so likely that sophisticated systems such as battleship cannons, missile defense mechanisms, and artificial satellites are built using it. People entrust their lives to technology based on this theory when they fly in an airplane.

A different kind of reasoning, called **diagnostic reasoning**, is used in medical diagnosis, car repair, and in various kinds of trouble shooting. Doctors, during their training, learn about illnesses and their symptoms in *if-then* statements of the form

If a patient has illness i, then he or she displays symptoms s_1, s_2, ...

However, doctors do not usually know the illness and then determine the symptoms. Instead they begin with the symptoms displayed or reported by the patient and try to diagnose the illness. For example, if Theresa goes to the infirmary complaining of a congested nose, chest pains, and overall lethargy, and if the nurse or doctor observes red eyes and a fever, the most reasonable diagnosis might be that Theresa has the flu. The reasoning is: If Theresa displays symptoms s_1, s_2, ..., then she has illness i. This diagnosis is based on the knowledge that if a person has the flu, then she displays Theresa's symptoms and maybe a few more such as an upset stomach and a cough. Thus, medical personnel usually employ converses of *if-then* statements whose truth values are not known beyond some level of certainty. Mathematically, this reasoning is unreliable; the converse may or may not be true. Theresa's complaints could also be the symptoms of bronchitis or allergy. But this is often the best medical personnel can do; and our confidence in medical personnel through this process is based on their training and previous experience.

Diagnostic reasoning involves using the converse of a true statement to make a conjecture about an illness, needed repair, or correction. If used in a proof, it would be called "converse error." However, doctors, auto mechanics, and others who use it seldom claim that their diagnoses are always correct. They know they are conjecturing. There is always the possibility that a wrong diagnosis may be made.

For instance, consider trying to determine the cause of a problem in the operation of a video cassette recorder. Here is a part of the Owner's Manual of a Sylvania Model VC3140 SL 01 VCR.

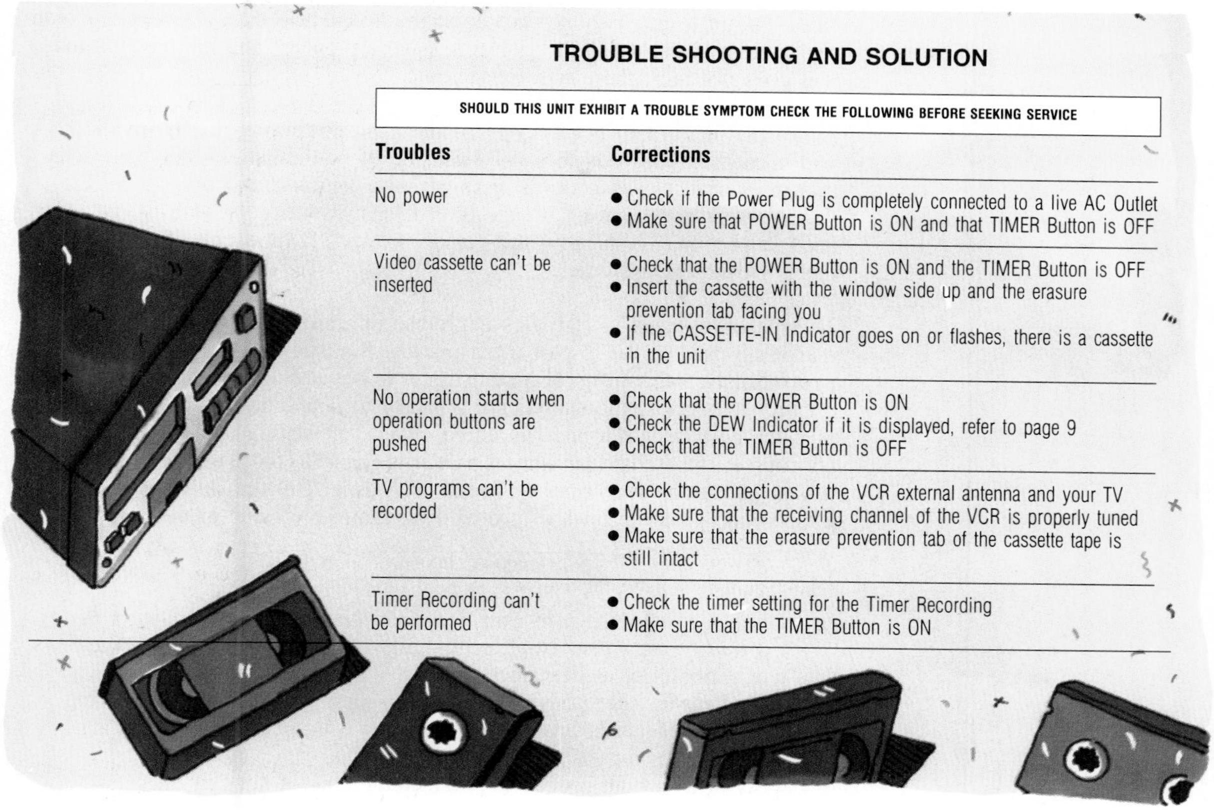

TROUBLE SHOOTING AND SOLUTION

SHOULD THIS UNIT EXHIBIT A TROUBLE SYMPTOM CHECK THE FOLLOWING BEFORE SEEKING SERVICE

Troubles	Corrections
No power	• Check if the Power Plug is completely connected to a live AC Outlet • Make sure that POWER Button is ON and that TIMER Button is OFF
Video cassette can't be inserted	• Check that the POWER Button is ON and the TIMER Button is OFF • Insert the cassette with the window side up and the erasure prevention tab facing you • If the CASSETTE-IN Indicator goes on or flashes, there is a cassette in the unit
No operation starts when operation buttons are pushed	• Check that the POWER Button is ON • Check the DEW Indicator if it is displayed, refer to page 9 • Check that the TIMER Button is OFF
TV programs can't be recorded	• Check the connections of the VCR external antenna and your TV • Make sure that the receiving channel of the VCR is properly tuned • Make sure that the erasure prevention tab of the cassette tape is still intact
Timer Recording can't be performed	• Check the timer setting for the Timer Recording • Make sure that the TIMER Button is ON

You know that if the POWER button is in the OFF position, then there is no power. The first row of this trouble-shooting manual reasons from the converse: If there is no power, it may be that the POWER button is in the OFF position.

Similarly, the operating instructions to the VCR (elsewhere in the manual) indicate that if (a) the POWER button is OFF or (b) the TIMER button is ON or (c) the cassette is placed with the window side down or (d) there is a cassette already in the machine, then a video cassette cannot be inserted. So the trouble-shooting section of the manual states the converse: If a video cassette cannot be inserted, then perhaps the POWER button is OFF or the TIMER button is ON or the cassette is being placed with its window side up or there is a cassette already in the machine.

Making Connections
Many manufacturing firms have quality control departments which use probabilistic reasoning to maintain a certain level of confidence in the firm's product. A certain number of items from a production lot are tested. If the defective rate is x%, then one would expect to find no more than y defective items among the sampled items. If more than y defective items were found, then it is likely that the defective rate would be greater than x%.

ADDITIONAL EXAMPLES
Here are some additional questions you may ask regarding trouble shooting the VCR.

1. What should you check if the VCR does not react to the pushing of operation buttons?
Check that the POWER Button is ON, the DEW indicator is off, and that the TIMER Button is OFF.

2. What can you not do if the CASSETTE-IN indicator is on or is flashing?
You cannot insert another cassette into the VCR.

Concluding the inverse of an if-then statement is *inverse error* when used in mathematics, but there are times in everyday occurrences when you may be expected to reason this way. When a friend says, "If you want to go with us, be at my house at 6 P.M.," the friend may want you to think "If I don't want to go, I should not be at the house at 6 P.M." When a person says, "If it rains, we aren't going to have a picnic," the inference is that if it does not rain, there will be a picnic.

All of this may disturb you. Why is the same reasoning that is invalid within mathematics possibly valid outside of mathematics? The answer is that in everyday conversation, *if* sometimes means *if and only if*. So the friend's sentence means "If (and only if) you want to go with us, be at my house at 6 P.M." Similarly, the person means "If (and only if) it rains, there will not be a picnic."

The reasoning used in statistics is precise in a slightly different way from reasoning used in the rest of mathematics. Statistics makes use of **probabilistic reasoning**. If a statistician tosses a coin 100 times and obtains 100 heads, the statistician is likely to conclude that the coin is *not fair*: the coin does not have an equal chance of turning up heads or tails. Of course, the statistician knows that it is *possible* for a fair coin to be tossed 100 times and come up heads each time. But the statistician also knows, applying the laws of probability, that the chance of this happening is only $\frac{1}{2^{100}}$, or about $\frac{1}{1267651000000000000000000000000}$. This leads the statistician to reason as follows.

> Let p: *the coin is fair*, and let q: *100 consecutive tosses were heads.*
> *p and q* has a very small probability of occurring (applying the laws of probability as described above).
> That means that *not (p and q)* has a high probability of occurring.
> By De Morgan's Laws, *not (p and q)* is equivalent to *(not p) or (not q)*.
> So, with a high probability, either the coin is not fair or 100 consecutive heads were not heads.
> But the 100 heads did occur.
> Therefore, (with only a tiny amount of doubt) the coin is not fair.

The statistician could not have concluded that the coin is not fair with the same level of confidence after tossing it 5 times and getting heads each time. That is because when q is *5 consecutive heads occur*, the probability of *p and q* is $\frac{1}{2^5} = \frac{1}{32}$, a much greater probability than getting 100 heads in a row. Thus, the statistician would look for stronger evidence and use a larger number of trials. In other words, by tossing the coin so many times the statistician reduces the chances for error, or a wrong conclusion.

Although very few people would demand so low a probability of error when deciding on the fairness of a coin, when people's lives are at stake,

Summary

In this chapter you have studied the logic of statements the reasoning processes used in mathematics and the everyday world. Universal statements assert that all members of a set have a certain property, and existential statements assert that at least one member of a set has a certain property. To write the negation of a statement, you express what it would mean for the statement to be false. The negation of a universal statement is existential, and the negation of an existential statement is universal.

The three words *and, or,* and *not* are important in the study of logic, and have applications to the design and evaluation of computer logic networks. De Morgan's Laws can be used to negate logical expressions containing *and* and *or*.

Conditional statements, symbolized $p \Rightarrow q$, are false under only one set of circumstances: when the antecedent is true and the consequent is false. As was the case with existential and universal statements, conditionals can be expressed in a variety of forms.

A conditional and its contrapositive are logically equivalent, that is, they always have the same truth values. But a conditional and its converse may have different truth values, as may a conditional and its inverse.

An argument consists of premises and a conclusion. An argument is said to be valid if any argument of its form that has true premises has a true conclusion. The Laws of Detachment (*modus ponens*), Indirect Reasoning (*modus tollens*), and Transitivity are valid forms of argument. Converse Error, Inverse Error, and Improper Induction are invalid forms of argument. The Method of Direct Proof of a Universal Statement involves the use of valid forms of argument to make a chain of deductions leading from the hypothesis to the conclusion of the conditional to be proved.

Vocabulary

Below are the most important terms and phrases for this chapter.
For the starred (*) terms you should be able to give a definition of the term.
For the other terms you should be able to give a general description and a specific example of each.

Lesson 1-1
$\forall$, $\exists$, quantifier
*universal statement
Law of Substitution
*existential statement
*counterexample

Lesson 1-2
*negation, *not p*
truth table

Lesson 1-3
**p or q*, **p and q*
exclusive *or*, inclusive *or*
logical expression
*logically equivalent expressions, $\equiv$
De Morgan's Laws

Lesson 1-4
input-output table
network of logic gates
functionally equivalent networks

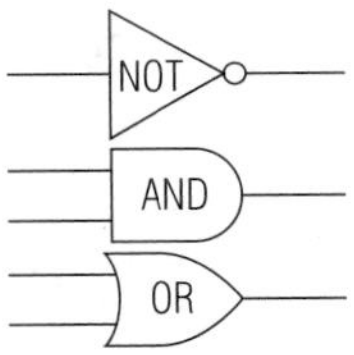

Lesson 1-5
*conditional statement, $p \Rightarrow q$
*antecedent, hypothesis
*consequent, conclusion
*negation of a simple (or universal) conditional
*contrapositive
Contrapositive Theorem
*converse, *inverse
**p* if and only if *q*, $p \Leftrightarrow q$
*biconditional
**p* only if *q*
sufficient condition
necessary condition

Lesson 1-6
argument, premises, $\therefore$
conclusion of an argument
*valid argument, valid conclusion
Law of Detachment, *modus ponens*
Law of Indirect Reasoning, *modus tollens*
Law of Transitivity

Lesson 1-7
*even integer, odd integer
Method of Direct Proof

Lesson 1-8
invalid argument
converse error, inverse error
improper induction

Lesson 1-9
inductive, deductive reasoning
diagnostic reasoning
concluding the inverse
probabilistic reasoning

SUMMARY

The Summary gives an overview of the entire chapter and provides an opportunity for students to consider the material as a whole. Thus, the Summary can be used to help students relate the various concepts presented in the chapter.

VOCABULARY

Terms, symbols, and properties are listed by lesson to provide a checklist of concepts a student must know. Emphasize to students that they should read the vocabulary list carefully before starting the Progress Self-Test. If students do not understand the meaning of a term, they should refer back to the individual lesson.

Definitions or descriptions of all terms in vocablulary lists may be found in the Glossary or the List of Theorems in the student text.

PROGRESS SELF-TEST

Whereas end-of-chapter materials may be considered optional in some texts, they should not be considered optional in UCSMP *Precalculus and Discrete Mathematics.* The Progress Self-Test provides the opportunity for feedback and correction; the Chapter Review provides additional opportunities for practice. It is at this point that the material "gels" for many students, allowing them to solidify skills and concepts before a test. In general, students' performance is markedly improved after doing these pages.

USING THE PROGRESS SELF-TEST

Assign the Progress Self-Test as a one-night assignment. Worked-out *solutions* for all questions are in the Selected Answers section of the student text. Encourage students to take the Progress Self-Test honestly, grade themselves, and then be prepared to discuss the test in class. (See page T35 for more information on the Progress Self-Test.)

Advise students to pay special attention to those Chapter Review questions (pages 74–78) which correspond to questions missed on the Progress Self-Test. A chart provided with the Selected Answers keys the Progress Self-Test questions to the lettered SPUR Objectives in the Chapter Review or to the Vocabulary. It also keys the questions to the corresponding lessons where the material is covered.

Progress Self-Test

See margin for answers not shown.

Take this test as you would take a test in class. Then check the test yourself using the solutions at the back of the book.

In 1 and 2, match each sentence with the best description of it:

(a) statement
(b) universal statement
(c) existential statement
(d) none of these

1. $4 + x \geq 17$ (d)

2. *There is a condor that has been born in captivity.* (c)

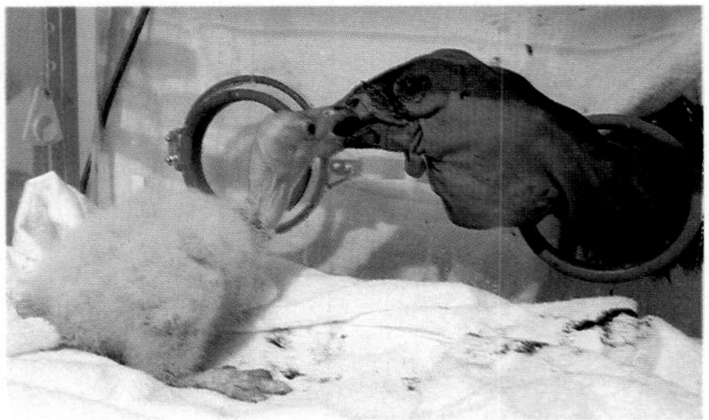

3. *True* or *false*? $\forall$ *mammals m, m is not a mosquito.* True

4. Identify the statements equivalent to *All Olympic medalists are tested for illegal drug use.*

(a) *No Olympic medalists are tested for illegal drug use.*
(b) *There exists an Olympic medalist who is tested for illegal drug use.*
(c) *Some Olympic medalist is tested for illegal drug use.*
(d) *For all Olympic medalists m, m is tested for illegal drug use.*
(e) *If an individual is an Olympic medalist, then the individual is tested for illegal drug use.*
(f) None of these sentences is equivalent to the original. (d) and (e)

5. *True* or *false*? $\forall$ *real numbers y,* $0 + y = y$. True

6. Write the negation of the statement in Question 5.

7. The following statement is true:

$$\forall \text{ real numbers } x, \sqrt{x^2} = |x|.$$

Use substitution to simplify $\sqrt{(-7c)^2}$, where c is a real number. $\sqrt{(-7c^2)} = |-7c|$

8. Rewrite $-8 \leq x < 12$ spelling out the implied *and* and *or*.

9. Write the negation of *The bald eagle is the national bird and "The Star-Spangled Banner" is the national anthem.*

10. A researcher has collected information on courses taken by students of various ethnic backgrounds. Because the researcher wants to look at the interaction of ethnicity and courses taken, student files have been coded as follows:

1 Caucasian
2 Black
3 Hispanic
4 Asian
5 Native American
6 Other

At this point the researcher just wants to look at the results for blacks and Hispanics. Consider the following program fragment:

```
 5 REM ENTRY PROGRAM A
10 INPUT C
15 IF NOT (C = 2 OR C = 3) THEN 10
20 REM BEGIN PROGRAM A
   .
   .
   .
```

Use De Morgan's Laws to change line 15 to an equivalent statement so that the output of the program remains the same.

11. Write the following statement as an *if-then* statement:
A person can be admitted to an R-rated movie only if the person is at least 17 years old.

12. Imagine that s is a particular real number. Consider the sentence

If $s < 4$ *then* $|s| < 4$.

Under what circumstances is this sentence false? if $s < 4$ and $|s| \geq 4$

13. *Multiple choice.* A conditional is logically equivalent to which of the following?
(a) its converse
(b) its inverse
(c) its contrapositive (c)

14. *Multiple choice.* Identify the sentence which is the contrapositive of
If you travel to Africa, then you need to have a malaria shot.
(a) *If you need to have a malaria shot, then you are traveling to Africa.*
(b) *If you do not need to have a malaria shot, then you are not traveling to Africa.*
(c) *If you are not traveling to Africa, then you do not need to have a malaria shot.* (b)

15. Write the following theorem in an equivalent way using two *if-then* statements:
Two lines are parallel if and only if, when cut by a transversal, corresponding angles have the same measure.

In 16 and 17, match the argument with the appropriate argument form and indicate whether the argument is valid or invalid.
(a) Law of Detachment (*modus ponens*)
(b) Law of Indirect Reasoning (*modus tollens*)
(c) Law of Transitivity
(d) Converse Error
(e) Inverse Error
(f) Improper Induction

16. $\forall$ *real numbers* x, $x^3 \geq 27 \Rightarrow x \geq 3$.
$\sqrt{8} < 3$.
$\therefore (\sqrt{8})^3 < 27$. (b); valid

17. *If* $\triangle ABC \cong \triangle DEF$, *then* $\angle ABC \cong \angle DEF$.
$\triangle ABC$ *is not congruent to* $\triangle DEF$.
$\therefore \angle ABC$ *is not congruent to* $\angle DEF$.
(e); invalid

18. Determine whether the argument is valid or invalid. Justify your reasoning.
If Pete hit three home runs in the game, then his team won. If Pete was not named most valuable player, then his team did not win. Pete hit three home runs in the game. Therefore, Pete was named most valuable player.

19. Use the method of Direct Proof to prove the following statement, or disprove it by giving a counterexample.
If n *is an odd integer, then* $n^2 - 1$ *is an even integer.*

In 20 and 21, consider the network below.

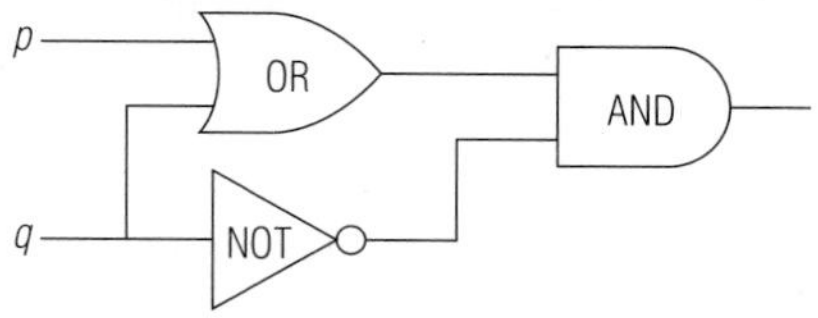

20. Write the logical expression that corresponds to the network. *(p or q) and ~q*

21. Write an input-output table for the network.

22. Write the truth table for *not (p or q)*.

23. Consider the universal statement
$\forall$ *real numbers* x *and* y,
$\sqrt{x + y} = \sqrt{x} + \sqrt{y}$.
Is the statement *true* or *false*? If true, explain why. If false, give a counterexample.

ADDITIONAL ANSWERS

6. ***$\exists$ a real number y, such that $0 + y \neq y$.***

8. x is greater than or equal to -8 and less than 12.

9. *The bald eagle is not the national bird or "The Star-Spangled Banner" is not the national anthem.*

10. IF NOT (C = 2) AND NOT (C= 3) THEN 10.

11. *If you can be admitted to a R-rated movie, then you are at least 17 years old.*

15. *If two lines cut by a transversal have corresponding angles with the same measure, then they are parallel. If two lines are parallel, then when cut by a transversal, corresponding angles have the same measure.*

18. valid; for example, use the Law of Detachment followed by the Law of Indirect Reasoning

19. Suppose that n is any odd integer. Thus there exists an integer r such that $n = 2r + 1$. Then $n^2 - 1 = (2r + 1)^2 - 1 = 4r^2 + 4r + 1 - 1 = 4r^2 + 4r = 2(2r^2 + 2r)$. Since $(2r^2 + 2r)$ is an integer, $n^2 - 1$ is an even integer by definition.

21. See Additional Answers in the back of this book.

22.

p	q	p or q	not (p or q)
T	T	T	F
T	F	T	F
F	T	T	F
F	F	F	T

23. False; counterexample: Let $x = 4$ and $y = 4$. $\sqrt{4 + 4} = \sqrt{8}$, $\sqrt{4} + \sqrt{4} = 4$. But $\sqrt{8} \neq \sqrt{4}$, so $\sqrt{4 + 4} \neq \sqrt{4} + \sqrt{4}$.

RESOURCES
- Chapter 1 Test, Form A
- Chapter 1 Test, Form B
- Quiz and Test Writer

CHAPTER REVIEW

The main objectives for the chapter are organized here into sections corresponding to the four main types of understanding this book promotes: Skills, Properties, Uses, and Representations. We call these the SPUR objectives. (See pages T38–39 for more information.)

Skills range from the carrying out of simple and complicated procedures for getting answers to the study of algorithms.

Properties range from the mathematical justifications for procedures and other theory to the writing of proofs.

Uses range from simple real-world applications of the mathematics to the modeling of real situations.

Representations range from graphs and diagrams to the invention of other metaphors to describe the mathematics.

Notice that the four types of understanding are not in increasing order of difficulty. There may be hard skills and easy representations; some uses may be easier than anything else; and so on.

USING THE CHAPTER REVIEW
Students should be able to answer questions like these with about 85% accuracy by the end of the chapter.

You may assign these questions over a single night to help students prepare for a test the next day, or you may assign the questions over a two-day period.

Chapter Review

Questions on SPUR Objectives

See margin for answers not shown below.

SPUR stands for **S**kills, **P**roperties, **U**ses, and **R**epresentations.
The Chapter Review questions are grouped according to the SPUR Objectives for this chapter.

SKILLS deal with the procedures used to get answers.

Objective A: *Identify forms of logical statements.* *(Lessons 1-1, 1-5)*

In 1–3, identify the sentence as an existential statement, a universal statement, or neither.

1. *For all baseball seasons s, s runs from April through October.* universal
2. *-5 is an integer.* neither
3. *∃ a positive integer n such that n is smaller than 1.* existential
4. Consider the statement
   If the ballet is The Nutcracker, then it is performed during the holiday season.
   Identify each statement below as its *contrapositive*, *converse*, or *inverse*.
   a. *If the ballet is not The Nutcracker, then it is not performed during the holiday season.* inverse
   b. *If the ballet is not performed during the holiday season, then it is not The Nutcracker.* contrapositive
   c. *If the ballet is performed during the holiday season, then it is The Nutcracker.*

Objective B: *Write logically equivalent forms of statements.* *(Lessons 1-1, 1-3, 1-5)*

In 5–7, rewrite each sentence as the appropriate universal or existential statement using *for all* or *there exists*, respectively.

5. *No country has landed people on Mars.*
6. *Every intelligence memo is read by at least one government official.*
7. *Every composite number has a positive integer factor other than 1 and the number itself.*
8. Rewrite the given statement in *if-then* form.
   Never practicing your piano lessons is a sufficient condition for not learning to play piano.
9. Determine which of the following is (are) equivalent to
   tan $\theta < 0$ is a necessary condition for θ to be in Quadrant II or IV.
   (a) *If tan $\theta < 0$ then θ is in Quadrant II or IV.*
   (b) *If θ is in Quadrant II or IV, then tan $\theta < 0$.*
   (c) *If tan $\theta \not< 0$, then θ is not in Quadrant II and not in Quadrant IV.*
   (d) *If θ is not in Quadrant II and not in Quadrant IV, then tan $\theta \not< 0$.*
   (e) None of the above is equivalent to the given conditional. (b) and (c)
10. *True* or *false*? *A student is allowed to take calculus only if the student has taken two years of algebra* can be rewritten as *If a student has taken two years of algebra, then the student is allowed to take calculus.* False
11. Write two *if-then* statements which together are equivalent to the statement below.
    Passing a state's bar exam is a necessary and sufficient condition for practicing law in that state.
12. **a.** Rewrite the inequality $|x| \le 3$ using a double inequality. $-3 \le x \le 3$
    b. Express the double inequality in part **a** by writing out the implied *and* and *or*.
13. Write an *if-then* statement that is logically equivalent to
    ∀ real numbers x, log $x > 0$ only if $x > 1$.
14. Given the statements
    p: Sue is wearing a blue sweater.
    q: Sue has brown eyes.
    Write the statement whose logical expression is given.
    a. *not p* **b.** *p and (not q)*

Objective C: *Write the negation of a logical statement. (Lessons 1-2, 1-3, 1-5)*

In 15–18, write the negation of the statement.

15. *No British bobby (police officer) carries a gun.*

16. *Every President is guarded by at least one Secret Service agent.*

17. *If a person wants to travel from the U.S. to Europe, then the person must fly or must travel by ship.*

18. *I'm Chevy Chase and you're not.*

19. *True* or *false*? The negation of $\forall$ *real numbers z, if* $z^2 > 1$ *then* $z > 1$ is $\exists$ *a real number z such that* $z^2 > 1$ *and* $z \not> 1$. True

20. Use De Morgan's Laws to write the negation of the logical expression *(not p) or q*. p and ~q

Objective D: *Determine the truth value of a statement. (Lessons 1-1, 1-2, 1-3, 1-5)*

In 21 and 22, determine the truth value of each logical expression given the following statements:

$p(x): x < -4$ $\quad q(x): x > 15$ $\quad r(x): x \le 6$

21. $q(7)$ *or* $\sim r(7)$ True

22. $\sim(p(7)$ *or* $q(7))$ *and* $r(7)$ False

In 23 and 24, determine the truth value of the given statement. If false, provide a counterexample.

23. *Every quadrilateral whose diagonals bisect each other is a rectangle.*

24. $\exists$ *a real number x such that* $x^2 < x$. True

25. Consider the statement

$\forall$ *real numbers x,* $\sin^2 x + \cos^2 x = 1$.

a. Write the negation of this statement.

b. Which is true: the statement or its negation? the statement

PROPERTIES deal with the principles behind the mathematics.

Objective E: *Identify properties of logical statements. (Lessons 1-1, 1-2, 1-3, 1-5)*

In 26 and 27, determine whether the given expression is a statement.

26. $7 + 3 = 12$ Yes **27.** $2x - 5 \ge 8$ No

28. *True* or *false*? The negation of an existential statement is another existential statement.

29. *True* or *false*? A conditional and its converse are logically equivalent. False

30. Imagine that *z* is a particular real number and consider the following statement:

If $z^2 > 1$ *then* $z > 1$.

Under what circumstances is this statement false? when $z^2 > 1$ and $z \le 1$

31. *True* or *false*? The negation of an *if-then* statement is another *if-then* statement. False

32. *True* or *false*? *p or q* is false only when both *p* and *q* are false. True

Objective F: *Use substitution to verify specific statements. (Lesson 1-1)*

33. The following statement is true:

$\forall$ *real numbers a and b,*

$(a + b)^3 = a^3 + 3a^2b + 3ab^2 + b^3$.

Use substitution to expand $(3x + 4)^3$.

34. Suppose you wanted to evaluate $(\sqrt{45} + \sqrt{20})^2$ without using a calculator.

a. What universal statement might you use?

b. Use your answer to part **a** to evaluate the given expression. a) $(a + b)^2 = a^2 + 2ab + b^2$ b) 125

If you assign the questions over two days, then we recommend assigning the *evens* for homework the first night so that students get feedback in the class the next day. Then assign the *odds* for the second night (the night before the test) so that students can use the answers provided in the book as a study aid.

EVALUATION

Two forms of a Chapter Test—Forms A and B—are provided in the Teacher's Resource File. For information on grading, see *General Teaching Suggestions: Grading* on page T43.

ASSIGNMENT RECOMMENDATION

We strongly recommend that you assign the reading and some questions from Lesson 2-1 for homework the evening of the test. It gives students work to do if they complete the test before the end of the period and keeps the class moving.

If you do not give assignments on the days of tests, you may cover one less *chapter* over the course of the year.

ADDITIONAL ANSWERS

4.c. converse

5. ***$\forall$ countries x, x has never landed people on Mars.***

6. ***$\forall$ intelligence memos x, $\exists$ a government official y, such that y reads x.***

7. ***$\forall$ composite numbers x, $\exists$ a positive integer y, such that $y \ne x$, $y \ne 1$, but y is a factor of x.***

8. ***If you never practice your piano lessons, you will not learn to play piano.***

11., 12.b., 13., 14., 15., 16., 17., 18., 23., 25.a., 28., 33. See the margin on page 76.

ADDITIONAL ANSWERS

11. ***If you pass a state's bar exam, then you may practice law in that state. If you practice law in a state, then you must pass that state's bar exam.***

12.b. x is greater than or equal to -3 and less than or equal to 3.

13. ***If log x > 0, then x > 1.***

14.a. ***Sue is not wearing a blue sweater.***
b. Sue is wearing a blue sweater, and she does not have brown eyes.

15. ***There is a British bobby that carries a gun.***

16. ***There is a president that is not guarded by a Secret Service agent.***

17. ***A person wants to travel from the U.S. to Europe and the person does not have to fly and the person does not have to travel by ship.***

18. ***I'm not Chevy Chase or you are.***

23. False; counter-example: rhombus

25.a. ***∃ a real number x, such that*** $\sin^2 x + \cos^2 x \neq 1$.

28. False

33. $(3x + 4)^3 = 27x^3 + 108x^2 + 144x + 64$

35. IV; invalid

43. ***All even numbers are real numbers; Law of Transitivity***

44. $|\pi + -13| \leq |\pi| + |-13|$; **Law of Substitution**

45. even; $a = 2k$ **for some integer** k, **so** $3a \cdot b = 3(2k)b = 2(3kb)$

46. odd; $(8s^2 + 4s + 3) = 2(4s^2 + 2s + 1) + 1$

47. using the same value k **for expressing** m **and** n

Objective G: *Determine whether arguments are valid or invalid. (Lessons 1-6, 1-8)*

In 35–38, tell whether the argument is valid or invalid. Support your answer with a reference to one or more of the following.

I. Law of Detachment (*modus ponens*)
II. Law of Indirect Reasoning (*modus tollens*)
III. Law of Transitivity
IV. Converse Error
V. Inverse Error
VI. Improper Induction

35. *If a polygon is a regular pentagon, then it has five sides.*
Polygon PENTA has five sides
∴ Polygon PENTA is a regular pentagon.

36. $\forall$ *real numbers* $x > 10 \Rightarrow x^2 > 100$.
$x^2 \leq 100$.
$\therefore x \leq 10$ II; valid

37. *If a quadrilateral is a square, then it is a rhombus.*
If a quadrilateral is a rhombus, then it is a parallelogram.
∴ If a quadrilateral is a square, then it is a parallelogram. III; valid

38. *If* θ *is in quadrant I, then* $\sin\theta > 0$.
θ *is not in quadrant I.*
$\therefore \sin\theta \leq 0$. V; invalid

39. Consider the argument

$$2^2 \cdot 2^3 = (2 \cdot 2)(2 \cdot 2 \cdot 2) = 2^5 = 2^{2+3}$$
$$2^3 \cdot 2^4 = (2 \cdot 2 \cdot 2)(2 \cdot 2 \cdot 2 \cdot 2) = 2^7 = 2^{3+4}$$
$$\therefore 2^n 2^m = 2^{n+m} \text{ for all integers } n, m.$$

a. Is the conclusion true? Yes
b. Is the argument valid? No

Objective H: *Use logic to prove or disprove statements. (Lessons 1-1, 1-3, 1-6, 1-7)*

40. Find a counterexample to show that the following statement is false. sample: $n = -1$
For all integers n, $\sqrt{n^2} = n$.

41. Consider the conditional
For all positive real numbers, y, a, and b, $y = ab \Rightarrow \log y = \log a + \log b$.
Is it true or false? If false, give a counterexample. True

42. *Multiple choice.* Which is a valid conclusion using all of the following premises?
If triangles ADC and ADB are congruent, then triangle ABC is isosceles.
If $\overline{AD}$ *is perpendicular to* $\overline{BC}$*, then triangles ADC and ADB are congruent.*
$\overline{AD}$ *is perpendicular to* $\overline{BC}$.

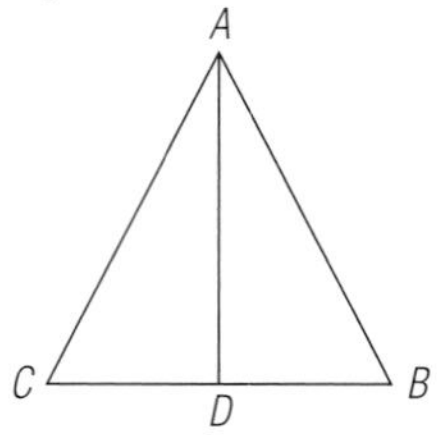

(a) *Triangles ADC and ADB are congruent.*
(b) *If* $\overline{AD}$ *is perpendicular to* $\overline{BC}$*, then triangle ABC is isosceles.*
(c) *Triangle ABC is isosceles.* (c)

In 43 and 44, form a valid conclusion from the given statements and justify your reasoning.

43. *All even numbers are integers.*
All integers are real numbers.

44. $\forall$ *real numbers a and b,*
$|a + b| \leq |a| + |b|$.
π *and* -13 *are real numbers.*

In 45 and 46, determine whether the given integer is even or odd. Justify your answer.

45. $3ab$, where a and b are integers and a is even

46. $8s^2 + 4s + 3$, where s is any integer

47. Find the error in this "proof" of the following theorem.
If m and n are even integers, then $m - n$ *is an even integer.*

Proof Because m is even there is an integer k such that $m = 2k$. Also, because n is even, $n = 2k$. It follows that $m - n = 2k - 2k = 0$, which is even. Thus, the difference between any two even integers is even.

In 48 and 49, either prove the given statement using a direct proof or disprove it by giving a counterexample.

48. *If m is any even integer, then* $m^2 = 4k$ *for some integer k.*

49. *If r and s are odd integers, then* $r \cdot s = 4k + 1$ *for some integer k.*

USES deal with applications of mathematics in real situations.

Objective I: *Determine the truth of quantified statements outside of mathematics.* *(Lessons 1-1, 1-2, 1-3, 1-5)*

In 50–56, refer to the table below.

At a summer camp, students participate in three types of activities: sports, arts and crafts, and nature identification. The activities of four campers are summarized below.

	sports				arts and crafts			nature identification		
camper	swimming	volleyball	canoeing	hiking	pottery	wood-working	jewelry design	bird watching	botany	entomology
Kenji	√		√			√				√
Jennifer		√		√	√		√	√	√	
Ruby				√		√		√	√	√
Oscar	√	√	√					√		

Use the table to determine if each statement is true or false. Justify your answers.

50. *∀ campers c in the sample, c participates in a sports activity.*

51. *∃ a camper c in the sample such that ∀arts and crafts activities y, c participates in y.*

52. *Some camper participates in every sports activity.*

53. *No camper fails to participate in arts and crafts.*

54. *For every camper c in the sample, if c participates in nature identification, then c participates in arts and crafts.*

55. *∃ a camper c in the sample such that c participates in entomology or c participates in hiking.*

56. *Some camper participates in jewelry design and in swimming.*

Objective J: *Determine whether or not a logical argument outside of mathematics is valid.* *(Lessons 1-6, 1-8)*

In 57–60, tell whether the argument is valid or invalid. Support your answer with a reference to one or more of the following.

I. Law of Detachment (*modus ponens*)
II. Law of Indirect Reasoning (*modus tollens*)
III. Law of Transitivity
IV. Converse Error
V. Inverse Error

57. *If you dial a number which is busy or has no answer, you may dial the number again by just pressing REDIAL. You just dialed a number by pressing REDIAL. Therefore, you must have dialed a number which was busy or had no answer.* **IV; invalid**

58. *If Sharon skates, then she wins the gold medal. Sharon is skating. Thus, she will win the gold medal.* **I; valid**

59. *When it fails to rain within 30 days, there is a drought. There was not a drought. Thus, there was rain within a 30-day period.* **II; valid**

60. *The ski lift runs when the ski resort is open. When the ski lift runs, hundreds of skiers try the slopes. Therefore, if hundreds of skiers are not trying the slopes, then the ski resort is not open.* **II and III; valid**

48. Suppose *m* is any even integer. Thus there exists an integer *s*, such that $m = 2s$. Then $m^2 = 4s^2 = 4k$, where $k = s^2$ is an integer by closure properties.

49. Counterexample: Let $r = 3$ and $s = 1$. $r \cdot s = 3 \cdot 1 = 3 = 4k + 1 \therefore k = \frac{1}{2}$, but $\frac{1}{2}$ is not an integer.

50. True; all campers participate in a sports activity.

51. False; no camper participates in all the arts and crafts activities.

52. False; no camper participates in all the sports activities.

53. False; Oscar does not participate in an arts and crafts activity.

54. False; Oscar participates in nature identification, but he does not participate in an arts and crafts activity.

55. True; Kenji participates in entomology but not hiking.

56. False; no camper participates in both jewelry design and swimming.

ADDITIONAL ANSWERS

62. No, $x = 3, y = 1, z = 1$ is a set of values for which the two statements have different values. Therefore, the two statements are not equivalent.

63. See Additional Answers in the back of this book.

64.a. *(p and q) or (q and r)*

65.c. See Additional Answers in the back of this book.

d. $\sim(p \text{ and } \sim q) \equiv (\sim p)$ or $\sim(\sim q) \equiv (\sim p)$ or q

66.

p	q	p or q
T	F	T
T	T	T
F	F	F
F	T	T

67.

p	q	p ⇒ q
T	F	F
T	T	T
F	F	T
F	T	T

68.

p	q	r	p ⇒ q	(p ⇒ q) ⇒ r
T	T	T	T	T
T	T	F	T	F
T	F	T	F	T
T	F	F	F	T
F	T	T	T	T
F	T	F	T	F
F	F	T	T	T
F	F	F	T	F

69., 70. See Additional Answers in the back of this book.

Objective K: *Read and interpret computer programs using if-then statements. (Lessons 1-3, 1-5)*

In 61 and 62, refer to the following programs which are identical except for line 20:

Program A:

```
 5  REM PROGRAM A
10  INPUT X, Y, Z
20  IF X > 2 OR (Y > 2 AND Z > 2)
    THEN PRINT "TRUE"
    ELSE PRINT "FALSE"
30  END
```

Program B:

```
 5  REM PROGRAM B
10  INPUT X, Y, Z
20  IF (X > 2 OR Y > 2) AND Z > 2
    THEN PRINT "TRUE"
    ELSE PRINT "FALSE"
30  END
```

61. Trace each of the following inputs through both programs and predict the outputs:

a. X = 3, Y = 1, Z = 3 **A:True; B:True**

b. X = 1, Y = 1, Z = 3 **A:False; B:False**

c. X = 3, Y = 1, Z = 1 **A:True; B:False**

62. Are the two statements $x > 2$ *or* $(y > 2$ *and* $z > 2)$ and $(x > 2$ *or* $y > 2)$ *and* $z > 2$ equivalent? Explain why or why not.

REPRESENTATIONS deal with pictures, graphs, or objects that illustrate concepts.

Objective L: *Translate logic networks into logical expressions and input-output tables and determine output signals. (Lesson 1-4)*

63. Suppose the logic gate G has the following input-output table:

p	q	output
1	1	0
1	0	0
0	1	1
0	0	1

Write an input-output table for the following network.

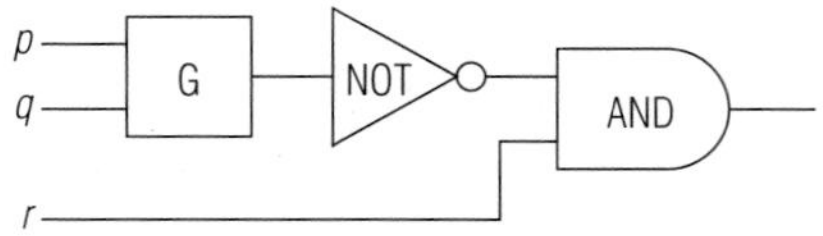

64. a. Write the logical expression that corresponds to the network below.

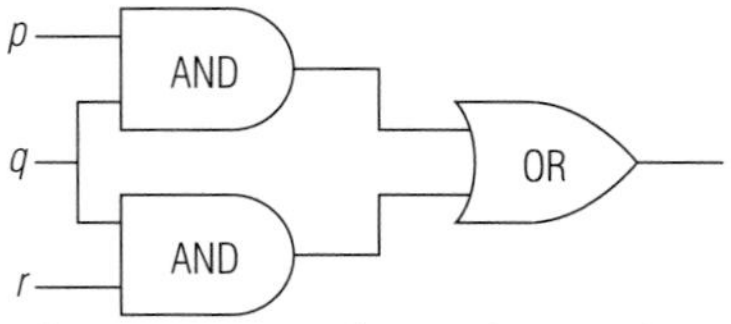

b. Suppose *p* carries a signal of 1, *q* carries a signal of 0, and *r* carries a signal of 1. What output signal will the network produce? **0**

65. In **a** and **b**, write the logical expression that corresponds to the network.

a. ***not (p and not q)***

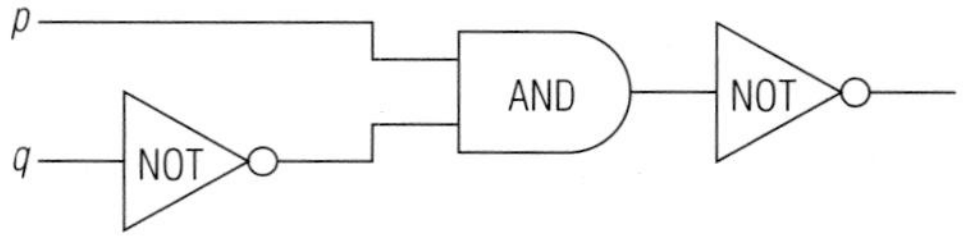

b. ***(not p) or q***

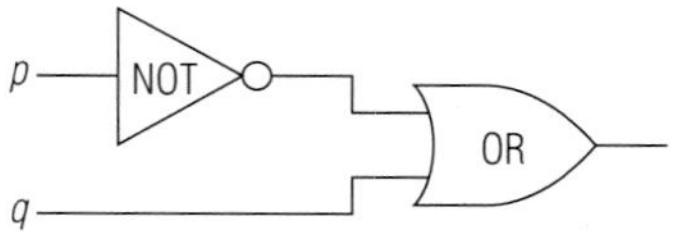

c. Use input-output tables to show that the networks in parts **a** and **b** are functionally equivalent.

d. Use De Morgan's Laws to show that the logical expressions obtained in parts **a** and **b** are equivalent.

Objective M: *Write truth tables for logical expressions. (Lessons 1-3, 1-5)*

In 66–69, write the truth table for the given logical expression.

66. *p or q*

67. $p \Rightarrow q$

68. $(p \Rightarrow q) \Rightarrow r$

69. *p and (q or r)*

70. Use a truth table to show that *(not p) or (not q) ≡ not (p and q)*.

CHAPTER 2 ■ ANALYZING FUNCTIONS

OBJECTIVES ■ CHAPTER 2

The objectives listed here are the same as in the Chapter 2 Review on pages 146–150 of the student text. The Progress Self-Test on page 145 and the tests in the Teacher's Resource File cover these objectives. For recommendations regarding the handling of this end-of-chapter material, see the notes in the margin on the corresponding pages of this Teacher's Edition.

OBJECTIVES FOR CHAPTER 2 (Organized into the SPUR Categories—Skills, Properties, Uses, and Representations)	**Progress Self-Test Questions**	**Chapter Review Questions**	**Teacher's Resource File**	
			Lesson Master*	**Chapter Test Forms A & B**
SKILLS				
A Determine relative minima and maxima of a function and intervals on which it is increasing or decreasing.	4	1–8	2-3, 2-4, 2-8	1, 3
B Rewrite exponential and logarithmic expressions and equations.	7	9–15	2-7	5, 6, 7
PROPERTIES				
C Identify the domain, range, and minimum and maximum values of functions.	1, 2	16–26	2-1, 2-2	2
D Determine the end behavior of a function.	3d, 5c	27–36	2-4, 2-5, 2-6, 2-7, 2-8	4, 10, 11
USES				
E Use trigonometric, exponential, and logarithmic functions as models.	9	37–42	2-6, 2-7, 2-8	9
F Solve maxima-minima problems.	6	43–44	2-2	12
REPRESENTATIONS				
G Analyze a function from its graph.	3, 5, 8	45–55	2-2, 2-3, 2-4, 2-5, 2-6, 2-7, 2-8	8

*** The masters are numbered to match the lessons.**

OVERVIEW 2

OVERVIEW ■ CHAPTER 2

This chapter has two parts. In the first part, some broad ideas about functions are reviewed. In the second part, these ideas are applied to analyze exponential, logarithmic, and trigonometric functions. It is assumed that much of the material is review.

This chapter is not about obtaining values of a particular function. It is assumed that students know how to obtain function values, either by using its definition or with the aid of a calculator. Instead, the focus of this chapter is on properties of the function itself: its usual domain and the corresponding range; its maximum or minimum values (if any); the intervals over which it is increasing or decreasing; its end behavior; the broad shape of its graph; the kinds of situations it models; and its special properties.

Other differences between this treatment and what students might have seen in previous courses are that the discussions are more formal, and, as befits a review, more material is covered in each lesson.

It is assumed that students have access to function graphing technology throughout this chapter and for the remainder of this book, both for classwork and assignments. Either function graphing calculators or computers with a function graphing program will suffice. We call either an *automatic grapher.*

More than any other chapter in this book, the ease or difficulty of this chapter and the time it takes to cover it will depend on the background and ability of your students. Students who have covered UCSMP *Advanced Algebra* and *Functions, Statistics, and Trigonometry* will have seen many of the ideas here twice before; for them, the lesson-a-day pace is quite reasonable.

PERSPECTIVES 2

PERSPECTIVES ■ CHAPTER 2

The Perspectives provide the rationale for the inclusion of topics or approaches, provide mathematical background, and make connections with other lessons and within UCSMP.

2-1

GRAPHS OF FUNCTIONS

Lesson 2-1 reviews vocablulary and notation that is used throughout the book. Although the definition of a function as a set of ordered pairs and the definition of a function as a correspondence (or mapping) between two sets are equivalent definitions, the mapping definition used in this lesson is more common in higher mathematics.

The terminology of automatic graphers is also reviewed. Students are expected to know how to specify a window for a graph, enter a rule for a function, and obtain approximate coordinates of any point on the graph they obtain.

Automatic graphers plot the values of a given function at a finite number of points selected from its domain. These plotted functions are examples of *discrete functions.* The discussion of discrete functions and their continuous counterparts is begun in this lesson and interwoven throughout the chapter.

2-2

RANGE, MAXIMA, AND MINIMA

To graph a function over a given domain, its range must be known. In many cases, the range may be determined from knowing the maximum and/or minimum values. This terminology is reviewed in this lesson and applied to the classic problem of determining the cylinder of minimum surface area for a given volume. The solution to this problem is approximated by graphing.

2-3

INCREASING AND DECREASING FUNCTIONS

The emphasis in this lesson is on translating the visual picture of a function increasing (or decreasing) on an interval into its algebraic counterpart and then working with that counterpart.

2-4

SEQUENCES AND LIMITS

Sequences are a special type of discrete function. Discussed in this lesson are arithmetic, geometric, harmonic, and alternating sequences, as well as several other specific sequences. The limit of the *n*th term of a numerical sequence is defined in preparation for the next lesson.

2-5

END BEHAVIOR

The end behavior of a function is a description of what happens to the values of the function as $x \to +\infty$ and as $x \to -\infty$. If $\lim_{x \to +\infty} f(x) = L$ or if $\lim_{x \to -\infty} f(x) = L$, then $y = L$ is a horizontal asymptote to the graph of the function. For some functions, the values of $f(x)$ increase (or decrease) without bound and we call the limit $+\infty$ (or $-\infty$). For still other functions, such as the sine function, no such limit exists.

Introduced in this lesson are the power functions, those functions f with equations of the form $f(x) = ax^n$, where n is an integer. Their end behavior forms the basis for the end behavior of polynomial functions to be discussed in Chapter 5. The power functions provide perhaps the simplest examples of even and odd functions.

2-6

ANALYZING EXPONENTIAL FUNCTIONS

The information about functions conveyed in the titles of Lessons 2-1, 2-2, 2-3, and 2-5 constitutes a significant part of what it means to analyze a real function. Also important in the analysis are the maximum or minimum values of the function, the situations in which it is employed as a model, and its special properties.

This lesson uses the exponential functions, defined by $f: x \rightarrow ab^x$ for all real numbers x, as a vehicle for discussing what it means to *analyze* a function. These functions should be quite familiar to students with previous UCSMP experience; their properties, including the applications to growth, are discussed in three previous courses.

2-7

ANALYZING LOGARITHM FUNCTIONS

The logarithm functions to the base b, defined by $y = \log_b x$, are analyzed in this lesson. The special cases $b = 10$ (common logarithm function) and $b = e$ (natural logarithm function) are emphasized. The Change of Base Theorem is deduced and applied in the solving of exponential equations of the form $a^x = b$. The properties of logarithms are applied to a discussion of the decibel logarithmic scale. (The inverse relationship between the exponential function $f: x \rightarrow b^x$ and the logarithm function $g: x \rightarrow \log_b x$ is discussed in Chapter 3.)

2-8

ANALYZING THE SINE AND COSINE FUNCTIONS

Both the right triangle definitions and the unit circle definitions of the sine and cosine are given in this lesson and the equivalence of the two definitions for acute angles is demonstrated. The unit circle definitions are used more often in this course because this is the definition that is most suited for calculus. Also included in this lesson are the notions of periodicity and of even and odd functions. The tangent, cotangent, secant, and cosecant functions are the subject of Lesson 5-6. Lesson 2-8 assumes that students bring with them a strong knowledge of trigonometry.

2-9

WHAT IS INFINITY?

This reading lesson covers various uses of the word *infinity* and the symbol ∞, starting from the meaning as found in the discussion of limits in Lessons 2-4 and 2-5. The work of Georg Cantor regarding the cardinalities of the positive integers, the positive rationals, and the reals is discussed.

DAILY PACING CHART 2

DAILY PACING CHART ■ CHAPTER 2

Every chapter of UCSMP *Precalculus and Discrete Mathematics* includes lessons, a Progress Self-Test, and a Chapter Review. For optimal student performance, the self-test and review should be covered. (See *General Teaching Suggestions: Mastery* on page T35 of this Teacher's Edition.) By following the pace of the Full Course given here, students can complete the entire text by the end of the year. Students following the pace of the Minimal Course spend more time when there are quizzes and on the Chapter Review and will generally not complete all of the chapters in this text.

When chapters are covered in full (the recommendation of the authors), then students in the Minimal Course can cover 11 chapters of the book. For more information on pacing, see *General Teaching Suggestions: Pace* on page T34 of this Teacher's Edition.

DAY	MINIMAL COURSE	FULL COURSE
1	2-1	2-1
2	2-2	2-2
3	2-3	2-3
4	Quiz (TRF); Start 2-4.	Quiz (TRF); 2-4
5	Finish 2-4.	2-5
6	2-5	2-6
7	2-6	Quiz (TRF); 2-7
8	Quiz (TRF); Start 2-7.	2-8
9	Finish 2-7.	2-9
10	2-8	Progress Self-Test
11	2-8	Chapter Review
12	2-9	Chapter Test (TRF)
13	Progress Self-Test	
14	Chapter Review	
15	Chapter Review	
16	Chapter Test (TRF)	

TESTING OPTIONS

TESTING OPTIONS

- ■ Quiz for Lessons 2-1 through 2-3
- ■ Quiz for Lessons 2-4 through 2-6
- ■ Chapter 2 Test, Form A
- ■ Chapter 2 Test, Form B
- ■ Chapter 2 Test, Cumulative Form

A Quiz and Test Writer is available for generating additional questions, additional quizzes, or additional forms of the Chapter Test.

INDIVIDUAL DIFFERENCES

PROVIDING FOR INDIVIDUAL DIFFERENCES

The student text is written for, and tested with, average students. It also has been successfully used with better and more poorly prepared students.

The Lesson Notes often include Error Analysis and Alternative Approach features to help you with those students who need more help. Students of all abilities often learn from their peers and may benefit from small group work referenced as appropriate throughout the Notes. A blackline Lesson Master (in the Teacher's Resource File), keyed to the chapter objectives, is provided for each lesson to allow more practice. (However, since it is important to keep up with the daily pace, you are not expected to use all of these masters. Again, refer to the suggestions for pacing on page T34.) Extension activities are provided in the Lesson Notes for those students who have completed the particular lesson in a shorter amount of time than is expected, even in the Full Course.

Analyzing Functions

The Mississippi River is about 2340 miles long, and it flows at an average rate of about 2 miles per hour.

This caption may bring to your mind an image of Tom Sawyer or Huckleberry Finn rafting down the river. On a raft, how far would they float downstream in a given amount of time? The correspondence between time and the distance traveled is an example of a *function*. Functions are so important in mathematics that various ways have been developed to describe them. They can be described by:

a *table:*

Time (hr)	**Distance** (mi)
10	20
1	2
403	806
35	70
⋮	⋮
p	$2p$

a *graph:*

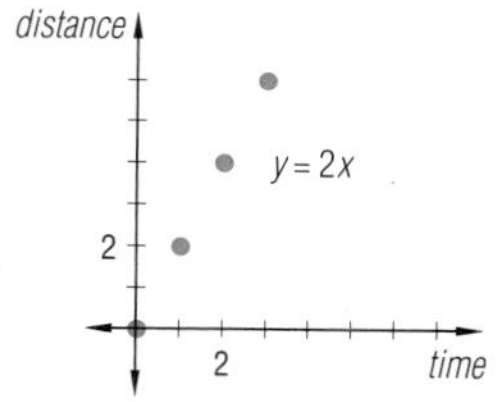

a *rule:*

Let f be the function mapping the time traveled x onto the distance traveled. Then
$f(x) = 2x$ ($f(x)$ notation)
f: $x \rightarrow 2x$ (mapping notation)

an arrow *diagram:*

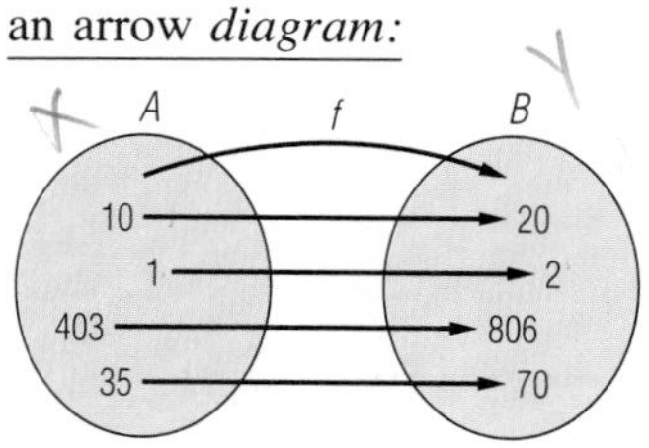

The arrow diagram makes it explicit that two sets are involved in a function, the set of possible times and the set of possible miles traveled.

The function f is an example of one of the **elementary functions**, that are the basic functions from which more complicated functions may be constructed. The elementary functions include linear, quadratic, power, polynomial, trigonometric, exponential, and logarithmic functions.

CHAPTER 2

For students with strong backgrounds, this should be a one-day-per-lesson chapter. For less well-prepared or weaker students, Lessons 2-7 and 2-8 are two-day lessons. We recommend that 12–16 days be spent on this chapter: 9 to 12 days on the lessons and quizzes; 1 day for the Progress Self-Test; 1 or 2 days for the Chapter Review; and 1 day for a Chapter Test. (See the Daily Pacing Chart on page 79D.)

USING PAGE 79
This chapter contains much needed review. Use this page to determine how much your students know about functions. Note which functions are classified as elementary functions.

Here is another example. Suppose A is the set of the top six countries in the gold medal standings in the 1988 Winter Olympics and B is the set of integers from 1 to 11. Define a function $f: A \rightarrow B$ by the following rule: For each country x in A, $f(x)$ = the number of gold medals won by x in the 1988 Winter Olympics.

table:

Country	Gold medals earned
Austria	3
East Germany	9
Finland	4
Netherlands	3
Switzerland	5
Soviet Union	11

arrow diagram:

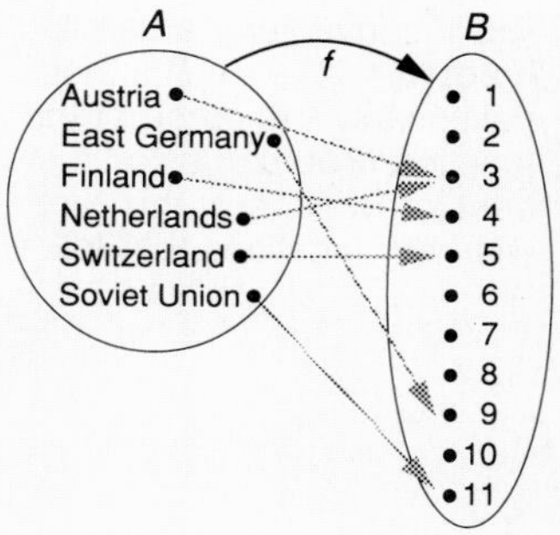

LESSON 2-1

RESOURCES
- Lesson Master 2-1
- Teaching Aid 4 can be used with **Questions 18 and 21**.

OBJECTIVE

C Identify the domain, range, and minimum and maximum values of functions.

TEACHING NOTES

Spend some time discussing the definition of function and the determination of the largest possible domain for a real function. Spend the rest of the time on the technology. Go through the logistics of using an automatic grapher with students. (See pages T40–42 and T52 for general suggestions regarding the use of automatic graphers.)

First, make certain that students interpret correctly the definition of function. Note that not every element of B needs to be used. We say that every function f maps A *into* B. If every element of B is used, then we say that f maps A *onto* B.

Though the word *range* is not defined formally until the next lesson, students have heard it before, and you may find it convenient to use it here. In $f: A \to B$, B is not necessarily the range of the function. You can assess student knowledge of this fact with the questions that follow.

LESSON

2-1 Graphs of Functions

real numbers- set of numbers that can be represented by decimals 0, 1, -7, 2.34, π, √5

> **Definition**
>
> A **function** f from a set A to a set B is a correspondence in which each element in A corresponds to exactly one element of B.

The **domain** of a function from set A to set B is A, and in the case of the Mississippi River rafting function f, A is the set of nonnegative real numbers less than or equal to 1170 (half the length of the Mississippi). This causes the graph of the function (below at the left) to be a segment.

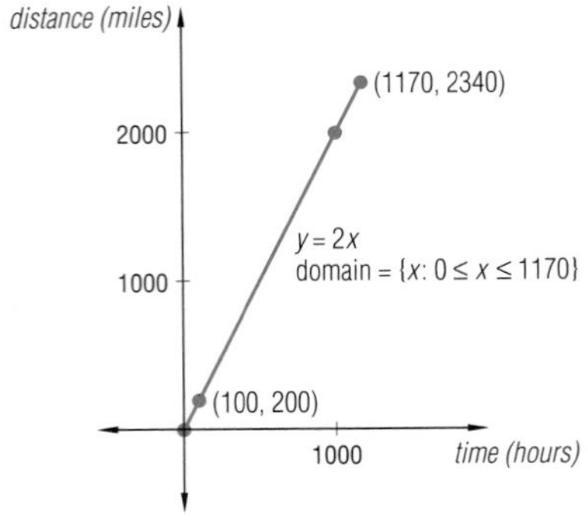

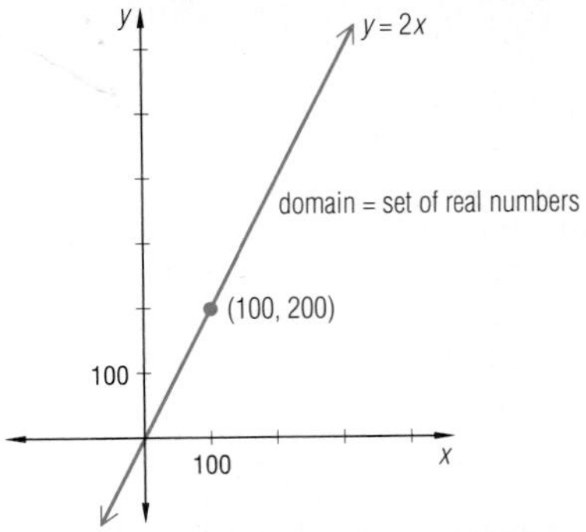

If the domain of this function is enlarged to be the set of all real numbers, then the graph of $y = 2x$ is a line, as graphed above at the right.

For a function f, the function values $y = f(x)$ depend on the values of x. For this reason y is called the **dependent variable** and x is called the **independent variable**. A function whose independent and dependent variables have only real number values is called a **real function**.

If you are given a rule for a real function, you can assume that unless a different domain is specified, the domain is the set of all real numbers for which the function rule yields a real number.

Example 1 Find the domain for the real function $h: x \to \frac{1}{\sqrt{x+3}}$.

Solution The fraction $\frac{1}{\sqrt{x+3}}$ is not defined when its denominator equals zero. This occurs when $x = -3$. Also $\sqrt{x+3}$ is not a real number when $x + 3 < 0$, or, in other words, when $x < -3$. Thus $\frac{1}{\sqrt{x+3}}$ is not a real number if $x \le -3$. For $x > -3$, however, $\frac{1}{\sqrt{x+3}}$ is a real number. Therefore, the domain of h is

$$\{x: x > -3\}$$

read as "the set of all real numbers x such that x is greater than -3."

80

A real function can always be graphed on a coordinate plane. Its **graph** is the set of all points (x, y) such that x is in the domain of f and $y = f(x)$.

The value of $f(x)$ is the **directed distance** or **height of the graph** from the horizontal axis at x. When $f(x) < 0$ the point $(x, f(x))$ is below the horizontal axis. When $f(x) > 0$ the point $(x, f(x))$ is above the horizontal axis. In the graph shown here, $f(x_1) < 0$ and $f(x_2) > 0$.

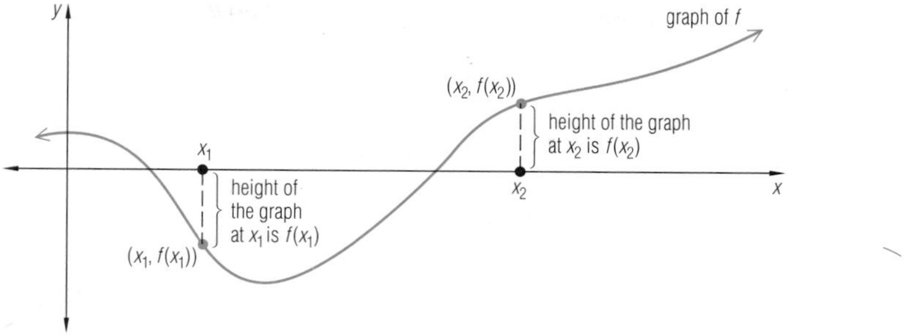

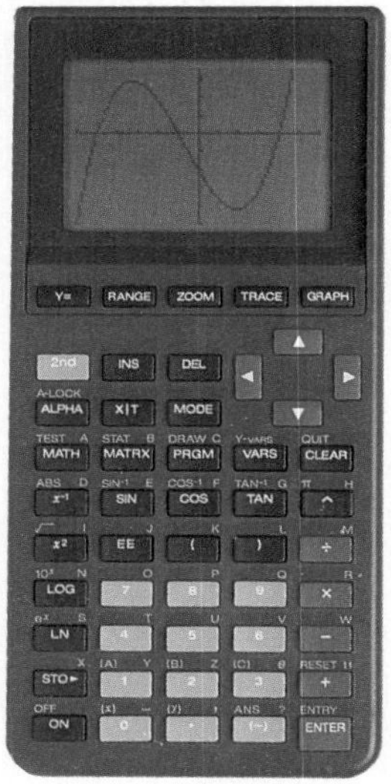

So much information can be obtained from graphs of functions that there are calculators and programs for personal computers that display such graphs. Graphics calculators and computer graphing programs work in similar ways, and so we call them both **automatic graphers** and do not distinguish between them. In this course, we assume you have access to some automatic grapher.

Of course, no grapher is completely automatic. Each has particular keys you must press. Here we discuss what you need to know in order to use any automatic grapher. Consult the calculator's or the function grapher's documentation, or ask your teacher or friends for more specific instructions.

The part of the coordinate grid that is displayed on the screen is called a **window**. The appearance of the graph depends on the choice of the window. (If you pick an inappropriate window, you may miss the graph entirely.) The screen at the right displays a window in which

$$-2 \le x \le 12$$
and $$-3 \le y \le 7.$$

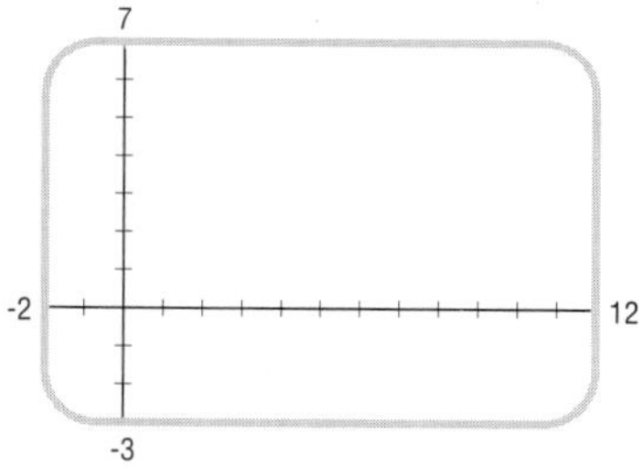

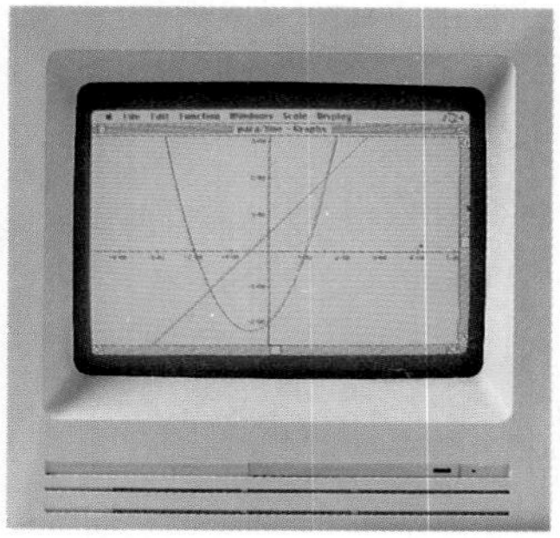

On some automatic graphers, the tick marks on the x- and y-axes are omitted. Usually you need to choose the x-values at each end of the window. Some graphers automatically adjust the y-values so that your graph will fit, but often you also need to choose the y-values. If you don't choose values, the grapher will usually select a **default window**, that is, a window that is used whenever you don't specify the intervals on which to plot x and y. Note that the set of x-values shown by the window is not necessarily the same as the domain of the function. It is only the set of x-values over which the function is currently being displayed.

Suppose a function $f: A \to B$ is specified by an arrow diagram. *True* or *false*?
(1) Every element of A must have an arrow coming out of it. (True)
(2) Every element of B must have an arrow pointing to it. (False)
(3) An element of A can have two arrows coming out of it point to different elements of B. (False)
(4) There can be an element of B that has two arrows point to it from different elements of A. (True)

While most mathematics textbooks (this one included) distinguish carefully between the domain and the range of a function, many automatic graphers show only *range* as a key or menu selection even for the domain variable. Point this out to students.

Some automatic graphers, if given a new function but not a new range, will use as the default whatever range was used last. This can be quite annoying, for example, if the last user had zoomed in on a particular point on some graph and so has a very small window.

Remind students that although the graphs of some functions on (calculator or computer) screens may look continuous, the graphs consist only of finitely many dots. This means that when the function defined for all real values is plotted with an automatic grapher, the display shows only the graph of a discrete function that approximates the given function in the selected window.

On almost all graphers, the equation or function f to be graphed must be entered as a formula for y in terms of x or as a rule for $f(x)$.

$y = 3x^2$	and	$f(x) = \frac{5}{9}(x - 32)$	can be handled.
$x = 4y$	and	$x + f(x) = 17$	cannot be handled.

On many graphers you enter equations by using the keys *, /, and ^ to indicate multiplication, division and powering, respectively. Automatic graphers generally follow the standard rules for order of operations. For instance, to graph $y = 3x^2$ or $f(x) = \frac{5}{9}(x - 32)$ you may need to enter

Y = 3*X^2 or Y = (5/9)*(X − 32).

The steps needed to graph an equation with an automatic grapher are:

1. Solve the equation you wish to graph for the dependent variable in terms of the independent variable. Most graphers will use y or $f(x)$.
2. Enter the function into your grapher.
3. Determine a suitable window and key it in.
4. Give instructions to graph.

Most automatic graphers allow you to plot many graphs simultaneously. This enables you to compare the graphs and see relationships among them.

Example 2 **a.** Graph $f(x) = ax^2$ when $a = \frac{1}{2}$, 1, 2, and 3.

b. What happens to the graph as a gets larger?

Solution

a. The question asks you to graph four functions. We identify the functions as f_1, f_2, f_3, and f_4.

$f_1(x) = \frac{1}{2}x^2$

$f_2(x) = x^2$

$f_3(x) = 2x^2$

$f_4(x) = 3x^2$

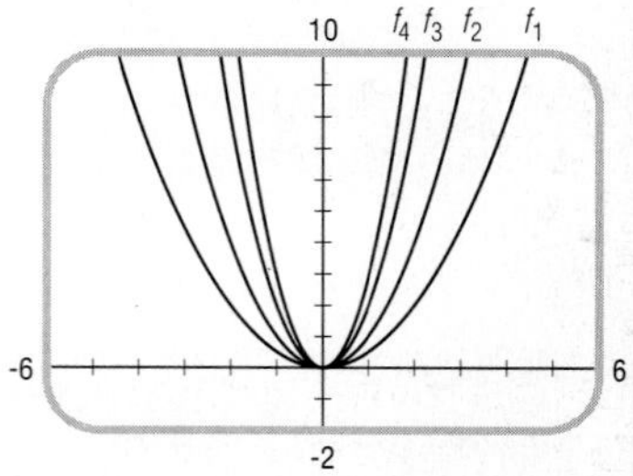

Each equation is already solved for $f(x)$. Enter them one at a time, following the instructions for your grapher. We use the window $-6 \le x \le 6$, $-2 \le y \le 10$ knowing that these parabolas are symmetric to the y-axis, and that the y-values are nonnegative.

b. The parabola that looks widest is $f_1(x) = \frac{1}{2}x^2$. The thinnest parabola is $f_4(x) = 3x^2$. As the value of a increases, the parabola looks thinner and thinner.

ADDITIONAL EXAMPLES

1.a. Find the domain for the real function $g: x \to \sqrt{9 - x^2}$.
$\{x: -3 \le x \le 3\}$

b. Give the domain of f for the real function
$f: x \to \frac{5}{\sqrt{9 - x^2}}$
$\{x: -3 < x < 3\}$

You could further extend this idea by comparing the domains of
$h: x \to \frac{1}{\sqrt{x + 3}}$,
$f: x \to \sqrt{x + 3}$,
and $g: x \to \frac{\sqrt{x + 1}}{\sqrt{x + 3}}$.
The domain of f is determined by solving $x + 3 \ge 0$. The domain of h is the same as the domain of f except that x cannot be equal to -3. The domain of g is the intersection of the solution sets to $x + 1 \ge 0$ and $x + 3 > 0$. Now consider
$d: x \to \frac{x - 1}{\sqrt{x + 3}}$.
In the rule for d, the numerator plays no restrictive role in determining the domain.

2.a. Graph $f(x) = (x - a)^2$ when a = -1, 0, 1, and 2, using the window $-4 \le x \le 5$ and $-1 \le y \le 10$.
The four functions are
$f_{-1}(x) = (x + 1)^2$,
$f_0(x) = x^2$,
$f_1(x) = (x - 1)^2$,
$f_2(x) = (x - 2)^2$.

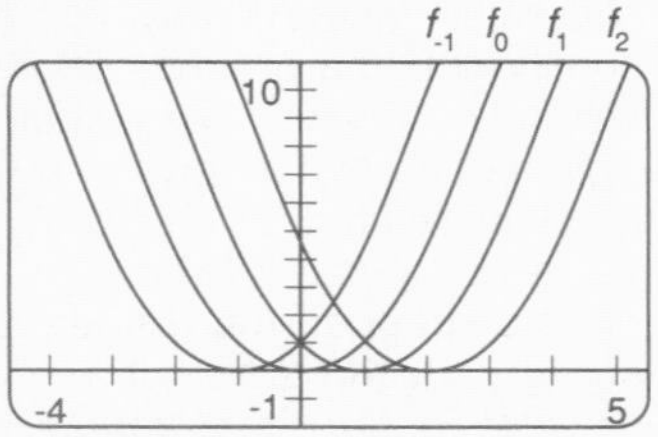

b. What happens to the graph as a increases?
As a increases, the graph is translated to the right.

LESSON

2-2

Range, Maxima, and Minima

Designing and producing containers utilize maximum and minimum techniques. (See pages 88 and 89.)

When graphing a function, whether by hand or with an automatic grapher, you normally know its domain, the set of possible values of the independent variable. It is usually more difficult to determine the set of all values taken on by the dependent variable. This set is called the **range** of the function. Symbolically, the range of a function f: $A \rightarrow B$ is the set of all elements y in B such that $\exists$ x with $f(x) = y$.

You can estimate the range of a function from its graph.

Example 1 Estimate the range of the quadratic function with equation $y = 4x^2 - 3x + 5$.

parabola - a plane curve formed by the intersection of a cone and a plane parallel to one of its sides

Solution It is understood that the domain of this function is the set of real numbers. Because this is a quadratic function and the coefficient of x^2 is positive, you should know that its graph is a parabola that opens upward. With the window $-10 \leq x \leq 10$ and $-10 \leq y \leq 10$, the graph resembles what is shown here. The range is the set of all values taken on by y, which seems to be approximately $\{y: y \geq 4.75\}$, the set of numbers greater than or equal to 4.75.

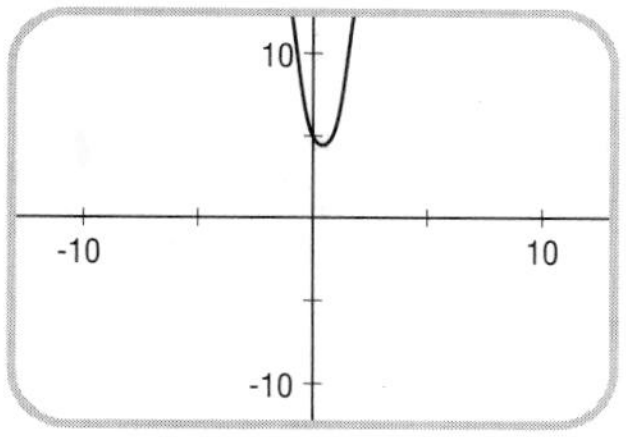

Many automatic graphers have a **trace key.** This key activates a cursor that indicates a point on the graph along with its approximate coordinates, and can move, pixel by pixel, along the graph. Using the trace function on our automatic grapher, we found that the smallest value of y was about 4.45 (when $x \approx 0.315$). That gives an approximate range of $\{y: y \geq 4.45\}$. Zooming in, we found an even more accurate smallest value to be 4.44.

Caution: If you do not know the shape of the graph of a function, do not work from as little information as is given in the graph of Example 1. Pick a new window that shows more of the graph.

An exact range for a quadratic function can be obtained from the coordinates of the vertex of the parabola, which in turn can be found by completing the square.

LESSON 2-2

RESOURCES

- Lesson Master 2-2
- Teaching Aid 5 displays the graph of the minimization example of the lesson.

OBJECTIVES

C Identify the domain, range, and minimum and maximum values of functions.
F Solve maxima-minima problems.
G Analyze a function from its graph.

TEACHING NOTES

You may find it convenient to use the word *extremum* (plural: *extrema*) to refer either to a maximum or minimum value. Whereas the range of a function is the set of all possible values of the function, extrema of a function may be associated with the entire domain, or they may be *relative extrema* associated with a particular interval. (Relative extrema are discussed in the next lesson.) Obviously, the extrema of a function restricted to a particular interval might not be the extrema for the function over its entire domain.

Making Connections The problem of minimizing a cylinder's surface area for a given volume is important and should be examined in detail. For calculus students, framing maximum/minimum problems in terms of functions is often the most difficult step in solving such problems. Other examples of this type of problem which you may want to discuss include minimizing the surface area of containers other than cylinders, minimizing cost functions, or maximizing profit functions. Any calculus text contains an abundance of typical problems, but does not solve them the way we do. Here is an additional example you might use.

A box with a square base and no lid is to have a volume of 3000 in^3. Find the dimensions of the box with the given volume that has the least surface area. (Let x be a side of the base, y be the height of the box, and A be its surface area.

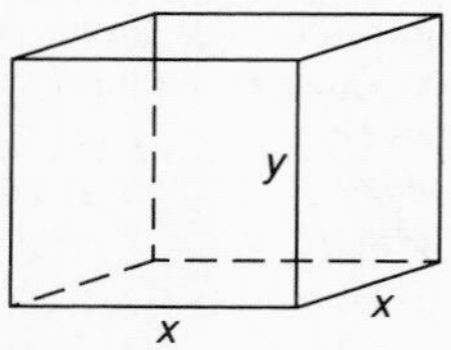

Then $3000 = x^2y$,

so $y = \frac{3000}{x^2}$.

Then $A = x^2 + 4xy$

$= x^2 + \frac{12000}{x}$

With an automatic grapher, the minimum area is when $x \approx 18.2$ in. Substituting into the second equation yields $y \approx 9.1$ in.)

Making Connections
One advantage of using calculus to solve maximum/minimum problems such as these is that the calculus approach produces exact solutions while the graphical approach yields only approximations to those solutions. More important, however, is that calculus provides a formal logical framework into which the solutions to these and whole classes of similar problems can be described and analyzed accurately.

Making Connections
The function f defined by $f(r) = 2\pi r^2 + \frac{116}{r}$ is a rational function, of the type to be discussed in Chapter 5. You might ask students to describe the appearance of a can with $r = 0.1$ or $r = 100$. As r gets close to zero, the term $\frac{116}{r}$ dominates $f(r)$, so

Example 2 Give the exact range of the function of Example 1.

Solution $y = 4x^2 - 3x + 5$

$$y = 4\left(x^2 - \frac{3}{4}x + \_\_\right) + 5 - \_\_$$

The expression in parenthesis will be a perfect square if the blank is filled by the square of half the coefficient of x, $\left(\frac{3}{8}\right)^2$, or $\frac{9}{64}$. This adds $4 \bullet \frac{9}{64}$ to that side (the expression having been multiplied by 4), so the same amount must be subtracted.

$$y = 4\left(x^2 - \frac{3}{4}x + \frac{9}{64}\right) + 5 - \frac{9}{16}$$

$$y = 4\left(x - \frac{3}{8}\right)^2 + \frac{71}{16}$$

The smallest value occurs when $x = \frac{3}{8}$, in which case $y = \frac{71}{16} = 4.4375$. Thus the vertex is $\left(\frac{3}{8}, \frac{71}{16}\right)$, and the exact range is $\{y: y \geq 4.4375\}$.

Caution: On some automatic graphers, the word "range" refers to the values of both the independent and dependent variables currently displayed by the window. Our usage is more common within mathematics.

For the function of Examples 1 and 2, 4.4375 is the *minimum value*, and it occurs at $x = \frac{3}{8}$.

Definition

m is the **minimum value** of a real function f with domain A if and only if $\exists c$ in A such that $f(c) = m$ and $\forall$ x in A, $m \leq f(x)$.
m is the **maximum value** of a real function f with domain A if and only if $\exists c$ in A such that $f(c) = m$ and $\forall$ x in A, $m \geq f(x)$.

For example, the maximum value of the sine function is 1 and the minimum value is -1, as a graph verifies. That is, $\forall x$, $-1 \leq \sin x \leq 1$.

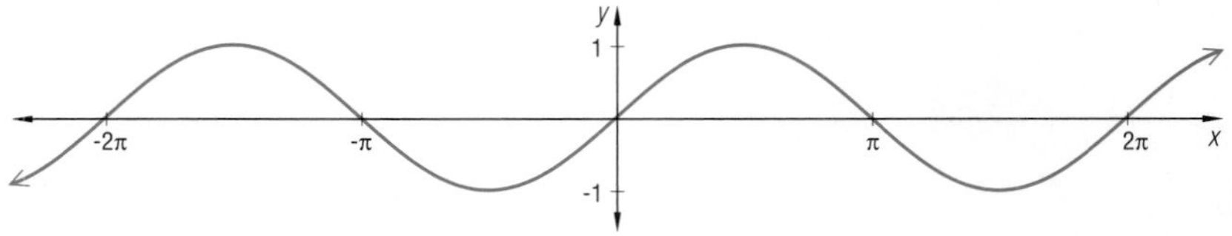

Here is a design problem that is rather complicated but leads to a typical kind of function whose maximum or minimum value is of interest.

A container manufacturer wants a cylindrical metal can that will hold one quart of paint with a little extra room for air space. One liquid quart occupies a volume of 57.749 in.^{3}, so the manufacturer decides to design a can with a volume of 58 in.^{3}. Of course, the manufacturer wants to keep the cost of material at a minimum. The amount of sheet metal required to construct the can is equal to the total surface area A of the top, bottom, and sides of the can. The problem, then, is to find the dimensions r and h which give a volume of 58 in.^{3} and a minimum value for A.

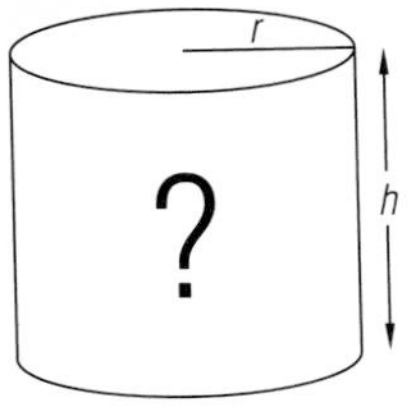

In order to do this we develop a formula for A as a function of a single variable. The total surface area A of the can is the sum of the areas of two circles (the top and bottom of the can), and the area of the rectangle that is formed when the side of the can is slit and unrolled. Thus, as shown here,

$$A = 2\pi r^2 + 2\pi rh.$$

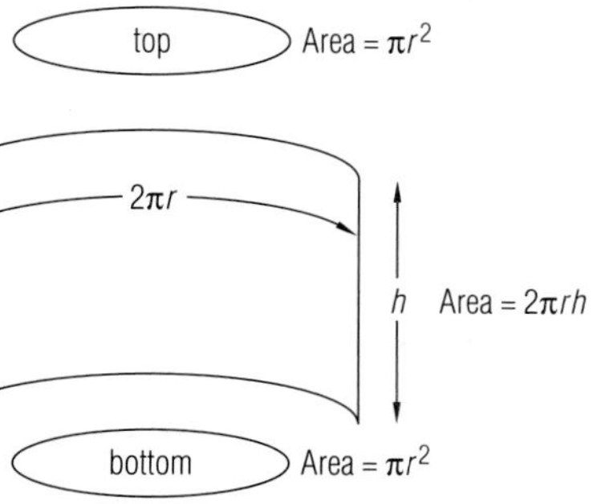

Now, to write A as a function of one variable only, the radius, we must find a way to write h in terms of r. The volume V of a cylinder of radius r and height h is given by $V = \pi r^2 h$. Because $V = 58$ in.^{3}, it follows that $58 = \pi r^2 h$, and so $h = \frac{58}{\pi r^2}$ expresses h as a function of r. Substitute $\frac{58}{\pi r^2}$ for h in the area formula.

$$\begin{aligned} A &= 2\pi r^2 + 2\pi r\left(\frac{58}{\pi r^2}\right) \\ &= 2\pi r^2 + \frac{116}{r} \end{aligned}$$

The area is now stated as a function of the radius, so we write

$$A = f(r) = 2\pi r^2 + \frac{116}{r}.$$

A graph can be used to determine the value of r which would make the area $f(r)$ of sheet metal as small as possible. However, simply entering the formula

$$f(r) = 2\pi r^2 + \frac{116}{r}$$

into an automatic grapher may result in a view of part of the graph that does not contain the minimum value. You may even get a blank screen. The values for the graphing window must be set so that the window contains the part of the graph which contains the minimum. To determine appropriate window values, it is helpful to compute some sample values of $f(r)$.

$$f(1) = 2\pi + 116 \approx 122 \qquad f(2) = 8\pi + 58 \approx 83,$$
$$f(3) = 18\pi + \frac{116}{3} \approx 95 \qquad f\left(\frac{1}{2}\right) = \frac{\pi}{2} + 232 \approx 234.$$

$r = 0$ is a vertical asymptote. What has happened is that the radius of the can is small but the height is very large, and the amount of material needed is quite great. (The mathematical idea is discussed in Lesson 5-3.)

As r becomes larger and larger, the term $2\pi r^2$ dominates, and the graph of f is indistinguishable from the graph of $y = 2\pi r^2$. In this case, the radius is quite large and since the can has a fixed volume, its height is very small; thus the amount of material needed is just about the same as that needed for the two circles comprising the top and bottom of the can. (The mathematical idea is discussed in Lesson 5-5.)

You might point out that because the formula $A = 2\pi r^2 + 2\pi rh$ has two independent variables (r and h), its graph is 3-dimensional and cannot be analyzed fully with a 2-dimensional coordinate system. Thus we must use our knowledge of the actual volume needed to write $\pi r^2 h = 58$, solve for h in terms of r, and use this in the formula for area to get a formula with only one independent variable.

Technology Students may have difficulty knowing how to determine accuracy when tracing and zooming in. When tracing a function, have students pay attention to what decimal place is changing in the y-coordinate. From this, the accuracy of the approximation can be determined. If the change is occurring in the third decimal place, then the estimate is accurate to two decimal places.

ADDITIONAL EXAMPLES

1. Estimate the range of the function f with $f(x) = -4x^2 - 3x + 5$.

Sample: Using a window of $-2 \le x \le 0$ and $4 \le y \le 6$ and the trace function, the maximum value appears to be approximately $y = 5.56$. Thus, the range is approximately $\{y: y \le 5.56\}$.

ADDITIONAL EXAMPLES

2. Give the exact range of the function from Example 1.

Let $y = -4x^2 - 3x + 5$.
Completing the square,
$$y = -4\left(x^2 + \tfrac{3}{4}x + \tfrac{9}{64}\right) + 5 + \tfrac{9}{16}$$
$$y = -4\left(x + \tfrac{3}{8}\right)^2 + \tfrac{89}{16}.$$
Thus, the exact maximum value of the function is $y = \frac{89}{16} = 5.5625$, and the exact range is $\{y: y \le 5.5625\}$.

NOTES ON QUESTIONS

Question 11: Some students may think that if the area of the rectangle is fixed, then the perimeter is also a constant for all lengths and widths. A spreadsheet or calculator generated table of values will help dispel this misconception. For instance, with an area of 2000, the following are possible.

Length	Width	Perimeter
500	4	1008
400	5	810
200	10	420
100	20	240
…	…	…

Question 18: This question provides an opportunity to see if students have remembered anything from their previous study of logarithms.

Question 19: Discuss with students what factors other than economy of material are considered in designing a container. Appearance, ease of handling, ease of appropriate use, and stackability on store shelves are additional considerations. A sphere is the most economical shape for a container with a given volume, but a gallon of milk packed in a sphere would roll off the breakfast table.

ADDITIONAL ANSWERS

8. minimum value: -0.1

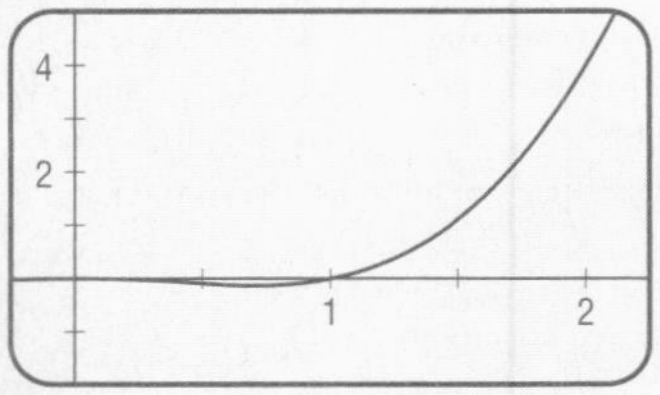

These values indicate that the scale on the y-axis should reach at least 100. With a suitable choice of values on the horizontal axis, the graph of f looks like the one shown.

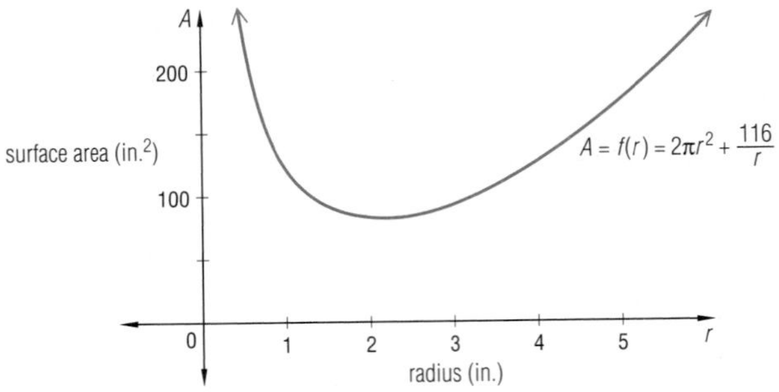

By using the trace key and zooming in, we found the coordinates of the minimum point on the curve to be about (2.1, 82.9); that is, the minimum value of A is approximately 82.9, and it occurs when $r \approx 2.1$. Therefore, the best choice for the radius of the can is approximately 2.1 inches and, since $h = \frac{58}{\pi r^2}$, the corresponding height for the can is given by

$$h \approx \frac{58}{\pi(2.1)^2} \approx 4.2 \text{ inches}$$

The value of A when $r \approx 2.1$ gives some additional information: approximately 82.9 square inches of sheet metal is required to construct this can.

Questions

Covering the Reading

1. Consider the function $f: x \rightarrow y$ given by the rule $y = 2x$. Find the range of this function if its domain is
   a. the set of positive integers **the set of positive even integers**
   b. the set of positive real numbers **the set of positive real numbers**
   c. the set of all real numbers. **the set of real numbers**

2. A number t is in the range of a function f if and only if $\exists$ x in the domain of f such that __?__. **$f(x) = t$**

3. M is the maximum value of a function g if and only if $\forall$ z in the domain of g, __?__. **$g(z) \le M$**

4. Consider the quadratic function with equation $y = 3x^2 + 4x + 10$.
   a. Estimate the range using an automatic grapher.
   b. Calculate the exact range. **{real numbers y: $y \ge 8\frac{2}{3}$}**
   a) sample: {real numbers y: $y \ge 8.7$}

5. What is the mean of the maximum and minimum values of the sine function? **0**

6. Consider a cylindrical can with height 4.2 inches and radius 2.1 inches.
   a. What is its volume? **$V \approx 58.2$ in.³**
   b. What is its surface area? **$A \approx 83.1$ in.²**
   c. What property does a can with these dimensions possess? **These dimensions come close to minimizing that volume's surface area.**

(6b) 2·3.14·(2.1)² = 27.69
2·3.14·2.1·4.2 = 55.39
27.69 + 55.39 = 83.1
V = πr²h
a) 3.14·(2.1)²·4.2

90

7. Suppose that a cylindrical can is to hold a liter of oil. Recall that 1 liter = 1000 cm^3.

a. Give a formula for the surface area of this can in terms of its radius. $S(r) = 2\pi r^2 + \frac{2000}{r}$

b. Approximately what radius and height (cm) would use the least amount of material to make the can? $r \approx 5.4$ cm; $h \approx 10.9$ cm

Applying the Mathematics

In 8 and 9, graph the function defined by the given rule over the indicated domain. Use the graph to estimate (to the nearest tenth) the minimum value of the function over this domain.

8. $f(x) = x^3 - x^2$, domain $\{x: 0 \le x \le 2\}$ See margin.

9. $g(x) = x \cos x$, domain $\{x: -10 \le x \le 10\}$ See margin.

10. The manufacturer discussed in the text is also commissioned to make cylindrical cans with volume 58 $in.^3$ and no tops because these will be closed with plastic lids. Once again the manufacturer is interested in finding the dimensions which require the least amount of metal for each can. Determine approximate values for the optimum dimensions. radius $\approx$ 2.6 in.; height $\approx$ 2.7 in.

11. A rancher is planning to fence in a rectangular area of 2000 m^2 for his cattle to graze. He is interested in determining the least amount of fencing needed. Perform the following steps to obtain an approximate answer to this question.

a. Determine a formula which expresses the length, l, of a 2000 m^2 piece of land in terms of its width, w. $l = \frac{2000}{w}$

b. Use part **a** to express the perimeter, P, of a 2000 m^2 rectangular piece of land in terms of its width. $P(w) = 2w + \frac{4000}{w}$

c. Evaluate the function, $P(w)$, in part **b** at $w = 40$, $w = 45$, and $w = 50$.

d. Use an automatic grapher to approximate the coordinates of the lowest point on the graph of $P(w)$ in the first quadrant.

e. What is the least amount of fencing that the rancher needs?

c–e) See margin.

12. Consider the function g whose entire graph is given at right.

a. Identify the domain and range of g.

b. Is g a discrete function? Explain.

c. Identify the minimum and maximum values of g.

See margin.

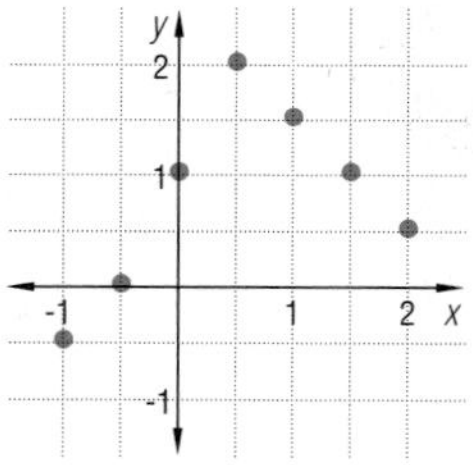

13. Let h be the function whose entire graph is shown at right.

a. Identify the domain and range of h.

b. Is h a discrete function? Explain why or why not.

c. Give the minimum and maximum values of h.

d. Find all x such that $h(x) = -2$.

e. On what interval(s) is $h(x) < 0$?

See margin.

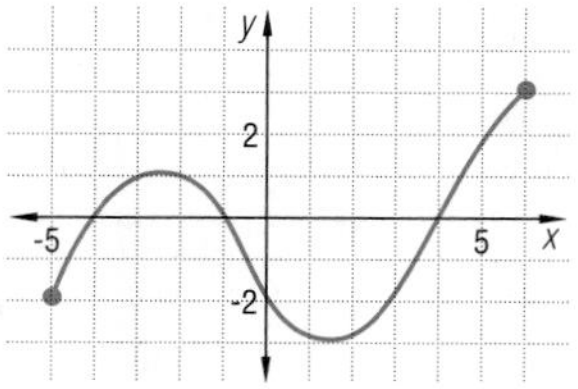

9. See Additional Answers in the back of this book.

11.c. $P(40) = 180$
$P(45) = 178.9$
$P(50) = 180$
d. (44.7, 178.9)
e. 178.9 m

12., 13. See Additional Answers in the back of this book.

NAME ______

LESSON **MASTER 2–2**
QUESTIONS ON **SPUR** OBJECTIVES

PROPERTIES *Objective C* *(See pages 146–150 for objectives.)*

1. Let $f(x) = 2x^2 + 14x + 21.5$.

a. Complete the square to find the exact range.

$f(x) = 2(x + 3.5)^2 - 3$; $\{y: y \ge -3\}$

b. Check your answer to part **a** by using an automatic grapher. Sketch the graph on the axes at the right.

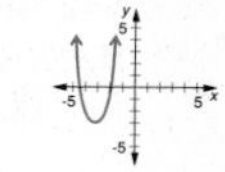

2. What is the range of the tangent function? set of all real numbers

3. Let g be the function defined by $g(x) = x^2$, where the domain is the set of integers. What is the range? set of all nonnegative perfect squares

USES *Objective F*

4. A canning factory is going to cut up pineapples for processing. A pineapple is roughly cylindrical. It will be sliced in half lengthwise to test for freshness. Then the half cylinder will be trimmed to form a rectangular solid, as shown at the right. The maximum volume of pineapple is obtained when the area of the rectangular cross-section is a maximum. Assume the pineapple has a radius of 3 inches.

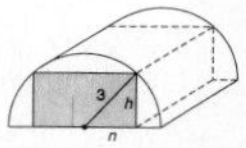

a. Use the Pythagorean Theorem to express n, half the base of the rectangle, in terms of h. $n = \sqrt{9 - h^2}$

b. Write an equation to express A, the area of the rectangle, in terms of h. $A = 2h\sqrt{9 - h^2}$

16 *Continued* *Precalculus and Discrete Mathematics* © Scott, Foresman and Company

NAME ______
Lesson MASTER 2-2 (page 2)

c. Use an automatic grapher to graph your answer to part **b**. Sketch the graph on the axes at the right.

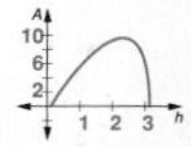

d. Find the dimensions of the rectangular cross-section that produce the maximum area. $h \approx 2.1$ in., $2n \approx 4.2$ in.

e. What is the maximum area? ≈ 8.8 in.

REPRESENTATIONS *Objective G*

5. The entire graph of the function f is given at the right.

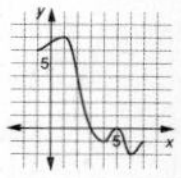

a. What is the domain of f? $-1 \le x \le 7$

b. What is the range of f? $-2 \le y \le 7$

c. Estimate the maximum and minimum values. max = 7 at x = 1, min = 2 at x = 6

d. On what interval(s) is $f(x)$ nonpositive? $3 \le x \le 7$

e. Is f a discrete function? Why or why not? No, the domain includes all reals between -1 and 7, and this is not a discrete set.

6. Sketch a graph with all of the following characteristics.

(a) The domain is $1 \le x \le 5$ for real numbers x.

(b) The range is $-3 \le y \le 3$ for real numbers y.

(c) $f(2)$ is the minimum value and $f(4)$ is the maximum value.

Answers will vary, but must have $f(2) = -3$ and $f(4) = 3$.

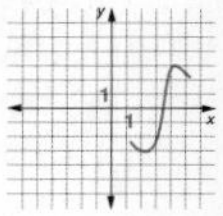

7. Use an automatic grapher to estimate the maximum or minimum values of $f(x) = x^3 + x^2 - 17x + 5$ on the interval $-5 \le x \le 5$. max = 70 (at x = 5), min $\approx$ -17 (at x $\approx$ 2)

Precalculus and Discrete Mathematics © Scott, Foresman and Company 17

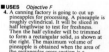

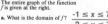

FOLLOW-UP

MORE PRACTICE
For more questions on SPUR Objectives, use *Lesson Master 2-2,* shown on page 91.

EXTENSION
Suppose that a can manufacturer wants to know the dimensions of the cylindrical can of maximum volume that can be constructed from 82.9 square inches of sheet metal.
(1) How does this problem differ from the minimization problem solved in this lesson? (The surface area is given and the volume must be maximized.)
(2) Express the volume of the can as a function of the radius.
($V \approx 41.5r - \pi r^3$)
(3) Use an automatic grapher to approximate the dimensions of the can of maximum volume.
($r \approx 2.1$ in., $h \approx 4.2$ in.^{2}, maximum volume ≈ 58 in.^{3})

The exact solutions for the minimization problem and the maximization problem in this extension can be shown to be identical. Such pairs of maximum and minimum problems are called *dual problems.* A class of mathematical results called *duality theorems* provide conditions under which the solutions to dual problems are identical.

ADDITIONAL ANSWERS
19. sample:
radius: 1.25 inches;
height: 5.75 inches;
volume: ≈ 28.23 cubic inches;
surface area: ≈ 54.98 square inches;
optimal height: ≈ 3.30 inches;
optimal radius: ≈ 1.65 inches
Economical dimensions are not used to obtain this volume.

Review

14. A camera store uses the chart below to determine the discount earned for quantity purchases. Let R be the set of numbers of rolls of film purchased and let P be the set of percent discounts.

Number of rolls of film purchased	Percent discount
0-2	0
3-6	5
7-12	8
13-20	10
21 and over	15

See below.

a. For all r in R, let $f(r)$ = percent discount on r rolls. Is $f: R \rightarrow P$ a function? If so, is it a discrete function? Explain your answers.
b. For all p in P, let $g(p)$ = number of rolls receiving discount p. Is $g: P \rightarrow R$ a function? If so, is it a discrete function? Explain your answers. No, each number in P corresponds to more than one in R.
c. If possible, evaluate.
i. $f(7)$ 8 **ii.** $g(10)$ not possible **iii.** $f(23)$ 15 *(Lesson 2-1)*

15. Consider the real function g described by the rule $g(y) = \frac{\sqrt{y+1}}{\sqrt{y-1}}$.
a. For what real values of y is the numerator defined? $y \geq -1$
b. For what real values of y is the denominator a nonzero real number? $y > 1$
c. Use the results of **a** and **b** to find the domain of g.
(Lesson 2-1) c) {real numbers y: $y > 1$}

16. Write the negation of the following statement about the range of a function $f: A \rightarrow B$.
$\forall y$ *in B,* $\exists x$ *in A with* $f(x) = y$. *(Lesson 1-2)*
$\exists y$ in B such that $\forall x$ in A, $f(x) \neq y$.

17. Let f be the function with rule $f(x) = 2x + 3$, and consider the conditional
If n is an integer, then f(n) is an odd integer.
a. Write the negation of the conditional. See below.
b. Which is true, the original conditional or its negation?
c. If the original conditional is true, prove it. Otherwise, find a counterexample that shows that its negation is true.
(Lessons 1-7, 1-5) b) the original conditional; c) See below.

18. Evaluate without using a calculator. *(Previous course)*
a. $\log_5 25$ 2 **b.** $\log_5 \frac{1}{25}$ -2 **c.** $\log_5 1$ 0

Exploration

19. Find a soda can and determine its radius, height, volume, and surface area. Using the idea of this lesson, determine whether its dimensions are economical in the sense that the surface area is the smallest possible for the given volume. See margin.

14. a) Yes, each element in R corresponds to exactly one element in P. Yes, the domain is a discrete set, the set of nonnegative integers.
17. a) *n is an integer, and f(n) is not an odd integer.*
c) Suppose n is any integer. Then $f(n) = 2n + 3 = 2n + 2 + 1 = 2(n + 1) + 1$. By the closure property of addition, $(n + 1)$ is an integer; and $f(n)$ is odd, according to the definition of odd.

92

LESSON

2-3 Increasing and Decreasing Functions

Here is a table and graph of total energy consumption (coal, natural gas, petroleum, hydroelectric power, nuclear electric power, and geothermal) in the United States from 1954–1988. (The data are from the 1988 Energy Information Administration Annual Energy Review. Consumption is given in quadrillion BTUs.)

Year	Consumption
1954	35.27
1955	38.82
1956	40.38
1957	40.48
1958	40.35
1959	42.14
1960	43.80
1961	44.46
1962	46.53
1963	48.32
1964	50.50
1965	52.68
1966	55.66
1967	57.57
1968	61.00
1969	64.19
1970	66.43
1971	67.89
1972	71.26
1973	74.28
1974	72.54
1975	70.55
1976	74.36
1977	76.29
1978	78.09
1979	78.90
1980	75.96
1981	73.99
1982	70.84
1983	70.50
1984	74.06
1985	73.96
1986	74.26
1987	76.77
1988	79.94

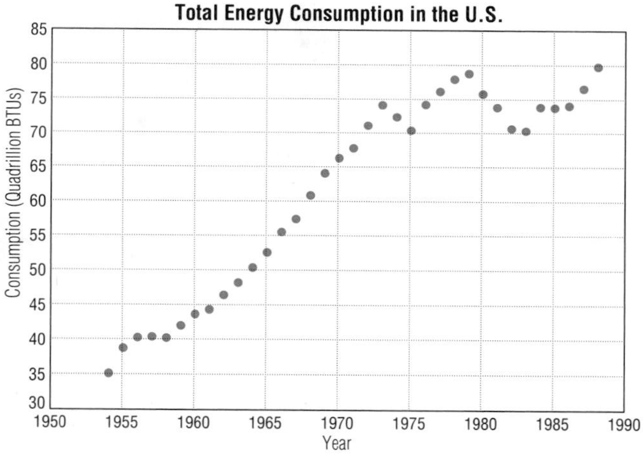

Think of these ordered pairs as determining a function E mapping years into quadrillions of BTUs: For each year x, $E(x)$ = the energy used in year x (in quadrillions of BTUs). For instance, $E(1985) = 73.96$. E is a discrete function that is defined by the table. The domain of E is the set of integers from 1954 to 1988.

By examining either the table or the graph of E, you can tell when the total energy consumption in the U.S. increased from one year to the next. For instance, since $E(1985) < E(1986)$, energy consumption increased from 1985 to 1986.

We also say that energy consumption increased during the *time intervals* 1954–1957, 1958–1973, 1975–1979, and 1985–1988. There are various types of *intervals*. All have graphs that are segments or rays (or possibly discrete segments or rays) with or without their endpoints.

LESSON 2-3

RESOURCES
- Lesson Master 2-3
- Quiz for Lessons 2-1 through 2-3
- Teaching Aid 6 displays the data and graph of total U.S. energy consumption.
- Teaching Aid 7 displays the definitions of increasing and decreasing functions.
- Teaching Aid 8 can be used with **Questions 11 and 12**.
- Computer Master 4

OBJECTIVES

A Determine relative minima and maxima of a function and intervals on which it is increasing or decreasing.

G Analyze a function from its graph.

TEACHING NOTES

In this book, the terms *open, half-open, closed,* and *infinite* are seldom used as they apply to intervals, but many people like to use them and we know of no difficulty this causes unless it is overdone. On the other hand, we often use the phrases *between a and b* (for an open interval) and *from a to b* (for a closed interval).

Discuss the definitions of *increasing function* and *decreasing function.* Notice the use of the universal quantifier to express with precision the fact that a function is increasing or decreasing on a set S. Increasing on an interval means increasing *everywhere* on that interval.

What we call *increasing* is sometimes called *strictly increasing.* Some textbooks also provide the following definitions.

f is nondecreasing or monotonic decreasing on S if and only if $\forall\ x_1, x_2 \in S$, if $x_1 < x_2$, then $f(x_1) \le f(x_2)$.

f is nonincreasing or monotonic decreasing on S if and only if $\forall\ x_1, x_2 \in S$, if $x_1 < x_2$, then $f(x_1) \ge f(x_2)$. Every constant function is both nonincreasing and nondecreasing.

It is not obvious to all students that for a continuous function, a relative maximum occurs when the function stops increasing and begins decreasing. Use the discrete example at the beginning of the lesson to point this out. Then ask what corresponding statement can be made about the place where a relative minimum occurs.

Technology The **Example** shows how an automatic grapher can be used to find approximate endpoints of intervals where a function is increasing or decreasing. For this purpose, all students should be able to make use of the trace function on an automatic grapher.

For the proof on this page, mastery is not expected, but it is instructive in its use of the definition of *increasing* and in its logic. In the proof, we must let x_1 and x_2 be *any* real numbers in order for the sufficient condition of an increasing function to be satisfied.

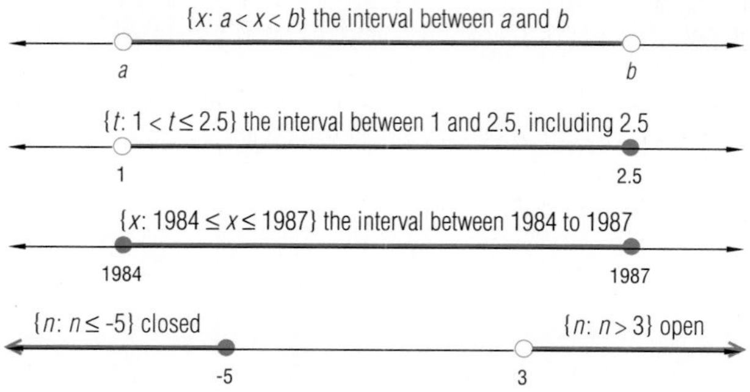

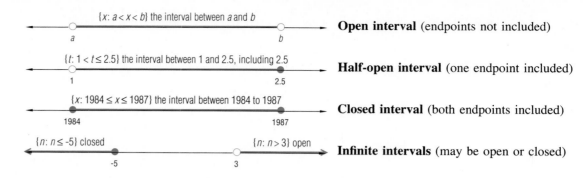

Open interval (endpoints not included)

Half-open interval (one endpoint included)

Closed interval (both endpoints included)

Infinite intervals (may be open or closed)

Returning to the energy consumption function, note that although $E(1984) = 74.06$ and $E(1987) = 76.77$, the function E is not increasing on the interval from 1984 to 1987 because the values of E went down from 1984 to 1985. In order for a function to be increasing on a set, whenever *any* two domain values are chosen from that set, the greater of the two domain values must yield a greater range value. Graphically, this means that as you trace the graph with a pencil from left to right, your pencil continually moves upward.

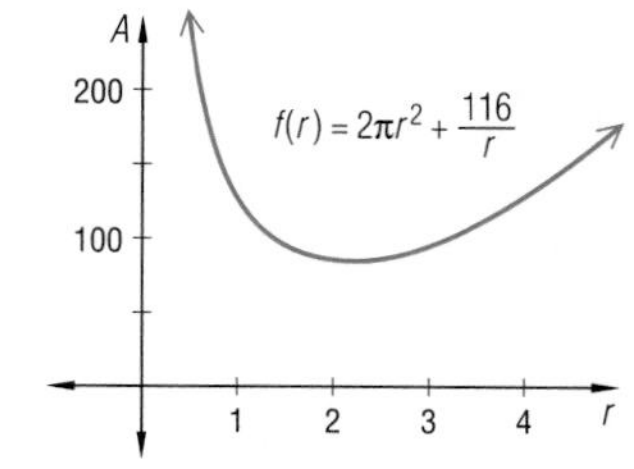

The surface area function $f(r) = 2\pi r^2 + \frac{116}{r}$ considered in the can design problem that we solved in the last lesson is a nondiscrete real function whose domain is the set of all positive real numbers. From the graph of f it appears that the function f is decreasing on the interval $0 < r < 2.1$ (approximately) and that f is increasing on the interval $r > 2.1$ (approximately) in its domain. However, although $f(1) \approx 122.28$ and $f(3) \approx 95.22$, the function f is not decreasing on the interval $1 < r < 3$ because, for example, $f(2) \approx 83.13 < f(3)$. These ideas are embodied in the following definitions.

Definitions

Suppose f is a real function and S is an interval in the domain of f.

f* is increasing on *S
if and only if $\forall\ x_1$ and x_2 in S, if $x_1 < x_2$ then $f(x_1) < f(x_2)$.

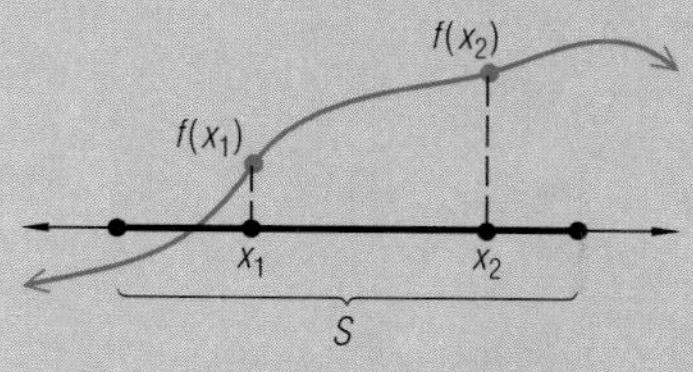

f* is decreasing on *S
if and only if $\forall\ x_1$ and x_2 in S, if $x_1 < x_2$ then $f(x_1) > f(x_2)$.

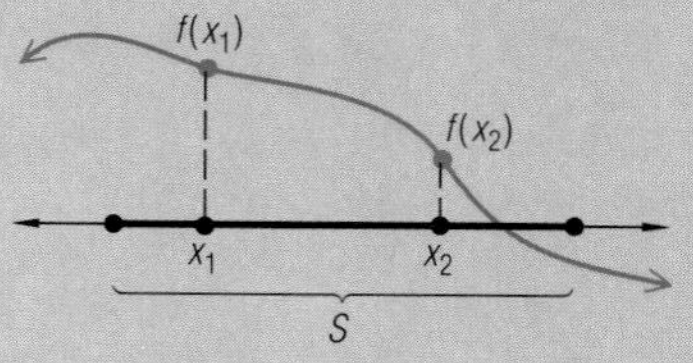

94

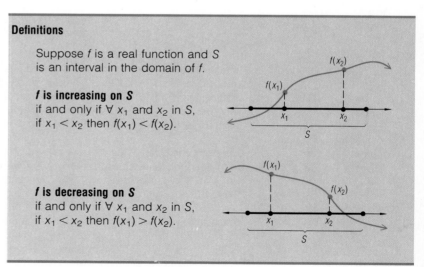

Knowing when energy consumption increases or decreases may indicate whether conservation measures are working or not, how much energy might be needed in the future, what it might cost, and so on. Knowing when a function increases or decreases can also indicate any maximum or minimum values. Conversely, knowing maximum or minimum values can help determine where a function increases or decreases. The example shows how you can do this with an automatic grapher.

Example Let f be the function defined by the rule $f(x) = x^3 + x^2 - x - 1$. Use an automatic grapher to estimate the intervals within the interval $\{x: -3 \le x \le 3\}$ on which f is increasing and on which f is decreasing.

Solution Graph the function for $-3 \le x \le 3$. A plot will look like the one shown at the right.

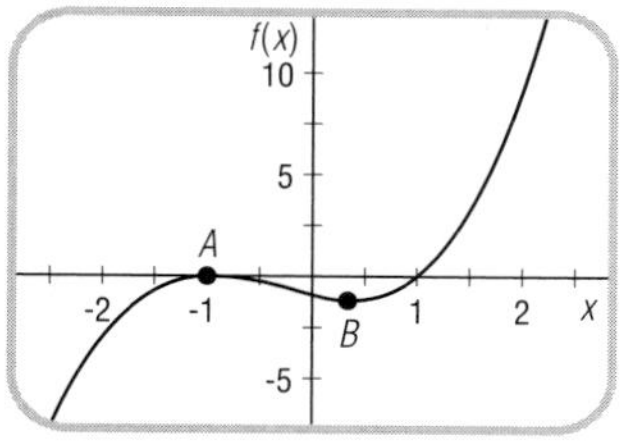

Trace along the curve from left to right. When x is -3, y is approximately -16. As you move from $x = -3$ to point A where $x = -1$, the y values increase, and so the graph goes up. Therefore, f is increasing on the interval $-3 \le x \le -1$. As you move left to right from point A (at $x = -1$) to point B (at $x \approx 0.3$), the values of f decrease, and so the graph goes down. Therefore, f is decreasing on the approximate interval $-1 \le x \le 0.3$. As you move to the right of B, the y values again get larger and the graph goes up, so f is also increasing on the interval $0.3 \le x \le 3$. Thus approximate intervals on which f is increasing are

$$-3 \le x \le -1 \quad \text{and} \quad 0.3 \le x \le 3$$

and the interval on which f is decreasing is $-1 \le x \le 0.3$.

Points like A and B in the graph of the example are significant points in the application of functions to real world problems. The value 0 of the function f at $x = -1$ is greater than or equal to the value of f at any other point near $x = -1$. For instance, for all x in the interval $-2 < x < 0$, $f(-1) \ge f(x)$. Therefore, we say that f has a **relative maximum** at $x = -1$, and that the relative maximum value is 0. The word *relative* is important because there are obviously values of the function outside this interval that are larger than $f(-1)$. For instance, $f(1.5) > f(-1)$. Notice that a relative maximum can be identified by locating a point such that the function is increasing on the left of that point and decreasing on the right.

Similarly, to locate a **relative minimum**, find a point where the function is decreasing on the left and increasing on the right. In the example, a relative minimum occurs at B, where $x \approx .3$. The relative minimum value is approximately $f(.3) \approx -1.2$.

Error Analysis

x^3+x^2-x-1

$-27+9+3-1$

-16

...increment might improve understanding of how to find the extrema.

ADDITIONAL EXAMPLE
Let g be the function defined by $g(x) = x^3 - 4x^2 + 2x + 3$ for all real numbers x. Use an automatic grapher to estimate where within the interval $\{x: -1 \le x \le 4\}$ the function g is increasing and where g is decreasing.
Graphing the function for $-1 \le x \le 4$ yields the following:

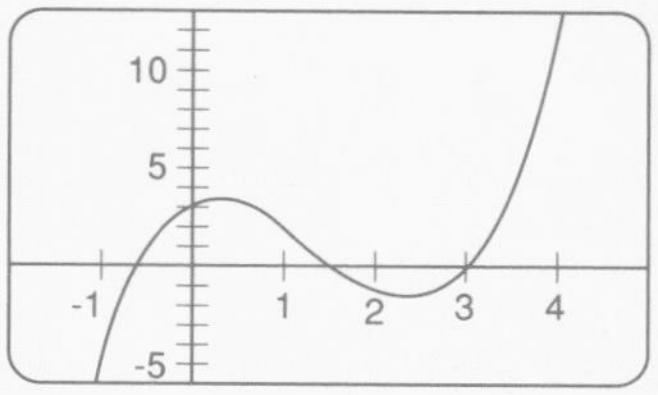

Using the trace key and moving from left to right, we estimate that g is increasing when $-1 \le x \le .3$ or $2.4 \le x \le 4$ and decreasing when $.3 \le x \le 2.4$.

NOTES ON QUESTIONS

Question 14: A kilowatt-hour is literally 1000-watt hours, the amount of energy used by a 1000-watt bulb when lit for one hour.

ADDITIONAL ANSWERS

11.a. $1954 \le x \le 1957$, $1958 \le x \le 1970$, $1975 \le x \le 1980$, $1986 \le x \le 1988$
b. $1972 \le x \le 1975$, $1980 \le x \le 1983$, $1984 \le x \le 1986$
c. 41.65, 62.07, 62.42, 64.76, 65.81, 65.88
d. 35.13, 38.81, 61.29, 59.86, 61.19, 64.25

12.a. $\forall$ integers x_1 and x_2 in $\{x: 1972 \le x \le 1976\}$, if $x_1 < x_2$, then $f(x_1) > f(x_2)$.
b. $\exists$ integers x_1 and x_2 in $\{x: 1972 \le x \le 1976\}$, such that $x_1 < x_2$ and $f(x_1) \le f(x_2)$.
c. $x_1 = 1975$ and $x_2 = 1976$

17.a.

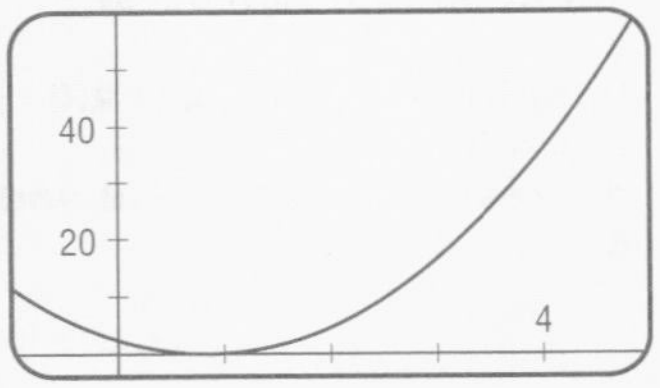

In 11 and 12, refer to the total U.S. energy production (in quadrillions of BTUs) for the years 1954–1988 given below. Let f be the function mapping years to production.

Year	Production	Year	Production	Year	Production
1954	35.13	**1966**	52.17	**1978**	61.10
1955	38.73	**1967**	55.04	**1979**	63.80
1956	41.21	**1968**	56.81	**1980**	64.76
1957	41.65	**1969**	59.10	**1981**	64.42
1958	38.81	**1970**	62.07	**1982**	63.89
1959	40.60	**1971**	61.29	**1983**	61.19
1960	41.49	**1972**	62.42	**1984**	65.81
1961	41.99	**1973**	62.06	**1985**	64.78
1962	43.58	**1974**	60.84	**1986**	64.25
1963	45.85	**1975**	59.86	**1987**	64.82
1964	47.72	**1976**	59.89	**1988**	65.88
1965	49.34	**1977**	60.22		

11. **a.** Identify all intervals of length ≥ 2 years over which f is increasing.
b. Identify all intervals of length ≥ 2 years over which f is decreasing.
c. Identify all relative maximum values in the interval $1954 \le x \le 1988$.
d. Identify all relative minimum values in the interval $1954 \le x \le 1988$.
See margin.

12. Note that $f(1972) \ge f(1976)$.
a. Using mathematical symbols, write what it would mean for f to be decreasing on the interval $1972 \le x \le 1976$.
b. Write the negation of the statement you wrote in part **a**.
c. Show that f is not decreasing on the interval $1972 \le x \le 1976$ by finding two numbers which satisfy the statement you wrote in part **b**.
See margin.

13. The height (in meters) t seconds after a certain freely falling object is thrown vertically up is given by the formula $h(t) = 1 + 25t - 4.9t^2$.
a. For what values of t is the function h decreasing? $t \ge 2.55$ sec
b. What is the physical meaning of the answer to part **a**? At $t = 2.55$ sec, the object reaches maximum height and starts to descend.

Review

14. One kilowatt-hour, the amount of electricity needed to keep a 100-watt bulb lit for 10 hours, is 3,412 BTUs. Convert the amount of energy used in the U.S. in 1988 into kilowatt-hours. *(Previous course)* 19.308 trillion kilowatt-hours

15. The values of the function E in this lesson are in quadrillions of BTUs. What is a quadrillion? *(Previous course)* 1,000,000,000,000,000 or 1×10^{15}

16. Explain why the function E of this lesson is a discrete function. *(Lesson 2-1)* Its domain is a discrete set, {integers x: $1954 \le x \le 1988$}.

17. **a.** Sketch a graph of the quadratic function defined by $f(x) = 3x^2 - 5x + 2$ with domain $-1 \le x \le 5$. See margin.
b. Give the range of f. *(Lesson 2-2)* $\{y: -\frac{1}{12} \le y \le 52\}$

18. A box is to be constructed with a square base and no top. It must hold a volume of 10 cubic meters but be made with a minimum of materials. Let s be the length of a side of the base and let h be the height of the box. a) $h = \frac{10}{s^2}$; b) $A(s) = \frac{40}{s} + s^2$

a. Find a formula for h in terms of s.

b. Find a formula for the surface area of the box as a function of s.

c. Estimate the value of s which minimizes the surface area. ≈ 2.7 m

d. What are the approximate dimensions and surface area of the box with minimal surface area? *(Lesson 2-2)*
2.7 m × 2.7 m × 1.4 m; S.A. ≈ 22.1 m²

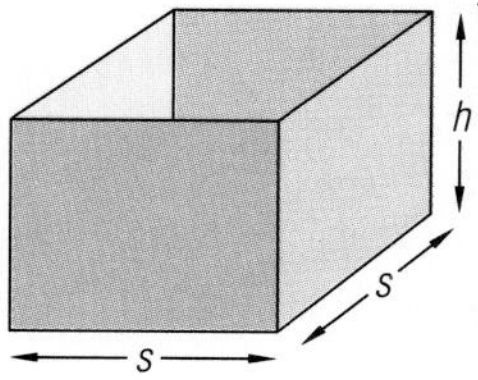

19. Consider the following line of a computer program.

```
100 IF (X < Y) AND (Y < Z) THEN PRINT "IN ORDER"
    ELSE PRINT "OUT OF ORDER"
```

Fill in the blank below so that a computer language that does not allow the ELSE command would run the program correctly:

```
100 IF (X < Y) AND (Y < Z) THEN PRINT "IN ORDER"
101 IF _?_ THEN PRINT "OUT OF ORDER"
```

(Lessons 1-5, 1-3) sample: (X > = Y) OR (Y > = Z)

In 20-22, use the laws of exponents to simplify. *(Previous course)*

20. $8^{30} \cdot 2^{15}$ 2^{105} **21.** $(ab)^2(ab)^3$ $(ab)^5$

22. $\dfrac{3m^4 \cdot 4m^3}{2m^5 \cdot 5m^2}$ 1.2

Exploration

23. According to the table in this lesson, energy consumption in the U.S. decreased in the middle 1970s and then again in the early 1980s. What were the reasons for these decreases? sample: increased effort at conservation in both decades; increased cost of fuel in the middle 70s

FOLLOW-UP

MORE PRACTICE

For more questions on SPUR Objectives, use *Lesson Master 2-3,* shown below.

EXTENSION

You may wish to invite students to prove the following theorem:

If $g(x) = \frac{1}{x}$, then g is decreasing on the set of positive real numbers.

(It needs to be shown that if $x_1 < x_2$, then $\frac{1}{x_1} > \frac{1}{x_2}$. Begin with $x_1 < x_2$, and divide both sides by x_1x_2 (a positive number) to obtain $\frac{1}{x_2} < \frac{1}{x_1}$. So $\frac{1}{x_1} > \frac{1}{x_2}$, which means that $g(x_1) < g(x_2)$.)

EVALUATION

A quiz covering Lessons 2-1 through 2-3 is provided in the Teacher's Resource File.

NAME ______

LESSON **MASTER 2–3**
QUESTIONS ON **SPUR** OBJECTIVES

■SKILLS *Objective A (See pages 146–150 for objectives.)*

1. The chart at the right shows the amount of gold produced in the U.S. between 1970 and 1985. Let $G(x)$ be the amount of gold produced in year x.

Year	U.S. Gold Production in 1,000 oz
1970	1,743
1975	1,052
1979	964
1980	970
1981	1,379
1982	1,466
1983	2,003
1984	2,085
1985	2,475

a. Describe the longest interval over which G is increasing. $1979 \le x \le 1985$

b. Describe the longest interval over which G is decreasing. $1970 \le x \le 1979$

c. What is the relative minimum? 964

d. What is the relative maximum? 1,743 and 2,475

e. Solve $G(x) = 2{,}003$. $x = 1983$

2. Let $f(x) = -2(x + 1)^2 + 3$. Give the interval(s) on which f is a. decreasing and b. increasing.

a. $x > -1$ b. $x < -1$

■REPRESENTATIONS *Objective G*

3. Use an automatic grapher to estimate the relative maximum and relative minimum of the function $g(x) = x^3 + x^2 - 17x + 15$ on the interval $-5 \le x \le 3$.
rel. max. ≈ 48.5 at x ≈ -2.7, rel. min. ≈ -7 at x ≈ 2.1

4. The graph of function h is given at the right.

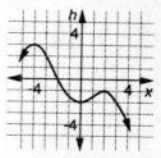

a. Describe any intervals where h is increasing.
$x < -4$, $0 < x < 2$

b. Find any relative maximum or relative minimum values of h. rel max: 3 (at x = -4) and -1 (at x = 2); rel min: -2 (at x = 0)

5. Graph $y = |x + 4|$ over $-10 \le x \le 10$. Describe all intervals on which the function is decreasing.
$x < -4$

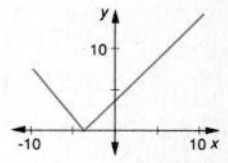

18 *Precalculus and Discrete Mathematics* © Scott, Foresman and Company

LESSON 2-4

RESOURCES

■ Lesson Master 2-4

OBJECTIVES

A Determine relative minima and maxima of a function and intervals on which it is increasing or decreasing.

D Determine the end behavior of a function.

G Analyze a function from its graphs.

TEACHING NOTES

The sequence 9, 13, 17, ... , 4*n* + 5, ... can be thought of as an ordered set. To describe the order, we say the first term is 9; the second term is 13; the third term is 17; ... ; the *n*th term is 4*n* + 5; ... . From this there is a natural correspondence 1 → 9, 2 → 13, 3 → 17, with the general rule $n \to 4n + 5$. That correspondence is a function, and the original sequence is seen to contain the range elements of the function. As another example, the "sequence" of Superball heights 2, 1.8, 1.62, ... is just the range of the function (1, 2), (2, 1.8), (3, 1.62), ...

In some books, the sequence with rule $s_n = \ldots$ is called $\{s_n\}$. We prefer the single letter *s* because it is a simple extension of function notation. Students have learned the $y = f(x)$ notation for functions. Tell them that while the definition for a sequence *g* could be written as $g(n) = 2(0.9)^{n-1}$, writing $g_n = 2(0.9)^{n-1}$ is more customary. Were we to write it as $g(n)$ ... , then we could call

LESSON

2-4 Sequences and Limits

You are familiar with *sequences* such as

$$\frac{1}{2}, \frac{2}{3}, \frac{3}{4}, \frac{4}{5}, \frac{5}{6}, \ldots .$$

The numbers are called the **terms** of the sequence. They are often written in order as above and referred to as the 1st term, 2nd term, 3rd term, and so on.

The above sequence was created using the rule

$$s_n = \frac{n}{n+1}, \ \forall \text{ positive integers } n.$$

This sequence can be thought of as a function mapping each positive integer to the corresponding term of the sequence:

$$f: n \to \frac{n}{n+1}, \ \forall \text{ positive integers } n.$$

$f(1) = \frac{1}{2}, f(2) = \frac{2}{3}, f(3) = \frac{3}{4}, \ldots$. Just as we name the function *f*, so the sequence whose terms are s_n is named by the single letter *s*.

Sequences may have finitely or infinitely many terms. A **finite sequence** is a function whose domain is the set of nonnegative integers from *a* to *b*. An **infinite sequence** is a function whose domain is the set of integers greater than or equal to *a*, a fixed integer. Normally, $a = 1$ or *a* is the smallest integer for which the formula for the sequence is defined, unless you are told otherwise. You can think of a finite sequence as one that has a last term, and an infinite sequence as one that goes on forever. For instance, if a Superball is dropped from a height of 2 meters and each time it bounces up to 90% of its previous height, then the consecutive heights form the sequence

$$2, 1.8, 1.62, 1.458, 1.3122, \ldots .$$

In theory, the ball bounces forever and so the sequence is infinite. But in practice the ball eventually stops and the sequence of heights is finite.

The sequence of Superball heights can be defined by giving a formula for the *n*th term in terms of *n*: for all $n \geq 1$,

$$g_n = 2(0.9)^{n-1}.$$

This formula is called an **explicit formula** for the sequence *g*. The sequence can also be defined by giving the first term and a formula for the $(n + 1)$st term in terms of previous terms

$$\begin{cases} g_1 = 2 \\ g_{n+1} = 0.9g_n \ \forall n \geq 1, \end{cases}$$

which is called a **recursive formula** for the sequence.

The sequence g is an example of a *geometric sequence*, one of several important types of sequences. In general, a **geometric sequence** is a sequence in which $\forall\ n > 1$, $g_n = rg_{n-1}$, where r is a constant. That is, each term beyond the first is r (the **constant ratio**) times the preceding term. In a geometric sequence,

$$g_2 = rg_1$$
$$g_3 = rg_2 = r^2 g_1$$
$$g_4 = rg_3 = r^3 g_1$$

and so the nth term has the explicit formula

$$g_n = r^{n-1} g_1.$$

In the Superball example above, the first term g_1 is 2 meters and the constant ratio r is 0.9.

If adding instead of multiplying is used to get the next term, then an *arithmetic sequence* is formed. An **arithmetic sequence** is a sequence in which $\forall\ n > 1$, $a_n = a_{n-1} + d$, where d is a constant. That is, each term beyond the first is d (the **constant difference**) greater than the preceding one. In an arithmetic sequence,

$$a_2 = a_1 + d$$
$$a_3 = a_2 + d = a_1 + 2d$$
$$a_4 = a_3 + d = a_1 + 3d$$

and so the nth term has the explicit formula

$$a_n = a_1 + (n - 1)d.$$

An example is the increasing sequence of odd positive integers 1, 3, 5, 7, 9, ... in which the first term is 1, the constant difference is 2, and so the nth term is $a_n = 1 + (n - 1)2 = 2n - 1$.

Fibonacci

Names are given to many other sequences. A **harmonic sequence** is one whose terms are reciprocals of an arithmetic sequence. The simplest harmonic sequence is the infinite sequence $1, \frac{1}{2}, \frac{1}{3}, \frac{1}{4}, \ldots$ of the reciprocals of the positive integers, in which $\forall\ n$, $a_n = \frac{1}{n}$. The **Fibonacci sequence** (named after Leonardo of Pisa, also known as Fibonacci, a great mathematician of the early 13th century) is the infinite sequence 1, 1, 2, 3, 5, 8, 13, 21, ..., in which each term is the sum of the two previous terms. A recursive formula for this sequence is

$$\begin{cases} a_1 = 1 \\ a_2 = 1 \\ a_{n+1} = a_n + a_{n-1}\ \forall \text{ integers } n \geq 2. \end{cases}$$

$A_{2+1} = A_2 + A_{2-1}$
$A_3 = 1+1$
$A_3 = 2$

Some other particular sequences are mentioned in the questions.

the sequence g, so why not do so even though we use g_n?

In many books, the domain of a sequence is restricted to the set of positive integers or the set of integers from 1 to n. In this book we allow any set of integers from a to b to be the domain of a sequence.

Note that the arrow (→) in $\lim_{n\to\infty} a_n$ limit notation has a different meaning than the arrow in the mapping form of function notation $f: x \to f(x)$.

Point out that in $\lim_{n\to\infty} (2n - 1) = +\infty$, the symbol $+\infty$ is not a number.

Making Connections
Mastery of the concept of the limit of a sequence provides a firm foundation for understanding end behavior of functions, the subject of the next lesson.

A rigorous conception of the limit of a sequence $\lim_{n\to\infty} a_n = L$ is challenging. To illustrate the idea that "L is the limit if we can get as close to L as we wish by selecting sufficiently large values of n" may be made clearer by thinking of the following game:

"Let $a_n = \frac{n}{n+1}$. I think that $\lim_{n\to\infty} a_n = 1$. You challenge me with a target interval around 1. I will then have to come up with an integer N so that a_n is within that target for all $n \geq N$."

This game can be played either on a number line or coordinate plane.

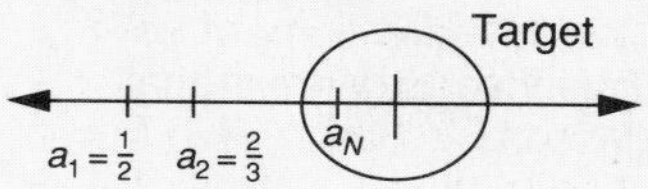

Choose a_n so that a_n is within the target for all $n \geq N$.

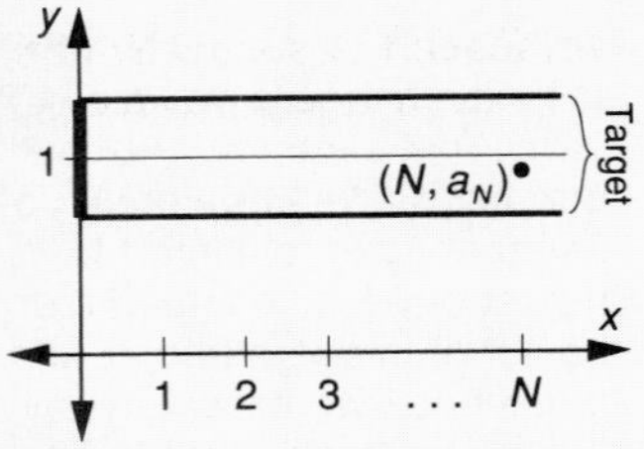

Find N so that for all $n \geq N$, a_n is within the target. Another way of putting it: $\lim_{n \to +\infty} s_n = L$ means that given any integer $k \geq 1$, there is an integer N such that s_n and L agree to at least k decimal places for all $n \geq N$.

Making Connections
Students who have studied from UCSMP *Advanced Algebra* or *Functions, Statistics, and Trigonometry* have seen recursive definitions for sequences and should be familiar with the language and notation used in this lesson.

For other students, recursion may be a new idea. You may wish to define a few arithmetic and geometric sequences both explicitly and recursively. Discuss commonalities and differences in these definitions. Point out that the explicit definition is preferable if one needs the thousandth term of a sequence. Also, mention that some sequences are more easily defined recursively than explicitly. The Fibonacci sequence is an example.

The Fibonacci sequence will be discussed in more detail as the solution to one of the two famous problems of Lesson 7-1. A more rigorous treatment of recursive and explicit definitions of sequences is given in that lesson.

The terms of a sequence need not be numbers. For example, suppose P is a point not on a number line l. $\forall$ positive integers n, let s_n be the line containing P and the point on l with coordinate n. The terms of the sequence s are lines. The first few terms of s are drawn here.

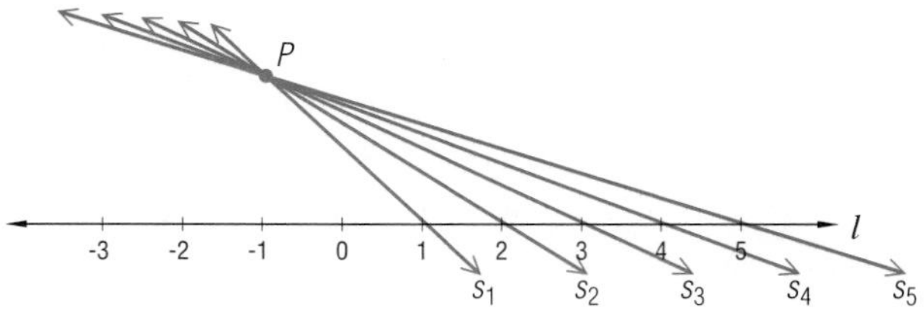

Notice that as n gets larger, these lines get closer and closer to the line parallel to l through P.

Since the set of positive integers is a discrete set, every sequence is a discrete function. If the terms of the sequence are real numbers, then the sequence is a real function and can be graphed.

Example Sketch the graph of the sequence a defined by the rule $a_n = \frac{n}{n+1}$ $\forall$ positive integers n.

Solution Make a table of values and plot the points but do not connect them. In the table and on the graph, we show the first seven terms of the sequence. Note that the larger n is, the closer the value of $a_n = \frac{n}{n+1}$ is to 1. Thus the height of the graph of f gets closer and closer to 1 as n gets larger and larger. It can be made as close to 1 as desired by going out far enough; that is, by taking n large enough.

n	$a_n = \frac{n}{n+1}$
1	$\frac{1}{2}$
2	$\frac{2}{3}$
3	$\frac{3}{4}$
4	$\frac{4}{5}$
5	$\frac{5}{6}$
6	$\frac{6}{7}$
7	$\frac{7}{8}$

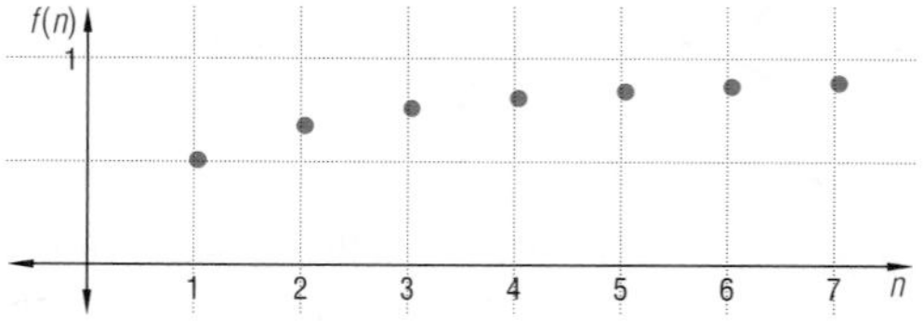

In both the Example and the sequence s of lines that precedes it, the terms of the sequence get closer and closer to some value as the number of the term increases. Terms can be found that are as close to that value as desired by going out far enough in the sequence. This value is called the *limit* of the sequence. For the Example, the limit of the sequence is the number 1. We write

$$\lim_{n \to \infty} a_n = 1,$$

read "the limit of a-sub-n as n approaches infinity is 1," or, more specifically,

$$\lim_{n \to \infty} \frac{n}{n+1} = 1.$$

For the sequence s of lines, the limit can be thought of as the line parallel to l through P. Calling that line e, we write

$$\lim_{n \to \infty} s_n = e.$$

In the rest of this discussion, only numerical sequences are considered, although in the questions you are asked to consider some sequences from geometry.

When a formula for the nth term of a sequence is known, a computer can be programmed to print as many terms of the sequence as you wish. The printout can suggest or confirm the existence of a limit for a sequence. For instance, here is a program to print the first 100 terms of the sequence of the Example:

```
10  FOR N = 1 TO 100
20      LET TERMN = N/(N + 1)
30      PRINT N, TERMN
40  NEXT N
50  END
```

One computer prints

```
1       0.5
2       0.6666666666
3       0.75
.       .
.       .
98      0.9898989899
99      0.99
100     0.9900990099
```

If line 10 is changed to FOR N = 950 TO 1000, then the last three lines printed are

```
.       .
.       .
998     0.9989989989
999     0.999
1000    0.9990009990
```

The printout exhibits the basic property of a limit of a sequence: all terms after a certain term are as close to the limit as you would want. For the above sequence, after the 99th term all terms of the sequence are within .01 of the limit. After the 999th term, all terms are within .001 of the limit. For numerical sequences, here is a precise definition of the idea of limit.

Definition

A number L is the limit of a numerical sequence a_n ⇔ for any positive number p (regardless how small) there is a term in the sequence beyond which all terms of the sequence are within p of L.

ADDITIONAL EXAMPLES

1. Sketch the graph of the sequence b defined by the rule $b_n = \frac{n+1}{n}$ ∀ positive integers n.

Make a table of values and plot the points. Note that this sequence has the same limit as that of the Example in the text. By writing b_n as $1 + \frac{1}{n}$, students should see why the limit is 1.

n	$b_n = \frac{n+1}{n}$
1	$2 = 2$
2	$\frac{3}{2} = 1.5$
3	$\frac{4}{3} \approx 1.33$
4	$\frac{5}{4} = 1.25$
5	$\frac{6}{5} = 1.2$
6	$\frac{7}{6} \approx 1.17$
7	$\frac{8}{7} \approx 1.14$

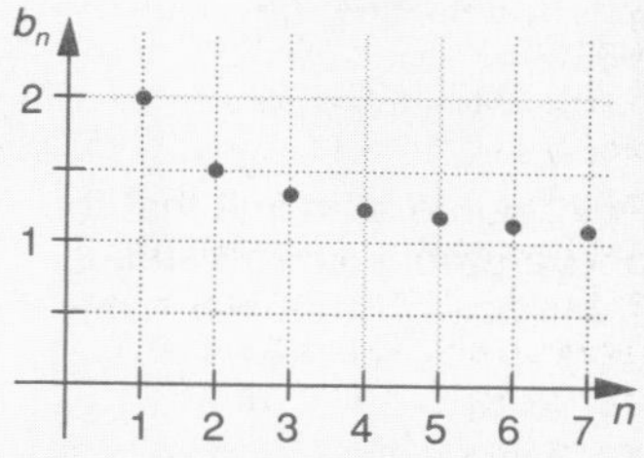

2. Find $\lim_{n \to \infty} \left(1 + \frac{1}{n}\right)^n$.

Note that $\frac{n+1}{n} = 1 + \frac{1}{n} > 1$. Thus, the Example in the text shows that $\lim_{n \to \infty} \left(1 + \frac{1}{n}\right) = 1$. On the other hand, for any constant $a > 1$, $\lim_{n \to \infty} a^n = +\infty$. Although $\left(1 + \frac{1}{n}\right) > 1$, it does not follow that $\lim_{n \to \infty} \left(1 + \frac{1}{n}\right)^n = +\infty$ or that $\lim_{n \to \infty} \left(1 + \frac{1}{n}\right)^n = 1$. In fact, $\lim_{n \to \infty} \left(1 + \frac{1}{n}\right)^n = e$, the base of the natural logarithms.

ADDITIONAL EXAMPLES

3. Let S_1 be the line through an arbitrary point *P*, not on a number line, and the point with coordinate 8 on the number line. For $n > 1$, let S_{n+1} be the line through point *P* and through the point on the number line with a coordinate that comes from multiplying the coordinate for S_n by $-\frac{1}{2}$.

a. Picture S_1, S_2, S_3, and S_4.

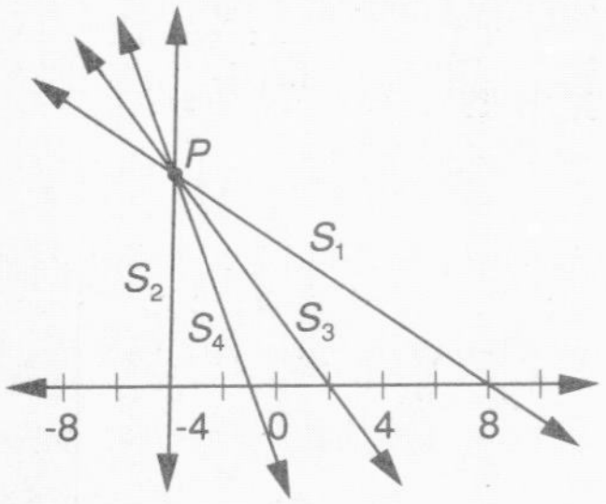

b. Describe $\lim_{n\to\infty} S_n$.

the line through P and the point on the number line with coordinate 0

Point out to students that the above problem generates a sequence of lines, not numbers.

4. Find the limit as $n \to \infty$ of $g_n = (-1)^n\left(\frac{n}{n+1}\right)$.

For even values of *n*, $\lim_{n\to\infty} g_n = 1$. For odd values of *n*, $\lim_{n\to\infty} g_n = -1$. Thus, Thus, $\lim_{n\to\infty} g_n$ does not exist.

As an example, the Superball geometric sequence g: 2, 1.8, 1.62, 1.458, 1.3122, . . . , in which $g_n = 2(.9)^{n-1}$, has the limit 0. We say that the terms *approach* 0. This means that no matter how close you choose p to 0, there is a certain point in the sequence beyond which the numbers are all within p of 0. For instance, suppose that you want terms to be within .01 of the limit 0. Then the terms would have to be between -.01 and .01. You can check with a calculator that $g_{52} = 0.0092768$, and all terms of g after this term are also positive numbers less than 0.01. You can write

$$\lim_{n\to\infty} 2(.9)^{n-1} = 0.$$

Now consider the sequence w of powers of -1: -1, 1, -1, 1, -1, . . . , with $w_n = (-1)^n$. The terms switch back and forth from one number to another so they never remain as close as you might wish to either number. We say "the limit does not exist."

When terms in a sequence alternate in sign, the sequence is called an **alternating sequence**. Both the harmonic sequence and the alternating harmonic sequence h: 1, $-\frac{1}{2}$, $\frac{1}{3}$, $-\frac{1}{4}$, . . . , with $h_n = (-1)^{n+1}\frac{1}{n}$, have limit 0. For instance, beyond the millionth term all terms are within .000001 of the limit 0, though some are greater than 0 and some are less than 0. You can write

$$\lim_{n\to\infty} \frac{1}{n} = 0 \text{ and } \lim_{n\to\infty} (-1)^{n+1}\frac{1}{n} = 0.$$

Two other expressions involving limit notation are commonly used. When for any number you might pick, beyond a certain term the terms of a sequence are larger than that number, then we say that the terms of the sequence *increase without bound* and that the limit of the sequence is $+\infty$ (read "positive infinity"). For example, the terms of the arithmetic sequence a: 1, 3, 5, 7, 9, 11, . . ., with $a_n = 2n - 1$, increase without bound, and you can write

$$\lim_{n\to\infty} (2n - 1) = +\infty.$$

Similarly, when for any number you might pick, beyond a certain term all terms of a sequence are less than that number, then we say that the terms of the sequence *decrease without bound* and that the limit of the sequence is $-\infty$ (read "negative infinity"). For instance, the terms of the sequence d: -2, -4, -6, -8, -10, -12, . . . , with $d_n = -2n$, decrease without bound, and you can write

$$\lim_{n\to\infty} (-2n) = -\infty.$$

The notions of limits all have graphical interpretations. They are discussed in the next lesson.

Questions

Covering the Reading

1. An infinite sequence is a function with domain __?__.
the set of all integers greater than or equal to a fixed integer

2. Write $\lim_{n\to\infty} b_n = 5$ in words.
The limit of b_n as n approaches infinity is 5.

104

In 3–6, a formula for the nth term of a sequence is given.

a. Identify the sequence as arithmetic, geometric, harmonic, alternating, or none of these.
b. Give the first five terms of the sequence.
c. Identify the sequence as increasing, decreasing, or neither.
d. Find $\lim_{n\to\infty} a_n$.

3. $a_n = \frac{n}{n-1}\ \forall\ n > 1$ **See margin.**

4. $a_n = 3^n$ **See margin.**

5. $a_n = (-1)^n \cdot \frac{2}{n}$ **See margin.**

6. $\begin{cases} a_1 = 6 \\ a_{n+1} = a_n - 4\ \forall\ n \geq 1 \end{cases}$ **See margin.**

7. a. Graph the first six terms of the alternating harmonic sequence $1, -\frac{1}{2}, \frac{1}{3}, \ldots$. **See margin.**

b. What is the earliest term of this sequence that is within .01 of the limit of the sequence? **the 101st term**

8. A ball is dropped from a height of 10 feet and each time bounces back to 80% of its previous height.

a. Give a recursive formula for the sequence of successive heights.
b. Give an explicit formula for the sequence of successive heights.
c. After how many bounces will the ball remain less than 1 foot high? **a)** $\begin{cases} g_0 = 10 \\ g_{n+1} = (.8)g_n\ \forall n \geq 1 \end{cases}$ **b)** $g_n = 10(.8)^n$**; c) 11 bounces**

Applying the Mathematics

9. Given a line segment $\overline{AB}$, a sequence is defined as follows.
P_1 is the midpoint of $\overline{AB}$.
P_n is the midpoint of $\overline{AP_{n-1}}\ \forall\ n > 1$.

a. Is this a recursive or an explicit definition? **recursive**
b. Draw the first four terms of the sequence. **See margin.**
c. What is $\lim_{n\to\infty} P_n$? ***A***

10. Let A be a fixed point on a fixed circle C with radius r. $\forall$ positive integers n, let R_n be the regular polygon with $n + 2$ sides inscribed in C with one vertex at A. Let p_n and a_n be the perimeter and area of R_n.

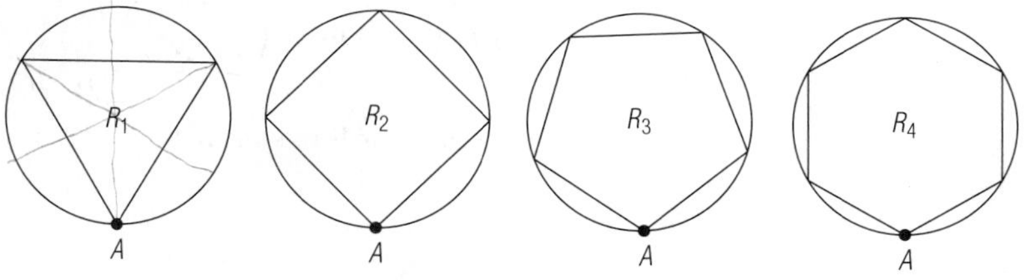

a. Find formulas for p_4 and a_4 in terms of r. **See margin.**
b. Draw R_6. **See margin.** **c.** What is $\lim_{n\to\infty} R_n$? ***C***
d. What is $\lim_{n\to\infty} p_n$? $2\pi r$ **e.** What is $\lim_{n\to\infty} a_n$? πr^2

NOTES ON QUESTIONS

Questions 3-5: You may wish to ask students to give a recursive definition for each.

(3. $a_2 = 2; a_{n+1} = \frac{n^2-1}{n^2} a_n$
4. $a_1 = 3; a_{n+1} = 3a_n$
5. $a_1 = -2; a_{n+1} = -\frac{n}{n+1} a_n$)

Question 6: You may wish to ask students to give an explicit formula for this sequence. ($a_n = 10 - 4n$)

Question 22: If students calculate the limits of R_n and R_n^2, they should realize that the limits differ by the integer 1. This will indicate that the limit is the solution to the equation $x^2 - x = 1$.

ADDITIONAL ANSWERS

3.a. none of these
b. starting with $n = 2$: $2, \frac{3}{2}, \frac{4}{3}, \frac{5}{4}, \frac{6}{5}$
c. decreasing
d. 1

4.a. geometric
b. 3, 9, 27, 81, 243
c. increasing
d. ∞

5.a. alternating
b. $-2, 1, -\frac{2}{3}, \frac{1}{2}, -\frac{2}{5}$
c. neither
d. 0

6.a. arithmetic
b. 6, 2, -2, -6, -10
c. decreasing
d. $-\infty$

7.a.

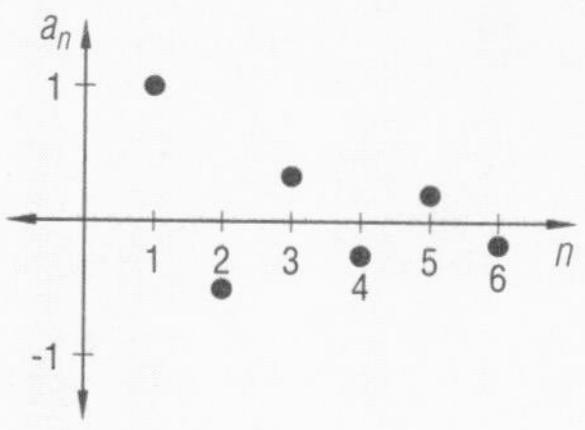

9.b., 10.a., b. See the margin on page 106.

FOLLOW-UP

MORE PRACTICE
For more questions on SPUR Objectives, use *Lesson Master 2-4,* shown on page 107.

EXTENSION
You may wish to use the program below to generate terms for the sequence given in the Example:

```
10 LET N = 1
20 LET TERMN =
   N/(N + 1)
30 PRINT N, TERMN,
   1 - TERMN
40 GOTO 20
```

Have students enter this program and observe how 1 − TERMN approaches 0 as N increases. What is the smallest value of N for which TERMN = 1 and 1 − TERMN = 0? (Answers will vary.)
Given the *n*th term should never equal 1, why does the computer say it does? (The computer must round infinite decimals.)

PROJECTS
The projects for Chapter 2 are described on page 142. **Projects 4 and 6** are related to the content of this lesson.

ADDITIONAL ANSWERS
9.b.

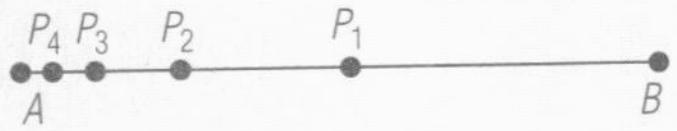

10.a. $p_4 = 6r$;
$a_4 = \frac{3\sqrt{3}}{2} r^2$, or $\approx 2.60\, r^2$
b.

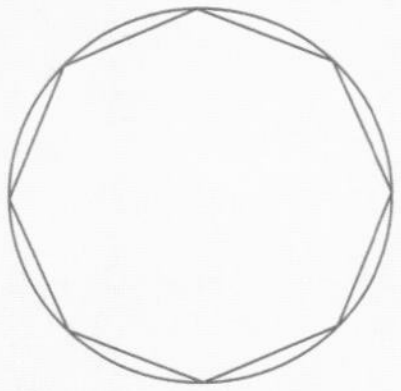

15.b. increasing: $0.23 \le x \le 5$; decreasing: $-1 \le x \le 0.23$

---

11. Give a formula for the *n*th term s_n of an increasing sequence such that $\lim_{n\to\infty} s_n = 2$. sample: $s_n = \frac{2n-1}{n}$

12. With a calculator or computer, explore the limit as $n\to\infty$ of the sequence s with $s_n = \frac{3n^2 + 10n + 6}{4n^2 - 30n + 2}$. The limit is $\frac{3}{4}$.

13. Consider the sequence $1, \frac{1}{4}, \frac{1}{9}, \frac{1}{16}, \frac{1}{25}, \ldots$ whose *n*th term is $\frac{1}{n^2}$.

a. This sequence is the same as the function f with what rule and what domain? domain: {integers n: $n > 0$}; rule: $f(n) = \frac{1}{n^2}$

b. What is $\lim_{n\to\infty} f(n)$? 0

Review

14. *Multiple choice.* On which of the following intervals is the function graphed at the right decreasing?

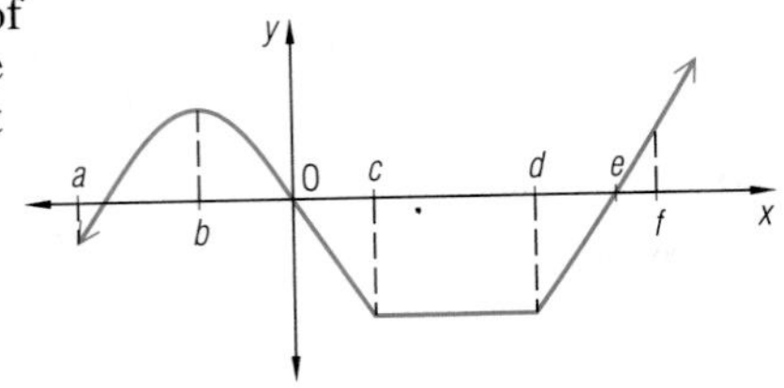

(a) $a \le x \le c$
(b) $b \le x \le c$
(c) $b \le x \le d$
(d) $0 \le x \le e$
(Lesson 2-3) (b)

15. Let f be the real function defined by $f(x) = x^2 - \sqrt{x+1}$.

a. What is the domain of f? {x: $x \ge -1$}

b. Estimate the intervals on which f is increasing and on which f is decreasing. (Limit your search to $x \le 5$.) See margin.

c. Estimate any relative minimum or maximum values that f may have. *(Lessons 2-3, 2-1).* relative minimum: ≈ -1.06

16. A rectangular garden is to be placed along a wall with a fence surrounding the other three sides. What is the largest area that can be fenced off with 20 feet of fence? (Hint: Write a formula for the area as a function of the width of the garden.)
(Lesson 2-2) 50 sq ft

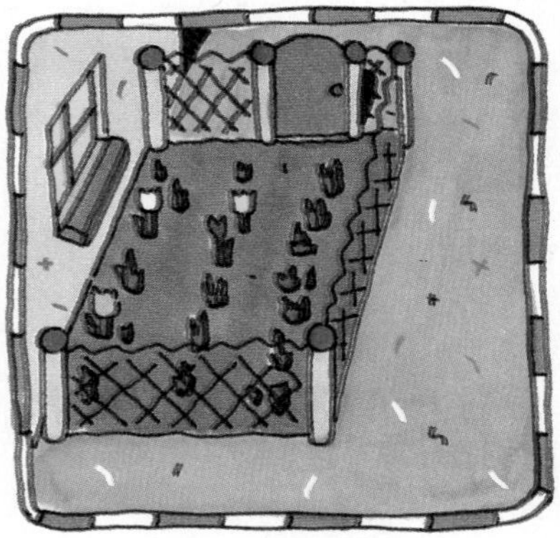

17. a. On one set of axes, graph $f(x) = \sin ax$ over the interval $-2\pi \le x \le 2\pi$ for $a = 1, 2,$ and 3. See margin.

b. How does the graph change as a increases?

c. How do the minimum and maximum values change as a increases?
(Lesson 2-1) They do not change.
b) The period decreases.

18. A stairway light is controlled by two switches–one at each end of the stairs. If the switches are both up or both down, then the light is on. Otherwise, it is off.

a. Make a table showing whether the light is on or off for every possible combination of switch positions. (Let "on" = 1 and "off" = 0.) See below.

b. Explain how this situation is a physical representation of the logical expression $p \Leftrightarrow q$. *(Lesson 1-5)* See below.

19. Prove: $\forall\ n$, if n is an odd integer, then n^2 is an odd integer. *(Lesson 1-7)* See below.

20. The height of a cylinder varies directly as its volume and inversely as the square of the radius of its base.

a. Give a formula relating height, volume, and radius. $V = \pi r^2 h$

b. If the radius is doubled but the volume is kept constant, what happens to the height? *(Previous course)*
It is one-fourth of the original value.

21. To the nearest degree, what are the measures of the three angles in a triangle whose sides have lengths 5, 12, and 13? *(Previous course)*
23°, 67°, 90°

Exploration

22. $\forall\ n$, let F_n be the nth term of the Fibonacci sequence, and let $R_n = \frac{F_n}{F_{n-1}}$ for $n \geq 2$. Explore the sequence R_n in an attempt to determine $\lim_{n\to\infty} R_n$. You may find it helpful to write a computer program that calculates R_n and $R_n{}^2$ for many values of n. ≈ 1.618

18. a)

switch 1	switch 2	light
1	1	1
1	0	0
0	1	0
0	0	1

b)

p	q	$p \Leftrightarrow q$
T	T	T
T	F	F
F	T	F
F	F	T

If 1 corresponds to T and 0 to F, it is apparent that the two truth tables are equivalent. Hence, the stairway light situation is a physical representation of $p \Leftrightarrow q$.

19. Suppose that n is any odd integer. According to the definition of odd, there exists an integer k such that $n = 2k + 1$. Then $n^2 = (2k + 1)^2 = 4k^2 + 4k + 1 = 2(2k^2 + 2k) + 1$. By closure properties, $(2k^2 + 2k)$ is an integer and n^2 is odd, by definition of odd.

17.a.

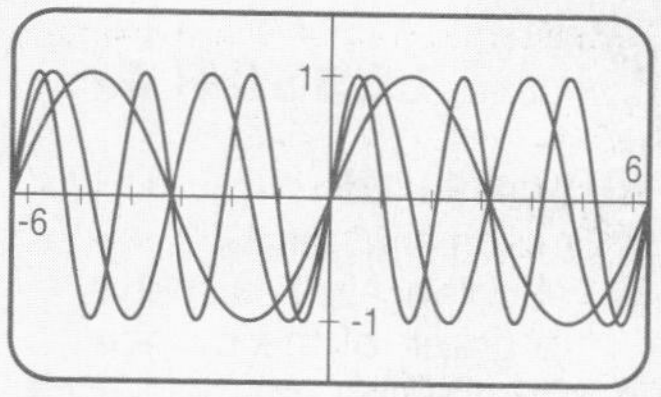

NAME ______

LESSON **MASTER 2–4**
QUESTIONS ON **SPUR** OBJECTIVES

■ **SKILLS** *Objective A (See pages 146–150 for objectives.)*
In 1 and 2, a formula for a sequence is given. a. Give the first four terms of the sequence, and b. tell if the sequence is increasing, decreasing, or neither.

1. $a_n = 5 + (-1)^n$
a. 4, 6, 4, 6
b. neither

2. $b_n = \frac{2n+1}{n^2}$
a. $3, \frac{5}{4}, \frac{7}{9}, \frac{9}{16}$
b. decreasing

3. Give an example of a decreasing arithmetic sequence.
sample: 5, 3, 1, -1, . . . ; that is, $c_n = 7 - 2n$

In 4–6, a. tell if each sequence is arithmetic, geometric, harmonic, or none of these, and b. tell if it is increasing, decreasing, or neither.

4. $a_n = (1.01)^n$ a. geometric b. increasing
5. $b_n = \frac{1}{n}$ a. harmonic b. decreasing
6. $c_n = 3n - 7$ a. arithmetic b. increasing

■ **PROPERTIES** *Objective D*

7. Find $\lim_{n\to\infty} a_n$ if $\begin{cases} a_1 = 8 \\ a_{k+1} = \frac{a_k}{10}\ \forall\ k \geq 1 \end{cases}$. 0

8. Use limit notation to describe the limit of $s_n = \left(-\frac{1}{2}\right)^{2n+1}$ as n increases without bound. $\lim_{n\to\infty} \left(-\frac{1}{2}\right)^{2n+1} = 0$

9. Let $\begin{cases} b_1 = 5 \\ b_{k+1} = \frac{1}{2} b_k\ \forall\ k \geq 1 \end{cases}$.
a. What is the limit of b_n as $n \to \infty$? 0
b. For what value of n is b_n within .001 of the limit? 14

NAME ______
Lesson MASTER 2-4 (page 2)

■ **REPRESENTATIONS** *Objective G*
In 10 and 11, graph the first six terms of the sequence and find its limit as $n \to \infty$.

10. $a_n = \frac{2n^2 + 1}{n^2}$

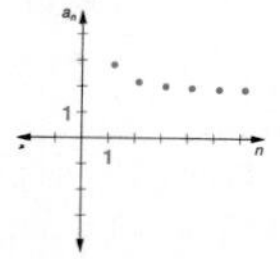

$\lim_{n\to\infty} a_n = 2$

11. $b_n = 3 + \frac{(-1)^n}{n}$

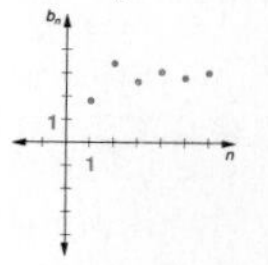

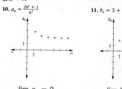

$\lim_{n\to\infty} b_n = 3$

LESSON 2-5

RESOURCES
- Lesson Master 2-5
- Teaching Aid 5 displays the graph of the cylinder minimization example from Lesson 2-2.
- Teaching Aid 9 can be used with **Questions 6, 7, and 16**.
- Computer Master 4

OBJECTIVES

D Determine the end behavior of a function.
G Analyze a function from its graph.

TEACHING NOTES

You might begin with the sequence $a_n = \frac{n}{n+1}$ from the Example in Lesson 2-4. Because the domain is a discrete set, the graph consists of discrete points. What if we now defined a function f by $\forall$ $x \in$ positive reals, $f(x) = \frac{x}{x+1}$. Sketch the graph of $y = f(x)$. How does it compare to the graph in the Example of Lesson 2-4? (It contains all the points of that graph and it is a decreasing function.) What is $\lim_{x \to +\infty} f(x)$? (1)

When discussing the **Example** of this lesson, help students to see that $\lim_{r\to\infty}\left(2\pi r^2 + \frac{116}{r}\right) = \lim_{r\to\infty} 2\pi r^2$ by having them use an automatic grapher to do the following: With a domain of $0 \le x \le 10$ and range of $0 \le y \le 500$, graph both $y = 2\pi r^2 + \frac{116}{x}$ and $y = 2\pi x^2$

LESSON

2-5 End Behavior

The fairness of a coin cannot be tested by a single toss.

The **end behavior** of a function f is a description of what happens to the values $f(x)$ of the function as x grows larger and larger in magnitude; that is, as $x \to \infty$ and as $x \to -\infty$. As an example, think of testing whether a coin is fair or not. The first few tosses may not tell you much, for even an unfair coin could give you a head and then a tail, or a fair coin could give you a few heads in a row. More important is what happens in the long run. After thousands of tosses, will the percentage of heads be near 50 or not? Put symbolically, if $P(n)$ is the relative frequency of heads after n tosses, what you are testing is whether $\lim_{n\to\infty} P(n) = .5$. You are asking about the end behavior of the function P.

We simulated the tossing of a coin 100 times. Below is a graph of P for $1 \le n \le 100$. Does $\lim_{n\to\infty} P(n) = .5$? It appears that this conjecture is reasonable.

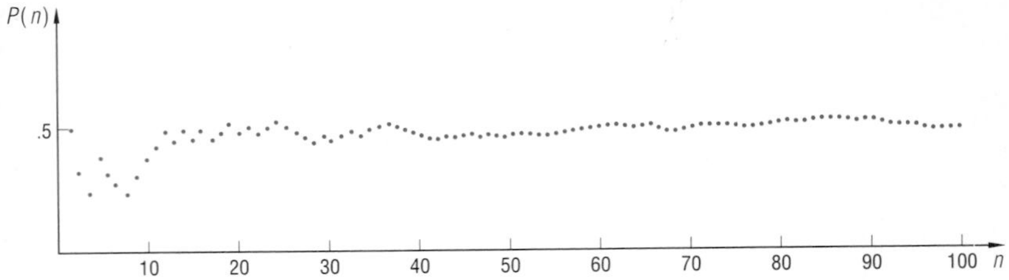

The behavior of a function as $x \to \infty$ (or as $x \to -\infty$) falls into one of three categories. (1) The values of $f(x)$ can increase or decrease without bound, that is $f(x) \to \infty$ or $f(x) \to -\infty$; (2) the values of $f(x)$ can approach some real number L; or (3) the values of $f(x)$ can follow neither of these patterns.

Exponential functions illustrate the first two types of end behavior. Shown is a graph of the function with equation $f(x) = 2^x$.

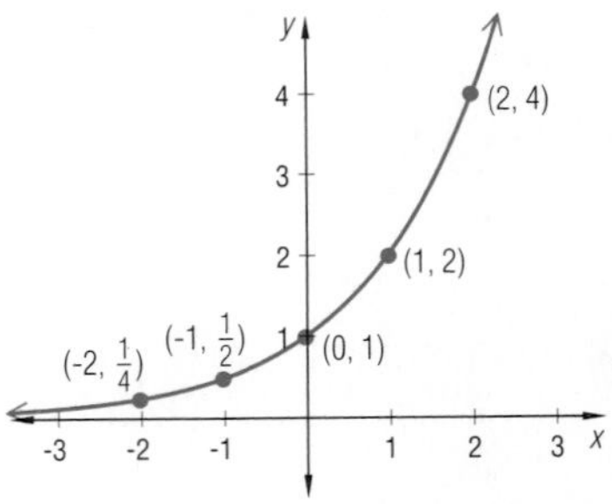

The first type of end behavior is clearly seen from the graph. As x gets larger and larger, 2^x does also; f is an increasing function. On a typical graph, when $x = 100$, the values of $f(x)$ are so large that they cannot be shown. (Written in base 10, 2^{100} has over 30 digits.) Since the values of 2^x increase without bound as x increases without bound, we write

$$\lim_{x\to\infty} 2^x = \infty.$$

the limit of 2 to the xth power as x approaches infinity is infinity asymptote- a straight line approached by a given curve as one of the variables in the equation of the curve approaches infinitely.

In general, $\lim_{x \to \infty} f(x) = \infty$ means that no matter how large a number you pick, $f(x)$ eventually becomes and stays larger than that number when x is large enough.

The graph of $f(x) = 2^x$ also demonstrates the second type of end behavior discussed. When x gets smaller and smaller (that is, x is farther and farther to the left of the origin), 2^x gets smaller, but not without bound. The values of 2^x get closer and closer to zero. For instance, when $x = -100$, $2^x \approx 9.31 \cdot 10^{-10} = 0.000000000931$. The value of 2^x can be made as close to zero as desired by making x small enough. So we write

$$\lim_{x \to -\infty} 2^x = 0.$$

The x-axis is a **horizontal asymptote** for the function $f(x) = 2^x$ because the graph of f gets closer and closer to this horizontal line as $x \to -\infty$.

In general, $\lim_{x \to -\infty} f(x) = L$ means that for any number p (no matter how small), $f(x)$ eventually remains within p of L when x is small enough. If $\lim_{x \to -\infty} f(x) = L$, the line $y = L$ is a horizontal asymptote of the function $y = f(x)$.

The sine function exemplifies a third type of end behavior, the absence of a limit. As x increases or decreases without bound, the values of the sine function continue to oscillate between 1 and -1.

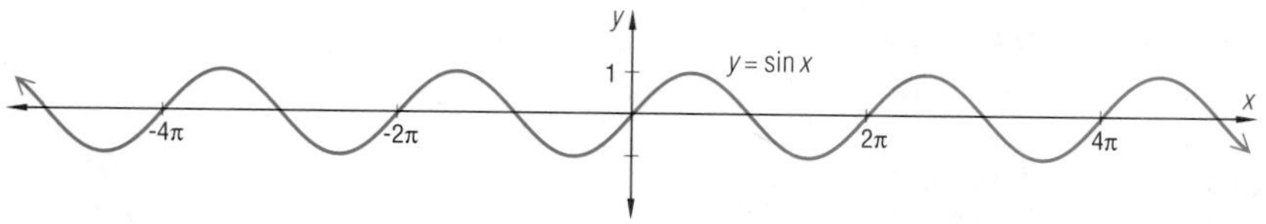

We write $\quad \lim_{x \to \infty} \sin x$ **does not exist**

and $\quad \lim_{x \to -\infty} \sin x$ **does not exist.**

Often the end behavior of one function is best described as being similar to the end behavior of another simpler function.

Example Describe the behavior of the function $f(r) = 2\pi r^2 + \frac{116}{r}$ as r increases without bound.

Solution A graph of this function was given in Lesson 2-2, but it only covers the interval $0 \le r \le 6$. That is not nearly enough to get an idea of the end behavior. However, algebraic analysis is quite easy. Examine the formula for $f(r)$. As r gets larger and larger, $\frac{116}{r}$ gets closer and closer to 0. Specifically, the values of $f(116, r)$ are 116 times the values of $\frac{1}{r}$. But $\lim_{r \to \infty} \frac{1}{r} = 0$, as you learned from the study of the harmonic sequence. Consequently, the term $\frac{116}{r}$ contributes nothing to the end behavior. Thus $\lim_{r \to \infty} f(r) = \lim_{r \to \infty} (2\pi r^2 + \frac{116}{r}) = \lim_{r \to \infty} 2\pi r^2$. You might write:

and compare. Then change the domain to $0 \le x \le 100$ and the range to $0 \le y \le 50000$ and graph both again. The two graphs appear to coincide. *Teaching Aid 5* from Lesson 2-2 may be helpful again here.

Error Analysis Some students think that every function is either even or odd. To dispel this false view, use examples such as:
Let $f(x) = x^2 + x$.
Then $f(3) = 11$.
$f(-3) = 6$.
So $f(-3) \ne f(3)$ and $f(-3) \ne -f(3)$. Thus, f is neither even nor odd.

Small Group Work Give each group a bag containing pieces of paper of two different colors. Have them draw five pieces with replacement and calculate the percentage of times one of the colors is drawn. Repeat five more times, combining the results with the first five draws and calculate the percentages again. This process can be repeated until 50 or more draws have been made. A graph can be drawn illustrating the limiting behavior of the percentage.

ADDITIONAL EXAMPLES

1. In Question 18 of Lesson 2-3, the total surface area A of the box in terms of s, one of its sides, is given by $A(s) = s^2 + \frac{40}{s}$. Describe the end behavior as s increases without bound.
As s gets larger and larger, $\frac{40}{s}$ approaches 0. Specifically, $\lim_{s \to \infty} \frac{1}{s} = 0$ and $\lim_{s \to \infty} \frac{40}{s} = 40(\lim_{s \to \infty} \frac{1}{s}) = 40(0) = 0$. Thus, $\lim_{s \to \infty} A(s) = \lim_{s \to \infty} (s^2 + \frac{40}{s}) = \lim_{s \to \infty} s^2 + \lim_{s \to \infty} \frac{40}{s} = \lim_{s \to \infty} s^2 = \infty$.

ADDITIONAL EXAMPLES

2. Because sin x oscillates between -1 and 1 as x increases without bound, $\lim_{x\to\infty} \sin x$ does not exist. What is the end behavior for f when $f(x) = \frac{1}{x} \sin x$ as x increases without bound?

$\lim_{x\to\infty} \frac{1}{x} \sin x = 0$

This can be verified by graphing *f*.

3. Prove that $\forall x$, $g(x) = 5x^4 - 8x^2$ is an even function.

We need to show that $g(-x) = g(x)\ \forall x$.

$g(-x) = 5(-x)^4 - 8(-x)^2$
$= 5x^4 - 8x^2$
$= g(x)$

Therefore, $g(-x) = g(x)\ \forall x$, and so g is an even function.

Symmetries of graphs provide a powerful means of recognizing whether a function might be even or odd. Graphs of even functions are reflection-symmetric with respect to the y-axis. Graphs of odd functions are rotation-symmetric with respect to the origin (point-symmetric). Urge students to use graphs to help conjecture whether a function is even or odd.

The end behavior of the function f as r increases without bound is thus the same as the end behavior of the function $y = 2\pi x^2$, whose value increases without bound as x increases without bound.
Therefore, $\lim_{r\to\infty} f(r) = \infty$.

Check Recall how the function f was derived. Increasing the radius r while keeping the volume of the cylinder constant means that the cylinder is getting wider and shorter. Its surface area becomes closer and closer to the sum of the areas of its bases, or $2\pi r^2$, as r becomes larger and larger without bound.

The solution in the above example relies on knowing the end behavior of a quadratic function of the form $y = ax^2$. More generally, it helps to know the end behavior of the **power functions**, those functions with equations of the form $y = ax^n$. Specifically, when x is positive, the larger x is, the larger its powers. For instance, $\lim_{x\to\infty} x^3 = \lim_{x\to\infty} x^4 = \infty$. However, when x is negative, its odd powers are negative, but its even powers are positive. So $\lim_{x\to-\infty} x^4 = \infty$ but $\lim_{x\to-\infty} x^3 = -\infty$.

The figures below summarize the end behavior of the power functions

$$f(x) = ax^n$$

where n is a positive integer and a is a positive real number:

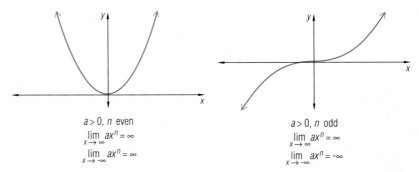

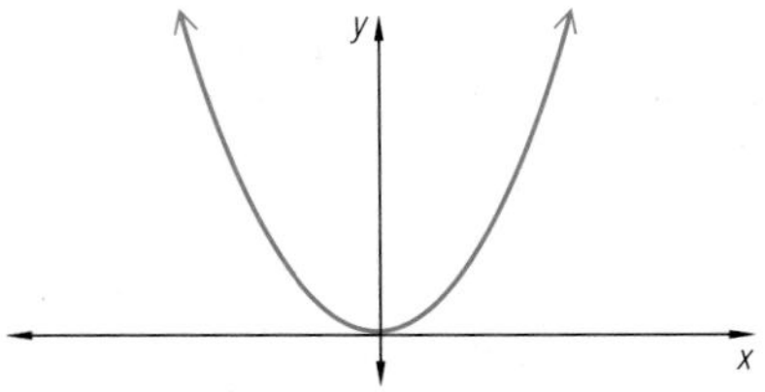

$a > 0$, n even
$\lim_{x\to\infty} ax^n = \infty$
$\lim_{x\to-\infty} ax^n = \infty$

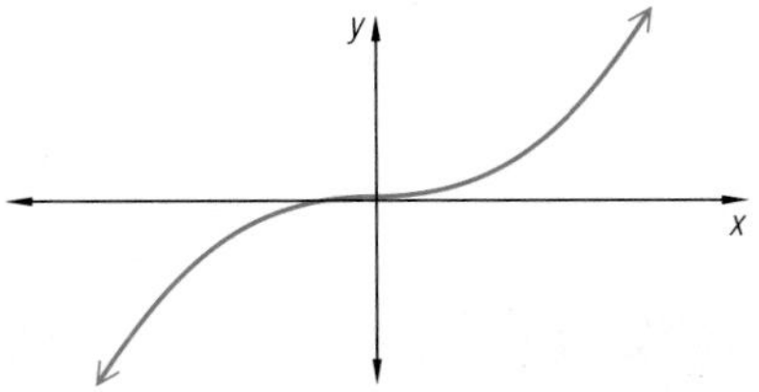

$a > 0$, n odd
$\lim_{x\to\infty} ax^n = \infty$
$\lim_{x\to-\infty} ax^n = -\infty$

If $a < 0$, then the graphs are reflected over the x-axis and the end behavior changes accordingly.

Notice that the graphs of the power functions possess symmetry. The even power functions are reflection-symmetric with respect to the y-axis. This is due to the fact that when n is even, $(-x)^n = x^n\ \forall$ real numbers x. Thus, if f is an even power function, $f(-x) = f(x)\ \forall$ real numbers x.

The odd power functions are rotation-symmetric with respect to the origin. That is, the graph of an odd power function coincides with its image under a 180° rotation about the origin. This is because when n is odd, $(-x)^n = -(x^n)\ \forall\ x$. As a result, if f is an odd power function, then $f(-x) = -f(x)\ \forall$ real numbers x.

Other functions are called *even* or *odd* if they possess either of these properties of the even and odd power functions.

Definition

A real function is an **even function** if and only if $\forall$ x, $f(-x) = f(x)$.
A real function is an **odd function** if and only if $\forall$ x, $f(-x) = -f(x)$.

Among the familiar functions, there are examples of even and odd functions other than the power functions. Since $\forall$ x, $|-x| = |x|$, the absolute value function is an even function. Since $\forall$ x, $\sin(-x) = -\sin x$, the sine function is an odd function.

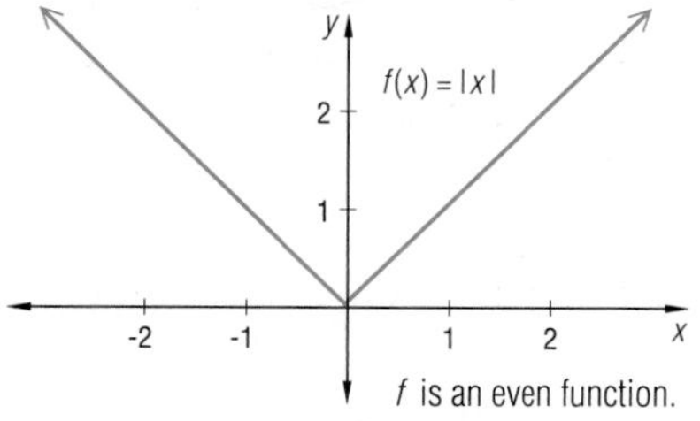

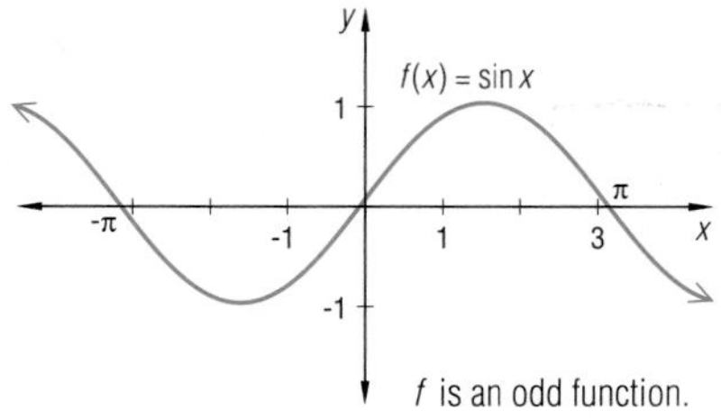

If a function is odd or even and you know its end behavior as $x \to \infty$, then you can rotate or reflect the right side of its graph to determine its end behavior as $x \to -\infty$.

Questions

Covering the Reading

In 1 and 2, consider the function P at the beginning of the lesson.

1. Is P a discrete function? Justify your answer.
Yes, the domain is a discrete set, the set of positive integers.

2. If the coin is fair, give an equation for an asymptote of P. $y = 0.5$

3. Consider the function g: $x \to 3^{-x}$.

a. Sketch a graph of g for x in the interval $-2 \le x \le 2$. **See margin.**

b. Explain why $\lim_{x\to\infty} 3^{-x} = 0$. **See margin.**

c. What is $\lim_{x\to-\infty} 3^{-x}$? ∞

d. Does g have a horizontal asymptote? If so, give an equation for it. $y = 0$

4. a. Explain why $\lim_{z\to\infty} \frac{4000}{z^2} = 0$. **See margin.**

b. Describe the end behavior of the function $f(z) = \frac{4000}{z^2} - 12z^2$.
as $z \to \infty$, $f(z) \to -\infty$; as $z \to -\infty$, $f(z) \to -\infty$

5. a. Let $f(x) = \cos x$. What is $\lim_{x\to\infty} \cos x$? **The limit does not exist.**

b. Is f even, odd, or neither? Justify your answer.
even; $\forall$ x, $\cos(-x) = \cos(x)$

NOTES ON QUESTIONS

Question 3: Have students use an automatic grapher and, for part **b**, graph $y = 3^{-x}$ with window $3 \le x \le 7$, $0 \le y \le .05$, and again with window $7 \le x \le 17$, $0 \le y \le .0001$ to see how rapidly $y = 3^{-x}$ approaches 0 as x gets larger.

For part **b**, it may be helpful to rewrite the function as g: $x \to \frac{1}{3^x}$.

For part **c**, remind students that if $x < 0$, then $-x > 0$. Even at this level, some students may still think that "$-x$" always represents a negative number.

Question 5: Invite students to sketch a graph of $y = \cos x$ to verify the answers.

ADDITIONAL ANSWERS

3.a.

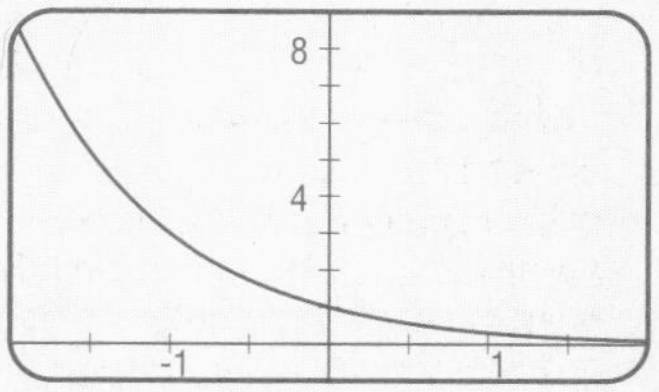

b. As the values of x increase without bound, the values of 3^{-x} become closer and closer to 0.

4.a. As the values of z increase without bound, the values of $\frac{4000}{z^2}$ become closer and closer to 0.

NOTES ON QUESTIONS
Making Connections for Question 13: Some students assume that because $\lim_{x\to\infty}\left(1+\frac{1}{x}\right)=1$, therefore $\lim_{x\to\infty}\left(1+\frac{1}{x}\right)^x=1$ also. They are surprised by what they see. It is counterintuitive. This question should be discussed, as it helps prepare students for the discussion of *e* in Lesson 2-7.

ADDITIONAL ANSWERS
8.a. sample:

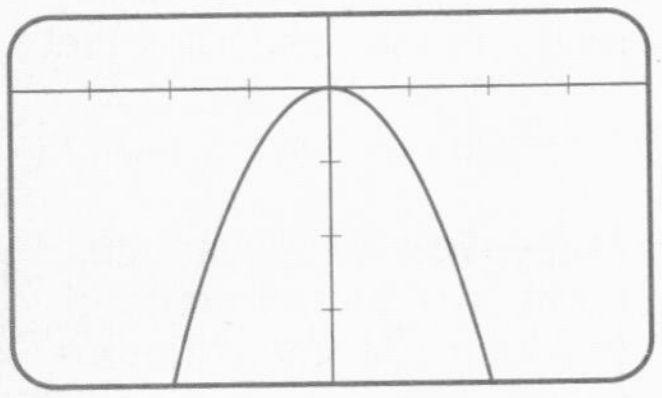

9.

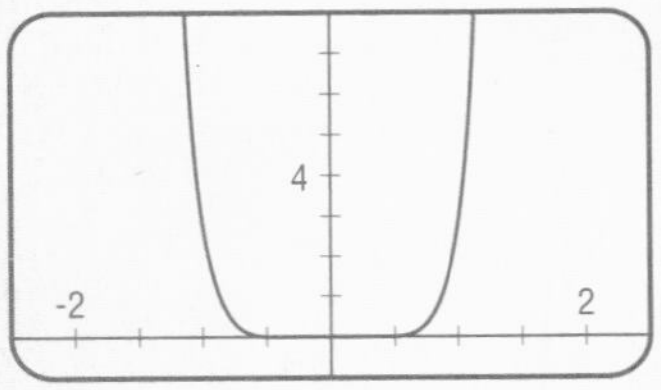

as x → ∞, y → ∞;
as x → -∞, y → ∞

10.

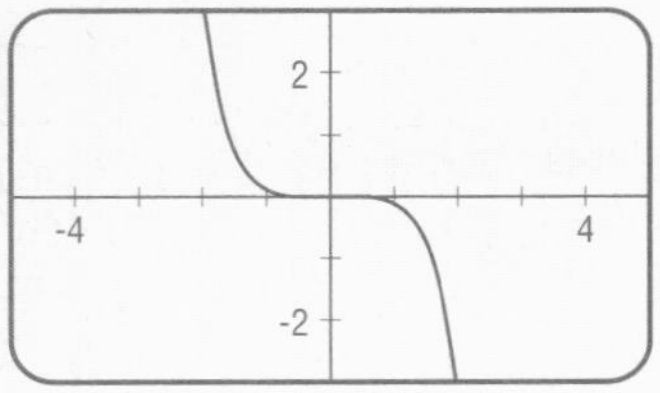

as x → ∞, y → -∞;
as x → -∞, y → ∞

11.a.

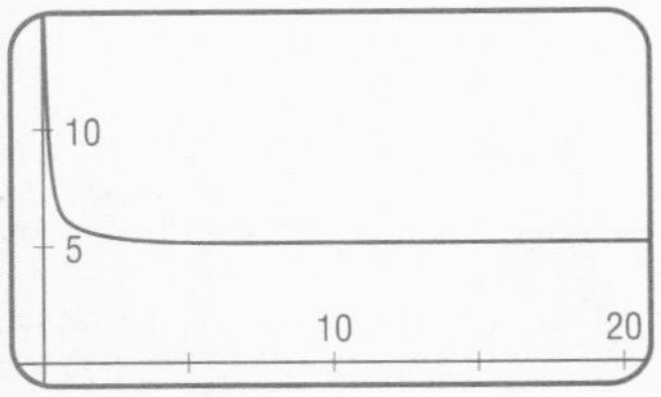

In 6 and 7, specify whether the function whose graph is given appears to be even, odd, or neither.

6. odd

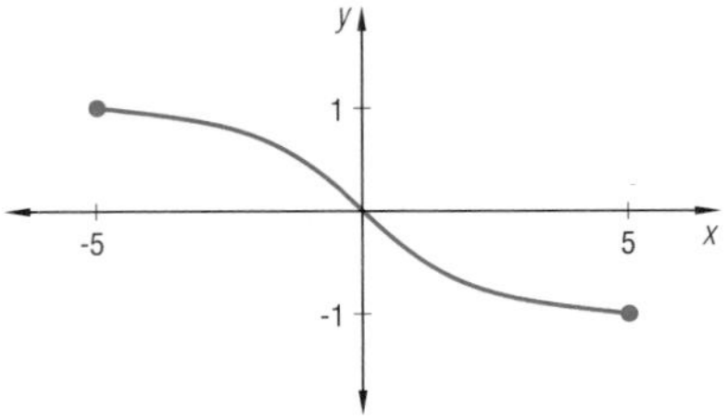

7. neither

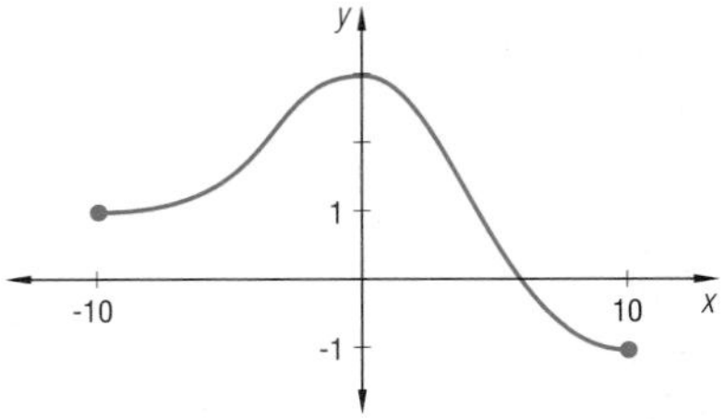

8. Suppose f is an even function and $\lim_{x\to-\infty} f(x) = -\infty$.

a. Sketch a possible graph for such a function. **See margin.**

b. What is $\lim_{x\to\infty} f(x)$? $-\infty$

9. Sketch a graph of $y = 3x^8$ and describe its end behavior. **See margin.**

Applying the Mathematics

10. Sketch a graph of $y = -\frac{1}{10}x^5$ and describe its end behavior. **See margin.**

11. Let $f(x) = 5 + \frac{1}{x}$.

a. Sketch a graph of f over the interval $0 < x \le 20$. **See margin.**

b. Find $\lim_{x\to\infty} f(x)$. 5

c. Give an equation of the horizontal asymptote. $y = 5$

d. Find the values of x needed for $f(x)$ to be within 0.1 of the limit you found in part **b**. $x > 10$

12. a. If h is an odd function and $\lim_{x\to\infty} h(x) = 3$, find $\lim_{x\to-\infty} h(x)$. -3

b. Sketch a graph of such a function. **See margin.**

13. Consider the function f defined by the formula $f(x) = \left(1+\frac{1}{x}\right)^x$.

a. Complete the table below to examine the behavior of f as x gets larger and larger.

x	10	100	1000	10,000	100,000	1,000,000
$\left(1+\frac{1}{x}\right)^x$	**2.59374**	**2.70481**	**2.71692**	**2.71815**	**2.71827**	**2.71828**

b. Use the results from **a** to approximate an equation for a horizontal asymptote to the graph of f. **y = 2.71828 or y = 2.71829 or y = e**

c. Use an automatic grapher to check your results from **a** and **b**. **See margin.**

Review

In 14 and 15, identify the sequence as arithmetic, geometric, harmonic, or none of these, and find its limit as $n\to\infty$.

14. $a_n = (-2)^n$ **geometric; the limit does not exist**

15. $b_n = -(2^n)$ *(Lesson 2-4)* **geometric;** $-\infty$

112

16. Let f be the function with domain $\{x: -5 \le x \le 5\}$ graphed here.

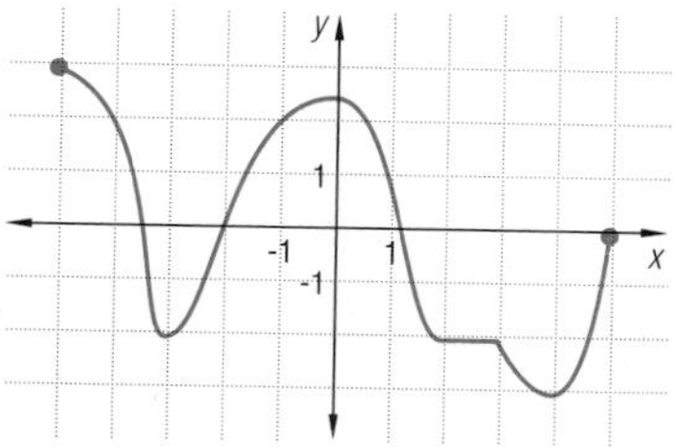

a. Identify the interval(s) on which f is increasing. See below.
b. Identify the interval(s) on which f is decreasing. See below.
c. Identify the interval(s) on which f is constant. $2 \le x \le 3$
d. Estimate the relative maximum and minimum values of f and tell where they occur. See below.
e. Give the range of f. $\{y: -3 \le y \le 3\}$
f. Find $f(2.5)$. -2
g. Estimate the value(s) of x for which $f(x) = -1$. -3.3; -2.3; 1.5; 4.8
h. Estimate the value(s) of x for which $f(x) \le -2$. *(Lessons 2-3, 2-2, 2-1)*
$x = -3$ and $2 \le x \le 4.5$

17. Find the exact range of the function g given by $g(x) = -3x^2 + 4x - 2$. *(Lesson 2-2)* See below.

18. Calculate $\left(\frac{9}{4}\right)^x$ for $x = 2, \frac{3}{2}, 1, \frac{1}{2}, 0, -\frac{1}{2}, -1, -\frac{3}{2}$, and -2. *(Previous course)*
$\frac{81}{16}, \frac{27}{8}, \frac{9}{4}, \frac{3}{2}, 1, \frac{2}{3}, \frac{4}{9}, \frac{8}{27}, \frac{16}{81}$

In 19 and 20, use the laws of exponents to simplify. *(Previous course)*

19. $4^{1/5} \cdot 8^{1/5}$ 2

20. $\frac{x^{1/2}y^{-3/4}}{x^{-2/3}\,y^{1/4}}$ $\frac{x^{7/6}}{y}$

21. Which is larger, $81^{-1/4}$ or $81^{-1/2}$? *(Previous course)* $81^{-1/4}$

22. Deduce a valid conclusion from the following premises, and identify the argument form. *(Lesson 1-6)*
∀ real numbers x and y, $x < y \Rightarrow e^x < e^y$
$e^n \ge e^5$ *for a particular value of n.*
$n \ge 5$; Law of Indirect Reasoning

Exploration

23. Suppose $g(x) = \frac{1}{f(x)}$ for all x with $f(x) \ne 0$. What is the relationship between the end behavior of f and that of g? See below.

16. a) $-3 \le x \le 0$, $4 \le x \le 5$ b) $-5 \le x \le -3$, $0 \le x \le 2$, $3 \le x \le 4$
d) relative maxima: -2, occurs at $2 \le x \le 3$; 2.5, occurs at $x = 0$; 3 at $x = -5$; 0 at $x = 5$
relative minima: -2, occurs at $x = -3$ and $2 \le x \le 3$; -3, occurs at $x = 4$

17. $\{y: y \le -\frac{2}{3}\}$

23. If as $x \to \infty$, $g(x) \to \infty$ or $-\infty$, then as $x \to \infty$, $f(x) \to 0$.
If as $x \to \infty$, $g(x) \to L \ne 0$, then as $x \to \infty$, $f(x) \to \frac{1}{L}$.
If as $x \to \infty$, $g(x) \to 0$, then as $x \to \infty$, $f(x) \to \infty$ or $-\infty$ depending on the function g. Similar end behavior is exhibited when $x \to -\infty$.

FOLLOW-UP

MORE PRACTICE

For more questions on SPUR Objectives, use *Lesson Master 2-5,* shown below.

ADDITIONAL ANSWERS

12.b. sample:

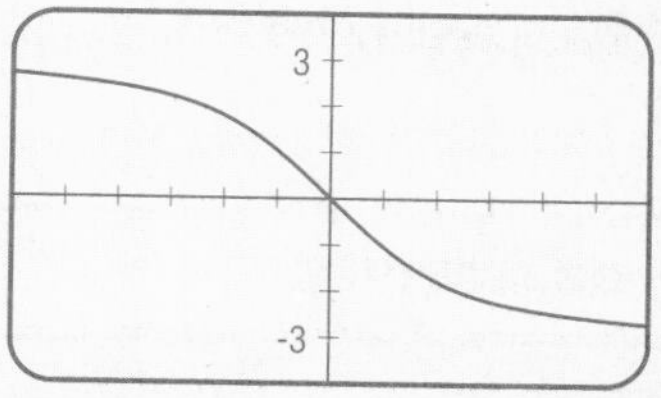

13.c.

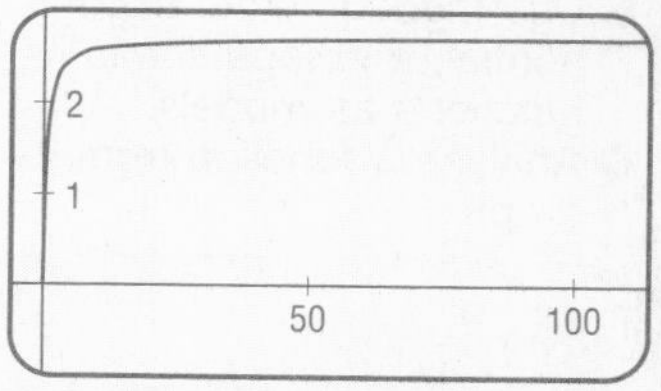

NAME ______

LESSON **MASTER 2–5**
QUESTIONS ON SPUR OBJECTIVES

PROPERTIES *Objective D (See pages 146–150 for objectives.)*

1. Consider the function $f(x) = \frac{x}{x+2}$.

a. Complete the table below to give decimal approximations for function values as x becomes large.

x	10	100	1,000	10,000
$\frac{x}{x+2}$	.8333	.9804	.9980	.9998

b. What is $\lim_{x\to\infty} f(x)$? 1
c. Write an equation of the horizontal asymptote of f. $y = 1$

2. a. Let $g(x) = \frac{\sin x}{x}$. Find $\lim_{x\to\infty} g(x)$ and $\lim_{x\to-\infty} g(x)$. 0; 0
b. For what values of x is $g(x)$ within .001 of the limit? $x < -1{,}000$ or $x > 1{,}000$
c. Is g even, odd, or neither? even

3. If h is an even function and $\lim_{x\to\infty} h(x) = -3$, find $\lim_{x\to-\infty} h(x)$. -3

4. Describe the end behavior of $y = -6x^3$.
$\lim_{x\to\infty} (-6x^3) = -\infty$, $\lim_{x\to-\infty} (-6x^3) = \infty$

REPRESENTATIONS *Objective G*

5. The function f is graphed at the right.

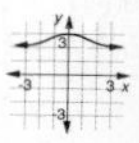

a. Is f even, odd, or neither? even
b. Describe its end behavior.
$\lim_{x\to-\infty} f(x) = 2$, $\lim_{x\to\infty} f(x) = 2$
c. Write the equation of the horizontal asymptote. $y = 2$

6. Sketch the graph of an odd function f that has a relative maximum at $x = -2$ such that $\lim_{x\to\infty} f(x) = -1$.

sample:

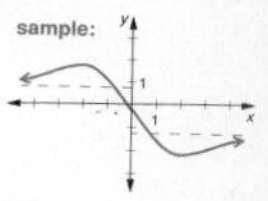

 21

LESSON 2-6

RESOURCES
- Lesson Master 2-6
- Quiz for Lesson 2-4 through 2-6
- Teaching Aid 10 displays the list of tasks concerning the analysis of a function.
- Computer Master 5

OBJECTIVES

D Determine the end behavior of a function.
E Use trigonometric, exponential, and logarithmic functions as models.
G Analyze a function from its graph.

TEACHING NOTES

The eight tasks on the first page of this lesson can be applied to any real-valued function whose domain is a subset of the real numbers. Exponential functions are taken first because their behavior is not trivial, yet it is still reasonably easy to describe.

In this lesson, the familiar laws of exponents are thought of as properties of exponential functions. Of course, they are properties of the *values* of these functions, but we use the function language just as we do when we think of relationships between sines and cosines as properties of the trigonometric functions. In the next lesson, the properties of logarithms are viewed as properties of logarithm functions.

LESSON

2-6 Analyzing Exponential Functions

In this and the next two lessons of this chapter, three types of functions—exponential, logarithmic, and trigonometric—are analyzed. We expect that you have studied these functions before, so that some of the material is familiar to you. By "analyzing" we mean responding *in mathematical language* to the following tasks concerning a particular numerical function f:

1. Domain. Identify the values for which f is defined.
2. Range. Describe the set of all possible values of f.
3. Increasing or Decreasing. Identify the intervals on which $f(x)$ increases and the intervals on which $f(x)$ decreases.
4. Maxima or Minima. Find the largest or smallest value of $f(x)$ on any interval.
5. End Behavior. Describe what happens to $f(x)$ as $x \to \infty$ and as $x \to -\infty$.

A good graph of the function can help you with all five of these tasks. Consequently, it is important to be able to:

6. Graph. Graph f with an automatic grapher. Sketch a graph of f without such a grapher.

These tasks mean very little unless you know what makes the function important.

7. Model. What types of situations might lead to a function like f? How is f normally used in those situations?
8. Properties. What special properties does f have? Are any of these unique to the function f or functions like it?

Now consider these eight tasks relative to the **exponential function with base b**, where $b \neq 0$, defined by

$$f: x \to b^x, \ \forall x.$$

The domain of an exponential function is normally the set of all real numbers. But if the domain is the set of positive integers, a geometric sequence is formed. For instance, the nth term of the sequence

$$\frac{2}{3}, \frac{4}{9}, \frac{8}{27}, \frac{16}{81}, \ldots$$

is given by the formula $f(n) = \left(\frac{2}{3}\right)^n$ $\forall$ positive integers n. The common ratio $\frac{2}{3}$ of the sequence is the base of the exponential function.

114

Values of exponential functions are related by the familiar **laws of exponents**. These can be proved from the properties of addition and multiplication of real numbers, but the proofs are quite difficult.

For all real numbers x and y,

$b^x \cdot b^y = b^{x+y}$	(Product of Powers)
$\frac{b^x}{b^y} = b^{x-y}$	(Quotient of Powers)
$(b^x)^y = b^{xy}$	(Power of a Power)

From these laws, other relationships can be deduced which help in calculating powers.

For all positive integers m and n and positive real numbers b,

$b^0 = 1$

$b^{-n} = \frac{1}{b^n}$

$b^{1/n}$ = the positive nth root of $b = \sqrt[n]{b}$

$b^{m/n}$ = the positive nth root of $b^m = \sqrt[n]{b^m}$

= the mth power of the positive nth root of $b = (\sqrt[n]{b})^m$

The laws of exponents imply that products, quotients, and powers of powers are all themselves powers of the same base. For instance, multiplication, division, or taking a power on the integral powers of 2 (shown below) results in another integral power of 2.

... 0.015625, 0.03125, 0.0625, 0.125, 0.25, 0.5, 1, 2, 4, 8, 16, 32, ...

In Lesson 2-5 it was noted that the exponential function $y = 2^x$ is an increasing function throughout its domain. This property is true of all exponential functions whose bases are greater than 1, and is illustrated in the graph shown below. That is, if $b > 1$ and $x_1 < x_2$, then $b^{x_1} < b^{x_2}$. A general proof is beyond the scope of this course, but a proof for special cases is not difficult.

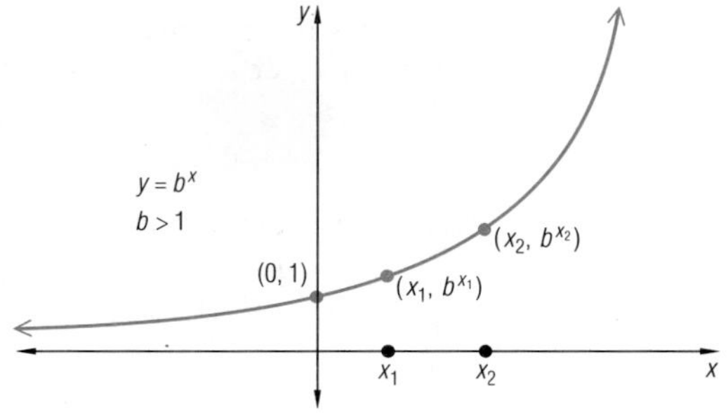

Making Connections In the Additional Examples to Lesson 2-4, $\lim_{n\to\infty}\left(1 + \frac{1}{n}\right)^n$ was explored. Let $h = \frac{1}{n}$ and the result is a limit description of e, $e = \lim_{h\to 0}(1 + h)^{1/h}$. This limit can be explored using a calculator. As $h \to 0$, the base $1 + h$ approaches 1 while the exponent $\frac{1}{h}$ becomes larger and larger; however, the value of $(1 + h)^{1/h}$ gets closer and closer to the value of e, as the following table shows.

h	$1 + h$	$\frac{1}{h}$
0.1	1.1	10
-.1	.9	-10
.01	1.01	100
-.01	.99	-100
.001	1.001	1000
-.001	.999	-1000

$(1 + h)^{1/h}$	$\approx$	e
1.1^{10}	$\approx$	2.5937
$.9^{-10}$	$\approx$	2.8680
1.01^{100}	$\approx$	2.7048
$.99^{-100}$	$\approx$	2.7320
1.001^{1000}	$\approx$	2.7170
$.999^{-1000}$	$\approx$	2.7196

The limit description of e will be used later in the text in discussions of the growth property of the function f: $x \to e^x$. This limit description also provides a means to compute decimal approximations to e; consequently, be sure to cover it.

Alternate Approach
Here is an alternate proof of **Example 1**: $1 < b$, $b < b^2$, $b^2 < b^3$, $b^3 < b^4$, $b^4 < b^5$, $b^5 < b^6$ by repeated multiplications of both sides by the positive number b. Thus, by transitivity, $b^4 < b^6$. This proof is a little less mysterious but involves more steps.

Example 1 Prove: If $b > 1$, then $b^4 < b^6$.

Proof Start with the given inequality,

$$1 < b.$$

Since b is positive, multiplying both sides by b does not change the sense of the inequality. So

$$b < b^2.$$

Multiplying both sides by b again yields

$$b^2 < b^3.$$

By the transitive property of inequality,

$$b < b^3.$$

Now multiply both sides by b^3. Since b^3 is positive (do you see why?), the sense of the inequality is not changed. Thus

$$b^4 < b^6.$$

When the domain of $f\colon x \rightarrow b^x$ is a set of integers, as in a geometric sequence, the base b may be positive or negative, and b^x is always a real number. However, when x is a real number, if the base b is negative, then b^x is not always a real number. For instance, there is no real number equal to $(-9)^{1/2}$. Hence, when the domain of $f(x) = b^x$ is the set of all real numbers, the base b must be positive.

Exponential functions are neither even nor odd functions. But the exponential functions $f(x) = b^x$ and $g(x) = \left(\frac{1}{b}\right)^x$ are related. Using the laws of exponents, for all real numbers x,

$$g(x) = \left(\frac{1}{b}\right)^x = (b^{-1})^x = b^{-x} = f(-x).$$

Thus the graph of g is the reflection image of the graph of f over the y-axis. This is pictured at right for $b = \frac{2}{3}$. The graph of $y = \left(\frac{2}{3}\right)^x$ is the reflection image of the graph of $y = \left(\frac{3}{2}\right)^x$ over the y-axis. The graph of $y = \left(\frac{2}{3}\right)^x$ at right verifies that it is a decreasing function. This is the case for all exponential functions in which the base $b < 1$. Notice that, since $\forall\, b$, $b^0 = 1$, the graphs of all exponential functions contain the point $(0, 1)$.

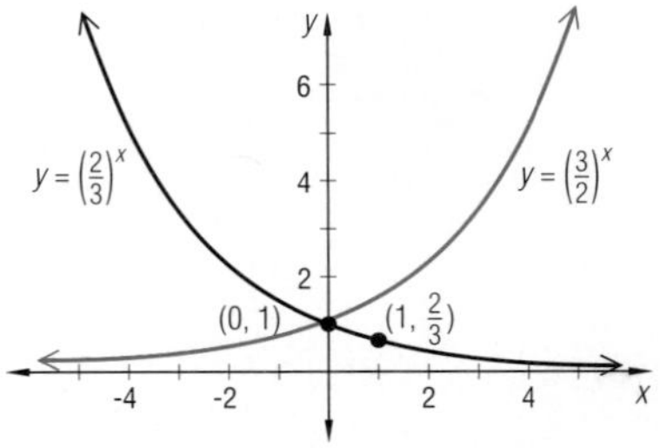

In the last lesson we noted the end behavior of $y = 2^x$. The end behavior of all exponential functions with base $b > 1$ is similar.

$$\lim_{x\to\infty} b^x = \infty \text{ and } \lim_{x\to-\infty} b^x = 0 \qquad \text{for } b > 1.$$

In the questions, you are asked to consider the end behavior of $y = b^x$ when $b < 1$.

116

The values of an exponential function with a positive base are positive numbers. This fact, combined with its end behavior, suggests that the range of an exponential function is the set of positive real numbers. Since exponential functions are either increasing or decreasing, their minimum value on a finite interval will always be at one end of the interval, their maximum value at the other end. However, they have no maxima or minima over the set of *all* real numbers.

Three bases are more commonly used than any others. Base 2 is found in computer science applications. Base 10 is the foundation of our decimal system and scientific notation. The irrational base $e = \lim_{x \to \infty} \left(1 + \frac{1}{x}\right)^x =$ 2.718281828459045... is the most commonly used base in more advanced mathematics and in applications. The reason that e is so important is that, as x increases, the values of

$$y = e^x$$

increase at a rate exactly equal to the value of the function. You will study this idea in more detail in Chapter 9.

One of the most important applications of the number e occurs in the Continuous Change Model that you may have used before to study the growth of bacteria cultures, the decay of radioactive substances, or continuously compounded interest.

> **Continuous Change Model**
>
> If a quantity grows or decays continuously at an annual rate r, the amount $A(t)$ after t years is given by
> $$A(t) = Be^{rt}$$
> where $B = A(0)$.

In the case of continuous growth, r is positive. In the case of continuous decay, r is negative. The requirement that the quantity increases or decreases *continuously* is important. When this is not satisfied, the model becomes only an approximation.

Example 2 When Rhonda started high school, her parents gave her a certificate of deposit for \$5,000 to help pay her college expenses four years later. If the bank pays an annual rate of 8.6% compounded continuously, how much will Rhonda have when starting college?

Solution Because the investment is growing continuously at a constant rate, we can apply the Continuous Change Model. In this case, $B = 5000$, $r = .086$, and $t = 4$. So the amount Rhonda will have four years later is

$$A(4) = 5000 \bullet e^{(.086)(4)}$$
$$\approx 7052.89$$

(.086)(4)
5,000 • e^.344 = 7,052.89

or nearly \$7053.

Making Connections
Students who have studied from UCSMP *Advanced Algebra* or *Functions, Statistics, and Trigonometry* should be familiar with the Continuous Change Model. Remind students that in the Continuous Change Model, a value for r that is greater than 0 is synonymous with growth. A value for r that is less than 0 is synonymous with decay.

ADDITIONAL EXAMPLES

1. Prove: If $0 < b < 1$, then $b^2 > b^3 > 0$.
Multiply all sides by b^2. This does not change the senses of the inequality because b^2 is positive. The result is $0 < b^3 < b^2$, which is equivalent to what was desired.

2. Suppose that shortly after Rhonda's birth, her parents decided to open a savings account for her with an initial principle that would ensure having \$50,000 for her college education 18 years later. Assuming that the bank pays an annual interest rate of 8.6% compounded continuously, what must be the initial deposit of principle?
Use the continuous rate of change model with $t = 18$, $r = .086$, and $A(t) = 50{,}000$. Thus, $50{,}000 = Be^{(.086)(18)}$ or $50{,}000 = Be^{1.548}$. From this, $B = \frac{50000}{e^{1.548}} \approx 10{,}633.64$. Notice that only a linear equation needs to be solved; the variable is not in the exponent.

NOTES ON QUESTIONS
Question 5: You might ask students to set up equations whose solutions would answer the following questions. This will help to motivate the need to solve exponential equations, discussed in the next lesson.
(1) If the annual rate is only 5%, what initial deposit is needed to provide the same balance at age 62? ($Be^{(.05 \cdot 32)} = 5000e^{(.07 \cdot 32)}$)
(2) If the initial deposit is \$4000, what annual rate is needed to provide the same balance at age 62? ($4000e^{32r} = 5000e^{(.07 \cdot 32)}$)
(3) If the initial deposit is \$6,000, at what age will this person be able to retire with the same amount as the \$5000 deposit would have provided at age 62? ($6000e^{.07t} = 5000e^{(.07 \cdot 32)}$)

Small Group Work for Question 8: Students may be interested in investigating this model a little more. Each group could make a table that shows the predicted value given by the model for each year, discuss why some of the discrepancies occur, and then see if they can come up with a better model.

Error Analysis for Question 8: Be sure students understand how t is defined: in year y, $t = y - 1790$. The most common error would be for students to use the value of y itself, which will result in function values that are far too large.

ADDITIONAL ANSWERS
2.a.

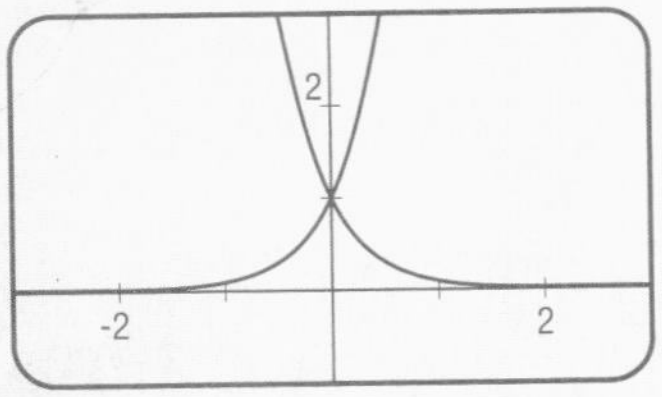

The following example summarizes the analysis of exponential functions with bases greater than 1.

Example 3 Analyze the exponential function with base b, where $b > 1$.

Solution Here is a summary of the analysis of $y = b^x$ where $b > 1$ as you might be expected to write it.

Usual domain: set of real numbers.
Range: set of positive real numbers.
Increasing or decreasing: increasing over its entire domain.
Maxima or minima: none.
End behavior: $\lim_{x \to \infty} b^x = \infty$; $\lim_{x \to -\infty} b^x = 0$.
Graph: (See page 115).
Model: growth and decay, compound interest. When $b = e$, the function models continous growth.
Special Properties: Values of the function are related by the laws of exponents.

Questions

Covering the Reading

1. *Multiple choice.* Which of the following could be the graph of $y = .6^x$?
(a)

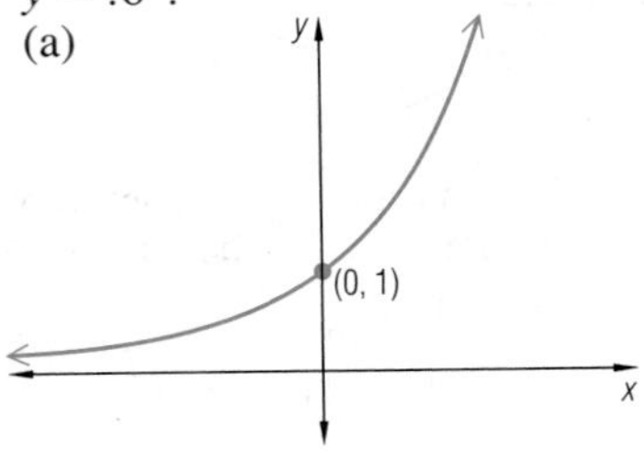

(b)

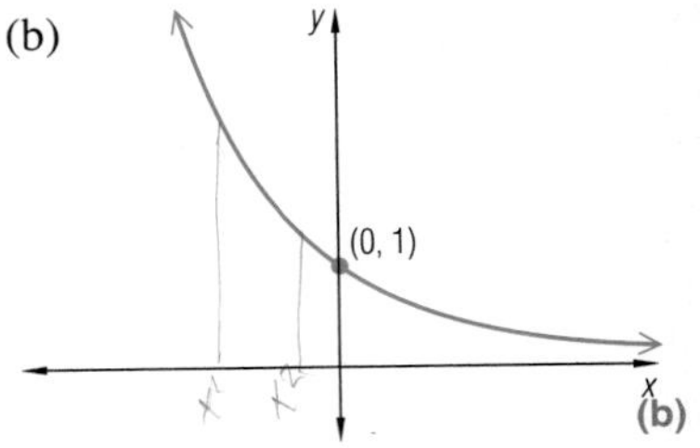

(b)

2. a. Sketch the graphs of f: $x \to 10^x$ and g: $x \to \left(\frac{1}{10}\right)^x$ on the same axes. See margin.
b. How are the graphs related? The graph of g is the reflection image of the graph of f over the y-axis.

3. Give a formula for the exponential function determined by the sequence 3, 9, 27, 81, 243, ... $y = 3^x$, ∀ positive integers x

4. Multiply the smallest and largest powers of 2 shown in this lesson. What power of 2 results? $\frac{1}{2}$, the -1 power of 2

5. If a person puts \$5000 in a retirement account at age 30, paying an annual rate of 7% compounded continously, how much will it grow to by retirement at age 62 if no additional deposits or withdrawals are made? \$46,966.66

$5000 \cdot e^{.07 \cdot 32} =$

118

6. **a.** *True* or *false?* $\exists$ *real numbers x and y such that* $x < y$ *and* $\left(\frac{3}{4}\right)^x < \left(\frac{3}{4}\right)^y$. False

b. Justify your answer to part **a**. See margin.

7. Analyze the exponential function with base b, $0 < b < 1$, following the form of Example 3. See margin.

Applying the Mathematics

8. Use the following table, which lists the population of the United States at each 10-year census between 1790 and 1980 (in millions).

Year	Population	Year	Population	Year	Population
1790	3.93	1860	31.44	1930	122.76
1800	5.31	1870	39.82	1940	131.67
1810	7.24	1880	50.16	1950	150.70
1820	9.64	1890	62.95	1960	179.32
1830	12.87	1900	76.00	1970	203.30
1840	17.07	1910	91.97	1980	226.55
1850	23.19	1920	105.71	1990	250.37

a. Use the Continuous Change Model to obtain a formula for the population $P(t)$ of the United States t years after 1790, based on an annual growth rate of 2.96% from that time. $P(t) = 3.93e^{(.0296)t}$

b. Compute the predicted population of the United States for the year 2000 based on the formula for $P(t)$ obtained in part **a**. $\approx$ 1.97 billion

c. Calculate the population predicted by the formula $P(t)$ in part **a** for the years

i. 1850, 23.21 million

ii. 1900, and 101.97 million

iii. 1950. 447.9 million

For which period is the Continuous Change Model a reasonably accurate predictor of the actual population of the United States? for the period from 1790 to 1850

9. **a.** Graph $f(x) = e^x$ and $g(x) = 3e^x$ on the same axes. See margin.

b. *True* or *false?* $\forall$ x, $g(x) > f(x)$. True

10. Consider the situation outlined in Example 2. As mentioned in the lesson, the Continuous Change Model is at best a good approximation if the quantity under consideration is changing in discrete increments. For these cases, the exact formula is

$$A(t) = B\left(1 + \frac{r}{n}\right)^{nt},$$

where n is the number of times the quantity is increased (or decreased) in a year. Determine the amount that would accumulate in Rhonda's account in four years if the interest is compounded with the indicated frequency.

a. yearly compounding $6954.87

b. monthly compounding $7044.25

c. daily compounding $7052.61

Question 10: You can use the fact that $\lim_{h \to \infty} (1 + h)^{1/h} = e$ to show that the Continuous Change Model is the limit of what you get if you invested something at compound interest as the number n of compounding periods goes to infinity. If a principal of B is invested at an annual rate of r, then the amount $A(t)$ after t years is given by the formula $A(t) = B\left(1 + \frac{r}{n}\right)^{nt} = B\left(\left(1 + \frac{r}{n}\right)^{n/r}\right)^{rt}$ which approaches Be^{rt} as $n \to \infty$.

ADDITIONAL ANSWERS

6.b. The function f, with $f(x) = \left(\frac{3}{4}\right)^x$ is decreasing over the set of real numbers. So, from the definiton of a decreasing function, if $x < y$, then $f(x) > f(y)$ for all x and y.

7. usual domain: set of real numbers;
range: set of positive real numbers;
increasing or decreasing: decreasing over its entire domain;
maxima or minima: none;
end behavior: $\lim_{x \to \infty} b^x = 0$, $\lim_{x \to -\infty} b^x = \infty$.
graph:

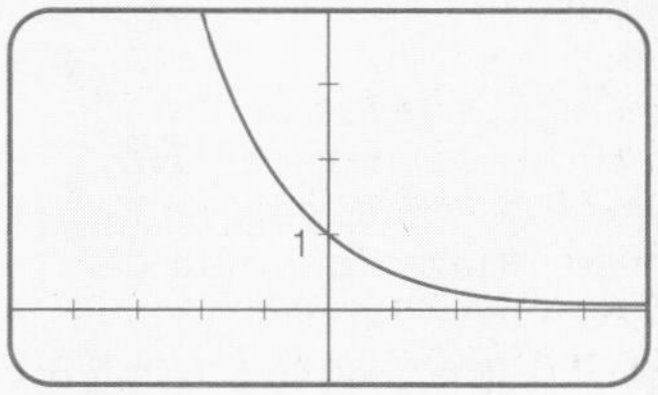

model: decay;
special properties: values of the function related by the laws of exponents

9.a.

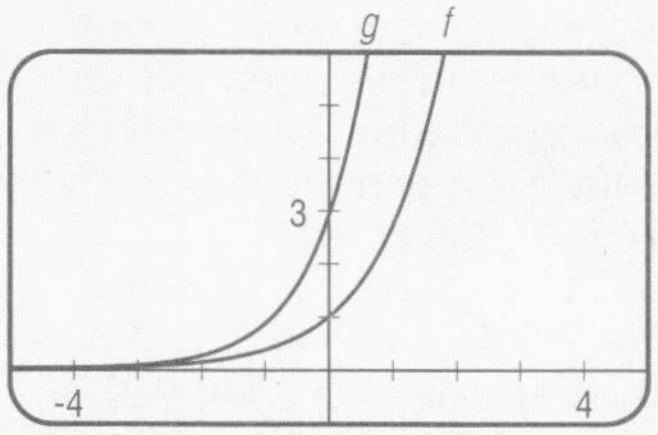

NOTES ON QUESTIONS
Making Connections for Question 13: Students who have studied *Functions, Statistics, and Trigonometry* should be familiar with this function.

Question 23: You may invite students to calculate $\frac{1}{0!} + \frac{1}{1!} + \frac{1}{2!} + \frac{1}{3!} + \ldots + \frac{1}{n!}$ and $\left(1 + \frac{1}{n}\right)^n$ for various values of n. Which formula provides a better approximation for e? $\left(\frac{1}{0!} + \frac{1}{1!} + \frac{1}{2!} + \frac{1}{3!} + \ldots + \frac{1}{n!}\right)$

FOLLOW-UP

MORE PRACTICE
For more questions on SPUR Objectives, use *Lesson Master 2-6,* shown on page 121.

EXTENSION
Some automatic graphers and statistical software, given ordered pairs of data that are appropriate to the Continuous Change Model, will calculate the exponential curve of best fit. Both the increase in new car prices and depreciation of used cars can be approximated by this model. As a project, have students gather data on the price increase of a new car model over the past 10 years (or the decrease in price of a used car), graph the data, and use a calculator or software to determine the curve of best fit.

Note that if $y = b^x$, then $\log y = (\log b)x$; that is, $\log y$ is a multiple of x. This observation provides the following criterion for deciding if data fit an exponential model $y = b^x$. Given data $\{(x_1, y_1), (x_2, y_2), \ldots, (x_k, y_k)\}$, plot the points $\{(x_1, \log y_1), (x_2, \log y_2), \ldots (x_k, \log y_k)\}$. If these plotted points generally lie along a line through the origin, then the given data will fit reasonably well with an exponential function.

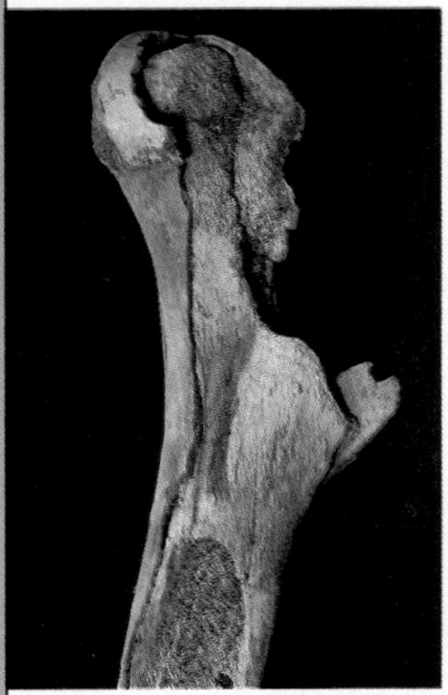
Wooly mammoth leg bone

11. One of the best examples of continuous decrease in nature is radioactive decay. The radioactive isotope carbon 14, used in carbon dating of organic findings, disintegrates at the very slow annual rate $r = .000124$. If 1.2 mg of carbon 14 started decaying in the bones of an animal that died 20,000 years ago, how much of it would be detected today? **about 0.1 mg**

12. Prove that if $b < 1$, the function $f: x \rightarrow b^x$ with domain the set of integers is a decreasing function. **See page 121.**

13. The **standard normal distribution** in statistics is described by a bell-shaped curve whose equation is

$$y = \frac{1}{\sqrt{2\pi}} e^{-x^2/2}$$

a. Use an automatic grapher with the window $-4 \le x \le 4$, $0 \le y \le 0.4$ to graph this function. **See page 121.**

b. On what interval(s) is the function increasing? decreasing?
c. Find any relative maximum or minimum values. **rel. max.: $\frac{1}{\sqrt{2\pi}}$**
d. Describe the apparent end behavior. **See page 121.**
e. Estimate the range. **$\{y: 0 < y \le 0.4\}$**
b) See p. 121.

Review

14. Is $f: t \rightarrow t^2 + 2t^3$ an even function, an odd function, or neither? *(Lesson 2-5)* **neither**

15. If $\forall n$, $a_n = \frac{3n + 5}{n + 1}$, determine $\lim\limits_{n \to \infty} a_n$. *(Lesson 2-4)* **3**

16. Let g be the function defined by the rule $g(x) = x^3 + 3x^2 - x - 3$.
a. Sketch a graph of g. **See page 121.**
b. For what values of x is $g(x) = 0$? **$x = 1$, $x = -1$, and $x = -3$**
c. For what values of x is $g(x) > 0$? **$-3 < x < -1$ and $x > 1$**
d. For what values of x is $g(x) < 0$? *(Lesson 2-1)*
$x < -3$ and $-1 < x < 1$

In 17-19, use the laws of exponents to simplify. *(Previous course)*

17. $\frac{x^5y^3}{x^2y^6} \cdot \frac{(xy)^2}{x^2y}$ **$\frac{x^3}{y^2}$**

18. $\frac{6(n + 1)^3}{4(n + 1)^4}$ **$\frac{3}{2(n + 1)}$**

19. $x^0 + \frac{3}{x^{-2}} + x^2$ **$1 + 4x^2$**

20. Determine when the function f with $f(x) = 2x^2 + 8x - 9$ is increasing. *(Lesson 2-3)* **$x \ge -2$**

21. Give the exact value. *(Previous course)*
a. $\sin 30°$ **$\frac{1}{2}$** **b.** $\cos \frac{\pi}{6}$ **$\frac{\sqrt{3}}{2}$** **c.** $\sin \frac{\pi}{4}$ **$\frac{\sqrt{2}}{2}$**

6. Evaluate $\log_2 n$ for $n = 1, 2, 3, 4, 5, 6, 7, 8$, and 9. **See margin.**

7. Analyze the logarithm function with base $b > 1$. (Use the form of Example 3 of Lesson 2-6.) **See margin.**

8. Prove the Logarithm of a Quotient Theorem by using the Law of Exponents $\frac{b^r}{b^s} = b^{r-s}$. **See margin.**

9. Solve for x: $\log_x 625 = 2$. **$x = 25$**

In 10 and 11, express the logarithm in terms of $\log_b n$ and $\log_b w$.

10. $\log_b \frac{5n^2}{w}$ **$\log_b 5 + 2\log_b n - \log_b w$**

11. $\log_b \sqrt[4]{nw^3}$ **$\frac{1}{4}\log_b n + \frac{3}{4}\log_b w$**

12. Solve $15^t = 30$. **$t \approx 1.256$**

In 13 and 14, refer to the situation in Example 5.

13. By how much would the loudness level be increased if there were three machines instead of one? **≈ 4.77 decibels**

14. How many machines, each with a loudness level of 70 decibels, would be necessary to surpass the 90 dB limit set by the EPA? **101 machines**

Applying the Mathematics

15. Explain what it would mean for a sound to have a negative loudness (as measured in decibels). **See margin.**

16. Africa's population during the decade 1980–1990 grew at the annual rate of 2.9%, the highest of any continent. Determine the number of years (to the nearest year) the population will take to double if this growth rate continues. **$t = 24$ years**

17. **a.** Compute $\log_{12} 98$. **≈ 1.845**
    b. Compute $\log_{98} 12$. **≈ 0.542**
    c. Generalize the result you found in parts **a** and **b**. **$\log_a b = \frac{1}{\log_b a}$**

18. Use the definition and properties of logarithms to solve the equation $\log(x - 15) + \log x = 2$. **$x = 20$**

19. Radiocarbon, the radioactive isotope carbon 14, has a half-life of about 5730 years. Using the Continuous Change Model, this means that if B grams is the initial amount, then the amount after 5730 years is $\frac{B}{2}$ grams, where $\frac{B}{2} = Be^{5730r}$ for a certain constant r.
    a. Solve the above equation for r to determine the annual rate at which carbon 14 decays. **$r \approx -1.21 \times 10^{-4} = -0.000121$**
    b. Suppose the skull of a prehistoric relative of the rhinoceros is discovered and analyses show that it contains only 20% of the carbon 14 that bones of living animals do. Use the result of part **a** to approximate the age of the skull. **≈ 13,000 years old**

15. Loudness in decibels is $.10 \log (I \cdot 10^{12})$.
If sound has a negative loudness, then
$I \cdot 10^{12} < 1$,
so $I < 10^{-12}$.
But since the weakest intensity that can still be heard is
$I = 10^{-12}$ watts/m², the sound cannot be heard.

NAME ____

LESSON MASTER 2–7
QUESTIONS ON **SPUR** OBJECTIVES

■SKILLS *Objective B (See pages 146–150 for objectives.)*
In 1–3, evaluate each expression.

1. $\log_2 8^2$ **6**
2. $\log_b \sqrt[3]{b}$ **$\frac{1}{3}$**
3. $\log_{10} \sqrt{1{,}000}$ **$\frac{3}{2}$**

4. Express $\log_b \frac{n^3}{2w^4}$ *in terms of* $\log_b n$ and $\log_b w$.
$3 \log_b n - (\log_b 2 + 4 \log_b w)$

5. Let $\log_k 2 = .231$ and $\log_k 3 = .367$.
   a. Use properties of logarithms to find $\log_k 6$. **.598**
   b. Find the value of k and use it to check your answer to part **a**. **$k \approx 20$; $20^{.598} \approx 6$**

6. Solve: $\log_3 x = 6$ **729**

■PROPERTIES *Objective D*
7. Explain why $\lim_{x \to -\infty} \log_b x = 0$ is false for any $b > 0$ with $b \neq 1$.
$\log_b x$ is undefined for $x < 0$, so the limit is also undefined.

■USES *Objective E*
8. The sales of a certain new product increase over time according to $s(t) = 105 + 200 \log (5t + 1)$ where $s(t)$ is the annual sales (in thousands of dollars) after t years. About how many years does it take for the annual sales to reach $500,000?
≈ 18.7 years

NAME ____

9. One formula that relates the average weight w in pounds of girls to their average height h in inches is
$\log w = -2.86625 + 2.72158 \log h$.
   a. Predict the weight of a girl who is 63 inches tall. **107.3 lb**
   b. Assume that a child's weight is normal if it is in a range of ±5% of the average. Give the range of normal weights for a girl who is 63 inches tall. **$101.9 \le w \le 112.7$**

■REPRESENTATIONS *Objective G*
10. a. Graph the function $g(x) = \log_2 x$.

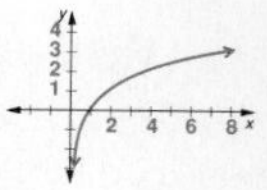

   b. Give the domain and range of g.
   domain: $\{x : x > 0\}$,
   range: set of all reals
   c. Use the Change of Base Theorem to express g in terms of $\log_{10} x$.
   $g(x) = \frac{\log_{10} x}{\log_{10} 2}$ or $g(x) = \log_2 10 \cdot \log_{10} x$
   d. Compare the graph of g to the graph of $f(x) = \log_{10} x$.
   The height of the graph of g is $\log_2 10 \approx 3.3$ times the height of f; that is, the graph of g results from the scale change $S_{1, \log_2 10}$ applied to the graph of f.

NOTES ON QUESTIONS
Question 20: A function is increasing if and only if the slope of the line through any two of its points is positive. This result shows that the inverse of an increasing function is increasing, and thus shows that the logarithm function with base $b > 1$ is increasing (since it is the inverse of the exponential function with base b).

Question 28: This result is related to the fact that for an integer $N \lfloor \log N \rfloor + 1$ is the number of decimal digits in N. Students may be interested to know that $\lfloor \log_2 N \rfloor + 1$ is the number of *binary* digits in the binary expansion of N.

FOLLOW-UP

MORE PRACTICE
For more questions on SPUR Objectives, use *Lesson Master 2-7,* shown on page 127.

EXTENSION
Have students use an automatic grapher to investigate the rate of growth of different types of functions. For example, the graphs of $y = 2^x$, $y = x^2$, $y = x$, $y = \sqrt{x}$, and $y = \log_2 x$ on a common set of axes would show the differences between exponential, quadratic, and linear rates of growth.

PROJECTS
The projects for Chapter 2 are described on page 142. **Project 5** is related to the content of this lesson.

ADDITIONAL ANSWERS
26.a. ***If the set S is finite, then S is a discrete set.***
b. ***If the set S is a discrete set, then S is finite.***
c. Counterexample: ***The set of integers is a discrete set, but it is not finite.***

20. Let l be the line containing the points (a, b) and (c, d), and let l' be its image under a reflection over the line $y = x$. Thus, l' contains the points (b, a) and (d, c). Prove that the slope of l' is the reciprocal of the slope of l. See below.

Review

In 21–24, refer to the graph at right. *True* or *false*? *(Lesson 2-3)*

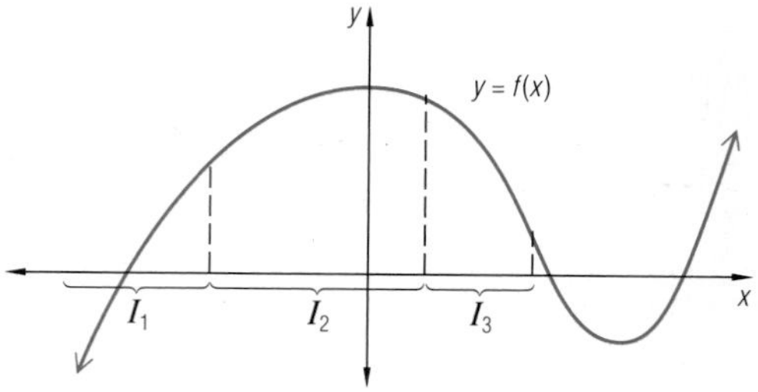

21. $\forall x_1$ and x_2 in I_1, if $x_1 < x_2$ then $f(x_1) < f(x_2)$. True

22. f is increasing on I_2. False

23. $\exists x_1$ and x_2 in I_2 such that $x_1 < x_2$ and $f(x_1) < f(x_2)$. True

24. $\forall x_1$ and x_2 in I_3, if $x_1 < x_2$ then $f(x_1) < f(x_2)$. False

25. Refer to the sequence h below.

$$\frac{2}{2}, \frac{3}{4}, \frac{4}{6}, \frac{5}{8}, \frac{6}{10}, \frac{7}{12}, \ldots$$

a. Find an explicit formula for h_n. $h_n = \frac{n+1}{2n}$
b. Write the decimal forms of h_{10}, h_{100}, and h_{1000}. See below.
c. What does $\lim_{n \to \infty} h_n$ appear to be? *(Lesson 2-4)* 0.5

26. Consider the statement *The set S being discrete is a necessary condition for S to be finite.*
a. Rewrite the statement in *if-then* form.
b. Write the converse of the statement in **a**.
c. Is the converse true? If so, explain. If not, give a counterexample. *(Lessons 2-1, 1-5)* See margin.

27. In logic, the word *but* means *and*.
a. Write the negation of the following statement using *but*:
$x^2 \neq 4$ *or* $x = 2$. x^2 is equal to 4 but x is not equal to 2.
b. Assume the statement in **a** is false. What is x? *(Lesson 1-3)*
$x = -2$

Exploration

28. a. Find the common logarithms of the following numbers

2385	≈ 3.377
238.5	≈ 2.377
23.85	≈ 1.377
2.385	≈ 0.377
0.2385	≈ -0.623

b. Generalize the results of what you found in **a**.
c. Relate the generalization to scientific notation.
b) $\log\left(\frac{1}{10}a\right) = \log \frac{1}{10} + \log a = \log a - 1$; c) $\log(a \cdot 10^n) = \log a + n$

20. For two points (x_1, y_1) and (x_2, y_2) on a line L, slope is defined as $\frac{y_2 - y_1}{x_2 - x_1}$. Therefore, the slope of $l = \frac{d-b}{c-a}$ and the slope of $l' = \frac{c-a}{d-b}$. Since $\frac{1}{\text{slope of } l'} = \frac{1}{\frac{c-a}{d-b}} = \frac{d-b}{c-a} =$ slope of l, the slope of l' is the reciprocal of the slope of l.

25. b) $h_{10} = .55$; $h_{100} = .505$; $h_{1000} = .5005$

128

LESSON

2-8

Analyzing the Sine and Cosine Functions

You were probably introduced to sines and cosines through definitions related to the ratios of the sides of a right triangle.

Right Triangle Definitions of the Sine and Cosine

Given an acute angle θ in a right triangle:

$$\textbf{sine of } \theta = \sin \theta = \frac{\text{side opposite } \theta}{\text{hypotenuse}}$$

$$\textbf{cosine of } \theta = \cos \theta = \frac{\text{side adjacent to } \theta}{\text{hypotenuse}}$$

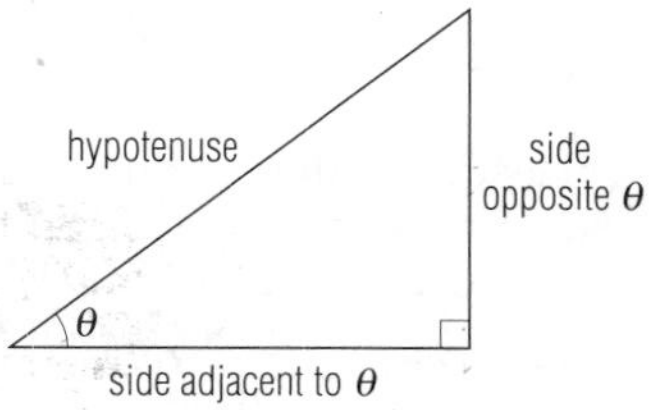

Values of the sine and cosine can be obtained in various ways. If you know the angle measures, use a calculator. Calculators measure angles in at least two ways, degrees and radians, so be certain the calculator is set to the mode you want. Degrees and radians are related by the conversion formula

$$\pi \text{ radians} = 180 \text{ degrees}.$$

So, dividing by π, $\quad 1 \text{ radian} \approx 57°.$

If you know the sides of a right triangle, as at left below, you can use the definition to find $\sin \theta$ or $\cos \theta$. For the special angles 30°, 45°, and 60° (or, equivalently, $\frac{\pi}{6}$, $\frac{\pi}{4}$, and $\frac{\pi}{3}$ radians), you should know the exact values (see Questions 3-6).

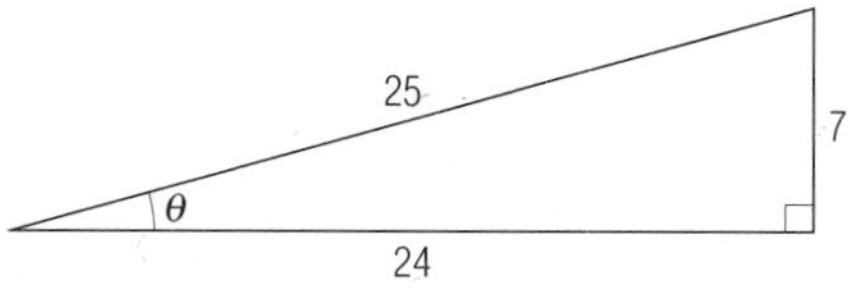

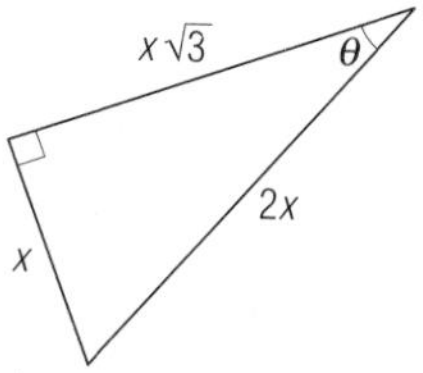

$\sin \theta = \frac{7}{25} = 0.28$ $\qquad$ $\sin 30° = \sin \frac{\pi}{6} = \frac{1}{2} = 0.5$

$\cos \theta = \frac{24}{25} = 0.96$ $\qquad$ $\cos 30° = \cos \frac{\pi}{6} = \frac{\sqrt{3}}{2} \approx 0.866$

LESSON 2-8

RESOURCES

- Lesson Master 2-8
- Teaching Aid 12 displays the definitions of the sine and cosine.
- Computer Master 5

OBJECTIVES

A Determine relative minima and maxima of a function and intervals on which it is increasing or decreasing.
D Determine the end behavior of a function.
E Use trigonometric, exponential, and logarithmic functions as models.
G Analyze a function from its graph.

TEACHING NOTES

This lesson reviews the trigonometric functions. There is a lot of information in this lesson, and if your students have not had a great deal of trigonometry, they may need to spend an extra day on this lesson.

Students should be able to use either the right triangle definition or the unit circle definition to find values of trigonometric functions at special angles. They also should be able to analyze the sine and cosine functions.

The formula π radians = 180 degrees is the springboard for conversion between radians and degrees. Some students, however, prefer to use a proportion:

$$\frac{\text{degree measure}}{\text{radian measure}} = \frac{180}{\pi}$$

This will help in **Question 2**.

In the unit circle definition of the trigonometric functions, assume that the magnitude θ

is measured in radians unless degrees are explicitly specified.

Because radian measure is more common than degree measure in advanced mathematics, try to wean students away from converting radians to degrees.

Some students have seen the Law of Sines in the form $\frac{a}{\sin A} = \frac{b}{\sin B} = \frac{c}{\sin C}$. Many need to be reminded that any two of these ratios can form a proportion to solve problems.

The Law of Cosines, as given in this lesson, finds side c of ΔABC. Ask students to write the other two versions of the Law of Cosines for ΔABC.

Error Analysis Warn students that pressing the "clear" or "all clear" key on some calculators will set them into degree mode. When problems require radians, this will produce errors.

Small Group Work Students could brainstorm in small groups to come up with as many examples of phenomena that are periodic as they can think of. Some examples are days of the week, phases of the moon, hours on an analog watch, and so so. Have students give the period for each.

Making Connections You may want to ask this question: What is the image of the point $(r, 0)$ under the rotation with magnitude θ about the origin? (The image is $(r \cdot \cos\theta, r \cdot \sin\theta)$; this fact will be used in Chapter 8 in the introduction of polar coordinates $[r, \theta]$.)

At this point, you might want to review the graphs of functions of the form $f(x) = A\cos(B(x - C)) + D$. It is not necessary to do this now since the transformations of graphs are studied in Lesson 3-9. However, you may want

Note: When angle measures are written without units, radian measure is understood. So, $\sin\left(\frac{\pi}{6}\right)$ is understood to be $\sin\left(\frac{\pi}{6}\text{ radians}\right)$.

The above ratios define the corresponding two trigonometric functions

$$\text{sine: } \theta \rightarrow \sin\theta \text{ and cosine: } \theta \rightarrow \cos\theta$$

on the interval $0 < \theta < 90°$ in degree measure, or $0 < \theta < \frac{\pi}{2}$ in radian measure. To extend the domain to the set of all real numbers, it is customary to give a second definition of these values using the *unit circle*. The **unit circle** is the circle with center (0, 0) and radius 1.

Unit Circle Definitions of the Sine and Cosine

($\cos\theta$, $\sin\theta$) is the image of the point (1, 0) under the rotation of θ about the origin.

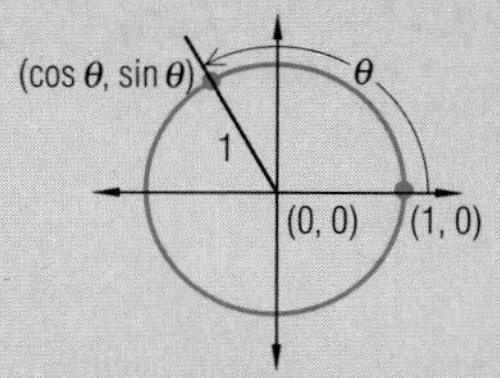

If the magnitude θ is in radians, and $0 < \theta < 2\pi$, then θ is the length of the arc on the unit circle from (1, 0) counterclockwise to ($\cos\theta$, $\sin\theta$).

From the unit circle definitions, any sine or cosine value can be found or estimated. For instance, to find $\cos\frac{7\pi}{6}$, rotate (1, 0) by $\frac{7\pi}{6}$ about the origin and use the first coordinate of the image. You could also use a calculator. These methods show

$$\cos\frac{7\pi}{6} = -\frac{\sqrt{3}}{2} \approx -0.866.$$

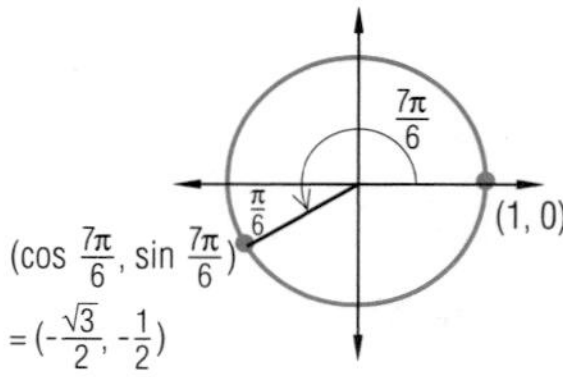

The second coordinate of the image is $\sin\frac{7\pi}{6}$.

$$\sin\frac{7\pi}{6} = -\frac{1}{2} = -0.5$$

To avoid confusion, it is important that the unit circle definitions of the trigonometric values agree with the right triangle definitions. To prove this, recall that for a given θ, the right triangle ratios have the same values regardless of the triangle chosen (because such triangles are all similar). In particular, consider a right triangle whose acute angle θ is placed at (0, 0) with one side along the x-axis and the other side (the hypotenuse) intersecting the unit circle at (x, y), as shown at right. Then, by the right triangle definition,

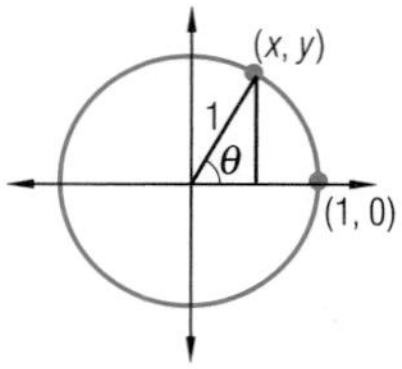

$$\sin\theta = \frac{\text{side opposite } \theta}{\text{hypotenuse}} = \frac{y}{1} = y,$$

which is exactly what is required by the unit circle definition. You can use a similar analysis to show that the unit circle and right triangle definitions agree for the cosine.

In general, rotations about the origin that differ by multiples of 2π yield the same image point (x, y) on the circle. Hence,

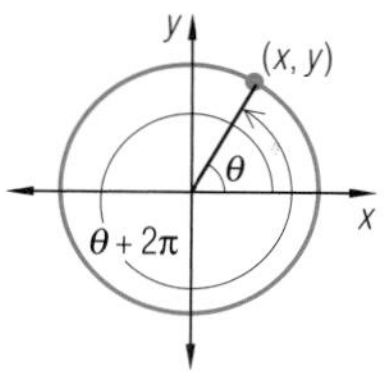

$$\cos(\theta + 2\pi n) = \cos\theta$$

and

$$\sin(\theta + 2\pi n) = \sin\theta$$

$\forall\ \theta$ and $\forall$ integers n. For example,

$\cos\frac{7\pi}{6} = \cos\left(\frac{7\pi}{6} + 2\pi n\right)$ and $\sin\frac{7\pi}{6} = \sin\left(\frac{7\pi}{6} + 2\pi n\right)$ $\forall$ integers n. Rotations can also be expressed in degrees. Since π radians $= 180°$, $\frac{7\pi}{6}$ radians $= 210°$. Thus, $\cos\frac{7\pi}{6} = \cos 210° = \cos(210 + 360n)°$ $\forall$ integers n.

While the Greek letter θ is frequently used in trigonometry to denote angles, the variable x is preferred when discussing the trigonometric functions as real functions. Thus we write $\cos(x + 2\pi n) = \cos x$ and $\sin(x + 2\pi n) = \sin x$ $\forall\ x$ and $\forall$ integers n. This means that sine and cosine are *periodic functions*.

Definition

A real function f is **periodic** if and only if there is a positive number p such that $f(x + p) = f(x)$ for all x in the domain of f. The smallest p with this property is the **period** of f.

Periodic functions cycle through the same values repeatedly. The argument above the definition proves that the period of the sine and cosine function is 2π or less. Calculation of values proves that the period is not less than 2π. So it is exactly 2π.

Since a point can be rotated as much as one wishes in either a positive (counterclockwise) or negative (clockwise) direction, the domain of cosine: $x \to \cos x$ and sine: $x \to \sin x$ is the set of real numbers. And, since both coordinates of points on the unit circle range from -1 to 1, the range of each function is $\{y: -1 \le y \le 1\}$. These properties can be seen in the graphs below. Notice that, because of periodicity, the graphs are **translation-symmetric**, that is, each graph coincides with its own image under a translation.

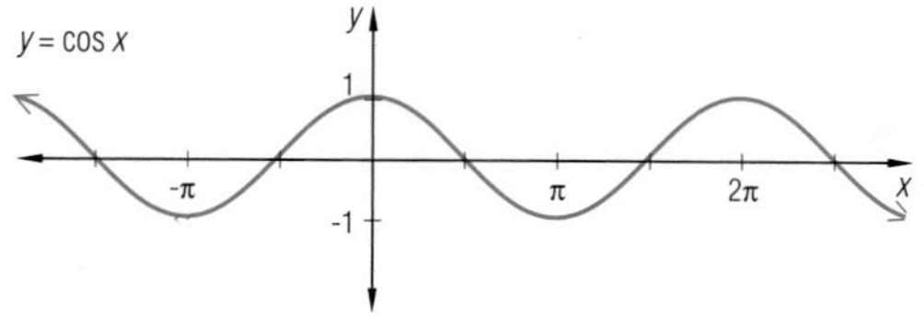

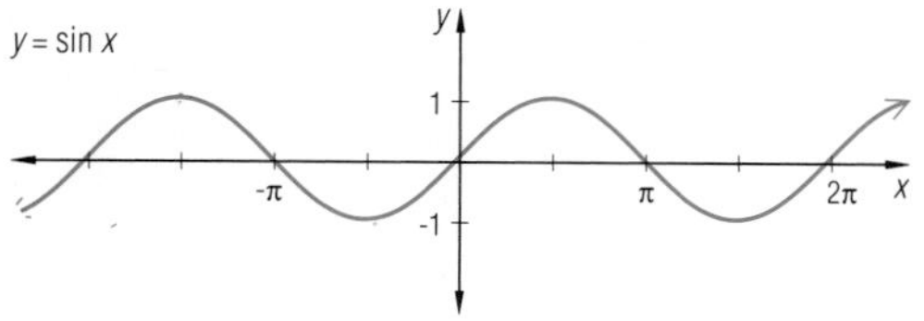

The graphs of the two trigonometric functions above suggest that the cosine is an even function, while the sine is an odd function. These results can be proved.

to use an automatic grapher to graph some of these functions and have students determine the period from the graph. Also, emphasize that the period is the *smallest* p such that $f(x + p) = f(x)$ for all x in the domain of f.

ADDITIONAL EXAMPLES

1. Give the exact values.

a. $\cos\frac{3\pi}{2}$

0

b. $\sin\frac{3\pi}{2}$

-1

2. Describe the maxima and minima of the cosine function.

The cosine function achieves its maximum value of 1 when $x = 2k\pi$ for any integer k and its minimum value of -1 when $x = (2k + 1)\pi$ for any integer k.

NOTES ON QUESTIONS
Questions 3-7: A unit circle may be very helpful in doing these questions successfully.

Question 12: The most difficult part of this question requires describing the intervals over which the sine function is increasing or decreasing. Use the result of Question 11 for the former; then note that the sine function is either increasing or decreasing throughout its domain, so at all other times it is decreasing.

Making Connections for Question 16: Part **a** anticipates work with trigonometric identities in Chapter 6. The unit circle is a major assistance again.

Question 18: Because the measures of two sides and an included angle are given, the SAS Triangle Congruence Theorem guarantees that the measures of the unknown sides and angles are uniquely determined. Remind students of this fact.

Question 21: There is the implicit assumption that for each additional meter below the surface, the light intensity is only 90% of what it was at the previous meter.

Question 27: Setting tick marks on the x-axis at multiples of $\frac{\pi}{4}$ or $\frac{\pi}{2}$ would make the period of this graph easier to determine. It crosses the x-axis at about -2.7 and, in a similar fashion, at 3.6; this suggests a period of 6.3, which is about 2π. Some students might be able to prove that the period is 2π.

Theorems

The cosine function is an even function.
The sine function is an odd function.

Proof The points $(\cos x, \sin x)$ and $(\cos(-x), \sin(-x))$ are the images of $(1, 0)$ under rotations of x and $-x$. But they are reflection images of each other over the x-axis. So for all x,

$$\cos(-x) = \cos x$$
$$\text{and } \sin(-x) = -\sin x,$$

which indicates that the cosine function is even and the sine function is odd.

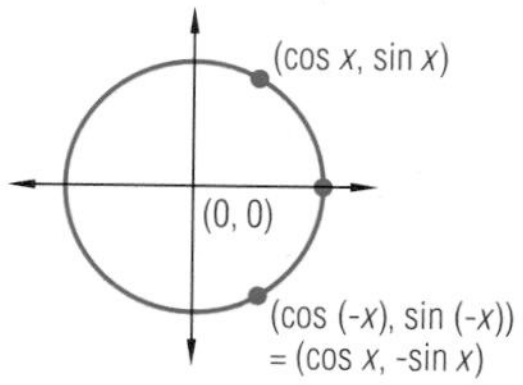

Unlike the exponential and logarithmic functions, the sine and cosine functions increase on some intervals and decrease on others. For instance, on the interval $0 \le x \le \frac{\pi}{2}$, so important in right triangle trigonometry, the sine function is increasing while the cosine function is decreasing. On the interval $\frac{\pi}{2} \le x \le \pi$, both functions are decreasing. (Check these by looking at the graphs.)

These functions also have a different end behavior from that of exponential or logarithmic functions. Neither function approaches a particular value as $x \to \infty$ or $x \to -\infty$. This means, for example, that $\lim_{x \to \infty} \cos x$ does not exist.

The sine and cosine functions are called **trigonometric functions** because they describe relationships between sides and angles in triangles. (The word "trigonometry" is from the Greek word trigonometria, meaning "triangle measure.") We expect that you have seen the following two relationships before.

Theorems

In any $\triangle ABC$ with sides a, b, and c opposite angles A, B, and C:

Law of Sines
$$\frac{\sin A}{a} = \frac{\sin B}{b} = \frac{\sin C}{c}$$

Law of Cosines
$$c^2 = a^2 + b^2 - 2ab \cos C$$

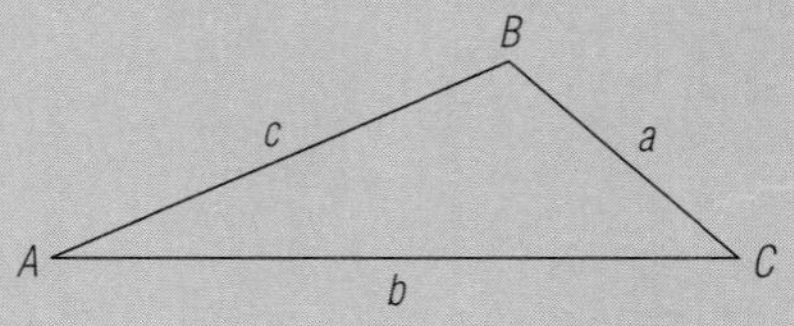

Functions based on sine and cosine functions are almost perfect models of sound waves and phenomena based on rotations, such as the times of sunrise and sunset (with period = 1 year). Think of two people twirling a rope at 1 revolution per second. Focus on a fixed point P in the middle of the rope and observe its height h above the ground. At time $t = 0$, P is on the ground; when $t = 0.25$ seconds, P reaches a height of 3.5 feet; after another 0.25 seconds, P reaches a maximum height of 7 feet. Another 0.25 seconds later, P again is at a height of 3.5 feet, and then returns to its initial position on the ground in a total time of 1 second.

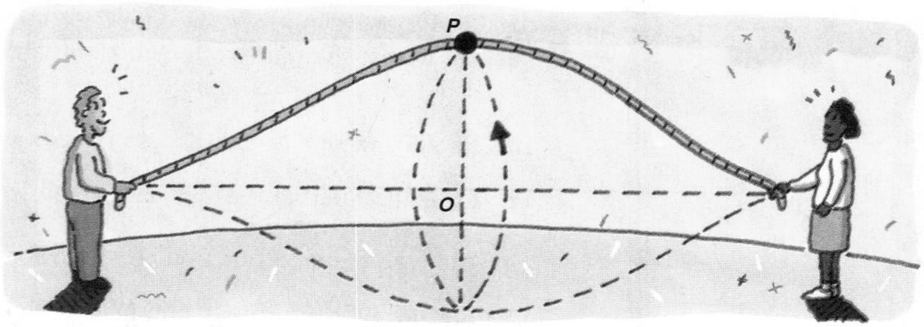

The function which maps time to the height of point P is periodic and can be described using a trigonometric model. Below is a sketch of height vs. time over an interval of 2 seconds. The period is 1 because the graph repeats itself each second. You can verify that $h = -3.5 \cos(2\pi t) + 3.5$ gives the height h of P at time t.

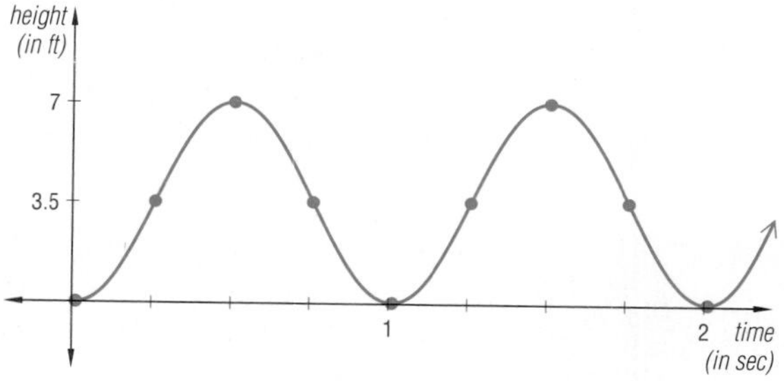

Questions

Covering the Reading

1. Use right triangle MON shown at right. Give the exact value of:

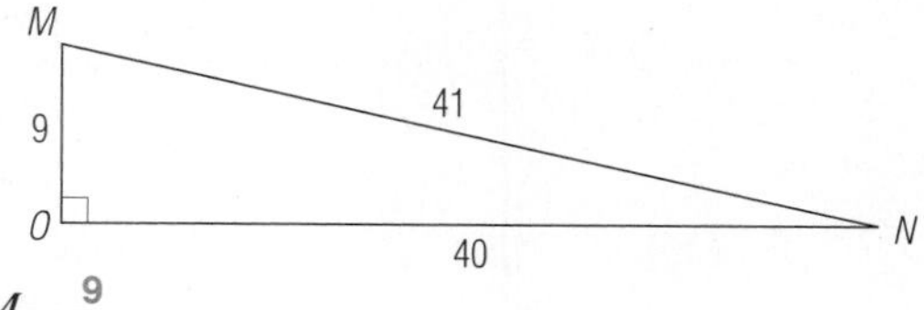

a. $\sin M$ $\frac{40}{41}$ **b.** $\cos M$ $\frac{9}{41}$

c. $\cos N$ $\frac{40}{41}$ **d.** $\sin N$ $\frac{9}{41}$

2. a. $\frac{\pi}{3}$ radians = __?__ degrees. 60 π/3 press radians then degree

b. 135° = __?__ radians. $\frac{3\pi}{4}$ 2.356 135 set to degrees then to radians

FOLLOW-UP

MORE PRACTICE
For more questions on SPUR Objectives, use *Lesson Master 2-8,* shown on page 135.

PROJECTS
The projects for Chapter 2 are described on page 142. **Project 1** is related to the content of this lesson.

In 3–7, give the exact value of cos x and sin x when x has the given value.

3. $\frac{\pi}{3}$ **4.** $\frac{3\pi}{4}$ **5.** 300° **6.** 90° **7.** 8π

3–7) See margin.

8. If $\theta = 148°$, give the coordinates of point P on the unit circle

a. exactly (in terms of trigonometric functions); (cos 148°, sin 148°)

b. to the nearest hundredth. (-0.85, 0.53)

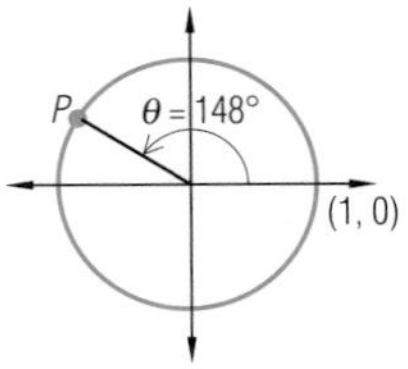

9. Show that the unit circle and right triangle definitions for cos θ agree if $0 < \theta < \frac{\pi}{2}$. See margin.

10. Explain why $\forall x$, $\cos(x + 4\pi) = \cos x$. Since the period of cos x is 2π, $\cos(x + 2\pi n) = \cos x$. Let $n = 2$, then $\cos(x + 4\pi) = \cos x$.

11. Over the interval $0 \le x \le 2\pi$, when is the sine function increasing? when $0 \le x \le \frac{\pi}{2}$ and $\frac{3\pi}{2} \le x \le 2\pi$

12. Analyze the sine function. Use the form of Example 3 of Lesson 2-6. See margin.

13. For the twirling rope mentioned in this lesson, verify that the equation $h = -3.5\cos(2\pi t) + 3.5$ gives the height h of point P on the rope at $t = 0$, 0.25, 0.5, 0.75, and 1 sec. See margin.

Applying the Mathematics

14. A plane at an altitude of 10,000 ft is seen directly overhead. One minute later while at the same altitude it is seen at an angle of 24° from the horizontal.

a. How far has the plane traveled? 22,460 ft

b. Find the plane's speed in feet per minute and in miles per hour. 22,460 ft/min ≈ 255.2 mph

15. a. Sketch a right triangle with angle θ such that $\cos\theta = \frac{3}{5}$.

b. Find $\sin\theta$. $\frac{4}{5}$

a) See margin.

16. a. Prove: For all real numbers x, $\sin^2 x + \cos^2 x = 1$. See margin.

b. If $\cos x = \frac{-2}{7}$, use the result in part **a** to find $\sin x$. $\pm\frac{3\sqrt{5}}{7}$

c. Find a formula for $\sin x$ in terms of $\cos x$. $\sin x = \pm\sqrt{1 - \cos^2 x}$

17. If $\frac{3\pi}{2} < x < 2\pi$ and $\cos x = \frac{3}{5}$, find $\sin x$. $-\frac{4}{5}$

18. Find the missing side and angles in triangle ABC. side $BC \approx 8$; m∠$B \approx 90°$; m∠$C \approx 36.87°$

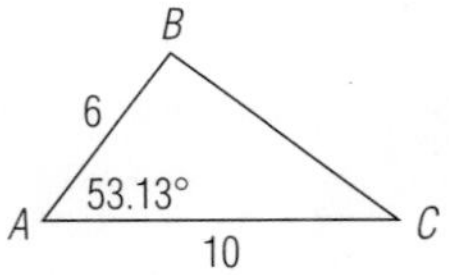

19. A bird watcher spots a sandhill crane nest about 32° north of east. Then he walks east for 500 m, and determines that the nest is at 16° west of north. How far is he from the nest at this point? ≈ 276 m

Review

20. Let b be a positive real number such that $b \neq 1$. Simplify. *(Lesson 2-7)*

a. $\log_b b^x$ x **b.** $b^{\log_b x}$ $(x > 0)$ x

ADDITIONAL ANSWERS

3. $\cos x = \frac{1}{2}$; $\sin x = \frac{\sqrt{3}}{2}$

4. $\cos x = -\frac{\sqrt{2}}{2}$; $\sin x = \frac{\sqrt{2}}{2}$

5. $\cos x = \frac{1}{2}$; $\sin x = -\frac{\sqrt{3}}{2}$

6. $\cos x = 0$; $\sin x = 1$

7. $\cos x = 1$; $\sin x = 0$

9. Consider a right triangle whose acute angle θ is at (0, 0) with one side along the x-axis and the other side intersecting the unit circle at (x, y). By the right triangle definition,
$\cos\theta = \frac{\text{adjacent side}}{\text{hypotenuse}} = \frac{x}{1} = x$, which corresponds to the unit circle definition.

12. usual domain: set of real numbers; range: $-1 \le y \le 1$; relative maximum: 1; relative minimum: -1; increasing: $-\frac{\pi}{2} + 2\pi n \le x \le \frac{\pi}{2} + 2\pi n$, $\forall$ integers n; decreasing: $\frac{\pi}{2} + 2\pi n \le x \le \frac{3\pi}{2} + 2\pi n$, $\forall$ integers n; end behavior: limits do not exist.
graph:

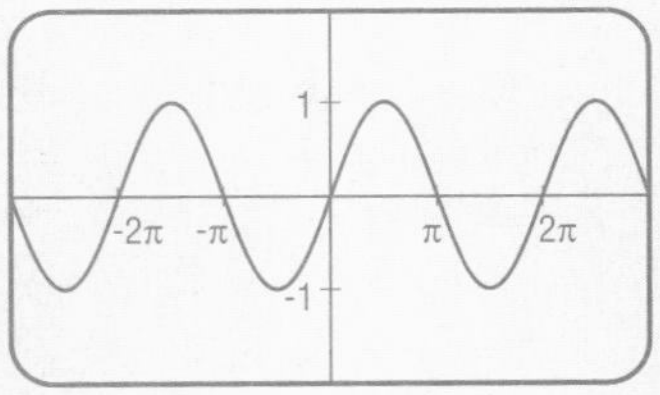

model: sound and light waves, phenomena based on rotations; special properties: odd function, periodic function

13. $h(0) = 0$ ft; $h(0.25) = 3.5$ ft;
$h(0.5) = 7.0$ ft; $h(0.75) = 3.5$ ft;
$h(1) = 0$ ft

15.a.

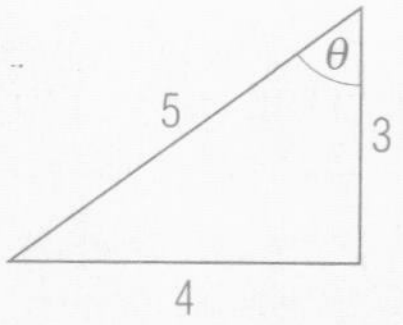

21. Pollution in a lake affects the ability of light to penetrate the surface. Suppose the intensity of light one meter below the surface is 90% that of the intensity at the surface.

a. Write an equation that expresses the proportion $p(x)$ of light hitting the surface that is found at a depth of x meters. $p(x) = .9^x$

b. What proportion of light hitting the surface is found at a depth of 3.8 meters? $\approx 67\%$

c. At what depth is the intensity only 43% of that hitting the surface? *(Lessons 2-7, 2-6)* ≈ 8.0 m

22. *Multiple choice.* On which interval is the function graphed at right increasing?

(a) $-a \le x \le a$

(b) $\frac{a}{2} \le x \le a$

(c) $-\frac{a}{4} \le x \le 2a$

(d) $-\frac{a}{4} \le x \le 0$ *(Lesson 2-3)* **(d)**

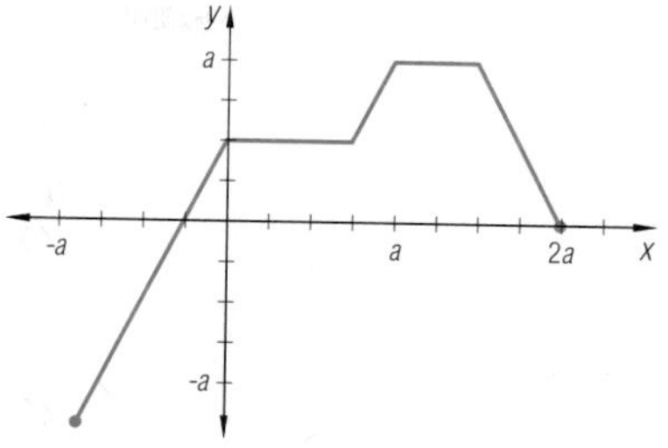

23. Use properties of logarithms to solve the equation $2 \log_4 x - \log_4 3 = \frac{1}{2}$. *(Lesson 2-7)* $x = \sqrt{6}$

In 24–26, simplify using the laws of exponents. *(Previous course)*

24. $\frac{(x + y)^{-3}}{(x + y)^{-5}}$ $(x + y)^2$ **25.** $\left(\frac{p}{q}\right)^2\left(\frac{q}{p}\right)^3$ $\frac{q}{p}$ **26.** $\frac{m^3 + n^3}{m^2 + n^2}$ cannot be simplified

Exploration

27. At right is a graph of $y = \sin 2x + \cos x$. Explore other combinations of sine and cosine functions and sketch the three most interesting combinations you find. For each one, estimate the period p, and count the number of relative maxima and minima that occur in an interval of length p. **Answers will vary.**

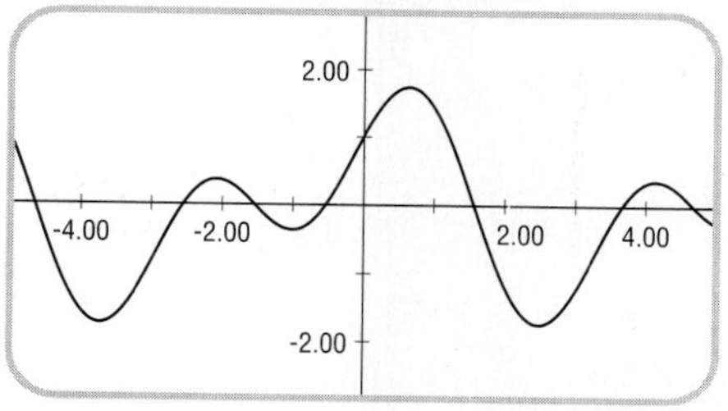

16.a. The equation of the unit circle is $x^2 + y^2 = 1$. By the unit circle definition of sine and cosine, $(\cos x, \sin x)$ is on the circle. Hence, $\cos^2 x + \sin^2 x = 1$.

NAME ____________

LESSON **MASTER 2-8**
QUESTIONS ON **SPUR** OBJECTIVES

■ PROPERTIES *Objective D (See pages 146–150 for objectives.)*

1. a. Graph $f(x) = 1 - \cos(.01x^2 - 1)^2$ on an automatic grapher using the viewing window $-10 \le x \le 10$, $-1 \le y \le 1$. Sketch the graph on the axes at the right.

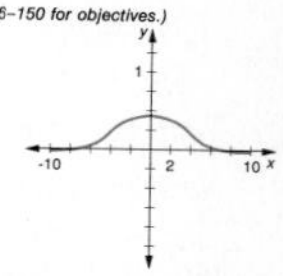

b. Describe the end behavior of f, based on your graph from part **a**.
It appears that $\lim_{x \to -\infty} f(x) = 0$ and $\lim_{x \to \infty} f(x) = 0$.

c. Graph f using the window $-20 \le x \le 20$, $-5 \le y \le 5$. Sketch the graph on the axes at the right.

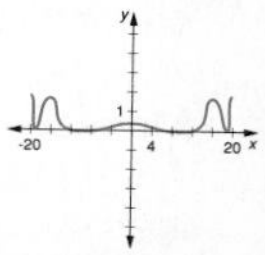

d. Does your graph from part **c** agree with your answer to part **b**? Explain why or why not.
No, although the graph approaches the x-axis near $x = \pm 8$, it then moves away from it past $x = \pm 12$ and oscillates indefinitely.

■ USES *Objective E*

2. An office building is topped by a radio antenna. An observer on the ground 300 feet from the building finds that the angle α to the top of the antenna is 56°. The angle β to the top of the building is 52°. About how tall is the antenna? ≈ 60.8 ft

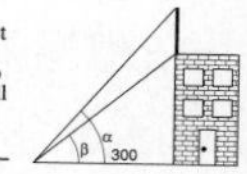

26 *Continued* *Precalculus and Discrete Mathematics* © Scott, Foresman and Company

NAME ____________
Lesson MASTER 2-8 (page 2)

3. The motion of a point P on a Ferris wheel is approximated by $f(t) = 50 \sin(1.5\pi t) + 58$, where t is the time in minutes and $f(t)$ is the height in feet.

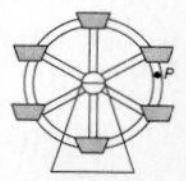

a. Graph f.

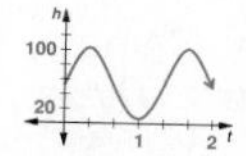

b. What is the diameter of the Ferris wheel? 100 ft

c. The lowest point on the wheel is how many feet above the ground? 8 ft

d. How long does it take for the Ferris wheel to make one revolution? $\frac{4}{3}$ min

■ REPRESENTATIONS *Objective G*
In 4–6, tell whether each function is odd, even, or neither.

not mirror image → **4.** $f(x) = \sin x$ odd

mirror image → **5.** $g(x) = |\sin x|$ even **6.** $h(x) = \sin |x|$ even

7. Consider the function $f(x) = -\cos x$ on the interval $0 \le x \le 2\pi$.

a. Over what interval is f decreasing? $\pi \le x \le 2\pi$

b. Where do any maximum and minimum values occur?
maximum at $x = \pi$, minimum at $x = 0$ and $x = 2\pi$

Hit the trace button for the decreasing and for maximum and minimum

Precalculus and Discrete Mathematics © Scott, Foresman and Company **27**

LESSON

2-9

What is Infinity?

Is space infinite in size? Does it go on forever, or does it have a boundary?

In this chapter you have seen the expressions $\lim_{x \to \infty} f(x)$ and $\lim_{n \to \infty} a_n$. Perhaps these expressions confused you. Can a number approach infinity? Is infinity itself a number? This question has been a subject of discussion since at least as long ago as the time of Zeno, a Greek philosopher of the 6th century B.C. However, the mathematical resolution to this problem has come about only within the past 200 years.

First let us resolve the meaning of infinity as used in the expressions "$\lim_{x \to \infty} f(x)$" and "$\lim_{n \to \infty} a_n$." In this case, "approaches infinity" is merely a shorthand for "gets larger and larger without bound," and "approaches negative infinity" is short for "gets smaller and smaller without bound." There is no number that is being approached. The sentence $\lim_{x \to \infty} 2^x = \infty$ means that, whatever large *number* L you might pick, there is a number h such that $\forall\, x$ greater than h, $2^x > L$.

Graphically $\lim_{x \to \infty} f(x) = \infty$ means that however high a horizontal line $y = L$ you might draw, there is a vertical line $x = h$ so that all points on the graph of the function to the right of that line are also above the horizontal line.

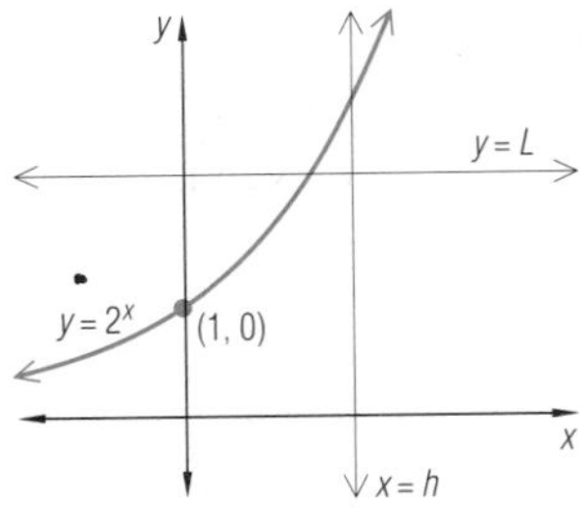

A different meaning of the idea of infinity, the meaning "goes on forever," is found in *infinite decimals*. Going on forever means that wherever you are, you are not at the end, because there is always more to follow. This meaning also is related to limits. For instance, an infinite decimal (one that goes on forever) is the limit of a sequence of finite decimals. The number

$$.\overline{32} = .32323232323232323232323232323232\ldots$$

is the limit of the sequence .32, .3232, .323232, .32323232, ..., which happens to be $\frac{32}{99}$.

In the above uses, infinity is a figure of speech, describing things without bound or things that continue forever, but infinity is not a number.

LESSON 2-9

TEACHING NOTES

For some students, it is worth saying that "x approaches negative infinity" means both that "x gets smaller and smaller without bound" and "the absolute value of x gets larger and larger without bound." Some people say "gets larger and larger in the negative direction." Choose the phrase with which you are most comfortable.

Mentioning that .99999 ... = 1 is almost guaranteed to lead to some lively discussion. Do not say this unless you are prepared to spend some time explaining why.

Students may enjoy the following limerick.

There was a young fellow
from Trinity
Who took $\sqrt{\infty}$
But the number of digits
Gave him the fidgets;
He dropped Math and took
up Divinity.

[Source: Gamow, George, *One, Two, Three ... Infinity,* Bantam Books]

Cantor used a Hebrew letter to represent cardinality because virtually all of the Greek letters (the traditional symbols to use) had some interpretation in mathematics or physics. He choose $\aleph_0$ because, being Jewish, he was quite familiar with the Hebrew alphabet. Aleph is the first letter of that alphabet; bet is the second. The Hebrew "Aleph-bet," may have predated the Greek "alpha-beta" from which the English word "alphabet" is derived.

However, there is a sense in which there is a number called infinity. To the question: "How many integers are there?" an answer is "There are infinitely many."

The theory of the number of elements in infinite sets was developed by a German mathematician, Georg Cantor, in 1895. The number of elements in a set is called its **cardinality**. The cardinality of {8, 5, 11} is 3; Cantor called the cardinality of the set of positive integers $\aleph_0$ (aleph-null), using the first letter of the Hebrew alphabet.

There are many brilliant ideas in Cantor's theory. Normally we think of deciding whether two sets have the same cardinality by counting their elements. This would not work for infinite sets, because you would never finish counting. Cantor realized that sometimes people use a different strategy; in deciding whether there are enough chairs for students, a teacher may simply ask students to sit in chairs and see if there are any students or chairs left. This led Cantor to an alternate definition of equal cardinality.

Definition

Two sets *A* and *B* have the same cardinality
$\Leftrightarrow$ there is a 1-1 correspondence between their elements.

For instance, the set {2, 4, 6, 8, 10, 12, 14, ...} of positive even integers has cardinality $\aleph_0$ because there is a 1-1 correspondence between the set of positive even integers and the set of positive integers:

$$\begin{array}{cccccc} 2 & 4 & 6 & 8 & 10 & 12\ldots \\ \updownarrow & \updownarrow & \updownarrow & \updownarrow & \updownarrow & \updownarrow \\ 1 & 2 & 3 & 4 & 5 & 6\ldots \end{array}$$

This means that infinite sets can have the same cardinality as some of their subsets, a fact noticed also by Galileo. If the elements of an infinite set can be listed in an order, then there is an automatic 1-1 correspondence between the set and the set of positive integers; the number of the element in the list is its corresponding integer. Such sets are called **countably infinite**. A set that is either finite or countably finite is called **countable**. Notice that a set is a discrete set if and only if it is countable.

Surprises arise from this definition. Some sets we think of as much bigger than the set of integers turn out to be countable. For instance, the positive rational numbers can be listed in order of the sum of their numerator and denominator when they are written in lowest terms. Here is the beginning of such a list. The numbers are ordered by the sum of their numerator and denominator; if two numbers have the same sum, the one with the smaller numerator is placed first.

$$\frac{1}{1}, \frac{1}{2}, \frac{2}{1}, \frac{1}{3}, \frac{3}{1}, \frac{1}{4}, \frac{2}{3}, \frac{3}{2}, \frac{4}{1}, \frac{1}{5}, \frac{5}{1}, \frac{1}{6}, \frac{2}{5}, \frac{3}{4}, \frac{4}{3}, \frac{5}{2}, \frac{6}{1}, \ldots$$

Making Connections You might ask students to try to give a graphical description of what is meant by such statements as $\lim_{x \to 0} \log_b x = -\infty$ and $\lim_{x \to \infty} f(x) = 1$. (Whichever large number *L* you choose, $\exists$ a value *h* such that $\forall\, x < h$, $\log_b x < -L$. Whichever small positive value *L* you choose, $\exists$ a value *h* such that whenever $x > h$, $|f(x) - 1| < L$.)

Alternate Approach Another way to show the countability of the set of rational numbers is shown below.

Every rational number can be written in the form $\frac{a}{b}$, where *a* and *b* are integers, and all these numbers can be put in an array, with $\frac{a}{b}$ in the *a*th column and *b*th row.

1	2	3	4	5	6	7	...
$\frac{1}{2}$	$\frac{2}{2}$	$\frac{3}{2}$	$\frac{4}{2}$	$\frac{5}{2}$	$\frac{6}{2}$	$\frac{7}{2}$	...
$\frac{1}{3}$	$\frac{2}{3}$	$\frac{3}{3}$	$\frac{4}{3}$	$\frac{5}{3}$	$\frac{6}{3}$	$\frac{7}{3}$	...
$\frac{1}{4}$	$\frac{2}{4}$	$\frac{3}{4}$	$\frac{4}{4}$	$\frac{5}{4}$	$\frac{6}{4}$	$\frac{7}{4}$	...
$\frac{1}{5}$	$\frac{2}{5}$	$\frac{3}{5}$	$\frac{4}{5}$	$\frac{5}{5}$	$\frac{6}{5}$	$\frac{7}{5}$	...
$\frac{1}{6}$	$\frac{2}{6}$	$\frac{3}{6}$	$\frac{4}{6}$	$\frac{5}{6}$	$\frac{6}{6}$	$\frac{7}{6}$	...
...	...	...	...	...	...	...	...

All the positive rational numbers may now be ordered according to the following scheme: in the above array, draw a continuous, broken line that goes through all the numbers in the array. Starting at 1, go horizontally to the next place on the right, obtaining 2, then diagonally down to the left to $\frac{1}{2}$, then vertically down one place to $\frac{1}{3}$, diagonally up to $\frac{2}{2}$, and so on as shown on the following page.

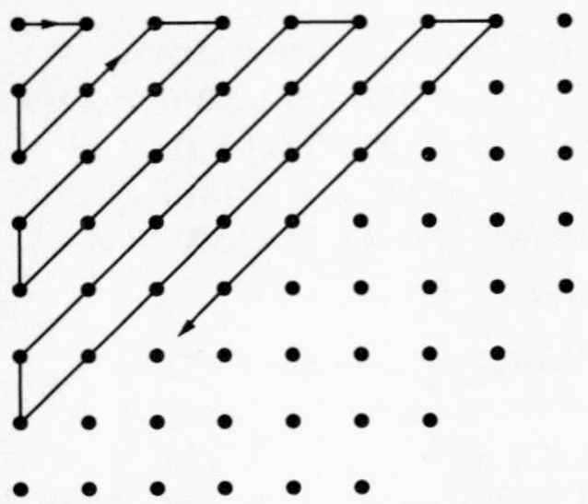

The sequence 1, 2, $\frac{1}{2}$, $\frac{1}{3}$, $\frac{2}{2}$, 3, 4, $\frac{3}{2}$, $\frac{2}{3}$, $\frac{1}{4}$, $\frac{1}{5}$, $\frac{2}{4}$, $\frac{3}{3}$, $\frac{4}{2}$, 5, ... contains the rational numbers in the order in which they occur along the broken line. Now delete all those numbers $\frac{a}{b}$ in this sequence for which a and b have a common factor, so that each rational number r will appear exactly once and in its simplest form. This results in the sequence 1, 2, $\frac{1}{2}$, $\frac{1}{3}$, 3, 4, $\frac{3}{2}$, $\frac{2}{3}$, $\frac{1}{4}$, $\frac{1}{5}$, 5, ... which contains each positive rational number once and only once. This shows that the set of all positive rational numbers is countable.

NOTES ON QUESTIONS

Question 11: There is a tacit assumption in this question that time will extend infinitely into the future.

ADDTIONAL ANSWERS

5. Consider a countably infinite hotel in which all the rooms are filled. Thus, there are $\aleph_0$ people in the hotel. Suppose 100 new people want to check in. Move the person currently in room 1 to room 101, the person in room 2 to room 102, and in general, move the person in room *n* to room *n* + 100. Then, put the new guests in rooms 1-100. Therefore, $\aleph_0 + 100 = \aleph_0$.

Since every positive rational number must appear somewhere on this list, the set of positive rational numbers is countable and its cardinality is $\aleph_0$.

Through an ingenious construction, Cantor proved that not all sets are countable. Specifically, the set of real numbers is **uncountable**. That is, no 1-1 correspondence is possible between the set of reals and the set of positive integers. His proof relies on the fact that every real number can be written as a decimal. It uses indirect reasoning, as follows:

Suppose there were a list of all the real numbers. Perhaps it would start as follows

1st	3.**1**497852345...
2nd	2.0**0**00000000...
3rd	687.88**5**5885588...
4th	3.141**5**925635...
5th	18.7500**0**00000...
6th	0.00000**0**0286...

and so on, where each decimal is thought of as an infinite decimal. (If the decimal terminates, write zeros for all of its remaining digits.) Now we show that there is a real number r that is not on the list. Let r be any number whose 1st decimal place is different from the 1st decimal place in the 1st number, whose 2nd decimal place is different from the 2nd decimal place in the 2nd number, and so on. One such number is $r = 0.216611...$ (The digits that are different from those in r are shown above in boldface.) Since r is a real number that differs from every number on the list, the list does not contain all real numbers. Since this argument can be used with any list of real numbers, no list can include all the reals.

Consequently, the number of real numbers is infinite, but this is a different infinity from $\aleph_0$. Often the letter ***c*** (for continuum) is used to represent the cardinality of the reals. c is larger than $\aleph_0$.

Cantor went on to prove that there are infinities larger than c, and indeed, that there is an infinite sequence of infinities each larger than the previous!

Cantor also developed an arithmetic of these infinities, called **transfinite arithmetic**. As an example of transfinite arithmetic, consider a hotel with a countably infinite number of rooms that are all filled. There are thus $\aleph_0$ people in the hotel. Now suppose a new person wants to check in. Is there space? Of course there is. Simply move the person currently in room 1 to room 2, the person in room 2 to room 3, and in general move the person in room n to room $n + 1$. Then put the new guest in room 1. This example illustrates the transfinite arithmetic fact

$$\aleph_0 + 1 = \aleph_0.$$

Clearly, transfinite arithmetic does not operate in the same way as typical arithmetic. In general, what holds for finite sets does not necessarily hold for infinite sets, and vice versa.

138

Questions

Covering the Reading

In 1–3, give the cardinality of the set

1. the set of negative integers $\aleph_0$

2. the set of rational numbers $\aleph_0$

3. the set of positive real numbers c

4. Give the next ten numbers in the list of positive rational numbers begun on page 137. $\frac{1}{7}, \frac{3}{5}, \frac{5}{3}, \frac{7}{1}, \frac{1}{8}, \frac{2}{7}, \frac{4}{5}, \frac{5}{4}, \frac{7}{2}, \frac{8}{1}$

5. Use the infinite hotel to explain why $\aleph_0 + 100 = \aleph_0$. See margin.

6. Show that the set of positive odd integers has cardinality $\aleph_0$. See margin.

7. Use the result of Question 6 to aid in demonstrating that $\aleph_0 + \aleph_0 = \aleph_0$. See margin.

8. Suppose the list of rational numbers in this lesson were used to try to list all the real numbers. Give the first six decimal places of a real number that is not on the list. sample: 0.31141

9. Draw two segments of different lengths. Show how to form a 1-1 correspondence between their points. (In so doing, you have proved that any two segments have the same cardinality.) See margin.

10. How does this diagram help to prove that the number of points on an open semicircle is uncountable? See margin.

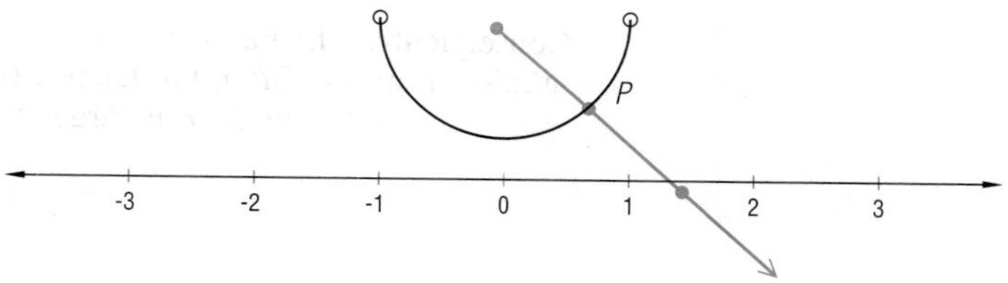

11. What assumption about time underlies the following words from the hymn "Amazing Grace," written by John Newton? Explain your answer.

When we've been there ten thousand years
Bright shining as the sun,
We've no less days to sing God's praise
Than when we've first begun.

Time has cardinality c, and c + 10,000 = c.

Review

12. Analyze the function $y = 2\cos(3x)$. Use the format of Example 3 of Lesson 2-6. *(Lesson 2-8)* See margin.

6. The set of positive odd integers has a cardinality of $\aleph_0$ because there is a 1-1 correspondence between the set of positive odd integers and the set of positive integers.

1	3	5	7	9	11
↑	↑	↑	↑	↑	↑
1	2	3	4	5	6

7. Consider the set of positive odd integers and the set of positive even integers. Both sets have a cardinality of $\aleph_0$. Combining the two sets into one set forms the set of positive integers which also has a cardinality of $\aleph_0$. Therefore, $\aleph_0 + \aleph_0 = \aleph_0$.

9. Take any two line segments, such as $\overline{AB}$ and $\overline{CD}$ below. Extend a line through *A* and *C*, then another through *B* and *D*. Since $\overline{AB}$ and $\overline{CD}$ have different lengths, these lines intersect at some point *P*. To establish a 1-1 correspondence, pair each point *E* on $\overline{AB}$ with the intersection of $\overleftrightarrow{PE}$ and $\overline{CD}$.

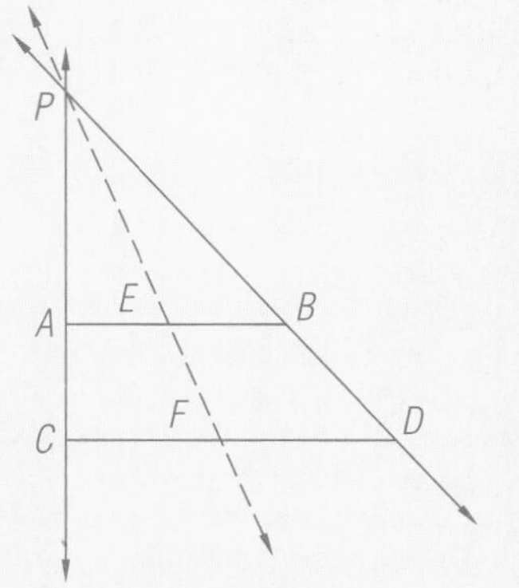

10. A 1-1 correspondence can be established between the points on the semicircle and the points on the number line as follows: let *P* be any point on the semicircle. Draw a ray from the center of the semicircle through *P*, then pair *P* with the point of intersection of that ray and the number line.

12. See the margin on page 140.

NOTES ON QUESTIONS

Questions 15-17: These questions can be interpreted in terms of graphs by asking whether there exists a horizontal line that intersects the graph of the key function in more than one point. For the exponential and logarithm functions (Questions 15 and 16), the answer is *no* (unless $b = 1$, the single counterexample that makes Question 15 false); for the cosine function (Question 17), the answer is *yes.*

ADDITIONAL ANSWERS

12. usual domain: set of real numbers; range: $-2 \le x \le 2$; relative maximum: 2; relative minimum: -2; decreasing: $\frac{2n\pi}{3} \le x \le \frac{(2n+1)\pi}{3}$, ∀ integers n; increasing: $\frac{(2n-1)\pi}{3} \le x \le \frac{2n\pi}{3}$, ∀ integers n; end behavior: the limit does not exist. graph:

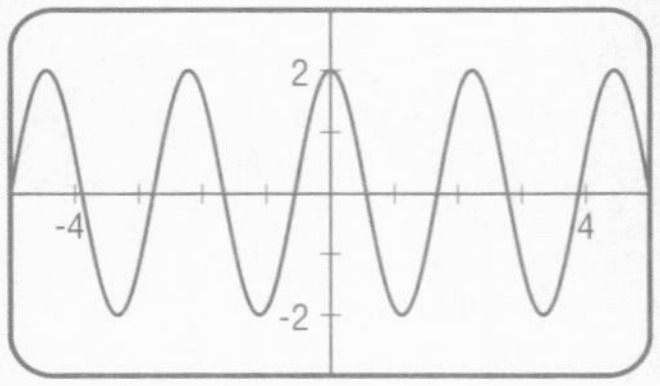

model: sound waves, phenomena based on rotation; special properties: even function, periodic function

14.a. Let a and b be positive real numbers both unequal to 1. Setting $x = b$ in the Change of Base Theorem, $\log_b a \cdot \log_a b = \log_b b = 1$. Thus, $\log_a b = \frac{1}{\log_b a}$.

18.a.

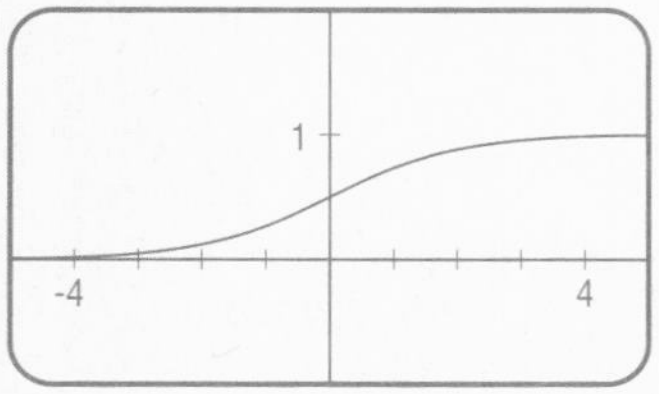

13. In triangle ABC, angle B is 60°, angle C is 50°, and side b is 10 yards. Find the length of side a. *(Lesson 2-8)* 10.85 yards

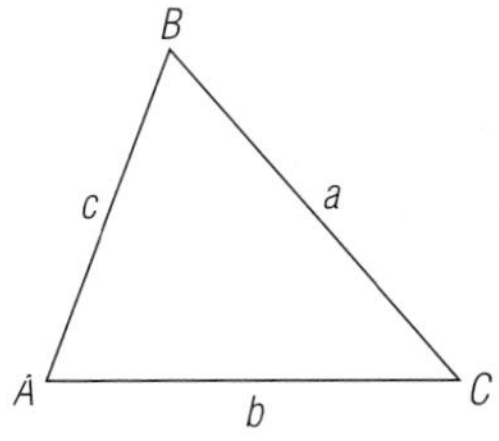

14. a. Use the change of base theorem to prove

$$\log_a b = \frac{1}{\log_b a}$$

for any positive numbers, a and b, both unequal to 1.

b. Use the result in **a** to evaluate $\log_{17} e$. *(Lesson 2-7)* $\log_{17} e \approx .3530$

a) See margin.

In 15–17, *true* or *false? (Lessons 2-6, 2-7, 2-8)*

15. ∀ real numbers x and y and positive real number b, $b^x = b^y \Rightarrow x = y$. False

16. ∀ positive real numbers x, y, and b, $\log_b x = \log_b y \Rightarrow x = y$. True

17. ∀ real numbers x and y, $\cos x = \cos y \Rightarrow x = y$. False

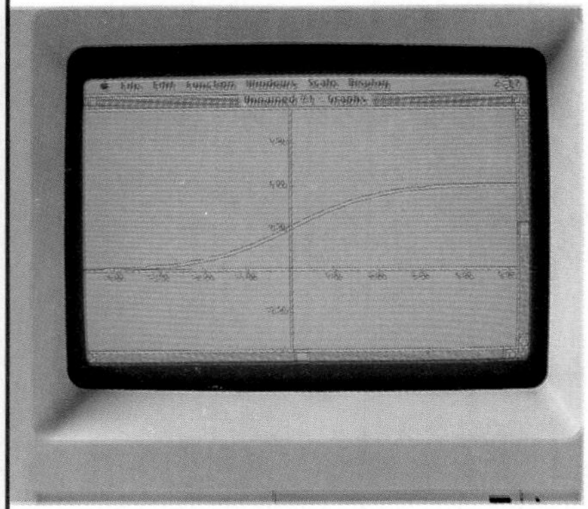

18. Use an automatic grapher to answer the following questions about the function defined by $f(x) = \frac{e^x}{1 + e^x}$.

a. Print out or sketch the graph of the function. See margin.

b. On what interval(s) is f increasing and on what interval(s) is f decreasing? increasing over entire domain

c. Find any relative maxima or minima. none

d. Describe the end behavior of f. $\lim_{x \to \infty} f(x) = 1$; $\lim_{x \to -\infty} f(x) = 0$

e. Estimate the range of f. $\{y: 0 < y < 1\}$

f. Does f appear to be even, odd, or neither? neither

(Lessons 2-2, 2-3, 2-5)

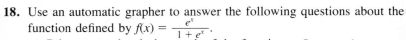

19. Prove: If the first term of an arithmetic sequence is odd, and the constant difference is even, then all terms of the sequence are odd. (Hint: use the explicit formula for an arithmetic sequence.) *(Lessons 1-7, 2-4)* See margin.

20. Consider the sequence given by the formula:

$$a_n = 4 + \frac{(-1)^n}{n}$$

a. Give the domain and formula for this sequence. See margin.

b. Graph the first ten terms of the sequence. See margin.

c. If $\lim_{n \to \infty} a_n = L$, find L. 4

d. How large must n be so that a_n is within .02 units of L? 50

(Lesson 2-4)

140

21. On a piano, the note that is n keys above "middle A" (the A above middle C) has frequency $f(n) = 440 \cdot 2^{n/12}$. (A negative value for n means the key is below middle A.)
  a. Is f a discrete function?
  b. Determine the location of the note with frequency 329.6.
  (Lessons 2-1, 2-7) **5 keys below "middle A"**
  a) Yes, the domain is a discrete set, a set of 88 integers.

22. In a study of memory, a sample of people studied a topic for 3 hours and were then tested on it monthly. The function defined by

$$f(x) = 24 - 10\log(x + 1)$$

approximates the group's average score $f(x)$ after x months. Sketch a graph of the function. *(Lesson 2-7)* **See below.**

23. Sketch the graph of an even function h which is decreasing on the interval $2 \le x \le 4$, has a relative minimum at $x = 1$, and for which $\lim_{x \to \infty} h(x) = -5$. *(Lessons 2-3, 2-5)* **See below.**

24. Refer to the function g with domain $-4 \le x \le 4$, graphed at right.
  a. Estimate $g(-2)$. **-3**
  b. Estimate $g(1)$. **5**
  c. Estimate x such that $g(x) = 0$. **-1.5 and 2.5**
  d. Estimate x such that $g(x) \ge 2$. **$-1 \le x \le 2$**
  e. Estimate the range of g.
  (Lessons 2-1, 2-2, 2-3)
  $-4 \le y \le 5$

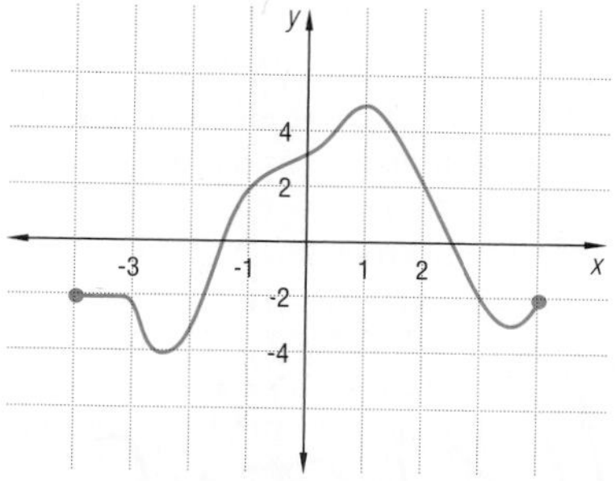

Exploration

25. Georg Cantor has lent his name to a set called the *Cantor set*. Find out what this set is and describe its special properties. **The Cantor perfect set is the subset of the real interval [0, 1] consisting of all numbers of the form $\sum_{i=1}^{\infty} \frac{e_i}{3^i}$, where e_i is 0 or 2. Geometrically, it may be described as follows: one removes from [0, 1] its middle third interval $(\frac{1}{3}, \frac{2}{3})$, then the middle thirds of $[0, \frac{1}{3}]$, $[\frac{2}{3}, 1]$, and so on. Among the properties of this set is that it is nowhere dense in the real line but does have cardinality c.**

22.

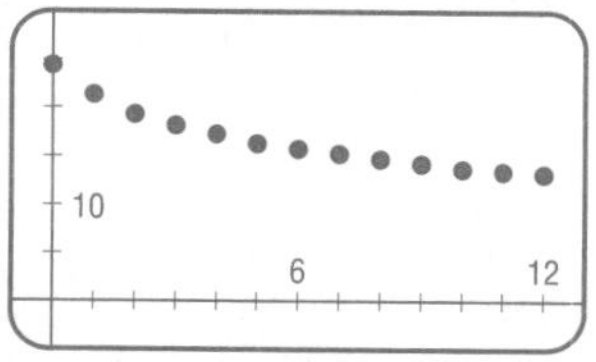

23. sample:

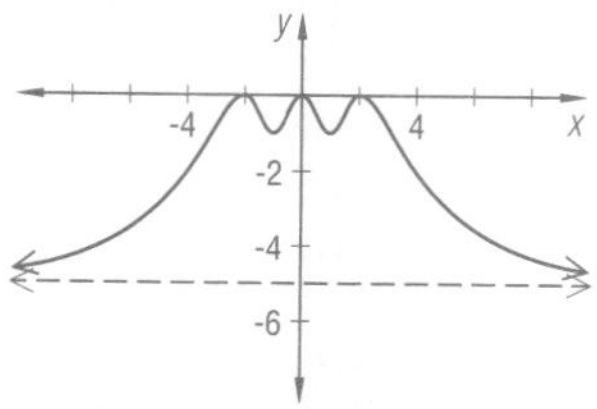

FOLLOW-UP

PROJECTS
The projects for Chapter 2 are described on page 142. **Project 3** is related to the content of this lesson.

ADDITIONAL ANSWERS
19. Let the first term of the arithmetic sequence, a_1, be any odd integer. There exists an integer r such that $a_1 = 2r + 1$ by definition of odd. By definition of even, there exists an integer s such that the constant difference, d, is 2s. The explicit form of an arithmetic sequence is $a_n = a_1 + (n - 1)d$. So $a_n = (2r + 1) + (n - 1)2s = 2r + 1 + 2sn - 2s = 2(r + sn - s) + 1$, which is odd since $r + sn - s$ is an integer.

20.a. domain: set of positive integers; $f(x) = 4 + \frac{(-1)^x}{x}$
b.

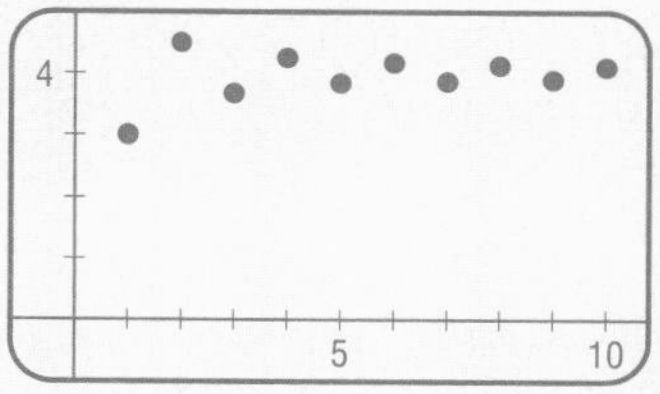

PROJECTS

The projects provide an opportunity for extended work on topics related to the content of the chapter. (See pages 70 and T37 for more information on the purposes of projects.) Some or all of these projects may be suitable for **small group work**. Solutions for the projects can be found at the end of the Additional Answers section in the back of this book.

Projects

1. Some functions have more than one independent variable. Graphing functions of the form $z = f(x, y)$ yields three-dimensional objects. Use software capable of graphing in three dimensions to do the following.
   a. Sketch the graph of the functions below.
   $$f(x, y) = x^2 + y^2$$
   $$f(x, y) = \sqrt{x^2 + y^2}$$
   $$f(x, y) = \sin(x + 2y)$$
   b. Experiment with some other functions with two independent variables, and see what shapes you get.
   c. Find or formulate definitions for *relative maximum*, *relative minimum*, and *saddle point* for such functions.

2. **a.** When Rhonda begins college she has \$7053 in an account paying 8.6% annual interest compounded continuously. She plans to use the money to help pay her tuition at the beginning of each of her 4 years of college, taking out an equal amount each year and leaving the remainder in the account. If she plans to use all the money, how much should she withdraw each year?
   b. The situation in **a** is similar to that of a person planning for retirement. Suppose a person retires with d dollars in an account earning r% annual interest compounded continuously. The money is to be withdrawn in equal amounts each month for y years, at which time it will be gone. How much should each monthly withdrawal be? Write a computer program to answer this question. Have tables printed showing monthly balances for two different situations.

3. Lesson 2-9 discusses George Cantor's work on infinity and transfinite numbers. Find out more and write an essay on transfinite arithmetic. Discuss the relationship between $\aleph_0$, $\aleph_1$, $\aleph_2$ and describe some sets with these numbers as their cardinalities. The following sources may be helpful: *The Main Stream of Mathematics*, by Edna Kramer; *What is Mathematics?* by Richard Courant and Herbert Robbins; *One, Two, Three ... Infinity*, by George Gamow.

4. A sequence of snowflakes is formed in the following manner. An equilateral triangle with area 1 is drawn. Then, an equilateral triangle is drawn outward on the middle third of each side. Then, an equilateral triangle is drawn outward on the middle. This is repeated forever. Let $A(n)$ be the area of the figure obtained at the nth step of this procedure.

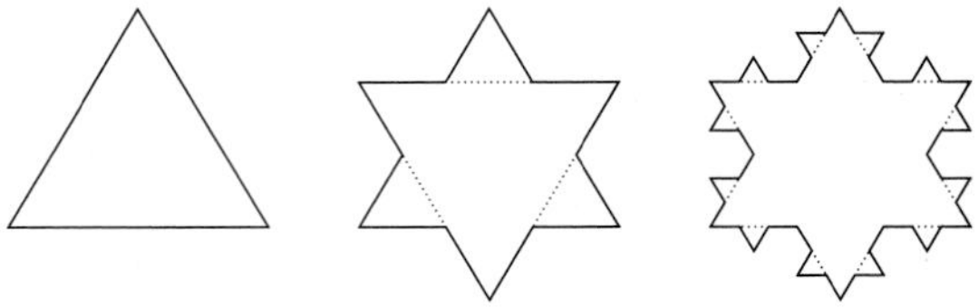

$A(1) = 1 \quad A(2) = 1 + 3 \cdot \frac{1}{9} = \frac{4}{3} \quad A(3) = ?$

   a. Write a computer program to calculate $A(n)$ for $n = 1, 2, 3, \ldots, 100$.
   b. Graph A for $n \le 10$.
   c. Describe the behavior of A as $n \to +\infty$. What does this mean in terms of the snowflake?
   d. Find a formula for $A(n)$ and use it to explain the behavior you noticed in part **c**.
   e. Make up a similar sequence using a geometric figure of your own choice, and repeat the above investigation for it.

5. Investigate how a logarithmic scale is used to measure some quantity such as star brightness, earthquake intensity, or acidity. Find out what physical characteristics underlie the quantity, how the quantity is measured, how it is "transformed" to a logarithmic scale, and why or how the logarithmic scale is more useful than the original scale. What is the Weber-Fechner law in physiology, and how does it relate to these issues?

6. Find out what figurate numbers are. Report on the recursive and explicit formulas for sequences of figurate numbers, and on the relationships that exist among these sequences.

142

Summary

A function f is a correspondence from a set A (its domain) to a set B which assigns to every element of the domain exactly one element of B. If the domain of f can be put into a 1-1 correspondence with a subset of the set of integers, then f is a discrete function. A sequence is a discrete function because its domain is the set of positive integers from a to b or the set of all positive integers greater than or equal to a fixed integer. The set of elements of B that the function uses is its range. If the domain and range of f both consist of real numbers, then f is a real function.

Given a formula for a function, an automatic grapher can quickly give its graph in a given window. It can also zoom in on parts of the graph and obtain the approximate coordinates of points on the graph.

Many properties of a function f can be inferred from its graph and proved algebraically. We call the investigation of these properties "analyzing the function." It includes studying the following:

1. the domain of f.
2. the range of f.
3. the intervals on which f is increasing and decreasing.
4. maximum and minimum values of f. If a value of $f(x)$ is a maximum or minimum only over a particular interval, it is called a relative maximum or minimum.
5. the end behavior of f; that is, the limit of $f(x)$ as x becomes larger or smaller without bound. There are three possibilities: If $f(x)$ can be made as close to some number L as desired by taking x large enough, we write $\lim_{x \to \infty} f(x) = L$. If $f(x)$ can be made as large (or small) as desired by taking x large enough, we write $\lim_{x \to \infty} f(x) = \infty$ $\left(\text{or } \lim_{x \to \infty} f(x) = -\infty\right)$. If neither of these is the case, then we say $\lim_{x \to \infty} f(x)$ does not exist. $\lim_{x \to -\infty} f(x)$ is defined analogously. Since an infinite sequence is a function defined on the set of positive integers, its end behavior is defined in the same way.
6. the shape of the graph.
7. situations which are modeled by f.
8. special properties unique to f or functions of the same type.

The graph of a function also can confirm whether a function is even or odd. Graphs of even functions are reflection-symmetric with respect to the y-axis; graphs of odd functions are rotation-symmetric with respect to the origin.

Exponential functions arise from situations of constant growth. When the base of an exponential function f is greater than 1, f is increasing, $\lim_{x \to -\infty} f(x) = 0$, and $\lim_{x \to \infty} f(x) = \infty$.

Logarithmic functions are inverses of exponential functions. When the base of a logarithmic function g is greater than 1, g is increasing and $\lim_{x \to \infty} g(x) = \infty$.

The trigonometric functions sine and cosine can be defined in terms of right triangles or in terms of the unit circle. As real functions, they are periodic and their range is the interval $-1 \le y \le 1$. They are alternately increasing and decreasing on intervals of length π, and have relative maxima and minima at regular intervals. The limits of sine and cosine as $x \to \infty$ and as $x \to -\infty$ do not exist.

SUMMARY

The Summary gives an overview of the entire chapter and provides an opportunity for students to consider the material as a whole. Thus, the Summary can be used to help students relate the various concepts presented in the chapter.

RESOURCES
- Chapter 2 Test, Form A
- Chapter 2 Test, Form B
- Chapter 2 Test, Cumulative Form
- Teaching Aid 13 can be used with Chapter Review **Questions 45-49**.
- Quiz and Test Writer

CHAPTER REVIEW

The main objectives for the chapter are organized here into sections corresponding to the four main types of understanding this book promotes: Skills, Properties, Uses, and Representations. We call these the SPUR objectives.

USING THE CHAPTER REVIEW
Students should be able to answer questions like these with about 85% accuracy by the end of the chapter. (See pages 74 and T38–39 for more information.)

Chapter Review

Questions on SPUR Objectives

See margin for answers not shown below.

SPUR stands for **S**kills, **P**roperties, **U**ses, and **R**epresentations.
The Chapter Review questions are grouped according to the SPUR Objectives for this chapter.

SKILLS deal with the procedures used to get answers.

Objective A: *Determine relative minima and maxima of a function and intervals on which it is increasing or decreasing.* (*Lessons 2-3, 2-4, 2-8*)

In 1 and 2, refer to the table below which gives total enrollment (in thousands) in U.S. public elementary and secondary schools. Let E be the function mapping year to enrollment.

Year	Enrollment
1900	15,503
1910	17,814
1920	21,578
1930	25,678
1940	25,434
1950	25,111
1960	36,087
1970	45,619
1980	41,645
1987	40,024

Source: *World Almanac and Book of Facts* 1989.

1. a. Find the intervals on which E is increasing and on which E is decreasing.
b. Note that $E(1900) < E(1940)$. Is E increasing on the interval $1900 \le x \le 1940$? Justify your answer.

2. a. Find all relative minimum and maximum values of E and give the years at which they occur.
b. Choose one of the relative minima from part **a**, and find an interval (containing the year in which the relative minimum occurs) on which E is greater than or equal to that minimum value.
$1930 \le x \le 1950$

3. Consider the sequence $s_n = \sin \frac{n\pi}{2}$.
a. Find s_n for all integers n in the interval $0 \le n \le 8$. 0, 1, 0, -1, 0, 1, 0, -1, 0
b. Find an interval of longest possible length on which s is increasing. $3 \le n \le 5$
c. Find an interval of longest possible length on which s is decreasing. $1 \le n \le 3$
d. Find the first three values of n at which s has a relative maximum. $n = 1, 5, 9$
e. Find the first three values of n at which s has a relative minimum. $n = 3, 7, 11$

4. Consider the function $f(t) = -3t^2 + 4t + 1$.
a. Find the intervals on which f is increasing and on which f is decreasing.
b. Find any relative minimum and maximum values of f. relative maximum: $\frac{7}{3}$

In 5–7, **a.** identify the sequence as geometric, arithmetic, harmonic, or none of these; and **b.** determine whether it is increasing, decreasing, or neither.

5. $a_n = 5 - 2n \ \forall n$ a) arithmetic; b) decreasing

6. $c_n = \cos n\pi \ \forall n$ a) none of these; b) neither

7. $\begin{cases} b_1 = 1 \\ b_{k+1} = \frac{k}{k+1} b_k \ \forall \ k > 1 \end{cases}$ a) harmonic b) decreasing

8. Fill in the blanks. If a function g is decreasing on an interval $a \le x \le b$ and increasing on the interval $b \le x \le c$, then g has a __?__ at b.
relative minimum

146

Objective B: *Rewrite exponential and logarithmic expressions and equations.* *(Lesson 2-7)*

In 9–13, use the definition or properties of logarithms to rewrite the equation and solve.

9. $\log_2 8 = x$ $2^x = 8, x = 3$

10. $\log_2 x = 8$ $2^8 = x, x = 256$

11. $\log_b 9 = 2$ $b^2 = 9, b = 3$

12. $6^t = 42$ $t = \log_6 42, t = 2.086$

13. $\log_3 2z - \log_3 5 = -2$ $3^{-2} = \frac{2z}{5}, z = \frac{5}{18}$

In 14 and 15, rewrite the expression in terms of $\log_{10} N$, $\log_{10} M$, and $\log_{10} P$.

14. $\log_{10} \frac{N^2 \bullet M^3}{P}$ $2\log_{10}N + 3\log_{10}M - \log_{10}P$

15. $\log_{10} \left(\frac{N \bullet M^2}{P^3}\right)^{1/2}$ $\frac{1}{2}\log_{10}N + \log_{10}M - \frac{3}{2}\log_{10}P$

PROPERTIES deal with the principles behind the mathematics.

Objective C: *Identify the domain, range, and minimum and maximum values of functions.* *(Lessons 2-1, 2-2)*

In 16 and 17, refer to the table below which gives student prices for a series of symphony orchestra concerts based on the row assignment. Let R be the set of row numbers and S be the set of series prices.

Row Number	Series Price
1-2	$18.00
3-13	$56.25
14-25	$45.00

16. For each s in S, let $f(s)$ = a particular row number associated with series price s. Is $f: S \rightarrow R$ a function? Why or why not? If it is a function, tell whether it is a discrete function and identify its domain, range, minimum value, and maximum value.

17. For each r in R, let $g(r)$ = the series price associated with row r. Is $g: R \rightarrow S$ a function? Why or why not? If it is a function, tell whether it is a discrete function and identify its domain, range, minimum value, and maximum value.

18. For $0 \le s \le 80$, the equation $f(L) = s = 2\sqrt{5L}$ relates the speed s of a car in miles per hour and the length L of the skid marks in feet if the brakes have been fully applied.

a. Is f a discrete function? Explain your answer.

b. Identify the domain. $\{L: 0 \le L \le 320\}$

c. Identify the range of f. $\{s: 0 \le s \le 80\}$

In 19–21, a function f is defined according to the indicated rule. Identify the domain.

19. $f(p) = 4p^3 - 2p^2 + 3p + 1$ all real numbers

20. $f(r) = \frac{10r}{\sqrt{30 + 3r}}$ $\{r: r > -10\}$

21. $f(z) = 3\sqrt{z - 7}$ $\{z: z \ge 7\}$

In 22–24, suppose g is defined by $g(x) = 2x - 1$ and has the indicated domain. Find the minimum and maximum values of g (if they exist) and give the range of g.

22. {-1, -5, 0, 5, 10}

23. the set of positive integers

24. the set of real numbers

25. Find the range of $h: t \rightarrow -4t^2 + 6t - 7$ if the domain is

a. the interval $-1 \le t \le 2$ $\{y: -17 \le y \le -\frac{19}{4}\}$

b. the set of real numbers $\{y: y \le -\frac{19}{4}\}$

26. A function k is defined by $k(x) = x^3 + 1$ on the domain $5 \le x \le 10$. What is the range?
$\{y: 126 \le y \le 1001\}$

Objective D: *Determine the end behavior of a function.* *(Lessons 2-4, 2-5, 2-6, 2-7, 2-8)*

27. Describe the end behavior of $f: x \rightarrow 2e^x$.

28. Describe the end behavior of the cosine function.   No limits exist as $x \rightarrow \infty$ or as $x \rightarrow -\infty$.

ADDITIONAL ANSWERS

1.a. increasing: $1900 \le x \le 1930$, $1950 \le x \le 1970$; decreasing: $1930 \le x \le 1950$, $1970 \le x \le 1987$

b. No, E is not increasing over the entire interval, because E is decreasing over the interval $1930 \le x \le 1940$.

2.a. relative minimum value: 25,111 in 1950; relative maximum value: 25,678 in 1930 and 45,619 in 1970

4.a. increasing: $x \le \frac{2}{3}$; decreasing: $x \ge \frac{2}{3}$

16. No, each element in S corresponds to more than one element in R.

17. Yes, each element in R corresponds to exactly one element in S. discrete function; domain: $\{r: 1 \le r \le 25$, r an integer$\}$; range: {18.00, 56.25, 45.00}; minimum value: $18.00; maximum value: $56.25

18.a. No, L can be any real number between 0 and 320.

22. minimum value: -11; maximum value: 19; range: {-11, -3, -1, 9, 19}

23. minimum value: 1; maximum value: none; range: set of positive odd integers

24. minimum value: none; maximum value: none; range: set of real numbers

27. As $x \rightarrow \infty$, $f(x) \rightarrow \infty$; as $x \rightarrow -\infty$, $f(x) \rightarrow 0$.

ADDITIONAL ANSWERS

33.a. $\lim_{x\to\infty} f(y) = -4$;
$\lim_{x\to\infty} f(y) = -4$
b. $y = -4$

34.a. $\lim_{x\to\infty} g(y) = -\infty$;
$\lim_{x\to-\infty} g(y) = -\infty$
b. none

36.a. $\lim_{x\to\infty} f(x) = 0$;
$\lim_{x\to-\infty} f(x) = \infty$
b. $\lim_{x\to\infty} f(x) = 1$; $\lim_{x\to-\infty} f(x) = 1$
c. $\lim_{x\to\infty} f(x) = \infty$; $\lim_{x\to-\infty} f(x) = 0$

43.b. $f(p) = 150p - 10p^2 - 500$

ADDITIONAL ANSWERS
(For Questions on p. 150)
47.a. increasing: $x \le -4$; $-3 \le x \le -2$, $2 \le x \le 3$, $x \ge 4$; decreasing: $-4 \le x \le -3$, $-1 \le x \le 1$, $3 \le x \le 4$
b. relative minima: 1.5 occurs at $x = -3$, -3 occurs at $x = 4$, 2 for all x such that $-2 \le x \le -1$, and -2 for all x such that $1 \le x \le 2$; relative maxima: 3 occurs at $x = -4$, -1.5 occurs at $x = 3$, 2 for all x such that $-2 \le x \le -1$, and -2 for all x such that $1 \le x \le 2$
c. odd

51. sample:

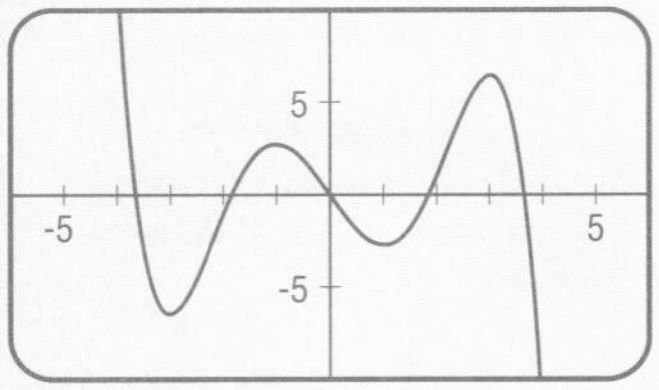

52.a.

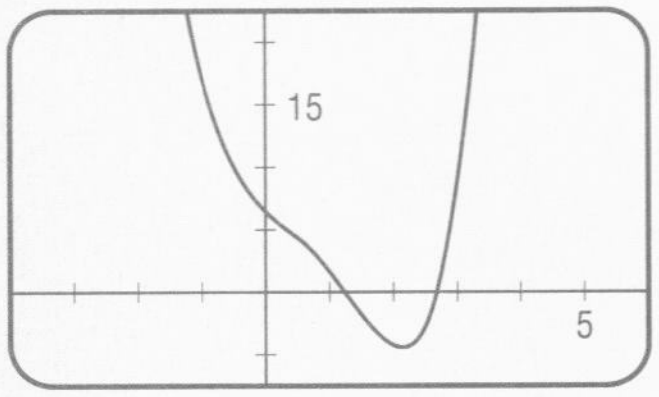

29. Consider the sequence a given in Question 5.
a. Find the values of n for which
i. $a_n < -100$ $n \ge 53$
ii. $a_n < -1000$ $n \ge 503$
b. Find $\lim_{x\to\infty} a_n$. $-\infty$

In 30 and 31, find the limit of the sequence given in

30. Question 6; The limit does not exist.

31. Question 7. 0

32. Let $h(x) = x \sin\left(\frac{1}{x}\right)$.
a. Use a calculator to estimate $\lim_{x\to\infty} h(x)$. 1
b. For what values of x is $h(x)$ within .001 of that limit? $|x| \ge 13$
c. Estimate $\lim_{x\to-\infty} h(x)$. 1
d. Find the equations of all horizontal asymptotes. $y = 1$

In 33 and 34, **a.** describe the end behavior of the given function, and **b.** give the equations of any horizontal asymptotes.

33. $f(y) = -4 - \frac{1}{y^3}$

34. $g(y) = -4y^4 - \frac{1}{y^3}$

35. Suppose the function h is odd and $\lim_{z\to\infty} h(z) = 10$. Find $\lim_{z\to\infty} h(z)$. -10

36. Describe the end behavior of the function $f(x) = b^x$ for
a. $b = .99$.
b. $b = 1.00$.
c. $b = 1.01$.

USES deal with applications of mathematics in real situations.

Objective E: *Use trigonometric, exponential, and logarithmic functions as models.*
(Lessons 2-6, 2-7, 2-8)

37. A sheet of paper is 0.1 mm thick. When it is folded in half once, the folded sheet is 0.2 mm thick. Folding it in half again produces a stack 0.4 mm thick. Define the function g so that $g(n)$ is the thickness when folded n times.
a. Is g discrete? Yes
b. Find a formula for $g(n)$. $2^n(.1)$
c. How thick is the stack after 7 folds? 12.8 mm
d. How many folds are required to obtain a stack that at least reaches the moon, which is 380,000 km from the earth? 42

38. a. At what interest rate, compounded continuously, will \$1500 grow to \$2500 in 5 years? 10.22%
b. At what rate of simple interest, added each year, would \$1500 grow to \$2500 in 5 years? (Simple interest adds the same amount, a percent of the original investment, each year.) 13.33%

39. The pH of a chemical solution is given by
$\text{pH} = -\log x$
where x is the $[H_3O^+]$ ion concentration in moles per liter.
a. What is the pH of a solution whose $[H_3O^+]$ concentration is $2.7 \cdot 10^{-12}$ moles per liter? ≈ 11.6
b. What $[H_3O^+]$ concentration corresponds to a pH of 7? 1.0×10^{-7} moles/liter

40. Each time a liquid is passed through a filter, 73% of the impurities are removed. If a liquid starts out 60% pure, how many times must it be filtered to obtain a liquid that is at least 99.9% pure? (Hint: Study the percentage of impurity instead of the percentage of purity.) 5

41. Near an airport, suppose that an overhead airplane creates a noise level of 60 dB. How many such planes could be overhead simultaneously without surpassing the "annoyance level" of 65 dB?
(Use Loudness $= 10 \log_{10}(I \cdot 10^{12})$.) 3 planes

3-4

CHUNKING

Chunking is a term used in psychology to describe how the brain often puts together individual pieces to form a whole; for instance, as you read these sentences you are reading chunks of letters rather than individual letters. From at least as early as the time students have had to consider a two-digit numeral as a single number, they have been chunking. In this lesson, chunking is applied to an algebraic expression by thinking of it as a single variable.

3-5

APPROXIMATING SOLUTIONS TO EQUATIONS

Automatic graphers have eased the difficulty of finding approximate real solutions to equations. The underlying theory involves two ideas whose formal treatment requires calculus: continuity and the Intermediate Value Theorem. These ideas are applied informally in this lesson and are used to justify the Bisection Method for obtaining more accurate solutions.

3-6

THE LOGIC OF SOLVING INEQUALITIES

Given a graph of a continuous real function, it is very easy to determine whether the function is increasing or decreasing on an interval. One would think there could be no important applications. This lesson shows that view to be incorrect: If g were increasing on an interval containing $f(x)$ and $h(x)$, and if $f(x) < h(x)$, then $g(f(x)) < g(h(x))$. If g were decreasing, the sense of the inequality would be reversed. Thus, increasing and decreasing functions provide a general framework for discussing how to solve inequalities.

3-7

THE TEST POINT METHOD FOR SOLVING INEQUALITIES

The Test Point Method involves solving the inequality $f(x) < g(x)$ by examining values of x on intervals whose endpoints are the zeros of the function $h = f - g$. The intuition of this method comes from looking at the graph of h. Its proof involves the Intermediate Value Theorem. The execution of the Test Point Method uses the logic of *and* and *or.* This lesson brings together many of the topics that students have studied thus far in this course.

3-8

ABSOLUTE VALUE EQUATIONS AND INEQUALITIES

Another application of *and-or* logic appears in the solving of absolute value sentences. The solutions to $|x| = a$, $|x| > a$, and $|x| < a$ are applied to the more general situations $|f(x)| = g(x)$, $|f(x)| < g(x)$, and $|f(x)| > g(x)$.

3-9

GRAPHS, TRANSFORMATIONS, AND SOLUTIONS

The Graph Transformation Theorems deal with the effects on the graph of a sentence in x and y by the replacements $x - h$ and $y - k$ (translation) and by the replacements $\frac{x}{a}$ and $\frac{y}{b}$. These widely applicable theorems are used in this lesson to graph offspring of parent functions and to review the amplitude, period, and phase shift of trigonometric functions. This lesson is review for students who have studied from UCSMP *Functions, Statistics, and Trigonometry.*

3-10

SOME UNUSUAL GRAPHS

When the Fract function $F(x) = x - \lfloor x \rfloor$ and the absolute value function $A(x)$ are composed with other functions, graphs of those other functions may repeat, or parts may be reflected over the x-axis or y-axis. This reading lesson shows students how they can use the Fract and absolute value functions to create a variety of interesting graphs.

DAILY PACING CHART ■ CHAPTER 3

Every chapter of UCSMP *Precalculus and Discrete Mathematics* includes lessons, a Progress Self-Test, and a Chapter Review. For optimal student performance, the self-test and review should be covered. (See *General Teaching Suggestions: Mastery* on page T35 of this Teacher's Edition.) By following the pace of the Full Course given here, students can complete the entire text by the end of the year. Students following the pace of the Minimal Course spend more time when there are quizzes and on the Chapter Review and will generally not complete all of the chapters in this text.

When chapters are covered in full (the recommendation of the authors), then students in the Minimal Course can cover 11 chapters of the book. For more information on pacing, see *General Teaching Suggestions: Pace* on page T34 of this Teacher's Edition.

DAY	MINIMAL COURSE	FULL COURSE
1	3-1	3-1
2	3-2	3-2
3	3-3	3-3
4	Quiz (TRF); Start 3-4.	Quiz (TRF); 3-4
5	Finish 3-4.	3-5
6	3-5	3-6
7	3-6	3-7
8	3-7	Quiz (TRF); 3-8
9	Quiz (TRV); Start 3-8.	3-9
10	Finish 3-8.	3-10
11	3-9	Progress Self-Test
12	3-10	Chapter Review
13	Progress Self-Test	Chapter Test (TRF)
14	Chapter Review	Comprehensive Test (TRF)
15	Chapter Review	
16	Chapter Test (TRF)	
17	Comprehensive Test (TRF)	

TESTING OPTIONS

TESTING OPTIONS

- Quiz for Lessons 3-1 through 3-3
- Quiz for Lessons 3-4 through 3-7
- Chapter 3 Test, Form A
- Chapter 3 Test, Form B
- Chapter 3 Test, Cumulative Form
- Comprehensive Test, Chapters 1–3

A Quiz and Test Writer is available for generating additional questions, additional quizzes, or additional forms of the Chapter Test.

INDIVIDUAL DIFFERENCES

PROVIDING FOR INDIVIDUAL DIFFERENCES

The student text is written for, and tested with, average students. It also has been successfully used with better and more poorly prepared students.

The Lesson Notes often include Error Analysis and Alternate Approach features to help you with those students who need more help. Students of all abilities often learn from their peers and may benefit from small group work referenced as appropriate throughout the Notes. A blackline Lesson Master (in the Teacher's Resource File), keyed to the chapter objectives, is provided for each lesson to allow more practice. (However, since it is important to keep up with the daily pace, you are not expected to use all of these masters. Again, refer to the suggestions for pacing on page T34.) Extension activities are provided in the Lesson Notes for those students who have completed the particular lesson in a shorter amount of time than is expected, even in the Full Course.

Functions, Equations, and Inequalities

CHAPTER 3

For many years you have solved equations and inequalities. You know methods for getting exact solutions to linear equations, quadratic equations, equations of the form $a^x = b$, and some others. These methods can be combined to find exact solutions to equations like

$$y^{10} - 5y^6 + 4y^2 = 0,$$

for which you might think no method exists.

Yet for some sentences, there are no theorems or formulas that give exact solutions. For example, consider the following situation.

> The teachers in one school earn \$25,000 per year plus \$1,000 for each year of teaching experience. The salaries in another school start at \$23,000 per year and increase 4% for each year of teaching experience. After how many years (if ever) do teachers in the second school earn more than teachers in the first?

The question leads to solving an inequality relating arithmetic and geometric sequences.

$$23{,}000(1.04)^x > 25{,}000 + 1000x$$

Although there is no formula for solving this inequality, there are algebraic and graphical ways to determine the solutions.

In this chapter, the logic you encountered in Chapter 1 and the function ideas of Chapter 2 are combined to derive powerful methods for solving equations and inequalities like these, to examine why these methods work, and to explain why some other methods work some of the time but not always.

CHAPTER 3

We recommend 14–17 days be spent on this chapter: 10 to 12 days on the lessons and quizzes; 1 day for the Progress Self-Test; 1 or 2 days for the Chapter Review; 1 day for a Chapter Test, and 1 day for the Comprehensive Test, which can serve as a quarterly exam. (See the Daily Pacing Chart on page 151D).

USING PAGE 151

Both of the equations on this page are found in questions in the chapter. At this time, you may ask students if they can find any solutions to the first equation (0, 1, and -1 are three of them), and if so, can they find them all. For the inequality, note the situation from which it is derived, and ask if anyone has a feeling regarding the answer (no pencils or calculators allowed). (The answer is that after 13 years, teachers in the second school will earn more.)

Point out to students that in this chapter they will review the solving of equations and inequalities studied in previous years, and they also will learn some more general methods.

LESSON 3-1

RESOURCES
■ Lesson Master 3-1

OBJECTIVES

A Solve equations by applying a function to each side, taking into account nonreversible steps.
F Analyze the reversibility of steps used in solving equations and inequalities.
J Apply equation-solving techniques to real-world problems.

TEACHING NOTES

After students have read the text, you may want to synthesize the key vocabulary and points of the lesson. There are only two new terms: *1-1 function* and *reversible.* Ask students what these have to do with each other. Be sure to discuss the key idea that follows the definition of 1-1 function on page 154.

Making Connections
You may have to connect the informal treatment students have seen of equation-solving to the more formal treatment given in this lesson. Most students do not equate the colloquial "squaring of both sides" with the formalism of "applying the squaring function."

Emphasize also the reversibility of steps idea. Point out that just one nonreversible step among many reversible steps makes possible the loss of zeros or the introduction of extraneous zeros.

You may want to introduce the term *extraneous solution.*

LESSON

3-1

The Logic of Equation-Solving

If you were asked to find all real numbers x that satisfy the equation

$$4 - x = \sqrt{x - 2},$$

you might write the following, and give explanations as shown here.

$4 - x = \sqrt{x-2}$	Given
$(4 - x)^2 = x - 2$	Square both sides.
$16 - 8x + x^2 = x - 2$	Expand the binomial square.
$x^2 - 9x + 18 = 0$	Add $-x + 2$ to each side and simplify.
$(x - 3)(x - 6) = 0$	Factor.
$x = 3$ or $x = 6$	Zero Product Property

This work explains what you did to go from one equation to the next, but it does not tell what "$x = 3$ or $x = 6$" has to do with the original equation. The connection is found by examining the underlying logic. The six lines written above are shorthand for a sequence of five *if-then* statements.

Step 1 If $4 - x = \sqrt{x-2}$, then $(4 - x)^2 = x - 2$.
Step 2 If $(4 - x)^2 = x - 2$, then $16 - 8x + x^2 = x - 2$.
Step 3 If $16 - 8x + x^2 = x - 2$, then $x^2 - 9x + 18 = 0$.
Step 4 If $x^2 - 9x + 18 = 0$, then $(x - 3)(x - 6) = 0$.
Step 5 If $(x - 3)(x - 6) = 0$, then $x = 3$ or $x = 6$.

By the Law of Transitivity, the following conclusion can be drawn.

If $4 - x = \sqrt{x-2}$, then $x = 3$ or $x = 6$.

Note that this conclusion does not imply that 3 and 6 are solutions of the given equation. It only implies that *If a real number satisfies the equation $4 - x = \sqrt{x-2}$, then it must be either 3 or 6.* That is, the conclusion restricts the possible solutions to those two numbers. In fact, the followin; check shows that 3 is a solution, but 6 is not.

Check:	Does $4 - 3 = \sqrt{3-2}$?	Yes, so 3 is a solution.
	Does $4 - 6 = \sqrt{6-2}$?	No, so 6 is not a solution.

What has gone wrong? In order for 3 and 6 both to be solutions, the converse of the above conclusion must be true. That is, we must know *If $x = 3$ or $x = 6$, then $4 - x = \sqrt{x-2}$.* This converse will be true if the converses of each of the if-then statements in Steps 1–5 above are true for any real number x. The converses of the if-then stements in Steps 2–5 are true for any real number x. Consequently, for any such number the following if-and-only-if statements are true:

Step 2 $(4 - x)^2 = x - 2$ if and only if $16 - 8x + x^2 = x - 2$.
Step 3 $16 - 8x + x^2 = x - 2$ if and only if $x^2 - 9x + 18 = 0$.
Step 4 $x^2 - 9x + 18 = 0$ if and only if $(x - 3)(x - 6) = 0$.
Step 5 $(x - 3)(x - 6) = 0$ if and only if $x = 3$ or $= 6$.

152

Therefore, by the Law of Transitivity:
x satisfies the equation $(4 - x)^2 = x - 2$ if and only if $x = 3$ or $x = 6$.

On the other hand, the converse of the if-then statement in Step 1 is false for $x = 6$, because 6 satisfies the second equation but not the first.

These observations are often summarized by saying that Steps 2–5 are *reversible,* while Step 1 is *nonreversible*. In general, **reversible** steps in solving an equation or inequality correspond to if-and-only if statements. Equations related by reversible steps have the same solutions and are called **equivalent equations**. **Nonreversible** steps correspond to true if-then statements for which the converse is false. In the example above, squaring both sides of $4 - x = \sqrt{x - 2}$ proved to be a nonreversible step.

Squaring both sides of an equation may be a reversible step or it may be nonreversible. Thus, whenever you are trying to solve an equation and you square both sides, you need to check your answer to make sure that each value you finally obtain actually satisfies the given equation.

On the other hand, some kinds of steps are *always* reversible. For instance, adding or subtracting a number to or from both sides of an equation, and multiplying or dividing both sides by a nonzero number are always reversible steps. Also, applying an algebraic property, such as the distributive property or the Zero Product Property or a law of exponents, to simplify one or both sides of an equation is always a reversible step.

Note that multiplying both sides of an equation by an expression that can have the value zero may be a nonreversible step. For instance, attempting to solve $\frac{1}{x} + 3 = \frac{1}{x}$ by multiplying both sides by x gives the equation $1 + 3x = 1$ for which 0 is a solution. Yet, 0 does not work in the original equation, so that equation has no solution. In the language of if-then statements, *for any real number x, if* $\frac{1}{x} + 3 = \frac{1}{x}$, *then* $1 + 3x = 1$ is true, but its converse is false.

Example 1 Find all real numbers x that satisfy $\frac{1}{15 - 5x} = \frac{1}{x^2 - 4x + 3}$.

Solution Let x be any real number.

$\frac{1}{15 - 5x} = \frac{1}{x^2 - 4x + 3}$	
$\Rightarrow x^2 - 4x + 3 = 15 - 5x$	Multiply by $(15 - 5x)(x^2 - 4x + 3)$.
$\Leftrightarrow x^2 + x - 12 = 0$	Add $5x - 15$ to both sides.
$\Leftrightarrow (x + 4)(x - 3) = 0$	Factor the left side.
$\Leftrightarrow x = -4$ or $x = 3$.	Zero Product Property

Now by the Law of Transitivity, the only possible numbers that could satisfy the given equation are -4 and 3, but because the first step may not be reversible, both must be checked in the original equation:

Does $\frac{1}{15 - 5 \cdot (-4)} = \frac{1}{(-4)^2 - 4 \cdot (-4) + 3}$? Yes, so $x = -4$ is a solution.

Does $\frac{1}{15 - 5 \cdot 3} = \frac{1}{3^2 - 4 \cdot 3 + 3}$? No, because both denominators equal 0 . Thus, $x = 3$ is not a solution, and the only solution is $x = -4$.

Extraneous solutions satisfy an equation in one of the intermediate steps but fail to satisfy the original equation because some nonreversible process was used to find a solution. However, if you use this term, stress that an extraneous solution is not a solution!

Students are asked to derive the equation for **Example 4** in **Question 17.** The value $h = 0$ does not satisfy the problem because it asks for the satellite height for which the *farthest* observable distance is $2h$.

ADDITIONAL EXAMPLES

1. Solve $\sqrt{x + 14} = x + 2$.

$\sqrt{x + 14} = x + 2$
$\Rightarrow x + 14 = (x + 2)^2$
$\Leftrightarrow x + 14 = x^2 + 4x + 4$
$\Leftrightarrow x^2 + 3x - 10 = 0$
$\Leftrightarrow (x + 5)(x - 2) = 0$
$\Leftrightarrow x = -5$ or $x = 2$

Thus, -5 and 2 are possible solutions. The only solution is $x = 2$.

In Example 1, the symbol $\Rightarrow$ is used on the second line because multiplying both sides of an equation by $(15 - 5x)(x^2 - 4x + 3)$ may be nonreversible, since that expression may have the value 0. The symbol $\Leftrightarrow$ is used in each of the later lines because those steps are always reversible.

Many steps in equation solving involve applying a function to both sides of an equation. For instance, we might square both sides of an equation, cube both sides, take the logarithm of both sides, and so forth. In each case, any solution of the given equation is also a solution of the transformed equation. This is because a function assigns only one value to each element in its domain.

Theorem

Suppose that g is a real function whose domain is the set of all real numbers. If a real number is a solution of an equation $u = v$, then it is also a solution of the equation $g(u) = g(v)$.

The requirement that g is defined for all real numbers is *necessary* in order to have a reversible step, as the logarithm function in Example 3 below shows. However, as we have seen from the squaring function that opened this lesson, this requirement is not *sufficient*. To assure reversibility, g must be a *one-to-one function*.

Definition

A function g is a **one-to-one** or **1–1 function** if and only if for all u and v in the domain of g,

$$g(u) = g(v) \Rightarrow u = v.$$

Thus, if a function g is 1–1, then for all u and v in the domain of g,

$$u = v \Leftrightarrow g(u) = g(v).$$

This, when the domain of g is the set of all real numbers, makes applying g to both sides of an equation a reversible step.

One-to-one real functions can be recognized from their graphs. To see how, note that because a conditional statement and its contrapositive are equivalent, it follows from the definition that when g is one-to-one, $u \neq v \Rightarrow g(u) \neq g(v)$. That is, g sends any two different domain values to two unequal range values. This means that no horizontal line intersects the graph of g more than once.

On the next page are drawn the squaring and cubing functions each with domain the set of real numbers. The squaring function $g: x \rightarrow x^2$ is not 1–1. For instance, $7 \neq -7$, but $g(7) = g(-7) = 49$. Many horizontal lines intersect its graph more than once. Thus, squaring both sides of an equation is not always a reversible step.

154

y, has values on both sides not 1-1

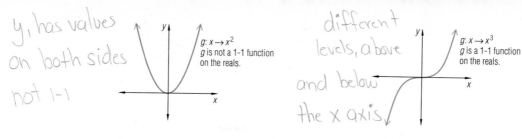

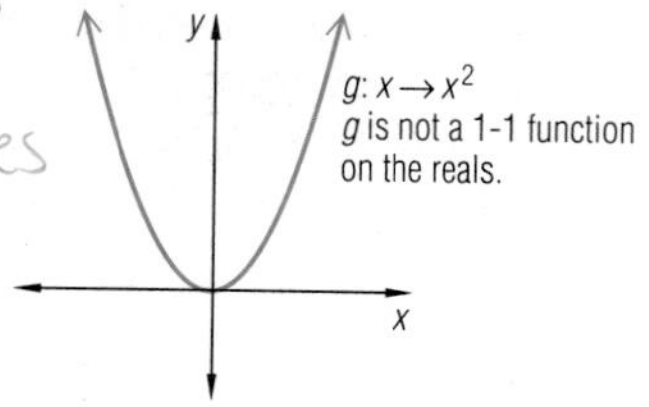

different levels, above and below the x axis

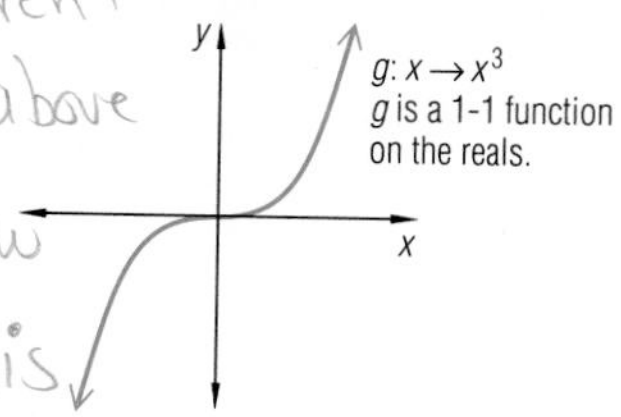

In contrast, the cubing function $g: x \to x^3$ is a 1–1 function on the set of all real numbers, so cubing both sides of an equation is always a reversible operation.

Example 2 Solve $\sqrt[3]{5n} = 2$.

Solution Cubing both sides is a reversible operation.

$$\sqrt[3]{5n} = 2$$
$$\Leftrightarrow \quad 5n = 8$$
$$\Leftrightarrow \quad n = \frac{8}{5}$$

Check Because all steps are reversible, the check is not required as an integral part of the solution. However, it is always a good idea to confirm the arithmetic.

$\sqrt[3]{5 \cdot \frac{8}{5}} = \sqrt[3]{8} = 2$. It checks.

You can see that knowing whether a function is 1–1 is quite valuable. Properties of functions studied in the last chapter can be helpful in deciding whether a function is 1–1.

Theorem

Suppose that g is a real function. If g is increasing throughout its domain, or if g is decreasing throughout its domain, then g is a 1–1 function.

Proof To show that $g(x) = g(y) \Rightarrow x = y$, it is sufficient to prove the contrapositive: if $x \neq y$, then $g(x) \neq g(y)$. Here is a proof for all increasing functions.

If $x \neq y$, then either (1) $x < y$ or (2) $x > y$.

(1) If $x < y$, then $g(x) < g(y)$, since g is an increasing function.
(2) If $x > y$, then $g(x) > g(y)$, since g is an increasing function.

Thus, in all cases when $x \neq y$, $g(x) \neq g(y)$. Since the contrapositive is true, the theorem is true.
The proof for decreasing functions is left to you in the Questions.

ADDITIONAL EXAMPLES

2. Solve $\sqrt[5]{4n} = 3$.
$\sqrt[5]{4n} = 3 \Leftrightarrow 4n = 3^5$
$\Leftrightarrow 4n = 243$
$\Leftrightarrow n = \frac{243}{4}$

3. Solve $\log(t^3 + 3t^2 + 5t - 7) = 3\log(t + 1)$.
$\log(t^3 + 3t^2 + 5t - 7) = 3\log(t + 1)$
$\Leftrightarrow \log(t^3 + 3t^2 + 5t - 7) = \log(t + 1)^3$
$\Leftrightarrow t^3 + 3t^2 + 5t - 7 = (t + 1)^3$
$\Leftrightarrow t^3 + 3t^2 + 5t - 7 = t^3 + 3t^2 + 3t + 1$
$\Leftrightarrow 5t - 7 = 3t + 1$
$\Leftrightarrow t = 4$

ADDITIONAL EXAMPLES

4. Find all values of x satisying $\frac{1}{2x^2 + 7x - 3} = \frac{1}{x^2 + 12x + 11}$.

For all values of x for which neither denominator is zero, we can multiply both sides by the product of the denominators.

$x^2 + 12x + 11 = 2x^2 + 7x - 3$

$x^2 - 5x - 14 = 0$

$(x - 7)(x + 2) = 0$

$x = 7$ or $x = -2$

Thus, 7 and -2 are candidates for solutions.

When $x = -2$, does $\frac{1}{2(-2)^2 + 7(-2) - 3} = \frac{1}{(-2)^2 + 12(-2) + 11}$?

$-\frac{1}{9} = -\frac{1}{9}$? Yes

When $x = 7$, does $\frac{1}{2(7)^2 + 7(7) - 3} = \frac{1}{(7)^2 + 12(7) + 11}$?

$\frac{1}{144} = \frac{1}{144}$? Yes

5. Find the height of the satellite for which the farthest observable distance is $11h$.

With $d = 11h$ in the equation, then $\forall h \geq 0$,

$11h = \sqrt{7920h + h^2}$

$\Rightarrow 121h^2 = 7920h + h^2$

$\Leftrightarrow 120h^2 - 7920h = 0$

$\Leftrightarrow 120h(h - 66) = 0$

$\Leftrightarrow h = 0$ or $h = 66$

The only reasonable solution is $h = 66$. However, the lifespan of a satellite at this height would be very short!

From this theorem, it follows that all logarithmic functions and all exponential functions are one-to-one functions. However, they are only one-to-one for values in their domains. The following example shows how this technical point can affect the solving of an equation.

Example 3 A student solved the equation

$$\log(x^2) = \log(-x)$$

in the following way.

Suppose that x is a real number.

Step 1 $\log(x^2) = \log(-x) \Leftrightarrow x^2 = -x$ Because the logarithm function is one-to-one

Step 2 $x^2 = -x \Leftrightarrow x = 0$ or $x = -1$ By solving the quadratic equation

So, the given equation has two solutions, 0 and -1.

a. Find the error in the student's work.
b. Modify the student's work to correct the error.

Solution

a. $x = 0$ is not a solution of the given equation because log 0 is undefined. Although the logarithm function is one-to-one, step 1 is in error because $x = 0$ is not in the domain of the logarithm function.

b. To correct the error, make the following changes in steps 1 and 2.

Step 1 $\log(x^2) = \log(-x) \Rightarrow x^2 = -x$ Because the logarithm function is one-to-one

Step 2 $x^2 = -x \Leftrightarrow x = 0$ or $x = -1$ By solving a quadratic equation

So, the given equation has at most those two *possible* solutions, 0 and -1.

Add a step 3 to check the possible solutions.

Step 3 Check: For $x = -1$, $\log((-1)^2) = \log(-(-1))$ so -1 is a solution of the given equation. For $x = 0$, $\log(0)$ is undefined, so 0 is not a solution.
$\therefore$ -1 is the only solution to the given equation.

Example 4 A satellite is in a circular orbit above the surface of the Earth. Let h and d be the distances in miles from the satellite to the nearest and farthest points on the Earth's surface that are visible from the satellite. (The earth can be regarded as a sphere of radius 3960 miles.) The equation

$$d = \sqrt{7920h + h^2}$$

relates d and h. (You are asked to derive this equation in the Questions.) Find the height h of the satellite for which the farthest observable distance is $2h$.

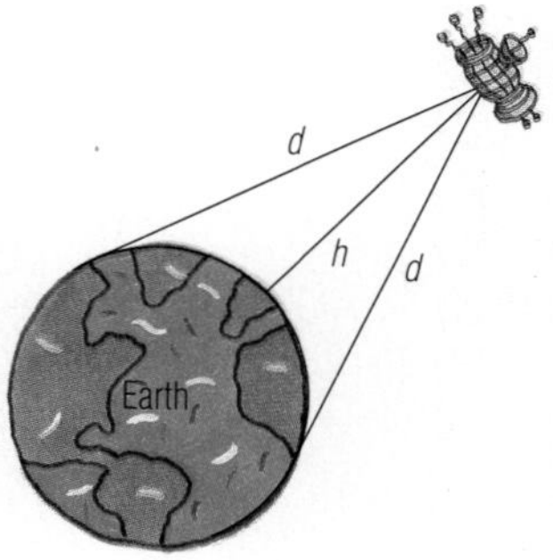

156

Solution Suppose h is a height for which $d = 2h$ in the equation. Then $\forall\ h \geq 0$,

	$2h = \sqrt{7920h + h^2}$	
$\Rightarrow$	$4h^2 = 7920h + h^2$	Applying $g: x \rightarrow x^2$ to both sides
$\Leftrightarrow$	$3h^2 - 7920h = 0$	Addition Property of Equality
$\Leftrightarrow$	$3h(h - 2640) = 0$	Distributive Property (factoring)
$\Leftrightarrow$	$h = 0$ or $h = 2640$.	Zero Product Theorem

Thus, the statement

If $2h = \sqrt{7920h + h^2}$ *then* $h = 0$ *or* $h = 2640$.

is true. But because the converse may not be true, you must determine if $h = 0$ or $h = 2640$ satisfies the original equation. Both of the equalities

$$2 \cdot 0 = \sqrt{7920 \cdot 0 + 0^2} \text{ and } 2 \cdot 2640 = \sqrt{7920 \cdot 2640 + (2640)^2}$$

do hold (you should check this), which indicates that the step of squaring both sides did not introduce any extra values in this case. So,

$2h = \sqrt{7920h + h^2}$ *if and only if* $h = 0$ *miles or* $h = 2640$ *miles.*

However, although $h = 0$ miles satisfies the original equation, it fails to satisfy the physical conditions of the problem. A satellite cannot be in orbit 0 miles above the earth. Only at a height of 2640 miles can one see twice as far to the horizon as to the nearest point.

Questions

Covering the Reading

1. Show that for any real number x, $2x + 5 = 3x - 2 \Leftrightarrow x = 7$. **See margin.**
2. When is a reasoning step reversible? **when the step is a biconditional**
3. In an equation involving real numbers, tell whether the reasoning step is always reversible.
   a. adding the same number to both sides **Yes**
   b. squaring both sides **No**
   c. multiplying both sides by 3 **Yes**
   d. multiplying both sides by 0 **No**
4. *True* or *false?* Nonreversible steps in the solution process of an equation may lead to values of the variable that fail to satisfy the equation. **True**
5. Explain why cubing both sides of an equation is a reversible step. **See margin.**
6. Consider the solution of the equation $\sqrt{y - 3} = 3 - y$ below. Replace the ? with a $\Leftrightarrow$ for a reversible step or with $\Rightarrow$ for a nonreversible step.

a.	$\sqrt{y - 3} = 3 - y$	?	$y - 3 = (3 - y)^2$	$\Rightarrow$
b.		?	$y - 3 = 9 - 6y + y^2$	$\Leftrightarrow$
c.		?	$y^2 - 7y + 12 = 0$	$\Leftrightarrow$
d.		?	$(y - 3)(y - 4) = 0$	$\Leftrightarrow$
e.		?	$y - 3 = 0$ or $y - 4 = 0$	$\Leftrightarrow$
f.		?	$y = 3$ or $y = 4$.	$\Leftrightarrow$

NOTES ON QUESTIONS

Question 3: You might ask students to give the result in function language if the given equation is $f(x) = g(x)$.
((a) $f(x) + a = g(x) + a$;
(b) $h(f(x)) = h(g(x))$ where $h(x) = x^2$;
(c) like (b) but where $h(x) = 3x$;
(d) $0 = 0$)

Question 7: This problem has no solution. This can be demonstrated using a graphing utility to graph $f(x) = \sqrt{2x^2 + 3x + 1}$ and $g(x) = x - 1$ on the same coordinate axes. Since the graphs do not intersect there is no solution.

Question 14: The student could correct the solution by indicating on the second line that the division can be done only if $4x - 1 \neq 0$, so there still is a possibility that $4x - 1 = 0$.

Question 18: Point out that satisfying this property implies that h is a 1-1 function.

Questions 20 and 21: The algorithm described is simply two applications of the strategy used in Example 2 or Example 4. It is possible to begin Question 21 by squaring both sides immediately, but the calculations become a little complicated. The solution can be checked by noting that the graph of $f(x) = \sqrt{2x + 3} + \sqrt{7 - x}$ never intersects the line $y = 1$.

Small Group Work for Question 23: Students should be able to provide a correct proof of the theorem.

ADDITIONAL ANSWERS

1. See Additional Answers in the back of this book.

5. The function $f: x \rightarrow x^3$ is a 1-1 function.

ADDITIONAL ANSWERS

7.b. See Additional Answers in the back of this book.

c. Neither of these is an actual solution. They arose because the first step is not reversible, thus the solution sets of the original and the final equations are different. There is no solution to this equation.

12. False, $g(x) = x^6$ is not a 1-1 function by definition since, for example, $g(1) = g(-1)$, but $1 \neq -1$. Thus, taking the sixth power of both sides of an equation is not a reversible operation.

14. When $x = \frac{1}{4}, \frac{1}{4x-1} = \frac{1}{0}$, which is not defined. So the third line does not follow from the second line. A correct third line should insert $4x - 1 \neq 0$. A correct fourth line is $x^2 = 1$ or $4x - 1 = 0$.

17.a.

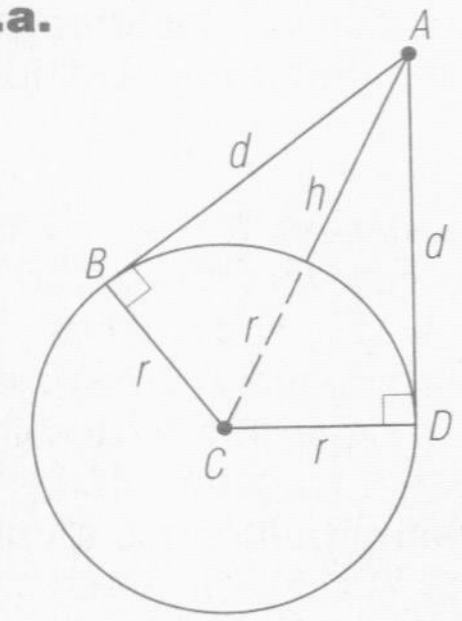

$\overline{AB}$ is tanget to $\odot$ C at B, so $\triangle ABC$ is a right triangle.
So $r^2 + d^2 = (r + h)^2$

$$r^2 + d^2 = r^2 + 2rh + h^2$$
$$d^2 = 2rh + h^2$$
$$d = \sqrt{2rh + h^2}$$
$$d = \sqrt{7920h + h^2}$$

18. See Additional Answers in the back of this book.

7. a. What function is being applied to both sides of an equation when you square both sides? $f: x \to x^2$

b. Apply that function as a first step in solving $\sqrt{2x^2 + 3x + 1} = x - 1$. See margin.

c. Which of the candidates for solution are actually solutions? Explain how any nonsolutions arose. See margin.

In 8–10, solve.

8. $\log x = \log (x^2 - 30)$ $x = 6$

9. $\frac{1}{y^2 - 9} = \frac{1}{y - 3}$ $y = -2$

10. $\sqrt[3]{v} = \sqrt[3]{2v - 2}$ $v = 2$

11. Refer to Example 4. At what altitude does d equal four times h? 528 miles

Applying the Mathematics

12. *True* or *false?* Taking the 6th power of both sides of an equation is always a reversible operation. Justify your answer. See margin.

13. Find all real numbers x such that $\sqrt{x^2 + 1} + 1 = x$. (Hint: Start by subtracting 1 from both sides to isolate the radical expression.) no real solution

14. A student attempted to solve the equation $4x^3 - x^2 = 4x - 1$ as follows.

$$x^2(4x - 1) = 4x - 1$$
$$\frac{1}{4x - 1} \cdot x^2(4x - 1) = (4x - 1) \cdot \frac{1}{4x - 1}$$
$$x^2 = 1$$
$$x = 1 \text{ or } x = -1.$$

The solution $\frac{1}{4}$ to the original equation has been lost. What happened? See margin.

In 15 and 16, solve.

15. $e^{x^2} = e^{2x}$ $x = 0$ or $x = 2$

16. $\log_2(3x + 5) = \log_2(4x + 9)$ no real solution

17. Refer to Example 4.

a. Derive the equation $d = \sqrt{7920h + h^2}$.

b. At what satellite altitude is the observed distance 1000 miles greater than the height of the satellite? (Assume h must be between 50 and 500 miles.)

a) See margin. b) 168.92 miles

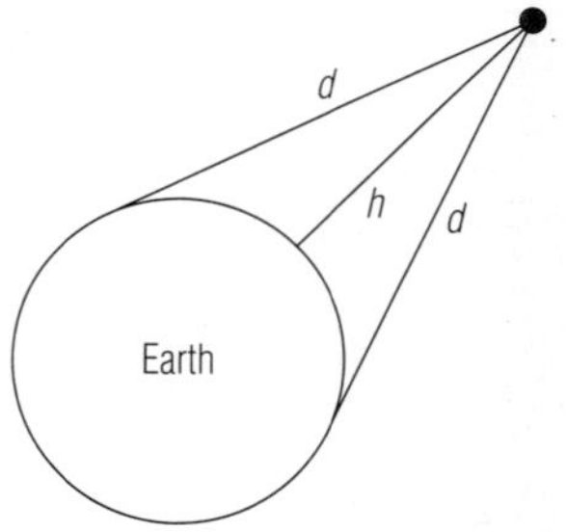

18. Prove: If h is a real function that is decreasing throughout its domain, then, for all u and v in its domain, $h(u) = h(v) \Rightarrow u = v$. See margin.

19. a. Find the error in this "solution" of $(2z + 1)^2 = z^2$.

$(2z + 1)^2 = z^2 \Rightarrow 2z + 1 = z$

$\Rightarrow \quad z = -1$. See margin.

b. Correct this "solution" to find all real numbers for which $(2z + 1)^2 = z^2$. See margin.

20. The equation $\sqrt{x - 3} = 3 - \sqrt{x}$ contains two square roots. Use the following procedure to find all real numbers that satisfy this equation.

a. Square both sides. $x - 3 = 9 - 6\sqrt{x} + x$

b. Simplify to obtain an equation in which there is a radical on one side of the equation, and all other terms are on the other side. $6\sqrt{x} = 12$

c. Square both sides of the equation obtained in part **b**. $36x = 144$

d. Solve the resulting equation. $x = 4$

e. The solutions of the given equation (if any) are found among the solutions of the equation in part **d**. Determine if any of the solutions in part **d** are solutions to the given equation. Does $\sqrt{4 - 3} = 3 - \sqrt{4}$? Does $\sqrt{1} = 3 - 2$? Yes

21. Apply the technique of Question 20 to solve

$\sqrt{2x + 3} + \sqrt{7 - x} = 1$.

(Hint: First rewrite the equation so that there is a radical expression on each side.) no real solution

Review

22. Prove or disprove: *If m and n are even integers then $m \cdot n = 4k$ for some integer k.* *(Lesson 1-7)* See margin.

23. Find the error in this "proof" of the following theorem.

If m and n are any odd integers, then $m + n$ is an even integer.

Proof Suppose m and n are any odd integers. If $m = 1$ and $n = 3$, then $m + n = 4$ which is even. If $m = 5$ and $n = 13$, then $m + n = 18$, which is even. Thus the sum of any two odd integers is even. *(Lesson 1-8)* See margin.

24. Consider a circle with points P and Q_1. Let l be a sequence of lines defined by

$\begin{cases} l_1 = \text{line } PQ_1 \\ l_{n+1} = \text{line } PQ_{n+1} \end{cases}$

where Q_{n+1} is the midpoint of arc PQ_n.

What is $\lim_{n \to \infty} l_n$? *(Lesson 2-4)*

The line through P that is tangent to the circle.

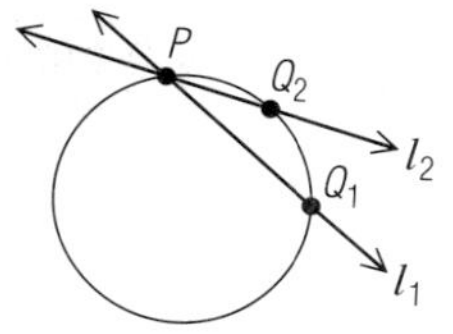

25. Let f and g be functions defined by $f(x) = 4x + 2$ and $g(x) = x^2 + 1$. Find

a. $f(g(2))$ 22 **b.** $g(f(5))$. 485 *(Previous course)*

Exploration

26. Make up an equation different from any in this lesson in which the use of nonreversible steps leads to potential solutions that do not satisfy the original equation.

Sample: $\frac{1}{\sqrt{4x + 2}} = \frac{1}{2x - 3} \Leftrightarrow 2x - 3 = \sqrt{4x + 2}$

FOLLOW-UP

MORE PRACTICE

For more questions on SPUR Objectives, use *Lesson Master 3-1,* shown below.

EXTENSION

Refer to **Example 4.** As values of h increase, how does d compare to h? That is, what is

$\lim_{h \to \infty} \frac{\sqrt{7920h + h^2}}{h}$?

(The limit is 1; h becomes closer and closer to d.)

19.a. The first step is nonreversible.

b. $(2z + 1)^2 = z^2$

$\Leftrightarrow 4x^2 + 4z + 1 = z^2$

$\Leftrightarrow 3z^2 + 4z + 1 = 0$

$\Leftrightarrow (3z + 1)(z + 1) = 0$

$\Leftrightarrow 3z = -1$ or $z = -1$

$\Leftrightarrow z = -\frac{1}{3}$ or $z = -1$

22., 23. See Additional Answers in the back of this book.

$x^2 - 13 = 243$

$x^2 = 256$

$x = \sqrt{256}$

$x = \pm 16$

NAME ______

LESSON **MASTER 3–1**

QUESTIONS ON **SPUR** OBJECTIVES

■ SKILLS *Objective A (See pages 217–220 for objectives.)*

In 1–8, find all real solutions.

1. $\sqrt[5]{x^2 - 13} = 3$ — $x = \pm 16$

2. $\sqrt{y + 5} = y - 7$ — $y = 11$

3. $\sqrt{x - 3} + 2 = \sqrt{2x + 1}$ — $x = 4$ or 12

4. $\frac{1}{z^2 + 2z} = \frac{1}{5z}$ — $z = 3$

5. $5^{3a - 3} = 25^a$ (Hint: $25 = 5^2$) — $a = 3$

6. $e^{2x^2 - 6x} = e^{4x - 12}$ — $x = 2$ or 3

7. $\ln (t^2 - 10) = \ln (3t + 30)$ — $t = 8$ or -5

8. $\ln (y^2 - 14) = \ln (6y + 13)$ — $y = 9$

■ PROPERTIES *Objective F*

In 9–11, tell whether the step is always reversible.

9. taking the square root of both sides — no

10. subtracting a variable from both sides — yes

11. dividing both sides by a variable — no

■ USES *Objective J*

12. A baseball is thrown straight up with an initial velocity of 44 feet per second. After how many seconds is it 33 feet above the ground? (Hint: Use $h = -\frac{1}{2}gt^2 + v_0t$, where h is the height in feet, t is the time in seconds, g is the acceleration due to gravity in feet/second^2, 9.8 ft/s^2, and v_0 is the initial velocity in feet/second.)

It never reaches 33 ft.

32ft/s

and so $g(k)$ is the one and only solution to the given equation, provided that $g(k)$ is in the domain of f.

LESSON 3-2

RESOURCES
- Lesson Master 3-2
- Teaching Aid 14 can be used with **Questions 12 and 13**.
- Computer Master 6

LESSON

3-2

An equation in the variable x can be considered to be of the form

$$f(x) = h(x).$$

For instance, in the equation

$$\sqrt{x-2} = 4 - x$$

Error Analysis The notation f^{-1} could mean $\frac{1}{f}$, that is, the function whose formula is the reciprocal of the formula of f, but by convention it does not. Since it is convention and not logic or a pattern that has decided what the symbol f^{-1} represents, this symbol can be quite confusing to students. Remind students that f^{-1} is not "f to the negative one power" and is not $\frac{1}{f}$.

Students often want to sketch the graph of the inverse function as a 90° rotation rather than a reflection about the line $y = x$. A good strategy is to graph $y = x$, locate two or more points on the graph of f, and plot the corresponding points of f^{-1} before sketching the entire graph of f^{-1}.

Making Connections Students have seen many examples of inverse functions as shown on page 163. The graphical relationship between inverse functions (at the bottom of this page) also is one that students should have seen before. This visual display reinforces the fact that if (a, b) is on the graph of f, then (b, a) must be on the graph of f^{-1}.

Point out that $f(x) = x^2$ is not a 1-1 function on the set of real numbers and does not have an inverse according to the theorem on page 163.

ADDITIONAL EXAMPLES

1. Suppose $f(x) = 3x - 2$ and $g(x) = x^3 + 1$ for all real numbers x.

a. Determine a formula for $f \circ g(x)$.
$f \circ g(x) = 3x^3 + 1$

b. Determine a formula for $g \circ f(x)$.
$g \circ f(x) = 27x^3 - 54x^2 + 36x - 7$

c. Compare the results in parts **a** and **b**. What does this tell you about commutativity of composition?
$f \circ g \neq g \circ f$
One counterexample proves that composition of functions is not a commutative operation.

You have probably already used this procedure to solve equations, although you may not have thought of it in terms of function composition.

Example 2 Solve: $x^{2/3} = 64$.

Solution It is understood that $x \geq 0$ because a noninteger power of x is being taken. This equation is of the form $f(x) = k$, where $f(x) = x^{2/3}$ for all nonnegative real numbers x, and $k = 64$. Let g be the function with rule $g(x) = x^{3/2}$, and apply g to each side. The function g satisfies $g(f(x)) = x$ for all x in the domain of f. Thus

$$g(x^{2/3}) = g(64)$$
$$(x^{2/3})^{3/2} = 64^{3/2}$$
$$x = 512.$$

In Example 2, the function g is called the *inverse* of the function f. Here is a general definition.

Definition

f and g are **inverse functions,** written $\mathbf{f = g^{-1}}$ or $\mathbf{g = f^{-1}}$, if and only if $f \circ g(x) = x$, for all x in the domain of g, and $g \circ f(x) = x$, for all x in the domain of f.

In general, if f has an inverse f^{-1}, and if

$$f(a) = b,$$

then $$f^{-1} \circ f(a) = f^{-1}(b).$$

So $$a = f^{-1}(b).$$

Thus if f is a real function and (a, b) is on the graph of f, then (b, a) is on the graph of f^{-1}. Recall that (a, b) is the reflection image of (b, a) over the line $y = x$. Thus, when a real function f has an inverse f^{-1}, the graphs of f and f^{-1} are reflection images of one another over the line $y = x$. Consequently, the domain of f^{-1} is the range of f, and vice versa.

There is a simple criterion for the existence of an inverse function.

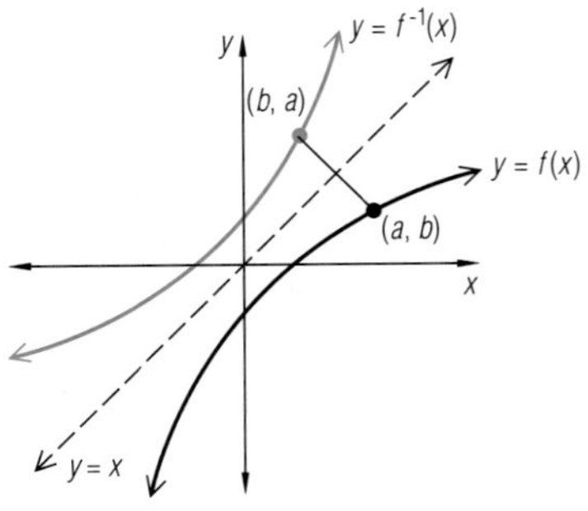

Theorem

A function has an inverse if and only if it is a 1–1 function.

Proof If f has an inverse f^{-1}, and $f(x) = f(y)$, then $f^{-1} \circ f(x) = f^{-1} \circ f(y)$, so $x = y$, and so f is a 1–1 function. If f is a 1–1 function, then switching the components of its ordered pairs yields a function, its inverse.

Some of the reversible operations that are commonly used in solving equations may be considered as functions that have inverses. When you add k to both sides of an equation, it is as if you are applying the function f: $x \to x + k$ to both sides. You can reverse the step by subtracting k.

Adding and Subtracting k:

Let f: $x \to x + k$ $\forall$ real numbers x.
g: $x \to x - k$ $\forall$ real numbers x.

Then $\forall$ real numbers x, $f(g(x)) = f(x - k) = (x - k) + k = x$. (You should verify that $g(f(x)) = x$.) Hence, f and g are inverses.

Multiplying and Dividing by $k \neq 0$:

Let f: $x \to kx$ $\forall$ real numbers x.
g: $x \to \frac{x}{k}$ $\forall$ real numbers x.

(You should verify that $\forall$ real numbers x, $f(g(x)) = g(f(x)) = x$ and so f and g are inverses.)

You are familiar with other examples of inverse functions.

Exponential and logarithmic functions:

Let b be a positive number, $b \neq 1$, and
let f: $x \to b^x$ $\forall$ real numbers x
g: $x \to \log_b x$ $\forall$ positive real numbers x.

Then $f(g(x)) = f(\log_b x) = b^{\log_b x} = x$, from the definition of log and $g(f(x)) = g(b^x) = \log_b(b^x) = x$, from the definition of log. So f and g are inverse functions.

A function may have no inverse over its entire domain, but may have an inverse over a restricted domain. For instance, let

$$f: x \to x^2 \quad \text{and} \quad g: x \to \sqrt{x}.$$

f and g are not inverses if their domains are the set of real numbers. They are inverses if their domains are the set of nonnegative real numbers, because $\sqrt{x^2} = x$ and $(\sqrt{x})^2 = x$ for all nonnegative x but not for all real x.

The sine and cosine functions defined over the set of real numbers are not 1–1 so they do not have inverses. Yet, if their domains are restricted so that the functions are 1–1, then they do. These inverses are discussed in Chapter 6.

The composite of two inverse functions f and f^{-1} maps each number in its domain onto itself and is sometimes called the **identity** function and denoted by the symbol I. For all x in its domain, $I(x) = x$.

2. Solve $x^{5/7} = 243$.
2187

NOTES ON QUESTIONS

Question 13: The graph of the inverse function will look only like the reflection image of the graph of the function if the scales on the axes are the same.

Question 16: The following is an alternative way of determining a formula for the inverse of $f(x) = \frac{2}{3}x + 6$. To get $f(x)$ from x, multiply x by $\frac{2}{3}$ and then add 6. Thus, to get f^{-1}, perform inverse operations from right to left: subtract 6, then divide by $\frac{2}{3}$. Thus, $f^{-1}(x) = \frac{x - 6}{2/3} = \frac{2}{3}(x - 6)$.

FOLLOW-UP

MORE PRACTICE

For more questions on SPUR Objectives, use *Lesson Master 3-2*, shown on page 165.

EXTENSION

Having students write a complicated function as a composite of two or more functions (as in **Question 24**) is a useful exercise and one that often gives many calculus students difficulty. Here are two examples you might wish to use as **small group work**.

(1) Let f be the function defined by $f(x) = \sin x^2$. If $f = g \circ h$, what could g and h be?
(We could have $g(x) = \sin x$ and $h(x) = x^2$.)

(2) Let p be the function defined by $p(x) = \log(x - 5)^{1/3}$. Suppose p is the composite of three functions. What could those functions be?
(It could be that $p = \log \circ h \circ r$ where r: $x \to x - 5$, h: $x \to x^{1/3}$, and log: $x \to \log x$.)

ADDITIONAL ANSWERS

2. sample: $x = \frac{\pi}{2}$;

$f \circ g(x) = \cos(x^3) = \cos\frac{\pi^3}{8}$
$\approx -.7424$;
$g \circ f(x) = (\cos x)^3 = \left(\cos\frac{\pi}{2}\right)^3 = 0$;
$\therefore f \circ g(x) \neq g \circ f(x)$

4. $f \circ g(x) = \dfrac{1}{x^2 - 2 + \frac{1}{x^2}}$;
domain: $\{x: x \neq -1, 0, 1\}$

5.a. the set of nonnegative real numbers

b.

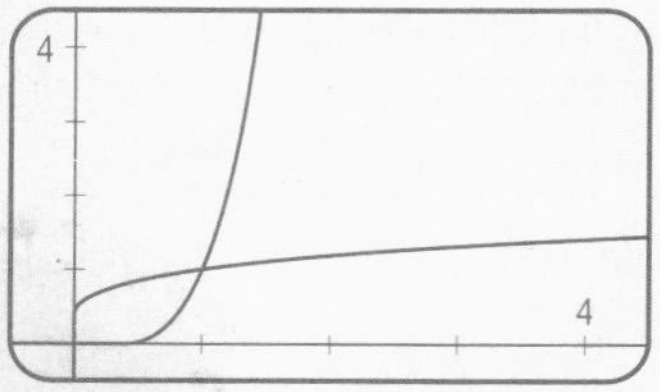

6.

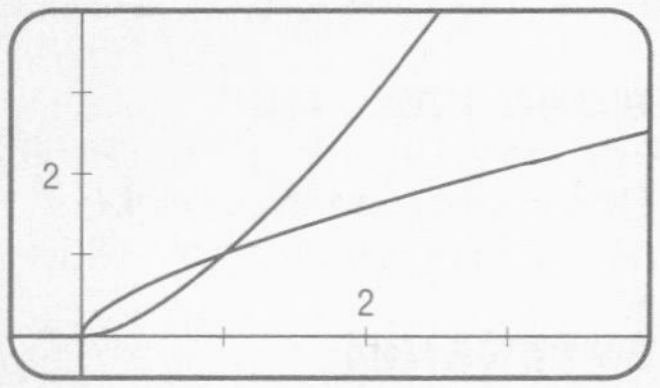

7. $f \circ g(x) = f(g(x)) = k\left(\frac{x}{k}\right) = x$;
$g \circ f(x) = g(f(x)) = \frac{(kx)}{k} = x$;
$\therefore f \circ g = g \circ f = I$

13.

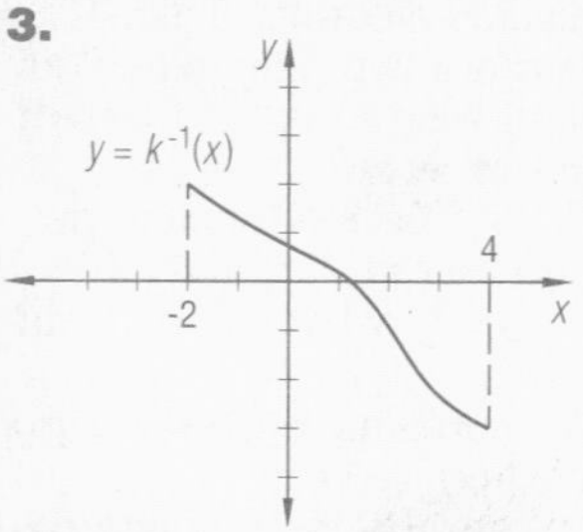

Questions

Covering the Reading

1. Let $h(x) = -2x + 3$ and $k(x) = \frac{1}{2}x^4$ for all real numbers x.
a. Give a formula for $h \circ k(x)$. $-x^4 + 3$
b. Compute $h \circ k(7)$. -2398
c. Give a formula for $k \circ h(x)$. $\frac{1}{2}(-2x+3)^4$
d. Compute $k \circ h(7)$. 7320.5
e. Is $h \circ k = k \circ h$? No

2. Let f: $x \to \cos x$ and g: $x \to x^3$ $\forall$ real numbers x. Show that $\exists$ a real number x such that $f \circ g(x) \neq g \circ f(x)$. **See margin.**

In 3 and 4, two real functions, f and g, are defined by formulas. Find a formula for the composite function $f \circ g$ and give its domain.

3. $f(x) = 3x + 1$ $\quad g(x) = 2x^3$
$f \circ g(x) = 6x^3 + 1$; domain: the set of real numbers

4. $f(x) = \dfrac{1}{x^2 - 4}$ $\quad g(x) = x + \dfrac{1}{x}$ **See margin.**

5. a. For what domain are f and g, as defined below, inverses?
$f(x) = x^4$ $\quad g(x) = \sqrt[4]{x}$ **See margin.**
b. Graph these functions over a part of that domain containing the origin. **See margin.**

6. Graph the inverse functions $f(x) = x^{3/2}$ and $g(x) = x^{2/3}$ for all nonnegative x. **See margin.**

7. Let k be a nonzero real number. Let f: $x \to kx$ $\forall x$ and g: $x \to \frac{x}{k}$ $\forall x$. Verify that $f \circ g = g \circ f = I$. **See margin.**

In 8–10, use inverse functions to solve.

8. $x^{3/4} = 4096$ $\quad x = 65536$ $\qquad$ **9.** $\sqrt{x^2 + 2} = 3$ $\quad x = \sqrt{7}$ or $x = -\sqrt{7}$

10. $e^y = 7000$ $\quad y \approx 8.854$ $\qquad$ **11.** $3 \log_2 x = 1.5$ $\quad x = \sqrt{2}$

Applying the Mathematics

12. Explain why the function h graphed below cannot have an inverse.

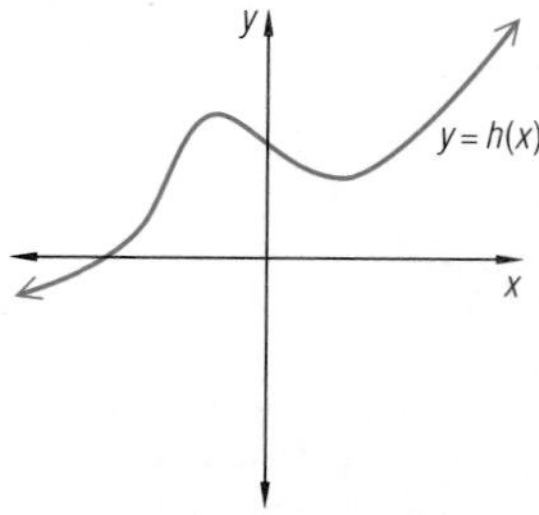

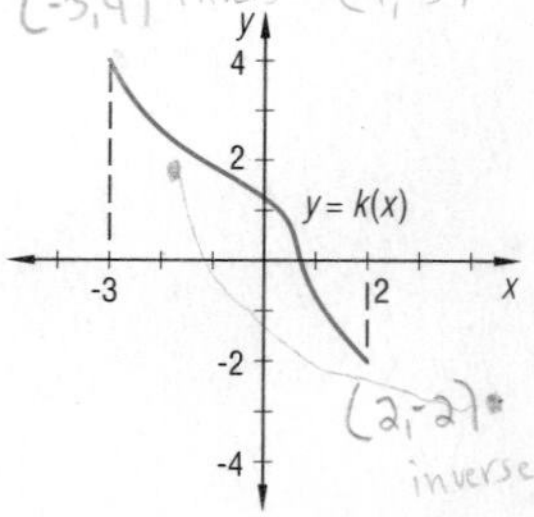

h is not a 1–1 function.

13. Explain why the function k graphed above has an inverse and sketch the graph of k^{-1}. **k is a 1–1 function. See margin.**

14. Suppose f and g are the real functions defined by

$$f(x) = -x^2 \qquad g(x) = \log_{10} x.$$

a. Find the formula and the domain for the function $f \circ g$.
b. Explain why the composite function $g \circ f$ is undefined.
a,b) See margin.

15. Explain why every increasing function has an inverse. An increasing function is a 1–1 function, and all 1–1 functions have inverses.

16. Give a formula for the inverse of $f\colon x \to \frac{2}{3}x + 6$. $f^{-1}(x) = \frac{3}{2}x - 9$

17. *Multiple choice.* Suppose that for all real numbers x,
$f(x) = e^x$, $g(x) = \sqrt{x^2 - 1}$, and $h(x) = e^{\sqrt{x^2 - 1}}$. Which is true?
(a) $f = g \circ h$ (b) $h = f \circ g$ (c) $g = h \circ f$
(d) $f = h \circ g$ (e) $h = g \circ f$ (f) $g = f \circ h$ (b)

Review

18. Solve: $\frac{3}{\sqrt{t - 5}} = \frac{1}{\sqrt{t}}$. *(Lesson 3-1)* no real solutions

19. a. Is $f\colon x \to \log_5 x$ a 1–1 function? Explain your answer.
b. Solve $\log_5 (x + 3) = 2 \log_5 (x + 1)$ *(Lesson 3-1)* $x = 1$
a) Yes, $f(x)$ is increasing over its entire domain, the positive reals.

20. In $\triangle ABC$, $AB = 15$, $BC = 21$ and $AC = 28$.
a. Find the measure of $\angle C$. $\approx 31.8°$
b. Find the length of the altitude drawn from $\angle B$. ≈ 11.1
(Lesson 2-8)

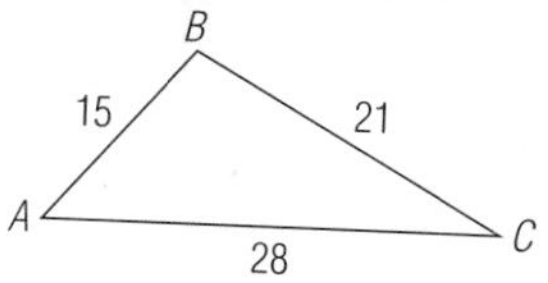

21. At the right is a part of the graph of the real function $f\colon x \to \frac{2x - 7}{x - 4}$.
a. What is the domain of f?
b. Identify the x- and y-intercepts.
c. Use limit notation to describe the end behavior of f.
(Lessons 2-1, 2-4, Previous course)
a) $\{x\colon x \neq 4\}$; b,c) See margin.

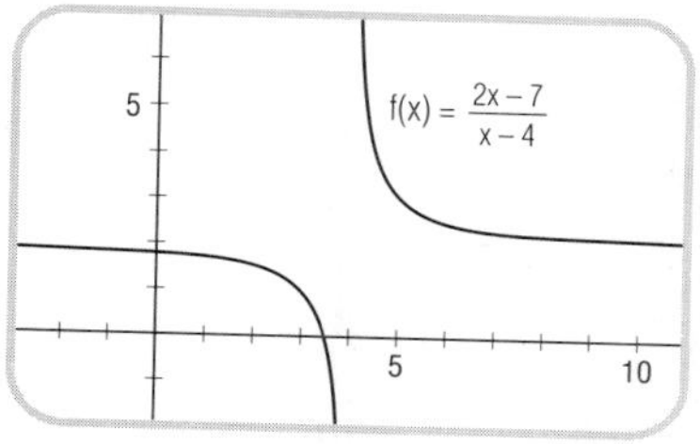

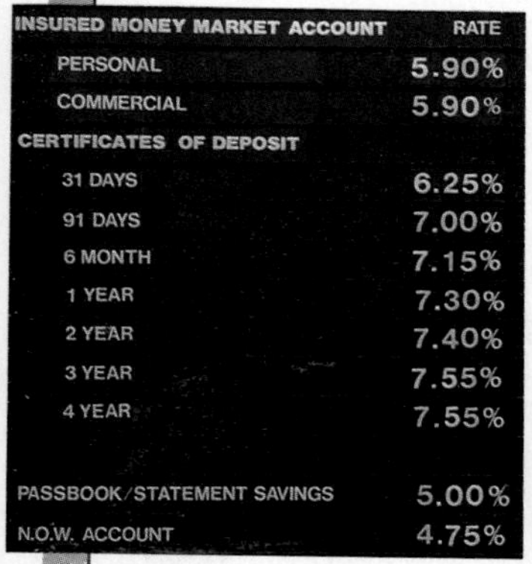

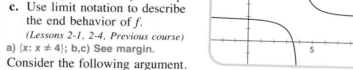

22. Consider the following argument.
If the interest rate drops, investment drops.
Investment has dropped.
∴ *The interest rate has dropped.*
a. Write the form of this argument. See margin.
b. Is the argument valid or invalid? *(Lessons 1-6, 1-8)* invalid

23. Prove that if n is an odd integer, then $n^2 + 1$ is even. *(Lesson 1-7)*
See margin.

Exploration

24. Suppose $f(x) = \frac{1}{e^{x^2}}$ for all real numbers x.
a. Express f as the composite of three simpler functions.
b. Graph f over a reasonable domain.
c. Explain how each of the three simpler functions contributes something to the graph.
See margin.

14.a. $f \circ g(x) = -(\log_{10} x)^2$; domain: $\{x\colon x > 0\}$
b. $g \circ f(x) = \log_{10} -(x^2)$. Since the log of a nonpositive number is undefined, and $-x^2 \le 0\ \forall$ real x, $g \circ f$ is always undefined.

21.b. x-intercept: $(\frac{7}{2}, 0)$; y-intercept: $(0, \frac{7}{4})$
c. $\lim_{x\to\infty} f(x) = 2$; $\lim_{x\to-\infty} f(x) = 2$

22.a. $p \Rightarrow q$
q
$\therefore p$

23. Suppose n is any odd integer. By definition, there exists an integer k such that $n = 2k + 1$. Then $n^2 + 1 = (2k + 1)^2 + 1 = 4k^2 + 4k + 2 = 2(2k^2 + 2k + 1)$. Since $(2k^2 + 2k + 1)$ is an integer by closure properties, $n^2 + 1$ is even by definition.

24.a. $f = f_1 \circ f_2 \circ f_3$ when $f_1(x) = \frac{1}{x}$, $f_2(x) = e^x$, and $f_3(x) = x^2$.
b., c. See Additional Answers in the back of this book.

NAME ______

LESSON **MASTER 3–2**
QUESTIONS ON SPUR OBJECTIVES

■ **SKILLS** *Objective A* *(See pages 217–220 for objectives.)*
In 1–4, solve each equation.

1. $e^x = 280$ $x = \ln 280 \approx 5.63$
2. $y^{2/5} = 16$ $y = 1024$
3. $(\sqrt{2t + 5})^3 = 27$ $t = 2$
4. $8 + \log_5 m = 133$ $m = 5^{125}$

■ **SKILLS** *Objective B*
5. Let $f(x) = 5x + 2$, $g(x) = x^2$, and $h(x) = \frac{1}{x}$. Give a simplified formula for each function.
a. $f \circ g$ $5x^2 + 2$
b. $g \circ f$ $25x^2 + 20x + 4$
c. $h \circ f$ $\frac{1}{5x + 2}$
d. Does $g \circ h = h \circ g$? yes

6. If r and s are real functions such that $r(x) = \sqrt{x}$ and $s(x) = \log x$, give the domains of $r \circ s$ and $s \circ r$.
domain of $r \circ s$ is $\{x\colon x \ge 1\}$; domain of $s \circ r$ is $\{x\colon x > 0\}$

■ **PROPERTIES** *Objective G*
7. If $f(x) = \frac{x - 5}{7}$, give a formula for f^{-1}. $f^{-1}(x) = 7x + 5$

8. Prove that $g(x) = \sqrt[3]{x + 5}$ and $h(x) = x^3 - 5$ are inverses.
$(g \circ h)(x) = g(x^3 - 5) = \sqrt[3]{(x^3 - 5) + 5} = \sqrt[3]{x^3} = x$; $(h \circ g)(x) = h(\sqrt[3]{x + 5}) = (\sqrt[3]{x + 5})^3 - 5 = x + 5 - 5 = x$

■ **REPRESENTATIONS** *Objective M*
9. Given the graph of f at the right, graph the inverse of f on the same grid.

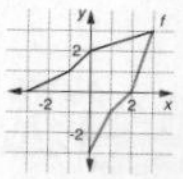

LESSON 3-3

RESOURCES

- Lesson Master 3-3
- Quiz for Lessons 3-1 through 3-3
- Teaching Aid 15 displays the graph of $P(x) = R(x) - C(x)$.
- Teaching Aid 16 displays the graphs of the **Example** and **Question 9**.

OBJECTIVES

B Describe the sum, difference, product, quotient, and composite of two given functions.
K Use operations on functions to find formulas which model realistic situations.
M Graph functions obtained from other functions by function operations or inverses.

TEACHING NOTES

When graphing two functions and their sum, product, difference, or quotient, use different colors if possible. Ask students to use different colors in their work also.

When discussing the washing machine problem that begins the lesson, you might want to discuss the fact that the cost per day of production, $C(x)$, also can be considered as the sum of two functions: a constant function F that gives the fixed costs $F(x) = 15{,}000$; and a function I that gives the cost per item for components $I(x) = 125x$. It might be helpful for students to graph both F and I on the same coordinate axes, find values of the functions for several specific

LESSON

3-3 Arithmetic Operations on Functions

Composition is not the only way to combine functions. The following situation naturally leads to subtraction of functions. It also shows how mathematics can help in making business decisions.

Suppose that an automated assembly line has the capacity to fill orders for up to 1000 automatic washing machines per day to be bought by a nationwide chain of department stores. For each day that the assembly line is in operation, there are fixed costs of \$15,000 for line operators, equipment depreciation, and maintenance. The cost of the components used to assemble each washing machine is \$125. The manufacturer wishes to consider selling the washing machines to the department store chain at a price of \$225 minus a sales incentive discount of $\$d$ per washing machine, where d is 8% of the number of washing machines bought per day by the chain.

Several discrete functions are important for an economic analysis of this manufacturing process. If x is the number of washing machines manufactured per day, then

$C(x)$ = the **total cost** per day for manufacturing x washing machines
$S(x)$ = the **selling price** per washing machine
$R(x)$ = the **total revenue** from the sale of one day's production of washing machines
$P(x)$ = the **total profit** from the sale of one day's production of washing machines.

Using the information given about the assembly line, the following formulas for these functions can be written.

$C(x)$ is the sum of the fixed costs of \$15,000 and the component costs $125x$:
$$C(x) = 15{,}000 + 125x.$$

$S(x)$ is the base price, \$225, minus the incentive discount $.08x$:
$$S(x) = 225 - .08x.$$ reimburse store and people

$R(x)$ is the product of the number of washing machines produced and the selling price per washing machine:
$$R(x) = x \bullet S(x) = 225x - .08x^2.$$

$P(x)$ is equal to the total revenue $R(x)$ minus the total cost $C(x)$:
$$\begin{aligned} P(x) &= R(x) - C(x) \\ &= (225x - .08x^2) - (15{,}000 + 125x) \\ &= 100x - 15{,}000 - .08x^2. \end{aligned}$$

These formulas indicate that it is sometimes useful to subtract functions to obtain new functions. Here $C(x)$ is subtracted from $R(x)$ to obtain $P(x)$. We say that $R - C = P$. It is also possible to add, multiply, and divide functions. These operations are defined by telling what the resulting function does with each element of its domain.

Definitions

Suppose that f and g are real functions defined on a set S of real numbers. Then $\boldsymbol{f + g}$, $\boldsymbol{f - g}$, $\boldsymbol{f \cdot g}$, and $\boldsymbol{\frac{f}{g}}$ are the functions defined $\forall x$ in S:

$$(f + g)(x) = f(x) + g(x)$$
$$(f - g)(x) = f(x) - g(x)$$
$$(f \cdot g)(x) = f(x) \cdot g(x)$$
$$\left(\frac{f}{g}\right)(x) = \frac{f(x)}{g(x)}, \text{ provided } g(x) \neq 0.$$

These operations can be interpreted graphically. Consider again the total cost function C, the total revenue function R, and the total profit function P related by the formula

$$P = R - C.$$

Because C is a linear function with

$$C(x) = 15{,}000 + 125x,$$

its graph is a straight line. The graph of

$$R(x) = 225x - .08x^2$$

is a parabola opening downward. With the window $-50 \leq x \leq 1200$ and $-10{,}000 \leq y \leq 160{,}000$, the graphs for C and R are pictured below.

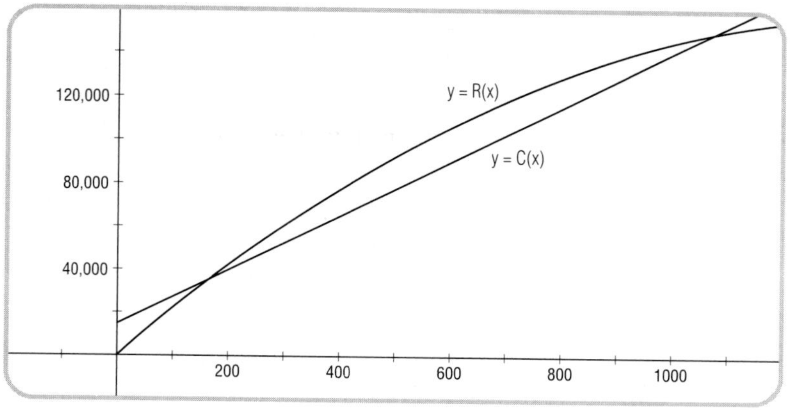

Since $P(x) = R(x) - C(x)$, the graph of P can be found by subtracting the y-coordinates of points on the graph of C from those on the graph of R with the same x-coordinate. This subtraction procedure is described by arrows in the figure on the next page.

values of x, and show that values of C can be obtained by adding ordinates of points on the graph of F to those on the graph of I.

You also may want to have students find several other points on the graphs of R and C on this page. Then have them show that subtracting ordinates yields points on the graph of P. Doing such computations for a number of points should suggest to students the reasonableness of this approach as a technique for sketching graphs.

The graph on page 168 merits some discussion with students. The graphs of functions R and C intersect twice. After the curves intersect beyond $x = 1000$, the revenue curve is below the cost curve. This means that continuing the 8% discount for production beyond this second intersection point begins to cost the manufacturer money.

Making Connections
Students may wonder why the arithmetic operations are so important. Sums are applied here and in the breaking up of sound waves into their pure tones. Subtraction is employed in Lesson 3-5 in the theory of solving equations. Applications of products of functions are found in many places, including this lesson and the next one, and in Lesson 6-8 in the discussion of phase-amplitude analysis. Division is basic to the construction and analysis of rational functions from polynomial functions.

Error Analysis Students may find the various operations with functions confusing and make errors in interpreting the notation. Continue to stress the following:

$f \circ g$ is a new function: the composite of f with g.

$f \cdot g$ is a new function: the product of f with g.

$f \cdot g(x)$ is the value of $f \cdot g$ at x.

$f(x) \cdot g(x)$ is the formula for calculating the value of $f \cdot g$ at x.

ADDITIONAL EXAMPLE

For all x, let h be defined by $h(x) = x^2 \cos x$. Give some properties of h.

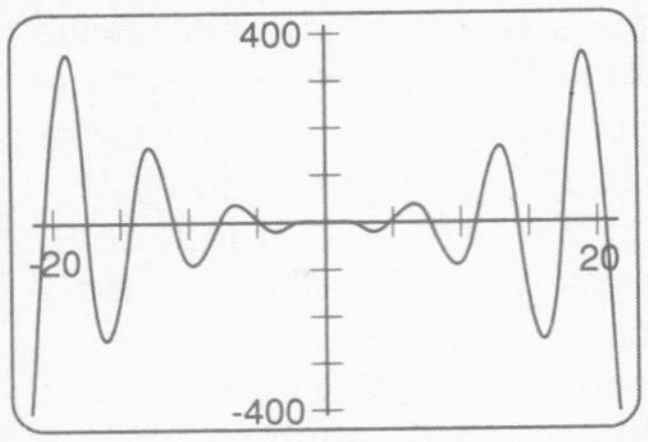

Above is a graph of *h*. The domain of *h* is the set of all real numbers, as is its range. The function *h* seems to be an even function with relative maxima and minima that get larger as *x* is farther from the origin. This can be explained by the fact that x^2 increases without bound as $x \to \pm\infty$ and the cosine varies between -1 and 1, so their product varies more and more as $|x|$ increases.

NOTES ON QUESTIONS

Question 7: To sketch $f + g$, students may either wish to add ordinates or start with points on the graph of f and add the corresponding values on the graph of g.

To sketch $f - g$, you may suggest that students sketch $-g$ (just reflect g over the x-axis) and then add $f + (-g)$. This approach reinforces the notion that subtraction is "adding the opposite."

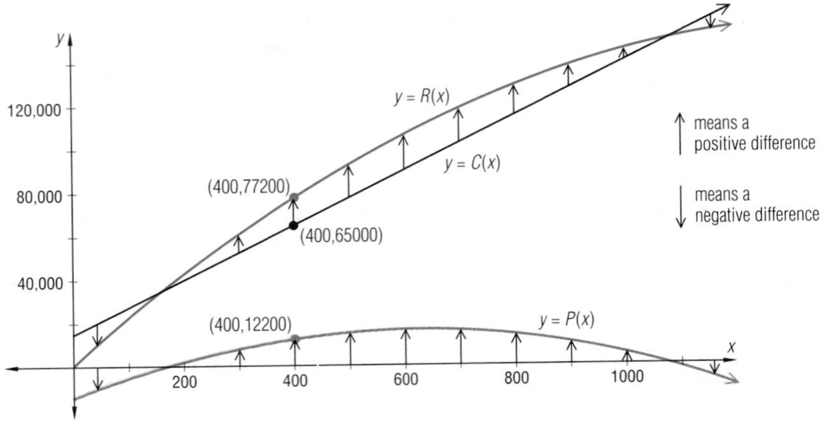

For instance, at $x = 400$, the y-coordinate of the point on the graph of P is the difference of the y-coordinates on the graphs of R and C: $77{,}200 - 65{,}000 = 12{,}200$. That means that, when 400 washing machines are sold, there will be a profit of \$12,200 for the day.

Notice that the function $P = R - C$ is increasing on part of the interval $0 \le x \le 1000$ and is decreasing on another part of the interval. Because of this incentive discount, the profit does not increase as more items are sold because in the long run the revenue increases at a slower rate than the cost. Consequently, this is not a particularly wise incentive plan.

The graphs of the sum, the product, or the quotient of two functions can be found by adding, multiplying, or dividing y-coordinates, respectively. Sometimes the result is quite unusual. In the following example, the sine function is multiplied by the identity function I: $x \to x$.

Example Let h be the function defined by $h(x) = x \sin x$ for all real numbers x. Give some properties of h.

Solution We show the three functions $I(x) = x$, $f(x) = \sin x$, and $h(x) = x \sin x$ on the interval $-12 \le x \le 12$.

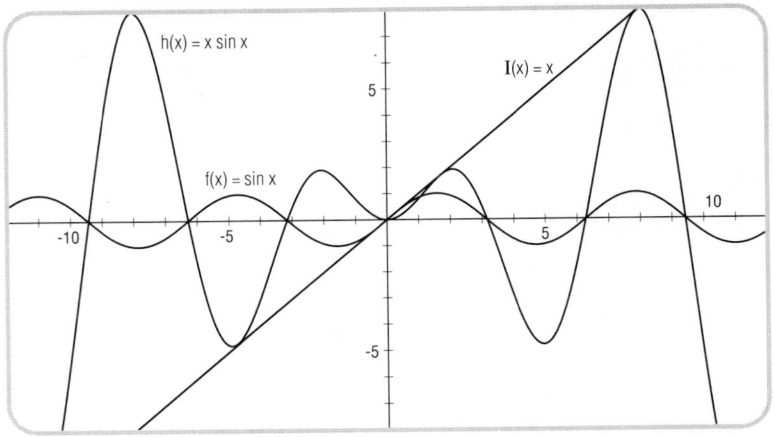

168

The domain of h is the set of real numbers. Since $h = I \cdot f$, the graph of h intersects the x-axis at all values of x for which $I(x) = 0$ or $f(x) = \sin x = 0$. h seems to be an even function with relative maxima and minima that get larger as x gets farther from the origin.

Caution: Do not confuse function composition and multiplication.

$$(f \circ g)(x) = f(g(x))$$
$$(f \cdot g)(x) = f(x) \cdot g(x)$$

When $f(x) = \sin x$ and $g(x) = x$, as in the example, $f \circ g(x) = \sin x$ but $f \cdot g(x) = x \sin x$.

Questions

Covering the Reading

1. a. In the situation at the beginning of the lesson, find the total cost, selling price, total revenue, and total profit from one day's production of 700 washing machines.
b. Find a number of washing machines to produce in a day that gives a greater profit than your answer to part **a**. What is the total cost and total revenue for producing that number of washing machines?
See margin.

2. *True* or *false?* If $f(x) = 3x^2$, $g(x) = 3x^3$, and $h(x) = x$,
a. $\frac{f}{g} = h$ False **b.** $\frac{g}{f} = h$ False **c.** $h \cdot f = g$ True **d.** $g - f = h$ False

3. Let $f(x) = x \cos x$. Give some properties of f. See margin.

4. Let $f(x) = x^2 - \frac{1}{x}$ and $g(x) = x^2 + \frac{1}{x}$. Compute and simplify formulas for **a.** $f + g$, **b.** $f - g$, **c.** $f \cdot g$, and **d.** $\frac{f}{g}$. In each case, find the largest set S of real numbers on which the function is defined.
See margin.

5. Let $f(x) = e^x$ and $g(x) = e^{-x}$.
a. Use an automatic grapher to sketch f, g, $f - g$ and $f \cdot g$, for $-5 \le x \le 5$. See margin.
b. Describe the results you get.
Sample: $f \cdot g$ is the line $y = 1$. $f - g$ is an odd, increasing function.

6. Let $h(x) = x^2$ and $m(x) = x + 1$. -1, because make y zero
a. Graph h, m, and $h + m$ without an automatic grapher.
b. Describe $h + m$. a) See margin. b) $h + m$ is a parabola that opens upward with vertex $(-\frac{1}{2}, \frac{3}{4})$

7. Trace the graphs of f and g shown, then sketch the graphs of $f + g$ and $f - g$.
See margin.

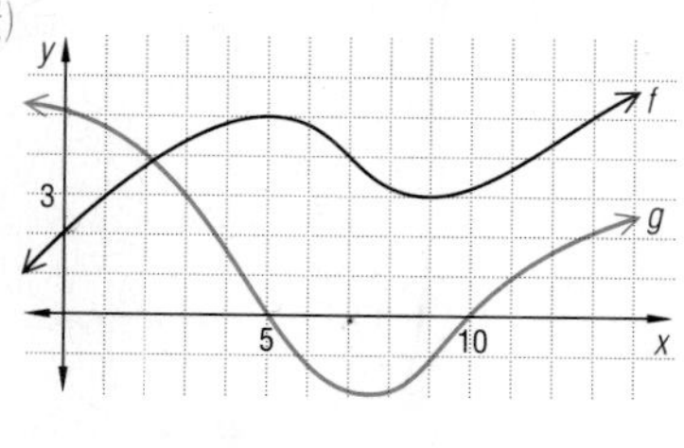

ADDITIONAL ANSWERS

1.a. $C(700) = \$102{,}500$
$S(700) = \$169$
$R(700) = \$118{,}300$
$P(700) = \$15{,}800$
b. sample: 625 machines;
$C(625) = \$93{,}125$
$R(625) = \$109{,}375$
$P(625) = \$16{,}250$

3. The domain of f is the set of real numbers. Let $h(x) = \cos x$, then, since $f = I \cdot h$, whenever $h(x) = \cos x = 1$, $f(x) = I(x) = x$, so the range of f is the set of real numbers. f crosses the x-axis whenever $I(x) = 0$ or $h(x) = \cos x = 0$. f is an odd function with relative maxima and minima that get larger as x is farther from the origin.

4.a. $2x^2$, $\{x: x \ne 0\}$
b. $-\frac{2}{x}$, $\{x: x \ne 0\}$
c. $x^4 - \frac{1}{x^2}$, $\{x: x \ne 0\}$
d. $\frac{x^3 - 1}{x^3 + 1}$, $\{x: x \ne 0 \text{ or } -1\}$

5.a.

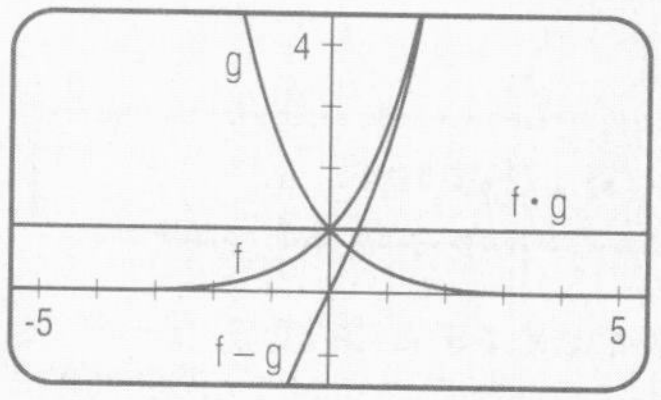

6.a.

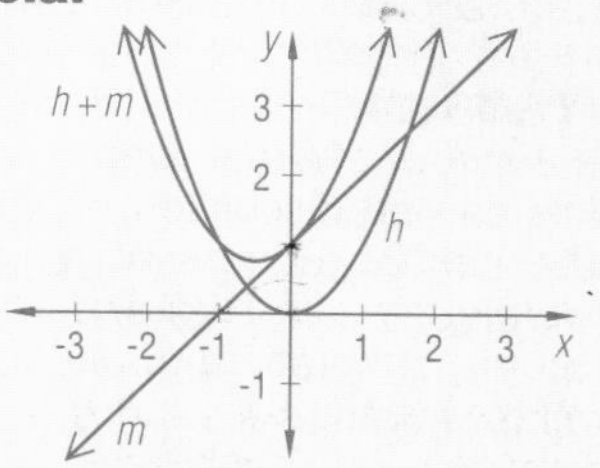

7.

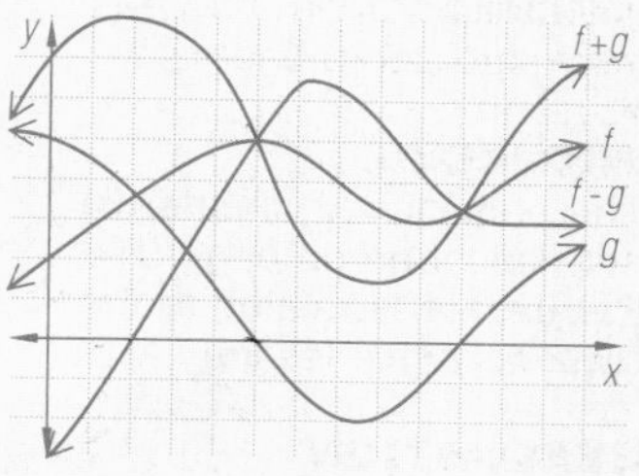

NOTES ON QUESTIONS
Question 8: Did any students notice that $f(x) = \frac{1}{2}\sin 2x$?

Error Analysis for Question 22: Students may be bothered by the first sentence of the question, because the profit function does increase on part of the interval. When we speak of a function increasing on an interval, we mean that it is increasing on the entire interval.

FOLLOW-UP

MORE PRACTICE
For more questions on SPUR Objectives, use *Lesson Master 3-3,* shown on page 171.

EXTENSION
The concept of breakeven points can be discussed. These can be found exactly by setting $C(x) = R(x)$ and using the quadratic formula or other equation-solving techniques. They can be found approximately using an automatic grapher that has a trace and zoom capability.

PROJECTS
The projects for Chapter 3 are described on page 214. **Project 4** is related to the content of this lesson.

EVALUATION
A quiz covering Lessons 3-1 through 3-3 is provided in the Teacher's Resource File.

Applying the Mathematics

8. **a.** Sketch the graphs of $y = \sin x$ and $y = \cos x$, then use multiplication of y-coordinates at key points (points where either function is -1, 0, or 1) to sketch the graph of $f(x) = \sin x \cos x$. Check your graph with an automatic grapher. See margin.
b. What do the range, amplitude, and period of f appear to be? range: $\{y: -.5 \le y \le .5\}$; amplitude: .5; period: π

9. Consider the graphs of f and $f + g$ given below.

* try to figure out what g is, because that's how you get f+g *

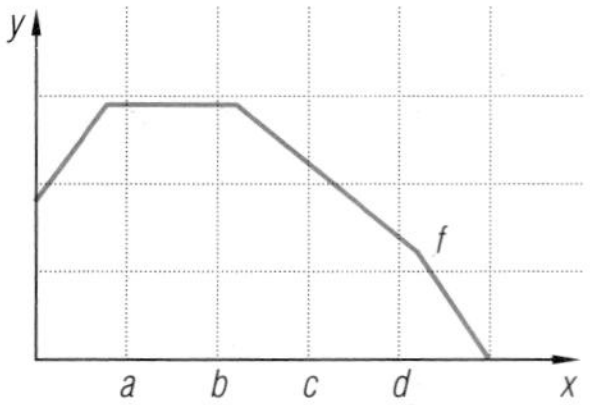

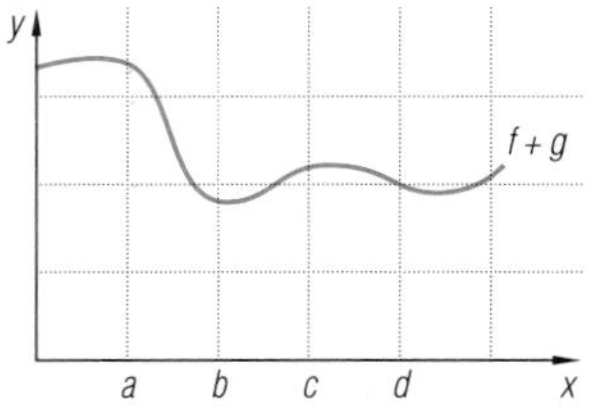

Determine whether $g(x)$ is positive, negative, or zero when
a. $x = a$. positive **b.** $x = b$. negative **c.** $x = c$. zero **d.** $x = d$. positive

10. **a.** Find the coordinates of the vertex of the graph of the function P in this lesson. (625, 16250)
b. Does the vertex of the graph of the function R lie directly above the vertex of the graph of P? No

11. The function h defined by $h(x) = x \sin x$ is an even function that is the product of two odd functions. Give an example of another even function that is the product of two odd functions. See margin.

12. Disprove: *For all real functions f and g, if f and g are increasing functions, then f • g is an increasing function.* See margin.

13. At time $t = 0$ a cargo boat is in a harbor with its deck level with the pier. It is being loaded with grain, which causes the boat to sink 0.1 meter every minute. At the same time the wave action raises and lowers the boat according to the formula $h = \sin(.5t)$ where h = height in meters after t seconds.
a. Write a formula for the function s, where $s(t)$ is the height above or below the pier after t seconds. $s(t) = \sin(.5t) - \frac{.1}{60}t$
b. s is the sum of a sine wave and what other kind of function?
c. Graph s, for $0 \le t \le 300$. b) linear; c) See margin.

Review

14. Let f and g be functions defined by

$$f: x \rightarrow x^2 + 1 \text{ and } g: x \rightarrow \sqrt{x - 1}.$$

a. Find a rule for $g \circ f$. **b.** Identify the domain of $g \circ f$.
c. Find a rule for $f \circ g$. **d.** Identify the domain of $f \circ g$.
e. Are f and g inverses? *(Lesson 3-2)* **a) $g \circ f(x) = |x|$; b) the set of real numbers; c) $f \circ g(x) = x$; d) $\{x: x \geq 1\}$; e) No**

In 15–18, solve. *(Lessons 3-2, 3-1, 2-7)*

15. $x - \sqrt{x + 14} = -2$. **$x = 2$**

16. $a^{-1/2} = 6$. **$a = \frac{1}{36}$**

17. $10^x = 4128$ **$x \approx 3.616$**

18. $e^t = 16.3$ **$t \approx 2.791$**

19. Suppose f is a function defined on the set of natural numbers such that $\forall\ n$, $f(n) > 3$ and $\lim_{n \to \infty} f(n) = 3$. Sketch a possible graph of f. *(Lesson 2-5)* **See margin.**

20. Write the negation of each statement.
a. $\forall$ *satellites s, s is a military spy satellite.*
b. $\exists$ *a person p such that p is a leader of a trade union and p has received the Nobel Peace Prize.* *(Lessons 1-2, 1-3)* **a) $\exists$ a satellite s such that s is not a military spy satellite. b) See margin.**

21. Factor $x^4 + 6x^2 + 8$. *(Previous course)* **$(x^2 + 4)(x^2 + 2)$**

Exploration

22. The profit function in this lesson has an unwise property: it is not increasing on the interval $0 \leq x \leq 1000$. Modify the incentive discount plan to get a profit function that is increasing for all x on this interval. **Sample: Change the sales incentive discount to 5%, yielding $P(x) = 100x - 15{,}000 - .05x^2$.**

23. Does Question 11 exemplify a more general property? That is, is the product of two odd functions always an even function? Justify your answer. **Let f and g be two odd functions, and let $h = f \cdot g$. Then for all x in the domain of h**

$$\begin{aligned} h(x) &= f(x) \cdot g(x) \\ &= (-f(-x)) \cdot (-g(-x)) \quad \text{Definition of odd function} \\ &= f(-x) \cdot g(-x) \\ &= h(-x). \end{aligned}$$

So h is an even function, by the definition of even function, and the conjecture is true.

ADDITIONAL ANSWERS
8.a., 11., 12., 13.c., 19.
See Additional Answers in the back of this book.

20.b. ***$\forall$ people p, p is not the leader of a trade union or p has not received the Nobel Peace Prize.***

NAME ____________

LESSON MASTER 3–3
QUESTIONS ON **SPUR** OBJECTIVES

■ **SKILLS** *Objective B (See pages 217–220 for objectives.)*
For 1 and 2, formulas for f and g are given. Write simplified formulas for a. $f + g$, b. $f \circ g$, and c. $\frac{f}{g}$.

1. $f(x) = e^x$ and $g(x) = e^{x+2}$
   a. $(f + g)(x) = e^x + e^{x+2} = (1 + e^2)e^x$
   b. $(f \circ g)(x) = e^{e^{x+2}}$
   c. $\frac{f}{g}(x) = \frac{1}{e^2}$

2. $f(x) = x^3 - 5x$ and $g(x) = \frac{1}{x}$
   a. $(f + g)(x) = x^3 - 5x + \frac{1}{x}$
   b. $(f \circ g)(x) = \frac{1}{x^3} - \frac{5}{x}$
   c. $\frac{f}{g}(x) = x^4 - 5x^2$

3. Using the formulas given in Question 2, give the domains of $\frac{f}{g}$ and $\frac{g}{f}$.
   $\{x : x \neq 0\}$; $\{x : x \neq 0, x \neq \sqrt{5}, \text{ and } x \neq -\sqrt{5}\}$

■ **USES** *Objective K*

4. Working at its slowest speed, a machine can produce 250 magnetic disks per hour with a 98% efficiency rate (2% are defective). For each step up in speed, the machine can produce 10 more disks per hour, but its efficiency rate drops .5%.
   a. Write a formula for n, where $n(x)$ is the total number of disks produced hourly when the speed is increased x steps. $n(x) = 250 + 10x$
   b. Write a formula for $e(x)$, the efficiency rate when the speed is raised x steps. $e(x) = .98 - .005x$
   c. Let $w(x)$ be the number of nondefective disks produced per hour when the speed is raised x steps. How are w, n, and e related? $w(x) = n(x)e(x)$
   d. Write a formula for $w(x)$ in terms of x. $w(x) = 245 + 8.55x - .05x^2$
   e. Graph the function w on the grid at the right.

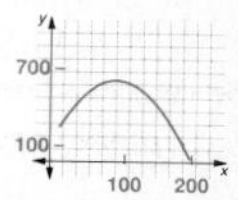

NAME ____________
Lesson MASTER 3–3 (page 2)

   f. At how many steps above the slowest speed should the machine be set to produce the maximum number of nondefective disks per hour? 85 or 86
   g. What is the maximum number of nondefective disks that the machine can produce per hour? 610

■ **REPRESENTATIONS** *Objective M*

5. The functions f and g are graphed at the right. On the grids below, sketch the graphs of
   a. $f + g$. b. $f \cdot g$.

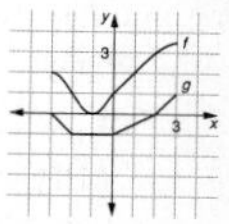

a.

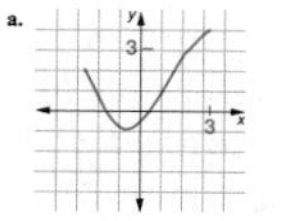

b.

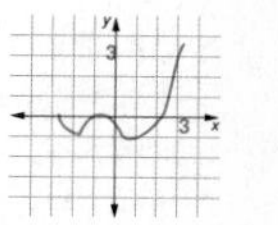

6. The graphs of two functions f and $f + g$ are shown at the right. Tell if $g(x)$ is positive, negative, or zero when

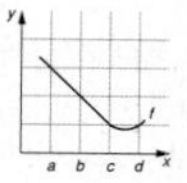

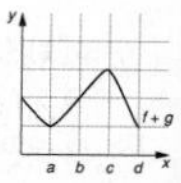

   a. $x = a$. negative
   b. $x = b$. zero
   c. $x = c$. positive
   d. $x = d$. zero

LESSON 3-4

RESOURCES
- Lesson Master 3-4
- Teaching Aid 16 (same as in Lesson 3-3)

OBJECTIVE

C Find zeros of functions and solutions to equations using factoring or chunking.

TEACHING NOTES

Chunking is, in theory, simply the use of the Substitution Property from Lesson 1-1. In **Example 1**, you may wish to make the chunking explicit by substituting u for e^x as follows:

$$e^{2x} - e^x - 6 = 0$$
$$\Leftrightarrow \quad (e^x)^2 - e^x - 6 = 0$$
$$\Leftrightarrow \quad u^2 - u - 6 = 0$$
$$\Leftrightarrow \quad (u - 3)(u + 2) = 0.$$

Now the chunking idea is used again. Both $u - 3$ and $u + 2$ can be thought of as single numbers whose product is zero. Thus, one or the other is zero.

$$\Leftrightarrow \quad u - 3 = 0 \text{ or } u + 2 = 0$$
$$\Leftrightarrow \quad u = 3 \text{ or } u = \text{-}2$$

Now, let $e^x = u$.
$e^x = 3$ or $e^x = \text{-}2$ (impossible)
Thus, $x = \ln 3$.

Chunking is made explicit in **Example 3** also. But it is not necessary always to do the substitution. Students should get in the habit of seeing an expression and thinking of it as a chunk, as is the case in **Example 4**.

LESSON

3-4 Chunking

Chunking refers to thinking of an expression as if it were a single variable. Chunking is a very powerful method with wide applicability.

Example 1

Solve the equation $e^{2x} = e^x + 6$.

Solution Think of e^x as a chunk. The equation in e^x is a quadratic equation which you know how to solve.

$$e^{2x} - e^x - 6 = 0$$
$$\Leftrightarrow \quad (e^x)^2 - e^x - 6 = 0$$
$$\Leftrightarrow \quad (e^x - 3)(e^x + 2) = 0$$
$$\Leftrightarrow \quad e^x - 3 = 0 \text{ or } e^x + 2 = 0$$
$$\Leftrightarrow \quad e^x = 3 \text{ or } e^x = \text{-}2$$
$$\Leftrightarrow x = \ln 3 \approx 1.0986 \text{ or impossible}$$

(When x is real, e^x is always positive, so the equation $e^x + 2 = 0$ has no real solutions.)

Check We use the window $0 \le x \le 2$, $0 \le y \le 12$, and graph $y = e^{2x}$ and $y = e^x + 6$. The x-coordinate of the point of intersection is the solution to $e^{2x} = e^x + 6$.

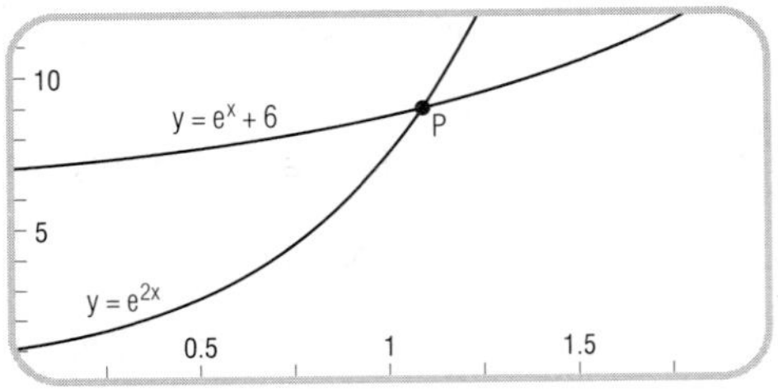

The trace function shows that the x-coordinate of P is approximately 1.10 . This is very close to ln 3.

The equation in Example 1 is of the form
$$f(x) = g(x).$$
To solve such an equation, it is often useful to rewrite it as an equation of the form
$$h(x) = 0$$
which has the same solutions. This replacement is always possible since you can take h to be the function $f - g$ defined by
$$h(x) = f(x) - g(x)$$
for all x in the domains of both f and g.

172

One advantage of working with $h(x) = 0$ is that it may be possible to factor $h(x)$, as it was in Example 1. Then the Zero Product Property can be used. Recall that when $\exists\, c$ such that $h(c) = 0$, then c is called a **zero** of h. Zeros of h that are real numbers are called **real zeros**. In the language of functions, the Zero Product Property is:

If there exists c such that $h(c) = 0$ and $h = f \cdot g$, then either $f(c) = 0$ or $g(c) = 0$.

Stated another way: For all functions f, g, and h, if $h = f \cdot g$, then $\forall\, x$,

$$h(x) = 0 \Leftrightarrow f(x) = 0 \text{ or } g(x) = 0.$$

Thus if an expression for a function is factored, the task of finding its zeros is reduced to finding the zeros of the factors.

Example 2 Find the zeros of the function p defined by
$$p(x) = (x^2 - 3)(5x + 8)^2.$$

Solution Here $p = f \cdot g$, where $f(x) = x^2 - 3$ and $g(x) = (5x + 8)^2$. So a real number x is a zero of p if and only if

$$\begin{array}{lrcr} & f(x) = 0 & \text{or} & g(x) = 0 \\ \Leftrightarrow & x^2 - 3 = 0 & \text{or} & (5x + 8)^2 = 0 \\ \Leftrightarrow & x^2 = 3 & \text{or} & 5x + 8 = 0 \\ \Leftrightarrow & x = \pm\sqrt{3} & \text{or} & x = -\frac{8}{5}. \end{array}$$

Therefore, p has three zeros, $\sqrt{3}$, $-\sqrt{3}$, and $-\frac{8}{5}$.

Notice in Example 2 that $5x + 8$ could be considered as a chunk. Like any real number, its square is zero if and only if it equals zero. Sometimes work is made easier by using substitution to replace a chunk with a single variable. Once the resulting equation has been solved, the chunk is substituted back.

Example 3 Find all real solutions to $(m + 3)^4 - 3 = m$.

Solution This looks difficult to solve, but notice that if 3 is added to both sides, then the variable m appears only in the expression $m + 3$. Suppose m is a real number. Then

$$\begin{array}{lrl} & (m + 3)^4 - 3 = m & \\ \Leftrightarrow & (m + 3)^4 = m + 3 & \\ \Leftrightarrow & x^4 = x & \text{Where } x = m + 3 \text{ (the chunk)} \\ \Leftrightarrow & x^4 - x = 0 & \\ \Leftrightarrow & x(x^3 - 1) = 0 & \\ \Leftrightarrow & x = 0 \text{ or } x^3 - 1 = 0 & \text{Zero Product Property} \\ \Leftrightarrow & x = 0 \text{ or } x = 1 & \\ \Leftrightarrow & m + 3 = 0 \text{ or } m + 3 = 1 & \text{Substituting back } m + 3 \text{ for } x \\ \Leftrightarrow & m = -3 \text{ or } m = -2. & \end{array}$$

Check If $m = -3$, $(m + 3)^4 - 3 = (-3 + 3)^4 - 3 = 0 - 3 = -3 = m$
If $m = -2$, $(m + 3)^4 - 3 = (-2 + 3)^4 - 3 = 1 - 3 = -2 = m$

ADDITIONAL EXAMPLES

1. Solve the equation $e^{4x} = 13e^{2x} - 36$.

Think of e^{2x} as a chunk.
$(e^{2x})^2 = 13e^{2x} - 36$
$\Leftrightarrow (e^{2x})^2 - 13e^{2x} + 36 = 0$
$\Leftrightarrow (e^{2x} - 4)(e^{2x} - 9) = 0$
Now chunk again, thinking of e^{2x} as $(e^x)^2$.
$((e^x)^2 - 4)((e^x)^2 - 9) = 0$
$\Leftrightarrow (e^x)^2 - 4 = 0$ or $(e^x)^2 - 9 = 0$
$\Leftrightarrow (e^x + 2)(e^x - 2) = 0$
or $(e^x + 3)(e^x - 3) = 0$
$\Leftrightarrow e^x = 2$ or $e^x = 3$
$\Leftrightarrow x = \ln 2 \approx .6931$
or $x = \ln 3 \approx 1.0986$
$e^x = -2$, $e^x = -3$ do not have real solutions.

2. Find the zeros of the function q defined by $q(x) = (4x^2 - 5)(3x - 7)^2$.

Here, $q = f \cdot g$ where $f(x) = 4x^2 - 5$ and $g(x) = (3x - 7)^2$. So $q(x) = 0$ if and only if
$f(x) = 0$ or $g(x) = 0$
$\Leftrightarrow 4x^2 - 5 = 0$ or $(3x - 7)^2 = 0$
$\Leftrightarrow x^2 = \frac{5}{4}$ or $3x = 7$
$\Leftrightarrow x = \pm\frac{\sqrt{5}}{2}$ or $x = \frac{7}{3}$
Therefore, q has three zeros: $\frac{\sqrt{5}}{2}$, $-\frac{\sqrt{5}}{2}$, and $\frac{7}{3}$.

3. Find all real solutions to $(m + 5)^4 - 5 = m$.

$(m + 5)^4 - 5 = m$
$\Leftrightarrow (m + 5)^4 = m + 5$
Let $x = m + 5$
$\Leftrightarrow x^4 = x$
$\Leftrightarrow x^4 - x = 0$
$\Leftrightarrow x(x^3 - 1) = 0$
$\Leftrightarrow x = 0$ or $x^3 - 1 = 0$
$\Leftrightarrow x = 0$ or $x = 1$
(Substitute back $m + 5$ for x.)
$\Leftrightarrow m = -5$ or $m = -4$.

ADDITIONAL EXAMPLES
4. Solve $y^8 - 8y^5 + 12y^2 = 0$.
Factor out y^2 on the left side:
$y^2(y^6 - 8y^3 + 12) = 0$
$\Leftrightarrow y^2 = 0$ or $y^6 - 8y^3 + 12 = 0$
$\Leftrightarrow y = 0$ or $y^6 - 8y^3 + 12 = 0$.
Let $x = y^3$ so $x^2 - 8x + 12 = 0$
or $(x - 2)(x - 6) = 0$
or $x = 2$ or $x = 6$.
Replacing x by y^3 means
$y^3 = 2$ or $y^3 = 6$
or $y = \sqrt[3]{2}$ or $y = \sqrt[3]{6}$
Thus, the three real solutions are $0, \sqrt[3]{2}, \sqrt[3]{6}$.

NOTES ON QUESTIONS
Error Analysis for Question 11: This equation may confuse some students, who will replace *T* by 1512. Remind them what *E* and *T* represent. Then you may have to suggest chunking. What is true if $T = 100$? (You are at sea level.)

Question 18: Students may wonder where the equation for *c* comes from. For the car, the position is given by the product of half the acceleration and the square of the time. Half the acceleration is 20. Because the runner started 2 seconds before the car, the time that the car has traveled is $t - 2$.

The next example combines chunking and factoring to solve the equation given on the first page of this chapter.

Example 4 Solve $y^{10} - 5y^6 + 4y^2 = 0$.

Solution By factoring the left side,
$$y^2(y^8 - 5y^4 + 4) = 0$$
So either $y^2 = 0$ or $y^8 - 5y^4 + 4 = 0$. The first equation gives 0 as a solution. Concentrate now on the second equation, and think of y^4 as a chunk. The equation is quadratic in y^4.
$$\Leftrightarrow \quad (y^4)^2 - 5(y^4) + 4 = 0$$
$$\Leftrightarrow \quad (y^4 - 4)(y^4 - 1) = 0$$
$$\Leftrightarrow \quad y^4 = 4 \text{ or } y^4 = 1$$
$$\Leftrightarrow \quad y = \sqrt[4]{4} \text{ or } y = -\sqrt[4]{4} \text{ or } y = \sqrt[4]{1} = 1 \text{ or } y = -\sqrt[4]{1} = -1.$$
There are thus five real solutions: 0, 1, – 1, $\sqrt[4]{4}$, and $-\sqrt[4]{4}$. You can check that there is no error; each of these satisfies the given equation.

You sometimes have to use quite a bit of creativity and experience to find the exact values of the zeros of a given function by algebraic methods. In fact, algebraic methods sometimes fail completely. In the next lesson you will learn how to approximate zeros of functions when an exact solution is not obvious.

Questions

Covering the Reading

$8^{\frac{1}{2}}$

In 1 and 2, **a.** what expression might be thought of as a chunk? **b.** Solve.

1. $4 \cdot 8^{2x} = 4 \cdot 8^x - 1$
a) 8^x; b) $x = -\frac{1}{3}$

2. $(2d - 5)^3 + 5 = 2d$
a) $2d - 5$; b) $d = \frac{5}{2}$, $d = 2$, and $d = 3$

3. Define: zero of a function.
A zero of a function $h(x)$ is a number c such that $h(c) = 0$.

4. State the Zero Product Property in the language of **a.** numbers; **b.** functions. See margin.

5. Finding the solutions to $3x = 2^x + 1$ is equivalent to finding the zeros of what function? $h(x) = 3x - 2^x - 1$

In 6–8, find the zeros of the function defined by the given rule.

6. $f(x) = x(x + 1)(x + 2)$ $x = 0$, $x = -1$, and $x = -2$

7. $g(t) = (t^2 - 1)^3$ $t = 1$ and $t = -1$

8. $h(x) = x^3 + 2x^2 - 2x$ $x = 0$, $x = -1 - \sqrt{3}$, and $x = -1 + \sqrt{3}$

In 9 and 10, find all real solutions.

9. $3x^6 - 20x^3 = 32$
$x = 2$ and $x = -\left(\frac{4}{3}\right)^{1/3} \approx -1.1$

10. $y^{10} = 2y^5$
$y = 0$ and $y = 2^{1/5} \approx 1.15$

174

Applying the Mathematics

11. At higher altitudes, water boils at lower temperatures. The equation $E \approx 1000(100 - T) + 580(100 - T)^2$ relates the temperature T (in °C) at which water boils to the elevation E meters above sea level. To the nearest degree T, find the boiling point of water in Albuquerque, about 1512 meters above sea level. 99°C

In 12–15, use chunking to solve the given equation.

12. $(\log_2 x)^2 - 10(\log_2 x) = -16$
[Hint: Use the substitution $u = \log_2 x$.] $x = 4$ or $x = 256$

13. $2t^4 + 3t^2 = 2$ $t = \pm\sqrt{\frac{1}{2}}$

14. $2^{2x-1} = 10$ $x \approx 2.161$

15. $22 - 3\left(\frac{3n+5}{4}\right)^2 = 5\left(\frac{3n+5}{4}\right)$ $n = 1$ or $n = -\frac{59}{9}$

16. Find the coordinates of the point of intersection of the graphs of $g(x) = e^{-x}$ and $f(x) = e^{-2x} - 2$ by using the substitution $u = e^{-x}$. (-.693, 2)

17. Find all zeros of the function h in the interval $0 \le x \le 2\pi$, where $h(x) = 2\cos^2 x - 3\cos x - 2$. $\frac{2\pi}{3}, \frac{4\pi}{3}$

18. A runner travels at a speed of 20 ft/sec. Two seconds after the runner begins, a car leaves the same location and heads in the same direction with a constant acceleration of 40 ft/sec². As a result, the distance in feet traveled by the car t seconds $t \ge 2$ after the runner starts is given by $c(t) = 20(t - 2)^2$.

a. Write a formula for the distance $r(t)$ traveled by the runner after t seconds. $r(t) = 20t$

b. Write the equation which expresses the condition that the car has caught up with the runner. $20t = 20(t - 2)^2, t \ge 2$

c. Write the equation from part **b** in the form $h(t) = 0$.

d. What is the meaning of $h(t)$?

e. Solve the equation from part **c**. c) $20t^2 - 100t + 80 = 0, t \ge 2$;
d) When $h(t) = 0$, the runner and the car are at the same location.
e) $t = 4$ seconds

Review

19. a. Is the function $f: x \rightarrow 2^x$ a 1–1 function? Yes

b. Solve $\left(\frac{1}{2}\right)^{x-1} = 4^{(x^2)}$ by first writing each base as a power of 2. *(Lesson 3-1)* $x = \frac{1}{2}$ or $x = -1$

20. Let $f(x) = \frac{1}{2 - 4x}$ and $g(x) = 2x - 1$.

a. Find and simplify the formula for $f \bullet g$. $f \bullet g = -\frac{1}{2}$

b. What is the domain of $f \bullet g$? *(Lesson 3-3)* $\{x: x \neq \frac{1}{2}\}$

21. Let $f(x) = x$ and $g(x) = x^2 - 8x + 16$ and let $h = f \bullet g$.

a. Graph h on an automatic grapher. Describe some properties of h.

b. Describe how special values of functions f and g determine the graph of h. *(Lesson 3-3)* See margin.

ADDITIONAL ANSWERS

4.a. For all numbers a and b, $ab = 0 \Leftrightarrow a = 0$ or $b = 0$.

b. For all functions f, g, h, if $h = f \cdot g$, then $\forall x$, $h(x) = 0 \Leftrightarrow f(x) = 0$ or $g(x) = 0$.

21.a.

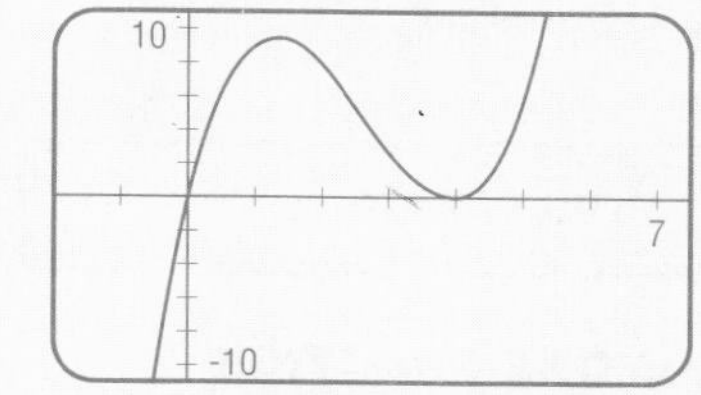

b. The zeros of f and the zeros of g will also be zeros of $f \cdot g$. So the x-intercepts of $f \cdot g$ will be the x-intercepts of f and the x-intercepts of g.

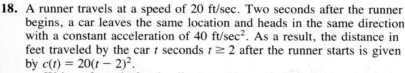

NAME ____________

LESSON **MASTER 3–4**
QUESTIONS ON **SPUR** OBJECTIVES

■ **SKILLS** *Objective C (See pages 217–220 for objectives.)*

1. Consider the equation $(\log_7 (x + 4))^2 - 8\log_7 (x + 4) + 15 = 0$.

a. Write the equation resulting from the substitution $u = \log_7 (x + 4)$. $u^2 - 8u + 15 = 0$

b. Solve the equation you found in part **a**. $u = 3$ or $u = 5$

c. Substitute $u = \log_7 (x + 4)$ back into your results from part **b** and solve for x.

$x = 339$ or $x = 16{,}803$

In 2 and 3, find the zeros of the function.

2. $g(x) = x^2 - 36$ — $x = \pm 6$

3. $h(x) = x^3 - 7x^2 - 8x$ — $x = 0$ or $x = 8$ or $x = -1$

In 4 and 5, find all solutions.

4. $(3k + 8)^3 - 2 = 3k + 6$ — $k = -3$ or $k = -\frac{7}{3}$ or $k = -\frac{8}{3}$

5. $0 = x^4 - 7x^2 + 12$ — $x = \pm 2$ or $x = \pm\sqrt{3}$

6. Give exact values of the zeros of $f(x) = 2\sin^2 x + \sin x - 1$ on $0 \le x \le 2\pi$.

$x = \frac{\pi}{6}$ or $x = \frac{5\pi}{6}$ or $x = \frac{3\pi}{2}$

7. Approximate the solutions of $5^{2x} + 18 = 9 \cdot 5^x$ to the nearest tenth. $x \approx .7$ or $x \approx 1.1$

8. a. Multiply $(n^2 + 7n + 30)$ by $(n - 4)$. $n^3 + 3n^2 + 2n - 120$

b. When spheres are stacked in a triangular pyramid with n layers, $s(n) = \frac{1}{6}n^3 + \frac{1}{2}n^2 + \frac{1}{3}n$ gives the number of spheres used. Use part **a** to help you find the number of layers that can be made with 20 spheres. 4

32

NOTES ON QUESTIONS
Question 23: Some students may wish to estimate the answer to part **a** by using an automatic grapher, but they should be able to obtain the vertex of the parabola exactly.

FOLLOW-UP

MORE PRACTICE
For more questions on SPUR Objectives, use *Lesson Master 3-4,* shown on page 175.

EXTENSION
Give students a challenge of coming up with an equation that looks difficult to solve but is, like **Example 3**, a reasonably simple equation with an unexpected or complicated chunk. This activity is suitable for **small group work**.
(Sample: $\sqrt[6]{x} + 2 = \sqrt[3]{x}$; here, $x = 64$; solutions are found by letting $y = \sqrt[6]{x}$.)

ADDITIONAL ANSWERS
23.b. At $t = .625$ seconds, the object reaches its maximum height, changes direction, and comes back down.

26.

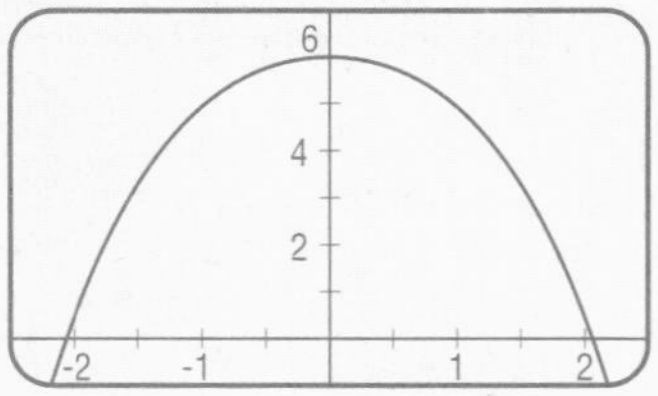

22. Use an automatic grapher to estimate the range of f defined by $f(x) = x^4 - 2x^2 + 3x + 1$. *(Lesson 2-2)* Sample: $\{y: y \geq -3.4\}$

23. The height h of an object thrown upward from a height of 6 ft with a velocity of 20 ft/sec is given by the equation $h(t) = -16t^2 + 20t + 6$, where t is the time in seconds.

a. Determine the interval over which the function is increasing.

b. What is the physical meaning of your answer to part **a**? *(Lesson 2-3)*

a) $0 \leq t \leq .625$; b) See margin.

24. Write a logical expression that describes the network below. *(Lesson 1-4)* *(p and q) or (not r)*

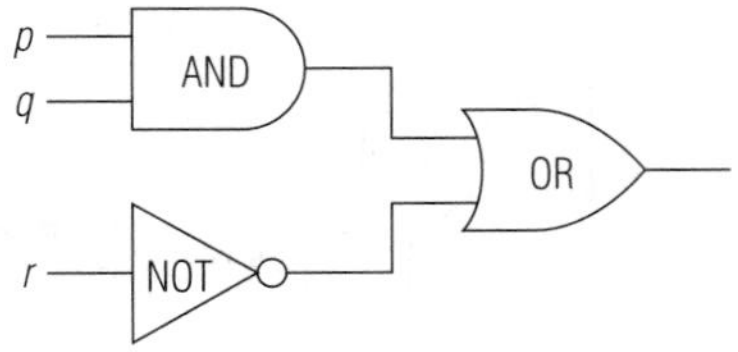

25. Find all solutions to the equation between 0 and 2π.

a. $\sin x = \frac{1}{2}$ $\frac{\pi}{6}, \frac{5\pi}{6}$

b. $\sin^2 x = \frac{1}{4}$ *(Previous course)* $\frac{\pi}{6}, \frac{5\pi}{6}, \frac{7\pi}{6}, \frac{11\pi}{6}$

Exploration

26. The curve in the design of the Gateway Arch in St. Louis, the catenary, was selected not only for its beauty but also because it is the best shape to support its own weight. This is the same type of curve as the graph of the function f defined by

$$f(x) = 8 - (e^x + e^{-x}).$$

Give some properties of f. Sample: f is symmetric to the y-axis, has y-intercept 6 and x-intercepts approximately -2.1 and 2.1. The graph is shown in the margin.

LESSON

3-5

Approximating Solutions to Equations

In Lesson 3-3, the total cost of manufacturing x washing machines in a day was given by $C(x)$, and the total revenue from selling those machines was $R(x)$. The solutions to the equation

$$C(x) = R(x)$$

are the *break-even values*, the values of x for which the revenue equals the cost. There are two ways to find the break-even values graphically. One way is to graph

$$y = C(x) \text{ and } y = R(x)$$

and look for the x-coordinates of any points of intersection of the graphs. The other way is to subtract one function from the other, graph the difference, and look for the zeros of the difference function. Recall that the profit function P is defined by

$$P(x) = R(x) - C(x).$$

C, R, and P are graphed here

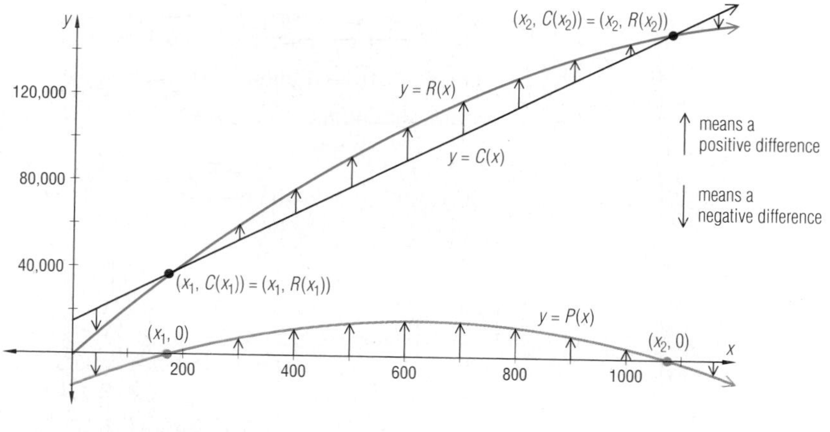

Notice that the zeros of the function P are the same as the x-coordinates of the points of intersection of C and R. This is because $\forall\, x$, $P(x) = 0 \Leftrightarrow R(x) - C(x) = 0 \Leftrightarrow R(x) = C(x)$. The break-even values are exactly those values for which the profit is zero.

LESSON 3-5

RESOURCES

■ Lesson Master 3-5

OBJECTIVES

D Use the Intermediate Value Theorem and the Bisection Method to locate or approximate zeros of a function.

H Identify continuous functions and their properties.

J Apply equation-solving techniques to real-world problems.

O Use graphs to approximate zeros of functions and solve equations and inequalities.

TEACHING NOTES

The two graphs at the top of page 178 provide a visual representation of the algebraic techniques of Lesson 3-3. Students can see that the x-coordinates of the intersection points of two curves correspond to the zeros of a new function formed by subtracting the two original functions.

We do not provide a formal definition of *continuous*. The intuitive notion of an unbroken curve on some interval provides students with a sufficient understanding of the term for this course. In particular, students need to understand the Intermediate Value Theorem so they can determine appropriate intervals where zeros can be expected. You might ask students to determine whether or not the Intermediate Value Theorem would apply to the functions on the next page.

(1) $f(x) = x^3 + x^2 + 4$ on the interval $-3 \leq x \leq 3$ (yes)
(2) $g(x) = \log x$ on the interval $-1 \leq x \leq 10$ (no)
(3) $h(x) = e^x - 2$ on the real numbers (yes)
(4) $p(x) = \tan x$ on the interval $\frac{5\pi}{4} \leq x \leq \frac{7\pi}{4}$ (no)

Problems such as those above also serve to review many of the ideas from earlier lessons of this chapter and give students an opportunity to discuss the domains of these functions.

Technology For **Example 2**, use an automatic grapher to show the graphs of $y = 2^x$, $y = 3x^2$, and $y = 2^x - 3x^2$ on one screen. Point out that at the points where $y = 2^x$ and $y = 3x^2$ intersect, $y = 2^x - 3x^2$ has its zeros. Many students prefer to find zeros of $h(x) = f(x) - g(x)$ rather than the x-value of the intersection of $f(x)$ and $g(x)$.

A spreadsheet can generate a table similar to that for **Example 2** very easily. Only columns for x and $2^2 - 3x^2$ are needed. For intervals with sign changes, students may use the spreadhseet to "zoom in" on zeros. For example, to zoom in on a zero(s) in the $7 < x < 8$ interval, have the spreadsheet calculate $f(x)$ for $7 < x < 8$ where x values are incremented by .1. Then repeat on an appropriate interval with x increments of .01, and so on.

You also may apply the Bisection Method when zooming in with an automatic grapher. Be sure to change both x and y intervals to keep the graph in perspective.

In general, the real solutions of the equation $f(x) = g(x)$ are the x-coordinates of the intersection points of the two graphs $y = f(x)$ and $y = g(x)$. They are also the values of x at which the graph of $f - g$ intersects the x-axis.

The x-coordinates of P_1 and P_2, namely c_1 and c_2, are the solutions to $f(x) = g(x)$.

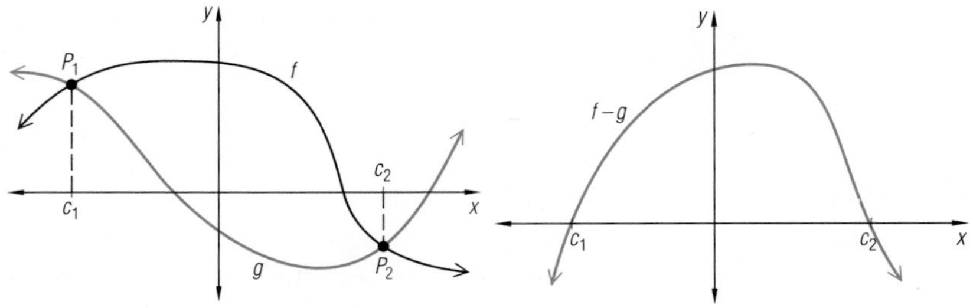

This suggests two graphical approaches to solving equations:
(1) plot f and g and find the approximate values of x where the graphs intersect
or (2) graph $f - g$ and find the approximate values of x where the graph of $f - g$ crosses the x-axis.

In the case of the washing machines, $P(x) = 100x - 15{,}000 - .08x^2$, and graphing shows $P(174) \approx 0$ and $P(1076) \approx 0$.

For the above example, you can get a better and better solution by zooming in on the graph, and an exact solution by using the quadratic formula. Equations that can be solved by applying a formula like the quadratic formula are rather special. Consequently, we would now like to put forth a method that works to approximate solutions to many kinds of equations. For this we need to use the concept of *continuity*. Informally, a function f is **continuous on an interval** I if every point in I is in the domain of f and if the graph of f on I is an unbroken curve. Although this description of the term *continuous* is not a precise definition (because the term *unbroken curve* has not been defined), it is a useful test in simple situations in deciding whether a function is continuous.

Note that if the interval I contains a point that is not in the domain of f, then f is not continuous on I. For example, the reciprocal function $f(x) = \frac{1}{x}$ is not continuous on the interval $-2 < x < 3$ because f is not defined at $x = 0$ but 0 is in that interval.

However, a real function f may be defined at all points of an interval and still may not be continuous on that interval.

Example 1 Sketch a graph to decide if the **floor function** ***F***, defined by
$F(x) = \lfloor x \rfloor =$ the greatest integer $\leq x$,
is continuous on the interval $-2 \leq x \leq 2$.

$-2 \leq x \leq 2$
greatest integer
$f(x) = f(-2) = -2$
$f(-1) = -1$
$f(-1.1) = -2$

178

Solution Using the definition of $\lfloor x \rfloor$, if $-2 \le x < -1$, then $F(x) = -2$; if $-1 \le x < 0$, then $F(x) = -1$; if $0 \le x < 1$, then $F(x) = 0$; if $1 \le x < 2$, then $F(x) = 1$; and finally, $F(2) = 2$. A graph of F is given at right. The graph of F is broken at $x = -1, 0, 1$, and 2 in the interval $-2 \le x \le 2$. Therefore, F is not continuous on this interval.

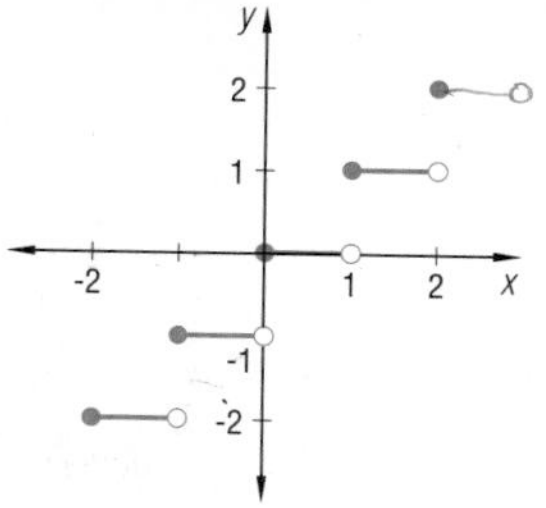

In Example 1 $F(-2) = -2$ and $F(2) = 2$ but many real numbers between -2 and 2 which are not attained as values of the function. For instance, there is no real number c in the interval $-2 \le x \le 2$ such that $F(c) = \frac{1}{2}$. F "skips" over some values between $F(-2)$ and $F(2)$. The next theorem, whose proof is beyond the scope of this course, states that such "skipping" is not possible for a continuous function. It says that if a continuous function takes two distinct values on some interval, then it must take all values between these two.

The Intermediate Value Theorem

Suppose that f is a continuous real function on an interval I, and that a and b are on I. Then for every real number y_0 between $f(a)$ and $f(b)$, there is at least one real number x_0 between a and b such that $f(x_0) = y_0$.

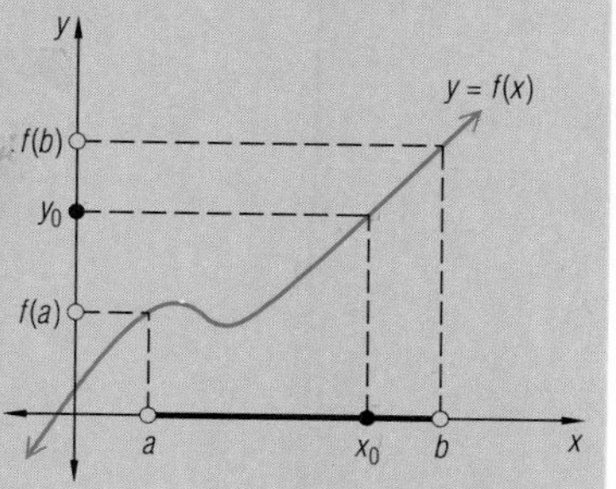

Your experience with the graphs of power, exponential, logarithmic, and trigonometric functions should suggest that these functions are continuous on any interval that is entirely contained in their domains. By methods beyond the scope of this book, this can be proved. It can also be proved that if f and g are continuous functions on an interval I, then

(1) the functions $f + g$, $f - g$, and $f \cdot g$ are continuous on I

and (2) the function $\frac{f}{g}$ is continuous on I provided that $g(x) \neq 0$ for any x in I.

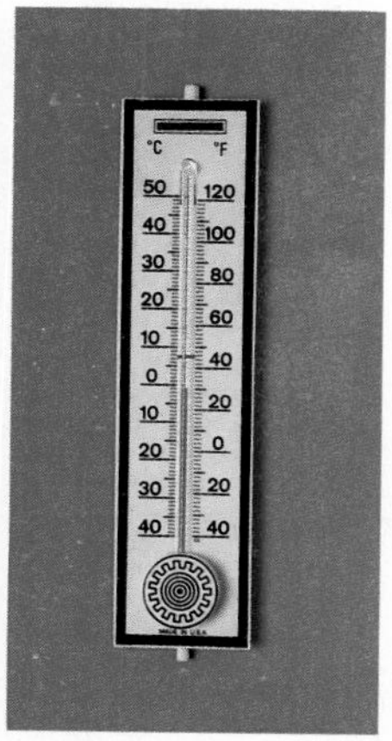

Temperature readings over time can be modeled by a continuous function.

For instance, since the natural logarithm and cosine functions $f(x) = \ln x$ and $g(x) = \cos x$ are continuous on the interval $x > 0$ of positive real numbers, so are

their sum $s(x) = \ln x + \cos x$,

their difference $d(x) = \ln x - \cos x$,

and their product, $p(x) = (\ln x)(\cos x)$.

However, their quotient $q(x) = \frac{\ln x}{\cos x}$ is not continuous on the interval of positive real numbers $x > 0$ because there are many positive real numbers for which $\cos x = 0$. Yet, on the smaller interval $0 < x < \frac{\pi}{2}$, the function q is continuous.

Students probably will need help with the Bisection Method in order to understand both the process involved and the stopping condition. Go through **Example 2** carefully. Discuss the stopping criteria. You might then ask students to apply the technique to estimate the zero that is in the interval $7 \le x \le 8$ to within .1. (7.3) In **Question 7b**, students are asked to approximate the zero between -1 and 0.

In using the Bisection Method, numbers with many decimal places can arise very quickly. Remind students to use their calculator's memory feature(s) when doing calculations. If using bisection to locate a zero between a and b, it helps to draw a long coordinate line with a and b as its x-axis endpoints. As successive midpoints and values are calculated, students should keep track of these on this coordinate system.

Error Analysis Some students assume that the inverse of the following statement is true: If $f(a)$ and $f(b)$ have opposite signs, then there would be at least one zero c between a and b. You can dispel this misconception with a table and graph of $f(x) = 9x^2 - 45x + 56$. $f(2) = 2$ and $f(3) = 2$, but the zeros of f are $x = \frac{7}{3}$ and $x = \frac{8}{3}$.

ADDITIONAL EXAMPLES

1. In 1990, the postal service charged 25 cents for up to a one-ounce first class letter and 20 cents for each additional ounce or fraction thereof.

a. Write a formula for the cost of sending a first class letter as a function of its weight (in ounces).

$f(x) = 25 + 20\lfloor x - 1 \rfloor$

ADDITIONAL EXAMPLES
b. Sketch the graph to decide if the function mapping weight to cost is continuous on the interval of weights x, where $0 < x \leq 6$.

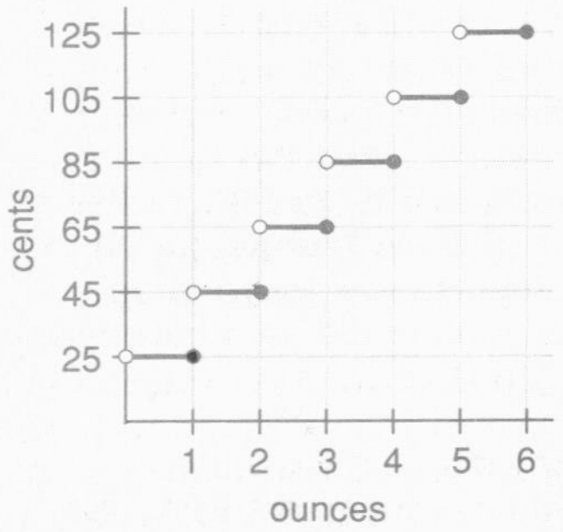

The function is not continuous on $0 < x \leq 6$.

2.a. Use the Intermediate Value Theorem to show that there must be a solution to the equation $\sin x = .4 \ln x$ between 2 and 3.
Remind students to have their calculators in radian mode. Let $f(x) = \sin x - .4 \ln x$. $f(2) = \sin 2 - .4 \ln 2 \approx .6320$ $f(3) = \sin 3 - .4 \ln 3 \approx -.2983$ Thus, $f(2) > 0$ and $f(3) < 0$. Since both $\sin x$ and $\ln x$ are continuous on $2 < x < 3$, we can apply the Intermediate Value Theorem. Since $f(2)$ and $f(3)$ have opposite signs, $\exists$ c such that $2 < c < 3$ and $f(c) = 0$.
b. How many solutions does the equation have?
The automatic grapher suggests that there are three solutions. (Use $0 \leq x \leq 20$ and $-2 \leq y \leq 2$.)

The Intermediate Value Property has the following special case, an important tool for solving equations and inequalities:

If f is a continuous function on the interval $a \leq x \leq b$ and if one of the values $f(a)$ or $f(b)$ is positive while the other is negative, then f has at least one zero between a and b.

Example 2 Given the equation $2^x = 3x^2$.
a. Use the Intermediate Value Theorem to show that there must be a solution between 0 and 1.
b. How many solutions does the equation have?

Solution
a. Let h be the function obtained from subtracting $3x^2$ from 2^x:
$$h(x) = 2^x - 3x^2.$$
Compute $h(0)$ and $h(1)$:
$$h(0) = 2^0 - 3(0)^2 = 1 \quad h(0) \text{ is positive.}$$
$$h(1) = 2^1 - 3(1)^2 = -1 \quad h(1) \text{ is negative.}$$
The function h is continuous because it is the difference of an exponential function and a power function, both of which are continuous. Because $h(0)$ and $h(1)$ have opposite signs, the Intermediate Value Theorem guarantees that there is some number c between 0 and 1 such that $h(c) = 0$. This value of c is a zero of the function h and a solution to the equation.
b. An automatic grapher suggests that the function h has at least three zeros between -2 and 8. Even without an automatic grapher, you can compute some values of 2^x and $3x^2$ with a calculator and determine which is greater. When the sign of $h(x)$ changes, it means there is a solution in that interval.

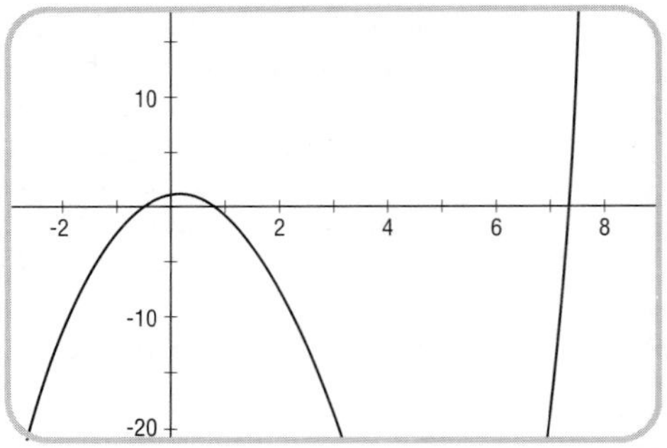

x	-2	-1	0	1	2	3	4	5	6	7	8
$h(x) = 2^x - 3x^2$	-11.75	-2.5	1	-1	-8	-19	-32	-43	-44	-19	64
sign of $h(x) = 2^x - 3x^2$	–	–	+	–	–	–	–	–	–	–	+

You can see from this table that there is a real solution in each of the intervals
$$-1 < x < 0, \quad 0 < x < 1, \quad \text{and} \quad 7 < x < 8$$
because $2^x - 3x^2$ has different signs at the endpoints of each of these intervals. Thus $2^x = 3x^2$ has at least three real solutions.

180

The Intermediate Value Theorem is behind the following general technique for approximating solutions to $f(x) = g(x)$. Suppose f and g are real functions and you want to find the approximate locations of real solutions of the equation

$$f(x) = g(x).$$

Let h be the function defined by $h(x) = f(x) - g(x)$. You are looking for values of x for which $h(x) = 0$. Find intervals $a \le x \le b$ that lie entirely in the domains of both f and g and are such that $h(a)$ and $h(b)$ have different signs. That is, at one endpoint the value of h is positive and at the other endpoint the value of h is negative. If h is a continuous function on $a \le x \le b$, the Intermediate Value Theorem guarantees that the value of h must pass through every intermediate value between $h(a)$ and $h(b)$. In particular, since 0 is between $h(a)$ and $h(b)$, there must be at least one number c in the interval $a \le x \le b$ such that $h(c) = 0$. Thus, the Intermediate Value Theorem allows you to find intervals containing zeros of continuous functions. By finding smaller and smaller such intervals, zeros can be approximated to any desired level of accuracy.

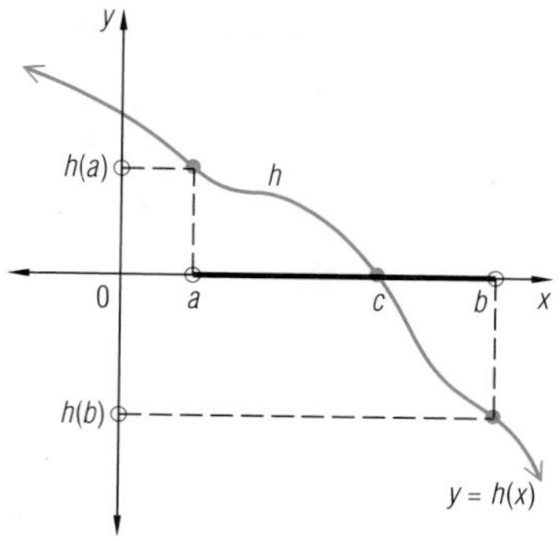

$h(b) < 0 < h(a) \Rightarrow \exists\ c$ between a and b such that $h(c) = 0$

One way to get zeros with more accuracy is by using the **Bisection Method**. This method is like the strategy used in this child's game: "I am thinking of a number between 0 and 100. I will tell you if your guess is too large or too small. Guess my number in as few tries as possible." The best strategy is to split the interval in half (bisect it) and first guess 50. Then depending on the answer to the first guess, you either guess 25 (bisecting the interval between 0 and 50) or you guess 75 (bisecting the interval between 50 and 100). And so on.

For a solution to $2^x - 3x^2 = 0$ between 0 and 1, first test the midpoint 0.5. Since $h(0.5) \approx .6642 > 0$ and $h(1) < 0$, there is a solution between 0.5 and 1. So bisect that interval and calculate $h(0.75)$. Since $h(0.75) \approx -.00571 < 0$ and $h(0.5) > 0$, there is a solution between 0.5 and 0.75. The next number to try would be 0.625. Repeating this method, you can get as close to a zero as you wish.

The Bisection Method is fairly easy to implement on a computer. Writing a BASIC program that executes the algorithm is one of the projects at the end of this chapter.

NOTES ON QUESTIONS

Questions 3 and 4: The domain is a critical part of the criterion for continuity. Theorems that specify continuity in their hypotheses must be applied carefully (if they are applied at all) to functions that are not continuous over all real numbers.

Question 9b: Of course, f has the same zero as the solution to $2x^3 - 3x + 2 = 5$.

Making Connections for Question 10b: You may wish to put this in the language of absolute value, in preparation for Lesson 3-8. The goal is to find a value of x such that $|\sin x - 0.3x| < 0.01$.

Question 20: Ask students to keep zooming in as they keep track of the count of zeros. Eventually the answer will be clear. Challenge them to explain why.
(As x approaches 0, $\frac{1}{x}$ gets larger and larger and so its sine vacillates between 1 and -1 faster and faster.)

FOLLOW-UP

MORE PRACTICE
For more questions on SPUR Objectives, use *Lesson Master 3-5,* shown on page 183.

EXTENSION
Students can use the Bisection Method with words. Have students form **small groups**. One student selects a word in a dictionary. Another chooses a word approximately in the middle of the dictionary and asks if the student's word is alphabetically before or after it. A third student splits the half into halves and asks again. This continues until someone has identified the student's word.

PROJECTS
The projects for Chapter 3 are described on page 214. **Project 3** is related to the content of this lesson.

ADDITIONAL ANSWERS
7.b. $h(.625) = .37034 > 0$. Since $h(.75) < 0$ and $h(.5) > 0$, the solution is between .625 and .75 and is therefore closer to .75 than to .5.

18. Suppose m is any even integer and k is any integer. By definition, there exists an integer r such that $m = 2r$. Thus, $m \cdot k = 2r(k) = 2(rk)$. Since rk is an integer by closure properties, the product of m and any integer is even by definition.

Questions

Covering the Reading

1. a. Approximate the break-even values for the washing machine situation. $x \approx 174$ and $x \approx 1076$

b. Where can these values be found on the graph of the functions C, R, and P? They can be found where the graphs of R and C intersect and where the graph of P intersects the x-axis.

2. Examine the graphs of f and g at right, where $f(x) = \log_2 x$ and $g(x) = (x - 2)^3$.

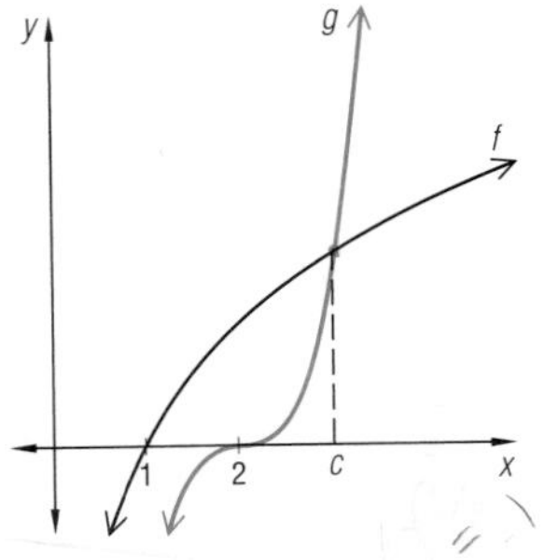

a. Write an equation for which c is a solution.

b. Write a formula for a function which has c as a zero.

a) $\log_2 x = (x - 2)^3$; b) $h(x) = \log_2 x - (x - 2)^3$

3. a. Is the floor function continuous on the interval $1 \le x \le 1.5$? Yes

b. Is the floor function continuous on the interval $1.5 \le x \le 2.5$? No

4. *Multiple choice.* A function is graphed at right. On which interval is the function continuous?

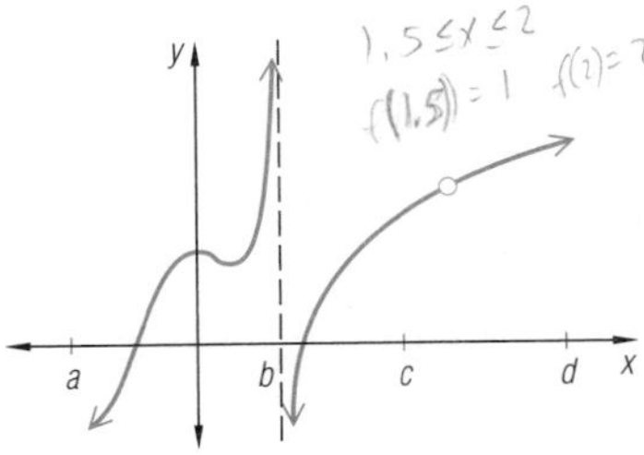

(a) $a \le x \le 0$
(b) $a \le x \le b$
(c) $b \le x \le c$
(d) $c \le x \le d$ (a)

5. Assume f is a continuous function on the interval $-2 \le x \le 3$. Some values of f are given in the table below.

x	-2	-1	0	1	2	3
$f(x)$	-4	2	10	1	3	-5

a. Find two consecutive integers such that $\exists$ some number x_0 between them with $f(x_0) = 7.5$. -1 and 0 (or 0 and 1)

b. The function f has at least how many zeros? 2

c. Between what two consecutive integers can you be sure that f has a zero? -2 and -1 (or 2 and 3)

6. Rewrite the consequent of the Intermediate Value Theorem using the $\forall$ and $\exists$ symbols. $\forall$ real numbers y_0 between $f(a)$ and $f(b)$, $\exists$ at least one real number x_0 between a and b such that $f(x_0) = y_0$.

7. For the equation $2^x - 3x^2 = 0$, a solution was found in the interval $0.5 \le x \le 0.75$.

a. Which endpoint seems to be a better approximation to the zero?

b. Verify your answer to part **a** by doing one further bisection.

a) $x = .75$; b) See margin.

Applying the Mathematics

8. Use the Intermediate Value Theorem to find an interval between two consecutive integers that contains a zero of the function h defined by $h(x) = e^x - 5x^2 + 3$. $1 \le x \le 2$ (or $-1 \le x \le 0$ or $4 \le x \le 5$)

9. a. Use the Intermediate Value Theorem to find an interval between two integers that contains a solution to the equation $2x^3 - 3x + 2 = 5$. sample: $1 \le x \le 2$

b. Use an automatic grapher to find an interval of length 0.1 that contains a zero of the function $f(x) = 2x^3 - 3x - 3$. sample: $1.5 \le x \le 1.6$

10. a. Determine the number of solutions of the equation $\sin x = .3x$ in the interval $-\pi < x < \pi$. 3 solutions

b. Find a positive value of x so that the difference between $\sin x$ and $.3x$ is less than .01. sample: 2.35

11. Is the given function continuous on the interval $-2 \le x \le 2$? If not, tell why not.

a. $f(x) = |x|$ Yes **b.** $g(x) = \frac{x}{|x|}$ No, g is not defined at $x = 0$.

12. Let $f(x) = \frac{x}{x-2}$. Then $f(1) = -1$ and $f(3) = 3$. Yet f does not have a zero between 1 and 3. Why does the Intermediate Value Theorem not apply? *f is not defined on the entire interval $1 \le x \le 3$.*

13. A hot baby bottle is set on a table to cool. Its temperature (in degrees Fahrenheit) after x minutes is given by

$$f(x) = 80e^{-0.555x} + 70.$$

When will the drink reach body temperature (98.6°)? after approximately 1.9 minutes

Review

14. Find all zeros on the interval $0 \le x \le \frac{\pi}{2}$ of the function h defined by $h(x) = 4\sin^3 x - \sin x$. *(Lesson 3-4)* $0, \frac{\pi}{6}$

15. By thinking of $\sqrt{n}$ as a chunk, solve $5\sqrt{n} = n - 300$. $n = 400$

16. Describe the end behavior of the function $f: x \to x^2$. *(Lesson 2-4)* See below.

17. Let p be the statement $\forall x$, *if 2 is a factor of x, then 6 is a factor of x.*

a. Describe exactly what it means for p to be false.

b. Give a counterexample to show p is false. sample: $x = 4$

c. Is the converse of p true or false? *(Lesson 1-5)* True

a) $\exists$ x such that 2 is a factor of x and 6 is not a factor of x.

18. Prove that the product of an even integer and any integer is even. *(Lesson 1-7)* See margin.

19. Graph the solution to ($x > -5$ and $x < 2$) or ($x > 4$ and $x < 8$). *(Previous course)* See margin.

Exploration

20. How many zeros does the function $f(x) = \sin \frac{1}{x}$ have between 0 and 1? an infinite number: $\{\frac{1}{\pi}, \frac{1}{2\pi}, \frac{1}{3\pi}, \ldots\}$

16. $\lim_{x \to \infty} x^2 = \infty$; $\lim_{x \to -\infty} x^2 = \infty$

19.

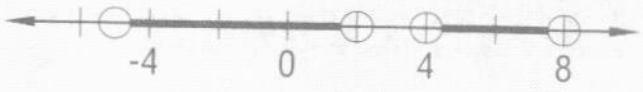

NAME ____

LESSON **MASTER 3–5**
QUESTIONS ON **SPUR** OBJECTIVES

■ **SKILLS** *Objective D (See pages 217–220 for objectives.)*

1. The function f is continuous on the interval $-3 \le x \le 3$. The chart below gives some values of f.

x	-3	-2	-1	0	1	2	3
$f(x)$	5	3	1	-.5	-2	1	9

Find consecutive integers a and b such that

a. $\exists x_1$ such that $a < x_1 < b$ and x_1 is a zero of f. $a = -1, b = 0$ or $a = 1, b = 2$

b. $\exists x_2$ such that $a < x_2 < b$ and $f(x_2) = -1$. $a = 0, b = 1$ or $a = 1, b = 2$

2. Consider the equation $\cos x = \frac{3}{2^{x^2}}$. Use the Intermediate Value Theorem to

a. find an interval of length 1 that contains a solution. sample: $7 < x < 8$

b. find an interval of length .1 that contains a solution. sample: $7.8 < x < 7.9$

■ **PROPERTIES** *Objective H*

3. Use the diagram at the right to fill in the blanks.

According to the Intermediate Value Theorem, if it is known that g is a continuous function and y_0 is between $g(k)$ and $g(j)$, then $\exists$ x_0 between j and k) such that $g(x_0) = y_0$.

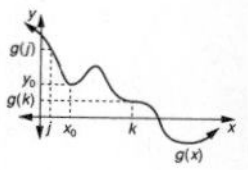

In 4–6, determine whether the function is continuous on the interval $0 \le x \le \pi$. If not, tell where the function is discontinuous.

4. $f(x) = x$ yes

5. $g(x) = \frac{1}{x-2}$ no; discontinuous at $x = 2$

6. $h(x) = \frac{1}{\cos x}$ no; discontinuous at $x = \frac{\pi}{2}$

 Continued 33

NAME ____
Lesson MASTER 3-5 (page 2)

7. *Multiple choice.* A function is graphed at the right. On which interval is the function *not* continuous?

(a) $b \le x \le c$
(b) $0 \le x \le c$
(c) $0 \le x \le d$
(d) $c \le x \le d$ (a)

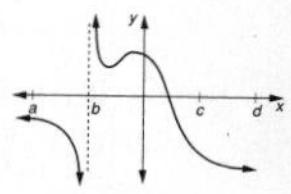

■ **USES** *Objective J*

8. For some cardiac tests, a dye is injected into a vein. The concentration of the dye can be approximated by $f(x) = -.003x^4 + .07x^3 - .026x^2 + .8x$, where $f(x)$ is the percent of dye in the bloodstream after x seconds. After about how many seconds does half the dye remain in the bloodstream? about 22 sec

■ **REPRESENTATIONS** *Objective O*

9. Refer to the graphs of $f(x) = e^x$ and $g(x) = -x^2 + 2x + 4$ below.

a. Write an equation for which c is a solution. $e^x = -x^2 + 2x + 4$

b. Write a formula for a function h for which c is a zero. $h(x) = e^x + x^2 - 2x - 4$

c. Find an interval of length .1 that contains c. $1.5 < c < 1.6$

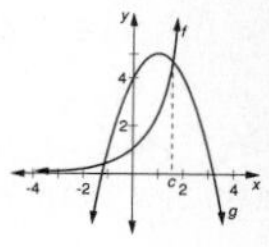

34

LESSON 3-6

RESOURCES
■ Lesson Master 3-6

OBJECTIVES

E Solve inequalities algebraically.
F Analyze the reversibility of steps used in solving equations and inequalities.
L Use inequalities to solve real-world problems.

TEACHING NOTES

The theorems on applying increasing or decreasing functions to solve inequalities (pages 185 and 186) are quite elegant and powerful. Students must be reminded that while applying an increasing function preserves the sense of the inequality, applying a decreasing function reverses it.

You might point out that the familiar procedure of multiplying both sides of an inequality by a positive or a negative number is a special case of these theorems in which $g(x) = kx$ and $k \neq 0$. When k is positive, g is increasing; when k is negative, g is decreasing. For example, suppose $f(x) = -3x$ and $h(x) = 12$, and we wish to solve $f(x) > h(x)$, that is, $-3x > 12$. Let $g(x) = -\frac{1}{3}x$, a decreasing function. So $g(f(x)) < g(h(x))$. That is, $-\frac{1}{3}h(x)$. Thus, $x < -4$.

LESSON

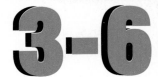

3-6 The Logic of Solving Inequalities

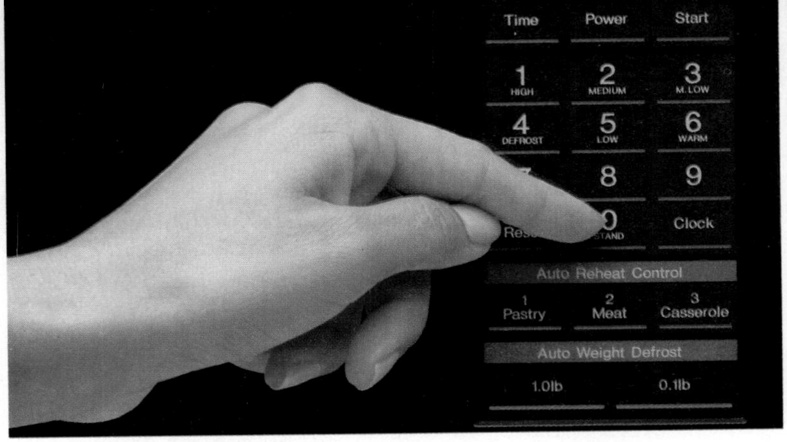

Inequalities are used to determine microwave cooking times. (See Example 2.)

Solving inequalities is a little more complicated than solving equations. Not only does the solver have to do the same thing to both sides, but the *sense* of the inequality (whether it is < or >) has to be taken into account. In general, adding the same quantity to both sides or multiplying both sides by a positive quantity is a reversible step that preserves the sense of the inequality.

However, to solve
$$-5x < 100$$
if you multiply both sides by $-\frac{1}{5}$ then you must reverse the sense of the inequality. The given inequality implies
$$x > -20.$$
Conversely, if $x > -20$, then multiplying both sides by -5 gives $-5x < 100$. Thus the two inequalities are equivalent.
$$-5x < 100 \Leftrightarrow x > -20.$$
This is an example of the fact that multiplication of both sides of an inequality by a negative number is a reversible step that reverses the sense of the inequality.

If the situation is this complicated with an operation as simple as multiplication, you might expect even greater problems trying to solve an inequality like
$$2^x < 3.$$
However, this inequality can be solved using by some rather simple ideas closely related to those we discussed with equations in Lessons 3-1 and 3-2 and based on properties of increasing and decreasing functions.

Theorem

Let f be a real function
(1) If f is increasing on its entire domain, then f has an inverse function f^{-1} which is increasing on its entire domain.
(2) If f is decreasing on its entire domain, then f has an inverse function f^{-1} which is decreasing on its entire domain.

Proof We prove part (1) and leave part (2) for you. Suppose f is increasing on its entire domain. Then f is a 1–1 function (by a theorem of Lesson 3-1), and so f has a inverse function f^{-1} (by a theorem of Lesson 3-2). Now suppose

$$u < v.$$

Then because f is increasing

$$f(u) < f(v).$$

We need to show that $f^{-1}(u) < f^{-1}(v)$. Let $f^{-1}(u) = a$ and $f^{-1}(v) = b$. (Thus we need to show that $a < b$.) By the definition of inverse, $f(a) = u$ and $f(b) = v$, and by substitution into the top inequality,

$$f(a) < f(b).$$

Now, since f is increasing we must have

$$a < b$$

because otherwise if $a > b$, then $f(a) > f(b)$ and if $a = b$ then $f(a) = f(b)$. Thus, since $a < b$, substituting again,

$$f^{-1}(u) < f^{-1}(v),$$

which is the conclusion needed to demonstrate that f^{-1} is increasing.

Notice that in the proof of the above theorem, another statement about the increasing function f was proved, namely that if $f(a) < f(b)$, then $a < b$.

Example 1 Find all real numbers x that satisfy $2^x < 3$.

Solution This inequality has the form $f(x) < k$, so it is natural to try to apply the inverse of f to each side. The inverse here is the logarithm function to the base 2. Since the logarithm function with base $b > 1$ is an increasing function, if $0 < m < n$, then $\log m < \log n$, and conversely. This means that taking $\log_2$ of both sides preserves the sense of the inequality and the step is reversible.

$$2^x < 3$$
$$\Leftrightarrow \quad \log_2(2^x) < \log_2 3$$
$$\Leftrightarrow \quad x < \log_2 3$$

The argument of Example 1 can be generalized and provides a basis for solving other inequalities. Any increasing function can be applied to both sides of an inequality without changing the sense of the inequality. Furthermore, the step is reversible.

Theorem

If g is an increasing function on an interval, then $\forall x$ such that $f(x)$ and $h(x)$ are in that interval, $f(x) < h(x) \Leftrightarrow g(f(x)) < g(h(x))$.

Proof From the definition of an increasing function, $\forall$ a and b on the interval on which g is increasing, $a < b \Rightarrow g(a) < g(b)$. The $\Rightarrow$ direction of the theorem follows by substituting $f(x)$ for a and $h(x)$ for b. Now because g is an increasing function, it has an inverse g^{-1} which is also increasing. Applying g^{-1} to both sides of $g(f(x)) < g(h(x))$ yields the $\Leftarrow$ direction.

Examples 1-3 apply the theorems on increasing and decreasing functions. In **Example 1**, $g(x) = \log_2 x$, and so g is an increasing function. In **Example 2**, $g(x) = x^{1/0.585}$, an increasing function. In **Example 3**, the function being applied to both sides is $g^{-1}(x) = x^{1/5}$, a decreasing function.

The formula of **Example 2** can be derived as follows: We are looking for a function $f: A \rightarrow t$ such that $\frac{f(2A)}{f(A)} = 1.5$ for all $A > 0$. In general, a function f that has the property that $\frac{f(bx)}{f(x)} = m$ for all $x > 0$ is the function $f(x) = x^p$, where $p = \log_b m$. Here is a proof: With this f, $\frac{f(bx)}{f(x)} = \frac{2(bx)^p}{x^p} = b^p$. But since the ratio must be constant, $b^p = m$, from which $p = \log_b m$. In Example 2, $b = 2$ and $m = 1.5$, and so $\log_b m = \log_2 1.5 \approx 0.585$.

In **Example 4**, stress the logic used, especially the roles of "and" and "or." The need for considering all cases is vital. Tell students that since there are no real numbers x such that $x > 0$ and $x < -1$, the word "impossible" denotes that the solution set is empty.

Do not overlook the visual check for **Example 4**. For some students, the connection between inequalities and graphs is most persuasive. This graphical approach provides a useful tie-in with the next lesson.

ADDITIONAL EXAMPLES

1. Solve $5^x < 7$.
$5^x < 7$
$\Leftrightarrow \log_5 5^x < \log_5 7$
$\Leftrightarrow x < \log_5 7$
So $x < 1.2091$ approximately.

ADDITIONAL EXAMPLES

2. In the same microwave oven, one potato cooks in 4 minutes. If you are hosting a dinner party and guests are arriving in one hour, at most, how many potatoes can be cooked in one hour?

$4w^{0.585} \le 60$

$(w^{0.585})^{1/0.585} \le (15)^{1/0.585}$

$w \le 15^{1/0.585}$

$w \le 102.43$

At most, 102 potatoes could be cooked in one hour—but will they fit?

3. Find all solutions to $\sin x > 0.5$ between $-\frac{\pi}{2}$ and $\frac{\pi}{2}$.

Because sin x is increasing on the interval $-\frac{\pi}{2} < x < \frac{\pi}{2}$, applying $\sin^{-1}$ preserves the sense of the inequality.

$\sin x > 0.5$

$\Leftrightarrow \sin^{-1}(\sin x) > \sin^{-1}(0.5)$

$\Leftrightarrow x > \frac{\pi}{6}$

Since we assumed $-\frac{\pi}{2} < x < \frac{\pi}{2}$, the solution is $\frac{\pi}{6} < x < \frac{\pi}{2}$.

4. Can the sum of a number and that number raised to the sixth power ever be negative?

Let x be such a number.

$x^6 + x < 0$

$\Leftrightarrow x(x^5 + 1) < 0$

$\Leftrightarrow [(x < 0) \text{ and } (x^5 + 1 > 0)]$

or $[(x > 0) \text{ and } (x^5 + 1 < 0)]$

$\Leftrightarrow (x < 0 \text{ and } x > -1)$

or $(x > 0 \text{ and } x < -1)$

(impossible)

$-1 < x < 0$

Thus, the answer is yes when $-1 < x < 0$.

Example 2 According to one microwave cookbook, if you double the amount of a particular food to be cooked, you should multiply the cooking time by 1.5. From this, it can be deduced that amount A and cooking time t are related by the formula $t = kA^{0.585}$, where k is a constant determined by the particular food being cooked. If one hot dog can be cooked in 40 seconds, at most how many hot dogs can be cooked in 2 minutes?

Solution First we need to find k in the formula $t = kA^{0.585}$. Since $t = 40$ seconds when $A = 1$, we find $k = 40$. If the time is less than 2 minutes, which is 120 seconds, the inequality to solve is

$$40A^{0.585} \le 120.$$

Then divide both sides by 40 to isolate the power.

$$A^{0.585} \le 3$$

Now take the $\frac{1}{0.585}$ power of each side. All positive power functions are increasing on the set of positive reals, so this will not change the sense of the inequality.

$$(A^{0.585})^{1/0.585} \le 3^{1/0.585}$$

$$A \le 6.54$$

At most 6 hot dogs could be cooked in 2 minutes.

Check From the given information, 1 hot dog could be cooked in 40 seconds; 2 in 1.5 times as long, or 60 seconds; 4 in 1.5 times as long or 90 seconds; 8 in l.5 times as long or 135 seconds. So 6 is reasonable for 2 minutes. Caution: Microwave ovens and foods vary. Do not apply this formula without confirming it for the oven and food being used.

The same ideas apply to decreasing functions. By definition, a function f is decreasing if and only if whenever $x_1 < x_2$, then $f(x_1) > f(x_2)$. Already this tells us: decreasing functions cause the sense of the inequality to be switched.

Theorem

If g is a decreasing function on an interval, then $\forall$ x such that $f(x)$ and $h(x)$ are in that interval, $f(x) < h(x) \Leftrightarrow g(f(x)) > g(h(x))$.

Example 3 Find all positive solutions to $x^{-5} > 32$.

Solution Think of this inequality as being of the form $g(x) > k$, where $g: x \to x^{-5}$. Here $g^{-1}: x \to x^{-1/5}$, and so we take both sides to the $-\frac{1}{5}$ power. In this case, g is a decreasing function, and so is its inverse. Thus, when applying g^{-1} to both sides the sense of the inequality is reversed. Thus an inequality equivalent to the given inequality is

$$(x^{-5})^{-1/5} < 32^{-1/5}$$

That is, $$x < \frac{1}{2}.$$

186

Check Since the inequalities are equivalent, a check is not logically necessary, but it is still wise to check work. Let x be a value satisfying $x < \frac{1}{2}$, say 0.1. Then $x^{-5} = 100{,}000$, which works in the original inequality.

When a function is not increasing or decreasing, then the solving of $f(x) < k$ requires different techniques. Some of those techniques are discussed in the next lesson.

Almost all inequalities can be solved by methods related to the methods you have learned for solving equations. A method related to factoring has to do with the products of positive and negative numbers. If the product of two numbers is positive, then either the numbers are both positive or both negative. If the product of two numbers is negative, then one of the numbers must be positive, the other negative. We call this the Sign of the Product Property.

Thus, if the formula for a function is factored or can be factored, the problem of solving inequalities involving that formula can be reduced to solving inequalities involving the factors.

Example 4 Can the sum of a real number and its square ever be negative?

Solution The question requires solving $x^2 + x < 0$.
Factoring is easy.

$\Leftrightarrow \quad x(x + 1) < 0$

By the Sign of the Product Property one of the factors must be positive, the other negative.

$\Leftrightarrow (x < 0 \text{ and } x + 1 > 0) \text{ or } (x > 0 \text{ and } x + 1 < 0)$

$\Leftrightarrow \quad (x < 0 \text{ and } x > -1) \text{ or } (x > 0 \text{ and } x < -1)$

$\Leftrightarrow \quad -1 < x < 0 \text{ or impossible}$

$\Leftrightarrow \quad -1 < x < 0$

Thus the sum of a number and its square is negative if and only if the number is between 0 and -1.

Check A graph of $f(x) = x^2 + x$ shows that $f(x)$ is negative when $-1 < x < 0$ and at no other time.

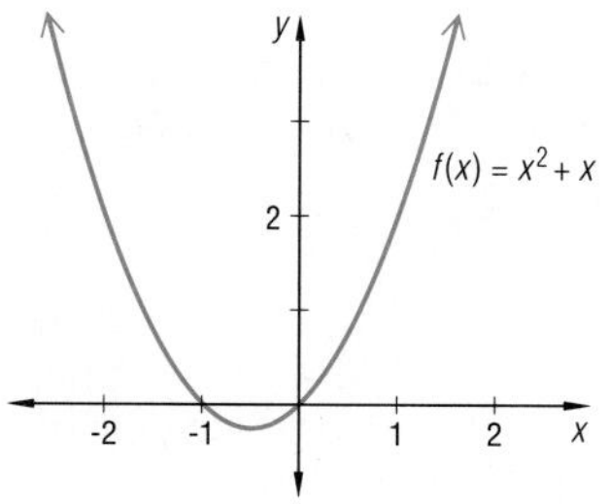

NOTES ON QUESTIONS

Question 2: After obtaining $x < \log_3 6$, some students may be tentative about applying the Change of Base Theorem. Reassure them that simplifying one side of an inequality does not affect the sense of the inequality.

Error Analysis for Question 3: Many students will apply $\log_{0.5}$ to both sides without reversing the sense of the inequality. To show that $f(x) = \log_{0.5}x$ is a decreasing function (for $x > 0$), you may use an automatic grapher to graph $y = \frac{\log_{10} x}{\log_{10} 0.5}$ or $y = \frac{\ln x}{\ln 0.5}$. Alternatively, rewrite $(0.5)^y$ as $(2^{-1})^y = (2)^{-y}$ and solve $2^{-y} \geq 16$ by taking $\log_2$ of both sides.

Question 10: The solution follows the pattern of the proof of the second theorem of the lesson.

Error Analysis for Question 14: Students may be tempted to "cross multiply" and then solve the resulting linear inequality. Point out that this easily can result in the wrong solution, since it assumes that the quantities t and $8t - 40$ must be positive. The correct solution is obtained by subtracting one fraction from another and comparing the resulting fraction with 0, using the Sign of the Product Property.

NOTES ON QUESTIONS
Small Group Work for Question 23: Since $y = 10 - x$, one can explore by considering the point of intersection of the graphs of the functions $f(x) = x^{10-x}$ and $g(x) = 1000x$. Does the point of intersection occur at a relative maximum of f? You may want to have students work in small groups. Even after several zooms with an automatic grapher, it is not easy to see that the answer is no. This question begs for an algebraic solution, but that is inaccessible until calculus. (Analyzing the derivative of the function $f(x) = x^{10-x}$ will suffice.)

ADDITIONAL ANSWERS
10. Suppose $u > v$. Then because f is decreasing, $f(u) > f(v)$. Let $f^{-1}(u) = a$ and $f^{-1}(v) = b$. By the definition of inverse, $f(a) = u$ and $f(b) = v$, and by substitution into the first inequality, $f(a) < f(b)$. Now since f is decreasing, we must have $a > b$, so $f^{-1}(u) > f^{-1}(v)$. Therefore, f^{-1} is decreasing.

17. $h(x)$ is continuous on the interval $2 \le x \le 3$. By the Intermediate Value Theorem, for every real number y_0 between $h(2)$ and $h(3)$ there exists at least one real number x_0 such that $f(x_0) = y_0$. Since $h(2) \approx -.693$ and $h(3) \approx 3.901$, then 0 is between $h(2)$ and $h(3)$, hence the function has a zero between 2 and 3.

20. domain: the set of real numbers;
range: $0 < y \le 1$;
maximum: (0, 1);
minima: none;
increasing: $x < 0$;
decreasing: $x > 0$;
end behavior:
$\lim_{x\to\infty} g(x) = 0$, $\lim_{x\to-\infty} g(x) = 0$

Questions

Covering the Reading

1. *Multiple choice*. Suppose a function is applied to both sides of an inequality. What property of the function determines whether the sense of the inequality should be changed or remains the same?
(a) whether the function is 1–1 or not
(b) whether the function is increasing or decreasing
(c) whether the function has an inverse or not
(d) whether the function has a maximum or minimum value **(b)**

In 2–5, solve.

2. $3^x < 6$ **$x < \log_3 6 \approx 1.631$**

3. $(0.5)^y \ge 16$ **$y \le -4$**

4. $9x^{-3} \ge 18$ **$0 < x \le \frac{1}{\sqrt[3]{2}}$**

5. $t^{2.5} < 100$ **$0 \le t < 6.3$**

6. Refer to Example 2. If 2 pieces of fried chicken take 4 minutes to cook in the microwave, at most how many pieces could be cooked in 12 minutes? **13**

7. Find all values of x such that $(3x - 2)(x - 5) > 0$. **$x > 5$ or $x < \frac{2}{3}$**

8. Can the sum of twice a real number and half its square ever be negative? **Yes, if and only if the number is between -4 and 0.**

Applying the Mathematics

9. Use the graph of the function f at right to describe the solutions to $f(x) < 0$ on $0 \le x \le 7$. **$2 < x < 6$**

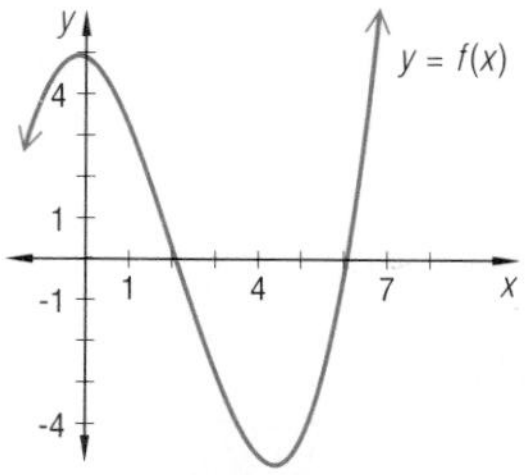

10. Prove: If a real function f is decreasing on its entire domain, then f has an inverse function f^{-1} which is decreasing on its entire domain. **See margin.**

11. Can the sum of a real number and its fourth power ever be negative? **Yes, if and only if the number is between -1 and 0.**

12. Find all values of x such that $7x^2 - 11x + 4 < 0$. **$\frac{4}{7} < x < 1$**

13. The common logarithm of an integer is less than 6. What are the possible values of the integer? **$0 < x < 1{,}000{,}000$**

14. Find all real numbers t such that $\frac{1}{t} > \frac{5}{8t - 40}$. **$t > \frac{40}{3}$ or $0 < t < 5$**

15. The half-life of carbon 14, used to date ancient artifacts, is about 5730 years. If between 20% and 30% of the carbon 14 originally in an artifact remains, about how old is the artifact? **10,000 to 13,000 years old**

Review

16. Given the function h defined by $h(x) = 2x^3 + 5x^2 - 2$.
a. Use an automatic grapher to determine three intervals between successive integers in which the zeros of h lie.
b. Find an interval of length 0.25 that includes the largest zero.
(Lesson 3-5) a) -3 and -2; -1 and 0; 0 and 1; b) sample: $.5 \le x \le .75$

17. Use the Intermediate Value Theorem to explain why the function h defined by $h(x) = x^2 - \ln x - 4$ must have a zero between 2 and 3. *(Lesson 3-5)* See margin.

18. Solve over the set of real numbers: $5x^6 + x^3 - 4 = 0$. *(Lesson 3-4)*
See below

19. **a.** Consider the function $h: x \rightarrow x^4$. Is h a 1–1 function when its domain is the set of all real numbers? No
b. Is h a 1–1 function when its domain is $\{x: x \ge 3\}$? Yes
c. Solve $\sqrt[4]{x-3} = \frac{1}{5}$. *(Lesson 3-1)* $x = 3.0016$

20. Analyze the exponential function $g: x \rightarrow 2^{-x^2}$. Discuss the domain, range, maximum and minimum values, intervals on which the function is increasing or decreasing, and end behavior. *(Lesson 2-6)*
See margin.

21. *Multiple choice.* Which of the following statements is true?
(a) $\forall$ positive real numbers x, $\log x < 0$.
(b) $\exists$ a positive real number x such that $\log x < 0$.
(Lesson 1-1, 2-7) (b)

22. **a.** Write the contrapositive of the following statement:
If $x > 0$ then $6x > 1$. *If $6x \le 1$, then $x \le 0$.*
b. Compare the truth values of the contrapositive and of the statement in part **a**. *(Lesson 1-5)* They have the same truth value; both are false.

Exploration

23. Explore the following conjecture. When x and y are positive numbers whose sum is 10, then the largest value of x^y occurs when $x^y = 1000x$. See margin.

24. Consider the function f with $f(x) = \dfrac{1}{1 - e^{(-1/x)}}$.
a. Explain why f is not continuous on the interval $-2 \le x \le 2$.
b. Graph f on $-2 \le x \le 2$ with an automatic grapher. Does the graph support your answer to part **a**?
c. Use the graph from part **b** to explain why you would expect $f(x) = \frac{1}{2}$ to have no solution on $-2 \le x \le 2$.
d. Prove that $f(x) = \frac{1}{2}$ has no solution in the reals.
See margin.

18. $x = -1$ and $x = \left(\frac{4}{5}\right)^{1/3} \approx 0.928$

FOLLOW-UP

MORE PRACTICE

For more questions on SPUR Objectives, use *Lesson Master 3-6,* shown below.

ADDITIONAL ANSWERS

23. If $x + y = 10$, then $x^y = x^{10-x}$. Graph $f(x) = x^{10-x}$ and $g(x) = 1000x$ as shown below. The largest value of f occurs when $x \approx 4.134$, in which case $x^{10-x} \approx 4127$ (which is not $1000x$). So the conjecture is false.

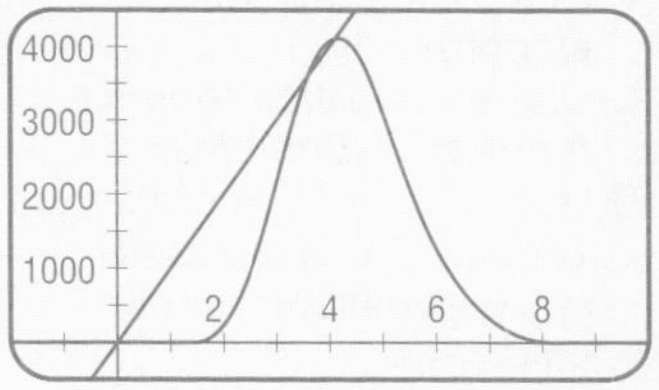

24. See Additional Answers in the back of this book.

NAME ______

LESSON **MASTER 3–6**
QUESTIONS ON **SPUR** OBJECTIVES

■ SKILLS *Objective E (See pages 217–220 for objectives.)*
In 1–6, solve each inequality.

1. $4^x < 1024$ — $x < 5$
2. $\frac{1}{y-5} \ge 3.2$ — $5 < y \le 5.3125$
3. $(a + 10)(a - 3) < 0$ — $-10 < a < 3$
4. $3b^2 + 10b \le b^2 - b - 5$ — $-5 \le b \le -\frac{1}{2}$
5. $2x^2 - 2 \ge -x^2 - x$ — $x \le -1$ or $x \ge -\frac{2}{3}$
6. $\log(2x - 8) > \log(6x + 9)$ — no solution
7. When can the cube of a number be greater than twice its square? — when the number is greater than 2
8. Find all solutions to $\cos x < \frac{1}{2}$ on the interval $0 \le x \le 2\pi$. — $\frac{\pi}{3} < x < \frac{5\pi}{3}$

■ PROPERTIES *Objective F*

9. Since $f: x \rightarrow x^{35}$ is an increasing function, if $\sqrt[5]{x-3} < \sqrt[7]{2-x}$, what can you conclude? — $(x-3)^7 < (2-x)^5$

■ USES *Objective L*

10. The formula $R = e^{21.4b}$ relates the percent risk R of an automobile accident to the percent b of the blood alcohol level of the driver. At what value of b does a driver have at least a 50% chance of having an accident? — when the blood alcohol level is over .1828%

 35

LESSON 3-7

RESOURCES

- Lesson Master 3-7
- Quiz for Lessons 3-4 through 3-7
- Teaching Aid 17 provides the graphs for **Examples 1 and 2**.

OBJECTIVES

- **E** Solve inequalities algebraically.
- **L** Use inequalities to solve real-world problems.
- **O** Use graphs to approximate zeros of functions and solve equations and inequalities.

TEACHING NOTES

For a worthwhile discussion, display the graph for **Example 1**. Give visual meaning to the inequality by extending vertical segments from points *A* and *B* to the *x*-axis. Ask students for the significance of the intersections of these segments with the *x*-axis. Also, ask for the solutions to $x + 2 < 15x^2 - x - 6$. Elicit from students that on a graph, ">" means above and "<" means below. Students need to have these connections made.

For **Example 1**, also point out that for very large values of $|x|$, the $15x^2$ of $15x^2 - x - 6$ dominates and thus $15x^2 - x - 6 > x + 2$. For values of $|x|$ near 0, the -6 becomes the dominant value of $15x^2 - x - 6$, and thus it is near zero where $x + 2 > 15x^2 - x - 6$ becomes true.

LESSON

3-7

The Test Point Method for Solving Inequalities

Just as with equations, graphing is a powerful method for approximating solutions to inequalities of the form

$$f(x) > g(x) \text{ or } f(x) < g(x)$$

when *f* and *g* are real functions.

Example 1 Approximate the solutions of the inequality
$$x + 2 > 15x^2 - x - 6.$$

Solution Graph the functions

$$f(x) = x + 2 \text{ and } g(x) = 15x^2 - x - 6$$

using the window $-2 \le x \le 2$, $-8 \le y \le 8$. We know that the graphs of *f* and *g* are a line and a parabola.

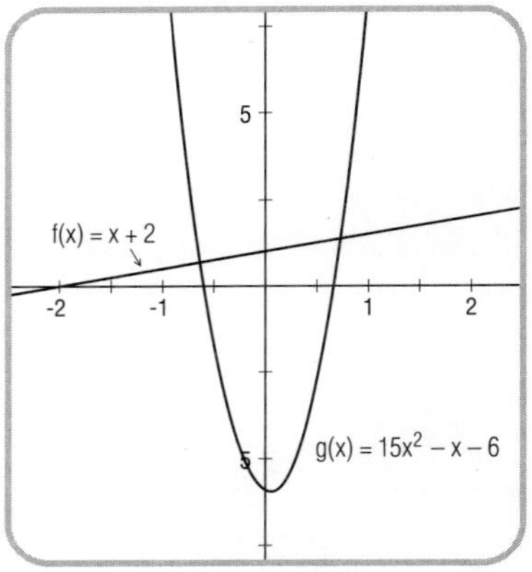

By zooming in on the two intersection points *A* and *B* of these graphs, their *x*-coordinates are approximated as $x \approx -.66$ and $x \approx .80$. Note from the graphs that $x + 2 > 15x^2 - x - 6$ for values of *x* where the line is above the parabola. This is when *x* is between the two points found above. Therefore, an approximate solution of the given inequality is the set of all real numbers *x* such that $-.66 < x < .80$.

The problem in Example 1 can also be solved algebraically by using the Sign of the Product Property.

Example 2 Describe the exact solutions to the inequality of Example 1.

Solution Rewrite the given inequality with 0 on one side.

$$x + 2 > 15x^2 - x - 6$$
$$\Leftrightarrow \quad 0 > 15x^2 - 2x - 8$$
$$\Leftrightarrow \quad 0 > (5x - 4)(3x + 2)$$

The product on the right side is negative if and only if one of the factors is positive and the other is negative.

$$\Leftrightarrow (5x - 4 < 0 \text{ and } 3x + 2 > 0) \text{ or } (5x - 4 > 0 \text{ and } 3x + 2 < 0)$$

$$\Leftrightarrow \quad (5x < 4 \text{ and } 3x > -2) \text{ or } (5x > 4 \text{ and } 3x < -2)$$

$$\Leftrightarrow \quad \left(x < \tfrac{4}{5} \text{ and } x > -\tfrac{2}{3}\right) \text{ or } \left(x > \tfrac{4}{5} \text{ and } x < -\tfrac{2}{3}\right)$$

$$\Leftrightarrow \quad -\tfrac{2}{3} < x < \tfrac{4}{5} \quad \text{or} \quad \text{impossible}$$

Thus, $x + 2 > 15x^2 - x - 6$ if and only if $-\frac{2}{3} < x < \frac{4}{5}$. This agrees well with the approximation $-.66 < x < .80$ found in Example 1.

The method of solution used in Example 2 is analogous to that used in Lessons 3-4 and 3-5 to solve equations. The inequality $f(x) > g(x)$ was rewritten in the form $h(x) < 0$, where $h = g - f$. Then $h(x) < 0$ was solved.

Below is a graph of h. Notice that the key values $-\frac{2}{3}$ and $\frac{4}{5}$ in the solution to Example 2 are the zeros of h.

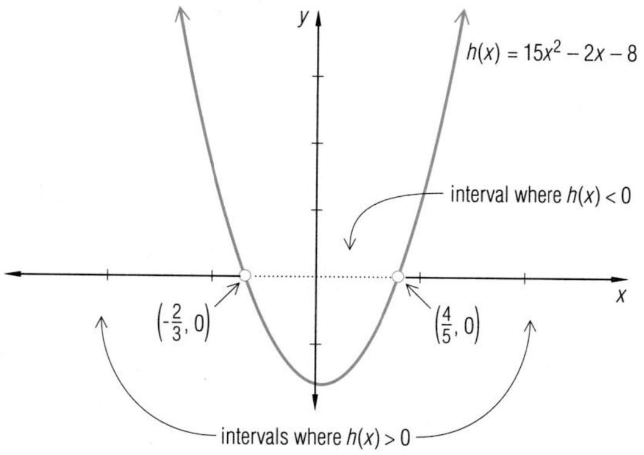

Between these zeros, all values of the function are negative, illustrating the following general theorem.

Function Inequality Theorem

Suppose that f is a continuous real function with zeros a and b. If f has no zeros between a and b, then either

$f(x) > 0$ for all x between a and b

or

$f(x) < 0$ for all x between a and b.

Proof We prove the contrapositive. The negation of the consequent of the theorem is: There exist x_1 and x_2 between a and b such that $f(x_1) \leq 0$ and $f(x_2) \geq 0$. If so, then by the Intermediate Value Theorem, f has a zero at or between x_1 and x_2. Then f would have a zero between a and b. This is the negation of the antecedent, thus proving the contrapositive.

Error Analysis Be sure to discuss **Example 2** with students, since a common error is to consider only one case to determine the solution. Students need to understand that the solutions from the different cases are connected by the word *or*.

In the proof of the Function Inequality Theorem, point out the power of the logic learned in Chapter 1. You may wish to write out the theorem on the chalkboard, and then write its contrapositive below; both are lengthy.

The Test Point Method for Solving Inequalities when the function is continuous is a very useful technique. Many students will be more successful with this technique than with the algebraic technique of solving cases. Also, if there are more than two factors to be considered, the test point method usually is more efficient.

In **Example 3**, point out to students that it is not necessary to evaluate the function in each interval. You simply need to determine whether the function is positive or negative in that interval. This is often easy to accomplish when the function is factored. Simply determine whether each factor is positive or negative and then use the rules for multiplying positive and negative numbers to determine whether the product is positive or negative.

In **Example 3**, you also may want to discuss the use of the Test Point Method to make a rough sketch of the function. This is hinted at in the diagrams accompanying the various steps of the solution. Because the function keeps the same sign in a given interval, you know whether it lies completely above or below the axis in that interval. Because of the zeros, you can predict that turns must occur. This type of analysis allows students to check the reasonableness of a more accurate result obtained by using an automatic grapher.

ADDITIONAL EXAMPLES

1. Approximate the solutions of the inequality $5x^2 + 6x - 23 < -x^2 + 5x + 12$.

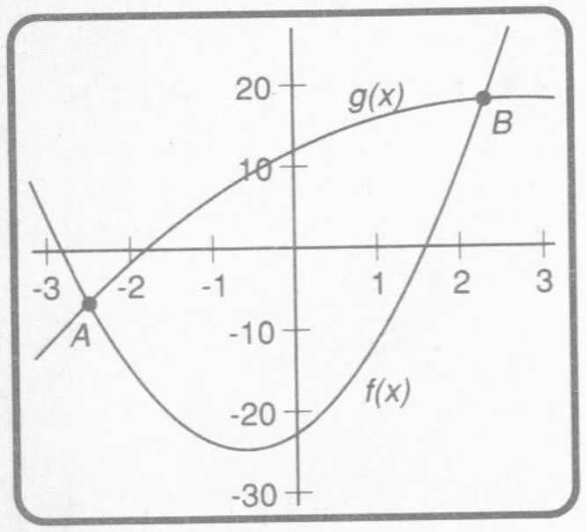

The solution is approximately $-2.50 < x < 2.33$, since in this interval the graph of $f(x)$ is below that of $g(x)$.

The Function Inequality Theorem implies that the solution set to an inequality of the form $f(x) < 0$ or $f(x) > 0$ is a union of intervals whose endpoints are zeros of f. For instance, for the function pictured below, the solutions to $f(x) > 0$ are in the intervals $a_1 < x < a_2$ or $a_3 < x < a_4$ or $a_4 < x$. The solutions to $f(x) < 0$ are in the intervals $x < a_1$ or $a_2 < x < a_3$.

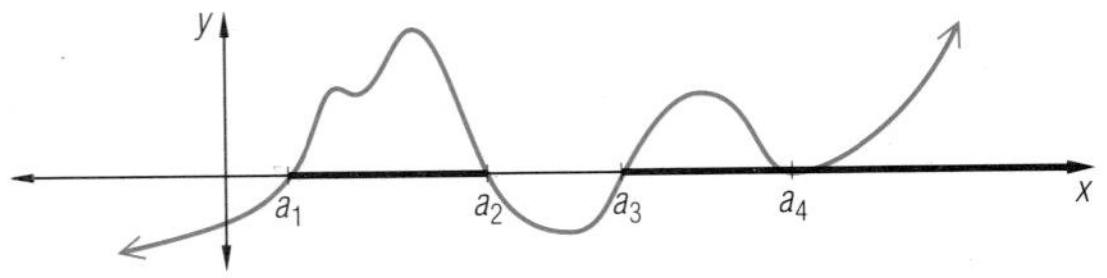

Thus, if you know the zeros of f, and you have checked that f is continuous, you only have to determine whether the interval between two successive zeros (or the interval to the left of the smallest zero or the interval to the right of the largest zero) is part of the solution set to $f(x) < 0$ or part of the solution set to $f(x) > 0$. This determination can be made by testing a value in the interval. The procedure of solving inequalities in this way is called the *Test Point Method*.

Test Point Method for Solving $f(x) > 0$ for Continuous f	
Step 1:	Find all real zeros of the function f by solving $f(x) = 0$ and arrange them in increasing order: $$a_1 < a_2 < \ldots < a_n.$$
Step 2:	For each pair a_i and a_{i+1} of consecutive zeros of f, evaluate f at a convenient test point c between a_i and a_{i+1}. If $f(c) > 0$, then the interval $a_i < x < a_{i+1}$ is part of the solution set of $f(x) > 0$; otherwise, *no* point in this interval is in the solution set.
Step 3:	Evaluate f at a convenient test point $c < a_1$. If $f(c) > 0$, then all real numbers x such that $x < a_1$ belong to the solution set of $f(x) > 0$; otherwise, none of these numbers belong to the solution set. The same procedure can be used to test all real $x > a_n$.
To solve $f(x) < 0$, replace $f(x) > 0$ and $f(c) > 0$ in steps 2 and 3 by $f(x) < 0$ and $f(c) < 0$.	

Examples 1 and 2 involved the inequality $0 > 15x^2 - 2x - 8$. To solve by the Test Point Method, let $h(x) = 15x^2 - 2x - 8$. h is continuous on the reals because $h(x)$ is a polynomial. In Example 2, the zeros were found to be $a_1 = -\frac{2}{3}$ and $a_2 = \frac{4}{5}$. We note these on a number line. These points determine the intervals $x < -\frac{2}{3}$, $-\frac{2}{3} < x < \frac{4}{5}$, and $x > \frac{4}{5}$. Now, choose test points on each interval determined by $-\frac{2}{3}$ and $\frac{4}{5}$.

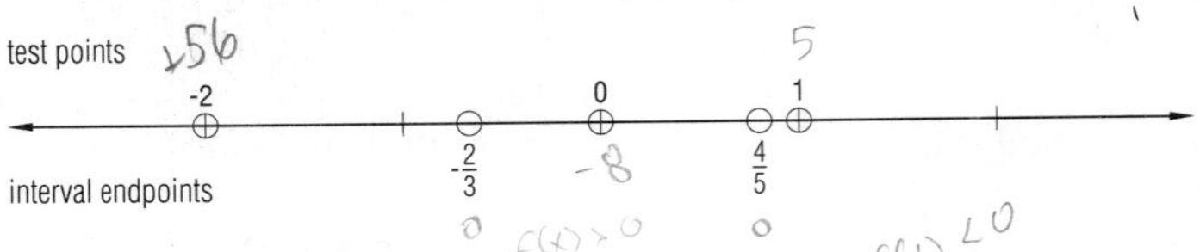

On $x < -\frac{2}{3}$ we choose -2. Since $h(-2) = 15(-2)^2 - 2(-2) - 8 = 56 > 0$, all values of $h(x)$ for $x < -\frac{2}{3}$ are positive. On $-\frac{2}{3} < x < \frac{4}{5}$, we choose 0. Since $h(0) = -8 < 0$, all values of $h(x)$ for $-\frac{2}{3} < x < \frac{4}{5}$ are negative. On $x > \frac{4}{5}$, we choose 1. Since $h(1) = 15(1)^2 - 2(1) - 8 = 5 > 0$, all values of $h(x)$ for $x > \frac{4}{5}$ are positive. The solution to $h(x) < 0$ is the set of values on the intervals on which $h(x)$ is negative, namely the single interval $-\frac{2}{3} < x < \frac{4}{5}$.

Example 3 Apply the Test Point Method to solve the inequality

$$2x^4 - 3x^3 - 6x^2 < x^4 - 2x^3.$$

Solution First rewrite the inequality so that 0 is on one side:

$$x^4 - x^3 - 6x^2 < 0.$$

In other words, solve $h(x) < 0$, when

$$h(x) = x^4 - x^3 - 6x^2.$$

Notice that h is continuous, which is necessary when applying the Test Point Method.

Step 1: Find the zeros of h. In this case it can be done by factoring:

$$x^4 - x^3 - 6x^2 = 0$$
$$x^2(x - 3)(x + 2) = 0$$

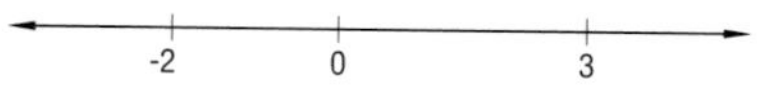

So the zeros of h are $a_1 = -2$, $a_2 = 0$, and $a_3 = 3$. These points determine four open intervals: $x < -2$, $-2 < x < 0$, $0 < x < 3$, and $x > 3$.

Step 2: Choose a test point in the interval $-2 < x < 0$, say $x = -1$, and evaluate $h(-1)$:
$(-1)^4 - (-1)^3 - 6(-1)^2 = -4 < 0$.
Since $h(-1) < 0$, each value in the interval $-2 < x < 0$ is a solution. For a test point in $0 < x < 3$, we choose $x = 1$, and evaluate $h(1)$:
$(1)^4 - (1)^3 - 6(1)^2 = -6 < 0$.
Since $h(1) < 0$, each value in $0 < x < 3$ is a solution.

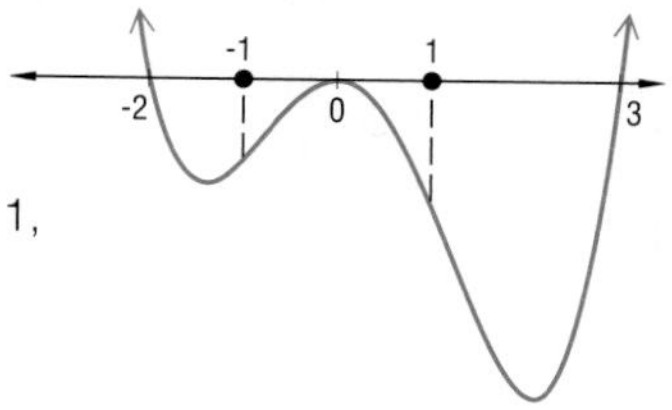

Step 3: For a test point in $x < -2$, we choose $x = -3$:
$h(-3) = (-3)^4 - (-3)^3 - 6(-3)^2 = 54 \not< 0$.
Since $h(-3) \not< 0$, the points in $x < -2$ are not solutions. For the interval $x > 3$, we choose $x = 4$:
$h(4) = (4)^4 - (4)^3 - 6(4)^2 = 96 \not< 0$.
Since $h(4) \not< 0$, the points in $x > 3$ are not solutions.

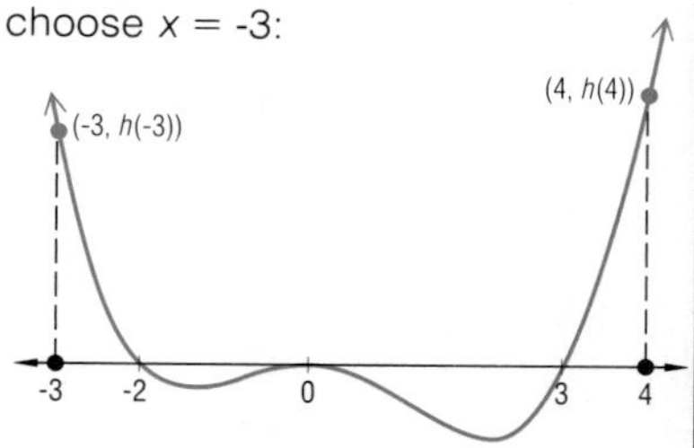

Therefore, the solution set of the inequality is the set of all real numbers x such that $-2 < x < 0$ or $0 < x < 3$.

2. Describe the exact solutions to
$5x^2 + 6x - 23 < -x^2 + 5x + 12$.
$-\frac{5}{2} < x < \frac{7}{3}$

3. Apply the Test Point Method to solve the inequality $8x^4 + 4x^3 - 17x^2 > 2x^4 + 3x^3 + 18x^2$
Rewrite as
$x^2(3x - 7)(2x + 5) > 0$.
$x < -\frac{5}{2}$ **or** $x > \frac{7}{3}$

NOTES ON QUESTIONS
Question 1: You might ask students what the union of the solutions to parts **a** and **b** represents. (the solution set to $(3x + 5)(x + 2) \neq 0$) What real numbers are not in this union? (the solutions to the equation $(3x + 5)(x + 2) = 0$)

Question 5: A graph of $f(x) = x^2(x - 1)(x + 4)$ will clarify the situation.

Error Analysis for Question 9: Students may come up with an incorrect solution if they forget that the domain of the ln function is the set of positive real numbers. Remind students that the solution set always will be a subset of the domain of the original function and to check their solution against it.

Question 10: A graph is again useful.

Question 12: Graphing can be used to answer the question, because the functions involved are continuous and increasing.

Question 13: This question should be discussed. Students may have seen the ceiling function before, and the function is applied later in this book.

FOLLOW-UP

MORE PRACTICE
For more questions on SPUR Objectives, use *Lesson Master 3-7,* shown on page 195.

EXTENSION
You may wish to play a "name that inequality" game with students. Use the number line to give "solutions," ask students to give a corresponding inequality. Ask for more than one inequality. Example:

Inequalities:
$(x + 4)(x + 1) < 0$,
$7(x + 4)(x + 1) < 0$

EVALUATION
A quiz covering Lessons 3-4 through 3-7 is provided in the Teacher's Resource File.

ADDITIONAL ANSWERS
4.a. p: f is a continuous function with zeros a and b and with no zeros between a and b.
q: f(x) > 0 for all x between a and b.
r: f(x) < 0 for all x between a and b.

Questions

Covering the Reading

1. a. Describe the values of x that satisfy $(3x + 5)(x + 2) > 0$.
   b. Describe the values of x that satisfy $(3x + 5)(x + 2) < 0$.
   a) $x < -2$ or $x > -\frac{5}{3}$; b) $-2 < x < -\frac{5}{3}$
2. a. Determine when the parabola $y = x^2$ is above the line $y = 2x + 1$.
   b. Determine when the parabola $y = x^2$ is below the line $y = 2x + 1$.
   a) $x < 1 - \sqrt{2}$ or $x > 1 + \sqrt{2}$
   b) $1 - \sqrt{2} < x < 1 + \sqrt{2}$

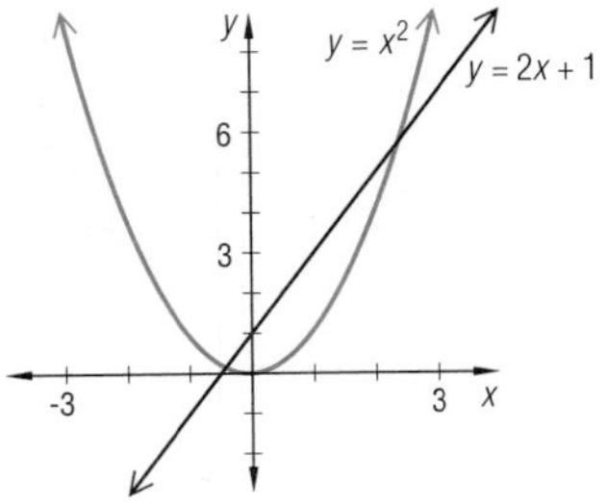

3. a. Use an automatic grapher to approximate the solutions of $x^3 - 6x^2 + 8x > x^2 - 4x$. $0 < x < 3$ or $x > 4$
   b. Find the exact solutions using the Test Point Method. $0 < x < 3$ or $x > 4$
4. The Function Inequality Theorem has the following form:
   If p, then (q or r).
   a. Identify p, q, and r. See margin.
   b. Give the form of its contrapositive. *If not q and not r, then not p.*
   c. Its contrapositive is a special case of what theorem? The Intermediate Value Theorem
5. Solve the inequality $c^2(c - 1)(c + 4) > 0$ using the Test Point Method. $c < -4$ or $c > 1$

Applying the Mathematics

6. The function $f(x) = x^3 - 4x$ has zeros -2 and 2. Both -1 and 1 are in the interval $-2 \le x \le 2$, but $f(-1) > 0$ and $f(1) < 0$. Explain why this does not contradict the Function Inequality Theorem. $f(x)$ also has a zero at $x = 0$ which is between -2 and 2, so the theorem does not apply.
7. The function $g(x) = \frac{x^2 - 9}{x}$ has zeros -3 and 3. Both -1 and 1 are in the interval $-3 \le x \le 3$ but $g(-1) > 0$ and $g(1) < 0$. Explain why this does not contradict the Function Inequality Theorem. (Hint: Graph g.) $g(x)$ is discontinuous at $x = 0$, so the theorem does not apply.
8. Find the exact solution set of $7^{(x^2)} < 7^{3x - 2}$. $\{x: 1 < x < 2\}$
9. Find the exact solution set of the inequality $\ln(x^2) > \ln(x + 2)$. $\{x: -2 < x < -1 \text{ or } x > 2\}$
10. A rocket's height in feet t seconds after take-off is given by $h(t) = -16t^2 + 300t + 20$. For how long is it higher than 1000 ft? approximately 10.3 seconds
11. Approximate the solutions to $\sin x > \frac{1}{2}x$. $x < -1.9$ or $0 < x < 1.9$
12. Use graphing to answer this question, first posed on page 151. The teachers in school A earn \$25,000 per year plus \$1,000 for each year of teaching experience. The salaries in school G start at \$23,000 per year and increase 4% for each year of teaching experience. When (if ever) do teachers in school G earn more than teachers in school A? after 13 years

Review

13. The **ceiling function** C is defined as

$$C(x) = \lceil x \rceil = \text{the least integer} \geq x.$$

a. Sketch a graph of C on the interval $-2 \leq x \leq 2$.

b. Is C continuous on this interval? Explain. *(Lesson 3-5)*
See margin.

14. In Lesson 3-4 the equation

$$E \approx 1000(100 - T) + 580(100 - T)^2$$

was used to find the temperature T(in °C) at which water boils at an elevation of E meters above sea level. To the nearest degree T, find the boiling point of water in Denver, about 1600 meters above sea level. *(Lesson 3-4)* **99°C**

15. Suppose f and g are functions defined by

$$f(x) = (\log x) + 5 \text{ and } g(x) = 10^{x-5}.$$

Show that f and g are inverse functions. *(Lessons 3-2, 2-6, 2-7)*
See margin.

16. Solve the equation $3^x = 9^{x+1}$. *(Lesson 3-1)* **x = -2**

17. Two forest ranger lookout stations are 50 miles apart. A ranger in one station spots a fire at an angle of 35° with the line connecting the two stations. The ranger in the other station spots a fire at an angle of 42° with the line connecting the two stations. How far is the fire from both stations? *(Lesson 2-8)* **The fire is approximately 34 miles from the first station and 29 miles from the second station.**

18. Write the negation of the statement

$$\forall \textit{ real numbers } x, \textit{ if } |x - 4| > 3, \textit{ then } x > 6 \textit{ or } x < 2.$$

(Lesson 1-3, 1-5) **$\exists$ x such that $|x - 4| > 3$ and $2 \leq x \leq 6$.**

Exploration

19. Find a function whose graph resembles that of f below.

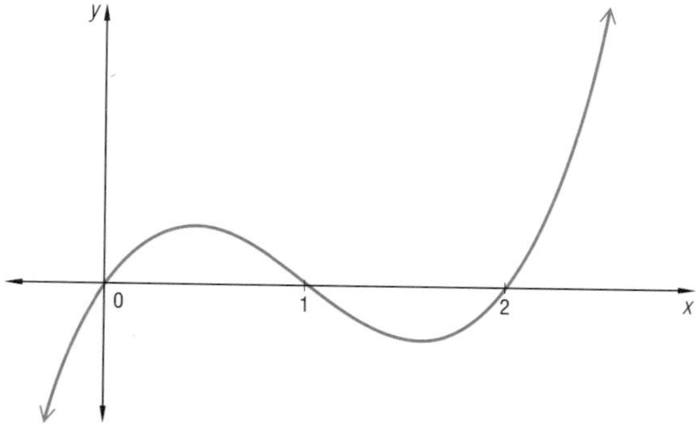

That is, the function should satisfy the following criteria.

$f(x) = 0$	$\Leftrightarrow$	$x = 0$	or	$x = 1$	or	$x = 2$
$f(x) > 0$	$\Leftrightarrow$	$0 < x < 1$	or	$x > 2$		
$f(x) < 0$	$\Leftrightarrow$	$x < 0$	or	$1 < x < 2.$		

sample: $f(x) = x^3 - 3x^2 + 2x$

13.a.

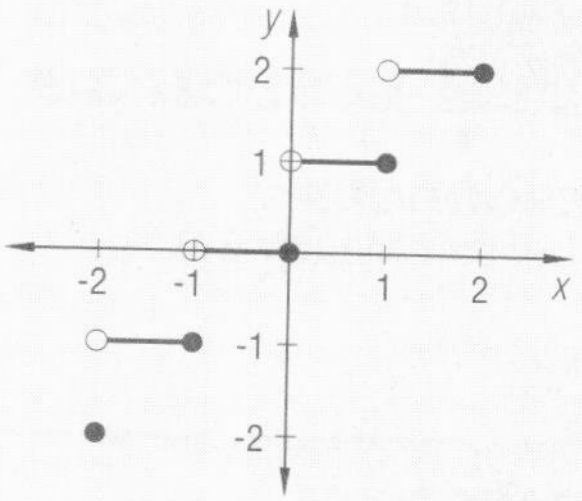

b. No, the graph of C is broken at x = -2, -1, 0, and 1 in the interval $-2 \leq x \leq 2$, thus it is not continuous by definition.

15. $f \circ g(x) = f(g(x)) = \log(10^{x-5}) + 5 = x - 5 + 5 = x$, $\forall$ real numbers x; $g \circ f(x) = g(f(x)) = 10^{(\log x + 5 - 5)} = 10^{\log x} = x$, $\forall$ positive real numbers x; $f \circ g = g \circ f = I$; $\therefore$ f and g are inverse functions by definition.

NAME ______

LESSON **MASTER 3–7**
QUESTIONS ON **SPUR** OBJECTIVES

■ SKILLS *Objective E (See pages 217–220 for objectives.)*
In 1–4, describe the solutions to the given inequality.

1. $3x^2 + 17x + 10 > 0$ **$x < -5$ or $x > -\frac{2}{3}$**

2. $x^2 - 4x < 77$ **$-7 < x < 11$**

3. $(x + 5)(x + 3)(2x - 1) > 0$ **$x > \frac{1}{2}$ or $-5 < x < -3$**

4. $x^3 + 2x^2 + x < 0$ **$x < 0$ and $x \neq -1$**

■ USES *Objective L*

5. Midville School has 3,700 students but is decreasing in enrollment by about 3% per year. Westville has 2,100 students but is gaining about 50 students per year. If these trends continue, when will Westville have a larger enrollment? **in about 11 years**

6. When a particular baseball is hit, its height y in feet after traveling x feet horizontally is given by $y = -.04x^2 + 1.4x + 3$. At the same time the baseball is hit, a football is kicked. Its path is described by $y = -.02x^2 + 1.1x$. For how many feet horizontally is the baseball higher than the football? **for about 21.9 ft**

■ REPRESENTATIONS *Objective O*

7. Let f and g be the functions graphed at the right.

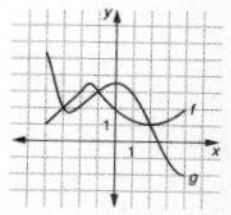

a. For which values of x is $f(x) > g(x)$? **$-3 < x < -1$, $2 < x < 4$**

b. When is $g(x) > f(x)$? **$-4 < x < -3$, $-1 < x < 2$**

8. Use a graph to determine when $\sin x > (x - 1)^2$ on the interval $0 \leq x \leq \pi$. **$.38 < x < 1.96$**

36 *Precalculus and Discrete Mathematics* © Scott, Foresman and Company

LESSON 3-8

RESOURCES

■ Lesson Master 3-8

OBJECTIVES

A Solve equations by applying a function to each side, taking into account nonreversible steps.
E Solve inequalities algebraically.
L Use inequalities to solve real-world problems.

TEACHING NOTES

Some students may have been taught to solve problems such as the one in **Example 1** by translating absolute values to square roots using the identity $|x| = \sqrt{x^2}$. They would begin with $|3x - 2| = 5x \Leftrightarrow \sqrt{(3x-2)^2} = 5x \Rightarrow (3x - 2)^2 = (5x)^2$, and so on. There is nothing incorrect with this method; its disadvantage, however, is that it introduces a quadratic inequality where none is needed.

Error Analysis The requirement in **Example 1** that $5x \geq 0$ is subtle. Some students will ignore it and then think that $x = -1$ is a solution. Stress with students the importance of checking their results by substitution in the original equation or inequality or by a quick sketch of its graph.

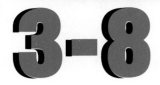

LESSON

3-8 Absolute Value Equations and Inequalities

In this lesson, the ideas of the preceding lessons are applied to equations and inequalities involving the absolute value function. The **absolute value** of x, written $|x|$, is defined as follows: if $x \geq 0$, then $|x| = x$; if $x < 0$, then $|x| = -x$.

From the definition, notice that $|x|$ is never negative, for when x is negative then its absolute value is its opposite, a positive number, and when x is nonnegative, its absolute value is x. This shows that $|x| = a$ has no solution if $a < 0$.

The equation $|x| = 5$ has two solutions, 5 and -5, a fact easily generalized and shown at right.

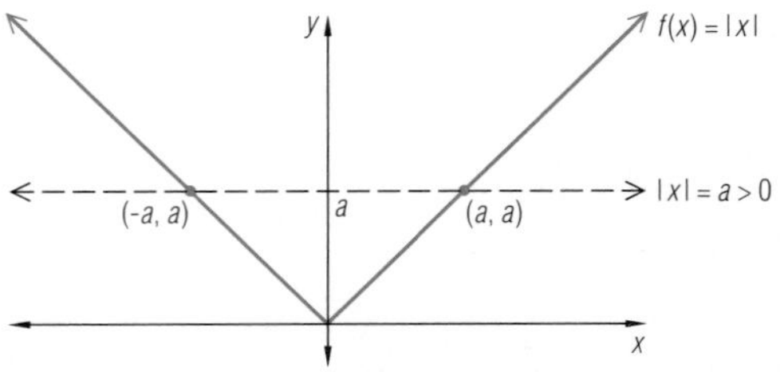

Theorem

$\forall\, a \geq 0,\ |x| = a \Leftrightarrow x = a$ or $x = -a$.

Proof When $x \geq 0$, $|x| = x$, so $|x| = a \Leftrightarrow x = a$.
When $x < 0$, $|x| = -x$, so $|x| = a \Leftrightarrow -x = a \Leftrightarrow x = -a$.

The above theorem is the basic theorem for solving absolute value equations. Yet, with it, rather complicated equations involving absolute values can be solved.

Example 1 Solve $|3x - 2| = 5x$.

Solution There is no solution if $5x < 0$, so we know $x \geq 0$. Use the theorem with $5x$ in place of a and $3x - 2$ in place of x. If $x \geq 0$,

$$\begin{aligned} |3x-2| = 5x &\Leftrightarrow 3x - 2 = 5x \text{ or } 3x - 2 = -5x \\ &\Leftrightarrow -2 = 2x \text{ or } -2 = -8x \\ &\Leftrightarrow x = -1 \text{ or } x = \tfrac{1}{4} \end{aligned}$$

However, when $x = -1$, $5x < 0$, so -1 is not a solution. $\frac{1}{4}$ is the only solution.

Check A graph of the two functions

$$y = |3x - 2|$$

and

$$y = 5x$$

verifies that there is exactly one point of intersection, when $x = \frac{1}{4}$.

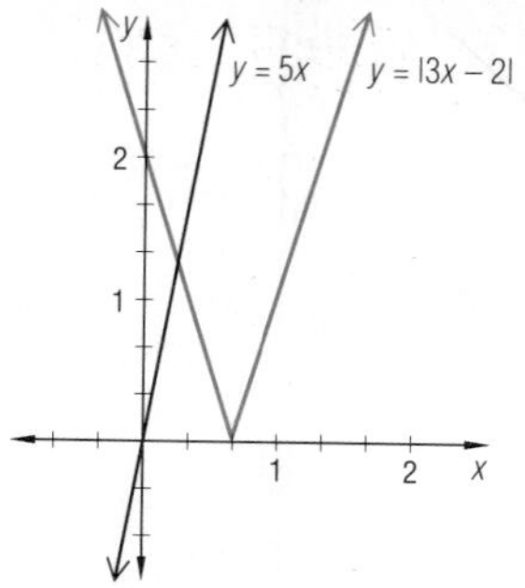

Recall the relation between absolute value and distance; $|x|$ is the distance from x to 0 on a number line. More generally, recall from geometry that $|x_1 - x_2|$ is the distance between two points on a number line with coordinates x_1 and x_2. In many applications, this distance is called the **absolute error** of x_1 from x_2.

For example, if the true value of a certain quantity is x and if x_m is the measured value obtained from an experiment, then the statement:

The absolute error of x_m from x is at most .01.

can be expressed as:

$$|x - x_m| \leq .01.$$

The geometric interpretation of this statement is that the distance from x to x_m on a number line is at most .01 unit.

To solve such inequalities, first note that if a is zero or negative, $|x| < a$ has no solution. The following theorem provides the basis for solving absolute value inequalities when a is positive.

Theorem

For all real numbers x and a with $a > 0$, $|x| < a$ if and only if $-a < x < a$.

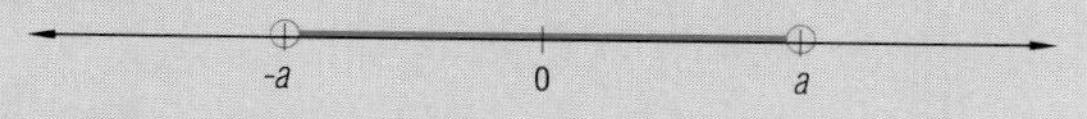

graph of solution set to $|x| < a$

Proof Let x and a be real numbers with $a > 0$. From the definition of $|x|$, either $|x| = x$ (when $x \geq 0$) or $|x| = -x$ (when $x < 0$). Thus

$$\begin{aligned} |x| < a &\Leftrightarrow (x < a \text{ and } x \geq 0) \quad \text{or} \quad (-x < a \text{ and } x < 0) \\ &\Leftrightarrow (x < a \text{ and } x \geq 0) \quad \text{or} \quad (x > -a \text{ and } x < 0) \\ &\Leftrightarrow 0 \leq x < a \quad \text{or} \quad -a < x < 0 \\ &\Leftrightarrow -a < x < a \end{aligned}$$

Notice that the above theorem shows how to change an absolute value inequality into a double inequality.

Making Connections
Example 2 is a type that should be familiar to students who have studied from UCSMP *Algebra* or *Advanced Algebra.* It is placed here to prepare students for **Example 3**. Just as $|x|$ is the distance between x and 0 on a number line, $|x_1 - x_2|$ is the distance between two points x_1 and x_2 on a number line. Here, $x_1 = a_n$ and $x_2 = L$ to get $|a_n - 2|$, which we want to be less than .01.

An algebraic proof for the Theorem following **Example 3** is another example of the power of logic. Suppose $a \geq 0$. Then using the definition of absolute value,
$|x| > a \Leftrightarrow (x > 0$ and $x > a)$ or $(x < 0$ and $-x > a)$
$\Leftrightarrow x > a$ or $(x < 0$ and $x < -a)$
$\Leftrightarrow x > a$ or $x < -a$. If $a < 0$, then $|x| > a$ is always true. But so is $x < -a$ or $x > a$; the former includes all negative values of x, and the latter includes all nonnegative values.

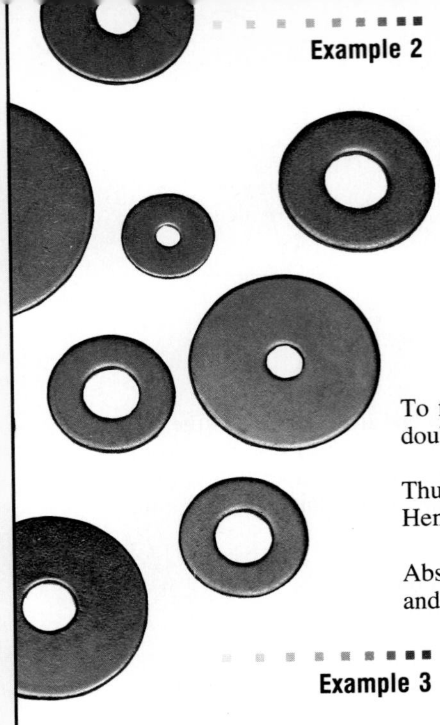

Example 2 A machine stamps out circular washers that are supposed to have a hole with diameter 0.8 cm. The absolute error can be no more than 0.03 cm. Let d be an acceptable diameter. Write inequalities that d must satisfy
a. using absolute value;
b. without using absolute value.

Solution
a. The fact that the absolute error can be no more than 0.03 cm indicates that d must satisfy $|d - 0.8| \leq 0.03$ cm.
b. According to the theorem, this inequality can be written as $-0.03 \text{ cm} \leq d - 0.8 \leq 0.03$ cm.

To find the range of acceptable diameters in Example 2, convert the double inequality into an *and* statement.

$$-0.03 \leq d - 0.8 \quad \text{and} \quad d - 0.8 \leq 0.03$$

Thus, $$0.77 \leq d \quad \text{and} \quad d \leq 0.83$$

Hence, $|d - 0.8| \leq 0.03$ cm is equivalent to $0.77 \text{ cm} \leq d \leq 0.83$ cm.

Absolute value inequalities arise when a sequence is approaching a limit and you wish to know when it gets as close as possible to a certain value.

Example 3 The sequence $a_n = \frac{2n+5}{n+3}$ has terms $\frac{7}{4}, \frac{9}{5}, \frac{11}{6}, \frac{13}{7}, \ldots$, and $\lim_{n\to\infty} a_n = 2$. For what values of n is a_n less than .01 away from the limit?

Solution Restated using absolute value, the question is to solve

$$|a_n - 2| < .01.$$

That is, $$\left|\frac{2n+5}{n+3} - 2\right| < .01.$$

From the theorem, this inequality can be rewritten as a double inequality without employing absolute value.

$$-.01 < \frac{2n+5}{n+3} - 2 < .01$$

By the meaning of the double inequality,

$$-.01 < \frac{2n+5}{n+3} - 2 \quad \text{and} \quad \frac{2n+5}{n+3} - 2 < .01$$

Solve the inequalities.

$$1.99 < \frac{2n+5}{n+3} \quad \text{and} \quad \frac{2n+5}{n+3} < 2.01$$

Multiply both sides of each inequality by $n + 3$. Since n is a positive integer, $n + 3$ is positive and so the sense of each inequality does not change.

$$1.99(n+3) < 2n+5 \quad \text{and} \quad 2n+5 < 2.01(n+3)$$
$$1.99n + 5.97 < 2n+5 \quad \text{and} \quad 2n+5 < 2.01n + 6.03$$
$$.97 < .01n \quad \text{and} \quad -1.03 < .01n$$
$$97 < n \quad \text{and} \quad -103 < n$$

Any integer greater than 97 is greater than -103. So $n > 97$. Thus all terms of the sequence after the 97th term are less than .01 away from 2.

198

Check The 97th term, $a_{97} = \frac{199}{100} = 1.99$ which is exactly .01 from 2.
The 98th term $a_{98} = \frac{201}{101} = 1.\overline{9900}$ which is less than .01 away from 2.

Suppose that x is a real number. Then the distance of x from the origin is greater than a if and only if x is to the left of $-a$ or to the right of a.

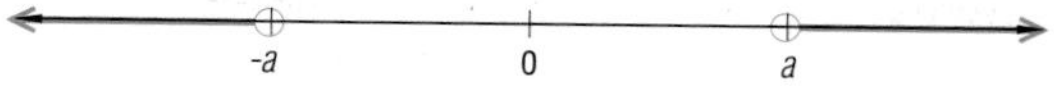

This geometric statement can be expressed algebraically as follows:

Theorem

For all real numbers x and a, $|x| > a$ if and only if $x < -a$ or $x > a$.

Example 4 Find and graph all real numbers x such that $|5 - 2x| > 4$.

Solution Let x be a real number. From the theorem,

$$\begin{aligned} |5 - 2x| > 4 &\Leftrightarrow 5 - 2x < -4 \quad \text{or } 5 - 2x > 4 \\ &\Leftrightarrow \quad -2x < -9 \quad \text{or} \quad -2x > -1 \\ &\Leftrightarrow \quad x > 4.5 \text{ or} \quad x < 0.5 \end{aligned}$$

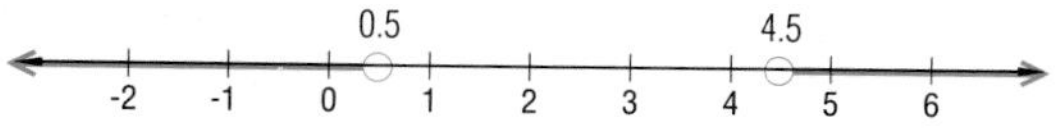

Check When $x = 4.5$ or $x = 0.5$, then $|5 - 2x| = 4$. This checks the endpoints of the intervals. Now check a point within each interval. Pick a value less than 0.5, for example, 0. It should work. $|5 - 2 \cdot 0| > 4$, so it does work. Pick a value between 0.5 and 4.5, for example, 3. $|5 - 2 \cdot 3| \not> 4$, so 3 doesn't work. And it shouldn't. Pick a value greater than 4.5, for example, 5. It should work and it does.

Questions

Covering the Reading

In 1–6, solve.

1. $|2v + 6| = 0$ $v = -3$
2. $|9 - 3x| = 11$ $x = -\frac{2}{3}$ or $x = \frac{20}{3}$
3. $|-y| = -6$ no solution
4. $|5n + 4| = 2n + 3$ $n = -1$ or $n = -\frac{1}{3}$
5. $|m + 6| = 6 + m$ $m \geq -6$
6. $|t| \geq 743$ $t \leq -743$ or $t \geq 743$

In 7–9, interpret each absolute value in terms of distance.

7. $|x| = 7$
x is 7 units away from the origin.
8. $|y - 11| = 2$ See margin.
9. $|a_n - 6| < .01$
The distance between a_n and 6 on a number line is less than .01 units.

ADDITIONAL EXAMPLES

1. Solve $|2x - 10| = 3x$.
$x = 2$

2. If the absolute error can be no more than 0.04 cm for oversize circular washers with a hole of diameter 1.4 cm, write inequalities that an acceptable diameter d must satisfy:
a. using absolute value;
$|d - 1.4| \leq 0.04$ cm
b. without using absolute value.
$-0.04 \text{ cm} \leq d - 1.4 \leq 0.04 \text{ cm}$

3. The sequence $b_n = \frac{5n + 2}{2n + 3}$ has terms $\frac{7}{5}, \frac{12}{7}, \frac{17}{9}, \frac{22}{11}, \frac{27}{13}, \ldots$, and $\lim_{n \to \infty} b_n = \frac{5}{2}$. For what values of n is b_n within .01 of its limit?
$n \geq 274$

4. Find all real numbers x such that $|7 - 4x| < 3$. Graph them.
$1 < x < 2.5$

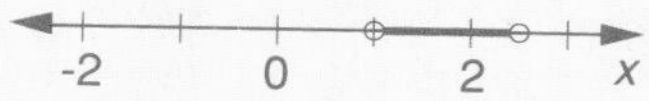

NOTES ON QUESTIONS

Error Analysis for Question 4: Reminiscent of Example 1, $2n + 3 \geq 0$ is a restriction on permissible values for n. In theory, students need to solve the system $|5n + 4| = 2n + 3$ and $2n + 3 \geq 0$.

Question 5: $6 + m \geq 0$ is required for permissible values of m.

ADDITIONAL ANSWERS

8. The distance between y and 11 on a number line is 2 units.

NOTES ON QUESTIONS
Error Analysis for Question 16: When solving this inequality, students may find that $n > -602$ appears as a possibility. Remind students that the domain of n is the set of positive integers. Thus, $n > -602$ cannot be sensible. Students who got that answer made the mistake of not being careful with multiplication by a negative number.

Question 27: This question is in preparation for Lesson 3-10. Ask students to select some trigonometric functions.

ADDITIONAL ANSWERS
19.b.

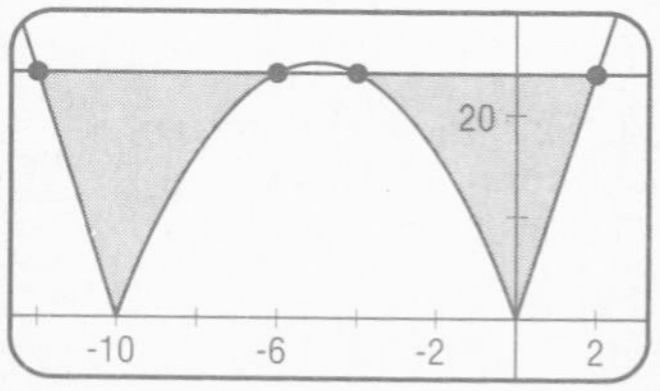

10. *Multiple choice.* Which graph below represents the solution set of the inequality $|x| < 4$? (a)

(a)
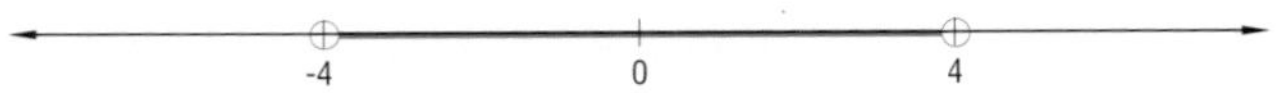

(b)

(c)

(d)
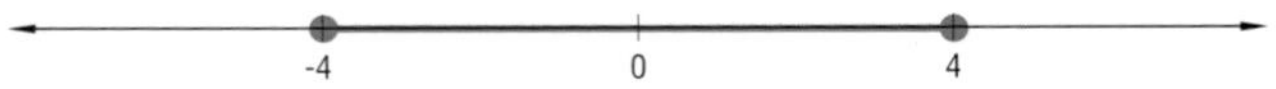

11. *Multiple choice.* Which of the following is equivalent to the inequality $|2x + 1| > 2$?
(a) $-2 > 2x + 1 > 2$
(b) $2x + 1 < -2$ or $2x + 1 > 2$
(c) $2x + 1 > -2$ or $2x + 1 < 2$
(d) $2x + 1 > -2$ and $2x + 1 > 2$ (b)

In 12 and 13, write an absolute value inequality to describe the situation.

12. When the thermostat is set for $k°$, the actual room temperature T may vary at most 3° from $k°$. $|T - k| \leq 3$

13. A manufacturer makes spindles that are supposed to be 15.7 cm long. Parts are rejected if their length is off by more than 0.5 cm.
rejected lengths L satisfy $|L - 15.7| > 0.5$

14. Solve $|7t + 9| > 1$. $t > -\frac{8}{7}$ or $t < -\frac{10}{7}$

15. Solve $|3x - 14| < 12x + 2$ $x > \frac{4}{5}$

16. For the sequence $s_n = \frac{3n - 6}{4 + 2n}$, $\lim_{n \to \infty} s_n = 1.5$. For what values of n is s_n less than .01 from its limit? $n > 598$

Applying the Mathematics

17. Use an absolute value inequality to describe the interval graphed below.

$|x - 57| < 2$

X-57<2 X-57>-2
X<59 X>55

18. If a poll percent is within 3% of the true percent p, and the poll percent is 31%, describe the possible values of p:
a. using a double inequality; $.28 \leq p \leq .34$
b. using absolute value. $|.31 - p| \leq .03$

19. a. Solve $|x^2 + 10x| \leq 24$ algebraically. $-12 \leq x \leq -6$ or $-4 \leq x \leq 2$
b. Verify your solution with an automatic grapher. See margin.

200

20. *True* or *false?* For all functions f and g, if there is a real number x such that $f(x) < g(x)$, then there is a real number x such that $|f(x)| < |g(x)|$. Justify your answer. False, counterexample: let $f: x \to -2$ and $g: x \to -1$. $\forall x,\ f(x) < g(x)$, but $\forall x,\ |f(x)| \geq |g(x)|$.

21. One of De Morgan's Laws was used in this lesson. Where? It was used in the proof of the theorem: $|x| > a$ if and only if $x < -a$ or $x > a$.

22. **a.** Is the absolute value function a 1–1 function? No

b. On what interval is the absolute value function increasing? $x \geq 0$

Review

23. **a.** Is $f: x \to 2^x$ a 1–1 function? Yes

b. Find the exact solution set of $2^{(x^2)} > 2^{x+6}$ $x < -2$ or $x > 3$

c. Let $g: x \to 2^{(x^2)}$ and $h: x \to 2^{x+6}$. Graph g and h on the same axes and use the graph to check your answer to part **b**. *(Lessons 3-1, 3-7)* See margin.

24. Let $h(x) = \sqrt{2x+3}$ and $g(x) = x$.

a. Find a formula for $(h - g)(x)$. $(h - g)(x) = \sqrt{2x+3} - x$

b. What is the domain of $h - g$? $\{x: x \geq -\frac{3}{2}\}$

c. Find the zeros of $h - g$. *(Lessons 3-4, 3-1, 2-1)* $x = 3$

25. **a.** Sketch a graph of the sequence p defined by the rule $p_n = \dfrac{3n+4}{n+1}$ for all positive integers n. See margin.

b. Find $\lim_{n \to \infty} p_n$. *(Lesson 2-4)* 3

26. Prove: *If m and n are even integers and p is an odd integer, then mn−p is an odd integer.* *(Lesson 1-7)* See below.

Exploration

27. Choose a variety of functions f and graph both $y = |f(x)|$ and $y = f(x)$. How do the graphs compare? They are alike when $f(x)$ is positive. All parts of the graph of $y = f(x)$ below the x-axis are reflected over the x-axis in the graph of $y = |f(x)|$.

26. Suppose m and n are any even integers and p is any odd integer. By definition there exists integers r, s, and t such that $m = 2r$, $n = 2s$, and $p = 2t + 1$. Thus $m \cdot n - p = (2r)(2s) - (2t + 1) = 4rs - 2t - 1 = 2(2rs - t) - 1$. Since $(2rs - t)$ is an integer by closure properties, $m \cdot n - p$ is odd by definition.

FOLLOW-UP

MORE PRACTICE

For more questions on SPUR Objectives, use *Lesson Master 3-8,* shown below.

23.c.

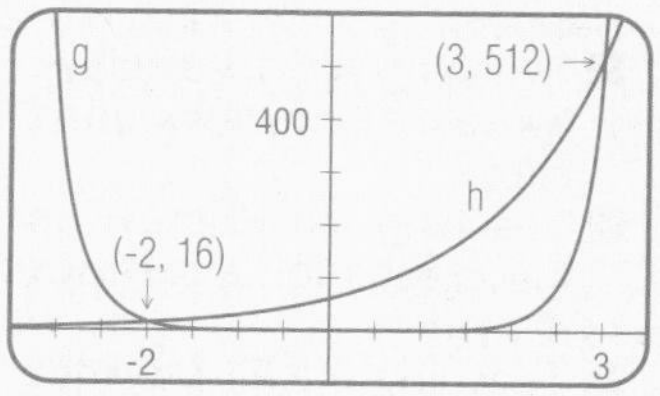

25.a.

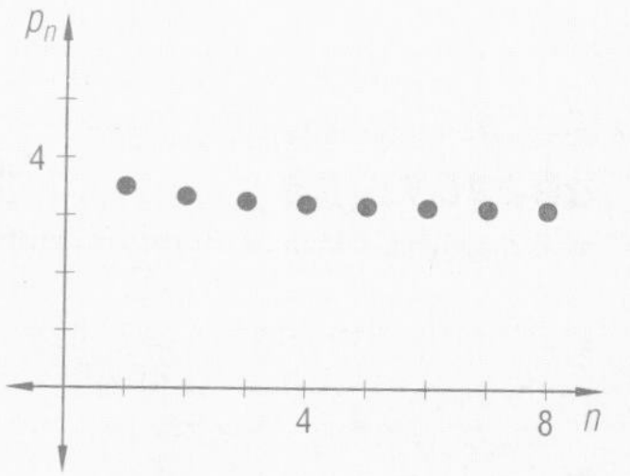

NAME ________

LESSON **MASTER 3–8**
QUESTIONS ON **SPUR** OBJECTIVES

■ **SKILLS** *Objective A (See pages 217–220 for objectives.)*

1. Interpret $|t - 15| = 9$ in terms of distance. The distance between t and 15 is 9.

In 2–5, solve the equation.

2. $|7 + 2q| = 19$ $q = 6$ or $q = -13$
3. $|5 - x| = -6$ no solution
4. $|2b - 3| = 17$ $b = 10$ or $b = -7$
5. $|4m + 21| = 2m + 7$ no solution

■ **SKILLS** *Objective E*

In 6 and 7, write an absolute value inequality to describe each interval.

6. $|x| \leq 6$
7. $|x - 25| < 3$

8. *Multiple choice.* Which of the following is equivalent to the inequality $|3x - 25| < 10$? (c)

(a) $3x - 25 < -10$ and $3x - 25 < 10$
(b) $3x - 25 < -10$ or $3x - 25 < 10$
(c) $3x - 25 > -10$ and $3x - 25 < 10$
(d) $3x - 25 > -10$ or $3x - 25 < 10$

In 9 and 10, solve the inequality.

9. $|z - 12| > 2$ $z < 10$ or $z > 14$
10. $|10y - 17| \leq 13$ $.4 \leq y \leq 3$

■ **USES** *Objective L*

11. A contractor estimates that a construction job will cost $27,500. Suppose the actual cost is within $1,500 of the estimate. Describe this situation

a. with absolute value. $|c - 27{,}500| \leq 1{,}500$
b. with a double inequality. $26{,}000 \leq c \leq 29{,}000$

 37

LESSON 3-9

RESOURCES

- Lesson Master 3-9
- Teaching Aid 18 displays the graph of the average cost at U.S. private colleges.
- Teaching Aid 19 displays two transformations of a circle.
- Teaching Aid 20 provides the graphs for **Examples 1 and 2**.
- Teaching Aid 21 provides the graph for **Question 7**.

OBJECTIVES

I Identify the amplitude, period, and phase shift of trigonometric functions.

N Find an equation of a graph after a transformation.

TEACHING NOTES

The Graph Scale Change Theorem is sometimes difficult for students. While students understand that the scale change $S_{a,b}$ maps (x, y) to (ax, by), they may fail to understand why the image is described by replacing x with $\frac{x}{a}$ and y with $\frac{y}{b}$. Here is a proof. In a scale change, $x' = ax$ and $y' = by$. Then $\frac{x'}{a} = x$ and $\frac{y'}{b} = y$. Thus, we can substitute $\frac{x'}{a}$ for x and $\frac{y'}{b}$ for y. This describes the image. In practice, we drop the primes.

LESSON

3-9 Graphs, Transformations, and Solutions

According to *Peterson's Guides Annual Survey of Undergraduate Institutions*, the average cost of tuition, mandatory fees, and room and board at four-year private colleges in the U.S. in 1989–90 was \$11,051. Although this was the most recent information when this book was being readied for publication, it will be outdated as you read this paragraph. Still, you can use this value to estimate what the corresponding fees might be in a later year. Assuming a 4% annual increase due to inflation, then the pattern of costs forms a geometric sequence.

1989	starting point	11,051
1990	one year later	11,051(1.04)
1991	two years later	$11{,}051(1.04)^2$
	. . .	
$1989 + n$	n years later	$11{,}051(1.04)^n$.

A formula for the function mapping the numbers in the middle column onto the numbers in the right column is evident. Let n be the number of years after 1989 and C be the cost.

$$C = 11{,}051(1.04)^n.$$

A second and perhaps more useful formula would be to match the actual year number Y to the cost. This would be a function mapping the left column of the chart onto the right column. Because $Y = 1989 + n$, $n = Y - 1989$. Thus a formula matching year number to cost is

$$C = 11{,}051(1.04)^{Y-1989}.$$

Replacing n and Y by x yields two functions defined by the formulas

$$C = 11{,}051(1.04)^x.$$
$$C = 11{,}051(1.04)^{x-1989}.$$

The graph of the second function is 1989 units to the right of the first function, exhibiting the first of the two fundamental theorems relating graphs, transformations, and solutions to equations.

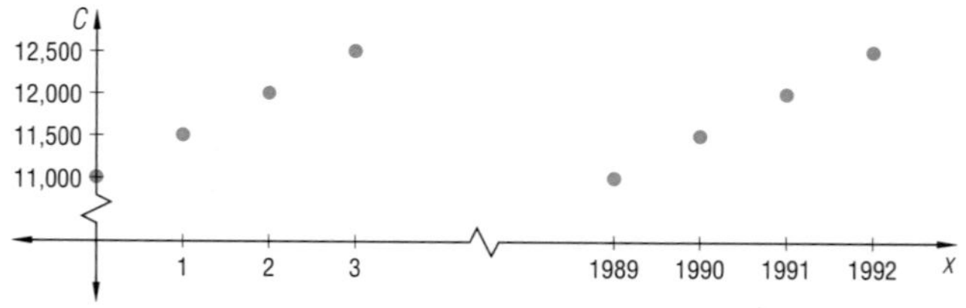

Graph Transformation Theorems

Translation For an equation relating real variables x and y, the following two processes yield the same graph for any specific values of h and k:

(1) replacing x by $x - h$ and y by $y - k$ in the equation and graphing the result;

(2) applying the translation $(x, y) \rightarrow (x + h, y + k)$ to the graph of the original relation.

EACH POINT

Scale Change For an equation relating real variables x and y, the following two processes yield the same graph for any specific nonzero values of a and b:

(1) replacing x by $\frac{x}{a}$ and y by $\frac{y}{b}$ in the equation and graphing the result;

(2) applying the scale change $(x, y) \rightarrow (ax, by)$ to the graph of the original relation.

Proof A proof is given here for the translation part in one dimension. You should attempt the corresponding proof for the scale change.

Any equation in the variable x is equivalent to one in the form $f(x) = c$. Replacing x by $x - h$ means that the new equation has the form $f(x - h) = c$. Let s be any real number. Then

$$s \text{ is a solution to } f(x) = c \Leftrightarrow f(s) = c$$
$$\Leftrightarrow f((s + h) - h) = c$$
$$\Leftrightarrow s + h \text{ is a solution to } f(x - h) = c.$$

The graph at left below pictures the circle $x^2 + y^2 = r^2$ and its translation image $(x - h)^2 + (y - k)^2 = r^2$ with center (h, k) and radius r. The graph at right pictures the circle and its scale change image resulting from the substitutions $\frac{x}{a}$ for x and $\frac{y}{b}$ for y. The image is an ellipse.

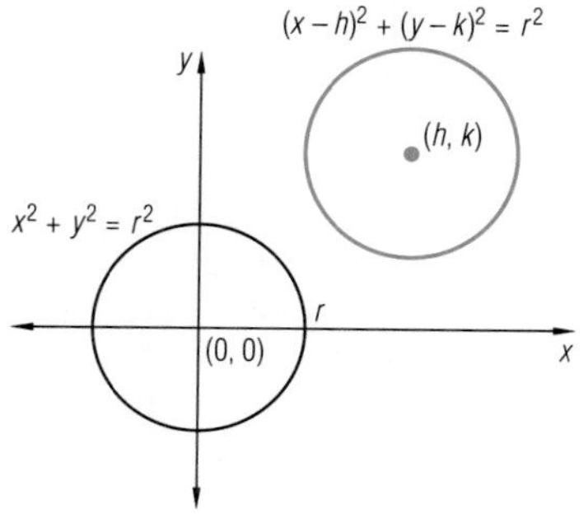

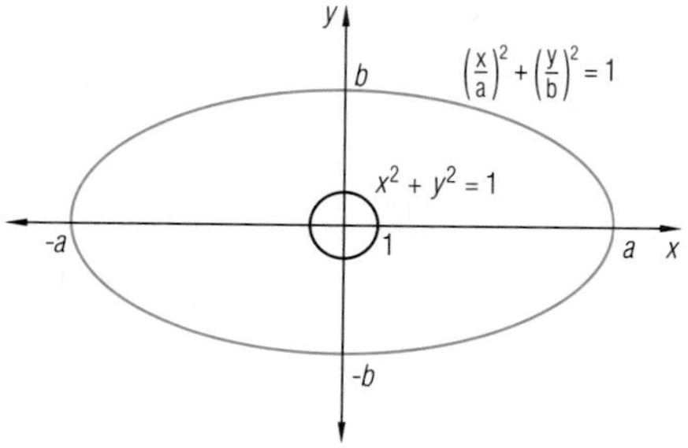

The Graph Transformation theorems generalize to include inequalities in 1, 2, or 3 dimensions. Translations and scale changes can be done simultaneously.

A similar discussion can be used to prove the Graph Translation Theorem. Let $x' = x + h$ and $y' = y + k$. Then $x' - h = x$ and $y' - k = y$. Appropriate substitution now gives the theorem.

For students new to UCSMP materials, the notation $S_{a,b}$ and $T_{h,k}$ probably will need to be explained. You can do this by using the graphs following the Graph Transformation Theorem. Note that the center (0, 0) of the circle $x^2 + y^2 = r^2$ has been moved to $(0 + h, 0 + k) = (h, k)$. For example, with $x^2 + y^2 = 9$ and $h = 6$ and $k = 5$, the center of the circle after translation is (6, 5).

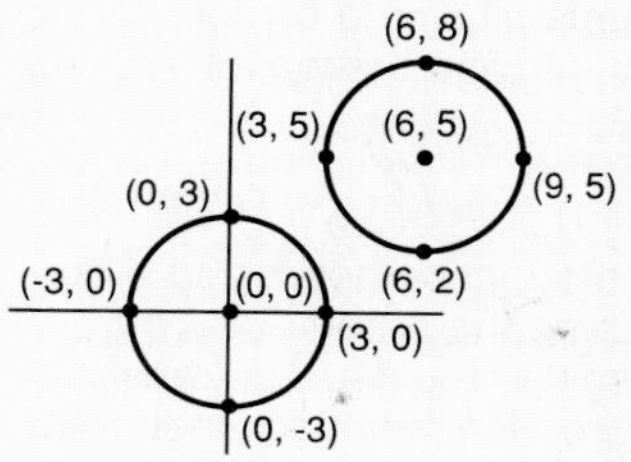

Point out that the coordinates of every point on the circle $(x - 6)^2 + (y - 5)^2 = 9$ are found by adding 6 to the x-coordinate and 5 to the y-coordinate of every point on the original circle.

For the graph of the ellipse, do an example with $a = 5$ and $b = 2$. Sketch both the original circle and its image after the scale change.

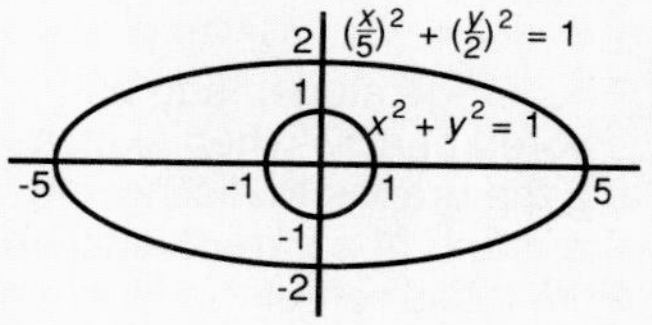

Point out that
$(0, 0) \rightarrow (0, 0)$
$(1, 0) \rightarrow (5, 0)$
$(0, 1) \rightarrow (0, 2)$
$(-1, 0) \rightarrow (-5, 0)$
$(0, -1) \rightarrow (0, -2)$
That is, $x_{new} = 5x_{old}$, and $y_{new} = 2y_{old}$.

You may wish to do some concrete examples with equations. For instance, the solutions to $x^2 - 2x - 8 = 0$ are $x = -2$ and $x = 4$. Now replace x by $x - 5$. The solutions to $(x - 5)^2 - 2(x - 5) - 8 = 0 \Leftrightarrow x^2 - 12x + 27 = 0$ are 5 greater: $x = 3$ and $x = 9$. Now replace x by $\frac{x}{5}$. The solutions to $\left(\frac{x}{5}\right)^2 - 2\left(\frac{x}{5}\right) - 8 = 0 \Leftrightarrow x^2 - 10x - 200 = 0 \Leftrightarrow (x + 10)(x - 20) = 0$ are 5 times as great: $x = -10$ and $x = 20$.

In reviewing **Examples 1 and 2**, stress the importance of form. One goal of mathematics is to transform the unfamiliar into the familiar. If a function can be expressed as $\frac{y - k}{b} = f\left(\frac{x - h}{a}\right)$, then the information needed to determine the parent function, scale change, and translation are all neatly packaged and waiting to be used.

Error Analysis Students may be confused in **Example 1** as to why the equation was not rewritten as $y - 1.3 = \frac{|x - 5|}{1/2}$, indicating a scale change of $S_{1/2, 1}$ rather than $S_{1,2}$. Point out that these scale changes are identical in this case, since they correspond to the same change in the slopes of the graph.

Encourage students to use colored pencils when sketching the graphs to distinguish the graph of a parent function from the graph of the function under consideration.

Example 1 Find all maximum or minimum points and symmetries of the graph of the function with equation $y = 2|x - 5| + 1.3$.

Solution This function can be compared to the absolute value function $y = |x|$, whose graph is known to be a right angle with vertex (0,0) opening upward and symmetric to the y-axis. Rewrite the given equation as

$$y - 1.3 = 2|x - 5|$$

and then $$\frac{y - 1.3}{2} = |x - 5|.$$

The graph of the absolute value function has been translated 5 units to the right and 1.3 units up. The right angle has been stretched by a scale factor of 2 in the y-direction, into an acute angle. Thus the graph is an acute angle with vertex (5, 1.3) opening upward and symmetric to the line $x = 5$. Consequently the graph of the function has a minimum point at (5, 1.3) and is symmetric to the line $x = 5$.

Check Below are the graphs of $y = |x|$ and $y = 2|x - 5| + 1.3$.

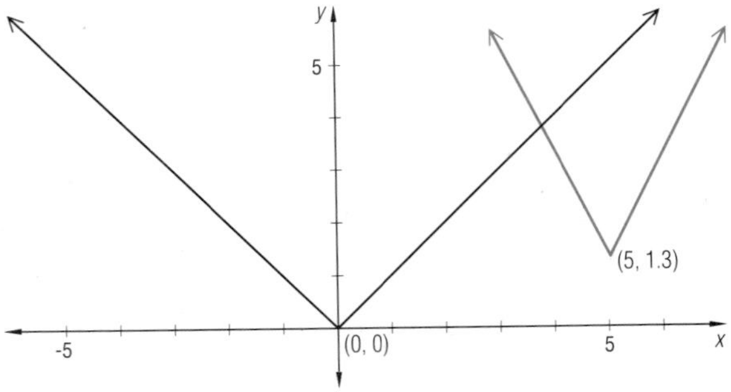

We denote the translation h units to the right and k units up by $T_{h,k}$, and the scale change of a in the x-direction and b in the y-direction by $S_{a,b}$. The translation $T_{0,0}$ and the scale change $S_{1,1}$ both represent the **identity transformation**; they do nothing to a figure. In this language, the graph of the function in Example 1 is the graph of $y = |x|$ transformed first by $S_{1,2}$, and then by $T_{5,1.3}$. In general, applying $S_{a,b}$ and then $T_{h,k}$ to the graph of a function f results in the graph of the function with equation

$$\frac{y - k}{b} = f\left(\frac{x - h}{a}\right), \text{ or equivalently, } y = bf\left(\frac{x - h}{a}\right) + k.$$

Translation and scale change images of the graphs of the sine or cosine functions are quite important because the key numbers have physical properties. A **sine wave** is an image of the graph of the sine or cosine function under a composite of scale changes or translations. Pure sound tones travel in sine waves. Applying a horizontal scale change affects the frequency or pitch of the tone (and the period of the function). Applying a vertical scale change affects the tone's volume or amplitude (and the range of the function). Applying a horizontal translation changes the time at which the tone is heard. For any sine wave, two special terms apply.

3-10

Some Unusual Graphs

Many homes are heated and appliances run by natural gas. Although natural gas is one of the safer energy sources, still there are accidents, and there tend to be more accidents in the winter months when more heat is needed. These data from *The World Almanac 1985* report the number of accidental deaths by gas poisoning in the U.S. for each month in 1980.

Month	Jan	Feb	Mar	Apr	May	Jun	Jul	Aug	Sep	Oct	Nov	Dec
Number	158	139	113	84	58	68	57	51	55	119	124	216

In the scatter plot of these data below, January is represented by 1, February by 2, and so on. The variables, month of year x and number of accidental deaths y, appear to be related.

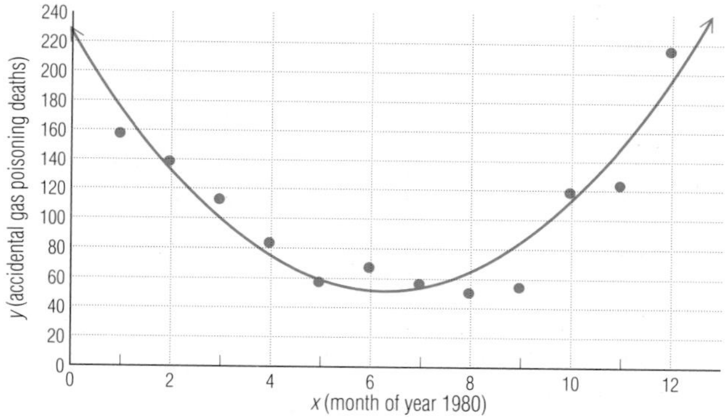

The points on this graph lie close to the parabola which is the graph of the function given by $f(x) = 4.4x^2 - 55.5x + 227.6$. However, since it is reasonable to assume that the same pattern of accidents would repeat each year, this equation is not a good model beyond the end of the year. A model for this situation might use a trigonometric function with a period of one year. But there is a way to construct an equation whose graph exactly repeats any section of a graph of an equation. We will show how to do this for f. Here is the kind of graph we would like.

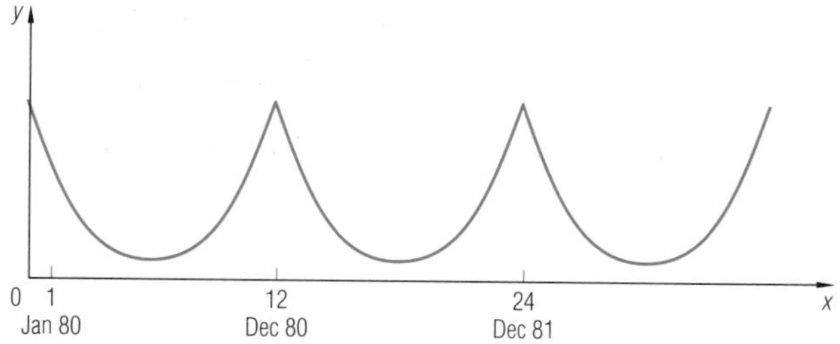

LESSON 3-10

RESOURCES

- Teaching Aid 21 provides the graph of periodic function g.

TEACHING NOTES

One of the wonders of mathematics is that operations defined for one purpose often have uses quite apart from (and seemingly quite unrelated to) that purpose. In this lesson, examples using two such operations are shown. The first gives an application of the floor (greatest integer) function to generate graphs of periodic functions. Remind students that their first encounter with the idea of the floor function was with rounding down in elementary school.

The second type of application uses the absolute value function, which students might have first thought of (incorrectly) as "drop the minus sign."

Making Connections
Have students refer again to the modeling of a twirling jump rope in Lesson 2-8, and compare its graph's smooth, rolling shape with the very different one shown at the bottom of this page. Discuss the similarities of the two cases.

In the function $g(x) = f\left(p \text{ Fract}\left(\frac{x}{p}\right)\right)$, note that p represents the period of the desired function. Adapt the argument on page 210 for $x = 15$ and then for a value other than $x = 15$.

Students may be fascinated by the equation for the square on page 212. By using a scale transformation, this equation can be transformed into an equation for a rhombus with diagonals $2a$ and $2b$: $\left|\frac{x}{a}\right| + \left|\frac{y}{b}\right| = 1$. This rhombus, with vertices at $(a, 0)$, $(-a, 0)$, $(0, b)$, and $(0, -b)$, is inscribed in the ellipse graphed in Lesson 3-9.

In 36 month graph. what is the 27 month fpart(27/12) = .25 March of the third year.

To find an equation for this graph, first consider the function h with formula $h(x) = x - \lfloor x \rfloor$, the difference of the identity and floor functions. For all positive x, this function subtracts the integer part of x from x, so what remains for positive x is the part to the right of the decimal point. For example, $h(\pi) = 3.14159\ldots - \lfloor 3.14159\ldots \rfloor = 3.14159\ldots - 3 = 0.14159\ldots$ Some automatic graphers have a key or name for this function, **Frac** or **Fract** or **F Part** (for fractional part). Here is a graph of Fract when the domain is the set of nonnegative real numbers. The graph resembles an infinite rake.

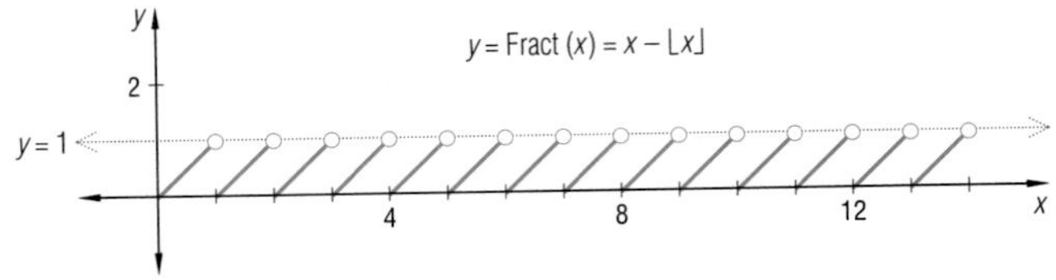

(When x is negative, a slightly different formula is needed to get the decimal part of the number. You are asked to explore this in the questions.)

Now let f be a function on the interval $0 \le x \le p$ and suppose we wish to find the function g whose graph repeats the graph of f. Then it can be proved that the desired function is given by

$$g(x) = f\left(p\text{Fract}\left(\frac{x}{p}\right)\right), \text{ for all } x \ge 0.$$

So if f is the quadratic function above, which we want to repeat with period 12, we should have

$$\begin{aligned} g(x) &= f\left(12\text{Fract}\left(\frac{x}{12}\right)\right) \\ &= 4.4\left(12\text{Fract}\left(\frac{x}{12}\right)\right)^2 - 55.5\left(12\text{Fract}\left(\frac{x}{12}\right)\right) + 227.6 \\ &= 633.6\left(\text{Fract}\left(\frac{x}{12}\right)\right)^2 - 666\text{Fract}\left(\frac{x}{12}\right) + 227.6 \end{aligned}$$

$12^2 \cdot 4.4 = 633.6$ $\quad 12 \cdot 55.5 = 666$

Here is a graph of g for $0 \le x \le 36$.

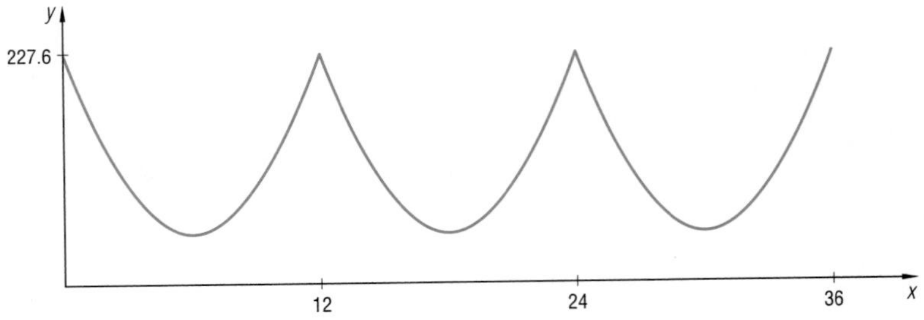

To see why this works, examine the formula for $g(x)$ in terms of $f(x)$ displayed at the beginning of this lesson, taking a particular value of x, say 15. Since months are counted from January of 1980, $x = 15$ represents March of 1981. We would like to somehow transform it to be 3, the value for March of 1980. The $\frac{x}{12}$ in the formula divides 15 by 12, resulting in $\frac{15}{12}$, or 1.25. The Fract function now ignores the 1, yielding .25 or $\frac{3}{12}$, which indicates that March is $\frac{3}{12}$ of the way into the interval. The multiplication by 12 now brings us back to the 3 we wanted, and the function f is applied. In this way, $g(15) = f(3)$. In general, for all x such that $0 \le x \le 12$, and for all nonnegative integers n, $g(x + 12n) = f(x)$, and so g is periodic with period 12 and looks like f over that period, which is exactly what was desired.

No vertical line can intersect the graph of a function in more than one point, but sometimes we want graphs that have two or more points on the same vertical line. With automatic graphers, these relations often must be graphed as the union of two or more functions. For example, to graph the circle

$$(x - h)^2 + (y - k)^2 = r^2,$$

first solve for y.

$$(y - k)^2 = r^2 - (x - h)^2$$

$$y - k = \pm\sqrt{r^2 - (x - h)^2}$$

and so $\quad y = \sqrt{r^2 - (x - h)^2} + k$ or $y = -\sqrt{r^2 - (x - h)^2} + k$.

Now let $f(x) = \sqrt{r^2 - (x - h)^2} + k$ and $g(x) = -\sqrt{r^2 - (x - h)^2} + k$ and graph f and g on the same axes. f is the upper semicircle and g is the lower semicircle. Shown here is the circle with center $(h, k) = (2,-3)$ and radius $r = 4$.

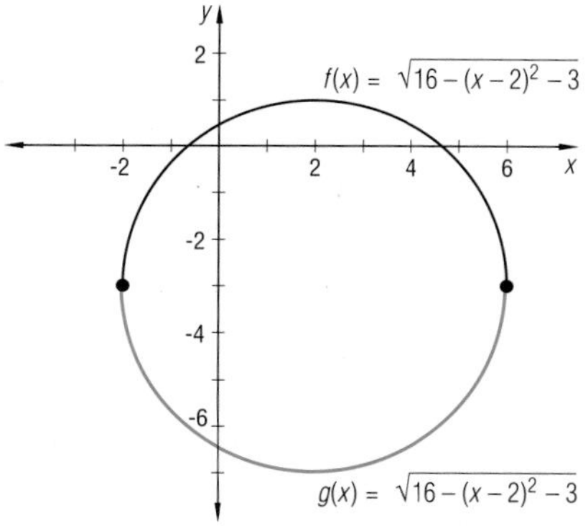

Relations based on the absolute value function can result in still other unusual graphs.

The graph of the relation $|y| = |f(x)|$ is the union of the graphs of $y = f(x) = x^2 - 1$ and $y = -f(x) = -x^2 + 1$, which is the union of the graph of f and its reflection image over the x-axis. At right is graphed $|y| = |x^2 - 1|$, which is the union of two parabolas.

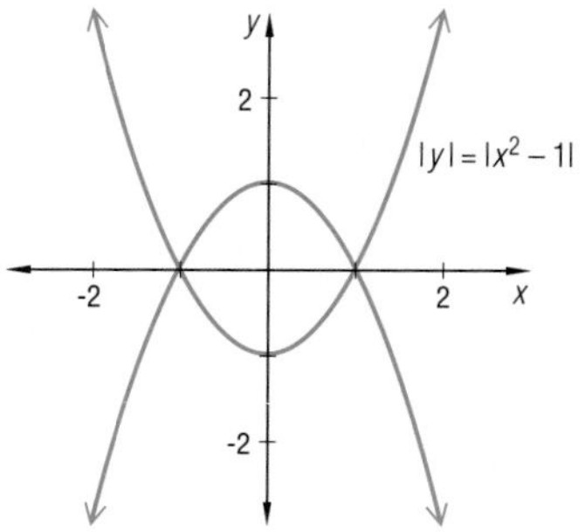

NOTES ON QUESTIONS
Question 5: As an extension, ask students to graph $y = \sqrt{16 - x^2}\sin(8x)$ and $y = -\sqrt{16 - x^2}\sin(8x)$. Invite them to explore functions of the form $y = B\sin(Ax)$, where B is a function rather than a constant.

FOLLOW-UP

EXTENSION
Have students graph and write equations for other periodic phenomena. Some ideas include:
(1) the inventory level of a particular item which is sold at a constant rate over a period of days and then restocked to its initial level;
(2) the power level of a rechargeable battery that decreases exponentially to some pre-set value and then recharges according to a linear function.

PROJECTS
The projects for Chapter 3 are described on page 214. **Projects 2** and **4** are related to the content of this lesson.

ADDITIONAL ANSWERS
3.a. ***h*** **subtracts the greatest integer less than or equal to x from x.**

5.a. $f(x) = \sqrt{16^2 - (x+7)^2} + 9$ or $\sqrt{-x^2 - 14x + 207} + 9$
b. $g(x) = -\sqrt{16^2 - (x+7)^2} + 9$ or $-\sqrt{-x^2 - 14x + 207} + 9$

6.

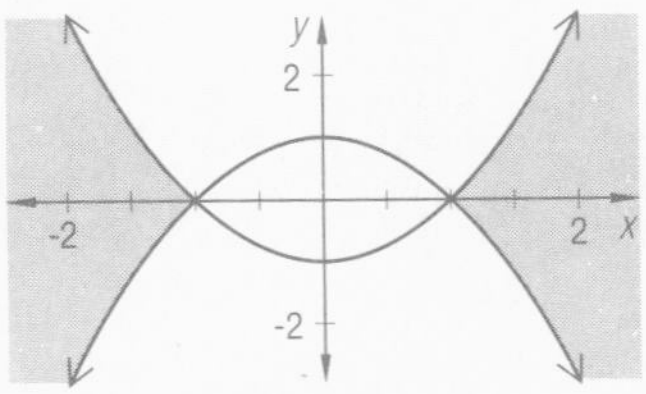

$|y| < x^2 - 1$

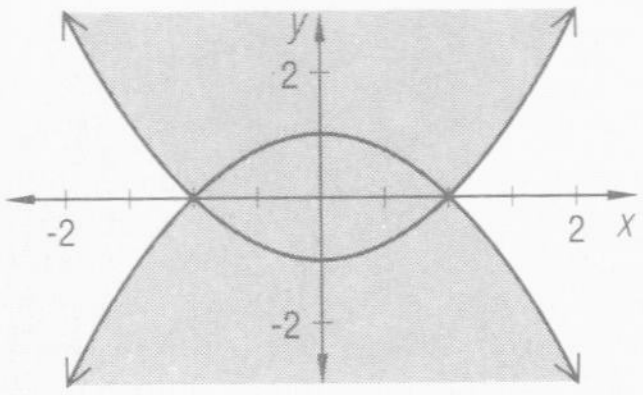

$|y| > x^2 - 1$

If instead of putting the absolute value on y, the absolute value is put on x, then the negative values of x yield the same function values as the positive values of x. For instance, at the right is the graph of the function $y = h(x) = e^{|x|}$. The function h is the composite of the exponential function with base e and the absolute value function. Notice that h is an even function; the graph is the union of the right half of the graph of the exponential function with its reflection image over the y-axis.

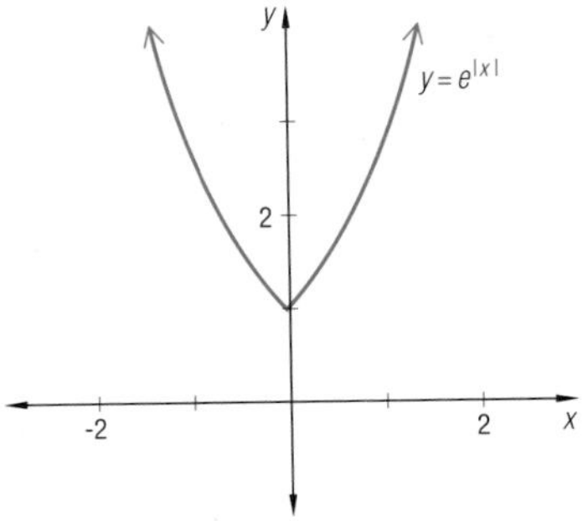

Putting absolute values over both variables results in graphs that are symmetric to both axes. For instance, the graph of the rather simple equation $|x| + |y| = 2$ is a square. The sides of the square are reflection or rotation images of that part of the graph of $x + y = 2$ that lies in the first quadrant. The graph of $|x| + |y| \leq 2$ is the square region shaded at right.

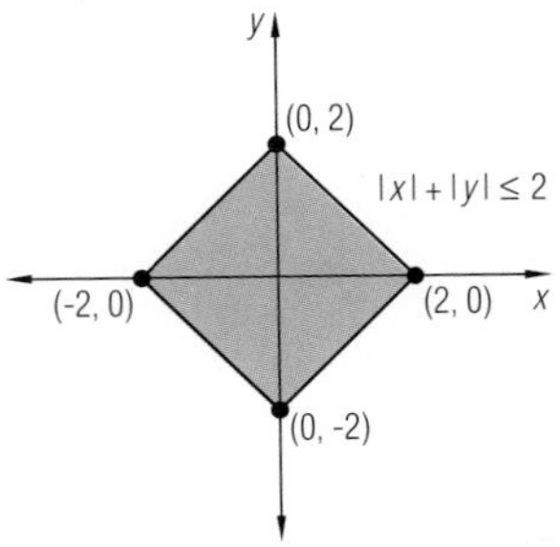

Graphs like these are used in all sorts of computer-aided design. With knowledge of their equations, the time it takes to draw them can be dramatically shortened.

Questions

Covering the Reading

1. The graph below repeats the graph of $y = \frac{1}{2}x^2$ for $0 \leq x < 3$ with period 3. What is an equation for the function that is graphed?

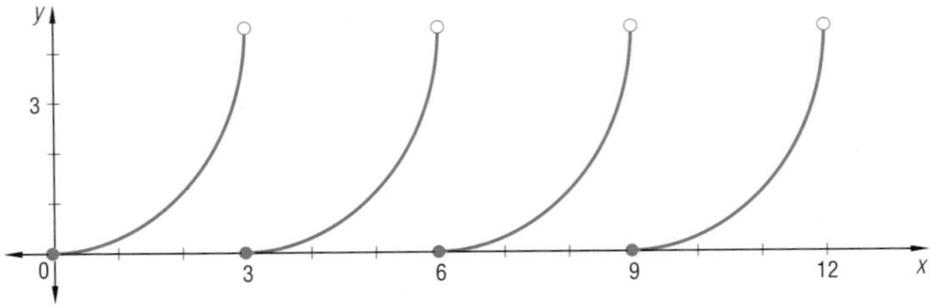

$f(x) = \frac{9}{2}\left(\text{Fract } \frac{x}{3}\right)^2$

212

2. Find a formula for the periodic sequence often heard in exercise classes: 1, 2, 3, 4, 1, 2, 3, 4, 1, 2, 3, 4, . . . (Sometimes this sequence feels infinite.) **$a_n = 4(\text{Fract } \frac{n}{4}) + 1$, $\forall$ nonnegative integers n**

3. When x is negative, the formula $h(x) = x - \lfloor x \rfloor$ does *not* give the fractional part of x.
   a. In words, describe what h does. **See margin.**
   b. Modify the formula so that when $x < 0$, $h(x)$ is the fractional part of x. **$h(x) = |x - \lceil x \rceil|$**

4. Suppose for all x such that $0 \le x \le 20$ and for all integers n, $g(x + 20n) = f(x)$. How are the graphs of g and f related? **g is a periodic function with a period of 20 units. g looks like f over its period.**

5. A circle has center (-7, 9) and radius 16.
   a. What is an equation for its upper semicircle? **See margin.**
   b. What is an equation for its lower semicircle? **See margin.**
   c. Do the endpoints of each semicircle satisfy these equations? **Yes**

6. The relation $|y| = |x^2 - 1|$ is graphed in this lesson. Describe the graphs of $|y| < x^2 - 1$ and $|y| > x^2 - 1$. **See margin.**

7. Graph the function with equation $y = \log |x|$. **See margin.**

8. What geometrical figure has the equation $|x| + 2|y| = 6$? **rhombus**

Review

9. Let $f(x) = x^2 + x$ and $g(x) = x + 7$. Give simplified formulas for $f \circ g$ and $f \bullet g$ to show that $f \circ g \ne f \bullet g$. *(Lessons 3-2, 3-3)*
   $f \circ g(x) = x^2 + 15x + 56$; $f \bullet g(x) = x^3 + 8x^2 + 7x$

10. If $f(x) = c$ for some constant c, how does the graph of a function g relate to the graph of $g + f$? *(Lesson 3-3)*
    $g + f$ is the graph of g translated up c units.

11. Use chunking to solve $e^{2x} - e^x = 30$. *(Lesson 3-4)* **$x = \ln 6 \approx 1.792$**

12. Let $f(x) = \sqrt{25 - x^2}$. *(Lessons 2-1, 2-2, 3-7)*
    a. Graph f. **See margin.**
    b. Give its domain and range.
    c. For what values of x is the graph of f above the graph of $g(x) = .5x + 3$?
    b) domain: $\{x: -5 \le x \le 5\}$; range: $\{y: 0 \le y \le 5\}$; c) $-4.97 < x < 2.57$

13. Recall that the formula $A = 1000e^{.08t}$ gives the amount A in a savings account in which a \$1000 initial investment earns 8% annual interest compounded continuously. Use this formula to find an equation which gives the time required to reach a certain amount A. *(Lesson 3-2)*
    $t = 12.5 \ln \frac{A}{1000}$

Exploration

14. **a.** Describe the graph of the function f, where $f(x) = |x| + |x - 2|$.
    b. Describe the graph of the function $g: x \to |x - h_1| + |x - h_2|$, where h_1 and h_2 are constants.
    c. Explore what happens if more than two absolute values are added and write up the results of your exploration.
    See margin.

7.

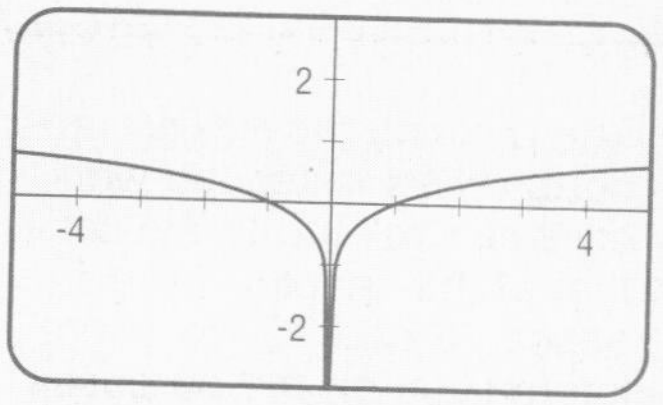

12.a.

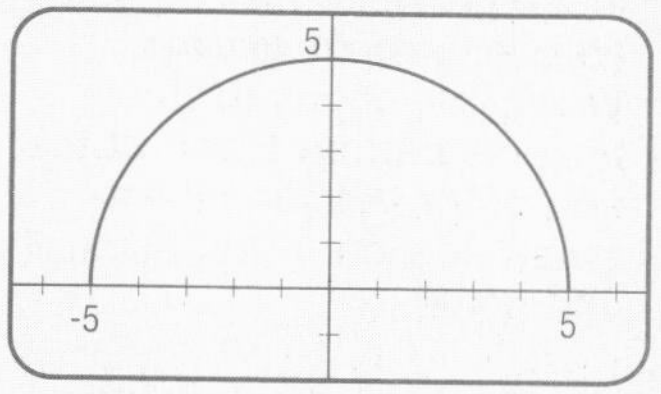

14.a. The graph, shown below, is a union of three connected line segments. When $x < 0$, it is the segment defined by $y = -2x + 2$; when $0 \le x \le 2$, it is the horizontal line $y = 2$; and when $x > 2$, it is the segment defined by $y = 2x - 2$.

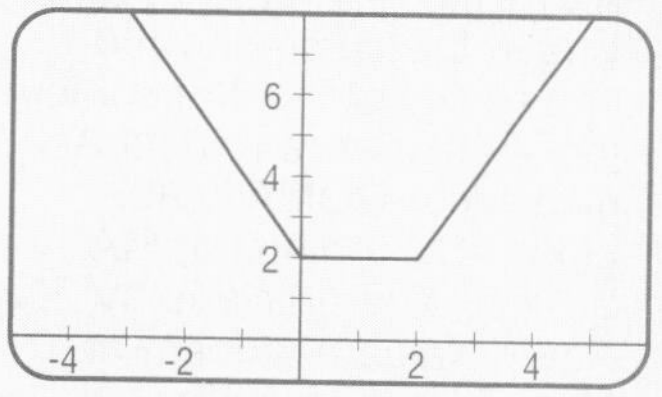

b. As in part a, the graph, shown below, is a union of 3 connected line segments.

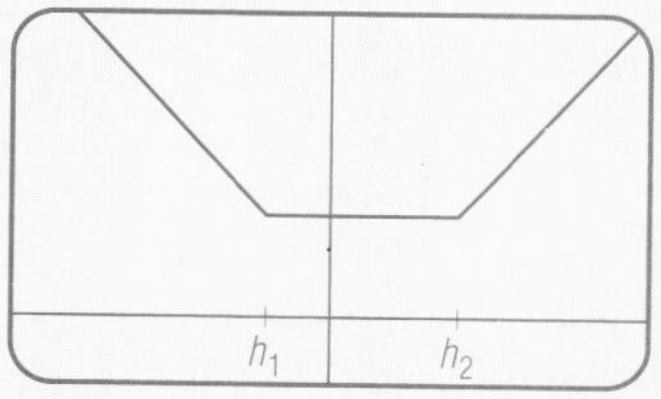

c. The graph of a function which is the sum of n different absolute values will be a union of n connected line segments. Shown below is what the graph might look like for $n = 4$.

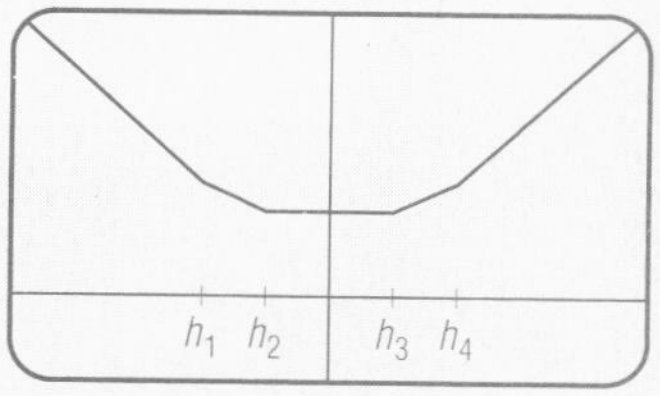

PROJECTS

The projects provide an opportunity for extended work on topics related to the content of the chapter. (See pages 70 and T37 for more information on the purposes of projects.) Some or all of these projects may be suitable for **small group work**. Solutions for the projects can be found at the end of the Additional Answers section in the back of this book.

For **Project 1**, students could use other data that are periodic; for example the length of daylight throughout the year in a location in the northern and southern latitudes, the number of visitors to a ski lodge, and almost anything else that is seasonal.

An article related to **Project 2** is in the summer 1990 newsletter of the Consortium for Mathematics and its Applications (COMAP, 60 Lowell St., Arlington, MA, 02174). The article is by Adam Kalai, a student at New Trier High School in Winnetka, IL.

Projects

1. Many weather conditions, such as temperature, vary throughout the year in a way that can be modeled by a sine function.

a. Use your library or the weather bureau to get information on average monthly high and low temperatures for your area. Graph the yearly cycle of these averages. Write sine functions h and l that approximate each graph.

b. Determine the transformation that transforms h into l and the one that transforms l into h.

c. Find a transformation that will change h into h', which models the high temperatures in degrees Celsius.

d. Graph the average high temperatures for three cities in various parts of the world. Explain differences and similarities.

e. To show how well your function h models the temperatures in your area, compare the real data with the approximate values. Graph the errors for each month. Are there any trends?

2. Many automatic graphers are not designed to graph functions that are defined piecewise. However, such graphs can be defined by manipulating the domain of the function.

a. How does the graph of $f(x) = x^2$ differ from those of $g(x) = x^2 + \frac{0}{|x| + x}$ and $h(x) = x^2 + \frac{0}{|x| - x}$? Why?

b. What is the domain of $f(x) = \frac{0}{|x - 5| + (x - 5)}$? If g is another function, describe the domain of $f + g$.

c. Describe a function f so that $f + g$ is undefined outside the interval $a \leq x \leq b$.

d. Determine an equation of a function whose graph is the segment connecting points (x_1, y_1) and (x_2, y_2).

e. Use restrictions to draw the following stick figure:

3. The BASIC program below executes the bisection method. Line 20 defines the function whose zero you wish to approximate.

```
 10 REM BISECTION METHOD
 20 DEF FN F(X) = X^3 – X^2 + X/3 – 1/27
 30 INPUT "ENTER ERROR BOUND.",E
 40 IF E > 0 THEN 70
 50 PRINT "ERROR BOUND MUST BE
    POSITIVE."
 60 GOTO 30
 70 INPUT "ENTER INTERVAL A < B
    CONTAINING A ZERO.",A,B
 80 IF A < B THEN 110
 90 PRINT "A MUST BE LESS THAN B."
100 GOTO 70
110 IF (FN F(A)*FN F(B)) < 0 THEN 150
120 PRINT "THE INTERVAL ";A;" < X <
    ";B;" MAY NOT CONTAIN A ZERO."
130 PRINT "F(";A;") AND F(";B;") HAVE
    THE SAME SIGN."
140 GOTO 70
150 LET T = (A + B)/2
160 IF (B – A) < (2*E) THEN 190
170 IF (FN F(A)*FN F(T)) > 0 THEN LET
    A = T ELSE LET B = T
180 GOTO 150
190 PRINT T,"APPROXIMATES A ZERO."
200 PRINT "F(";T;") = ";FN F(T)
210 END
```

a. What is the purpose of line 110?

b. Run the program with the function $f: x \rightarrow x^3 - x^2 + \frac{x}{3} - \frac{1}{27}$ given in line 20. Find the zero within 0.005. How close is the approximation you found to $\frac{1}{3}$, the true zero?

c. Repeat part **b** for the function: $x \rightarrow e^x - 10x^2$.

d. Repeat part **b** for several functions of your own choosing.

4. Using equations for semicircles and lines and restricting domains, give equations for each of the letters of the alphabet. Try to make all your letters the same size, and position them in some consistent fashion on the coordinate plane.

Summary

Some of the most useful tools of mathematics are the methods for solving equations and inequalities. In this chapter the logic you learned in Chapter 1 was applied to explore a variety of these methods. Be aware that taking nonreversible steps can cause you to lose solutions or gain extraneous ones.

Automatic graphers can be used to approximate solutions. Examining a graph can help you tell when two functions have equal values. When a function is continuous over an interval, the Intermediate Value Theorem allows you to locate small intervals in which a zero must lie by observing where the function changes signs. The Bisection Method helps to narrow the approximation to any level of accuracy. Inequalities can also be solved graphically.

Composites and inverses of functions can be used to solve equations or inequalities. If an equation is of the form $f(x) = h(x)$, then applying a function g to both sides can be done without affecting the solutions if g is a 1–1 function. If an increasing function g is applied to both sides of the inequality $f(x) < h(x)$, then the sense of the inequality is preserved. If a decreasing function g is applied to both sides, then the sense of the inequality is reversed. To solve $f(x) = k$ or $f(x) < k$, where k is a constant, then if f has an inverse f^{-1}, let $g = f^{-1}$ and apply g to both sides.

There are several techniques for special situations. Chunking can be used to recognize when an equation fits a simpler pattern. Some equations can be related to more basic ones by the properties of transformations (their graphs are also related). Inequalities involving continuous functions can be solved using the Test Point Method. Equations and inequalities involving absolute value can be solved by reducing them to equations and inequalities that do not involve absolute value.

Vocabulary

Below are the most important terms and phrases for this chapter.
For the starred (*) items you should be able to give a definition of the terms. For the other items you should be able to give a general description and a specific example of each.

Lesson 3-1
reversible operation
*1–1 function

Lesson 3-2
function composition, $\circ$
composite of g with f, $g \circ f$
*inverse functions, f^{-1}
identity function

Lesson 3-3
$g + f$, $g - f$, $g \cdot f$, $\frac{f}{g}$

Lesson 3-4
chunking
*zero of a function

Lesson 3-5
continuous on an interval
floor function, $F: x \rightarrow \lfloor x \rfloor$
Intermediate Value Theorem
Bisection Method

Lesson 3-6
Sign of the Product Property

Lesson 3-7
Function Inequality Theorem
ceiling function, $C: x \rightarrow \lceil x \rceil$
Test Point Method for Solving Inequalities

Lesson 3-8
absolute value
absolute error

Lesson 3-9
Graph Transformation Theorems
translation
scale change
identity transformation
sine wave
amplitude
phase shift

Lesson 3-10
Frac, Fract

SUMMARY

The Summary gives an overview of the entire chapter and provides an opportunity for students to consider the material as a whole. Thus, the Summary can be used to help students relate the various concepts presented in the chapter.

VOCABULARY

Terms, symbols, and properties are listed by lesson to provide a checklist of concepts a student must know. (See page 71 for more information.)

PROGRESS SELF-TEST

The Progress Self-Test provides the opportunity for feedback and correction. The Chapter Review provides additional opportunities for practice.

USING THE PROGRESS SELF-TEST

Assign the Progress Self-Test as a one-night assignment. (See pages 72 and T35 for more information.)

ADDITIONAL ANSWERS

5.

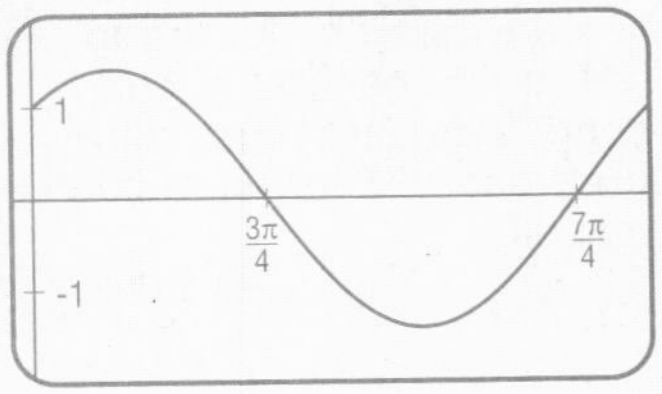

6. $(a \cdot b)(x) = 15x^3 + 12x^2 + 45x + 36$

8.b. the number of quarts of water in the body of a person who is x inches tall

14. No, the function has a break in it.

16.a. amplitude: 4; period: π; phase shift: $\frac{\pi}{4}$

b. $y = 4 \sin\left(2\left(x - \frac{\pi}{4}\right)\right)$

Progress Self-Test

See margin for answers not shown below.

Take this test as you would take a test in class. You will need an automatic grapher. Then check your test using the solutions at the back of the book.

In 1–3, find all real number solutions.

1. $\sqrt{8x + 12} = x - 1$. $x = 11$

2. $2 \log x = \log (6x - 8)$ $x = 2$ or $x = 4$

3. $|4z - 3| \le 7$ $-1 \le z \le \frac{5}{2}$

4. *Multiple choice.* Which procedure can result in a nonreversible step?
(a) subtracting the same number from both sides of an equation
(b) raising both sides of an equation to the fifth power
(c) raising both sides of an equation to the sixth power
(d) dividing both sides of an equation by a nonzero number (c)

5. Let $f(x) = \sin x$ and $g(x) = \cos x$. Sketch a graph of $f + g$ over the interval $0 \le x \le 2\pi$.

6. Let $a(x) = 5x + 4$ and $b(x) = 3x^2 + 9$. Give a simplified formula for $(a \bullet b)(x)$.

7. Let $f(x) = x^2$ and $g(x) = \sqrt{x}$. Write a simplified formula for $f \circ g$, and give its domain. $f \circ g(x) = x$; domain: $\{x: x \ge 0\}$

8. The function $W(x) = \frac{11(x - 40)}{2}$ gives the weight in pounds W of a person whose height is x inches. The function $Q(x) = .325x$ approximates the number of quarts of water Q in the body of a person who weighs x pounds. For the function $Q \circ W$:
a. what does x measure? height in inches
b. what does $Q \circ W(x)$ measure?

9. *Multiple choice.* Which of the intervals listed below must contain a zero of the function f defined by $f(x) = 2x^3 - 7x^2 + 2x - 7$?
(a) $-1 \le x \le 0$ (b) $1 \le x \le 2$
(c) $3 \le x \le 4$ (d) $4 \le x \le 5$ (c)

10. Use chunking to solve $(\sqrt[3]{x})^2 + 2\sqrt[3]{x} = 3$. $x = 1$ or $x = -27$

11. *True* or *false?* If a function is 1–1, then it has an inverse. True

12. Solve $x^2 - 8 > x + 22$. $x < -5$ or $x > 6$

13. Find the zeros of $f(x) = 2x(x - 1)(x + 3)^2$. 0, 1, -3

14. A function is graphed below. Is f a continuous function? Explain your answer.

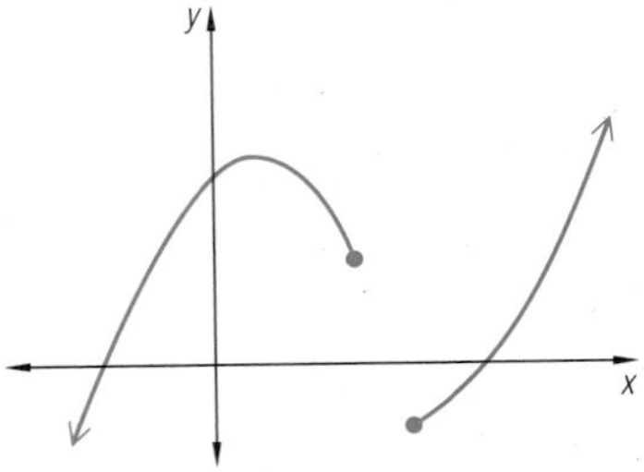

15. The formula
$$L = L_0\sqrt{1 - \left(\frac{v}{c}\right)^2}$$
relates the length L_0 of an object at rest to its length L when traveling at a velocity v. (c represents the speed of light.) Determine the velocity, as a function of the speed of light, for which the traveling length is one-third the rest length. $v = \frac{2\sqrt{2}}{3}c$

16. Given the sine wave graphed below
a. Identify its amplitude, period, and phase shift.
b. Write an equation for this wave.

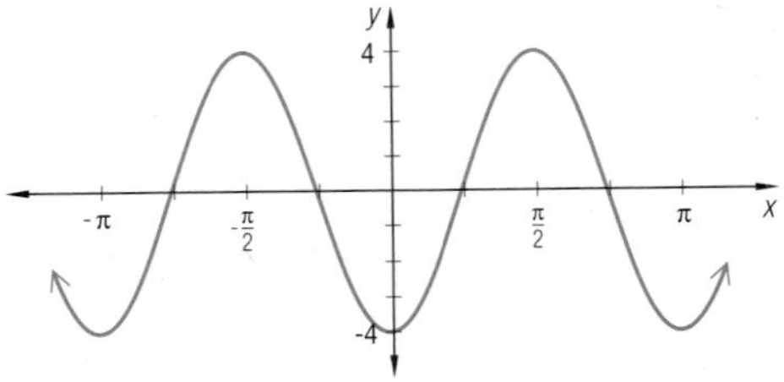

17. Approximate the zero of $f: x \rightarrow e^x - 3x$ to within 0.1 of the true value. ≈ 0.6 and 1.5

18. Radioactive carbon 14, used to date fossils, has a half-life of 5730 years. If A_0 is the original amount of carbon 14 in the bones, then the amount A left after t years is given by
$$A = A_0e^{-.00012t}$$
If less than 30% of the original carbon 14 content is found in the fossil, about how old is it? more than 10,000 years old

216

Chapter Review

Questions on SPUR Objectives

See margin for answers not shown below.

SPUR stands for **S**kills, **P**roperties, **U**ses, and **R**epresentations.
The Chapter Review questions are grouped according to the SPUR Objectives for this chapter.

SKILLS deal with the procedures used to get answers.

Objective A: *Solve equations by applying a function to each side, taking into account nonreversible steps. (Lessons 3-1, 3-2, 3-8)*

In 1–7, solve each equation or inequality for all real number solutions.

1. $\sqrt{3x + 10} = x$ $x = 5$
2. $\sqrt{y + 1} = 5 - y$ $y = 3$
3. $4 \log_5 t = \log_5 (2t^2 - 1)$ $t = 1$
4. $2 \log (x + 1) = \log (x + 13)$ $x = 3$
5. $\frac{1}{x + 5} = \frac{1}{x^2 + 3}$ $x = -1$ or $x = 2$
6. $\frac{1}{3x - 2} = \frac{1}{7x + 4}$ $x = -\frac{3}{2}$
7. $3^v = 9^{-4v + 1}$ $v = \frac{2}{9}$
8. $4^w = 8^{w - 1}$ $w = 3$
9. $\sqrt{z - 1} + 1 = \sqrt{z + 6}$ $z = 10$
10. $\sqrt{2p} - 4 = \sqrt{p - 14}$ $p = 50$ or $p = 18$
11. $(4x + 1)^2 = 49$ $x = \frac{3}{2}$ or $x = -2$
12. $|s + 2| = 3s$ $s = 1$
13. $|6t - 5| = t$ $t = 1$ or $t = \frac{5}{7}$

Objective B: *Describe the sum, difference, product, quotient and composite of two given functions. (Lessons 3-2, 3-3)*

In 14 and 15, find simplified formulas for the functions $f \cdot g$, $\frac{f}{g}$, $f \circ g$, and $g \circ f$, and identify their domains.

14. $f(x) = \ln x$ $g(x) = -x^3$
15. $f(x) = \frac{1}{x}$ $g(x) = x - 6$

In 16–18, for all $x \geq 0$ let $f(x) = \sqrt{x}$, $g(x) = x^3$, and $h(x) = x^2 + 6$. Identify the single function that is equal to:

16. $\frac{h}{f}$ $\frac{h}{f}(x) = \frac{x^2 + 6}{\sqrt{x}}$
17. $f \circ g$ $f \circ g(x) = \sqrt{x^3}$
18. $g - h$ $(g - h)(x) = x^3 - x^2 - 6$

Objective C: *Find zeros of functions and solutions to equations using factoring or chunking. (Lesson 3-4)*

19. Find the zeros of the function $f(x) = 3^{2x} + 3^x - 6$ by using the substitution $u = 3^x$ and solving for u first. $x \approx .631$

In 20–22, find the zeros of the given function.

20. $f(y) = (y + 2)(y + 1)^2(y - 3)$ -2, -1, 3
21. $h(z) = z^3 + 9z^2 + 20z$ -5, -4, 0
22. $f(x) = \cos^2 x - \cos x$ (for $0 \leq x \leq 2\pi$)

In 23 and 24, use chunking to solve each equation.

23. $(\log_5(x - 1))^2 - \log_5(x - 1) = 2$ $x = 26$ or $\frac{6}{5}$
24. $e^{4x} - 4e^{2x} = 12$ $x \approx .896$

Objective D: *Use the Intermediate Value Theorem and the Bisection Method to locate or approximate zeros of a function. (Lesson 3-5)*

25. Consider the function $f: x \rightarrow 2e^{-x^2} - 1$.
    a. Graph the function to determine how many real zeros it has. 2
    b. For each zero, find the two consecutive integers between which it is located.
26. a. How many solutions does the equation $\sin x = \frac{2}{5} + \ln x$ have? 1
    b. Between which two consecutive integers does each solution lie?
27. Use the Bisection Method to find an interval of length .25 that contains the largest real zero of f when $f(x) = \ln x - (x - 2)^2$. $3 < x < 3.25$
28. Use the Bisection Method to find an interval of length .25 that contains a solution to $x^4 = 3x^2 + 2$. sample: $1.75 \leq x \leq 2$

Objective E: *Solve inequalities algebraically. (Lessons 3-6, 3-7, 3-8)*

In 29–31, solve the inequality for all real solutions.

29. $x^3 + x^2 > 2x$
30. $|w + 2| \leq 3$ $-5 \leq w \leq 1$
31. $|3z + 1| > 2$ $z < -1$ or $z > \frac{1}{3}$

RESOURCES
- Chapter 3 Test, Form A
- Chapter 3 Test, Form B
- Chapter 3 Test, Cumulative Form
- Comprehensive Test, Chapters 1-3
- Quiz and Test Writer

CHAPTER REVIEW

The main objectives for the chapter are organized here into sections corresponding to the four main types of understanding this book promotes: Skills, Properties, Uses, and Representations. We call these the SPUR objectives.

USING THE CHAPTER REVIEW

Students should be able to answer questions like these with about 85% accuracy by the end of the chapter. (See pages 74 and T38–39 for more information.)

ADDITIONAL ANSWERS

14. $f \cdot g = -x^3 (\ln x)$, domain: $\{x: x > 0\}$; $\frac{f}{g} = \frac{\ln x}{-x^3}$, domain: $\{x: x > 0\}$; $f \circ g = \ln(-x^3)$, domain: $\{x: x < 0\}$; $g \circ f = -(\ln x)^3$, domain: $\{x: x > 0\}$

15. $f \cdot g = 1 - \frac{6}{x}$, domain: $\{x: x \neq 0\}$; $\frac{f}{g} = \frac{1}{x^2 - 6x}$, domain: $\{x: x \neq 0 \text{ and } x \neq 6\}$; $f \circ g = \frac{1}{x - 6}$, domain: $\{x: x \neq 6\}$; $g \circ f = \frac{1}{x} - 6$, domain: $\{x: x \neq 0\}$

22. $0, \frac{\pi}{2}, \frac{3\pi}{2}, 2\pi$

25.b. $-1 < x < 0$ and $0 < x < 1$

26.b. $1 < x < 2$

29. See the margin on page 218.

ADDITIONAL ANSWERS
26.b. between 1 and 2

29. $-2 < x < 0$ or $x > 1$

34.

35. (b), (c)

36.a. In the second step, both sides of the equation were divided by x^3, which is a valid step only for $x \neq 0$. However, $x = 0$ is a solution.
b. $x = 0$, $x = 2$, $x = -1$

37.b. No, it is not a reversible operation since it is not a 1-1 function.

40. $f \circ g(z) = (z^{5/3})^{3/5} = z$; $g \circ f(z) = (z^{3/5})^{5/3} = z$; $\therefore$ since $f \circ g = g \circ f = I$, f and g are inverse functions.

41. $h \circ m(x) = \log(10^{x-7}) + 7 = x$; $m \circ h(x) = 10^{\log x + 7 - 7} = x$, $\therefore$ since $h \circ m = m \circ h = I$, h and m are inverse functions.

42. If $g(x)$ is a decreasing function, then $g(x) < g(y)$ for all $x > y$ and $g(x) > g(y)$ for all $x < y$. Therefore, when $x \neq y$, $g(x) \neq g(y)$. So g is a 1-1 function, and therefore it has an inverse.

43.c. No, $g(x)$ is undefined at $x = -2$, thus it is not continuous. Therefore, the Intermediate Value Theorem cannot be applied in the interval $-3 \le x \le -1$.

48.d.

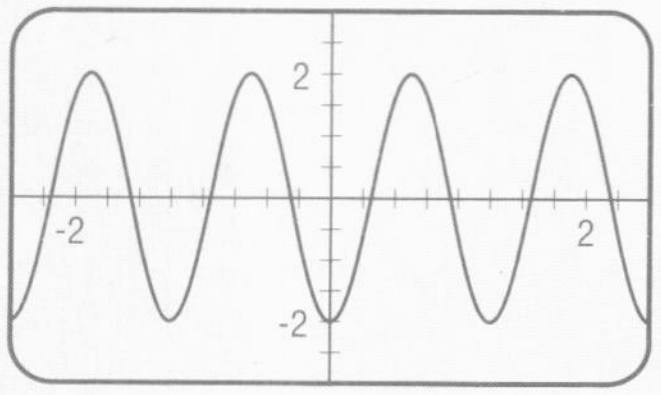

---

32. Solve $(3x + 1)(x - 2) > 0$
a. by using the Zero Product Theorem for Inequalities.
b. by using the Test Point Method.
a, b) $x < -\frac{1}{3}$ or $x > 2$

33. Solve $5^t < 5^{2t+1}$. $t > -1$

34. Solve $2x^2 + 4x < 70$ and graph the solutions on a number line. $-7 < x < 5$; See margin.

PROPERTIES deal with the principles behind the mathematics.

Objective F: *Analyze the reversibility of steps used in solving equations and inequalities.* (Lessons 3-1, 3-6)

35. *Multiple choice.* Which of the following procedures (a) to (e) can result in a nonreversible step?
(a) adding the same number to both sides of an equation
(b) squaring both sides of an equation
(c) dividing both sides of an equation by a quantity containing a variable
(d) multiplying both sides of an equation by a nonzero number
(e) All of the above are nonreversible steps.

36. a. Find the error in the solution of $x^5 - x^4 = 2x^3$.

$$x^5 - x^4 = 2x^3$$
$$\Leftrightarrow \quad x^2 - x = 2$$
$$\Leftrightarrow \quad x^2 - x - 2 = 0$$
$$\Leftrightarrow \quad (x - 2)(x + 1) = 0$$
$$\Leftrightarrow \quad x = 2 \text{ or } x = -1$$

b. Correct the solution to find all real numbers for which $x^5 - x^4 = 2x^3$.

37. a. Is the function $h: t \rightarrow t^4$ a 1–1 function? Explain. No, $g(2) = g(-2) = 16$, but $2 \neq -2$.
b. Use your answer to part **a** to determine whether or not raising both sides of an equation to the fourth power is a reversible operation?

38. In each step replace ? by $\Leftrightarrow$ for a reversible step or by $\Rightarrow$ for a nonreversible step.

$$\sqrt{2 - x} = x - 2$$

a. ? $\quad 2 - x = (x - 2)^2 \quad \Rightarrow$
b. ? $\quad 2 - x = x^2 - 4x + 4 \quad \Leftrightarrow$
c. ? $\quad x^2 - 3x + 2 = 0 \quad \Leftrightarrow$
d. ? $\quad (x - 2)(x - 1) = 0 \quad \Leftrightarrow$
e. ? $\quad x - 2 = 0$ or $x - 1 = 0 \quad \Leftrightarrow$
f. ? $\quad x = 2$ or $x = 1 \quad \Leftrightarrow$

39. *True* or *false?* If f is an increasing function, then applying f to both sides of $g(x) < h(x)$ preserves the sense of the inequality. True

Objective G: *Identify and prove properties of inverse functions.* (Lesson 3-2)

In 40 and 41, prove that the two functions are inverses.

40. $f: z \rightarrow z^{3/5}$ and $g: z \rightarrow z^{5/3}$

41. $h: x \rightarrow (\log x) + 7$ and $m: x \rightarrow 10^{x-7}\ \forall\ x > 0$.

42. Explain why every decreasing function has an inverse.

Objective H: *Identify continuous functions and their properties.* (Lesson 3-5)

43. Let g be the function defined by $g(x) = \frac{1}{x + 2}$.
a. Find $g(-3)$. -1 **b.** Find $g(-1)$. 1
c. Can the answers to parts **a** and **b** be applied to show g has a zero between -1 and -3? Explain.

44. Tell whether the function $f: x \rightarrow \lfloor 2x \rfloor$ is continuous on the given interval.
a. $1 \le x \le 2$ No **b.** $1 \le x \le 1\frac{1}{2}$ No
c. $1 \le x \le 1\frac{1}{4}$ Yes

45. Suppose h is a continuous function over the interval $a \le x \le b$. If $h(a) < 0$ and $h(b) > 0$, then what conclusion can you draw about a zero of h? h has a zero between a and b.

Objective I: *Identify the amplitude, period, and phase shift of trigonometric functions.* (Lesson 3-9)

In 46 and 47, determine the period of the function.

46. $f(x) = \sin 3x$ $\frac{2\pi}{3}$ **47.** $g(x) = \cos x$ 2π

In 48 and 49, a trigonometric function is defined.
a. Determine its amplitude.
b. Determine its period.
c. Determine its phase shift.
d. Graph the function and check your answers to **a**, **b**, and **c**.

48. $y = 2 \sin\left(5\left(x - \frac{\pi}{2}\right)\right)$ a) 2; b) $\frac{2\pi}{5}$; c) $\frac{\pi}{2}$

49. $y = 6 \cos(3x + \pi)$ a) 6; b) $\frac{2\pi}{3}$; c) $-\frac{\pi}{3}$

218

50. Given the cosine function graphed at the right.
  a. Identify the amplitude, period, and phase shift.
  b. Write an equation that defines the function graphed at the right. $y = 3\cos\left(\frac{x+\pi}{2}\right)$

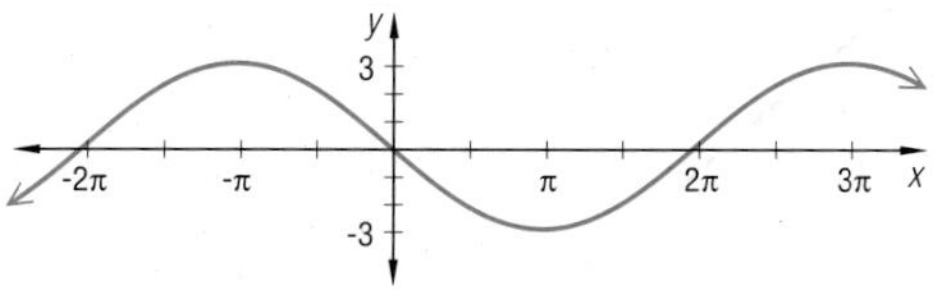

USES deal with applications of mathematics in real situations.

■ **Objective J:** *Apply equation-solving techniques to real world problems.* *(Lessons 3-1, 3-5)*

51. The formula

$$m = m_0\left(\frac{1}{\sqrt{1-\left(\frac{v}{c}\right)^2}}\right)$$

relates the mass of an object at rest, m_0, to its mass m when traveling at a velocity v. (c represents the speed of light.) Determine the velocity, as a function of the speed of light, for which the traveling mass is twice the rest mass. $v = \frac{\sqrt{3}}{2}c$

52. The weight w_h (in kg) of an object at altitude h (in km) above the Earth's surface satisfies

$$w_h = \left(\frac{r}{r+h}\right)^2 w_0$$

where w_0 is the weight at sea level and r is the radius of the Earth (about 6400 km). At what altitude would a person weigh $\frac{1}{4}$ of his sea level weight? $h = 6400$ km

53. Slowtown has a population of 3500 and is losing people at the rate of 150 per year. Boomtown has a population of 1500 and is growing 5% per year. In about how many years will the two towns have the same population? about 8 years

54. If a ball is dropped from a height of 50 feet, its height at any time t (in sec) is given by the equation $h(t) = -16t^2 + 50$ At what time is the ball at a height of 20 feet? $t \approx 1.369$ seconds

■ **Objective K:** *Use operations on functions to find formulas which model realistic situations.* *(Lesson 3-3)*

55. An assembly line has the capacity to fill orders for up to 750 television sets per day to be sold by a nationwide department store. For each day the assembly line is in operation, the fixed costs are $10,000 for line operators, equipment depreciation, and maintenance. The cost of the components used to assemble each television set is $55. The televisions are sold to the department store at a price of $150 minus a sales incentive discount of $$t$ per television, where t is 6% of the number of televisions per day for the chain.
  a. Find formulas for
    i. the total cost function C
    ii. the selling price function S
    iii. the total revenue function R
    iv. the total profit function P
  b. Sketch the graphs of functions C and R on the same coordinate axes.
  c. Use subtraction of y-coordinates to sketch the function P.

56. At time $t = 0$, an oil tanker is in a harbor with its deck level with the pier. It is being unloaded which causes the ship to rise .05 meters per minute. At the same time the wave action raises and lowers the ship according to the formula $h = .4\cos(.5t)$ where h is the height in meters after t minutes.
  a. Write a formula for s, where $s(t)$ is the height above or below the pier after t minutes. $s(t) = .4\cos(.5t) + .05t - .4$
  b. Graph s.
  c. Explain why this model can only be used over a short period of time.

49.d.

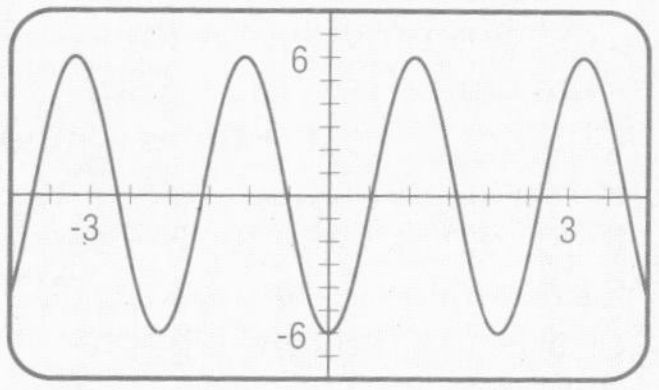

50.a. amplitude: 3; period: 4π; phase shift: -π

55.a. i. $C(x) = 10000 + 55x$
ii. $S(x) = 150 - .06x$
iii. $R(x) = 150x - .06x^2$
iv. $P(x) = 95x - 10000 - .06x^2$
b.

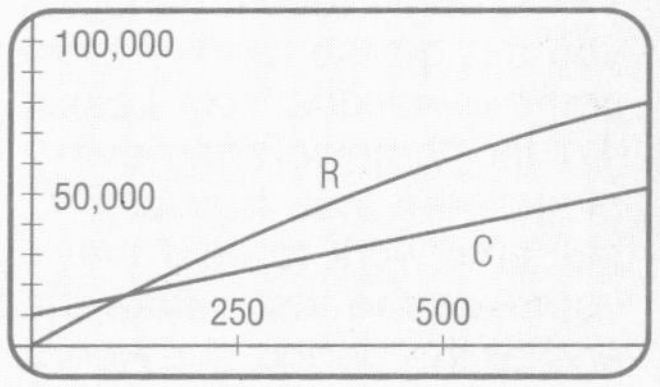

c.

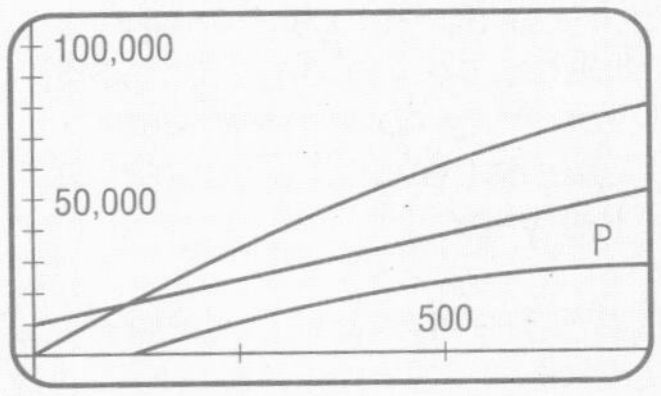

56.b.

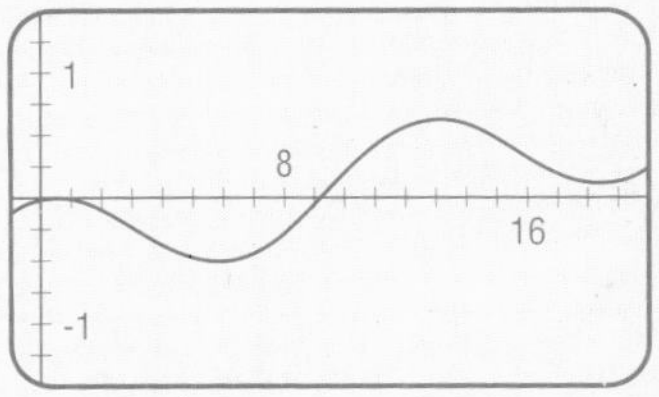

c. As soon as the ship is finished unloading, it will stop rising .05 meters per minute. After that time, the wave action will be the only cause of the ship's movement.

EVALUATION
Three forms of a Chapter Test are provided in the Teacher's Resource File. Chapter 3 Test, Forms A and B cover just Chapter 3. The third test is Chapter 3 Test, Cumulative Form. About 50% of this test covers Chapter 3, 25% covers Chapter 2, and 25% covers Chapter 1. A fourth test, Comprehensive Test, Chapters 1–3, that is primarily multiple choice in format, is also provided. For information on grading, see *General Teaching Suggestions: Grading* on page T43 in the Teacher's Edition.

ASSIGNMENT RECOMMENDATION
We strongly recommend that you assign the reading and some questions from Lesson 4-1 for homework the evening of the test. It gives students work to do if they complete the test before the end of the period and keeps the class moving.

If you do not give assignments on the days of tests, you may cover one less *chapter* over the course of the year.

ADDITIONAL ANSWERS
58.b. $.56 < p < .60$

59.

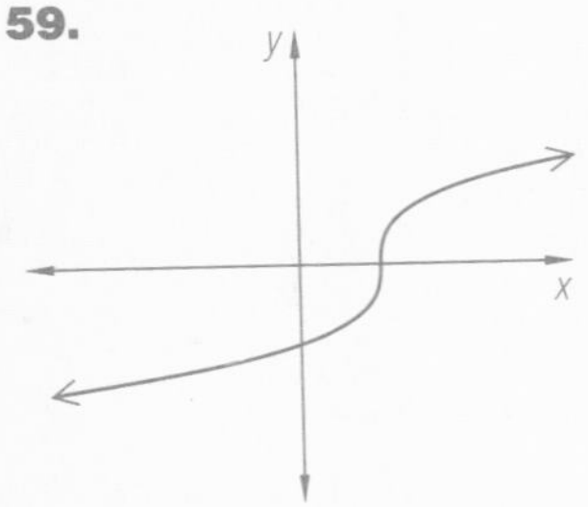

60. *f* is not a 1-1 function, so it does not have an inverse.

61., 62.–64., 65.b., 68., 69. See Additional Answers in the back of this book.

Objective L: *Use inequalities to solve real-world problems.* (Lessons 3-6, 3-7, 3-8)

57. Suppose the ball in Problem 54 is thrown upward with an initial velocity of 48 feet/sec. Then its height at any time t (in seconds) is given by the equation $h(t) = -16t^2 + 48t + 50$. Find the times when the ball is higher than 18 feet above the ground. $0 < t < 3.56$

58. A pollster reports that 58% of Americans favor a law that is pending in Congress. The error is less than 2%.
- **a.** Write an absolute value inequality that describes the true percentage p of Americans who are in favor of the legislation. $|p - .58| < .02$
- **b.** Solve your inequality in part **a** for p.

Representations deal with pictures, graphs, or objects that illustrate concepts.

Objective M: *Graph functions obtained from other functions by function operations or inverses.* (Lessons 3-2, 3-3)

In 59 and 60, a graph of a function f is given. If possible, sketch the graph of f^{-1}. If not possible, explain why not.

59.

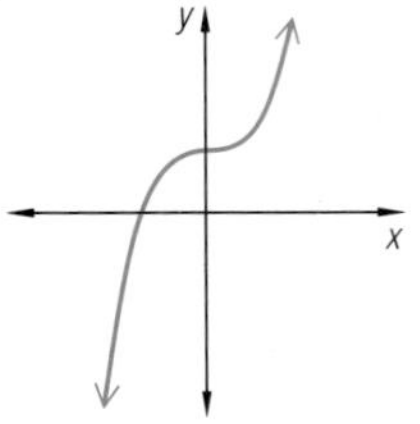

60.

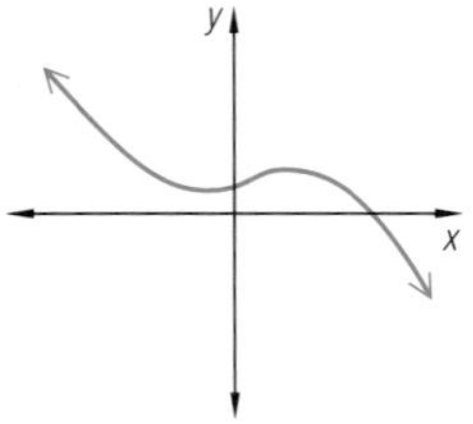

61. Graph $f: x \rightarrow \sin x$ and $g: x \rightarrow x$ on the same axes.
- **a.** Graph $f + g$.
- **b.** Graph $f \cdot g$.
- **c.** Use the properties of f and g to explain the differences in the graphs of $f + g$ and $f \cdot g$.

In 62–64, let $f(x) = \sin x$, $g(x) = \sin 2x$, $h(x) = \frac{1}{x}$, and $k(x) = x$. Sketch graphs of the function over the interval $0 < x < 2\pi$.

62. $h \cdot f$

63. $h - k$

64. $g + k$

Objective N: *Find an equation of a graph after a transformation.* (Lesson 3-9)

65. The circle $x^2 + y^2 = 1$ is transformed by $S_{3,1/3}$ and then $T_{2,-5}$.
- **a.** Find the equation of the image under $S_{3,1/3}$. $\frac{x^2}{9} + 9y^2 = 1$
- **b.** Find the equation of the final image.

In 66 and 67, find an equation for the graph below:

66. as an image of $y = \sin x$. $y = -\frac{1}{2} \sin 2x$

67. as an image of $y = \cos x$. $y = \frac{1}{2} \cos(2x + \frac{\pi}{2})$

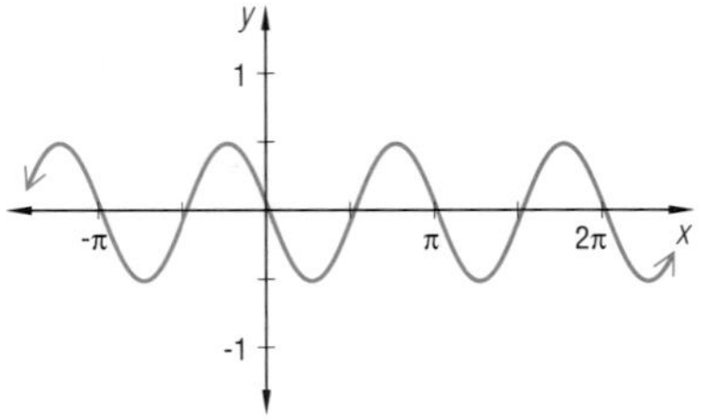

68. Find an equation for the image of the hyperbola $xy = 1$ under $S_{a,\, 1/a}$. Explain what has happened.

69. The function f defined by $f(x) = \sin 4x$ is transformed so that the amplitude is 3 times that of f and the period is half the period of f.
- **a.** Write the equation of the new function g.
- **b.** Graph the new function.

Objective O: *Use graphs to approximate zeros of functions and solve equations and inequalities.* (Lessons 3-5, 3-7)

70. The x-coordinates of the intersection point of the graphs $y = f(x)$ and $y = g(x)$ are solutions of which of the following equations?
(a) $f(x) = g(x)$
(b) $f(x) - g(x) = 0$
(c) $|g(x) - f(x)| = 0$ (a), (b), (c)

71. Find an interval of length .05 that contains a solution for the equation $x^3 + 3x^2 - 2 = x^2 - 4x + 5$. sample: $.975 \le x \le 1.0$

72. Approximate the zeros of the function $f(x) = e^{(x^2)} - 2$ to within .02 of their actual value. $x \approx -0.83$ or $x \approx 0.83$

220

CHAPTER 4 INTEGERS AND POLYNOMIALS

OBJECTIVES

The objectives listed here are the same as in the Chapter 4 Review on pages 283–286 of the student text. The Progress Self-Test on page 282 and the tests in the Teacher's Resource File cover these objectives. For recommendations regarding the handling of this end-of-chapter material, see the notes in the margin on the corresponding pages of this Teacher's Edition.

OBJECTIVES FOR CHAPTER 4 (Organized into the SPUR Categories—Skills, Properties, Uses, and Representations)	**Progress Self-Test Questions**	**Chapter Review Questions**	**Teacher's Resource File**	
			Lesson Masters*	**Chapter Test Forms A & B**
SKILLS				
A Use the Quotient-Remainder Theorem for integers or polynomials.	1c	1–7	4-2, 4-4	1, 2
B Use synthetic substitution to find values of polynomial functions.	5	8–12	4-5	16
C Divide polynomials.	2, 3	13–19	4-4, 4-5	4
D Determine the congruence of integers for a given modulus.	8	20–24	4-3	10
E Factor polynomials.	12	25–34	4-1, 4-8	3
PROPERTIES				
F Justify properties of factors of integers or factors of polynomials.	6, 16	35–49	4-1	6, 7
G Use the properties of congruence of integers for a given modulus to rewrite sentences.	9	50–54	4-3	11
H Use the Remainder Theorem, Factor Theorem, or Quotient-Remainder Theorem to describe characteristics of given polynomials.	4, 17	55–61	4-4, 4-6	5, 15
I Use proof by contradiction.	18	62–66	4-8	17
J Use the Factor Search Theorem and the Fundamental Theorem of Arithmetic in determining prime numbers and prime factorizations.	10, 11	67–73	4-8	13, 14
USES				
K Use the Quotient-Remainder Theorem to solve applied problems.	1a, b	74–77	4-2	1
L Use modular arithmetic to solve applied problems.	7	78–81	4-3	12
REPRESENTATIONS				
M Represent numbers in other bases and perform addition in base 2.	13, 14, 15	82–88	4-7	8, 9

*** The masters are numbered to match the lessons.**

OVERVIEW 4

OVERVIEW ■ CHAPTER 4

Integers and polynomials over a field (such as the field of real numbers or the field of complex numbers) have much in common. Structurally, with the operations of addition and multiplication, they are both rings. For each, there is a Quotient-Remainder Theorem. Some integers and some polynomials can be factored over particular domains; others cannot and are prime over that domain. When integers are represented in base 10, that representation is an abbreviation for a polynomial in the number 10. Thus, it is natural to discuss these ideas together.

Although polynomials are built up through the addition or subtraction of powers of a particular number, the key operation in this chapter is division. In Lesson 4-1, the terminology of factors is introduced for both integers and polynomials.

Lesson 4-2 considers integers only, and discusses what happens when one integer is not a factor of another. The result is that a remainder exists, following conditions of the Quotient-Remainder Theorem.

If a number is divided by n, each of the possible remainders gives rise to a congruence class modulo n. Lesson 4-3 discusses modular arithmetic.

Lessons 4-4 through 4-6 are devoted to the division of polynomials. Long division and the Quotient-Remainder Theorem for polynomials are discussed first. Then the process of synthetic division is described and analyzed. The last of these three lessons covers the Remainder and Factor Theorems.

Lesson 4-7 presents a different kind of connection between polynomials and integers, namely that the representation of an integer m in base x is a shorthand for the value of a polynomial in x.

Lesson 4-8 discusses prime numbers and prime polynomials over a field. The method of indirect proof is analyzed also and exemplified by Euclid's proof of the infinitude of the primes.

The use of polynomials to approximate other functions is the subject of Lesson 4-9.

In this chapter, students are expected to do proofs of divisibility. Some of the proofs involve various congruence classes, such as even and odd integers, or those of the form $3k + 1$. Others involve proving that, for all integer arguments, the values of a particular polynomial are divisible by a particular number. Students also are expected to begin developing some proficiency with indirect proof.

PERSPECTIVES 4

PERSPECTIVES ■ CHAPTER 4

The Perspectives provide the rationale for the inclusion of topics or approaches, provide mathematical background, and make connections with other lessons and within UCSMP.

4-1

FACTORS OF INTEGERS AND POLYNOMIALS

In this lesson, divisibility is discussed in two different contexts, namely, in the system of integers and in the system of polynomials. Students already may be familiar with many of the facts presented in the lesson. The topic of divisibility is presented carefully in order to prove the various theorems about divisibility and in order to give students the tools to write divisibility proofs.

4-2

THE QUOTIENT-REMAINDER THEOREM

The Quotient-Remainder Theorem for integers has been known by students since they studied division in elementary school. This lesson is devoted to explaining the theorem for negative as well as positive integers, and to showing how the theorem suggests a partitioning into sets of integers with equal remainders.

4-3

MODULAR ARITHMETIC

There are a number of reasons for discussing modular arithmetic in this lesson. (1) It is an immediate application of the Quotient-Remainder Theorem. (2) We use the language of modular arithmetic to describe properties of the trigonometric functions later in the book. (For example, the solutions to the equation $\sin x = \frac{1}{2}$ can be described as $x = \frac{\pi}{6} \pm 2n\pi$ for any integer n, or as $x \equiv \frac{\pi}{6} \pmod{2\pi}$.) (3) There are interesting problems that can be solved using congruences, and there are interesting applications. (4) Modular arithmetic is a topic that all college students of mathematics and computer science are expected to know. The definition and basic properties of congruence are presented, with applications to check digits and to finding the right-most digits of certain numbers.

4-4

DIVISION OF POLYNOMIALS

There are two ways to divide one integer by another, say 32 by 5. In *integer division,* the quotient is 6 and the remainder is 2. In *rational number division,* the quotient is 6.4. Similarly, there are two ways of dividing polynomials. There is *polynomial division,* in which $x^2 - 2$ divided by $x - 1$ is $x + 1$ with a remainder of -1. There is also *rational expression division*, in which the quotient of $x^2 - 2$ divided by $x - 1$ is $x + 1 + \frac{-1}{x - 1}$. The focus in this lesson is on polynomial division, for which there is a Quotient-Remainder Theorem analogous to the one for integers.

4-5

SYNTHETIC DIVISION

Synthetic division or synthetic substitution (the name depends on the use to which it is put) is discussed in this lesson. It provides important background for the Remainder Theorem of Lesson 4-6, and its explanation involves the nested form of polynomials often used in computer programs.

4-6

THE REMAINDER AND FACTOR THEOREMS

The Remainder Theorem was justified informally in Lesson 4-6; it is proved formally in this lesson. The Factor Theorem, which follows immediately from the Remainder Theorem, basically says that a polynomial contains each of its zeros in the form of a linear factor. The Factor Theorem is used to prove that a polynomial of degree n has at most n zeros, thus its graph can cross a horizontal line at most n times.

4-7

POLYNOMIAL REPRESENTATIONS OF INTEGERS

In this lesson, students learn how to represent integers in bases other than base 10. The first part of the lesson discusses the test for divisibility by 9 and the check called "casting out nines" in base 10. The second part of the lesson discusses other bases, particularly base 2 (binary) because it is the base used most often in computer science and other applications. Students should be able to convert from one base to another and should be able to add integers written in base 2.

4-8

PRIME NUMBERS AND PRIME POLYNOMIALS

This lesson formalizes the factoring of integers and polynomials. The infinitude of primes is proved indirectly, and the factorization of integers and polynomials into prime factors is explained. Some books use the term *irreducible polynomial* rather than *prime polynomial.* We use prime polynomial to emphasize the common features in both sets.

4-9

THE VARIETY OF POLYNOMIALS

A continuous function may be approximated over any closed interval, to any assigned degree of accuracy, by a polynomial function. This statement, known as the Weierstrass Approximation Theorem, is another reason polynomials are important. In this lesson, polynomial approximations (found by using power series) to familiar functions are displayed, and students are asked to explore the idea in the Questions.

DAILY PACING CHART ■ CHAPTER 4

Every chapter of UCSMP *Precalculus and Discrete Mathematics* includes lessons, a Progress Self-Test, and a Chapter Review. For optimal student performance, the self-test and review should be covered. (See *General Teaching Suggestions: Mastery* on page T35 of this Teacher's Edition.) By following the pace of the Full Course given here, students can complete the entire text by the end of the year. Students following the pace of the Minimal Course spend more time when there are quizzes and on the Chapter Review and will generally not complete all of the chapters in this text.

When chapters are covered in full (the recommendation of the authors), then students in the Minimal Course can cover 11 chapters of the book. For more information on pacing, see *General Teaching Suggestions: Pace* on page T34 of this Teacher's Edition.

DAY	MINIMAL COURSE	FULL COURSE
1	4-1	4-1
2	4-2	4-2
3	4-3	4-3
4	Quiz (TRF); Start 4-4.	Quiz (TRF); 4-4
5	Finish 4-4.	4-5
6	4-5	4-6
7	4-6	Quiz (TRF); 4-7
8	Quiz (TRF); Start 4-7.	4-8
9	Finish 4-7.	4-9
10	4-8	Progress Self-Test
11	4-9	Chapter Review
12	Progress Self-Test	Chapter Test (TRF)
13	Chapter Review	
14	Chapter Review	
15	Chapter Test (TRF)	

TESTING OPTIONS

TESTING OPTIONS

- Quiz for Lessons 4-1 through 4-3
- Quiz for Lessons 4-4 through 4-6
- Chapter 4 Test, Form A
- Chapter 4 Test, Form B
- Chapter 4 Test, Cumulative Form

A Quiz and Test Writer is available for generating additional questions, additional quizzes, or additional forms of the Chapter Test.

INDIVIDUAL DIFFERENCES

PROVIDING FOR INDIVIDUAL DIFFERENCES

The student text is written for, and tested with, average students. It also has been successfully used with better and more poorly prepared students.

The Lesson Notes often include Error Analysis and Alternate Approach features to help you with those students who need more help. Students of all abilities often learn from their peers and may benefit from small group work referenced as appropriate throughout the Notes. A blackline Lesson Master (in the Teacher's Resource File), keyed to the chapter objectives, is provided for each lesson to allow more practice. (However, since it is important to keep up with the daily pace, you are not expected to use all of these masters. Again, refer to the suggestions for pacing on page T34.) Extension activities are provided in the Lesson Notes for those students who have completed the particular lesson in a shorter amount of time than is expected, even in the Full Course.

Integers and Polynomials

CHAPTER 4

Caesar's troops, captured in stone, adorn the Roman Forum.

Perhaps the earliest examples of coded messages were those sent by drumbeat in heavily forested areas of Africa or the smoke signals of Native Americans. Early secret codes were simple substitutions of one letter for another. Julius Caesar is said to have used the following substitution scheme to send battle plans to his troops (using the Latin alphabet, of course).

A	B	C	D	E	F	G	H	I	J	K	L	M	N	O	P	Q	R	S	T	U	V	W	X	Y	Z
D	E	F	G	H	I	J	K	L	M	N	O	P	Q	R	S	T	U	V	W	X	Y	Z	A	B	C

The message "ATTACK" would be written by substituting the corresponding letters in the second row: "DWWDFN". In the mathematical language of this chapter, the position $m(x)$ of the encoded letter for the letter in position x of the alphabet would be

$$m(x) = x + 3 \pmod{26}.$$

Codes are still important in national defense. In fact, the largest employer of mathematicians in the United States is the National Security Agency, the part of the Department of Defense responsible for ensuring the security of U.S. communications.

The need for secure coding systems now extends beyond the military to the needs of business to keep records and electronic messages safe from tampering, and to ensure privacy. Modern codes are based on factoring very large integers into two primes. If the primes are each about 75 digits in length, their product will have 150 digits. At this time, prime codes based on 150-digit numbers seem to be relatively safe. The problem is to find 75-digit primes. This is not an easy task even with high-speed computers, so mathematicians have built new mathematical theories, techniques, and programs which extend many of the ideas you will study in this chapter.

CHAPTER 4

We recommend 12–15 days be spent on this chapter: 9 to 11 days on the lessons and quizzes; 1 day for the Progress Self-Test; 1 or 2 days for the Chapter Review; and 1 day for a Chapter Test. (See the Daily Pacing Chart on page 221D.)

USING PAGE 221
It may come as a surprise to students to learn that the National Security Agency is the largest employer of mathematicians in the United States. Many of these people work on the mathematical theory of codes. The interception and transcription of coded messages is essential for national defense. Early in World War II, British intelligence intercepted German-coded messages. These messages were constructed by a machine using a very complicated code that the Germans thought would be impossible to decipher. British mathematicians and cryptologists after much effort, broke the code. It is thought by many people that the outcome of the war might have been different had this code not been broken. (Delving into this story is an Exploration Question in Lesson 4-8.)

LESSON 4-1

RESOURCES
■ Lesson Master 4-1

OBJECTIVES

E Factor polynomials.
F Justify properties of factors of integers or factors of polynomials.

TEACHING NOTES

In the definition of factor, the letter d is chosen to suggest divisor, and the letter q to suggest quotient. In the Quotient-Remainder Theorem (Lesson 4-2), these letters are used again along with r to suggest remainder.

In the definition of factor for polynomials, you may have to remind students that the quotient $q(x)$ could be a polynomial of degree zero, that is, a constant polynomial. In **Example 1a**, it is also the case that $m - 2n$ is a factor of $3m - 6n$, because $3m - 6n = 3(m - 2n)$.

We expect that students are familiar with the idea of degree of a polynomial but are unfamiliar with nested form, which is discussed in Lesson 4-5. You may wish to tell students that the nested form is preferable to the power form in computer programs because computers usually evaluate multiplications more efficiently than they evaluate powers.

LESSON

4-1 Factors of Integers and Polynomials

Integers and polynomials do not look alike. Yet, as you will see throughout this chapter, they have many structural similarities. Consider division. You can always divide a given integer n by another integer d provided that $d \neq 0$. However, the result may not be an integer. If the result is an integer, d is said to be a *factor* of n.

For example, 8 is a factor of 56 because there is an *integer* q, namely 7, such that

$$56 = q \cdot 8.$$

Note also that 5 is a factor of -5 because $-5 = -1 \cdot 5$. However, 4 is not a factor of 58 because there is no integer q such that

$$58 = q \cdot 4.$$

There is a solution to this equation, but the solution is not an integer.

Definition

Suppose that n and d are integers and $d \neq 0$. ***d* is a factor of *n*** if and only if there is an integer q such that $n = q \cdot d$.

Other phrases with the same meaning are: ***n* is a multiple of *d*, *n* is divisible by *d***, and ***d* is a divisor of *n***. The letter q is chosen in the definition because q is the **quotient** when n is divided by d.

For instance, -117 is a multiple of 13 because $-117 = -9 \cdot 13$.

The definition specifically excludes the possibility that d is equal to 0 because division by 0 is not defined. Thus, 0 is not a factor of any integer.

Example 1 For all integers m and n:
a. Is 3 a factor of $3m - 6n$?
b. Is $7m^2n^3$ a multiple of mn^5?

Solution **a.** Yes, because $3m - 6n = 3(m - 2n)$, and $m - 2n$ is an integer for all integers m and n.
b. When $m = 2$ and $n = 3$, $7m^2n^3 = 756$ and $mn^5 = 486$. So, $7m^2n^3$ is not always a multiple of mn^5.

Definition

A function p is a **polynomial function** if there are $n + 1$ numbers $a_0, a_1, \ldots, a_n$ with $a_n \neq 0$ such that
$p(x) = a_nx^n + a_{n-1}x^{n-1} + \ldots + a_1x + a_0$ for all x.

222

$p(x)$ itself is a **polynomial**. The numbers $a_0, a_1, \ldots, a_n$ are the **coefficients** of $p(x)$, and n is the **degree** of $p(x)$. If all the a_i are from a particular set S, then $p(x)$ is a **polynomial over S**. For example,

$$p(x) = x^3 + 2x^2 - x - 2$$

is a polynomial of degree 3 over the set of integers. Two polynomials are **equal** if their corresponding coefficients are equal.

The form in which $p(x)$ is written above is called the **expanded form** or **power form** of $p(x)$. This polynomial could also be written in **factored form** as

$$p(x) = (x + 2)(x + 1)(x - 1),$$

which clearly shows the zeros of p to be -2, -1, and 1. $p(x)$ also could be written in **nested form** as

$$p(x) = x(x(x + 2) - 1) - 2,$$

which is a form quite convenient for calculation of its values.

From the definition, the degree of a **constant polynomial** $p(x) = k \neq 0$ is 0. It is convenient to think of the **zero function** defined by $p(x) = 0 \; \forall \; x$ as being a polynomial function (the **zero polynomial** function) with no assigned degree. Thus, horizontal lines other than the x-axis are graphs of polynomial functions of degree 0, oblique lines are graphs of polynomial functions of degree 1, and parabolas are graphs of polynomial functions of degree 2.

Example 2 Find the degree of

a. $(x^4 + 3x^2 + 8)(x^2 - x - 1)$

b. $(x^4 + 3x^2 + 8) + (x^2 - x + 1)$

c. $(x^4 + 3x^2 + 8) + (-x^4 + 2x^3 - 6)$.

Solution

a. The terms of the product polynomial are found by multiplying each term of the first by each term of the second. Here the highest degree term is the product x^6, the product of x^4 and x^2. Therefore the degree is 6.

b. The sum of the two polynomials is found by combining like terms. The highest degree term of the sum is x^4. Thus, the sum is of degree 4.

c. Again the sum is found by combining like terms. Here the x^4 terms add to 0 and the highest degree term with a nonzero coefficient is $2x^3$. Thus, the sum is of degree 3.

In Example 2**a**, the polynomials to be multiplied are of degree 4 and 2; the product is of degree 6. In 2**b**, the polynomials to be added are of degree 4 and 2, and the sum is of degree 4. In 2**c**, both polynomials are of degree 4 but the sum is of degree 3. These results suggest the following theorems, which will be applied in the next lesson.

Making Connections
The set S over which the polynomials are defined is often taken to be either the integers, the rational numbers, the real numbers, or the complex numbers. Students should note what set S is used in the statement of each theorem in later lessons of this chapter, since it is an important part of each result.

The 155-digit number factored by Arjen Lenstra and Mark Manasse (see page 224) is the product of three primes that are 7, 49, and 99 digits long. The number is equal to 2^{2^9} (or 2^{512}) + 1. This number is the ninth in a series of numbers devised by the French mathematician Pierre de Fermat over 300 years ago.

Some students may think that the Degree of a Sum Theorem should state that the degree of $p(x) + q(x)$ is just the larger of m and n. But, if the polynomials had the same degree and the leading coefficients of the polynomials were opposites, then the degree of the sum would be less than the degree of either polynomial. For instance, if $p(x) = 7x^4 - 8x + 5$, and $q(x) = -7x^4 + 3x^2 + 17$, then the degree of $p(x) + q(x)$ would be 2. In fact, if $q(x) = k - p(x)$, where $k \neq 0$, then the degree of $p(x) + q(x)$ would be 0.

You might ask: Why is there no Degree of a Difference Theorem? ($p(x) - q(x) = p(x) + (-q(x))$, which is the sum of polynomials of the same degrees as given. That is, any subtraction can be converted to an addition.)

When discussing the proof of the Transitive Property of Factor Theorem, remind students that they are expected to be able to write a proof such as those written in boldface. Discuss this proof in class, along with **Question 8**. Note that the closure property of multiplication is used in the proof although it has not been stated explicitly.

Example 4 provides an opportunity to recall what it means to find a counterexample to a universal statement, and to review the meaning of ∃ and ∀. Notice the form of the conjecture: ∀ a, b, c, if $p(a, b, c)$, then ($q(a, b, c)$ or $r(a, b, c)$). The negation is ∃ a, b, c such that $p(a, b, c)$ and $\sim(q(a, b, c)$ or $r(a, b, c))$. The last, by De Morgan's Laws, is equivalent to $\sim q(a, b, c)$ and $\sim r(a, b, c)$, which is the form of the negation in the solution.

Dr. Arjen Lenstra displays factored 155-digit number.

Theorems

Let $p(x)$ be a polynomial of degree m and $q(x)$ be a polynomial of degree n.

Degree of a Product: The degree of $p(x) \cdot q(x)$ is $m + n$.

Degree of a Sum: The degree of $p(x) + q(x)$ is less than or equal to the larger of m and n, or else $p(x) + q(x)$ is the zero polynomial.

The words associated with factoring integers are also used with polynomials. Notice the similarity of the definitions.

Definition

Suppose that $n(x)$ and $d(x)$ are polynomials and $d(x) \neq 0$. **$d(x)$ is a factor of $n(x)$** if and only if there exists a polynomial $q(x)$ such that $n(x) = q(x) \cdot d(x)$.

Example 3 Is $x + 15$ a factor of $x^2 - 225$?

Solution Yes, because $x^2 - 225 = (x + 15)(x - 15)$. Here $n(x) = x^2 - 225$ and $d(x) = x + 15$. The quotient polynomial $q(x) = x - 15$.

Now we return to properties of factors of integers. By division, you can verify that 13 is a factor of 1001, and 1001 is a factor of 364,364. Can you explain why 13 is then a factor of 364,364? Here is the general theorem and its proof.

Theorem (Transitive Property of Factors of Integers)

For all integers a, b, and c, if a is a factor of b, and b is a factor of c, then a is a factor of c.

Proof Suppose that a, b, and c are integers such that a is a factor of b and b is a factor of c. Then there are integers m and n such that

$$b = a \cdot m \text{ and } c = b \cdot n$$

by the definition of factor. It follows by substitution that

$$c = b \cdot n = (a \cdot m) \cdot n = a \cdot (m \cdot n)$$

because multiplication is associative. By closure of multiplication of integers, $m \cdot n$ is an integer; so by the definition of factor, a is a factor of c.

Another key property of factors has to do with the factor of a sum. Since 13 is a factor of both 1001 and 52, is it also a factor of their sum, 1053?

Theorem (Factor of a Sum of Integers)

For all integers a, b, and c, if a is a factor of b and a is a factor of c, then a is a factor of $b + c$.

Proof We begin the proof. You are asked to complete the proof in the questions.

Suppose a, b, and c are any integers such that a is a factor of b and a is a factor of c. By the definition of factor, there exist integers q and r such that

$$b = a \cdot q \text{ and } c = a \cdot r.$$

Example 4 Disprove the following conjecture.

∀ integers a, b, and c, if c is a factor of a • b, then c is a factor of a or c is a factor of b.

Solution Only one counterexample is needed to disprove the conjecture. The conjecture is false if and only if its negation is true. The negation of the given statement is

∃ integers a, b, and c such that c is a factor of a • b, and c is not a factor of a, and c is not a factor of b.

Try to think of integers that make this negation true. For instance, let $a = 2$ and $b = 6$. Then $ab = 12$. Can you find a factor of 12 which is not a factor of 2 and not a factor of 6? When you do, this is what you might write.

Let $a = 2$, $b = 6$, and $c = \square$. Note that $\square$ is a factor of $2 \cdot 6 = 12$ but $\square$ is not a factor of 2 and $\square$ is not a factor of 6. Thus c is a factor of $a \cdot b$, but c is not a factor of a and c is not a factor of b. Hence, the conjecture is false.

We leave it to you to find a number to put in the $\square$.

Factors of polynomials have properties that correspond to the properties of factors of integers.

Theorems

For all polynomials $a(x)$, $b(x)$, and $c(x)$,

Transitive Property of Factors of Polynomials If $a(x)$ is a factor of $b(x)$ and $b(x)$ is a factor of $c(x)$, then $a(x)$ is a factor of $c(x)$.

Factor of a Sum of Polynomials If $a(x)$ is a factor of $b(x)$ and $a(x)$ is a factor of $c(x)$, then $a(x)$ is a factor of $b(x) + c(x)$.

ADDITIONAL EXAMPLES

1. For all integers m and n:
a. is 17 a factor of 171177?
No
b. is $11m^5n^2$ a factor of m^4n^3?
No

2. Is $x - 7$ a factor of $x^2 - 49$? Explain your answer.
Yes; $x^2 - 49 = (x + 7)(x - 7)$.

3. Give the degree of each of the following:
a. $(x^6 - 7x^2 + 8x)(3x^4 + 5x^2 - 8)$
10
b. $(5y^4 - 7y^2 + 8y) - (5y^4 - 7y^2 + 11)$
1
c. $(8x^5 - 7x^3 + 58x) - (9x^5 - 7x^3 + 16x^2)$
5

4. Disprove the following conjecture:
∀ integers a, b, and c, if c is a factor of a and c is a factor of b, then c is a factor of a ÷ b.
A counterexample is $a = 100$, $b = 50$, $c = 10$. 10 is a factor of 100, and 10 is a factor of 50, but 10 is not a factor of $2 = \frac{100}{50}$.

5. Let $p(x) = x + 3$, $q(x) = x^2 + 7x + 12$, and $r(x) = x^2 - 5x - 24$.
a. Show that $p(x)$ is a factor of $q(x)$ and of $r(x)$.
$q(x) = (x + 3)(x + 4)$; $r(x) = (x + 3)(x - 8)$
b. Show that $q(x) + r(x)$ can be expressed as $p(x) \cdot$ (some polynomial).
$q(x) + r(x) = (x + 3)(2x - 4)$

NOTES ON QUESTIONS
Questions 1-4: Remind students to use the definition of factor correctly. For instance, it is incorrect to say that 11 is a factor of 132 because $\frac{132}{11} = 12$. The correct reason should be stated as "11 is a factor of 132 because $132 = 12 \cdot 11$."

Question 8: Discuss this question along with the proof of the Transitivity of Factors Theorem. You can call upon different students to supply the missing phrases.

Error Analysis for Question 11: Students may not see at first why this conjecture is false. As a hint, you could mention, for example, that -9 is a factor of 63. After students complete the exercise, challenge them to reword the conjecture to make it true. ($\forall$ integers a and b, if a is divisible by b and b is divisible by a, then $|a| = |b|$.)

Question 14: Some students will need to be reminded of the definition of factorial.

ADDITIONAL ANSWERS

6. ***If d is a factor of n, then there is an integer q such that n = q · d.***
If there is an integer q such that n = q · d, then d is a factor of n.
The first direction broke down b and c into its factors.
The second direction was used to show a was indeed a factor of c.

9. Then $b + c = (a \cdot q) + (a \cdot r) = a(q + r)$. Therefore, since $q + r$ is an integer by closure properties, a is a factor of $b + c$ by definition.

10.a. $q(x) = 2x^2 - 12x = 2x(x - 6) = 2x \cdot p(x)$; $r(x) = x^2 - 3x - 18 = (x + 3)(x - 6) = (x + 3) \cdot p(x)$
b. $q(x) + r(x) = (2x^2 - 12x) + (x^2 - 3x - 18) = 3x^2 - 15x - 18 = (3x + 3)(x - 6) = (3x + 3) \cdot p(x)$

Example 5 Let $p(x) = x - 2$, $q(x) = x^2 - 4$, and $r(x) = x^2 + 8x - 20$.
a. Show that $p(x)$ is a factor of $q(x)$ and of $r(x)$.
b. Show that $q(x) + r(x)$ can be expressed as $p(x) \cdot$(some polynomial).

Solution
a. $q(x) = x^2 - 4 = (x + 2)(x - 2)$
$r(x) = x^2 + 8x - 20 = (x + 10)(x - 2)$
Thus, $p(x)$ is a factor of both $q(x)$ and $r(x)$.
b. To show that $x - 2$ is a factor of $q(x) + r(x)$, first express $q(x)$ and $r(x)$ in factored form.

$$\begin{aligned} q(x) + r(x) &= (x^2 - 4) + (x^2 + 8x - 20) \\ &= (x + 2)(x - 2) + (x + 10)(x - 2) \\ &= [(x + 2) + (x + 10)](x - 2) \\ &= (2x + 12)(x - 2) \end{aligned}$$

Use $x - 2$ as a common factor.

Thus,
$q(x) + r(x) = p(x) \cdot (2x + 12)$.

Factors of polynomials are important for two major reasons. First, just as factors of integers help to understand how integers are related, factors of polynomials are important in understanding how polynomials are related. Second, factors of polynomials can be used to determine the zeros of the polynomials and thus help to solve polynomial equations and inequalities. Before we turn to these ideas, we will take a more detailed look at division of integers.

Questions

Covering the Reading

In 1–5, determine whether the statement is *true* or *false* and explain your answer.

1. 11 is a factor of 132.
   True, there exists an integer, 12, such that $11 \cdot 12 = 132$.
2. 17 is a multiple of 17.
   True, there exists an integer, 1, such that $17 \cdot 1 = 17$.
3. 2 is a factor of 0.
   True, there exists an integer, 0, such that $2 \cdot 0 = 0$.
4. For any integers n and m, 4 is a factor of $2n(2m + 6)$.
   True, $2n(2m + 6) = 4(nm + 3n)$.
5. $n - 6$ is a factor of $n^2 - 17n + 66$. True, $n^2 - 17n + 66 = (n - 6)(n - 11)$

6. Write the two *if-then* statements contained in the definition of *d is a factor of n*. Explain how each direction of the definition was used in the proof of the Transitive Property of Factors. See margin.

7. Find the degree of the polynomial.
   a. $(4y^{10} - 3y^8 + 4)(y^2 + 7y + 2)$ 12
   b. $(4y^{10} - 3y^8 + 4) + (y^2 + 7y + 2)$ 10
   c. $(4y^{10} - 3y^8 + 4) - (y^{10} - 5y^4 + 7)$ 10

8. Consider the following theorem:
Theorem (Factor of a Product)
For all integers m, n, and p, if m is a factor of n, then m is a factor of $n \cdot p$.
Fill in the blanks in the proof of this theorem.

Suppose m, n, and p are any integers such that m is __a.__ of n. By the definition of factor, there exists an integer q such that __b.__ . Then

$n \cdot p =$ __c.__ $\cdot p$ By substitution
$= m$ __d.__ By the Associative Property

Therefore, by the definition of factor and because __e.__ is an integer, __f.__ . a) a factor; b) $n = q \cdot m$; c) $q \cdot m$; d) $(q \cdot p)$; e) $q \cdot p$; f) m is a factor of $n \cdot p$.

9. Complete the proof of the Factor of the Sum of Integers Theorem. See margin.

10. Let $p(x) = x - 6$, $q(x) = 2x^2 - 12x$, and $r(x) = x^2 - 3x - 18$.
   a. Show that $p(x)$ is a factor of both $q(x)$ and $r(x)$.
   b. Show that $q(x) + r(x)$ can be expressed as $p(x) \cdot$ (some polynomial).
   c. What theorem is illustrated by parts **a** and **b**?
   a, b) See margin. c) Factor of a Sum Theorem

Applying the Mathematics

11. Consider the conjecture:
   ∀ integers a and b, if a is divisible by b and b is divisible by a, then a = b.
   a. Disprove this conjecture by giving a counterexample.
   b. Write the negation of this conjecture.
   See margin.

12. Use the proof of the Transitive Property for integer factors to prove the Transitive Property for polynomial factors. See margin.

13. $p(x) = x^3 + 1$ and $r(x) = x^7 + 2x^6 + x^4 - x^3 - 3$. If $p(x) \cdot q(x) = r(x)$, what is the degree of $q(x)$? 4

14. If m is any integer that is greater than 3, must $m!$ be divisible by 3? Why or why not? See margin.

15. Prove: *For any integer n, the sum of the three consecutive integers n, n + 1, and n + 2 is divisible by 3.* See margin.

16. a. State the converse of the Factor of a Sum of Integers Theorem.
   b. Is the converse *true* or *false*? Justify your answer.
   See margin.

17. Write the polynomial $p(x) = (8x - 1)(4x + 3)x + (8x - 1)(4x + 3)5$
   a. in factored form. $(8x - 1)(4x + 3)(x + 5)$
   b. in expanded form. $32x^3 + 180x^2 + 97x - 15$

18. Express $x^4 - 81$ as a product of three polynomials.
   $(x^2 + 9)(x - 3)(x + 3)$

11.a. Counterexample: Let $a = 1$ and $b = -1$. Then a is divisible by b and b is divisible by a, but $a \neq b$.
b. ∃ *integers a and b such that a is divisible by b and b is divisible by a and a ≠ b.*

12. Suppose that $a(x)$, $b(x)$, and $c(x)$ are polynomials such that $a(x)$ is a factor of $b(x)$ and $b(x)$ is a factor of $c(x)$. Then there are polynomials $p(x)$ and $q(x)$ such that $b(x) = a(x) \cdot p(x)$ and $c(x) = b(x) \cdot q(x)$ by the definition of a factor. It follows that $c(x) = b(x) \cdot q(x) = a(x) \cdot p(x) \cdot q(x)$. $(p(x) \cdot q(x))$ is a polynomial by closure properties, so $a(x)$ is a factor of $c(x)$ by definition.

14. Yes, $m! = m \cdot m - 1 \cdot \ldots \cdot 3 \cdot 2 \cdot 1 = 3(m \cdot m - 1 \cdot \ldots \cdot 2 \cdot 1)$. By closure properties, $(m \cdot m - 1 \cdot \ldots \cdot 2 \cdot 1)$ is an integer, so 3 is a factor of $m!$ by definition.

15., 16. See the margin on page 228.

NAME ______________________

LESSON **MASTER 4–1**
QUESTIONS ON **SPUR** OBJECTIVES

■ **SKILLS** *Objective E (See pages 283–286 for objectives.)*
1. Show that $x + 5$ is a factor of $x^2 + x - 20$.
   $x^2 + x - 20 = (x + 5)(x + 4)$
2. If $5x^2 - 8x - 21 = (x - 3) \cdot p(x)$, what is $p(x)$? $5x + 7$

■ **PROPERTIES** *Objective F*
In 3–5, determine whether the statement is *true* or *false* and explain your answer.
3. -6 is a factor of 720.
   True, because 720 = -6(-120) and -120 is an integer.
4. 21 is a factor of 98.
   False, because ∃ no integer k such that $98 = 21k$.
5. $2x + 7y$ is a factor of $4x^2 - 49y^2$.
   True, because $4x^2 - 49y^2 = (2x + 7y)(2x - 7y)$.
6. Prove: *If a, b, and c are integers such that a is a factor of b and b^2 is a factor of c, then a is a factor of c.*
   If a is a factor of b and b^2 is a factor of c, then ∃ integers k and m such that $b = ak$ and $c = b^2m$. Thus, by substitution, $c = b^2m = (ak)^2m = a(ak^2m)$. Since ak^2m is an integer, a is a factor of m.
7. Prove: *If a(x), b(x), and c(x) are polynomials such that a(x) is a factor of b(x) and a factor of c(x), then a(x) is a factor of b(x) − c(x).*
   If $a(x)$ is a factor of $b(x)$ and of $c(x)$, then ∃ polynomials $k(x)$ and $m(x)$ such that $b(x) = a(x)k(x)$ and $c(x) = a(x)m(x)$. Thus, $b(x) - c(x) = a(x)k(x) - a(x)m(x) = a(x)[k(x) - m(x)]$. Since $k(x) - m(x)$ is a polynomial, $a(x)$ is a factor of $b(x) - c(x)$.
8. Consider the conjecture *If a and b are integers and a is a factor of b, then a is a factor of $(b + 1)^2 - 1$.*
   a. Find values for a and b to illustrate this conjecture. sample: $a = 3$, $b = 6$; then $(b + 1)^2 - 1 = 48$, and 3 is a factor of 48.
   b. Prove the conjecture. If a is a factor of b, then ∃ an integer k such that $b = ak$. Then $(b + 1)^2 - 1 = b^2 + 2b = (ak)^2 + 2(ak) = a(ak^2 + 2k)$. Since $ak^2 + 2k$ is an integer, a is a factor of $(b + 1)^2 - 1$.

NOTES ON QUESTIONS
Question 23: An interesting project is to research perfect, abundant, and deficient numbers. Students interested in history may enjoy this activity.

FOLLOW-UP

MORE PRACTICE
For more questions on SPUR Objectives, use *Lesson Master 4-1,* shown on page 227.

EXTENSION
You might want to ask students to think about the parallel between the degree of a product of polynomials and the logarithm of a product of integers, extending the idea of **Question 24**. The reason for this parallel is that for an integer N, log N is approximately the number of decimal digits in N. In other words, log N is approximately the degree of N when N is written as a polynomial in powers of 10. More exactly, for an integer N, the number of decimal digits in N is $\lfloor \log N \rfloor + 1$, and the degree of N when written as a polynomial in powers of ten is $\lfloor \log N \rfloor$. If you choose to discuss this now, you will want to recall it in Lesson 4-7. At that time, you can relate logarithms to other bases and integers written in different bases.

ADDITIONAL ANSWERS
15. Suppose n is any integer. Then $n + (n + 1) + (n + 2) = 3n + 3 = 3(n + 1)$. Since $n + 1$ is an integer by closure properties, the sum of any three consecutive integers is divisible by 3.

16.a. For all integers, a, b, and c, if a is a factor of $b + c$, then a is a factor of b and a is a factor of c.
b. False; counterexample: Let $a = 3$, $b = 2$, and $c = 1$. Then a is a factor of $b + c$ but is not a factor of either b or c.

Review

19. Consider the graph of $f(x) = (x - 2)^2 + 3$ at the right.

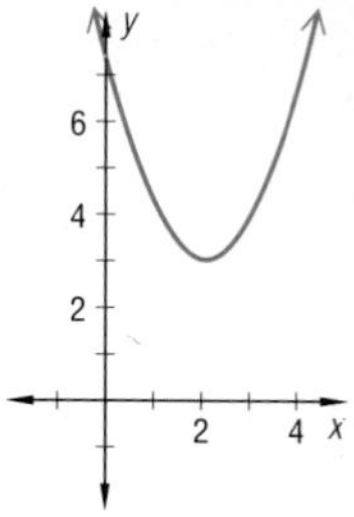

a. Identify the domain and range of f.
b. Over what interval is the function increasing? $x \geq 2$
c. Is this function even, odd, or neither?
d. This graph can be obtained from the parent graph $y = x^2$ by what translation? $T_{2,3}$
(Previous course, Lessons 2-1, 2-2, 2-5, 3-9)
a) See below. c) neither

20. Analyze the function f defined by $f(x) = 2^{1-x}$. Discuss its domain, range, end behavior, and any intervals on which it is increasing or decreasing. *(Lessons 2-1, 2-2, 2-3, 2-5, 2-6)* See below.

21. Let a, b, and c be integers. Write the negation of the following statement using DeMorgan's Law:
b is divisible by a, or c is divisible by a.
(Lesson 1-3)
b is not divisible by a, and c is not divisible by a.

22. a. Write a logical expression that corresponds to the following network. ((not q) or p) and (not r)

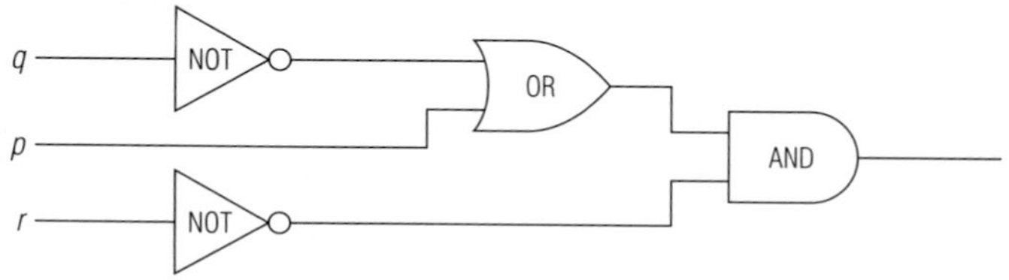

b. Find the output if $p = 0$, $q = 1$, and $r = 0$. *(Lesson 1-4)* 0

Exploration

23. A positive integer n is a **perfect number** if n equals the sum of its *proper divisors*, those divisors less than n. Thus, 6 is a perfect number because $6 = 1 + 2 + 3$; 1, 2, and 3 are the proper divisors of 6. Find the next perfect number. $28 = 1 + 2 + 4 + 7 + 14$

24. The degree of a polynomial has much in common with the number of digits of an integer. What theorems about integers correspond to the Degree of a Product and Degree of a Sum Theorems? Let the number of digits in two integers be m and n. Then the number of digits in the product of the two integers is $m + n$ or $m + n - 1$. The number of digits in the sum of the two integers is less than or equal to the larger of $(m + 1)$ and $(n + 1)$.

19. a) domain: the set of real numbers; range: $\{y: y \geq 3\}$
20. domain: the set of real numbers
range: $y > 0$
end behavior: $\lim_{x \to \infty} f(x) = 0$; $\lim_{x \to -\infty} f(x) = \infty$
decreasing over entire domain

LESSON

4-2

The Quotient-Remainder Theorem

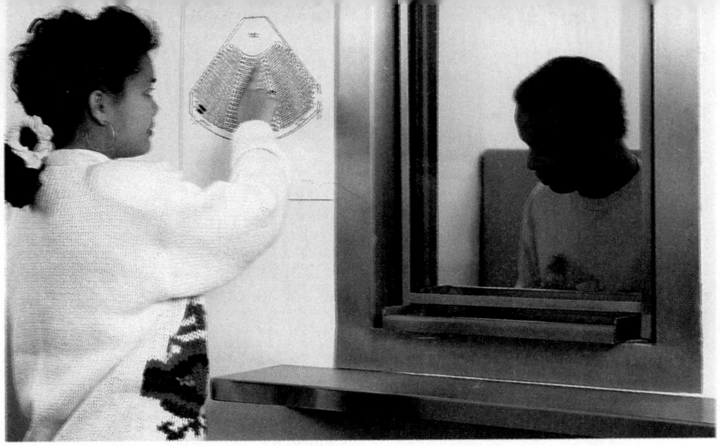

Lesson 4-1 examined the division of one integer or polynomial by another when no remainder is left. This lesson explores division that has a nonzero remainder. Consider the following situation.

> Tickets to a certain concert are first offered to ticket brokers in lots of 200. Only tickets not bought by the brokers are sold directly to the general public. The concert hall has 4566 seats. What is the maximum number of tickets that can be purchased by brokers? What is the minimum number of tickets that will be offered for sale to the general public?

To answer this question, divide 4566 by 200 to get a quotient of 22 and a remainder of 166.

```
         22     quotient
  200)4566
      400
      ---
      566
      400
      ---
      166       remainder
```

Thus, the maximum number of tickets that can be purchased by brokers is $22 \cdot 200 = 4400$, and the minimum number of tickets to be sold directly to the general public is 166.

Another way to describe this division is to say that 4566 equals 22 groups of 200 with 166 left over:

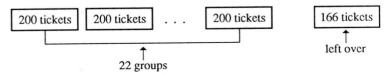

200 tickets	200 tickets	. . .	200 tickets		166 tickets

↑ 22 groups ↑ left over

Or, you could write

$$4566 = 22 \cdot 200 + 166$$

↑ 22 groups of 200 ↑ 166 left over

The number left over (166) is less than the size of a group (200) because if 200 or more were left over, another group could be formed.

This idea is generalized in the Quotient-Remainder Theorem. This theorem says that when *any* integer n (even a negative one) is divided by *any* positive integer d, there is an **integer quotient** q and a nonnegative **integer remainder** r that is less than d.

LESSON 4-2

RESOURCES

■ Lesson Master 4-2

OBJECTIVES

A Use the Quotient-Remainder Theorem for integers or polynomials.

K Use the Quotient-Remainder Theorem to solve applied problems.

TEACHING NOTES

Because students have known about quotients and remainders for a long time, they may need to be convinced of the importance of the Quotient-Remainder Theorem. The reasons for discussing it are given in the last paragraph of the lesson.

The restriction that $0 \le r < d$ in the Quotient-Remainder Theorem is important not only because it guarantees the uniqueness of q and r, but also because it provides the basis for the modular arithmetic discussed in the next lesson. Discuss **Examples 1 and 2** in class, paying special attention to Example 2. It is a good idea to have students work several more similar examples in class. (See the Additional Examples that follow.)

Technology Many versions of BASIC allow variables to be declared as integer variables and integer division to be performed on such variables. Check your version of BASIC to see how integer arithmetic is handled. In the version available to us, the integer division of 5 by 2 gives the correct quotient of 2, but in integer division of -5 by 2 gives the incorrect quotient of -2. Ask students who have a background in programming to write a program using real number divison along with the function INT to compute the integers q and r in the Quotient-Remainder Theorem. One such program is shown below.

```
10 INPUT "Please enter the
   integer N.";N
20 INPUT "Please enter the
   integer D.";D
30 LET Q=INT(N/D)
40 LET R=N−Q*D
50 PRINT
   N;"=";Q;"*";D;"+";R
60 END
```

The partitioning of the integers into three subsets on page 231 is important for Lesson 4-3. These subsets become the equivalence classes for arithmetic modulo 3.

Example 3 may seem so obvious to students that they think it does not need to be proved. Its form, however, is identical to that used in **Question 10,** whose result is not so obvious.

Quotient-Remainder Theorem (for Integers)

If n is an integer and d is a positive integer, then there exist unique integers q and r such that

$$n = q \cdot d + r \text{ and } 0 \le r < d.$$

It follows immediately from the Quotient-Remainder Theorem that if n is an integer and d is a positive integer,

d is a factor of n if and only if (the remainder r obtained by applying the Quotient-Remainder Theorem to n and d equals zero.)

Note that the Quotient-Remainder Theorem gives the result of an *integer division* as a pair of integers: a quotient and a remainder (which might be zero). Thus it expresses the result of integer division in terms of the integers. Of course, you can also think of the integers n and d as real numbers and form their real number quotient. The result of the *real number division* of n by d is the single real number $\frac{n}{d}$. Notice the distinction:

Integer division of 26 by 3 gives a quotient of 8 and a remainder of 2. *Real number division* of 26 by 3 gives the real number $\frac{26}{3}$, which equals $8.\overline{6}$.

The restriction $0 \le r < d$ in the Quotient-Remainder Theorem guarantees that the values of q and r will be unique. That is, for any given choice of n and d, there is only one pair of integers q and r such that $n = q \cdot d + r$ and $0 \le r < d$. For instance, if $n = 4566$ and $d = 200$, then each of the equations

$$4566 = 20 \cdot 200 + 566$$
$$4566 = 21 \cdot 200 + 366$$
$$4566 = 22 \cdot 200 + 166$$

is true. But only the third equation (with $q = 22$ and $r = 166$) satisfies the restriction $0 \le r < d$, that is, that the remainder be less than the divisor.

Example 1 Use a calculator to determine integers q and r that satisfy the conditions of the Quotient-Remainder Theorem when $n = 8714$ and $d = 73$.

Solution If you perform the division 8714 ÷ 73 on a calculator, a number like 119.369863 will appear on your display. This tells you that q, the integer quotient of 8714 divided by 73, is 119. What is the remainder r?

To find r, note that because

$n = q \cdot d + r,$

$r = n - q \cdot d.$

So in this case,

$r = 8714 - 119 \cdot 73 = 27.$

Thus $q = 119$ and $r = 27$.

Check You should check whether $n = d \cdot q + r$ and whether $0 \le r < d$. Does $8714 = 119 \cdot 73 + 27$? Yes. Is $0 \le 27 < 73$? Yes.

When n is negative, you must take extra care to obtain the correct values for q and r.

Example 2 If $n = -16$ and $d = 3$, find values of q and r so that $n = q \bullet d + r$ and $0 \le r < d$.

Solution If you divide -16 by 3 on a calculator, you obtain a number like -5.333333333 on your display. You may be tempted to conclude that $q = -5$ (by ignoring the decimal part). Unfortunately, this conclusion is wrong.

You need to find integers q and r for which

$$-16 = q \bullet 3 + r \text{ and } 0 \le r < 3.$$

Because of the second condition, *r must be nonnegative*. Therefore, $q \bullet 3$ must be less than -16. This means you must take $q = -6$. Then

$$\begin{aligned} -16 &= q \bullet 3 + r \\ &= (-6) \bullet 3 + r \\ &= -18 + r, \end{aligned}$$

and so $r = 2$.

Observe that if you had taken $q = -5$, then when you solved for r, you would have obtained

$$-16 = (-5) \bullet 3 + r$$

for which $r = -1$.

But this value of r does not satisfy the condition

$$0 \le r < 3.$$

The distinction between the quotient in real number division and the quotient in integer division can be made using the floor function. Dividing an integer n by a nonzero integer d gives the quotient

$q = \frac{n}{d}$ in real number division

$q = \left\lfloor \frac{n}{d} \right\rfloor$ in integer division.

In Example 2, $n = -16$ and $d = 3$, so $q = \left\lfloor \frac{-16}{3} \right\rfloor = \left\lfloor -5\frac{1}{3} \right\rfloor = -6$.

Because the remainder r must be nonnegative, the pattern of remainders when dividing by 3 remains the same for both positive integers and negative integers.

n	-16	-15	-14	-13	-12	-11	-10	-9	-8	-7	-6	-5	-4	-3	-2	-1	0	1	2	3	4	5	6	7	8	9 ...
q	-6	-5	-5	-5	-4	-4	-4	-3	-3	-3	-2	-2	-2	-1	-1	-1	0	0	0	1	1	1	2	2	2	3 ...
r	2	0	1	2	0	1	2	0	1	2	0	1	2	0	1	2	0	1	2	0	1	2	0	1	2	0 ...

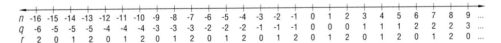

In this way, the Quotient-Remainder Theorem leads to the fact that the integers can be partitioned into three subsets.

remainder 0 when divided by 3: ..., -15, -12, -9, -6, -3, 0, 3, 6, 9, 12, ...

remainder 1 when divided by 3: ..., -14, -11, -8, -5, -2, 1, 4, 7, 10, 13, ...

remainder 2 when divided by 3: ..., -13, -10, -7, -4, -1, 2, 5, 8, 11, 14, ...

ADDITIONAL EXAMPLES

In Additional Examples 1 and 2 below, determine integers q and r that satisfy the conditions of the Quotient-Remainder Theorem for the given values of n and d.

1. $n = 6784$, $d = 93$
$q = 72$, $r = 88$; $6784 = 72 \cdot 93 + 88$ and $0 \le 88 < 93$

2. $n = -38$, $d = 5$
$q = -8$, $r = 2$; $-38 = (-8)5 + 2$ and $0 \le 2 < 5$

3. Prove that no integer is both even and odd.
Suppose n is such a number. Then the uniqueness required by the Quotient-Remainder Theorem (for integers) is violated with $d = 2$.

NOTES ON QUESTIONS

Question 2: This question relates the Quotient-Remainder Theorem to the familiar long division algorithm. In Lesson 4-4, long division of polynomials is based on the Quotient-Remainder Theorem for polynomials.

Question 9: The answers to the divisions in the various parts of this question are not the point of the question. For each part which has an answer of "integer," you could ask if $\lfloor \frac{n}{d} \rfloor$ or $\lceil \frac{n}{d} \rceil$ is the better answer. For instance, in part **a**, the answer is $\lfloor \frac{n}{d} \rfloor$. If the question asked, "How many buses are needed," the answer could be either $\lfloor \frac{n}{d} \rfloor$ or $\lfloor \frac{n}{d} \rfloor + 1$, depending on the requirement that all students be bused.

Making Connections for Question 10: To prepare students for Lesson 4-3, ask the class to prove that any integer N can be written in one of the forms $4k$, $4k + 1$, $4k + 2$, or $4k + 3$. You can then discuss the case when $d = 5$ and when $d = 6$. Seek the generalization that "Given d, any integer N can be written in one of forms dk, $dk + 1$, $dk + 2$, ..., $dk + (d - 1)$."

For division by 2, there are only two possible remainders, and the two corresponding subsets are the familiar sets of even and odd integers. Here is how this can be proved.

Example 3 Prove that every integer is either even or odd.

Solution **Suppose n is any integer.** (To show that n is either even or odd, apply the Quotient-Remainder Theorem using 2 in place of d. That is, think of 2 as the divisor.) **The Quotient-Remainder Theorem guarantees the existence of integers q and r such that**

$$n = q \cdot 2 + r \quad \text{and } 0 \le r < 2.$$

But because r is an integer and $0 \le r < 2$, the only possible values for r are 0 and 1. In other words, either

$$n = q \cdot 2 + 0 \quad \text{or} \quad n = q \cdot 2 + 1.$$

Rewriting these equations shows that either

$n = 2q$, **which means that n is even,**

or $n = 2q + 1$, **which means that n is odd.**

Though it looks simple, and though it may seem unnecessary, the Quotient-Remainder Theorem is important for three reasons. (1) It enables a division of integers (used in many computer languages) that is expressed entirely in terms of integers without the use of decimals or fractions. You encountered this in elementary school. (2) By considering the remainders when dividing by n, it partitions the integers into n distinct subsets. This leads to the useful technique of modular arithmetic introduced in the next lesson. (3) There is a similar theorem for polynomials that is discussed in Lesson 4-4.

Questions

Covering the Reading

1. Ms. Smith wants to make copies of a map to a picnic location but finds she has only \$1.30 in her pocket, and copies cost 3¢ apiece. To determine how many copies she can make, she performs the division at the right.

$$\begin{array}{r} 43 \\ 3\overline{)130} \\ \underline{12} \\ 10 \\ \underline{9} \\ 1 \end{array}$$

   a. Provide a meaning for 43 and 1 in the context of this problem. See below.
   b. Match the numbers 3, 43, 130, and 1 with n, q, r, and d in the Quotient-Remainder Theorem.
   c. Rewrite the long division shown at the right in the form utilized in the Quotient-Remainder Theorem. $130 = 43 \cdot 3 + 1$

   b. $n = 130$, $q = 43$, $r = 1$, $d = 3$

2. Find and correct the error in the integer division at the right. Use the Quotient-Remainder Theorem to explain why the error is, in fact, an error. See below.

$$\begin{array}{r} 6 \\ 8\overline{)60} \\ \underline{48} \\ 12 \end{array}$$

1. a) Ms. Smith can make 43 copies, and she will have 1¢ left over.
2. In the problem, $r = 12$ and $d = 8$. However, from the Quotient-Remainder Theorem, r must be less than d. Therefore q needs to be larger. The correct division is $60 = 7 \cdot 8 + 4$.

In 3–6, for the pair of integers n and d, find **a.** the real number quotient and **b.** integers q and r that satisfy the conditions of the Quotient-Remainder Theorem.

3. $n = 62$, $d = 5$
a) 12.4; b) $q = 12$, $r = 2$

4. $n = -29$, $d = 4$
a) -7.25; b) $q = -8$, $r = 3$

5. $n = 1063$, $d = 38$
a) 27.97368...; b) $q = 27$, $r = 37$

6. $n = 78{,}312$, $d = 66$
a) 1186.545...; b) $q = 1186$, $r = 36$

7. Use a number line to show the q and r values of the Quotient-Remainder Theorem for an integer n with $-10 \leq n \leq 15$ and $d = 4$. See margin.

8. An integer n is divided by 13.
a. What are the possible remainders? See margin.
b. Find 3 different values for n for which the remainder is 4.
c. Find a value for n for which the remainder is 0. sample: $n = 13$
d. Find a value for n for which the integer quotient is 2.
b) sample: $n = 17$, $n = 30$, $n = 43$; d) sample: $n = 27$

Applying the Mathematics

9. The senior class of a high school is planning a picnic by a river. Many questions about this trip can be answered by division. Which interpretation of n divided by d, integer or real number, would be more appropriate in the given situation?
a. How many buses will be filled if there are n people going and each bus can hold d persons? integer
b. If the picnic site is n miles away and the buses can go d miles an hour, how long will it take? real number
c. If each person drinks n bottles of soda and there are d bottles in a case, how many full cases of soda will be drunk? integer
d. If the temperature is rising n degrees per minute and it will take d minutes to get to the picnic site, how much warmer will it be there than it was when the bus ride began? neither, not a division problem
e. If students want to go on rafts, each raft has a weight capacity of n kg, and each student weighs about d kg, how may students can go on each raft? integer

10. Fill in the blanks below to prove that any integer n can be written in one of the three forms

$$n = 3q,\ n = 3q + 1,\ \text{or } n = 3q + 2$$

for some integer q.

Proof: Suppose n is any integer. Apply the Quotient-Remainder Theorem using 3 in place of d. Then the Quotient-Remainder Theorem guarantees the existence of integers q and r such that

$n =$ __a.__ and $0 \leq r < 3$. a) $q \cdot 3 + r$

But because r is an integer and $0 \leq r < 3$, the only possible values for r are __b.__, __c.__, and __d.__ b) 0; c) 1; d) 2

In other words, we know that either

$n =$ __e.__ or $n =$ __f.__ or $n =$ __g.__ .

Rewriting these equations shows that either
$n = 3q$ or $n = 3q + 1$ or $n = 3q + 2$.
e) $q \cdot 3 + 0$; f) $q \cdot 3 + 1$; g) $q \cdot 3 + 2$

ADDITIONAL ANSWERS
7. See Additional Answers in the back of this book.

8.a. 0, 1, 2, 3, 4, 5, 6, 7, 8, 9, 10, 11, 12

NOTES ON QUESTIONS
Question 18: It is implicit that the statement is meant to be universal, not existential. That is, it is true for *any* even integers *a* and *b*.

FOLLOW-UP

MORE PRACTICE
For more questions on SPUR Objectives, use *Lesson Master 4-2,* shown on page 235.

EXTENSION
You might want to extend **Question 28** to have students write the BASIC program given above. You might also have students explore some of the integer division and remainder functions available in other computer languages such as Pascal or Logo.

ADDITIONAL ANSWERS
11. See Additional Answers in the back of this book.

11. Use the result of Question 10 to explain why exactly one of every three consecutive integers is divisible by 3. See margin.

12. Martin gets a bonus point for each \$2500 worth of computer sales he makes. Last month his computer sales totaled \$22,150.
a. How many points did Martin earn? 8 points
b. On how much in sales did Martin fail to earn points? \$2150
c. Relate your answers to parts **a** and **b** to the Quotient-Remainder Theorem.
$22150 = 8 \cdot 2500 + 2150$; $n = 22150$, $q = 8$, $d = 2500$, and $r = 2150$

13. Use your calculator to find the quotient and remainder when -18,743 is divided by 436. $q = -43$, $r = 5$

14. When n is divided by d, the quotient is 7 and the remainder is 2.
a. Find values for n and d that satisfy the above requirement.
b. Are your answers to part **a** unique? Explain.
a) sample: $n = 23$, $d = 3$; b) No, for each d there exists a different n.

15. If m is any positive integer, must $(2m)!$ be divisible by $m!$? Why or why not? Yes, $(2m)! = 1 \cdot 2 \cdot 3 \cdot \ldots \cdot m \cdot (m+1) \cdot \ldots \cdot 2m = m![(m+1) \cdot \ldots \cdot 2m]$. Therefore, by definition $(2m)!$ is divisible by $m!$.

In 16–18, determine whether each statement is *true* or *false*. Justify your answer.

16. 93 is divisible by 5. False, 93 divided by 5 yields a remainder of 3.

17. $27x^2 + 9x + 12$ is divisible by 3, for any integer x.
True, $27x^2 + 9x + 12 = 3(9x^2 + 3x + 4)$.

18. $3a^2b$ is divisible by 24 if a and b are even integers.
True, if $a = 2k$ and $b = 2m$, then $3a^2b = 3(2k)^2(2m) = 24(k^2m)$.

Review

19. Prove: $\forall$ *integers a and b, if a is divisible by b, then a^2 is divisible by b^2*. *(Lesson 4-1)* See page 235.

In 20–22, write the expression as a polynomial in expanded form. *(Lesson 4-1)*

20. $(3x^2 + 4x - 8)(x^3 - 1)$ $3x^5 + 4x^4 - 8x^3 - 3x^2 - 4x + 8$

21. $(z^6 + z^3 + 1)(z^2 + z + 1)$ $z^8 + z^7 + z^6 + z^5 + z^4 + z^3 + z^2 + z + 1$

22. $\left(y + \frac{1}{2}\right)^3 - \left(y - \frac{1}{2}\right)^3$ $3y^2 + \frac{1}{4}$

23. The formula $m = m_0\left(\dfrac{1}{\sqrt{1 - \frac{v^2}{c^2}}}\right)$

relates the mass of an object at rest m_0 to its mass m when traveling at velocity v. c is the speed of light. At what percentage of the speed of light should an object be moving to have its mass at rest tripled? *(Lesson 3-1)* 94.3%

24. a. Sketch the graph of the sequence t defined by the rule $t_n = \frac{3n}{n+2}$ for all positive integers n. See page 235.
b. Find $\lim_{n \to \infty} t_n$. *(Lesson 2-4)* 3

25. Lookouts on two naval vessels, about 2 miles apart, spot a disabled boat off the port side. One lookout spots the boat at an angle of 42° with the line connecting the naval vessels. The other lookout spots the boat at an angle of 28° with the same line. **a.** Which naval vessel is closer to the boat? **b.** How much closer is it? *(Lesson 2-8)*
a) The vessel that spotted the boat at an angle of 42° is closer. b) ≈ 0.4 mi

26. Write the contrapositive of the following statement:

If the temperature stays below 28°F, the citrus crop is ruined.
(Lesson 1-5) *If the citrus crop is not ruined, then the temperature has not stayed below 28°F.*

Exploration

27. The Quotient-Remainder Theorem can be generalized to include the case when $d < 0$. We modify the theorem as follows:

If n and $d \neq 0$ are integers, then there exist unique integers q and r such that $n = q \cdot d + r$ and $0 \le r < |d|$.

For each n, d pair, find the corresponding q, r pair.

a. $n = 17 \quad d = 3$ **b.** $n = 17 \quad d = -3$
c. $n = -17 \quad d = 3$ **d.** $n = -17 \quad d = -3$

a) $q = 5$, $r = 2$; b) $q = -5$, $r = 2$; c) $q = -6$, $r = 1$; d) $q = 6$, $r = 1$

28. In BASIC, the command INT (N/D) gives the integer quotient when n is divided by d. Write a BASIC command that will give the integer remainder. N – (INT(N/D)*D)

19. Let a and b be any integers such that a is divisible by b. Then, by definition of divisible, $a = q \cdot b$ for some integer q. Then $a^2 = (q \cdot b)^2 = q^2 \cdot b^2$. Since q^2 is an integer by closure of multiplication, b^2 is a factor of a^2, and thus a^2 is divisible by b^2.

24. a)

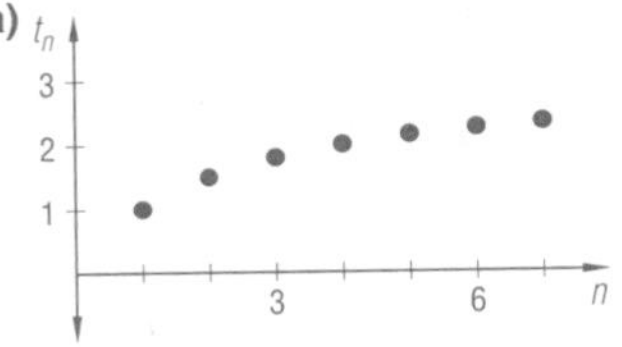

NAME ______

LESSON **MASTER 4–2**
QUESTIONS ON SPUR OBJECTIVES

■ **SKILLS** *Objective A (See pages 283–286 for objectives.)*
In 1–4, values for n and d are given. Find values for q and r as defined in the Quotient-Remainder Theorem.

1. $n = 138$, $d = 23$ — $q = 6$, $r = 0$
2. $n = 15$; $d = 40$ — $q = 0$, $r = 15$
3. $n = -72$; $d = 13$ — $q = -6$, $r = 6$
4. $n = 247{,}291$; $d = 718$ — $q = 344$, $r = 299$

In 5–7, when n is divided by d, the quotient is 11 and the remainder is 5.

5. Can d be 4? Explain your answer. No, because the remainder must be less than d.

6. If $n = 93$, find the value of d. 8

7. If $d = 10$, find the value of n. 115

In 8–10, suppose n is an integer and n is divided by 9.

8. How many different remainders are possible? 9

9. Give two positive values of n for which $r = 7$. sample: $n = 16, 25$

10. Give two negative values of n for which $r = 7$. sample: $n = -2, -11$

11. Explain why every integer m can be written in exactly one of the following forms (where k is an integer):
$m = 6k$, $m = 6k + 1$, $m = 6k + 2$, $m = 6k + 3$, $m = 6k + 4$, $m = 6k + 5$
Let $d = 6$ in the Quotient-Remainder Theorem. Then ∃ integers k and r such that $m = 6k + r$ where $0 \le r < 6$. Thus, r is one of the following: 0, 1, 2, 3, 4, 5. That is, exactly one of the following is true: $m = 6k$, $m = 6k + 1$, $m = 6k + 2$, …, $m = 6k + 5$.

40 *Continued*

NAME ______
Lesson MASTER 4-2 (page 2)

■ **USES** *Objective K*

12. A particular brand of pen is sold in three ways: individually (for \$4 each), in boxes of 12 (for \$45 per box), and in cases of 100 pens (for \$320 per case). Suppose n pens are bought using c cases, b boxes, and i individual pens, in such a way that the total cost is minimized.

a. What are the values of c, b, and i if $n = 88$? $c = 0$, $b = 7$, $i = 4$

b. What are the values of c, b, and i if $n = 362$? $c = 3$, $b = 5$, $i = 2$

c. Are your answers to parts **a** and **b** unique? Why or why not?
Yes, if a case (or box) is replaced by boxes or individual pens, the cost will be greater, and therefore not minimized.

 41

LESSON 4-3

RESOURCES
- Lesson Master 4-3
- Quiz for Lessons 4-1 through 4-3

OBJECTIVES

D Determine the congruence of integers for a given modulus.
G Use the properties of congruence of integers for a given modulus to rewrite sentences.
L Use modular arithmetic to solve applied problems.

TEACHING NOTES

Students should understand that every integer is an element of exactly one of the congruence classes *R*1, *R*2, or *R*0, as shown on this page. Remind students of Question 10 from Lesson 4-2 during this discussion.

The Congruence Theorem defines $a \equiv b \pmod{m}$ if and only if m is a factor of $a - b$. This approach has the advantage of extending directly to equivalence classes of real numbers. To make the theorem more meaningful to students, have them choose two elements from the set *R*2 and check that their difference is divisible by 3.

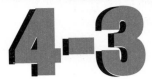

LESSON

4-3 Modular Arithmetic

Karl Friedrich Gauss

In the chapter opener, we mentioned that codes can be constructed using *modular arithmetic*.

This arithmetic was developed by the great German mathematician Karl Friedrich Gauss (1777–1855). Gauss described the technique in his book *Disquisitiones Arithmeticae* (''Inquiries About Arithmetic''), published in 1801, in which he laid the foundation for the modern theory of numbers. Before showing you an application to codes, we describe the idea.

Recall from Lesson 4-2 that integer division by 3 partitions the integers into three sets whose elements are determined by their remainders under this division. One set is the set of integers with remainder 1, a second set is the set of integers with remainder 2, and the third set is the set of integers with remainder 0. Let us call these sets *R*1, *R*2, and *R*0.

You can check that

$$R1 = \{\dots, -5, -2, 1, 4, 7, 10, 13, 16, 19, 22, \dots\}.$$

Check: $22 = 7 \cdot 3 + 1$ and $-5 = -2 \cdot 3 + 1$. In general, by the Quotient-Remainder Theorem, *R*1 consists of all the integers of the form $3k + 1$ for some integer k.

The elements of *R*2 all have the form $3k + 2$ for some integer k.

$$R2 = \{\dots, -4, -1, 2, 5, 8, 11, 14, 17, 20, 23, \dots\}$$

*R*0 is the set of all multiples of 3. These have the form $3k$, for some integer k.

$$R0 = \{\dots, -6, -3, 0, 3, 6, 9, 12, 15, 18, 21, \dots\}$$

The Quotient-Remainder Theorem implies that every integer is an element of exactly one of the sets *R*1, *R*2, or *R*0.

We call 3 the **modulus** or **mod** for these sets. If two integers a and b are in the same set, we say that a and b are **congruent modulo 3,** and we write $a \equiv b \pmod 3$, read ''a is congruent to b mod 3.'' For instance, both -2 and 19 are in the first set, so $-2 \equiv 19 \pmod 3$. Because all multiples of 3 are in the third set, $420 \equiv -15 \equiv 0 \pmod 3$. We call the sets *R*0, *R*1, and *R*2 **congruence classes modulo 3.** More generally, we have the following definition.

Definition

Let a and b be integers and let m be a positive integer. ***a* is congruent to *b* modulo *m*, denoted $a \equiv b \pmod{m}$,** if and only if a and b have the same integer remainder when they are divided by m.

For a given modulus, it is often useful to find the smallest positive number congruent to a given number. This equals the remainder obtained when the given number is divided by the modulus. For instance, let the modulus be 7, the number of days in a week. Now identify Sunday as 1, Monday as 2, Tuesday as 3, and so on. What day of the week is 100 days after a Tuesday? This question can be answered by finding the smallest positive number congruent to 100 + 3 and translating that back to a day of the week. All numbers congruent to 103 modulo 7 have the same remainder when divided by 7. Integer division of 103 by 7 gives a quotient of 14 and a remainder of 5. So 100 days after a Tuesday is the 5th day of the week, Thursday.

There is another way to determine when two numbers are congruent modulo m. If two numbers a and b are in the same congruence class modulo m, then there exist integers q_1, q_2, and r such that

$$a = mq_1 + r$$

and

$$b = mq_2 + r.$$

The key idea is that the remainder r is the same in both equations. So when the equations are subtracted, r disappears.

$$a - b = m(q_1 - q_2)$$

Because $q_1 - q_2$ is an integer, m is a factor of $a - b$. So if two integers are congruent mod m, then m is a factor of their difference. Conversely, it can be shown that if m is a factor of the difference $a - b$ between two integers a and b, then a and b are congruent modulo m. This leads to the following theorem.

Congruence Theorem

∀ integers a and b and positive integers m, $a \equiv b \pmod{m}$ if and only if m is a factor of $a - b$.

For instance, consider the modulus 12. By the congruence theorem, two integers are congruent modulo 12 if their difference is divisible by 12. Thus $31 \equiv 7 \pmod{12}$ because $31 - 7 = 24$ and 24 is divisible by 12. By repeatedly adding or subtracting 12, you can find all the integers that are congruent to 7 modulo 12.

$$7, 19, 31, 43, \ldots \text{ and } -5, -17, -29, -41, \ldots$$

The modulus 12 explains *clock arithmetic*. For instance, 7 hours after 6:00 is 1:00 because $6 + 7 \equiv 1 \pmod{12}$.

Now let us turn to a modern application of codes. When people enter or copy information, mistakes are made. In today's world, with so much information stored in computers, an error of just a single digit or number may mean that a company does not realize that you paid your bill, or that you will receive an item different from the one you ordered by mail, or that the wrong emergency vehicle is dispatched to an accident. To catch errors, many identification numbers (such as credit card numbers or billing numbers) have extra digits called *check digits*. The check digit is a code number determined in some mathematical way from the other digits in the identification number. Almost all check digits are calculated using modular arithmetic.

The starting point for generalizing the definition of congruence to noninteger values of m is the Congruence Theorem. For noninteger values of m, $a \equiv b \pmod{m}$ if and only if $a - b$ is an integer multiple of m. For instance, $\frac{17\pi}{4} \equiv \frac{\pi}{4} \pmod{2\pi}$ because $\frac{17\pi}{4} - \frac{\pi}{4} = 4\pi$, which is an integer multiple of 2π.

The idea of multiples of nonintegers will bother some students, which can lead to a lively class discussion.

In the day-of-week discussion on this page, some students may suggest the following alternative. Dividing 100 by 7 leaves a remainder of 2. Two days after Tuesday is Thursday. Thus, the solution is Thursday.

Alternate Approach
When working with mod m, encourage students to replace any integer by the smallest nonnegative integer to which it is congruent, whenever possible. This greatly simplifies calculations mod m. **Example 2** shows the power of this technique. It may be even clearer to students if they find the last 2 digits of 9^{10} by hand. In this case, they are working (mod 100) and can discard all but the last two digits. After this exercise, Example 2 should be easier to follow.

Technology Symbol manipulator programs, such as *Mathematica* or *Derive,* display all the decimal places of numbers like 17^9, and thus provide an alternate way of doing questions such as those in **Example 2**.

ADDITIONAL EXAMPLES

1. Using the ISBN check digit formula, as given in Example 1 of this lesson, fill in the correct check digit for 0-688-03118- __?__.
Opposite of Sum of products = -212 ≡ 8 mod 11. Thus, the check digit is 8.

2. Find the last three digits of 19^9.
779

Example 1 Books published since 1972 are assigned ten-digit International Standard Book Numbers (ISBN). The first 9 digits give information; the last digit is a check digit. The check digit is obtained by multiplying the first nine digits by 10, 9, 8, 7, 6, 5, 4, 3, and 2, respectively. The opposite of the sum of these products must be congruent to the check digit modulo 11. (A check digit must be a single character, and the symbol X is used for the check digit value of 10.) Fill in the correct check digit for 0-07-062341-___.

Solution Calculate the sum of the products.
$0 \cdot 10 + 0 \cdot 9 + 7 \cdot 8 + 0 \cdot 7 + 6 \cdot 6 + 2 \cdot 5 + 3 \cdot 4 + 4 \cdot 3 + 1 \cdot 2 = 128$
Now determine the smallest positive integer congruent to -128 mod 11. Doing integer division of -128 by 11, the quotient is -12 and the remainder is 4. Thus the check digit should be 4.

Addition, subtraction, and multiplication behave in a familiar way with respect to congruence modulo m. Consider the following congruences mod 3.

$$11 \equiv 8 \pmod 3$$
$$7 \equiv 19 \pmod 3$$

Add the numbers on the two sides and another true congruence is obtained

$$18 \equiv 27 \pmod 3.$$

Similarly, by subtraction $4 \equiv -11 \pmod 3$.
Also, by multiplication $77 \equiv 152 \pmod 3$.
The following theorem generalizes these results.

Theorem

Let a, b, c, and d be any integers and let m be a positive integer. If $a \equiv b \pmod m$ and $c \equiv d \pmod m$, then

$a + c \equiv b + d \pmod m$ **(Addition Property of Congruence)**
$a - c \equiv b - d \pmod m$ **(Subtraction Property of Congruence)**
and $ac \equiv bd \pmod m$. **(Multiplication Property of Congruence)**

Proof (Addition Property of Congruence):
Suppose $a \equiv b \pmod m$ and $c \equiv d \pmod m$.
By the Congruence Theorem, the hypothesis becomes
m is a factor of $a - b$ and m is a factor of $c - d$.
Thus there exist integers k_1 and k_2 such that
$a - b = k_1 m$ and $c - d = k_2 m$.

Example 1 When $x^8 - x + 4$ is divided by $x^3 + 2$, the quotient is $x^5 - 2x^2$. Find the remainder and its degree.

Solution Let $n(x) = x^8 - x + 4$, $d(x) = x^3 + 2$, $q(x) = x^5 - 2x^2$, and let $r(x)$ be the remainder polynomial. Then

$n(x) = q(x) \cdot d(x) + r(x)$.

So $r(x) = n(x) - q(x) \cdot d(x)$ — By subtracting $q(x) \cdot d(x)$ from both sides

$= (x^8 - x + 4) - (x^5 - 2x^2)(x^3 + 2)$ — By substitution

$= (x^8 - x + 4) - (x^8 - 2x^5 + 2x^5 - 4x^2)$

$= x^8 - x + 4 - x^8 + 4x^2$

$= 4x^2 - x + 4$.

Thus $r(x) = 4x^2 - x + 4$. The degree of $r(x)$ is 2, which is smaller than 3, the degree of the divisor.

In Example 1, the quotient was given and the remainder was found. If the quotient is not known, it and the remainder can be computed from $p(x)$ and $d(x)$ by a long division procedure that is essentially the same as the one that you learned for dividing integers. We first illustrate the procedure and then relate the procedure to the Quotient-Remainder Theorem.

To divide $p(x) = 6x^3 - 9x^2 + 8x + 1$ by $d(x) = 2x + 1$, proceed as follows:

Step 1:

$$\begin{array}{r} 3x^2 \phantom{{}- 9x^2 + 8x + 1} \\ 2x + 1 \overline{\smash{)}\,6x^3 - 9x^2 + 8x + 1} \\ \underline{6x^3 + 3x^2} \phantom{{}+ 8x + 1} \\ -12x^2 + 8x + 1 \end{array}$$

Think: $\frac{6x^3}{2x} = 3x^2$. This is the first term in the quotient.

Now multiply $3x^2$ by $2x + 1$. Subtract from $6x^3 - 9x^2 + 8x + 1$.

Step 2:

$$\begin{array}{r} 3x^2 - 6x \phantom{{}+ 8x + 1} \\ 2x + 1 \overline{\smash{)}\,6x^3 - 9x^2 + 8x + 1} \\ \underline{6x^3 + 3x^2} \phantom{{}+ 8x + 1} \\ -12x^2 + 8x + 1 \\ \underline{-12x^2 - 6x} \phantom{{}+ 1} \\ 14x + 1 \end{array}$$

Think: $\frac{-12x^2}{2x} = -6x$. This is the second term in the quotient.

Multiply $-6x$ by $2x + 1$. Subtract from $-12x^2 + 8x + 1$.

Step 3:

$$\begin{array}{r} 3x^2 - 6x + 7 \\ 2x + 1 \overline{\smash{)}\,6x^3 - 9x^2 + 8x + 1} \\ \underline{6x^3 + 3x^2} \phantom{{}+ 8x + 1} \\ -12x^2 + 8x + 1 \\ \underline{-12x^2 - 6x} \phantom{{}+ 1} \\ 14x + 1 \\ \underline{14x + 7} \\ -6 \end{array}$$

Think: $\frac{14x}{2x} = 7$. This is the third term in the quotient.

Multiply 7 by $2x + 1$. Subtract from $14x + 1$.
The degree of -6 is less than the degree of $2x + 1$, so the division is finished.

You may wish to review the following characteristics of polynomial long division:
(1) Both dividend and divisor should be written in decreasing powers of the variable.
(2) "Missing" powers should be represented by 0 times the variable to that power.
(3) Quotient terms should be placed over like terms in the dividend.
(4) Unlike integer division, keep bringing down the entire leftover dividend.
(5) Avoid adding unlike terms.
(6) The division is over when the degree of the divisor is more than the degree of the remaining dividend.

Having quotient terms placed over like terms in the dividend parallels long division in arithmetic. It also clearly indicates when the division is over, since there is no room for any quotient term after the constant. Some books instruct students to place quotient terms as far left as possible. That has the advantage of being similar to the recording used in synthetic division. Either way is acceptable; the important thing is that the work be well-organized and legible.

Error Analysis When students subtract in long division, instruct them not to physically change the signs of the subtrahend. While this may ease the subtraction for some students, it then becomes very difficult to check or review the computations. One cannot tell if what was written was one of the intermediate products or if it was the opposite of that product.

You may want to discuss **Example 2** in class. Students frequently have difficulty when the dividend polynomial has terms with zero coefficients, or when there is a fraction in the quotient. Note the use of the Quotient-Remainder Theorem in the check.

The lesson concludes with a justification for the long division algorithm based on the Quotient-Remainder Theorem. To help students appreciate this justification, you may want them to carry out the integer division using only repeated applications of the Quotient-Remainder Theorem for integers. Then you can discuss the parallels on page 245.

The quotient is $q(x) = 3x^2 - 6x + 7$ and the remainder is $r(x) = -6$. You can check that the division is correct by observing that

$$\begin{aligned} p(x) &= q(x) \cdot d(x) + r(x) \\ 6x^3 - 9x^2 + 8x + 1 &= (3x^2 - 6x + 7)(2x + 1) + (-6). \end{aligned}$$

By dividing both sides of this equation by $2x + 1$, the result of the long division procedure may be written in the rational form

$$\frac{6x^3 - 9x^2 + 8x + 1}{2x + 1} = 3x^2 - 6x + 7 + \frac{-6}{2x + 1}.$$

If some of the coefficients in the polynomial $p(x)$ are zero, you can use the long division scheme but you need to fill in the zero coefficients for all the missing powers of the variable.

Example 2 Divide $p(x) = 6x^5 - x^4 + x + 1$ by $d(x) = 2x^2 + x$ using the long division procedure.

Solution In the polynomial $p(x)$, the coefficients of x^3 and x^2 are zero. So in using the long division procedure, it is helpful to write the polynomial as $p(x) = 6x^5 - x^4 + 0x^3 + 0x^2 + x + 1$. The division is shown below.

$3x^3 - 2x^2 + x - \frac{1}{2}$	
$2x^2 + x \overline{)6x^5 - x^4 + 0x^3 + 0x^2 + x + 1}$ $\underline{6x^5 + 3x^4}$	Think: $\frac{6x^5}{2x^2} = 3x^3$. Multiply $3x^3$ by $2x^2 + x$ and subtract.
$-4x^4 + 0x^3 + 0x^2 + x + 1$ $\underline{-4x^4 - 2x^3}$	Think: $\frac{-4x^4}{2x^2} = -2x^2$. Multiply $-2x^2$ by $2x^2 + x$ and subtract.
$2x^3 + 0x^2 + x + 1$ $\underline{2x^3 + x^2}$	Think: $\frac{2x^3}{2x^2} = x$. Multiply x by $2x^2 + x$ and subtract.
$-x^2 + x + 1$ $\underline{-x^2 - \frac{1}{2}x}$	Think: $\frac{-x^2}{2x^2} = -\frac{1}{2}$. Multiply $-\frac{1}{2}$ by $2x^2 + x$ and subtract.
$\frac{3}{2}x + 1$	The degree of $\frac{3}{2}x + 1$ is less than the degree of $2x^2 + x$, so the division is complete.

Therefore $q(x) = 3x^3 - 2x^2 + x - \frac{1}{2}$ and $r(x) = \frac{3}{2}x + 1$.

Check

Does $p(x) = q(x) \cdot d(x) + r(x)$?

$6x^5 - x^4 + x + 1 \stackrel{?}{=} \left(3x^3 - 2x^2 + x - \frac{1}{2}\right)\left(2x^2 + x\right) + \left(\frac{3}{2}x + 1\right)$

$6x^5 - x^4 + x + 1 \stackrel{?}{=} 6x^5 + 3x^4 - 4x^4 - 2x^3 + 2x^3 + x^2 - x^2 - \frac{1}{2}x + \frac{3}{2}x + 1$

$6x^5 - x^4 + x + 1 \stackrel{?}{=} 6x^5 - x^4 + x + 1$. Yes.

Written in its rational form, the division of Example 2 shows that

$$\frac{6x^5 - x^4 + x + 1}{2x^2 + x} = 3x^3 - 2x^2 + x - \frac{1}{2} + \frac{\frac{3}{2}x + 1}{2x^2 + x}.$$

244

Why does the long division procedure work, and how does this procedure relate to the Quotient-Remainder Theorem? To answer this question, first consider the integer division: $4369 \div 9$.

$$\begin{array}{r} 485 \\ 9\overline{)4369} \\ \underline{3600} \\ 769 \\ \underline{720} \\ 49 \\ \underline{45} \\ 4 \end{array}$$

Quotient-Remainder Theorem Version

$$4369 = 9 \cdot 400 + 769$$
$$769 = 9 \cdot 80 \quad + 49$$
$$49 = 9 \cdot 5 \quad + 4$$
$$\therefore 4369 = 9 \cdot 400 + 9 \cdot 80 + 9 \cdot 5 + 4$$
$$= 9 \cdot 485 + 4$$

Consider now how the theorem relates to the first example of long division of polynomials in this lesson, $p(x) = 6x^3 - 9x^2 + 8x + 1$ divided by $d(x) = 2x + 1$.

$$\begin{array}{r} 3x^2 - 6x + 7 \\ 2x + 1\overline{)6x^3 - 9x^2 + 8x + 1} \\ \underline{6x^3 + 3x^2 \qquad\qquad\quad} \\ -12x^2 + 8x + 1 \\ \underline{-12x^2 - 6x \quad\;\;} \\ 14x + 1 \\ \underline{14x + 7} \\ -6 \end{array}$$

Quotient-Remainder Theorem Version

$$6x^3 - 9x^2 + 8x + 1 = (2x + 1) \cdot 3x^2 \quad + (-12x^2 + 8x + 1)$$
$$-12x^2 + 8x + 1 = (2x + 1) \cdot (-6x) \quad + (14x + 1)$$
$$14x + 1 = (2x + 1) \cdot 7 \quad + (-6)$$
$$\therefore 6x^3 - 9x^2 + 8x + 1 = (2x + 1) \cdot 3x^2 + (2x + 1) \cdot (-6x) + (2x + 1) \cdot 7 + (-6)$$
$$= (2x + 1)(3x^2 - 6x + 7) + (-6)$$

Thus, both with integers and polynomials the long division procedure involves repeated application of the Quotient-Remainder Theorem.

Questions

Covering the Reading

1. Pair the analogous terms: rational expression division, integer division, polynomial division, rational number division. See below.

2. When $x^5 + x^3 - x + 3$ is divided by $x^2 - 1$, the quotient is $x^3 + 2x$. Find the remainder and its degree. $r(x) = x + 3$; degree: 1

In 3 and 4, use long division to find the quotient $q(x)$ and the remainder $r(x)$ for the given polynomial $p(x)$ divided by the given polynomial $d(x)$.

3. $p(x) = 3x^2 + 2x + 4$, $d(x) = x - 2$ $q(x) = 3x + 8$; $r(x) = 20$

4. $p(x) = x^5 - 3x^3 + x + 1$, $d(x) = x^3 - 1$ $q(x) = x^2 - 3$; $r(x) = x^2 + x - 2$

1. integer division and polynomial division; rational expression division and rational number division

ADDITIONAL EXAMPLES

1. When $x^8 - x^4 + 3x^2 - 5$ is divided by $x^2 - 2$, the quotient is $x^6 + 2x^4 + 3x^2 + 9$. Find the remainder and its degree.
remainder 13, which has degree 0

2. Divide $p(x) = 6x^6 - 20x^5 - 7x^4 + 9x^3 - x^2 - 2x + 7$ by $d(x) = 3x^2 + 2x$ using the long division method.
quotient = $2x^4 - 8x^3 + 3x^2 + x - 1$, remainder = 7

NOTES ON QUESTIONS

Question 1: It is not vital that students remember these terms; the analogy is useful, however.

Questions 3-11: Advise students to check their work using the Quotient-Remainder Theorem. This is probably analogous to the way they learned to check long division arithmetic problems in elementary school.

5. Use long division to show that

$$h(x) = \frac{x^5 - x^3 - 2x}{x^2 - 3}$$

can be rewritten as

$$h(x) = x^3 + 2x + \frac{4x}{x^2 - 3}.$$

See page 247.

Applying the Mathematics

6. Use long division to find the quotient and remainder when $p(x) = 5x^5 - x^3 + 3x^2 - 1$ is divided by $d(x) = 2x^3 - x + 1$. **See margin.**

In 7 and 8, a function is defined by the indicated formula. Find another formula for the function using long division.

7. $f(x) = \frac{x^2 + 4x - 21}{x - 3}$ $f(x) = x + 7$ 8. $h(x) = \frac{14x - 11}{2x + 1}$ $h(x) = 7 - \frac{18}{2x + 1}$

9. Recall that $d(x)$ is a factor of $p(x)$ if $p(x) = q(x) \cdot d(x)$ for some polynomial $q(x)$. That is, dividing $p(x)$ by $d(x)$ gives a remainder of zero. Use long division of polynomials to show that $x - 2$ is a factor of $5x^3 - 4x^2 - 10x - 4$. $5x^3 - 4x^2 - 10x - 4 = (5x^2 + 6x + 2)(x - 2)$

10. Show that $x^2 + x + 1$ is a factor of

$$x^4 - x^2 - 2x - 1$$

and find another factor. $x^4 - x^2 - 2x - 1 = (x^2 - x - 1)(x^2 + x + 1)$. So, $x^2 - x - 1$ and $x^2 + x + 1$ are factors of $x^4 - x^2 - 2x - 1$.

11. Find the quotient and remainder when $x^3 + y^3$ is divided by $x + y$. (Hint: Rewrite $x^3 + y^3$ to show zero coefficients for the missing powers of the variables. The missing terms are of the form $x^n y^m$ for positive integer values of m and n less than 3, and $m + n = 3$.) $q(x) = x^2 - xy + y^2$; $r(x) = 0$

Review

12. Name 4 elements in each of the congruence classes modulo 5. *(Lesson 4-3)* Sample: $R0$: 0, 5, 10, 15; $R1$: 1, 6, 11, 16; $R2$: 2, 7, 12, 17; $R3$: 3, 8, 13, 18; $R4$: 4, 9, 14, 19

13. Find the smallest positive integer that makes the congruence true.
a. $x \equiv 87 \pmod{15}$ 12 **b.** $y \equiv -3 \pmod{7}$ *(Lesson 4-3)* 4

14. Find the last two digits of 15^{10}. *(Lesson 4-3)* 25

15. Use the Quotient-Remainder Theorem to show that every integer m can be written in one of the following forms:

$$m = 4k,\ m = 4k + 1,\ m = 4k + 2, \text{ or } m = 4k + 3$$

for some integer k. *(Lesson 4-2)* See page 247.

16. Write in factored form:
$(x - 5)^2 (x + 3)^4 (x + 7)^3 - (x - 5)^3 (x + 3)^4 (x + 7)$. *(Lesson 4-1)*
$(x - 5)^2(x + 3)^4(x + 7)(x^2 + 13x + 54)$

17. Decide if the following argument is valid or invalid. Justify your answer.

If an integer n is divisible by 6, then 2n is divisible by 4.
2 • 6144 is divisible by 4.
∴ 6144 is divisible by 6. *(Lessons 1-4, 1-5)* invalid, converse error

NOTES ON QUESTIONS
Making Connections for Question 5: In Chapter 5, students will learn that this answer means that the end behavior of the graph of h is the same as the end behavior of the quotient polynomial q, where $q(x) = x^3 + 2x$. You can anticipate that discussion by asking students to graph h and q. The distance between h and q for any value of x is $\frac{r(x)}{d(x)}$; this distance approaches 0 as $x \to +\infty$. This approach provides an alternate way to check any long division of polynomials.

Question 11: It is possible to avoid writing down the missing terms in $x^3 + y^3$ if enough room is left for the intermediate products in the long division. If students write down the missing terms, it should be by decreasing powers of x. Otherwise it will be difficult to keep track of the subtractions.

If students have difficulty with writing $x^3 + y^3$ as $x^3 + 0x^2y + 0xy^2 + y^3$, practice with $x^4 + y^4$.

Question 18: In part **c**, the actual value equals x and $s = L = \sqrt[3]{540}$. Some students may get $s \approx 8.14$ and $h \approx 8.15$.

Question 21: This exploration shows one of the few times in which direct substitution into variable expressions does not parallel all of the arithmetic. The reason is that the remainder polynomials can have negative and noninteger values. If a student were to pick a small even number for substitution into the second example, the numerical quotient and remainder would agree with their polynomial values. In all cases, the Quotient-Remainder Theorem works.

ADDITIONAL ANSWERS
6. $q(x) = \frac{5}{2}x^2 + \frac{3}{4}$;
$r(x) = \frac{1}{2}x^2 + \frac{3}{4}x - \frac{7}{4}$

18. A gravel bin in the shape of a box with a square top is to be constructed from sheet steel to hold 20 cubic yards of gravel when it is filled level to the top.

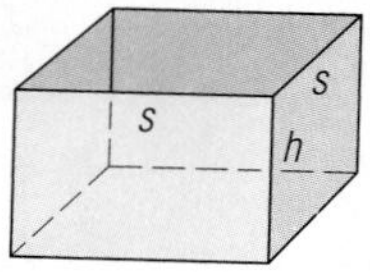

a. If each of the four sides of the bin has width s feet and height h feet, express the height as a function of the width. $h = \frac{540}{s^2}$

b. Express the total number of square feet of sheet steel needed to construct the bin and top as a function of s. Call your function A.

c. Use an automatic grapher to graph the function A and use this graph to estimate the dimensions of the bin that requires the least sheet steel for its construction.

d. Analyze the behavior of A as a function of s as $s \to 0$ and as $s \to +\infty$. (Consider only nonnegative s values.) *(Lessons 2-2, 2-5)*
b, c, d) See below.

19. Find all real number solutions to:

a. $x^2 + x - 1 = 0$. $x = \frac{-1 \pm \sqrt{5}}{2}$

b. $x^2 + x + 1 = 0$. *(Previous course)* no real numbers

20. Solve $|3t - 7| \geq 20$. *(Lesson 3-8)* $t \leq -\frac{13}{3}$ or $t \geq 9$

Exploration

21. Check each of the two long division examples in this lesson by substituting a number for x and performing the numerical division. Describe what happens. See margin.

5. Let $f(x) = x^5 - x^3 - 2x$. By long division, $f(x) = (x^3 + 2x)(x^2 - 3) + 4x$. So, dividing by $x^2 - 3$, $\frac{f(x)}{x^2 - 3} = h(x) = x^3 + 2x + \frac{4x}{x^2 - 3}$.

15. Suppose m is any integer. When m is divided by 4, the four possible remainders are 0, 1, 2, and 3. Then, by the Quotient-Remainder Theorem, there exists an integer k such that $m = 4k$, or $m = 4k + 1$, or $m = 4k + 2$, or $m = 4k + 3$.

18. b) $A(s) = \frac{2160}{s} + 2s^2$

c)

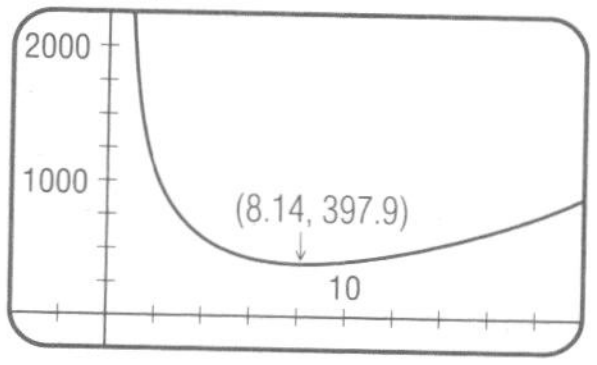

$s \approx 8.14$ feet
$h \approx 8.14$ feet

d) $\lim_{s \to 0^+} A(s) = \infty$; and $\lim_{s \to \infty} A(s) = \infty$

FOLLOW-UP

MORE PRACTICE

For more questions on SPUR Objectives, use *Lesson Master 4-4*, shown below.

ADDITIONAL ANSWERS

21. See Additional Answers in the back of this book.

NAME ______

LESSON **MASTER 4–4**
QUESTIONS ON **SPUR** OBJECTIVES

■ **SKILLS** *Objective A (See pages 283–286 for objectives.)*

1. If $p(x) = q(x)d(x) + r(x)$, $p(x) = 3x^3 - 2x$, and $d(x) = x^2 - 3$, find $q(x)$ and $r(x)$.

$q(x) =$ 3x $r(x) =$ 7x

2. a. Write an equation in the form $p(x) = q(x)d(x) + r(x)$ based on the long division at the right.

$$\begin{array}{r} 2x - 4 \\ 3x^2 + 5x \overline{)6x^3 - 2x^2 + 0x + 3} \\ 6x^3 + 10x^2 \quad\quad\quad\; \\ \hline -12x^2 + 0x \quad\; \\ -12x^2 - 20x \quad\; \\ \hline 20x + 3 \end{array}$$

$6x^3 - 2x^2 + 3 =$
$(2x - 4)(3x^2 + 5x) + (20x + 3)$

b. Simplify the expression on the right side of the equation from part **a** to show that it in fact equals the left side.
$6x^3 + 10x^2 - 12x^2 - 20x + 20x + 3 = 6x^3 - 2x^2 + 3$

3. Find k and m so that $-12x^2 + mx - 5 = (4x - 1)(kx + 2) - 3$ ∀ real numbers x.

$k =$ -3 $m =$ 11

■ **SKILLS** *Objective C*
In 4–8, find the quotient $q(x)$ and the remainder $r(x)$ when the polynomial $p(x)$ is divided by the polynomial $d(x)$.

4. $p(x) = 4x^3 + 7x - 4$
$d(x) = 2x + 1$

$q(x) =$ $2x^2 - x + 4$
$r(x) =$ -8

5. $p(x) = x^5 - x^3 + 2$
$d(x) = x - 1$

$q(x) =$ $x^4 + x^3$
$r(x) =$ 2

Precalculus and Discrete Mathematics © Scott, Foresman and Company *Continued* **43**

NAME ______
Lesson MASTER 4–4 (page 2)

6. $p(x) = 18x^3 - 13x^2 - 4x + 1$
$d(x) = 9x - 2$

$q(x) =$ $2x^2 - x - \frac{2}{3}$
$r(x) =$ $-\frac{1}{3}$

7. $p(x) = 4x^4 - 6x^3 + 6x^2 + 4$
$d(x) = 2x^2 + x + 1$

$q(x) =$ $2x^2 - 4x + 4$
$r(x) =$ 0

8. $p(x) = x^5 + 5x^4 - 7x^3 - 15x^2 - 16x + 44$
$d(x) = x^2 + 4x - 11$

$q(x) =$ $x^3 + x^2 - 4$ $r(x) =$ 0

9. What is the length of a rectangle whose width is $x + 4$ units and whose area is $2x^3 + 10x^2 + 9x + 4$ square units? $2x^2 + 2x + 1$ units

■ **PROPERTIES** *Objective H*

10. Suppose that polynomials $m(x)$ and $n(x)$ have degrees 6 and 2, respectively. If $m(x)$ is divided by $n(x)$, what do you know about the degrees of the quotient and remainder polynomials?
The degree of the quotient polynomial is 4, and the degree of the remainder polynomial is 0 or 1, or the remainder is 0.

11. Suppose that a polynomial $p(x)$ is divided by a polynomial $d(x)$, and that the quotient polynomial has degree 4 while the remainder polynomial has degree 2. What do you know about the degrees of polynomials $d(x)$ and $p(x)$?
The degree of polynomial $d(x)$ is at least 3, and the degree of polynomial $p(x)$ is at least 7.

44 *Precalculus and Discrete Mathematics* © Scott, Foresman and Company

LESSON 4-5

RESOURCES
- Lesson Master 4-5
- Teaching Aid 24 displays examples of synthetic division and long division.
- Computer Master 8

OBJECTIVES

B Use synthetic substitution to find values of polynomial functions.
C Divide polynomials.

TEACHING NOTES

Some books do the synthetic division by $x - c$ using $-c$ as the multiplier and then subtracting. This has the advantage of being more like polynomial long division but the disadvantage of being more difficult to carry out.

The synthetic substitution scheme is much harder to explain than to do. If students work out the examples as they read, they should have little difficulty learning the algorithm. Be sure to point out the necessity of including zero coefficients when using synthetic substitution.

During your discussion of synthetic substitution and synthetic division, emphasize that the algorithm is the same, but the information obtained from the scheme is different.

Making Connections
The fact that a polynomial is completely characterized by its coefficients is an important one. A similar idea is

LESSON

4-5 Synthetic Division

Suppose you were asked to evaluate the polynomial

$$p(x) = 3x^3 + x^2 - 8x - 5$$

at $x = 6.2$. Substituting 6.2 for x gives

$$p(6.2) = 3(6.2)^3 + (6.2)^2 - 8(6.2) - 5.$$

To evaluate this polynomial (in expanded form) with the aid of some calculators, you could use the powering key as indicated below.

3 [×] 6.2 [y^x] 3 [+] 6.2 [x^2] [−] 8 [×] 6.2 [−] 5 [=]

You should verify that this key sequence yields $p(6.2) = 698.824$.

Another technique that can be used to evaluate polynomials arises from looking at the polynomial in nested form. The nested form of $p(x)$ can be obtained from its expanded form by successive factorization as shown here.

$$p(x) = \underbrace{3x^3 + x^2 - 8x}_{\text{factor out } x} - 5$$

$$= (\underbrace{3x^2 + x}_{\text{factor out } x} - 8)x - 5$$

$$= ((3x + 1)x - 8)x - 5$$

In nested form, $p(6.2) = ((3(6.2) + 1)6.2 - 8)6.2 - 5$.
On some calculators, this can be computed using the key sequence below.

3 [×] 6.2 [+] 1 [=] [×] 6.2 [−] 8 [=] [×] 6.2 [−] 5 [=]

The key [=] must be used to complete the evaluation of each expression inside parentheses. You should verify that this key sequence also yields $p(6.2) = 698.824$.

A big advantage of nested form is that it permits you to evaluate a polynomial using only the operations of multiplication and addition. This difference can be significant because powering functions on a calculator or computer use the identity

$$y^x = e^{x \ln y}$$

and the built-in exponential and logarithmic functions with base e. Though for most computations there is little difference between the speed or accuracy of computations carried out these two ways, discrepancies may occur in a long calculation involving large numbers or carried out to extreme accuracy.

When $p(x)$ is in nested form, the evaluation of $p(x)$ at a specific value of x, say $x = 6.2$, follows a pattern:

Step 1: Multiply 3 by 6.2 and add 1.
Step 2: Multiply the sum by 6.2 and add -8.
Step 3: Multiply the sum by 6.2 and add -5.

This sequence of steps can be written in a convenient scheme. Write the specific value of the variable at the far left. Then write the coefficients of the polynomial for descending powers of the variable.

6.2	3	1	-8	-5

Now bring down the leading coefficient, 3. Multiply by 6.2, add to 1, and repeat the process. The work is shown below.

6.2	3	1	-8	-5
		18.6	121.52	703.824
	3	19.6	113.52	$698.824 = p(6.2)$

The above scheme for evaluating a polynomial for some value of the variable is called **synthetic substitution**. It makes use of the fact that when a polynomial in a single variable is arranged in descending powers with each degree included so that an nth degree polynomial has $n + 1$ terms, the polynomial is completely characterized by its coefficients. Thus, in using synthetic substitution, only the coefficients of the polynomial, including 0 for any missing terms, need to be written.

Example 1 Given the polynomial $p(x) = 7x^4 - 4x^3 + 2x + 6$.
a. Write $p(x)$ in nested form.
b. Use synthetic substitution to evaluate $p(3)$.

Solution **a.** You should notice that the coefficient of the x^2 term is 0. So, you must first write $p(x)$ as

$$p(x) = 7x^4 - 4x^3 + 0x^2 + 2x + 6.$$

Then proceed to repeatedly factor out one power of x.

$$p(x) = \underbrace{7x^4 - 4x^3 + 0x^2 + 2x}_{\text{factor out } x} + 6$$

$$= (\underbrace{7x^3 - 4x^2 + 0x}_{\text{factor out } x} + 2)x + 6$$

$$= ((\underbrace{7x^2 - 4x}_{\text{factor out } x} + 0)x + 2)x + 6$$

$$= (((7x - 4)x + 0)x + 2)x + 6$$

$$= (((7x - 4)x)x + 2)x + 6$$

exploited when representing integers in different bases (Lesson 4-7).

To emphasize the connection between synthetic substitution and long division, write the long division for $3x^3 + x^2 - 8x - 3$ divided by $x - 2$ on the chalkboard and then erase all the variables, leaving only the coefficients. Compare the numbers in the remaining skeleton to the synthetic substitution problem.

Point out to students that when they read the coefficients for the quotient from the last row of the synthetic division scheme that they know the degree of the quotient must be one less than the degree of the dividend. You may also want to point out that the remainder must be equal to $p(x) = q(x) \cdot (x - c) + r(x)$ and substituting c for x. This anticipates the Remainder and Factor Theorems of Lesson 4-6.

Technology You may wish to set up a synthetic division demonstration using a spreadsheet. Divide a polynomial $p(x)$ by $x - c$ synthetically (having preset the spreadsheet to do the necessary multiplications and additions). Using different values of c, students will see different $q(x)$'s and $p(c)$'s being generated. Because a value of $p(c)$ is generated each time, you may use these calculations to locate intervals between consecutive integers that contain zeros of $p(x) = 0$. There is also a theorem (not presented in this book) that if for a given $c > 0$, the coefficients of $q(x)$ are all positive, then there are no zeros of $p(x)$ greater than c.

ADDITIONAL EXAMPLES

1. Let $f(x) = 5x^4 + 7x^3 - 8x^2 + 2x + 1$.

a. Write $f(x)$ in nested form.

$f(x) = (((5x + 7)x - 8)x + 2)x + 1$

b. Use synthetic substitution to find $f(2)$.

2	5	7	-8	2	1
		10	34	52	108
	5	17	26	54	109

So $f(2) = 109$.

2. Use synthetic division to divide $p(x) = 8x^5 + 2x^4 - x^3 - 26x^2 + 5$ by $d(x) = x - \frac{3}{2}$.

3/2	8	2	-1	-26	0	5
		12	21	30	6	9
	8	14	20	4	6	14

$q(x) = 8x^4 + 14x^3 + 20x^2 + 4x + 6$, $r(x) = 14$

3. Use synthethic substitution to divide $3x^4 - 8x^3 + 13x^2 - 3x + 6$ by $3x - 2$. (This extra example is similar to Questions 10 and 11.)

Rewrite $3x - 2$ as $3\left(x - \frac{2}{3}\right)$.

2/3	3	-8	13	-3	6
		2	-4	6	2
	3	-6	9	3	8

So $q(x) = x^3 - 2x^2 + 3x + 1$, $r(x) = 8$.

b. Write 3 at the far left and then list the coefficients of descending powers of the variable. Be sure to list the 0 coefficient of the x^2 term. Then follow the scheme as illustrated before.

3	7	-4	0	2	6
		21	51	153	465
	7	17	51	155	471 = $p(3)$

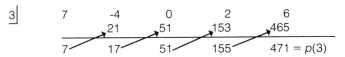

Generally, the arrows are not written in the scheme. With this convention, the work would look like that below.

3	7	-4	0	2	6
		21	51	153	465
	7	17	51	155	471

Thus, $p(3) = 471$.

It turns out that, if $p(x)$ is a polynomial and c is any number, then there is an important and surprising relationship between the synthetic substitution used to evaluate $p(c)$ and the division of $p(x)$ by $x - c$. We examine this relationship by considering a specific instance. Suppose $p(x) = 3x^3 + x^2 - 8x - 5$ and $c = 2$.

Synthetic substitution to find $p(2)$:

2	3	1	-8	-5	
		6	14	12	1st row
					2nd row
	3	7	6	(7) = $p(2)$	3rd row

Division of $p(x)$ by $x - 2$:

$$\begin{array}{r|rrrr} & 3x^2 & + 7x & + 6 & \leftarrow \text{quotient} \\ \hline x - 2 & 3x^3 + x^2 & - 8x & - 5 & \\ & 3x^3 - 6x^2 & & & \\ \hline & 7x^2 & - 8x & & \\ & 7x^2 & - 14x & & \\ \hline & & 6x & - 5 & \\ & & 6x & - 12 & \\ \hline & & & (7) & = \text{remainder} \end{array}$$

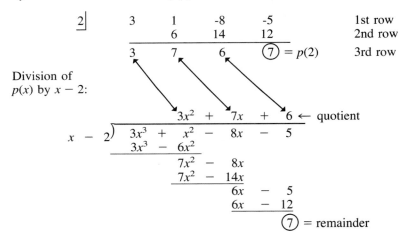

Synthetic substitution exhibits the quotient and the remainder! The last entry in the third row of the synthetic substitution procedure is the same as the remainder in the division procedure. The other numerical entries in the third row are the coefficients of the quotient, whose degree is one less than the degree of $p(x)$.

To understand why this relationship works in this case, we return to the synthetic substitution and attach appropriate powers of x to the coefficients so that the first row represents $p(x)$. Next, we attach powers of x to the coefficients in the second and third rows so that all terms in a given column have the same power of x.

2	$3x^3$	$1x^2$	$-8x$	-5	1st row
		$6x^2$	$14x$	12	2nd row
	$3x^3$	$7x^2$	$6x$	7	3rd row

Adding the first and the second rows together yields the third row:
$(3x^3 + 1x^2 - 8x - 5) + (6x^2 + 14x + 12) = 3x^3 + 7x^2 + 6x + 7$.

This equation can be expressed in an equivalent form as the work below indicates. We first substitute $p(x)$ for $3x^3 + 1x^2 - 8x - 5$ and then solve for $p(x)$.

$$p(x) + 2(3x^2 + 7x + 6) = x(3x^2 + 7x + 6) + 7$$
$$\Leftrightarrow \quad p(x) = x(3x^2 + 7x + 6) - 2(3x^2 + 7x + 6) + 7$$
$$\Leftrightarrow \quad p(x) = (3x^2 + 7x + 6)(x - 2) + 7$$

The final equation has the form of the Quotient-Remainder Theorem for Polynomials, where $x - 2$ is the divisor, $3x^2 + 7x + 6$ is the quotient, and 7 is the remainder.

Although we have shown this relationship between synthetic substitution and division by a polynomial of the form $x - c$ for only one specific instance, it is valid for all polynomials $p(x)$ and all values of c. To find the quotient and remainder for the division of $p(x)$ by $d(x) = x - c$, simply perform synthetic substitution to evaluate $p(c)$. When synthetic substitution is used for this purpose it is called, appropriately, **synthetic division**.

Example 2 Use synthetic division to divide $p(x) = 6x^5 + x^4 + x^3 - 3x^2 + 9$ by $d(x) = x + \frac{1}{2}$.

Solution Observe that the coefficient of x in $p(x)$ is zero; hence, write the polynomial as $p(x) = 6x^5 + x^4 + x^3 - 3x^2 + 0x + 9$. Also, $d(x) = x - \left(-\frac{1}{2}\right)$ so $c = -\frac{1}{2}$. Then the synthetic division algorithm yields the following result.

$-\frac{1}{2}$	6	1	1	-3	0	9
		-3	1	-1	2	-1
	6	-2	2	-4	2	8
	coefficients of $q(x)$					remainder

The polynomial $q(x)$ has degree one less than the degree of $p(x)$; so $q(x)$ is of degree 4. Using the values in the third row, the quotient and remainder can be written as

$$q(x) = 6x^4 - 2x^3 + 2x^2 - 4x + 2 \text{ and } r(x) = 8.$$

NOTES ON QUESTIONS

Questions 5-7: Remind students to use the Quotient-Remainder Theorem to check their work.

Question 8: The result depends upon the calculator. Students who obtain the same answer to parts **a** and **b** should be encouraged to try the problem again with a value of x containing more decimal places.

Questions 10 and 11: Emphasize that the synthetic division scheme only works for divisors of the form $x - c$. Encourage students to factor a constant from the given divisor to get one of the proper form. After dividing synthetically, they then should use the Quotient-Remainder Theorem to get the correct quotient.

Question 13: Generalize the result. The function M_k is periodic with period k.

ADDITIONAL ANSWERS
3. The $x \cdot x \cdot x \cdot x \cdot x$ calculation may result in greater accuracy, since the powering key is evaluated by using the exponential and logarithmic functions.

7.a. $q(x) = 2x^3 + 3x^2 - 4x + 5$; $r(x) = 0$

9.

1	1	0	0	0	0	-1
		1	1	1	1	1
	1	1	1	1	1	0

Since the remainder is 0, $x - 1$ is a factor of $x^5 - 1$.

13.a.

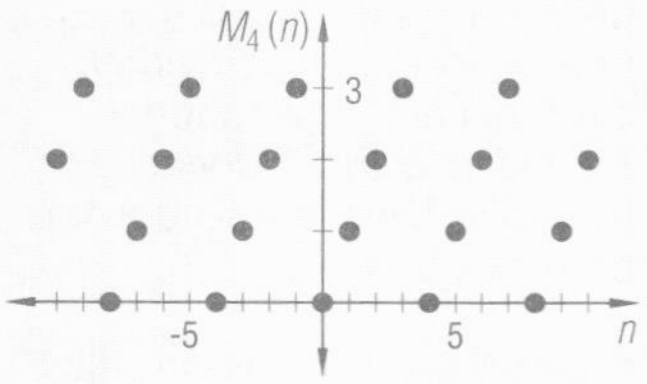

14. Counterexample: Let $a = 2$, $b = 6$, and $c = 4$. Then $a + b = 8$, which is divisible by c. However, neither a nor b is divisible by c.

15.a., c. See Additional Answers in the back of this book.

17.

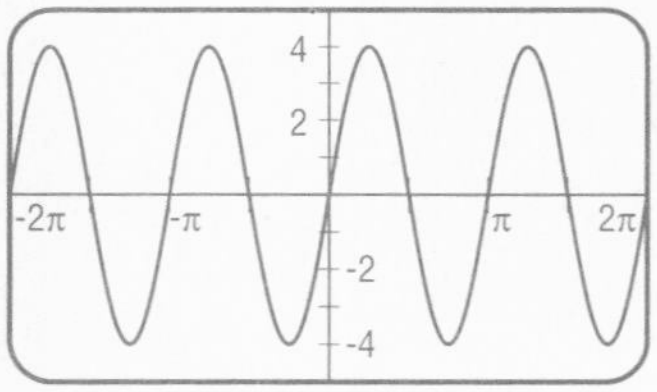

19. See Additional Answers in the back of this book.

Questions

Covering the Reading

1. Let $p(x) = 2x^3 - 4x^2 + x - 2$ and $c = 2$.
  a. Write $p(x)$ in nested form. $p(x) = ((2x - 4)x + 1)x - 2$
  b. Use synthetic substitution to find $p(c)$. 0
  c. Refer to the nested form of the polynomial in part **a**. Write a key sequence needed to use this form to evaluate $p(x)$ at $x = 7.54$. See below.

2. Repeat Questions **1a** and **1b** when $p(x) = -5x^3 + x^2 - x + 1$ and $c = 1.1$. a) $((-5x + 1)x - 1)x + 1$; b) -5.545

3. On a calculator, why might it be better to evaluate x^5 as $x \cdot x \cdot x \cdot x \cdot x$ rather than using the powering key? See margin.

4.

-3	2	0	-9	17	3
		-6	18	-27	30
	2	-6	9	-10	33

This synthetic division was done to divide $p(x)$ by $d(x)$.
  a. What is $p(x)$? **b.** What is $d(x)$? $x + 3$
  c. What is the quotient? **d.** What is the remainder? 33
  e. What value of what polynomial has been found? $p(-3)$
  a) $2x^4 - 9x^2 + 17x + 3$; c) $2x^3 - 6x^2 + 9x - 10$

In 5 and 6, use synthetic division to find the quotient $q(x)$ and remainder $r(x)$ when $p(x)$ is divided by $d(x)$.

5. $p(x) = 3x^3 - x^2 + x - 2$, $d(x) = x - 1$ $q(x) = 3x^2 + 2x + 3$; $r(x) = 1$

6. $p(x) = 2x^5 - x^3 + 4x - 3$, $d(x) = x + 1$
$q(x) = 2x^4 - 2x^3 + x^2 - x + 5$; $r(x) = -8$

Applying the Mathematics

7. a. Use synthetic division to divide $p(x) = 2x^4 - 3x^3 - 13x^2 + 17x - 15$ by $d(x) = x - 3$. Identify the quotient, $q(x)$, and the remainder, $r(x)$. See margin.
  b. Write your solution to part **a** in the form used in the Quotient-Remainder Theorem for Polynomials.
  c. *True* or *false?*
    i. $x - 3$ is a factor of $p(x)$. **ii.** $q(x)$ is a factor of $p(x)$. True
    iii. $p(3) = 0$ True **iv.** $q(3) = 0$ False
  b) $p(x) = (2x^3 + 3x^2 - 4x + 5)(x - 3) + 0$; c) i. True

8. Compare the values that you get with a calculator when you evaluate the polynomial
$$p(x) = x^4 - 3x^3 + 5x^2 - 7x + 9$$
at $x = 2.3456789$ by:
  a. using the power key to compute values of x^2, x^3, and x^4;
  b. using the nested form of p. samples: a) 11.646354; b)11.646354

9. Use synthetic division to show that $x - 1$ is a factor of $x^5 - 1$. See margin.

1. c.
2 [×] 7.54 [−] 4 [=] [×] 7.54 [+] 1 [=] [×] 7.54 [−] 2 [=]

In 10 and 11, use synthetic division to find the quotient $q(x)$ and the remainder $r(x)$ when the given polynomial $p(x)$ is divided by $d(x)$.

10. $p(x) = 3x^3 - 2x^2 + 2x - 1$, $d(x) = -x + 1$
(Hint: Write $d(x) = -(x - 1)$. After dividing, multiply the quotient by -1.) $q(x) = -3x^2 - x - 3$; $r(x) = 2$

11. $p(x) = 4x^4 - 2x^2 + 3x + 1$, $d(x) = 2x + 1$
(Hint: Write $d(x) = 2\left(x + \frac{1}{2}\right)$. Divide $p(x)$ by $x + \frac{1}{2}$, then multiply the quotient by an appropriate number.) $q(x) = 2x^3 - x^2 - \frac{1}{2}x + \frac{7}{4}$; $r(x) = -\frac{3}{4}$

Review

12. Use long division to find the quotient $q(x)$ and the remainder $r(x)$ when $p(x) = 6x^4 - x^3 + 2x^2 - 3x + 1$ is divided by $d(x) = x^2 + x + 1$. *(Lesson 4-4)* $q(x) = 6x^2 - 7x + 3$; $r(x) = x - 2$

13. Let $M_4(n)$ be the smallest nonnegative integer congruent to n modulo 4.
a. Graph the function M_4 for integers between -10 and 10.
b. What is the period of this function? *(Lessons 4-3, 2-8)*
a) See margin. b) 4

14. Prove or disprove the following conjecture:
∀ integers a, b, and c, if a + b is divisible by c, then a is divisible by c and b is divisible by c. *(Lesson 4-1)* See margin.

15. A neighborhood group decides to sell candy bars to raise money for playground equipment.
a. The group is charged 75¢ for each of the first 1500 candy bars they buy, and 60¢ for all additional candy bars. Write a function C that describes the group's cost as a function of the number n of candy bars. See margin.
b. The group sells the candy bars for \$1.25 each. Write another function R that describes the group's revenue as a function of the number of candy bars. $R(n) = 1.25n$
c. Write a profit function P for this situation. See margin.
d. If the group wants to earn \$3000, what is the minimum number of candy bars that must be sold? *(Lesson 3-3)* 4962

16. Solve: $\log_{25}(x + 2) - \log_{25}x = \frac{1}{2}$. *(Lesson 2-7)* $x = \frac{1}{2}$

17. Identify the period, amplitude, and phase shift of $f(x) = 4\cos 2\left(x - \frac{\pi}{4}\right)$. Sketch a graph of the function. *(Lesson 3-9)*
See below.

18. a. Between what two consecutive integers must a zero of $f(x) = x^3 - 3x^2 + 4x + 2$ lie? -1 and 0
b. Find an interval of length 0.1 that contains the zero. *(Lesson 3-5)* Sample: $-.4 \le x \le -.3$

Exploration

19. Write a computer program that will let you divide a polynomial of degree at most seven by $x - c$. See margin.

17. period: π; amplitude: 4; phase shift: $\frac{\pi}{4}$; see margin for graph.

FOLLOW-UP

MORE PRACTICE
For more questions on SPUR Objectives, use *Lesson Master 4-5,* shown below.

EXTENSION
You might extend **Question 9** by asking students to use synthetic division to show that certain polynomials are *not* factors of other polynomials. For instance, show that $x + 1$ is not a factor of $x^4 + 1$. This anticipates the Factor Theorem.

PROJECTS
The projects for Chapter 4 are described on page 279. **Project 5** is related to the content of this lesson.

NAME ____________

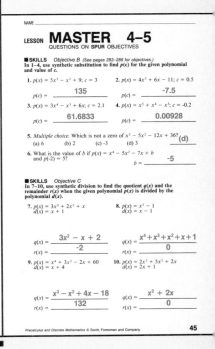

LESSON **MASTER 4–5**
QUESTIONS ON **SPUR** OBJECTIVES

■ **SKILLS** *Objective B* *(See pages 283–286 for objectives.)*
In 1–4, use synthetic substitution to find $p(c)$ for the given polynomial and value of c.

1. $p(x) = 5x^3 - x^2 + 9$; $c = 3$
$p(c) =$ 135

2. $p(x) = 4x^3 + 6x - 11$; $c = 0.5$
$p(c) =$ -7.5

3. $p(x) = 3x^4 - x^3 + 6x$; $c = 2.1$
$p(c) =$ 61.6833

4. $p(x) = x^5 + x^4 - x^3$; $c = -0.2$
$p(c) =$ 0.00928

5. *Multiple choice.* Which is not a zero of $x^3 - 5x^2 - 12x + 36$? (d)
(a) 6 (b) 2 (c) -3 (d) 3

6. What is the value of b if $p(x) = x^4 - 5x^2 - 7x + b$ and $p(-2) = 5$?
$b =$ -5

■ **SKILLS** *Objective C*
In 7–10, use synthetic division to find the quotient $q(x)$ and the remainder $r(x)$ when the given polynomial $p(x)$ is divided by the polynomial $d(x)$.

7. $p(x) = 3x^3 + 2x^2 + x$
$d(x) = x + 1$
$q(x) =$ $3x^2 - x + 2$
$r(x) =$ -2

8. $p(x) = x^5 - 1$
$d(x) = x - 1$
$q(x) =$ $x^4 + x^3 + x^2 + x + 1$
$r(x) =$ 0

9. $p(x) = x^4 + 3x^3 - 2x + 60$
$d(x) = x + 4$
$q(x) =$ $x^3 - x^2 + 4x - 18$
$r(x) =$ 132

10. $p(x) = 2x^3 + 5x^2 + 2x$
$d(x) = 2x + 1$
$q(x) =$ $x^2 + 2x$
$r(x) =$ 0

 45

LESSON 4-6

RESOURCES
- Lesson Master 4-6
- Quiz for Lessons 4-4 through 4-6
- Teaching Aid 25 can be used with **Question 9**.
- Computer Master 8

OBJECTIVE

H Use the Remainder Theorem and the Factor Theorem to describe characteristics of given polynomials.

TEACHING NOTES

Making Connections In the UCSMP series, the Factor Theorem is discussed in *Algebra, Advanced Algebra,* and in *Functions, Statistics, and Trigonometry.* The Remainder Theorem is discussed only in the last book mentioned above.

In *Advanced Algebra,* a completely different proof of the Factor Theorem is given, using the Graph Translation Theorem: If $x - c$ is a factor of a polynomial $p(x)$, then $p(x) = (x - c)q(x)$. Therefore, $p(c) = (c - c)q(c) = 0 \cdot q(c) = 0$. If 0 is a solution to a polynomial equation $f(x) = 0$, then the graph of the function $f: x \to f(x)$ contains (0, 0) and the constant term of the polynomial is 0. Therefore, x is a factor of $f(x)$. Any polynomial function p with a zero c contains $(c, 0)$, so it is the image

LESSON

4-6

The Remainder and Factor Theorems

In Lesson 4-5, you saw an example of the following theorem, known as the *Remainder Theorem.* This theorem states a fundamental relationship between polynomial division and evaluating a polynomial.

The Remainder Theorem

If a polynomial $p(x)$ of degree $n \geq 1$ is divided by $x - c$, where c is any number, then the remainder is the constant $p(c)$. That is,

$$p(x) = q(x)(x - c) + p(c).$$

Proof Since $p(x)$ is divided by $x - c$, the Quotient-Remainder Theorem guarantees the existence of polynomials $q(x)$ and $r(x)$ such that

$$p(x) = q(x)(x - c) + r(x)$$

and either degree of $r(x) <$ degree of $(x - c)$ or $r(x) = 0$ for all x. Because the degree of $x - c$ is 1, it follows that either the degree of $r(x)$ is 0, or $r(x)$ is itself 0. In either case, $r(x)$ has a constant value, call it R. Thus for all x,

$$p(x) = q(x)(x - c) + R.$$

Since this equation is true for all x, it must be true when $x = c$. So

$$p(c) = q(c)(c - c) + R$$
$$p(c) = R.$$

This says that the remainder R is the constant $p(c)$. That is, for all x,

$$p(x) = q(x)(x - c) + p(c).$$

Example 1 Let $p(x) = x^5 - 3x^4 + 8x^2 - 9x + 27$.
Find the remainder when $p(x)$ is divided by $x + 2$.

Solution 1 In this case, $x - c = x + 2$, so $c = -2$. Then, according to the Remainder Theorem, the remainder is $p(-2)$.

$$p(-2) = (-2)^5 - 3(-2)^4 + 8(-2)^2 - 9(-2) + 27$$
$$= -32 - 48 + 32 + 18 + 27$$
$$= -3$$

Solution 2 You could also use synthetic substitution with $c = -2$. Be careful to identify 0 as the coefficient of the x^3 term.

-2	1	-3	0	8	-9	27
		-2	10	-20	24	-30
	1	-5	10	-12	15	-3 ↑ $p(-2)$

The remainder is -3 when $p(x)$ is divided by $x + 2$. That is, $p(x) = q(x)(x + 2) + -3$.

254

The linear polynomial $x - c$ is a factor of $p(x)$ if and only if $p(x) = q(x) \bullet (x - c)$. Thus, "$x - c$ is a factor of $p(x)$" means that $p(x) = q(x)(x - c) + 0$. But this is the statement that the remainder when $p(x)$ is divided by $x - c$ is zero. Now, by the Remainder Theorem, this remainder must be $p(c)$. This reasoning proves the following very powerful theorem.

The Factor Theorem

For all polynomials $p(x)$, $x - c$ is a factor of $p(x)$ if and only if $p(c) = 0$, that is, if and only if c is a zero of the polynomial function.

Example 2 Show that $x - 1$ is a factor of $x^7 - 1$.

Solution Apply the Factor Theorem with $c = 1$ and $p(x) = x^7 - 1$. Since $p(1) = 1^7 - 1 = 0$, $x - 1$ is a factor of $p(x)$.

The Factor Theorem can also be applied when the factor is linear but not of the form $x - c$.

Example 3 Use synthetic substitution to show that $2x + 3$ is a factor of the polynomial $p(x)$ defined by

$$p(x) = 2x^4 - x^3 - 4x^2 - x - 6.$$

Solution The problem asks to show that $p(x) = (2x + 3)q(x)$, where $q(x)$ is a polynomial. This is equivalent to asking to show that $p(x) = \left(x + \frac{3}{2}\right) \bullet 2q(x)$, where $q(x)$ is a polynomial. That is, $2x + 3$ is a factor of $p(x) \Leftrightarrow x + \frac{3}{2}$ is a factor of $p(x)$. Now, by the Factor Theorem, $x + \frac{3}{2}$ is a factor of $p(x) \Leftrightarrow p\left(-\frac{3}{2}\right) = 0$. Use synthetic substitution to evaluate $p\left(-\frac{3}{2}\right)$.

$-\frac{3}{2}$	2	-1	-4	-1	-6
		-3	6	-3	6
	2	-4	2	-4	0

Since $p\left(-\frac{3}{2}\right) = 0$, $x + \frac{3}{2}$ is a factor of $p(x)$. Thus $2x + 3$ is a factor of $p(x)$.

If $p(x)$ is a polynomial, then zeros of the function $x \to p(x)$ are also called **zeros of the polynomial** $p(x)$. The Remainder and Factor Theorems are useful tools for determining exact values of zeros of a polynomial. Suppose you are able to show that c is a zero of $p(x)$. Then $x - c$ is a factor of $p(x)$. To find other zeros of $p(x)$, divide $p(x)$ by $x - c$ to obtain the quotient polynomial $q(x)$. Other zeros of $p(x)$ can then be found by finding the zeros of $q(x)$. This is usually an easier task because $q(x)$ has degree one less than $p(x)$. This procedure is justified by the following theorem.

of such a function f under the translation c units to the right. By the Graph Translation Theorem, an equation for p can be found by using $x - c$ in place of x in the formula for $f(x)$ and so $x - c$ is a factor of $p(x)$.

Example 1 is a simple application of the Remainder Theorem. **Example 2** shows the power of the Factor Theorem; factors of polynomials can be obtained without dividing and sometimes with little substitution.

If your students have studied geometric series, you might point out that the finite geometric series $1 + x + x^2 + x^3 + x^4 + x^5 + x^6 = \frac{x^7 - 1}{x - 1}$. The quotient in **Example 2** is this series.

Example 3 uses the fact that when $a \neq 0$, $q(x)$ is a factor of $p(x)$ if and only if $aq(x)$ is a factor of $p(x)$.

The idea of the theorem at the top of page 256 is that dividing $p(x)$ by $x - c_1$ "divides out" the zero c_1 from $p(x)$. Any zeros that are still left must be zeros of the original polynomial $p(x)$. This result can sometimes be used to find zeros of a polynomial if some of the zeros are already known, as **Example 4** illustrates.

Alternate Approach To clarify the theorem on this page, you might want to carry out the solutions to a polynomial equation in its entirety. For instance, let $p(x) = x^4 - 4x^3 - x^2 + 16x - 12$. The zero 1 is easy to find. Dividing by $x - 1$, $p(x) = (x - 1)(x^3 - 3x^2 - 4x + 12)$. Another zero of the quotient polynomial is 2, so after division, $x^3 - 3x^2 - 4x + 12 = (x - 2)(x^2 - x - 6)$. The new quotient polynomial has -2 as a zero, and dividing gives $x^2 - x - 6 = (x + 2)(x - 3)$. Thus, the original polynomial $p(x) = (x - 1)(x - 2)(x + 2)(x - 3)$.

It is not uncommon for one solution to an equation to be used to find other solutions. For instance, with trigonometric equations, if you know one solution to $\sin x = k$, then another solution can be found by using $\sin(\pi - x) = \sin x$, and all others can be found by adding 2π to these solutions.

In general, a problem of the type in **Example 5** is solved by looking for the maximum number of times that some horizontal line intersects the graph of $y = p(x)$.

Since the prime factorization of a polynomial is unique (see page 271), and since a polynomial of degree n can have no more than n linear factors, then it can have no more than n zeros. In order to prove that it has exactly n zeros, we need to know that every polynomial has at least one zero. This theorem, whose proof is beyond the scope of this course, is discussed in Chapter 8. This should assuage those students who are unhappy with the "imprecision" of the "Number of Zeros of a Polynomial" Theorem.

Theorem

If c_1 is a zero of a polynomial $p(x)$ and if c_2 is a zero of the quotient polynomial $q_1(x)$ obtained when $p(x)$ is divided by $x - c_1$, then c_2 is a zero of $p(x)$.

Proof Start with the antecedent of the theorem. Since c_1 is a zero of a polynomial $p(x)$, then by the Factor Theorem

$$p(x) = q_1(x)(x - c_1).$$

Since c_2 is a zero of the quotient $q_1(x)$ that is obtained when $p(x)$ is divided by $(x - c_1)$, by the Factor Theorem there is a polynomial $q_2(x)$ with

$$q_1(x) = q_2(x)(x - c_2).$$

So, by substitution for $q_1(x)$,

$$p(x) = q_2(x)(x - c_2)(x - c_1).$$

Thus $(x - c_2)$ is a factor of $p(x)$. Applying the Factor Theorem once again, c_2 is a zero of $p(x)$.

Because you know the Quadratic Formula, you only need to know one solution to a cubic equation in order to obtain the other two. As Example 4 shows, you can solve such an equation even when the solutions are not real numbers.

Example 4 Given that $-\frac{1}{2}$ is a solution to $2x^3 + 5x^2 + 6x + 2 = 0$, find the remaining solutions.

Solution Let $p(x) = 2x^3 + 5x^2 + 6x + 2$. From the given, $-\frac{1}{2}$ is a zero of this function. Thus, by the Factor Theorem, $x + \frac{1}{2}$ is a factor of $p(x)$. To find the quotient polynomial $q(x)$ when $p(x)$ is divided by $x + \frac{1}{2}$, you could use synthetic division or long division. We use synthetic division.

$$\begin{array}{r|rrrr} -\frac{1}{2} & 2 & 5 & 6 & 2 \\ & & -1 & -2 & -2 \\ \hline & 2 & 4 & 4 & 0 \end{array}$$

Thus $q(x) = 2x^2 + 4x + 4$. The zeros of $q(x)$ can be found by using the Quadratic Formula:

$$x = \frac{-4 \pm \sqrt{4^2 - 4(2)(4)}}{4} = \frac{-4 \pm \sqrt{-16}}{4} = \frac{-4 \pm 4i}{4} = -1 \pm i.$$

Consequently, the zeros of $p(x)$ are $-\frac{1}{2}$, $-1 + i$, and $-1 - i$.

Now imagine beginning with a polynomial $p(x)$ of degree n and repeatedly dividing by $x - c$ for each zero c of $p(x)$. Each division reduces the degree of the current polynomial by 1. So the process of repeated division can have at most n steps (in which case it would end with a polynomial of degree 0). This reasoning justifies the following theorem.

256

Theorem (Number of Zeros of a Polynomial)

A polynomial of degree n has at most n zeros.

This theorem provides a test of the possible shape of a graph of a polynomial equation $y = p(x)$ where $p(x)$ has real coefficients. Because a polynomial with real coefficients has at most n zeros, its graph can cross the x-axis at most n times. In fact, it cannot cross any horizontal line $y = k$ more than n times. This can be proved rather easily.

Theorem

Let $p(x)$ be a polynomial of degree $n \geq 1$ with real coefficients. The graph of $y = p(x)$ can cross any horizontal line $y = k$ at most n times.

Proof Suppose that $p(x)$ is a polynomial with degree $n \geq 1$. The points of intersection of the graph of $y = p(x)$ and the horizontal line $y = k$ are the solutions of the equation $p(x) = k$. In other words, they are numbers x such that $p(x) - k = 0$. This means that the number of points of intersection equals the number of zeros of the polynomial $p(x) - k$. Let $q(x) = p(x) - k$. Because the polynomials $p(x)$ and $q(x)$ only differ by the constant k, the degree of $q(x)$ is the same as the degree of $p(x)$. That is, the degree of $q(x)$ is n. By the theorem on the number of zeros of a polynomial, q has at most n zeros. It follows that the graph of $y = p(x)$ can cross the horizontal line $y = k$ at most n times.

Example 5 Suppose $p(x)$ is a polynomial whose graph is shown at the right. What is the smallest possible value for the degree of $p(x)$?

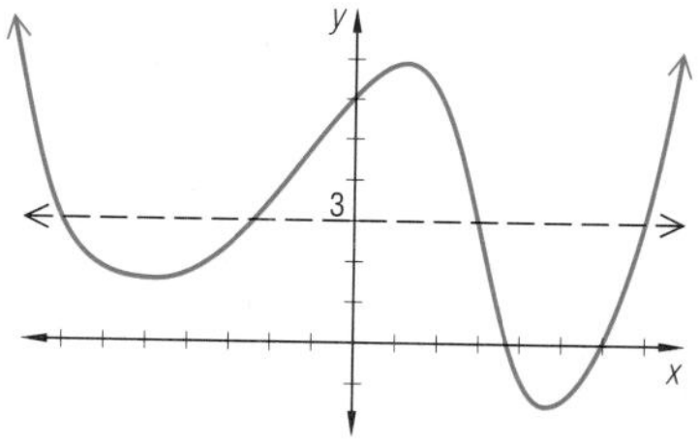

Solution The graph of $y = p(x)$ crosses some horizontal lines, such as the line $y = 3$, four times. Hence, the degree of $p(x)$ must be at least 4.

Some students quickly (too quickly) jump to the conclusion that the degree of a polynomial indicates how many "wiggles" a polynomial has. It indicates only the *maximum* number of wiggles. For instance, the graph of $y = x^{10}$ has only one wiggle; it looks quite a bit like the flattened parabola $y = x^2$.

Technology Some calculators have the capability of evaluating expressions for a given value of the variable. This calculator feature could be used in this section to verify results. Also, **Questions 4, 5, 7, 11, 12, and 17** can be solved using this feature.

ADDITIONAL EXAMPLES

1. Find the remainder when $p(x) = x^6 - 7x^4 - 18x^2 + 5x - 3$ is divided by $x - 3$.
12

2. For any positive integer n, show that $x - 1$ is a factor of $p(x) = x^{2n} - 1$.
For all n, $p(1) = 1^{2n} - 1 = 0$, so by the Factor Theorem, $x - 1$ is a factor of $x^{2n} - 1$.

3. Use synthetic substitution to show that $3x - 2$ is a factor of $p(x) = 3x^4 + 4x^3 - 19x^2 + 7x + 2$.

2/3	3	4	-19	7	2
		2	4	-10	-2
	3	6	-15	-3	0

Since $p\left(\frac{2}{3}\right) = 0$, $x - \frac{2}{3}$ is a factor of $p(x)$.
$\therefore$ $3x - 2$ is also a factor of $p(x)$.

4. Given that $\frac{2}{3}$ is a solution to $3x^3 - 11x^2 - 15x + 14 = 0$, find the other solutions.
$\frac{3 + \sqrt{37}}{2}, \frac{3 - \sqrt{37}}{2}$

5. What is the smallest possible degree for the polynomial function whose graph is shown below?

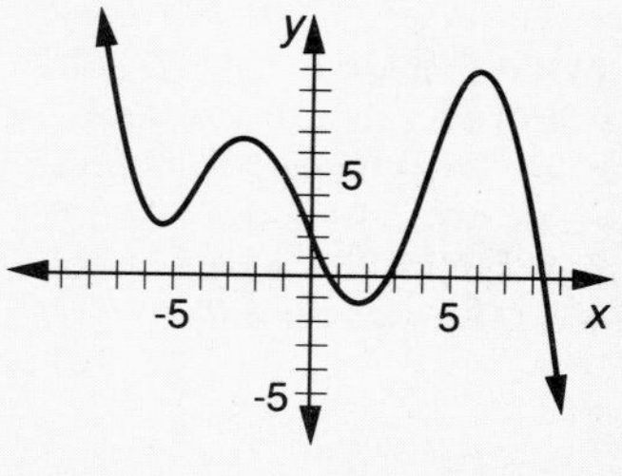

The line $y = 4$ intersects the graph 5 times, so the degree is at least 5.

NOTES ON QUESTIONS
Question 7: Remind students that after using synthetic division with -1 (or 3) as the divisor, they should use the quotient polynomial (not the original) for the next synthetic division. Because each division results in a polynomial of one degree less, two synthetic divisions (using -1 and 3) will produce a quotient polynomial of degree two. The quadratic formula then provides the remaining zeros.

Question 11: This question extends the idea of **Example 2** to the Remainder Theorem: If $a \neq 0$, then dividing $p(x)$ by $q(x)$ gives the same remainder as dividing $p(x)$ by $aq(x)$, because
$$p(x) = q(x)\,d(x) + R = aq(x)\frac{d(x)}{a} + R.$$

Question 13a: Students have known the result when $n = 2$ for many years. The general result is far more powerful.

Question 15: This question proves the uniqueness part of a theorem useful in modeling and in numerical analysis: Given $n + 1$ points in the xy-plane, if there is no polynomial of degree $n - 1$ that contains them, then there exists a unique polynomial of degree n which contains them. For instance, 3 points, no 2 of them collinear, determine a parabola with a vertical line of symmetry.

ADDITIONAL ANSWERS
9. (b) and (e); There are horizontal lines the graphs pass through more than 4 times, but the polynomial is only of degree 4.

Questions

Covering the Reading

1. A polynomial $p(x)$ is divided by $x + 4$ to obtain a polynomial $q(x)$ with degree 3, and a remainder of -5.
a. What is the degree of $p(x)$? 4
b. Give a value of x for which $p(x) = -5$. -4

In 2 and 3, use the Remainder Theorem to find the remainder when $p(x)$ is divided by $q(x)$.

2. $p(x) = 3x^5 - x^3 + 4x - 3$, $q(x) = x + 1$ -9

3. $p(x) = x^3 + x^2 - x + 2$, $q(x) = x - 13$ 2355

4. Without dividing, how can you tell that $t - 5$ has to be a factor of $t^4 - 5t - 600$? $(5)^4 - 5(5) - 600 = 0$, so by the Factor Theorem, $t - 5$ is a factor of $t^4 - 5t - 600$.

5. Determine if $p(x)$ has $3x + 1$ as a factor.
a. $p(x) = 9x^4 - 19x^2 - 2$ No **b.** $p(x) = 9x^4 - 19x^2 + 2$ Yes

6. Given that -2 is a zero of the polynomial $p(y) = 6y^3 + 11y^2 - 17y - 30$, find the remaining zeros of $p(y)$. $y = \frac{5}{3}$ and $y = -\frac{3}{2}$

7. a. Show that -1 and 3 are zeros of the polynomial $p(x) = x^4 - 5x^2 - 10x - 6$.
b. Find the remaining zeros of $p(x)$.

8. A polynomial $p(x)$ is divided by $x - 2$ to obtain a polynomial $q(x)$. with a remainder of 0. $q(x)$ has degree 5 and a zero at $x = -3$.
a. Give two zeros of $p(x)$. $x = 2$ and $x = -3$
b. Give two factors of $p(x)$. $x - 2$ and $x + 3$
c. What is the maximum number of zeros that $p(x)$ can have? 6
d. What is the remainder when $q(x)$ is divided by $x + 3$? 0

9. Which of the following *cannot* be the shape of the graph of a polynomial of degree 4? Explain your answer. See margin.

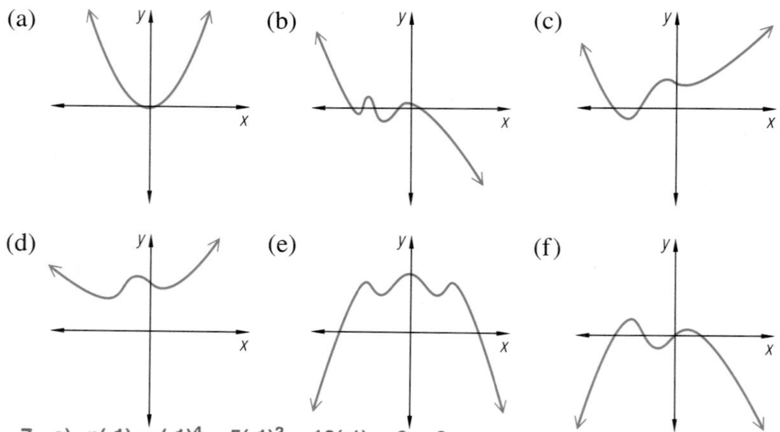

7. a) $p(-1) = (-1)^4 - 5(-1)^2 - 10(-1) - 6 = 0$; $p(3) = (3)^4 - 5(3)^2 - 10(3) - 6 = 0$; b) $x = -1 + i$ and $x = -1 - i$

258

Applying the Mathematics

10. Suppose $p(t) = t^4 - 2t^2 - t + k$ and 2 is a zero of p.
  a. Use synthetic division to find k. $k = -6$
  b. Check your answer by evaluating $p(2)$.
  $p(2) = (2)^4 - 2(2)^2 - 2 - 6 = 0$

11. Use the Remainder Theorem to find the remainder when $2x^3 - 3x^2 + 4x - 6$ is divided by $3x - 2$. $-\frac{110}{27}$

12. Refer to Question 1. If 2 is a zero of q, find $p(2)$. $p(2) = 0$

13. **a.** Suppose that c and d are real numbers and that n is any positive integer. Use the Factor Theorem to prove that $c - d$ is a factor of $c^n - d^n$. (Hint: Consider the polynomial $p(x) = x^n - d^n$.)
  b. Use the result from part **a** to prove that, if $n > 1$, $4^n - 1$ is never prime.
  See margin.

14. **a.** If $p(x)$ is a polynomial of degree 3, explain why the graph of $y = p(x)$ cannot cross the line $y = x$ more than 3 times. (Hint: Consider the polynomial $p_1(x) = p(x) - x$ and use the idea in the proof of the last theorem in this lesson.) See margin.
  b. Generalize part **a**. Sample: The graph of a polynomial of degree n can cross the line $y = x$ at most n times.

15. Prove that if the graphs of two polynomials $p_1(x)$ and $p_2(x)$ of degree n intersect at more than n points, then $p_1(x)$ and $p_2(x)$ are identical. (Hint: Consider the polynomial $p(x) = p_1(x) - p_2(x)$ and use the theorem on the number of zeros of a polynomial.) See margin.

16. The zeros of a fourth degree polynomial function p are 0, -4, $\frac{2}{3}$, and $\frac{1}{5}$. Write a formula for $p(x)$ in the form $a_4x^4 + a_3x^3 + a_2x^2 + a_1x + a_0$, where the a_i are all integers. $p(x) = 15x^4 + 47x^3 - 50x^2 + 8x + 0$

Review

17. Use synthetic division to find the quotient and remainder when
$$x^4 + 7x^3 - 2x + 1$$
is divided by $x + 3$. *(Lesson 4-5)*
$q(x) = x^3 + 4x^2 - 12x + 34$; $r(x) = -101$

18. Find the smallest nonnegative integer that makes the congruence true.
  a. $x \equiv 11 \pmod 4$ **b.** $y \equiv 250 \pmod{14}$ **c.** $z \equiv -84 \pmod{11}$
  (Lesson 4-3) a) 3; b) 12; c) 4

19. Factor: $(2x + y)(x + 3y)^2(x - y)^3 + (x - 4y)(x + 3y)(x - y)^4$.
  (Lesson 4-1) $(x + 3y)(x - y)^3(3x^2 + 2xy + 7y^2)$

20. Solve the inequality $(2x + 1)(x - 4) < 0$
  a. algebraically **b.** graphically. *(Lessons 3-6, 3-7)*
  $-\frac{1}{2} < x < 4$ $-\frac{1}{2} < x < 4$

21. Use the Intermediate Value Theorem to find an interval between two consecutive integers that contains a zero of the function g defined by $g(x) = x^5 - 4x^2 + 6$. *(Lesson 3-5)* $-2 < x < -1$

22. Find the transformations that will transform the graph of $y = x^3$ into the graph of $y = \frac{1}{4}(x + 6)^3 - 2$. *(Lesson 3-9)*
  $S_{1,1/4}$ then $T_{-6,-2}$

13.a. Given $p(x) = x^n - d^n$. The Factor Theorem tells us that $x - d$ is a factor of $p(x)$ if and only if $p(d) = 0$. $p(d) = d^n - d^n = 0$, so $x - d$ is a factor of $x^n - d^n$. If $x = c$, then $c - d$ is a factor of $c^n - d^n$.
b. For all $n > 1$, $4^n - 1 = 4^n - 1^n$, which has $(4 - 1)$ or 3 as a factor by part a. Since $4^n - 1 > 3$ for $n > 1$ and 3 is a factor of $4^n - 1$, then $4^n - 1$ is not prime.

14.a. Let $p_1(x) = p(x) - x$. When $p(x) = x$, $p_1(x) = 0$. Since $p_1(x)$ and $p(x)$ differ only by a first degree term, and since $p(x)$ is a polynomial of degree 3, $p_1(x)$ has degree 3 also. $p_1(x)$ has at most 3 zeros, so $p(x) = x$ at most three times.

15. Consider $p(x) = p_1(x) - p_2(x)$. The degree of this polynomial is at most n. But $p(x)$ has more than n zeros. However, a polynomial of degree n has at most n zeros, so $p(x)$ must be the zero polynomial. So $p(x) = 0$, and $p_1(x) - p_2(x) = 0$. Hence, $p_1(x) = p_2(x)$ for all x.

NAME ______

LESSON **MASTER 4–6**
QUESTIONS ON **SPUR** OBJECTIVES

■ **PROPERTIES** *Objective H (See pages 283–286 for objectives.)*

1. If $x^2 - 4$ is a factor of a given polynomial $p(x)$, find the value(s) of c for which $p(c) = 0$. -2, 2

2. If 3, 7, and -2 are the zeros of a cubic polynomial $c(x)$, give a possible formula for $c(x)$.
$(x - 3)(x - 7)(x + 2) = x^3 - 8x^2 + x + 42$, or a multiple thereof

3. Suppose $p(t) = t^4 - 3t^2 - t + 14$. Find $p(3)$ by dividing $p(t)$ by $d(t) = t - 3$.
$p(3) =$ 65

4. If -1 is a zero of $p(x) = x^3 - 5x^2 + 2x + 8$, find the remaining zeros of $p(x)$.
2, 4

5. *Multiple choice.* If the graph of a polynomial $p(x)$ crosses the x-axis exactly two times, then (c)
(a) $p(x)$ is of degree 2
(b) $p(x)$ is of degree greater than 2
(c) $p(x)$ is of degree 2 or more
(d) $p(x)$ may have more than two real zeros

6. *True or false?* If $t(c) = 0$ and $p(x) = q(x)t(x)$, then $p(c) = 0$. true

7. If $p(x)$ and $q(x)$ are polynomials such that $p(x) = q(x)(x + 3) - 2$, then you can conclude that $p($-3$) =$ -2.

46

FOLLOW-UP

MORE PRACTICE
For more questions on SPUR Objectives, use *Lesson Master 4-6,* shown on page 259.

PROJECTS
The projects for Chapter 4 are described on page 279. **Project 4** is related to the content of this lesson.

EVALUATION
A quiz covering Lessons 4-4 through 4-6 is provided in the Teacher's Resource File.

23. Let $f: x \rightarrow \sqrt{x+2}$ and $g: x \rightarrow \frac{1}{x^2}$.

a. Identify the domains of f and g. f: {x: x ≥ -2}; g: {x: x ≠ 0}

b. Find a rule for $g \circ f$. $g \circ f(x) = \frac{1}{x+2}$

c. Identify the domain of $g \circ f$. {x: x > -2}

d. Explain why f has an inverse f^{-1} but g does not.

(Lessons 3-1, 3-2) **f is a 1-1 function, but g is not.**

24. a. Write the negation of the following statement:

$\exists$ *x such that* $\log_2 x < 0$. $\forall$ x, $\log_2 x \geq 0$

b. Which is true: the statement or its negation?

(Lessons 1-1, 1-2) **the statement**

Exploration

25. A polynomial is of degree 9.

a. What is the largest number of relative maxima that it can have? 4

b. What is the largest number of relative minima that it can have? 4

c. Find a polynomial of degree 9 that illustrates your answer to either part **a** or part **b**.

sample: $p(x) = (x-1)(x-2)(x-3)(x-4)(x-5)(x-6)(x-7)(x-8)(x-9)$

LESSON

4-7

Polynomial Representation of Integers

The polynomial $a_nx^n + a_{n-1}x^{n-1} + \ldots + a_2x^2 + a_1x + a_0$ is a polynomial *in x*, but when you use synthetic substitution or synthetic division, you ignore the powers of x and just write the coefficients

$$a_n \quad a_{n-1} \quad \ldots \quad a_2 \quad a_1 \quad a_0.$$

This procedure may seem artificial, but it is something you have been doing ever since you learned to write whole numbers. When an integer, say 3407, is written as a decimal, the numbers 3, 4, 0, and 7 are the coefficients of a polynomial *in 10*.

$$3407 = 3 \cdot 10^3 + 4 \cdot 10^2 + 0 \cdot 10^1 + 7 \cdot 10^0$$

You write down only the coefficients and you mentally register that 3 is the thousands digit, 4 is the hundreds digit, and so on, and that the value of the number is the sum of the products of the coefficients with the place values. More generally, if an integer in base 10 has thousands digit T, hundreds digit h, tens digit t, and units digit u, then its value equals the polynomial *in 10*

$$T \cdot 10^3 + h \cdot 10^2 + t \cdot 10 + u.$$

This is often written as $1000T + 100h + 10t + u$, which disguises the powers of ten.

Base 10 has some special properties. Recall that

$$10 \equiv 1 \pmod 9.$$

That is, the remainder when 10 is divided by 9 is 1. Applying the Multiplication Property of Congruence,

$$10^2 \equiv 1 \pmod 9.$$

That is, the remainder when 100 is divided by 9 is 1. Repeatedly multiplying both sides by 10 or 1, which are congruent mod 9, yields

$\forall$ nonnegative integers n, $\quad 10^n \equiv 1 \pmod 9$.

Multiplying both sides by a_n, for any n,

$\forall$ nonnegative integers n, $\quad a_n \cdot 10^n \equiv a_n \pmod 9$.

For instance, $5000 \equiv 5 \pmod 9$, as you can check. With this machinery, you can prove a divisibility test you may have learned many years ago.

Example 1 Prove: An integer is divisible by 9 if and only if the sum of its digits in base 10 is divisible by 9.

Solution Remember that a number is divisible by 9 if and only if it is congruent to 0 (mod 9). When the number is written in base 10, it has the following value.

$$a_n10^n + a_{n-1}10^{n-1} + \ldots + a_210^2 + a_110 + a_0$$

Since $a_n \cdot 10^n \equiv a_n$ and $a_{n-1} \cdot 10^{n-1} \equiv a_{n-1}$, and so on, by the Addition Property of Congruence,

$$a_n10^n + a_{n-1}10^{n-1} + \ldots + a_210^2 + a_110 + a_0$$
$$\equiv a_n + a_{n-1} + \ldots + a_2 + a_1 + a_0 \pmod 9.$$

So the number is congruent to the sum of its digits (mod 9). That means that either both number and sum are congruent to 0 or neither is. So the number being divisible by 9 guarantees that the sum of its digits is also divisible by 9, and vice versa.

LESSON 4-7

RESOURCES

- Lesson Master 4-7
- Teaching Aid 26 displays the table of integers in base 10 and base 2.
- Teaching Aid 27 shows the network and truth table for the half-adder.
- Computer Master 9

OBJECTIVE

M Represent numbers in other bases and perform addition in base 2.

TEACHING NOTES

Example 1 presents a proof of a divisibility test that students may have known for some time. Point out that the result (and the proof using modular arithmetic) extends to other bases. A number is divisible by $b - 1$ if and only if the sum of its digits in its base b representation is divisible by $b - 1$.

Students may find working in other bases confusing. You may wish to represent the same number in bases 2, 3, 4, 5, and 10. Discuss which digits are allowed for each base. Note the similarity of place values in various bases. All share a right-most place value of one. In the hexadecimal system (base 16), commonly used in computer languages, extra digits needed to be invented. Ask why!

Making Connections Recall the discussion in the notes for Lesson 4-1 about $\log_b N$ approximating the number of digits in the base *b* representation of *N*. This fact helps explain why the logarithm to base 2 is so important in computer science.

Students may be interested to know that computers store all kinds of data in binary form. Occasionally, programmers need to see the contents of certain memory locations. When the contents of memory are printed, they are represented in hexadecimal notation. The programmer knows that two hexadecimal digits are equal to one byte and therefore is able to determine the contents of memory.

Technology Some advanced scientific and graphing calculators can represent numbers and perform arithmetic on them in binary, octal, and hexadecimal. If you are using such calculators in your class, consult the owner's manual to see how to work with binary and hexadecimal numbers.

Because an integer written in base 10 is congruent to the sum of its digits (mod 9), you can substitute the sum of the digits for the integer and use that sum (mod 9) to check computations. This procedure, called **casting out nines**, dates back to the Middle Ages. Here is how a multiplication problem can be checked. Does $569 \cdot 277 = 157613$?

Original problem	Sum of digits	mod 9
569	20	2
× 277	16	× 7
157613	23	14

The multiplication at left is wrong if 157613 is not congruent (mod 9) to the product of 2×7. Since the sum of the digits of 157613 is 23, it is congruent to 14, as required. Caution: As with the use of check digits, casting out nines is not a perfect check. If a random error is made, one out of nine times casting out nines will not sense it.

Computers do not use the digits 0 through 9 that we use to represent numbers. Rather, in digital computers, numbers are represented by electronic components called **bits** that exist in one of two states, on or off, and are denoted by 1 or 0, respectively.

When numbers are represented using just the digits 0 and 1, we say they are written in **base 2** or **binary notation**. In base 2, a number is written as a polynomial *in 2*, and the allowable coefficients are 0 and 1. The notation 10100_2 (read "one zero one zero zero base two") is evaluated as follows:

$$10100_2 = 1 \cdot 2^4 + 0 \cdot 2^3 + 1 \cdot 2^2 + 0 \cdot 2^1 + 0 \cdot 2^0.$$

Whenever a number is written in a base other than our ordinary base 10 system, a subscript is used to indicate the base. Also, commas are *not* used in numbers in any base other than 10. The integers from 0 to 16 in base 2 are given in the following table.

Base 10 Integer	Base 2 Place Values					Base 2 Integer
	2^4 16	2^3 8	2^2 4	2^1 2	2^0 1	
0	0	0	0	0	0	0_2
1	0	0	0	0	1	1_2
2	0	0	0	1	0	10_2
3	0	0	0	1	1	11_2
4	0	0	1	0	0	100_2
5	0	0	1	0	1	101_2
6	0	0	1	1	0	110_2
7	0	0	1	1	1	111_2
8	0	1	0	0	0	1000_2
9	0	1	0	0	1	1001_2
10	0	1	0	1	0	1010_2
11	0	1	0	1	1	1011_2
12	0	1	1	0	0	1100_2
13	0	1	1	0	1	1101_2
14	0	1	1	1	0	1110_2
15	0	1	1	1	1	1111_2
16	1	0	0	0	0	10000_2

In general, the digits in the base b representation of a number are the coefficients of a polynomial in b whose value is that number.

Definition

A number is written in base b **notation,**

$$(d_n\, d_{n-1} \ldots d_1\, d_0)_b,$$

if and only if each digit d_i is a particular integer from 0 through $b - 1$, and the number equals

$d_n \bullet b^n + d_{n-1} \bullet b^{n-1} + \ldots + d_1 \bullet b^1 + d_0 \bullet b^0$.

Example 2 illustrates how the definition can be used to find the base 10 representation of a number expressed in another base.

Example 2 Find the base 10 representation for the number 101011_2.

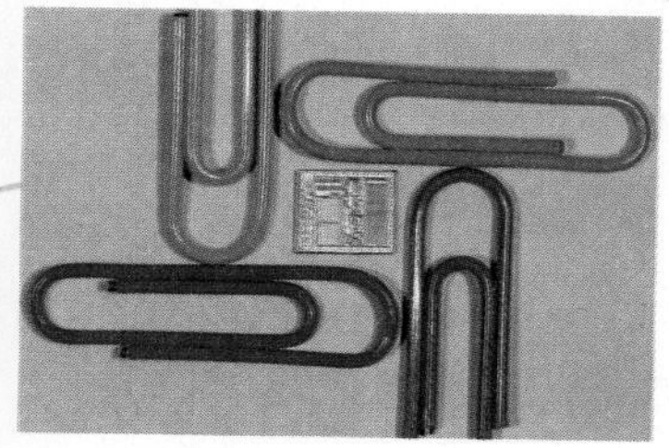
Computer chip compared to size of paper clip

Solution Use the definition and simplify. This number is written in base 2, so $b = 2$ in the definition.

$$\begin{aligned} 101011_2 &= 1 \bullet 2^5 + 0 \bullet 2^4 + 1 \bullet 2^3 + 0 \bullet 2^2 + 1 \bullet 2^1 + 1 \bullet 2^0 \\ &= 32 + 0 + 8 + 0 + 2 + 1 \\ &= 43 \end{aligned}$$

Computers store numbers in binary form. Each bit of computer memory stores one binary digit (0 or 1). Bits are organized into groups of eight called **bytes**. Example 2 shows that 43_{10} can be stored in one computer byte as 00101011.

To change a number in base 10 to another base, you can use division. Example 3 shows how to do this.

Example 3 Write 397 in base 2.

Solution Write down the integer powers of 2 that are less than 397. Begin with the largest power of 2 in the list. There is one 256 in 397. So the leftmost digit is 1. Then 397 − 256 = 141. There is one 128 in 141 so the next digit is also 1. 141 − 128 = 13. There are no 64s, 32s, or 16s in 13. So the next three digits are all 0s. There is one 8 in 13 so the next digit is 1. 13 − 8 = 5. There is one 4 in 5 and 5 − 4 = 1. There is no 2 in 1. Then the number is 110001101_2.

$2^8 = 256$
$2^7 = 128$
$2^6 = 64$
$2^5 = 32$
$2^4 = 16$
$2^3 = 8$
$2^2 = 4$
$2^1 = 2$
$2^0 = 1$

Base 10 notation is more compact than binary, but base 2 has a computational advantage. You only need four facts to add any binary numbers. These facts are $0 + 0 = 0_2$; $0 + 1 = 1_2$; $1 + 1 = 10_2$; and $1 + 1 + 1 = 11_2$. Otherwise, the general procedure for adding numbers in base 2 is similar to that used in base 10.

Students may wonder about other arithmetic operations on binary integers. The subtraction algorithm familiar to students can be mimicked to subtract binary integers, but computers perform the operation in quite a different way. Multiplication as done by computers is even more difficult. However, the special case of multiplying by powers of 2 has properties like those of multiplying by 10 in base 10. To multiply by 2, you simply "shift left one bit." For example, the result of multiplying 101100101_2 by 2 is 1011001010_2.

Error Analysis The table of base 10 and base 2 integers (preceding the base b notation definition) should be explained carefully. Many students will incorrectly call 10_2 ten and 100_2 one hundred. Do not let students be careless. It will greatly interfere with their understanding.

Making Connections
Regarding the half-adder network example, reassure students that they will not be expected to become computer designers in this course. The purpose of this example is to tie together concepts in number representation and logic with computer design.

Example 4 Perform the base 2 addition at right.

$$\begin{array}{rrrrrr} & & 1 & 1 & 1 & 0_2 \\ + & & & 1 & 1 & 1_2 \\ \hline \end{array}$$

Solution Just as in base 10, addition is usually done from right to left. In the 2^0 or 1s place: $0 + 1 = 1 = 1_2$, so write the 1 in the 2^0 place of the sum row.

$$\begin{array}{rrrrrr} & & 1 & 1 & 1 & 0_2 \\ + & & & 1 & 1 & 1_2 \\ \hline & & & & & 1_2 \end{array}$$

In the 2^1 or 2s place: $1 + 1 = 2 = 10_2$, so write the 0 in the 2s place of the sum row and carry the 1 to the top of the 2^2 place column. (The carried 1 represents 2 groups of 2 which equals 1 group of 4, or 2^2.)

$$\begin{array}{rrrrrr} & & & \mathbf{1} & & \\ & & 1 & 1 & 1 & 0_2 \\ + & & & 1 & 1 & 1_2 \\ \hline & & & \boxed{1} & 0 & 1_2 \end{array}$$

In the 2^2 or 4s place: $1 + 1 + 1 = 3 = 11_2$, so write a 1 in the 2^2 place of the sum row and carry the other 1 to the 2^3 place column. (The carried 1 represents 2 groups of 4 which equals 1 group of 8, or 2^3.)

$$\begin{array}{rrrrrr} & & \mathbf{1} & \mathbf{1} & & \\ & & 1 & 1 & 1 & 0_2 \\ + & & & 1 & 1 & 1_2 \\ \hline & & \boxed{1} & 1 & 0 & 1_2 \end{array}$$

In the 2^3 or 8s place: $1 + 1 = 2 = 10_2$, so write a 0 in the 2^3 place and carry the other 1 to the 2^4 place column. (The carried 1 represents 2 groups of 8 which equals 1 group of 16, or 2^4.)

$$\begin{array}{rrrrrr} & \mathbf{1} & \mathbf{1} & \mathbf{1} & & \\ & & 1 & 1 & 1 & 0_2 \\ + & & & 1 & 1 & 1_2 \\ \hline & \boxed{1} & 0 & 1 & 0 & 1_2 \end{array}$$

In the 2^4 or 16s place: Bring the one down.

$$\begin{array}{rrrrrr} & \mathbf{1} & \mathbf{1} & \mathbf{1} & & \\ & & 1 & 1 & 1 & 0_2 \\ + & & & 1 & 1 & 1_2 \\ \hline & 1 & 0 & 1 & 0 & 1_2 \end{array}$$

The final sum is 10101_2.

Check In base 10, the addends are 14 and 7. So the sum should be 21. Check that $10101_2 = 21_{10}$:

$$1 \cdot 2^4 + 0 \cdot 2^3 + 1 \cdot 2^2 + 0 \cdot 2^1 + 1 = 21.$$

In the early days of computers, programmers had to use numbers written in binary format to code their machines and to enter data. These bases are important because they are powers of 2, and it is easy to convert numbers in them to and from base 2.

You now have the tools to understand the mechanisms by which a computer performs addition. Digital computers use several networks, one of which is a **half-adder**. A half-adder takes two binary digits as input and produces the two digits of their sum as output. (The sum $1_2 + 0_2$ can be thought of as having a two-digit value by writing $1_2 + 0_2 = 01_2$.)

A half-adder network is shown below.

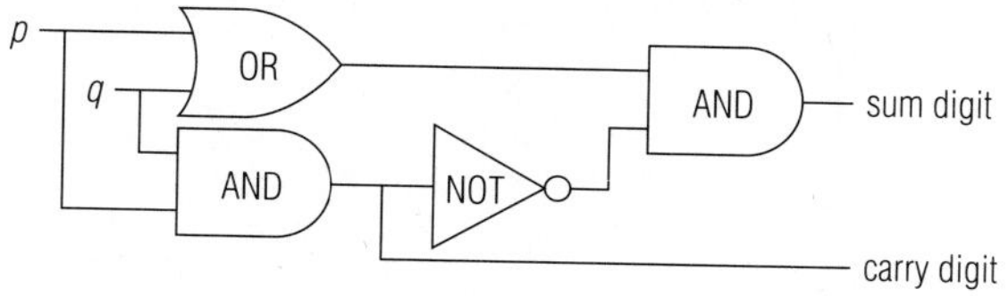

p and q represent the digits to be added; and they are in State 0 or State 1. Trace through the network and determine the output when 1 and 0 are added. (This is the second row of the table.) Then $p = 1$ and $q = 0$. Notice that the network gives two outputs—the sum digit and the carry digit. What is the sum digit? The OR gate outputs a 1 and sends it to the upper AND gate. Along the lower path, the AND gate outputs a 0 which is reversed to a 1 by the NOT gate. Because the input to the upper AND gate is two 1s, its output, the sum digit, is 1. What about the carry digit? The output of the lower AND gate is 0, so 0 is the carry digit. The other rows are completed in a similar way.

p	q	Sum digit	Carry digit
1	1	0	1
1	0	1	0
0	1	1	0
0	0	0	0

Notice that the half-adder network can be viewed as the combination of two networks. The *carry digit* network is the simpler one and corresponds to the logical expression *p and q*. The *sum digit* network is more complicated, but, as you can check for yourself, corresponds to (*p or q*) and (*not*(*p and q*)).

The half-adder can only be used to compute the sum of the ones digits of two binary numbers. But when adding the other digits you must take carrying into account. This requires the use of a network called a *full-adder* that is capable of adding three binary digits. We leave the discussion of this network to a computer science course.

Questions

Covering the Reading

1. Write the number $6 \cdot 10^9 + 2 \cdot 10^5 + 3 \cdot 10^2$ in base 10. 6,000,200,300

2. What is the smallest nonnegative integer congruent to 395 (mod 9)? 8

3. Indicate how casting out nines can show that $2947 \times 6551 \neq 19295797$. See margin.

4. *Multiple choice*. What is the correct way to read 10_2?
   (a) ten
   (b) ten base two
   (c) one zero two
   (d) one zero base two (d)

5. Why is it incorrect to write the base 2 representation of 8 as 200_2? 2 is not a digit in base 2.

ADDITIONAL EXAMPLES

1. Prove: An integer is divisible by $b - 1$ if and only if the sum of its digits in base b is divisible by $b - 1$.
Refer to Example 1 and the paragraph that precedes it. Replace 10 by *b* and 9 by $b - 1$.

2. Give the base 10 representation for the number 11011011_2.
219

3. Write 289 in base 2.
100100001_2

4. Perform the base 2 addition below.

$$\begin{array}{r} 1\ 0\ 1\ 1\ 0\ 1\ 1_2 \\ +\quad 1\ 0\ 1\ 1\ 1\ 0_2 \\ \hline \end{array}$$

10001001_2

ADDITIONAL ANSWERS

3. original problem:

$$\begin{array}{r} 2947 \\ \times\ 6551 \\ \hline 19295797 \end{array}$$

sum of digits: 22, 17, 49

mod 9:

$$\begin{array}{r} 4 \\ \times\ 8 \\ \hline 32 \end{array}$$

The sum of the digits of the product is 49. This should be equivalent to 32(mod 9). But $49 \not\equiv 32 \pmod 9$, so the multiplication is incorrect.

NOTES ON QUESTIONS

Question 9: Ask students to calculate $\log_2 72$ and compare it to the answers to parts **a** and **b.** (For any positive integer x, $\log_2 x$ is one less than the number of digits in the base 2 representation of x.)

Question 15: Ask students to think of what you get by adding one to 11111_2.

Question 16: Ask students if they can tell how an integer written in base 3 is even or odd. (Because of the result in Additional Example 1, the number is even if and only if the sum of its digits is even.)

Question 17: As in Question 9, suggest $\log_2$ as being helpful.

ADDITIONAL ANSWERS

19. A number is divisible by 3 if and only if it is congruent to 0(mod 3). In base 10, the number has the value: $a_n 10^n + a_{n-1}10^{n-1} + \ldots + a_0$. By the addition property of congruence, $a_n \cdot 10^n + a_{n-1} \cdot 10^{n-1} + \ldots + a_0 \equiv a_n + a_{n-1} + \ldots + a_0 \pmod 3$. Either the value of the number and the sum of its digits are congruent to 0(mod 3) or neither is.

20.

-2/5	5	22	53	18
		-2	-8	-18
	5	20	45	0

29. Suppose m is any even integer and n is any odd integer. By definition, there exist integers r and s such that $m = 2r$ and $n = 2s + 1$. Then $m \cdot n = 2r(2s + 1) = 4rs + 2r = 2(2rs + 4)$. Since $(2rs + 4)$ is an integer by closure properties, $m \cdot n$ is even by definition.

In 6–8, find the base 10 representation of the number.

6. 111_2 7 **7.** 110010_2 50 **8.** 1011011_2 91

9. a. How many digits are in the base 2 representation of 72? (Hint: Write integer powers of 2 as in Example 3.) 7

b. Find the base 2 representation of 72. 1001000_2

c. Check your result in part **b** by expanding the base 2 representation with appropriate powers of 2. $1 \cdot 2^6 + 0 \cdot 2^5 + 0 \cdot 2^4 + 1 \cdot 2^3 + 0 \cdot 2^2 + 0 \cdot 2^1 + 0 \cdot 2^0 = 2^6 + 2^3 = 64 + 8 = 72$

10. Find the binary representation of 150. 10010110_2

11. a. How many addition facts do you need to add in base 2? 4

b. How many addition facts do you need to add in base 10? 100

In 12 and 13, perform the indicated additions:

12. $10111_2 + 11010_2$ 110001_2 **13.** $11011_2 + 1011_2$ 100110_2

14. Refer to the half-adder network at the end of the lesson. Let $p = 0$ and $q = 0$. Trace through the network and determine the sum and carry digits. Show that your results agree with the entries in the fourth row of the input-output table. 0 out of the OR gate, so 0 out of the AND gate at the sum digit. 0 out of the first AND gate, so 0 as the carry digit.

Applying the Mathematics

15. *Multiple choice.* Which of the following equals the base 10 representation of 11111_2?
(a) 2^6 (b) $2^6 - 1$ (c) 2^5 (d) 2^{5-1} (e) $2^5 - 1$ (e)

16. How can you tell just by looking at a base 2 representation whether the number is even or odd?
The number is odd if the last digit is 1 and even if the last digit is 0.

17. In 1985, the total federal debt was about 1,827,000,000,000 dollars. This takes 13 digits to write. How many digits are in the base 2 representation? 41

18. a. What digits are allowed in the base 3 representation of an integer?

b. Determine the base 10 representation for 1201_3. 46

c. Find the base 3 representation for 46. 1201_3

a) 0, 1, 2

19. Prove: *An integer is divisible by 3 if and only if the sum of its digits in base 10 is divisible by 3.* (Hint: Use mod 3.) See margin.

Review

20. Use synthetic substitution to show that $5x + 2$ is a factor of the polynomial $p(x) = 5x^3 + 22x^2 + 53x + 18$. *(Lesson 4-6)*
See margin.

21. Given $p(x) = x^4 + 2x^3 - 5x^2 + 1$.

a. Write $p(x)$ in nested form. $x(x(x(x + 2) - 5)) + 1$

b. Use synthetic division to find $p(-2.3)$. *(Lesson 4-5)* -21.7999

22. Use synthetic division to find the quotient $q(x)$ and remainder $r(x)$ when $p(x) = 2x^3 - 3x^2 + 4x - 6$ is divided by $d(x) = x - 1.25$. *(Lesson 4-5)* $q(x) = 2x^2 - .5x + 3.375$; $r(x) = -1.78125$

23. Use long division to find the quotient and remainder when $p(x) = (x^2 + x + 1)^2$ is divided by $3x^2 - x$. *(Lesson 4-4)*
$q(x) = \frac{1}{3}x^2 + \frac{7}{9}x + \frac{34}{27}$; $r(x) = \frac{88}{27}x + 1$

266

24. Find the smallest positive integer that makes the congruence true.
a. $x \equiv -12 \pmod 5$ **3** **b.** $y \equiv 784 \pmod{11}$ *(Lesson 4-3)* **3**

25. Use your calculator to find the integer quotient and integer remainder when 64,942 is divided by 98. *(Lesson 4-2)*
quotient: 662; remainder: 66

26. The equation $d = .044v^2 + 1.1v$ gives the approximate distance d in feet that it takes to stop a car traveling at v miles per hour. At what speed(s) is the stopping distance 50 feet? *(Lesson 3-5)* **≈ 23.5 mph**

27. Find the phase shift, period, and amplitude for the function g defined by $g(x) = 7\cos(8x - \pi)$. *(Lessons 2-8, 3-9)*
phase shift: $\frac{\pi}{8}$; period: $\frac{\pi}{4}$; amplitude: 7

28. Which letter on the unit circle at the right could stand for the following? *(Lesson 2-8)*

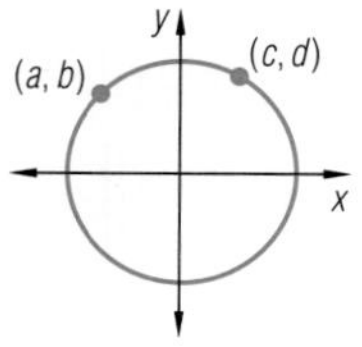

a. $\cos\left(\frac{3\pi}{4}\right)$ **a** **b.** $\sin\left(\frac{\pi}{3}\right)$ **d**

c. $\cos 60°$ **c** **d.** $\sin\left(\frac{3\pi}{4}\right)$ **b**

29. Prove that the product of an even integer and an odd integer is an even integer. *(Lesson 1-7)* **See margin.**

30. Determine whether the statement is logically equivalent to:
If figure ABC is a right triangle, then it has exactly two acute angles.
a. *If ABC has exactly two acute angles, then it is a right triangle.*
b. *If ABC does not have exactly two acute angles, then it is not a right triangle.* **Yes**
c. *If ABC is not a right triangle, then it does not have exactly two acute angles.* *(Lesson 1-5)* **No; a) No**

31. *True* or *false*? If false, give a counterexample.
$\forall$ *real numbers* x, $|x| = x$. *(Lesson 1-1)*
False, let x = -1. Then $|x| = |-1| = 1 \neq -1$.

Exploration

32. Use a computer book to find out how hexadecimal numbers are used to represent ASCII codes used by computers. Write the ASCII code for
a. 253 **32 35 33** **b.** 7655 **37 36 35 35** **c.** the letter A. **41**

FOLLOW-UP

MORE PRACTICE
For more questions on SPUR Objectives, use *Lesson Master 4-7,* shown below.

EXTENSION
Ask students to write addition and multiplication tables for base 2:

+	0	1
0	0	1
1	1	10

•	0	1
0	0	0
1	0	1

Compare these tables to the base 10 tables that students learned in elementary school.

NAME

LESSON **MASTER 4–7**
QUESTIONS ON **SPUR** OBJECTIVES

■ **REPRESENTATIONS** *Objective M (See pages 283–286 for objectives.)*
In 1 and 2, find the base-2 representation of the number.
1. 40 101000_2 2. 129 10000001_2

In 3 and 4, find the base-10 representation of the number.
3. 110110_2 54 4. 1000001_2 65

In 5 and 6, perform the following steps:
a. Find the binary representation of r.
b. Find the binary representation of t.
c. Use your answers to parts a and b to find the base-2 representation of $r + t$.
d. Convert your answer from part c into base 10.

5. $r = 23$, $t = 24$
a. 10111_2
b. 11000_2
c. 101111_2
d. 47

6. $r = 62$, $t = 7$
a. 111110_2
b. 111_2
c. 1000101_2
d. 69

7. Express 94 in base n where
a. $n = 2$. 1011110_2 b. $n = 8$. 136_8 c. $n = 4$. 1132_4

 47

LESSON 4-8

RESOURCES

- Lesson Master 4-8
- Computer Master 9

OBJECTIVES

E Factor polynomials.
I Use proof by contradiction.
J Use the Factor Search Theorem and the Fundamental Theorem of Arithmetic in determining prime numbers and prime factorizations.

TEACHING NOTES

In the Infinitude of Primes Theorem, the integer $n + 1$ obtained in the proof (1 more than the product of the first m primes) is pointed out as being either prime or having a prime factor larger than any prime on the list. Students may think that the latter posibility does not occur and that $n + 1$ is prime for all m. For instance, $2 \cdot 3 + 1$, $2 \cdot 3 \cdot 5 + 1$, $2 \cdot 3 \cdot 5 \cdot 7 + 1$, and $2 \cdot 3 \cdot 5 \cdot 7 \cdot 11 + 1$ are all prime. However, $2 \cdot 3 \cdot 5 \cdot 7 \cdot 11 \cdot 13 + 1 = 30031 = 59 \cdot 509$. Let n_p be 1 more than the product of primes less than or equal to p. Then, among the primes less than 11,213, it is known that n_p is prime only for $p = 2, 3, 5, 7, 11$ (the cases mentioned above), and 31, 379, 1019, 1021, and 2657.

On most calculators, the number of digits in $2^{216091} - 1$ cannot be found by taking its common logarithm. However, subtracting 1 does not change the number of digits in this number (it is not an integer

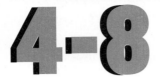

LESSON

4-8 Prime Numbers and Prime Polynomials

Recall the definition of *prime number*.

Definition

An integer $n > 1$ is **prime** if and only if 1 and n are the only positive integer factors of n.

For example, 53 is prime because its only positive integer factors are 1 and 53. But 54 is not prime because 6 is a factor of 54. If n is not prime, then either $n \leq 1$ or $n = d \cdot q$, where both d and q are positive integers greater than 1. With $n = 54$, d and q could be 6 and 9 or 3 and 18 or 2 and 27. Notice that the factors d and q are smaller than n and they themselves either are prime or they can be factored.

Any decreasing sequence of positive integers must be finite. If you continue the process of factoring a number, one of the factors must eventually be prime because the factors in each new pair are smaller than their product and all are positive. This proves:

Prime Factor Theorem

Every integer greater than 1 has a prime factor.

Prime numbers have long been important in pure mathematics. Until recently, it was thought that prime numbers had few practical applications. However, as we pointed out in the opening of this chapter, with the advent of electronic computing, prime numbers have been used to develop codes that are very difficult to break.

As computers have become faster and mathematicians have developed better algorithms for factoring large numbers, larger and larger primes are needed. A prime number larger than any given known prime always exists because it has long been established that there are infinitely many primes. The earliest known proof of the infinitude of primes is in Book 9 of the 13 books of Euclid's *Elements,* which was written about 250 B.C. The basic idea of Euclid's proof is to assume that there are only finitely many primes, which is equivalent to assuming that there is a largest prime. Then, from this assumption, we deduce a contradiction. The general form of the proof is called a *proof by contradiction.*

In general, **proof by contradiction** works in the following way: Let s be the statement you want to prove. You reason from *not s* until you deduce a statement of the form *p and* (*not p*), which is a contradiction. From this you conclude that *not s* is false, which means that s is true. The Latin name for this form of argument is **reductio ad absurdum**, literally meaning "reducing to an absurdity." The following theorem guarantees that Proof by Contradiction is a valid form of argument.

Theorem (Validity of Proof by Contradiction)

The following form of argument is valid.

If not s then (p and (not p))

∴ *s*.

Proof To prove the validity of this argument, we must prove that the conditional $(\sim s \Rightarrow (p \text{ and } \sim p)) \Rightarrow s$ is always true. This is shown by the truth table below.

s	p	~s	~p	p and ~p	~s ⇒ (p and ~p)	(~s ⇒ (p and ~p)) ⇒ s
T	T	F	F	F	T	T
T	F	F	T	F	T	T
F	T	T	F	F	F	T
F	F	T	T	F	F	T

The proof given below is a version of the one presented by Euclid, but it has been put into today's mathematical language and notation.

Infinitude of Primes Theorem

There are infinitely many prime numbers.

Proof Let

s: There are infinitely many prime numbers.

Assume

not s: There are finitely many prime numbers.

Because there is a finite number, say m, of primes, they can be listed from smallest to largest:

$$p_1 = 2, p_2 = 3, p_3 = 5, \ldots, p_m.$$

Multiply these primes to get a new number

$$n = p_1 p_2 p_3 \bullet \ldots \bullet p_m.$$

Now consider the next consecutive integer $n + 1$. Since $n + 1$ is larger than the assumed largest prime p_m, it is not prime. By the Prime Factor Theorem, $n + 1$ must have a prime factor, say p. Then p, being prime, must be on the list of all primes given at the start of the proof. Now n is the product of all the primes on the list, and so p is a factor of n. Also, p is a factor of $n + 1$. Hence p is a factor of the difference $(n + 1) - n$, which equals 1. Because the only positive integer factor of 1 is 1, p must equal 1. But p is a prime number and all prime numbers are greater than 1. Thus we have arrived at a contradiction: $p = 1$ and $p > 1$. Hence the assumption *not s* is false. So s is true: there are infinitely many prime numbers.

power of 10), and $\log 2^{216091} = 216091(\log 2) \approx 65049.87\ldots$, from which the number of digits in the base 10 representation is 65050.

You may have to elaborate on the implications of the Prime Factor Theorem for testing for primes. It may help to go back to the search for prime factors of 391, this time testing prime divisors only.

The efficiency of the Factor Search Theorem can be illustrated by factoring a number such as $21021 = 3 \cdot 7^2 \cdot 11 \cdot 13$. By the Factor Search Theorem, you first have to check for prime factors up to $\sqrt{21021} \approx 144$. After finding that 3 is a factor, however, the problem is reduced to factoring 7007. Now it is necessary only to check factors up to $\sqrt{7007} \approx 83$. Continuing in this way, only a small number of primes need to be checked.

Although we define polynomials as prime only over the reals, you could replace *real* by *rational* or, for that matter, by any field.

The only polynomials that are prime over the complex numbers are polynomials of degree 1, because any quadratic can be factored into linear factors if complex coefficients are allowed. For instance, $x^2 + 1 = (x - i)(x + i)$

Because there is no natural order relation on the set of polynomials, there is no standard prime factorization; the factors may be given in any order.

Technology Invite students familiar with programming to write programs that will test if a number n is prime by testing up to and including $n - 1$, $\frac{n}{2}$, and $\sqrt{n}$. Enter a large test integer for each and compare computing times.

ADDITIONAL EXAMPLES
The proofs of the Infinitude of Primes Theorem and the Factor Search Theorem are indirect. Some additional statements you might have students prove using indirect proofs are:

1. There is no largest even integer.
s: There is no largest even integer. Assume *not* s: There is a largest even integer. Suppose this largest integer is $2m$. The even number smaller than $2m$ is $2m - 2$. $(2m) + (2m - 2)$ is an integer by the closure property. Furthermore, $(2m) + (2m - 2) = 2(2m - 1)$ is an even integer that is larger than $2m$. Hence, *not* s is false. Therefore, s is true.

2. If a function f is increasing on a set S and if $f(x_1) = f(x_2)$, then $x_1 = x_2$.
s: $x_1 = x_2$: *not* s: $x_1 \neq x_2$; p: $f(x)$ is increasing; q: $f(x_1) = f(x_2)$. Assume $x_1 \neq x_2$. Then $x_1 < x_2$ or $x_1 > x_2$. Since $f(x)$ is increasing on S, if $x_1 < x_2$, then $f(x_1) < f(x_2)$, and if $x_1 > x_2$, then $f(x_1) > f(x_2)$. Hence, $f(x_1) \neq f(x_2)$. Therefore, if $f(x_1) = f(x_2)$, then $x_1 = x_2$.

The largest prime known (as of 1990) is $2^{216091} - 1$, discovered by David Slowinski. Because $\log(2^{216091} - 1) = 65049.87\ldots$, this number is larger than 10^{65049} and has 65,050 digits in base 10. The universe is not old enough for all integers between 1 and $2^{216091} - 1$ to have been checked for factors, even if a trillion trillion trillion numbers could be checked each second. Obviously Slowinski knew some efficient testing techniques.

A first step toward efficiency is found by using the Prime Factor Theorem. That theorem tells us that, if an integer $n > 1$ is not prime, then it has a prime factor. So, for instance, to test whether 391 is prime, we need test only whether 391 is divisible by a prime number. However, it is not necessary to test *all* the prime numbers less than 391. Why? Suppose that d is a prime factor of 391. Then $391 = d \cdot q$ and $q > 1$. If both d and q are greater than $\sqrt{391}$, then $d \cdot q > 391$. But that's impossible. Thus, if 391 is not prime, at least one of its factors must be less than or equal to $\sqrt{391}$. In general, to test whether n is prime, you need only search the prime numbers less than or equal to $\sqrt{n}$. All this is summarized in the following theorem.

Factor Search Theorem

If an integer n has no prime factors between 1 and $\sqrt{n}$ inclusive, then n is prime.

Thus, to test whether 391 is prime, it is necessary only to test its divisibility by primes between 1 and $\sqrt{391} \approx 19.77$, that is, the numbers 2, 3, 5, 7, 11, 13, 17, and 19.

Prime numbers are often thought of as the *building blocks* of the positive integers because it can be proved that every integer $n > 1$ is either prime or can be written as a product of prime factors. For example, 29 and 41 are primes while $26 = 2 \cdot 13$ and $36 = 2 \cdot 2 \cdot 3 \cdot 3 = 2^2 \cdot 3^2$ are products of prime factors.

If $n > 1$ is not a prime number, then a representation of n as a product of primes is called a **prime factorization** of n. For example,

$$2 \cdot 2 \cdot 3 \cdot 7, \quad 7 \cdot 2 \cdot 3 \cdot 2, \quad 2^2 \cdot 3 \cdot 7$$

are all prime factorizations of 84. The following theorem states that the only way two prime factorizations of an integer can differ from each other is in the order in which the factors are written. This result is so important that it is called the *Fundamental Theorem of Arithmetic*. The proof of the complete result is beyond the level of this course, but parts of the theorem are proved in Chapter 7.

Fundamental Theorem of Arithmetic

Suppose that n is an integer and that $n > 1$. Then either n is a prime number or n has a prime factorization which is unique except for the order of the factors.

Thus, according to this theorem, if you first notice that 11 is a factor of 1001, and someone else first notices that 7 is a factor, the ultimate prime factorizations you both get will be the same.

Among the different prime factorizations of an integer $n > 1$ that is not prime, there is one and only one in which:

(a) all like factors are combined using exponents;

(b) the prime factors are arranged in increasing order of magnitude.

This factorization is called the **standard prime factorization** of n. For example, the standard prime factorization of 60 is $60 = 2^2 \cdot 3 \cdot 5$ and that of 231 is $231 = 3 \cdot 7 \cdot 11$.

Now consider polynomials. Like integers, polynomials can be factored. For instance,

$$\begin{aligned} 4x^4 - 56x^2 + 180 &= 4(x^4 - 14x^2 + 45) \\ &= 4(x^2 - 9)(x^2 - 5) \\ &= 4(x + 3)(x - 3)(x^2 - 5). \end{aligned}$$

A polynomial is not considered to be *completely* factored unless all its factors are prime. How can you determine if a polynomial is prime? After all, $x^2 - 5$ could be factored into $(x + \sqrt{5})(x - \sqrt{5})$.

Definition

A polynomial $p(x)$ with degree $n \geq 1$ is **prime over the real numbers** if and only if the only polynomial factors of $p(x)$ with real coefficients and leading coefficient 1 are constants or constant multiples of $p(x)$.

So, factored into prime factors over the real numbers,

$$4x^4 - 56x^2 + 180 = 4(x + 3)(x - 3)(x + \sqrt{5})(x - \sqrt{5}).$$

You may recall from earlier work that, if a, b, and c are real numbers, then $ax^2 + bx + c$ is factorable if and only if $b^2 - 4ac \geq 0$. Thus, over the real numbers, $x^2 + 1$ is a prime polynomial, as is $3x^2 - 9x + 12$.

A natural question to ask is whether there is a theorem for polynomials like the Fundamental Theorem of Arithmetic for integers. That is, is the prime factorization of a polynomial over the real numbers unique except for order? For example, consider the polynomial $p(x) = x^8 - 1$. Because $p(1) = 0$, $x - 1$ is a factor. The quotient is $q_1(x) = x^7 + x^6 + x^5 + x^4 + x^3 + x^2 + x + 1$ and there is no remainder.

$$x^8 - 1 = (x - 1)(x^7 + x^6 + x^5 + x^4 + x^3 + x^2 + x + 1) = (x - 1)\, q_1(x)$$

But $p(-1) = 0$ also, so $x + 1$ is also a factor.

$$x^8 - 1 = (x + 1)(x^7 - x^6 + x^5 - x^4 + x^3 - x^2 + x - 1) = (x + 1)\, q_2(x)$$

If there is unique factorization into primes, then $q_1(x)$ and $q_2(x)$ must each be factorable. This is, in fact, what happens.

$$\begin{aligned} x^8 - 1 &= (x - 1)q_1(x) \\ &= (x - 1)(x + 1)(x^6 + x^4 + x^2 + 1) \\ &= (x - 1)(x + 1)(x^2 + 1)(x^4 + 1) \end{aligned}$$

3. If p is prime and n^2 is divisible by p, then n is divisible by p.
Sample: $p = 13$, $n^2 = 8281$ (divisible by 13); $n = 91$ is divisible by 13. s: n is divisible by p. not s: n is not divisible by p.
r: p is a prime.
q: n^2 is divisible by p.
Assume n is not divisible by p. Then n^2 is not divisible by p since $n^2 = n \cdot n$. Hence, n is divisible by p by the Law of Indirect Reasoning: If (*not* s) then (*not* q), q, therefore s.
(You may need to do **3.** above for specific values of p first. This result will be used in Lesson 5-7 in proofs that certain numbers are irrational.)

NOTES ON QUESTIONS

Question 16: Have students think of x^2 as a chunk.

Question 17: Some students may notice first that 6 is a factor; others may notice first that the polynomial is quadratic in x^4 (that is, with x^4 as a chunk).

Question 24: There are other ways to show that the numbers are unequal. For instance, begin the multiplication. Since 9×89 does not end in 11, the multiplication must be incorrect. As a curiosity, you might ask students to multiply 12345679 by 9. The result may surprise them.

ADDITIONAL ANSWERS

2.c. $n + 1$ is an integer by closure properties since both n and 1 are integers. This contradicts part b, thus the assumption is false and it is true that there is no largest integer.

13. Every even number is divisible by 2, thus it always has a factor of 2 and cannot be prime.

14. Assume that there is a largest multiple of 5, N. By the closure of the integers, there exists a number $N + 5$. Now 5 divides N and 5 divides 5, so by the Factor of a Sum Theorem, 5 divides $N + 5$. Hence, $N + 5 > N$ and $N + 5$ is a multiple of 5. This contradicts the initial assumption that N is the largest multiple of 5. The assumption is false, so there is no largest multiple of 5.

15. Assume that there is a smallest positive real number, S. Consider $\frac{1}{2}S$. Since the real numbers are closed under multiplication, $\frac{1}{2}S$ is a real number. $\frac{1}{2}S < S$ and positive, so the assumption is invalid, and therefore there is no smallest positive real number.

19. The Fundamental Theorem of Arithmetic states that if a number is not prime it has a unique prime factorization. 1,000,000,000 has a prime factorization of $2^9 \cdot 5^9$; therefore, 11 is not a prime factor.

22. It is sufficient to prove the contrapositive: $\forall$ *integers, n, if n is even, then n^2 is even.* Suppose n is any even integer. Thus, there exists an integer r such that $n = 2r$ by the definition of even. Then $n = (2r)^2 = 4r^2 = 2(2r^2) \cdot 2r^2$ is an integer by closure properties, so n is even. Since the contrapositive is true, the original statement is true, thus if n^2 is odd, then n is odd.

Factoring $q_2(x)$ yields the same factorization of $x^8 - 1$.

$$\begin{aligned} x^8 - 1 &= (x + 1)\ q_2(x) \\ &= (x + 1)(x - 1)(x^6 + x^4 + x^2 + 1) \\ &= (x + 1)(x - 1)(x^2 + 1)(x^4 + 1) \text{ as above.} \end{aligned}$$

However, neither is a prime factorization over the reals, because $x^4 + 1 = (x^2 + \sqrt{2}x + 1)(x^2 - \sqrt{2}x + 1)$. A prime factorization of $x^8 - 1$ over the reals is thus

$$(x + 1)(x - 1)(x^2 + 1)(x^2 + \sqrt{2}x + 1)(x^2 - \sqrt{2}x + 1)$$

and it seems to be a unique factorization. In general, there is unique factorization of a polynomial into prime polynomials over the reals as long as constant multiples are ignored.

Unique Factorization Theorem for Polynomials

Suppose that $p(x)$ is a polynomial with integer coefficients. Then either $p(x)$ is prime over the real numbers or $p(x)$ has a factorization into polynomials prime over the reals which is unique except for the order of the factors or multiplications by real constants.

In the next chapter, you will see that the factorization of polynomials helps in understanding the behavior of the functions known as rational functions.

Questions

Covering the Reading

1. You know $1 \neq 2$. Suppose you can prove the following.
   If there are unicorns, then 1 = 2.
   What conclusion can you draw? **There are no unicorns.**

2. Consider the statement: *There is no largest positive integer.*
   a. To write a proof by contradiction, with what assumption should you start the proof? **There exists a largest positive integer.**
   b. If you let n be the largest positive integer, what can you say about $n + 1$? **$n + 1$ cannot be an integer.**
   c. Use your results from parts **a** and **b** to complete the proof by contradiction. **See margin.**

3. Write the first 15 prime numbers.
   2, 3, 5, 7, 11, 13, 17, 19, 23, 29, 31, 37, 41, 43, 47

4. Refer to the proof of the Infinitude of Primes Theorem.
   a. What number n was constructed from the primes?
   b. What was the contradiction derived from the assumption?
   a) the product of all the primes; b) $p = 1$ and $p > 1$

5. *True* or *false?* If n is an integer and p is a prime factor of $n + 1$, then p is not a factor of n. **True**

6. a. To determine if 783 is prime, what is the largest number that must be tested to see if it is a factor of 783? $\sqrt{783}$
b. Is 783 prime? No

7. Give the standard prime factorization of 480. $2^5 \cdot 3 \cdot 5$

In 8–10, give a prime factorization of the polynomial.

8. $3x^2 - 33x - 72$ $3(x^2 - 11x - 24)$
9. $9y^{100} - y^{98}$ $y^{98}(3y + 1)(3y - 1)$
10. $z^2 - 17$ $(z - \sqrt{17})(z + \sqrt{17})$

11. John noticed that $x^2 + 1$ is a factor of $p(x) = x^4 + x^3 + x^2 + x$. Joan noticed that -1 is a zero of p. Jan noticed only that x is a factor. Put all this information together to factor $p(x)$. $p(x) = x(x + 1)(x^2 + 1)$

Applying the Mathematics

12. *True* or *false?* The only common positive integer factor of two consecutive integers is 1. True

13. Every prime number greater than 2 is an odd number. Explain why. See margin.

In 14 and 15, give a proof by contradiction.

14. There is no largest multiple of 5. See margin.

15. There is no smallest positive real number. See margin.

In 16–17, give a prime factorization over the reals of the polynomial.

16. $x^4 - 13x^2 + 36$ $(x + 3)(x - 3)(x + 2)(x - 2)$
17. $6 - 12y^4 + 6y^8$ $6(y^2 + 1)^2(y - 1)^2(y + 1)^2$

18. One factor of $x^3 - 39x + 70$ is $x - 2$. Use division to determine the other factor. Then express $x^3 - 39x + 70$ as the product of three prime factors over the reals. $(x - 2)(x - 5)(x + 7)$

19. Use the Fundamental Theorem of Arithmetic to argue that 11 is not a factor of 1,000,000,000. See margin.

Review

20. Write the base 10 representation of 101100_2. *(Lesson 4-7)* 44

21. *Multiple choice.* The base 2 representation of 257 is
(a) 1111111_2 (b) 11111111_2
(c) 100000001_2 (d) 10000001_2 *(Lesson 4-7)* (c)

22. Prove: $\forall$ *integers n, if n^2 is odd, then n is odd.*
(Hint: Consider the contrapositive.) *(Lessons 1-5, 1-7)* See margin.

23. Provide a proof of the following statement used in the proof of the Infinitude of Primes Theorem:
If p is a factor of a and p is a factor of b, then p is a factor of a − b. *(Lesson 4-1)* See margin.

24. a. Use modular arithmetic to show that $123456789 \times 9 \neq 1111111111$. See margin.
b. What is the correct product? *(Lesson 4-3)* 1111111101

23., 24. See Additional Answers in the back of this book.

NAME ____________

LESSON MASTER 4–8
QUESTIONS ON SPUR OBJECTIVES

■ **SKILLS** *Objective E (See pages 283–286 for objectives.)*
In 1–6, factor into prime factors over the real numbers.

1. $6y^2 + 17y + 5$ $(2y + 5)(3y + 1)$
2. $25n^4 - 30n^2 + 9$ $(5n^2 - 3)^2$
3. $x^4 - 2x^2 - 8$ $(x - 2)(x + 2)(x^2 + 2)$
4. $7x^2 - 49x + 42$ $7(x - 6)(x - 1)$
5. $p(x) = 2x^3 + 7x^2 - 14x + 5$, given that $x - 1$ is a factor of $p(x)$ $(2x - 1)(x - 1)(x + 5)$
6. $(x^2 + xy)(3x^2 + y^2) - (x^2 + xy)(2x^2 + 2y^2)$ $x(x - y)(x + y)^2$

■ **PROPERTIES** *Objective I*
7. *True* or *false?* In a proof by contradiction, the assumption which begins the proof is proved to be true. false

48 *Continued*

NAME ____________
Lesson MASTER 4–8 (page 2)

8. Consider the statement *There is no largest real number less than one.*
a. To write a proof by contradiction, with what assumption would you start?
Suppose ∃ a largest real number less than 1.
b. Complete the proof.
Let that largest real number be r. Let s be the real number halfway between r and 1, that is, $s = \frac{r+1}{2}$. Then $r < s < 1$, so that s is a real number less than 1. But s is larger than r, contradicting the statement that r is the largest such number. Thus, the assumption must be false, so there is no largest real number less than 1.
c. What does this tell you about the number $0.\overline{9}$? Justify your answer.
$0.\overline{9} = 1$, because if $0.\overline{9} < 1$, then $0.\overline{9}$ would be the largest real number less than 1.

■ **PROPERTIES** *Objective J*
9. a. List the numbers that must be tested as factors to determine whether 907 is prime.
2, 3, 5, 7, 11, 13, 17, 19, 23, 29
b. Is 907 prime? yes

10. a. *Multiple choice.* Which prime factorization is not equivalent to the others?
(a) $2 \cdot 2 \cdot 2 \cdot 3 \cdot 11 \cdot 19$ (b) $2^3 \cdot 2 \cdot 3 \cdot 19 \cdot 11$
(c) $19 \cdot 2^3 \cdot 3 \cdot 11$ (d) $2^3 \cdot 3 \cdot 11 \cdot 19$ (b)
b. Which of the above represents a standard prime factorization? (d)

In 11 and 12, give the standard prime factorization of the number.
11. 360 $2^3 \cdot 3^2 \cdot 5$ 12. 1938 $2 \cdot 3 \cdot 17 \cdot 19$

13. *True* or *false?* In the context of the Fundamental Theorem of Arithmetic, $3 \cdot 5 \cdot 7$ and $7 \cdot 5 \cdot 3$ would be considered distinct factorizations. false

 49

FOLLOW-UP

MORE PRACTICE
For more questions on SPUR Objectives, use *Lesson Master 4-8,* shown on page 273.

EXTENSION
Because of the Factor Theorem, if a quadratic with real coefficients had real zeros, then it can be factored over the reals. For instance, the zeros of $3x^2 - 2x - 4 = 0$ are $\frac{1 \pm \sqrt{13}}{3}$, which implies that $3x^2 - 2x - 4 = 3\left(x - \frac{1 - \sqrt{13}}{3}\right)\left(x - \frac{1 + \sqrt{13}}{3}\right)$. Ask students to make up other quadratics with irrational zeros and to give their prime factorizations over the reals.

PROJECTS
The projects for Chapter 4 are described on page 279. **Projects 3 and 4** are related to the content of this lesson.

EVALUATION
Alternative Assessment
Have students investigate $2n + 1$ when n is a prime. Ask them to conjecture as to whether or not $2n + 1$ is prime. Can they find a counterexample?

ADDITIONAL ANSWERS
29.c.

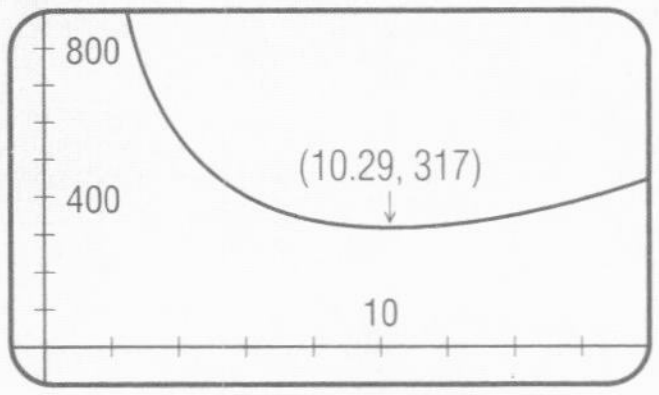

$s \approx 10.29$ inches;
$h \approx 5.14$ inches

31. See Additional Answers in the back of this book.

25. Residential electricity is called AC for "alternating current" because the direction of current flow alternates through a circuit. Below is a graph of current (in amperes) as a function of time (in seconds).

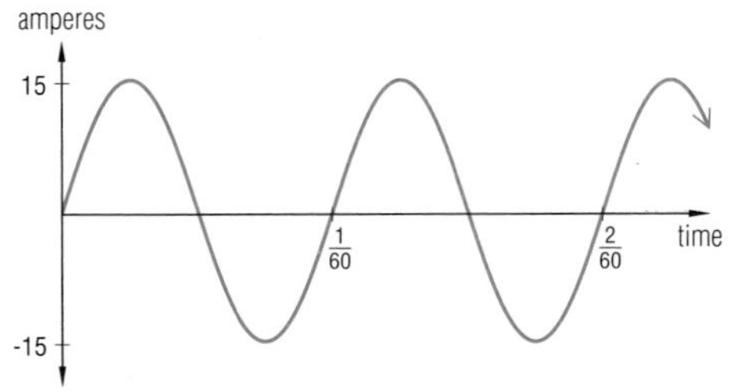

a. What type of function has a graph like this one? sine function
b. Identify the period and amplitude of this graph.
c. Write an equation for current as a function of time.
d. Find the current produced at 0.10 seconds. *(Lessons 2-8, 3-9)*

b) period: $\frac{1}{60}$ seconds; amplitude: 15; c) $c(t) = 15\sin(120\pi t)$; d) 0 amperes

26. Is the sine function even, odd, or neither? *(Lesson 2-8)* odd

27. Solve: $x^2 - 3x = (x - 3)(x^2 - 2)$. *(Lesson 3-1)* $x = -1$, $x = 2$, and $x = 3$

28. Solve: $\sqrt{3x - 8} = \frac{1}{2}x$. *(Lesson 3-1)* $x = 8$ and $x = 4$

29. A manufacturer needs to construct a box with a square base, no top, and a volume of 544 in.^{3} and wants to use as little material as possible in constructing the box.
a. Express the height of the box as a function of the length of the side of the base. $h = \frac{544}{s^2}$
b. Express the surface area of the box as a function of the length of the base. $A(s) = s^2 + \frac{2176}{s}$
c. Use an automatic grapher to find the dimensions of the box that would require the least amount of material for construction. *(Lesson 2-2)* See margin.

Exploration

30. Pick a rate (primes per second) at which you think a computer can test a number to determine whether it is prime. With the rate you pick, how many years would it have taken to test all primes less than the square root of the largest prime number known? How fast a rate would be needed to test the number $10^{100,000} + 1$ if the computer was going all day and night for a century? Answers will vary.

31. One of the most famous cases of code breaking involves the *Enigma*, used by the Germans during World War II. Use the library to research this device, the ULTRA project that broke the code, the English mathematician Alan Turing, and how the work done by Turing and others on the project set the stage for the post-war development of digital computers. See margin.

LESSON

4-9

The Variety of Polynomials

One reason for the importance of polynomial functions is that linear functions, quadratic functions, the power functions of the form $f\colon x \to x^n$, and the sum of the first n terms of the finite geometric sequence defined by $g_n = gr^{n-1}$ are all polynomials. Yet you have not seen many applications of other polynomials, and you may wonder why other polynomials of degree greater than 2 are important.

A major use of higher degree polynomial functions is to approximate other continuous functions. At the left below are the graphs of several functions. At the right are graphs of polynomial functions that approximate them.

$f_1(x) = e^x$

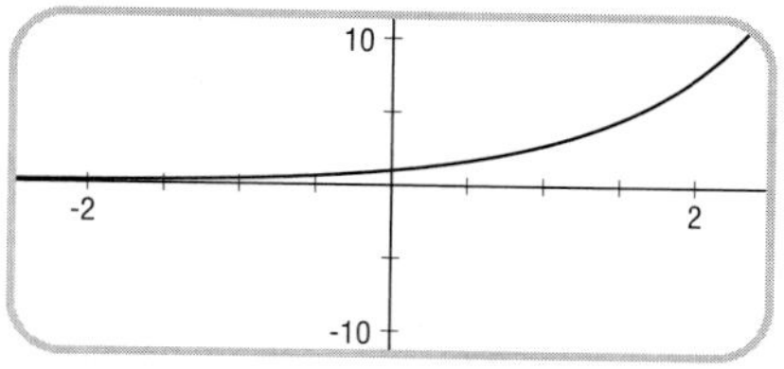

$p_1(x) = \frac{1}{24}x^4 + \frac{1}{6}x^3 + \frac{1}{2}x^2 + x + 1$

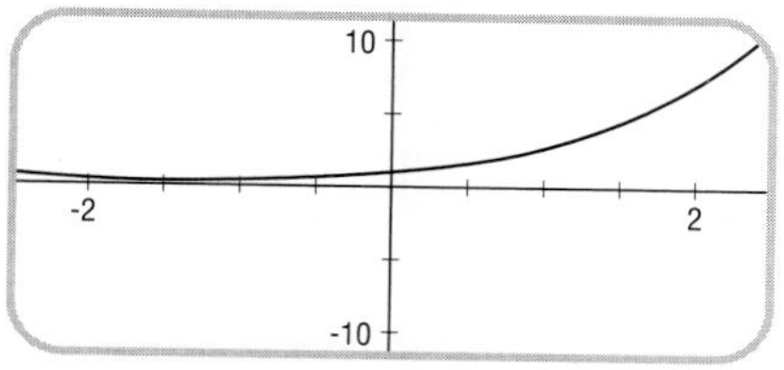

$f_2(x) = \cos x$

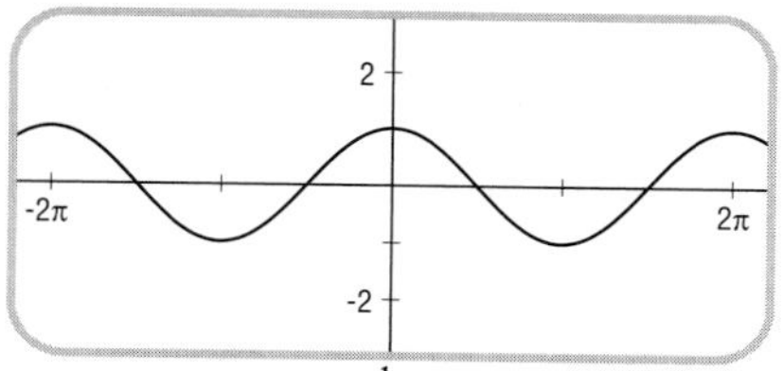

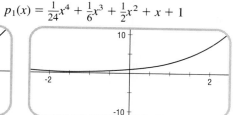

$p_2(x) = \frac{1}{40320}x^8 - \frac{1}{720}x^6 + \frac{1}{24}x^4 - \frac{1}{2}x^2 + 1$

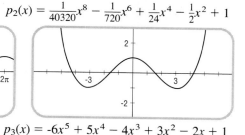

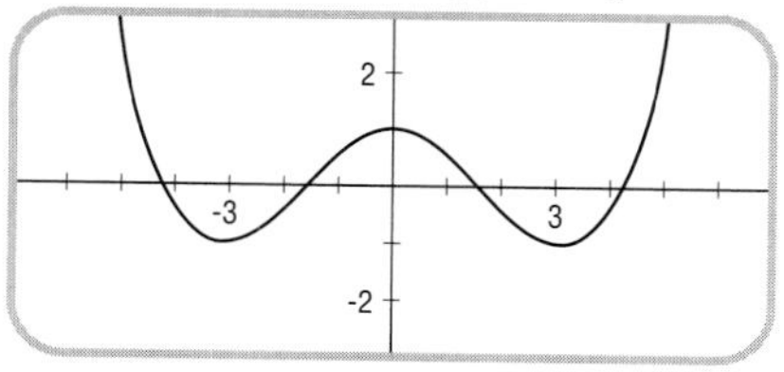

$f_3(x) = \frac{1}{(1 + x)^2}$

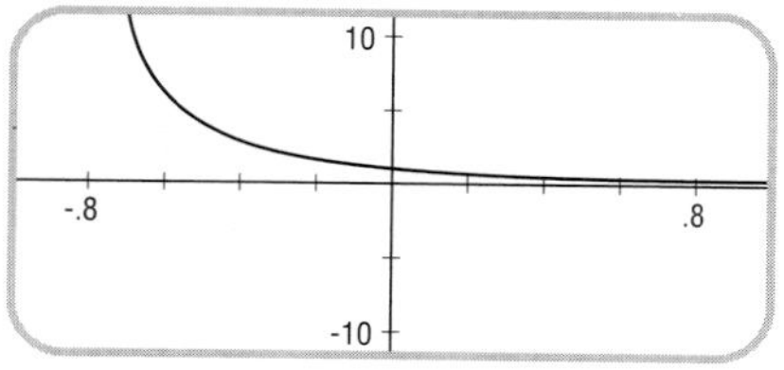

$p_3(x) = -6x^5 + 5x^4 - 4x^3 + 3x^2 - 2x + 1$

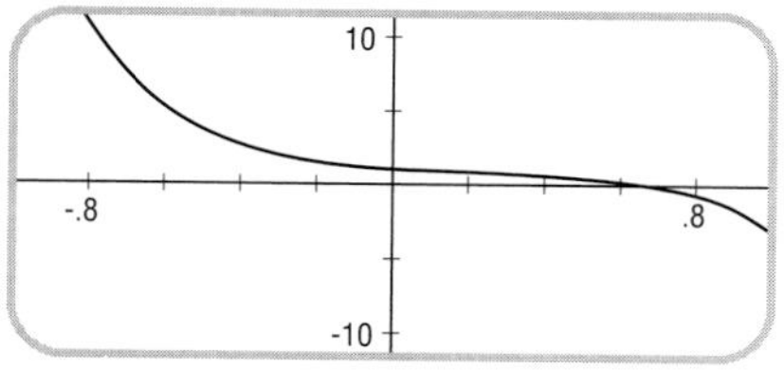

$f_4(x) = \ln(x + 1)$

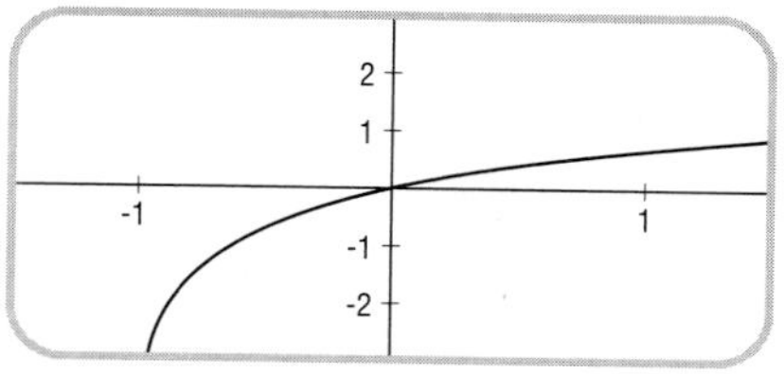

$p_4(x) = \frac{1}{5}x^5 - \frac{1}{4}x^4 + \frac{1}{3}x^3 - \frac{1}{2}x^2 + x$

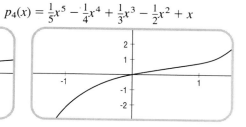

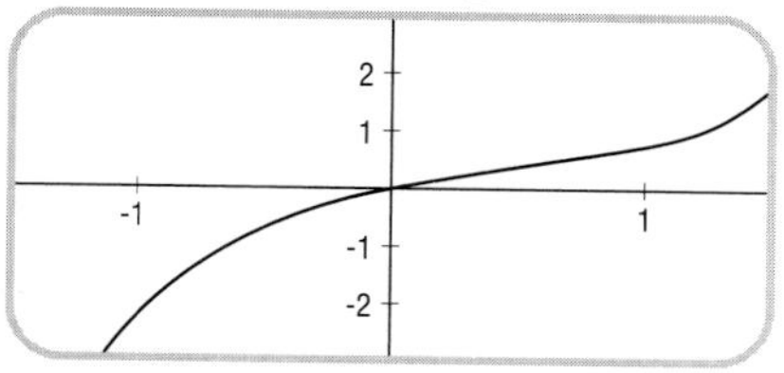

LESSON 4-9

RESOURCES

■ Teaching Aid 28 shows a polynomial approximation to $f_1(x) = e^x$.

TEACHING NOTES

You might want to widen the domain and regraph at least one of the four pairs of functions on this page to show that the polynomial approximation is good only on a limited subset of the domain of each function. Dividing the domain into relatively small intervals and determining an approximating polynomial for each interval is one way around this limitation. The topic of *splining* is appropriate in this regard. It involves fitting cubic polynomials to a function sliced into intervals—in effect, creating a piecewise defined polynomial approximation function.

Graphing both the function and its polynomial approximation on one screen can be very effective, especially with a color monitor and different colors for the two graphs.

Making Connections
Give students six or seven points that are approximately ina parabolic pattern to plot. Have them sketch a parabola through the points that they think best fit the data. Now pick three points on this parabola and estimate the coordinates. The equation of the parabola through these three points can now be found. Students can calculate the distance from the points plotted to the parabola and find the sum. They can compare the sums with each other and find the parabola with the smallest sum.

NOTES ON QUESTIONS
Technology for Questions 1-3: A spreadsheet or an automatic grapher will be very helpful for these questions. Software such as DERIVE allows you to have a graphing window and a table window on the screen simultaneously. The TI-81 graphing calculator, in TRACE mode, allows you to jump vertically between two graphs, displaying both coordinates on the screen simultaneously. It also allows you to calculate the value of the difference of functions for stored values of *x*.

ADDITIONAL ANSWERS
2.a. $p_4(x) = x - \frac{1}{2}x^2 + \frac{1}{3}x^3 - \frac{1}{4}x^4 + \frac{1}{5}x^5 - \frac{1}{6}x^6 + \frac{1}{7}x^7$

2.b. ≈ 0.066

3.a.

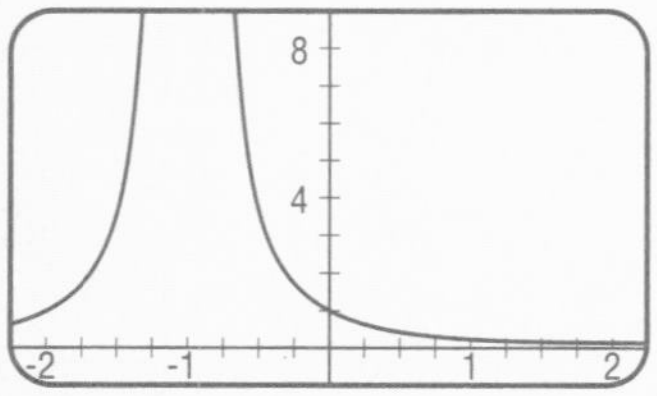

b. Since $f_3(x)$ approaches infinity near $x = -1$, there may not be a polynomial function which can approximate f_3 near $x = -1$.

Each pair of graphs is close to congruent over the selected domains. This can be tested by examining the difference of the two functions in the pair. Here is the graph of $h(x) = e^x - p_1(x)$. Notice that all the values of h are quite near zero.

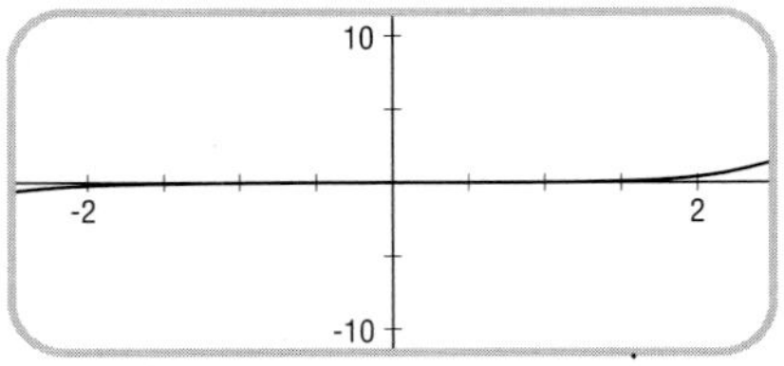

Karl Weierstrauss

Polynomial approximations to many functions, including those graphed above, were discovered by mathematicians in the 1700s. If you study calculus, you will very likely learn how to arrive at these approximations. (Like many things in mathematics, finding such approximations is not difficult if you know how.) Some 200 years later, Karl Weierstrass (1815–1897), a German mathematician, proved that for *any* continuous function f on *any* interval $a \le x \le b$, there is a polynomial function P such that $|P(x) - f(x)|$ is smaller than any assigned accuracy you might specify for all x. In less formal language, this means that any continuous function can be approximated, on any interval $a \le x \le b$, as close as you like, by a polynomial function. The theorem is known appropriately as the **Weierstrass Approximation Theorem**. Of course, one might need a polynomial with many terms, but still the theorem reflects the extraordinary variety of polynomials.

Questions

Covering the Reading

In Questions 1–3, refer to the functions mentioned in the lesson.

1. What is the largest value of $|p_2(x) - f_2(x)|$ on the interval $-\pi \le x \le \pi$?
approximately 0.024

2. **a.** The polynomial $p_4(x)$ could be written as
$p_4(x) = x - \frac{1}{2}x^2 + \frac{1}{3}x^3 - \frac{1}{4}x^4 + \frac{1}{5}x^5$.
Extend the pattern of the formula to create a 7th degree polynomial that better approximates $f_4(x)$ on the interval $-0.5 \le x \le 1$.
b. What is the farthest that $f_4(x)$ gets from $p_4(x)$ on this interval?
See margin.

3. **a.** Graph $f_3(x)$ for $-2 \le x \le 2$.
b. Explain why there might be no good polynomial function approximation to $f_3(x)$ on this interval.
See margin.

4. **a.** Graph the functions $f(x) = \sin x$ and $P(x) = -\frac{1}{6}x^3 + x$ on the interval $-0.5 \le x \le 0.5$. See margin.
b. What is the farthest that $f(x)$ gets from $P(x)$? 0.0003

276

Review

5. Write the standard prime factorization of 177,750. *(Lesson 4-8)*
$2 \cdot 3^2 \cdot 5^3 \cdot 79$

6. Give a prime factorization of $x^6 - 1$.
(Hint: Use $a^3 + b^3 = (a + b)(a^2 - ab + b^2)$.) *(Lesson 4-8)*
$(x + 1)(x - 1)(x^2 - x + 1)(x^2 + x + 1)$

7. a. Every integer can be expressed in the form $6n$, $6n + 1$, $6n + 2$, $6n + 3$, $6n + 4$, or $6n + 5$ where n is an integer. Show that if $n > 1$, then $6n$, $6n + 2$, $6n + 3$, and $6n + 4$ do not represent prime numbers. **See margin.**

b. Explain why a number of the form $6n + 5$ can also be written as $6k - 1$ for some integer k.

c. Parts **a** and **b** together establish that every prime number greater than 3 is one more or less than a multiple of 6. Show that this property is true for the prime numbers between 40 and 60.
(Lessons 4-8, 4-2) **See margin.**
b) $6n + 5 = 6n + 6 - 1 = 6(n + 1) - 1 = 6k - 1$, for some integer k.

8. Use a proof by contradiction to prove the following:
If ABC is a triangle, then it does not have two obtuse angles.
(Lesson 4-8) **See margin.**

9. Perform the addition. $1\,0\,1\,1\,1_2 + 1\,0\,1\,1_2$ *(Lesson 4-7)* **100010_2**

10. Given $n = 11{,}024$ and $d = 13$. Use a calculator to find the integer quotient q and integer remainder r determined by the Quotient-Remainder Theorem. *(Lesson 4-2)* **$q = 848$, $r = 0$**

11. A sine wave is shown below.

a. Find its period and amplitude. **period: .002; amplitude: 100**

b. Write an equation that expresses y as a function of t.
(Lessons 2-8, 3-9) **$y = 100\sin(1000\pi t)$**

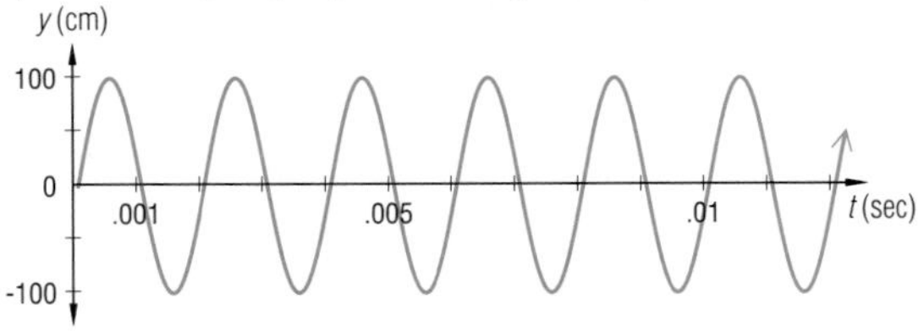

12. A lighthouse and Coast Guard rescue boat are located 25 miles apart. One evening, the lighthouse operator sees a ship in distress at an angle of 42° with the rescue boat. At the same time, the Coast Guard boat picks up a distress signal at a bearing of 65° from the lighthouse. Find the distance of the ship from the Coast Guard rescue boat.
(Lesson 2-8)
approximately 17.5 miles

4.a.

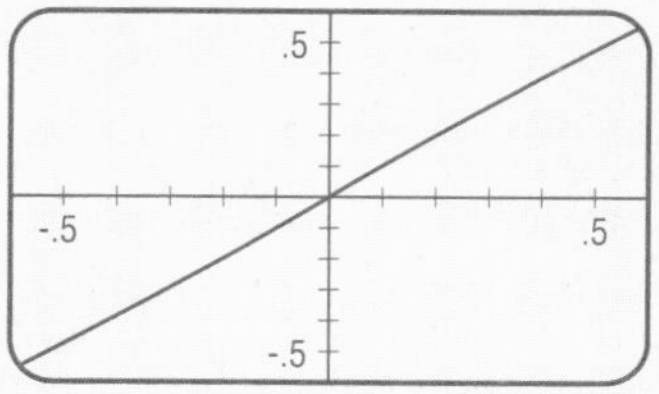

7.a. $6n = 3(2n)$, so 3 is a factor and $6n$ is not prime. $6n + 2 = 2(3n + 1)$, so 2 is a factor and $6n + 2$ is not prime. $6n + 3 = 3(2n + 1)$, so 3 is a factor and $6n + 3$ is not prime. $6n + 4 = 2(3n + 2)$, so 2 is a factor and $6n + 4$ is not prime.
c. $41 = (6 \cdot 7) - 1$
$43 = (6 \cdot 7) + 1$
$47 = (6 \cdot 8) - 1$
$53 = (6 \cdot 9) - 1$
$59 = (6 \cdot 10) - 1$

8. Assume: *ABC* is a triangle and has two obtuse angles. Let $\angle A$ and $\angle B$ be the two obtuse angles and $\angle C$ be the third angle. The sum of the measures of the angles is $m\angle A + m\angle B + m\angle C$. Since $m\angle A > 90°$ and $m\angle B > 90°$, $m\angle A + m\angle B + m\angle C > 180°$. Therefore, *ABC* cannot be a triangle since its angles sum to more than 180°. Hence, the assumption is false, and triangle *ABC* does not have two obtuse angles.

In 13 and 14, write the form of the argument and decide if it is valid or invalid. *(Lessons 1-6, 1-8)*

13. *If a prime number n has the form $2^q - 1$, where q is an integer, then q is a prime number.*
11 is a prime number.
$\therefore n = 2^{11} - 1 = 2047$ is prime.
$p(x) \Rightarrow q(x)$; $q(11)$; $\therefore p(11)$; invalid, converse error

14. *If a prime number n has the form $2^q - 1$, where q is an integer, then q is a prime number.*
$n = 2^{127} - 1$ is prime.
$\therefore$ 127 is a prime number.
$p(x) \Rightarrow q(x)$; $p(127)$; $\therefore q(127)$; valid, Law of Detachment

15. Refer to the graph of f at the right.

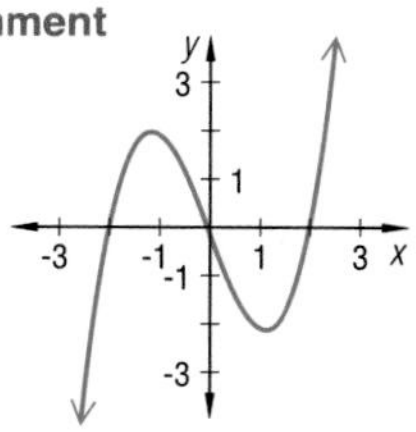

a. Identify the intervals on which f is increasing. $x \leq -1, x \geq 1$

b. Is f a 1-1 function? Explain.

c. Does f have an inverse function? If so, sketch a graph of the inverse. If not, explain why not. See below.

(Lessons 2-2, 3-2)

b) No, $f(0) = f(2)$, but $0 \neq 2$

Exploration

16. The denominators of the coefficients of $p_1(x)$ in this lesson are the factorials. By using polynomials of higher and higher degree, see how close you can get to $f_1(x) = e^x$ on the interval $-5 \leq x \leq 5$. sample: if $p(x) = 1 + x + \frac{x^2}{2!} + \frac{x^3}{3!} + \ldots + \frac{x^{12}}{12!}$, then $|e^x - p(x)| \leq 0.3$ for all x in the interval $-5 \leq x \leq 5$.

15. c) No, f is not 1–1, therefore it cannot have an inverse function.

278

Projects

1. Karl Friedrich Gauss (1777-1855) introduced the theory of congruences modulo d in a book entitled *Disquisitiones arithmeticae*. Use an encyclopedia, biography, or a history of mathematics book to learn more about the discoveries made by this remarkable mathematician. Write a short report about what you find.

2. **a.** Any solution to the equation $ax \equiv b \pmod d$ must be congruent to one of the integers $0, 1, 2, \ldots, d-1$. To determine whether any solutions exist, find the greatest common factor (GCF) of a and d. Let $\text{GCF}\{a, d\} = k$. The fact is that if b is divisible by k, then there are k solutions in the set $\{0, 1, 2, \ldots, d-1\}$. If b is not divisible by k, then there are no solutions. For instance, the equation $6x \equiv 10 \pmod 8$ has two solutions in the set $\{0, 1, 2, \ldots, 7\}$ because $\text{GCF}\{6, 8\} = 2$ and 10 is divisible by 2. The solutions are $x = 3$ and $x = 7$. Determine the number of solutions that exist in each case and then find the solutions, if any.
   i. $5x \equiv 8 \pmod{12}$ **ii.** $3x \equiv 12 \pmod 9$
   iii. $4x \equiv 5 \pmod 6$ **iv.** $6x \equiv 24 \pmod{12}$
   Identify those numbers d which yield exactly one solution for every equation $ax \equiv b \pmod d$, $a \not\equiv 0$.
   b. Take a value of d in part **a**. Explore the solution of quadratic equations $ax^2 + bx + c \equiv 0 \pmod d$. Establish criteria for the solvability of quadratics in this system.

3. Prime numbers and modular arithmetic are frequently-used tools in the study of cryptography. Look in an index of periodicals and find one or more articles dealing with this topic and write a report. Some possible journals are *Scientific American*, *Science*, and *Byte*. Although there are many articles, here are four possibilities:
   ''The Mathematics of Public-Key Cryptography'' by Martin E. Hellman in *Scientific American*, August 1979, pp. 146-152+.
   ''Cryptography: On the Brink of a Revolution?'' by Gina Bari Kolata in *Science*, August 19, 1977, pp. 747-748.
   ''Implementing Cryptographic Algorithms on Microcomputers'' by Charles Kluepfel in *Byte*, October 1984, pp. 126-27+.
   ''Secret Encryption with Public Keys'' by Linda L. Deneen in *UMAP: The Journal of Undergraduate Mathematics and Its Application*, Spring 1987, pp. 9-29.

4. $x^2 - 3x + 2$ factors over the integers as $(x-2)(x-1)$. $x^2 - 3x - 2$ is not factorable over the integers. What percent of quadratic expressions
   $x^2 + ax + b \qquad -9 \le a \le 9,\ -9 \le b \le 9$
   are factorable over the integers? Answer this question one of two ways.
   a. By using a computer to test all 361 quadratics.
   b. By using a random number table to pick values for a and b in twenty quadratics, then estimating the percent of factorables in the sample.

5. Approximate all zeros of a cubic equation using the following algorithm, which merges bisection (Lesson 3-5), synthetic substitution (Lesson 4-5), and the quadratic formula. Given $p(x) = ax^3 + bx^2 + cx + d$:
   a. Use synthetic substitution to evaluate $p(t)$ for values of t until you find a value t_1 that gives a positive remainder, and a value t_2 that gives a negative remainder.
   b. Use bisection to generate a sequence of values $t_3, t_4, \ldots$ which causes the remainders to approach zero. (Stop when you get within, say .001 of zero, that is, $|p(t_n)| < .001$.)
   c. The value t_n is an approximation to one of the zeros. The remaining numbers on the bottom line of the synthetic division are the coefficients of a quadratic $q(x)$, called the reduced polynomial. Use the quadratic formula to estimate the real zeros, if any. These will be zeros of $p(x)$.
   Write a computer program to do this, or use a calculator to follow the algorithm to solve $5x^3 + 4x^2 - 551x + 110 = 0$. (Hint: Try to find the first solution between 2 and -2.)

PROJECTS

The projects provide an opportunity for extended work on topics related to the content of the chapter. (See pages 70 and T37 for more information on the purposes of projects.) Some or all of these projects may be suitable for **small group work**. Solutions for the projects can be found at the end of the Additional Answers section in the back of this book.

Projects 1-3 are related to modular arithmetic.

Project 2 provides an algorithm for finding the greatest common factor of two numbers without first finding their prime factorizations.

Projects 4 and 5 require access to a computer.

SUMMARY

The Summary gives an overview of the entire chapter and provides an opportunity for students to consider the material as a whole. Thus, the Summary can be used to help students relate the various concepts presented in the chapter.

Summary

The operation of division is fundamental in mathematics. This chapter extends your knowledge of integer division to division of polynomials, to the development of modular arithmetic, to the solution of polynomial equations, and to the examination of the theory of prime numbers and also of prime polynomials.

Just as $n = q \cdot d$ means that q and d are factors of n, so $n(x) = q(x) \cdot (x)$ means that the polynomials $q(x)$ and $d(x)$ are factors of $n(x)$. When an integer n is divided by d, there is a unique integer quotient q and remainder r such that $n = q \cdot d + r$ and $0 \leq r < d$. Similarly, when a polynomial $n(x)$ is divided by $d(x)$, there is a unique polynomial quotient $q(x)$ and remainder $r(x)$ such that $n(x) = q(x) \cdot d(x) + r(x)$ and either the degree of $r(x)$ is nonzero and less than the degree of $d(x)$ or $r(x)$ is the zero polynomial. These are the Quotient-Remainder theorems for integers and polynomials.

You can find the quotient polynomial $q(x)$ by long division. The algorithm for polynomial division is organized like the long division algorithm for whole number division that you learned in elementary school. Other algorithms that are related to long division are synthetic substitution and division. Synthetic substitution provides a fast way to evaluate a polynomial. The final value in the synthetic substitution to find $p(c)$ is the remainder when $p(x)$ is divided by the degree one polynomial $d(x) = x - c$.

Pick a particular divisor d, and group numbers together whose remainders are equal when divided by d. This produces an arithmetic modulo d.

Electronic computing devices operate using binary logic, as described in Chapter 1, and apply that logic to perform arithmetic on numbers represented in binary (base 2) notation. The Quotient-Remainder theorem gives a method for changing any base 10 representation of a number into a different base. Addition in different bases follows place-value algorithms similar to those of base 10 arithmetic.

Prime numbers are those positive integers greater than 1 that are not divisible by any positive integer other than 1 and themselves. The Fundamental Theorem of Arithmetic asserts that the primes are the building blocks of the set of integers: any integer greater than 1 is either prime or has a prime factorization that is unique except for the order of the factors. Polynomials also have unique prime factorizations except for order and multiplication by real constants. Determining whether a number is prime is very difficult for large numbers, but it is possible to prove that there are an infinite number of primes.

The proof of the infinitude of primes uses a method of indirect reasoning called proof by contradiction. To prove a statement s using proof by contradiction, start by assuming *not s* and show that this assumption leads to a contradiction.

280

Vocabulary

Below are the most important terms and phrases for this chapter.
For the starred (•) items you should be able to give a definition of the terms.
For the other items you should be able to give a general description and a specific example of each.

Lesson 4-1
*factor, divisor
multiple
quotient
nested form
zero polynomial
degree of a product
degree of a sum
Transitive Property of Factors
Factor of a Sum Theorem
Factor of a Product Theorem

Lesson 4-2
Quotient-Remainder Theorem (for Integers)

Lesson 4-3
modulus, mod
≡, congruent to
*congruent modulo *m*
congruence classes
Congruence Theorem
properties of congruence

Lesson 4-4
Quotient-Remainder Theorem (for Polynomials)

Lesson 4-5
synthetic substitution
synthetic division

Lesson 4-6
zero of a polynomial
Remainder Theorem
Factor Theorem
Number of Zeros of a Polynomial Theorem

Lesson 4-7
base 2 representation
binary notation
*base *b* notation
half-adder

Lesson 4-8
proof by contradiction
*prime
Infinitude of Primes Theorem
Factor Search Theorem
prime factorization
Fundamental Theorem of Arithmetic
standard prime factorization
Unique Factorization for Polynomials

Lesson 4-9
Weierstrass Approximation Theorem

VOCABULARY

Terms, symbols, and properties are listed by lesson to provide a checklist of concepts a student must know. (See page 71 for more information.)

PROGRESS SELF-TEST

The Progress Self-Test provides the opportunity for feedback and correction; the Chapter Review provides additional opportunities for practice.

USING THE PROGRESS SELF-TEST

Assign the Progress Self-Test as a one-night assignment. (See pages 72 and T35 for more information.)

ADDITIONAL ANSWERS

1.c. $q = 18153$, $r = 6$; $145230 = 18153(8) + 6$

5. 47.625

6. Let n, p, and m be integers such that n is a factor of p and n is a factor of m. By definition, there exists integers r and s such that $p = r \cdot n$ and $m = s \cdot n$. Then, $p \cdot m = r \cdot n \cdot s \cdot n = (r \cdot s \cdot n)n$. Since $r \cdot s \cdot n$ is an integer by closure properties, n is a factor of $p \cdot m$ by definition.

16. Counterexample: Let $a = 5$ and $b = 8$. Then b is a factor of $(5)^2 - 1 = 24$, but b is not a factor of $5 - 1 = 4$ or $5 + 1 = 6$.

18. Assume there is a largest even integer, N. $N = 2k$ for some integer k. Consider $N + 2$. Since the integers are closed under addition, $N + 2$ is an integer. Further, since $N + 2 = (2k) + 2 = 2(k + 1)$, $N + 2$ is even. $N + 2 > N$, so N is not the largest even integer. This contradicts the initial assumption. Therefore, the original statement is true; there does not exist a largest even integer.

Progress Self-Test

See margin for answers not shown below.

Take this test as you would take a test in class. Then check the test yourself using the solutions at the back of the book.

1. A bottling company packages bottles of soda in 8-packs. If 145,230 bottles of soda are processed in one day, determine
   a. the number of 8-packs that can be packaged. 18,153
   b. the number of bottles left unpackaged. 6
   c. Relate your answers to parts **a** and **b** to the Quotient-Remainder Theorem.

2. Use long division to find the quotient $q(x)$ and the remainder $r(x)$ when $p(x) = 3x^4 - 19x^3 + 36x^2 - 32x + 15$ is divided by $d(x) = x^2 - 4x$.
$q(x) = 3x^2 - 7x + 8$; $r(x) = 15$

In 3 and 4, refer to the synthetic division scheme below for $p(x) = 2x^3 - 3x^2 - 10x + 6$ divided by $x - 4$.

4⌋	2	-3	-10	6
		8	20	40
	2	5	10	46

3. The remainder is ___?___. 46

4. *True* or *false?* $x - 4$ is a factor of $2x^3 - 3x^2 - 10x + 6$. False Explain

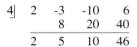

5. Use synthetic substitution to find $p(c)$ if $p(x) = 7x^3 - 2x^2 + 11x + 12$ and $c = 1.5$.

6. Prove: *If n, p, and m are integers, n is a factor of p, and n is a factor of m, then n is a factor of p • m.*

7. There are 2^{50} different ways to answer a true-false test with 50 questions. Find the last four digits of 2^{50}. 2624

8. Give the smallest positive integer solution to $y \equiv 151 \pmod{11}$. 8

9. Write a statement about factors which expresses $a \equiv 5 \pmod{n}$. n is a factor of $(a - 5)$.

10. Give the standard prime factorization of 819. $3^2 \cdot 7 \cdot 13$

Ch review #14, 58, 72.

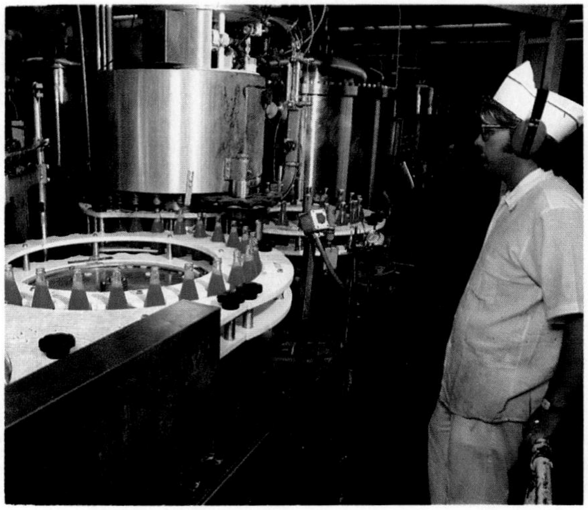

11. *True* or *false? For all positive integers m, if m is not divisible by any prime number less than $\sqrt{m}$, then m is prime.* False

12. Give the prime factorization of $7x^4 - 3x^3 - 4x^2$ over the set of polynomials with real coefficients. $x^2(7x + 4)(x - 1)$

13. Write the base 2 representation of 33. 100001_2

14. Add: $101_2 + 110_2$. Give your answer in base 2. 1011_2

15. Write the base 10 representation of 312_5. 82

16. Find a counterexample to show that the following statement is false.
For all integers a and b, if b is a factor of $a^2 - 1$, then b is a factor of either $a + 1$ or $a - 1$.

17. If $x - 3$ is a factor of $p(x)$ then $p(3) =$ ___?___. 0

18. Use a proof by contradiction to prove the following statement:
There is no largest even integer.

282

See margin for answers not shown below.

Chapter Review

Questions on SPUR Objectives

SPUR stands for **S**kills, **P**roperties, **U**ses, and **R**epresentations.
The Chapter Review questions are grouped according to the SPUR Objectives for this chapter.

SKILLS deal with the procedures used to get answers.

Objective A: *Use the Quotient-Remainder Theorem for integers or polynomials.* *(Lesson 4-2, 4-4)*

In 1 and 2, given the pair of integers n and d, find integers q and r as determined by the Quotient-Remainder Theorem.

1. $n = 81$, $d = 15$ **2.** $n = 47$, $d = 9$

In 3 and 4, for each pair of integers n and d use your calculator to determine integers q and r that satisfy the Quotient-Remainder Theorem.

3. $n = 7865$, $d = 94$ $q = 83$, $r = 63$

4. $n = -15758$, $d = 45$ $q = -351$, $r = 37$

5. Use a calculator to help in finding an exact remainder for $1739541 \div 859$. 66

6. When $x^7 + 2x^5 - x^3 + x^2 + 7$ is divided by $x^2 + 2$, the quotient is $x^5 - x + 1$. Find the remainder and its degree.

7. Find a real number a that makes the following true: $y^2 - 12y + 9 = (y + a)(y - 7) - 26$. $a = -5$

Objective B: *Use synthetic substitution to find values of polynomial functions.* *(Lesson 4-5)*

In 8–10 use synthetic substitution to find $p(c)$ for the given polynomial and value of c.

8. $p(x) = 8x^4 - 2x^3 + 11x - 10$, $c = 1.3$ 22.7548

9. $p(x) = 4x^6 - 7x^4 + 2x^2 - 3$, $c = -0.5$ -2.875

10. $p(x) = x^5 - 2x^2 + 7x$, $c = -0.8$ -7.20768

11. *True* or *false?* Refer to the synthetic substitution below. If $p(x) = 3x^3 + 8x^2 + 7x + 6$, then $p(2) = 76$. True

2⌋	3	8	7	6
		6	28	70
	3	14	35	76

12. Use synthetic substitution to show that -5 is a zero of $p(x) = x^4 - 3x^3 - 33x^2 + 39x + 20$.

Objective C: *Divide polynomials.* *(Lessons 4-4, 4-5)*

In 13–19, find the quotient $q(x)$ and the remainder $r(x)$ when the polynomial $p(x)$ is divided by the polynomial $d(x)$.

13. $p(x) = 3x^3 + 8x^2 - 33x + 18$, $d(x) = x + 5$

14. $p(x) = 35x^3 - 19x^2 + 11x - 9$, $d(x) = 5x^2 + 3x + 2$

15. $p(x) = x^5 + 7x^4 + 12x^3 + 5$, $d(x) = x^2 + 4x$

16. $p(x) = 6x^3 - 13x^2 - 6x + 6$, $d(x) = 3x^2 + x - 1$

17. $p(x) = x^4 - 7x^3 + 3x^2 - 4x + 9$, $d(x) = x + 1$

18. $p(x) = 5x^3 + x - 4$, $d(x) = x - 6$

19. $p(x) = x^5 + x^4 - x^3 + x^2 - x - 3$, $d(x) = x + \frac{1}{2}$

Objective D: *Determine the congruence of integers for a given modulus.* *(Lesson 4-3)*

In 20–23, give the smallest positive integer that makes the congruence true.

20. $x \equiv 214 \pmod 7$ 4

21. $y \equiv -1 \pmod{13}$ 12

22. $z \equiv 1000 \pmod{11}$ 10

23. $t \equiv -482 \pmod{25}$ 18

24. Consider the congruence classes $R0$, $R1$, $R2$, …, $R11$ for integers modulo 12. If you add an element from $R9$ to an element in $R5$, which set contains the sum? $R2$

Objective E: *Factor polynomials.* *(Lessons 4-1, 4-8)*

In 25–32, factor completely into prime polynomials over the integers.

25. $5x^2 - 5y^2$ $5(x - y)(x + y)$

26. $9t^2 + 45t + 54$ $9(t + 2)(t + 3)$

27. $3x^3 + x^2 - 10x$ $x(3x - 5)(x + 2)$

28. $4y^4 - 37y^2 + 9$ $(2y - 1)(2y + 1)(y - 3)(y + 3)$

29. $18v^4 + 60v^2 + 50$ $2(3v^2 + 5)^2$

RESOURCES
- Chapter 4 Test, Form A
- Chapter 4 Test, Form B
- Chapter 4 Test, Cumulative Form
- Quiz and Test Writer

CHAPTER REVIEW

The main objectives for the chapter are organized here into sections corresponding to the four main types of understanding this book promotes: Skills, Properties, Uses, and Representations. We call these the SPUR objectives.

USING THE CHAPTER REVIEW

Students should be able to answer questions like these with about 85% accuracy by the end of the chapter. (See pages 74 and T38–39 for more information.)

ADDITIONAL ANSWERS

1. $q = 5$, $r = 6$

2. $q = 5$, $r = 2$

6. $r(x) = 2x + 5$; degree $= 1$

12.

-5⌋	1	-3	-33	39	20
		-5	40	-35	-20
	1	-8	7	4	0

Since the remainder is 0, -5 is a zero of $p(x)$.

13. $q(x) = 3x^2 - 7x + 2$; $r(x) = 8$

14. $q(x) = 7x - 8$; $r(x) = 21x + 7$

15. $q(x) = x^3 + 3x^2$; $r(x) = 5$

16. $q(x) = 2x - 5$; $r(x) = x + 1$

17. $q(x) = x^3 - 8x^2 + 11x - 15$; $r(x) = 24$

18., **19.** See the margin on page 284.

ADDITIONAL ANSWERS

18. $q(x) = 5x^2 + 30x + 181$; $r(x) = 1082$

19. $q(x) = x^4 + \frac{1}{2}x^3 - \frac{5}{4}x^2 + \frac{13}{8}x - \frac{29}{16}$; $r(x) = -\frac{67}{32}$

31. $(2x + 1)(2x^2 - 4x - 9)$

32. $(7t - 2)(8t + 1) \cdot (-13t^2 - 66t + 11)$

33. $7(x - 3) \cdot \left(x + \frac{3 - \sqrt{37}}{14}\right)\left(x - \frac{3 - \sqrt{37}}{14}\right)$

36. False, there does not exist an integer q such that $156 = q \cdot 9$.

37. True, a and b may be represented by $2k$ and $2m$, respectively, since they are both even integers. Thus, $2a + 2b = 2(2k) + 2(2m) = 4(k + m)$, so 4 is a factor by definition.

38. Suppose n is any odd integer. Then, by definition, there exists an integer r such that $n = 2r + 1$. Then, $n^2 + n = (2r + 1)^2 + (2r + 1) = 4r + 4r + 1 + 2r + 1 = 4r^2 + 6r + 2 = 2(2r^2 + 3r + 1)$. Since $2r^2 + 3r + 1$ is an integer by closure properties, $n^2 + n$ is divisible by 2.

41. Let a, b, c, and d be integers such that $a = b - c$ and a and c are divisible by d. Then there exist integers r and s such that $a = d \cdot r$ and $c = d \cdot s$ by definition. Then, $b = a + c = d \cdot r + d \cdot s = d(r + s)$. Since $(r + s)$ is an integer by closure properties, b is divisible by d.

42. Counterexample: Let $a = 3$. Then $a^3 - 1 = 26$, but 26 is not divisible by 4.

43. Let $2k + 1$ and $2k + 3$ represent two consecutive odd integers, Then, $(2k + 1)(2k + 3) = 4k^2 + 8k + 3$. Adding one gives $4k^2 + 8k + 4 = 4(k^2 + 2k + 1)$. $k^2 + 2k + 1$ is an integer for all integers k, so 4 is a factor.

30. $w^4 - z^4$ $(w - z)(w + z)(w^2 + z^2)$

31. $(2x + 1)^2(x - 4) + (2x + 1)(3x - 5)$

32. $(7t - 2)^2(8t + 1)(5t - 6) - (7t - 2)(8t + 1)^2(6t + 1)$

33. Given that $x - 3$ is a factor of $p(x) = 7x^3 - 18x^2 - 10x + 3$, factor $p(x)$ completely over the reals.

34. Given that $2x + 5$ is a factor of $f(x) = 24x^3 + 34x^2 - 59x + 15$, factor $f(x)$ completely. $(2x + 5)(3x - 1)(4x - 3)$

PROPERTIES deal with the principles behind the mathematics.

Objective F: *Justify properties of factors of integers or factors of polynomials. (Lesson 4-1)*

In 35–37, determine whether the statement is true or *false*. Explain your answer.

35. *18 is a factor of 90.* True, $18 \cdot 5 = 90$.

36. *156 is divisible by 9.*

37. *$2a + 2b$ is divisible by 4 if a and b are even integers.*

38. Prove: *For all odd integers n, $n^2 + n$ is divisible by 2.*

39. *True* or *false? If n is a nonzero integer, then 0 is divisible by n.* True

40. *Multiple choice.* Which of the following expressions are divisible by 6 if a is divisible by 3?

(a) $6 + a$ (b) $2a$ (c) $4a$
(d) $3a^2$ (e) $6 - 2a$ (b), (c), (e)

In 41 and 42, prove the conjecture or find a counterexample to disprove it.

41. *For all integers a, b, c, and d, if $a = b - c$ and a and c are divisible by d, then b is divisible by d.*

42. *For all integers a, if a is odd then $a^3 - 1$ is divisible by 4.*

43. Prove that the number one greater than the product of any two consecutive odd integers is divisible by 4.

In 44–47, determine whether the given integer is even or odd. Justify your answer.

44. -86 **45.** 215

46. $3ab$, where a and b are integers and a is even

47. $8s^2 + 4s + 3$ where s is any integer

48. If k is an odd integer and $n = 2k - 1$, is n even or odd? Explain.

49. Find the error in this "proof" of the following theorem.

If m and n are even integers, then $m - n$ is an even integer.

"Proof: Because m is even there is an integer k such that $m = 2k$. Also, because n is even, $n = 2k$. It follows that $m - n = 2k - 2k = 0$, which is even. Thus, the difference of any two even integers is even."

Objective G: *Use the properties of congruence of integers for a given modulus to rewrite sentences. (Lesson 4-3)*

50. If $x \equiv y \pmod{11}$ what can you conclude about $x - y$? $x - y$ is divisible by 11.

In 51 and 52, rewrite in the language of congruences.

51. x is a multiple of 5. $x \equiv 0 \pmod 5$

52. If $\sin x = -\frac{1}{2}$, then $x = \frac{7\pi}{6} + 2\pi k$ or $x = \frac{11\pi}{6} + 2\pi k$ for some integer k.

53. If $c \equiv 5 \pmod{11}$ and $d \equiv 7 \pmod{11}$, what congruence statement can you write for the following: **a.** $c - d$ **b.** cd

54. In a listing of students in a school, the tops of the columns start with the 1st, 13th, 25th, 37th, 49th, ... students. Let x stand for the number of a student. Write a congruence that indicates that student x is at the top column of a page. $x \equiv 1 \pmod{12}$

Objective H: *Use the Remainder Theorem, Factor Theorem, or Quotient-Remainder Theorem to describe characteristics of given polynomials. (Lessons 4-4, 4-6)*

55. In the Quotient-Remainder Theorem, given polynomials $n(x)$ and $d(x)$, if $n(x) = q(x) \cdot d(x) + r(x)$, what are the restrictions on $r(x)$?

56. If the degree of $p(x)$ is 5, the degree of $q(x)$ is d, and the degree of their product is 10, find d. $d = 5$

57. Use long division to show that $x^4 - 3$ is a factor of $x^6 + 2x^5 - 7x^4 - 3x^2 - 6x + 21$.

58. *Multiple choice*. When $p(x) = 4x^5 - 7x^3 + 2x^2 + 1$ is divided by $x + 5$, the remainder is given by which of the following?
(a) $p(x + 5)$ (b) $p(x - 5)$
(c) $p(5)$ (d) $p(-5)$. (d)

59. If 7 is a zero of $p(x)$, then _?_ is a factor of $p(x)$. $x - 7$

60. If 2 is a zero of $p(x) = x^3 - 4x^2 + 3x + 2$, find the other zeros. $x = 1 + \sqrt{2}$ and $x = 1 - \sqrt{2}$

61. If p is a polynomial of degree 7, then p can have at most _?_ different zeros. 7

Objective I: *Use proof by contradiction.* *(Lesson 4-8)*

In 62 and 63, write the assumption that you would make to begin the proof of the given statement by contradiction.

62. *There is no smallest integer.*

63. *For all primes p, if p is a factor of n^2, then p is a factor of n.*

64. Use proof by contradiction to prove the statement in Question 62.

65. If n_p is one more than the product of all the primes from 2 to p, inclusive, what is n_{11}? 2311

66. Give Euclid's proof of the infinitude of the primes.

Objective J: *Use the Factor Search Theorem and the Fundamental Theorem of Arithmetic in determining prime numbers and prime factorizations.* *(Lesson 4-8)*

67. *True* or *false*? *To determine if 653 is prime, you only need to check if 653 is divisible by the prime numbers between 1 and 26.* True

In 68–70, determine if the given integer is prime. If it is not prime, write a prime factorization of it.

68. 653 prime 69. 551 $19 \cdot 29$

70. 8430 $2 \cdot 3 \cdot 5 \cdot 281$

71. *Multiple choice*. The standard prime factorization of 2352 is:
(a) $2 \cdot 2 \cdot 2 \cdot 2 \cdot 3 \cdot 7 \cdot 7$
(b) $7^2 \cdot 3 \cdot 2^4$
(c) $2^4 \cdot 3 \cdot 7^2$
(d) $2^2 \cdot 2^2 \cdot 3 \cdot 7^2$ (c)

72. How many primes must be tested to determine whether 227 is prime? 6

73. Only primes less than what number must be tested to determine whether 1,000,001 is prime? 1001

USES deal with applications of mathematics in real situations.

Objective K: *Use the Quotient-Remainder Theorem to solve applied problems.* *(Lesson 4-2)*

74. A computer printer prints 66 lines per page. A document with 5789 lines is printed.
a. How many pages are needed? 88
b. How many lines are on the last page? 47
c. Write an equation in the form
$$5789 = pq + r$$
to describe the situation.

75. Once around a track is 400 meters. If a runner wants to run 5.5 miles, find the number of complete laps and the number of meters that still must be run. Use the fact that 1 mile is about 1609 meters. 22 laps and 49.5 meters

76. Elizabeth is a frequent flyer with an airline company and earns a free ticket for every 40,000 miles credited to her account. Last year 387,500 miles were credited to her account.
a. How many free tickets did she earn? 9
b. How many credited miles can be applied to the next free ticket? 27500 miles
c. Relate your answers in parts **a** and **b** to the Quotient-Remainder Theorem.

77. How many months and how many days per month would give a 365-day year months of the same length?

44. -86 = 2(-43), so -86 is even.

45. 215 = 2(107) + 1, so 215 is odd.

46. Let $a = 2k$. Then $3ab = 6kb = 2(3kb)$, so $3ab$ is even.

47. $8s^2 + 4s + 3 = 2(4s^2 + 2s + 1) + 1$. Let $k = 4s^2 + 2s + 1$, then $8s^2 + 4s + 3 = 2k + 1$. So $8s^2 + 4s + 3$ is odd.

48. Odd; let $k = 2m + 1$. Then $2k - 1 = 2(2m + 1) - 1 = 2(2m) + 1$.

49. The same integer k should not be used for both m and n.

52. If $\sin x = -\frac{1}{2}$, then $x \equiv \frac{7\pi}{6} \pmod{2\pi}$ or $x \equiv \frac{11\pi}{6} \pmod{2\pi}$.

53.a. $c - d \equiv -2 \pmod{11}$
b. $cd \equiv 35 \pmod{11}$

55. $0 \le$ degree of $r(x) <$ degree of $d(x)$ or $r(x) = 0$

57. See Additional Answers in the back of this book.

62. *There is a smallest integer.*

63. *p is a prime, p is a factor of n^2, and p is not a factor of n.*

64. Assume there exists a smallest integer, P. Consider $P - 1$. By the closure of integers under addition, $P - 1$ is an integer. $P - 1$ is also smaller than P, which contradicts the initial assumption. Therefore, a smallest integer does not exist.

66., 74.c. See the margin on page 286.

76.c., 77. See Additional Answers in the back of this book.

EVALUATION
Three forms of a Chapter Test are provided in the Teacher's Resource File. Chapter 4 Test, Forms A and B cover just Chapter 4. The third test is Chapter 4 Test, Cumulative Form. About 50% of this test covers Chapter 4, 25% covers Chapter 3, and 25% covers previous chapters. For information on grading, see *General Teaching Suggestions: Grading* on page T43 in the Teacher's Edition.

ASSIGNMENT RECOMMENDATION
We strongly recommend that you assign the reading and some questions from Lesson 5-1 for homework the evening of the test. It gives students work to do if they complete the test before the end of the period and keeps the class moving.

If you do not give assignments on the days of tests, you may cover one less *chapter* over the course of the year.

ADDITIONAL ANSWERS
66. Assume there are finitely many prime numbers. They can be listed from smallest to largest: $p_1 = 2, p_2 = 3, \ldots, p_m$. Multiply these: $n = p_1p_2p_3 \cdots p_m$. Consider the integer $n + 1$. Since $n + 1$ is larger than the assumed largest prime p_m, it is not prime. By the Prime Factor Theorem, $n + 1$ must have a prime factor, p, on the list of primes. Then p is a factor of both n and $n + 1$. It is also a factor of the difference $(n + 1) - n = 1$. Therefore, p must equal 1. But p is a prime number, so p cannot equal 1. This contradiction proves that the original assumption is false. Hence, there are infinitely many prime numbers.

74.c. $5789 = 87 \cdot 66 + 47$

Objective L: *Use modular arithmetic to solve applied problems. (Lesson 4-3)*

78. There are 5^{30} ways to answer a 30-question multiple choice test with 5 choices for each question. Write the last 4 digits of this number. 5625

79. A calculator writes 7^{19} in scientific notation. Find the last 3 digits of this number. 143

In 80 and 81, find the check digit for the ISBN number.

80. 0-939765-04-__?__ 7

81. 0-7216-1983-__?__ 5

REPRESENTATIONS deal with pictures, graphs, or objects that illustrate concepts.

Objective M: *Represent numbers and perform addition in base 2. (Lesson 4-7)*

In 82 and 83, write the base 2 representation of the number.

82. 65 1000001_2 **83.** 165 10100101_2

In 84 and 85, write the base 10 representation of the number.

84. 11111_2 31 **85.** 101010_2 42

In 86 and 87, perform the indicated addition and then verify your answer by finding the base 10 representations of the numbers.

86. $10101_2 + 11010_2$ 101111_2; $21 + 26 = 47$

87. $11111_2 + 11111_2$

88. *Multiple choice.* Which of the following is evenly divisible by 4?
(a) 1001_2 (b) 1010_2
(c) 1011_2 (d) 1100_2 (d)

87. 111110_2; $31 + 31 = 62$

CHAPTER 5 RATIONAL NUMBERS AND RATIONAL FUNCTIONS

OBJECTIVES

The objectives listed here are the same as in the Chapter 5 Review on pages 341–344 of the student text. The Progress Self-Test on page 340 and the tests in the Teacher's Resource File cover these objectives. For recommendations regarding the handling of this end-of-chapter material, see the notes in the margin on the corresponding pages of this Teacher's Edition.

OBJECTIVES FOR CHAPTER 5 (Organized into the SPUR Categories—Skills, Properties, Uses, and Representations)	Progress Self-Test Questions	Chapter Review Questions	Teacher's Resource File	
			Lesson Masters*	Chapter Test Forms A & B
SKILLS				
A Simplify rational expressions.	1, 2, 3	1–13	5-1	1, 2
B Identify numbers as rational or irrational.	4	14–20	5-1, 5-2	3
C Simplify expressions involving radicals.	6	21–26	5-2	6
D Find values of trigonometric functions.	11	27–34	5-6	12
E Solve rational equations.	10	35–39	5-7	8
PROPERTIES				
F Prove properties of rational and irrational numbers.	5	40–47	5-1, 5-2	4, 15
G Use limit notation to describe the behavior of rational functions.	7, 8	48–57	5-3, 5-4, 5-5	9, 10
H Classify discontinuities as essential or removable.	9	58–61	5-4, 5-6	7
USES				
I Apply rational expressions and rational equations.	3	62–65	5-1, 5-7	5
REPRESENTATIONS				
J Graph quotients of polynomials or trigonometric functions.	8, 12	66–70	5-3, 5-4, 5-5, 5-6	7, 14
K Relate the limit of a function to its graph, and find equations for its asymptotes.	14	71–73	5-3, 5-4, 5-5	7, 11
L Use right triangles to find values of trigonometric functions.	13	74–77	5-6	13

*** The masters are numbered to match the lessons.**

OVERVIEW 5

OVERVIEW ■ CHAPTER 5

The preceding chapter compared systems of integers and polynomials, thus enhancing students' understanding of both. In this chapter, the system of rational numbers is used to build understanding of rational functions. Irrational numbers and some nonrational functions are used for contrast.

It is useful to think of the following pairs of terms as analogous:
(1) *rational expression*
(an indicated quotient of two polynomials) and
fraction
(an indicated quotient of two integers);
(2) *rational function*
(a function that can be expressed as the quotient of two polynomial functions) and
rational number
(a number that can be written as the quotient of two integers);
(3) *nonrational function*
(a real function that is not a rational function) and
irrational number
(a real number that is not a rational number).

This chapter can be thought of as having three parts. The first two lessons provide the groundwork for the study of rational functions. In Lesson 5-1, there is work with operations on rational expressions, which is analogous to operations on fractions. In Lesson 5-2, irrational numbers are defined. Some irrational numbers appear in fractions, of course, and the operations with fractions can be used with them.

The next three lessons explore rational functions themselves. The simplest rational functions (other than polynomial functions) are the reciprocals of power functions, that is, those functions with rules of the form $f(x) = \frac{k}{x^n}$. These are studied in Lesson 5-3. In Lessons 5-4 and 5-5, the behavior of rational functions is discussed.

The last three lessons apply the ideas of the earlier parts of the chapter. In Lesson 5-6, the tangent, cotangent, secant, and cosecant functions are nonrational functions, which are defined as reciprocals or quotients of sine and cosine functions. Thus, some of the ideas of reciprocal functions and rational functions (such as vertical asymptotes) apply to these functions in the same way that ideas about fractions can be used with irrational numbers. Lesson 5-7 covers the solving of equations of the form $f(x) = g(x)$, where both are rational expressions. These kinds of equations appear naturally from discussions of rational functions. Lesson 5-8 discusses unit fractions (analogous to reciprocals of the power functions) and also their surprising relationship to the determination of those regular polygons that will tessellate and to the determination of regular polyhedra.

The mathematics of this chapter is beautiful but was relatively inaccessible to students before the advent of automatic graphers. Today, the behavior of rational functions can be studied in detail.

PERSPECTIVES 5

PERSPECTIVES ■ CHAPTER 5

The Perspectives provide the rationale for the inclusion of topics or approaches, provide mathematical background, and make connections with other lessons and within UCSMP.

5-1

RATIONAL NUMBERS AND RATIONAL EXPRESSIONS

This lesson provides practice in the manipulation of rational expressions. The mathematical underpinnings of these manipulations relate to work with fractions. The rules for adding, multiplying, and simplifying fractions are similar to those for rational expressions. In working with rational expressions, however, there are two additional concerns: (1) there may be values of the variable for which an expression is undefined (the domain of the expression); and (2) two expressions are equal if and only if they are equal for every value of the variable for which they are defined (the notion of identity).

5-2

IRRATIONAL NUMBERS

This lesson reviews some ideas about irrational numbers. The proof that $\sqrt{2}$ is irrational, and proofs about the results of combining rational and irrational numbers, are used as models to practice the writing of indirect proofs. The decimal representations of rational and irrational numbers also are discussed. The lesson closes with work on rationalizing denominators of fractions representing irrational numbers.

5-3

RECIPROCALS OF THE POWER FUNCTIONS

The functions $f: x \to \frac{k}{x}$, whose graphs are hyperbolas, and the functions $f: x \to \frac{k}{x^2}$, which represent inverse square laws, are familiar examples of the reciprocals of the power functions $f: x \to \frac{k}{x^n}$. Other than the polynomial functions, these are the simplest rational functions. An analysis of their behavior near asymptotes and also their end behavior helps students to understand more complicated rational functions.

5-4

RATIONAL FUNCTIONS

A rational function is a function f such that $f(x) = \frac{p(x)}{q(x)}$ for all x in its domain, and p and q are polynomial functions. The behavior of rational functions is highly dependent on the zeros of q, for at these points the function is not continuous. The lesson analyzes two types of discontinuities: removable discontinuities, for which the zero of q is also a zero of p and the function can be made continuous by the insertion of a single point; and essential discontinuities, for which there is a vertical asymptote to the graph at the zero.

5-5

END BEHAVIOR OF RATIONAL FUNCTIONS

The end behavior of a rational function $f: x \to \frac{p(x)}{q(x)}$ is the same as the end behavior of the function defined as the quotient of the leading terms of the polynomials p and q. If the degree of p is greater than the degree of q, then the end behavior is the same as that of a polynomial. If the degree of p equals the degree of q, then there is a horizontal asymptote to the graph. If the degree of p is less than the degree of q, then the end behavior is the same as that of one of the reciprocals of the power functions studied in Lesson 5-3. Automatic graphers are used to picture these end behaviors.

5-6

THE TAN, COT, SEC, AND CSC FUNCTIONS

The properties of the sine and cosine functions are combined with those of the rational functions to review the basic properties of the four functions discussed in this lesson. Although these functions are not rational functions, they share some properties with the functions studied in Lessons 5-3 and 5-5. The secant and cosecant functions are reciprocals of the cosine and sine functions, and the tangent and cotangent functions are quotients of the sine and cosine. All of these functions have essential discontinuities at the zeros of either the sine or the cosine functions.

5-7

RATIONAL EQUATIONS

Throughout the chapter, students have been reviewing algebraic manipulations with rational expressions. These skills now are applied to the solution of equations involving rational expressions. As with some of the equations in Chapter 3, a step used in solving an equation may not be reversible, so all potential solutions must be checked to see if they are actual solutions.

5-8

TESSELLATING POLYGONS AND REGULAR POLYHEDRA

In this lesson, arithmetic and algebra are applied to solve two geometry problems.

As the artist Maurits Escher has shown so dramatically, tessellations can be constructed from figures of many varieties and forms; however, only three regular polygons tessellate: the equilateral triangle, square, and hexagon. In this lesson, the proof of this fact is shown to be equivalent to the solving of the Diophantine equation $\frac{1}{n} + \frac{1}{s} = \frac{1}{2}$. There are only five regular polyhedra: the regular tetrahedron, cube, regular octahedron, regular dodecahedron, and regular icosahedron. The proof that there are only five regular polyhedra is shown to be equivalent to the solving of the related Diophantine equation $\frac{1}{n} + \frac{1}{s} > \frac{1}{2}$.

This reading lesson not only covers some beautiful mathematics related to the solving of rational equations, but also it illustrates the use of one branch of mathematics to solve significant problems in another branch.

DAILY PACING CHART 5

DAILY PACING CHART ■ CHAPTER 5

Every chapter of UCSMP *Precalculus and Discrete Mathematics* includes lessons, a Progress Self-Test, and a Chapter Review. For optimal student performance, the self-test and review should be covered. (See *General Teaching Suggestions: Mastery* on page T35 of this Teacher's Edition.) By following the pace of the Full Course given here, students can complete the entire text by the end of the year. Students following the pace of the Minimal Course spend more time when there are quizzes and on the Chapter Review and will generally not complete all of the chapters in this text.

When chapters are covered in full (the recommendation of the authors), then students in the Minimal Course can cover 11 chapters of the book. For more information on pacing, see *General Teaching Suggestions: Pace* on page T34 of this Teacher's Edition.

DAY	MINIMAL COURSE	FULL COURSE
1	5-1	5-1
2	5-2	5-2
3	5-3	5-3
4	Quiz (TRF); Start 5-4.	Quiz (TRF); 5-4
5	Finish 5-4.	5-5
6	5-5	5-6
7	5-6	Quiz (TRF); 5-7
8	Quiz (TRF); Start 5-7.	5-8
9	Finish 5-7.	Progress Self-Test
10	5-8	Chapter Review
11	Progress Self-Test	Chapter Test (TRF)
12	Chapter Review	
13	Chapter Review	
14	Chapter Test (TRF)	

TESTING OPTIONS

TESTING OPTIONS

- ■ Quiz for Lessons 5-1 through 5-3
- ■ Quiz for Lessons 5-4 through 5-6
- ■ Chapter 5 Test, Form A
- ■ Chapter 5 Test, Form B
- ■ Chapter 5 Test, Cumulative Form

A Quiz and Test Writer is available for generating additional questions, additional quizzes, or additional forms of the Chapter Test.

INDIVIDUAL DIFFERENCES

PROVIDING FOR INDIVIDUAL DIFFERENCES

The student text is written for, and tested with, average students. It also has been successfully used with better and more poorly prepared students.

The Lesson Notes often include Error Analysis and Alternate Approach features to help you with those students who need more help. Students of all abilities often learn from their peers and may benefit from small group work referenced as appropriate throughout the Notes. A blackline Lesson Master (in the Teacher's Resource File), keyed to the chapter objectives, is provided for each lesson to allow more practice. (However, since it is important to keep up with the daily pace, you are not expected to use all of these masters. Again, refer to the suggestions for pacing on page T34.) Extension activities are provided in the Lesson Notes for those students who have completed the particular lesson in a shorter amount of time than is expected, even in the Full Course.

Rational Numbers and Rational Functions

CHAPTER 5

This chapter uses rational functions in situations such as weightlessness. (American astronaut John F. Young)

As you saw in the last chapter, the systems of integers and polynomials over the reals have much the same structure. In each system, the operations of addition, subtraction, and multiplication are closed. Any integer is either prime or can be factored into a product of primes; any polynomial is either irreducible or can be factored into a product of irreducible polynomials.

$$80 = 2^4 \cdot 5 \qquad x^4 + x^3 + 4x^2 + 9x + 5 = (x+1)^2(x^2 - x + 5)$$

There is a Quotient-Remainder Theorem that enables a division problem to be treated without leaving the system.

$$80 = 29 \cdot 2 + 22 \qquad x^4 + x^3 + 4x^2 + 9x + 5 = (x^2 + 6)(x^2 + x - 2) + 3x + 17$$

However, these systems are not closed with respect to division. The quotient of two elements in the system may not be in the system.

$$\frac{80}{29}, \text{ not an integer} \qquad \frac{x^4 + x^3 + 4x^2 + 9x + 5}{x^2 + 6}, \text{ not a polynomial}$$

The quotient of two integers is a *rational number*; the quotient of two polynomial functions is a *rational function*. Rational numbers arise in measurements and as ratios. Yet not all real numbers can be expressed as quotients of integers. The *irrational* numbers $\frac{\sqrt{3}}{2}$ and π are common in geometry. Similarly, not all real functions are quotients of polynomial functions. The *nonrational* expression $\frac{\sin \theta}{\cos \theta}$ gives rise to an important trigonometric function. Some rational and nonrational functions behave quite differently from those functions discussed in preceding chapters.

CHAPTER 5

This is another one of those chapters for which it is tempting to spend a great deal of time. Few students are as facile with rational expressions as their teachers would like them to be. Since the material from this chapter is reviewed continually, we recommend a lesson a day pace, with a possible extra day or two for quizzes. Allowing a day for the Progress Self-Test, 1 or 2 days for the Review, and a day for a Chapter test, this chapter should take 11–14 days to cover. (See the Daily Pacing Chart on page 287D.)

USING PAGE 287

Clarify the terms to be used in this chapter. You may wish to write the terms in two columns on the chalkboard:

Chapter 4 terms

arithmetic	algebra
*integer (in base b)	*polynomial
integer	polynomial function

Chapter 5 terms

arithmetic	algebra
*(simple) fraction	*rational expression
rational number	rational function
irrational number	nonrational function

The starred (*) terms refer to how a mathematical idea is represented. The other terms are independent of the representation.

LESSON 5-1

RESOURCES
- Lesson Master 5-1

OBJECTIVES

A Simplify rational expressions.
B Identify numbers as rational or irrational.
F Prove properties of rational and irrational numbers.
I Apply rational expressions and rational equations.

TEACHING NOTES

The greatest difficulty students have with understanding the definition of rational number is that a number is rational regardless of the way it is written. The number $\frac{9}{4}$ is rational in any other written form, including 2.25, $\sqrt{\frac{81}{16}}$, 1.5^2, $\frac{9\pi}{4\pi}$, and $2 + \frac{1}{4}$. On the other hand, a rational expression is a written form. The expression $\sqrt{(x+5)^4}$ is not a rational expression even though it is identically equal to $(x + 5)^2$ (for all real values of x).

Example 1 proves that the set of rational numbers is closed under addition. Question 19 in Lesson 5-2 asks students to prove that the set of rational numbers also is closed under multiplication. Since both operations for rational numbers satisfy the associative, commutative, and distributive properties, and have identities and inverses, the set of rational numbers is a field. As a result, the set is closed under subtraction and division (except for division by zero).

LESSON

5-1 Rational Numbers and Rational Expressions

A number that can be written as a ratio of two integers is said to be *rational*.

Definition

A real number r is **rational** if and only if there exist integers a and b ($b \neq 0$) such that $r = \frac{a}{b}$.

Rational numbers are not always written as fractions. However, if you can express a given number as a ratio of two integers, that number must be rational. For example, since $-0.283 = \frac{-283}{1000}$, there is a way to write -0.283 as a ratio of integers, and so -0.283 is rational. $1\frac{3}{5}$ is rational because $1\frac{3}{5} = \frac{8}{5}$. 0 is rational because 0 can be written as $\frac{0}{1}$. On the other hand, $\frac{\sqrt{2}}{2}$ is a ratio, but not a ratio of two integers. You will see in the next lesson that $\frac{\sqrt{2}}{2}$ is not a rational number.

The rational numbers are **closed** with respect to the basic operations arithmetic, addition and multiplication. That is, the sum or product of any two rational numbers is rational.

Example 1 Show that the sum of any two rational numbers is a rational number.

Solution Suppose that r and s are any two rational numbers. Then there exist integers a, b, c, and d with $b \neq 0$ and $d \neq 0$ such that $r = \frac{a}{b}$ and $s = \frac{c}{d}$. So

$$r + s = \frac{a}{b} + \frac{c}{d}$$

$$= \frac{a}{b} \cdot \frac{d}{d} + \frac{b}{b} \cdot \frac{c}{d} \quad \text{Multiplication Property of 1}$$

$$= \frac{ad}{bd} + \frac{bc}{bd} \quad \text{Multiplication of Fractions}$$

$$= \frac{ad + bc}{bd} \quad \text{Distributive Property}$$

Because sums and products of integers are integers, $ad + bc$ and bd are integers. Furthermore, $bd \neq 0$ because neither b nor d is 0. Hence $\frac{ad + bc}{bd}$ is a ratio of integers, meaning that $r + s$ is a rational number.

Just as the quotient of two integers is a rational number, so the quotient of two polynomials is a *rational expression*.

Definition

An algebraic expression $r(x)$ is said to be a **rational expression** if and only if it has the form $r(x) = \frac{p(x)}{q(x)}$, where $p(x)$ and $q(x)$ are polynomials (with $q(x)$ not the zero polynomial).

You saw in Chapter 4 that the arithmetic of polynomials is similar to that of integers. A similar analogy holds between the arithmetic of rational expressions and the arithmetic of fractions.

For instance, to add two rational expressions, you can use the same rules that you use to add fractions.

Example 2 A group of N students ordered one sausage pizza and one pepperoni pizza of the same size. They divided the sausage pizza evenly. Since two students did not want any pepperoni pizza, they divided that one into $N - 2$ pieces. Suppose a student ate one piece of each kind of pizza. Expressed as a fraction of one whole pizza, how much pizza did that student eat?

Solution One piece of the sausage pizza is $\frac{1}{N}$ of one pizza; one piece of the pepperoni pizza is $\frac{1}{N-2}$ of one pizza. So the student ate a total of

$$\frac{1}{N} + \frac{1}{N-2}$$

of one whole pizza. Since $N > 2$, these expressions are defined for all values of N possible in this situation. Add the rational expressions.

$$\frac{1}{N} + \frac{1}{N-2} = \frac{1}{N} \bullet \frac{N-2}{N-2} + \frac{1}{N-2} \bullet \frac{N}{N} \quad \text{Multiplication Property of 1}$$

$$= \frac{N-2}{N^2-2N} + \frac{N}{N^2-2N} \quad \text{Multiplication of fractions}$$

$$= \frac{2N-2}{N^2-2N} \quad \text{Distributive Property}$$

Thus the total amount eaten by the student is $\frac{2N-2}{N^2-2N}$ of one whole pizza.

In Example 2, the domain of N is the set of integers greater than 2. But the algebraic properties hold for all values of N except for those values for which the expressions are undefined. The above steps show that, $\forall$ real numbers N with $N \neq 0$ and $N \neq 2$, $\frac{1}{N} + \frac{1}{N-2} = \frac{2N-2}{N^2-2N}$. That is, when N is replaced by any real number other than 0 or 2, the expressions on both sides have the same value. Such an equation is called an *identity*.

Definitions

An **identity** is an equation that is true for all values of the variables for which both sides are defined. The set of all such values is called the **domain of the identity.**

The analogy between rational numbers and rational expressions is used to explain the techniques for manipulating rational expressions. Facility with such manipulations is one of the objectives of this lesson. The theme of logic also is maintained by using the manipulations in the context of proving identities (see **Question 11**).

Technology In **Example 2**, to check whether the following is an identity,
$\frac{1}{N} - \frac{1}{N-2} = \frac{2N-2}{N^2-2N}$,
have a computer (program or spreadsheet) generate a table of values for $N = 1, 2, 3, \ldots$ and evaluate the expression on each side of the identity. Remind students that the domain is the set of integers $N > 2$.

Making Connections
Students who have studied from other UCSMP texts may have had less experience with algebraic manipulations. However, our research indicates that teachers think that almost all students can benefit from additional practice such as that provided in the text.

In **Example 5**, manipulative skills are used to simplify the difference quotient developed in Lesson 2-1. Besides providing a reason for the manipulation, this example prepares students for calculus.

Error Analysis In simplifications like those found in **Example 4**, if you see $\frac{a^2 + 5a}{a^2 + 10a} = \frac{5a}{10a} = \frac{1}{2}$, then you have a student who "slashes" identical addends. Discuss $\frac{15 + 5}{15 + 10}$; does it equal $\frac{20}{25}$ or $\frac{1}{2}$?

In **Example 4b**, remind students to determine restrictions using the original problem and not the product in simplified form.

Two ways of dividing fractions are given on page 291. Another method for dividing fractions is to multiply both the numerator and denominator by the reciprocal of the divisor.

ADDITIONAL EXAMPLES

1. Show that the quotient of two nonzero rational numbers is a rational number.
Any two nonzero rational numbers can be denoted as $\frac{a}{b}$ and $\frac{c}{d}$, where a, b, c, and d are integers with $abcd \neq 0$. Then $\frac{a}{b} \div \frac{c}{d} = \frac{a}{b} \cdot \frac{d}{c} = \frac{ad}{bc}$. The fraction $\frac{ad}{bc}$ is a rational number because the nonzero integers are closed under multiplication.

2. For a party, Mrs. Ortiz orders one vegetarian and two meat submarine sandwiches. Of the N students at the party, three would eat only vegetarian subs and two would eat only meat subs. Mrs. Ortiz sliced the vegetarian sub into $N - 2$ equal sandwiches and each meat sub into $N - 3$ equal sandwiches. If Horacio eats one of each kind of sandwich, what fraction of one whole submarine sandwich has he eaten?
$\mathbf{\frac{2N - 5}{(N - 2)(N - 3)}}$

The domain of the identity $\frac{1}{N} - \frac{1}{N-2} = \frac{2N-2}{N^2 - 2N}$ is the set of real numbers N such that $N \neq 0$ and $N \neq 2$. These conditions are called **restrictions** on N. You have already learned that paying attention to restrictions on variables can keep you from making errors when solving equations. Later in this chapter you will see that the domain of an identity may signal critical features of the graph of a function.

As with a numerical fraction, a rational expression is **in lowest terms** if its numerator and denominator have no common factor other than 1. To check whether $\frac{2N-2}{N^2-2N}$ from Example 2 is in lowest terms, factor the numerator and denominator: $\frac{2N-2}{N^2-2N} = \frac{2(N-1)}{N(N-2)}$. Since they have no factor in common other than 1, the rational expression is in lowest terms.

If an expression is not in lowest terms, it can be simplified. You have simplified rational expressions many times in previous courses. For example, if $b \neq 0$, then $\frac{b^7}{b^5} = \frac{b^2 \cdot \cancel{b^5}}{\cancel{b^5}} = b^2$. However be very cautious. Eliminating identical *factors* from both numerator and denominator is allowed because it is in essence dividing both by the same nonzero number. But you may not eliminate identical *terms*. In Example 3, the x^2 and the 4 cannot be eliminated.

Example 3 Simplify $\frac{x^2 + 4x + 4}{x^2 - 4}$.

Solution First note that the expression is defined for all real numbers x such that $x^2 - 4 \neq 0$. Since $x^2 - 4 = (x - 2)(x + 2)$, the restriction that $x^2 - 4 \neq 0$ is equivalent to $x \neq 2$ and $x \neq -2$. When $x \neq 2$ and $x \neq -2$,

$$\frac{x^2 + 4x + 4}{x^2 - 4} = \frac{(x+2)(x+2)}{(x-2)(x+2)}$$

$$= \frac{(x+2)\cancel{(x+2)}}{(x-2)\cancel{(x+2)}} \quad \text{because } x \neq -2 \text{ and so } x + 2 \neq 0$$

$$= \frac{x+2}{x-2}.$$

Thus the equation $\frac{x^2 + 4x + 4}{x^2 - 4} = \frac{x+2}{x-2}$ is an identity whose domain is the set of all real numbers except 2 and -2.

Before multiplying rational expressions, first try to find identical factors in the numerators and denominators.

Example 4 **a.** Multiply: $\frac{a^2 + 5a + 6}{a^2 - 3a - 10} \cdot \frac{a^2 - 25}{a + 3}$.
b. State any restrictions on the variable a.

Solution **a.** Factor the numerators and denominators and multiply.

$$\frac{a^2 + 5a + 6}{a^2 - 3a - 10} \cdot \frac{a^2 - 25}{a + 3} = \frac{\cancel{(a+3)}\cancel{(a+2)}}{\cancel{(a-5)}\cancel{(a+2)}} \cdot \frac{\cancel{(a-5)}(a+5)}{\cancel{(a+3)}} = a + 5$$

b. The denominators in the original problem must not be zero. So $a \neq 5$, $a \neq -2$, and $a \neq -3$.

290

Check The value of the original expression should equal the value of the answer for all possible values of a. For example, when $a = 2$, the original expression has value $\frac{20}{-12} \bullet \frac{-21}{5} = 7$, which is the value of $a + 5$ when $a = 2$.

When the numerator or denominator of a fraction includes a fraction, the original fraction is called a **complex fraction.** (This is a different use of the word "complex" than occurs in the term *complex number*.)

To simplify complex numerical fractions such as $\frac{\frac{2}{3}}{\frac{4}{5}}$, you may use one of two methods. One method is to replace the division $\frac{a}{b}$ by the equivalent multiplication $a \bullet \frac{1}{b}$.

Method (1): $\frac{\frac{2}{3}}{\frac{4}{5}} = \frac{2}{3} \bullet \frac{1}{\frac{4}{5}} = \frac{2}{3} \bullet \frac{5}{4} = \frac{10}{12} = \frac{5}{6}$.

A second method is to multiply both numerator and denominator by a number designed to clear the fractions. In this case multiply by 15, since 15 is divisible by the denominators 3 and 5.

Method (2): $\frac{\frac{2}{3}}{\frac{4}{5}} = \frac{\frac{2}{3}}{\frac{4}{5}} \bullet \frac{15}{15} = \frac{10}{12} = \frac{5}{6}$

In Example 5, a complex fraction occurs in an expression for slope. Method (1) above is used to simplify it.

Example 5 A graph of $f(x) = \frac{1}{x^2}$ for $x > 0$ is shown at right. What is the slope of the segment connecting the points $(x, f(x))$ and $(x + h, f(x + h))$?

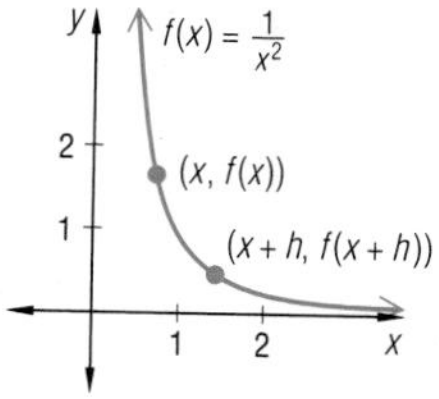

Solution The slope is

$$\frac{f(x+h) - f(x)}{(x+h) - x} = \frac{\frac{1}{(x+h)^2} - \frac{1}{x^2}}{h} \quad \text{if } h \neq 0,\ x + h \neq 0,\ x \neq 0$$

$$= \left(\frac{1}{(x+h)^2} - \frac{1}{x^2}\right) \bullet \frac{1}{h}$$

$$= \left(\frac{x^2}{(x+h)^2x^2} - \frac{(x+h)^2}{(x+h)^2x^2}\right) \bullet \frac{1}{h}$$

$$= \frac{x^2 - (x^2 + 2xh + h^2)}{x^2(x+h)^2h}$$

$$= \frac{-2xh - h^2}{hx^2(x+h)^2}$$

$$= \frac{-2x - h}{x^2(x+h)^2}.$$

3. Simplify $\frac{2x^2 - 5x - 3}{x^2 - 9}$.
$\frac{2x + 1}{x + 3}$

4.a. Multiply: $\frac{a^2 + 6a + 8}{a^2 - 5a - 14} \bullet \frac{a - 7}{a^2 - 16}$.
$\frac{1}{a - 4}$

b. State restrictions on a.
$a \neq 7, a \neq -2, a \neq 4$, and $a \neq -4$

5. A graph of $f(x) = x^2$ for $x > 0$ is shown below. What is the slope of the line connecting the point, $(x, f(x))$ and $(x + h, f(x + h))$?

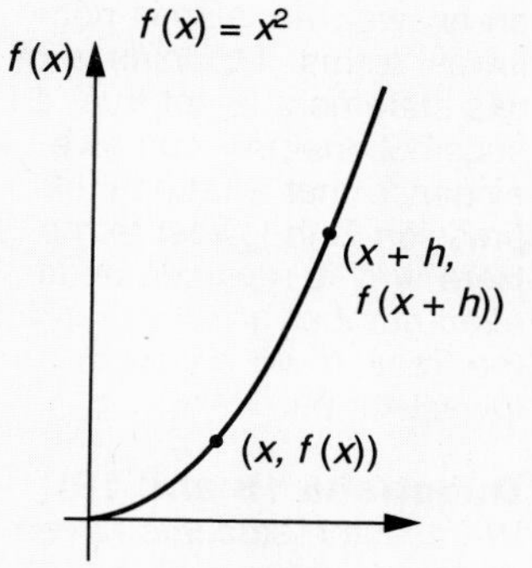

$2x + h$

NOTES ON QUESTIONS
Error Analysis for Question 6: Some students include zeros of the numerator. Remind them that while division by zero is not defined, division into zero always produces a quotient of zero. Using a calculator to divide by zero and into zero may be helpful.

Questions 8 and 9: You might invite students to do part **b** first, since restrictions depend upon the original expression.

Making Connections for Question 10: As with **Example 5**, this question anticipates the study of the derivative in Chapter 10. Doing some of these manipulations now decreases the quantity of new material to be mastered later.

NOTES ON QUESTIONS
Technology for Question 11: Students may wish to verify this identity by using an automatic grapher and graphing the left side with various values of *a*. Point out that for a given graph, the same number must replace all occurrences of *a* in the expression.

Question 17: If a rational expression with integer coefficients is not in lowest terms, then all but a finite number of integer substitutions will yield a fraction for an answer that also is not in lowest terms. The inverse of this statement is not true, so substitutions only can give evidence that a rational expression is in lowest terms. Here, if *N* is replaced by an even number greater than 2, the fraction will not be in lowest terms.

Questions 18 and 19: We assume students have done examples such as these before. If they have not, try simpler examples:

(1) Write $.\overline{3}$ as a ratio of two integers.
$10x = 3.333\ldots$
$x = .333\ldots$
$9x = 3$
$x = \frac{3}{9}$ or $\frac{1}{3}$
(2) Repeat (1) for $.\overline{27}$.
$100x = 27.272727\ldots$
$x = .272727\ldots$
$99x = 27$
$x = \frac{27}{99}$ or $\frac{3}{11}$
(3) For a stimulating discussion, repeat (1) for $.\overline{9}$.
$10x = 9.999\ldots$
$x = .999\ldots$
$9x = 9$
$x = 1 \therefore .\overline{9} = 1$

Question 20: Encourage students to solve this problem by graphing. Let *x* represent miles driven and *y* represent total rental cost.

ADDITIONAL ANSWERS
4., 5.c. See Additional Answers in the back of this book.

The restrictions are $h \neq 0$, $x \neq 0$, and $x + h \neq 0$. But $x + h \neq 0$ means that $x \neq -h$. Therefore the slope of the segment between $(x, f(x))$ and $(x + h, f(x + h))$ is

$$\frac{-2x - h}{x^2(x + h)^2} \text{ if } h \neq 0,\ x \neq 0 \text{ and } x \neq -h.$$

Check Because identities are true for all values of the variables for which both sides are defined, when you check you almost always have a large choice of values to substitute for the variables. Let $x = 2$ and $h = 3$. This yields the points (2, .25) and (5, .04) on the graph. The slope of the segment containing them is $\frac{.04 - .25}{5 - 2}$ or -.07. The final expression gives the value $\frac{-2 \cdot 2 - 3}{4 \cdot 25} = \frac{-7}{100}$, which checks.

Questions

Covering the Reading

In 1–3, determine whether each statement is *true* or *false*. If false, give a counterexample.

1. *Every rational number is an integer.* **False, counterexample: $\frac{2}{3}$ is a rational number, but it is not an integer.**
2. *All integers are rational numbers.* **True**
3. *The sum of any two rational expressions is a rational expression.* **True**
4. Show that the difference of any two rational numbers is a rational number. **See margin.**
5. An organization has 400 envelopes to stuff and mail.
   a. If this job is split among *N* people, how many envelopes will each person have to stuff and mail? $\frac{400}{N}$
   b. If two more people help, how many fewer envelopes will each person have to stuff and mail? $\frac{800}{N^2 + 2N} = \frac{800}{N(N + 2)}$
   c. Check your answers to parts **a** and **b** by letting $N = 5$. **See margin.**
6. If *x* can be any real number, what is the domain of the expression $\frac{x + 6}{2x - 5}$? $\{x: x \neq \frac{5}{2}\}$
7. A student did the following simplification.
$$\frac{x^2 - 5\not{x}}{\not{x}} = x^2 - 5$$
   a. Explain what the student did wrong. **See margin.**
   b. Simplify $\frac{x^2 - 5x}{x}$ correctly. $x - 5$

In 8 and 9, **a.** simplify and **b.** state any restrictions on the variables.

8. $\frac{2}{x - 5} + \frac{3}{x^2 - 25}$ **See margin.**
9. $\frac{2y + 6}{y^2 - 2y - 24} \cdot \frac{y - 6}{y + 3}$ **See margin.**

10. The graph of $f(x) = \frac{1}{x}$ for $x > 0$ is shown at the right. Write the slope of $\overline{AB}$ as a simple fraction.
$\frac{-1}{x^2 + hx}$, $x \neq 0$, and $x \neq -h$

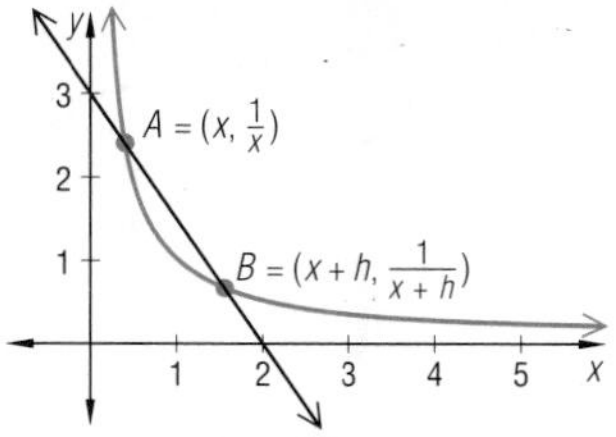

Applying the Mathematics

11. a. Show that $\frac{4-a}{x(x-a)} + \frac{x-4}{x^2-ax} = \frac{1}{x}$ is an identity by transforming the expression on the left side into that on the right. See margin.
b. What is the domain of the identity? $\{x: x \neq 0 \text{ and } x \neq a\}$

In 12–15, write the complex fraction as a simple fraction and give all restrictions on the variable.

12. $\dfrac{\frac{3}{4}}{\frac{2}{x}}$ $\frac{3x}{8}$, $x \neq 0$

13. $\dfrac{\frac{3}{y+1}}{\frac{2}{y}}$ $\frac{3y}{2y+2}$, $y \neq -1$, $y \neq 0$

14. $\dfrac{2 + \frac{1}{x^2}}{1 - \frac{1}{x^2}}$ $\frac{2x^2+1}{x^2-1}$, $x \neq -1$, $x \neq 1$, and $x \neq 0$

15. $1 - \dfrac{1}{1 + \frac{a}{1-a}}$ a, $a \neq 1$

16. Grade point averages (GPA) are often computed by assigning 4 points to every A, 3 to every B, 2 to every C, 1 to every D, and 0 to every F, and then dividing by the number of grades assigned. Let a, b, c, d, and f represent the number of each of these grades made by a certain student.
a. Write an expression that can be used to compute the student's GPA.
b. Must your answer to part **a** represent a rational number? Explain.
a) GPA = $\frac{a(4) + b(3) + c(2) + d(1) + f(0)}{a + b + c + d + f}$; b) See margin.

17. In Example 2, the rational expression $\frac{2N-2}{N^2-2N}$ has the value $\frac{12}{35}$ when $N = 7$. Explain how this value gives evidence that the rational expression is in lowest terms. 12 and 35 do not have any common factors.

Review

In 18 and 19, show that the repeating decimal is a rational number by writing it as a ratio of two integers. *(Previous course)*

18. $.\overline{148}$ See margin.

19. $.\overline{012345679}$ See margin.

20. One car rental company charges $39.95 a day plus 30¢ per mile after 100 miles. Another company charges $32.95 a day plus 35¢ per mile after 100 miles. A traveler plans to rent a car for 5 days and drive over 100 miles. For what total mileage is the second rental plan cheaper? *(Lesson 3-7)* less than 800 miles

7.a. The student eliminated identical terms, not identical factors.

8.a. $\frac{2x + 13}{x^2 - 25}$
b. restrictions: $x \neq -5$ and $x \neq 5$

9., 11.a., 16.b., 18., 19. See Additional Answers in the back of this book.

NAME ____________

LESSON **MASTER 5–1**
QUESTIONS ON **SPUR** OBJECTIVES

SKILLS *Objective A (See pages 341–344 for objectives.)*
In 1–6, a. simplify and b. state any restrictions on the variables.

1. $\frac{3}{p^2+6p+5} - \frac{4}{p^2+2p-15}$
a. $\frac{-p-13}{(p-3)(p+1)(p+5)}$
b. $p \neq -5, -1, 3$

2. $\frac{2r-3}{2r+3} + \frac{7}{r}$
a. $\frac{2r^2+11r+21}{r(2r+3)}$
b. $r \neq 0, -\frac{3}{2}$

3. $\frac{4x^2-13x-12}{2x^2+3x-2} \div \frac{x-4}{2x^2-5x+2}$
a. $\frac{4x^2-5x-6}{x+2}$
b. $x \neq -2, \frac{1}{2}, 2, 4$

4. $\dfrac{\frac{1}{x} + \frac{3}{x^2}}{\frac{9}{x^2} - 1}$
a. $\frac{1}{3-x}$
b. $x \neq 0, -3, 3$

5. $\frac{4}{y^2-4} + \frac{6}{2y^2-7y+6}$
a. $\frac{14y}{(y+2)(y-2)(2y-3)}$
b. $y \neq -2, 2, \frac{3}{2}$

6. $\frac{5}{n^2+n-12} - \frac{2}{n^2+6n+8}$
a. $\frac{3n+16}{(n-3)(n+2)(n+4)}$
b. $n \neq -4, -2, 3$

7. a. Prove that the equation $\frac{4x}{4x-3} + \frac{3}{3-4x} = 1$ is an identity.
$\frac{4x}{4x-3} + \frac{3}{3-4x} = \frac{4x}{4x-3} + \frac{-3}{4x-3} = \frac{4x-3}{4x-3} = 1$

b. What is the domain of the identity?
all real numbers except $\frac{3}{4}$

NAME ____________
Lesson MASTER 5–1 (page 2)

SKILLS *Objective B*
In 8–11, show that the given number is rational by expressing it as a ratio of two integers.

8. $\sqrt{\frac{1}{4}}$ $\frac{1}{2}$
9. $0.4\overline{18}$ $\frac{414}{990}$
10. 3.232 $\frac{3232}{1000}$
11. $3.23\overline{2}$ $\frac{2909}{900}$

PROPERTIES *Objective F*
12. *Prove:* If a rational number is divided by a nonzero rational number, the result is a rational number.
Suppose r and s are rational, $s \neq 0$. Then $\exists$ integers a, b, c, d with $b, c, d \neq 0$, such that $r = \frac{a}{b}$, $s = \frac{c}{d}$. So $\frac{r}{s} = \frac{\frac{a}{b}}{\frac{c}{d}} = \frac{a}{b} \cdot \frac{d}{c} = \frac{ad}{bc}$. Since ad and bc are integers, and neither b nor c is 0, $\frac{r}{s}$ is rational.

In 13 and 14, determine whether the statement is *true* or *false*. Justify your answer.

13. All whole numbers are rational numbers.
True; any whole number n can be written as $\frac{n}{1}$, and n and 1 are integers.

14. The reciprocal of any nonzero rational number is rational.
True; the rational number $\frac{a}{b}$ (where a, b are nonzero integers) has reciprocal $\frac{b}{a}$ which is rational.

USES *Objective I*
15. It takes 100 experienced construction workers N days to construct a large office building. It will take 80 novice workers $2N - 30$ days to construct the same building.

a. Write an expression for the fraction of work done on a single day by the 100 experienced workers. $\frac{1}{N}$

b. Write an expression for the fraction of work done on a single day by the 80 novice workers. $\frac{1}{2N-30}$

c. Write and simplify a rational expression that represents the fraction of work done on a single day by both groups of workers working together. $\frac{3N-30}{2N^2-30N}$

d. What restrictions must be placed on N? $N \neq 0, 15$

NOTES ON QUESTIONS
Question 27: Because the two graphs will coincide on the same screen, some students do not believe that the second function was graphed. A technique you can use to convince students that the graphs coincide is to graph g with a slight "offset." Depending upon your selection of range, try $y = \frac{2N - 2}{N^2 - 2N} + k$ where k could be .2 or 1. This new graph will be "parallel" to the others and convincing to students.

FOLLOW-UP

MORE PRACTICE
For more questions on SPUR Objectives, use *Lesson Master 5-1,* shown on page 293.

EXTENSION
Question 15 is reminiscent of continued fractions. A small research project on using continued fractions to approximate mathematical constants or functions may be appropriate. Ask students to find x, the value of the following infinite continued fraction:

$$x = 1 + \cfrac{1}{1 + \cfrac{1}{1 + \ldots}}.$$

$\left(x = 1 + \frac{1}{x} \text{ and } x > 0, \text{ so } x = \frac{1 + \sqrt{5}}{2}.\right)$

ADDITIONAL ANSWERS
22.a. domain: the set of real numbers; range: $0 < y \le 4$

23.b. Suppose n is any odd integer. By definition, there exists an integer r such that $n = 2r + 1$. Then $n^2 = (2r + 1)^2 = 4r^2 + 4r + 1 = 2(2r^2 + 2r) + 1$. Since $2r^2 + 2r$ is an integer by closure properties, n^2 is an odd integer by definition.

27. See Additional Answers in the back of this book.

21. Use chunking to solve for x: $e^{2x} - 5e^x + 6 = 0$. *(Lesson 3-4)*
$x = \ln 2$ or $x = \ln 3$

22. Refer to the graph of g at the right.

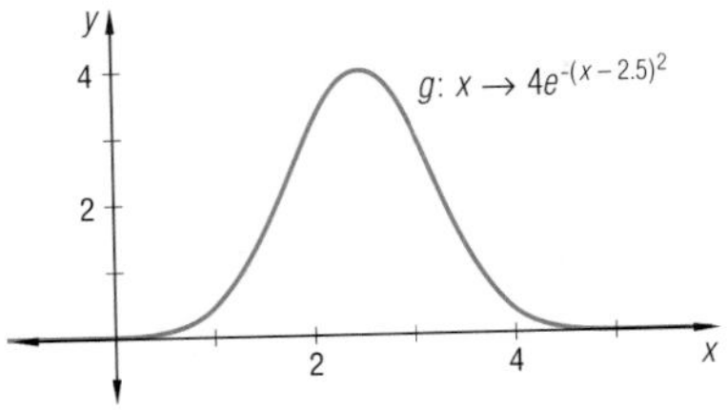

a. Identify the domain and range of the function.
b. Identify the maximum value of the function. $y = 4$
c. Identify the interval on which the function is increasing.
(Lessons 2-3, 2-2, 2-1)
a) See margin. c) $x \le 2.5$

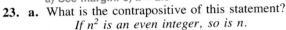

23. a. What is the contrapositive of this statement?
If n^2 is an even integer, so is n.
b. Prove the contrapositive. See margin.
c. Is the original statement true? *(Lessons 1-7, 1-5)* Yes
a) If n is not an even integer, then n^2 is not an even integer.

24. Give the negation of the universal conditional $\forall\, x$, *if $p(x)$, then $q(x)$.* *(Lesson 1-5)* $\exists\, x$ *such that $p(x)$ and not $q(x)$.*

In 25 and 26, simplify without using a calculator. *(Previous course)*

25. $(7\sqrt{2})^2$ 98

26. $\sqrt{7500} - \sqrt{363}$ $39\sqrt{3}$

Exploration

27. Graph the functions f and g where
$$f\colon N \to \frac{1}{N} + \frac{1}{N - 2}$$
and $$g\colon N \to \frac{2N - 2}{N^2 - 2N}.$$
Describe what happens. See margin.

294

LESSON

5-2

Irrational Numbers

The diagram at the right shows an isosceles right triangle with sides of lengths 1, 1, and x. By the Pythagorean Theorem,

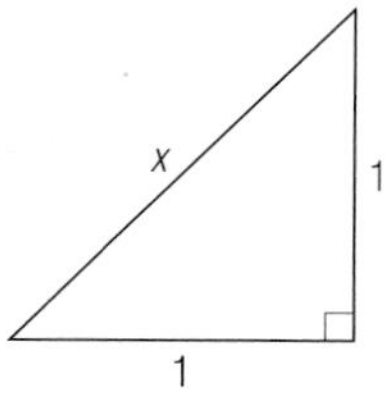

$$x^2 = 1^2 + 1^2 = 2,$$

and so

$$x = \sqrt{2}.$$

Pythagoras, after whom the Pythagorean Theorem is named, led a group of philosophers in Greece during the sixth century B.C. For many years, the Pythagoreans believed that all natural phenomena could be explained in terms of the natural numbers 1, 2, 3, ... One part of their belief was that any number could be expressed as a ratio of two natural numbers. In the triangle pictured above, this would mean that the length of the hypotenuse, which is $\sqrt{2}$, could be expressed as $\frac{a}{b}$ for some natural numbers a and b.

The Greek stamp above illustrates a simple case of the Pythagorean Theorem.

However, despite their belief that $\sqrt{2}$ should be rational, some ingenious Pythagorean mathematicians proved otherwise. They proved that $\sqrt{2}$ could not be written as a ratio of two natural numbers. Such numbers were called **irrational** because they could not be obtained as ratios of integers. This discovery shook the very foundations of their belief system and created a major crisis in their group. Proclus, a Greek mathematician who lived about 1000 years later, wrote, "It is told that those who first brought out the irrationals from concealment into the open perished in a shipwreck, to a man."

The most common proof that $\sqrt{2}$ is irrational is an indirect proof, specifically, a proof by contradiction. It makes use of the fact that if the square of an integer is even, then the integer itself must be even, which was to be proved in Question 23 of Lesson 5-1. The proof also makes use of the fact that any fraction can be written in lowest terms, with no factors greater than 1 common to both the numerator and the denominator.

Theorem (The Irrationality of $\sqrt{2}$)

$\sqrt{2}$ is irrational.

Analysis The statement to be proved is
s: *$\sqrt{2}$ is irrational.*

The proof starts by assuming $\sim s$; in other words, by assuming
$\sim s$: *$\sqrt{2}$ is a rational number.*

Then a contradiction is deduced. The existence of this contradiction will imply that the assumption is false and hence that s is true.

LESSON 5-2

RESOURCES

■ Lesson Master 5-2

OBJECTIVES

B Identify numbers as rational or irrational.
C Simplify expressions involving radicals.
F Prove properties of rational and irrational numbers.

TEACHING NOTES

To help students understand the proof that $\sqrt{2}$ is irrational, review the Proof by Contradiction Theorem in Lesson 4-8. Draw a vertical line on the chalkboard. To its left, write the form of the proof by contradiction argument as is presented in that lesson. To its right, write the corresponding steps of the proof that $\sqrt{2}$ is irrational. Remind students why a is an even number if a^2 is even.

A terminating decimal can always be made a repeating decimal by affixing an infinite number of 0's to the right. For this reason, fact (3) on page 296 can be replaced by the statement: If a real number x has a repeating decimal expansion, then it is a rational number.

The application of *modus tollens* to show that π is irrational can be presented as follows: The first premise, $p(x) \Rightarrow q(x)$, is what Lambert proved: *If x is a nonzero rational number, then tan x is irrational.* Let $x = \frac{\pi}{4}$. Then the second premise, $\sim q(c)$, is that *tan* $\frac{\pi}{4}$ *is rational.* This is so because tan $\frac{\pi}{4}$ = 1. Therefore, we can conclude $\sim p(c)$, that is, that $\frac{\pi}{4}$ *is irrational or zero.* Since we know $\frac{\pi}{4} \neq 0$, $\frac{\pi}{4}$ must be irrational.

Using the fact that $\frac{\pi}{4}$ is irrational to conclude that π is irrational requires the following theorem: The quotient (or product) of an irrational and a rational number is irrational. This can be proved by the method of **Example 1**.

The creation of an irrational number $x = 0.1010010001 \ldots$ that is not a square root illustrates a method of creating an infinite number of decimal representations of irrational numbers. Some students, however, confuse a pattern with a repeating cycle. Stress that they are not necessarily the same. The construction of an irrational number as a decimal expansion relies on creating a pattern that is not just a block of repeating digits. For variety, consider $x = 0.1234567891011121314 \ldots$, which is irrational, but with a different pattern than $0.1010010001 \ldots$.

In **Example 2**, to convince students that multiplying a number of the form $c + \sqrt{d}$ by its conjugate will always eliminate the square root, show the actual multiplication: $(c + \sqrt{d})(c - \sqrt{d}) = c^2 + c\sqrt{d} - c\sqrt{d} + \sqrt{d^2} = c^2 + d$.
You may wish to extend this to $(a\sqrt{b} + c\sqrt{d}) \cdot (a\sqrt{b} - c\sqrt{d})$ in order to prepare students for **Questions 12 and 13**.

Proof **Assume $\sqrt{2}$ is a rational number. Then $\sqrt{2}$ can be expressed as a ratio of integers $\frac{a}{b}$ where $b \neq 0$ and $\frac{a}{b}$ is written in lowest terms. That is, a and b have no factors in common greater than 1.**

$$\sqrt{2} = \frac{a}{b}$$

$$\Rightarrow \quad 2 = \frac{a^2}{b^2} \quad \text{Square both sides.}$$

$$\Rightarrow \quad 2b^2 = a^2 \quad \text{Multiply both sides by } b^2.$$

But since $2b^2 = a^2$, by the definition of even, a^2 is even. Hence, as seen earlier, a is even. By the definition of even, then, $a = 2k$ for some integer k. Now we show that b must also be even.

$$2b^2 = (2k)^2 \quad \text{Substitute } 2k \text{ for } a.$$

$$\Rightarrow \quad 2b^2 = 4k^2 \quad \text{Laws of exponents}$$

$$\Rightarrow \quad b^2 = 2k^2 \quad \text{Divide both sides by 2.}$$

But since $b^2 = 2k^2$, b^2 is even; thus b is even. Now it has been proved that both a and b are even, and so they must have a common factor of 2. Consequently, on the one hand a and b have no factors in common, and on the other hand, they have a common factor of 2. This contradiction shows that the assumption that $\sqrt{2}$ is a rational number is false. Therefore, $\sqrt{2}$ is not a rational number.

Proofs of the irrationality of other square roots of primes p can follow the same pattern. In each proof you would need to make use of the following theorem:

If n^2 is divisible by p, then n is divisible by p.

The irrationality of $\sqrt{3}$, $\sqrt{5}$, $\sqrt{6}$, $\sqrt{7}$, $\sqrt{8}$, $\sqrt{10}$, $\sqrt{11}$, $\sqrt{12}$, $\sqrt{13}$, $\sqrt{14}$, $\sqrt{15}$, and $\sqrt{17}$ was proved by Theodorus of Cyrene around 390 B.C. (You will be asked to prove the irrationality of some of these square roots in the Questions.) In general, it can be proved that for any integer n, if n is not a perfect square, then $\sqrt{n}$ is irrational. On the other hand, of course, if n is a perfect square, then $\sqrt{n}$ is an integer and thus rational.

Recall the following two general facts from previous courses:
(1) Any terminating decimal number can be expressed as a ratio of two integers.
(2) Any infinite repeating decimal number can be expressed as a ratio of two integers.
Therefore:
(3) If a real number x has a terminating or repeating decimal expansion, then x is a rational number.

Because $\sqrt{2}$ is irrational, it follows from (3) (by *modus tollens*) that its decimal expansion is neither terminating nor repeating. The value 1.4142136 that you might see when you use a calculator is simply an approximation.

In 1761, Johann Heinrich Lambert proved that if x is a nonzero rational number, then e^x and tan x are irrational. Thus $e^1 = e$ is irrational. Also, using *modus tollens*, since tan $\frac{\pi}{4} = 1$, a rational number, then $\frac{\pi}{4}$ is irrational. So π is irrational.

296

Gregory Chudnovsky

David Chudnovsky

The values of $\frac{22}{7}$ and 3.14 that you may have used for π are just approximations. Even your calculator only gives an approximation. Currently at least 1,011,196,691 decimal places of π are known, based on results in 1989 by Gregory and David Chudnovsky at Columbia University in New York.

Recall from Lesson 2-9 that Cantor proved that the set of rational numbers is countable and the set of real numbers is uncountable. Thus there are more irrational numbers than rational numbers. You can generate irrational numbers quite easily by using the converse of (3) on the previous page: If x is rational, then its decimal expansion is either terminating or repeating. (See Question 18.) It follows that irrational numbers are exactly those numbers whose decimal expansions neither terminate nor repeat.

For instance, let $x = 0.1010010001\ldots$

The number of 0s between successive 1s keeps increasing by 1.

The decimal expansion of x neither terminates nor repeats; hence x must be an irrational number.

Proof by contradiction is often used to prove properties involving irrational numbers. Example 1 illustrates such a proof.

Example 1 Prove by contradiction: The sum of a rational number and an irrational number is an irrational number.

Solution Let p be the statement to be proved. Write p formally as follows.

p: ∀ real numbers r and l, if r is rational and l is irrational, then r + l is irrational.

Start the proof by assuming $\sim p$. In other words, assume

~p: ∃ real numbers r and l such that r is rational and l is irrational and r + l is rational.

Then reason to a contradiction. Here is what you could write.

Proof **Assume the negation of the statement to be proved. That is, assume there is a rational number *r* and an irrational number *l* whose sum is a rational number, call it *s*. Then**

$$r + l = s$$
$$l = s - r.$$

As was shown in Lesson 5-1, Question 4, the difference of two rational numbers is also rational. Since *s* and *r* are rational, *l* is rational. This contradicts the part of the assumption that states *l* is irrational. Hence, the assumption as a whole is false; this means the original statement is true. That is, the sum of any rational number and any irrational number is an irrational number.

Error Analysis Some students may use $2 + \sqrt{3}$ as its own conjugate and write $(2 + \sqrt{3})(2 + \sqrt{3}) = 4 + 3 = 7$. They assume that the square roots add to zero. Have students compare the general cases $(x + y)^2 = x^2 + 2xy + y^2$ and $(x + y) \cdot (x - y) = x^2 - y^2$ to see why this does not work. (Students may check that $\frac{4}{5 - \sqrt{7}} = \frac{10 + 2\sqrt{7}}{9}$ by keying both expressions into a calculator and comparing results.)

ADDITIONAL EXAMPLES

1. Prove by contradiction: The product of a nonzero rational number and an irrational number is irrational.

Assume there is a nonzero rational number *r* and an irrational number *t* such that *rt* = *s* is rational. Then $t = \frac{r}{s}$. But the quotient of two nonzero rational numbers is a rational number. This contradicts the assumption that *t* is irrational. So, the assumption is false. Thus, the product of a nonzero rational number and an irrational number is irrational.

2. Rationalize the denominator of $\frac{2 + \sqrt{3}}{7 + \sqrt{5}}$.

$\frac{14 - 2\sqrt{5} + 7\sqrt{3} - \sqrt{15}}{44}$

NOTES ON QUESTIONS

Questions 1-3: A Venn diagram can be used to picture the relations among the sets.

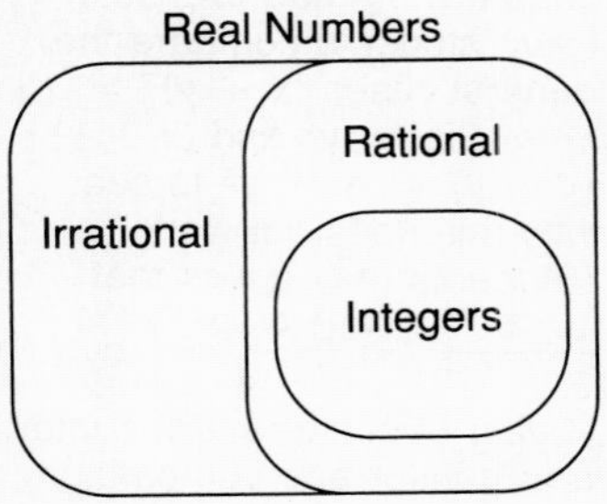

Question 5: Students should be able to generalize the proof to prove that for any prime number p, $\sqrt{p}$ is irrational.

Question 10: Point out to students that rational expressions can have irrational values, but only when the variables are irrational.

Question 12: The resulting expression can be checked with a calculator; its approximate value is 1.850.

Questions 14 and 15: Point out that "any two" can mean that the same number is used twice.

Making Connections for Question 16: Some students may have seen this result before. It is important and should be discussed.

Question 18: Although this question is review, it is directly related to the content of this lesson.

Note that when a rational *expression* is evaluated, the result may be an irrational number. The reason is that irrational numbers can be substituted for the variables in such expressions. For instance,

$$\frac{x+3}{x^2-1}$$

is a rational expression. But if you substitute $x = \sqrt{2}$ into it, you obtain

$$\frac{(\sqrt{2})+3}{(\sqrt{2})^2-1} = \frac{\sqrt{2}+3}{2-1} = \sqrt{2}+3.$$

This number is irrational by the result of Example 1 because it is the sum of $\sqrt{2}$, which is irrational, and 3, which is rational.

Sometimes, it is necessary to be able to rewrite rational expressions involving square roots. Such work often uses the Difference of Squares identity

$$(x+y)(x-y) = x^2 - y^2 \quad \forall \text{ real numbers } x \text{ and } y.$$

For instance: $(\sqrt{11}+\sqrt{3})(\sqrt{11}-\sqrt{3}) = 11 - 3 = 8.$

In general, if c is an integer or the square root of an integer and d is an integer, then $c + \sqrt{d}$ and $c - \sqrt{d}$ are called **conjugates**, and their product is the integer $c^2 - d$. Here $\sqrt{11}+\sqrt{3}$ and $\sqrt{11}-\sqrt{3}$ are conjugates and their product is the integer 8.

Multiplying the numerator and denominator of a fraction by the conjugate of the denominator removes the square root(s) from the denominator. This process is called *rationalizing the denominator*.

Example 2 Rationalize the denominator of $\frac{4}{5-\sqrt{7}}$.

Solution Multiply both numerator and denominator by $5+\sqrt{7}$, the conjugate of $5-\sqrt{7}$.

$$\begin{aligned}\frac{4}{5-\sqrt{7}} &= \frac{4}{5-\sqrt{7}} \cdot \frac{5+\sqrt{7}}{5+\sqrt{7}}\\ &= \frac{20+4\sqrt{7}}{25-7}\\ &= \frac{20+4\sqrt{7}}{18}\\ &= \frac{2(10+2\sqrt{7})}{2 \cdot 9}\\ &= \frac{10+2\sqrt{7}}{9}.\end{aligned}$$

Before the days of calculators and computers, rationalizing the denominator helped in evaluating expressions involving square roots. It was much easier to approximate $\frac{10+2\sqrt{7}}{9}$ (requiring division by 9) than to approximate $\frac{4}{5-\sqrt{7}}$ (requiring division by an approximation to an infinite decimal). Today, calculators and computers do the arithmetic, and rationalizing the denominator is ordinarily done to prepare the way for further computations.

298

Questions

Covering the Reading

In 1–3, tell whether the statement is *true* or *false*.

1. $\exists$ *a real number that is irrational.* True

2. *Some irrational numbers can be expressed as ratios of integers.* False

3. *Some rational expressions have irrational values.* True

4. To use a proof by contradiction, what is the first assumption you make?
Assume the negation of the original statement.

5. Use the proof of the irrationality of $\sqrt{2}$ as a model to complete the steps below in the proof that $\sqrt{3}$ is irrational.

a. Let $\sqrt{3} = \frac{a}{b}$ where a and b are integers with $b \neq 0$ and $\frac{a}{b}$ is in lowest terms. Square both sides and solve for a^2. $a^2 = 3b^2$

b. What does your result in part **a** imply about a^2? about a?

c. Let $a = 3k$ for some integer k. Substitute $3k$ for a in part **a** and solve the result for b^2. $b^2 = 3k^2$

d. What does your result in part **c** imply about b^2? about b?

e. Show that the results from parts **b** and **d** lead to a contradiction. Use this contradiction to finish the proof. b) a^2 and a have a factor of 3. d) b^2 and b have a factor of 3. e) See margin.

6. a. Is $\sqrt{4}$ a rational number? Why or why not? Yes, $\sqrt{4} = \frac{2}{1}$.

b. Is $\sqrt{4}$ an irrational number? Why or why not? No, 2 is a rational number.

7. Determine whether each decimal expansion represents a rational or an irrational number. Explain.

a. .313113111311113 . . . **b.** $.\overline{31}$ **c.** .31

The number of 1s between successive 3s keeps increasing by 1.

See margin.

8. *True* or *false?* e^2 is irrational. True

9. Use Example 1 as a model to prove the following statement: *The difference of any rational number and any irrational number is an irrational number.* See margin.

10. a. Evaluate the expression $\frac{5 - x}{x^2 - 2}$ when $x = \sqrt{3}$. $5 - \sqrt{3}$

b. Is the result a rational or an irrational number? Explain. See margin.

11. a. Give the conjugate of $5 + \sqrt{3}$. $5 - \sqrt{3}$

b. Rationalize the denominator of $\frac{10}{5 + \sqrt{3}}$. $\frac{25 - 5\sqrt{3}}{11}$

c. Check your answer to part **b** by finding decimal approximations to each fraction. $\frac{10}{5 + \sqrt{3}} \approx 1.4854315$, $\frac{25 - 5\sqrt{3}}{11} \approx 1.4854315$

ADDITIONAL ANSWERS

5.e. Both a and b have a common factor of 3. However, from the beginning of the proof, $\frac{a}{b}$ is assumed in lowest terms, thus having no common factors. There is a contradiction, so the negated statement is false, and the original statement ($\sqrt{3}$ is irrational) must be true.

7.a. irrational; The decimal expansion neither terminates nor repeats.
b. rational; The decimal expansion repeats.
c. rational; The decimal expansion terminates.

9. Assume the negation of the original statement is true. Thus, there is a rational number r and an irrational number i whose difference is a rational number d. Then, $r - i = d$ and $r - d = i$. However, by the closure properties of rational numbers, i is rational because the difference of two rational numbers is a rational number. Thus, there is a contradiction, so the assumption must be false, and the original statement is therefore true.

10.b. irrational; $\sqrt{3}$ is an irrational number, and the difference of an irrational number and a rational number is another irrational number.

In 12 and 13, rationalize the denominator.

12. $\dfrac{3\sqrt{5}+2}{3\sqrt{5}-2}$ $\frac{49 + 12\sqrt{5}}{41}$

13. $\dfrac{4x}{x\sqrt{6}-2x\sqrt{3}}$ $\frac{2\sqrt{6} + 4\sqrt{3}}{-3}$

Applying the Mathematics

In 14 and 15, determine whether the statement is *true* or *false*. If false, give a counterexample.

14. *The sum of any two irrational numbers is an irrational number.* **See margin.**

15. *The product of any two irrational numbers is an irrational number.* **See margin.**

16. Recall the Quadratic Formula $x = \dfrac{-b \pm \sqrt{b^2 - 4ac}}{2a} \Leftrightarrow ax^2 + bx + c = 0$ and $a \neq 0$. If a, b, and c are all rational numbers, under what conditions will the values of x be

a. rational? **See margin.** **b.** irrational? **See margin.**

17. Prove: The reciprocal of $\sqrt{7} + \sqrt{6}$ is $\sqrt{7} - \sqrt{6}$. **See margin.**

Review

18. The long division at the right converts the rational number $\frac{29}{54}$ into a decimal. The remainder at each step of the division has been circled. **a-c) See margin.**

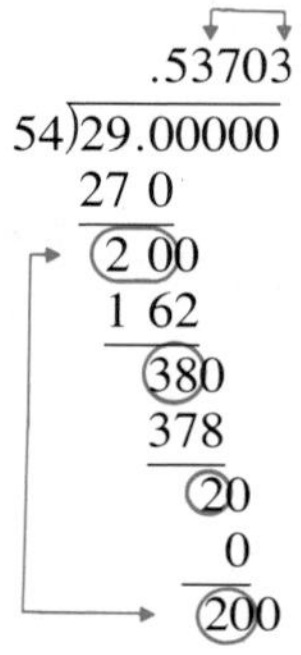

a. Use the Quotient-Remainder Theorem to explain why, after 54 steps (or less), some integer (here 20) must repeat as the remainder.

b. What effect does the repetition of a remainder have on the decimal expansion of $\frac{29}{54}$?

c. Generalize this long division to explain why the decimal expansion of every rational number is terminating or repeating.

d. What is the decimal equal to $\frac{29}{54}$? *(Lessons 5-1, 4-2)* $\frac{29}{54} = 0.5\overline{370}$

19. Prove: *If a rational number is multiplied by a rational number, the result is a rational number.* *(Lesson 5-1)* **See margin.**

20. The electrical circuit at the right is a parallel circuit with resistances R_1 and R_2. The total resistance R is given by the formula

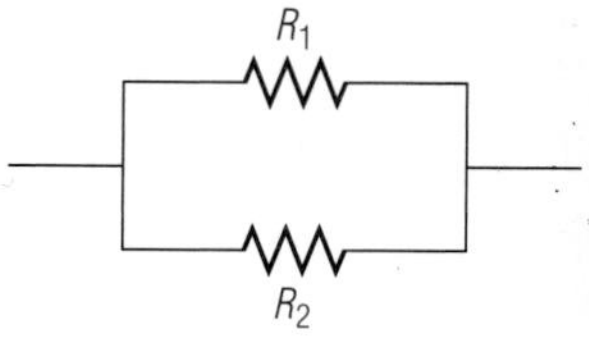

$$R = \frac{1}{\frac{1}{R_1} + \frac{1}{R_2}}.$$

Find a simple fraction that expresses the value of R. *(Lesson 5-1)*
$R = \frac{R_1 R_2}{R_1 + R_2}$

FOLLOW-UP

MORE PRACTICE

For more questions on SPUR Objectives, use *Lesson Master 5-2,* shown on page 301.

EXTENSION

In some situations, rationalizing the numerator is appropriate. Define a function f by the rule $f(x) = \sqrt{x}$ for all $x > 0$. In calculus, limits such as the following must be computed:

$$\lim_{h\to 0} \frac{f(2+h) - f(2)}{h}.$$

What is the limit?

$$\frac{f(2+h) - f(2)}{h} = \frac{\sqrt{2+h} - \sqrt{2}}{h} = \frac{\sqrt{2+h} - \sqrt{2}}{h} \cdot \frac{(\sqrt{2+h} + \sqrt{2})}{(\sqrt{2+h} + \sqrt{2})} = \frac{1}{\sqrt{2+h} + \sqrt{2}}$$

Therefore,

$$\lim_{h\to 0} \frac{f(2+h) - f(2)}{h} = \lim_{h\to 0} \frac{1}{\sqrt{2+h} + \sqrt{2}} = \frac{1}{2\sqrt{2}}.$$

PROJECTS

The projects for Chapter 5 are described on page 338. **Projects 1 and 3** are related to the content of this lesson.

ADDITIONAL ANSWERS

14. False; counterexample: π and $3 - \pi$ sum to 3, which is rational.

15. False; counterexample: Let $a = \sqrt{2}$ and $b = \sqrt{8}$. Then, $a \cdot b = \sqrt{2} \cdot \sqrt{8} = \sqrt{16} = 4$, but 4 is a rational number.

16.a. when $b^2 - 4ac$ is zero or a perfect square
b. when $b^2 - 4ac > 0$ and not a perfect square

300

In 21 and 22, **a.** simplify; **b.** state any restrictions on the variables; and **c.** check your answer. *(Lesson 5-1)*

21. $\frac{3}{x} + \frac{3x}{x-4}$ **See below.**

22. $\frac{t+1}{t-1} - \frac{t-1}{t+1}$ **See below.**

In 23 and 24, solve over the set of real numbers. *(Lessons 3-7, 3-1)*

23. $(y + 1)(2y - 5) > 0$ **$y < -1$ or $y > \frac{5}{2}$**

24. $\sqrt{x - 1} + 1 = 0$ **no solution**

25. *Multiple choice.* Which equation could generate the graph at the right?
(a) $y = 2^x$
(b) $y = \log_2 x$
(c) $y = 2x$
(d) $y = 2\sqrt{x}$
(Lessons 2-7, 2-6, Previous course) **(b)**

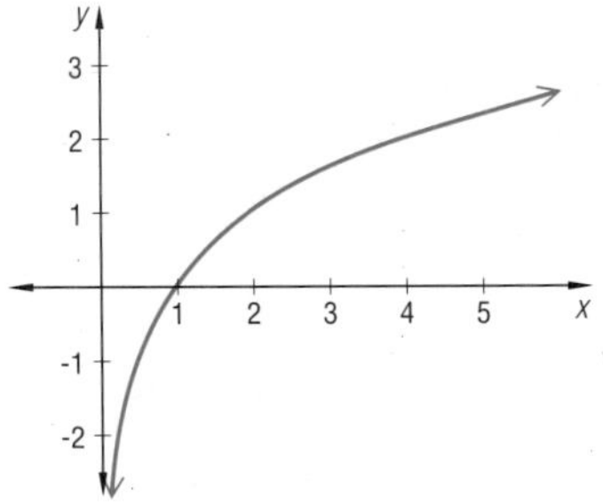

Exploration

26. In Question 17, the conjugate of a number of the form $\sqrt{x} + \sqrt{y}$ was found to be its reciprocal. In general, when will this happen? **when $|x - y| = 1$**

21. a) $\frac{3(x^2 + x - 4)}{x(x-4)}$; **b) restrictions: $x \neq 0$, $x \neq 4$**
c) Let $x = 2$ in the original and simplified expressions.
$\frac{3}{2} + \frac{3(2)}{2-4} = -\frac{3}{2}$, **and** $\frac{3(2^2 + 2 - 4)}{2(2-4)} = -\frac{6}{4}$, **which checks.**

22. a) $\frac{4t}{t^2 - 1}$; **b) restrictions: $t \neq 1$, $t \neq -1$**
c) Let $t = 3$ in the original and simplified expressions.
$\frac{3+1}{3-1} - \frac{3-1}{3+1} = \frac{3}{2}$, **and** $\frac{4 \cdot 3}{3^2 - 1} = \frac{12}{8}$, **which checks.**

17., 18.a.-c., 19. See Additional Answers in the back of this book.

NAME ______

LESSON **MASTER 5–2**
QUESTIONS ON **SPUR** OBJECTIVES

■ **SKILLS** *Objective B (See pages 341–344 for objectives.)*

1. If a real number is irrational, then its decimal representation is either **terminating** or **repeating**.

In 2–5, identify each number as rational or irrational. Justify your answer.

2. $\frac{1}{2 - \sqrt{3}}$ **irrat.; since $\sqrt{3}$ is irrational.**

3. $\sqrt{841}$ **rational; $\sqrt{841} = 29$**

4. 2.1212212221 . . . **irrat.; neither repeating nor terminating**

5. $-5.1\overline{37}$ **rational; it is a repeating decimal**

■ **SKILLS** *Objective C*

In 6–8, rationalize the denominator.

6. $\frac{11\sqrt{2}}{1 - \sqrt{2}}$ **$-22 - 11\sqrt{2}$**

7. $\frac{-5\sqrt{3}}{\sqrt{7} - \sqrt{3}}$ **$\frac{-5\sqrt{21} + 15}{4}$**

8. $\frac{\sqrt{a} - \sqrt{b}}{\sqrt{a} + \sqrt{b}}$ **$\frac{a + b - 2\sqrt{ab}}{a - b}$**

9. Rationalize the *numerator* in the expression $\frac{\sqrt{3} + \sqrt{2}}{2}$. **$\frac{1}{2(\sqrt{3} - \sqrt{2})}$**

■ **PROPERTIES** *Objective F*

10. Prove that $\sqrt{17}$ is irrational. **Assume $\sqrt{17}$ is rational, then ∃ integers a, b ($b \neq 0$) with $\sqrt{17} = \frac{a}{b}$ in lowest terms. Thus, $17 = \frac{a^2}{b^2}$ and $17b^2 = a^2$. So 17 is a factor of a^2 and thus of a. Then $a = 17k$ and $17b^2 = (17k)^2 = 289k^2$, and $b^2 = 17k^2$. Therefore, 17 is a factor of b^2 and thus of b, contradicting the assumption that $\frac{a}{b}$ is in lowest terms. Therefore, $\sqrt{17}$ is irrational.**

52 *Continued* *Precalculus and Discrete Mathematics* © Scott, Foresman and Company

NAME ______
Lesson MASTER 5–1 (page 2)

11. a. *True* or *false*? The quotient of two irrational numbers is irrational. **false**

b. If true, prove the statement. Otherwise give a counterexample.
$\sqrt{2}$ and $\sqrt{8}$ are irrational, but $\frac{\sqrt{2}}{\sqrt{8}} = \frac{\sqrt{2}}{2\sqrt{2}} = \frac{1}{2}$, which is rational.

12. Prove that if a and b are integers and a is not a perfect square, then $\sqrt{a} + b$ is irrational.
Suppose $\sqrt{a} + b$ is rational. Then since $-b$ is rational (since b is), $(\sqrt{a} + b) + (-b) = \sqrt{a}$ is rational, contradicting the fact that $\sqrt{a}$ is irrational (since a is not a perfect square). Thus, $\sqrt{a} + b$ must be irrational.

Precalculus and Discrete Mathematics © Scott, Foresman and Company 53

LESSON 5-3

RESOURCES
- Lesson Master 5-5
- Quiz for Lessons 5-1 through 5-3

OBJECTIVES

G Use limit notation to describe the behavior of rational functions.
J Graph quotients of polynomial or trigonometric functions.
K Relate the limit of a function to its graph, and find equations for its asymptotes.

TEACHING NOTES

Alternate Approach Students should have encountered the word *asymptote* in earlier courses. The initial discussion of asymptotes begins with a function that is quite familiar to UCSMP students. If your students have never studied the graph of $y = \frac{k}{x}$, or are unfamiliar with the term *hyperbola,* then you may want to begin by graphing a number of equations of this form with different values of k. To do this, a table of values is helpful. Consider the part of the graph in the first quadrant that is near the y-axis. The table below gives some values of $\frac{1}{x}$ for positive values of x near zero.

x	1	0.5	0.1	0.01
$\frac{1}{x}$	1	2	10	100

x	0.001	0.0001	10^{-10}
$\frac{1}{x}$	1000	10,000	10^{10}

As x gets closer to 0 from the right, the function values get larger and larger; they

LESSON

5-3

Reciprocals of the Power Functions

At the right is the familiar graph of the function f defined by $f(x) = \frac{1}{x}$. Both the domain and the range of f are the set of nonzero real numbers. The graph is a hyperbola with two branches, one in the first quadrant and one in the third quadrant.

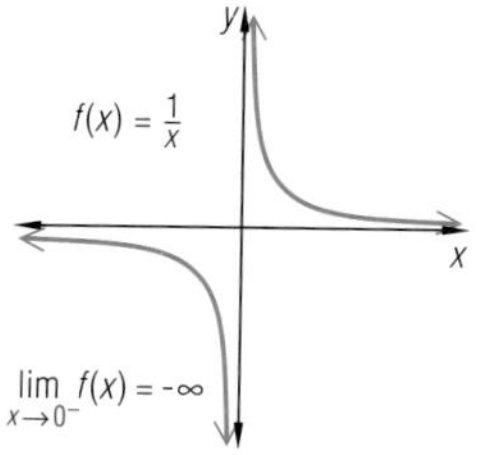

The function f is different in two respects from the polynomial and other functions analyzed so far. First, it is not continuous on any interval that contains $x = 0$ but is continuous everywhere else. Second, as x gets closer and closer to 0, the graph of f gets closer and closer to the y-axis, which is a *vertical asymptote* to it.

The behavior of this function near $x = 0$ can be used to analyze many other functions and their graphs, so it will be analyzed in detail. Consider the part of the graph in the first quadrant near the y-axis, in blue at the right. As x approaches 0 *from the right*, the function values get larger and larger. For instance, when $x = 0.5$, $f(x) = 2$; when $x = 0.05$, $f(x) = 20$; and when $x = 0.0001$, $f(x) = 10{,}000$. The values of the function can be made as large as desired simply by taking x close enough to 0. You are familiar with this idea; it is the idea of limit. The only difference here is that x is approaching a particular value (0) from a particular direction (the right). We write

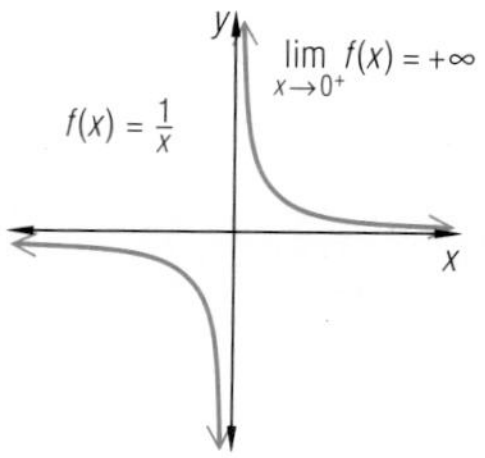

$$\lim_{x \to 0^+} \frac{1}{x} = +\infty,$$

which is read "the limit of $\frac{1}{x}$ as x approaches 0 from the right is positive infinity."

The behavior of the part of the graph of f that is near the y-axis in the third quadrant can be analyzed in a similar way. Note that as x gets closer to 0 *from the left*, the values of $\frac{1}{x}$ are negative and get smaller and smaller (though larger and larger in absolute value). They can be made as small as desired by taking x close enough to 0. For instance, to make $\frac{1}{x} < -25{,}000$, choose a value of x between -0.00004 and 0. We write

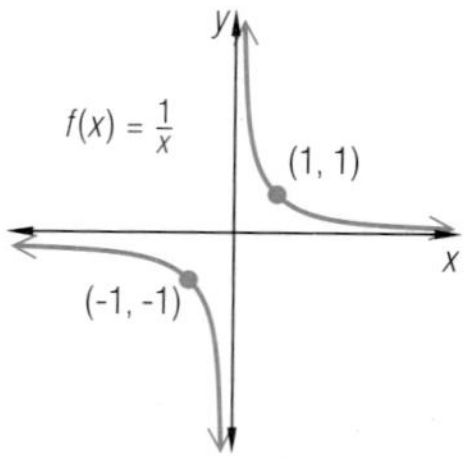

$$\lim_{x \to 0^-} \frac{1}{x} = -\infty,$$

which is read "the limit of $\frac{1}{x}$ as x approaches 0 from the left is negative infinity."

Notice the two new symbols in this limit notation.

$x \to a^-$	$x \to a^+$
x approaches a from the left.	x approaches a from the right.

This notation is used to define a vertical asymptote.

Definition

The line $x = a$ is a **vertical asymptote** for a function f if and only if $\lim_{x \to a^+} f(x) = +\infty$ or $\lim_{x \to a^+} f(x) = -\infty$ or $\lim_{x \to a^-} f(x) = +\infty$ or $\lim_{x \to a^-} f(x) = -\infty$.

Example 1 At the right is graphed the function h defined for all real numbers $x \neq 4$ by the rule $h(x) = \frac{1}{x-4}$.

a. Determine an equation for the vertical asymptote.

b. Use limit notation to describe the behavior of the function near the vertical asymptote.

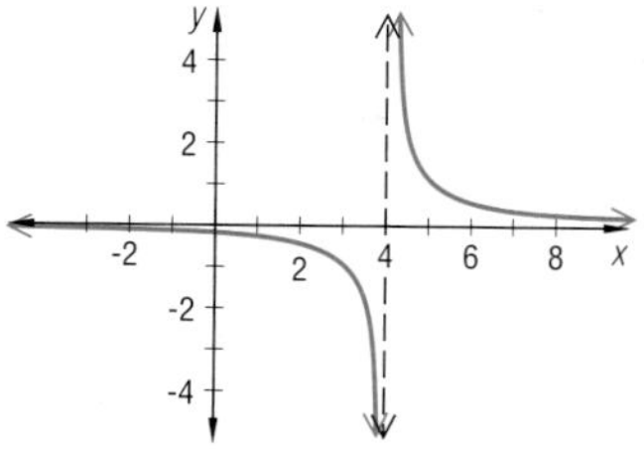

Solution By the Graph Translation Theorem the graph of $h(x) = \frac{1}{x-4}$ is the image of the graph of $f(x) = \frac{1}{x}$ under a translation that slides all points four units to the right; that is, under the translation $T_{4,0}$.

a. $f(x) = \frac{1}{x}$ has a vertical asymptote at $x = 0$. The translation shifts the asymptote for the image four units to the right. Hence, an equation for the vertical asymptote to the graph of h is $x = 4$.

b. The domain of h consists of all real numbers except $x = 4$. So h is not defined at $x = 4$. The behavior of the function near $x = 4$ is:

$$\lim_{x \to 4^+} h(x) = +\infty \text{ and } \lim_{x \to 4^-} h(x) = -\infty.$$

Many physical phenomena obey *inverse square laws*. That is, one quantity varies as the reciprocal of the square of another. (Do not confuse this inverse with a function inverse.) For instance, the intensity of light I varies inversely as the square of the distance d of the observer from it. That is $I = \frac{k}{d^2}$. A graph of the function $f: d \to \frac{k}{d^2}$ is shown here.

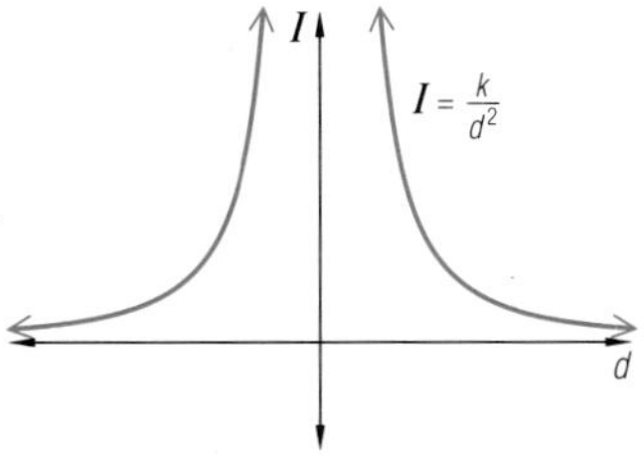

Think of the light source as being at the origin. If you travel along the x-axis from left to right, then the graph illustrates quite well what happens to the intensity of the light. When you are far left from the origin, the light is dim and the intensity is low. As you near the origin from the left, the light becomes more intense until at the origin it is blinding. Then as you continue past the origin to the right, the light again becomes less

increase without bound. This means that the value of the function can be made as large as desired by taking x close enough to 0. (For example, to obtain $f(x) > 1000$, take $0 < x < .001$.) We describe this behavior by writing $\lim_{x \to 0^+} \left(\frac{1}{x}\right) = +\infty$.

The behavior of the part of the graph of $f(x) = \frac{1}{x}$ that is near the y-axis in the third quadrant can be analyzed in a similar way. The table below gives values of the function for negative values of x near zero.

x	-1	-0.1	-0.01	-0.001
$\frac{1}{x}$	-1	-10	-100	-1000

x	-0.0001	-10^{-10}
$\frac{1}{x}$	-10,000	-10^{10}

We describe this behavior by writing $\lim_{x \to 0^-} \left(\frac{1}{x}\right) = -\infty$.

You might wish to note the variety of ways in which a reciprocal can be calculated:

(1) using a calculator 1/x key;

(2) using the arithmetic of fractions, that is, $\frac{1}{\frac{1}{10{,}000}} = 1 \cdot \frac{10{,}000}{1} = 10{,}000$;

(3) using negative exponents; $\frac{1}{10^{-4}} = 10^4 = 10{,}000$;

(4) using long division, $0.0001\overline{)1.0000}$ with quotient 1 0000.

In using limit notation, make sure students do not confuse this notation with function mapping notation. The notation $x \to 4$ means x *approaches* 4, while the notation $f: x \to 4$ means the function f *maps* x to the number 4, that is, $f(x) = 4$.

ADDITIONAL EXAMPLES

1. Consider the graph of the function f with rule $f(x) = \frac{1}{x-3} + 2$, where $x \neq 3$.

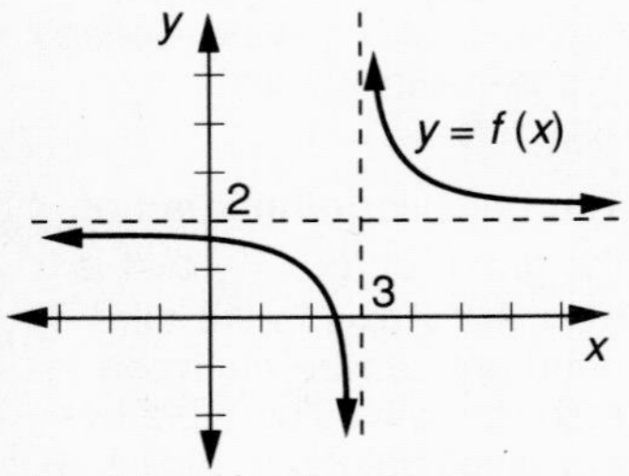

a. Determine an equation for the vertical asymptote.
$x = 3$

b. Use limit notation to describe the behavior of the function near the vertical asymptote.
$\lim_{x\to3^+} f(x) = +\infty$; and
$\lim_{x\to3^-} f(x) = -\infty$

2.a. Describe the behavior of the function $g: x \to \frac{5}{x^4}$ as x approaches 0.
$\lim_{x\to0^-} g(x) = +\infty$, and
$\lim_{x\to0^+} f(x) = +\infty$; and

b. Graph g.

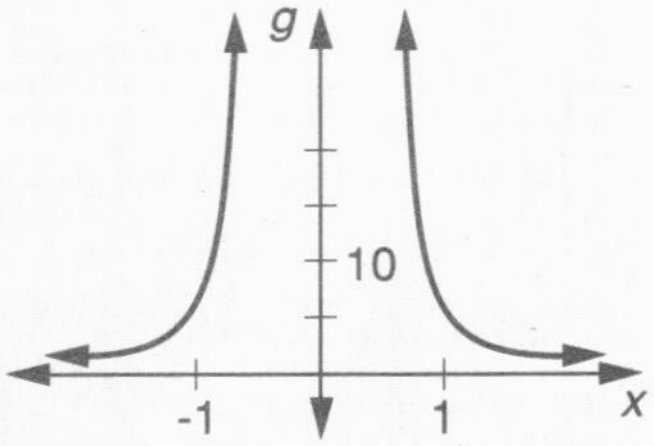

NOTES ON QUESTIONS
Questions 1-5: Discuss these questions so that students understand the symbolism and vocabulary in the lesson.

Making Connections for Question 6d: These are ideas that students will see more formally in a calculus course. Make sure students can complete this problem algebraically.

intense. The function is not continuous at the origin. The behavior at the origin can be described by

$$\lim_{d\to0^-} \frac{k}{d^2} = +\infty \qquad \lim_{d\to0^+} \frac{k}{d^2} = +\infty.$$

The functions f: $x \to y$ with $y = \frac{k}{x}$ and $y = \frac{k}{x^2}$ are the simplest examples of the reciprocals of the power functions. Like the power functions themselves, reciprocals of the odd power functions all have similar behavior, and reciprocals of the even power functions also behave similarly. However, the reciprocals of the odd powers and the reciprocals of the even powers behave differently from each other.

Example 2 **a.** Describe the behavior of the function f: $x \to \frac{-3}{x^5}$ as x approaches 0.
b. Graph f.

Solution

a. As x approaches 0 from the left, x^5 is negative and becomes closer and closer to 0 itself. Thus $f(x) = \frac{-3}{x^5}$ increases without bound in the positive direction. So $\lim_{x\to0^-} f(x) = +\infty$. As x approaches 0 from the right, x^5 is positive and itself approaches 0. Thus $f(x) = \frac{-3}{x^5}$ is negative and its magnitude increases without bound. So $\lim_{x\to0^+} f(x) = -\infty$.

b. To graph this function, you need to know its end behavior: $\lim_{x\to+\infty} f(x) = 0$ and $\lim_{x\to-\infty} f(x) = 0$. Using the end behavior and the behavior near the vertical axis, you can see that the graph resembles a reflection image of the hyperbola that opened this lesson, but its curve hugs the horizontal axis more tightly and the vertical axis less tightly.

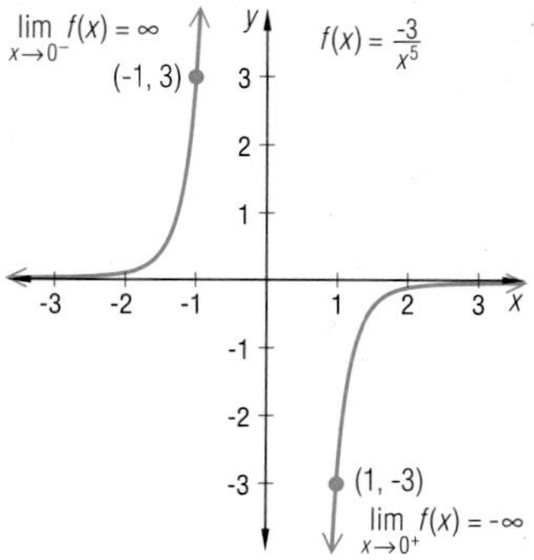

You can check the graphs of Examples 1 and 2 with an automatic grapher, but be aware that some automatic graphers do not handle vertical asymptotes well. For this reason, you should be sure to dot or color any asymptotes when you sketch such a graph on paper.

Questions

Covering the Reading

In 1–3, refer to the notation $\lim_{x\to3^+} f(x) = +\infty$.

1. *True* or *false?* The limit notation indicates that as $x \to 3^+$ the function values get larger and larger, increasing without bound. **True**

304

2. *Multiple choice.* Which of the following is the correct way to read "$x \rightarrow 3^+$"?
   (a) x is to the right of 3
   (b) x is to the left of 3
   (c) x approaches 3
   (d) x approaches 3 from the right
   (e) x approaches 3 from the left (d)

3. Sketch a graph of a function f with this property. See margin.

4. Explain in words and draw a graph of a function f with the indicated property.
   a. $\lim_{x \to 0^-} f(x) = -\infty$ **b.** $\lim_{x \to 0^-} f(x) = +\infty$
   See margin.

5. Consider the function h defined by the rule $h(x) = \frac{1}{x+6}$.
   a. The graph of h can be considered as the translation image of $f: x \rightarrow \frac{1}{x}$ under what translation? $T_{-6,\,0}$
   b. Determine the equation for any vertical asymptote to the graph of h.
   c. Describe the behavior of function h near the vertical asymptote using limit notation. b) $x = -6$; c) $\lim_{x \to -6^+} h(x) = +\infty$ and $\lim_{x \to -6^-} t(x) = -\infty$

6. Consider the parent function at the right defined by the rule $f(x) = \frac{1}{x^2}$.

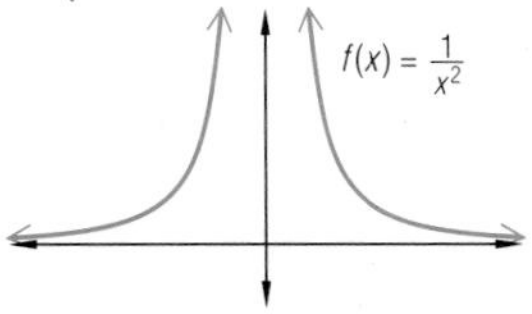

   a. *True* or *false?* $\lim_{x \to 0^+} f(x) = +\infty$. True
   b. *True* or *false?* $\lim_{x \to 0^-} f(x) = -\infty$. False
   c. Write an equation for the vertical asymptote to the graph of f.
   d. Determine the values of x that make
   i. $f(x) > 400$ $-.05 < x < .05$ **ii.** $f(x) > 250{,}000$. $-.002 < x < .002$
   c) $x = 0$

7. **a.** Describe the behavior of the function $g: t \rightarrow \frac{5}{t^4}$ as t approaches 0.
   b. Describe the end behavior of g. $\lim_{t \to +\infty} g(t) = 0$ and $\lim_{t \to -\infty} g(t) = 0$
   c. Graph g. See margin.
   a) $\lim_{t \to 0^+} g(t) = +\infty$ and $\lim_{t \to 0^-} g(t) = +\infty$

Applying the Mathematics

8. Consider the function f defined by $f(x) = \frac{1}{x^3}$.
   a. Give its domain and range of f. domain: $\{x: x \neq 0\}$; range: $\{y: y \neq 0\}$
   b. Indicate when f is increasing and when f is decreasing.
   c. Describe its end behavior using limit notation.
   d. Describe its behavior as $x \rightarrow 0^+$ and $x \rightarrow 0^-$ using limit notation.
   e. Tell whether f is odd or even. odd
   f. Graph f. b) decreasing over its entire domain as x goes from $-\infty$ to $+\infty$; c) $\lim_{x \to +\infty} f(x) = 0$ and $\lim_{x \to -\infty} f(x) = 0$; d,f) See margin.

9. On the average, Mars is about 1.5 times as far from the Sun as Earth. Compare the brightness of the Sun as seen from Mars with its brightness as seen from Earth. The sun's brightness seen from Earth is 2.25 times the brightness seen from Mars.

10. **a.** Graph h, where $h(y) = \frac{1}{y+6} - 3$. See margin.
    b. To what graph of this lesson is the graph of h congruent? the graph of $h(x) = \frac{1}{x-4}$ given in Example 1.

ADDITIONAL ANSWERS
3., 4., 7.c., 8.d., f., 10.a.
See Additional Answers in the back of this book.

NAME ______

LESSON **MASTER 5–3**
QUESTIONS ON **SPUR** OBJECTIVES

■ **PROPERTIES** *Objective G (See pages 341–344 for objectives.)*

1. Refer to the graph of f at the right.

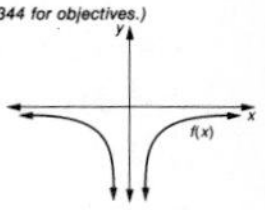

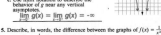

   a. Explain in words what happens to $f(x)$ as x approaches zero.
   As x approaches zero from the right or left, f(x) decreases without bound.
   b. Use limit notation to describe the behavior of f as x approaches zero for positive values of x. $\lim_{x \to 0^+} f(x) = -\infty$
   c. Use limit notation to describe the behavior of f as x approaches zero for negative values of x. $\lim_{x \to 0^-} f(x) = -\infty$
   d. Describe, in words, the end behavior of f.
   As x increases or decreases without bound, f(x) approaches zero.
   e. Describe, using limit notation, the end behavior of f.
   $\lim_{x \to -\infty} f(x) = \lim_{x \to \infty} f(x) = 0$

2. Consider the functions g and h where $g(x) = \frac{2}{x-3}$ and $h(x) = \frac{2}{x}$.
   a. What transformation maps the graph of h to the graph of g? $T_{3,0}$
   b. Describe, using limit notation, the behavior of g near any vertical asymptotes.
   $\lim_{x \to 3^+} g(x) = \infty$, $\lim_{x \to 3^-} g(x) = -\infty$
   c. Describe, using limit notation, the end behavior of g.
   $\lim_{x \to -\infty} g(x) = \lim_{x \to \infty} g(x) = 0$

■ **REPRESENTATIONS** *Objective J*

3. Let f be the function $f(x) = \frac{1}{x+1}$.

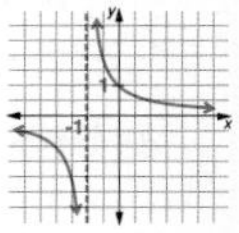

   a. Graph f.
   b. Write an equation for the vertical asymptote to the graph of f.
   $x = -1$

NAME ______

4. **a.** Graph the function g where $g(x) = \frac{-1}{x^4}$.

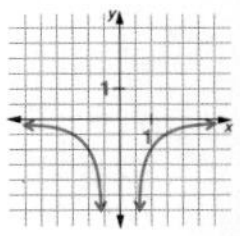

   b. Use limit notation to describe the end behavior of g.
   $\lim_{x \to -\infty} g(x) = \lim_{x \to +\infty} g(x) = 0$
   c. Use limit notation to describe the behavior of g near any vertical asymptotes.
   $\lim_{x \to 0^-} g(x) = \lim_{x \to 0^+} g(x) = -\infty$

5. Describe, in words, the difference between the graphs of $f(x) = \frac{1}{x^{2k}}$ and $g(x) = \frac{1}{x^{2k+1}}$ for positive integers k.
   The graph of f lies above the x-axis and is symmetric with respect to the y-axis, while the graph of g lies in Quadrants 1 and 3 and is symmetric with respect to the origin. Both graphs have the x-axis as a horizontal asymptote, f approaching it from above and g from below.

■ **REPRESENTATIONS** *Objective K*

In 6 and 7, a. write an equation for the graph's vertical asymptote and b. use limit notation to describe the behavior of the function near the vertical asymptote.

6.

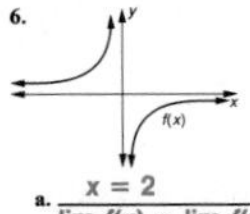

   a. $x = 2$
   b. $\lim_{x \to 2^-} f(x) = \lim_{x \to 2^+} f(x) = \infty$

7.

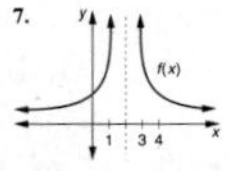

   a. $x = 0$
   b. $\lim_{x \to 0^-} f(x) = \lim_{x \to 0^+} f(x) = -\infty$

8. Let f be the function defined by $f(x) = \frac{1}{x^n}$ such that $\lim_{x \to 0^-} f(x) = \infty$.

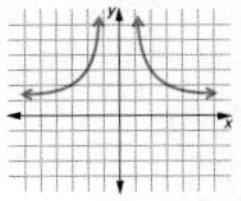

   a. Sketch a possible graph of f.
   b. Is n even or odd? even
   c. Find $\lim_{x \to 0^+} f(x)$. ∞

Review

11. Prove or disprove: *The difference of two irrational numbers is an irrational number.* *(Lesson 5-2)* See margin.

12. Rationalize the denominator: $\dfrac{5}{\sqrt{12} + \sqrt{3}}$. *(Lesson 5-2)* $\frac{5\sqrt{3}}{9}$

13. **a.** Write the negation:
The decimal expansion of x is terminating or the decimal expansion of x is repeating.

b. Write the contrapositive:
If x is a rational number, then the decimal expansion of x is terminating or repeating.

c. Use the result in part **b** to determine whether the given number is rational or irrational. *(Lessons 5-2, 1-5, 1-3)*

i. $.\overline{78}$ **ii.** .7878878887... (The number of 8s between successive 7s keeps increasing by 1.)

See margin.

In 14 and 15, **a.** simplify and **b.** state any restrictions on the variables. *(Lesson 5-1)*

14. $\dfrac{7}{3 - x} + \dfrac{11}{x - 3}$ a) $\frac{4}{x-3}$; b) $x \neq 3$

15. $\dfrac{a^2 + 2a}{3a^2 + 5a - 2} \cdot \dfrac{6a^2 + 13a - 5}{4a^2 + 20a + 25}$. $\frac{a}{2a+5}$; b) $a \neq \frac{1}{3}$, $a \neq -2$, $a \neq -\frac{5}{2}$

16. Use synthetic division to determine whether $x + 7$ is a factor of $3x^3 + 5x^2 - 14x + 2$. Justify your answer. *(Lessons 4-6, 4-5)*
No, because $3x^3 + 5x^2 - 14x + 2 = (x + 7)(3x^2 - 16x + 98) - 684$

17. Prove: *∀ integers a, b, and c, if a is a factor of b, and a is a factor of b + c, then a is a factor of c.* (Hint: $c = (b + c) - b$). *(Lesson 4-1)*
See margin.

In 18 and 19, **a.** write a rule for the given function if $f\colon t \to t^2 - 1$ and $g\colon t \to t - 1$. **b.** State the domain of the function. *(Lessons 3-3, 3-2, 2-1)*

18. $f \circ g$ a) $f \circ g = t^2 - 2t$; b) domain: the set of real numbers

19. $\dfrac{f}{g}$ a) $\frac{f}{g} = t + 1$; b) domain: $\{t\colon t \neq 1\}$

Exploration

20. Some automatic graphers use lines to connect the points they actually compute so that the resulting graph looks continuous. When a function has vertical asymptotes, this can cause problems. The graph at the right shows the result of using such an automatic grapher with $f(x) = \dfrac{-2}{x + 1} + 2$.

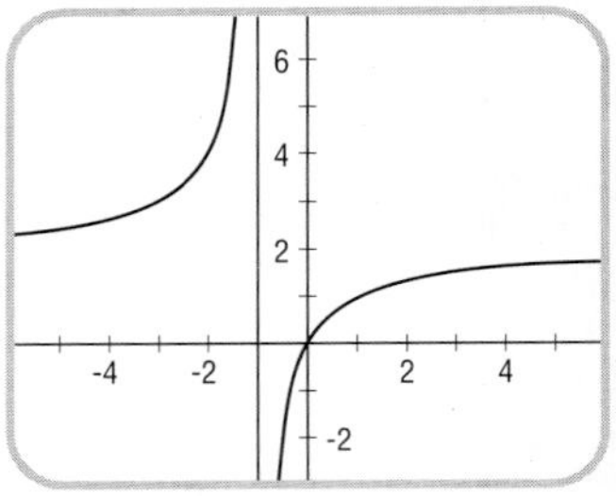

a. Use the techniques you learned in the lesson to determine the behavior of f as $x \to -1^+$ and as $x \to -1^-$. If it is appropriate, use limit notation to describe the behavior.

b. Use an automatic grapher to sketch a graph of f with a wider range than is shown here. Does your graph have the same problem as the graph shown here?

a) $\lim_{x \to -1^+} f(x) = -\infty$ and $\lim_{x \to -1^-} f(x) = +\infty$; b) See margin.

NOTES ON QUESTIONS
Technology for Question 20: Use this question as a springboard to discuss limitations of technology. Some automatic graphers display asymptotes as if they were part of a graph. Few, if any, show points of discontinuity. Large domain and range value settings hide local behavior while small ones reveal little about global or end behavior.

Make certain students see that the vertical line that appears on the graph was meant to connect the highest point on the left branch with the lowest point on the right branch. Automatic graphers may sometimes give a deceiving result. Hence, students must not rely only on technology but must apply mathematical understanding to interpret results correctly.

FOLLOW-UP

MORE PRACTICE
For more questions on SPUR Objectives, use *Lesson Master 5-3,* shown on page 305.

EXTENSION
Ask students where the graph of the power function $y = x^n$, $n \in \{1, 2, 3, \ldots\}$, intersects the graph of its reciprocal function.
(if n is odd, at (1, 1) and (-1, -1); if n is even, at (1, 1) and (-1, 1))

PROJECTS
The projects for Chapter 5 are described on page 338. **Project 2** is related to the content of this lesson.

EVALUATION
A quiz covering Lessons 5-1 through 5-3 is provided in the Teacher's Resource File.

ADDITIONAL ANSWERS
11., 13., 17., 20.b. See Additional Answers in the back of this book.

LESSON

5-4

Rational Functions

When a car gets better gas mileage, it is more fuel efficient and less expensive to run. But how much will be saved? If gas mileage increases from 15 $\frac{\text{mi}}{\text{gal}}$ to 25 $\frac{\text{mi}}{\text{gal}}$, will that save more, less, or the same amount of money as if mileage increases from 10 $\frac{\text{mi}}{\text{gal}}$ to 20 $\frac{\text{mi}}{\text{gal}}$?

Clearly the amount saved depends on how many miles the car is driven and on the cost per gallon of gas. Let's suppose that a gallon of gas costs \$1.25 and a car is driven 10,000 miles in a year. The problem can be analyzed using unit analysis.

$$\frac{\text{cost}}{\text{year}} = \frac{\text{miles}}{\text{year}} \cdot \frac{\text{cost}}{\text{mile}}$$

$$= \frac{\text{miles}}{\text{year}} \cdot \frac{\text{cost}}{\text{gal}} \cdot \frac{\text{gal}}{\text{mile}}$$

where the last unit is the reciprocal of miles per gallon.

Thus if C is the total cost in a year, then at 25 $\frac{\text{mi}}{\text{gal}}$,

$$C = 10{,}000\ \frac{\text{mi}}{\text{yr}} \cdot 1.25\frac{\$}{\text{gal}} \cdot \frac{1\ \text{gal}}{25\ \text{mi}}$$

$$= \$500.$$

Although any comparison of costs could be done by arithmetic using the method described above, it is more efficient to use algebra. Then the question can be answered in general.

Example 1 Suppose a car is driven d miles a year and gas costs g dollars a gallon. If fuel efficiency increases from $x\frac{\text{mi}}{\text{gal}}$ to $(x + a)\frac{\text{mi}}{\text{gal}}$, what is the amount saved in a year?

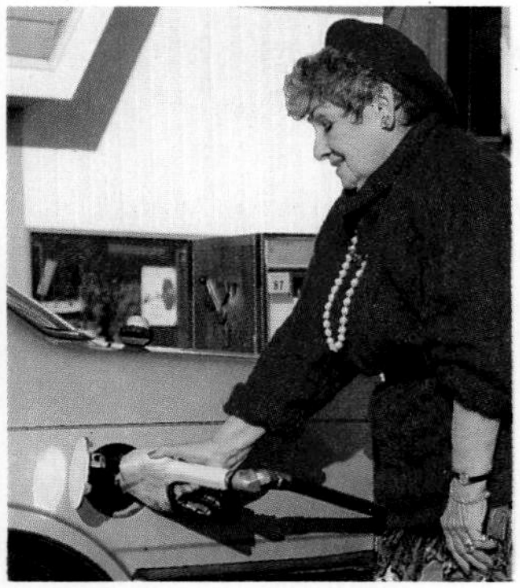

Solution Following the analysis above, the cost for a year at $x\frac{\text{mi}}{\text{gal}}$ is $\frac{dg}{x}$. The cost for a year at $(x + a)\ \frac{\text{mi}}{\text{gal}}$ is $\frac{dg}{x + a}$. Thus the amount saved in a year is $\frac{dg}{x} - \frac{dg}{x + a}$.

By substituting into the answer to Example 1, the amount saved can be determined for any values of d, g, x, and a. For instance, to determine the amount S saved in an increase in fuel efficiency from 15 $\frac{\text{mi}}{\text{gal}}$ to 25 $\frac{\text{mi}}{\text{gal}}$ when a car is driven 10,000 miles and a gallon of gas costs \$1.25, set $d = 10{,}000$, $g = \$1.25$, $x = 15$, and $a = 10$. Then

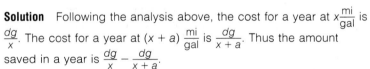

$$S = \frac{10000 \cdot 1.25}{15} - \frac{10000 \cdot 1.25}{25} = \frac{12500}{15} - \frac{12500}{25} \approx 833 - 500 = \$333.$$

LESSON 5-4

RESOURCES

- Lesson Master 5-4
- Teaching Aid 29 displays the graph of $S(x) = \frac{10}{x(x + 10)}$.
- Teaching Aid 30 displays three rational functions.
- Computer Master 10

OBJECTIVES

G Use limit notation to describe the behavior of rational functions.
H Classify discontinuities as essential or removable.
J Graph quotients of polynomial or trigonometric functions.
K Relate the limit of a function to its graph, and find equations for its asymptotes.

TEACHING NOTES

This lesson has three parts. The first two pages motivate the definition of *rational function.* Then examples and properties of rational functions are discussed. The last part is devoted to a discussion of singularities.

Making Connections
Dimensional analysis, the arithmetic of units, is a powerful way of composing a function to model a problem for which no formula exists. While having units come out properly is no guarantee of correctness, incorrect units in the final answer are a very strong indicator that the answer is wrong. Students with previous UCSMP experience should be familiar with these ideas.

Making Connections In the text following **Example 1**, we first let S be the amount saved. Then, to emphasize that S depends on x, the fuel efficiency of the car in miles per gallon, the variable S becomes $S(x)$. Students may have seen this done with geometry formulas such as $A = \pi r^2$, which when written as a function, becomes $A(r) = \pi r^2$.

The fact that $S(x)$ decreases as x increases illustrates the law of diminishing returns. While a change in gas efficiency from 10 mpg to 20 mpg is an increase of 100%, a change from 100 mpg to 110 mpg is only 10%. In each case, the absolute change is 10 mpg, but the relative changes are much different.

We assert that the formulas $r(x) = \frac{x^2 + 5}{\cos x + 2}$ and $s(x) = \frac{\sqrt{x - 1} + 3}{x^2 - 9}$ do not define rational functions, but we do not prove this. Though $r(x)$ and $s(x)$ are not rational functions, they can be estimated by rational functions over any closed interval in which they are defined.

In **Example 2**, the discontinuity is removable because 4 is a zero (and thus $x - 4$ is a factor) of multiplicity 1 of both the denominator and numerator. If 4 were a zero of the denominator and not the numerator, there would be an essential discontinuity. Note that this discontinuity can be removed only by defining $g(4) = 6$. If you defined $g(4) = 3$, then the function would still not be continuous at $x = 4$.

Technology In **Example 3**, there is also an essential discontinuity at $x = -2$. To dramatize the vertical asymptotes $x = -2$ and $x = 2$ of this example, you might use the trace feature of an automatic grapher. With a domain of $-6 \le x \le 6$ and range of

To determine whether increasing fuel efficiency by 10 $\frac{\text{mi}}{\text{gal}}$ has more effect when the original efficiency is low or high, let $a = 10$. Since d and g are constants, for simplicity set $dg = k$. Now

$$S = \frac{k}{x} - \frac{k}{x + 10} = \frac{k(x + 10)}{x(x + 10)} - \frac{kx}{x(x + 10)} = \frac{10k}{x(x + 10)}.$$

A reasonable domain for the fuel efficiency x for this situation is $0 \le x \le 50$. A graph of the function $f\colon x \rightarrow S$ when $k = 1$ is given here. This graph is interesting when $x < 0$, so we show it for $-20 \le x \le 25$.

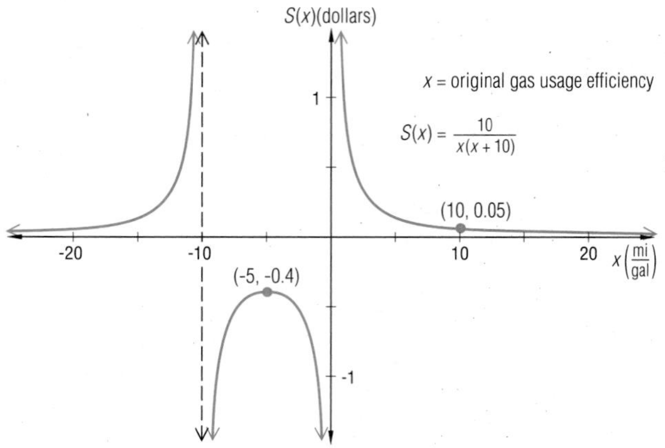

Notice that, when $x > 0$, as x increases S decreases. Thus the savings from increasing efficiency by $10\frac{\text{mi}}{\text{gal}}$ are greater when the original miles per gallon x is poor (when x is small) than when it is good (when x is large). You can verify this by substitution. For the particular situation described above, $k = dg = 12500$, so let

$$S(x) = \frac{125{,}000}{x(x + 10)} = \text{amount saved when fuel efficiency increases from } x \text{ to } (x + 10)\ \frac{\text{mi}}{\text{gal}}.$$

Then $$S(10) = \frac{125{,}000}{10 \bullet 20} = \$625 \text{ savings}$$

$$S(15) = \frac{125{,}000}{15 \bullet 25} \approx \$333 \text{ savings}.$$

Thus an increase from 10 $\frac{\text{mi}}{\text{gal}}$ to 20 $\frac{\text{mi}}{\text{gal}}$ saves almost twice as much as an increase from 15 $\frac{\text{mi}}{\text{gal}}$ to 25 $\frac{\text{mi}}{\text{gal}}$.

The function S is an example of a *rational function*, a function whose formula can be written as a rational expression.

Definition

A function f is a rational function if and only if there are polynomials $p(x)$ and $q(x)$ with $f(x) = \frac{p(x)}{q(x)}$ for all x in the domain of f.

308

All polynomial functions are rational functions: for example, if f is defined by $f(x) = x^2 - 2x - 15$, then $f(x) = \frac{x^2 - 2x - 15}{1}$, and since 1 is a constant polynomial function, f is rational. Also, reciprocals of nonzero polynomial functions are rational. However, the formulas $r(x) = \frac{x^2 + 5}{\cos x + 2}$ and $s(x) = \frac{\sqrt{x-1} + 3}{x^2 - 9}$ do not define rational functions because it is not possible to find any polynomial functions whose quotients equal either of these. (Proving this is beyond the scope of this book.)

Because the sum, difference, product, and quotient of two rational expressions can be written as rational expressions, the sum, difference, product, and quotient of two rational functions are rational functions. For instance, when $g(x) = \frac{9}{x + 2}$ and $h(x) = x^3$ then $f = g + h$ is a rational function because

$$\begin{aligned} f(x) = g(x) + h(x) &= \frac{9}{x+2} + x^3 \\ &= \frac{9}{x+2} + \frac{x^3(x+2)}{x+2} \\ &= \frac{x^4 + 2x^3 + 9}{x+2}. \end{aligned}$$

Thus $f(x)$ is a quotient of polynomials and so f is a rational function. Notice that f is defined for all real numbers x such that $x \neq -2$.

The domain of a rational function has a great bearing on its shape. If

$$f(x) = \frac{p(x)}{q(x)},$$

where $p(x)$ and $q(x)$ are polynomials, then the domain of f is at most the set of all real numbers x such that $q(x) \neq 0$. At points where $q(x) = 0$, f is undefined and its graph has one of these features:

(i) a vertical asymptote, as when $x = 2$ or $x = -2$ in h below,

or, (ii) a hole, as when $x = 4$ in g below.

A function f is said to have a **removable discontinuity** at a point x if the graph of f has a hole at x but is otherwise continuous on an interval around x. The discontinuity is called removable because the function could be made continuous by redefining its value at that one point. On the other hand, defining the value of the function at a single point cannot remove the discontinuity at a vertical asymptote, as you can see in the graph of h. A discontinuity that cannot be removed by insertion of a single point is called an **essential discontinuity**.

$f(x) = \frac{x^3}{3} - \frac{x^2}{2} - 6x$

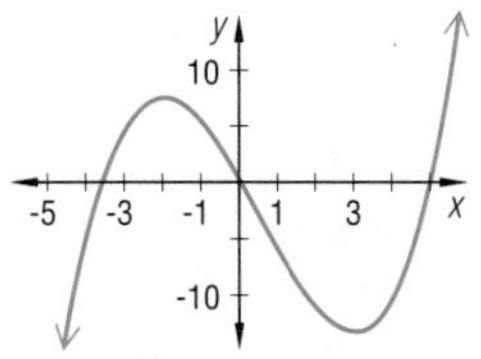

no discontinuities

$g(x) = \frac{x^2 - 2x - 8}{x - 4}$

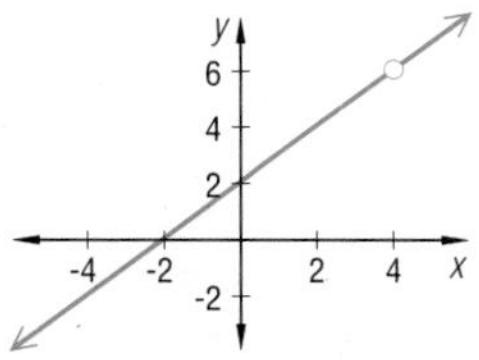

one removable discontinuity

$h(x) = \frac{3}{4 - x^2}$

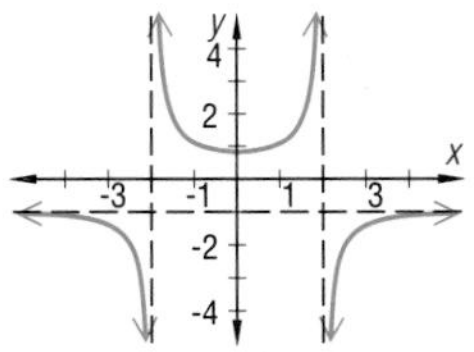

two essential discontinuities

$-10 \leq y \leq 10$, have students note carefully the values of $h(x)$ as they trace from $x = -6$ to $x = 6$. Special attention should be paid to values near $x = -2$ and $x = 2$. The jumps are somewhat spectacular as one "crosses" these two values.

A spreadsheet or short BASIC program also can describe what is happening to the function h in **Example 3**. Numerically, zoom in near $x = -2$ and $x = 2$ to show students the numerical behavior of function h near its asymptotes. The following is a sample BASIC program:

```
10 FOR X = 1 TO 3 STEP .1
20 PRINT X, 3/(4 - X^2)
30 NEXT X
```

By using smaller STEP values, the behavior of h can be seen closer and closer to the asymptote.

Error Analysis In order to conclude that $x = c$ is a vertical asymptote based on the Discontinuity Theorem, emphasize that two conditions must be true: (1) $p(c) \neq 0$ and (2) $q(c) = 0$. Some students focus exclusively on solving $q(x) = 0$ and assume that all of its zeros determine vertical asymptotes. A few students may use zeros of $p(x)$ for vertical asmyptotes. A student who can correctly sketch the graphs of $y = \frac{1}{x - 2}$ and $y = \frac{x^2 - 4}{x - 2}$ probably knows the difference between an asymptote and a hole in a graph.

Mention the use of the Factor Theorem related to the Discontinuity Theorem. If $p(c) = q(c) = 0$, then $x - c$ is a factor of both $p(x)$ and $q(x)$.

ADDITIONAL EXAMPLES

1. Suppose a local train travels m miles between two cities at an average speed of x miles per hour and an express train covers the same distance at an average speed of $x + a$ miles per hour. How much time is saved by taking the express train instead of the local train?

$\frac{m}{x} - \frac{m}{x+a}$ or $\frac{ma}{x(x+a)}$

2.a. Determine the location(s) of the removable discontinuities of the function f with $f(x) = \frac{2x^3 - 18x}{x^2 - 9}$.

at $x = 3$ and $x = -3$

b. How can these discontinuities be removed?

Define $f(3) = 6$ and $f(-3) = -6$.

3. Show that $g(x) = \frac{4}{x^2 - x - 6}$ has essential discontinuities at $x = -2$ and at $x = 3$.

$\lim_{x \to -2^-} g(x) = \infty$

$\neq \lim_{x \to -2^+} g(x) = -\infty$;

$\lim_{x \to 3^-} g(x) = -\infty$

$\neq \lim_{x \to 3^+} g(x) = \infty$

NOTES ON QUESTIONS

Questions 7 and 8: Urge students first to decide analytically whether the discontinuity is removable or essential, and then use an automatic grapher to test their answers.

Making Connections for Questions 9-11: Remind students of the methods used in Chapter 3 for analyzing when a function is positive or negative. This will make the analysis of behavior near vertical asymptotes simpler.

Question 12: The numerator and denominator of the rational expression in the formula could be multiplied by any nonzero real number.

The following examples illustrate how a study of the discontinuities of a function can be used to sketch its graph.

Example 2 Determine the location of the removable discontinuity of the function g above.

Solution A discontinuity exists at $x = 4$ because this is where the function is undefined. Notice that the formula for g can be reduced; $\frac{x^2 - 2x - 8}{x - 4} = \frac{(x+2)(x-4)}{x-4} = x + 2$, provided $x \neq 4$. Thus, the graph of g is the line $y = x + 2$ everywhere except at $x = 4$ where it has a hole. Since $x + 2 = 6$ when $x = 4$, this hole is located at $(4, 6)$. If we were to define $g(4) = 6$, then the hole would be filled in. Thus the discontinuity is removable.

Caution! If a function has a removable discontinuity and you graph it with an automatic grapher, the graph may look as if it has no hole. When you sketch the graph on paper, be sure to indicate the hole.

Example 3 Show that the function h above has an essential discontinuity at $x = 2$.

Solution As $x \to 2^+$, $4 - x^2$ is negative, so $h(x) = \frac{3}{4 - x^2}$ is negative. And since $4 - x^2 \to 0$ as $h \to 2^+$, $\lim_{x \to 2^+} h(x) = -\infty$. Thus it is impossible to define $h(2)$ so as to make h continuous. Hence h has a discontinuity at $x = 2$ that is essential.

The function g in Example 2 has a hole at $x = 4$. Since $x - 4$ is a factor of both the numerator and denominator of $g(x)$, both of those are zero when $x = 4$. This idea can be used to distinguish between holes and asymptotes.

Discontinuity Theorem

Given a rational function f with $f(x) = \frac{p(x)}{q(x)}$, where $p(x)$ and $q(x)$ are polynomials over the reals, if $\exists$ a real number c such that $q(c) = 0$ but $p(c) \neq 0$, then f has an essential discontinuity when $x = c$ and the line $x = c$ is a vertical asymptote to its graph.

If both $q(c) = 0$ and $p(c) = 0$, then both $p(x)$ and $q(x)$ have a factor $x - c$. You can then factor out $x - c$ from the numerator and denominator of $f(x)$. If the resulting rational function is defined when $x = c$, then there is a removable discontinuity at $x = c$. This is what happened in Example 2. If the resulting function is still not defined at $x = c$, analyze it for discontinuities.

Finally, note that a nonrational function may have an essential discontinuity without an asymptote. For instance, the floor function $f: x \to \lfloor x \rfloor$ has an essential discontinuity at each integer, but it has no asymptotes.

Questions

Covering the Reading

In 1 and 2, refer to Example 1.

1. Suppose a person drives 15,000 miles a year and a gallon of gas costs \$1.40 on average. How much would this person save by driving a car that gets $30\frac{\text{mi}}{\text{gal}}$ over one that gets $20\frac{\text{mi}}{\text{gal}}$? **\$350**

2. a. Graph the function f: $x \rightarrow S$, where S is the amount saved by increasing gas efficiency from $x\frac{\text{mi}}{\text{gal}}$ to $(x + 15)\frac{\text{mi}}{\text{gal}}$. (Let $dg = 1$.)

b. Describe $\lim\limits_{x\to\infty} f(x)$. $\lim\limits_{x\to\infty} f(x) = 0$

c. What real world significance does the answer to **b** have?

a,c) See margin.

In 3–5, tell whether the function defined by the given formula is a rational function. If it is, determine its domain.

3. $g(x) = \frac{x^2}{x + 1} - \frac{x}{x + 1}$ **Yes, {x: x ≠ -1}**

4. $f(z) = \sqrt{x^2 + 1}$ **No**

5. $h(x) = -3x^4 + 2x^3 - 7x^2 + x$ **Yes, domain: the set of real numbers**

6. The function f: $x \rightarrow \frac{1}{x^2}$ graphed at the right has a discontinuity at $x = 0$. Is this discontinuity removable or essential? **essential**

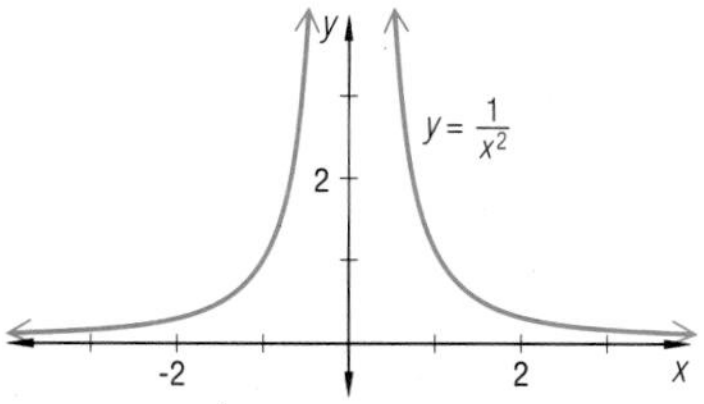

7. The function g defined by $g(u) = \frac{3u + 30}{u + 10}$ has a discontinuity at $u = -10$.

a. Is this discontinuity removable or essential? **removable**

b. Graph g. **See margin.**

8. *True* or *false?* The function f defined by $f(t) = \frac{t^2 - t}{(t - 1)^2}$ has a removable discontinuity at $t = 1$. **False**

Applying the Mathematics

In 9–11, **a.** determine the domain of the function f; **b.** classify its discontinuities; **c.** find equations of its vertical asymptotes; and **d.** sketch a rough graph of the function. (Mark holes or asymptotes wherever they occur.)

9. $f(u) = \frac{u + 2}{3u + 5}$ **See margin.**

10. $f(x) = \frac{2x^2 - 1}{x^2 - 1}$ **See margin.**

11. $f(k) = \frac{2k + 3}{k^2 + 9}$ **See margin.**

12. Write an equation for the function f whose graph is the line with a hole in it shown at the right. $y = \frac{x^2 - 6x + 8}{2x - 8}$

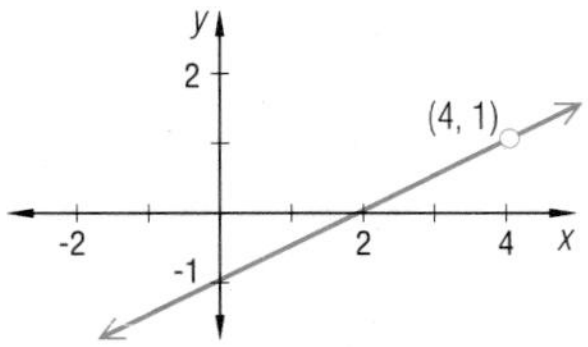

Question 16: Ask students for a function f that satisfies the given condition. $\left(\text{sample: } f\text{: } x \rightarrow \frac{-1}{x - 7}\right)$

Question 21b: The Bisection Method is appropriate for this question.

Question 27a: Have students predict the vertical asymptotes before using an automatic grapher. Did any students sketch the component rational functions and then add their graphs?

ADDITIONAL ANSWERS

2.a.

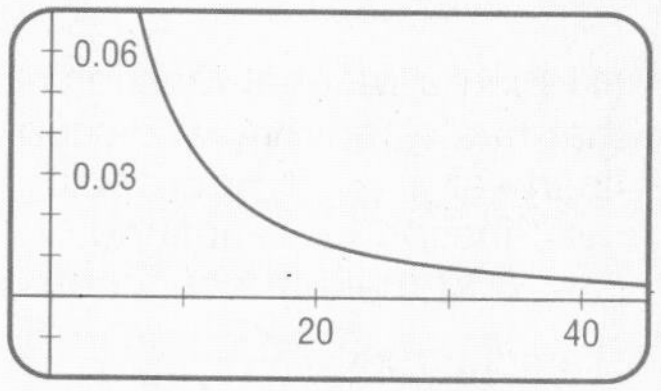

c. The greater the initial gas mileage, the less the savings for an increase of 15 mpg.

7.b.

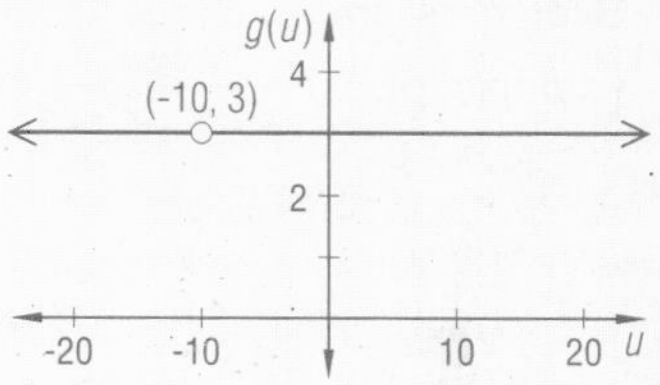

9.a. $\{u\text{: } u \neq -\frac{5}{3}\}$

b. essential

c. $u = -\frac{5}{3}$

d.

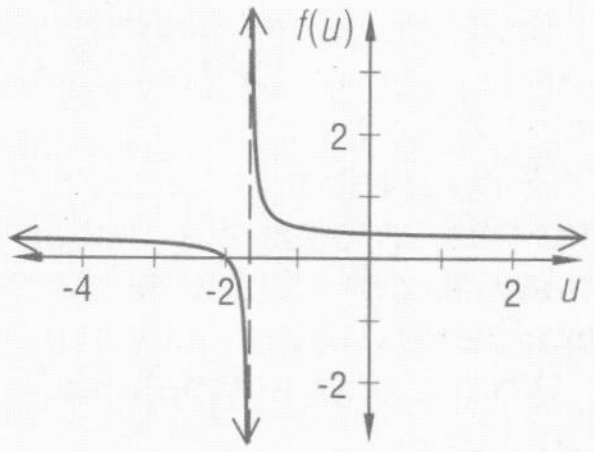

10., 11. See Additional Answers in the back of this book.

FOLLOW-UP

MORE PRACTICE

For more questions on SPUR Objectives, use *Lesson Master 5-4,* shown on page 313.

PROJECTS

The projects for Chapter 5 are described on page 338. **Project 5** is related to the content of this lesson.

EVALUATION

Alternate Assessment

Students could do these problems in **small groups** or individually, perhaps working at the chalkboard.

(1) Find a rational function that has a removable discontinuity at $x = 3$. How could the function be redefined at $x = 3$ to make it continuous? (sample: $f(x) = \frac{x^2 - x - 6}{x - 3}$; redefine $f(3) = 5$ to make f continuous.)

(2) Find a rational function that has essential discontinuities at $x = -4$ and $x = 1$. Sketch its graph. $\left(\text{sample: } g(x) = \frac{x}{x^2 + 3x - 4}\right)$

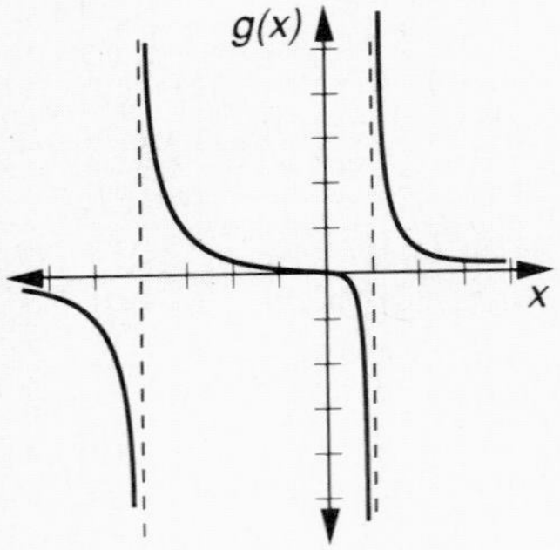

(3) Find a rational function that has an essential discontinuity at $x = -2$ and a removable discontinuity at $x = 2$. What are its asymptotes? (sample: $h(x) = \frac{x - 2}{x^2 - 4}$; horizontal asymptote: x-axis; vertical asymptote: $x = -2$)

13. Given the function g defined by

$$g(x) = \begin{cases} \frac{x^2 + 4x - 5}{x - 1} & \text{if } x \neq 1 \\ 6 & \text{if } x = 1. \end{cases}$$

Explain why g is a continuous function. **See margin.**

14. Consider the rational function defined by $f(x) = \frac{1}{x}$. Let g be the function whose graph is obtained by applying $T_{2,-3}$ to the graph of f.

a. Write a formula for g. **$g(x) = \frac{1}{x-2} - 3$**

b. Check your answer to part **a** by using an automatic grapher to graph g. **See margin.**

Review

15. Consider the function h defined by $h(x) = \frac{5}{x^4}$.

a. Give the domain and range of h. **See below.**

b. Describe the behavior of h as $x \to 0$. **See below.**

c. Describe the end behavior of h using limit notation. **See below.**

d. Prove that h is even.

e. Graph the function. *(Lessons 5-3, 2-5, 2-1)* **See margin.**

d) $\forall$ x, $h(-x) = \frac{5}{(-x)^4} = \frac{5}{x^4} = h(x)$.

16. Explain, in words, the meaning of $\lim_{x \to 7^+} f(x) = -\infty$. *(Lesson 5-3)*

As x approaches 7 from the right, f(x) approaches negative infinity.

17. Rationalize denominators and simplify: $1 - \frac{x}{1 + \sqrt{3}} + \frac{x}{1 - \sqrt{3}}$. *(Lesson 5-2)* **$1 - x\sqrt{3}$**

18. Prove that $\sqrt{17}$ is irrational. *(Lesson 5-2)* **See margin.**

19. Express the following sum as a single fraction. Assume that $a \neq b$.

$$\frac{9}{a - b} + \frac{2}{b - a} + \frac{7}{4a - 4b}$$

(Hint: Use the identity $x - y = -1(y - x)$.) *(Lesson 5-1)* **$\frac{35}{4(a - b)}$**

20. Determine whether the value of each expression below is a rational or an irrational number. *(Lessons 5-1, 2-6)*

a. $\left(1 + \frac{1}{1000}\right)^{1000}$ **rational** **b.** $\lim_{n \to \infty}\left(1 + \frac{1}{n}\right)^n$ **irrational**

21. a. Use the Intermediate Value Theorem to show that the equation $\cos x = 0.2x^2$ has a solution in the interval from 1 to 1.5.

b. Find an interval of length .01 that contains this solution. *(Lesson 3-5)* **a) See margin. b) $1.25 \le x \le 1.26$**

22. Consider the function defined by $f(x) = 3x^2 - 5x + 9$. Use an automatic grapher to approximate the coordinates of the minimum point of the graph f, to the nearest tenth. *(Lesson 2-2)* **(.8, 6.9)**

15. a) domain: $\{x: x \neq 0\}$; range: $\{y: y > 0\}$

b) $\lim_{x \to 0^+} h(x) = +\infty$; $\lim_{x \to 0^-} h(x) = +\infty$

c) $\lim_{x \to \infty} h(x) = 0$; $\lim_{x \to -\infty} h(x) = 0$

23. Write the negation of the statement. *(Lessons 1-5, 1-3)*
  a. *If the duplicating machine works, then it is Sunday.* See below.
  b. *The country is in Southeast Asia, and it is not the Philippines.*
  The country is not in Southeast Asia, or it is the Philippines.

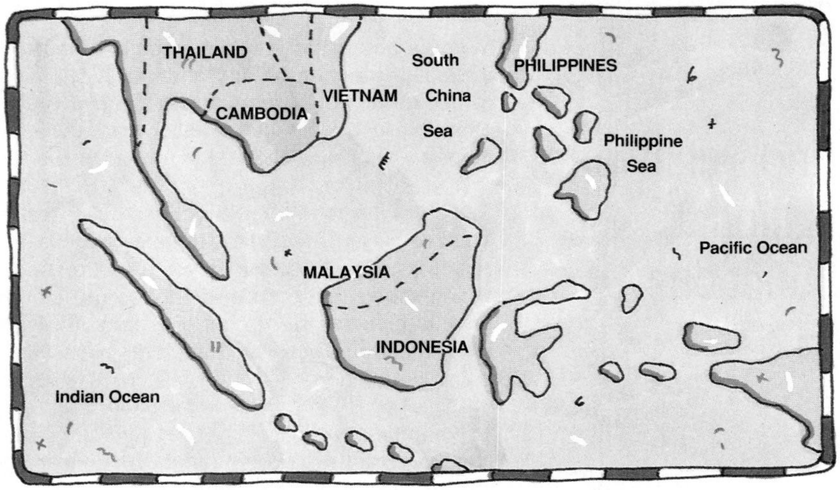

24. Find the base 10 representation of 111001_2. *(Lesson 4-7)* 57

25. Use long division to find the quotient and remainder when $p(x) = x^5 + 3x^4 + 3x^3 + x^2 + 4x - 10$ is divided by $d(x) = x^2 + 1$. *(Lesson 4-4)* $q(x) = x^3 + 3x^2 + 2x - 2$; $r(x) = 2x - 8$

26. Solve over the set of real numbers. *(Lesson 3-8)*
  a. $|5x + 2| = 9$ b. $|5x + 2| \geq 9$
  $x = \frac{7}{5}$ or $x = \frac{-11}{5}$ $x \leq \frac{-11}{5}$ or $x \geq \frac{7}{5}$

Exploration

27. a. Graph f when $f(x) = \frac{1}{x-1} + \frac{1}{x^2} + \frac{1}{x+1}$. See below.
  b. Experiment with other functions that have discontinuities. Copy the equations and graphs of the ones that you find most interesting.
  Answers will vary.

23. a) *The duplicating machine works, and it is not Sunday.*
27. a)

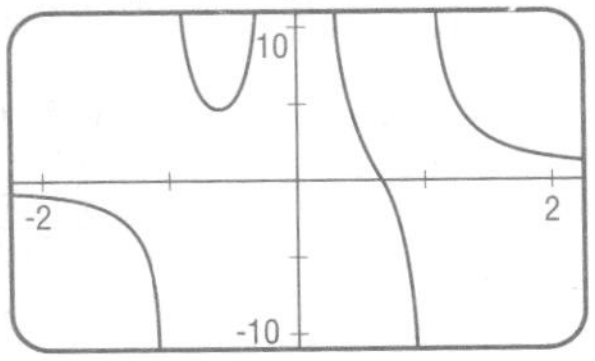

ADDITIONAL ANSWERS
13., 14.b., 15.e., 18., 21.a. See Additional Answers in the back of this book.

NAME ____________

LESSON **MASTER 5–4**
QUESTIONS ON **SPUR** OBJECTIVES

■ **PROPERTIES** *Objective G (See pages 341–344 for objectives.)*

1. Consider the function g defined by $g(y) = \frac{-2y}{4y^2 - 9}$.
  a. Identify the vertical asymptotes of the graph of g. $y = -\frac{3}{2}, y = \frac{3}{2}$
  b. Use limit notation to describe the behavior of g near the asymptotes.
  $\lim_{y \to -\frac{3}{2}^-} g(y) = \infty$, $\lim_{y \to -\frac{3}{2}^+} g(y) = -\infty$,
  $\lim_{y \to \frac{3}{2}^-} g(y) = \infty$, $\lim_{y \to \frac{3}{2}^+} g(y) = -\infty$,
  c. Use limit notation to describe the end behavior of g.
  $\lim_{y \to \infty} g(y) = \lim_{y \to -\infty} g(y) = 0$

2. Given $f(x) = \frac{x+4}{x^2 + 2x - 8}$, find
  a. $\lim_{x \to -4^-} f(x)$. $-\frac{1}{6}$ b. $\lim_{x \to -4^+} f(x)$. $-\frac{1}{6}$
  c. $\lim_{x \to 2^-} f(x)$. $-\infty$ d. $\lim_{x \to 2^+} f(x)$. ∞

■ **PROPERTIES** *Objective H*

3. Given $f(x) = \frac{x+2}{x^2 - x - 6}$, f has a(n) **removable** discontinuity at $x = -2$ and a(n) **essential** discontinuity at $x = 3$.

In 4 and 5, classify any discontinuities as essential or removable, and at each removable discontinuity, redefine the function to make it continuous.

4. $f(a) = \frac{2a^2 + a - 3}{a + 1}$
  essential at $a = -1$

5. $f(b) = \frac{2b^2 + b - 3}{b - 1}$
  removable at $b = 1$; redefine $f(1) = 5$

6. Find a rule for a function that has a removable discontinuity at $x = -2$ and no essential discontinuities.
  sample: $f(x) = \frac{(x+1)(x+2)}{x+2} = \frac{x^2 + 3x + 2}{x + 2}$

56 *Continued* *Precalculus and Discrete Mathematics* © Scott, Foresman and Company

NAME ____________
Lesson MASTER 5–4 (page 2)

■ **REPRESENTATIONS** *Objective J*
In 7–10, sketch a graph of the function, indicating holes at the removable discontinuities and dashed lines for any asymptotes.

7. $g(x) = \frac{2x^2 + 7x}{x + 3}$

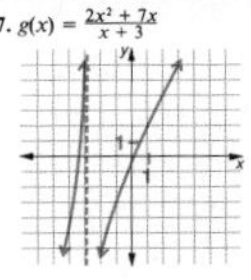

8. $f(x) = \frac{x^2 + 2x}{x^3 - 4x}$

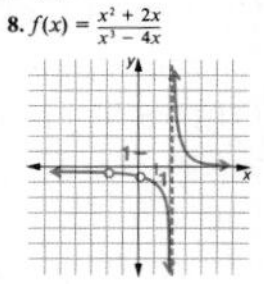

9. $h(x) = \frac{x^2 + 5x - 6}{x + 6}$

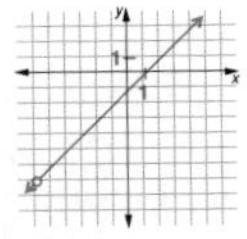

10. $f(x) = \frac{x - 2}{x^2 - 4}$

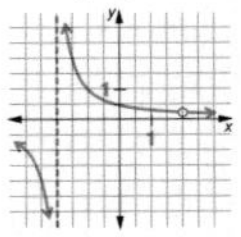

■ **REPRESENTATIONS** *Objective K*
In 11 and 12, a. find equations for any vertical asymptotes, b. use limit notation to describe the behavior of the function near any asymptote, and c. use limit notation to describe the behavior of the function near any removable discontinuity.

11. the function in Question 8
  a. $x = 2$
  b. $\lim_{x \to 2^-} f(x) = -\infty$, $\lim_{x \to 2^+} f(x) = \infty$
  c. $\lim_{x \to -2} f(x) = -\frac{1}{4}$, $\lim_{x \to 0} f(x) = -\frac{1}{2}$

12. the function graphed at the right
  a. $x = -3$, $x = 4$
  b. $\lim_{x \to -3^-} f(x) = \infty$, $\lim_{x \to -3^+} f(x) = -\infty$
  $\lim_{x \to 4^-} f(x) = \infty$, $\lim_{x \to 4^+} f(x) = \infty$
  c. $\lim_{x \to 1} f(x) = 2$

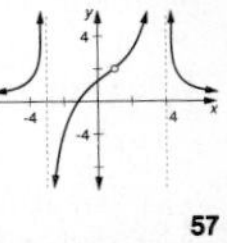

Precalculus and Discrete Mathematics © Scott, Foresman and Company 57

LESSON 5-5

RESOURCES
- Lesson Master 5-5
- Teaching Aid 31 shows the end behavior of rational functions table.
- Computer Master 11

OBJECTIVES

G Use limit notation to describe the behavior of rational functions.
J Graph quotients of polynomial or trigonometric functions.
K Relate the limit of a function to its graph, and find equations for its asymptotes.

TEACHING NOTES

In discussing the end behavior of rational functions, there are three possible levels of presentation and understanding. At the most general level, there is the question of the limiting value of the function as $x \to \infty$ or as $x \to -\infty$. There are only four possibilities for such a limit: (1) such a value may exist; (2) the function may become larger than any number chosen, so that the limit is said to be $+\infty$; (3) the function may become smaller than any number chosen, so that the limit is said to be $-\infty$; or (4) the function may vacillate and there is no limit at all. In the case of rational functions, all but (4) can occur.

At a more specific level, we are concerned with the speed with which a given function approaches its end behavior. The speed is described by identifying a simpler function whose end behavior is the same as the given function.

LESSON

5-5

End Behavior of Rational Functions

With an automatic grapher, you can rather easily determine the behavior of a rational function on any interval. The Discontinuity Theorem enables you to determine any asymptotes of the function, and so tells you if it will have unusual behavior far away from the x-axis. Still, neither of these helps to determine the end behavior of the function.

Since graphs of rational functions have many possible shapes, it would seem to be a difficult task to determine the end behaviors of all rational functions. This lesson is devoted to explaining and proving a theorem which shows that the task is not so difficult at all. The theorem states that the end behavior of any rational function is the same as that of one of three functions: a power function, a constant function, or the reciprocal of a power function.

Polynomial functions, in some ways the simplest of the rational functions, are analyzed first. (Polynomial functions are to rational functions as integers are to rational numbers.)

Consider the polynomial function p defined by

$$p(x) = 3x^4 - 5x^3 + 8x^2 - 20x + 16.$$

Factor out x^4. This complicates the expression for $p(x)$ but simplifies the analysis of its end behavior.

$$p(x) = \left(3 + \frac{-5}{x} + \frac{8}{x^2} + \frac{-20}{x^3} + \frac{16}{x^4}\right) x^4$$

The end behavior of p is the behavior $p(x)$ as $|x| \to \infty$. With a calculator, you can see how close $3 + \frac{-5}{x} + \frac{8}{x^2} + \frac{-20}{x^3} + \frac{16}{x^4}$ is to 3 when $|x|$ is large.

x	$3 + \frac{-5}{x} + \frac{8}{x^2} + \frac{-20}{x^3} + \frac{16}{x^4}$	$p(x)$	$3x^4$	$\frac{p(x)}{3x^4}$
50	2.90304	1.814×10^7	$1.875 \cdot 10^7$	0.9675
100	2.95078	2.951×10^8	$3.000 \cdot 10^8$	0.9837
500	2.99003	1.869×10^{11}	$1.875 \cdot 10^{11}$	0.9968
1000	2.99501	2.995×10^{12}	$3.000 \cdot 10^{12}$	0.9983
10,000	2.99950	2.999×10^{16}	$3.000 \cdot 10^{16}$	0.9998

As $|x|$ gets larger, $p(x) = \left(3 + \frac{-5}{x} + \frac{8}{x^2} + \frac{-20}{x^3} + \frac{16}{x^4}\right)x^4$ does not get closer and closer to $3x^4$, but the ratio $\frac{p(x)}{3x^4}$ of these polynomials gets closer and closer to 1. Thus the end behavior of p is the same as the end behavior of the power function $3x^4$. That is,

$$\lim_{x \to +\infty} p(x) = +\infty \text{ and } \lim_{x \to -\infty} p(x) = +\infty.$$

In general, suppose that p is a polynomial function of degree n defined by

$$p(x) = a_nx^n + a_{n-1}x^{n-1} + \ldots + a_1x + a_0.$$

When $x \neq 0$, this polynomial can be rewritten or transformed into the following equivalent (and unusual) form by factoring out x^n.

$$p(x) = \left(a_n + \frac{a_{n-1}}{x} + \frac{a_{n-2}}{x^2} + \ldots + \frac{a_1}{x^{n-1}} + \frac{a_0}{x^n}\right)x^n$$

As x gets farther and farther from the origin, the values of the terms

$$\frac{a_{n-1}}{x}, \frac{a_{n-2}}{x^2}, \ldots, \frac{a_1}{x^{n-1}}, \frac{a_0}{x^n}$$

get closer and closer to 0. In fact $a_n + \frac{a_{n-1}}{x} + \frac{a_{n-2}}{x^2} + \ldots + \frac{a_1}{x^{n-1}} + \frac{a_0}{x^n}$ can be made as close as one wishes to a_n by taking x sufficiently far to the left or right of the origin. Hence when $|x|$ is large, the ratio of the values of

$$p(x) = \left(a_n + \frac{a_{n-1}}{x} + \frac{a_{n-2}}{x^2} + \ldots + \frac{a_1}{x^{n-1}} + \frac{a_0}{x^n}\right)x^n$$

and $a_n x^n$ come arbitrarily close to the values of 1.

Therefore, far to the left or right of the origin, the graph of p will behave much like the graph of the power function q defined by $q(x) = a_n x^n$. Consequently the graphs of p and q display the same end behavior even though they may be quite different for values of x relatively close to the origin. It follows that you can determine the end behavior of any polynomial function by examining its degree and the sign of its leading coefficient!

The polynomial $p(x) = a_n x^n + a_{n-1}x^{n-1} + \ldots + a_1 x + a_0$, is customarily written in decreasing order of the powers of its terms. So the term $a_n x^n$ is called the **leading term**. The analysis above shows that the end behavior of a polynomial function is the same as the end behavior of its leading term.

The procedure used to determine the end behavior of polynomial functions can be extended to any rational functions. Notice how many techniques you have seen for determining end behavior.

Example 1 Describe the end behavior of the function g defined by the formula

$$g(x) = \frac{2x^2 + 1}{x^2 - 1}$$

Solution 1 A calculator or computer can generate a table of values of $g(x)$ for large values of $|x|$. From the last row, it seems that $\lim_{x \to +\infty} g(x) = 2$.

x	± 2	± 5	± 10	± 100	± 1000
$g(x) = \frac{2x^2 + 1}{x^2 - 1}$	3	2.125	2.0303	2.0003	2.000003

Since g is an even function, it is symmetric to the y-axis. From all this, it appears that $\lim_{x \to -\infty} g(x) = \lim_{x \to +\infty} g(x) = 2$.

A third level of understanding which is between the first two, is to note a class of functions whose end behavior is similar to the behavior of a given function. For instance, an algorithm might be said to take "polynomial time," which means that as the complexity of the problem n increases, the end behavior of the number of steps required for the algorithm is $p(n)$, where $p(n)$ is a polynomial in n.

On page 314, note that even though p and $q: x \to 3x^4$ have the same behavior, the values of $p(x)$ and $q(x)$ do not get closer and closer to each other as $x \to \infty$. In fact, $p(x) - q(x) = -5x^3 + 8x^2 - 20x + 16$, so the absolute value of their difference becomes larger and larger as $|x| \to \infty$. In general, two functions p and q have the same end behavior if and only if $\lim_{x \to +\infty} \frac{p(x)}{q(x)} = \lim_{x \to -\infty} \frac{p(x)}{q(x)}$. You might relate this to the tossing of a fair coin. Generally, as the coin is tossed more, the absolute difference between the numbers of heads and tails tends to increase, but the ratio of heads to tails approaches 1.

When we write $p(x) \approx a_n x^n$, for example, we do not mean that the absolute difference $|p(x) - a_n x^n|$ is small; in fact, it becomes arbitrarily large as x grows in magnitude. Instead, we mean that this difference relative to the absolute value of the polynomial, $\frac{|p(x) - a_n x^n|}{|p(x)|}$, becomes small. This can be seen by noting that this relative difference is the absolute value of a rational function whose numerator has smaller degree than its denominator. Thus, the relative difference approaches zero as x grows in magnitude.

In **Example 2**, however, the function h does get closer and closer to the line $y = \frac{1}{3}x + 4$.

Making Connections
Polynomial functions and rational functions have a symbiotic relationship with regard to end behavior. The behavior of the simplest rational functions (the reciprocals of the power functions) is critical in determining the end behavior of polynomial functions. In turn, the end behavior is needed for analyzing the end behavior of more complicated rational functions.

To determine the end behavior of a polynomial, the polynomial is written as a product of its leading power and the rest of the polynomial. This will be new for most students. Go through **Examples 1 and 2** in detail. By factoring out x to the highest power, emphasize that $p(x)$ has been rewritten in a way that facilitates analyzing its behavior as $|x| \to \infty$, because when f is the reciprocal of a power function, $\lim_{x\to\infty} f(x) = \lim_{x\to-\infty} f(x) = 0$. The key point to make is that when $|x|$ is large, the leading term's behavior dominates the behavior of the entire polynomial function.

You might point out that the method of Solution 1 of **Example 1** is quite limited because you may not be able to see the limit from the table of values.

In discussing the method of Solution 2 of **Example 1**, emphasize that the divisor is the highest power of the independent variable in the denominator. Then apply the distributive property to divide, in effect, each term of the expression by x^2. First, analyze the numerator and

Solution 2 Divide the numerator and denominator of the formula by the highest power of the independent variable that appears in the denominator. In this case, divide by x^2.

$$g(x) = \frac{2x^2+1}{x^2-1} = \frac{\frac{2x^2+1}{x^2}}{\frac{x^2-1}{x^2}} = \frac{2+\frac{1}{x^2}}{1-\frac{1}{x^2}}.$$

For values of x far to the left or to the right of the origin, $\frac{1}{x^2}$ is negligibly small, and so the values of $g(x)$ become arbitrarily close to $\frac{2}{1} = 2$. In limit notation, this is written

$$\lim_{x\to+\infty} g(x) = \lim_{x\to+\infty}\left(\frac{2x^2+1}{x^2-1}\right) = \lim_{x\to-\infty}\left(\frac{2x^2+1}{x^2-1}\right) = 2.$$

Therefore, the line $y = 2$ is a horizontal asymptote for the function.

Solution 3 Use long division. Divide $2x^2 + 1$ by $x^2 - 1$.

$$\begin{array}{r} 2 \\ x^2-1\overline{)2x^2+1} \\ \underline{2x^2-2} \\ 3 \end{array}$$

The long division shows that $g(x) = \frac{2x^2+1}{x^2-1} = 2 + \frac{3}{x^2-1}$. Since $x^2 - 1$ grows without bound as $x \to +\infty$, $\lim_{x\to+\infty}\left(\frac{3}{x^2-1}\right) = 0$. Thus $\lim_{x\to+\infty} g(x) = \lim_{x\to-\infty}\left(2 + \frac{3}{x^2-1}\right) = 2$. Similarly, $\lim_{x\to-\infty} g(x) = 2$.

Check A graph of g is shown at the right. The graph confirms that the end behavior is described by $\lim_{x\to+\infty} g(x) = \lim_{x\to-\infty} g(x) = 2$.

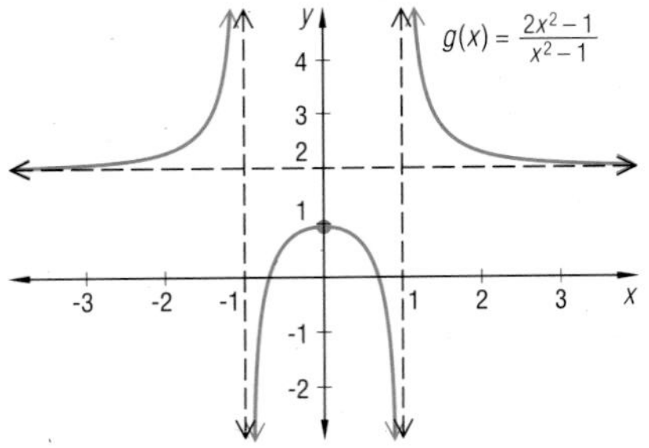

Solution 3 to Example 1 illustrates how long division can be used to express a function in a form that permits the behavior as $x\to+\infty$ and as $x\to-\infty$ to be quickly determined. This can be done with any rational function in which the degree of the numerator is greater than or equal to the degree of the denominator.

Example 2 Let h be the function defined by $h(x) = \frac{x^2+14x+12}{3x+6}$.

a. Rewrite $h(x)$ using long division.

b. Use the result from part **a** to discuss the behavior of h as $x\to+\infty$ and as $x\to-\infty$.

316

Solution

a.

$$\begin{array}{r} \tfrac{1}{3}x + 4 \\ 3x+6\overline{)\ x^2 + 14x + 12} \\ \underline{x^2 + 2x} \\ 12x + 12 \\ \underline{12x + 24} \\ -12 \end{array}$$

Therefore,

$$h(x) = \frac{1}{3}x + 4 + \frac{-12}{3x + 6}.$$

b. Notice that as $x \to +\infty$ and as $x \to -\infty$, $\frac{-12}{3x + 6}$ gets arbitrarily close to 0. It follows that the values of $h(x)$ get arbitrarily close to the values of $\frac{1}{3}x + 4$. That is,

$$\lim_{x \to +\infty} h(x) = \lim_{x \to +\infty} \left(\frac{1}{3}x + 4\right) = +\infty,$$

$$\lim_{x \to -\infty} h(x) = \lim_{x \to -\infty} \left(\frac{1}{3}x + 4\right) = -\infty.$$

In Example 2, the line $y = \frac{1}{3}x + 4$ is called an **oblique asymptote** to the graph of h. Below, the function h and its oblique asymptote, are plotted. At the left the x- and y-axes have the same scale. The center and right graphs show the effects of zooming along the x-axis. Notice that for very large and very small values of x, the graph of h gets closer and closer to the line $y = \frac{1}{3}x + 4$.

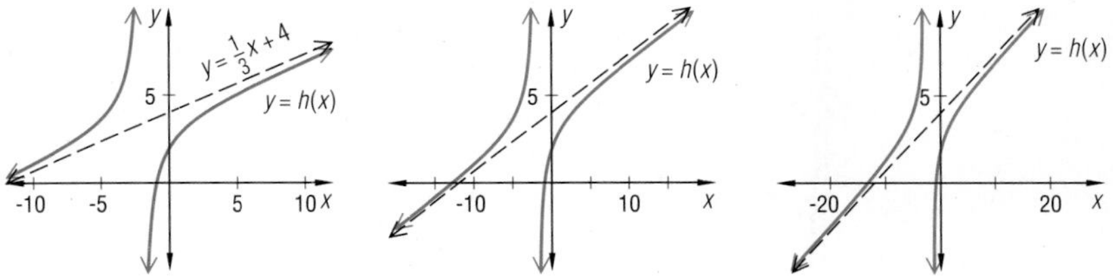

From the polynomial example that began this lesson and from Example 1, perhaps you have already conjectured the general pattern. When a rational function is written as a quotient of polynomials, its end behavior is found by dividing the leading terms of the polynomials.

Theorem (End Behavior of Rational Functions)

Suppose $f(x) = \frac{a_m x^m + \ldots + a_1 x + a_0}{b_n x^n + \ldots + b_1 x + b_0}$ $\forall$ real numbers x for which the denominator is nonzero, where the a_i and b_i are real numbers $\forall$ i, $a_m \neq 0$, and $b_n \neq 0$. Then the end behavior of f is the same as the end behavior of the function g defined by $g(x) = \frac{a_m}{b_n} x^{m-n}$ $\forall$ real numbers x.

denominator of $\dfrac{2 + \frac{1}{x^2}}{1 - \frac{1}{x^2}}$ separately as $|x| \to \infty$, then divide to get the quotient. Implicit in this work is the theorem that the limit of a quotient is the quotient of the limits.

Error Analysis Some students may extend incorrectly the method of Solution 2 of **Example 1** to analyzing the end behavior of polynomial functions by writing $f(x) = 5x^4 + 3x^2 + 7x + 8$ as $f(x) = 5 + \frac{3}{x^2} + \frac{7}{x^3} + \frac{8}{x^4}$ and concluding that $\lim_{|x| \to \infty} f(x) = 5$. The use of a table of values showing several values of x^4 will help to convince students that the factor x^4 cannot be eliminated.

In Solution 3 to **Example 2**, implicit in $\lim_{|x| \to \infty} \left(2 + \frac{2}{x^2 - 1}\right) = 2$, is the theorem that "the limit of a sum is the sum of the limits." Some students have difficulty understanding $\lim_{|x| \to \infty} 2 = 2$.

Making Connections In **Example 2**, the quotient is a function whose end behavior is identical to that of the given rational function. Earlier examples of polynomial division could be interpreted in terms of nonlinear asymptotes of rational functions. For instance, the division following Example 1 of Lesson 4-4 gave $\frac{6x^3 - 9x^2 + 8x + 1}{2x + 1} = 3x^2 - 6x + 7 + \frac{-6}{2x + 1}$. Thinking of the fraction as a formula for a rational function, its end behavior is the same as that of the parabola $y = 3x^2$, and in fact, it gets closer and closer to the parabola $y = 3x^2 - 6x + 7$.

ADDITIONAL EXAMPLES

1. Describe the end behavior of the function f defined by the formula $f(x) = \dfrac{5x^4 + 3x^2 + 8}{2x^4 + 7x^2 + 1}$.

Using a calculator or computer:

x	1	5	10
$f(x)$	1.6	2.2496	2.4302

x	100	1000
$f(x)$	2.4993	2.499993

So it appears that $\lim_{x \to +\infty} f(x) = 2.5$. Because f is an even function, $\lim_{x \to -\infty} f(x) = 2.5$.

$$\lim_{x \to \infty} \frac{5x^4 + 3x^2 + 8}{2x^4 + 7x^2 + 1} = \lim_{x \to \infty} \frac{5 + \frac{3}{x^2} + \frac{8}{x^4}}{2 + \frac{7}{x^2} + \frac{1}{x^2}} = \frac{5}{2} \text{ or } 2.5.$$

2. Let h be the function defined by $h(x) = \dfrac{3x^2 + 5x - 9}{2x + 6}$.

a. Rewrite h using long division.

$h(x) = \frac{3}{2}x - 2 + \dfrac{3}{2x + 6}$

b. Describe the behavior of h as $x \to +\infty$ and as $x \to -\infty$.

$\lim_{x \to \infty} h(x) = \infty$; $\lim_{x \to -\infty} h(x) = -\infty$

The graph of h approaches the line $y = \frac{3}{2}x - 2$.

3. Describe the end behavior of the function $m(h) = 100\left(\dfrac{1740}{1740 + h}\right)^2$.

The end behavior is the same as the reciprocal of an inverse square function, so $\lim_{h \to \infty} m(h) = 0$; $\lim_{h \to -\infty} m(h) = 0$.

This formula gives the weight of an object that is h kilometers above the surface of our moon.

4. Describe the end behavior of the function f, where $f(x) = \dfrac{10(x + 3)^2}{(x - 4)^3}$.

It is the same as the function g, where $g(x) = \frac{10}{x}$, so $\lim_{x \to +\infty} f(x) = 0$. $\lim_{x \to -\infty} f(x) = 0$.

Proof Rewrite the formula for $f(x)$ by factoring x^m from the numerator and factoring x^n from the denominator.

$$f(x) = \frac{\left(a_m + \frac{a_{m-1}}{x} + \frac{a_{m-2}}{x^2} + \ldots + \frac{a_1}{x^{m-1}} + \frac{a_0}{x^m}\right) x^m}{\left(b_n + \frac{b_{n-1}}{x} + \frac{b_{n-2}}{x^2} + \ldots + \frac{b_1}{x^{n-1}} + \frac{b_0}{x^n}\right) x^n}$$

As $|x|$ gets larger and larger, all the terms in the parentheses approach zero except a_m and b_n, which remain constant. Thus

$$\lim_{x \to +\infty} f(x) = \lim_{x \to +\infty} \frac{a_m x^m}{b_n x^n} = \lim_{x \to +\infty} \frac{a_m}{b_n} x^{m-n} = \lim_{x \to +\infty} g(x)$$

and $\lim_{x \to -\infty} f(x) = \lim_{x \to -\infty} \frac{a_m x^m}{b_n x^n} = \lim_{x \to -\infty} \frac{a_m}{b_n} x^{m-n} = \lim_{x \to -\infty} g(x).$

Example 3 Describe the end behavior of the function w defined by

$$w(h) = 150\left(\frac{4000}{4000 + h}\right)^2.$$

Solution Write the formula as a quotient of polynomials. Notice that the polynomial in the denominator is expanded and written in decreasing order of the exponents.

$$w(h) = \frac{150 \cdot 4000^2}{h^2 + 8000h + 4000^2}$$

The numerator is a constant polynomial $p(h) = 150 \cdot 4000^2$. The denominator is a quadratic expression $q(h) = h^2 + 8000h + 4000^2$. The end behavior of w is the same as the end behavior of

$$g: h \to \frac{150 \cdot 4000^2}{1} \cdot \frac{1}{h^2} = \frac{150 \cdot 4000^2}{h^2}.$$

This is an inverse square function whose end behavior you should know. $\lim_{h \to +\infty} w(h) = \lim_{h \to +\infty} g(h) = 0$ and $\lim_{h \to -\infty} w(h) = \lim_{h \to -\infty} g(h) = 0$.

Dr. Anna L. Fisher in weightlessness training

The function of Example 3 may seem exotic, but it is not. If an astronaut weighs 150 lb on the surface of the Earth, $w(h)$ is the astronaut's weight h miles above the surface. Since $\lim_{h \to +\infty} w(h) = 0$ as the astronaut goes farther and farther from the surface, the astronaut's weight approaches zero, which is called weightlessness.

The four specific functions in this lesson illustrate the major cases of the End Behavior of Rational Functions Theorem.

function rule	m	n	type	end behavior
$p(x) = 3x^4 - 5x^3 + 8x^2 - 20x + 16$	3	0	$m > n$	like the power function $f(x) = 3x^4$
$h(x) = \dfrac{x^2 + 14x + 12}{3x + 6}$	2	1	$m > n$ $m = n + 1$	like the linear function $f(x) = \frac{1}{3}x + 4$
$g(x) = \dfrac{2x^2 + 1}{x^2 - 1}$	2	2	$m = n$	like the constant function $f(x) = 2$
$w(h) = 150\left(\dfrac{4000}{4000 + h}\right)^2$	0	2	$m < n$	like the reciprocal of a power function $f(h) = \dfrac{150 \cdot 4000^2}{h^2}$

318

Questions

Covering the Reading

1. Fill in the parentheses: $4x^3 - 11x^2 + 6x - 2 = (?)\,x^3$. $4 - \frac{11}{x} + \frac{6}{x^2} - \frac{2}{x^3}$

2. Consider the polynomial function p with $p(x) = 6 - 4x^5 + 2x^3$.
a. What is the leading term of $p(x)$? $-4x^5$
b. The end behavior of p is like that of which power function?
$f(x) = -4x^5$

3. Consider the graph of the rational function r with $r(x) = \dfrac{2x^2 + 9x - 11}{6x^2 + 3}$.
a. Describe all asymptotes. $y = \frac{1}{3}$
b. Describe the end behavior of r
like the function $f(x) = \frac{1}{3}$; $\lim\limits_{x\to+\infty} r(x) = \frac{1}{3}$, $\lim\limits_{x\to-\infty} r(x) = \frac{1}{3}$

4. Give the end behavior of the function f with $f(k) = \dfrac{2k + 3}{k^2 + 9}$. (Note: You were asked to analyze this function in Question 11 of Lesson 5-4.)
like the function $g(k) = \frac{2}{k}$; $\lim\limits_{k\to+\infty} f(k) = 0$, $\lim\limits_{k\to-\infty} f(k) = 0$

5. Give the end behavior of

$$g: x \to \frac{7x^8 - 12x^7 + 6x^6 - 9x^5 + 4x^3 - 12x + 1000}{7x^4 + 6x^3 - 2x^2}.$$

like the function $f(x) = x^4$; $\lim\limits_{x\to+\infty} g(x) = +\infty$, $\lim\limits_{x\to-\infty} g(x) = +\infty$

6. Use long division to find the oblique asymptote for the graph of the function f with equation $f(x) = \dfrac{6x^2 - 8x + 4}{2x + 1}$. $y = 3x - \frac{11}{2}$

7. Use long division to show that the end behavior of $h: v \to \dfrac{4v^3 + 3v^2 + 8v - 2}{2v - 1}$ is like the end behavior of a polynomial function of degree 2. See margin.

8. a. If you weigh 150 lb on the surface of the Earth, how much will you weigh in an airplane 7 miles above the surface? $\approx$ 149.5 lb
b. An astronaut who weighs 150 lb on the surface of the Earth has what weight in a space shuttle 120 miles above the surface?
c. When astronauts went to the moon, how much did the earth's gravitation contribute to their weight halfway there, about 225,000 miles above the surface of the Earth? $\approx$ 0.05 lb
b) $\approx$ 141.4 lb

Applying the Mathematics

9. A more general formula for the weight w of an object in pounds h miles above the surface of a planet is $w(h) = w(0)\left(\dfrac{r}{r + h}\right)^2$, where r is the radius of a planet. If an object weighed 100 lb on the surface of Mars, whose radius is about 2000 miles, graph what its weight would be at all altitudes above the surface. See margin.

10. Give a formula for a function f that has a vertical asymptote at $x = 5$, contains the point (1, 2), and has end behavior like the function $g: x \to \dfrac{3}{x^3}$. sample: $f(x) = \dfrac{3x - 11}{x^3(x - 5)}$

ADDITIONAL ANSWERS
7., 9. See Additional Answers in the back of this book.

NAME ______________________

LESSON **MASTER 5–5**
QUESTIONS ON **SPUR** OBJECTIVES

■ PROPERTIES *Objective G (See pages 341–344 for objectives.)*
In 1–4, a. find a constant function, linear function, power function, or reciprocal of a power function whose end behavior resembles that of the given function, b. use limit notation to describe the end behavior, and c. use limit notation to describe the behavior near any vertical asymptotes.

1. $f(x) = \dfrac{3x^2 + 2x + 1}{5x - 2x^3}$
a. $y = -\frac{3}{2x}$
b. $\lim\limits_{x\to\infty} f(x) = 0$, $\lim\limits_{x\to-\infty} f(x) = 0$
c. $\lim\limits_{x\to-\sqrt{\frac{5}{2}}^-} f(x) = \lim\limits_{x\to0^+} f(x) = \lim\limits_{x\to\sqrt{\frac{5}{2}}^-} f(x) = \infty$, $\lim\limits_{x\to-\sqrt{\frac{5}{2}}^+} f(x) = \lim\limits_{x\to0^-} f(x) = \lim\limits_{x\to\sqrt{\frac{5}{2}}^+} f(x) = -\infty$

2. $g(x) = \dfrac{2x^2 - 5}{5x^2 + 2}$
a. $y = \frac{2}{5}$
b. $\lim\limits_{x\to-\infty} g(x) = \lim\limits_{x\to\infty} g(x) = \frac{2}{5}$
c. no vertical asymptotes

3. $h(x) = \dfrac{x^2 + x - 1}{x - 1}$
a. $y = x + 2$
b. $\lim\limits_{x\to-\infty} h(x) = -\infty$, $\lim\limits_{x\to\infty} h(x) = \infty$
c. $\lim\limits_{x\to1^-} h(x) = -\infty$, $\lim\limits_{x\to1^+} h(x) = \infty$

4. $f(t) = t^5 + 12t^3 - 4t - \frac{2}{t}$
a. $y = x^5$
b. $\lim\limits_{t\to-\infty} f(t) = -\infty$, $\lim\limits_{t\to\infty} f(t) = \infty$
c. no vertical asymptotes

■ REPRESENTATIONS *Objective J*
In 5–7, a. find the x- and y-intercepts of the graph of the function and b. sketch the graph.

5. the function in Question 1
a. no x- or y-intercepts
b.
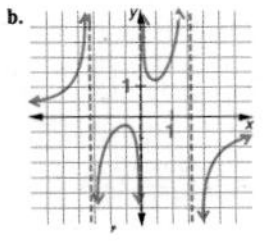

6. the function in Question 2
a. x-int.: $\pm\sqrt{\frac{5}{2}}$; y-int.: $-\frac{5}{2}$
b.
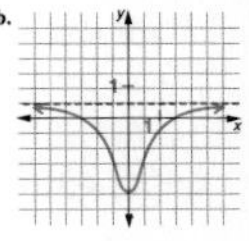

58 *Continued* *Precalculus and Discrete Mathematics* © Scott, Foresman and Company

NAME ______________________
Lesson MASTER 5–5 (page 2)

7. the function in Question 3
a. x-intercepts: $\approx$ -1.5, $\approx$ 0.7
y-intercept: 1
b.
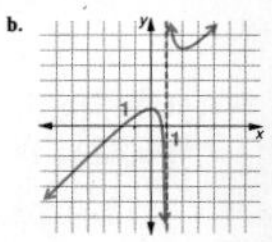

■ REPRESENTATIONS *Objective K*
In 8–10, find an equation for each asymptote of the graph of the function and state whether the asymptote is vertical, horizontal, or oblique.

8. the function in Question 2
$y = \frac{2}{5}$, horizontal

9. the function in Question 3
$x = 1$, vertical
$y = x + 2$, oblique

10. the function graphed at the right
$x = -2$, vertical; $x = 2$, vertical;
$y = \frac{1}{2}x + 1$, oblique

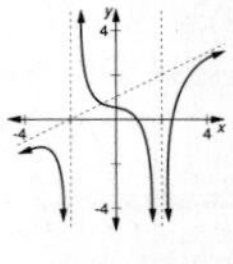

11. Sketch the graph of a function with asymptotes $y = -x + 4$ and $x = 3$, $\lim\limits_{x\to3^-} f(x) = -\infty$, and $\lim\limits_{x\to3^+} f(x) = \infty$.

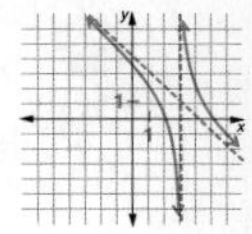

Precalculus and Discrete Mathematics © Scott, Foresman and Company 59

NOTES ON QUESTIONS

Error Analysis for Question 4: For automatic graphing, the selection of a domain and range is critical. Have students try $-30 \le x \le 30$ and $-0.5 \le y \le 0.75$. Many students think of an asymptote as a line that a graph approaches but never touches. They are surprised to see the graph of $f(x) = \frac{2x + 3}{x^2 + 9}$ crossing its horizontal asymptote $y = 0$ at the point (-1.5, 0). This problem provides an excellent opportunity to expand students' understanding of asymptotes.

Technology for Questions 6 and 7: Have students use an automatic grapher to graph these functions. Suggest settings of $-100 \le x \le 100$ and $-500 \le y \le 500$. The end behaviors will be illustrated nicely. Stress that the oblique asymptote describes the end behavior of the function. It is not exactly the same as the function when $|x|$ is large. Demonstrate by evaluating $f(x)$ and y for large values of x.

Question 13: This function is interesting because it has one removable and one essential discontinuity. Using a window of $-5 \le x \le 5$ and $-5 \le y \le 5$ and the trace mode, some calculators display no y-value for $x = 3$. The hole at $\left(3, \frac{5}{7}\right)$ is actually visible because one pixel on the display screen is not activated.

FOLLOW-UP

MORE PRACTICE

For more questions on SPUR Objectives, use *Lesson Master 5-5,* shown on page 319.

ADDITIONAL ANSWERS

11.a., 13.b., 14., 18.a., 20. See Additional Answers in the back of this book.

11. In Mr. Ease's class, all points on quizzes and tests are combined to determine a student's final grade. Viola only got 3 of 20 points possible on the first quiz. Suppose Viola does not miss any point for the rest of the year. Let x be the number of points after the first quiz and $r(x)$ be the total fraction of points Viola has earned.

a. Graph r. See margin.

b. What does the end behavior of r tell you about this situation? See below.

12. Let s be the sequence defined by $s_n = \frac{7n + 11}{4n - 3}$. Find $\lim_{n \to +\infty} s_n$. $\frac{7}{4}$

Review

13. Consider the real function T defined by $T(z) = \frac{z^2 - z - 6}{z^2 + z - 12}$.

a. Determine the location of the removable discontinuity of function T.

b. Show that T has an essential discontinuity at $z = -4$. *(Lesson 5-4)*

a) $z = 3$; b) See margin.

14. Prove that $\sqrt{7}$ is irrational. *(Lesson 5-2)* See margin.

15. Consider the function $f: x \to \frac{4}{x}$. Find the slope of the segment joining $\left(x, \frac{4}{x}\right)$ to $\left(x + h, \frac{4}{x + h}\right)$. *(Lesson 5-1)* $\frac{-4}{x(x + h)}$, $x \ne 0$, $h \ne 0$, and $x \ne -h$

In 16 and 17, write in factored form. *(Lesson 4-8)*

16. $(x + 11)^2(x - 5)^3(x + 2) + (x + 11)^3(x - 5)^2(x + 2)^2$

$(x + 11)^2(x - 5)^2(x + 2)(x^2 + 14x + 17)$

17. $(3t - 7y)^4(8t + 5y)^3 - (3t - 7y)^3(8t + 5y)^4(t + y)$

$(3t - 7y)^3(8t + 5y)^3(-8t^2 + 3t - 7y - 13ty - 5y^2)$

18. Refer to the graph of a cosine function at the right.

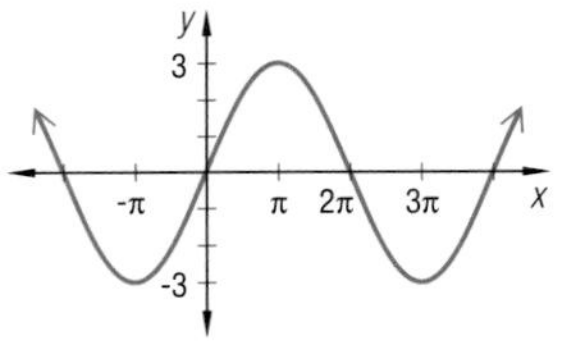

a. Identify the amplitude, period, and phase shift.

b. Write a cosine equation that can represent the graph. *(Lesson 3-9)*

a) See margin. b) $y = 3\cos\left(\frac{x - \pi}{2}\right)$

19. Evaluate without using a calculator. *(Lesson 2-8, Previous course)*

a. $\sin\left(\frac{\pi}{4}\right)$ $\frac{\sqrt{2}}{2}$ **b.** $\cos\left(\frac{2\pi}{3}\right)$ $-\frac{1}{2}$

c. $\sin\left(\frac{3\pi}{2}\right)$ -1 **d.** $\cos\left(\frac{7\pi}{6}\right)$ $-\frac{\sqrt{3}}{2}$

Exploration

20. A ball that is tossed upward from ground level to a height of 10 meters and then bounces to 75% of its height on each succeeding bounce will travel a total of $\frac{80(4^n - 3^n)}{4^n}$ meters by the time of its $(n + 1)$st bounce. Determine the end behavior of this function using one or more of the techniques of Example 1 and interpret the physical meaning of what you have determined. See margin.

11. b. As the number of total points increases, Viola's grade approaches 100%.

LESSON

5-6

The Tan, Cot, Sec, and Csc Functions

You can create a rational function by using the reciprocal of a polynomial function or by dividing one polynomial function by another. Similarly, new trigonometric functions can be obtained by using the reciprocal of the sine function or cosine function or by dividing one of these functions by the other. These four functions are not rational functions but they can be analyzed using many of the techniques you have seen in the previous lessons. Their names should be familiar to you from previous courses.

Definitions

For any real number x:

the **tangent** of x = **tan** $x = \frac{\sin x}{\cos x}$ provided $\cos x \neq 0$.

the **cotangent** of x = **cot** $x = \frac{\cos x}{\sin x}$ provided $\sin x \neq 0$.

the **secant** of x = **sec** $x = \frac{1}{\cos x}$ provided $\cos x \neq 0$.

the **cosecant** of x = **csc** $x = \frac{1}{\sin x}$ provided $\sin x \neq 0$.

Each of these definitions defines a function whose domain is the set of all real numbers except those for which the denominator is 0.

From the definitions, values of each function can be calculated. For instance,

$$\cot \frac{\pi}{6} = \frac{\cos \frac{\pi}{6}}{\sin \frac{\pi}{6}} = \frac{\frac{\sqrt{3}}{2}}{\frac{1}{2}} = \sqrt{3} \approx 1.732 \text{ and } \sec 68° = \frac{1}{\cos 68°} \approx 2.669.$$

Also, from the definitions, some values of the functions can be obtained from a consideration of right triangles. For instance, in the triangle shown here,

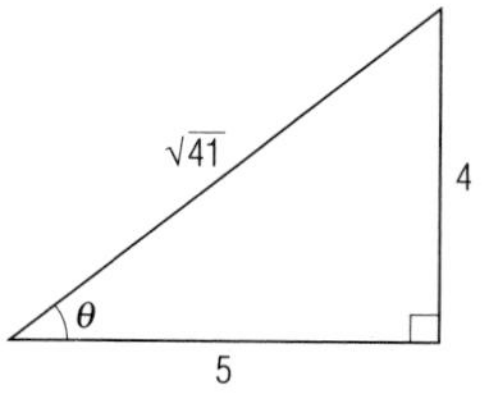

$$\tan \theta = \frac{\sin \theta}{\cos \theta} = \frac{\frac{\text{side opposite } \theta}{\text{hypotenuse}}}{\frac{\text{side adjacent to } \theta}{\text{hypotenuse}}} = \frac{\text{side opposite } \theta}{\text{side adjacent to } \theta} = \frac{4}{5}.$$

With rational functions, when a value of x causes the denominator to be 0, there is a discontinuity. The same is true for these trigonometric functions. Consider the cosecant function. Because $\sin x = 0$ when x is a multiple of π, the cosecant function is not defined when x is a multiple of π.

LESSON 5-6

RESOURCES
- Lesson Master 5-6
- Quiz for Lessons 5-4 through 5-6
- Teaching Aid 32 displays the graphs of sine, cosecant, cosine, and secant functions.
- Teaching Aid 33 displays the graphs of sine, cosine, tangent, and cotangent functions.

OBJECTIVES

D Find values of trigonometric functions.
H Classify discontinuities as essential or removable.
J Graph quotients of polynomial or trigonometric functions.
L Use right triangles to find values of trigonometric functions.

TEACHING NOTES

To assist students in understanding the graph of $y = \csc x$, you might begin with a discussion of the trend in values of $\frac{1}{x}$ as x itself increases and as x decreases. Then use a calculator or computer to prepare a table as shown below (x is in radians, not degrees).

x	$\sin x$	$\csc x$
0	0	not defined
.1	.09983	10.01669
.2	.19867	5.03349
⋮	⋮	⋮
1.5	.99749	1.00251
$\frac{\pi}{2}$	1	1
1.6	.99957	1.00043
1.7	.99166	1.00841
⋮	⋮	⋮
3.1	.04158	24.04964
π	0	not defined

Graph both $y = \sin x$ and $y = \csc x$ simultaneously. This also can be done quite effectively on some calculators by setting the graphing mode to "Simul" (simultaneous).

Error Analysis Because $x^{-1} = \frac{1}{x}$, some students may assume that the $\sin^{-1}$, $\cos^{-1}$, and $\tan^{-1}$ keys on a calculator are for reciprocals of these functions. Point out that this is not the case. To stress this point, you may wish to calculate $\sin^{-1} x$ and $\frac{1}{\sin x}$ for several values of x.

To help students understand the behavior of the $y = \tan x$ graph, you might begin with a brief analysis of signs and values of sine and cosine. Sketch the graphs of $y = \sin x$, $y = \cos x$, and $y = \tan x$ on the chalkboard when developing the following charts.

$y = \sin x$

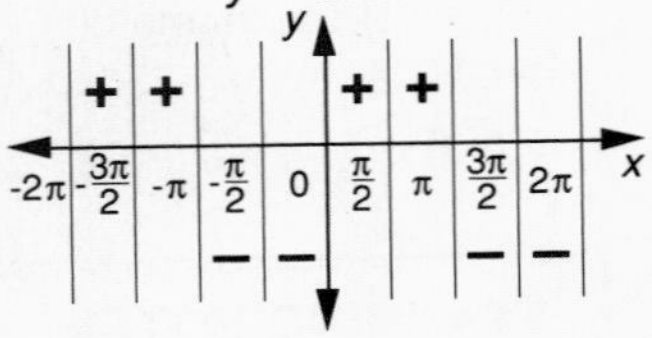

$y = \cos x$

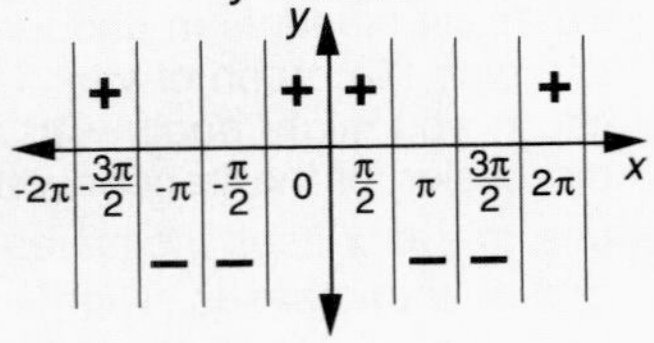

Now, think of dividing positive and negative numbers,

$y = \dfrac{\sin x}{\cos x} = \tan x$

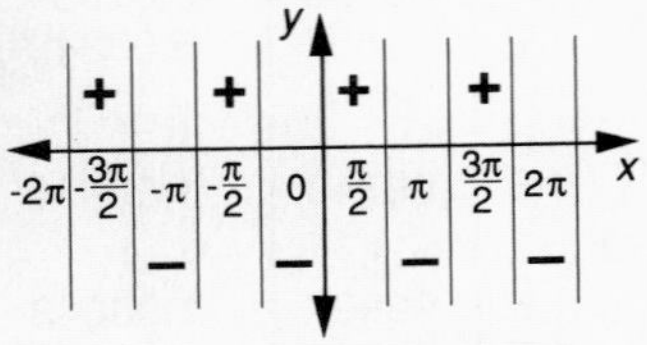

Discuss values of $\frac{\sin x}{\cos x}$ as $\sin x$ and $\cos x$ increase or decrease.

At each value of x where $\sin x = 0$ there is a vertical asymptote to the cosecant function. Notice also that the two functions have the same values when $\sin x = 1$ or when $\sin x = -1$. This is because the numbers 1 and -1 equal their reciprocals. The information that we have determined so far is shown below.

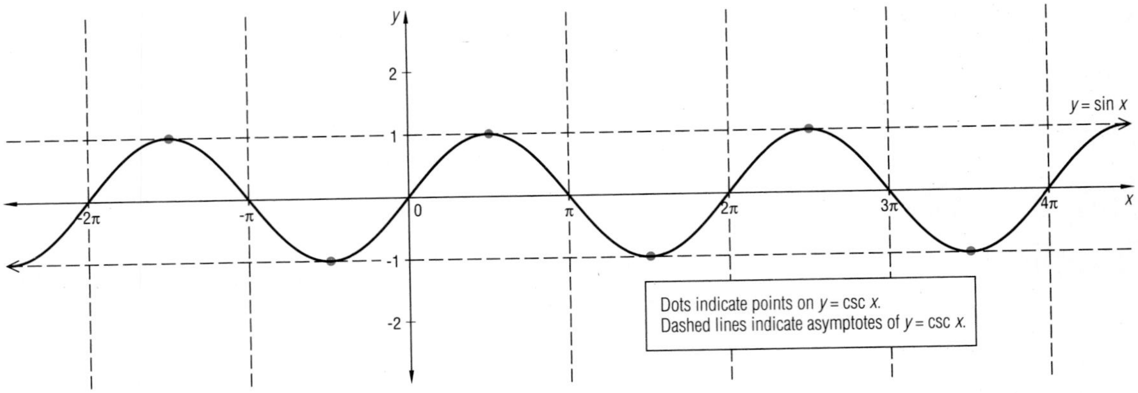
y = sin x
Dots indicate points on y = csc x.
Dashed lines indicate asymptotes of y = csc x.

When a number is positive, its reciprocal is positive, and when a number is negative so is its reciprocal. Thus the sine and cosecant are positive at the same time and negative at the same time. As the sine gets closer to zero, the cosecant gets farther away from zero. And since the sine is periodic with period 2π, the cosecant is periodic also. This information allows the rest of a sketch of the graph to be made.

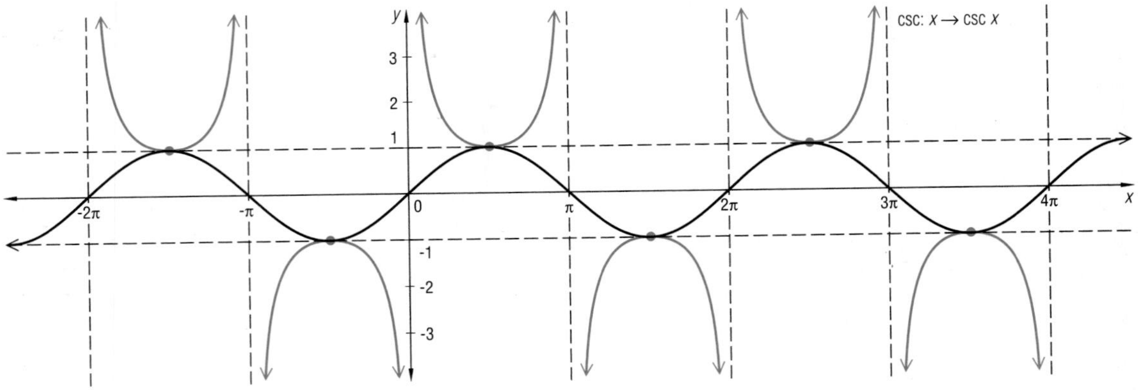
csc: x → csc x

The graph of the secant function is related in a similar way to the graph of the cosine.

The graphs of the tangent and cotangent functions have quite a different shape from any graphs you have yet seen in this book. First, notice that since $\tan x = \dfrac{\sin x}{\cos x}$, the tangent is not defined when $\cos x = 0$. Second, by the definition, $\tan x = 0$ if and only if $\sin x = 0$. Third, the tangent is positive when the sine and cosine have the same sign and it is negative when the sine and cosine have opposite signs. The graph of the tangent function exhibits all of these properties.

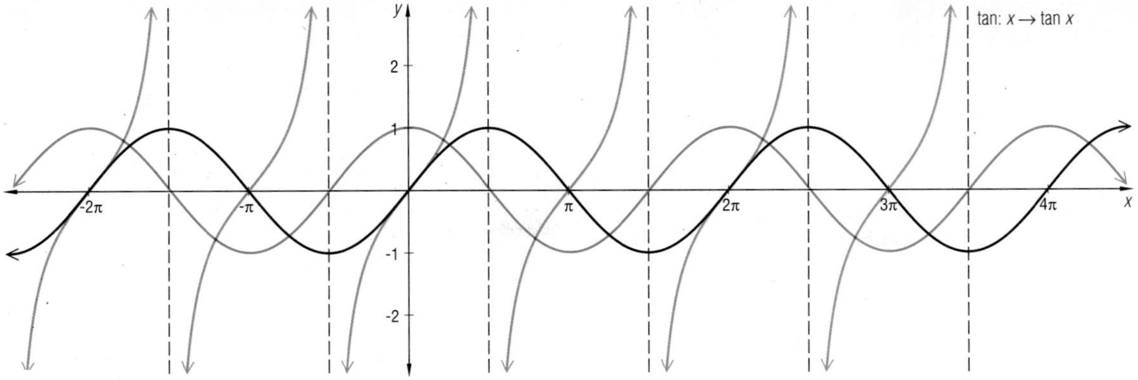

A surprise is that the graph of the tangent function has period π rather than 2π. This is proved in the next chapter.

Questions

Covering the Reading

1. If $\sin x = .8$ and $\cos x = .6$, give the values of $\tan x$, $\cot x$, $\sec x$, and $\csc x$. $\tan x = \frac{4}{3}$; $\cot x = \frac{3}{4}$; $\sec x = \frac{5}{3}$; $\csc x = \frac{5}{4}$

2. Describe the function value in terms of the sides of the right triangle as named in the drawing.
a. $\tan \theta$ **b.** $\cot \theta$
c. $\sec \theta$ **d.** $\csc \theta$
See margin.

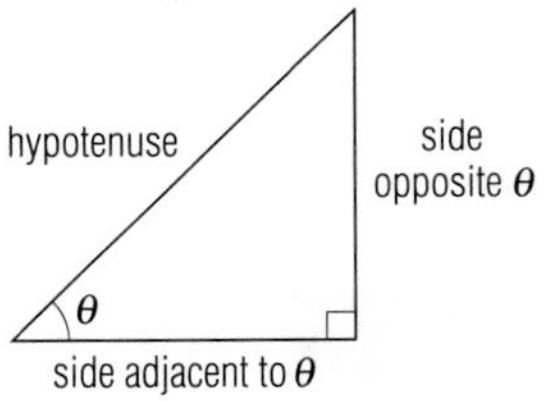

3. A graph of the cosecant function is in this lesson. Identify four points on this graph with $0 \le x \le 2\pi$. sample: $\left(\frac{\pi}{4}, \sqrt{2}\right), \left(\frac{\pi}{2}, 1\right), \left(\frac{3\pi}{4}, \sqrt{2}\right), \left(\frac{3\pi}{2}, -1\right)$

4. a. For what values of x is $\sec x$ not defined? $x = \frac{\pi}{2} + n\pi$, n an integer
b. Graph the secant function $y = \sec x$. See margin.

5. Identify four points on the graph of $y = \tan x$ with $0 \le x \le \frac{\pi}{2}$. See margin.

6. a. For what values of x is $\cot x$ not defined? $x = n\pi$, n an integer
b. Graph the cotangent function $y = \cot x$. See margin.

7. Consider the six trigonometric functions sine, cosine, tangent, cotangent, secant, and cosecant.
a. Give their periods. See margin.
b. Which are even functions, which odd, which neither? See margin.
c. Which are rational functions? none
d. Which are continuous functions over the set of all real numbers? sin and cos
e. Make a table of values of these functions for the domain values $\frac{\pi}{6}$, $\frac{\pi}{4}$, and $\frac{\pi}{3}$. See margin.

ADDITIONAL EXAMPLES
Students are asked to graph $y = \sec x$ and $y = \cot x$ in **Questions 4 and 6** of this lesson. They also are appropriate Additional Examples.

NOTES ON QUESTIONS
Question 2: For students familiar with the SOHCAHTOA (SineOppositeHypotenuse, CosineAdjacentHypotenuse, TangentOppositeAdjacent) mnemonic for remembering the right triangle definitions of sin, cos, and tan, CHOSHACAO is the corresponding mnemonic for csc, sec, and cot relative to a right triangle.

Questions 4 and 6: Have students do the same careful planning and analysis in graphing these functions as they did when graphing $y = \csc x$ and $y = \tan x$.

Question 7: Be sure that students understand that if a function is odd (even), then its reciprocal function is also odd (even).

ADDITIONAL ANSWERS
2.a. $\tan \theta = \frac{\text{side opposite } \theta}{\text{side adjacent to } \theta}$
b. $\cot \theta = \frac{\text{side adjacent to } \theta}{\text{side opposite } \theta}$
c. $\sec \theta = \frac{\text{hypotenuse}}{\text{side adjacent to } \theta}$
d. $\csc \theta = \frac{\text{hypotenuse}}{\text{side opposite } \theta}$

4.b.

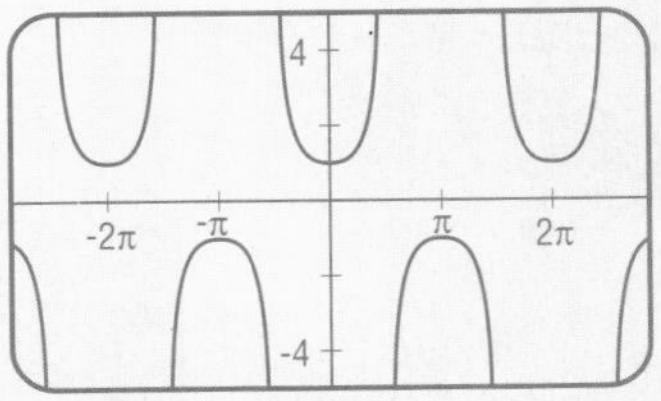

5. sample: $(0, 0), \left(\frac{\pi}{6}, \frac{\sqrt{3}}{3}\right), \left(\frac{\pi}{4}, 1\right), \left(\frac{\pi}{3}, \sqrt{3}\right)$

6.b., 7.a., b., e. See Additional Answers in the back of this book.

NOTES ON QUESTIONS

Question 8: A major historical reason for the use of the cotangent, secant, and cosecant functions was to avoid having to divide by decimals. This question shows how any division by sin x can be converted into a multiplication by another trigonometric function.

Question 11b: Students also could verify this formula for equilateral triangles (n = 3) and regular hexagons (n = 6).

Question 20: Assume that the two peaks are the same height.

Question 21: This question gives the origin of the names tangent, cotangent, secant, and cosecant.

ADDITIONAL ANSWERS

9. $\tan\left(-\frac{\pi}{4}\right) = -1$; $\cot\left(-\frac{\pi}{4}\right) = -1$; $\sec\left(-\frac{\pi}{4}\right) = \sqrt{2}$; $\csc\left(-\frac{\pi}{4}\right) = -\sqrt{2}$

10.

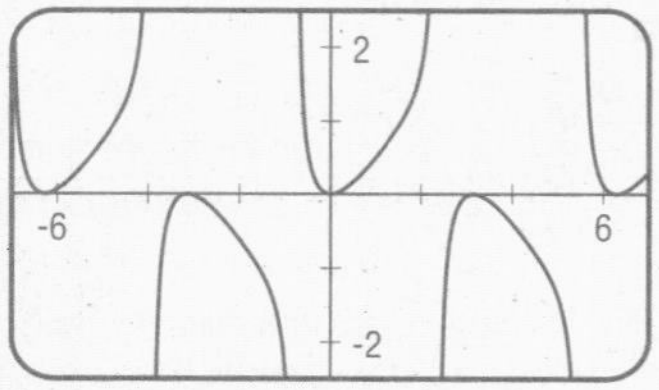

11.a. The area of the triangle ABO is $\frac{1}{2}(AB)h$. So, the area of the regular n-gon is the sum of the areas of n congruent triangles, $n\frac{1}{2}(AB)h$. $m\angle AOB = \frac{2\pi}{n}$. The altitude h splits $\angle AOB$ into two smaller angles measuring $\frac{\pi}{n}$. $\tan\left(\frac{\pi}{n}\right) = \frac{\frac{1}{2}AB}{h}$. So, $AB = 2h\tan\frac{\pi}{n}$. Hence, the area of the n-gon is $n\frac{1}{2}\left(2h\tan\frac{\pi}{n}\right)h = nh^2\tan\frac{\pi}{n}$.

11.b., 15.a., 18.b. See Additional Answers in the back of this book.

Applying the Mathematics

8. *Multiple choice.* If p and q are real numbers and $p = \frac{q}{\sin x}$, then $p =$
(a) $q \tan x$ (b) $q \cot x$
(c) $q \sec x$ (d) $q \csc x$ (d)

9. Give the values of tan x, cot x, sec x, and csc x when $x = -\frac{\pi}{4}$. See margin.

10. Graph the function f defined by $f(x) = \frac{\sin x}{1 + \cot x}$ for $-2\pi \le x \le 2\pi$. See margin.

11. Suppose that $\overline{AB}$ is a side of a regular n-gon inscribed in circle O and suppose that h is the distance from O to $\overline{AB}$.

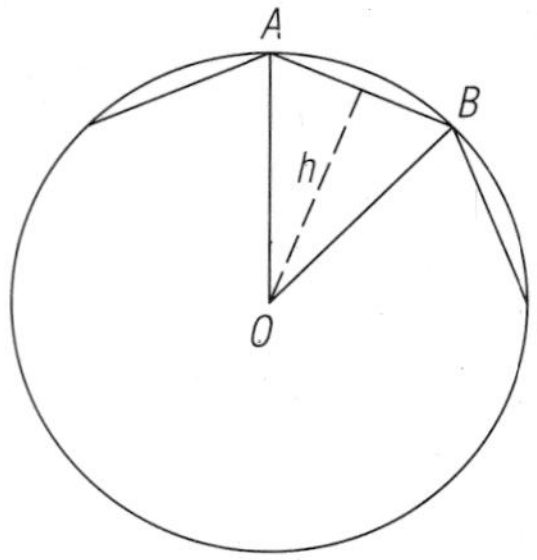

a. Prove that the area of the polygon is $nh^2 \tan\left(\frac{\pi}{n}\right)$.
b. Check the formula for the case where $\overline{AB}$ is a side of a square in a circle with radius 6. a,b) See margin.

12. Here is a proof that the tangent function is an odd function. Justify each conclusion.
a. For all real numbers x in the domain of the tangent function,
$\tan(-x) = \frac{\sin(-x)}{\cos(-x)}$ definition of tangent
b. $= \frac{-\sin x}{\cos x}$ sine is an odd function and cosine an even function
c. $= -\tan x$ definition of tangent
d. $\therefore$ the tangent function is an odd function.
transitive property of equality and definition of odd function

Review

13. The end behavior of $w: z \rightarrow \frac{6z^4 - 4z^3 + 2z + 1}{11z^5 + 2}$ is the same as the end behavior of what simpler function? *(Lesson 5-5)* $f: z \rightarrow \frac{6}{11z}$

14. Use limit notation to describe the end behavior of $g: y \rightarrow \frac{7y^3 + 2y - 2}{8y^3 + 5y^2 + 11y}$. *(Lesson 5-5)* $\lim_{y \to +\infty} g(y) = \frac{7}{8}$; $\lim_{y \to -\infty} g(y) = \frac{7}{8}$

15. a. Sketch a graph of $f(x) = \frac{5x + 1}{x - 2}$. See margin.
b. Use limit notation to describe the behavior of the function near $x = 2$.
c. Use limit notation to describe the end behavior of f.
(Lessons 5-5, 5-4) b) $\lim_{x \to 2^+} f(x) = +\infty$; $\lim_{x \to 2^-} f(x) = -\infty$; c) $\lim_{x \to +\infty} f(x) = 5$; $\lim_{x \to -\infty} f(x) = 5$

16. Rationalize the denominator: $\frac{6}{\sqrt{10} + 5}$. *(Lesson 5-2)* $\frac{10 - 2\sqrt{10}}{5}$

324

17. Find the smallest nonnegative integer that satisfies each congruence.
a. $y \equiv 17 \pmod 2$ 1 **b.** $t \equiv 1000 \pmod{11}$ *(Lesson 4-3)* 10

18. a. Solve over the set of real numbers:
$$\log_5(x^2 + 2) < \log_5(3x + 6).\quad -1 < x < 4$$
b. The function f: $x \rightarrow \log_5 x$ is an increasing function. How was this fact used in answering part **a**? *(Lesson 3-6)* See margin.

19. Use chunking to solve $e^{2x} - 4e^x + 3 = 0$. *(Lesson 3-4)*
$x = \ln 3$ or $x = 0$

20. One mountain peak is 10 miles due east of another. A hiker locates the first mountain peak at an angle of 65° W of N, and the second at an angle of 32° N of E. How far is the hiker from each mountain peak? *(Lesson 2-8)*
≈ 6.32 miles from the first and ≈ 5.04 miles from the second

Exploration

21. Use the diagram of a unit circle at the right. All six trigonometric values of θ can be represented by lengths of segments in this unit circle diagram. Match the values in the left column with the appropriate lengths in the right column.

a. $\cos \theta$	(i) AC	iv
b. $\sin \theta$	(ii) BF	i
c. $\tan \theta$	(iii) DE	ii
d. $\cot \theta$	(iv) OA	iii
e. $\sec \theta$	(v) OD	vi
f. $\csc \theta$	(vi) OF	v

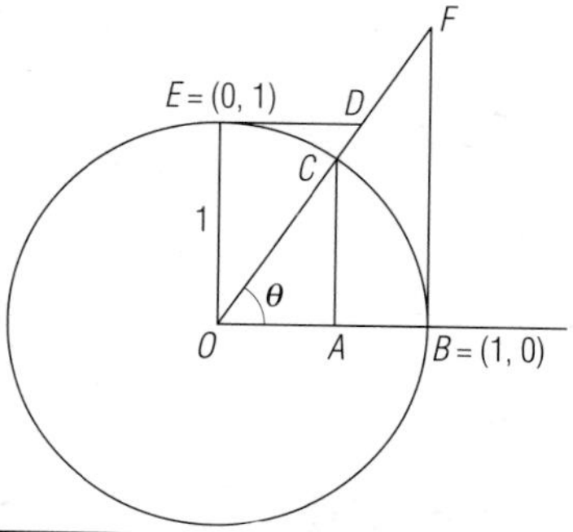

FOLLOW-UP

MORE PRACTICE
For more questions on SPUR Objectives, use *Lesson Master 5-6,* shown below.

EXTENSION
The cosine, cotangent, and cosecant are *cofunctions.* They got their names because for all x in radians, function (x) = cofunction $\left(\frac{\pi}{2} - x\right)$ and function (x) = cofunction $(90 - x)$ for degrees. (**Question 21** shows this graphically.) Thus, $\sin(x) = \cos\left(\frac{\pi}{2} - x\right)$, and so on. In any identity involving only the single argument x, you can switch $\sin x$ with $\cos x$, $\tan x$ with $\cot x$, and $\sec x$ with $\csc x$ to obtain another (possibly different) identity. Have students find other identities and their cofunction counterpart identities.

EVALUATION
A quiz covering Lessons 5-4 through 5-6 is provided in the Teacher's Resource File.

NAME ________

LESSON **MASTER 5–6**
QUESTIONS ON **SPUR** OBJECTIVES

■**SKILLS** *Objective D (See pages 341–344 for objectives.)*

1. If $\sin x = \frac{-\sqrt{3}}{2}$ and $\cos x = -\frac{1}{2}$, find the value of $\tan x$, $\sec x$, $\cot x$, and $\csc x$.
$\sqrt{3}$ -2 $\frac{\sqrt{3}}{3}$ $-\frac{2\sqrt{3}}{3}$

In 2–7, find the indicated value.

2. $\cot \frac{7\pi}{6}$ $\sqrt{3}$ 3. $\sec -\frac{3\pi}{2}$ undefined
4. $\csc \frac{\pi}{2}$ 1   5. $\tan \pi$ 0
6. $\sec \frac{3\pi}{4}$ $-\sqrt{2}$ 7. $\tan -\frac{\pi}{4}$ -1

■**PROPERTIES** *Objective H*

8. Consider the cotangent function f: $x \rightarrow \cot x$. For what values of x does this function have essential discontinuities? $x = n\pi$, where n is an integer

9. What transformation can be applied to the graph of the function $\csc x$ so that the image has essential discontinuities at $x = \frac{n\pi}{2}$ ∀ integers n? $S_{\frac{1}{2},1}$

■**REPRESENTATIONS** *Objective J*

10. a. Sketch a graph of $y = \cot x$ for $-2\pi < x < 2\pi$.
b. On the same axes, sketch a graph of $y = \csc x$.
c. For which values of x is $\cot x > \csc x$?
$(2n - 1)\pi < x < 2n\pi$, where n is an integer

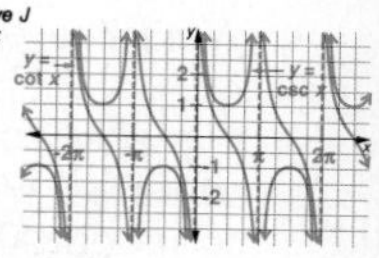

■**REPRESENTATIONS** *Objective L*
In 11–14, use the right triangle below to find the indicated value.

11. $\cot \theta$ $\frac{\sqrt{2}}{2}$ 12. $\csc \phi$ $\sqrt{3}$
13. $\sec \theta$ $\sqrt{3}$ 14. $\tan \phi$ $\frac{\sqrt{2}}{2}$

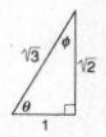

60

LESSON 5-7

RESOURCES
- Lesson Master 5-7
- Teaching Aid 34 displays the graphical solution of $\frac{1}{x} = \frac{1}{x+1} + \frac{1}{x+2}$.

OBJECTIVES

E Solve rational equations.
I Apply rational expressions and rational equations.

TEACHING NOTES

It is better to approach the determination of the least common denominator (LCD) in a systematic way rather than by stating an arbitrary rule. Focus on the reason for finding the LCD: this one expression must have enough factors so that when each fraction in an equation is multiplied by it, each product is a polynomial.

Error Analysis A common error students make is forgetting to multiply *every* term in the rational equation by the LCD. This is especially true in a problem like **Example 2**, where one term (the 1) is not a fraction. Working some additional examples and practice problems should help students to avoid this error.

It is easy for students to get confused when solving rational equations. This confusion stems from the fact that solving such equations involves multiplying both sides by a variable quantity, which is not a reversible step.

LESSON

5-7 Rational Equations

An equation of the form $f(x) = g(x)$, where $f(x)$ and $g(x)$ are rational expressions or combinations of rational expressions, is called a **rational equation.** For example, the equations

$$\frac{x^2-1}{x+2} = \frac{3x+2}{x-4}, \qquad t + \frac{1}{t} = t^2 - 1,$$

$$\frac{y+1}{y-1} + \frac{1}{y+1} = \frac{2y^2-1}{y+3}, \qquad z^2 = 2z - 1$$

are rational equations. In this lesson, you will review techniques for solving certain rational equations.

Example 1 is motivated by equations such as

$$\frac{1}{3} + \frac{1}{6} = \frac{1}{2} \quad \text{and} \quad \frac{1}{5} + \frac{1}{20} = \frac{1}{4},$$

in which the reciprocal of a positive integer is the sum of the reciprocals of two other positive integers.

Example 1 Are there any rational numbers x, $x + 1$, and $x + 2$ such that the reciprocal of the smallest number is the sum of the reciprocals of the others?

Solution The problem asks you to solve

$$\frac{1}{x} = \frac{1}{x+1} + \frac{1}{x+2}.$$

To do this, multiply both sides by the least common multiple of the denominators. This is called the least common denominator (LCD). The LCD here is $x(x + 1)(x + 2)$.

$$x(x+1)(x+2) \bullet \frac{1}{x} = x(x+1)(x+2)\left(\frac{1}{x+1} + \frac{1}{x+2}\right)$$
$$(x+1)(x+2) = x(x+2) + x(x+1)$$
$$x^2 + 3x + 2 = x^2 + 2x + x^2 + x$$

Add $-x^2 - 3x$ to each side to obtain

$$2 = x^2.$$

So $x = \sqrt{2}$ or $x = -\sqrt{2}$. Hence, the only real numbers that satisfy the condition are

$$\sqrt{2},\ \sqrt{2}+1, \text{ and } \sqrt{2}+2$$
$$\text{and } -\sqrt{2},\ -\sqrt{2}+1, \text{ and } -\sqrt{2}+2.$$

None of these are rational, so the answer to the question posed is no.

The equation in Example 1 reduced to $2 = x^2$. Things do not always work out so nicely! When a rational equation is difficult to solve, you can approximate the solution as close as you wish by graphing and using what you know about rational functions. In Example 1, let $f(x) = \frac{1}{x}$ and $g(x) = \frac{1}{x+1} + \frac{1}{x+2}$.

326

The graph at the right shows that f and g have two points of intersection P_1 and P_2 with x-coordinates at about 1.41 and -1.41, which approximate the exact values $\sqrt{2}$ and $-\sqrt{2}$. Since $\lim_{x \to -\infty} g(x) = 0$ and $\lim_{x \to -\infty} f(x) = 0$, if you did not solve the equations algebraically you would have to check in some other way that the graphs of f and g do not intersect at some value of x less than $-\sqrt{2}$ or greater than $\sqrt{2}$.

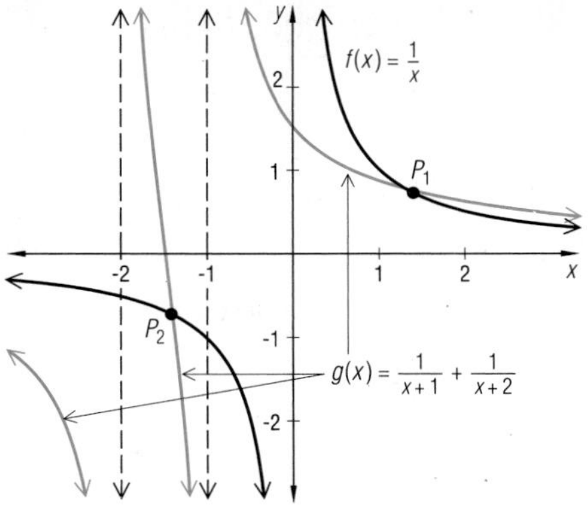

In Example 1, both sides of the original equation were multiplied by an expression containing a variable. Multiplying both sides of an equation by an expression which contains a variable is not a reversible step when the value of the expression is zero. Therefore, when you solve a rational equation by the method illustrated above you must check that the values obtained are, in fact, solutions. The graph above provided a check.

Example 2 Solve $1 + \frac{8}{x^2 - 16} = \frac{1}{x - 4}$.

Solution To find the LCD, first factor all denominators:

$$1 + \frac{8}{(x+4)(x-4)} = \frac{1}{x-4}$$

The LCD is $(x + 4)(x - 4)$. Now multiply both sides by the LCD.

$$(x+4)(x-4)\left(1 + \frac{8}{(x+4)(x-4)}\right) = (x+4)(x-4)\left(\frac{1}{x-4}\right)$$

$\Rightarrow \quad (x+4)(x-4) + 8 = x + 4$ — Distribute on the left side and multiply on the right side.

$\Leftrightarrow \quad x^2 - 16 + 8 = x + 4$

$\Leftrightarrow \quad x^2 - x - 12 = 0$ — Collect like terms.

$\Leftrightarrow \quad (x - 4)(x + 3) = 0$ — Factor.

$\Leftrightarrow \quad x - 4 = 0 \text{ or } x + 3 = 0$ — Zero Product Property

$\Leftrightarrow \quad x = 4 \text{ or } x = -3$

This analysis shows that *if* x is a solution of the given equation then $x = 4$ or $x = -3$. But $x = 4$ in the original equation makes the denominators zero. Thus 4 is *not* a solution. If you substitute -3 for x in the original equation, you get

$$1 + \frac{8}{(-3)^2 - 16} = \frac{1}{-3-4} \Leftrightarrow 1 + \frac{8}{-7} = -\frac{1}{7}$$

which is true. Therefore, -3 is the only solution to this equation.

The following is true for all a, b, and c (even for $c = 0$):
$a = b \Rightarrow ac = bc$.
However, the converse is true only for $c \neq 0$:
$ac = bc$ and $c \neq 0 \Rightarrow a = b$,
or, alternatively,
$ac = bc \Rightarrow c = 0$ or $a = b$.

Thus, in **Examples 1 and 2**,
$\frac{1}{x} = \frac{1}{x+1} + \frac{1}{x+2}$
$\Rightarrow x = \sqrt{2}$ or $x = -\sqrt{2}$,
and $1 + \frac{8}{x^2 - 16} = \frac{1}{x - 4}$
$\Rightarrow x = 4$ or $x = -3$.
The converses of the above implications are not necessarily true. Ask students to find the nonreversible step in **Example 1**. (It is the step in which both sides of the equation are multiplied by the LCD).

ADDITIONAL EXAMPLES

1. Notice that $\frac{1}{2} + \frac{1}{3} = 1 - \frac{1}{2 \cdot 3}$. Are there any other consecutive integers that fit this pattern? In other words, are there any other consecutive integers n and $n + 1$ such that $\frac{1}{n} + \frac{1}{(n+1)} = 1 - \frac{1}{(n(n+1))}$?
No; the above equation leads to $n = 2$ or $n = -1$, but -1 is not in the domain of the equation.

2. Solve $\frac{2x - 1}{x + 3} + \frac{1}{x^2 + x - 6} = \frac{4 - x}{x - 2}$.
$x = 3$ or $x = -1$

ADDITIONAL EXAMPLES

3. Solve $\frac{3}{x+2} + \frac{12}{x^2-4} = \frac{1}{2-x}$.

There is no solution.

4. Consider the situation of **Example 3**. If upstream flooding increased the current's rate by 2 mph, what would the still water speed of the boat have to be to make the same round trip in 15 minutes?

$\frac{6}{r+5} + \frac{6}{r-5} = \frac{1}{4}$

$\Rightarrow r^2 - 48r - 25 = 0$

$\Rightarrow r = 24 + \sqrt{601} \approx$ **48.52 mph**

With s as the speed of the river, the equation of Additional Example 3 and that of Example 3 are of the form $r^2 - 48r - s^2 = 0$.

As with all problem solving, if a problem reduces to solving an equation, you must check potential solutions back in the original problem, not just in the equation.

Example 3 The current in a river was estimated to be 3 mph. A speedboat went downstream 6 miles and came 6 miles back in 15 minutes. What was the average speed of the speedboat in still water?

Solution Let r be the average speed (in miles per hour) of the speedboat in still water. Then going downstream the speedboat travels at a rate $r + 3$, and returning it travels at a rate $r - 3$. Since rate $= \frac{\text{distance}}{\text{time}}$ in general, time $= \frac{\text{distance}}{\text{rate}}$. Substitution of the given information for d and r indicates that the time going is $\frac{6}{r+3}$ and the time returning is $\frac{6}{r-3}$. Since the total time is $\frac{1}{4}$ hour,

$$\frac{6}{r+3} + \frac{6}{r-3} = \frac{1}{4}.$$

Multiply both sides by the LCD, which is $4(r + 3)(r - 3)$.

$$\Rightarrow \quad 4(r+3)(r-3)\left(\frac{6}{r+3} + \frac{6}{r-3}\right) = \frac{1}{4} \cdot 4(r+3)(r-3)$$

$$\Rightarrow \quad 24(r-3) + 24(r+3) = (r+3)(r-3)$$

$$\Leftrightarrow \quad 24r - 72 + 24r + 72 = r^2 - 9$$

$$\Leftrightarrow \quad 0 = r^2 - 48r - 9$$

$$\Leftrightarrow \quad r = \frac{48 \pm \sqrt{2304 + 36}}{2}$$

$$\Leftrightarrow \quad = \frac{48 \pm \sqrt{2340}}{2}$$

$$\Leftrightarrow \quad = 24 \pm \sqrt{585}$$

So. $\quad r = 24 + \sqrt{585} \approx 48.2$ or $r = 24 - \sqrt{585} \approx -0.2$.

Since we assume r is positive (the speedboat was going forward) the solution -0.2 to the equation is not considered a solution to the problem. Therefore, the average speed of the speedboat was about 48.2 miles per hour.

Rational equations appear in applications from many different fields such as economics, optics, acoustics, and electricity. You will be asked to solve problems from these fields in the questions of this and later lessons.

328

Questions

Covering the Reading

1. Consider the rational equation $\frac{2x+9}{x^2-25} + \frac{1}{x+5} = \frac{x+11}{4x+20}$.
   a. Find the least common denominator. 4(x + 5)(x − 5)
   b. Without solving the equation, give two numbers that are not solutions to the equation. x = 5 and x = -5

2. Determine whether or not the equations are equivalent.
   a. $\frac{x}{x-1} + \frac{3}{x-1} = \frac{5}{x-1}$; $x + 3 = 5$. Yes
   b. $\frac{x}{x-2} + \frac{3}{x-2} = \frac{5}{x-2}$; $x + 3 = 5$. No, the first equation is undefined when x = 2, but x = 2 is a solution to the second equation.

3. a. Find all real numbers x, $x + 2$, and $x + 4$ such that the reciprocal of the smallest number is the sum of the reciprocals of the other two. $\{\sqrt{8}, \sqrt{8}+2, \sqrt{8}+4\}$ and $\{-\sqrt{8}, -\sqrt{8}+2, -\sqrt{8}+4\}$
   b. Estimate all solutions to the nearest hundredth.
   c. Are any of the solutions rational? No
   b) {2.83, 4.83, 6.83} and {-2.83, -0.83, 1.17}

In 4–7, solve.

4. $\frac{4}{2x} + \frac{2}{3x} = \frac{1}{6}$ x = 16

5. $\frac{3}{x+1} + 2 = \frac{5}{x+1}$ x = 0

6. $\frac{12}{x-2} - \frac{6}{x-3} = 1$ x = 5 or x = 6

7. $\frac{t+5}{t+1} + \frac{3}{t-3} = \frac{-16}{t^2-2t-3}$ t = -4

8. Refer to Example 3.
   a. If there were no current, how fast would the speedboat have been going? ≈ 48.2 mph
   b. Did the current affect how far the speedboat could go in 15 minutes? Yes, the current decreases the length of travel about 0.05 miles every 15 minutes.

9. A sightseeing plane travels into a 60 mph headwind and returns, going 150 miles out and coming 150 miles back in 2 hours.
   a. How fast would the plane have gone in still air? ≈ 171 mph
   b. How far could the plane have gone in 2 hours had there been no wind? ≈ 342 miles

Applying the Mathematics

10. In an electrical circuit, the equivalent resistance, R, of two resistors, R_1 and R_2, connected in parallel is given by the equation

$$\frac{1}{R} = \frac{1}{R_1} + \frac{1}{R_2}.$$

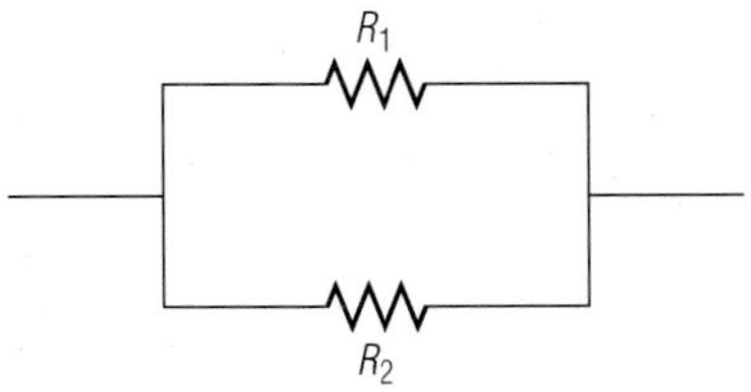

   a. If $R_1 = 5$ ohms and $R_2 = 6$ ohms, find R. $R \approx 2.73$ ohms
   b. If R is known to be 4 ohms and R_1 is 6 ohms, find R_2. $R_2 = 12$ ohms

11. In $\triangle ABC$, AB is one foot longer than BC. Determine exact values for the lengths of BC and AB. $BC = \sqrt{2} + 1$; $AB = \sqrt{2} + 2$

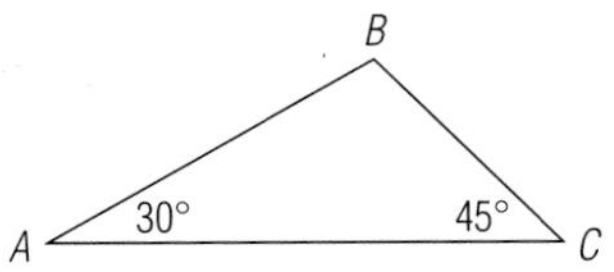

NOTES ON QUESTIONS
Question 2: This is a different sort of question, and it may confuse some students. Go back to the definition of "equivalent equations" as equations having the same solutions.

Questions 4-7: Remind students to check potential solutions in the original rational equations.

Question 11: This question requires the use of the Law of Sines. Encourage students to rationalize the denominators in their answers.

NOTES ON QUESTIONS

Question 12: Dimensional analysis is helpful to solve this problem.

$\text{hours} \cdot \frac{\text{parts}}{\text{hour}} = \text{parts}$

Since 120 parts are needed in all,

$x \text{ hours} \cdot 120 \frac{\text{parts}}{\text{hour}} +$

$x \text{ hours} \cdot 240 \frac{\text{parts}}{\text{hour}} =$

120 parts.

Question 17g: Generalize the result of this question: to introduce a removable discontinuity at $x = a$ for a function $f: x \rightarrow y$, simply define $g(x) = f(x) \cdot \frac{x - a}{x - a}$.

Question 21: There are three ways to approach this problem: (1) using logarithms; (2) using an automatic grapher; and (3) using a calculator and the guess-and-check method.

ADDITIONAL ANSWERS

15. $\tan \frac{5\pi}{6} = -\frac{\sqrt{3}}{3}$;

$\cot \frac{5\pi}{6} = -\sqrt{3}$;

$\sec \frac{5\pi}{6} = -\frac{2\sqrt{3}}{3}$;

$\csc \frac{5\pi}{6} = 2$

17.f.

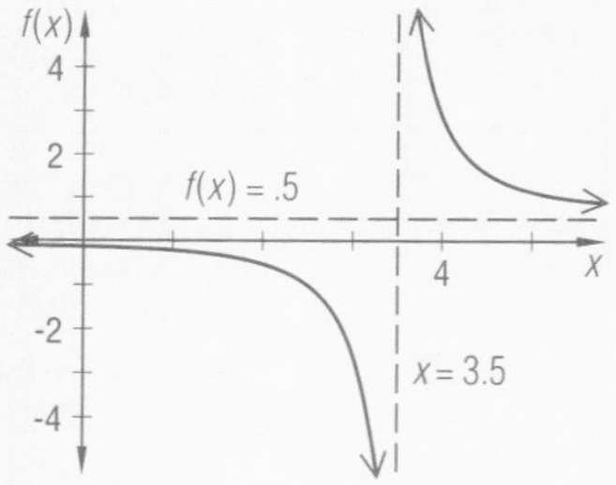

g. $g(x) = \frac{x^2 - 2x - 15}{2x^2 - 17x + 35}$

19.a. amplitude: 4; period: $\frac{2\pi}{3}$; phase shift: $\frac{\pi}{3}$

12. A manufacturing company has an old machine that produces 120 parts per hour. If the company buys a new machine that makes twice as many parts per hour, how long will it take the two machines to make 120 parts if they work together? (Hint: You may wish to let x = the number of hours the machines run to make the 120 parts.)
20 minutes

13. Suppose that you travel 10 miles at r miles per hour, 20 miles at 4 mph less than r, and 30 miles at 4 mph faster than r. Write your average speed as a rational expression in r. $\frac{3(r^3 - 16r)}{3r^2 - 2r - 8}$

Review

14. If $\sin x = 0.3$ and $\cos x$ is positive, give the values of $\tan x$, $\cot x$, $\sec x$, and $\csc x$. *(Lesson 5-6, 2-8)* $\tan x \approx 0.31$; $\cot x \approx 3.18$; $\sec x \approx 1.05$; $\csc x = \frac{10}{3}$

15. Give the values of $\tan x$, $\cot x$, $\sec x$, and $\csc x$ when $x = \frac{5\pi}{6}$. *(Lesson 5-6)* **See margin.**

16. Let f be the rational function defined by $f(x) = \frac{2x^2 - x - 3}{x - 2}$. The end behavior of f is like the end behavior of what simpler function? *(Lesson 5-5)* $2x + 3$

17. Consider the function f defined by

$$f(x) = \frac{x + 3}{2x - 7}.$$

a. Determine where f is undefined. $x = \frac{7}{2}$

b. Identify any discontinuities and state whether they are removable or essential. **essential discontinuity at** $x = \frac{7}{2}$

c. Find equations for any vertical asymptotes to the graph of f.

d. Find equations for any horizontal asymptotes to the graph of f.

e. Find the coordinates of the x- and y-intercepts of the graph of f.

f. Sketch the graph of f. **See margin.**

g. Write an equation for a new function, g, whose graph is identical to that of f except that g has a removable discontinuity at $x = 5$. *(Lessons 5-5, 5-4, 2-1)* c) $x = \frac{7}{2}$; d) $y = \frac{1}{2}$; e) x-int.: $(0, -\frac{3}{7})$; y-int.: $(-3, 0)$; g) See margin.

18. a. Simplify $\frac{16x^2 - 4}{8x + 4}$. $2x - 1$

b. Give any restrictions on x. *(Lesson 5-1)* $x \neq -\frac{1}{2}$

19. Refer to the graph at the right.

a. Identify the amplitude, period, and phase shift of this sine curve.

b. Use the results from part **a** to write an equation that describes the graph. *(Lesson 3-9)*

a) See margin. b) $y = 4 \sin (3x - \pi)$

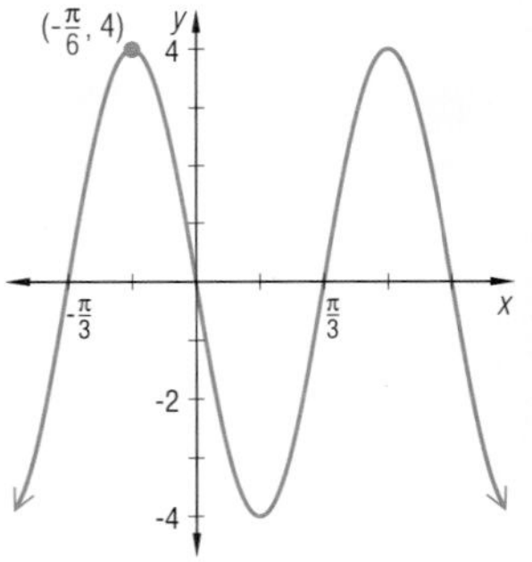

330

20. The system of star magnitudes is based on a logarithmic scale. For an increase in brightness by a factor of 100, the magnitude decreases by 5. For instance, a star with a magnitude of 3.6 appears $\frac{1}{100}$ as bright as the brightest nighttime star visible from Earth, Sirius, with a magnitude of -1.4. The Sun's magnitude is -26.8. How many times brighter than Sirius is the Sun? *(Lesson 2-7)*
$\approx 1.45 \times 10^{10}$ **times brighter**

21. If a deposit doubles in value in 10 years when compounded quarterly, find the interest rate. *(Previous course)* $\approx$ **7.0%**

22. What can be concluded using all these statements? *(Lesson 1-6)*
If the legislature passes this bill, the governor will veto it.
If the governor vetoes this bill, the legislature will override the veto.
If the legislature overrides the governor's veto, the bill will become law. **If the legislature passes this bill, the bill will become law.**

Texas legislators vote yes *by holding up two fingers and* no *by holding up one.*

Exploration

23. Consider the equation

$$\frac{ax - 1}{x^2 - 1} = \frac{1}{x - 1}.$$

a. Find three values of a for which this equation has no solution.
b. In case the equation does have a solution, write an expression for the solution in terms of a. $x = \frac{2}{a - 1}$
c. Find a value for a so that the solution is $x = 4$. $a = \frac{3}{2}$
a) $a = 1$, $a = 3$, and $a = -1$

FOLLOW-UP

MORE PRACTICE

For more questions on SPUR Objectives, use *Lesson Master 5-7,* shown below.

NAME ______

LESSON **MASTER 5–7**
QUESTIONS ON **SPUR** OBJECTIVES

■SKILLS *Objective E (See pages 341–344 for objectives.)*
In 1–6, solve each equation.

1. $\frac{1}{y - 3} = \frac{y}{y + 4} + \frac{7y}{(y + 4)(y - 3)}$ **$y = 1$**

2. $\frac{8}{t + 3} + \frac{t}{t - 3} = \frac{-2}{t^2 - 9}$ **$t = \frac{-11 \pm \sqrt{209}}{2}$**

3. $\frac{r}{r - 1} + \frac{r - 1}{r} = 1$ **$r = \frac{1 \pm i\sqrt{3}}{2}$**

4. $2 + \frac{20}{s^2 - s - 6} = \frac{-3}{s - 3}$ **$s = \frac{-1 \pm i\sqrt{111}}{4}$**

5. $\frac{z + 3}{z - 6} + \frac{z}{3z - 4} = \frac{2z^2 - 3}{3z^2 - 22z + 24}$ **$z = \frac{1 \pm \sqrt{73}}{4}$**

6. $\frac{x}{2x - 1} = \frac{6}{5x + 7} + \frac{x^2 - x + 1}{10x^2 + 9x - 7}$ **$x = \frac{1}{2} \pm i$**

■USES *Objective I*

7. A tired swimmer takes 30 minutes longer to return to shore from a pier as it took to swim from shore to the pier. The pier is 1 km from shore and the swimmer's average speed returning is 3 km/h less than the swimmer's average speed going out to the pier. What is the swimmer's average speed going out to the pier?
about 4.4 km/h

8. In an electrical circuit, if two resistors with resistances R_1 and R_2 are connected in parallel, then the equivalent resistance, R, is found using the equation $\frac{1}{R} = \frac{1}{R_1} + \frac{1}{R_2}$. If R_1 is 5 ohms less than R_2 and R is one-third of R_2, what are the values of R_1, R_2, and R?
$R_1 = 5$ ohms, $R_2 = 10$ ohms, $R = \frac{10}{3}$ ohms

 61

LESSON 5-8

RESOURCES

■ Teaching Aid 35 displays the five regular polyhedra.

TEACHING NOTES

Finding equivalent problems in different contexts is exciting and important because it opens the possibility of different methods of solution. When Descartes invented analytic geometry, he thought that every geometry problem would be able to be solved because he could now find an algebraic equivalent. Perhaps he had too much faith in algebra, but he was correct in the sense that many problems that are very difficult to solve geometrically have easier algebraic solutions.

Equivalent problems may not look at all alike. **Problem 1**, **Problem 2**, and the problems of **Questions 8 and 9** are all equivalent. Point out to students that equivalent problems are not like alternate solutions to a problem. They are alternate problems to solve.

LESSON

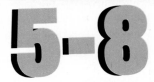

5-8 Tessellating Polygons and Regular Polyhedra

To follow this lesson, you need to know the meanings of a few terms. A **unit fraction** is a fraction of the form $\frac{1}{n}$, where n is a positive integer. Thus $\frac{1}{2}, \frac{1}{3}, \frac{1}{4}, \ldots$ are the unit fractions. A **regular polygon** is a convex polygon whose sides are all the same length and whose angles are all the same measure. Equilateral triangles and squares are regular polygons, and there can be regular polygons with any number of sides. Since the sum of the measures of the angles of a convex polygon is $(n - 2)180°$, each angle of a regular polygon has measure $\frac{(n-2)180°}{n}$.

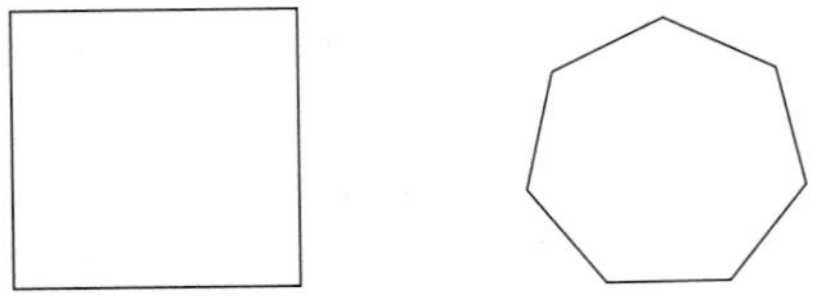

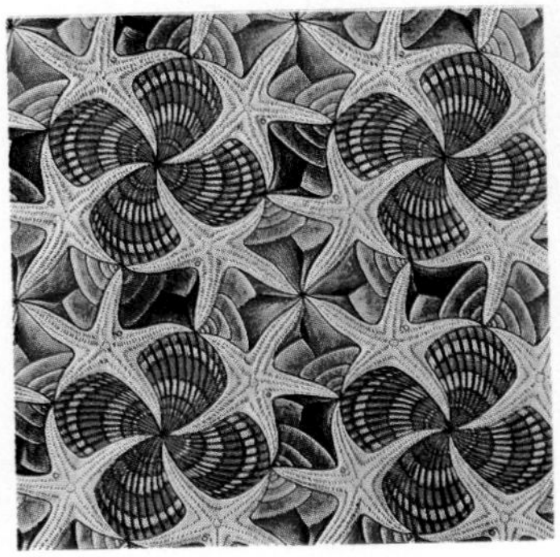

A **tessellation** is a covering of the plane with congruent copies of the same region, with no holes and no overlaps. There are many possible shapes for that region; each is called a **fundamental region** for the tessellation. One of the many tessellations of the famous Dutch artist Maurits Escher (1898-1973) is shown at the left.

It would not seem that unit fractions, regular polygons, and tessellations are related, but they are. Their relationship arises from trying to solve the following problem:

Problem 1: Which regular polygons can be fundamental regions for a tessellation?

From floor tiles and designs, and perhaps from earlier mathematics courses, you may be familiar with tessellations having equilateral triangles, squares, and regular hexagons as fundamental regions. They are drawn here. So the question is: Are there any others?

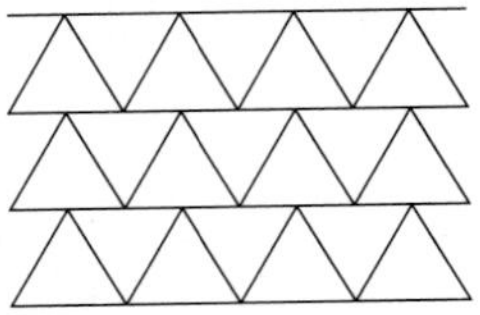

equilateral triangle tessellation

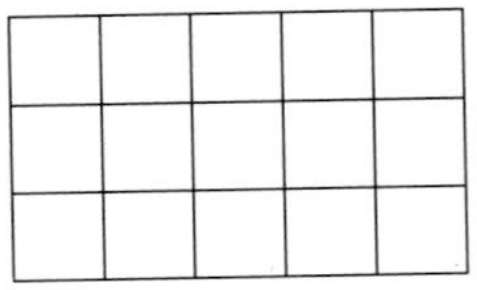

square tessellation

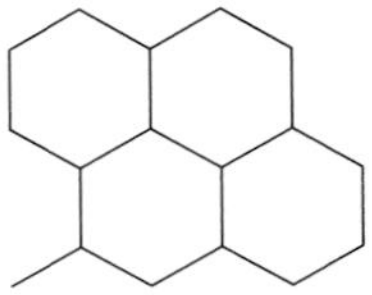

regular hexagon tessellation

Now let us consider a problem involving unit fractions. It seems to be an unrelated digression, but it is not.

332

The method of solving **Problem 2** is one of considering and exhausting all possible cases. There are exactly three ways to write $\frac{1}{2}$ as the sum of two positive unit fractions: $\frac{1}{3} + \frac{1}{6}$, $\frac{1}{4} + \frac{1}{4}$, and $\frac{1}{6} + \frac{1}{3}$. This implies $(n, s) =$ (3, 6), (4, 4), or (6, 3). Some students will fail to make the connection.

Problem 2: Write $\frac{1}{2}$ as the sum of two unit fractions.

Restated, this problem asks for a pair of positive integers n and s such that $\frac{1}{2} = \frac{1}{n} + \frac{1}{s}$. You can see that at least one of n and s must be small, because the sum of fractions like $\frac{1}{100}$ and $\frac{1}{120}$ is much less than $\frac{1}{2}$. In fact, at least one of $\frac{1}{n}$ and $\frac{1}{s}$ must be as great as $\frac{1}{4}$. (Here is a short proof by contradiction: If both are less than $\frac{1}{4}$, then adding them yields a sum that is less than $\frac{1}{2}$, and that contradicts the condition of the problem.) Thus

$$\text{either } \frac{1}{n} \geq \frac{1}{4} \quad \text{or} \quad \frac{1}{s} \geq \frac{1}{4}.$$

Suppose $\frac{1}{n} \geq \frac{1}{4}$. Then, multiplying both sides by $4n$, $4 \geq n$.
Since n is a positive integer, $n = 1, 2, 3,$ or 4. To find s, substitute these values for n.

when $n = 1$:	$\frac{1}{2} = \frac{1}{1} + \frac{1}{s}$	This implies $s = -2$, which is not allowed.
when $n = 2$:	$\frac{1}{2} = \frac{1}{2} + \frac{1}{s}$	This implies $\frac{1}{s} = 0$, which is impossible.
when $n = 3$:	$\frac{1}{2} = \frac{1}{3} + \frac{1}{s}$	This implies $s = 6$.
when $n = 4$:	$\frac{1}{2} = \frac{1}{4} + \frac{1}{s}$	This implies $s = 4$.

Because n and s can be interchanged in the original equation, the same reasoning can be applied again. Therefore, when $s = 3$, then $n = 6$, and when $s = 4$, then $n = 4$ (which was found the first time). There are thus three integer solutions to the equation $\frac{1}{2} = \frac{1}{n} + \frac{1}{s}$:

$$(n, s) = (3, 6) \quad (n, s) = (4, 4) \text{ or } \quad (n, s) = (6, 3).$$

So $\frac{1}{2} = \frac{1}{3} + \frac{1}{6}$ or $\frac{1}{2} = \frac{1}{4} + \frac{1}{4}$ or $\frac{1}{2} = \frac{1}{6} + \frac{1}{3}$. The work on Problem 2 enables us to solve Problem 1 to which we now return.

Think of what happens at the vertex P of a polygon in a tessellation. Either P is the vertex of other polygons (as in the square or regular hexagon tessellations drawn on page 332) or the vertex lies on a side of another polygon (as in the equilateral triangle tessellation drawn on page 332).

The vertex can lie on a side of another polygon only when an angle of the regular polygon is a factor of 180°, because all the angles formed must be congruent. There are no factors of 180° between 90° and 180°, so this can only occur for squares and equilateral triangles.

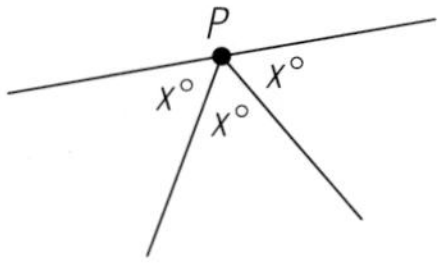

To help explain why there are only three possibilities for regular polygons fitting around a vertex, a table of measures of each angle of a regular n-gon will help.

n	$\frac{180(n-2)}{n}$
3	60
4	90
5	108
6	120

When there are more than 6 sides, the measure of an angle is too large to allow three polygons to meet.

For **Problem 3**, the argument that each angle of a face polygon must be less than $\frac{360}{s}$ may not be intuitively obvious. You can have students build straw or pipe cleaner models or bring in models of the regular polyhedra to explain this point.

NOTES ON QUESTIONS

Questions 1 and 2: Each of these tessellations is different than the one shown in the lesson. Invite students to do tessellations that combine two regular polygons with different numbers of sides. (Strips of equilateral triangles and squares can form a tessellation. If one tries to tessellate with octagons, there are holes into which squares fit.)

Question 8: This question provides a third problem equivalent to solving $\frac{1}{2} = \frac{1}{n} + \frac{1}{s}$.

Question 11: Invite students to write a new function with a graph identical to that of $y = h(x)$, but with two exceptions: removable discontinuities at $x = -3$ and at $x = 7$.
$\left(f(x) = \frac{(x+2)(x+3)(x-7)}{(x^2-25)(x+3)(x-7)}\right)$

The only other possibility is that s regular n-gons meet at a vertex P. The figure will look something like what is shown at the right. Since there are s congruent angles with vertex P, and the sum of the measures of these angles is 360°, each has measure $\frac{360°}{s}$. And since these are angles of regular n-gons, each angle must also have measure $\frac{(n-2)180°}{n}$. Thus n and s are integer solutions to the equation

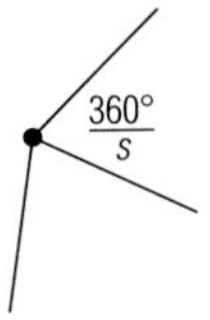

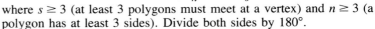

$$\frac{(n-2)180°}{n} = \frac{360°}{s}$$

where $s \geq 3$ (at least 3 polygons must meet at a vertex) and $n \geq 3$ (a polygon has at least 3 sides). Divide both sides by 180°.

$$\frac{n-2}{n} = \frac{2}{s}$$

Rewrite the fraction on the left side as the difference of two fractions.

$$1 - \frac{2}{n} = \frac{2}{s}$$

Divide both sides by 2.

$$\frac{1}{2} - \frac{1}{n} = \frac{1}{s}$$

Adding $\frac{1}{n}$ to both sides yields the equation of Problem 2. That equation has only three solutions $(n, s) = (3, 6)$, $(n, s) = (4, 4)$, and $(n, s) = (6, 3)$. Each of those solutions determines a geometric configuration.

$(n, s) = (3, 6)$ means that the polygons are triangles (3-gons) with 6 meeting at each vertex.
$(n, s) = (4, 4)$ means that the polygons are squares (4-gons) with 4 meeting at each vertex.
$(n, s) = (6, 3)$ means that the polygons are hexagons (6-gons) with 3 meeting at each vertex.

So, at most three regular polygons can be fundamental regions for a tessellation. The drawings on page 332 show that all three tessellations are possible.

Now consider regular polyhedra, again recalling some terminology. A **regular polyhedron** (plural **polyhedra**) is a convex polyhedron whose faces are all congruent regular polygons. The ancient Greeks knew of five different regular polyhedra. They are pictured here.

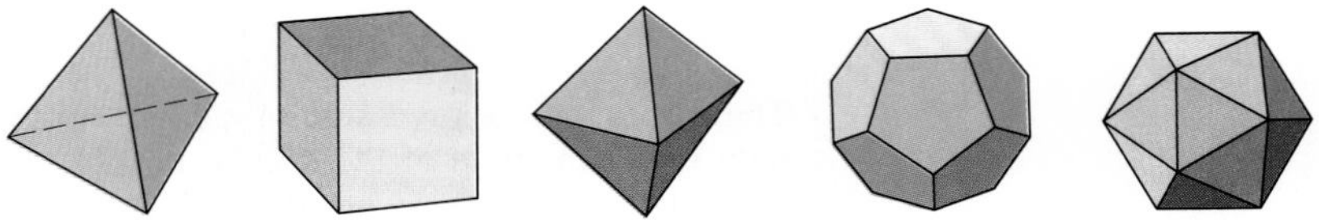

regular tetrahedron (4 faces) | cube (6 faces) | regular octahedron (8 faces) | regular dodecahedron (12 faces) | regular icosahedron (20 faces)

The question is: Do regular polyhedra exist that were unknown to the Greeks?

334

Problem 3: How many regular polyhedra are there?

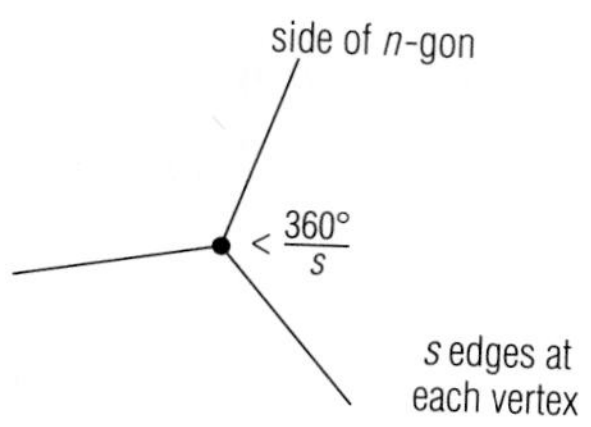

This problem is related to the first two in the following way. Think of looking at a regular polyhedron from above one of its vertices. You see a number of edges (call it s) of the polyhedron radiating from the vertex, forming equal angles. The edges are not all in the same plane, so the sum of the measures of these angles must be less than 360°, and thus the measure of each angle must be less than $\frac{360°}{s}$. If the angle is an interior angle of an n-gon, then its measure is, as before, $\frac{(n-2)180°}{n}$. So Problem 3 reduces to finding integers n and s such that $n \geq 3$ and $s \geq 3$ and

$$\frac{(n-2)180°}{n} < \frac{360°}{s}.$$

This sentence is identical to the equation already solved in Problem 2 except that there is a < sign where there was an = sign. By the same steps used in solving the equation, the inequality is equivalent to

$$\frac{1}{2} < \frac{1}{n} + \frac{1}{s}.$$

Again both n and s must be small because otherwise the sum of their reciprocals will not be greater than $\frac{1}{2}$. But by the conditions of the problem, $n \geq 3$ and $s \geq 3$. Because $(n, s) = (3, 6)$ makes the sum of the reciprocals equal to $\frac{1}{2}$, when $n = 3$, s can only be 3, 4, or 5.

$$\frac{1}{2} < \frac{1}{3} + \frac{1}{3} \qquad \frac{1}{2} < \frac{1}{3} + \frac{1}{4} \qquad \frac{1}{2} < \frac{1}{3} + \frac{1}{5}$$

When $n = 4$, s can only be 3.

$$\frac{1}{2} < \frac{1}{4} + \frac{1}{3}$$

When $n = 5$, s can only be 3, because $\frac{1}{5} + \frac{1}{4}$ is not large enough.

$$\frac{1}{2} < \frac{1}{5} + \frac{1}{3}$$

When $n \geq 6$, there is no value of s that works. Consequently, there are only 5 pairs of numbers (n, s) that satisfy the inequality $\frac{1}{2} < \frac{1}{n} + \frac{1}{s}$ with $n \geq 3$ and $s \geq 3$. They are (3, 3), (3, 4), (3, 5), (4, 3), and (5, 3). Each solution refers to a particular regular polyhedron. For instance, $(n, s) = (3, 4)$ means 4 triangles (3-gons) meet at each vertex of the polyhedron. This is the regular octahedron, with 8 faces. In Question 7, you are asked to match each of the other solutions with its corresponding polyhedron.

So the Greeks knew of all the regular polyhedra. There are no others.

Problem 1 is a geometry problem and Problem 2 is an arithmetic problem, yet they can both be solved using the same algebra. Problems that can be solved using the same mathematics are called *equivalent problems*. Equivalent problems demonstrate the power, unity, and internal consistency of mathematics.

Question 12: A visit to a well-equipped physics lab will provide the equipment needed to demonstrate the relationship $\frac{1}{o} + \frac{1}{i} = \frac{2}{r}$. Ask students to design an experiment to approximate the radius of the spherical mirror.

FOLLOW-UP

EXTENSION

The ancient Egyptians strove to express all fractions as sums of unit fractions. Students interested in history may wish to explore the mathematics of ancient Egypt.

There are many interesting problems related to unit fractions. For instance, ask students to find a general pattern in the following:

$\frac{1}{12} + \frac{1}{4} = \frac{1}{3}$

$\frac{1}{20} + \frac{1}{5} = \frac{1}{4}$

Is your generalization true? $\left(\frac{1}{n(n+1)} + \frac{1}{n+1} = \frac{1}{n}\right.$ is a true generalization.)

You could ask students the following: In how many ways can 1/3 be written as the sum of two unit fractions? (3) What about 1/4? (5) This could be extended into a project or **small group work**.

ADDITIONAL ANSWERS

1. sample:

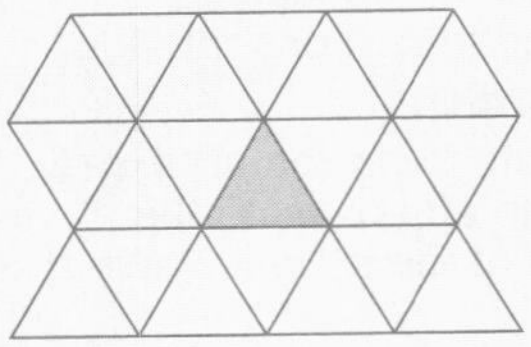

2. sample:

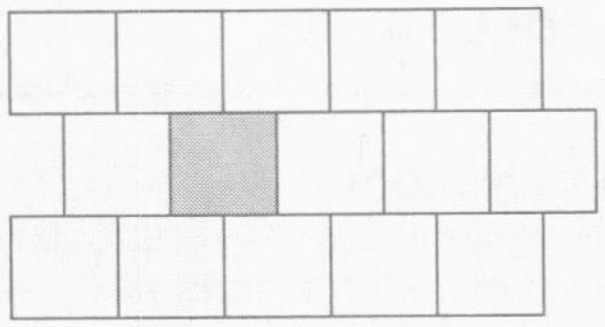

3. $\left(\frac{1}{3}, \frac{1}{3}\right), \left(\frac{1}{3}, \frac{1}{4}\right), \left(\frac{1}{3}, \frac{1}{5}\right), \left(\frac{1}{3}, \frac{1}{6}\right), \left(\frac{1}{3}, \frac{1}{7}\right), \left(\frac{1}{3}, \frac{1}{8}\right), \left(\frac{1}{4}, \frac{1}{3}\right), \left(\frac{1}{4}, \frac{1}{4}\right), \left(\frac{1}{5}, \frac{1}{3}\right), \left(\frac{1}{6}, \frac{1}{3}\right), \left(\frac{1}{7}, \frac{1}{3}\right), \left(\frac{1}{8}, \frac{1}{3}\right)$

8. The area of a rectangle is its length times its width, ℓw. Its perimeter is then $2\ell + 2w$. By setting these equal, the problem is to find positive integers ℓ and w such that $\ell w = 2\ell + 2w$. Dividing both sides by $2\ell w$ (which is nonzero) yields $\frac{\ell w}{2\ell w} = \frac{2\ell + 2w}{2\ell w}$, which simplifies to $\frac{1}{2} = \frac{1}{w} + \frac{1}{\ell}$. This is equivalent to the equation in Problem 2, which is also equivalent to Problem 1.

9. Let n be an integer greater than 2. Suppose $(n - 2)$ is a factor of $2n$. Then $2n = (n - 2)k$ for some integer k. Then $\frac{2n}{n-2} = k$, since $n \neq 2$. By taking the reciprocal, $\frac{n-2}{2n} = \frac{1}{k}$. So $\frac{1}{2} - \frac{1}{n} = \frac{1}{k}$, or $\frac{1}{2} = \frac{1}{k} + \frac{1}{n}$. This is equivalent to the equation in Problem 2, which is also equivalent to Problem 1.

Questions

Covering the Reading

1. Draw a tessellation of the plane by equilateral triangles in which each vertex of each triangle is a vertex of five others. See margin.

2. Draw a tessellation of the plane by squares in which each vertex of every square is a vertex of only one other square. See margin.

3. Find all pairs of unit fractions whose denominators are greater than or equal to 3 and whose sum is greater than $\frac{5}{11}$. See margin.

4. Solve for t: $\frac{1}{t} + \frac{1}{3} = \frac{1}{2}$. $t = 6$

5. Let a_n = the measure of an interior angle of a regular n-gon, for $n \geq 3$.
- **a.** Give a formula for a_n in terms of n.
- **b.** Compute the first five terms of the sequence a.
- **c.** What is $\lim_{n \to \infty} a_n$. (Hint: Think of the sequence as a discrete rational function.) a) $a_n = \frac{(n-2)180°}{n}$; b) 60°, 90°, 108°, 120°, $128\frac{4}{7}$°; c) 180°

6.
- **a.** How many regular polyhedra are there? 5
- **b.** How many of these were known to the ancient Greeks? all 5

7. Match each solution (n, s) to the inequality $\frac{1}{n} + \frac{1}{s} < \frac{1}{2}$ with its corresponding regular polyhedron.
- **a.** (3, 3) **b.** (3, 4) **c.** (3, 5) regular icosahedron
- **d.** (4, 3) cube **e.** (5, 3) regular dodecahedron

a) regular tetrahedron b) regular octahedron

8. Consider this problem: Find all rectangles with integer sides whose area and perimeter are numerically equal. Show that this problem is equivalent to Problems 1 and 2 in this lesson. See margin.

9. Consider this problem: For which positive integers $n > 2$ is $n - 2$ a factor of $2n$? Show that this problem is equivalent to Problems 1 and 2 in this lesson. See margin.

Review

10. An insurance company has a machine that can write a batch of checks in 2 hours. The company buys a new machine capable of writing the same batch of checks in $\frac{3}{4}$ hour. If both machines work simultaneously, how long will it take the company to write a batch of checks? *(Lesson 5-7)* $\frac{6}{11}$ of an hour

11. Given the function h defined by $h(x) = \frac{x + 2}{x^2 - 25}$.
- **a.** Identify the domain of h. $\{x: x \neq 5 \text{ and } x \neq -5\}$
- **b.** Describe the behavior of h near any vertical asymptotes.
- **c.** Describe the end behavior of h using limit notation.
- **d.** Identify the x- and y-intercepts of h.
- **e.** Sketch a graph of h.

(Lessons 5-5, 5-4, 2-1) b), d), e) See margin. c) $\lim_{x \to -\infty} h(x) = 0$; $\lim_{x \to +\infty} h(x) = 0$

336

12. The diagram below shows a spherical mirror with center C and radius r. A light ray from a point on an object O reflects off the mirror and passes through an image point I. The relationship between the image distance i the object distance o and the radius r can be approximated by

$$\frac{1}{o} + \frac{1}{i} = \frac{2}{r}.$$

Suppose a spherical mirror has radius 5 cm.

a. If an object is located at a distance of 15 cm from the mirror, find the image distance. 3 cm

b. Where should the object be placed so that the image distance is twice the object distance? *(Lesson 5-7)* 3.75 cm from the mirror

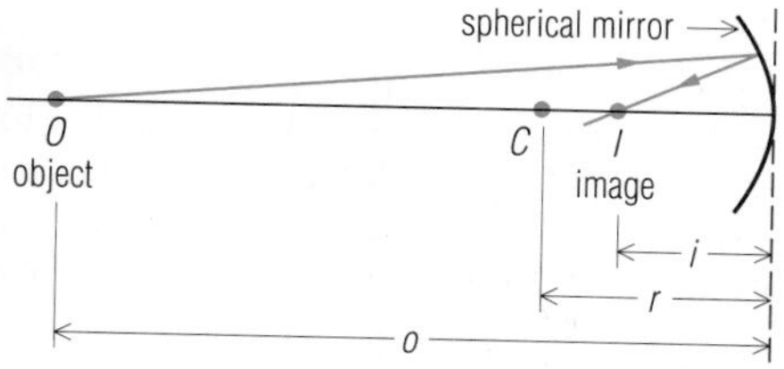

13. Find tan x, cot x, sec x, and csc x if $x = \frac{5\pi}{4}$. *(Lesson 5-6)* See margin.

14. Rationalize the denominator: $\frac{20}{\sqrt{10} - 6}$. *(Lesson 5-2)* $-\frac{10\sqrt{10} + 60}{13}$

15. **a.** Show that $\frac{5x^2 + 3x - 2}{7x^2 + 8x + 1} \bullet \frac{7x^2 - 27x - 4}{5x^2 + 8x - 4} = \frac{x - 4}{x + 2}$ is an identity.

b. State any restrictions on x. *(Lesson 5-1)*
a), b) See margin.

16. Prove: $\forall$ *integers p and q, if 4 is a factor of p and q is even, then 8 is a factor of p • q.* *(Lessons 4-1, 1-7)* See margin.

17. Find the standard prime factorization of 4388. *(Lesson 4-8)* $2^2 \bullet 1097$

18. The relative error in a measurement is usually reported as a percent and given by $\left|\frac{x - x_m}{x}\right|$ where x is the true value of a quantity and x_m is its measured value. Suppose a machine makes ball bearings whose diameter should be 2.8 cm. If the relative error can be at most 2%, find the acceptable values for the measured diameter. *(Lessons 5-7, 3-8)*
2.744 cm ≤ x ≤ 2.856 cm

Exploration

19. What are the *semi-regular polyhedra?* See margin.

11.b. near $x = 5$: $\lim_{x \to 5^-} h(x) = -\infty$, $\lim_{x \to 5^+} h(x) = +\infty$; near $x = -5$: $\lim_{x \to -5^-} h(x) = -\infty$, $\lim_{x \to -5^+} h(x) = +\infty$

d. x-intercept: (-2, 0); y-intercept: $\left(0, -\frac{2}{25}\right)$

e.

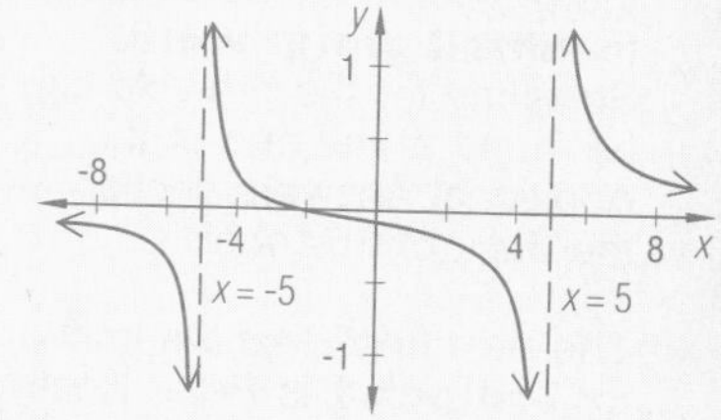

13. $\tan \frac{5\pi}{4} = 1$; $\cot \frac{5\pi}{4} = 1$; $\sec \frac{5\pi}{4} = -\sqrt{2}$; $\csc \frac{5\pi}{4} = -\sqrt{2}$

15.a. $\frac{5x^2 + 3x - 2}{7x^2 + 8x + 1} \cdot \frac{7x^2 - 27x - 4}{5x^2 + 8x - 4} = \frac{(5x - 2)(x + 1)}{(7x + 1)(x + 1)} \cdot \frac{(7x + 1)(x - 4)}{(5x - 2)(x + 2)} = \frac{x - 4}{x + 2}$

b. $x \neq -\frac{1}{7}$, $x \neq -1$, $x \neq \frac{2}{5}$, and $x \neq -2$

16. Suppose p and q are integers such that 4 is a factor of p, and q is even. By definition, there exist integers r and s such that $p = 4r$ and $q = 2s$. Then $pq = (4r)(2s) = 8(rs)$. Since rs is an integer by closure properties, 8 is a factor of pq by definition.

19. If two or more different regular polygons are allowed as faces, there are 13 polyhedra for which each vertex is surrounded by the same arrangement of polygons. These are called semi-regular polyhedra, and they are convex.

Projects

PROJECTS

The projects provide an opportunity for extended work on topics related to the content of the chapter. (See pages 70 and T37 for more information on the purposes of projects.) Some or all of the projects may be suitable for **small group work**. Solutions for the projects can be found at the end of the Additional Answers section in the back of this book.

The best book that we know for **Project 1** is Peter Beckmann's *A History of Pi* (Golem Press, 1970, 1971).

Project 2 points out that the graph of the quotient of two linear functions is a hyperbola; this provides a way to find the asymptotes of the hyperbola.

Project 3 could be extended into a discussion of continued fractions.

The Rational Zero Theorem in **Project 4** is found in some mathematics texts; you may wish to mention it to your students.

Once data have been collected, the mathematics of **Project 5** are relatively easy to do with a spreadsheet.

1. In Lesson 5-2, it was mentioned that π was shown to be irrational in 1761. However, there have been attempts in the United States to legislate a value for π that is rational. Find some references that review the history of π. Write a short report on π's history. Include information about the state(s) that tried to legislate a value for π and about some of the attempts to find π's decimal expansion.

2. Let $f(x) = \frac{1}{x}$. Show that any function g of the form $g(x) = \frac{ax + b}{cx + d}$ has f as its parent function by finding an appropriate scale change and translation under which g is the image of f. Test your generalization for at least 5 specific cases of your own choosing for f and g.

3. a. Approximate the value of:

i. $\sqrt{2 + \sqrt{2}}$

ii. $\sqrt{2 + \sqrt{2 + \sqrt{2}}}$

iii. $\sqrt{2 + \sqrt{2 + \sqrt{2 + \ldots}}}$.

b. Approximate the value of $\sqrt{6 + \sqrt{6 + \sqrt{6 + \ldots}}}$.

c. Approximate the value of $x = \sqrt{a + \sqrt{a + \sqrt{a + \ldots}}}$ for various other values of a.

d. Use the data from parts **a–c** to conjecture a relationship between x and a.

e. Use the fact that $x = \sqrt{a + x}$ to prove the relationship you conjectured in part **d**.

Herbert Bayer's 1966 representation of infinity

4. a. Suppose that a_0, a_1, a_2, a_3 are integers with $a_3 \neq 0$ and that $p(x)$ is the polynomial defined by

$$p(x) = a_3x^3 + a_2x^2 + a_1x + a_0.$$

Suppose that m and k are positive integers with no common factors.

i. Show that the equation $p\left(\frac{m}{k}\right) = 0$ can be rewritten in the form $m[a_3m^2 + a_2mk + a_1k^2] = -a_0k^3$.

ii. Use the result of part **i** to show that m must be a factor of the constant coefficient a_0 of p.

iii. Show that the equation $p\left(\frac{m}{k}\right) = 0$ can also be rewritten in the form

$$a_3m^3 = -k(a_2m^2 + a_1mk + a_0k^2).$$

iv. Use the result of part **iii** to show that k must be a factor of the leading coefficient a_3 of p.

b. i. Generalize the results in parts **ii** and **iv** above to prove the following result.

Rational Zero Theorem Suppose that $p(x)$ is a polynomial with integer coefficients: $p(x) = a_nx^n + a_{n-1}x^{n-1} + \ldots + a_1x + a_0$, $a_n \neq 0$. If $r = \frac{m}{k}$ is a rational number in lowest terms that is a zero of $p(x)$, then m is a factor of the constant a_0 and k is a factor of the leading coefficient a_n.

ii. List all possible candidates for rational zeros of the polynomial $p(x) = 6x^4 - 7x^3 + 8x^2 - 7x + 2$.

iii. Determine if p has any rational zeros and then find all remaining zeros of p.

5. Research to find the gas mileage obtained by current cars and estimate a cost per gallon in your area. Make a table of amounts a person could save in a year in gas costs by driving some particular cars rather than others; have the table include a variety of miles driven in a year.

Summary

Chapter 4 discussed properties of integers and extended these properties to polynomials. This chapter continues this theme by considering rational numbers and relating their properties to the behavior of rational functions. Irrational numbers and nonrational functions are also treated.

Rational numbers are those real numbers that can be expressed as a ratio of integers $\frac{a}{b}$ where $b \neq 0$. Any infinite repeating decimal can be expressed as a ratio of integers. Hence, any finite decimal or any infinite repeating decimal is a rational number.

Those real numbers that cannot be expressed as ratios of integers are called irrational numbers. For any integer n, $\sqrt{n}$ is irrational if and only if n is not a perfect square. Any decimal expansion that neither terminates nor repeats represents an irrational number.

A rational function f is a function that can be written as $f(x) = \frac{p(x)}{q(x)}$ where $p(x)$ and $q(x)$ are polynomials and $q(x)$ is not zero. Rough sketches of rational functions can be drawn by considering the end behavior of the function as well as the behavior of the function near vertical asymptotes. Discontinuities, or breaks in the graph, may be either removable or essential.

The end behavior of the rational function f with $f(x) = \frac{a_m x^m + \ldots + a_1 x + a_0}{b_n x^n + \ldots + b_1 x + b_0}$ is the same as the end behavior of the function g with $g(x) = \frac{a_m}{b_n} x^{m-n}$. If $m > n$ then the end behavior of the rational function is the same as the end behavior of a power function. If $m < n$, then the end behavior is the same as that of a reciprocal of a power function. If $m = n$, then $y = \frac{a_m}{b_n}$ is a horizontal asymptote to the graph of the function.

Four trigonometric functions, defined in terms of sine and/or cosine, were discussed in this chapter. These nonrational functions are

$$\tan x = \frac{\sin x}{\cos x} \text{ for } \cos x \neq 0$$
$$\cot x = \frac{\cos x}{\sin x} \text{ for } \sin x \neq 0$$
$$\sec x = \frac{1}{\cos x} \text{ for } \cos x \neq 0$$
$$\csc x = \frac{1}{\sin x} \text{ for } \sin x \neq 0.$$

The graphs of these functions can be obtained by considering where the sine and cosine functions are positive, negative, or zero.

Rational equations can be solved by multiplying both sides of the equation by the least common denominator. The same logic of equation solving employed in Chapter 3 applies to solving rational equations.

Vocabulary

For the starred (*) terms you should be able to give a definition of the term.
For the other terms you should be able to give a general description and a specific example of each.

Lesson 5-1
*rational number
*rational expression
identity; domain of an identity
restrictions
complex fraction

Lesson 5-2
irrational number
*conjugates

Lesson 5-3
vertical asymptote
$x \to a^-$, $x \to a^+$

Lesson 5-4
*rational function
*removable discontinuity
*essential discontinuity
Discontinuity Theorem

Lesson 5-5
leading term
end behavior of rational function

Lesson 5-6
oblique asymptote
*tangent, tan x
*cotangent, cot x
*secant, sec x
*cosecant, csc x

Lesson 5-7
rational equation

Lesson 5-8
unit fraction
regular polygon
tessellation
fundamental region
regular polyhedron

SUMMARY

The Summary gives an overview of the entire chapter and provides an opportunity for students to consider the material as a whole. Thus, the Summary can be used to help students relate the various concepts presented in the chapter.

VOCABULARY

Terms, symbols, and properties are listed by lesson to provide a checklist of concepts a student must know. (See page 71 for more information.)

ADDITIONAL ANSWERS
(For Progress Self-Test Questions on p. 340)

8.a. $\lim_{x \to 2^+} g(x) = +\infty$

b. $\lim_{x \to 2^-} g(x) = -\infty$

c. $\lim_{x \to +\infty} g(x) = 3$; $\lim_{x \to -\infty} g(x) = 3$

d.

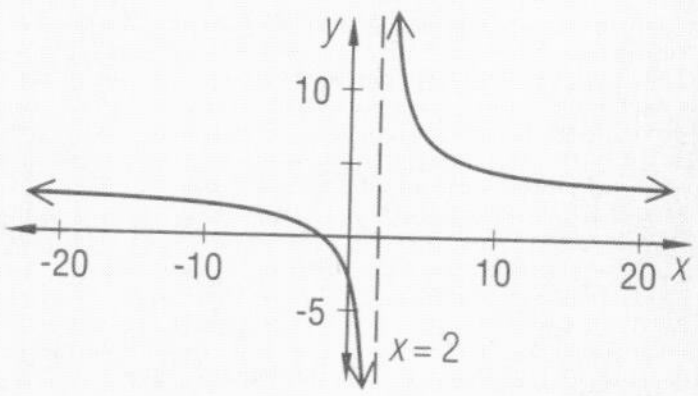

9. essential discontinuity

12.

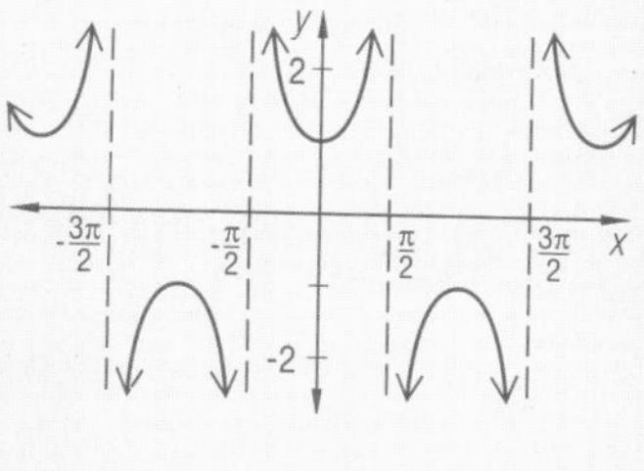

PROGRESS SELF-TEST

The Progress Self-Test provides the opportunity for feedback and correction; the Chapter Review provides additional opportunities for practice.

USING THE PROGRESS SELF-TEST
Assign the Progress Self-Test as a one-night assignment. (See pages 72 and T35 for more information.)

ADDITIONAL ANSWERS

1.a. $\frac{2x(4x+9)}{(x+1)(x+2)(x+3)}$
b. $x \neq -3, x \neq -2$, and $x \neq -1$

2.a. $\frac{2t^2-t-1}{3t^2-2t-5} \cdot \frac{3t^2+7t-20}{t^2+3t-4} = \frac{(2t+1)(t-1)}{(3t-5)(t+1)} \cdot \frac{(3t-5)(t+4)}{(t-1)(t+4)} = \frac{2t+1}{t+1}$
b. $t \neq \frac{5}{3}, t \neq -1, t \neq 1$, and $t \neq -4$

5. Assume the negation is true, that $\sqrt{11}$ is rational. By definition, there exist integers a and b, with $b \neq 0$, such that $\sqrt{11} = \frac{a}{b}$, where $\frac{a}{b}$ is in lowest terms. Then, $11 = \frac{a^2}{b^2} \Rightarrow a^2 = 11b^2$. Thus, a^2 has a factor of 11. And a has a factor of 11, because if a is an integer and a^2 is divisible by a prime, then a is divisible by that prime. Therefore, let $a = 11k$ for some integer k. Then $11b^2 = (11k)^2 \Rightarrow b^2 = 11k^2$. So b^2 and b have a factor by 11 by similar argument. Thus, a and b have a common factor of 11. This is a contradiction since $\frac{a}{b}$ is in lowest terms. Hence, the assumption must be false, and so $\sqrt{11}$ is irrational.

7. As x approaches 2 from the left, $f(x)$ approaches positive infinity.

8., 9., 12. See the margin on page 339.

Progress Self-Test

See margin for answers not shown below.

Take this test as you would take a test in class. Then check the test yourself using the solutions at the back of the book.

1. a. Write as a single fraction in lowest terms: $\frac{6x}{(x+3)(x+1)} + \frac{2x}{(x+2)(x+1)}$.
b. State any restrictions on the variable.

2. a. Simplify. $\frac{2t^2-t-1}{3t^2-2t-5} \cdot \frac{3t^2+7t-20}{t^2+3t-4}$
b. State any restrictions on the variable.

3. The diagram below shows a converging lens with focal length f.

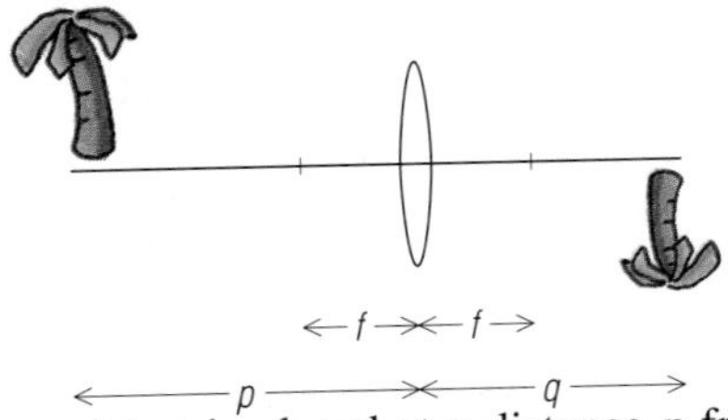

If an object is placed at a distance p from the center of the lens, its image is located at a distance q from the center, and the distance p, q, and f are related by the equation
$$f = \frac{1}{\frac{1}{p} + \frac{1}{q}}.$$
Express f as a simple fraction. $f = \frac{pq}{p+q}$

4. Identify each of the following numbers as rational (R) or irrational (I).
a. $\frac{10}{6}$ R **b.** $\sqrt{24}$ I
c. $\sqrt{49} + 3$ R **d.** $7.\overline{63}$ R
e. .4646646664666664... (pattern continues) I

5. Prove that $\sqrt{11}$ is irrational.

6. Rationalize the denominator: $\frac{8}{\sqrt{10} - \sqrt{5}}$. $\frac{8(\sqrt{10}+\sqrt{5})}{5}$

7. Write in words: $\lim_{x \to 2^-} f(x) = +\infty$.

8. Consider the function $g: x \to \frac{3x+6}{x-2}$.
a. Describe the behavior of g as $x \to 2^+$.
b. Describe the behavior of g as $x \to 2^-$.
c. Use limit notation to describe the end behavior of g.
d. Sketch a graph of g.

9. For the function of Problem 8, is $x = 2$ an essential discontinuity, a removable discontinuity, or neither?

10. Solve for y: $\frac{2}{y-1} + \frac{3y}{y+4} = 3$. $y = 2$

11. Find the indicated value.
a. $\tan \frac{2\pi}{3}$ $-\sqrt{3}$ **b.** $\csc \frac{\pi}{4}$ $\sqrt{2}$

12. Sketch a graph of $y = \sec x$ over the interval $-2\pi \leq x \leq 2\pi$.

13. Find $\tan \alpha$ in the triangle below.

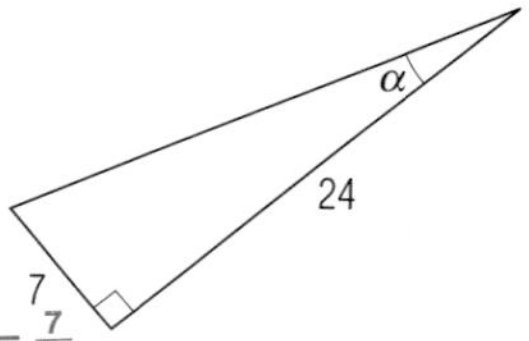

$\tan \alpha = \frac{7}{24}$

14. Refer to the graph of the function h below. The dashed lines are asymptotes. Evaluate the following:

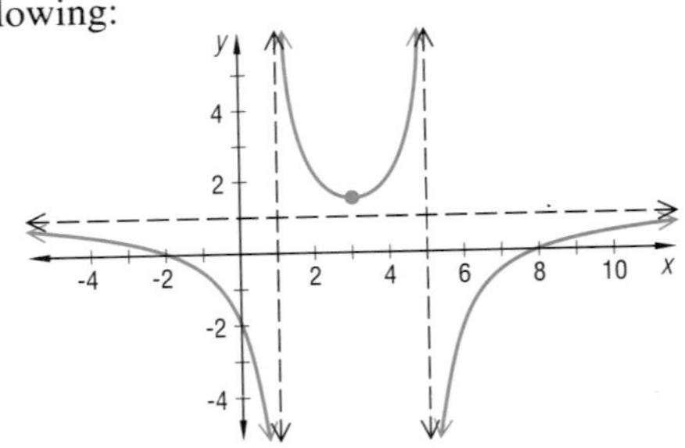

a. $\lim_{x \to 1^-} h(x)$ $-\infty$ **b.** $\lim_{x \to 5^+} h(x)$ $-\infty$
c. $\lim_{x \to -\infty} h(x)$ 1 **d.** $\lim_{x \to +\infty} h(x)$ 1

See margin for answers not shown below.

Chapter Review

Questions on **SPUR** Objectives

SPUR stands for **S**kills, **P**roperties, **U**ses, and **R**epresentations.
The Chapter Review questions are grouped according to the SPUR Objectives for this chapter.

SKILLS deal with the procedures used to get answers.

Objective A: *Simplify rational expressions.* *(Lesson 5-1)*

In 1-6, **a.** simplify. **b.** State any restrictions on the variables.

1. $\frac{x+5}{3x-2} + \frac{4x+3}{2-3x}$ a) -1; b) $x \neq \frac{2}{3}$

2. $\frac{9}{y^2+7y+12} + \frac{4}{y^2+8y+15}$

3. $\frac{z-3}{2(z+1)(z-2)} \cdot \frac{6z-12}{(z-3)(z+4)}$

4. $\frac{2p^2+p-1}{p^2+6p+5} \div \frac{6p^2+p-2}{p^2+11p+30}$

5. $\frac{t+2}{t-5} - \frac{3t+1}{t+4}$ a) $\frac{-2t^2+20t+13}{(t-5)(t+4)}$; b) $t \neq 5$ and $t \neq -4$

6. $\frac{1}{r+1} - \frac{1}{r-1}$ a) $\frac{-2}{r^2-1}$; b) $r \neq 1$ and $r \neq -1$

In 7 and 8, write the expression in lowest terms.

7. $\frac{6x^2-27x-15}{2x^2-11x-6}$ $\frac{3(x-5)}{(x-6)}$ **8.** $\frac{z^3-3z^2-4z}{z^4-z^3-12z^2}$ $\frac{z+1}{z(z+3)}$

In 9 and 10, write as a simple fraction and state any restrictions on the variables.

9. $\dfrac{\frac{1}{3}-\frac{3}{a^2}}{\frac{1}{3a}+\frac{1}{a^2}}$ $a-3$, $a \neq -3$ and $a \neq 0$

10. $\dfrac{\frac{x^2+7x+12}{x^2-25}}{\frac{x^2-3x-18}{x^2-3x-10}}$

In 11–12, **a.** show that the first expression simplifies to the second expression, and **b.** state any restrictions on the variables.

11. $\frac{z^2-z-2}{z^2-4z-5} \cdot \frac{z-5}{z^2+z-6}$, $\frac{1}{z+3}$

12. $\frac{5}{y^2-9} + \frac{7}{y^2-2y-3}$, $\frac{12y+26}{(y+1)(y+3)(y-3)}$

13. Show that the equation is an identity.
$$\frac{\frac{1}{x}+\frac{2}{x^2}}{1-\frac{4}{x^2}} = \frac{1}{x-2}$$

Objective B: *Identify numbers as rational or irrational.* *(Lessons 5-1, 5-2)*

In 14–20, identify each number as rational or irrational. Justify your reasoning.

14. e^5

15. -7

16. .01001000100001... (pattern continues)

17. 0 rational, since 0 is an integer

18. $9.42\overline{67}$ rational, equals $\frac{3733}{396}$

19. $\sqrt{36} + \frac{8}{3}$ rational, equals $\frac{26}{3}$

20. $\frac{\sqrt{3}}{4}$ irrational, $\sqrt{3}$ is not an integer.

Objective C: *Simplify expressions involving radicals.* *(Lesson 5-2)*

In 21–24, rationalize the denominator to simplify the expression.

21. $\frac{5}{3-\sqrt{5}}$ $\frac{15+5\sqrt{5}}{4}$ **22.** $\frac{12}{4+\sqrt{6}}$ $\frac{24-6\sqrt{6}}{5}$

23. $\frac{4\sqrt{3}}{\sqrt{2}+\sqrt{3}}$ $12-4\sqrt{6}$ **24.** $\frac{-9\sqrt{2}}{\sqrt{5}-\sqrt{2}}$ $-3\sqrt{10}-6$

In 25 and 26, rationalize the numerator.

25. $\frac{3-\sqrt{2}}{5}$ $\frac{7}{15+5\sqrt{2}}$

26. $\frac{\sqrt{x+h}-\sqrt{x}}{h}$

RESOURCES
- Chapter 5 Test, Form A
- Chapter 5 Test, Form B
- Chapter 5 Test, Cumulative Form
- Quiz and Test Writer

CHAPTER REVIEW

The main objectives for the chapter are organized here into sections corresponding to the four main types of understanding this book promotes: Skills, Properties, Uses, and Representations. We call these the SPUR objectives.

USING THE CHAPTER REVIEW
Students should be able to answer questions like these with about 85% accuracy by the end of the chapter. (See pages 74 and T38-39 for more information.)

ADDITIONAL ANSWERS

2.a. $\frac{13y+61}{(y+4)(y+5)(y+3)}$
b. $y \neq -4, y \neq -5$, and $y \neq -3$

3.a. $\frac{3}{(z+1)(z+4)}$
b. $z \neq -1, z \neq 2, z \neq 3$, and $z \neq -4$

4.a. $\frac{p+6}{3p+2}$
b. $p \neq \frac{1}{2}, p \neq -\frac{2}{3}, p \neq -1, p \neq -5$, and $p \neq -6$

10. $\frac{x^2+6x+8}{x^2-x-30}$; $x \neq 6, x \neq -3, x \neq 5, x \neq -2$, and $x \neq -5$

11.a. $\frac{z^2-z-2}{z^2-4z-5} \cdot \frac{z-5}{z^2+z-6} = \frac{(z+1)(z-2)}{(z+1)(z-5)} \cdot \frac{z-5}{(z-2)(z+3)} = \frac{1}{z+3}$
b. $z \neq -1, z \neq 5, z \neq 2$, and $z \neq -3$

12., 13., 14., 15., 16., 26. See the margin on page 342.

ADDITIONAL ANSWERS

12.a. $\frac{5}{y^2-9}+\frac{7}{y^2-2y-3}$
$=\frac{5}{(y+3)(y-3)}+\frac{7}{(y+1)(y-3)}$
$=\frac{5(y+1)+7(y+3)}{(y+3)(y-3)(y+1)}$
$=\frac{5y+5+7y+21}{(y+3)(y-3)(y+1)}$
$=\frac{12y+26}{(y+3)(y-3)(y+1)}$
b. $y \neq -3$, $y \neq -1$, and $y \neq 3$

13. $\frac{\frac{1}{x}+\frac{2}{x^2}}{1-\frac{4}{x^2}}=\frac{x+2}{x^2-4}=$
$\frac{x+2}{(x+2)(x-2)}=\frac{1}{x-2}$;
$x \neq 0$, $x \neq -2$, and $x \neq 2$

14. irrational, since e^x is irrational for any nonzero rational number x

15. rational, because -7 is an integer

16. irrational, the decimal expansion neither terminates nor repeats

26. $\frac{1}{\sqrt{x+h}+\sqrt{x}}$

27. $\tan x=-\frac{3}{4}$; $\sec x=\frac{5}{4}$; $\cot x=-\frac{4}{3}$; $\csc x=-\frac{5}{3}$

43. Assume the negation is true, that $\sqrt{13}$ is rational. By definition, there exist integers a and b, with $b \neq 0$, such that $\sqrt{13}=\frac{a}{b}$, where $\frac{a}{b}$ is in lowest terms. Then, $13=\frac{a^2}{a^2} \Rightarrow a^2=13b^2$. Thus, a^2 has a factor of 13. And a has a factor of 13, because if a is an integer and a^2 is divisible by a prime, then a is divisible by that prime. Therefore, let $a=13k$ for some integer k. Then, $13b^2=(13k)^2 \Rightarrow b^2=13k^2$. So b^2 and b have a factor of 13 by similar argument. Thus, a and b have a common factor of 13. This is a contradiction since $\frac{a}{b}$ is in lowest terms. Hence, the assumption must be false, and so $\sqrt{13}$ is irrational.

Objective D: *Find values of trigonometric functions.* (Lesson 5-6)

27. If $\sin x = -.6$ and $\cos x = .8$, find the values of $\tan x$, $\sec x$, $\cot x$, and $\csc x$.

28. Let $\sec x = 2$.
 a. Find $\cos x$. $\frac{1}{2}$
 b. Find x if $0 \le x \le \pi$. $x=\frac{\pi}{3}$

In 29–34, find the indicated value.

29. $\tan \frac{3\pi}{4}$ -1
30. $\sec \frac{5\pi}{6}$ $\frac{-2\sqrt{3}}{3}$
31. $\csc \frac{3\pi}{2}$ -1
32. $\cot \frac{\pi}{6}$ $\sqrt{3}$
33. $\tan \pi$ 0
34. $\sec \pi$ -1

Objective E: *Solve rational equations.* (Lesson 5-7)

In 35–39, solve each equation.

35. $1+\frac{12}{x^2-4}=\frac{3}{x-2}$ $x = 1$

36. $\frac{m}{m-1}-\frac{2}{m+3}=\frac{-m+5}{(m-1)(m+3)}$ no solutions

37. $\frac{3}{t}+\frac{2}{t+1}=4$ $t=-\frac{3}{4}$ and $t = 1$

38. $\frac{2}{v+1}-\frac{v-1}{v}=\frac{1}{v^2+v}$ $v = 2$

39. $\frac{5}{y+1}-\frac{7}{y-2}=8$ $y=\frac{1}{2}$ and $y=\frac{1}{4}$

PROPERTIES deal with the principles behind the mathematics.

Objective F: *Prove properties of rational and irrational numbers.* (Lessons 5-1, 5-2)

In 40–43, show that each number is a rational number by expressing it as a ratio of integers $\frac{a}{b}$, where $b \neq 0$.

40. .893 $\frac{893}{1000}$
41. $2.\overline{47}$ $\frac{245}{99}$
42. $3.98\overline{1}$ $\frac{3583}{9000}$
43. Show that $\sqrt{13}$ is irrational.

In 44–47, *true* or *false*. Justify your answer.

44. *Every integer is a rational number.*
45. *The product of any two rational numbers is rational.*
46. *The product of any two irrational numbers is irrational.*
47. Prove: *If p is a rational number and q is an irrational number, then p − q is irrational.*

Objective G: *Use limit notation to describe the behavior of rational functions.* (Lessons 5-3, 5-4, 5-5)

48. *Multiple choice.* The notation $x \to -4^+$ is read
 (a) x approaches positive four from the left.
 (b) x approaches positive four from the right.
 (c) x approaches negative four from the left.
 (d) x approaches negative four from the right. (d)

49. Write in words: $\lim_{x \to -\infty} f(x) = 4$.

50. Consider the function $f: x \to \frac{1}{x+2}$.
 a. What transformation maps $g: x \to \frac{1}{x}$ onto the function f? $T_{-2,0}$
 b. Find
  i. $\lim_{x \to -2^+} f(x)$ $+\infty$
  ii. $\lim_{x \to -2^-} f(x)$ $-\infty$
  iii. $\lim_{x \to -\infty} f(x)$ 0
  iv. $\lim_{x \to +\infty} f(x)$ 0

51. Consider the function h defined by the rule $h(x)=\frac{2x+5}{x-6}$.
 a. Use limit notation to describe the behavior of h as $x \to 6^+$.
 b. Use limit notation to describe the behavior of h as $x \to 6^-$.
 c. Use limit notation to describe the end behavior of h.

52. Consider the function f defined by $f(x)=\frac{3x^2-11}{x^2-4}$.
 a. Where would you expect to find vertical asymptotes for f? $x = 2$ and $x = -2$
 b. Use limit notation to describe the behavior of f near the value(s) indicated in part **a**.
 c. Use limit notation to describe the end behavior of f.
 d. Sketch a graph of f.

In 53 and 54, find the oblique asymptote of the graph of the function.

53. $f\colon x \rightarrow \dfrac{3x^3 + 2x^2 + x}{4x^2 + 2x + 1}$ $y = \frac{3}{4}x + \frac{1}{8}$

54. $g\colon t \rightarrow \dfrac{18t^2 - 3t + 1}{2t + 5}$ $y = 9t - 24$

In 55–57, find a constant function, a power function, or a reciprocal of a power function that has the same end behavior as the given function.

55. $h(y) = \dfrac{7y^4 - 8y^2 - 2y + 1}{3y^2 + 7y + 2}$ $h(y) = \frac{7}{3}y^2$

56. $p(t) = \dfrac{11t^5 - 5t^4 + 6t^3 + 2t + 1}{9t^5 + 4t^2 - 7t + 2}$ $p(t) = \frac{11}{9}$

57. $q(z) = \dfrac{z^5 + 4z^2 + 2}{4z^6 + z - 1}$ $q(z) = \frac{1}{4z}$

Objective H: *Classify discontinuities as essential or removable.* *(Lesson 5-4, 5-6)*

58. The function $f\colon x \rightarrow \dfrac{x + 5}{x^2 - 25}$ is undefined at $x = 5$ and at $x = -5$.

a. At which of these values is there an essential discontinuity? 5

b. At which of these values is there a removable discontinuity? -5

c. Redefine the function at the value in part **b** so that the discontinuity is removed.

59. Construct a rule for a function that has a removable discontinuity at $x = 4$ and that has no essential discontinuities.

60. Consider the tangent function $f\colon x \rightarrow \tan x$. For what values of x does this function have essential discontinuities?

61. *True* or *false?* The cosecant function has essential discontinuities at $y = n\pi$ for all integers n. True

USES deal with applications of mathematics in real situations.

Objective I: *Apply rational expressions and rational equations.* *(Lessons 5-1, 5-7)*

62. To score a batch of computerized tests, two computers work on the task. Working alone the first computer would take c hours and the second computer would take h hours.

a. What expression represents the portion of the task done by the first computer each hour? $\frac{1}{c}$

b. What expression represents the portion of the task done by the second computer each hour? $\frac{1}{h}$

c. Write a single fraction that represents the portion of the task completed each hour if both computers work together. $\frac{c+h}{ch}$

d. If c is 4 hours and h is 6 hours, how long does it take the two computers to do the job when working together? 2.4 hr

63. The velocity v of a wave is given by the formula $v = \dfrac{\lambda}{\frac{1}{f}}$ where λ is the wavelength and f is the frequency of the wave. (λ is the distance between two consecutive points in corresponding positions and f is the number of waves passing a given point per time period.) Find a simple expression for the velocity of the wave. $v = f\lambda$

64. Suppose a person drives 10,000 miles a year and a gallon of gas costs \$1.15 on average. How much would this person save in a year by driving a car that gets 25 $\frac{\text{mi}}{\text{gal}}$ over one that gets 18 $\frac{\text{mi}}{\text{gal}}$? ≈ \$179

65. A plane, capable of flying at 500 miles per hour, makes a 3000-mile trip. Because of a headwind, the return flight takes 1.5 hours longer than the original flight which had the advantage of a tailwind. Find the speed of the wind, assuming it is the same on both trips. 61.6 mph

44. True, every integer a can be expressed as $\frac{a}{1}$.

45. True, if $\frac{a}{b}$ and $\frac{c}{d}$ are two rational numbers, where $b \neq 0$ and $d \neq 0$, then $\frac{a}{b} \cdot \frac{c}{d} = \frac{ac}{bd}$. ac and bd are integers and $bd \neq 0$ since $b \neq 0$ and $d \neq 0$. Hence, $\frac{ac}{bd}$ is rational.

46. False, counter-example: $\sqrt{2} \cdot \sqrt{2} = 2$

47. Assume the negation is true. Thus, the difference of a rational number p and an irrational number q is a rational number r. Then $p - q = r$. So $p - r = q$. However, by the closure property of rational numbers, the difference between two rational numbers is another rational number. Hence, there is a contradiction, and so the assumption is false, which proves the original statement.

49. The limit of $f(x)$ as x decreases without bound is 4.

51.a. $\lim_{x \to 6^+} h(x) = +\infty$

b. $\lim_{x \to 6^-} h(x) = -\infty$

c. $\lim_{x \to +\infty} h(x) = 2$; $\lim_{x \to -\infty} h(x) = 2$

52.b. $\lim_{x \to 2^+} f(x) = +\infty$; $\lim_{x \to 2^-} f(x) = -\infty$; $\lim_{x \to -2^+} f(x) = -\infty$; $\lim_{x \to -2^-} f(x) = +\infty$

c. $\lim_{x \to +\infty} f(x) = 3$; $\lim_{x \to -\infty} f(x) = 3$

d.

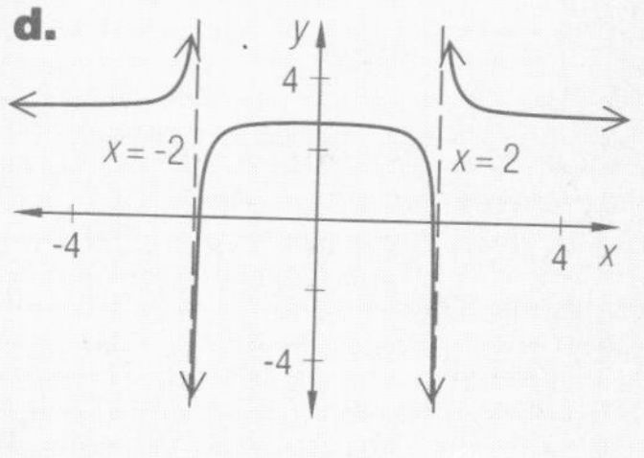

58.c., 59., 60. See Additional Answers in the back of this book.

EVALUATION
Three tests are provided for this chapter in the Teacher's Resource File. Chapter 5 Test, Forms A and B cover just Chapter 5. The third test is Chapter 5 Test, Cumulative Form. About 50% covers Chapter 5, 25% covers Chapter 4, and 25% covers previous chapters. For information on grading, see *General Teaching Suggestions: Grading* on page T43 in the Teacher's Edition.

ASSIGNMENT RECOMMENDATION
We strongly recommend that you assign the reading and some questions from Lesson 6-1 for homework the evening of the test. It gives students work to do if they complete the test before the end of the period and keeps the class moving.

If you do not give assignments on the days of tests, you may cover one less *chapter* over the course of the year.

ADDITIONAL ANSWERS
66.a. $x = 1$ and $x = -1$
b. $x = 1$ and $x = -1$: essential
c. $y = 2$
d. x-intercept: none; y-intercept: -3
e. $\lim_{x \to +\infty} h(x) = 2$; $\lim_{x \to -\infty} h(x) = 2$
f.

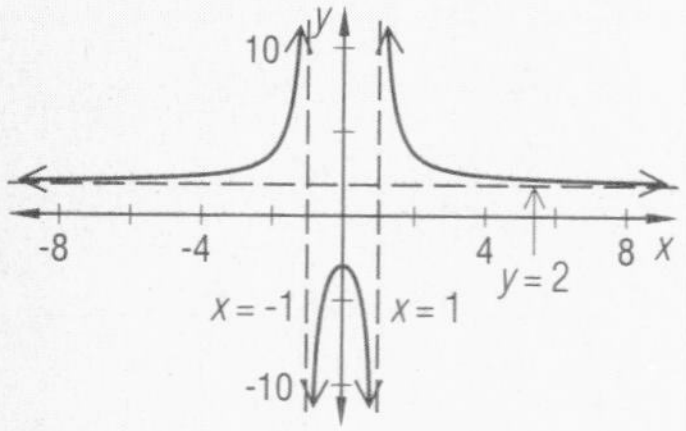

67.a. $x = 3$
b. $x = 3$: essential
c. none
d. x-intercept: $-\sqrt[3]{6} \approx -1.8$; y-intercept: -2
e. $\lim_{x \to +\infty} f(x) = \infty$; $\lim_{x \to -\infty} f(x) = \infty$

67.f., 68.-73. See Additional Answers in the back of this book.

REPRESENTATIONS deal with pictures, graphs, or objects that illustrate concepts.

Objective J: *Graph quotients of polynomial or trigonometric functions. (Lessons 5-3, 5-4, 5-5, 5-6)*

In 66–68, use the given function.

- **a.** For what value(s) is the function undefined?
- **b.** Identify any discontinuities as essential or removable.
- **c.** Find an equation for any horizontal asymptotes.
- **d.** Find the x- and y-intercepts of the function.
- **e.** Describe the end behavior of the function.
- **f.** Sketch a graph of the function.

66. $h: x \to \dfrac{2x^2 + 3}{x^2 - 1}$

67. $f: x \to \dfrac{x^3 + 6}{x - 3}$

68. $g: x \to \dfrac{x + 4}{x^2 + 4x}$

69. Sketch a graph of $y = \tan x$.

70. Sketch a graph of $y = \csc x$.

Objective K: *Relate the limit of a function to its graph, and find equations for its asymptotes. (Lessons 5-3, 5-4, 5-5)*

71. Consider a function f with the following properties:

$\lim_{x \to 5^+} f(x) = -\infty$ $\lim_{x \to +\infty} f(x) = -2$

$\lim_{x \to 5^-} f(x) = -\infty$ $\lim_{x \to -\infty} f(x) = -2$

- **a.** Construct a possible graph of f.
- **b.** Write equations for its horizontal and vertical asymptotes.

72. Consider the function f graphed below.
- **a.** Use limit notation to describe the behavior of the function near $x = 4$.
- **b.** Use limit notation to describe the behavior of the function near $x = -3$.
- **c.** Use limit notation to describe the behavior of the function near the horizontal asymptote.
- **d.** Find equations of all asymptotes.

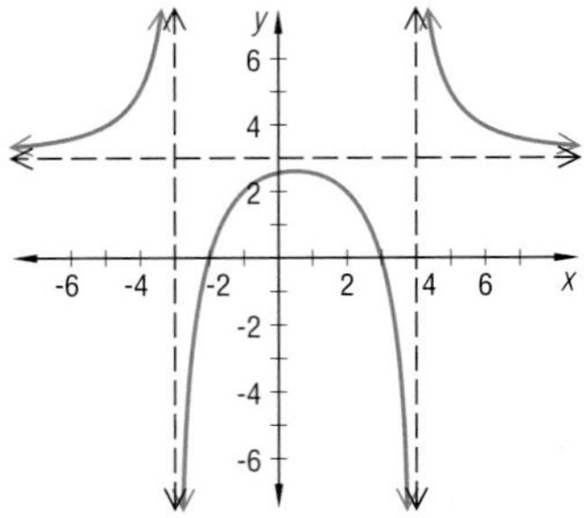

73. Graph the function f defined by $f(x) = \dfrac{-1}{(x + 1)^2}$ and write equations for its horizontal and vertical asymptotes.

Objective L: *Use right triangles to find values of trigonometric functions. (Lesson 5-6)*

In 74–77, use the triangle below to find the indicated value.

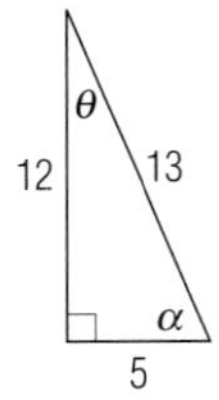

74. $\tan \alpha$ $\frac{12}{5}$ **75.** $\sec \theta$ $\frac{13}{12}$

76. $\csc \alpha$ $\frac{13}{12}$ **77.** $\cot \theta$ $\frac{12}{5}$

CHAPTER 6 TRIGONOMETRIC IDENTITIES AND EQUATIONS

OBJECTIVES

The objectives listed here are the same as in the Chapter 6 Review on pages 396–398 of the student text. The Progress Self-Test on page 395 and the tests in the Teacher's Resource File cover these objectives. For recommendations regarding the handling of this end-of-chapter material, see the notes in the margin on the corresponding pages of this Teacher's Edition.

OBJECTIVES FOR CHAPTER 6 (Organized into the SPUR Categories—Skills, Properties, Uses, and Representations)	Progress Self-Test Questions	Chapter Review Questions	Teacher's Resource File	
			Lesson Masters*	Chapter Test Forms A & B
SKILLS				
A Without a calculator, use trigonometric identities to express values of trigonometric functions in terms of rational numbers and radicals.	1, 2, 5, 6	1–12	6-2, 6-3, 6-4, 6-5	1, 4, 6, 8
B Evaluate inverse trigonometric functions with or without a calculator.	3, 4	13–17	6-6	11, 12
C Solve trigonometric equations and inequalities algebraically.	10, 11	18–25	6-7	9
PROPERTIES				
D Prove trigonometric identities and identify their domains.	8, 9	26–38	6-2, 6-3, 6-4, 6-5	2, 3, 5
USES				
E Solve problems using inverse trigonometric functions.	14	39–41	6-6	14
F Use trigonometric equations and inequalities to solve applied problems.	13	42–44	6-7	13
REPRESENTATIONS				
G Use an automatic grapher to test proposed trigonometric identities.	7	45–50	6-1, 6-2	7
H Use graphs to solve trigonometric equations and inequalities.	12	51–53	6-7	10

*** The masters are numbered to match the lessons.**

OVERVIEW ■ CHAPTER 6

This chapter continues the development of trigonometry that is present in UCSMP materials. If students have studied from *Functions, Statistics, and Trigonometry,* they will have seen all the material in the chapter, including the formulas, but will not have had experience proving identities.

Although the content of the chapter is standard for a precalculus course, the approach taken is somewhat different. In the first lesson, for example, graphing technology is used to conjecture that an equation might be an identity, and also is used to find a counterexample to a proposed identity. Proof techniques are introduced in Lesson 6-2 as a means to validate conjectures. Several proof techniques are discussed so that students can choose the one most appropriate for a given situation.

The next three lessons apply proof techniques to deduce the formulas for $\cos(\alpha \pm \beta)$ and $\sin(\alpha \pm \beta)$. These formulas are then used to obtain the formulas for $\cos 2x$ and $\sin 2x$. The last of these identities is used to derive an equation for the range of a projectile. This equation is one that students may have seen in a physics course, but perhaps without its derivation.

The inverse trigonometric functions were mentioned briefly in Lesson 3-8 in the context of function composition. Lesson 6-6 discusses these functions in more detail and provides a careful treatment of their development. Students are led through a discussion of the need to restrict the domain of the trigonometric functions so that their inverses are functions, and also are provided with the criteria needed to make these restrictions.

Lesson 6-7 applies the sentence-solving techniques from earlier chapters to equations and inequalities involving trigonometric functions. Students are expected to solve these equations over specific intervals as well as over the set of real numbers. Graphing technology is used again to provide a first stage of analysis for solving inequalities.

The content of Lesson 6-8 provides a powerful application of the identity for the cosine or sine of the sum of two angles. Students are introduced to several situations that are modelled by equations of the form $f(t) = a \cos ct + b \sin ct$.

Proofs of trigonometric identities are found in all lessons of the chapter after Lesson 6-2, and students are expected to gain some facility with them. Although this content is traditional, some people have suggested that it should be deemphasized. We feel that the amount of emphasis in this text is appropriate, for the following reasons: (1) proving relationships between values of functions is a skill useful in later mathematics courses; (2) the key identities have some very nice applications; and (3) some identities are very beautiful and can motivate students to want to find others.

PERSPECTIVES ■ CHAPTER 6

The Perspectives provide the rationale for the inclusion of topics or approaches, provide mathematical background, and make connections with other lessons and within UCSMP.

6-1

GRAPHS OF IDENTITIES

Since students have been introduced to identities in Chapter 5, the ideas of this lesson are not new. To test whether $f(x) = g(x)$ is true for all x, one can graph $y = f(x)$ and $y = g(x)$ on some appropriate interval to see if the graphs are identical. The key point is that regardless of how the graphs look, the graphs by themselves do not constitute a proof. However, if the graphs are different, then a value of x can be found at which $f(x) \neq g(x)$. Be sure to emphasize the difference between identities and other equations, a difference that some students find hard to understand.

6-2

PROVING IDENTITIES

There are no new concepts introduced in this lesson, but the context is different enough that students will think the material is new. Four techniques are given for proving identities. (1) Rewrite one side using definitions, known identities, and algebraic properties until it equals the other side. This is a commonly used technique. (2) Rewrite each side independently until expressions are obtained that are equal. This is a very good technique that should be used more frequently by students. (3) Begin with a known identity and transform it until the desired identity appears. This technique is used in the next lesson to deduce the formula for $\cos(\alpha + \beta)$. (4) Transform both sides of the equation using reversible steps until an equation known to be an identity appears. This technique is probably the most commonly used one, although some teachers do not like to use it because you begin with what is supposed to be proved.

6-3

FORMULAS FOR $\cos(\alpha + \beta)$ AND $\cos(\alpha - \beta)$

Students who have studied from *Functions, Statistics, and Trigonometry* have seen these formulas proved using matrices for rotations. In this book, a proof of the formula for $\cos(\alpha + \beta)$ is given by using circular functions. Then, using a process that is repeated in later lessons, the identity for $\cos(\alpha + \beta)$ is used to obtain an exact cosine value for an argument that is not a multiple of $\frac{\pi}{4}$ or $\frac{\pi}{6}$ but rather a sum of a combination of both. A formula for $\cos(\alpha - \beta)$, and the identity $\cos\left(\frac{\pi}{2} - x\right) = \sin x$ also are derived from the formula for $\cos(\alpha + \beta)$.

6-4

FORMULAS FOR $\sin(\alpha \pm \beta)$ AND $\tan(\alpha \pm \beta)$

This lesson follows naturally from the previous one. Using the identities for $\cos\left(\frac{\pi}{2} - x\right)$ and $\cos(\alpha - \beta)$, an identity for $\sin(\alpha + \beta)$ is deduced. This leads to identities for $\sin(\alpha - \beta)$ and $\sin(\pi - x)$. By dividing the formulas for the sine by the corresponding formulas for the cosine, the identities for the tangent function are derived.

6-5

FORMULAS FOR $\cos 2x$ AND $\sin 2x$

It is natural to let $\alpha = \beta = x$ in the formulas for $\cos(\alpha + \beta)$ and $\sin(\alpha + \beta)$, and thus derive formulas for $\cos 2x$ and $\sin 2x$. The formulas for $\cos 2x$ and $\sin 2x$ are sometimes called *double angle identities*.

In the previous lessons, two reasons for having identities are discussed: to prove other identities, and to find exact values of circular functions for new arguments. In this lesson, one additional reason is discussed, namely to simplify a formula.

6-6

INVERSE TRIGONOMETRIC FUNCTIONS

In preparation for the solving of equations, this lesson discusses the inverses of the sine, cosine, and tangent functions. In order to define inverse functions, the domains of the given functions have to be restricted. Criteria for the domains are given which force a particular choice of domain for each function: for the sine, $-\frac{\pi}{2} \le x \le \frac{\pi}{2}$; for the cosine, $0 \le x \le \pi$; and for the tangent, $-\frac{\pi}{2} < x < \frac{\pi}{2}$. A major objective of the lesson is to evaluate the inverse functions; that is, to obtain exact values or approximations to them for any x in their domains.

6-7

SOLVING EQUATIONS AND INEQUALITIES INVOLVING TRIGONOMETRIC FUNCTIONS

In this lesson, trigonometric equations, inverse functions, and inequalities are considered. To solve an equation such as $\sin x = k$, students first find the single value $\sin^{-1} k$, and then use identities to find other solutions. The solutions to $\sin x = k$ are combined with graphical techniques to solve sentences such as $\sin x < k$.

6-8

PHASE-AMPLITUDE ANALYSIS AND VIBRATIONS

When a weight hanging on a spring is set in motion (assuming no friction), after t seconds its height is given by $f(b) = a \cos ct + b \sin ct$. This formula can be rewritten as the single cosine function $\sqrt{a^2 + b^2} \cos (ct - \theta)$, where θ depends only on a and b, indicating that the sum of the cosine and sine function with the same argument is a cosine function. Thus, the motion of the weight over time is described by a sine wave.

If there is friction, then there is a dampening force, and $f(t) = e^{-kt}(a \cos ct + b \sin ct)$. This lesson describes both models of the motion of the weight on a spring, and in so doing provides a classical application of some of the identities and inverse functions studied in the chapter.

DAILY PACING CHART ■ CHAPTER 6

Every chapter of UCSMP *Precalculus and Discrete Mathematics* includes lessons, a Progress Self-Test, and a Chapter Review. For optimal student performance, the self-test and review should be covered. (See *General Teaching Suggestions: Mastery* on page T35 of this Teacher's Edition.) By following the pace of the Full Course given here, students can complete the entire text by the end of the year. Students following the pace of the Minimal Course spend more time when there are quizzes and on the Chapter Review and will generally not complete all of the chapters in this text.

When chapters are covered in full (the recommendation of the authors), then students in the Minimal Course can cover 11 chapters of the book. For more information on pacing, see *General Teaching Suggestions: Pace* on page T34 of this Teacher's Edition.

DAY	MINIMAL COURSE	FULL COURSE
1	6-1	6-1
2	6-2	6-2
3	6-3	6-3
4	Quiz (TRF); Start 6-4.	Quiz (TRF); 6-4
5	Finish 6-4.	6-5
6	6-5	6-6
7	6-6	Quiz (TRF); 6-7
8	Quiz (TRF); Start 6-7.	6-8
9	Finish 6-7.	Progress Self-Test
10	6-8	Chapter Review
11	Progress Self-Test	Chapter Test (TRF)
12	Chapter Review	Comprehensive Test (TRF)
13	Chapter Review	
14	Chapter Test (TRF)	
15	Comprehensive Test (TRF)	

TESTING OPTIONS

TESTING OPTIONS

- ■ Quiz for Lessons 6-1 through 6-3
- ■ Quiz for Lessons 6-4 through 6-6
- ■ Chapter 6 Test, Form A
- ■ Chapter 6 Test, Form B
- ■ Chapter 6 Test, Cumulative Form
- ■ Comprehensive Test, Chapters 1–6

A Quiz and Test Writer is available for generating additional questions, additional quizzes, and additional forms of the Chapter Test.

INDIVIDUAL DIFFERENCES

PROVIDING FOR INDIVIDUAL DIFFERENCES

The student text is written for, and tested with, average students. It also has been successfully used with better and more poorly prepared students.

The Lesson Notes often include Error Analysis and Alternate Approach features to help you with those students who need more help. Students of all abilities often learn from their peers and may benefit from small group work referenced as appropriate throughout the Notes. A blackline Lesson Master (in the Teacher's Resource File), keyed to the chapter objectives, is provided for each lesson to allow more practice. (However, since it is important to keep up with the daily pace, you are not expected to use all of these masters. Again, refer to the suggestions for pacing on page T34.) Extension activities are provided in the Lesson Notes for those students who have completed the particular lesson in a shorter amount of time than is expected, even in the Full Course.

Trigonometric Identities and Equations

CHAPTER 6

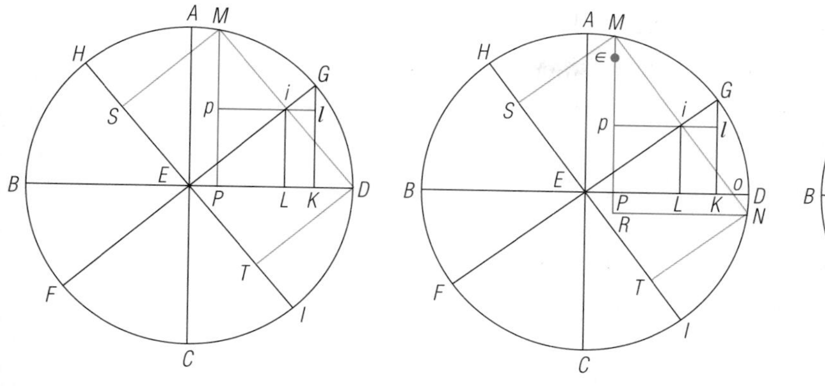

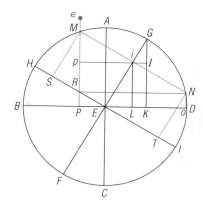

How would you multiply two numbers like 5,842,497 and 1,950,903 without a calculator?

That was the problem facing Renaissance astronomers, who had few aids that would simplify the computations resulting from their studies. Relief came from a surprising area of mathematics: trigonometry. In 1593, the Italian Christopher Clavius published a book on astronomy that described, and proved, a method for using sines and cosines to change a difficult multiplication problem into one which used addition and subtraction. Clavius's proof required heavy use of geometry and consideration of three cases. His drawings, shown above, indicate the complexity of his arguments. The method applies a *trigonometric identity*

$$2 \sin x \sin y = \cos(x - y) - \cos(x + y)$$

together with tables of trigonometric values. Such tables were available in the sixteenth century. For instance, for the numbers at the top of the page, use $0.5842497 \approx \sin 35°45'$ and $0.1950903 \approx \sin 11°15'$.

Now, with $x = 35°45'$ and $y = 11°15'$, the identity yields

$$\begin{aligned} 2 \cdot \sin 35°45' \cdot \sin 11°15' &= \cos(35°45 - 11°15') - \\ &\quad \cos(35°45' + 11°15') \\ &= \cos 24°30' - \cos 47° \\ &\approx 0.2279630. \end{aligned}$$

Thus $0.5842497 \cdot 0.1950903 \approx 0.1139815$. To multiply the corresponding integers, all that is left is to put the decimal point in the proper place.

In this chapter you will develop some fundamental properties of trigonometric functions from which you will be able to prove the identity $2 \sin x \sin y = \cos(x - y) - \cos(x + y)$ as a homework question! We hope you will be as pleased with this material as Clavius was when he wrote "these things are entirely new and full of pleasure and satisfaction."

CHAPTER 6

If students have seen these ideas previously, this chapter can be done at a lesson-a-day pace. Otherwise, you will have to add a day for the formulas and a day for identities. With a day for the Progress Self-Test, 1 or 2 days for the Chapter Review, and one day for a Chapter test, this chapter should take 11–14 days. The Comprehensive Test, covering Chapters 1–6, which can serve as an end-of-semester test, adds another day. (See the Daily Pacing Chart on page 345D.)

USING PAGE 345

It is not well known today that before logarithms were invented, trigonometric identities were used to simplify computations. The identity $2 \sin x \sin y = \cos(x - y) - \cos(x + y)$ was used for this purpose. Tell students they will prove this identity in Lesson 6-3.

You might ask students for simpler values of x and y to verify the identity. For instance, if $x = 0$ and $y = \pi$, then the left side is $2 \sin 0 \sin \pi = 0$, and the right side is $\cos(-\pi) - \cos \pi = 0$.

LESSON 6-1

RESOURCES
- Lesson Master 6-1
- Teaching Aid 36 can be used with **Questions 7-10**.

OBJECTIVE

G Use an automatic grapher to test proposed trigonometric identities.

TEACHING NOTES

This lesson provides students with a visual understanding of the meaning of an identity. Students need access to an automatic grapher throughout the lesson.

Before beginning the lesson, you may want to review the six basic trigonometric functions. Students should be able to provide a rough sketch of each function and identify its domain and range. Facility with the basic functions is essential if students are to formulate conjectures about possible identities. Appendix A in the back of this book can serve as a reference for this review.

To show that **Example 2** is not an identity, students can simply calculate the value of each side for $x = \frac{\pi}{6}$. A real number x for which $f(x) \neq g(x)$ is not a counterexample unless x is in the domains of both f and g. For example, -2 is not a counterexample for $\log x^4 = 4 \log x$ since -2 is not in the domain of $\log x$.

LESSON

6-1 Graphs of Identities

Consider the equations

$$\text{(a) } \sin^2 x + \cos^2 x = 1 \quad \text{and} \quad \text{(b) } \tan x = \frac{\sin x}{\cos x}.$$

Equation (a) is true for all real numbers x and equation (b) is true for all real numbers x for which $\cos x \neq 0$. Recall from Lesson 5-1 that an identity is an equation that is true for all values of the variables for which both sides are defined. The set of all such values is called the **domain of the identity.** Equations (a) and (b) are, therefore, examples of identities.

Identities need not involve the trigonometric functions; $2(x + 3) = 2x + 6$, $\log (t^3) = 3 \log t$, and $\frac{1}{N} + \frac{1}{N-2} = \frac{2N-2}{N^2-2N}$ are identities. Also, identities can involve more than one variable; $2(x + y) = 2x + 2y$ is yet another identity.

In mathematics, identities are useful for simplifying complicated expressions and for solving equations and inequalities. But it is important to know that an equation is indeed an identity before using it as one. In the next lesson you will learn to prove identities. In this lesson you will learn to test equations using graphs to see if they might be identities.

Example 1 Graph the function defined by the equation $y = \cos^2 x - \sin^2 x$ for $-2\pi \leq x \leq 2\pi$. Find another equation that appears to produce the same graph, and conjecture a possible identity.

Solution It is helpful to use an automatic grapher to graph this function.

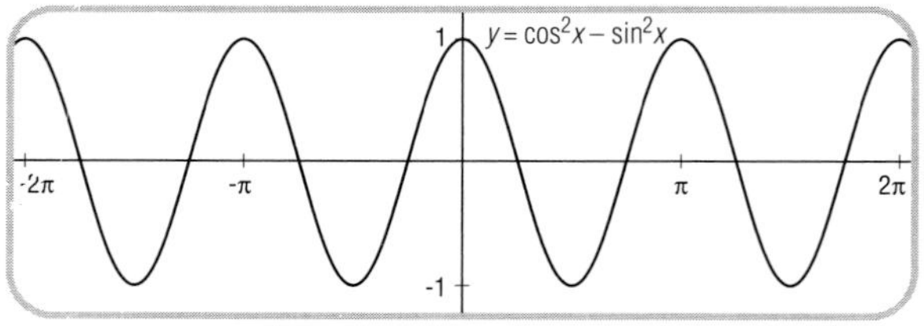

The result looks like a cosine curve with period π. An equation for that curve is $y = \cos 2x$. The graph of $y = \cos 2x$ with the same scale is shown below.

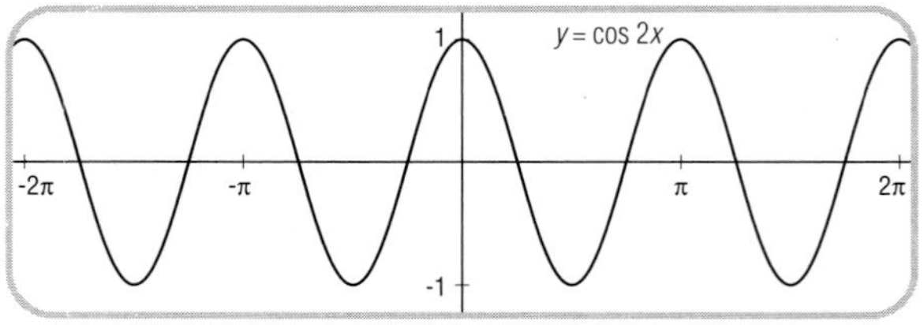

346

The graphs seem to be identical. From this evidence, the equation

$$\cos^2 x - \sin^2 x = \cos 2x$$

seems to hold true for all x in the interval $-2\pi \le x \le 2\pi$. Because the sine and cosine are periodic, it seems reasonable to conjecture that this equation holds for all real numbers x. Thus this equation appears to be an identity.

Be cautious, however, when drawing conclusions from graphs. Even the most powerful computer grapher cannot show the behavior at *every* point on the graph. A graph can only *suggest* identities, not prove them.

A single counterexample shows that a given equation is not an identity; if $f(a) \neq g(a)$ for a single real number a, then the equation $f(x) = g(x)$ is not an identity.

Example 2 Use an automatic grapher to help decide whether

$$(\cos x + \sin x)^2 = \cos^2 x + \sin^2 x$$

is an identity. If it is not an identity, find a value of x for which the equation is not true.

Solution On the same set of axes, graph the functions with equations $y = (\cos x + \sin x)^2$ and $y = \cos^2 x + \sin^2 x$.

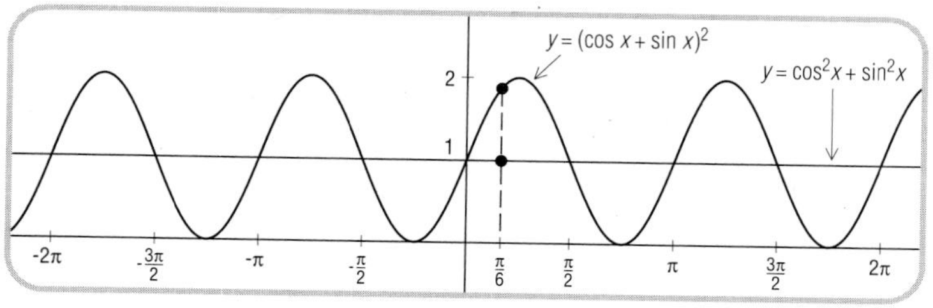

It is apparent that the two graphs are quite different from each other. For instance, the graphs have different values at $\frac{\pi}{6}$. You can check this as follows.

$$\begin{aligned}\left(\cos\frac{\pi}{6} + \sin\frac{\pi}{6}\right)^2 &= \left(\frac{\sqrt{3}}{2} + \frac{1}{2}\right)^2 \\ &= \left(\frac{\sqrt{3}+1}{2}\right)^2 \\ &= \frac{3 + 2\sqrt{3} + 1}{4} \\ &= \frac{4 + 2\sqrt{3}}{4} \\ &= 1 + \frac{\sqrt{3}}{2},\end{aligned}$$

whereas

$$\left(\cos\frac{\pi}{6}\right)^2 + \left(\sin\frac{\pi}{6}\right)^2 = \left(\frac{\sqrt{3}}{2}\right)^2 + \left(\frac{1}{2}\right)^2 = \frac{3}{4} + \frac{1}{2} = 1.$$

Since the equation is not true when $x = \frac{\pi}{6}$, it is not an identity.

If available, use an automatic grapher that allows you to select different colors for graphs. Ask students to use different colored pens or pencils when sketching more than one graph on a single coordinate system.

It should be relatively easy for students to see that equations such as those in **Example 2** are not identities because the two graphs do not coincide for many values of the independent variable. To show two graphs that are not identical but are almost identical for part of the domain, consider $\cos x = 1 - \frac{x^2}{2!} + \frac{x^4}{4!} - \frac{x^6}{6!}$, when $-1 < x < 1$.

This example makes the point that a graph can only *suggest* an identity. To show that a sentence is true for all values in its domain, a logical argument is needed.

ADDITIONAL EXAMPLES

1. Find another equation that appears to produce the same graph as $y = 1 - 2\sin^2 x$ for $-2\pi \le x \le 2\pi$, and conjecture a possible identity.
sample: $y = \cos 2x$; $1 - 2\sin^2 x = \cos 2x$

2. Is each equation given below an identity? If not, find a value of x for which the equation is not true.

a. $\sin\left(x + \frac{\pi}{6}\right)\sin\left(x - \frac{\pi}{6}\right) = \sin^2 x - \sin^2 \frac{\pi}{6}$
It seems to be an identity.

b. $(\cos x - \sin x)^2 = \cos^2 x - \sin^2 x$
It is not an identity; let $x = \frac{\pi}{6}$. Then the left side is $\frac{(\sqrt{3} - 1)^2}{4}$ and the right side is $\frac{1}{2}$.

ADDITIONAL EXAMPLES

3. Use an automatic grapher to help decide whether sec x sin x = tan x is an identity. If so, identify its domain. If not, find a counterexample.
It seems to be an identity. The domain is the set of all reals x such that $x \neq \frac{k\pi}{2}$, where k is an odd number; that is, the set of all reals x such that $x \not\equiv \frac{\pi}{2} \pmod{\pi}$ or $x \neq \frac{\pi}{2} + k\pi$, where k is an integer.

NOTES ON QUESTIONS

Questions 7-10: Students should verify their answers using an automatic grapher.

Making Connections for Question 13: You might ask students for what interval $4x(1 - x)$ is a good approximation for $\sin \pi x$. Remind them of the Weierstrass Approximation Theorem (Lesson 4-9).

Question 15: This question points out that to check a proposed identity in two variables by graphing in 2-space, one of the variables must be held constant. The proposed identity can then be graphed as before. However, students should be cautioned against choosing just one value for the variable being held constant. Several different values should be used to guard against the possibility that the sentence is true in special cases. There are 3-dimensional graphers in which the left side could be graphed as $z = \sin(x + y) + \sin(x - y)$ and the right side as $x = 2 \sin x \cos y$. Three software packages that will do this are *Derive, Mathematica,* and *Theorist.*

Although the graph in Example 2 does not suggest an identity, the graph can be used to help solve the equation $(\cos x + \sin x)^2 = \cos^2 x + \sin^2 x$. The solutions are the x-coordinates of the points of intersection of the graphs. The graphs suggest that the solutions are at $x = \frac{k\pi}{2}$ for all integers k. In Lesson 6-7, the algebraic solution to this equation is discussed.

One limitation of automatic graphers is that they may not show gaps in the graph of the function at values for which the function is undefined.

Example 3 **a.** Use an automatic grapher to help decide whether

$$\cos x \tan x = \sin x$$

is an identity.

b. If it seems to be an identity, identify its domain; if it is not an identity, give a counterexample.

Solution

a. Below is the graph of $y = \cos x \tan x$.

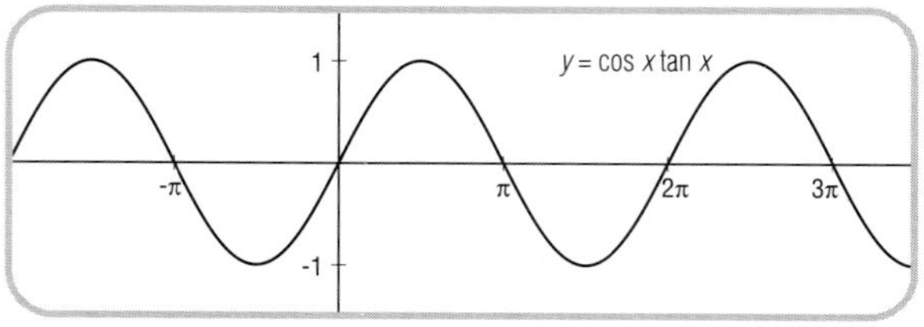

Below is the graph of $y = \sin x$.

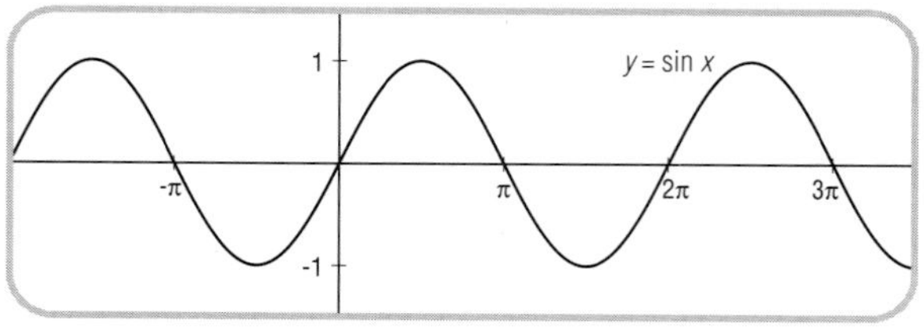

It appears as if the two graphs are identical. This suggests that the proposed identity is, in fact, an identity.

b. The domain of the identity is the set of values for which both sides are defined. The sine and cosine are defined for all real numbers. However, tan x is undefined when $x = \frac{\pi}{2} \pm k\pi$, for all integers k. Thus, the domain of the identity is the set of all real numbers x such that $x \neq \frac{\pi}{2} + k\pi$. Written in modular arithmetic, the restriction is $x \not\equiv \frac{\pi}{2} \pmod{\pi}$. Notice that you could not tell this restriction from the graphs in part **a**.

Since in this chapter the equations and identities you will be working with will be trigonometric, it will be necessary for you to remember the domains of the basic functions. Refer to pages 130 and 321 if you need to.

348

Questions

Covering the Reading

1. What is an identity? **an equation which is true for all values of the variable for which both sides are defined**

In 2–4, give the domain of the identity.

2. $\cot x = \frac{\cos x}{\sin x}$ **$x \neq 0 \pmod{\pi}$**

3. $\sin^2 x = 1 - \cos^2 x$ **all real numbers**

4. $\log t^3 = 3\log t$ **$t > 0$**

5. What identity did Christopher Clavius use to convert multiplications into additions and subtractions? **$2 \sin x \sin y = \cos(x - y) - \cos(x + y)$**

6. Use the information in the Chapter Opener (and not your calculator) to estimate 5,842,497 • 1,950,903. **≈ 11,398,150,000,000**

In 7–10, a graph drawn by an automatic grapher is shown. What identity is suggested by the graph? (Hint: Think about the graphs of the six basic trigonometric functions.)

7. $y = \sin^2 x + \cos^2 x$

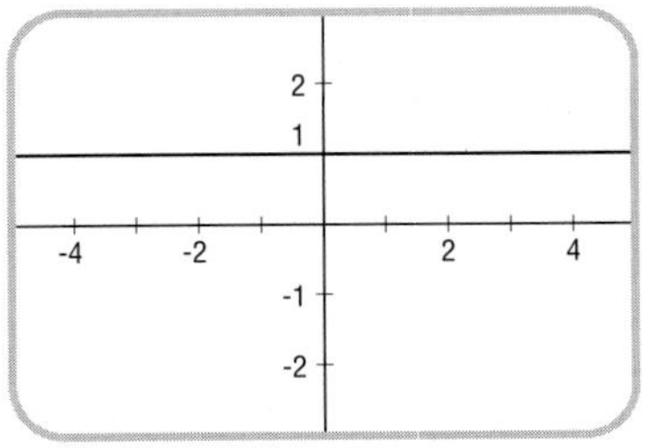

$\sin^2 x + \cos^2 x = 1$

8. $y = \csc x \tan x \sin x$

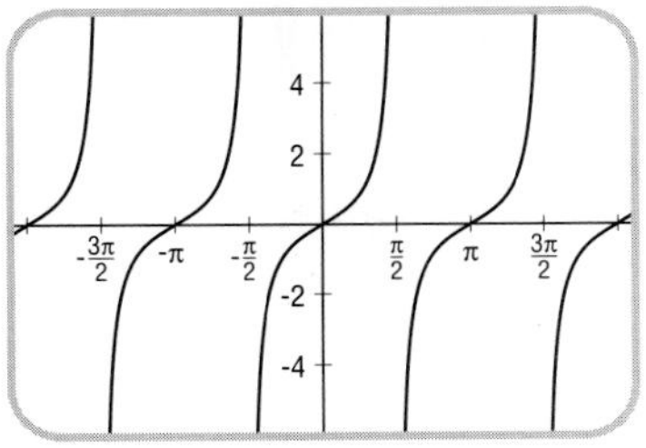

$\csc x \tan x \sin x = \tan x$

9. $y = \tan x \csc x$

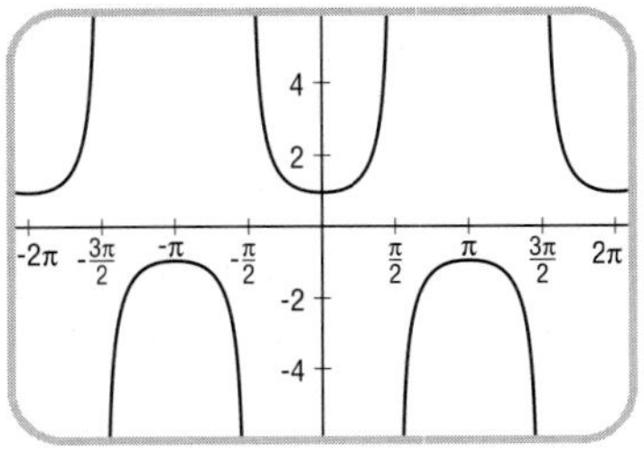

$\tan x \csc x = \sec x$

10. $y = \cos\left(\frac{3\pi}{2} + x\right)$

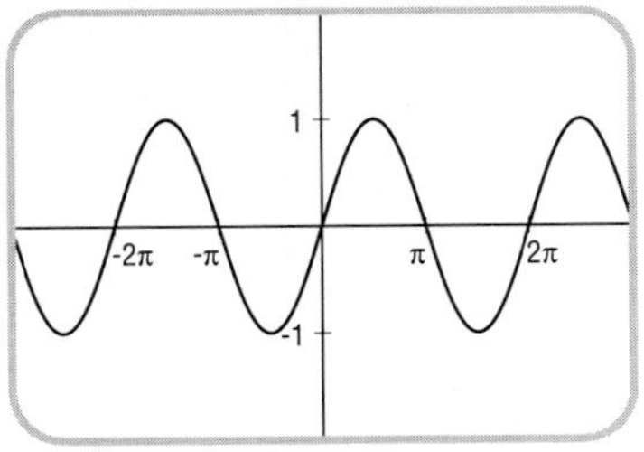

$\cos\left(\frac{3\pi}{2} + x\right) = \sin x$

In 11 and 12, **a.** graph the two sides of the equation as separate functions on the same set of axes. **b.** Use the graph to help decide whether the equation appears to be an identity. If it does not appear to be an identity, show it is not by giving a counterexample. If it appears to be an identity, specify its domain.

11. $\sec^2 x - \tan^2 x = 1$ **a) See margin. b) identity, domain: $x \neq \frac{(2n+1)\pi}{2}$, $\forall$ integers n**

12. $\sin\left(\frac{\pi}{2} - x\right) = \sin\left(x - \frac{\pi}{2}\right)$ **See margin.**

Question 26: Ask students to graph $y = \sin x$ and $y = x - \frac{x^3}{6} + \frac{x^5}{120}$ on the intervals $-1 \le x \le 1$, $-2 \le x \le 2$, $-3 \le x \le 3$, and $-4 \le x \le 4$. What conclusions can be arrived at from this activity? (The function with equation $y = x - \frac{x^3}{6} + \frac{x^5}{120}$ can be used to obtain approximate values for the sine function, provided x is close to 0.) In fact, the sequence of functions defined as follows yields better and better approximations to the sine function near $x = 0$.

$f_1(x) = \frac{x^1}{1!}$

$f_1(x) = \frac{x^1}{1!} - \frac{x^3}{3!}$

$f_1(x) = \frac{x^1}{1!} - \frac{x^3}{3!} + \frac{x^5}{5!}$

$f_1(x) = \frac{x^1}{1!} - \frac{x^3}{3!} + \frac{x^5}{5!} - \frac{x^7}{7!}$

$f_1(x) = \frac{x^1}{1!} - \frac{x^3}{3!} + \frac{x^5}{5!} - \frac{x^7}{7!} + \frac{x^9}{9!}$

ADDITIONAL ANSWERS

11.a.

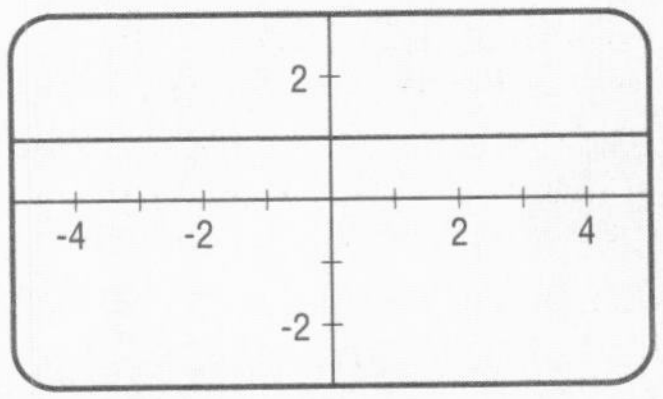

12.a.

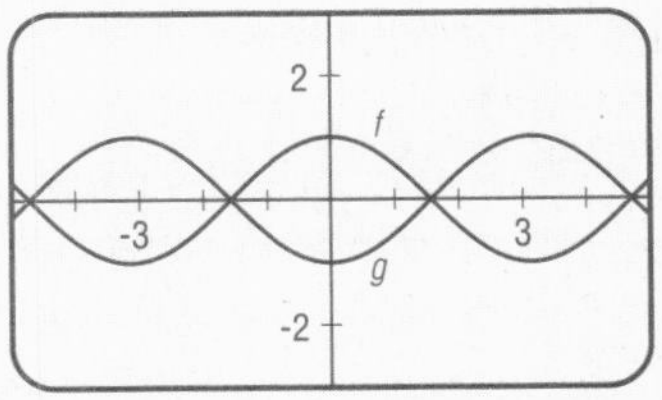

$f(x) = \sin\left(\frac{\pi}{2} - x\right)$; $g(x) = \sin\left(x - \frac{\pi}{2}\right)$

b. This is not an identity. Sample counterexample: Let $x = 0$. Then $\sin\left(\frac{\pi}{2} - 0\right) = 1$, but $\sin\left(0 - \frac{\pi}{2}\right) = -1$.

ADDITIONAL ANSWERS

13.a. It is not an identity.

b. Sample: Let $x = 0.2$. Then $\sin(0.2\pi) = 0.588$, but $4 \cdot 0.2 \cdot (1 - 0.2) = 0.64$.

14.a. $\sin\left(2 \cdot \frac{\pi}{4}\right) = 1$, and $2 \sin\frac{\pi}{4} \cos\frac{\pi}{4} = 2 \cdot \frac{\sqrt{2}}{2} \cdot \frac{\sqrt{2}}{2} = 1$; $\sin\left(2 \cdot \frac{\pi}{3}\right) = \frac{\sqrt{3}}{2}$, and $2 \sin\frac{\pi}{3} \cos\frac{\pi}{3} = 2 \cdot \frac{\sqrt{3}}{2} \cdot \frac{1}{2} = \frac{\sqrt{3}}{2}$; $\sin(2 \cdot \pi) = 0$, and $2 \sin \pi \cos \pi = (2)(0)(-1) = 0$.

b.

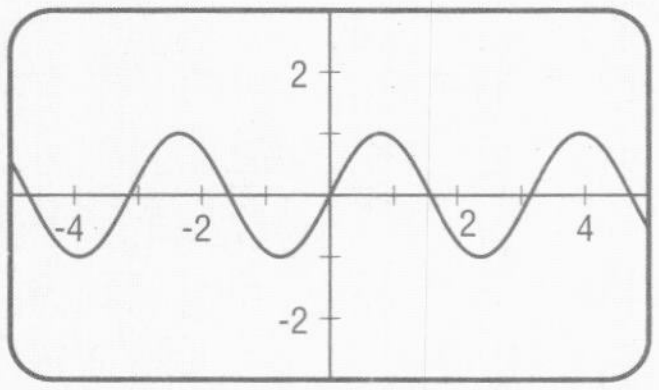

15.a. the graph of $f(\alpha) = \sin(\alpha + 0) + \sin(\alpha - 0)$ and $g(\alpha) = 2 \sin \alpha \cos 0$

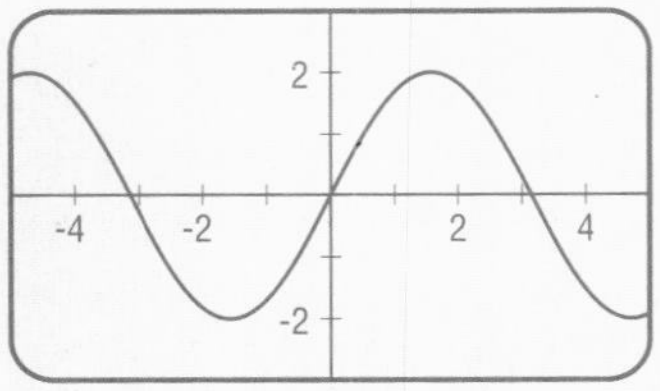

b. the graph of $f(\alpha) = \sin\left(\alpha + \frac{\pi}{6}\right) + \sin\left(\alpha - \frac{\pi}{6}\right)$ and $g(\alpha) = 2 \sin \alpha \cos \frac{\pi}{6}$

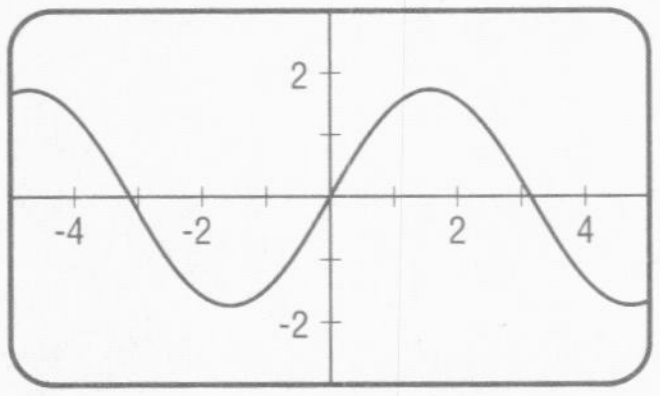

c. See Additional Answers in the back of this book.

Applying the Mathematics

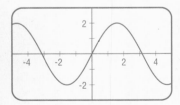

13. a. Use an automatic grapher to conjecture whether or not $4x(1 - x) = \sin \pi x$ is an identity for $0 \le x \le 1$.

b. By substituting a particular value for x, test your conjecture.
a,b) See margin.

14. Consider the proposed identity: $\sin 2\theta = 2 \sin \theta \cos \theta$.

a. Check that the expressions on the two sides of the equation give the same result for $\theta = \frac{\pi}{4}$, $\theta = \frac{\pi}{3}$, and $\theta = \pi$. See margin.

b. Use an automatic grapher to graph the two equations $y = \sin 2\theta$ and $y = 2 \sin \theta \cos \theta$. Do they seem to have identical graphs? Yes. See margin.

15. Graphs can be used to explore whether

$$\sin(\alpha + \beta) + \sin(\alpha - \beta) = 2 \sin \alpha \cos \beta$$

may be an identity, even though it has two variables. Graph each side of the proposed identity as a function of α when

a. $\beta = 0$. **b.** $\beta = \frac{\pi}{6}$. **c.** $\beta = \frac{\pi}{2}$. a,b,c) See margin.

d. Do the graphs appear to coincide for the tested values of α and β?
Yes

Review

In 16–18, find the exact value. *(Lessons 5-6, 2-8)*

16. $\cos\left(\frac{5\pi}{6}\right)$ $-\frac{\sqrt{3}}{2}$ **17.** $\sin\left(-\frac{\pi}{3}\right)$ $-\frac{\sqrt{3}}{2}$ **18.** $\tan\left(\frac{5\pi}{3}\right)$ $-\sqrt{3}$

19. Given that $2t - 1$ is a factor of $h(t) = 2t^3 - t^2 - 8t + 4$, find all zeros of h. *(Lessons 4-6, 4-5, 3-4)* $\frac{1}{2}$, ± 2

20. Use the language of congruences modulo π to express the domain of the cotangent function and the secant function. *(Lesson 4-3)*
cot: $\{x : x \not\equiv 0 \pmod{\pi}\}$; sec: $\{x : x \not\equiv 0 \pmod{\frac{\pi}{2}}\}$

In 21 and 22, rules are given to define functions f and g. In each case, compute and simplify formulas for $f + g$, $f - g$, $f \cdot g$, and $\frac{f}{g}$. Find the domains for these functions. *(Lesson 3-3)*

21. $f(x) = 2x + 1 - \frac{1}{x}$; $g(x) = x^2 + \frac{1}{x}$
See margin.

22. $f(t) = e^t$; $g(t) = e^{-t}$
See margin.

23. The circles at the right are concentric and $m\angle AOB = m\angle COD$. Use triangle congruence to prove that $AB = CD$.
(Previous course) See margin.

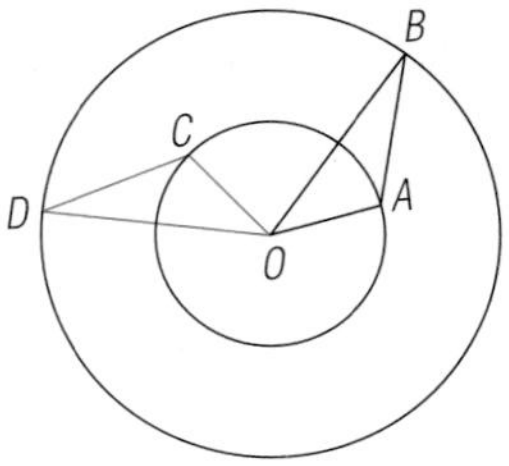

24. *Multiple choice.* The local high school marching band sponsored a sale of gift wrap. Gift wrap is shipped in cartons of 12 boxes each. If the band has taken orders for b boxes of birthday gift wrap, which of the following gives the number of cartons the band must order? (Recall that $f(x) = \lfloor x \rfloor$ is the floor function and $g(x) = \lceil x \rceil$ is the ceiling function.) *(Lesson 3-5, Previous course)*

(a) $\left\lfloor \frac{b}{12} \right\rfloor$ (b) $\left\lfloor \frac{12}{b} \right\rfloor$ (c) $\left\lceil \frac{b}{12} \right\rceil$ (d) $\left\lceil \frac{12}{b} \right\rceil$ (c)

25. An angle is said to be "in Quadrant I" if it has a radian measure θ in the interval $0 < \theta < \frac{\pi}{2}$. Use an interval to describe the radian measures between 0 and 2π of angles in each of the other three quadrants.
(Previous course)
See margin.

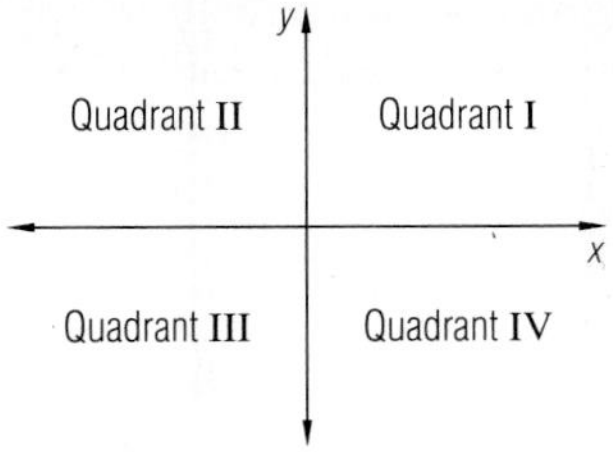

Exploration

26. a. Recall from Lesson 4-9 that $\sin x \approx x - \frac{x^3}{6} + \frac{x^5}{120}$ for some values of x. Write and run a computer program to generate the values for the table below to determine how close an approximation this is.

x	$\sin x$	$x - \frac{x^3}{6} + \frac{x^5}{120}$	$\left\|\sin x - (x - \frac{x^3}{6} + \frac{x^5}{120})\right\|$
-3.0	-0.141	-0.525	0.384
-2.5	-0.598	-0.710	0.111
-2.0	-0.909	-0.933	0.024
-1.5	-0.997	-1.001	0.003
-1.0	-0.841	-0.842	0.0002
-0.5	-0.479	-0.479	0.0000
0	0.000	0.000	0.0000
0.5	0.479	0.479	0.0000
1.0	0.841	0.842	0.0002
1.5	0.997	1.001	0.003
2.0	0.909	0.933	0.024
2.5	0.598	0.710	0.111
3.0	0.141	0.525	0.384

b. Based on the table in part **a**, determine an interval on which the approximation is accurate to within 0.05. $-2 \le x \le 2$

FOLLOW-UP

MORE PRACTICE
For more questions on SPUR Objectives, use *Lesson Master 6-1,* shown below.

EXTENSION
You might define a *near-identity* to be an equation of the form $f(x) = g(x)$, where $|f(x) - g(x)| < .01$ for every value of x. Then ask, for the equations of **Questions 13 and 26**, over what intervals are the equations near identities? ($0 < x < .01$, $.4 < x < .6$, $.99 < x < 1$)

ADDITIONAL ANSWERS
21. $(f + g)(x) = x^2 + 2x + 1, x \neq 0;$
$(f - g)(x) = 2x + 1 - x^2 - \frac{2}{x}, x \neq 0;$
$(f \cdot g)(x) = 2x^3 + x^2 - x + 2 + \frac{1}{x} - \frac{1}{x^2}, x \neq 0;$
$\left(\frac{f}{g}\right)(x) = \frac{2x - 1}{x^2 - x + 1}, x \neq 0, -1$

22., 23., 25. See Additional Answers in the back of this book.

NAME ____________

LESSON **MASTER 6–1**
QUESTIONS ON SPUR OBJECTIVES

■ **REPRESENTATIONS** *Objective G (See pages 396–398 for objectives.)*
In 1–6, use an automatic grapher to determine whether the equation appears to be an identity. If so, conjure the domain of the identity. If not, give a counterexample.

1. $1 + \cos^2 x = 2 \sin^2 \left(x + \frac{\pi}{2}\right)$
no; sample: $x = \frac{3\pi}{2}$

2. $\tan y = \cot \left(\frac{\pi}{2} - y\right)$
yes; $\{x : x \neq \frac{(2k + 1)\pi}{2}$ where k is an integer$\}$

3. $\sin x \cdot \tan x = \sec x + \cos x$
no; sample: $x = 0$

4. $\sin x \cdot \tan x = \sec x - \cos x$
yes; $\{x : x \neq \frac{(2k + 1)\pi}{2}$ where k is an integer$\}$

5. $\sin 3z = 3 \sin z \cos z$
no; sample: $z = \frac{\pi}{2}$

6. $\csc u = -\csc (\pi + u)$
yes; $\{u : u \neq k\pi$ where k is an integer$\}$

7. Use an automatic grapher to decide whether $\tan (\alpha + \beta) = \tan \alpha + \tan \beta$, for $\alpha = \frac{\pi}{4}$, $\alpha = \frac{\pi}{2}$, and a value of your choice for α.
no, no, no

8. *True* or *false*? Identities can be proved using an automatic grapher.
false

LESSON 6-2

RESOURCES

■ Lesson Master 6-2

OBJECTIVES

A Without a calculator, use trigonometric identities to express values of trigonometric functions in terms of rational numbers and radicals.

D Prove trigonometric identities and identify their domains.

G Use an automatic grapher to test proposed trigonometric identities.

TEACHING NOTES

The four techniques discussed in this lesson all use the same identity. By using the same problem four different times, students can compare and contrast the different methods of proof.

In Technique 1, we write "Left side" and then what it equals. This helps students focus on the idea of beginning with one side of the equation and transforming it until it equals the other side, at which time "Right side" is written. These words are for recording only and are not part of the proof.

In Technique 2, known identities are used to transform both sides simultaneously. Insist, however, that students draw a vertical line between the two sides until they are able to show the two sides that yield the same result. The vertical line reminds students that they cannot add a quantity to both sides or multiply both sides by a quantity.

LESSON 6-2

Proving Identities

The equation

$$\sin x = \frac{2 \sin^3 x + \sin 2x \cos x}{4 \sin^2 x + 2 \cos 2x}$$

is an identity because it is true for all real numbers x. This means that if the right side occurred in a formula, you could replace it with $\sin x$. But how would you know that these expressions are equal for all x?

In the last lesson you saw that graphing can suggest that an equation is an identity. In this lesson, you will learn basic algebraic techniques for proving identities. Later in this chapter, you will learn techniques that can be applied to derive the particular identity above.

There are a few general principles that you should keep in mind when proving identities. First, there is no new or special logic for proving identities. *Identities are equations and all the logic that was discussed with equation-solving applies to them.* Second, you will utilize some of the techniques that you have used in doing earlier proofs in this book. For instance, you will often make use of definitions to rewrite expressions.

In this lesson, you will see various ways to prove the same identity:

$$1 + \tan^2 x = \sec^2 x.$$

You may find yourself using these techniques again and again in this chapter, so you should read the proofs carefully.

Technique 1: Rewrite one side using definitions, known identities, and algebraic properties until it equals the other side.

Proof We start with the left side of the proposed identity and rewrite it until it is transformed into the right side.

For all real numbers x for which $\tan x$ is defined,

Left side	$= 1 + \tan^2 x$	
	$= 1 + \dfrac{\sin^2 x}{\cos^2 x}$	Definition of tangent
	$= \dfrac{\cos^2 x}{\cos^2 x} + \dfrac{\sin^2 x}{\cos^2 x}$	Forming a common denominator
	$= \dfrac{\cos^2 x + \sin^2 x}{\cos^2 x}$	Adding fractions with a common denominator
	$= \dfrac{1}{\cos^2 x}$	Pythagorean Identity
	$= \sec^2 x$	Definition of secant
	= Right side.	

Therefore,

$1 + \tan^2 x = \sec^2 x$ By the Transitive Property of Equality.

352

Technique 2: Rewrite each side *independently* until expressions are obtained that are known to be equal.

When using this technique, since you cannot be sure that the proposed identity actually is an identity until you have finished, you should not write an equal sign between the two sides until the very end. We draw a vertical line between the two sides as a reminder of this restriction. A proof will, therefore, have the following form.

$$\begin{array}{c|c} A & B \\ = \ldots & = \ldots \\ = \ldots & = \ldots \\ = E & = E \end{array}$$
$$\therefore A = B$$

The steps of the derivation must show that $A = E$ and that $B = E$. The fact that $A = B$ then follows by the symmetric and transitive properties of equality.

Proof For all real numbers x for which $\tan x$ and $\sec x$ are defined,

	$1 + \tan^2 x$	$\sec^2 x$	
Def. of tangent	$= 1 + \frac{\sin^2 x}{\cos^2 x}$	$= \frac{1}{\cos^2 x}$	Def. of secant
Adding fractions	$= \frac{\cos^2 x + \sin^2 x}{\cos^2 x}$		
Pythagorean Identity	$= \frac{1}{\cos^2 x}$		

$\therefore$ $1 + \tan^2 x = \sec^2 x$ Transitive Property of Equality

Technique 3: Begin with a known identity and transform it until the desired identity appears.

A natural identity to use is the Pythagorean Identity $\sin^2 x + \cos^2 x = 1$. Because $\tan x = \frac{\sin x}{\cos x}$, the quantity $\tan x$ can be introduced into the Pythagorean Identity by dividing both sides by $\cos^2 x$.

Proof **For all real numbers x,**

	$\sin^2 x + \cos^2 x = 1$	Pythagorean Identity
$\Rightarrow$	$\frac{\sin^2 x}{\cos^2 x} + \frac{\cos^2 x}{\cos^2 x} = \frac{1}{\cos^2 x}$	Provided $\cos x \neq 0$
$\Rightarrow$	$\tan^2 x + 1 = \sec^2 x$	Definition of tangent, secant

This shows that the equation $\tan^2 x + 1 = \sec^2 x$ holds for all x for which $\cos x \neq 0$.

To be successful with Technique 3, students must be able to choose the appropriate identity to start the proof. Most students lack the experience to make the proper choice. Students should not become discouraged if their choice for a starting identity fails to lead to the desired result. Point out that mathematicians often try many paths before finding the one that is successful in obtaining new knowledge.

Point out to students that the last line of Technique 4 is the first line of Technique 3. Mathematicians will sometimes do a proof "backwards." They begin with Technique 4 and apply reversible steps until an identity is found. For publication, they begin with the identity derived by Technique 4. This should explain how, in **Example 3**, $\sin^2 x + \cos^2 x = 1$ was a natural identity to use.

Invite students to start a notebook of trigonometric identities. New identities can be listed as they are proved. Students should keep a record of how each identity was proved. Then these identities and techniques of proof can be used to prove more identities. Each identity should be accompanied by all restrictions on its domain.

ADDITIONAL EXAMPLE
Use the identity $\cot^2 x + 1 = \csc^2 x$ (from **Questions 1-3**) for class discussion, having students attempt the proof by each of the four strategies. Students could put their proofs on the chalkboard for the class to critique.

NOTES ON QUESTIONS

Question 10: To help students determine signs of the solutions, the following diagram may be helpful.

II sin (csc)	all I
III tan (cot)	cos IV (sec)

The trigonometric functions whose values are positive within each quadrant are labeled.

Question 11: Ask students why the graphs of both equations are never below the x-axis. (Squares of real numbers are nonnegative.)

Question 13: In this question, students classify each trigonometric function as odd or even. This can be helpful for determining whether a given equation is an identity for all real numbers x. If one side is a formula for an odd function and the other for an even function, they cannot be equal for all x.

Question 20: Ask students why geologists do not drill above the point of the oil-bearing rock near the surface. (because they would run out of oil more quickly)

Question 23: The exploration question introduces students to hyperbolic sines and cosines. Some scientific calculators contain a "hyp" key that is used in conjunction with the "cos" and "sin" keys to evaluate the hyperbolic functions. Graphics calculators also can be used to graph these functions. The definitions provide a useful vehicle for reviewing the exponential functions:
$\cosh x = \frac{e^x + e^{-x}}{2}$ and
$\sinh x = \frac{e^x - e^{-x}}{2}$.

Technique 4: Transform both sides of the equation to be proved using reversible steps until an equation known to be an identity appears.

When using this technique you must keep the domain of the identity in mind, and you must be careful to check the reversibility of each step.

Proof For all real numbers x for which $\cos x \neq 0$,

	$\tan^2 x + 1 = \sec^2 x$	
$\Leftrightarrow$	$\frac{\sin^2 x}{\cos^2 x} + 1 = \frac{1}{\cos^2 x}$	Definitions of tangent and secant
$\Leftrightarrow$	$\cos^2 x\left(\frac{\sin^2 x}{\cos^2 x} + 1\right) = \cos^2 x \cdot \frac{1}{\cos^2 x}$	Multiplication Property of Equality (reversible since $\cos x \neq 0$)
$\Leftrightarrow$	$\sin^2 x + \cos^2 x = 1$	Pythagorean Identity

Because $\tan x = \frac{\sin x}{\cos x}$ and $\sec x = \frac{1}{\cos x}$, both $\tan x$ and $\sec x$ are undefined when $\cos x = 0$. It follows that the domain of the identity is the set of all x for which $\cos x \neq 0$. But $\cos x = 0$ for all numbers of the form $\frac{\pi}{2} + n\pi$ where n is an integer. Consequently, the domain is the set of all x such that $x \neq \frac{\pi}{2} + n\pi$ for any integer n; that is, the set of all x such that $x \not\equiv \frac{\pi}{2} \pmod{\pi}$.

When using technique 4, it is essential that you explicitly indicate the reversibility of the steps. You must take care when multiplying both sides of the equation by an expression containing a variable. Unless you have explicitly excluded values of the variable that make the expression equal 0 (as we did above by noting $\cos x \neq 0$), the step will not be reversible.

When doing proofs, unless you are asked to use a particular technique, you can use any of these ways.

All the above proofs at some step use the Pythagorean Identity: $\sin^2 x + \cos^2 x = 1$. At this point, you are expected to know that identity, the definitions of the trigonometric functions, and the identities $\sin(-x) = -\sin x$ and $\cos(-x) = \cos x$. In the next two lessons, these properties are employed to deduce some additional very powerful and important identities.

Questions

Covering the Reading

1. **a.** Use Technique 1 to prove that $\cot^2 x + 1 = \csc^2 x$ for all real numbers x for which both sides are defined. **See margin.**
   b. What is the domain of the identity in part **a**? $x \neq n\pi$, $\forall$ integers n
2. Use Technique 2 to prove the identity in Question 1. **See margin.**
3. Use Technique 3 or 4 to derive the identity of Question 1. (Identify which technique you use.) **See margin.**

4. The following is an identity: $\sin x \cot x = \cos x$.
   a. Give the domain of the identity. $x \neq n\pi$, $\forall$ integers n
   b. Prove the identity. See margin.

Applying the Mathematics

In 5-8, prove the identity and give its domain.

5. $\cos x \tan x = \sin x$ See margin.

6. $\tan x \cot x = \cos^2 x + \sin^2 x$ See margin.

7. $\csc^2 x \sin x = \frac{\sec^2 x - \tan^2 x}{\sin x}$ See margin.

8. $\tan x + \cot x = \sec x \csc x$ See margin.

9. Use an automatic grapher or calculator to determine whether $\sin^2 x (\cot^2 x + 1) = 1$ might be an identity. If it is, prove it; if it is not, give a counterexample. See margin.

10. Suppose x is in the interval $\frac{3\pi}{2} < x < 2\pi$ and $\tan x = -\frac{3}{8}$. Use trigonometric identities to determine
    a. $\cot x$ $-\frac{8}{3}$ b. $\sec x$ $\frac{\sqrt{73}}{8}$ c. $\sin x$. $\frac{-3}{\sqrt{73}}$

11. Here is a graph of $y = \tan^2 x$. Use the identity $\tan^2 x + 1 = \sec^2 x$ to sketch a graph of $y = \sec^2 x$. See margin.

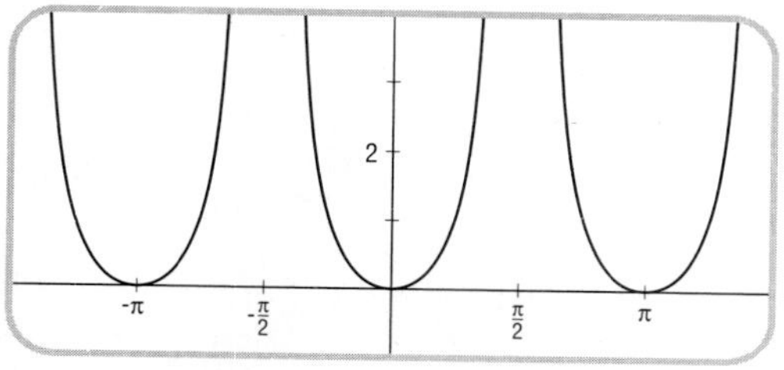

Review

12. Solve $\frac{1}{x} + \frac{4}{x^2} = \frac{x + 4}{5x}$. *(Lesson 5-7)* $x = 5$ or $x = -4$

13. Let f be one of the six trigonometric functions sin, cos, tan, cot, sec, csc.
    a. For which of these functions is $f(-x) = -f(x)$ for all x?
    b. For which of these functions is $f(-x) = f(x)$ for all x? cos x, sec x
    (Lessons 5-6, 2-8) a) sin x, tan x, csc x, cot x

14. Give the period of the function $f: x \rightarrow y$ defined by the given equation. *(Lessons 3-9, 2-8)*
    a. $y = \sin x$ 2π b. $y = \sin 2x$ π c. $y = 2 \sin x$ 2π

15. Simplify and state any restrictions on the variable. *(Lesson 5-1)*

$$\frac{\frac{3}{x} - \frac{9}{x^2}}{\frac{1}{3x} - 1} \qquad \frac{9(x-3)}{x(1-3x)},\ x \neq 0,\ x \neq \frac{1}{3}$$

FOLLOW-UP

MORE PRACTICE

For more questions on SPUR Objectives, use *Lesson Master 6-2,* shown below.

ADDITIONAL ANSWERS

1.a., 2., 3., 4.b., 5-9., 11.
See Additional Answers in the back of this book.

NAME ________________

LESSON **MASTER 6–2**
QUESTIONS ON **SPUR** OBJECTIVES

■SKILLS *Objective A (See pages 396–398 for objectives.)*

1. Suppose x is in the interval $\frac{\pi}{2} < x < \pi$ and $\sin x = \frac{\sqrt{2}}{\sqrt{3}}$. Use trigonometric identities to find each value.
   a. $\sec x$ $-\sqrt{3}$ b. $\tan x$ $-\sqrt{2}$
   c. $\cos x$ $-\frac{\sqrt{2}}{2}$ d. $\csc x$ $\frac{\sqrt{3}}{\sqrt{2}}$

2. If $\sin \alpha = -.8$ and $\tan \alpha = 2.1$, find $\sec \alpha$. ≈ -2.63

■PROPERTIES *Objective D*
In 3–5, prove the identity and specify its domain.

3. $\sec x \cdot \cos x = \sin x \csc x$

$\sec x \cos x$	$\sin x \csc x$
$\left(\frac{1}{\cos x}\right) \cos x$	$\sin x \left(\frac{1}{\sin x}\right)$

$1 = 1$
$\{x: x \neq \frac{k\pi}{2}, k \text{ is an integer}\}$

4. $\frac{1}{1 + \cos x} + \frac{1}{1 - \cos x} = 2 \csc^2 x$

L.S. $= \frac{(1 - \cos x) + (1 + \cos x)}{1 - \cos^2 x}$
$= \frac{2}{\sin^2 x}$
$= 2 \csc^2 x$
$\{x: x \neq k\pi, k \text{ is an integer}\}$

5. $\cot x \cdot \sin x \cdot \sec x = \cos^2 x + \sin^2 x$

$\cot x \sin x \sec x$	$\cos^2 x + \sin^2 x$
$\left(\frac{\cos x}{\sin x}\right) \sin x \left(\frac{1}{\cos x}\right)$	1

$1 = 1$
$\{x: x \neq \frac{k\pi}{2} \text{ where } k \text{ is an integer}\}$

6. Fill in the blank to make an identity: $\cos x \cdot \cot x + \sin x =$ $\csc x$.

■REPRESENTATIONS *Objective G*
In 7 and 8, use an automatic grapher to conjecture whether the equation is an identity. If so, prove it and identify its domain. If not, give a counterexample.

7. $\cot 2x + \cot x = \cot 3x$
   no; sample: $x = \frac{\pi}{4}$

8. $\frac{\cos x \cot x}{1 - \sin x} = 1 + \csc x$ yes;

$\frac{\cos x \cot x}{1 - \sin x}$	$1 + \csc x$
$\frac{\cos^2 x}{(1 - \sin x) \sin x}$	$1 + \frac{1}{\sin x}$
$\frac{1 - \sin^2 x}{(1 - \sin x) \sin x}$	$\frac{\sin x + 1}{\sin x}$
$\frac{1 + \sin x}{\sin x}$	

$\{x: x \neq k\pi \text{ and } x \neq \frac{(4k + 1)\pi}{2} \text{ where } k \text{ is an integer}\}$

ADDITIONAL ANSWERS

17.b. The set in part a is the set of all integers with remainder of 3 when divided by 7.

18. Let m be any even integer. There exists an integer k such that $m = 2k$ by definition of even. $m^2 = (2k)^2 = 4k^2$. Since k^2 is an integer, m^2 is divisible by 4 by definitions.

23.a. hyperbolic cosine; $\cosh x = \frac{1}{2}(e^x + e^{-x})$; hyperbolic sine: $\sinh x = \frac{1}{2}(e^x - e^{-x})$

b. samples: $\cosh^2 x - \sinh^2 x = 1$; $\frac{\sinh x}{\cosh x} = \tanh x$; $1 - \tanh^2 x = \text{sech}^2 x$

16. Consider the function f graphed at the right. Can f be a polynomial function of degree 4? Explain your answer. *(Lesson 4-6)* No, there is a horizontal line which crosses the graph 5 times, hence the degree must be greater than 4.

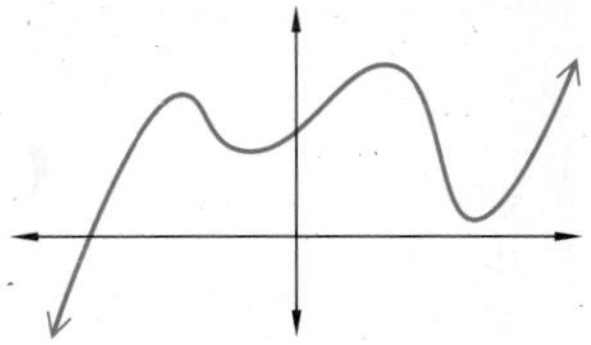

17. **a.** Here is a congruence class mod 7: {... , 3, 10, 17, 24, 31, ...}. Describe this congruence class algebraically.
b. Relate your description in part **a** to the results for division by 7 as expressed in the Quotient-Remainder Theorem. *(Lessons 4-3, 4-2)*
a) $\{n: n = 3 + 7q$ for some integer $q\}$; b) See margin.

18. Prove the following statement:
$\forall$ *even integers m, m^2 is divisible by 4.* *(Lessons 4-1, 1-7)*
See margin.

19. Let (a, b) be the image of $(1, 0)$ under a rotation of θ with center $(0,0)$.
a. Write a and b in terms of θ. $a = \cos\theta$, $b = \sin\theta$
b. Give the distance between $(1, 0)$ and (a, b).
(Lesson 2-8, Previous course)
$d = \sqrt{(a-1)^2 + b^2}$ or $\sqrt{2 - 2\cos\theta}$

20. Oil company geologists have determined that there is a rich strip of oil-bearing rock starting 100 ft underground and descending at 28° to the horizontal, as pictured below. To get the optimal amount of crude oil from this source, they decide to build the well 2500 ft away from the initial spot. How deep do they need to drill?
(Lesson 2-8) ≈ 1429 feet

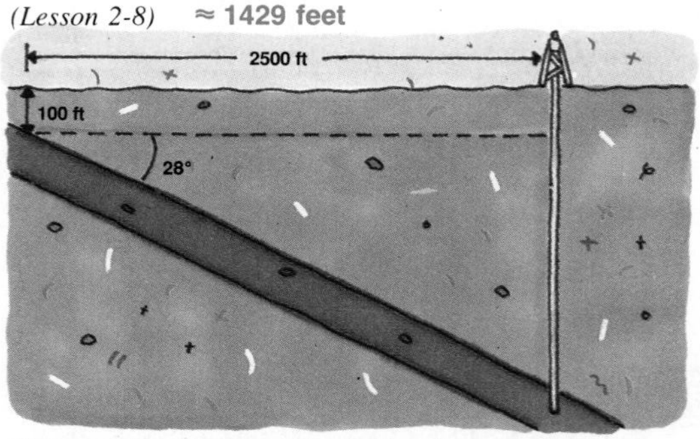

21. Determine whether the following argument is valid or invalid. Justify your reasoning. *(Lessons 1-8, 1-6)*
If Mr. Gutierrez is at the door, then the man at the door is short and has dark hair.
The man at the door is short but does not have dark hair.
Therefore, it is not Mr. Gutierrez. valid, Law of Indirect Reasoning

22. *Multiple choice.* $\forall$ all real numbers x and y, $x - y =$
(a) $y - x$ (b) $-(y - x)$ (c) $-x + y$
(d) $-(x - y)$ (e) $-(x + y)$. *(Previous course)* (b)

Exploration

23. There are functions called the *hyperbolic cosine* and *hyperbolic sine* functions.
a. Using other references, find how they are defined.
b. Find one or more hyperbolic identities analogous to the trigonometric identities mentioned in this lesson. a,b) See margin.

356

LESSON

6-3

Formulas for $\cos(\alpha+\beta)$ and $\cos(\alpha-\beta)$

In Lesson 2-8, calculation of exact values of the sine and cosine functions for the domain values 0, $\frac{\pi}{6}$, $\frac{\pi}{4}$, $\frac{\pi}{3}$, and $\frac{\pi}{2}$ was reviewed. In this lesson, you will see how to compute cosines for sums and differences of these domain values. For example, because

$$\frac{5\pi}{6} = \frac{\pi}{2} + \frac{\pi}{3} \quad \text{and} \quad \frac{\pi}{12} = \frac{\pi}{3} - \frac{\pi}{4},$$

you will be able to compute $\cos \frac{5\pi}{6}$ and $\cos \frac{\pi}{12}$ using formulas for $\cos(\alpha + \beta)$ and $\cos(\alpha - \beta)$.

You might hope that $\cos(\alpha + \beta)$ would equal $\cos \alpha + \cos \beta$ for all real numbers α and β, but one counterexample will prove that this proposed identity is false. For example, if $\alpha = 0$ and $\beta = \frac{\pi}{3}$,

$$\cos(\alpha + \beta) = \cos\left(0 + \frac{\pi}{3}\right) = \cos \frac{\pi}{3} = \frac{1}{2},$$

but

$$\cos \alpha + \cos \beta = \cos 0 + \cos \frac{\pi}{3} = 1 + \frac{1}{2} = \frac{3}{2}.$$

Therefore $\cos(\alpha + \beta) = \cos \alpha + \cos \beta$ is *not* an identity. The identity for $\cos(\alpha + \beta)$ is a little more complicated.

Theorem (Identity for $\cos(\alpha + \beta)$)

For all real numbers α and β,
$$\cos(\alpha + \beta) = \cos \alpha \cos \beta - \sin \alpha \sin \beta.$$

Proof The proof we give here applies when $\alpha > 0$, $\beta > 0$, and $\alpha + \beta < \pi$. Variations of this proof can be done to prove the identity for other values of α and β. Recall from Lesson 2-8 that the image of the point (1, 0) on the unit circle under a rotation of θ has coordinates $(\cos \theta, \sin \theta)$. The diagram shown below displays the images of $P = (1, 0)$ under rotations of α, $\alpha + \beta$, and $-\beta$. These images are labeled Q, R, and S, respectively. Notice that the first coordinate of R is $\cos(\alpha + \beta)$, which is the left side of the proposed identity.

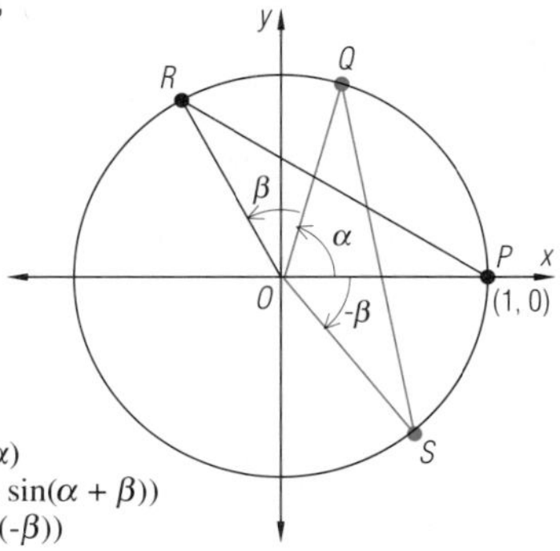

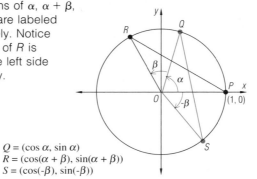

$Q = (\cos \alpha, \sin \alpha)$
$R = (\cos(\alpha + \beta), \sin(\alpha + \beta))$
$S = (\cos(-\beta), \sin(-\beta))$

LESSON 6-3

RESOURCES

- Lesson Master 6-3
- Quiz for Lessons 6-1 through 6-3
- Teaching Aid 37 displays the proof of identity for $\cos(\alpha + \beta)$.

OBJECTIVES

A Without a calculator, use trigonometric identities to express values of trigonometric functions in terms of rational numbers and radicals.

D Prove trigonometric identities and identify their domains.

TEACHING NOTES

This lesson gives students the tools they need to find exact values of the trigonometric functions for many other arguments.

Making Connections
The proof of the identity for $\cos(\alpha + \beta)$ is based on the unit circle and the distance formula. In some books, the identity for $\cos(\alpha - \beta)$ is proved first. We chose our method because the addition identity is more natural to try to prove first. There is also an elegant proof of this identity using matrices. The proof yields the identities for $\cos(\alpha + \beta)$ and $\sin(\alpha + \beta)$ simultaneously. The beginning of the proof is outlined in Question 12 of Lesson 6-4. Students who have used other UCSMP books should be familiar with matrix operations, and those who have

used *Functions, Statistics, and Trigonometry* will have seen the matrix proof in that text.

In the proof of the Cosine of a Sum Theorem, the recognition that $\cos^2(\alpha + \beta) + \sin^2(\alpha + \beta)$ is a form of $\cos^2 x + \sin^2 x$ allows a simplifying substitution to be made. Encourage students to look for disguised identities.

Examples 1 and 2 show two immediate applications of identities in two variables. By substituting numbers for both variables, a value may be obtained that has not been seen before. By substituting a number for just one variable, an identity in one variable may appear.

Error Analysis In finding $\cos \frac{7\pi}{12}$ using known values, many students have difficulty determining a useful sum equal to $\frac{7\pi}{12}$. Here are two suggestions for helping students overcome this difficulty. (1) Ask them to write $\frac{7\pi}{12}$ as the sum of two fractions, *both* of which reduce. Thus, possible numerators must be nontrivial factors of 12 (2, 3, 4, 6). Ask which pair(s) of these also add to 7. (2) Write $\frac{\pi}{6} = \frac{2\pi}{12}, \frac{\pi}{4} = \frac{3\pi}{12}$, and $\frac{\pi}{3} = \frac{4\pi}{12}$ on the chalkboard and ask which pair adds to $\frac{7\pi}{12}$.

If $x + y = \frac{\pi}{2}$, then x and y can be thought of as measures of two complementary angles. If $f(x) = g(y)$, then we say that f and g are **co**-functions because f (one angle) $= g$(its **co**mplement). For example, the sine and **co**sine functions are **co**functions. Ask students to write the pairs of cofunctions using complete names of the trigonometric functions.

The aim is to describe this expression in terms of sines and cosines of α and β. This suggests using the coordinates of Q and S. Notice also that $m\angle ROP = \alpha + \beta$ and $m\angle QOS = \alpha + \beta$. Thus, by the SAS Triangle Congruence Theorem, $RP = QS$.

Now translate $RP = QS$ into cosines and sines using the coordinates of these points and the distance formula, then solve for $\cos(\alpha + \beta)$.

$$\sqrt{(\cos(\alpha+\beta) - 1)^2 + (\sin(\alpha+\beta) - 0)^2} = \sqrt{(\cos\alpha - \cos(-\beta))^2 + (\sin\alpha - \sin(-\beta))^2}$$

$\cos(-x) = \cos(x)$ and $\sin(-x) = -\sin(x)\ \forall$ real x.

$$\sqrt{(\cos(\alpha+\beta) - 1)^2 + (\sin(\alpha+\beta))^2} = \sqrt{(\cos\alpha - \cos\beta)^2 + (\sin\alpha + \sin\beta)^2}$$

Square both sides.

$$(\cos(\alpha+\beta) - 1)^2 + (\sin(\alpha+\beta))^2 = (\cos\alpha - \cos\beta)^2 + (\sin\alpha + \sin\beta)^2$$

Expand the squared terms.

$$\cos^2(\alpha+\beta) - 2\cos(\alpha+\beta) + 1 + \sin^2(\alpha+\beta) = \cos^2\alpha - 2\cos\alpha\cos\beta + \cos^2\beta + \sin^2\alpha + 2\sin\alpha\sin\beta + \sin^2\beta$$

Rearrange the sums.

$$(\cos^2(\alpha+\beta) + \sin^2(\alpha+\beta)) - 2\cos(\alpha+\beta) + 1 = (\cos^2\alpha + \sin^2\alpha) + (\cos^2\beta + \sin^2\beta) - 2\cos\alpha\cos\beta + 2\sin\alpha\sin\beta$$

Pythagorean Identity

$$1 - 2\cos(\alpha+\beta) + 1 = 1 + 1 - 2\cos\alpha\cos\beta + 2\sin\alpha\sin\beta$$

Subtract 2.

$$-2\cos(\alpha+\beta) = -2\cos\alpha\cos\beta + 2\sin\alpha\sin\beta$$

Divide by -2.

$$\cos(\alpha+\beta) = \cos\alpha\cos\beta - \sin\alpha\sin\beta$$

Example 1 illustrates a situation in which the formula for the cosine of a sum is used to determine a value of the cosine function.

Example 1 Express $\cos\left(\frac{7\pi}{12}\right)$ in terms of rational numbers and radicals.

Solution Write $\frac{7\pi}{12}$ as a sum of two numbers for which you know exact values of the sine and cosine.

$$\frac{7\pi}{12} = \frac{\pi}{4} + \frac{\pi}{3}$$

Then apply the identity $\cos(\alpha + \beta) = \cos\alpha\cos\beta - \sin\alpha\sin\beta$ with $\alpha = \frac{\pi}{4}$ and $\beta = \frac{\pi}{3}$.

$$\begin{aligned}\cos\left(\frac{7\pi}{12}\right) &= \cos\left(\frac{\pi}{4} + \frac{\pi}{3}\right)\\ &= \cos\frac{\pi}{4}\cos\frac{\pi}{3} - \sin\frac{\pi}{4}\sin\frac{\pi}{3}\\ &= \frac{\sqrt{2}}{2}\bullet\frac{1}{2} - \frac{\sqrt{2}}{2}\bullet\frac{\sqrt{3}}{2}\\ &= \frac{\sqrt{2} - \sqrt{6}}{4}\end{aligned}$$

Check Use a calculator: $\cos\left(\frac{7\pi}{12}\right) \approx -.259$ and $\frac{\sqrt{2} - \sqrt{6}}{5} \approx -.259$. So it checks.

358

A formula for $\cos(\alpha - \beta)$ can be obtained easily from the formula for $\cos(\alpha + \beta)$ as follows: For all real numbers α and β,

$$\begin{aligned}\cos(\alpha - \beta) &= \cos(\alpha + (-\beta)) \\ &= \cos\alpha\cos(-\beta) - \sin\alpha\sin(-\beta) \\ &= \cos\alpha\cos\beta - \sin\alpha(-\sin\beta) \\ &= \cos\alpha\cos\beta + \sin\alpha\sin\beta.\end{aligned}$$

Replace β by $-\beta$ in the identity for $\cos(\alpha + \beta)$.
$\forall x$, $\cos(-x) = \cos x$ and $\sin(-x) = -\sin x$.

This proves the following theorem.

Theorem (Identity for $\cos(\alpha - \beta)$)

For all real numbers α and β,
$$\cos(\alpha - \beta) = \cos\alpha\cos\beta + \sin\alpha\sin\beta.$$

Sines and cosines are related in many ways. In a right triangle ABC with right angle at C, $\angle A$ and $\angle B$ are complements because their measures add to 90°. From the right triangle definitions of sine and cosine, $\cos A = \sin B$. That is, the cosine of an angle equals its complement's sine. More generally, if x and y are real numbers with $x + y = \frac{\pi}{2}$, then $y = \frac{\pi}{2} - x$ and from the formula for $\cos(\alpha - \beta)$ it can be proved that $\cos y = \sin x$. The next example shows this.

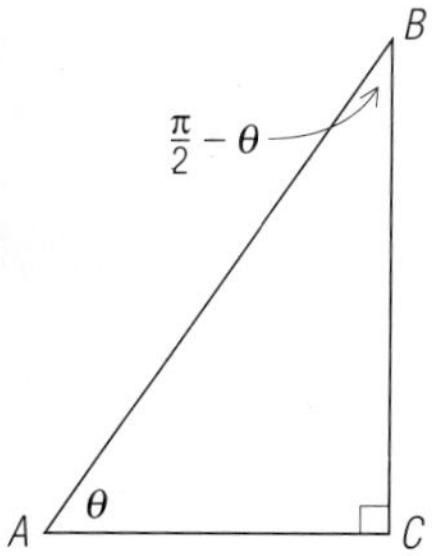

Example 2 Prove the following identity: $\forall$ real numbers x, $\cos\left(\frac{\pi}{2} - x\right) = \sin x$.

Solution Transform the left side using the formula for $\cos(\alpha - \beta)$ with $\alpha = \frac{\pi}{2}$ and $\beta = x$. Then use the exact values of $\cos\frac{\pi}{2}$ and $\sin\frac{\pi}{2}$.

Left side $= \cos\left(\frac{\pi}{2} - x\right)$
$= \cos\frac{\pi}{2}\cos x + \sin\frac{\pi}{2}\sin x$ — Formula for $\cos(\alpha - \beta)$
$= 0 \cdot \cos x + 1 \cdot \sin x$ — $\cos\frac{\pi}{2} = 0$, $\sin\frac{\pi}{2} = 1$
$= \sin x$
= Right side

The identity in Example 2 is called a **cofunction identity**. In the next lesson, it is applied to derive a formula for $\sin(\alpha + \beta)$.

In this and the next few lessons, many identities will be proved. You do not need to memorize all of them. If you memorize the key identities, you can quickly derive the others. One of the identities you should memorize is the formula for the cosine of a sum,
$$\cos(\alpha + \beta) = \cos\alpha\cos\beta - \sin\alpha\sin\beta,$$
because its derivation is so lengthy.

We use α and β when there are two variables to make it easier when there will be one variable. For instance, with $\cos(\alpha + \beta) = \cos\alpha\cos\beta - \sin\alpha\sin\beta$, replace α by π and replace β by x. Then $\cos(\pi + x) = \cos\pi\cos x - \sin\pi\sin x$.

ADDITIONAL EXAMPLES

1. Express $\cos\left(\frac{5\pi}{12}\right)$ in terms of rational numbers and radicals.
$\cos\frac{5\pi}{12} = \cos\left(\frac{2\pi}{12} + \frac{3\pi}{12}\right) = \cos\left(\frac{\pi}{6} + \frac{\pi}{4}\right) = \frac{\sqrt{6} - \sqrt{2}}{4}$

2. Prove the following identity: $\cos(\pi - x) = -\cos(x)$.
$\cos(\pi - x) = \cos\pi \cdot \cos(x) + \sin\pi \cdot \sin(x) = -1 \cdot \cos(x) + 0 \cdot \sin(x) = -1\cos(x) = -\cos(x)$

NOTES ON QUESTIONS

Question 4: Students are expected to use the identity $\cos\left(\frac{\pi}{2} - x\right) = \sin x$ and translate it to degrees. But they could use the right triangle definition of sine and cosine, or the unit circle definition and symmetry.

Question 7: Suggest using Technique 1 of Lesson 6-2. Have students work on the right side of the identity.

Question 11: Since $\frac{17\pi}{12} = \frac{\pi}{6} + \frac{5\pi}{4} = \frac{\pi}{4} + \frac{7\pi}{6} = \frac{2\pi}{3} + \frac{3\pi}{4}$, find out which pair was used the most by students and why.

Question 19d: With the substitution $x = \frac{\theta}{2}$, this identity becomes $\cos\frac{\theta}{2} = \pm\sqrt{\frac{1 + \cos\theta}{2}}$, a half-angle formula.

ADDITIONAL ANSWERS

1. *OP, OQ, OR, OS* are all radii, so they are equal. $m\angle POR = \alpha + \beta = m\angle QOS$. So, $\triangle ROP \cong \triangle QOS$ (Side-Angle-Side congruence). Hence, $PR = QS$.

5. $\cos\left(\frac{3\pi}{2} + x\right) =$
$\cos\frac{3\pi}{2} \cdot \cos x - \sin\frac{3\pi}{2} \cdot \sin x =$
$0 - (-1) \cdot \sin x = \sin x$

6. $\cos\left(\frac{\pi}{4} + \frac{\pi}{4}\right) = \cos\frac{\pi}{2} = 0;$
$\cos\frac{\pi}{4} \cdot \cos\frac{\pi}{4} - \sin\frac{\pi}{4} \cdot \sin\frac{\pi}{4} =$
$\frac{\sqrt{2}}{2} \cdot \frac{\sqrt{2}}{2} - \frac{\sqrt{2}}{2} \cdot \frac{\sqrt{2}}{2} = \frac{1}{2} \cdot \frac{1}{2} = 0$
So, $\cos\left(\frac{\pi}{4} + \frac{\pi}{4}\right) =$
$\cos\frac{\pi}{4} \cdot \cos\frac{\pi}{4} - \sin\frac{\pi}{4} \cdot \sin\frac{\pi}{4}$.

8.a. The formula becomes $\cos 0 = \cos^2 \alpha + \sin^2 \alpha$ or $1 = \cos^2 \alpha + \sin^2 \alpha$.
b. The formula becomes $\cos(\alpha + \alpha) = \cos(2\alpha) = \cos \alpha \cos \alpha - \sin \alpha \sin \alpha = \cos^2 \alpha - \sin^2 \alpha$.

10.a.

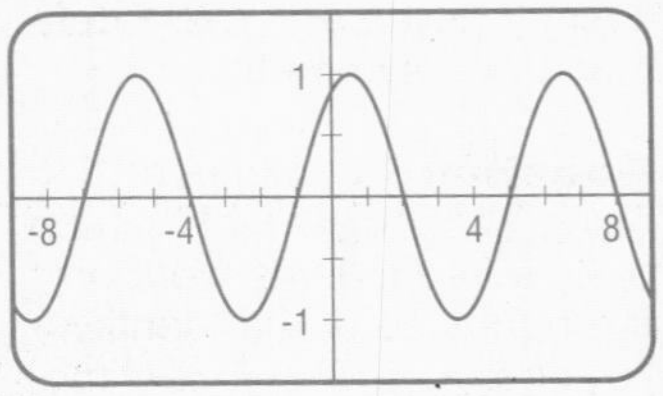

c. $\cos\left(x - \frac{\pi}{6}\right) = \cos\frac{\pi}{6} \cdot \cos x + \sin\frac{\pi}{6} \cdot \sin x = \frac{\sqrt{3}}{2} \cos x + \frac{1}{2} \sin x$

12. See Additional Answers in the back of this book.

14. $\frac{\csc x}{\sec x} = \frac{\frac{1}{\sin x}}{\frac{1}{\cos x}} = \frac{\cos x}{\sin x} =$
$\cot x, x \neq 0 \left(\text{mod } \frac{\pi}{2}\right)$

15., 17.d. See Additional Answers in the back of this book.

Questions

Covering the Reading

1. Explain why $RP = QS$ in the beginning of the proof of the identity for $\cos(\alpha + \beta)$. See margin.

2. Give formulas for $\cos(x + y)$ and $\cos(x - y)$.
$\cos(x + y) = \cos x \cos y - \sin x \sin y$; $\cos(x - y) = \cos x \cos y + \sin x \sin y$

3. Express $\cos 15°$ in terms of rational numbers and radicals. (Hint: $15 = 45 - 30$.) $\frac{\sqrt{6} + \sqrt{2}}{4}$

4. If $0 < x < 90°$ and $\sin x = \cos 15°$, what is x? 75°

5. Prove: $\forall$ *real numbers x*, $\cos\left(\frac{3\pi}{2} + x\right) = \sin x$. See margin.

6. Verify the identity for $\cos(\alpha + \beta)$ when $\alpha = \frac{\pi}{4}$ and $\beta = \frac{\pi}{4}$.
See margin.

Applying the Mathematics

Christopher Clavius

7. Prove Christopher Clavius's identity shown on page 345 of this chapter. $\cos(x - y) - \cos(x + y) =$
$\cos x \cos y + \sin x \sin y - (\cos x \cos y - \sin x \sin y) = 2 \sin x \sin y$

8. **a.** What happens in the formula for $\cos(\alpha - \beta)$ when $\alpha = \beta$?
b. What happens in the formula for $\cos(\alpha + \beta)$ when $\alpha = \beta$?
a,b) See margin.

9. Consider the cofunction identity from Example 2. Substitute $\frac{\pi}{2} - x$ for x and simplify the result. $\sin\left(\frac{\pi}{2} - x\right) = \cos x$

10. **a.** Use an automatic grapher to graph the function f defined by $f(x) = \frac{\sqrt{3}}{2} \cos x + \frac{1}{2} \sin x$. See margin.
b. Based on the graph, conjecture a formula for f of the form $f(x) = \cos(x \pm k)$. $f(x) = \cos\left(x - \frac{\pi}{6}\right)$
c. Prove your conjecture.
See margin.

11. Express $\cos \frac{17\pi}{12}$ in terms of rational numbers and radicals. $\frac{\sqrt{2} - \sqrt{6}}{4}$

Review

12. Use an automatic grapher to conjecture whether or not the equation below is an identity. If it is, prove it; if not, find a counterexample.
$\cos x \cot x = \csc x - \sin x$ *(Lessons 6-2, 6-1)* See margin.

13. Suppose α is in the interval $\pi < \alpha < \frac{3\pi}{2}$ and $\cot \alpha = 0.6$. Find $\sec \alpha$. *(Lesson 6-2)* $\frac{-\sqrt{34}}{3}$

In 14 and 15, prove the identity and give its domain. *(Lesson 6-2)*

14. $\frac{\csc x}{\sec x} = \cot x$
See margin.

15. $\frac{1}{1 + \sin x} + \frac{1}{1 - \sin x} = 2 \sec^2 x$
See margin.

16. Evaluate without using a calculator.
a. $\cos \pi$ -1 **b.** $\sin \pi$ 0 **c.** $\tan \pi$ 0 **d.** $\cot \pi$ undefined
e. $\sec \pi$ -1 **f.** $\csc \pi$ undefined *(Lessons 5-6, 2-8)*

17. Let $h: x \rightarrow \frac{4x^2 + 23}{x^2 + 3}$.

a. Use division of polynomials to rewrite $h(x)$ as the sum of a polynomial and a rational expression. $h(x) = 4 + \frac{11}{x^2 + 3}$

b. Determine the end behavior of the function h. Write your results using limit notation. $\lim_{x \to +\infty} h(x) = 4$ and $\lim_{x \to -\infty} h(x) = 4$

c. Does h have any discontinuities? If so, identify them as essential or removable. If not, explain why not.

d. Use an automatic grapher to graph h. Is the graph consistent with your results in part **b**? *(Lessons 5-5, 5-4, 4-4)* Yes. See margin.

c) No, h is defined for all real numbers.

18. Multiply the matrices.

a. $[3 \;\; 4]\begin{bmatrix} -1 & 9 \\ 6 & 0 \end{bmatrix}$ $[21 \;\; 27]$

b. $\begin{bmatrix} \frac{1}{2} & -2 \\ 5 & 8 \end{bmatrix}\begin{bmatrix} 2 & 10 \\ \frac{1}{2} & 7 \end{bmatrix}$ $\begin{bmatrix} 0 & -9 \\ 14 & 106 \end{bmatrix}$

(Previous course)

Exploration

19. Consider the equation $\sqrt{\frac{1 + \cos 2x}{2}} = \cos x$.

a. Use an automatic grapher to graph each side as a function of x for $-\pi \le x \le \pi$. See below.

b. For what values of x in the above domain does the equation seem to be true? $-\frac{\pi}{2} \le x \le \frac{\pi}{2}$

c. Check your answer to part **b** by evaluating both sides of the equation for

i. $x = \frac{\pi}{4}$ **ii.** $x = \frac{3\pi}{4}$. See below.

d. Modify the equation to make it an identity. $\sqrt{\frac{1 + \cos 2x}{2}} = |\cos x|$

19. a)

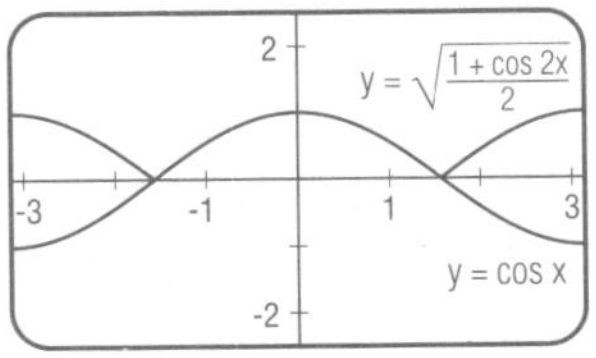

c) i. $\sqrt{\frac{1 + \cos 2(\frac{\pi}{4})}{2}} = \frac{\sqrt{2}}{2}$ and $\cos \frac{\pi}{4} = \frac{\sqrt{2}}{2}$

ii. $\sqrt{\frac{1 + \cos 2(\frac{3\pi}{4})}{2}} = \frac{\sqrt{2}}{2}$ but $\cos \frac{3\pi}{4} = -\frac{\sqrt{2}}{2}$

FOLLOW-UP

MORE PRACTICE

For more questions on SPUR Objectives, use *Lesson Master 6-3,* shown below.

EVALUATION

A quiz covering Lessons 6-1 through 6-3 is provided in the Teacher's Resource File.

NAME ____________

LESSON **MASTER 6–3**
QUESTIONS ON **SPUR** OBJECTIVES

■ SKILLS *Objective A (See pages 396–398 for objectives.)*

1. Suppose $0 < \alpha < \frac{\pi}{2} < \beta < \pi$, $\cos \alpha = .75$, and $\sin \beta = .3$. Find the following values.

a. $\sin \alpha$ .66 b. $\cos \beta$ -.95

c. $\cos(\alpha + \beta)$ -.91 d. $\cos(\alpha - \beta)$ -.52

In 2–4, express the following in terms of rational numbers and radicals.

2. $\cos\left(\frac{\pi}{4} + \frac{\pi}{6}\right)$ $\frac{\sqrt{6} - \sqrt{2}}{4}$

3. $\cos \frac{11\pi}{12}$ $\frac{-\sqrt{2} - \sqrt{6}}{4}$

4. $\cos \frac{\pi}{9} \cos \frac{2\pi}{9} - \sin \frac{\pi}{9} \sin \frac{2\pi}{9}$ $\frac{1}{2}$

5. Find the cosine of $\frac{\pi}{12}$:

a. by writing $\frac{\pi}{12}$ as $\frac{\pi}{4} - \frac{\pi}{6}$. $\frac{\sqrt{6} + \sqrt{2}}{4}$

b. by writing $\frac{\pi}{12}$ as $\frac{\pi}{3} - \frac{\pi}{4}$. $\frac{\sqrt{6} + \sqrt{2}}{4}$

6. Simplify $\cos\left(\frac{\pi}{2} + x\right) + \cos\left(\frac{\pi}{2} - x\right)$. 0

64 *Continued* *Precalculus and Discrete Mathematics* © Scott, Foresman and Company

NAME ____________
Lesson MASTER 6–3 (page 2)

■ PROPERTIES *Objective D*
In 7–9, prove the identity and specify the domain.

7. $-2 \sin x \sin y = \cos(x + y) - \cos(x - y)$
Right side = cos x cos y − sin x sin y − (cos x cos y + sin x sin y)
= -2 sin x sin y
{all real numbers}

8. $\cos(\pi - x) = -\cos x$
Left side = cos π cos x + sin π sin x
= (-1) cos x + (0) sin x
= -cos x
{all real numbers}

9. $\sin x = \cos\left(x + \frac{3\pi}{2}\right)$
Right side = $\cos x \cos \frac{3\pi}{2} - \sin x \sin \frac{3\pi}{2}$
= (cos x)(0) − (sin x)(-1)
= sin x
{all real numbers}

Precalculus and Discrete Mathematics © Scott, Foresman and Company 65

LESSON 6-4

RESOURCES
■ Lesson Master 6-4

OBJECTIVES

A Without a calculator, use trigonometric identities to express values of trigonometric functions in terms of rational numbers and radicals.
D Prove trigonometric identities and identify their domains.

TEACHING NOTES

Since we have identities for $\cos(\alpha + \beta)$ and $\cos(\alpha - \beta)$, it is natural to want to find similar identities for the other common trigonometric functions.

In the first paragraph of this lesson, replacing x by $\frac{\pi}{2} - x$ in $\sin x = \cos\left(\frac{\pi}{2} - x\right)$ to obtain $\sin\left(\frac{\pi}{2} - x\right) = \cos\left(\frac{\pi}{2} - \left(\frac{\pi}{2} - x\right)\right)$ may be difficult for some students to understand. You may wish to begin with $\sin \alpha = \cos\left(\frac{\pi}{2} - \alpha\right)$ and then replace α by $\frac{\pi}{2} - x$.

Point out the two applications of cofunctions in proving the Sine of a Sum Theorem. Some students will need help in going from the second to the third line of the proof. Write $\cos(x - y) = \cos x \cos y + \sin x \sin y$ and tell them to replace x by $\frac{\pi}{2} - \alpha$ and y by β. Then simplify the resulting expression.

LESSON

6-4

Formulas for $\sin(\alpha \pm \beta)$ and $\tan(\alpha \pm \beta)$

Because of the cofunction identity $\sin x = \cos\left(\frac{\pi}{2} - x\right)$, formulas for the cosine can be used to obtain values of the sine function. To find $\sin x$, just find the cosine of $\frac{\pi}{2} - x$. Also, replacing x by $\frac{\pi}{2} - x$ on each side yields another cofunction identity.

$$\sin\left(\frac{\pi}{2} - x\right) = \cos\left(\frac{\pi}{2} - \left(\frac{\pi}{2} - x\right)\right) = \cos x$$

These two cofunction identities help in deducing an identity for $\sin(\alpha + \beta)$.

Theorem (Identity for $\sin(\alpha + \beta)$)

For all real numbers α and β,
$$\sin(\alpha + \beta) = \sin \alpha \cos \beta + \cos \alpha \sin \beta.$$

Proof Begin with the left side and transform it to equal the right side.

$\sin(\alpha + \beta) = \cos\left(\frac{\pi}{2} - (\alpha + \beta)\right)$	Cofunction Identity
$= \cos\left(\left(\frac{\pi}{2} - \alpha\right) - \beta\right)$	
$= \cos\left(\frac{\pi}{2} - \alpha\right) \cos \beta + \sin\left(\frac{\pi}{2} - \alpha\right) \sin \beta$	Formula for $\cos(x - y)$
$= \sin \alpha \cos \beta + \cos \alpha \sin \beta$	Cofunction identities

You should memorize this identity for $\sin(\alpha + \beta)$.

The theorem below is a corresponding identity for $\sin(\alpha - \beta)$. It can be proved by replacing β in $\sin(\alpha + \beta)$ by $-\beta$. You are asked to write this proof in the Questions.

Theorem (Identity for $\sin(\alpha - \beta)$)

For all real numbers α and β,
$$\sin(\alpha - \beta) = \sin \alpha \cos \beta - \cos \alpha \sin \beta.$$

Example 1 Prove the identity $\sin(\pi - x) = \sin x$.

Solution Use the formula for $\sin(\alpha - \beta)$, with $\alpha = \pi$ and $\beta = x$. For all real numbers x,

$$\begin{aligned}\sin(\pi - x) &= \sin \pi \cos x - \cos \pi \sin x \\ &= 0 \cdot \cos x - (-1) \sin x \\ &= \sin x.\end{aligned}$$

The identity in Example 1 has a geometric significance. When $0 < x < \pi$, x and $\pi - x$ are measures of supplementary angles. So the sines of supplements are equal. Graphically, the identity shows that for all x, $(x, \sin x)$ and $(\pi - x, \sin(\pi - x))$ lie on the same horizontal line.

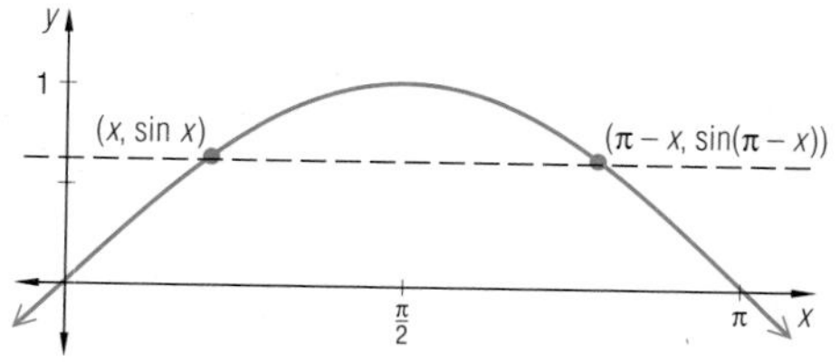

The identities for the tangent of a sum and the tangent of a difference can be derived from the corresponding identities for the sine and cosine.

Theorem (Identities for $\tan(\alpha + \beta)$ and $\tan(\alpha - \beta)$)

For all real numbers α and β such that $\tan \alpha$, $\tan \beta$, $\tan(\alpha + \beta)$, and $\tan(\alpha - \beta)$ are defined,

$$\tan(\alpha + \beta) = \frac{\tan \alpha + \tan \beta}{1 - \tan \alpha \tan \beta}$$

and

$$\tan(\alpha - \beta) = \frac{\tan \alpha - \tan \beta}{1 + \tan \alpha \tan \beta}.$$

Proof Suppose that α and β are real numbers such that $\cos(\alpha + \beta)$, $\cos \alpha$, and $\cos \beta$ are nonzero. Then

$$\tan(\alpha + \beta) = \frac{\sin(\alpha + \beta)}{\cos(\alpha + \beta)} \qquad \underline{\ ?\ }$$

$$= \frac{\sin \alpha \cos \beta + \cos \alpha \sin \beta}{\cos \alpha \cos \beta - \sin \alpha \sin \beta} \qquad \underline{\ ?\ }$$

$$= \frac{\dfrac{\sin \alpha \cos \beta}{\cos \alpha \cos \beta} + \dfrac{\cos \alpha \sin \beta}{\cos \alpha \cos \beta}}{\dfrac{\cos \alpha \cos \beta}{\cos \alpha \cos \beta} - \dfrac{\sin \alpha \sin \beta}{\cos \alpha \cos \beta}} \qquad \text{Divide numerator and denominator by } \cos \alpha \cos \beta.$$

$$= \frac{\tan \alpha + \tan \beta}{1 - \tan \alpha \tan \beta}. \qquad \underline{\ ?\ }$$

You are asked to prove the identity for $\tan(\alpha - \beta)$ in the Questions.

You might use a unit circle to make the identity of **Example 1** reasonable.

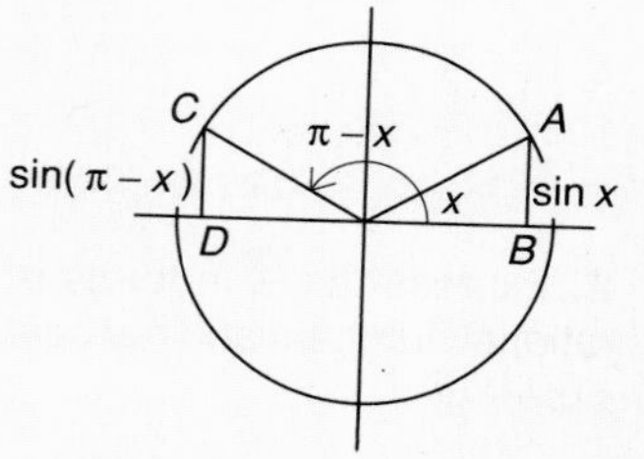

By symmetry, $\sin x = \sin(\pi - x)$.

In the proof of $\tan(\alpha + \beta) = \frac{\tan \alpha + \tan \beta}{1 - \tan \alpha \tan \beta}$, many students do not understand why $\frac{\sin \alpha \cos \beta + \cos \alpha \sin \beta}{\cos \alpha \cos \beta - \sin \alpha \sin \beta}$ was divided by $\cos \alpha \cos \beta$. Because $\tan x = \frac{\sin x}{\cos x}$, it is necessary to create fractions with only sines in numerators and corresponding cosines in denominators. Because the $\cos \alpha \cos \beta$ term in the lower left of the expression lacks any sines, it cannot be transformed into tangents and therefore must be "eliminated" by dividing by $\cos \alpha \cos \beta$. A wonderful result is that all the remaining terms are transformed into tangents and the identity is proved.

Be sure to discuss the last paragraph of this lesson. The finding or creation of functions L that would satisfy $L(ab) = L(a) + L(b)$ greatly facilitated the work of astronomers.

ADDITIONAL EXAMPLES

1. Prove the identity $\sin\left(\frac{3\pi}{2} + x\right) = -\cos x$.

$\sin\left(\frac{3\pi}{2} + x\right) = \sin\frac{3\pi}{2}\cos x + \cos\frac{3\pi}{2}\sin x = (-1)\cos x + 0(\sin x) = -\cos x$

2. Express $\tan\frac{7\pi}{12}$ in terms of rational numbers and radicals.

sample:

$\tan\frac{7\pi}{12} = \tan\left(\frac{3\pi}{12} + \frac{4\pi}{12}\right) = \tan\left(\frac{\pi}{4} + \frac{\pi}{3}\right) = \frac{1 + \sqrt{3}}{1 - \sqrt{3}}$

NOTES ON QUESTIONS

Question 3: Students will need to remember that $\sin(-x) = -\sin x$ and $\cos(-x) = \cos x$.

Question 11: Students should be able to do this proof in two ways. (1) $\tan(\alpha - \beta) = \frac{\sin(\alpha - \beta)}{\cos(\alpha - \beta)} = \ldots$, as in the proof for $\tan(\alpha + \beta)$ in the lesson. (2) $\tan(\alpha - \beta) = \tan(\alpha + (-\beta))$ and then apply the $\tan(x + y)$ identity.

Making Connections for Question 12: Students who used *Functions, Statistics, and Trigonometry* will have seen this proof in that book.

Question 13: In expanding $\cos\left(x + \frac{\pi}{2}\right)$ using the Cosine of a Sum Theorem, students will be proving the identity as they find it.

ADDITIONAL ANSWERS

1.a.

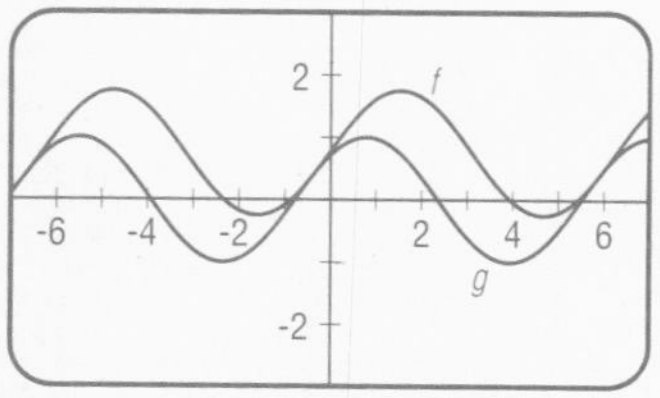

$f(\alpha) = \sin\alpha + \sin\frac{\pi}{4}$;
$g(\alpha) = \sin\left(\alpha + \frac{\pi}{4}\right)$

Example 2 Express $\tan\frac{11\pi}{12}$ in terms of rational numbers and radicals.

Solution First rewrite $\frac{11\pi}{12}$ as a sum of two numbers whose tangents you know: $\frac{11\pi}{12} = \frac{3\pi}{4} + \frac{\pi}{6}$.

Then $\tan\frac{11\pi}{12} = \tan\left(\frac{3\pi}{4} + \frac{\pi}{6}\right)$

$= \dfrac{\tan\frac{3\pi}{4} + \tan\frac{\pi}{6}}{1 - \tan\frac{3\pi}{4}\cdot\tan\frac{\pi}{6}}$

$= \dfrac{-1 + \frac{\sqrt{3}}{3}}{1 - (-1)\left(\frac{\sqrt{3}}{3}\right)}$ Substituting for $\tan\frac{3\pi}{4}$ and $\tan\frac{\pi}{6}$

$= \dfrac{-3 + \sqrt{3}}{3 + \sqrt{3}}$ Simplifying the complex fraction

$= \dfrac{-3 + \sqrt{3}}{3 + \sqrt{3}}\cdot\dfrac{3 - \sqrt{3}}{3 - \sqrt{3}}$ Rationalizing the denominator

$= \dfrac{-12 + 6\sqrt{3}}{6}$ Multiplying fractions

$= -2 + \sqrt{3}$. Simplifying to lowest terms

The sum and difference formulas for the sine, cosine, and tangent lead to a variety of identities involving these functions. Some of these are in the Questions, some are in the next lesson, and still more are to be found in Chapter 8. For instance, from the formula for $\tan(\alpha + \beta)$, you can prove that for all x, $\tan(x + \pi) = \tan x$.

Relationships among the values of a function tell you how the function behaves. For instance, the identity $\tan(x + \pi) = \tan x$ indicates that the tangent function is periodic. The identity $\log(xy) = \log x + \log y$ for $x > 0$ and $y > 0$ indicates that the log function converts a multiplication into an addition. The real number property $a^{x+y} = a^x \cdot a^y$ can be thought of as an identity involving the values of the exponential function $f: x \rightarrow a^x$ that shows how powers with the same base can be multiplied.

Questions

Covering the Reading

1. Consider the equation $\sin(\alpha + \beta) = \sin\alpha + \sin\beta$.
   a. Let $\beta = \frac{\pi}{4}$ and graph each side of the equation as a function of α.
   b. Does the equation appear to be an identity?
   a) See margin. b) not an identity
2. Write formulas for $\sin(x + y)$ and $\sin(x - y)$.
   $\sin(x + y) = \sin x\cos y + \cos x\sin y$; $\sin(x - y) = \sin x\cos y - \cos x\sin y$
3. Substitute $-\beta$ for β in the identity for $\sin(\alpha + \beta)$ to derive the identity for $\sin(\alpha - \beta)$.
   $\sin(\alpha + (-\beta)) = \sin\alpha\cos(-\beta) + \cos\alpha\sin(-\beta) = \sin\alpha\cos\beta - \cos\alpha\sin\beta$

364

4. Prove: ∀ *real numbers x*, $\sin(\frac{\pi}{2} + x) = \cos x$. See margin.

5. Supply the missing reasons for each step in the proof of the identity for $\tan(\alpha + \beta)$. Step 1: Definition of tangent; Step 2: Identities for $\sin(\alpha + \beta)$ and $\cos(\alpha + \beta)$; Step 4: Definition of tangent

In 6–8, find the value in terms of rational numbers and radicals.

6. $\sin \frac{7\pi}{12}$ $\frac{\sqrt{6} + \sqrt{2}}{4}$ 7. $\tan 345°$ $\sqrt{3} - 2$ 8. $\sin 105°$ $\frac{\sqrt{6} + \sqrt{2}}{4}$

Applying the Mathematics

9. a. Prove: ∀ *real numbers x*, $\tan(x + \pi) = \tan x$. See margin.
   b. What does the theorem in part **a** indicate about the period of the tangent function? The period is no larger than π.

10. In the identity for $\tan(\alpha + \beta)$, let $\alpha = \beta$. What happens? See margin.

11. Prove the identity for $\tan(\alpha - \beta)$. See margin.

12. An elegant alternate proof of two of the identities of this and the previous lesson is possible using matrices. Assume that for any real number θ,
$$\begin{bmatrix} \cos\theta & -\sin\theta \\ \sin\theta & \cos\theta \end{bmatrix}$$
is the matrix for R_θ, where R_θ is the rotation of θ about the origin. Translate the transformation identity $R_{\theta+\phi} = R_\theta \cdot R_\phi$ into matrix form and then multiply the matrices. Which two identities result? See margin.

Review

13. Find and prove an identity for $\cos\left(x + \frac{\pi}{2}\right)$. *(Lesson 6-3)* See margin.

14. Express $\cos \frac{23\pi}{12}$ in terms of rational numbers and radicals. *(Lesson 6-3)* See margin.

15. Use an automatic grapher to determine whether $\sin x = \sqrt{1 - \cos^2 x}$ could be an identity. If so, prove it. If not, find a counterexample. *(Lessons 6-2, 6-1)* See margin.

16. A small grain combine can harvest 100 acres in 24 hours, a medium combine can do the same job in 18 hours, and a large combine can do the job in 12 hours. If all three combines are used at the same time, how long will it take to harvest 800 acres? *(Lesson 5-7)* ≈ 44.3 hours

17. Tell whether the number is rational or irrational. *(Lessons 5-2, 5-1)*
   a. $.8\overline{7}$ rational b. $\sqrt{15}$ irrational c. $\frac{\sqrt{36}}{9}$ rational

18. Express in factored form:
$(8x + 3)^2(5x - 7)^3 + (8x + 3)(5x - 7)^2$. *(Lesson 4-8)*
$(8x + 3)(5x - 7)^2((8x + 3)(5x - 7) + 1)$

4. $\sin\left(\frac{\pi}{2} + x\right) = \sin\frac{\pi}{2}\cos x + \cos\frac{\pi}{2}\sin x = 1 \cdot \cos x + 0 \cdot \sin x = \cos x$

9.a., 10., 11., 13., 14., 15. See the margin on page 366.

12. See the Additional Answers in the back of this book.

NAME ____________

LESSON **MASTER 6–4**
QUESTIONS ON **SPUR** OBJECTIVES

■ **SKILLS** *Objective A (See pages 396–398 for objectives.)*
In 1–4, express the following in terms of rational numbers and radicals.

1. $\sin\left(\frac{2\pi}{3} - \frac{\pi}{4}\right)$ $\frac{\sqrt{6} + \sqrt{2}}{4}$
2. $\tan \frac{7\pi}{12}$ $\frac{1 + \sqrt{3}}{1 - \sqrt{3}}$
3. $\frac{\tan \frac{3\pi}{16} + \tan \frac{\pi}{16}}{1 - \tan \frac{3\pi}{16} \tan \frac{\pi}{16}}$ 1
4. $\sin \frac{13\pi}{12}$ $\frac{\sqrt{2} - \sqrt{6}}{4}$
5. Given that $\sin \alpha = \frac{2}{3}$, find $\sin(\pi - \alpha)$. $\frac{2}{3}$

In 6–9, suppose r is in the interval $0 < r < \frac{\pi}{2}$ with $\cos r = \frac{3}{5}$ and s is in the interval $\pi < s < \frac{3\pi}{2}$ with $\sin s = -\frac{1}{4}$. Use this information to find each value.

6. $\sin(s - r)$ $\frac{4\sqrt{15} - 3}{20}$
7. $\tan(r + s)$ $\frac{4\sqrt{15} + 3}{3\sqrt{15} - 4}$
8. $\sin(\pi + s - r)$ $\frac{3 - 4\sqrt{15}}{20}$
9. $\tan(\pi + s - r)$ $\frac{3 - 4\sqrt{15}}{3\sqrt{15} + 4}$

66 *Continued* *Precalculus and Discrete Mathematics* © Scott, Foresman and Company

NAME ____________
Lesson MASTER 6–4 (page 2)

■ **PROPERTIES** *Objective D*
In 10–13, prove the identity and specify its domain.

10. $\sin\left(\frac{3\pi}{2} + x\right) = -\text{cox } x$
Left side $= \sin\frac{3\pi}{2}\cos x + \cos\frac{3\pi}{2}\sin x$
$= (-1)\cos x + (0)\sin x$
$= -\cos x$
{all real numbers}

11. $\tan x = -\tan(\pi - x)$
Right side $= -\left(\frac{\tan\pi - \tan x}{1 + \tan\pi\tan x}\right)$
$= -\frac{0 - \tan x}{1 + (0)\tan x}$
$= \tan x$
$\left\{x: x \neq \frac{(2k+1)\pi}{2} \text{ where } k \text{ is an integer}\right\}$

12. $\sin(x + y)\cos(x - y) = \sin x \cos x + \sin y \cos y$
Left side $= (\sin x \cos y + \cos x \sin y)(\cos x \cos y + \sin x \sin y)$
$= \sin x \cos x \cos^2 y + \sin^2 x \sin y \cos y + \cos^2 x \sin y \cos y + \cos x \sin x \sin^2 y$
$= \sin x \cos x(\cos^2 y + \sin^2 y) + (\sin^2 x + \cos^2 x)\sin y \cos y$
$= \sin x \cos x + \sin y \cos y$
{all real numbers}

13. $\sin(x + y) + \cos(x - y) = (\sin x + \cos x)(\sin y + \cos y)$
(Compare this identity to the one in Question 12.)
Left side $= \sin x \cos y + \cos x \sin y + \cos x \cos y + \sin x \sin y$
$= \sin x(\cos y + \sin y) + \cos x(\sin y + \cos y)$
$= (\sin x + \cos x)(\cos y + \sin y)$
{all real numbers}
(Note that replacing multiplication with addition in the identity in Question 12 (except for in $(x + y)$) yields the identity in Question 13!)

Precalculus and Discrete Mathematics © Scott, Foresman and Company 67

FOLLOW-UP

MORE PRACTICE
For more questions on SPUR Objectives, use *Lesson Master 6-4,* shown on page 365.

EXTENSION
You may wish to have students use an automatic 3-D grapher to examine identities that involve two variables. For example, to verify $\sin(x + y) = \sin x \cos y + \cos x \sin y$, graph $z_1 = \sin(x + y)$ and $z_2 = \sin x \cos y + \cos x \sin y$. You may even graph $z_3 = z_1 - z_2$. *Mathematica, Derive,* and *Theorist* are software packages that do 3-D graphics.

PROJECTS
The projects for Chapter 6 are described on page 393. **Projects 3 and 5** are related to the content of this lesson.

ADDITIONAL ANSWERS

9.a. $\tan(x + \pi) = \dfrac{\tan x + \tan \pi}{1 - \tan x \tan \pi} = \dfrac{\tan x + 0}{1 - (\tan x) \cdot 0} = \tan x$

10. The identity becomes $\tan 2\beta = \dfrac{2 \tan \beta}{1 - \tan^2 \beta}$.

11. $\tan(\alpha - \beta) = \tan(\alpha + (-\beta)) = \dfrac{\tan \alpha + \tan(-\beta)}{1 - \tan \alpha \tan(-\beta)} = \dfrac{\tan \alpha - \tan \beta}{1 + \tan \alpha \tan \beta}$

13. $\cos\left(x + \frac{\pi}{2}\right) = \cos x \cos \frac{\pi}{2} - \sin x \sin \frac{\pi}{2} = (\cos x) \cdot 0 - (\sin x) \cdot 1 = -\sin x$

14. $\dfrac{\sqrt{6} + \sqrt{2}}{4}$

15. Counterexample: $\sin \frac{3\pi}{2} = -1$, but $\sqrt{1 - \cos^2 \frac{3\pi}{2}} = 1$.

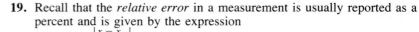

19. Recall that the *relative error* in a measurement is usually reported as a percent and is given by the expression

$$\left|\frac{x - x_m}{x}\right|$$

where x is the expected value of a quantity and x_m is its measured value. Suppose a machine cuts steel bars whose desired length is 27.5 cm. If the relative error can be at most 1.5%, find the acceptable values for the measured length. *(Lesson 3-8)* $27.0875 \text{ cm} \le x_m \le 27.9125 \text{ cm}$

20. Use chunking to find all real numbers that satisfy $e^{2x} - 4e^x - 12 = 0$. *(Lessons 3-4, 2-6)* $x = \ln 6 \approx 1.79$

21. A semicircular tunnel through a mountain has a radius of 15 feet. A truck with a height of 12 feet needs to pass through the tunnel. If the truck straddles the center line of the road when passing through the tunnel, what is the maximum allowable width for the truck? *(Previous course)* 18 ft

Exploration

22. a. Find an identity for $\sin(\alpha + \beta + \gamma)$ in terms of $\sin \alpha$, $\sin \beta$, $\sin \gamma$, $\cos \alpha$, $\cos \beta$, $\cos \gamma$.
b. Use this identity to predict a similar identity for $\cos(\alpha + \beta + \gamma)$.
c. Determine whether your prediction is true. a,b,c) See below.

22. a) $\sin(\alpha + \beta + \gamma) = \sin \alpha \cos \beta \cos \gamma + \cos \alpha \sin \beta \cos \gamma + \cos \alpha \cos \beta \sin \gamma - \sin \alpha \sin \beta \sin \gamma$

b) Sample: $\cos(\alpha + \beta + \gamma) = \cos \alpha \cos \beta \cos \gamma - \sin \alpha \sin \beta \cos \gamma - \sin \alpha \cos \beta \sin \gamma + \cos \alpha \sin \beta \sin \gamma$

c) Proof: (Chunk $(\alpha + \beta)$ and apply cosine and sine of sum identities.)

$$\begin{aligned}\cos[(\alpha + \beta) + \gamma] &= \cos(\alpha + \beta)\cos \gamma - \sin(\alpha + \beta)\sin \gamma \\ &= (\cos \alpha \cos \beta - \sin \alpha \sin \beta)\cos \gamma \\ &\quad - (\sin \alpha \cos \beta + \cos \alpha \sin \beta)\sin \gamma \\ &= \cos \alpha \cos \beta \cos \gamma - \sin \alpha \sin \beta \cos \gamma \\ &\quad - \sin \alpha \cos \beta \sin \gamma - \cos \alpha \sin \beta \sin \gamma\end{aligned}$$

366

LESSON

6-5

Formulas for cos 2*x* and sin 2*x*

In Example 1 of Lesson 6-1 you saw that the graph of $y = \cos^2 x - \sin^2 x$ looks like the graph of $y = \cos 2x$.

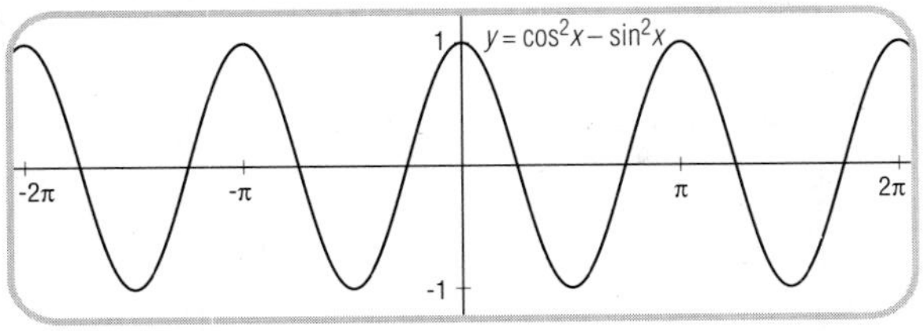

In other words, the equation $\cos^2 x - \sin^2 x = \cos 2x$ appears to be an identity. From the identities for $\cos(\alpha + \beta)$ and $\sin(\alpha + \beta)$ you can prove this identity as well as some others.

Theorem (Identities for cos 2*x*)

For all real numbers x:

$$\cos 2x = \cos^2 x - \sin^2 x \quad (1)$$
$$\cos 2x = 2\cos^2 x - 1 \quad (2)$$
$$\cos 2x = 1 - 2\sin^2 x. \quad (3)$$

Proof (1) This identity is a special case of the identity for $\cos(\alpha + \beta)$. For all real numbers α and β,

$$\cos(\alpha + \beta) = \cos\alpha\cos\beta - \sin\alpha\sin\beta.$$

Now let $\alpha = x$ and $\beta = x$. Then

$$\cos(x + x) = \cos x \cos x - \sin x \sin x.$$

That is, $\cos 2x = \cos^2 x - \sin^2 x$.

(2) This identity follows from (1) using the Pythagorean Identity $\cos^2 x + \sin^2 x = 1$, in the form $\sin^2 x = 1 - \cos^2 x$.

Left side $= \cos 2x$
$= \cos^2 x - \sin^2 x$ From (1)
$= \cos^2 x - (1 - \cos^2 x)$ $\sin^2 x = 1 - \cos^2 x$
$= \cos^2 x - 1 + \cos^2 x$
$= 2\cos^2 x - 1$
= Right side

The proof of (3) is left for you to do.

Example 1 If $\cos x = -\frac{3}{5}$, find $\cos 2x$.

Solution You could solve this problem using either (1), (2), or (3), but by choosing (2), you can avoid computing $\sin x$.

$$\cos 2x = 2\cos^2 x - 1 = 2\left(-\frac{3}{5}\right)^2 - 1 = 2 \cdot \frac{9}{25} - 1 = \frac{18}{25} - 1 = -\frac{7}{25}$$

LESSON 6-5

RESOURCES

- Lesson Master 6-5
- Teaching Aid 37 displays the drawing for **Example 3**.
- Computer Master 12

OBJECTIVES

A Without a calculator, use trigonometric identities to express values of trigonometric functions in terms of rational numbers and radicals.

D Prove trigonometric identities and identify their domains.

TEACHING NOTES

Stress that since the double angle identities $\cos 2x = \cos^2 x - \sin^2 x$ and $\sin 2x = 2\sin x \cos x$ can be derived quickly from the formulas for $\cos(x + y)$ and $\sin(x + y)$, they do not need to be memorized. Also, show how $\cos 2x$ identities (2) and (3) follow from (1).

Example 2 uses the Pythagorean Identity and the double inequality $\frac{\pi}{2} < x < \pi$. This inequality needs to be transformed to $\pi < 2x < 2\pi$ in order to analyze the sign of $\sin 2x$.

Making Connections A formula that physics students should know is derived in **Example 3**. The horizontal and vertical components of the velocity should be understandable to students if you draw a right triangle whose hypotenuse is v_0. Students should recognize that the formula for the height is the standard quadratic equation for the motion of a projectile thrown into the air: $h = -\frac{1}{2}gt^2 + v_0t + h_0$, where g is the acceleration due to gravity, v_0 is the initial velocity in the vertical direction, and h_0 is the initial height from which the object is thrown.

Go over **Example 4** carefully. Students find it unusual to use a double-angle formula to determine a half angle. Tell them why $\cos 2x = 2\cos^2 x - 1$ was chosen to help calculate $\cos\frac{\alpha}{2}$. Some students will need to see the following steps explicitly.
$\cos 2x = 2\cos^2 x - 1$.
Let $2x = \alpha$, so $x = \frac{\alpha}{2}$.
$\cos 2\left(\frac{\alpha}{2}\right) = 2\cos^2\left(\frac{\alpha}{2}\right) - 1$,
$\cos \alpha = 2\cos^2\left(\frac{\alpha}{2}\right) - 1$,
and so on.

Also, $\cos\alpha = \frac{3}{5} \Rightarrow \cos\alpha > 0 \Rightarrow 0 < \alpha < \frac{\pi}{2}$ or $\frac{3\pi}{2} < \alpha < 2\pi$. Since we want $\frac{\alpha}{2}$, divide these double inequalities by 2. $0 < \frac{\alpha}{2} < \frac{\pi}{4} \Rightarrow \cos\frac{\alpha}{2} > 0$ and $\frac{3\pi}{4} < \frac{\alpha}{2} < \pi \Rightarrow \cos\frac{\alpha}{2} < 0$. For this reason, the sign of $\cos\frac{\alpha}{2}$ cannot be determined.

Technology Have students check the results to problems such as those in **Example 5**. A scientific calculator can be used to approximate the value of $\sin\frac{\pi}{8}$. Students can approximate the exact value on the calculator to see if the results appear to be equal, within a normal round-off error. Such a check should determine incorrect signs as well as computational errors introduced while using the identities.

A special case of the formula for $\sin(\alpha + \beta)$ leads to the next theorem. Its proof is like proof (1) of the previous theorem and is left for you to do.

Theorem (Identity for sin 2*x*)

For all real numbers x,

$$\sin 2x = 2\sin x\cos x.$$

Example 2 If $\cos x = -\frac{3}{5}$ and $\frac{\pi}{2} < x < \pi$, find $\sin 2x$.

Solution The formula $\sin 2x = 2\sin x\cos x$ can be used, but you need to know $\sin x$. If $\cos x = -\frac{3}{5}$ and $\frac{\pi}{2} < x < \pi$, x is in the 2nd quadrant, so $\sin x$ is positive. Now $\sin^2 x + \cos^2 x = 1$, so $\sin^2 x + \left(-\frac{3}{5}\right)^2 = 1$, and thus $\sin^2 x = \frac{16}{25}$, from which $\sin x = \frac{4}{5}$.
Thus $\sin 2x = 2 \cdot \frac{4}{5} \cdot -\frac{3}{5} = -\frac{24}{25}$.

One use for identities is to simplify formulas, as Example 3 illustrates.

Example 3 A projectile is launched from ground level with an initial speed of v_0 feet per second at an angle θ above the horizontal. The following is a drawing of this situation showing the components of the initial velocity vector.

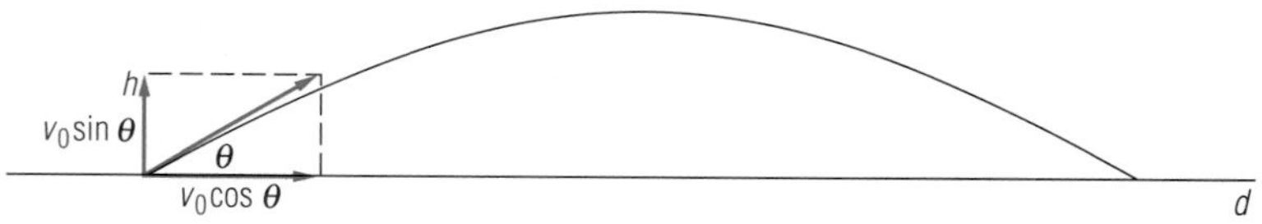

Let $h(t)$ be the height in feet of the projectile t seconds after it is launched, and let $d(t)$ be the horizontal distance of the projectile from the launch site. If only the force of gravity on the projectile is considered and other forces such as air resistance are ignored, then

$$d(t) = (v_0\cos\theta)t$$
$$h(t) = -16t^2 + (v_0\sin\theta)t.$$

a. How long will the projectile remain airborne?
b. The range R is the horizontal distance traveled by the projectile. (Do not confuse this with the range of a function.) Find a formula for R in terms of v_0 and θ.
c. What is the range when the projectile is launched at 120 ft/sec at an angle of 30°?

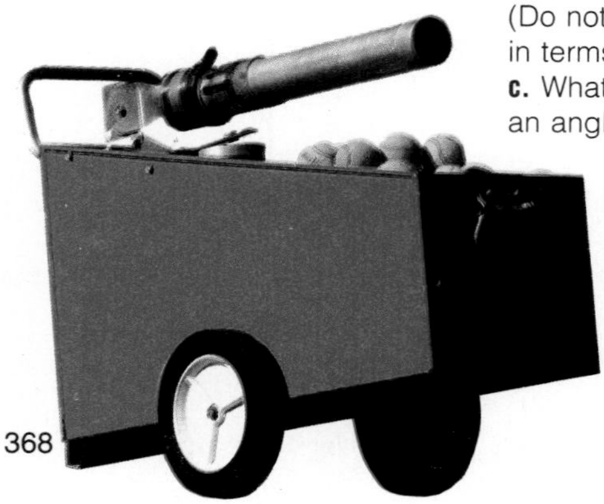

368

Solution

a. The length of time the projectile is airborne is the positive value of t for which $h(t) = 0$. Thus, substituting for $h(t)$,

$$-16t^2 + (v_0 \sin\theta)t = 0$$
$$(-16t + v_0 \sin\theta)t = 0 \qquad \text{Factoring}$$
$$-16t + v_0 \sin\theta = 0 \quad \text{or} \quad t = 0 \qquad \text{Zero Product Property}$$
$$t = \frac{v_0 \sin\theta}{16} \quad \text{or} \quad t = 0.$$

The projectile is on the ground at 0 seconds and $\frac{v_0 \sin\theta}{16}$ seconds, so it remains airborne for $\frac{v_0 \sin\theta}{16}$ seconds.

b. The range R is simply the value of $d(t)$ when $t = \frac{v_0 \sin\theta}{16}$. In other words,

$$R = (v_0 \cos\theta)t$$
$$= (v_0 \cos\theta)\left(\frac{v_0 \sin\theta}{16}\right)$$
$$= \frac{{v_0}^2 \cos\theta \sin\theta}{16} \text{ feet.}$$

This formula can be simplified using the identity $\sin 2\theta = 2\cos\theta \sin\theta$.

$$R = \frac{{v_0}^2\, 2\cos\theta \sin\theta}{32}$$
$$= \frac{{v_0}^2 \sin 2\theta}{32} \text{ feet}$$

c. Here $v_0 = 120$ ft/sec and $\theta = 30°$. So

$$R = \frac{(120)^2 \sin(2 \cdot 30)}{32}$$
$$\approx 389.7 \text{ feet.}$$

The formulas for $\cos 2x$ and $\sin 2x$ are sometimes called **double angle identities**. From the formulas for $\cos 2x$, half angle identities involving $\cos\frac{x}{2}$ and $\sin\frac{x}{2}$ can be deduced. Recall the Bisection Method for finding solutions to equations. That method utilizes the fact that any real number can be located, to within any accuracy you desire, by successively splitting intervals in two. This same idea was used hundreds of years ago to obtain values of sines and cosines. If you knew how to find $\cos\left(\frac{x}{2}\right)$, you could substitute $\alpha + \beta$ for x to calculate $\cos\left(\frac{\alpha+\beta}{2}\right)$, the cosine of the mean of α and β, for any α and β. The same is true for $\sin\left(\frac{x}{2}\right)$. Examples 4 and 5 show how this can be done.

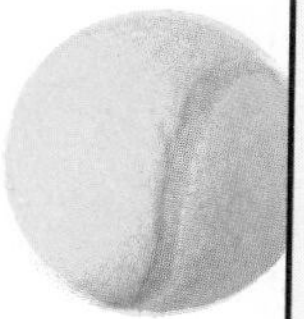

For **Example 5,** you may need to explain why $\cos 2x = 1 - 2\sin^2 x$ was chosen to find $\sin\frac{\pi}{8}$. One way to proceed is as follows: Let $x = \frac{\pi}{8}$. Then $\cos 2x = 1 - 2\sin^2 x$ becomes $\cos 2\left(\frac{\pi}{8}\right) = 1 - 2\sin^2\frac{\pi}{8}$. So $\cos\frac{\pi}{4} = 1 - 2\sin^2\frac{\pi}{8}$ and so on.

If you have time, invite students to use $\cos 2x = 2\cos^2 x - 1$ and chunking $\left(\text{let } x = \frac{\theta}{2}\right)$ to derive $\cos\frac{\theta}{2} = \pm\sqrt{\frac{1+\cos\theta}{2}}$. Many texts call this a half-angle formula.

Similarly, you can use $\cos 2x = 1 - 2\sin^2 x$ to develop $\sin\frac{\theta}{2} = \pm\sqrt{\frac{1-\cos\theta}{2}}$.

ADDITIONAL EXAMPLES

1. If $\sin x = \frac{12}{13}$, find $\cos 2x$.
$-\frac{119}{169} \approx -0.70414$

2. If $\sin x = \frac{12}{13}$ and $\frac{\pi}{2} < x < \pi$, determine $\sin 2x$.
$-\frac{120}{169}$

3. An object is shot from a cannon with an initial velocity of 50 feet per second. The cannon barrel makes an angle of 30° with the ground. (Ignore air resistance.)
a. How long will the object remain in the air?
$\frac{50(\sin 30°)}{16} \approx 1.56$ sec
b. What is the maximum height that it attains in flight?
≈ 9.77 feet
c. What horizontal distance does it travel?
≈ 67.7 feet

4. If $\cos\theta = \frac{5}{13}$ and $0 < \theta < \frac{\pi}{2}$, find $\cos\frac{\theta}{2}$.
$\frac{3}{\sqrt{13}}$

ADDITIONAL EXAMPLES

5. Write the value of $\sin \frac{\pi}{12}$ in terms of rational numbers and radicals.

sample:

$\sin \frac{\pi}{12} = \sin\left(\frac{1}{2} \cdot \frac{\pi}{6}\right) = \sqrt{\frac{1 - \cos \frac{\pi}{6}}{2}} = \frac{\sqrt{2 - \sqrt{3}}}{2}$

NOTES ON QUESTIONS

Question 7: Some students prefer to write $\cos\left(\frac{1}{2}x\right)$ rather than $\cos\left(\frac{x}{2}\right)$. To determine the sign of $\cos \frac{x}{2}$, note that $\frac{\pi}{2} < \frac{x}{2} < \frac{3\pi}{4}$; thus $\frac{x}{2}$ is in quadrant II, and therefore $\cos \frac{x}{2}$ is negative.

Question 11: Invite students to derive identities for $\sin 4x$ in three ways:
(1) $\sin 4x = \sin(3x + x) = \ldots$
(2) $\sin 4x = \sin 2(2x) = \ldots$
(3) $\sin 4x = \sin(2x + 2x) = \ldots.$

Question 12: Another way to show that these expressions are equal is to square both sides.

Question 13: Many students will need some guidance in doing this problem. The goal is to be able to do parts **c** and **d**. To do this, parts **a** and **b** are needed.

Question 21: Just as there are three identities for $\cos 2x$ in terms of $\cos x$ or $\sin x$, there are many identities for $\cos 3x$ and $\cos 4x$. You may wish to make the restriction that these identities be in terms of $\cos x$ alone. Then the answer is unique.

ADDITIONAL ANSWERS

3.a. $\cos\left(2 \cdot \frac{\pi}{6}\right) = \cos^2\left(\frac{\pi}{6}\right) - \sin^2\left(\frac{\pi}{6}\right) = \frac{3}{4} - \frac{1}{4} = \frac{1}{2}$; $\sin\left(2 \cdot \frac{\pi}{6}\right) = 2 \sin\left(\frac{\pi}{6}\right) \cos\left(\frac{\pi}{6}\right) = 2 \cdot \frac{1}{2} \cdot \frac{\sqrt{3}}{2} = \frac{\sqrt{3}}{2}$
b. $\cos\left(2 \cdot \frac{\pi}{6}\right) = \cos \frac{\pi}{3} = \frac{1}{2}$; $\sin\left(2 \cdot \frac{\pi}{6}\right) = \sin \frac{\pi}{3} = \frac{\sqrt{3}}{2}$

Example 4 If $\cos \alpha = \frac{3}{5}$, find $\cos \frac{\alpha}{2}$.

Solution Begin with the identity

$$\cos 2x = 2 \cos^2 x - 1.$$

Let $\alpha = 2x$, so $\frac{\alpha}{2} = x$.

$$\cos \alpha = 2 \cos^2\left(\frac{\alpha}{2}\right) - 1$$

Since $\cos \alpha = \frac{3}{5}$,

$$\frac{3}{5} = 2 \cos^2\left(\frac{\alpha}{2}\right) - 1.$$

Now solve for $\cos \frac{\alpha}{2}$. (Think of $\cos \frac{\alpha}{2}$ as a chunk.)

$$\frac{8}{5} = 2 \cos^2\left(\frac{\alpha}{2}\right)$$

$$\frac{4}{5} = \cos^2\left(\frac{\alpha}{2}\right)$$

$$\pm\sqrt{\frac{4}{5}} = \cos \frac{\alpha}{2}$$

Thus $$\cos \frac{\alpha}{2} = \pm\frac{2}{\sqrt{5}} = \pm\frac{2\sqrt{5}}{5}.$$

There are thus two possible values of $\cos \frac{\alpha}{2}$. Without additional information determining the sign of $\cos \frac{\alpha}{2}$, a unique solution cannot be found.

Example 5 Write the value of $\sin \frac{\pi}{8}$ in terms of rational numbers and radicals.

Solution $\frac{\pi}{8}$ is half of $\frac{\pi}{4}$, and the cosine and sine of $\frac{\pi}{4}$ are known. This suggests using the formula

$$\cos 2x = 1 - 2 \sin^2 x, \text{ with } \frac{\pi}{8} \text{ in place of } x.$$

$$\cos 2\left(\frac{\pi}{8}\right) = 1 - 2 \sin^2\left(\frac{\pi}{8}\right)$$

$$\cos \frac{\pi}{4} = 1 - 2 \sin^2\left(\frac{\pi}{8}\right)$$

$$\frac{\sqrt{2}}{2} = 1 - 2 \sin^2\left(\frac{\pi}{8}\right)$$

Now the problem is reduced to solving an equation for $\sin \frac{\pi}{8}$.

$$\frac{\sqrt{2}}{2} - 1 = -2 \sin^2\left(\frac{\pi}{8}\right)$$

$$-\frac{\sqrt{2}}{4} + \frac{1}{2} = \sin^2\left(\frac{\pi}{8}\right)$$

$$\frac{2 - \sqrt{2}}{4} = \sin^2\left(\frac{\pi}{8}\right)$$

$$\pm\sqrt{\frac{2 - \sqrt{2}}{4}} = \sin \frac{\pi}{8}$$

370

As in Example 4, at this point there are two possible solutions. But $0 < \frac{\pi}{8} < \frac{\pi}{2}$, so $\sin \frac{\pi}{8}$ is positive. Therefore,

$$\sin \frac{\pi}{8} = \sqrt{\frac{2-\sqrt{2}}{4}} = \frac{\sqrt{2-\sqrt{2}}}{2}.$$

Check A calculator indicates that both $\frac{\sqrt{2-\sqrt{2}}}{2}$ and $\sin \frac{\pi}{8}$ are approximately equal to 0.38268.

Covering the Reading

1. Prove identity (3) for $\cos 2x$.
$\cos 2x = \cos^2 x - \sin^2 x = (1 - \sin^2 x) - \sin^2 x = 1 - 2\sin^2 x$

2. Prove the identity $\sin 2x = 2 \sin x \cos x$.
$\sin 2x = \sin(x + x) = \sin x \cos x + \cos x \sin x = 2\sin x \cos x$

3. a. Use the identities in this lesson to find $\cos\left(2 \cdot \frac{\pi}{6}\right)$ and $\sin\left(2 \cdot \frac{\pi}{6}\right)$.

b. Check that your answers agree with the known values for $\cos \frac{\pi}{3}$ and $\sin \frac{\pi}{3}$. a,b) See margin.

4. If $\sin x = \frac{5}{13}$ and $0 < x < \frac{\pi}{2}$, find
a. $\cos 2x$ $\frac{119}{169}$ **b.** $\sin 2x$. $\frac{120}{169}$

5. If $\cos x = \frac{1}{4}$ and $\frac{3\pi}{2} < x < 2\pi$, find
a. $\cos 2x$ $-\frac{7}{8}$ **b.** $\sin 2x$. $-\frac{\sqrt{15}}{8}$

6. Refer to Example 3.
a. What is the range of a projectile launched with an initial speed of 90 ft/sec at an angle 45° above the horizontal? ≈ 253 ft
b. How long will it be in the air? ≈ 4 sec

7. If $\cos y = -\frac{12}{13}$ and $\pi < y < \frac{3\pi}{2}$, find $\cos \frac{y}{2}$. $\frac{-1}{\sqrt{26}}$

8. Express $\sin \frac{3\pi}{8}$ in terms of rational numbers and radicals. $\frac{\sqrt{2+\sqrt{2}}}{2}$

Applying the Mathematics

9. Graph $f: x \rightarrow 2\cos^2 x - 1$ and $g: x \rightarrow 1 - 2\sin^2 x$ on the same axes. Explain the results you get.
See margin. Both f and g are the function $x \rightarrow \cos 2x$.

10. Refer to Example 3.
a. If the velocity v_0 is fixed, find the angle θ which maximizes the range. $\frac{\pi}{4}$
b. What is the maximum range when $v_0 = 120$ ft/sec? 450 ft
c. When $v_0 = 120$ ft/sec and θ is chosen to maximize the range, what is the greatest height attained by the projectile? 112.5 ft

11. Prove the identity $\sin 3x = 3 \sin x - 4 \sin^3 x$. (Hint: Express $3x$ as $2x + x$.) See margin.

9., 11. See Additional Answers in the back of this book.

NAME ______

LESSON **MASTER 6–5**
QUESTIONS ON SPUR OBJECTIVES

■ **SKILLS** *Objective A (See pages 396–398 for objectives.)*

1. Given $0 < x < \frac{\pi}{2}$ and $\cos x = \frac{8}{17}$, find $\sin 2x$ and $\sin \frac{x}{2}$. $\frac{240}{289}$; $\frac{3}{\sqrt{34}}$

In 2 and 3, express the following in terms of rational numbers and radicals.

2. $\cos \frac{\pi}{8}$ $\frac{\sqrt{\sqrt{2}+2}}{2}$

3. $\sin \frac{\pi}{12}$ $\frac{\sqrt{2-\sqrt{3}}}{2}$

4. a. Use the identity for $\cos(\alpha - \beta)$ to find $\cos \frac{7\pi}{12}$. $\frac{\sqrt{2}-\sqrt{6}}{2}$

b. Use the identity $\cos 2\alpha = 2\cos^2 \alpha - 1$ to find $\cos \frac{7\pi}{12}$. $\frac{-\sqrt{2-\sqrt{3}}}{2}$

c. Show that your answers to parts **a** and **b** are equal.

$$\left(\frac{\sqrt{2}-\sqrt{6}}{4}\right)^2 = \frac{2+6-2\sqrt{12}}{16} = \frac{1}{2} - \frac{\sqrt{3}}{4}$$

$$\left(\frac{-\sqrt{2-\sqrt{3}}}{2}\right)^2 = \frac{2-\sqrt{3}}{4} = \frac{1}{2} - \frac{\sqrt{3}}{4}$$

Since both answers are negative and their squares are equal, they are equal.

68 *Continued* *Precalculus and Discrete Mathematics* © Scott, Foresman and Company

NAME ______
Lesson MASTER 6–5 (page 2)

■ **PROPERTIES** *Objective D*

5. *Multiple choice.* Which expression equals $\cos 2x$ for all x? (d)
(a) $1 - 2\cos^2 x$ (b) $\sin^2 x - \cos^2 x$
(c) $2\sin^2 x - 1$ (d) $2\cos^2 x - 1$

In 6–9, prove the identity and specify its domain.

6. $(\sin x + \cos x)^2 = \sin 2x + 1$
Left side $= \sin^2 x + \cos^2 x + 2 \sin x \cos x$
$= 1 + \sin x$
{all real numbers}

7. $1 - \tan^2 \theta = \frac{2 \tan \theta}{\tan 2\theta}$
Right side $= \dfrac{2\tan\theta}{\frac{2\tan\theta}{1-\tan^2\theta}}$
$= 2\tan\theta \cdot \frac{1-\tan^2\theta}{2\tan\theta}$
$= 1 - \tan^2\theta$
$\left\{\theta : \theta \neq \frac{k\pi}{4} \text{ where } k \text{ is an integer}\right\}$

8. $4\sin^2\alpha\cos^2\alpha + (\cos^2\alpha - \sin^2\alpha)^2 = 1$
Left side $= (2\sin\alpha\cos\alpha)^2 + (\cos 2\alpha)^2$
$= (\sin 2\alpha)^2 + (\cos 2\alpha)^2$
$= 1$
{all real numbers}

9. $2\sin\beta\csc 2\beta = \sec\beta$
Left side $= 2\sin\beta\left(\frac{1}{\sin 2\beta}\right)$
$= 2\sin\beta\left(\frac{1}{2\sin\beta\cos\beta}\right)$
$= \frac{1}{\cos\beta}$
$= \sec\beta$
$\left\{\beta : \beta \neq \frac{k\pi}{2} \text{ where } k \text{ is an integer}\right\}$

Precalculus and Discrete Mathematics © Scott, Foresman and Company 69

FOLLOW-UP

MORE PRACTICE
For more questions on SPUR Objectives, use *Lesson Master 6-5,* shown on page 371.

PROJECTS
The projects for Chapter 6 are described on page 393. **Projects 3 and 4** are related to the content of this lesson.

EVALUATION
Alternative Assessment
(1) The text does not try to define or work with inverses for the secant, cosecant, or cotangent functions. Why do you think this is so? (Sample: Since these three functions are reciprocals of the other three trigonometric functions, it is somewhat redundant to work with inverses for them as well.)

(2) Find the following:
(a) $\cos(\sec^{-1} 2)$ $\left(\frac{1}{2}\right)$
(b) $\sec^{-1} 0$ (undefined)
(c) $\sec^{-1} 1$ (0)
(d) $\sec^{-1} \tan\left(-\frac{\pi}{4}\right)$ (π)

ADDITIONAL ANSWERS
13.e. Does $\sin^2(x + y) + \cos^2(x + y) = 1$?
Does $\left(\frac{3 + 8\sqrt{2}}{15}\right)^2 + \left(\frac{4 - 6\sqrt{2}}{15}\right)^2 = 1$?
Does $\frac{137 + 48\sqrt{2}}{225} + \frac{88 - 48\sqrt{2}}{225} = 1$?
Does $\frac{225}{225} = 1$? Yes

14. See Additional Answers in the back of this book.

21.a. $\cos 3x = 4\cos^3 x - 3\cos x$; $\cos 4x = 8\cos^4 x - 8\cos^2 x + 1$
b. Sample: For $\cos(nx)$, the leading terms are always $2^{n-1}\cos^n x$.

12. Prove that $\sqrt{\frac{\sqrt{3} + 2}{4}} = \frac{\sqrt{2} + \sqrt{6}}{4}$ in the following way.
a. Find a value for $\cos \frac{\pi}{12}$ by thinking of $\frac{\pi}{12}$ as $\frac{\pi}{3} - \frac{\pi}{4}$ and using the identity for $\cos(\alpha - \beta)$. $\frac{\sqrt{6} + \sqrt{2}}{4}$
b. Find a value for $\cos \frac{\pi}{12}$ by thinking of $\frac{\pi}{6}$ as twice $\frac{\pi}{12}$ and using an identity for $\cos 2x$. $\frac{\sqrt{2 + \sqrt{3}}}{2}$
c. Why do parts **a** and **b** prove the desired result? Since the answers to parts a and b both represent $\cos \frac{\pi}{12}$, they are equal.

Review

13. Suppose $\pi < x < \frac{3\pi}{2}$, $\frac{\pi}{2} < y < \pi$, $\sin x = -\frac{4}{5}$, and $\cos y = -\frac{1}{3}$.
a. Find $\cos x$. $-\frac{3}{5}$
b. Find $\sin y$. $\frac{2\sqrt{2}}{3}$
c. Find $\cos(x + y)$. $\frac{3 + 8\sqrt{2}}{15}$
d. Find $\sin(x + y)$. $\frac{4 - 6\sqrt{2}}{15}$
e. Check your answers to parts **c** and **d** using the Pythagorean Identity. *(Lessons 6-4, 6-3, 2-8)* See margin.

14. Prove the identity $\tan^2\theta(\cot^2\theta + \cot^4\theta) = \csc^2\theta$ and specify its domain. *(Lesson 6-2)* See margin.

15. In a circle of radius 5, the point (5, 0) is rotated x radians counterclockwise to the point $P = (-3, 4)$. Find
a. $\cos x$ $-\frac{3}{5}$
b. $\sin x$ $\frac{4}{5}$
c. $\tan x$ $-\frac{4}{3}$
d. x. ≈ 2.214
(Lessons 5-6, 2-8)

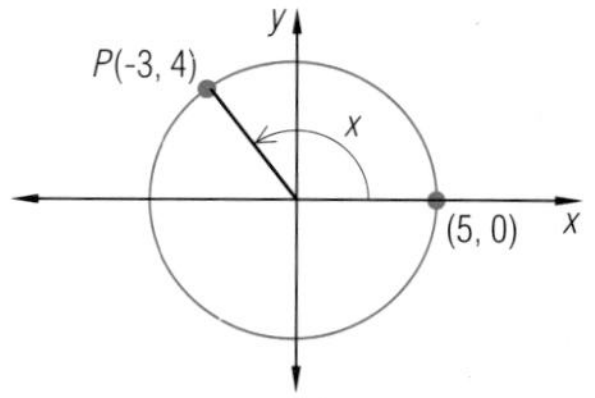

16. If $x - 5$ is a factor of a polynomial $p(x)$ then $p(5) =$ __?__. *(Lesson 5-2)* 0

17. Solve: $t^4 - 13t^2 + 36 = 0$. *(Lesson 3-4)* $t = \pm 3, \pm 2$

18. A function f is defined by the formula $y = f(x) = 3x + 2$. What is a formula for f^{-1}? *(Lesson 3-2)* $f^{-1}(x) = \frac{x - 2}{3}$

19. Determine whether the following argument is valid or invalid. Justify your reasoning.
If this coffee is a mild blend, then it is from Colombia.
If this coffee is popular in the United States, then it is a mild blend.
This coffee is from Colombia.
$\therefore$ *This coffee is a mild blend.*
(Lessons 1-8, 1-6) invalid, converse error

20. a. Write the negation of
$\forall$ *real numbers x,* $\sqrt{x - 2}$ *is a real number.*
b. Which is true: the statement or its negation? the negation
(Lessons 1-2, 1-1) a) $\exists$ a real number x such that $\sqrt{x - 2}$ is not real.

Exploration

21. a. Find identities for $\cos 3x$ and for $\cos 4x$.
b. Is there a pattern in the identities for $\cos 2x$, $\cos 3x$, and $\cos 4x$? If so, describe the pattern. See margin.

372

LESSON

6-6

Inverse Trigonometric Functions

You have already used the calculator keys $\boxed{\sin^{-1}}$, $\boxed{\cos^{-1}}$, and $\boxed{\tan^{-1}}$. These keys give values of the inverses of the sine, cosine, and tangent functions. In this lesson, these inverses are analyzed in some detail. But first some ideas about functions and their inverses are reviewed.

Because a function $f: x \rightarrow y$ cannot map a particular value of x onto more than one value of y, the graph of a real function cannot intersect any vertical line more than once. This is the basis for the Vertical Line Test for a function. In Lesson 3-2, it was shown that if a real function f has an inverse f^{-1}, then the graph of $y = f^{-1}(x)$ is the reflection image, over the line $y = x$, of the graph of $y = f(x)$.

When a vertical line is reflected about $y = x$, a horizontal line is obtained. It follows that if a real function f has an inverse f^{-1} that is a function, f must pass the Horizontal Line Test: that is, the graph of f cannot intersect any horizontal line more than once.

Now recall the graphs of the cosine, sine, and tangent functions.

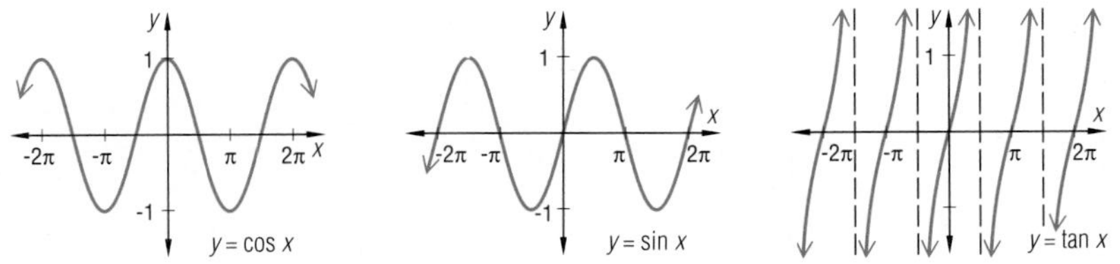

Notice that all three functions fail the Horizontal Line Test miserably! There are horizontal lines that intersect each graph at infinitely many points. For instance, if $y = \frac{1}{2}$, there are infinitely many values of x for which $\sin x = \frac{1}{2}$.

In order to obtain an inverse of any of these functions, the domain must be restricted so that for each y there is only one x with $f(x) = y$. But how should the domain be restricted? The following criteria for the choice of the restricted domains have become generally accepted:

1. The domain should include the angles between 0 and $\frac{\pi}{2}$ because these are the measures of the acute angles in a right triangle.
2. On the restricted domain, the function should take on all values in the range exactly once.
3. If possible, the function should be continuous on the restricted domain.

These criteria force the following choices for the domains of the restricted cosine, sine, and tangent functions.

LESSON 6-6

RESOURCES

- Lesson Master 6-6
- Quiz for Lessons 6-4 through 6-6
- Teaching Aid 38 provides the graphs of inverse functions for the cosine, sine, and tangent.

OBJECTIVES

B Evaluate inverse trigonometric functions with or without a calculator.
E Solve problems using inverse trigonometric functions.

TEACHING NOTES

In Lesson 3-8, there was a brief discussion of the inverse trigonometric functions. This lesson provides a more thorough discussion of these functions. The criteria at the bottom of this page provide the basis for restricting the domains of the trigonometric functions so that their inverses are also functions. Students should understand these criteria and how they are used in the definitions on page 374.

In discussing the criteria on restricting domains, emphasize that the criteria are applied to original functions in order to create new restricted functions. The restricted functions are the source for the inverse functions. A restriction on the domain of the original function becomes a restriction on the range of the inverse.

Concerning the criteria used to determine the restricted

domains, point out the following to students. Criterion 1 does not limit the domain to angles between 0 and $\frac{\pi}{2}$ (as some students assume). Criterion 2 is needed so that the function *f* passes the horizontal line test and thus has an inverse function. Criterion 3 implies that the domain must be continuous.

Some students may wonder why the restricted interval for the sine is from $\frac{-\pi}{2}$ to $\frac{\pi}{2}$ rather than from 0 to $\frac{\pi}{2}$ and then from $\frac{3\pi}{2}$ to 2π. Point out that the former choice meets criterion 3 while the latter does not.

Note that the range for the inverse tangent is the open interval $\frac{-\pi}{2} < \theta < \frac{\pi}{2}$, whereas the range for the sine is the closed interval with the same endpoints. You might ask students why there is this difference. (The tangent is undefined at both $\frac{-\pi}{2}$ and at $\frac{\pi}{2}$.)

Error Analysis You may need to remind students once again that $\sin^{-1} x$ means "the inverse of the sine function" and not $\frac{1}{\sin x}$. You might mention that some books use the notation arcsin *x* or Arcsin *x* instead of $\sin^{-1} x$ to avoid this confusion. Sin^{-1} is now more popular, however, because of the existence of calculators. Some books also use arcsin to denote the *relation* that is inverse to the sine function and Arcsin to denote the *function* that is inverse to the restricted sine function.

Alternate Approach In this book, the method used to determine the ranges of the inverse trigonometric functions focuses on the original trigonometric functions. As an alternate approach, you could

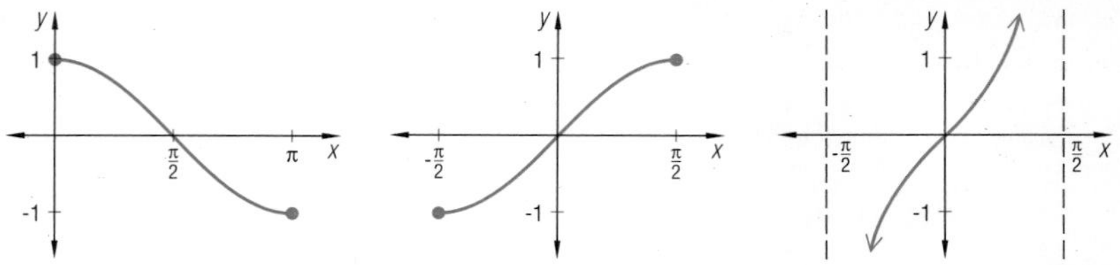

$y = \cos x$, $0 \le x \le \pi$ | $y = \sin x$, $-\frac{\pi}{2} \le x \le \frac{\pi}{2}$ | $y = \tan x$, $-\frac{\pi}{2} < x < \frac{\pi}{2}$

A restriction on the domain of a function becomes a restriction on the range of its inverse. Thus, to find the inverse function for $y = \cos x$ on the interval $0 \le x \le \pi$, first form the inverse relation $x = \cos y$ and then restrict y so that $0 \le y \le \pi$. This inverse function is denoted $\cos^{-1}$ and called the **inverse cosine function**. (Read $\cos^{-1}$ as "the number whose cosine is.") Repeating this procedure for the restricted sine and tangent functions gives the **inverse sine function** and the **inverse tangent function**, which are denoted by $\sin^{-1}$ and $\tan^{-1}$, respectively. These results are summarized below.

Definitions

$\forall$ *x* in the interval $-1 \le x \le 1$,
$\cos^{-1} x$ is the unique number θ in the interval $0 \le \theta \le \pi$ such that $\cos \theta = x$.

$\forall$ *x* in the interval $-1 \le x \le 1$,
$\sin^{-1} x$ is the unique number θ in the interval $-\frac{\pi}{2} \le \theta \le \frac{\pi}{2}$ such that $\sin \theta = x$.

$\forall$ real numbers *x*,
$\tan^{-1} x$ is the unique number θ in the interval $-\frac{\pi}{2} < \theta < \frac{\pi}{2}$ such that $\tan \theta = x$.

From the definition,

$$y = \cos^{-1} x \Leftrightarrow \cos y = x \text{ and } 0 \le y \le \pi.$$

(In the Questions, you are asked to write similar statements for $\sin^{-1}$ and $\tan^{-1}$.)

Example 1 A rectangular picture 4 feet high is hung on a wall so that the bottom edge is at your eye level. Your view of this picture is determined by the angle θ formed by the lines of sight from your eye to the top and bottom edges of the picture.

a. How does the angle θ depend on the distance *d* between you and the wall?
b. Find θ when $d = 8$ ft.

374

Solution

a. From the given information, $\tan \theta = \frac{4}{d}$. By the definition of $\tan^{-1}$,

$$\theta = \tan^{-1} \frac{4}{d}.$$

b. Substitute 8 for d in the answer to part **a**.

$$\theta = \tan^{-1} \left(\frac{4}{8}\right) = \tan^{-1} .5$$

$$\approx 26.6° \text{ or } .46 \text{ radians}$$

To graph the inverse trigonometric functions, use the fact that the graph of $y = f(x)$ and the graph of $y = f^{-1}(x)$ are reflection images of each other about the line $y = x$. Hence to find the graphs of the inverse trigonometric functions, first graph the corresponding trigonometric function over its restricted domain. Then reflect each point across the line $y = x$. The resulting curve is the graph of the inverse. This procedure is illustrated in Example 2.

Example 2 At the right is graphed $y = \cos x$ over the restricted domain $0 \le x \le \pi$. Graph $y = \cos^{-1} x$.

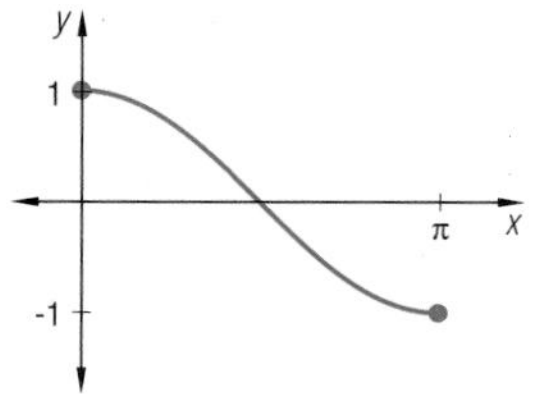

Solution The graph of $y = \cos x$ restricted to $0 \le x \le \pi$ is the black curve in the figure on the right. The blue curve is the reflection image of the black curve over the line $y = x$. The blue curve is the graph of $y = \cos^{-1} x$.

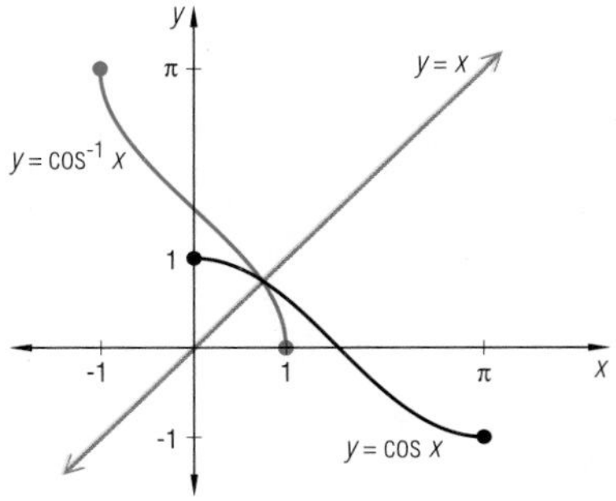

Example 3 **a.** Find the domain and range of $\cos^{-1}$.
b. Is $\cos^{-1}$ increasing or decreasing?

Solution

a. The domain of $\cos^{-1}$ is the range of the (restricted) cosine function. Therefore, the domain of $\cos^{-1}$ is $-1 \le x \le 1$. By definition of $\cos^{-1}$, its range is $0 \le y \le \pi$, the domain of the restricted cosine function.

b. As x increases from -1 to 1, as you move from left to right along the curve $y = \cos^{-1}x$, the y-values decrease. Thus, the inverse cosine function is a decreasing function. Another argument: since the cosine function is decreasing on this interval, so is its inverse, because the inverse of any decreasing function is decreasing.

change that focus to the inverse. Then, for example, you would begin with a graph of $y = \sin x$ for $-2\pi \le x \le 2\pi$. When this graph is reflected over the line $y = x$, the result is not a function. Ask students what restrictions on the range of the image would result in the image being a function. The range of the image is identical to the domain of the original function; thus applying criteria similar to those given in the text, the same results follow.

We do not encourage students to memorize the graphs of $y = \cos^{-1} x$, $y = \sin^{-1} x$, or $y = \tan^{-1} x$. Rather, students should use the fact that these graphs can be obtained by reflecting the restricted portion of the trigonometric function about the line $y = x$.

We have deliberately avoided discussion of $y = \sec^{-1} x$, $y = \csc^{-1} x$, or $y = \cot^{-1} x$. In practice, these three functions are seldom used.

You will probably need to go over **Example 4** carefully. The notation $\cos\left(\tan^{-1} \frac{b}{a}\right)$ is confusing to some students. Translating the problem into a triangle with sides labeled appropriately is a new strategy. Point out that calling $\tan^{-1} \frac{b}{a}$ "θ" is arbitrary; it could have been called α or β or anything else. Some students may try to do this problem algebraically. Knowing that $1 + \tan^2 \theta = \sec^2 \theta$, they replace $\tan \theta$ by $\frac{b}{a}$ and solve for $\sec \theta$. Then, because $0 < \theta < \frac{\pi}{2}$ and $\sec \theta > 0$, they use $\cos \theta = \frac{1}{\sec \theta}$ to get the final result.

ADDITIONAL EXAMPLES

1.a. Every morning at camp, Benny plays the bugle while the flag is being raised. He follows Old Glory with his eyes as it is being pulled skyward. If the flagpole is 32 feet tall, Benny's eyes are 4 feet above the ground, and he is standing d feet from the flagpole, at what angle θ with the horizontal is Benny's head tilted when the flag reaches the top of pole?
$\theta = \tan^{-1}\left(\frac{28}{d}\right)$

b. If $d = 100$ feet, what is θ?
$\theta \approx .273$ radians or $15.64°$

2. Graphed below is $y = \cos x$ over the restricted domain $-\pi \le x \le 0$. Why was this not chosen to be the restricted cosine function from which $\cos^{-1} x$ would be defined?

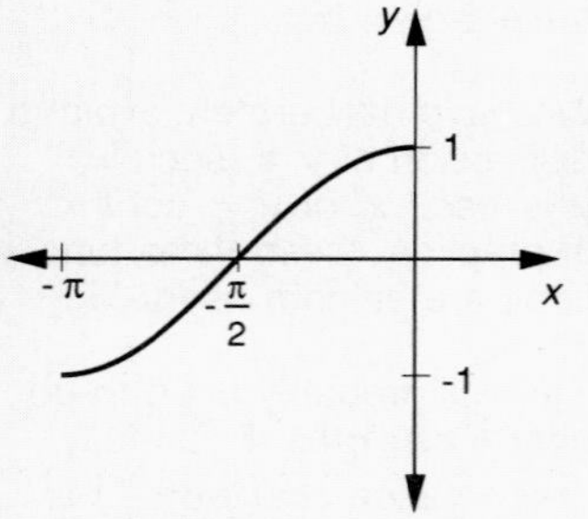

It excludes the right triangle values $0 < x \le \frac{\pi}{2}$ from its range.

3.a. Find the domain and range of $\sin^{-1} x$.
domain = $\{x: -1 \le x \le 1\}$;
range = $\left\{y: -\frac{\pi}{2} \le y \le \frac{\pi}{2}\right\}$

b. Is $\sin^{-1} x$ increasing or decreasing?
increasing

4. If $a > 0$ and $b > 0$, write $\csc\left(\tan^{-1} \frac{b}{a}\right)$ in terms of a and b.
$\csc\left(\tan^{-1} \frac{b}{a}\right) = \frac{\sqrt{a^2 + b^2}}{b}$

NOTES ON QUESTIONS

Questions 5-7: If students know the sin, cos, and tan of the integer multiples of $\frac{\pi}{6} = 30°$ and $\frac{\pi}{4} = 45°$, these questions can be answered quickly.

You can find the graphs of $y = \sin^{-1} x$ and $y = \tan^{-1} x$ in a similar way. In the Questions you will be asked to verify that the graphs look like the ones shown below.

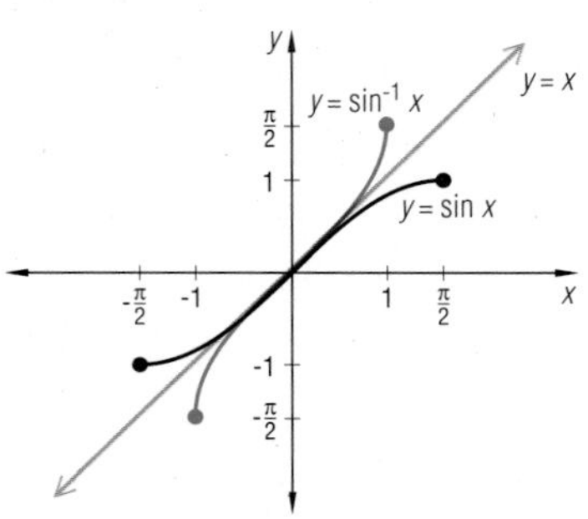

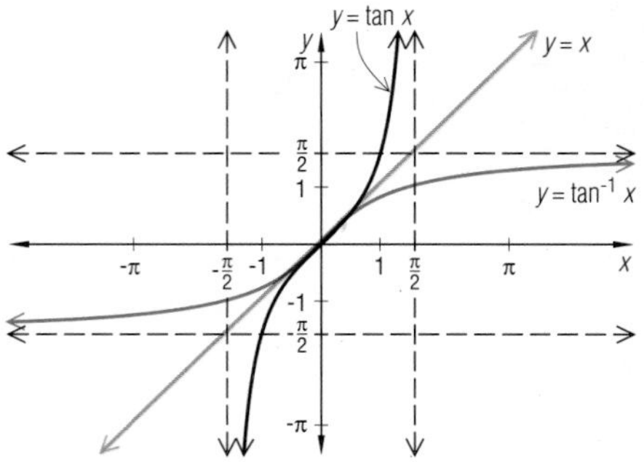

Inverse trigonometric functions, like other functions, can be added, subtracted, multiplied, divided, or composed with other functions. For instance, the expression $\cos(\sin^{-1} 0.5)$, read "the cosine of the number whose sine is 0.5," is obtained by composing the cosine and inverse sine functions. In the diagram at the left below, θ is a number whose sine is 0.5. The right diagram shows that $\cos \theta = \cos(\sin^{-1} 0.5) = \frac{\sqrt{3}}{2}$.

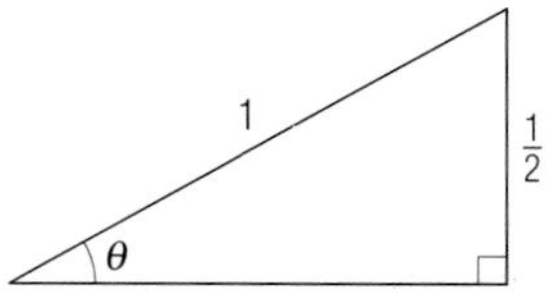

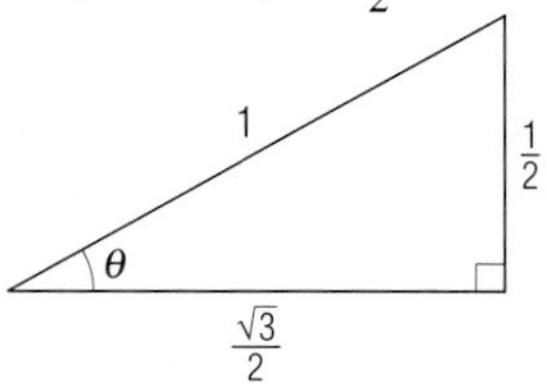

Example 4 If $a > 0$ and $b > 0$, write $\cos\left(\tan^{-1} \frac{b}{a}\right)$ in terms of a and b.

Solution By definition of cosine and $\tan^{-1}$, $\cos\left(\tan^{-1} \frac{b}{a}\right)$ is the cosine of the number θ whose tangent is $\frac{b}{a}$. Also, θ has to be in the interval $-\frac{\pi}{2} < \theta < \frac{\pi}{2}$. Because a and b are positive, θ, a, and b can be pictured as in the triangle at the right, with $\tan \theta = \frac{b}{a}$.

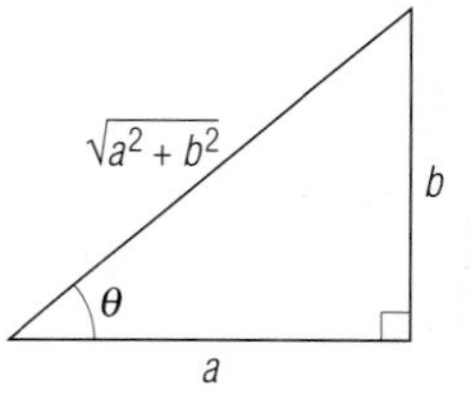

The hypotenuse of this triangle is $\sqrt{a^2 + b^2}$. You can see from the diagram that $\cos\left(\tan^{-1}\left(\frac{b}{a}\right)\right) = \cos \theta = \frac{\text{side adjacent to } \theta}{\text{hypotenuse}} = \frac{a}{\sqrt{a^2 + b^2}}$.

Check Pick values for a and b, say $a = 3$ and $b = 2$. Then

$$\tan^{-1} \frac{b}{a} = \tan^{-1} \frac{2}{3} \approx .588.$$

Then $$\cos\left(\tan^{-1} \frac{2}{3}\right) \approx \cos .588 \approx .832.$$

Now substitute 3 for a and 2 for b in the answer.

$$\frac{a}{\sqrt{a^2 + b^2}} = \frac{3}{\sqrt{2^2 + 3^2}} \approx \frac{3}{\sqrt{13}} \approx .832.$$ It checks.

376

Questions

Covering the Reading

1. Refer to the statement about $\cos^{-1}$ immediately following the definitions of the inverse cosine, inverse sine, and inverse tangent functions. Fill in the blanks.

a. $y = \sin^{-1}x \Leftrightarrow$ __?__ and __?__. $x = \sin y$ and $-\frac{\pi}{2} \le y \le \frac{\pi}{2}$

b. $y = \tan^{-1}x \Leftrightarrow$ __?__ and __?__. $x = \tan y$ and $-\frac{\pi}{2} < y < \frac{\pi}{2}$

2. Refer to Example 1. Let h be the height of the picture. Suppose an individual stands 6 feet away from the wall.

a. Write an equation that describes how θ depends on h.

b. Find the viewing angle for a picture that is 8 ft from top to bottom. a) $\theta = \tan^{-1}\left(\frac{h}{6}\right)$; b) $\approx 53°$

3. a. Complete the table of values below.

points on $y = \sin x$	$\left(-\frac{\pi}{2}, \ \right)$	$\left(-\frac{\pi}{3}, \ \right)$	$\left(-\frac{\pi}{4}, \ \right)$	$\left(-\frac{\pi}{6}, \ \right)$	$(0, \ )$	$\left(\frac{\pi}{6}, \ \right)$	$\left(\frac{\pi}{4}, \ \right)$	$\left(\frac{\pi}{3}, \ \right)$	$\left(\frac{\pi}{2}, \ \right)$
corresponding points on $y = \sin^{-1} x$									

b. Graph $y = \sin x$ and $y = \sin^{-1} x$ on the same axes.
See margin.

4. Refer to the graph of $y = \tan^{-1} x$.

a. Find the domain. all real numbers

b. Find the range. $-\frac{\pi}{2} < y < \frac{\pi}{2}$

c. Is the function increasing or decreasing? increasing

d. What are $\lim_{x \to +\infty} \tan^{-1}x$ and $\lim_{x \to -\infty} \tan^{-1}x$? $\frac{\pi}{2}$; $-\frac{\pi}{2}$

In 5–7, use your knowledge of the values of the trigonometric functions for integer multiples of $\frac{\pi}{6} = 30°$ and $\frac{\pi}{4} = 45°$ to find the value: **a.** in radians; and **b.** in degrees.

5. $\sin^{-1}\left(-\frac{\sqrt{2}}{2}\right)$ a) $-\frac{\pi}{4}$; b) $-45°$ **6.** $\cos^{-1}(0)$ a) $\frac{\pi}{2}$; b) $90°$ **7.** $\tan^{-1}(-\sqrt{3})$ a) $-\frac{\pi}{3}$; b) $-60°$

In 8–10, use your calculator to compute the value (in radians).

8. $\cos^{-1}\left(\frac{\sqrt{3}}{3}\right)$ 0.955 **9.** $\sin^{-1}(-.9)$ -1.120 **10.** $\tan^{-1}(2)$ 1.107

11. a. How is the expression $\sin\left(\cos^{-1}\frac{3}{5}\right)$ read?

b. Evaluate $\sin\left(\cos^{-1}\frac{3}{5}\right)$. $\frac{4}{5}$

a) the sine of the number whose cosine is $\frac{3}{5}$

12. Evaluate $\cos(\tan^{-1}(-1))$. $\frac{\sqrt{2}}{2} \approx .707$

In 13 and 14, draw an appropriate triangle and evaluate the expression.

13. $\sin\left(\tan^{-1}\frac{b}{a}\right)$ See margin. **14.** $\sin\left(\cos^{-1}\frac{x}{3}\right)$, $x > 0$ See margin.

ADDITIONAL ANSWERS

3.a. See Additional Answers in the back of this book.

b.

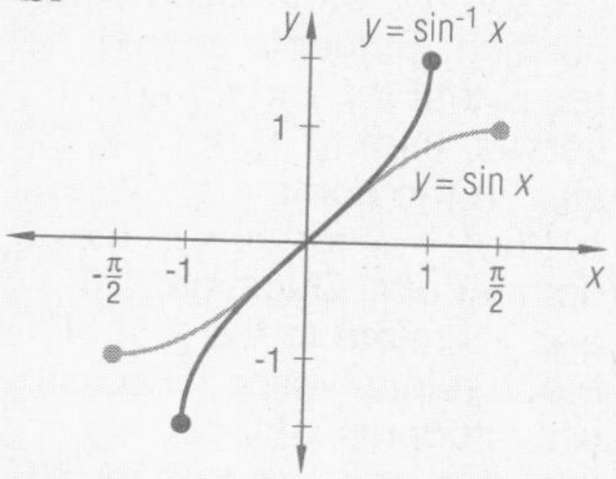

13.

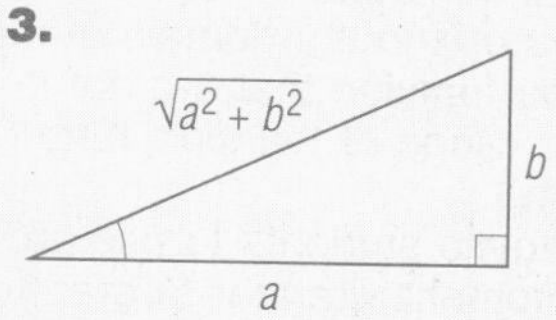

$\sin\left(\tan^{-1}\frac{b}{a}\right) = \frac{b}{\sqrt{a^2 + b^2}}$

14.

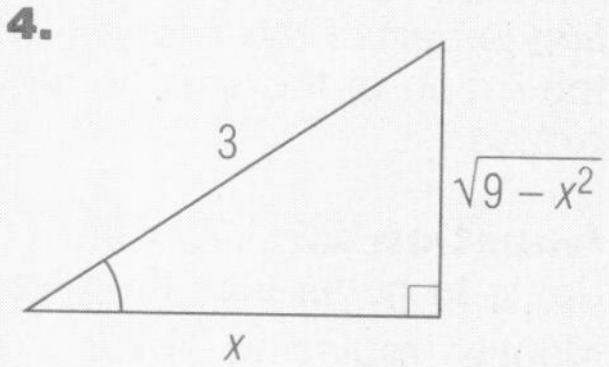

$\sin\left(\cos^{-1}\frac{x}{3}\right) = \frac{\sqrt{9 - x^2}}{3}$

NAME ______________________

LESSON **MASTER 6–6**
QUESTIONS ON **SPUR** OBJECTIVES

■ SKILLS *Objective B* *(See pages 396–398 for objectives.)*

1. Let $f(x) = \tan^{-1}(x)$.

a. Within what interval must x lie? x can be any real number.

b. Within what interval must $f(x)$ lie? $-\frac{\pi}{2} < f(x) < \frac{\pi}{2}$

In 2–5, compute without using a calculator.

2. $\sin^{-1}\left(\frac{-\sqrt{3}}{2}\right)$ $-\frac{\pi}{3}$ 3. $\tan(\tan^{-1} 0)$ 0

4. $\cos(\sin^{-1} 1)$ 0 5. $\sin\left(\tan^{-1}\left(\tan\frac{\pi}{6}\right)\right)$ $\frac{1}{2}$

In 6 and 7, use ΔABC to evaluate the expression.

6. $\sin\left(\tan^{-1}\left(\frac{a}{b}\right)\right)$ $\frac{a}{c}$

7. $\cos^{-1}\left(\frac{a}{c}\right)$ ϕ

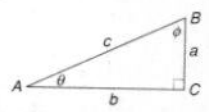

■ USES *Objective E*

8. Snell's Law of Refraction states that when a light ray passes from one medium into a second one, its path bends so that the angles α and β (shown in the diagram) obey $\frac{\sin\alpha}{\sin\beta} = C$, where C is a constant called the *index of refraction*. If the index of refraction is 1.24, and a ray of light strikes the liquid at an angle of incidence of $\alpha = 65°$, find the angle of refraction β. about 47.0°

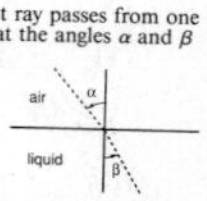

9. Suppose that the sun, Earth, and Jupiter are positioned as shown in the diagram at the right.

a. If Jupiter is x km from the sun and Earth is 150 million km from the sun, give an expression for the angle θ formed by the sight lines from Earth to the other two bodies. $\theta = \tan^{-1}\left(\frac{x}{1.5 \cdot 10^8}\right)$

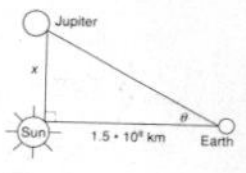

b. If $x = 7.8 \cdot 10^8$ km, find θ. about 79.1°

70 *Precalculus and Discrete Mathematics* © Scott, Foresman and Company

NOTES ON QUESTIONS
Making Connections for Questions 15 and 16: These questions relate to Lesson 3-2 on function composition. Students should recall that if *f* and f^{-1} are inverses, then $f(f^{-1}(x)) = x$ for all *x* in the domain of f^{-1}, and $f^{-1}(f(x)) = x$ for all *x* in the domain of *f*. Questions 15 and 16 relate to the first of these relationships. Discuss with students why $\sin^{-1}(\sin x) = x$ is not true for all real numbers. (There are values of *x* in the domain of the sine function that are not in the range of the $\sin^{-1}$ function.)

Invite students to use an automatic grapher to graph $y = \sin(\sin^{-1} x)$ and $y = \sin^{-1}(\sin x)$ for $-2\pi \le x \le 2\pi$, $-2\pi \le y \le 2\pi$. Relate the graphs to the above discussion. (The graph of the first is part of the line $y = x$; the graph of the second is a zigzag.)

Question 23: Ask students to generalize the given identity, replacing $\frac{\pi}{3}$ by β. ($\sin(\alpha + \beta) + \sin(\alpha - \beta) = 2 \sin\alpha \cos\beta$; in this question, $\cos\beta = \frac{1}{2}$.)

FOLLOW-UP

MORE PRACTICE
For more questions on SPUR Objectives, use *Lesson Master 6-6, shown on page 377.*

EVALUATION
A quiz covering Lessons 6-4 through 6-6 is provided in the Teacher's Resource File.

ADDITIONAL ANSWERS
19. $\theta = \tan^{-1}\left(\frac{h}{10}\right)$, where h = altitude (in miles)

20., 23. See Additional Answers in the back of this book.

Applying the Mathematics

In 15 and 16, compute without using a calculator.

15. $\sin\left(\sin^{-1}\left(-\frac{\sqrt{2}}{2}\right)\right)$ $-\frac{\sqrt{2}}{2}$ **16.** $\tan(\tan^{-1}(1.2))$ 1.2

17. Use the Pythagorean Identity to determine $\cos(\sin^{-1} .6)$ without using a calculator. .8

18. Use the Law of Sines to approximate the value of θ, in degrees, for the triangle pictured at the right. ≈ 74°

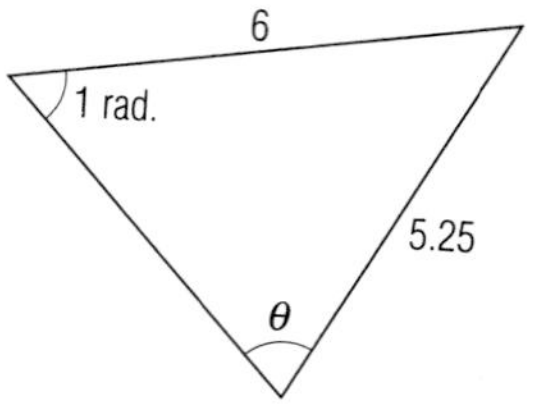

19. A radar tracking station is located 10 miles from a rocket launching pad. If a rocket is launched straight upward, express the angle of elevation of the rocket from the tracking station as a function of the altitude (in miles) of the rocket. See margin.

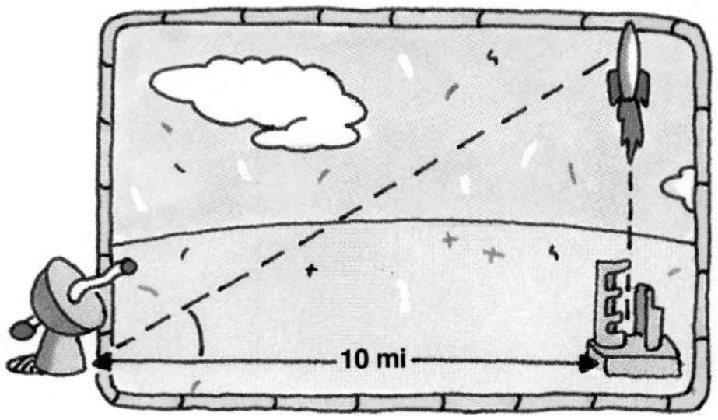

Review

20. Express $\sin\left(\frac{5\pi}{8}\right)$ in terms of rational numbers and radicals. *(Lesson 6-5)* See margin.

21. Suppose that $\sin x = \frac{2}{3}$ and $\frac{\pi}{2} < x < \pi$. Find $\sin 2x$ and $\cos 2x$. *(Lesson 6-5)* $\sin 2x = -\frac{4\sqrt{5}}{9}$, $\cos 2x = \frac{1}{9}$

22. Prove: ∀ *real numbers x such that* $\cos x \neq 0$, $\frac{\sin 2x}{\cos x} = 2 \sin x$. *(Lesson 6-5)* $\frac{\sin 2x}{\cos x} = \frac{2 \sin x \cos x}{\cos x} = 2 \sin x$

23. Show that ∀ real numbers α, $\sin\left(\alpha + \frac{\pi}{3}\right) + \sin\left(\alpha - \frac{\pi}{3}\right) = \sin\alpha$. *(Lesson 6-4)* See margin.

24. Consider the following identity: $\tan x \cdot \csc x = \sec x$.
a. What is the domain of the identity? $x \neq 0 \pmod{\frac{\pi}{2}}$
b. Prove the identity. $\tan x \cdot \csc x = \frac{\sin x}{\cos x} \cdot \frac{1}{\sin x} = \frac{1}{\cos x} = \sec x$
(Lesson 6-2)

25. Find ten positive integers satisfying $y \equiv 4 \pmod{11}$. *(Lesson 4-3)* sample: 15, 26, 37, 48, 59, 70, 81, 92, 103, 114

Exploration

26. a. Use an automatic grapher to guess the values of *x* for which $\sin(\cos^{-1}x) = \cos(\sin^{-1}x)$. $-1 \le x \le 1$
b. Prove your conjecture in part **a** by finding an expression which contains no trigonometric functions and is equal to both sides of the equation for the values of *x* you specified.
c. Find and prove another identity (valid only on an interval you specify) which contains one trigonometric and one inverse trigonometric function on each side of the equation.
b) $\sin(\cos^{-1}x) = \sqrt{1 - x^2} = \cos(\sin^{-1}x)$

c) Sample: $\tan(\cot^{-1}x) = \cot(\tan^{-1}x)$ for $x > 0$.

LESSON

6-7

Solving Equations and Inequalities Involving Trigonometric Functions

"There goes Archimedes with his confounded lever again."

In this lesson the logic of solving equations and inequalities as well as algebraic and graphical methods for finding solutions are applied to trigonometric equations and inequalities. The first example illustrates how to compute and describe the numbers that give rise to a particular value of a trigonometric function.

Example 1 Find and describe all real numbers x such that $\sin x = \frac{3}{4}$.

Solution One solution of this equation is $x = \sin^{-1}\left(\frac{3}{4}\right)$.
Use a calculator to find that $x \approx .84806$ radians $\approx 48.59°$.
By the definition of the inverse sine function, this is the only solution of the given equation in the interval $-\frac{\pi}{2} \le x \le \frac{\pi}{2}$. However, because $\sin(\pi - x) = \sin x$ for all real numbers x, another solution to the equation is $\pi - x \approx (\pi - .84806)$ radians ≈ 2.29353 radians $\approx 131.41°$.
This second solution lies in the interval from $\frac{\pi}{2}$ to π.

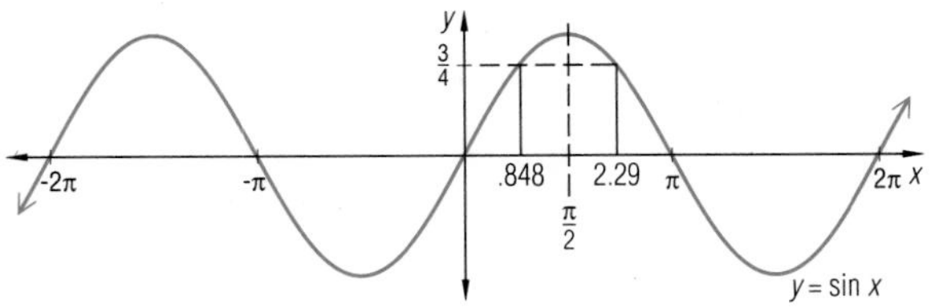

You can see from the graph that these are the only two solutions of the given equation in the interval $0 \le x \le 2\pi$. Because the sine function is periodic with period 2π, the solutions to the given equation are numbers of the form

$$x = .84806 + 2n\pi \text{ or } x = 2.29353 + 2n\pi, \text{ for all integers } n.$$

Another way of writing this is with modular arithmetic.

$$x \equiv .84806 \pmod{2\pi} \text{ or } x \equiv 2.29353 \pmod{2\pi}$$

LESSON 6-7

RESOURCES
- Lesson Master 6-7

OBJECTIVES

C Solve trigonometric equations and inequalities algebraically.
F Use trigonometric equations and inequalities to solve applied problems.
H Use graphs to solve trigonometric equations and inequalities.

TEACHING NOTES

Technology The use of an automatic grapher can enhance the discussion of this lesson. A visual understanding of the meaning of a solution should prove helpful when shifting the focus to algebraic techniques. It also should help students realize that, when solved over the reals, trigonometric equations frequently have infinitely many solutions.

You can use the following graphs as an aid in finding all values of x for $0 \le x \le 2\pi$ that have given sine, cosine or tangent values.

Two numbers with same sine are x and $\pi - x$.

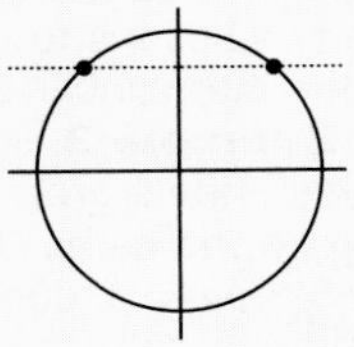

Two numbers with same cosine are x and $2\pi - x$.

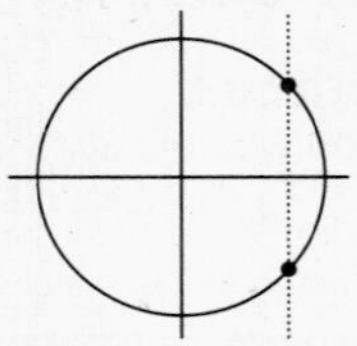

Two values with the same tangent are x and $\pi + x$.

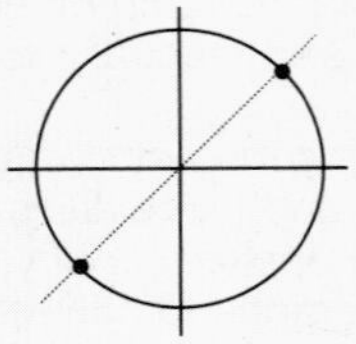

Alternate Approach
There is a jump in the level of difficulty from **Example 1** to **Example 2**. You may wish to increase the complexity of these problems more gradually in the following way. Using as the domain $0 \le x \le 2\pi$, ask students to solve:

(a) $\sin x = 1$; $\left(x = \frac{\pi}{2}\right)$

(b) $\sin x = \frac{1}{2}$; $\left(x = \frac{\pi}{6} \text{ or } x = \frac{5\pi}{6}\right)$

(c) $\sin x - 1 = 0$; (same as (a))

(d) $2 \sin x - 1 = 0$; (same as (b))

(e) $(\sin x - 1)(2 \sin x - 1) = 0$; (same as (a) and (b) combined)

(f) $\sin^2 x - 3 \sin x + 1 = 0$. (same as (e))

Point out that solving quadratic trigonometric equations often involves going from step (f) to steps (a) and (b).

Students often do not know how many solutions to expect in the interval $0 \le x \le 2\pi$. For **Example 2**, suggest that students graph the following on the same coordinate system: $y = 2\cos^2 x + \cos x$ and $y = 1$ for $0 \le x \le 2\pi$ and $-3 \le y \le 3$.

The next example involves the solving of a quadratic equation in $\cos x$.

Example 2 Solve $2\cos^2 x + \cos x = 1$
a. over the interval $0 \le x \le 2\pi$,
b. over the set of all real numbers.

Solution
a. Rewrite the equation as $2\cos^2 x + \cos x - 1 = 0$. If you think of "$\cos x$" as a chunk, you see that the left side has the form $2q^2 + q - 1$ which equals $(2q - 1)(q + 1)$. Thus the given equation implies

$$(2\cos x - 1)(\cos x + 1) = 0.$$

By the Zero Product Property,

$$2\cos x - 1 = 0 \quad \text{or} \quad \cos x + 1 = 0$$

$$\cos x = \frac{1}{2} \quad \text{or} \quad \cos x = -1.$$

Now find the solutions in the interval $0 \le x \le 2\pi$, which is one period of the cosine function.

$$x = \frac{\pi}{3} \text{ or } x = \frac{5\pi}{3} \quad \text{or} \quad x = \pi$$

Check that these values satisfy the given equation. When $x = \frac{\pi}{3}$, $2\cos^2\left(\frac{\pi}{3}\right) + \cos\left(\frac{\pi}{3}\right) = 2\left(\frac{1}{2}\right)^2 + \frac{1}{2} = 1$, as required. The checks of the other values are left to you.

b. By the result of part **a**, together with the fact that the cosine function has period 2π, you can conclude that the solutions are all x of the form

$$x = \frac{\pi}{3} + 2\pi n, \quad x = \frac{5\pi}{3} + 2\pi n, \quad \text{or } x = \pi + 2\pi n = (2n + 1)\pi,$$

where n is an integer. You could also write

$$x \equiv \frac{\pi}{3} \pmod{2\pi} \text{ or } x \equiv \frac{5\pi}{3} \pmod{2\pi} \text{ or } \quad x \equiv \pi \pmod{2\pi}.$$

Examples 1 and 2 illustrate a general two-step strategy for finding all real number solutions of a trigonometric equation.

Step 1: Solve the given equation over a restricted domain equal to the largest fundamental period of the given trigonometric functions.

Step 2: Use periodicity properties of the trigonometric functions, together with the solutions from Step 1, to find all solutions.

At times you may need to use an identity to rewrite one side of the equation in a form which can be factored.

Example 3 Find the solution set for the equation $\tan x + \sec^2 x - 1 = 0$.

Solution Use the identity $1 + \tan^2 x = \sec^2 x$ to substitute for $\sec^2 x$. The resulting equation is quadratic in $\tan x$ and can be solved over the restricted domain $-\frac{\pi}{2} < x < \frac{\pi}{2}$.

$$\tan x + (1 + \tan^2 x) - 1 = 0$$
$$\Leftrightarrow \quad \tan^2 x + \tan x = 0$$
$$\Leftrightarrow \quad \tan x\,(1 + \tan x) = 0$$
$$\Leftrightarrow \quad \tan x = 0 \quad \text{or} \quad 1 + \tan x = 0$$
$$\tan x = -1$$
$$\Leftrightarrow \quad x = 0 \qquad x = -\frac{\pi}{4}$$

Because $\tan x$ has period π, it follows that the exact solution set is the set of all real numbers of the form

$$x = 0 + n\pi = n\pi \quad \text{or} \quad x = -\frac{\pi}{4} + n\pi$$

for all integers n. That is,

$$x \equiv 0 \pmod{\pi} \quad \text{or} \quad x \equiv -\frac{\pi}{4} \pmod{\pi}.$$

Graphing can help solve trigonometric inequalities.

Example 4 **a.** Use an automatic grapher to approximate the solutions to

$$2\sin^2 x + \sin x - 1 \geq 0$$

in the interval $0 \leq x \leq 2\pi$.

b. Find the exact solutions to the inequality.

Solution

a. Use an automatic grapher to plot the function

$$f(x) = 2\sin^2 x + \sin x - 1$$

on the interval $0 \leq x \leq 2\pi$. Then use the trace function on the grapher to estimate the solution set for the inequality. A reasonable estimate is that x can be any value in the interval $.52 \leq x \leq 2.62$, or may be the single value 4.71, or in a small interval around 4.71.

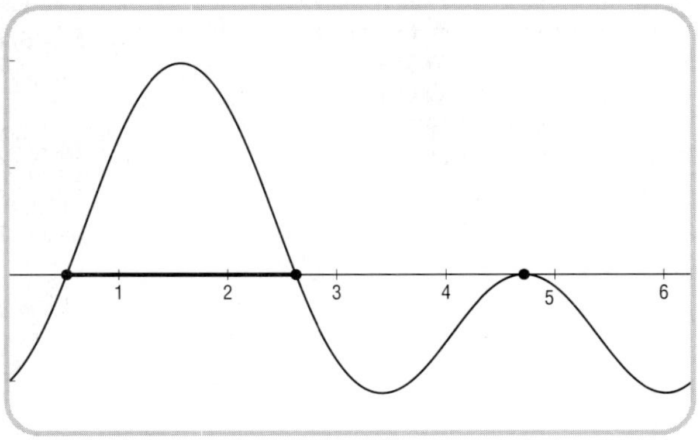

Students will see that the period of $y = 2\cos^2 x + \cos x$ is 2π and that there are three solutions over the desired interval.

Chunking ideas are used in this lesson to help factor trigonometric equations. Furthermore, in **Example 3**, students should see that identities often must be used before attempting a solution. Remind students that it is not acceptable to divide by a variable or by an expression containing a variable. There is the possibility that solutions may be lost.

In **Examples 1-4**, all of the quadratic equations are factorable and they all have standard solutions. Students may mistakenly think that all trigonometric equations have these kinds of solutions. Show students one equation that does not have a standard solution; for example, $10\sin^2 x + 2\sin x - 1 = 0$. The quadratic formula is needed and leads to the following solutions (to the nearest .001 radians) in the interval $0 \leq x \leq 2\pi$: .234, 2.908, 3.588, and 5.837.

In **Example 4**, the combination of quadratics, trigonometric functions, and inequalities can be intimidating. Note that first the related equation is considered. This equation is easier than the one in **Example 3**. If need be, substitute t for $\sin x$ and solve $2t^2 + t - 1 = 0$.

Example 5 uses the formula for the range of a projectile that was developed in Lesson 6-5. Students may not realize that the picture of the path of the cannonball and the graph are not showing the same thing. The picture relates vertical distance to horizontal distance, showing a single flight path. The graph relates horizontal distance to the angle formed by the cannon.

ADDITIONAL ANSWERS

1. Find and describe all real numbers x such that

a. $\cos x = \frac{1}{2}$;

$x = \frac{\pi}{3} + 2n\pi$ or $x = \frac{5\pi}{3} + 2n\pi$

b. $\cos x = \frac{3}{4}$.

$x \approx .7227342478 + 2n\pi$ or $x \approx 5.560451059 + 2n\pi$

2. Solve $4 \sin^2 x - 1 = 0$

a. over the interval $0 \le x \le 2\pi$;

$x = \frac{\pi}{6}, \frac{5\pi}{6}, \frac{7\pi}{6}, \frac{11\pi}{6}$

b. over the set of all real numbers.

$x = \frac{\pi}{6} + 2n\pi$ or $x = \frac{11\pi}{6} + 2n\pi$

3. Find the solution set for the equation $\cot x + \csc^2 x = 1$.

$x = \frac{3\pi}{4} + n\pi$ or $x = \frac{\pi}{2} + n\pi$

4.a. Use an automatic grapher to approximate the solutions to $4 \cos^2 x - 1 \ge 0$ on the interval $0 \le x \le 2\pi$.

$0 \le x \le 1.05$ or $2.09 \le x \le 4.19$ or $5.24 \le x \le 6.28$

b. Find the exact solutions to the inequality.

$0 \le x \le \frac{\pi}{3}$ or $\frac{2\pi}{3} \le x \le \frac{4\pi}{3}$ or $\frac{5\pi}{3} \le x \le 2\pi$

5. The formula $H = \frac{v_0^2 \sin^2 \theta}{64}$ estimates the maximum height achieved by the person launched by the Rocket Launcher. Given that the person's initial velocity is 50 ft/sec and that the person needs to go above a banner 20 ft above the ground and below a supportive beam of the tent above the banner, which is 30 feet above the ground, find the values of θ ($0 < \theta < 90°$) for a safe flight.

$45.69 < \theta < 61.21°$

b. To determine the exact solution set of the inequality, first solve the equation $2 \sin^2 x + \sin x - 1 = 0$ by noting that the left side has the form $2q^2 + q - 1 = 0$ which can be factored.

$$(2 \sin x - 1)(\sin x + 1) = 0$$

$$\Leftrightarrow 2 \sin x - 1 = 0 \quad \text{or } \sin x + 1 = 0$$

$$\Leftrightarrow \quad \sin x = \frac{1}{2} \quad \text{or} \quad \sin x = -1$$

$$\Leftrightarrow x = \frac{\pi}{6} \text{ or } x = \frac{5\pi}{6} \quad \text{or} \quad x = \frac{3\pi}{2}$$

The function $y = 2 \sin^2 x + \sin x - 1$ is continuous so the Test Point Method can be used. With that you should find that $2 \sin^2 x + \sin x - 1 \ge 0$ for x in the interval $\frac{\pi}{6} \le x \le \frac{5\pi}{6}$ and for the single value $x = \frac{3\pi}{2}$.

Check Since $\frac{\pi}{6} \approx .52$, $\frac{5\pi}{6} \approx 2.62$, and $\frac{3\pi}{2} \approx 4.71$, the values in part **b** agree with the estimates found in part **a** from graphing.

The following example uses the results on projectile motion developed in Example 3 of Lesson 6-5.

Example 5 In recent years, the Ringling Brothers Circus featured an act called the Rocket Launcher. A person is launched from a rocket with an initial velocity of 64 ft/sec. The formula

$$R = \frac{v_0^2}{32} \sin 2\theta$$

estimates the horizontal distance R (in feet) traveled when the person is launched at an angle of θ with an initial speed of v_0 ft/sec. Find the values of θ so that the person will travel a horizontal distance greater than 100 ft, the distance to the nearest part of the safety net.

Solution Given $v_0 = 64$ ft/sec, the solution set consists of all values of θ such that

$$100 < \frac{(64)^2}{32} \sin 2\theta$$

or, equivalently,

$$100 < 128 \sin 2\theta.$$

382

The situation requires that θ lie in the interval $0 < \theta < 90°$. The colored part of the graph below indicates those points on the graph of $y = 128 \sin 2\theta$ whose y-coordinates are greater than $y = 100$.

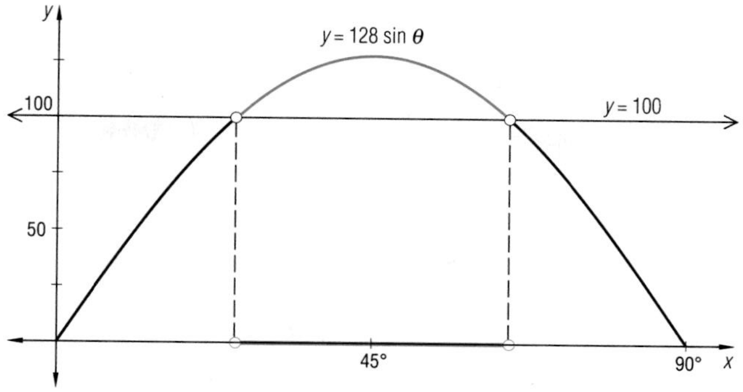

To find the corresponding x-coordinates, solve the equation

$$100 = 128 \sin 2\theta$$

or $$.78125 \approx \sin 2\theta.$$

A calculator shows $2\theta \approx 51.38°$, making $\theta \approx 25.69°$,

but since $\sin x = \sin(\pi - x) = \sin(180° - x)$,

the other solution is $2\theta \approx 128.62°$.

So the solutions to the equation are

$$\theta \approx 25.69° \quad \text{or} \quad \theta \approx 64.31°.$$

Thus for the inequality $25.69° < \theta < 64.31°$, the person will travel greater than 100 feet when shot at an angle between 25.69° and 64.31°.

Questions

Covering the Reading

In 1 and 2, describe all real numbers x satisfying the equation.

1. $\sin x = 1$ See margin.

2. $\tan x = 1$ See margin.

3. Find all solutions (to the nearest thousandth) to $\cos x = \frac{2}{3}$:
a. over the interval $0 \le x \le 2\pi$,
b. over the set of real numbers.
See margin.

4. Check whether the values $\frac{5\pi}{3}$ and π are solutions to the equation of Example 2. See margin.

5. Solve the equation $2 \sin^2 x - 5 \sin x - 3 = 0$ for x:
a. over the interval $0 \le x \le 2\pi$, $\frac{7\pi}{6}, \frac{11\pi}{6}$
b. over the set of real numbers. See margin.

6. Use an appropriate trigonometric identity to help find all real numbers x such that $\cos^2 x - \sin x + 1 = 0$. $\frac{\pi}{2} + 2\pi n$, n an integer

NOTES ON QUESTIONS

Question 1: In this and other questions in which the domain is the set of real numbers, allow solutions to be presented in a variety of ways. Some students may prefer modular arithmetic; others may prefer adding $\pm 2\pi n$ or some other indication of period to a basic solution; still others may wish to combine the algebra to obtain $x = \left(2n + \frac{1}{2}\right)\pi$; and others may prefer set notation $\{x: x = \frac{\pi}{2} + 2\pi n\}$. The important criteria are clarity and completeness.

Question 6: To facilitate factoring, students should use the strategy of rewriting the equation so that it contains only one type of trigonometric function. Ask if it is simpler to rewrite $\cos^2 x$ in terms of $\sin x$ or to rewrite $\sin x$ in terms of $\cos x$. (The former is easier because no square root is involved.)

ADDITIONAL ANSWERS

1. $\frac{\pi}{2} + 2\pi n$, n an integer, or equivalently $x \equiv \frac{\pi}{2} \pmod{2\pi}$

2. $\frac{\pi}{4} + \pi n$, n an integer or $x \equiv \frac{\pi}{4} \pmod{\pi}$

3.a. 0.841 and 5.442
b. $\pm 0.841 + 2n\pi$, n an integer

4. $2\cos^2 \frac{5\pi}{3} + \cos \frac{5\pi}{3} = 2\left(\frac{1}{2}\right)^2 + \frac{1}{2} = \frac{1}{2} + \frac{1}{2} = 1$ as required. $2\cos^2 \pi + \cos \pi = 2(-1)^2 + -1 = 2 - 1 = 1$ as required. Hence, $\frac{5\pi}{3}$ and π are solutions.

5.b. $x = \frac{7\pi}{6} + 2n\pi$ or $x = \frac{11\pi}{6} + 2n\pi$, where n is any integer, or equivalently $x \equiv \frac{7\pi}{6} \pmod{2\pi}$ or $x \equiv \frac{11\pi}{6} \bmod 2\pi$.

NOTES ON QUESTIONS
Question 24: This question is worth discussion time in class because it points to the power of chunking as an equation-solving technique.

FOLLOW-UP

MORE PRACTICE
For more questions on SPUR Objectives, use *Lesson Master 6-7,* shown on page 385.

PROJECTS
The projects for Chapter 6 are described on page 393. **Projects 1 and 2** are related to the content of this lesson.

ADDITIONAL ANSWERS
16., 17.b. See Additional Answers in the back of this book.

18.d.

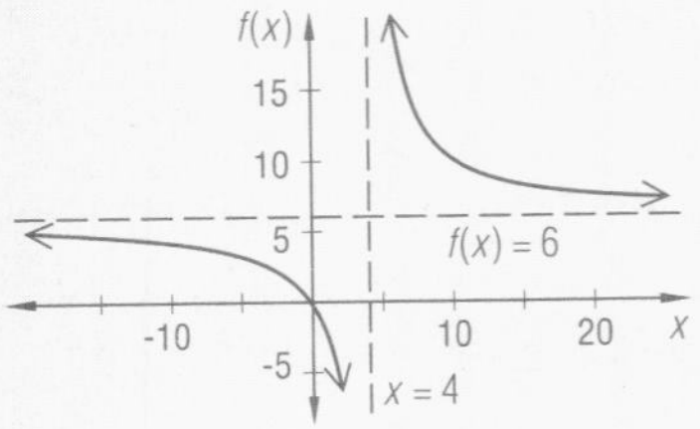

21.

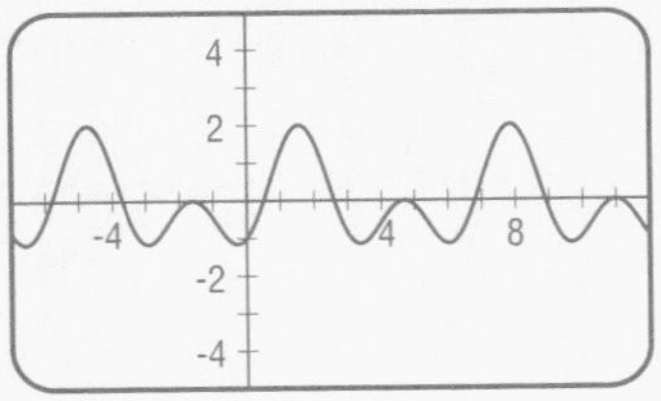

22.a.

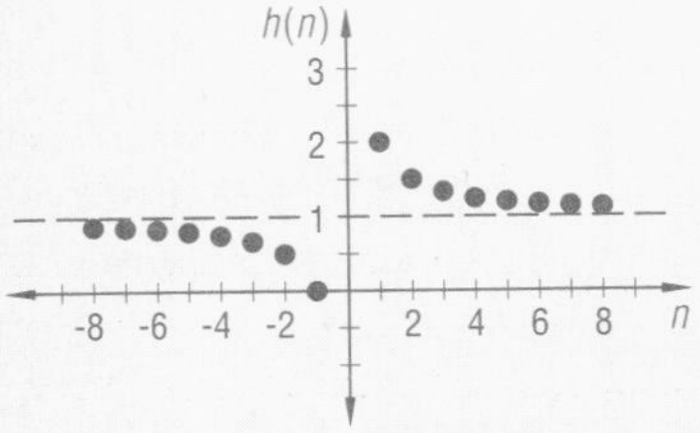

7. Use the result of Example 1 to find an approximate solution set for the inequality $\sin x \le \frac{3}{4}$ over the interval $0 \le x \le 2\pi$.
$0 \le x \le .848$ or $2.29 \le x \le 2\pi$

In 8 and 9, find the exact solutions to the inequality over the interval $0 \le x \le 2\pi$. (Hint: refer to Example 2.) 9. $\frac{\pi}{3} \le x \le \frac{5\pi}{3}$

8. $\tan x \ge \sqrt{3}$ $\frac{\pi}{3} \le x < \frac{\pi}{2}, \frac{4\pi}{3} \le x < \frac{3\pi}{2}$ 9. $2\cos^2 x + \cos x - 1 \le 0$

10. Refer to Example 5. Find all angle values θ so that the person will travel a horizontal distance greater than 30 feet. $6.78° < \theta < 83.22°$

Applying the Mathematics

11. Find exact values of all real solutions of $\sin x \tan x + \cos x = 2$.
$x = \frac{\pi}{3} + 2\pi n$ or $x = \frac{5\pi}{3} + 2\pi n$, n an integer

12. To solve the inequality $\sin x + \cos x \ge 0$ over the interval $0 \le x \le 2\pi$, a student divided by $\cos x$ to obtain $\tan x + 1 \ge 0$.
    a. Is this a reversible step? No
    b. Solve the original inequality. $0 \le x \le \frac{3\pi}{4}, \frac{7\pi}{4} \le x \le 2\pi$

13. A *nautical mile* was originally defined as the length of a minute of arc of a meridian. Because the Earth flattens at the poles, the number of feet in this nautical mile varied with latitude. Let y be the number of feet in the original nautical mile at latitude θ degrees. Then y and θ are approximately related by the equation $y = 6077 + 31 \cos 2\theta$. At what latitude is the original nautical mile equal to
    a. 6080 feet, called an *Admiralty mile*; $\approx 42.22°$
    b. 6076.115 feet (1852 meters), the nautical mile now used officially in the United States and internationally? $\approx 41.40°$

Review

14. Suppose $\theta = \sin^{-1}\left(\frac{2}{\sqrt{5}}\right)$.
    a. Approximate θ to the nearest hundredth of a degree. $\theta \approx 63.43°$
    b. Find the exact value of $\cos \theta$. *(Lesson 6-6)* $\cos \theta = \frac{1}{\sqrt{5}}$

15. Evaluate without a calculator. *(Lesson 6-6)*
    a. $\sin^{-1}\left(-\frac{\sqrt{2}}{2}\right)$ $-\frac{\pi}{4}$ b. $\tan^{-1}\sqrt{3}$ $\frac{\pi}{3}$ c. $\cos^{-1}\left(-\frac{1}{2}\right)$ $\frac{2\pi}{3}$

16. Prove the identity $\sin x = \frac{2\sin^3 x + \sin 2x \cos x}{4\sin^2 x + 2\cos 2x}$ mentioned at the beginning of Lesson 6-2. *(Lessons 6-5, 6-2)* See margin.

17. Given the equation $\frac{\tan x}{\sin x} + \frac{1}{\cos x} = 2 \sec x$.
    a. For what values of x are both sides of this equation defined?
    b. Prove that the equation is an identity. *(Lesson 6-2)* See margin.
    a) for all real numbers $x \ne \frac{n\pi}{2}$, where n is any integer

18. Given the function defined by $f(x) = \frac{6x + 1}{x - 4}$. *(Lessons 5-7, 5-5, 5-4)*
    a. *True* or *false? f* is a rational function. True
    b. Determine the equations of all vertical asymptotes. $x = 4$
    c. Find $\lim_{x \to +\infty}\left(\frac{6x + 1}{x - 4}\right)$. 6
    d. Sketch a graph of the function. See margin.

384

19. Find the base 10 representation of 11101_2. *(Lesson 4-7)* 29

20. Solve: $|2t - 5| > 3$. *(Lesson 3-8)* $t < 1$ or $t > 4$

21. Plot the function f from Example 4 over a longer interval than $0 \le x \le 2\pi$. *(Lesson 2-8)* See margin.

a. What is the period of f? 2π

b. Explain why f has this period. Values of the function repeat every interval of length 2π; that is, $f(x + 2\pi) = f(x)$, $\forall$ real numbers x.

22. a. Sketch the graph of the function h defined by

$$h(n) = \frac{n+1}{n} \ \forall \text{ nonzero integers } n.$$ See margin.

b. Is h a discrete function? Why or why not?

c. Find $\lim_{n \to +\infty} h(n)$. *(Lessons 2-4, 2-1)* 1

b) Yes, its domain is the set of nonzero integers, a discrete set.

23. a. Rewrite the following statement in *if-then* form:
Having one right angle is a sufficient condition for a parallelogram to be a rectangle.

b. Is the conditional *true* or *false*? *(Lesson 1-5, Previous course)*

a) *If a parallelogram has one right angle, then it is a rectangle.* b) True

24. a. Use the result of Question 5 to find all real solutions of:

i. $2 \sin^2 3x - 5 \sin 3x - 3 = 0$ $\frac{7\pi}{18} + \frac{2\pi n}{3}, \frac{11}{18} + \frac{2\pi n}{3}$

ii. $2 \sin^2(x + \pi) - 5 \sin(x + \pi) - 3 = 0$ $\frac{\pi}{6} + 2\pi n, \frac{5\pi}{6} + 2\pi n$

iii. $2 \sin^2\left(\frac{1}{x}\right) - 5 \sin\left(\frac{1}{x}\right) - 3 = 0.$ $\left(\frac{1}{\frac{7\pi}{6} + 2\pi n}\right), \left(\frac{1}{\frac{11\pi}{6} + 2\pi n}\right)$

b. If f is a function, how are the solutions of $2 \sin^2 x - 5 \sin x - 3 = 0$ related to those of $2 \sin^2[f(x)] - 5 \sin[f(x)] - 3 = 0$?

b) The solutions of $2 \sin^2[f(x)] - 5 \sin[f(x)] - 3 = 0$ are all numbers x such that $f(x)$ is equal to the solutions of $2 \sin^2 x - 5 \sin x - 3 = 0$.

NAME ____________

LESSON MASTER 6–7
QUESTIONS ON SPUR OBJECTIVES

■SKILLS *Objective C (See pages 396–398 for objectives.)*
In 1–4, solve without a calculator over the interval $0 \le x \le 2\pi$.

1. $\csc x = \frac{2\sqrt{3}}{3}$

$x = \frac{\pi}{3}$ or $\frac{2\pi}{3}$

2. $\tan x \le -\sqrt{3}$

$\frac{\pi}{2} < x \le \frac{2\pi}{3}$ or $\frac{3\pi}{2} < x \le \frac{5\pi}{3}$

3. $\sec 2x = 2$

$x = \frac{\pi}{6}, \frac{5\pi}{6}, \frac{7\pi}{6}$, or $\frac{11\pi}{6}$

4. $(\sin x)(2 \sin x - 1) < 0$

$0 < x < \frac{\pi}{6}$ or $\frac{5\pi}{6} < x < \pi$

In 5 and 6, solve over the reals using a calculator or automatic grapher, if necessary.

5. $\sec^2 x + \csc^2 x = \sec^2 x \csc^2 x$

$\left\{x: x \ne \frac{k\pi}{2} \text{ where } k \text{ is an integer}\right\}$

6. $7 \sin^2 \phi + 10 \sin \phi + 3 = 0$

$\sin \phi = -1$ or $-\frac{3}{7}$
Therefore,
$\phi = \frac{3\pi}{2} + 2k\pi$ or
$\phi = -.44 + 2k\pi$ or
$\phi = 3.58 + 2k\pi$
where k is an integer

NAME ____________

■USES *Objective F*

7. Two jugglers toss bowling pins to each other with an initial velocity v_0 of 35 ft/s. The horizontal distance R that the bowling pins travel is approximated by the equation $R = \frac{v_0^2}{32} \sin 2\theta$. At what angle should the pins be thrown if the jugglers are standing 30 ft apart?

about 25.8°

■REPRESENTATIONS *Objective H*

8. Use an automatic grapher to approximate to the nearest tenth all solutions to $\tan x + \sin x = 1.7$ over the interval $0 \le x \le 2\pi$.

$x \approx 0.8$ or ≈ 4.3

9. Refer to the graph below of $g(x) = x$ and $h(x) = \tan^2 x$.

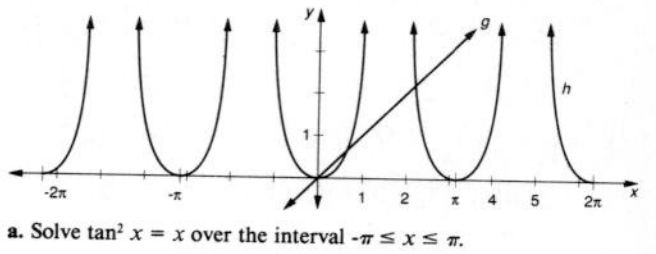

a. Solve $\tan^2 x = x$ over the interval $-\pi \le x \le \pi$.

$x = 0$ or $x \approx .7$ or $x \approx 2.2$

b. Solve $\tan^2 x < x$ over the interval $-\pi \le x \le \pi$.

$0 < x < .7$ or $2.2 < x \le \pi$

LESSON 6-8

RESOURCES

- Teaching Aid 39 displays the graphs of vibrating strings.

TEACHING NOTES

Making Connections In relation to mechanical vibration and the collapse of the Tacoma Narrows Bridge, you may wish to mention the phenomenon called natural frequency. Perhaps your classroom has venetian blinds that begin vibrating loudly when the windows are open. Why does a flag sometimes flutter in a strong wind? Why do soldiers break cadence when crossing a bridge? These examples all relate to natural frequency and mechanical vibration.

Students may wonder about the source of the equation $f(t) = a \cos ct + b \sin ct$. Explain that it comes from solving a type of equation that arises in calculus called a differential equation, which describes the mechanical vibration of a spring.

Review the proof of the Phase-Amplitude Theorem. It is quite straightforward but rich in diagram, definition, and identity applications. Students need to realize that the techniques studied in Chapter 3 for finding the amplitude, period, and phase shift of trigonometric functions are still applicable in this situation.

LESSON

6-8 Phase-Amplitude Analysis and Vibrations

Mechanical vibration is a very important and commonplace physical phenomenon. The pleasant sounds produced by many musical instruments result from the vibrations of tightly stretched strings or wooden reeds. Vibrations can also be destructive. The collapse of the Tacoma Narrows Bridge in 1940 and the failure of aircraft components due to metal fatigue are examples of disastrous side effects of mechanical vibrations.

Although mechanical vibrations can be very complicated, there is a simple physical model that mathematicians, engineers, and scientists use to study their most basic features. It consists of a weight w hanging from the lower end of a spring whose upper end is attached to a support. If the weight is hanging motionless from the spring, the downward pull of gravity on the weight is exactly counterbalanced by the upward pull on the weight by the stretched spring, and we say that the weight-spring system is in **equilibrium position**. If the weight is disturbed from its equilibrium position by, for example, giving it a shove upward or pulling it downward and releasing it, then it will bounce up and down around its equilibrium position.

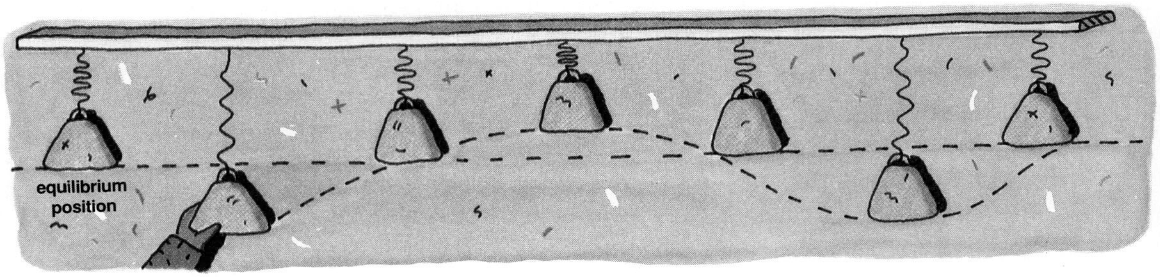

Suppose $f(t)$ represents the distance of the weight above or below the equilibrium position t seconds after the weight was disturbed, with $f(t) > 0$ when the weight is above its equilibrium position and $f(t) < 0$ when it is below that position.

If the spring is assumed to be perfectly elastic and only the forces of gravity and of the spring are considered, then f can be described by the equation

$$f(t) = a \cos ct + b \sin ct$$

where a, b, and c are constants, $c \neq 0$, and at least one of a or b is nonzero. The values of a, b, and c are determined by the spring, weight, and initial disturbance.

For instance, the motion of a weight-spring system could be modeled by

$$f(t) = 3 \cos 2t + 4 \sin 2t$$

where $f(t)$ is measured in inches.

386

The graph of $y = f(t)$ looks like this. You can use the trace function on the grapher to verify that when t is 8 seconds, $f(t)$ is approximately equal to -4. That means, the weight is approximately 4 inches below its equilibrium position 8 seconds after the weight is initially disturbed. The formula verifies this.

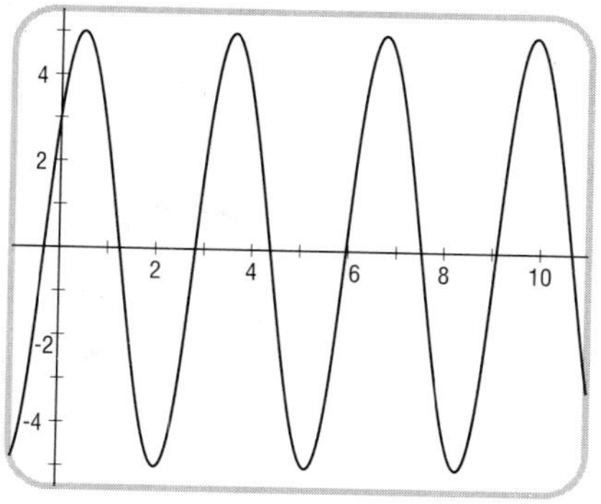

The graph of $y = f(t)$ has the general shape of a graph of a sine or cosine function; its amplitude is approximately 5 and its period is about 3.1. The identities for $\cos(\alpha \pm \beta)$ can be applied to show that the graph of $y = f(t)$ is just a translation image of the graph of $y = 5 \cos 2t$. It is as easy to prove a general identity as it is to work with specific functions.

Phase-Amplitude Theorem

Let a, b, and c be real numbers with $c \neq 0$ and $a^2 + b^2 \neq 0$. Then, for all t,

$$a \cos ct + b \sin ct = \sqrt{a^2 + b^2} \cos(ct - \theta),$$

a cosine function with period $\frac{2\pi}{c}$, amplitude $\sqrt{a^2 + b^2}$, and phase shift $\frac{\theta}{c}$, where

$$\cos \theta = \frac{a}{\sqrt{a^2 + b^2}}, \sin \theta = \frac{b}{\sqrt{a^2 + b^2}}, \text{ and } 0 < \theta < 2\pi.$$

Proof The figure at the right relates θ, a, and b if $0 < \theta < \frac{\pi}{2}$. Notice that $\cos \theta = \frac{a}{\sqrt{a^2 + b^2}}$ and $\sin \theta = \frac{b}{\sqrt{a^2 + b^2}}$. (Similar figures can be drawn if θ is in one of the other three quadrants.)

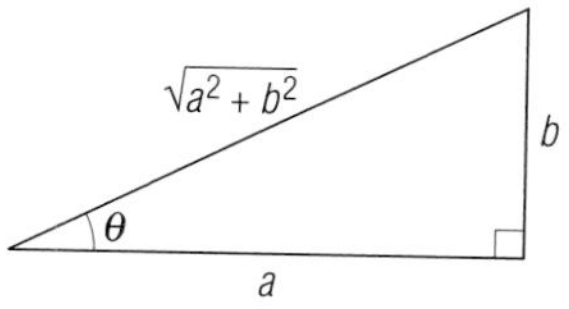

By the formula for the cosine of a difference,

$$\sqrt{a^2 + b^2}\cos(ct - \theta) = \sqrt{a^2 + b^2}(\cos ct \cos \theta + \sin ct \sin \theta).$$

Now substitute for $\cos \theta$ and $\sin \theta$.

$$= \sqrt{a^2 + b^2}\left((\cos ct) \bullet \frac{a}{\sqrt{a^2 + b^2}} + (\sin ct) \bullet \frac{b}{\sqrt{a^2 + b^2}}\right)$$

$$= a \cos ct + b \sin ct$$

This proves the theorem for the case $0 < \theta < \frac{\pi}{2}$. Proofs for the other possible values of θ are similar.

Error Analysis Students may calculate the phase shift incorrectly, misreading the sign or taking the phase shift to be θ rather than $\frac{\theta}{c}$. If this error occurs, emphasize the usefulness of factoring out c to obtain $\cos c\left(t - \frac{\theta}{c}\right)$, and remind students that $(t - k)$ represents a phase shift of k, while $(t + k)$ represents a phase shift of $-k$.

Small Group Work If you have access to springs, you might try the following class experiment. Have a student hold a spring near the top of the chalkboard. Have two other students serve as recorders. Mark the bottom of the spring at its equilibrium position. Pull the spring down by a certain amount and record the amount of displacement from equilibrium. Release the spring, with students marking the top and bottom positions of the oscillations and the times at which those top and bottom displacements occur. (Record times as the number of seconds from release of the spring.) Have students plot displacement versus time and observe that the resulting graph appears to be a sine curve. By attempting to duplicate the situation pictured on the middle of page 386, students should be convinced of the reasonableness of the model.

Making Connections
Discuss why multiplication by $e^{-(1/8)t}$ acts as a dampening influence on the graph. Such a discussion provides a good opportunity to review exponential functions. If students graph $y = e^{-(1/8)t}$, remind them that $\lim_{t \to \infty} e^{-(1/8)t} = 0$. Hence, when this quantity serves as a multiplier, the values of $f(t)$ get smaller and smaller in magnitude as t takes on larger and larger values.

Thus if

$$f(t) = a \cos ct + b \sin ct,$$

then f can be written also as

$$f(t) = \sqrt{a^2 + b^2} \cos(ct - \theta),$$

or

$$f(t) = \sqrt{a^2 + b^2} \cos c\left(t - \frac{\theta}{c}\right).$$

The *amplitude* of the graph of $y = f(t)$ is $\sqrt{a^2 + b^2}$ and its *period* is $\frac{2\pi}{c}$.

Also $\frac{\theta}{c}$ is the *phase shift* needed to map the cosine function onto the graph of $y = f(t)$. The rewritten form makes it easy to determine the amplitude and phase shift of the function f, which is how the theorem got its name.

Now the function $f(t) = 3 \cos 2t + 4 \sin 2t$ can be analyzed. Just apply the Phase-Amplitude Theorem with $a = 3$, $b = 4$, and $c = 2$.

The amplitude is $\sqrt{a^2 + b^2} = \sqrt{3^2 + 4^2} = \sqrt{25} = 5$.

The period is $\frac{2\pi}{c} = \frac{2\pi}{2} = \pi$.

The phase shift is $\frac{\theta}{c}$, where

$$\cos \theta = \frac{a}{\sqrt{a^2 + b^2}} = \frac{3}{5} \text{ and}$$

$$\sin \theta = \frac{b}{6\sqrt{a^2 + b^2}} = \frac{4}{5}.$$

So $0 < \theta < \frac{\pi}{2}$. In particular, $\theta \approx .927 \text{ radian} \approx 53.13°$.

Consequently, $\frac{\theta}{c} \approx \frac{.927}{2} \approx .464$ and

$$f(t) \approx 5 \cos(2(t - .464)).$$

The amplitude, period, and phase shift of f are labeled on the graph below.

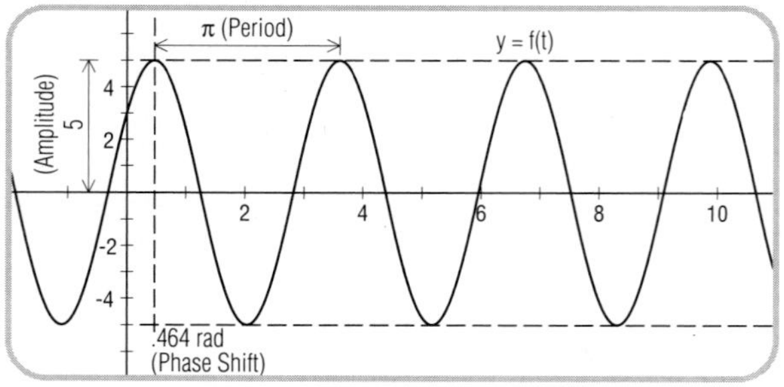

Real springs are not perfectly elastic; frictional forces resist the motion of the weight. These cause the spring to eventually stop moving. If these are taken into account, then the resulting mathematical model is

$$f(t) = e^{-kt}(a \cos ct + b \sin ct)$$

where a, b, c, and k are constants such that $k > 0$, $c \neq 0$, and at least one of a or b is nonzero.

388

For example, after an initial disturbance, the weight-spring system discussed earlier in the lesson might oscillate according to the formula

$$h(t) = e^{-t/8}(3 \cos 2t + 4 \sin 2t)$$

where $h(t)$ is the distance in inches of the weight from its equilibrium position t seconds after the weight is initially disturbed.

Notice that h can be written as the product $g \cdot f$ where g is the exponential function $g: t \rightarrow e^{-t/8}$ and f is the previous weight-spring function $f: t \rightarrow 3 \cos 2t + 4 \sin 2t$.

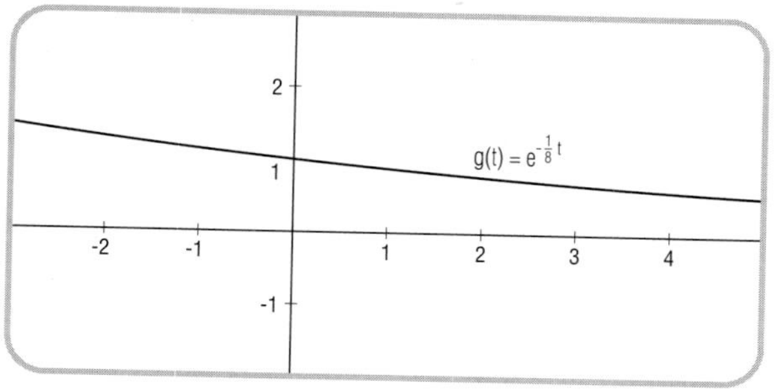

Because the amplitude of f is 5, its values lie between -5 and 5. Hence, for $t > 0$, the values of the product function h must lie between $-5e^{-t/8}$ and $5e^{t/8}$. Therefore, the graph of h oscillates as a cosine function that is bounded above by $y = 5e^{-t/8}$ and below by $y = -5e^{-t/8}$.

Below (in black) is the graph of h as done with an automatic grapher on the interval $0 \leq t \leq 10$. From this, you can see that after the initial disturbance, the weight starts upward from a point 3 inches above the equilibrium position and then oscillates. The time between successive "crests" of the oscillation is π, but the amplitude of oscillation dies out with the exponentials $y = 5e^{-t/8}$ and $y = -5e^{-t/8}$.

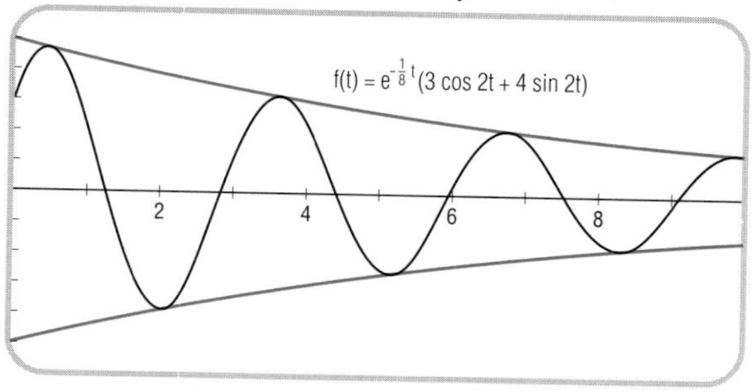

ADDITIONAL EXAMPLE
Rewrite $f(t) = -2\sqrt{2} \cos 5t - 2\sqrt{2} \sin 5t$ as a single cosine function and identify the period, amplitude, and phase shift.

$f(t) = 4 \cos\left(5t - \frac{5\pi}{4}\right)$, period $= \frac{2\pi}{5}$, amplitude $=$ 4, phase shift $= \frac{\pi}{4}$

NOTES ON QUESTIONS

Question 6: This question presents an excellent application of the Phase-Amplitude Theorem in solving trigonometric equations. You may wish to have students solve $\sqrt{3} \sin t + \cot t < \sqrt{2}$ as well.

Question 10: In general, since squaring is involved, reversibility of steps is an issue, and potential solutions must be checked. However, in this case, the expression in the radical is always positive, so there is no problem with the domain of the cosine function. The problem is with the range.

Question 13: There is an identity for $\tan \frac{\alpha}{2}$ in terms of $\cos \alpha$, namely:

$\tan \frac{\alpha}{2} = \pm \sqrt{\frac{1 - \cos \alpha}{1 + \cos \alpha}}$,

where the sign depends on the quadrant in which $\frac{\alpha}{2}$ is located.

Questions

Covering the Reading

In 1 and 2, $f(t)$ represents the distance in inches from the equilibrium position of the weight in a weight-spring system t seconds after an initial disturbance.

a. Use an automatic grapher to graph f.

b. Estimate the position of the weight 1 second after the disturbance.

1. $f(t) = 5 \cos 3t + 12 \sin 3t$ a) See margin. b) ≈ -3.26

2. $f(t) = e^{-t/3}[2\sqrt{2} \cos t - 2\sqrt{2} \sin t]$ a) See margin. b) ≈ -0.610

3. *True* or *false?* The Phase-Amplitude Theorem asserts that the sum of a sine and a cosine function can always be written as a single cosine function. False

4. Use the Phase-Amplitude Theorem to rewrite the function f defined in Question 1 as a single cosine function. Give its amplitude, period, and phase shift. $f(t) = 13\cos(3t - 1.176)$; amplitude: 13; period: $\frac{2\pi}{3}$; phase shift: ≈ 0.392

5. *Multiple choice.* The function $f(t) = 2 \sin t + 2 \cos t$ can be rewritten as

(a) $f(t) = 2 \cos \left(t + \frac{\pi}{4}\right)$

(b) $f(t) = 2 \cos \left(t - \frac{\pi}{4}\right)$

(c) $f(t) = 2\sqrt{2} \cos \left(t + \frac{\pi}{4}\right)$

(d) $f(t) = 2\sqrt{2} \cos \left(t - \frac{\pi}{4}\right)$. (d)

6. The purpose of this problem is to solve the equation $\sqrt{3} \sin t + \cos t = \sqrt{2}$ for $0 \le t \le 2\pi$.

a. Write the left side of the equation as a single cosine function using the Phase-Amplitude theorem. $2\cos\left(t - \frac{\pi}{3}\right)$

b. Divide both sides of the new equation by 2, and use chunking to determine that $t - \frac{\pi}{3}$ must be $\frac{\pi}{4}$ or $-\frac{\pi}{4}$. See margin.

c. Solve for t. $t = \frac{7\pi}{12}$ or $t = \frac{\pi}{12}$

7. Describe the motion of the weight in Question 2 in your own words as precisely as you can. See margin.

Review

8. Solve each equation over the set of real numbers.

a. $\sin x = \frac{1}{2}$ **b.** $\sin 2x = \frac{1}{2}$ **c.** $\sin \frac{x}{2} = \frac{1}{2}$ *(Lesson 6-7)*

a,b,c) See margin.

9. Solve the inequality $3 \tan^2 x \ge 1$ for $-\pi \le x \le \pi$. *(Lessons 6-7, 3-7)* $-\frac{5\pi}{6} \le x \le -\frac{\pi}{6}$ or $\frac{\pi}{6} \le x \le \frac{5\pi}{6}$, but $x \ne \frac{\pi}{2}$ and $x \ne -\frac{\pi}{2}$

10. Solve over the interval $0 \le x \le 2\pi$: $\sqrt{2 - \cos x} = \cos x$. *(Lessons 6-7, 3-4, 3-1)* $x = 0$ or $x = 2\pi$

390

11. The bearing of a ship is the angle measured clockwise from due north. Suppose a ship travels 20 miles north in 1 hour. Find a formula for the bearing θ as a function of the easterly distance traveled, d. *(Lesson 6-6)*
$\theta = \tan^{-1}\left(\frac{d}{20}\right)$

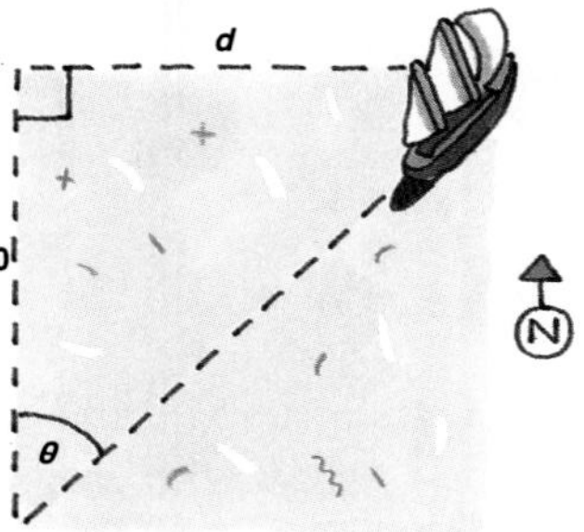

In 12–14, express the given trigonometric value in terms of rational numbers and radicals. *(Lessons 6-5, 6-4, 6-3)*

12. $\sin \frac{7\pi}{8}$ $\frac{\sqrt{2-\sqrt{2}}}{2}$ **13.** $\tan \frac{\pi}{12}$ $\frac{\sqrt{3}-1}{1+\sqrt{3}}$ **14.** $\cos \frac{\pi}{12}$ $\frac{\sqrt{2}+\sqrt{6}}{4}$

15. Prove the following statement: $\forall$ *real numbers x and y,* $\sin(x + y) + \sin(x - y) = 2 \sin x \cos y$. *(Lesson 6-4)* See margin.

16. An AM radio station transmits a **carrier wave** which is a sine curve whose frequency is assigned to that station. Suppose the carrier has equation $w(t) = a \sin ct$. To transmit the sound wave $s(t)$, the carrier wave's amplitude a is modulated by multiplying it by $1 + s(t)$. This is called **amplitude modulation** (AM). The transmitted signal then has equation $f(t) = a(1 + s(t)) \sin ct$.

a. Use identities for $\cos(\alpha + \beta)$ and $\cos(\alpha - \beta)$ to rewrite $\frac{1}{2}[\cos(\alpha - \beta) - \cos(\alpha + \beta)]$.

b. Suppose the sound wave to be transmitted is given by $s(t) = b \sin dt$, so that the signal transmitted by the radio station has equation $f(t) = a(1 + b \sin dt) \sin ct$. Use the result of part **a** to show that the signal transmitted by the radio station can be written as
$$f(t) = a \sin ct + \frac{ab}{2} \cos[(c - d)t] - \frac{ab}{2} \cos[(c + d)t].$$
(Lessons 6-4, 6-3)

a) $\frac{1}{2}(\cos(\alpha - \beta) - \cos(\alpha + \beta)) = \sin\alpha \sin\beta$; b) See margin.

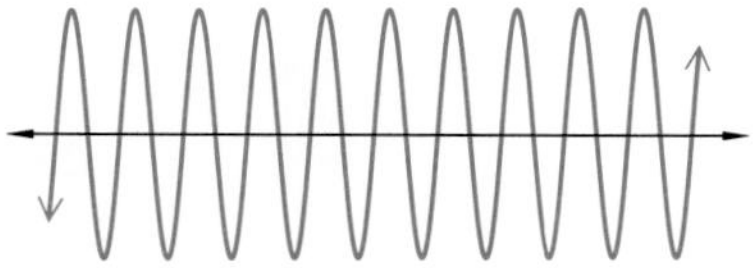

carrier wave $c(t) = a \sin ct$

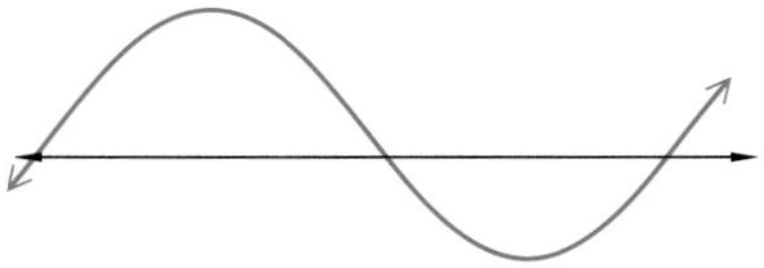

sound wave $s(t)$

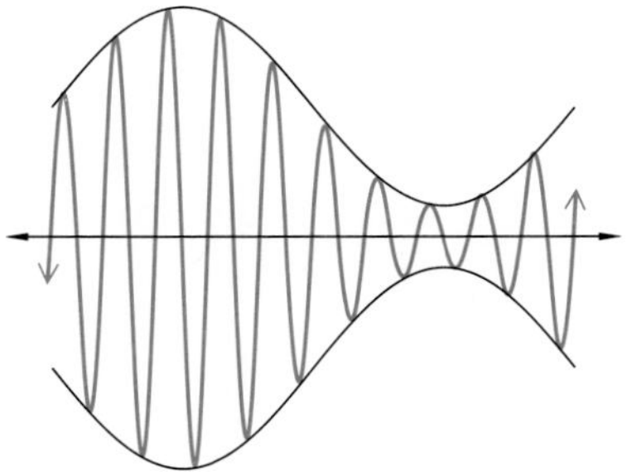

transmitted signal
$f(t) = a(1 + s(t)) \sin ct$

ADDITIONAL ANSWERS

1.a.

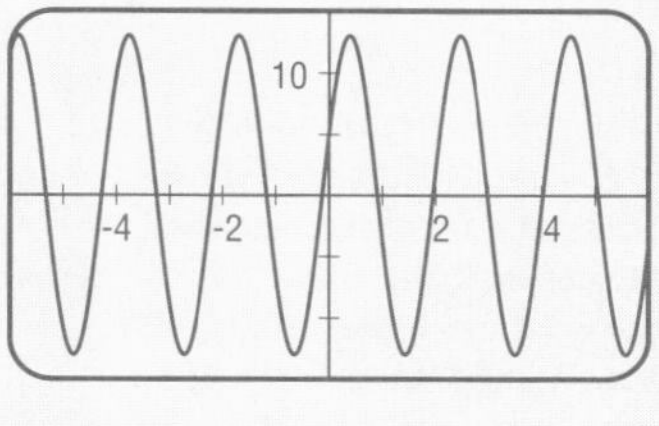

2.a.

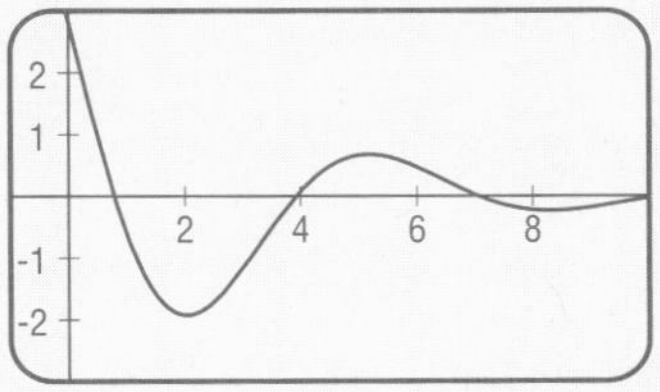

6.b. $\cos\left(t - \frac{\pi}{3}\right) = \frac{\sqrt{2}}{2}$, so $t - \frac{\pi}{3} = \frac{\pi}{4}$ or $-\frac{\pi}{4}$, since $-\frac{\pi}{3} \leq t - \frac{\pi}{3} \leq \frac{5\pi}{3}$.

7. The weight oscillates with a maximum amplitude of 4 which decreases by a factor of $e^{-t/3}$ after t seconds. The weight is at the equilibrium position at $t = \frac{\pi}{4}$ seconds.

8.a. $\frac{\pi}{6} + 2n\pi$ and $\frac{5\pi}{6} + 2n\pi$, n an integer

b. $\frac{\pi}{12} + n\pi$ and $\frac{5\pi}{12} + n\pi$, n an integer

c. $\frac{\pi}{3} + 4n\pi$ and $\frac{5\pi}{3} + 4n\pi$, n an integer

15., 16.b. See Additional Answers in the back of this book.

ADDITIONAL ANSWERS
19. See Additional Answers in the back of this book.

21. The Tacoma Narrows Bridge collapsed as a result of complicated oscillations caused by high winds and the bridge's design. The oscillations can be modeled by sums of sine and cosine functions.

17. **a.** Use an identity to simplify.
  i. $\cos(\pi - x)$ -cos x
  ii. $\cos(\pi + x)$ -cos x
  iii. $\cos(2\pi - x)$ cos x

b. Relate your answers to the diagram at the right. *(Lesson 6-3)*
The x-coordinates of A and D are cos x, while those of B and C are -cos x.
But D = (cos(2π − x), sin (2π − x)),
B = (cos(π − x), sin (π − x)),
C = (cos (π + x), sin (π + x))

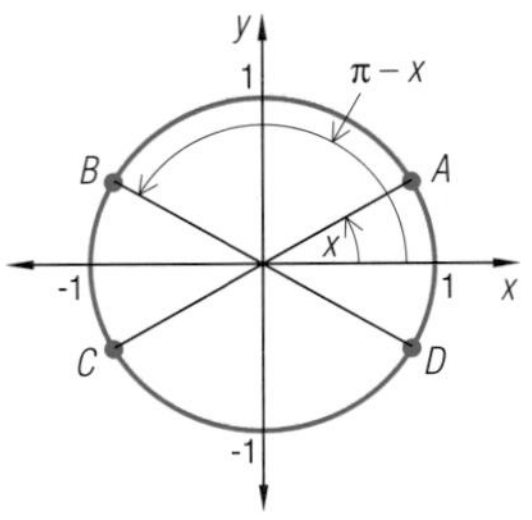

In 18 and 19, use an automatic grapher to conjecture whether or not the given equation is an identity. If it is, prove it; if not, give a counterexample. *(Lessons 6-2, 6-1)*

18. $\cos \pi x = 1 - 4x^2$
not an identity; Counterexample: let x = 1. cos (π • 1) = -1, but 1 − 4 • 1² = -3.

19. $\sin^2 x(\sec^2 x + \csc^2 x) = \sec^2 x$ See margin.

20. Refer to the graph of g at the right.
a. For what values of x is $g(x) > 0$? x > -2
b. For what values is the function decreasing? 0 ≤ x ≤ 1
c. Identify any relative minimum or relative maximum values. *(Lessons 2-3, 2-2, 2-1)*
relative minimum: 2;
relative maximum: 3

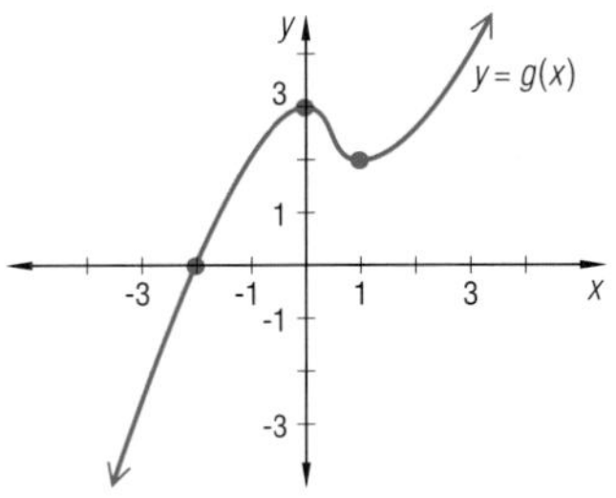

21. Find out about the Tacoma Narrows Bridge disaster. See margin.

Projects

1. In Lessons 6-5 and 6-7 there are situations dealing with projectiles. An object projected from ground level with an initial velocity v (in meters per second) at an angle θ to the horizontal will land $R = \frac{v^2 \sin 2\theta}{9.8}$ meters away. This formula only applies if the points of release and landing are at the same horizontal level. This assumption is often not satisfied. For instance, in the shot-put an athlete propels a shot, a heavy metal sphere, and tries to maximize the horizontal distance it travels. A shot-putter releases the shot at about shoulder level, not at ground level. It can be shown that when the shot is released at a height h meters above the ground with a velocity of v meters per second at an angle θ with the horizontal its horizontal displacement R is given by

$$R = \frac{v^2 \sin 2\theta}{19.6} + \frac{v^2 \cos \theta}{9.8}\sqrt{\sin^2\theta + \frac{19.6h}{v^2}}.$$

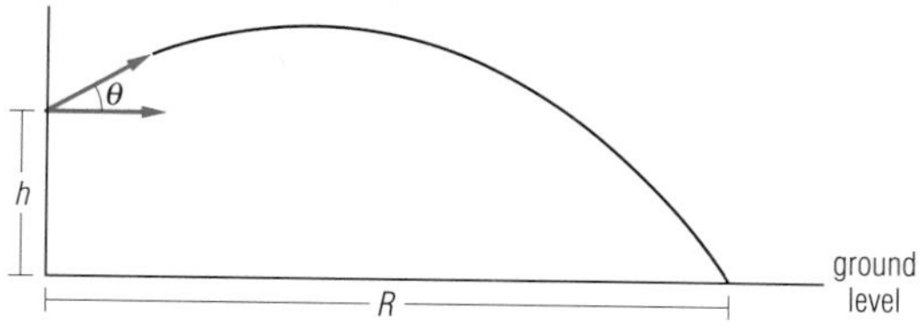

When the release velocity and release height are given, R is solely a function of θ.

   a. World-class shot-putters can easily achieve a release velocity of 13 meters per second. Suppose for such a shot-putter $h = 2$ m. Use an automatic grapher to graph R over the interval from 20° to 70°. What is the maximum distance possible for this shot-putter?

   b. Consider a shorter athlete who releases the shot from a height of 1.7 m at the same velocity. Redo part **a** for this shot-putter.

   c. Compare the data you find in parts **a** and **b** with data from shot-putters in
   i. your school **ii.** Olympic events.

2. Refer to Question 42 in the Chapter Review.

   a. Determine the acceleration due to gravity at your latitude.

   b. Using an identity for $\sin 2\theta$, rewrite this equation as a function in the variable $\sin \theta$.

   c. Using the quadratic formula, find the latitude at which $g = 9.8$ m/s^2. How well does your answer agree with that found in **a** of Question 42?

   d. Through other sources, find out what further modifications may be necessary to determine g at high altitudes and at very low altitudes.

3. Define "pseudotrigonometric" functions for all real numbers x as follows:

   $\text{psin } x = 2^{x-1} - 2^{-x-1}$ $\qquad \text{ptan } x = \frac{2^{2x} - 1}{2^{2x} + 1}$

   $\text{pcos } x = 2^{x-1} + 2^{-x-1}$ $\qquad \text{psec } x = \frac{2^{x+1}}{2^{2x} + 1}.$

   a. Prove the following identities:
   i. "Pseudo-Pythagorean identity": $\text{pcos}^2 x - \text{psin}^2 x = 1$
   ii. Quotient identity: $\frac{\text{psin } x}{\text{pcos } x} = \text{ptan } x$
   iii. Reciprocal identity: $\frac{1}{\text{pcos } x} = \text{psec } x$

   b. Write formulas for pcot x and pcsc x, and prove some identities which involve them and are similar to those in part **a**.

   c. Prove the addition identities:
   i. $\text{psin}(x + y) = \text{psin } x \text{ pcos } y + \text{pcos } x \text{ psin } y$
   ii. $\text{pcos}(x + y) = \text{pcos } x \text{ pcos } y + \text{psin } x \text{ psin } y$
   iii. $\text{ptan}(x + y) = \frac{\text{ptan } x + \text{ptan } y}{1 + \text{ptan } x \text{ ptan } y}.$

   d. Prove other pseudotrigonometric identities that are analogous to trigonometric identities you have seen in the chapter.

4. In this chapter, the sines and cosines of several acute angles were expressed in terms of rational numbers and radicals. Use the sum and difference formulas as well as the formulas for $\sin 2x$ and $\cos 2x$ to express $\sin x°$, where x is an integer between 0 and 90, in terms of rational numbers and radicals for as many values of x as possible.

5. Prove the following incredible identity:
   In *any* acute or obtuse $\triangle ABC$,
   $\tan A + \tan B + \tan C = \tan A \tan B \tan C$.

PROJECTS

The projects provide an opportunity for extended work on topics related to the content of the chapter. (See pages 70 and T37 for more information on the purposes of projects.) Some or all of these projects may be suitable for **small group work**. Solutions for the projects can be found at the end of the Additional Answer section in the back of this book.

If students are planning to complete **Project 1**, you might add one more part to the project. Students might develop some sort of handicapping scheme that could be used to put competitors of different heights on a more equal footing.

Project 3 may be too easy if students explored the hyperbolic functions in Lesson 6-2. This project simply replaces e by 2.

In **Project 4**, x can be any integer multiple of 3.

Here are two other identities like those found in Project 5:
$\sin A + \sin B + \sin C = 4 \cos \frac{A}{2} \cos \frac{B}{2} \cos \frac{C}{2}$;
$\cos A + \cos B + \cos C = 1 + 4 \sin \frac{A}{2} \sin \frac{B}{2} \sin \frac{C}{2}$.

SUMMARY

The Summary gives an overview of the entire chapter and provides an opportunity for students to consider the material as a whole. Thus, the Summary can be used to help students relate the various concepts presented in the chapter.

VOCABULARY

Terms, symbols, and properties are listed by lesson to provide a checklist of concepts a student must know. (See page 71 for more information.)

Summary

The emphasis of this chapter was on proving trigonometric identities and showing that other equations are not identities. One important tool for doing this is an automatic grapher. While a graph cannot prove that an equation is an identity, it can indicate when an equation is not an identity. When a graph suggests that an equation is an identity, other techniques are needed to provide an algebraic proof. These techniques include transforming one side of the identity to obtain the other, transforming each side of the identity separately to obtain identical expressions, using a previously known identity to derive the desired identity, and working with reversible steps from the identity to be proved to a known identity.

Several important identities were proved in this chapter. These form the basis for proofs of many other identities. You should memorize the following identities if you have not already done so.

$$\cos(\alpha + \beta) = \cos\alpha\cos\beta - \sin\alpha\sin\beta$$
$$\sin(\alpha + \beta) = \sin\alpha\cos\beta + \cos\alpha\sin\beta$$

The following identities are used frequently. You should either learn them by heart or be able to derive them quickly when you need them.

$$\cos 2\alpha = \cos^2\alpha - \sin^2\alpha$$
$$\sin 2\alpha = 2\sin\alpha\cos\alpha$$
$$\tan(\alpha + \beta) = \frac{\tan\alpha + \tan\beta}{1 - \tan\alpha\tan\beta}$$

By replacing β by $-\beta$, you can obtain formulas for $\cos(\alpha - \beta)$, $\sin(\alpha - \beta)$, and $\tan(\alpha - \beta)$.

These identities can be used to express values of trigonometric functions in terms of rational numbers and radicals. They are also used in phase-amplitude analysis to rewrite the sum of a sine and cosine function that have the same period as a single sine or cosine function. Such analysis is used in modeling the motion of the weight in a spring system and other systems involving vibration.

By restricting the domains of the trigonometric functions, it is possible to define their inverses in such a way that the inverses are also functions. In particular,

$\cos^{-1} x$ is the unique number θ in the interval $0 \le \theta \le \pi$ whose cosine is x;

$\sin^{-1} x$ is the unique number θ in the interval $-\frac{\pi}{2} \le \theta \le \frac{\pi}{2}$ whose sine is x;

$\tan^{-1} x$ is the unique number θ in the interval $-\frac{\pi}{2} < \theta < \frac{\pi}{2}$ whose tangent is x.

The graphs of the inverse trigonometric functions are obtained by reflecting the graph of the restricted trigonometric function about the line $y = x$.

Trigonometric equations and inequalities can be solved using identities and algebraic techniques with which you are already familiar. They can then be applied to solve problems.

Vocabulary

For the starred (*) items you should be able to give a definition of the term.
For the other items you should be able to give a general description and a specific example of each.

Lesson 6-1
*domain of an identity

Lesson 6-3
identity for $\cos(\alpha + \beta)$
identity for $\cos(\alpha - \beta)$
cofunction identity,
$\cos\left(\frac{\pi}{2} - x\right) = \sin x$

Lesson 6-4
identity for $\sin(\alpha + \beta)$
identity for $\sin(\alpha - \beta)$
identity for $\tan(\alpha + \beta)$
identity for $\tan(\alpha - \beta)$

Lesson 6-5
identities for $\cos 2x$
identities for $\sin 2x$

Lesson 6-6
*$\cos^{-1} x$
*$\sin^{-1} x$
*$\tan^{-1} x$

Lesson 6-8
equilibrium position
Phase-Amplitude Theorem

Progress Self-Test

See margin for answers not shown below.

Take this test as you would take a test in class. You will need an automatic grapher. Then check the test yourself using the solutions at the back of the book.

For Questions 1–6, do not use a calculator.

1. Suppose $\pi < \alpha < \frac{3\pi}{2}$ and $\cos\alpha = \frac{x}{3}$. Find $\sin\alpha$.

2. *Multiple choice.*
$\cos\frac{\pi}{3}\cos\frac{\pi}{6} + \sin\frac{\pi}{3}\sin\frac{\pi}{6} =$
(a) $\cos\left(\frac{\pi}{3}+\frac{\pi}{6}\right)$ (b) $\cos\left(\frac{\pi}{3}-\frac{\pi}{6}\right)$
(c) $\sin\left(\frac{\pi}{3}+\frac{\pi}{6}\right)$ (d) $\sin\left(\frac{\pi}{3}-\frac{\pi}{6}\right)$. (b)

3. Evaluate $\sin(\cos^{-1}\frac{1}{2})$. $\frac{\sqrt{3}}{2}$

4. Determine $\cos(\tan^{-1}\frac{2}{3})$. $\frac{3}{\sqrt{13}}$

In 5 and 6, use an appropriate trigonometric identity to express the following in terms of rational numbers and radicals.

5. $\cos\frac{7\pi}{12}$ $\frac{\sqrt{2}-\sqrt{6}}{4}$ **6.** $\sin\frac{\pi}{12}$ $\frac{\sqrt{6}-\sqrt{2}}{4}$

7. Use an automatic grapher to determine whether $\tan(x+\frac{\pi}{2}) = \tan(-x)$ appears to be an identity. If it does, prove that it is indeed an identity. If not, find a counterexample.

In 8 and 9, prove the identity and determine its domain.

8. $\cos x + \tan x \sin x = \sec x$

9. $\frac{\sin(\alpha+\beta)}{\cos\alpha\cos\beta} = \tan\alpha + \tan\beta$

10. Solve $2\sin^2 x - \sin x - 1 = 0$
a. over the interval $0 \le x \le 2\pi$,
b. over the set of real numbers.

In 11 and 12, refer to the graph below.

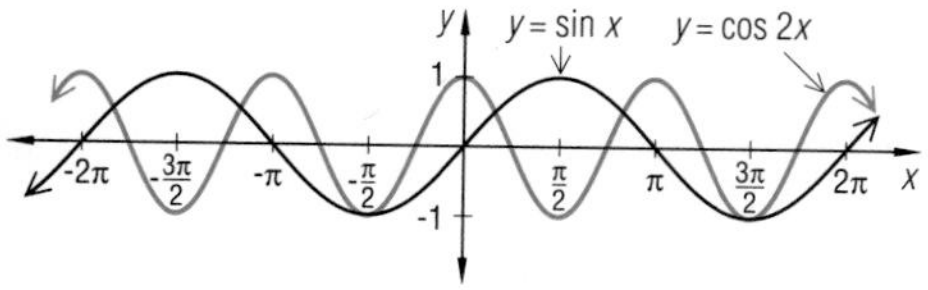

11. In the interval $0 \le x \le \frac{\pi}{2}$, determine exactly where $\cos 2x = \sin x$. $x = \frac{\pi}{6}$

12. On the interval $-2\pi \le x \le 0$, determine where $\sin x \le \cos 2x$. $0 \ge x \ge -\frac{7\pi}{6}$ or $-\frac{11\pi}{6} \ge x \ge -2\pi$

13. When light travels from one medium to another, the angle that the light ray makes with the vertical changes. This bending of light where the two mediums intersect is called *refraction* and is governed by Snell's law,

$$n_1 \sin\theta_1 = n_2 \sin\theta_2,$$

where n_1 and n_2 are the *indices of refraction* in the two mediums. Suppose light in air ($n = 1.0$) traveling at an angle of 20° with the vertical hits water ($n = 1.33$). In the water, the light ray will make what angle with the vertical? ≈ 14.9°

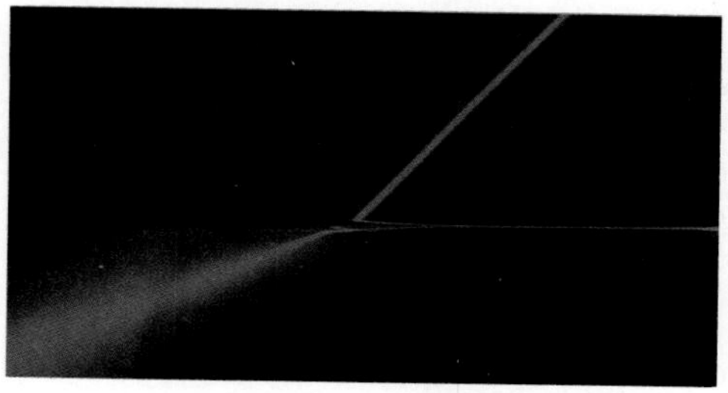

14. A 150-ft radio tower is held in place by two guy wires. Find a formula for the angle θ the guy wire makes with the ground in terms of its distance d from the base of the tower.
$\theta = \tan^{-1}\left(\frac{150}{d}\right)$

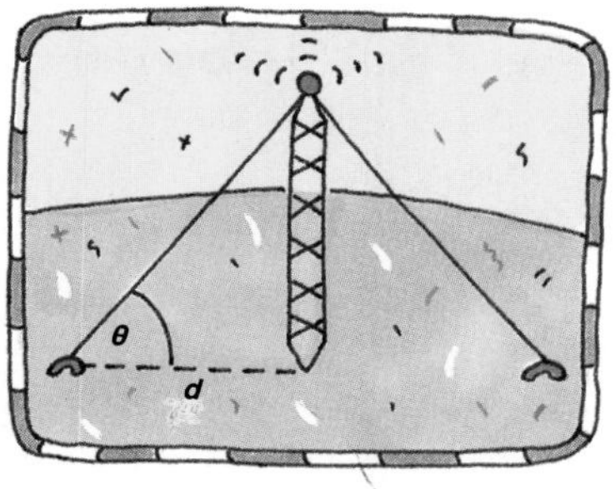

PROGRESS SELF-TEST

The Progress Self-Test provides the opportunity for feedback and correction; the Chapter Review provides additional opportunities for practice.

USING THE PROGRESS SELF-TEST

Assign the Progress Self-Test as a one-night assignment. (See pages 72 and T35 for more information.)

ADDITIONAL ANSWERS

1. $-\sqrt{1-\left(\frac{x}{3}\right)^2}$ or $-\frac{1}{3}\sqrt{9-x^2}$

7. not an identity; Counterexample: Let $x = -\frac{\pi}{6}$. Then $\tan\left(-\frac{\pi}{6}+\frac{\pi}{2}\right) = \sqrt{3}$, but $\tan\left(-\frac{\pi}{6}\right) = \frac{-1}{\sqrt{3}}$.

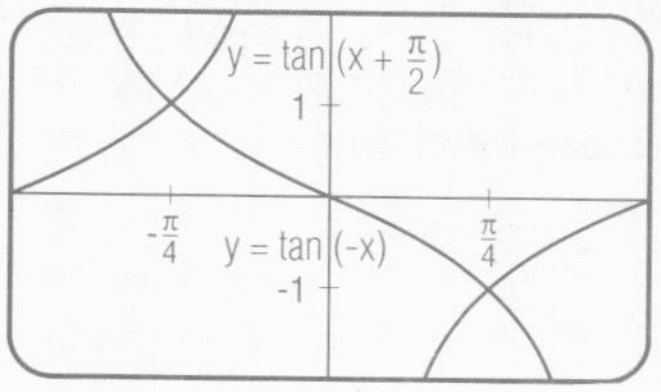

8., 9. See Additional Answers in the back of this book.

10.a. $x = \frac{7\pi}{6}, \frac{11\pi}{6}$, or $\frac{\pi}{2}$

b. $x = \frac{7\pi}{6} + 2\pi n, \frac{11\pi}{6} + 2\pi n, \frac{\pi}{2} + 2\pi n$, n an integer

RESOURCES
- Chapter 6 Test, Form A
- Chapter 6 Test, Form B
- Chapter 6 Test, Cumulative Form
- Comprehensive Test, Chapters 1–6
- Quiz and Test Writer

CHAPTER REVIEW

The main objectives for the chapter are organized here into sections corresponding to the four main types of understanding this book promotes: Skills, Properties, Uses, and Representations. We call these the SPUR objectives.

USING THE CHAPTER REVIEW
Students should be able to answer questions like these with about 85% accuracy by the end of the chapter. (See pages 74 and T38–39 for more information.)

ADDITIONAL ANSWERS

4. $\frac{\sqrt{2+\sqrt{2}}}{2}$

5. $\frac{\sqrt{2+\sqrt{2}}}{2}$

6. $2-\sqrt{3}$

8. $-\frac{4\sqrt{2}+\sqrt{21}}{15}$

9. $\frac{2\sqrt{42}+2}{15}$

Chapter Review

Questions on SPUR Objectives

See margin for answers not shown below.

SPUR stands for **S**kills, **P**roperties, **U**ses, and **R**epresentations.
The Chapter Review questions are grouped according to the SPUR Objectives for this chapter.

SKILLS deal with the procedures used to get answers.

Objective A: *Without a calculator, use trigonometric identities to express values of trigonometric functions in terms of rational numbers and radicals. (Lessons 6-2, 6-3, 6-4, 6-5)*

In 1–3, suppose x is in the interval $\pi < x < \frac{3\pi}{2}$ and $\sin x = \frac{-3}{8}$. Use trigonometric identities to find each value.

1. $\cos x$ $-\frac{\sqrt{55}}{8}$ **2.** $\tan x$ $\frac{3}{\sqrt{55}}$ **3.** $\csc x$ $-\frac{8}{3}$

In 4–6, use appropriate identities to express the following in terms of rational numbers and radicals.

4. $\sin \frac{3\pi}{8}$ **5.** $\cos \frac{\pi}{8}$ **6.** $\tan \frac{\pi}{12}$

7. *Multiple choice.*
$\cos \frac{\pi}{4} \cos \frac{\pi}{6} - \sin \frac{\pi}{4} \sin \frac{\pi}{6} =$
(a) $\sin\left(\frac{\pi}{4}+\frac{\pi}{6}\right)$ (b) $\cos\left(\frac{\pi}{4}+\frac{\pi}{6}\right)$
(c) $\sin\left(\frac{\pi}{4}-\frac{\pi}{6}\right)$ (d) $\cos\left(\frac{\pi}{4}-\frac{\pi}{6}\right)$. (b)

In 8–11, suppose x is in the interval $\frac{\pi}{2} < x < \pi$ with $\cos x = -\frac{1}{3}$ and y is in the interval $0 < y < \frac{\pi}{2}$ with $\sin y = \frac{2}{5}$. Use this information to find each value.

8. $\cos(x + y)$ **9.** $\sin(x - y)$
10. $\sin 2x$ $-\frac{4\sqrt{2}}{9}$ **11.** $\sin \frac{x}{2}$ $\frac{\sqrt{6}}{3}$

12. a. Use the identity for $\sin(x - y)$ to find $\sin \frac{5\pi}{12}$. $x = \frac{2}{3}\pi$, $y = \frac{\pi}{4}$, $\sin \frac{5\pi}{12} = \frac{\sqrt{6}+\sqrt{2}}{4}$
b. Use the identity $\cos 2x = 1 - 2 \sin^2 x$ to find $\sin \frac{5\pi}{12}$.
c. Show that your answers to parts **a** and **b** are equal.

Objective B: *Evaluate inverse trigonometric functions with or without a calculator. (Lesson 6-6)*

In 13–16, find the exact value without using a calculator.

13. $\sin^{-1}\left(\frac{-\sqrt{2}}{2}\right)$ $-\frac{\pi}{4}$ **14.** $\cos^{-1} 1$ 0
15. $\sin(\tan^{-1} 1)$ $\frac{\sqrt{2}}{2}$ **16.** $\sin\left(\sin^{-1}\left(\frac{-1}{2}\right)\right)$ $-\frac{1}{2}$

17. Draw an appropriate triangle to determine $\sin\left(\cos^{-1} \frac{2}{5}\right)$. $\frac{\sqrt{21}}{5}$

Objective C: *Solve trigonometric equations and inequalities algebraically. (Lesson 6-7)*

In 18–20, solve over the interval $0 \le x \le 2\pi$.

18. $\cos x = \frac{1}{2}$ $x = \frac{\pi}{3}, \frac{5\pi}{3}$ **19.** $\tan x = -1$ $x = \frac{3\pi}{4}, \frac{7\pi}{4}$

20. $\sin x = \frac{-\sqrt{3}}{2}$ $x = \frac{4\pi}{3}, \frac{5\pi}{3}$

21. Solve over the set of real numbers: $(\sin x + 1)(\tan x - 1) = 0$.

22. Solve
$$2\sin^2 x - \cos x - 1 = 0$$
a. over the interval $0 \le x \le 2\pi$, $x = \frac{\pi}{3}, \pi, \frac{5\pi}{3}$
b. over the set of real numbers.

23. Use an automatic grapher to determine the values of x in the interval $0 \le x \le 2\pi$ that satisfy $.8 \le \cos x$.

24. Find the exact solutions of the inequality
$$\cos x\,(\cos x - 2) < 0$$
over the interval $0 \le x \le 2\pi$.

25. Solve $\cos 2x + 7 \sin x - 4 > 0$ over the interval $0 \le x \le 2\pi$. $\frac{\pi}{6} < x < \frac{5\pi}{6}$

PROPERTIES deal with the principles behind the mathematics.

Objective D: *Prove trigonometric identities and identify their domains. (Lessons 6-2, 6-3, 6-4, 6-5)*

In 26–30, complete each blank so that the resulting equation is an identity.

26. $\cos^2 x + \sin^2 x =$ __?__ 1

27. $\cos x \sin y - \sin x \cos y =$ __?__ $\sin(y - x)$

28. $\sin 2x =$ __?__ $2 \sin x \cos x$

29. $\cos^2 x - 1 =$ __?__ $-\sin^2 x$

30. $\sin(x + y) =$ __?__ $\sin x \cos y + \cos x \sin y$

In 31–38, prove the identity and identify its domain.

31. $\cos\left(\frac{3\pi}{2} + x\right) = \sin x$

32. $\sec x \cot x = \csc x$

33. $\sin\left(\frac{\pi}{2} + x\right) = \cos x$

34. $\frac{1}{1 + \cos \alpha} + \frac{1}{1 - \cos \alpha} = 2 \csc^2 \alpha$

35. $\cos(\alpha - \beta) - \cos(\alpha + \beta) = 2\sin \alpha \sin \beta$

36. $\sec x + \cot x \csc x = \sec x \csc^2 x$

37. $\cos 4x = \cos^4 x - 6\cos^2 x \sin^2 x + \sin^4 x$

38. $\tan^2 x = \frac{1 - \cos 2x}{1 + \cos 2x}$

USES deal with applications of mathematics in real situations.

Objective E: *Solve problems using inverse trigonometric functions. (Lesson 6-6)*

39. A child 3 ft tall flies a kite on a 200-foot straight string. Find a formula for the angle θ that the string makes with the horizontal in terms of the height of the kite above the ground.

40. A weight is placed at the end of a spring so that, at rest, the weight is 6 inches from the top of its stand. It is then shoved upward so that the spring is compressed 3 inches. When released, the weight oscillates up and down with a displacement of ± 3 inches from its equilibrium position. If the time for one complete oscillation is .4 sec, then the equation $d = 3 \cos 5\pi t$ gives the distance, in inches, of the spring from its equilibrium position. Solve this equation for t.

41. A ship travels on a bearing of θ degrees, where θ is measured clockwise from due north. When the ship has traveled 100 miles north of its original position, describe its bearing in terms of its distance x east of its original position.

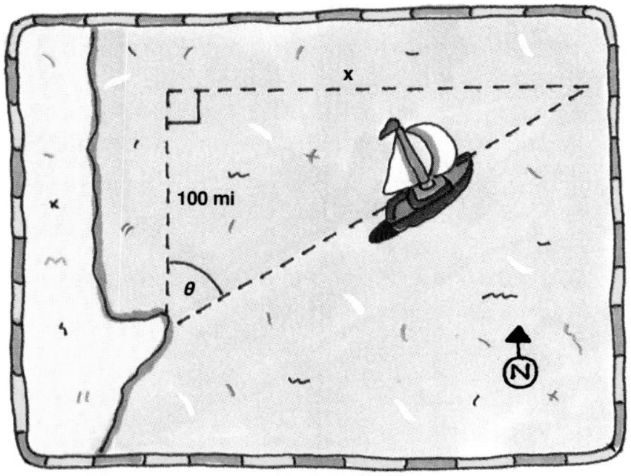

Objective F: *Use trigonometric equations and inequalities to solve applied problems. (Lesson 6-7)*

42. In many situations, the value 9.8 m/sec^2 is used for acceleration due to gravity. Actually, the equation $g = 9.78049(1 + 0.005288 \sin^2 \theta - 0.000006 \sin^2 2\theta)$ estimates the acceleration g due to gravity (in m/sec^2) at sea level as a function of the latitude θ in degrees.

a. Use an automatic grapher to estimate the latitude at which g is 9.8 m/sec^2.

b. For what latitudes is the acceleration due to gravity greater than 9.81 m/sec^2?

12.b. $x = \frac{\sqrt{\sqrt{3} + 2}}{2}$

c. Does $\left(\frac{\sqrt{\sqrt{3} + 2}}{2}\right)^2 = \left(\frac{\sqrt{6} + \sqrt{2}}{4}\right)^2$?

Does $\frac{\sqrt{3} + 2}{4} = \frac{8 + 4\sqrt{3}}{16}$?

Does $\frac{\sqrt{3} + 2}{4} = \frac{2 + \sqrt{3}}{4}$?

Yes

17.

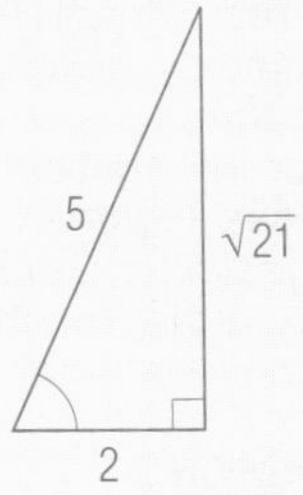

21. $x = -\frac{\pi}{2} + 2\pi n, \frac{\pi}{4} + \pi n$, n an integer

22.b. $x = \frac{\pi}{3} + 2\pi n$, $(2n + 1)\pi$, or $\frac{5\pi}{3} + 2\pi n$, n an integer

23. $0 \le x \le 0.644$ or $5.640 \le x \le 2\pi$, approximately

24. $0 \le x < \frac{\pi}{2}$, $\frac{3\pi}{2} < x \le 2\pi$

31.–38. See Additional Answers in the back of this book.

39. $\theta = \sin^{-1}\left(\frac{h - 3}{200}\right)$, where h = height of the kite above the ground in feet.

40. $t = \frac{1}{5\pi} \cos^{-1}\left(\frac{d}{3}\right)$

41. $\theta = \tan^{-1}\left(\frac{x}{100}\right)$

42.a. $\theta = \pm 38.0°$

b. $\theta < -49.1°$ or $\theta > 49.1°$, approximately

EVALUATION
Three forms of a Chapter Test are provided in the Teacher's Resource File. Chapter 6 Test, Forms A and B cover just Chapter 6. The third test is Chapter 6 Test, Cumulative Form. About 50% of this test covers Chapter 6, 25% covers Chapter 5, and 25% covers previous chapters. A fourth test, Comprehensive Test, Chapters 1–6, that is primarily multiple choice in format, is also provided. For information on grading, see *General Teaching Suggestions: Grading* on page T43 in the Teacher's Edition.

ASSIGNMENT RECOMMENDATION
We strongly recommend that you assign the reading and some questions from Lesson 7-1 for homework the evening of the test. It gives students work to do if they complete the test before the end of the period and keeps the class moving.

ADDITIONAL ANSWERS

43. $\approx 34.8°$ or $55.2°$

44. $22.3° \le \theta \le 67.7°$, approximately

45. See Additional Answers in the back of this book.

46. not an identity; Counterexample: Let $x = 0$. Then $\cos(2 \cdot 0) = 1$, but $2\cos(0) = 2$.

47., 49. See Additional Answers in the back of this book.

50.a.

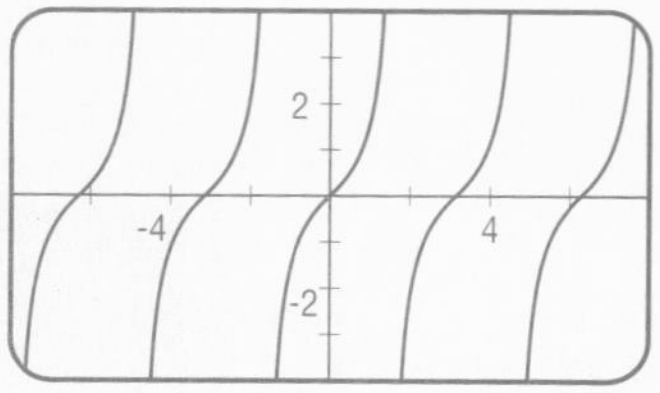

In 43 and 44, a quarterback throws a football with an initial velocity of 64 ft/sec. The range is approximated by the equation from Lesson 6-5:

$$R = \frac{{v_0}^2}{32}\sin 2\theta.$$

43. If the quarterback wants to make a 40 yard pass (120 ft), at approximately what angle should the football be thrown?

44. For what angle values is the range more than 30 yards (90 ft)?

REPRESENTATIONS deal with pictures, graphs, or objects that illustrate concepts.

Objective G: *Use an automatic grapher to test proposed trigonometric identities.* (Lessons 6-1, 6-2)

In 45–47, use an automatic grapher to determine whether the proposed identity appears to be an identity. If it does, prove it algebraically. If not, give a counterexample.

45. $1 + \cot^2 x = \csc^2 x$

46. $\cos 2x = 2\cos x$

47. $\tan(\pi + \gamma) = \tan \gamma$

48. Use an automatic grapher to determine over what domain $\cos x$, with x in radians, can be approximated by
$$f(x) = 1 - \frac{x^2}{2!} + \frac{x^4}{4!} - \frac{x^6}{6!}$$
to within .01. approximately $-2 \le x \le 2$

49. How can you use an automatic grapher to determine whether the proposed identity $\sin(\alpha + \beta) = \sin\alpha\cos\beta - \cos\alpha\sin\beta$ is true? (Do not actually use the automatic grapher, just describe the procedure.)

50. **a.** Graph the function $f(x) = \sin x \sec x$.
b. What single trigonometric function has a similar graph? $y = \tan x$
c. What identity is suggested? $\sin x \sec x = \tan x$

Objective H: *Use graphs to solve trigonometric equations and inequalities.* (Lesson 6-7)

51. **a.** Use an automatic grapher to find all solutions to $\sin x - \cos x = \frac{1}{2}$ (to the nearest tenth) over the interval $0 \le x \le 2\pi$. $x \approx 1.1$ or $x \approx 3.6$
b. Use your answer to part **a** to solve $\sin x - \cos x < \frac{1}{2}$. $0 \le x < 1.1,\ 3.6 < x < 2\pi$

52. Refer to the graph below of $y = 1.5$ and $y = \cos^2 x + 1$.
a. Solve $\cos^2 x + 1 = 1.5$ over the interval from -2π to 2π. $x = \pm\frac{\pi}{4}, \pm\frac{3\pi}{4}, \pm\frac{5\pi}{4}, \pm\frac{7\pi}{4}$
b. Solve the inequality $\cos^2 x + 1 < 1.5$ over the interval $0 < x < \pi$. $\frac{\pi}{4} < x < \frac{3\pi}{4}$

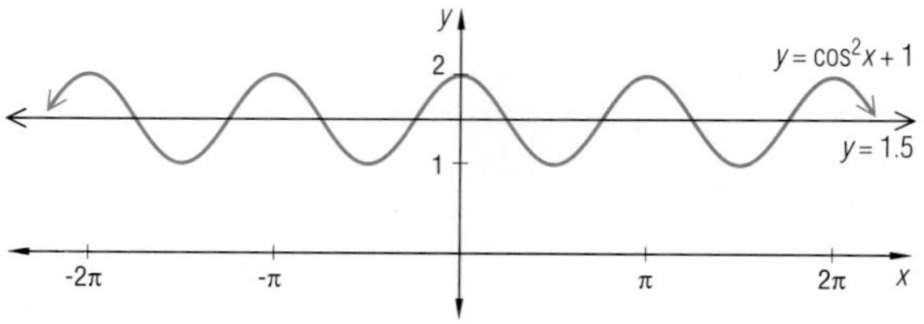

53. Solve $\tan x \le 0.8$ over the interval $0 \le x < \frac{\pi}{2}$. $0 \le x \le 0.675$

398

CHAPTER 7 RECURSION AND MATHEMATICAL INDUCTION

OBJECTIVES

The objectives listed here are the same as in the Chapter 7 Review on pages 462–464 of the student text. The Progress Self-Test on page 461 and the tests in the Teacher's Resource File cover these objectives. For recommendations regarding the handling of this end-of-chapter material, see the notes in the margin on the corresponding pages of this Teacher's Edition.

OBJECTIVES FOR CHAPTER 7 (Organized into the SPUR Categories—Skills, Properties, Uses, and Representations)	Progress Self-Test Questions	Chapter Review Questions	Teacher's Resource File: Lesson Masters*	Teacher's Resource File: Chapter Test Forms A & B
SKILLS				
A Determine terms of a sequence defined either explicitly or recursively.	1a	1–7	7-1	1
B Conjecture explicit formulas for recursively defined sequences.	1b	8–12	7-2	2
C Use summation notation to write sums.	3	13–18	7-3	4, 5
D Rewrite sums recursively.	4	19–21	7-3	6
E Evaluate a finite or infinite geometric series.	6	22–25	7-6	7
PROPERTIES				
F Prove that a recursively defined sequence has a particular explicit formula.	5	26–27	7-3, 7-4	9
G Prove statements using mathematical induction.	8, 9, 10	28–33	7-4, 7-5, 7-7	8, 10, 11
USES				
H Use recursive thinking to solve problems.	2	34–36	7-1	3
I Execute algorithms on sets of numbers.	11	37–38	7-8	13
REPRESENTATIONS				
J Interpret computer programs which calculate terms of sequences.	7	39–40	7-2, 7-3, 7-6	12

*** The masters are numbered to match the lessons.**

OVERVIEW ■ CHAPTER 7

The previous two chapters have dealt primarily with precalculus topics. This chapter returns to topics in discrete mathematics, although mathematical induction is often a topic in precalculus courses. Almost all of the material of this chapter is specifically mentioned in the National Council of Teachers of Mathematics *Curriculum and Evaluation Standards for School Mathematics* as part of the core curriculum for college-bound students.

There are a number of reasons why the content of this chapter is important. Recursive and inductive thinking is used to solve problems, to write instructions or computer programs, and to view the ways in which mathematical objects are related logically. Mathematical induction is a powerful tool for writing proofs in many branches of mathematics. The ideas of recursion and induction are connected in this chapter in a way that points out their similarity.

The first two lessons of the chapter discuss recursion in the context of sequences. Students are expected to use the recursive definition of a sequence to generate the terms of the sequence and to conjecture an explicit formula for the sequence. In both lessons, students study two classic problems involving recursion, the Tower of Hanoi and the Fibonacci sequence.

Lesson 7-3 reviews summation notation. While this notation may be new to some students, it will be familiar to students who have studied from *Functions, Statistics, and Trigonometry.*

Lessons 7-4 through 7-8 are involved with the principle of mathematical induction. It is not uncommon for students to see induction problems in the context of sums only, as discussed in Lesson 7-4. Since this approach gives students too narrow a focus of the power of this proof technique, we have presented other situations that are proved by mathematical induction. Do not expect students to master the use of induction in all of the contexts presented, but a discussion of each situation can help students to grasp the wide applicability of induction and gives them time to master its application to summations and divisibility.

Lesson 7-6 discusses the sum of the terms of a geometric sequence, including what it means to sum the terms of an infinite geometric sequence. The formula for the sum of the first n terms of a geometric sequence is proved using mathematical induction.

Lessons 7-7 discusses strong mathematical induction. This form of induction is applied to prove the number of moves required to assemble a jigsaw puzzle and to prove that every integer is either prime or a product of primes.

The last two lessons of the chapter discuss algorithms. In Lesson 7-8, the ideas of recursion and iteration are used to define computer sorting algorithms. In the last lesson, the efficiency of such algorithms is studied, a topic of importance to users of computer time.

Students who are planning to take or who are now taking Advanced Placement Computer Science will find the content of this chapter to be very valuable. The topics of recursion, sorting techniques, and analysis of efficiency of algorithms are all significant in that course.

PERSPECTIVES ■ CHAPTER 7

The Perspectives provide the rationale for the inclusion of topics or approaches, provide mathematical background, and make connections with other lessons and within UCSMP.

7-1

TWO FAMOUS PROBLEMS

The topic of recursion is familiar to students who have studied from previous UCSMP courses. Recursive definitions for sequences are given in Chapter 1 of UCSMP *Advanced Algebra* and also are found in *Functions, Statistics, and Trigonometry.* In those texts, it is assumed that the recursive definition results in a unique sequence; that is, there is no Recursion Principle that gives criteria under which a recursive definition for a sequence yields a unique sequence. We also introduce and use the Recursion Principle in Lesson 7-1 because it is needed to understand mathematical induction.

Many students are probably familiar with the Fibonacci sequence; it is discussed in many elementary and high school mathematics textbooks. The Fibonacci sequence is perhaps the most well-known example of a recursion sequence. The Tower of Hanoi problem will probably not be familiar to many students.

7-2

RECURSIVE AND EXPLICIT FORMULAS

Some sequences, like the Tower of Hanoi sequence, are more easily defined recursively than explicitly. Then, when an explicit formula is conjectured for the sequence, there is an obvious question: how can you tell if that formula is correct? This lesson gives one method. If the explicit formula generates a sequence with the same initial values and recursive relation as that of the given sequence, then by the Recursion Principle, they must be the same sequence.

In this rather straightforward way, students are introduced to the two parts of a mathematical induction proof informally.

7-3

SUMMATION NOTATION

This lesson completes the prerequisites for understanding the mathematical induction proofs introduced in the next lesson. We expect that students have seen and used the Σ symbol before. A new idea is to express $\sum_{i=1}^{n+1}$ in terms of $\sum_{i=1}^{n}$ and the $(n + 1)$th term. Zeno's paradox of Achilles and the tortoise is used as an example for presenting a sum.

7-4

THE PRINCIPLE OF MATHEMATICAL INDUCTION

The Principle of Mathematical Induction is, in some sense, a statement that modus ponens can be used infinitely many times simultaneously. Because it involves the infinite, and other logical principles do not, it is independent of them. Thus, in rigorous developments of mathematics from first principles, some postulate equivalent to the Principle of Mathematical Induction is needed. Some books assume the *well-ordering principle,* that every nonempty set of positive integers has a least element, and then deduce mathematical induction. We choose to go directly to mathematical induction because the background has been developed in preceding lessons. The analogy between recursive definitions and mathematical induction is drawn and the Recursion Principle is proved. In this lesson, mathematical induction is used to prove a formula for the sum of the first n integers and for the sum of a particular sequence of reciprocals.

7-5

FACTORS AND MATHEMATICAL INDUCTIONS

Students have been introduced to two methods for proving divisibility theorems: (1) use the definition of factor and properties of divisibility; and (2) use modular arithmetic. A third method is proof by mathematical induction. The purpose of this lesson is to practice mathematical induction in this context.

7-6

GEOMETRIC SERIES AND THE RESOLUTION OF ZENO'S PARADOX

In the preceding lessons, mathematical induction was used in proving rather specific results. In this lesson, the general problem of evaluating a finite geometric series is discussed; part of the solution to this problem involves a mathematical induction proof. Then, by taking the limit, infinite geometric series can be evaluated, and Zeno's paradox from Lesson 7-3 can be resolved.

It is possible that students have seen some of the formulas in this lesson before; however, the use of mathematical induction probably is new.

7-7

STRONG MATHEMATICAL INDUCTION

The Strong Form of Mathematical Induction is akin to defining the nth term of a sequence recursively in terms of any of the preceding terms. That is, in the strong form, the truth of a proposition $S(x)$ for any values of $x \leq k$ can be used to deduce the truth of the proposition $S(k + 1)$. Mathematically, the two forms of mathematical induction are equivalent; each can be proved from the other. In practice, each form is called "mathematical induction," but for instructional purposes, it is best to separate them.

7-8

SORTING ALGORITHMS

Sorting is one of the most basic operations that can be performed on data. Examples of sorting are alphabetizing, numerical ordering, ordering by dates, sizing, or coding. Two sorting algorithms are discussed in this lesson: Bubblesort, an iterative algorithm in which the same procedure is repeated again and again, and Quicksort, a recursive algorithm that calls upon a smaller version of itself. Because Quicksort is a recursive algorithm, mathematical induction can be used to prove that it does the job it claims to do.

The purpose of this lesson is to exhibit these algorithms and compare their properties.

7-9

EFFICIENCY OF ALGORITHMS

This lesson begins with a discussion of the ideas of the size of a problem and the efficiency of an algorithm for solving that problem. The efficiency of two algorithms for computing x^n are compared, and then the efficiency of Bubblesort and Quicksort are compared.

If you have never seen these ideas before, you may be surprised that series, logarithms, and the floor function are involved. The algorithms for computing x^n will be seen again in Lesson 8-6 when computing powers of complex numbers.

DAILY PACING CHART 7

DAILY PACING CHART ■ CHAPTER 7

Every chapter of UCSMP *Precalculus and Discrete Mathematics* includes lessons, a Progress Self-Test, and a Chapter Review. For optimal student performance, the self-test and review should be covered. (See *General Teaching Suggestions: Mastery* on page T35 of this Teacher's Edition.) By following the pace of the Full Course given here, students can complete the entire text by the end of the year. Students following the pace of the Minimal Course spend more time when there are quizzes and on the Chapter Review and will generally not complete all of the chapters in this text.

When chapters are covered in full (the recommendation of the authors), then students in the Minimal Course can cover 11 chapters of the book. For more information on pacing, see *General Teaching Suggestions: Pace* on page T34 of this Teacher's Edition.

DAY	MINIMAL COURSE	FULL COURSE
1	7-1	7-1
2	7-2	7-2
3	7-3	7-3
4	Quiz (TRF); Start 7-4.	Quiz (TRF); 7-4
5	Finish 7-4.	7-5
6	7-5	7-6
7	7-6	Quiz (TRF); 7-7
8	Quiz (TRF); Start 7-7.	7-8
9	Finish 7-7.	7-9
10	7-8	Progress Self-Test
11	7-9	Chapter Review
12	Progress Self-Test	Chapter Test (TRF)
13	Chapter Review	
14	Chapter Review	
15	Chapter Test (TRF)	

TESTING OPTIONS

TESTING OPTIONS

- ■ Quiz for Lessons 7-1 through 7-3
- ■ Quiz for Lessons 7-4 through 7-7
- ■ Chapter 7 Test, Form A
- ■ Chapter 7 Test, Form B
- ■ Chapter 7 Test, Cumulative Form

A Quiz and Test Writer is available for generating additional questions, additional quizzes, or additional forms of the Chapter Test.

INDIVIDUAL DIFFERENCES

PROVIDING FOR INDIVIDUAL DIFFERENCES

The student text is written for, and tested with, average students. It also has been successfully used with better and more poorly prepared students.

The Lesson Notes often include Error Analysis and Alternate Approach features to help you with those students who need more help. Students of all abilities often learn from their peers and may benefit from small group work referenced as appropriate throughout the Notes. A blackline Lesson Master (in the Teacher's Resource File), keyed to the chapter objectives, is provided for each lesson to allow more practice. (However, since it is important to keep up with the daily pace, you are not expected to use all of these masters. Again, refer to the suggestions for pacing on page T34.) Extension activities are provided in the Lesson Notes for those students who have completed the particular lesson in a shorter amount of time than is expected, even in the Full Course.

Recursion and Mathematical Induction

Before reading even one more sentence, try to determine a pattern for the numbers in the table above. Use that pattern to find the missing number.

There are two quite different ways of working on this question. One way is to work *across*, seeing how the numbers in each row are related, and then determining what number is related to 7. This is what you often do when you find a rule for a function.

The second way is to work *down*, seeing how the numbers in the right column are related, and using the number 63 and possibly previous numbers to determine a number that might go in the blank.

The second kind of thinking is called *recursive thinking* or *recursion*. Recursive thinking means using previous elements in a sequence in a prescribed way to generate later ones. Computers, because of their capacity to repeat operations very quickly, are often programmed in ways that use recursion.

If you use recursive thinking on the above table, you can figure out a way to generate successive entries in the right column without referring to the left column. Observe that each number in the right column, after the first number, is one more than twice the number just above it.

In this chapter, you will learn how recursive thinking can be used in a variety of situations. You will also be introduced to *mathematical induction*, a method of proof that is related to recursive thinking. The methods you learn are used to tackle and solve important problems in mathematics, computer science, many other fields, and games.

CHAPTER 7

It is critical in this chapter to keep the pace going. Although mathematical induction takes a while to master, it is best to work on that mastery after a variety of contexts have been given. Otherwise, students learn form rather than substance.

With 9–11 days on the lessons and quizzes, 1 day for the Progress Self-Test, 1 or 2 days for the Chapter Review, and 1 day for a Chapter Test, this chapter should take 12–15 days. (See the Daily Pacing Chart on page 399D.)

USING PAGE 399
The sequence presented on this page comes from the Tower of Hanoi problem which is discussed in the first lesson. In determining the number that fits in the blank, some students might recognize that all the numbers in the second column are 1 less than a power of 2. Their guess would be $2^7 - 1 = 127$, which is correct. Others may use a method of finite differences, seeing that the differences between successive numbers in the right column are powers of 2. Then they will add 64 to the 63 to get 127.

Ask your students: How many of you worked across to find the missing term? How many worked down? Usually more students work down. This points out the naturalness of recursive thinking.

LESSON 7-1

RESOURCES
- Lesson Master 7-1
- Teaching Aid 40 displays the calendar of the Fibonacci sequence found in the Teaching Notes.

OBJECTIVES

A Determine terms of a sequence defined either explicitly or recursively.
H Use recursive thinking to solve problems.

TEACHING NOTES

As a class demonstration, do the Tower of Hanoi problem for a given number of objects and count the number of required moves.

You can begin with a quarter. Place it on pile *A* and ask how many moves are needed to get it to pile *C*. (1) Now place a quarter and nickel on pile *A*. Ask students to describe and count the moves now needed. Write this as $1 + 1 + 1 = 3$. Now place a quarter, nickel, and penny on pile *A* and ask the same question again. Summarize as: move 2 coins from *A* to *B*, 1 coin from *A* to *C*, and 2 coins from *B* to *C*. Write this as $3 + 1 + 3 = 7$. Now place 4 coins on pile *A* and repeat as: move 3 coins from *A* to *B*, 1 from *A* to *C*, and 3 from *B* to *C*. Write this as $7 + 1 + 7 = 15$. In effect, start small and keep increasing the complexity of the problem. When finished, have students do the same process at their desks.

LESSON

7-1 Two Famous Problems

In this lesson you will read about two famous problems whose solutions involve recursive thinking. The first was invented by the French mathematician Edouard Lucas in 1883, and it has come to be known as the **Tower of Hanoi.** (Hanoi is now the capital of Vietnam.) Here is the story made up by Lucas.

According to legend, at the time of Creation, 64 golden disks, each with a small central hole, were placed on one of the three golden needles in a temple in Hanoi. No two of the disks were the same size, and they were placed on the needle in such a way that no larger disk was on top of a smaller disk.

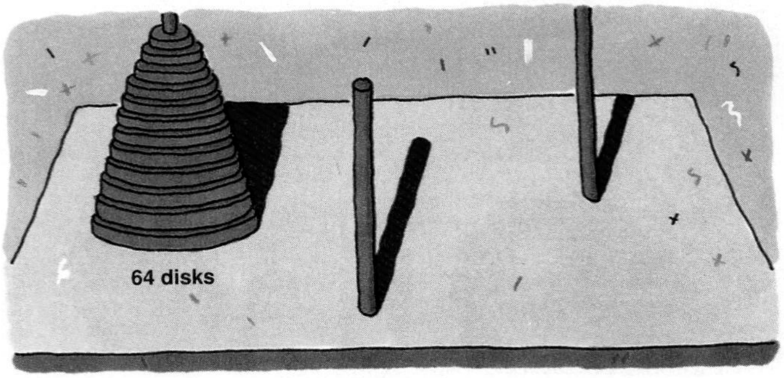

The Creator ordained that the monks of the temple were to move all 64 disks, one by one, to one of the other needles, never placing a larger disk on top of a smaller disk. When all the disks are stacked on the other needle, the world will end, the faithful will be rewarded, and the unfaithful will be punished. If the monks work very rapidly, moving one disk every second, how long will it be until the end of the world?

To solve this problem, imagine that you know the solutions to smaller problems of the same type, and then figure out a way to use those solutions to solve the given problem. The aim is to reduce the given problem repeatedly to obtain problems that are so small that their solutions are obvious. In this case, for example, you could start by supposing you know the length of time needed to transfer a tower of 63 disks. Then you can figure out the time needed to transfer 64 disks. But in order to figure out the time needed to transfer 63 disks, you would need to know the time it takes to transfer 62 disks. This reasoning would proceed until you have reduced the problem to finding the time it takes to transfer one disk. To be specific, you could reason as follows.

To transfer all 64 disks to a second needle, the monks first must have moved the top 63 disks, one by one, to the third needle (Step 1). Then they can move the bottom disk to the second needle (Step 2). Finally they can transfer the 63 disks on top of the bottom disk on the second needle (Step 3). The diagram on page 401 illustrates these steps.

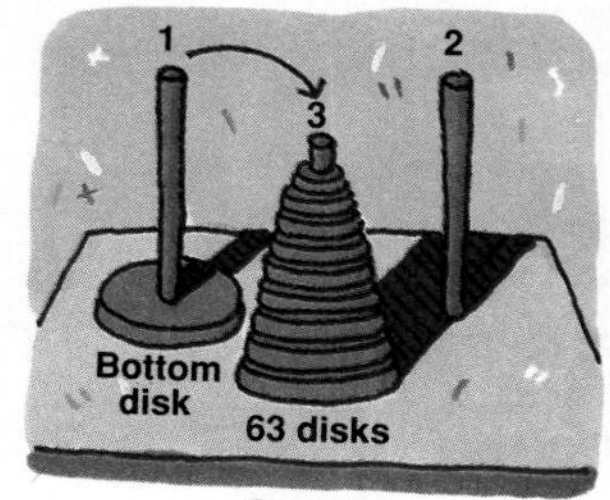

Step 1

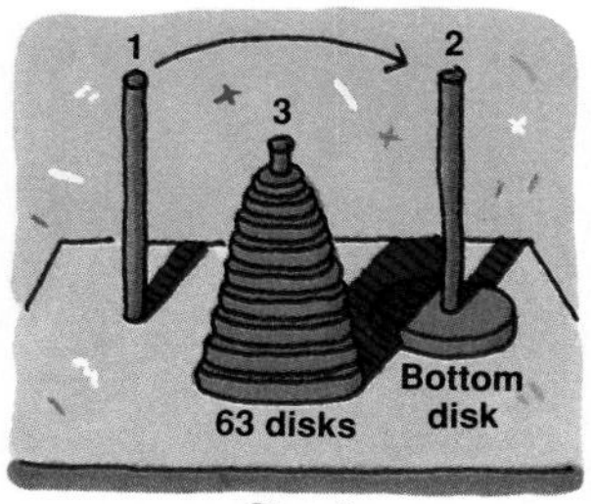

Step 2

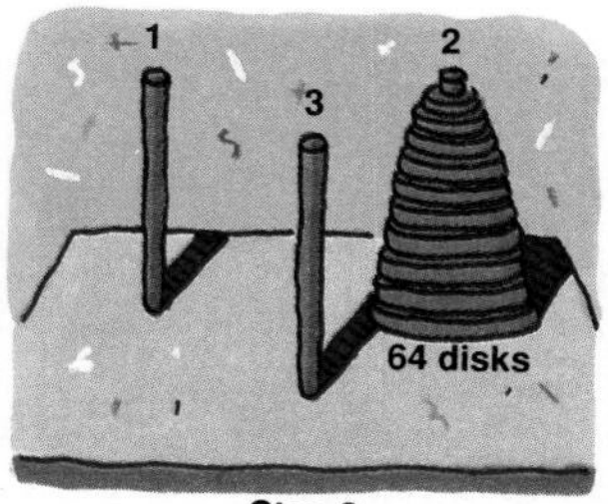

Step 3

Thus to transfer 64 disks, they need to transfer 63 disks twice (Steps 1 and 3) and move the 64th disk once (Step 2).

For each integer $k \geq 1$, let T_k be the number of moves needed to shift k disks from one needle to another according to the rules. The thinking above is recursive because T_{64} is described in terms of T_{63}. Specifically, T_{64}, the number of moves needed to move all 64 disks from one needle to another, is given by

$$\begin{aligned} T_{64} &= \underset{\text{Step 1}}{T_{63}} + \underset{\text{Step 2}}{1} + \underset{\text{Step 3}}{T_{63}} \\ &= 2T_{63} + 1. \end{aligned}$$

The same reasoning can be used to derive a relationship between the number of moves needed to move $k + 1$ disks and the number needed to move k disks for *any* integer $k \geq 1$:

$$\begin{aligned} T_{k+1} &= \underset{\text{Step 1}}{T_k} + \underset{\text{Step 2}}{1} + \underset{\text{Step 3}}{T_k} \\ &= 2T_k + 1. \end{aligned}$$

Since the monks move one disk each second, T_{k+1} also represents the number of seconds needed to transfer $k + 1$ disks. Hence, the problem of finding the number of seconds needed to transfer a tower of $k + 1$ disks has been reduced to the problem of finding the number of seconds needed to transfer a tower of k disks for each integer $k \geq 1$. The problem of transferring a single disk is obvious: just one second is required. In other words, $T_1 = 1$. Thus,

$$\begin{cases} T_1 = 1 \\ T_{k+1} = 2T_k + 1, \text{ for each integer } k \geq 1. \end{cases}$$

These two equations constitute a *recursive definition* for the sequence T, which is called the **Tower of Hanoi sequence.**

Definitions

A **recursive definition for a sequence** consists of two statements:
(1) a specification of one or more initial terms of the sequence, called **initial conditions;**
(2) an equation that expresses each of the other terms of the sequence in terms of the previous terms of the sequence, called a **recurrence relation.**

Not all initial conditions and recurrence relations determine a unique sequence. For instance, suppose $a_1 = 5$ and $a_{n+1} = 2a_{n-1}$ for $n > 1$. The sequence 5, 5, 10, 10, 20, 20, ... , the sequence 5, 6, 10, 12, 20, 24, ... , and many other sequences satisfy these conditions. However, the Tower of Hanoi recursive definition and the recursion formulas for arithmetic and geometric sequences you saw in Lesson 2-4 are examples of a type of recursion that always defines a unique sequence.

One way to solve the Tower of Hanoi problem would be to continue the sequence of calculations begun in the **Example** until T_{64} has been computed. The value of T_{64} would then equal the number of seconds from the creation until the end of the world. Another way to solve the Tower of Hanoi problem is to find an explicit formula for T_n and substitute the value $n = 64$. We show how to do this in the next lesson.

The solution to the problem posed in the text, namely, the time required to move 64 disks, should not be discussed until Question 1 of Lesson 7-2.

Subscript notation may confuse some students. To help them, you can build on the presentation in Lesson 2-4. Remind students that T_k is equivalent to $T(k)$. For students who have difficulty understanding what T_{k+1} means, ask them what integer comes after k. T_{k+1} represents the term following T_k. An explicit formula for T_n will be developed in Lesson 7-2.

In the **Example**, students need to see that in each line the subscript of T is different from (one more than) the value of k. Remind them that the second column is at T_{k+1}, not T_k.

Be sure to discuss the derivation of the formula for the Fibonacci sequence. Students need to understand the lag time before a pair of rabbits starts producing new rabbits.

Keeping track of infertile, fertile, and baby rabbits is a difficult problem. A calendar as described below may be helpful. Assume that all births take place at 12:01 AM on the first of each month, beginning on January 1.

	A	B	C	D
January 1	1	0	0	1
February 1	0	1	0	1
March 1	1	0	1	2
April 1	1	1	1	3
May 1	2	1	2	5
June 1	3	2	3	8

A = number of pairs of rabbits born today
B = number of pairs of rabbits one month old
C = number of pairs of rabbits $\geq$ 2 months old
D = total number of pairs of rabbits

$A_n = B_{n-1} + C_{n-1}$

Go over the differences between recursive and explicit formulas for sequences. Students need to understand this terminology for Lesson 7-2.

Recursion Principle

Suppose that a recurrence relation defines x_{n+1} in terms of x_n and n for each integer $n \geq 1$. Then there is exactly one sequence defined by this recurrence relation and the initial condition $x_1 = a$.

Consequently, since the sequence is uniquely defined, its terms can be determined.

Example Compute the first seven terms of the Tower of Hanoi sequence.

Solution T_1 is given by the initial condition.

$T_1 = 1$

The terms after T_1 are determined by the recurrence relation

$T_{k+1} = 2T_k + 1.$

$k = 1$: $T_2 = 2 \cdot T_1 + 1 = 2 \cdot 1 + 1 = 3$

$k = 2$: $T_3 = 2 \cdot T_2 + 1 = 2 \cdot 3 + 1 = 7$

$k = 3$: $T_4 = 2 \cdot T_3 + 1 = 2 \cdot 7 + 1 = 15$

$k = 4$: $T_5 = 2 \cdot T_4 + 1 = 2 \cdot 15 + 1 = 31$

$k = 5$: $T_6 = 2 \cdot T_5 + 1 = 2 \cdot 31 + 1 = 63$

$k = 6$: $T_7 = 2 \cdot T_6 + 1 = 2 \cdot 63 + 1 = 127$

This is the sequence which opened the chapter.

One way to solve the Tower of Hanoi puzzle is to continue the sequence of calculations begun in Example 1 until T_{64} has been computed. You are asked to do this, using a computer, in the Questions. The value of T_{64} would then equal the number of seconds predicted by Lucas's legend from creation until the end of the world. Another way to solve the Tower of Hanoi problem would be to find an explicit formula for T_n and substitute the value $n = 64$; this is done in the next lesson.

The second famous problem in this lesson is almost 800 years old, and was invented by the Italian mathematician Fibonacci, whose given name was Leonardo of Pisa (1170?–1250?). Fibonacci was the greatest European mathematician of his time. In 1202 his book *Liber Abaci (The Book of Counting)* introduced the Hindu-Arabic system of numerals and decimal notation to a Europe that used Roman numerals. One of the problems in *Liber Abaci* gives rise to a sequence called the *Fibonacci sequence* that is useful in describing phenomena as varied as patterns of sunflower and daisy petals, seeds, mollusk shells, and spiral galaxies. Here is Fibonacci's problem.

> Suppose that a certain breed of rabbit is infertile during its first month of life, but after two months and after every month thereafter each male-female pair of rabbits produces one additional male-female pair. Starting with one newly-born male-female pair and assuming that no rabbit dies, how many rabbit pairs will there be at the beginning of the 7th month?

This problem, too, can be solved using recursive thinking. Let F_n = the number of rabbit pairs at the beginning of the nth month. From the given information, the number of pairs at the beginning of the 1st month is

$$F_1 = 1.$$

At the beginning of the 2nd month, that pair, call it $P1$, has not yet produced offspring, so

$$F_2 = 1.$$

These are the initial conditions.

Now by the beginning of the 3rd month, $P1$ has produced offspring which we will call $P2$. There are now two pairs, so

$$F_3 = 2.$$

By the beginning of the 4th month, $P1$ produces more offspring (call them $P3$) but $P2$ is not yet fertile, so

$$F_4 = 3.$$

By the 5th month, both $P1$ and $P2$, the pairs alive in the 3rd month, are fertile, so

$$F_5 = F_4 + 2 = 5.$$

By the beginning of the 6th month, the three pairs alive in the 4th month are fertile and producing, so

$$F_6 = F_5 + 3 = 5 + 3 = 8.$$

By the beginning of the 7th month, the five pairs alive in the 5th month are fertile, so

$$F_7 = F_6 + 5 = 8 + 5 = 13.$$

In general, notice that the *increase* during each month is equal to the number of rabbit pairs that were alive at the beginning of the previous month (for they are the rabbit pairs that can produce new pairs). That is,

$$\begin{bmatrix}\text{the number}\\ \text{of rabbit pairs}\\ \text{at the beginning}\\ \text{of month } n+1\end{bmatrix} = \begin{bmatrix}\text{the number}\\ \text{of rabbit pairs}\\ \text{at the beginning}\\ \text{of month } n\end{bmatrix} + \begin{bmatrix}\text{the number}\\ \text{of rabbit pairs}\\ \text{at the beginning of}\\ \text{month } n-1\end{bmatrix}$$

This yields the recurrence relation

$$F_{n+1} = F_n + F_{n-1} \qquad \text{for all integers } n \geq 2.$$

Thus a recursive definition of F is:

$$\begin{cases} F_1 = 1 \\ F_2 = 1 \\ F_{n+1} = F_n + F_{n-1} & \text{for all integers } n \geq 2. \end{cases}$$

Now any terms of F can be found. The initial conditions are:

$$F_1 = F_2 = 1.$$

From the recurrence relation,

$$F_3 = F_2 + F_1 = 1 + 1 = 2$$
$$F_4 = F_3 + F_2 = 2 + 1 = 3$$
$$F_5 = F_4 + F_3 = 3 + 2 = 5$$

and so on. The sequence F is called the **Fibonacci sequence.** Its terms are called **Fibonacci numbers**.

ADDITIONAL EXAMPLES

1. List the first ten terms of the sequence defined by $p_k = \lfloor \ln k \rfloor$, for integers $k \geq 1$.
0, 0, 1, 1, 1, 1, 1, 2, 2, 2

This sequence is defined explicitly and affords an opportunity to review logarithms and to discuss when the values of the sequence change. This kind of sequence also is used in evaluating the efficiency of an algorithm.

2. The factorial function can be defined as follows: If $n = 0$, then $n! = 1$.
If $n > 0$, then $n! = n \cdot (n - 1)!$. Generate 0!, 1!, . . . , 10!
1, 1, 2, 6, 24, 120, 720, 5040, 40320, 326880, 3628800

NOTES ON QUESTIONS
Question 2: Assume 1 move per second and no mistakes.

Questions 7 and 8: Note how much easier it is to generate the terms using the recursive definition than using the explicit formula on this page.

Question 15: Emphasize that it is possible to find more than one explicit formula which can generate the same first few terms of a sequence. This is the reason we do not give several terms of a sequence and ask students to generate an explicit formula, unless there is additional knowledge such as the sequence is arithmetic or geometric.

Question 18: Some students may think that the discount after two weeks is 20% of the initial price. Remind them that the price after a 10% discount is 90% of what it was. Elicit that the sequence of prices is geometric. As an extension to this question ask students which of the following is better for the consumer: a 20% discount followed by a 10% discount applied to the reduced price or a 10% discount followed by a 20% discount applied to the reduced price? (They are the same.)

ADDITIONAL ANSWERS
6. sample: 5, 8, 10, 16, 20, 32, 40, 64

11. The initial conditions, defined by the first two terms of the sequences, are different. $x_1 = 1, x_2 = 1$ for the first sequence, and $x_1 = 1, x_2 = 0$ for the second.

15.c. The first four terms of each sequence are identical, but the fifth terms are different. One must be cautious when generalizing because many different sequence formulas may generate the same first few terms.

A variant of the recursion principle explains why the definition of the Fibonacci sequence determines that sequence and no other.

Recursion Principle (Variant)

Suppose that a recurrence relation defines x_{n+1} in terms of x_n, x_{n-1}, and n for each integer $n \geq 2$. Then there is exactly one sequence defined by this recurrence relation and the initial conditions $x_1 = a$ and $x_2 = b$.

This variant itself can be generalized to apply to sequences in which the nth term is defined in terms of the previous k terms, where $k < n$, and the initial conditions specify the values of x_1 through x_k. The Recursion Principle and these variants can be proved using the Principle of Mathematical Induction, which you will encounter in Lesson 7-4.

It turns out that there is also an explicit formula for the Fibonacci sequence, although its derivation is beyond the scope of this book. It can be shown that for each integer $n \geq 1$,

$$F_n = \frac{1}{\sqrt{5}}\left(\frac{1+\sqrt{5}}{2}\right)^n - \frac{1}{\sqrt{5}}\left(\frac{1-\sqrt{5}}{2}\right)^n.$$

Perhaps the most surprising thing about this formula is that all its values are integers. In the Questions you are asked to use this formula to check that $F_{13} = 233$.

Questions

Covering the Reading

1. Compute T_7 and T_8, the seventh and eighth terms of the Tower of Hanoi sequence. $T_7 = 127$; $T_8 = 255$

2. Using the recursive definition and a calculator, find the number of moves required to transfer 15 disks in the Tower of Hanoi problem. How many hours would it take? 32,767 moves, 9.1 hours

3. Find two different sequences that both satisfy the following recurrence relation:

$$s_n = 3s_{n-1} + 2 \quad \text{for all integers } n \geq 2.$$

This shows that it is necessary to specify initial conditions when defining a sequence recursively.
samples: 2, 8, 26, 80, 242, . . . and 5, 17, 53, 161, 485, . . .

In 4 and 5, you may wish to refer to Lesson 2-4 for definitions of arithmetic and geometric sequences.

4. a. Write the first five terms of an arithmetic sequence with initial term 3 and common difference 4. 3, 7, 11, 15, 19

b. Write the initial condition and the recurrence relation for this sequence. $a_1 = 3$, $a_{k+1} = a_k + 4$, $\forall$ integers $k \geq 1$

c. Write an explicit formula for this sequence.
$a_n = 4n - 1$, $\forall$ integers $n \geq 1$

404

5. a. Write the first five terms of a geometric sequence with initial term 3 and common ratio 4. **3, 12, 48, 192, 768**

b. Write the initial condition and the recurrence relation for this sequence. **$a_1 = 3$, $a_{k+1} = 4a_k$, $\forall$ integers $k \geq 1$**

c. Write an explicit formula for this sequence. **$a_n = 3(4)^{n-1}$, $\forall$ integers $n \geq 1$**

6. Give a sequence other than the two mentioned in the lesson that satisfies the initial condition $a_1 = 5$ and the recurrence relation $a_{n+1} = 2a_{n-1}$. **See margin.**

7. Compute the first thirteen terms of the Fibonacci sequence using the recursive definition. **1, 1, 2, 3, 5, 8, 13, 21, 34, 55, 89, 144, 233**

8. Use a calculator to compute the thirteenth term of the Fibonacci sequence, F_{13}, using the explicit formula given in the lesson. **$F_{13} = 233$**

Applying the Mathematics

In 9 and 10, write the first six terms of the recursively defined sequence.

9. $\begin{cases} z_1 = 5 \\ z_2 = 2 \\ z_{k+1} = 3z_k + z_{k-1} \text{ for } k \geq 2 \end{cases}$ **5, 2, 11, 35, 116, 383**

10. $\begin{cases} p_1 = 1 \\ p_{k+1} = (k+1) \cdot p_k \text{ for } k \geq 1 \end{cases}$ **1, 2, 6, 24, 120, 720**

11. Verify that the constant sequence defined by
$x_n = 1$ for all n
and the alternating sequence 1, 0, 1, 0, ... defined by
$x_n = 1$ if n is odd and $x_n = 0$ if n is even
both satisfy the recurrence relation $a_{n+1} = a_{n-1}$ and the initial condition $a_1 = 1$. Why does this not contradict the Recursion Principle?
See margin.

In 12–14, tell whether the definition of the sequence is recursive or explicit, and write the first six terms of the sequence.

12. $s_n = \sin\left(\frac{n\pi}{2}\right)$ **explicit; 1, 0, -1, 0, 1, 0**

13. $a_n = \begin{cases} 1 \text{ if } n \text{ is even} \\ -1 \text{ if } n \text{ is odd} \end{cases}$ **explicit; -1, 1, -1, 1, -1, 1**

14. $\begin{cases} v_1 = 1 \\ v_{k+1} = 1 + v_{\lceil (k+1)/2 \rceil} \text{ for } k \geq 1 \end{cases}$
(Recall that $\lceil \ \rceil$ denotes the ceiling function.) **recursive; 1, 2, 3, 3, 4, 4**

15. a. Write the first five terms of the sequence defined by the explicit formula $a_n = n$. **1, 2, 3, 4, 5**

b. Repeat part **a** for the sequence defined by $t_n = n^4 - 10n^3 + 35n^2 - 49n + 24$. **1, 2, 3, 4, 29**

c. Compare the answers to parts **a** and **b**. What is the point of this question? **See margin.**

16. There is a sequence L called the *Lucas sequence* named after Edouard Lucas, the Tower of Hanoi problem inventor. L has the initial conditions $L_1 = 1$ and $L_2 = 3$ and the same recurrence relation as the Fibonacci sequence. Give the first ten terms of L.
1, 3, 4, 7, 11, 18, 29, 47, 76, 123

NAME ____________

LESSON **MASTER 7–1**
QUESTIONS ON **SPUR** OBJECTIVES

■ SKILLS *Objective A (See pages 462–464 for objectives.)*
In 1–6, write the first five terms of the sequence defined by the given formula.

1. $d_n = n^3 + 4n^2 - 3n + 1$ **3, 19, 55, 117, 211**

2. $\begin{cases} a_1 = 4 \\ a_{k+1} = -3a_k - 7 \ \forall \ k \geq 1 \end{cases}$ **4, -19, 50, -157, 464**

3. $a_n = 2\left\lceil \frac{n}{3} \right\rceil$ **2, 2, 2, 4, 4**

4. $b_n = (-1)^n \cdot \frac{n(n+1)}{2}$ **-1, 3, -6, 10, -15**

5. $f_n = \begin{cases} \cos n\pi, & n \text{ even} \\ \sin n\pi, & n \text{ odd} \end{cases}$ **0, 1, 0, 1, 0**

6. $c_n = 3\left\lfloor \frac{n}{3} \right\rfloor$ **0, 0, 3, 3, 3**

7. Let a be the sequence defined by $a_n = (-2)^{n+1}$.

a. Which term is greater: a_7 or a_8? **a_7**

b. Given two consecutive integers, how can one tell which will yield the greater term of the sequence?
The odd integer will yield the greater term.

8. Let b be the sequence defined by
$\begin{cases} b_1 = 2 \\ b_{k+1} = 3b_k - 2 \ \forall \ k \geq 1. \end{cases}$ Find b_6. **244**

9. Which of the sequences in Questions 1–7 are defined recursively? **Questions 2, 7**

10. If 5, a, b, c, d, 14 are consecutive terms of an arithmetic sequence, find a, b, c, and d. **6.8, 8.6, 10.4, 12.2**

 Continued 73

NAME ____________
Lesson MASTER 7–1 (page 2)

■ USES *Objective H*

10. A leading brand of soap advertises that it is $99\frac{44}{100}$% pure. Suppose that in the manufacturing process, the soap is initially 80% pure, and is then put through a series of chemical processes to purify it further, each step removing 40% of the existing impurities. Let A_n represent the percent of impurities present after the nth step.

a. Write a recursive definition for the sequence A. **$\begin{cases} A_1 = .12 \\ A_{k+1} = .6A_k \ \forall \ k \geq 1 \end{cases}$**

b. Find A_3. **.0432**

c. Find n so that the soap is at least $99\frac{44}{100}$% pure. **7**

11. A phone company suggests that you call two friends, with instructions for each of them to call two of their friends. If this process is continued, how many friends will receive a call as the seventeenth in the line of calls? **131,072**

12. Suppose an economist suggests that the government adopt an income tax system consisting of income brackets of equal size, and in which the tax rate increases by a constant amount from one bracket to the next. A possible tax table for this system is given below.

Income Bracket	Annual Income (in dollars)	Tax Rate (in percent)
1	0 – 20,000	12
2	20,001 – 40,000	16.5
3	40,001 – 60,000	21
4	60,001 – 80,000	25.5

a. If the tax rate in the nth income bracket is R_n, write a recursive definition for R_n. **$\begin{cases} R_1 = .12 \\ R_{k+1} = R_k + .045 \ \forall \ k \geq 1 \end{cases}$**

b. Find R_8. **.435**

c. If the maximum tax rate were 48 percent, what level of income would be taxed at this rate?
Bracket 9: $160,001 and above

d. Suppose the tax rates change so that the lowest rate is 9 percent, with an increase of 5.1 percent from one bracket to the next. At what rate would someone who earned $136,000 annually be taxed? Is this more or less than under the originally proposed system? **39.6%, more**

74

FOLLOW-UP

MORE PRACTICE
For more questions on SPUR Objectives, use *Lesson Master 7-1,* shown on page 405.

EXTENSION
Archimedes (287–212 B.C.) developed recursion formulas for both the areas and perimeters of inscribed and circumscribed regular polygons: If a_n and A_n are respectively the areas of regular polygons of n sides inscribed in and circumscribed about a circle, then $a_{2n} = \sqrt{a_n A_n}$ and

$$A_{2n} = \frac{2A_n a_{2n}}{A_n + a_{2n}}.$$

If p_n and P_n are perimeters of regular polygons inscribed in and circumscribed about a circle, then $P_{2n} = \frac{2P_n p_n}{P_n + p_n}$ and $p_{2n} = \sqrt{p_n P_{2n}}$.

Archimedes used these formulas to develop the best approximation to π known at that time. A challenge for students is to prove these recursion formulas. Then, beginning with equilateral triangles or hexagons, they should use the formulas to approximate π.

ADDITIONAL ANSWERS

18.c. $P_n = (0.9)^{n-1}x$, $\forall$ integers $n \geq 1$

19., 23. See Additional Answers in the back of this book.

24. sample:
```
10 REM TOWER OF HANOI
20 TERM = 1
30 FOR K = 2 TO 64
40    TERM = TERM *2 + 1
50 NEXT K
60 PRINT TERM
70 END
```

17. The U.S. Postal Service in 1990 charged 25¢ for the first ounce of first class mail. Any additional ounce or fraction thereof cost 20¢. Consider this table of postages according to weight.

Weight not exceeding (oz)	Postage cost (cents)
1	25
2	45
3	65
4	85
5	105
6	125
7	145
8	165

a. Fill out the rest of the second column in the table.
b. Write an explicit formula for the sequence C of postage costs.
c. Write a recursive definition for the sequence C of postage costs.
b) $C_n = 20n + 5$, $\forall$ integers $n \geq 1$; c) $C_1 = 25$, $C_{k+1} = C_k + 20$, $\forall$ integers $k \geq 1$

18. A certain discount store makes the following advertisement:

> We offer low prices to move merchandise. Each week that an item remains on our racks, we will discount the price by 10% of the current price.

Assume that the first week an item is on the shelf, its price P is x dollars.
a. Write the sequence $P_1, P_2, \ldots, P_6$ that gives the price of an item for each of the first six weeks that an item is on the racks.
b. Write a recursive definition for the sequence defined in part **a**.
c. Write an explicit formula for the sequence defined in part **a**.
d. Use either **b** or **c** to determine after how many weeks the price is roughly $.25x$. a) x, $0.9x$, $(0.9)^2x$, $(0.9)^3x$, $(0.9)^4x$, $(0.9)^5x$; b) $P_1 = x$, $P_{k+1} = (0.9)P_k$, $\forall$ integers $k \geq 1$; c) See margin. d) during week 14

Review

19. On the same set of axes, sketch the graphs of $y = \tan x$ and $y = \cot x$ for $-2\pi \leq x \leq 2\pi$, and explain how the graphs are related. *(Lesson 5-6)* See margin.

20. Find numbers x and y such that $x \equiv 5 \pmod 7$ and $y \equiv 3 \pmod 7$.
a. For the x and y you found, to what number (mod 7) is xy congruent? 1
b. Must everyone get the same answer you got for part **a?** Yes *(Lesson 4-3)*

21. Solve for x: $5(2^x - 1) = 10$. *(Lesson 3-2)* $x \approx 1.585$

22. Expand. *(Previous Course)*
a. $(x + 1)^2$ $x^2 + 2x + 1$
b. $(x + y)^2$ $x^2 + 2xy + y^2$
c. $\left(x + \frac{1}{x}\right)^2$ $x^2 + 2 + \frac{1}{x^2}$

Exploration

23. a. Construct a model of the Tower of Hanoi problem with four disks.
b. Describe the steps needed to transfer a tower of 3 disks from one needle to another.
c. Describe the steps needed to transfer a tower of 4 disks from one needle to another.
See margin.

24. Write a computer program that uses the recursive definition of the Tower of Hanoi sequence to calculate T_{64}. See margin.

LESSON

7-2

Recursive and Explicit Formulas

It is easy to square an integer whose units digit (in base 10) is 1, such as 41. You know the square of the previous integer: $40^2 = 1600$. Then add 40 and 41 to 1600. The resulting sum, 1681, is the square of 41. This works because for all real numbers x,

$$(x + 1)^2 = x^2 + 2x + 1$$
$$= x^2 + x + (x + 1).$$

Specifically, when $x = 40$, $\quad 41^2 = 40^2 + 40 + 41.$

The kind of thinking exemplified in this mental arithmetic is recursive, in the sense that a particular value in a sequence (in this case the sequence of squares) is used to obtain the next value. You can use this thinking to obtain a recursive formula for the sequence of squares of the positive integers 1, 4, 9, 16, 25, ...

Example 1 Find a recursive formula for the sequence with explicit formula $a_n = n^2$.

Solution A recursive formula has to have a value for one or more initial terms and a recurrence relation for a_{n+1} in terms of previous terms. The first term of this sequence is 1. Since $a_n = n^2$, by substitution of $n + 1$ for n,

$$a_{n+1} = (n + 1)^2.$$

Now expand this power as was done above.

$$a_{n+1} = n^2 + 2n + 1$$

Substituting a_n for n^2 gives the recurrence relation $a_{n+1} = a_n + 2n + 1$. So a recursive formula for the sequence is

$$\begin{cases} a_1 = 1 \\ a_{n+1} = a_n + 2n + 1 \quad \text{for all integers } n > 1. \end{cases}$$

Check Let $n = 1$ in the recurrence relation. Then $a_{n+1} = a_2 = a_1 + 2 \cdot 1 + 1 = 1 + 2 + 1 = 4 = 2^2$, as it should be.

Recursive formulas are frequently used in computer programs to print out values of sequences. Here are two programs to print out the first 100 terms of the sequence of squares. One uses an explicit formula, the other a recursive one.

```
5  REM EXPLICIT
   FORMULA
10 FOR N = 1 TO 100
20   TERM = N * N
30   PRINT TERM
40 NEXT N
50 END
```

```
5  REM RECURSIVE
   FORMULA
10 TERM = 1
20 FOR N = 1 TO 100
30   PRINT TERM
40   TERM = TERM + 2*N + 1
50 NEXT N
60 END
```


LESSON 7-2

RESOURCES
- Lesson Master 7-2
- Computer Master 13

OBJECTIVES

B Conjecture explicit formulas for recursively defined sequences.
J Interpret computer programs which calculate terms of sequences.

TEACHING NOTES

Examples 1 and 2 have students find a recursive formula for a sequence given an explicit definition for that sequence. **Examples 3 and 4** treat the problem of finding an explicit formula for a sequence with a given recursive definition.

Students should see that the recurrence relation for the sequence of squares defined by $a_n = n^2$ in **Example 1** is just a special case of the sentence $(x + 1)^2 = x^2 + 2x + 1$, which is true for all real numbers x. Give students the square of a two-digit integer x (say $45^2 = 2025$) and ask them for the square of $x + 1$. ($46^2 = 2025 + 45 + 46 = 2116$)

Making Connections Students who studied from UCSMP *Advanced Algebra* or *Functions, Statistics, and Trigonometry* will have distinguished programs that define sequences recursively from those that define them explicitly. The following two programs generate the same sequence. The first uses an explicit definition to generate

100 terms of the sequence defined by $a_n = n^2 + 5n + 4$.

```
10 FOR N = 1 TO 100
20     AN = N∧2 + 5*N + 4
30     PRINT A
40 NEXT N
50 END
```

The second uses a recursive definition to generate terms of the sequence defined by $a_{n+1} = a_n + 2n + 4$.

```
10 A = 10
15 PRINT A
20 FOR I = 2 TO 100
30     A = A + 2 * I + 4
40     PRINT A
50 NEXT I
60 END
```

The sequence printed begins 10, 18, 28, 40, 54, ... Notice that the differences of consecutive terms of this sequence form a linear sequence with slope 2; that is how the recursion relation in line 30 of the second program was found.

The sums S_1, S_2, S_3, ... in **Example 2** form a sequence of partial sums, a concept defined in Lesson 7-6. To help students understand the solution, you may wish to write the following:

$S_5 = 1 + 2 + 3 + 4 + 5$
$S_6 = 1 + 2 + 3 + 4 + 5 + 6$
(the part $1 + 2 + 3 + 4 + 5$ is S_5)

Thus, $S_6 = S_5 + 6$, and for $n = 5$, $S_{n+1} = S_n + (n + 1)$.

To lead into the conclusion of **Example 3** that $T_n = 2^{n-1}$ explicitly, you may wish to prove that the sequence defined recursively as

$$\begin{cases} a_1 = 2 \\ a_{n+1} = 2a_n \text{ for } n \geq 1 \end{cases}$$

has explicit definition $a_n = 2^n$. Be sure to write 2, 4, 8, 16, 32, . . . on the chalkboard. Seeing powers of two may

Sums lend themselves to recursion, since if you have $k + 1$ terms to be added and you know the sum of the first k of them, you need only add the last term to get the grand total. Any set of n terms to be added can be considered as the n terms $a_1, a_2, \ldots, a_n$ of a sequence. Here is part of a computer program in BASIC that uses recursion to find the sum of n numbers $A(1), A(2), \ldots, A(N)$ which have already been entered into the computer's memory. Line 110 inputs the number of terms to be added so that the sum stops at the proper place. Line 120 is the initial condition and line 140 is the recurrence relation which gives the new sum.

```
      :
110  INPUT N
120  SUM = A(1)
130  FOR K = 2 TO N
140     SUM = SUM + A(K)
150  NEXT K
160  PRINT SUM
170  END
```

The computer program contains the idea behind the solution to the next example.

Example 2 Let S_n = the sum of the integers from 1 to n. For example, $S_5 = 1 + 2 + 3 + 4 + 5$. Find a recursive definition for S.

Solution For the initial condition, S_1 is the sum of the integers from 1 to 1, or simply 1. To develop the recurrence relation, note that for example,

$$S_6 = (1 + 2 + 3 + 4 + 5) + 6 = S_5 + 6.$$

In general,

S_{n+1} = the sum of the integers from 1 to $n + 1$
= the sum of the integers from 1 to n, plus the integer $n + 1$
$= S_n + (n + 1)$.

Thus a recursive definition for S is

$$\begin{cases} S_1 = 1 \\ S_{n+1} = S_n + n + 1 \quad \text{for all integers } n \geq 1. \end{cases}$$

Sometimes it is easier to have a recursive formula for a sequence, while at other times it is easier to have an explicit formula. For instance, with the Tower of Hanoi sequence in the last lesson, a recursive formula was used to calculate T_6, the number of moves necessary to transfer six disks. But the Tower of Hanoi problem requires the calculation of T_{64}. For this purpose, an explicit formula is most efficient.

The most basic technique for conjecturing an explicit formula for a sequence that has been defined recursively is to list consecutive terms until you see a pattern emerging.

408

Example 3 The first six terms of the Tower of Hanoi sequence are 1, 3, 7, 15, 31, and 63. What explicit formula for the nth term of T do these terms satisfy?

Solution Each of these terms is one less than a power of 2.

$T_1 = 1 = 2^1 - 1$
$T_2 = 3 = 2^2 - 1$
$T_3 = 7 = 2^3 - 1$
$T_4 = 15 = 2^4 - 1$
$T_5 = 31 = 2^5 - 1$
$T_6 = 63 = 2^6 - 1$

Thus an explicit formula for the Tower of Hanoi sequence appears to be $T_k = 2^k - 1$.

You can prove that the explicit formula $T_k = 2^k - 1$ for the Tower of Hanoi sequence is correct by using the Recursion Principle. Recall that the recursive formula for T is

$$\begin{cases} T_1 = 1 \\ T_{k+1} = 2T_k + 1 \quad \text{for each integer } k \geq 1. \end{cases}$$

According to the Recursion Principle, this definition determines exactly one sequence.

Example 4 Prove that $T_k = 2^k - 1$ is an explicit formula for the Tower of Hanoi sequence.

Solution Let S be the sequence defined by the explicit formula $S_n = 2^n - 1$. By the Recursion Principle, if S satisfies the recursive definition of the Tower of Hanoi sequence T, then S is the Tower of Hanoi sequence.
$S_1 = 2^1 - 1$, so S has the same initial condition as T.

For all $n \geq 1$, $S_{n+1} = 2^{n+1} - 1$	Substituting $n + 1$ for n in the formula for S
$= 2^1 \cdot 2^n - 1$	Laws of exponents
$= 2 \cdot 2^n - 2 + 2 - 1$	Subtracting 2 and adding 2
$= 2 \cdot (2^n - 1) + 1$	Distributive Property
$= 2 \cdot S_n + 1.$	Substituting using the formula for S_n

So S satisfies the same recurrence relation as T. Hence, $T_k = 2^k - 1$ for all integers $k \geq 1$.

Thus according to the Tower of Hanoi story made up by Edouard Lucas, if the monks work to transfer one disk a second, the world will end in $T_{64} = 2^{64} - 1$ seconds.

help students recognize the explicit formula for T_n from the table:

n	T_n
1	1
2	3
3	7
4	15
5	31

To prove that $S_n = 2^n - 1$ is the correct explicit formula for the Tower of Hanoi problem, **Example 4** shows that this explicit formula leads to a sequence that has the same recursive definition. The Recursion Principle guarantees that the explicit formula will generate the identical sequence to that of the recursive formula.

ADDITIONAL EXAMPLES

1. Find a recursive formula for the sequence with explicit formula $a_n = n^3$.
$a_1 = 1$; $a_{n+1} = a_n + 3n^2 + 3n + 1$ for all integers $n \geq 1$

2. Let S_n = sum of the squares of the integers from 1 to n. For example, $S_4 = 1^2 + 2^2 + 3^2 + 4^2$. Find a recursive definition for S.
$S_1 = 1$; $S_{n+1} = S_n + (n+1)^2$ for all integers $n \geq 1$

An extension is to let S_n = sum of the mth powers of the integers from 1 to n. For example, $S_4 = 1^m + 2^m + 3^m + 4^m$. Then a recursive definition for S is $S_1 = 1$; $S_{n+1} = S_n + (n + 1)^m$ for all intergers $n \geq 1$.

3. Suppose
$$\begin{cases} a_1 = 2 \\ a_{n+1} = 3a_n + 2. \end{cases}$$
Conjecture an explicit formula for a_n.
$a_n = 3^n - 1$ for all integers $n \geq 1$

ADDITIONAL EXAMPLES

4. Prove that the explicit formula in Additonal Example 3 is correct.

Let s be the sequence defined by $s_n = 3^n - 1$. Then

$s_1 = 3^1 - 1 = 2$;

$s_{n+1} = 3^{n+1} - 1$

$= 3 \cdot 3^n - 3 + 3 - 1$

$= 3(3^n - 1) + 2$

$= 3s_n + 2$

$\therefore s_{n+1} = 3s_n + 2$

Thus, by the Recursion Principle, since s satisfies the same recursive definition as a, the sequences are identical.

NOTES ON QUESTIONS

Question 1: By the method of **Example 1**, students find the recurrence relation by writing $S_{n+1} = 4(n + 1) + 7 = 4n + 4 + 7 = 4n + 7 + 4 = S_n + 4$. Suggest that writing a few terms and inspecting the results may be easier: $S_1 = 11$, $S_2 = 15$, $S_3 = 19$, $S_4 = 23$, . . . , so each term is 4 more than its predecessor. Thus, $S_{n+1} = S_n + 4$.

Question 6: Relate the results to the periodicity of the sine function. (This sequence picks key values of the function, and ignores all values in between.) A graph will be useful.

Question 9: This question allows students to focus on obtaining a recursive definition from a computer program and then generating an explicit formula.

Question 11: Alternately subtracting and adding the terms of this sequence gives $e^x - 1$.

Question 12: Even though c_1 and c_2 are not given, the value of any term can be found as long as $n \geq 2$ and the values of two consecutive terms c_n and c_{n-1} are given.

In general, if the recursive definition of a sequence satisfies the conditions of the Recursion Principle or its variant, then you can prove that an explicit formula is correct if you can show that the formula satisfies

(1) the initial condition or conditions, and

(2) the recurrence relation

of the recursive formula.

Questions

Covering the Reading

In 1 and 2, find a recursive formula for the sequence S with the given explicit formula.

1. $S_n = 4n + 7$ **2.** $S_n = n^2 - 1$ See margin.

$S_1 = 11$, $S_{k+1} = S_k + 4$, $\forall$ integers $k \geq 1$

3. According to the Tower of Hanoi legend, how many *years* after Creation will the world end? ≈ 585 billion years

In 4 and 5, **a.** Write the first six terms of the sequence t. **b.** Find an explicit formula that works for the six terms. **c.** Use the Recursion Principle and the method of Example 4 to prove that the explicit formula is correct.

4. $\begin{cases} t_1 = 1 \\ t_{k+1} = t_k + 2k - 1 \end{cases}$ for all integers $k \geq 1$ See margin.

5. $\begin{cases} t_1 = 3 \\ t_{k+1} = t_k + 4 \end{cases}$ for all integers $k \geq 1$ See margin.

6. a. Write a computer program to evaluate and print out the first fifty terms of the sequence defined by $a_n = \sin\left(\frac{\pi}{2} n\right)$, beginning with a_1.

b. Describe the sequence in words. a) See margin. b) The odd terms alternate between -1 and 1; the even terms are all zero.

Applying the Mathematics

In 7 and 8, let $a_1, a_2, a_3, \ldots$ be the sequence defined recursively by the given formulas. Write the first six terms of the sequence and determine an explicit formula that is suggested by the pattern of numbers you obtain.

7. $\begin{cases} a_1 = 1 \\ a_{k+1} = \frac{a_k}{1 + a_k} \end{cases}$ $\forall$ integers $k \geq 1$ $1, \frac{1}{2}, \frac{1}{3}, \frac{1}{4}, \frac{1}{5}, \frac{1}{6}$; $a_n = \frac{1}{n}$, $\forall$ integers $n \geq 1$

8. a. $\begin{cases} a_1 = 1 \\ a_{k+1} = -a_k + 2 \end{cases}$ $\forall$ integers $k \geq 1$ 1, 1, 1, 1, 1, 1; $a_n = 1$, $\forall$ integers $n \geq 1$

b. $\begin{cases} a_1 = 1 \\ a_{k+1} = -a_k + 3 \end{cases}$ $\forall$ integers $k \geq 1$ See margin.

c. $\begin{cases} a_1 = 1 \\ a_{k+1} = -a_k + c \end{cases}$ $\forall$ integers $k \geq 1$, where c is a constant See marg

410

9. Consider the computer program below.

```
10 TERM = 2
20  PRINT TERM
30  FOR I = 2 TO 100
40     TERM = TERM * 3 + 2
50     PRINT TERM
60  NEXT I
70  END
```

a. Write a recursive definition for the sequence printed by the computer. $a_1 = 2$; $a_{k+1} = 3a_k + 2$, $\forall$ integers $k \geq 1$

b. Find the values of enough terms to conjecture an explicit formula for the sequence. (Hint: Each value is 1 less than something.)

c. Prove that the explicit formula is correct using the Recursion Principle and the method of Example 4. See margin.

b) $a_n = 3^n - 1$, $\forall$ integers $n \geq 1$, conjectured from 2, 8, 26, 80, ...

Review

10. a. Explain why there is no sequence that satisfies the recursive definition

$$\begin{cases} x_1 = 0 \\ x_{n+1} = \frac{1}{x_n} \text{ for all integers } n \geq 1 \end{cases}$$

x_2 does not exist; it would require division by 0.

b. Why does this not contradict the Recursion Principle? *(Lesson 7-1)* x_n is not defined for all n.

11. For any real number x, define a sequence u by the formula $u_n = \frac{x^n}{n!}$ for all integers $n \geq 1$.

a. Is this an explicit or recursive definition? explicit

b. Write the first five terms of the sequence. $x, \frac{x^2}{2}, \frac{x^3}{6}, \frac{x^4}{24}, \frac{x^5}{120}$ *(Lesson 7-1)*

12. Suppose a sequence c satisfies the recurrence relation

$$c_{n+1} = 2c_n + c_{n-1} \ \forall \ n \geq 2,$$

and suppose $c_5 = 4$ and $c_6 = 7$. Find c_8. *(Lesson 7-1)* 43

13. Consider the function f defined by $f(x) = \frac{2x^2 + 4}{x^2 - 9}$.

a. Use limit notation to describe the end behavior of f. See margin.

b. Classify any discontinuities as removable or essential.

c. Find the equations of all asymptotes to the graph of f. *(Lessons 5-5, 5-4)* $x = 3$, $x = -3$, $y = 2$

b) There are essential discontinuities at $x = 3$ and $x = -3$.

In 14 and 15, rewrite the expression as a single fraction in factored form. *(Lesson 5-1)*

14. $\left[\frac{k(k+1)}{2}\right]^2 + (k+1)^3$ See margin.

15. $\frac{n}{n+1} + \frac{1}{(n+1)(n+2)}$ See margin.

16. a. Sketch the graph of an even function g such that $\lim_{x \to \infty} g(x) = -\infty$ and g is increasing on the interval $0 \leq x \leq 3$. See margin.

b. What is the smallest number of relative maxima that g can have? *(Lessons 2-5, 2-3, 2-2)* 2

ADDITIONAL ANSWERS

2. $S_1 = 0$, $S_{k+1} = S_k + 2k + 1$, $\forall$ integers $k \geq 1$

4.a. 1, 2, 5, 10, 17, 26
b. $t_n = n^2 - 2n + 2$
c. $t_1 = 1^2 - 2(1) + 2 = 1$, so the initial condition is met. $t_{n+1} = (n+1)^2 - 2(n+1) + 2 = n^2 + 2n + 1 - 2n - 2 + 2 = (n^2 - 2n + 2) + 2n - 1 = t_n + 2n - 1$, so the recursive relationship is satisfied. Therefore, the explicit formula is correct.

5.a. 3, 7, 11, 15, 19, 23
b. $t_n = 4n - 1$
c. $t_1 = 4(1) - 1 = 3$, so the initial condition is met. $t_{n+1} = 4(n+1) - 1 = 4n + 4 - 1 = t_n + 4$, so the recursive relationship is satisfied. Therefore, the explicit formula is correct.

6.a., 8.b., c., 9.c., 13.a., 14. See Additional Answers in the back of this book.

15., 16.a. See the margin on page 412.

NAME ____________

LESSON MASTER 7-2
QUESTIONS ON SPUR OBJECTIVES

■SKILLS *Objective B (See pages 462–464 for objectives.)*
In 1–4, a. write the first five terms of the sequence and b. determine an explicit formula that is suggested by the pattern of numbers in part a.

1. $\begin{cases} a_1 = 2 \\ a_{k+1} = 3a_k \ \forall \ k \geq 1 \end{cases}$ a. 2, 6, 18, 54, 162 b. $a_n = 2 \cdot 3^{n-1}$

2. $\begin{cases} d_1 = -1 \\ d_{k+1} = d_k + 5 \ \forall \ k \geq 1 \end{cases}$ a. -1, 4, 9, 14, 19 b. $d_n = 5n - 6$

3. $\begin{cases} a_1 = 1 \\ a_{k+1} = (k+1)a_k \ \forall \ k \geq 1 \end{cases}$ a. 1, 2, 6, 24, 120 b. $a_n = n!$

4. $\begin{cases} x_1 = 3 \\ x_2 = 5 \\ x_{k+1} = x_k + 2x_{k-1} - 2 \ \forall \ k \geq 1 \end{cases}$ (Hint: Subtract 1 from each term to see the pattern.) a. 3, 5, 9, 17, 33 b. $x_n = 2^n + 1$

5. Conjecture an explicit formula for S_n, the sum of the first n positive odd integers. $S_n = n^2$

■REPRESENTATIONS *Objective J*

6. a. List the terms generated by the program at the right. 0, 10, 30, 60, 100, 150, 210

```
10 FOR N=0 TO 6
20    C=5N*(N+1)
30    PRINT C
40 NEXT N
```

b. Does the program use a recursive or explicit formula? What is this formula? explicit; $c_n = 5n(n+1)$

7. Complete the program at the right so it uses a recursive formula to generate the first seven terms of the sequence defined by $a_n = \frac{n}{2} + 3$ $\forall \ n \geq 1$.

```
10 TERM = 3.5
20 PRINT TERM
30 FOR K = 2 TO 7
40    TERM = TERM+.5
50    PRINT TERM
60 NEXT K
```

 75

NOTES ON QUESTIONS
Question 18: Invite students to use the method of **Example 4** to prove that their explicit formulas are correct.

FOLLOW-UP

MORE PRACTICE
For more questions on SPUR Objectives, use *Lesson Master 7-2,* shown on page 411.

EXTENSION
Suppose the Creator had added one additional disk to the original pile in the Tower of Hanoi problem, making 65 disks. How would that affect the monks' time of labor? ($T_{65} = 2^{65} - 1 = 2T_{64} + 1$; hence, it will take 1 second more than twice as long.)

PROJECTS
The projects for Chapter 7 are described on pages 458–459. **Project 1** is related to the content of this lesson.

ADDITIONAL ANSWERS

15. $\frac{n+1}{n+2}$, for $n \neq -1$

16.a. sample:

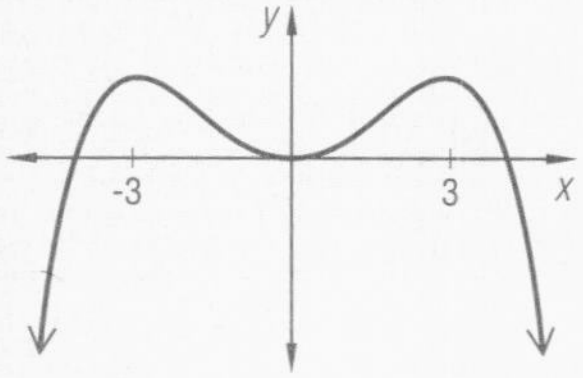

18.a.i. $\frac{1}{3}, \frac{4}{9}, \frac{13}{27}, \frac{40}{81}$
$a_n = \frac{1}{2}\left(1 - \left(\frac{1}{3}\right)^n\right)$,
$\forall$ integers $n \geq 1$

ii. $\frac{1}{4}, \frac{5}{16}, \frac{21}{64}, \frac{85}{256}$
$a_n = \frac{1}{3}\left(1 - \left(\frac{1}{4}\right)^n\right)$,
$\forall$ integers $n \geq 1$

b. $a_n = \frac{1}{c-1} \cdot \left(1 - \left(\frac{1}{c}\right)^n\right)$;
yes, the formula works for all noninteger numbers $c \neq 0$ or 1.

17. Consider the universal statement
$\forall$ *real numbers* x, $p(x) \Rightarrow q(x)$.

a. Under what circumstances will the universal statement be false?

b. Suppose $p(x)$ is $(x + 3)^2 = 49$, and $q(x)$ is $x = 4$. Is
$\forall$ *real numbers* x, $p(x) \Rightarrow q(x)$
true or *false?* *(Lessons 1-3, 1-1)* False

a) if there exists an x such that $p(x)$ is true and $q(x)$ is false

Exploration

18. a. List the first four terms of each sequence and conjecture an explicit formula.

i. $\begin{cases} a_1 = \frac{1}{3} \\ a_{k+1} = a_k + \frac{1}{3^{k+1}} \text{ for all integers } k \geq 1 \end{cases}$

ii. $\begin{cases} a_1 = \frac{1}{4} \\ a_{k+1} = a_k + \frac{1}{4^{k+1}} \text{ for all integers } k \geq 1 \end{cases}$

b. Conjecture an explicit formula for the sequence defined by

$\begin{cases} a_1 = \frac{1}{c} \\ a_{k+1} = a_k + \frac{1}{c^{k+1}} \end{cases}$ for all integers $k \geq 1$, and constant integer $c \neq 0$.

Does your formula work for noninteger values of c?
See margin.

19. There are stacking puzzles other than the Tower of Hanoi. Try this one by Henry Ernest Dudeney, which he called Transferring the Counters. Divide a sheet of paper into six compartments, as shown in the illustration, and place a pile of fifteen counters, numbered consecutively 1, 2, 3, . . . , 15 downwards, in compartment *A*. The puzzle is to transfer the complete pile, in the fewest possible moves, to compartment *F*. You can move the counters one at a time to a compartment, but may never place a counter on one that bears a smaller number than itself. Thus, if you place 1 on *B* and 2 on *C*, you can then place 1 on 2, but not 2 on 1.

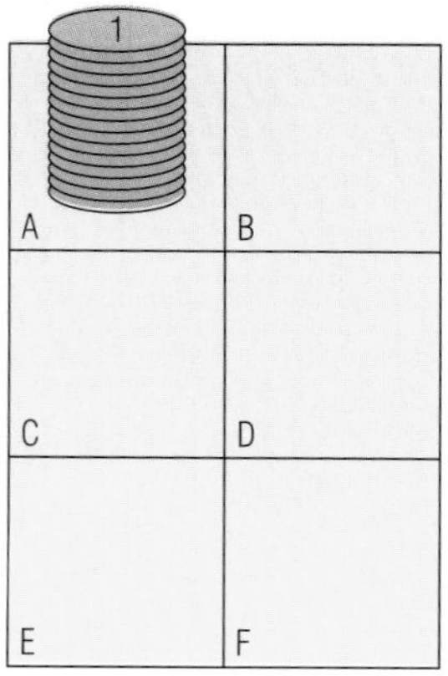

Make a pile of five counters (1 to 5) on *B* in 9 moves. Make a pile of four (6 to 9) on *C* in 7 moves. Make a pile of three (10 to 12) on *D* in 5 moves. Make a pile of two (13 and 14) on *E* in 3 moves. Place one (15) on *F* in 1 move. Replace 13 and 14 on *F* in 3, 10 to 12 on *F* in 5, 6 to 9 in 7, and 1 to 5 in 9 moves. That makes 49 moves in all.

412

LESSON

7-3

Summation Notation

The Greek philosopher Zeno of Elea lived around 450 B.C. and is famous for his paradoxes. One of them involves a runner who is trying to go from point A to point B. Zeno pointed out that the runner would first have to go half the distance, then half the remaining distance, and so forth. Suppose the runner travels at a constant speed, taking one minute to go half the distance from A to B. It will then take $\frac{1}{2}$ minute to go half of the remaining distance, $\frac{1}{4}$ minute to go half of the new remaining distance, and so forth, as shown below.

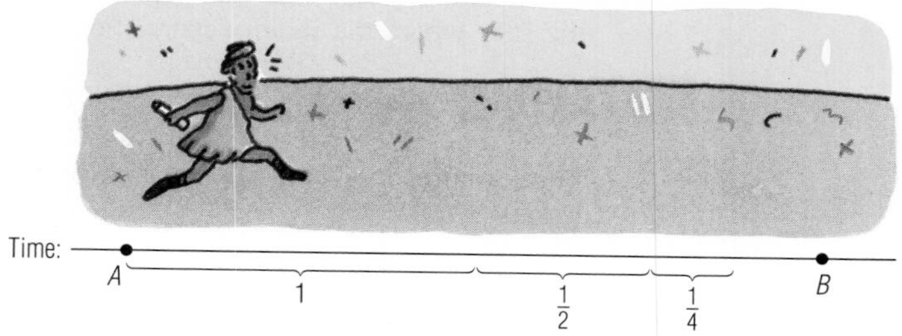

Zeno observed that the total time required for the runner to reach point B would have to equal the sum of an *infinite* collection of positive numbers:

$$1 + \frac{1}{2} + \frac{1}{4} + \frac{1}{8} + \dots .$$

Zeno argued that such an infinite sum could not equal a finite number, and therefore the runner could never reach point B. This argument is called **Zeno's paradox** because obviously runners do get from point A to point B.

Summation notation, or *sigma notation*, is very convenient for concisely representing sums such as that arising from Zeno's paradox. This notation uses the symbol $\sum$, the Greek letter sigma. Sigma is used because it is the Greek version of S, the first letter of the word *sum*. The numbers to be added are thought of as consecutive terms of a sequence.

Definition (Summation notation)

Suppose m and n are integers with $m < n$. Then

$$\sum_{i=m}^{n} a_i = a_m + a_{m+1} + \dots + a_n.$$

The symbol $\sum_{i=m}^{n} a_i$ is read "the sum of the numbers a_i from i equals m to i equals n." The expression $a_m + a_{m+1} + \dots + a_n$ is called the **expanded form** of $\sum_{i=m}^{n} a_i$, and the variable i is called the **index** of this sum.

LESSON 7-3

RESOURCES
- Lesson Master 7-3
- Quiz for Lessons 7-1 through 7-3

OBJECTIVES

C Use summation notation to write sums.
D Rewrite sums recursively.
F Prove that a recursively defined sequence has a particular explicit formula.
J Interpret computer programs which calculate terms of sequences.

TEACHING NOTES

Did Zeno actually think that the runner would not get from point A to point B? Zeno knew that the runner got from point A to point B. He was showing that the formal logic and mathematics of his time were insufficient for dealing with the infinite, in this case, with an infinite sum.

A paradox of Zeno related to the paradox of this lesson is sometimes called The Dichotomy. In order to traverse a distance, the runner has to go halfway. To go halfway, the runner has to go a quarter of the way. To go a quarter of the way, the runner has to go an eighth of the way, and so on. Since the runner has to satisfy infinitely many of these conditions, the runner will never get started!

Alternate Approach If the symbol Σ is new for your students, you might introduce an intermediate definition:

$\sum_{i=1}^{n} = a_1 + a_2 + \ldots + a_n$.

You can even begin more slowly than this. Start with a few examples in expanded form first. Point out that there is a pattern in the expansion that can be expressed with a variable whose values increase by 1. The final step is expression in summation notation. Some examples are:

$1 + 2 + 3 + 4 + 5 = \sum_{i=1}^{5}$;

$\frac{1}{2} + \frac{1}{4} + \frac{1}{8} + \frac{1}{16} + \frac{1}{32} = \frac{1}{2^1} + \frac{1}{2^2} + \frac{1}{2^3} + \frac{1}{2^4} + \frac{1}{2^5} = \sum_{i=1}^{5} \frac{1}{2^i}$.

Point out that the purpose of summation notation is to simplify an expression, not to make it more complicated.

It is important for students to understand that the index variable is the only variable for which values are substituted when the expanded form is written. That is,

$\sum_{i=1}^{4} i = 1 + 2 + 3 + 4 + 10$,

$\sum_{i=1}^{4} (i + c) = (1 + c) + (2 + c) + (3 + c) + (4 + c) = 10 + 4c$,

$\sum_{i=5}^{10} k = k + k + k + k + k + k = 6k$

In the second and third expressions above, c and k are not indices of summation and thus do not assume the range of integer values over which the sum is computed.

Example 1 Write $\sum_{i=1}^{4} 3i$ in expanded form and find the value of this sum.

Solution In this example, $a_i = 3i$ for each integer i. $\sum_{i=1}^{4} 3i$ is the sum of all the numbers $3i$ as i takes on integer values from 1 to 4. Consequently,

$$\begin{aligned}\sum_{i=1}^{4} 3i &= 3 \cdot 1 + 3 \cdot 2 + 3 \cdot 3 + 3 \cdot 4 \quad \text{Expanded form}\\ &= 3 + 6 + 9 + 12\\ &= 30.\end{aligned}$$

A given sum can be expressed in summation notation in more than one way. The index i can be replaced by any other variable without changing the meaning of the sum as long as the replacement is made every place in the sum where the index occurs. For example,

$$\sum_{i=2}^{5} 2^i = \sum_{n=2}^{5} 2^n = 2^2 + 2^3 + 2^4 + 2^5.$$

Furthermore, this sum can be denoted with different upper and lower index values by

$$\sum_{i=1}^{4} 2^{i+1}$$

because $\quad 2^{1+1} + 2^{2+1} + 2^{3+1} + 2^{4+1} = 2^2 + 2^3 + 2^4 + 2^5.$

The next example clarifies the meaning of summation notation when the lower or upper value of the index variable is itself a variable.

Example 2 Consider the sum $\sum_{k=1}^{n} \frac{n+1}{n+k}$.

a. Write the expanded form of the sum for $n = 1$.
b. Write the expanded form of the sum for $n = 4$, and approximate its value.
c. Write the expanded form of the sum for arbitrary n.

Solution
a. The index variable k varies from 1 to 1 so it takes on only the single value 1:

$$\sum_{k=1}^{1} \frac{1+1}{1+k} = \frac{1+1}{1+1}.$$

b. $$\begin{aligned}\sum_{k=1}^{4} \frac{4+1}{4+k} &= \frac{4+1}{4+1} + \frac{4+1}{4+2} + \frac{4+1}{4+3} + \frac{4+1}{4+4}\\ &= \frac{5}{5} + \frac{5}{6} + \frac{5}{7} + \frac{5}{8} \approx 3.173.\end{aligned}$$

c. The index variable k varies from 1 to n:

$$\begin{aligned}\sum_{k=1}^{n} \frac{n+1}{n+k} &= \frac{n+1}{n+1} + \frac{n+1}{n+2} + \frac{n+1}{n+3} + \ldots + \frac{n+1}{n+n}\\ &= 1 + \frac{n+1}{n+2} + \frac{n+1}{n+3} + \ldots + \frac{n+1}{2n}.\end{aligned}$$

In summation notation, the sum of the integers from 1 to n, $1 + 2 + 3 + \ldots + n$, can be written as $\sum_{i=1}^{n} i$. You saw a recursive definition for this sum S_n in the last lesson.

$$\begin{cases} S_1 = 1 \\ S_{n+1} = S_n + n + 1 \end{cases}$$

In summation notation, the recurrence relation is

$$\sum_{i=1}^{n+1} i = \sum_{i=1}^{n} i + (n + 1).$$

Notice that, in general,

$$\begin{aligned} \sum_{i=m}^{k+1} a_i &= a_m + a_{m+1} + \ldots + a_k + a_{k+1} \\ &= [a_m + a_{m+1} + \ldots + a_k] + a_{k+1} \\ &= \left(\sum_{i=m}^{k} a_i\right) + a_{k+1}. \end{aligned}$$

This recurrence relation is satisfied by any sum. Replacing $\sum_{i=m}^{k+1} a_i$ with $\left(\sum_{i=m}^{k} a_i\right) + a_{k+1}$ is referred to as "writing the sum recursively."

Questions

Covering the Reading

In 1 and 2, **a.** write the expanded form of the given sum and **b.** evaluate the sum.

1. $\sum_{k=0}^{3} (k^2 + 3k - 2)$ a) $-2 + 2 + 8 + 16$; b) 24

2. $\sum_{n=-4}^{-1} 2^n$ a) $2^{-4} + 2^{-3} + 2^{-2} + 2^{-1}$; b) $\frac{15}{16}$

3. *True* or *false?* $\sum_{p=3}^{8} 4p = \sum_{k=3}^{8} 4k$. Justify your answer in one or two sentences.
True, the only difference is that different letters are used for the indices.

In 4 and 5, write using summation notation.

4. $1^2 + 2^2 + 3^2 + \ldots + 43^2$ $\sum_{i=1}^{43} i^2$

5. $\frac{1}{2} + \frac{1}{3} + \frac{1}{4} + \ldots + \frac{1}{k+1}$ $\sum_{j=1}^{k} \frac{1}{j+1}$

6. Note that the terms of the sequence of times in Zeno's Paradox are powers of $\frac{1}{2}$. Use summation notation to write the sum of
a. the first 5 terms **b.** the first n terms.
See margin.

7. Refer to Example 2.
a. Find the value of the sum if $n = 3$. $\frac{37}{15} \approx 2.47$
b. Write $\sum_{k=n-1}^{n+1} \frac{n+1}{n+k}$ in expanded form. $\frac{n+1}{2n-1} + \frac{n+1}{2n} + \frac{n+1}{2n+1}$

ADDITIONAL EXAMPLES

1. Write $\sum_{i=1}^{4} (4i + i^2)$ in expanded form and find the value of this sum.
$\sum_{i=1}^{4} (4i + i^2) = (4 \cdot 1 + 1^2) + (4 \cdot 2 + 2^2) + (4 \cdot 3 + 3^2) + (4 \cdot 4 + 4^2) = 5 + 12 + 21 + 32 = 70$

2. Consider the sum $\sum_{k=1}^{n} \frac{n(k+1)}{(k+2)}$.
a. Write the expanded form of the sum for $n = 1$.
$\sum_{k=1}^{1} \frac{n(k+1)}{(k+2)} = \frac{1(1+1)}{(1+2)}$
b. Write the expanded form of the sum for $n = 4$ and approximate its value.
$\sum_{k=1}^{4} \frac{n(k+1)}{(k+2)} = \frac{4(1+1)}{(1+2)} + \frac{4(2+1)}{(2+2)} + \frac{4(3+1)}{(3+2)} + \frac{4(4+1)}{(4+2)} \approx 12.2$
c. Write the expanded form of the sum for arbitrary n.
$\sum_{k=1}^{n} \frac{n(k+1)}{(k+2)} = \frac{n(1+1)}{(1+2)} + \frac{n(2+1)}{(2+2)} + \frac{n(3+1)}{(3+2)} + \ldots + \frac{n(n+1)}{(n+2)}$

ADDITIONAL ANSWERS

6.a. $\sum_{k=0}^{4} \left(\frac{1}{2}\right)^k$

b. $\sum_{j=0}^{n-1} \left(\frac{1}{2}\right)^j$

NOTES ON QUESTIONS

Questions 9b and 11: This discussion will help prepare students for the inductive step in proofs using the Principle of Mathematical Induction.

Question 10: This question outlines skills needed to prove that $p(n)$ is true for all integers $n \geq 1$ by the Principle of Mathematical Induction.

Question 13: This question illustrates a general property for summation notation:

$\sum_{i=1}^{n} [af(i) + bg(i)] = a\sum_{i=1}^{n} f(i) + b\sum_{i=1}^{n} g(i)$

Question 14: Students who expand and regroup the terms will find this question quite easy. It is a special case of the general form: $(a_1 - a_2) + (a_2 - a_3) + (a_3 - a_4) + \ldots + (a_n - a_{n+1}) = a_1 - a_{n+1}$; that is, $\sum_{i=1}^{k+1} (a_i - a_{i+1}) = a_1 - a_{k+1}$.

ADDITIONAL ANSWERS

9.a. $\sum_{i=1}^{1} a_i = 1$, $\sum_{i=1}^{2} a_i = 5$, $\sum_{i=1}^{3} a_i = 14$, $\sum_{i=1}^{4} a_i = 30$

b. $\sum_{i=1}^{k+1} i^2 = 1^2 + 2^2 + \ldots + k^2 + (k+1)^2 = (1^2 + 2^2 + \ldots + k^2) + (k+1)^2 = \left(\sum_{i=1}^{k} i^2\right) + (k+1)^2$

10.a.–d., 11., 14., 15., 16.; 19. See Additional Answers in the back of this book.

8. a. Write $\sum_{i=1}^{7} i$ in expanded form and evaluate.

b. Check your answer to part **a** by using the theorem for the sum of the first n integers. $S_7 = \frac{7(7+1)}{2} = 28$

c. Find the sum of the integers from 1 to 100. 5050

d. Find the sum of the integers from 1 to $k + 2$. $\frac{(k+2)(k+3)}{2}$

a) $1 + 2 + 3 + 4 + 5 + 6 + 7 = 28$

9. Consider the sequence defined by $a_i = i^2$.

a. Compute $\sum_{i=1}^{n} a_i$ for $n = 1, 2, 3$, and 4.

b. Explain why $\sum_{i=1}^{k+1} i^2 = \left(\sum_{i=1}^{k} i^2\right) + (k+1)^2$ for each integer $k \geq 1$. See margin.

Applying the Mathematics

10. Let $p(n)$ be the sentence: $\sum_{i=1}^{n} (2i - 1) = n^2$.

a. Show that $p(1)$ is true. See margin.

b. Write $p(k)$. See margin.

c. Write $p(k + 1)$. See margin.

d. Assuming that $p(k)$ is true, show that $\sum_{i=1}^{k+1} (2i - 1) = k^2 + [2(k+1) - 1]$. See margin.

e. Assuming that $p(n)$ is true for all integers $n \geq 1$, find $1 + 3 + 5 + 7 + \ldots + 101$. 2601

11. Write $\sum_{i=1}^{k+1} i(i-1)$ in terms of $\sum_{i=1}^{k} i(i-1)$. See margin.

12. Express the sum in Question 1 in summation notation with an index j that varies from 1 to 4. See below.

13. Determine whether or not $\sum_{i=1}^{4} (i^3 + 5i) = \sum_{i=1}^{4} i^3 + 5\sum_{i=1}^{4} i$. It does.

14. Show that $\sum_{k=1}^{5} \left(\frac{1}{k} - \frac{1}{k+1}\right) = 1 - \frac{1}{6}$. (Hint: This problem is easier if you do not simplify the terms.) See margin.

15. Find a sequence a with $\sum_{n=1}^{4} (a_n^2) \neq \left(\sum_{n=1}^{4} a_n\right)^2$ See margin.

16. Consider the computer program below.

```
10 SUM = 0
20 FOR K = -5 TO 5
30 SUM = SUM + .1*(K/10)^2
40 NEXT K
50 PRINT SUM
60 END
```

In summation notation, write the expression this program evaluates. See margin.

12. $\sum_{j=1}^{4} (j^2 + j - 4)$

416

Review

17. Consider the sequence defined recursively by

$$\begin{cases} a_1 &= \frac{1}{2} \\ a_{k+1} &= \frac{k+1}{k+2}\, a_k \ \forall \text{ integers } k \geq 1. \end{cases}$$

Write the first six terms of the sequence and conjecture an explicit formula for a_n. *(Lesson 7-2)* $\frac{1}{2}, \frac{1}{3}, \frac{1}{4}, \frac{1}{5}, \frac{1}{6}, \frac{1}{7}$; $a_n = \frac{1}{n+1}$, $\forall$ integers $n \geq 1$

18. Newton's Law of Cooling states that the change in temperature of an object per unit of time is equal to a constant multiplied by the temperature difference between the object and the surrounding air. Suppose that a cake is removed from a 325° oven and placed in a 75° room. The cake's temperature decreases each minute by an amount equal to .05 times the difference in temperature between the cake and the room at the beginning of the minute. Let T_k be the temperature of the cake at the beginning of the kth minute.

a. Write a recursive definition for the sequence $T_1, T_2, T_3, \ldots$.

b. Use the result from part **a** to find the cake's temperature (to the nearest degree) at the beginning of the fifth minute. 279°

(Lesson 7-1) a) $T_1 = 325$, $T_{k+1} = 0.95T_k + 3.75$, $\forall$ integers $k \geq 1$

19. Prove: $\forall$ *real numbers x and y,*

$$\sin^2 x - \sin^2 y = \cos^2 y - \cos^2 x.$$ *(Lesson 6-2)* See margin.

20. Write the following expression as a single fraction in factored form:

$$\frac{k}{2k+1} + \frac{1}{(2k-1)(2k+1)}.$$ *(Lesson 5-1)* $\frac{2k^2 - k + 1}{(2k+1)(2k-1)}$

21. Let $f(k) = k + 1$ and $g(k) = \frac{k(k+1)}{2}$ for all real numbers k. Write a formula for $g \circ f$. *(Lesson 3-2)* $(g \circ f)(k) = \frac{(k+1)(k+2)}{2}$

Exploration

22. Find out something about a paradox of Zeno different from the paradox mentioned in this lesson.

Sample: the Arrow Paradox states that an arrow never moves, because at each instant the arrow is in a fixed position. Another of Zeno's paradoxes, known as the Paradox of Achilles and the Tortoise, is frequently summarized as follows. Achilles, who could run 10 yards per second, competed against a tortoise which ran 1 yard per second. In order to make the race more fair, the tortoise was given a headstart of 10 yards. Zeno's argument, that Achilles could never pass the tortoise, was based on the "fact" that whenever Achilles reached a certain point where the tortoise had been, the tortoise would have moved ahead of that point.

FOLLOW-UP

MORE PRACTICE

For more questions on SPUR Objectives, use *Lesson Master 7-3,* shown below.

EVALUATION

A quiz covering Lessons 7-1 through 7-3 is provided in the Teacher's Resource File.

NAME ____________

LESSON MASTER 7–3
QUESTIONS ON **SPUR** OBJECTIVES

■ SKILLS *Objective C (See pages 462–464 for objectives.)*

1. If $n = 3$, find $\sum_{k=-1}^{n} k^2 + 3k$. 30

In 2 and 3, write using summation notation.

2. $7 + 14 + 21 + 28 + 35 + 42$ $\sum_{j=1}^{6} 7j$

3. $-\frac{2}{(k+1)^2} - \frac{1}{(k+1)} + 0 + (k+1) + 2(k+1)^2$ $\sum_{j=-2}^{2} j(k+1)^j$

4. a. Rewrite the equation $1^3 + 2^3 + 3^3 + 4^3 + \ldots + n^3 = (1 + 2 + 3 + 4 + \ldots + n)^2$ using summation notation. $\sum_{j=1}^{n} j^3 = \left(\sum_{j=1}^{n} j\right)^2$

b. Show that the equation is true for $n = 5$. Both sides yield 225.

■ SKILLS *Objective D*

5. a. Express $\sum_{j=0}^{11} (j^4 - j)$ in terms of $\sum_{j=0}^{10} (j^4 - j)$. $\sum_{j=0}^{10} (j^4 - j) + (11^4 - 11)$

b. Given that $\sum_{j=0}^{10} (j^4 - j) = 25{,}278$, find $\sum_{j=0}^{11} (j^4 - j)$. 39,908

6. Let $S(k)$ be the statement: $\sum_{j=1}^{k} j^3 = \frac{1}{4}k^2(k+1)^2$.

a. Find $\sum_{j=1}^{5} j^3$ and show that $S(5)$ is true.
$\sum_{j=1}^{5} j^3 = 225$; $\frac{1}{4}(5^2)(5+1)^2 = 225$; thus, $S(5)$ is true.

b. Rewrite $\sum_{j=1}^{k+1} j^3$ in terms of $\sum_{j=1}^{k} j^3$. $\sum_{j=1}^{k} j^3 + (k+1)^3$

c. Use your answers to parts **a** and **b** to find $\sum_{j=1}^{6} j^3$. $225 + 6^3 = 441$

d. Use the answer to part **c** to determine if $S(6)$ is true.
$\frac{1}{4}(6^2)(6+1)^2 = 441 = \sum_{j=1}^{6} j^3$, so $S(6)$ is true.

76 *Continued* *Precalculus and Discrete Mathematics* © Scott, Foresman and Company

NAME ____________
Lesson MASTER 7–3 (page 2)

■ PROPERTIES *Objective F*

7. Prove that the sequence defined by $\begin{cases} a_1 = \frac{1}{2} \\ a_{k+1} = \frac{a_k}{k+2} \ \forall\, k \geq 1 \end{cases}$ has explicit formula $a_n = \frac{1}{(n+1)!}$.

$a_1 = \frac{1}{(1+1)!} = \frac{1}{2!} = \frac{1}{2}$; $a_{k+1} = \frac{1}{(k+1+1)!} = \frac{1}{(k+2)!} = \frac{1}{(k+2)(k+1)!} = \frac{1}{k+2} \cdot \frac{1}{(k+1)!} = \frac{1}{k+2} \cdot a_k = \frac{a_k}{k+2}$

8. Prove that the sequence defined by $\begin{cases} x_1 = 3 \\ x_{k+1} = \frac{x_k}{4} \ \forall\, k \geq 1 \end{cases}$ has explicit formula $x_n = \frac{3}{4^{n-1}}$.

$x_1 = \frac{3}{4^{1-1}} = 3$; $x_{k+1} = \frac{3}{4^{k+1-1}} = \frac{3}{4^k} = \frac{3}{4 \cdot 4^{k-1}} = \frac{1}{4} \cdot \frac{3}{4^{k-1}} = \frac{x^k}{4}$

■ REPRESENTATIONS *Objective J*

9. Consider the computer program below.

```
10 INPUT N
20 SUM = 0
30 FOR K = 4 TO N
40   TERM = K/(K + 1)
50   SUM = SUM + TERM
60 NEXT K
70 PRINT SUM
```

a. Use summation notation to express the sum computed by this program. $\sum_{k=4}^{n} \frac{k}{k+1}$

b. If 7 is input for N, what output is generated? 3.36547619

10. How would you change the program in Question 9 so it would compute $\sum_{k=n}^{15} \frac{n-k}{k}$?

```
30 FOR K = N TO 15
40   TERM = (N – K)/K
```

Precalculus and Discrete Mathematics © Scott, Foresman and Company 77

LESSON 7-4

RESOURCES
- Lesson Master 7-4

OBJECTIVES

F Prove that a recursively defined sequence has a particular explicit formula.
G Prove statements using mathematical induction.

TEACHING NOTES

In order to understand mathematical induction, students need (1) to apply the $S(n)$ notation of propositions, and (2) to realize the significance of the two parts of the induction proof. In this book, students have seen the notation of (1) since Lesson 1-1, and their work with recursive definitions makes (2) quite a bit more reasonable. Still, the form of mathematical induction proofs is different from other proofs students have encountered, and the algebraic manipulation also is different. Thus, mathematical induction is difficult for students when they study it the first time.

Improper induction is the invalid argument form by which a generalization is made from specific examples. The Principle of Mathematical Induction is the valid argument form studied in this lesson. A proof that uses the Principle of Mathematical Induction is called a *mathematical induction* proof, or a proof by mathematical induction. All of these terms are usually just called *induction.*

LESSON

7-4 The Principle of Mathematical Induction

Induction is the use of inductive reasoning to arrive at a generalization. As was pointed out in Lesson 1-9, induction is not a valid method of proof but it is very useful in coming up with conjectures. Example 3 in Lesson 7-2 used induction to conjecture that the Tower of Hanoi sequence T had the explicit formula $T_n = 2^n - 1$. Then deduction from the Recursion Principle was used to prove that this formula did in fact hold for all integers $n \geq 1$.

Mathematical induction is not the same as induction. *Mathematical induction* refers to a valid proof argument form that is closely related to recursion.

The idea of mathematical induction is often illustrated as follows: Imagine a row of dominoes lined up in such a way that for each integer $k \geq 1$, if the kth domino falls over then it causes the $(k + 1)$st domino to fall over also. Now what happens if you push the first domino over? All the dominoes will fall.

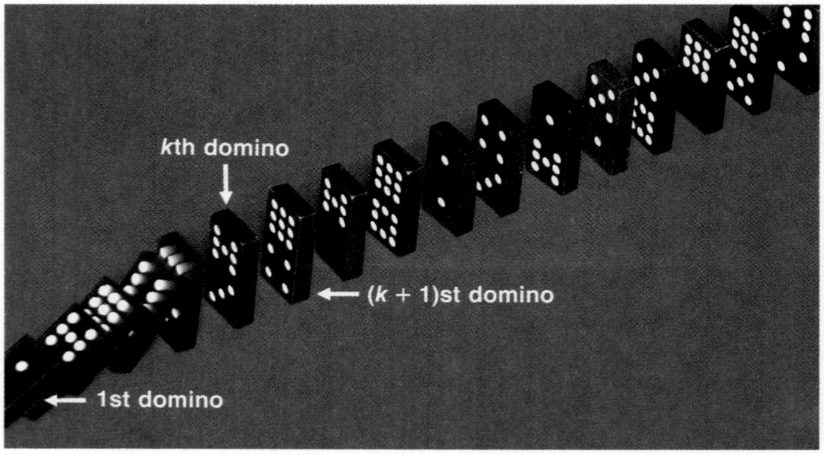

Principle of Mathematical Induction

> Suppose that for each positive integer n, $S(n)$ is a sentence in n. If
> (1) $S(1)$ is true, and
> (2) for all integers $k \geq 1$, the assumption that $S(k)$ is true implies that $S(k + 1)$ is true,
>
> then $S(n)$ is true for all positive integers n.

Verification of (1) (pushing the first domino) is called the **basis step** of the induction, and verification of (2) (if the kth domino falls, so will the $(k + 1)$st) is called the **inductive step.** Note that (2) is a universal conditional statement. It is proved with a direct proof.

You

suppose that $S(k)$ is true for a particular but arbitrarily chosen integer $k \geq 1$,

and then you

show that the supposition that $S(k)$ is true implies that $S(k+1)$ is true.

The supposition that $S(k)$ is true is called the **inductive assumption.**

Example 1 Use the Principle of Mathematical Induction to prove that the sum of the first n positive odd integers is n^2. That is, prove that

$1 + 3 + 5 + ... + (2n - 1) = n^2$ *for all integers* $n \geq 1$,

or in summation notation,

$$\sum_{i=1}^{n} (2i - 1) = n^2 \text{ for all integers } n \geq 1.$$

Solution First identify $S(n)$. In this case, $S(n)$ is $\sum_{i=1}^{n} (2i - 1) = n^2$.

(1) Basis step: Show that $S(1)$ is true.

When $n = 1$, the left side of $S(1)$ is $2 \cdot 1 - 1$, which is just 1, and the right side is 1^2. Hence the two sides of the equation are equal when $n = 1$, and so $S(1)$ is true.

(2) Inductive step: Show that the assumption that $S(k)$ is true implies that $S(k + 1)$ is true.

Suppose that $S(k)$ is true for a particular but arbitrarily chosen integer $k \geq 1$. That is, suppose

$$S(k): \sum_{i=1}^{k} (2i - 1) = k^2.$$

This is the inductive assumption.

Now show that $S(k + 1)$ is true, where

$$S(k+1): \sum_{i=1}^{k+1} (2i - 1) = (k + 1)^2.$$

Begin with the left side of $S(k + 1)$, and use algebra and the inductive assumption to transform it into the right side of $S(k + 1)$.

$$\sum_{i=1}^{k+1} (2i - 1) = \left(\sum_{i=1}^{k} (2i - 1)\right) + 2(k + 1) - 1$$ Write the sum recursively.

$$= \left(\sum_{i=1}^{k} (2i - 1)\right) + 2k + 1$$ Simplify.

$$= k^2 + (2k + 1)$$ Use the inductive assumption (the supposition that $S(k)$ is true).

$$= (k + 1)^2$$ Factor.

Therefore, for all integers $k \geq 1$, if $S(k)$ is true, then $S(k + 1)$ is true.

Thus, from (1) and (2) above, by the Principle of Mathematical Induction, $S(n)$ is true for all integers $n \geq 1$.

The basis step of a mathematical induction proof provides a starting point, a particular instance for which $S(n)$ is known to be true. The inductive statement is a universal conditional statement. The inductive assumption, that $S(k)$ is true for some arbitrarily chosen $k \geq 1$, is used to show that $S(k + 1)$ is true. Once the universal conditional statement is known to be true, it can be used, together with the basis step, to create a series of simultaneous *modus ponens* arguments. In this way, $S(n)$ is shown to be true for all positive integers greater than or equal to the value of n used in the basis step.

You can compare the Principle of Mathematical Induction to climbing an infinite staircase to the stars. To prove that you will be able to make that infinite climb, you must do two things: (1) show that you can climb the first step, and (2) assuming that you can climb the kth step, show that you will be able to climb step $k + 1$ also.

A large set of dominos can be arranged so that pushing one will topple the entire set. At times, people have arranged thousands of dominos in this way. The weakness of the domino analogy is that it takes time for all the dominos to fall. When the Principle of Mathematical Induction is applied, the truth of $S(n)$ for all integers $n \geq 1$ is established simultaneously.

Students may have difficulty understanding the inductive step. It bothers them that we assume the formula to be true for $n = k$. Relate this step to the domino experiment. It is akin to placing the dominos carefully. If we place the dominos so that if the kth domino falls over, the $(k + 1)$th domino falls over also. Then all the dominos will fall if we just topple the first domino. Pushing over the first domino is akin to showing the statement $S(1)$ is true.

After completing a proof by mathematical induction, be sure to summarize why the proof is complete. Students often do not know when they have completed such a proof.

The formula for **Example 1** can be illustrated as shown below.

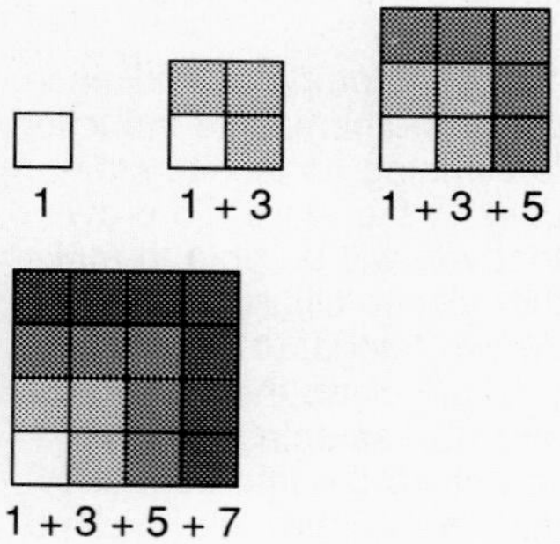

Notice that in **Examples 1-3** we explicitly write the statement $S(n)$. At that point, it has not been established that $S(n)$ is true for any value of n. Before beginning the proof, write (1) the statement $S(1)$ to be used in the basis step, (2) the inductive assumption $S(k)$, and (3) the conclusion $S(k + 1)$ to be proved from the inductive assumption.

In **Examples 1-3**, the bold type illustrates what students are expected to write in their proofs. Point out that the recursive use of summation is used in Examples 1 and 2 to rewrite the statement for $S(k + 1)$ so that the inductive assumption can be used.

In the last two lessons, you have proved that certain explicit formulas satisfy the recursive definitions for sequences. Proving by mathematical induction is quite similar to defining using recursion. Compare the processes.

Mathematical induction proof	Recursive definition
$S(n)$ is a proposition. Basis step: $S(1)$ is true. Inductive step: $S(k + 1)$ is proved from $S(k)$. Result: $S(n)$ is true for all integers $n \geq 1$.	*a_n is a term of a sequence.* Initial condition: a_1 has a particular value. Recurrence relation: a_{k+1} is defined in terms of a_k. Result: a_n is uniquely defined for all integers $n \geq 1$.

The Recursion Principle asserts that, under certain circumstances, a recursive definition uniquely defines a sequence. Because of the similarity between mathematical induction proofs and recursive definitions, it is natural to use mathematical induction to prove the Recursion Principle. Here is the principle restated and with its mathematical induction proof.

Recursion Principle

Suppose that a recurrence relation defines x_{n+1} in terms of x_n and n for each integer $n \geq 1$. Then there is exactly one sequence X defined by this recurrence relation and the initial condition $x_1 = a$.

Proof Suppose that there is a second sequence Y in which $y_1 = a$ and y_{n+1} is defined by the same recurrence relation as x_{n+1}. We need to show that $x_n = y_n$ for all positive integers n. Thus here $S(n)$ is the statement $x_n = y_n$.

(1) Basis step: Show that $S(1)$ is true.
$S(1)$ is the statement $x_1 = y_1$. It is true because $x_1 = a$ and $y_1 = a$. That is, both sequences have the same initial condition.

(2) Inductive step: Show that the assumption that $S(k)$ is true implies that $S(k + 1)$ is true.
$S(k)$ is the statement that $x_k = y_k$. Since the sequences have the same recurrence relation and since x_{k+1} is defined in terms of x_k and k, and y_{k+1} is defined in terms of y_k and k, $x_{k+1} = y_{k+1}$. Thus $S(k + 1)$ is true.

By the Principle of Mathematical Induction, $S(n)$ is true for every positive integer n. Consequently, the sequences X and Y are identical. So there cannot be two different sequences defined by this recurrence relation and initial condition.

Although the correctness of an explicit formula for a recursively defined sequence can be proved using the Recursion Principle, it is very common to use mathematical induction to achieve the same purpose.

Here is Example 4, Lesson 7-2, recast in the language of mathematical induction.

Example 2 The Tower of Hanoi sequence was defined recursively as follows:

$$\begin{cases} T_1 = 1 \\ T_{k+1} = 2T_k + 1 \text{ for all integers } k \geq 1. \end{cases}$$

Use mathematical induction to prove that an explicit formula for the sequence is

$$T_n = 2^n - 1 \text{ for all integers } n \geq 1.$$

Solution In this case, $S(n)$ is the explicit formula $T_n = 2^n - 1$.

(1) Basis step: Show that $S(1)$ is true.

When $n = 1$, the formula is $T_1 = 2^1 - 1 = 1$. This agrees with the recursive definition of the sequence. Hence $S(1)$ is true.

(2) Inductive step: Show that the assumption that $S(k)$ is true implies that $S(k + 1)$ is true.

Suppose that $S(k)$ is true for a particular but arbitrarily chosen integer $k \geq 1$. That is, suppose

$S(k)$: $T_k = 2^k - 1.$ Inductive assumption

Now deduce $S(k + 1)$, where

$S(k + 1)$: $T_{k+1} = 2^{k+1} - 1.$

Transform the left side of $S(k + 1)$ into the right side using the inductive assumption.

$T_{k+1} = 2T_k + 1$	The recursive definition, of the sequence
$= 2(2^k - 1) + 1$	Substitute from the inductive assumption.
$= 2^{k+1} - 2 + 1$	Distribute.
$= 2^{k+1} - 1$	Simplify.

Therefore, for all integers $k \geq 1$, if $S(k)$ is true, then $S(k + 1)$ is true.

Thus, from (1) and (2) above, by the Principle of Mathematical Induction, $S(n)$ is true for all integers $n \geq 1$.

Above, the Principle of Mathematical Induction was stated as a way to deduce that a proposition $S(n)$ is true for all natural numbers 1, 2, 3, … . However, the set of integers for which $S(n)$ is defined or true need not start at 1. To prove $S(n)$ is true for all integers $\geq m$, begin by proving that $S(m)$ is true, and then prove that for all integers $k \geq m$, the assumption that $S(k)$ is true implies that $S(k + 1)$ is true. Example 3 illustrates this technique.

Alternate Approach For some students the use of Σ-notation in **Example 1** is overwhelming. Some initial expansions will be helpful. Remind students to use $S(n)$ as the basis for each expansion.

$S(n)$ is $\sum_{i=1}^{n} (2i - 1) = (2 \cdot 1 - 1) + (2 \cdot 2 - 1) + (2 \cdot 3 - 1) + \ldots +, (2 \cdot n - 1) = n^2$.

We are to prove that $S(n)$ is true for all integers $n \geq 1$.

$S(1)$ is $\sum_{i=1}^{1} (2 \cdot i - 1) = 2 \cdot 1 - 1 = 1^2$ (which is true).

$S(k)$ is $\sum_{i=1}^{k} (2i - 1) = (2 \cdot 1 - 1) + (2 \cdot 2 - 1) + (2 \cdot 3 - 1) + \ldots + (2k - 1) = k^2$ (which is assumed true in the inductive assumption).

$S(k + 1)$ is $\sum_{i=1}^{k+1} (2i - 1) = (2 \cdot 1 - 1) + (2 \cdot 2 - 1) + (2 \cdot 3 - 1) + \ldots + (2k - 1) + (2(k + 1) - 1) = (k + 1)^2$ (which must be proved from the inductive assumption). The left side of $S(k + 1)$ is

$= k^2 + (2(k + 1) - 1)$

$= k^2 + 2k - 1$

$= (k + 1)^2$.

The substitution of most of the expansion of $S(k + 1)$ with k^2 is quite effective.

To prove the inductive step to be true, many students benefit from the following:

(1) Expand $S(k)$'s summation expression.

(2) Expand $S(k + 1)$'s summation expression.

(3) Replace most of the expansion of $S(k + 1)$'s summation expression with the corresponding explicit expression of $S(k)$ (since this is assumed true).

(4) Simplify the result of step 3 to show equality with the explicit side of $S(k + 1)$.

Example 2 shows how mathematical induction can be used to prove that the explicit formula conjectured for the Tower of Hanoi sequence is correct. Because the sequence is defined recursively, it is easy to see how $S(k + 1)$ follows from $S(k)$.

Example 3 helps students realize that the basis step is not necessarily a verification that $S(1)$ is true. In this particular instance, $n = 2$ is the first integer for which the given statement is defined.

ADDITIONAL EXAMPLES

1. Use the Principle of Mathematical Induction to prove that for all integers $n \geq 1$, $\sum_{i=1}^{n} \frac{1}{i(i+1)} = \frac{n}{n+1}$.

Let $S(n)$ be the equation above. When $n = 1$, both sides equal 1/2, so $S(1)$ is true. Suppose that $S(k)$ is true for a particular but arbitrarily chosen integer $k \geq 1$. Then $\sum_{i=1}^{k} \frac{1}{i(i+1)} = \frac{k}{k+1}$. Now begin with the left side of $S(k + 1)$: $\sum_{i=1}^{k+1} \frac{1}{i(i+1)}$ $= \sum_{i=1}^{k} \frac{1}{i(i+1)} + \frac{1}{(k+1)(k+2)}$ $= \frac{k}{k+1} + \frac{1}{(k+1)(k+2)}$ (by the inductive assumption) $= \frac{k^2+2k+1}{(k+1)(k+2)}$ (adding rational expressions). This indicates that $S(k + 1)$ is true. Now, applying the Principle of Mathematical Induction, $S(n)$ is true for all $n \geq 1$.

Example 3 Use mathematical induction to prove

$$\frac{1}{2 \cdot 1} + \frac{1}{3 \cdot 2} + \frac{1}{4 \cdot 3} + \frac{1}{5 \cdot 4} + \ldots + \frac{1}{n(n-1)} = \frac{n-1}{n}$$

for all integers $n \geq 2$; that is,

$$\sum_{i=2}^{n} \frac{1}{i(i-1)} = \frac{n-1}{n} \text{ for all integers } n \geq 2.$$

Solution Note that the summation begins with $i = 2$. In fact, $\frac{1}{i(i-1)}$ is undefined for $i = 0$ and $i = 1$.

The statement $S(n)$ is $\sum_{i=2}^{n} \frac{1}{i(i-1)} = \frac{n-1}{n}$.

(1) Basis step: Show that $S(2)$ is true.

When $n = 2$, the left side of $S(2)$ is

$$\sum_{i=2}^{2} \frac{1}{i(i-1)} = \frac{1}{2(2-1)} = \frac{1}{2}$$

and the right side is $\frac{2-1}{2} = \frac{1}{2}$ also. Hence $S(2)$ is true.

(2) Inductive step: Show that the assumption that $S(k)$ is true implies that $S(k + 1)$ is true.

Suppose that $S(k)$ is true for a particular but arbitrarily chosen integer $k \geq 2$, where

$$S(k)\colon \sum_{i=2}^{k} \frac{1}{i(i-1)} = \frac{k-1}{k}. \qquad \text{This is the inductive assumption.}$$

Now show that $S(k + 1)$ is true, where

$$S(k+1)\colon \sum_{i=2}^{k+1} \frac{1}{i(i-1)} = \frac{(k+1)-1}{k+1}.$$

Begin with the left side of $S(k + 1)$ and use algebra and the inductive assumption to transform it into the right side of $S(k + 1)$:

$$\sum_{i=2}^{k+1} \frac{1}{i(i-1)} = \left(\sum_{i=2}^{k} \frac{1}{i(i-1)}\right) + \frac{1}{(k+1)(k+1-1)} \qquad \text{Write the sum recursively.}$$

$$= \frac{k-1}{k} + \frac{1}{(k+1)k} \qquad \text{Substitute from the inductive assumption.}$$

$$= \frac{(k+1)(k-1)+1}{(k+1)k} \qquad \text{Find a common denominator and add.}$$

$$= \frac{k^2-1+1}{(k+1)k} \qquad \text{Simplify the numerator.}$$

$$= \frac{k}{k+1} \qquad \text{Simplify.}$$

$$= \frac{(k+1)-1}{k+1} \qquad \text{Rewrite in the form of } S(k+1).$$

Therefore, for all integers $k \geq 2$, $S(k)$ implies $S(k+1)$.

Thus, from (1) and (2) above by the Principle of Mathematical Induction, $S(n)$ is true for all integers $n \geq 2$.

Questions

Covering the Reading

1. Fill in the missing steps in the following proof of the formula for the sum of the first n integers: See margin.

$1 + 2 + 3 + \ldots + n = \frac{n(n+1)}{2}$ *for all integers* $n \geq 1$.

Proof Let $S(n)$ be the equation $\sum_{i=1}^{n} i = \frac{n(n+1)}{2}$.

(1) Show that $S(1)$ is true.
When $n = 1$, the left side of the equation is 1 and the right side is __a.__ which also equals 1. Hence the equation is true for $n = 1$ and so $S(1)$ is true.

(2) Show that the assumption that $S(k)$ is true implies that $S(k+1)$ is true.
Suppose $S(k)$ is true for a particular but arbitrarily chosen integer $k \geq 1$, where
$S(k)$: __b.__ . This is the inductive assumption.

We must show that $S(k+1)$ is true, where
$S(k+1)$: __c.__ .
Begin with the left side of $S(k+1)$ and use the inductive assumption to transform it to obtain the right side of $S(k+1)$.

__d.__ $= \sum_{i=1}^{k} i + (k+1)$	Write the sum recursively.
$= \frac{k(k+1)}{2} + (k+1)$	__e.__
$= \frac{k(k+1)}{2} + \frac{2(k+1)}{2}$	Equivalent fractions property
= __f.__	Add fractions with a common denominator.
= __g.__	Factor out $k + 1$.
$= \frac{(k+1)((k+1)+1)}{2}$	Rewrite.

Thus, if the inductive assumption $S(k)$ is true, then $S(k+1)$ is true.

From (1) and (2) above, you can conclude that $S(n)$ is true $\forall$ integers $n \geq 1$ by __h.__ .

2. A sequence t is defined recursively by
$\begin{cases} t_1 = 2 \\ t_{n+1} = 3t_n + 2. \end{cases}$
Prove that $t_n = 3^n - 1$ for all integers $n \geq 1$ using mathematical induction.
$S(n)$ is the statement that $t_n = 3^n - 1$. $S(1)$ is the statement that $t_1 = 3^1 - 1 = 2$. This is true by the recursive definition. Suppose $S(k)$ is true. Then $t_{k+1} = 3t_k + 2$ (from the definition of the sequence) $= 3(3^k - 1) + 2$ (from the inductive assumption) $= 3^{k+1} - 1$. Thus, $S(k+1)$ is true. By the Principle of Mathematical Induction, $S(n)$ is true for all $n \geq 1$.

3. Use mathematical induction to prove that $5 + 7 + 9 + \ldots + [5 + 2(n-1)] = n(n+4)$ for all integers $n \geq 1$.
Let $S(n)$ be the above equation. $S(1) = 5 = 1 \cdot (1 + 4)$, so $S(1)$ is true. Suppose $S(k)$ is true. Then $5 + 7 + 9 + \ldots + [5 + 2(k-1)] = k(k+4)$. Add $5 + 2k$ to each side so that the left side becomes the left side of $S(k+1)$. Then $5 + 7 + 9 + \ldots + [5 + 2k] = k(k+4) + 5 + 2k = k^2 + 6k + 5 = (k+1)(k+4+1)$. Thus, $S(k) \Rightarrow S(k+1)$ and by the Principle of Mathematical Induction, $S(n)$ is true for all $n \geq 1$.

ADDITIONAL ANSWERS

1.a. $\frac{1(1+1)}{2} = 1$

b. $1 + 2 + 3 + \ldots + k = \frac{k(k+1)}{2}$

c. $1 + 2 + 3 + \ldots + k + k + 1 = \frac{(k+1)(k+2)}{2}$

d. $1 + 2 + 3 + \ldots + k + k + 1$

e. Use inductive assumption.

f. $\frac{k(k+1) + 2(k+1)}{2}$

g. $\frac{(k+1)(k+2)}{2}$

h. the Principle of Mathematical Induction

NOTES ON QUESTIONS
Question 3: This question can serve as a model for proofs by mathematical induction. Be sure to focus on the steps as related to mathematical induction.

Questions 4-6: These questions are designed to help students learn some of the steps needed for mathematical induction without going through complete proofs.

Question 5: The inequality in this question shows that statements of the form $S(n)$ do not have to be addition formulas or equations.

Question 13: Inform students that adding and subtracting the same number, as is done in part **c,** is not a trick but a common mathematical procedure.

FOLLOW-UP

MORE PRACTICE
For more questions on SPUR Objectives, use *Lesson Mater 7-4,* shown on page 425.

EXTENSION
Example 1, **Question 2**, and **Question 3** are examples of arithmetic series. Challenge students to prove by mathematical induction that $\sum_{i=1}^{n} [a + (i - 1)d] = \frac{n[2a + (n - 1)d]}{2}$, where a and d are real numbers. Have students give the specific values of a and d for which the above formula becomes Example 1, Question 2, and Question 3, respectively.

PROJECTS
The projects for Chapter 7 are described on pages 458–459. **Project 2** is related to the content of this lesson.

2. Refer to Example 1, which proves the following statement:
$1 + 3 + 5 + \ldots + (2n - 1) = n^2$ *for all integers* $n \geq 1$.
a. Identify the basis step. $1 = 1^2$
b. Write the inductive assumption. See margin.

3. Consider the equation
$$S(n): \sum_{i=1}^{n} 2i = n(n + 1) \text{ for all integers } n \geq 1.$$
a. Show that $S(1)$ is true. See margin.
b. Write $S(k)$ and $S(k + 1)$. See margin.
c. Show that for any integer $k \geq 1$, $S(k) \Rightarrow S(k + 1)$. See margin.
d. What can you conclude from the results of parts **a–c**?
$S(n)$ is true $\forall$ integers $n \geq 1$.

In 4–6, $S(n)$ is given. **a.** Write $S(1)$, $S(3)$, and $S(5)$. **b.** Which of these are true and which are false? **c.** Write $S(k + 1)$.

4. $S(n)$: 3 is a factor of $n^3 + 2n$. See margin.

5. $S(n)$: $n^2 + 4 < (n + 1)^2$. See margin.

6. $S(n)$: $\sum_{i=1}^{n} (3i - 2) = \frac{n(3n - 1)}{2}$. See margin.

Applying the Mathematics

7. Consider the equation
$$S(n): 1^2 + 2^2 + 3^2 + \ldots + n^2 = \frac{n(n + 1)(2n + 1)}{6}.$$
The steps below use mathematical induction to prove $S(n)$ true for all integers $n \geq 1$.
a. Write and verify the basis step. $S(1)$: $1^2 = \frac{1(2)(3)}{6}$, so $S(1)$ is true.
b. Show that, for any integer $k \geq 1$, $S(k) \Rightarrow S(k + 1)$. See margin.
c. Complete the proof. See margin.

8. **a.** Write out the first 5 terms of the sequence defined recursively by
$$\begin{cases} a_1 = 1 \\ a_{k+1} = 3a_k + 1 \text{ for all integers } k \geq 1. \end{cases}$$
1, 4, 13, 40, 121
b. Write out the first five terms of the sequence defined explicitly by the formula
$$b_n = \frac{3^n - 1}{2} \text{ for all integers } n \geq 1.$$
1, 4, 13, 40, 121
c. Use mathematical induction to verify that the explicit formula describes the same sequence as the recursive formula. (Hint: Follow the model of Example 2.) See margin.

9. Consider the sequence 1, 7, 17, 31, ... defined recursively by
$$\begin{cases} a_1 = 1 \\ a_{k+1} = a_k + 4k + 2 \text{ for all integers } k \geq 1. \end{cases}$$
Use mathematical induction to prove that an explicit formula for this sequence is
$$a_n = 2n^2 - 1 \quad \text{for all integers } n \geq 1.$$
See margin.

10. Use mathematical induction to prove that $S(n)$ is true for all integers $n \geq 1$ where

$$S(n): \sum_{i=1}^{n} (3i - 2) = \frac{n(3n - 1)}{2}.$$

(Hint: Follow the steps of Question 3.) See margin.

Review

11. Use summation notation to rewrite the sum. *(Lesson 7-3)*

$$\frac{1}{n} + \frac{2}{n} + \frac{3}{n} + \ldots + \frac{n}{n} \quad \sum_{i=1}^{n} \frac{i}{n}$$

12. **a.** Write the sum $\sum_{i=-3}^{2} (i^2 + i)$ in expanded form.
    b. Find the value of the sum in part **a**. *(Lesson 7-3)* 16
    a) $(-3)^2 + -3 + (-2)^2 + -2 + (-1)^2 + -1 + (0)^2 + 0 + (1)^2 + 1 + (2)^2 + 2$

13. **a.** Given that $x - y$ is a factor of $x^4 - y^4$, explain why $x - y$ is a factor of $x^5 - xy^4$. See margin.
    b. Factor $xy^4 - y^5$. $xy^4 - y^5 = y^4(x - y)$
    c. Use parts **a** and **b** to explain why $x - y$ is a factor of $x^5 - xy^4 + xy^4 - y^5$. *(Lesson 4-1)* See margin.

14. Recall that $\lfloor x \rfloor$ denotes the floor of x, the greatest integer that is less than or equal to x. Consider the sequence defined by:

$$\begin{cases} a_1 = 1 \\ a_k = a_{\lfloor (k+1)/2 \rfloor} + a_{\lfloor k/2 \rfloor} & \forall \text{ integers } k \geq 2. \end{cases}$$

    a. Write the first six terms of the sequence. 1, 2, 3, 4, 5, 6
    b. Conjecture an explicit formula for the sequence. *(Lessons 7-1, 7-2)*
    Sample: $a_n = n$, for all integers $n \geq 1$.

15. **a.** Graph $y = \cos x$ on the interval $0 \leq x \leq \pi$. See margin.
    b. Graph the line $y = x$ on the same set of axes. See margin.
    c. Form the graph of $y = \cos^{-1} x$ on the interval $-1 \leq x \leq 1$ by reflecting the graph of $y = \cos x$ across the line $y = x$. See margin.
    d. Evaluate without using a calculator.
    i. $\cos^{-1}\left(\frac{\sqrt{2}}{2}\right)$ $\frac{\pi}{4}$ **ii.** $\cos^{-1}\left(\frac{1}{2}\right)$ $\frac{\pi}{3}$ **iii.** $\cos^{-1}\left(\frac{-\sqrt{3}}{2}\right)$ $\frac{5\pi}{6}$

(Lessons 6-6, 2-8)

16. *True* or *false?* If k is an odd integer, then 2 is a factor of $3k^2 + k + 2$. *(Lesson 4-1)* True

Exploration

17. Consider the computer program in BASIC below.

```
10 INPUT N
20 PRINT N
30 N = N + 1
40 GO TO 20
```

Explain how this program models the Principle of Mathematical Induction. The program contains an initial condition in line 10 and a recurrence relation in line 30. Given an infinite amount of time and computer memory, it would print all the integers greater than N – 1.

ADDITIONAL ANSWERS
2.b., 3.a.–c., 4.–6., 7.b., c., 8.c., 9., 10., 13.a., c., 15. a.–c. See Additional Answers in the back of this book.

NAME ______________________

LESSON **MASTER 7–4**
QUESTIONS ON **SPUR** OBJECTIVES

■ **PROPERTIES** *Objective F (See pages 462–464 for objectives.)*

1. Prove that the sequence defined by $\begin{cases} a_1 = 2 \\ a_{k+1} = a_k + k + 1 \ \forall\, k \geq 1 \end{cases}$ has explicit formula $a_n = \frac{n^2}{2} + \frac{n}{2} + 1$.

Let $S(n)$: $a_n = \frac{n^2}{2} + \frac{n}{2} + 1$. Show $S(1)$ is true: $a_1 = \frac{1^2}{2} + \frac{1}{2} + 1 = 2$, so $S(1)$ is true. Assume $S(k)$ is true $\left(a_k = \frac{k^2}{2} + \frac{k}{2} + 1\right)$ for a positive integer k. Show $S(k + 1)$ is true:

$$a_{k+1} = a_k + k + 1 = \frac{k^2}{2} + \frac{k}{2} + 1 + k + 1$$
$$= \frac{k^2 + k + 2 + 2k + 2}{2}$$
$$= \frac{k^2 + 2k + 1 + k + 1 + 2}{2} = \frac{(k + 1)^2}{2} + \frac{k + 1}{2} + 1$$

So $S(k + 1)$ is true. Therefore, by induction, $S(n)$ is true ∀ positive integers n.

2. Identify the basis step and the inductive step in your proof for Question 1.

basis: The verification that $S(1)$ is true

inductive: Proving that $S(k)$ is true ⇒ $S(k + 1)$ is true

3. Find an explicit formula for the sequence defined by $\begin{cases} a_1 = 2 \\ a_{k+1} = 3a_k \ \forall\, k \geq 1. \end{cases}$ Use mathematical induction to prove that your formula is correct.

$a_n = \frac{2}{3} \cdot 3^n$ ∀ positive integers n. Proof: Let $S(k)$: $a_k = \frac{2}{3} \cdot 3^k$. Show $S(1)$ is true: $a_1 = \frac{2}{3} \cdot 3^1 = 2$, so $S(1)$ is true. Assume $S(k)$ is true for a positive integer k; that is, $a_k = \frac{2}{3} \cdot 3^k$. Show $S(k + 1)$ is true:
$a_{k+1} = \frac{2}{3} \cdot 3^{k+1} = \frac{2}{3}(3^k \cdot 3) = 3\left(\frac{2}{3} \cdot 3^k\right) = 3a_k$
So $S(k + 1)$ is true. Therefore, by induction, $S(n)$ is true ∀ positive integers n.

78 *Continued* *Precalculus and Discrete Mathematics* © Scott, Foresman and Company

NAME ______________________
Lesson MASTER 7–4 (page 2)

■ **PROPERTIES** *Objective G*
In 4–6, use mathematical induction to prove that $S(n)$ is true for all positive integers n, or find a counterexample.

4. $S(n)$: $3 + 9 + 15 + \ldots + (6n - 3) = 3n^2$
Show $S(1)$ is true: $3 = 3 \cdot 1^2$. Assume $S(k)$ is true for a positive integer k: $3 + 9 + 15 + \ldots + (6k - 3) = 3k^2$. Show $S(k + 1)$ is true:

$$3 + 9 + 15 + \ldots + (6(k + 1) - 3)$$
$$= 3 + 9 + 15 + \ldots + (6k - 3) + (6(k + 1) - 3)$$
$$= 3k^2 + 6k + 3$$
$$= 3(k^2 + 2k + 1)$$
$$= 3(k + 1)^2$$

So $S(k + 1)$ is true. Therefore, by induction, $S(n)$ is true ∀ positive integers n.

5. $S(n)$: $\sum_{i=0}^{n} 2^i = 2^{n+1} - 1$
Show $S(1)$ is true: $\sum_{i=0}^{1} 2^i = 1 + 2 = 3 = 2^{1+1} - 1$.
Assume $S(k)$ is true for a positive integer k:
$\sum_{i=0}^{k} 2^i = 2^{k+1} - 1$. Show $S(k + 1)$ is true:

$$\sum_{i=0}^{k+1} 2^i = \sum_{i=0}^{k} 2^i + 2^{k+1} = 2^{k+1} - 1 + 2^{k+1}$$
$$= 2 \cdot 2^{k+1} - 1 = 2^{(k+1)+1} - 1$$

So $S(k + 1)$ is true. Therefore, by induction, $S(n)$ is true ∀ positive integers n.

6. $S(n)$: $(a + b)^n = a^n + b^n$
False; sample: Let $a = b = n = 2$. Then $(a + b)^n = (2 + 2)^2 = 16$, but $a^n + b^n = 2^2 + 2^2 = 8$.

Precalculus and Discrete Mathematics © Scott, Foresman and Company 79

LESSON 7-5

RESOURCES

- Lesson Master 7-5

OBJECTIVE

G Prove statements using mathematical induction.

TEACHING NOTES

In this lesson, students see that the same steps are used in mathematical induction proofs regardless of the content of the proofs. Students still need to verify the basis step, and still need to prove that $S(k+1)$ is true if the inductive assumption $S(k)$ is true. In this lesson, to prove that $S(k+1)$ is true, students need to use the properties of divisibility reviewed at the beginning of the lesson.

Ask students for numerical examples of both the Factor of a Sum Theorem and the Transitive Property of Factors. Students need to understand these theorems thoroughly.

Making Connections In Lessons 1-6 and 4-1, students completed proofs about factors using the definition of a factor and the basic properties of divisibility. In this lesson, students return to factor proofs but use mathematical induction instead. While induction is not necessary for all the proofs, it is a very powerful technique that can be applied to situations in which other techniques are not applicable.

In **Example 1,** students need to expand an expression of the form $(k+1)^3$.

LESSON

7-5 Factors and Mathematical Induction

Mathematical induction is a widely applicable method of proof. In the last lesson, you saw that it can be used to prove a wide variety of formulas involving sums. Mathematical induction can also be used to prove divisibility properties.

Recall from Chapter 4 that if m and n are integers, then m is a *factor* of n if and only if $n = m \cdot k$ for some integer k. In Chapter 4 this idea was used to prove some fundamental properties of divisibility, which are repeated below.

Factor of a Sum Theorem

For all integers a, b, and c, if a is a factor of b and a is a factor of c, then a is a factor of $b + c$.

Transitive Property of Factors

For all integers a, b, and c, if a is a factor of b and b is a factor of c, then a is a factor of c.

Example 1 applies these properties in a proof by mathematical induction.

Example 1 Prove that for every positive integer n, 3 is a factor of $n^3 + 2n$.

Solution **Let $S(n)$ be the sentence: *3 is a factor of $n^3 + 2n$.***

(1) Basis step: Show that $S(1)$ is true.
$S(1)$ is the statement: *3 is a factor of $1^3 + 2 \cdot 1$.*
Since $1^3 + 2 \cdot 1 = 3$, and 3 is a factor of 3, $S(1)$ is clearly true.

(2) Inductive step: Show that if $S(k)$ is true, then $S(k+1)$ is true for all integers $k \geq 1$.
Assume that $S(k)$ is true for some particular but arbitrarily chosen positive integer k. That is, assume

$S(k)$: *3 is a factor of $k^3 + 2k$.* This is the inductive assumption.

Now it must be shown that $S(k+1)$ is true, where
$S(k+1)$: *3 is a factor of $(k+1)^3 + 2(k+1)$.*
Expanding $(k+1)^3$ and $2(k+1)$, the expression in $S(k+1)$ becomes
$$k^3 + 3k^2 + 3k + 1 + 2k + 2$$
which can be regrouped as
$$(k^3 + 2k) + 3k^2 + 3k + 3.$$

426

By the inductive assumption, 3 is a factor of $k^3 + 2k$. The remaining terms in the sum all have a factor of 3. Then, by the Factor of a Sum Theorem, 3 is a factor of $(k^3 + 2k) + 3k^2 + 3k + 3$, and so $S(k + 1)$ is true.

From (1) and (2) using the Principle of Mathematical Induction,

$S(n)$: *3 is a factor of $n^3 + 2n$*

is true for all integers $n \geq 1$.

In the next example, some manipulation must be done before the factorization needed in the proof becomes evident.

Example 2 Prove that for every positive integer n,

3 is a factor of $4^n - 1$.

Solution **Let $S(n)$: *3 is a factor of $4^n - 1$.***

(1) Show that $S(1)$ is true.

$S(1)$ is the statement *3 is a factor of $4^1 - 1$*. Since $4^1 - 1 = 3$, $S(1)$ is true.

(2) Show that the assumption that $S(k)$ is true implies that $S(k + 1)$ is true.

Assume that $S(k)$ is true for some positive integer k. That is, assume

$S(k)$: *3 is a factor of $4^k - 1$.*

It must be shown that this assumption implies $S(k + 1)$ is true where

$S(k + 1)$: *3 is a factor of $4^{k+1} - 1$.*

Transform the expression in $S(k + 1)$ so that it contains the expression in $S(k)$. This can be done using the fact that you can subtract 1 from a number and then add 1 without changing the value of the number.

$4^{k+1} - 1 = 4(4^k) - 1$	Properties of Exponents
$= 4(4^k - 1 + 1) - 1$	$4^k = 4^k - 1 + 1$
$= 4(4^k - 1) + 4(1) - 1$	Distributive Property
$= 4(4^k - 1) + 3$	

The inductive assumption states that 3 is a factor of $4^k - 1$. Since $4^k - 1$ is a factor of $4(4^k - 1)$, then by the Transitive Property of Factors, 3 is then a factor of $4(4^k - 1)$. Since 3 is also a factor of itself, then by the Factor of a Sum Theorem, 3 is a factor of $4(4^k - 1) + 3$. Therefore, 3 is a factor of $4^{k+1} - 1$, and so $S(k + 1)$ is true.

From (1) and (2) above, using the Principle of Mathematical Induction, 3 is a factor of $4^n - 1$ for all positive integers n.

Alternate Approach
You may wish to show an alternate proof for **Example 1** that uses properties of divisibility. Proof: By the Quotient-Remainder Theorem, every integer is of the form $3k$, $3k + 1$, or $3k + 2$. If $n = 3k$, then $n^3 + 2n = (3k)^3 + 2(3n) = 3(9k^3 + 2n)$, which is divisible by 3. If $n = 3k + 1$, then $n^3 + 2n = (3k + 1)^3 + 2(3k + 1) = 27k^3 + 27k^2 + 9k + 1 + 6k + 2 = 3(9k^3 + 9k^2 + 5k + 1)$, which is divisible by 3. If $n = 3k + 2$, then $n^3 + 2n = (3k + 2)^3 + 2(3k + 2) = 27k^3 + 54k^2 + 36k + 8 + 6k + 4 = 3(9k^3 + 18k^2 + 14k + 4)$, which is divisible by 3.

In **Examples 2 and 3**, the hardest part for students is knowing what term needs to be added and subtracted to obtain $S(k + 1)$ in the proper form. You may need to give students a hint to guide them in adding and subtracting the proper term.

Some students respond well to using the definition of factor explicitly. In the inductive step of the proof of **Example 2**, use the following: 3 divides $4^k - 1$ means there exists an integer m such that $4^k - 1 = 3m$. Then the proof becomes

$$\begin{aligned} 4^{k+1} &= 4(4^k) - 1 \\ &= 4(4^k - 1 + 1) - 1 \\ &= 4(4^k - 1) + 4(1) - 1 \\ &= 4(3m) + 3 \\ &= 3(4m + 1). \end{aligned}$$

Because $4m + 1$ is an integer, it is divisible by 3. Thus, $4^{k+1} - 1$ also is divisible by 3.

Mathematical proofs often involve considerable creativity. Even though the structure of each proof by mathematical induction is the same, sometimes creativity is needed to make connections between the expressions in $S(k)$ and $S(k + 1)$.

ADDITIONAL EXAMPLES

1. Prove that for every positive integer n, 5 is a factor of $n^5 + 4n$.

$S(1)$ is the statement that 5 is a factor of $1^5 + 4(1) = 5$, which is true. Assume $S(k)$: 5 is factor of $k^5 + 4k$. Now consider the expression in $S(k + 1)$: $(k + 1)^5 + 4(k + 1) = k^5 + 5k^4 + 10k^3 + 10k^2 + 5k + 1 + 4(k + 1) = k^5 + 4k + 5(k^4 + 2k^3 + 2k^2 + k + 1)$, and since by the inductive assumption, $k^5 + 4k$ is divisible by 5, the entire expression is divisible by 5. Thus, $S(k) \Rightarrow S(k + 1)$ and by the Principle of Mathematical Induction, $S(n)$ is true for all integers $n \geq 1$.

2. Prove that for every positive integer n, 5 is a factor of $6^n - 1$.

$S(1)$ is that 5 is a factor of $6^1 - 1 = 5$, which is true. Assume $S(k)$. Now $6^{k+1} - 1 = 6(6^k) - 1 = 6(6^k - 1) + 5 = 6(5m) + 5$ (by the inductive assumption) $= 5(6m + 1)$, so the expression in $S(k + 1)$ is divisible by 5, and thus $S(k + 1)$ is true if $S(k)$ is true. By the Principle of Mathematical Induction, $S(n)$ is true for all integers $n \geq 1$.

3. Use the Principle of Mathematical Induction to prove that for all positive integers n, $\left(1 + \frac{3}{1}\right)\left(1 + \frac{5}{4}\right) \dots \left(1 + \frac{2n + 1}{n^2}\right) = (n + 1)^2$.

$S(1)$ is $\left(1 + \frac{3}{1}\right) = (1 + 1)^2$, which is true. If $S(k)$ is true, then the left-hand side of $S(k + 1)$ can be written as $(k + 1)^2\left(1 + \frac{2(k + 1) + 1}{(k + 1)^2}\right)$, which equals $(k + 2)^2 = [(k + 1) + 1]^2 = (k + 2)^2$. Thus, $S(k + 1)$ is true if $S(k)$ is true. By the Principle of Mathematical Induction, $S(n)$ is true for all integers $n \geq 1$.

You may have viewed the adding and subtracting of 1 in the middle of Example 2 as a trick. But there is a saying among some problem solvers: If an idea is used once, it is a trick; but if it is used twice, then it is a *technique*. In Lesson 7-2 we added 2 and subtracted 2 in proving the explicit formula for the Tower of Hanoi sequence.

In Example 3, the technique of adding and subtracting a number is used to show that $x - y$ is a factor of $x^n - y^n$ for all n. Here xy^k is subtracted and then added again.

Example 3 Use the Principle of Mathematical Induction to prove that, when $x \neq y$, $x - y$ is a factor of $x^n - y^n$ for each positive integer n.

Solution Here

$$S(n)\text{: } x - y \text{ is a factor of } x^n - y^n.$$

(1) Show that $S(1)$ is true.
$S(1)$ is the statement $x - y$ *is a factor of* $x^1 - y^1$, which is true when $x \neq y$.

(2) Show that the assumption that $S(k)$ is true implies that $S(k + 1)$ is true.

Assume that $S(k)$ is true for some integer $k \geq 1$. That is, assume
$S(k)$: $x - y$ *is a factor of* $x^k - y^k$
is true.

It must be shown that this assumption implies that $S(k + 1)$ is true, where

$S(k + 1)$: $x - y$ *is a factor of* $x^{k+1} - y^{k+1}$.

Begin with $x^{k+1} - y^{k+1}$ and add and subtract xy^k so that the expression becomes a sum of terms clearly divisible by $x - y$.

$$x^{k+1} - y^{k+1} = x^{k+1} - xy^k + xy^k - y^{k+1}$$
$$= x(x^k - y^k) + y^k(x - y) \qquad \text{Factor.}$$

Certainly $x - y$ is a factor of $y^k(x - y)$. By the inductive assumption, $x - y$ is a factor of $x^k - y^k$. Since $x^k - y^k$ is a factor of $x(x^k - y^k)$, by the Transitive Property of Factors, $x - y$ is a factor of $x(x^k - y^k)$. According to the Factor of a Sum Theorem, since $x - y$ is a factor of each term, $x - y$ is a factor of the sum. So $x - y$ is a factor of $x^{k+1} - y^{k+1}$.

From (1) and (2), using the Principle of Mathematical Induction, $S(n)$ is true for all positive integers n.

The result in Example 3 shows, for instance, that because $5 = 7 - 2$, 5 is a factor of $7^n - 2^n$ for each natural number n. It also shows that for any integer $a \neq 1$, $a - 1$ is a factor of $a^n - 1$ for every natural number n. Note that Example 2 is the special case of this when $a = 4$.

Questions

Covering the Reading

1. Show that when $n = 3$, 16 is a factor of $5^n - 4n - 1$.

$5^3 - 4 \cdot 3 - 1 = 112 = 16 \cdot 7$, so 16 is a factor.

2. Show that if 2 is a factor of $n^2 + n$, then 2 is a factor of $(n + 1)^2 + (n + 1)$. (Hint: Expand $(n + 1)^2 + (n + 1)$ and regroup.)

See margin.

3. Consider the sentence:

$S(n)$: *2 is a factor of* $n^2 - n + 2$.

a. Show that $S(1)$, $S(13)$, and $S(20)$ are all true. See margin.

b. Write $S(k)$ and $S(k + 1)$. See margin.

c. Expand the expression in $S(k + 1)$ from part **b** and simplify the expanded expression. $k^2 + k + 2$

d. Use appropriate regrouping to show that for all integers k, $S(k) \Rightarrow S(k + 1)$. See margin.

e. What can you conclude from the results of parts **a–d**?

2 is a factor of $n^2 - n + 2$ $\forall$ integers $n \geq 1$.

4. Consider

$S(n)$: *5 is a factor of* $6^n - 1$.

Prove that $S(n)$ is true for every positive integer n, using the following steps.

a. Show that $S(1)$ is true. Since 5 is a factor of $6^1 - 1 = 5$, $S(1)$ is true.

b. Write the inductive assumption $S(k)$. $S(k)$: 5 is a factor of $6^k - 1$.

c. Write $S(k + 1)$. $S(k + 1)$: 5 is a factor of $6^{k+1} - 1$.

d. Use the fact that $6^{k+1} - 1 = 6(6^k - 1) + 5$ to show that $\forall$ integers $k \geq 1$, $S(k) \Rightarrow S(k + 1)$. See margin.

e. Finish the proof. See margin.

In 5 and 6, use the result of Example 3 to show that the statement is true.

5. 6 is a factor of $9^n - 3^n$ for all positive integers n. See margin.

6. 12 is a factor of $13^n - 1$ for all positive integers n. See margin.

Applying the Mathematics

7. Let $S(n)$ be the sentence: *3 is a factor of* $n^3 + 14n + 3$. Use the Principle of Mathematical Induction to prove that $S(n)$ is true for all positive integers $n \geq 1$. See margin.

8. Prove that 6 is a factor of $n^3 + 11n$ for all positive integers n. Use the theorem: $\forall$ *positive integers n, 6 is a factor of* $3n(n + 1)$.

See margin.

9. Given $S(n)$: *8 is a factor of* $12^n - 8^n$.

a. Determine whether $S(1)$, $S(2)$, and $S(3)$ are true.

b. $12^n - 8^n$ is in the form $x^n - y^n$ as in Example 3. However, 8 is not a factor of $12 - 8$. What, if anything, can you conclude about $S(n)$ based on the result of that example? See margin.

c. Use the factorization

$$12^n - 8^n = 4^n \cdot 3^n - 4^n \cdot 2^n = 4^n(3^n - 2^n)$$

to prove that $S(n)$ is true for all integers $n \geq 2$. (You do not need to use mathematical induction.) See margin.

a) $S(1)$ is false. $S(2)$ and $S(3)$ are true.

ADDITIONAL ANSWERS
2., 3.a., b., d., 4.d., e., 5.–8., 9.b., c. See Additional Answers in the back of this book.

NAME ______

LESSON **MASTER 7–5**
QUESTIONS ON **SPUR** OBJECTIVES

■ **PROPERTIES** *Objective G (See pages 462–464 for objectives.)*

1. Let $P(n)$ be a statement in n. What can be concluded in each of the following instances?

a. $P(1)$ is true and $P(k) \Rightarrow P(k + 1)$.
$P(n)$ is true $\forall$ positive integers n.

b. $P(1)$ is true and $P(k)$ does not imply $P(k + 1)$.
$P(1)$ is true.

c. $P(6)$ is true and $P(k) \Rightarrow P(k + 1)$.
$P(n)$ is true $\forall$ integers $n \geq 6$.

In 2 and 3, use mathematical induction to prove that the statement is true.

2. 4 is a factor of $5^n - 1$ $\forall$ positive integers n.

Let $S(n)$: 4 is a factor of $5^n - 1$. Show $S(1)$ is true: $5^1 - 1 = 4$ which has 4 as a factor. Assume $S(k)$ is true for a positive integer k: 4 is a factor of $5^k - 1$. Show $S(k + 1)$ is true:
$5^{k+1} - 1 = 5(5^k) - 1 = 5(5^k - 1 + 1) - 1$
$= 5(5^k - 1) + 5 \cdot 1 - 1 = 5(5^k - 1) + 4$
Since 4 is a factor of $5^k - 1$ and of 4, 4 is a factor of $5(5^k - 1) + 4$, and $S(k + 1)$ is true. Therefore, by induction, $S(n)$ is true $\forall$ positive integers n.

3. 3 is a factor of $n^3 + 8n$ $\forall$ positive integers n.

Let $S(n)$: 3 is a factor of $n^3 + 8n$. Show $S(1)$ is true: $1^3 + 8(1) = 9$ which has 3 as a factor. Assume $S(k)$ is true for a positive integer k: 3 is a factor of $k^3 + 8k$. Show $S(k + 1)$ is true:
$(k + 1)^3 + 8(k + 1) = k^3 + 3k^2 + 3k + 1 + 8k + 8$
$= (k^3 + 8k) + 3(k^2 + k + 3)$
Since 3 is a factor of $k^3 + 8k$ and of $3(k^2 + k + 3)$, 3 is a factor of their sum, so $S(k + 1)$ is true. Therefore, by induction, $S(n)$ is true $\forall$ positive integers n.

80 *Continued* *Precalculus and Discrete Mathematics* © Scott, Foresman and Company

NAME ______
Lesson MASTER 7–5 (page 2)

4. **a.** Show that $x + y$ is a factor of $x^3 + y^3$.
$x^3 + y^3 = (x + y)(x^2 - xy + y^2)$, therefore $x + y$ is a factor of $x^3 + y^3$.

b. Use mathematical induction to prove that $x + y$ is a factor of $x^{2n-1} + y^{2n-1}$ $\forall$ positive integers n. (Hint: In the inductive step, add and subtract x^2y^{2k-1}.)

Let $S(n)$: $x + y$ is a factor of $x^{2n-1} + y^{2n-1}$. Show $S(1)$ is true: $x^{2 \cdot 1 - 1} + y^{2 \cdot 1 - 1} = x + y$, and $x + y$ is a factor of itself. Assume $S(k)$ is true for a positive integer k: $x + y$ is a factor of $x^{2k-1} + y^{2k-1}$. Show $S(k + 1)$ is true:
$x^{2(k+1)-1} + y^{2(k+1)-1} = x^{2k+1} + y^{2k+1}$
$= x^{2k+1} + x^2y^{2k-1} - x^2y^{2k-1} + y^{2k+1}$
$= x^2(x^{2k-1} + y^{2k-1}) - y^{2k-1}(x^2 - y^2)$
Since $x + y$ is a factor of $x^{2k-1} + y^{2k-1}$ and of $x^2 - y^2$, it is a factor of $x^2(x^{2k-1} + y^{2k-1}) - y^{2k-1}(x^2 - y^2)$, so $S(k + 1)$ is true. Therefore, by induction, $S(n)$ is true $\forall$ positive integers n.

Precalculus and Discrete Mathematics © Scott, Foresman and Company 81

NOTES ON QUESTIONS
Question 11: For this question, help students see that it is often easier to get results into the desired form if factors are not expanded. That is, it is much easier to simplify

$$\sum_{i=1}^{k+1} i^3 = \left[\frac{k(k+1)}{2}\right]^2 + (k+1)^3$$
$$= \frac{k^2(k+1)^2}{4} + \frac{4(k+1)^3}{4}$$
$$= \frac{(k+1)^2[k^2 + 4(k+1)]}{4}$$

by leaving quantities factored.

Question 18: Ask students why $n(n + 1)(n + 2)$ is an example of such a polynomial. (because either n, $n + 1$, or $n + 2$ is divisible by 3)

FOLLOW-UP

MORE PRACTICE
For more questions on SPUR Objectives, use *Lesson Master 7-5*, shown on page 429.

ADDITIONAL ANSWERS
11. See Additional Answers in the back of this book.

12. Let $S(n)$: $T_n = 2^n + 3$. $T_1 = 2^1 + 3 = 5$ which agrees with the recursive definition, hence $S(1)$ is true. Assume $S(k)$: $T_k = 2^k + 3$ is true for some integer k. Show $S(k + 1)$: $T_{k+1} = 2^{k+1} + 3$ is true. $T_{k+1} = 2T_k - 3 = 2(2^k + 3) - 3 = 2^{k+1} + 6 - 3 = 2^{k+1} + 3$. Since $S(1)$ is true and $S(k) \Rightarrow S(k + 1)$, by the Principle of Mathematical Induction, $S(n)$ is true for all $n \geq 1$. Hence, the explicit formula correctly defines the sequence.

14., 15., 16.a. See Additional Answers in the back of this book.

10. Using results from this lesson, prove in one sentence that $\forall$ positive integers $n \geq 1$, 3 is a factor of $2^{2n} - 1$. Sample: By Example 3 with $x = 2^2$ and $y = 1$, $x - y = 3$ is a factor of $x^n - y^n = 2^{2n} - 1$.

Review

11. Use mathematical induction to prove: $\sum_{i=1}^{n} i^3 = \left[\frac{n(n+1)}{2}\right]^2$ is true for all integers $n \geq 1$. *(Lesson 7-4)* See margin.

12. Prove: The sequence 5, 7, 11, 19, 35, ... , defined recursively by
$$\begin{cases} T_1 = 5 \\ T_{k+1} = 2T_k - 3 \quad \forall \text{ integers } k \geq 1, \end{cases}$$
has explicit formula $T_n = 2^n + 3$ $\forall$ integers $n \geq 1$. *(Lesson 7-4)* See margin.

13. Determine whether or not the equation below is true:
$$\sum_{j=-2}^{3} \frac{1}{3^j} = \sum_{j=-2}^{3} 3^{j-1}.$$ *(Lesson 7-3)* True

In 14 and 15, show that the equation is an identity.

14. $\cos^4 x - \sin^4 x = \cos 2x$ *(Lesson 6-5)* See margin.

15. $\frac{a^4 - b^4}{a - b} = a^3 + a^2b + ab^2 + b^3$ *(Lesson 4-4)* See margin.

16. Define a function f by $f(x) = (.7)^x$. *(Lesson 2-6)*
- **a.** Use an automatic grapher to graph f on the interval $-2 \leq x \leq 20$.
- **b.** What is $\lim_{x \to +\infty} (.7)^x$? 0
- **c.** Find the smallest integer n such that $(.7)^n < .01$. 13
- **d.** Use the answer to part **b** to find $\lim_{x \to +\infty} \frac{1 - (.7)^x}{1 - .7}$. $\frac{10}{3}$

a) See margin.

17. A bacteriologist places 30 bacteria organisms in a vial and counts the organisms at 1-hour intervals, observing that the population triples each hour. Let t be the number of hours the organisms have been in the vial. Find a rule for a function f that gives the number of bacteria in terms of t. *(Lesson 2-6)* $f(t) = 30 \cdot 3^t$

Exploration

18. Find several polynomials in n (like the ones in Example 1 or Question 7) which have 3 as a factor for all integers $n \geq 1$. samples: $n^4 + 2n^2$, $n^3 + 14n$, $n(n + 1)(n + 2) = n^3 + 3n^2 + 2n$

430

LESSON

7-6

Geometric Series and the Resolution of Zeno's Paradox

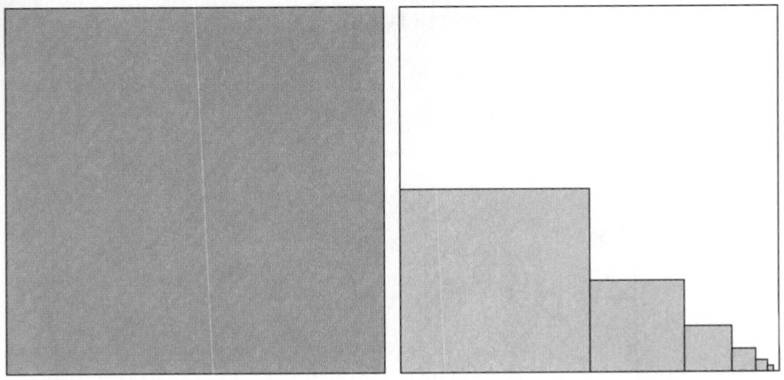

How does this drawing suggest that an infinite *number of terms can have a* finite *sum?*

The indicated sum of consecutive terms of a sequence is called a **series**. If the sequence is finite, then the sum

$$\sum_{i=1}^{n} a_i = a_1 + a_2 + \ldots + a_n$$

is called a **finite series**. If the sequence is infinite, then there are infinitely many terms to add. The sum is denoted by

$$\sum_{i=1}^{\infty} a_i = a_1 + a_2 + \ldots$$

and the expression on either side of the equation is called an **infinite series**.

For example, if

$$a_i = \left(\frac{1}{2}\right)^i$$

then

$$\sum_{i=0}^{10} a_i = \sum_{i=0}^{10} \left(\frac{1}{2}\right)^i = 1 + \frac{1}{2} + \frac{1}{4} + \frac{1}{8} + \ldots + \frac{1}{1024}$$

is a finite series, while

$$\sum_{i=0}^{\infty} a_i = \sum_{i=0}^{\infty} \left(\frac{1}{2}\right)^i = 1 + \frac{1}{2} + \frac{1}{4} + \frac{1}{8} + \ldots$$

is an infinite series. In fact, this is the infinite series from Zeno's paradox in Lesson 7-3.

In stating his paradox about the impossibility of moving from point A to point B, Zeno argued that the value of an infinite series must be infinite. Of course, he knew you could get from point A to point B, but he wanted to demonstrate that the logic and mathematics of his time could not deal with this situation. In this lesson, you will see how mathematicians today deal with such an infinite summation. To do so, first consider finite series.

A finite **geometric series** is the sum of consecutive terms of a geometric sequence. The simplest such series has first term 1 and is the sum of consecutive powers of the constant ratio r. There is a simple formula for this sum.

LESSON 7-6

RESOURCES
- Lesson Master 7-6
- Quiz for Lessons 7-4 through 7-6
- Computer Master 14

OBJECTIVES

E Evaluate a finite or infinite geometric series.
J Interpret computer programs which calculate terms of sequences.

TEACHING NOTES

Summation notation is a way of compactly expressing the terms of a series. In this lesson, we define series and focus our attention on geometric series.

In the drawing at the top of this page, there is an infinite number of squares suggested by the figure on the right. However, the sum of their areas is finite, since it is less than the area of the square on the left. This is another counterexample to Zeno's assumption that the value of an infinite series must be infinite.

In the Sum of the First n Powers Theorem, students may ask why the last term on the left side is not r^n. Point out that we wish to have a formula for the first n terms; then rewrite the left side as $r^0 + r^1 + r^2 + \ldots + r^{n-1}$, which has exactly n terms. Also, write the series in summation notation $\sum_{i=1}^{n} r^{i-1}$.

Making Connections It is possible that students have seen the formula for the evaluation of a finite geometric series. In earlier UCSMP courses, students may have seen the following proof:
$S = a + ar + ar^2 + ... + ar^{n-1}$;
$-rS = - ar - ar^2 - ... - ar^{n-1} - ar^n$;
$S - rS = a - ar^n$;
$S(1 - r) = a(1 - r^n)$;
$S = \frac{a(1 - r^n)}{1 - r}$ if $r \neq 1$.
This proof is not rigorous because an infinite distributive property is used to get from S to $-rS$, and because there is an infinite subtraction. This is why mathematical induction is used in this lesson.

Since Zeno's paradoxes require treatment of geometric series, it may not be surprising for students to learn that The Evaluation of a Finite Geometric Series Theorem was known at the time of Euclid and is included as Proposition 35 of Book IX of his *Elements*. Of course, since the ancient Greeks did not have the algebra of today, it is not expressed there as we express it. The following is a translation found in Carl Boyer's *A History of Mathematics* (New York: John Wiley, 1968), pp. 127–128: "If as many numbers as we please be in continued proportion, and there be subtracted from the second and the last numbers equal to the first, then as the excess of the second is to the first, so will the excess of the last be to all those before it." In algebraic terms, this means that for a geometric sequence g with first term g_1 and common ratio r,
$\frac{g_{n+1} - g_1}{S_n} = \frac{g_2 - g_1}{g_1}$,
which in turn is equivalent to
$S_n = \frac{g_1 - g_1 r^n}{1 - r}$.

Theorem (Sum of the First *n* Powers)

If $r \neq 1$, then $\sum_{i=0}^{n-1} r^i = 1 + r + r^2 + ... + r^{n-1} = \frac{1 - r^n}{1 - r}$
$\forall$ integers $n \geq 1$.

Proof Our proof uses mathematical induction. Here $S(n)$ is the equation

$1 + r + r^2 + ... + r^{n-1} = \frac{1 - r^n}{1 - r}$, where r is a real number and $r \neq 1$.

(1) Basis step: $S(1)$ is the equation $1 = \frac{1 - r^1}{1 - r}$. When $r \neq 1$, the right side equals 1, so $S(1)$ is true.

(2) Inductive step: Assume $S(k)$ is true. That is, assume $1 + r + r^2 + ... + r^{k-1} = \frac{1 - r^k}{1 - r}$. We must show that $S(k + 1)$ is true, where $S(k + 1)$ is the equation $1 + r + r^2 + ... + r^k = \frac{1 - r^{k+1}}{1 - r}$. Begin with the left side and show that it equals the right side.

$1 + r + r^2 + ... + r^{k-1} + r^k = \frac{1 - r^k}{1 - r} + r^k$	Inductive assumption
$= \frac{1 - r^k}{1 - r} + \frac{(1 - r)r^k}{1 - r}$	Common denominator
$= \frac{1 - r^k + r^k - r^{k+1}}{1 - r}$	Adding fractions
$= \frac{1 - r^{k+1}}{1 - r}$	Opposites add to zero.

Therefore, $S(k + 1)$ is true if $S(k)$ is true.

Since $S(1)$ is true and, for all integers $k \geq 1$, $S(k) \Rightarrow S(k + 1)$, then by the Principle of Mathematical Induction, $S(n)$ is true for all integers $n \geq 1$.

Example 1 Find the sum $1 + 3 + 9 + 27 + ... + 3^{10}$.

Solution Here $r = 3$ and $n = 11$. So the sum is $\frac{1 - 3^{11}}{1 - 3} = \frac{3^{11} - 1}{2}$. A calculator shows this number to be 88,573.

Multiplying both sides of the formula for the Sum of the First *n* Powers by the number *a* results in a formula for the sum of the terms of any finite geometric sequence.

Theorem (Evaluation of a Finite Geometric Series)

If *a* is any real number and *r* is any real number other than 1, then for all integers $n \geq 1$,

$$a + ar + ar^2 + ... + ar^{n-1} = a\left(\frac{1 - r^n}{1 - r}\right).$$

Example 2 In Zeno's paradox as given in Lesson 7-3, after going half the distance, then half the remaining distance, and so forth for n repetitions, the runner will have traveled the following number of minutes:

$$1 + \frac{1}{2} + \frac{1}{4} + \frac{1}{8} + \ldots + \frac{1}{2^{n-1}}.$$

a. Simplify this sum.
b. Calculate the sum for $n = 5$, 10, and 20.

Zeno of Elea

Solution

a. Use the formula for the Evaluation of a Finite Geometric Series, with $a = 1$ and $r = \frac{1}{2}$.

$$1 + \frac{1}{2} + \frac{1}{4} + \frac{1}{8} + \ldots + \frac{1}{2^{n-1}} = \frac{1 - \left(\frac{1}{2}\right)^n}{1 - \frac{1}{2}} = \frac{1 - \frac{1}{2^n}}{\frac{1}{2}} = 2\left(1 - \frac{1}{2^n}\right)$$

b. For $n = 5$, the sum equals $2\left(1 - \frac{1}{2^5}\right) = 1.9375$.
For $n = 10$, the sum equals $2\left(1 - \frac{1}{2^{10}}\right) \approx 1.9980$.
For $n = 20$, the sum equals $2\left(1 - \frac{1}{2^{20}}\right) \approx 1.999998$.

The expressions calculated in Example 2b are the sums of the first five, ten, and twenty terms of the infinite geometric series $1 + \frac{1}{2} + \frac{1}{4} + \frac{1}{8} + \ldots$. Such sums are called *partial sums*. Given any sequence of numbers $a_1, a_2, a_3, \ldots$, the **partial sums** of the sequence are the numbers

$$a_1, \quad a_1 + a_2, \quad a_1 + a_2 + a_3, \quad a_1 + a_2 + a_3 + a_4, \ldots .$$

Denote the sequence of partial sums by $S_1, S_2, S_3, \ldots$. Then

$$S_1 = a_1 = \sum_{i=1}^{1} a_i$$

$$S_2 = a_1 + a_2 = \sum_{i=1}^{2} a_i$$

$$S_3 = a_1 + a_2 + a_3 = \sum_{i=1}^{3} a_i$$

$$S_4 = a_1 + a_2 + a_3 + a_4 = \sum_{i=1}^{4} a_i$$

$$\vdots$$

$$S_n = a_1 + a_2 + a_3 + \ldots + a_n = \sum_{i=1}^{n} a_i$$

$$\vdots$$

In Example 2b above, the partial sums appear to get closer and closer to 2 as n gets larger and larger. If the sequence of partial sums $S_1, S_2, S_3, \ldots$ approach a finite limit as n gets larger and larger without bound, then the infinite sum is defined to equal this limit.

Encourage students to be careful in their language when talking about infinite summation. Students need to grasp the concept that the infinite sum is the limit of the sequence of partial sums. This notion of the limit of the sequence of partial sums is an important concept in calculus that will be used in the study of the integral in Chapter 11.

Have students express the last sentence of the text on page 435 in if-then language: *If the limit of the sequence of partial sums exists, then the terms of the sequence approach zero.*

The text shows that the converse is not true. To convince students that $\sum_{i=1}^{\infty} \frac{1}{i} = 1 + \frac{1}{2} + \frac{1}{3} + \frac{1}{4} + \ldots + \frac{1}{n} + \ldots$ has no finite limit, suggest grouping:

$$1 + \underbrace{\frac{1}{2} + \frac{1}{3} + \frac{1}{4}}_{>1} + \underbrace{\frac{1}{5} + \frac{1}{6} + \frac{1}{7} + \frac{1}{8} + \frac{1}{9} + \frac{1}{10}}_{>1} + \ldots$$

We can group the terms of $\sum_{i=1}^{\infty} \frac{1}{i}$ into clusters whose sums are all 1 or more. Thus, $\sum_{i=1}^{\infty} \frac{1}{i}$ has no finite limit.

ADDITIONAL EXAMPLES

1.a. Find the sum $1 + 5 + 25 + 125 + \ldots + 5^{10}$.

$\frac{1 - 5^{11}}{1 - 5} = 12{,}207{,}031$

b. Let's say I give you 1 penny today, double the amount tomorrow, and keep doubling every day for one 30-day month. How many pennies would you have received from me?

$\frac{1 - 2^{30}}{1 - 2} = 1{,}073{,}741{,}823$ **pennies which is equal to \$10,737,418.23.**

2.

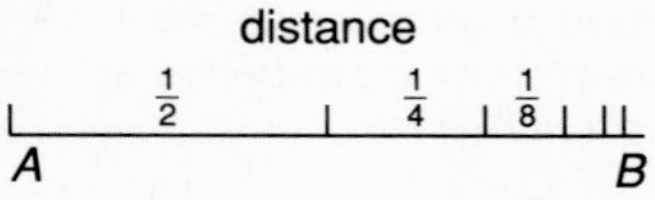

Zeno's runner of Lesson 7-3, after reaching a midpoint, always has half the remaining distance to run. For n repetitions, the fraction of distance from A to B that he has run is $D_n = \frac{1}{2} + \frac{1}{4} + \frac{1}{8} + \ldots + \frac{1}{2^n}$

a. Simplify the sum above.

$D_n = 1 - \frac{1}{2^n}$ **or** $\frac{2^n - 1}{2^n}$

b. Calculate the sum for $n = 5$, 10, and 20.

$D_5 = \frac{31}{32} = 0.96875$,
$D_{10} = \frac{1023}{1024} \approx 0.9990234375$,
$D_{20} \approx \frac{1{,}048{,}575}{1{,}048{,}576} \approx 0.9999990463$

3. Evaluate the infinite geometric series
$\frac{1}{2} + \frac{1}{4} + \frac{1}{8} + \ldots + \frac{1}{2^n} + \ldots$

1

This result is consistent with Additional Example 2. In an infinite number of repetitions, the runner will have covered the entire distance from A to B.

Definition

Let $a_1, a_2, a_3, \ldots$ be an infinite sequence of numbers and let $S_n = \sum_{i=1}^{n} a_i$. Then $\sum_{k=1}^{\infty} a_k = \lim_{n \to +\infty} S_n = \lim_{n \to +\infty} \sum_{k=1}^{n} a_k$ provided this limit exists and is finite.

Thus it appears that $1 + \frac{1}{2} + \frac{1}{4} + \frac{1}{8} + \ldots = \sum_{k=1}^{\infty} \frac{1}{2^{k-1}} = 2.$

The proof of this depends on the following important property of real numbers with absolute value less than 1. If $|b| < 1$, then $\lim_{n \to +\infty} b^n = 0$. This agrees with what we found in Lesson 2-6 to be the end behavior of the exponential function $f(x) = b^x$ when $0 < b < 1$:

$$\lim_{n \to +\infty} b^x = 0.$$

That is, the graph of f approaches the x-axis as x becomes larger and larger without bound. Thus when $0 < r < 1$, $\lim_{n \to +\infty} r^n = 0$.

Since for any integer n, $|r^n| = |(-r)^n|$, it is also the case that $\lim_{n \to +\infty} r^n = 0$ when $-1 < r < 0$. These properties enable infinite geometric series to be evaluated.

Theorem (Evaluation of an Infinite Geometric Series)

If a is any real number and r is a real number with $0 < |r| < 1$, then $\sum_{k=0}^{\infty} ar^k = \frac{a}{1-r}$.

Proof Suppose a is any real number and r is a real number with $0 < |r| < 1$.

$\sum_{k=0}^{\infty} ar^k = \lim_{n \to +\infty} \sum_{k=0}^{n} ar^k$	Definition of infinite summation
$= \lim_{n \to +\infty} \left[a\left(\frac{1-r^n}{1-r}\right)\right]$	Evaluation of a Finite Geometric Series
$= \lim_{n \to +\infty} \left[\left(\frac{a}{1-r}\right)(1 - r^n)\right]$	Rewriting
$= \frac{a}{1-r} \lim_{n \to +\infty} (1 - r^n)$	Factoring out the constant $\frac{a}{1-r}$
$= \frac{a}{1-r}(1 - 0)$	$\lim_{n \to +\infty} r^n = 0$
$= \frac{a}{1-r}$	

434

With the theorem on page 434, the infinite series arising from Zeno's paradox can be evaluated.

Example 3 Evaluate the infinite series $\sum_{k=0}^{\infty} \left(\frac{1}{2}\right)^k = 1 + \frac{1}{2} + \frac{1}{4} + \frac{1}{8} + \ldots + \frac{1}{2^n} + \ldots$.

Solution This is an infinite geometric series with first term $a = 1$ and constant ratio $r = \frac{1}{2}$. By the Evaluation of an Infinite Geometric Series Theorem, the sum is $\frac{a}{1-r} = \frac{1}{1-\frac{1}{2}} = 2$.

This result agrees with common sense and resolves Zeno's paradox. If it takes the runner one minute to go halfway from A to B, it takes two minutes to go all the way.

In many cases the sequence of partial sums of a given infinite series does not approach a finite limit. For instance, consider the following sum of terms of a geometric sequence with $r > 1$:

$$1 + 2 + 4 + 8 + \ldots + 2^n + \ldots .$$

Since the terms of the geometric sequence become larger and larger, the sequence of corresponding partial sums approaches infinity, not a finite limit. A necessary condition for the limit of partial sums to exist and be finite is that the terms of the sum approach zero. But this is not sufficient. For the infinite harmonic series

$$\sum_{i=1}^{\infty} \frac{1}{i} = 1 + \frac{1}{2} + \frac{1}{3} + \frac{1}{4} + \frac{1}{5} + \ldots + \frac{1}{n} + \ldots,$$

$\lim_{n \to +\infty} \frac{1}{n} = 0$, but the partial sums of the series have no finite limit.

NOTES ON QUESTIONS
Making Connections for Question 4: This question helps to prepare students for understanding the epsilon-delta proofs that they may see in calculus. Solving such inequalities requires to use of logarithms.

Question 8: This question introduces the definitions of convergent and divergent series.

Question 11: Ask students to write out the first five terms of both the sequence (the changing values of TERM) and the partial sums of the series (the changing values of SUM) associated with this question.

Question 19: Writing $\sum_{k=1}^{\infty} \frac{6}{k^2}$ as $6 \sum_{k=1}^{\infty} \frac{1}{k^2}$ and evaluating $6\left(\frac{1}{1^2} + \frac{1}{2^2} + \frac{1}{3^2} + \frac{1}{4^2} + \ldots\right)$ will be easier.

Questions

Covering the Reading

In 1 and 2, find the sum.

1. $1 + 2.5 + 6.25 + 15.625 + \ldots + (2.5)^7$ ≈ 1016.6

2. $\sum_{i=0}^{n-1} \frac{3}{2^i}$ $6 - 6\left(\frac{1}{2}\right)^n$

3. Given the sequence $a_1, a_2, a_3, \ldots$ with $a_k = \frac{1}{k}$ for each integer $k \geq 1$. Let $S_1, S_2, S_3, \ldots$ denote the sequence of corresponding partial sums. Find S_3, S_4, and S_5. $S_3 = \frac{11}{6}, S_4 = \frac{25}{12}, S_5 = \frac{137}{60}$

4. In the lesson, it was noted that if $0 < r < 1$, then $\lim_{n \to +\infty} r^n = 0$.

a. If $r = \frac{1}{2}$, how large must n be to ensure that $r^n < .001$? $n \geq 10$

b. Calculate r^n for this value of n. ≈ 0.00098

c. If $r = .99$, how large must n be to ensure that $r^n < .001$? $n \geq 688$

5. a. Find the sum $\frac{4}{3} + \frac{4}{3^2} + \frac{4}{3^3} + \ldots + \frac{4}{3^{20}}$. ≈ 2.00000

b. Find $\sum_{k=0}^{\infty} \frac{4}{3^k}$. 6

6. Find the sum of the infinite geometric series

$100 + 80 + 64 + \frac{256}{5} + \frac{1024}{25} + \ldots$. 500

7. Is the value of $\sum_{k=0}^{\infty} \left(\frac{10}{9}\right)^k$ finite? Why or why not?
No, it diverges since the ratio $r = \frac{10}{9} > 1$ and $\lim_{k \to \infty} \left(\frac{10}{9}\right)^k = +\infty \neq 0$.

Applying the Mathematics

8. An infinite series is called **convergent** if its sequence of partial sums has a finite limit, **divergent** otherwise.

a. Give an example of a convergent series. See margin.

b. Give an example of a divergent series. See margin.

9. a. Explain why, for all real numbers a and r with $r \neq 1$,

$a\left(\frac{1 - r^n}{1 - r}\right) = a\left(\frac{r^n - 1}{r - 1}\right)$. $a\left(\frac{1 - r^n}{1 - r}\right) = a \cdot \frac{-1}{-1} \cdot \frac{1 - r^n}{1 - r} = a\left(\frac{r^n - 1}{r - 1}\right)$

b. Evaluate the left and right sides of the equation in part **a** for $a = 1$ and for both $r = 2$ and $r = \frac{1}{2}$. See margin.

c. Which side of the equation is easier to evaluate when $r > 1$?

d. Which side of the equation is easier to evaluate when $r < 1$?
c) the right; d) the left

10. Find the sum $p + 3p + 3^2p + \ldots + 3^{n+1}p$. $p\left(\frac{3^{n+2} - 1}{3 - 1}\right) = \frac{p}{2}(3^{n+2} - 1)$

11. Consider the computer program below.

```
10 TERM = 3
20 SUM = 3
30 FOR K = 2 TO 25
40      TERM = TERM * .5
50      SUM = SUM + TERM
60 NEXT K
70 PRINT SUM
80 END
```

a. Write a recursive definition for the sequence whose terms are stored in the variable TERM. See margin.

b. Write an explicit formula for the sequence. $a_n = 3\left(\frac{1}{2}\right)^{n-1}$

c. Use summation notation to write the sum that is calculated by the program. See margin.

d. Use the formula for the sum of the terms of a finite geometric series to find the sum in part **c**. ≈ 5.99999982

e. Suppose line 30 is modified by changing 25 to larger and larger integers, running the program each time. The printed answer will get closer and closer to what number? 6

f. Run the program to confirm your answers to parts **d** and **e**.
The result confirms d and e.

12. Let $a_1, a_2, a_3, \ldots$ be a sequence and $S_1, S_2, S_3, \ldots$ be the corresponding sequence of partial sums. Explain why for all integers $k \geq 1$, $S_{k+1} = S_k + a_{k+1}$. See margin.

FOLLOW-UP

MORE PRACTICE

For more questions on SPUR Objectives, use *Lesson Master 7-6,* shown on page 437.

EXTENSION

Discuss what happens to an infinite geometric series when the common ratio is -1. Euler thought that the sum $S = 1 - 1 + 1 - 1 + 1 - 1 + \ldots$ was $\frac{1}{2}$, arguing as follows:
Grouping one way, $(1 - 1) + (1 - 1) + (1 - 1) + \ldots = 0$.
Grouping a second way, $1 - 1 + 1 - 1 + 1 - 1 + \ldots = 1 + (-1 + 1) + (-1 + 1) + (-1 + 1) + \ldots = 1$.
The average is $\frac{1}{2}$.
Two arguments confirm Euler's answer: Let $S = 1 - 1 + 1 - 1 + 1 - 1 + \ldots$. Then $S = 1 - (1 - 1 + 1 - 1 + 1 - 1 + \ldots)$, so $S = 1 - S \Rightarrow 2S = 1 \Rightarrow S = \frac{1}{2}$. Ignoring the stipulation that $|r| < 1$ in the formula for a infinite geometric sequence,

$$S = \frac{a}{1 - r} = \frac{1}{1 - (-1)} = \frac{1}{2}.$$

Today, however, we would say that the series has no limit.

PROJECTS

The projects for Chapter 7 are described on pages 458–459. **Projects 3 and 5** are related to the content of this lesson.

EVALUATION

A quiz covering Lessons 7-4 through 7-6 is provided in the Teacher's Resource File.

ADDITIONAL ANSWERS

8.a. sample: $\sum_{n=0}^{\infty} \frac{1}{5^n}$

b. sample: $\sum_{n=0}^{\infty} \left(\frac{3}{2}\right)^n$

13. a. Write a recursive definition for the sequence S, where $S_n = \sum_{i=1}^{n} ar^{i-1}$, the sequence of partial sums of the geometric series $\sum_{n=1}^{\infty} ar^{n-1}$.

b. Use the Recursion Principle to prove: $S_n = \frac{a(1 - r^n)}{1 - r}$ when $r \neq 1$.
See margin.

Review

14. Prove: For every positive integer n, 3 is a factor of $n^3 + 14n$. *(Lesson 7-5)* **See margin.**

15. Consider the equation $S(n)$: $\sum_{i=1}^{n} [i(i + 1)] = \frac{n(n + 1)(n + 2)}{3}$.

a. Verify that $S(3)$ is true.

b. Use mathematical induction to prove that $S(n)$ is true for all integers $n \geq 1$. *(Lesson 7-4)*
See margin.

16. Suppose Dolores owes \$8000 on a new car in January. In February, she is charged .8% interest on the loan, so she owes $8000 + .008 \cdot 8000 = 8064$ dollars. However, she makes a payment of \$400, so after her February payment she owes \$7664. This process continues; each month she is charged .8% interest and she makes a \$400 payment. Let A_k represent the amount Dolores owes after her month k payment (where month 1 is January).

a. Find a recursive formula for the sequence $A_1, A_2, A_3, \ldots$.

b. Use the recursive formula and a calculator to see how much she owes in May. *(Lesson 7-1)* **\$6639.79**

a) $A_1 = 8000$; $A_{k+1} = A_k(1.008) - 400$ for all integers $k \geq 1$

17. a. Graph $y = \tan\left(\theta - \frac{\pi}{2}\right)$ on the interval $-\pi \leq \theta \leq \pi$. How is this graph related to the parent graph $y = \tan \theta$? **See margin.**

b. Solve the equation $\tan\left(\theta - \frac{\pi}{2}\right) \leq 1$ on the interval $-\pi \leq \theta \leq \pi$. *(Lessons 6-7, 5-6, 3-9)* **$-\pi < \theta \leq -\frac{\pi}{4}$ or $0 < \theta \leq \frac{3\pi}{4}$**

18. Solve for q: $\frac{1}{25} + \frac{1}{q} = \frac{1}{10}$. *(Lesson 5-7)* **$\frac{50}{3}$**

Exploration

19. a. Use a computer or programmable calculator to approximate $\sum_{k=1}^{\infty} \frac{6}{k^2}$.

b. Find the square root of your answer to part **a**. **3.141**

c. What do you think the exact value of the answer to part **b** is? **π**

a)9.867

20. Define a sequence by $a_k = \frac{(-1)^{k+1} \cdot 4}{2k - 1}$.

a. Find the first six terms of the sequence. **$4, -\frac{4}{3}, \frac{4}{5}, -\frac{4}{7}, \frac{4}{9}, -\frac{4}{11}$**

b. Write a program to print the corresponding partial sums S_n for $n = 1, 2, 3, \ldots, 100$. Note that the terms of S alternately increase and decrease. **See margin.**

c. Use your program to approximate $\sum_{k=1}^{\infty} \frac{(-1)^{k+1} \cdot 4}{2k - 1}$.
A good approximation can be obtained by averaging two consecutive partial sums, such as S_{10000} and S_{10001}. **3.141594**

d. What do you think the exact value of the answer to part **c** is? **π**

9.b., 11.a., c., 12.-15., 17.a., 20.b. See Additional Answers in the back of this book.

NAME ______

LESSON **MASTER 7–6**
QUESTIONS ON **SPUR** OBJECTIVES

■ **SKILLS** *Objective E (See pages 462–464 for objectives.)*
In 1–3, a. find the value of the series for $n = 4$ and b. find the limit of the series as $n \to \infty$.

1. $\sum_{k=0}^{n} \frac{2}{7^k}$	2. $\sum_{k=1}^{n} 4^k$	3. $\sum_{k=1}^{n} c(.9)^k$
a. 2.333	a. 340	a. 3.095c
b. $2\frac{1}{3}$	b. ∞	b. 9c

4. Let b be the sequence defined by $\begin{cases} b_1 = 2 \\ b_{k+1} = \frac{3}{4}b_k \ \forall\, k \geq 1. \end{cases}$
Let S_n be the nth partial sum of the sequence.

a. Find a formula for S_n. $S_n = 8\left[1 - \left(\frac{3}{4}\right)^n\right]$

b. Find S_6. 6.576

c. Find $\lim_{n\to\infty} S_n$. 8

5. Give an example of a series which converges and whose seventh term is greater than $\frac{1}{2}$. sample: $\sum_{i=1}^{\infty} (.95)^i$

6. Consider the finite geometric series $a + 2a + 4a + \ldots + 512a$.

a. Use sigma notation to express the series. $\sum_{n=0}^{9} 2^n a$

b. If the value of the series is 613.8, find a. .6

82 *Continued*

NAME ______
Lesson MASTER 7–6 (page 2)

■ **REPRESENTATIONS** *Objective J*
7. Consider the computer program below.

```
10  TERM = 3
20  PRINT TERM
30  SUM = TERM
40  FOR K = 2 TO 15
50     TERM = 2 * (TERM)/3
60     PRINT TERM
70     SUM = SUM + TERM
80  NEXT K
90  PRINT SUM
```

a. Use summation notation to write the sum that is calculated by the program. sample: $\sum_{n=1}^{15} 3\left(\frac{2}{3}\right)^{n-1}$

b. Use the formula for the sum of the terms of a finite geometric series to find the sum in part **a**. ≈ 8.98

c. What would the sum approach if the 15 in line 40 were changed to a larger and larger number? 9

8. What changes would have to be made to the program in Question 7 to compute the partial sums of the sequence defined by $a_n = 2(.3)^n$?
sample: 10 TERM = .6
50 TERM = 3 * (TERM)/10

 83

LESSON 7-7

RESOURCES

- Lesson Master 7-7
- Teaching Aid 41 displays the statements of the original form and strong form of mathematical induction.

OBJECTIVE

G Prove statements using mathematical induction.

TEACHING NOTES

The strong form of mathematical induction is not always found in texts at this level. The examples and theorems in the lesson show the power of the strong form of the Principle of Mathematical Induction. Students are led from the seemingly trivial theorem about jigsaw puzzles to a statement that is almost the Fundamental Theorem of Arithmetic.

Small Group Work You might consider doing an experiment with students related to the jigsaw puzzle problem. Bring in several pieces of a puzzle and have students count the number of moves needed to assemble different numbers of those pieces. Students might be able to conjecture the relationship that is proved in **Example 1**. Implicit in Example 1 is the assumption that no mistakes are made in building the puzzle.

LESSON

7-7

Strong Mathematical Induction

When defining a sequence recursively, sometimes the recurrence relation requires that you know more than one previous term (as in the Fibonacci sequence, in which a_{n+1} is defined in terms of a_n and a_{n-1}). Similarly, the inductive step in some mathematical induction proofs requires an inductive assumption that more than $S(k)$ is true. This modified inductive assumption results in what is called the *strong form of mathematical induction*, or *strong mathematical induction*.

Principle of Mathematical Induction (Strong Form)

Suppose that for each positive integer n, $S(n)$ is a sentence in n. If
(1) $S(1)$ is true, and
(2) for all integers $k \geq 1$, the assumption that $S(1)$, $S(2)$, ... , $S(k-1)$, $S(k)$ are all true implies that $S(k+1)$ is also true,
then $S(n)$ is true for all integers $n \geq 1$.

The Strong Form of Mathematical Induction given above differs from the original form in Lesson 7-4 only in the inductive step. In the inductive step in the Strong Form, you are allowed to use the assumption that *all* of the statements $S(1)$, $S(2)$, ... , $S(k-1)$, $S(k)$ are true in order to prove that $S(k+1)$ is true. It can be shown that these two forms of mathematical induction are logically equivalent in the sense that the validity of either form implies the validity of the other. (The proof is beyond the scope of this course.)

In this lesson there are three proofs which utilize strong mathematical induction. The first proof involves assembling a jigsaw puzzle. The process can be thought of as a game in which a *move* consists of putting together two *blocks*, each consisting of one or more individual puzzle pieces, to form a larger block. Before reading on, try to answer these two questions: If the puzzle has n pieces, how many moves are needed? Does the number of moves depend on how you put the puzzle together?

To many people, the answers to these questions are not obvious. Although there are many different ways to assemble a given jigsaw puzzle, the *number* of moves that are needed is always the same, and it is remarkably easy to find.

Example 1 Prove: ∀ positive integers n, the number of moves necessary to assemble any jigsaw puzzle with n pieces is $n - 1$.

Solution **Let $S(n)$ be the sentence: Any assembly of a jigsaw puzzle with n pieces requires $n - 1$ moves.**

(1) Show that $S(1)$ is true.

$S(1)$: Any assembly of a jigsaw puzzle with 1 piece requires 0 moves.

A jigsaw puzzle with only one piece is already assembled, and so no moves are required for its assembly. Therefore $S(1)$ is true. (Since a two-piece puzzle requires 1 move for its assembly, $S(2)$ is true also.)

(2) Show that the assumption that $S(1)$, $S(2)$, ..., $S(k)$ are all true implies that $S(k + 1)$ is true.

Assume that for some integer $k \geq 1$, $S(1)$, $S(2)$, ... , $S(k - 1)$, $S(k)$ are all true. This means if a jigsaw puzzle has k or fewer pieces, then the number of moves required for any assembly of that puzzle is 1 fewer than the number of pieces.

Show that this assumption implies that $S(k + 1)$ is true:

$S(k + 1)$: Any assembly of a jigsaw puzzle with $k + 1$ pieces requires k moves.

Now a jigsaw puzzle with $k + 1$ pieces has been assembled by some finite sequence of moves. The last move in this sequence put together two blocks. If r and s are the numbers of pieces in these two blocks, then $r + s$ is the total number of pieces in the puzzle. That is, $r + s = k + 1$. Hence both r and s are integers from 1 to k, inclusive.

These two final blocks of pieces can be thought of as completed jigsaw puzzles with r and s pieces, and so by the inductive assumption, these two blocks required $r - 1$ and $s - 1$ moves for their assembly. Therefore, the total number of moves required for the assembly of the puzzle with $k + 1$ pieces is

$$\underbrace{(r-1)+(s-1)}_{\text{moves to assemble two blocks}} + \overset{\text{move to put the two blocks together}}{1}$$

$$= r + s - 1$$

$$= (k + 1) - 1 \quad \text{because } r + s = k + 1$$

$$= k.$$

This shows that $S(k + 1)$ is true, which completes the inductive step.

Therefore, by the Strong Form of the Principle of Mathematical Induction, $S(n)$ is true for every positive integer n.

Making Connections
The first paragraph following the proof of **Example 1** is an important variant of strong induction. The original definition of strong induction specifies $n = 1$. The corresponding idea with the original form of mathematical induction is found in **Example 3** and Additional Example 3 of Lesson 7-4. In essence, you do not have to push over the first domino to begin induction. You may begin at the ath domino.

Some students may note that **Example 2** is basically a repeated application of a theorem that the sum of any three odd numbers is also odd.

Most of the proofs that students are expected to complete using the Strong Form of Induction will relate to properties of recursively defined sequences, such as **Example 2**. Students need to realize that in such situations the basis step may involve verifying the property for more than one specific statement.

By now students should understand the basics of mathematical induction proofs even though they may still have some problems in writing such proofs. There are two more lessons in the chapter that provide opportunities for further practice.

The reason that strong mathematical induction is needed in the proof of the theorem on page 441 is that the factors of a given number are not one less than the number but may be any number less than that number. Thus, it is necessary to connect $S(k + 1)$ to any of the $S(i)$, $i = 1$ to k, that precede it.

ADDITIONAL EXAMPLES

1. The "We're All Siblings Society" is sponsoring a hands-across-your-town campaign. Their goal is to have citizens in each community form a hand-holding chain linking together diverse groups. Between each two people there will be a flag. Assuming that no person in the chain holds hands with himself (herself), how many flags will be needed when there are n people in the chain? Prove the result.

$n - 1$; Let $S(n)$ be the number of flags needed for a chain of n people. $S(1) = 0$ and $S(2) = 1$. Assume for some integer k, $S(1)$ through $S(k)$ are all true. Now a chain of $k + 1$ people is formed by joining chains of r and s people, where $r + s = k + 1$ and since $r \le k$ and $s \le k$, by the inductive assumption, these chains hold $r - 1$ and $s - 1$ flags. One more flag is necessary to join them, so for $k + 1$ people, $1 + (r - 1) + (s - 1) = (r + s) - 1 = (k + 1) - 1 = k$ flags are needed. Thus, by strong induction, $S(n)$ is true for all integers $n \ge 1$.

2. Consider the sequence defined recursively by

$$\begin{cases} a_1 = 2 \\ a_2 = 4 \\ a_{k+1} = 7a_k + a_{k-1} \end{cases}$$

for $k \ge 2$.
Prove that all the terms of this sequence are even.

$S(n)$: a_n is even. $S(1)$ and $S(2)$ are true from the initial conditions of the sequence. Assume for some integer k, $S(1)$ through $S(k)$ are all true. By this inductive assumption, there are integers r and s such that $a_{k+1} = 7a_k + a_{k-1} = 7(2r) + (2s) = 2(7r + s)$. Thus, a_{k+1} is even, and $S(k + 1)$ is true. By strong induction, $S(n)$ is true for all integers $n \ge 1$.

As with the original form of mathematical induction, strong mathematical induction does not have to begin with $n = 1$. Also, the basis step may require establishing the proof of $S(n)$ for more than one particular value of n. So in the basis step it may be necessary to show that for some particular integers a and b with $a \le b$,

$S(a)$, $S(a + 1)$, $S(a + 2)$, ..., $S(b)$ are true.

Then in the inductive step, it is assumed that $k \ge a$, and $S(a)$, $S(a + 1)$, $S(a + 2)$, ..., $S(b)$ are true.

Consider the sequence defined recursively by

$$\begin{cases} a_1 = 1 \\ a_2 = 1 \\ a_3 = 1 \\ a_{k+1} = a_k + a_{k-1} + a_{k-2} \text{ for } k \ge 3. \end{cases}$$

Using successive substitution to calculate a_4, a_5, a_6, and a_7 yields

$$a_4 = a_3 + a_2 + a_1 = 1 + 1 + 1 = 3$$
$$a_5 = a_4 + a_3 + a_2 = 3 + 1 + 1 = 5$$
$$a_6 = a_5 + a_4 + a_3 = 5 + 3 + 1 = 9$$
$$a_7 = a_6 + a_5 + a_4 = 9 + 5 + 3 = 17.$$

A reasonable conjecture would be that every term in the sequence is odd.

Strong mathematical induction can be used to prove this conjecture. The inductive step uses the recurrence relation $a_{k+1} = a_k + a_{k-1} + a_{k-2}$, but this relation is only valid for $k \ge 3$. Thus, the inductive step (showing that $S(1)$, ..., $S(k)$ are true implies $S(k + 1)$ is true) can only be done for $k \ge 3$. This is like showing that in a row of dominoes, each falling domino, starting with the third one, causes the next one to fall. You must still show that the first three dominoes fall. Thus, in the proof which follows, the basis step shows that $S(1)$, $S(2)$, and $S(3)$ are true.

Example 2 Prove that all of the terms of the sequence defined above are odd.

Solution The following is to be proved for all natural numbers n.
$S(n)$: *a_n is an odd integer.*

(1) Show that $S(1)$, $S(2)$, and $S(3)$ are true.

$S(1)$, $S(2)$, and $S(3)$ are the statements that a_1, a_2, and a_3 are odd integers. The initial conditions $a_1 = 1$, $a_2 = 1$, and $a_3 = 1$ imply that they are. So $S(1)$, $S(2)$, and $S(3)$ are true.

(2) Show that the assumption that $S(1)$, $S(2)$, ... , $S(k - 1)$, $S(k)$ are all true for some integer $k \ge 3$ implies that $S(k + 1)$ is true.

Assume that a_1, a_2, ... , a_{k-1}, a_k are all odd integers. — This is the inductive assumption.

We must show that a_{k+1} is an odd integer where $a_{k+1} = a_k + a_{k-1} + a_{k-2}$. From the definition of *odd,* an integer m is odd if and only if it can be written as $m = 2t + 1$ for some integer t. Thus, since a_k, a_{k-1}, and a_{k-2} are all odd, there are integers p, q, and r such that

$$a_k = 2p + 1,\ a_{k-1} = 2q + 1,\ a_{k-2} = 2r + 1.$$

440

Substituting into the recurrence relation

$$a_{k+1} = a_k + a_{k-1} + a_{k-2}$$
$$a_{k+1} = (2p + 1) + (2q + 1) + (2r + 1)$$
$$= 2(p + q + r + 1) + 1.$$

Since $p + q + r + 1$ is an integer, it follows that a_{k+1} has the form of an odd integer. Thus, if $S(1)$, $S(2)$, ..., $S(k)$ are all true, $S(k + 1)$ is true.

Therefore, by the Strong Form of the Principle of Mathematical Induction, the statement

$S(n)$: a_n is an odd integer

is true for all positive integers n.

Some important theorems can be proved most easily by using strong mathematical induction. Here it is used to prove a statement that is almost the Fundamental Theorem of Arithmetic.

Theorem

Every positive integer $n \geq 2$ is either a prime or a product of primes.

(The Fundamental Theorem of Arithmetic also states that the prime factorization is unique.)

Proof Let $S(n)$ be the sentence *n is a prime or a product of primes*. The Strong Form of the Principle of Mathematical Induction is used, starting with $n = 2$.

(1) Show that $S(2)$ is true.
$S(2)$ is the statement *2 is a prime or a product of primes*. Clearly $S(2)$ is true because 2 is a prime.

(2) Show that the assumption that $S(2)$, $S(3)$, ..., $S(k)$ are all true implies that $S(k + 1)$ is true.

Assume that $S(2)$, $S(3)$, ... $S(k - 1)$, $S(k)$ are all true. That is, assume that

Every integer from 2 to k, inclusive, is either a prime or a product of primes.

It needs to be shown that this assumption implies that $S(k + 1)$ is true, where

$S(k + 1)$: *$k + 1$ is a prime or a product of primes.*

Now, there are two possibilities: either $k + 1$ is prime or it is not prime. If $k + 1$ is prime, then $S(k + 1)$ is true. If $k + 1$ is not prime, then $k + 1 = a \cdot b$ where a and b are positive integers greater than 1 and less than $k + 1$. Therefore, a and b are integers from 2 to k, inclusive, and so the inductive assumption implies that both a and b are either primes or products of primes. Consequently, $k + 1 = a \cdot b$ is also a product of primes, and so $S(k + 1)$ is true in this case also.

It follows that $S(n)$ must be true for all integers $n \geq 2$ by the Strong Form of the Principle of Mathematical Induction.

NOTES ON QUESTIONS

Question 4: Compare this question to **Example 2**.

Questions 5 and 6: You may wish to use one of these as a class exercise and assign the other for homework. Invite students to work on one of these in class, write their proofs on the chalkboard, and then discuss the correctness of their proofs.

Question 10: Students should recognize that the sequence is a geometric one.

Question 12: Ask: What if the manufacturer switches to cylindrical cans with radii equal to their heights? (If the radius is r, $P(x) = 0.15\pi r^3 - .50\pi r^2$. A can with radius $\frac{10}{3}$ inches yields a net income of zero; smaller cans yield a loss while larger cans yield a profit.)

Technology for Question 14: Students may notice that $a_{k+1} = a_k + a_{k-1}$ for $k \geq 2$ is the recurrence relation for a generalized Fibonacci sequence. Invite students to write a program that will ask for the input of the first two terms of such a sequence. The program will then generate as many terms as desired. It will be useful to check hypotheses.

ADDITIONAL ANSWERS

1. In the Strong Form of Mathematical Induction, the assumption that each of $S(1), S(2), \dots, S(k)$ are true is used to show that $S(k+1)$ is true. In the original form, only the assumption that $S(k)$ is true is used to prove $S(k+1)$ is true.

2.a. Sample: Combine 1 and 2, then join 3, then join 4; combine 1 and 4, then join 2, then join 3.

3. The 7-piece and 13-piece blocks needed 6 and 12 steps, respectively, by the inductive assumption. That is 18 steps; joining them is the 19th step.

4.b.iii. $a_1, a_2, \dots, a_k$ are all even integers. Prove that a_{k+1} is an even integer.
iv. $a_{k+1} = a_k + a_{k-1} + a_{k-2}$. a_k, a_{k-1}, and a_{k-2} are even integers by the inductive assumption. Hence, $a_k + a_{k-1}$ is an even integer by the Sum of Two Even Integers Theorem. By the same theorem, $a_k + a_{k-1} + a_{k-2}$ is then also an even integer. And so a_{k+1} is an even integer.

5.b. $a_1 = 5$ and $a_2 = 15$ are multiples of 5. Assume $a_1, a_2, \dots, a_k$ are all multiples of 5. Show that a_{k+1} is a multiple of 5. Now $a_{k+1} = a_k + a_{k-1}$. 5 is a factor of a_k and a_{k-1}, so it is a factor of their sum, a_{k+1}. Hence, by the Strong Form of Mathematical Induction, a_n is a multiple of 5 $\forall\, n \geq 1$.

6. $a_1 = 3$ and $a_2 = 5$ are odd integers. Assume $a_1, a_2, \dots, a_k$ are all odd integers. $a_{k+1} = a_{k-1} + 2a_k$, where $a_{k-1} = 2q + 1$ and $a_k = 2r + 1$ for some integers q and r. Then $a_{k+1} = 2q + 1 + 2(2r + 1) = 2q + 4r + 3 = 2(q + 2r + 1) + 1$ so a_{k+1} is odd. Hence, by the Strong Form of Mathematical Induction, every term in the sequence is an odd integer.

Questions

Covering the Reading

1. Explain the difference between the Strong Form of Mathematical Induction and the original form. See margin.

2. A 4-piece jigsaw puzzle is shown. One way to put it together is to combine 1 and 4, then combine 2 and 3, then join the blocks.

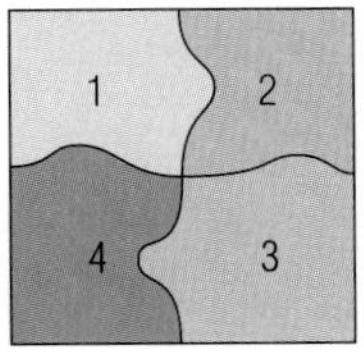

a. Describe two other ways of putting the puzzle together. See margin.
b. How many steps are needed for each of your ways? 3

3. The last move in putting together a 20-piece jigsaw puzzle was to combine a block of 7 pieces and a block of 13 pieces. Explain how strong mathematical induction tells you that the last move was the 19th move made. See margin.

4. A sequence is defined recursively as follows:

$$\begin{cases} a_1 = 2 \\ a_2 = 2 \\ a_3 = 4 \\ a_{k+1} = a_k + a_{k-1} + a_{k-2} \text{ for } k \geq 3. \end{cases}$$

a. Compute the first five terms of this sequence. 2, 2, 4, 8, 14
b. Complete the steps below to prove, using the Strong Form of Mathematical Induction, that the terms of this sequence are all even integers.
i. Write the statement $S(n)$. $S(n)$: a_n is an even integer.
ii. Verify the basis step. $a_1 = 2$, $a_2 = 2$, $a_3 = 4$ are even integers.
iii. Write the inductive assumption and the statement that must be proved assuming it. See margin.
iv. Use **iii** and the Sum of Two Even Integers Theorem to complete the inductive step. See margin.
v. What can you conclude from the results of parts **ii–v**? All the terms in the sequence are even integers.

Applying the Mathematics

5. Consider the sequence defined recursively as follows:

$$\begin{cases} a_1 = 5 \\ a_2 = 15 \\ a_{k+1} = a_k + a_{k-1} \text{ for } k \geq 2. \end{cases}$$

a. Write the first four terms of the sequence. 5, 15, 20, 35
b. Use the Strong Form of Mathematical Induction to prove that every term of the sequence is a multiple of 5. (Hint: Use the Factor of a Sum Theorem.) See margin.

6. Consider the sequence defined recursively as follows:

$$\begin{cases} a_1 = 3 \\ a_2 = 5 \\ a_{k+1} = a_{k-1} + 2a_k \text{ for } k \geq 2. \end{cases}$$

Prove that every term of the sequence is an odd integer. See margin.

7. Recall, from the Questions in Lesson 7-1, that the *Lucas sequence L* has the same recurrence relation as the Fibonacci sequence F but it has different initial conditions. It is described recursively as follows:

$$\begin{cases} L_1 = 1 \\ L_2 = 3 \\ L_{k+1} = L_k + L_{k-1} \text{ for integers } k \geq 2. \end{cases}$$

Complete the steps below to show that $S(n)$: $L_n = F_{n+1} + F_{n-1}$ is true for all integers $n \geq 2$, using strong mathematical induction.

a. Show that $S(2)$ and $S(3)$ are true. See margin.

b. Write the inductive assumption.

c. Write $S(k+1)$. $S(k+1)$: $L_{k+1} = F_{k+2} + F_k$

d. Use the inductive assumption and the recursive definition of the Lucas sequence to rewrite L_{k+1} as a sum of four terms of the Fibonacci sequence. $L_{k+1} = L_k + L_{k-1} = (F_{k+1} + F_{k-1}) + (F_k + F_{k-2})$

e. Regroup the result from part **d** and use the recursive definition of the Fibonacci sequence to finish the inductive step. See margin.

b) $\forall$ integers j such that $2 \leq j \leq k$, $L_j = F_{j+1} + F_{j-1}$

Review

8. Find

a. $\sum_{k=0}^{\infty}\left[3 \cdot \left(\frac{2}{3}\right)^k\right]$ 9 **b.** $3 + \sum_{k=0}^{\infty} 2 \cdot \left(\frac{2}{3}\right)^k$. 9

(Lesson 7-6)

9. Use mathematical induction to prove that $n^3 + 3n^2 + 2n$ is divisible by 3 for all integers $n \geq 1$. *(Lesson 7-5)* See margin.

10. Given the sequence defined recursively as follows:

$$\begin{cases} a_1 = 4 \\ a_{k+1} = 3 \cdot a_k \text{ for } k \geq 1. \end{cases}$$

Prove that $a_n = 4 \cdot 3^{n-1}$ is an explicit formula for the sequence. *(Lesson 7-4)* See margin.

11. **a.** Use an automatic grapher to graph $y = \csc x \sec x - \cot x$.

b. Use the graph in part **a** and your knowledge of the graphs of trigonometric functions to conjecture an identity: $\csc x \sec x - \cot x =$ _?_. tan x

c. Prove the identity in part **b**. a, c) See margin.

(Lessons 6-2, 6-1)

12. A manufacturer produces cubical boxes with lids. The boxes cost 25 cents per square inch of surface area to make. For each box sold the manufacturer receives 15 cents per cubic inch of volume.

a. Assume that the manufacturer is able to sell every box made. Express the net income received for each box as a polynomial function $P(x)$ where x is the length of a side in inches.

b. What size box yields a net income of zero? 10″ × 10″ × 10″

c. Do larger boxes yield a positive profit? Yes

d. Does the manufacturer incur a loss if smaller boxes are made? Yes

(Lessons 3-7, 3-3, 3-1) a) $P(x) = 0.15x^3 - 1.5x^2$

7.a. $S(2)$: $L_2 = F_3 + F_1$. $3 = 2 + 1$, so $S(2)$ is true. $S(3)$: $L_3 = F_4 + F_2$. $4 = 3 + 1$, so $S(3)$ is true.
e. $L_{k+1} = (F_{k-1} + F_{k-2}) + (F_{k+1} + F_k) = F_{k+2} + F_k$

9., 10., 11.a., c. See Additional Answers in the back of this book.

NAME ____________________

LESSON **MASTER 7–7**
QUESTIONS ON **SPUR** OBJECTIVES

■PROPERTIES *Objective G (See pages 462–464 for objectives.)*

1. The Strong Form of Mathematical Induction differs from the original form only in the __________ step.

In 2–4, use the Strong Form of Mathematical Induction.

2. Consider the sequence defined recursively by
$$\begin{cases} a_1 = 3 \\ a_2 = -12 \\ a_{k+1} = a_k - 4a_{k-1} \ \forall \text{ integers } k \geq 2. \end{cases}$$
Prove that every term of the sequence is a multiple of 3.
Let $S(n)$: a_n is a multiple of 3. Show $S(1)$ and $S(2)$ are true: $a_1 = 3$ and $a_2 = -12$ are both multiples of 3. Assume $S(1), S(2), \ldots, S(k)$ are all true for a positive integer k. Show $S(k+1)$ is true: $a_{k+1} = a_k - 4a_{k-1}$. Since $S(k)$ and $S(k-1)$ are true, a_k and a_{k-1} are multiples of 3. Therefore, $a_k - 4a_{k-1}$ is a multiple of 3. Thus, $S(k+1)$ is true. Therefore, by strong induction, $S(n)$ is true $\forall$ positive integers n.

3. Consider the sequence defined recursively by
$$\begin{cases} a_1 = 2 \\ a_2 = 4 \\ a_{k+1} = 2a_k - 3a_{k-1} \ \forall \text{ integers } k \geq 2. \end{cases}$$
Prove that every term of the sequence is an even integer.
Let $S(n)$: a_n is even. Show $S(1)$ and $S(2)$ are true: $a_1 = 2$ and $a_2 = 4$ are both even. Assume $S(1), S(2), \ldots, S(k)$ are all true for a positive integer k. Show $S(k+1)$ is true: $a_{k+1} = 2a_k - 3a_{k-1}$. Since $S(k)$ and $S(k-1)$ are true, a_k and a_{k-1} are even. Therefore, $2a_k - 3a_{k-1}$ is even. Thus, $S(k+1)$ is true. Therefore, by strong induction, $S(n)$ is true $\forall$ positive integers n.

84 *Continued*

NAME ____________________
Lesson MASTER 7–7 (page 2)

4. Let c be a fixed integer. Prove that every term of the sequence defined by
$$\begin{cases} b_1 = c \\ b_2 = 2c \\ b_{k+1} = 2b_k + b_{k-1} \ \forall \text{ integers } k \geq 2 \end{cases}$$
is divisible by c.
Let $S(n)$: b_n is divisible by c. Show $S(1)$ and $S(2)$ are true: b_1 and b_2 are divisible by c since $b_1 = c$ and $b_2 = 2c$. Assume $S(1), S(2), \ldots, S(k)$ are all true for a positive integer k. Show $S(k+1)$ is true: $b_{k+1} = 2b_k + b_{k-1}$. Since $S(k)$ and $S(k-1)$ are true, b_k and b_{k-1} are divisible by c. Therefore, $2b_k + b_{k-1}$ is divisible by c. Thus, $S(k+1)$ is true. Therefore, by strong induction, $S(n)$ is true $\forall$ positive integers n.

 85

FOLLOW-UP

MORE PRACTICE
For more questions on SPUR Objectives, use *Lesson Master 7-7,* shown on page 443.

PROJECTS
The projects for Chapter 7 are described on pages 458–459. **Project 2** is related to the content of this lesson.

ADDITIONAL ANSWERS
14.c. a_1 and a_2 are multiples of m. Assume a_1, a_2, ..., a_k are multiples of m. $a_{k+1} = a_k + a_{k-1}$. Because a_{k-1} and a_k are multiples of m, their sum, a_{k+1}, is also a multiple of m by the Factor of a Sum Theorem. By the Strong Form of Mathematical Induction, every term in the sequence is a multiple of m.

15.a. Suppose that a recurrence relation defines x_{n+1} in terms of x_n, x_{n-1}, ..., x_1, and n for each integer $n \geq 1$. Then there is exactly one sequence defined by this recurrence relation and the initial condition $x_1 = a$.
b. Suppose there is a second sequence y for which $y_1 = a$ and y_{n+1} is defined by the same recurrence relation as x_{n+1}. Let $S(n)$ be the statement $x_n = y_n$. (1) Because $x_1 = a$ and $y_1 = a$, $x_1 = y_1$. So, $S(1)$ is true. (2) Assume $S(1)$, $S(2)$, ..., $S(k)$ are true. Then $x_1 = y_1$, $x_2 = y_2$, ..., $x_k = y_k$. The sequences have the same recurrence relation; y_{k+1} is defined in terms of y_1, ..., y_k and k; and x_{k+1} is defined in terms of x_1, x_2, ..., x_k and k; so $y_{k+1} = x_{k+1}$. Thus, $S(k + 1)$ is true. By (1), (2), and the Strong Form of Mathematical Induction, $S(n)$ is true for all $n \geq 1$. Therefore, the Adapted Recursion Principle is proven.

13. Solve $\sqrt{(x-1)(x+5)} = \sqrt{x+5}$ over the reals. *(Lesson 3-1)* $x = -5, 2$

Exploration

14. Consider a sequence a that satisfies the recurrence relation $a_{k+1} = a_k + a_{k-1}$ $\forall$ integers $k \geq 2$.
   a. Of what form must a_1 and a_2 be so that every term of the sequence a_1, a_2, a_3, ... is a multiple of 7?
   b. Of what form must a_1 and a_2 be so that every term is a multiple of m? **a_1 and a_2 must be multiples of m.**
   c. Prove that if a_1 and a_2 are of the form specified in part **b**, then every term is a multiple of m. See margin.
   a) a_1 and a_2 must be multiples of 7.

15. a. Adapt the Recursion Principle from Lesson 7-1 so that it parallels strong mathematical induction.
   b. Use strong mathematical induction to prove your adaptation of the principle.
   See margin.

444

LESSON

7-8

Sorting Algorithms

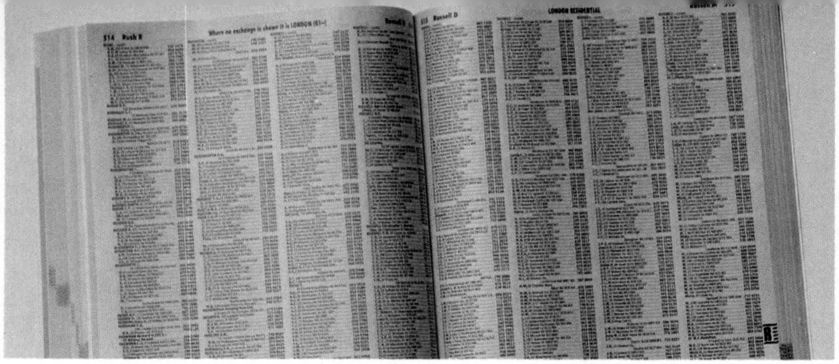

Computers sort listings for telephone books.

Often computers are employed to arrange or sort a given list of items in some desired order. Arranging a list of names in alphabetical order or a list of numbers in increasing order can be a tedious job if the lists are very long. **Sorting algorithms** provide the computer with the necessary instructions to carry out this task. This lesson describes and compares two computer sorting algorithms. It also shows how the Strong Form of Mathematical Induction can be used to verify that an algorithm does the job it is supposed to do.

Suppose you are given a list of n numbers to arrange in increasing order. One algorithm for accomplishing this task is called *Bubblesort*. Bubblesort makes successive passes through the list *from the bottom to the top*. In each pass, successive pairs of adjacent numbers are compared; if the number on the bottom is larger than the one on top, the two are exchanged; otherwise, they are left as is. One result of the first pass is the largest number "bubbles up" to the top; a result of the second pass is the next largest number ends up second to the top. After each pass n, the n top numbers in the list are called the *sorted section* of the list; the rest make up the *unsorted section*.

Example 1 Apply the Bubblesort algorithm to arrange the list 7, 3, 11, 1, 9, 4 in increasing order.

Solution Imagine the numbers in the list written in a column starting at the bottom and going up to the top. For example, the list 7, 3, 11, 1, 9, 4 would appear as

4
9
1
11
3
7

The effect of the first pass is shown on page 446. In each case, the numbers in the blue boxes are compared. At the end of the first pass, the number in the top, yellow box is in its correct position. Therefore, it is now in the sorted section of the list.

LESSON 7-8

RESOURCES

- Lesson Master 7-8
- Teaching Aid 42 can be used with **Example 1**.
- Teaching Aid 43 can be used with **Example 2**.

OBJECTIVE

I Execute algorithms on sets of numbers.

TEACHING NOTES

Students may wonder why they are asked to use one of the text algorithms to sort a small list of numbers. We use a small list for illustrative purposes only. Clearly, such algorithms would be used in practice only with long lists of numbers to be sorted. Such long lists are not practical for understanding the mechanics of the algorithms.

Computers also use similar algorithms to alphabetize lists. Because a computer assigns numerical values to the letters of the alphabet, a program similar to that for the sorting algorithms can be used to arrange lists alphabetically.

Students will enjoy and benefit from a live application of Bubblesort. Ask ten students to write their last names in large letters on a sheet of paper. Invite them to come up to the front of the room, mix them randomly, and have them sort themselves using the Bubblesort Algorithm. Place the unsorted list of students one step away from the front of the room and the sorted list against the wall. Repeat this activity with another group of students.

You may have 12 or more students go through a live demonstration of Quicksort. Give each student a card with a unique number on it. Have them line up in a random order. Then let them sort themselves in ascending order using the Quicksort Algorithm.

Alternate Approach
Have students arrange themselves in pairs. Give one student in each pair 20 3″ × 5″ index cards and ask that person to number the cards 1 to 20. Have students shuffle the cards and spread them out in a row on the floor in front of them. Select either Bubblesort or Quicksort. As one student keeps rereading the algorithm, the other student applies it until the cards are in order. Then have students reverse their roles. Students will enjoy this activity and develop a good understanding of both algorithms.

The writing of the list of numbers in Bubblesort from bottom to top will confuse some students. The reason it is done this way is that the name "Bubblesort" is an apt description of what occurs during the sorting process when the larger numbers "bubble up" through the list.

Some versions of Bubblesort will terminate when a pass through the remaining list produces no switches. That list is already in order.

In the use of the Quicksort Algorithm in **Example 2**, double subscripts are used to name the lists. Students may need some help in interpreting such lists. Encourage students to read from the right-most subscript to the left-most one. For instance, $(L_\ell)_r$ is the right sublist of the left sublist from the previous step.

Pass 1:

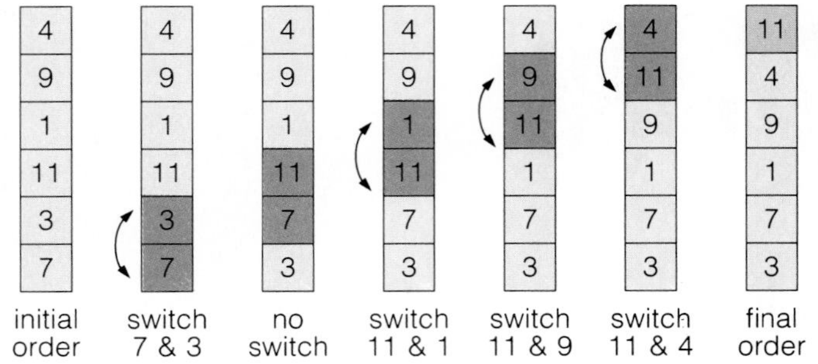

At the end of the first pass the numbers in the list are in the order

3, 7, 1, 9, 4, 11.

For each succeeding pass, the unsorted numbers are compared. The successive comparisons are shown in the blue boxes. Those in the yellow boxes are in the sorted section of the list.

Pass 2:

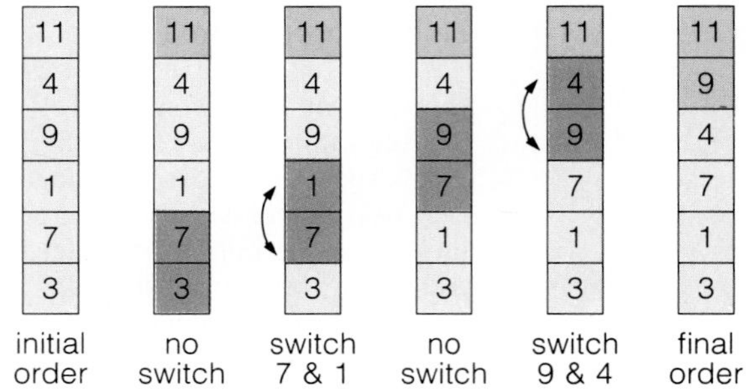

At the end of the second pass the numbers in the list are in the order

3, 1, 7, 4, 9, 11.

Pass 3:

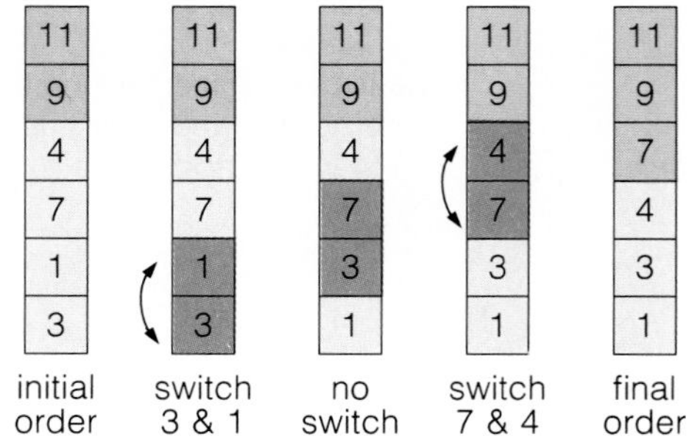

At the end of the third pass the numbers in the list are in the order

1, 3, 4, 7, 9, 11,

which is in increasing order.

You can stop here if you are executing the algorithm by hand. However, if you program a computer for the algorithm, it will continue to execute the last two passes without performing any interchanges between boxes.

The general algorithm can be described as follows:

Bubblesort Algorithm

To arrange a list L of numbers in increasing order:

Step 1: If L contains only one number (or no numbers), then L is already sorted, so stop. Otherwise, let the "unsorted section of the list" be the entire list and the "sorted section of the list" be empty.

Step 2: Make a pass through the unsorted section of L from bottom to top. Compare pairs of successive numbers and switch the numbers in the pair if the bottom number of the pair is greater than the top one. (At the end of each pass, the greatest of the unsorted numbers is at the top of the unsorted section of L.) This number becomes the bottom of the sorted section of L.

Step 3: If the unsorted section of L contains more than one number, repeat step 2; otherwise stop.

Conference participants are sorted alphabetically.

Bubblesort is an example of an **iterative algorithm**. The term iterative refers to the fact that the same steps are repeated over and over, that is, they are **iterated**. The Bisection Method studied in Lesson 3-5 is another example of an iterative algorithm because the same steps are repeated on smaller and smaller intervals.

Another sorting algorithm, called *Quicksort,* is a **recursive algorithm**, in the sense that during its execution it makes use of, or "calls on," itself. Recall how the Tower of Hanoi puzzle was solved by assuming that a smaller similar problem could be solved. Quicksort sorts a list by assuming that smaller lists can be sorted. It arranges a list of real numbers in increasing order by performing the following steps:

Point out to students that in each sublist, the first number will be placed in the correct position relative to the numbers in that sublist. The "divide and conquer" strategy of Quicksort keeps approximately halving the size of the lists used, thus continually making the process easier.

Making Connections
Sorting Algorithms is a mandatory topic in Advanced Placement Computer Science (APCS) courses. If your school offers this course, ask your APCS teacher if he or she has a program that visually shows a variety of sorting algorithms. Students will be astounded at the time difference required by Bubblesort (slow) versus Quicksort (fast) to sort a large list of numbers.

Students who are interested in sorting techniques and recursion will find a wealth of information in *Algorithms + Data Structures = Programs,* by Niklaus Wirth (Prentice-Hall, 1976). The author is the creator of both Pascal and Modula II. The book is a classic in its field.

The University of Toronto Library has an excellent film (also available in VCR format) called "Sorting Out Sorts." It provides excellent visual examples of a wide variety of sorting techniques along with analyses of efficiency, which will be discussed in the next lesson.

ADDITIONAL EXAMPLES

1. Repeat the instructions for Example 1 using the list 6, 5, 4, 2, 3, 1.

Reading top to down, here are the steps:

Pass 1:
1, 3, 2, 4, 5, 6
1, 3, 2, 4, 6, 5
1, 3, 2, 6, 4, 5
1, 3, 6, 2, 4, 5
1, 6, 3, 2, 4, 5
6, 1, 3, 2, 4, 5

Pass 2:
6, 1, 3, 2, 5, 4
6, 1, 3, 5, 2, 4
6, 1, 5, 3, 2, 4
6, 5, 1, 3, 2, 4

Pass 3:
6, 5, 1, 3, 4, 2
6, 5, 1, 4, 3, 2
6, 5, 4, 1, 3, 2

Pass 4:
6, 5, 4, 1, 3, 2 (stays same)
6, 5, 4, 3, 1, 2

Pass 5:
6, 5, 4, 3, 2, 1

Ask students if a list in exactly reverse order to that desired is the worst case for Bubblesort to handle. (Yes)

2. Apply Quicksort to the list L = 6, 5, 4, 2, 3, 1.
L_ℓ = 5, 4, 2, 3, 1; f = 6; L_r is empty so is sorted. Sort 5, 4, 2, 3, 1; here L_ℓ = 4, 2, 3, 1; f = 5; L_r is empty so is sorted. Sort 4, 2, 3, 1; here L_ℓ = 2, 3, 1; f = 4; L_r is empty so is sorted. Sort 2, 3, 1; here L_ℓ = 1; f = 2; L_r = 3, and they are sorted. The sorted list is 1, 2, 3, 4, 5, 6.

Quicksort Algorithm

To arrange a list L of distinct numbers in increasing order:

Step 1: If L contains only one number (or no numbers), then L is already sorted, so stop. Otherwise continue with Steps 2 through 4.

Step 2: Divide L into two *sublists* by comparing the first number f in L with the remaining numbers. Place those numbers that are less than f in a *left sublist* L_ℓ and those that are greater than f in a *right sublist* L_r. Then place f between L_ℓ and L_r. The list is now L_ℓ, f, L_r.

Step 3: Use Quicksort to sort L_ℓ.

Step 4: Use Quicksort to sort L_r.

The action of the Quicksort algorithm on the original list of Example 1 is described below.

Example 2 Apply the Quicksort algorithm to the list L = 7, 3, 11, 1, 9, 4.

Solution

Step 1: The list L contains more than one number, so continue to Step 2.

Step 2: $f = 7$, so compare 7 with the remaining numbers and group them into two sublists:
  L_ℓ = 3, 1, 4 and L_r = 11, 9.
  L now looks like
  3, 1, 4, 7, 11, 9.

Step 3: Use Quicksort to sort L_ℓ = 3, 1, 4.
  Step 1: L_ℓ contains more than one number, so continue.
  Step 2: $f = 3$, so divide L_ℓ into
    $(L_\ell)_\ell = 1$ and $(L_\ell)_r = 4$.
    L_ℓ is now 1, 3, 4.
  Step 3: Use Quicksort to sort $(L_\ell)_\ell$.
    Step 1: $(L_\ell)_\ell$ contains only one number, so it is already sorted.
  Step 4: Use Quicksort to sort $(L_\ell)_r$.
    Step 1: $(L_\ell)_r$ contains only one number, so it is already sorted.
  Now L_ℓ is sorted, and L looks like
    1, 3, 4, 7, 11, 9.

Step 4: Use Quicksort to sort L_r = 11, 9.
  Step 1: L_r contains more than one number, so continue.
  Step 2: $f = 11$, so divide L_r into
    $(L_r)_\ell = 9$ and $(L_r)_r$ is empty.
    L_r is now 9, 11.

Step 3: Use Quicksort to sort $(L_r)_\ell$.

Step 1: $(L_r)_\ell$ contains only one number, so it is already sorted.

Step 4: Use Quicksort to sort $(L_r)_r$.

Step 1: $(L_r)_r$ contains no numbers, so it is already sorted.

Now L_r is sorted, and L looks like

1, 3, 4, 7, 9, 11.

Now L is sorted.

Notice that step 1 is the basis step of mathematical induction. In steps 3 and 4, the algorithm calls on itself.

The diagram below illustrates the action of the Quicksort algorithm on the list given in Example 2.

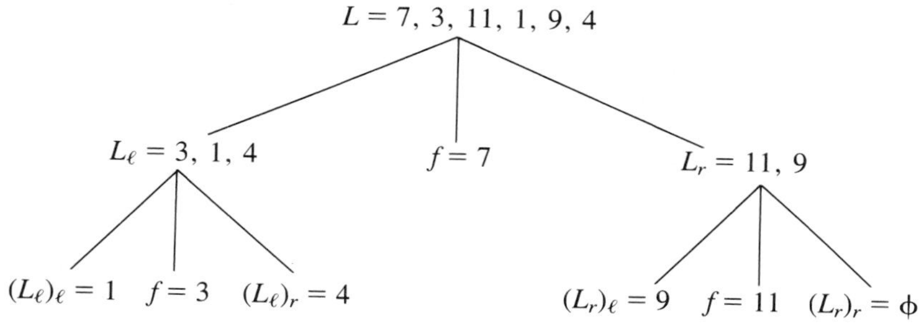

You can read the sorted list from the diagram. Start at the bottom left. Read the ends of the branches from left to right. If there are no branches down, move up and to the right for the next branch.

Based on the descriptions of these algorithms and the preceding examples, it is reasonable to believe that either of these algorithms will always succeed in arranging any list of n numbers in increasing order. Strong mathematical induction can be used to *prove* that the Quicksort algorithm does what it is supposed to do.

Quicksort Theorem

For each integer $n \geq 1$, the Quicksort algorithm arranges any list of n distinct real numbers in increasing order.

Proof Let $S(n)$ be the sentence: *The Quicksort algorithm arranges any list of n distinct real numbers in increasing order.*
The inductive step of the proof uses Steps 3 and 4 of the algorithm, which are executed only if $n \geq 2$. Therefore, the basis step must show that both $S(0)$ and $S(1)$ are true.

(1) Show $S(0)$ and $S(1)$ are true.
Any list containing no number or only one number is already arranged in increasing order and Quicksort leaves it alone. Therefore, $S(0)$ and $S(1)$ are true.

NOTES ON QUESTIONS

Questions 5, 6, 8, and 9: Require students to list all the intermediate steps when performing the given algorithms; otherwise, there is no way of knowing that the algorithm has been applied properly.

Question 8: Every time a number smaller than the first number in the list is encountered, exchange that number with the first number.

Question 16: You could repeat the instructions for equilateral triangles. Use the area formula $A = \frac{s^2\sqrt{3}}{4}$. Or you could repeat by focusing on the sum of perimeters rather than areas.

ADDITIONAL ANSWERS

5.

5, -7, 1.5, -1, 13, 6	initial order
-7, 1.5, -1, 5, 6, 13	after first pass
-7, -1, 1.5, 5, 6, 13	after second pass

6. See Additional Answers in the back of this book.

8.a.

7, 9, 4, 6, -4, 5, 0 initial order
4, 7, 9, 6, -4, 5, 0
-4, 4, 7, 9, 6, 5, 0

b. Apply the Filterdown algorithm to the list. The smallest number is now in front. Apply Filterdown to the sublist which includes all but the first number. Apply Filterdown to the sublists which are successively one element smaller. Continue until the list is exhausted.

13. a_1 and a_2 are even. Assume $a_1, a_2, \ldots, a_{k-1}, a_k$ are all even. Show that a_{k+1} is even. $a_{k+1} = c \cdot a_k + a_{k-1} = c(2p) + 2r = 2(cp + r)$ for some integers p and r. Since $cp + r$ is an integer by closure properties, a_{k+1} is even. Therefore, a_n is even for all integers $n \geq 1$ by the Strong Form of Mathematical Induction.

14. $S(1)$: 6 is a factor of $1^3 + 5 \cdot 1$. 6 is a factor of 6, so $S(1)$ is true. Assume $S(k)$: 6 is a factor of $k^3 + 5k$ is true. Show $S(k + 1)$: 6 is a factor of $(k + 1)^3 + 5(k + 1)$ is true. $(k + 1)^3 + 5(k + 1) = k^3 + 3k^2 + 3k + 1 + 5k + 5 = (k^3 + 5k) + 3k^2 + 3k + 6 = (k^3 + 5k) + 3k(k + 1) + 6$. 6 is a factor of $k^3 + 5k$ by the inductive assumption, a factor of $3k(k + 1)$ by the given hint, and a factor of 6. Hence, 6 is a factor of $(k^3 + 5k) + 3k(k + 1) + 6$ by the Factor of a Sum Theorem. By mathematical induction, 6 is a factor of $n^3 + 5n$ for all integers $n \geq 1$.

(2) Show that the assumption that $S(0), \ldots, S(k)$ are true implies that $S(k + 1)$ is true.
Assume that statements $S(0), \ldots, S(k)$ are true for some integer $k \geq 1$; that is, assume that the Quicksort algorithm arranges any list of up to k distinct numbers in increasing order. It must be shown that Quicksort can then arrange any list of $k + 1$ numbers in increasing order.
Let L be any list of $k + 1$ distinct real numbers. Since $k \geq 1$, Step 2 of the Quicksort algorithm applied to the list L forms 3 groups: the first number f of L, the left sublist L_ℓ, and the right sublist L_r. Because L_ℓ and L_r both contain k or fewer numbers, the inductive assumption implies that Steps 3 and 4 of the Quicksort algorithm will arrange them in increasing order. Once the numbers of L_ℓ are arranged in increasing order to the left of f and the numbers of L_r are arranged in increasing order to the right of f, the entire list L is in increasing order. Therefore, $S(k + 1)$ is true assuming that $S(0), S(1), \ldots, S(k)$ are true.

Hence, by strong mathematical induction, $S(n)$ is true for all integers $n \geq 0$.

Quicksort was invented in 1960 by C.A.R. Hoare. As you have seen throughout your study of mathematics and again saw in the case of sorting numbers into increasing order, more than one algorithm may be available to solve a given problem. The question arises as to which algorithm is best. This question is the subject of the next lesson.

Questions

Covering the Reading

In 1 and 2, the given algorithm is applied to the list 2, 5, 3, 8.
a. Determine the first two numbers compared.
b. Specify the action taken as a result of that comparison.

1. Bubblesort **a) 2 and 5; b) 2 and 5 are not exchanged.**

2. Quicksort **a) 2 and 5; b) 5 is placed in L_r**

3. Write the list 9, 1, 6, 4 after the first pass of Bubblesort. **1, 6, 4, 9**

4. Suppose Quicksort is applied to the list 9, 1, 6, 4. What are the results of Step 2 of Quicksort? **$f = 9$, $L_\ell = \{1, 6, 4\}$, $L_r = \theta$**

In 5 and 6, use the specified algorithm to arrange the list 5, -7, 1.5, -1, 13, 6 in increasing order. Show intermediate results of each step.

5. Bubblesort **See margin.**

6. Quicksort **See margin.**

Applying the Mathematics

7. Suppose that no interchanges are necessary during a given pass in the Bubblesort algorithm. Explain why the list at that pass must already be in increasing order. **If no interchanges are necessary, adjacent numbers are in order. Hence, by the Transitive Property, the entire list is in order.**

8. The following algorithm finds the smallest number in a list of n distinct numbers. Compare the first number in the list successively to the remaining numbers. When a number smaller than the first is encountered, place it in front of the rest of the list and continue making successive comparisons with it. The number that is in the first position in the list after the last comparison is the smallest number in the list. This is called the *Filterdown algorithm*.

a. Apply the Filterdown algorithm to the list 7, 9, 4, 6, -4, 5, 0, and write the intermediate list after each rearrangement of the list.

b. Construct an algorithm for arranging a list in increasing order that uses Filterdown as a ''subalgorithm.''

See margin.

9. Construct a list of 6 numbers for which the Bubblesort algorithm will require interchanges at each of the 5 passes. sample: 6, 5, 4, 3, 2, 1

In 10 and 11, when calculating efficiencies of sorting algorithms, the maximum number of *comparisons* necessary for a list of n numbers is usually considered. Based on this definition, determine which one of the Bubblesort and Quicksort algorithms is more efficient for sorting the indicated list.

10. the list of Example 1 Quicksort

11. the list of Questions 1 and 2 Quicksort

12. Fill in the blanks below to describe an algorithm which moves a tower of disks from needle A to needle B in the Tower of Hanoi puzzle. (You may want to refer back to Lesson 7-1.)

Tower of Hanoi Algorithm

Step 1: If the tower consists of 1 disk, __a.__ and stop. Otherwise, let k be the number of disks in the tower and continue with Steps 2–4. move the disk to needle *B*

Step 2: Use __b.__ to move the top $k - 1$ disks from needle A to needle C. the Tower of Hanoi Algorithm for $k - 1$ disks

Step 3: Move the remaining disk __c.__ . to needle *B*

Step 4: Use __d.__ to move $k - 1$ disks __e.__ . d) the Tower of Hanoi Algorithm for $k - 1$ disks; e) from needle *C* to needle *B*

Review

13. Let c be any integer and define a sequence recursively by

$$\begin{cases} a_1 = 2 \\ a_2 = 2 \\ a_{k+1} = ca_k + a_{k-1} \text{ for integers } k \geq 2. \end{cases}$$

Use strong mathematical induction to prove:

$\forall$ *integers* $n \geq 1$, a_n *is even*. *(Lesson 7-7)* See margin.

14. Prove that 6 is a factor of $n^3 + 5n$ for all integers $n \geq 1$. (Hint: In the inductive step, use the fact that 6 is a factor of $3n(n + 1)$ for all integers n.) *(Lesson 7-5)* See margin.

15. Prove: *For all integers* $n \geq 1$,

$$\frac{1}{1 \cdot 2} + \frac{1}{2 \cdot 3} + \frac{1}{3 \cdot 4} + \dots + \frac{1}{n(n+1)} = \frac{n}{n+1};$$

that is, $\displaystyle\sum_{i=1}^{n} \frac{1}{i(i+1)} = \frac{n}{n+1}$. *(Lesson 7-4)* See margin.

15. Prove S(n): $\displaystyle\sum_{i=1}^{n} \frac{1}{i(i+1)} = \frac{n}{n+1}$ **for all integers $n \geq 1$.**

S(1): $\displaystyle\sum_{i=1}^{1} \frac{1}{i(i+1)} = \frac{1}{1+1}$ **is true.**

Assume S(k): $\displaystyle\sum_{i=1}^{k} \frac{1}{i(i+1)} = \frac{k}{k+1}$ **is true. Show that**

S(k + 1): $\displaystyle\sum_{i=1}^{k+1} \frac{1}{i(i+1)} = \frac{k+1}{k+2}$ **is true.** $\displaystyle\sum_{i=1}^{k+1} \frac{1}{i(i+1)} =$

$\displaystyle\sum_{i=1}^{k} \frac{1}{i(i+1)} + \frac{1}{(k+1)(k+2)} = \frac{k}{k+1} + \frac{1}{(k+1)(k+2)} = \frac{k(k+2)}{(k+1)(k+2)} + \frac{1}{(k+1)(k+2)} = \frac{(k+1)^2}{(k+1)(k+2)} = \frac{k+1}{k+2}$. **So by mathematical induction, S(n) is true $\forall$ integers $n \geq 1$.**

NAME ______________________

LESSON **MASTER 7–8**
QUESTIONS ON **SPUR** OBJECTIVES

■ USES *Objective I (See pages 462–464 for objectives.)*
In 1–5, use the specified algorithm to arrange the given list in increasing order. Show all intermediate steps.

1. -7, 3, 0, 1 (Bubblesort)

1		3
0	⟶	1
3		0
-7		-7

2. 9, 4, -2, 10, 8 (Quicksort)

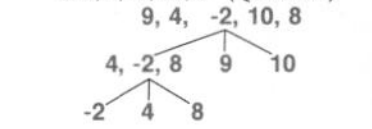

-2, 4, 8, 9, 10

3. 2, -3, .5, -7, -6 (Quicksort)

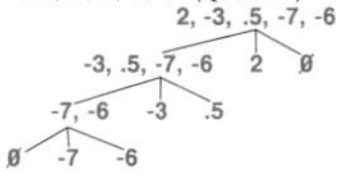

-7, -6, -3, .5, 2

4. 1, 3, 5, 7, 2, 4, 6 (Bubblesort)

6		7		7		7
4		6		6		6
2		4		5		5
7	⟶	2	⟶	4	⟶	4
5		5		2		3
3		3		3		2
1		1		1		1

5. -3, -2, -1, 7, 1, 2, 3 (Quicksort)

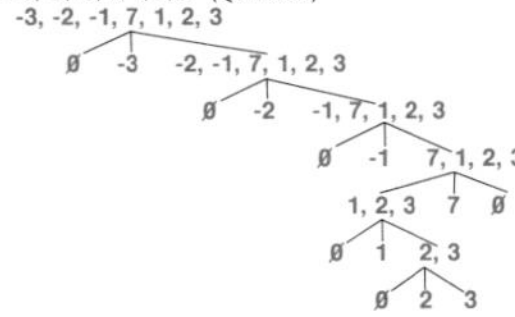

-3, -2, -1, 1, 2, 3, 7

6. For a list of four numbers, what is the maximum number of passes needed for the Bubblesort algorithm to arrange them in increasing order? 3

7. Is the Quicksort algorithm recursive or iterative? recursive

86

FOLLOW-UP

MORE PRACTICE
For more questions on SPUR Objectives, use *Lesson Master 7-8,* shown on page 451.

EXTENSION
Sorting is a major topic in computer science. Below are the names of a dozen sorting algorithms.

1. Insertion Sort
2. Selection
3. Exchange (Bubblesort) Sort
4. Tree Sort
5. Heapsort
6. Quicksort
7. Shellsort
8. Mergesort
9. Shakersort
10. Natural Mergesort
11. Balanced Mergesort
12. Polyphase Sort

Interested students or **small groups** of students could explore these algorithms as a project, comparing their efficiencies (see Lesson 7-9) and appropriateness of application.

ADDITIONAL ANSWERS

17.a. $R = \frac{R_1R_2}{R_1 + R_2}$

b. $\frac{70}{17} \approx 4.12$ **ohms**

19. The argument is valid. The three premises are: (1) $p \Rightarrow q$, (2) $q \Rightarrow r$, (3) $\sim r$. By the Transitive Property and statements (1) and (2), $p \Rightarrow r$. Then by modus tollens and (3), $\sim p$. This is the conclusion, and so the argument is valid.

16. The figure at the right is formed from an infinite sequence of squares where the side of each square is half as long as the side of the preceding square. Let A_k be the area of square k, and let $A_1 = 1$.

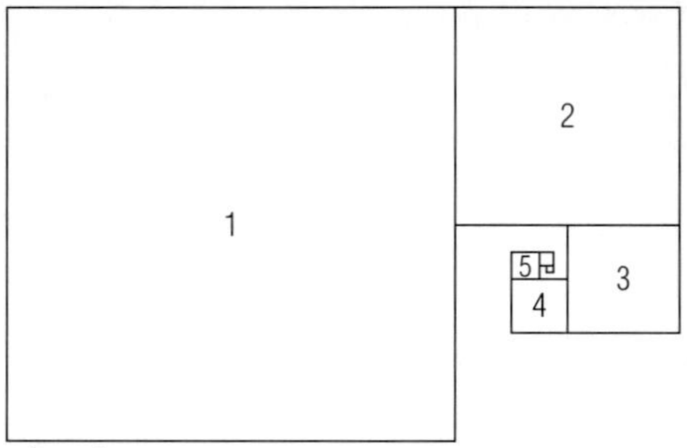

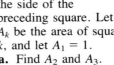

a. Find A_2 and A_3.
b. Write an explicit formula for A_k.
c. Find a formula for $\sum_{k=1}^{n} A_k$, the sum of the areas of the first n squares.
d. Find the total area of the figure. *(Lesson 7-6)* **c)** $\frac{4}{3}(1 - (\frac{1}{4})^n)$ **d)** $\frac{4}{3}$
a, b) See below.

17. The equivalent resistance R of two resistors R_1 and R_2 connected in parallel is given by the equation

$$\frac{1}{R} = \frac{1}{R_1} + \frac{1}{R_2}.$$

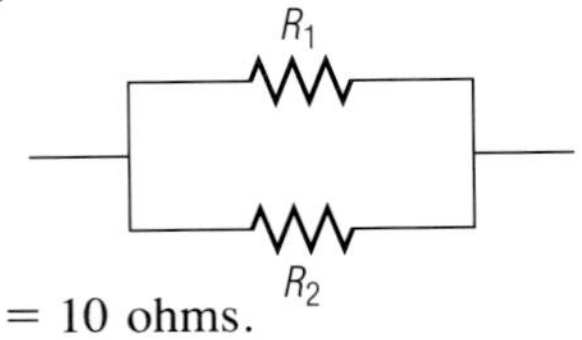

a. Solve this equation for R.
b. Find R if $R_1 = 7$ ohms and $R_2 = 10$ ohms.
(Lesson 5-7) **See margin.**

18. Find integers q and r such that $57 = 4q + r$ and $0 \le r < 4$. *(Lesson 4-2)*
$q = 14, r = 1$

19. Determine whether the argument below is valid or invalid. Justify your reasoning.
If it is Thanksgiving, then dinner includes turkey.
If dinner includes turkey, then dinner includes potatoes.
Dinner does not include potatoes.
$\therefore$ *It is not Thanksgiving.* *(Lessons 1-8, 1-6)* **See margin.**

Exploration

20. Consider a list of 5 letters: a, b, c, d, e. Arrange these letters in an order that requires 4 passes of the Bubblesort algorithm to rearrange them in alphabetical order. How many such orderings are there? **e, d, c, b, a; there are 24 such orderings. The only restriction is that a must be the last letter in the list.**

16. a) $A_2 = \frac{1}{4}$, $A_3 = \frac{1}{16}$
b) $A_k = \frac{1}{4^{k-1}}$, for $k \ge 1$

LESSON

7-9

Efficiency of Algorithms

As was mentioned in the last lesson, several different algorithms may be available to solve a given problem. How do you choose among these algorithms? Sometimes the best choice might be the simplest one to do by hand or the simplest one to program on a computer. This may be the case for ''small'' problems that do not require a large number of calculations or comparisons. However, ''large'' problems might force you to consider other features of the algorithm, such as how much computer time or computer memory it would require. Such features are measures of the *efficiency* of an algorithm.

In this lesson, the **efficiency** of an algorithm is the maximum number $E(n)$ of *significant operations* necessary for the algorithm to solve the given problem if the problem is of *size n*. This is not a precise definition of efficiency until size and significant operations are defined for the given problem. The two ideas of *size* and *significant operations* are usually easy to identify in specific problems. For instance, consider the familiar algorithm for adding two multi-digit positive integers.

''...ere's a throwback for you. He ...cks all his answers with pencil and ...er.''

A reasonable way to measure the size of an addition problem would be to use the maximum number of digits in either of the two positive integers that are added. Using this measure, addition of 372 and 457, for example, is a problem of size 3. Now apply the algorithm.

Given Problem		Step 1		Step 2		Step 3
				1		
372		372		372		372
+457	→	457	→	457	→	457
		9		29		829

Each step of the addition algorithm involves the addition of single-digit numbers to produce a digit in the sum, and sometimes a ''carry'' digit of 1. It is reasonable to regard these single-digit additions as the significant operations for this algorithm and the ''carries'' as insignificant because they are by-products of the additions.

With these meanings of size and significant operations, for the addition algorithm,

$$E(3) = 3$$

because at most 3 single-digit additions are required to add two 3-digit numbers.

Algorithms can differ greatly in their efficiency. Consider now the problem of computing x^n for some real number x and some positive integer n. Compare these two algorithms.

First Algorithm for computing x^n: repeated multiplication. Multiply x repeatedly $n - 1$ times. That is, x^n is the nth term of the sequence defined recursively by

$$a_1 = x \text{ and } a_{k+1} = x \cdot a_k \text{ for } k \geq 1.$$

LESSON 7-9

TEACHING NOTES

Efficiency of algorithms is an area of interest to any user of computer time. Cost requirements demand that the most efficient algorithm be used if all other factors are equal.

For the two algorithms for calculating powers described on pages 453–454, students can get a concrete feel for the efficiency of the algorithms by performing calculations on their calculators. Have students compute 3^{12} by multiplying by 3 repeatedly. Then have students compute 3^2, 3^4, and 3^8 by squaring. Make sure students understand that 3^4 is obtained by squaring 3^2 and that 3^8 is obtained by squaring 3^4. Now 3^{12} can be computed as $3^4 \cdot 3^8$. Students should keep track of the number of computations needed in each case. Such an experiment may help them understand the importance of efficiency in dealing with extremely large powers.

Making Connections
Your computer science students may have studied the measure of efficiency of algorithms in computer science classes. A function with the same overall end behavior as the function E used in this lesson is called the "Big O" (the letter O), for Order of efficiency. For instance, because the end behavior of Bubblesort is like that of a quadratic polynomial, the efficiency of Bubblesort is said to be of the order n^2, written as $O(n^2)$. For Quicksort, the efficiency is of the order $O(n \log_2 n)$. You can use

these values to have students determine the respective time multiplier, if the number of items to be sorted is doubled for each.
For Bubblesort, $E(2n) \approx O((2n)^2) = 4n^2 \approx 4(E(n))$.
For Quicksort, $E(2n) \approx O(2n(\log_2 (2n))) = O(2n + 2n \log_2 n) \approx 2n + 2E(n)$. Thus, for Bubblesort, the needed time is quadrupled. For Quicksort, it is slightly more than doubled.

NOTES ON QUESTIONS
Questions 1-4: For each question, you may wish to discuss with students the best-case and worst-case scenarios for calculating the efficiency of an algorithm.
Question 1: The efficiency of the algorithm is the same regardless of the given numbers. **Question 2:** The efficiency is best if there are no carries; worst if there is a carry in every digit. **Questions 3 and 4:** The efficiency $E(n) = n - 1$ for all n in the repeated multiplication algorithm; for the sums of powers, $E(n)$ is best if n is itself a power of 2.

Question 5: Compare the efficiency of the Bubblesort alogrithm for answering this question to the efficiency of Quicksort. As the number of items to be sorted increases tenfold repeatedly, ask students how the efficiency of Bubblesort compares to that of Quicksort? (Bubblesort's efficiency is multiplied by 100; Quicksort's by about 33.)

This algorithm requires $n - 1$ multiplications. For instance, to calculate 7^{32}, it would take 31 multiplications:

$$7 \cdot 7 = 7^2;\ 7 \cdot 7^2 = 7^3;\ 7 \cdot 7^3 = 7^4; \text{ and so on.}$$

If the size of the problem is taken to be n and the significant operations are the required multiplications, then the efficiency of this algorithm is

$$E(n) = n - 1$$

because it requires $n - 1$ multiplications.

The second algorithm is more difficult to explain but it is surprisingly efficient. It is based on the fact that certain powers can be calculated quickly. With it, 7^{32} can be calculated in only five multiplications:

$$7 \cdot 7 = 7^2;\ 7^2 \cdot 7^2 = 7^4;\ 7^4 \cdot 7^4 = 7^8;\ 7^8 \cdot 7^8 = 7^{16};\ 7^{16} \cdot 7^{16} = 7^{32}.$$

Second Algorithm for computing x^n: writing n as a sum of powers of 2. Write n as a sum of powers of 2. (This is equivalent to writing n in base 2.) Calculate x^n for those powers by repeatedly squaring. Then multiply the powers.

For example, to calculate x^{43}, note that $43 = 101011_2 = 2^5 + 2^3 + 2^1 + 2^0 = 32 + 8 + 2 + 1$. So

$$x^{43} = x^{32} \cdot x^8 \cdot x^2 \cdot x.$$

The computation of x^{43} by this algorithm requires 5 multiplications to produce

$$x^2, x^4, x^8, x^{16}, x^{32}$$

and then 3 more to combine powers for a total of 8 multiplications.

Notice that computing x^n for any integer n between $2^5 = 32$ and $2^6 - 1 = 63$ would require the 5 multiplications necessary to produce

$$x^2, x^4, x^8, x^{16}, x^{32}$$

and then at most 5 additional multiplications to form x^n.

Of course, the second algorithm is more complicated to describe than the first. However, it is more efficient. The highest power of 2 needed is 2^p, where

$$2^p \leq n < 2^{p+1}.$$

To solve for p, take the logarithm to the base 2 of each part:

$$p \leq \log_2 n < p + 1.$$

Therefore, the exponent p is the greatest integer less than or equal to $\log_2 n$; that is,

$$p = \lfloor \log_2 n \rfloor.$$

Once p has been determined, p multiplications are needed to compute

$$x^2, x^4, x^8, \ldots, x^{(2p)}$$

and then at most p additional multiplications to compute x^n. Therefore, because the multiplications are taken as the only significant operations, the efficiency of the second algorithm is

$$\begin{aligned} E(n) &= p + p \\ &= 2p \\ &= 2 \lfloor \log_2 n \rfloor. \end{aligned}$$

For large values of n, $E(n)$ for the second algorithm is much smaller than $E(n)$ for the first; the second algorithm is much more efficient.

For example, for x^{1000}, the first algorithm requires 999 multiplications while the second requires at most

$$2\lfloor \log_2 1000 \rfloor = 2 \cdot 9 = 18$$

multiplications. This is the sense in which the second algorithm is more efficient than the first.

For the Bubblesort and Quicksort algorithms considered in the last lesson, a natural interpretation for the size n of a problem is the number n of items in the list. The significant operations are the comparisons of numbers in the list. We will define the efficiency $E(n)$ of these algorithms to be the number of comparisons necessary to arrange the list.

For Bubblesort, a formula for $E(n)$ is not difficult to derive. Recall that to sort a list of n real numbers, Bubblesort makes (at most) $n - 1$ passes through the list. In the first pass, it makes $n - 1$ comparisons; in the second pass, it makes $n - 2$ comparisons; and so on to the $(n - 1)$st pass, in which it makes only one comparison. Thus the total number of comparisons it makes is

$$E(n) = \underset{\text{Pass 1}}{(n-1)} + \underset{\text{Pass 2}}{(n-2)} + \ldots + \underset{\text{Pass }(n-2)}{2} + \underset{\text{Pass }(n-1)}{1}.$$

Thus, $E(n)$ is the sum of the first $n - 1$ natural numbers. From Lesson 7-4, this sum is known to be $\frac{(n-1)[(n-1)+1]}{2}$, which is $\frac{n(n-1)}{2}$.

For Quicksort, the formula for $E(n)$ is more difficult to derive. It can be shown that the efficiency of the Quicksort algorithm is approximately $n(\log_2 n - 2)$, which is significantly better than that of Bubblesort. The efficiency of algorithms translates into the computer time required for execution of the algorithm. Suppose that a computer is capable of executing one million significant operations per second. The following table lists approximate execution times for the various efficiency functions $E(n)$ seen in this lesson and for various values of n.

Algorithm	$E(n)$	$n = 100$	$n = 1000$	$n = 10{,}000$	$n = 1{,}000{,}000$
x^n repeated multiplication	$n - 1$	.0001 sec	.001 sec	.01 sec	1 sec
x^n n as sum of powers of 2	$2\lfloor \log_2 n \rfloor$	.000012 sec	.000018 sec	.000026 sec	.000038 sec
Bubblesort n numbers	$\frac{n(n-1)}{2}$	.005 sec	.5 sec	50 sec	139 hours!
Quicksort n numbers	$n(\log_2 n - 2)$	.0005 sec	.008 sec	.11 sec	18 seconds

Since computers often have to calculate large powers and sort lengthy lists, improved algorithms are valuable—they save time and thereby save money.

Questions

Covering the Reading

In 1 and 2, refer to the addition of 372 and 457, as shown in the lesson.

1. Using the same meanings for size and significant operations, what is the efficiency of the addition algorithm for adding two positive integers with at most n digits each? $E(n) = n$

Question 10b: Ask students to generalize their indicated sum to any finite arithmetic sequence a_n. Then ask them to derive an explicit formula for $\sum_{i=1}^{n} [a_1 + (i - 1)d]$.

Question 16: Selection sort is an iterative algorithm designed to find the kth smallest number of a set of numbers. Selection sort works by finding the smallest element, then the second smallest element (by finding the smallest of the elements remaining after the smallest is omitted), and so forth. One special case of selection sort occurs in the determination of the median of a set of numbers.

Merge sort is a recursive algorithm which works as follows. The first half of the list of numbers is sorted using Merge sort, then the second half is sorted. Finally, the two halves are merged into a single sorted list.

ADDITIONAL ANSWERS
3.b. repeated multiplication: 9999; sum of powers of 2: 14

7. See Additional Answers in the back of this book.

8. Let $S(n)$: $a_n = 3n^2 + 5n - 3$. For $n = 1$, the formula yields $a_1 = 3 \cdot 1^2 + 5 \cdot 1 - 3 = 5$. This agrees with the initial condition, so $S(1)$ is true. Assume $S(k)$: $a_k = 3k^2 + 5k - 3$ is true for some positive integer k. Show that $S(k+1)$: $a_{k+1} = 3(k+1)^2 + 5(k+1) - 3$ is true. $a_{k+1} = 3(k+1)^2 + 5(k+1) - 3 = (3k^2 + 5k - 3) + 6k + 8 = a_k + 6k + 8$. This agrees with the recursive formula, so $S(k+1)$ is true. Hence, by mathematical induction, $S(n)$ is true $\forall$ integers $n \geq 1$, and so the explicit formula for the sequence is $3n^2 + 5n - 3$.

**9. $S(n)$: $5^n - 4n - 1$ is divisible by 16. $S(1)$: $5^1 - 4 \cdot 1 - 1$ is divisible by 16. $S(1)$ is true since 0 is divisible by 16. Assume $S(k)$: $5^k - 4k - 1$ is divisible by 16 is true for some positive integer k. Show that 16 is a factor of $5^{k+1} - 4(k+1) - 1$.
$5^{k+1} - 4(k+1) - 1$
$= 5(5^k - 4k - 1 + 4k + 1) - 4(k+1) - 1$
$= 5(5^k - 4k - 1) + 20k + 5 - 4(k+1) - 1$
$= 5(5^k - 4k - 1) + 16k$
16 is a factor of $5^k - 4k - 1$ (by the inductive assumption) and a factor of $16k$. Hence, by the Factor of a Sum Theorem, 16 is also a factor of $5(5^k - 4k - 1) + 16k$. Therefore, by mathematical induction, $5^n - 4n - 1$ is divisible by 16 for all positive integers n.**

2. If each carry is also considered to be a significant operation, compute the efficiency. $E(n) = 2n - 1$

3. Compare the efficiencies of the two algorithms of this lesson for calculating
a. x^{10} **b.** x^{10000}. See margin.
a) repeated multiplication: 9; sum of powers of 2: 4

4. The second algorithm discussed in this lesson for computing x^n requires successive comparisons of 2, 2^2, 2^3, etc., with n until a positive p is found so that
$$2^p \leq n < 2^{p+1}.$$
a. How many comparisons are needed for this? p
b. If these comparisons are also regarded as significant operations for the algorithm, compute the efficiency $E(n)$ of the second algorithm. $3\lfloor \log_2 n \rfloor$

5. What is the efficiency of the Bubblesort algorithm if it is used to arrange 1000 numbers in increasing order? $E(10) = 499{,}500$

6. Suppose the efficiency of an algorithm is given by the formula
$$E(n) = n^3.$$
If a computer can execute one million significant operations per second, calculate the execution time for
a. $n = 100$ 1 second **b.** $n = 1000$ **c.** $n = 1{,}000{,}000$.
b) 1000 seconds or ≈ 16.67 minutes; c) 10^{12} seconds or ≈ 31,710 years

Review

7. Apply the Quicksort algorithm to arrange the following list in increasing order: 7, -3, 2, -6, 10, 5. Show the intermediate results of each step. *(Lesson 7-8)* See margin.

8. A sequence is defined recursively as follows:
$$\begin{cases} a_1 = 5 \\ a_{k+1} = a_k + 6k + 8 \quad \forall k \geq 1. \end{cases}$$
Show that an explicit formula for the sequence is $a_n = 3n^2 + 5n - 3$. *(Lessons 7-4, 7-2)* See margin.

9. Use mathematical induction to show that 16 is a factor of $5^n - 4n - 1$ for all integers $n \geq 1$. (Hint: In the inductive step, use the fact: $5^{k+1} = 5(5^k - 4k - 1 + 4k + 1) = 5(5^k - 4k - 1) + 20k + 5$.) *(Lesson 7-5)* See margin.

10. In an auditorium, each row of seats contains 5 more seats than the row in front of it. The first row has 34 seats and there are 27 rows.
a. If a_n is the number of seats in row n, write an explicit formula for a_n. $a_n = 34 + 5(n - 1)$
b. Use summation notation to describe the total number of seats in the auditorium. *(Lessons 7-3, 7-2)* $\sum_{n=1}^{27} (34 + 5(n - 1))$

11. If the a_i, b_i, and c are real numbers, show that
$\sum_{i=0}^{4} (a_i + cb_i) = \sum_{i=0}^{4} a_i + c\sum_{i=0}^{4} b_i$ *(Lesson 7-3)* See margin.

12. Consider the function h defined by $h(x) = \frac{5x + 3}{x - 4}$.

a. Use division to express this function rule in another form.

b. Identify any horizontal or oblique asymptotes to the graph and give the appropriate equations. $y = 5$

c. Give the equation of any vertical asymptote. $x = 4$

d. Sketch a graph of the function. *(Lessons 5-5, 5-4)* See margin.

a) $h(x) = 5 + \frac{23}{x - 4}$

13. a. Write a logical expression to describe the network below.

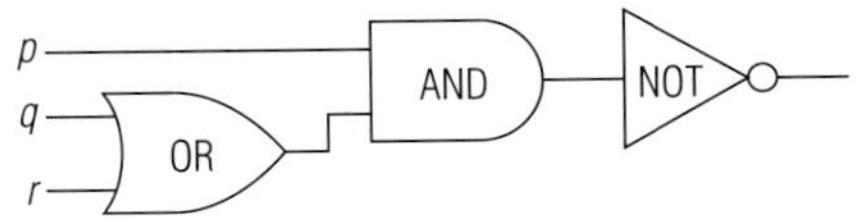

b. Give a set of input values for p, q, and r which will yield an output value of 1. *(Lesson 1-4)* sample: $p = 0$, $q = 1$, $r = 0$

a) not(p and (q or r))

14. Compute. *(Previous Course)*

a. $(6 - 4i) + (-5 + i)$ $1 - 3i$ **b.** $(6 - 4i) - (-5 + i)$ $11 - 5i$ **c.** $(6 - 4i) \cdot (-5 + i)$ $-26 + 26i$

Exploration

15. An algorithm for multiplying two positive integers with at most n digits each is illustrated in the following diagram for $n = 2$.

			2					
34	34	34	34	34	34	34	34	34
×52→	52→	52→	52→	52→	52→	52→	52→	52
	8	68	68	68	68	68	68	68
			0	170	170	170	170	170
					8	68	768	1768

a. Take the size of a given multiplication to be the maximum number of digits in either of the two numbers being multiplied. What is the size of the problem in the illustration above? 2

b. What is the size of the problem if 48 is multiplied by 186? 3

c. If single-digit multiplications are taken to be significant operations, show that there are at most n^2 multiplications. See margin.

d. Also take the single-digit additions of the partial products to be significant operations. You must determine the maximum number of columns that must be added. The first partial product can have at most $n + 1$ digits so there are at most $n + 1$ columns. Each row below it can have at most 1 additional column of digits. Use this information to determine the maximum number of single-digit additions. $2n$ column additions

e. Use parts **c** and **d** to determine a formula for the efficiency of the algorithm. Verify that the formula works for the problem above. See margin.

16. Two other sorting algorithms are Selection and Merge. Look up Selection sort and Merge sort in a computer science book and find formulas for their efficiency $E(n)$. Compare their efficiency with that of Bubblesort and Quicksort for various values of n. See margin.

11. $\sum_{i=0}^{4} (a_i + cb_i) = a_0 + cb_0 + a_1 + cb_1 + \ldots + a_4 + cb_4 = a_0 + a_1 + \ldots + a_4 + cb_0 + cb_1 + \ldots + cb_4 = (a_0 + a_1 + \ldots + a_4) + c(b_0 + b_1 + \ldots + b_4) = \sum_{i=0}^{4} a_i + c\sum_{i=0}^{4} b_i$

12.d.

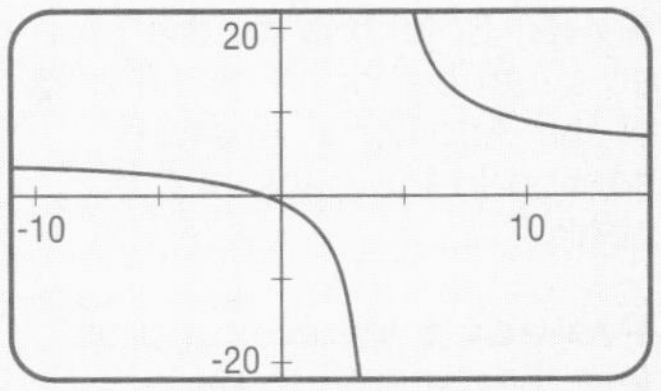

15.c. Each of the n digits of the second number are multiplied by up to n digits of the first number. Hence, there are at most n^2 multiplications.

e. $n^2 + 2n$

For the problem, $n = 2$, so the efficiency should be 8. All four multiplication steps shown, must be carried out, as well as four column additions. 4 + 4 = 8, so the algorithm checks.

16. See Additional Answers in the back of this book.

Projects

PROJECTS

The projects provide an opportunity for extended work on topics related to the content of the chapter. (See pages 70 and T37 for more information on the purposes of projects.) Some or all of these projects may be suitable for **small group work**. Solutions for the projects can be found at the end of the Additional Answers section in the back of this book.

Project 1 involves a problem that seems to have started at MIT in the 1960s. After a while, regardless of the *N* you begin with, it seems that the sequence ends in 1 (which itself generates the sequence 4, 2, 1), but no one has proved this. The number of iterations that it takes for a sequence to get to 1 has been studied and forms some nice patterns.

Projects 2 and 4 both relate to the Fibonacci sequence and various properties of the terms of that sequence.

1. A famous unsolved problem can be explored using the computer program shown below. Let N be a positive integer.

```
10 INPUT N
15 PRINT "START WITH"; N
20 LET X = N
25 IF X = 1 THEN END
30 IF X/2 = INT(X/2) THEN GOTO 50
35 X = 3 * X + 1
40 PRINT X
45 GOTO 25
50 X = X/2
55 PRINT X
60 GOTO 25
```

a. Describe what this program does in general.

b. Investigate the outputs of this program for a variety of values of N.

As of 1990, no one knew whether this program always generates the final value of X that you found or whether it takes on new values indefinitely for some N or whether it returns to some previously taken value and then loops.

"This must be Fibonacci's."

2. Below are some interesting sums involving the terms of the Fibonacci sequence. It would be helpful to have a list of the first 12 terms of the sequence which, as you recall, is defined as

$$\begin{cases} F_1 = 1 \\ F_2 = 1 \\ F_{k+1} = F_k + F_{k-1} \text{ for integers } k \geq 2. \end{cases}$$

In **a–c**, **i.** fill in the blanks;
ii. based on the results, form a conjecture, $S(n)$, about the sum in the nth row;
iii. show that $S(n)$ is true for some n other than those shown;
iv. use the Principle of Mathematical Induction to prove that $S(n)$ is true for all natural numbers.

a.

$$\begin{aligned} 1^2 &= 1 \cdot 1 \\ 1^2 + 1^2 &= 1 \cdot 2 \\ 1^2 + 1^2 + 2^2 &= 2 \cdot 3 \\ 1^2 + 1^2 + 2^2 + 3^2 &= 3 \cdot \underline{\ ?\ } \\ 1^2 + 1^2 + 2^2 + 3^2 + 5^2 &= \underline{\ ?\ } \\ 1^2 + 1^2 + 2^2 + 3^2 + 5^2 + 8^2 &= \underline{\ ?\ } \end{aligned}$$

b.

$$\begin{aligned} 1 + 1 &= 2 \\ 1 + 1 + 2 &= 4 \\ 1 + 1 + 2 + 3 &= \underline{\ ?\ } \\ 1 + 1 + 2 + 3 + 5 &= \underline{\ ?\ } \\ 1 + 1 + 2 + 3 + 5 + 8 &= \underline{\ ?\ } \end{aligned}$$

c.

$$\begin{aligned} 1^2 + 1^2 &= 2 \\ 1^2 + 2^2 &= \underline{\ ?\ } \\ 2^2 + 3^2 &= \underline{\ ?\ } \\ 3^2 + 5^2 &= \underline{\ ?\ } \\ 5^2 + 8^2 &= \underline{\ ?\ } \end{aligned}$$

d. Investigate some sums of your own choosing. Prove that any patterns you find are valid for all natural numbers.

458

3. A square has sides of length s. An infinite sequence of squares is formed by placing the vertices of each one at the midpoints of the sides of the previous one.

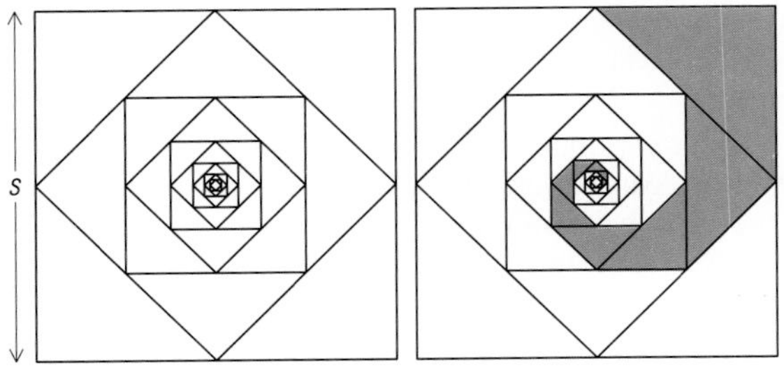

a. Find the sum of the perimeters of the squares, if possible.
b. Find the area of the shaded region (shown above at the right), if possible.
c. Form a similar sequence using some other regular polygon, and try to repeat parts **a** and **b**.

4. Let $M = \begin{bmatrix} 1 & 1 \\ 1 & 0 \end{bmatrix}$.
a. Calculate M^2, M^3, M^4, M^5, and M^6.
b. Make a conjecture which relates M^n to the Fibonacci sequence for integers $n \geq 2$.
c. Prove your conjecture by induction.
d. Find the first six terms of a sequence that is analogously related to the powers of the matrix $\begin{bmatrix} 2 & 1 \\ 1 & 0 \end{bmatrix}$.
e. Find a recursive definition for the sequence you found in part **d.**
f. Generalize by finding a recursive definition for the sequence related to the powers of $\begin{bmatrix} a & 1 \\ 1 & 0 \end{bmatrix}$, where a is any real number.

5. Suppose a car is purchased with \$5,000 borrowed at an annual interest rate of 12% (that is, a monthly rate of 1%). The loan is paid off in monthly installments of \$166 each. In other words, each month after the car is purchased, the amount owed (balance) is equal to the balance in the previous month, plus 1% of that balance in interest, minus the payment of \$166. Let A_n be the amount owed n months after the car is purchased.
a. Calculate A_1, A_2, and A_3.
b. Write a recursive definition for the sequence A.
c. Show that
$A_3 = 1.01^3 \cdot 5000 - 1.01^2 \cdot 166 - 1.01 \cdot 166 - 166.$
(Hint: Find a numeric expression for A_1, but do not calculate its value. Then substitute it into the recurrence relation to find an expression for A_2. Repeat this one more time to find an expression for A_3.)
d. Show that $A_3 =$
$5000 \cdot 1.01^3 - 166(1.01^0 + 1.01^1 + 1.01^2)$.
e. Conjecture a similar expression for A_n.
f. Use your knowledge of geometric series to rewrite the expression in parentheses in part **e** to obtain an explicit formula for A_n.
g. Use your formula from part **f** to find the amount owed on the car after 3 years.
h. Generalize the above by using the same steps to find an explicit formula for the balance after n months when \$$B$ is borrowed at a monthly interest rate of r.
i. Suppose you know B and r, and you wish to calculate the amount p of the monthly payment so that the loan will be paid off after n months. (This is called *amortizing the loan over n months.*) To find an equation for p in terms of B, r, and n, set the explicit formula equal to 0, and solve for p.
j. Estimate how much money you would need to borrow to buy a car of your choosing, and find out the typical interest rate for a car loan. Use the formula from part **i** to calculate your monthly payment in order to pay off the loan in 3 years. How much would you pay altogether over the 3 years? How much more than the actual amount borrowed would this be?

Project 3 is a kind of construction once considered just a game, but now is used in forming fractals.

Project 5 applies geometric sequences to the study of annuities.

SUMMARY

The Summary gives an overview of the entire chapter and provides an opportunity for students to consider the material as a whole. Thus, the Summary can be used to help students relate the various concepts presented in the chapter.

VOCABULARY

Terms, symbols, and properties are listed by lesson to provide a checklist of concepts a student must know. (See page 71 for more information.)

ADDITIONAL ANSWERS

2. $r_1 = 5$, $r_k = 3r_{k-1}$ for $k \geq 2$

4. $\sum_{i=1}^{k+1} i^2 = \left(\sum_{i=1}^{k} i^2\right) + (k+1)^2$

5. Let $S(n)$: $c_n = 3 \cdot 2^n - 3$ for integers $n \geq 1$. $S(1)$: $c_1 = 3 \cdot 2^1 - 3$. This matches the recursive definition since $c_1 = 3$, so $S(1)$ is true. Assume $S(k)$: $c_k = 3 \cdot 2^k - 3$ is true for an arbitrary integer k. Show that $S(k+1)$: $c_{k+1} = 3 \cdot 2^{k+1} - 3$ is true. From the recursive definition, $c_{k+1} = 2c_k + 3 = 2(3 \cdot 2^k) - 3 = 3 \cdot 2^{k+1} - 3$. Therefore, $S(k+1)$ is true. Hence, by mathematical induction, $S(n)$ is true for all $n \geq 1$, and so the explicit formula $c_n = 3 \cdot 2^n - 3$ yields the correct definition of the sequence.

7.b. $b_1 = 1$; $b_{j+1} = 2b_j + 5$ for $j \geq 1$

Summary

A recursive definition for a sequence consists of an initial condition or conditions and a recurrence relation that describes how to obtain a term of the sequence from one or more preceding terms. After using the recursive definition to calculate the first several terms of a sequence, one may see a pattern for an explicit formula for the sequence. If correct, the explicit formula can be proved true by showing that it satisfies the recursive definition and satisfies the conditions of the Recursion Principle.

A series is an indicated sum of consecutive terms of a sequence. Summation notation is a shorthand way of indicating the sum. $\sum_{i=m}^{n} a_i$ means $a_m + a_{m+1} + \ldots + a_{n-1} + a_n$. For a sequence $a_1, a_2, a_3, \ldots$, the nth partial sum S_n is defined by $S_n = \sum_{i=1}^{n} a_i$. If the sequence of partial sums $S_1, S_2, S_3, \ldots$ has a limit, then the value of the infinite series $\sum_{i=1}^{\infty} a_i$ is defined to be $\lim_{n \to +\infty} S_n$. Infinite series help to resolve one of Zeno's Paradoxes. The value of an infinite geometric series with first term a and common ratio r is $\frac{a}{1-r}$ for $0 < |r| < 1$.

An important proof technique, mathematical induction, can be used to show that some statements $S(n)$ are true for all positive integers n. To use induction, you first show that $S(1)$ is true. Then you show that if the inductive assumption $S(k)$ is true, then $S(k+1)$ is also true. The Strong Form of Mathematical Induction is a variant of this technique; its inductive assumption is that $S(1)$, $S(2)$, ..., $S(k)$ are all true. The two forms of the Principle of Mathematical Induction can be used to prove theorems about sums, factors, recursively-defined sequences, and algorithms.

Some computer algorithms are iterative; they repeat the same procedure again and again to solve a problem. Others are recursive; they solve a problem by reducing it to smaller and smaller versions of itself. Two algorithms for sorting lists are Bubblesort, which is iterative, and Quicksort, which is recursive. Quicksort is the more efficient algorithm.

Vocabulary

For the starred (*) terms you should be able to give a definition of the term.
For the other terms you should be able to give a general description and a specific example of each.

Lesson 7-1
Tower of Hanoi problem
Tower of Hanoi sequence
*recursive definition for a sequence, initial condition(s), recurrence relation
*Fibonacci sequence, Fibonacci numbers
Recursion Principle

Lesson 7-3
Zeno's Paradox
*summation notation, $\sum_{i=m}^{n} a_i$,
index, expanded form

Lesson 7-4
*Principle of Mathematical Induction
*basis step, inductive step
*inductive assumption
sum of the first n positive integers

Lesson 7-6
series, finite series, infinite series
*geometric series
sum of the first n powers
value of a finite geometric series
partial sum(s)
*infinite sum, $\sum_{k=1}^{\infty} a_k$
value of an infinite geometric series
convergent, divergent

Lesson 7-7
*Principle of Strong Mathematical Induction

Lesson 7-8
sorting algorithm
Bubblesort, Quicksort
iterate
iterative algorithm
recursive algorithm

Lesson 7-9
size of a problem
significant operations
efficiency of an algorithm

460

Polar Coordinates and Complex Numbers

CHAPTER 8

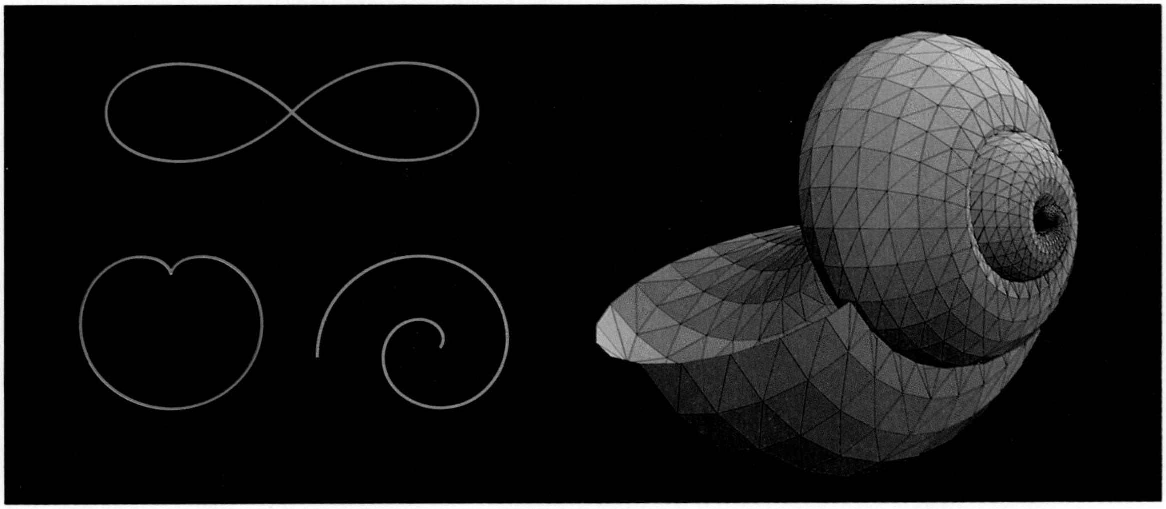

Your first use of complex numbers such as i and $2 + 3i$, and the use that came first historically, was probably as solutions to equations like $x^2 + 6 = 0$ that have no real solutions. Mathematicians of the 16th, 17th, and 18th centuries knew of no other uses for complex numbers and were uncomfortable with these objects. Euler's discovery of the relationship $e^{i\pi} + 1 = 0$, and Gauss's proof in 1797 of the Fundamental Theorem of Algebra, made mathematicians realize that complex numbers play a unique role in mathematics. Casper Wessel in that same year, and Jean Robert Argand in 1806, showed that complex numbers can be associated with points in the plane. This led to using complex numbers to represent geometric and physical objects such as a chambered nautilus and quantities such as forces, velocities, and flows.

Complex numbers today have a variety of uses, all stemming from their algebraic origins and their geometric interpretations. They are associated with transformations, they can be used to describe some beautiful geometric figures, and they can have the properties of vectors. They have become important in many areas of mathematics and science, including electronics and electrical, hydraulic, and aeronautical engineering. A new area of mathematics called *dynamical systems* has arisen since 1900 and, within the past 15 years, the words *chaos* and *fractal* have entered the mathematician's vocabulary, producing new applications for complex numbers.

In this chapter, two graphical representations of complex numbers are used: rectangular coordinates and polar coordinates. Polar coordinates, however, have applications far beyond complex numbers. Some plane curves such as those pictured above have very complicated rectangular coordinate equations but rather simple polar coordinate equations.

CHAPTER 8

We recommend 13–16 days for this chapter. 10 to 12 days on the lessons and quizzes, 1 day for the Progress Self-Test, 1 or 2 days for the Chapter Review, and 1 day for a Chapter test. An extra day may be needed for a quiz, for work in a computer lab, or for extended discussion of problems. (See the Daily Pacing Chart on page 465D.)

USING PAGE 465

Discuss Euler's Theorem that $e^{i\pi} + 1 = 0$. Point out that it contains the most important constants of mathematics (e, i, π, 0, and 1), and operations of additon, multiplication, and exponentation. Rewritten as $e^{i\pi} = -1$, we see the astounding fact that a transcendental number (e) raised to an imaginary power ($i\pi$) equals a real number (-1).

There are many wonderful theorems in this chapter that connect algebra and geometry.

LESSON 8-1

RESOURCES
- Lesson Master 8-1

OBJECTIVES

A Express complex numbers in $a + bi$ and rectangular form.
B Perform operations with complex numbers.
F Prove or verify properties of complex numbers.
H Use complex numbers to solve AC circuit problems.
I Graph complex numbers.

TEACHING NOTES

Some students think the imaginary part of the complex numbers $a + bi$ is bi. This would cause difficulty for interpreting the imaginary part of (a, b).

Emphasize to students that when working with radicals, negative signs must be removed from the radicands by writing the expressions in terms of $\sqrt{-1}$. This should be done before performing any manipulations, as **Example 1** illustrates.

Alternate Approach
After defining i, students can work **Example 1** in the following way: $\sqrt{-25} = \sqrt{25}\sqrt{-1} = 5i$ and $\sqrt{-16} = \sqrt{16}\sqrt{-1} = 4i$. Then $\sqrt{-25} \cdot \sqrt{-16} = 5i \cdot 4i = 20i^2 = 20(-1) = -20$.

LESSON
8-1
History and Basic Properties of Complex Numbers

The development of complex numbers had its origin in the search for methods to solve polynomial equations. Prior to the 1500s, quadratic equations without real solutions, such as

$$x^2 + 1 = 0,$$

were regarded as "unsolvable." The quadratic formula

$$x = \frac{-b \pm \sqrt{b^2 - 4ac}}{2a}$$

had been used (in a more primitive notation) long before 1500 to solve quadratic equations

$$ax^2 + bx + c = 0$$

for cases in which $b^2 - 4ac$ is positive. However, Girolamo Cardano, a 16th century Italian mathematician, seems to have been the first to treat the quadratic formula expressions

$$-\frac{b}{2a} + \frac{\sqrt{b^2 - 4ac}}{2a} \text{ and } -\frac{b}{2a} - \frac{\sqrt{b^2 - 4ac}}{2a}$$

as numbers even when $b^2 - 4ac$ is negative. He did this by adopting the following algebraic rules for operating with the square roots of negative numbers.

1) $\sqrt{-1} \cdot \sqrt{-1} = -1$
2) For any positive real number d, $\sqrt{-d} = \sqrt{d}\,\sqrt{-1}$.
3) Multiplication is associative and commutative.

Using these rules, he could carry out computations as in Example 1.

Example 1 Multiply $\sqrt{-25} \cdot \sqrt{-16}$.

Solution Simplify each square root first by using rule 2 above.

$$\sqrt{-25} = \sqrt{25}\,\sqrt{-1} = 5\sqrt{-1}$$
$$\sqrt{-16} = \sqrt{16}\,\sqrt{-1} = 4\sqrt{-1}$$

Then apply rules 1 and 3.

$$\sqrt{-25} \cdot \sqrt{-16} = 5\sqrt{-1} \cdot 4\sqrt{-1} = 20\sqrt{-1}\,\sqrt{-1} = -20$$

René Descartes

Square roots of negative numbers were later given the name *imaginary numbers* by René Descartes in the early 1600s, and, in 1777, Leonhard Euler introduced the symbol i for $\sqrt{-1}$. An **imaginary number** is defined to be a number of the form bi, where b is a real number and the **imaginary unit** i satisfies $i^2 = -1$.

To deal with the quadratic formula expressions that involved real and imaginary numbers, Cardano applied his rules for imaginary numbers

466

Leonhard Euler

together with the algebraic rules for combining binomials to ''numbers'' of the form

$$a + bi$$

where a and b are real numbers. For example, to add $3 + 2i$ to $-1 + 5i$, he added ''like'' terms of these binomials.

$$(3 + 2i) + (-1 + 5i) = (3 + (-1)) + (2 + 5)i = 2 + 7i$$

Similarly, to multiply these two numbers, he used the product formula for binomials and the property $i^2 = -1$.

$$(3 + 2i) \cdot (-1 + 5i) = -3 - 2i + 15i + 10i^2$$
$$= (-3 - 10) + (-2 + 15)i = -13 + 13i$$

In the 1830s, Gauss introduced the modern name, *complex numbers*, for these numbers.

Definitions

A **complex number** is a number that can be written in the form $a + bi$ where a and b are real numbers and $i^2 = -1$. The **real part** of $a + bi$ is a and the **imaginary part** is b.

The variable z is commonly used to represent a complex number. If $z = 2 - 3i$, the real part of z is 2 and the imaginary part of z is -3. Any real number a can be expressed as $a + 0i$, and any imaginary number bi can be expressed as $0 + bi$. Consequently, real and imaginary numbers can be regarded as complex numbers.

Equality for complex numbers is defined as follows:

$$a + bi = c + di \text{ if and only if } a = c \text{ and } b = d.$$

Thus, the complex numbers $2 + 3i$ and $3 + 2i$ are not equal even though they are both constructed from the same two real numbers, 2 and 3.

Addition and multiplication of two complex numbers, $z = a + bi$ and $w = c + di$, are defined by using the corresponding operations for binomials and the property $i^2 = -1$.

Definitions

Let a, b, c, and d be real numbers. Then for the complex numbers $z = a + bi$ and $w = c + di$:

$z + w = (a + c) + (b + d)i$ **(Addition)**

$zw = (ac - bd) + (ad + bc)i.$ **(Multiplication)**

With these customary definitions of addition and multiplication, the complex numbers satisfy the **field properties**. Specifically, addition and multiplication are closed, commutative, and associative. There is an identity ($0 = 0 + 0i$) for addition and a different identity ($1 = 1 + 0i$) for multiplication. Every complex number $z = a + bi$ has an additive inverse ($-z = -a - bi$) and every nonzero complex number z has a multiplicative inverse $\frac{1}{z} = \frac{1}{a + bi}$. Finally, multiplication is distributive over addition.

Error Analysis If students try to apply the Square Root of a Product Theorem in **Example 1**, they will obtain $\sqrt{-25} \cdot \sqrt{-16} = \sqrt{400} = 20$, which is incorrect. Stress that they must simplify the radicals first. Then they can apply the familiar rules to the square roots of nonnegative real numbers.

You may wish to list the three cases concerning square roots of products. If $a \geq 0$ and $b \geq 0$, then $\sqrt{a}\sqrt{b} = \sqrt{ab}$. If either $a < 0$ or $b < 0$, but not both, then $\sqrt{a}\sqrt{b} = \sqrt{ab}$. If both $a < 0$ and $b < 0$, then rewrite both with $\sqrt{-1}$ or i forms to get $-\sqrt{ab}$.

Students may have difficulty believing that a single variable z can represent a complex number, because they typically see complex numbers written with two variables ($a + bi$). The use of graphing helps students to see that a complex number is a single entity.

We assume without proof that all of the field axioms apply to the set of complex numbers; in fact, these axioms are used to motivate the definitions of the operations on complex numbers. Note that the set of complex numbers is not ordered; that is, given two complex numbers, you cannot say that one is less than or greater than the other.

Students should not memorize the definitions of addition and multiplication of complex numbers. Instead, they should use the distributive property twice, remember that $\sqrt{-1}\sqrt{-1} = -1$, and combine like terms. The operations are completely consistent with what one would expect.

Making Connections In Lesson 5-2, students learned that to rationalize a denominator of the form $a + \sqrt{b}$ (b is positive), one could multiply both numerator and denominator by $a - \sqrt{b}$. The process of rewriting the quotient in **Example 2** extends this procedure to instances where b is negative. Specifically, the given fraction equals $\frac{6 + 7\sqrt{-1}}{8 + 5\sqrt{-1}}$ or $\frac{6 + \sqrt{-49}}{8 + \sqrt{-25}}$, and both numerator and denominator are multiplied by $8 - 5\sqrt{-1}$ or $8 - \sqrt{-25}$.

Technology Some calculators have operations on complex numbers built in. Regarding **Example 3**, some symbol manipulation software, *Derive* among them, ask the user to select the domain for factoring polynomials. Allowable choices include factoring over polynomials with rational, radical, and complex number coefficients. Thus, depending upon the domain of coefficients, $x^2 + 5$ could be prime or could be factored as $(x + i\sqrt{5})(x - i\sqrt{5})$. This idea will be discussed again in Lesson 8-9.

To assist students in proving theorems such as those implied by **Example 3** and **Questions 16 and 20**, suggest that they begin as follows. If the theorem involves only one complex number, represent it as $z = a + bi$. If two complex numbers are involved, represent one as $z_1 = a + bi$ and the other as $z_2 = c + di$ (or $z_1 = a_1 + b_1 i$, $z_2 = a_2 + b_2 i$).

With the definitions of addition and multiplication given above, the complex numbers form a field, the *field of complex numbers*.

The consequences of the complex number system forming a field are that all of the familiar operations (addition, subtraction, multiplication, division, and integer powers) can be done with complex numbers just as they are done with real numbers, in the manner used by Cardano.

Subtraction and division of complex numbers are defined as they are with real numbers: For all complex numbers z and w, $z - w = z + (-w)$ and, when $w \neq 0$, $\frac{z}{w} = z \cdot \frac{1}{w}$. As one consequence of these definitions, rational expressions involving complex numbers can be dealt with in the same way as rational expressions involving real numbers. For instance, if a fraction has a complex number in its denominator, a process akin to rationalizing the denominator can be used to rewrite the number in $a + bi$ form.

Example 2 Express the quotient $\frac{6 + 7i}{8 + 5i}$ in $a + bi$ form.

Solution. Think of the denominator as $8 + 5\sqrt{-1}$. Its conjugate is $8 - 5\sqrt{-1}$, or $8 - 5i$. Multiply both the numerator and denominator by $8 - 5i$.

$$\frac{6 + 7i}{8 + 5i} = \frac{6 + 7i}{8 + 5i} \cdot \frac{8 - 5i}{8 - 5i} \qquad \text{Because } \frac{8 - 5i}{8 - 5i} = 1$$

$$= \frac{48 - 30i + 56i - 35i^2}{64 - 25i^2}$$

$$= \frac{48 + 26i + 35}{64 + 25} \qquad \text{Because } i^2 = -1$$

$$= \frac{83 + 26i}{89}$$

$$= \frac{83}{89} + \frac{26}{89}i$$

Note that $8 + 5i$ and $8 - 5i$ are a pair of complex numbers of the form $a + bi$ and $a - bi$. The numbers in such a pair are called *complex conjugates* of one another.

Definition

Let $z = a + bi$ be a complex number. Its **complex conjugate,** denoted $\bar{z} = \overline{a + bi}$, is the complex number with the same real part and the opposite imaginary part. That is,

$$\bar{z} = \overline{a + bi} = a - bi.$$

The product of a complex number and its complex conjugate is always real, as shown in the next example.

468

Example 3 Prove: $\forall$ *complex numbers* z, $z \cdot \bar{z}$ *is a real number.*

Solution Let $z = a + bi$, where a and b are real numbers. Then $\bar{z} = a - bi$.

$$\begin{aligned} z \cdot \bar{z} &= (a + bi)(a - bi) \\ &= a^2 - abi + abi - b^2i^2 \\ &= a^2 - b^2(-1) \qquad \text{Because } i^2 = -1 \\ &= a^2 + b^2 \end{aligned}$$

Since a and b are real, $a^2 + b^2$ is also real. Thus, $z \cdot \bar{z}$ is a real number.

The result of Example 3 provides a way to factor the polynomial $x^2 + y^2$ over the field of complex numbers. For all real numbers x and y, $x^2 + y^2 = (x + yi)(x - yi)$. Polynomials that are prime over the set of real numbers can be factored if complex coefficients are allowed.

Complex numbers were viewed as a mathematical curiosity without application outside of mathematics until Wessel and Argand independently developed a graphical representation for complex numbers.

To graph the complex number $z = a + bi$, write it as the ordered pair (a, b), where the first coordinate represents the real part and the second coordinate represents the imaginary part; this ordered pair is the **rectangular form** of z. Thus, the complex numbers $z = 3 - 5i$ and $w = -4 + 6i$ can be represented as the ordered pairs (3, -5) and (-4, 6), respectively.

In this way, complex numbers can be identified with points in a plane just as real numbers can be identified with points on a number line. This plane, called the **complex plane**, is constructed much like the ordinary coordinate plane. The horizontal axis is the **real axis**, and the vertical axis perpendicular to it is the **imaginary axis**.

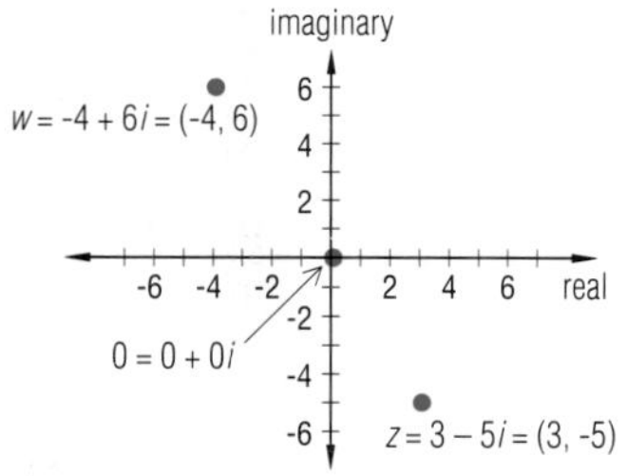

Complex Plane

Notice that the origin in the complex plane represents the complex number $0 = 0 + 0i$. Further, because all real numbers $a + 0i$ have coordinates $(a, 0)$, real numbers are on the real axis. Similarly all imaginary numbers $0 + bi$ are plotted on the imaginary axis because they have coordinates $(0, b)$. Finally, the point in the complex plane that represents the complex conjugate $\bar{z} = a - bi$ is the reflection image over the real axis of the point representing $z = a + bi$.

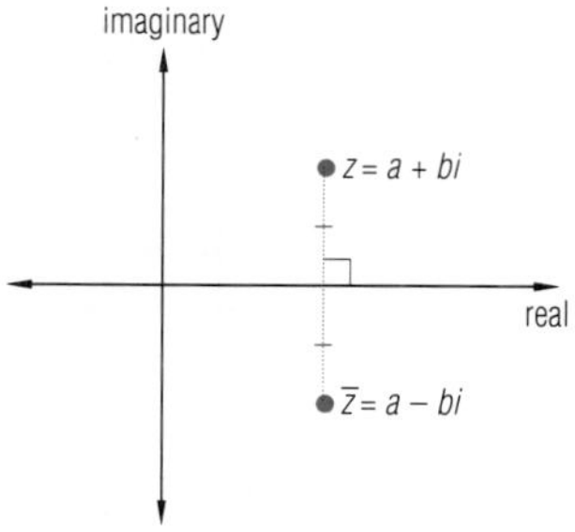

ADDITIONAL EXAMPLES

1. Multiply $\sqrt{-36}\sqrt{-9}$.
-18

2. Express the quotient $\frac{7 + 2i}{3 - 5i}$ in $a + bi$ form.
$\frac{11}{34} + \frac{41}{34}i$

3. Prove: For all complex numbers z, $z + \bar{z}$ is a real number equal to twice the real part of z.
Let $z = a + bi$. Then $\bar{z} = a - bi$, and so $z + \bar{z} = a + bi + a - bi = 2a$.

4. If the total voltage in a circuit is 20 amps and the current is $2 + i$ amps, find the impedance.
$\frac{20}{2 + i} = 8 - 4i$

NOTES ON QUESTIONS

Question 6: Some mathematicians do not consider the set of real numbers to be subset of the set of complex numbers, but a set with operations isomorphic to a subset of the set of complex numbers. The question is philosophical: whether isomorphism means that two sets are identical but just represented differently; or whether it means they are different sets.

Question 9: In this question, students are applying the theorem $\overline{z \cdot w} = \overline{z} \cdot \overline{w}$. Ask if the same holds true for an irrational number such as $5 + 3\sqrt{2}$ and its conjugate. (Yes)

Question 13: This question introduces the geometric interpretation of addition of complex numbers. This topic is covered fully in Lesson 8-3.

Question 17: This question anticipates some of the work in Lesson 8-8. Ask students to use synthetic division to determine the other solution of the polynomial equation.

Question 18: This question foreshadows the determination of complex solutions that is investigated in Lesson 8-8. You may wish to ask students to find the solutions of $x^2 + 9 = 0$ and $x^2 + 4 = 0$.

Question 20: Because the statement is a biconditional, proofs of two if-then statements are required.

Question 25: This question anticipates some of the work of Lesson 8-8.

Complex numbers are commonly used in electronics when dealing with alternating current (AC) to represent *voltage*, *current*, and *impedance*. **Voltage** refers to the electrical potential between two points in an electrical circuit and is measured in volts. **Current**, which is measured in amps, refers to the rate of flow of electrical charge through a circuit. **Impedance**, which is measured in ohms, refers to the opposition to the flow of current caused by components called resistors, coils, and capacitors. The total impedance in a circuit is represented by a complex number whose real part indicates the opposition to current flow due to resistors, and whose imaginary part indicates the opposition due to coils and capacitors. For example, suppose a circuit contains resistors with impedance 8 ohms, and coils and capacitors with impedance -3 ohms. Then the total impedance in the circuit is given by $8 - 3i$.

Impedance is related to voltage and current according to Ohm's Law

$$I = \frac{V}{Z},$$

where I is current, V is voltage, and Z is impedance. Also, if two circuits are connected in *series*—that is, in such a way that current flows through one circuit and then through the other—then the total impedance and total voltage are found by adding the impedances and adding the voltages of the individual circuits, respectively. This is illustrated in Example 4.

Example 4 Suppose two AC circuits are connected in series, one with an impedance of $-5 + 7i$ ohms and the other with an impedance of $8 - 13i$ ohms.

a. Find the total impedance.

b. If the total voltage is 15 volts, find the current.

Solution

a. The total impedance is the sum of the individual impedances, or

$$(-5 + 7i) + (8 - 13i) = 3 - 6i.$$

b. Substituting into Ohm's Law yields

$$I = \frac{V}{Z} = \frac{15}{3 - 6i} = \frac{15(3 + 6i)}{(3 - 6i)(3 + 6i)} = \frac{45 + 90i}{45} = 1 + 2i.$$

The current is $1 + 2i$ amps.

The interpretation of a complex number as current is explained in Question 13 of Lesson 8-3.

Questions

Covering the Reading

1. Who was the first to work with complex numbers expressed in the given notation, and in what century did this happen?

a. $a + bi$ **b.** $a + b\sqrt{-1}$ **c.** (a, b)

a) Leonhard Euler; 18th; b) Cardano; 16th; c) Wessel; 18th

2. Rewrite using i notation.

a. $\sqrt{-64}$ $8i$ **b.** $\sqrt{-20}$ $\sqrt{20}\, i = 2\sqrt{5}\, i$

3. Simplify each product.
   a. $\sqrt{-9} \cdot \sqrt{-49}$ -21
   b. $\sqrt{-8} \cdot \sqrt{-2}$ -4

4. Identify the real and imaginary parts of the complex number $8 - 7i$.
   real = 8; imaginary = -7

5. Rewrite $10 + \sqrt{-225}$ in $a + bi$ form. $10 + 15i$

6. a. *True* or *false? Every real number can be expressed as a complex number.*   True
   b. Is the converse true?   No

In 7 and 8, perform the indicated operations and write the result in $a + bi$ form.

7. $(7 - 6i) + (-4 + 2i)$ $3 - 4i$

8. $\dfrac{14 + 6i}{7}$ $2 + \frac{6}{7}i$

9. Let $z = 2 - i$ and $w = 3 + 2i$.
   a. Compute $z \cdot w$ and write the answer in $a + bi$ form. $8 + i$
   b. Write the complex conjugate of your answer to part **a**. $8 - i$
   c. Find $\bar{z}$. $2 + i$
   d. Find $\bar{w}$. $3 - 2i$
   e. Multiply your answers to parts **c** and **d**. Write the result in $a + bi$ form. Compare with your answer to part **b**. $8 - i$; they are equal.

10. Express $\dfrac{5i}{4 - 3i}$ in $a + bi$ form. $-\frac{3}{5} + \frac{4}{5}i$

11. Rewrite each complex number as an ordered pair and graph it in the complex plane.
   a. $12 + 8i$
   b. $-4 + 7i$
   c. -3
   d. $-7i$
   a-d) See margin.

12. Two AC circuits with impedances of $9 - i$ ohms and $-5 + 3i$ ohms are connected in series.
   a. Find the total impedance. $4 + 2i$ ohms
   b. If the total voltage is 6 volts, find the current. $\frac{6}{5} - \frac{3}{5}i$

Applying the Mathematics

13. Let $z = 7 + 2i$ and $w = -2 - 8i$.
   a. Compute $z + w$. $5 - 6i$
   b. Plot 0, z, w, and $z + w$ in the complex plane.   See margin.
   c. The points you have plotted in part **b** are vertices of a quadrilateral. What kind of quadrilateral is it? How do you know?
   parallelogram; Sample: slopes of opposite sides are equal.

14. Determine the values for the real numbers x and y that make the following equation true: $12 + xi = (3 + y) + 7i$. $x = 7, y = 9$

15. In an AC circuit, the voltage across the resistors is 15 volts, and the voltage across the coils and capacitors is 18 volts, so that the total voltage is given by $15 + 18i$ volts. The current is 6 amps.
   a. Find the impedance. $\frac{5}{2} + 3i$ ohms
   b. If the circuit actually consists of two circuits connected in series, and the impedance of one of those circuits is $2.5 - 2i$ ohms, find the impedance of the other. $5i$ ohms

ADDITIONAL ANSWERS
11., 13.b. See Additional Answers in the back of this book.

NAME ____________

LESSON **MASTER 8–1**
QUESTIONS ON **SPUR** OBJECTIVES

■SKILLS *Objective A (See pages 534–536 for objectives.)*
In 1 and 2, rewrite the complex number as an ordered pair.

1. $3 + 5i$ (3, 5)
2. $-i - 2$ (-2, -1)

In 3 and 4, rewrite the complex number in $a + bi$ form.

3. (-5, 0) $-5 + 0i$
4. $(\frac{1}{4}, -1)$ $\frac{1}{4} - i$

5. If the real part and the imaginary part of a complex number are 7 and 15, respectively, write the number in $a + bi$ form. $7 + 15i$

■SKILLS *Objective B*
In 6–10, perform the indicated operation. Write the result in $a + bi$ form.

6. $\sqrt{-4} \cdot \sqrt{-81}$ -18
7. $\frac{6 - 2\sqrt{9}}{3}$ 0
8. $(12 - 3i) - i(5 + 7i)$ $19 - 8i$
9. $\frac{5i}{1 - i} + 3 - 4i$ $\frac{1}{2} - \frac{3}{2}i$
10. $(5 - 8i)(2 + 7i)$ $66 + 19i$

In 11 and 12, express the solutions in $a + bi$ form.

11. $z^2 = -45$ $\pm 3\sqrt{5}i$
12. $-3 + 2i + w = 8 - 9i$ $11 - 11i$

■PROPERTIES *Objective F*
13. Let $z = 6 - 2i$. Verify that $z \cdot \bar{z}$ is a real number.
$(6 - 2i)(6 + 2i) = 36 + 4 = 40$

14. Let $z = a + bi$. Prove that $\bar{\bar{z}} = z$.
$\bar{z} = a - bi$, so $\bar{\bar{z}} = a - (-bi) = a - bi = z$

 Continued

NAME ____________
Lesson MASTER 8–1 (page 2)

15. Prove that for all complex numbers u, v, and w, $(u + w)v = uv + wv$.
Let $u = a + bi$, $v = c + di$, and $w = e + fi$. Then $(u + w)v = [(a + e) + (b + f)i](c + di) = (ac + ec - bd - fd) + (ad + ed + bc + fc)i$. Also, $uv + wv = [(ac - bd) + (bc + ad)i] + [(ce - df) + (cf + de)i] = (ac - bd + ce - df) + (bc + ad + cf + de)i$. Therefore, $(u + w)v = uv + wv$

■USES *Objective H*
16. If the voltage in an AC circuit is 120V and the current is $6 + 3i$ amps, find the impedance. $16 - 8i$ ohms

17. Two AC circuits with impedances of $9 + 16i$ ohms and $-6 + 8i$ ohms are connected in series.
   a. Find the total impedance. $3 + 24i$ ohms
   b. If the total voltage is 10 volts, find the current. $\frac{2}{39} - \frac{16}{39}i$ amps

■REPRESENTATIONS *Objective I*
In 18 and 19, let $A = 0$, $B = -2i$, and $C = -4 + 2i$.

18. a. Graph A, B, and C on the same complex plane.
   b. Find the area of ΔABC.   4

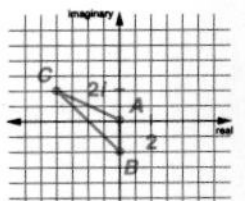

19. Let f be the function defined by $f(z) = (-2 + i)z$. Graph the points $f(A)$, $f(B)$, and $f(C)$ in the same complex plane.

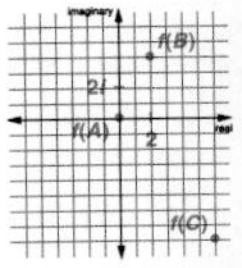

FOLLOW-UP

MORE PRACTICE
For more questions on SPUR Objectives, use *Lesson Master 8-1*, shown on page 471.

EXTENSION
Small groups can examine the counterparts of **Question 16** for the operations with imaginary numbers. Then ask the same questions for nonimaginary, nonreal complex numbers. (Imaginary numbers are not closed under any of the operations; nor are the nonimaginary, nonreal complex numbers.)

PROJECTS
The projects for Chapter 8 are described on pages 529 and 530. **Projects 1 and 6** are related to the content of this lesson.

ADDITIONAL ANSWERS
16. For two imaginary numbers mi and ni, $mi + ni = (m + n)i$. Since m and n are real, $m + n$ is real, and $(m + n)i$ is an imaginary number.

17. $(1 + i)^2 - 2(1 + i) + 2$
$= 1 + i + i + i^2 - 2 - 2i + 2$
$= 1 + i + i - 1 - 2 - 2i + 2$
$= 0$
So $1 + i$ is a solution of $z^2 - 2z + 2 = 0$.

19.b. $z \cdot \frac{1}{z} =$
$(5 - 2i)\left(\frac{5}{29} + \frac{2}{29}i\right)$
$= \frac{25}{29} + \frac{10}{29}i - \frac{10}{29}i - \frac{4}{29}i^2$
$= \frac{25}{29} + \frac{4}{29}$
$= 1$

20. Let $z = a + bi$. Then $\bar{z} = a - bi$. By definition of equality for complex numbers, $a + bi = a - bi$ if and only if $a = a$ and $b = -b$, then $2b = 0$ and $b = 0$. So $z = a + bi = a + 0i$, and z is a real number.

21., 24.b., c., 25.a., c., d. See Additional Answers in the back of this book.

16. Prove or disprove: *The sum of any two imaginary numbers is an imaginary number.* See margin.

17. Show that $1 + i$ is a solution of $z^2 - 2z + 2 = 0$. See margin.

18. Find all four solutions of the equation $x^4 + 13x^2 + 36 = 0$ by factoring. $\pm 3i, \pm 2i$

19. **a.** If $z = 5 - 2i$, find $\frac{1}{z}$ and express the result in $a + bi$ form. $\frac{5}{29} + \frac{2}{29}i$
b. Check by multiplying $z \cdot \frac{1}{z}$. See margin.

20. If z is a complex number, prove that $z = \bar{z}$ if and only if z is a real number. See margin.

Review

21. Use proof by contradiction to prove that there is no smallest integer. *(Lesson 4-8)* See margin.

22. The conversion formula between the two most common measures of temperature, Celsius and Fahrenheit, is $C = \frac{5}{9}(F - 32)$.
The numerical values for the same temperature on the two scales is usually quite different, for example, $90°\ F \approx 32°\ C$.
a. There is one temperature whose numerical value on both scales is the same. Find this temperature. -40°
b. Suppose that a reading in degrees Fahrenheit is considered close to its Celsius reading if the readings differ by no more than 5°. Find the range of temperatures whose Fahrenheit and Celsius readings are close to each other. *(Lessons 3-8, 3-1)* $-51.25°F \le t \le -28.75°F$

23. Use the definition of absolute value to fill in the blanks:
a. $x > 5$ or $x < -5$ if and only if $|x|$ _?_. > 5
b. $-4 < x < 2$ if and only if $|x + 1|$ _?_. < 3
(Lesson 3-8)

24. Refer to the increasing function f graphed at the right.
a. Give the domain and range of $y = f(x)$.
b. Reflect the graph of $y = f(x)$ across the line $y = x$. See margin.
c. Find the domain and range of the function represented by the reflection image. See margin.
d. How are f and the function in part **c** related?
(Lessons 3-2, 2-2, 2-1) d) They are inverses.
a) domain: $\{x: 2 \le x \le 5\}$; range: $\{y: 1 \le y \le 2\}$

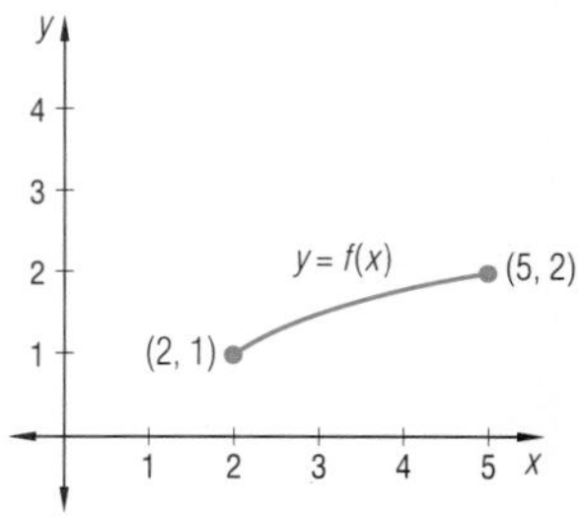

Exploration

25. **a.** Solve the quadratic equation $2x^2 + 3x + 5 = 0$. Call your solutions z_1 and z_2. See margin.
b. How are z_1 and z_2 related? They are complex conjugates.
c. Compute $z_1 + z_2$ and $z_1 z_2$. How are the sum and product related to the coefficients of the given equation? See margin.
d. Generalize parts **b** and **c** and, if possible, prove your generalization.
See margin.

26. Generalize Question 19. If $z = a + bi$, then $\frac{1}{z} = \frac{a}{a^2 + b^2} - \frac{b}{a^2 + b^2}i$.

472

LESSON

8-2

Polar Coordinates

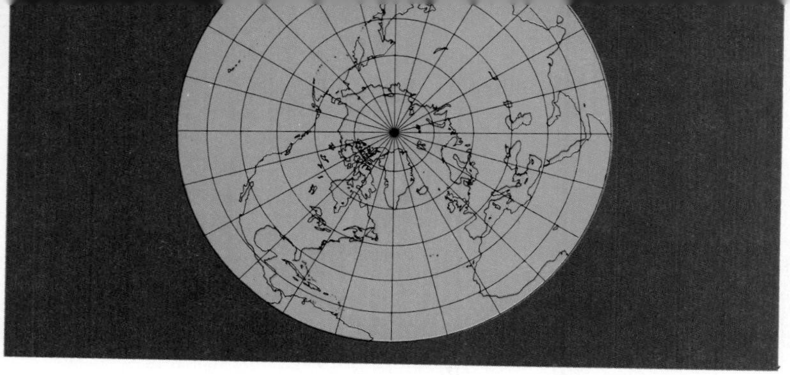

People had observed since ancient times that stars at night rotated around a point in the sky; that point was called a *pole*, and the nearby bright star was called Polaris. When, in 1543, Copernicus published his theory that the Earth orbits around the sun once a year and spins on its axis once each day, the points on the Earth which intersect the axis were naturally called the north and south poles.

Maps of polar regions, like the one of the north pole shown above, show lines radiating from the pole. These lines, which on the Earth are longitude lines, are identified by their degree measure from some reference line. On the map, the circles of latitude are concentric. *Polar coordinates* in the plane use a grid similar to that found in the map.

To construct a polar coordinate system in a plane, first select a point O to be the **pole** of the system. Then select a line through O to be the **polar axis**. Coordinatize this line so that O has coordinate 0. Usually the polar axis is drawn to be horizontal, as shown below. Any point P in the plane has **polar coordinates** $[r, \theta]$ if and only if under a rotation of θ about the pole O, P is the image of the point with coordinate r.

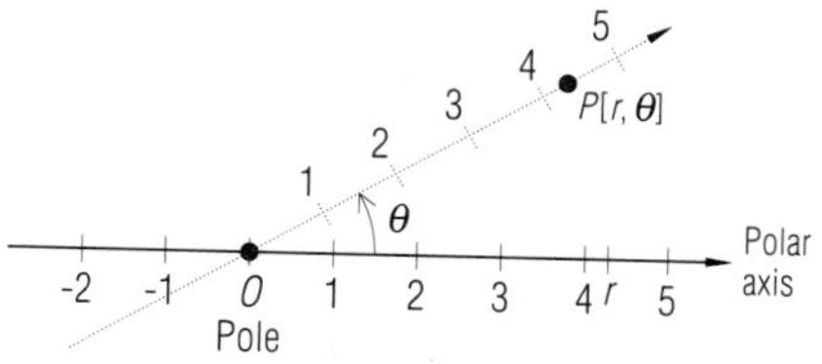

In this picture, it looks as if P is the image of the point with coordinate 4.3 under a rotation of 30°, so $r \approx 4.3$ and $\theta \approx 30° = \frac{\pi}{6}$, and P has approximate polar coordinates [4.3, 30°] or $[4.3, \frac{\pi}{6}]$. The brackets [ ] are used to distinguish polar from rectangular coordinates.

Every rotation of θ yields the same image as a rotation of $\theta + 2\pi n$ (in radians) or $\theta + 360°n$ (in degrees), where n is any integer. So unlike the situation with rectangular coordinates, a point P has more than one polar coordinate representation. For the point P above, three other representations are $[4.3, \frac{13\pi}{6}]$, [4.3, -330°], and $[4.3, -\frac{11\pi}{6}]$. Since points on the polar axis can have negative coordinates, r can be negative, as Example 1 shows.

LESSON 8-2

RESOURCES

- Lesson Master 8-2
- Teaching Aid 44 provides a large polar grid for demonstration purposes.
- Teaching Aid 45 provides smaller polar grids for duplication purposes.

OBJECTIVES

A Express complex numbers in *a* + *bi*, rectangular, polar, and trigonometric form.
C Convert between polar and rectangular coordinate representations of points.
I Graph complex numbers.

TEACHING NOTES

Jakob Bernoulli, a member of the famous family of brilliant Swiss mathematicians and scientists, introduced the concept of polar coordinates in 1691. Students may wish to read about this extraordinary family and their many contributions to both mathematics and science.

The primary objectives of this lesson are to help students gain facility with plotting points on a polar grid and with making conversions between the rectangular and polar representations of a point in the plane. Both of these skills require practice. Point out that brackets are used for polar form and parentheses for rectangular form.

Some students find it difficult to plot points $[r, \theta]$ if $r < 0$. Suggest that they view the polar axis as a spinner. You may fasten a thin, coordinatized stick with a thumbtack to the pole of a sheet of polar graph paper. Rotating the stick by the angle θ and using its coordinatization for r will help students understand polar plotting.

In discussing the theorem on this page, remind students that "$\forall$ integers n" allows n to be either positive or negative. Some books use the notation $\theta \pm 2\pi k$, where k is a positive integer, to emphasize that any integer multiple of 2π can be added to or subtracted from θ.

If some students are bothered by the existence of more than one pair of polar coordinates for the same point, introduce the analogy with fraction and decimal representations of rational numbers. (See the Perspectives for this lesson on page 465B.)

Making Connections
With Cartesian coordinates, the point (2, 5) is the intersection of the line $x = 2$ with the line $y = 5$. In polar coordinates, with $r > 0$, students may think of the point $\left[4, \frac{\pi}{6}\right]$ as the intersection of the circle of radius $r = 4$ with the ray $\theta = \frac{\pi}{6}$.

The conversion of the rectangular coordinates of a point to polar coordinates by use of the Rectangular to Polar conversion formulas $r^2 = x^2 + y^2$ and $\tan \theta = \frac{y}{x}$ offers an opportunity to remind students that the equation $\tan \theta = c$ always has infinitely many solutions because the tangent function is periodic. However, if students use the inverse tangent key on their calculator to solve this equation, only the solution in the interval $-\frac{\pi}{2} < \theta < \frac{\pi}{2}$ is displayed because the arctangent function selects this value from among the infinite number of solutions to $\tan \theta = c$.

Example 1 **a.** Plot the point $Q = [-5, \frac{3\pi}{4}]$.
b. Give a different polar representation $[r, \theta]$ for Q in which $r > 0$.

Solution
a. Rotate the polar axis $\frac{3\pi}{4}$ and look for the point with coordinate -5.

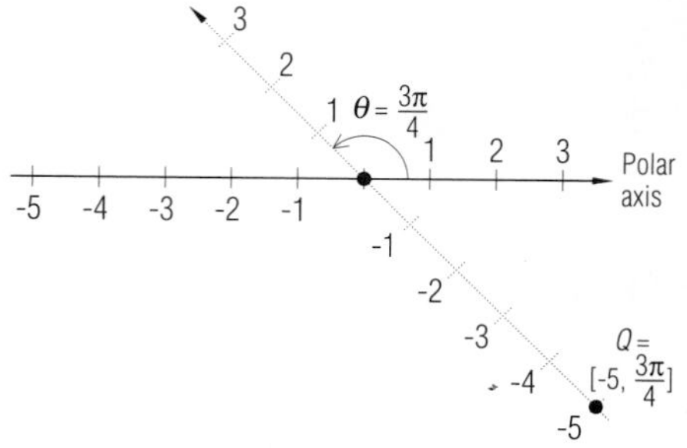

b. A rotation of $-\frac{\pi}{4}$ will bring the polar axis to the same position, but points on the axis will have coordinates opposite those from a rotation of $\frac{3\pi}{4}$. So $Q = [5, -\frac{\pi}{4}]$ is another possible representation for point Q.

In any polar coordinate system, the pole O is assigned the polar coordinate representations $[0, \theta]$ for *any* number θ. Thus, $[0, 0]$, $[0, 30°]$, and $[0, \frac{\pi}{2}]$ are three of the infinitely many polar coordinate representations of the pole.

The possible polar coordinates for a point other than the pole are summarized in the following theorem.

Theorem

For any particular values of r and θ, the following polar coordinate representations name the same point.

a. $[r, \theta]$
b. $[r, \theta + 2\pi n]$, $\forall$ integers n
c. $[-r, \theta + (2n + 1)\pi]$, $\forall$ integers n

The **polar grid** pictured at the right is very helpful for plotting points and sketching curves in polar coordinates and is commonly used.

Each of the concentric circles in the grid represents a value of r, and each ray from the pole represents a value of θ. When plotting using polar coordinates, you should identify the positive polar axis with an arrow and put a scale on it to indicate values of r.

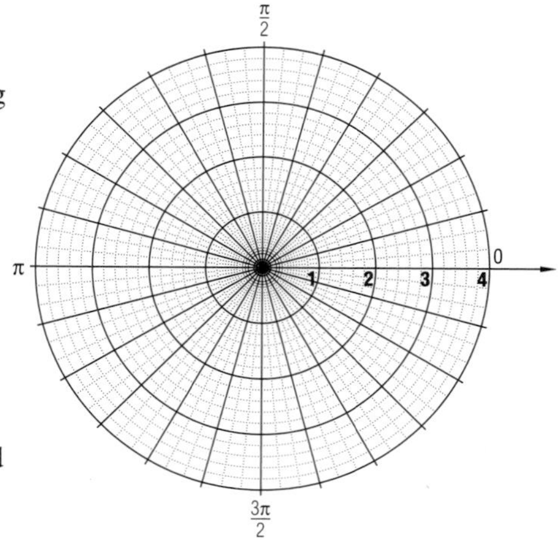

474

Example 2 On the same polar grid,

a. plot the points $Q_1 = [-2.4, \frac{\pi}{2}]$ and $Q_2 = [2.1, -\frac{7\pi}{3}]$;

b. sketch all solutions $[r, \theta]$ to the equation $r = 3$;

c. sketch all solutions $[r, \theta]$ to the equation $\theta = \frac{5\pi}{12}$.

Solution

a. The points are plotted at the right.

b. The equation $r = 3$ describes the circle of radius 3 centered at the pole. This circle is drawn in blue.

c. $\theta = \frac{5\pi}{12}$ is the line obtained by rotating the polar axis by $\frac{5\pi}{12}$. This line is drawn in orange.

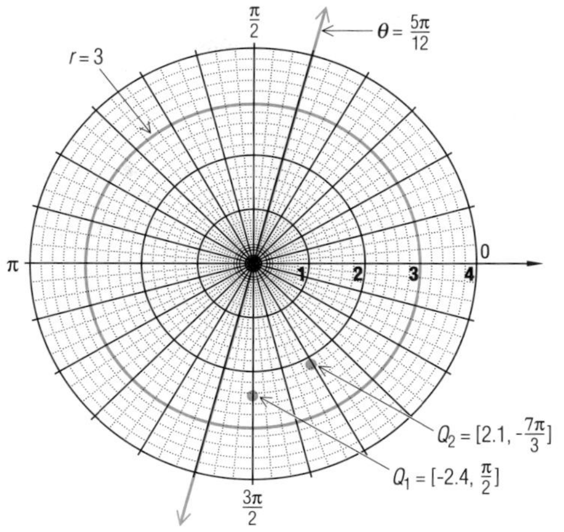

Often polar and rectangular coordinate systems are superimposed on the same plane. Then the polar axis coincides with the x-axis, and the pole is the origin. When this is done, you can use trigonometry to find the unique rectangular coordinate representation for any point whose polar coordinate representation is known.

Example 3 Find the rectangular coordinates of the point $P = [3, \frac{5\pi}{6}]$.

Solution First plot P as shown at the right.

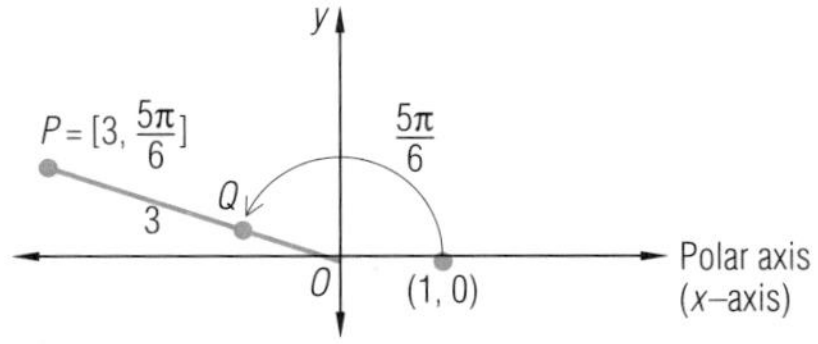

By definition of the cosine and sine, the point $Q = [1, \frac{5\pi}{6}]$, where the unit circle intersects $\overline{OP}$, has rectangular coordinates $\left(\cos\frac{5\pi}{6}, \sin\frac{5\pi}{6}\right)$. Because $P = [3, \frac{5\pi}{6}]$, the point P is three times as far from the origin as Q, so its rectangular coordinates are $\left(3\cos\frac{5\pi}{6}, 3\sin\frac{5\pi}{6}\right)$. Exact values of these are known: $\cos\frac{5\pi}{6} = -\frac{\sqrt{3}}{2}$ and $\sin\frac{5\pi}{6} = \frac{1}{2}$.

Alternate Approach You may wish to use a reference triangle to verify the Polar-Rectangular Conversion Theorem. For simplicity, select θ such that $0 < \theta < \frac{\pi}{2}$. Draw a segment from the pole to $[r, \theta]$ and segments from $[r, \theta]$ to $[r, 0]$, and $[0, 0]$ to $[r, \theta]$ to create a right ΔOAB as pictured below.

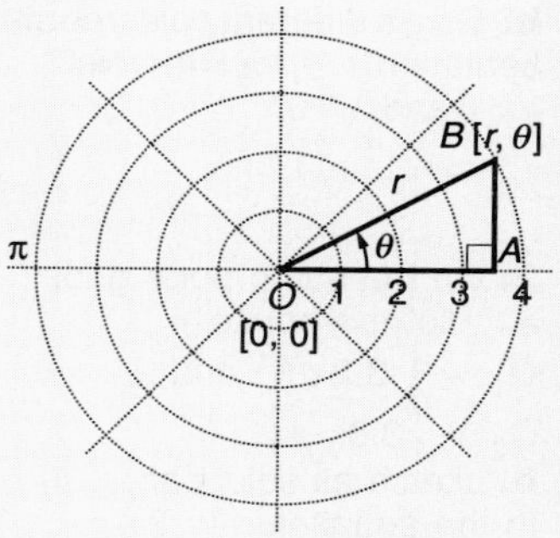

From right triangle trigonometry $\cos\theta = \frac{OA}{OB}$ and $\sin\theta = \frac{AB}{OB}$, or $\cos\theta = \frac{x}{r}$ and $\sin\theta = \frac{y}{r}$, or $x = r\cos\theta$ and $y = r\sin\theta$.

Technology Many scientific calculators can do conversions between rectangular and polar coordinates. For example, one calculator has the shift register keys Pol(x, y) and Rec(r, θ) for this purpose. Another calculator has these conversions included in the MATH menu. Ask your students if their calculators have this capacity.

ADDITIONAL EXAMPLES

1.a. Plot the point $P = \left[2, -\frac{\pi}{4}\right]$.

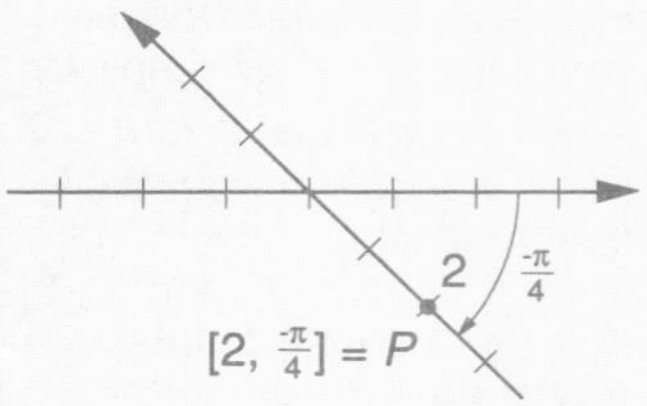

b. Give a different polar representation $[r, \theta]$ for P in which $r > 0$ and $\theta > 0$.
sample: $\left[2, \frac{7\pi}{4}\right]$

2. On the same polar grid,
a. plot the points
$Q_1 = \left[-3.2, \frac{3\pi}{4}\right]$ and
$Q_2 = \left[3.8, \frac{7\pi}{6}\right]$;
b. sketch all solutions $[r, \theta]$ to the equation $r = 2.4$;
c. sketch all solutions $[r, \theta]$ to the equation $\theta = \frac{13\pi}{12}$.

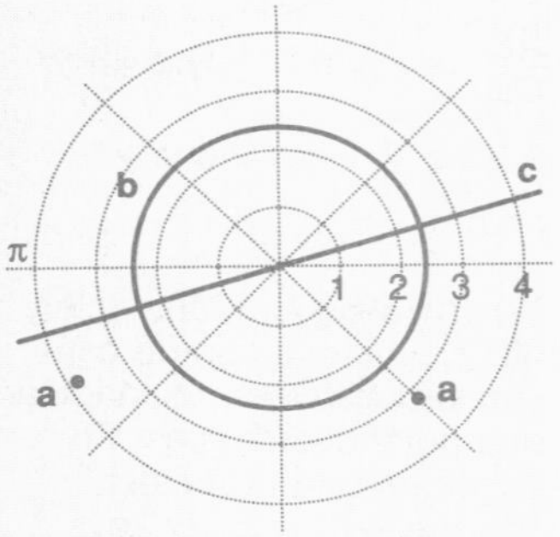

3. Find the rectangular coordinates of the point P with polar coordinates $\left[5, \frac{\pi}{6}\right]$.
$\left(5\cos\frac{\pi}{6}, 5\sin\frac{\pi}{6}\right) = \left(\frac{5\sqrt{3}}{2}, \frac{5}{2}\right)$

Students with scientific calculators that can do polar to rectangular conversions should check their solutions with the calculator.

4. Find three different polar coordinate representations for the point whose rectangular coordinates are (5, -12).
samples: $\left[13, \tan^{-1}\left(-\frac{12}{5}\right)\right] \approx$ **[13, -67.4°]**, $\approx$ **[13, 292.6°]**, $\approx$ **[-13, 112.6°]**

Thus

$$P = \underbrace{[3, \tfrac{5\pi}{6}]}_{\text{polar}} = \underbrace{\left(3\cos\tfrac{5\pi}{6}, 3\sin\tfrac{5\pi}{6}\right) = \left(-\tfrac{3\sqrt{3}}{2}, \tfrac{3}{2}\right)}_{\text{rectangular}}.$$

Check Both $P = [3, \frac{5\pi}{6}]$ and $P = \left(-\frac{3\sqrt{3}}{2}, \frac{3}{2}\right)$ are in the 2nd quadrant, closer to the x-axis than the y-axis.

The procedure used in Example 3 generalizes to the following result.

Polar-Rectangular Conversion Theorem

If $[r, \theta]$ is a polar coordinate representation of a point P, then the rectangular coordinates (x, y) of P are given by
$$x = r\cos\theta \quad \text{and} \quad y = r\sin\theta.$$

Proof The argument of Example 3 applies to all points but the pole. It is left for you to generalize. If P is the pole, then $P = (0, 0)$. The theorem holds because
$$x = 0 = 0 \cdot \cos\theta \quad \text{and} \quad y = 0 = 0 \cdot \sin\theta$$
for any θ.

Translating from rectangular to polar coordinates is also possible. To obtain a general formula, solve the system
$$\begin{cases} x = r\cos\theta \\ y = r\sin\theta \end{cases}$$
for r and θ.

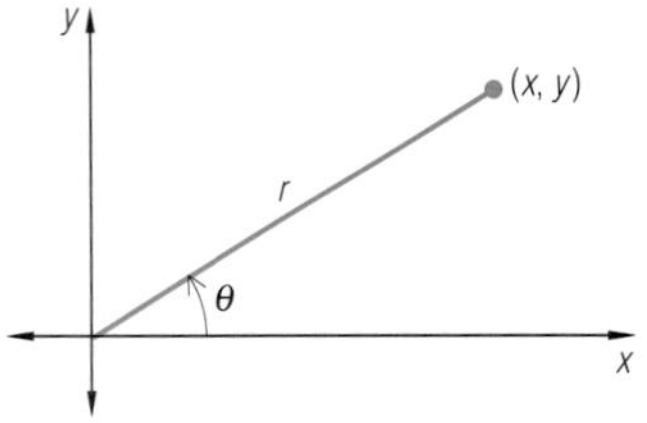

To solve the system for θ, notice that
$$\frac{y}{x} = \frac{r\sin\theta}{r\cos\theta} = \tan\theta.$$
There are many possible values of θ, leading to the many possible polar representations. For a given point, the values of θ differ by multiples of π (or 180°).

To solve the above system for r, notice that
$$x^2 + y^2 = (r\cos\theta)^2 + (r\sin\theta)^2 = r^2(\cos^2\theta + \sin^2\theta) = r^2 \cdot 1 = r^2.$$
So $\quad r = \pm\sqrt{x^2 + y^2}.$

This confirms that r can be positive or negative. The choice for r (and for θ) depends on the quadrant in which the point is located.

476

Example 4 Find 3 different polar coordinate representations for the point whose rectangular coordinates are (-3, 4).

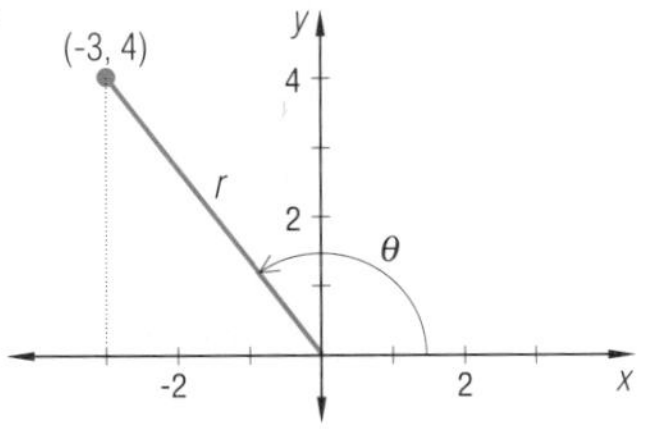

Solution $r = \pm\sqrt{(-3)^2 + 4^2} = \pm 5$ and $\tan \theta = -\frac{4}{3}$. A calculator shows

$$\theta = \tan^{-1}\left(-\frac{4}{3}\right) \approx -53°.$$

So $\theta \approx -53° + 180°n$.

Since the point is in the second quadrant, one representation is [-5, -53°]. From this obtain the other polar coordinate representations: [5, 127°], [-5, 307°], and so on.

The general relationships that govern the conversions between rectangular and polar coordinates are summarized below.

Polar to Rectangular	Rectangular to Polar
$x = r \cos \theta$	$r^2 = x^2 + y^2$
$y = r \sin \theta$	$\tan \theta = \frac{y}{x}$

In Lesson 8-1, the complex number $z = a + bi$ was plotted in the complex plane by writing it in rectangular form as the ordered pair (a, b). Polar coordinates provide a way of representing complex numbers $a + bi$ as $[r, \theta]$, where $r = \sqrt{a^2 + b^2}$ and $\tan \theta = \frac{b}{a}$. Both these ways are quite useful, as you will see throughout the rest of this chapter.

Questions

Covering the Reading

1. Plot all the points on the same polar grid.
   a. [2, 300°] **b.** [-4, 100°] **c.** $[0, \frac{\pi}{6}]$ **d.** $[1, \frac{5\pi}{3}]$
   a-d) See margin.
2. Suppose $P = [2, \frac{7\pi}{6}]$. Give another polar coordinate representation for P:
   a. with $r = 2$, $\theta \neq \frac{7\pi}{6}$; sample: $[2, \frac{19\pi}{6}]$
   b. with $r = -2$; sample: $[-2, \frac{\pi}{6}]$
   c. with $\theta < 0$. sample: $[2, -\frac{5\pi}{6}]$
3. Explain why the points are polar coordinate representations of $[4, \frac{\pi}{3}]$.
   a. $[4, -\frac{5\pi}{3} + 2k\pi]$ for any integer k
   b. $[-4, \frac{4\pi}{3} + 2k\pi]$ for any integer k a,b) See margin.

NOTES ON QUESTIONS

Question 2: Elicit enough responses so that students can see the patterns.

Question 4a: Using $r = \sqrt{x^2 + y^2}$ and $r = 2$, ask students to convert the polar equation $r = 2$ to its corresponding rectangular form. They should get $x^2 + y^2 = 4$, which is the equation of the circle with center at the origin and radius 2. Point out that $r = 2$ is a much simpler equation than $x^2 + y^2 = 4$. In the equation $r = 2$, θ can be any real number because θ does not appear in it.

Question 4b: The polar equation of any line through the pole is of the form $\theta = c$. The rectangular form for such a line is $y = mx$. Ask for the relationship between m and c. ($m = \tan c$)

Question 8: Ask students to write expressions that will generate all possible polar coordinates for these two points. Plotting the points on polar graph paper is helpful.

Making Connections for Questions 11, 12, and 16: These questions introduce the polar representation of complex numbers discussed in the next lesson.

Making Connections for Questions 14 and 15: Refer students to Question 4. Polar equations and graphs are studied extensively in Lessons 8-4 and 8-5.

Question 18: You may summarize part **a** as "the conjugate of a sum is the sum of the conjugates."

ADDITIONAL ANSWERS
1., 3. See Additional Answers in the back of this book.

ADDITIONAL ANSWERS
4.a., b.

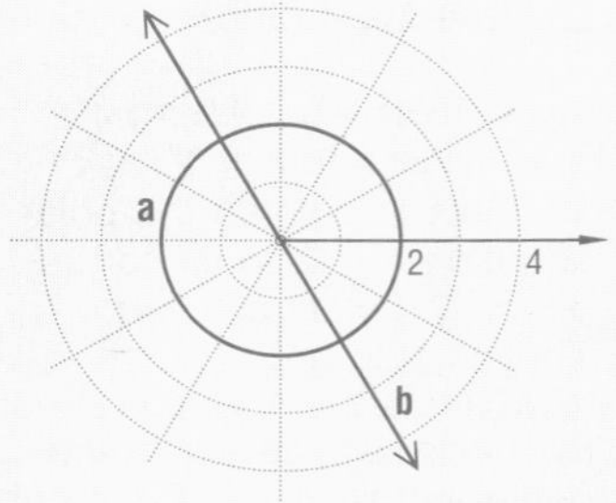

5. Given any point $P = [r, \theta]$. First plot P and $Q = [1, \theta] = (\cos\theta, \sin\theta)$.

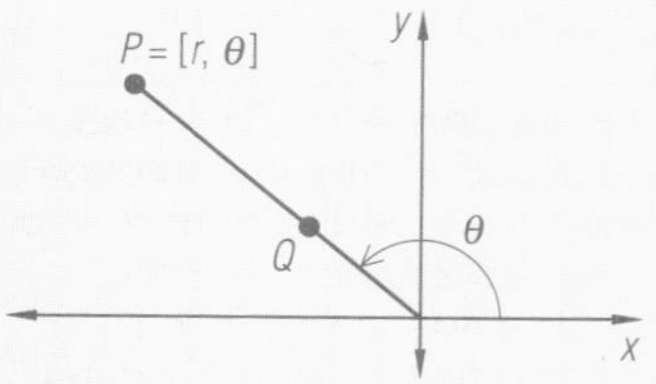

Because $[r, \theta]$ is r times as far from the origin as Q, its rectangular coordinates are $(r\cos\theta, r\sin\theta)$. Thus, the rectangular coordinates of P are given by $x = r\cos\theta$ and $y = r\sin\theta$.

14. $P_1 = [3, 0°]$, $P_2 = [3, 30°]$, $P_3 = [3, 90°]$, $P_4 = [3, 120°]$, $P_5 = [3, 195°]$, $P_6 = [3, 240°]$, $P_7 = [3, 285°]$; $r = 3$

15. $Q_1 = \left[-4, \frac{\pi}{6}\right]$, $Q_2 = \left[-3, \frac{\pi}{6}\right]$, $Q_3 = \left[-2, \frac{\pi}{6}\right]$, $Q_4 = \left[-1, \frac{\pi}{6}\right]$, $Q_5 = \left[0, \frac{\pi}{6}\right]$, $Q_6 = \left[1, \frac{\pi}{6}\right]$, $Q_7 = \left[2, \frac{\pi}{6}\right]$, $Q_8 = \left[3, \frac{\pi}{6}\right]$, $Q_9 = \left[4, \frac{\pi}{6}\right]$; $\theta = \frac{\pi}{6}$

18.a. $z + w = (6 - 3i) + (2 + 4i) = 8 + i$, so $\overline{z + w} = 8 - i$. $\overline{z} = 6 + 3i$ and $\overline{w} = 2 - 4i$. So $\overline{z} + \overline{w} = (6 + 3i) + (2 - 4i) = 8 - i$. $\therefore \overline{z + w} = \overline{z} + \overline{w}$.
b., c. See Additional Answers in the back of this book.

4. On the same polar grid, sketch all solutions to these equations.
a. $r = 2$ **b.** $\theta = \frac{2\pi}{3}$ a,b) See margin.

5. Prove the Polar-Rectangular Conversion Theorem for positive values of r by generalizing the argument of Example 3. See margin.

In 6 and 7, find the rectangular coordinates for the point P whose polar coordinates are given.

6. $[4, \frac{3\pi}{2}]$ (0, -4) **7.** [2.3, -42°] ≈ (1.7, -1.5)

In 8 and 9, give one pair of polar coordinates for each (x, y) pair.

8. (5, 2) ≈ $[\sqrt{29}, 21.8°]$ **9.** (-2, -3) ≈ $[-\sqrt{13}, 56.3°]$

Applying the Mathematics

10. Let $P = [3, \frac{\pi}{6}]$. State one pair of polar coordinates for the image of P under the given transformation.
a. reflection over the polar axis sample: $[3, \frac{11\pi}{6}]$
b. reflection over the line $\theta = \frac{\pi}{2}$ sample: $[3, \frac{5\pi}{6}]$
c. rotation of $\frac{2\pi}{3}$ radians about the origin sample: $[3, \frac{5\pi}{6}]$

11. Example 4 could be interpreted as getting polar coordinate representations of what complex number? $-3 + 4i$

12. Give polar coordinates for each complex number.
a. (1, 0) **b.** i **c.** $1 + i$
a) sample: [1, 0]; b) sample: [1, 90°]; c) sample: $[\sqrt{2}, 45°]$

13. When the coordinates of P are written in polar form, $\theta = \frac{\pi}{6}$. When the coordinates of P are written in rectangular form, $x = 5$. Find polar and rectangular coordinates for P. $[\frac{10\sqrt{3}}{3}, \frac{\pi}{6}]$, $(5, \frac{5\sqrt{3}}{3})$

In 14 and 15, the rays are equally spaced around the pole. Give polar coordinates for each point on the graph. Then write an equation, using r and/or θ, that the coordinates satisfy. See margin.

14.

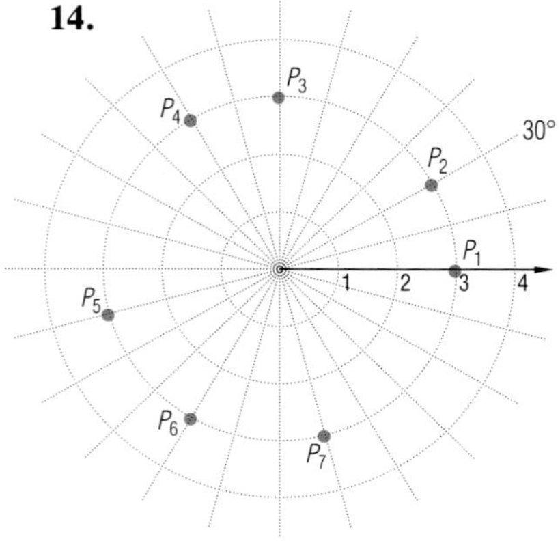

15.

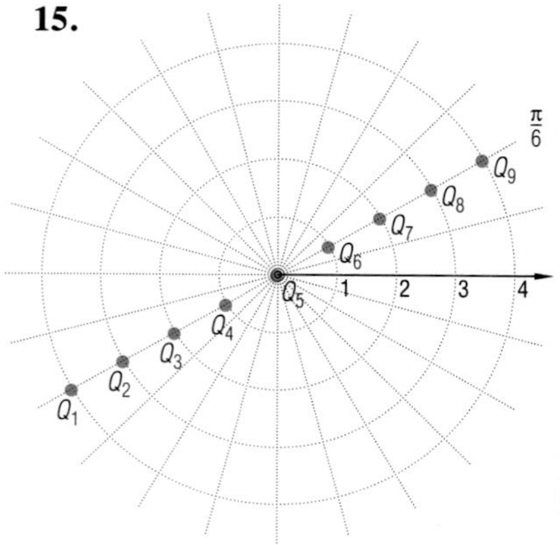

478

16. What complex number is represented by the point P_6 in Question 14?
$-1.5 + (-1.5\sqrt{3})i$

Review

17. If the voltage in an AC circuit is $10 - 15i$ volts and the impedance is $-4 + 6i$ ohms, find the current. *(Lesson 8-1)* $-\frac{5}{2}$ amps

18. Let $z = 6 - 3i$ and $w = 2 + 4i$. Verify the following algebraic properties of the complex conjugate for z and w. *(Lesson 8-1)*
a. $\overline{z + w} = \overline{z} + \overline{w}$ **b.** $\overline{z \cdot w} = \overline{z} \cdot \overline{w}$
c. $\overline{\left(\frac{z}{w}\right)} = \frac{\overline{z}}{\overline{w}}$ a, b, c) See margin.

19. Does $\sum_{k=0}^{\infty} (.9)^k$ have a finite limit? If so, find the limit. If not, explain why not. *(Lesson 7-6)* Yes, 10

20. Suppose a person drives 8,000 miles per year with a car that averages 26 miles/gallon. Assume that gasoline costs \$1.30/gallon.
a. Find a single fraction that gives the amount of money this person would save in a year by improving the gas mileage by h miles/gallon. $\frac{400h}{26 + h}$
b. By how much must the gas mileage improve to save \$80? 6.5 mpg
(Lessons 5-7, 5-1)

21. Prove that the points (-1, 0), (-5, -5), (-11, -5), and (-7, 0) are vertices of a parallelogram. *(Previous course)* See margin.

Exploration

22. On Earth, where are the north and south magnetic poles?

The north magnetic pole lies just north of North America and west of Greenland and is about latitude 76° N and longitude 101° W on Bathurst Island in Canada. The south magnetic pole is in Antarctica and is about latitude 66° S and longitude 140° E, just off the coast of Antarctica due south of Australia. The location of both magnetic poles vary over time.

FOLLOW-UP

MORE PRACTICE
For more questions on SPUR Objectives, use *Lesson Master 8-2*, shown below.

PROJECTS
The projects for Chapter 8 are described on pages 529–530. **Project 4** is related to the content of this lesson.

21. Let the vertices in clockwise order be $A(-1, 0)$, $B(-5, -5)$, $C(-11, -5)$, and $D(-7, 0)$. The slope of $\overline{AD} = 0$; the slope of $\overline{BC} = 0$. The slope of $\overline{AB} = \frac{-5 - 0}{-5 - (-1)} = \frac{5}{4}$; the slope of $\overline{DC} = \frac{-5 - 0}{-11 - (-7)} = \frac{5}{4}$. Since $ABCD$ is composed of two pairs of parallel lines, it is a parallelogram.

NAME ____________

LESSON **MASTER 8–2**
QUESTIONS ON **SPUR** OBJECTIVES

■ **SKILLS** *Objective C (See pages 534–536 for objectives.)*
In 1 and 2, give one pair of polar coordinates for each (x, y) pair.

1. $(1, \sqrt{3})$ $[2, 60°]$ 2. $(-5, -12)$ $[13, 247°]$

3. If $P = \left[r, \frac{7\pi}{4}\right] = (5, y)$, $r =$ $5\sqrt{2}$ and $y =$ -5.

4. Suppose $P = \left[10, \frac{\pi}{4}\right]$. Give another polar coordinate representation for P:
a. with $r < 0$; sample: $\left[-10, \frac{5\pi}{4}\right]$
b. with $r > 0$, $\theta < 0$. sample: $\left[10, -\frac{7\pi}{4}\right]$

In 5 and 6, find the rectangular coordinates for the point P whose polar coordinates are given.

5. $[5\sqrt{2}, 225°]$ $(-5, -5)$ 6. $\left[-4, -\frac{\pi}{3}\right]$ $(-2, 2\sqrt{3})$

■ **REPRESENTATIONS** *Objective I*
7. Plot the following points on the polar grid at the right.
a. $[1, 45°]$ b. $\left[-4, -\frac{\pi}{2}\right]$
c. $\left[0, -\frac{3\pi}{2}\right]$ d. $\left[-3, -\frac{2\pi}{3}\right]$

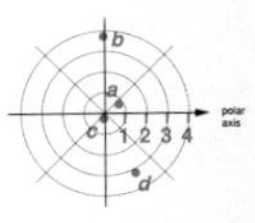

8. On the polar grid, sketch all solutions to the equation $r = 3$.

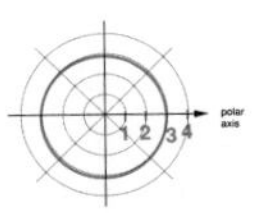

9. Give two polar representations of the point P graphed below.
sample: $[3, -60°]$, $[3, 300°]$

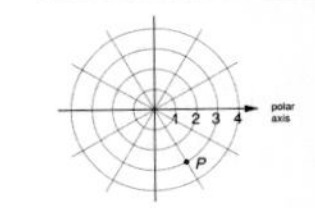

 89

LESSON 8-3

RESOURCES
- Lesson Master 8-3
- Quiz for Lessons 8-1 through 8-3
- Teaching Aid 46 displays **Example 1** and the Geometric Addition Theorem.
- Teaching Aid 47 displays the Geometric Multiplication Theorem.

OBJECTIVES

B Perform operations with complex numbers.
F Prove or verify properties of complex numbers.
I Graph complex numbers and verify the Geometric Additon and Geometric Multiplication Theorems.

TEACHING NOTES

By the end of this lesson, students will have been introduced to four ways of writing complex numbers: binomial form $a + bi$; rectangular coordinates form (a, b); trigonometric form $r(\cos\theta + i\sin\theta)$; and polar form $[r, \theta]$. (A fifth common form is the exponential $re^{i\theta}$.) In the last two forms, $r \geq 0$ so that nth roots can be obtained later without confusion. It is a good idea to summarize these forms and their names at this point.

Point out that it is rather easy to multiply and divide complex numbers if they are in polar form; it is somewhat more difficult if they are in trigonometric form (only because there is more to write); and it is even more difficult if they are in $a + bi$ form.

LESSON

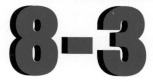

8-3 The Geometry of Complex Numbers

In Lesson 8-1, you learned to add, subtract, multiply, and divide complex numbers by treating complex numbers $a + bi$ as binomials and using the property $i^2 = -1$. These arithmetic operations can be interpreted geometrically using rectangular and polar coordinate systems in the complex plane.

Consider first the addition of complex numbers. Recall that if $z = a + bi$ and $w = c + di$, then their sum is

$$z + w = (a + c) + (b + d)i.$$

Since $z = (a, b)$ and $w = (c, d)$, in rectangular coordinates this addition is

$$(a, b) + (c, d) = (a + c, b + d).$$

There is a simple geometric interpretation for this addition.

Example 1 Suppose that $z = 4 + 6i$ and that $w = 3 - i$. Plot the points z, w, and $z + w$ in the complex plane and verify that the points representing z, 0, w, and $z + w$ are consecutive vertices of a parallelogram.

Solution In this case,

$$z + w = (4 + 6i) + (3 - i) = 7 + 5i.$$

Let
$z = 4 + 6i = (4, 6) \quad = Z,$
$0 = 0 + 0i = (0, 0) \quad = O,$
$w = 3 - i = (3, -1) \quad = W,$ and
$z + w = 7 + 5i = (7, 5) \quad = P.$

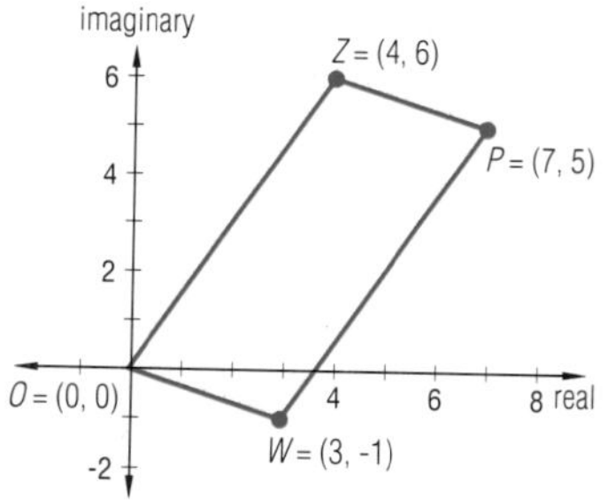

Plotting these four points results in the figure shown above. To prove that $ZOWP$ is a parallelogram, show that the diagonals bisect each other. The midpoint of $\overline{OP}$ is $\left(\frac{0+7}{2}, \frac{0+5}{2}\right) = \left(\frac{7}{2}, \frac{5}{2}\right)$, and the midpoint of $\overline{ZW}$ is $\left(\frac{4+3}{2}, \frac{6-1}{2}\right) = \left(\frac{7}{2}, \frac{5}{2}\right)$. Therefore, $ZOWP$ is a parallelogram.

The midpoint calculations in Example 1 can be carried out for any nonzero complex numbers $z = (a, b)$, $w = (c, d)$, and $z + w = (a + c, b + d)$. The result is the following theorem, whose proof is left as Question 2.

Geometric Addition Theorem

Let $z = a + bi$ and $w = c + di$ be two complex numbers that are not collinear with (0, 0). Then the point representing $z + w$ is the fourth vertex of a parallelogram with consecutive vertices $z = a + bi$, 0, and $w = c + di$.

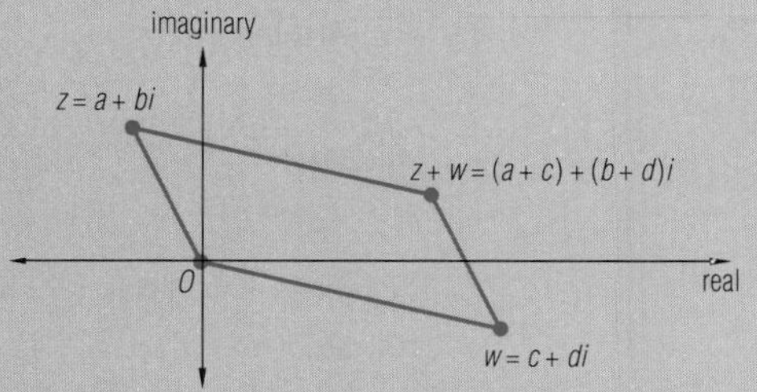

If you have studied vectors in a science class or a previous mathematics course, you may recognize that the rules for adding complex numbers are the same as those for vectors. In this book, vectors are studied in Chapter 12.

Another geometric interpretation of addition of complex numbers uses the language of transformations. Adding a complex number $z = (a, b)$ to a complex number w applies the translation a units horizontally and b units vertically to w.

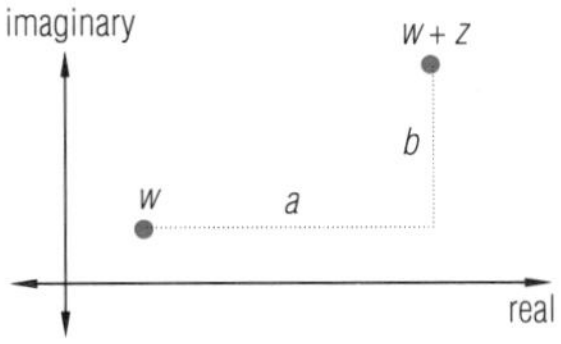

Multiplication of complex numbers can also be interpreted geometrically, but it is easier with polar coordinates. Let $[r, \theta]$ with $r \geq 0$ be the polar coordinates for (a, b) or $a + bi$. Then, from Lesson 8-2, you know that

$$r = \sqrt{a^2 + b^2} \text{ and } \tan\theta = \frac{b}{a}.$$

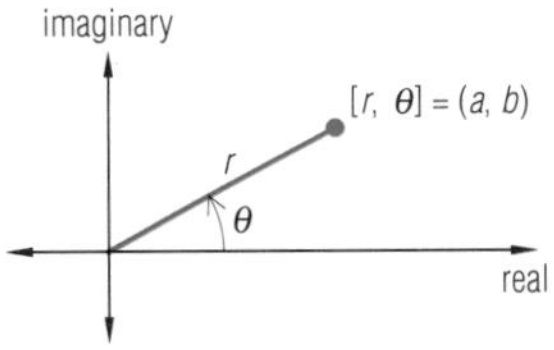

Since the period of the tangent function is π, there are always two values of θ between 0 and 2π that satisfy $\tan\theta = \frac{b}{a}$. The correct value is obtained by examining the quadrant in which (a, b) is located.

Example 2 Find the polar coordinates $[r, \theta]$ of $2 - 3i$ with $r \geq 0$ and $0 \leq \theta \leq 2\pi$.

Solution (2, -3) is a point in the 4th quadrant.

$r = \sqrt{2^2 + (-3)^2} = \sqrt{13}$

$\tan\theta = \frac{-3}{2} = -1.5$, so $\theta \approx -.983$ and a polar coordinate representation is $[\sqrt{13}, -.983]$. But that value of θ is not between 0 and 2π. Add 2π to get $\theta \approx -.983 + 2\pi \approx 5.3$. Thus, approximate polar coordinates for $2 - 3i$ are $[\sqrt{13}, 5.3]$.

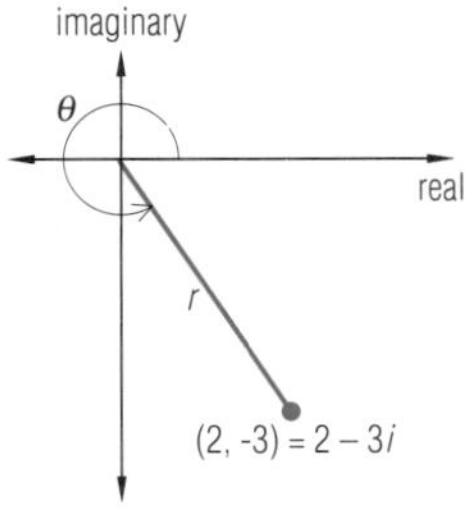

However, $a + bi$ form is easiest for addition and subtraction and relates better to work with polynomials.

Point out that the proof of the Geometric Multiplication Theorem begins with the rectangular representation of complex numbers, then converts the numbers to trigonometric form, and then expresses the final result in rectangular form. Each form has advantages and disadvantages in terms of doing calculations or proofs. Point out also that recognizing the expanded forms of $\sin(\theta + \phi)$ and $\cos(\theta + \phi)$ is crucial in completing the proofs. Hidden and proved in this proof is a theorem for multiplying complex numbers written in trigonometric form: $[r(\cos\theta + i\sin\theta)] \cdot [s(\cos\phi + i\sin\phi)] = rs[\cos(\theta + \phi) + i\sin(\theta + \phi)]$. Emphasize that to multiply two complex numbers, multiply their absolute values and add their arguments.

ADDITIONAL EXAMPLES

1. If $z = -2 + 3i$ and $w = 5 - 2i$, show that the graphs of z, O, w, and $z + w$ are vertices of a parallelogram.

Let $P = z + w = (3, 1)$. Now $Z = (-2, 3)$, $O = (0, 0)$, and $W = (5, -2)$.

Slope of $\overline{OZ} = \frac{3 - 0}{-2 - 0} = -\frac{3}{2}$; slope of $\overline{WP} = \frac{1 - -2}{3 - 5} = -\frac{3}{2}$; therefore, $\overline{OZ} \parallel \overline{WP}$.

Slope of $\overline{ZP} = \frac{1 - 3}{3 - -2} = -\frac{2}{5}$; slope of $\overline{OW} = \frac{-2 - 0}{5 - 0} = -\frac{2}{5}$; therefore, $\overline{ZP} \parallel \overline{OW}$. ZOWP is a parallelogram.

ADDITIONAL EXAMPLES

2. Find the polar coordinates $[r, \theta]$ of $z = -5 - 2i$ with $r \geq 0$ and $0 \leq \theta \leq 2\pi$.

$\left[\sqrt{29}, \pi + \tan^{-1}\left(\frac{2}{5}\right)\right] \approx [\sqrt{29}, 3.522]$

If possible, have students check this result with a scientific calculator that has a polar to rectangular conversion built-in.

3. Write the complex number $-5\sqrt{3} + 5i$ in trigonometric form.

$10\left(\cos\frac{5\pi}{6} + i \sin\frac{5\pi}{6}\right)$

NOTES ON QUESTIONS

Questions 3-5: Ask students to evaluate the sines and cosines of the trigonometric form and then simplify. The result should be in the form $a + bi$. This is a good check. You might also ask students to write each number in rectangular form.

Question 6: These exercises will be easier if students express each one in $[r, \theta]$ form.

Question 12: Students should also be able to explain why the product of these numbers is a real number. (The sum of their arguments is zero.)

Making Connections for Question 14: Be sure to make connections between this question and the Geometric Multiplication Theorem. Have students draw an accurate diagram in order to measure and compare arguments and modula of points and their images. Ask students what is equal to the ratio of similitude in the problem and why this is so.

Error Analysis for Question 20: Students who answer $\frac{5\pi}{8}$ have neglected to consider the range of the inverse sine function, that is, the domain of the restricted sine function.

For any complex number z, the distance from z to the origin is called the **absolute value** or **modulus** of z, and is written $|z|$. For any complex number $z = [r, \theta]$, $|z| = |r|$. This generalizes the idea of absolute value of real numbers. If $z = a + bi$, then $|z| = \sqrt{a^2 + b^2}$. When $z = [r, \theta]$ and $r \geq 0$, the form $[r, \theta]$ is called a **polar form** of z. A complex number has infinitely many polar forms, but if $z = [r_1, \theta_1] = [r_2, \theta_2]$, with $r_1 \geq 0$ and $r_2 \geq 0$, then $|z| = r_1 = r_2$ and $\theta_1 \equiv \theta_2 \pmod{2\pi}$. That is, θ_1 and θ_2 differ by a multiple of 2π.

If $[r, \theta]$ is a polar form of a complex number z, then θ is called an **argument** of z. Notice that if θ is an argument of z, the numbers $\theta \pm 2k\pi$ for any integer k are also arguments of z.

In Lesson 8-2, conversion from polar coordinates to rectangular coordinates was shown.

$$(a,b) = [r,\theta] \Rightarrow a = r\cos\theta \text{ and } b = r\sin\theta.$$

Substituting,

$$\begin{aligned} a + bi &= r\cos\theta + (r\sin\theta)i \\ &= r(\cos\theta + i\sin\theta). \end{aligned}$$

If $r > 0$, then the expression $r(\cos\theta + i\sin\theta)$ is called a **trigonometric form of the complex number**. Like polar coordinates, the trigonometric form denotes a complex number in terms of its absolute value and argument. But unlike polar coordinates, a complex number in trigonometric form is still in $a + bi$ form. This makes the trigonometric form quite useful.

Example 3 Write the complex number $-2 - 2\sqrt{3}\,i$ in trigonometric form.

Solution First find r and θ.

$$r = \sqrt{(-2)^2 + (-2\sqrt{3})^2} = \sqrt{4 + 12} = 4$$

$$\tan\theta = \left(\frac{-2\sqrt{3}}{-2}\right) = \sqrt{3}$$

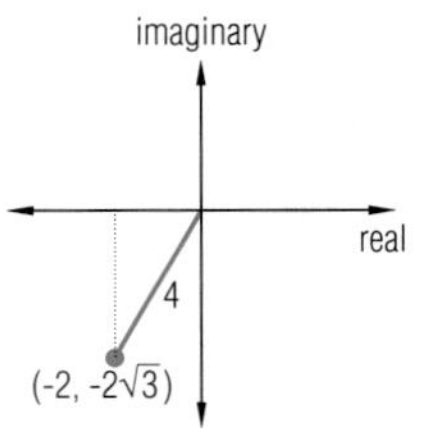

Since $-2 - 2\sqrt{3}\,i$ is in the 3rd quadrant,

$$\theta = \frac{\pi}{3} + \pi = \frac{4\pi}{3}.$$

Therefore, $-2 - 2\sqrt{3}\,i = 4\left(\cos\frac{4\pi}{3} + i\sin\frac{4\pi}{3}\right)$.

The trigonometric form of a complex number is not unique. You can add $2\pi n$ to the argument for any integer n. So,

$$\forall \text{ integers } n,\ -2 - 2\sqrt{3}\,i = 4\left(\cos\left(\frac{4\pi}{3} + 2\pi n\right) + i\sin\left(\frac{4\pi}{3} + 2\pi n\right)\right).$$

You have now seen four ways of writing complex numbers.

$a + bi$ form	rectangular form	polar form	trigonometric form
$a + bi$ =	(a, b) =	$[r, \theta]$ =	$r(\cos\theta + i\sin\theta)$

real part, imaginary part, absolute value, argument

482

The trigonometric form of complex numbers makes it possible to deduce the following geometric interpretation of multiplication of complex numbers.

Geometric Multiplication Theorem

Let z and w be complex numbers. If $z = [r, \theta]$ and $w = [s, \phi]$, then $zw = [rs, \theta + \phi]$. That is, multiplying a complex number z by w applies to z the composite of a size change of magnitude s and a rotation of ϕ about the origin.

Proof If $z = [r, \theta]$ and $w = [s, \phi]$, then in trigonometric form $z = r(\cos\theta + i\sin\theta)$ and $w = s(\cos\phi + i\sin\phi)$. So

$$z \bullet w = r(\cos\theta + i\sin\theta) \bullet s(\cos\phi + i\sin\phi)$$
$$= rs[(\cos\theta\cos\phi - \sin\theta\sin\phi) + i(\cos\theta\sin\phi + \sin\theta\cos\phi)].$$

Now apply the trigonometric identities for $\cos(\theta + \phi)$ and $\sin(\theta + \phi)$.

$$z \bullet w = rs[\cos(\theta + \phi) + i\sin(\theta + \phi)]$$

Finally, translate back into polar form.

$$z \bullet w = [rs, \theta + \phi]$$

For example, if $z = [4, 132°]$ and $w = [3, 95°]$, then a polar form of the product $z \bullet w$ is

$[4 \bullet 3, 132° + 95°] = [12, 227°]$.

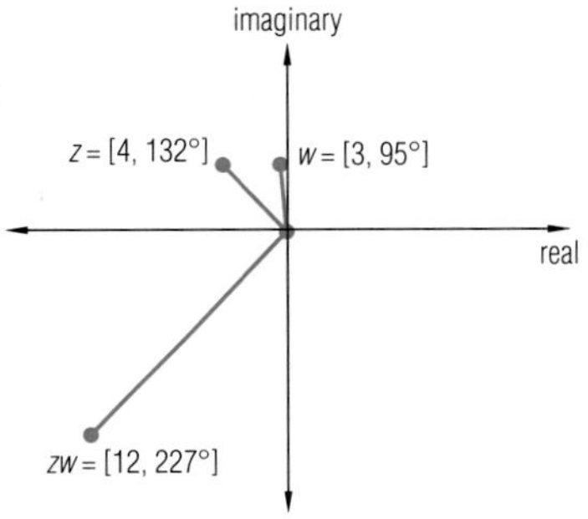

In general, multiplication of $z = [r, \theta]$ by $w = [s, \phi]$:

(1) produces a size change on z of magnitude s, because the size (absolute value) of z is r and the size of the product is rs; and

(2) rotates z through ϕ units because the argument of z is θ and the argument of the product is $\theta + \phi$.

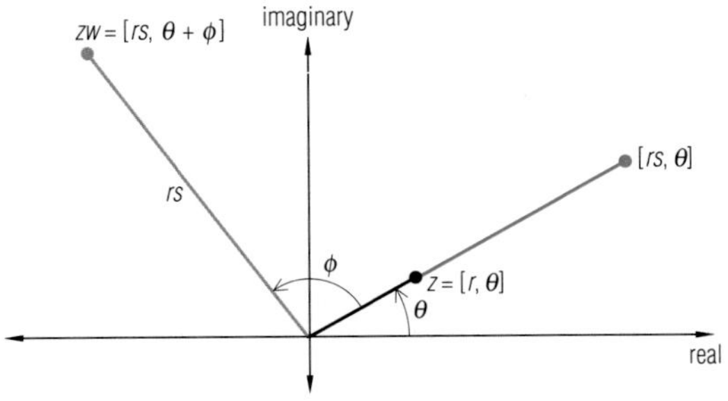

ADDITIONAL ANSWERS
(For Questions on p. 484)

3.a. $\left[3\sqrt{2}, \frac{3\pi}{4}\right]$

b. $3\sqrt{2}\left(\cos\frac{3\pi}{4} + i\sin\frac{3\pi}{4}\right)$

4.a. $\left[\frac{\sqrt{3}}{3}, 30°\right]$

b. $\frac{\sqrt{3}}{3}(\cos 30° + i\sin 30°)$

5.a. $\approx [5, 127°]$

b. $\approx 5(\cos 127° + i\sin 127°)$

6.a.–c.

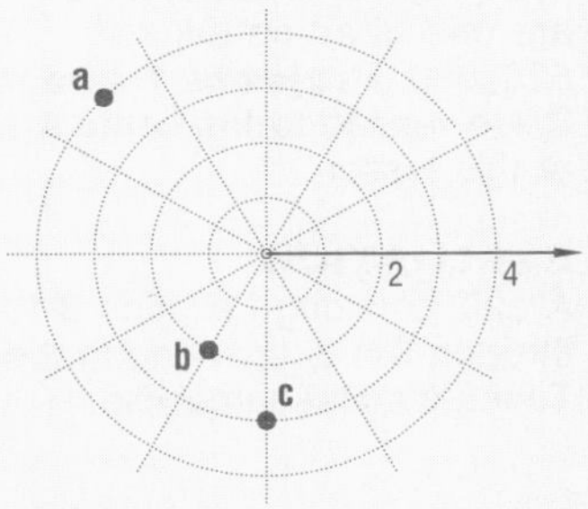

10.

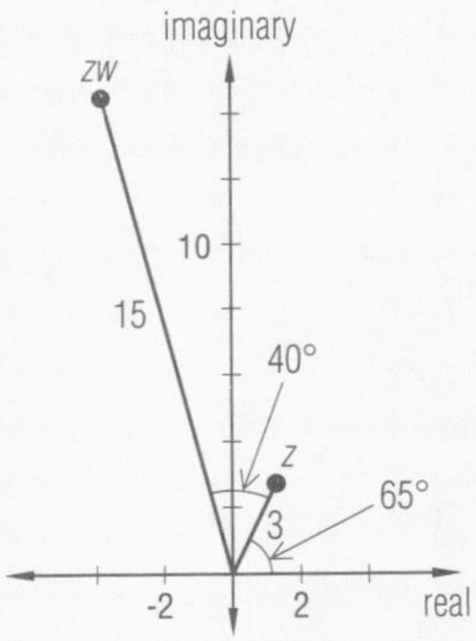

12. $\bar{z}$ **is the reflection image of** z **with respect to the real axis, so the fourth vertex of the parallelogram,** $z + \bar{z}$**, will be on the real axis, and** $z + \bar{z}$ **is a real number.**

13.b.

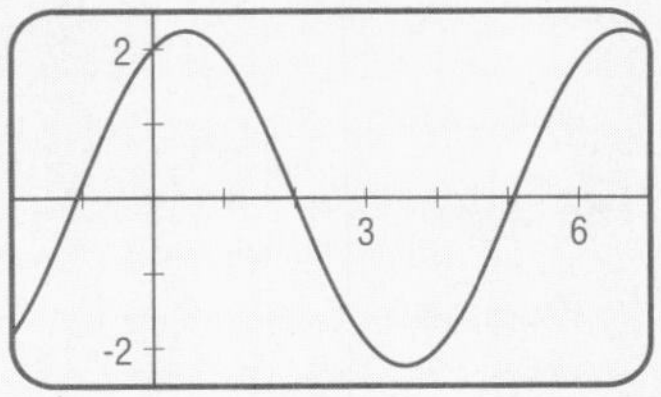

FOLLOW-UP

MORE PRACTICE
For more questions on SPUR Objectives, use *Lesson Master 8-3,* shown on page 485.

EXTENSION
In **Questions 23 and 24**, invite students to prove their theorems. Illustrations should accompany the proofs.

PROJECTS
The projects for Chapter 8 are described on pages 529–530. **Projects 1 and 2** are related to the content of this lesson.

EVALUATION
A quiz covering Lessons 8-1 through 8-3 is provided in the Teacher's Resource File.

ADDITIONAL ANSWERS

1. $z = 7 + 3i = (7, 3) = Z$; $o = 0 + 0i = (0, 0) = O$, $w = 4 - 9i = (4, -9) = W$; $z + w = 11 - 6i = (11, -6) = P$
The slope of $\overline{OZ} = \frac{3-0}{7-0} = \frac{3}{7}$; the slope of $\overline{WP} = \frac{-6-(-9)}{11-4} = \frac{3}{7}$; and so $\overline{OZ} \parallel \overline{WP}$.
The slope of $\overline{ZP} = \frac{-6-3}{11-7} = -\frac{9}{4}$; the slope of $\overline{OW} = \frac{-9-0}{4-0} = -\frac{9}{4}$; and so $\overline{ZP} \parallel \overline{OW}$. Therefore, the figure is a parallelogram.

2. $z = a + bi = (a, b) = Z$; $o = 0 + 0i = (0, 0) = O$; $w = c + di = (c, d) = W$; $z + w = (a + c) + (b + d)i = (a + c, b + d) = P$
The slope of $\overline{OZ} = \frac{b-0}{a-0} = \frac{b}{a}$; the slope of $\overline{WP} = \frac{b+d-d}{a+c-c} = \frac{b}{a}$; and so $\overline{OZ} \parallel \overline{WP}$.
The slope of $\overline{ZP} = \frac{b+d-b}{a+c-a} = \frac{d}{c}$; the slope of $\overline{OW} = \frac{d-0}{c-0} = \frac{d}{c}$; and so $\overline{ZP} \parallel \overline{OW}$. Therefore, the figure is a parallelogram.

Questions

Covering the Reading

1. Verify the Geometric Addition Theorem for the complex numbers $z = 7 + 3i$ and $w = 4 - 9i$. See margin.

2. Prove the Geometric Addition Theorem for the complex numbers $z = a + bi$ and $w = c + di$, when $z \neq 0$ and $w \neq 0$. See margin.

In 3–5, write each number **a.** in polar form; **b.** in trigonometric form.

3. $-3 + 3i$ **4.** $\frac{1}{2} + \frac{1}{2\sqrt{3}}i$ **5.** $-3 + 4i$
3-5) See margin.

6. Graph the numbers on the same complex plane.
a. $4\left(\cos \frac{3\pi}{4} + i \sin \frac{3\pi}{4}\right)$
b. $2\left(\cos\left(-\frac{2\pi}{3}\right) + i \sin\left(-\frac{2\pi}{3}\right)\right)$
c. $3\left(\cos \frac{3\pi}{2} + i \sin \frac{3\pi}{2}\right)$ a-c) See margin.

In 7–9, find zw. Express results in the form of the given numbers.

7. $z = [3, 150°]$ $w = [2, 60°]$ [6, 210°]

8. $z = 10\left(\cos \frac{11\pi}{12} + i \sin \frac{11\pi}{12}\right)$ $w = 5\left(\cos \frac{\pi}{4} + i \sin \frac{\pi}{4}\right)$
$50\left(\cos \frac{7\pi}{6} + i \sin \frac{7\pi}{6}\right)$

9. $z = 2 + 3i$ $w = -4 + i$ $-11 - 10i$

10. Illustrate the multiplication of $z = 3(\cos 65° + i \sin 65°)$ by $w = 5(\cos 40° + i \sin 40°)$ with a diagram showing the appropriate size transformation and rotation. See margin.

11. Find the modulus and an argument θ for the imaginary number $-3i$.
modulus: 3, argument: 270°

Applying the Mathematics

12. Explain geometrically why $z + \bar{z}$ is a real number for any complex number z. See margin.

13. The AC circuit in Example 4 of Lesson 8-1 has a current of $w = 1 + 2i$ ohms. This can be interpreted as follows. If an oscilloscope were connected to the circuit in order to graph the rate of flow of electrons over time, a sine curve would result. The amplitude of the sine curve would be equal to $|w|$, and the phase shift would be equal to $-\theta$ where θ is the argument of w.
a. Find $|w|$ and $-\theta$. $|w| = \sqrt{5}$, $-\theta \approx -63.4°$
b. Graph the sine curve with this amplitude and phase shift and period 2π. See margin.

14. Consider the triangle in the complex plane with vertices $A = 3 + i$, $B = 2 - i$, and $C = 2 - 3i$.
a. Multiply each of the vertices by $z = 1 + 2i$ to obtain points A', B', and C'. $A' = 1 + 7i$, $B' = 4 + 3i$, $C' = 8 + i$
b. Show that the triangles ABC and $A'B'C'$ are similar. See margin.
c. What is the ratio of similitude? $\sqrt{5}$

484

15. Consider the triangle in the complex plane with vertices at $E = 4$, $F = 3i$, and $G = 2 + 5i$.

a. Add the complex number $1 + i$ to each of the vertices E, F, and G to obtain points E', F', and G'. $E' = 5 + i$, $F' = 1 + 4i$, $G' = 3 + 6i$

b. Show that $\triangle E'F'G'$ is congruent to $\triangle EFG$. (Hint: Find lengths of corresponding sides). See below.

c. Describe the geometric transformation which maps $\triangle EFG$ onto $\triangle E'F'G'$. $T_{1,1}$

16. a. Show that $|12 - 15i| = 3|4 - 5i|$.

b. Generalize part **a** and prove your generalization. a,b) See below.

Review

17. a. Graph all solutions to the equation $r = 4$ on a polar grid.

b. What equation in x and y do all such solutions satisfy? *(Lesson 8-2)* a) See margin. b) $x^2 + y^2 = 16$

18. Simplify $(18 - \sqrt{-100}) + (-4 + 3\sqrt{-36})$. *(Lesson 8-1)* $14 + 8i$

19. Prove or disprove: *For all complex numbers, the sum of the number and its complex conjugate is a real number.* *(Lesson 8-1)* See below.

20. Find the exact value of $\sin^{-1}\left(\sin \frac{5\pi}{8}\right)$. *(Lesson 6-6)* $\frac{3\pi}{8}$

21. Consider the graph at the right.

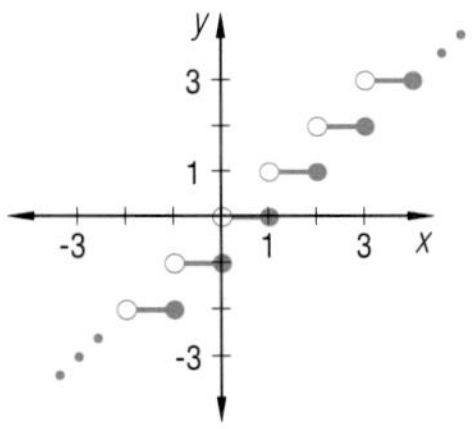

a. Does the graph represent a function? Yes

b. If so, identify its domain and range. If not, why not? *(Lessons 2-2, 2-1)*

b) domain: the set of real numbers; range: the set of integers

22. Complete the square to find the center and radius of the circle with equation $x^2 + 2x + y^2 - 4y = 4$. *(Previous course)*

center = (-1, 2), radius = 3

Exploration

23. What could be a Geometric Subtraction Theorem? See margin.

24. What could be a Geometric Division Theorem? See margin.

15. b) $EF = 5$, $EG = \sqrt{29}$, and $FG = 2\sqrt{2}$, while $E'F' = 5$, $E'G' = \sqrt{29}$, and $F'G' = 2\sqrt{2}$. So $\triangle EFG \cong \triangle E'F'G'$ by the SSS Congruence Theorem.

16. a) $|12 - 15i| = \sqrt{12^2 + (-15)^2} = \sqrt{144 + 225} = \sqrt{369} = 3\sqrt{41}$, and $3|4 - 5i| = 3\sqrt{4^2 + (-5)^2} = 3\sqrt{16 + 25} = 3\sqrt{41}$.

b) $\forall$ real numbers c and complex numbers $z = a + bi$, $|cz| = |c||z|$. Proof: $|cz| = |ca + cbi| = \sqrt{c^2a^2 + c^2b^2} = |c|\sqrt{a^2 + b^2} = |c||a + bi| = |c||z|$.

19. Let $z = a + bi$, where a and b are real numbers. Then $\bar{z} = a - bi$. $z + \bar{z} = (a + bi) + (a - bi) = 2a$. Since a is real, $2a$ is real. So for all complex numbers, the sum of the number and its complex conjugate is a real number.

3.-5., 6, 10., 12., 13.b. See the margin on page 483.

14.b., 17.a., 23., 24. See Additional Answers in the back of this book.

NAME ______

LESSON **MASTER 8–3**
QUESTIONS ON **SPUR** OBJECTIVES

■ SKILLS *Objective A (See pages 534–536 for objectives.)*
In 1–4, the complex number is written in either binomial, rectangular, polar, or trigonometric form. Write it in the other three forms.

1. 10
(10, 0)
[10, 0°]
$10(\cos 0° + i \sin 0°)$

2. $[-3, \frac{3\pi}{2}]$
$3i$
(0, 3)
$-3\left(\cos \frac{3\pi}{2} + i \sin \frac{3\pi}{2}\right)$

3. $(7\sqrt{3}, -7)$
$7\sqrt{3} - 7i$
[14, 330°]
$14(\cos 330° + i \sin 330°)$

4. $4[\cos(-120°) + i \sin(-120°)]$
$-2 - 2\sqrt{3}i$
$(-2, -2\sqrt{3})$
[4, -120°]

In 5 and 6, find the modulus and the argument θ ($0 \le \theta < 2\pi$) of the complex number.

5. $15 - 4i$
modulus: ≈ 15.5
$\theta \approx 6.02$

6. $a - ai$ where $a > 0$
modulus: $a\sqrt{2}$
$\theta = \frac{7\pi}{4}$

■ SKILLS *Objective B*
7. Let $z = \sqrt{3}(\cos 25° + i \sin 25°)$ and $w = \sqrt{2}(\cos 60° + i \sin 60°)$.

a. Find z^2. $3(\cos 50° + i \sin 50°)$

b. Find $z \cdot w$. $\sqrt{6}(\cos 85° + i \sin 85°)$

In 8 and 9, solve for z, writing the answer in polar form.

8. $z = [3, 115°] \cdot [4, 290°]$
[12, 45°]

9. $[8, 50°] \cdot z = [6, 172°]$
$[\frac{3}{4}, 122°]$

■ PROPERTIES *Objective F*
10. Let z be a complex number with argument θ. Show that $z \cdot [1, -2\theta] = \bar{z}$.
Let $z = [r, \theta]$. Then $z \cdot [1, -2\theta] = [r, \theta] \cdot [1, -2\theta] = [r, -\theta] = \bar{z}$.

90 *Continued* *Precalculus and Discrete Mathematics* © Scott, Foresman and Company

NAME ______
Lesson MASTER 8-3 (page 2)

11. What does multiplying a complex number by a real number do to its polar representation?
Let r be the real number. If $r > 0$, the argument remains the same. If $r < 0$, the argument increases by π. In each case, the modulus is multiplied by $|r|$.

■ REPRESENTATIONS *Objective I*
12. Let $G = 0$, $H = -2 - 3i$, $I = 3 - i$, and $J = H + I$.

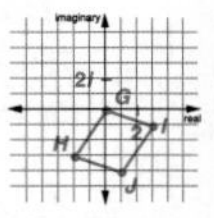

a. Sketch quadrilateral $GHJI$ in the complex plane.

b. What kind of figure is $GHJI$? parallelogram

c. Prove your answer to part b.
Slope of $\overline{GH} = \frac{3}{2}$ = slope of $\overline{IJ}$, so $\overline{GH} \parallel \overline{IJ}$. Slope of $\overline{GI} = -\frac{1}{3}$ = slope of $\overline{HJ}$, so $\overline{GI} \parallel \overline{HJ}$. Therefore, $GHJI$ is a parallelogram.

13. Illustrate the multiplication of $z = 2(\cos 75° + i \sin 75°)$ by $w = 4(\cos 45° + i \sin 45°)$ with a diagram showing the appropriate size transformation and rotation.

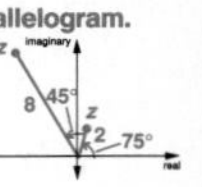

14. a. Graph $\triangle ABC$ in the complex plane where $A = 0$, $B = 3 - 4i$, and $C = -2 - i$.

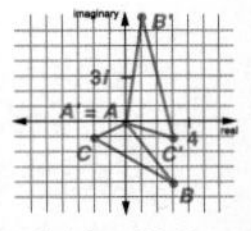

b. Let A', B', and C' be the images of A, B, and C under the transformation $f: z \to (-1 + i)z$. Graph $\triangle A'B'C'$.

c. Find the ratio of similitude of $\triangle A'B'C'$ to $\triangle ABC$ and relate it to the Geometric Multiplication Theorem.
$-1 + i = [\sqrt{2}, \frac{3\pi}{4}]$; Multiplication by $-1 + i$ multiplies the modulus by $\sqrt{2}$. The ratio of similitude is $\sqrt{2}$.

d. How does the argument of C' compare to that of C? Use the Geometric Multiplication Theorem.
Multiplying by $[\sqrt{2}, \frac{3\pi}{4}]$ increases the argument by $\frac{3\pi}{4}$.

Precalculus and Discrete Mathematics © Scott, Foresman and Company 91

LESSON 8-4

RESOURCES
- Lesson Master 8-4
- Teaching Aid 48 displays the table and graph for **Example 1**.
- Teaching Aid 49 displays the table and graphs for **Example 3**.
- Teaching Aid 50 displays the graphs for **Example 4**.

OBJECTIVE

J Sketch graphs of polar equations.

TEACHING NOTES

Sketching the polar graph of an equation $r = f(\theta)$ by constructing a table of values and then plotting the points is easy for most students. The second method discussed in this lesson uses the graph of the rectangular equation to get the polar graph, and thus takes advantage of students knowledge of rectangular graphs.

In **Example 1**, some students may notice a pattern in the right column of the table: $\sqrt{0}$, $\sqrt{1}$, $\sqrt{2}$, $\sqrt{3}$, $\sqrt{4}$. This is just happenstance.

Alternate Approach
Some students may have difficulty plotting those points for which $r < 0$. To help, suggest a transparent ruler. To plot the point $\left[-\sqrt{2}, \frac{5\pi}{4}\right]$, place the ruler along the $\theta = \frac{5\pi}{4}$ ray. The point will be $\sqrt{2}$ (≈ 1.4) from the origin on the ray opposite $\theta = \frac{5\pi}{4}$.

LESSON

8-4

Polar Equations and Their Graphs

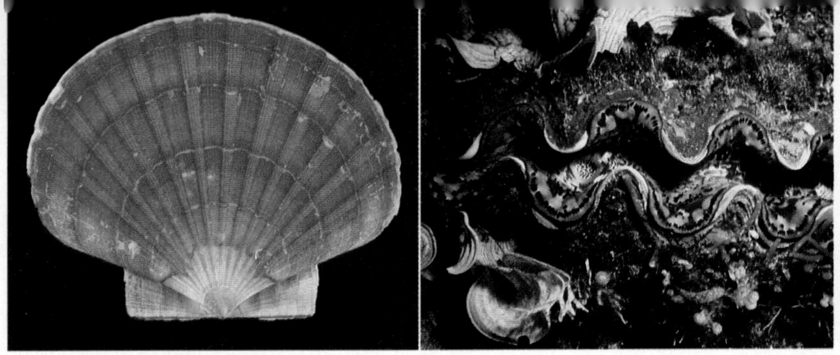

Polar graphs model shapes in nature.

Polar coordinates have a variety of applications. In the last lesson, you saw a connection between complex numbers and polar coordinates, and in Lesson 8-2, you saw that circles centered at the origin have very simple equations in polar coordinates. Some other circles also have simple **polar equations**, as Examples 1 and 2 show.

Example 1 Sketch the graph of the set of points $[r, \theta]$ that satisfy the polar equation $r = 2 \sin \theta$.

Solution Construct a table of ordered pairs $[r, \theta]$ that satisfy the equation. Then plot these points on a polar grid and connect successive points with a smooth curve.

θ	$r = 2 \sin \theta$
0	0
$\pi/6$	1
$\pi/4$	$\sqrt{2}$
$\pi/3$	$\sqrt{3}$
$\pi/2$	2
$2\pi/3$	$\sqrt{3}$
$3\pi/4$	$\sqrt{2}$
$5\pi/6$	1
π	0
$7\pi/6$	-1
$5\pi/4$	$-\sqrt{2}$
$4\pi/3$	$-\sqrt{3}$
$3\pi/2$	-2
$5\pi/3$	$-\sqrt{3}$
$7\pi/4$	$-\sqrt{2}$
$11\pi/6$	-1
2π	0

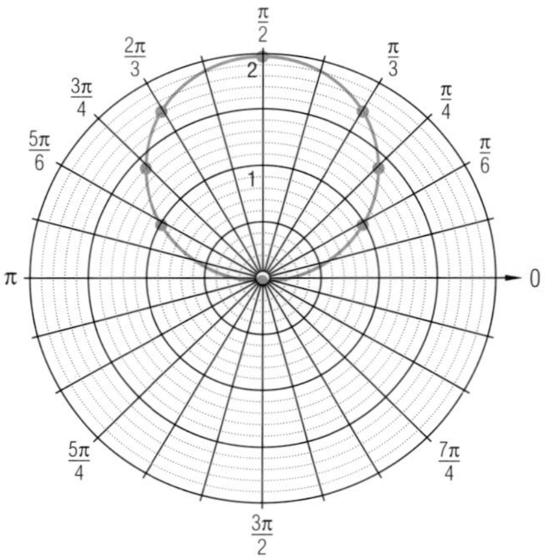

Notice that when $\pi < \theta < 2\pi$, r is negative and so these points coincide with points plotted when $0 \le \theta \le \pi$. This indicates it would have been sufficient to plot points $[r, \theta]$ when $0 \le \theta \le \pi$. Also, since the sine function has period 2π, no other values of θ need be considered.

The graph in Example 1 can be proved to be a circle by applying the conversion formulas between rectangular and polar coordinates.

Example 2 Prove that the graph of $r = 2 \sin \theta$ is a circle.

Solution The idea is to show that the polar equation is equivalent to a rectangular equation whose graph is known. There are three conversion formulas:
$r^2 = x^2 + y^2$, $y = r \sin \theta$, and $x = r \cos \theta$.
The first two of these are useful here.

$[r, \theta]$ is a point on $r = 2 \sin \theta$

$\Rightarrow$ $\quad r^2 = 2r \sin \theta$

$\Rightarrow$ $\quad x^2 + y^2 = 2y$ $\quad$ **Substituting using the two formulas**

This is known to be an equation for a circle. To find its center and radius, complete the square.

$$x^2 + y^2 - 2y = 0$$
$$x^2 + (y^2 - 2y + 1) = 1 \qquad \textbf{Complete the square on } y.$$
$$x^2 + (y - 1)^2 = 1$$

This verifies that all points of the polar graph of $r = 2 \sin \theta$ lie on the circle with center (0, 1) and radius 1. To prove that no point on the circle is missing from the polar graph, reverse the steps. Suppose (x, y) lies on this circle. Then $x^2 + (y - 1)^2 = 1$, and so $2y = x^2 + y^2$. Let $[r, \theta] = (x, y)$. Then $r \sin \theta = y$ and $r^2 = x^2 + y^2$, so $2r \sin \theta = r^2$. That is, $r = 2 \sin \theta$ or $r = 0$. Since $[0, 0]$ satisfies $r = 2 \sin \theta$, the graph of $x^2 + (y - 1)^2 = 1$ is identical to the graph of $r = 2 \sin \theta$.

This procedure can be used to show that when a is any nonzero real number, the polar graphs of the equations $r = a \cos \theta$ and $r = a \sin \theta$ are circles.

Example 3 shows that a seemingly minor change in the polar equation considered in Example 1 can result in a quite different polar graph.

Example 3 Sketch the graph of the polar equation $r = 1 + 2 \sin \theta$.

Solution Construct a table of values for this equation.

θ	0	$\frac{\pi}{4}$	$\frac{\pi}{2}$	$\frac{3\pi}{4}$	π	$\frac{5\pi}{4}$	$\frac{3\pi}{2}$	$\frac{7\pi}{4}$	2π
r	1	$1 + \sqrt{2}$	3	$1 + \sqrt{2}$	1	$1 - \sqrt{2}$	-1	$1 - \sqrt{2}$	1

Notice that r is negative when $\theta = \frac{5\pi}{4}$, $\theta = \frac{3\pi}{2}$, and $\theta = \frac{7\pi}{4}$. Also, $r = 0$ for those values of θ for which $1 + 2 \sin \theta = 0$; that is, when $\sin \theta = -\frac{1}{2}$. Now in the interval $0 \le \theta \le 2\pi$, $\sin \theta = -\frac{1}{2}$ when $\theta = \frac{7\pi}{6}$ or $\frac{11\pi}{6}$. Thus the value of r is negative throughout the interval $\frac{7\pi}{6} < \theta < \frac{11\pi}{6}$. When successive points in the table are plotted and connected with a smooth curve, a graph like the one shown on the next page occurs.

Error Analysis Students may think because the graphs of $r = 2 \sin \theta$ or $r = 1 + 2 \sin \theta$ fail the vertical line test, these euqations are not functions. Functions in polar coordinates are of the form $f: \theta \rightarrow r$, and their graphs do not need to satisfy either a vertical or horizontal line test. Stress that even though ordered pairs are written $[r, \theta]$, θ is the independent variable and r is the dependent variable. Thus, we write $r = f(\theta)$. In polar coordinates, the independent variable is listed second, whereas in rectangular coordinates, the independent variable x is first.

There are advantages and disadvantages associated with the fact that a point in the plane has many polar representations. The main disadvantage is that extra care is necessary when information about polar curves is derived from their equations. For example, the circles given by the polar equations
$r = \sin \theta$
$r = \cos \theta$
intersect at the pole and at the point $\left[\frac{\sqrt{2}}{2}, \frac{\pi}{4}\right]$.

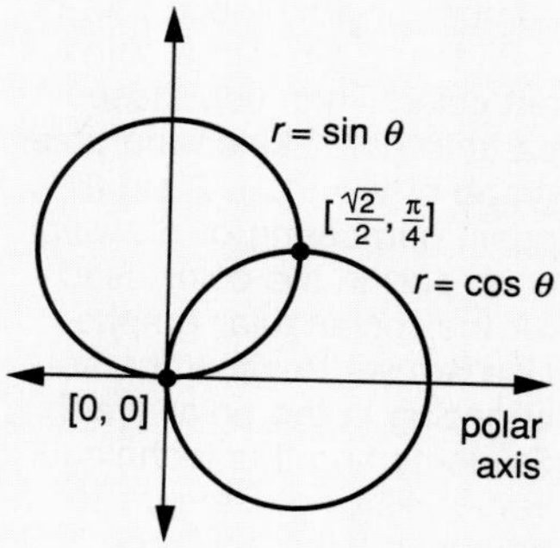

However, only the point $\left[\frac{\sqrt{2}}{2}, \frac{\pi}{4}\right]$ satisfies the given polar equations simultaneously. The pole is on both circles because one of its polar representations, $[0, 0]$, satisfies $r = \sin \theta$ while another, $\left[0, \frac{\pi}{2}\right]$, satisfies $r = \cos \theta$. The nonuniqueness of a polar representation, however, also is the reason why the polar coordinate system provides effective graphs of some equations, and this advantage outweighs the disadvantages involved.

As polar graphs become more complicated, students may lose track of the sequencing of points to be plotted. Suggest that they take the table of ordered pairs used in **Example 1** and divide it into four parts: $0 \le \theta < \frac{\pi}{2}$, $\frac{\pi}{2} \le \theta < \pi$, $\pi < \theta \le \frac{3\pi}{2}$, and $\frac{3\pi}{2} < \theta \le 2\pi$. Have students use a different color pencil for each part of the table and the same color for sketching the graph. If students do this for an easy problem like Example 1, they will build a useful technique for succeeding with the more complicated problems in the next lesson.

Alternate Approach
You may wish to embellish upon the method of using a rectangular graph of $r = 1 + 2 \sin \theta$ to assist in sketching the polar graph. Divide the rectangular graphs into intervals of width $\frac{\pi}{2}$ and either number or draw each part of $y = 1 + 2 \sin \theta$ in a different color. Then use these parts to help sketch the polar graph of $y = 1 + 2 \sin \theta$, again numbering or drawing each part in the color used for the rectangular graphs. Use arrows to show the sequencing in the polar graph. Students find this technique very satisfying. Be sure to emphasize that the rectangular graph is for assistance only; it is not the polar graph.

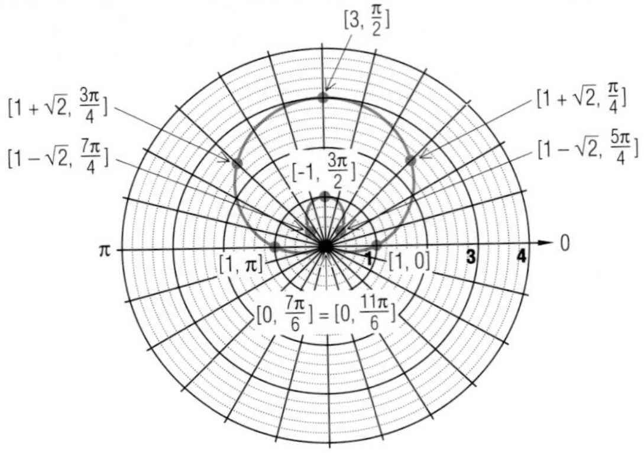

The polar graph of $r = 1 + 2 \sin \theta$ constructed in Example 3 is a type of curve known as a **limaçon** (pronounced lim′ ə son′). Limaçon is an Old French word for "snail." The polar graphs of equations of the form

$$r = a + b \cos \theta$$

or

$$r = a + b \sin \theta,$$

where a and b are nonzero real numbers, are all limaçons. To have an inside loop like the limaçon in Example 3, r must be negative for some interval of values of θ; otherwise, the limaçon simply has a "dimple" like the graph of $r = 3 + 2 \sin \theta$ displayed below.

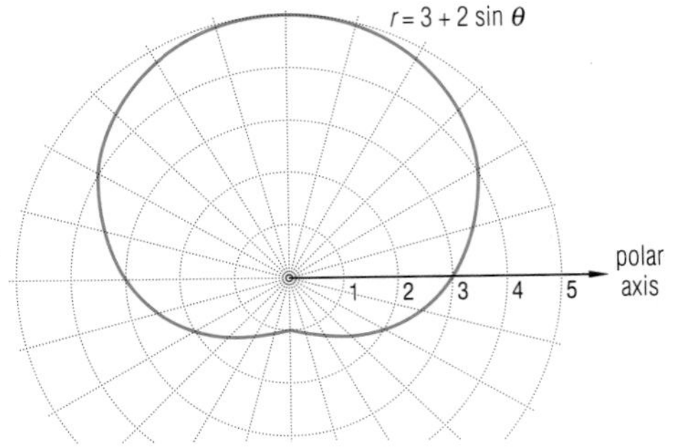

The rectangular graph and the polar graph of a given equation $r = f(\theta)$ are usually strikingly different. For example, while the polar graph of $r = 1 + 2 \sin \theta$ on $0 \le \theta \le 2\pi$ is the limaçon of Example 3, the rectangular graph of this equation is a sine curve of amplitude 2 translated 1 unit up the vertical r-axis.

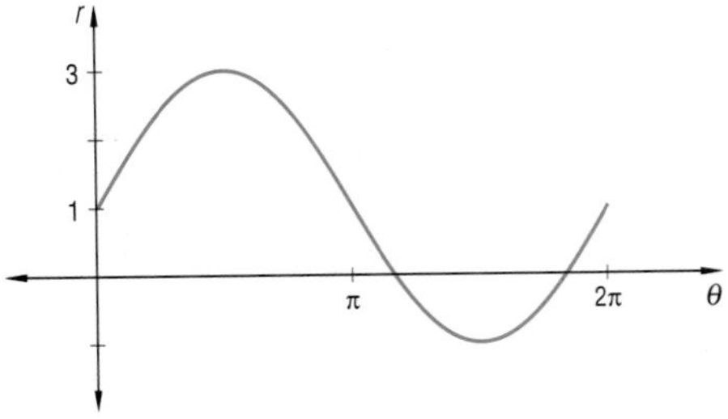

488

Although the polar and rectangular graphs of an equation $r = f(\theta)$ have very different appearances, there is a close geometric relationship between the two that is helpful for graphing polar equations. This relationship is illustrated in the figures below.

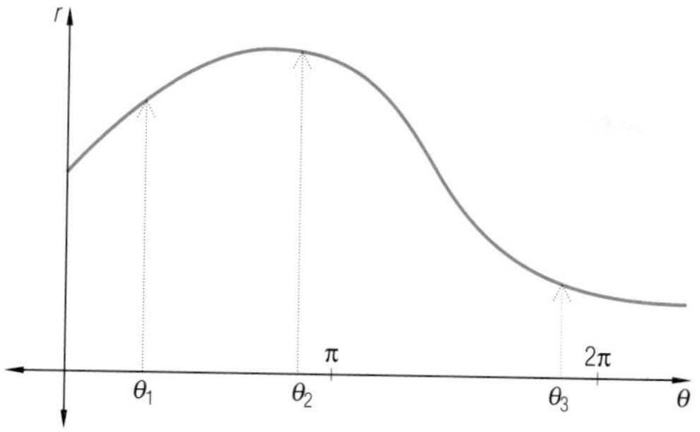

Rectangular graph of $r = f(\theta)$

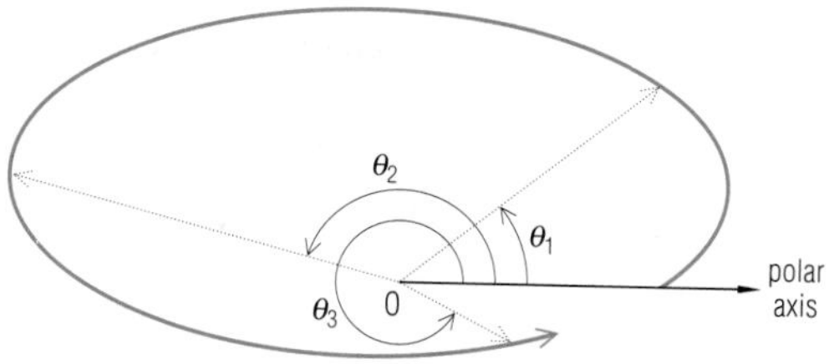

Polar graph of $r = f(\theta)$

The vertical arrows in the rectangular graph correspond to the radial arrows in the polar graph as follows: For each value of θ, the length of the vertical arrow from the θ-axis to the rectangular graph of $r = f(\theta)$ equals the length of the segment from the pole to the polar graph at an angle θ to the polar axis. Below, this relationship between the rectangular and polar graphs is illustrated for the equation $r = 1 + 2 \sin \theta$.

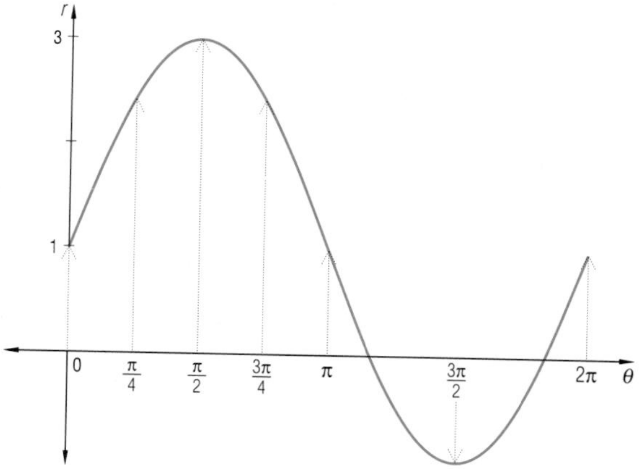

Rectangular graph of
$r = 1 + 2 \sin \theta$
$0 \le \theta \le 2\pi$

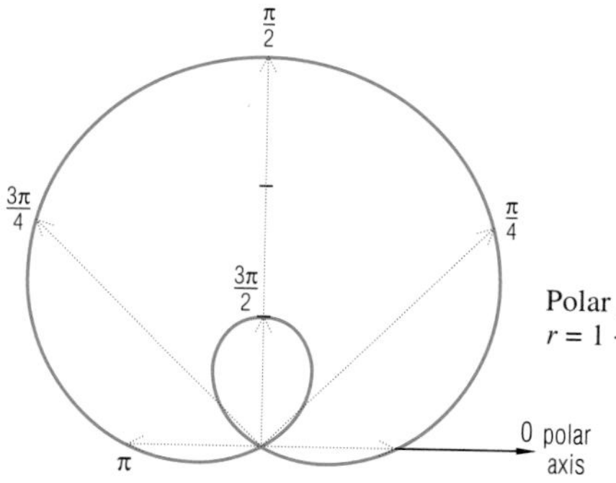

Polar graph of
$r = 1 + 2 \sin \theta$

Some automatic graphers plot polar graphs directly; others do not. If yours does not, you can still plot a polar graph by taking advantage of the relationship described above. One way to do this is to store a program in a grapher's memory that transforms the rectangular graph of an equation to the polar graph. Your teacher or the manual for the grapher may provide instructions for doing this. However, an automatic grapher can help you to sketch polar graphs even if it is not programmed to produce polar plots. The following example shows you how to do this.

ADDITIONAL EXAMPLES

1. Sketch the polar equation $r = 2 \cos \theta$.

The graph contains the following points:
$[2, 0], \left[\sqrt{3}, \frac{\pi}{6}\right], \left[\sqrt{2}, \frac{\pi}{4}\right], \left[1, \frac{\pi}{3}\right], \left[0, \frac{\pi}{2}\right], \left[-1, \frac{2\pi}{3}\right], \left[-\sqrt{2}, \frac{3\pi}{4}\right], \left[-\sqrt{3}, \frac{5\pi}{6}\right], [-2, \pi], \left[-\sqrt{3}, \frac{7\pi}{6}\right], \left[-\sqrt{2}, \frac{5\pi}{4}\right], \left[-1, \frac{4\pi}{3}\right], \left[0, \frac{3\pi}{2}\right], \left[1, \frac{5\pi}{3}\right], \left[\sqrt{2}, \frac{7\pi}{4}\right], \left[\sqrt{3}, \frac{11\pi}{6}\right], [2, 2\pi].$

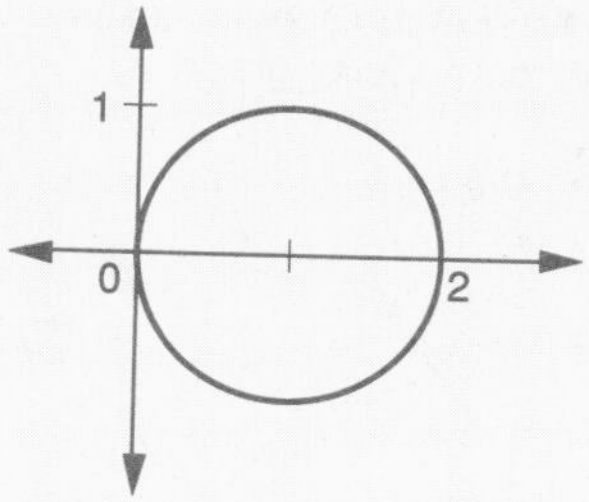

2. Prove that the polar graph of $r = 2 \cos \theta$ is a circle.

Use the substitutions $r = \sqrt{x^2 + y^2}$; $\cos \theta = \frac{x}{\sqrt{x^2 + y^2}}$. **Then,** $r = 2 \cos \theta$ **becomes** $\sqrt{x^2 + y^2} = \frac{2x}{\sqrt{x^2 + y^2}}$ **or** $(x - 1)^2 + y^2 = 1$, **which is a circle with center at (1, 0) and radius equal to 1.**

3. Sketch the graph of the polar equation $r = 1 + \sin \theta$. Some points on the graph are:

$[1, 0], \left[1\frac{1}{2}, \frac{\pi}{6}\right], \left[\frac{2 + \sqrt{2}}{2}, \frac{\pi}{4}\right], \left[\frac{2 + \sqrt{3}}{2}, \frac{\pi}{3}\right], \left[2, \frac{\pi}{2}\right], \left[\frac{2 + \sqrt{3}}{2}, \frac{2\pi}{3}\right], \left[\frac{2 + \sqrt{2}}{2}, \frac{3\pi}{4}\right], \left[1\frac{1}{2}, \frac{5\pi}{6}\right], [1, \pi], \left[\frac{1}{2}, \frac{7\pi}{6}\right], \left[\frac{2 - \sqrt{2}}{2}, \frac{5\pi}{4}\right], \left[\frac{2 - \sqrt{3}}{2}, \frac{4\pi}{3}\right], \left[0, \frac{3\pi}{2}\right], \left[\frac{2 - \sqrt{3}}{2}, \frac{5\pi}{3}\right], \left[\frac{2 - \sqrt{2}}{2}, \frac{7\pi}{4}\right], \left[\frac{1}{2}, \frac{11\pi}{6}\right], [1, 2\pi].$

(The graph for this example is shown on the next page.)

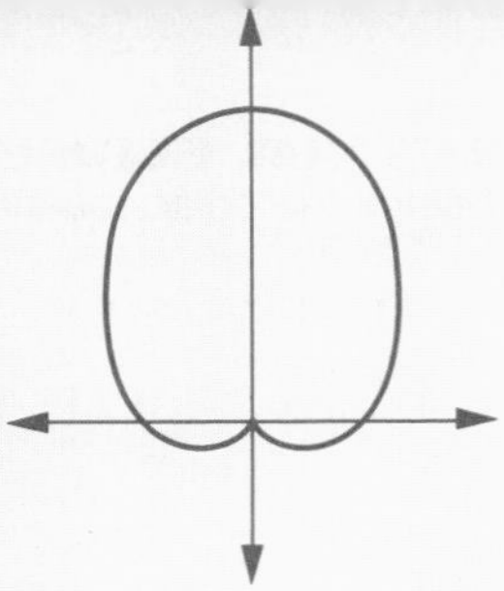

$r = 1 + \sin \theta$ is a cardioid.

4. Use the rectangular graph of $r = \sin \theta - 1$ on $0 \le \theta \le 2\pi$ to sketch its polar graph. The rectangular graph is:

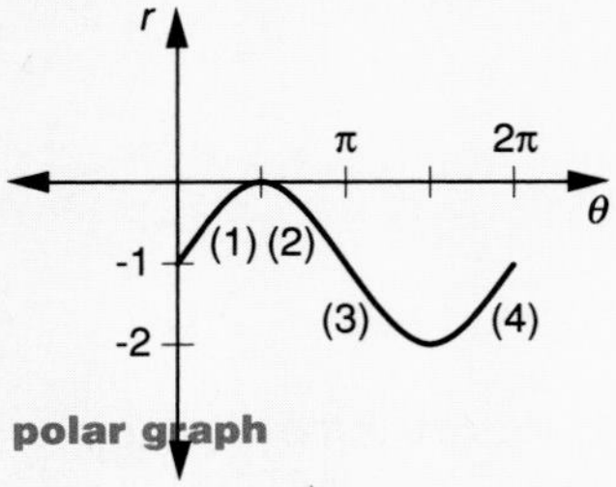

polar graph

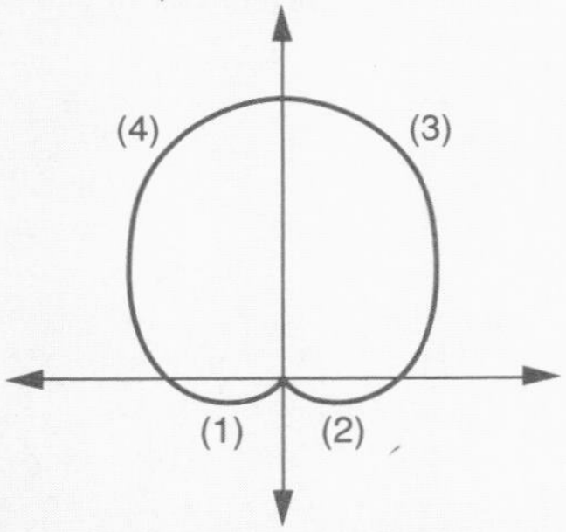

This curve is called a cardioid.

NOTES ON QUESTIONS
Question 7: This curve is the simplest Archimedian spiral. The general form is discussed in the next lesson.

Question 14: Students should check their answer by looking to see that ΔABC and its image are congruent.

Example 4 Use the rectangular graph of $r = 3 + 2 \cos \theta$ on $0 \le \theta \le 2\pi$ to sketch its polar graph.

Solution Because this is an equation of the form $r = a + b \cos \theta$, the polar graph will be a limaçon. Use a grapher to plot the rectangular graph of the equation.

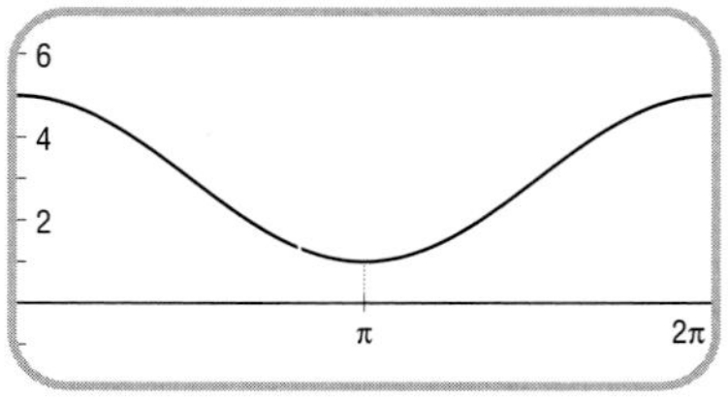

Rectangular graph of $r = 3 + 2 \cos \theta$; $0 \le \theta \le 2\pi$

Notice that as θ increases from 0 to 2π, the value of r decreases from its maximum value of 5 at $\theta = 0$ to its minimum value of 1 at $\theta = \pi$, and then it increases to its maximum value of 5 again at $\theta = 2\pi$. Consequently, in the corresponding polar graph, the lengths of the segments from the pole decrease from a maximum length of 5 at $\theta = 0$ to the minimum length of 1 at $\theta = \pi$, and then increase to a maximum length of 5 at $\theta = 2\pi$. Thus the polar graph of $r = 3 + 2 \cos \theta$ can be sketched as below.

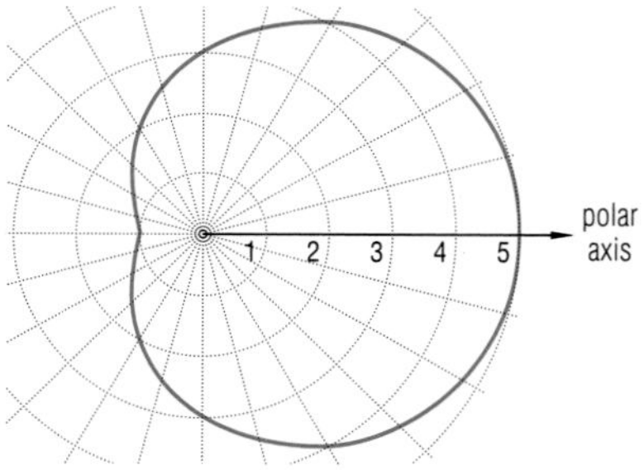

Polar graph of $r = 3 + 2 \cos \theta$, $0 \le \theta \le 2\pi$

If you had sketched the graph of the equation in Example 4 on the interval $0 \le \theta \le \pi$, you would have obtained only the upper half of the limaçon and your graph would have had the shape shown here.

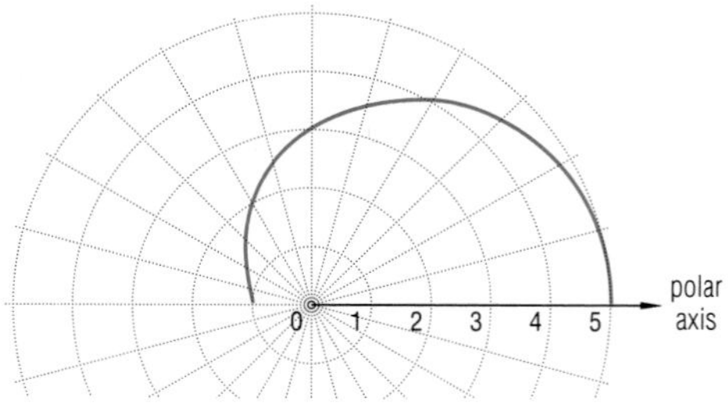

490

However, if you had sketched the same equation on an interval larger than $0 \le \theta \le 2\pi$, or on any interval whose length is greater than or equal to 2π radians, your polar graph would have looked like that in Example 4. This would have happened because the cosine function is periodic with period 2π, so the values of $r = 3 + 2\cos\theta$ repeat every 2π radians.

It would be tempting to generalize the preceding observations and state the following conjecture: If $r = f(\theta)$ is periodic with period p, then the complete polar graph of $r = f(\theta)$ can be obtained by graphing $r = f(\theta)$ on any interval of length p. As you will see in the next lesson, this conjecture is false.

Questions

Covering the Reading

In 1 and 2, make a table of values and sketch the polar graph of the equation on the interval $0 \le \theta \le 2\pi$.

1. $r = 3\sin\theta$ **See margin.** **2.** $r = 2\cos 3\theta$ **See margin.**

3. Prove that the curve in Question 1 is a circle, and find its center and radius. **See margin.**

In 4 and 5, plot the rectangular graph of the given equation. Then use the rectangular graph of the equation to sketch the polar graph.

4. $r = 1 + \sin\theta$, for $0 \le \theta \le 2\pi$ **See margin.**

5. $r = 2 + 3\cos\theta$, for $0 \le \theta \le 2\pi$ **See margin.**

Applying the Mathematics

6. **a.** Convert the polar equation $r(\sin\theta + \cos\theta) = 1$ to rectangular coordinates by using the polar-rectangular conversion formulas.
b. Sketch the polar graph of the equation. **a) $y + x = 1$; b) See margin.**

In 7 and 8, determine whether or not the function is periodic. If it is, find an interval $\theta_1 \le \theta \le \theta_2$ that determines the complete graph of $r = f(\theta)$.

7. $r = \theta$
not periodic

8. $r = 2\cos\theta + \sin 2\theta$
periodic, sample: $0 \le \theta \le 2\pi$

9. Suppose that a and b are positive numbers.
a. Sketch rectangular graphs of $r = a + b\cos\theta$ when $a < b$ and when $a > b$.
b. Sketch a polar graph of the equation in part **a** for each case.
c. Explain why the polar graphs for the two cases have the following properties:
i. $a < b$; a limaçon with an inner loop **See margin.**
ii. $a > b$; a limaçon that does not include the pole. **$r > 0$ for all θ.**
a, b) See margin.

ADDITIONAL ANSWERS
1.–5, 6.b., 9.a., b., c. i.
See Additional Answers in the back of this book.

NAME ____

LESSON **MASTER 8–4**
QUESTIONS ON SPUR OBJECTIVES

■ REPRESENTATIONS *Objective J* *(See pages 534–536 for objectives.)*
In 1–6, sketch the graph of the polar equation and identify the type of curve obtained.

1. $r = 5$ **circle**

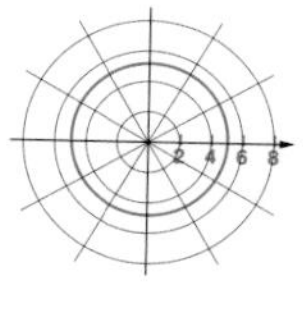

2. $\theta = 60°$ **line**

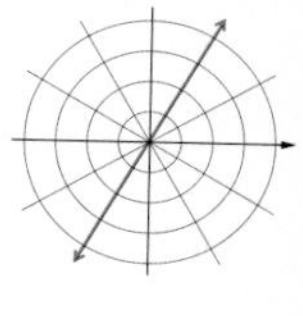

3. $r\sin\theta = -3$ **line**

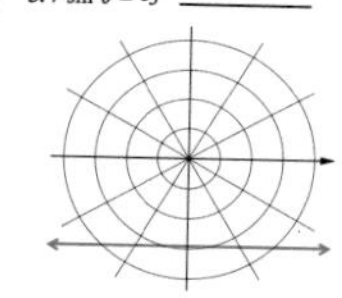

4. $r = 2 + 2\cos\theta$ **cardioid**

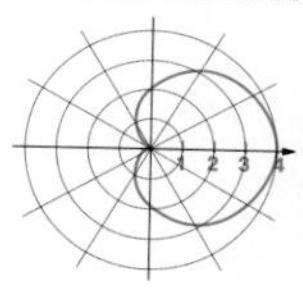

NAME ____
Lesson MASTER 8–4 (page 2)

5. a. Sketch the rectangular graph of the equation $r = 1 + 3\sin\theta$.

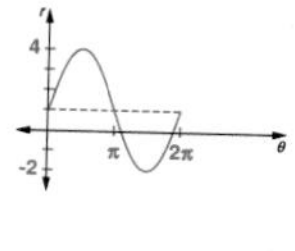

b. Use the rectangular graph in part **a** to sketch its polar graph.

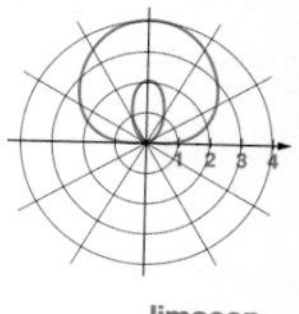

c. Identify the type of curve that results in part **b**. **limaçon**

6. a. Sketch the graph of $r = 5\cos\theta$.

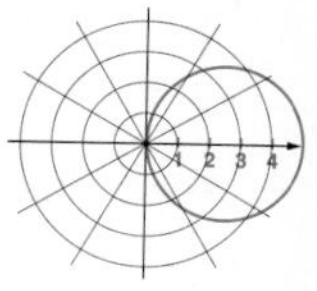

b. Prove that the graph is a circle and find its center.

$\sqrt{x^2 + y^2} = 5 \cdot \frac{x}{\sqrt{x^2 + y^2}}$
$x^2 + y^2 = 5x$
$x^2 - 5x + y^2 = 0$
$x^2 - 5x + \frac{25}{4} + y^2 = \frac{25}{4}$
$\left(x - \frac{5}{2}\right)^2 + y^2 = \left(\frac{5}{2}\right)^2$
center: $\left(\frac{5}{2}, 0\right)$

NOTES ON QUESTIONS
Technology for Question 19: Some graphing facilities that do not produce polar graphs directly can be programmed to accomplish this. For example, the following program for one calculator can be stored and recalled to produce the polar graph of $r = \cos 3\theta$ on the interval $0 \le \theta \le 2\pi$.

```
CLS:
RAD:
0 → T:
RANGE -1.3, 1.3, 1, -1, 1, 1:
LBL 1:
COS(3T) → R:
REC(R, T):
PLOT I, J:
LINE:
T + .1 → T:
T ≤ 2π ⇒ GOTO 1:
```

With another calculator, to graph $r = 1 + 2 \sin \theta$, press MODE and then select PARAMETRIC. Press Y =. For X_{IT} =, enter (1 + 2 sin T) cos T and for Y_{IT} =, enter (1 + 2 sin T) sin T. Press ZOOM and select STANDARD. You may then adjust the display window using ZOOM or RANGE. You may also TRACE through the curve.

FOLLOW-UP

MORE PRACTICE
For more questions on SPUR Objectives, use *Lesson Master 8-4,* shown on page 491.

PROJECTS
The projects for Chapter 8 are described on pages 529–530. **Project 3** is related to the content of this lesson.

ADDITIONAL ANSWERS
10., 11.a., 12., 14.b., 15.b., 17. See Additional Answers in the back of this book.

10. The polar graph of $r = 1 + \cos \theta$, or any equation of the form $r = a \pm a \cos \theta$ or $r = a \pm a \sin \theta$, is a curve known as a **cardioid**. Carefully graph this curve and use the graph to explain how the curve got its name. **See margin.**

11. a. Without graphing, explain how the polar graph of $r = k \sin \theta$ is related to the polar graph of $r = \sin \theta$. **See margin.**

b. Describe the polar graph of $r = k \sin \theta$. **It is a circle with radius $\frac{k}{2}$ and center $\left(0, \frac{k}{2}\right)$**

Review

12. Illustrate the multiplication of $z = 2(\cos 150° + i \sin 150°)$ by $w = 5(\cos 40° + i \sin 40°)$ with a diagram showing the appropriate size transformation and rotation. *(Lesson 8-3)* **See margin.**

13. Suppose Z and W are the points in the plane that represent the complex numbers $z = 2 - i$ and $w = -1 + 3i$, respectively. Then the distance $ZW = 5$. Let $v = -\frac{7}{2} - 8i$, and suppose points Z' and W' represent $z + v$ and $w + v$, respectively.

a. What are the coordinates of Z' and W'? $Z' = \left(-\frac{3}{2}, -9\right)$, $W' = \left(-\frac{9}{2}, -5\right)$

b. What transformation maps Z to Z' and W to W'? $T_{-7/2,\,-8}$

c. Find $Z'W'$. *(Lesson 8-3)* **5**

14. a. Consider $\triangle ABC$ with vertices $A = [2, 30°]$, $B = [1, 135°]$, and $C = [4, 270°]$. Rotate the triangle 60° about the origin and give polar coordinates for the new vertices.

b. Graph $\triangle ABC$ and its image on a polar grid. *(Lesson 8-2)*
a) $A' = [2, 90°]$, $B' = [1, 195°]$, $C' = [4, 330°]$; b) See margin.

15. Let $z = \frac{1}{2} + \frac{\sqrt{3}}{2}i$ and $w = \frac{1}{z}$.

a. Write w in $a + bi$ form. $\frac{1}{2} - \frac{\sqrt{3}}{2}i$

b. Compare $\bar{z}$ with w. **See margin.**

c. Write z and w in polar form. How do they compare? *(Lessons 8-2, 8-1)* **$z = [1, 60°]$, $w = [1, -60°]$; their arguments are opposites.**

16. Let $z = -3 + 4i$, $v = 2 - i$, and $w = -3i$. Find the value of each expression below and write your answer in $a + bi$ form.

a. $\frac{v + w}{z}$ $-.88 + .16i$ **b.** $\frac{v}{z}$ $-.4 - .2i$ **c.** $\frac{w}{z}$ $-.48 + .36i$

d. Add the answers to parts **b** and **c**, and compare the sum to the answer to part **a**. *(Lesson 8-1)*
The sum, $-.88 + .16i$, is equal to the answer to part a.

17. If x and y are real numbers, graph $\{(x, y)\colon x^2 + y^2 = 0\}$. *(Previous course)* **See margin.**

18. Write in summation notation:
$1 \cdot 3 + 2 \cdot 5 + 3 \cdot 7 + 4 \cdot 9 + \ldots + 10 \cdot 21$. *(Lesson 7-3)* $\sum_{k=1}^{10} [k(2k + 1)]$

Exploration

19. Learn how to use an automatic grapher to graph polar equations. Then graph several of the polar equations in this lesson. **Answers will vary.**

492

LESSON

8-5

Rose Curves and Spirals

Celandine poppies

Polar coordinates are especially useful for describing two beautiful categories of curves: *rose curves* and *spirals*.

Example 1 Sketch the polar graph of $r = \sin 2\theta$ for $0 \le \theta \le 2\pi$.

Solution The rectangular graph of $r = \sin 2\theta$ for $0 \le \theta \le 2\pi$ is given below.

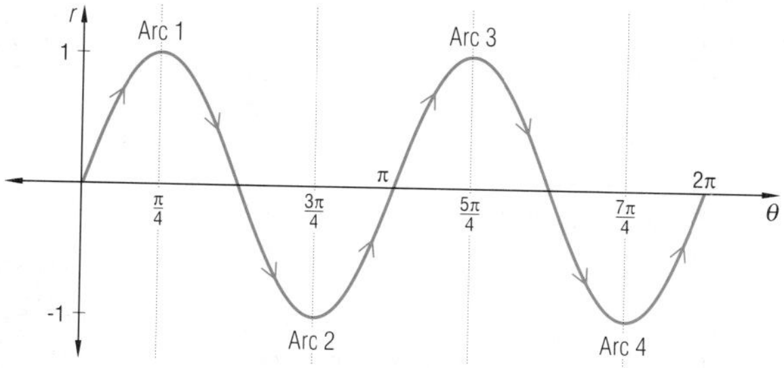

The θ-axis divides this graph into 4 congruent arcs, each symmetric to one of the dotted vertical lines where θ is an odd multiple of $\frac{\pi}{4}$. Each of these arcs contributes one of the four congruent loops in the polar graph below.

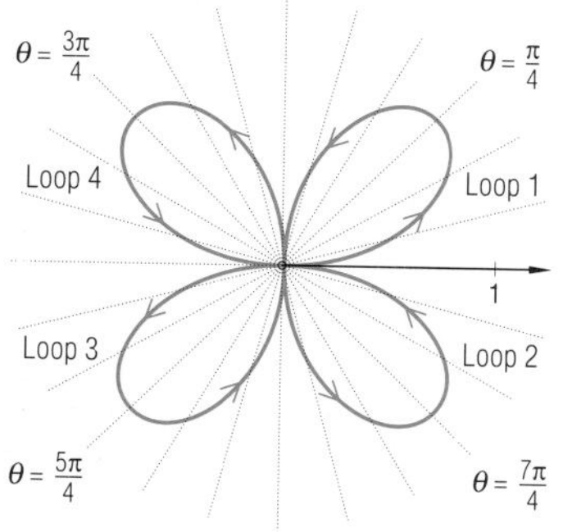

Because of the periodicity of the sine function, values of θ outside $0 \le \theta \le 2\pi$ produce no new points for the graph.

LESSON 8-5

RESOURCES

- Lesson Master 8-5
- Teaching Aid 51 can be used with **Example 1**.
- Teaching Aid 52 can be used with **Example 2**.
- Teaching Aid 53 can be used with **Questions 6 and 7**.
- Computer Master 15

OBJECTIVE

J Sketch graphs of polar equations.

TEACHING NOTES

You might wish to discuss the symmetry properties of polar graphs. You can ask the following questions:
(1) Suppose $r = f(\theta)$ and for all θ, $f(-\theta) = f(\theta)$. What kind of symmetry does the graph of f possess? (symmetry to the line containing polar axis)
(2) Suppose $r = f(\theta)$ and for all θ, $f(\pi - \theta) = f(\theta)$. What kind of symmetry does the graph of f possess? (symmetry to the line $\theta = \frac{\pi}{2}$)
(3) Suppose $r = f(\theta)$ and for all θ, $f(\pi + \theta) = f(\theta)$. What kind of symmetry does the graph of f possess? (point symmetry to the pole)
However, there is one caution: The converses of these sufficient conditions are not true. For instance, the polar graph of $r = \sin 2\theta$ for $0 \le \theta \le 2\pi$ is symmetric to the vertical line $\theta = \frac{\pi}{2}$ even though $f(\theta) \ne f(\pi - \theta)$.

At the top of this page, the phrase "the number of leaves is determined by n" may need some explanation. If n is even, there are $2n$ leaves. If n is odd, there are n leaves.

The relationship between the spirals mentioned on page 495 has an analog for rectangular coordinates: If you plot $(x, \log y)$ in place of (x, y) for the exponential function $y = ab^x$, the resulting rectangular graph is a straight line instead of an exponential curve. It is this fact that underlies the use of semilog paper to plot experimental data that are expected to fit an exponential model.

The nautilus (a relative of the squid) may have the most beautiful shell of all the sea animals. Like natural beehives, the regularity of these shells is astounding; they look as if they had been crafted by machine.

ADDITIONAL EXAMPLES

1. Sketch the polar graph of $r = 2 \cos 3\theta$ for $0 \le \theta \le 2\pi$.

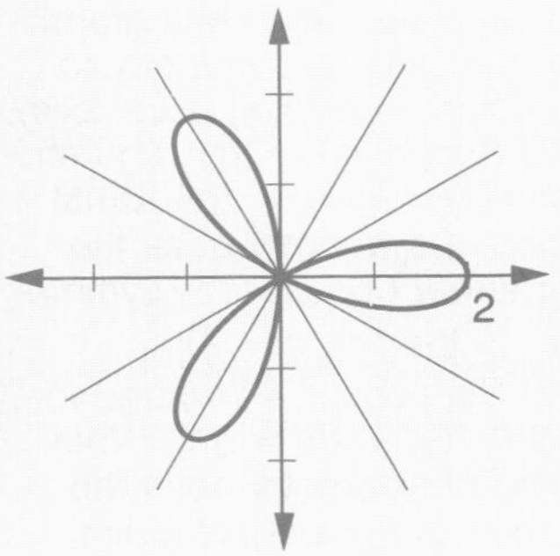

2. Sketch the polar graph of $r = \pi + \frac{\theta}{2}$ for $0 \le \theta \le 4\pi$.

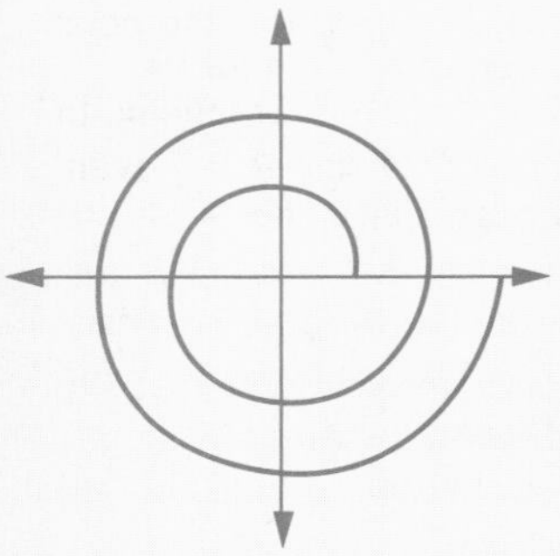

The beautiful polar graph in Example 1 is called a *four-leafed rose*. In general, the polar graphs of equations of the form

$$r = a\cos(n\theta), \quad a > 0,\ n \text{ a positive integer,}$$
$$\text{or} \quad r = a\sin(n\theta), \quad a > 0,\ n \text{ a positive integer,}$$

are called **rose curves**. The length of each leaf or petal is a and the number of leaves is determined by n. For example, the polar graph of the equation

$$r = 2\cos 3\theta$$

is a 3-leafed rose and each leaf has length 2.

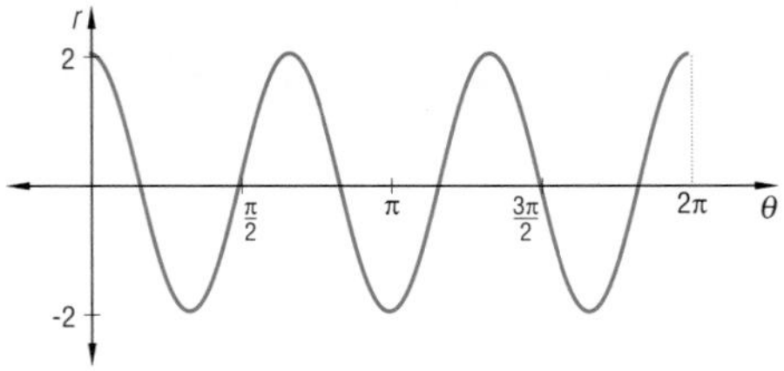

Rectangular graph of
$r = 2\cos 3\theta$
$0 \le \theta \le 2\pi$

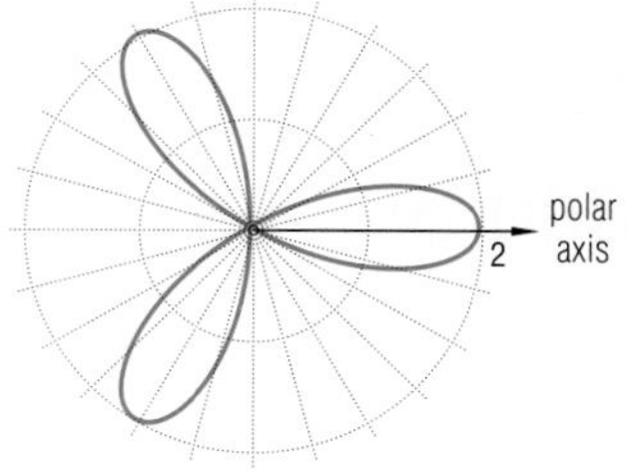

Three-leafed rose
Polar graph of $r = 2\cos 3\theta$
(covered twice in $0 \le \theta \le 2\pi$)

In determining the points that satisfy $r = 2\cos 3\theta$, notice that values of r repeat with period $\frac{2\pi}{3}$ but the complete polar graph of $r = f(\theta)$ requires the entire interval $0 \le \theta \le \pi$. Therefore, the conjecture stated at the end of the last lesson is false. You sometimes need to consider a larger interval than the period of the trigonometric function to obtain the complete polar graph.

Example 2 Sketch the polar graphs of the following equations on the same polar coordinate grid:
$r = \theta + 1, \quad 0 \le \theta \le 2\pi;$ and $r = 2^\theta, \quad 0 \le \theta \le 2\pi.$

Solution First sketch the rectangular graphs of the two equations on the same rectangular coordinate graph.

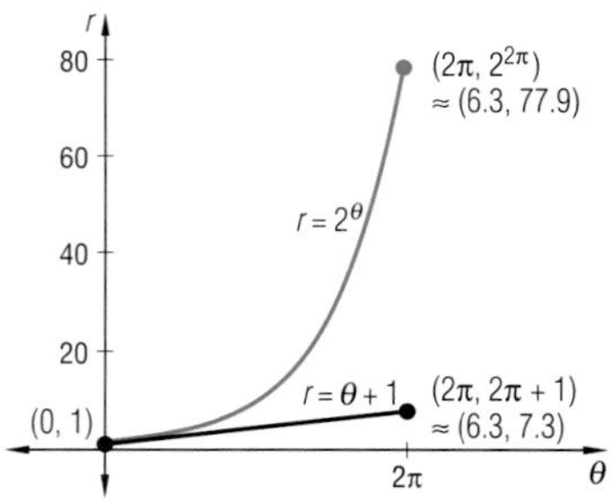

Both functions are increasing on the interval $0 \le \theta \le 2\pi$ and both have the value 1 at $\theta = 0$. But the exponential function $r = 2^\theta$ increases much more rapidly than the linear function $r = \theta + 1$. The rectangular graphs of these equations transform to the polar graphs displayed below.

Thus, both polar graphs are sections of spirals, but the polar graph of $r = 2^\theta$ expands as a spiral much more rapidly than that of $r = \theta + 1$.

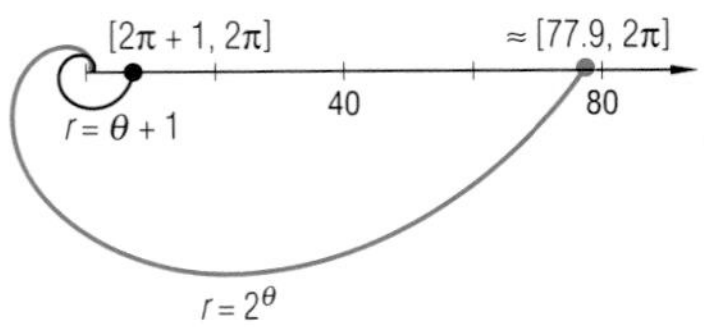

494

The polar graphs of $r = \theta + 1$ and, more generally,

$$\{[r, \theta]: r = a\theta + b\},$$

where a is positive and b is nonnegative, are called **spirals of Archimedes.** They have the appearance of a coil of rope or hose with a constant distance between successive coils.

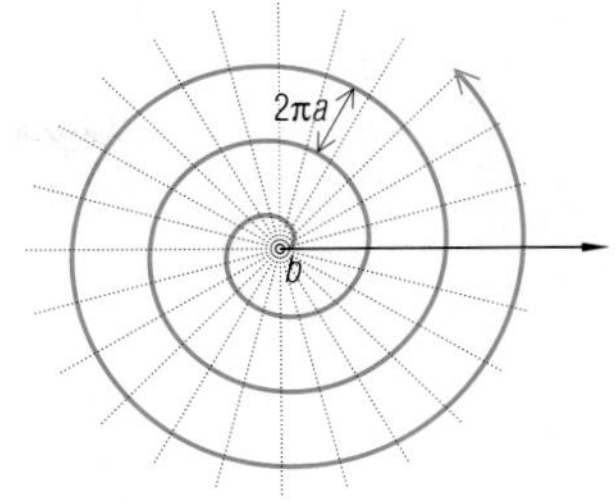

Spiral of Archimedes $r = a\theta + b$

In contrast, the polar graphs of $r = 2^{\theta}$ and, more generally,

$$\{[r, \theta]: r = ab^{\theta}\},$$

where $a > 0$ and $b > 1$, are called **logarithmic spirals.** The distance between successive coils of a logarithmic spiral is not constant as with spirals of Archimedes. Rather, this distance d has the value a when $\theta = 0$, approaches $+\infty$ as θ approaches $+\infty$, and approaches 0 as θ approaches $-\infty$.

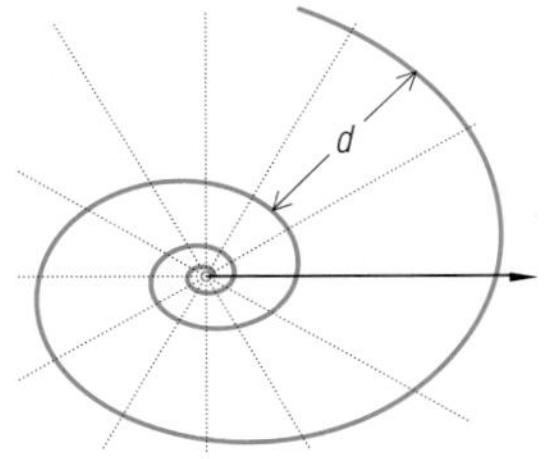

Logarithmic spirals derive their name from the fact that

$$\begin{aligned} & r = ab^{\theta} \\ \Leftrightarrow \quad & \log r = \log(ab^{\theta}) \\ \Leftrightarrow \quad & \log r = \log a + (\log b)\theta, \end{aligned}$$

which is of the form $\log r = A + B\theta$.

Thus, if you plot $[\log r, \theta]$ in place of $[r, \theta]$ for the equation $r = ab^{\theta}$, its polar graph is a spiral of Archimedes instead of a logarithmic spiral.

One very remarkable geometric property of the logarithmic spiral $r = ab^{\theta}$ is that the angle ϕ between the tangent line to the polar graph and the line from the pole has the same measure at any point of the spiral.

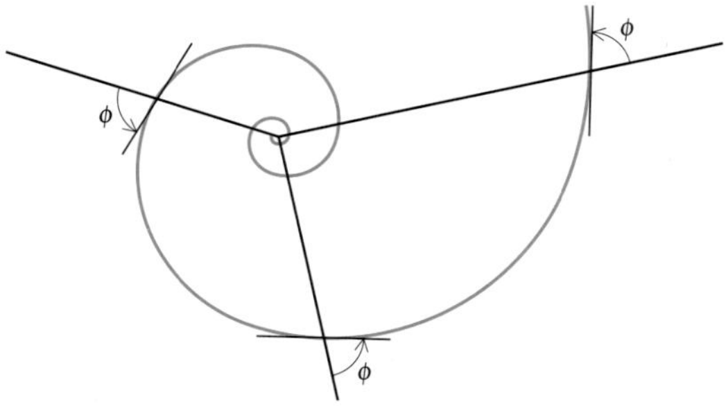

NOTES ON QUESTIONS

Question 10: This question shows that polar graphs can be used to test for identities.

Question 11: In general, replacing θ by $\theta - \phi$ in an equation of a function $r = f(\theta)$ rotates a graph ϕ about the origin.

Question 12: Although the graphs intersect at [0, 0], these coordinates do not satisfy $r = 4 \cos \theta$. However, $\left[0, \frac{\pi}{2}\right]$ does.

FOLLOW-UP

MORE PRACTICE

For more questions on SPUR Objectives, use *Lesson Master 8-5,* shown on page 497.

EXTENSION

You may wish to take an extra day to introduce students to other families of curves such as the three that follow. You might have students work in **small groups**, trying different values of *a*, and then seeing the effects of changing *a* on the graphs.

cissoid: $r = 2a \tan \theta \sin \theta$

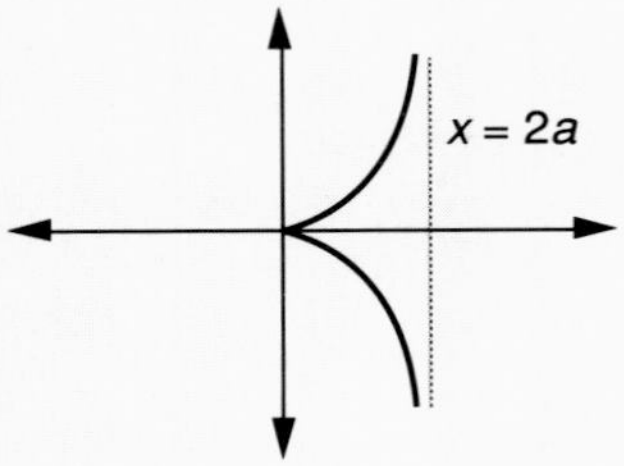

lemniscate: $r^2 = a^2 \cos 2\theta$

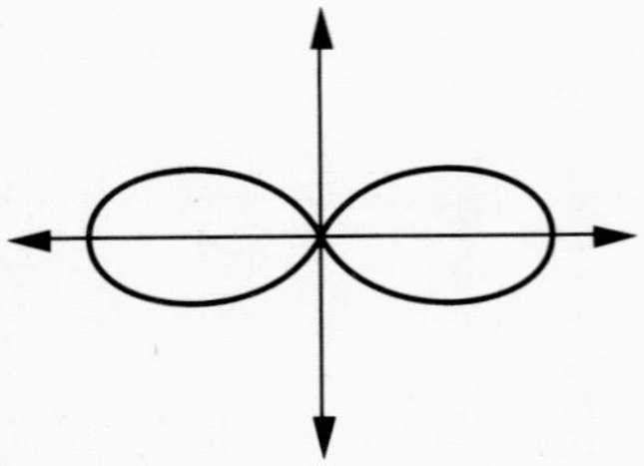

lituus: $r^2 = \frac{a^2}{\theta}$

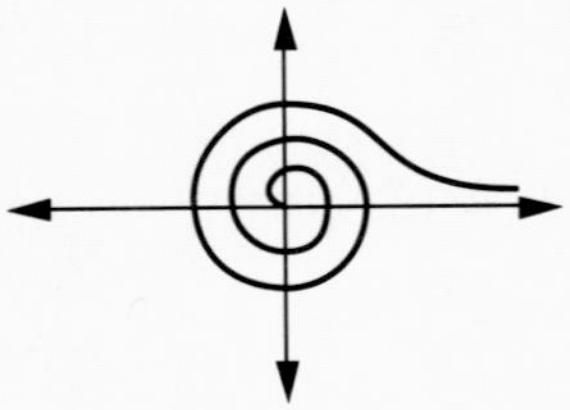

PROJECTS

The projects for Chapter 8 are described on pages 529–530. **Project 5** is related to the content of this lesson.

One consequence of this property is that any two regions bounded by two lines intersecting at a fixed angle at the pole and two consecutive intercepted arcs of the spiral are similar.

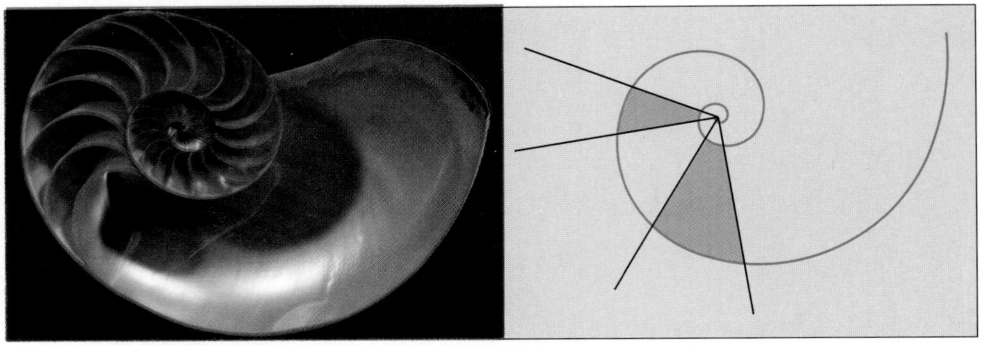

The shells of some sea creatures are shaped like logarithmic spirals for this reason. As the creature grows, the shell compartment expands in a way that allows the creature to retain its shape. Shown here is the shell of a chambered nautilus, cut in half to show the compartments.

Questions

Covering the Reading

In 1–3, sketch the polar graph.

1. $r = 4 \cos 5\theta$ **2.** $r = 3 \sin 2\theta$ **3.** $r = \cos 6\theta$
1-3) See margin.

4. Give polar coordinates for five points on the graph of $r = \theta + 1$.
sample: $[1, 0]$, $[\frac{\pi + 6}{6}, \frac{\pi}{6}]$, $[\frac{\pi + 4}{4}, \frac{\pi}{4}]$, $[\frac{\pi + 2}{2}, \frac{\pi}{2}]$, $[\pi + 1, \pi]$

5. Give polar coordinates for five points on the graph of $r = 2^\theta$.
sample: $[1, 0]$, $[2^{\pi/6}, \frac{\pi}{6}]$, $[2^{\pi/4}, \frac{\pi}{4}]$, $[2^{\pi/2}, \frac{\pi}{2}]$, $[2^\pi, \pi]$

Applying the Mathematics

6. Give an equation for the rose curve graphed below.

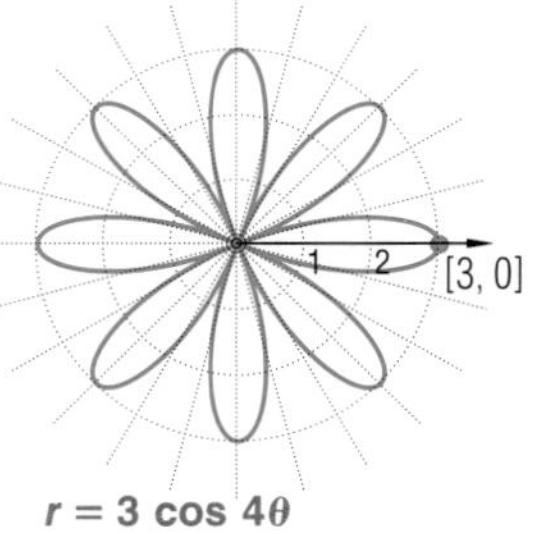

$r = 3 \cos 4\theta$

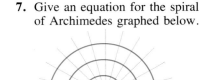

7. Give an equation for the spiral of Archimedes graphed below.

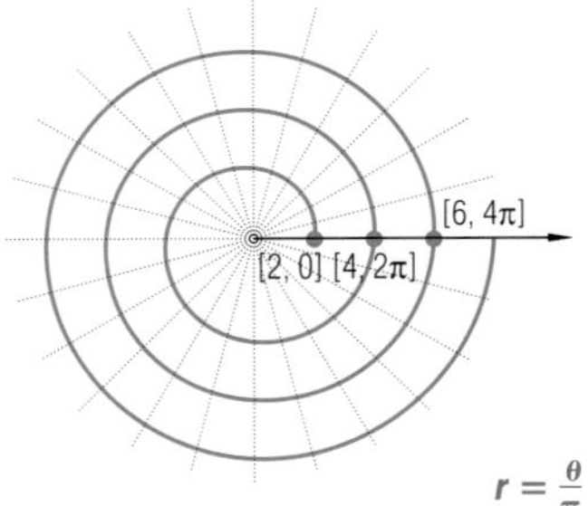

$r = \frac{\theta}{\pi} + 2$

8. a. What is an equation for a five-leafed rose with leaves of length 2?
b. Graph this curve. a) $r = 2 \cos 5\theta$ or $r = 2 \sin 5\theta$; b) See margin.

9. a. Give the polar graph of $r = 2 \sin 3\theta$. See margin.
b. Give equations for all lines of symmetry for the graph in part **a**.
See margin.

Review

10. a. Use an automatic grapher to plot the rectangular graphs of the equations $r = \cos\theta + \sin\theta$ and $r = \sqrt{2}\cos\left(\theta - \frac{\pi}{4}\right)$.

b. Verify that the equations in part **a** are equivalent by using an appropriate identity.

c. Sketch the polar graph of $r = \cos\theta + \sin\theta$ and verify that it is a circle.

(Lessons 8-4, 6-3) a-c) See margin.

11. Consider the equations $r = 1 + 2\cos\theta$ and $r = 1 + 2\cos\left(\theta - \frac{\pi}{4}\right)$.

a. Describe the geometric relationship between the rectangular graphs of these equations.

b. Describe the geometric relationship between the polar graphs of these equations. The 2nd graph is a rotated image of the 1st by $\frac{\pi}{4}$.

c. Sketch the polar graphs of both equations on the same polar grid. *(Lessons 8-4, 3-9)* a, c) See margin.

12. a. On a polar grid, sketch the curves $r = 4\cos\theta$ and $r = 4\sin\theta$.

b. Find the points of intersection of the two curves. *(Lessons 8-4, 6-7)*
a) See margin. b) $[0, 0]$, $\left[2\sqrt{2}, \frac{\pi}{4}\right]$

13. a. A size change of magnitude $\frac{1}{2}$ followed by a rotation of $\frac{4\pi}{3}$ about the origin is equivalent to multiplication by what complex number?

b. Write your answer to part **a** in $a + bi$ form. $-\frac{1}{4} - \frac{\sqrt{3}}{4}i$

c. Use your answer to part **b** to find the images of the points $u = -2$, $v = 6$, $w = 6 + 4i$, and $z = -2 + 4i$ under the transformations described.

d. Graph the points u, v, w, and z and their images. *(Lessons 8-3, 8-2)*
a, c, d) See margin.

14. In an AC circuit, the current is $-4 + 3i$ amps and the impedance is $2 + 4i$ ohms.

a. Find the voltage. $-20 - 10i$ volts

b. Suppose the circuit is connected in series with another circuit with voltage $11 + 8i$ volts. Find the total voltage. *(Lesson 8-1)*
$-9 - 2i$ volts

15. Use an automatic grapher to approximate the solutions to $\sin x + 2\cos x > 2$ for $0 \le x \le 2\pi$. *(Lesson 6-7)*
$0 < x < 0.92$

16. If f and g are functions defined by $f(x) = \frac{2x - 2}{x + 1}$ and $g(x) = \frac{8 - x}{x^2 - 1}$, find rules for $f + g$ and $f \cdot g$. *(Lessons 5-1, 3-3)*
$(f + g)(x) = \frac{2x^2 - 5x + 10}{x^2 - 1}$; $(f \cdot g)(x) = \frac{16 - 2x}{(x + 1)^2}$, $x \neq 1$

Exploration

17. Find out something about the chambered nautilus other than what is stated in this lesson. The chambered nautilus is also known as the pearly nautilus. It has a smooth, coiled shell 15–25 cm in diameter, consisting of 30–36 chambers; it lives in the outermost chamber.

ADDITIONAL ANSWERS
1.-3., 8.b., 9., 10., 11.a., c., 12.a., 13.a., c., d. See Additional Answers in the back of this book.

NAME ____________

LESSON **MASTER 8–5**
QUESTIONS ON **SPUR** OBJECTIVES

■REPRESENTATIONS *Objective J* *(See pages 534–536 for objectives.)*
In 1–3, sketch the graph of the polar equation and identify the type of curve obtained.

1. $r = 5\cos 2\theta$ rose curve

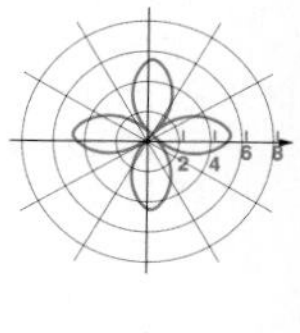

2. $r = \frac{3}{2}\theta + 2$ spiral of Archimedes

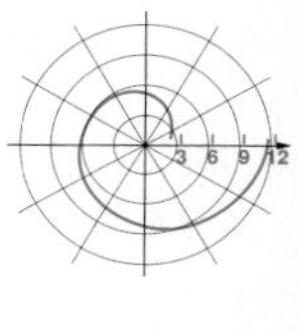

3. $r = 5^\theta$ logarithmic spiral

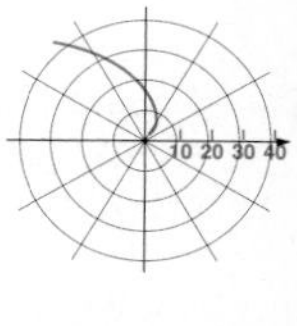

NAME ____________
Lesson MASTER 8–5 (page 2)

4. a. Graph the three-leafed rose curve $r = 6\sin 3\theta$.

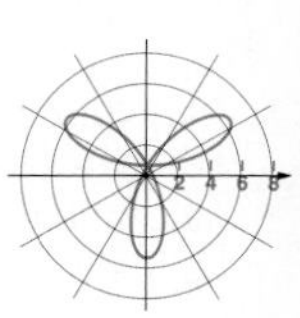

b. Does it have any reflection symmetries? Prove that your answer is correct.
Reflection-symmetric over the line $\theta = \frac{\pi}{2}$; If $[r, \theta]$ is on the graph, then $r = 6\sin 3\theta$.
$6\sin 3(\pi - \theta) =$
$6\sin(3\pi - 3\theta) =$
$6\sin(\pi - 3\theta) = -6\sin(3\theta - \pi) = 6\sin(3\theta)$;
Therefore, $[r, \pi - \theta]$ is on the graph.

5. a. Write an equation for a five-leafed rose with leaves of length 2.5 which is reflection-symmetric over the polar axis. $r = 2.5\cos 5\theta$

b. Graph this curve.

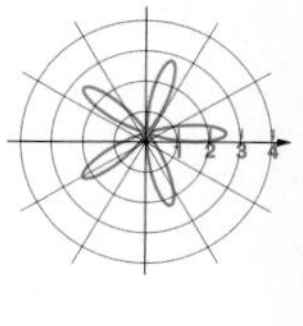

LESSON 8-6

RESOURCES
- Lesson Master 8-6
- Teaching Aid 54 displays the graphs of powers of complex numbers.

OBJECTIVES

D Find powers and roots of complex numbers.
F Prove or verify properties of complex numbers.
K Use DeMoivre's Theorem to graph powers and roots of complex numbers.

TEACHING NOTES

DeMoivre's Theorem is so simple that students do not always appreciate how astounding it is. Point out that in $a + bi$ form, the first form known to mathematicians, this theorem is not obvious at all. It is only in the trigonometric and polar forms that the pattern is evident.

To reinforce understanding of the theorem and its proof by induction, do a series of applications of the Geometric Multiplication Theorem. For example, to calculate $[3, 20°]^4$, you can do
$[3, 20°] \cdot [3, 20°] = [9, 40°]$
$[3, 20°] \cdot [9, 40°] =]27, 60°]$
$[3, 20°] \cdot [27, 60°] = [81, 80°]$.
But $[81, 80°] = [3^4, 4 \cdot 20°]$.
Thus, $[3, 20°]^4 = [3^4, 4 \cdot 20°]$.

Small Group Work You may wish to divide your class into four groups to calculate and simplify $(1 - i)^{10}$ using a different algorithm. All groups should write down their work.

LESSON

8-6 Powers of Complex Numbers

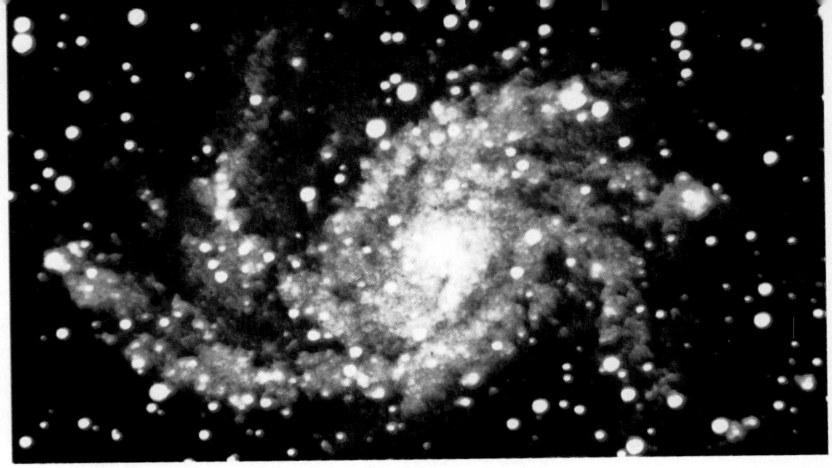

Spiral galaxy in Cepheus

Lessons 8-1 and 8-3 covered the algebraic and geometric properties of the two basic operations in the complex number field: addition and multiplication. Now we consider the taking of positive integer powers of a complex number. The ability to find these powers provides all of the computational equipment needed to obtain values of polynomial functions involving complex numbers. Also, there is an interesting relationship between the powers of a complex number and some of the curves you studied in the last lesson.

Recall that in Lesson 7-9, two algorithms were given for the calculation of x^n for a given real number x and positive integer n. The first, using repeated multiplication, required $n - 1$ main steps. The second required thinking of the integer n as a sum of powers of 2 but was more efficient because it required fewer steps.

Abraham DeMoivre

Similarly, there is more than one way to calculate z^n for a given complex number z and positive integer n. For instance, one way to calculate z^9 when $z = -1 + \sqrt{3}\,i$ is to repeatedly square z to obtain z^2, z^4, and z^8, and then multiply z^8 by z to arrive at z^9. For most values of z and n, however, there is a more efficient way to obtain z^n. This way is based on the Geometric Multiplication Theorem, and utilizes a simple formula known as DeMoivre's (pronounced di mwavs′) Theorem, named after its discoverer, Abraham DeMoivre (1667–1754).

DeMoivre's Theorem extracts a price for its efficiency. The complex number must be written in polar or trigonometric form in order to apply the theorem easily.

DeMoivre's Theorem

(Polar Form)	For all positive integers n, if $z = [r, \theta]$, then $z^n = [r^n, n\theta]$.
(Trigonometric form)	For all positive integers n, if $z = r(\cos\theta + i\sin\theta)$, then $z^n = r^n(\cos n\theta + i\sin n\theta)$.

Proof As is often the case when trying to show that a mathematical statement is true for all positive integers, mathematical induction is used. Given $z = [r, \theta]$, the truth of $S(n)$ must be proved for all positive integers n, where $S(n)$ is $z^n = [r^n, n\theta]$. First show that $S(1)$ is true.

$S(1)$: $\quad z^1 = [r^1, 1 \cdot \theta]$.

Since $z = [r, \theta]$, $S(1)$ is true.

Second, show that for any positive integer k, the assumption $S(k)$ is true implies that $S(k + 1)$ is true.

Here $\quad S(k)$: $\quad z^k = [r^k, k\theta]$,

and $\quad S(k + 1)$: $z^{k+1} = [r^{k+1}, (k + 1)\theta]$.

If $S(k)$ is true, then multiplying both sides by z yields

$$z \cdot z^k = [r, \theta] \cdot [r^k, k\theta].$$

$$z^{k+1} = [r \cdot r^k, \theta + k\theta] \quad \text{Geometric Multiplication Theorem}$$

$$= [r^{k+1}, (k + 1)\theta]$$

Thus, $S(k + 1)$ is true. Therefore, by the Principle of Mathematical Induction, $S(n)$ is true for all positive integers n.

The next example shows how to use DeMoivre's Theorem to compute the power mentioned earlier.

Example 1 Compute z^9 when $z = -1 + \sqrt{3}i$.

Solution 1 Find a trigonometric form for z: if $z = r(\cos\theta + i\sin\theta)$,

then $$r = \sqrt{(-1)^2 + (\sqrt{3})^2} = \sqrt{4} = 2.$$

$$\tan\theta = \frac{\sqrt{3}}{-1} \text{ and } \theta \text{ is in the second quadrant.}$$

Therefore, $\theta = \frac{2\pi}{3}$, and so a trigonometric form of z is

$$z = 2\left(\cos\frac{2\pi}{3} + i\sin\frac{2\pi}{3}\right).$$

Apply DeMoivre's Theorem with $r = 2$, $\theta = \frac{2\pi}{3}$, and $n = 9$.

$$z^9 = 2^9\left[\cos\left(9 \cdot \frac{2\pi}{3}\right) + i\sin\left(9 \cdot \frac{2\pi}{3}\right)\right]$$

Because $2^9 = 512$ and $9 \cdot \frac{2\pi}{3} = 6\pi \equiv 0 \pmod{2\pi}$, it follows that a simpler trigonometric form of z^9 is

$$z^9 = 512[\cos(0) + i\sin(0)] = 512.$$

That is, $(-1 + \sqrt{3}\,i)^9$ is the real number 512.

Solution 2 Use a polar form for z. If $z = [r, \theta]$, then find r and θ as in Solution 1. So a polar form of z is

$$z = \left[2, \frac{2\pi}{3}\right].$$

Apply DeMoivre's Theorem to find

$$z^9 = \left[2^9, 9 \cdot \frac{2\pi}{3}\right] = [512, 6\pi] = [512, 0] = 512.$$

The geometry of the sequence of successive powers of a complex number is quite beautiful.

Group A: Expand $(1 - i)^{10}$ by rewriting it as $(1 - i)(1 - i) \ldots (1 - i)$ and multiplying:

Group B: Express $(1 - i)^{10}$ in trigonometric form as $[\sqrt{2}(\cos(-45°) + i\sin(-45°))]^{10}$ and evaluate using repeated applications of $[r(\cos\theta + i\sin\theta)] \cdot [s(\cos\phi + i\sin\phi)] = [rs(\cos\theta + \phi) + i\sin(\theta + \phi)]$.

Group C: Do repeated applications of the Geometric Multiplication Theorem for $([\sqrt{2}, -45°])^{10}$.

Group D: Use DeMoivre's Theorem to evaluate $[2, -45°]^{10}$.

If some students know the Binomial Theorem, have them constitute a fifth group to calculate $(1 - i)^{10}$.

(All groups should get $-32i$ as the answer.)

In Solution 2 of **Example 1**, some students may have difficulty understanding why $[512, 6\pi] = [512, 0]$. Remind them that in Lesson 8-2 they learned that $[r, \theta] = [r, \theta + 2\pi n]$, for all integers n. Here, $r = 512$, $\theta = 0$ and $n = 3$.

In the discussion on pages 501 and 502, some students will need to be reminded that there are infinite number of complex numbers whose absolute value is 1, unlike real numbers where there are only two such numbers. All of these complex numbers lie on the unit circle. Like real numbers, complex numbers w and z satisfy the relation $|zw| = |z||w|$; thus, if two complex numbers have an absolute value of 1, so does their product. For this reason, the power of any complex number with absolute value 1 also has an absolute value of 1. Another way of seeing this is from DeMoivre's Theorem: $[1, \theta]^n = [1, n\theta]$. A generalization is found in **Question 13**.

ADDITIONAL EXAMPLES

1. Compute z^{11} when $z = \frac{1}{2} + \frac{\sqrt{3}}{2} i$.

$z = \cos\frac{\pi}{3} + i \sin\frac{\pi}{3} = \left[1, \frac{\pi}{3}\right]$, so $z^{11} = \left[1^{11}, \frac{11\pi}{3}\right] = \left[1, \frac{5\pi}{3}\right]$.

2. Suppose $z = \left[1.2, \frac{\pi}{2}\right]$. Compute the successive powers z^2 through z^9, plot them in the complex plane, and draw a smooth curve through the points.

With r to two decimal places, $z = \left[1.2, \frac{\pi}{2}\right]$, $z^2 = [1.44, \pi]$, $z^3 = \left[1.73, \frac{3\pi}{2}\right]$, $z^4 = [2.07, 0]$, $z^5 = \left[2.49, \frac{\pi}{2}\right]$, $z^6 = [2.99, \pi]$, $z^7 = \left[3.58, \frac{3\pi}{2}\right]$, $z^8 = [4.30, 0]$, $z^9 = \left[5.16, \frac{\pi}{2}\right]$.

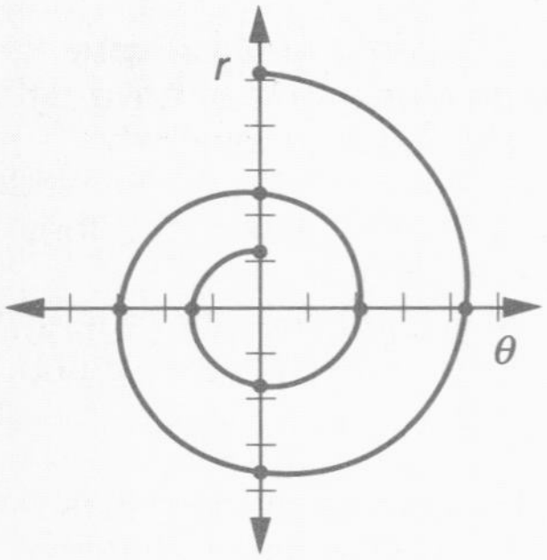

Example 2 Suppose that $z = [1.1, \frac{\pi}{3}]$.

a. Use DeMoivre's Theorem to compute the successive powers: $z^2, z^3, z^4, z^5, z^6, z^7, z^8, z^9$.

b. Plot these points on a polar grid and draw a smooth curve through the successive points.

Solution

a. Here are the values of these complex numbers with r to two decimal places.

$$z^2 = [1.21, \tfrac{2\pi}{3}]$$
$$z^3 = [1.33, \tfrac{3\pi}{3}] = [1.33, \pi]$$
$$z^4 = [1.46, \tfrac{4\pi}{3}]$$
$$z^5 = [1.61, \tfrac{5\pi}{3}]$$
$$z^6 = [1.77, \tfrac{6\pi}{3}] = [1.77, 0]$$
$$z^7 = [1.95, \tfrac{7\pi}{3}] = [1.95, \tfrac{\pi}{3}]$$
$$z^8 = [2.14, \tfrac{8\pi}{3}] = [2.14, \tfrac{2\pi}{3}]$$
$$z^9 = [2.36, \tfrac{9\pi}{3}] = [2.36, \pi]$$

b. Shown here are the graphs of these points on a polar grid.

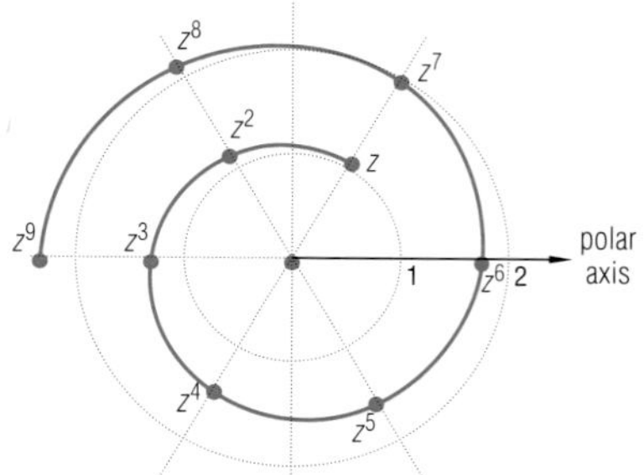

The curve looks like a spiral. In fact, all the powers of $z = [1.1, \frac{\pi}{3}]$ lie on the logarithmic spiral $r = (1.1)^{3\theta/\pi}$. You can check this by substituting integer multiples of $\frac{\pi}{3}$ for θ and obtaining values of r. For example, for $\theta = \frac{4\pi}{3}$, $r = (1.1)^{(3/\pi)\bullet(4\pi/3)} = 1.1^4$. Thus, $z^4 = [1.1^4, \frac{4\pi}{3}]$ lies on the spiral.

In Example 2, the powers of $z = [r, \theta]$ are successively farther and farther from the origin because r, the absolute value of z, satisfies $r > 1$. Consequently, as n increases, so does r^n. When $r < 1$, the powers spiral inward. For instance, to the right is the graph of the first through ninth powers of w where $w = .9\left(\cos\left(\frac{\pi}{4}\right) + i \sin\left(\frac{\pi}{4}\right)\right)$.

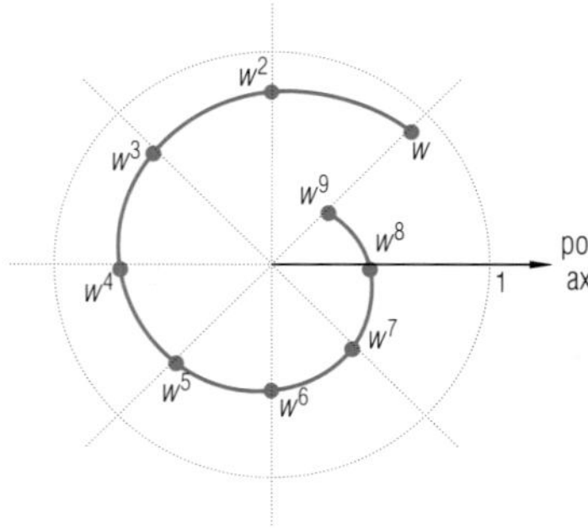

If z is a complex number with $|z| = 1$, then $|z^n| = 1$ for all positive integers n. (You are asked to prove a more general form of this in Question 13.) This means that the graphs of the sequence of successive powers of z,

$$z, z^2, z^3, \ldots, z^k, \ldots,$$

all lie on the circle of radius 1 centered at the origin (the unit circle).

For some choices of z this sequence of points is periodic. For example, if $z = i$, then $z, z^2, z^3, \ldots$ is the sequence

$$i, -1, -i, 1, i, -1, -i, 1, \ldots$$

with period 4, since the terms repeat every 4 terms. As another example, if $z = 1$, then $z, z^2, z^3, \ldots$ is the constant sequence with period 1 and all terms equal to 1.

There are values of z with $|z| = 1$ for which all terms of the sequence $z, z^2, z^3, \ldots$ are distinct points on the unit circle in the complex plane. For example, if $z = \cos 1 + i \sin 1$ (that is, the argument of z is 1 radian), then by DeMoivre's Theorem,

$$z^n = \cos n + i \sin n.$$

In this case, it can be shown that for all n, the values of z^n are different. The diagrams below show the graphs of the first 10, 50, and 500 points of this sequence.

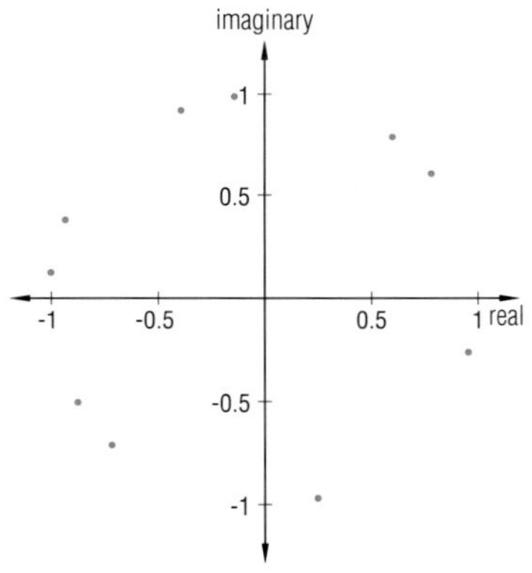

$\{z^n: 1 \leq n \leq 10\}$

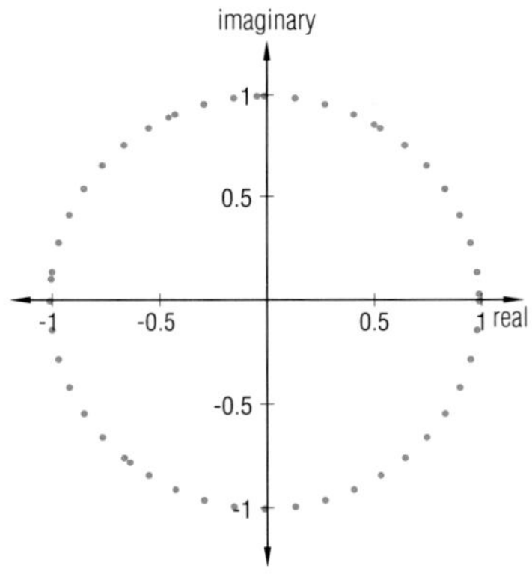

$\{z^n: 1 \leq n \leq 50\}$

Not all points on the circle are included in the infinite sequence of powers of z, but it can be shown, using techniques similar to those in Lesson 8-10, that given any point on the circle, there is some power of z that is as close as you wish to that point.

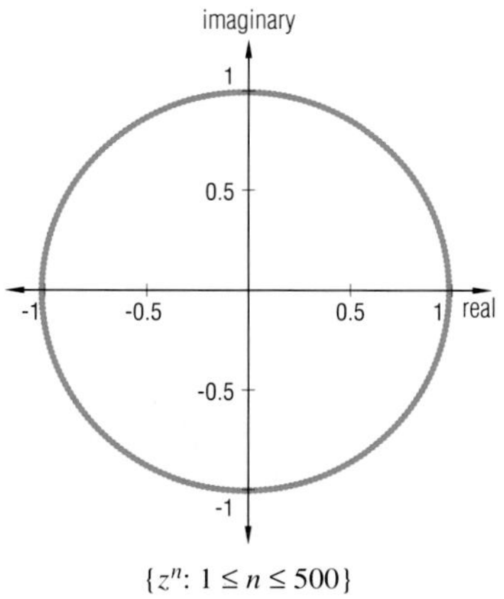

$\{z^n: 1 \leq n \leq 500\}$

NOTES ON QUESTIONS
Questions 6, 7, 9, 15, and 16: Students need polar graph paper for these problems.

Question 11: This question anticipates Lesson 8-7.

Question 12: Is it obvious to students that all points on the graph $z = \left[1.1, \frac{\pi}{3}\right]^n$ lie on exactly one of three lines? Ask them why. (The arguments of the powers are multiples of $\frac{\pi}{3}$. All these multiples lie on one of three lines.)

Question 13: Note that each side is a real number. You might want to check this by asking students to calculate $|z|^2$ and $|z^2|$ when $z = 3 + 4i$. ($z^2 = -7 + 24i$; $|z|^2 = |z^2| = 25$)

Question 17: Remind students that many proofs concerning properties of complex numbers begin with "Let $z = a + bi$."

FOLLOW-UP

MORE PRACTICE
For more questions on SPUR Objectives, use *Lesson Master 8-6,* shown on page 503.

EXTENSION
In the example given to check **Question 13** in the Notes on Questions, $a = 3$, $b = 4$, and $|z| = 5$. If $z^2 = c + di$, then $c = -7$, $d = 24$, and $|z^2| = 25$. Now 3, 4, and 5 are sides of a right triangle and so are 7, 24, and 25; these numbers are called Pythagorean triples because all three are integers. Can powers of complex numbers be used to generate more Pythagorean triples? Have students explore this idea. (Yes; if *a, b,* and *c* are integer sides of a right triangle and $z = a + bi$, then for any n, if $z^n = c + di$, then c, d, and $|z^n|$ form a Pythagorean triple.)

ADDITIONAL ANSWERS
6.

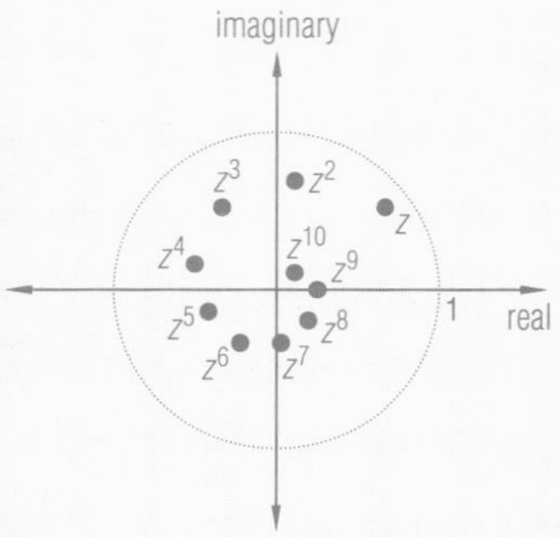

7.

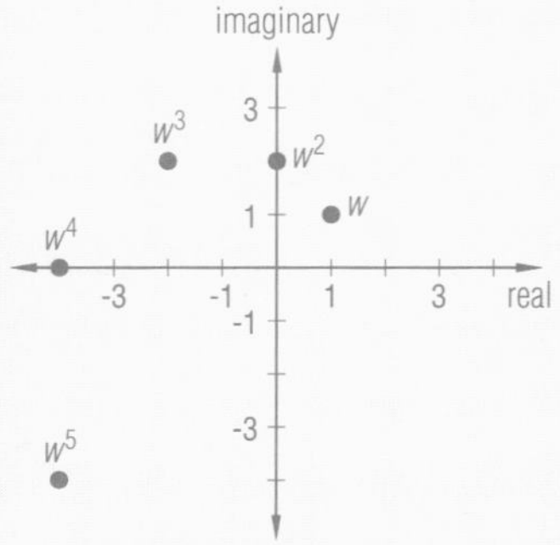

Questions

Covering the Reading

In 1–3, use DeMoivre's Theorem to compute the power. Write your answer in the same form as the base.

1. $[3, \frac{\pi}{5}]^4$ $[81, \frac{4\pi}{5}]$

2. $[6(\cos \frac{4\pi}{7} + i \sin \frac{4\pi}{7})]^3$ $216(\cos\frac{12\pi}{7} + i \sin \frac{12\pi}{7})$

3. $(2 + 2\sqrt{3}\, i)^5$ $512 - 512\sqrt{3}i$

In 4 and 5, do the graphs of the sequence of numbers get closer to or farther from the origin?

4. z^1, z^2, z^3, z^4, z^5 when $z = \frac{1}{3}\left(\cos\left(\frac{\pi}{7}\right) + i \sin\left(\frac{\pi}{7}\right)\right)$ closer

5. w^1, w^2, w^3, w^4, w^5 when $w = [5, \frac{2\pi}{11}]$ farther

6. Graph $z^1, z^2, \ldots, z^{10}$ when $z = [.85, 40°]$. See margin.

7. Graph w^1, w^2, w^3, w^4, and w^5 when $w = 1 + i$. See margin.

8. Give the polar coordinates of the points w^3, w^6, and w^9 from the graph following Example 2. $w^3 = [.729, \frac{3\pi}{4}]$; $w^6 \approx [.53, \frac{3\pi}{2}]$; $w^9 \approx [.39, \frac{\pi}{4}]$

9. Graph the first 10 positive integer powers of $[1, \frac{\pi}{6}]$. See margin.

Applying the Mathematics

10. a. Calculate $(2 + 2\sqrt{3}\, i)^4$ by thinking of it as $((2 + 2\sqrt{3}\, i)^2)^2$.
b. Check your work by converting $2 + 2\sqrt{3}\, i$ to polar form and using DeMoivre's Theorem.
See margin.

11. A fourth root of a certain complex number is $5(\cos 45° + i \sin 45°)$. Write the complex number in $a + bi$ form. -625

12. Refer to Example 1.
a. Compute z^7, z^{13}, z^{19}, and z^{25}. See margin.
b. Show that the argument for each of the powers in part **a** is of the form $\frac{2\pi}{3} + 2n\pi$, for some integer n. See margin.
c. The powers z^1, z^7, z^{13}, z^{19}, and z^{25} all lie along the line $\theta = \frac{2\pi}{3}$ and start the loops of the spirals. Find the distances from the origin to the graphs of z^1, z^7, z^{13}, z^{19}, and z^{25}.
2; 128; 8192; 524288; 33,554,432

13. Let z be any complex number. Use either mathematical induction or DeMoivre's Theorem to prove that for all positive integers n, $|z^n| = |z|^n$. See margin.

14. Find an equation for the logarithmic spiral on which the powers in Question 5 lie. $r = 5^{11\theta/2\pi}$

Review

In 15 and 16, **a.** sketch the graph of the equation, and **b.** classify the graph as a limaçon, rose curve, spiral of Archimedes, or logarithmic spiral. *(Lessons 8-5, 8-4)*

15. $r = 3 \sin 4\theta$
a) See margin. b) rose curve

16. $r = 1 - \cos \theta$
a) See margin. b) limaçon

502

17. Prove that if z is any complex number, then $z - \bar{z}$ is an imaginary number. *(Lesson 8-1)* Let $z = a + bi$. Then $\bar{z} = a - bi$ and $z - \bar{z} = a + bi - (a - bi) = a + bi - a + bi = 0 + 2bi$ which is an imaginary number.

18. Graph the function $y = 3 \sin(3x + \pi)$ and identify its amplitude, period, and phase shift. *(Lesson 3-9)*
See margin. amplitude = 3, period = $\frac{2\pi}{3}$, phase shift = $-\frac{\pi}{3}$

19. *Multiple choice.* The end behavior of $p(x) = 7x^5 - 4x^3 + 2x^2 - 5x + 9$ can be described by which of the following? *(Lesson 2-5)*

(a) $\lim\limits_{x \to -\infty} p(x) = -\infty$ and $\lim\limits_{x \to +\infty} p(x) = -\infty$

(b) $\lim\limits_{x \to -\infty} p(x) = -\infty$ and $\lim\limits_{x \to +\infty} p(x) = +\infty$

(c) $\lim\limits_{x \to -\infty} p(x) = +\infty$ and $\lim\limits_{x \to +\infty} p(x) = -\infty$

(d) $\lim\limits_{x \to -\infty} p(x) = +\infty$ and $\lim\limits_{x \to +\infty} p(x) = +\infty$ (b)

20. *Multiple choice.* Determine the statement that is the correct negation of
If a person is going 15 miles per hour over the speed limit when stopped by the police, then the person gets a ticket.

(a) *If a person is not going 15 miles per hour over the speed limit when stopped by the police, then the person does not get a ticket.*
(b) *If a person is going 15 miles per hour over the speed limit when stopped by the police, then the person does not get a ticket.*
(c) *There is a person who was going 15 miles per hour over the speed limit when stopped by the police and did not get a ticket.*
(d) *There is a person who was not going 15 miles per hour over the speed limit when stopped by the police and did not get a ticket.*

(Lesson 1-5) (c)

21. Solve each equation over the set of real numbers. *(Previous course)*
a. $x^3 = 13$ **b.** $x^3 = -13$ **c.** $x^4 = 13$ **d.** $x^4 = -13$
a) $\sqrt[3]{13}$; b) $-\sqrt[3]{13}$; c) $\sqrt[4]{13}$, $-\sqrt[4]{13}$; d) no real solution

Exploration

22. DeMoivre proved many theorems that are now known by other mathematicians' names. One of these is Stirling's Formula. Find this theorem in a probability or statistics book and explain what it does.
Stirling's formula gives an approximation for $n!$. $n! \approx \sqrt{2\pi n} \cdot \left(\frac{n}{e}\right)^n$

9., 10., 12.a., b., 13., 15.a., 16.a., 18. See Additional Answers in the back of this book.

NAME ____________

LESSON **MASTER 8–6**
QUESTIONS ON **SPUR** OBJECTIVES

■ SKILLS *Objective D (See pages 534–536 for objectives.)*
In 1–3, use DeMoivre's Theorem to compute the power. Write your answer in the same form as the base.

1. $(3 + 3i)^5$ — $-972 - 972i$
2. $\left[\sqrt{2}\left(\cos \frac{3\pi}{4} + i \sin \frac{3\pi}{4}\right)\right]^4$ — $4(\cos \pi + i \sin \pi)$
3. $\left[2, \frac{\pi}{2}\right]^{10}$ — $[1024, \pi]$

4. A sixth root of a certain complex number z is $2(\cos 30° + i \sin 30°)$. Write z in $a + bi$ form. — $-64 + 0i$

■ PROPERTIES *Objective F*
5. Let $z = [3, -12°]$.
   a. Write a polar representation of z^5. — $[243, -60°]$
   b. Use your answer to part **a** to find a polar representation for $(z^5)^4$. — $[243^4, -240°]$
   c. Verify that $(z^5)^4 = z^{20}$.
   $z^{20} = [3, -12°]^{20} = [3^{20}, -240°] = [(3^5)^4, -240°] = [243^4, -240°] = (z^5)^4$

6. Let $z = [r, \theta]$ and $w = [s, \phi]$, and let n be a positive integer.
   a. Write polar representations for z^n, w^n, and $z \cdot w$.
   $[r^n, n\theta], [s^n, n\phi], [rs, \theta + \phi]$
   b. Use your answer to part **a** to prove that $z^n \cdot w^n = (z \cdot w)^n$.
   $z^n \cdot w^n = [r^n, n\theta] \cdot [s^n, n\phi] = [r^n s^n, n\theta + n\phi] = [(rs)^n, n(\theta + \phi)]$; $(z \cdot w)^n = [rs, \theta + \phi]^n = [(rs)^n, n(\theta + \phi)]$; Therefore, $z^n \cdot w^n = (z \cdot w)^n$.

NAME ____________
Lesson MASTER 8–6 (page 2)

■ REPRESENTATIONS *Objective K*
7. a. Graph z^1, z^2, z^3, z^4, z^5, and z^6 when $z = \frac{3}{4}\left(\cos \frac{\pi}{3} + i \sin \frac{\pi}{3}\right)$.
   b. Is the sequence of points getting closer to or farther from the origin?
   closer

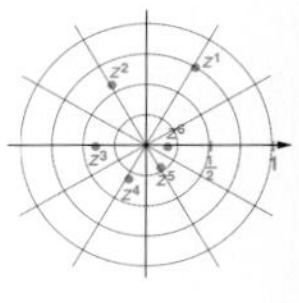

8. Give the polar coordinates of the points w^1, w^2, w^3, w^4, and w^5 where the first four are graphed below.

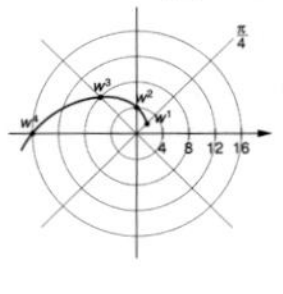

$w^1 = \left[2, \frac{\pi}{4}\right]$
$w^2 = \left[4, \frac{\pi}{2}\right]$
$w^3 = \left[8, \frac{3\pi}{4}\right]$
$w^4 = [16, \pi]$
$w^5 = \left[32, \frac{5\pi}{4}\right]$

LESSON 8-7

RESOURCES
- Lesson Master 8-7
- Quiz for Lessons 8-4 through 8-7
- Teaching Aid 55 can be used with **Example 2** and displays the Geometric *n*th Roots Theorem.
- Teaching Aid 56 displays the *n*th roots of real numbers.

OBJECTIVES

- **D** Find powers and roots of complex numbers.
- **K** Use DeMoivre's Theorem to graph powers and roots of complex numbers.

TEACHING NOTES

The material in this lesson often causes students difficulty when approached using trigonometric form only. However, the content is made easier by the use of polar form and by the large number of diagrams indicating the position of the roots as vertices of a regular polygon centered at the origin. Emphasize that the angles between consecutive *n*th roots are equal. Work through **Example 1** before the proof of the theorem. Note how much easier **Example 2** is because the Complex *n*th Roots Theorem is used.

Point out the following difference between finding roots of complex numbers and real numbers. For real numbers, if you know r is a square root of a given number, then so is $-r$, because $r^2 = (-r)^2$. Thus, you find one square root and use it to find the others. For complex numbers, the idea is

LESSON

8-7

Roots of Complex Numbers

Roots of complex numbers are defined in the same way as roots of real numbers. For instance, since $(2 - 7i)^2 = -45 - 28i$, we say that $2 - 7i$ is a square root of $-45 - 28i$. Since $(-1 + \sqrt{3}\,i)^9 = 512$, as you saw in the last lesson, $-1 + \sqrt{3}\,i$ is a 9th root of 512. In general, if z and w are complex numbers and if n is an integer greater than or equal to 2, then **z is an nth root of w** if and only if $z^n = w$. Although real numbers have 0, 1, or 2 real roots (depending on both the number and the value of n), nonzero complex number always have exactly n nth roots. The nth roots of complex numbers have a simple structure, and their geometry is exquisitely elegant.

Example 1 a. Find the cube roots of $27i$. b. Plot them in the complex plane.

Solution **a.** Let $z = [s, \phi]$ represent a cube root of $27i$. By the definition of cube root, z must satisfy the equation

$$z^3 = 27i.$$

In polar form, this equation is

$$[s, \phi]^3 = [27, \tfrac{\pi}{2}].$$

So, by DeMoivre's Theorem,

$$[s^3, 3\phi] = [27, \tfrac{\pi}{2}].$$

This means that a complex number $z = [s, \phi]$ satisfies $z^3 = 27i$ if and only if $s^3 = 27$ and $3\phi \equiv \frac{\pi}{2} (\text{mod } 2\pi)$.

Therefore, $s = \sqrt[3]{27} = 3$. Since 3ϕ and $\frac{\pi}{2}$ may differ by integral multiples of 2π, $3\phi = \frac{\pi}{2} + k \cdot 2\pi$, where k is any integer.

Solving for ϕ, $\quad \phi = \frac{\pi}{6} + k \cdot \frac{2\pi}{3}$, where k is any integer.

Thus the cube roots of $27i$ have the polar form $[3, \frac{\pi}{6} + \frac{2\pi k}{3}]$, where k is any integer. Replacing k with 0, 1, and 2 gives the roots z_0, z_1, and z_2 as shown here in polar, trigonometric, and $a + bi$ forms.

$$k = 0:\ z_0 = [3, \tfrac{\pi}{6}] = 3\left(\cos \tfrac{\pi}{6} + i \sin \tfrac{\pi}{6}\right) = \tfrac{3\sqrt{3}}{2} + \tfrac{3}{2}i$$

$$k = 1:\ z_1 = [3, \tfrac{\pi}{6} + \tfrac{2\pi}{3}] = 3\left(\cos \tfrac{5\pi}{6} + i \sin \tfrac{5\pi}{6}\right) = -\tfrac{3\sqrt{3}}{2} + \tfrac{3}{2}i$$

$$k = 2:\ z_2 = [3, \tfrac{\pi}{6} + \tfrac{4\pi}{3}] = 3\left(\cos \tfrac{3\pi}{2} + i \sin \tfrac{3\pi}{2}\right) = -3i$$

All other values of k yield numbers equal to z_0, z_1, or z_2. For instance, when $k = 3$,

$$z_3 = [3, \tfrac{\pi}{6} + \tfrac{6\pi}{3}] = 3\left(\cos\left(\tfrac{\pi}{6} + 2\pi\right) + i \sin\left(\tfrac{\pi}{6} + 2\pi\right)\right)$$
$$= 3\left(\cos \tfrac{\pi}{6} + i \sin \tfrac{\pi}{6}\right)$$

which, due to the periodicity of cosine and sine, is equal to z_0. Therefore, $27i$ has exactly 3 complex cube roots: z_0, z_1, and z_2.

b. These roots are plotted below.

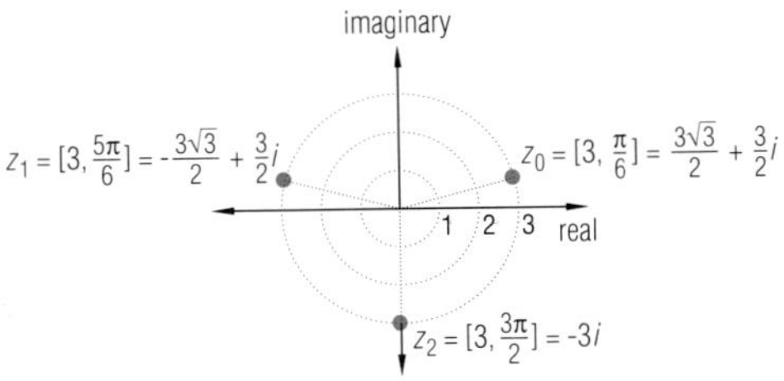

Check If z_1 is a cube root, it must satisfy the equation $z^3 = 27i$. Rewrite $z_1 = -\frac{3\sqrt{3}}{2} + \frac{3}{2}i = -\frac{3}{2}(\sqrt{3} - i)$. Now substitute.

$$\begin{aligned}\left(-\tfrac{3}{2}\left(\sqrt{3} - i\right)\right)^3 &= \left(-\tfrac{3}{2}\right)^3\left(\sqrt{3} - i\right)^3 \\ &= -\tfrac{27}{8}\left(\sqrt{3} - i\right)\left(3 - 2\sqrt{3}\,i - 1\right) \\ &= -\tfrac{27}{8}\left(\sqrt{3} - i\right)\left(2 - 2\sqrt{3}\,i\right) \\ &= -\tfrac{27}{8} \bullet \left(-8i\right) \\ &= 27i\end{aligned}$$

The procedure outlined in Example 1 can be used to find all nth roots of any complex number $[r, \theta]$, with $r > 0$, as follows. Let $z = [s, \phi]$ be an nth root of $[r, \theta]$, where $s > 0$. Then $z^n = [s^n, n\phi] = [r, \theta]$. Since the absolute values of these polar forms of z^n are equal, $s^n = r$, and so $s = \sqrt[n]{r}$. The arguments $n\phi$ and θ may differ by integral multiples of 2π, so

$$n\phi = \theta + k \bullet 2\pi, \text{ for } k \text{ an integer,}$$

$$\text{or } \phi = \frac{\theta}{n} + k \bullet \frac{2\pi}{n}, \text{ for } k \text{ an integer.}$$

Thus one nth root of $[r, \theta]$ is $z = [\sqrt[n]{r}, \frac{\theta}{n}]$. The others are found by adding multiples of $\frac{2\pi}{n}$ to the argument.

This proves the following theorem:

Complex *n*th Roots Theorem

Let z be a complex number with $z = [r, \theta] = r(\cos\theta + i\sin\theta)$ and $r > 0$.

(Polar Form) The n nth roots of $[r, \theta]$ are $[\sqrt[n]{r}, \frac{\theta}{n} + k \bullet \frac{2\pi}{n}]$, where $k = 0, 1, 2, \ldots,$ and $n - 1$.

(Trigonometric Form) The n nth roots of $r(\cos\theta + i\sin\theta)$ are $\sqrt[n]{r}\left(\cos\left(\frac{\theta}{n} + k \bullet \frac{2\pi}{n}\right) + i\sin\left(\frac{\theta}{n} + k \bullet \frac{2\pi}{n}\right)\right)$, where $k = 0, 1, 2, \ldots,$ and $n - 1$.

to determine one of the nth roots $[r, \theta]$. Then the others are found by repeatedly adding $\frac{2\pi}{n}$ to the argument, because $[r, \theta]^n = \left[r, \theta + \frac{2\pi}{n}\right]^n$.

The Complex nth Roots Theorem should remind students of DeMoivre's Theorem. Write the two theorems (polar and trigonometric forms) next to each other and make comparisons.

Making Connections
The Geometric nth Roots Theorem is certainly one of the great results of elementary mathematics. Review with students how many ideas it relates: regular polygons from geometry; powers and roots of complex numbers from algebra; and the graphing that connects them.

Point out how beautifully the Geometric nth Roots Theorem explains why there are different numbers of real nth roots of a real number x, depending on whether x is positive or negative and n is odd or even.

Error Analysis When generating all of the nth roots of a complex number, students may start with a correct solution $[r, \theta]$ but make mistakes in adding $\frac{2\pi k}{n}$, or use $k\phi$ itself to get the other roots. If they consistently check themselves by calculating the argument given by $k = n$ to see if they are back to $[r, \theta]$ where they began, both of these errors can be avoided.

ADDITIONAL EXAMPLES

1.a. Find the cube roots of 8 and plot them in the complex plane.

$[2, 0], \left[2, \frac{2\pi}{3}\right], \left[2, \frac{4\pi}{3}\right]$

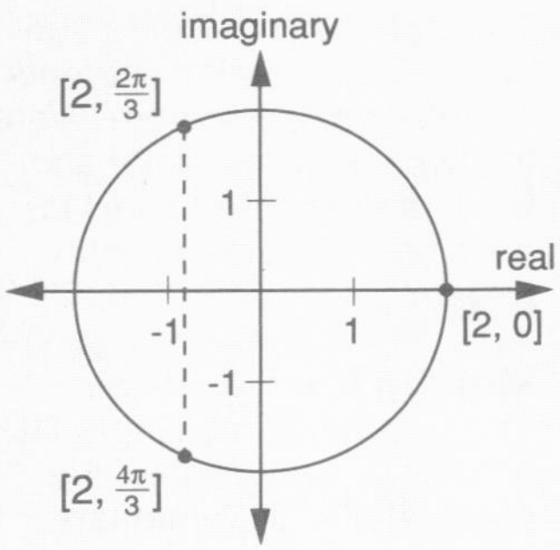

b. Find the cube roots of $4 + 4\sqrt{3}\,i$ and plot them in the complex plane.

$\left[2, \frac{\pi}{9}\right], \left[2, \frac{7\pi}{9}\right], \left[2, \frac{13\pi}{9}\right]$

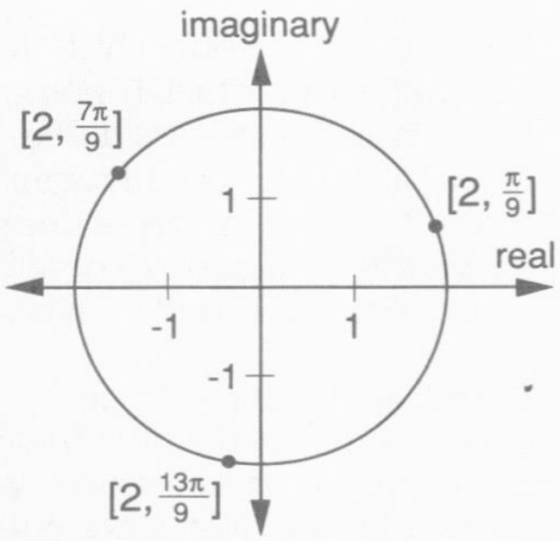

2. Find the 6th roots of $32 + 32\sqrt{3}\,i$. Express the roots in both trigonometric and polar form, and graph them in the complex plane.

$z_0 = [2, 10°]$
$= 2(\cos 10° + i \sin 10°)$
$z_1 = [2, 70°]$
$= 2(\cos 70° + i \sin 70°)$
$z_2 = [2, 130°]$
$= 2(\cos 130° + i \sin 130°)$
$z_3 = [2, 190°]$
$= 2(\cos 190° + i \sin 190°)$
$z_4 = [2, 250°]$
$= 2(\cos 250° + i \sin 250°)$
$z_5 = [2, 310°]$
$= 2(\cos 310° + i \sin 310°)$

(The graph is shown at the top of the next page.)

Example 2 Find the 5th roots of $16 + 16\sqrt{3}\,i$. Express the roots in both trigonometric and polar form, and graph them in the complex plane.

Solution Let $[r, \theta] = 16 + 16\sqrt{3}\,i$. Then $r = \sqrt{16^2 + (16\sqrt{3})^2} = \sqrt{256 + 768} = 32$ and $\tan\theta = \frac{16\sqrt{3}}{16} = \sqrt{3}$, so $\theta = 60°$. Hence, $[r, \theta] = [32, 60°]$, and so one 5th root is $\left[\sqrt[5]{32}, \frac{60°}{5}\right]$, which is $[2, 12°]$. In trigonometric form, this root is $2(\cos 12° + i \sin 12°)$. The other four roots also have an absolute value of 2 and are spaced $\frac{360°}{n} = \frac{360°}{5} = 72°$ apart.

$k = 1$: $[2, 84°] = 2(\cos 84° + i \sin 84°)$

$k = 2$: $[2, 156°] = 2(\cos 156° + i \sin 156°)$

$k = 3$: $[2, 228°] = 2(\cos 228° + i \sin 228°)$

$k = 4$: $[2, 300°] = 2(\cos 300° + i \sin 300°)$

Notice that if 72° is added once again to find one more root, $2(\cos 372° + i \sin 372°)$ would result, but this is the same as the first root, $2(\cos 12° + i \sin 12°)$.

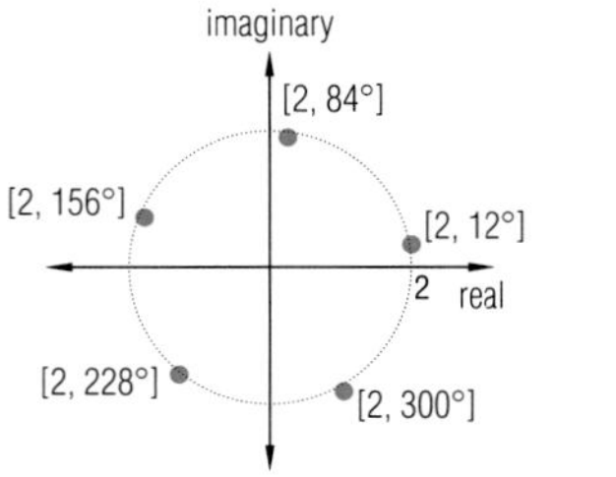

Notice that the three cube roots from Example 1 lie on the circle with radius 3 centered at the origin, and z_0, z_1, and z_2 are spaced $\frac{2\pi}{3}$ radians apart on that circle. Thus, z_0, z_1, and z_2 are the vertices of an equilateral triangle. In Example 2, the five 5th roots are the vertices of a regular pentagon centered at (0, 0). These examples are instances of the following gorgeous theorem.

Geometric *n*th Roots Theorem

When graphed in the complex plane, the nth roots of any nonzero complex number z are the vertices of a regular n-gon whose center is at (0, 0).

Proof Refer to the Polar Form of the Complex nth Roots Theorem. Each succeeding value of k adds $\frac{2\pi}{n}$ to the argument while keeping the absolute value the same. This means that the nth roots of a number are all the same distance from the origin and are spaced $\frac{2\pi}{n}$ apart. Thus, the n roots determine congruent arcs on a circle. These congruent arcs determine congruent chords, the sides of a regular n-gon.

506

Real numbers can be considered as complex numbers, so the Complex and Geometric nth Root theorems can be applied to picture the nth roots of real numbers. This picture can, in turn, be used to explain why real numbers may have 0, 1, or 2 real nth roots, depending on n and the sign of the real number.

If w is a positive real number, then one of its nth roots is $\sqrt[n]{w}$, which lies on the positive part of the real axis. Since the other nth roots are vertices of a regular n-gon centered at the origin, there will be another vertex on the real axis only if n is even. This confirms that for a positive number, there are exactly two real nth roots if n is even and exactly one real nth root if n is odd.

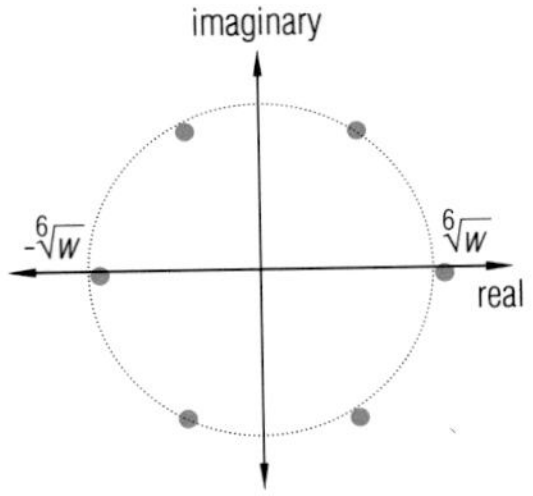

6th roots of $w > 0$

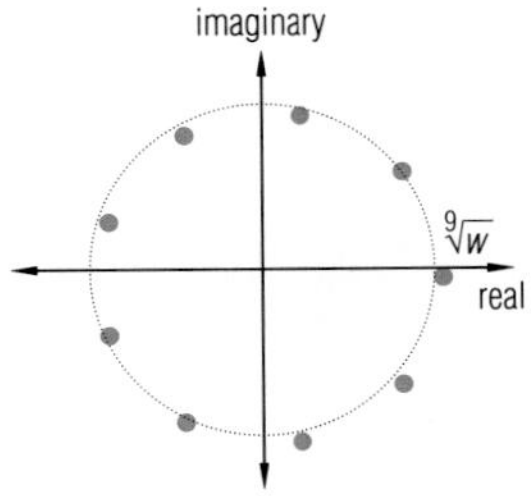

9th roots of $w > 0$

If w is a negative real number, then $w = [-w, \pi]$ in polar form (note that $-w$ is a positive number), and its n complex nth roots are $[\sqrt[n]{-w}, \frac{\pi}{n} + \frac{2\pi k}{n}]$, where $k = 0, 1, 2, \ldots,$ and $n - 1$. For the nth root to be real, its argument must be a multiple of π. The arguments of these nth roots equal $\pi(\frac{2k+1}{n})$, which is a multiple of π only if $n = 2k + 1$; that is, only if n is odd and $k = \frac{n-1}{2}$. This confirms that for a negative number, there is no real nth root when n is even and exactly one real nth root when n is odd.

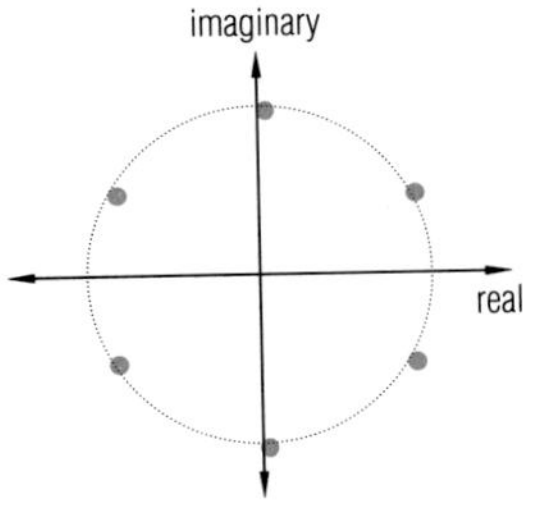

6th roots of $w < 0$

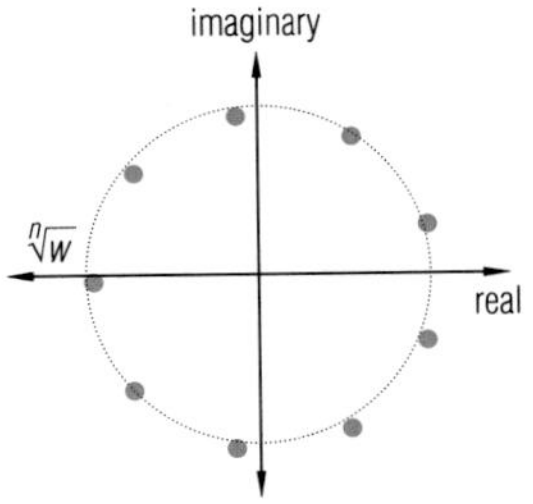

9th roots of $w < 0$

Questions

Covering the Reading

1. Refer to Example 1. The cube roots are described by the polar coordinates $[3, \frac{\pi}{6} + \frac{2\pi k}{3}]$ for k an integer. Show that $k = 4$ gives one of the three roots found in the example. See margin.

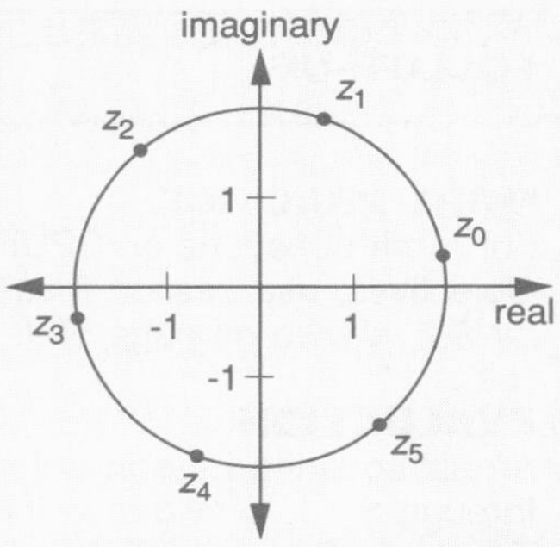

NOTES ON QUESTIONS
Making Connections for Questions 6-8, 12, and 13: These questions anticipate the content of the next two lessons and should be discussed.

Question 10: If the positive integer powers of $1 + i$ are graphed, then they lie on a spiral and (16, 0) is the 8th point on the spiral.

Question 11: This question asks students to generalize a pattern. Encourage students to make as precise a statement as possible about this pattern. From their statement, the exact location of each vertex should be determined.

Question 15: The general result of this question is startling: the sum of the nth roots of any complex number is 0. (This is proved most easily by noting that, from the Factor Theorem, the sum of the zeros of a polynomial $x^n + a_{n-1}x^{n-1} + \ldots + a_1x + a_0$ is equal to $-a_{n-1}$. In the special case of the polynomial $x^n - z$, the zeros are the nth roots of z, and $a_{n-1} = 0$.)

Question 16: This is a geometric sequence whose first term is 1 and whose common ratio is the nonreal number [.7, 80°].

ADDITIONAL ANSWERS
1. For $k = 4$, $\left[3, \frac{\pi}{6} + \frac{2\pi k}{3}\right] = \left[3, \frac{17\pi}{6}\right] = \left[3, \frac{5\pi}{6}\right]$, which is z_1.

FOLLOW-UP

MORE PRACTICE
For more questions on SPUR Objectives, use *Lesson Master 8-7,* shown on page 509.

EVALUATION
A quiz covering Lesson 8-4 through 8-7 is provided in the Teacher's Resource File.

ADDITIONAL ANSWERS
2. $z_0 \approx 1.93 + .52i$;
$z_1 \approx .52 + 1.93i$;
$z_2 = -\sqrt{2} + i\sqrt{2}$;
$z_3 \approx -1.93 - .52i$;
$z_4 \approx -.52 - 1.93i$;
$z_5 = \sqrt{2} - i\sqrt{2}$

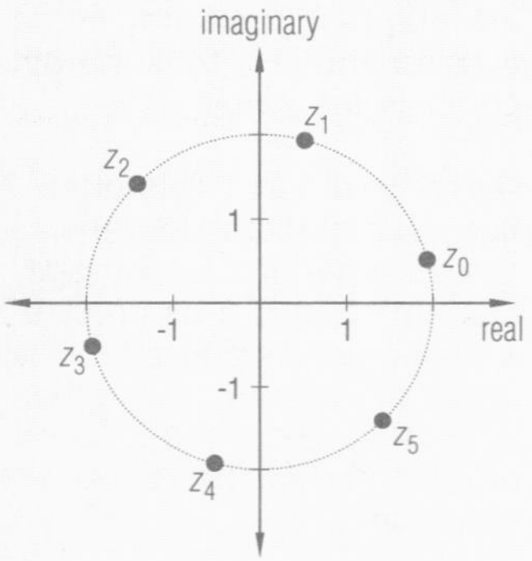

3. $z_0 = 3[\cos(63°) + i\sin(63°)]$;
$z_1 = 3[\cos(135°) + i\sin(135°)]$;
$z_2 = 3[\cos(207°) + i\sin(207°)]$;
$z_3 = 3[\cos(279°) + i\sin(279°)]$;
$z_4 = 3[\cos(351°) + i\sin(351°)]$

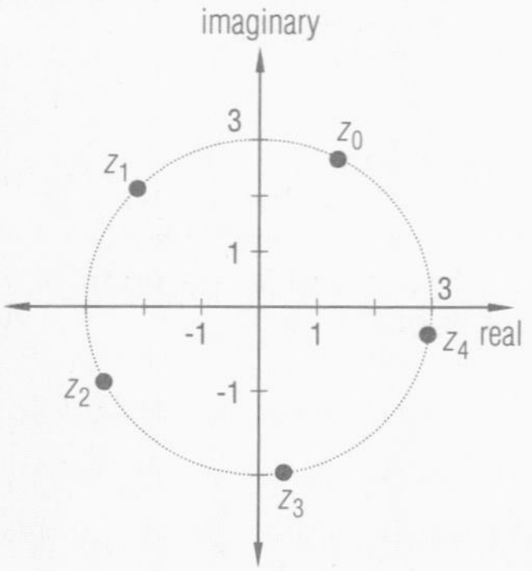

In 2 and 3, find and plot the roots in the complex plane. Write the roots in the same form as the given number.

2. the sixth roots of $64i$ See margin.

3. the fifth roots of $z = 243[\cos(315°) + i\sin(315°)]$ See margin.

4. Refer to Example 2. Show that $[2, 228°]$ is a fifth root of $16 + 16\sqrt{3}i$ by raising the root to the fifth power. $[2, 228°]^5 = [2^5, 5 \cdot 228°] = [32, 1140°] = [32, 60°] = 32(\cos 60° + i\sin 60°) = 16 + 16\sqrt{3}\,i$

5. a. The fourth roots of $8 + 8\sqrt{3}i$ form the vertices of what figure?
b. Find those fourth roots and graph them in the complex plane.
a) square; b) See margin.

In 6–8, give **a.** the number of real and **b.** the number of nonreal solutions to the equation. (You do not have to solve the equation.)

6. $z^5 = 10$ a) 1; b) 4 **7.** $z^9 = -4$ a) 1; b) 8 **8.** $z^{164} = 1000$ a) 2; b) 162

Applying the Mathematics

9. A cube root of a certain complex number is $9(\cos 60° + i\sin 60°)$. Find the complex number and its other cube roots.
-729; -9, $9(\cos 300° + i\sin 300°)$

10. a. $1 + i$ is an 8th root of what real number? 16
b. What are its other 8th roots? $\sqrt{2}\,i$, $-1 + i$, $-\sqrt{2}$, $-1 - i$, $-\sqrt{2}\,i$, $1 - i$, $\sqrt{2}$

11. a. Graph the 4th roots of 1. See margin.
b. Graph the 5th roots of 1. See margin.
c. Describe the graph of the nth roots of 1. They are the vertices of a regular n-gon, with center at the origin, and a vertex at (1, 0).

In 12 and 13, solve the equation over the set of complex numbers. Express the solutions in $a + bi$ form and plot them in the complex plane. See margin.

12. a. $x^3 = 8$ **13. a.** $z^4 = 81$
b. $z^3 = -8$ **b.** $z^4 = -81$

14. If the center of the stop sign at the right is (0, 0) and the distance from a vertex of the sign to the center is 1 foot, then the vertices of the sign are the 8th roots of what number? -1

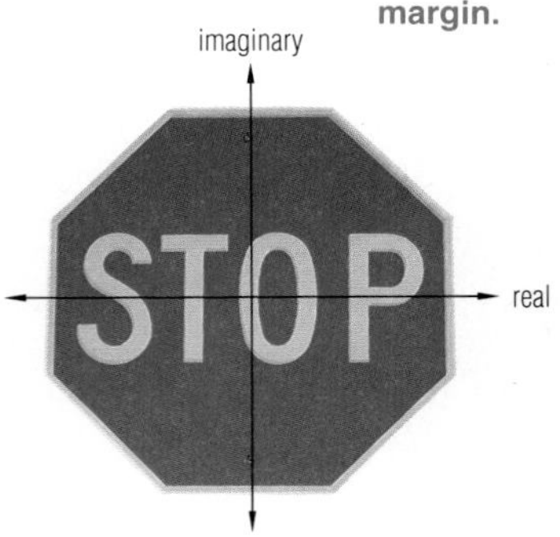

15. Show that the sum of the 3 cube roots of 8 is 0. The 3 cube roots of 8 are 2, $-1 + \sqrt{3}\,i$, $-1 - \sqrt{3}\,i$; their sum is $2 + (-1 + \sqrt{3}\,i) + (-1 - \sqrt{3}\,i) = 0$.

Review

16. Define a sequence by $a_n = [.7, 80°]^n$.
a. Write the first eight terms in polar form. See margin.
b. Graph the first six terms. See margin.
c. What is $\lim_{n\to\infty} a_n$? *(Lesson 8-6)* 0

17. Sketch the polar graphs of the equations $r = 3\theta$ and $r = 3^\theta$ on the same polar grid, and identify their shapes. *(Lesson 8-5)* See margin.

18. Prove that for all complex numbers z, $\overline{z^n} = (\overline{z})^n$. (Hint: Write z in polar form.) *(Lesson 8-6)* See margin.

19. Given the sequence defined recursively by

$$\begin{cases} a_1 = 3 \\ a_{k+1} = 2a_k - 1 \text{ for integers } k \geq 1. \end{cases}$$

a. Conjecture an explicit formula for the sequence. (Hint: It involves powers of 2.) $a_n = 2^n + 1$

b. Use mathematical induction to prove your conjecture. *(Lessons 7-4, 7-2)* See below.

20. Consider the function f defined by

$$f(x) = \frac{6x^2 + 5x}{3x - 2}.$$

a. Find equations for all asymptotes. $x = \frac{2}{3}$, $y = 2x + 3$

b. Use limit notation to describe the end behavior of f and its behavior near any vertical asymptotes. *(Lessons 5-5, 5-4)* See below.

21. Let $p(x) = x^3 + x^2 - 10x + 8$.

a. Given that $p(1) = 0$, name a factor of $p(x)$. $x - 1$

b. Given that $x + 4$ is a factor of $p(x)$, name a zero of $p(x)$. -4

c. At most, how many more zeros can $p(x)$ have? 1

d. Suppose $p(x)$ is a factor of a polynomial $q(x)$. Name two zeros of $q(x)$. Justify your answer. *(Lessons 4-6, 4-1)*
1 and -4, by the Transitive Property of Factors

Exploration

22. Write a computer program that will determine the n nth roots of any complex number $a + bi$. See margin.

5.b., 11.a., b., 12., 13., 16.a., b., 17., 18., 22. See Additional Answers in the back of this book.

NAME

LESSON MASTER 8–7
QUESTIONS ON SPUR OBJECTIVES

■ **SKILLS** *Objective D (See pages 534–536 for objectives.)*
In 1 and 2, find the roots. Express them in the same form as the given number.

1. cube roots of $[125, \pi]$

$\left[5, \frac{\pi}{3}\right], [5, \pi], \left[5, \frac{5\pi}{3}\right]$

2. seventh roots of $128\left(\cos \frac{\pi}{3} + i \sin \frac{\pi}{3}\right)$

$2\left(\cos \frac{\pi}{21} + i \sin \frac{\pi}{21}\right), 2\left(\cos \frac{7\pi}{21} + i \sin \frac{7\pi}{21}\right),$
$2\left(\cos \frac{13\pi}{21} + i \sin \frac{13\pi}{21}\right), 2\left(\cos \frac{19\pi}{21} + i \sin \frac{19\pi}{21}\right),$
$2\left(\cos \frac{25\pi}{21} + i \sin \frac{25\pi}{21}\right), 2\left(\cos \frac{31\pi}{21} + i \sin \frac{31\pi}{21}\right),$
$2\left(\cos \frac{37\pi}{21} + i \sin \frac{37\pi}{21}\right)$

3. The fifth power of some complex number z is 243. Find all possible values of z.

$[3, 0°], [3, 72°], [3, 144°], [3, 216°], [3, 288°]$

In 4–6, solve the equation over the set of complex numbers. Express the solutions in $a + bi$ form.

4. $x^6 = -1$

$\pm i, \frac{\pm\sqrt{3} \pm i}{2}$

5. $z^8 = 16$

$\pm\sqrt{2}, \pm\sqrt{2}i, \pm 1 \pm i$

6. $(z - 1)^8 = 16$ (Hint: Use your answer to Question 5.)

$1 \pm \sqrt{2}, 1 \pm \sqrt{2}i, 2 \pm i, \pm i$

NAME
Lesson MASTER 8–7 (page 2)

7. A sixth root of a certain complex number is $\sqrt{2}(\cos 108° + i \sin 108°)$. Find the complex number and its other sixth roots.

The number is $8(\cos 288° + i \sin 288°)$. The other roots are: $\sqrt{2}(\cos 48° + i \sin 48°)$, $\sqrt{2}(\cos 168° + i \sin 168°)$, $\sqrt{2}(\cos 228° + i \sin 228°)$, $\sqrt{2}(\cos 288° + i \sin 288°)$, and $\sqrt{2}(\cos 348° + i \sin 348°)$.

■ **REPRESENTATIONS** *Objective K*
8. Graph the sixth roots of -1 on a complex plane. (Refer to Question 4.)

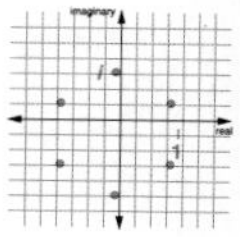

9. a. The seventh roots of $128\left(\cos \frac{\pi}{3} + i \sin \frac{\pi}{3}\right)$ form the vertices of what figure? (Refer to Question 2.) regular heptagon

b. Graph the figure.

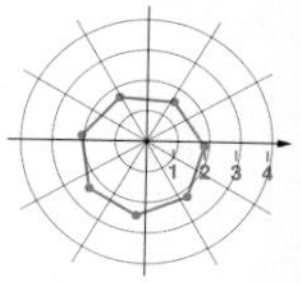

19\. b) $S(n): a_n = 2^n + 1$. $S(1)$: $a_1 = 2^1 + 1 = 3$, so $S(1)$ is true. Assume $S(k)$: $a_k = 2^k + 1$. Then

$$\begin{aligned} a_{k+1} &= 2a_k - 1 \\ &= 2(2^k + 1) - 1 \\ &= 2^{k+1} + 2 - 1 \\ &= 2^{k+1} + 1 \end{aligned}$$

So $S(k) \Rightarrow S(k+1)$. Therefore by mathematical induction, $S(n)$ is true for all positive integers n.

20\. b) $\lim_{x \to -\infty} f(x) = -\infty$; $\lim_{x \to \infty} f(x) = \infty$; $\lim_{x \to 2/3^+} f(x) = \infty$; $\lim_{x \to 2/3^-} f(x) = -\infty$

LESSON 8-8

RESOURCES
- Lesson Master 8-8

OBJECTIVES

E Find all zeros, and their multiplicities, of a given polynomial.
G Use the properties of polynomials to find or describe their zeros.

TEACHING NOTES

This lesson focuses on the number of zeros of a polynomial function. The first theorem shows that every polynomial of *odd* degree with *real* coefficients has at least one *real* zero. The Fundamental Theorem of Algebra is much more general; it says that every polynomial of *any* degree ≥ 1 with *complex* coefficients has at least one zero. A consequence of the generality of the theorem is that the zero may be nonreal.

As the preceding paragraph indicates, caution students to read this lesson very carefully. It is easy to overlook critical occurrences of the words real and complex.

Implicit in the proof of the first theorem is the fact that polynomial functions are everywhere continuous. The sign of the coefficient of the lead term determines the end behavior of $p(x)$. A few graphs will help show why this must be true.

LESSON

8-8 The Number of Zeros of a Polynomial

The number of solutions to an equation or inequality can be as important as the values of the solutions themselves. For instance, for a particular value of k, the number of nonnegative integer solutions (d, q) to $10d + 25q = k$ is the number of ways to make change of k cents using only dimes and quarters. When a quadratic equation is used to model the flight of a ball, the number of solutions to the equation can tell you how many times the ball reaches a certain height. If a ball reaches a height twice, then it cannot be the maximum height it reaches. If it reaches a height exactly once, that may be its maximum height.

In the last lesson, you saw that although real numbers may have 0, 1, or 2 real nth roots, *every* nonzero complex number has n complex nth roots. Thus, when n is a positive integer and $k \neq 0$, determining the number of solutions to the equation $x^n = k$ is simpler in the set of complex numbers than it is in the set of real numbers. In the set of complex numbers, the equation $x^n = k$ always has n solutions.

You also saw in Lesson 4-6 that every polynomial of degree n has *at most* n zeros. It is also easy to prove that certain polynomials have at *least* one zero.

Theorem

Every polynomial of odd degree with real coefficients has at least one real zero.

Proof Recall that, when $p(x)$ is a polynomial of odd degree with real coefficients, the end behavior of the function p is such that as $x \rightarrow +\infty$ and $x \rightarrow -\infty$, $p(x)$ approaches $+\infty$ in one case and $-\infty$ in the other. thus, because p is continuous on any interval the Intermediate Value Theorem implies that the graph of p must intersect the x-axis (at least once), and so $p(x)$ must have a real zero.

The above theorem cannot be extended to polynomials of even degree. It is possible for polynomials of even degree with real coefficients not to have any real zeros. However, in 1797, Gauss proved the following important generalization.

Fundamental Theorem of Algebra

Every polynomial of degree $n \geq 1$ with real or nonreal complex coefficients has at least one complex zero.

510

Gauss' proof was based on geometric properties of the graphs of the real and imaginary parts of the equation

$$p(z) = 0$$

as $z = x + yi$ varies over the complex plane. He gave three other proofs of the theorem during his lifetime because he was searching for a proof that was entirely algebraic. To this day, all known proofs of the Fundamental Theorem of Algebra use geometry in some essential way, and all are beyond the scope of this course.

At the time Gauss proved this theorem, it was known that all polynomial equations of the 2nd, 3rd, or 4th degree could be solved using complex numbers. Some thought that perhaps new numbers, different from complex numbers, might be needed for equations of higher degree. Gauss's result implies that no other numbers are needed.

Sometimes the zeros of a polynomial $f(x)$ are considered to occur more than once. For example, the zeros of $p(x) = (x - 1)(x + 2)^5(x - 3)$ are 1, -2, and 3, but there is a sense in which the zero -2 occurs 5 times. The zero -2 is said to have *multiplicity* 5.

Definitions

Suppose c is a zero of a polynomial $p(x)$ of degree at least 1. The largest positive integer m such that $(x - c)^m$ is a factor of $p(x)$ is called the **multiplicity** of the zero, c, of $p(x)$. Zeros of multiplicity 1 are called **simple zeros** of $p(x)$.

Example 1 Find all zeros and their corresponding multiplicities for the polynomial $p(x) = x^4 + 4x^3 - 5x^2$.

Solution The terms in the expression defining $p(x)$ have x^2 as the largest common factor:

$$p(x) = x^2(x^2 + 4x - 5).$$

The remaining quadratic factor $x^2 + 4x - 5$ of $p(x)$ can be factored

$$x^2 + 4x - 5 = (x - 1)(x + 5).$$

Therefore, a complete factorization of $p(x)$ is

$$p(x) = x^2(x - 1)(x + 5).$$

Thus applying the Factor Theorem and the definition of multiplicity for zeros, $x = 0$ is a zero of multiplicity 2, while $x = 1$ and $x = -5$ are simple zeros of $p(x)$. Thus, although $p(x)$ has only three distinct zeros, it is said that $p(x)$ has four zeros, counting multiplicities.

In Example 1, the polynomial $p(x)$ of degree four had four zeros, counting multiplicities. In general, if multiplicities are counted, does a polynomial of degree $n \geq 1$ always have n zeros? The answer depends on the number system in which the zeros are allowed to exist. For instance, if only integer zeros are allowed, the polynomial $f(x) = 2x + 1$, which has degree 1, has no zeros. But if x can be any real number, then $f(x)$ has exactly one zero, $-\frac{1}{2}$. Similarly, the polynomial $h(x) = x^2 + 1$ has no real zeros but has exactly two complex zeros, i and $-i$. Is there a polynomial of degree $n \geq 1$ that has fewer than n complex zeros even after counting multiplicities?

Each of the first two theorems has a limitation. For the first theorem, note that polynomials of *even* degree and real coefficients may have no real zeros: consider $p(x) = x^2 + 1$. (The zeros are i and $-i$.) For the Fundamental Theorem of Algebra, the polynomial must be of degree greater than or equal to one. A constant polynomial such as $f(x) = 4$ may have no zero. Again, a graph will help.

The Fundamental Theorem of Algebra could have been stated more succinctly as: Every nonconstant polynomial with complex coefficients has at least one complex zero. We added "real or noreal" for emphasis.

The concept of multiplicity builds on the idea (derived from the Factor Theorem) that a polynomial contains each of its zeros in the form of a linear factor. If the same linear factor appears more than once, then the number of times it appears is the multiplicity of the corresponding zero. Once students have this concept, they can show easily that the number of zeros that a polynomial has (counting multiplicities) is equal to its degree. You may wish to go through the proof of this theorem; it is a straightforward use of mathematicial induction.

ADDITIONAL EXAMPLES

1. Find all zeros and their corresponding multiplicities for the polynomial $p(x) = x^5 - 2x^3 + x$.
$p(x) = x(x - 1)^2(x + 1)^2$; -1 has multiplicity 2, 0 has multiplicity 1, and 1 has multiplicity 2.

ADDITIONAL EXAMPLES

2.a. Give a polynomial of degree 4 that has zeros -1 (multiplicity 2), 2, and 0.
$p(x) = x(x - 2)(x + 1)^2 = x^4 - 3x^2 - 2x$, or any nonzero multiple of $p(x)$

b. Give a polynomial of degree 4 that has zeros $2 + i$, $2 - i$, -1, and 1.
$p(x) = (x + 1)(x - 1)(x - 2 - i) \cdot (x - 2 + i) = x^4 - 4x^3 + 4x^2 + 4x - 5$

NOTES ON QUESTIONS

Technology for Question 1: Use an automatic grapher with domain set to $-5 \le x \le 5$ and range set to $-60 \le y \le 60$ to help students verify their conclusions.

Question 2: Remind students that all real numbers can be considered as complex numbers.

Question 5: Point out that the quadratic formula is valid for any quadratic polynomial, even one with nonreal coefficients.

Question 8: A diagram may prove to be a useful reminder and guide.

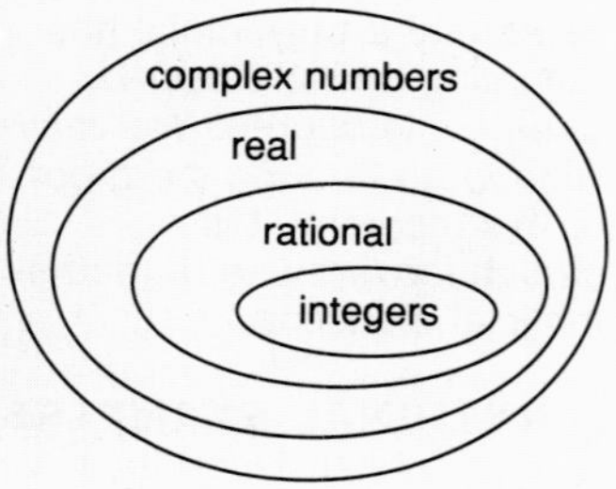

Question 12: Generalize parts **a** and **b**. The degree of the sum is the larger of the degrees of $p(x)$ and $q(x)$. The degree of the product is the sum of the degrees of $p(x)$ and $q(x)$. In part **c**, ask students for an example in which the number of zeros of the quotient is 5, 4, 3, 2, 1, and 0. (Begin with a polynomial $p(x)$ that has 5 distinct zeros. Every time a zero of $q(x)$ is a zero of $p(x)$, then there is one less zero in the quotient.)

Surprisingly, the answer to this question is *no*. The complex number system is large enough to provide n zeros for any polynomial of degree n, even those with nonreal complex coefficients.

Theorem

If $p(x)$ is any polynomial of degree n with real or complex coefficients, then $p(x)$ has exactly n real or complex zeros provided that each zero of multiplicity m is counted m times.

Proof If $p(x)$ has degree zero, then it is of the form $p(x) = k$ with $k \neq 0$, and so $p(x)$ has "zero zeros"; the theorem holds.

For polynomials of degree $n \ge 1$, the proof uses mathematical induction.
Let $S(n)$: *Every polynomial of degree n has exactly n complex zeros, provided that each zero of multiplicity m is counted m times.*
$S(1)$ is true because when $a \neq 0$, $p(x) = ax + b$ has one zero, $\frac{-b}{a}$.
Now we must show that for any positive integer k, $S(k)$ is true $\Rightarrow S(k + 1)$ is true. Suppose $S(k)$ is true. That is, suppose every polynomial of degree k has exactly k complex zeros allowing for multiplicities. Now consider any polynomial $p(x)$ of degree $k + 1$. By the Fundamental Theorem of Algebra, $p(x)$ has a zero, call it z_1. Then, by the Factor Theorem, there exists a polynomial $q(x)$ such that
$$p(x) = (x - z_1)\, q(x).$$
Since the degree of $(x - z_1)$ is 1, the degree of $q(x)$ must be k by the Degree of a Product Theorem.

Furthermore, the zeros of $q(x)$ are zeros of $p(x)$, and since $q(x)$ is of degree k, by the inductive assumption it has k zeros. Thus $p(x)$ has at least $k + 1$ zeros: z_1 and the k zeros of $q(x)$. Since it is known from Lesson 4-6 that $p(x)$ has no more than $k + 1$ zeros, $p(x)$ must have exactly $k + 1$ zeros. Thus $S(k) \Rightarrow S(k + 1)$ for all $k \ge 1$ and so $S(n)$ is true for all positive integers n.

By the Factor Theorem, if z_1 is a zero of $p(x)$, then $(x - z_1)$ is a factor of $p(x)$. Thus you can always find a polynomial with certain given zeros.

Example 2 Give a polynomial of degree 3 that has the zeros 3, i, and -4.

Solution Since a polynomial of degree 3 has exactly 3 zeros, there are no other zeros. Let $p(x)$ be a polynomial with these zeros. Then $x - 3$, $x - i$, and $x + 4$ are factors of $p(x)$. Thus one possibility is that
$$p(x) = (x - 3)(x - i)(x + 4).$$
In standard form,
$$p(x) = (x^2 - 3x - ix + 3i)(x + 4)$$
$$= x^3 + (1 - i)x^2 + (-12 - i)x + 12i.$$
All other possible polynomials are complex multiples of $p(x)$.

Questions

Covering the Reading

1. a. Describe the end behavior of $p(x) = 8x^5 - 20x^4 + 3x - 1$.
   b. Explain why this end behavior forces $p(x)$ to have at least one real zero.
   See margin.
2. How many complex zeros, counting multiplicities, does the polynomial $f(x) = 3x^{11} - \frac{1}{2}x^4 + ix^2 - 2$ have? 11

In 3–6, find all zeros and their multiplicities for the given polynomial.

3. $f(x) = x^6 - 4x^5 - 5x^4$
   zeros: 0 (with multiplicity 4), -1, 5
4. $g(x) = x - \frac{x^3}{6}$
   zeros: 0, $\pm\sqrt{6}$, all with multiplicity 1
5. $q(x) = 2x^2 - ix + 1$
   zeros: i, $-\frac{i}{2}$, both with multiplicity 1
6. $p(x) = (x^2 + 1)^2$
   zeros: $\pm i$, both with multiplicity 2
7. Give a polynomial of degree 3 that has the zeros 6, $4i$, and 0.
   sample: $p(x) = (x - 6)(x - 4i)x$

Applying the Mathematics

8. How many zeros (counting multiplicities) does the polynomial $p(x) = (x^2 + 2)(x^2 - 2)(x + 2)^2(2x - 1)$ have in the set of
   a. integers? 2
   b. rational numbers? 3
   c. real numbers? 5
   d. complex numbers? 7
9. A polynomial $p(x)$ has the following zeros: $1 + i$, $1 - i$, -1, and 3. The 3 has multiplicity 2, and the other zeros have multiplicity 1.
   a. What is the degree of $p(x)$? 5
   b. Write a possible formula for $p(x)$ in factored form.
   $p(x) = (x - 1 - i)(x - 1 + i)(x + 1)(x - 3)^2$
10. Given $h(x) = x^4 - 6x^3 + 18x^2 - 32x + 24$, use the fact that 2 is a zero of $h(x)$ with multiplicity 2 to find all the zeros and their multiplicities.
    zeros: 2 (with multiplicity 2), $1 \pm \sqrt{5}\,i$
11. A particular sixth degree polynomial has at least one simple zero and at least one zero of multiplicity 2. Give the number and possible multiplicities of the remaining zeros. See margin.
12. If, counting multiplicities, a polynomial $p(x)$ has 5 zeros and a polynomial $q(x)$ has 7 zeros, give the number of zeros of:
    a. $r(x)$, when $r(x) = p(x) + q(x)$ 7
    b. $s(x)$, when $s(x) = p(x) \cdot q(x)$ 12
    c. $t(x)$, when $t(x) = \frac{p(x)}{q(x)}$. ≤ 5

Review

13. Suppose $z = -\sqrt{2} + \sqrt{2}\,i$ is an 8th root of some number w.
    a. Find w. 256
    b. Find the other 8th roots of w (in polar form), and graph them on a polar grid. *(Lesson 8-7, 8-6)* See margin.
14. Let $z = [r, \theta]$, where $r \geq 0$, and let n be a nonnegative integer. Give an expression in polar form for:
    a. z^n $[r^n, n\theta]$
    b. $\bar{z}$ $[r, -\theta]$
    c. $(\bar{z})^n$ $[r^n, -n\theta]$
    d. $\overline{(z^n)}$. $[r^n, -n\theta]$
    e. Complete this sentence. Parts **c** and **d** show that, for any nonnegative integer n, the nth power of the conjugate of a complex number equals __?__. *(Lesson 8-6)*
    the conjugate of the nth power of the complex number

Making Connections for Question 14: The equality of parts **c** and **d** ($(\bar{z})^n = \overline{(z^n)}$) will be used in proving the Conjugate Zeros Theorem of Lesson 8-7.

ADDITIONAL ANSWERS

1.a. $\lim_{x \to -\infty} p(x) = -\infty$, and $\lim_{x \to \infty} p(x) = \infty$

b. Since $p(x) < 0$ for a small enough real number x, $p(x) > 0$ for a large enough real number x, and p is continuous, the Intermediate Value Theorem ensures there exists a real number c such that $p(c) = 0$.

11. It has three more zeros. There can be three more simple zeros, or one zero with multiplicity 3, or one simple zero and one zero with multiplicity 2.

13.b. See Additional Answers in the back of this book.

NAME ______

LESSON **MASTER 8–8**
QUESTIONS ON SPUR OBJECTIVES

■ **SKILLS** *Objective E (See pages 534–536 for objectives.)*
In 1–3, find all zeros and their multiplicities for the given polynomial.

1. $f(x) = 2x^5 + x^4 - 3x^3$
   zeros: 0, 1, $-\frac{3}{2}$
   mult.: 3, 1, 1
2. $g(x) = (x - 1)^4(x + 3)^2(x^2 + 1)^5$
   zeros: 1, 3, i, $-i$
   mult.: 4, 2, 5, 5
3. $p(r) = r^3 + 3r^2 - 9r + 5$ given that 1 is a zero of $p(r)$
   zeros: 1, -5; mult.: 1, 1
4. Give a polynomial of degree 3 that has the zeros 3, i, and $-i$.
   $p(x) = (x - 3)(x^2 + 1)$

■ **PROPERTIES** *Objective G*

5. A polynomial $g(x)$ has the following zeros: 2, 1, i, $-i$, and $\frac{1}{2}$. The 1 has multiplicity 3, and the other zeros have multiplicity 1.
   a. What is the degree of $g(x)$? 7
   b. Write a possible formula for $g(x)$ in factored form.
   sample: $g(x) = (x - 2)(x - 1)^3(x^2 + 1)(2x - 1)$
6. According to the Fundamental Theorem of Algebra, $p(x) = 6x^{12} - x^4 + 10x^3 + 5$ has exactly 12 complex zeros, counting multiplicities.
7. Suppose $p(x)$ and $g(x)$ are polynomials such that $\frac{p(x)}{q(x)}$ is a polynomial with 5 zeros (counting multiplicities), and $p(x) + q(x)$ has 7 zeros (counting multiplicities).
   a. What is the degree of $p(x)$? 7
   b. What is the degree of $q(x)$? 2
   c. How many zeros does $p(x)q(x)$ have (counting multiplicities)? 9

NOTES ON QUESTIONS
Making Connections for Question 16b: The general pattern is described easily with modular arithmetic:
$i^x = i^y \Leftrightarrow x \equiv y \pmod 4$. Thus, $i^{4k} = i^0$, $i^{4k+1} = i$, $i^{4k+2} = i^2$, and $i^{4k+3} = i^3$.

Question 21: This example is difficult to carry out without making errors. You might ask a student to program the formula into a computer.

FOLLOW-UP

MORE PRACTICE
For more questions on SPUR Objectives, use *Lesson Master 8-8,* shown on page 513.

ADDITIONAL ANSWERS
15. $(2 + i)^2 - 4(2 + i) + 5 = 4 + 4i - 1 - 8 - 4i + 5 = 0$; $(2 - i)^2 - 4(2 + i) + 5 = 4 - 4i - 1 - 8 - 4i + 5 = 0$

15. Show that $2 + i$ and $2 - i$ are both solutions to the equation $x^2 - 4x + 5 = 0$. *(Lesson 8-1)*
See margin.

16. **a.** Evaluate i^n for $n = 0, 1, 2, \ldots, 8$. $1, i, -1, -i, 1, i, -1, -i, 1$

b. Generalize the results of part **a** to write formulas for
$$i^{4k},\ i^{4k+1},\ i^{4k+2},\ i^{4k+3}$$
for any nonnegative integer k.

c. Use the result of part **b** to evaluate
i. i^{59} $-i$ **ii.** i^{18} -1 **iii.** i^{60}. 1

(Lesson 8-1)
b) $i^{4k} = 1$, $i^{4k+1} = i$, $i^{4k+2} = -1$, and $i^{4k+3} = -i$

17. Let $p(x)$ be the polynomial $p(x) = 3x^4 - 2x^3 + x$.

a. Find a polynomial of the form $f(x) = ax^n$ whose behavior is similar to that of $p(x)$ for large x. $f(x) = 3x^4$

b. Rewrite $p(x)$ to justify your claim in part **a**. $f(x) = 3x^4(1 - \frac{2}{3x} + \frac{1}{3x^3})$

c. Describe the end behavior of $p(x)$ using limit notation. *(Lesson 5-5)*
$\lim_{x \to -\infty} p(x) = +\infty$, $\lim_{x \to +\infty} p(x) = +\infty$

18. The commands MOD and DIV in the computer language Pascal can be used to determine the results in the Quotient-Remainder Theorem.
n DIV d computes the integer quotient for $n \div d$.
n MOD d computes the integer remainder for $n \div d$.

a. Evaluate the following.
i. 157 DIV 11 14 **ii.** 218 MOD 7 1

b. Write an expression in Pascal that gives the number of complete weeks in n days. n DIV 7

c. Carolyn's father is baking muffins and packaging them a dozen at a time. Assume that he bakes m muffins. If Carolyn gets all the leftovers, write an expression in Pascal that gives the number of muffins she gets. m MOD 12

(Lesson 4-2)

19. Use the Intermediate Value Theorem to find an interval between two consecutive integers where the function f, defined by $f(t) = 12t^3 + 28t^2 - 173t - 252$, has a zero. *(Lesson 3-5)*
any one of: $-5 < x < -4$; $-2 < x < -1$; or $3 < x < 4$

20. Suppose $p(x)$ is a polynomial of odd degree and $\lim_{x \to +\infty} p(x) = -\infty$. What is $\lim_{x \to -\infty} p(x)$? *(Lesson 2-5)* ∞

Exploration

21. Given the cubic polynomial $f(x) = x^3 + px^2 + qx + r$, let
$$A = \sqrt[3]{-\frac{(2p^3 - 9pq + 27r)}{54} + \sqrt{\frac{(2p^3 - 9pq + 27r)^2}{2916} + \frac{(3q - p^2)^3}{729}}}$$
and
$$B = \sqrt[3]{-\frac{(2p^3 - 9pq + 27r)}{54} - \sqrt{\frac{(2p^3 - 9pq + 27r)^2}{2916} + \frac{(3q - p^2)^3}{729}}}.$$
Cardano showed that the zeros of $f(x)$ are
$$A + B - \frac{p}{3},\ -\frac{A + B}{2} - \frac{p}{3} + \frac{A - B}{2}\sqrt{-3},\ \text{and}\ -\frac{A + B}{2} - \frac{p}{3} - \frac{A - B}{2}\sqrt{-3}.$$
Use these formulas to find the zeros of the polynomial $f(x) = x^3 + 2x^2 + 3x + 1$.
$\approx -0.4301, -0.7849 \pm 1.307i$

LESSON

8-9

Nonreal Zeros of Polynomials with Real Coefficients

Graphing polynomials (similar to those in this chapter) contributes to the production of this Landsat computer mapping. (Columbia Plateau, Washington State)

The last lesson showed that a polynomial of degree n has n zeros. This lesson starts with a related question: If all the coefficients of the polynomial are real numbers, how many of these n zeros are real and how many are nonreal?

For example, the polynomial $p(x)$ of degree 2, defined by

$$p(x) = x^2 + 1,$$

has no real zeros. It does, however, have two nonreal complex zeros, i and $-i$. In general, all polynomials *with real coefficients* satisfy the following important property.

Conjugate Zeros Theorem

Let $p(x)$ be a polynomial with real coefficients. If $z = a + bi$ is a zero of $p(x)$, then its complex conjugate $\bar{z} = a - bi$ is also a zero of $p(x)$.

This theorem is often paraphrased as follows: *Nonreal zeros of a polynomial with real coefficients always occur in complex conjugate pairs.* Before proving the theorem, the following example illustrates its use.

Example 1 Consider the polynomial $p(x)$ defined by $p(x) = x^4 + 2x^3 + 3x^2 + 2x + 2$.

a. Verify that i is a zero of $p(x)$.

b. Find the remaining zeros of $p(x)$ and their multiplicities.

Solution **a.** Recall that $i^2 = -1$, $i^3 = -i$, and $i^4 = 1$. Consequently,

$$\begin{aligned} p(i) &= i^4 + 2i^3 + 3i^2 + 2i + 2 \\ &= 1 - 2i - 3 + 2i + 2 \\ &= 0. \end{aligned}$$

LESSON 8-9

RESOURCES

- Lesson Master 8-9
- Teaching Aid 57 can be used with **Example 3**.
- Teaching Aid 58 displays the graphs and zeros of $P_c(x) = 4x^4 + 8x^3 - 3x^2 - 9x + c$.

OBJECTIVES

E Find all zeros, and their multiplicities, of a given polynomial.

G Use the properties of polynomials to find or describe their zeros.

TEACHING NOTES

Before the days of automatic graphers, it was not easy to see the real zeros of polynomials with real coefficients, though there are algebraic methods for determining bounds on them and there are algebraic ways to approximate them. Nowadays, it is relatively easy to determine the real zeros. This leads naturally to the question of seeing the complex zeros, the subject of this lesson.

The usefulness of the Conjugate Zeros Theorem is demonstrated by **Example 1** before the theorem is proved.

Review the proof of the Conjugate Zeros Theorem. If complex numbers z and w are conjugates, so are $p(z)$ and $p(w)$. Since any real number equals its conjugate, if $p(z)$ is a real number, then $p(w)$ is the same real number. In this case, the real number is 0.

The proof puts together simple properties to produce a powerful result.

Alternate Approach
In **Example 1**, students also may use synthetic division twice to derive the polynomial $x^2 + 2x + 2$. The first synthetic division $x - i$ yields a polynomial of degree 3 with complex coefficients and a remainder of 0. Dividing this reduced polynomial synthetically by $x + i$ yields a quotient of $x^2 + 2x + 2$ and a remainder of 0. The quadratic formula can be used to find the remaining zeros, as in the example.

The solution to **Example 3c** states that the graph of p crosses the x-axis at simple real zeros but just touches it at the real zero of multiplicity 2. A proof requires calculus, but here is the idea. Suppose that c is a simple real zero of a polynomial function p, so that $p(x) = q(x)(x - c)$, where $x - c$ is not a factor of the polynomial $q(x)$. Then, for values of x close enough to c, $q(x)$ remains close to some constant a so $p(x) \approx a(x - c)$. The graph of $y = a(x - c)$ is a straight line crossing the x-axis at $(c, 0)$, and the graph of p behaves roughly in the same way.

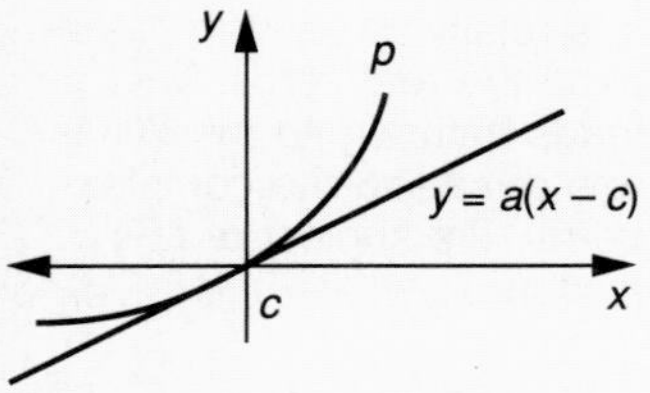

Suppose that d is a real zero of p with multiplicity 2, so that $p(x) = r(x)(x - d)^2$, where $x - d$ is not a factor of $r(x)$. Then, for values of x close enough to d, $r(x)$ remains close to some constant b so $p(x) \approx b(x - d)^2$. The graph of $y = b(x - d)^2$ is a parabola just touching the x-axis at its vertex, $(d, 0)$,

b. Because $z = i$ is a zero of $p(x)$, it follows that $\overline{z} = -i$ is also a zero of $p(x)$ by the Conjugate Zeros Theorem. The Factor Theorem implies that $(x - i)$ and $(x + i)$ are factors of $p(x)$. Thus, $(x - i)(x + i) = x^2 + 1$ is a factor of $p(x)$. Divide $p(x)$ by $x^2 + 1$ to obtain the other factor.

$$\begin{array}{r}
x^2 + 2x + 2 \\
x^2 + 1 \overline{)\,x^4 + 2x^3 + 3x^2 + 2x + 2} \\
\underline{x^4 + x^2 } \\
2x^3 + 2x^2 + 2x + 2 \\
\underline{2x^3 + 2x } \\
2x^2 + 2 \\
\underline{2x^2 + 2} \\
0
\end{array}$$

The remaining zeros of $p(x)$ are the zeros of the quotient polynomial

$$q(x) = x^2 + 2x + 2.$$

These can be found using the quadratic formula:

$$x = \frac{-2 \pm \sqrt{2^2 - 4 \cdot 2}}{2} = -1 \pm i.$$

Therefore, $p(x)$ has four simple zeros:

$$i, -i, -1 + i, \text{ and } -1 - i.$$

The proof of the Conjugate Zeros Theorem relies on the following algebraic properties of complex conjugates (see Question 20 of Lesson 8-1, Question 18 of Lesson 8-2, and Question 18 of Lesson 8-7): For all complex numbers z and w, and positive integers m,

(1) $\overline{z + w} = \overline{z} + \overline{w}$
(2) $\overline{z \cdot w} = \overline{z} \cdot \overline{w}$
(3) $\overline{(z^m)} = (\overline{z})^m$
(4) $z = \overline{z}$ if and only if z is a real number.

To prove the Conjugate Zeros Theorem, let

$$p(x) = a_n x^n + a_{n-1} x^{n-1} + \ldots + a_1 x + a_0,$$

where the a_i are all real numbers. First, we show that

$$\overline{p(z)} = p(\overline{z})$$

for any complex number z. The argument is as follows:

$$\overline{p(z)} = \overline{a_n z^n + a_{n-1} z^{n-1} + \ldots + a_1 z + a_0}$$

$$= \overline{a_n z^n} + \overline{a_{n-1} z^{n-1}} + \ldots + \overline{a_1 z} + \overline{a_0}$$ Property (1) applied repeatedly

$$= \overline{a_n}(\overline{z})^n + \overline{a_{n-1}}(\overline{z})^{n-1} + \ldots + \overline{a_1}(\overline{z}) + \overline{a_0}$$ Properties (2) and (3)

$$= a_n(\overline{z})^n + a_{n-1}(\overline{z})^{n-1} + \ldots + a_1(\overline{z}) + a_0$$ Property (4)

$$= p(\overline{z}).$$

516

Now it is easy to prove that complex zeros of polynomials with real coefficients occur in conjugate pairs. If z is a complex zero of p, then

$$0 = p(z).$$

Thus
$$\overline{0} = \overline{p(z)}$$

But $\overline{0} = 0$, so
$$0 = \overline{p(z)} = p(\overline{z}).$$

Therefore, if z is a complex zero of p, $\overline{z}$ is also a zero of p. This completes the proof of the Conjugate Zeros Theorem.

Example 2 Find a polynomial of smallest degree with real coefficients that has zeros 1, -2, and -1 + i.

Solution Let $p(x)$ be such a polynomial. Since it has real coefficients and -1 + i is a zero, -1 − i must be a zero of $p(x)$. Therefore $p(x)$ must have 4 zeros, and must therefore have degree 4. By the Factor Theorem, a possible formula for $p(x)$ is

$$\begin{aligned} p(x) &= (x - 1)(x + 2)[x - (-1 + i)][x - (-1 - i)] \\ &= (x - 1)(x + 2)(x + 1 - i)(x + 1 + i) \\ &= (x^2 + x - 2)(x^2 + 2x + 1 - i^2) \\ &= (x^2 + x - 2)(x^2 + 2x + 2) \\ &= x^4 + 3x^3 + 2x^2 - 2x - 4. \end{aligned}$$

Real zeros of a polynomial $p(x)$ with real coefficients occur at points along the x-axis where the graph of

$$y = p(x)$$

crosses or touches the x-axis. The nonreal zeros of $p(x)$ do not appear on the x-axis because only real values of the variable are on the x-axis. However, using the Conjugate Zeros Theorem, information about the nonreal zeros can be inferred from knowledge of the real zeros.

Example 3 The curve pictured below is the graph of $y = p(x)$ for a polynomial $p(x)$ of degree 4 with real coefficients. The x-axis has been removed.

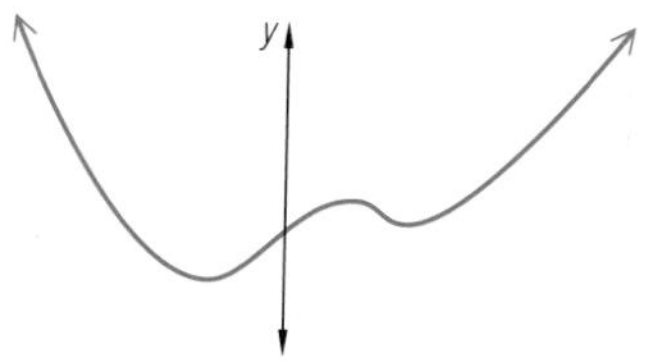

Reinsert the horizontal axis at a position that is consistent with the additional information provided about $p(x)$. Also, describe the nonreal zeros.

a. $p(x)$ has no real zeros.
b. $p(x)$ has one real zero.
c. $p(x)$ has exactly two real zeros and both are simple zeros.
d. $p(x)$ has two simple real zeros and one real zero of multiplicity 2.

and the graph of p behaves roughly in the same way.

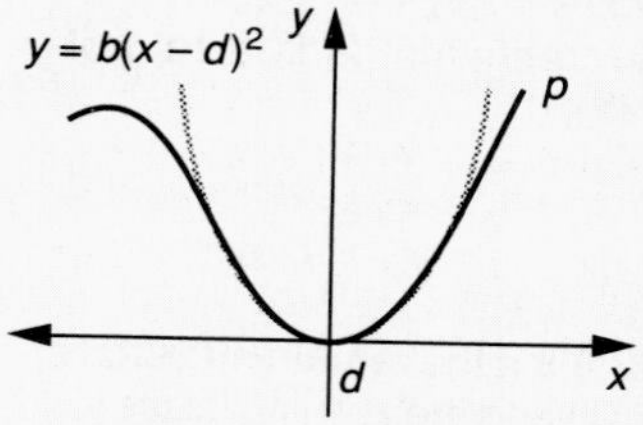

On this page and page 518, two important properties of polynomials are employed to provide a rough way of graphing nonreal zeros of real polynomial functions. The first is the fact that the zeros of a polynomial vary continuously as a function of the coefficients, a result which is not trivial. The second property is the Conjugate Zeros Theorem. The first property implies that when the constant coefficient in the polynomial graphed on this page is increased by a tiny amount, the real zero of multiplicity 2 must change only slightly. It can no longer be real since the graph of the function no longer intersects the x-axis, as shown on page 518. Then the Conjugate Zeros Theorem implies that its conjugate must also be a zero. Thus, the two zeros are symmetric about the real axis and are close to the old real zero.

Technology The family of polynomials p_c shown on these pages can be illustrated using *Mathematica* software. A *Mathematica* program can be written that will show the polynomial function's graph for different values of c along with a graph in the complex plane of corresponding zeros of $p_c(x)$. Seeing the pattern of zeros as c changes is quite remarkable and instructive.

ADDITIONAL EXAMPLES

1. Let $p(x) = x^6 - 5x^5 + 10x^4 - 20x^3 + 24x^2$.

a. Verify that $2i$ is a zero of $p(x)$.

$p(2i) = (2i)^6 - 5(2i)^5 + 10(2i)^4 - 20(2i)^3 + 24(2i)^2 = -64 - 5(32i) + 160 + 160i - 96 = 0$

b. Find the remaining zeros of $p(x)$ and their multiplicities.

0, multiplicity 2;
$-2i$, multiplicity 1;
2, multiplicity 1;
3, multiplicity 1

2. Find a polynomial with real coefficients that has zeros -1, 3, and $2 + 3i$.

$p(x) = (x + 1)(x - 3) \cdot (x - 2 - 3i)(x - 2 + 3i) = x^4 - 6x^3 + 18x^2 - 14x - 39$

3. The curve below is the graph of $y = p(x)$ for a polynomial $p(x)$ of degree 5 with real coefficients. Insert the horizontal axis at a position consistent with the additional information provided about $p(x)$. Also, describe the nonreal zeros.

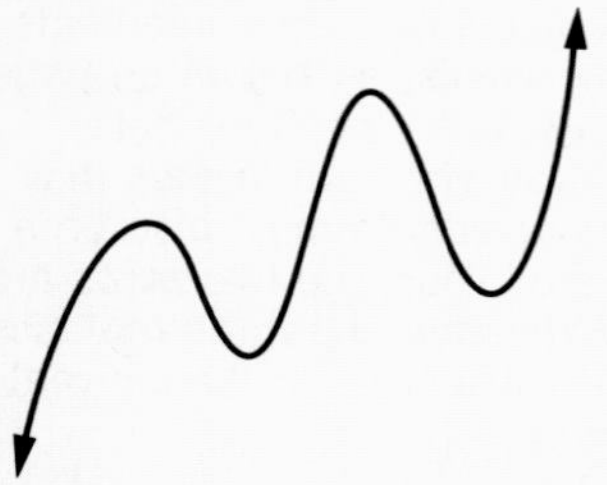

a. $p(x)$ has no real zeros.
impossible

b. $p(x)$ has exactly 1 real zero.

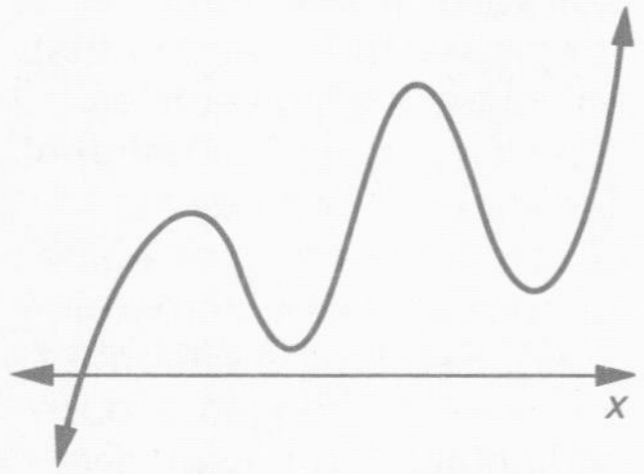

The other 4 zeros are complex in conjugate pairs.

Solution

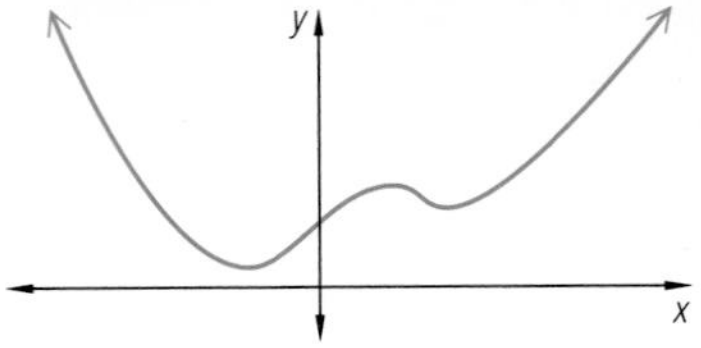

a. If $p(x)$ has no real zeros, the graph of $y = p(x)$ cannot intersect the x-axis. So the x-axis must lie below the graph. Since $p(x)$ has degree 4, it must have 4 zeros. Because they are all nonreal and the coefficients of $p(x)$ are real, they must be in complex conjugate pairs. Therefore, they are either 2 pairs of conjugate zeros, each of multiplicity 1, or a single pair of conjugate zeros, each of multiplicity 2.

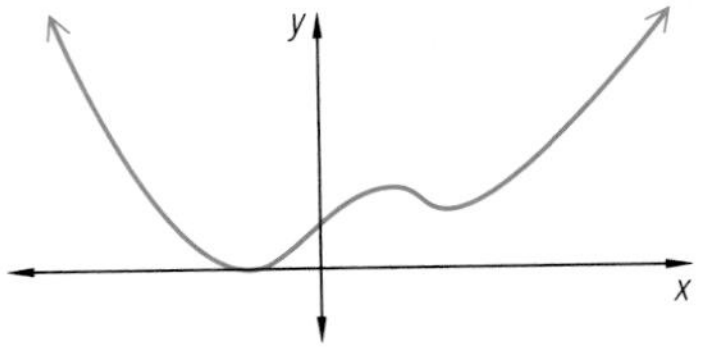

b. If $p(x)$ has one real zero, it intersects the x-axis at only one point. It must have 2 conjugate complex zeros, so its real zero must have multiplicity 2.

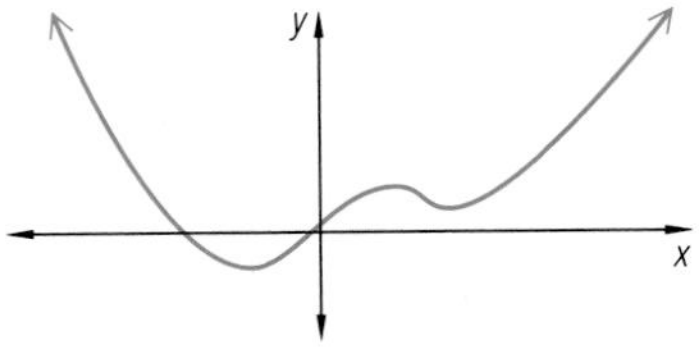

c. If $p(x)$ has only two real, simple zeros, the graph of $y = p(x)$ will cross the x-axis exactly twice. The other 2 zeros of $p(x)$ must be a complex conjugate pair.

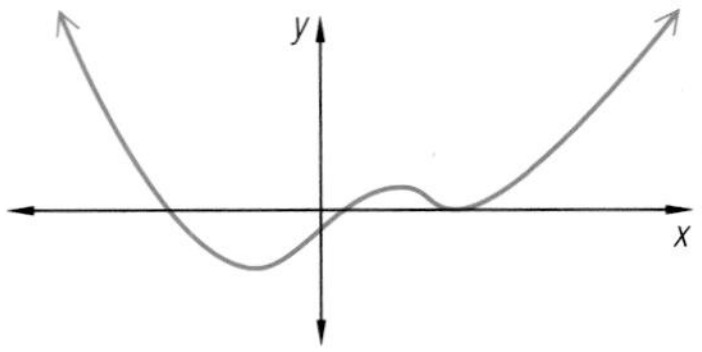

d. If $p(x)$ has two simple real zeros and one real zero of multiplicity 2, the graph of $y = p(x)$ will cross the x-axis at two of the zeros, and just touch the x-axis at the third. Since $p(x)$ can have no more than 4 zeros counting multiplicities, it can have no nonreal zeros.

The use of graphs can show the effect of changes in a polynomial's constant coefficient on its complex zeros. It is a consequence of continuity that if the coefficients of $p(x)$ are changed only slightly, then the locations of the zeros of $p(x)$ change only slightly also. For example, if you vary the polynomial $p(x)$ by increasing its constant coefficient, the graph of p will slide up the y-axis without changing its shape. As this happens, its zeros change. For example, shown below are graphs, on the interval $-2 \le x \le 2$, of the six polynomial functions defined by

$$p_c(x) = 4x^4 + 8x^3 - 3x^2 - 9x + c$$

for c = -4, -2, 0, 2, 4, and 6.

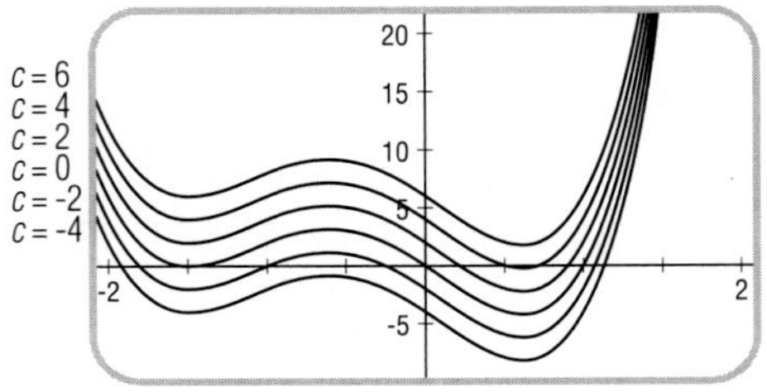

$p(x) = 4x^4 + 8x^3 - 3x^2 - 9x + c$

518

The lowest graph corresponds to $c = -4$ and the highest corresponds to $c = 6$, with the intermediate graphs increasing in vertical position with the value of c. Since each $p_c(x)$ has degree 4, each function has a total of 4 zeros. From their graphs, it appears that

p_{-4} has two simple real zeros (and thus two complex conjugate zeros),
p_{-2} has four simple real zeros,
p_0 has two simple real zeros and one real zero of multiplicity two (near -1.5),
p_2 has two simple real zeros (and thus two complex conjugate zeros),
p_4 has two simple real zeros that are very close to each other (and two complex conjugate zeros),
p_6 has no real zeros, so it must have two pairs of complex conjugate zeros.

The complex zeros of the $p_c(x)$ are displayed as dots in the complex plane below, with arrows to show how they move as the value of c increases.

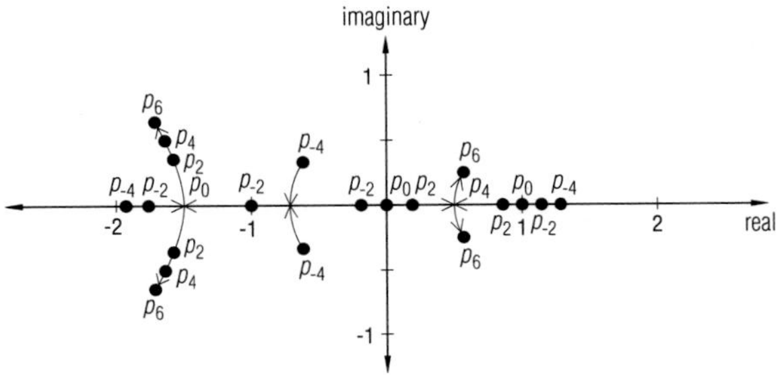

Read this graph by first finding the four zeros of p_{-4}. Notice that as c increases, two distinct real zeros merge into a real zero of multiplicity 2, and then split into complex conjugate zeros as c continues to increase. Automatic graphers have made it possible to track the movements of complex zeros more easily than ever before.

Questions

Covering the Reading

1. Suppose $p(x)$ is a polynomial with real coefficients such that $p(3 + 2i) = 4 - 7i$ and $p(2 - 3i) = 0$. Evaluate
a. $p(3 - 2i)$; $4 + 7i$ **b.** $p(2 + 3i)$. 0

2. Let $p(x) = x^2 + 3x - 2$.
a. Find $p(4 + 7i)$. $-23 + 77i$ **b.** Find $p(4 - 7i)$. $-23 - 77i$

3. Consider the polynomial $p(x) = x^2 + 4ix + 5$.
a. Show that the zeros of $p(x)$ are i and $-5i$. See margin.
b. The zeros are not complex conjugates of each other. Does this contradict the Conjugate Zeros Theorem? Why or why not?
No, the coefficients of p are not all real numbers.

c. $p(x)$ has exactly 3 real zeros; one is simple and the other has multiplicity 2.

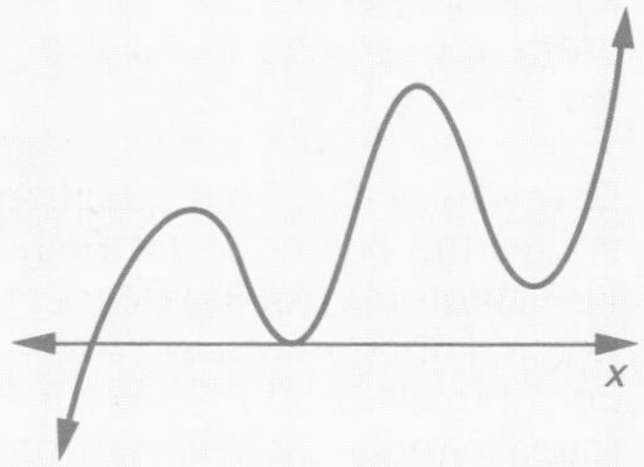

The other 2 zeros are a complex conjugate pair.

d. $p(x)$ has exactly 3 real zeros; all are simple zeros.

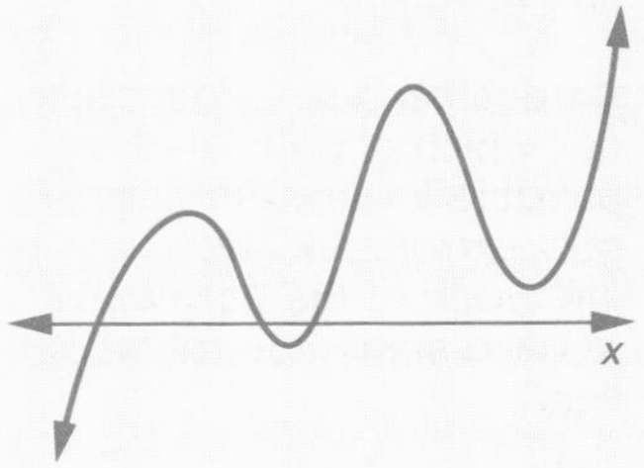

The other 2 zeros are a complex conjugate pair.

e. $p(x)$ has exactly 4 real, simple zeros.
impossible

f. $p(x)$ has 5 real, simple zeros.

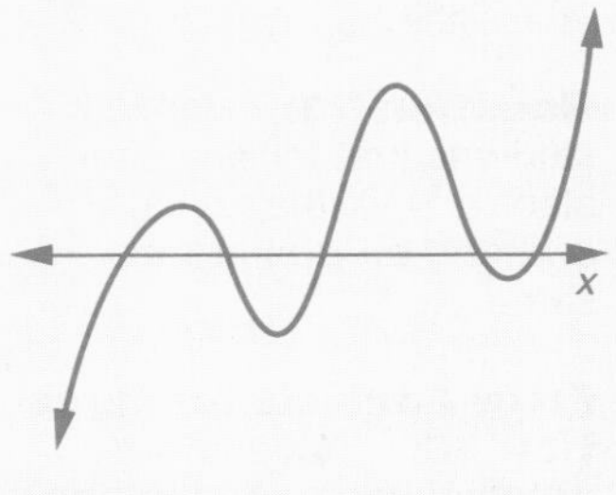

There are no other zeros.

NOTES ON QUESTIONS
Questions 1 and 2:
These questions use a generalization of the Conjugate Zeros Theorem found in the lesson but not stated explicitly as a theorem: Given complex nonreal numbers w and z, if $p(w) = z$, then $p(\overline{w}) = \overline{z}$.

ADDITIONAL ANSWERS
3.a. $p(i) = i^2 + 4i^2 + 5 = -1 - 4 + 5 = 0$; $p(-5i) = 25i^2 - 20i^2 + 5 = -25 + 20 + 5 = 0$

NOTES ON QUESTIONS

Question 5: Students should be able to see that the Conjugate Zeros Theorem applies and that there are exactly four zeros.

Question 6: This question shows the power of the Factor Theorem, Conjugate Zeros Theorem, division of polynomials, and the Quadratic Formula when they work together.

Question 7: All the polynomials that work are nonzero multiples of the sample answer given.

Question 10: You might ask which of parts **a–d** are possible for any 5th degree polynomial and which require the graph of the polynomial to have some special properties.

Making Connections for Question 11: These zeros are the 4th roots of 1. It is easy to verify that their graphs are vertices of a square centered at the origin, confirming the Geometric *n*th Roots Theorem of the previous lesson.

Question 13: Remind students that for any constant $c \neq 0$, the zeros of $p(x)$ and $c \cdot p(x)$ are the same.

Error Analysis for Question 16: Students who assume incorrectly that the coefficients of $q(x)$ are reals may get results contradicting the given information; their 5th degree polynomial $p(x)$ will have more than 5 zeros.

Technology for Question 22: This requires the use of a spreadsheet or computer program.

ADDITIONAL ANSWERS
8., 9., 10., 12.b., 17., 18.a., b., 21., 22. See Additional Answers in the back of this book.

4. *True* or *false*? *There is a polynomial with real coefficients that has exactly one nonreal zero.* False

5. Two zeros of the polynomial $p(x) = x^4 - 9x^3 + 50x^2 - 49x + 41$ are $\frac{1 + i\sqrt{3}}{2}$ and $4 + 5i$. Find the remaining zeros of $p(x)$. $\frac{1 - i\sqrt{3}}{2}$, $4 - 5i$

6. One of the zeros of $p(x) = x^4 - 7x^2 + 4x + 20$ is $2 + i$. Find the remaining zeros. $2 - i$, -2 (with multiplicity 2)

7. Find a polynomial of smallest degree with real coefficients that has zeros -3 and $2 - i$. sample: $p(x) = x^3 - x^2 - 7x + 15$

In 8 and 9, trace the curve in Example 3.

8. Insert an x-axis so that $p(x)$ has four real simple zeros. See margin.

9. Sketch a different solution to Example 3d. See margin.

10. The curve at the right is the graph of a polynomial $p(x)$ of degree 5 with real coefficients, with the x-axis removed. For each of the following, sketch the graph and reinsert the horizontal axis so that the given condition is satisfied. Also, describe the nonreal zeros of $p(x)$.

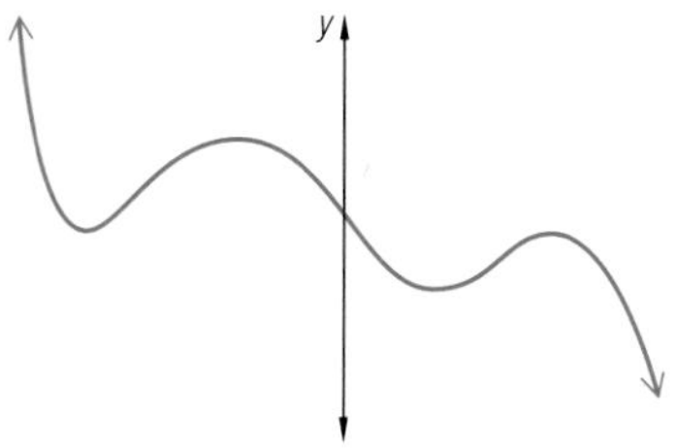

a. $p(x)$ has exactly one real simple zero.
b. $p(x)$ has exactly three real simple zeros.
c. $p(x)$ has exactly three real zeros, two of which have multiplicity two and one of which is a simple zero.
d. $p(x)$ has exactly two real zeros, one of multiplicity two and the other a simple zero. a-d) See margin.

Applying the Mathematics

11. Find all zeros of $p(x) = x^4 - 1$. 1, -1, i, $-i$

12. a. Find a polynomial $p(x)$, with real coefficients and of the lowest degree possible, that has the two zeros 2 and $1 - 3i$.
b. How does the answer to part **a** change if the requirement that $p(x)$ have real coefficients is dropped?
a) sample: $p(x) = x^3 - 4x^2 + 14x - 20$; b) See margin.

13. Find a polynomial of degree 3 with integer coefficients whose zeros include $\frac{1}{2}i$ and $\frac{2}{3}$. sample: $p(x) = 12x^3 - 8x^2 + 3x - 2$

14. Consider the polynomials of the form $p(x) = x^2 + c$, where c is a real number.
a. For each value of c given below, indicate the zeros of $p(x)$ in the complex plane.
i. $c = -1$ 1, -1 **ii.** $c = 0$ 0 **iii.** $c = 1$ i, $-i$
b. Imagine the graph of $p(x)$ sliding upwards as c varies from -1 to 1. Describe what happens to the location of the zeros of $p(x)$ as this occurs. They get closer together until they coincide at c = 0, then they split apart in opposite directions along the imaginary axis.

Review

15. Given that 1 is a zero of $p(x) = x^5 - x^4 + 6x^3 - 6x^2 + 9x - 9$, find all zeros of $p(x)$ and their multiplicities. *(Lesson 8-8)*

zeros: $\pm\sqrt{3}i$ (each with multiplicity 2), 1

16. Suppose $p(x)$ and $q(x)$ are polynomials, $p(x)$ has degree 5, and $p(x) = (x + 2)^2\, q(x)$.

a. What is the degree of $q(x)$? 3

b. If $3i$ is a zero of $q(x)$ with multiplicity 2, and -2 is a simple zero of $q(x)$, give all zeros of $p(x)$ and their multiplicities. *(Lesson 8-8)*

-2 (with multiplicity 3), $3i$ (with multiplicity 2)

17. Plot the solutions to $x^4 = 6 + 2i$ in the complex plane. *(Lesson 8-7)*

See margin.

18. Let $z = 4 + i$ and $w = 2 + 3i$.

a. Sketch a graph which verifies the Geometric Addition Theorem for $z + w$. See margin.

b. On the same set of axes, also graph $\bar{z}$, $\bar{w}$, and $\bar{z} + \bar{w}$. See margin.

c. Explain how your diagram illustrates the property that $\bar{z} + \bar{w} = \overline{z + w}$.

(Lessons 8-3, 8-1)

$\bar{z} + \bar{w}$ is the reflection image of $z + w$ over the real axis, so $\overline{z + w} = \bar{z} + \bar{w}$.

19. Let $p(x) = 2x^3 - 3x^2 + 4$.

a. Use the Remainder Theorem to find $p\left(-\frac{1}{2}\right)$. 3

b. Find $q(x)$ so that $p(x) = q(x)\left(x + \frac{1}{2}\right) + p\left(-\frac{1}{2}\right)$. *(Lessons 4-6, 4-5)*

$2x^2 - 4x + 2$

20. Write the logical expression associated with the following network. *(Lesson 1-4)*

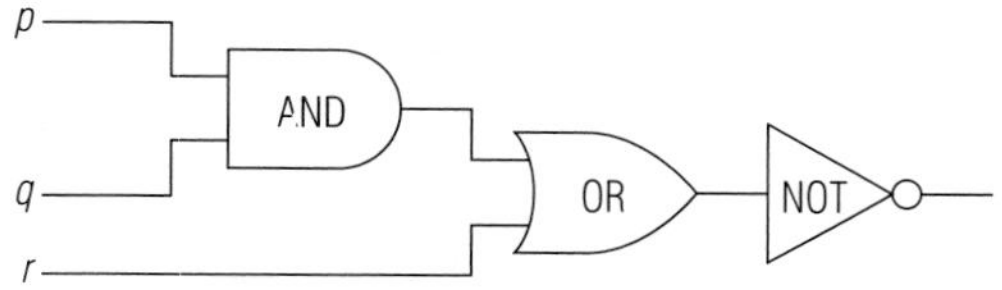

not ((*p* and q) or *r*)

Exploration

21. a. Use the technique of Question 14**b** with an automatic grapher to study the zeros of $p(x) = x^4 - 4x^2 + c$ for $c = -5$, $c = 0$, and $c = 3$.

b. Continue the analysis of the polynomial for $c > 3$. Good values to try are $c = 4$ and $c = \frac{25}{4}$. For $c = \frac{25}{4}$, one of the zeros is $\frac{3 + i}{2}$.

a,b) See margin.

22. The polynomial

$$p(x) = x^4 - 5x^2 + x + c$$

has a graph similar to the one displayed in Example 3. Use a computer to find the zeros of $p(x)$ for various values of c between -5 and 5. Then use the idea discussed at the end of the lesson to graph the zeros of $p(x)$ for these values of c. See margin.

FOLLOW-UP

MORE PRACTICE

For more questions on SPUR Objectives, use *Lesson Master 8-9,* shown below.

EVALUATION

Alternate Assessment

(1) Find all roots of each polynomial below, with their multiplicities.

(a) $p(x) = x^4 - 6x^3 + 14x^2 - 30x + 45$ (3 multiplicity 2; $\pm\sqrt{5}i$, both multiplicity 1)

(b) $q(x) = x^4 - 10x^2 + 9$ (± 1, ± 3, all multiplicity 1)

(c) $v(x) = x^4 + 16$ ($\sqrt{2} \pm i\sqrt{2}$; $-\sqrt{2} \pm i\sqrt{2}$, all multiplicity 1)

(d) $t(x) = x^3 + 5x^2 - 14x$ (0, 2, -7, each multiplicity 1)

(2) Find the sum of each set of roots in Question 1 above. ((a) 6; (b) 0; (c) 0; (d) -5)

(3) Conjecture how the sum of the roots of a polynomial of degree n is related to $\frac{a_{n-1}}{a_n}$. $\left(\text{sum} = -\frac{a_{n-1}}{a_n}\right)$

NAME ______

LESSON **MASTER 8–9**
QUESTIONS ON **SPUR** OBJECTIVES

■ **SKILLS** *Objective E (See pages 534–536 for objectives.)*

1. a. Find the zeros and corresponding multiplicities of $p(x) = (2x^2 + ix - 5)^2$.

$\frac{-i \pm \sqrt{39}}{4}$, each with multiplicity 2

b. Does your answer to part **a** contradict the Conjugate Zeros Theorem? Explain.

No; Although the zeros are not in conjugate pairs, the coefficients of $p(x)$ are not all real.

2. If $-i$ is a zero of $p(x) = x^4 - 5x^3 + 5x^2 - 5x + 4$, find the remaining zeros. i, 4, 1

3. Two of the zeros of the polynomial $f(t) = t^4 - 4t^3 + 9t^2 - 16t + 20$ are $2 - i$ and $2i$. Find the remaining zeros of $f(t)$. $2 + i$, $-2i$

■ **PROPERTIES** *Objective G*

4. Does there exist a polynomial $p(x)$ with real coefficients which has exactly five zeros: $2i$, $-2i$, $3 + i$, $3 - i$, and $11i$? Justify your answer.

No, by the Conjugate Zeros Theorem, $-11i$ must also be a zero.

5. What is the smallest possible degree of a polynomial $p(x)$ with real coefficients if $1 + 2i$, $7 - 5i$, and 2 are zeros of $p(x)$? 5

6. Use the Conjugate Zeros Theorem to prove: Every polynomial of odd degree with real coefficients has at least one real zero.

By the Conjugate Zeros Theorem, nonreal zeros must always come in conjugate pairs, so there must be an even number of nonreal zeros. Thus, if a polynomial has odd degree, it must have an odd number of zeros, so at least one must be real.

7. Find a polynomial of smallest degree with real coefficients that has zeros $2 - 3i$ and 0.

$p(x) = (x - 2 + 3i)(x - 2 - 3i)x = x^3 - 4x^2 + 13x$

8. Suppose $p(x)$ is a polynomial with real coefficients such that $p(x) = (x + 2 + i)q(x)$ where $q(x)$ is a polynomial. Give a factor of $q(x)$. $(x + 2 - i)$

 101

LESSON 8-10

TEACHING NOTES

Technology In first entering 2 and then repeatedly pressing the COS key, students should realize that they are using their calculators and recursion to solve $\cos(x) = x$ for its one real zero. They may wish to use an automatic grapher to check their solutions.

It is critical that students understand the definiton of *orbit.* Because $O(d)$ uses function notation, they may think that it represents a single number. Stress that $O(d)$ is a *sequence,* a sequence of values that begins with d, $f(d)$, $f(f(d))$, $f(f(f(d)))$, … . Indeed, f is a function mapping D into D; O is a function mapping D into the set of sequences over D.

Like other sequences, orbits may have a single limiting value as in the square root example that begins the lesson. They may be periodic, as is the case with the system whose function is $f(x) = \frac{1}{1-x}$ immediately following the definition and also with the example on page 524. The latter example uses complex numbers in polar form. The initial point on the unit circle (for the dynamical system $f: D \to D$ where $f(z) = z^2$ and D is the unit circle) is chosen so that the argument is a rational multiple of 360°. The orbit will be periodic for any initial point with this characteristic.

LESSON

8-10

Discrete Dynamical Systems

Some interesting things can happen when you enter a number in your calculator and then press one of the function keys repeatedly. For example, enter the number 2 and then press the square root key, $\boxed{\sqrt{x}}$, 9 times. (On some calculators, this will produce an error message. You may need to use $\boxed{\sqrt{x}}$ $\boxed{=}$ to take the square root of the previous result.) The following series of numbers will be displayed (rounded to three decimal places): 2.000, 1.414, 1.189, 1.091, 1.044, 1.022, 1.011, 1.005, 1.003, 1.001. These numbers are approximations to the first ten terms of the sequence defined recursively by

$$\begin{cases} x_0 = 2 \\ x_{k+1} = \sqrt{x_k} \text{ for integers } k \geq 0. \end{cases}$$

As k gets larger and larger, the corresponding x_k values appear to get closer and closer to 1; that is, apparently

$$\lim_{k \to \infty} x_k = 1.$$

Begin again with the number 2 but this time use the cosine function key, $\boxed{\cos}$, with your calculator set for radian measure. The following numbers will be displayed, again rounded to three decimal places:

2.000, -.416, .915, .610, .820, .683, .776, .714, .756, .728.

These are approximations to the first ten terms of the sequence defined recursively by

$$\begin{cases} y_0 = 2 \\ y_{k+1} = \cos y_k \text{ for integers } k \geq 0. \end{cases}$$

The numbers y_k seem to be "settling down" on some number between .7 and .8 as k gets larger and larger, but the exact value of this limit is much less obvious.

Finally, if you again begin with the number 2 but use the squaring function key, $\boxed{x^2}$, the following numbers (again rounded) may be displayed: 2, 4, 16, 256, 65536, $4.295 \cdot 10^9$, $1.8447 \cdot 10^{19}$, $3.4028 \cdot 10^{38}$, $1.1579 \cdot 10^{77}$, *overflow*. These numbers are values or estimates of the first ten terms of the sequence defined by

$$\begin{cases} z_0 = 2 \\ z_{k+1} = (z_k)^2 \text{ for integers } k \geq 0. \end{cases}$$

This is a sequence whose terms increase without bound as k gets larger and larger.

The preceding calculator examples are illustrations of *discrete dynamical systems*. A **discrete dynamical system** is a set D together with a function

$$f: D \to D$$

that is repeatedly applied from D into itself. For example, in the square root sequence given earlier, the set D is the set of all nonnegative real numbers. The function f is defined for all x in D by

$$f(x) = \sqrt{x}.$$

Because $\sqrt{x} \geq 0$ for all $x \geq 0$, f is a function from D into itself. The sequence 2.000, 1.414, 1.189, 1.091, ... is called the *orbit with initial point 2* for this system.

When a small change in the value of d produces dramatic changes in $O(d)$, as shown on page 525, it is said that the system is exhibiting chaotic behavior. In this example, an initial point is chosen in radian measure that is not a rational multiple of 2π.

Definition

Let a set D and a function f: $D \to D$ constitute a discrete dynamical system. The sequence $a_0, a_1, a_2, \ldots$ defined by

$$\begin{cases} a_0 = d \\ a_{k+1} = f(a_k) \text{ for integers } k \geq 0 \end{cases}$$

is the **orbit with initial point *d***, written ***O*(*d*)**, for f on D.

One of the fundamental problems concerning discrete dynamical systems is to determine how the orbit $O(d)$ varies with the initial point d. In the case of the discrete dynamical system given by the square root function on the set D of nonnegative real numbers, all orbits with nonzero initial points are sequences with limits equal to 1.

However, not all orbits in dynamical systems have limits. For instance, let f be a function defined by

$$f(x) = \frac{1}{1-x}.$$

The orbit with initial point $\frac{3}{4}$ is the sequence obtained as follows.

$$a_0 = \frac{3}{4}$$

$$a_1 = f(a_0) = f\left(\frac{3}{4}\right) = \frac{1}{1-\frac{3}{4}} = 4$$

$$a_2 = f(a_1) = f(4) = \frac{1}{1-4} = -\frac{1}{3}$$

$$a_3 = f(a_2) = f\left(-\frac{1}{3}\right) = \frac{1}{1+\frac{1}{3}} = \frac{3}{4}$$

$$a_4 = f(a_3) = f\left(\frac{3}{4}\right) = \frac{1}{1-\frac{3}{4}} = 4$$

Notice that $a_4 = a_1$, so the orbit begins to repeat. This orbit is periodic. Its *period*, the number of terms from one appearance of a term to its next appearance, is 3. The values of the orbit do not approach a limit. You are asked to verify that this would happen with initial points other than 0 and 1 in the Questions.

Technology You might wish to approach the examples in this lesson using a computer. A sample program for generating the orbits is given below. Input A is the argument of the initial point on the unit circle (in radians), N is the number of points you want, and the function is the squaring function.

```
10 INPUT A, N
20 FOR K = 1 TO N
30 PRINT COS(A), SIN(A)
40 IF A > 2*PI THEN A =
     A - 2*PI
50 LET A = 2*A
60 NEXT K
70 END
```

Start with the values in the lesson, A = 18° = .3415926 radians, and then with A = .33333 radians to see the difference in behavior. If you can graph the coordinates, the students are given a visual representation of what is going on. The computer instructions for graphing differ too much from machine to machine for us to give them here. You should be aware that roundoff error will eventually cause difficulties in the output, both numerical and graphical.

The orbits of the squaring function on the set of nonnegative real numbers are more complicated. Their behavior depends on the initial point d. If $d > 1$, the orbit $O(d)$ is a sequence whose terms increase without bound. For instance, $O(2) = 2, 4, 16, 256, 65536, 4294967296, \ldots$. If $0 \leq d < 1$, the orbit with initial point d has limit equal to zero. $O(.5) = .5, .25, .0625, .0039063, .0000153, 2.3283 \cdot 10^{-10}, \ldots$, which seems to approach 0 very rapidly. If $d = 1$, the orbit with initial point d is a constant sequence with each term equal to 1.

The orbits of the squaring function are extremely sensitive to changes in the initial point d when d is near 1. If the initial point d is changed from 1 to slightly less than 1, the orbit changes from a constant sequence whose terms all equal 1 to a sequence with limit equal to zero. If d is changed from 1 to a number slightly larger than 1, the orbit changes from a constant sequence whose terms all equal 1 to a sequence whose terms increase without bound. For example, the 15th term of the orbit $O(1.001)$ is approximately 1.29397×10^7, a very large positive number.

Now consider the squaring function defined over a domain of nonreal complex numbers, in particular, the domain D of all points of the unit circle in the complex plane. Each z in D has absolute value equal to 1. Therefore, $z = [1, \theta]$ for some θ, and $f(z) = [1, \theta]^2 = [1, 2\theta]$ because squaring a complex number doubles its argument.

Suppose $d = [1, 18°]$. Then the orbit is given by the sequence

$$\begin{aligned}
a_0 &= d &&= [1, 18°]\\
a_1 &= f([1, 18°]) &&= [1, 36°]\\
a_2 &= f([1, 36°]) &&= [1, 72°]\\
a_3 &= f([1, 72°]) &&= [1, 144°]\\
a_4 &= f([1, 144°]) &&= [1, 288°]\\
a_5 &= f([1, 288°]) &&= [1, 576°] = [1, 216°]\\
a_6 &= f([1, 216°]) &&= [1, 432°] = [1, 72°].
\end{aligned}$$

Note that $a_6 = a_2$. It follows that

$$\begin{aligned}
a_7 &= f(a_6) = f(a_2) = a_3\\
a_8 &= f(a_7) = f(a_3) = a_4
\end{aligned}$$

and so on. Therefore, the orbit $O(d)$ is periodic, with period 4: $O(d) = [1, 18°], [1, 36°], [1, 72°], [1, 144°], [1, 288°], [1, 216°], \ldots$, and then the last four of these numbers are repeated over and over. The six different points in the orbit are plotted at the right. The orange curve in the diagram describes the order of points in the orbit of d.

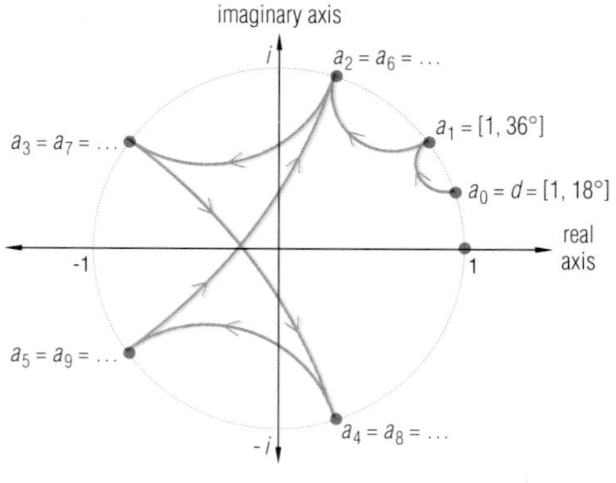

The orbit of $d = [1, 18°]$

524

The orbit with $d = [1, 18°]$ consists of only finitely many different points because the argument θ of the complex number is a *rational multiple* of one revolution; that is,

$$18° = \frac{1}{20}(360°).$$

For any other complex number d on the unit circle whose argument is a rational multiple of 360° or 2π radians, it can be shown that the orbit with initial point d eventually repeats the same points, and therefore consists of only a finite number of distinct points.

However, if the argument of d is not a rational multiple of 2π radians, then it can be shown that the orbit is not periodic; it consists of infinitely many distinct points which have no limit. For instance, let

$$d = [1, \tfrac{1}{3} \text{ radian}].$$

Since each squaring doubles the argument, the terms of the orbit are $[1, \frac{1}{3}]$, $[1, \frac{2}{3}]$, $[1, \frac{4}{3}]$, $[1, \frac{8}{3}]$, $[1, \frac{16}{3}]$, ... Hence, an explicit formula for the sequence is

$$a_n = [1, \tfrac{2^n}{3}] \text{ for integers } n \geq 1.$$

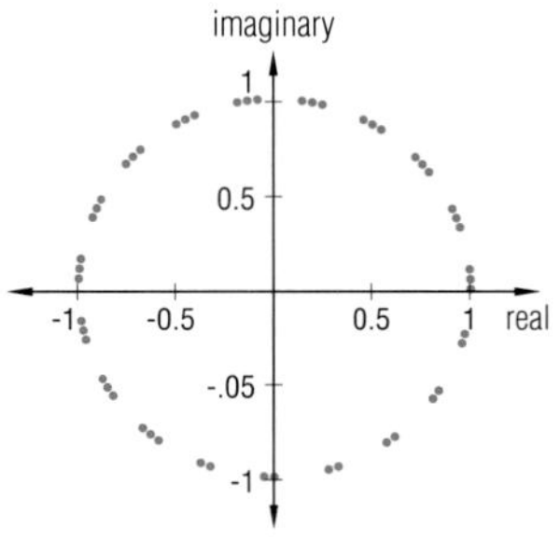

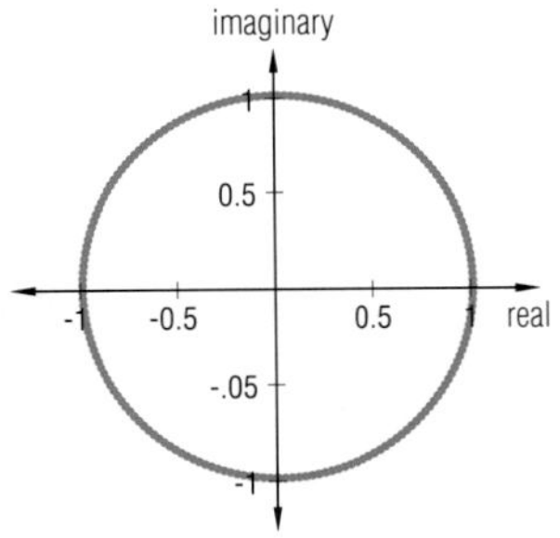

The following proof by contradiction shows that no two terms of the orbit are equal. Suppose that n and m were distinct positive integers with $a_n = a_m$. Then the arguments of a_n and a_m would differ by a multiple of 2π. That is, there would be an integer k such that

$$\frac{2^n}{3} = \frac{2^m}{3} + 2\pi k.$$

Solving this equation for π,

$$\pi = \frac{1}{2k}\left(\frac{2^n}{3} - \frac{2^m}{3}\right).$$

Because the right side of the preceding equation is a rational number and π is irrational, this last statement is false. Thus, the assumption that $a_n = a_m$ leads to a contradiction. Consequently, all terms of the orbit of d are distinct complex numbers. The graphs at the left show the first 50 terms (top) and the first 250 terms (bottom) of the orbit.

It can be proved that when the argument of d is not a rational multiple of 2π, the points in the orbit $O(d)$ are *dense* on the unit circle in the sense that any arc of this circle, no matter how short, contains infinitely many points of the orbit! For any initial point with an argument that *is* a rational multiple of 2π, a small change in the initial point can give an argument that is an irrational multiple of 2π, resulting in a dense orbit. For any initial point with an argument that is an irrational multiple of 2π, a small change in the initial point can give an argument that is a rational multiple of 2π, resulting in an orbit with a finite number of points. In this way, for any initial point on the unit circle, the orbit is extremely sensitive to small changes from the initial point. Mathematicians refer to the type of behavior exhibited by orbits in this example as **chaos**.

These ideas are new. Analysis of discrete dynamical systems began only in the early 1960s with mathematical models for weather forecasting. As you know, predicting the weather is tricky business. The complicated

The final discussion of chaos points out that mathematics is a continually developing subject. At this point, it is not known whether chaos will develop into a subject of great importance, but it is certainly a very interesting area of current research.

Students interested in reading more on this subject will enjoy the book *Chaos: Making a New Science,* by James Gleick. This book was a bestseller in 1989. It is mentioned in the Projects.

NOTES ON QUESTIONS
Question 1: Some students will conjecture that the limit of the sequence of values is 0. With a sufficient number of iterations, the calculator displays the number 1. Ask why this is the case.

Question 5: This question is deceptive in that the argument 1 radian would seem to divide anything including 2π; but it does not because there is no rational number q such that $2\pi = 1q$. This question should be related to **Example 4**.

Questions 7 and 8: Graphically, a fixed point of f is a value of x where the graph of $y = f(x)$ intersects the line $y = x$.

Question 19: Project 8 gives students additional opportunities to explore the function $f(z) = z^2 + c$.

interaction of temperature, moisture conditions, air pressures, and winds frequently confounds and frustrates professional meteorologists as well as people with ordinary interests in such things as planning a picnic or raising a garden. However, our interest in predicting the weather goes far beyond curiosity, picnics, and gardens. The ability to predict the weather with some degree of accuracy over reasonable lengths of time could save lives lost in severe storms, vastly improve farm economy through better planning, and facilitate travel and communication.

The development of high speed, large memory digital computers that began in the 1950s gave scientists of that day strong reason to hope that the problem of predicting weather could finally be solved in a way that would be quite adequate for most practical purposes. After all, they already knew the most important variables affecting the weather and they could apply the laws of physics to obtain equations relating these variables. Thus, they had the tools necessary to construct a reasonably realistic mathematical model for the weather system. To be sure, such a model would involve many variables and very complicated equations, but a computer large enough and fast enough to analyze the model seemed only a few years away.

Edward Lorenz

Preliminary work on a computer model of the weather system was already underway at the Massachusetts Institute of Technology in 1960. There, the meteorologist Edward Lorenz used a vacuum-tube digital computer to analyze a rather crude mathematical model of the weather system. He varied the model until he identified twelve equations relating temperature, atmospheric pressure, and wind velocity that seemed to reflect the essential features of the real weather patterns that he had studied for many years. Lorenz entered initial weather data and the computer printed out a row of numbers once each minute that corresponded to the successive one-day forecasts of this mathematical model.

However, one day in 1961, Lorenz accidently discovered a feature of his model that had profound consequences not only for weather forecasting but also for the study of other complex systems in economics, physics, chemistry, and mathematics. In attempting to repeat part of a computer run, he reentered the data, but used approximations to three decimal places instead of the six-decimal place-values used previously. He expected the use of these approximations to have little noticeable effect on the results of the run. At first, the output for the two runs matched closely, as expected. But then they began to deviate from one another more and more until the two predictions were radically different.

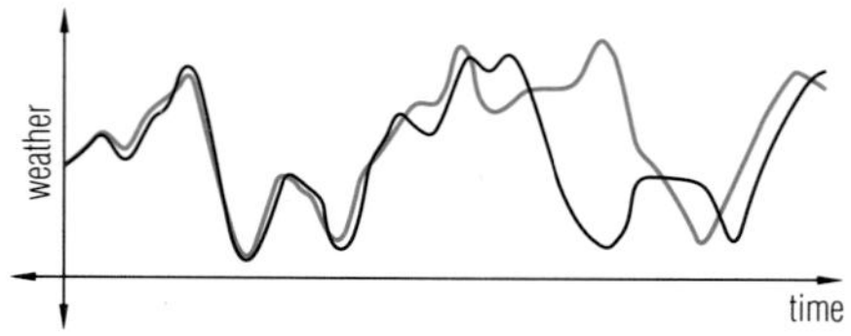

After careful consideration, Lorenz realized that the differences in the outputs of the two runs were not due to a computer error but rather that they revealed a disturbing intrinsic property of his mathematical model: *small differences in the initial conditions for the system could result in radically different predictions.*

In this lesson, you have observed a similar phenomenon for discrete dynamical systems. Small changes in the initial point on the unit circle result in radically different orbits generated by the squaring function. The chaotic behavior that Lorenz observed in his weather model indicates that unpredictability may be an intrinsic characteristic of our weather system rather than an annoying result of the simplifying assumptions that underlie any mathematical model of it. Many other phenomena, such as chemical reactions, the flow of fluids, the flow of highway traffic, economic growth, and the beating of the heart, also exhibit chaotic behavior. The study of chaos in dynamical systems is in its infancy and may provide keys to understanding these diverse phenomena.

Questions

Covering the Reading

1. Enter .5 in your calculator and press the square root key, $\boxed{\sqrt{x}}$, repeatedly. Describe what happens and make a conjecture. See margin.

2. Compute the orbit of $d = -\frac{1}{2}$ for the dynamical system defined by D = set of real numbers not equal to 1 and $f(x) = \frac{1}{1-x}$. $-\frac{1}{2}, \frac{2}{3}, 3, -\frac{1}{2}, \frac{2}{3}, 3, \ldots$

3. Compute the first four terms in the orbit of $d = 7$ for the dynamical system defined by D = set of nonnegative real numbers and $f(x) = x^2$. 7, 49, 2401, 5764801

In 4 and 5, consider the dynamical system defined by D = the unit circle in the complex plane and $f(z) = z^2$.

4. Compute and graph the orbit of $d = [1, 10°]$. See margin.

5. Compute and graph the first six terms of the orbit of [1, 1 radian]. See margin.

6. In this lesson, we observed that if you enter the number 2 in your calculator and repeatedly press the cosine function key, the numbers displayed by your calculator seem to be approaching some number x between .7 and .8.

a. Assuming that these numbers do converge to some number x, explain why it must be true that $\cos x = x$. See margin.

b. Use part **a** to compute x to three decimal places. 0.739

In 7 and 8, a **fixed point** of a dynamical system with function f is a value of x for which $f(x) = x$.

7. Find all fixed points for the function f when $f(x) = x^2$. 1, 0

FOLLOW-UP

PROJECTS

The projects for Chapter 8 are described on pages 529–530. **Projects 7 and 8** are related to the content of this lesson.

ADDITIONAL ANSWERS

1. To 4 digits, the sequence is .5000, .7071, .8409, .9170, … . The sequence approaches 1.

4. [1, 10°], [1, 20°], [1, 40°], [1, 80°], [1, 160°], [1, 320°], [1, 280°], [1, 200°], [1, 40°], [1, 80°], …

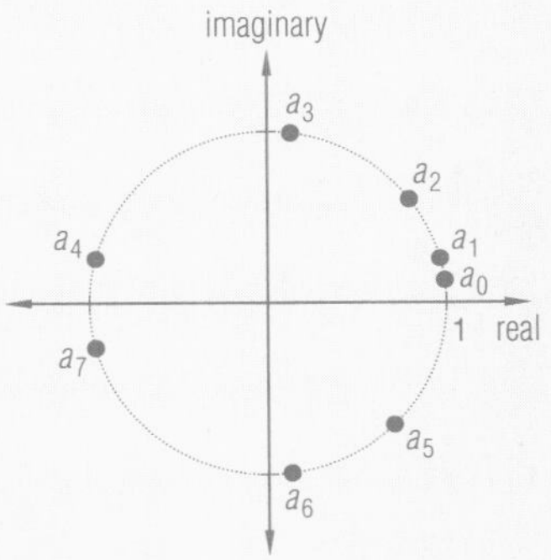

5. [1, 1], [1, 2], [1, 4], [1, 8], [1, 16], [1, 32]

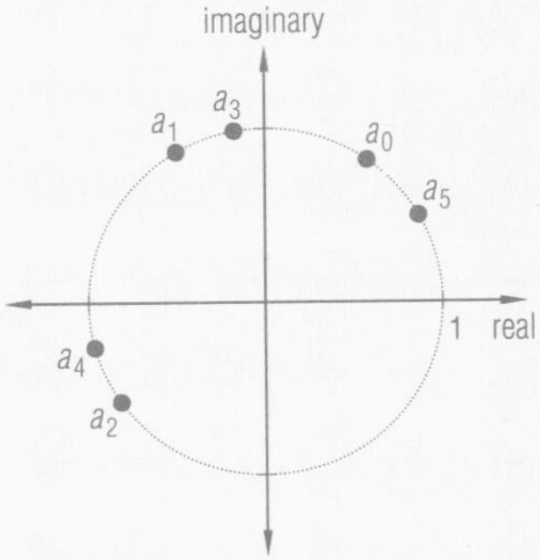

6.a. Let a_k be the kth number obtained (by pressing the cosine key k times). Then a_k is approaching x for larger and larger k. But $a_k = \cos a_{k-1}$, and $\cos a_{k-1}$ approaches $\cos x$ since the cosine function is continuous. Therefore, $x = \cos x$.

ADDITIONAL ANSWERS

9. For all $x \neq 1$, let $a_0 = x$. Then $a_1 = \frac{1}{1-x}$. And if $x \neq 0$,

$$a_2 = \frac{1}{1 - \frac{1}{1-x}} = \frac{1-x}{1-x-1} = \frac{1-x}{-x} = \frac{x-1}{x},$$

and $a_3 = \frac{1}{1 - \frac{x-1}{-x}}$

$$= \frac{x}{x-(x-1)} = \frac{x}{1} = x.$$ So $a_3 = a_0$, hence the period is 3.

10.

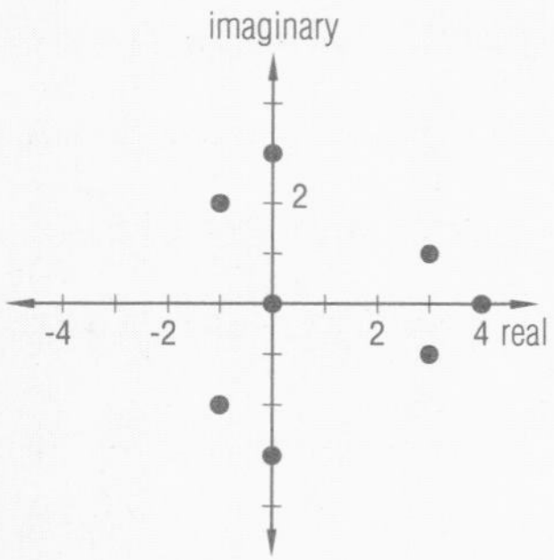

13. $z_0 = \sqrt[4]{7}\left(\cos\frac{\pi}{5} + i \sin\frac{\pi}{5}\right)$;

$z_1 = \sqrt[4]{7}\left(\cos\frac{7\pi}{10} + i \sin\frac{7\pi}{10}\right)$;

$z_2 = \sqrt[4]{7}\left(\cos\frac{6\pi}{5} + i \sin\frac{6\pi}{5}\right)$;

$z_3 = \sqrt[4]{7}\left(\cos\frac{17\pi}{10} + i \sin\frac{17\pi}{10}\right)$

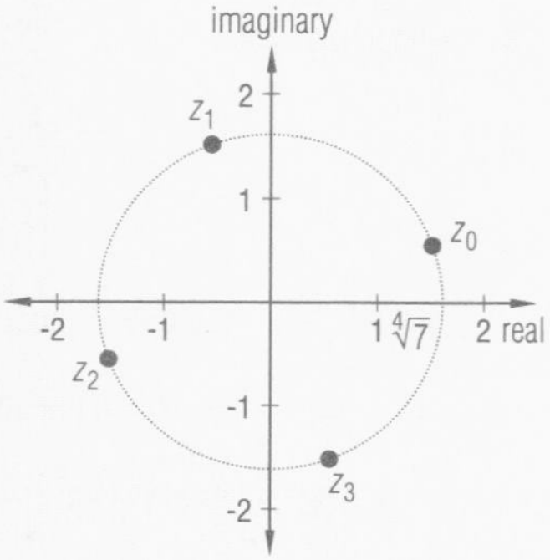

14.b., 15.a., 18., 19. See Additional Answers in the back of this book.

8. Suppose that f is the function defined for all real numbers x by $f(x) = \sin x$ and that D is the set of all real numbers x between $-\pi$ and π.
   a. Find a fixed point of this system. **0** **b.** Are there any others? **No**

9. Prove that, in the dynamical system in Question 2, the period of every number is no more than 3. **See margin.**

Review

10. The graph at the right shows five of the zeros of a polynomial of degree 8 with real coefficients. Copy and complete the graph so it shows all of the zeros. *(Lessons 8-9, 8-1)* **See margin.**

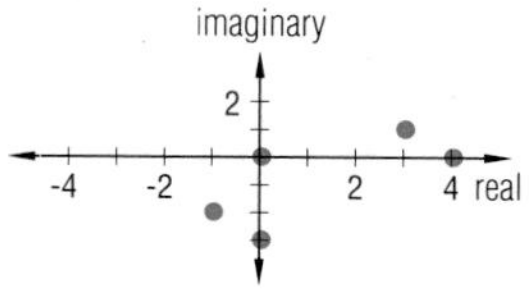

11. Write a formula for a fourth degree polynomial with real coefficients whose zeros include 0, $1 - 2i$, and 2. *(Lesson 8-9)*
    Sample: $x^4 - 4x^3 + 9x^2 - 10x$

12. Find all zeros and their multiplicities for the polynomial $p(x) = (x + 2)^2(x^4 - 2x^2 - 8)$. *(Lesson 8-8)*
    zeros: -2 (with multiplicity 3), 2, $\pm 2i$

13. Find the fourth roots of $7\left(\cos\frac{4\pi}{5} + i \sin\frac{4\pi}{5}\right)$ and graph the roots in a complex plane. *(Lesson 8-7)* **See margin.**

14. Let $z = [3, 40°]$ and $w = [2, 70°]$.
    a. Find a polar representation for
    i. zw **ii.** z^5 **iii.** w^5.
    b. Use your answers to part **a** to show that $(zw)^5 = z^5 \cdot w^5$.
    (Lessons 8-6, 8-3) **ai) sample: [6, 110°]; aii) sample: [243, 200°]; aiii) sample: [32, 350°]; b) See margin.**

15. **a.** Sketch the polar graph of the equation $r = 1 + 2\cos\theta$. **See margin.**
    b. Give equations for any lines of symmetry. *(Lessons 8-5, 8-4)*
    $\theta = 0°$

16. Find a recursive definition for the sequence with explicit formula $t_n = 5 - 4n$ for integers $n \geq 1$. *(Lessons 7-1, 7-2)* $\begin{cases} t_1 = 1 \\ t_{n+1} = t_n - 4 \text{ for } n \geq 1 \end{cases}$

17. The period T (in seconds) of a pendulum is given by $T = 2\pi\sqrt{\frac{\ell}{g}}$, where ℓ is the length of the pendulum (in feet) and g is the gravitational acceleration (32 ft/sec²). Find the length of a pendulum whose period is 12 seconds. *(Lesson 3-1)* **$\frac{1152}{\pi^2} \approx 116.7$ feet**

18. Prove: If *m is any odd integer, then $m^2 + m - 3$ is an odd integer.* *(Lesson 1-7)* **See margin.**

Exploration

19. If c is a fixed complex number and f is the function defined for all complex numbers z by $f(z) = z^2 + c$, then use what you know about the geometry of addition and multiplication to explain how the point $f(z)$ can be located from the point z in the complex plane. (Note: This function is used to define the Mandelbrot set which is important in the study of fractals.) **See margin.**

Projects

1. Four representations of complex numbers are in this chapter: $a + bi$, rectangular, polar, and trigonometric.
Show that a fifth representation, the matrix $\begin{bmatrix} a & -b \\ b & a \end{bmatrix}$ for $a + bi$, satisfies the following properties of the complex numbers.
   a. The rules for complex addition and multiplication hold.
   b. The modulus of $a + bi$ is related to the determinant of its matrix.
   c. The additive and multiplicative inverses of a complex number are related to the additive and multiplicative inverses of this matrix.
   d. The matrix form has special matrices which correspond to 1 (the multiplicative identity) and 0 (the additive identity).
   e. The argument of a complex number is related to its matrix representation.

2. Report on one of the following applications of complex numbers.
   a. acoustics
   b. electricity
   c. higher mathematics, such as differential equations

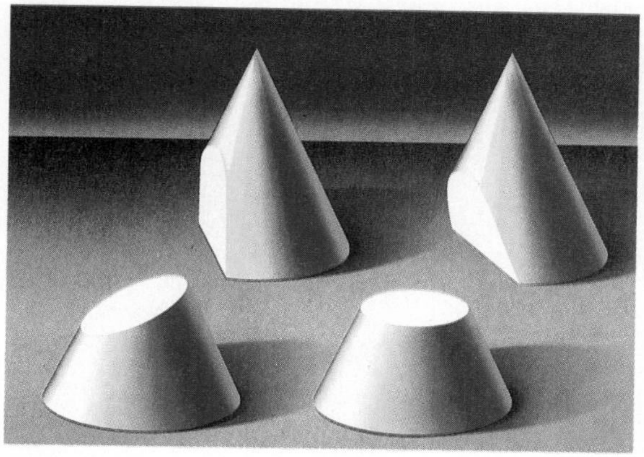

3. The conic sections have polar equations. Write a report describing these equations and the role played by the eccentricity of the conic. Use analytic geometry, calculus, or trigonometry books as aids in your research.

4. There are coordinate systems which extend the idea of polar coordinates to three dimensions. Find out about and describe at least two of these systems. (Books on mapmaking are one source.)

Northern maidenhair fern, a spiral in nature

5. Spirals occur frequently in nature. From a library, copy pictures of chambered nautiluses, sunflowers, and other natural objects to show your class, and explain their mathematical properties.

6. The Irish mathematician William Rowan Hamilton attempted to reconcile the diverse phenomena of eighteenth century physics with a new mathematical system he called *quaternions*. This system extended the ordered pair concept to ordered 4-tuples. Find out how he defined addition and multiplication, and what the relationships among his ''quaternionic units'' are.

7. Write a report on chaotic phenomena found in fields outside mathematics. Describe what makes the phenomena chaotic, and when, how, and by whom their chaotic nature was discovered. A good source of information is *Chaos: Making a New Science*, by James Gleick.

PROJECTS

The projects provide an opportunity for extended work on topics related to the content of the chapter. (See pages 70 and T37 for more information on the purposes of projects.) Some or all of these projects may be suitable for **small group work**. Solutions for the projects can be found at the end of the Additional Answers section in the back of this book.

A variety of projects are given on this page and the next. Of these, **Projects 1 and 3** are the most mathematical, and **Project 8** requires the most knowledge of computers. **Projects 2 and 6** may require the use of a substantial amount of mathematics depending on how they are done. **Projects 4, 5, and 7** require using the ideas of the chapter.

8. Let c be a complex number and consider the dynamical system defined by D = set of complex numbers and $f(z) = z^2 + c$.

a. Write a computer program which will input values for c and d and print out the first several terms in the orbit $O(d)$ while also graphing them.

b. Select a few values for c, and for each one, use your program to investigate $O(d)$ for different values of d. Values of c and d with $|c| < 2$ and $|d| < 2$ work the best. Describe the behavior of the orbits you get. Some interesting values to try for c are $c = 0$, $-.4 + .2i$, $-.12 + .74i$, $-.39054 - .58679i$, -1.25, $.32 + .043i$, and $-.11 + .67i$.

c. The program below inputs a value of c, then tests values of d to determine whether the terms of $O(d)$ attain an arbitrarily large distance from the origin. If not, the point d is plotted on the screen. In the program, variables ending in R pertain to real parts, those ending in I pertain to imaginary parts.

```
10 NC = 319
20 NR = 199
30 LR = -1.5
40 LI = -1.5
50 UR = 1.5
60 UI = 1.5
70 INPUT "ENTER C AS ORDERED
     PAIR:"; CR,CI
80 SCREEN 1
90 FOR DR = LR TO UR STEP
     (UR-LR)/NC
100    FOR DI = LI TO UI STEP
       (UI-LI)/NR
110      K = 0
120      ZR = DR
130      ZI = DI
140        K = K + 1
150        TEMP = ZR*ZR - ZI*ZI + CR
160        ZI = 2*ZR*ZI + CI
170        ZR = TEMP
180        DS = ZR*ZR + ZI*ZI
190        IF DS < 100 AND K < 30 THEN
           GO TO 140
200      IF DS < 100 THEN PSET
         ((DR-LR)/(UR-LR)*NC,(UI-DI)/
         (UI-LI)*NR)
210  NEXT DI
220 NEXT DR
230 END
```

Note that lines 10 and 20 assume that pixels are numbered from 0 to 319 from left to right and from 0 to 199 from top to bottom. Lines 30 through 60 set the viewing window so the lower left is $-1.5 - 1.5i$ and the upper right is $1.5 + 1.5i$. Line 80 sets the screen to graphics mode, and line 200 plots points. These lines may have to be modified for your particular computer and for the window you want. The program will plot a region corresponding to the points d for which the first 30 terms of $O(d)$ stay within 10 units of the origin. The edge of the region is called the **Julia set** of the function $f(z) = z^2 + c$.

Run the program to see the Julia sets for some of the values of c you investigated in part **b**. Also try $c = i$, $-.194 + .6557i$, and $.11031 - .67037i$. Note that in interpreted BASIC, the program can take a long time to run.

d. Find out what the Mandelbrot set is, and write a program to draw it. Some good resources are *The Beauty of Fractals*, by Heinz-Otto Peitgen and Peter H. Richter, and *Chaos, Fractals, and Dynamics*, by Robert L. Devaney.

This fractal landscape was automatically generated by a computer from the decimal expansion of pi.

530

Summary

Complex numbers are numbers that can be written in the form $a + bi$, where a and b are real numbers and $i^2 = -1$. With normal polynomial addition and multiplication they form a field.

Four forms of writing complex numbers were discussed: $a + bi$ form, rectangular form (a, b), polar form $[r, \theta]$, trigonometric form $r(\cos\theta + i\sin\theta)$. They are related by the equations $r^2 = a^2 + b^2$, $a = r\cos\theta$, and $b = r\sin\theta$. The graph of a complex number on the complex plane is shown here.

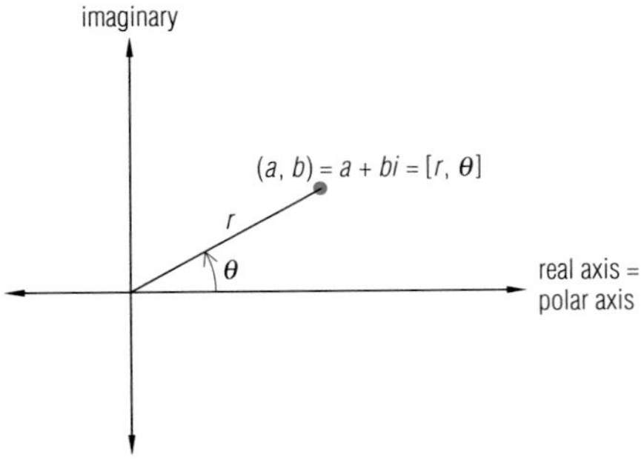

Adding $w = (a, b)$ to z applies the translation $(x, y) \to (x + a, y + b)$ to z. The graphs in the complex plane of w, 0, z, and $z + w$ are consecutive vertices of a parallelogram. Multiplying z by $w = [s, \phi]$ applies to z a composite of a size change of magnitude s and a rotation of ϕ about the origin. If $z = [r, \theta]$ and $w = [s, \phi]$, then $zw = [rs, \theta + \phi]$.

A polar equation $r = f(\theta)$ can be graphed on a polar grid by finding sets of ordered pairs $[r, \theta]$ that satisfy the equation. A second method involves plotting the rectangular graph of $r = f(\theta)$, then using the height of the graph at various angles θ to make the polar graph. The height becomes the distance from the pole at the angle θ.

Polar graphs are often quite beautiful. Here are some equations and the names of their graphs (a is a nonzero real number, n is a positive integer: $\sin\theta$ can be replaced by $\cos\theta$.)

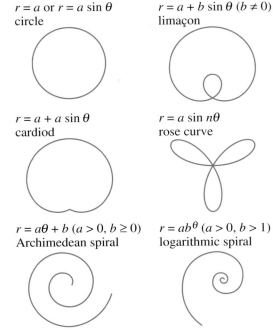

$r = a$ or $r = a\sin\theta$
circle

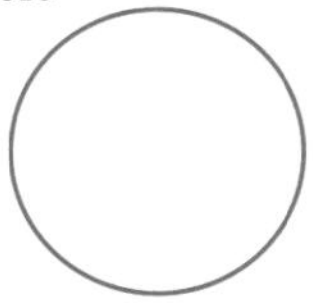

$r = a + b\sin\theta\ (b \neq 0)$
limaçon

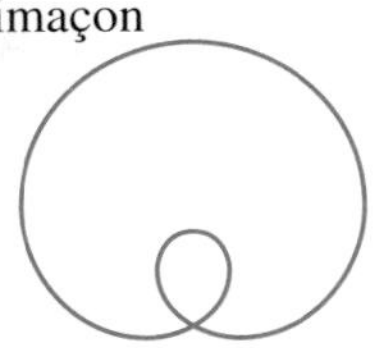

$r = a + a\sin\theta$
cardiod

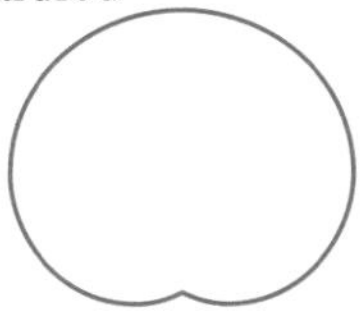

$r = a\sin n\theta$
rose curve

$r = a\theta + b\ (a > 0, b \geq 0)$
Archimedean spiral

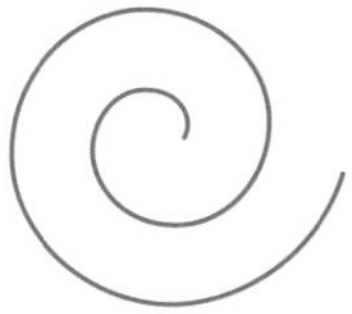

$r = ab^{\theta}\ (a > 0, b > 1)$
logarithmic spiral

Polar form is most convenient for DeMoivre's Theorem: If $z = [r, \theta]$, then $z^n = [r^n, n\theta]$. When graphed in the complex plane, the powers of z lie on a logarithmic spiral. From DeMoivre's Theorem, one can prove that the nth roots of z are given by $[\sqrt[n]{r}, \frac{\theta}{n} + k \cdot \frac{2\pi}{n}]$ for $k = 0, 1, 2, \ldots, n - 1$. When graphed in the complex plane, the nth roots of z form the vertices of a regular n-gon centered at $(0, 0)$.

The Fundamental Theorem of Algebra states that every polynomial $p(x)$ of degree at least 1, over the set of complex numbers, has at least one zero in the set of complex numbers. In fact, the number of zeros it has is equal to its degree, provided that multiplicities are counted. In the case that all coefficients of $p(x)$ are real, whenever a nonreal number is a zero of $p(x)$, so is its conjugate.

The study of chaos in discrete dynamical systems has only developed since the 1960s. Its promise is a better understanding of many different phenomena, including weather forecasting.

SUMMARY

The Summary gives an overview of the entire chapter and provides an opportunity for students to consider the material as a whole. Thus, the Summary can be used to help students relate the various concepts presented in the chapter.

VOCABULARY

Terms, symbols, and properties are listed by lesson to provide a checklist of concepts a student must know. (See page 71 for more information.)

ADDITIONAL ANSWERS

5.

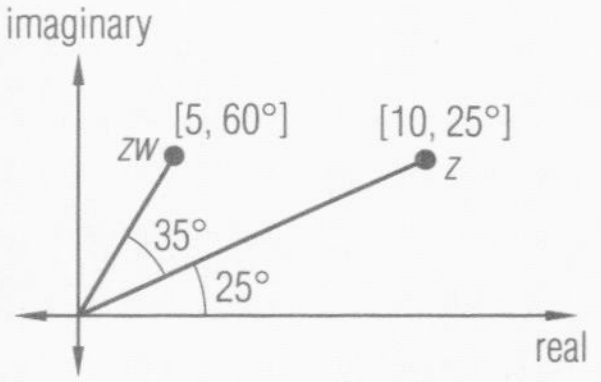

6b.

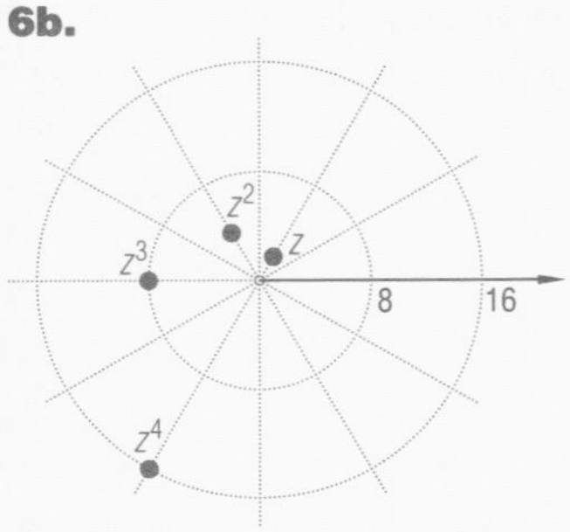

Vocabulary

For the starred (*) terms you should be able to give a definition of the term.
For the other terms you should be able to give a general description and a specific example of each.

Lesson 8-1
*imaginary number
*imaginary unit, i
*complex number, real part, imaginary part
*equality for complex numbers
*addition, multiplication of complex numbers
field properties
field of complex numbers
*complex conjugates
rectangular form
complex plane
real axis
imaginary axis
current
impedance

Lesson 8-2
pole
polar axis
polar coordinates, $[r, \theta]$
polar grid
Polar-Rectangular Conversion Theorem

Lesson 8-3
Geometric Addition Theorem
*absolute value, modulus, $|z|$
polar form
*argument
trigonometric form
Geometric Multiplication Theorem

Lesson 8-4
polar equation
limaçon
cardioid

Lesson 8-5
rose curve, leaves, petals
spiral of Archimedes
logarithmic spiral

Lesson 8-6
DeMoivre's Theorem

Lesson 8-7
*nth root of a complex number
Complex nth Roots Theorem
Geometric nth Roots Theorem

Lesson 8-8
Fundamental Theorem of Algebra
*multiplicity of a zero
*simple zero

Lesson 8-9
Conjugate Zeros Theorem

Lesson 8-10
discrete dynamical system
orbit with initial point d, $O(d)$
dense
chaos, chaotic behavior
fixed point

Progress Self-Test

See margin for answers not shown below.

Take this test as you would take a test in class. Then check the test yourself using the solutions at the back of the book.

1. Let $z = 8 - 5i$ and $w = -2 + 3i$. Calculate the following.
  a. $z - w$ $10 - 8i$ **b.** zw $-1 + 34i$
  c. $\frac{z}{w}$ $-\frac{31}{13} - \frac{14}{13}i$ **d.** $\overline{w}$ $-2 - 3i$

2. Express $[8, \frac{5\pi}{6}]$ using rectangular coordinates.

3. Let $z = 4\sqrt{3} - 4i$.
  a. Write z in rectangular coordinate form.
  b. Find the absolute value of z. 8
  c. Find the argument θ of z if $0 \le \theta < 2\pi$.
  d. Write z in polar form. $[8, \frac{11\pi}{6}]$

4. Suppose an AC circuit has voltage $-3 - 8i$ volts and current 6 amps. Use Ohm's Law $I = \frac{V}{Z}$ to find the impedance in ohms.

5. Illustrate the multiplication of $z = 10(\cos 25° + i \sin 25°)$ by $w = \frac{1}{2}(\cos 35° + i \sin 35°)$ with a diagram that verifies the Geometric Multiplication Theorem.

6. Let $z = 2\left(\cos \frac{\pi}{3} + i \sin\frac{\pi}{3}\right)$.
  a. Calculate z^1, z^2, z^3, and z^4.
  b. Graph them in the complex plane.
  c. Is the sequence of numbers getting closer to or farther from 0? farther

7. **a.** Calculate the fifth roots of $243i$.
  b. Plot them in the complex plane.

8. Find all zeros and their multiplicities for the polynomial
$$p(x) = (x^4 - 81)(x^2 + 6x + 9).$$

9. Let $z = a + bi$.
  a. Find a formula for z^2. $a^2 - b^2 + 2abi$
  b. Find a formula for $(\overline{z})^2$. $a^2 - b^2 - 2abi$
  c. Use parts **a** and **b** to show that $\overline{(z^2)} = (\overline{z})^2$ $\forall$ complex numbers z.

10. Suppose $p(x)$ is a fifth degree polynomial with real coefficients and with zeros 2, $5i$, and $1 - i$.
  a. Find all other zeros of $p(x)$. $-5i$, $1 + i$
  b. Write a possible formula for $p(x)$ in factored form.

11. Sketch the polar graph of the equation $r = 2 + 2 \sin \theta$.

12. **a.** Sketch the polar graph of the equation $r = 3 \cos 4\theta$.
  b. Identify the type of curve you get.

2. $(-4\sqrt{3}, 4)$
3. a) $(4\sqrt{3}, -4)$; c) $\frac{11\pi}{6}$
4. $-\frac{1}{2} - \frac{4}{3}i$ ohms
6. a) $z^1 = 2(\cos \frac{\pi}{3} + i \sin \frac{\pi}{3})$, $z^2 = 4(\cos \frac{2\pi}{3} + i \sin \frac{2\pi}{3})$, $z^3 = 8(\cos \pi + i \sin \pi)$, $z^4 = 16(\cos \frac{4\pi}{3} + i \sin \frac{4\pi}{3})$
7. a) [3, 18°], [3, 90°], [3, 162°], [3, 234°], [3, 306°]
8. zeros: -3 (with multiplicity 3), 3, $\pm 3i$
9. c) $\overline{z^2} = \overline{a^2 - b^2 + 2abi} = a^2 - b^2 - 2abi = (\overline{z})^2$
10. b) $p(x) = (x - 2)(x - 5i)(x + 5i)(x - 1 - i)(x - 1 + i)$
12. b) eight-leafed rose curve

PROGRESS SELF-TEST

The Progress Self-Test provides the opportunity for feedback and correction; the Chapter Review provides additional opportunities for practice.

USING THE PROGRESS SELF-TEST

Assign the Progress Self-Test as a one-night assignment. (See pages 72 and T35 for more information.)

7.b.

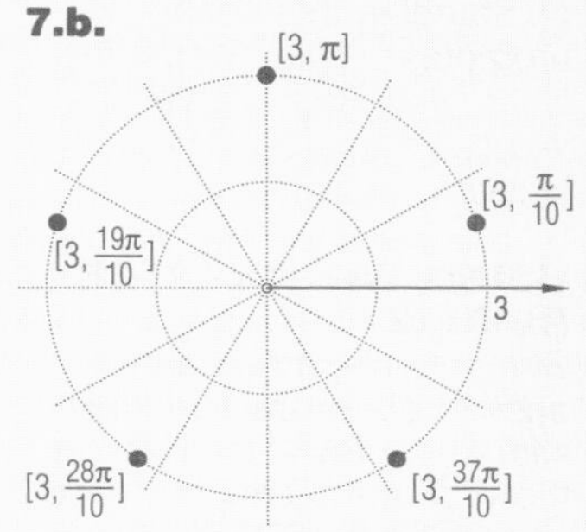

11.

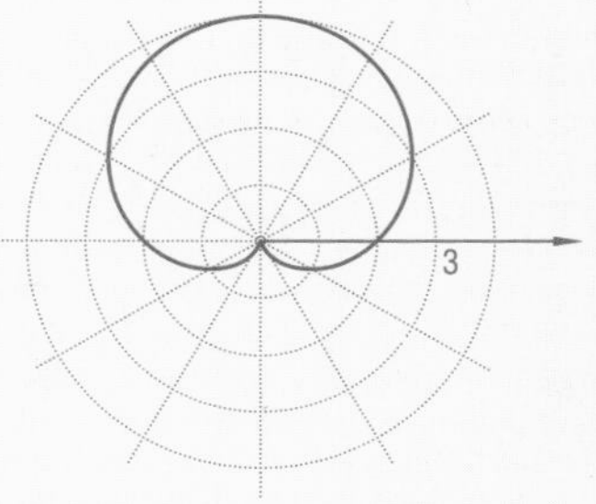

12.a.

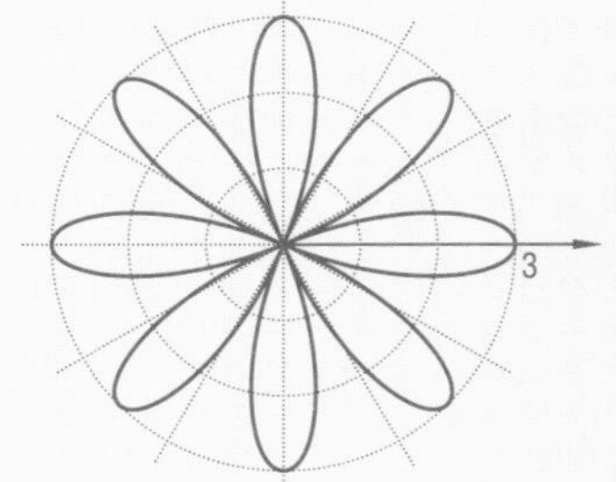

RESOURCES
- Chapter 8 Test, Form A
- Chapter 8 Test, Form B
- Chapter 8 Test, Cumulative Form
- Quiz and Test Writer

CHAPTER REVIEW

The main objectives for the chapter are organized here into sections corresponding to the four main types of understanding this book promotes: Skills, Properties, Uses, and Representations. We call these the SPUR objectives.

USING THE CHAPTER REVIEW
Students should be able to answer questions like these with about 85% accuracy by the end of the chapter. (See pages 74 and T38–39 for more information.)

ADDITIONAL ANSWERS

1. $(6\sqrt{3}, -6)$, $\left[12, \frac{11\pi}{6}\right]$, $12\left(\cos \frac{11\pi}{6} + i \sin \frac{11\pi}{6}\right)$

2. $4\sqrt{2} - 4\sqrt{2}\,i$, $(4\sqrt{2}, -4\sqrt{2})$, $\left[8, \frac{7\pi}{4}\right]$

3. $2.5 \cos 35° + 2.5i \sin 35°$, $(2.5 \cos 35°, 2.5 \sin 35°)$, $2.5(\cos 35° + i \sin 35°)$

4. $(-4, 0)$, $[4, \pi]$, $4(\cos \pi + i \sin \pi)$

5. $-7 + 5i$, $\approx [\sqrt{74}, 144°]$, $\approx \sqrt{74}(\cos 144° + i \sin 144°)$

6. $-\frac{1}{2}i$, $\left(0, -\frac{1}{2}\right)$, $\frac{1}{2}\left(\cos \frac{3\pi}{2} + i \sin \frac{3\pi}{2}\right)$

7. $|z| = 25$, $\theta \approx 163.7°$

8. $|z| = b$, $\theta = \frac{\pi}{2}$

9.c. $\left[4, \frac{\pi}{2} + 2n\pi\right]$, n an integer.

17. 125

21. $\sqrt{21}(\cos 62° + i \sin 62°)$

Chapter Review

Questions on SPUR Objectives

SPUR stands for **S**kills, **P**roperties, **U**ses, and **R**epresentations.
The Chapter Review questions are grouped according to the SPUR Objectives for this chapter.
See margin for answers not shown below.

SKILLS deal with the procedures used to get answers.

Objective A: *Express complex numbers in $a + bi$, rectangular, polar, and trigonometric form.* *(Lessons 8-1, 8-2)*

In 1–6, the complex number is written in either $a + bi$, rectangular, polar, or trigonometric form. Write it in the other three forms.

1. $6\sqrt{3} - 6i$ **2.** $8\left(\cos \frac{7\pi}{4} + i \sin \frac{7\pi}{4}\right)$

3. $[2.5, 35°]$ **4.** -4

5. $(-7, 5)$ **6.** $\left[-\frac{1}{2}, \frac{5\pi}{2}\right]$

7. Find the absolute value and argument θ for the complex number $-24 + 7i$, with $0 \le \theta < 360°$.

8. Given any imaginary number $0 + bi$ with $b > 0$, find the absolute value r and argument θ if θ is restricted to the interval $0 \le \theta < 2\pi$.

9. Suppose $P = \left[-4, \frac{3\pi}{2}\right]$.
- **a.** Give polar coordinates for P with $r \ne -4$. sample: $\left[4, \frac{\pi}{2}\right]$
- **b.** Give polar coordinates for P with $r > 0$ and $\theta < 0$. sample: $\left[4, -\frac{3\pi}{2}\right]$
- **c.** Give the general form of the polar coordinates for P with $r > 0$.

Objective B: *Perform operations with complex numbers.* *(Lessons 8-1, 8-3)*

In 10 and 11, rewrite each expression in $a + bi$ form.

10. $\sqrt{-48}$ $4\sqrt{3}\,i$

11. $\frac{4 - \sqrt{-12}}{6}$ $\frac{2}{3} - \frac{\sqrt{3}}{3}i$

12. Express the solutions to $x^2 = -80$ in $a + bi$ form. $\pm 4\sqrt{5}\,i$

In 13–16, perform the indicated operation and write the result in $a + bi$ form.

13. $(7 + 3i) + (8 - 6i)$ $15 - 3i$

14. $(4 - 9i)^2$ $-65 - 72i$

15. $\frac{3 - 5i}{10 + 2i}$ $\frac{5}{26} - \frac{7}{13}i$

16. $(10 + 3i)(8 - 4i)$ $92 - 16i$

17. Find the product of $2 + 11i$ and its conjugate.

18. Express in $a + bi$ notation: $(3 + 9i)(4 - i) - \frac{8 + 2i}{1 + i}$. $16 + 36i$

19. Evaluate i^9. i

In 20 and 21, find $z \cdot w$ and express results in the same form as that of the given numbers.

20. $z = [10, 150°]$, $w = [2, 40°]$ $[20, 190°]$

21. $z = \sqrt{3}(\cos 50° + i \sin 50°)$, $w = \sqrt{7}(\cos 12° + i \sin 12°)$

22. Find z so that $z \cdot [3, 80°] = [24, 300°]$.

Objective C: *Convert between polar and rectangular coordinate representations of points.* *(Lesson 8-2)*

23. Suppose $P = \left[1, \frac{\pi}{3}\right]$ and $Q = \left[-1, \frac{2\pi}{3}\right]$.
- **a.** Determine rectangular coordinates for P and Q.
- **b.** *True* or *false*? *P and Q are reflection images of each other over the polar axis.*

In 24 and 25, give one pair of polar coordinates for each (x, y) pair.

24. $(-6, 7)$ **25.** $(\sqrt{3}, -1)$ $\left[2, \frac{11\pi}{6}\right]$

26. If $P = \left[r, \frac{5\pi}{6}\right] = (x, 2)$, solve for r and x.

Objective D: *Find powers and roots of complex numbers.* *(Lessons 8-6, 8-7)*

27. Express $[1.7, \frac{\pi}{3}]^6$ in $a + bi$ form. $\approx 24.1 + 0i$

28. Find $(-1 + 3i)^4$. $28 + 96i$

In 29–31, find the indicated roots of the given number. Express the roots in the same form as the given number.

29. cube roots of $[64, \frac{\pi}{2}]$ $[4, \frac{\pi}{6}], [4, \frac{5\pi}{6}], [4, \frac{3\pi}{2}]$

30. sixth roots of $729\left(\cos\frac{\pi}{6} + i\sin\frac{\pi}{6}\right)$

31. tenth roots of 1024 $2\cos\frac{\pi n}{5} + 2i\sin\frac{\pi n}{5}$, for $n = 0, 1, 2, ..., 9$

32. One cube root of z is $\frac{5\sqrt{2}}{2} + \frac{5\sqrt{2}}{2}i$.

a. Write z in $a + bi$ form. $-\frac{125\sqrt{2}}{2} + \frac{125\sqrt{2}}{2}i$

b. Write the other two cube roots of z in polar form. $[5, \frac{11\pi}{12}], [5, \frac{19\pi}{12}]$

33. a. According to DeMoivre's Theorem, $[r, \theta]^n = \underline{\ ?\ }$. $[r^n, n\theta]$

b. Restate DeMoivre's Theorem in trigonometric form.

34. If $3(\cos 30° + i \sin 30°)$ is a fifth root of w, write w in polar form.
$243(\cos 150° + i \sin 150°) = [243, 150°]$

Objective E: *Find all zeros, and their multiplicities, of a given polynomial.* *(Lessons 8-8, 8-9)*

In 35 and 36, find all zeros and the corresponding multiplicities for the given polynomial.

35. $p(x) = x^4 - 6x^3 + 9x^2$

36. $p(x) = x^4 - 5x^2 - 36$

37. If $1 - i$ is a zero of $p(x) = x^3 + x^2 - 4x + 6$, find the remaining zeros. $1 + i, -3$

38. a. Find all zeros and their multiplicities for the polynomial $p(x) = 4x^2 - 4xi - 1$.

b. Does the answer to part **a** contradict the Conjugate Zeros Theorem? Explain why or why not.

PROPERTIES deal with the principles behind the mathematics.

Objective F: *Prove or verify properties of complex numbers.* *(Lessons 8-1, 8-3, 8-6)*

In 39 and 40, let $z = 3 + 2i$, $v = -1 + 3i$, and $w = 4 - i$.

39. Verify that $\overline{z - w} = \overline{z} - \overline{w}$.

40. Verify that $z(v + w) = zv + zw$.

41. Prove that if z and w are imaginary numbers, then zw is a real number.

42. Let $z = [r, \theta]$ and $w = [r, -\theta]$. Show that z and w are complex conjugates by writing them in $a + bi$ form.

43. Let $z = [r, \theta]$ and let n and m be positive integers.

a. Use DeMoivre's Theorem to write z^n, z^m, and z^{n+m} in polar form.

b. Use the answers to part **a** to show that $z^n \bullet z^m = z^{n+m}$.

Objective G: *Use the properties of polynomials to find or describe their zeros.* *(Lessons 8-8, 8-9)*

44. $p(x) = 5x^7 - 9x^4 + 2ix^3 - 6$ has exactly $\underline{\ ?\ }$ complex zeros, counting multiplicities. 7

45. *True* or *false? Every polynomial of odd degree $n \geq 1$ with real coefficients has at least one real zero.* True

46. If $5 + 2i$ is a zero of a polynomial $q(x)$ with real coefficients, then $\underline{\ ?\ }$ must also be a zero. $5 - 2i$

47. Without doing any computation, explain why 3, -2, and $2i$ cannot all be zeros of the polynomial $p(x) = -3x^3 + 9x^2 - 12x + 36$.

48. Suppose that $q(x)$ is a fourth degree polynomial with real coefficients, $q(4) = 0$, $q(4i) = 0$, and $q(3 - i) = 1 + 2i$.

a. Counting multiplicities, how many real zeros and how many nonreal zeros does $q(x)$ have? 2 real, 2 nonreal

b. Give the value of:

i. $q(-4)$; cannot be determined

ii. $q(-4i)$; 0

iii. $q(3 + i)$. $1 - 2i$

49. Suppose $p(x)$ and $q(x)$ are polynomials such that $p(x) = (x - z)q(x)$ and z is a zero of $q(x)$. Then what do you know about the multiplicity of z as a zero of $p(x)$?

22. $[8, 220°]$

23.a. $P = \left(\frac{1}{2}, \frac{\sqrt{3}}{2}\right)$, $Q = \left(\frac{1}{2}, -\frac{\sqrt{3}}{2}\right)$
b. True

24. sample: $\approx [\sqrt{85}, 130.6°]$

26. $r = 4$, $x = -2\sqrt{3}$

30. $3\left(\cos\left(\frac{\pi}{36} + \frac{\pi n}{3}\right) + i\sin\left(\frac{\pi}{36} + \frac{\pi n}{3}\right)\right)$ for $n = 0, 1, 2, \dots, 5$

33.b. $(r(\cos\theta + i\sin\theta))^n = r^n(\cos n\theta + i\sin n\theta)$

35. zeros: 0, 3 (both with multiplicity 2)

36. zeros: ± 3, $\pm 2i$ (all with multiplicity 1)

38.a. $\frac{i}{2}$ with multiplicity 2
b. No, the Conjugate Zeros Theorem does not apply if the coefficients of the polynomial are not all real numbers.

39. $z - w = -1 + 3i$, so $\overline{z - w} = -1 - 3i$. $\overline{z} - \overline{w} = (3 - 2i) - (4 + i) = -1 - 3i$. So $\overline{z - w} = \overline{z} - \overline{w}$.

40. $z(v + w) = (3 + 2i) \cdot (3 + 2i) = 5 + 12i$, and $zv + zw = (-9 + 7i) + (14 + 5i) = 5 + 12i$, so $z(v + w) = zv + zw$.

41. If $z = 0 + bi$ and $w = 0 + di$, then $zw = bdi^2 = -bd = -bd + 0i$, which is a real number.

42. $z = r\cos\theta + (r\sin\theta)i$ and $w = r\cos(-\theta) + (r\sin(-\theta))i = r\cos\theta - (r\sin\theta)i$. So z and w are complex conjugates.

43.a. $z^n = [r^n, n\theta]$; $z^m = [r^m, m\theta]$; $z^{n+m} = [r^{n+m}, (n + m)\theta]$
b. $z^n \cdot z^m = [r^n, n\theta] \cdot [r^m, m\theta] = [r^n \cdot r^m, n\theta + m\theta] = [r^{n+m}, (n + m)\theta] = z^{n+m}$

47., 49. See the margin on page 536.

OVERVIEW ■ CHAPTER 9

The purpose of the chapter is to help students become comfortable with the idea of the derivative. We would like students to have a solid understanding of rates of change and how these rates relate to the derivative.

Shortcut formulas for finding the derivative are purposely omitted, with the exception of the formula for finding the derivative of a quadratic. If students spend time working with the concept of the derivative first, then the shortcut formulas will be more meaningful when studied in calculus.

This chapter has received the highest ratings by pilot teachers of any chapter in the book, so we encourage you not to skip it. There are two major reaosns for the high rating. The approach taken is intuitive rather than formal, and thus your students can concentrate on the major ideas. The material is relatively easy for most students, so it takes some of the potential fear and mystery out of calculus.

The evidence is that students who have been introduced to the ideas of calculus, rather than just the techniques, do better in a calculus course. There is an analogy with algebra; if a student is to be successful in algebra, it is best for that student to have experience with the ideas of algebra before learning manipulative techniques.

The first three lessons lead to the study of the first derivative. In Lesson 9-1, rates of change are reviewed and related to the difference quotient in Chapter 2. These ideas are extended in Lesson 9-2 to finding the limit of the difference quotient as $\Delta x \to 0$. The second lesson also introduces the term *derivative*, and all derivatives are computed only at a particular point. Lesson 9-3 culminates the study of the first derivative by treating it as a function.

Lesson 9-4 introduces the second derivative, using the concepts of acceleration and deceleration. Again, there is an emphasis on rate of change, but this time it is on the rate of change of a rate.

Lesson 9-5 discusses the use of derivatives to find maximum and minimum points of graphs. Some of the maximum and minimum problems that students approximated earlier by using an automatic grapher can now be solved exactly by using the derivative. The derivative also provides sophisticated techniques for proving properties that students studied previously.

The final lesson of the chapter considers derivatives of exponential functions, which are important functions in the solution of differential equations. Arguments are given to demonstrate that the values of the derivative of an exponential function are proportional to the values of the function itself.

If you have access to software such as *The Math Exploration Toolkit* or the *Calculus Toolkit*, you may want to use it frequently throughout this chapter. In class activities, using some form of large scale monitor or overhead projector, you can graph a number of secant lines and give students a visual interpretation of the mathematical ideas. If you do not have either software package, you may still be able to accomplish the same results provided your graphing software can graph a function and several lines on the same coordinate system.

PERSPECTIVES ■ CHAPTER 9

The Perspectives provide the rationale for the inclusion of topics or approaches, provide mathematical background, and make connections with other lessons and within UCSMP.

9-1

DIFFERENCE QUOTIENTS AND RATES OF CHANGE

Throughout this lesson, there is an emphasis on helping students connect *rate of change* to *slope* and to the *difference quotient*. The lesson opens with a discrete function that maps the number of a day of a year to its length at 50° N latitude. The utility of this situation is that everyone knows that the length of a day changes, getting longer in summer and shorter in winter (in the Northern Hemisphere); however, most people are unaware of how fast the length of a day changes. Geometrically, the average rate of change in the length of a day from one time of a year to another is the slope of the line containing these points of the function.

Another example considered in this lesson is the classic one of the height of a projectile over time. The rate of change in this case is the average velocity of the projectile. This example is used to introduce the difference quotient because it involves a continuous function, and therefore the denominator of the quotient can be made as small as one wishes.

9-2

THE DERIVATIVE AT A POINT

The projectile example of the previous lesson is examined in some detail in this lesson. Now, the limit of the difference quotient is taken as the denominator approaches 0.

What was the average velocity becomes (in the limit) the instantaneous velocity. What was the slope of a secant line to the graph of the function now becomes (in the limit) the slope of the line tangent to the graph at a particular point. That slope is the derivative of the function at the point.

A variety of examples of the derivative are given. Using a graph of the distance a car travels, from a starting point over a period of time, students are asked to draw tangents to estimate the slope. From a formula for a quadratic function, the derivative can be calculated by using its definition. Students also are asked to compute the instantaneous rate of change of the volume of a sphere in relation to its changing radius.

9-3

THE DERIVATIVE FUNCTION

In the previous lesson, students computed the derivative at a particular point. This lesson expands on that idea by computing the derivative at every point of the function. This step leads to the consideration of the derivative as a function itself.

Examples from the previous lesson are extended in this lesson. The derivative of the projectile height function (its velocity function) is shown to be a linear function. This is generalized to obtain the only formula of the chapter: the derivative of a quadratic function. The derivative of the volume function for a sphere is found to be its surface area function.

9-4

ACCELERATION AND DECELERATION

The goal of this lesson is to discuss the second derivative, which is the derivative of a derivative. Again, the first example is discrete, this time using population growth. The rate of change of the rate of change of a population yields an acceleration or deceleration of population growth.

The continuous example discussed is again the projectile height function. Now there can be instantaneous acceleration, and for the projectile function the instantaneous acceleration is shown to be constant and equal to the acceleration due to gravity.

9-5

USING DERIVATIVES TO ANALYZE GRAPHS

This lesson brings together and extends the geometric ideas about the derivative from earlier lessons and also the definitions of increasing and decreasing functions from Lesson 2-3. The key ideas are not proved but presented visually, using what is known about derivatives as slopes of tangents to curves. These key ideas are that a function whose derivative exists on an interval is (1) increasing on that interval if the derivative is positive, (2) decreasing on that interval if the derivate is negative, and (3) either has a relative minimum or maximum or "flattens out" where the derivative is zero.

From the ideas above, the coordinates of the vertex of the graph of a quadratic function can be deduced. This is a result students have seen before, but this time it completes nicely the projectile example that has been used throughout the chapter.

9-6

DERIVATIVES OF EXPONENTIAL FUNCTIONS

The informal approach taken in this chapter has allowed a variety of functions to be considered, including sine, log, and some polynomial functions. Perhaps surprising is the fact that the derivates of these familiar functions also are familiar functions. Now, in this lesson, the functions f of the form $f(x) = ab^x$ are considered and shown to have derivative f', where $f'(x) = (\ln b)f(x) = a(\ln b)b^x$. This means that the function f grows at a rate proportional to its values and that all of its derivatives are multiples of itself. Because f satisfies the differential equation $f' = kf$, exponential functions are fundamental in solving a variety of these kinds of equations.

DAILY PACING CHART ■ CHAPTER 9

Every chapter of UCSMP *Precalculus and Discrete Mathematics* includes lessons, a Progress Self-Test, and a Chapter Review. For optimal student performance, the self-test and review should be covered. (See *General Teaching Suggestions: Mastery* on page T35 of this Teacher's Edition.) By following the pace of the Full Course given here, students can complete the entire text by the end of the year. Students following the pace of the Minimal Course spend more time when there are quizzes and on the Chapter Review and will generally not complete all of the chapters in this text.

When chapters are covered in full (the recommendation of the authors), then students in the Minimal Course can cover 11 chapters of the book. For more information on pacing, see *General Teaching Suggestions: Pace* on page T34 of this Teacher's Edition.

DAY	MINIMAL COURSE	FULL COURSE
1	9-1	9-1
2	9-2	9-2
3	9-3	9-3
4	Quiz (TRF); Start 9-4.	Quiz (TRF); 9-4
5	Finish 9-4.	9-5
6	9-5	9-6
7	9-6	Progress Self-Test
8	Progress Self-Test	Chapter Review
9	Chapter Review	Chapter Test (TRF)
10	Chapter Review	Comprehensive Test (TRF)
11	Chapter Test (TRF)	
12	Comprehensive Test (TRF)	

TESTING OPTIONS

- Quiz for Lessons 9-1 through 9-3
- Chapter 9 Test, Form A
- Chapter 9 Test, Form B
- Chapter 9 Test, Cumulative Form
- Comprehensive Test, Chapters 1–9

A Quiz and Test Writer is available for generating additional questions, additional quizzes, or additional forms of the Chapter Test.

PROVIDING FOR INDIVIDUAL DIFFERENCES

The student text has been written for, and tested with, average students. It also has been successfully used with better and more poorly prepared students.

The Lesson Notes often include Error Analysis and Alternate Approach features to help you with those students who need more help. Students of all abilities often learn from their peers and may benefit from small group work referenced as appropriate throughout the Notes. A blackline Lesson Master (in the Teacher's Resource File), keyed to the chapter objectives, is provided for each lesson to allow more practice. (However, since it is important to keep up with the daily pace, you are not expected to use all of these masters. Again, refer to the suggestions for pacing on page T34.) Extension activities are provided in the Lesson Notes for those students who have completed the particular lesson in a shorter amount of time than is expected, even in the Full Course.

The Derivative in Calculus

CHAPTER 9

Assembly line procedures affect the cost of items produced.

Algebra can be described as the study of variables and their properties. Geometry can be described as the study of visual patterns. Calculus is often divided into two parts: differential calculus and integral calculus. This chapter concentrates on differential calculus, which can be briefly described as *the study of rates of change in functions*. The ideas in differential calculus help to determine whether functions are increasing or decreasing and how fast they are doing so.

Statements involving rates of change appear in a variety of places.

In physics: The rate of change of the temperature of a body with respect to time is proportional to the difference between the temperature of the body and that of its surroundings. (Newton's Law of Cooling)

In biology: Under ideal growth conditions, the rate of growth of a colony of bacteria is proportional to the number of bacteria at the time.

In business: As the number of items being produced increases, the cost of producing each item tends to decrease, but the rate of decrease diminishes.

In the news: "The decrease in the unemployment rate for June was the smallest decrease this year."

Calculus is so important that it is required of all college students majoring in the sciences, engineering, or economics, and in many business schools. Differential calculus provides the mathematical tool to describe rates of change precisely and compute them efficiently. That tool is the *derivative*. This chapter focuses on the meaning, interpretation, and significance of the derivative. In a later chapter, you will study some ideas of integral calculus.

CHAPTER 9

We recommend 10–12 days for this chapter: 6 to 7 days for the lessons and quiz; 1 day for the Progress Self-Test; 1 or 2 days for the Chapter Review; and 1 day for a Chapter test. There is also a Comprehensive Test, which can serve as a quarterly exam. (See the Daily Pacing Chart on page 537D.)

USING PAGE 537
Four statements involving rates of change are given. Newton's Law is related to the derivative. The biological law relates to the derivative of an exponential function, discussed in Lesson 9-6. The last two laws deal with rates of rates, namely accelerations and decelerations, which are discussed in Lesson 9-4.

You might ask students what they know about calculus. If they seem fearful, point out that most things people do not know seem difficult, but once learned, they seem much easier.

LESSON 9-1

RESOURCES
- Lesson Master 9-1
- Teaching Aid 59 displays the table and graph for the length-of-day example.
- Teaching Aid 60 displays the difference quotient.
- Teaching Aid 61 can be used with **Questions 5-9 and 21**.

OBJECTIVES

A Compute average rates of change in functions.
D Find rates of change in real situations.
G Relate average rate of change to secant lines of graphs of functions.

TEACHING NOTES

When the same idea has different terms associated with it, the idea usually arose in a variety of different contexts. The major idea of this lesson has three names associated with it: (1) average rate of change, (2) slope, and (3) difference quotient. *Average rate of change* involves an application of the idea. Average velocity is just a special case. *Slope* relates to a graph. When the function is continuous, the use of the term slope refers to the slope of a secant line to a curve. *Difference quotient* refers to the algebraic way of expressing the idea that is useful in calculus.

Be sure that students include appropriate units of measurement for the numerical value of the average rate of change and for the slope.

LESSON

9-1

Difference Quotients and Rates of Change

Photo: Bill Brooks/Masterfile   Photo: Brian Sytnyk/Masterfile

Winnipeg Legislative Building (spring and winter)

This lesson reviews some ideas about rate of change and introduces terms and notation that are needed to make them precise. Consider the following example of rate of change familiar to everyone except those who live on the equator.

All through the year the length of a day from sunrise to sunset changes. Here is a table of lengths of days in 1989 for places at 50° north latitude, the latitude of Vancouver and Winnipeg in Canada, Frankfurt in West Germany, and Prague, the capital of Czechoslovakia. Days are numbered beginning with January 1, 1989.

Length of day (in minutes) on first day of month at 50° north latitude

Date	Jan 1	Feb 1	Mar 1	Apr 1	May 1	Jun 1	Jul 1	Aug 1	Sep 1	Oct 1	Nov 1	Dec 1
Day of year	1	32	60	91	121	152	182	213	244	274	305	335
Length	490	560	658	775	882	964	977	914	809	699	583	504

Let $f(x)$ be the length of day x. Then f is periodic with period one year, because each year the lengths of days are about the same. The graph of f has much the same appearance as a sine curve.

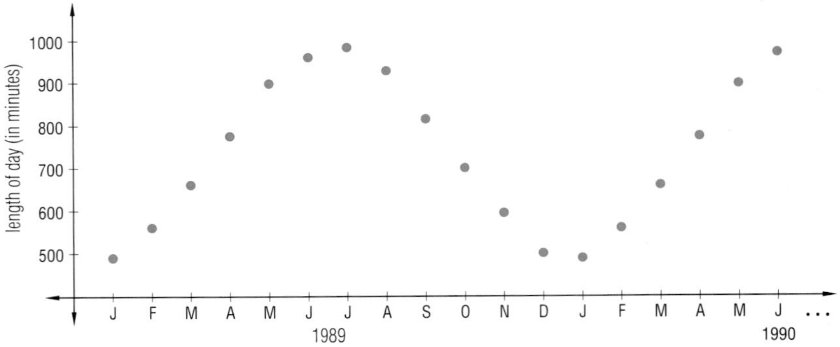

From the graph, you can see that the length of a day does not change evenly. In March and April the length is increasing rapidly, but in June the length is increasing slowly to its peak. In September and October the length is decreasing rapidly, but the decrease slows in December.

To discuss these changes, some new notation is useful. Change is found by subtraction; the change in days from day x_1 to day x_2 is $x_2 - x_1$. This difference is indicated as $\boldsymbol{\Delta x}$, read "delta x." (Δ is the upper case Greek letter Delta, which corresponds to D, the first letter in *Difference*.) The change in length of day from $f(x_1)$ to $f(x_2)$ is indicated by $\boldsymbol{\Delta y}$, read "delta y." That is, $\Delta y = f(x_2) - f(x_1) = y_2 - y_1$. For instance, from April 1st (day 91) to May 1st (day 121), $\Delta x = 121 - 91 = 30$ days, and $\Delta y = 882 - 775 = 107$ minutes.

The *average rate of change* from x_1 to x_2 is found by dividing Δy by Δx. To find how fast the length of day has been changing from April 1st to May 1st, divide $\Delta y = 107$ minutes by $\Delta x = 30$ days to get

$$\frac{\Delta y}{\Delta x} = \frac{107 \text{ minutes}}{30 \text{ day}} \approx 3.6 \frac{\text{minutes}}{\text{day}}.$$

This means that on the average in April, each day is about 3.6 minutes longer than the previous day. You should recognize that the average rate of change is the slope of the line through (91, 775) and (121, 882). In fact, if $y = f(x)$, then

$$\frac{\Delta y}{\Delta x} = \frac{y_2 - y_1}{x_2 - x_1} = \frac{f(x_2) - f(x_1)}{x_2 - x_1}.$$

Definition

Let f be a function defined at x_1 and x_2, with $x_1 \neq x_2$. The **average rate of change** of f from x_1 to x_2 is the slope of the line through $(x_1, f(x_1))$ and $(x_2, f(x_2))$.

The length-of-day function is a discrete function, but the definition applies to all real functions. The figure below shows the graph of a continuous function f, along with two distinct points $P_1 = (x_1, f(x_1))$ and $P_2 = (x_2, f(x_2))$ on its graph. Recall that in geometry a **secant line to a circle** is a line that intersects the circle at two distinct points. Similarly, when a function is continuous, a line passing through two distinct points on the graph of the function is called a **secant line for the graph of the function.** Thus the average rate of change of the function f from x_1 to x_2 equals the slope m_{sec} of the secant line through P_1 and P_2.

$$m_{\text{sec}} = \frac{f(x_2) - f(x_1)}{x_2 - x_1} = \frac{\Delta y}{\Delta x}$$

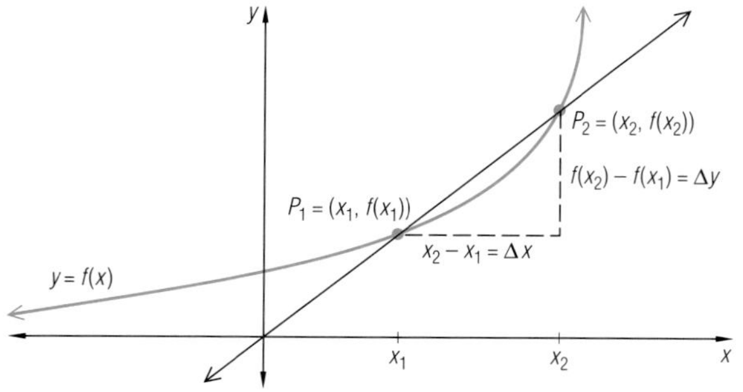

Attention to the language and notation in this lesson will help students throughout the chapter.

Go through the substitutions that transform the familiar formula $\frac{y_2 - y_1}{x_2 - x_1}$ into the unfamiliar formula $\frac{f(x_1 + \Delta x) - f(x_1)}{\Delta x}$. First substitute $f(x)$ for y in each case, then let $x_2 = x_1 + \Delta x$.

Example 1 relates velocity to the average rate of change of directed distance with respect to time. This connection is important because Lesson 9-2 begins with a discussion of instantaneous velocity. Help students see that the units for rate of change, in this case ft/sec, are appropriate units for velocity.

Relate the calculated velocities to their respective secant lines. The average velocity of 0 from $t = 20$ to $t = 30$ bears special mention. Point out that the graph is of time versus height, not the path of the projectile. Its path is vertical.

Error Analysis In problems such as **Example 1**, some students make errors evaluating $\dfrac{h(t_2) - h(t_1)}{t_2 - t_1}$. Students should follow the textbook's method of explicitly calculating $h(t_2)$ and $h(t_1)$ and then substituting these values into the expression for slope.

Example 2 reviews the manipulative skills needed to find the derivative. In this example, students evaluate the difference quotient over the interval t to $t + \Delta t$. In the next lesson, we use this same computation to find the limit as $\Delta t \to 0$. Show students that the formula derived in Example 2 can be used to evaluate the average rate of change when t and Δt are known. This is done in **Example 3** in preparation for Lesson 7-3, where the derivative is treated as a function.

Error Analysis In problems such as **Example 2**, many students evaluate expressions of the form $\dfrac{h(t + \Delta t) - h(t)}{\Delta t}$ incorrectly. Ask students what simpler expressions are needed to evaluate the given expression. They should respond

Example 1 A projectile is propelled into the air from ground level with an initial velocity of 800 ft/sec. If only the effect of the Earth's gravity is considered, its height (in feet) after t seconds is given by the function $h(t) = 800t - 16t^2$. This function is graphed below for $0 \le t \le 40$. Several secant lines are also drawn. Find the average rate of change of height with respect to time over the following intervals.

a. $10 \le t \le 20$ **b.** $20 \le t \le 30$ **c.** $30 \le t \le 40$

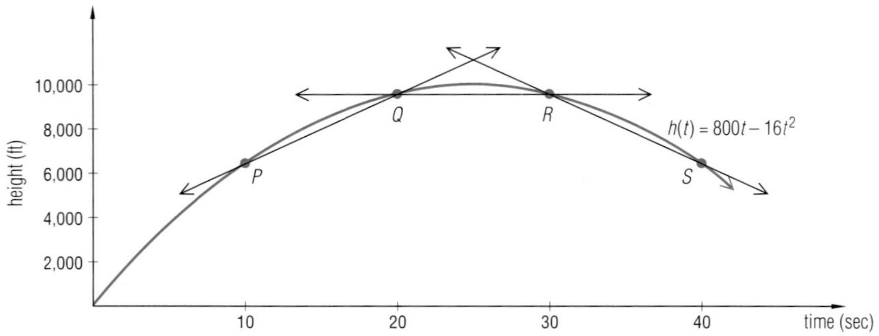

Solution In this example, the function is called h (for height) instead of f and the independent variable is called t (for time) instead of x. The average rate of change of height with respect to time over the interval $t_1 \le t \le t_2$ is $\dfrac{h(t_2) - h(t_1)}{t_2 - t_1}$.

a. At $t_1 = 10$, $h(t_1) = 800(10) - 16(10)^2 = 6400$.
At $t_2 = 20$, $h(t_2) = 800(20) - 16(20)^2 = 9600$.

Thus, the average rate of change is

slope of $\overleftrightarrow{PQ} = \dfrac{h(t_2) - h(t_1)}{t_2 - t_1} = \dfrac{9600 - 6400}{20 - 10} = 320$ ft/sec.

b. At $t_1 = 20$, $h(t_1) = 9600$, and
at $t_2 = 30$, $h(t_2) = 800(30) - 16(30)^2 = 9600$.

Thus, the average rate of change is

slope of $\overleftrightarrow{QR} = \dfrac{9600 - 9600}{30 - 20} = 0$ ft/sec.

Since $\overleftrightarrow{QR}$ has slope 0, $\overleftrightarrow{QR}$ is horizontal.

c. At $t_1 = 30$, $h(t_1) = 9600$, and at $t_2 = 40$, $h(t_2) = 6400$.

Thus, the average rate of change is

slope of $\overleftrightarrow{RS} = \dfrac{6400 - 9600}{40 - 30} = -320$ ft/sec.

The average rate of change of an object's directed distance (or the position of an object on a line) from a fixed point over a time interval is called its **average velocity** over that interval. The results of Example 1 thus calculate average velocities of the projectile. Comparing parts **a** and **c** shows that the projectile travels the same distance during both 10-second intervals: 3200 feet. However, from $t = 10$ to $t = 20$ it travels in a positive direction (up), so its average velocity is positive (320 ft/sec). From $t = 30$ to $t = 40$ it travels in a negative direction (down), so its average velocity is negative (-320 ft/sec).

540

What does the average velocity of zero in part **b** mean? It does not mean that the projectile is stopped in midair, like a cartoon character, for 10 seconds. It simply means that the projectile's *average* velocity from $t = 20$ to $t = 30$ is 0 because its height is the same at both times. In actuality, in this time interval the projectile first moves up and then back down.

The concept of average rate of change is an important one in mathematics and a general formula is useful. One such formula is

$$\begin{pmatrix}\text{the average rate of change}\\ \text{in } f \text{ from } x_1 \text{ to } x_2\end{pmatrix} = \frac{f(x_2) - f(x_1)}{x_2 - x_1}.$$

This formula can be rewritten in terms of f, x_1, and Δx.

Since $x_2 - x_1 = \Delta x$, then $x_2 = x_1 + \Delta x$ and so $f(x_2) = f(x_1 + \Delta x)$.

The denominator is Δx. By substitution,

$$\begin{pmatrix}\text{the average rate of change}\\ \text{in } f \text{ from } x_1 \text{ to } x_2\end{pmatrix} = \frac{\Delta y}{\Delta x} = \frac{f(x_1 + \Delta x) - f(x_1)}{\Delta x}.$$

The quantity $\frac{f(x_1 + \Delta x) - f(x_1)}{\Delta x}$ is called the **difference quotient** of f over the interval from x_1 to $(x_1 + \Delta x)$.

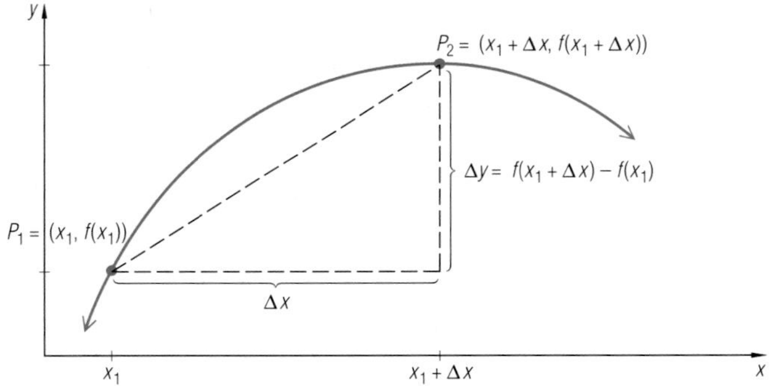

The difference quotient has an interpretation in the diagram above. In traveling from P_1 to P_2, the x-coordinate changes by Δx and the y-coordinate by $\Delta y = f(x_1 + \Delta x) - f(x_1)$. Thus when $x_2 = x_1 + \Delta x$ and $y = f(x)$,

$$\begin{pmatrix}\text{average rate of change}\\ \text{of } f \text{ from } x_1 \text{ to } x_2\end{pmatrix} = \begin{pmatrix}\text{slope of line through}\\ (x_1, y_1) \text{ and } (x_2, y_2)\end{pmatrix} = \begin{pmatrix}\text{difference quotient}\\ \text{of } f \text{ from } x_1 \text{ to } x_2\end{pmatrix}.$$

$$\frac{\Delta y}{\Delta x} = \frac{y_2 - y_1}{x_2 - x_1} = \frac{f(x_2) - f(x_1)}{x_2 - x_1} = \frac{f(x_1 + \Delta x) - f(x_1)}{\Delta x}$$

Furthermore, if y is the directed distance of an object at time x, then these quotients give the average velocity of the object over the interval from x_1 to x_2.

with $h(t + \Delta t)$ and $h(t)$. On the chalkboard, write $h(t) = 800t - 16t^2$ and circle it. Then write $h(t + \Delta t) = 800(t + \Delta t) - 16(t + \Delta t)^2$, reminding students that each t in $h(t)$ is replaced by $t + \Delta t$ for $h(t + \Delta t)$. Expand and simplify the expression for $h(t + \Delta t)$ and circle it with a different color chalk. Next write $\frac{h(t + \Delta t) - h(t)}{\Delta t}$ and, using different colors of chalk for the two components of the numerator, replace $h(t + \Delta f)$ and $h(t)$ by their circled expansions. Then proceed to simplify. Invite students to evaluate $\frac{h(t + \Delta t) - h(t)}{\Delta t}$ for another $h(t)$. Remind students that the denominator Δt comes from $(t + \Delta t) - t$.

Point out to students that the length-of-day example is discrete, so Δx cannot be smaller than 1. But in **Examples 2 and 3**, the function is continuous, so Δx can be as small as we wish. These ideas will help prepare students for Lesson 9-2.

ADDITIONAL EXAMPLES

1. Farmer MacDonald tries to scare away some chicken hawks that are circling his henhouse by pointing his rifle straight up and shooting. A bullet leaves the muzzle of his rifle at a velocity of 960 feet per second. Find the average rate of change of the height of the bullet with respect to time over the given intervals. Assume the height at time t is given by $h(t) = 960t - 16t^2$.

a. $10 \le t \le 20$
480 ft/sec
b. $20 \le t \le 30$
160 ft/sec
c. $30 \le t \le 40$
-160 ft/sec
d. $25 \le t \le 35$
0 ft/sec

ADDITIONAL EXAMPLES

2. Refer to the function h with $h(t) = 960t - 16t^2$ in Additional Example 1. Find a formula for the difference quotient given the average rate of change of h for each interval t to $t + \Delta t$.

$\mathbf{\dfrac{h(t + \Delta t) - h(t)}{\Delta t} =}$
$960 - 32t - 16\Delta t$

3. Use the formula from Additional Example 2 to find the projectile's average velocity from t to $t + \Delta t$ for $t =$ 5 and

a. $\Delta t = 1$;
784 ft/sec

b. $\Delta t = .5$;
792 ft/sec

c. $\Delta t = .1$.
798.4 ft/sec

NOTES ON QUESTIONS

Questions 5-9: As an extension, **small groups** of students may enjoy the following activity. Draw a graph (as below) and ask students to make up a story of an automobile trip consistent with the graph.

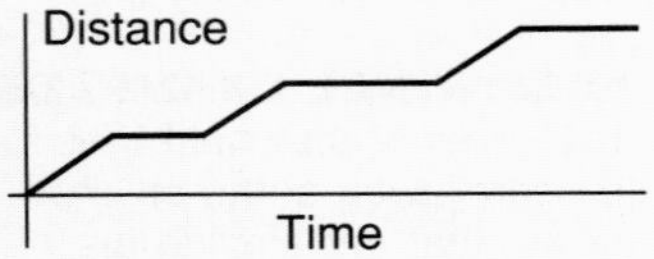

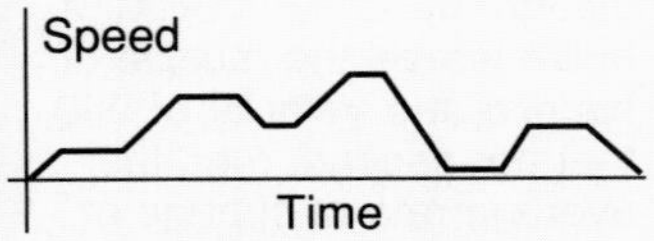

Ask students to interpret positive, negative, and zero slopes. Given one of these graphs, ask students to sketch the other.

Example 2 Refer to the function h defined in Example 1. Find a formula for the difference quotient giving the average rate of change of h for each interval t to $t + \Delta t$.

Solution In this case t plays the role of x_1, and Δt the role of Δx in the difference quotient. The computation is tedious but straightforward.

$$\frac{h(t + \Delta t) - h(t)}{\Delta t)} = \frac{[800(t + \Delta t) - 16(t + \Delta t)^2] - [800t - 16t^2]}{\Delta t}$$

$$= \frac{[800t + 800\Delta t - 16(t^2 + 2t\Delta t + (\Delta t)^2)] - [800t - 16t^2]}{\Delta t}$$

$$= \frac{800t + 800\Delta t - 16t^2 - 32t\Delta t - 16(\Delta t)^2 - 800t + 16t^2}{\Delta t}$$

$$= \frac{800(\Delta t) - 32t(\Delta t) - 16(\Delta t)^2}{\Delta t}$$

$$= 800 - 32t - 16\Delta t$$

In calculating average rates of change, continuous functions differ from discrete functions in the following fundamental way: for any continuous function, the value of Δx (or, in the case of Example 2, Δt) can be made as close to zero as you wish.

Example 3 Use the formula from Example 2 to find the projectile's average velocity from t to $t + \Delta t$ for $t = 5$ and

a. $\Delta t = 1$ (that is, from 5 to 6 seconds);

b. $\Delta t = .5$ (that is, from 5 to 5.5 seconds);

c. $\Delta t = -.1$ (that is, from 4.9 to 5 seconds).

Solution From Example 2, the average velocity equals the difference quotient, $800 - 32t - 16\Delta t$.

a. For $t = 5$ and $\Delta t = 1$, the average velocity is $800 - 32(5) - 16(1) = 624$ ft/sec.

b. For $t = 5$ and $\Delta t = .5$, the average velocity is $800 - 32(5) - 16(.5) = 632$ ft/sec.

c. For $t = 5$ and $\Delta t = -.1$, the average velocity is $800 - 32(5) - 16(-.1) = 641.6$ ft/sec.

In Example 3, you can see that the average velocity seems to be getting closer and closer to 640 ft/sec as Δt gets closer and closer to 0. This limit idea is studied further in the next lesson.

Questions

Covering the Reading

In 1 and 2, consider the length-of-day data given at the beginning of this lesson.

1. Find the average rate of change in the length of day from August 1st to September 1st. ≈ -3.39 minutes/day

2. From February 1st to March 1st, on the average each day is _?_ minutes _?_ than the previous day. 3.5; longer

3. Finish this sentence. The slope of the secant line through $(x_1, f(x_1))$ and $(x_1 + \Delta x,$ _?_ $)$ equals the average rate of change of f from _?_ to _?_. $f(x_1 + \Delta x)$; x_1; $x_1 + \Delta x$

4. If $y = f(x)$, give three expressions equal to $\frac{\Delta y}{\Delta x}$. $\frac{y_2 - y_1}{x_2 - x_1}, \frac{f(x_2) - f(x_1)}{x_2 - x_1}, \frac{f(x_1 + \Delta x) - f(x_1)}{\Delta x}$

In 5–9, refer to the function graphed at the right.

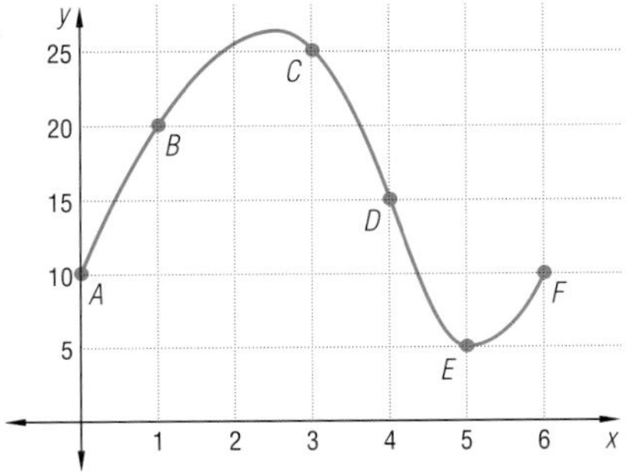

5. In going from B to C, find Δx and Δy. $\Delta x = 2$, $\Delta y = 5$

6. Between which two named points is $\Delta y = -10$? C to D or D to E

7. Find the average rate of change of the function from D to F. ≈ -2.5

8. Find the average rate of change of the function over the interval $0 \le x \le 3$. 5

9. Between which two named points is the average rate of change of the function zero? A and F

10. Find the average velocity of the projectile in Example 1 over the indicated time interval. See margin.
a. from 10 to 11 **b.** from 10 to 10.3 **c.** from 9.99 to 10

11. A rocket is propelled vertically into the air from a height of 20 ft with an initial velocity of 480 ft/sec. If only the effect of gravity is considered, then its height (in feet) after t seconds is given by the equation $h(t) = 480t - 16t^2 + 20$.
a. Sketch a graph of height vs. time. See margin.
b. Find the average velocity from $t = 4$ to $t = 12$. Include appropriate units with your answer. 224 ft/sec
c. Find a formula for the average velocity from $t = 4$ to $t = 4 + \Delta t$. $352 - 16\Delta t$ ft/sec

Question 16: This is an important question to discuss in class. Students need to relate the various statements in part **b** to the different parts of the difference quotient. For those statements that could be either true or false, have students give you a particular situation that makes the statement true and a particular situation that makes the statement false. Of course, the situations must be in agreement with the data given in the problem.

Question 17: Use this problem to introduce Lesson 9-2.

Question 22: Every constant in the formula should be explained. The 250 is half the difference (in minutes) between the shortest and longest days. The 733 is the average length of a day. (It is slightly more than 12 hours because the sun has width.) The $\frac{2\pi}{365}$ is the period. The 80 is the phase shift to get the function to have its maximum on June 21st. The data for part **b** can be found in some almanacs, which contain the information for latitudes that are multiples of 10°.

ADDITIONAL ANSWERS
10.a. 464 ft/sec
b. 475.2 ft/sec
c. 480.16 ft/sec

11.a.

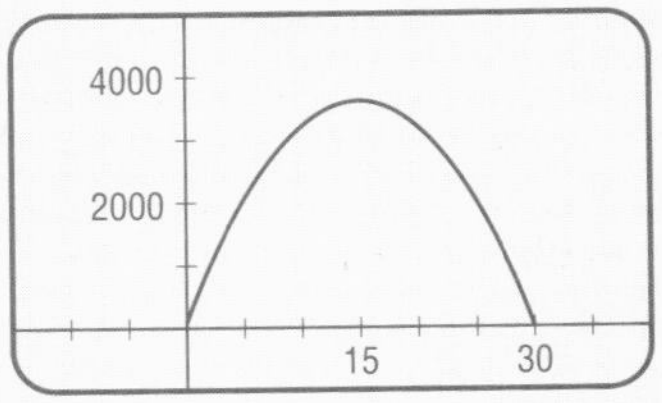

ADDITIONAL ANSWERS

14.b. The slope of the line through any two points on the graph of *f* is always $\frac{1}{2}$.

15.a.

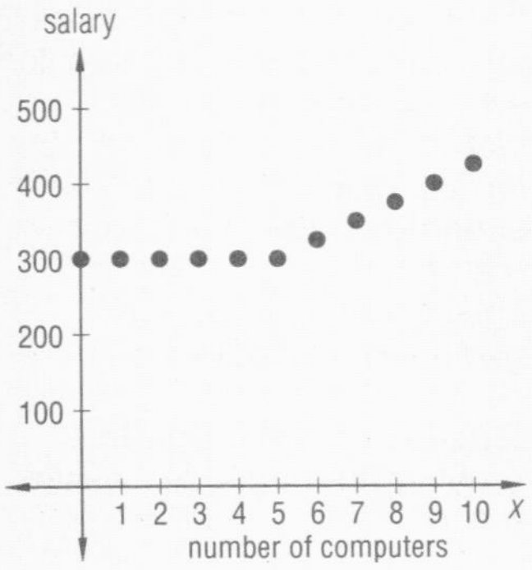

n	S
0	300
1	300
2	300
3	300
4	300
5	300
6	325
7	350
8	375
9	400
10	425

Applying the Mathematics

12. Suppose P and Q are two points on the curve $y = x^2 - x + 2$. If the x-coordinate of P is 1 and the x-coordinate of Q is 4, find the slope of the secant line $\overleftrightarrow{PQ}$. 4

13. Determine the average rate of change of the function f defined by $f(x) = x^2$ over the interval $-1 \le x \le 2$. 1

14. a. Calculate the difference quotient, $\frac{f(x + \Delta x) - f(x)}{\Delta x}$, for the function f defined by $f(x) = \frac{1}{2}x - 5$. $\frac{1}{2}$

b. Give a geometric interpretation for your answer to part **a**. See margin.

15. Tina has a job selling portable computers. Her weekly salary is $300 plus a $25 commission for each computer that she sells after the first five.

a. Construct a table of values and a graph with number of computers sold as the independent variable and salary as the dependent variable. Assume Tina sells at most 10 computers. See margin.

b. If the number of computers she sells increases from 5 to 8, what is her average salary increase per computer sold? Include appropriate units with your answer. $25 /computer

c. If the number of computers she sells increases from 3 to 7, what is her average salary increase per computer sold? Include appropriate units with your answer. $12.50 /computer

16. A meteorologist records temperatures at half-hour intervals from the beginning to the end of the shift. At the end of the shift, the meteorologist reports that the average rate of change of temperature with respect to time over the entire shift is .4° F/hr, and that the temperature over the same time period increased by 3° F.

a. If the shift began at 6:00 A.M., at what time did it end?

b. With regard to the above situation, determine whether each of the following statements must be *true*, must be *false*, or could be either *true* or *false*.

i. Over the entire shift, the recorded temperatures steadily increased.

ii. A linear model with a slope of .4 is an accurate model for the data.

iii. The temperature never decreased during the shift.

iv. If a continuous curve is drawn through the data points, then the slope of the secant line determined by the temperatures at the beginning and end of the shift will be positive.

a) 1:30 P.M.; b) i, ii, iii could be true or false; iv is true.

17. In Example 3, the average velocity $800 - 32(5) - 16(\Delta t) = 640 - 16(\Delta t)$ was computed for smaller and smaller values of Δt.

a. Find the value of $640 - 16(\Delta t)$ when $\Delta t = .001$. 639.984 ft/sec

b. What is $\lim_{\Delta t \to 0} (640 - 16\Delta t)$? 640 ft/sec

c. If t is fixed, what is $\lim_{\Delta t \to 0} (800 - 32t - 16\Delta t)$? $800 - 32t$ ft/sec

Review

18. Solve $\sin x > \frac{1}{2}$ over the interval $0 \le x \le \frac{\pi}{2}$. *(Lesson 6-7)* $\frac{\pi}{6} < x \le \frac{\pi}{2}$

19. Assume $x \neq 0$, $x \neq \frac{1}{3}$, and $x \neq -\frac{1}{3}$. Write the following expression as a simple fraction. *(Lesson 5-1)*

$$\frac{\frac{3}{x} - \frac{1}{3}}{\frac{1}{x^2} - 9} \quad \frac{9x - x^2}{3 - 27x^2}$$

20. *Multiple choice.* If $\sqrt{3}$ is a zero of a polynomial $p(x)$, then
(a) $\sqrt{3}$ is a factor of $p(x)$.
(b) $x + \sqrt{3}$ is a factor of $p(x)$.
(c) $x - \sqrt{3}$ is a factor of $p(x)$.
(d) $p(-\sqrt{3}) = 0$. (c)
(Lesson 4-6)

21. Refer to the graph of the function f with $y = f(x) = x^2 - 9$ at the right.
a. For what values of x is $y > 0$?
b. For what values of x is $y < 0$?
c. What are the zeros of the function?
d. What is the range of the function?
(Lessons 3-7, 3-1, 2-2, 2-1)
a) $x > 3$ or $x < -3$; b) $-3 < x < 3$;
c) -3, 3; d) $\{y: y \geq -9\}$

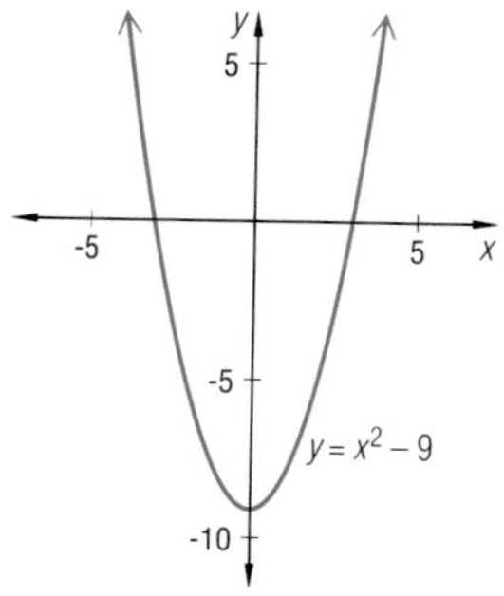

Exploration

22. The values of the length-of-day function in this lesson can be approximated by using the formula

$$f(x) = 250 \sin\left(\frac{2\pi}{365}(x - 80)\right) + 733, \text{ where } x = 1 \text{ for January 1.}$$

a. How close to the data given is the value given by $f(x)$, for
i. January 1st? **ii.** November 1st?
b. Use this formula to approximate the lengths of the longest and shortest days of the year at latitude 50° N.
c. Find data for the lengths of days where you live, and search for a sine function that approximates these values.

a i) $f(1) \approx 488.5$; that is very close to the actual value of 490.
a ii) $f(305) \approx 566$; that is 17 minutes less than the actual value, or within 3% of the actual value.
b) For the shortest day, Dec. 21, $x = 355$, and for the longest day, June 21, $x = 172$. $f(355) \approx 483$ minutes and $f(172) \approx 983$ minutes.
c) Answers will vary depending on students' latitude.

FOLLOW-UP

MORE PRACTICE

For more questions on SPUR Objectives, use *Lesson Master 9-1,* shown on below.

EXTENSION

The opening example on the length of a day may surprise students in that the data fit a sine curve. You might have students call the weather bureau to get even more detailed data for an entire year. A large plot could be made on paper for a bulletin board. As a project, students might compare these data with those for a city near the Arctic Circle and a city near the equator, giving the information in tabular and graphical form, and estimating the length of day by a formula such as that found in **Question 22**.

PROJECTS

The projects for Chapter 9 are described on page 577. **Project 3** is related to the content of this lesson.

NAME ____________

LESSON **MASTER 9–1**
QUESTIONS ON **SPUR** OBJECTIVES

SKILLS *Objective A (See pages 581–584 for objectives.)*

1. Find the average rate of change in $f(x) = 2x^3 + 5x - 4$ from $x = -2$ to $x = 5$. **43**

2. Find the average rate of change in $h(x) = x^2 + 25$ over the interval $-3 \leq x \leq 3$. **0**

3. Let $g(x) = \frac{1}{2}t^2 + 6t$.
 a. Find the average rate of change in g from t to $t + \Delta t$. **$t + 6 + \frac{1}{2}\Delta t$**
 b. Use your answer to part **a** to find the average rate of change in g from 0 to 2. **7**

USES *Objective D*

4. A ball is thrown upward from a height of 4.5 feet with an initial velocity of 21 feet per second. If only the effect of gravity is considered, then its height (in feet) after t seconds is given by the equation $u(t) = -16t^2 + 21t + 4.5$.
 a. Find a formula for the average velocity from $t = \frac{1}{2}$ to $t = \frac{1}{2} + \Delta t$. **$5 - 16\Delta t$**
 b. Use the formula from part **a** to find the average velocity from $t = \frac{1}{2}$ to $t = 4\frac{1}{2}$. Include appropriate units with your answer. **-59 ft/sec**

REPRESENTATIONS *Objective G*

5. Refer to the graph of f at the right.

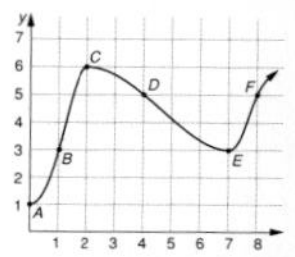

 a. Find the average rate of change in f from A to C. **$\frac{5}{2}$**
 b. Over what interval is the average rate of change in f zero? **$1 \leq x \leq 7$**
 c. Over what interval is the average rate of change $\frac{2}{3}$? **$4 \leq x \leq 7$**
 d. Find the average rate of change in f over the interval $0 \leq x \leq 8$. **$\frac{1}{2}$**

6. Suppose P and Q are points on the graph of the function g, with $P = (-4, b)$ and $Q = (6, 1)$. If the average rate of change in g from $x = -4$ to $x = 6$ is -2, find b. **$b = 21$**

LESSON 9-2

RESOURCES
- Lesson Master 9-2
- Teaching Aid 62 can be used with **Questions 3 and 6**.
- Teaching Aid 63 can be used with **Questions 10, 11, and 12**.
- Computer Master 16

OBJECTIVES

B Use the definition of derivative to compute derivatives.
D Find rates of change in real situations.
E Use derivatives to find velocity of a moving object.
H Estimate derivatives by finding slopes of tangent lines.

TEACHING NOTES

Recall the three key terms of the previous lesson and point out that each has a corresponding term in this lesson. Average rate of change becomes *instantaneous rate of change.* As a special case, average velocity becomes instantaneous velocity. The slope of the secant line to the graph of a continuous function becomes the *slope of the tangent line to that graph.* The difference quotient becomes the *derivative.*

You can begin the lesson by referring back to Example 3 in Lesson 9-1. Have students evaluate the average velocity from t to $t + \Delta t$ for $t = 5$ and then for successively smaller values of Δt, namely, $\Delta t = .01, .001, .0001$, and $.00001$. Or you can begin with Question

LESSON

9-2 The Derivative at a Point

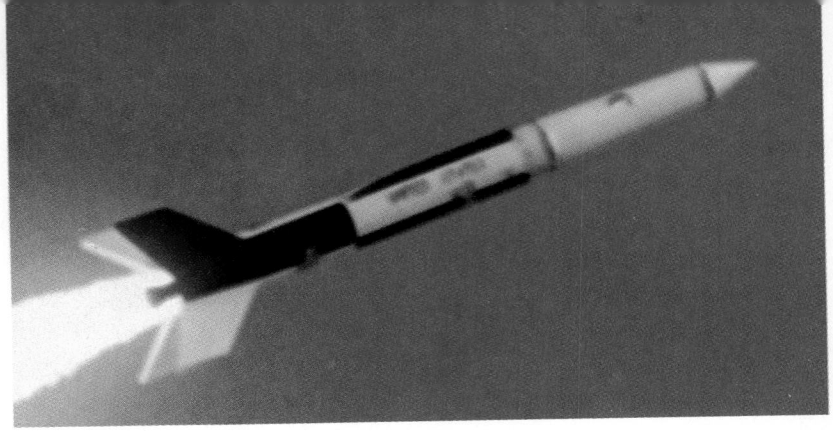

Derivatives can be used to calculate instantaneous velocities of projectiles, such as the prototype of the Mercury rocket used in the 1950s and 1960s.

Consider again the motion of the projectile in the examples of Lesson 9-1. After t seconds the projectile is $h(t) = 800t - 16t^2$ feet above ground level. It follows that between times $t = 0$ and $t = 40$ seconds, the projectile's average velocity is $\frac{h(40) - h(0)}{40 - 0} = 160$ ft/sec. But is this how fast the projectile is moving at each moment of its flight? No. At the moment of its launch it has its maximum velocity, and it climbs more and more slowly until its velocity is zero as it reaches its maximum height. Then it begins to fall back to Earth, dropping faster and faster, its velocity decreasing from zero through negative values, until it hits the Earth.

To determine the velocity of the projectile at a particular instant of time, say exactly 5 seconds after it is launched, you could approximate the answer by computing the projectile's average velocity over the interval from 5 to 6 seconds after launch. You could obtain a more accurate approximation by computing the average velocity over the interval from 5 to 5.5 seconds or even from 4.9 to 5 seconds after launch. In Example 3 of Lesson 9-1 these values were found to be 624 ft/sec, 632 ft/sec, and 641.6 ft/sec, respectively. You can imagine that this process of calculating average velocities over smaller and smaller time intervals could be continued forever. The numbers obtained would come closer and closer to the projectile's velocity at exactly 5 seconds after launch. For this reason, the term *instantaneous velocity* is defined to mean the limit approached by these numbers as the lengths of the time intervals approach zero.

Definition

Suppose an object is moving so that at each time t it is at position $f(t)$ along a line. Then,

$$\begin{pmatrix}\textbf{instantaneous velocity}\\ \text{of the object at time } t\end{pmatrix} = \lim_{\Delta t \to 0}\begin{pmatrix}\text{average velocity of the object}\\ \text{between times } t \text{ and } t + \Delta t\end{pmatrix}$$
$$= \lim_{\Delta t \to 0}\frac{f(t + \Delta t) - f(t)}{\Delta t},$$

provided this limit exists and is finite.

546

Example 1 For the projectile from Example 1 of Lesson 9-1, what is the instantaneous velocity at time $t = 5$ seconds?

Solution Use the definition of instantaneous velocity with $t = 5$.

$$\begin{pmatrix}\text{instantaneous velocity}\\ \text{at time } t = 5\end{pmatrix} = \lim_{\Delta t \to 0}\begin{pmatrix}\text{average velocity between}\\ \text{times } t = 5 \text{ and } t = 5 + \Delta t\end{pmatrix}$$
$$= \lim_{\Delta t \to 0} \frac{f(5 + \Delta t) - f(5)}{\Delta t}$$
$$= \lim_{\Delta t \to 0}(800 - 32 \cdot 5 - 16\Delta t) \quad \text{From Example 2 of Lesson 7-1}$$
$$= \lim_{\Delta t \to 0}(640 - 16\Delta t)$$
$$= 640 \text{ ft/sec}$$

Just as the idea of average velocity has a geometric interpretation in terms of slopes of secant lines, the idea of instantaneous velocity also has a geometric interpretation. Let $f(x)$ be the distance, from some reference point, of a moving object at time x seconds, and consider the graph of f. Take an interval of time from x seconds to $x + \Delta x$ seconds. You saw in Lesson 9-1 that

$$\begin{pmatrix}\text{average velocity from}\\ \text{time } x \text{ to time } x + \Delta x\end{pmatrix} = \frac{f(x + \Delta x) - f(x)}{\Delta x}$$
$$= \begin{pmatrix}\text{the slope of the secant line though the}\\ \text{points } (x, f(x)) \text{ and } (x + \Delta x, f(x + \Delta x))\end{pmatrix}.$$

Now imagine computing average velocities for smaller and smaller values of Δx. Each average velocity equals the slope of a secant line through $P = (x, f(x))$ and a second point $Q = (x + \Delta x, f(x + \Delta x))$ on the graph of f, where Q gets closer and closer to P. For instance, the drawing below shows a sequence of secant lines through P and points Q_1, Q_2, Q_3, and Q_4 that get closer and closer to P. If the graph of f is the kind of smooth curve shown in the drawing, then the secant lines will come closer and closer to a line that just skims the graph at the point P. This line is defined to be the **tangent line** to the graph at P. The slopes of the secant lines get closer and closer to the slope of this tangent line.

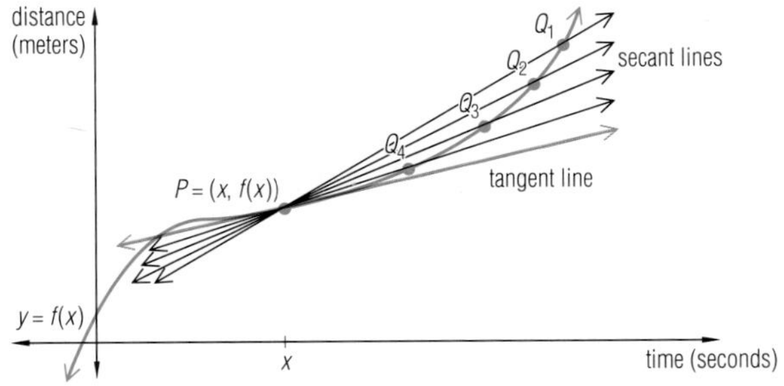

$$\begin{pmatrix}\text{the slope of the tangent}\\ \text{line at } (x, f(x))\end{pmatrix} = \lim_{\Delta x \to 0}\begin{pmatrix}\text{the slope of the secant line through}\\ (x, f(x)) \text{ and } (x + \Delta x, f(x + \Delta x))\end{pmatrix}$$
$$= \lim_{\Delta x \to 0} \frac{f(x + \Delta x) - f(x)}{\Delta x}$$
$$= \lim_{\Delta x \to 0}(\text{average velocity from time } x \text{ to } x + \Delta x)$$
$$= \text{instantaneous velocity at time } x$$

17 of Lesson 9-1. Such computations provide a concrete basis for the results obtained in **Example 1** of this lesson.

Making Connections You might ask students what kind of velocity (average or instantaneous) is displayed by a car's speedometer. (instantaneous) Compare this to the velocity calculated after completing a 200-mile car trip in 3 hours and 46 minutes. (average)

Technology If you have the appropriate software, graph the function f from **Example 1**. Then try to graph a line that is tangent to the curve when $t = 5$. Next graph several secant lines joining point P to other points Q on the curve. Make sure that you choose points Q that get closer and closer to point P. Students should see that the secant lines get closer and closer to the tangent line. The tangent line should have a slope that is very close to the value of the derivative at point P. The accuracy of the results depends on how accurately you estimated the tangent line.

In discussing the definition of derivative, point out that f' is called the derivative function because it is "derived from" function f. Mention that $f'(x)$ is read as "f prime of x." You may wish to mention that some calculus texts use the notation Df and others use $\frac{dy}{dx}$.

Students should understand the derivative in the following three ways: algebraically from its definition; geometrically as the slope of a tangent line; and in an applied way, as an instantaneous rate of change.

Example 1 emphasizes the application aspect of the derivative, while **Example 3** emphasizes calculation skills.

The visual relationship of the derivative to the slope of the tangent line is the idea of **Example 2**. Example 4 combines an application with calculation skills.

Some students may ask if the work of **Example 3** can be generalized to an instantaneous velocity formula, as was done in Example 2 of Lesson 9-1 for average velocity. The answer, of course, is yes. The derivative function is the topic of Lesson 9-3. These students may simply replace occurrences of 3 with x in the work of Example 3 and take the limit as $\Delta x \to 0$ as the final step.

ADDITIONAL EXAMPLES

1. In Additional Example 1 of Lesson 9-1, the average velocity of the bullet from time t to time $t + \Delta t$ was given by $\frac{h(t + \Delta t) - h(t)}{\Delta t} = 960 - 32t - 16\Delta t$. What is the instantaneous velocity at time $t = 30$ seconds?
0 ft/sec; the bullet must be at its peak then, going neither up nor down.

2. Repeat **Example 2** on page 548 when $t = 3$ hours.
A tangent to the graph at (3, 80) seems to go through (1, 50). Thus, the instantaneous velocity is about 15 mph.

3.a. Using $f(x) = x^2 + 4$ again, calculate $f'(-3)$, $f'(-2)$, $f'(-1)$, $f'(0)$, $f'(1)$, and $f'(2)$.

x	-3	-2	-1	0	1	2
f'(x)	-6	-4	-2	0	2	4

b. Make a conjecture about the function f'.
It seems that $f'(x) = 2x$. Also, for the values of x considered here, $f'(-x) = -f'(x)$, so f' may be an odd function.

So instantaneous velocity can be interpreted geometrically as the slope of a tangent line. Notice that the tangent line at P may intersect the curve at a second point different from P, as in the above graph, in which there is a second point of intersection to the left of P.

Example 2 The graph at the right shows the distance d (in miles) of a car from its starting point t hours after it begins a trip. Estimate the instantaneous velocity of the car at time $t = 1$ hour by sketching the tangent line to the graph at $t = 1$ and finding an approximate value for its slope.

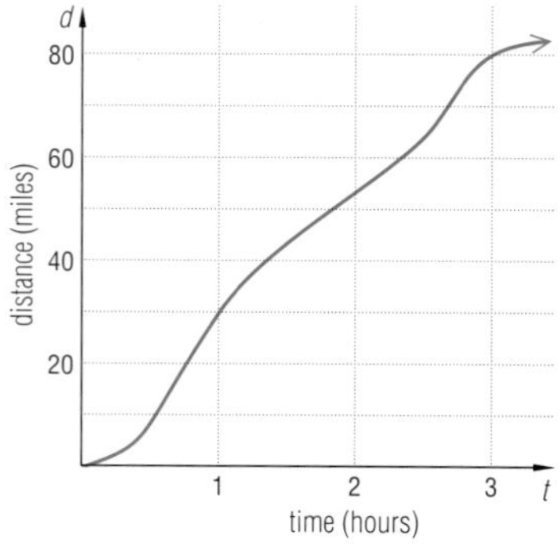

Solution From the graph, you can see that after one hour of travel, the car is 30 miles from its starting point. Sketch the tangent line at the point (1, 30) by drawing a line which skims the curve at that point. It appears that this line also passes through the point (2, 70). Therefore, the slope of the line is approximately

$$\frac{(70 - 30) \text{ miles}}{(2 - 1) \text{ hours}} = 40 \text{ mph}.$$

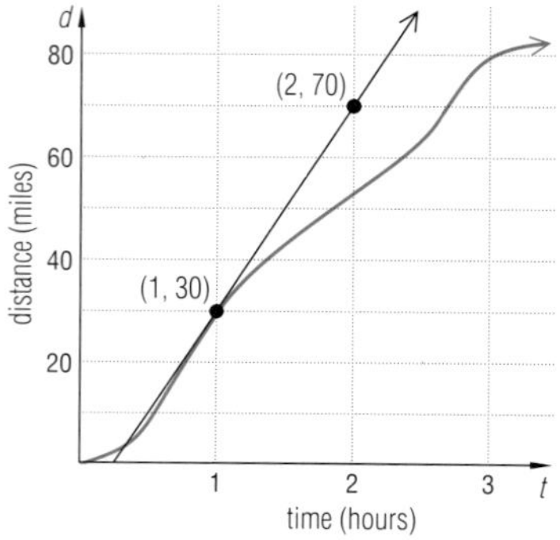

It follows that the instantaneous velocity of the car at time $t = 1$ hour is approximately 40 mph.

In the discussion above, the quantity $\lim\limits_{\Delta x \to 0} \frac{f(x + \Delta x) - f(x)}{\Delta x}$ appears both in the definition of instantaneous velocity and in the calculation of the slope of a tangent line. This remarkable quantity arises naturally in many other different settings, and so it has been given a special name. It is called the *derivative of f at x*.

Definition

The **derivative of a real function f at x,** denoted **$f'(x)$,** is given by

$$f'(x) = \lim_{\Delta x \to 0} \frac{f(x + \Delta x) - f(x)}{\Delta x},$$

provided this limit exists and is finite.

548

The definition asserts that a derivative is a limit of difference quotients. With discrete functions, for any value of x, values of Δx can be found so that $x + \Delta x$ is between two domain values and $f(x + \Delta x)$ is consequently not defined, so there is no limit. Thus discrete functions do not have derivatives. A function must be continuous and its graph must be "smooth" at a point to have a derivative at that point. It can be shown that all polynomial, exponential, logarithmic, and trigonometric functions defined on the real numbers have derivatives at every point in their domains.

Example 3 Let f be the function defined by the rule $f(x) = x^2 + 4$ for all real numbers x. Find $f'(3)$, the derivative of f at $x = 3$.

Solution Use the definition of derivative with $x = 3$.

$$f'(3) = \lim_{\Delta x \to 0} \frac{f(3 + \Delta x) - f(3)}{\Delta x} \quad \text{Definition of derivative}$$

$$= \lim_{\Delta x \to 0} \frac{((3 + \Delta x)^2 + 4) - (3^2 + 4)}{\Delta x} \quad \text{Definition of } f$$

$$= \lim_{\Delta x \to 0} \frac{(9 + 6(\Delta x) + (\Delta x)^2 + 4) - (9 + 4)}{\Delta x} \quad \text{Expanding}$$

$$= \lim_{\Delta x \to 0} \frac{6(\Delta x) + (\Delta x)^2}{\Delta x}$$

$$= \lim_{\Delta x \to 0} (6 + \Delta x)$$

$$= 6$$

Check Sketch the tangent line to the graph of $y = f(x)$ at the point $(3, f(3))$. The slope of this line seems approximately equal to 6.

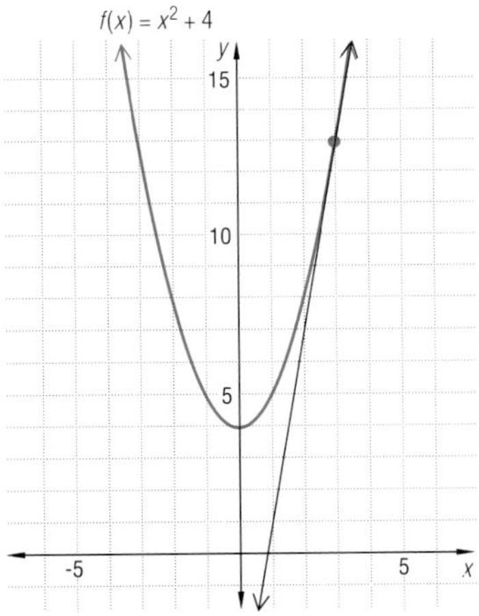

Velocity is only one of many kinds of rates of change. It applies when x represents time and $y = f(x)$ represents the position of an object in one dimension such as height or distance. If x and y are any quantities related by an equation $y = f(x)$, then for a particular value of x, the **instantaneous rate of change of f at x** is defined to be the derivative of f at x. This means

$$\begin{pmatrix}\text{instantaneous rate of} \\ \text{change of } f \text{ at } x\end{pmatrix} = f'(x) = \lim_{\Delta x \to 0} \frac{f(x + \Delta x) - f(x)}{\Delta x}.$$

When there is no confusion, the instantaneous rate of change or instantaneous velocity is sometimes referred to as the rate of change or velocity at x.

4. The volume of a cube with side s is given by the formula $V(s) = s^3$. Calculate $V'(2)$ and $V'(4)$ and interpret your answers.
$V'(2) = 12$; $V'(4) = 48$; The instantaneous rate of change of the volume of a continuously expanding cube when the side has length 4 is 4 times that when the side has length 2.

NOTES ON QUESTIONS
Questions 2, 6, 10, 11, and 13: These questions are important for understanding the geometry of the derivative. Be flexible with the answers, as accuracy depends on how well the tangent lines are drawn.

Making Connections for Questions 10-12: The reason calculus is used in many different areas is that the derivative has meaning in a variety of situations. In these questions, water level in a reservoir, a roller coaster, and sounds in music are discussed. In this lesson, the derivative was applied to a projectile, and in the previous lesson an example about the length of a day was discussed. These situations all change over time, and the derivative gives a mathematical way of dealing with change over time.

Question 14: The rate of change for a linear function is constant, that is, it is the slope.

Question 15: The average rate of change in this question means the amount that the measure of the angle increases as the number of sides increase. This amount gets smaller as n gets larger. There is no meaning to the instantaneous rate of change for this situation because the function is discrete.

ADDITIONAL ANSWERS

2.a. $y = -\frac{5}{2}x - 2$

b.

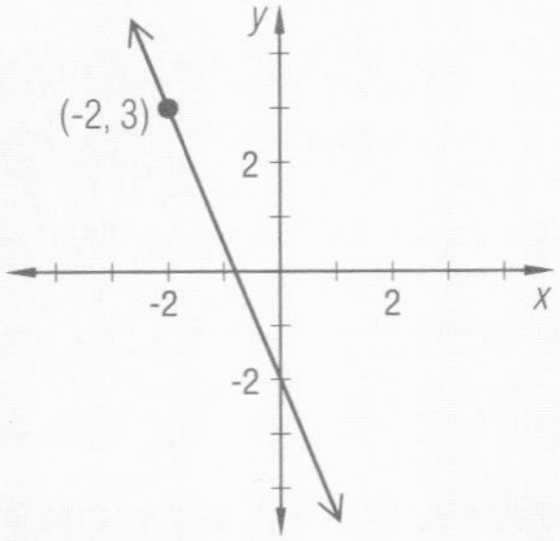

c. sample:

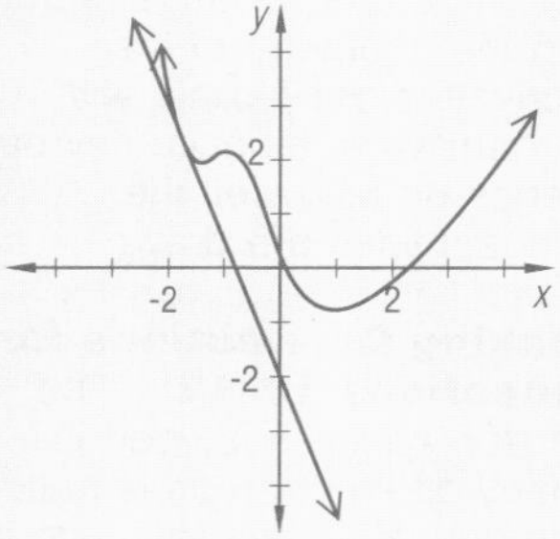

Imagine a spherical balloon being blown up so that the radius is increasing at a steady rate. Example 4 asks for the instantaneous rate of change in the sphere's volume. You need to expand the cube of a binomial: $(a + b)^3 = a^3 + 3a^2b + 3ab^2 + b^3$.

Example 4 Recall that the volume V of a sphere with radius r is given by $V(r) = \frac{4}{3}\pi r^3$.

a. Compute $V'(1)$, the instantaneous rate of change of V at $r = 1$.

b. Compute $V'(2)$, the instantaneous rate of change of V at $r = 2$.

Solution

a. From the definition of instantaneous rate of change,

$$\begin{aligned}
V'(1) &= \lim_{\Delta x \to 0} \frac{V(1 + \Delta x) - V(1)}{\Delta x} \\
&= \lim_{\Delta x \to 0} \frac{\frac{4}{3}\pi(1 + \Delta x)^3 - \frac{4}{3}\pi}{\Delta x} \\
&= \lim_{\Delta x \to 0} \frac{\frac{4}{3}\pi(1 + 3\Delta x + 3(\Delta x)^2 + (\Delta x)^3) - \frac{4}{3}\pi}{\Delta x} \\
&= \lim_{\Delta x \to 0} \frac{\frac{4}{3}\pi(3\Delta x + 3(\Delta x)^2 + (\Delta x)^3)}{\Delta x} \\
&= \lim_{\Delta x \to 0} \frac{4}{3}\pi(3 + 3\Delta x + (\Delta x)^2) \\
&= \lim_{\Delta x \to 0} \frac{4\pi(3)}{3} \\
&= 4\pi.
\end{aligned}$$

b. Similarly, $$V'(2) = \lim_{\Delta x \to 0} \frac{\frac{4}{3}\pi(2 + \Delta x)^3 - \frac{4}{3}\pi \cdot 8}{\Delta x}.$$

You should do the algebra to obtain

$$V'(2) = 16\pi.$$

Physically, Example 4 could be interpreted as follows: If you want to blow up a balloon so that its radius increases steadily, you will need to put in 4 times as much air at the instant when the radius is 2 units as you will at the instant when the radius is 1 unit.

Questions

Covering the Reading

1. For the projectile in Example 1, the average velocity between times $t = 10$ and $t = 10 + \Delta t$ is $480 - 16\Delta t$ ft/sec. Find the instantaneous velocity at time $t = 10$ seconds. 480 ft/sec

2. a. Find an equation for the line ℓ through (-2, 3) with slope $\frac{-5}{2}$.

b. Graph ℓ.

c. Graph a function that has ℓ as a tangent line at $x = -2$.
See margin.

550

3. Use the graph at the right that shows the distance traveled by a driver going through a turn.

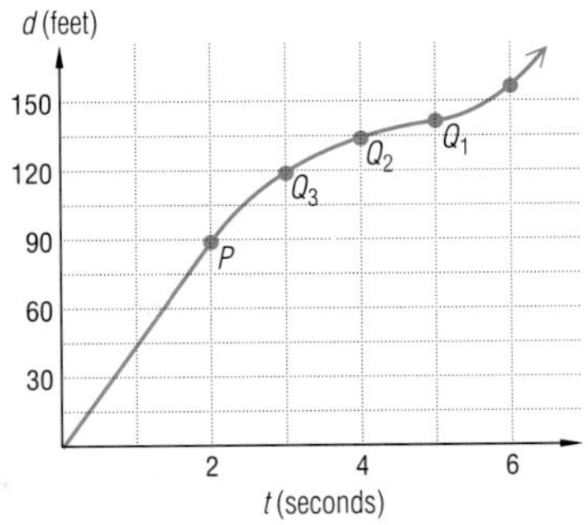

a. Estimate the slopes of the secant lines $\overleftrightarrow{PQ_1}$, $\overleftrightarrow{PQ_2}$, $\overleftrightarrow{PQ_3}$. (Include units with your answer.)

b. Trace the graph and sketch the tangent line to the curve at P. Find the slope of the line you drew. (Include units with your answer.) See margin.

c. Estimate the instantaneous velocity of the driver at $t = 2$ seconds.

a) Samples: 17.5 ft/sec, 22.5 ft/sec, 30 ft/sec; c) Sample: 40 ft/sec

4. Give the definition of a derivative at a point.
See margin.

5. *True* or *false? Suppose a function f has a derivative at every point. Then the slope of the tangent line to the graph of f at x is f′(x).*
True

6. The figure below shows the graph of an even function f.

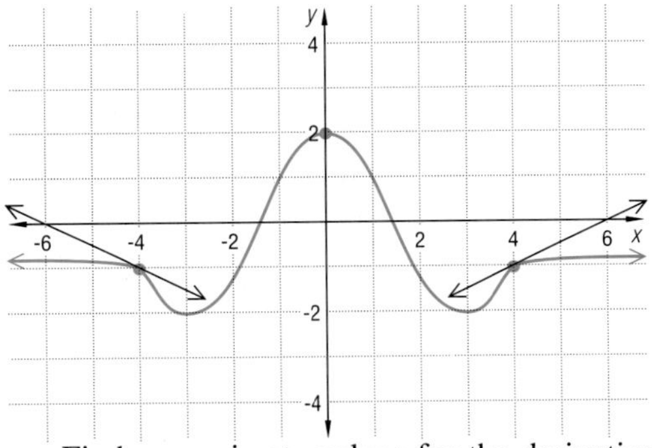

a. Find approximate values for the derivatives of f at $x = -4$, $x = 0$, and $x = 4$.

b. Write the equations of the tangent lines to the graph at $x = -4$, $x = 0$, and $x = 4$. a) $f'(-4) \approx -\frac{1}{2}$, $f'(0) = 0$, $f'(4) \approx \frac{1}{2}$;
b) See margin.

7. Explain why discrete functions do not have derivatives.
For a discrete function, you cannot find values of x so that $\Delta x \to 0$.

8. Let $f(x) = .5x^2$.

a. Graph f over $0 \le x \le 4$. See margin.

b. Sketch secants through $(1, f(1))$ and $(1 + \Delta x, f(1 + \Delta x))$ for $\Delta x = 2$, 1, and 0.5. See margin.

c. Find the slopes of the three secants in part **b**. 2, 1.5, 1.25

d. Use the definition of the derivative of f at x to find the slope of the tangent to f at $x = 1$ algebraically. 1

e. Sketch the tangent to f at $x = 1$. See margin.

9. Fill in the algebraic steps missing in Example 4, part **b**. See margin.

3.b.

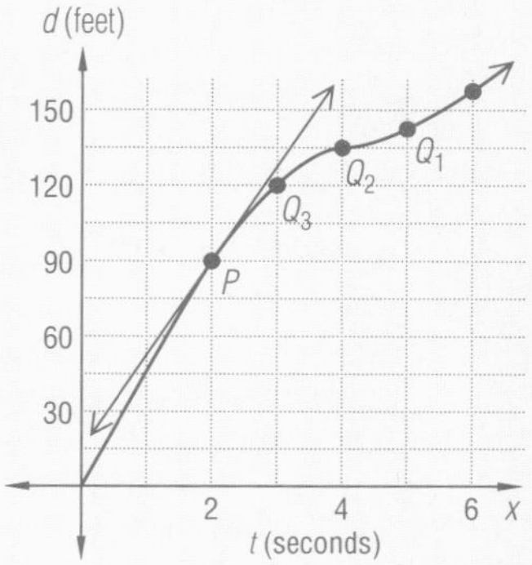

slope ≈ 40 ft/sec

4. The derivative of a real function f at a point x is
$$\lim_{\Delta x \to 0} \frac{f(x + \Delta x) - f(x)}{\Delta x},$$
provided this limit exists and is finite.

6.b.

value of x	equation
-4	$y = -\frac{1}{2}x - 3$
0	$y = 2$
4	$y = \frac{1}{2}x - 3$

8.a., b.

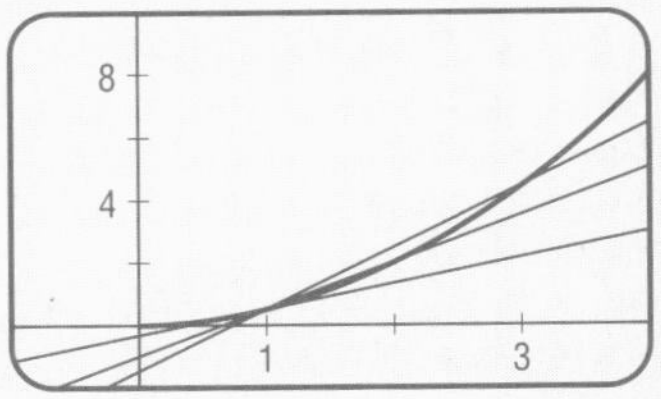

e.

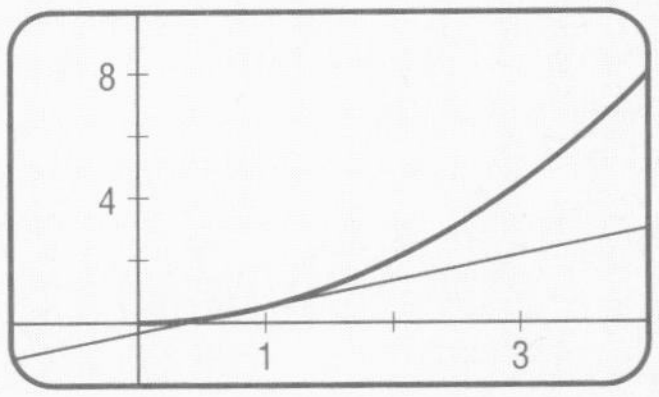

9. See Additional Answers in the back of this book.

ADDITIONAL ANSWERS
10.b. At 7:05, the water level is falling at about 1.3 ft/min; at 7:15, the water level is rising at about 1 ft/min.

12.a.i. during the attack and slope times
ii. during the decay and release time
iii. sample: at the break point and during the sustain time
b.i. sound getting louder
ii. sound getting softer
iii. constant volume

13.c. $A'(0)$ is $\lim_{\Delta x \to 0} \frac{A(0 + \Delta x) - A(0)}{\Delta x}$; this limit does not exist because it has different values when approaching zero from the right and from the left.

14.b. The average rate of change is the slope of the secant line through $x = 2$ and $x = 2 + \Delta x$, but f is a line. Hence, the slope of the secant line is the slope of f.

16. Let S be the sequence defined by $S_n = 2n^2 + 1$, $\forall$ positive integers n. Show S satisfies the recursive definition:
$T_1 = 3$; $T_{k+1} = T_k + 4k + 2$, $\forall$ positive integers $k \geq 1$. $S_1 = 2 \cdot 1^2 + 1 = 3$; so S has the same initial condition as T. For all n, $S_{n+1} = 2(n + 1)^2 + 1$
$= 2(n^2 + 2n + 1) + 1$
$= 2n^2 + 4n + 2 + 1$
$= (2n^2 + 1) + 4n + 2$
$= S_n + 4n + 2$.
So, S satisfies the same recurrence relation as T. Since in this relation S_{n+1} is defined in terms of S_n by the Recursion Principle, they are the same sequence.

22.a. degrees Fahrenheit/minute
b.i. The oven is heating up slowly.
ii. The oven is heating up quickly.
iii. The oven is cooling off.
iv. The oven is maintaining a constant temperature.

Applying the Mathematics

10. The water level $f(t)$ in a reservoir at various times t is graphed below.
a. Estimate the derivative of f at 7:05 and at 7:15.
b. What does the derivative mean in each case? **See margin.**
a) $f'(7{:}05) \approx -\frac{9}{7} \approx -1.3$ ft/min; $f'(7{:}15) \approx 1$ ft/min

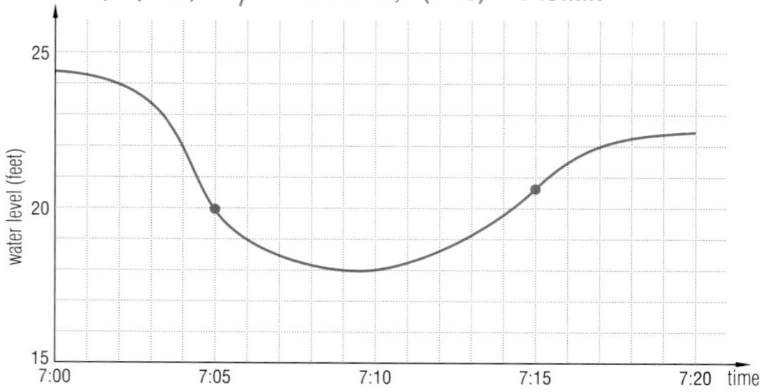

11. Let h be the height (in feet) of a car on a roller coaster t seconds after a ride begins, and suppose t and h are related by the equation $h = f(t)$ graphed at the right. By drawing tangents, approximate the instantaneous rate of change of height with respect to time at $t = 30$ seconds, $t = 60$ seconds, and $t = 100$ seconds. At which of these times is the height of the roller coaster changing the fastest? $h'(30) \approx 1$ ft/sec; $h'(60) \approx -.7$ ft/sec; $h'(100) \approx \frac{1}{2}$ ft/sec; The fastest change is at $t = 30$.

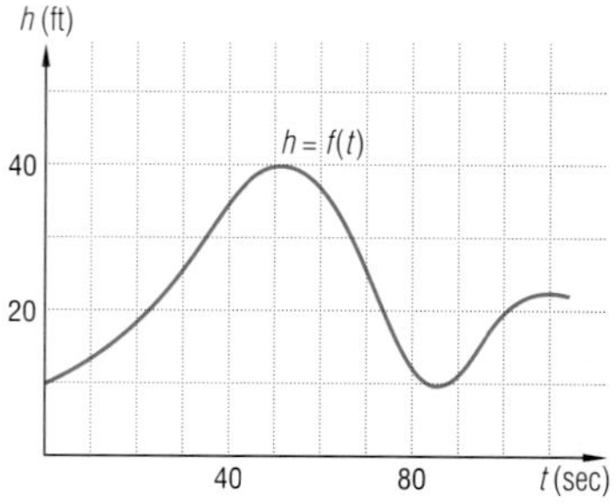

12. Many musical synthesizers allow the user to control the *envelope* of created sounds. The envelope determines the volume of the tone from the time that a key is first played until the time the tone dies out. The graph below shows the envelope for a typical sound.

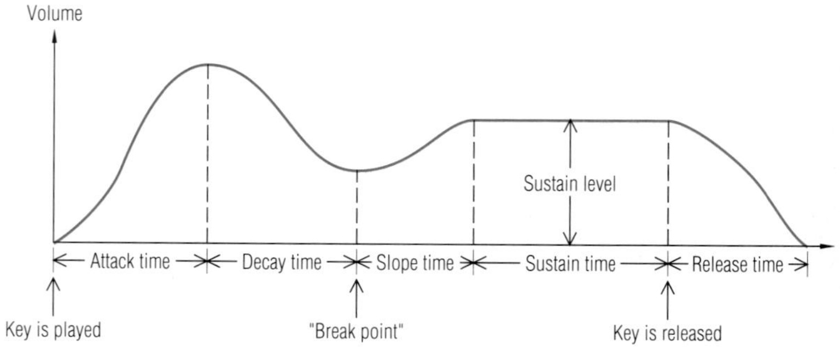

a. Describe two different times at which the derivative of the volume with respect to time is
i. positive **ii.** negative **iii.** zero.
b. Describe what you would hear when the derivative is
i. positive **ii.** negative **iii.** zero.
a, b) See margin.

552

13. Consider the absolute value function A with $A(x) = |x|$.

a. Calculate

$$\lim_{\Delta x \to 0+} \frac{A(0 + \Delta x) - A(0)}{\Delta x}.$$ 1

b. Calculate

$$\lim_{\Delta x \to 0-} \frac{A(0 + \Delta x) - A(0)}{\Delta x}.$$ -1

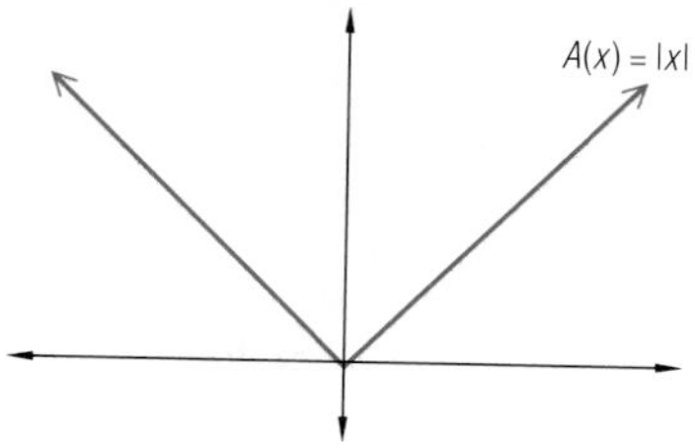

c. Explain why the results of parts **a** and **b** imply that $A'(0)$ does not exist. See margin.

Review

14. Given the function f with $f(x) = 3x - 4$.

a. Calculate the average rate of change of the function from $x = 2$ to $x = 2 + \Delta x$. 3

b. Interpret your answer to part **a** in terms of slopes of secant lines from $x = 2$ to $x = 2 + \Delta x$. *(Lesson 9-1)* See margin.

15. Recall that the sum of the interior angles of an n-gon is $180(n - 2)$. Let $f(n)$ be the measure of *one* angle of a regular n-gon.

a. Write a formula for $f(n)$ in terms of n. $f(n) = \frac{180(n-2)}{n}$

b. Find $f(20)$ and $f(24)$. $f(20) = 162$, $f(24) = 165$

c. What is the average rate of change of f from $n = 20$ to $n = 24$? *(Lesson 9-1)* $\frac{3}{4}$

16. Prove that the explicit formula,

$a_n = 2n^2 + 1$ $\forall$ positive integers n,

and the recursive formula,

$$\begin{cases} a_1 = 3 \\ a_{k+1} = a_k + 4k + 2 \quad \forall \text{ integers } k \geq 1, \end{cases}$$

define the same sequence. *(Lessons 7-4, 7-2)* See margin.

In 17–21, evaluate without using your calculator. *(Lessons 6-6, 6-2, 2-8, 2-7)*

17. $\sin \frac{7\pi}{6}$ -0.5

18. $\cos^2\left(\frac{5\pi}{12}\right) + \sin^2\left(\frac{5\pi}{12}\right)$ 1

19. $\tan^{-1}\sqrt{3}$ $\frac{\pi}{3}$

20. $\log_2 \frac{1}{16}$ −4

21. $\ln e^{\pi}$ π

Exploration

In 22–24, imagine a function describing the given situation.

a. Give a reasonable unit for the derivative.

b. Describe when the derivative would be

i. small and positive. ii. large and positive.

iii. negative. iv. zero.

22. baking (temperature of oven over time) See margin.

23. weightlifting (height of bar over time) See margin.

24. jogging (distance over time) See margin.

FOLLOW-UP

MORE PRACTICE

For more questions on SPUR Objectives, use *Lesson Master 9-2*, shown below.

23., 24. See Additional Answers in the back of this book.

NAME

LESSON **MASTER 9–2**

QUESTIONS ON **SPUR** OBJECTIVES

■SKILLS *Objective B (See pages 581–584 for objectives.)*

In 1–3, find the derivative of the function at the given point.

1. $f(x) = 2x^2 - x + 5$; (0, 5) -1

2. $g(x) = 20x + 17$; (.6, 29) 20

3. $h(y) = -15$; (18, -15) 0

4. Let $f(x) = -x^3 + 3x$. Use the definition of derivative to compute $f'(1)$ and $f'\left(\frac{3}{2}\right)$. $f'(1) = 0$, $f'\left(\frac{3}{2}\right) = -\frac{15}{4}$

■USES *Objective D*

5. Suppose that the profit (in cents per pound) a grocer makes in a day from selling hamburger at a price of s cents per pound is given by $P(s) = -s^2 + 320s - 5000$.

a. Find the derivative when $s = 120$. 80

b. What does your answer to part **a** mean?

sample: At the price of $1.20/lb, each increase of 1¢/lb in the price results in an increase of 80¢ in profit.

■USES *Objective E*

6. Suppose the distance in meters that a car has travelled at time t (in seconds) is given by $s(t) = 4t^2 + 3t$. Find its instantaneous velocity at time $t = \frac{1}{4}$. 5

NAME

Lesson MASTER 9-2 (page 2)

7. A ball is dropped from the roof of a house 25 feet high. The height (in feet) of the ball above the ground at time t seconds is given by $h(t) = -16t^2 + 25$.

a. What is the instantaneous velocity of the ball at time $t = .5$ seconds? -16 ft/sec

b. At what time t does the ball hit the ground? $t = 1.25$ sec

c. What is the instantaneous velocity of the ball at the moment it hits the ground? -40 ft/sec

■REPRESENTATIONS *Objective H*

8. Refer to the graph of f at the right. Estimate $f'(x)$ for each value of x given below.

a. $x = 0$ 0

b. $x = 2$ 2

c. $x = -3$ 0

d. $x = -4$ -2

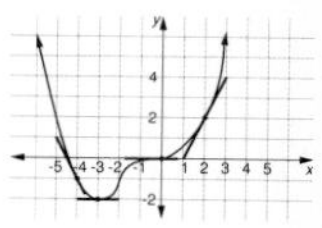

LESSON 9-3

RESOURCES
- Lesson Master 9-3
- Quiz for Lessons 9-1 through 9-3
- Teaching Aid 64 can be used with **Example 2**.
- Teaching Aid 65 can be used with **Questions 6, 7, and 11**.
- Computer Master 17

OBJECTIVES

B Use the definition of derivative to compute derivatives.
D Find rates of change in real situations.
E Use derivatives to find the velocity and acceleration of a moving object.
H Estimate derivatives by finding slopes of tangent lines.

TEACHING NOTES

In this lesson, the limit of the difference quotient will be obtained for any value of x.

The result that the derivative of the sine is the cosine may surprise students. The proof is better left to a calculus course. Instead, we present this result as simply a statement of fact. Use an automatic grapher to display $y = \sin x$ and $y = \cos x$ on one screen. Then discuss the slope of the lines tangent to $y = \sin x$ at many points in the interval $0 \le x \le 2\pi$. Relate each slope to the corresponding point on the curve $y = \cos x$.

Technology The four functions shown on page 555

LESSON

9-3 The Derivative Function

Here is a graph of the sine function f defined by $f(x) = \sin x$.

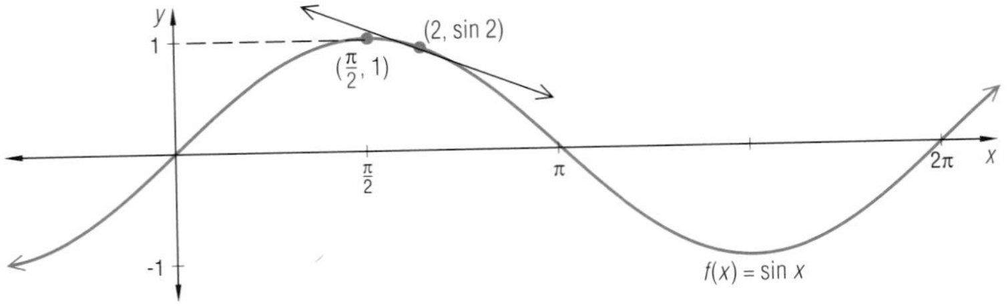

It is easy to see that the value of the derivative of f at $x = \frac{\pi}{2}$ is zero because the tangent to the graph of f at $\left(\frac{\pi}{2}, \sin \frac{\pi}{2}\right)$ is horizontal, with a slope of 0. What about other points? For instance, consider the point at $x = 2$. The tangent to the graph at (2, sin 2) has a negative slope. An exact value of the slope is the derivative of f at $x = 2$. By definition,

$$f'(2) = \lim_{\Delta x \to 0} \frac{\sin(2 + \Delta x) - \sin 2}{\Delta x}.$$

To estimate this limit, evaluate the difference quotient for some small values of Δx. (You should check the following computations. Make certain your calculator is set to radians.)

$$\Delta x = 0.1: \quad \frac{\sin(2 + 0.1) - \sin 2}{0.1} \approx \frac{0.8632 - 0.9093}{0.1} = \frac{-0.0461}{0.1} = -0.461$$

$$\Delta x = 0.01: \quad \frac{\sin(2 + 0.01) - \sin 2}{0.01} \approx \frac{0.9051 - 0.9093}{0.01} = \frac{-0.00421}{0.01} = -0.421$$

$$\Delta x = 0.001: \quad \frac{\sin(2 + 0.001) - \sin 2}{0.001} \approx \frac{0.908881 - 0.909297}{0.001} = \frac{-0.000417}{0.001} = -0.417$$

$$\Delta x = 0.0001: \quad \frac{\sin(2 + 0.0001) - \sin 2}{0.0001} \approx \frac{0.9092558 - 0.9092974}{0.0001} = \frac{-0.0000416}{0.0001} = -0.416$$

It seems that the slope of the tangent is about -0.416. But what number is it exactly? Could it be $-\frac{5}{12} = -0.41\overline{6}$? It turns out that it is not $-\frac{5}{12}$, but rather $\cos 2 = -0.4161 \ldots$! Now look back at the slope at $x = \frac{\pi}{2}$. It is 0, which is $\cos \frac{\pi}{2}$.

In fact, the slopes of tangents to the graph of $f(x) = \sin x$ are positive when $0 \le x < \frac{\pi}{2}$, and on this interval the cosine function is positive. The slopes are negative when $\frac{\pi}{2} < x < \frac{3\pi}{2}$, and on that interval the cosine function is negative. If you proceed to approximate the derivative of the sine function at any number x, you will find values approaching $\cos x$. The cosine function is the *derivative* of the sine function.

Some calculator and computer programs enable you to see this graphically. Below are graphs of four functions. The black curve is the cosine function. The other three curves are functions defined by the difference quotients for the sine function,

$$f_{\Delta x}(x) = \frac{\sin(x + \Delta x) - \sin x}{\Delta x},$$

when $\Delta x = 1$, $\Delta x = 0.5$, and $\Delta x = 0.1$. You can see that the limit of $f_{\Delta x}$ as $\Delta x \to 0$ seems to be the cosine function. That is, for all x, the exact value of the derivative of the sine function at x is $\cos x$.

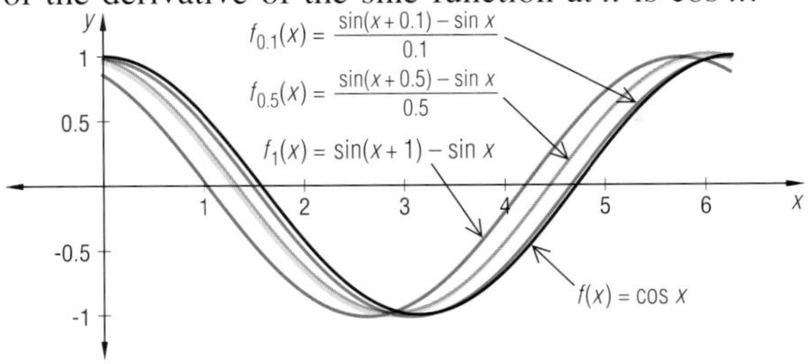

For $f(x) = \sin x$, since $f'(x) = \cos x$, $\cos x$ is called the *derivative function* of $\sin x$.

Definition

Suppose that f is a function that has a derivative $f'(x)$ at each point x in the domain of f. Then the function f': $x \to f'(x)$ for all x in the domain of f is called the **derivative function of *f*.**

For the projectile discussed in Lessons 9-1 and 9-2, the derivative at each time is the (instantaneous) velocity of the projectile at that time. So the derivative function is a velocity function. We return to this application in the following example.

Example 1 Suppose the height (in feet) of a projectile t seconds after launch is given by $h(t) = 800t - 16t^2$. Find a formula for its velocity t seconds after launch.

Solution The velocity at each time t is the instantaneous rate of change of the position function h at t. This is the limit of the difference quotient as the change in t approaches 0; that is, as $\Delta t \to 0$. In other words, it is the derivative of the position function.

$$h'(t) = \text{velocity at time } t = \lim_{\Delta t \to 0} \frac{h(t + \Delta t) - h(t)}{\Delta t}$$

$$= \lim_{\Delta t \to 0} (800 - 32t - 16\Delta t) \qquad \text{From Lesson 9-1, Example 2}$$

$$= (800 - 32t) \text{ ft/sec}$$

For example, after 30 seconds the velocity is $800 - 32 \cdot 30 =$ -160 ft/sec, which means that the projectile is moving downward at a speed of 160 ft/sec. This is about 110 miles per hour.

give a graphic display of the cosine function as the limit of the difference quotients of the sine function. Some automatic graphers have the capability of showing such displays. In general, to approximate the derivative of a function f, enter a new function

$$g(x) = \frac{f(x + \Delta x) - f(x)}{\Delta x},$$

where Δx is some very small number, say 0.001. (Larger values of Δx are used in the textbook so that the graphs can be distinguished.) Then, for most well-behaved functions, g is a very good approximation to f'.

Making Connections
Make the connection between the instantaneous rate of change function of **Example 1** in this lesson and the average rate of change function of Example 2 in Lesson 9-1. All we have done here is to take the limit as $\Delta t \to 0$.

Example 2 continues to emphasize the geometric interpretation of the derivative. You will probably need to discuss this example in detail. After working through the proof of the theorem on the derivative of a quadratic, you may want to do another example like that of Example 2 but starting with a quadratic function. Show that the type of analysis done in Example 2 leads to the graph of a straight line as required by the theorem.

In doing a problem such as **Example 2**, it is useful to have the graph of f' directly below that of f or to graph f' on the same coordinate system as f.

Ask students to translate the graph of $y = f(x)$ in **Example 2** either upward or downward (vertically only). Ask: How does the graph of the derivative of the translated function compare to that of the original? This exercise should help students see that the constant term in a function is not reflected in the derivative.

Making Connections
Many students have seen the conclusion of the Derivative of a Quadratic Function Theorem in another form in a previous course. They may have been taught that the vertex of the graph of $y = ax^2 + bx + c$ has first coordinate $x = -\frac{b}{2a}$. If you set $2ax + b = 0$ and solve for x, you obtain the same result. This is done in Lesson 9-5.

The theorem on the derivative of a quadratic is the only shortcut formula for finding the derivative that is introduced. Emphasize to students that when questions ask them to find the derivative by using the definition, they must take the limit of the difference quotient. If no specific instructions are given, then then may use the theorem for the derivative of a quadratic function.

Making Connections
Example 3 of this lesson continues Example 4 of the previous lesson. Students may be very surprised to learn that the derivative of the volume formula is the surface area formula. In general, the derivative of a volume formula for a class of similar figures is a multiple of the surface area formula for that class.

ADDITIONAL EXAMPLES
1. Suppose that the height (in feet) of a projectile t seconds after lunch is given by $h(t) = 100 + 960t - 16t^2$. Find a formula for its velocity t seconds after launch.
$h'(t) = 960 - 32t$ ft/sec

From this formula we can see that its initial velocity was 960 ft/sec. The decrease in velocity over time is 32 ft/sec, which is due to gravity. In general, if $h(t) = h_0 + v_0t - 16t^2$, then $h'(t) = v_0 - 32t$ ft/sec, and so h_0 = initial height and v_0 = initial velocity.

Given a graph of a function f, you can make a rough sketch of the derivative function f' as follows. Imagine moving from left to right along the graph of f. For each value of x, draw or imagine the tangent line to the graph at the point $(x, f(x))$. Estimate its slope $f'(x)$, and plot the point $(x, f'(x))$ on the graph of f'. Example 2 shows how this is done.

Example 2 A function f is graphed below. Use the graph of f to estimate $f'(x)$ when x = -5, -2, 2, and 6. Sketch a graph of f' for values of x between -5 and 6.

Solution Draw tangent lines to the graph of f at the indicated values of x.

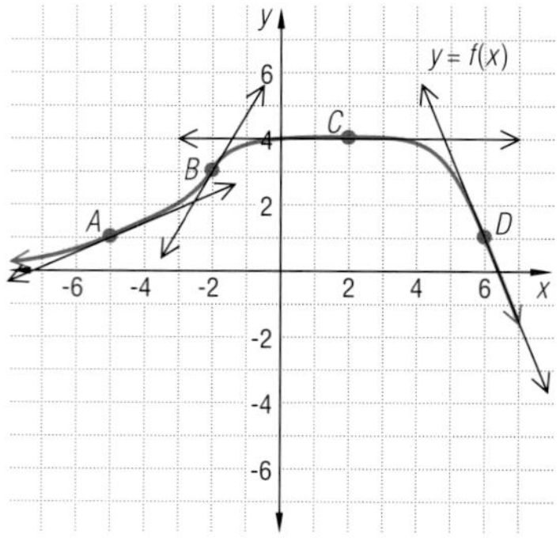

At the point $A = (-5, f(-5))$, the tangent line has a slope of about $\frac{1}{2}$. Therefore the point $A' = (-5, \frac{1}{2})$ is on the graph of f'.

The slope of the tangent at the point $B = (-2, f(-2))$ is about 2. Therefore $B' = (-2, 2)$ is on the graph of f'. Between $x = -5$ and $x = -2$ the graph of f rises more and more steeply, indicating that the graph of f' increases from $\frac{1}{2}$ to 2.

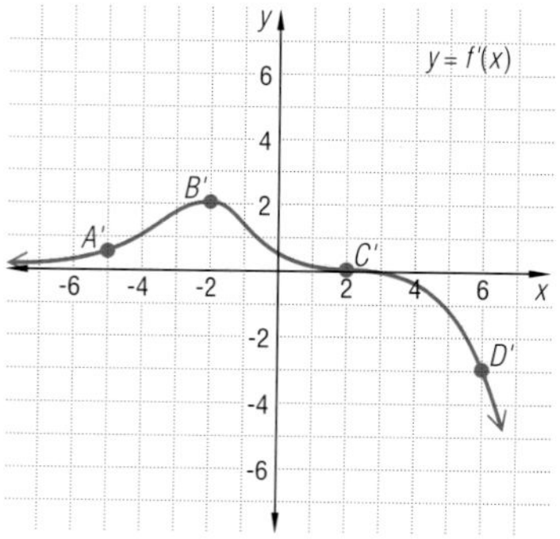

At $C = (2, f(2))$ the tangent is horizontal, with slope 0. Therefore $C' = (2, 0)$ is on the graph of f'. Since the graph of f flattens out rapidly between $x = -2$ and $x = 2$, the graph of f' decreases quickly to 0 at $x = 2$.

At $D = (6, f(6))$ the slope of the tangent line is about -3. Therefore $D' = (6, -3)$ is on the graph of f'. The graph of f decreases more and more steeply as x moves from left to right between $x = 2$ and $x = 6$. Therefore between $x = 2$ and $x = 6$, the graph of f' decreases from 0 to -3.

In general, it is not always as easy as was in Example 1 to find a formula for the derivative of a function. However, there are computer programs that calculate derivatives, and some time is spent in most calculus classes calculating derivatives. In Example 1, the derivative of the quadratic function $h(t) = 800t - 16t^2$ was calculated. The ideas used in that calculation can be applied to any quadratic function.

Theorem (Derivative of a Quadratic Function)

If $f(x) = ax^2 + bx + c$, where a, b, and c are real numbers and $a \neq 0$, then $f'(x) = 2ax + b$ for all real numbers x.

Proof Apply the definition of derivative at each point x.

$$f'(x) = \lim_{\Delta x \to 0} \frac{f(x + \Delta x) - f(x)}{\Delta x}$$

Substitute for $f(x)$.
$$= \lim_{\Delta x \to 0} \frac{a(x + \Delta x)^2 + b(x + \Delta x) + c - (ax^2 + bx + c)}{\Delta x}$$

Expand and distribute.
$$= \lim_{\Delta x \to 0} \frac{a(x^2 + 2x\Delta x + (\Delta x)^2) + bx + b\Delta x + c - ax^2 - bx - c}{\Delta x}$$

Combine like terms.
$$= \lim_{\Delta x \to 0} \frac{\cancel{ax^2} + 2ax\Delta x + a(\Delta x^2) + \cancel{bx} + b\Delta x + \cancel{c} - \cancel{ax^2} - \cancel{bx} - \cancel{c}}{\Delta x}$$

Δx is a factor of each term in the numerator.
$$= \lim_{\Delta x \to 0} \frac{2ax\Delta x + a(\Delta x)^2 + b\Delta x}{\Delta x}$$

$$= \lim_{\Delta x \to 0} (2ax + a\Delta x + b)$$

Take the limit.
$$= 2ax + b$$

For the function $h(t) = 800t - 16t^2$, $a = -16$, $b = 800$, and $c = 0$. The theorem states that the derivative function will be $h'(t) = 2at + b = -32t + 800$, which is exactly what was found in Example 1.

The formula for finding the derivative of a quadratic function can be extended to linear functions ($a = 0$ and $b \neq 0$) and to constant functions ($a = 0$ and $b = 0$) since the proof above does not depend on the values of a, b, or c. For example, the derivative of the linear function $f(x) = 0x^2 + 3x + 2$ is $f'(x) = 2 \cdot 0 \cdot x + 3 = 3$, and the derivative of the constant function $g(x) = 5 = 0x^2 + 0x + 5$ is $g'(x) = 0x + 0 = 0$. These results agree with the slopes of these functions.

The next example generalizes Example 4 of the previous lesson.

Example 3 The volume $V(r)$ and the surface area $S(r)$ of a sphere of radius r are given by the formulas

$$V(r) = \frac{4}{3}\pi r^3 \text{ and } S(r) = 4\pi r^2.$$

Verify that the surface area is the derivative function of the volume. That is, verify that $V'(r) = S(r)$ for all r.

Solution From the definition of the derivative,

$$V'(r) = \lim_{\Delta r \to 0} \frac{V(r + \Delta r) - V(r)}{\Delta r}$$

$$= \lim_{\Delta r \to 0} \frac{\frac{4}{3}\pi(r + \Delta r)^3 - \frac{4}{3}\pi r^3}{\Delta r} \text{ for all } r.$$

Now there is only algebra to do, just as in Example 4 of Lesson 9-2. You should do the work to find that $V'(r) = 4\pi r^2$. Since $S(r) = 4\pi r^2$ for all r, $V'(r) = S(r)$.

2. The graph of $f(x) = x^2 + 4$ from Example 3 of Lesson 9-2 is given below. Sketch the graph of $y = f'(x)$.

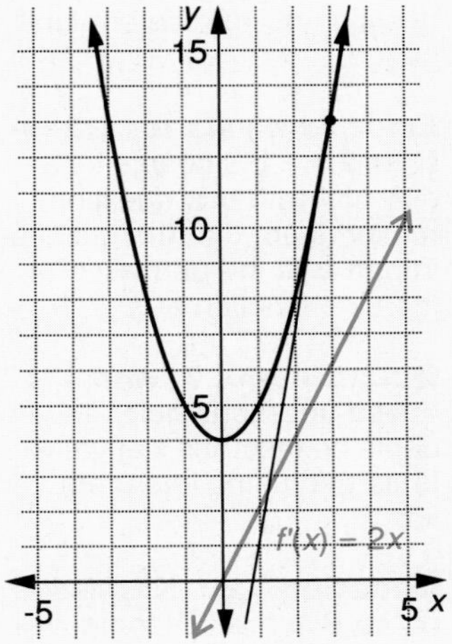

It is the line $y = 2x$.

3. A rectangular region adjacent to a building is to be enclosed with 120 feet of fencing. The area of the region is given by $A(w) = 120w - 2w^2$. Calculate the instantaneous rate of change of this function when $w = 20$ feet. What is the meaning of the number that you get?

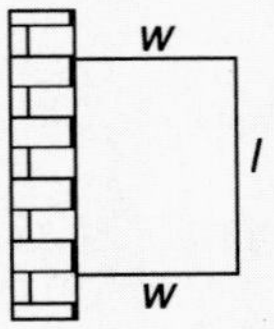

$A'(20) = 40$ square feet per foot. This means that when the width is 20, then a small increase in width will produce a small square unit increase of 40 times the numerical amount in area.

Example 3 verifies that the instantaneous rate of change of the volume of a sphere at a given radius exactly equals its surface area for that radius. That is, as the radius increases (or decreases), you can think of the volume changing by the addition (or peeling away) of very thin layers of surface.

Questions

Covering the Reading

1. a. For $0 \le x \le 2\pi$, when is the slope of the tangent to $f(x) = \sin x$ positive? $0 \le x < \frac{\pi}{2}, \frac{3\pi}{2} < x \le 2\pi$

b. For $0 \le x \le 2\pi$, when is the cosine function positive? $0 \le x < \frac{\pi}{2}, \frac{3\pi}{2} < x \le 2\pi$

2. a. Estimate $\lim\limits_{\Delta x \to 0} \frac{\sin(3 + \Delta x) - \sin 3}{\Delta x}$ to three decimal places by using very small positive values of Δx. $\approx -.990$

b. According to what is said in this lesson, the answer to part **a** should be very close to what value of which trigonometric function? $\cos 3 \approx -.98999$

3. If the projectile in Example 1 is propelled upward with an initial velocity of 600 ft/sec, then its height is given by $h(t) = 600t - 16t^2$.

a. Find the velocity function. $v(t) = h'(t) = 600 - 32t$

b. What is the projectile's velocity at 6 seconds? 408 ft/sec

In 4 and 5, **a.** use the definition of derivative to find the derivative of each of the functions. **b.** Check your answer to part **a** using the theorem for the derivative of a quadratic function or its extension.

4. $f(x) = 3x + 1$ See margin.

5. $g(x) = 5x^2 + 2x$ See margin.

6. Refer to the graph of the function f at the right. Indicate whether the value of f' for the given value of x is positive, negative, or zero.
a. a negative **b.** b zero **c.** c positive **d.** d zero

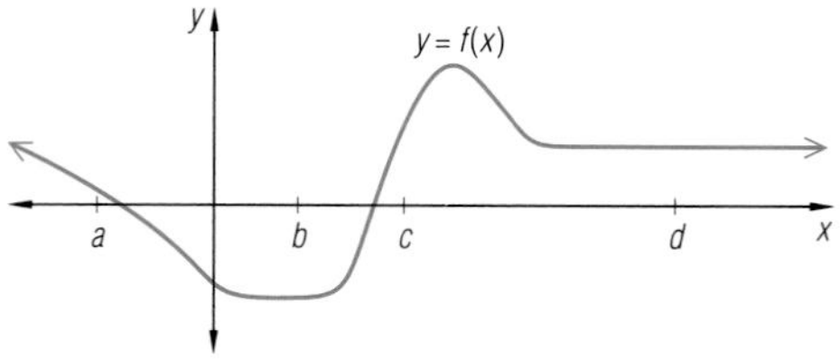

7. The function k is graphed at the right. Estimate the values of k' when $x = -6$, -3, -1, and 2, and use this information and the graph of k to sketch a graph of k'. See margin.

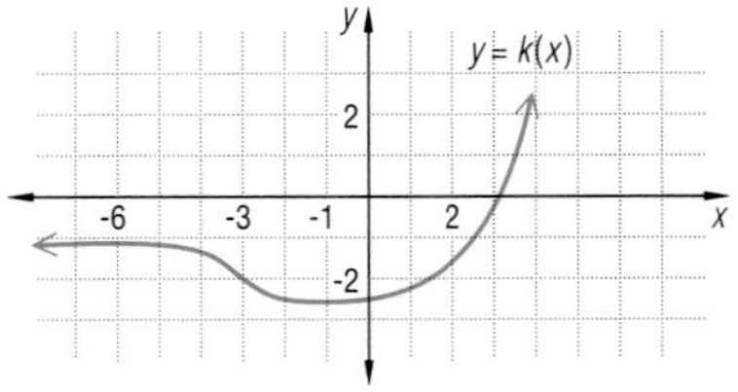

NOTES ON QUESTIONS

Question 1: The entire idea of this question is for students to see that the two parts are asking the same question using different language.

Error Analysis for Question 2b: If students have difficulty with this question, remind them that the quantity in part **b** is the derivative of the sine function at $x = 3$.

Questions 6, 7, and 11: These questions help students connect the derivative to its geometric interpretation.

Question 12: Knowledge of the Binomial Theorem will help students with this question.

Question 15: The unit of $A'(x)$ is square inches per inch.

Question 17: The connection between these functions usually is introduced in the other direction, with the natural log appearing as the integral of the reciprocal function. Either way, the result should be a surprise.

ADDITIONAL ANSWERS

4.a. $f'(x) = \lim\limits_{\Delta x \to 0} \frac{3(x + \Delta x) - 3x}{\Delta x} = \lim\limits_{\Delta x \to 0} \frac{3\Delta x}{\Delta x} = 3$

b. $f(x) = 0x^2 + 3x + 1$, so $f'(x) = 2 \cdot 0x + 3 = 3$.

5. See Additional Answers in the back of this book.

7. $k'(-6) = 0$, $k'(-3) \approx -1$, $k'(-1) \approx \frac{1}{10}$, $k'(2) \approx 1$

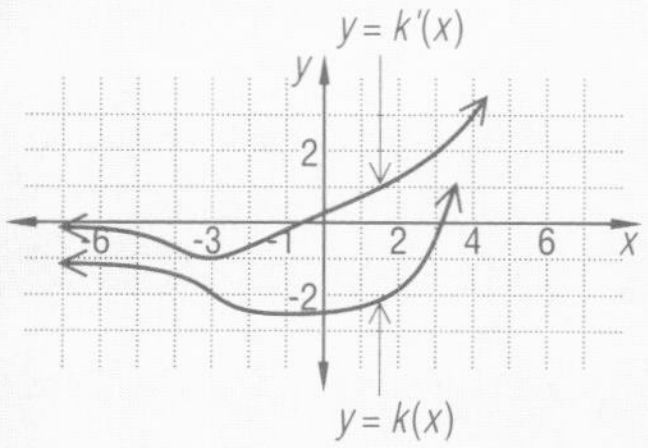

In 8 and 9, use the extension of the theorem for the derivative of a quadratic function to find the derivative of the given function.

8. $f(x) = -6x + 4$ $f'(x) = -6$ **9.** $f(t) = 7$ $f'(t) = 0$

10. **a.** If $g(x) = 4x^2 - 3$, find $g'(x)$. $g'(x) = 8x$

b. What is the value of $g'(5)$?   40

c. Use an automatic grapher to graph g. Check your answer to part **b** by estimating the slope of the tangent line to the graph of g at the point $(5, g(5))$.   See margin.

11. At the right is a graph of $f(x) = -x^2 + 2x + 3$. Some tangent lines to the graph are also shown.

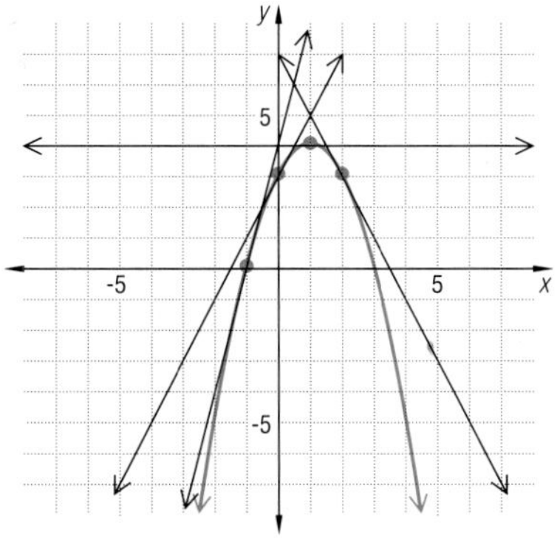

a. Use the graph to fill in approximate values for the table below.   Samples:

x	$f'(x)$
-1	4
0	2
1	0
2	-2

b. Find the derivative of f using the theorem for the derivative of a quadratic function. $f'(x) = -2x + 2$

c. Check your answers to part **a** by evaluating this derivative at the given values of x. $f'(-1) = 4$, $f'(0) = 2$, $f'(1) = 0$, $f'(2) = -2$

12. Supply the missing algebraic steps in Example 3.   See margin.

Applying the Mathematics

13. Plot $f(x) = \dfrac{\cos(x + 0.1) - \cos x}{0.1}$ and $g(x) = -\sin x$ on the same screen. Make a conjecture based on the results.   See margin.

14. **a.** Let $f(x) = x^2 - 2$ and $g(x) = x^2 + 1$. Use the theorem of this lesson to find f' and g'. $f'(x) = 2x$, $g'(x) = 2x$

b. On the same set of axes, sketch graphs of f and g. Choose three values of x and draw tangent lines to each graph at those three points.   See margin.

c. The functions f and g differ by a constant. How are their derivatives related? Explain why this relationship holds.   See margin.

15. A flyer is to be made by putting a white border of width x inches around an 8″ by 10″ photograph, as shown at the right. The area of the border is given by $A(x) = 4x^2 + 36x$ square inches.

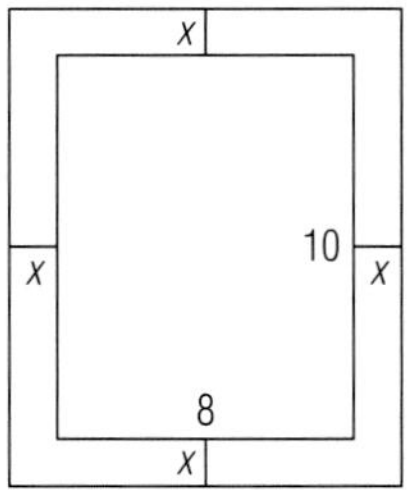

a. Find the instantaneous rate of change of A with respect to x when $x = 2$ inches.

b. What is the meaning of the number you get?   See margin.

a) $A'(2) = 52$ square inches per inch

10.c.

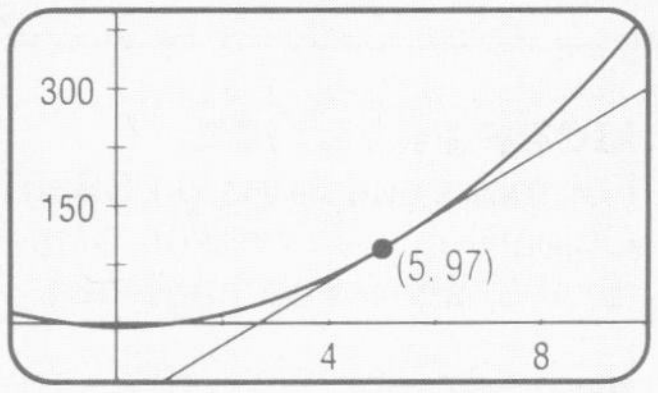

The slope at (5, 97) is about 40.

12. See Additional Answers in the back of this book.

13.

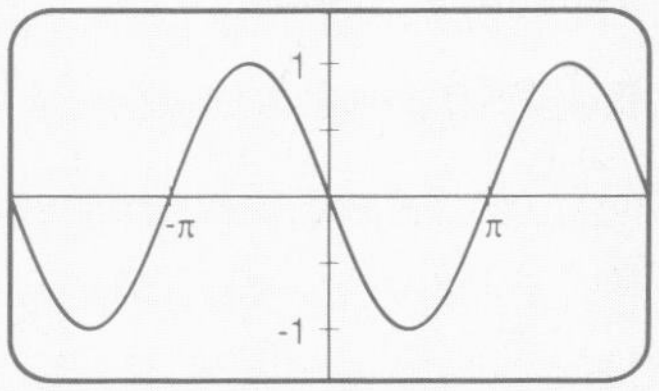

f and g are almost identical

If $h(x) = \cos x$, then $h'(x) = -\sin x$.

14.b.

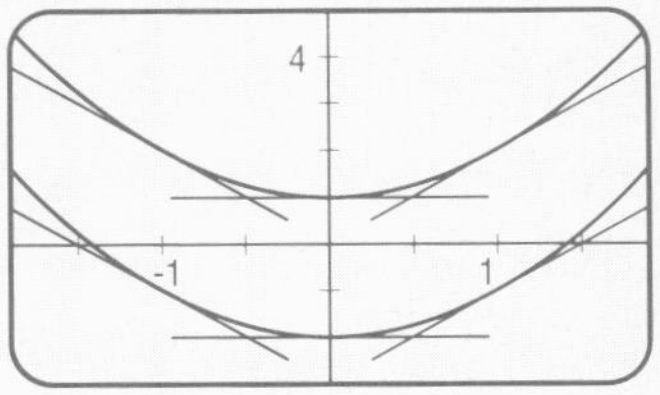

Shows tangents at $x = -1$, 0, and 1.

c. $f'(x) = g'(x)$ for all x. The derivative of a constant is zero, so if $h(x) = x^2 + c$, then $h'(x) = 2x + 0 = 2x$ for any real number c.

15.b. If the value is 2 inches of width, then the area of the border is increasing at the rate of 52 square inches of border for an inch of width.

FOLLOW-UP

MORE PRACTICE
For more questions on SPUR Objectives, use *Lesson Master 9-3,* shown on page 561.

PROJECTS
The projects for Chapter 9 are described on page 577. **Projects 2 and 4** are related to the content of this lesson.

EVALUATION
A quiz covering Lessons 9-1 through 9-3 is provided in the Teacher's Resource File.

ADDITIONAL ANSWERS
17.a.

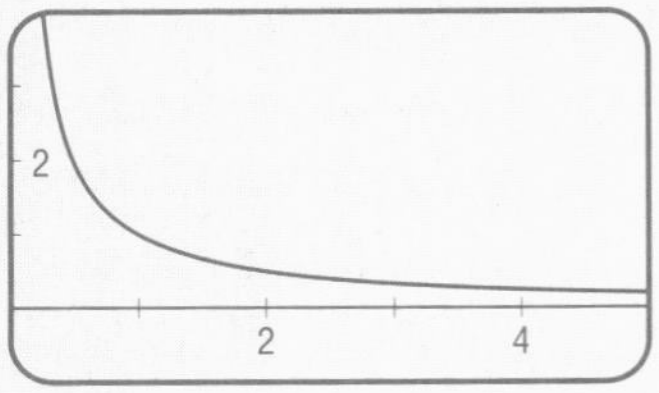

22.a.

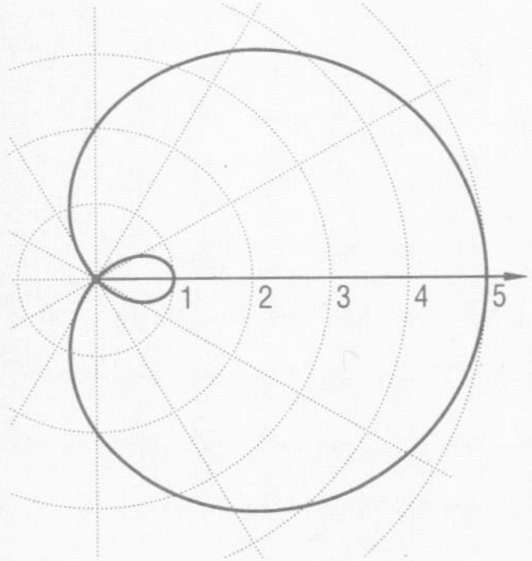

b. limaçon

23. $r = 10\cos(2\theta)$

16. The diagram at the right shows part of a race track. A driver travels at top speed in the straight parts but must slow down in the turns.

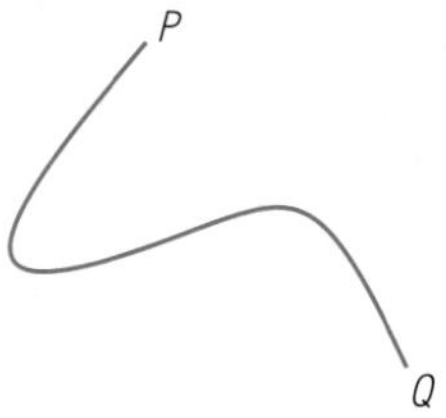

a. If the driver goes from point P to point Q, which of the graphs below is the better description of distance as a function of time? **i**

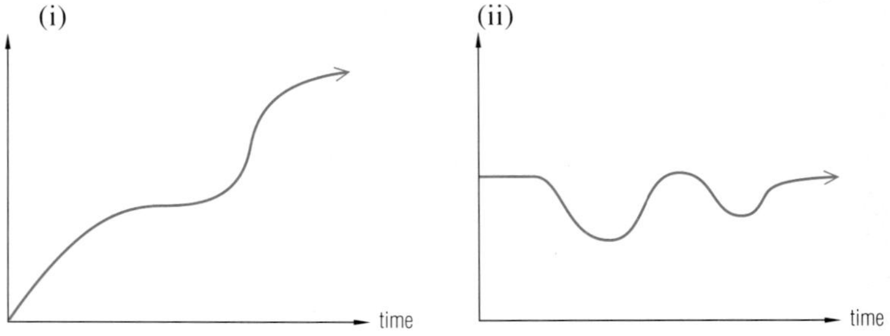

b. What does the graph you did not choose describe? **Velocity as a function of time**

17. a. Graph $f(x) = \frac{\ln(x + .001) - \ln x}{.001}$ for $0 < x \le 5$. **See margin.**

b. The graph in part **a** should resemble the graph of a familiar function. Use it to conjecture a formula for the derivative of the natural log function. **If $f(x) = \ln x$, then $f'(x) = \frac{1}{x}$.**

18. For what values of x is the derivative of $f(x) = -3x^2 - 18x + 7$ positive? **$x < -3$**

Review

19. *True* or *false*? If a function f has a derivative function at the point $(x_0, f(x_0))$, then the instantaneous rate of change of a function f at the point is the slope of the tangent line at that point. *(Lesson 9-2)* **True**

20. The function f is graphed here, along with tangent lines at $x = -4$, $x = 0$, and $x = 4$.

a. Based on the symmetry of the graph, what kind of function is f?

b. Estimate the derivative at

i. $x = -4$ **$\approx \frac{3}{2}$**

ii. $x = 0$ **≈ -1**

iii. $x = 4$. **$\approx \frac{3}{2}$**

(Lessons 9-2, 2-5)

a) an odd function; that is, $f(-x) = -f(x)$

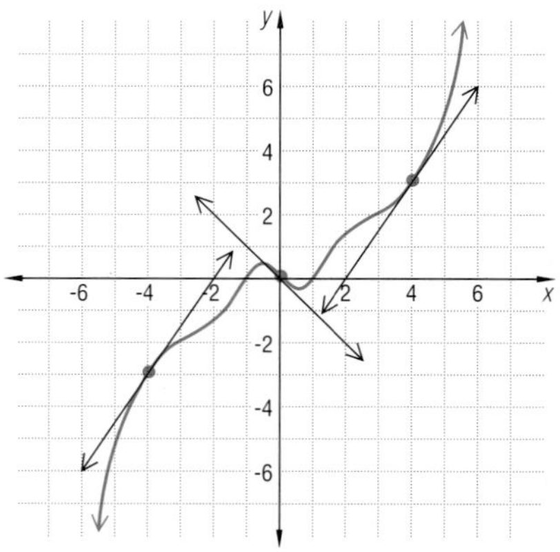

560

21. At the right is graphed $f: x \to -x^2$. Find the average rate of change of f from $x = 0$ to $x = 2$. *(Lesson 9-1)* **-2**

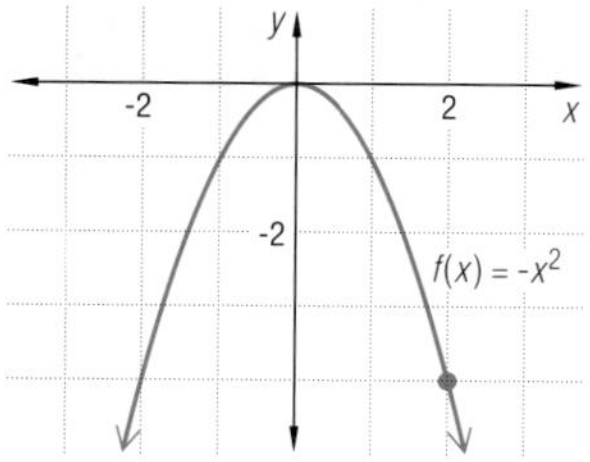

22. **a.** Graph $r = 2 + 3 \cos \theta$ on a polar grid.
b. Identify the curve that results.
(Lessons 8-5, 8-4)
See margin.

23. Find a polar equation for a four-leafed rose curve that intersects the polar axis at $[10, 0°]$. *(Lesson 8-5)* **See margin.**

24. Snell's Law describes the behavior of light when it travels from one medium into another. It states that $n_1 \sin \theta_1 = n_2 \sin \theta_2$, where n_1 and n_2 are the indices of refraction for the first and second media, and θ_1 and θ_2 are the angles from the vertical at which the light crosses the boundary between the media. Suppose light in air (index of refraction = 1.0) traveling at an angle of 60° with the vertical enters a diamond (index of refraction = 2.42). In the diamond, what angle will the light ray make with the vertical? *(Lessons 6-7, 6-6)* **≈ 21°**

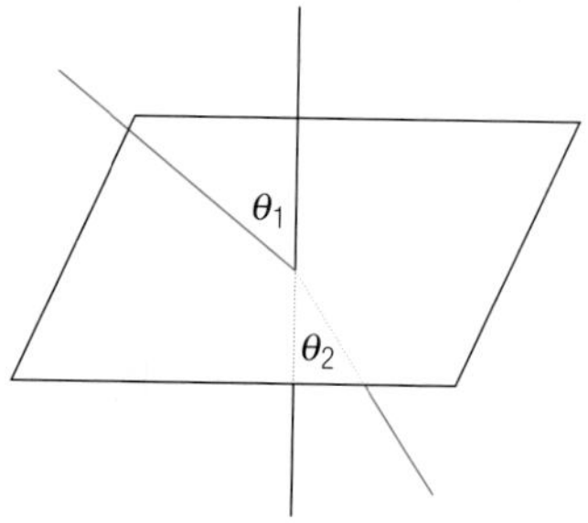

25. Describe the end behavior of the function $g(x) = \dfrac{2x^2 - 2x - 1}{x - 2}$.
(Lesson 5-5) **As $x \to -\infty$, $g(x) \to -\infty$; as $x \to +\infty$, $g(x) \to +\infty$.**

Exploration

26. Obtain information about the derivative function for the function f defined by $f(x) = x^3$:
a. by using the difference quotient to approximate values of f';
b. by using the definition of derivative; and
c. by using an automatic grapher and carefully obtaining slopes of tangents. **a,b,c) If $f(x) = x^3$, then $f'(x) = 3x^2$.**

NAME __________

LESSON **MASTER 9–3**
QUESTIONS ON **SPUR** OBJECTIVES

SKILLS *Objective B (See pages 581–584 for objectives.)*
In 1–4, find the derivative of the function whose formula is given.

1. $f(x) = -6x$ **-6**
2. $p(x) = 2x^2 - 6$ **4x**
3. $q(x) = \frac{1}{2}x^2 - 4x + 650$ **x − 4**
4. $g(x) = 620$ **0**

USES *Objective D*
5. If \$500 is invested at an interest rate of 6% compounded continuously, the amount in the account after t years is $A(t) = 500e^{.06t}$ dollars. The derivative of A is $A'(t) = 30e^{.06t}$.
a. Find the amount in the account after 10 years. **\$911**
b. Find $A'(10)$. **\$55**
c. What does the answer to part **b** mean? **sample: After 10 years, the amount is increasing by \$55 per year.**

6. A piece of metal expands by heat so that the surface area at time t (in minutes) is $A(t) = \frac{\pi}{10}t^2 + \frac{1}{6}t + 3$.
a. Find the surface area at $t = 10$ minutes. $4\frac{2}{3} + 10\pi$
b. Find the derivative of $A(t)$. $A'(t) = \frac{\pi}{5}t + \frac{1}{6}$
c. Find the instantaneous rate of change of the surface area when $t = 10$ minutes. $2\pi + \frac{1}{6}$
d. Calculate $A'(2)$. $\frac{2}{5}\pi + \frac{1}{6}$
e. What is the initial rate of change of the surface area (that is, at $t = 0$)? $\frac{1}{6}$

Precalculus and Discrete Mathematics © Scott, Foresman and Company *Continued* **105**

NAME __________
Lesson MASTER 9–3 (page 2)

USES *Objective E*
7. A particle moves so that the distance s traveled in meters at time t seconds is given by $s(t) = t^2 + 3t - 2$.
a. Find the average velocity between 2 and 3 seconds. **8 m/sec**
b. What is the instantaneous velocity of the particle at time $t = 6$? **15 m/sec**
c. What is the initial velocity of the particle? **3 m/sec**

REPRESENTATIONS *Objective H*
8. The function g is graphed at the right.

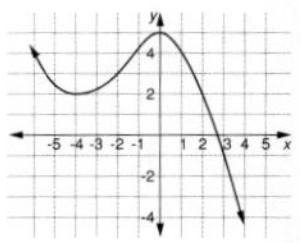

a. Estimate the values of $g'(x)$ when $x = -5, -4, -2, 0,$ and 2.
$g'(-5) = -1$, $g'(-4) = 0$, $g'(-2) = 1$, $g'(0) = 0$, $g'(2) = -3$

b. Use this information and the graph of g to sketch a graph of g'.

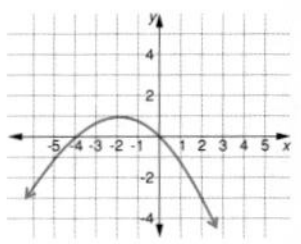

106 *Precalculus and Discrete Mathematics* © Scott, Foresman and Company

LESSON 9-4

RESOURCES
■ Lesson Master 9-4

OBJECTIVES

D Find rates of change in real situations.
E Use derivatives to find the velocity and acceleration of a moving object.

TEACHING NOTES

This lesson applies the ideas of the first three lessons to the first derivative. The example on this page shows that taking the rate of change of a rate of change has meaning. Point out also that the business example from the opening page of the chapter is about a rate of a rate.

Some students may be familiar with the fact that acceleration tells how fast velocity is changing. Thus, it is natural that acceleration is the derivative of the velocity. The discussion provides a natural lead-in to acceleration being the second derivative of distance from a starting point.

Point out that the use of the term *average acceleration* applies whether the function is continuous or discrete. The term *instantaneous acceleration,* however, can be used only for a continuous function with first and second derivatives.

You might wish to extend the graph-sketching activities of the first three lessons. Give students a time versus distance graph of a car trip.

LESSON

9-4 Acceleration and Deceleration

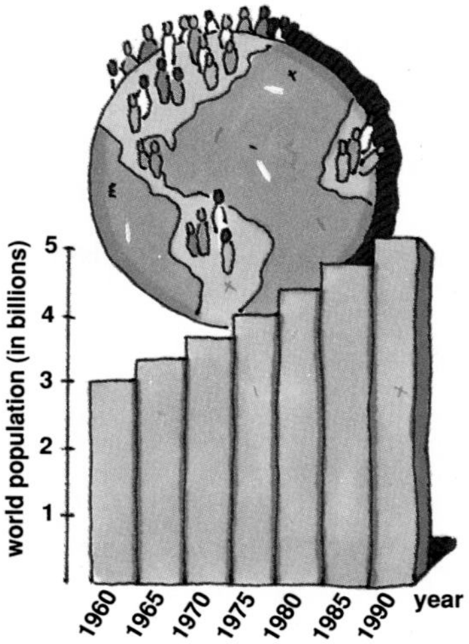

Below are estimates of the world population in recent years, from the United States Bureau of the Census. In the right column are average rates of change of population per year between the successive values in the middle column. For instance, the average rate of change of population between 1975 and 1980 is calculated as

$$\frac{4{,}473{,}000{,}000 - 4{,}103{,}000{,}000}{1980 - 1975} = 74{,}000{,}000 \ \frac{\text{people}}{\text{year}}.$$

Year	World Population	Average rate of change during previous 5 years $\left(\frac{\text{people}}{\text{year}}\right)$
1960	3,049,000,000	
1965	3,358,000,000	61,800,000
1970	3,721,000,000	72,600,000
1975	4,103,000,000	76,400,000
1980	4,473,000,000	74,000,000
1985	4,882,000,000	81,800,000
1990 (est.)	5,315,000,000	86,600,000

The world population has not been constant. Because the population is growing, the average rates of change are positive. The rate of change has also not been constant. For most of the time periods, the rate of change has been growing. It is said that the growth is *accelerating*. **Acceleration** is the rate of change of a rate. Average accelerations are given in the rightmost column below. The unit in that column is $\left(\frac{\text{people}}{\text{year}}\right)$ per year. For instance, the number 760,000 means that in the years from 1970 to 1975, the amount of population increase per year itself increased by an average of 760,000 each year. It was calculated using the average rate of change numbers for 1970 and 1975.

$$\frac{76{,}400{,}000 \ \frac{\text{people}}{\text{year}} - 72{,}600{,}000 \ \frac{\text{people}}{\text{year}}}{1975 - 1970} = 760{,}000 \ \frac{\left(\frac{\text{people}}{\text{year}}\right)}{\text{year}}$$

Year	World Population	Average rate of change	Average rate of change of average rate of change
1960	3,049,000,000		
1965	3,358,000,000	61,800,000	
1970	3,721,000,000	72,600,000	2,160,000
1975	4,103,000,000	76,400,000	760,000
1980	4,473,000,000	74,000,000	-480,000
1985	4,882,000,000	81,800,000	1,560,000
1990 (est.)	5,315,000,000	86,600,000	960,000

From 1975 to 1980, the average rate of change of population lessened. It was 76,400,000 people per year from 1970 to 1975; it became 74,000,000 people per year from 1975 to 1980. It is said that the population increase *decelerated*. When acceleration is negative, it is called **deceleration**.

You are, of course, familiar with the terms acceleration and deceleration as used with buses or cars. These words have a similar meaning to that in the population growth example on page 562. A car accelerates when its velocity (rate of change of position) increases. That is, the rate of change of velocity is what is called acceleration. For instance, if a car goes from 20 mph to 60 mph in 5 seconds, then its average acceleration—its average rate of change of velocity—is $\frac{40 \text{ mph}}{5 \text{ sec}}$ or 8 mph per second. A car is said to decelerate when its velocity decreases.

There is a difference between the world population and car acceleration examples. Population data are discrete whereas the positions of a car are continuous. So the instantaneous acceleration for population cannot be defined, but instantaneous acceleration for the motion of an object can be defined.

The **instantaneous acceleration** $a(t)$ of a projectile or other object at time t is defined to be the instantaneous rate of change of its velocity with respect to time at time t. To compute it, take the derivative of the velocity function v at t:

$$a(t) = v'(t) = \lim_{\Delta t \to 0} \frac{v(t + \Delta t) - v(t)}{\Delta t}.$$

If time is measured in seconds and distance in feet, then the units for acceleration are $\frac{\text{ft/sec}}{\text{sec}}$, the same as the units of the difference quotient $\frac{v(t + \Delta t) - v(t)}{\Delta t}$. The unit is written as feet per second per second or ft/sec^2.

Example **a.** Find the instantaneous acceleration $a(t)$ of a projectile whose position at time t is given by $h(t) = 800t - 16t^2$.
b. Determine the instantaneous acceleration of the projectile at $t = 20$ sec.

Solution

a. By definition, the acceleration $a(t)$ is the derivative of the velocity function at t.

Since $v(t) = 800 - 32t$, from Lesson 9-3, Example 1
then $v'(t) = -32$.
$\therefore$ for all t, $a(t) = -32$ ft/sec^2.

b. Since $a(t) = -32$ for *all* values of t, the acceleration at $t = 20$ sec is -32 ft/sec^2.

In general, position, velocity, and acceleration are related as follows:

velocity = derivative of position
acceleration = derivative of velocity
= derivative of (derivative of position).

Another way to state the relationship between position and acceleration is the following:

Let $s(t)$ represent the position of an object at time t. Then
velocity = $v(t) = s'(t)$, and
acceleration = $a(t) = v'(t) = (s')'(t)$.

Because acceleration is a derivative of a derivative, acceleration is said to be the *second derivative* of position with respect to time, written $a(t) = s''(t)$.

Have them sketch plausible corresponding velocity and acceleration graphs. Ask: How is a slowdown in velocity reflected in the acceleration graph?

The **Example** completes the explanation of how the height of a projectile is modeled by a quadratic polynomial. The coefficient of the square term is half of the acceleration.

The graphs on page 564 should be discussed in class. They continue the emphasis on understanding ideas from calculus geometrically. The slopes of the tangents to the curve of $s(t)$, lead to the graph of $s'(t)$. Likewise, the slopes of the tangents to the graph of $s'(t)$ lead to the graph of $s''(t)$. Students should observe that the graph of $s''(t)$ is in agreement with the theorem from the last lesson. The derivative of a quadratic ($s'(t)$, in this case) is a linear function.

ADDITIONAL EXAMPLES

1.a. Using the height function $h(t) = 960t - 16t^2$, find the acceleration of the bullet at any time t.
$a(t) = -32$ ft/sec^2
b. Determine the acceleration of the bullet at time $t = 4$ seconds.
$a(4) = -32$ ft/sec^2

2. When an object moves so that its position at time t seconds is given by $s(t) = \frac{2}{3}t^3 - 8t^2 + 5t + 7$ m/sec, then the velocity of the object is given by $v(t) = 2t^2 - 16t + 5$.
a. Find a formula for the acceleration of the object.
$a(t) = 4t - 16$
b. At time $t = 60$ seconds, is the object speeding up or slowing down?
speeding up

Given an arbitrary function f, the derivative function f' of f is called the **first derivative** of f. The **second derivative** of f is the derivative of the first derivative and is denoted f''.

The functions h, v, and a of the Example are graphed below.

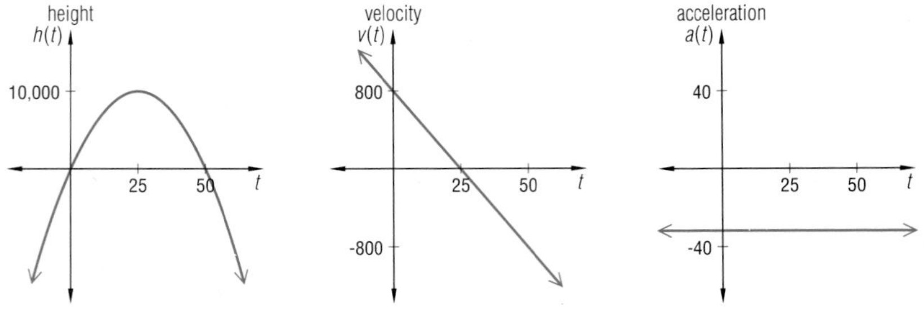

$h(t) = 800t - 16t^2$ ft $\qquad$ $v(t) = 800 - 32t$ ft/sec $\qquad$ $a(t) = -32$ ft/sec^2

The equation $h(t) = 800t - 16t^2$ in the Example is of the form $h(t) = v_0t - \frac{1}{2}gt^2$, the equation for the height of a projectile at time t if v_0 is the initial velocity and g is the acceleration due to gravity. You can see that v_0 is, in fact, the value of the first derivative $v(t)$ when $t = 0$, and g is -32 ft/sec^2, the second derivative of $v(t)$ for all t. This is the constant acceleration due to the gravity of Earth. When the acceleration is constant, people usually do not speak of the *instantaneous acceleration at time t*, but simply *the acceleration*.

Questions

Covering the Reading

In 1–3, use the world population data in this lesson.

1. What does the number 61,800,000 in the third column mean, and how was it calculated? See margin.

2. What is the average rate of change of the average rate of change of population from 1975 to 1980? **-480,000 $\frac{\text{people}}{\text{year}}$ per year**

3. If the average acceleration of the world population remained the same from 1985 to 1990 as it was from 1980 to 1985, what would the population be in 1990? **5,330,000,000 people**

4. A car went from 0 to 60 mph in 6 seconds. What was its average acceleration during this time interval? **10 mph/sec**

5. *True* or *false? If the acceleration of a moving object is negative, then its velocity is decreasing.* **True**

6. *True* or *false? If the velocity of a moving object is negative, then the acceleration of the object is negative.* **False**

ADDITIONAL EXAMPLES

3. The Handy Box Corp. is making open top boxes from 24-in. by 12-in. rectangles of cardboard by cutting out congruent squares in all 4 corners and folding the resulting flaps up to form an open top box. Given that each edge of a cutout square is of length x inches, determine:

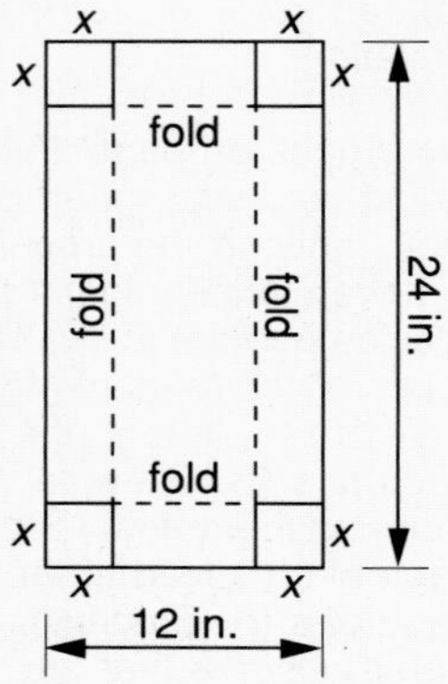

a. a formula $V(x)$ for the volume of a box in terms of x.
$V(x) = 4x^3 - 72x^2 + 288x$

b. Given that $V'(x) = 12x^2 - 144x + 288$, find a formula for the acceleration of volume as a function of x.
$V''(x) = 24x - 144$

c. What is the rate of change of volume at $x = 1$?
$V'(1) = 156$ in.^{3} per in.

d. What is the "acceleration" at $x = 5$? What does this negative acceleration imply?
$V''(5) = -24$ in.^{3} per in.^{2}; the volume is changing at a slower rate.

e. Graph $y = V(x)$, $y = V'(x)$, and $y = V''(x)$ on the same coordinate system.
Use an automatic grapher with settings $0 \le x \le 6$ and $-200 \le y \le 400$.

ADDITIONAL ANSWERS

1. 61,800,000 is the average rate of change in world population for the years 1960 to 1965. It is found by taking one-fifth of the difference between the 1965 and 1960 populations.

10. (graphs), 12. See Additional Answers in the back of this book.

564

7. a. Is the height of the projectile in the Example increasing or decreasing at time $t = 3$ seconds? increasing
   b. Is its velocity increasing or decreasing at this time? decreasing
   c. Is its acceleration increasing or decreasing at this time?
   Neither; it is always -32 ft/sec^2.

8. Let $s(t) = 10 + 15t - 4.9t^2$ be the height (in meters) of an object at time t (in seconds).
   a. Find the velocity $v(t)$ at time t. $v(t) = s'(t) = 15 - 9.8t$ m/sec
   b. Find the acceleration $a(t)$ at time t. $a(t) = v'(t) = s''(t) = -9.8$ m/sec^2

Applying the Mathematics

In 9 and 10, a hot object is placed on a table to cool. The temperature (in degrees Fahrenheit) of the object after x minutes is given by

$$f(x) = 80e^{-0.555x} + 70.$$

9. What units should be used to measure the second derivative of f?
   $\frac{\text{degree Fahrenheit}}{\text{min}}$ per minute or degrees Fahrenheit/min^2

10. For this function, the first derivative is $f'(x) = -44.4e^{-0.555x}$ and the second derivative of f is $f''(x) = 24.642e^{-0.555x}$. Use an automatic grapher to graph f, f', and f'' over the interval $0 \le x \le 10$. (Use a separate set of axes for each function.) See margin.
    a. How fast is the rate of cooling changing at time $t = 2$ minutes?
    b. How fast is the temperature changing at time $t = 2$ minutes?
    c. At what time during the first 10 minutes is the rate of cooling changing the fastest? at the beginning; $t = 0$
    d. At what time during the first 10 minutes is the rate of cooling changing the slowest? at the end; $t = 10$
    a) $\approx$ 8.12 degrees/min; b) $\approx$ -14.63 degrees/min^2

11. Refer to the world population data in this lesson. If the average acceleration of the world population remained the same from 1990 to 2025 as it was from 1985 to 1990, what would the population be in 2025? 9,018,000,000

Review

12. A function f is graphed at the right along with some tangents to the graph. Estimate $f'(x)$ at the points shown and sketch a graph of f'. *(Lesson 9-3)*
    $f'(1) \approx -2$, $f'(4) \approx 1$, $f'(6) \approx \frac{1}{2}$
    See margin.

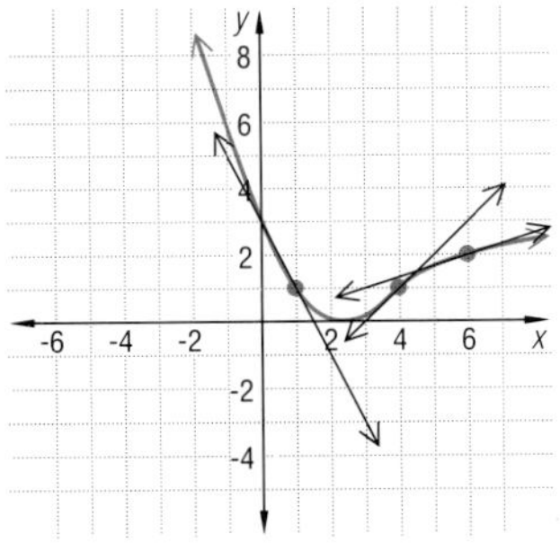

13. It can be shown that the derivative of the natural logarithm function is the reciprocal function. That is, the derivative of $g(x) = \ln x$ is $g'(x) = \frac{1}{x}$ for all $x > 0$. What is the slope of the tangent line to the graph of g at the point $(2, g(2))$? *(Lesson 9-3)* $\frac{1}{2}$

NOTES ON QUESTIONS

Questions 5 and 6: The distinction between these two statements is important in understanding the geometric interpretation of acceleration. The statement in Question 6 is false because an object could have negative velocity and positive acceleration, as, for example, an object coming down to earth at a faster and faster pace.

Questions 9 and 10: These questions preview exponential functions in Lesson 9-6. At this point, you might just mention that exponential functions have the property that their derivatives are exponential with the same base.

Question 12: This question again requires students to graph the first derivative by investigating the graph of the function. You might wish to do this problem in class and then extend it by asking: What would the graph of the second derivative look like? (Because the tangent line to the first derivative has constant slope, the graph of the second derivative is a horizontal line.)

NAME ______

LESSON **MASTER 9–4**
QUESTIONS ON **SPUR** OBJECTIVES

USES *Objective D (See pages 581–584 for objectives.)*

1. A pot of water is being heated on a stove. The temperature (in degrees Fahrenheit) of the water after t minutes is given by $F(t) = 212 - 140e^{-.5t}$. For this function, $F'(t) = 70e^{-.5t}$ and $F''(t) = -35e^{-.5t}$.
   a. What units should be used to measure $F'(t)$? degrees/min
   b. What units should be used to measure $F''(t)$? degrees/min^2
   c. How fast is the temperature changing at time $t = 2$ minutes? 25.8 degrees/min
   d. How fast is the rate of heating changing at time $t = 2$ minutes? -12.9 degrees/min^2
   e. How fast is the temperature changing at time $t = 3$ minutes? 15.6 degrees/min

USES *Objective E*

2. A kite which is 125 feet high is rising vertically. The height of the kite after t minutes is given by $h(t) = 2t^2 + 6t + 125$.
   a. How fast is the string being played out when $t = 2$ minutes? 14 ft/min
   b. What is the vertical acceleration of the kite at $t = 10$ minutes? 4 ft/min^2
   c. Is the velocity of the kite changing at a constant rate? yes

3. A ball thrown directly upward with a speed of 96 ft/sec moves according to the equation $h(t) = 96t - 16t^2$, where h is the height of the ball in feet and t is the time in seconds after the ball is thrown.
   a. Find the velocity of the ball 2 seconds after it is thrown. 32 ft/sec
   b. For how many seconds does the ball rise? 3 sec
   c. What is the highest height reached by the ball? 144 ft
   d. Find its acceleration at time t. -32 ft/sec^2

4. If the position of a particle at time t is given by $s(t) = (2t + 3)^2$, find its velocity and acceleration at time $t = 0$. $v(0) = 12$, $a(0) = 8$

 107

NOTES ON QUESTIONS
Question 18: Remind students that English words do not always convey the same ideas as their mathematical counterparts. For instance, an acceleration in housing starts seems to mean only an increase in the number of starts, which mathematically refers to the first derivative. But if housing starts means a rate (perhaps a percentage of new to old buildings), then the acceleration refers to the second derivative.

FOLLOW-UP

MORE PRACTICE
For more questions on SPUR Objectives, use *Lesson Master 9-4,* shown on page 565.

EXTENSION
Students might want to research the populations of some countries since 1975. Thirteen European countries have had a zero growth rate, meaning that births and deaths are in equilibrium. Discuss the implications of this situation on the world population.

ADDITIONAL ANSWERS
14.b.

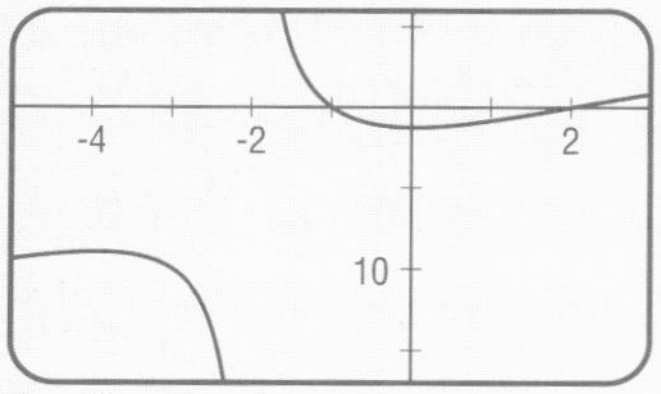

removable discontinuity at x = 1, essential discontinuity at x = -2

15. See Additional Answers in the back of this book.

14. Let $f\colon x \to \dfrac{x^3 - 2x^2 - x + 2}{x^2 + x - 2}$ be a real function.

a. What is the domain of f? **all real numbers except x = -2 or x = 1**

b. Use an automatic grapher to graph f. Classify the discontinuities of f as removable or essential. **See margin.**

c. Write equations for any horizontal or vertical asymptotes. **x = -2**

(Lessons 5-5, 5-4, 2-1)

15. Prove the identity $\sin^2 x = \dfrac{1 - \cos 2x}{2}$.
(Hint: Begin with an identity for cos $2x$.) *(Lesson 6-5)* **See margin.**

16. Given that $\sum_{j=1}^{10} j^3 = 3025$, find:

a. $\sum_{j=1}^{11} j^3$ **4356** **b.** $\sum_{j=1}^{10} (2j)^3$. *(Lesson 7-3)* **24,200**

17. Match each graph with the appropriate parent function.

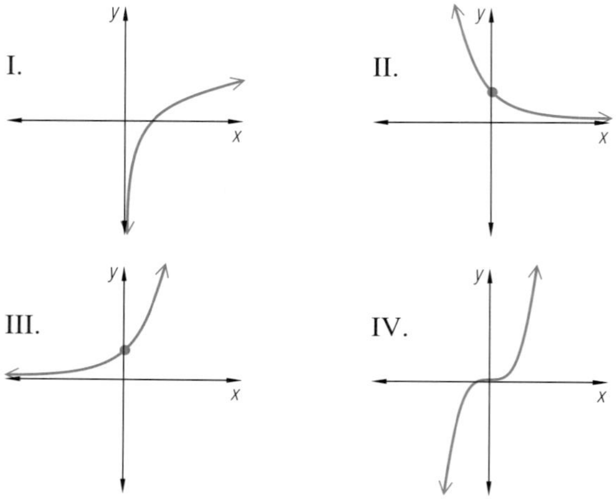

a. $y = e^x$ **III** **b.** $y = x^3$ **IV**

c. $y = \left(\frac{1}{3}\right)^x$ **II** **d.** $y = \ln x$ **I**

(Lessons 2-7, 2-6, Previous course)

Exploration

18. What do the following mean?

a. an acceleration in housing starts (real estate)

b. a deceleration in unemployment (labor)

c. the acceleration principle (economics)

d. the acceleration due to gravity (physics)

Sample:
a) There were more houses begun (to be built) in the current period than in the previous period. b) The unemployment level started dropping more slowly. c) The principle that an increase in the demand for a finished product will create a greater demand for capital goods d) Two bodies attract each other, and that force causes them to come together at an ever-increasing velocity.

LESSON

9-5

Using Derivatives to Analyze Graphs

In Lesson 9-3 it was stated that the derivative of the sine function is the cosine function. Now compare the graphs of these functions in a new way. When the value of the cosine function (in blue) is positive, as it is between $x = -\frac{\pi}{2}$ and $x = \frac{\pi}{2}$, the sine function (in black) is increasing. When the value of the cosine function is negative, as it is between $x = \frac{\pi}{2}$ and $x = \frac{3\pi}{2}$, the sine function is decreasing.

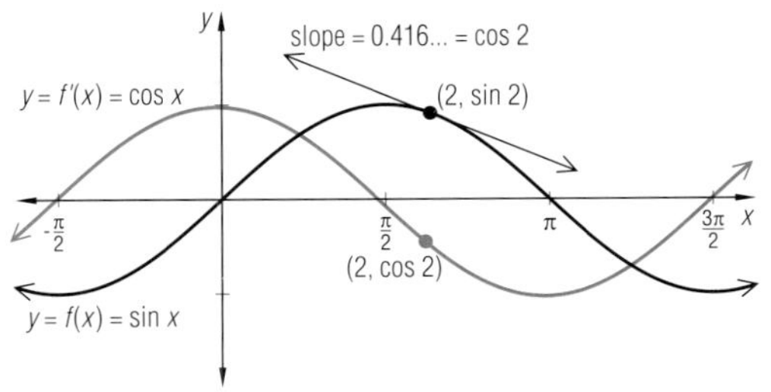

This is because the values of the derivative function are the slopes of lines tangent to the graph of the function. When the slopes of tangents to the graph of a function are positive, the function is increasing. When the slopes of tangents to the graph of a function are negative, the function is decreasing.

These important facts are summarized in the theorem below.

Theorem

Suppose f is a function whose derivative function f' exists for all x in the interval $a < x < b$.

(1) If $f'(x) > 0$ $\forall$ x in the interval $a < x < b$, then f is increasing on the interval.

(2) If $f'(x) < 0$ $\forall$ x in the interval $a < x < b$, then f is decreasing on the interval.

Example 1 Prove that $f(x) = x^3 - 6x^2 + 9x - 5$ is decreasing on the interval $1 < x < 3$ using the above theorem and the fact that $f'(x) = 3x^2 - 12x + 9$.

Solution It would be very difficult to prove that f is decreasing on $1 < x < 3$ by the methods of Lesson 2-3. But, if the above theorem is used, then it is only necessary to show that $f'(x) < 0$ for all x in the interval $1 < x < 3$. This can be done using the Test Point Method from Lesson 3-7. Note that f' is a polynomial function on the real numbers, so it is continuous, and $f'(x) = 3(x^2 - 4x + 3) = 3(x - 3)(x - 1)$. Therefore $f'(1) = 0$ and $f'(3) = 0$ and 1 and 3 are consecutive zeros of f'. A convenient test point is $x = 2$. Since $f'(2) = 3(2)^2 - 12(2) + 9 = -3$, it follows that $f''(x) < 0$ for all x in the interval $1 < x < 3$. Therefore, by the above theorem, f is decreasing on $1 < x < 3$.

LESSON 9-5

RESOURCES
- Lesson Master 9-5
- Teaching Aid 66 can be used with **Questions 10-12**.

OBJECTIVES

C Use derivatives to identify properties of functions.
F Use derivatives to solve optimization problems.
I Determine properties of derivatives from the graph of a function.

TEACHING NOTES

You might wish to introduce this lesson by continuing to use the projectile idea. Suppose a ball is tossed into the air. At first, it is going up (positive velocity, positive derivative) and then coming down (negative velocity, negative derivative). The velocity function is continuous, so there is an intermediate value at which the velocity is zero. At that point, the ball is neither going up nor down; it is at its maximum height.

Making Connections The discussion about relative maxima and minima and the derivative provides an opportunity for reviewing logic from Chapter 1: If a function has a relative maximum or relative minimum value at a point and the derivative exists at that point, then the derivative at that point is zero. The graphs show that the converse is not true. Even if the derivative is zero at a point, there may not be a relative maximum or minimum at that point.

Technology For **Example 1**, use an automatic grapher to display the graphs of $y = f(x)$ and $y = f'(x)$ simultaneously. Point out how zeros and intervals where $y = f'(x)$ is positive or negative are related to the graph of $y = f(x)$.

Students may have proved the Parabola Vertex Theorem algebraically in an earlier course. Now, however, they can use the tools of calculus to provide a sophisticated proof of the vertex relationship. Remind students that $f'(x) = 2ax + b$ was the conclusion of the Derivative of a Quadratic Function Theorem in Lesson 9-3.

In **Example 2**, some students may need to be reminded that while the zero of $h'(t)$ provides the time at which the maximum height is reached, the height itself is calculated using $h(t)$.

In **Example 2**, some students may think that if the amount of fencing is fixed, then the area also is fixed. A table of lengths, widths, and corresponding areas will help students see that this is not the case.

Students sometimes think that optimization problems are all either obvious or contrived. Some students use optimization ideas when studying for tests. These students wish to optimize their total score, so they study long enough to get a good score but not so long that they do not have any free time.

ADDITIONAL ANSWERS

2. increasing: $x < \frac{3}{4}$; decreasing: $x > \frac{3}{4}$

3.a., b., d. See Additional Answers in the back of this book.

The previous theorem discusses nonzero derivatives. If the value of the derivative at a particular point is zero, then the situation is more complex. The point may be a relative maximum or minimum (as in the graph at the left) or the point may indicate where the graph is flat or momentarily "flattens out" (as in the graph at the right).

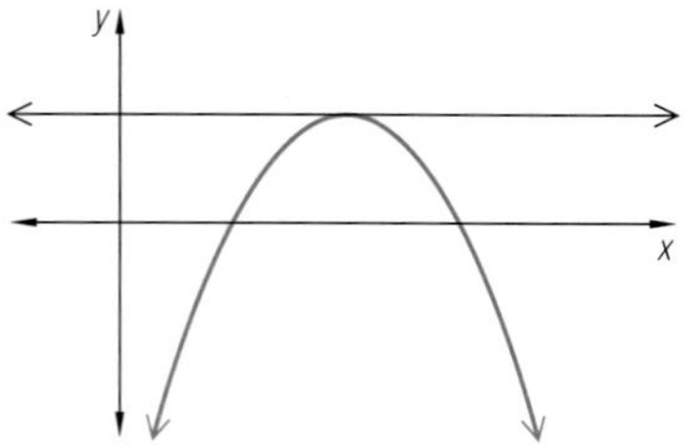

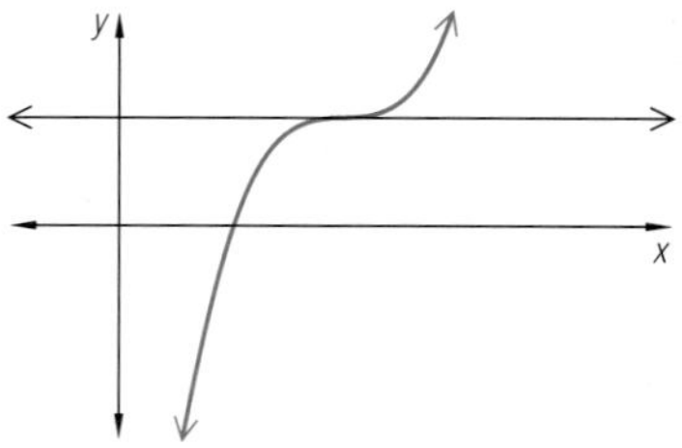

On the other hand, if there is a relative maximum or minimum at a point, and the derivative exists at that point, then it equals zero. This fact can be used to give an elegant proof, using calculus, of a result that you may have seen before.

Parabola Vertex Theorem

Let a, b, and c be real numbers with $a \neq 0$. Then the parabola that is the graph of $f(x) = ax^2 + bx + c$ has its vertex at the point where $x = -\frac{b}{2a}$.

Proof By the Derivative of a Quadratic Function Theorem, f has a derivative at each point, and hence at the vertex. The vertex of a parabola is a relative maximum or minimum. So, the derivative of f is zero at the vertex.

Since $\quad f'(x) = 2ax + b,$

at the vertex, $\quad 0 = 2ax + b.$

Thus $\quad x = -\frac{b}{2a}.$

Example 2 What is the maximum height reached by our favorite projectile, the one whose height at time t is given by $h(t) = 800t - 16t^2$?

Solution Use the above theorem. In $h(t) = 800t - 16t^2$, $a = -16$ and $b = 800$. So the maximum height occurs at time $t = -\frac{b}{2a} = -\frac{800}{-32} =$ 25 seconds. That height is $h(25)$, so substitute $t = 25$ into the formula for $h(t)$: $h(25) = 800 \cdot 25 - 16 \cdot 25^2 = 10{,}000$. The projectile thus reaches a maximum height of 10,000 feet.

The next example illustrates the use of the preceding theorem to solve what is called an **optimization problem**—one in which the value of one variable is sought to obtain the optimal, or most desirable, value of another.

Example 3 A rectangular region adjacent to a building is to be enclosed with 120 feet of fencing. What should the dimensions of the region be in order to maximize the enclosed area?

Solution In this case, the variable to be optimized is the area of the region. The area is a function of length and width, which are to be chosen. However, since the fencing is limited to 120 feet, choosing one dimension, say the width w, forces a choice for the other, the length $\ell = 120 - 2w$. Thus, the area can be described as a function of the width:

$$\begin{aligned} A(w) &= \ell w \\ &= (120 - 2w)w \\ &= 120w - 2w^2. \end{aligned}$$

Notice that the graph of A is a parabola which opens downward, so that A has a maximum at its vertex. Thus, the maximum area is obtained when the width is chosen to correspond to this vertex; that is, when

$$w = -\frac{b}{2a} = \frac{-120}{2(-2)} = 30.$$

Then $$\ell = 120 - 2(30) = 60.$$

Thus the optimal dimensions of the region are 30 ft x 60 ft. The maximum area is 1800 sq ft.

Check Try increasing or decreasing the width just a little and calculating the area. If $w = 30.1$, then $\ell = 59.8$ and $A = 1799.98$ sq ft. If $w = 29.9$, then $\ell = 60.2$ and $A = 1799.98$ sq ft. Since $1799.98 < 1800$, this checks that the value $w = 30$ yields a relative maximum.

Questions

Covering the Reading

1. The derivative g' of a function g is graphed at the right.

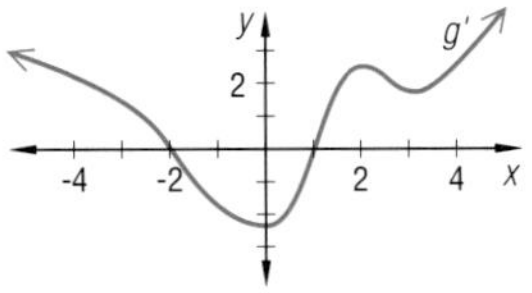

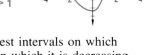

a. Where is g increasing? $x < -2$ or $x > 1$

b. Where is g decreasing? $-2 < x < 1$

2. Use the derivative to determine the largest intervals on which $f(x) = -2x^2 + 3x + 1$ is increasing and on which it is decreasing. See margin.

3. Use the fact that if $f(x) = -x^3 + 3x^2 - 3x$, then $f'(x) = -3x^2 + 6x - 3$.

a. Graph f and f'. See margin.

b. For what values of x does the graph of f' lie below the x-axis? What does this tell you about f? See margin.

c. Find $f'(1)$. 0

d. Does f have a relative maximum or minimum at $x = 1$? Explain. See margin.

4. Assume that the cosine function has a derivative at every point.

a. Find all values of x such that $h(x) = \cos x$ has a relative maximum or minimum at x. $x = n\pi$ for integers n

b. What does your answer to part **a** tell you about h'? $h'(n\pi) = 0$ for all integers n

ADDITIONAL EXAMPLES

1. Given that $f(x) = \frac{1}{3}x^3 - x^2 - 8x + 12$ and that $f'(x) = x^2 - 2x - 8$, determine the intervals on which the function f is increasing or decreasing.
f increases on $x \le -2$ or $x \ge 4$; f decreases on $-2 \le x \le 4$. An automatic grapher can easily verify this solution. Set $-6 \le x \le 10$ and $-50 \le y \le 50$ for a good view.

2. Using the height function from Additional Example 1 of Lesson 9-1, $h(t) = 960t - 16t^2$, what is the maximum height reached by a bullet shot straight up from farmer MacDonald's rifle?
14,400 feet, which is approximately 2.73 miles

3. Suppose that the rectangular region adjacent to the building is to be enclosed and divided into 3 separate yards using the 120 feet of fencing. What dimensions will result in a region of total maximum area?

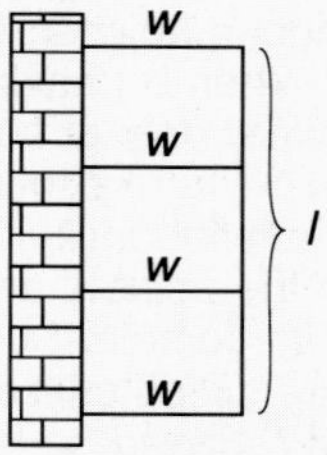

$A(w) = 120w - 4w^2$, so $A'(w) = 120 - 8w$. The maximum area is achieved if $w = 15$ ft and $\ell = 60$ ft. This area is 900 square feet.

NOTES ON QUESTIONS

Question 3: If you did exploration Question 26 of Lesson 9-3, then you can explain how $f'(x)$ is found from $f(x)$ without having to go through the calculation. This question relates the sign of the derivative to whether the function is increasing or decreasing and also relates the derivative to maximum and minimum values.

NOTES ON QUESTIONS
Questions 10-12:
These questions reinforce the geometric interpretation of the derivative.

ADDITIONAL ANSWERS
8.d., e., 9., 15.a., 17.b.
See Additional Answers in the back of this book.

FOLLOW-UP

MORE PRACTICE
For more questions on SPUR Objectives, use *Lesson Master 9-5,* shown on page 571.

PROJECTS
The projects for Chapter 9 are described on page 577. **Project 1** is related to the content of this lesson.

EVALUATION
Alternative Assessment
In the discussion preceding the Parabola Vertex Theorem there is the following result: A continuous function f has relative extrema (maxima and minima) only at points in its domain where either (1) $f'(x)$ is undefined, (2) $f'(x) = 0$, or (3) the point is an endpoint of its domain. Such points are called critical points of f. Based on this result, answer the following questions:

(1) If $f'(x) = 4x(x - 2)(x + 3)$:
(a) How many relative extrema might the function $f(x)$ have? (at most, 3 relative extrema)
(b) At what x-values could these relative extrema occur? (at $x = 0$, $x = 2$, $x = -3$)

5. If $f(x) = 2x^3 - x^2 + 1$, then $f'(x) = 6x^2 - 2x$. Is f increasing or decreasing on the interval $0 < x < \frac{1}{3}$? decreasing

6. A rock is thrown upwards so that its height (in feet) t seconds after being thrown is given by $h(t) = -16t^2 + 30t + 10$. How high does it go? $\approx$ 24 ft

7. Let P be a positive real number. Find the dimensions of the rectangle with maximal area that has perimeter P. $\ell = w = \frac{P}{4}$

Applying the Mathematics

8. Use the fact that if $f(x) = \frac{1}{3}x^3 - x^2 - 3x + 5$, then $f'(x) = x^2 - 2x - 3$.
a. Compute f''. $f''(x) = 2x - 2$
b. Determine the intervals on which f is increasing.
c. Determine the intervals on which f is decreasing.
d. Use f' to find possible relative minima and maxima of f.
e. Check your answers to parts **b–d** above by plotting f using an automatic grapher. d, e) See margin.
b) f is increasing for $x < -1$ or $x > 3$. c) f is decreasing for $-1 < x < 3$.

9. On a certain interval, as one travels along the graph of $y = f(x)$ from left to right, the slopes of the tangents to the curve are increasing. Must the function be increasing on that interval? Explain your answer. See margin.

Multiple choice. In 10–12, the graph of f' is given at the left. Tell which of the other graphs could be f.

10.
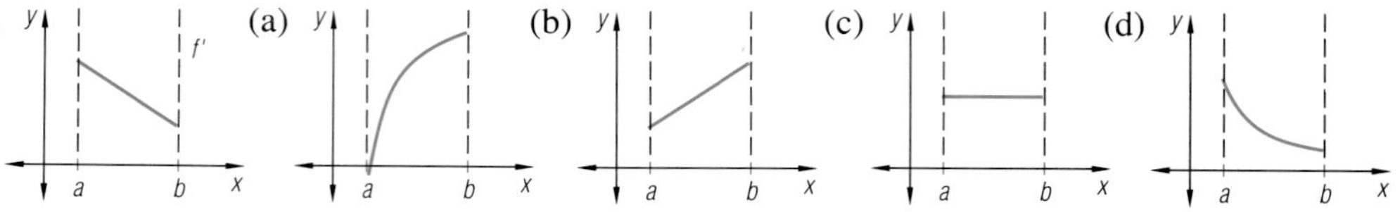

11.
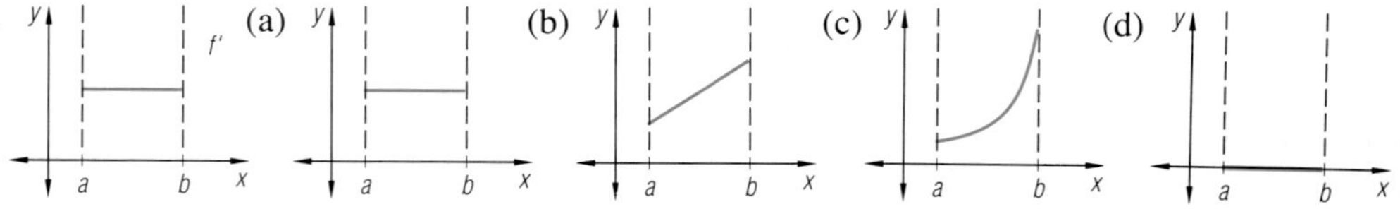

12.
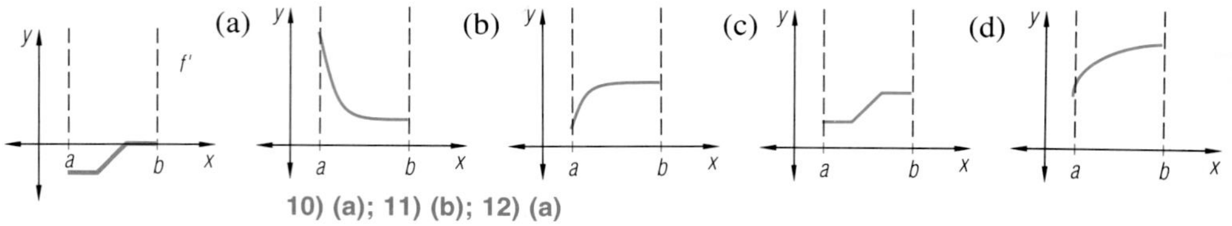

10) (a); 11) (b); 12) (a)

Review

13. A particle moves so that its position (in feet) after t seconds is given by $f(t) = 2t^2 + 3t - 1$. Find the instantaneous velocity of the particle at time $t = 3$ seconds. What is the acceleration at $t = 3$ seconds? *(Lessons 9-4, 9-3)* $v(3) = 15$ ft/sec; $a(3) = 4$ ft/sec^2

14. Let $f(x) = e^{-x^2}$. The derivative function of f is $f'(x) = -2xe^{-x^2}$. Find the slope of the tangent line to the graph of f at the point $(1, f(1))$. *(Lessons 9-3, 9-2)* $f'(1) = -2e^{-1} \approx -.74$

15. The table below gives the average price of a certain stock each month throughout a particular year.

month	Jan	Feb	Mar	Apr	May	Jun	Jul	Aug	Sep	Oct	Nov	Dec
price ($)	22	20	19	18.5	19	20.5	23	22.5	21.5	20	22	22.5

a. Draw a scatterplot of price vs. month.
b. Determine whether each statement is *true* or *false*.
   i. The average price of the stock is rising from April to July. True
   ii. The average price of the stock is falling from October to December. False
   iii. The rate at which the average price of the stock is falling increases from July to October. True
   iv. The rate at which the average price of the stock is rising decreases from April to July. *(Lesson 9-4)* False

a) See margin.

16. a. Let $f(x) = 3x^2 + 1$ for all real numbers x. Use the difference quotient to write an expression for the average rate of change of f over the interval from -1 to $-1 + \Delta x$. $-6 + 3\Delta x$
b. Use this expression to find the average rate of change of f over the interval $-1 < x < -0.9$. *(Lesson 9-1)* -5.7

17. a. Give the measure, to the nearest degree, of the acute angle between the lines with equations $y = 3x - 1$ and $y = 4x + 5$. $\approx 4.4°$
b. Graph the lines by hand and measure the angle to check your answer to part **a**. *(Lesson 6-4)* See margin.

18. Let $f(x) = e^x$ and $g(x) = \ln x$.
a. Find the formula for h if the graph of h is the image of the graph of f under $T_{2,3}$. $h(x) = e^{x-2} + 3$
b. Find the formula for k if the graph of k is the image of the graph of g under $T_{3,2}$. $k(x) = \ln(x - 3) + 2$
c. Find a formula for $h \circ k$. *(Lessons 3-9, 3-2)* $h \circ k(x) = x, x \geq 3$

19. What is the probability that at least one head will appear in two tosses of a fair coin? *(Previous course)* .75

20. Fill in the blank.
a. $6! = \underline{\ ?\ } \bullet 5!$ 6 b. $8! = 8 \bullet \underline{\ ?\ }$ 7!
c. $n! = n \bullet \underline{\ ?\ }$ for any positive integer n. $(n - 1)!$
(Previous course)

Exploration

21. In this lesson you learned a theorem which told you how to get information about a graph by using derivatives. The converse of each part of this theorem is false. Find counterexamples as explained below to show this.
a. Find a function f that is increasing on an interval but whose derivative is not positive on the interval.
b. Find a function f that is decreasing on an interval but whose derivative is not negative on the interval.

a) Sample: $f(x) = x^3$ is increasing on the set of reals but at $x = 0$, $f'(x) = 0$.
b) Sample: $f(x) = -x^3$ is decreasing on the set of reals, but at $x = 0$, $f'(x) = 0$.

(2) The graph of a function is shown below:

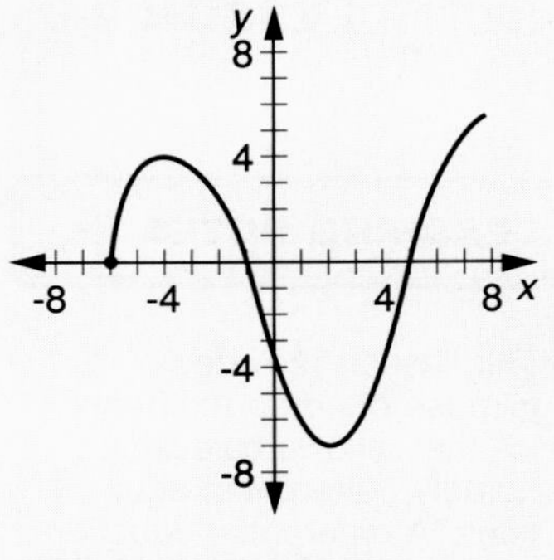

(a) What is the domain of g? ($\{x: x \geq -6\}$)
(b) What are the approximate x-values of its critical points? ($x = -6$, $x = 2$, $x = -4$)

(3) If $f(x) = \ln x$, $f'(x) = \frac{1}{x}$. Given its derivative, could the natural logarithm function have any relative extrema? (No; the domain of f is $\{x: x > 0\}$, and over this interval, $f'(x) = \frac{1}{x}$ is continuous and always positive. So no critical points exist, and there cannot be any relative extrema either.)

NAME ______

LESSON **MASTER 9–5**
QUESTIONS ON **SPUR** OBJECTIVES

PROPERTIES *Objective C (See pages 581–584 for objectives.)*

1. Suppose f is a function such that $f(x) = \frac{1}{3}x^3 - \frac{1}{2}x^2 - 2x$. Then $f'(x) = x^2 - x - 2$. Use the first derivative to find
a. the interval(s) on which f is increasing, $x < -1, x > 2$
b. the interval(s) on which f is decreasing, $-1 < x < 2$
c. the points at which f may have a relative maximum or minimum. $x = -1, 2$

2. Suppose $f(x) = 2x^3 + x - 620$. Then $f'(x) = 6x^2 + 1$. Is f increasing on the set of all real numbers? Justify your answer.
Yes; $f'(x) = 6x^2 + 1 > 0$ for all x. Therefore, f is an increasing function.

USES *Objective F*

3. What rectangle with the largest area can be formed using 80 ft of fencing?
20 ft × 20 ft square (area: 400 ft²)

REPRESENTATIONS *Objective I*

4. The derivative g' of a function g is graphed at the right.

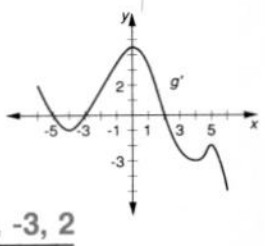

a. Where is g increasing? decreasing?
$-6 < x < -5, -3 < x < 2$;
$-5 < x < -3, 2 < x < 6$
b. Where may g have a relative maximum or minimum? $x = -5, -3, 2$

5. Consider the function f graphed at right.

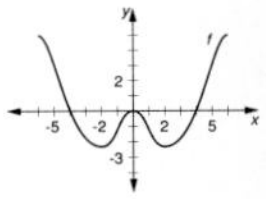

a. On what interval(s) is $f'(x)$ positive? $-2 < x < 0, 2 < x < 6$
b. On what interval(s) is $f'(x)$ negative? $-6 < x < -2, 0 < x < 2$
c. For what value(s) of x is $f'(x) = 0$? $x = -2, 0, 2$
d. Is $f''(x)$ greater than or less than 0 on the interval $-4 < x < -2$? $f''(x) > 0$

108 *Precalculus and Discrete Mathematics* © Scott, Foresman and Company

LESSON 9-6

TEACHING NOTES

This lesson provides a glimpse of some mathematics that follows calculus, namely, differential equations. A differential equation is an equation that involves a function and one or more of its derivatives. Differential equations is an important course for all engineers and physical scientists.

The simplest differential equations are those that ask for the function given its derivative. Finding such antiderivatives is a basic point of integral calculus. The next simplest are those in which the rate of change of a function is proportional to the values of the function; that is, equations of the form $f'(x) = kf(x)$. The solutions to these equations are exponential functions, the same exponential functions that students have seen before.

This lesson relates those prior experiences to derivatives by considering a situation involving bacteria growth, and generalizing from it the conclusion that the rate of change of the function at a given value of x is a fixed multiple of that value.

LESSON

9-6

Derivatives of Exponential Functions

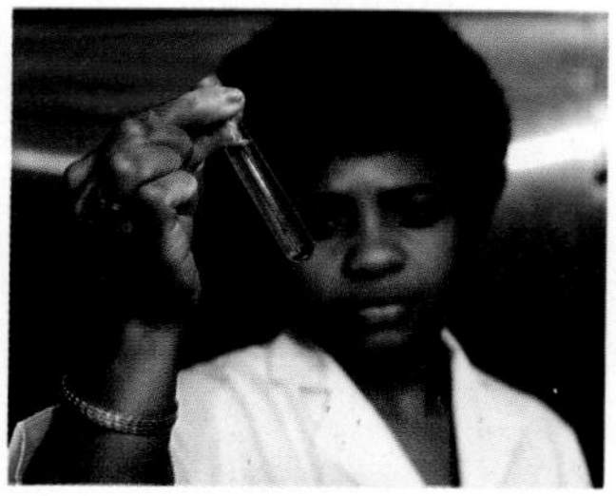

Exponential functions can be used to model bacterial growth.

In this chapter, you have encountered derivatives of a few elementary functions. Here is a quick summary.

Function f	Derivative f'	
sine	cosine	Lesson 9-3
quadratic	linear	Lesson 9-3
cubic	quadratic	Questions in Lessons 9-3, 9-5
natural log	reciprocal	Questions 13, 19 in Lesson 9-4

For each of these functions f, the function f' which maps x onto the derivative $f'(x)$, the instantaneous rate of change of f at x, is another familiar function.

Now consider exponential functions, those functions f of the form $f(x) = ab^x$ for all x, with $a > 0$, $b > 0$, and $b \neq 1$. From Lesson 2-6, recall that these functions model many growth situations. For instance, if a population of bacteria begins with 5 bacteria and doubles every 3 hours, then $f(x) = 5 \cdot 2^{x/3}$ gives the population after x hours.

For a fixed Δx, an exponential function increases or decreases by an amount proportional to the value of the function. In the example above, at 3 hours there are 10 bacteria, and they will increase by 10 in the next three hours. At 6 hours there are 20 bacteria, and they will increase by 20 in the next three hours.

If we let Δx go to zero, then the instantaneous rate of change at a point is obtained. It is also proportional to the value of the exponential function at that point. This can be seen as follows. Begin with

$$f'(x) = \lim_{x \to 0} \frac{f(x + \Delta x) - f(x)}{\Delta x}$$

$$= \lim_{x \to 0} \frac{ab^{x + \Delta x} - ab^x}{\Delta x} \qquad \text{Substitution}$$

$$= \lim_{x \to 0} \frac{ab^x(b^{\Delta x} - 1)}{\Delta x}.$$

Since ab^x is independent of Δx, it can be factored out of the limit. So

$$f'(x) = \left(\lim_{x \to 0} \frac{b^{\Delta x} - 1}{\Delta x}\right) ab^x.$$

It can be proved that for $b > 0$ and $b \neq 1$, the limit in the parentheses above exists and equals ln b. While a proof is beyond the scope of this book, a few calculations for small values of Δx shows that this is plausible. For example, examine the values below for $b = 2$ (as in the bacteria growth example).

Δx	$\frac{2^{\Delta x} - 1}{\Delta x}$
.1	0.7177
.01	0.6956
.001	0.6934
.0001	0.6932

572

Since $\ln 2 = 0.693147\ \ldots$, this result supports the conclusion that $\lim\limits_{x \to 0} \frac{2^{\Delta x} - 1}{\Delta x} = \ln 2$. Then when $f(x) = a \bullet 2^x$,

$$f'(x) = \left(\lim_{x \to 0} \frac{2^{\Delta x} - 1}{\Delta x}\right) \bullet a \bullet 2^x = (\ln 2)\ a \bullet 2^x = (\ln 2)\ f(x).$$

In general, when $f(x) = ab^x$, then

$$f'(x) = (\ln b)\ f(x).$$

In words, the derivative of an exponential function is the product of the natural log of the base times the original exponential function. Since $\ln b$ is a constant, every exponential function thus possesses the property that its instantaneous rate of change is proportional to its value. Moreover, when $b = e$, then $\ln b = \ln e = 1$. Consequently, the exponential function with base e, defined by $f(x) = e^x$, equals its derivative $f'(x)$ for all real numbers x!

That is, the instantaneous rate of change of this function at any point equals the value of the function at that point. Geometrically, at each point (a, e^a) on the graph of the function, the slope of the tangent $f'(a)$ equals $f(a) = e^a$.

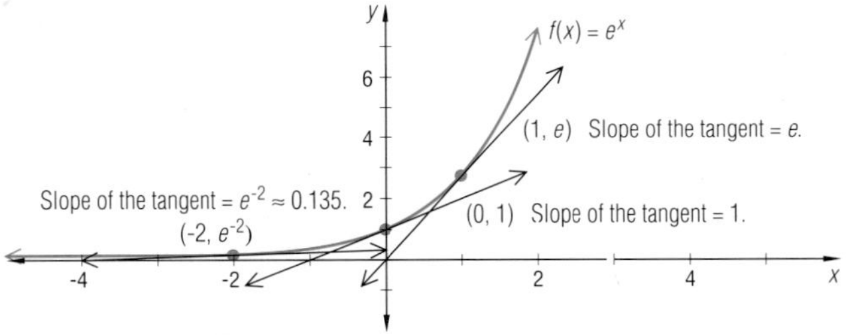

This can be confirmed by examining the graphs of functions of the form

$$f_{\Delta x}(x) = \frac{e^{x + \Delta x} - e^x}{\Delta x}$$

for values of Δx approaching zero. Below are graphed these functions for $\Delta x = 0.5$, 0.3, and 0.1, along with $f(x) = e^x$ itself.

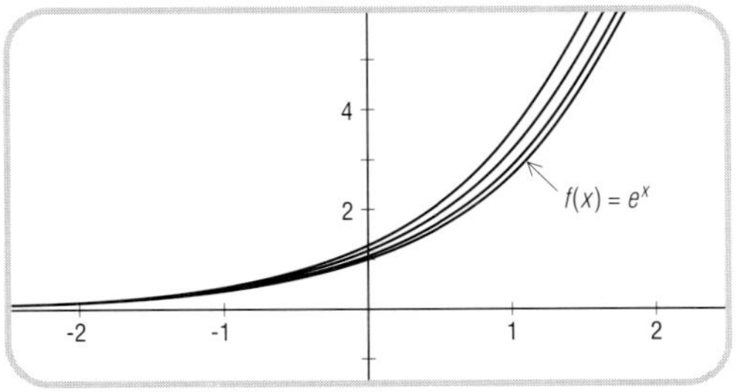

Since the exponential function with base e equals its first derivative, its first derivative also equals its second derivative. The significance of this property is that if a situation is modeled by the exponential function $f(x) = e^x$, then the velocity and acceleration for that situation are also modeled by that same equation.

Making Connections
Point out the properties of the number e. Students have seen e used in a variety of different ways: (1) in the formula $P = e^{rt}$ involving continuous compounding; (2) as the base for natural logarithms; (3) as a sum of an infinite series; (4) in the equation $e^{i\pi} = -1$; and now (5) as the base of the function $y = e^x$, which is fundamental to solving differential equations.

NOTES ON QUESTIONS

Question 4: Some students may skip this question because they think the derivation is long. Remind students that not all proofs are long.

Question 5: Like many questions throughout this chapter, this question would be inconceivable without an automatic grapher.

ADDITIONAL ANSWERS

1.b. $f(6) - f(3) = 20 - 10 = 10$; $f(3) - f(0) = 10 - 5 = 5$; so $f(6) - f(3)$ is double $f(3) - f(0)$.

4. Given $f(x) = ab^x \Rightarrow f'(x) = (\ln b)f(x)$, let $a = 1$ and $b = e$. Then, $f(x) = 1 \cdot e^x \Rightarrow f'(x) = (\ln e)f(x)$. Hence, $f(x) = e^x \Rightarrow f'(x) = f(x)$.

5.a.

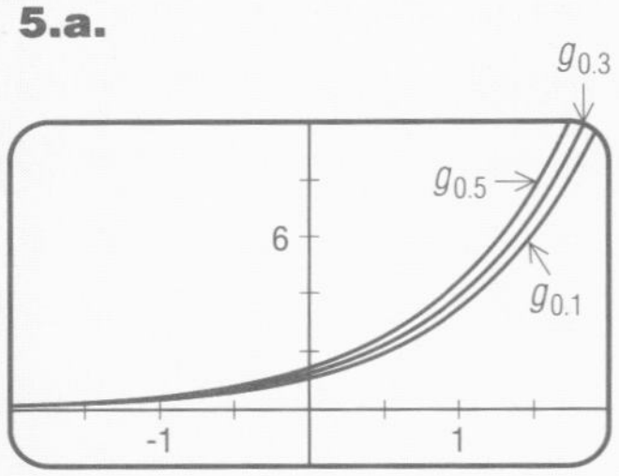

The importance of derivatives, such as those at the beginning of this lesson, is that many physical laws are statements involving quantities, rates of change of those quantities (for instance, velocity), and rates of change of rates of change of those quantities (for instance, acceleration due to gravity). These laws, then, can be succinctly stated in terms of a function and its first and second derivatives.

Given such a property, to determine exactly what function or functions might model the situation, one has to solve an equation involving functions and their derivatives. Finding derivatives is called **differentiation**, and such an equation is called a **differential equation**. Exponential functions with base e uniquely satisfy the simplest differential equation $f' = f$ and are fundamental in solving other, more complicated, differential equations. Differential equations are often used in the study of the physical world, enough so that almost all college programs of study for engineering and physical science require a course in differential equations, often taken just after calculus.

Questions

Covering the Reading

1. Consider the function $f(x) = 5 \cdot 2^{x/3}$ used in this lesson to model a bacteria population.
   a. Evaluate $f(4)$ and tell what it signifies.
   b. Show that $f(6) - f(3)$ is double $f(3) - f(0)$. See margin.
   a) $f(4) \approx 12.6$; there are about 13 bacteria at 4 hours.

2. **a.** Name four points on the graph of $f: x \rightarrow e^x$.
   b. Give the slopes of the tangent to the graph at each of the points you name in part **a**. 1, e, e^2, e^3
   a) sample: (0, 1), (1, e), (2, e^2), (3, e^3)

3. **a.** Calculate $\frac{e^{2+\Delta x} - e^2}{\Delta x}$ for $\Delta x = 0.1$, 0.01, and 0.001.
   b. Conjecture a value for $\lim\limits_{\Delta x \to 0} \frac{e^{2+\Delta x} - e^2}{\Delta x}$. $e^2 \approx 7.389$
   a) ≈ 7.77, ≈ 7.43, ≈ 7.39

4. Derive the statement $f(x) = e^x \Rightarrow f'(x) = f(x)$ from the statement $f(x) = ab^x \Rightarrow f'(x) = (\ln b)f(x)$. See margin.

5. **a.** Graph the function $g_{\Delta x}$ defined by $g_{\Delta x}(x) = \frac{3^{x+\Delta x} - 3^x}{\Delta x}$, for $\Delta x = 0.5$, $\Delta x = 0.3$, and $\Delta x = 0.1$. See margin.
   b. What function is $\lim\limits_{\Delta x \to 0} g_{\Delta x}$? $g': x \rightarrow (\ln 3)3^x$

6. Write Newton's Law of Cooling (from the first page of the chapter) as a differential equation, letting f be the function mapping time onto the temperature of a body and a_0 be the temperature of the surroundings.
   $f'(t) = k(f(t) - a_0)$

574

Review

7. Suppose an object's velocity at time t seconds is given by $v(t) = -2t^2 + 3t$. At what times is the object's velocity increasing? *(Lesson 9-5)* $t < \frac{3}{4}$ sec

8. A function f possesses the following characteristics: The graph of f contains $(-1, 7\frac{2}{3})$, $(1, 1\frac{2}{3})$, and $(3, -3)$; $f'(x) > 0$ on the intervals $x < -1$ and $x > 3$; and $f'(x) < 0$ on the interval $-1 < x < 3$.

a. On which interval(s) is f

i. increasing? $x < -1$ and $x > 3$ **ii.** decreasing? $-1 < x < 3$

b. Sketch a possible graph of f. *(Lesson 9-5)* See margin.

9. At the instant the driver of a car applies the brakes, the car is traveling at 55 mph. After 4 seconds, it is traveling at 25 mph, and 10 seconds after applying the brakes, it comes to a complete stop.

a. Find the average acceleration of the car over each of the two time intervals. -7.5 mph/sec, ≈ -4.2 mph/sec

b. Compare the answers to part **a**. Interpret your comparison in terms of the motion of the car. *(Lesson 9-4)* See margin.

10. The horizontal distance from a vertical plumb line in centimeters of a swinging ball at time t seconds is given by $s(t) = 5 \sin \pi t$, its velocity by $v(t) = 5\pi \cos \pi t$, and its acceleration by $a(t) = -5\pi^2 \sin \pi t$. Consider only values of t in the interval $0 \le t \le 2$. *(Lessons 9-4, 9-3)*

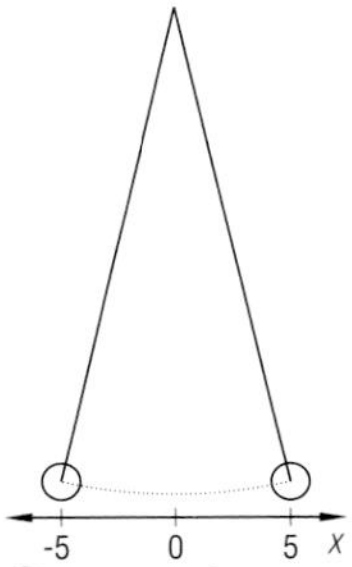

a. At what times is the ball moving the fastest?

b. Where is the ball at these times?

c. At what times it the velocity changing the most?

d. Where is the ball at these times? at the extremes

e. At what times is the velocity changing the least? See margin.

f. Where is the ball at the times of part **e**? vertical; over the zero mark

a) $t = 0$, 1, and 2 seconds; b) vertical; over the zero mark; c) See margin.

11. It is a fact that if $f(x) = 2x^4 - 3x^3 + 4$, then $f'(x) = 8x^3 - 9x^2$. Find the slope of the tangent to the graph of f at the point (1, 3). *(Lesson 9-3)* $f'(1) = -1$

12. At what point of the graph of $g(x) = 2x^2 + 6x - 4$ is the tangent line to the graph horizontal? *(Lesson 9-3)* (-1.5, -8.5)

13. In a controlled experiment, a penny is dropped from the top of the Sears Tower, which is 1454 feet tall. Its height (in feet) after t seconds is given by $h(t) = 1454 - 16t^2$. *(Lesson 9-2)*

a. When does it hit the ground? at about $t = 9.53$ seconds

b. How fast is it falling at the instant it hits the ground?

c. Convert your answer in part **b** to miles per hour. about 208 mph

b) about 305 ft/sec

14. Let $f(x) = -3x^2 - x$.

a. Find a formula for the slope of the secant line through the points $(2, f(2))$ and $(2 + \Delta x, f(2 + \Delta x))$. $m = -13 - 3\Delta x$

b. Use the answer to part **a** to find the slope of the tangent line to the graph of f at $x = 2$. *(Lessons 9-2, 9-1)* -13

8.b. sample:

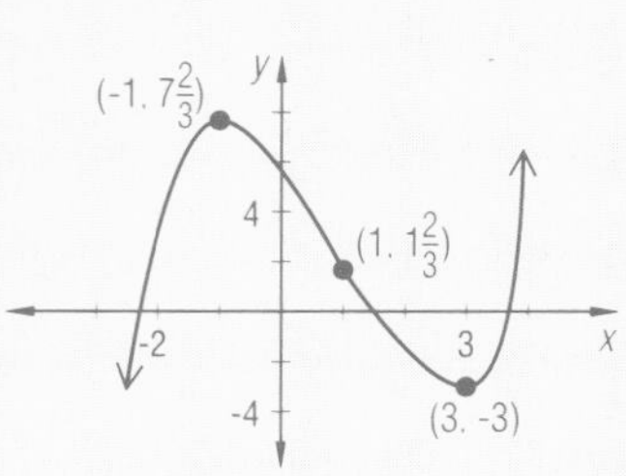

9.b. The car is decelerating more rapidly during the first four seconds of breaking than during the last six seconds.

10.c. $t = \frac{1}{2}$ **and** $\frac{3}{2}$ **seconds**

e. $t = 0$**, 1, and 2 seconds**

ADDITIONAL ANSWERS

15.a.

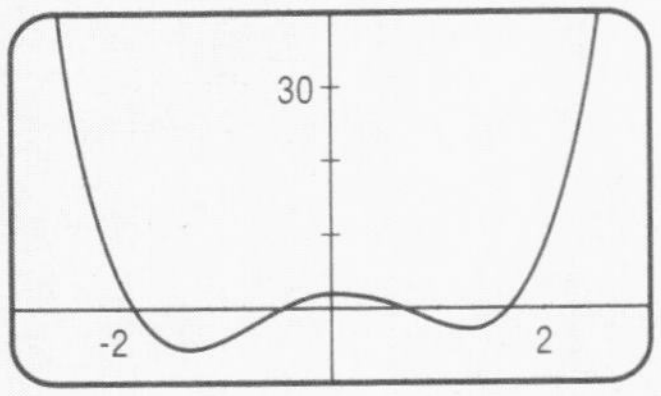

b. increasing: $x < -1.9$, $-.4 < x < .6$, $x > 1.8$; decreasing: $-1.9 < x < -.4$, $.6 < x < 1.8$

c. sample:

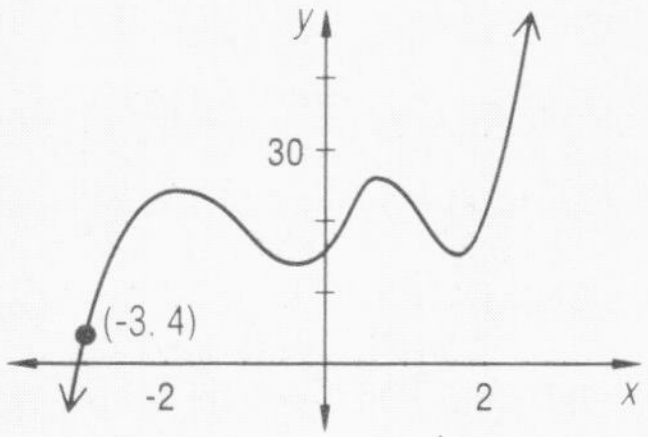

16. $n = 5$; $z = 32i$

17. sample:

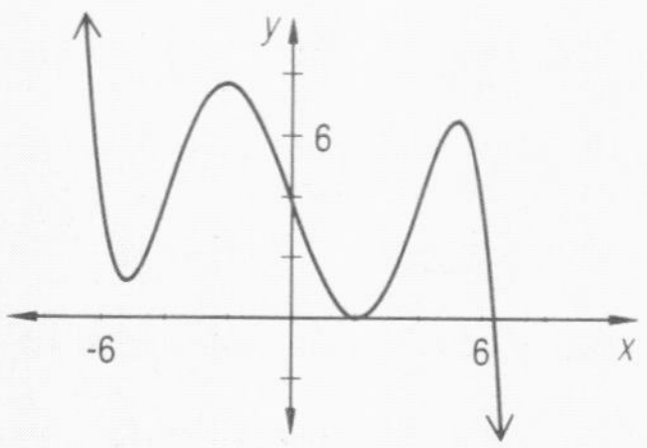

15. Suppose that $f'(x) = 2x^4 - 7x^2 + x + 3$.
  a. Graph f' on the interval $-3 \le x \le 3$.
  b. From your graph in part **a**, estimate when f is increasing and when f is decreasing.
  c. Sketch what the graph of f might be, given that $f(-3) = 4$.
(Lesson 9-5) **See margin.**

16. The vertices of the polygon here are the nth roots of z. What are n and z? *(Lesson 8-7)* **See margin.**

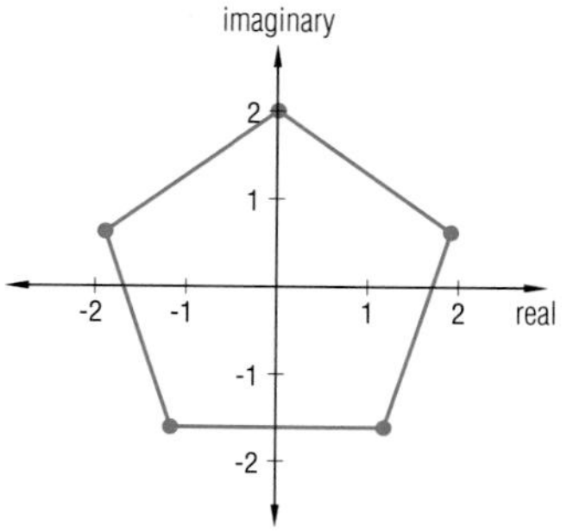

17. Sketch the graph of a fifth-degree polynomial function whose leading coefficient is negative, and which has one real zero of multiplicity 2 and one simple real zero. *(Lessons 8-7, 5-5)* **See margin.**

18. Write the base 2 representations of 8, 16, and 24. *(Lesson 4-7)* **1000, 10000, 11000**

Exploration

19. Find the value of b so that the exponential function f with $f(x) = b^x$ satisfies the differential equation $f' = 2f$. (That is, the slope of the tangent at each point of the graph of f is double the value of the function.) **$b = e^2$**

20. Find a physical law, not mentioned in this chapter, that can be described using a differential equation. **sample: Kirchoff's law for circuits: $E(t) = LI' + RI$ where E is a voltage source, L is an inductance, R is resistance, and I is current**

576

Projects

1. a. In this chapter you have seen many functions whose derivatives exist.
   i. Choose at least 5 of these functions and graph them using an automatic grapher.
   ii. Locate any relative maximum or relative minimum points for the function.
   iii. Evaluate the derivative at these relative minimum or relative maximum points.
   iv. Choose any one of the relative maximum or relative minimum points. Repeatedly zoom in on the function at that location. What appears to happen to the graph?
   v. Relate your result in part **iv** to your knowledge of the derivative and its value in part **iii**.

   b. Not all functions have a derivative at every value in the domain of the function. For instance, $f(x) = |x - 2| + 3$ has a minimum point at (2, 3) but the derivative does not exist there. Repeatedly zoom in on this part of the function. Compare your result to part **iv** above.

   c. Generalize your results from **a** and **b**.

2. Isaac Newton and Gottfried Leibniz are generally credited as the coinventors of calculus. Write a report on the particular contributions that each of these mathematicians made.

3. Use an almanac or some other source to find the time of sunrise where you live for the first day of each month of the year.
   a. Graph your data and draw a smooth curve that approximates it. What kind of curve do you get? Find a formula for the function you have graphed.
   b. Find the average rate of change in time of sunrise from month to month. Graph these data and try to find the formula for a function that fits these data.
   c. When is the time of sunrise changing fastest? When is it changing slowest?
   d. Compare your answers to parts **a–c** with the results of the analysis performed on the length-of-day function in this chapter.

4. *Newton's Method* is another technique that can be used to approximate the zeros of an equation. The technique uses equations of tangent lines to the curve at various points. Refer to the graph of f below. Let x_1 be the first approximation to the zero r. Construct the tangent line to the graph at $(x_1, f(x_1))$. Because its slope is $f'(x_1)$, the equation for this line is
$$y - f(x_1) = f'(x_1)(x - x_1).$$

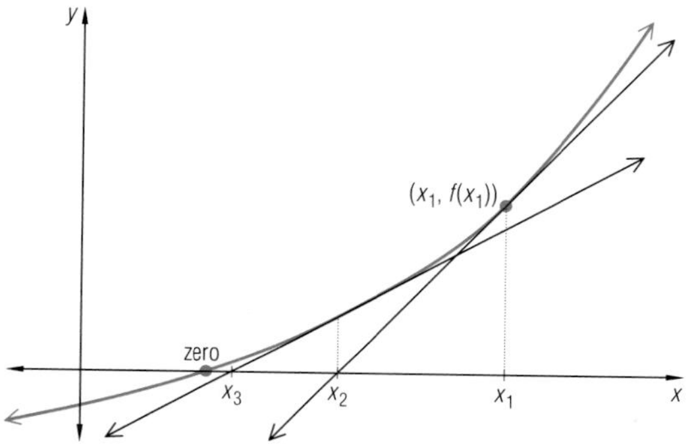

The x-intercept of this line is x_2, which becomes the next approximation to the zero.
   a. Find a formula for x_2.
   b. Draw the tangent to the graph at $(x_2, f(x_2))$ and write the equation of this tangent line.
   c. Find the x-intercept, x_3, of the tangent line in part **b**. This becomes the next approximation to the zero.
   d. Generalize the process described above to find a formula for the $(n + 1)$st approximation.
   e. Apply the above technique to approximate a zero to $f(x) = 2x^3 + 3x^2 + 1$. (The derivative of f is $f'(x) = 6x^2 + 6x$.) Compare your results using *Newton's Method* to an approximation found by using some other technique that you know.

PROJECTS

The projects provide an opportunity for extended work on topics related to the content of the chapter. (See pages 70 and T37 for more information on the purposes of projects.) Some or all of these projects may be suitable for **small group work**. Solutions for the projects can be found at the end of the Additional Answers section in the back of this book.

Project 1 involves the fact that functions are not always differentiable at all values in their domain, despite the fact that such has been true for the functions in this chapter.

Project 2 could be varied to examine the work of mathematicians whose work helped Newton and Leibniz, such as Cavelieri and Kepler.

Project 3 is related to the length-of-day work in Lesson 9-1. It is interesting to find the times of sunrise R and sunset S (ignoring daylight savings time) for an entire year; then the length-of-day function $L = S - R$. Putting S, R, and L on the same graph is most interesting.

Project 4 could be extended to have students develop a computer program to calculate the values to some desired level of accuracy.

SUMMARY

The Summary gives an overview of the entire chapter and provides an opportunity for students to consider the material as a whole. Thus, the Summary can be used to help students relate the various concepts presented in the chapter.

VOCABULARY

Terms, symbols, and properties are listed by lesson to provide a checklist of concepts a student must know. (See page 71 for more information.)

Summary

This chapter introduces the fundamental ideas of differential calculus, the branch of calculus concerned with rates of change.

The average rate of change of a function f over the interval from x_1 to x_2 equals the slope of the secant line through $(x_1, f(x_1))$ and $(x_2, f(x_2))$. Writing $x_2 - x_1$ as Δx, then the average rate of change is

$$\frac{\Delta y}{\Delta x} = \frac{y_2 - y_1}{x_2 - x_1} = \frac{f(x_2) - f(x_1)}{x_2 - x_1} = \frac{f(x_1 + \Delta x) - f(x_1)}{\Delta x}.$$

This last expression is called a difference quotient.

When f is a function giving the position of an object along a line relative to a starting point, then the average rate of change of f over an interval is the average velocity over that interval.

The limit as $\Delta x \to 0$ of the average rate of change or difference quotient of a function f is the derivative of f at x_1. Geometrically, this number is the limit of the slope of the secant line as $x_2 \to x_1$, and is the slope of the line tangent to the curve at the point (x_1, y_1). If f is a function giving the position of an object, then the derivative at a point is the instantaneous velocity at that point.

At the top of the next column, the left side contains quantities that can be calculated for all functions, including discrete functions. The corresponding limits on the right side can only be calculated if the derivative exists.

change between two points	limit as one point approaches the other
average rate of change	instantaneous rate of change
slope of secant	slope of tangent
difference quotient	derivative

Given a function f, the function mapping each number x to the derivative of f at x is the derivative function f'. A function is increasing on an interval when its derivative is positive at all points on that interval. A function is decreasing on an interval when its derivative is negative on that interval. When a smooth function has a relative maximum or minimum at a point, its derivative is zero at that point.

Acceleration is the rate of change of a rate of change. It is found by calculating the derivative of a derivative, or second derivative. If a function is a position function, then acceleration is given by the derivative of the velocity function. Negative acceleration is known as deceleration.

Exponential functions possess the property that their derivatives are proportional to their values. Further, when the base is e, the derivative is equal to the function itself. This fact makes exponential functions particularly useful in modeling phenomena in the sciences.

Vocabulary

For the starred (*) terms you should be able to give a definition of the term.
For the other terms you should be able to give a general description and a specific example of each.

Lesson 9-1
differential calculus
*Δx, Δy
*average rate of change
secant line to a circle
secant line for the graph of a function
velocity
*difference quotient

Lesson 9-2
*instantaneous velocity
tangent line to the graph of a function
*derivative of a function at a point
instantaneous rate of change of f at x

Lesson 9-3
*derivative function of f, f'
Derivative of a Quadratic Function Theorem

Lesson 9-4
acceleration
instantaneous acceleration
deceleration
first derivative, second derivative

Lesson 9-5
Parabola Vertex Theorem
optimization problem

Lesson 9-6
differentiation
differential equation

578

Progress Self-Test

See margin for answers not shown below.

Take this test as you would take a test in class. Then check the test yourself using the solutions at the back of the book.

1. The function f is graphed below.

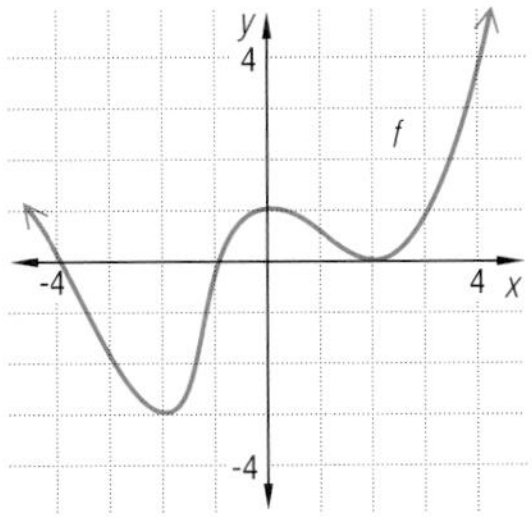

a. Find the average rate of change in f from $x = -4$ to $x = -2$. $-\frac{3}{2}$

b. Find the average rate of change in f from $x = -2$ to $x = 2$. $\frac{3}{4}$

c. Find an interval over which the average rate of change in f is $-\frac{1}{2}$.

d. Find an interval over which the average rate of change in f is 0.

2. The table below shows the percent of the U.S. labor force in farm-related occupations every ten years from 1940 to 1980.

Year	% of labor force in farm occupation
1940	17.4
1950	11.6
1960	6.0
1970	3.1
1980	2.2

a. Find the average rate of change per year in the interval

i. from 1940 to 1950; -.58% per year

ii. from 1970 to 1980. -.09% per year

b. Does the average rate of change appear to be dropping faster or slower as time goes on? slower

c. Is the acceleration positive or negative? positive

3. Let $f(x) = 4x^2 - 3$.

a. Find a formula for the average rate of change of f from 2 to $2 + \Delta x$. $16 + 4\Delta x$

b. Use your answer to **a** to find the average rate of change of f from 2 to 2.1. 16.4

4. The function f is graphed below, along with two tangent lines to its graph. Find:

a. $f'(-3)$ -2 **b.** $f'(1)$. 1

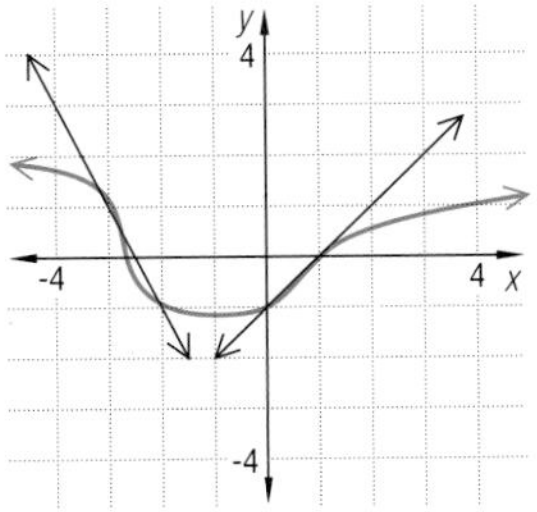

5. Let $g(x) = -2x^2 + x$.

a. Use the definition of derivative to find a formula for g'. $g'(x) = -4x + 1$

b. Find the slope of the line tangent to the graph of g at $x = 3$. -11

Today's farmers use equipment improved since the 1950s.

PROGRESS SELF-TEST

The Progress Self-Test provides the opportunity for feedback and correction; the Chapter Review provides additonal opportunities for practice.

USING THE PROGRESS SELF-TEST

Assign the Progress Self-Test as a one-night assignment. (See pages 72 and T35 for more information.)

ADDITIONAL ANSWERS

1.c. sample: $x = 0$ to $x = 2$

d. samples: $x = -4$ to $x = -1$, $x = -1$ to $x = 2$

ADDITIONAL ANSWERS
7.

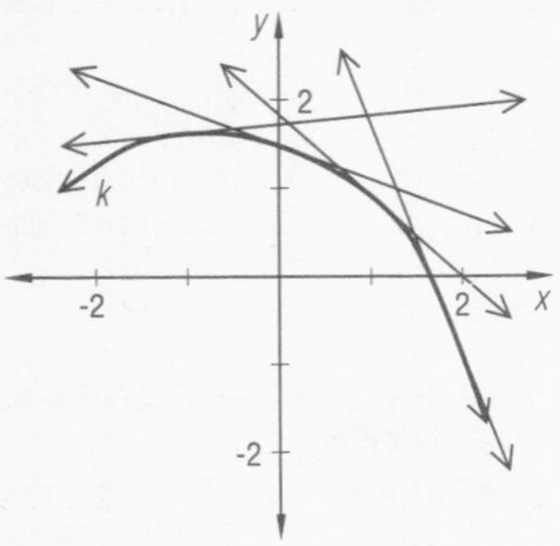

negative

8.d. 0 ft/sec

6. *Multiple choice*. The graph of f is shown below.

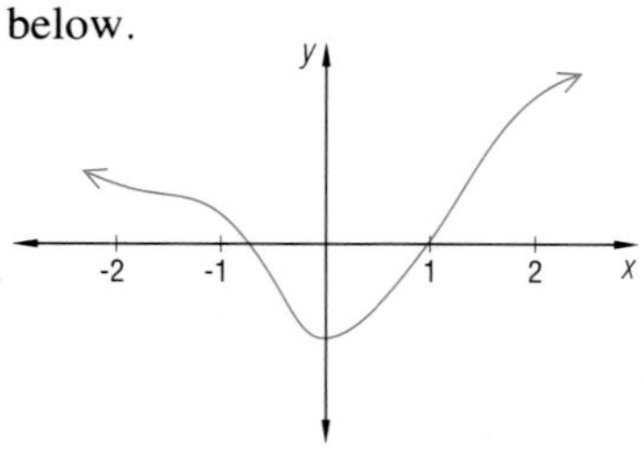

Which of the graphs below could be the graph of f'?

(a)

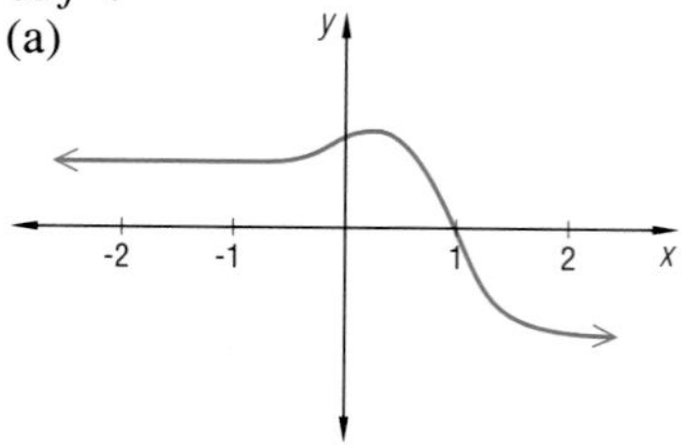

(b)

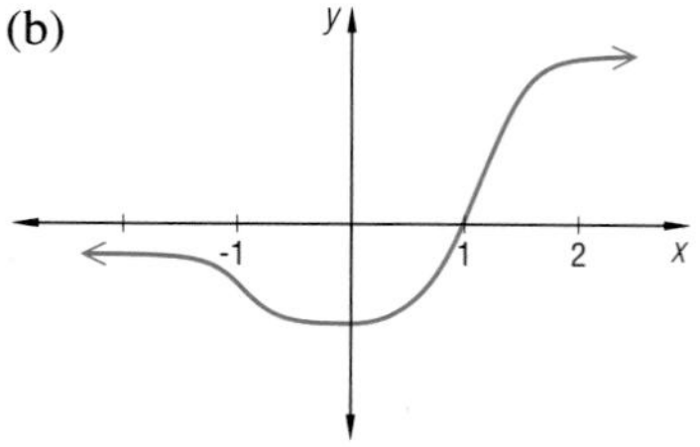

(c)

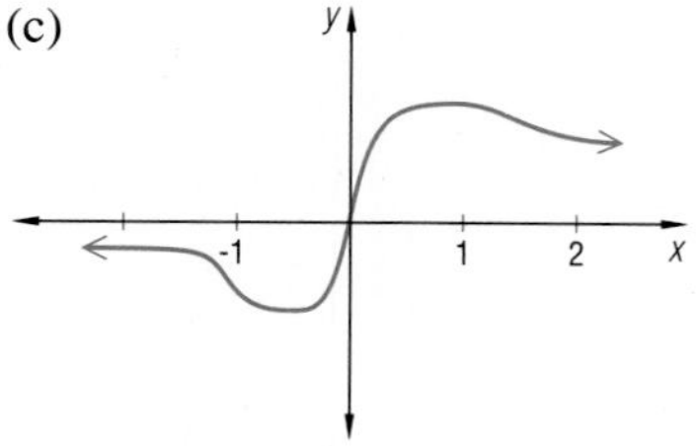

(c)

7. Copy the graph of the function k below and sketch tangent lines at $x = -1$, $x = 0$, $x = 1$, and $x = 2$. Is $k''(x)$ positive or negative on the interval $-1 < x < 2$?

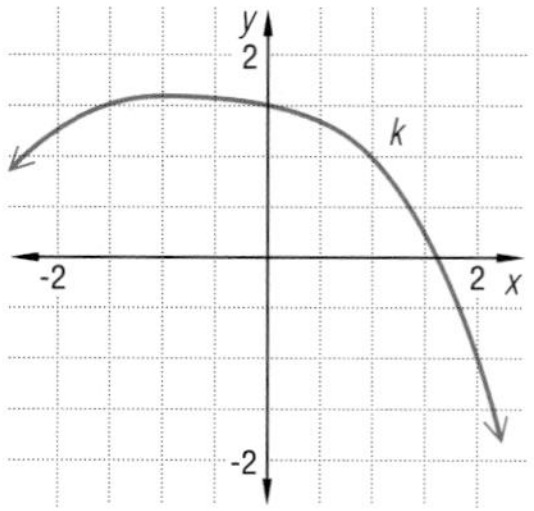

8. A ball is thrown so that its height (in feet) after t seconds is given by

$$h(t) = -16t^2 + 64t + 50.$$

a. Find the instantaneous velocity of the ball 1 second after it is thrown. **32 ft/sec**

b. Find the acceleration of the ball 1 second after it is thrown. What is the physical meaning of your answer?

c. Find the maximum height reached by the ball. **114 ft**

d. What is the instantaneous velocity of the ball when it reaches its maximum height?

9. Given that if $f(x) = e^{-x^2}$, then $f'(x) = -2xe^{-x^2}$. Find the interval(s) on which f is increasing and those on which f is decreasing. (Use the fact that $e^{-x^2} > 0$ for all x.) **f is increasing when $x < 0$. f is decreasing when $x > 0$.**

8. b. -32 ft/sec^2; acceleration due to gravity

580

Chapter Review

Questions on SPUR Objectives

SPUR stands for **S**kills, **P**roperties, **U**ses, and **R**epresentations.
The Chapter Review questions are grouped according to the SPUR Objectives for this chapter.
See margin for answers not shown below.

SKILLS deal with the procedures used to get answers.

Objective A: *Compute average rates of change in functions. (Lesson 9-1)*

1. Find the average rate of change in $f(x) = 3x^2 + 1$ from $x = -1$ to $x = 4$. 9
2. Let g be the function defined by $g(x) = x^3$.
   a. Calculate the average rate of change in g from 1 to $1 + \Delta x$. $3 + 3\Delta x + (\Delta x)^2$
   b. Use your answer to part **a** to find the average rate of change in g
   i. from 1 to 1.1;
   ii. from 1 to 1.01.
3. Let $k(t) = -2t^2 + 5t$.
   a. Find the average rate of change in k from t to $t + \Delta t$. $-4t + 5 - 2\Delta t$
   b. Use your answer to part **a** to find the average rate of change of k from 2 to 2.5. -4

Objective B: *Use the definition of derivative to compute derivatives. (Lessons 9-2, 9-3)*

4. Let f be the function defined by $f(x) = 2x^2$ $\forall$ real numbers x. Use the definition to compute $f'(1)$ and $f'(-1)$. $f'(1) = 4$, $f'(-1) = -4$
5. Given that $f(x) = -x^2 + x + 1$ for all real numbers x, find a formula for the instantaneous rate of change of f at x.

In 6–8, find the derivative of the function whose formula is given.

6. $f(x) = 2x$ $\quad f'(x) = 2$
7. $g(x) = -3x^2$ $\quad g'(x) = -6x$
8. $k(x) = -3x^2 + 2x$ $\quad k'(x) = -6x + 2$

PROPERTIES deal with the principles behind the mathematics.

Objective C: *Use derivatives to identify properties of functions. (Lesson 9-5)*

9. *True* or *false*? If the derivative of a function is 0 at a point, then the function has a relative minimum or maximum at that point. False
10. Let $f(x) = x + e^{-2x}$ for all real numbers x. The first derivative of f is $f'(x) = 1 - 2e^{-2x}$ and the second derivative is $f''(x) = 4e^{-2x}$. Is f increasing, decreasing, or neither on the interval $1 < x < 2$? increasing
11. Use the fact that the derivative of $f(x) = x^3 + 3x + 2$ is $f'(x) = 3x^2 + 3$ to prove that f is increasing on the set of all real numbers.
12. Suppose for a function f, $f''(x) < 0$ $\forall$ x in the interval $1 < x < 4$. Are the slopes of tangents to the graph of f increasing or decreasing on $1 < x < 4$? decreasing
13. Given $f(x) = x^3 - 1.5x^2 - 18x$ and $f'(x) = 3x^2 - 3x - 18$.
    a. Use the first derivative to find
    i. the interval(s) on which f is increasing, $x < -2$ or $x > 3$
    ii. the interval(s) on which f is decreasing, and $-2 < x < 3$
    iii. the points at which f may have a relative maximum or minimum.
    b. Check your answers with an automatic grapher.

RESOURCES
- Chapter 9 Test, Form A
- Chapter 9 Test, Form B
- Chapter 9 Test, Cumulative Form
- Comprehensive Test, Chapter 1–9
- Quiz and Test Writer

CHAPTER REVIEW

The main objectives for the chapter are organized here into sections corresponding to the four main types of understanding this book promotes: Skills, Properties, Uses, and Representations. We call these the SPUR objectives.

USING THE CHAPTER REVIEW
Students should be able to answer questions like these with about 85% accuracy by the end of the chapter. (See pages 74 and T38–39 for more information.)

ADDITIONAL ANSWERS
2.b.i. 3.31; ii. 3.0301

5. $f'(x) = -2x + 1$

11. $f'(x) = 3x^2 + 3$. Since $x^2 \ge 0$ for all real x, $f'(x) = 3x^2 + 3 \ge 0$ for all real numbers. Since the derivative is positive, the slopes of the tangents to the curve are all positive, and the function is increasing for all real numbers.

13.a.iii. $x = -2$ or $x = 3$
b.

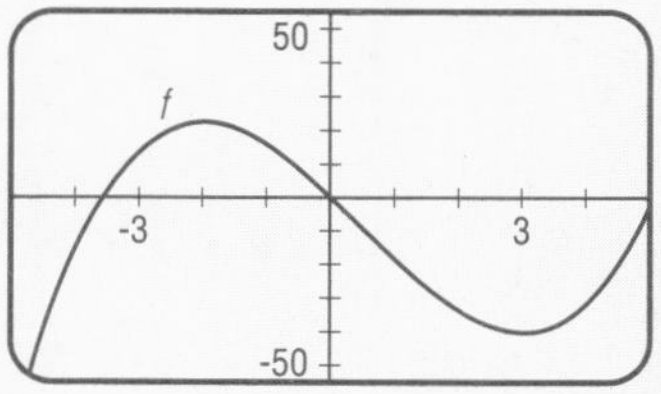

ADDITIONAL ANSWERS

14.a. 1 min/oz; an extra minute of cooking time is required for every additional ounce of potatoes.

b. $\frac{2}{3}$ min/oz; as weight increases, the rate of change of baking time needed decreases for every additional ounce of potatoes.

c. Potatoes weighing between 10 and 16 ounces need less cooking time per ounce than potatoes weighing between 4 and 6 ounces.

15.a. $A'(t)$: grams/day; $A''(t)$: grams/day^2

b. ≈-25.5 grams/day; at 7 days, the amount of radon present is decreasing by about 25.5 grams/day.

16.a.

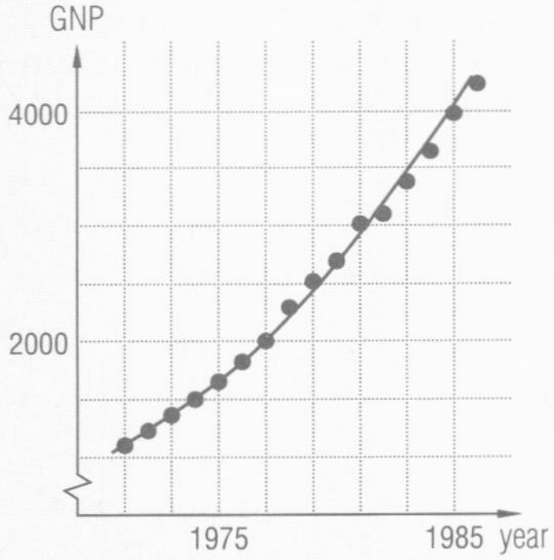

17.d. -80 ft/sec

18.b. slowing down

USES deal with applications of mathematics in real situations.

Objective D: *Find rates of change in real situations.* *(Lessons 9-1, 9-2, 9-3, 9-4)*

14. Suppose a 4 oz. potato takes 5 minutes to bake in a microwave oven, a 6 oz. potato takes 7 minutes to bake, 10 oz. of potatoes takes 10 minutes, and 16 oz. of potatoes take 14 minutes.

a. Find the average rate of change in baking time with respect to potato weight from 4 oz. to 6 oz. What is the meaning of your answer in terms of baking time?

b. Find the average rate of change from 10 oz. to 16 oz.

c. Compare your answers to parts **a** and **b** in terms of baking time.

d. Assume the average rate of change from 16 oz. to 20 oz. is the same as that from 10 oz. to 16 oz. How long would it take to bake 20 oz. of potatoes? $16\frac{2}{3}$ minutes

15. Radon gas is an odorless radioactive gas emitted naturally from the Earth. If 500 grams of radon is initially present, then the amount remaining after t days is $A(t) = 500e^{-0.182t}$ grams. The derivative of A is $A'(t) = -91e^{-0.182t}$ and the second derivative is $A''(t) = 16.562e^{-0.182t}$.

a. What units are appropriate for measuring $A'(t)$ and $A''(t)$, respectively?

b. Find $A'(7)$. What does the derivative mean in this case?

c. Is the radon decaying faster after 5 days or after 7 days? 5 days

16. The following table gives the yearly Gross National Product (GNP) of the United States for the years 1970–1988.

a. Make a scatterplot of the GNP (in billions of dollars) versus years after 1970. Sketch a reasonable curve to fit the data.

b. On the interval from 1974 to 1978, is the average rate of change of the GNP increasing or decreasing? increasing

c. During what year does it appear that the GNP changed fastest? 1983–1984

Year	Gross National Product (in billions of dollars)
1970	1015.5
1971	1102.7
1972	1212.8
1973	1359.3
1974	1472.8
1975	1598.4
1976	1782.8
1977	1990.5
1978	2249.7
1979	2508.2
1980	2732.0
1981	3052.6
1982	3166.0
1983	3405.7
1984	3772.2
1985	4014.9
1986	4231.6
1987	4524.3
1988	4880.6

Source: *Statistical Abstract of the United States, 1990.* U. S. Department of Commerce.

Objective E: *Use derivatives to find the velocity and acceleration of a moving object.* *(Lessons 9-2, 9-3, 9-4)*

17. A ball is dropped from a 100-foot tower. The height (in feet) of the ball above the ground at time t seconds is given by $h(t) = -16t^2 + 100$.

a. What is the instantaneous velocity of the ball at time $t = 2$ seconds? -64 ft/sec

b. What is the acceleration of the ball at time $t = 2$ seconds? -32 ft/sec^2

c. At what time t does the ball hit the ground? 2.5 sec

d. What is the instantaneous velocity of the ball at the moment it hits the ground?

18. A particle moves horizontally so that its position (in meters) to the right of its starting point at time t seconds is given by $f(t) = -3t^2 + 13t + 16$.

a. At time $t = 2$ seconds, is the particle moving to the left, to the right, or stationary? to the right

b. Is the particle speeding up or slowing down at time $t = 2$ seconds?

19. Below is the graph of the distance of a subway car from the beginning of its route over time.
  a. Estimate the velocity of the subway car at
     i. $t = 2$ **1 mi/min** ii. $t = 5$ **0 mi/min**
     iii. $t = 8$ **1 mi/min** iv. $t = 12$. **0 mi/min**
  b. What is happening to the subway car at $t = 5$ and $t = 12$? **It is stationary.**
  c. Sketch a rough graph of the subway car's velocity.
  d. When is the subway car's acceleration positive? When is it negative?

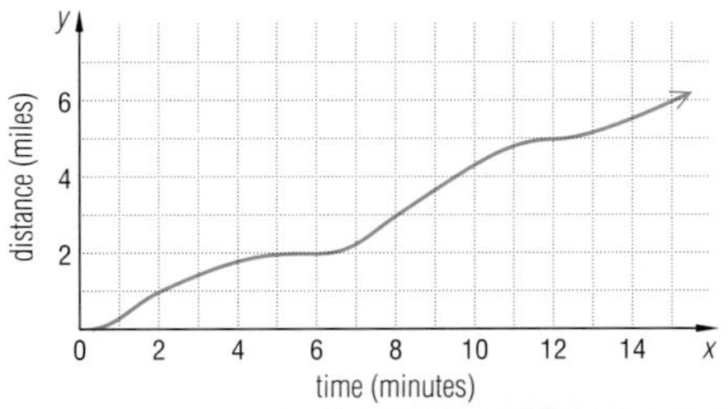

Objective F: *Use derivatives to solve optimization problems. (Lesson 9-5)*

20. An object is thrown upwards from a platform so that its height (in feet) t seconds after being thrown is given by $h(t) = -16t^2 + 50t + 80$.
  a. What is the maximum height it reaches?
  b. What is the height of the platform?

21. A triangular platform is to be built in the corner formed by two walls as shown below. The 4-meter outer edge of the platform forms an angle of θ with one of the walls. The area of the platform is to be maximized.

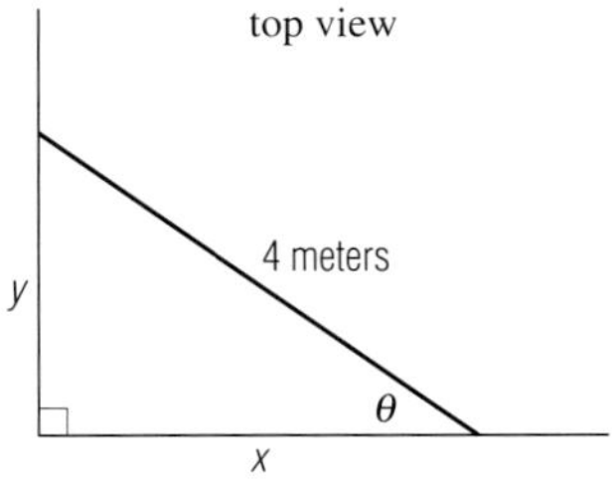

  a. Use trigonometry to show that the area can be written as a function of θ:
     $A(\theta) = 8 \sin \theta \cos \theta$
     $= 4 \sin 2\theta$.
  b. Using the fact that $A'(\theta) = 8 \cos 2\theta$, find θ in the interval $0 < \theta < \frac{\pi}{2}$ so that $A(\theta)$ is maximized. $\theta = \frac{\pi}{4}$

REPRESENTATIONS deal with pictures, graphs, or objects that illustrate concepts.

Objective G: *Relate average rate of change to secant lines of graphs of functions. (Lesson 9-1)*

22. Refer to the graph of f at the right.
  a. What is the average rate of change in f from $x = -4$ to $x = -1$? $-\frac{1}{3}$
  b. What is the average rate of change in f from $x = -1$ to $x = 1$? $\frac{3}{2}$
  c. With two of the five indicated points as endpoints, over what interval is the average rate of change in f zero?
  d. Over what interval is the average rate of change $\frac{2}{5}$? **sample: x = -4 to x = 1**

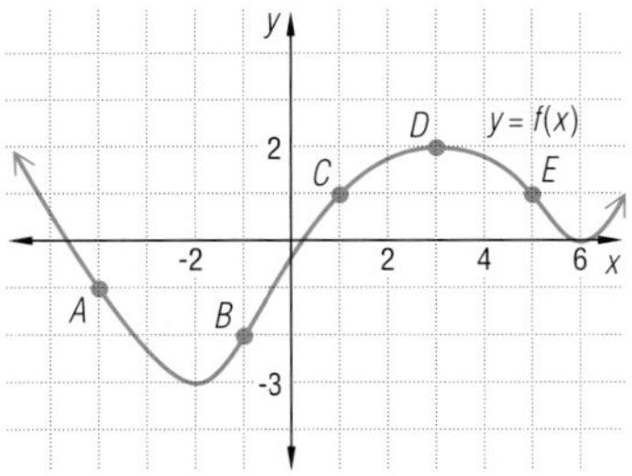

23. Suppose that U and V are points on the graph of the function g, and that $U = (-1, y)$ and $V = (5, -7)$. If the average rate of change in g from $x = -1$ to $x = 5$ is $-\frac{2}{3}$, find y. -3

EVALUATION
Three forms of a Chapter Test are provided in the Teacher's Resource File. Chapter 9 Test, Forms A and B cover just Chapter 9. The third test is Chapter 9 Test, Cumulative Form. About 50% of this test covers Chapter 9, 25% covers Chapter 8, and 25% covers previous chapters. A fourth test, Comprehensive Test, Chapters 1-9, that is primarily multiple choice in format, is also provided. For information on grading, see *General Teaching Suggestions: Grading* on page T43 in the Teacher's Edition.

ASSIGNMENT RECOMMENDATION
We strongly recommend that you assign the reading and some questions from Lesson 10-1 for homework the evening of the test. It gives students work to do if they complete the test before the end of the period and keeps the class moving.

19.c.

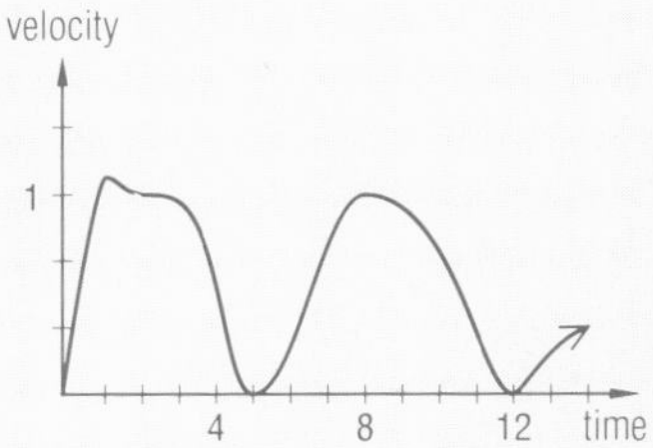

d. positive: $0 < x < 2$, $5 < x < 8$; negative: $2 < x < 5$, $8 < x < 12$

20.a. ≈ 119 ft
b. 80 ft

21.a. $A(\theta) = \frac{1}{2}xy = \frac{1}{2}(4 \cos \theta)(4 \sin \theta) = 8 \sin \theta \cos \theta = 4 (2 \sin \theta \cos \theta) = 4 \sin 2\theta$

22.c. x = 1 to x = 5

ADDITIONAL ANSWERS

25.a. $f'(-4) \approx -1$, $f'(-2) \approx 0$, $f'(0) \approx 2$, $f'(3) \approx 0$, $f'(5) \approx -3$

b.

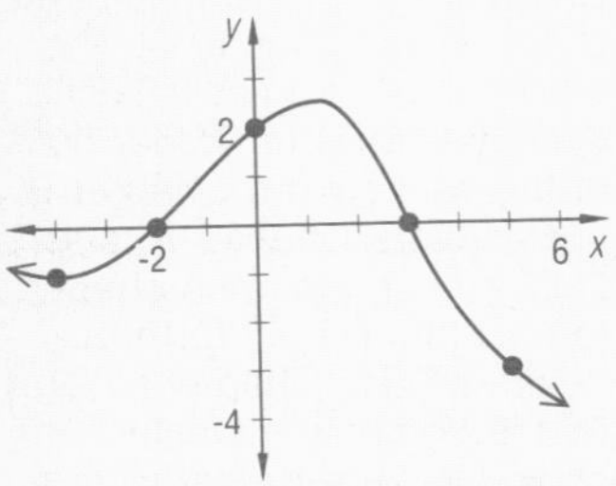

27. sample:

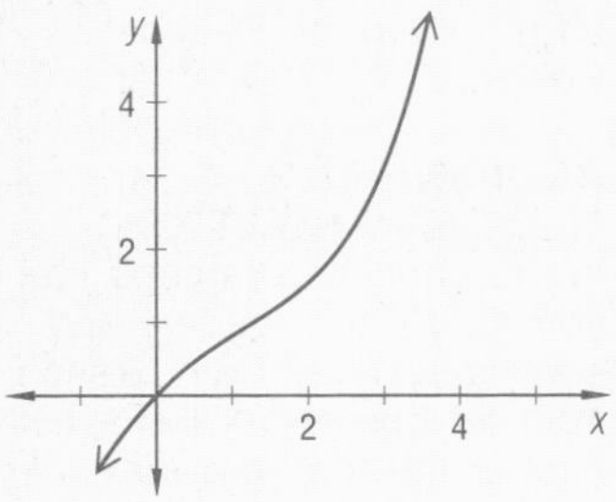

28.c. $x = -1$, $x = 3$

d. $f'(0) \approx 1.5$, $f'(3) \approx 0$, $f'(6) \approx -2$

29.a. increasing: $-5 < x < -3$, $1 < x < 5$; **decreasing:** $-3 < x < 1$

c. positive: $-1 < x < 5$; **negative:** $-5 < x < -1$

Objective H: *Estimate derivatives by finding slopes of tangent lines.* *(Lessons 9-2, 9-3)*

24. Refer to the graph of f below. Estimate $f'(x)$ when

a. $x = -2$; 3 **b.** $x = 1$; 0

c. $x = 5$. -1

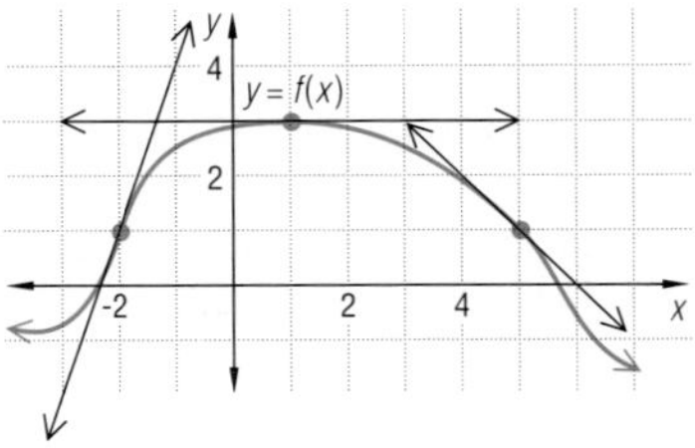

25. Refer to the graph of f below.

a. Estimate the values of $f'(-4)$, $f'(-2)$, $f'(0)$, $f'(3)$, and $f'(5)$.

b. Sketch a graph of f' over the interval $-6 \le x \le 6$.

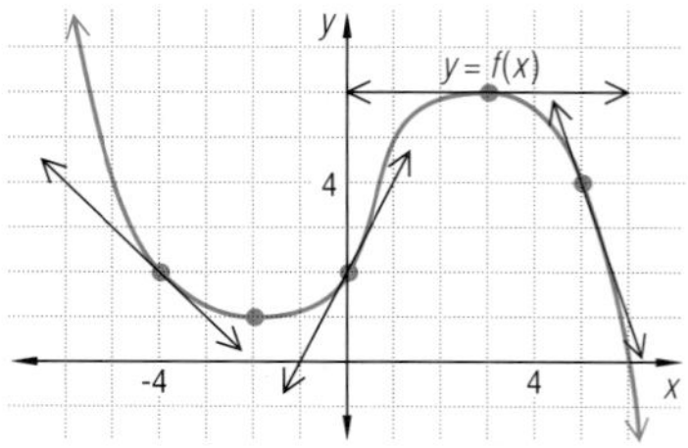

26. *Multiple choice.* The graph of f' is shown below. Which of the graphs below could be the graph of f? (a)

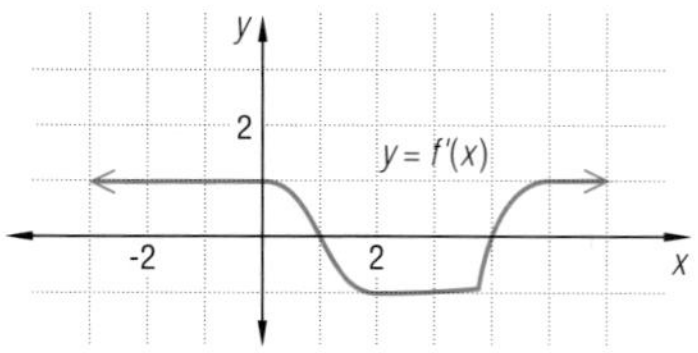

(a)

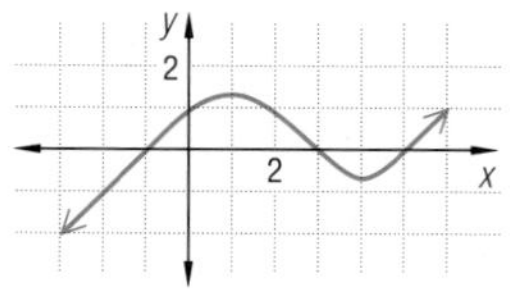

(b)

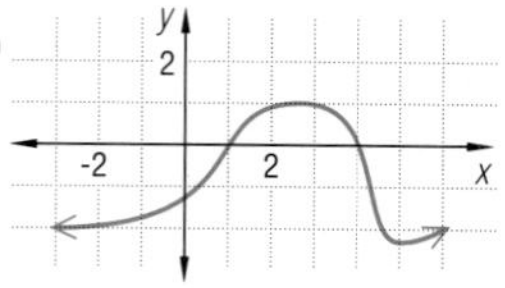

(c)

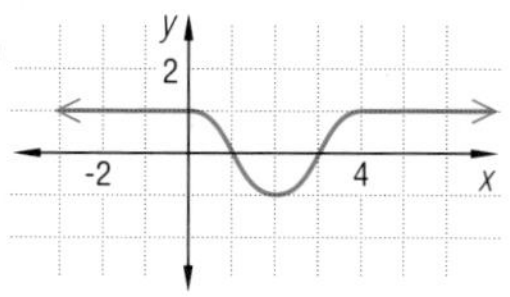

27. Sketch the graph of a function f such that $f'(0) = 1$, $f'(2) = 1$, and $f'(3) = 2$.

Objective I: *Determine properties of derivatives from the graph of a function.* *(Lesson 9-5)*

28. Use the graph of f below to estimate answers.

a. On what interval(s) is the derivative of f positive? $-1 < x < 3$

b. On what interval(s) is the derivative of f negative? $x < -1$, $x > 3$

c. Where is the derivative equal to 0?

d. Estimate $f'(0)$, $f'(3)$, and $f'(6)$.

e. Is $f''(x)$ positive, negative, or zero on the interval $0 \le x \le 7$? negative

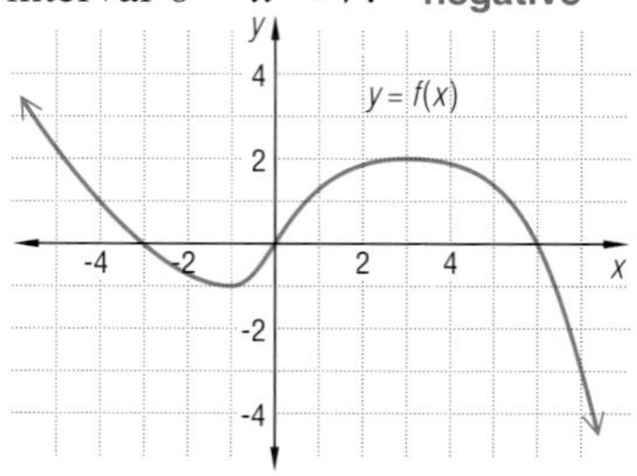

29. Use the graph of g' shown below to estimate answers.

a. Identify the intervals on which g is increasing and the intervals on which g is decreasing.

b. Where may g have a relative minimum or maximum? $x = -3$, $x = 1$

c. Identify the intervals on which g'' is positive and the intervals on which g'' is negative.

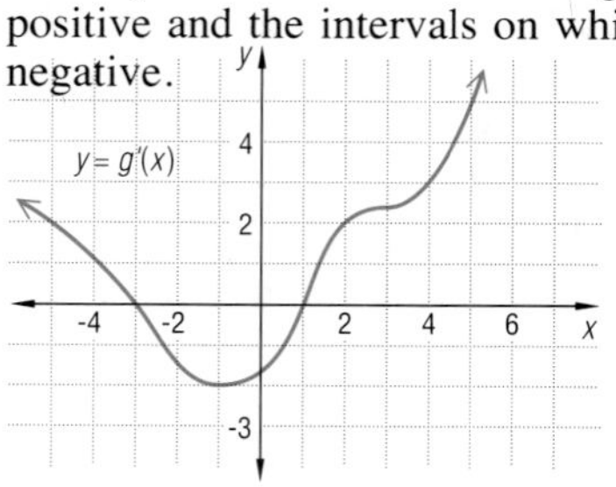

CHAPTER 10 COMBINATORICS

OBJECTIVES

The objectives listed here are the same as in the Chapter 10 Review on pages 636–638 of the student text. The Progress Self-Test on page 635 and the tests in the Teacher's Resource File cover these objectives. For recommendations regarding the handling of this end-of-chapter material, see the notes in the margin on the corresponding pages of this Teacher's Edition.

OBJECTIVES FOR CHAPTER 10 (Organized into the SPUR Categories—Skills, Properties, Uses, and Representations)	Progress Self-Test Questions	Chapter Review Questions	Teacher's Resource File	
			Lesson Masters*	Chapter Test Forms A & B
SKILLS				
A Describe the essential features of counting problems.	1a, 2a	1–5	10-1	1, 2
B Evaluate expressions indicating permutations or combinations.	6, 7	6–10	10-3, 10-4	5, 6
C Apply the Binomial Theorem to expand binomials or find specific terms.	11	11–14	10-5	13, 14
PROPERTIES				
D Use properties of permutations and combinations to prove identities.	15	15–17	10-3, 10-4	7, 15
E Apply the Binomial Theorem to deduce properties of sets.	13	18–19	10-6	16
USES				
F Use the Multiplication Counting Principle and permutations to solve counting problems.	4, 5, 8, 9	20–28	10-2, 10-3	4, 9–12
G Use combinations and the Binomial Theorem to solve counting problems.	1b, 2b, 10, 14	29–35	10-4, 10-6, 10-7	8
H Find binomial probabilities in realistic situations.	12, 16	36–37	10-6	17
REPRESENTATIONS				
I Use a possibility tree to determine the number of outcomes in a given situation.	3	38–40	10-2	3

*** The masters are numbered to match the lessons.**

OVERVIEW ■ CHAPTER 10

Most students taking *Precalculus and Discrete Mathematics* have studied permutations, combinations, and the Binomial Theorem before. In previous UCSMP courses, the material of Lessons 10-2 through 10-6 is covered in detail. In *Functions, Statistics, and Trigonometry,* it is covered in two chapters and mastery is expected.

The content of Chapter 10 is included in this book for a number of reasons. (1) The material is very important in all of mathematics and deserves to be seen twice by most students. (2) Combinatorics is an important part of discrete mathematics and belongs in a course that gives strong attention to that subject. (3) Counting problems are subtle and almost all students can benefit from a second exposure. (4) In previous courses, many of the major theorems were stated but not proved; in this book, they can all be proved using mathematical induction. (5) The counting ideas can be taken further than is customarily done at this level. In particular, combinations with repetition and multinomial coefficients are considered.

The problems presented in the chapter opener page represent the span of problems to be discussed in the chapter. The variety of the problems suggests the need for the first lesson. Lesson 10-1 is designed to help students focus on the essential features of a counting problem prior to studying those techniques needed to solve the problem.

Lessons 10-2, 10-3, and 10-4 cover the standard counting problems that require a knowledge of the Multiplication Counting Principle, permutations, or combinations. Then there are two lessons covering the Binomial Theorem. In Lesson 10-5, the Binomial Theorem is used to expand binomials raised to any positive integer power. In Lesson 10-6, the Binomial Theorem is related to counting problems and probability.

Lesson 10-7 discusses combinations which allow repetitions, a topic not found in most schoolbooks but one which completes the range of problems considered in this book. Lesson 10-8 connects permutations and combinations with the coefficients of $(a_1 + a_2 + \ldots + a_k)^n$, thus completing the connection between counting problems and powers of polynomials.

PERSPECTIVES ■ CHAPTER 10

The Perspectives provide the rationale for the inclusion of topics or approaches, provide mathematical background, and make connections with other lessons and within UCSMP.

10-1

WHAT EXACTLY ARE YOU COUNTING?

Many textbooks begin a discussion of combinatorics with permutations and combinations. After students have studied both ideas, problems are mixed together and students are asked to solve them using either permutations or combinations. We have chosen to begin the study of combinatorics somewhat differently. By having students focus first on the essential features of a problem, namely, whether repetitions are allowed and whether order is important, we hope to avoid some of the confusion that occurs when students learn techniques first and then try to apply them.

10-2

POSSIBILITY TREES AND THE MULTIPLICATION COUNTING PRINCIPLE

The Multiplication Counting Principle is discussed in four previous UCSMP textbooks. The treatment in this book is more formal in three ways: students have to complete proofs by induction presented in outline form; the Multiplication Counting Principle is stated more formally using subscript notation; and the language of strings is used. A string refers to an ordered set of symbols while a set is an unordered grouping of symbols. A corollary of the Multiplication Counting Principle is that n^r is the number of r-symbol strings that can be made from a set of n symbols. A set of n symbols means that the symbols themselves have no repetition.

10-3

PERMUTATIONS

A permutation is a string of length r from a collection of n symbols without repeating any of the symbols. Common notations for the number of permutations for a fixed n and r are ${}_nP_r$ and $P(n, r)$. We use the latter because it exhibits the permutation as a function of two variables, and also shows the key letters in the order they are read: Permutations of n things taken r at a time. The familiar formula

$P(n, r) = \frac{n!}{(n-r)!}$ is discussed in this lesson.

10-4

COMBINATIONS

A combination is a set of length r from a collection of n symbols without repeating any of the symbols. There are three common notations for combinations with a fixed n and r: ${}_nC_r$, $C(n, r)$ and $\binom{n}{k}$. In this book, the middle notation is used for the same reasons given for the permutation notation. However, the last notation also is used because it is the one most commonly used by mathematicians. In complicated expressions, it is the shortest to write and, thus, the clearest. In this lesson, the familiar computation formula $C(n, r) = \frac{n!}{r!(n-r)!}$ is derived, as are some of its special cases.

10-5

THE BINOMIAL THEOREM

The Binomial Theorem, $(x + y)^n = \sum_{k=0}^{n} \binom{n}{k} x^{n-k}y^k$, shows that the coefficients of the nth power of the binomial $x + y$ are the combinations $\binom{n}{k}$. When these combinations are written in order for increasing values of n, the array of positive integers known as Pascal's triangle appears. Thus, there are algebraic, combinatorial, and arithmetic ways, of determining binomial coefficients. This lesson uses the combinatorial and arithmetic ways, and gives students practice in expanding powers of binomials.

10-6

COUNTING AND THE BINOMIAL THEOREM

Suppose an experiment has two outcomes, success or failure, with a probability p of success. Then, if there are n repetitions (trials) of this experiment, the Multiplication Counting Principle tells us that there are 2^n permutations of successes and failures. Of these, $\binom{n}{k}$ have k successes, and thus the probability of k successes in n trials is $\frac{\binom{n}{k}}{2^n}$. This lesson develops the mathematics of this very important application.

10-7

COMBINATIONS WITH REPETITION

The number of combinations of k elements that can be constructed from a set of n elements with repetition allowed is $\binom{k+n-1}{k}$, or equivalently, $\binom{k+n-1}{n-1}$. This kind of counting problem determines (1) the number of terms in the expansion of a polynomial of n terms to the kth power, (2) the number of strings of n nonnegative integers whose sum is k, and (3) the number of ways of distributing k balls into n boxes. This lesson develops and applies the mathematics of these kinds of problems.

10-8

MULTINOMIAL COEFFICIENTS

The Binomial Theorem indicates that $(x + y)^n$ has $n + 1$ terms, each with variables of the form $x^{n-k}y^k$, and that the coefficient of the term $x^{n-k}y^k$ is $\binom{n}{k}$. A natural generalization is to the expansion of the multinomial $(x_1 + x_2 + \ldots + x_k)^n$. Each term in the expansion has variables of the form $x_1^{a_1}x_2^{a_2} \ldots x_k^{a_k}$, where $a_1 + a_2 + \ldots + a_k = n$. In Lesson 10-7, the number of terms was found to be $\binom{k+n-1}{k}$. In this lesson, the coefficient $\frac{n!}{a_1!a_2! \ldots a_k!}$ of each term is derived. This is the number of permutations from a set with n elements of k different kinds, a_1 of one kind, a_2 of a second kind, ..., and a_k of the kth kind.

DAILY PACING CHART ■ CHAPTER 10

Every chapter of UCSMP *Precalculus and Discrete Mathematics* includes lessons, a Progress Self-Test, and a Chapter Review. For optimal student performance, the self-test and review should be covered. (See *General Teaching Suggestions: Mastery* on page T35 of this Teacher's Edition.) By following the pace of the Full Course given here, students can complete the entire text by the end of the year. Students following the pace of the Minimal Course spend more time when there are quizzes and on the Chapter Review and will generally not complete all of the chapters in this text.

When chapters are covered in full (the recommendation of the authors), then students in the Minimal Course can cover 11 chapters of the book. For more information on pacing, see *General Teaching Suggestions: Pace* on page T34 of this Teacher's Edition.

DAY	MINIMAL COURSE	FULL COURSE
1	10-1	10-1
2	10-2	10-2
3	10-3	10-3
4	Quiz (TRF); Start 10-4.	Quiz (TRF); 10-4
5	Finish 10-4.	10-5
6	10-5	10-6
7	10-6	Quiz (TRF); 10-7
8	Quiz (TRF); Start 10-7.	10-8
9	Finish 10-7.	Progress Self-Test
10	10-8	Chapter Review
11	Progress Self-Test	Chapter Test (TRF)
12	Chapter Review	
13	Chapter Review	
14	Chapter Test (TRF)	

TESTING OPTIONS

TESTING OPTIONS

- Quiz for Lessons 10-1 through 10-3
- Quiz for Lessons 10-4 through 10-6
- Chapter 10 Test, Form A
- Chapter 10 Test, Form B
- Chapter 10 Test, Cumulative Form

A Quiz and Test Writer is available for generating additional questions, additional quizzes, or additional forms of the Chapter Test.

INDIVIDUAL DIFFERENCES

PROVIDING FOR INDIVIDUAL DIFFERENCES

The student text is written for, and tested with, average students. It also has been successfully used with better and more poorly prepared students.

The Lesson Notes often include Error Analysis and Alternate Approach features to help you with those students who need more help. Students of all abilities often learn from their peers and may benefit from small group work referenced as appropriate throughout the Notes. A blackline Lesson Master (in the Teacher's Resource File), keyed to the chapter objectives, is provided for each lesson to allow more practice. (However, since it is important to keep up with the daily pace, you are not expected to use all of these masters. Again, refer to the suggestions for pacing on page T34.) Extension activities are provided in the Lesson Notes for those students who have completed the particular lesson in a shorter amount of time than is expected, even in the Full Course.

Combinatorics

CHAPTER 10

While counting is usually a simple, everyday task, there are many counting problems for which sophisticated counting procedures are needed. Here are some examples.

Problem 1: The automobile license plates for a certain state list two letters followed by four digits from 0 to 9, inclusive. How many different license plates can be made?

Problem 2: A bridal party consists of 3 bridesmaids, 3 ushers, and the bride and groom. In how many ways can the bridal party be arranged in a row for portraits if the bride and groom always stand together?

Problem 3: A play in the Lotto game in the Illinois State Lottery consists of choosing a set of six different numbers from among the integers from 1 to 54, inclusive. What is the probability of winning in a single play?

Problem 4: A ski club has 80 members, 50 men and 30 women. A planning committee consisting of 5 men and 3 women is to be selected. How many different committees are possible?

Problem 5: Among the integers from 1000 to 9999 (inclusive), how many have at least one digit which is a 7 or a 5?

These and other types of counting problems are studied in a branch of mathematics called *combinatorics*. **Combinatorics** is the science of counting. From simple principles, general theorems can be deduced and the above problems and many others can be solved. Combinatorial problems arise in virtually all fields of endeavor.

CHAPTER 10

If students have studied permutations and combinations previously, this chapter can be taken at a lesson-a-day pace. With 8 to 10 days for the lessons and quizzes, 1 day for the Progress Self-Test, 1 or 2 days for the Chapter Review, and 1 day for a Chapter Test, the entire chapter should take 11–14 days. (See the Daily Pacing Chart on page 585D.)

If students have not studied the content of this chapter previously, then allow a couple of extra days in Lessons 10-3 through 10-6 and allow 16 days for the chapter.

USING PAGE 585
To determine how much of the content of this chapter your students may have studied previously, you might ask them to answer the questions on this page. Problem 1 is the easiest; some students studied this type of counting problem before algebra. Problem 2 involves permutations; Problem 3 applies combinations; and Problem 4 uses both combinations and the Multiplication Counting Principle. Problem 5 is a type that students probably have not seen before. All of these problems are discussed in the following lesson; and they all are solved in the chapter.

Make sure that students do not overlook the definition of *combinatorics*.

LESSON 10-1

RESOURCES
■ Lesson Master 10-1

OBJECTIVE

A Describe the essential features of counting problems.

TEACHING NOTES

The box at the top of this page illustrates the four major types of counting problems to be discussed in the chapter. In this lesson, students determine to which cell a given problem belongs. During the course of the chapter, students will learn the techniques needed to solve problems in any of the four cells. The appropriate techniques for each cell are indicated in the Chapter Summary on page 634.

Example 1 belongs in the upper left cell; **Examples 2 and 3** belong in the lower right cell; and **Example 4** is in the lower left cell.

It is important that students associate the word *string* with the concept of an ordered list of symbols. Hence, when the term string is used in the text or in the answers, it should be understood that order is important, whether or not it is explicitly stated.

LESSON

10-1

What Exactly Are You Counting?

The key to solving many counting problems is to identify the essential mathematical features of the items being counted. If you can recognize these features, you can reformulate a problem in a way that makes it easier to solve and that is free of details that do not affect the solution. How do you determine the essential features of a counting problem? Generally, the items to be counted are written down as lists of symbols. You must determine whether the order of the symbols is important, and you must also consider whether symbols can be repeated. One way to think about the essential features is to determine in which cell of the following diagram the problem would fit.

	repetition of symbols allowed	repetition of symbols not allowed
ordered symbols		
unordered symbols		

The focus of this lesson is to analyze some problems to determine the cell in which each problem belongs. Later lessons of the chapter focus on techniques for actually solving the problems.

Example 1 Identify the essential features of the items to be counted in Problem 1 of the chapter opener:

The automobile license plates for a certain state list two letters followed by four digits from 0 to 9, inclusive. How many different license plates can be made?

Solution In this state, a license number consists of six symbols. Each of the first two symbols can be any of the 26 letters in the alphabet and each of the last four symbols can be any of the 10 digits from 0 through 9.

Two license numbers such as

ET 1996 and **TE 6991**

are different even though both use the same six symbols. Thus, for a license number the order of the six symbols matters. Repetition of letters or digits is allowed because a letter or digit can be used more than once in a license number. So this problem belongs in the cell containing ordered symbols with repetition allowed.

When the symbols in a problem must be ordered, it is common to refer to the ordered list of symbols as a **string**. Then the license plate problem can be reformulated in the following way:

> Count the number of strings of six symbols in which the first two symbols are selected from the 26 letters of the alphabet (with repetition allowed) and the last four symbols are selected from the digits 0 through 9 (with repetition allowed).

The license plate problem can be solved using the Multiplication Counting Principle. This principle is discussed in the next lesson.

Example 2 Identify the essential features of the items to be counted in Problem 3 of the chapter opener:

> A play in the Lotto game in the Illinois State Lottery consists of choosing six different numbers from among the integers from 1 to 54, inclusive. What is the probability of winning in a single play?

Solution To know the probability, you must know how many different choices are possible. The order in which the six numbers are selected does not matter and the six numbers selected must all be different. Consequently, a play in the Lotto game fits into the cell of unordered symbols with no repetitions allowed.

When the symbols in a problem are unordered, with no repetitions allowed, the unordered list of symbols is referred to as a **set**.

Sets of six numbers selected without repetition from among the integers from 1 through 54 are the same as 6-element subsets of the set {1, 2, 3, ..., 53, 54}. Thus, the number of different choices for the Lotto problem can also be stated as follows:

> How many different 6-element subsets are there of the set of all integers from 1 through 54?

Many states have a lottery that consists of a variety of games other than the one with the million dollar jackpot. If your state has such a lottery, a natural extension of **Example 2** is to answer the question for some of your state's lottery games. Later in the chapter, you might actually determine the number of possible plays in each game.

Though some students may know techniques for solving some of these problems from previous courses, it is important to avoid discussing the actual solutions at this point. Exploration **Question 23** allows these students to display their knowledge. For those students who have not had a previous exposure to combinatorics, a discussion of solutions can be confusing and make the content appear more difficult than necessary.

We also recommend that you do not introduce the words permutation and combination at this time. If students are not familiar with these terms, their introduction now may cause students to focus on terminology rather than on the essential characteristics of the problems.

ADDITIONAL EXAMPLES

In 1–4, identify the essential features of the items to be counted. That is, are the symbols ordered or unordered? Is repetition allowed or not allowed?

1. the number of strings of four letters that can be made from the letters of CONQUEST
ordered symbols, no repetition

2. the number of subsets of four letters that can be formed from the letters of CONQUEST
unordered symbols, no repetition

ADDITIONAL EXAMPLES

3. the number of different three-scooped ice-cream cones that can be made if 31 flavors are available
ordered symbols with repetition

4. the number of different orders of three one-scoop ice-cream cones that can be made if 31 flavors are available
unordered symbols with repetition

NOTES ON QUESTIONS

For the Covering the Reading questions in this lesson, students may have to go to the chapter opener page to find an answer.

Question 2: Contrast the word *string* with the word *set*, which refers to an unordered collection.

Questions 3-9: It is not necessary to discuss the actual answers to the questions at this time. Instead, focus on the essential features so that the arithmetic does not get in the way of the concepts. If students wish to show you that they can answer some of the questions, have them do it on their homework papers as part of **Question 23**.

Question 11: This is a problem that can be solved later using combinations.

Questions 12: You might use this question as a lead-in to the next lesson; its solution should be known to many students.

Question 13: This kind of problem will be discussed again in Lesson 10-8.

The committee problem (Problem 4) stated in the chapter opener is similar in some essential ways to the Lotto problem but it has an important new ingredient.

Example 3 Identify the essential features of the items to be counted in the following problem.

> A ski club has 80 members, 50 men and 30 women. A planning committee consisting of 5 men and 3 women is to be selected. How many different committees are possible?

Solution The order in which the persons on the planning committee are selected or listed does not change the composition of the committee. Also, no person can be selected more than once for a given committee. Consequently, you can think of a committee as a subset of 5 men of the set of 50 male club members, together with a subset of 3 women of the set of 30 female members. Therefore, the committee problem can be reformulated as follows:

> How many ways is it possible to choose a 5-element subset of a set of 50 elements *and* a 3-element subset of a set of 30 elements?

Counting the "male" and "female" subcommittees of the committee is essentially the same counting problem as the Lotto problem. So these parts of the problem fit the "order doesn't matter, no repetition" cell. The new ingredient is that these two subcommittees are put together to form a committee. This feature requires taking a step beyond the diagram.

Example 4 Describe the essential features of the items to be counted in the following problem.

> At a school fair, a $1 ticket at the fishing pool allows the participant to choose three prizes from four bags, each of which contains a different type of prize. Any combination of selections is acceptable: all selections can be made from the same bag, two can be made from one bag and one from a different bag, or all three can be made from different bags. How many different selections are possible?

Solution Let the four bags be labeled *a*, *b*, *c*, and *d*. Selection of prizes can be thought of as choosing 3 of these labels, with repetition allowed. The order of the selection is unimportant. For instance, a selection of *abb* is equivalent to a selection of *bab* or of *bba*. So this problem is one of counting unordered symbols when repetition is allowed. A general technique for handling such problems is discussed in Lesson 10-7.

588

Questions

Covering the Reading

1. What is meant by finding the essential features of a counting problem? **See margin.**
2. Define: *string*. **A string is an ordered list of symbols.**

In 3–9, describe the essential features of each problem. That is, decide whether the list of symbols used to describe the items being counted consists of ordered or unordered symbols and whether repetition of symbols is allowed. (You do not have to compute the solution to these problems.)

3. A bicycle combination lock has four dial wheels with the digits from 0 through 9 on each wheel. How many different lock combinations are possible? **Ordered symbols; repetition is allowed.**
4. How many integers from 100 to 999, inclusive, have three different digits? **Ordered symbols; repetition not allowed.**
5. How many different code words can be constructed from the six letters in the word *STRING*? **Ordered symbols; repetition not allowed.**
6. There are 13 applicants to fill three sales positions in different departments of a large department store. In how many different ways can these openings be filled? **Unordered symbols; repetition not allowed.**
7. A club with 50 members has the following officers: president, vice-president, treasurer, and secretary. If all members of the club are eligible for all offices, how many different slates of officers are possible? **Ordered symbols; repetition not allowed.**
8. How many different four-digit house numbers can be constructed from the six brass numerals 1, 2, 3, 4, 5, 6? **Ordered symbols; repetition not allowed.**
9. During Halloween, Ms. May allows trick-or-treaters to choose 4 pieces of candy from any of 5 bowls set on a table. If there is only one kind of candy in each bowl and no bowl contains the same kind of candy as another bowl, how many different selections are possible? **Unordered symbols; repetition is allowed.**

Applying the Mathematics

10. Identify the essential features of Problem 2 on page 585. **Ordered symbols; repetition not allowed.**
11. Suppose that your home is 6 blocks north and 4 blocks west of your school.
    a. Draw two different street routes 10 blocks long that you can follow from where you live to your school. **See margin.**
    b. Describe the two street routes in part **a** as strings of 10 symbols selected from the letters *S* (South one block) and *E* (East one block).
    c. Identify the essential features of the problem of counting all street routes from where you live to your school.
    b) See margin. c) Ordered symbols; repetition is allowed.
12. A school club is conducting an election of officers. There are 3 candidates for president, 2 for vice-president, and 3 for secretary-treasurer. Identify the essential features of the problem of counting the number of different possible ballots. **Offices are ordered, and repetition is not allowed.**

ADDITIONAL ANSWERS

1. determining whether or not the order of symbols counts and whether or not repetition of symbols is allowed

11.a. samples:

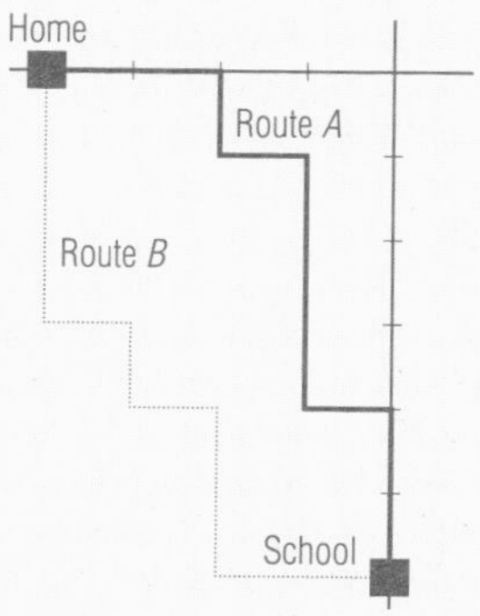

b. for the samples in part a, SSSESESSEE, EESESSSESS

NAME ______________

LESSON **MASTER 10–1**
QUESTIONS ON **SPUR** OBJECTIVES

■SKILLS *Objective A* *(See pages 636–638 for objectives.)*
In 1–5, describe the essential features of the problem.

1. The coach of a volleyball team chooses six starting players from a group of nine people. How many different starting teams are possible?
   unordered symbols, repetition not allowed
2. A department store offers a free monogramming service for customers who purchase a set of bath towels. If the monogram consists of three letters, how many different monograms are possible?
   ordered symbols, repetition allowed
3. The winner of a contest held by a record store is allowed to pick a total of 10 free albums from any of five different musical categories: reggae, classical, country, jazz, and rock. How many different selections are possible?
   unordered symbols, repetition allowed
4. Five married couples attend an opera performance together. If they wish to be seated in a row of 10 seats so that each person is sitting next to his or her spouse, how many different seating arrangements are possible?
   ordered symbols, repetition not allowed
5. How many numbers between 100 and 999 have at least one digit that is a seven?
   ordered symbols, repetition allowed
6. Ten Olympic swimmers compete for gold (1st place), silver (2nd place), and bronze (3rd place) medals in a 400-meter freestyle event. In how many different ways can the three medals be awarded?
   ordered symbols, repetition not allowed

 109

FOLLOW-UP

MORE PRACTICE
For more questions on SPUR Objectives, use *Lesson Master 10-1,* shown on page 589.

ADDITIONAL ANSWERS
14.a. Because the cosine function has period 2π, any interval of length 2π contains all possible values of the function.

b. $\frac{\pi}{2} < \theta < \frac{3\pi}{2}$

c.

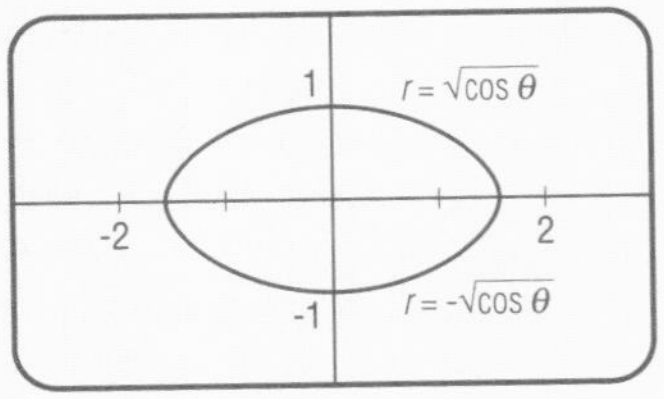

d.

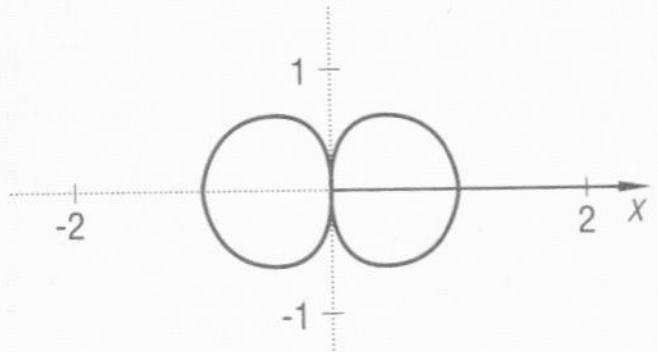

15. See Additional Answers in the back of this book.

13. Suppose that three different coins, a nickel, a dime, and a quarter, are to be placed in three different cups in a collection of 9 different cups. Identify the essential features of the problem of counting the number of different ways of distributing the three coins among the 9 cups.
Ordered symbols; repetition not allowed.

Review

14. Consider the polar equation
$$r^2 = \cos \theta.$$
a. Explain why the complete polar graph of this equation can be obtained by plotting it on the interval $0 \le \theta \le 2\pi$.
b. For what values of θ in the range $0 \le \theta \le 2\pi$ is r undefined?
c. Use an automatic grapher to plot the rectangular graphs of $r = \sqrt{\cos \theta}$ and $r = -\sqrt{\cos \theta}$ for those values of θ for which r is defined.
d. Use the rectangular graphs in part **c** to sketch the polar graph of $r^2 = \cos \theta$. *(Lesson 8-4)* **See margin.**

15. Use an automatic grapher to conjecture whether the equation below is an identity. If it is an identity, prove it. If not, give a counterexample.
$(\tan \theta)(\sin \theta + \cot \theta \cos \theta) = \sec \theta$ *(Lessons 6-2, 6-1)* **See margin.**

16. *Multiple choice.* If 3 is a zero of a polynomial $p(x)$, then
(a) 3 is a factor of $p(x)$.
(b) $x + 3$ is a factor of $p(x)$.
(c) $x - 3$ is a factor of $p(x)$.
(d) $p(-3) = 0$. **(c)**
(Lesson 4-6)

17. **a.** How many positive integers less than 1000 are divisible by 7?
b. How many positive integers less than an integer n are divisible by d? (Hint: Use the greatest integer or floor function.) *(Lesson 3-5)*
a) 142; b) $\lfloor \frac{n}{d} \rfloor$

18. For the function f graphed at the right, determine the interval(s) over which the function is decreasing. *(Lesson 2-3)*
$x < -1$, $x > 0$

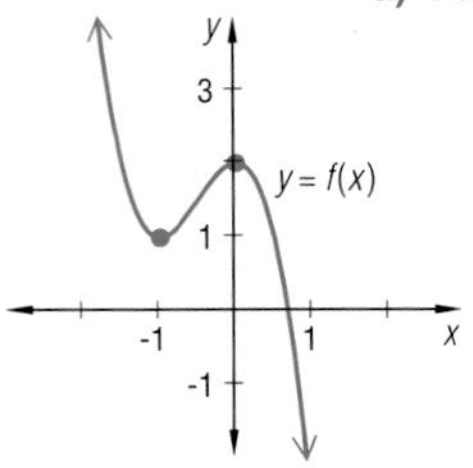

In 19 and 20, write the negation of each statement. *(Lesson 1-2)*

19. *For all positive integers n, n is prime.*
∃ a positive integer n that is not prime.

20. *There exist positive integers n and m such that nm = 11.*
∀ positive integers n and m, $nm \ne 11$.

In 21 and 22, expand each product. *(Previous course)*

21. $(x + y)^2$
$x^2 + 2xy + y^2$

22. $(2x - y)^3$
$8x^3 - 12x^2y + 6xy^2 - y^3$

Exploration

23. Pick one of the problems from Questions 3–9 in this lesson, and try to determine the actual solution to the problem. **(3) 10,000; (4) 648; (5) 720; (6) 286; (7) 5,527,200; (8) 360; (9) 625**

590

LESSON

10-2

Possibility Trees and the Multiplication Counting Principle

Many counting problems can be solved by taking advantage of the fact that the items to be counted can be selected or constructed through a succession of steps. For example, a string $s_1, s_2, s_3, \ldots, s_k$ of k symbols can be constructed by the following succession of k steps.

Step 1: Select the first element, s_1.
Step 2: Select the second element, s_2.
Step 3: Select the third element, s_3.
⋮
Step $k-1$: Select the $(k-1)$st element, s_{k-1}.
Step k: Select the kth element, s_k.

The following example illustrates this.

Example 1 In some competitions between two players or teams, the first to win three games wins the competition. Describe the possible outcomes of such a competition as a succession of steps.

Solution Call the players A and B. By the time five games have been played, either A or B must have won three games. So this competition is a succession of 3 to 5 games. The possible outcomes of the competition are the sequences of wins and losses described below.

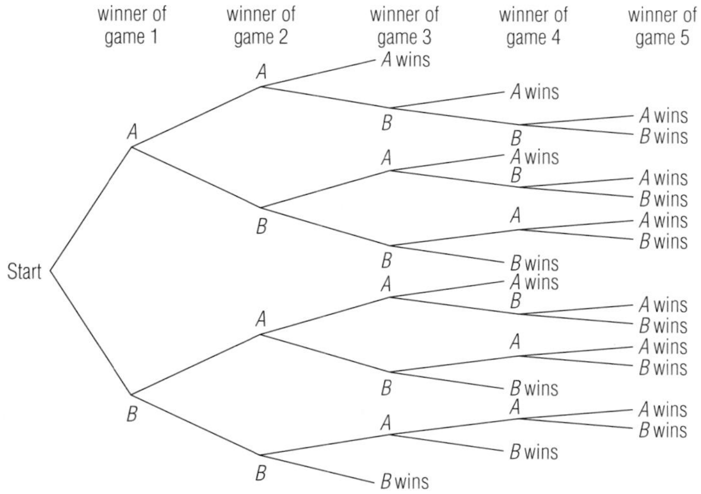

Because there are 20 ways to get "wins," the diagram shows that there are 20 possible outcomes for this competition. Each outcome is represented by a path from the point marked "start" to a point where either A or B wins the competition. Two outcomes require 3 games, 6 outcomes require 4 games, and 12 outcomes require 5 games.

LESSON 10-2

RESOURCES
- Lesson Master 10-2
- Teaching Aid 67 can be used with **Examples 1 and 2**.
- Computer Master 18

OBJECTIVES

F Use the Multiplication Counting Principle to solve counting problems.
I Use a possibility tree to determine the number of outcomes in a given situation.

TEACHING NOTES

This lesson emphasizes two types of problems—those that can be solved by constructing possibility trees and those that can be solved by using the Multiplication Counting Principle. Although we use possibility trees to lead to the Multiplication Counting Principle, students need to understand that not all tree problems can be solved by applying the counting principle.

Because the branches of the tree in **Example 1** are of different lengths, some students may have difficulty reading the tree to determine the total number of outcomes. Students need to see that each time a branch terminates, there is one outcome. You might also have students write out the sequence of team wins for each branch of the tree, similar to the code words for the tree in **Example 2**.

Small Group Work **Examples 3 and 4** relate naturally to **Project 1** on page 632. If students consider the population of their state and the number of license plate identifications found in these two problems, they may wonder how the Department of Motor Vehicles comes up with enough license plate identifications for all registered vehicles in the state. This could lead to an interesting project for a single student or for a **small group** of students.

The solution of **Example 5** applies a De Morgan Law with negations on both sides, namely: *p or q* ≡ *not ((not p) and (not q))*. The complexity of the law signals the possibility that students may have difficulty understanding it.

We assume that students are familiar with the following Addition Counting Principle: The number of elements in the union of two disjoint sets is the sum of the elements in each one of them. Symbolically, $A \cap B = \emptyset \Rightarrow N(A \cup B) = N(A) + N(B)$.

The word *complement* means "to fill out" or "complete." The use of the word with respect to angles and with respect to sets are variants of this meaning.

The diagram used in Example 1 to display the possible outcomes of the competition is called a **possibility tree.** Each game in each outcome of the competition is represented by a **branch point** or **node** of the tree; the first node is labeled "start." A node corresponds to a step in which several choices or results are possible. The segments between nodes are called **branches,** and the ends of the branches are called **leaves.** The number of possible outcomes of the playoff series, 20, is equal to the number of leaves in the possibility tree for the competition.

A possibility tree can be used to count strings if the number of strings is not too large.

Example 2 Use a possibility tree to count the number of different two-letter code words that can be constructed from the letters *A, B,* and *C* if repetition of letters is not allowed.

Solution There are three branches leaving the start node, representing the selection of the first letter. However, because repetition of letters is not allowed, the possibility tree has only two branches leaving each of those nodes, representing selection of the second letter.

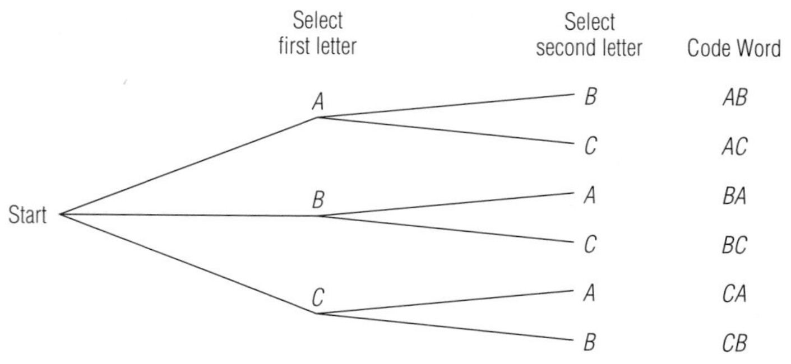

There are six leaves on this possibility tree, and so there are six possible code words if repetition is not allowed.

In the possibility tree of Example 2, in each step the same number of branches leads from each node. The number of leaves in such a possibility tree is the product of the numbers of branches leaving the nodes for each step of the tree. For instance, in Example 2, there are 3 branches to indicate selection of the first letter and 2 branches leaving each of those branches to indicate selection of the second letter. So the total number of leaves in the tree is $3 \cdot 2 = 6$.

This special feature of certain possibility trees illustrates the following general Multiplication Counting Principle, which you may have encountered, without proof, in previous courses.

592

> **The Multiplication Counting Principle**
>
> Suppose that strings result from a procedure which consists of k successive steps and that:
> the 1st step can be done in n_1 ways,
> the 2nd step can be done in n_2 ways,
> ⋮
> and the kth step can be done in n_k ways.
> Then the number of strings is
> $$n_1 \cdot n_2 \cdot \ldots \cdot n_k.$$

A proof of the Multiplication Counting Principle using mathematical induction is outlined in Question 15. The following example illustrates how this principle can be used to solve the license plate problem of the chapter opener.

Example 3 Compute the number of different 6-symbol license plate identifications that are possible if the first two symbols can be any of the 26 letters of the alphabet and the last four symbols can be any digit from 0 to 9.

Illinois employs handicapped people to make license plates.

Solution Think of constructing a 6-symbol license plate identification as a process of selecting six symbols, one at a time from left to right. There are 26 possible selections for each of the first two symbols because there are 26 letters of the alphabet. There are 10 possible selections for each of the last four symbols because there are 10 digits from 0 to 9. Thus, by the Multiplication Counting Principle, there are

26	•	26	•	10	•	10	•	10	•	10	= 6,760,000
1st letter		2nd letter		1st digit		2nd digit		3rd digit		4th digit	

different possible license plate identifications.

The answer to Example 3 could be written $26^2 \cdot 10^4$, where 26^2 is the number of two-letter strings with repetition and 10^4 is the number of four-digit strings with repetition. Those two powers are instances of the following corollary to the Multiplication Counting Principle.

> **Theorem**
>
> If repetition is allowed, the number of r-symbol strings that can be made from a set of n symbols is n^r.

Proof Each symbol in the string can be selected in n ways. There are r selections to be made. Thus the number of strings is $\underbrace{n \cdot n \cdot \ldots \cdot n}_{r \text{ factors}} = n^r$.

ADDITIONAL EXAMPLES

1. Describe the possible outcomes of a competiton in which the first team to win two games wins the series.
Call the competitors *A* and *B*. The possibilities are *A*-*A*, *A*-*B*-*A*, *A*-*B*-*B*, *B*-*B*, *B*-*A*-*A*, and *B*-*A*-*B*. Using a tree:

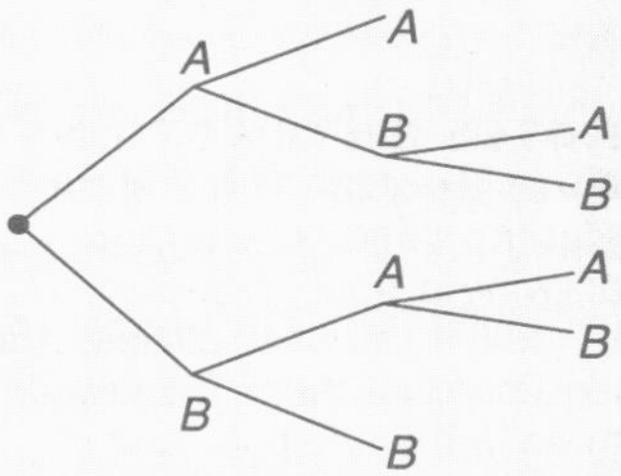

2. Use a possibility tree to count the number of ways to arrange the numbers 1, 2, 4, and 5 so that no number is next to a number that is one larger than it.

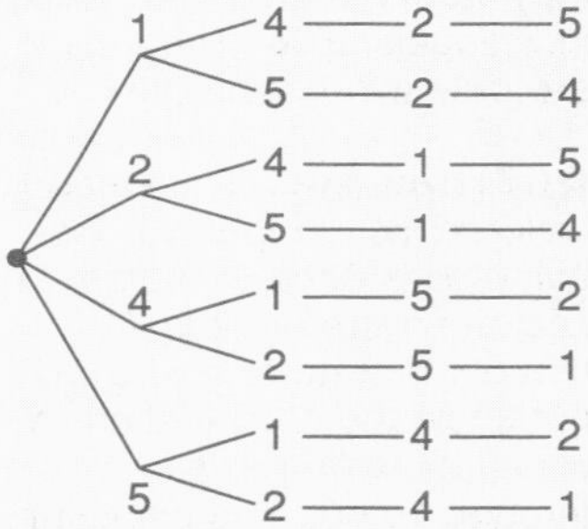

8 ways

3. A business wants a phone number that ends in 00. If there are three exchanges (first three numbers) that are available to the business's community, how many phone numbers ending in 00 are available?
There are ten choices for each of the remaining digits, so 300 numbers in all.

ADDITIONAL EXAMPLES
4. How many 7-digit phone numbers have no repeated digits? Assume that all numbers are possible.
604,800

5. Among the 3-digit integers, how many have at least one digit that is a 2 or a 4?
452

NOTES ON QUESTIONS
Question 3: Emphasize that the major reason for drawing a possibility tree is to exhibit the possibilities, not just to count them. To count them without listing, one would use the Multiplication Counting Principle.

Question 4: If the chairs were in a row, then they would be somehow different (left, center, right). If the chairs were in a circle, then they would be in some sense indistinguishable. Then there are only two possibilities.

Question 10: The kind of lock mentioned in this question also is used for some security systems. It is thought that no one who did not know the combination would be willing to go through 10,000 combinations.

Question 11: If the positions were indistinguishable, then the number of ways would be 286. (Divide 1716 by 3!.) This question can be used as a lead-in to the next lesson.

Example 4 Suppose that the license plate problem is modified by requiring that no symbol in the license plate be repeated. Determine the total number of different license plates that is possible in this case.

Solution Because no repetition is allowed, the number of letters and digits that can be selected is reduced by one each time a selection is made. Therefore, there are

26	•	25	•	10	•	9	•	8	•	7
1st letter		2nd letter		1st digit		2nd digit		3rd digit		4th digit

or 3,276,000 different license plates.

Sometimes it is rather complicated to count the items of a set that possess a particular property. In such cases it may be more efficient to count the items in the **complement**, that is, to count the items in the set that do not have the given property.

Example 5 Among the integers from 1000 through 9999, how many have at least one digit that is a 5 or a 7?

Solution The set S of all integers from 1000 through 9999 consists of all strings $d_1d_2d_3d_4$ of 4 digits in which repetition is allowed. Note that d_1 must be one of the digits from 1 through 9, while d_2, d_3, and d_4 can be any of the digits from 0 through 9. By the Multiplication Counting Principle, the set S contains 9000 integers.

The given problem asks you to count the integers in S which have one, two, three, or four digits of 5 or 7. But it is simpler to count those integers in S that contain neither a 5 nor a 7.

$$\begin{pmatrix}\text{number of integers}\\ \text{containing a 5 or 7}\end{pmatrix} = \begin{pmatrix}9000 - (\text{number of integers not}\\ \text{containing a 5 and not containing a 7})\end{pmatrix}$$

The integers not containing a 5 and not containing a 7 are strings of 4 digits where the first digit can be any of 1, 2, 3, 4, 6, 8, 9 and the other digits can be any of 0, 1, 2, 3, 4, 6, 8, 9. So the number of integers satisfying this condition is $7 \cdot 8 \cdot 8 \cdot 8 = 3584$. Thus the number of integers containing a 5 or 7 is $9000 - 3584 = 5416$.

Questions

Covering the Reading

1. Refer to Example 1. How many possible outcomes are there if Team A wins the first two games? 4

2. Modify the possibility tree in Example 2 to find the number of possible code words that can be formed if the letters can be repeated. See margin. 9 possible code words

3. Each runner in a 10-K race receives a T-shirt as a souvenir. The T-shirts are available in three colors: white, blue, and red; and four sizes: S, M, L, and XL. Draw and label a possibility tree to count the number of different T-shirts that are available. See margin. 12 different shirts

4. Use a possibility tree to count the number of different ways that Luis, Van, and Charisse can be seated in a row of three chairs. See margin. 6 different seatings

5. In the World Series for baseball, the first team to win four games wins the series. Suppose that the 2099 World Series is played between the Atlanta Braves and the Toronto Blue Jays and that the Braves win the first three games. Use a possibility tree to count the number of different ways that the series can be completed. See margin. 5 ways to complete the series

6. How many different four-digit numbers can be constructed from the digits 1, 2, 3, 4, 5, 6
   a. if repetition of digits is not allowed? 360
   b. if repetition is allowed? 1296

7. A grocery store carries n brands of liquid dish soap. Each brand offers four different sized bottles, small, medium, large, and economy, and each is available as either lemon-scented or pine-scented. How many different types of liquid dish soap bottles are available at the store? $8n$

8. Explain why it would not be easy to use a possibility tree to solve the license plate problem in Example 4. It would be impractical to represent all 3,276,000 possible outcomes.

9. How many of the integers from 1000 through 9999 have at least one digit which is a 5? 3168

Applying the Mathematics

10. **a.** A bicycle combination lock has four dial wheels with the digits from 0 through 9 on each wheel. How many different possible combinations are there for this type of lock? 10,000
    b. Generalize part **a** if there are w dial wheels and d symbols on each wheel. d^w

11. There are 13 applicants to fill three sales positions in different departments of a large department store. In how many different ways can these openings be filled? 1716

12. A computer program includes the following nested loop.

```
100 FOR I = 1 TO 7
200   FOR J = 1 TO 12
300     B(I) = B(I) − C(I,J) * B(J)
400   NEXT J
500 NEXT I
600 END
```

How many times is line 300 executed when the program is run? 84

13. **a.** Among the integers from 200 through 999, how many have at least one digit that is a 4 or a 5 or a 6? 555
    b. If one of the integers in the answer to part **a** is chosen, what is the probability that all 3 digits are the same? ≈ 0.0054

ADDITIONAL ANSWERS

2.

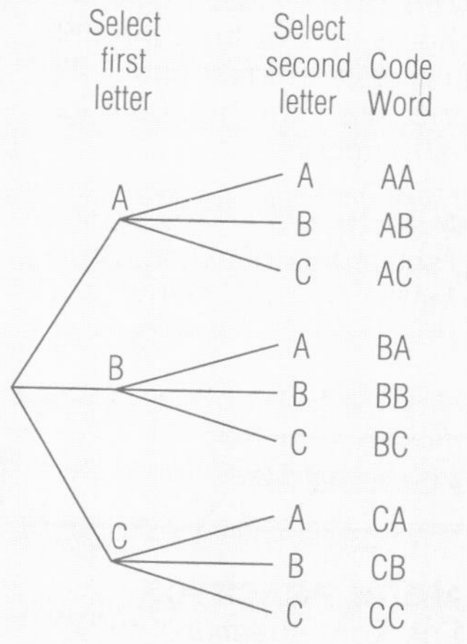

3.

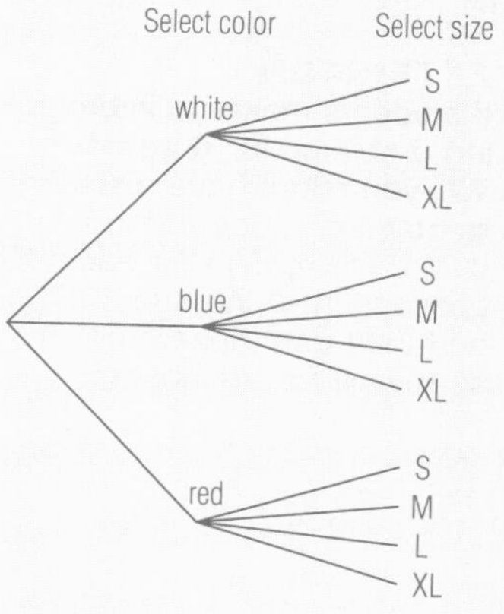

4.

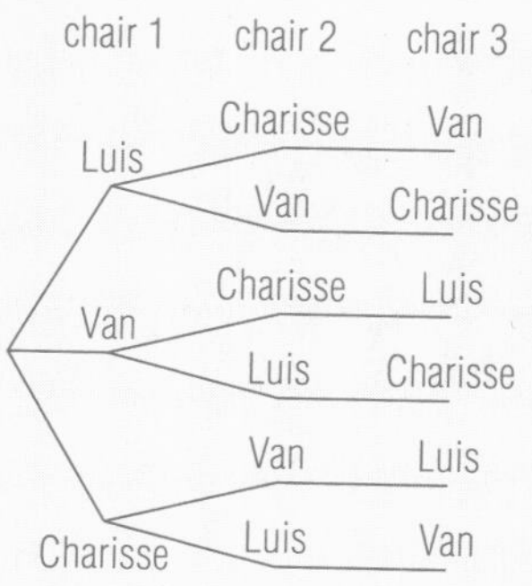

5.

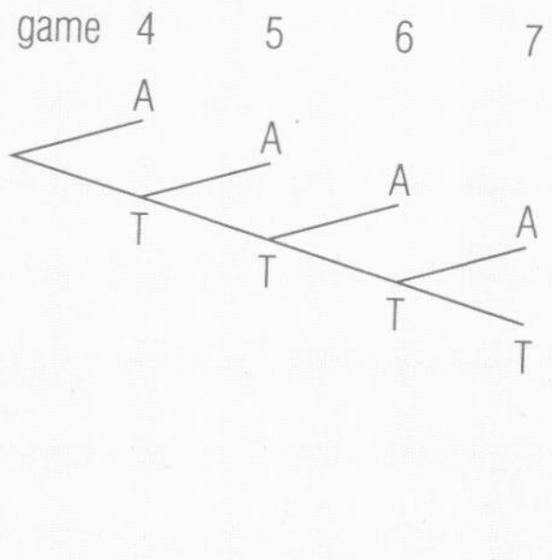

NOTES ON QUESTIONS

Question 13b: The choice here is from the 555 that have at least one 4, 5, or 6, not from the 800 original numbers.

Question 22: Be sure to discuss this question. Students can be very inventive.

FOLLOW-UP

MORE PRACTICE

For more questions on SPUR Objectives, use *Lesson Master 10-2*, shown on page 597.

EXTENSION

If students make up interesting problems for **Question 22**, you might have **small groups** get together to invent one problem for each of Lessons 10-2 through 10-7 and then use these problems as review for the chapter.

ADDITIONAL ANSWERS

14.d. Sample: The advertiser is using the term "endless" loosely.

14. a. If A is a set with n elements, use the Multiplication Counting Principle to prove that there are 2^n different subsets of A (counting the empty set and A itself as subsets of A). (Hint: Think of constructing subsets of A as follows: Arrange the elements of A in some order as a list. Then make n successive decisions, Yes or No, for the question: Do you put this element on the list in the subset?) **See page 597.**

b. Consider the excerpt from a cereal box reproduced here. How many combinations of fruits are possible? **15**

So many kinds of fruit, the possibilities are endless.
Delicious apples.
Sweet pineapples.
Chewy apricots.
Plump raisins.
So many kinds of fruit, it gives you a different fruit taste in every bite.
Try some and see.

c. What does the empty set in part **a** represent in the situation of part **b?**

d. Given the answer to part **b** is a finite number, why can the manufacturer claim "the possibilities are endless." **See margin.**

c) a spoonful of bran (and milk) only; no fruit

15. Take the following as an axiom: *If one step can be done in m ways and a second step can be done in n ways, then the two steps can be done in mn ways.* Use this axiom and the Principle of Mathematical Induction to prove the Multiplication Counting Principle. (Hint: If each item in the set to be counted is described by $k + 1$ successive steps, group the first k steps into a single new step.) **See page 597.**

Review

In 16 and 17, identify the essential features of the problem. You do not have to do the counting. *(Lesson 10-1)*

16. Each question on a 20-question multiple choice test has five options. Determine the number of ways of answering the test.
Ordered symbols; repetition is allowed.

17. A buffet table is divided into four sections. The first section contains a choice of salad, soup, or fruit plate. The second section contains peas, carrots, green beans, and two types of potatoes. The third section contains chicken, roast beef, pork chops, and fish. The last section contains five different desserts. If all the items within each section are arranged in a row, determine the number of different arrangements of items. **Ordered symbols; repetition is not allowed.**

18. Use the difference quotient to prove that for any linear function $f(x) = mx + b$, the average rate of change of f from x to $x + \Delta x$ is m. *(Lesson 9-1)* See below.

19. Use the formula $\sum_{i=1}^{n} i = \frac{n(n+1)}{2}$ to find the following.

a. $1 + 2 + 3 + \ldots + 70$ 2485

b. $\sum_{i=1}^{200} i - \sum_{i=1}^{100} i$ 15,050

c. the sum of all 4-digit positive integers 49,495,500

d. the sum of the integers from m to n *(Lesson 7-3)* $\frac{n}{2}(n+1) - \frac{m}{2}(m+1)$

In 20 and 21, solve without using a calculator. *(Lessons 3-2, 2-7)*

20. $3^x = 9^{4x+3}$ $-\frac{6}{7}$

21. $\log_5 x + \log_5 3 = 4$ $\frac{625}{3}$

Exploration

22. Invent a problem, different from the ones in this lesson, which requires the Multiplication Counting Principle for its solution. Answers will vary.

14. a) There are two choices for each element: include in the subset, or don't include it. Since there are n elements, by the Multiplication Counting Principle there are n factors of 2, or 2^n subsets.

15. $s(1)$ is true, because the number of ways the first step can be done is n_1. Assume $s(k)$, the number of ways to do the first k steps, is $n_1 \cdot n_2 \cdot \ldots \cdot n_k$. Let m be the number of ways to do the first k steps and let n represent the number of ways to do the $(k + 1)$st step. Then by the inductive hypothesis, $m = n_1 \cdot n_2 \cdot \ldots \cdot n_k$ and $n = n_{k+1}$ so $mn = n_1 \cdot n_2 \cdot \ldots \cdot n_k \cdot n_{k+1}$. So $S(k + 1)$, the number of ways to do the $(k + 1)$ steps is $n_1 \cdot n_2 \cdot \ldots \cdot n_k \cdot n_{k+1}$. Thus $S_{(n)}$ is true for all n.

18. $$\frac{\Delta y}{\Delta x} = \frac{f(x + \Delta x) - f(x)}{\Delta x} = \frac{m(x + \Delta x) + b - (mx + b)}{\Delta x} = \frac{mx + m\Delta x + b - mx - b}{\Delta x} = \frac{m\Delta x}{\Delta x} = m$$

PROJECTS
The projects for Chapter 10 are described on pages 632–633. **Project 1** is related to the content of this lesson.

NAME

LESSON MASTER 10–2
QUESTIONS ON **SPUR** OBJECTIVES

USES *Objective F (See pages 636–638 for objectives.)*

1. At a pharmacy, toothbrushes are available in c different colors, three different firmnesses (hard, medium, and soft), and two different sizes (adult and youth). How many different kinds of toothbrushes are available? 6c

2. How many numbers between 1,000 and 9,999 inclusive have at least one digit that is a seven? 3,168

3. If Roger owns 12 shirts, 4 belts, and 10 pairs of pants, how many more shirts must he buy in order to have at least 600 different outfits? 3

4. How many 6-letter codewords can be made from the letters F, G, H, I, J, K, and L:

a. if the letters can be repeated? 117,649

b. if the letters cannot be repeated? 5,040

5. A mother, a father, and their seven children plan to have a family portrait taken. If one parent is to stand at each end of the row, in how many ways can the family members be arranged? 10,080

6. a. How many 3-digit numbers contain at least one six? 252

b. What is the probability of choosing one of the numbers in part **a** which has exactly two sixes? $\frac{26}{252}$, or $\approx .103$

7. How many lines of output are produced by the following computer program? 120

```
10 FOR I = 0 TO 7
20  FOR J = 1 TO 3
30   FOR K = 2 TO 6
40    PRINT I * J * K
50   NEXT K
60  NEXT J
70 NEXT I
```


NAME

REPRESENTATIONS *Objective I*

8. Draw a possibility tree to count the number of different 3-letter codewords that can be made from the letters A, B, and C if only the C may be repeated. 11

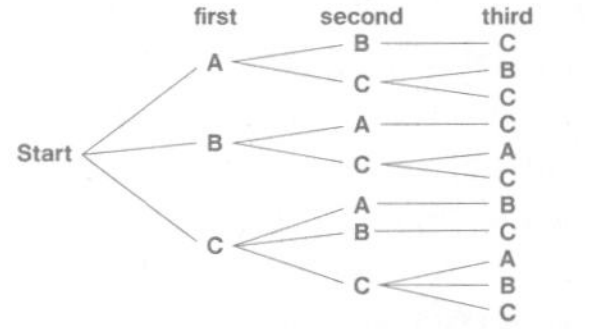

9. Jack, Patrice, Lydia, and Miguel are the finalists in a statewide science fair. First and second prizes will be awarded.

a. Draw a possibility tree to determine all possible distributions of the awards.

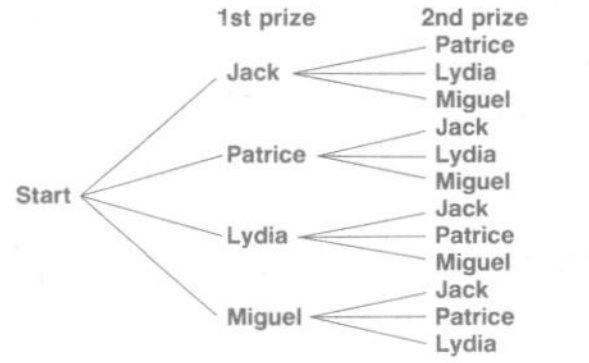

b. In how many of these distributions does Patrice receive 2nd prize? 3

LESSON 10-3

RESOURCES

- Lesson Master 10-3
- Quiz for Lessons 10-1 through 10-3
- Teaching Aid 68 provides the possibility tree for the permutations of "Math."

OBJECTIVES

B Evaluate expressions indicating permutations.
D Use properties of permutations to prove identities.
F Use the Multiplication Counting Principle and permutations to solve counting problems.

TEACHING NOTES

We assume that students are familiar with the factorial symbol. Make certain, however, that they can locate the factorial key on their calculators.

In this lesson, we consider only permutations of n distinct objects. In Lesson 10-8, students will learn about the number of permutations of n objects when two or more of those objects are the same.

Making Connections
Emphasize the connection between the Permutation Theorem and the Multiplication Counting Principle. The Multiplication Counting Principle, together with mathematical induction, is needed to prove the Permutation Theorem. In fact, you may want to do **Question 14**, which requires the proof, as part of the class discussion.

LESSON

10-3

Permutations

The word *math* is a string consisting of the four letters m, a, t, and h in which repetition does not occur. Other strings without repetition of these four letters are listed below, but none of them is a word in the English language.

amht *tamh* *atmh* *htma*

All of these strings are examples of *permutations*.

Definition

A **permutation** of the symbols $a_1, a_2, ..., a_n$ is a string of all these symbols without repetition.

For example,

1 2 3 3 1 2 2 3 1

are three different permutations of the numbers 1, 2, and 3.

You could use a possibility tree to list all of the permutations of the four letters m, a, t, h. Here is part of that tree.

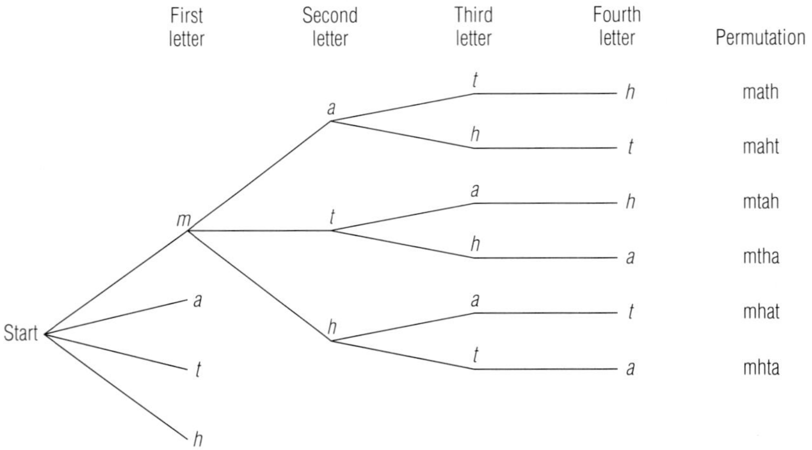

If completed, this possibility tree would be quite large. However, you can count the number of permutations of these four letters without listing them.

Any one of the four letters can be selected as the first letter. Any of the three remaining letters can be chosen as the second letter. Two letters remain available as the third letter and only one letter remains for the fourth letter. Therefore, by the Multiplication Counting Principle, there are

$$4 \cdot 3 \cdot 2 \cdot 1 = 4! = 24$$

different permutations of these four letters.

Here are the 24 permutations of the letters *m*, *a*, *t*, and *h* systematically listed. (Do you see the system?)

math	amth
maht	amht
mtah	atmh
mtha	athm
mhat	ahmt
mhta	ahtm
tmah	hmat
tmha	hmta
tamh	hamt
tahm	hatm
thma	htma
tham	htam

Notice that all of the permutations in a given section begin with the same first letter. The permutations in the section are obtained by following the first letter with all possible permutations of the remaining three letters. This recursive nature of permutations, along with mathematical induction, can be used to prove the following result. In Question 14 you are asked to supply a proof.

Permutation Theorem

There are $n!$ permutations of n different elements.

For example, there are $5! = 5 \cdot 4 \cdot 3 \cdot 2 \cdot 1 = 120$ permutations of the letters a, b, c, d, and e.

Example 1 In how many ways can 6 boys and 3 girls be arranged in a row if the three girls must always stand next to each other?

Solution You can think of arranging the boys and girls in the row as a two-step process.

Step 1 Select an arrangement for the boys and reserve a space for the girls' part of the row.

Step 2 Select an arrangement for the girls within their part of the row.

You can think of the girls' part of the row as a single unit G that is to be arranged with the six boys B_1, B_2, B_3, B_4, B_5, B_6 to complete the row. An arrangement of $\{G, B_1, B_2, B_3, B_4, B_5, B_6\}$ is a string of ordered symbols without repetition; it is a permutation, so there are 7! arrangements of the boys with a space for the girls. The arrangement of the three girls in their part of the row is also a string without repetition, thus there are 3! different ways to arrange the girls. Therefore, by the Multiplication Counting Principle, there are

$$7! \cdot 3! = 30{,}240$$

different ways to arrange the boys and girls in a row with the girls next to one another.

Now consider the word *logarithm*. Supppose you wanted to make four-letter strings using each of the letters in this word at most once. There are nine letters to choose from, so the first letter can be chosen in 9 ways. Because no letter can be used more than once, the second letter can be chosen in only 8 ways, the third in 7 ways, and the fourth in 6 ways. Hence, by the Multiplication Counting Principle, there are

$$9 \cdot 8 \cdot 7 \cdot 6 = 3024$$

different four-letter strings using the letters in *logarithm* only once. These strings are called *permutations of 9 elements taken 4 at a time*.

Error Analysis The situation of **Example 1** contains a slight variant of the standard permutation problem. The three girls must be treated as a single symbol for the purposes of obtaining the number of permutations of 7 objects (6 boys and 1 set of girls). The potential for error occurs when students fail to realize that the three girls, treated previously as a single object, can themselves be permuted. The final solution requires the product of the number of permutations of seven objects with the number of permutations of the girls. After discussing **Example 1**, you might want to discuss **Question 3** to help students focus on the differences involved when these 10 people are arranged with no restrictions versus restrictions.

Making Connections Students who have studied from *Functions, Statistics, and Trigonometry* probably used ${}_nP_r$ notation rather than $P(n, r)$. You can refer to the Perspectives to inform students why the notation $P(n, r)$ is used in this book. We have used different notations in the UCSMP books so that students will be familiar with the various symbols used for these ideas.

Technology In **Example 2**, help students see that

$$\frac{8!}{5!} = \frac{8 \cdot 7 \cdot 6 \cdot 5 \cdot 4 \cdot 3 \cdot 2 \cdot 1}{5 \cdot 4 \cdot 3 \cdot 2 \cdot 1} = 8 \cdot 7 \cdot 6$$

by dividing out common factors. Hence, it is not necessary to evaluate 8! on the calculator and then divide it by 5!. Division of factorials can be handled in this case because the numbers are small. When the numbers are greater than 69, many calculators cannot do factorial computations even in scientific notation. Thus, simplifying a computation before resorting to technology is important.

ADDITIONAL EXAMPLES

1. Four boy-girl couples go together to a dance.

a. If the pair in each couple stands next to each other for a picture, in how many ways can the eight people be arranged?

$4! \cdot 2^4 = 384$

b. If the pair in each couple stands next to each other and boys and girls alternate, in how many ways can the eight people be arranged?

$2 \cdot 4! = 48$

2. In a 1990 contest of a mathematics organization, three of the six people with the best scores came from the same school. The contest rules specify that if there were ties, three winners would be selected at random. It turns out that the three winners all came from the same school. What was the probability of this occurring?

$\frac{3!3!}{6!} = .05$

NOTES ON QUESTIONS

Question 3: Do not underestimate the difficulty of these types of questions. Advise students to proceed slowly and carefully, writing down instances of each type to make the possibilities clear. They can call the boys B1, B2, B3, B4, B5, and B6 and the girls G1, G2, and G3.

Questions 4 and 9: As a possible follow-up to Question 4, you may wish to ask: Are there more six-letter strings or seven-letter strings? (seven-letter strings) Are there more eight-letter strings than seven-letter strings? (Yes) Are there more nine-letter strings than eight-letter strings? (No, they are equal.) Ask students to try to generalize their results. (If $n > r > s$, then ${}_nP_r > {}_nP_s$. The same is not true for combinations.) Question 9 considers the case when $r = n - 1$.

Definition

Let S be a set with n (distinct) elements and let $r \leq n$. A **permutation of the n elements of S taken r at a time** is a string of r of the elements of S without repetition.

Both the symbols $\boldsymbol{P(n, r)}$ and ${}_n\boldsymbol{P}_r$ are used to denote the number of permutations of n elements taken r at a time. In the *logarithm* example above, you saw that $P(9,4) = {}_9P_4 = 9 \cdot 8 \cdot 7 \cdot 6 = 3024$.

There is a way to express $P(9, 4)$ as a quotient of factorials.

$$P(9, 4) = 9 \cdot 8 \cdot 7 \cdot 6 = \frac{9 \cdot 8 \cdot 7 \cdot 6 \cdot 5!}{5!} = \frac{9!}{5!} = \frac{9!}{(9-4)!}.$$

Similarly, $P(12, 2) = \frac{12 \cdot 11 \cdot 10!}{10!} = \frac{12!}{(12-2)!}$ and

$$P(100,13) = \frac{100!}{87!} = \frac{100!}{(100-13)!}.$$

The reasoning in the computations above and mathematical induction can be used to prove the following theorem.

***P(n, r)* Calculation Theorem**

The number $P(n, r)$ of permutations of n elements taken r at a time is given by

$$P(n, r) = \frac{n!}{(n-r)!}.$$

Example 2 There are 8 competitors in the final heat for the 800-meter race in a track meet. A large, a medium, and a small trophy will be awarded to the first-, second-, and third-place finishers. How many different ways can these trophies be awarded?

Solution The number of different ways to award the trophies equals the number of different ways that a first-, a second-, and a third-place finisher can be chosen from among the 8 competitors. This is the same as the number of 3-symbol strings that can be taken from a set of 8 symbols without repetition. Therefore, there are $P(8, 3)$ different ways to award the three trophies.

$$P(8, 3) = \frac{8!}{(8-3)!} = \frac{8!}{5!} = 8 \cdot 7 \cdot 6 = 336$$

Check Use the Multiplication Counting Principle: There are 8 ways to select the first place finisher, 7 ways to select the second place finisher, and 6 ways to select the third place finisher. Therefore, there are $8 \cdot 7 \cdot 6 = 336$ different ways to award the trophies.

Notice that if $r = n$ in the formula for $P(n, r)$, then

$$P(n, n) = \frac{n!}{(n-n)!} = \frac{n!}{0!}.$$

Because the Permutation Theorem states that there are $n!$ permutations of n different things, it follows that

$$n! = \frac{n!}{0!}.$$

This is one reason why 0! is defined to equal 1.

Questions

Covering the Reading

1. Refer to the possibility tree at the beginning of the lesson.
a. Complete the portion of the tree whose first letter is a.
b. Should you get the six permutations in the first section of the second column of the list found preceding the Permutation Theorem?
a) See margin. b) Yes

2. How many 6-digit numbers with all digits different can be constructed from the numbers 1, 2, 3, 4, 5, and 6? 720

3. In Example 1, the number of different ways to arrange 6 boys and 3 girls in a row with the girls next to one another was computed. Find the number of different ways to arrange the 6 boys and 3 girls in a row
a. with no restrictions; 362,880
b. if the boys must all stand next to each other; 17,280
c. if the boys must all stand next to each other and the girls must stand next to each other. 8640

4. How many six-letter strings can be made using letters of the word *logarithm* no more than once? 60,480

In 5 and 6, evaluate the expression.

5. $P(7, 3)$ 210 **6.** $P(9, 9)$ 362,880

7. How many 3-digit numbers with different digits can be constructed from the numbers 1, 2, 3, 4, and 5? 60

Applying the Mathematics

8. Find n so that $P(18, n) = \frac{18!}{15!}$. $n = 3$

9. a. Use the $P(n, r)$ Calculation Theorem to verify that $P(10, 9) = P(10, 10)$. See margin.
b. Generalize part **a**. $\forall$ integers $n \geq 2$ $P(n, n-1) = P(n, n)$

10. A combination for a padlock consists of three numbers from 0 to 35.
a. How many "combinations" are possible? 46,656
b. How many "combinations" consist of three different numbers? 42,840

11. Consider the integers from 1000 to 9999 which have only odd digits.
a. How many of these have all different digits? 120
b. How many have odd digits with possible digit repetitions? 625

Question 10: The word "combinations" is in quotes because this everyday use of the word is different from the mathematical use that students will see in the next lesson. There are some teachers who feel that the everyday use of the word should be changed, but an argument could be made that the mathematical use should be changed.

ADDITIONAL ANSWERS
1.a.

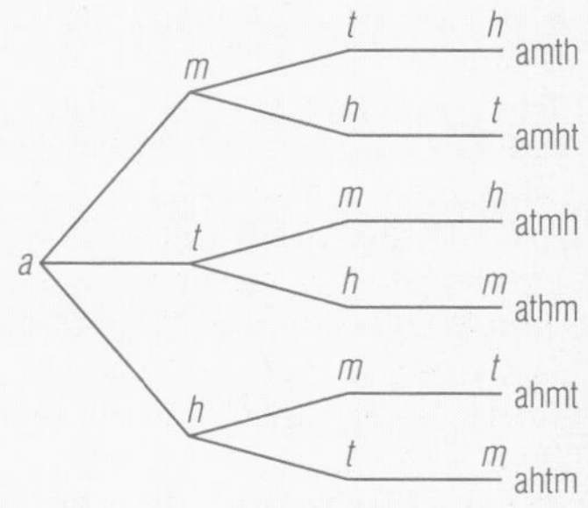

9.a. $P(10, 9) = \frac{10!}{(10-9)!} = \frac{10!}{1!} = \frac{10}{1} = \frac{10}{0!} = \frac{10!}{(10-0)!} = P(10, 10)$

NOTES ON QUESTIONS

Question 16: These seatings are known as *circular permutations.* In general, the number of circular permutations of n objects is $(n - 1)!$.

Question 17: The sample answer can be found by using the $P(n, r)$ Calculation Theorem. There also is another answer: $n = 12$, $r = 8$.

ADDITIONAL ANSWERS

14. For $n = 1$, the number of possible permutations is $1 = 1! = n!$ Assume the theorem is true for $n = k$. That is, there are $k!$ permutations of k different elements. Consider $n = k + 1$. There are $k!$ permutations of the first k elements by the inductive hypothesis. Then there are $k + 1$ places to insert the $(k + 1)$st element into any of these permutations. So by the Multiplication Counting Theorem, there are $k!(k + 1) = (k + 1)!$ permutations. Hence, by mathematical induction, the Permutation Theorem is true.

15.a. $(n - 2)! = (n - 2)(n - 3)(n - 4) \cdot (n - 5) \ldots (2)(1) = (n - 2)[(n - 3)(n - 4) \cdot (n - 5) \ldots (2)(1)] = (n - 2)(n - 3)!$

18.

12. A bridal party consists of 3 bridesmaids, 3 ushers, and the bride and groom. In how many ways can the bridal party be arranged in a row for portraits if the bride and groom always stand together? 10,080

13. Five boys and five girls are to give speeches in a contest. If boys and girls alternate, how many arrangements of speakers are possible? 28,800

14. Use mathematical induction and the Multiplication Counting Principle to prove the Permutation Theorem. (Hint: Each permutation of n elements can be constructed by selecting the first term and then following it by some permutation of the remaining $n - 1$ elements.) See margin.

15. **a.** Show that $(n - 2)! = (n - 2) \cdot (n - 3)!$. See margin.
b. Solve for n: $P(n, 3) = 5 \cdot P(n, 2)$. $n = 7$

16. Lin, Olivia, Aisha, Fred, and Helen are to be seated at a circular table. Two arrangements of these five people around the table are regarded as the *same seating arrangement* if and only if one is a rotation image of the other. That is, the two arrangements place the people in the same order counterclockwise around the table, even if the people sit in different chairs in the two seatings. The two seatings below are assumed to be the same.

How many different seating arrangements are possible? (Hint: Because only the *order* of people around the table and not the particular chair occupied by a person matters when counting seating arrangements, you can count all possible seating arrangements by selecting a particular chair for one of the people, say Lin, and then counting the number of ways that the remaining 4 people can be arranged in the other chairs.) 24

17. Find positive integers n and r such that $P(11, 6) \cdot P(5, 3) = P(n, r)$. Sample: $n = 11$, $r = 9$

Review

18. A salesperson must leave company headquarters in San Francisco to visit Denver, Minneapolis, and Phoenix, once each. Draw a possibility tree illustrating the number of different sales trips which are possible. How many are there? *(Lesson 10-2)* See margin. 6 trips

602

19. Rewrite in $a + bi$ form: $\frac{3 - 6i}{2 + 5i}$. *(Lesson 8-1)* $-\frac{24}{29} - \frac{27}{29}i$

20. Suppose that a computer program contains the following lines of code.

```
100 FOR I = 0 TO 11
110    FOR J = 1 TO 4
120       FOR K = 1 TO 10
130          PRINT I + J + K
140       NEXT K
150    NEXT J
160 NEXT I
170 END
```

How many lines of output are produced by this code? *(Lesson 10-2)* **480**

21. The lens equation

$$\frac{1}{p} + \frac{1}{q} = \frac{1}{f}$$

relates the object distance p, the image distance q, and the focal length f when a lens is used to concentrate or disperse light (as on a screen or photographic film). If an object is 20 cm away from a lens with a focal length of 8 cm, how far is the image from the lens? *(Lesson 5-7)* **$\frac{40}{3}$ cm**

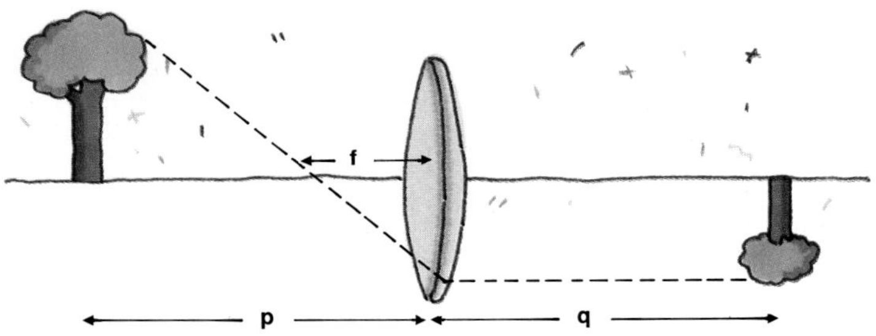

22. Find a formula for a polynomial function of lowest degree whose graph could be that at the right. *(Lesson 4-6)*
$f(x) = 2x^3 + 14x^2 - 12x - 144$

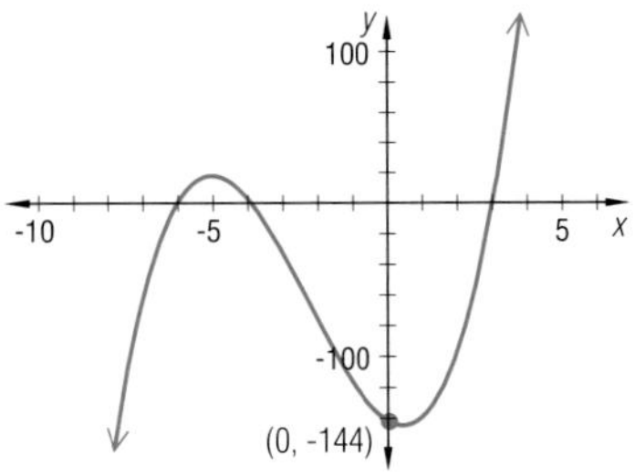

23. Determine the transformation that maps the graph of $y = \ln x$ onto the graph of $y = 2 \ln (x + 1) + 7$. *(Lesson 3-9)* **$(x, y) \rightarrow (x - 1, 2(y + 7))$**

Exploration

24. Choose a professional sport with more than one league or division, and determine the number of ways in which the final team standings could occur. How many of these would have your favorite team or teams in first place? **Answers may vary.**

FOLLOW-UP

MORE PRACTICE
For more questions on SPUR Objectives, use *Lesson Master 10-3*, shown below.

PROJECTS
The projects for Chapter 10 are described on pages 632–633. **Project 2** is related to the content of this lesson.

EVALUATION
A quiz covering Lessons 10-1 through 10-3 is provided in the Teacher's Resource File.

NAME ____________

LESSON MASTER 10–3
QUESTIONS ON **SPUR** OBJECTIVES

■ SKILLS *Objective B (See pages 636–638 for objectives.)*
In 1 and 2, evaluate the expressions.

1. $P(18, 4)$ **73,440** 2. $P(12, 3)$ **1,320**

3. $P(7, 4) \cdot P(3, 1) = P($**7**, **5**$)$

4. Find n so that $P(13, n) = \frac{13!}{4!}$. **9**

■ PROPERTIES *Objective D*
5. Verify that $P(7, 6) = P(7, 7)$ and generalize your result.
$P(7, 6) = \frac{7!}{(7-6)!} = \frac{7!}{1!} = \frac{7!}{1} = 7!$;
$P(7, 7) = \frac{7!}{(7-7)!} = \frac{7!}{0!} = \frac{7!}{1} = 7!$;
So $P(7, 6) = P(7, 7)$;
In general, $P(n, n-1) = P(n, n)$, $\forall$ integers $n \geq 2$.

6. If $(n - 4)! = (n - 4) \cdot x!$, what is x? **$n - 5$**

■ USES *Objective F*
7. Each player of a certain board game must choose a different token to move around the board. If there are 8 tokens, in how many different ways can 5 players choose them? **6,720**

8. Nine high school swim teams are competing in a 400-meter freestyle event. The pool contains 10 lanes. If the two swimmers from Lincoln High must swim in adjacent lanes and one swimmer from each of the other schools fill the remaining eight lanes, how many different arrangements of the swimmers are possible? **725,760**

112 *Precalculus and Discrete Mathematics* © Scott, Foresman and Company

LESSON 10-4

RESOURCES
■ Lesson Master 10-4

OBJECTIVES

B Evaluate expressions including permutations or combinations.
D Use properties of permutations and combinations to prove identities.
G Use combinations to solve counting problems.

TEACHING NOTES

Many states have lotteries. If your state has a lottery, now may be the time to discuss it in your classroom. Students can determine the number of different tickets for the games administered by the lottery commission. Some of these games may involve combinations; others may allow repetitions of numbers so that the Multiplication Counting Principle can be used to determine the number of outcomes.

At the bottom of this page, the relationship between combinations and permutations is discussed. It is important that students understand this relationship because of the connection between the computational formulas for these quantities. Students need to understand that each combination of r elements can itself be permuted in $r!$ ways.

LESSON
10-4
Combinations

In the Lotto game in the Illinois State Lottery, a player chooses a set of 6 different numbers from among the integers from 1 to 54, inclusive. The order in which the numbers are selected is unimportant; only the resulting set of 6 different numbers matters.

Example 2 of Lesson 10-1 pointed out that each play in the Lotto game can be thought of in either of the following two ways:

1. Choosing 6 numbers without repetition from among the integers from 1 to 54, inclusive.
2. Choosing a 6-element subset from the 54-element set $\{1, 2, 3, 4, \ldots, 52, 53, 54\}$.

Each choice is called a *combination* of 54 elements taken 6 at a time.

Definition

Let S be a set with n (distinct) elements and let $r \leq n$. A **combination of the n elements of S taken r at a time** is an r-element subset of S, or equivalently, an unordered set without repetition of r of these elements.

The difference between permutations and combinations of n elements taken r at a time is that permutations are ordered and combinations are unordered. For example,

312 451 123 145 231 345

are different permutations of the 5 numbers 1, 2, 3, 4, 5 taken 3 at a time. However, 312, 123, and 231 represent the same combination of the 5 numbers taken 3 at a time. So do 451 and 145.

The symbols $\boldsymbol{C(n, r)}$, $_n\boldsymbol{C}_r$, or $\binom{n}{r}$ are all commonly used to denote the number of combinations of n elements taken r at a time. $C(n, r)$, read "n choose r," is used to emphasize that C is a (discrete) function of the two variables n and r, and to emphasize that $C(n, r)$ is related to $P(n, r)$.

Specifically, it is possible to relate the number $P(n, r)$ of permutations of n elements taken r at a time to the number $C(n, r)$ of combinations of n elements taken r at a time. Consider the set $\{a, b, c, d, e\}$. The combinations of these 5 elements taken 3 at a time are listed below. (We write abc for $\{a, b, c\}$, etc.)

abc	acd	bcd	cde
abd	ace	bce	
abe	ade	bde	

Observe that each combination above contains 3 different, or distinct, letters. So for each combination, there are 3! permutations of its letters. Hence, the total number of permutations of 5 elements taken 3 at a time is

$$P(5, 3) = 3!\ C(5, 3).$$

This result can be generalized. There are $C(n, r)$ ways of choosing n elements taken r at a time. Then there are $r!$ ways of arranging each of those r-element sets. Therefore, the total number of permutations of n elements taken r at a time is, by the Multiplication Counting Principle, given by

$$P(n, r) = r!\ C(n, r).$$

By solving for $C(n, r)$ and using $P(n, r) = \frac{n!}{(n-r)!}$, the following formula, useful for computation, is derived.

$C(n, r)$ Calculation Theorem

The number $C(n, r)$ of combinations of n elements taken r at a time is given by $C(n, r) = \frac{n!}{r!(n-r)!}$.

Example 1 Evaluate $C(8, 3)$.

Solution Let $n = 8$ and $r = 3$ in the formula for $C(n, r)$.

$C(8, 3) = \frac{8!}{3!(8-3)!} = \frac{8!}{3!\ 5!}$

To compute this, rewrite the numerator: $\frac{8 \cdot 7 \cdot 6 \cdot 5!}{3!\ 5!} = \frac{8 \cdot 7 \cdot 6}{3 \cdot 2} = 56.$

Example 2 What is the probability of winning the Lotto game of the Illinois State Lottery in a single play?

Solution A play in the Lotto game consists of choosing six different numbers from among the integers from 1 to 54, inclusive. Thus, the number of different ways of playing Lotto is equal to the number of combinations of these 54 integers taken 6 at a time; that is, it is equal to $C(54, 6)$. By the $C(n, r)$ Calculation Theorem, $C(54, 6) = \frac{54!}{6!\ 48!}$.

Using the [x!] key on a calculator, 54 [x!] [÷] 6 [x!] [÷] 48 [x!] [=] displays 25,827,165. Assuming the combinations occur randomly, the probability of winning is $\frac{1}{25{,}827{,}165}$ or about 0.000000039.

Check $\frac{54!}{6!48!} = \frac{54 \cdot 53 \cdot 52 \cdot 51 \cdot 50 \cdot 49}{6 \cdot 5 \cdot 4 \cdot 3 \cdot 2 \cdot 1} = 25{,}827{,}165$

Technology All three examples in this lesson discuss the simplification of factorial expressions. Emphasize that cancellations can be used to simplify the work before using a calculator.

Students should not have any difficulty understanding the discussion of probability in **Example 2**. The definition applied should be familiar: If E is an event in a sample space S with $N(S)$ equally likely outcomes, and $N(E)$ is the number of outcomes in E, then the probability of E is $P(E) = \frac{N(E)}{N(S)}$.

Making Connections Have students distinguish between the situation of **Example 3** in this lesson and Example 1 of the previous lesson. Although they are similar and involve taking products of two computations, the essential difference is that the order of the people is unimportant in Example 3 but is important in Example 1. Hence, Example 3 relates to combinations while Example 1 relates to permutations.

ADDITIONAL EXAMPLES

1. Evaluate $\binom{12}{5}$.

792

2. A "Little Lotto" game consists of choosing five different numbers from the integers 1 to 35. What is the probability of winning this game?

$\frac{1}{324632}$, or about .000003 (about 3 in a million)

ADDITIONAL EXAMPLES

3. A test is to cover two chapters, one with ten lessons, the other with eight lessons. You have time to study only five lessons, so you decide to pick three from the longer chapter and two from the shorter chapter. How many different combinations of lessons could you choose?
3360

Some calculators have keys for computing $C(n, r)$ and $P(n, r)$. You should check to see if your calculator has these capabilities. The **check** for Example 2 illustrates a way to compute combinations if the values of the factorials exceed the operating range of your calculator.

The probability .000000039 of picking the winning numbers on one play of Lotto is not very encouraging news for Lotto players, even when the jackpot reaches 30 million dollars (as it sometimes does)! Although the Lotto game provides smaller prizes to players who match four or five of the winning numbers, it and other state lottery games are usually designed to return only about half of the receipts to the players in prize money. The other receipts go to advertise and conduct the game, and pay for projects funded by the state.

Notice that $C(8, 5) = \frac{8!}{5!(8-5)!} = \frac{8!}{5!\,3!} = \frac{8!}{(8-3)!\,3!}$, which equals $C(8, 3)$. Thus the number of ways of choosing 5 elements from 8 equals the number of ways of choosing 3 elements from 8. This is reasonable; each time you chose 5, you are "choosing" not to use the other 3. The generalization of this fact is one of the basic properties of combinations. Two basic properties are listed below, each given in two notations.

Theorem (Basic Properties of Combinations)

For all n and r for which $C(n, r)$ is defined:

a. $C(n, r) = C(n, n-r)$ $\quad \binom{n}{r} = \binom{n}{n-r}$;

b. $C(n, n) = C(n, 0) = 1$ $\quad \binom{n}{n} = \binom{n}{0} = 1$.

Proof

a. Here is an algebraic proof, relying on the $C(n, r)$ Calculation Theorem.

$$C(n, n-r) = \frac{n!}{(n-r)!(n-[n-r])!} \quad C(n, r) \text{ Calculation Theorem}$$

$$= \frac{n!}{(n-r)!\,r!} \quad \text{Because } n - [n-r] = r$$

$$= C(n, r) \quad C(n, r) \text{ Calculation Theorem}$$

A verbal proof can also be given, relying on the definition of $C(n, r)$. Each time you choose an r-element subset from a set of n elements, you leave behind the complementary subset of $n - r$ elements. Therefore, each combination of n elements taken r at a time corresponds to a combination of n elements taken $n - r$ at a time, making $C(n, r) = C(n, n-r)$.

b. Algebraic and verbal proofs are left to you as Question 9.

Problem 4 from the beginning of this chapter can be solved by identifying two combinations and then applying the Multiplication Counting Principle.

Example 3 A ski club has 80 members—50 men and 30 women. A planning committee consisting of 5 men and 3 women is to be selected. How many different committees are possible?

Solution Select the men and women separately. The 5 men can be selected in $\binom{50}{5}$ ways. The 3 women can be selected in $\binom{30}{3}$ ways. Using the Multiplication Counting Principle, the two selections can be made in $\binom{50}{5} \cdot \binom{30}{3}$ ways.

$$\binom{50}{5} \cdot \binom{30}{3} = \frac{50!}{5!\,45!} \cdot \frac{30!}{3!\,27!}$$

If you use the factorial or $C(n, r)$ key on a calculator, you may get the answer displayed in scientific notation.

$$\approx 8.6022 \times 10^9$$

To obtain an exact answer, you can combine some paper and pencil work with calculator work.

$$\frac{50}{5!\,45!} \cdot \frac{30!}{3!\,27!} = \frac{50 \cdot 49 \cdot 48 \cdot 47 \cdot 46 \cdot 45!}{5 \cdot 4 \cdot 3 \cdot 2 \cdot 1 \cdot 45!} \cdot \frac{30 \cdot 29 \cdot 28 \cdot 27!}{3 \cdot 2 \cdot 1 \cdot 27!}$$

$$= \frac{50 \cdot 49 \cdot 48 \cdot 47 \cdot 46 \cdot 30 \cdot 29 \cdot 28}{5 \cdot 4 \cdot 3 \cdot 2 \cdot 3 \cdot 2}$$

Since the result is an integer, there are always common factors in the numerator and denominator.

$$= \frac{\overset{10}{\cancel{50}} \cdot 49 \cdot \overset{\overset{4}{\cancel{12}}}{\cancel{48}} \cdot 47 \cdot \overset{23}{\cancel{46}} \cdot \overset{10}{\cancel{30}} \cdot 29 \cdot \overset{14}{\cancel{28}}}{\cancel{5} \cdot \cancel{4} \cdot \cancel{3} \cdot \cancel{2} \cdot \cancel{3} \cdot \cancel{2}}$$

$$= 10 \cdot 49 \cdot 4 \cdot 47 \cdot 23 \cdot 10 \cdot 29 \cdot 14$$

Hold off the multiplications by 10 until the end. Do the rest with a calculator.

$$= 86021656 \times 10^2$$

Thus, there are 8,602,165,600 different planning committees possible.

Questions

Covering the Reading

1. Describe the difference between permutations and combinations. See margin.
2. a. Give three different notations for the combinations of n elements taken r at a time. $C(n, r)$, $\binom{n}{r}$, ${}_nC_r$
   b. Write the $C(n, r)$ Calculation Theorem in each of these notations. See margin.
3. a. List all possible 4-element combinations of the set $\{a, b, c, d, e\}$.
   b. There are several possible permutations of the letters in each combination in part **a**. How many? 24
   c. Use the results from parts **a** and **b** to determine $P(5, 4)$. 120
   a) *abcd, abce, abde, acde, bcde*

In 4–6, evaluate the expressions.

4. $C(7, 3)$ 35 **5.** ${}_{12}C_5$ 792 **6.** $\binom{100}{3}$ 161,700

NOTES ON QUESTIONS
Question 1: Ask students for an example of two similar problems, one requiring permutations and one requiring combinations.

Question 2: This is an important question. Students may see any of these notations in a future course or on a test.

ADDITIONAL ANSWERS
1. A combination of elements of a set S is an unordered subset without repetition allowed. A permutation is a subset of S which is ordered.

2.b. $C(n, r) = \frac{n!}{r!(n-r)!}$

$\binom{n}{r} = \frac{n!}{r!(n-r)!}$

${}_nC_r = \frac{n!}{r!(n-r)!}$

NOTES ON QUESTIONS
Making Connections for Question 12: The generalization is a theorem on page 615 in Lesson 10-6.

Question 13: Point out that most of these "words" are not in the dictionary.

Question 16: If you choose to expand the discussion and work other problems with cards, you may need to explain a deck of cards to students. The result does not generalize easily. From a 52-card deck, the number of 50-card hands is smaller than either of these. From a 12-card deck, the number of 5-card hands equals the number of 7-card hands.

Project 5 on page 633 has students explore a variety of situations with cards and various card hands.

Questions 20 and 23: These questions review ideas that are in the next lesson. Be sure to discuss them.

Question 21b: The result is an interesting difference of two squares. You may wish to ask students to derive a similar identity for cosines.

ADDITIONAL ANSWERS

9.a. $C(n, n) = \frac{n!}{n!(n-n)!} = \frac{n!}{n!0!} = \frac{n!}{n! \cdot 1} = 1$

15.a.iii. $97 \cdot 96 \cdot 95 \cdot 94 = 83{,}156{,}160 = (24)(3{,}464{,}840) = (4!)(3{,}464{,}840)$

b. $C(n, 4) = \frac{n!}{4!(n-4)!} = \frac{n(n-1)(n-2)(n-3)}{4!}$, and $C(n, 4)$ is an integer. So, the product of any 4 consecutive positive integers n, $n - 1$, $n - 2$, and $n - 3$ is divisible by 4!.

7. The Winner Take All game in the state of Maryland requires a player to choose six different numbers from 1 to 35, inclusive.
a. How many different plays are possible in Winner Take All?
b. Determine the probability of winning Winner Take All with one play. a) 1,623,160 b) ≈.0000006

8. If $\binom{15}{8} = \binom{15}{x}$ and $x \neq 8$, what is a possible value of x? 7

9. a. Give an algebraic proof that $C(n, n) = 1$. See margin.
b. Give a verbal proof that $C(n, n) = 1$. If there are n objects, there is only one way to choose all n of them.

10. A math club has 40 members—15 men and 25 women. A planning committee consisting of 2 men and 3 women is to be selected. How many different committees are possible? 241,500

Applying the Mathematics

11. A test consists of 12 questions. Each student is required to choose 10 of these questions to answer and to omit the remaining 2 questions. How many different subsets of questions could be chosen? 66

12. a. Evaluate $\binom{4}{0} + \binom{4}{1} + \binom{4}{2} + \binom{4}{3} + \binom{4}{4}$. 16
b. Evaluate $\binom{5}{0} + \binom{5}{1} + \binom{5}{2} + \binom{5}{3} + \binom{5}{4} + \binom{5}{5}$. 32
c. Conjecture the value of
$\binom{6}{0} + \binom{6}{1} + \binom{6}{2} + \binom{6}{3} + \binom{6}{4} + \binom{6}{5} + \binom{6}{6}$. 64

13. There are 5 vowels and 21 consonants in the English alphabet. How many different 7-letter "words" can be formed that contain 4 different consonants and 3 different vowels? 301,644,000

14. Suppose that a company produces light bulbs in batches of 100. A sample of 3 light bulbs from each batch is selected at random for testing.
a. How many different test samples can be selected? 161,700
b. Suppose that there are 2 defective light bulbs in a certain batch. How many of the possible test samples will not contain either of the two defective bulbs? 152,096

15. a. Verify that each of the following products of four consecutive integers is divisible by 4!.
i. $10 \cdot 9 \cdot 8 \cdot 7$ $10 \cdot 9 \cdot 8 \cdot 7 = 5040 = (24)(210) = (4!)(210)$
ii. $33 \cdot 32 \cdot 31 \cdot 30$ See below.
iii. $97 \cdot 96 \cdot 95 \cdot 94$ See margin.
b. Use the formula for $C(n, 4)$ to write an elegant proof that the product of any four consecutive positive integers is divisible by 4!.
ii) $33 \cdot 32 \cdot 31 \cdot 30 = 982{,}080 = (24)(40{,}920) = (4!)(40{,}920)$; b) See margin.

16. In some varieties of poker, a poker hand consists of 5 cards selected from a 52-card deck. In other varieties, a poker hand has 7 cards. Which is greater, the number of different 7-card hands or the number of different 5-card hands? the number of different 7-card hands

608

Review

17. Consider the letters in the word *English*.
- **a.** How many permutations of these letters are there? 7!
- **b.** How many different five-letter strings are there using these letters if the letters can be used at most once? 2520
- **c.** How many different five-letter strings are there if letters can be used more than once? *(Lessons 10-3, 10-2)* 16,807

18. There are 50 contestants in a talent competition. Prizes will be awarded for first, second, and third place. How many different choices are possible for the top three finishers? *(Lessons 10-3, 10-2)*
117,600

19. Sketch the graph of a function whose derivative at $x = -1$ is -2 and whose derivative at $x = 5$ is 0.5. *(Lesson 9-2)* See margin.

20. Write $\sum_{j=0}^{4} a_j x^{4-j} y^j$ in expanded form. *(Lesson 7-3)*
$a_0x^4 + a_1x^3y + a_2x^2y^2 + a_3xy^3 + a_4y^4$

21.
- **a.** Prove that $\sin(\alpha + \beta)\sin(\alpha - \beta) = \sin^2\alpha\cos^2\beta - \cos^2\alpha\sin^2\beta$.
- **b.** Add and subtract $\sin^2\alpha\sin^2\beta$ from the right side of the identity above and use the Pythagorean Identity to simplify the result. *(Lesson 6-4)* See below.

22. Write the contrapositive of the following statement:
If the coat is green, then the pants are blue. *(Lesson 1-5)*
If the pants are not blue, then the coat is not green.

23. Expand $(x + y)^3$. *(Previous course)* $x^3 + 3x^2y + 3xy^2 + y^3$

Exploration

24. Find out how many flavors of ice cream are served at a nearby ice-cream parlor. Ignoring the order of the dips, how many different two-dip cones are possible
- **a.** if both dips can be the same;
- **b.** if both dips must be different?

Sample: Suppose there are 31 flavors.
a) $\frac{31}{2}(1 + 31) = 496$; b) $\frac{30}{2}(1 + 30) = 465$

21. a) $\sin(\alpha + \beta)\sin(\alpha - \beta)$
$= (\sin\alpha\cos\beta + \sin\beta\cos\alpha)(\sin\alpha\cos\beta - \sin\beta\cos\alpha)$
$= \sin^2\alpha\cos^2\beta - \sin\alpha\sin\beta\cos\beta\cos\alpha + \sin\alpha\sin\beta\cos\alpha\cos\beta - \sin^2\beta\cos^2\alpha$
$= \sin^2\alpha\cos^2\beta - \sin^2\beta\cos^2\alpha$
$= \sin^2\alpha\cos^2\beta - \cos^2\alpha\sin^2\beta$

b) $\sin^2\alpha\cos^2\beta - \cos^2\alpha\sin^2\beta$
$= \sin^2\alpha\cos^2\beta + \sin^2\alpha\sin^2\beta - \sin^2\alpha\sin^2\beta - \cos^2\alpha\sin^2\beta$
$= \sin^2\alpha(\cos^2\beta + \sin^2\beta) - \sin^2\beta(\sin^2\alpha + \cos^2\alpha)$
$= \sin^2\alpha - \sin^2\beta$

FOLLOW-UP

MORE PRACTICE
For more questions on SPUR Objectives, use *Lesson Master 10-4*, shown below.

PROJECTS
The projects for Chapter 10 are described on pages 632–633. **Project 5** is related to the content of this lesson.

19. sample:

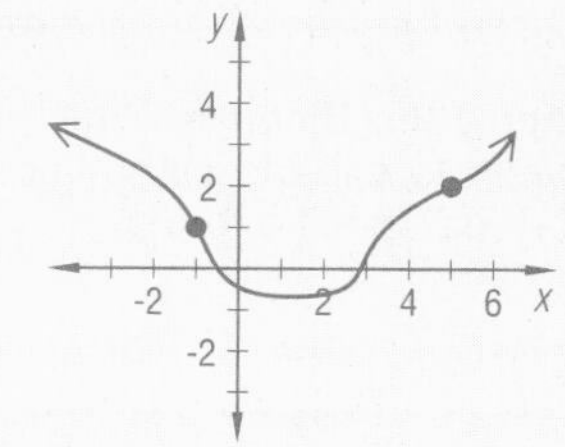

NAME ______

LESSON **MASTER 10–4**
QUESTIONS ON **SPUR** OBJECTIVES

■ SKILLS *Objective B (See pages 636–638 for objectives.)*
In 1 and 2, evaluate the expressions.

1. $C(12, 8)$ 495
2. $\binom{11}{6}$ 462
3. Find integers r and n such that $C(n, 2) = \frac{7!}{r! \cdot 2}$. $r = 5, n = 7$
4. $C(7, 2) \cdot 2! = P($7, 2$)$

■ PROPERTIES *Objective D*

5. Show that $\binom{10}{4} = \binom{10}{6}$.
$\binom{10}{4} = \frac{10!}{4!(10-4)!} = \frac{10!}{4!6!}$; $\binom{10}{6} = \frac{10!}{6!(10-6)!} = \frac{10!}{6!4!} = \frac{10}{4!6!}$;
So, $\binom{10}{4} = \binom{10}{6}$.
6. Prove that ∀ positive integers n, $C(n, n) = 1$.
By the $C(n, r)$ Calculation Theorem,
$C(n, n) = \frac{n!}{n!(n-n)!} = \frac{n!}{n!0!} = \frac{n!}{n!(1)} = \frac{n!}{n!} = 1$

■ USES *Objective G*

7. Suppose there are seven points in a plane, with no three points collinear. How many distinct triangles can be formed which have these points as vertices? 35
8. In a standard deck of 52 playing cards, how many different 4-card hands contain exactly one card from each of the four suits? 28,561
9. A choir director is to assemble a choir consisting of 4 girls and 5 boys. If 9 girls and 12 boys audition, how many different choirs are possible? 99,792
10. **a.** A questionnaire for a psychology experiment consists of 10 questions. Each one asks the subject to pick two words out of a group of five which he or she feels are most closely related. How many different responses are possible? 1×10^{10}
b. A second version of the questionnaire contains 15 questions, each with four words. Which version has a greater number of possible responses?
version with 15 questions (4.7×10^{11})

113

LESSON 10-5

RESOURCES
- Lesson Master 10-5
- Teaching Aid 69 displays the first 6 rows of Pascal's triangle.
- Teaching Aid 70 displays the first 10 rows of Pascal's triangle.

OBJECTIVE

C Apply the Binomial Theorem to expand binomials or find specific terms.

TEACHING NOTES

This is the first of two lessons on the Binomial Theorem. In this lesson, students use the theorem to expand a binomial to any positive integer power or to find a specific term of such an expansion without actually computing the entire expansion. Consequently, this is a very skill-oriented lesson. The applications of the theorem are contained in the next lesson.

Making Connections
Students who have studied from *Functions, Statistics, and Trigonometry* have covered this material in more depth than is given in this book. The topics of binomial probabilities and normal distributions as the limit of the binomial distributions are presented in the book mentioned above.

LESSON

10-5

The Binomial Theorem

Consider the expansions of the binomial powers shown below.

$$(x + y)^0 = 1$$
$$(x + y)^1 = 1x + 1y$$
$$(x + y)^2 = 1x^2 + 2xy + 1y^2$$
$$(x + y)^3 = 1x^3 + 3x^2y + 3xy^2 + 1y^3$$
$$(x + y)^4 = 1x^4 + 4x^3y + 6x^2y^2 + 4xy^3 + 1y^4$$
$$(x + y)^5 = 1x^5 + 5x^4y + 10x^3y^2 + 10x^2y^3 + 5xy^4 + 1y^5$$

In earlier courses you may have learned that the coefficients of these powers can be arranged in the following array.

	Row
1	0
1 1	1
1 2 1	2
1 3 3 1	3
1 4 6 4 1	4
1 5 10 10 5 1	5
	⋮

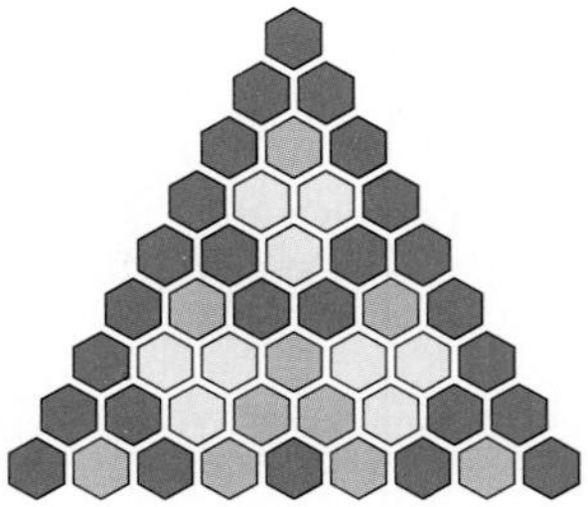

This array is known as **Pascal's triangle,** after the French mathematician and philosopher Blaise Pascal (1623–1662), and its entries are called **binomial coefficients.** Note that the first and last entries in each row are 1 and all other entries can be obtained as the sum of the two entries immediately above it to the left and right. (You are asked to prove this as Question 10.) For example, 10 in the 5th row is the sum of 6 and 4 in the 4th row. This gives a recursive way of constructing the triangle.

For small values of n, it is convenient to use recursion in Pascal's triangle to expand $(a + b)^n$.

Example 1 **a.** Find row 6 of Pascal's triangle.
b. Write $(a + b)^6$ in expanded form.

Solution

a. The sums of adjacent terms in row 5 determine the "inside" five elements of row 6.

Row 5: 1 5 10 10 5 1
Row 6: 1 6 15 20 15 6 1

The lines indicate the entries in row 5 which are added to obtain the entry in row 6.

b. The entries of row 6 are the seven coefficients in the expansion of $(a + b)^6$. The sum of the exponents in each term is 6. Also, the powers of a begin at a^6 and decrease to a^0, while the powers of b begin at b^0 and increase to b^6. Thus,

$$(a + b)^6 = a^6 + 6a^5b + 15a^4b^2 + 20a^3b^3 + 15a^2b^4 + 6ab^5 + b^6.$$

Pascal's triangle gives a recursive method for generating the powers of $x + y$, because to find the numbers in row k of the triangle, you must first find the numbers in row $k - 1$. But if you wanted to expand $(x + y)^{30}$, it is inefficient to construct 30 rows of Pascal's triangle. However, there is an explicit formula for $(x + y)^n$ which is quite elegant and is obtained by counting techniques. It is more efficient than recursion for finding the desired coefficients for larger values of n.

To develop the formula, first consider a special case $(x + y)^8$. By definition,

$$(x + y)^8 = (x + y)(x + y)(x + y)(x + y)(x + y)(x + y)(x + y)(x + y).$$

Each term in the expansion of $(x + y)^8$ can be found by choosing an x or y from each of the 8 binomials $(x + y)$. These eight xs and ys are multiplied. So each term is of the form $x^a y^b$, where $a + b = 8$ and a and b are integers from 0 to 8. Thus, when like terms are combined,

$$(x + y)^8 = \_\_\, x^8 + \_\_\, x^7y + \_\_\, x^6y^2 + \_\_\, x^5y^3 + \_\_\, x^4y^4 + \_\_\, x^3y^5 + \_\_\, x^2y^6 + \_\_\, xy^7 + \_\_\, y^8,$$

where each coefficient of $x^a y^b$ is the number of different ways xs and ys are chosen to obtain $x^a y^b$. There is only one way to obtain x^8: when all the choices are xs. So the coefficient of x^8 is 1. The choices that contribute to the x^7y term have 1 y chosen from the 8 binomials $(x + y)$. There are $\binom{8}{1} = 8$ ways to do that. So the coefficient of x^7y is 8. The choices which contribute to the x^6y^2 term have 2 ys chosen from the 8 binomials $(x + y)$. There are $\binom{8}{2} = 28$ ways of choosing 2 objects from 8, so the coefficient of x^6y^2 is 28. Continuing this process gives

$$(x + y)^8 = \binom{8}{0}x^8 + \binom{8}{1}x^7y + \binom{8}{2}x^6y^2 + \binom{8}{3}x^5y^3 + \binom{8}{4}x^4y^4 + \binom{8}{5}x^3y^5 + \binom{8}{6}x^2y^6 + \binom{8}{7}xy^7 + \binom{8}{8}y^8$$

$$= x^8 + 8x^7y + 28x^6y^2 + 56x^5y^3 + 70x^4y^4 + 56x^3y^5 + 28x^2y^6 + 8xy^7 + y^8.$$

This sum can be written more concisely using summation notation.

$$(x + y)^8 = \sum_{k=0}^{8} \binom{8}{k} x^{8-k} y^k$$

A more general result, obtained by replacing 8 with a positive integer n, is known as the *Binomial Theorem*. Omar Khayyam, perhaps the greatest mathematician of the 12th century, knew of it. In 1676, Isaac Newton generalized the Binomial Theorem to apply to any rational power of $(x + y)$. That result is beyond the scope of this book.

The proof given below of the Binomial Theorem is elegant and generalizes the example of $(x + y)^8$.

You might point out that Pascal's triangle can be considered a *two-dimensional sequence* because it involves two variables and its representation goes in two directions. The process in **Example 1** uses the recurrence relation for the sequence in **Question 12**. The explicit formula for the sequence is given by the Binomial Theorem.

Be sure to point out that $\binom{n}{r}$ is the rth term in the nth row only if the counting of both rows and terms begins with 0. (That is, the top row is the 0th row, as indicated on page 610, and the outside left diagonal of 1s is considered the 0th diagonal.) In **Example 3**, the counting begins with the more traditional 1. Because of this potential confusion, it is easier to write the exponents for the term first, and then write the coefficient that agrees with them, as is done in Example 3.

ADDITIONAL EXAMPLES

1.a. Find row 9 of Pascal's triangle.
1 9 36 84 126 126 84 36 9 1

b. Write $(a + b)^9$ in expanded form.
$\mathbf{a^9 + 9a^8b + 36a^7b^2 + 84a^6b^3 + 126a^5b^4 + 126a^4b^5 + 84a^3b^6 + 36a^2b^7 + 9ab^8 + b^9}$

ADDITIONAL EXAMPLES

2. Give the coefficient of x^6y^5 in the expansion of $(x + y)^{11}$.
462

3. Give the 4th term in the expansion of $(2b - 4a^2)^7$.
$\binom{7}{3}(2b)^4(-4a^2)^3 =$
$-35840b^4a^6$

NOTES ON QUESTIONS

Question 4b: You can prove this result for any power of $2a + b$ by letting $a = 1$ and $b = 1$. This proof will prepare students for the proof of the first theorem in the next lesson.

Questions 5 and 6: In these questions, the left-most term is called the 1st term.

Question 7: Students can check these results by rewriting $1 + i$ as $[\sqrt{2}, 45°]$ and using DeMoivre's Theorem.

Question 9: If students have difficulties with this question, have them write out the summation, first letting $k = 0$, then $k = 1$, then $k = 2$, and so on. In so doing, the answer becomes evident.

Question 12: You should go over this proof in class. Although the algebra is involved, it is important.

Technology for Question 17: One calculator gives 1.08972864 E10 as the answer. From this, the answer can be written as a decimal rather easily because the answer is divisible by 100. (10,897,286,400)

ADDITIONAL ANSWERS

1.a. 1, 7, 21, 35, 35, 21, 7, 1
c. $a^7 - 7a^6b + 21a^5b^2 - 35a^4b^3 + 35a^3b^4 - 21a^2b^5 + 7ab^6 - b^7$

The Binomial Theorem

For all positive integers n and numbers x and y,

$$(x+y)^n = \binom{n}{0}x^n + \binom{n}{1}x^{n-1}y + \binom{n}{2}x^{n-2}y^2 + \ldots + \binom{n}{k}x^{n-k}y^k + \ldots + \binom{n}{n-1}xy^{n-1} + \binom{n}{n}y^n$$

Or, equivalently,

$$= \sum_{k=0}^{n} \binom{n}{k}x^{n-k}y^k$$

Proof By definition,

$$(x+y)^n = \underbrace{(x+y)(x+y)(x+y)\ldots(x+y)}_{n \text{ factors}}.$$

Each term in the expansion is of the form x^ay^b, where a and b are nonnegative integers such that $a + b = n$, and one of x or y is selected from each of the n factors $(x + y)$. The term $x^{n-k}y^k$ is obtained by choosing y from any k of the n factors. Consequently, because there are $\binom{n}{k}$ ways of selecting k items from a set of n items, the coefficient of the term $x^{n-k}y^k$ in the expansion of $(x + y)^n$ is $\binom{n}{k}$.

Thus all of the coefficients in the expansion of $(x + y)^n$ are combinations of the form $\binom{n}{k}$, which means that Pascal's triangle is an array of combinations. The Binomial Theorem can be used to obtain any term in a binomial expansion.

Example 2 Give the coefficient of $x^{22}y^8$ in the expansion of $(x + y)^{30}$.

Solution From the Binomial Theorem, the expansion of $(x + y)^{30}$ has 31 terms, each of the form $\binom{30}{k}x^{30-k}y^k$ where $k = 0, 1, 2, \ldots, 30$. The term with $k = 8$ is

$$\binom{30}{8}x^{30-8}y^8 = \binom{30}{8}x^{22}y^8,$$

so the coefficient is $\binom{30}{8} = \frac{30!}{8!22!} = 5{,}852{,}925$.

It is customary to write the binomial theorem expansion of $(x + y)^n$ with the exponents of x in decreasing order, as is done above. Then $\binom{n}{0}x^n$ is the 1st term, $\binom{n}{1}x^{n-1}y$ is the 2nd term, $\binom{n}{2}x^{n-2}y^2$ is the 3rd term, and, in general, $\binom{n}{k}x^{n-k}y^k$ is the $(k + 1)$st term.

Example 3 Find the eighth term in the expansion of $(3p - 2)^{15}$.

Solution Look at the eighth term of $(x + y)^n$, where $n = 15$, $x = 3p$, and $y = -2$. The eighth term occurs when k, the index, is 7. So the eighth term is

$$\binom{15}{7}(3p)^{15-7}(-2)^7 = \frac{15!}{7!8!}(6561p^8)(-128)$$
$$= -5{,}404{,}164{,}480p^8.$$

Pascal investigated many of the properties of the binomial coefficients, and in fact invented mathematical induction in order to prove some of them! You will examine some of these properties in the Questions and in the next lesson.

Questions

Covering the Reading

1. Refer to Example 1.
   a. Use row 6 of Pascal's triangle to find row 7. See margin.
   b. Expand $(a + b)^7$.
   c. Expand $(a - b)^7$. See margin.
   b) $a^7 + 7a^6b + 21a^5b^2 + 35a^4b^3 + 35a^3b^4 + 21a^2b^5 + 7ab^6 + b^7$
2. What combination will yield the coefficient of the term $x^{20}y^{10}$ in the expansion of $(x + y)^{30}$? $\binom{30}{10}$
3. Write out the Binomial Theorem for the case $n = 9$. See margin.
4. a. Expand $(2a + b)^6$. See margin.
   b. Verify that the sum of the coefficients is a power of 3.
   $64 + 192 + 240 + 160 + 60 + 12 + 1 = 729 = 3^6$
5. Find the fourth term of $(s + 5r)^{15}$. $56875s^{12}r^3$
6. Find the tenth term of $(2x - 3y)^{13}$. $-225{,}173{,}520\ x^4y^9$

Applying the Mathematics

7. Use the Binomial Theorem to find:
   a. the 4th power of the complex number $1 + i$, -4
   b. the 8th power of the complex number $1 + i$. 16
8. Without computing the entire expansion of $(2a + b)^{25}$, find the coefficient of the term that has a^{12} as the power of a. $\approx 2.13 \times 10^{10}$
9. Write $\sum_{k=0}^{7}\left[\binom{7}{k}x^{7-k}2^k\right]$ as the power of a binomial. $(x + 2)^7$
10. Use the Binomial Theorem to explain why $\binom{n}{r} = \binom{n}{n-r}$. (Hint: Begin "$\binom{n}{r}$ is the coefficient of ....") See margin.
11. For all positive integers n, $\binom{n}{n} = 1$. How is the fact related to the makeup of Pascal's triangle?
   The outside right diagonal (the last term in each row) is 1.

3. $(x + y)^9 = \sum_{k=0}^{n}\binom{9}{k}x^{9-k}y^k = \binom{9}{0}x^9 + \binom{9}{1}x^8y + \binom{9}{2}x^7y^2 + \binom{9}{3}x^6y^3 + \binom{9}{4}x^5y^4 + \binom{9}{5}x^4y^5 + \binom{9}{6}x^3y^6 + \binom{9}{7}x^2y^7 + \binom{9}{8}xy^8 + \binom{9}{9}y^9$

4.a. $(2a + b)^6 = 64a^6 + 192a^5b + 240a^4b^2 + 160a^3b^3 + 60a^2b^4 + 12ab^5 + b^6$

10. $\binom{n}{r}$ **is the coefficient of** $x^{n-r}y^r$ **in the expansion of** $(x + y)^n$**. But** $(x + y)^n = (y + x)^n$**, so** $\binom{n}{r}$ **is also the coefficient of** $y^{n-r}x^r = x^{n-(n-r)}y^{n-r}$**, which is** $\binom{n}{n-r}$**. So** $\binom{n}{r} = \binom{n}{n-r}$**.**

12.b. Forming a least common denominator
c. Addition of fractions and distributive property

NAME ____________

LESSON **MASTER 10–5**
QUESTIONS ON **SPUR** OBJECTIVES

■**SKILLS** *Objective C (See pages 636–638 for objectives.)*
In 1–3, expand using the Binomial Theorem.

1. $(y + z)^6$
   $y^6 + 6y^5z + 15y^4z^2 + 20y^3z^3 + 15y^2z^4 + 6yz^5 + z^6$
2. $(3a - 2b)^5$
   $243a^5 - 810a^4b + 1080a^3b^2 - 720a^2b^3 + 240ab^4 - 32b^5$
3. $(x + 2)^4$
   $x^4 + 8x^3 + 24x^2 + 32x + 16$
4. What is the coefficient of the x^3y^4 term in the expansion of $(2x - 4y)^7$? 71,680
5. Find the tenth term of $(x - 3y)^{12}$. $-4{,}330{,}260x^3y^9$
6. a. What combination is the coefficient of the $x^{12}y^8$ term in the expansion of $(x + y)^{20}$? $\binom{20}{8}$ or $\binom{20}{12}$
   b. What is this coefficient? 125,970
7. Express as the power of a binomial: $\sum_{k=0}^{9}\binom{9}{k}y^{9-k}3^k$. $(y + 3)^9$
8. The coefficients of which terms in the expansion of $(x + y)^8$ are equal?
   1st and 9th, 2nd and 8th, 3rd and 7th, 4th and 6th

114

FOLLOW-UP

MORE PRACTICE
For more questions on SPUR Objectives, use *Lesson Master 10-5*, shown on page 613.

EXTENSION
The Binomial Theorem can be used to estimate powers of numbers just a little larger than 1, for example, the kinds of numbers that often appear in expressions of compound interest, such as $1.06^2 = (1 + .06)^2 = 1 + 2(.06) + (.06)^2 = 1.1236$. Notice that the last term does not change anything in the hundredths place. Students could explore other powers of these and similar numbers, noting how many terms of the binomial expansion are needed to get accuracy to the nearest hundredth or thousandth.

PROJECTS
The projects for Chapter 10 are described on pages 632-633. **Project 4** is related to the content of this lesson.

ADDITIONAL ANSWERS
12.b. See the margin on page 613.

18.d.

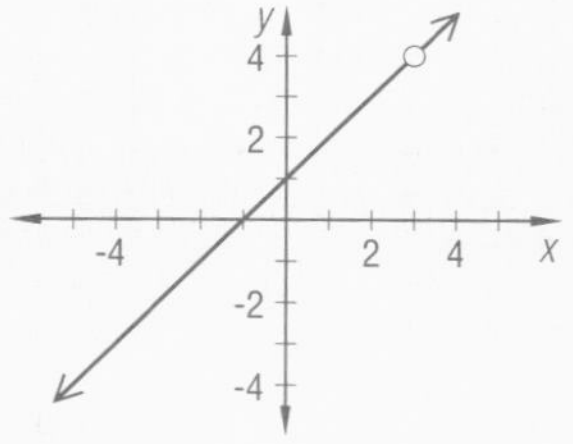

20.

row	sum of squares of elements
0	1
1	2
2	6
3	20

The sum of the squares for row *n* seems to be the middle element of row 2*n*. So the sum of the squares of the elements of the 12th row would be the middle element of row 24.

12. The fact that each "interior" entry in Pascal's triangle is the sum of the two entries immediately above it can be expressed as the identity $\binom{n}{r-1} + \binom{n}{r} = \binom{n+1}{r}$ for all positive integers n and r with $r \le n$.

Fill in the missing reasons in the following proof of this identity.

$$\binom{n}{r-1} + \binom{n}{r} = \frac{n!}{(r-1)!(n-r+1)!} + \frac{n!}{r!(n-r)!}$$ **a.** ______

$$= \frac{r \cdot n!}{r!(n-r+1)!} + \frac{n!(n-r+1)}{r!(n-r+1)!}$$ **b.** ______

$$= \frac{[r + (n-r+1)]n!}{r!(n-r+1)!}$$ **c.** ______

$$= \frac{(n+1)n!}{r!(n+1-r)!}$$ Property of Opposites

$$= \frac{(n+1)!}{r!(n+1-r)!}$$ Definition of **d.** ______

$$= \binom{n+1}{r}$$ **e.** ______

a) ${}_nC_r$ Calculation Theorem; b,c) See margin. d) $(n+1)!$ e) *nCr* Calculation Theorem

Review

In 13–15, evaluate the given expression. *(Lessons 10-4, 10-3)*

13. $C(9, 4)$ 126 **14.** $\binom{11}{6}$ 462 **15.** $P(93, 4)$ 70,073,640

16. A pizza place offers the following toppings: pepperoni, sausage, onions, green peppers, and mushrooms.
a. How many different kinds of pizzas with three different toppings can be ordered? 10
b. How many different kinds with any number of toppings (between none and all five) can be ordered? 32
(Lessons 10-4, 10-2)

17. Consider the following problem: How many 10-person batting orders can be formed from a softball team's 15-person roster?
a. State the essential features of the problem.
b. Solve the problem. *(Lessons 10-3, 10-1)* $\approx 1.0897 \times 10^{10}$
a) Ordered symbols; repetition is not allowed.

18. Given $f(x) = \frac{x^2 - 2x - 3}{x - 3}$.
a. Determine the values of x at which f is undefined. $x = 3$
b. Classify any discontinuities as essential or removable. removable
c. Find equations of any vertical asymptotes. none
d. Sketch a graph of f. *(Lessons 5-5, 5-4)* See margin.

19. *Multiple choice.* $\lim_{x \to -\infty} b^x = +\infty$ if
(a) $b > 1$ (b) $b = 1$ (c) $0 < b < 1$ (d) $b < 0$. *(Lesson 2-6)* (c)

Exploration

20. Find the sum of the squares of the elements of the first few rows of Pascal's triangle. Find a pattern to predict the exact value of the sum of the squares of the elements of the 12th row. See margin.

614

LESSON

10-6

Counting and the Binomial Theorem

Because of the Binomial Theorem, Pascal's triangle can be considered as a triangle of combinations.

$$\begin{array}{ccccccccc} & & & & 1 & & & & \\ & & & 1 & & 1 & & & \\ & & 1 & & 2 & & 1 & & \\ & 1 & & 3 & & 3 & & 1 & \\ 1 & & 4 & & 6 & & 4 & & 1 \\ & & & & \vdots & & & & \end{array} \qquad \begin{array}{ccccccccc} & & & & \binom{0}{0} & & & & \\ & & & \binom{1}{0} & & \binom{1}{1} & & & \\ & & \binom{2}{0} & & \binom{2}{1} & & \binom{2}{2} & & \\ & \binom{3}{0} & & \binom{3}{1} & & \binom{3}{2} & & \binom{3}{3} & \\ \binom{4}{0} & & \binom{4}{1} & & \binom{4}{2} & & \binom{4}{3} & & \binom{4}{4} \\ & & & & \vdots & & & & \end{array}$$

This explains why the entries in Pascal's Triangle are sometimes called binomial coefficients.

Notice that the sum of the elements in the third row of Pascal's Triangle is 2^3 and the sum in the fourth row is 2^4 (be sure to start with row 0 at the top). These facts are generalized in the following theorem.

Sum of Binomial Coefficients Theorem

$\forall$ integers $n \geq 0$,

$$\binom{n}{0} + \binom{n}{1} + \binom{n}{2} + \ldots + \binom{n}{k} + \ldots + \binom{n}{n} = 2^n.$$

That is, $$\sum_{k=0}^{n} \binom{n}{k} = 2^n.$$

Proof The proof is elegant. From the Binomial Theorem, for all numbers x and y and integers $n \geq 1$,

$$(x + y)^n = \sum_{k=0}^{n} \binom{n}{k} x^{n-k}y^k.$$

Since the Binomial Theorem is true for all numbers x and y, it is true when $x = 1$ and $y = 1$.

Thus $$(1 + 1)^n = \sum_{k=0}^{n} \binom{n}{k} 1^{n-k}1^k.$$

So $$2^n = \sum_{k=0}^{n} \binom{n}{k}.$$

The case when $n = 0$ is left for you to prove as Question 3.

LESSON 10-6

RESOURCES

- Lesson Master 10-6
- Quiz for Lessons 10-4 through 10-6
- Teaching Aid 69 displays Pascal's Triangle as a triangle of combinations.
- Computer Master 19

OBJECTIVES

E Apply the Binomial Theorem to deduce properties of sets.
G Use combinations and the Binomial Theorem to solve counting problems.
H Find binomial probabilities in realistic situations.

TEACHING NOTES

Making Connections In Question 14 of Lesson 10-2, students proved that there are 2^n subsets of a set with n elements. The result on this page provides another proof of the same result that ties in with Pascal's triangle and binomial coefficients.

The situations of **Examples 1 and 2** are known as *binomial experiments.* In any binomial experiment: (1) there is a series of n independent trials (independence in the trials means that the outcome of one trial is not affected by the outcome of any other trial); (2) each trial has two possible outcomes; and (3) the probability of each of the two outcomes remains constant from trial to trial.

Example 1 illustrates the fact that the binomial coefficient gives the number of ways that a particular event can occur in a series of *n* trials. In this type of situation, only the binomial coefficient is needed; the actual terms of the binomial are not important.

The situation in **Example 2** is, in theory, not precisely a binomial experiment because the total number of light bulbs produced is finite. Thus, if the probability of finding a first defective bulb truly is 1%, then the probability that the next bulb is defective is slightly less than 1% (because you already have found one of the 1% of bulbs that is defective). The actual distribution of good and defective bulbs is an example of a hypergeometric distribution. However, because the number of bulbs is large, the binomial probabilities are very close to the hypergeometric probabilities.

As in **Example 1**, the binomial coefficient in **Example 2** still gives the number of ways that the given situation can occur in *n* trials. But because the probability of the event is desired, the actual terms of the binomial are also of interest. The terms of the binomial are simply the probabilities of the two outcomes, in this case the probability of being acceptable and the probability of being defective. Students need to realize that the product. $(.99)^9(.01)^3$ gives the probability of getting a particular sequence of 12 bulbs of which 9 are good and 3 are defective. This result still must be multiplied by the binomial coefficient, which gives the total number of such sequences.

The identity $\sum_{k=0}^{n} \binom{n}{k} = 2^n$ has a combinatorial interpretation. Let S be a set with n elements. Because $\binom{n}{k}$ is the number of different subsets of S with k elements, it follows that the total number of different subsets of S is given by

$$\binom{n}{0} + \binom{n}{1} + \ldots + \binom{n}{k} + \ldots + \binom{n}{n}, \quad \text{or} \quad \sum_{k=0}^{n} \binom{n}{k}.$$

↑	↑	↑	↑
0-element subsets	1-element subsets	k-element subsets	n-element subsets

Earlier, in Question 14 of Lesson 10-2, the Multiplication Counting Principle was applied to show that there are 2^n subsets of S. So the Sum of Binomial Coefficients Theorem reflects the number of subsets of a set. This fact provides an alternate approach to solving certain counting problems. Suppose a coin is tossed 4 times. The following arrangements of *H*s and *T*s are possible.

HHHH	*HTHH*	*THHH*	*TTHH*
HHHT	*HTHT*	*THHT*	*TTHT*
HHTH	*HTTH*	*THTH*	*TTTH*
HHTT	*HTTT*	*THTT*	*TTTT*

There are a total of 16 arrangements. One way of calculating this total is to apply the theorem on page 605 to the number of 4-element strings of 2 elements with repetition, which gives 2^4 as the total.

An alternative method reformulates the problem in terms of subsets. A particular arrangement of *H*s and *T*s can be identified by indicating which of the tosses are *H*s. For instance, the arrangement *HTTH* can be indicated as {1, 4}, the arrangement *HHHT* as {1, 2, 3}, and the arrangement *TTTT* as the null set { }. Each arrangement corresponds to a different subset of {1, 2, 3, 4}, and the total number of arrangements is the total number of subsets of {1, 2, 3, 4}, which is 2^4.

Since the number of elements in a subset of {1, 2, 3, 4} corresponds to the number of *H*s in an arrangement, and the number of k-element subsets is $\binom{4}{k}$, the numbers of ways of obtaining k heads in 4 tosses is given by the elements in the 4th row of Pascal's triangle.

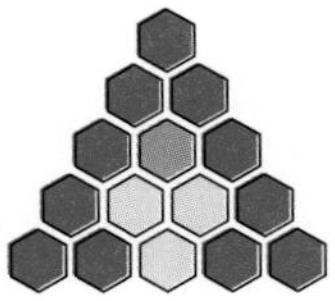

0 heads can occur in 1 way.	{1, 2, 3, 4} has 1 subset with 0 elements.
1 head can occur in 4 ways.	{1, 2, 3, 4} has 4 subsets with 1 element.
2 heads can occur in 6 ways.	{1, 2, 3, 4} has 6 subsets with 2 elements.
3 heads can occur in 4 ways.	{1, 2, 3, 4} has 4 subsets with 3 elements.
4 heads can occur in 1 way.	{1, 2, 3, 4} has 1 subset with 4 elements.

Tossing a coin repeatedly is an example of an experiment that involves a series of *trials*, each of which has two possible outcomes. In such an

616

experiment, the Binomial Theorem indicates the number of different ways that a particular event can happen. For example, in the series of n coin tosses, suppose the event is "obtaining exactly k tails" (and therefore $n - k$ heads). The number of ways the event can happen is the same as $\binom{n}{k}$, the number of ways to choose k tails from a total of n tosses. But $\binom{n}{k}$ is also the coefficient of the term $H^{n-k}T^k$ in the expansion of $(H + T)^n$. Notice that in this term, the exponent of T is k, the number of tails desired, and the exponent of H is $n - k$, the number of heads desired. Thus, if a coin is tossed n times, expanding $(H + T)^n$ gives a list of terms $\binom{n}{k} H^{n-k}T^k$, the coefficients of which are the number of ways to obtain $n - k$ heads and k tails.

Example 1 A coin is flipped six times. How many of the possible sequences of heads and tails have at least three tails?

Solution Using the discussion above, look at the terms in the expansion of $(H + T)^6$.

$$(H+T)^6 = \binom{6}{0}H^6 + \binom{6}{1}H^5T + \binom{6}{2}H^4T^2 + \binom{6}{3}H^3T^3 + \binom{6}{4}H^2T^4 + \binom{6}{5}HT^5 + \binom{6}{6}T^6$$

At least 3 tails means 3, 4, 5, or 6 tails. Therefore, the total number of ways to obtain at least 3 tails is the same as the sum of the coefficients of H^3T^3, H^2T^4, HT^5, and T^6:

$$\binom{6}{3} + \binom{6}{4} + \binom{6}{5} + \binom{6}{6} = 20 + 15 + 6 + 1 = 42.$$

Suppose that an unfair coin is used in the series of coin tosses above, so that the probability of obtaining a head from a single toss is $h = .7$ (thus the probability of a tail is $t = .3$). Suppose further that you wish to find the probability that in n tosses, you will obtain exactly k tails (and $n - k$ heads). Since each toss is independent of the others, the probability of obtaining a particular sequence consisting of k tails and $n - k$ heads is $h^{n-k}t^k = (.7)^{n-k}(.3)^k$. Since there are $\binom{n}{k}$ such sequences, all mutually exclusive, the total probability is $\binom{n}{k}h^{n-k}t^k$, which is one term of the expansion of $(h + t)^n$. For example, the probability of obtaining exactly 2 tails out of 6 tosses of the unfair coin is

$$\binom{6}{2}(.7)^{6-2}(.3)^2 = \frac{6!}{4!2!}(.7)^4(.3)^2 \approx .324.$$

In any two-outcome experiment with probabilities h and t for the outcomes, $h + t = 1$, and $\binom{n}{k}h^{n-k}t^k$ is called a **binomial probability**. An important application of binomial probabilities is in the calculation of probabilities involving samples taken from a very large population.

It may be useful to have students write a few of the different sequences that arise in the situation in **Example 2a**. Here are three possible ways to obtain 3 defective bulbs in 12 trials.
GGGGDGGDGGDG
DDDGGGGGGGGG
GDGGDGGGGGGD
By writing out a few of the different ways to obtain the desired result, students may see the need for multiplying the probabilities by the binomial coefficient.

ADDITIONAL EXAMPLES

1. In how many ways can a team win at least 10 of its first 15 games?
4944

2. If a team has won about 60% of its games in the past, what is the probability that it will win at least 10 of its first 15 games?
$\sum_{k=10}^{15} \binom{15}{k} (.6)^k(.4)^{15-k}$
$\approx .403$

NOTES ON QUESTIONS

Question 6: Ask students for other situations that are equivalent to this one. (samples: getting a disease if it is expected that 60% of people will get it; shooting baskets with a 60% probability of making one)

Question 12: Relate this proof, which involves raising (1 + -1) to the *n*th power, to the proof of the Sum of Binomial Coefficients Theorem, and to Question 4 of the previous lesson.

Error Analysis for Question 13: Some students will answer 32, forgetting that one of the subsets is the null set containing no letter.

Question 14: It is significant that de Méré's intuition was not accurate in this situation. It is a good example of how mathematics can clarify a situation that is not well understood.

Question 16: The answers given in the text are estimates. Some calculators will give the exact answer for part **a**, 1,294,365,618, from which, multiplying by 8, the exact answer for part **b** can be found, 10,354,924,944.

ADDITIONAL ANSWERS

1.b. {1, 2, 3, 4}, {1, 2, 3}, {1, 2}, {1}, { }, {1, 2, 4}, {1, 3}, {2}, {1, 3, 4}, {1, 4}, {3}, {2, 3, 4}, {2, 3}, {4}, {2, 4}, {3, 4}

Example 2 An assembly line that produces a large number of light bulbs is not operating well. If the probability is .01 that a particular light bulb from this assembly line is defective, what is the probability that a 12-bulb carton of light bulbs produced by this assembly line will contain

a. exactly 3 defective light bulbs?

b. more than one defective light bulb?

Solution Think of the 12-bulb carton as a series of 12 trials, each of which consists of selecting a light bulb. A trial has two possible outcomes—selecting a good bulb or selecting a defective bulb. Let g and d represent the probabilities of a bulb being good or defective, respectively. Here $g = .99$ and $d = .01$. Expand $(g + d)^{12}$.

$$(g + d)^{12} = \sum_{k=0}^{12} \binom{12}{k} g^{12-k} d^k$$

The probability that k bulbs in the carton are defective and $12 - k$ are good is given by the term $\binom{12}{k} g^{12-k} d^k$.

a. The probability that a 12-bulb carton will contain exactly 3 defective light bulbs is given by the term with $k = 3$.

$$\binom{12}{3} g^{12-3} d^3 = \binom{12}{3} (.99)^9 (.01)^3 = \frac{12!}{3!9!} (.99)^9 (.01)^3 \approx .000201$$

b. The probability that a 12-bulb carton will contain more than one defective light bulb is equal to 1 minus the probability of containing 0 or 1 defective light bulbs.

$$1 - \left(\binom{12}{0} g^{12-0} d^0 + \binom{12}{1} g^{12-1} d^1\right) = 1 - \left(\binom{12}{0} (.99)^{12} + \binom{12}{1} (.99)^{11} (.01)^1\right)$$
$$\approx .00617$$

The reciprocal of .000201 is about 4975; the reciprocal of .00617 is about 162. So about 1 in 5000 cartons will have 3 defective bulbs, and about 1 in 160 will have more than one defective bulb.

Questions

Covering the Reading

1. a. How many subsets are there of {1, 2, 3, 4}? 16

b. List them. See margin.

2. Compute the number of subsets of an 8-element set that contain

a. exactly 3 elements, 56 **b.** 3 or fewer elements. 93

In 3 and 4, a coin is flipped five times.

3. How many of the possible sequences of heads and tails have

a. 0 heads? 1 **b.** 1 head? 5 **c.** 2 heads? 10

d. 3 heads? 10 **e.** 4 heads? 5 **f.** 5 heads? 1

4. How many sequences of heads and tails are possible altogether? 32

5. Prove the sum of Binomial Coefficients Theorem for the case when $n = 0$. **See margin.**

6. Suppose a coin is weighted so that the probability of heads is .6. If the coin is tossed 10 times, give the probability of obtaining:
   a. 10 heads. **≈ .0060** b. 9 heads. **≈ .0403**
   c. 8 heads. **≈ .1209** d. 7 heads. **≈ .2150**

7. Refer to Example 2. What is the probability that the carton contains exactly 2 defective light bulbs? **≈ .00597**

8. Evaluate $\sum_{k=0}^{6} \binom{6}{k}$ without writing the terms and computing them. **64**

Applying the Mathematics

9. In a particular game at a fast-food restaurant, the probability that a given ticket will win a food prize is $\frac{1}{5}$. What is the probability that an individual will receive exactly four winning tickets in 10 plays of the game? **≈ 0.088**

10. A box contains only red and white balls of the same size. Suppose that there are twice as many red balls as white balls in the box. If you draw 5 balls from the box at random, one at a time, and return each ball to the box before drawing the next, what is the probability that you will draw at least three red balls? **≈ 0.790**

11. Suppose that $S = \{1, 2, 3, 4, 5, 6, 7, 8\}$ and that $S_1 = \{1, 2, 3, 4, 5, 6, 7\}$. Observe that every 5-element subset of S is either a 5-element subset of S_1 or a set consisting of 4 elements of S_1 and the number 8.
    a. Use this observation to explain why $\binom{7}{5} + \binom{7}{4} = \binom{8}{5}$.
    b. Prove the identity $\binom{n}{r-1} + \binom{n}{r} = \binom{n+1}{r}$ for all positive integers n and r with $r \le n$ by interpreting both sides of the identity in terms of counting r-element subsets of certain sets.
    See margin.

12. Prove that for each positive integer n,

$$\underbrace{\binom{n}{0} - \binom{n}{1} + \binom{n}{2} - \binom{n}{3} + \ldots \pm \binom{n}{n}}_{\text{positive and negative terms alternate}} = 0.$$

(Hint: Use the proof of the Sum of Binomial Coefficients Theorem as a guide.) **See margin.**

13. For a question on a test, a student may answer A, B, C, D, or E, or any combination of those choices. How many different answers are possible, assuming that every answer must include at least one letter? **31**

5. If $n = 0$, then $2^0 = 1$ and $\sum_{k=0}^{0} \binom{0}{k} = \binom{0}{0} = \frac{0!}{0!0!} = 1$, so $\sum_{k=0}^{n} \binom{n}{k} = 2^n$.

11.a. The number of 5-element subsets of S_1 plus the number of 4-element subsets of S_1 is the number of 5-element subsets of S.
b. Let S be a set of $n + 1$ elements and let S_1 be a set of n of these elements. Then every r-element subset of S is either an r-element subset of S_1 or an $(r - 1)$-element of S_1 along with the left-over element not in S_1. So $\binom{n+1}{r} = \binom{n}{r} + \binom{n}{r-1}$.

12. The expression $\binom{n}{0} - \binom{n}{1} + \binom{n}{2} - \binom{n}{3} + \ldots \pm \binom{n}{n}$ represents the coefficients of the expansion of $(x - y)^n$. Letting $x = y = 1$, $(x - y)^n = (1 - 1)^n = 0$, so the sum of the coefficients must be zero.

NAME ______

LESSON **MASTER 10–6**
QUESTIONS ON **SPUR** OBJECTIVES

■ **PROPERTIES** *Objective E (See pages 636–638 for objectives.)*
1. The number of 4-element subsets that can be formed from a set with 7 elements is given by the combination $\binom{7}{4}$.

2. How is ${}_7C_5 + {}_7C_6 + {}_7C_7$ related to the number of subsets that can be formed from a set?
It gives the number of subsets containing at least 5 elements that can be formed from a 7-element set.

■ **USES** *Objective G*
3. In how many ways is it possible to obtain at least five heads in 13 tosses of a coin? **7,099**

4. An "all-you-can-eat" buffet table contains 10 meat dishes. How many different combinations of meat dishes are possible, assuming that each customer chooses at least one? **1,023**

■ **USES** *Objective H*
5. If the "DON'T WALK" symbol on a pedestrian signal is lit for 40 seconds per minute, and the "WALK" symbol is lit for 20 seconds per minute, what is the probability that the signal will say "WALK" two out of six times that it is approached? **.329**

6. A soft drink manufacturer gives away prizes by printing the name of the prize on the inside of some of the cans. If 6% of the cans produced result in a prize, what is the probability that in a 24-can case,
   a. exactly three cans result in a prize? **.119**
   b. no more than one can results in a prize? **.573**

 115

FOLLOW-UP

MORE PRACTICE
For more questions on SPUR Objectives, use *Lesson Master 10-6*, shown on page 619.

PROJECTS
The projects for Chapter 10 are described on pages 632–633. **Project 6** is related to the content of this lesson.

EVALUATION
A quiz covering Lessons 10-4 through 10-6 is provided in the Teacher's Resource File.

ADDITIONAL ANSWERS
15. $128a^7 - 448a^6b + 672a^5b^2 - 560a^4b^3 + 280a^3b^4 - 84a^2b^5 + 14ab^6 - b^7$

18.a.

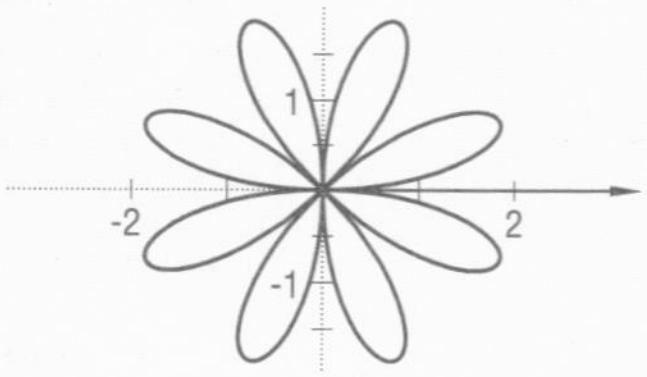

b. eight-leafed rose curve

19.

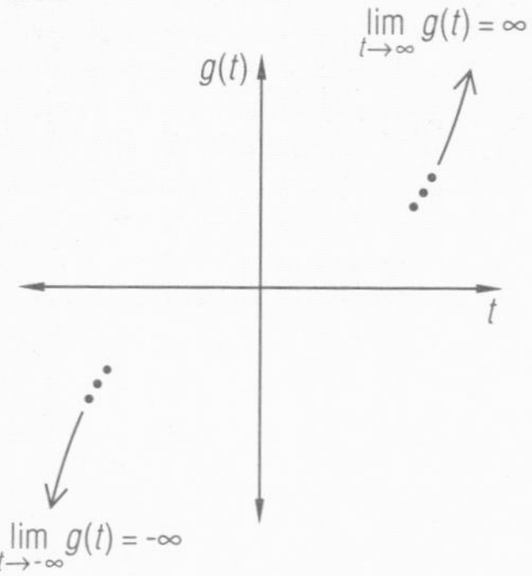

Blaise Pascal

14. Historians of mathematics usually give 1654 as the year in which the branch of mathematics called *probability theory* was born. That was the year in which a professional gambler, the Chevalier de Méré, posed several problems involving dice games to Pascal. One of the problems that de Méré discussed with Pascal was the following: Which is more likely to occur, (1) at least one 1 in a single roll of 4 dice, or (2) at least one pair of 1s in 24 rolls of two dice? Based on his experience as a gambler, de Méré believed that these outcomes were equally likely.

a. Think of a single roll of 4 dice as 4 successive rolls of a single die. Also, regard a roll of a single die as "good" if it comes up 1 and "defective" if it does not come up 1. Use the method of Example 2, part **b**, to compute the probability of getting at least one 1 in a single roll of 4 dice. $\approx .5177$

b. Find the probability of rolling a pair of 1s in a single roll of a pair of dice. $\approx .0278$

c. Compute the probability that out of 24 rolls of a pair of dice, at least one roll will result in a pair of 1s. $\approx .4914$

d. Was de Méré's belief correct? No

Review

15. Expand $(2a - b)^7$. *(Lesson 10-5)* See margin.

16. The student body of a small high school consists of 52 tenth-graders, 47 eleventh-graders, and 43 twelfth-graders.

a. A committee consisting of two representatives from each grade is to be formed. In how many ways can this committee be chosen?

b. A committee consisting of a president and vice-president for each class is to be formed. In how many ways can this committee be chosen? *(Lessons 10-4, 10-3)* See below.

17. Let $f(x) = x^3 - 3x^2$. Use the fact that $f'(x) = 3x^2 - 6x$ to find the interval(s) on which f is increasing. *(Lesson 9-5)* $x < 0$, $x > 2$

18. a. Sketch the polar graph of the equation $r = 2 \sin 4\theta$.

b. Identify the resulting curve.

See margin.

19. Describe the end behavior of the function g with $g(t) = 2t^3 + 5t^2 - 1$ pictorially and in limit notation. *(Lesson 5-5)*
See margin.

Exploration

20. Conduct an experiment to test the theoretical results obtained in Question 14. How well do your results agree with theory?
Answers may vary.

16. a) 1.2944×10^9; b) 1.035×10^{10}

LESSON

10-7

Combinations with Repetition

Consider the following:

Problem 1: How many terms are there in the expansion of $(x + y + z)^{13}$?

Problem 2: Find the number of strings of 3 nonnegative integers whose sum is 13. (Two such strings are 3, 5, 5 and 9, 0, 4.)

While these two problems appear to be quite different, they happen to be equivalent. The appropriate counting technique can be developed by again using the Binomial Theorem.

Consider a simpler variant of Problem 1, using 4 in place of 13. A term in the expansion of $(x + y + z)^4$ is obtained by selecting one of x or y or z from each of the four factors

$$(x + y + z)(x + y + z)(x + y + z)(x + y + z)$$

and then taking their product. The order in which the xs, ys, and zs are selected is not important and the same letter can be selected more than once. Therefore, each term in the expansion of $(x + y + z)^4$ corresponds to a collection of four letters selected from the three letters, x, y, z, *with repetition* allowed. For instance:

Term of $(x + y + z)^4$	Corresponding Collection
x^2yz	$xxyz$
yz^3	$yzzz$.

You can solve the problem of finding the number of distinct terms in the expansion of $(x + y + z)^4$ by listing all such collections, or (to save writing) by listing the variable part of all of the terms. They are: x^4, x^3y, x^3z, x^2y^2, x^2yz, x^2z^2, xy^3, xy^2z, xyz^2, xz^3, y^4, y^3z, y^2z^2, yz^3, z^4. There are 15 terms. Similarly, counting the terms in the expansion of $(x + y + z)^{13}$ is equivalent to counting the number of collections of 13 letters selected from 3 letters with repetition. However, this approach is impractical for $(x + y + z)^{13}$ because the number of terms is so large.

Note, however, that the sum of the powers of x, y, and z in each term of $(x + y + z)^4$ is 4. Similarly, the variable part of each term in the expansion of $(x + y + z)^{13}$ is of the form $x^my^nz^p$ where m, n, and p are nonnegative integers and $m + n + p = 13$. Therefore the problem of finding the number of terms in the expansion of $(x + y + z)^{13}$ is also equivalent to the problem of finding the number of sequences of 3 nonnegative integers which add to 13.

LESSON 10-7

RESOURCES

- Lesson Master 10-7
- Teaching Aid 71 can be used with **Questions 4 and 5**.

OBJECTIVE

G Use combinations and the Binomial Theorem to solve counting problems.

TEACHING NOTES

Because a set does not allow repeated elements, we use the term *collection* to refer to an unordered group of objects with repetition allowed. The collections given on the middle of this page are denoted in alphabetical order, for convenience only since *xyxz* refers to the same collection as *xxyz*.

The discussion on page 622 is important for understanding the theorem at the top of page 623. Students need to understand that if r objects are to be distributed among n boxes, then decisions are made only for $n - 1$ boxes. The decision for the nth box is then determined because there are exactly r objects to be distributed. So a particular allocation of the balls among the boxes can be shown by using $n - 1$ blanks as separators between the portions of balls in the different boxes. Hence, there are a total of $r + (n - 1)$ positions and r of these are to be taken. The number of ways this situation can occur is the combination given in the theorem.

Alternate Approach A different way to prove the theorem for $n = 3$, based on Pascal's triangle, is as follows: If all r balls are placed in the first box, there are zero balls to be placed in the other two boxes. This can be done in one way, just as there is one term in the 0th row of Pascal's triangle. If $r - 1$ balls are placed in the first box, there is one ball to be placed in either of the other two boxes. This can be done in one of two ways, just as there are two terms in the second row of Pascal's triangle.

In general, if $r - k$ balls are placed in the first box, then there are k balls to be placed in the other 2 boxes. These allocations, from $[k,0]$ $[k - 1, 1]$ to $[0, k]$ are just like the different number of terms of the form $a^s b^t$, where $s + t = k$ that appear in the kth row of Pascal's triangle, and there are $k + 1$ such terms.

Thus, the total number of ways of placing r balls in 3 boxes is $1 + 2 + \ldots + r = \frac{r(r+1)}{2}$. This proof is more directly connected with the idea of the trinomial expansion $(x + y + z)^r$, and might seem more straightforward to students. An argument using induction on n could generalize it.

Emphasize the equivalence of the following four problems: (1) the number of collections of r elements that can be constructed from a set of n elements (with repetition allowed); (2) the number of terms in the expansion of a polynomial of n terms to the rth power; (3) the number of strings of n nonnegative integers whose sum is r; and (4) the number of ways of distributing r balls into n boxes. Students who understand why these problems are equivalent have a deep understanding of this lesson.

Still another problem equivalent to these is the following. Think of the exponent 13 as thirteen identical balls. Think of each variable x, y, and z as a box. Each distribution of all thirteen balls into the three boxes can be identified with a term in the expansion of $(x + y + z)^{13}$. For instance, in the distribution pictured below,

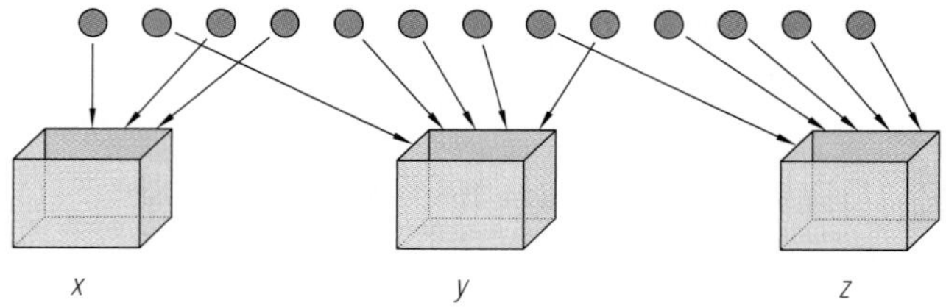

3 balls are placed in box x, 5 in box y, and 5 in box z. Therefore this distribution corresponds to the term $x^3y^5z^5$. Because the balls are identical, any distribution which places 3 balls in box x, 5 in box y, and 5 in box z also represents the term $x^3y^5z^5$. Therefore, the number of terms in the expansion of $(x + y + z)^{13}$ is the same as the number of ways of distributing 13 identical balls in 3 different boxes.

The essential feature of the balls-in-boxes distribution is that only the number of balls to be placed in each box is important. The order in which the balls are placed is not important. A distribution of 13 balls in three different boxes can therefore be described by the following sequence of steps.

Step 1: Decide how many balls are in the first box.
Step 2: Decide how many balls are in the second box.

Once these steps have been carried out, the number of balls in the third box is determined. The number of ways of carrying out these steps can be represented as follows.

Let a "❍" represent each ball and arrange 13 of them in a row. Insert separators, represented by blanks, between the balls in box 1 and box 2 and between the balls in box 2 and box 3. Thus

❍❍❍❍_❍❍❍❍❍_❍❍❍❍

Step 1: insert separator	Step 2: insert separator

represents placing 4 balls in box 1 and 5 in box 2, with 4 balls left over for box 3.

Thus there are 15 positions, 13 to be filled by a ball and 2 to remain blank. (The blanks can go on the ends; for example, starting with both blanks means box 3 gets all 13 balls.) Furthermore, the 13 is the number of balls, and the 2 is the number of boxes to choose from. Thus,

there are $\binom{15}{13}$ ways of taking 15 positions and choosing 13 to be filled with a ❍.

The preceding argument can be generalized as follows. Remember that a collection is like a combination in that the order of the elements does not matter, and think of $\binom{15}{13}$ in the example on page 622 as $\binom{13 + (3 - 1)}{13}$.

Theorem

Suppose that n and r are positive integers. The number of r-element collections that can be constructed from a set with n elements with repetitions allowed is given by

$$\binom{r + (n - 1)}{r}.$$

For instance, suppose that S is the set $\{a, b, c\}$ so that S has 3 elements. The number of 6-element collections that can be constructed from S, with repetitions allowed, is found by letting $n = 3$ and $r = 6$ in the theorem; that number is $\binom{6 + (3 - 1)}{6} = \binom{8}{6} = 28$. Two of these collections are *aabccc* and *bbbbbc*.

The expression $\binom{r + (n - 1)}{r}$ also gives the number of terms in the expansion of $(x_1 + \ldots + x_n)^r$. So, the expansion of $(x + y + z)^{13}$ has $\binom{13 + (3 - 1)}{13} = \binom{15}{13} = \frac{15 \cdot 14}{2} = 105$ terms.

Although the various representations used above are equivalent, usually one of them is more convenient or natural to use in a given application. The following problem illustrates a situation in which the balls-in-boxes model seems most appropriate.

Example How many integers from 0 to 9999 have the property that the sum of their digits is 8? (Two of these integers are 125 and 7001.)

Solution Every integer from 0 to 9999 can be expressed as a string of four digits from 0 through 9. Think of the positions in these strings of four digits as boxes B_1, B_2, B_3, B_4. Think of each such digit d as d identical balls. For example, the digit 3 corresponds to 3 balls and the digit 7 corresponds to 7 balls. With this interpretation of boxes and balls, every integer from 0 to 9999 whose digits add up to 8 corresponds to a distribution of 8 identical balls into 4 boxes. For example,

Integer With Digit Sum = 8	Distribution of 8 balls in 4 boxes
1 2 5	
7 0 0 1	

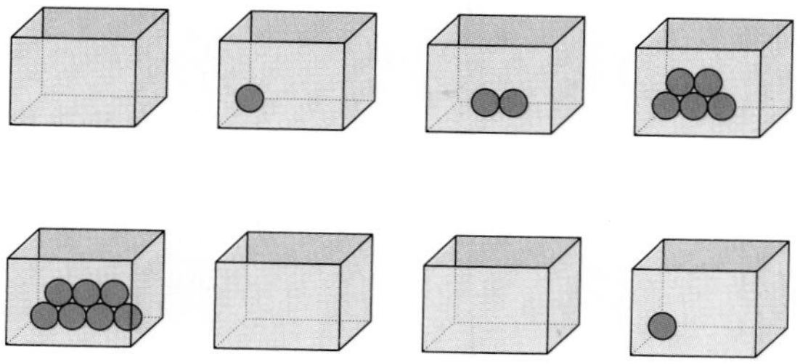

Ask students to state the balls-in-boxes equivalent to Problems 1 and 2. (How many ways can 13 balls be put into 3 boxes?) Then ask for the rewording of these two problems in the language of the theorem on this page. (How many 13-element collections can be constructed from a set of 3 elements with repetition allowed?)

The most difficult aspect of this lesson for students is to differentiate those problems requiring the use of the theorem for solution and those problems requiring the $C(n, r)$ Calculation Theorem from Lesson 10-4 for solution. Remind students that for typical combinations, repetition of elements is not allowed while for the balls and boxes interpretation repetition is allowed. Thus, if the context of the application allows for repetition of elements, then the techniques of this lesson are required.

ADDITIONAL EXAMPLES

1. How many terms are in the expansion of $(a + b + c + d)^{12}$?
$\binom{15}{12} = 455$

2. In how many ways can six new students to a school be added to the numbers currently in five homerooms?
$\binom{10}{6} = 210$

NOTES ON QUESTIONS
Questions 4 and 5: These questions test understanding of the balls-in-boxes model and should be covered carefully.

Questions 9–13: These questions sample a variety of situations that can be treated as combinations with repetition.

Question 12: Students may need help interpreting the situation of this problem in the language of this lesson. It is useful to think of the digits 1–6 as the boxes into which four balls, representing the four dice, are to be sorted. One box, or one digit, could get all four dice and a box could have no dice. In fact, on a given roll, there will be at least two digits which do not come up, that is, there will be at least two boxes with no dice in them.

Question 14: Treat this question as a binomial experiment. The actual distribution, as noted in the previous lesson, is hypergeometric.

Therefore, by the theorem, the number of integers from 0 to 9999 inclusive whose digit sum is 8 is

$$\binom{8 + (4 - 1)}{8} = \binom{11}{8} = \frac{11 \bullet 10 \bullet 9}{3 \bullet 2} = 165.$$

You will use both the collections with repetitions model and the balls-in-boxes model to solve a variety of counting problems in the Questions.

You have now encountered the four types of problems mentioned in the chart in Lesson 10-1: permutations (order matters) with and without repetition, and combinations (order doesn't matter) with and without repetition.

Questions

Covering the Reading

1. **a.** Expand $(x + y + z)^4$ to find all of the terms and their coefficients.
   b. Verify that the variable parts of the terms in part **a** are the ones listed in the lesson.
   See margin.
2. What is the answer to Problem 2 at the beginning of this lesson? 105
3. How many 7-element collections with repetitions allowed can be constructed using the letters x, y, and z? 36
4. Write a 12-symbol collection consisting of 8 Os and 4 blanks to represent each of the following distributions of 8 identical balls into 5 different boxes.

a.
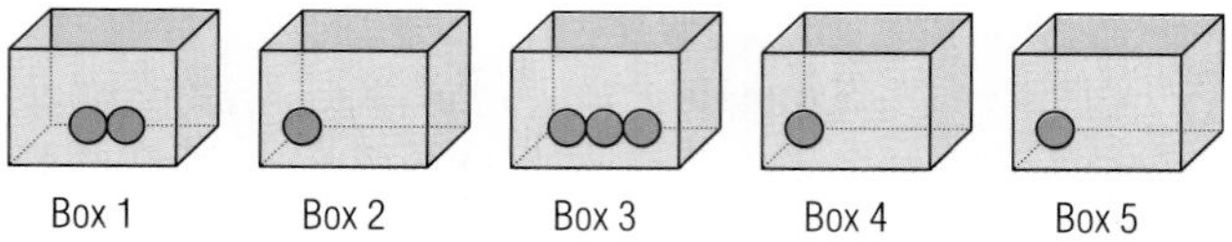

b.
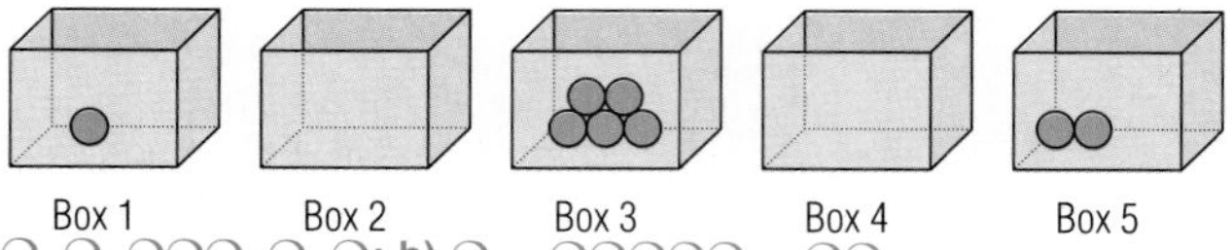

a) OO_O_OOO_O_O; b) O_ _OOOOO_ _OO

5. **a.** Suppose that 8 identical balls are to be placed into 5 different boxes. In each case, how many balls are in each box?
   i. O_O_O_OO_OOO 1, 1, 1, 2, 3
   ii. _OOO_OO_OOO_ 0, 3, 2, 3, 0
   iii. O_ _OOO_OOO_O 1, 0, 3, 3, 1
   b. In how many ways can 8 balls be placed into 5 boxes? 495
6. How many distinct terms are in the expansion of the trinomial power $(a + b + c)^9$? 55
7. How many distinct terms are in the expansion of $(x_1 + x_2 + x_3 + x_4 + x_5)^{11}$? 1365

8. How many integers from 1 to 9999 have the property that the sum of their digits is 6? **84**

Applying the Mathematics

9. The Tastee Donut Shoppe charges $3.00 for its Mix N' Match Selection that allows you to select a dozen doughnuts from among the following varieties: plain, maple frosted, chocolate, glazed, and French. How many different Mix N' Match Selections are possible? **1820**

10. In a game at a school fair, a $1 ticket at the fishing pool allows the participant to choose three prizes from four containers, each of which contains a different type of prize. Any combination of selections is acceptable. That is, all selections can be made from the same containers, two can be made from one container and one from a different container, or all three can be made from different containers. How many different selections are possible? **20**

11. Represent the problem of determining the number of terms in the expansion of $(w + x + y + z)^5$ as a problem of finding sequences of nonnegative integers with a given sum 5. **See margin.**

12. Suppose four identical dice are rolled. Define an outcome to be a particular collection of four numbers that come up on the four dice. For example, 2, 1, 5, 2 is the same outcome as 1, 2, 2, 5, but not the same as 2, 1, 6, 1 (even though the total is the same). How many different outcomes are possible? **126**

13. How many different solutions (w, x, y, z) of the equation $w + x + y + z = 20$ are there for which w, x, y, and z are all nonnegative integers? **1771**

Review

14. Suppose that 0.5% of the circuit boards produced on a certain assembly line are defective. If a sample of 100 circuit boards are selected at random to be tested, what is the probability that at least one of them will be defective? *(Lesson 10-6)* **≈ .394**

15. Find the coefficient of x^3y^5 in the expansion of $(2x + y)^8$. *(Lesson 10-5)* **448**

16. Find the number of committees which can be formed from a group of 25 people if each committee must contain at least 5 members, but no more than 8 members. *(Lesson 10-4)* **1,792,505**

17. Consider the word *boxes*.
   a. How many permutations of all the letters of *boxes* are there? **120**
   b. How many permutations of all letters of *boxes* end in a vowel? *(Lesson 10-3)* **48**

18. Let f be the function defined by $f(x) = x^3 + 4$. Find the average rate of change of f from $x = 2$ to $x = 3$. *(Lesson 9-1)* **19**

19. Evaluate $\sum_{k=1}^{\infty} 4\left(\frac{3}{5}\right)^{k-1}$. *(Lesson 7-6)* **10**

20. Write the argument form for *modus ponens*. *(Lesson 1-6)*
$((p \Rightarrow q)$ and $p) \Rightarrow q$

ADDITIONAL ANSWERS

1.a. $(x + y + z)^4 = x^4 + 4x^3y + 4x^3z + 6x^2y^2 + 12x^2yz + 6x^2z^2 + 4xy^3 + 12xy^2z + 12xyz^2 + 4xz^3 + y^4 + 4y^3z + 6y^2z^2 + 4yz^3 + z^4$

b. The 15 terms listed above are the same ones listed in the lesson, except for the addition of coefficients.

11. The problem of finding the number of terms in the expansion of $(w + x + y + z)^5$ is equivalent to the problem of finding the number of 4 nonnegative integers which add to 5.

NAME ____________

LESSON **MASTER 10–7**
QUESTIONS ON **SPUR** OBJECTIVES

■ **USES** *Objective G (See pages 636–638 for objectives.)*

1. How many terms are there in the expansion of $(m + n + p)^{11}$? **78**

2. How many positive integers less than 10,000 have digits which add to 16? **969**

3. The spinner for a game consists of a circle with five sections, colored red, blue, yellow, purple, and green. If an outcome is defined as a certain number of occurrences of each color, how many different outcomes are possible in seven spins? **330**

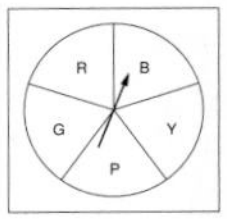

4. a. How many integers from 1 to 999 have the property that the sum of their digits is 9? **55**
   b. How many of these have exactly one five? **15**

FOLLOW-UP

MORE PRACTICE
For more questions on SPUR Objectives, use *Lesson Master 10-7*, shown on page 625.

PROJECTS
The projects for Chapter 10 are described on pages 632–633. **Project 3** is related to the content of this lesson.

21. A ship is on a bearing such that its position after 1 hour is 30 miles east of its starting position. Express the bearing θ as a function of the ship's northerly distance d from its original location. *(Lesson 6-6)*

$\theta = \tan^{-1} \frac{30}{d}$

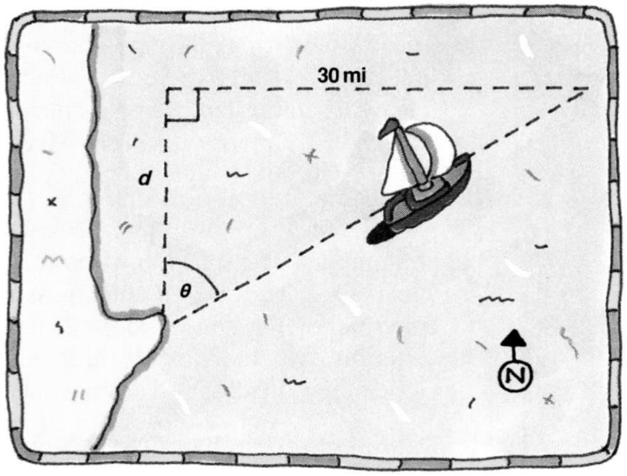

Exploration

22. a. Count the number of different distributions of 5 identical balls into 3 different boxes for which each box contains at least one ball. You can do this by writing down all such distributions.

b. Write a collection of ❍s and blanks, as in Question 5, to represent each of the distributions you found in part **a**.

c. Use the model of the collections of ❍s and blanks to count the number of different distributions of 15 identical balls into 7 different boxes for which each box contains at least one ball.

d. Try to generalize. Find a formula for the number of distributions of r identical balls into n different boxes for which each box contains at least one ball. **a) 1 1 3, 1 2 2, 1 3 1, 2 1 2, 2 2 1, 3 1 1; 6 distributions b) ❍_❍_❍❍❍, ❍_❍❍_❍❍, ❍_❍❍❍_❍, ❍❍_❍_❍❍, ❍❍_❍❍_❍, ❍❍❍_❍_❍; c) 3,003; d)** $\binom{r-1}{r-n}$

626

LESSON

10-8

Multinomial Coefficients

To expand a power of a trinomial or other polynomial, such as $(x + y + z)^{13}$, you need to know the variable part for each distinct term of the expansion and the coefficient of each term. The last lesson showed that there are 105 distinct terms in the expansion of $(x + y + z)^{13}$. Each term is of the form $x^a y^b z^c$, where a, b, and c are nonnegative integers and $a + b + c = 13$. But this is not enough to expand $(x + y + z)^{13}$ because the coefficients of the terms are still needed. These coefficients are called *multinomial coefficients*. In this lesson, you will learn how to determine these coefficients.

As with binomial coefficients, there is a combinatorial interpretation for multinomial coefficients. Write out $(x + y + z)^{13}$ as a product of 13 factors.

$$\underbrace{(x + y + z)(x + y + z)(x + y + z) \bullet \ldots \bullet (x + y + z)}_{13 \text{ factors}}$$

The expansion is found by choosing an x, y, or z from each factor, multiplying them, and adding the products. The sum of the exponents of x, y, and z is 13 in each term because there are 13 factors. Now consider a particular term, for instance, the one with variable part $x^6y^2z^5$. This term arises when x is chosen 6 times, y is chosen 2 times, and z is chosen 5 times. The coefficient of $x^6y^2z^5$ is the number of different ways in which this set of choices can be made.

To calculate this number, note that the 6 xs can be selected from the 13 factors in $\binom{13}{6}$ ways. There are now 7 factors remaining, so the 2 ys can be selected in $\binom{7}{2}$ ways. This leaves 5 factors from which to choose the 5 zs, and this can be done in only $\binom{5}{5} = 1$ way. Since the selections are made one after the other, by the Multiplication Counting Principle the total number of selections is the product of these three combinations, $\binom{13}{6} \bullet \binom{7}{2} \bullet \binom{5}{5}$.

In doing the actual calculation of this product, you see that many of the factorials appear in both the numerator and denominator. So what seems to be a difficult calculation can be simplified considerably.

$$\binom{13}{6}\binom{7}{2}\binom{5}{5} = \frac{13!}{6!7!} \bullet \frac{7!}{2!5!} \bullet \frac{5!}{5!0!} = \frac{13!}{6!2!5!}$$

Thus the coefficient of $x^6y^2z^5$ in the expansion of $(x + y + z)^{13}$ is $\frac{13!}{6!2!5!}$. You can see that the power of the polynomial and the exponents of the term appear as the factorials in the coefficient. The pattern is that simple.

LESSON 10-8

TEACHING NOTES

The word *multinomial* refers to any polynomial with more than one term; thus binomials and trinomials are multinomials. The *Multinomial Theorem* states that $(x_1 + x_2 + \ldots + x_k)^n = \sum \left(\frac{n!}{a_1!a_2! \ldots a_k!}\right) x_1^{a_1}x_2^{a_2} \ldots x_k^{a_k}$, where the summation ranges over all nonnegative integers a_i such that $a_1 + a_2 + \ldots + a_k = n$. We did not state the theorem using summation notation, but you should do so if you think it is appropriate for your class. When $k = 3$ the summation is equivalent to the Binomial Theorem.

The power of the multinomial theorem is shown at the bottom of page 628. Even with a small number of terms (3) and a small exponent (4), the expansion is not easy, but the power can be expanded by using the theorem at the top of the page and the theorem from the last lesson.

Point out that the text discusses the number of permutations of the word *proportion* using *all* of its letters. The number of permutations using some of the letters is a more difficult question.

Theorem

Let k and n be positive integers. In the expansion of $(x_1 + x_2 + \ldots + x_k)^n$, the coefficient of $x_1^{a_1}x_2^{a_2}\ldots x_k^{a_k}$ is $\frac{n!}{a_1!a_2!\ldots a_k!}$.

The general proof of this theorem requires mathematical induction. Here is a proof of the special case $k = 4$ without induction. The coefficient of $x_1^{a_1}x_2^{a_2}x_3^{a_3}x_4^{a_4}$ in the expansion of $(x_1 + x_2 + x_3 + x_4)^n$ is equal to the number of choices in the following counting problem. A set has n elements. You wish to choose a_1 of them. Then from the remaining $n - a_1$, you choose a_2. Then from the $n - a_1 - a_2$ that remain after this second choice, you choose a_3. Lastly, from the $n - a_1 - a_2 - a_3$ that remain after the third choice, you choose a_4. The number of ways to make these four selections is

$$\binom{n}{a_1}\cdot\binom{n-a_1}{a_2}\cdot\binom{n-a_1-a_2}{a_3}\cdot\binom{n-a_1-a_2-a_3}{a_4}$$

$$= \frac{n!}{a_1!(n-a_1)!}\cdot\frac{(n-a_1)!}{a_2!\,(n-a_1-a_2)!}\cdot\frac{(n-a_1-a_2)!}{a_3!\,(n-a_1-a_2-a_3)!}\cdot\frac{(n-a_1-a_2-a_3)!}{a_4!\,(n-a_1-a_2-a_3-a_4)!}$$

$$= \frac{n!}{a_1!a_2!a_3!a_4!}$$

since $(n - a_1 - a_2 - a_3 - a_4)! = 0! = 1$.

Notice now how what you have learned in this and the previous lessons makes it possible to expand any polynomial to any power. For instance, $(x + y + z)^4$ has 15 terms. The variable parts were listed in the last lesson. Now the coefficients of those terms can be determined. Here is the complete expansion.

$$\begin{aligned}(x+y+z)^4 &= \frac{4!}{4!}x^4 + \frac{4!}{3!1!}x^3y + \frac{4!}{3!1!}x^3z + \frac{4!}{2!2!}x^2y^2 + \frac{4!}{2!1!1!}x^2yz + \frac{4!}{2!2!}x^2z^2\\ &\quad + \frac{4!}{1!3!}xy^3 + \frac{4!}{1!2!1!}xy^2z + \frac{4!}{1!1!2!}xyz^2 + \frac{4!}{1!3!}xz^3 + \frac{4!}{4!}y^4\\ &\quad + \frac{4!}{3!1!}y^3z + \frac{4!}{2!2!}y^2z^2 + \frac{4!}{1!3!}yz^3 + \frac{4!}{4!}z^4\\ &= x^4 + 4x^3y + 4x^3z + 6x^2y^2 + 12x^2yz + 6x^2z^2 + 4xy^3\\ &\quad + 12xy^2z + 12xyz^2 + 4xz^3 + y^4 + 4y^3z + 6y^2z^2 + 4yz^3 + z^4\end{aligned}$$

A problem equivalent to the multinomial coefficient problem is the problem of determining the number of permutations of the n objects of a string when some of the objects are alike. (Until now, all the permutations you have seen have been from a set of different objects.) For instance, how many distinguishable permutations are there of the letters of *PROPORTION*?

628

You can think of this problem as being like determining the coefficient of $INO^3P^2R^2T$ in the expansion of $(I + N + O + P + R + T)^{10}$. That is, how many different arrangements are there of 1 I, 1 N, 3 Os, 2 Ps, 2 Rs, and a T? There are $\binom{10}{1}$ possible locations for the I. Then once the I has been located, there are $\binom{9}{1}$ locations for the N. Then there are $\binom{8}{3}$ locations for the three Os, $\binom{5}{2}$ locations for the two Ps, $\binom{3}{2}$ locations for the two Rs, and finally $\binom{1}{1}$ location for the T. The product is

$$\binom{10}{1}\binom{9}{1}\binom{8}{3}\binom{5}{2}\binom{3}{2}\binom{1}{1} = \frac{10!}{1!1!3!2!2!1!} = 151{,}200.$$

Some people prefer another way to analyze this problem. Let x be the number of distinguishable permutations of the letters of *PROPORTION*. If the three Os in *PROPORTION* are labeled as O_1, O_2, and O_3, then the number of distinguishable permutations is multiplied by 3!. Similarly, if the Ps are labeled P_1 and P_2, the number of permutations would be multiplied by 2!, and if the Rs are labeled R_1 and R_2, the number of permutations is again multiplied by 2! The result is the same as dealing with the number of permutations of the word $P_1R_1O_1P_2O_2R_2TIO_3N$, with 10 distinct letters. There are 10! permutations of $P_1R_1O_1P_2O_2R_2TIO_3N$, so $x \cdot 3! \cdot 2! \cdot 2! = 10!$ This implies that $x = \frac{10!}{3!2!2!}$, the same value as found by the multinomial coefficient analysis above. The letters that appear only once in *PROPORTION* do not affect the number of permutations; according to the analysis, each of them could be considered as multiplying the denominator by 1!, which does not change the value of the denominator.

Questions

Covering the Reading

1. Consider the expansion of $(x + y + z)^5$.
  a. How many terms are in this expansion? 21
  b. What is the coefficient of x^2y^3 in this expansion? 10

2. a. Expand $(m + n + p)^3$ completely. See margin.
  b. Apply your answer to part **a** to expand $(2x + y - 1)^3$.
  $8x^3 + 12x^2y - 12x^2 + 6xy^2 - 12xy + 6x + y^3 - 3y^2 + 3y - 1$

3. a. How many distinct permutations are there of the letters of the word *NONILLION*? 7560
  b. What is a nonillion? 10^{30}

4. A biology class has 12 microscopes. Four of these are identical and of one type, two are identical and of a second type, and six are identical and of a third type. In how many distinct ways can these microscopes be placed on 12 lab desks? 13,860

NOTES ON QUESTIONS

Question 2: The answer can be checked by multiplying $(2x + y - 1)(2x + y - 1) \cdot (2x + y - 1)$. Or you can let $x = 3$ and $y = 2$. The values of the given expression and the answer are both 343.

ADDITIONAL ANSWERS

2.a. $m^3 + 3m^2n + 3m^2p + 3mn^2 + 6mnp + 3mp^2 + n^3 + 3n^2p + 3np^2 + p^3$

NOTES ON QUESTIONS
Question 7a: We assume that a 6-digit "number" on a license plate can begin with zeros.

Question 9: It is assumed that there are at least eight jelly beans of each color.

Question 12: The answer given in scientific notation is the exact answer.

ADDITIONAL ANSWERS
5. The coefficient of $x_1^{a_1}x_2^{a_2}x_3^{a_3}$ in the expansion of $(x_1 + x_2 + x_3)^n$ is equal to the number of choices in the following counting problem. A set has n elements. You wish to choose a_1 of them. Then, from the remaining $n - a_1$, you choose a_2. Then, from the $n - a_1 - a_2$ that remains, you choose a_3. The number of ways to make the 3 selections is $\binom{n}{a_1}\binom{n-a_1}{a_2}\binom{n-a_1-a_2}{a_3} = \frac{n!}{a_1!(n-a_1)!} \cdot \frac{(n-a_1)!}{a_2!(n-a_1-a_2)!} \cdot \frac{(n-a_1-a_2)!}{a_3!(n-a_1-a_2-a_3)!} = \frac{n!}{a_1!a_2!a_3!}$, since $(n - a_1 - a_2 - a_3)! = 0! = 1$.

6. For $k = 2$, in the expansion of $(x + y)^n$, the coefficient of $x^{n-r}y^r$ is $\frac{n!}{(n-r)!r!} = \binom{n}{r}$, which is a restatement of the Binomial Theorem.

8.a. Think of the exponent 6 as 6 identical balls. Think of the variables x, y, z, and w as boxes. Each distribution of all six balls into the four boxes can be thought of as a term in the expansion of $(x + y + z + w)^6$.
b. Suppose that S is the set $\{x, y, z, w\}$. The number of 6-element collections selected from these 4 elements of S gives the number of terms in the expansion of $(x + y + z + w)^6$.

5. Prove the theorem of this lesson for the special case $k = 3$. You may assume that the problem is equivalent to a counting problem.
See margin.

6. Discuss the theorem of this lesson for the special case $k = 2$.
See margin.

7. While driving, Laronda noticed that the license plate of the car ahead of her had the number 226266. It thus had three occurrences of one digit and three occurrences of another.
   a. How many license plates with six digits have this property? (Hint: First you need to determine how many pairs of digits are possible.)
   b. What is the probability that a license plate with six randomly selected digits has this property? 0.0009
   a) 900

Review

8. Consider the problem of finding the number of terms in the expansion of $(x + y + z + w)^6$.
   a. Explain how this problem is equivalent to counting ways of distributing identical balls in boxes.
   b. Explain how this problem is equivalent to counting collections of symbols.
   c. How many terms are in the expansion of $(x + y + z + w)^6$?
   (Lesson 10-7) a, b) See margin. c) 84

9. A bowl contains red, orange, yellow, purple, and green jelly beans. In how many different ways can 8 jelly beans be taken from the bowl? *(Lesson 10-7)* 495

10. Suppose that the probability is 2% that a clover in a field is four-leafed. What is the probability that out of 40 clovers picked at random, at least two will be four-leafed? *(Lesson 10-6)* ≈ 0.19

11. A photo is to be taken of 5 tall people and 8 average-height people standing in a line. If the tall people are to stand together in the middle with 4 average-height people on each side, in how many different ways can the people be arranged? *(Lessons 10-4, 10-3, 10-2)* 4,838,400

630

12. A phone number consists of a three-digit area code, a three-digit exchange number, and then four more digits. The first digit of the area code cannot be 0 or 1, and the second digit of the area code must be 0 or 1. The first digit of the exchange number cannot be 0 or 1, and the exchange number cannot be 411, 911, or 555. If those are the only restrictions on telephone numbers, how many are possible? *(Lesson 10-2)*
1,275,200,000

13. Let $[r, \theta]$ and (x, y) represent the same point in the fourth quadrant. If $r = 6$ and $x = 4$, find θ and y. *(Lesson 8-2)*
$y \approx -4.47$; $\theta \approx 5.44$

14. Let c be an integer. Prove that every term of the sequence defined by

$$\begin{cases} a_1 = c \\ a_2 = c \\ a_{k+1} = a_k - 4a_{k-1} & \text{for integers } k \geq 2 \end{cases}$$

is divisible by c. *(Lesson 7-7)* See below.

15. When a polynomial $p(x)$ is divided by a polynomial $d(x)$, the quotient is $3x + 2$ and the remainder is $x + 2$.
a. What do you know about the degree of $d(x)$?
b. Find possible formulas for $p(x)$ and $d(x)$. *(Lesson 4-4)* See below.

Exploration

16. In the word *nonillion*, the subject of Question 3, every letter that appears occurs more than once. Find other English words with this property and calculate the number of distinct permutations of their letters. See below.

14. s(1): $a_1 = c$, so a_1 is divisible by c.
s(2): $a_2 = c$, so a_2 is divisible by c.
Assume s(k): a_k is divisible by c, for all $k \leq n$.
Then $a_{k+1} = a_k - 4a_{k-1}$
$= cq - 4cp$ for some integers p and q
$= c(q - 4p)$, where $q - 4p$ is an integer.
So a_{k+1} is divisible by c.
So $s(k) \Rightarrow s(k+1)$, and, therefore, every term is divisible by 3.

15. a) The degree of $d(x)$ is one less than the degree of $p(x)$.
b) $p(x) = (3x + 2)\, d(x) + x + 2$, so some samples are:

d(x)	p(x)
x	$3x^2 + 3x + 2$
x^2	$3x^3 + 2x^2 + x + 2$
$x + 5$	$3x^2 + 18x + 12$

16. samples:
deed: $\frac{4!}{2!2!} = 6$
noon: $\frac{4!}{2!2!} = 6$
tomtom: $\frac{6!}{2!2!2!} = 90$
deeded: $\frac{6!}{3!3!} = 20$

PROJECTS

The projects provide an opportunity for extended work on topics related to the content of the chapter. (See pages 70 and T37 for more information on the purposes of projects.) Some or all of these projects may be suitable for **small group work**. Solutions for the projects can be found at the end of the Additional Answers section in the back of this book.

Depending on the school requirements, the number of possible class schedules found in **Project 2** could be quite large.

To investigate the number of partitions of an integer for **Project 3**, students may, for example, try $n = 6$ and $n = 7$. Students can describe an iterative method or write a program to count the number of partitions for any integer n.

Projects

1. In Example 3 of Lesson 10-2 you learned that there are 6,760,000 different 6-symbol license plate identifications if the first two symbols can be any of the 26 letters of the alphabet and the last 4 symbols can be any of the digits 0 through 9. In 1987, the state of California had over 20 million vehicles registered for plates. Clearly the state of California must allow more types of license plates than those described in Example 3. Find out what types of license plates are used in your state and count the number of different license plates possible. In most states, certain license plate combinations are reserved for official vehicles and some letter combinations are not allowed. Find out about such restrictions in your state and estimate the number of license plate combinations available to private citizens. How many of these are actually in use in your state this year?

2. When you scheduled your classes for this year, you chose perhaps five classes to take during a seven- or eight-period school day. Some of your classes may have been required and others electives. Among all of the classes offered by your school, some, such as freshman English, you could not sign up for this year. Find out about the classes which were actually available to you and count the number of different schedules that you could have chosen.

3. The number 5 has the following positive integer *partitions.*

$1 + 1 + 1 + 1 + 1$
$1 + 1 + 1 + 2$
$1 + 1 + 3$
$1 + 2 + 2$
$1 + 4$
$2 + 3$
5

In general, a **partition** of a positive integer n is a finite sequence of positive integers whose sum is n. Investigate the number of partitions of any positive integer n.

4. You know that Pascal's triangle represents the coefficients of $(x + y)^n$. Actually, the variable portions can also be included in the triangle to obtain the figure at the right.

$$1$$
$$1x + 1y$$
$$1x^2 + 2xy + 1y^2$$

To illustrate the terms of $(x + y + z)^n$, a tetrahedron can be used. The top layer is a single 1. The second layer, which corresponds to $(x + y + z)^1$, is shown below at the left. The third layer is shown below at the right.

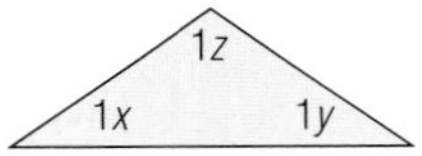

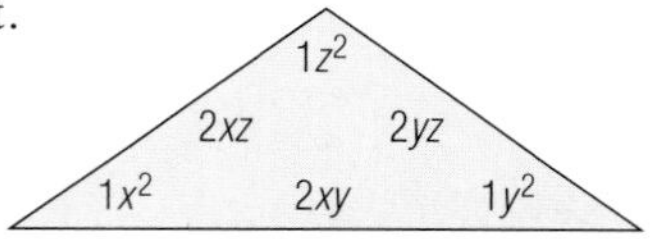

The variable portions of the fourth layer, representing $(x + y + z)^3$, are shown below.

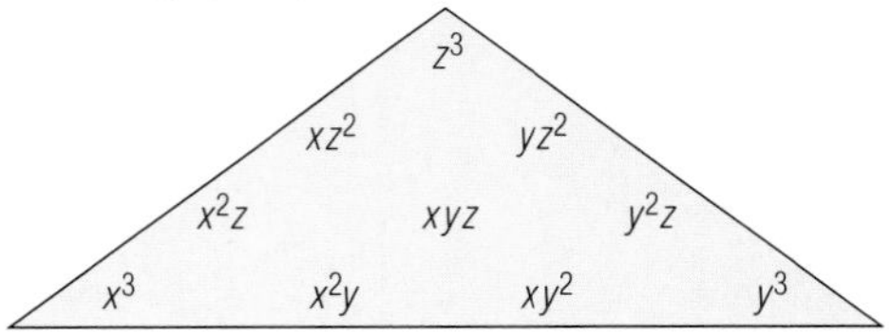

a. Fill in the coefficients of the terms of the fourth layer.
b. Find the fifth and sixth layers.
c. Describe some of the patterns that occur.
d. Make a 3-dimensional model of the first few layers.
e. Find Pascal's triangle in the tetrahedron. It appears several places. Explain why.

5. Combinations can be used to analyze card games such as poker and bridge. These games are played with a deck of 52 cards. The cards are divided into four *suits:* hearts ♥, diamonds ♦, clubs ♣, and spades ♠; and each suit has one card of each of the following ranks: 2, 3, 4, 5, 6, 7, 8, 9, 10, *J*, *Q*, *K*, *A*. A *poker hand* is any set of 5 cards dealt from a deck. A *bridge hand* consists of 13 cards from a deck. The order in which the cards in poker or bridge hands are dealt is unimportant. Many almanacs contain information on the number of possible poker and bridge hands of particular types and/or the odds against obtaining such hands. Pick one of these games. Find this information and use your knowledge of counting techniques to determine how these numbers were calculated.

6. Refer to Example 2 of Lesson 10-6. The BASIC program below simulates taking 10 samples, each consisting of a twelve-bulb carton of bulbs chosen at random. The probability of a bulb being defective is .01. The program represents each carton with a twelve-letter string in which the letter "G" represents a good bulb and the letter "D" represents a defective bulb.

```
10 RANDOMIZE
20 FOR SAMPLE = 1 TO 10
30    FOR I = 1 TO 12
40       IF RND(1) < .99 THEN PRINT "G";
            ELSE PRINT "D";
50    NEXT I
60    PRINT
70 NEXT SAMPLE
80 END
```

a. Enter and run the program. Approximately how many defective bulbs should you expect to find in the ten samples? How many appeared in the simulation?

b. Find the probability that a 12-bulb carton will contain exactly 2 defective bulbs. Run the program 10 times. (Use a different random number seed for each run.) How many cartons containing exactly 2 defective bulbs appeared? How many should you expect?

c. Modify the program to simulate an assembly line in which the probability that any one light bulb is defective is 5%. Repeat part **b** with the new program.

d. Run the program from part **c** and compute the quotient $\frac{\text{total number of defective bulbs}}{\text{total number of bulbs}}$. How does this compare with the probability that any one bulb is defective? Run the program several more times and keep track of the above quotient. Is it getting closer to the probability that any one bulb is defective? Explain your answer.

e. Repeat the "experiment" in part **d** with a partner. Have one of you determine the probability p that any one bulb is defective and modify the program to simulate it. Have the other person conduct the experiment in part **d** (without looking at a list of the program) and try to determine p experimentally.

Regarding **Project 5**, we also have seen odds for pinochle in almanacs, but that game is not as rich in combinations.

You might expand **Project 6** by having students contact the quality control department of a local company. Students might be able to talk with a quality control engineer and determine just how testing samples are selected for that industry.

SUMMARY

The Summary gives an overview of the entire chapter and provides an opportunity for students to consider the material as a whole. Thus, the Summary can be used to help students relate the various concepts presented in the chapter.

The box contained in the Chapter Summary is a handy reference tool for students thinking about the different types of counting problems. It is helpful for students to memorize its contents.

VOCABULARY

Terms, symbols, and properties are listed by lesson to provide a checklist of concepts a student must know. (See page 71 for more information.)

Summary

This chapter began with a discussion of various types of counting problems. Some counting problems can be considered as counting strings of symbols in which the order of the symbols distinguishes the strings. Other problems can be considered as counting sets of symbols in which order is not important. In determining the essential features of a problem, you must determine whether order is or is not important and whether repetition of symbols is or is not allowed. A possibility tree provides a visual representation of the various outcomes in a situation that involves a sequence of steps, each of which has several possible choices or results. By counting the number of "leaves" in the tree, the total number of possible outcomes can be obtained.

The Multiplication Counting Principle is one tool that can be used to count the number of different ways that something can occur. If k successive steps can be performed in $n_1, n_2, \ldots,$ and n_k ways, then the total number of ways all k steps can be performed is given by the product $n_1 \bullet n_2 \bullet \ldots \bullet n_k$.

Formulas useful for solving counting problems follow from the Multiplication Counting Principle. If a collection of r symbols is to be chosen from among n symbols, then the number of different collections is as given in the chart above at the right.

	repetition allowed	repetition not allowed
ordered symbols (permutations)	n^r	$P(n, r) = \dfrac{n!}{(n-r)!}$
unordered symbols (combinations)	$\dbinom{r + (n-1)}{r}$	$C(n, r) = \dbinom{n}{r} = \dfrac{n!}{r!(n-r)!}$

The number of permutations of n objects is $n!$. Permutations and combinations are related by the fact that $P(n, r) = r!\ C(n, r)$. Counting the number of terms in the expansion of $(x_1 + x_2 + \ldots x_r)^n$ and counting the number of ways n balls can be placed in r boxes are both equivalent to counting the number of n-element collections taken from r objects, with repetition allowed.

The Binomial Theorem provides an efficient means to expand $(x + y)^n$. The coefficient of $x^{n-k}y^k$ is given by $\dbinom{n}{k}$. The Binomial Theorem can help to compute the number of ways an event consisting of a sequence of trials can occur when each trial has two possible outcomes. It can also be used to compute the probability that an outcome occurs a certain number of times if the probability of one occurrence of that outcome is known.

Vocabulary

For the starred (*) terms you should be able to give a definition of the term.
For the other terms you should be able to give a general description and a specific example of each.

Lesson 10-1
combinatorics
essential features of a problem
string
set

Lesson 10-2
possibility tree
branch points, nodes
leaves
branches
Multiplication Counting Principle
complement

Lesson 10-3
*permutation
Permutation Theorem
*permutation of n elements taken r at a time
$P(n, r)$, ${}_nP_r$
$P(n, r)$ Calculation Theorem

Lesson 10-4
*combination of n elements taken r at a time
$C(n, r)$, $\dbinom{n}{r}$, ${}_nC_r$
$C(n, r)$ Calculation Theorem
Basic Properties of Combinations Theorem

Lesson 10-5
Pascal's triangle
binomial coefficients
Binomial Theorem

Lesson 10-6
Sum of Binomial Coefficients Theorem
trial
binomial probability

Lesson 10-8
multinomial coefficient

634

Progress Self-Test

See margin for answers not shown below.

Take this test as you would take a test in class. Then check the test yourself using the solutions at the back of this book.

In 1 and 2, **a.** describe the essential features of each problem. **b.** Solve the problem.

1. You wish to know how many different outcomes are possible when five identical dice are rolled.

2. A pizza parlor has the following toppings available: anchovy, pepperoni, sausage, black olive, double cheese, green pepper, mushroom, and onion. Thick or thin crust is offered. Determine the number of different 4-topping pizzas available.

3. A red die and a white die are rolled. Use a possibility tree to determine the number of rolls possible in which the red die is a five or six and the total showing on the two dice is less than ten. 7 possible rolls

4. Consider the computer program below.

```
10 FOR I = 1 TO 2
20   FOR J = 3 TO 8
30     FOR K = 10 TO 20
40         PRINT I + J + K
50     NEXT K
60   NEXT J
70 NEXT I
80 END
```

When the program is run, how many times is line 40 executed? 132

5. In how many ways can you answer a ten-question multiple-choice test if each question has four choices? 1,048,576

In 6 and 7, evaluate the expression.

6. $P(18, 4)$ 73,440 **7.** $\binom{8}{4}$ 70

8. A committee of twelve people is to choose a chairperson, assistant chairperson, and secretary from among the members of the committee. In how many ways can the selection be made? 1320

9. The alphabet has 21 consonants and 5 vowels. Suppose five-letter strings are to be created in which the first letter is a vowel.
a. How many such strings are there if the same letter can occur more than once?
b. How many are there if a letter can only occur once? 1,518,000

10. A caterer must choose three vegetable dishes from nine vegetable dishes which are available. In how many ways can this be done? 84

11. Find the fourth term in the expansion of $(2x - y)^9$. $-5376x^6y^3$

12. A coin is flipped seven times. How many of the possible sequences of heads and tails have exactly 4 tails? 35

13. How many subsets of a set with 11 elements are there that contain 3 or fewer members of that set? 232

14. A soda machine offers a choice of cola, diet cola, lemon-lime soda, and grape soda. If 7 cans are to be purchased, in how many different ways can this be done? 120

15. Prove that for positive integers n and r with $r \le n$, $P(n + 1, r + 1) = (n + 1)P(n, r)$.

16. Suppose that the probability is 3% that a Snacko does not have enough cream filling. If a box of 10 Snackos is purchased, what is the probability that exactly two Snackos are inadequately filled? ≈ 0.0317

PROGRESS SELF-TEST

The Progress Self-Test provides the opportunity for feedback and correction; the Chapter Review provides additional opportunities for practice.

USING THE PROGRESS SELF-TEST

Assign the Progress Self-Test as a one-night assignment. (See pages 72 and T35 for more information.)

ADDITIONAL ANSWERS

1.a. unordered symbols; repetition allowed
b. 252

2.a. unordered symbols; repetition not allowed
b. 140

3.

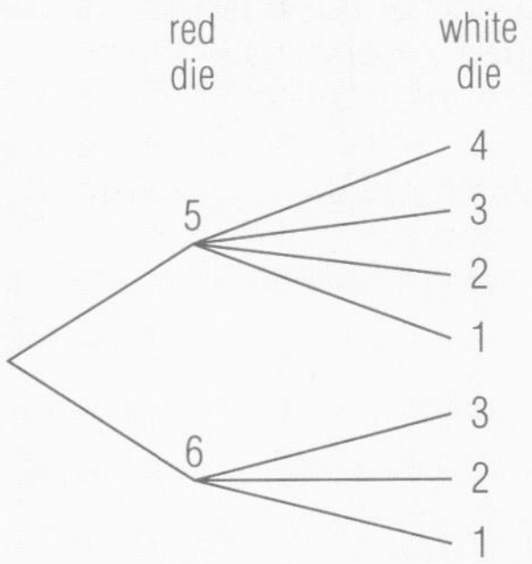

9.a. 2,284,880

15. $P(n + 1, r + 1) = \frac{(n + 1)!}{(n + 1 - (r + 1))!} = \frac{(n + 1)!}{(n - r)!}$; $(n + 1) \cdot P(n, r) = (n + 1)\frac{n!}{(n - r)!} = \frac{(n + 1)!}{(n - r)!}$. So, $P(n + 1, r + 1) = (n + 1) \cdot P(n, r)$.

Chapter Review

Questions on **SPUR** Objectives

RESOURCES
- Chapter 10 Test, Form A
- Chapter 10 Test, Form B
- Chapter 10 Test, Cumulative Form
- Quiz and Test Writer

CHAPTER REVIEW

The main objectives for the chapter are organized here into sections corresponding to the four main types of understanding this book promotes: Skills, Properties, Uses, and Representations. We call these the SPUR objectives.

USING THE CHAPTER REVIEW
Students should be able to answer questions like these with about 85% accuracy by the end of the chapter. (See pages 74 and T38–39 for more information.)

ADDITIONAL ANSWERS

1. unordered symbols; repetition allowed

2. unordered symbols; repetition not allowed

3. ordered symbols; repetition not allowed

4. unordered symbols; repetition not allowed

5. ordered symbols; repetition allowed

11. $x^8 + 8x^7y + 28x^6y^2 + 56x^5y^3 + 70x^4y^4 + 56x^3y^5 + 28x^2y^6 + 8xy^7 + y^8$

12. $16a^4 - 96a^3b + 216a^2b^2 - 216ab^3 + 81b^4$

13. -10240

15. $C(n, r) = \dfrac{n!}{r!(n-r)!} = \dfrac{n!}{(n-(n-r))!(n-r)!} = C(n, n-r)$

SPUR stands for **S**kills, **P**roperties, **U**ses, and **R**epresentations.
The Chapter Review questions are grouped according to the SPUR Objectives for this chapter.
See margin for answers not shown below.

SKILLS deal with the procedures used to get answers.

Objective A: *Describe the essential features of counting problems.* *(Lesson 10-1)*

In 1–5, describe the essential features of the problem. You do not have to solve these problems.

1. An ice cream shop carries 32 different flavors. How many different double cones (2 dips) can be made?

2. At the school cafeteria, a standard lunch offers a choice of salad or fruit cup, two of three vegetable dishes, one of four entrees, and one of two desserts. How many different lunches can a student choose?

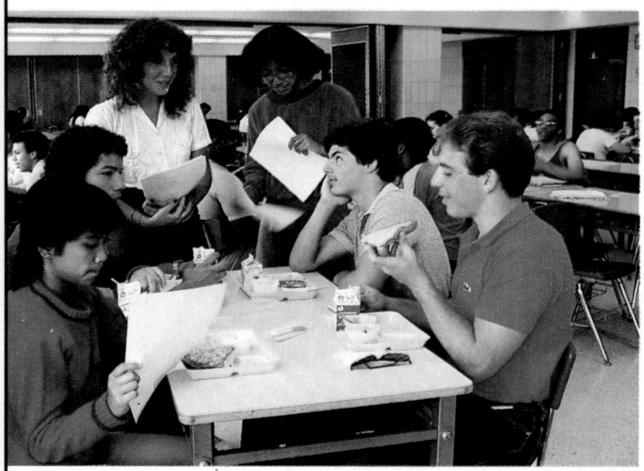

3. How many different 4-letter codewords can be constructed from different letters in the words *mind power*?

4. A company plans to hire 8 new employees with 4 men and 4 women. If 15 men and 7 women apply for the positions, in how many ways can the company hire its new employees?

5. How many four-digit integers have digits that alternate between even and odd when read from left to right?

Objective B: *Evaluate expressions indicating permutations or combinations.* *(Lessons 10-3, 10-4)*

In 6–10, evaluate the expression.

6. $P(17, 3)$ 4080 **7.** $C(14, 9)$ 2002

8. $C(18, 6)$ 18,564 **9.** $P(9, 6)$ 60,480

10. $\binom{9}{7}$ 36

Objective C: *Apply the Binomial Theorem to expand binomials or find specific terms.* *(Lesson 10-5)*

In 11 and 12, expand using the Binomial Theorem.

11. $(x + y)^8$ **12.** $(2a - 3b)^4$

13. Without computing the entire expansion of $(4a - 2b)^6$, find the coefficient of a^3b^3.

14. Find the fifth term of $(x + 5y)^9$. $78750x^5y^4$

PROPERTIES deal with the principles behind the mathematics.

Objective D: *Use properties of permutations and combinations to prove identities.* *(Lessons 10-3, 10-4)*

15. Show that $C(n, r) = C(n, n - r)$ for all positive integers n and r with $r \leq n$.

16. Show that $P(n, n) = n!$. $P(n, n) = \frac{n!}{(n-n)!} = \frac{n!}{0!} = n!$

17. Provide an explanation in terms of combinations to justify the relationship $P(n, r) = r!\,C(n, r)$.

Objective E: *Apply the Binomial Theorem to deduce properties of sets.* *(Lesson 10-6)*

18. How can the Binomial Theorem be used to find the number of different subsets that can be formed from a 10-element set?

19. What binomial coefficient, expressed as a combination, gives the number of 4-element subsets that can be formed from a 20-element set? $\binom{20}{4}$

USES deal with applications of mathematics in real situations.

Objective F: *Use the Multiplication Counting Principle and permutations to solve counting problems.* *(Lessons 10-2, 10-3)*

20. **a.** How many 6-letter strings can be created from the 26 letters of the English alphabet, if a letter can be used more than once?

b. How many 6-letter strings are possible if the first letter must be *d* and the last letter must be *t*, with no repetition allowed? 255,024

21. At a ceremony, seven people are to be seated in a row on the podium. How many different seatings are possible? 5040

22. A combination lock on a briefcase has three dials with the digits from 0 through 9 on each dial. How many different possible "combinations" are there for this lock? 1000

In 23–25, 15 old books and 4 old bookmarks are to be displayed in a row at an art exhibit.

23. How many different displays are possible?

24. How many displays are possible if the 4 bookmarks must be displayed together?

25. How many displays are possible if only 7 of the books and 2 of the bookmarks are to be displayed?

26. How many integers from 10,000 through 99,999 have at least one digit that is odd?

27. Consider the computer program below.

```
10 FOR I = 3 TO 5
15    FOR J = 6 TO 10
20       FOR K = 4 TO 11
25          B = I*J*K
30          PRINT B
35       NEXT K
40    NEXT J
50 NEXT I
60 END
```

How many numbers does line 30 print? 120

28. Consider the computer program below.

```
10 FOR I = 1 TO 4
20    FOR J = 2 TO 7
30       PRINT I, J
40    NEXT J
50    FOR K = 1 TO 3
60       PRINT I, K
70    NEXT K
80 NEXT I
90 END
```

How many lines of data are printed? 36

Objective G: *Use combinations and the Binomial Theorem to solve counting problems.* *(Lessons 10-4, 10-6, 10-7)*

29. An ice-cream shop has 20 flavors.

a. How many triple cones (3 dips) are possible if no flavor can be repeated? (Do not count different arrangements of the same flavors as different.) 1140

b. How many triple cones are possible if flavors can be repeated? 8000

30. Solve the problem in Question 4. 47,775

31. For a literature class, a teacher must choose four novels from a list of ten. In how many ways can this choice be made? 210

32. A dance company has 50 dancers. In how many ways can the company choose 23 of these dancers for a production?

EVALUATION

Three forms of a Chapter Test are provided in the Teacher's Resource File. Chapter 10 Test, Forms A and B cover just Chapter 10. The third test is Chapter 10 Test, Cumulative Form. About 50% of this test covers Chapter 10, 25% covers Chapter 9, and 25% covers previous chapters. A fourth test, Comprehensive Test, Chapters 1–10, that is primarily multiple choice in format, is also provided. For information on grading, see *General Teaching Suggestions: Grading* on page T43 in the Teacher's Edition.

ASSIGNMENT RECOMMENDATION

We strongly recommend that you assign the reading and some questions from Lesson 11-1 for homework the evening of the test. It gives students work to do if they complete the test before the end of the period and keeps the class moving.

17. For each of the $C(n, r)$ combinations of n objects taken r at a time, there are $r!$ arrangements. So $C(n, r) \cdot r! = P(n, r)$, or $P(n, r) = r!\, C(n, r)$.

18. Add the coefficients; $\sum_{k=0}^{10} \binom{10}{k} = 2^{10}$.

20.a. 308,915,780

23. $\approx 1.2165 \times 10^{17}$

24. $\approx 5.0215 \times 10^{14}$

25. $\approx 1.40 \times 10^{10}$

26. 87,500

32. ${}_{50}C_{23} \approx 1.0804 \times 10^{14}$

ADDITIONAL ANSWERS
34.a. C(2000, 50)
b. C(620, 12) · C(580, 12) · C(450, 13) · C(350, 13)

38.

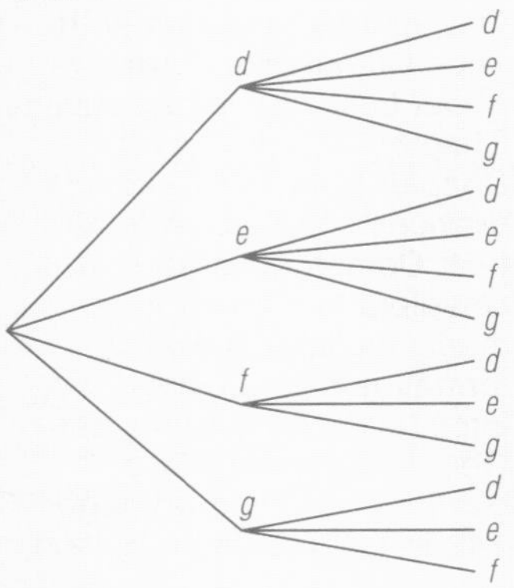

39.

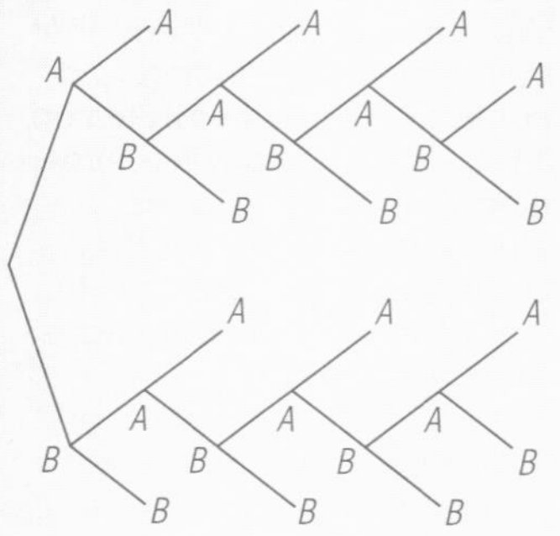

40.

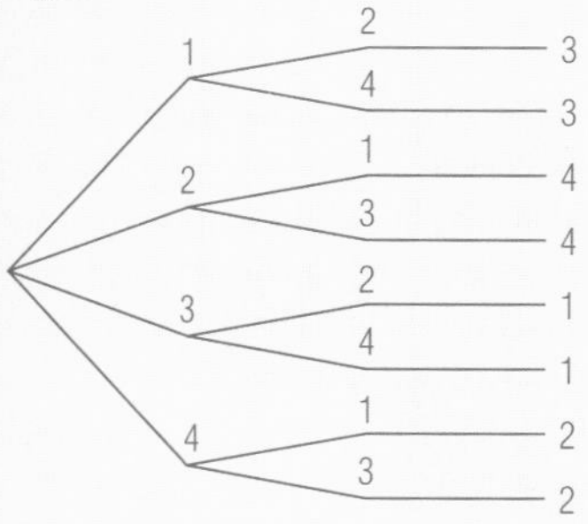

33. In how many ways is it possible to obtain at least two heads in a sequence of five coin tosses? **26**

34. Student council members are to be selected from the general student body. Suppose a high school has 2000 students: 620 freshmen, 580 sophomores, 450 juniors, and 350 seniors. Leave your answers to the following questions in $C(n, r)$ form.

a. How many 50-member councils can be formed from the whole student body?

b. How many 50-member councils can be formed if there must be 12 freshmen, 12 sophomores, 13 juniors, and 13 seniors in the council?

35. A Mix-n-Match selection of 10 cookies is to be made from three varieties: oatmeal, peanut butter, and sugar. How many different selections are possible? **66**

Objective H: *Find binomial probabilities in realistic situations.* *(Lesson 10-6)*

36. A jar contains a very large number of purple jelly beans and green jelly beans. There are three times as many purple ones as green ones. If an individual chooses 5 beans from the jar, estimate the probability that 3 of the beans will be green? **≈ 0.0879**

37. Suppose the probability is 0.2% that a pencil sharpener produced at a factory is defective.

a. If the quality control department randomly selects 100 sharpeners for testing, find the probability that exactly two of those sharpeners are defective. **≈ 0.0163**

b. Find the probability that at least one sharpener is defective. **≈ 0.181**

c. If the quality control department selects 500 sharpeners, find the probability that at least one of those sharpeners is defective. **≈ 0.6325**

REPRESENTATIONS deal with pictures, graphs, or objects that illustrate concepts.

Objective I: *Use a possibility tree to determine the number of outcomes in a given situation.* *(Lesson 10-2)*

38. Make a possibility tree to determine the number of two-letter strings that can be created from the letters *d*, *e*, *f*, and *g* if *d* and *e* can be repeated but *f* and *g* cannot. **14 strings**

39. In a tournament, two teams play each other until one team wins two in a row or a total of four games. Use a possibility tree to count the number of different possible outcomes of this tournament. **14 outcomes**

40. Three-digit numbers are to be made from the digits 1, 2, 3, and 4, with no repetition allowed. Use a possibility tree to count the number of three-digit numbers that can be made if digits must alternate even-odd-even or odd-even-odd. **8 numbers**

638

CHAPTER 11 GRAPHS AND CIRCUITS

OBJECTIVES

The objectives listed here are the same as in the Chapter 11 Review on pages 694–698 of the student text. The Progress Self-Test on pages 692–693 and the tests in the Teacher's Resource File cover these objectives. For recommendations regarding the handling of this end-of-chapter material, see the notes in the margin on the corresponding pages of this Teacher's Edition.

OBJECTIVES FOR CHAPTER 11 (Organized into the SPUR Categories—Skills, Properties, Uses, and Representations)	Progress Self-Test Questions	Chapter Review Questions	Teacher's Resource File: Lesson Masters*	Teacher's Resource File: Chapter Test Forms A & B
SKILLS				
A Draw graphs given sufficient information.	9	1–5	11-2, 11-3	4
PROPERTIES				
B Identify parts of graphs and types of graphs.	1, 2, 3, 4, 5, 6, 7, 11	6–11	11-2, 11-3, 11-4, 11-5	3, 10
C Determine whether there exists a graph containing vertices with given degrees.	14	12–16	11-3	8
D Determine whether a graph has an Euler circuit.	8	17–20	11-4	11, 12
USES				
E Use graphs to solve scheduling and probability problems.	15	21–23	11-1	13
F Use the Total Degree of a Graph Theorem and its corollaries to solve handshake problems.	16	24–26	11-3	5
G Solve application problems involving circuits.	13	27–28	11-1, 11-4	1
H Use stochastic matrices to make long-term predictions.	17	29–30	11-6	9
REPRESENTATIONS				
I Convert between the picture of a graph or directed graph, and its adjacency matrix.	10	31–35	11-2	2, 6
J Use the powers of the adjacency matrix of a graph to find the number of walks of a given length from a given starting vertex to a given ending vertex.	12	36–39	11-5	7

*** The masters are numbered to match the lessons.**

OVERVIEW ■ CHAPTER 11

This chapter is an introduction to the study of graphs (also called networks) and circuits, which consist of points and the segments or arcs that connect them. This branch of mathematics is known as *graph theory*. The chapter discusses graphs and circuits from three viewpoints: (1) as objects of study in themselves, (2) as tools for interpreting applications and solving problems, and (3) as links among mathematical ideas. Most discrete mathematics courses at the college level include the study of graphs, circuits, probability, and matrices. Students who complete this chapter will have covered the prerequisites for further work in discrete mathematics.

A generation ago, only a few colleges taught courses in graph theory. In recent years, however, the study of graphs has become a part of discrete mathematics. One reason is that a diverse collection of problems (including many practical problems) can be modeled by graphs. Another reason is that computer programs contain branches and loops (which are features of graphs), and their logical structures are most often described by graphs. A third reason is that the mathematics is very interesting and is still being discovered.

The chapter opener and Lesson 11-1 discuss the use of graphs to solve problems. The examples include scheduling problems, puzzles, and conditional probability, and introduce some of the terminology associated with graphs.

Lesson 11-2 provides a formal definition of graph and relates certain graphs to matrices. Lessons 11-3 and 11-4 build on those definitions and prove some simple theorems.

Lessons 11-5 and 11-6 continue the relationship between matrices and graphs. Powers of matrices give information about graphs that would be hard to acquire from a direct examination of a graph. Markov chains use ideas of large powers of probability matrices to show that the limiting effects of certain paths in a graph are predictable and stable.

Lesson 11-7 shows how theorems from the previous lessons can be employed to prove the famous relationship between the numbers of vertices V, edges E, and faces F of any polyhedron, $V - E + F = 2$.

PERSPECTIVES ■ CHAPTER 11

The Perspectives provide the rationale for the inclusion of topics or approaches, provide mathematical background, and make connections with other lessons and within UCSMP.

11-1

MODELING WITH GRAPHS

This is an informal lesson that introduces students to some of the kinds of problems they will encounter in the chapter and gives them practice in drawing graphs.

The first examples are problems that involve traversing a graph. The Königsberg bridge problem is a problem of traversing all the *edges* of a graph exactly once. A practical version of this problem is a snowplow driver having to traverse the streets of a city or town and wanting to do so efficiently. A problem of William Rowan Hamilton is to traverse all the *vertices* of a dodecahedron exactly once. A practical version of this problem is a mailman who has mailboxes at intersections of streets and wishes to be at each intersection exactly once for efficiency.

Although the snowplow and mailman problems might seem to be equivalent, they are not. Euler's solution to the Königsberg bridge problem is given in Lesson 11-4; there is no known general solution to Hamilton's type of problem.

A third use of graphs for solving problems that deal with efficiency is the scheduling of tasks to finish them in the shortest amount of time. Graphs used in scheduling problems have edges that only can be traversed in a particular direction; they are called *digraphs*. An algorithm for a solution to the scheduling problem is known and is given in this lesson.

A digraph, in which the vertices are labeled, is presented in the final example of graphs in this lesson and is called a probability tree. This kind of graph clarifies the computation of conditional probabilities.

11-2

THE DEFINITION OF GRAPH

A graph is defined formally by indicating its edges and the vertices for each edge. There are natural meanings for adjacent edges, adjacent vertices, isolated vertices, and loops. Parallel edges are two edges with the same endpoints, and a simple graph is one with no parallel edges and no loops.

Unlabeled graphs can be described by an adjacency matrix in which a_{ij} is the number of edges connecting vertex v_i to vertex v_j. This matrix description is applied in Lessons 11-5 and 11-6.

Students are expected to be able to convert between a geometric description of a graph (drawing), an algebraic description (definition), or an arithmetic description (matrix). We assume students have studied matrices in a previous course.

11-3

HANDSHAKE PROBLEMS

An edge of a graph can be considered a "handshake" between its vertices. The number of handshakes possible from a vertex is the number of edges, called the degree of the vertex. This lesson develops some theorems about the degree of a graph in order to apply them in the next lesson. Along the way, several "handshake problems" are solved.

11-4

THE KÖNIGSBERG BRIDGE PROBLEM

This lesson begins by introducing the terminology necessary to discuss traversing a graph. A *walk* is just a list of the vertices and edges traversed in order. A *path* is a walk that does not repeat edges, and a *circuit* is a path that starts and ends at the same vertex. The Königsberg bridge problem involves an Euler circuit, a circuit in which no edge is repeated.

The key to the solution of the Könisberg bridge problem is the theorem that every vertex of a graph with an Euler circuit has even degree. A sufficient condition for an Euler circuit also is given. This enables students to determine exactly which graphs have Euler circuits and solves the traversing problem for edges.

11-5

MATRIX POWERS AND WALKS

This lesson uses the matrix description of a graph to determine the number of walks from one vertex to another. The length of a walk is the number of edges in it. The method for finding how many walks there are of a given length is elegant: If A is the adjacency matrix for a graph, then the elements a_{ij} of A^n indicate the number of walks of length n from a_i to a_j.

11-6

MARKOV CHAINS

The adjacency matrices of Lesson 11-2 become probability matrices in this lesson. Instead of the a_{ij} element of the matrix representing the number of paths from vertex i to vertex j, the element a_{ij} is the probability of moving from vertex i to vertex j. A Markov chain is a situation that can be described by these transition probabilities.

In a matrix for a Markov chain, the elements of any row are nonnegative and add to 1, that is, the matrix is a stochastic matrix. If A is a stochastic matrix, so is A^n. Furthermore, the elements of A^n represent the probability of moving from vertex i to vertex j in exactly n transitions. So, for example, if the probabilities of particular changes in weather from one day to the next are considered constant, the probability of a particular kind of weather some number of days from now can be calculated.

11-7

EULER'S FORMULA

Euler is famous for many contributions to mathematics, and his name is associated with a number of formulas and theorems. The formula referred to in this lesson is $V - E + F = 2$ for the vertices, edges, and faces of any polyhedron. Euler's formula is explored for figures other than polyhedra, and it is proved for polyhedra using theorems from earlier lessons.

DAILY PACING CHART ■ CHAPTER 11

Every chapter of UCSMP *Precalculus and Discrete Mathematics* includes lessons, a Progress Self-Test, and a Chapter Review. For optimal student performance, the self-test and review should be covered. (See *General Teaching Suggestions: Mastery* on page T35 of this Teacher's Edition.) By following the pace of the Full Course given here, students can complete the entire text by the end of the year. Students following the pace of the Minimal Course spend more time when there are quizzes and on the Chapter Review and will generally not complete all of the chapters in this text.

When chapters are covered in full (the recommendation of the authors), then students in the Minimal Course can cover 11 chapters of the book. For more information on pacing, see *General Teaching Suggestions: Pace* on page T34 of this Teacher's Edition.

DAY	MINIMAL COURSE	FULL COURSE
1	11-1	11-1
2	11-2	11-2
3	11-3	11-3
4	11-4	11-4
5	Quiz (TRF); Start 11-5.	Quiz (TRF); 11-5
6	Finish 11-5.	11-6
7	11-6	11-7
8	11-7	Progress Self-Test
9	Progress Self-Test	Chapter Review
10	Chapter Review	Chapter Test (TRF)
11	Chapter Review	
12	Chapter Test (TRF)	

TESTING OPTIONS

TESTING OPTIONS

- Quiz for Lessons 11-1 through 11-4
- Chapter 11 Test, Form A
- Chapter 11 Test, Form B
- Chapter 11 Test, Cumulative Form

A Quiz and Test Writer is available for generating additional questions, additional quizzes, or additional forms of the Chapter Test.

INDIVIDUAL DIFFERENCES

PROVIDING FOR INDIVIDUAL DIFFERENCES

The student text is written for, and tested with, average students. It also has been successfully used with better and more poorly prepared students.

The Lesson Notes often include Error Analysis and Alternate Approach features to help you with those students who need more help. Students of all abilities often learn from their peers and may benefit from small group work referenced as appropriate throughout the Notes. A blackline Lesson Master (in the Teacher's Resource File), keyed to the chapter objectives, is provided for each lesson to allow more practice. (However, since it is important to keep up with the daily pace, you are not expected to use all of these masters. Again, refer to the suggestions for pacing on page T34.) Extension activities are provided in the Lesson Notes for those students who have completed the particular lesson in a shorter amount of time than is expected, even in the Full Course.

Graphs and Circuits

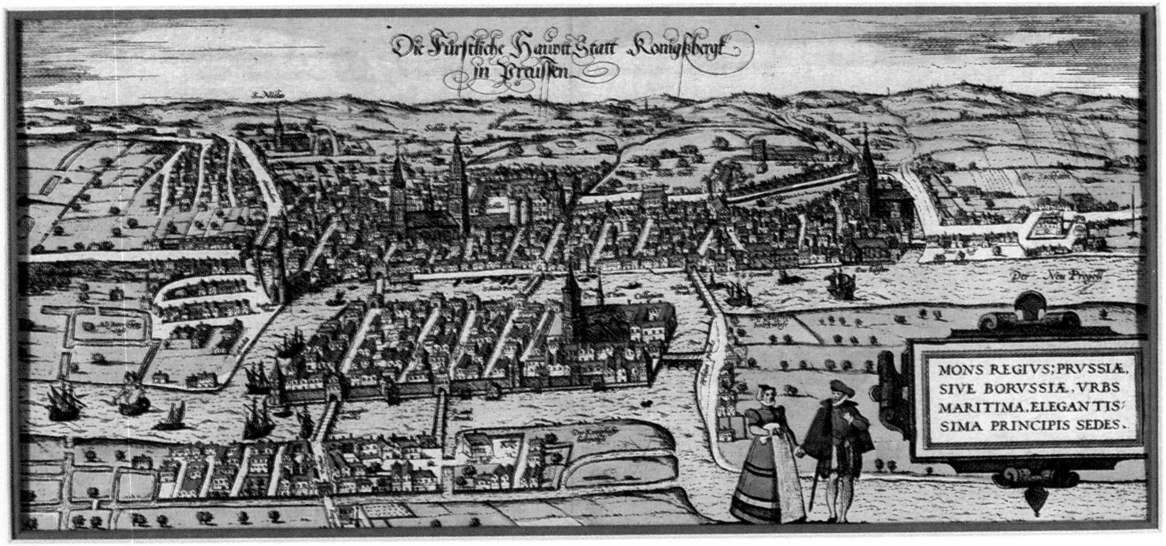

This historic etching, picturing Königsberg, predates Euler. Note that only six bridges existed and were shown; the seventh was built later.

The city of Kaliningrad in the Soviet Union is situated where two branches of the Pregol'a River come together. In 1736, this city was called Königsberg and was a part of East Prussia. At that time, parts of Königsberg were on the banks of the river, another part was on a large island in the middle, and a final part was between the two branches of the river. Seven bridges connected these four parts of the city. An unsolved problem of the time was:

> Is it possible for a person to walk around the city crossing each bridge exactly once, starting and ending at the same point?

This problem is now known as the **Königsberg bridge problem**. It became famous because it was the subject of a research paper written in that year by the great mathematician Leonhard Euler. Take a minute or two to see if you can find such a walk.

To solve the Königsberg bridge problem, Euler constructed a simple and helpful geometric model of the situation called a *graph*, and his paper is usually acknowledged to be the origin of the subject called **graph theory**. (As you will see, these graphs are not the same as the graphs of functions or relations.) In the two-and-a-half centuries since his solution of that problem, graphs have been used to solve a wide variety of problems. In this chapter, you will be introduced to a selection of those problems.

CHAPTER 11

This chapter should be done at a lesson-per-day pace. With a possible extra day for the quiz, 1 day for the Progress Self-Test, 1 or 2 days for the Chapter Review, and 1 day for the Chapter Test, allow 10–12 days for the chapter. You may need an extra day or two if your students have never studied matrices.

USING PAGE 639

If your students have studied from UCSMP *Geometry*, they will be familiar with the Königsberg bridge problem. Ask students if they have seen this problem before. If they have not, have them copy the picture onto a sheet of paper and try to find a path for the person. Although students may become convinced that no path is possible, few students can formulate an argument to prove that no path is possible.

LESSON 11-1

RESOURCES
- Lesson Master 11-1
- Teaching Aid 72 displays the graphs of Euler and Hamilton.
- Teaching Aid 73 can be used with **Example 2**.
- Teaching Aid 74 can be used with **Example 3** and **Question 14a**.

OBJECTIVES

E Use graphs to solve scheduling and probability problems.
G Solve application problems involving circuits.

TEACHING NOTES

In almost every lesson of this chapter, important terminology is introduced. While the number of terms per lesson is not large, the total is, so students need to learn the vocabulary in each lesson as they study it. In this lesson, the terms *vertex*, *edge*, *equivalent graphs*, and *digraphs* are introduced.

The idea most likely to cause difficulties for students is the distinction between the problems of Euler and Hamilton. Both are problems of traversing, but in Euler's problem, students should think of bridges as edges, and in Hamilton's problem, they should think of mailboxes as vertices.

The scheduling problem of **Example 2** is a fundamental problem in any business in which time is critical. At a muffler shop that promises quick service, for example,

LESSON

11-1

Modeling with Graphs

To solve the Königsberg bridge problem of the preceding page, Euler observed that, for this problem, each land mass could be represented by a point since it is possible to walk from any part of a land mass to any other part without crossing a bridge. The bridges could be thought of as arcs joining pairs of points. Thus the situation of Königsberg could be represented by the following geometric model consisting of four points and seven arcs.

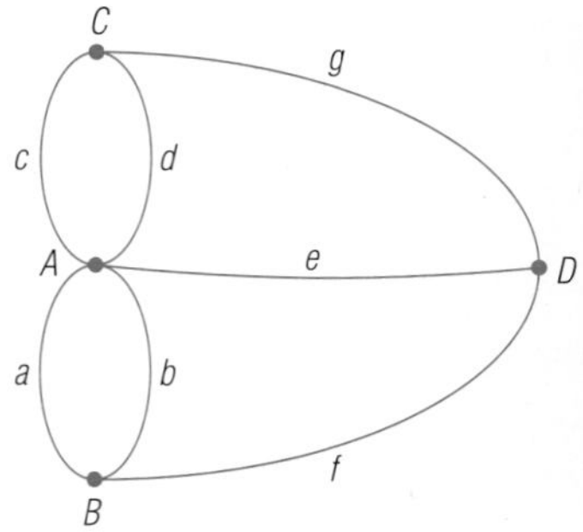

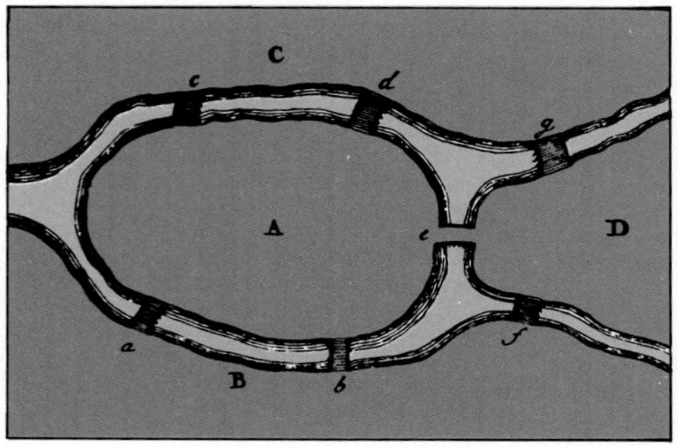

This type of geometric model is called a graph. The four points are the *vertices* and the seven arcs are the *edges* of the graph. In terms of this graph, the Königsberg bridge problem can be stated:

> Is it possible to trace this graph with a pencil, traveling each edge exactly once, starting and ending at the same vertex, without picking up the pencil?

Euler's solution to the Königsberg bridge problem is given in Lesson 11-4.

Another famous puzzle for which a graph is a helpful model was invented in 1859 by the Irish mathematician Sir William Rowan Hamilton (1805-1865). The puzzle consisted of a wooden block in the shape of a regular dodecahedron, as shown here.

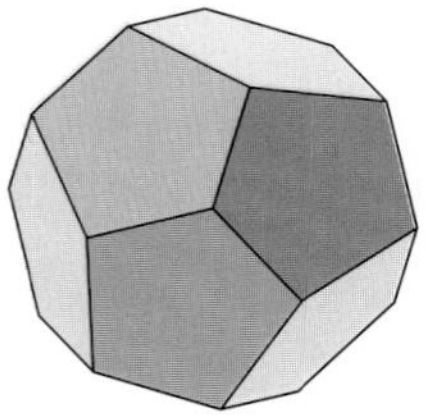

This polyhedron has 12 regular pentagons as its faces, 30 edges, and 20 vertices. Hamilton marked each vertex of the block with the name of a city, and the object of the puzzle was to find a travel route along the edges of the block that would visit each of the cities once and only once. A small pin protruded from each vertex so that the player could mark a route by wrapping string around each pin in order as its city was visited.

By thinking of the polyhedron as transparent, as shown below at the left, you can count to determine that Hamilton's problem involves a graph with 20 vertices and 30 edges in which 3 edges meet at each vertex. The graph below at the right shows the same relationships of vertices and edges in a 2-dimensional diagram; the two graphs are **equivalent**.

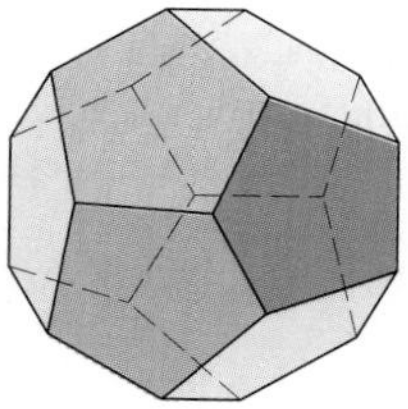

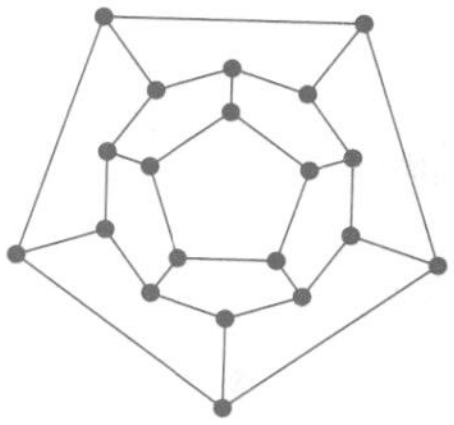

Hamilton's puzzle can be stated in terms of either graph as the following problem:

> Is it possible to trace this graph with a pencil, traveling through each *vertex* exactly once, without picking up the pencil?

In fact, Hamilton also sold the 2-dimensional version of his puzzle since it was easier to work with. Notice the similarity between the problems of Euler and Hamilton; in both the goal is to traverse all objects of one kind in the graph exactly once; in Euler's problem the objects are the edges; in Hamilton's the objects are the vertices.

Several practical situations give rise to problems like Hamilton's.

Example 1 One mailbox is located at each intersection in a city and a postal truck is required to collect the mail from all the boxes in the 42-block region in the map at the right.

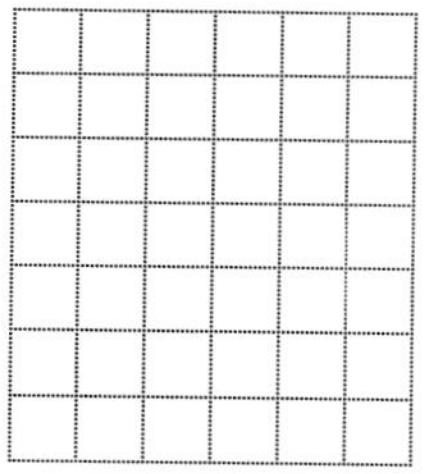

Can the driver plan a pick-up route that begins and ends at the same place and allows collection of the mail from all of the boxes without passing one that has already been collected?

Solution The problem can be modeled by a graph, with the mailboxes being the vertices and the streets being the edges. Experimentation gives several suitable routes, such as the one pictured here.

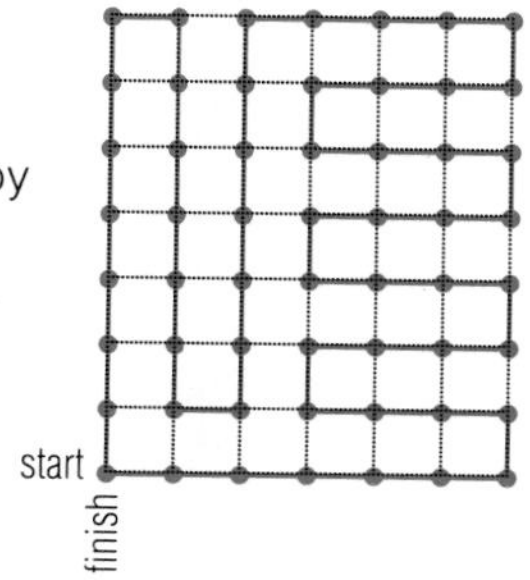

All of the examples you have seen so far result from situations that themselves are geometric. The next example is quite different; it shows the use of graphs in scheduling a complex task. First, some background information is necessary.

speed is based on an efficiency of schedule. Another example is an airplane that has landed and is being prepared for another flight. It needs to be serviced, its luggage and passengers unloaded and loaded, and its cabin cleaned in the shortest amount of time possible.

Making Connections Example 3 connects the mathematics of probability with graphs. In constructing the graph, students apply the formula for conditional probability, $P(A \cap B) = P(A) \cdot P(B \text{ given } A)$, when they complete the probabilities of the "leaves" of the tree. For example, $P(\text{Campus and Favor}) = P(\text{Campus} \cdot P(\text{Favor given Campus}) = .6 \cdot .7 = .42$.

The work with conditional probability in this lesson prepares students for Markov chains in Lesson 11-6.

ADDITIONAL EXAMPLES

1. A mailbox is located at each intersection in the downtown section of a city and a postal truck is required to collect the mail from all 45 boxes. Is there a route that begins and ends at the same point without passing a box from which the mail has already been collected?

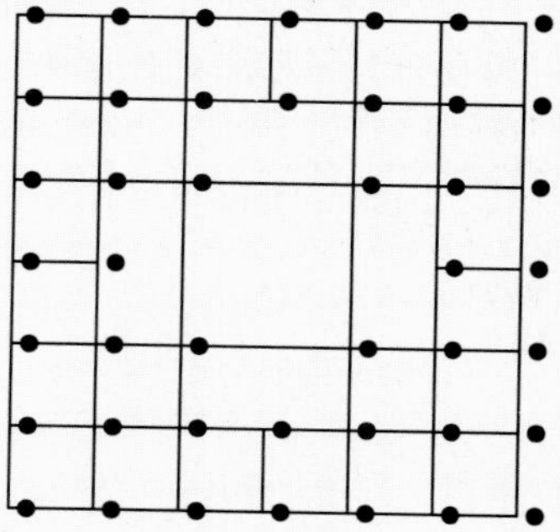

We have been unable to find such a route. This problem underscores the difficulty of such questions.

ADDITIONAL EXAMPLES

2. A scientific experiment involves the following tasks:

	Task	Time (month)
A	Choosing a project	1
B	Organizing a team	1
C	Obtaining funding	2
D	Setting up the equipment	0.5
E	Running procedures	6
F	Keeping records	5.5
G	Writing a report	1

The graph below shows the tasks, the time it takes to complete each task and the relationships among the tasks.

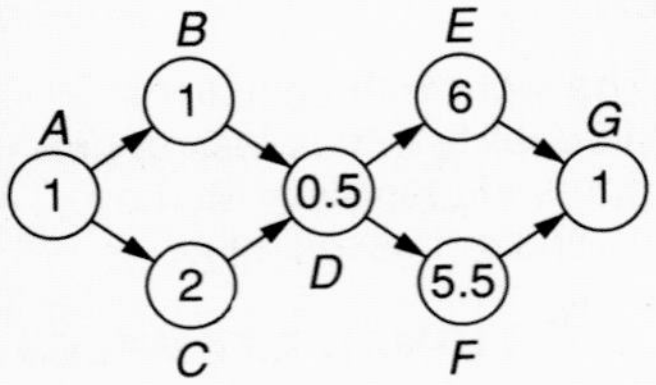

Use the graph above to determine the least number of months to complete the experiment.

The diagram below shows the total number of months to complete each part of the experiment. For instance, tasks *A*, *B*, and *D* take 1 + 1 + 0.5 = 2.5 months. Tasks *A*, *C*, and *D* take 1 + 2 + 0.5 months. So task *D* cannot be completed before 3.5 months. Thus, 3.5 is written below the circle for *D*.

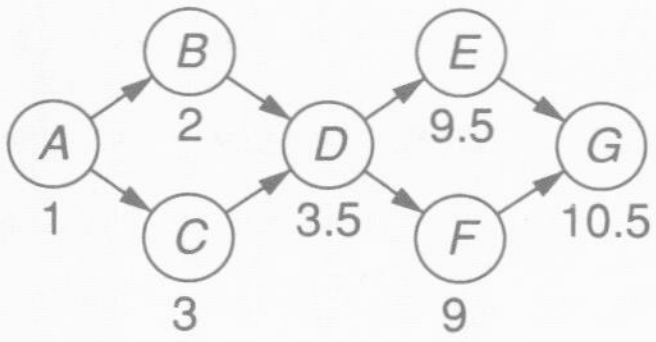

The total number of months needed to complete the experiment is 10.5.

Building a house is usually a team effort that involves specialists such as architects, excavators, concrete workers, framing carpenters, roofers, heating and air conditioning workers, electricians, plumbers, drywallers, finish carpenters, painters, and landscapers. Different specialists are often able to work at the same time provided that the work that must precede a particular specialist is completed before that specialist begins. By working simultaneously whenever possible, the house can be completed more quickly, and it is natural to wonder if there is some optimal way to schedule the various tasks for completion.

This can be done with a graph, but first the information to be graphed must be assembled. Here is a list of some of the tasks involved in building a house, along with the time they require and the tasks which must be completed prior to their beginning. (The list is simplified but the ideas are not.)

	Task	Time (days)	Immediately Preceding tasks
A	Preparing final house and site plans	3	none
B	Excavation and foundation construction	5	*A*
C	Framing and closing main structure	12	*B*
D	Plumbing	5	*C*
E	Wiring	3	*C*
F	Heating-cooling installation	7	*E*
G	Insulation and dry wall	9	*D, F*
H	Exterior siding, trim, and painting	15	*C*
I	Interior finishing and painting	7	*G*
J	Carpeting	3	*I*
K	Landscaping	4	*H*

It would take 73 working days to finish the house if only one task were done on any day. However, the following graph can help the builder decide which tasks can be done simultaneously in order to complete the job more quickly. The tasks are represented by vertices, drawn here as circles. The number of days needed for each job is indicated inside its circle. The arrows represent edges; when an arrow is drawn such as from vertex *A* to vertex *B*, it means that task *A* is a task immediately preceding *B* that must be completed before task *B* can begin.

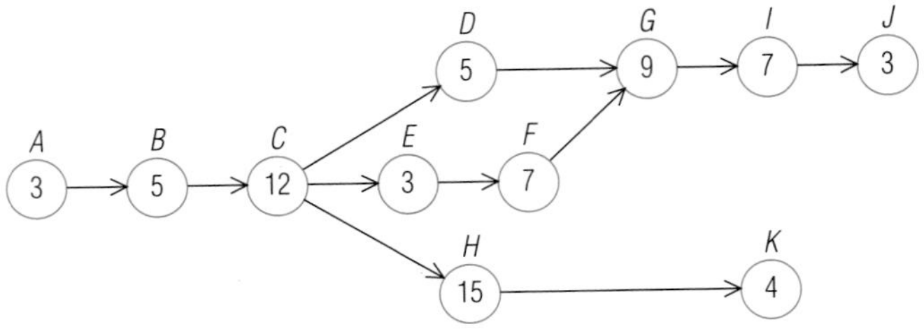

Example 2 Use the above graph to determine the least number of days to complete the house.

Solution From left to right along the graph, calculate for each task the least number of days in which it can be completed since the beginning of construction. Write this number beneath the circle representing the task. For instance, task *D* requires the completion of tasks *A*, *B*, and *C*, which takes 20 days. Since *D* requires 5 more days, write 25 beneath *D*. Through task *F* similarly requires 30 days. Since both *D* and *F* must be done before *G*, *G* cannot be started before 30 days, so will not be completed before 39 days.

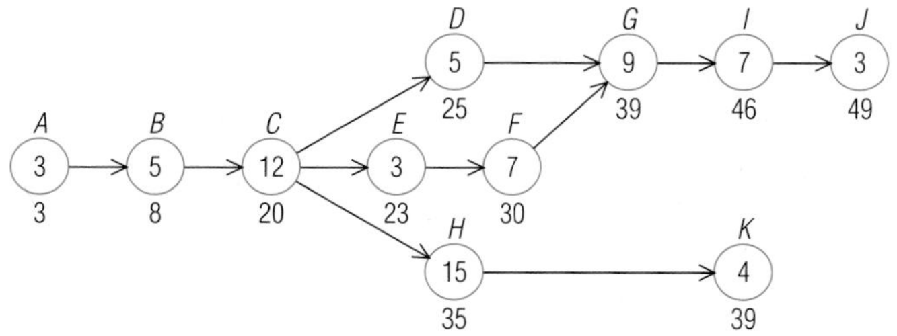

All the tasks will be completed when both *J* and *K* are done. Since *J* requires at least 49 days and *K* requires at least 39 days for its completion, the house can be completed in 49 days.

Notice that the algorithm used in Example 2 is recursive. Here is the general algorithm to calculate the number of days for a particular task:

If there are no prerequisite tasks, use the number of days required by this task alone.

Otherwise:
(1) Calculate the number of days for each prerequisite task by using this algorithm.
(2) Choose the largest of the numbers found in step (1), and add to it the number of days required by this task alone.

This algorithm is used to determine efficient job schedules for much more complex projects, and there exists computer software that will automatically create the graph and find the solution after the user inputs information like that given in Example 2.

The graphs preceding and in Example 2 are different from the previous graphs in this lesson in that you can travel along each edge in only one direction. Such graphs are called **directed graphs** or **digraphs**. The graph in the solution differs a second way: each vertex is labeled with a number. A different type of labeled digraph, called a *probability tree*, is useful for solving certain problems involving probabilities. In a **probability tree**, a vertex represents an event, and the edge leading from vertex *A* to vertex *B* is labeled with the probability that event *B* occurs if *A* occurs. The last vertex of each branch is called a *leaf* of the tree.

3. Suppose 75% of a class of students own automatic graphers. Suppose 80% of those who own graphers think they should be allowed on tests, and 90% of those who do not own graphers think they should not be allowed on tests.

a. Draw a graph of this situation.

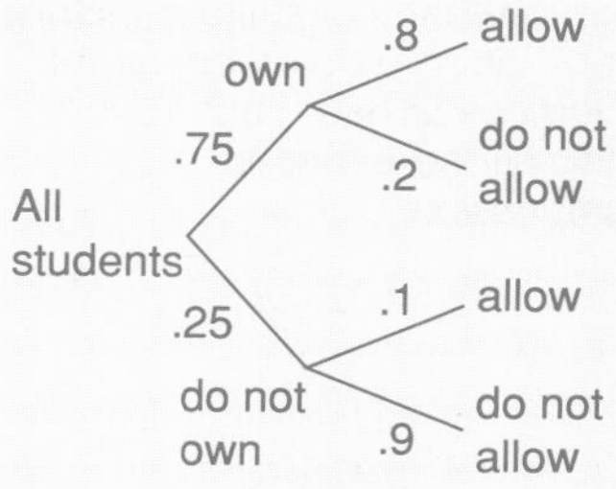

b. What is the probability that a student from the class favors the use of graphers on tests?
.625

c. If a randomly selected student opposes the use of graphers, what is the probability of that student not owning an automatic grapher?
.6

NOTES ON QUESTIONS
For the Covering the Reading questions in this lesson, students may have to go to the chapter opener page to find an answer.

Question 3: The differences between the problems of Hamilton and Euler can be seen easily by considering a graph that looks like a picture of a cube. Traversing all the vertices is easy, but traversing all the edges is impossible.

ADDITIONAL ANSWERS
1.a.

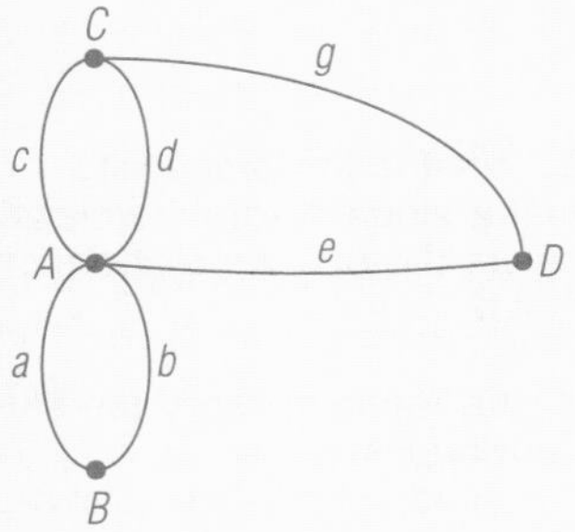

2. sample

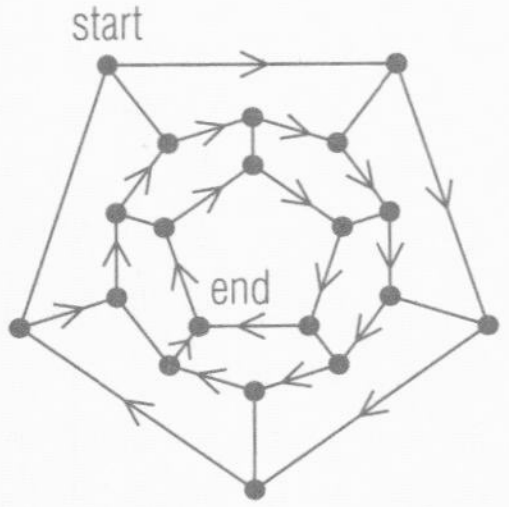

4. sample:

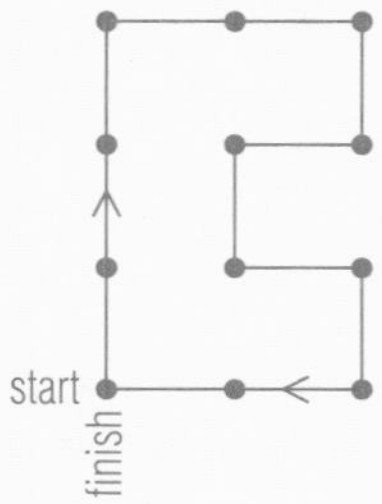

Example 3 60% of the students in a college live on campus. 70% of those who live on campus favor a tuition increase to pay for improved student health services. 40% of those who live off campus favor this increase. Draw a graph to determine the following.
a. What is the proportion of students who favor this increase to pay for improved student health services?
b. If a randomly selected student favors the tuition increase, what is the probability that the student lives on campus?

Solution
a. Draw a graph showing the division of students by residence: Campus or Off-Campus. Since 60% of the students live on campus, 40% must live off campus. Then break down each group by their opinions on the increase. The proportion of students who are both on campus and favor the increase is $.6 \cdot .7 = .42$. The proportion of students who are on campus and don't favor the increase is $.6 \cdot .3 = .18$. Similarly, you can compute the proportion of students in the two branches leading from the Off-Campus vertex.

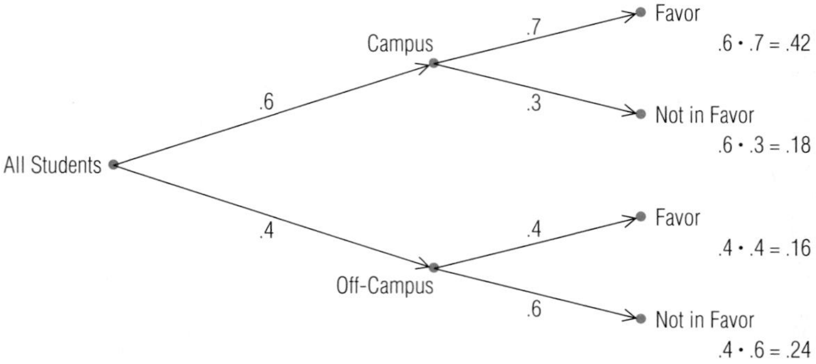

Notice that the proportions in the right column add to 1.00; all students are represented by the leaves of the tree. The proportion of students who favor the tuition increase is .42 + .16 or 58% of the student body.
b. 58% of the student body are in favor of the increase, while 42% are in favor *and* live on campus. Therefore, given that a student favors the tuition increase, there is a $\frac{.42}{.58} \approx 72\%$ probability that the student lives on campus.

Questions

Covering the Reading

1. The seventh bridge of Königsberg to be built connected the land masses B and D in the drawing on page 639.
a. Draw a graph of the Königsberg bridge problem when the city had only 6 bridges. See margin.
b. Show that there is a path around the city crossing each bridge exactly once by listing the land masses and bridges, in order. Does your path begin and end at the same point?
sample: *A*, *a*, *B*, *b*, *A*, *e*, *D*, *g*, *C*, *c*, *A*, *d*, *C*; no

2. Copy the graph for Hamilton's dodecahedron. Show that there is a solution to Hamilton's problem. See margin.

3. In Euler's problem, each __?__ is traversed exactly once, and in Hamilton's problem, each __?__ is traversed exactly once.
edge, vertex

In 4 and 5, copy the graph. Think of the vertices of the graph as mailboxes and the edges as streets, and draw a path that will allow a driver to collect the mail from each box without passing one whose mail has already been collected, starting and finishing at the same point.

4. 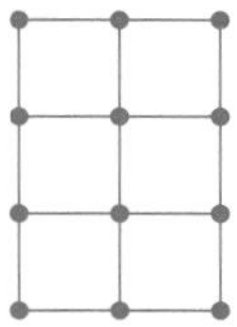

5. 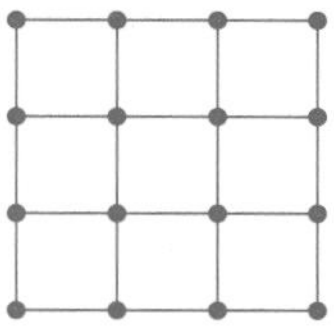

4,5) See margin.

6. In the situation of Example 2, a new task L, the moving in of appliances, taking one day, and requiring the completion of task I, is added to the schedule.
   a. Draw a new graph including task L. See margin.
   b. How does this affect the time to finish the house? does not affect it

In 7–9, refer to Example 3.

7. If there are 2000 students at the college, how many live off campus *and* favor a tuition increase for improved student health services?
320 students

8. What is the proportion of students who don't favor a tuition increase for improved student health services? .42

9. If a student selected at random doesn't favor a tuition increase for improved student health services, what is the probability that he or she lives on campus? ≈ .43

10. 80% of the customers of a store are female. The probability that a female wears contact lenses is 6%. The probability that a male wears contact lenses is 5%.
    a. Draw a probability tree and label its edges with the appropriate probabilities to represent this situation. See margin.
    b. What proportion of customers of this store are male and don't wear contact lenses? 19%
    c. What proportion wear contacts? 5.8%

Applying the Mathematics

11. A common children's puzzle is to traverse the edges of the drawing at the right exactly once. Is this puzzle more like Euler's problem or Hamilton's problem? Euler's

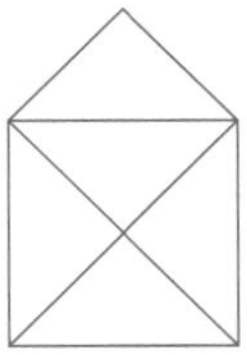

5. sample

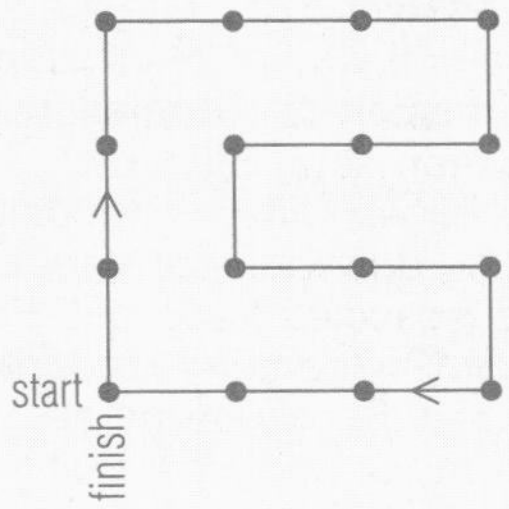

6.a. See Additional Answers in the back of this book.

10.a.

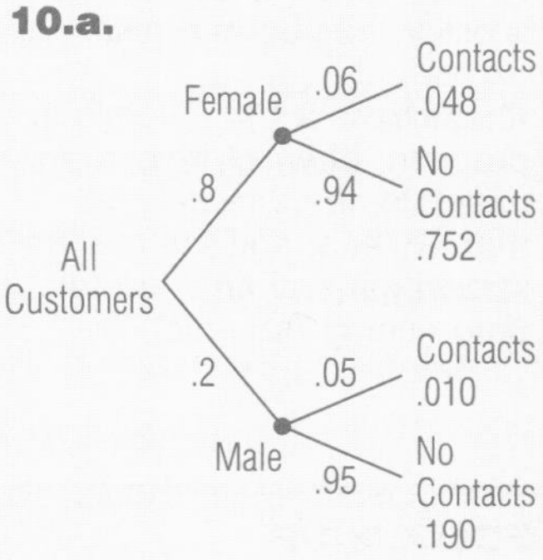

NOTES ON QUESTIONS
Making Connections for Question 13: Graphs are a natural way of organizing information about relations between sets. Students' graphs will differ depending on the definitions of trapezoid and isosceles trapezoid. For UCSMP *Geometry* students, trapezoids include parallelograms, and isosceles trapezoids include rectangles.

Question 14: This question is not difficult to do once a graph has been drawn.

Question 21b: Hamilton's problem, of which this is an example, is more difficult than Euler's, and there is no known general criterion for determining traversability.

FOLLOW-UP

MORE PRACTICE
For more questions on SPUR Objectives, use *Lesson Master 11-1*, shown on page 647.

PROJECTS
The projects for Chapter 11 are described on pages 689–690. **Projects 2, 3, and 4** are related to the content of this lesson.

ADDITIONAL ANSWERS
13. sample:

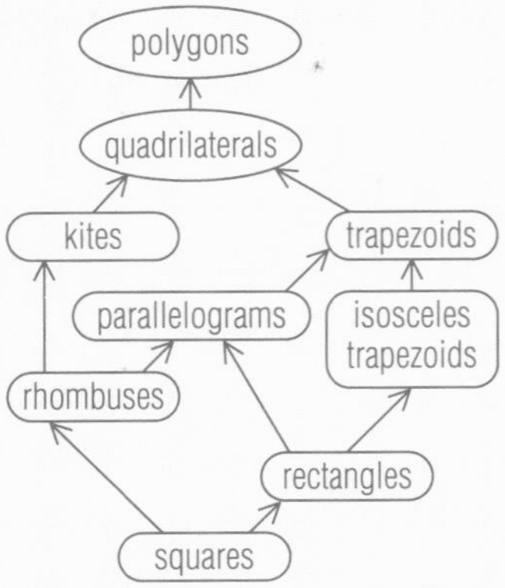

12. In Queenstown there is a river with two islands and bridges connecting the islands to the shore, as depicted at the right. Is there a path traversing each bridge exactly once? **Yes**

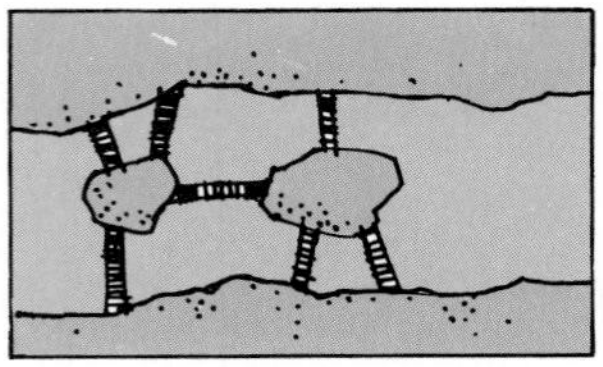

13. Draw a digraph whose vertices represent the following nine types of figures: isosceles trapezoids, kites, parallelograms, polygons, quadrilaterals, rectangles, rhombuses, squares, and trapezoids. Draw an arrow connecting vertex X to vertex Y if and only if all Xs are Ys, but do not connect X to Z if X is connected to Y, and Y is connected to Z.
See margin.

14. An oil company has used expensive test drilling to determine what areas of its land will be worth developing as oil fields. Only 10% of the test wells are gushers which indicate oil deposits worth developing. Test wells are extremely expensive to drill, so the company has been studying the use of less expensive seismic tests on a plot to predict whether drilling is worthwhile. Seismic tests indicate whether the underlying strata have a closed structure, open structure, or no structure. Over a period of time, the company performs seismic tests in conjunction with drilling. The table below shows the proportions of gushers and of nongushers that are associated with each structure. For example, 60% of the gushers came from closed structures.

	Closed	Open	No Structure
Gusher	.60	.30	.10
Not a gusher	.20	.30	.50

a. Find the probabilities v, w, x, y, and z in the following probability tree. **See page 647.**

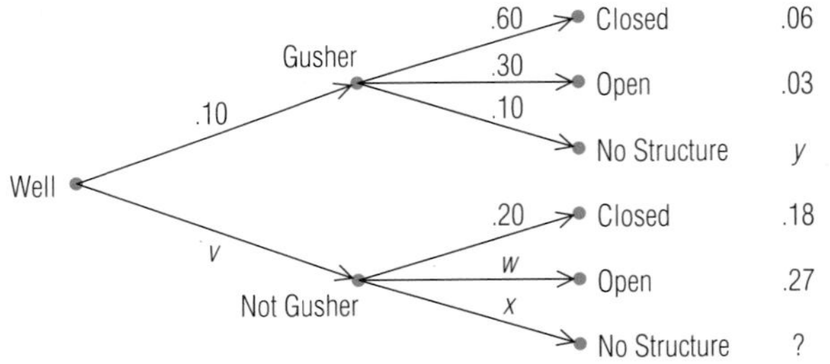

b. What is the probability that the underlying strata of a randomly selected well has a closed structure? **.24**

c. Suppose the company conducts a seismic test on one of its properties and finds a closed structure. According to the data above, what is the probability that drilling there would produce a gusher? **.25**

d. Suppose a seismic test indicates an open structure. What is the probability that drilling there would produce a gusher? **.10**

646

15. The possibility tree of Example 1 in Lesson 10-2 is a graph.
a. How many vertices does this graph have? 39
b. How many edges does this graph have? 38

Review

16. a. Sketch the graph of a function whose first derivative is negative on the interval $-5 < x < -2$ and positive on the interval $-2 < x < 3$.
b. Identify the location of a relative minimum or maximum. Which is it? *(Lesson 9-5)* **a) See margin. b) There is a relative minimum at x = -2.**

17. Consider the matrix at the right.

$$\begin{bmatrix} 5 & 1 & 0 & 3 \\ 2 & 0 & 8 & 0.5 \\ -3 & 4 & 6 & 10 \end{bmatrix}$$

a. What are its dimensions? 3×4
b. If a_{ij} is the element in the ith row and jth column, what is a_{23}? 8
c. Calculate $\sum_{i=1}^{3} a_{ii}$. *(Lesson 7-3, Previous course)* 11

18. Express in terms of tan x. *(Lesson 6-4)*
a. $\tan\left(x + \frac{\pi}{2}\right)$ -cot x **b.** $\tan(x + \pi)$ tan x **c.** $\tan\left(x + \frac{3\pi}{2}\right)$ -cot x

19. Suppose p is a polynomial function whose graph is shown at the right. What is the smallest possible value for the degree of p? *(Lesson 4-7)* 7

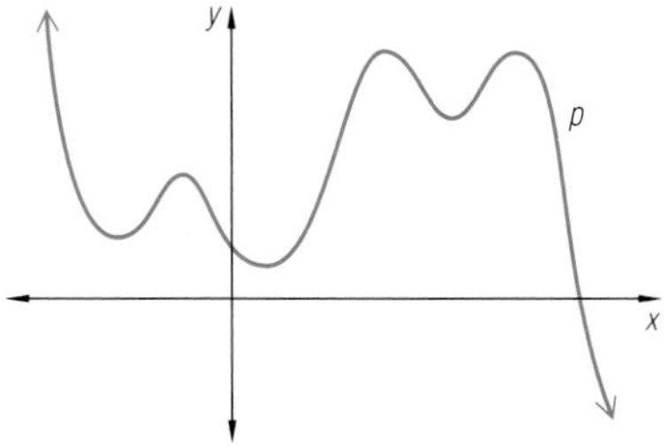

20. Solve $\sqrt{3y + 1} = 4$. Indicate which steps are reversible and which are not. *(Lesson 3-1)* **See below.**

Exploration

21. Example 1 and Questions 4 and 5 ask for the traversing of all the vertices of a rectangular array exactly once, going horizontally or vertically, and beginning and ending at the same point. In size, these arrays are 7 by 8, 3 by 4, and 4 by 4.
a. Find a rectangular array that cannot be traversed in this way.
b. Find a criterion that seems to distinguish the dimensions of those arrays that can be traversed from those which cannot be traversed.
a) samples: 3×3, 5×5; b) One dimension must be even for an array to be traversed.

14. $v = .90$; $w = .30$; $x = .50$; $y = .01$; $z = .45$

20. $\sqrt{3y + 1} = 4$
1. $3y + 1 = 16$ nonreversible
2. $3y = 15$ reversible
3. $y = 5$ reversible

16.a. sample

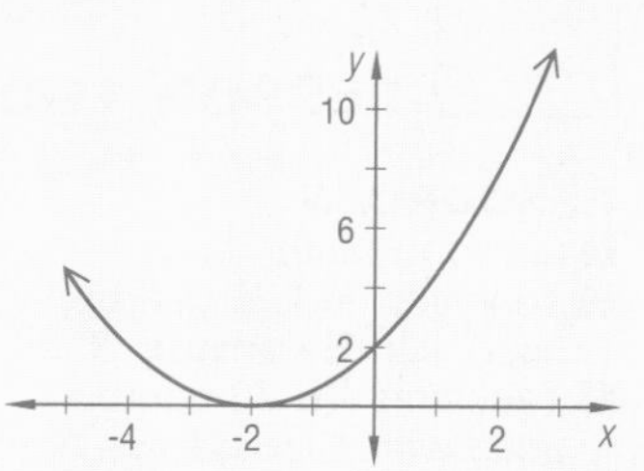

NAME ____

LESSON **MASTER 11-1**
QUESTIONS ON **SPUR** OBJECTIVES

■ **USES** *Objective E (See pages 694–698 for objectives.)*
In 1–4, suppose that a reporter conducts a poll of voters regarding a plan to construct a new sports stadium using city funds. Of those polled, 77% are people who vote regularly, while 23% vote only occasionally. Of the regular voters, 42% are in favor of the plan to construct the stadium. Among the occasional voters, 54% favor its construction.

1. Draw a probability tree to represent this situation.

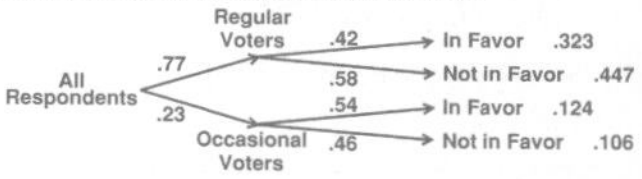

2. If 1,500 voters respond, how many are regular voters not in favor of the plan? about 670

3. Which category contains the largest percentage of respondents? regular voters not in favor of the plan

4. What is the probability that a randomly chosen voter who is not in favor of the plan is a regular voter? .808

5. Suppose the process of conducting an archaeological dig involves the following tasks:

Task	Description	Time Required (Days)	Prerequisite Tasks
A	choose a site	1	none
B	organize a team	3	none
C	travel to site	1	A, B
D	set up equipment	2	C
E	dig	8	D
F	keep journal of findings	9	D
G	write a report of findings	12	E, F

a. Sketch a directed graph to represent the situation.

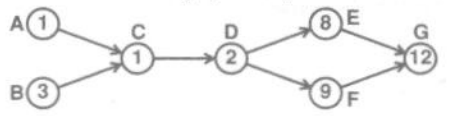

b. What is the minimal time required to conduct the entire dig? 27 days

NAME ____
Lesson MASTER 11-1 (page 2)

■ **USES** *Objective G*
In 6 and 7, the vertices of the graph at the right represent traffic lights and the edges represent streets.

6. Draw a path that will allow a maintenance person to pass each light exactly once.

sample:

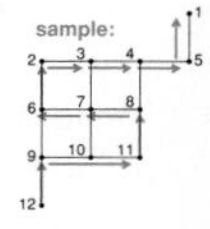

7. Suppose light 8 and the part of the street connecting it to light 7 are removed (so lights 4 and 11 are connected directly). Would a path satisfying Question 6 still be possible? yes

LESSON 11-2

RESOURCES
- Lesson Master 11-2
- Teaching Aid 75 can be used with **Example 3**.
- Teaching Aid 76 can be used with **Questions 5, 6, 7, and 10**.
- Teaching Aid 77 can be used with **Questions 11, 12, 14, and 15**.

OBJECTIVES

A Draw graphs given sufficient information.
B Identify parts of graphs and types of graphs.
I Convert between the picture of a graph or directed graph, and its adjacency matrix.

TEACHING NOTES

Point out to students that there are three ways to describe a graph: (1) geometrically, by drawing a picture; (2) algebraically, by naming vertices, naming edges, and giving the edge-endpoint function; and (3) arithmetically, by writing a matrix that shows how many edges there are from any vertex v_i to any vertex v_j.

In **Example 1**, an algebraic description of a graph is given and students are asked to give a geometric description. Students may wonder why the algebraic description is needed. With algebra, a problem involving graphs can be stated in computer language and can be treated in a computer program; also, theorems can be deduced about graphs.

LESSON

11-2

The Definition of Graph

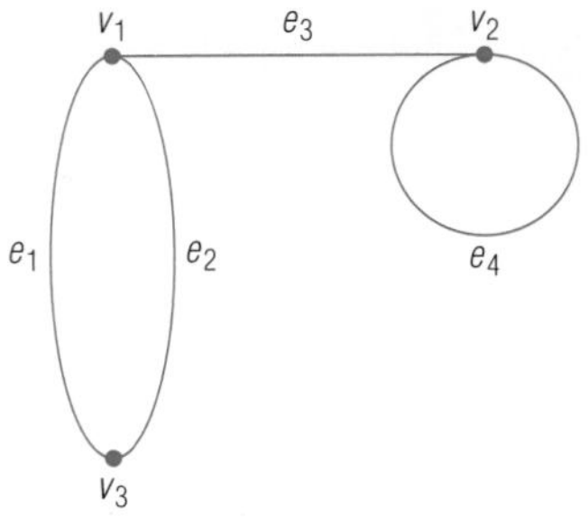

In the last lesson you saw examples of problems that could be solved using graphs, but no definition of *graph* was given. In order to deduce general properties of graphs, a definition is needed. Consider the graph at the right.
(1) Its vertices are v_1, v_2, and v_3.
(2) Its edges are e_1, e_2, e_3, and e_4.
(3) The endpoints of each edge are given by the following table.

edge	endpoints
e_1	$\{v_1, v_3\}$
e_2	$\{v_1, v_3\}$
e_3	$\{v_1, v_2\}$
e_4	$\{v_2\}$

This table describes an *edge-endpoint function*.

To specify a graph it is necessary to provide the kind of information given in (1)–(3) above, with words or a picture. Here is a formal definition of *graph*.

Definition

A **graph** G consists of
1. a finite set of **vertices,**
2. a finite set of **edges,**
3. a function (the **edge-endpoint function**) that maps each edge to a set of either one or two vertices (the **endpoints** of the edge).

The essential feature of an edge is that it *connects* or *joins* its endpoints; its particular shape—curve or segment—is not important. Two vertices connected by an edge are **adjacent vertices**. Two edges with a common endpoint are called **adjacent edges**. For instance, e_3 and e_2 above are adjacent edges. Caution: edges in graphs, unlike edges in polyhedra, have no points other than their endpoints.

Note that vertices v_1 and v_3 are connected by more than one edge. When this occurs the edges are said to be **parallel**. (This is a different meaning for ''parallel'' than that associated with lines.) Also, edge e_4 joins vertex v_2 to itself. Such an edge is called a **loop**. Note that in a graph there must be a vertex at each end of each edge.

Example 1 Draw a picture of the graph G defined as follows.

1. set of vertices: $\{v_1, v_2, v_3, v_4, v_5\}$
2. set of edges: $\{e_1, e_2, e_3, e_4, e_5\}$
3. edge-endpoint function:

edge	endpoints
e_1	$\{v_1, v_2\}$
e_2	$\{v_1, v_4\}$
e_3	$\{v_1, v_4\}$
e_4	$\{v_5\}$
e_5	$\{v_4, v_5\}$

Solution It is often convenient to start the graph by placing the vertices as though they were consecutive vertices of a convex polygon. Since there are 5 vertices, begin with a pentagon.

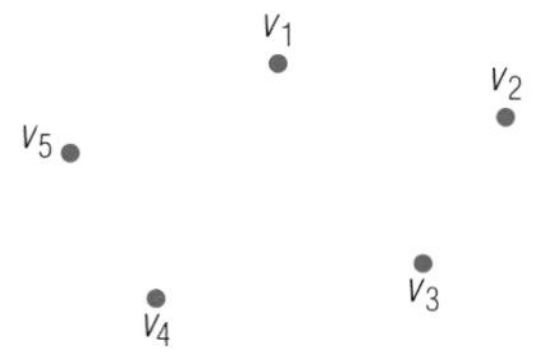

Then fill in edges as specified by the edge-endpoint function table.

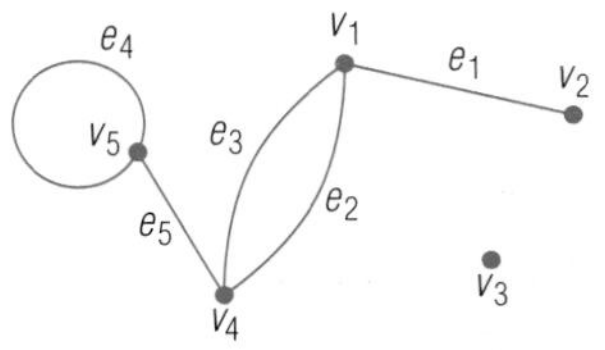

In Example 1, vertex v_3 is not the endpoint of any edge. It is called **isolated**. The definition of graph allows isolated vertices. Although all edges must have endpoints, a vertex need not be the endpoint of an edge.

A picture of the graph from the house-building problem in Lesson 11-1 is repeated here with the vertices shown by dots.

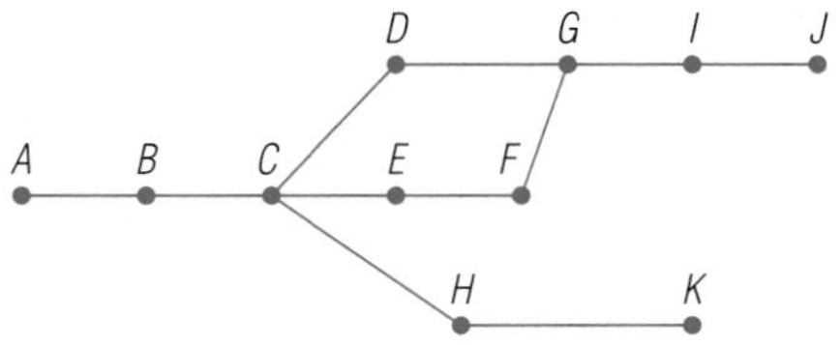

In this graph, there are no loops or parallel edges. Such a graph is called *simple*.

The terms *isolated*, *parallel*, and *simple* are easier for students to interpret in a setting involving a picture of a graph than by identifying the components of the definition. In terms of the edge-endpoint function, an isolated vertex is a vertex that does not appear in the table defining the function; parallel edges are those that map onto the same pair of vertices. The edge-endpoint function listing of a simple graph does not have single vertices in the endpoints column, and each pair of vertices will occur only once.

Example 3 asks for a translation of the geometric form into the arithmetic form (a matrix). Stress to students that before writing a matrix for a graph, the vertices should be ordered. If the graph is directed, then the rows represent the origins of arrows and the columns represent their destinations. **Example 4** translates a matrix to a picture.

At this point, students do not need to know how to perform operations on matrices. By Lesson 11-5, multiplication of matrices must be defined. See the note on **Question 20**.

The adjacency matrix for a graph is symmetric unless the graph is a directed graph. The type of check in **Example 3** can be modified to check **Example 4**. The number of edges of an undirected graph equals the sum of the elements on or above the diagonal of the matrix. Thus, the graph in Example 4 has 5 edges.

ADDITIONAL EXAMPLES

1. Draw a picture of the graph G with vertices v_1, v_2, v_3, and v_4; edges e_1, e_2, e_3, e_4, and e_5, and the following edge-endpoint function:
e_1 has endpoints $\{v_1, v_2\}$,
e_2 has endpoints $\{v_3\}$,
e_3 has endpoints $\{v_1, v_3\}$,
e_4 has endpoints $\{v_1, v_4\}$,
e_5 has endpoints $\{v_3, v_4\}$.

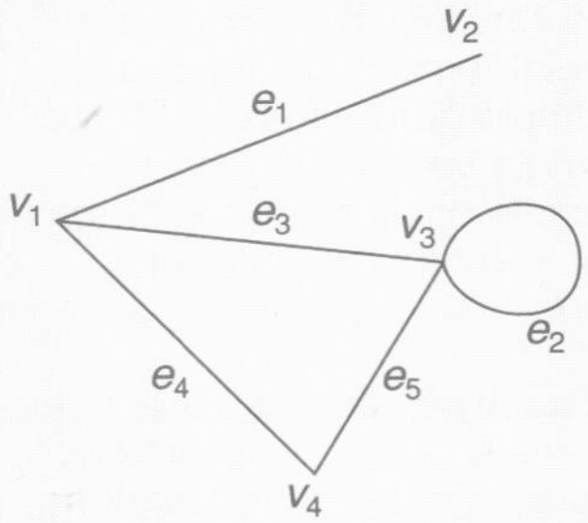

2. Draw a simple graph with 5 vertices and 7 edges.
sample:

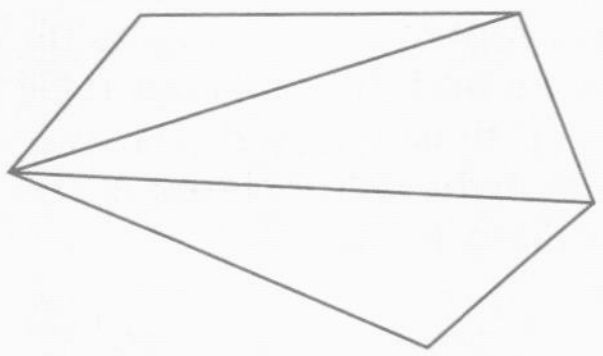

Definition

A graph is **simple** if and only if it does not have loops and it does not have parallel edges.

In a simple graph with vertices v and w, if edge $\{v, w\}$ exists it is unique, since there is at most one edge joining any two of its vertices. For instance, consider the simple graph with vertices a, b, c, d and edges $\{a, c\}$, $\{b, d\}$, and $\{b, c\}$. Two pictures of this graph are shown below. In the left picture, the pictures of edges $\{a, c\}$ and $\{b, d\}$ intersect but the edges do not (because they have no points other than their endpoints). Such intersections in pictures are called **crossings**. The figure at the right illustrates the same graph but it avoids crossings.

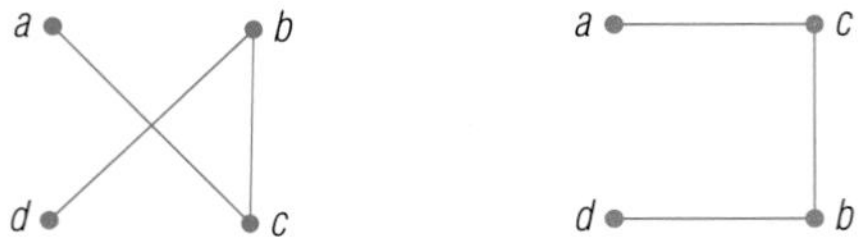

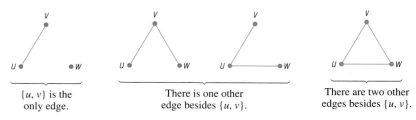

Example 2 Draw all simple graphs with vertices $\{u, v, w\}$ if one of the edges is $\{u, v\}$.

Solution

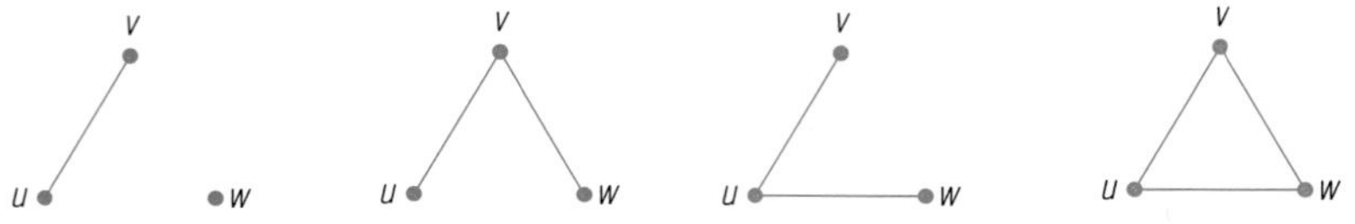

$\{u, v\}$ is the only edge.

There is one other edge besides $\{u, v\}$.

There are two other edges besides $\{u, v\}$.

Recall from Lesson 11-1 that it is sometimes useful to add direction to each edge of a graph. The resulting digraph is pictured like other graphs except that its edges are drawn with arrows. The formal definition of *digraph* is the same as the definition of *graph* except that the edge-endpoint function sends each edge to an *ordered pair* of vertices. (That would be indicated in an edge-endpoint table by, say (v_1, v_3) rather than $\{v_1, v_3\}$.)

For instance, some group-behavior studies investigate the influence one person has on another in a social setting. The directed graph pictured below shows such a set of influence relationships.

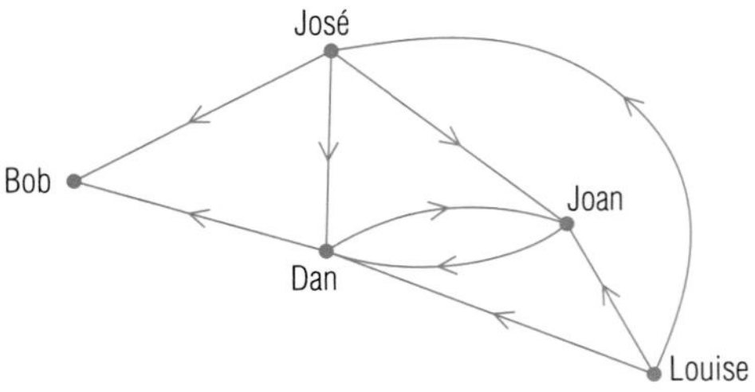

From the arrows you can see that, for example, José influences Bob, Dan, and Joan; and Louise influences José, Dan, and Joan. The edge connecting Louise and Joan is said to go *from* the vertex for Louise *to* the vertex for Joan.

It is natural to want to describe a graph numerically. It is somewhat surprising that this can be done using a matrix.

Definition

The **adjacency matrix** M for a graph with vertices $v_1, v_2, \ldots, v_n$ is the $n \times n$ matrix in which $\forall\ i, j$, the element in the ith row and jth column is the number of edges from vertex v_i to vertex v_j.

The next two Examples illustrate how to construct adjacency matrices for directed and undirected graphs.

Example 3 Write the adjacency matrix for the directed graph of influence relationships pictured above.

Solution Because there are 5 vertices, the adjacency matrix will have 5 rows and 5 columns. Label the rows and columns with the vertex names. To fill in the entry in the ith row and jth column, just count the number of edges from v_i to v_j. For instance, there are no edges from v_1 to v_1, so the entry in row 1, column 1 is 0. There is one edge from v_2 to v_3, so the entry in row 2, column 3 is 1. The entire adjacency matrix is given below.

	Bob v_1	José v_2	Dan v_3	Joan v_4	Louise v_5
Bob = v_1	0	0	0	0	0
José = v_2	1	0	1	1	0
Dan = v_3	1	0	0	1	0
Joan = v_4	0	0	1	0	0
Louise = v_5	0	1	1	1	0

Check In the matrix for a directed graph, each edge appears once in the matrix, so the sum of all elements of the matrix equals the number of edges of the graph. Here that number is 9, which checks.

3. Suppose A, B, and C are three cities. Every day there are 2 nonstop flights from A to B, 3 from B to C, and 1 from A to C. There are 2 nonstop flights from B to A, 2 from C to B, and 1 from C to A. Write the adjacency matrix for this information.

$$\begin{bmatrix} 0 & 2 & 1 \\ 2 & 0 & 3 \\ 1 & 2 & 0 \end{bmatrix}$$

4. Draw a picture of a directed graph that has the following adjacency matrix.

$$\begin{bmatrix} 1 & 2 & 1 & 0 \\ 0 & 0 & 1 & 3 \\ 0 & 0 & 2 & 0 \\ 1 & 1 & 0 & 0 \end{bmatrix}$$

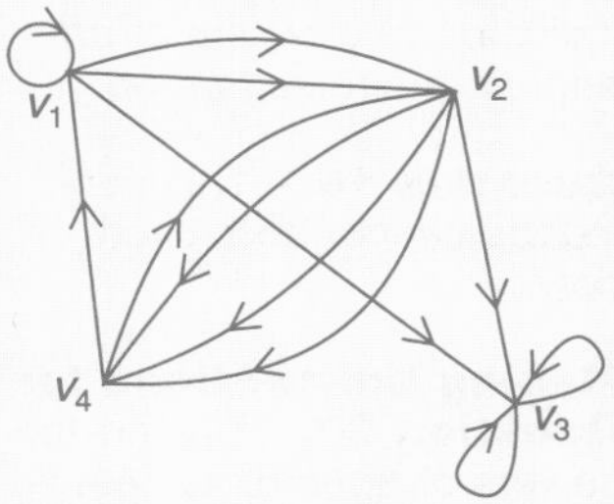

NOTES ON QUESTIONS
Question 12: The criterion for determining whether two pictures represent the same graph is simple. Do the graphs have the same algebraic definition? In this question, because no other information is known, one assumes that the vertices are interchangeable. But that might not be the case in other situations.

Question 14: It is important to discuss this question. Point out the use of parentheses rather than braces for the endpoints to indicate ordered pairs. The ordering of the endpoints is needed to indicate the direction of an edge in a directed graph.

Question 16: This conjecture cannot be proved now.

Making Connections for Question 20: Finding the powers of a matrix is needed in Lessons 11-5 and 11-6. Even students who have multiplied matrices may never have squared them. Anticipate the notation to be used: If the given matrix in this question is named A, the answer to part **a** is A^2.

ADDITIONAL ANSWERS

2.

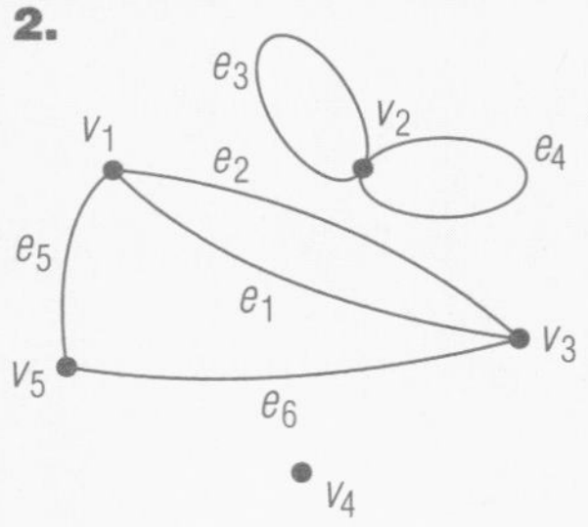

5.

$$\begin{array}{c} \\ v_1 \\ v_2 \\ v_3 \\ v_4 \end{array}\begin{array}{c} \begin{array}{cccc} v_1 & v_2 & v_3 & v_4 \end{array} \\ \begin{bmatrix} 0 & 1 & 0 & 0 \\ 0 & 0 & 1 & 1 \\ 0 & 0 & 1 & 1 \\ 1 & 1 & 0 & 0 \end{bmatrix} \end{array}$$

Example 4 Draw a picture of a graph (*not* directed) that has the following adjacency matrix.

$$\begin{array}{c} \\ v_1 \\ v_2 \\ v_3 \end{array}\begin{array}{c} \begin{array}{ccc} v_1 & v_2 & v_3 \end{array} \\ \begin{bmatrix} 1 & 2 & 0 \\ 2 & 2 & 0 \\ 0 & 0 & 0 \end{bmatrix} \end{array}$$

Solution Draw vertices v_1, v_2, and v_3 and connect them by edges as indicated in the matrix. For example, the 2 in the first row and second column indicates that two edges should go from v_1 to v_2. Since the graph is not directed, these two edges also go from v_2 to v_1, agreeing with the 2 in the second row and first column of the matrix. Note that when the graph is completed there is one loop at v_1 and two at v_2, there are parallel edges joining v_1 and v_2, and v_3 is an isolated vertex.

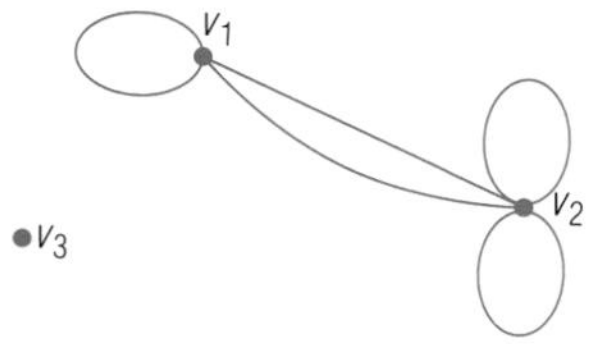

Questions

Covering the Reading

1. Refer to the figure at the right.
   a. Does the figure represent a graph? **Yes**
   b. If it does, tell how many edges and how many vertices it has. If it does not, explain why not.
   4 edges, 4 vertices

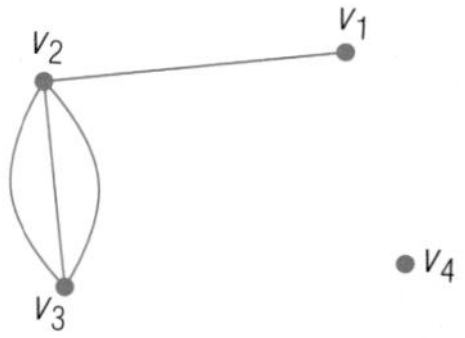

2. Draw a picture of the graph G defined as follows.
   (1) set of vertices: $\{v_1, v_2, v_3, v_4, v_5\}$
   (2) set of edges: $\{e_1, e_2, e_3, e_4, e_5, e_6\}$
   (3) edge-endpoint function: **See margin.**

edge	endpoints
e_1	$\{v_1, v_3\}$
e_2	$\{v_1, v_3\}$
e_3	$\{v_2\}$
e_4	$\{v_2\}$
e_5	$\{v_1, v_5\}$
e_6	$\{v_3, v_5\}$

652

3. Use the graph G of Question 2.
   a. Are edges e_1 and e_5 adjacent? **Yes**
   b. Are vertices v_1 and v_2 adjacent? **No**
   c. Identify all isolated vertices. **v_4, v_2**
   d. Identify all parallel edges. **e_3 and e_4, e_1 and e_2**
   e. Identify all loops. **e_3, e_4**

4. *True* or *false?* The directed graph following Example 2 shows that Bob influences Dan. **False**

5. Write the adjacency matrix for the directed graph pictured below. **See margin.**

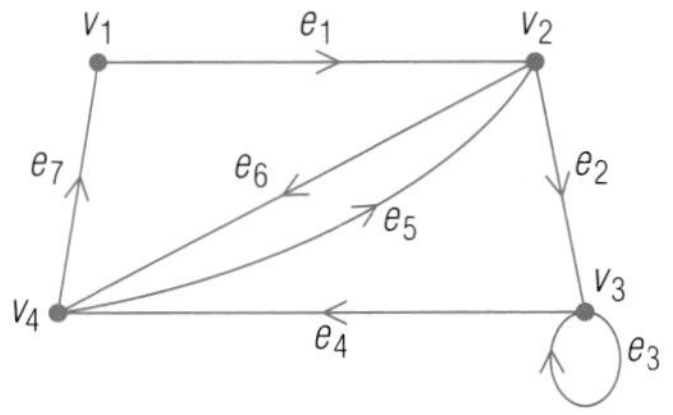

6. Draw a directed graph with the adjacency matrix shown at the right. **See margin.**

$$\begin{bmatrix} 1 & 2 & 0 & 1 \\ 0 & 0 & 1 & 0 \\ 1 & 3 & 0 & 0 \\ 0 & 1 & 0 & 2 \end{bmatrix}$$

7. Find the adjacency matrix for the graph pictured below. **See margin.**

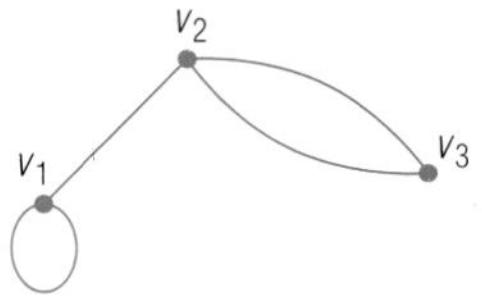

8. Give the numbers of vertices and edges of the graph in Example 2 of Lesson 11-1. **11 vertices, 11 edges**

9. Draw all simple graphs with vertices $\{a, b, c, d\}$, one edge $\{a, b\}$, and two other edges. **See margin.**

Applying the Mathematics

10. a. Does the adjacency matrix at the right represent a simple graph? **No**
    b. Explain your answer. **See margin.**

$$\begin{bmatrix} 0 & 1 & 2 \\ 1 & 0 & 1 \\ 2 & 1 & 0 \end{bmatrix}$$

11. Construct an edge-endpoint function table for the graph pictured below. **See margin.**

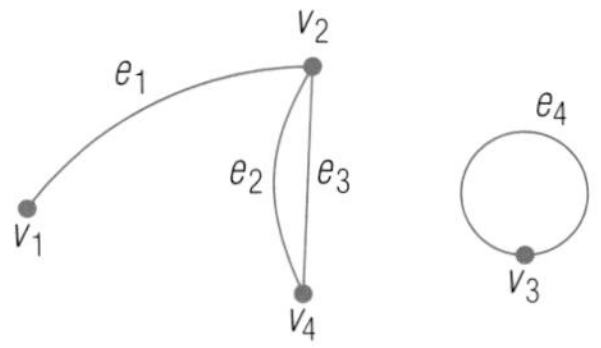

6.

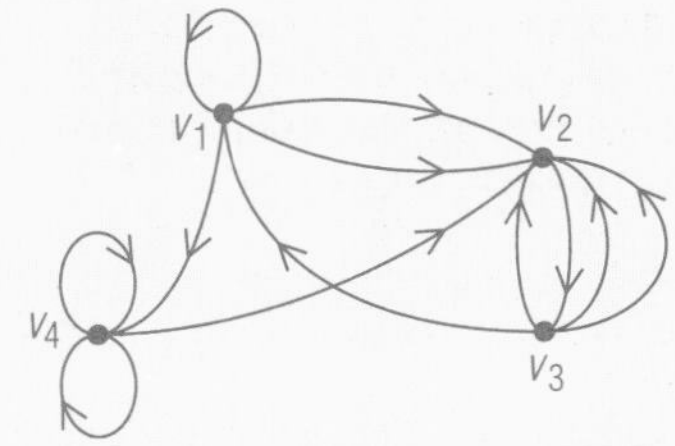

7.

$$\begin{array}{c} \\ v_1 \\ v_2 \\ v_3 \end{array} \begin{array}{c} \begin{array}{ccc} v_1 & v_2 & v_3 \end{array} \\ \begin{bmatrix} 1 & 1 & 0 \\ 1 & 0 & 2 \\ 0 & 2 & 0 \end{bmatrix} \end{array}$$

9.

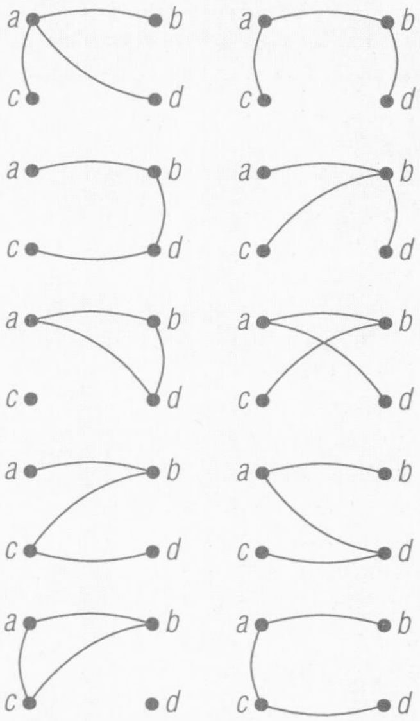

10.b. There are two edges between v_1 and v_3. Those edges are parallel, so the graph is not simple.

11.

edge	endpoint
e_1	$\{v_1, v_2\}$
e_2	$\{v_2, v_4\}$
e_3	$\{v_2, v_4\}$
e_4	$\{v_3\}$

12. Do the following two pictures represent the same graph? Why or why not? See margin.

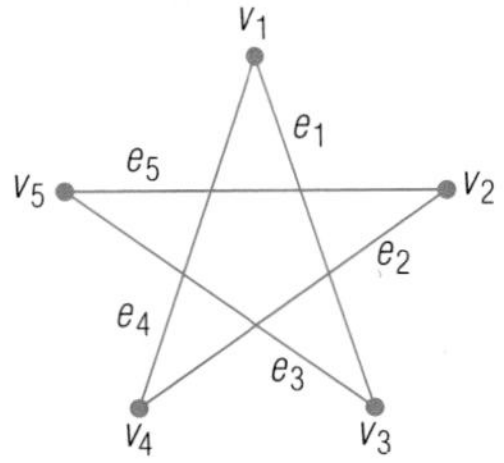

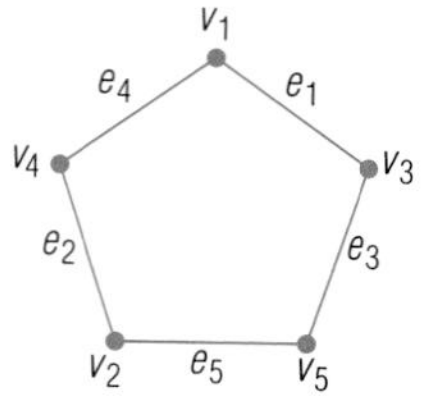

13. a. Write the negation of the following statement:
∀ *graphs G, if G does not have any loops, then G is simple.*

b. Is the negation you obtained in part **a** *true* or *false*? Justify your answer. a) ∃ a graph G such that G does not have any loops and G is not simple. b) True: G could have parallel edges and no loops.

14. The edge-endpoint function for a directed graph sends each edge to an ordered pair of vertices. For instance, edge e might be sent to (u, v). When a picture of this graph is drawn, there will be an arrow pointing from u to v to show that the edge e goes from u to v. Draw a picture of the directed graph defined as follows.

vertices: $\{v_1, v_2, v_3, v_4\}$
edges: $\{e_1, e_2, e_3, e_4, e_5, e_6\}$
edge-endpoint function: See margin.

edge	endpoints
e_1	(v_1, v_2)
e_2	(v_1, v_3)
e_3	(v_3, v_1)
e_4	(v_3, v_3)
e_5	(v_2, v_4)
e_6	(v_3, v_4)

15. A product P is being test-marketed against 3 leading brands A, B, and C. Below, vertices A, B, C, and P represent the products and an arrow is drawn from vertex x to vertex y if the tester prefers x to y. Explain why there is an inconsistency in the tester's preferences.

See margin.

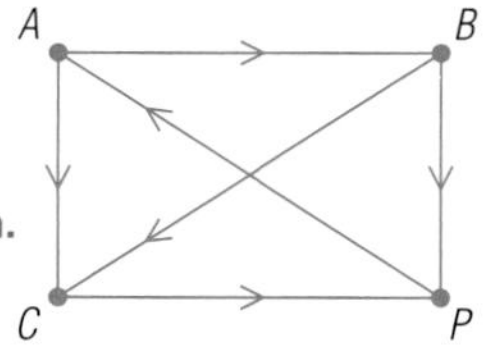

Review

16. a. How many vertices does the possibility tree shown at the right have? 18

b. How many edges does it have? 17

c. Make a conjecture about the number of vertices and edges in a possibility tree, based on parts **a** and **b**, and on Question 14 of Lesson 11-1. *(Lesson 11-1)* $V = E + 1$

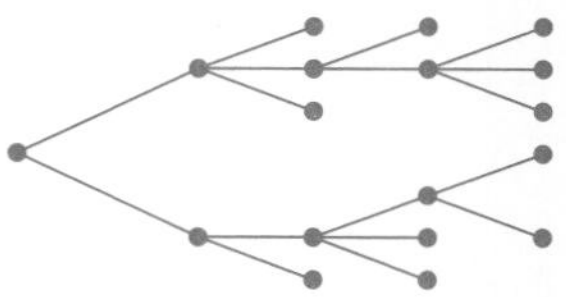

ADDITIONAL ANSWERS

12. Yes, they have the same lists of vertices and edges and the same edge-endpoint function.

14.

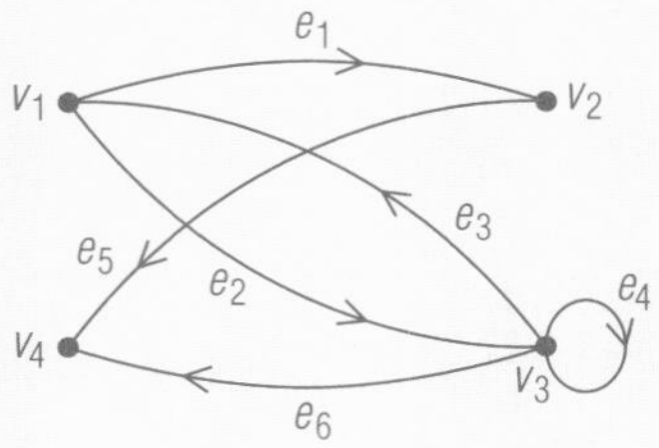

15. The tester prefers *A* to *C*, and prefers *C* to *P*. But in a direct comparison of *A* and *P*, the tester prefers *P*.

17.a.

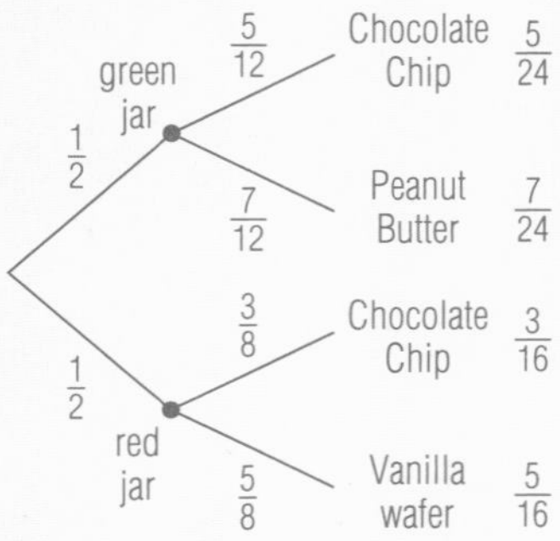

20.a. $\begin{bmatrix} \cos 2\theta & -\sin 2\theta \\ \sin 2\theta & \cos 2\theta \end{bmatrix}$

b. $\begin{bmatrix} \cos \theta & -\sin \theta \\ \sin \theta & \cos \theta \end{bmatrix}^n = \begin{bmatrix} \cos(n\theta) & -\sin(n\theta) \\ \sin(n\theta) & \cos(n\theta) \end{bmatrix}$

654

17. The green cookie jar in the kitchen contains five chocolate chip cookies and seven peanut butter cookies. The red cookie jar contains three chocolate chip cookies and five vanilla wafers. In the middle of the night little Freddy sneaks into the kitchen. He doesn't turn on the light for fear of waking his parents. He puts his hand in one of the jars at random and pulls out a cookie. *(Lesson 11-1)*

   a. Draw a probability tree and label its edges with probabilities to represent this situation. See margin.

   b. What is the probability that he gets a chocolate chip cookie? $\approx .396$

   c. When he bites into the cookie, he finds it is chocolate chip. What is the probability that it came from the red jar? $\approx .474$

18. Let $p(x)$ be a polynomial with real coefficients such that $p(2 - i) = 0$ and $p(-3) = 0$. Find a possible formula for $p(x)$. *(Lesson 8-9)*
Sample: $p(x) = x^3 - x^2 - 7x + 15$

19. Find the quotient and remainder when $p(x) = 6x^4 - 7x^2 + 3x + 1$ is divided by $d(x) = x + 7$. *(Lesson 4-4)*
quotient: $6x^3 - 42x^2 + 287x - 2006$; remainder; 14,043

20. **a.** Multiply the matrix $\begin{bmatrix} \cos\theta & -\sin\theta \\ \sin\theta & \cos\theta \end{bmatrix}$ by itself and simplify the result.

   b. Generalize the result in part **a.** *(Previous course)*
   See margin.

Exploration

21. **a.** Consider simple graphs that have four vertices $\{a, b, c, d\}$ and at least the edges $\{a, b\}$ and $\{b, c\}$. How many such graphs have exactly

   i. 2 edges? 1 **ii.** 3 edges? 4 **iii.** 4 edges? 6
   iv. 5 edges? 4 **v.** 6 edges? 1

   b. Consider simple graphs which have four vertices $\{a, b, c, d\}$ and at least the edges $\{a, b\}$, $\{b, c\}$, and $\{c, d\}$. How many such graphs have exactly

   i. 3 edges? 1 **ii.** 4 edges? 3
   iii. 5 edges? 3 **iv.** 6 edges? 1

   c. What do the sequences of answers to parts **a** and **b** suggest?

   d. How is Example 2 and its answer similar to this problem?

   c) Pascal's triangle; binomial coefficients; d) 1, 2, 1 is row 3 of Pascal's triangle.

FOLLOW-UP

MORE PRACTICE
For more questions on SPUR Objectives, use *Lesson Master 11-2*, shown below.

NAME

LESSON **MASTER 11–2**
QUESTIONS ON SPUR OBJECTIVES

■ **SKILLS** *Objective A (See pages 694–698 for objectives.)*
In 1–3, draw a graph with the specified characteristics.

1. two vertices and three edges
sample:

2. an isolated vertex, two loops, and three edges
sample:

3. two parallel edges, three vertices, and a loop
sample:

4. Draw the graph G defined as follows:
set of vertices: $\{v_1, v_2, v_3, v_4, v_5\}$;
set of edges: $\{e_1, e_2, e_3, e_4\}$
edge-endpoint function:

edge	endpoints
e_1	$\{v_1\}$
e_2	$\{v_2, v_4\}$
e_3	$\{v_3, v_5\}$
e_4	$\{v_1, v_3\}$

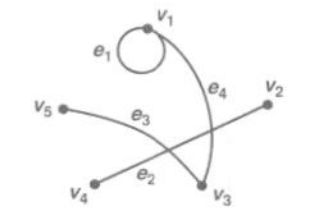

or

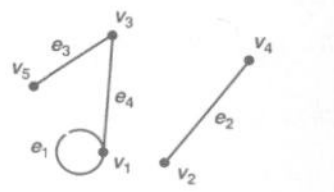

■ **PROPERTIES** *Objective B*
5. A graph consists of a finite set of vertices, a finite set of edges, and a function that maps each edge to a set of either one or two vertices.

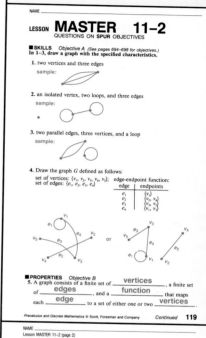

NAME
Lesson MASTER 11–2 (page 2)

6. Use the graph at the right.

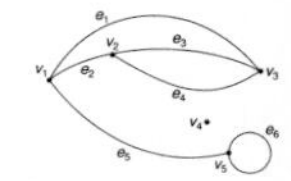

   a. Identify any isolated vertices. v_4

   b. Identify all edges adjacent to e_3. e_1, e_2, e_4

   c. Identify any loops. e_6

   d. Give the edge-endpoint function table for the graph.

edge	endpoints
e_1	$\{v_1, v_3\}$
e_2	$\{v_1, v_2\}$
e_3	$\{v_2, v_3\}$
e_4	$\{v_2, v_3\}$
e_5	$\{v_1, v_5\}$
e_6	$\{v_5\}$

■ **REPRESENTATIONS** *Objective I*
7. Write the adjacency matrix for the directed graph below.

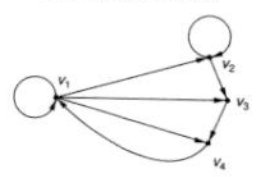

$$\begin{array}{c} \\ v_1 \\ v_2 \\ v_3 \\ v_4 \end{array} \begin{array}{c} \begin{array}{cccc} v_1 & v_2 & v_3 & v_4 \end{array} \\ \begin{bmatrix} 1 & 1 & 1 & 1 \\ 0 & 1 & 1 & 0 \\ 0 & 0 & 0 & 1 \\ 1 & 0 & 0 & 0 \end{bmatrix} \end{array}$$

8. What does the sum of the components of the matrix for a directed graph represent? number of edges of the graph

9. Draw an undirected graph with the following adjacency matrix.

$$\begin{array}{c} \\ v_1 \\ v_2 \\ v_3 \end{array} \begin{array}{c} \begin{array}{ccc} v_1 & v_2 & v_3 \end{array} \\ \begin{bmatrix} 0 & 0 & 1 \\ 0 & 2 & 0 \\ 1 & 0 & 2 \end{bmatrix} \end{array}$$

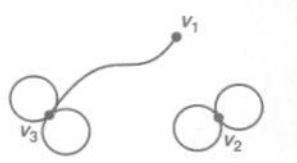

LESSON 11-3

RESOURCES

■ Lesson Master 11-3

OBJECTIVES

A Draw graphs given sufficient information.
B Identify parts of graphs and types of graphs.
C Determine whether there exists a graph containing vertices with given degrees.
F Use the Total Degree of a Graph Theorem and its corollaries to solve handshake problems.

TEACHING NOTES

The handshake problem that begins this lesson is solved using combinations. Graph theory often applies combinatorics.

Handshake problems are paradigms for many graph problems because they have the same structure. For instance, the problem at the top of page 657 has the same structure as the following: Is it possible to connect 47 points by segments so that each point is connected to 9 others? Is it possible to devise a chess tournament with 47 entrants so that each entrant would play exactly 9 games?

LESSON

11-3

Handshake Problems

Suppose n people are at a party. If each person shakes hands with every other person, how many handshakes are required?

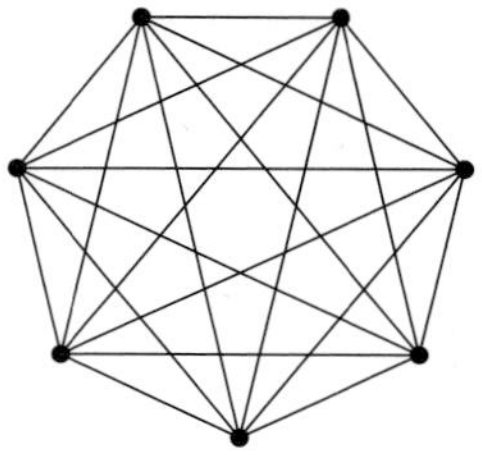

This *handshake problem* can be represented by a graph in the following way. Represent the n people by vertices and join two vertices by an edge if the corresponding people shake hands. Since each person shakes hands once with each other person, every pair of vertices is joined by exactly one edge. A graph with this property is called a **complete graph**. Here is a picture of the complete graph with 7 vertices.

You can see that a complete graph can be pictured as the union of a polygon with its diagonals. Thus the number of edges in the complete graph with n vertices equals the total number of sides and diagonals in an n-gon, which in turn is the answer to the handshake problem.

There are a number of ways to solve the handshake problem. One way is to use combinations. Note that there are as many handshakes (edges) as there are ways to choose 2 people (vertices) to shake hands out of a group of n. This number is $\binom{n}{2}$, which equals $\frac{n(n-1)}{2}$. For the above graph, when $n = 7$, there are 21 handshakes, corresponding to the total of 7 sides and 14 diagonals for a heptagon.

Many other problems are equivalent to handshake problems. For instance, replacing handshakes by games and people by teams converts any handshake problem into a problem involving games and teams. The solution to the above handshake problem implies that $\frac{n(n-1)}{2}$ games are required for each of n teams to play each of the other teams exactly once.

Here is a different handshake problem, one which seems more difficult.

> Forty-seven people attend a social gathering. During the course of the event, various people shake hands. Is it possible for each person to shake hands with exactly nine other people?

The concept of *degree*, together with properties of even and odd integers, can help to solve this problem.

Definitions

If v is a vertex of a graph G, the **degree of v**, denoted **deg(v)**, equals the number of edges that have v as an endpoint, with each edge that is a loop counted twice. The **total degree of G** is the sum of the degrees of all the vertices of G.

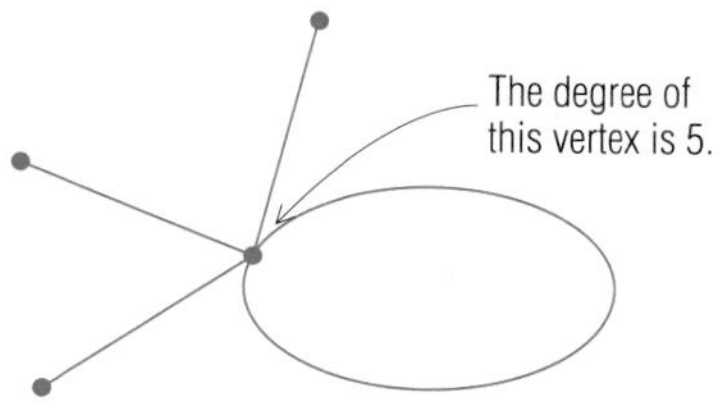

Consider the graph G pictured at the right. Its vertices have the following degrees:

$\deg(v_1) = 3$
$\deg(v_2) = 4$
$\deg(v_3) = 3$.

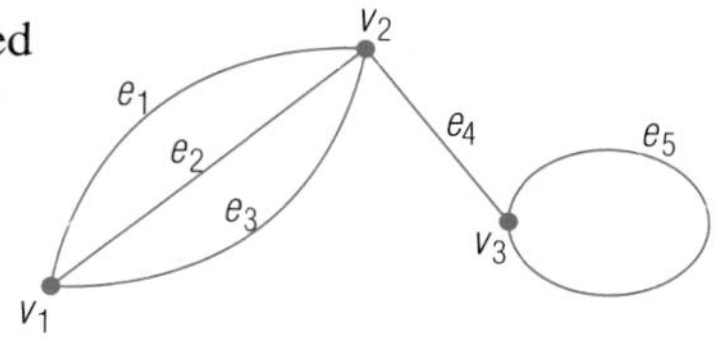

Thus the total degree of G is 10, which is twice the number of edges. Is this always the case? The answer is yes. The reason is that each edge of a graph contributes 2 to the total degree whether or not the edge is a loop. For instance, in the graph pictured above,

Edge	contributes	1 to the degree of	and	1 to the degree of
e_1		v_1		v_2
e_2		v_1		v_2
e_3		v_1		v_2
e_4		v_2		v_3
e_5, a loop, contributes 2 to the degree of v_3.				

This argument proves the following theorem.

Theorem (The Total Degree of a Graph)

The total degree of any graph equals twice the number of edges in the graph.

Alternate Approach To illustrate The Total Degree of a Graph Theorem and its use in **Example 1**, you might want to simplify the problem to one that involves fewer people, for example five people each shaking hands with exactly three others. Then you can have five students try to demonstrate the handshaking at the front of the room.

Students should go through the proof of **Example 2b** step by step.

ADDITIONAL EXAMPLES

1. In a group of 10 people, can each person shake hands with exactly 3 others?
Yes; call the people *A*, *B*, *C*, *D*, *E*, *F*, *G*, *H*, *I*, and *J*. Draw a polygon *ABCDEFGHIJ* and its diagonals *AF*, *BG*, *CH*, *DI*, and *EJ*. Now, considered as a graph, each vertex is connected to three others.

2. Draw a graph or show that no such graph exists.
a. a graph with three vertices of degrees 1, 2, and 3

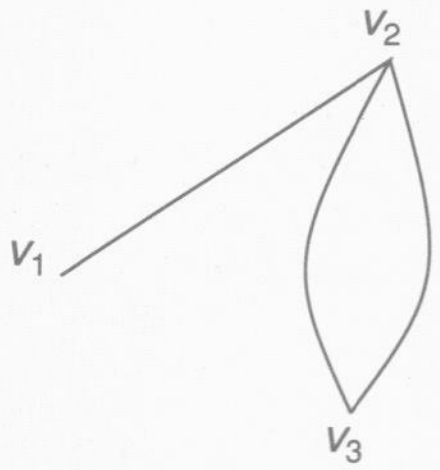

b. a simple graph with three vertices of degrees 1, 2, and 3
This cannot be done. The vertex with degree 3 has 3 edges from it, but there are only 2 other vertices, so it would be connected to one of the other vertices by 2 edges.

The theorem has corollaries, which you are asked to prove in the Questions.

Corollaries

1. Total Degree is Even: The total degree of any graph is an even positive integer.
2. Number of Odd Vertices is Even: Every graph has an even number of vertices of odd degree.

The second corollary helps to answer the question about handshakes at a social gathering.

Example 1 In a group of 47 people, can each person shake hands with exactly 9 other people?

Solution The answer is no. To see why, assume that each of the 47 people could shake hands with exactly 9 others. Represent each person as the vertex of a graph, and draw an edge joining each pair of people who shake hands. To say that a person shakes hands with 9 other people is equivalent to saying that the degree of the vertex representing that person is 9. The graph would then have 47 vertices, each of which would have degree 9. So it would have an odd number of vertices of odd degree. But this contradicts the second corollary, so the assumption must be false. Thus it is impossible for each of 47 people to shake hands with exactly 9 others.

Example 2 In parts **a** and **b**, draw the specified graph, or show that no such graph exists.
a. a graph with three vertices of degrees 2, 2, and 0
b. a simple graph with three vertices of degrees 2, 2, and 0

Solution
a. This combination of degrees is not forbidden by the Total Degree of a Graph Theorem because the total degree of the graph would be 4, which is even. The number of edges in the graph would be half the total degree, or 2. If you experiment by drawing three vertices connected in various ways by two edges, you quickly find that each of the graphs below satisfies the given properties.

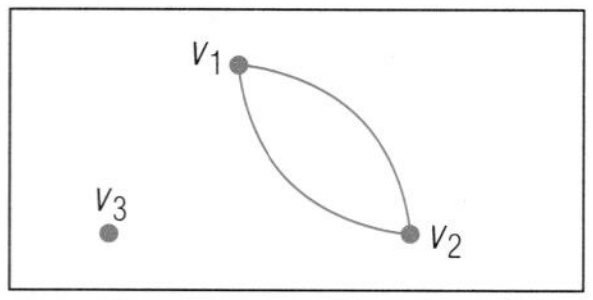

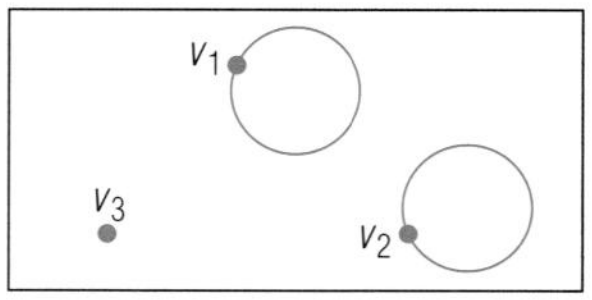

b. Neither graph in part **a** is simple. If you continue to experiment by shifting the positions of the edges in the graphs above, you continually come up with graphs that either do not have the given degrees or are not simple. At a certain point, you would probably conjecture that no such graph exists. Proof by contradiction is a natural approach to use to prove this conjecture.

658

To prove:
There is no simple graph with three vertices of degrees 2, 2, and 0.

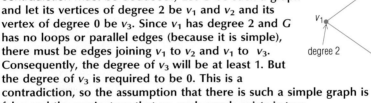

Proof (by contradiction): Assume that there is a simple graph with three vertices of degrees 2, 2, and 0. (A contradiction must be deduced.) **Let G be such a graph and let its vertices of degree 2 be v_1 and v_2 and its vertex of degree 0 be v_3. Since v_1 has degree 2 and G has no loops or parallel edges (because it is simple), there must be edges joining v_1 to v_2 and v_1 to v_3. Consequently, the degree of v_3 will be at least 1. But the degree of v_3 is required to be 0. This is a contradiction, so the assumption that there is such a simple graph is false and the conjecture that no such graph exists is true.**

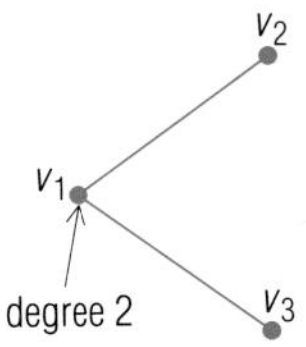

The result of Example 2b can be put into the language of handshakes. In a group of 3 people in which no pair shakes hands twice, it is impossible for 2 people to shake hands with 2 others and the third person not to shake any hands at all.

Questions

Covering the Reading

1. Consider the graph G pictured at the right.
a. Find the degree of each vertex.
b. What is the total degree of G?
a) See margin. b) 8

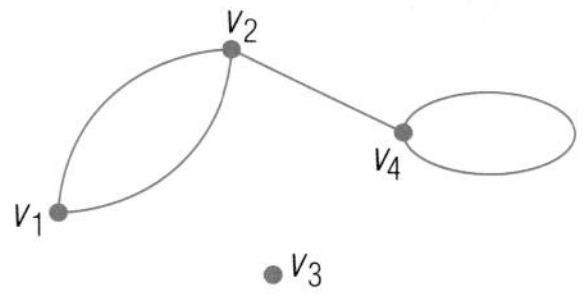

2. Explain why the following statement is *false*: The degree of a vertex equals the number of edges that have the vertex as an endpoint.
See margin.

3. Consider the graph pictured at the right. Fill in the table below for this graph. See margin.

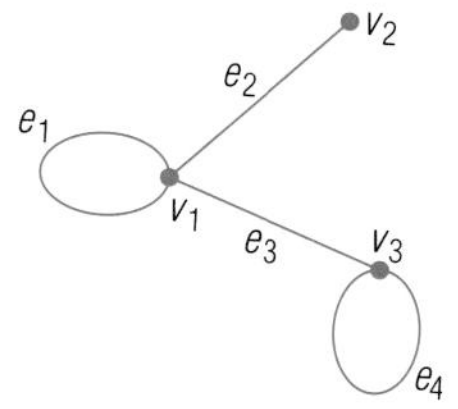

Edge	contribution
e_1	contributes **a.** ______ to the degree of **b.** ______ .
e_2	contributes **c.** ______ to the degree of **d.** ______ , and **e.** ______ to the degree of **f.** ______ .
e_3	**g.** ______
e_4	**h.** ______

NOTES ON QUESTIONS

Question 2: This question is tricky because the only counterexamples are loops.

Question 26: This question continues the review of matrix multiplication. If needed, here are two additional questions.

(a) $\begin{bmatrix} 1 & 5 \\ 2 & -1 \end{bmatrix} \cdot \begin{bmatrix} 4 \\ 8 \end{bmatrix}$ $\left(\begin{bmatrix} 44 \\ 0 \end{bmatrix}\right)$

(b) $\begin{bmatrix} 1 & 2 & -1 \\ 0 & 1 & 1 \\ -1 & 0 & 1 \end{bmatrix} \cdot \begin{bmatrix} 3 & 1 & 8 \\ 2 & 0 & 1 \\ 1 & 3 & 2 \end{bmatrix}$

$\left(\begin{bmatrix} 6 & -2 & 8 \\ 3 & 3 & 3 \\ -2 & 2 & -6 \end{bmatrix}\right)$

Question 27: A corollary to this statement is that there is a simple graph with m edges and n vertices for every pair of positive integers m and n such that $m \le \frac{n(n-1)}{2}$. (If an edge is deleted from a simple graph, the graph is still simple.)

ADDITIONAL ANSWERS

1.a. $\deg(v_1) = 2$; $\deg(v_2) = 3$; $\deg(v_3) = 0$; $\deg(v_4) = 3$

2. The statement does not include the case when edges are loops which are counted twice.

3.a. 2
b. v_1
c. 1
d. v_2
e. 2
f. v_1
g. contributes 1 to the degree of v_2 and 1 to the degree of v_1
h. contributes 2 to the degree of v_3

FOLLOW-UP

MORE PRACTICE

For more questions on SPUR Objectives, use *Lesson Master 11-3*, shown on page 661.

EXTENSION

Students can consider variants of the handshake problem. In a committee of 10 people, can subcommittees of 3 people be organized so that each person is in exactly 1 subcommittee? (No) Can subgroups of 3 people be organized so that each person is in exactly 2 subcommittees? (No) What numbers of subcommittees must a person be on so that each person could be on the same number of subcommittees? (3, 6, or 9)

ADDITIONAL ANSWERS

5.

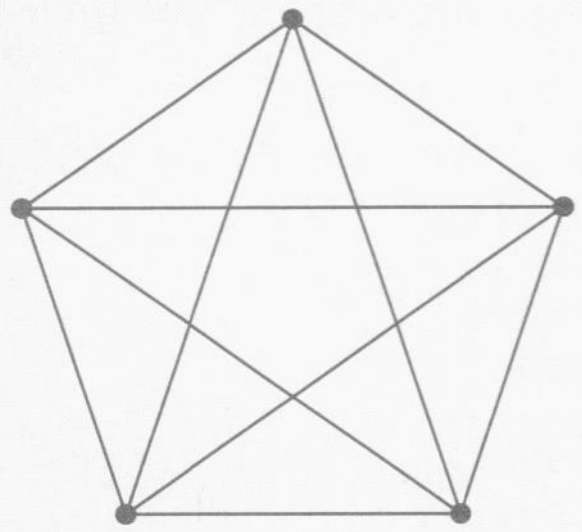

6.b.

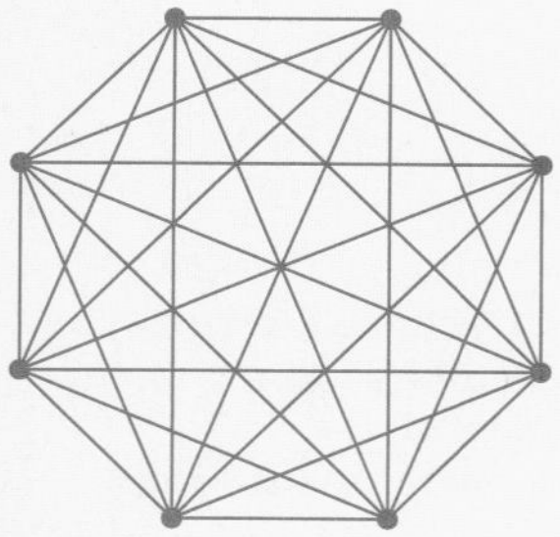

9. See Additional Answers in the back of this book.

10. The total degree of any graph equals twice the number of edges in the graph.

---

4. How many handshakes are needed for 23 people at a party if each person is to shake hands with every other person? 253

5. Draw a complete graph with 5 vertices. See margin.

6. At a family reunion, 8 cousins wish to reminisce with each other, two at a time.

a. How many conversations are needed? 28

b. Verify your answer to part **a** with a graph. See margin.

c. Explain how this problem is equivalent to a handshake problem.
It is equivalent to 8 people shaking hands with each other.

7. What is the total degree of the graph given at the beginning of this lesson? 42

8. What is the total degree of the second graph of this lesson? 8

9. Prove that, in a group of 9 people, it is impossible for every person to shake hands with exactly 3 others. See margin.

10. Correct this false statement. *The number of edges in any graph equals twice the total degree of the graph.* See margin.

11. How many different graphs are there that illustrate Example 2, part **a**? 2

Applying the Mathematics

12. Prove or disprove. *A graph must have an odd number of vertices of even degree.* See margin.

In 13–16, either draw a graph with the specified properties or explain why no such graph exists.

13. a graph with 10 vertices of degrees 1, 1, 1, 2, 3, 3, 3, 4, 5, 6
Impossible; it can't have an odd number of odd vertices.

14. a graph with 4 vertices of degrees 1, 1, 3, and 3 See margin.

15. a simple graph with 4 vertices of degrees 1, 1, 2, and 2 See margin.

16. a simple graph with 4 vertices of degrees 1, 1, 3, and 3 See margin.

17. Let G be a simple graph with n vertices.

a. What is the maximum degree of any vertex of G? $n - 1$

b. What is the maximum total degree of G? $n(n - 1)$

c. What is the maximum number of edges of G? $\frac{n(n-1)}{2}$

18. Use the answer to the handshake problem at the beginning of the lesson to deduce an expression for the number of diagonals of an n-gon. $\frac{n(n-3)}{2}$

19. At a party, the first guest to arrive shakes hands with the host. The second guest shakes hands with the host and the first guest, and so on.

a. If there are n guests, how many handshakes are there in all?

b. Relate part **a** to the first handshake problem of this lesson.
See margin.

20. Explain why the Total Degree is Even corollary follows immediately from the Total Degree of a Graph Theorem.
"Twice the number of edges" must be an even number.

21. Explain how the Number of Odd Vertices is Even Corollary follows from the Total Degree of a Graph Theorem. See margin.

660

Review

22. The numbers by the arcs show the number of non-stop flights between those cities when this question was written. Write the adjacency matrix for the graph of non-stop flights between the indicated cities. List the cities alphabetically for the rows and columns. *(Lesson 11-2)*

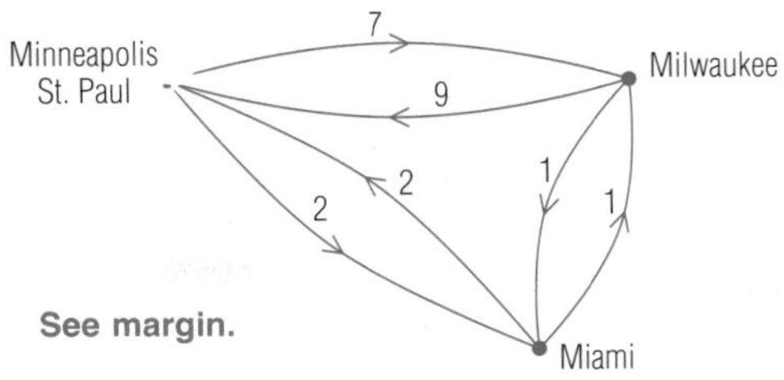

See margin.

23. Suppose that assembling a computer consists of the following steps.

	Step	Time required (hrs)	Prerequisite steps
A	Assemble memory & CPU chips	2	
B	Assemble I/O port components	3	
C	Assemble computer circuit board	3	*A*, *B*
D	Assemble disk drive	3	
E	Assemble computer	2	*C*, *D*
F	Assemble picture tube	5	
G	Assemble monitor circuit board	3	
H	Assemble monitor	4	*F*, *G*
I	Assemble key mechanism	3	
J	Assemble keyboard circuit board	2	
K	Assemble keyboard	1	*I*, *J*
L	Package computer and peripherals	1	*E*, *H*, *K*

Draw a digraph to determine the minimum time required for the entire assembling process. *(Lesson 11-1)* **See margin.**

24. Solve over the complex numbers: $20x^4 + 11x^2 - 3 = 0$. (Hint: Use chunking.) *(Lesson 3-4)* $x = \frac{\sqrt{5}}{5}, \frac{-\sqrt{5}}{5}, \frac{i\sqrt{3}}{2}, \frac{-i\sqrt{3}}{2}$

25. The graph of a function g is shown at the right. Approximate the values of x for which
 a. $g(x) > 0$, $x < -3$, $1 < x < 3$
 b. $g(x) < 0$, $-3 < x < 1$, $x > 3$
 c. g is increasing. $-1 < x < 2$
 (Lessons 2-3, 2-1)

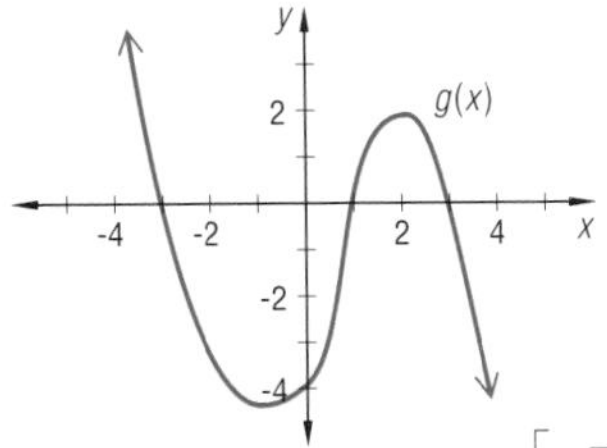

26. If $A = \begin{bmatrix} 1 & 2 \\ 3 & 5 \end{bmatrix}$, calculate $A \cdot A$ (which is A^2). *(Previous course)* $\begin{bmatrix} 7 & 12 \\ 18 & 31 \end{bmatrix}$

Exploration

27. Consider the statement: *If G is a simple graph with m edges and n vertices, then* $m \le \frac{n(n-1)}{2}$.
 a. Write its contrapositive.
 b. Prove or disprove either the statement or its contrapositive.
 See margin.

12. counterexample:

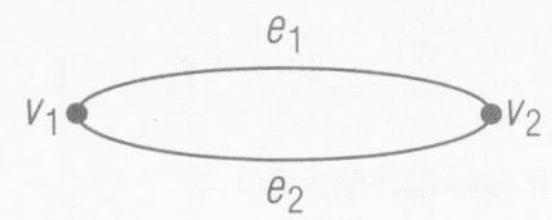

14., 15., 16., 19., 21., 22., 23., 27. See Additional Answers in the back of this book.

NAME ______

LESSON **MASTER 11–3**
QUESTIONS ON **SPUR** OBJECTIVES

■ **SKILLS** *Objective A (See pages 694–698 for objectives.)*

1. Draw a graph with four vertices of the following degrees: 1, 3, 3, and 5.

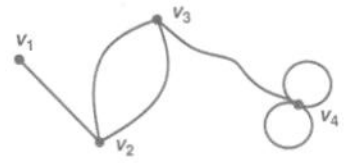

2. Draw a simple graph with three vertices of degrees 1, 2, and 1.

■ **PROPERTIES** *Objective B*

3. *Fill in the blank with the word* even *or* odd. Every graph has an **even** number of vertices of **odd** degree.

4. A graph has four edges. What is its total degree? **8**

5. Refer to the graph below.
 a. Give the degree of each vertex. **v_1: 4; v_2: 4; v_3: 6; v_4: 0**
 b. Give the total degree of the graph. **14**

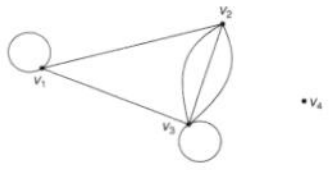

Precalculus and Discrete Mathematics © Scott, Foresman and Company *Continued* 121

NAME ______
Lesson MASTER 11–3 (page 2)

■ **PROPERTIES** *Objective C*
In 6 and 7, either draw a graph with the given properties or explain why no such graph exists.

6. a graph with five vertices of degrees 1, 2, 2, 3, and 5
 The graph is not possible since there would be an odd number of vertices of odd degree.

7. a graph with six vertices of degrees 1, 1, 2, 3, 4, and 5

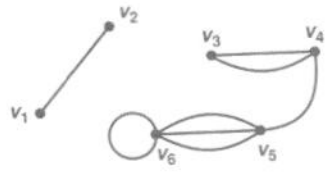

■ **USES** *Objective F*

8. Thirty-three parents meet at a graduation ceremony. Is it possible for each parent to shake hands with exactly seven other parents? Explain.
 No; Represent the situation as a graph. The 33 vertices would each have degree 7, which contradicts the corollary that states every graph has an even number of vertices of odd degree.

9. The instructor of a Spanish class requests that each one of the 12 students in the class meet with three other students, one at a time, to practice speaking Spanish.
 a. Draw a graph to represent this situation.

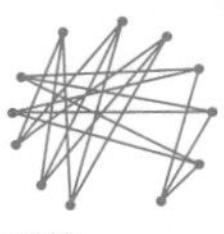

 b. How many different pairings result? **18**

122 *Precalculus and Discrete Mathematics* © Scott, Foresman and Company

LESSON 11-4

RESOURCES

- Lesson Master 11-4
- Quiz for Lessons 11-1 through 11-4
- Teaching Aid 78 can be used with **Questions 1, 2, 5, and 6**.
- Teaching Aid 79 can be used with **Questions 7, 8, and 12**.

OBJECTIVES

- **B** Identify parts of graphs and types of graphs.
- **D** Determine whether a graph has an Euler circuit.
- **G** Solve application problems involving circuits.

TEACHING NOTES

Students who have seen the Königsberg bridge problem before may remember the rule for determining whether a graph is traversable. A proof is given in this lesson that requires precise definitions of the terms *walk*, *path*, *circuit*, and *Euler circuit*. The table at the bottom of page 663 shows the successive restrictions on the terms from *walk* to *Euler circuit*.

The concept of *connectedness* is needed to give a sufficient condition for the existence of an Euler circuit. Guide students through the definition of connectedness on page 665. We have made the definition compact by using symbolic logic notation. You might ask students to write the definition without symbols.

LESSON

11-4

The Königsberg Bridge Problem

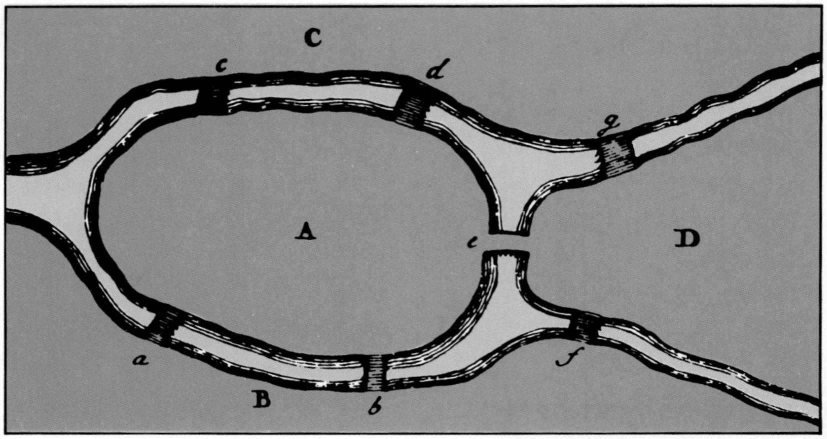

The beginning of this chapter posed the Königsberg bridge problem:

> In the city of Königsberg, is it possible for a person to walk around the city crossing each bridge exactly once, starting and ending at the same point?

It was pointed out in Lesson 11-1 that this problem is equivalent to the question of whether it is possible to trace the graph below without picking up your pencil, traversing each edge exactly once and starting and ending at the same vertex.

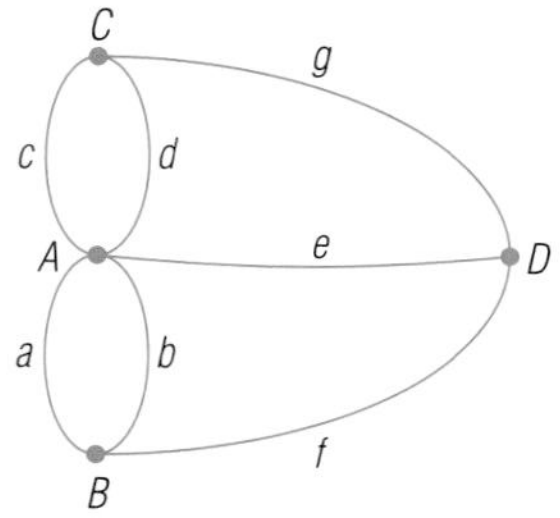

Now we are ready to present Euler's solution. The following definitions help to formulate the problem more precisely and provide a language for much of the subject of graph theory.

Definitions

Suppose that G is a graph and v and w are vertices of G.
A **walk from v to w** is an alternating sequence of adjacent vertices and edges beginning with v and ending with w.

A **path from v to w** is a walk from v to w in which no edge is repeated.

A **circuit** is a path that starts and ends at the same vertex.

An **Euler circuit** is a circuit that contains every edge and vertex of G.

With this language, the Königsberg bridge problem can be restated as follows: *Does the graph at the beginning of this lesson have an Euler circuit?*

To see the differences between these terms, consider the graph G below.

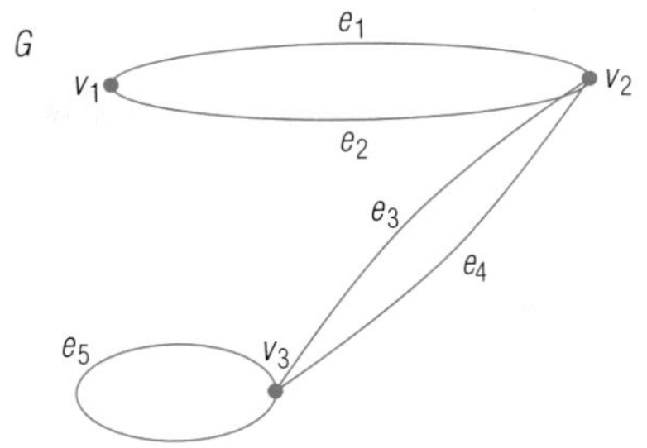

The following is a walk from v_3 to v_1:

$$v_3\ e_5\ v_3\ e_3\ v_2\ e_4\ v_3\ e_3\ v_2\ e_1\ v_1.$$

Of course, this is a very inefficient way to get from v_3 to v_1, but it *is* an alternating sequence of adjacent vertices and edges that starts at v_3 and ends at v_1. This walk, however, is not a path, since it repeats the edge e_3. (When the context prevents any confusion, a walk can be described by listing only the edges. In the above case, the walk is $e_5\ e_3\ e_4\ e_3\ e_1$.)

The following are some other paths from v_3 to v_1.

$e_3\ e_2$ (that is, $v_3\ e_3\ v_2\ e_2\ v_1$)
$e_5\ e_3\ e_2$
$e_4\ e_1$

Since a circuit must start and end at the same point and cannot have a repeated edge, the following are circuits from v_1 to v_1.

$e_2\ e_3\ e_4\ e_1$
$e_1\ e_2$

Since an Euler circuit must contain every edge and vertex of the graph, the following is an Euler circuit in the graph of G, starting at v_1.

$e_1\ e_4\ e_5\ e_3\ e_2$

The table below shows the differences among walks, paths, circuits, and Euler circuits.

	repeated edge?	starts and ends at the same point?	includes every edge and vertex?
walk	allowed	allowed	allowed
path	no	allowed	allowed
circuit	no	yes	allowed
Euler circuit	no	yes	yes

Euler proved the following theorem. It gives a necessary condition for a graph to have an Euler circuit, and it solves the Königsberg bridge problem.

Making Connections
The Circuits and Connectedness Theorem is used in the proof of Euler's Formula $V - E + F = 2$ in Lesson 11-7. This theorem is critical in reducing a figure with $n + 1$ edges to one that has n edges so that an induction hypothesis can be used. It may seem as if the theorem is so obvious that it could not have any place in a nonobvious proof, but it does.

ADDITIONAL EXAMPLES

1. Tell whether or not the graph is connected.

a.

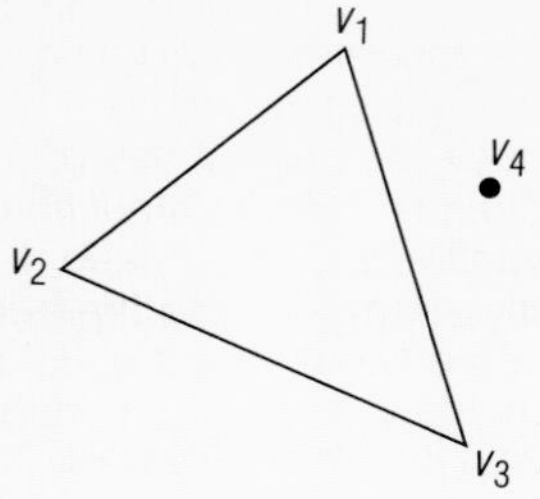

No

b.

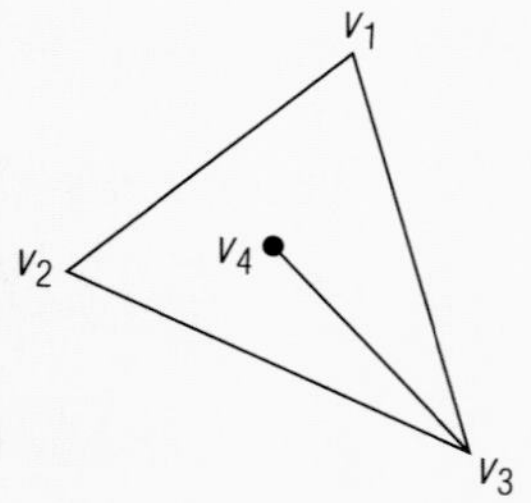

Yes

c.

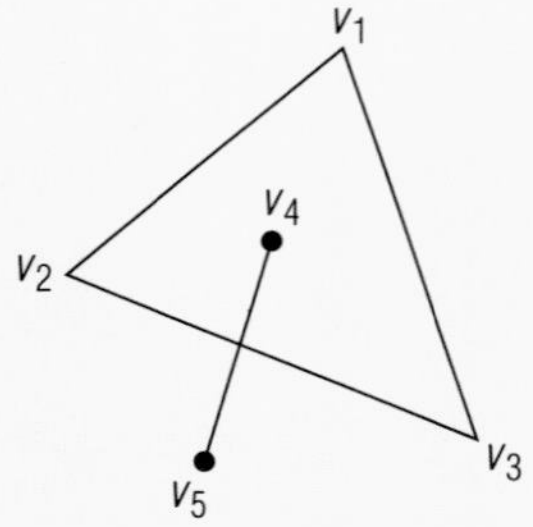

No

Euler Circuit Theorem

If a graph has an Euler circuit, then every vertex of the graph has even degree.

Proof Suppose G is any graph that has an Euler circuit. To show that every vertex of G has even degree, we take any particular but arbitrarily chosen vertex v and show that the degree of v must be even. Either v is or is not the vertex where the Euler circuit starts and stops. If v is not, then each time the circuit enters v on one edge, it must leave v by a different edge. Thus the edges having v as an endpoint must occur in entry/exit pairs. This idea is illustrated by the diagram below.

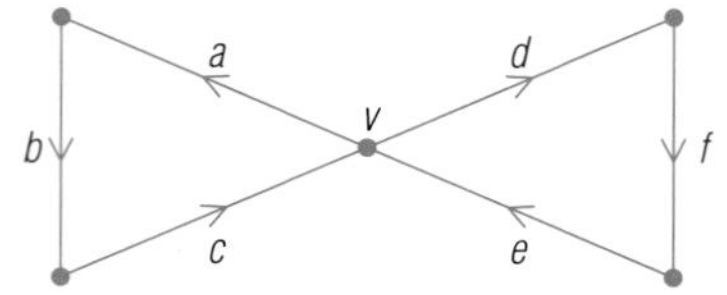

In an Euler circuit, each time v is entered by one edge, it is exited by another edge.

If v is the initial vertex of the Euler circuit, then the first edge of the circuit (that leads out of v) can be paired with the last edge of the circuit (that leads into v). Also, as above, any other edges having v as an endpoint occur in entry/exit pairs.

It follows that regardless of whether or not the circuit starts and ends at v, all the edges having v as an endpoint can be divided into pairs. Therefore, the number of such edges is even, which is what was to be shown.

Now examine the graph of the Königsberg bridges. All four vertices have odd degree. Since one vertex of odd degree is enough to preclude there being an Euler circuit, the answer to the question of the Königsberg bridge problem is "No." The Königsberg bridge graph has no Euler circuit.

Does the converse of the Euler Circuit Theorem hold true? Is it true that if every vertex of a graph has even degree, then the graph has an Euler circuit? The answer is no. The following 4-vertex counterexample shows a graph in which every vertex has even degree yet there is no Euler circuit.

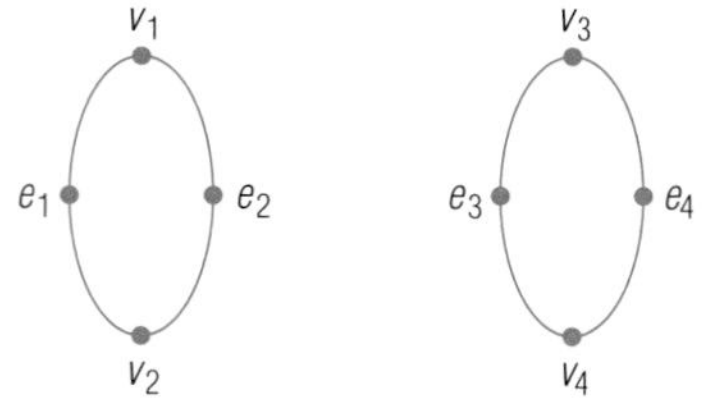

However, the converse to the Euler Circuit Theorem is true under certain conditions. To discuss these conditions requires the concept of *connectedness*. Roughly speaking, a graph is *connected* if it is possible to travel from any vertex to any other vertex along adjacent edges.

664

> **Definitions**
>
> Suppose G is a graph. Two vertices v and w in G are **connected vertices** if and only if there is a walk in G from v to w. G is a **connected graph** if and only if $\forall$ vertices v and w in G, $\exists$ a walk from v to w.

Example 1 Tell whether or not each graph is connected.

a.

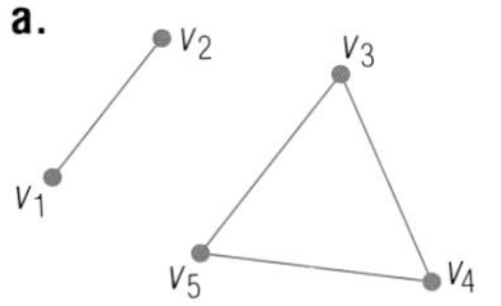

b.

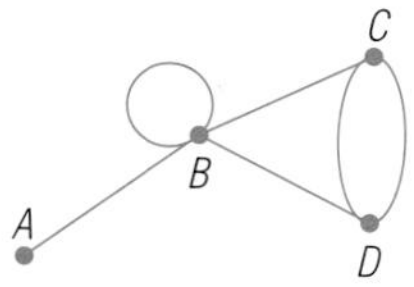

c.

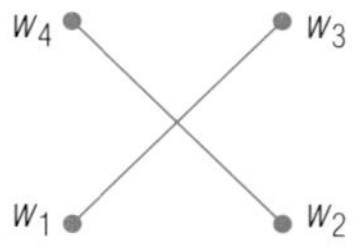

Solution

a. The graph is not connected. It is impossible to find a walk from v_1 to v_3, for instance.

b. The graph is connected; each vertex is connected to each other vertex by a walk.

c. The graph is not connected. For instance, there is no walk from w_1 to w_2. This can also be shown by redrawing the graph without crossings as follows.

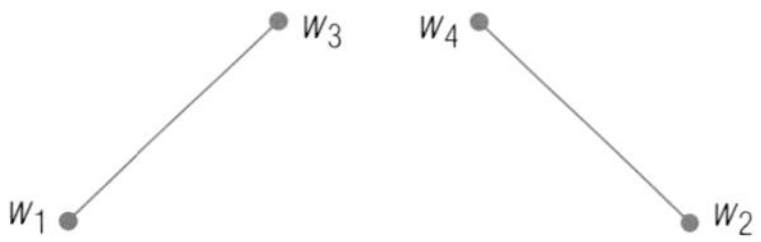

With the idea of connectedness, a sufficient condition for an Euler circuit can be given.

> **Theorem (Sufficient Condition for an Euler Circuit)**
>
> If a graph G is connected and every vertex of G has even degree, then G has an Euler circuit.

The proof is omitted because it is quite long.

Example 2 Does the following graph have an Euler circuit? If so, find such a circuit.

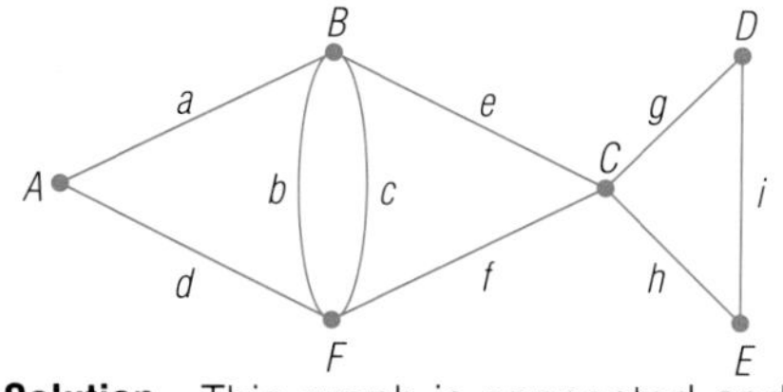

Solution This graph is connected and every vertex has even degree: deg(A) = deg(D) = deg(E) = 2, deg(B) = deg(C) = deg(F) = 4. Hence this graph has an Euler circuit. One such circuit starting at A is $a\ b\ c\ e\ g\ i\ h\ f\ d$.

2. Does the following graph have an Euler circuit? If so, find such a circuit. If not, why not?

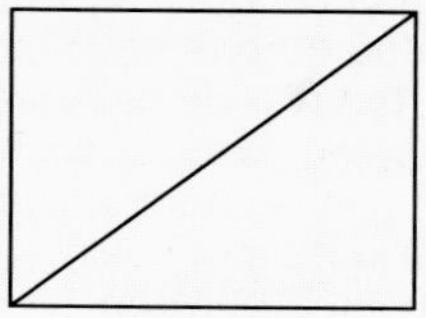

No; it has two vertices of odd degree.

NOTES ON QUESTIONS
Question 4: Emphasize that crossings do not give any indication of whether a graph is connected or not. All of these graphs have crossings, but two of them are not connected. And, of course, there can be a connected graph without crossings, for example, a polygon.

Question 17: The percentages give a false impression of the accuracy of the test in the sense that there are many more false positives that one would like. This is why it is critical that tests be accurate far more than even 98% of the time.

Now consider a connected graph which contains a circuit.

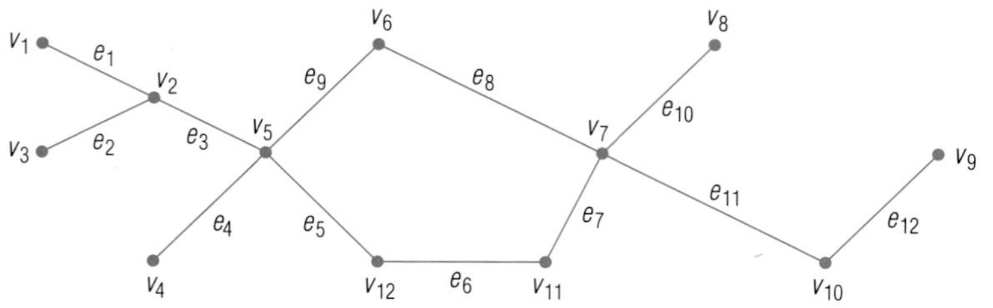

This graph has the following circuit starting at v_6: $e_9\ e_5\ e_6\ e_7\ e_8$. Suppose one edge is removed from the circuit. For instance, suppose edge e_6 is removed. Is the resulting graph connected?

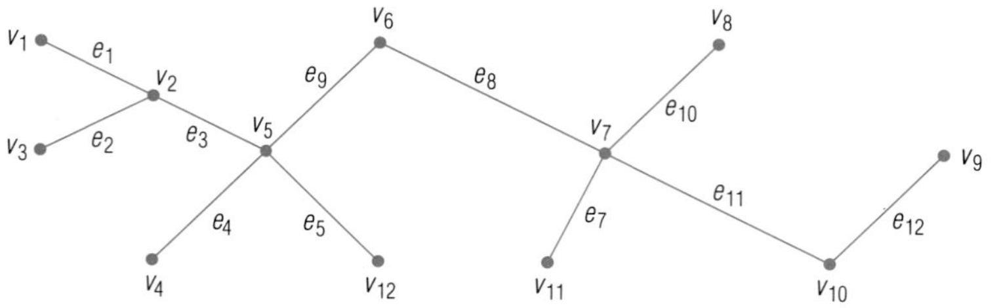

The answer, of course, is yes. For instance, although the direct connection from v_{12} to v_{11} has been removed, the indirect connection obtained by traveling the other way around the circuit $e_5\ e_9\ e_8\ e_7$ remains. By the same reasoning, any other pair of vertices connected using the original edge e_6 can also be connected by using this indirect route.

This discussion can be formalized to prove the following theorem which seems obvious but has a nonobvious application in Lesson 11-7.

Theorem (Circuits and Connectedness)

If a connected graph contains a circuit and an edge is removed from the circuit, then the resulting graph is also connected.

Questions

Covering the Reading

In 1 and 2, refer to the graph pictured below. A walk is given. **a.** Is it a path? **b.** Is it a circuit? **c.** Is it an Euler circuit?

1. $v_5\ e_7\ v_1\ e_8\ v_5\ e_7\ v_1$
a) No; b) No; c) No

2. $e_4\ e_6\ e_7\ e_2$
a) Yes; b) Yes; c) No

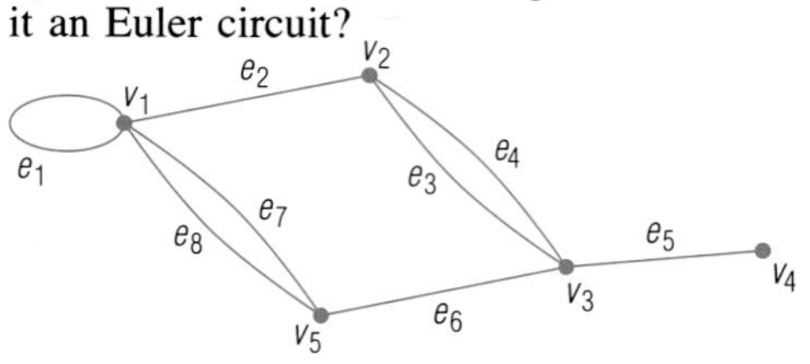

3. a. Write the contrapositive of the Euler Circuit Theorem. See margin.
b. How is the contrapositive used to test whether a given graph has an Euler circuit? **The presence of an odd vertex is sufficient to show that a graph cannot have an Euler circuit.**

4. Tell whether the graph pictured is connected.

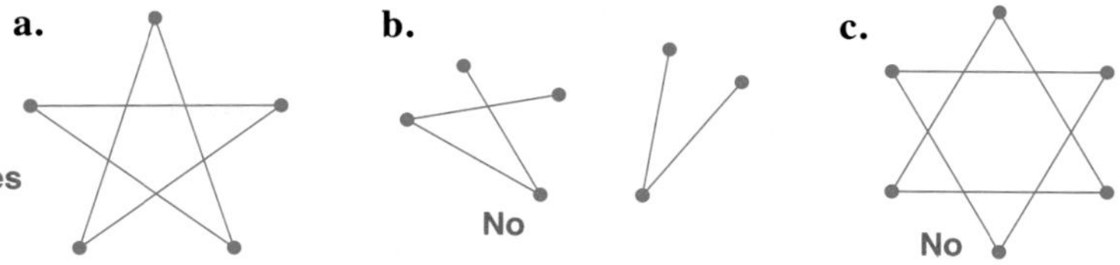

a. Yes b. No c. No

5. In the graph pictured below, the walk $e_1\ e_2\ e_3\ e_4\ e_5$ connects v_1 and v_9.

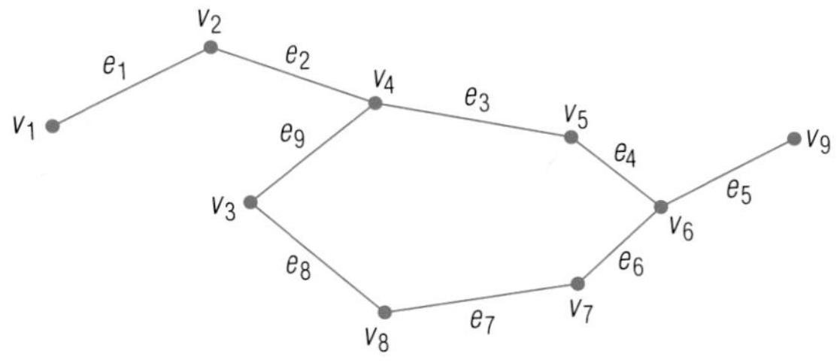

Now consider the graph obtained by removing edge e_3. Describe a walk in this graph that connects v_1 and v_9. $e_1\ e_2\ e_9\ e_8\ e_7\ e_6\ e_5$

6. Does the following graph have an Euler circuit? If so, describe the circuit. **Yes, sample from vertex *A*: *a b f g h i c e d j***

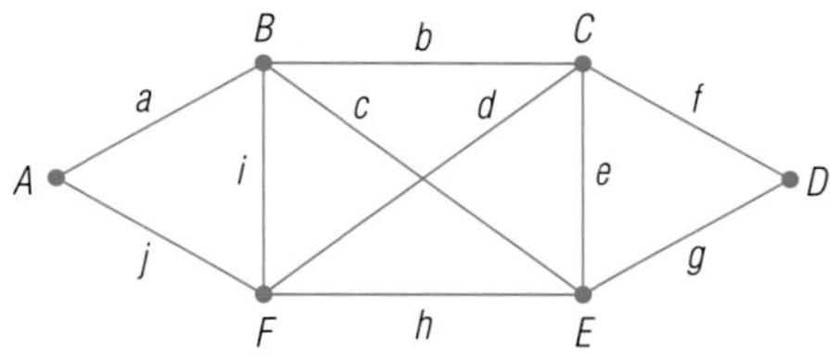

Applying the Mathematics

7. a. In the graph pictured below, list each edge that could be removed individually without disconnecting the graph. $e_1, e_2, e_3, e_4, e_5, e_6$
b. What is the maximum number of edges that can be removed at the same time without disconnecting the graph? 2

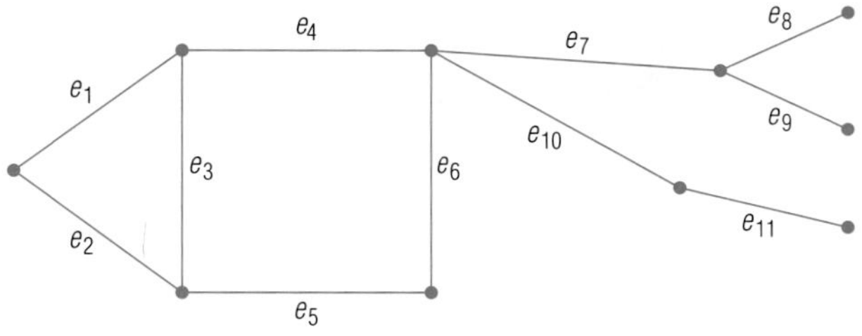

FOLLOW-UP

MORE PRACTICE
For more questions on SPUR Objectives, use *Lesson Master 11-4*, shown on page 669.

PROJECTS
The projects for Chapter 11 are described on pages 689–690. **Projects 1 and 5** are related to the content of this lesson.

EVALUATION
A quiz covering Lessons 11-1 through 11-4 is provided in the Teacher's Resource File.

Alternative Assessment
Have each student draw a graph that represents his or her walk through the school building on a typical day of school. The point at which students enter the building and each classroom, lunchroom, and so on, would be vertices and the hallways would be edges. Based on their graphs, have students answer the following questions:
(1) What is your usual route through your graph?
(2) Is an Euler circuit possible? If so, list it.
(3) If an Euler circuit is not possible, where could additional edges be added so that an Euler circuit does exist?

ADDITIONAL ANSWERS
3.a. If at least one vertex of a graph has an odd degree, then the graph does not have an Euler circuit.

ADDITIONAL ANSWERS

10. Not necessarily; there is a connected graph (which must have an Euler circuit) and a nonconnected graph (which cannot).

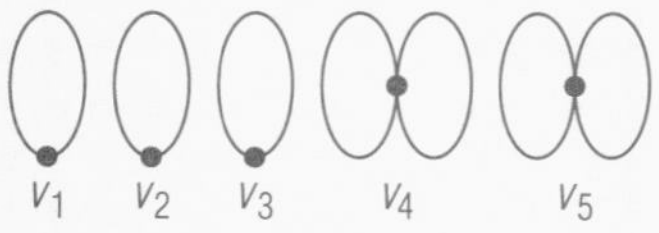

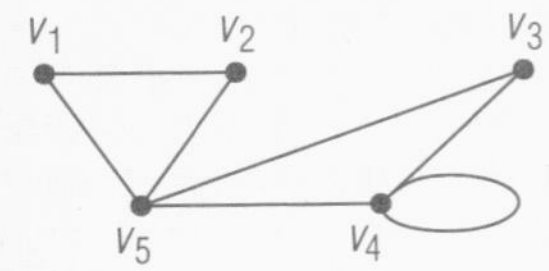

15. sample:

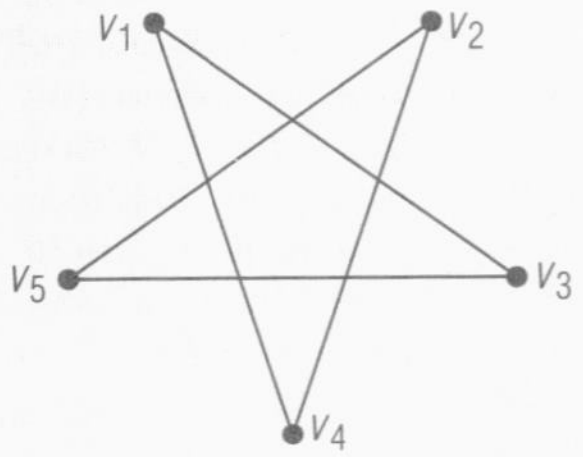

17.a.

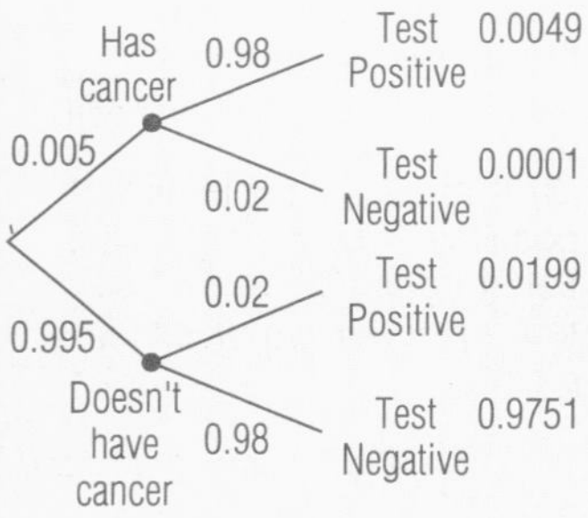

8. Consider the graph pictured below.
a. Find all paths from v_1 to v_3. **See page 669.**
b. Find five walks from v_1 to v_3 that are not paths. **See page 669.**
c. How many walks from v_1 to v_3 have no repeated vertex? **3**
d. Can you list all possible walks from v_1 to v_3? Explain.

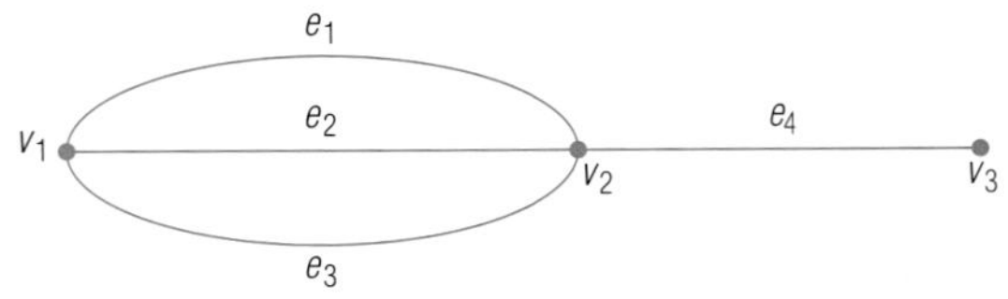

d) No, you can use pairs of e_1, e_2, and e_3 as many times as you wish.

9. If a graph has an Euler circuit, must the graph be connected? Explain your answer. **Yes, if the graph is not connected, there is no way a circuit could contain every vertex.**

10. Suppose G is a graph with five vertices of degrees 2, 2, 2, 4, and 4. Answer *yes*, *no*, or *not necessarily* to the question: Does G have an Euler circuit? Justify your answer. **See margin.**

11. Could a citizen of Königsberg have taken a walk around the city crossing each bridge exactly twice before returning to the starting point? Explain your answer. **See page 669.**

12. The city of Paris is situated along the banks of the Seine River and includes two islands in the river. Using the map below, determine if it is possible to take a walk around Paris starting and ending at the same point and crossing each bridge exactly once. **No**

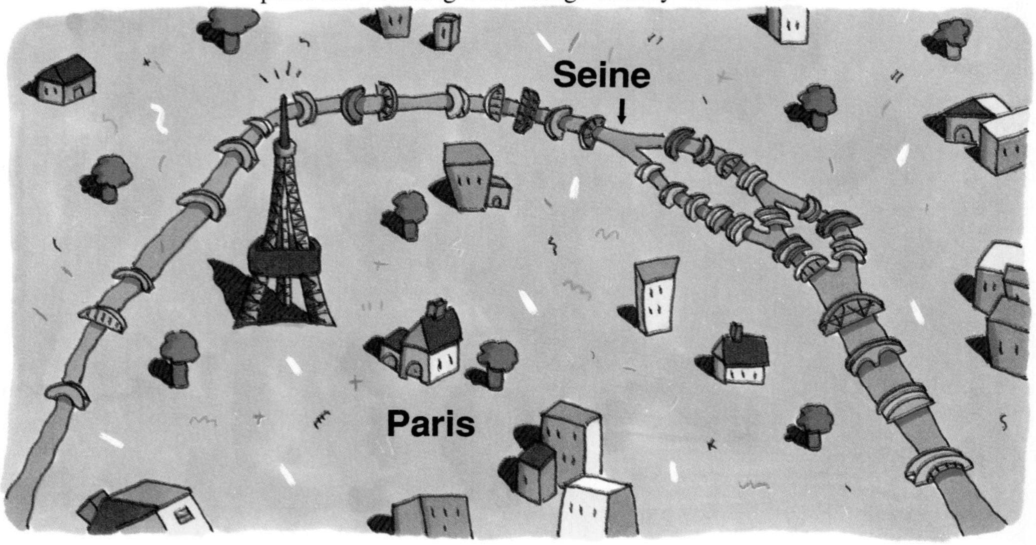

13. If a graph contains a walk from one vertex v to a different vertex w, must it contain a path from v to w? Explain. **Yes, if the walk repeats an edge, then there is a circuit. Remove edges from the graph until there is no circuit. Then connect v to w.**

668

Review

14. Who founded the subject of graph theory and in what century did he live? *(Lesson 9-1, Previous course)* **Leonhard Euler, eighteenth**

15. Show, with a graph, that it is possible for 5 people on a committee each to shake the hands of two others. *(Lesson 11-3)* **See margin.**

16. Write the adjacency matrix for the vertices of a tetrahedron. *(Lesson 11-2)* **See below.**

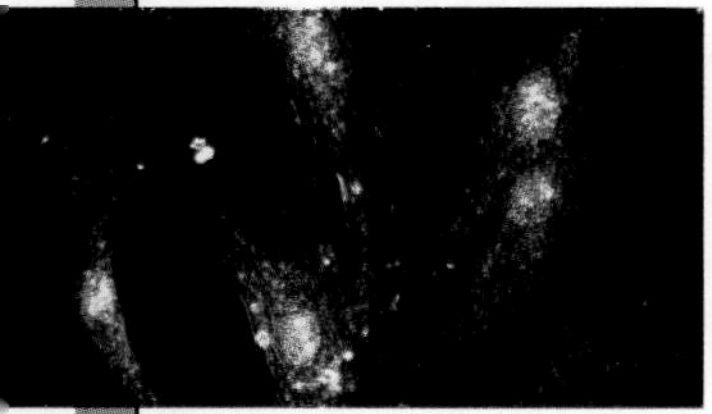

Normal cells from a person's lip

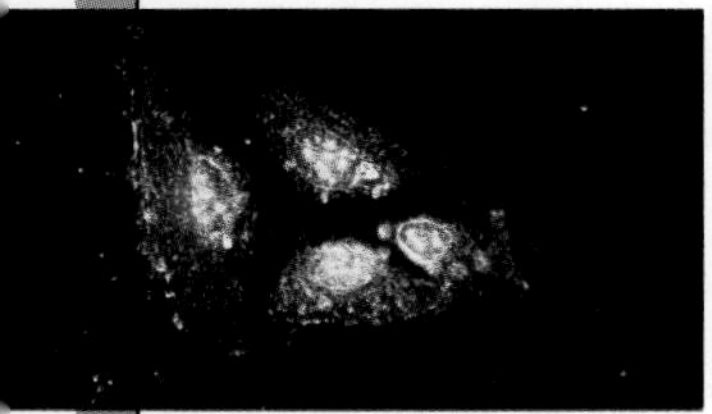

Cancer cells from a person's lip

17. Assume that there is a test for cancer which is 98 percent accurate; that is, if someone has cancer, the test will be positive (signifying a cancer) 98 percent of the time, and if one doesn't have it, the test will be negative (signifying no cancer) 98 percent of the time. Also assume that 0.5% of the population has cancer.

a. Draw a probability tree and label its edges with probabilities to represent this situation. **See margin.**

b. Imagine that you are tested for cancer. What is the probability that the test will be positive? **≈ 2.48%**

c. What is the probability that if the test is positive, then you have cancer? **≈ 19.76%**

d. What is the probability that if the test is positive, then you don't have cancer? *(Lesson 11-1)* **≈ 80.24%**

18. How many whole numbers less than 10,000 have the property that the sum of their digits is 7? *(Lesson 10-7)* **120**

19. Suppose $\frac{\pi}{2} < x < \pi$ and $\csc x = 5$. Find $\tan x$. *(Lesson 5-6)* **≈ -.204**

20. Find a transformation that will transform the graph of $y = e^x$ into the graph of $y = 3e^{x-5} + 4$. *(Lesson 3-9)* **$(x, y) \rightarrow (x + 5, 3y + 4)$**

Exploration

21. Let K_n denote the complete graph with n vertices. For what values of n does K_n have an Euler circuit? Justify your answer. **n must be an odd number. For the complete graph of an n-gon, the degree of each vertex is $n - 1$. That degree is even if n is odd.**

8. a) e_1e_4, e_2e_4, e_3e_4, $e_1e_2e_3e_4$, $e_1e_3e_2e_4$, $e_2e_1e_3e_4$, $e_2e_3e_1e_4$, $e_3e_1e_2e_4$, $e_3e_2e_1e_4$
b) samples: $e_1e_1e_1e_4$, $e_1e_2e_2e_4$, $e_1e_3e_3e_4$, $e_1e_2e_1e_4$, $e_1e_3e_1e_4$

11. Yes; think of replacing each bridge by two bridges. Such a walk exists by the sufficient condition for an Euler Circuit Theorem, since every vertex will have an even degree.

16. $\begin{bmatrix} 0 & 1 & 1 & 1 \\ 1 & 0 & 1 & 1 \\ 1 & 1 & 0 & 1 \\ 1 & 1 & 1 & 0 \end{bmatrix}$

NAME ______

LESSON **MASTER 11–4**
QUESTIONS ON **SPUR** OBJECTIVES

■ **SKILLS** *Objective A (See pages 694–698 for objectives.)*

1. Draw a graph with 4 vertices, one loop, and 2 parallel edges, and which has an Euler circuit. Identify the circuit.
sample: starting at v_1:
$e_1 e_6 e_4 e_5 e_3 e_2$

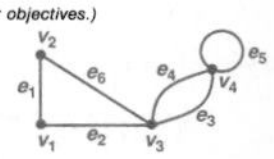

2. Consider the graph at the right below. Describe, if possible:

a. a walk from v_1 to v_4, **sample: $e_1 e_5 e_4$**

b. a path from v_4 to v_7, **sample: $e_4 e_6 e_7$**

c. an Euler circuit starting at v_1, **not possible**

d. a walk from v_1 to v_4 that is not a path. **sample: $e_1 e_8 e_7 e_6 e_5 e_2 e_3 e_5 e_4$**

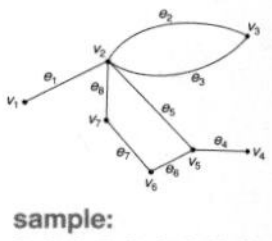

In 3–5, determine whether or not the graph is connected.

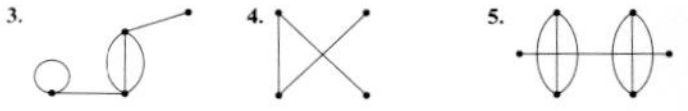

3. **yes** 4. **yes** 5. **no**

6. Consider the graph at the right.

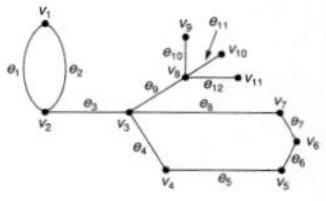

a. Give each edge whose removal (by itself) would keep the graph connected. **e_1, e_2, e_4, e_5, e_6, e_7, e_8**

b. What is the maximum number of edges that can be removed simultaneously while keeping the graph connected? **2**

 Continued 123

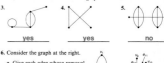

NAME ______
Lesson MASTER 11–4 (page 2)

■ **PROPERTIES** *Objective D*

7. Consider the statement: *If every vertex of a graph has even degree, then the graph has an Euler circuit.*

a. Write the converse of the statement.
If a graph has an Euler circuit, then every vertex of the graph has even degree.

b. Which is true, the statement or its converse? **converse**

c. What additional characteristic, if possessed by the graph, makes both statements true? **connectedness**

In 8 and 9, determine whether or not the graph described has an Euler circuit. Justify your answer.

8.

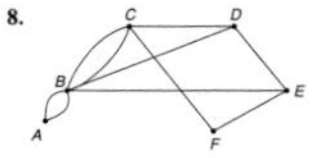

No; Vertices D and E have odd degree.

9. the graph with adjacency matrix

$$\begin{array}{c} \\ v_1 \\ v_2 \\ v_3 \\ v_4 \end{array} \begin{array}{c} \begin{array}{cccc} v_1 & v_2 & v_3 & v_4 \end{array} \\ \begin{bmatrix} 2 & 0 & 0 & 4 \\ 0 & 2 & 1 & 0 \\ 0 & 1 & 2 & 0 \\ 4 & 0 & 0 & 2 \end{bmatrix} \end{array}$$

No; Vertices v_2 and v_3 have odd degree.

■ **REPRESENTATIONS** *Objective G*

10. Suppose a pirate finds a treasure map like the one below. If the treasure is at one end of the points marked, is it possible for the pirate to search at every point, starting and ending at the same point and using each edge only once? If so, draw arrows on the map to show a possible route. If not, draw arrows on the map to show a route which starts and ends at the same point and passes through as many other points as possible.

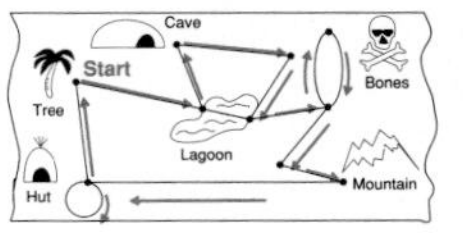

Not possible; skip edge shown across lagoon

124

LESSON 11-5

RESOURCES

- Lesson Master 11-5
- Teaching Aid 80 can be used with **Examples 1 and 2** and **Questions 2–4**.
- Teaching Aid 81 can be used with **Questions 8, 12, 15, and 18**.

OBJECTIVES

B Identify parts of graphs and types of graphs.

J Use the powers of the adjacency matrix of a graph to find the number of walks of a given length from a given starting vertex to a given ending vertex.

TEACHING NOTES

Making Connections
It may surprise students that matrix operations have an interpretation with graphs, since the matrix adjacency introduced in Lesson 11-2 seems to be merely a recording device. Furthermore, matrix multiplication tends to be viewed as an artificial operation by many people. Now, putting the two together, the powers of an adjacency matrix turn out to have a simple interpretation. This is a good opportunity to point out the unity of mathematics.

Although the double subscript notation a_{ij} is natural (it takes two numbers to indicate a row and a column) it can be confusing to students. Be certain to go over **Question 2**.

LESSON

11-5

Matrix Powers and Walks

In a matrix A, the element in the ith row and jth column is often denoted by a_{ij}. This notation is quite useful with adjacency matrices. For example, consider the directed graph pictured below along with its adjacency matrix.

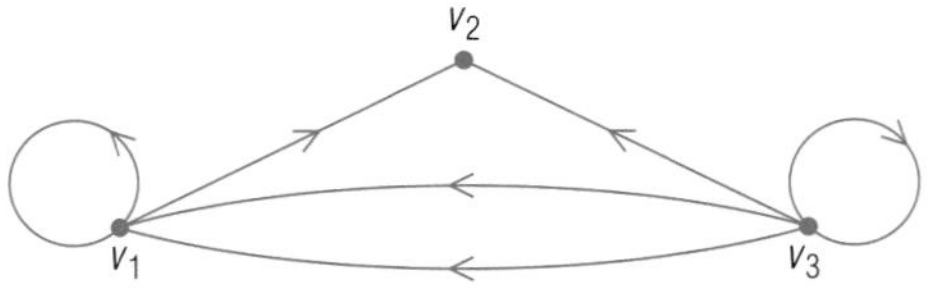

$$\begin{array}{c} \\ v_1 \\ v_2 \\ v_3 \end{array} \begin{array}{c} \begin{matrix} v_1 & v_2 & v_3 \end{matrix} \\ \begin{bmatrix} 1 & 1 & 0 \\ 0 & 0 & 0 \\ 2 & 1 & 1 \end{bmatrix} \end{array}$$

The **length** of a walk is defined to be the number of edges in the walk. Then the entry $a_{31} = 2$ can be interpreted as indicating that there are two walks of length 1 from v_3 to v_1. The first number of the subscript in a_{31} indicates the starting vertex; the second number, the ending vertex. The entry $a_{33} = 1$ indicates there is one walk of length 1 from v_3 to itself.

Now consider the undirected graph and its adjacency matrix pictured below.

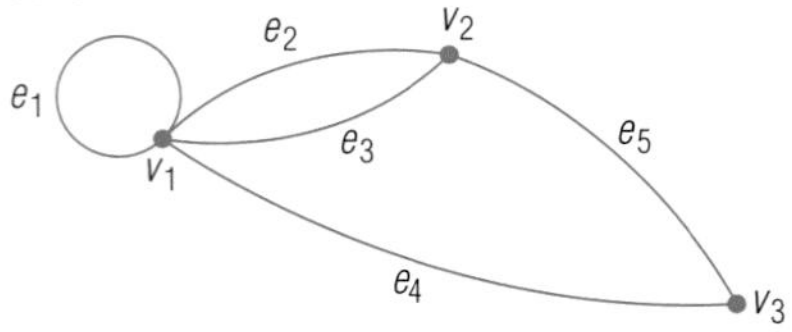

$$\begin{array}{c} \\ v_1 \\ v_2 \\ v_3 \end{array} \begin{array}{c} \begin{matrix} v_1 & v_2 & v_3 \end{matrix} \\ \begin{bmatrix} 1 & 2 & 1 \\ 2 & 0 & 1 \\ 1 & 1 & 0 \end{bmatrix} \end{array}$$

This matrix has an interesting characteristic. Its entries are symmetric around the main diagonal. For instance, $a_{12} = a_{21}$ and $a_{32} = a_{23}$. This is true since an edge from v_i to v_j also goes from v_j to v_i. Every matrix representation of an undirected graph will be a square matrix with this characteristic. Such a matrix is called **symmetric**.

Example 1 How many walks of length 2 are there from v_1 to v_3 in the above graph?

Solution A walk of length 2 from v_1 to v_3 will go through an "intermediate" vertex. There are two such walks with v_2 as the intermediate vertex (e_2e_5 and e_3e_5). There is one such walk with v_1 as the intermediate vertex (e_1e_4), and none with v_3 as the intermediate vertex. Thus, there are three walks of length 2 from v_1 to v_3.

The length of a walk has a connection with matrices. Let A be the adjacency matrix of the graph in Example 1.

$$A = \begin{array}{c} \\ v_1 \\ v_2 \\ v_3 \end{array} \begin{array}{c} \begin{matrix} v_1 & v_2 & v_3 \end{matrix} \\ \begin{bmatrix} 1 & 2 & 1 \\ 2 & 0 & 1 \\ 1 & 1 & 0 \end{bmatrix} \end{array}$$

Each entry a_{ij} gives the number of walks of length 1 from vertex v_i to vertex v_j. Thus $a_{21} = 2$ indicates that there are 2 walks of length 1 from vertex v_2 to vertex v_1. Since A is a square matrix, you can multiply it by itself.

$$A \bullet A = A^2 = \begin{bmatrix} 1 & 2 & 1 \\ 2 & 0 & 1 \\ 1 & 1 & 0 \end{bmatrix} \begin{bmatrix} 1 & 2 & 1 \\ 2 & 0 & 1 \\ 1 & 1 & 0 \end{bmatrix}$$

Let b_{ij} be the entry in row i, column j of A. Then b_{13} in A^2 is the product of row 1 and column 3 in A, by the rule for matrix multiplication.

$$b_{13} = [\ 1\ 2\ 1\ ] \begin{bmatrix} 1 \\ 1 \\ 0 \end{bmatrix} = 1 \bullet 1 + 2 \bullet 1 + 1 \bullet 0 = 3$$

Notice that this computation of b_{13} also computes the number of walks of length 2 from v_1 to v_3:

$$\begin{aligned} & \begin{bmatrix} \text{number of walks of length 1} \\ \text{from } v_1 \text{ to } v_1 \end{bmatrix} \times \begin{bmatrix} \text{number of walks of length 1} \\ \text{from } v_1 \text{ to } v_3 \end{bmatrix} \\ + & \begin{bmatrix} \text{number of walks of length 1} \\ \text{from } v_1 \text{ to } v_2 \end{bmatrix} \times \begin{bmatrix} \text{number of walks of length 1} \\ \text{from } v_2 \text{ to } v_3 \end{bmatrix} \\ + & \begin{bmatrix} \text{number of walks of length 1} \\ \text{from } v_1 \text{ to } v_3 \end{bmatrix} \times \begin{bmatrix} \text{number of walks of length 1} \\ \text{from } v_3 \text{ to } v_3 \end{bmatrix} \end{aligned}$$

The entire matrix A^2 is computed in the same way.

$$A^2 = \begin{bmatrix} 6 & 3 & 3 \\ 3 & 5 & 2 \\ 3 & 2 & 2 \end{bmatrix}$$

The entry in row 1 and column 3 of A^2 is 3, the number of walks of length 2 between v_1 and v_3 that was found in Example 1. Using similar reasoning, $b_{22} = 5$ indicates there are 5 walks of length 2 from v_2 to v_2. You should try to find these 5 walks.

The previous discussion is a special case of the following wonderful theorem, which can be proved using mathematical induction. As usual, $A^n = \underbrace{A \bullet A \bullet \ldots \bullet A}_{n \text{ factors}}$.

Theorem

Let G be a graph with vertices $v_1, v_2, \ldots, v_m$, and let n be a positive integer. Let A be the adjacency matrix for G. Then the element a_{ij} in A^n is the number of walks of length n from v_i to v_j.

The discussion on this page is designed to show that the connection between A^2 and walks of length 2 is not a coincidence; it can be predicted from the multiplications done to get each element of A^2. Briefly, to go from a vertex v_1 to a vertex v_3 with a walk of length 2, you must go from v_1 to an intermediate vertex v_i and then to v_3. Row 1 shows how to go from v_1 to v_i for all i, and the columns show how to go from v_i to v_3. Multiplying them gives the number of possible walks.

The computation of powers will not be difficult for students who have seen matrix multiplication before. If your students have not, you can define the power of a matrix using the explanation on this page.

The three missing walks from the answer to **Example 2** are the subject of **Question 6**.

ADDITIONAL EXAMPLES

1. Find the number of walks of length 2 from v_2 to v_1 in this graph.

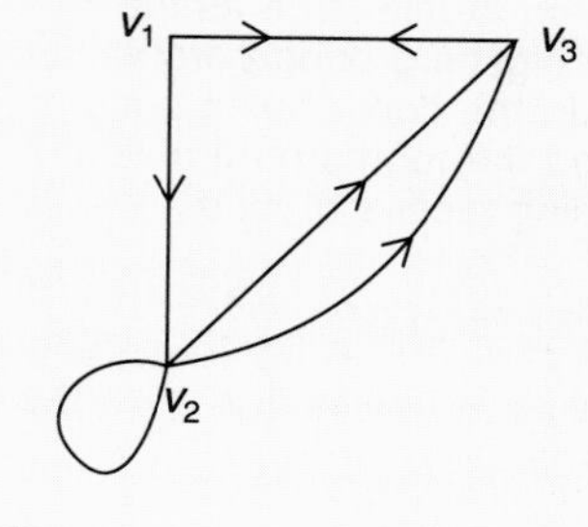

2
This can be verified by taking the second power of this adjacency matrix.

$$\begin{bmatrix} 0 & 1 & 1 \\ 0 & 1 & 2 \\ 1 & 0 & 0 \end{bmatrix}$$

ADDITIONAL EXAMPLES

2. In the graph of Additional Example 1, find the number of walks of length 3 from v_1 to v_3.

3

This should be done first by matrix multiplication. (Only one element of the matrix for the cube needs to be calculated.) Then it should be verified by examining the graph and seeing that there are three ways to go from v_1 to v_3 in a walk of length 3.

NOTES ON QUESTIONS

Question 8: This question can be answered both with and without matrices. Encourage students to use both approaches.

Question 9: You might wish to check a few elements of A^2 and A^3 by examining a graph for A.

Question 11: This kind of question reinforces the relationship between graphs and their adjacency matrices.

Question 18: Many students will need guidance with this question. Encourage them to think of it as the Königsberg bridge problem, with the doors being bridges and the rooms (and outside) being pieces of land.

ADDITIONAL ANSWERS

3. False; the matrix at the right represents a directed graph.

4. False; there is more than one walk from v_1 to v_2 in the graph at the right.

9. See Additional Answers in the back of this book.

10. $M^2 = \begin{bmatrix} 2 & 3 & 1 \\ 3 & 13 & 6 \\ 1 & 6 & 10 \end{bmatrix}$

$M^3 = \begin{bmatrix} 3 & 9 & 11 \\ 9 & 44 & 42 \\ 11 & 42 & 19 \end{bmatrix}$

Example 2 Determine the number of walks of length 3 between v_1 and v_2 in the graph of Example 1.

Solution The answer is given by the element a_{12} of A^3.

$$A^3 = \begin{bmatrix} 1 & 2 & 1 \\ 2 & 0 & 1 \\ 1 & 1 & 0 \end{bmatrix}^3 = \begin{bmatrix} 15 & 15 & 9 \\ 15 & 8 & 8 \\ 9 & 8 & 5 \end{bmatrix} \text{ and so } a_{12} = 15.$$

Thus, there are 15 walks of length 3 between v_1 and v_2.

Check The walks can be listed. Here are 12 of them. Which three are missing?

$e_1e_1e_2$ $e_1e_1e_3$ $e_2e_3e_2$ $e_3e_2e_3$ $e_2e_2e_2$ $e_3e_3e_3$ $e_1e_4e_5$
$e_2e_5e_5$ $e_3e_5e_5$ $e_4e_4e_3$ $e_4e_4e_2$ $e_2e_2e_3$

It can be proved that the powers of a symmetric matrix are symmetric. Since the matrix A in Example 1 is symmetric, its cube A^3 in Example 2 should also be symmetric. This provides another way of checking the multiplication in Example 2.

Questions

Covering the Reading

1. *True* or *false?* If in a graph there are 2 walks from v_2 to itself, then the entry in the second row, second column of the adjacency matrix for the graph will be 4. **False**

2. In the adjacency matrix of the graph at the right, the entry a_{32} = __?__ and a_{23} = __?__. **2, 0**

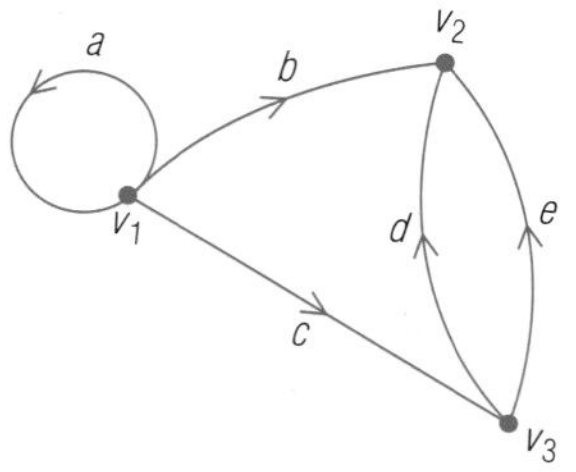

In 3 and 4, *true* or *false*? If *false*, rewrite the statement so it is *true*.

3. The matrix at the right represents an undirected graph. **See margin.**

$$\begin{array}{c|ccc} & v_1 & v_2 & v_3 \\ v_1 & 1 & 1 & 3 \\ v_2 & 1 & 1 & 2 \\ v_3 & 0 & 2 & 1 \end{array}$$

4. There is only one walk from v_1 to v_2 in the graph at the right. **See margin.**

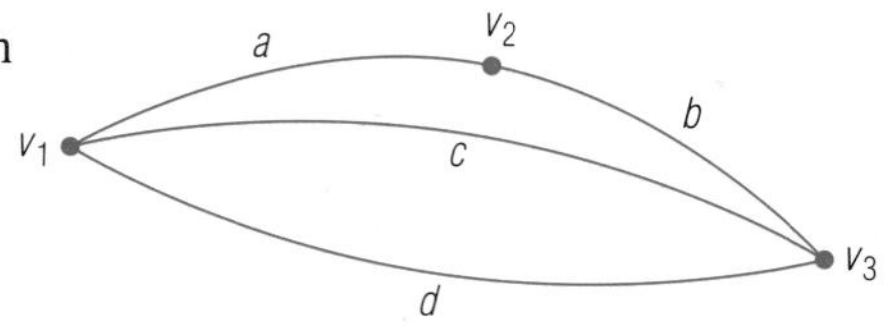

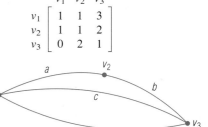

5. In Example 1, find the number of walks of length 2: **a.** from v_3 to v_2; **b.** from v_2 to v_2. **a) 2; b) 5**

672

6. In Example 2, find the 3 missing walks of length 3 from v_1 to v_2.
$e_2e_3e_3$, $e_3e_2e_2$, $e_3e_3e_2$

7. Describe all of the walks of length 3 from v_3 to v_3 in Example 1.
$e_4e_1e_4$, $e_4e_2e_5$, $e_4e_3e_5$, $e_5e_2e_4$, $e_5e_3e_4$

8. Find the number of walks of length 2 from v_3 to v_2 in the graph at the right.
e_1e_2, e_1e_5, e_4e_2, e_4e_5, e_6e_3

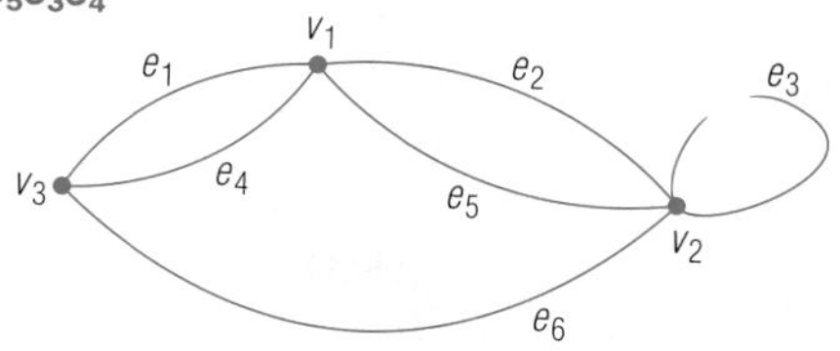

9. Verify the matrix for A^3 in Example 2. See margin.

10. If $M = \begin{bmatrix} 1 & 0 & 1 \\ 0 & 2 & 3 \\ 1 & 3 & 0 \end{bmatrix}$, , calculate M^2 and M^3. See margin.

Applying the Mathematics

11. If the entries on the main diagonal of the adjacency matrix for a graph are zero and the other entries are zero or one, then the graph is simple. Explain why. See margin.

12. Determine the total number of walks of length 3 for the graph given at the right.
105

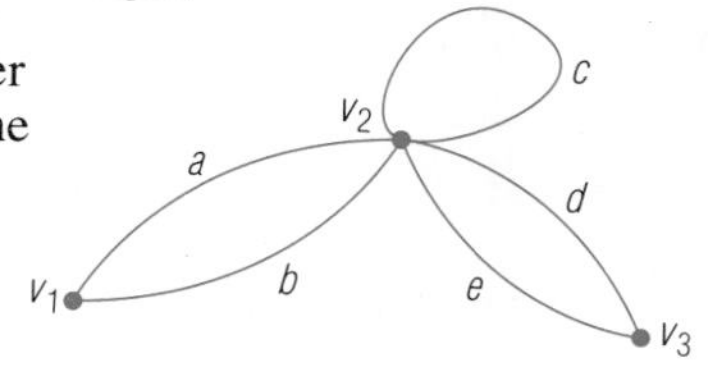

13. Refer to the graph on page 662. Determine the number of walks over 3 bridges of Königsberg that begin and end at C. 4

14. Consider the following true statement: *If a graph is not directed, then its adjacency matrix is symmetric.*
a. Write the converse.
b. Give a counterexample to show that the converse is false.
a,b) See margin.

Review

15. Does the following graph have an Euler circuit? If so, find it. If not, draw an edge that will make an Euler circuit possible. *(Lesson 11-4)* See margin.

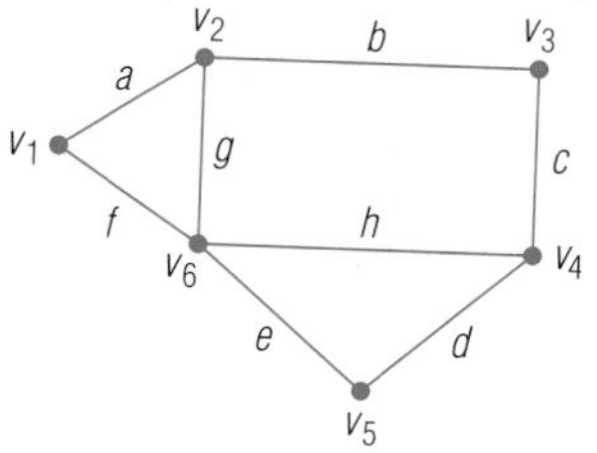

16. If there are 27 people at a party, is it possible for each one to shake hands with exactly 4 other people? *(Lesson 10-3)* Yes

17. Draw a graph which has the adjacency matrix at the right. *(Lesson 11-2)*
See margin.

	v_1	v_2	v_3	v_4
v_1	0	1	2	3
v_2	1	0	1	2
v_3	2	1	0	1
v_4	3	2	1	0

11. If the main diagonal is all zeros, there are no loops, and if all other entries are zero or one, there are no parallel edges, so the graph is simple.

14.a. If the adjacency matrix for a graph is symmetric, then its graph is not directed.
b. sample:

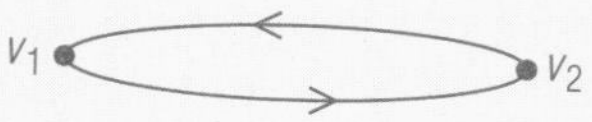

15. No, add edge *i*.

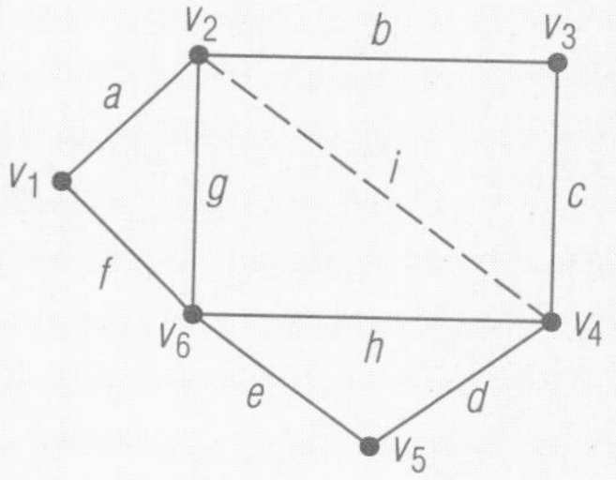

17.

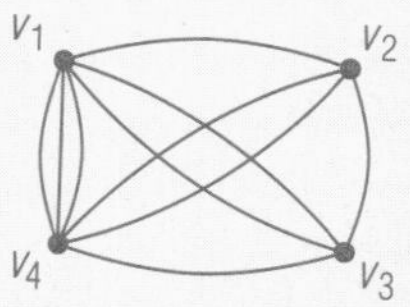

NAME ____________

LESSON **MASTER 11-5**
QUESTIONS ON SPUR OBJECTIVES

■ SKILLS *Objective A (See pages 694–698 for objectives.)*
1. Draw a simple graph with four vertices and three walks of length 2 from v_1 to v_2.
sample:

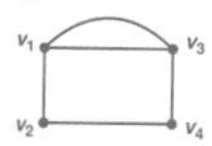

■ PROPERTIES *Objective B*
2. Refer to the graph at the right.

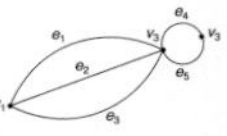

a. How many walks of length 3 are there from v_1 to v_2? 39
b. List four of these walks.
samples: $e_1e_4e_5$, $e_1e_5e_4$, $e_1e_2e_3$, $e_3e_2e_1$
c. What is the length of the longest path from v_1 to v_2? 5

■ REPRESENTATIONS *Objective J*
3. The adjacency matrix for an undirected graph is $\begin{bmatrix} 1 & 0 & 1 \\ 0 & 2 & 3 \\ 1 & 3 & 0 \end{bmatrix}$.
a. How many walks of length 2 go from v_1 to v_3? 1
b. How many walks of length 2 start from v_2? 22

4. a. Give the adjacency matrix for the directed graph at the right.

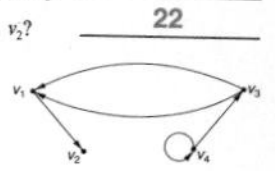

	v_1	v_2	v_3	v_4
v_1	0	1	0	0
v_2	0	0	0	0
v_3	2	0	0	0
v_4	0	0	1	1

b. How many walks of length 3 start at v_4? 6

FOLLOW-UP

MORE PRACTICE

For more questions on SPUR Objectives, use *Lesson Master 11-5*, shown on page 673.

ADDITIONAL ANSWERS

18.

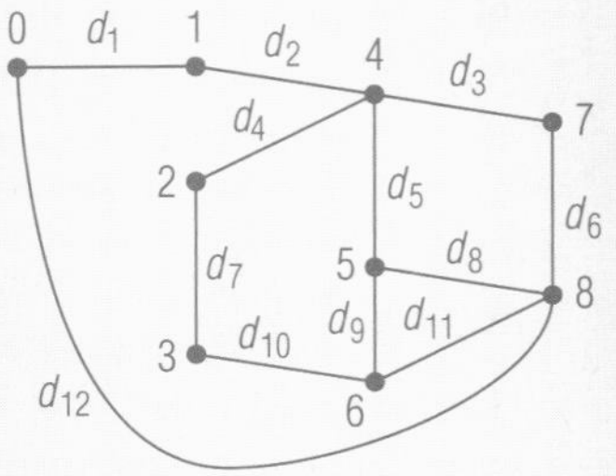

Add d_{13} between r_5 and r_6.

19.

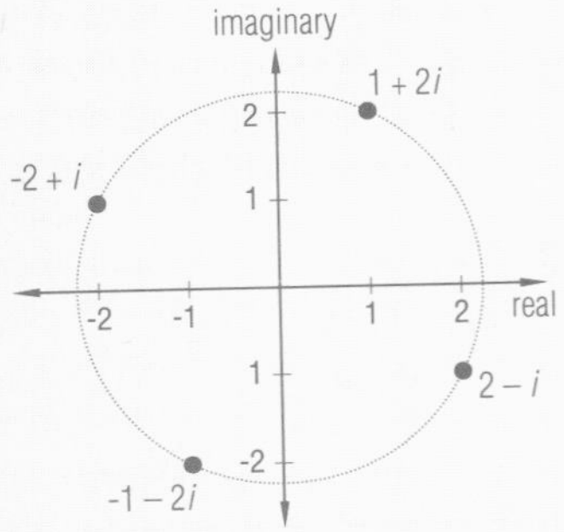

21. See Additional Answers in the back of this book.

22.b. There are an infinite number of solutions. For all real x, $y = \frac{4}{9}x + \frac{2}{3}$.

23. See Additional Answers in the back of this book.

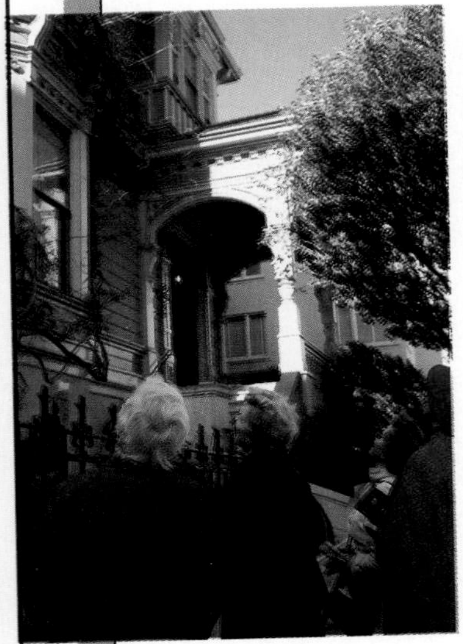

18. A house is open for public viewing. An outline of the floor plan is shown below. Is it possible to enter into room 1, pass through every interior doorway of the house exactly once, and exit from room 8? If so, how can this be done? If not, where could you put a new door to make such a tour possible? (Hint: Construct a graph to model this situation. Let 0 be the outside of the house.) *(Lesson 11-4)* **See margin.**

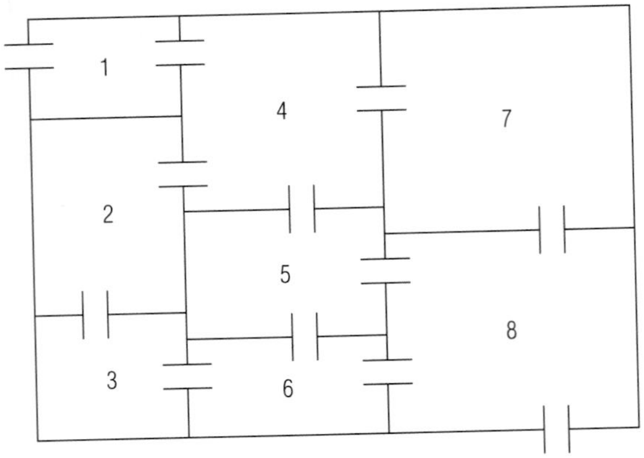

19. Suppose that $2 - i$ is the fourth root of some complex number z. Find and graph the other fourth roots. *(Lessons 8-7, 8-6)* **See margin.**

20. Simplify the expression $\left(\frac{1 + z^{-2}}{1 - z^{-2}}\right) \cdot (1 + z)$, and state restrictions on z. *(Lesson 5-1)* **$\frac{z^2 + 1}{z - 1}$ for $z \neq 1, -1$**

21. Show that the two computer logic networks given below are equivalent. *(Lessons 1-4, 1-3)* **See margin.**

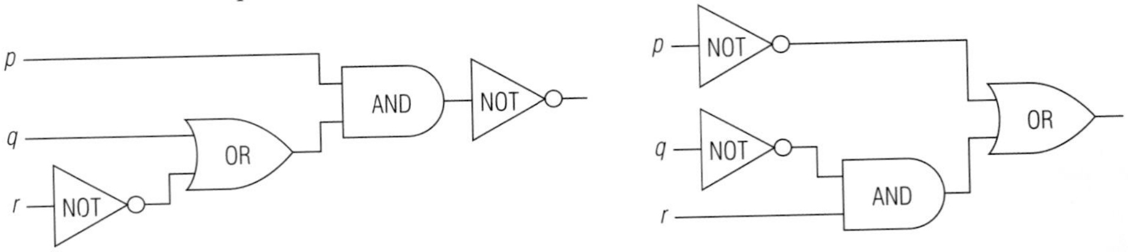

22. Solve the following systems. *(Previous course)*

a. $\begin{cases} 3x + 4y = 10 \\ 2x - y = 1 \end{cases}$ **$x = \frac{14}{11}, y = \frac{17}{11}$**

b. $\begin{cases} 8x = 18y - 12 \\ 24 = -16x + 36y \end{cases}$ **See margin.**

Exploration

23. Let $A = \begin{bmatrix} 0 & 1 & 0 \\ 0 & 0 & 1 \\ 1 & 0 & 0 \end{bmatrix}$

a. Calculate A^n for $n = 1, 2, 3, 4, \ldots$. Find a pattern in A^n.

b. What does this pattern tell you about the directed graph whose adjacency matrix is A?

c. Draw the graph to confirm your answer to part **b**.

a,b,c) See margin.

LESSON

11-6

Markov Chains

In this lesson, graphs and powers of matrices are combined with probability, limits, and systems of equations in a display of the interconnectedness of mathematics. The ideas of this lesson have wide applicability. We begin with an example involving weather.

Suppose that weather forecasters in a particular town have come up with data, represented in the following graph, concerning the probabilities of occurrence of mostly sunny days (S), rainy days (R), and mostly cloudy days (C).

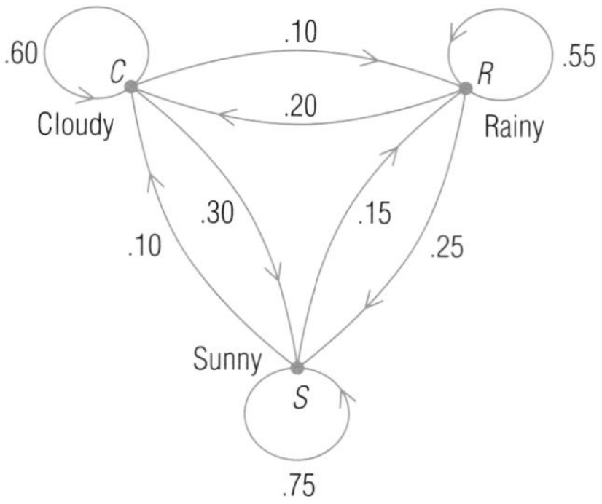

Interpret this directed graph as follows. The loop about point C, labeled .60, means that 60% of the time a cloudy day is followed by another cloudy day. The .10 by edge (C, R) means that 10% of cloudy days are followed by a rainy day. The .30 by edge (C, S) means that 30% of cloudy days are followed by a sunny day.

Now suppose today is cloudy. What is the weather likely to be two days from now?

LESSON 11-6

RESOURCES

- Lesson Master 11-6
- Teaching Aid 82 displays the Markov matrix for weather.
- Teaching Aid 83 can be used with **Questions 12, 13, and 14**.
- Computer Master 20

OBJECTIVE

H Use stochastic matrices to make long-term predictions.

TEACHING NOTES

A Markov chain is neither a chain nor a sequence. It is a *situation* in which enough information is given so that a sequence of states can be determined with their probabilities.

Making Connections Markov chain matrices are sometimes called *Markov matrices.* The fact that multiplying them gives probabilities of going from state n to state $n + 2$ is a combination of the idea of the previous lesson and the notion, first encountered in Lesson 11-1, that conditional probabilities can be represented by graphs.

Students should expect that the powers of the Markov weather matrix on page 676 gradually approach a limit in which all rows are the same. This merely represents the idea that weather n days from now is less and less dependent on today's weather as n increases and more and more dependent on the probabilities of a particular type of weather condition.

We have not considered all the possibilities of stochastic matrices in this lesson. For example, if a row has a 1 on the diagonal, then the 1 creates what is called an *absorbing state.* For example, a situation represented by

$$\begin{bmatrix} .6 & .4 \\ 0 & 1 \end{bmatrix}$$

will cause all objects to end in the second state (vertex). This is the absorbing state. If there are two rows with 1's, you can have oscillating states. For example, the matrix

$$\begin{bmatrix} 0 & 1 \\ 1 & 0 \end{bmatrix}$$

will never lead to a stable situation. Therefore, the Convergence of Powers Theorem explicitly excludes matrices with entries of zero.

To answer this question, represent the graph by a matrix T, where the rows and columns of T are labeled C, S, and R. The entries of T are probabilities that one type of weather on one day is followed by a particular type of weather the next day. For instance, $t_{23} = .15$ because t_{23} is in row S and column R and 15% of sunny days are followed by a rainy day.

$$\begin{array}{c} \\ C \\ S \\ R \end{array} \begin{array}{c} \begin{array}{ccc} C & S & R \end{array} \\ \begin{bmatrix} .60 & .30 & .10 \\ .10 & .75 & .15 \\ .20 & .25 & .55 \end{bmatrix} \end{array} = T$$

Notice that each element is nonnegative (since each is a probability), and that the entries in each row add to 1. A matrix with these properties is called a **stochastic matrix.** The matrix is indicated with the letter T to indicate that it contains the **transition probabilities** from one time period to the next.

In Lesson 11-5, you saw that the square of the adjacency matrix for a graph represents the number of walks of length two from one vertex to another. Here the square of T has a similar interpretation: the elements are the probabilities connecting weather two days apart.

$$T^2 = T \cdot T = \begin{bmatrix} .60 & .30 & .10 \\ .10 & .75 & .15 \\ .20 & .25 & .55 \end{bmatrix} \cdot \begin{bmatrix} .60 & .30 & .10 \\ .10 & .75 & .15 \\ .20 & .25 & .55 \end{bmatrix} = \begin{bmatrix} .41 & .43 & .16 \\ .165 & .63 & .205 \\ .255 & .385 & .36 \end{bmatrix}$$

Notice that the entries in each row still add up to 1, so T^2 is also a stochastic matrix. Reading across the first row of T^2 shows that if today is cloudy, there is a 41% chance that it will be cloudy two days from now, a 43% chance that it will be sunny, and a 16% chance of rain.

T^2 can be multiplied by itself to yield T^4, which indicates the probabilities of various types of weather occuring 4 days later. Similarly, $T^4 \cdot T^4 = T^8$ and $T^8 \cdot T^2 = T^{10}$. In general, each entry of T^k indicates the probability that one type of weather will be followed by a particular type k days later.

$$T^4 \approx \begin{bmatrix} .27985 & .50880 & .21135 \\ .22388 & .54678 & .22935 \\ .25988 & .49080 & .24933 \end{bmatrix}$$

$$T^8 \approx \begin{bmatrix} .24715 & .52432 & .22853 \\ .24466 & .52544 & .22990 \\ .24740 & .52295 & .22965 \end{bmatrix}$$

$$T^{10} \approx \begin{bmatrix} .24612 & .52458 & .22930 \\ .24563 & .52474 & .22962 \\ .24628 & .52426 & .22946 \end{bmatrix} \approx \begin{bmatrix} .25 & .52 & .23 \\ .25 & .52 & .23 \\ .25 & .52 & .23 \end{bmatrix}$$

676

The three rows of T^{10} are almost identical. This means that no matter what the weather is today, there is approximately a 25% chance of a cloudy day 10 days from now, a 52% chance of a sunny day, and a 23% chance of rain.

Weather is dependent on many factors. The key assumption in the model used here is that the probability of a certain type of weather tomorrow is only dependent on the weather today. When a situation can exist in only a finite number of states (above there are 3 states: C, S, and R), and the probability of proceeding from one state to the next depends only on the first state, then the situation is said to be an example of a **Markov chain.**

Markov chains are named after the Russian mathematician who first studied them, Andrei Andreevich Markov (1856–1922). Markov worked in a variety of areas of mathematics, with his greatest contributions being in the area of probability theory. He developed the concept of Markov chain from the theory of probability and applied it to a study of the distributions of vowels and consonants in Russian literature. His work is frequently considered to be the first research in *mathematical linguistics*, the mathematical study of language structure.

Recall that for the stochastic matrix T on the previous page, T^2 is also stochastic. In general, the kth power of any stochastic matrix is stochastic. This can be seen for the 2nd power of a 2×2 stochastic matrix as follows.

Because the entries in each row add to 1, the matrix has the form $\begin{bmatrix} x & 1-x \\ y & 1-y \end{bmatrix}$ where $0 \le x \le 1$ and $0 \le y \le 1$. Its square is

$$\begin{bmatrix} x & 1-x \\ y & 1-y \end{bmatrix} \bullet \begin{bmatrix} x & 1-x \\ y & 1-y \end{bmatrix} = \begin{bmatrix} x^2 + y - xy & 1 - x^2 - y + xy \\ xy + y - y^2 & 1 + y^2 - y - xy \end{bmatrix},$$

which is also stochastic.

Furthermore, if a stochastic matrix T has no zero entries, the rows of T^k will be nearly identical for large k. This indicates that over the long term the proportions of the occurrences of the different states stabilize. You saw this for T^{10} on page 676.

Theorem (Convergence of Powers)

Let T be an $n \times n$ stochastic matrix with no zero entries. Then $\lim_{k \to \infty} T^k$ is a stochastic matrix with n identical rows.

Stable populations occur in populations of plants and animals, as the following example illustrates.

ADDITIONAL EXAMPLE
In a particular desert climate, 80% of rainy days are followed by sunny days, and 95% of sunny days are followed by sunny days. Over the long run, what portion of days are sunny?
16/17

Example Consider a variety of rose that can have either a pale hue or a brilliant hue. It is known that seeds from a pale blossom yield plants of which 60% have pale flowers and 40% have brilliant flowers. Seeds from a brilliant flower yield plants of which 30% are pale and 70% are brilliant. After several generations of plants, what will be the proportion of pale and brilliant flowering plants?

Solution The transition matrix for this situation is

$$\begin{array}{ccc} & & \text{OFFSPRING} \\ & & \begin{array}{cc}\text{Pale} & \text{Brilliant}\end{array} \\ \text{FLOWER} & \begin{array}{r}\text{Pale} \\ \text{Brilliant}\end{array} & \begin{bmatrix} .6 & .4 \\ .3 & .7 \end{bmatrix} = T. \end{array}$$

Let a and b be the proportion of plants with pale and brilliant flowers, respectively, when the population stabilizes. Then, the proportion of the flowers of the next generation that are pale will be $.6a + .3b$ because .6 of those produced by the pale flowers are pale, and .3 of those produced by the brilliant ones are pale. But since the population has stabilized, the fraction of the next generation that is pale must still be a. This results in the equation

$$.6a + .3b = a.$$

Similarly, the fraction of flowers in the next generation that is brilliant is $.4a + .7b$. Again, since the population has stabilized, the fraction that is brilliant must be b.

$$.4a + .7b = b.$$

To solve the system of two equations, add $-a$ to both sides of the first equation, and $-b$ to both sides of the second:

$$\begin{cases} -.4a + .3b = 0 \\ .4a - .3b = 0 \end{cases}$$

This system has an infinite number of solutions. But if a and b are the proportions of pale and brilliant flowers, then it is also true that $a + b = 1$. Thus the following system must be satisfied.

$$\begin{cases} .4a - .3b = 0 \\ a + b = 1 \end{cases}$$

This has solution $(a, b) = \left(\frac{3}{7}, \frac{4}{7}\right) \approx (.43, .57)$. So when the population stabilizes, about 43% of the plants will have pale flowers and 57% will have brilliant flowers.

Check This matches the result obtained by calculating powers of T.

$$T^{10} \approx \begin{bmatrix} .42857 & .57143 \\ .42857 & .57143 \end{bmatrix} \approx \begin{bmatrix} \frac{3}{7} & \frac{4}{7} \\ \frac{3}{7} & \frac{4}{7} \end{bmatrix}.$$

After Markov published his theory, his techniques were adopted by scientists in a wide range of fields. Albert Einstein used these ideas to study the Brownian motion of molecules. Physicists have employed them in the

678

theory of radioactive transformation, nuclear fission detectors, and the theory of tracks in nuclear emulsions. Astronomers have used Markov theory to study fluctuations in the brightness of the Milky Way and in the spatial distribution of galaxies. Biologists have used Markov chains to describe population growth, evolution, molecular genetics, pharmacology, tumor growth, and epidemics. Sociologists have modeled voting behavior, geographical mobility, growth and decline of towns, sizes of businesses, changes in personal attitudes, and deliberations of trial juries with Markov chains.

Questions

Covering the Reading

In 1–4, consider the weather example.

1. **a.** If it is sunny today, what is the probability of rain tomorrow? 15%
b. If it is sunny today, what is the probability of sun tomorrow? 75%

2. **a.** If it is rainy today, what is the probability of rain two days from now? 36%
b. If it is sunny today, what is the probability of rain four days from now? ≈ 22.9%

3. Is T^{10} a stochastic matrix? Yes

4. **a.** In the matrix T^{10}, what does the number .24612 represent?
b. What is the significance of the fact that the rows of T^{10} are nearly identical? a) See margin. b) No matter what the weather is today, the probabilities for the weather in 10 days are about the same.

In 5–7, consider the rose example.

5. If a rose is brilliant, what is the probability that its offspring are brilliant? 0.7

6. If a rose is pale, what is the probability that its offspring are pale? 0.6

7. Using $T^{10} = \begin{bmatrix} \frac{3}{7} & \frac{4}{7} \\ \frac{3}{7} & \frac{4}{7} \end{bmatrix}$, calculate T^{20} and explain your result.

See margin.

Applying the Mathematics

8. At each four-month interval, two TV stations in a small town go through "ratings week." They try to offer special programs which will draw viewers from the other station. During each period, MBC (Markov Broadcasting Company) wins over 20% of SBS (Stochastic Broadcasting System) viewers, but loses 10% of its viewers to SBS.
a. Draw a graph (like that shown at the beginning of this lesson) to represent the movement of viewers between stations. See margin.
b. Write down the transition matrix. See margin.
c. Using the method of the rose example, find the long-term distribution of viewers watching each station. MBC: 67%; SBS: 33%

ADDITIONAL ANSWERS

4.a. the probability that it will be cloudy 10 days after a cloudy day

7. $T^{20} = \begin{bmatrix} \frac{3}{7} & \frac{4}{7} \\ \frac{3}{7} & \frac{4}{7} \end{bmatrix}$

In 20 generations, $\frac{3}{7}$ of the seeds will produce pale flowers, and $\frac{4}{7}$ brilliant flowers, no matter what seeds you start with.

8.a.

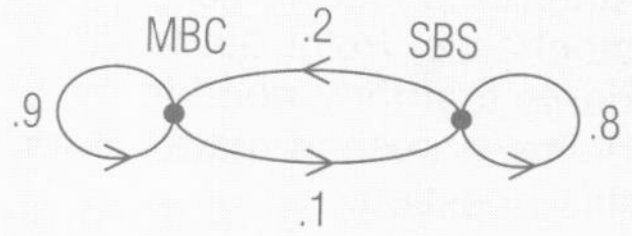

b.

	MBC	SBS
MBC	.9	.1
SBS	.2	.8

NOTES ON QUESTIONS
Making Connections for Question 9: The work of Galton led to some significant early results in modern statistics. His work is mentioned in conjunction with applications of ellipses in *Functions, Statistics, and Trigonometry.*

Question 10: Students may not realize that this result is not given in the text. There, it is shown that if A is stochastic then A^2 is stochastic. Point out to students that they need to define a second matrix B for this question.

Making Connections for Question 13: If paths containing circuits are included, the problem is more difficult but is a very nice example of a problem in permutations. There are $P(5, 5) + P(5, 3) + P(5, 1)$ paths from A to D and $P(3, 3) + P(3, 1)$ paths from D to G. Applying the Multiplication Counting Principle, the total number of possible paths is 1665.

ADDITIONAL ANSWERS
10. See Additional Answers in the back of this book.

11.b.

$$T^2 = \begin{bmatrix} .45 & .55 \\ .44 & .56 \end{bmatrix}$$

$$T^4 = \begin{bmatrix} .4445 & .5555 \\ .4444 & .5556 \end{bmatrix}$$

$$T^8 \approx \begin{bmatrix} .4444 & .5556 \\ .4444 & .5556 \end{bmatrix}$$

$$T^{16} \approx \begin{bmatrix} .4444 & .5556 \\ .4444 & .5556 \end{bmatrix}$$

c. $a = \frac{4}{9}, b = \frac{5}{9}$
d. Over the long term, the proportion of occurrences stabilize to *a* and *b*.

16.d. The time in part c is midway between the times in parts a and b.

9. The British scientist Sir Francis Galton studied inheritance by looking at distributions of the heights of parents and children. In 1886 he published data from a large sample of parents and their adult children showing the relation between their heights. The following matrix is based on his data. Since he had to use volunteers in his study, he could not be sure that his sample accurately reflected the English population.

		CHILD		
		Tall	Med	Short
PARENT	Tall	.53	.32	.15
	Med	.30	.34	.32
	Short	.15	.36	.53

$= T$

According to this matrix,

$$T^2 \approx \begin{bmatrix} .399 & .326 & .274 \\ .315 & .327 & .358 \\ .255 & .326 & .419 \end{bmatrix} \text{ and } T^{10} \approx \begin{bmatrix} .321 & .327 & .352 \\ .321 & .327 & .352 \\ .321 & .327 & .352 \end{bmatrix}.$$

a. What proportion of the children of tall parents were short? 15%
b. Use T^2 to tell what proportion of grandchildren of tall people were short. 27.4%
c. Use T^{10} to predict the approximate proportion of tall, medium, and short people in the population in the long run.
32.1% tall, 32.7% medium, 35.2% short

10. Prove for 2 × 2 matrices: *If A is stochastic and B is stochastic, the product AB must be stochastic.* See margin.

11. Consider the matrix T at the right.

$$\begin{bmatrix} .5 & .5 \\ .4 & .6 \end{bmatrix}$$

a. Is T stochastic? Yes
b. Calculate T^2, T^4, T^8, and T^{16}.
c. Find two numbers a and b such that $vT = v$ where $v = [a\ b]$ and $a + b = 1$.
d. What do a and b represent?
b-d) See margin.

Review

12. Find the total number of walks of length 4 which end at v_1 in the directed graph at the right. *(Lesson 11-5)* 24

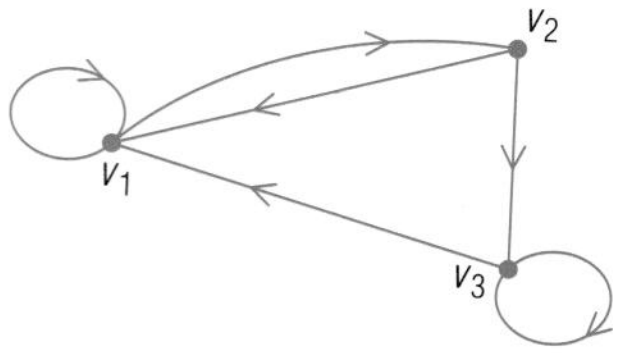

13. In the graph pictured below, determine the number of paths from A to G that contain no circuits. *(Lesson 11-4)* 10

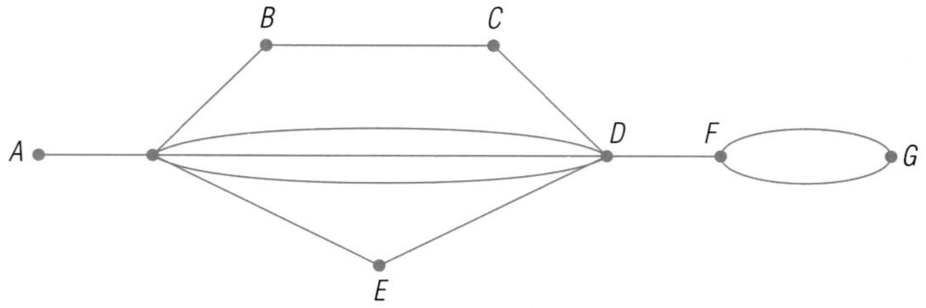

680

14. Consider the graph at the right.
  a. Does the graph have an Euler circuit? Justify your answer.
  b. What is the maximum number of edges that could be removed while keeping the graph connected? *(Lesson 11-4)*

a) Yes, because all vertices are even, and it is connected b) 4

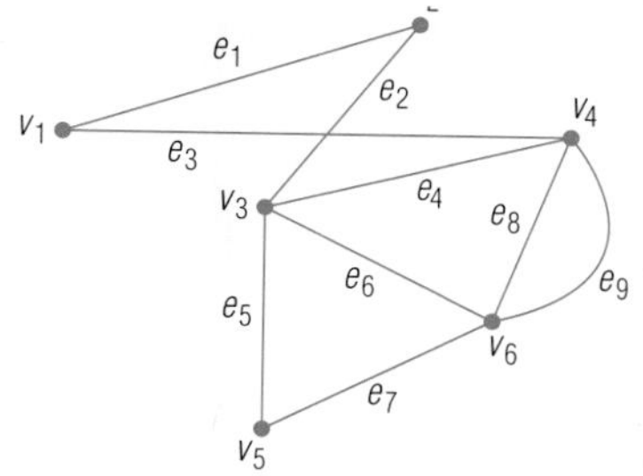

15. In a league of nine teams, is it possible for each team to play exactly seven other teams? Explain. *(Lesson 11-3)*
No, a graph cannot have an odd number of odd vertices.

16. Suppose the height (in feet) of an object t seconds after it is thrown is given by $h(t) = -16t^2 + 50t + 10$. *(Lessons 9-5, 9-4, 9-3)*
  a. Find the object's velocity 1 second after it is thrown. **$v(1) = 18$ ft/sec**
  b. When is the object's velocity the opposite of the velocity found in part **a**? **$t = \frac{68}{32} = 2.125$ sec**
  c. When does the object reach its maximum height? **$t = \frac{50}{32} = 1.5625$ sec**
  d. How are the times in parts **a**, **b**, and **c** related? **See margin.**
  e. When is the object's acceleration positive?
  Never, it is always -32 ft/sec^2.

17. Solve the inequality $2\sin^2 x + \cos^2 x < \frac{1}{4}$ for $0 \le x \le 2\pi$. *(Lesson 6-7)*
no solution

18. Use limit notation to describe the end behavior of the function given by $f(n) = \dfrac{2n^2 + 3n + 1}{6n^2}$. *(Lesson 5-5)* **$\lim_{n\to\infty} f(n) = \frac{1}{3}$; $\lim_{n\to-\infty} f(n) = \frac{1}{3}$**

19. Prove: *Exactly one of every four consecutive integers is divisible by 4.* (Hint: Use the Quotient-Remainder Theorem.) *(Lesson 4-2)*
See below.

Exploration

20. Find an example in a book or article that describes how a Markov chain is used in biology, linguistics, politics, geology, or physics.
See below.

19. Let n, $n + 1$, $n + 2$, and $n + 3$ represent the four consecutive integers. By the Quotient-Remainder Theorem, $n = 4q + r$ where q is an integer and $r = 0, 1, 2$, or 3. If $r = 0$, then n is divisible by 4. If $r = 1$, $n + 3 = (4q + 1) + 3 = 4(q + 1)$ is divisible by 4. If $r = 2$, $n + 2 = (4q + 2) + 2 = 4(q + 1)$ is divisible by 4. If $r = 3$, $n + 1 = (4q + 3) + 1 = 4(q + 1)$ is divisible by 4. So, exactly one of every four consecutive integers is divisible by 4.

20. A good source is *Markov Chains: Theory and Applications* by Dean Isaacson and Richard Madsen.

FOLLOW-UP

MORE PRACTICE
For more questions on SPUR Objectives, use *Lesson Master 11-6*, shown below.

EXTENSION
In general, the questions in this lesson do not require a computer program or technological support to compute high powers of matrices. We have provided the relevant power of a matrix and asked for interpretation of its elements. A good additional project for students is to write a computer program or develop a template on a spreadsheet that will compute powers of 3 × 3 matrices.

NAME ____________

LESSON **MASTER 11–6**
QUESTIONS ON **SPUR** OBJECTIVES

■ **USES** *Objective H (See pages 694–698 for objectives.)*

1. In a chemical experiment, molecules of a liquid are changing phase in a flask. It is known that from one minute to the next, 80% of the molecules remain liquid, while 20% become gaseous. At the same time, 40% of the gaseous molecules become liquid. The rest remain gaseous.
  a. Give T, the transition matrix.
  $T = \begin{matrix} & L & G \\ L & .8 & .2 \\ G & .4 & .6 \end{matrix}$

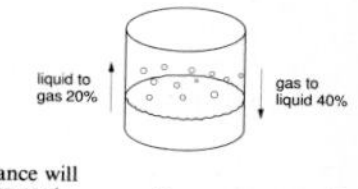

  b. After sufficient time, the substance will reach equilibrium so that the percent that is liquid and the percent that is gaseous become constant. Estimate these percents by calculating T^8. $T^8 \approx \begin{bmatrix} .6669 & .3331 \\ .6663 & .3337 \end{bmatrix}$
  c. Find the exact percents by solving a system of equations.
  67% liquid, 33% gas

2. At a particular time, it was found that 25% of adults drank Kool Cola, 20% drank Klassic Cola, and 55% preferred Fizzy Cola. The Kool Cola Company's surveys showed that people switched brands each month according to the directed graph below.

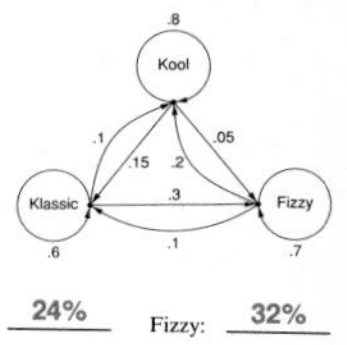

  a. Based on these data, what percent of the adults drank each brand one month after the initial data were gathered?
  Kool: **33%**
  Klassic: **21%**
  Fizzy: **46%**
  b. What percent would eventually drink each brand after a long period of time?
  Kool: **44%** Klassic: **24%** Fizzy: **32%**

126 *Precalculus and Discrete Mathematics* © Scott, Foresman and Company

LESSON 11-7

RESOURCES

- Teaching Aid 84 displays polyhedra and their planar graphs.
- Teaching Aid 85 shows the relationship among *V*, *E*, and *F* for various graphs.
- Teaching Aid 86 can be used with **Questions 5-8**, **11**, **and 17**.

TEACHING NOTES

Euler's Formula is sometimes called the Euler-Descartes formula. It is somewhat surprising that the Greeks did not know of the relationship $V - E + F = 2$ for polyhedra. Work through the table on this page. You may wish to consider other polyhedra also. In addition to **Questions 2-4**, ask students to give the values of *V*, *E*, and *F* for figures such as a tetrahedron ($V = 4, E = 6, F = 4$); a hexagonal prism ($V = 12, E = 18, F = 8$); and an icosahedron ($V = 12, E = 30, F = 20$).

The value of $V - E + F$ is called the *Euler characteristic* of a surface. The key to the proof of Euler's formula is the reduction of a graph with $k + 1$ edges to one that has k edges without changing the value of Euler's characteristic. That reduction is accomplished with the aid of the first two theorems of this lesson. These theorems thus have the role of *lemmas*, theorems that are stated and proved for the purpose of shortening the proof of a theorem that follows.

LESSON

11-7 Euler's Formula

A surprisingly simple relationship exists between the number of vertices, edges, and faces of a polyhedron. Although the ancient Greeks studied polyhedra extensively, no one seems to have known of this relationship until Descartes discovered it in the seventeenth century. Furthermore, it was not until the next century that it was proved by Leonhard Euler, whose name it now bears. In this lesson, Euler's formula is interpreted in terms of the ideas of graph theory that you have seen in this chapter. These ideas are combined along with mathematical induction to deduce the formula.

Consider the regular dodecahedron drawn below. Let V, E, and F be the number of its vertices, edges, and faces, respectively. In Lesson 11-1, it was noted that $V = 20$, $E = 30$, and $F = 12$.

In the table below are the corresponding values for several other polyhedra.

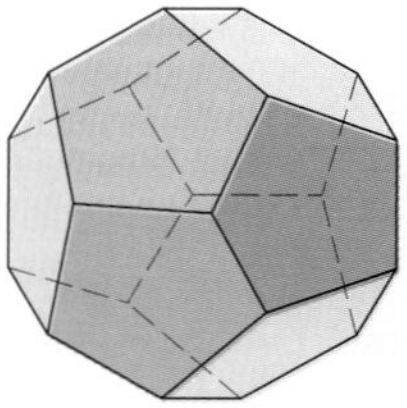
dodecahedron

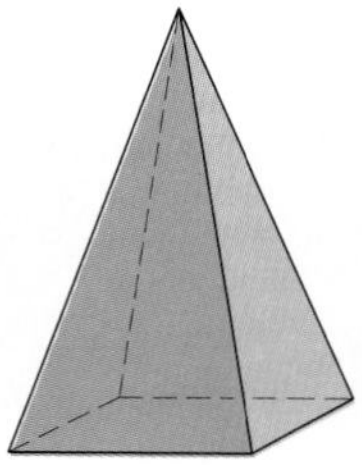
square pyramid

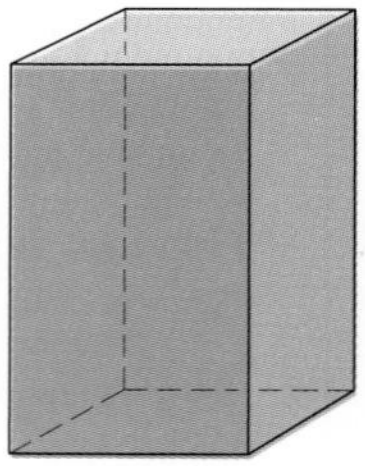
rectangular solid

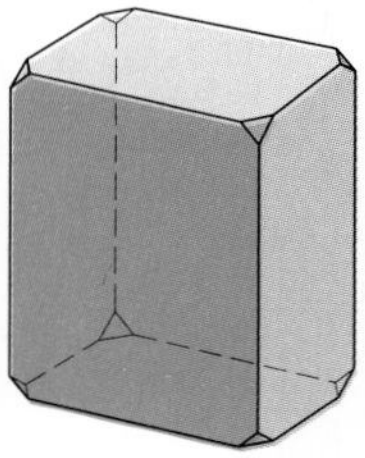
rectangular solid with truncated corners

polyhedron	V	E	F
dodecahedron	20	30	12
square pyramid	5	8	5
rectangular solid	8	12	6
rectangular solid with truncated corners	24	36	14

The relationship discovered by Descartes is that, for any polyhedron, $V - E + F = 2$. Euler reformulated this conjecture in terms of graphs. In Lesson 11-1, the dodecahedron was distorted into the graph shown on the next page at the left.

This distortion can be thought of in the following way: one face is removed to create a hole, allowing the remaining figure to be stretched and flattened. The same process was used to obtain graphs for the other polyhedra shown on the next page.

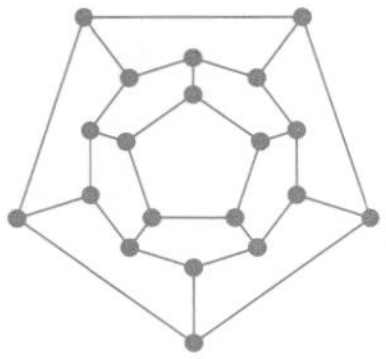

dodecahedron

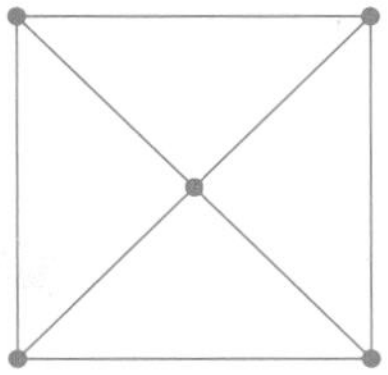

square
pyramid

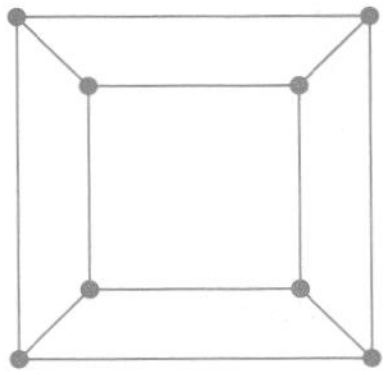

rectangular
solid

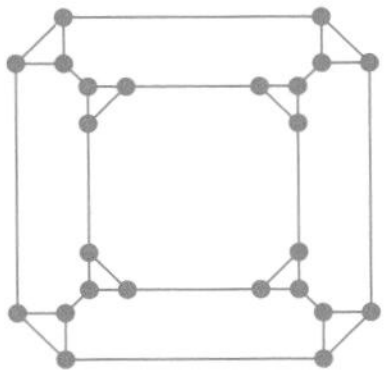

rectangular solid
with truncated
corners

Another way of visualizing this distortion is to imagine that each polyhedron is placed on a table top with a light source placed only slightly above it. If the faces of the polyhedron are transparent, but the edges are not, the light will project shadows of the edges and vertices onto the table top. These shadows will form the 2-dimensional graph.

In the distorting process, each face of the polyhedron becomes a region separated from the others by edges of the graph. The face that is removed for the flattening process (or that is on top for the shadowing process) corresponds to the region exterior to the graph and must be included in the count. These regions are called the **faces** of the graph. Since the distorting process does not change the number of vertices, edges, or faces, the relationship $V - E + F = 2$ holds for polyhedra if and only if it holds for the resulting graphs.

Note that these graphs are simple and connected, they have no crossings, no vertices of degree less than 3, and every edge is part of a circuit. A graph without crossings can always be obtained when the original polyhedron is **convex**; that is, when any two points in the polyhedron can be connected by a line segment lying completely inside it. This can be seen by noting that if a light source is placed close enough to the top of a transparent, convex polyhedron, the shadows of the edges will never cross.

The question arises as to whether all graphs satisfy $V - E + F = 2$. Two of the figures below show that under certain circumstances the value of $V - E + F$ can be different from 2.

The graph is not connected.

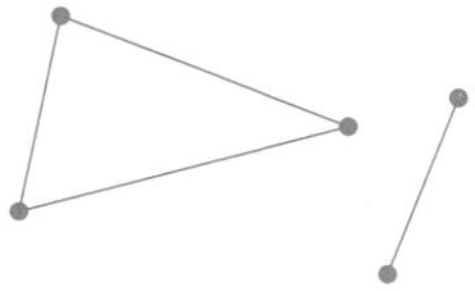

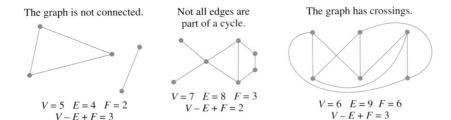

$V = 5 \quad E = 4 \quad F = 2$
$V - E + F = 3$

Not all edges are
part of a cycle.

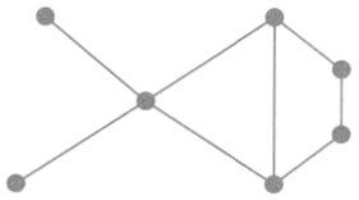

$V = 7 \quad E = 8 \quad F = 3$
$V - E + F = 2$

The graph has crossings.

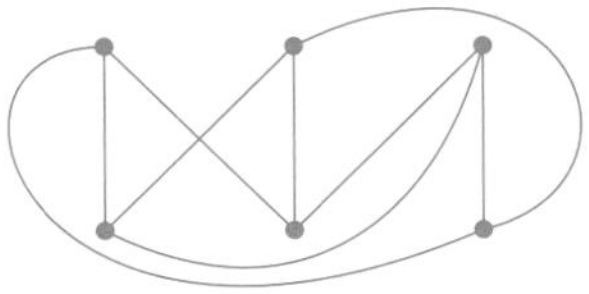

$V = 6 \quad E = 9 \quad F = 6$
$V - E + F = 3$

In the Questions, you will determine the values of $V - E + F$ for some graphs that are not simple.

In Imre Lakatos's book *Proofs and Refutations: The Logic of Mathematical Discovery* (Cambridge University Press, 1976), Euler's formula and its extensions are discussed in detail and presented as a dialogue between a teacher and his students.

If a polyhedral hole is drilled through a polyhedron, then the numbers of vertices, edges, and faces of the resulting surface satisfy $V - E + F = 0$. (See the Extension on page 688.)

It turns out that Euler's formula, $V - E + F = 2$, holds for any graph that is connected and has no crossings. We deduce this formula in two stages. First we prove that a graph can be changed without changing the value of $V - E + F$. For example, one edge of the graph shown below at the left is removed to obtain the graph at the right, but the value of $V - E + F$ remains the same because both E and F are decreased by 1.

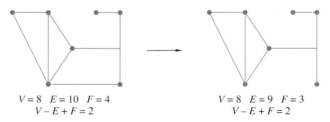

$V = 8 \quad E = 10 \quad F = 4$
$V - E + F = 2$

$V = 8 \quad E = 9 \quad F = 3$
$V - E + F = 2$

Theorem

Let G be a connected graph with no crossings, and let V, E, and F be the number of vertices, edges, and faces of G. The following alterations to G do not change the value of $V - E + F$:
(1) removing a vertex of degree 1 along with its adjacent edge, and
(2) removing an edge that is part of a circuit.

Proof (1) Let v be a vertex of degree 1, and let e be the edge emanating from it. (A vertex of degree 1 can only have 1 such edge.) Then, removing v and e decreases both V and E by 1 and does not affect F. But decreasing both V and E by 1 does not change the value of $V - E$, so the value of $V - E + F$ remains the same.

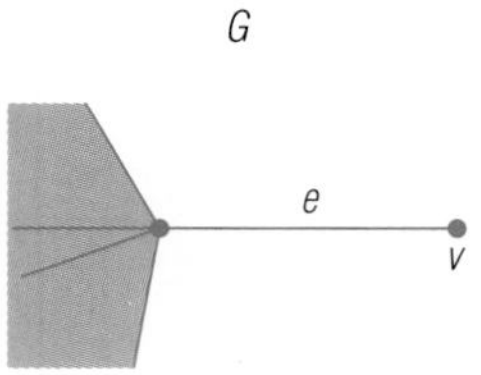

(2) Let e be an edge that is part of a circuit, and let f_1 and f_2 be the faces separated by e (as shown at the right). Removing e combines the two faces f_1 and f_2 into one. (Note that f_1 and f_2 could not have already been the same face, because otherwise e wouldn't have been part of a circuit.) Thus, removing e decreases both E and F by 1 and leaves V unchanged. But decreasing both E and F by 1 does not change $-E + F$, so $V - E + F$ remains the same.

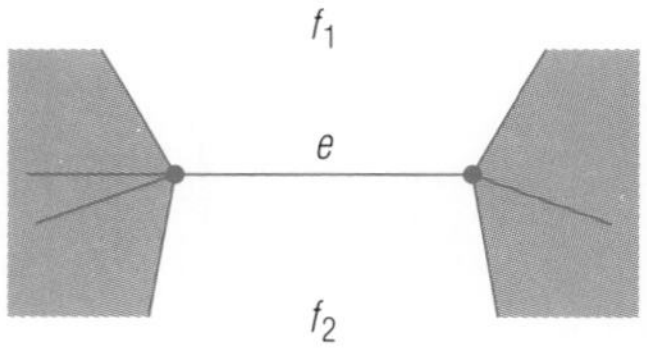

Part (2) of the preceding theorem gives information about graphs that have circuits. Now we prove a theorem that says part (1) of the preceding theorem can be applied to graphs that do not have circuits.

684

Theorem

Let G be a graph with at least one edge. If G has no circuits, then G has a vertex of degree 1.

Proof Choose any vertex of G and traverse a path from that vertex in the following way. Whenever you arrive at a vertex of degree greater than 1, leave the vertex via one of its other edges. (Note that this other edge will not have been used before because otherwise you would have traversed a circuit). Since there are only a finite number of vertices and none will ever be repeated (because there are no circuits), the path will eventually have to end. The ending vertex will have degree 1, because otherwise the path could continue.

For example, consider the graph G shown below. Starting at any vertex, if you follow any path eventually you will come to a vertex with degree 1.

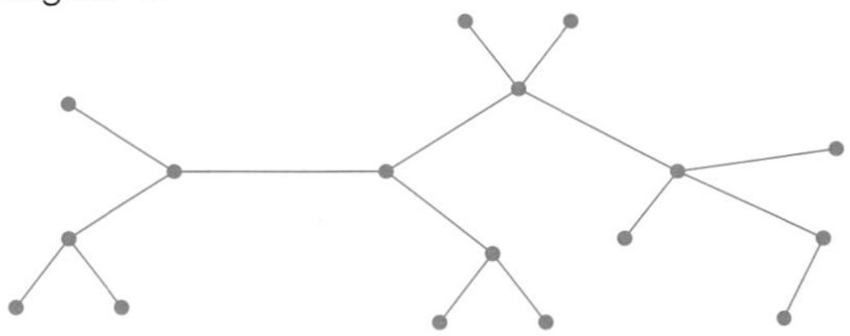

With these two theorems, Euler's formula can now be proved.

Theorem (Euler's Formula)

Let G be a connected graph with no crossings, and let V, E, and F be the number of vertices, edges, and faces of G. Then

$$V - E + F = 2.$$

Proof The proof uses induction. Let $S(n)$ be the statement that Euler's formula holds for any graph with n edges.

First, prove $S(0)$: Let G be a connected graph with 0 edges. The only such graph consists only of a single vertex. It has one face—the entire region surrounding it. For this graph, $V = 1$, $E = 0$, and $F = 1$. Thus, $V - E + F = 1 - 0 + 1 = 2$.

Now assume $S(k)$ is true for an arbitrarily chosen nonnegative integer k, and prove that $S(k + 1)$ is true. That is, assume that $V - E + F = 2$ for a connected graph with no crossings and with k edges. It must be proved that the formula holds for such a graph with $k + 1$ edges. Let G be such a graph. Either G has a circuit, or it doesn't.

NOTES ON QUESTIONS

Question 10: There are only two graphs of this type. You might ask students to draw all connected graphs with two edges. (There are three types, with 0 loops, 1 loop, and 2 loops.) Show that they too satisfy Euler's formula.

Question 14b: This is a well-known puzzle.

ADDITIONAL ANSWERS

2.a.

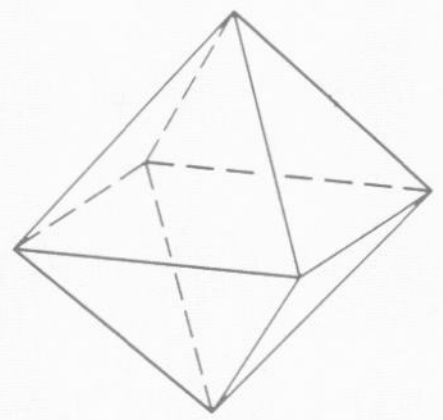

b.

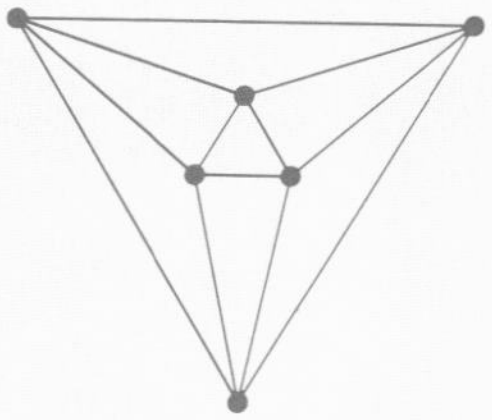

c. $V - E + F = 6 - 12 + 8 = 2$

3.a.

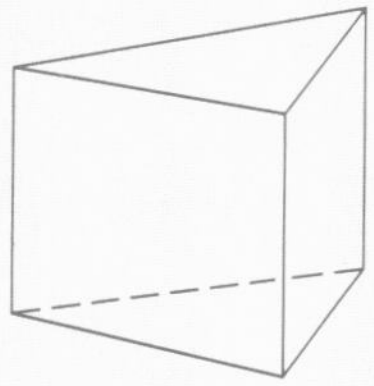

b.

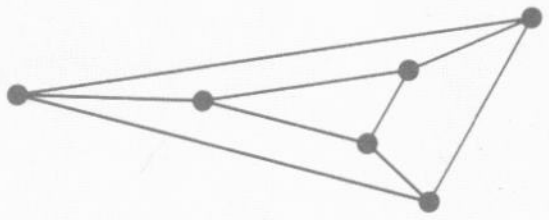

c. $V - E + F = 6 - 9 + 5 = 2$

If G has a circuit, remove one edge from the circuit. By part (2) of the theorem on page 684, this does not affect the value of $V - E + F$. The new graph is still connected, by the Circuits and Connectedness Theorem in Lesson 11-4, and it has only k edges. Therefore, by the inductive assumption, it satisfies $V - E + F = 2$. Since the value of $V - E + F$ is the same for both graphs, the formula also holds for G.

If G does not have a circuit, then by the theorem on page 685, it has a vertex of degree 1. Remove this vertex as well as its adjacent edge. By part (1) of the theorem on page 684, this does not affect the value of $V - E + F$. The new graph is connected and has only k edges, so by the inductive assumption, it satifies $V - E + F = 2$. Since $V - E + F$ is the same for both graphs, the formula also holds for G.

Thus, assuming that Euler's formula holds for a graph with k edges leads to the conclusion that it holds for a graph with $k + 1$ edges, whether it has a circuit or not. That is, assuming that $S(k)$ is true leads to the conclusion that $S(k + 1)$ is true. Since $S(0)$ is true and $S(k) \Rightarrow S(k + 1)$, $S(n)$ is true for all nonnegative integers n by the Principle of Mathematical Induction. In other words, Euler's formula holds for all connected graphs with no crossings.

So Euler's formula is true for polyhedra: In a polyhedron with V vertices, E edges, and F faces, $V - E + F = 2$, or as it is sometimes written, $V + F = E + 2$.

Questions

Covering the Reading

1. Show that the polyhedron drawn at the right satisfies Euler's formula.
$V - E + F = 9 - 16 + 9 = 2$

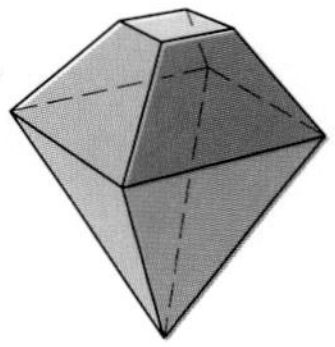

In 2–4, **a.** draw a polyhedron of the given type. **b.** Draw the graph obtained from it by the distorting process described in the lesson. **c.** Show that this graph satisfies Euler's formula. See margin.

2. regular octahedron **3.** triangular prism **4.** hexagonal pyramid

In 5 and 6, the graphs are not simple. Show that they still satisfy Euler's formula.

5.

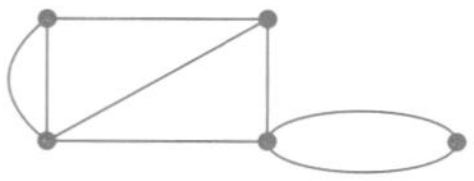

$V - E + F = 5 - 8 + 5 = 2$

6.

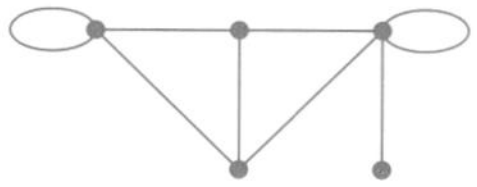

$V - E + F = 5 - 8 + 5 = 2$

686

In 7 and 8, the graph was obtained by altering the graph in Question 6. **a.** Tell what change was made. **b.** Explain why this change did not alter the value of $V - E + F$. See margin.

7. 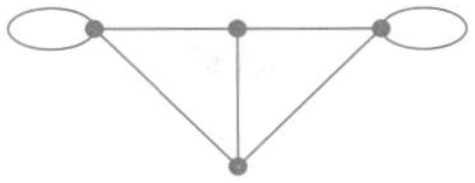

8. 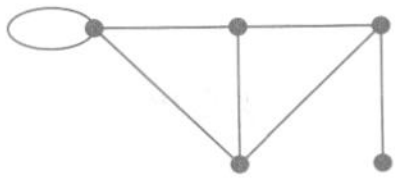

9. *True* or *false*? If G is a graph such that every vertex has degree greater than 1, then G has a circuit. If *true*, explain why. If *false*, give a counterexample. See margin.

10. Draw two different connected graphs, each of which has only one edge. Show that both satisfy Euler's formula. See margin.

11. Consider the graph with 7 edges shown at the right.

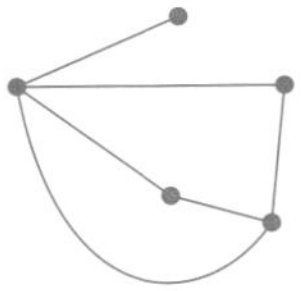

a. Draw a graph obtained from this one by performing one of the changes described in the theorem on page 684.

b. Suppose you know that $V - E + F = 2$ for all connected graphs with no crossings and with 6 edges. Based on your work in part **a**, what can you conclude about this graph? Justify your answer. a,b) See margin.

12. a. Show that Euler's formula does not hold for the graph at the right. $V - E + F = 6 - 6 + 8 = 8$

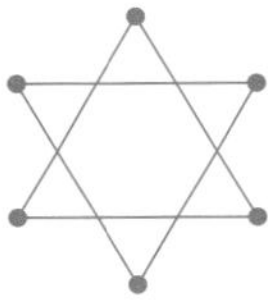

b. What condition of Euler's formula does this graph not satisfy? See margin.

13. Suppose a connected graph with no crossings has vertices of the following degrees: 1, 2, 2, 3, 5, 5.

a. How many edges does the graph have? 9

b. How many faces does it have? 5

14. Suppose that 3 houses are each to be connected to 3 utilities: electricity, gas, and telephone, and that the connecting lines are not to cross.

a. This situation may be represented as a graph in which vertices represent houses and utilities, and edges represent the connecting lines. How many vertices and edges must it have? See margin.

b. Try drawing a graph in the plane for this situation if there must be no crossings. impossible

c. Use Euler's formula to determine the number of faces the graph must have. 5

d. Explain why a face must have at least 4 edges. (Hint: it cannot have 3 edges, because that would mean that either two houses or two utilities would be connected to each other.)

e. Use part **d** along with the fact that an edge borders exactly 2 faces to explain why $4F \leq 2E$, or $2F \leq E$.

f. Use parts **c** and **e** to explain why it is impossible to connect 3 houses to 3 utilities without any lines crossing.

d–f) See margin.

4.a.

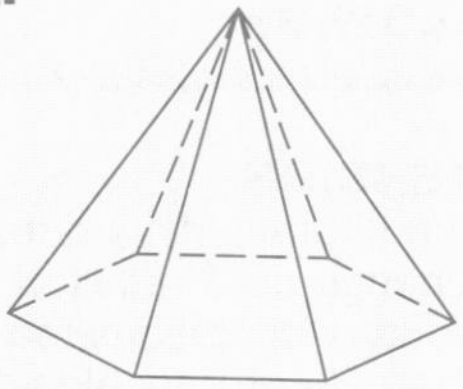

b.

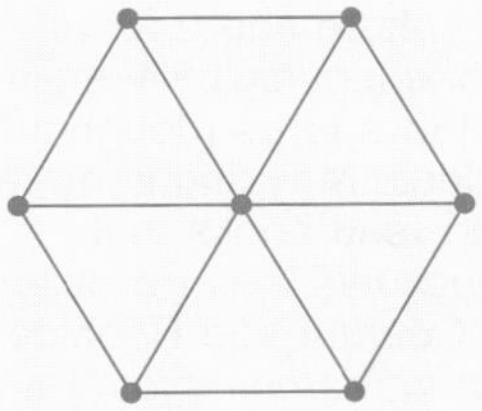

c. $V - E + F = 7 - 12 + 7 = 2$

7.a. The vertex of degree 1 and its adjacent edge was removed.
b. Both V and E were reduced by 1, so $V - E$ stayed the same. Since F did not change, $V - E + F$ did not change.

8.a. An edge was removed.
b. Both E and F were reduced by 1, so $-E + F$ was not changed. Since V did not change, $V - E + F$ did not change.

9. True, by the contrapositive of the second theorem of this lesson: Let G be a graph with at least one edge. If G has no vertex of degree 1, then G has a cycle.

10.

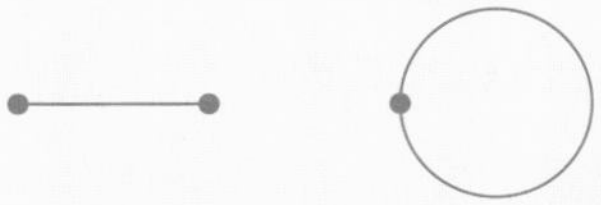

$V - E + F = 2 - 1 + 1 = 2$ $\qquad$ $V - E + F = 1 - 1 + 2 = 2$

11., 12.b., 14.a., d.–f. See Additional Answers in the back of this book.

FOLLOW-UP

EXTENSION
Euler's characteristic does not change as a result of continuous distortions, so it is an example of a topological property. It is 2 for any surface topologically equivalent to a sphere and 0 for any surface topologically equivalent to a torus (doughnut). Students may find it interesting to read books that discuss this idea (for example, Courant and Robbins, *What is Mathematics?*) and find out which figures have an Euler characteristic equal to 1.

ADDITIONAL ANSWERS
15.a.

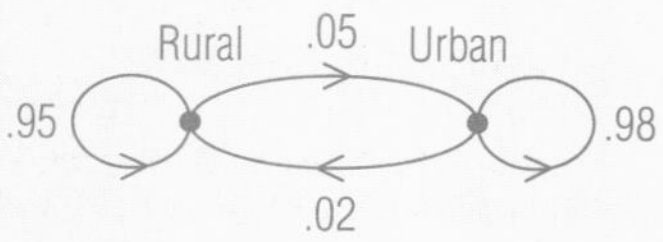

b.

	R	U
R	.95	.05
U	.02	.98

c. Urban ≈ 71%; Rural ≈ 29%

d.
$$\begin{aligned} .95a + .02b &= a \\ -5a + 2b &= 0 \\ 5a + 5b &= 5 \\ \hline 7b &= 5 \end{aligned}$$
$b = \frac{5}{7} \approx .714$; $a = \frac{2}{7} \approx .286$

16. 8

17.b. sample:

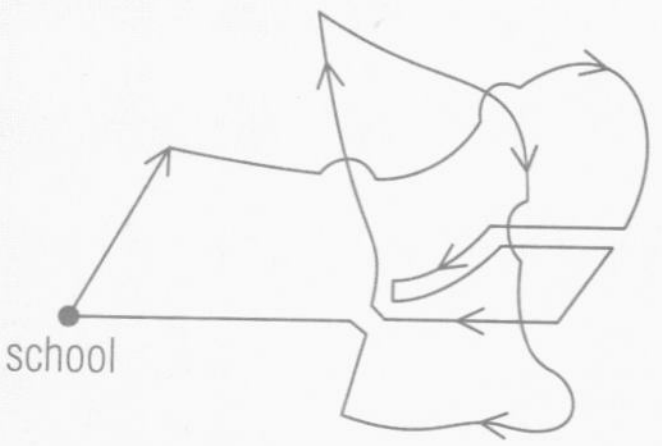

18., 20., 21.a., c., 22. See Additional Answers in the back of this book.

Review

15. Suppose that in a particular country, each year 5% of the rural population moves to urban areas, and 2% of the urban population moves to rural areas. *(Lesson 11-6)* See margin.
a. Sketch a digraph with edges labeled by numbers to represent the movement of the population.
b. Write the transition matrix T.
c. Find a large power of T and use it to predict the proportion of the population in rural and urban areas in the long run.
d. Solve a system of equations to check your prediction.

16. A digraph has the adjacency matrix given at the right. Find the number of paths of length 3 from v_3 to v_2. *(Lesson 11-5)* See margin.
$$\begin{bmatrix} 1 & 2 & 0 \\ 1 & 0 & 1 \\ 2 & 0 & 1 \end{bmatrix}$$

17. a. Can a school bus start and end at the school and pick up students along every section of road shown in the map at the right without repeating any road? Justify your answer.

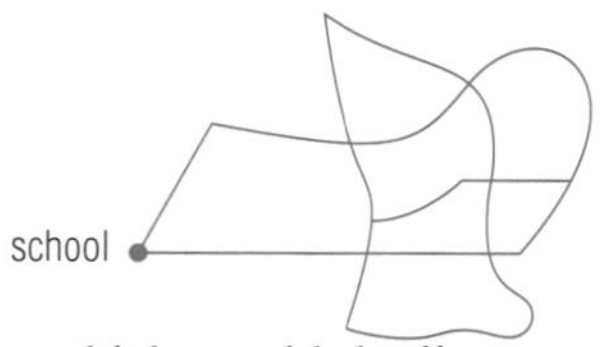

b. If so, draw the route the bus would follow. If not, draw a route which would duplicate as little of the roads as possible. *(Lesson 11-4)*
a) No, there are two vertices with odd degrees. b) See margin.

18. Determine whether there is a simple graph with four vertices: one of degree 1, and three of degree 3. Justify your answer. *(Lesson 11-3)* See margin.

19. If 6 dice are rolled, find the probability that exactly two 2s will come up. *(Lesson 10-6)* ≈ 20%

20. Prove that ∀ positive integers n, $\sum_{i=1}^{n} i(i+3) = \frac{n(n+1)(n+5)}{3}$.
(Lesson 7-4) See margin.

Exploration

21. a. Draw several graphs without crossings that are not connected, but have 2 separate components. See margin.
b. What is the value of $V - E + F$ for such graphs? 3
c. Repeat parts **a** and **b** for graphs with 3 separate components.
d. Conjecture the value of $V - E + F$ for graphs without crossings with n components, where n is a positive integer.
c) See margin. d) $n + 1$

22. Consider a regular polyhedron where each face is an n-sided regular polygon and m edges meet at each vertex. See margin.
a. Explain why $nF = 2E$. **b.** Explain why $mV = 2E$.
c. Solve the equations in parts **a** and **b** for F and V, respectively, and substitute into Euler's formula to obtain $\frac{1}{n} + \frac{1}{m} = \frac{1}{2} + \frac{1}{E}$.
d. Find all positive integer values for n and m for which the equation in part **c** gives reasonable values for E.
e. Use the results of part **d** to give a proof different from that in Lesson 5-8 that there are only five regular polyhedra: the tetrahedron, cube, octahedron, dodecahedron, and icosahedron.

Projects

1. The following unsolved problem was posed by G. Ringel and offered by Richard K. Guy in the December, 1989, issue of the *American Mathematical Monthly*, pp. 903–904. Consider the connected graph drawn below. Note that it has no circuits, and that all of its vertices have degree 1 or 3. Its 29 edges have been numbered from 1 to 29 in such a way that the sum of the numbers on the three edges leading into any vertex of degree 3 is always 45.

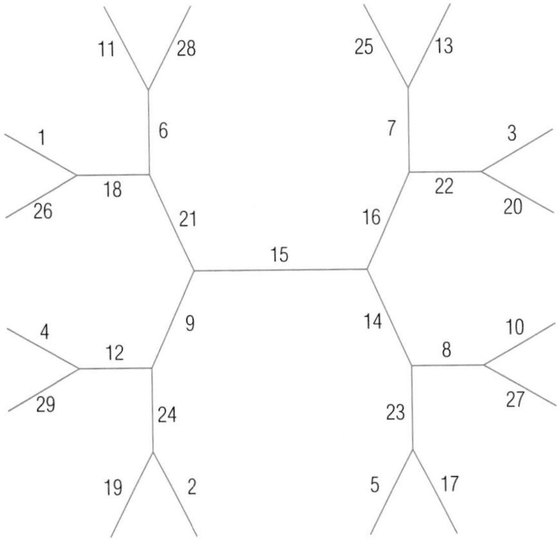

Ringel's conjecture is as follows:

> Suppose you are given any connected graph which has no circuits and all of whose vertices have degree 1 or 3. Let n be the number of edges in the graph. Then you can number the edges of the graph from 1 to n in such a way that the sum of the numbers on the edges leading into any vertex of degree 3 is a constant.

This conjecture has been neither proved nor disproved. Explore this problem by trying to verify the conjecture for smaller graphs. Is there a systematic way of numbering the edges? If so, can it be used to prove the conjecture? Or, can you find a counterexample?

2. Find out what *dynamic programming* is. Give some examples of problems that can be solved using dynamic programming and demonstrate their solutions. Describe the recursive algorithm that is used in solving these types of problems. For assistance, refer to books on artificial intelligence or discrete mathematics.

3. **a.** Suppose a salesperson wishes to travel to each city in the map below exactly once, starting and ending in New York, and using only the roads shown. The numbers on the roads indicate distances (in miles) between cities. Find the shortest route that the salesperson could use.

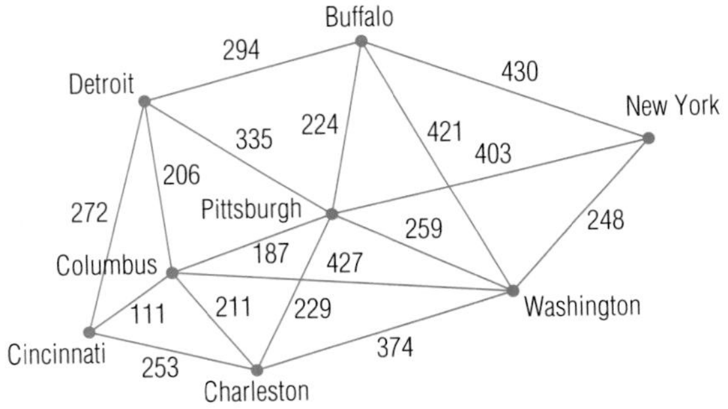

b. Explain why it would be impossible for the salesperson to visit each city below exactly once, using only the roads shown, and end at the starting point.

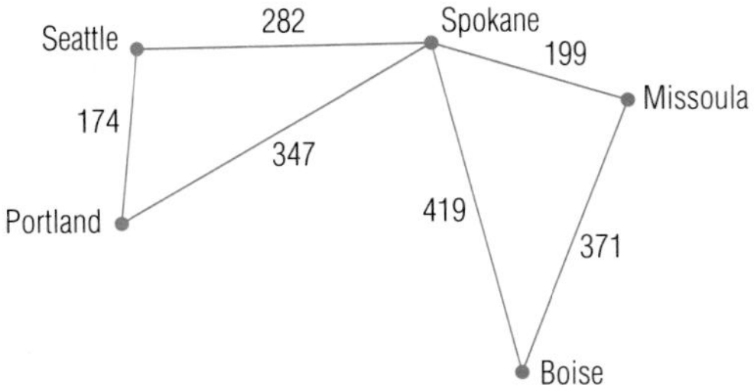

c. The general problem of parts **a** and **b** is known as the *traveling salesman problem*. Refer to a discrete mathematics book to find out about it. Write a report on your findings, along with your solutions to parts **a** and **b**.

PROJECTS

The projects provide an opportunity for extended work on topics related to the content of the chapter. (See pages 70 and T37 for more information on the purposes of projects.) Some or all of these projects may be suitable for **small group work**. Solutions for the projects can be found at the end of the Additional Answers section in the back of this book.

If any student is able to prove or disprove the conjecture stated in **Project 1**, you should write the editors of the *American Mathematical Monthly.* (If you write UCSMP, we will tell you whom to contact.)

The algorithm used in the solution of the scheduling problem of Lesson 11-1 is an example of dynamic programming. Students who enjoyed that problem might be interested in **Project 2**.

Regarding **Project 3**, in February 1991, it was announced that Donald L. Miller of the DuPont Co. and Joseph F. Penky of Purdue University had developed an algorithm that gives solutions to certain types of traveling salesman problems.

4. Find out some ways that graphs (also called *networks* or *nets*) are used in artificial intelligence, and write a report on them. Useful references are books on artificial intelligence, such as *Artificial Intelligence*, by Patrick Henry Winston, and *Elements of Artificial Intelligence*, by Steven L. Tanimoto.

5. A **tree** is a connected graph that has no circuits. Given a connected graph G, a **spanning tree** is a tree consisting of a subset of the edges of G but all of the vertices of G. Thus, by definition, it is a part of G which is simple and keeps all the vertices of G connected. For example, if G is the graph pictured below, then the graph above at the right is a spanning tree for G.

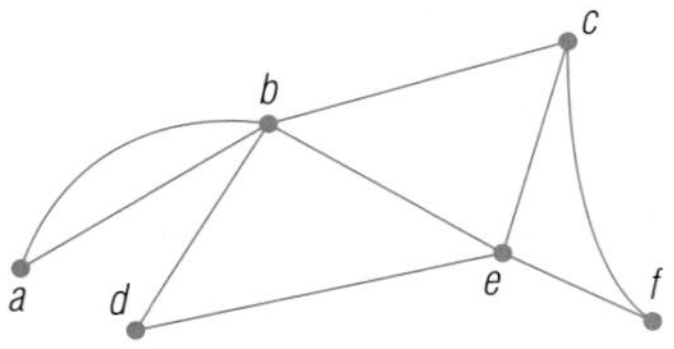

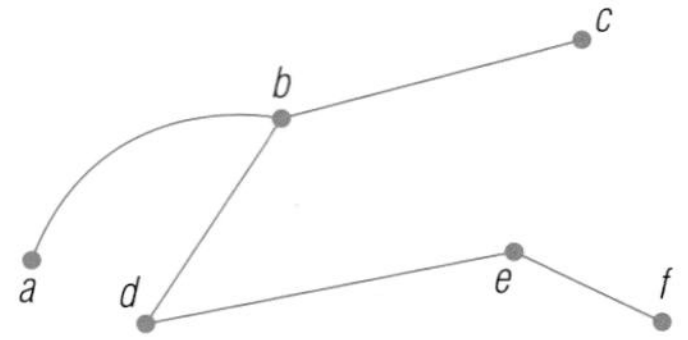

a. Find another spanning tree for G.

b. Describe a systematic algorithm that could be used to find a spanning tree given a connected graph, and demonstrate it on a graph of your choice.

c. Suppose each edge of the graph G is labeled with a number, as below. Find a spanning tree for G such that the sum of the labels on its edges is the smallest possible. Such a tree is called a **minimal spanning tree.**

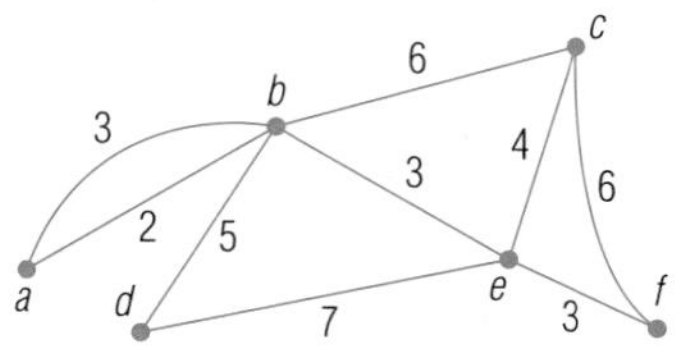

d. Find a systematic algorithm for finding a minimal spanning tree given a connected graph, and demonstrate it on a graph of your choice. (You might look up Kruskal's algorithm or Prim's algorithm in a book that discusses discrete mathematics.)

e. What would be some real-life applications for such an algorithm?

690

Summary

Problems in medicine, science, and business, as well as various puzzles, can be represented and solved using graphs. Directed graphs are useful for solving scheduling problems, and probability trees are helpful in situations involving the probability that one event occurs if another event occurs.

Because the total degree of any graph is twice the number of edges, it is even. Thus every graph has an even number of vertices of odd degree. These facts can sometimes be used to determine that certain types of graphs do not exist, providing solutions to a special class of problems referred to as handshake problems.

Euler proved that if a graph has an Euler circuit, then every vertex of the graph has even degree, and that if every vertex of a connected graph has even degree, then the graph has an Euler circuit. These results can be used to determine whether a given graph has an Euler circuit, and thus can be used to solve practical problems as well as puzzles such as the Königsberg bridge problem.

If an edge is removed from a circuit in a connected graph, the graph remains connected. This theorem helps prove Euler's formula: In any connected graph with no crossings, V vertices, E edges, and F faces, $V - E + F = 2$. This relation can be applied to any polyhedron, where V, E, and F are the number of vertices, edges, and faces of the polyhedron.

Every graph can be represented by an adjacency matrix that contains the numbers of edges from each vertex to each other vertex. The adjacency matrix for an undirected graph is always symmetric. The number of walks of length n from a given vertex to another given vertex can be obtained from the nth power of the adjacency matrix.

A Markov chain is a system involving a succession of changes from one state (or condition) to another, where the probability of moving to one state depends only on the previous state. It can be modeled by a stochastic matrix that contains those probabilities. It can be proved that for large values of n, the nth power of a stochastic matrix is approximately equal to a matrix in which every row is the same. This implies that in a Markov chain, after a long period of time, the probability of being in each state approaches a constant.

Vocabulary

For the starred (*) terms you should be able to give a definition of the term.
For the other terms you should be able to give a general description and a specific example of each.

Lesson 11-1
Königsberg bridge problem
graph theory
equivalent graphs
directed graph, digraph
probability tree

Lesson 11-2
*graph
vertex, edge
edge-endpoint function
endpoint
adjacent vertices
adjacent edges
parallel edges
loop, isolated vertex
*simple graph
crossing
adjacency matrix

Lesson 11-3
complete graph
*degree of a vertex, deg (v)
*total degree of a graph
Total Degree of a Graph Theorem

Lesson 11-4
*walk
*path
*circuit
*Euler circuit
Euler Circuit Theorem
connected vertices
*connected graph
Sufficient Condition for an Euler Circuit Theorem
Circuits and Connectedness Theorem

Lesson 11-5
length of a walk
symmetric matrix

Lesson 11-6
*stochastic matrix
transition probabilities
Markov chain
Convergence of Powers Theorem

Lesson 11-7
face of a graph
Euler's formula

SUMMARY

The Summary gives an overview of the entire chapter and provides an opportunity for students to consider the material as a whole. Thus, the Summary can be used to help students relate the various concepts presented in the chapter.

VOCABULARY

Terms, symbols, and properties are listed by lesson to provide a checklist of concepts a student must know. (See page 71 for more information.)

The Progress Self-Test provides the opportunity for feedback and correction; the Chapter Review provides additional opportunities for practice.

USING THE PROGRESS SELF-TEST
Assign the Progress Self-Test as a one-night assignment. (See pages 72 and T35 for more information.)

Save some class time for a discussion of **Question 15**. For one year, the state of Illinois required an AIDS test as a requirement for a marriage license. Finally, in light of evidence and an outcry that was predicted because of the answer to (c), the law was repealed. John Allen Paulos discusses this issue in his book *Innumeracy*.

ADDITIONAL ANSWERS
9.

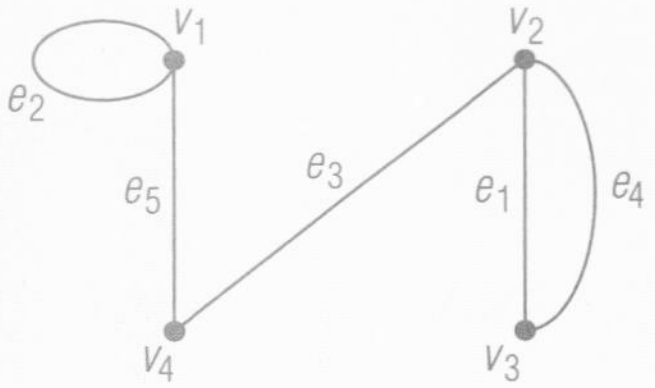

10.

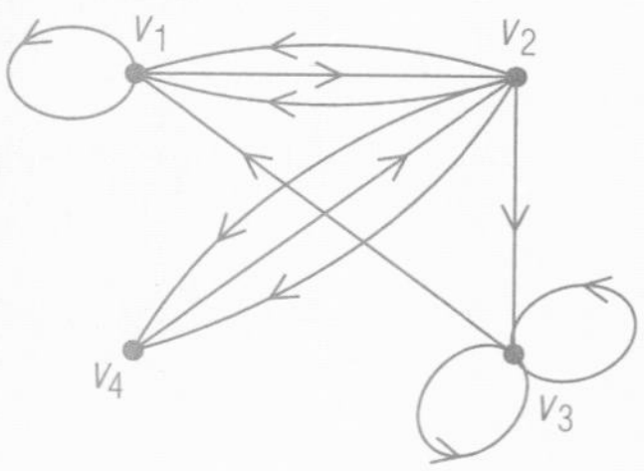

Progress Self-Test

See margin for answers not shown below.

Take this test as you would in class. Then check the test yourself using the solutions at the back of the book.

In 1–3, use the following graph.

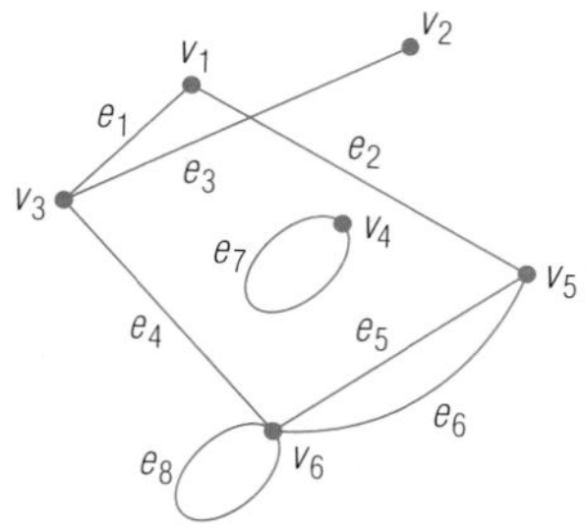

1. List all vertices adjacent to v_2. v_3

2. List all sets of parallel edges. e_5 and e_6

3. How can the graph be changed to make it simple? Remove e_7 and e_8 and either e_5 or e_6.

In 4–8, use the following graph.

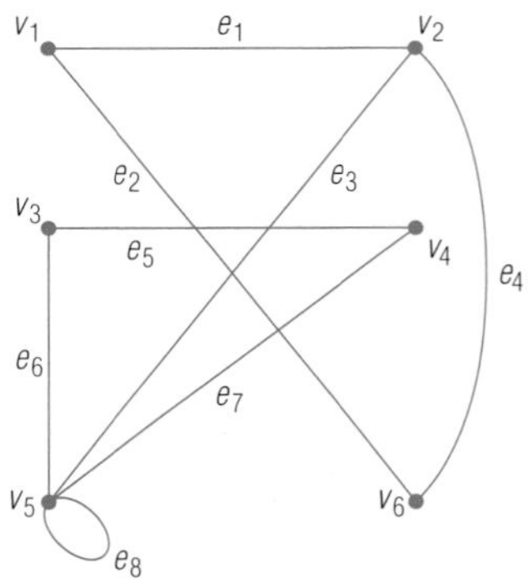

4. What is the degree of v_5? 5

5. What is the total degree of the graph? 16

6. Starting at v_3, is the walk $e_6\ e_8\ e_7\ e_5$:
a. a path? Yes
b. a circuit? Yes
c. an Euler circuit? No

7. List each edge which, if removed, would leave the graph connected. all except e_3

8. Does the graph have an Euler circuit? Justify your answer.

In 9 and 10, draw a graph satisfying the given conditions.

9. set of vertices: $\{v_1, v_2, v_3, v_4\}$
set of edges: $\{e_1, e_2, e_3, e_4, e_5\}$
edge-endpoint function

edge	endpoints
e_1	$\{v_2, v_3\}$
e_2	$\{v_1\}$
e_3	$\{v_2, v_4\}$
e_4	$\{v_2, v_3\}$
e_5	$\{v_1, v_4\}$

10. adjacency matrix: $\begin{bmatrix} 1 & 1 & 0 & 0 \\ 2 & 0 & 1 & 2 \\ 1 & 0 & 2 & 0 \\ 0 & 1 & 0 & 0 \end{bmatrix}$

In 11–13, consider the following graph of airline routes.

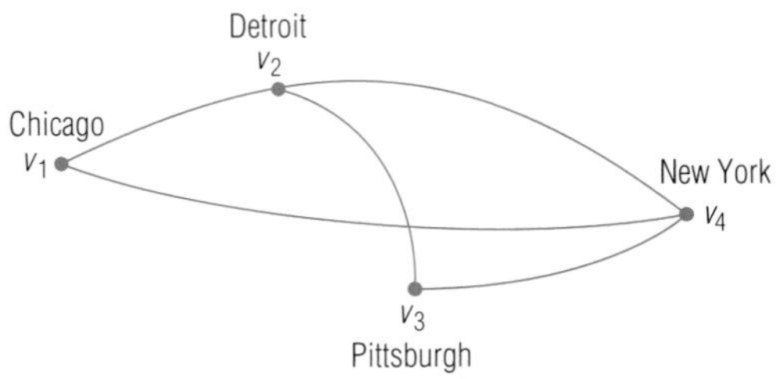

11. It is sometimes cheaper to fly through another city to get to your final destination. How many one-stop routes (i.e., walks of length 2) are there from Chicago to Pittsburgh? 2

12. a. Write the adjacency matrix for the graph.
b. Use the adjacency matrix to find the number of one-stop routes from Detroit to each of the other cities.

13. Is it possible for the airline to send a single plane so it covers every route exactly once and then is back where it begins? Justify your answer.

14. Does there exist a graph with 4 vertices of degrees 0, 1, 2, and 2? If so, draw one. If not, explain why not.

15. A drug manufacturer has developed a test for the AIDS virus (Acquired Immune Deficiency Syndrome). A study shows that the test registers positive for 98% of those who have the AIDS virus and for .2% of those who do not have the AIDS virus. Assume that .1% of the population has the AIDS virus.

a. Draw a graph and label its edges with probabilities to represent the situation.

b. Find the fraction of the population in each category: has the AIDS virus and tests positive; has the AIDS virus but tests negative; doesn't have the AIDS virus but tests positive; doesn't have the AIDS virus and tests negative.

c. What is the probability that a person who tests positive does not have the AIDS virus? ≈ 67%

16. Explain why it is impossible to set up a tournament in which each of nine teams plays exactly seven of the other teams.

8. No, there are vertices of odd degree.

12. a)

	v_1	v_2	v_3	v_4
v_1	0	1	0	1
v_2	1	0	1	1
v_3	0	1	0	1
v_4	1	1	1	0

b) The numbers of 1-stop routes to Chicago, Pittsburg, and New York are 1, 1, and 2, respectively.

13. No, this would be an Euler circuit, which is impossible since there are vertices with odd degree.

14. No, the total degree of a graph must be even.

15. b. has AIDS and tests positive: .00098
has AIDS but tests negative: .00002
doesn't have AIDS but tests positive: .001998
doesn't have AIDS and test negative: .997002

16. A graph cannot have an odd number of odd vertices.

17. A math teacher has the following policy regarding pop quizzes: If he gives one on one day, he will not give one the next day. However, if he does not give a pop quiz on a particular day, there is a 50-50 chance he will give one the next day.

a. Draw a graph for this situation, and label the edges with the correct probabilities.

b. Write down the corresponding stochastic matrix.

c. Find the long term probability that the teacher will give a pop quiz on any particular day. $\frac{1}{3}$

15.a.

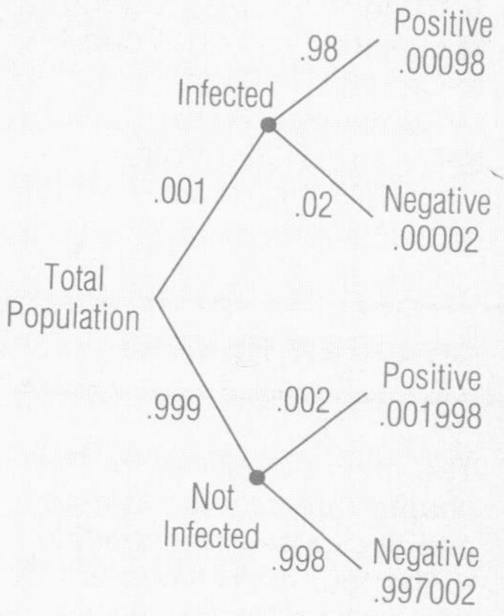

17.a.

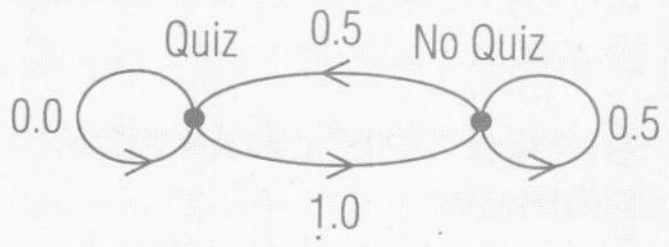

b.

	Q	NQ
Q	0.0	1.0
NQ	0.5	0.5

RESOURCES
- Chapter 11 Test, Form A
- Chapter 11 Test, Form B
- Chapter 11 Test, Cumulative Form
- Quiz and Test Writer

CHAPTER REVIEW

The main objectives for the chapter are organized here into sections corresponding to the four main types of understanding this book promotes: Skills, Properties, Uses, and Representations. We call these the SPUR objectives.

USING THE CHAPTER REVIEW
Students should be able to answer questions like these with about 85% accuracy by the end of the chapter. (See pages 74 and T38–39 for more information.)

ADDITIONAL ANSWERS
1.a. sample:

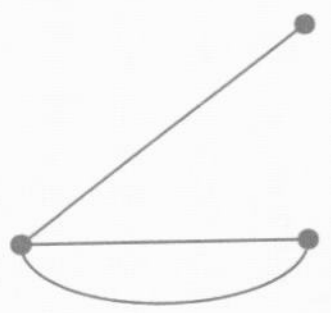

b. Yes, sample:

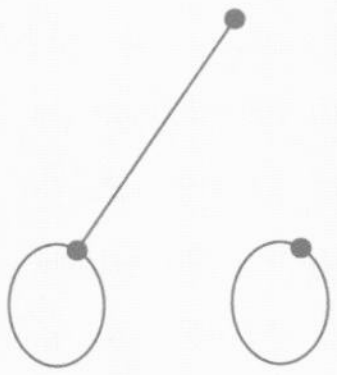

2. sample:

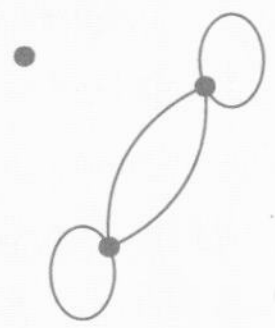

Chapter Review

Questions on SPUR Objectives

SPUR stands for **S**kills, **P**roperties, **U**ses, and **R**epresentations.
The Chapter Review questions are grouped according to the SPUR Objectives for this chapter.
See margin for answers not shown below.

SKILLS deal with the procedures used to get answers.

Objective A: *Draw graphs given sufficient information. (Lessons 11-2, 11-3)*

1. a. Draw a graph with three vertices and three edges.
b. Is it possible to draw a graph with three vertices and three edges with two edges which are not adjacent to each other? If so, do it. If not, explain why not.

2. Draw a graph with two loops, an isolated vertex, and two parallel edges.

3. Draw all simple graphs with three vertices.

4. Draw a simple graph with five vertices of the following degrees: 2, 3, 3, 4, and 4.

5. Draw the graph defined below.
set of vertices: $\{v_1, v_2, v_3, v_4, v_5\}$
set of edges: $\{e_1, e_2, e_3, e_4, e_5\}$
edge-endpoint function:

edge	endpoints
e_1	$\{v_1, v_3\}$
e_2	$\{v_1, v_4\}$
e_3	$\{v_2, v_4\}$
e_4	$\{v_1\}$
e_5	$\{v_1, v_3\}$

PROPERTIES deal with the principles behind the mathematics.

Objective B: *Identify parts of graphs and types of graphs. (Lessons 11-2, 11-3, 11-4, 11-5)*

6. Use the graph drawn below.
a. Identify all vertices adjacent to v_1. v_2, v_3, v_5
b. Identify all edges adjacent to e_5.
c. Identify any isolated vertices. none
d. Identify any parallel edges. e_1 and e_8
e. Identify any loops. e_6
f. *True* or *false?* If edge e_6 is removed, the graph is simple. False
g. Give the degree of each vertex.
h. Give the total degree of the graph. 16

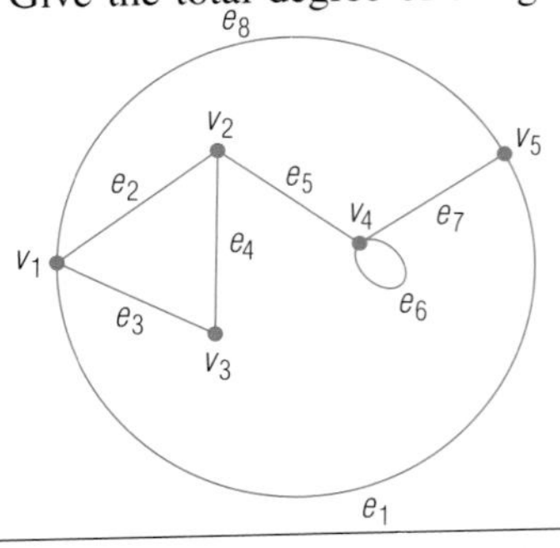

7. Consider the graph below.
a. Starting at v_2, is the walk $e_4\ e_7\ e_6\ e_5\ e_3$ **i.** a path? **ii.** a circuit? **iii.** an Euler circuit?
b. Starting at v_2, is the walk $e_4\ e_6\ e_5\ e_3\ e_1\ e_2\ e_3$ **i.** a path? **ii.** a circuit? **iii.** an Euler circuit?
c. Identify all paths from v_1 to v_3 and give their lengths.
d. Identify three circuits that go through v_1.

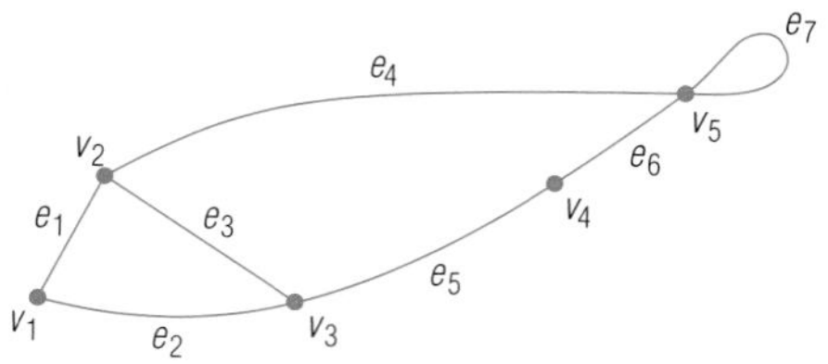

8. Consider the graph below.
 a. Identify an Euler circuit.
 b. Identify two circuits that are not Euler circuits.
 c. Identify a walk that is not a path.
 d. What is the minimum number of edges to remove so that the graph is no longer connected? **2**
 e. What is the maximum number of edges that can be removed at the same time while keeping the graph connected? List one such set of edges. **3: sample: e_1, e_3, e_6**

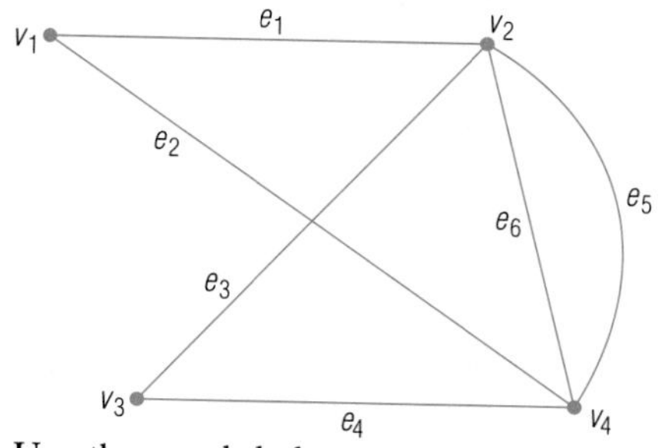

9. Use the graph below.
 a. *True* or *false?* There is exactly one walk from v_1 to v_2. **False**
 b. *True* or *false?* There is exactly one path from v_1 to v_2. **False**

10. Is the graph below connected? **Yes**

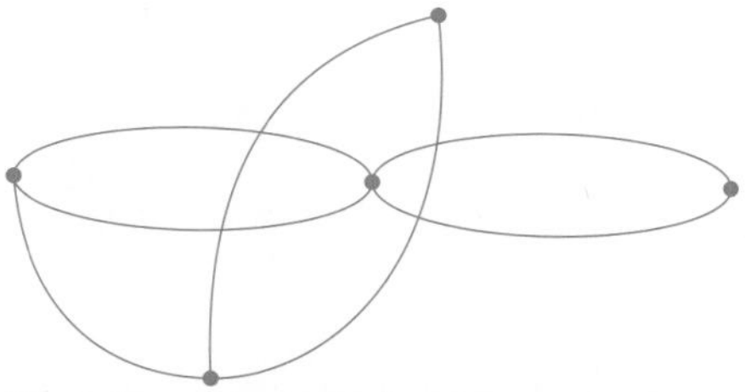

11. Give the edge-endpoint function table for the graph in Question 6.

Objective C: *Determine whether there exists a graph containing vertices with given degrees.* *(Lesson 11-3)*

In 12–15, either draw a graph with the given properties or show that no such graph exists.

12. a graph with 5 vertices of degrees 1, 2, 2, 3, and 5
13. a graph with 5 vertices of degrees 1, 2, 2, 3, and 0
14. a simple graph with 5 vertices of degrees 1, 2, 2, 3, and 0
15. a graph with 9 vertices of degrees 0, 1, 1, 1, 2, 2, 2, 3, and 3

16. Suppose that the sum of the entries in a matrix is odd. Can this matrix be the adjacency matrix of a graph? Explain your answer.

Objective D: *Determine whether a graph has an Euler circuit.* *(Lesson 11-4)*

In 17–20, determine, if possible, whether the graph has an Euler circuit. Justify your answer.

17.

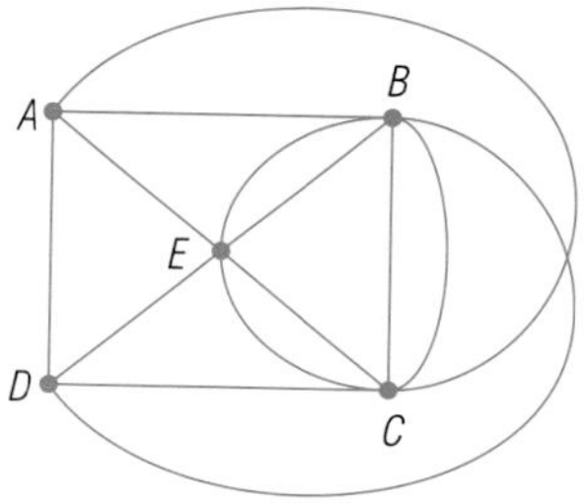

18.

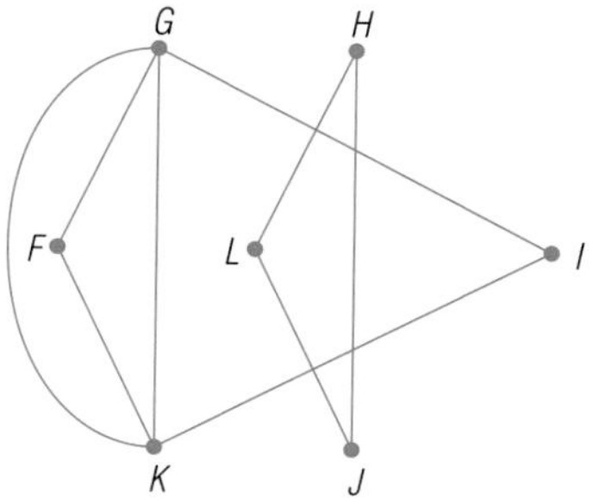

19. The graph whose adjacency matrix is $\begin{bmatrix} 1 & 1 & 1 \\ 1 & 0 & 2 \\ 1 & 2 & 0 \end{bmatrix}$

20. A graph with vertices of degrees 2, 2, 4, and 6

3.

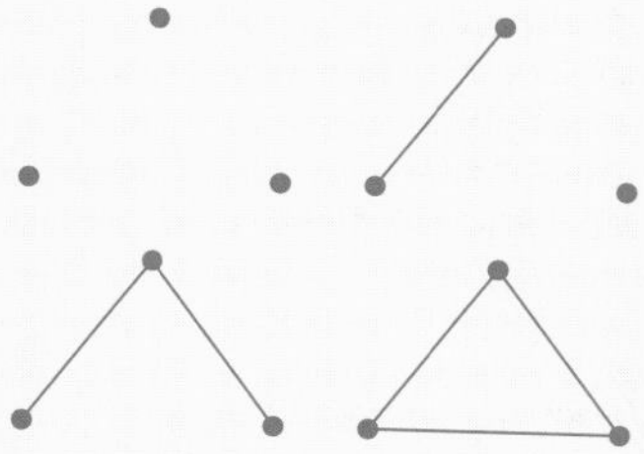

4. sample:

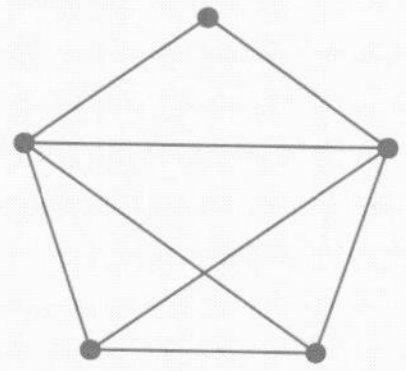

5.

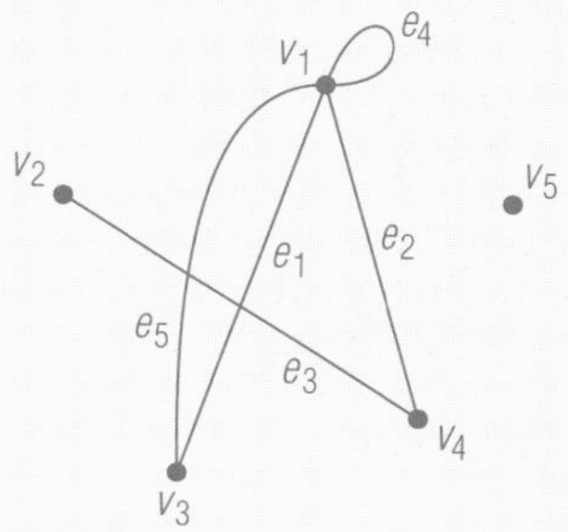

6.b. e_2, e_4, e_6, e_7
g. $deg(v_1) = 4$;
$deg(v_2) = 3$;
$deg(v_3) = 2$;
$deg(v_4) = 4$;
$deg(v_5) = 3$;

7.a. i. Yes; ii. Yes; iii. No
b. i. No; ii. No; iii. No
c.

length	path(s)
1	e_2
2	e_1e_3
3	none
4	$e_1e_4e_6e_5$
5	$e_1e_4e_7e_6e_5$
	$e_2e_3e_4e_6e_5$
	$e_2e_5e_6e_4e_3$
6	$e_2e_3e_4e_7e_6e_5$
	$e_2e_5e_6e_7e_4e_3$

d. samples: $e_1e_2e_5e_6e_4$; $e_3e_2e_1$; $e_4e_7e_6e_5e_2e_1$

8., 11., 12.-15., 16., 17.-20. See Additional Answers in the back of this book.

ADDITIONAL ANSWERS

21.a.

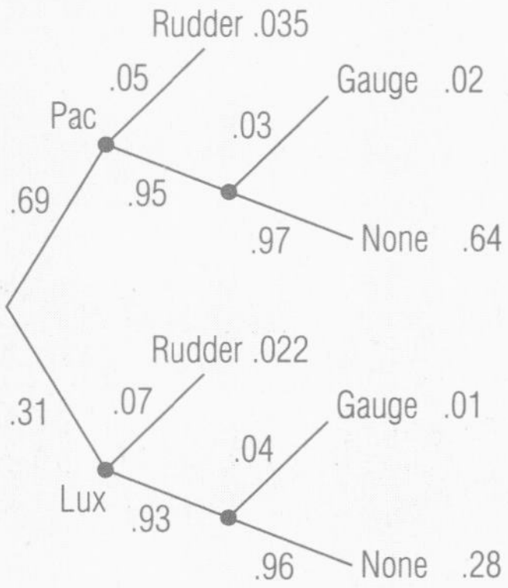

22.a.

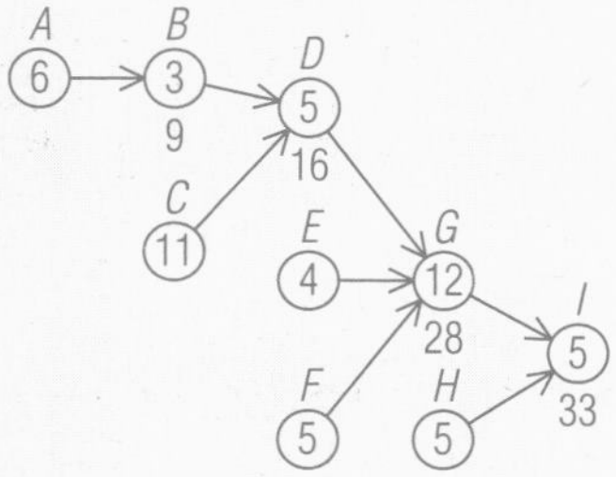

23.a.

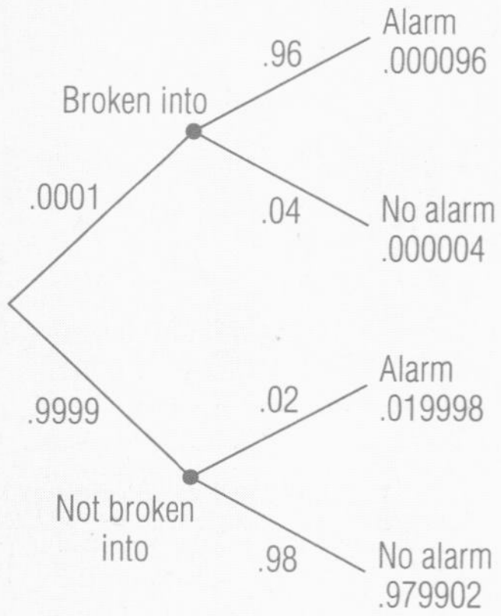

24. No; this situation may be represented as a graph with 25 vertices, each with 5 edges. This is not possible since a graph cannot have an odd number of odd vertices.

USES deal with applications of mathematics in real situations.

Objective E: *Use graphs to solve scheduling and probability problems.* *(Lesson 11-1)*

21. Oiler Motorboats manufactures two models of motorboat: a compact model called the Pac, and a luxury model called the Lux. In 1990, 69% of the boats sold were Oiler Pacs, and 31% were Oiler Luxes. Since then, 5% of the owners of an Oiler Pac have had to replace the rudder, and, of the others, 3% have had to replace the fuel gauge. The rest have needed no repairs. 7% of the owners of an Oiler Lux have had to replace the rudder, and, of the others, 4% had to replace the fuel gauge. The rest have needed no repairs.

a. Draw a probability tree to represent this situation, labeling edges with the proper probabilities.

b. If a 1990 Oiler was brought in for rudder replacement, what is the probability that it was a Pac? ≈ **61.4%**

22. Suppose the process of assembling a car at a particular plant can be broken down into the following tasks.

	Task Description	Time required (hrs)	Prerequisite tasks
A	Assemble body	6	
B	Paint exterior	3	*A*
C	Assemble engine	11	
D	Install engine	5	*B*, *C*
E	Assemble water pump	4	
F	Assemble carburetor	5	
G	Install fuel, exhaust, electrical, cooling systems	12	*D*, *E*, *F*
H	Assemble interior parts	5	
I	Install interior	5	*G*, *H*

a. Sketch a directed graph to represent the situation.

b. What is the minimal time required to assemble a car? **33 hours**

23. Suppose that at any given time in a particular city, the probability that a given car is being broken into is .01%. Also suppose that a Car-Safe alarm system installed on a car sounds 96% of the time that the car is broken into, but also sounds 2% of the time that the car is not being broken into.

a. Draw a probability tree to represent the situation.

b. Find the probability that the car is really being broken into when the alarm sounds. ≈ **0.48%**

Objective F: *Use the Total Degree of a Graph Theorem and its corollaries to solve handshake problems.* *(Lesson 11-3)*

24. In a class of 25 students, is it possible for each student to shake hands with exactly five other students? Justify your answer.

25. Six authors are writing a textbook, each one writing a different part. In order to maintain some unity in the book, they decide that each author should show the part he or she has written to three other authors. They want to do this in the following way: Each author will make three copies of what he or she has written, then trade each copy with a different author. Is this possible? Justify your answer.

26. From 1970 to 1975, the National Football League had two conferences each with 13 teams. If the league office had decided that every team should play 11 games in its own conference, each against a different team, would this have been possible? Justify your answer.

696

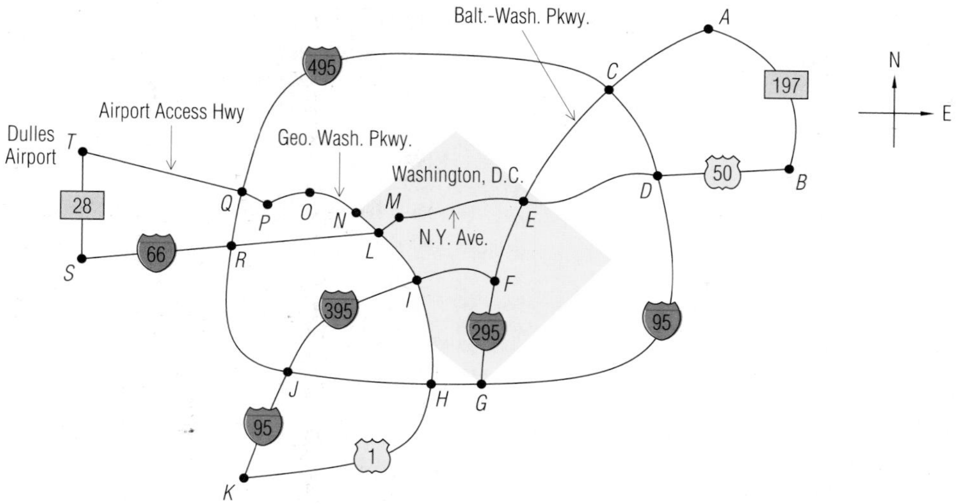

Objective G: *Solve application problems involving circuits.* (*Lessons 11-1, 11-4*)

27. A map of the Washington, DC area is shown above.

a. Explain why it is impossible to travel each road exactly once and return to where you started.

b. What one section of road (that is, what one edge of the graph) can be removed to make such a trip possible?

28. Consider the map of a section of a city shown below.

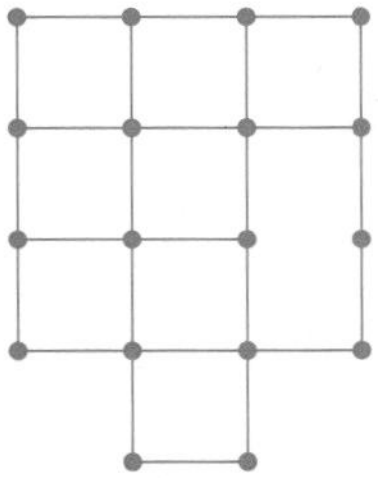

a. Each corner (indicated by a dot) is a recycling pickup point. Is there a route which a truck could follow which would begin and end at the same place and go past each of the other pickup points exactly once? If so, find it. If not, explain why not.

b. Is there a route which a street cleaner could follow which would begin and end at the same place and travel every section of road exactly once? If so, find it. If not, explain why not.

Objective H: *Use stochastic matrices to make long-term predictions.* (*Lesson 11-6*)

29. Some friends like to go bowling on Tuesdays. If they go on a particular Tuesday, there is a 40% chance they will go bowling the next Tuesday. Otherwise, there is a 75% chance that they will bowl the following Tuesday.

a. Draw a directed graph representing the situation.

b. Find T, the transition matrix.

c. Estimate how often the friends bowl on average over a long period of time by calculating T^8.

d. Find how often the friends bowl on average over a long period of time by solving a system of equations.

30. In a certain state, it was found that 60% of the daughters of women registered to vote as Democrats also register as Democrats, 15% register as Republicans, and the rest register as Independents. 70% of the daughters of Republicans are Republicans, 20% are Democrats, and the rest are Independents. 50% of the daughters of Independents are Independents, 30% are Democrats, and the rest are Republicans. Assume this pattern continues over many generations. What percentage of women will be registered in each group?

25. Yes; below is a sample graph:

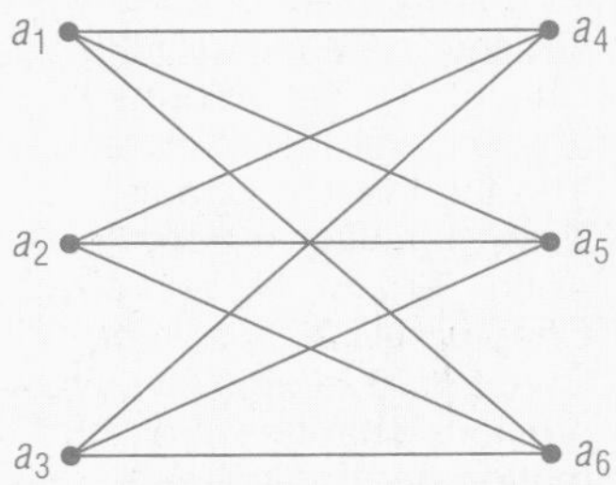

26. No, a graph cannot have an odd number of odd vertices.

27.a. Vertices *F* and *G* have odd degree, so there is not an Euler circuit.
b. the edge between *F* and *G*

28.a. Yes, sample:

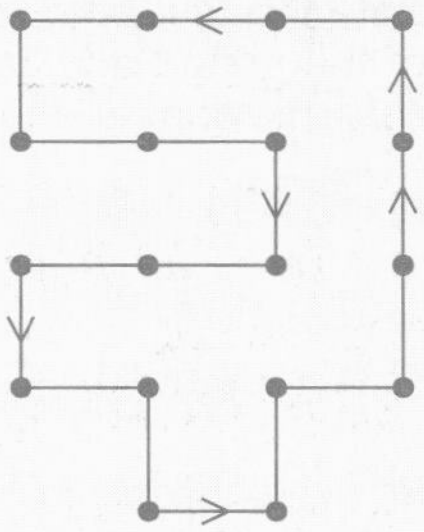

b. No, two of the vertices have odd degree, so no Euler circuit is possible.

29.a.

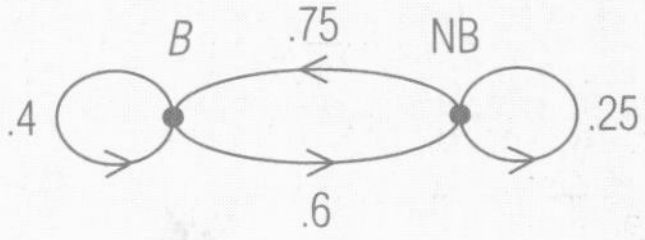

b. $\begin{array}{c} \\ B \\ NB \end{array}\begin{array}{c} \begin{array}{cc} B & NB \end{array} \\ \begin{bmatrix} .4 & .6 \\ .75 & .25 \end{bmatrix} \end{array}$

c. $T^8 \approx \begin{bmatrix} .5557 & .4443 \\ .5554 & .4446 \end{bmatrix}$
They bowl on about 56% of the Tuesdays.
d. $\approx$ **56%**

30. 38% Democrat, 36% Republican, 26% Independent

EVALUATION
Three forms of a Chapter Test are provided in the Teacher's Resource File. Chapter 11 Test, Forms A and B cover just Chapter 11. The third test is Chapter 11 Test, Cumulative Form. About 50% of this test covers Chapter 11, 25% covers Chapter 10, and 25% covers previous chapters. For information on grading, see *General Teaching Suggestions: Grading* on page T43 in the Teacher's Edition.

ASSIGNMENT RECOMMENDATION
We strongly recommend that you assign the reading and some questions from Lesson 12-1 for homework the evening of the test. It gives students work to do if they complete the test before the end of the period and keeps the class moving.

ADDITIONAL ANSWERS
31.

$$\begin{array}{c} \\ v_1 \\ v_2 \\ v_3 \\ v_4 \end{array} \begin{array}{c} \begin{array}{cccc} v_1 & v_2 & v_3 & v_4 \end{array} \\ \begin{bmatrix} 0 & 1 & 2 & 0 \\ 0 & 0 & 1 & 1 \\ 0 & 0 & 2 & 1 \\ 2 & 0 & 0 & 0 \end{bmatrix} \end{array}$$

32. The diagonal has zeros, and all other entries are either zeros or ones.

33.a.

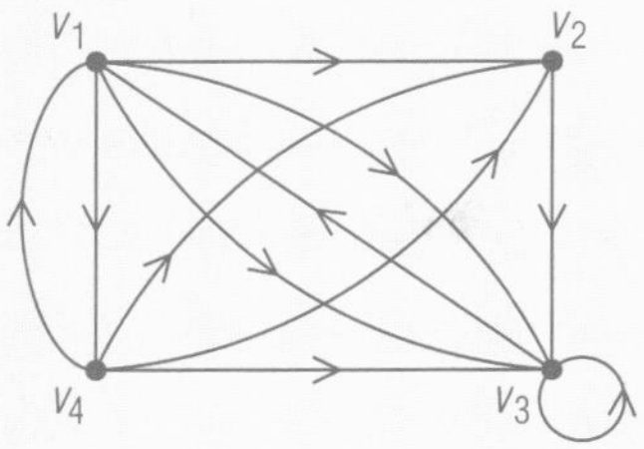

b. No, the matrix is not symmetric.

34., 37.a., 38.a., 39.c. See Additional Answers in the back of this book.

REPRESENTATIONS deal with pictures, graphs, or objects that illustrate concepts.

Objective I: *Convert between the picture of a graph or directed graph, and its adjacency matrix. (Lesson 11-2)*

31. Write the adjacency matrix for the directed graph shown below.

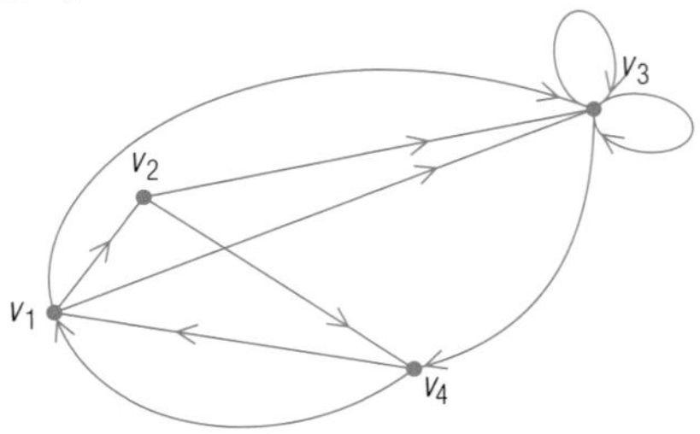

32. How can you tell from its adjacency matrix whether or not a graph is simple?

33. Consider the matrix shown below.

$$\begin{bmatrix} 0 & 1 & 2 & 1 \\ 0 & 0 & 1 & 0 \\ 1 & 0 & 1 & 0 \\ 1 & 2 & 1 & 0 \end{bmatrix}$$

a. Draw a directed graph whose adjacency matrix is the matrix given above.

b. Could the matrix above be the adjacency matrix of a graph that is undirected? If so, draw the graph. If not, explain.

34. Draw a graph (undirected) whose adjacency matrix is given below.

$$\begin{bmatrix} 0 & 2 & 1 \\ 2 & 2 & 0 \\ 1 & 0 & 0 \end{bmatrix}$$

35. In the adjacency matrix of the directed graph below,
a_{13} = __**a.**__ 0
and a_{22} = __**b.**__ . 0

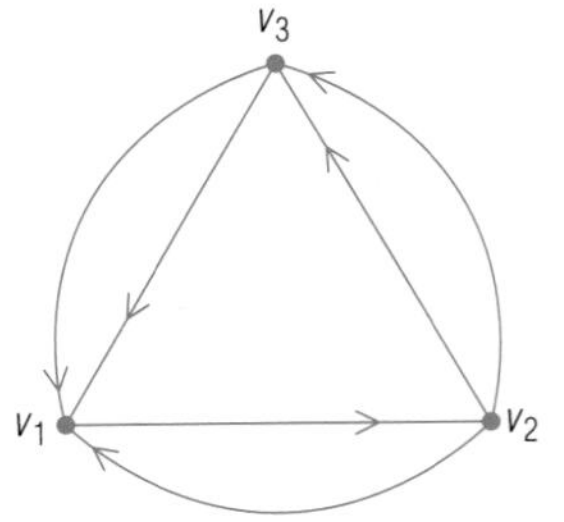

Objective J: *Use the powers of the adjacency matrix of a graph to find the number of walks of a given length from a given starting vertex to a given ending vertex. (Lesson 11-5)*

36. The adjacency matrix for a graph is $\begin{bmatrix} 2 & 1 & 2 \\ 1 & 0 & 1 \\ 2 & 1 & 0 \end{bmatrix}$. How many walks of length 2 go from v_2 to v_3? 2

37. a. Give the adjacency matrix for the graph below.

b. How many walks of length 3 are there which start at v_1? 39

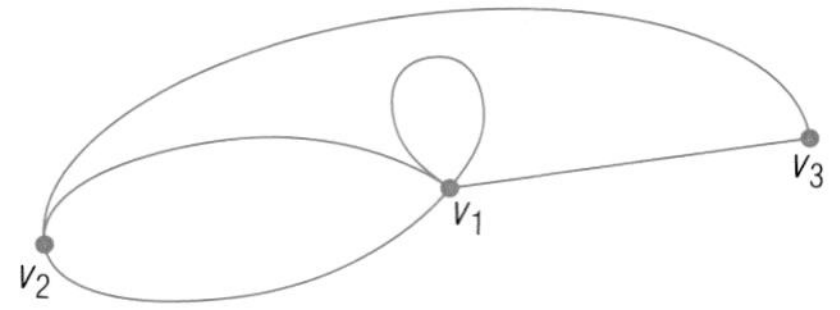

38. a. Give the adjacency matrix for the directed graph below.

b. How many walks of length 3 are there which start at v_1? 9

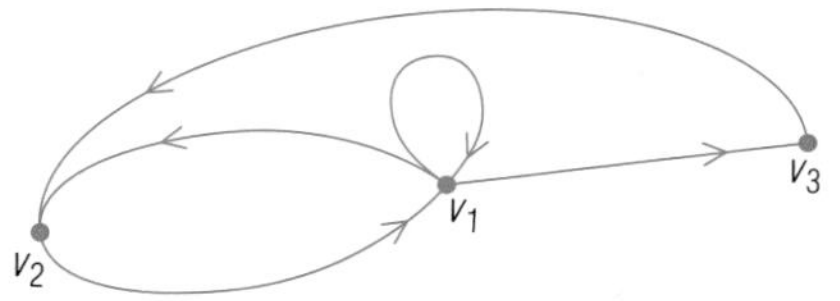

39. Consider the matrix

$$A = \begin{bmatrix} 0 & 1 & 1 & 1 \\ 0 & 0 & 1 & 1 \\ 0 & 0 & 0 & 1 \\ 0 & 0 & 0 & 0 \end{bmatrix}.$$

A has the property that

$$A^4 = \begin{bmatrix} 0 & 0 & 0 & 0 \\ 0 & 0 & 0 & 0 \\ 0 & 0 & 0 & 0 \\ 0 & 0 & 0 & 0 \end{bmatrix}.$$

a. *True* or *False*? $A^n = \begin{bmatrix} 0 & 0 & 0 & 0 \\ 0 & 0 & 0 & 0 \\ 0 & 0 & 0 & 0 \\ 0 & 0 & 0 & 0 \end{bmatrix}$ for all $n \geq 4$. True

b. What does your answer to part **a** imply about walks in the directed graph with adjacency matrix A? There are no walks of length 4 or more.

c. Confirm your answer to part **b** by drawing the directed graph.

CHAPTER 12 VECTORS

OBJECTIVES

The objectives listed here are the same as in the Chapter 12 Review on pages 756–758 of the student text. The Progress Self-Test on page 755 and the tests in the Teacher's Resource File cover these objectives. For recommendations regarding the handling of this end-of-chapter material, see the notes in the margin on the corresponding pages of this Teacher's Edition.

OBJECTIVES FOR CHAPTER 12 (Organized into the SPUR Categories—Skills, Properties, Uses, and Representations)	Progress Self-Test Questions	Chapter Review Questions	Teacher's Resource File	
			Lesson Masters*	Chapter Test Forms A & B
SKILLS				
A Find the magnitude and direction of two-dimensional vectors.	1	1–4	12-1	1
B Find sums, opposites, scalar products, and dot products of two-dimensional vectors.	2, 3, 4	5–16	12-2, 12-3, 12-4	2
C Find sums, lengths, scalar products, dot products, and cross products of vectors in 3-space.	6, 7, 8	17–24	12-6	5, 6
D Find the measure of the angle between two vectors.	5	25–27	12-4, 12-6	3
PROPERTIES				
E Prove or disprove statements about vector operations.	11	28–32	12-2, 12-3, 12-4, 12-6	7
F Identify parallel and orthogonal vectors.	12	33–38	12-3, 12-4, 12-6	4
USES				
G Use vectors in a plane to decompose motion or force into *x*- and *y*-components.	13b	39–40	12-1	14
H Use addition of vectors in a plane to solve problems involving forces or velocities.	13a	41–44	12-2	15
REPRESENTATIONS				
I Represent two-dimensional vectors in their component or polar representation, or as directed segments.	10	45–48	12-1, 12-3	8
J Represent addition, subtraction, and scalar multiplication of two-dimensional vectors graphically.	14, 15	49–50	12-2, 12-3	9
K Geometrically interpret three-dimensional vectors and their operations.	9	51–53	12-6	10
L Represent lines in a plane using vector or parametric equations.	16	54–59	12-3, 12-4	13
M Represent lines, planes, and spheres in 3-space using parametric, vector, or coordinate equations.	17, 18	60–67	12-5, 12-7	11, 12

*** The masters are numbered to match the lessons.**

OVERVIEW ■ CHAPTER 12

Three fundamental uses of vectors are presented in this chapter. First, vectors describe forces and motion in the physical world. Second, vectors provide a means to describe lines and planes algebraically in a way that is consistent and independent of the dimension of space. Third, vectors are a useful tool for solving systems. The first two uses are studied in detail; the third use is introduced briefly in the last lesson of the chapter. Vectors as elements of vector spaces are not discussed in this book.

The chapter begins by considering vectors in two dimensions. Lesson 12-1 introduces these vectors as quantities that have magnitude and direction. Two methods are used to describe them, namely, polar and component representations. Lesson 12-2 introduces vector addition and subtraction by means of problems involving combinations of forces and motions. In Lesson 12-3, vectors lead to new ways to describe lines in a plane, that is by vector equations and parametric equations. The dot product is introduced in Lesson 12-4 to calculate the angle between two vectors and to deal with perpendicularity.

Vectors in two dimensions can be extended to three dimensions (and higher). Lesson 12-5 develops the foundation for this work by discussing coordinates of points and the distance between points in 3-space. In Lesson 12-6, the basic properties of vectors are extended to three dimensions, and the operation of cross-product is introduced to find a vector that is perpendicular to two others. This makes it possible to use vectors in Lesson 12-7 to deduce equations for lines and planes in space. Lesson 12-8 looks back at the familiar topic of systems of linear equations, and uses the concepts in this chapter to connect the algebra of linear systems with three variables and the geometry of planes in space.

PERSPECTIVES ■ CHAPTER 12

The Perspectives provide the rationale for the inclusion of topics or approaches, provide mathematical background, and make connections with other lessons and within UCSMP.

12-1

VECTORS IN A PLANE

A vector is normally drawn as an arrow whose direction indicates the direction of the vector and whose length indicates the magnitude of the vector. Symbolically, a vector may be described as follows: (1) by a boldface letter **v** or by a letter with an arrow above it, such as $\vec{v}$; (2) by indicating its initial point P and endpoint Q, perhaps with the symbol $\overrightarrow{PQ}$, or (3) by indicating its endpoint if the initial point is taken as the origin (the standard position of the vector). Each of these types of descriptions can be used for vectors in the plane or in space, but description (3) will have two coordinates if the vector is a plane vector, and three coordinates if the vector is in 3-space.

The arrow representation for a vector is used at the beginning of the lesson. The first symbolic description is with polar coordinates, since they too involve magnitude and direction. The translation into rectangular coordinates gives a vector in terms of its horizontal and vertical components. Finding the length of a vector is easy in polar coordinates and is done by using the Pythagorean Theorem in rectangular coordinates.

12-2

ADDING AND SUBTRACTING VECTORS

This lesson discusses how to add vectors represented as arrows (use the parallelogram law) or when their polar representations are given (translate into rectangular coordinates).

The importance of vectors in applications has been due primarily to the fact that they can represent forces and that addition gives the resultant force. The applications in this lesson involve the calculation of the true velocity of an airplane (by adding the wind velocity to its instrument velocity) and the calculation of the direction a volleyball will travel (by adding the forces exerted by two players on it). More recently, vectors have been used in applications in social science and statistics (see Project 4 on page 752).

Subtraction of vectors is defined by adding the opposite; the questions in this lesson explore some of the properties of addition and subtraction.

12-3

PARALLEL VECTORS AND EQUATIONS OF LINES

The algebraic description of lines using vectors is a parametric description; that is, both the x- and y-coordinates are expressed in terms of a third variable t, the parameter. This description uses the second basic operation with vectors, scalar multiplication.

Multiplying a vector $\vec{v}$ by the real number k (the scalar) yields a vector $k\vec{v}$ whose length is k times the length of $\vec{v}$ and whose direction is the same as that of $\vec{v}$ if $k > 0$ and opposite that of $\vec{v}$ if $k < 0$. Since scalar multiples of vectors all go in the same or opposite directions, they lie on parallel lines. Thus, any point on a given line can be thought of as the sum of (1) a vector from the origin to a point on the line, and (2) a scalar multiple of a vector parallel to the line. The scalars by which the vector is multiplied are the values of the parameter.

This lesson defines scalar multiplication and uses it to yield parametric equations for a line in the plane. The same idea will be used in Lesson 12-7 to develop parametric equations for a line in space.

12-4

THE DOT PRODUCT AND THE ANGLE BETWEEN TWO VECTORS

A third binary operation on vectors is the dot product, also called the *inner product*. The result of this operation is a real number that has the following properties. (1) When the dot product of two nonzero vectors is divided by the product of the lengths of the vectors, the quotient is the cosine of the angle between the vectors. (2) When the dot product of two nonzero vectors is zero, the vectors are orthogonal (perpendicular). (3) The dot product of a vector with itself is the square of the length of the vector. As with the properties in the previous lesson, these properties all extend to three dimensions.

12-5

THREE-DIMENSIONAL COORDINATES

This lesson should be review for all students. It covers graphing in 3-space, equations for the planes containing two axes and for planes parallel to the axes, the distance formula, and the equation of a sphere.

12-6

VECTORS IN 3-SPACE

This lesson and the next one review the first half of the chapter. Addition, scalar multiplication, and the dot product are extended to three-dimensional vectors, and the same properties that were proved for these operations in 2-space are shown to be true in 3-space.

There is one operation on 3-space vectors that is not an extension of a 2-space operation, namely, the cross product (or vector product) of two vectors. The cross product of two vectors is the vector that is orthogonal to each of them and whose magnitude may be interpreted as the area of the parallelogram whose sides are the given vectors. The cross product can be used to obtain a vector equation of a plane. (See Question 8 of Lesson 12-7 or Question 20 of Lesson 12-8.)

12-7

LINES AND PLANES IN 3-SPACE

The same process that was used to determine parametric equations of a line in 2-space is extended in this lesson to lines in 3-space.

Although the parametric equations for a line extend quite nicely from 2-space to 3-space, the 3-dimensional extension of the linear equation $ax + by = c$ is not a line but a plane, $ax + by + cz = d$. This result is proved in the lesson by using the idea that there is exactly one plane perpendicular to a line m and passing through a particular point P. All vectors in the plane through P are perpendicular to m. Thus, the dot product of a vector parallel to m and a vector in the plane is 0. The result is that the vector (a, b, c) is perpendicular to any plane with equation $ax + by + cz = d$.

12-8

THE GEOMETRY OF SYSTEMS OF LINEAR EQUATIONS IN THREE VARIABLES

Since an equation of the form $ax + by + cz = d$ is a plane, solving a system of linear equations in three variables means finding the intersection of planes. Although students already have solved systems that have exactly one solution, this lesson considers all other possibilities.

A system of two linear equations in three variables may represent (1) planes that intersect in a line, or (2) parallel or coincident planes. In the first case, parametric equations for the line can be found. In the second case, there is no solution or the planes are, in fact, the same plane.

A system of three linear equations in three variables may represent (1) planes that have exactly one point in common; (2) three parallel planes; (3) two parallel planes and the third plane not parallel; (4) pairs of planes intersecting along parallel lines; (5) planes that have a line in common; or (6) planes that are all coincident. Students are asked to explore these situations.

DAILY PACING CHART ■ CHAPTER 12

Every chapter of UCSMP *Precalculus and Discrete Mathematics* includes lessons, a Progress Self-Test, and a Chapter Review. For optimal student performance, the self-test and review should be covered. (See *General Teaching Suggestions: Mastery* on page T35 of this Teacher's Edition.) By following the pace of the Full Course given here, students can complete the entire text by the end of the year. Students following the pace of the Minimal Course spend more time when there are quizzes and on the Chapter Review and will generally not complete all of the chapters in this text.

When chapters are covered in full (the recommendation of the authors), then students in the Minimal Course can cover 11 chapters of the book. For more information on pacing, see *General Teaching Suggestions: Pace* on page T34 of this Teacher's Edition.

DAY	MINIMAL COURSE	FULL COURSE
1	12-1	12-1
2	12-2	12-2
3	12-3	12-3
4	Quiz (TRF); Start 12-4.	Quiz (TRF); 12-4
5	Finish 12-4.	12-5
6	12-5	12-6
7	12-6	Quiz (TRF); 12-7
8	Quiz (TRF); Start 12-7.	12-8
9	Finish 12-7.	Progress Self-Test
10	12-8	Chapter Review
11	Progress Self-Test	Chapter Test (TRF)
12	Chapter Review	
13	Chapter Review	
14	Chapter Test (TRF)	

TESTING OPTIONS

TESTING OPTIONS

- Quiz for Lessons 12-1 through 12-3
- Quiz for Lessons 12-4 through 12-6
- Chapter 12 Test, Form A
- Chapter 12 Test, Form B
- Chapter 12 Test, Cumulative Form

A Quiz and Test Writer is available for generating additional questions, additional quizzes, or additional forms of the Chapter Test.

INDIVIDUAL DIFFERENCES

PROVIDING FOR INDIVIDUAL DIFFERENCES

The student text is written for, and tested with, average students. It also has been successfully used with better and more poorly prepared students.

The Lesson Notes often include Error Analysis and Alternate Approach features to help you with those students who need more help. Students of all abilities often learn from their peers and may benefit from small group work referenced as appropriate throughout the Notes. A blackline Lesson Master (in the Teacher's Resource File), keyed to the chapter objectives, is provided for each lesson to allow more practice. (However, since it is important to keep up with the daily pace, you are not expected to use all of these masters. Again, refer to the suggestions for pacing on page T34.) Extension activities are provided in the Lesson Notes for those students who have completed the particular lesson in a shorter amount of time than is expected, even in the Full Course.

Vectors

CHAPTER 12

Grand Canyon National Park, Arizona

An ordered pair (x, y) may stand for many different things.

(4, -5)	location on the coordinate plane
(36°N, 112°W)	point in the Grand Canyon
(a, b)	the complex number $a + bi$
95-86	scores in a basketball game
2:3	the ratio of 2 to 3
[17.6 8.8]	1×2 matrix

Since two numbers are involved, ordered pairs are generally thought of as two-dimensional. All but the complex number example can be extended to three dimensions using *ordered triples* (x,y,z).

(10, 0, -19.2)	location on a three-dimensional coordinate system
(48°50′N, 2°20′E, 100m)	a point in Paris 100 meters above ground
(32, 28, 12)	scores in a wrestling meet with three teams
$1:1:\sqrt{2}$	extended ratio of sides in a triangle
[6.7 -2.9 8.0]	1×3 matrix

You can extend these ideas to involve still more numbers.

In this chapter, you will encounter *vectors*, which involve yet another use of ordered pairs and triples, and you will learn certain operations on them. Historically, the primary application of vectors has been to represent physical forces. Furthermore, the operation called addition of vectors corresponds to the combining of physical forces. In recent years, many applications for vectors have been found in business; information is often conveniently stored in ordered pairs, triples, and longer sequences of numbers. Vectors also have applications to geometry. They provide a very nice way to describe lines and planes in both two and three dimensions.

CHAPTER 12

We recommend 11-14 days for this chapter: 8 to 10 days for the lessons and quizzes, 1 day for the Progress Self-Test, 1 or 2 days for the Chapter Review, and 1 day for the Chapter Test. (See the Daily Pacing Chart on page 699D.)

If time is a problem, the discussion of cross product in Lesson 12-6 and the entire Lesson 12-8 could be skipped.

USING PAGE 699

Another name for ordered pair is *2-tuple*; another name for ordred triple is *3-tuple*. All of the ideas in this chapter can be extended to *n-tuples*, that is, strings with *n* objects. Point out to students that there are 4-dimensional coordinate systems (though they are difficult to picture); for example, points can be located in space with time being the 4th dimension.

Some of your students may have studied physics. If so, they should have worked with vectors. These students may be able to help in explaining some of the ideas in this chapter to others in the class.

LESSON 12-1

RESOURCES
- ■ Lesson Master 12-1
- ■ Teaching Aid 87 can be used when discussing the length of a vector.

OBJECTIVES

A Find the magnitude and direction of two-dimensional vectors.
G Use vectors in a plane to decompose motion or force into *x*- and *y*-components.
I Represent two-dimensional vectors in their component or polar representation, or as directed segments.

TEACHING NOTES

Point out to students that, although in common English usage, "velocity" and "speed" are synonyms, to physical scientists, they do not mean the same thing. Speed is the magnitude of velocity; velocity includes direction as well. To say you are traveling at 50 mph indicates a speed; to say you are traveling north at 50 mph indicates a velocity.

Much of the work in this lesson centers on the two coordinate representations for vectors: *polar form* and *component form.* Square brackets are used for polar representations and parentheses for component form. This notation helps students to differentiate between the two forms. Students may see some books in which vectors in component form are represented with brackets. Remind them that a sentence such as

LESSON

12-1

Vectors in a Plane

Many quantities can be described by specifying a *direction* and a *magnitude*. For instance, when a weather forecaster says, "The winds are from the southwest at 15 miles per hour," the forecaster is describing the velocity of the winds by giving their direction ("from the southwest") and their magnitude ("15 miles per hour"). You could represent this quantity by drawing an arrow 15 units long pointing in a northeasterly direction relative to given compass headings. (Note: this arrow does not represent a ray; its length is finite.)

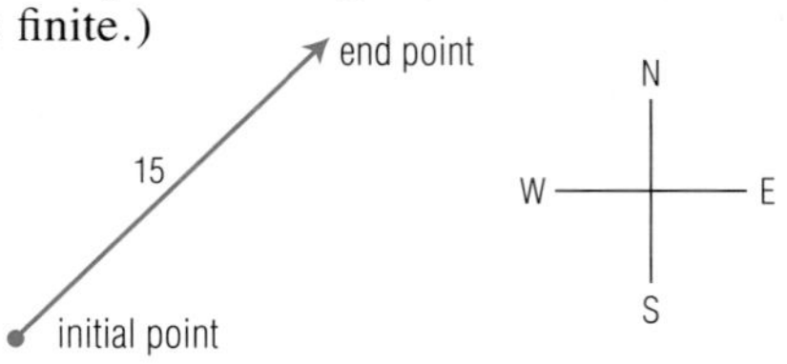

An arrow representing this quantity can begin at any point. The starting position of the arrow (its initial point) does not matter; any other arrow with the same magnitude (or *length*) and direction represents the same wind velocity. Velocity is an example of a vector quantity.

Definition

A **vector** is a quantity that can be characterized by its direction and its magnitude.

Vectors are named by letters with arrows above them, as in $\vec{u}$ and $\vec{v}$, or with boldface, as in $\boldsymbol{u}$ and $\boldsymbol{v}$.

If all the vectors under consideration lie in a single plane, then the vectors are called **plane vectors** or **two-dimensional vectors**. Descriptions of plane vectors can take several forms. The arrow describing wind velocity given above can also be represented by the ordered pair [15, 45°]. That ordered pair [magnitude, direction] is called the *polar representation* of a vector.

Definitions

The **polar representation** of a two-dimensional vector $\vec{v}$ with positive **magnitude** r and **direction** θ measured counterclockwise from the polar x-axis is $[r, \theta]$.

The arrow for $\vec{v}$ that joins the pole to the point $[r, \theta]$ is in **standard position.** Any arrow in the plane that is parallel to, in the same direction as, and the same length as the arrow in standard position is also a geometric representation of $\vec{v}$.

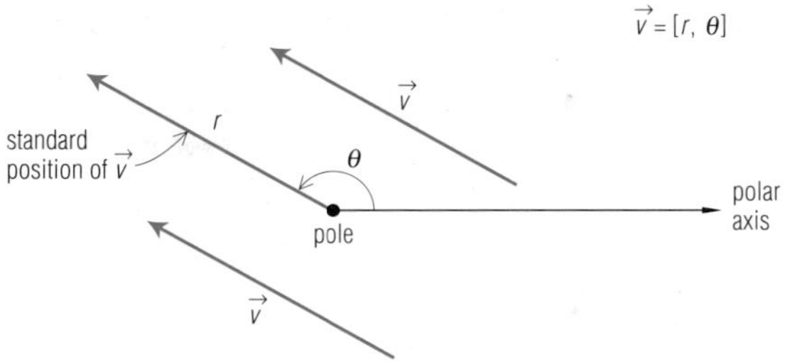

Trigonometry can be used to change the polar representation of a vector into a more familiar form.

Example 1 A ship's velocity is represented by [12, 82°], where the first component is measured in miles per hour and the second is the number of degrees north of east that the ship travels.
a. Draw the standard position arrow representing the velocity.
b. Describe the ship's movement each hour in terms of a number of miles east and a number of miles north.

Solution
a.

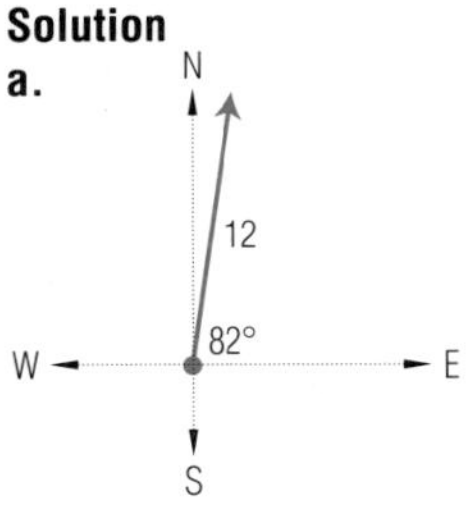

b. Draw a right triangle whose hypotenuse is the arrow and whose legs are parallel to the axes, as shown at the right. Then use the definitions of sine and cosine.

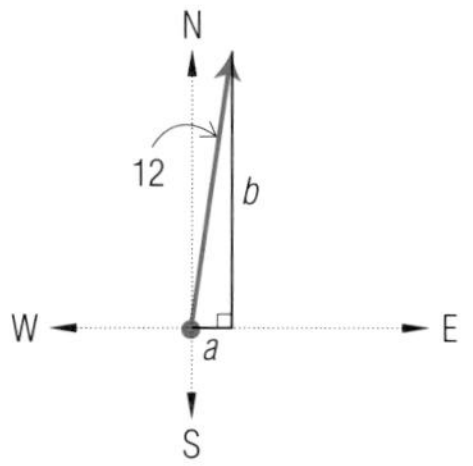

$\frac{a}{12} = \cos 82°$ $\quad\quad$ $\frac{b}{12} = \sin 82°$

$a = 12 \cos 82° \approx 1.67$ $\quad\quad$ $b = 12 \sin 82° \approx 11.88$

Each hour the ship's new position is approximately 1.67 miles to the east and 11.88 miles to the north of its old position.

In Example 1, the ordered pair $(12 \cos 82°, 12 \sin 82°) \approx (1.67, 11.88)$ describes the ship's motion. It also gives the rectangular coordinates of the endpoint of the arrow in standard position. This ordered pair is the *component representation* of the vector.

$[6, 45°] = (3\sqrt{2}, 2\sqrt{2})$ does not mean that the corresponding listings are equal, but that the vectors are equal. Point out that with vectors (as with complex numbers), in $[r, \theta]$, r is nonnegative, whereas with polar coordinates r can be negative.

The diagram explaining the formula for $|\vec{u}|$ is important. When looking at this diagram, students should think of several ways that the legs of the right triangle are used: (1) in applying the Pythagorean Theorem to find the length of $\vec{u}$; (2) in determining the tangent of the direction angle of $\vec{u}$ and thus finding the size of the angle; and (3) in determining the slope of the line containing the vector. Point out to students that a similar drawing can be made for vectors that are not in standard position.

ADDITIONAL EXAMPLES

1. An airplane is flying at a constant altitude at 600 km/hr in a direction 27° W of N. Express the airplane's velocity:
a. with an arrow;

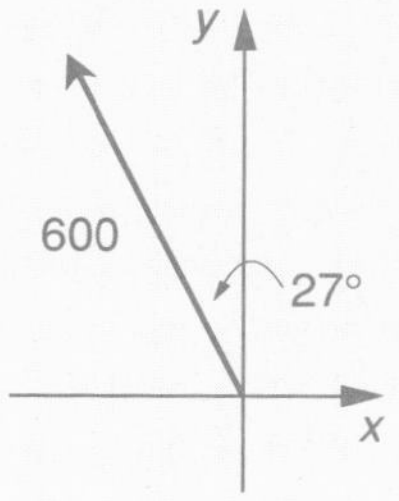

b. as a vector in polar coordinates;
[600, 117°]
c. as a vector in rectangular coordinates.
$(600 \cos 117°, 600 \sin 117°) \approx (-272, 535)$

ADDITIONAL EXAMPLES

2. The arrow from (13, 6) to (-27, 42) represents a plane vector $\vec{v}$.

a. Find the length of $\vec{v}$.
$\sqrt{2896} \approx 53.8$

b. What is the endpoint of the standard position arrow for $\vec{v}$?
(-40, 36)

c. Find the direction of $\vec{v}$.
$\approx 138°$

3. A vector $\vec{v}$ represents a force of 8 pounds that is being exerted at an angle of $\frac{2\pi}{3}$ with the positive x-axis. Find the x- and y-components of $\vec{v}$.
x-component of $\vec{v}$ is -4.
y-component of $\vec{v}$ is $4\sqrt{3}$.

Definitions

The **component representation** of a plane vector $\vec{u}$ is the ordered pair (u_1, u_2), the rectangular coordinates of the endpoint of the standard position arrow for $\vec{u}$. The numbers u_1 and u_2 are the ***x*-component** and ***y*-component** of $\vec{u}$, respectively, or the **horizontal** and **vertical components** of $\vec{u}$, respectively.

Example 2 The arrow from (-1, 2) to (3, 5) represents a plane vector $\vec{v}$.
a. Find the length and direction of $\vec{v}$.
b. Draw the standard position arrow for $\vec{v}$.

Solution

a. Using the diagram below and the Pythagorean Theorem, the length of $\vec{v}$ is $\sqrt{4^2 + 3^2} = 5$.

The direction of $\vec{v}$ is given by the angle θ between the positive x-axis and the vector. From the diagram, $\tan\theta = \frac{3}{4}$ and $0° < \theta < 90°$, so $\theta = \tan^{-1}(0.75) \approx 36.9°$.

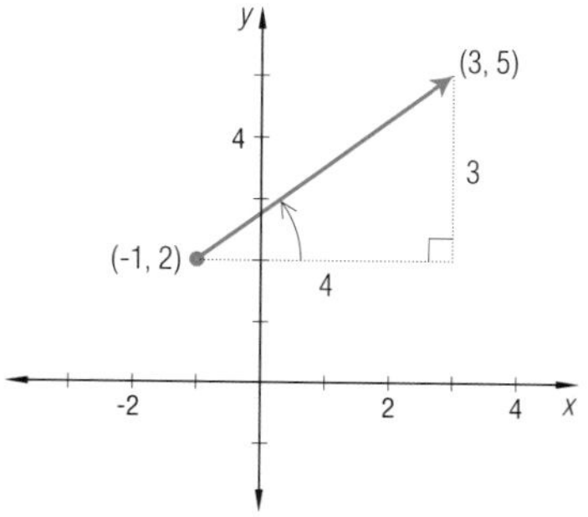

b. The translation from (-1, 2) to (3, 5) is a translation of 4 units horizontally and 3 units vertically. Thus, the standard position arrow for $\vec{v}$ is drawn from the origin to the point (4, 3).

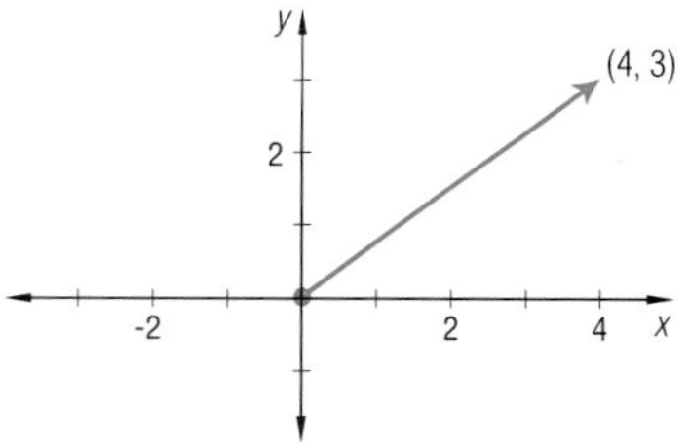

The vector quantity in Example 2 can be described either by an arrow, by the ordered pair (4, 3), or by its approximate polar representation [5, 36.9°].

Notice that in Example 2, (4, 3) = (3 − (-1), 5 − 2). In general, if $\vec{u} = (u_1, u_2)$ is the vector from (a, b) to (c, d), then $(u_1, u_2) = (c - a, d - b)$. That is, the horizontal component of $\vec{u}$ is $c - a$, and the vertical component of $\vec{u}$ is $d - b$.

Example 3 A vector $\vec{u}$ represents a force of 5 pounds that is being exerted at an angle of $\frac{5\pi}{6}$ with the positive x-axis. Find the x- and y-components of $\vec{u}$.

Solution The arrow in standard position for $\vec{u}$ is shown in the diagram below. The coordinates (u_1, u_2) of the endpoint of this arrow are

$$u_1 = 5\cos\frac{5\pi}{6} = 5\cdot\left(-\frac{\sqrt{3}}{2}\right) = -\frac{5\sqrt{3}}{2}$$

and

$$u_2 = 5\sin\frac{5\pi}{6} = 5\cdot\left(\frac{1}{2}\right) = \frac{5}{2}.$$

Therefore, $\left(-\frac{5\sqrt{3}}{2}, \frac{5}{2}\right)$ is the component representation of $\vec{u}$. The x-component of $\vec{u}$ is $-\frac{5\sqrt{3}}{2}$, and the y-component of $\vec{u}$ is $\frac{5}{2}$.

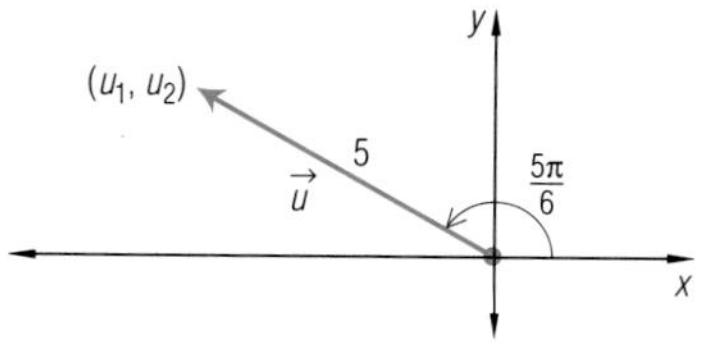

The symbol $|\vec{u}|$ denotes the length of $\vec{u}$. ($|\vec{u}|$ is sometimes called the **norm** of $\vec{u}$.) In the polar representation $\vec{u} = [r, \theta]$, $|\vec{u}| = r$. In its component representation $\vec{u} = (u_1, u_2)$, $|\vec{u}|$ can be easily determined using the Pythagorean Theorem.

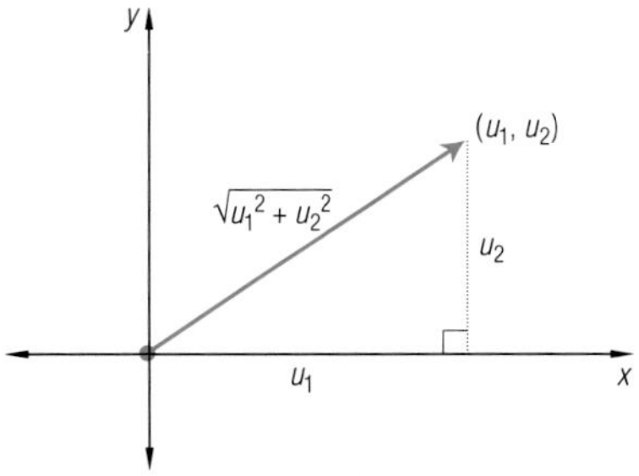

Theorem

If $\vec{u} = (u_1, u_2)$, then

$$|\vec{u}| = \sqrt{u_1^2 + u_2^2}.$$

The following relationship between the polar and component representations of a vector generalizes Examples 1 and 3.

Theorem

For all plane vectors $\vec{u}$, let θ be the angle between the horizontal and the vector, measured counterclockwise. Then,

$$[|\vec{u}|, \theta] = (|\vec{u}|\cos\theta, |\vec{u}|\sin\theta).$$

The point (0, 0) in the plane corresponds to the **zero vector** $\vec{0}$. The zero vector has length zero, and it can have any direction.

NOTES ON QUESTIONS

Question 3: Mathematically, vectors are very much like translations; addition of vectors is isomorphic to composition of translations.

Question 4: Students should be able to take a vector given in any form, as an arrow, in polar form, or in component form, and change it to the other two forms.

Question 12: All unit vectors have the form $(\cos\theta, \sin\theta)$.

Questions 13 and 14: Pictures can be used to help explain why there are two answers.

Question 15: This question previews the idea of a scalar multiple of a vector, covered in Lesson 12-3.

Question 21: Graphing the sum of two vectors, to be discussed in the next lesson, involves finding the fourth vertex of a parallelogram. In particular, the sum of the vectors (2, 7) and (-3, 6) is (-1, 13). You could use this question to introduce the next lesson.

ADDITIONAL ANSWERS

1.

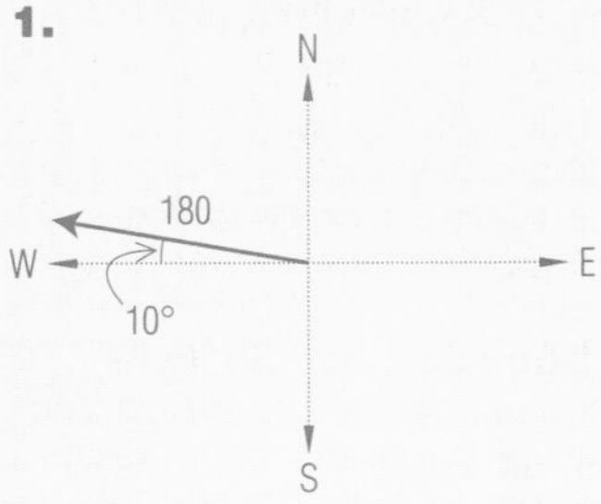

2.

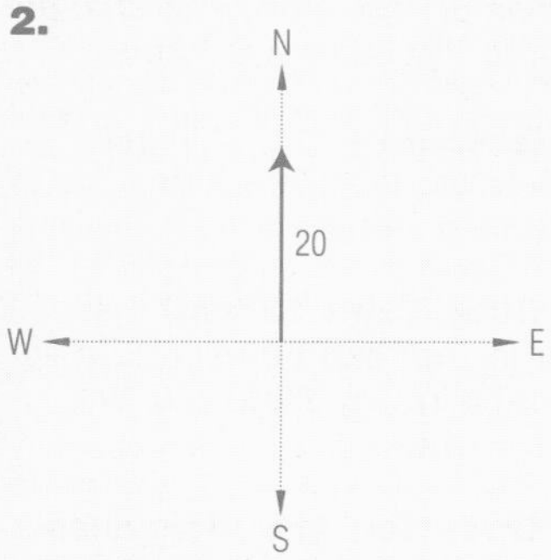

3.

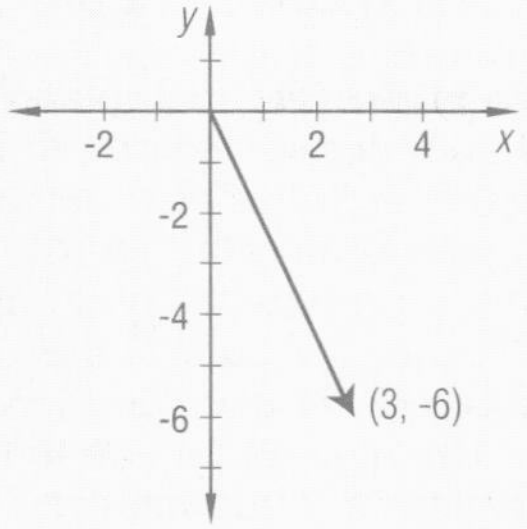

6.b.

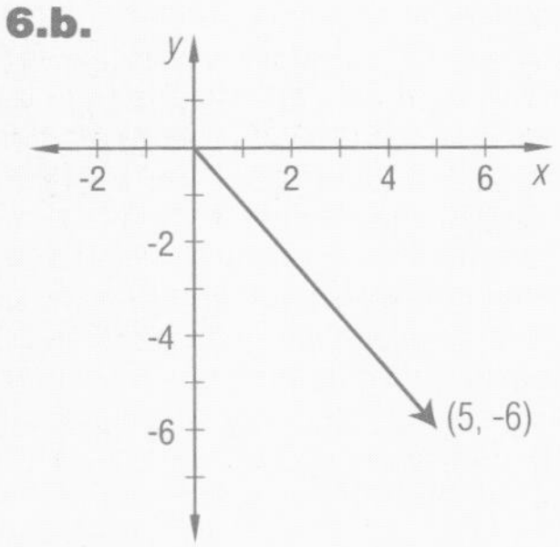

7.b., 9., 11., 16.a., 19. See Additional Answers in the back of this book.

Questions

Covering the Reading

In 1–3, draw an arrow to represent each vector.

1. the velocity of a plane flying at a speed of 180 miles per hour in the direction 10° north of west **See margin.**

2. the velocity of a ship traveling north at 20 knots **See margin.**

3. the translation of the graph of a function 3 units horizontally and -6 units vertically **See margin.**

4. A car is driving east-southeast (that is, 22.5° south of east) at a speed of 55 miles per hour. For this velocity, give

a. the polar representation, and **[55, -22.5°]**

b. the component representation.
(55 cos(-22.5°), 55 sin(-22.5°)) ≈ (50.8, -21.0)

5. Give the component representation of a vector with length 1 and direction 218° from the polar axis. **(cos 218°, sin 218°) ≈ (-0.788, -0.616)**

6. a. Give the component representation of the standard position arrow for the vector shown at the right. **(5, -6)**

b. Sketch the vector in standard position. **See margin.**

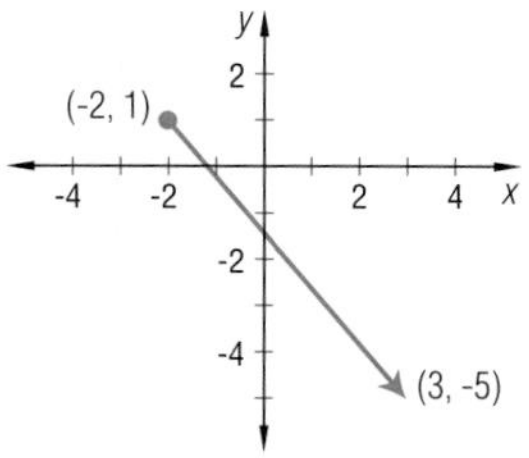

7. a. Find the length and direction of the plane vector represented by the arrow joining the point (3, 4) to (7, -1).

b. Sketch the standard position vector. **See margin.**
a) length = $\sqrt{41}$, direction ≈ 308.7°

8. Find $|\vec{v}|$ when $\vec{v} = (-6, 11)$. **$\sqrt{157}$ ≈ 12.5 units**

9. Sketch the vector described by (8 cos 10°, 8 sin 10°). **See margin.**

10. Give the polar representation of the vector (-16, 19).
$\left[\sqrt{617}, \tan^{-1}\left(\frac{-19}{16}\right)\right] \approx$ [24.8, 130°]

Applying the Mathematics

11. Prove that the arrow joining (-1, 2) to (4, -1) represents the same vector as the arrow joining (3, -2) to (8, -5). **See margin.**

12. A **unit vector** is a vector whose length is 1. Give the components of the unit vector with the same direction as the vector of Example 1.
(cos 82°, sin 82°) ≈ (0.139, 0.990)

13. Find the directions of all nonzero plane vectors whose x-components are 3 times their y-components. **$\tan^{-1}\frac{1}{3}$ ≈ 18.4° or 198.4°**

14. $\vec{v}$ is the vector from (3, 8) to (7, k) and $|\vec{v}| = 2\sqrt{13}$. Find the possible values of k. **$k = 2$ or $k = 14$**

704

15. Let $\vec{v} = [15, 45°]$. Suppose $\vec{w}$ has the same direction as $\vec{v}$, and $|\vec{w}| = \frac{1}{3}|\vec{v}|$. Give the polar and component representations of $\vec{w}$.
$\vec{w} = [5, 45°] = \left(\frac{5\sqrt{2}}{2}, \frac{5\sqrt{2}}{2}\right)$

Review

16. a. Sketch the graph of a function f with all of the following properties:

(i) $\lim_{x \to \infty} f(x) = 2$.
(ii) f is an odd function.
(iii) $f'(x) > 0$ for $0 < x < 2$.
(iv) $f'(x) < 0$ for $2 < x < 4$.

b. Identify all points where f must have a relative minimum or maximum. *(Lessons 9-5, 2-5)* **a) See margin.**
b) relative maximum at x = 2; relative minimum at x = -2

17. At the beginning of month 1, the buyer of a car owes $8,000 on the car. At the beginning of each month thereafter, 0.8% interest is added to the amount owed, and the owner then pays $200 toward the loan. Let A_k represent the amount owed during month k.

a. Write a recursive definition for the sequence $A_1, A_2, A_3, \ldots$.
b. Find the amount owed at the beginning of month 5. **≈ 7449.44**
(Lesson 7-1) **a)** $A_1 = 8000$; $A_2 = 7864$; $A_k = 1.008A_{k-1} - 200$ for $k > 1$.

18. a. *Multiple choice.* Which expression equals $\sin \frac{7\pi}{12}$?

(i) $\sin \frac{\pi}{3} \sin \frac{\pi}{4} - \cos \frac{\pi}{3} \cos \frac{\pi}{4}$

(ii) $\sin \frac{\pi}{3} \cos \frac{\pi}{4} + \cos \frac{\pi}{3} \sin \frac{\pi}{4}$

(iii) $\sin \frac{\pi}{3} \cos \frac{\pi}{4} - \cos \frac{\pi}{3} \sin \frac{\pi}{4}$

(iv) $\cos \frac{\pi}{3} \cos \frac{\pi}{4} - \sin \frac{\pi}{3} \sin \frac{\pi}{4}$ **(ii)**

b. Which two of the other three choices are equal to each other? *(Lessons 6-4, 6-3)* **(i) and (iii)**

Settling tank at water reclamation facility

19. Prove that if 3 is a factor of p and 6 is a factor of q, then 3 is a factor of $p + q$. *(Lesson 4-1)* **See margin.**

20. Each time a liquid is passed through a filter, 65% of the impurities are removed. If 30% of a liquid consists of impurities (so that it is 70% pure), how many times must it be filtered so that only 0.5% of it consists of impurities (that is, it is 99.5% pure)? *(Lesson 2-6)* **10 times**

21. Three vertices of a parallelogram (not necessarily in order) are (0, 0), (2, 7), and (-3, 6). Find all possible locations for the fourth vertex. *(Previous course)* **(-1, 13), (-5, -1), (5, 1)**

Exploration

22. Is there a vector whose polar and coordinate representations consist of the same numbers? In other words, is there a vector $\vec{v} = (x, y) = [r, \theta]$ such that $x = r$ and $y = \theta$? If so, find its components. If not, prove such a representation cannot exist.
Any vector of the form (r, 0) with r positive has polar form [r, 0].

FOLLOW-UP

MORE PRACTICE

For more questions on SPUR Objectives, use *Lesson Master 12-1*, shown below.

NAME ____

LESSON **MASTER 12–1**
QUESTIONS ON **SPUR** OBJECTIVES

■ **SKILLS** *Objective A (See pages 756–758 for objectives.)*
In 1 and 2, find the magnitude and direction of the given vector.

1. (2, -3) **$\sqrt{13}$, 303.7**
2. (-5, -12) **13, 247.4°**

3. Find a polar representation of the vector (12, 16). **[20, 53.1°]**

■ **USES** *Objective G*

4. A plane's velocity is represented by [550, 160°], where the magnitude is measured in miles per hour and the direction is in degrees counter-clockwise from due east.

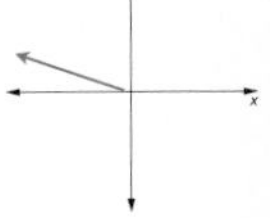

a. Sketch the vector for the velocity.
b. Give the vector in component form. **(-517, 188)**
c. Interpret the components. **Each hour the plane flies 517 miles to the west and 188 miles to the north.**

5. A fish is caught at the end of a 15-meter fishline. Its angle with the horizontal is 37°. Assume the fishline is taut.

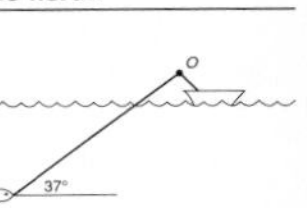

a. Write a polar representation for the fish's position, using the end of the fishing rod (point O) as the origin. **[15, 217°]**
b. Compute a component representation. **(-12, -9)**
c. Interpret the components. **The fish is 9 meters below the tip of the fishing rod and 12 meters horizontally away from the tip.**

Precalculus and Discrete Mathematics © Scott, Foresman and Company *Continued* 127

NAME ____
Lesson MASTER 12–1 (page 2)

■ **REPRESENTATIONS** *Objective I*

6. Suppose $A = (1, -2)$ and $B = (x, y)$ are points in a plane and $\vec{v}$ is the vector from A to B. If $\vec{v} = \left[6, \frac{\pi}{3}\right]$, find the coordinates of B. **$(4, 3\sqrt{3} - 2)$**

7. Find the component representation of [3, 210°] and sketch the vector. **$\left(-\frac{3}{2}\sqrt{3}, -\frac{3}{2}\right)$**

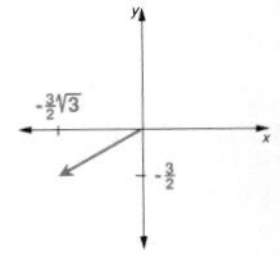

8. a. Find the component representation of the vector shown at the right. **(-4, -5)**
b. Sketch the vector in standard position.

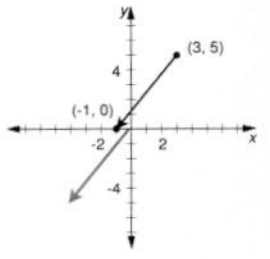

128 *Precalculus and Discrete Mathematics* © Scott, Foresman and Company

LESSON 12-2

RESOURCES
- Lesson Master 12-2
- Teaching Aid 88 can be used with **Example 1**.
- Teaching aid 89 can be used with **Questions 4-6**.

OBJECTIVES

A Find sums and opposites of two-dimensional vectors.
E Prove or disprove statements about vector operations.
H Use addition of vectors in a plane to solve problems involving forces or velocities.
J Represent addition and subtraction of two-dimensional vectors graphically.

TEACHING NOTES

In your discussions of plane vectors, remind students that everything that is done with vectors can be interpreted by using different forms: arrows, polar form, or component form. Then ask: Which form is used first in this lesson? (polar form, on page 706) Which form is used next? (component form, used for the definition of addition at the top of page 707) Addition using arrows is given after the component definition.

The theorem following the definition of the opposite of a vector shows both polar and component representations of the opposite; the arrow form is used next.

LESSON

12-2 Adding and Subtracting Vectors

Consider an airplane whose instrument panel indicates that its airspeed (that is, the speed of the airplane relative to the surrounding air) is 200 miles per hour and that its compass heading (that is, the direction in which the airplane is pointing) is due northeast. However, suppose there is a steady 50-mph wind blowing from the south. Because of the wind, the plane's airspeed and compass heading do not give the true direction and speed of the plane as measured by a control tower. The true velocity $\vec{v}$ of the plane is found by combining the instrument panel velocity (airspeed and compass heading) $\vec{p} = [200, 45°]$ with the wind velocity $\vec{w} = [50, 90°]$.

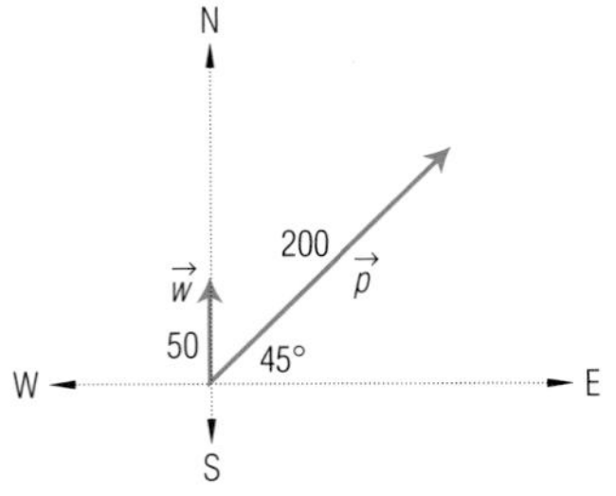

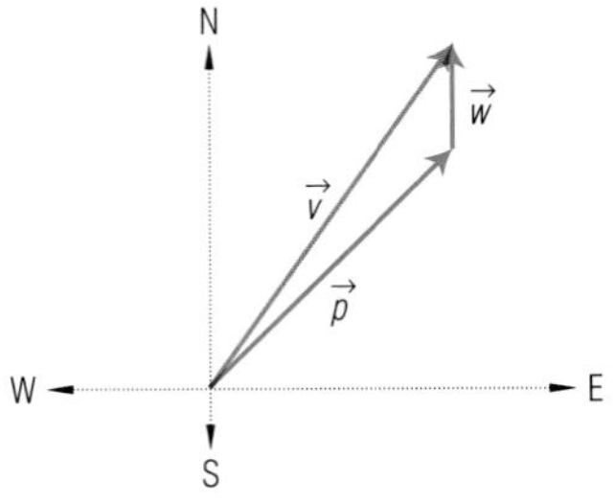

To find the true velocity, break the wind and instrument panel velocities into their horizontal (east-west) and vertical (north-south) components. The sums of corresponding components are the components of the true velocity.

$$\vec{p} = [200, 45°] = (200 \cos 45°, 200 \sin 45°) = (100\sqrt{2}, 100\sqrt{2})$$

$$\vec{w} = [50, 90°] = (0, 50)$$

Therefore, the true velocity of the plane is

$$\vec{v} = (100\sqrt{2}, 50 + 100\sqrt{2}).$$

The true speed of the plane is the length of $\vec{v}$:

$$|\vec{v}| = \sqrt{(100\sqrt{2})^2 + (50 + 100\sqrt{2})^2} \approx 238 \text{ miles per hour.}$$

The direction θ satisfies

$$\tan \theta = \frac{50 + 100\sqrt{2}}{100\sqrt{2}} \approx 1.35,$$

$$\therefore \theta \approx 53.5°.$$

Therefore, the plane is traveling at about 238 miles per hour in the direction 53.5° north of east.

This example exhibits the rule for adding vectors: add their x-components to get the x-component of the sum and add their y-components to get the y-component of the sum.

Definition

If $\vec{u} = (u_1, u_2)$ and $\vec{v} = (v_1, v_2)$, then the **sum of $\vec{u}$ and $\vec{v}$**, written $\vec{u} + \vec{v}$, is the vector $(u_1 + v_1, u_2 + v_2)$.

There are two equivalent ways to picture the sum of two vectors. In the diagram at the left below, the second arrow begins at the end of the first arrow. In the diagram at the right, both vectors are placed in standard position.

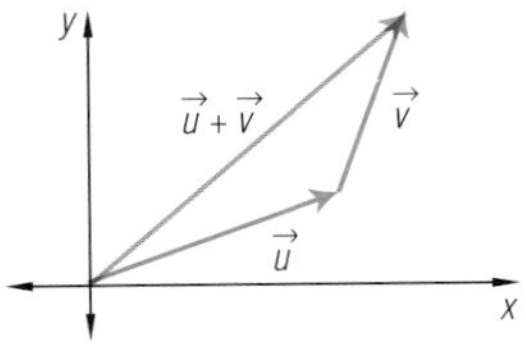

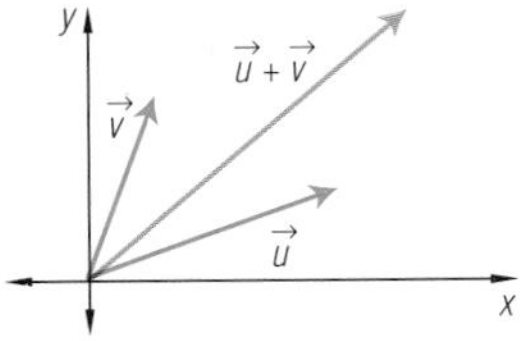

Either way, the two vectors represent sides of a parallelogram. The sum vector is a diagonal of the parallelogram!

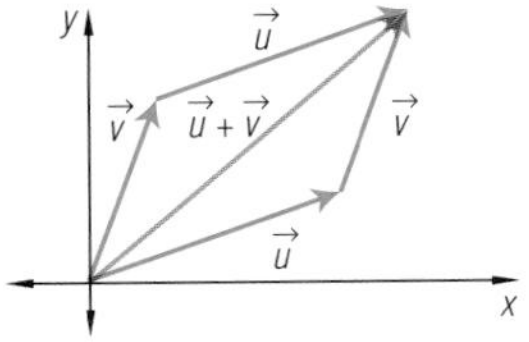

Curling event in the Olympics: friction causes the objects to stop.

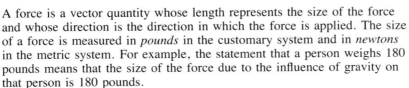

The procedure that was used to compute the true velocity of the plane from its instrument velocity and the wind velocity is often used with other types of vector quantities. *Forces* are among the most important of these. A **force** is an influence that changes the motion of an object. Anytime an object speeds up, slows down, stops, starts to move, or changes direction, one or more forces must be at work. For example, you might give an object a push to start it sliding. Then the friction of the object with the floor will cause it to stop. Your initial push and the influence of friction are both forces.

A force is a vector quantity whose length represents the size of the force and whose direction is the direction in which the force is applied. The size of a force is measured in *pounds* in the customary system and in *newtons* in the metric system. For example, the statement that a person weighs 180 pounds means that the size of the force due to the influence of gravity on that person is 180 pounds.

If two forces, $\vec{f}$ and $\vec{g}$, influence the motion of an object at the same time, then the sum of these two forces

$$\vec{h} = \vec{f} + \vec{g}$$

is a single force that produces the same resulting change in the motion of the object as the two forces $\vec{f}$ and $\vec{g}$ combined. For this reason, the force $\vec{h}$ is called the **resultant force** for the forces $\vec{f}$ and $\vec{g}$.

Subtraction with arrows is given on page 709. Subtraction with vectors given in component form can be discussed with **Question 14**. Subtraction with vectors given in polar form is the essence of **Question 15**.

In the discussion at the beginning of the lesson, the airplane's true velocity is the resultant of two forces $\vec{p}$ and $\vec{w}$ that act simultaneously, as shown in the first diagram. However, the sum can be envisioned by imagining $\vec{p}$ occurring first, then $\vec{w}$, as the second diagram illustrates. A third drawing for the situation could show $\vec{w}$ happening first, then $\vec{p}$. By combining all three drawings, students can see a parallelogram diagram like the one found on this page.

Alternate Approach
One reason for studying vectors is that they can be used to solve many familiar problems. In the airplane velocity example, the Law of Cosines can be used to find the plane's speed and direction. Showing students this alternate approach would point out the convenience of vector methods and will help to prepare them for Lesson 12-4 on the dot product and the angle between vectors.

When discussing **Example 1**, point out that in the original drawing the arrows terminate at the ball, representing the forces exerted on it. The diagram was then redrawn with the arrows eminating from the ball. In a way, they now show the result of each force on the ball's motion.

Technology **Example 2** can be extended by asking students to draw an arrow diagram to show that $(\vec{v} - \vec{w}) + \vec{w} = \vec{v}$. Another extension would be to ask students to write a computer program to give the sum of two vectors that are entered in polar form. A student familiar with graphics could also have the program draw the arrows on the screen.

You may want to instruct your students that, unless directed otherwise, vector answers should be written in the same form (polar or component) as the information given.

ADDITIONAL EXAMPLES

1. Suppose that two volleyball players hit a ball at the same time, and the force of the first hit was 40 lb in the direction -35°, as in **Example 1**, but the force of the second hit was 40 lb in the direction 35°. Show that the ball would go in the 0° direction as expected and find the magnitude of the force on it.

If $\vec{v}$ = [40, 35°] and $\vec{w}$ = [40, -35°], then $\vec{v} + \vec{w}$ = (40 cos 35° + 40 cos (-35°), 40 sin 35° + 40 sin (-35°)) = (80 cos 35°, 0), a vector in the direction 0° with magnitude ≈ 66 lb.

2. Let $\vec{u}$ = (2, 7) and $\vec{v}$ = (3, 5).

a. Draw $\vec{u}$, $\vec{v}$, and $\vec{u} - \vec{v}$.

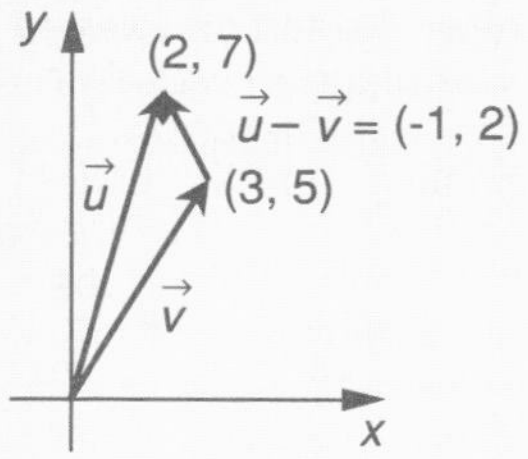

b. Give the components of $\vec{u} - \vec{v}$.

(-1, 2)

Example 1 Two volleyball players hit the ball at the same time. The forces they exert are represented by the arrows drawn at the right. Describe the vector representing the total force exerted on the ball.

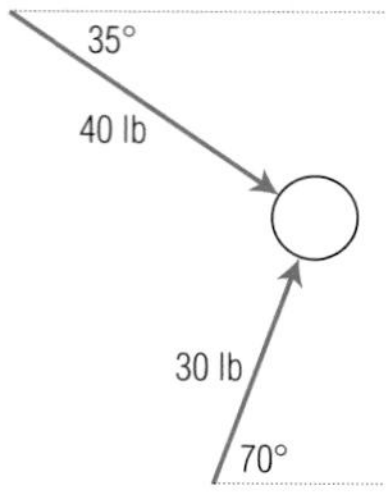

Solution First show each force as a vector in standard position. (Notice that one vector has direction -35°.) Each component of the total force on the ball will be the sum of the corresponding components of the two force vectors. So express each vector in component form.

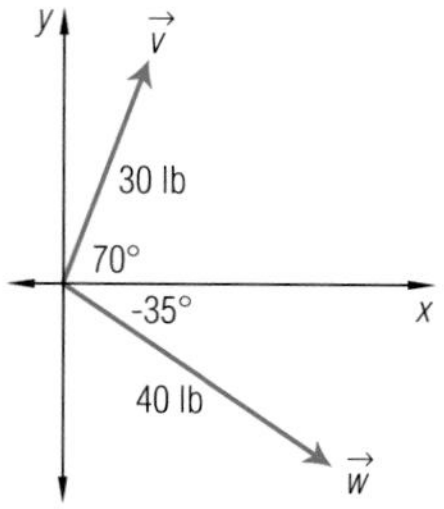

$\vec{v}$ = (30 cos 70°, 30 sin 70°)
≈ (10, 28)

$\vec{w}$ = (40 cos(-35°), 40 sin(-35°))
≈ (33, -23)

The total force in the *x*-direction is approximately 10 + 33 = 43 lb. The total force in the *y*-direction is approximately 28 + -23 = 5 lb. Thus, the resulting force has components of about (43, 5). That resultant is shown in orange at the right.

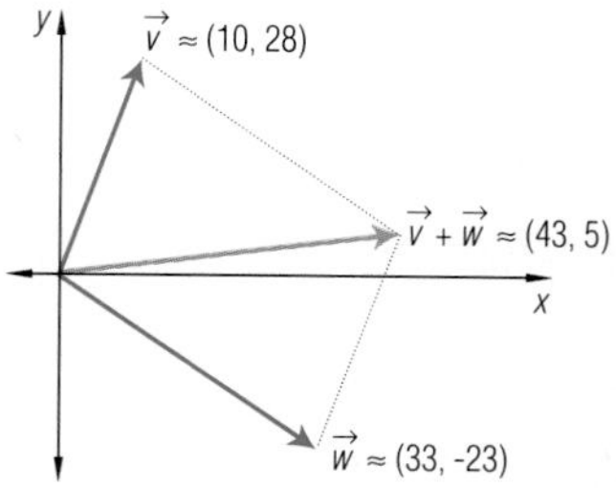

Check The origin and the endpoints of $\vec{v}$, $\vec{w}$, and $\vec{v} + \vec{w}$ in standard position should be vertices of a parallelogram, which they are.

If it makes sense to add two vectors, is it possible to subtract them? The answer is "Yes." As is commonly done in mathematical systems, subtraction is defined as adding the opposite, so the *opposite of a vector* must be defined. The definition is what you might expect.

Definition

The vector **$-\vec{v}$**, the **opposite of $\vec{v}$**, is the vector with the same magnitude as $\vec{v}$ and direction opposite to $\vec{v}$.

Part **a** of the following theorem follows immediately from the definition of $-\vec{v}$. In the Questions you are asked to prove part **b**.

708

Theorem

a. If $\vec{v}$ is a vector with polar representation $[r, \theta°]$, then $-\vec{v} = [r, 180° + \theta°]$.
b. If $\vec{v}$ has component representation (v_1, v_2), then $-\vec{v} = (-v_1, -v_2)$.

Opposite vectors are quite important in the study of forces. One of Newton's laws is (in his words): ''To every action there is always opposed an equal reaction, or the mutual reactions of two bodies upon each other are always equal and directed to contrary directions.'' This means that the forces in interactions between two bodies can be represented by opposite vectors.

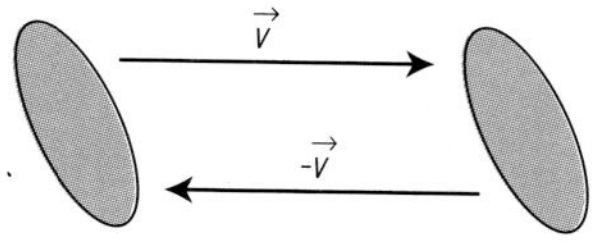

Using the definition of opposite vectors, it is possible to define vector subtraction.

Definition

The **difference $\vec{u} - \vec{v}$** is the vector $\vec{u} + (-\vec{v})$.

Example 2 The vectors $\vec{v}$ and $\vec{w}$ are shown at the right. Sketch **a.** $\vec{v} - \vec{w}$, **b.** $\vec{w} - \vec{v}$.

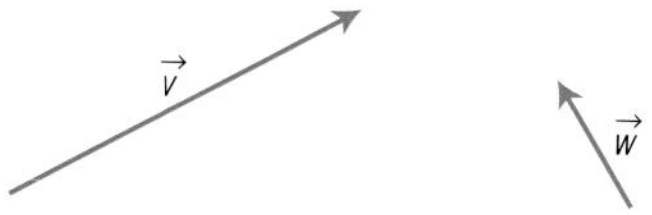

Solution First sketch the opposites of the two vectors.

a. $\vec{v} - \vec{w} = \vec{v} + -\vec{w}$, so sketch $-\vec{w}$ as done above, and add it to $\vec{v}$ using a parallelogram. The result is shown at the right.

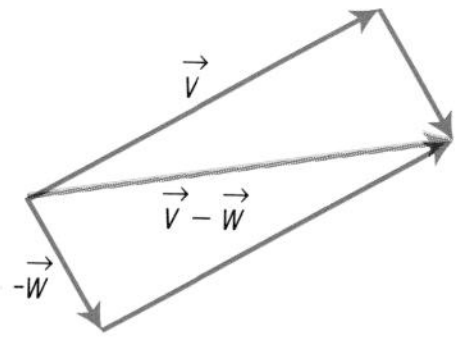

b. $\vec{w} - \vec{v} = \vec{w} + -\vec{v}$, so sketch $-\vec{v}$ and add it to $\vec{w}$ as shown.

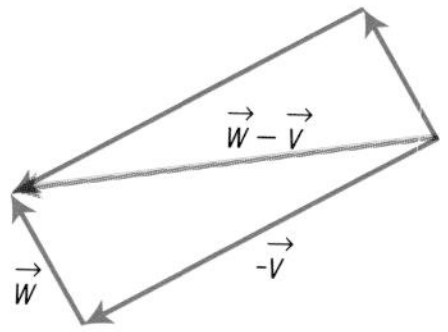

Notice that $\vec{v} - \vec{w}$ is the opposite of $\vec{w} - \vec{v}$. Addition and subtraction of vectors have many of the same properties as addition and subtraction of real numbers.

NOTES ON QUESTIONS

Questions 3 and 4: The arrows are not drawn to scale. Encourage students to draw accurate diagrams to check the reasonableness of their answers.

Question 6: Students should begin by drawing the two vectors beginning at the barge.

Question 12: Students may have a hard time getting started. It may help if you give a hint to let $\vec{v} = (v_1, v_2) = [r, \theta]$, so that $v_1 = r \cos \theta$ and $v_2 = r \sin \theta$. Then use the definition of opposite.

Question 14a: Generalize this question to give a rule for subtracting vectors given their components: If $\vec{s} = (a, b)$ and $\vec{t} = (c, d)$, then $\vec{s} - \vec{t} = (a - c, b - d)$.

Question 16: This is an important question to discuss. In the next lesson on scalar multiplication, this property will indicate that the magnitude of a scalar product is the product of the absolute value of the scalar and the magnitude of the original vector. Note the different, but related, use of the absolute value symbol for real numbers and for vectors.

Question 17: The circular arrangement of chairs indicates that even if everyone gets up and moves one chair to the left, for example, they will be in different chairs, but the seating is the *same*.

FOLLOW-UP

MORE PRACTICE
For more questions on SPUR Objectives, use *Lesson Master 12-2*, shown on page 711.

EXTENSION
In **Question 5**, the river is increasing the speed of the boat. You might ask students under what conditions the river slows the boat down rather than speeds it up.
(if the angle to the downstream bank is greater than 90°)
Relate this to the parallelogram diagram. Under what conditions is the length of the diagonal of a parallelogram less than the length of its sides?
(if the vertex has angle greater than 90°)
Can it be equal to one side or both?
(The diagonal equals the length of one side when the sides of the parallelogram are related by $\frac{1}{2}a = b \cos(180 - \theta)$ for $\theta > 90°$ and

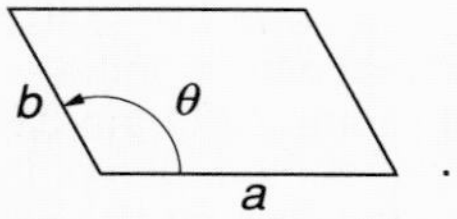

sample: $a = 6$, $b = 5$, $\theta = 180 - \tan^{-1}\frac{4}{3}$)
What is the maximum length?
(The maximum length is less than the sum of the two sides.)

PROJECTS
The projects for Chapter 12 are described on pages 751–752. **Projects 1 and 2** are related to the content of this lesson.

ADDITIONAL ANSWERS
3.a. magnitude: ≈ 43.3 lb; direction: ≈ 6.6° N of E

5. speed: $\sqrt{556} \approx 23.6$ mph; direction: $\theta = \tan^{-1}\left(\frac{10\sqrt{3}}{16}\right) \approx 47°$ N of E

Questions

Covering the Reading

In 1 and 2, find $\vec{u} + \vec{w}$.

1. $\vec{u} = (18, -3)$ and $\vec{w} = (-6, -11)$ (12, -14)

2. $\vec{u} = [9, 55°]$ and $\vec{w} = [3, 140°]$ ≈ [9.73, 72.9°]

3. Consider the resultant vector in Example 1.
a. Give its magnitude and direction. See margin.
b. What is the physical meaning of the answer to part **a**?
The resultant force is 43.3 lb in the direction 6.6° N of E.

4. Find the resultant force $\vec{h}$ for the forces $\vec{f}$ and $\vec{g}$ shown at the right.
71.96 lb in the direction of 7.18° N of E

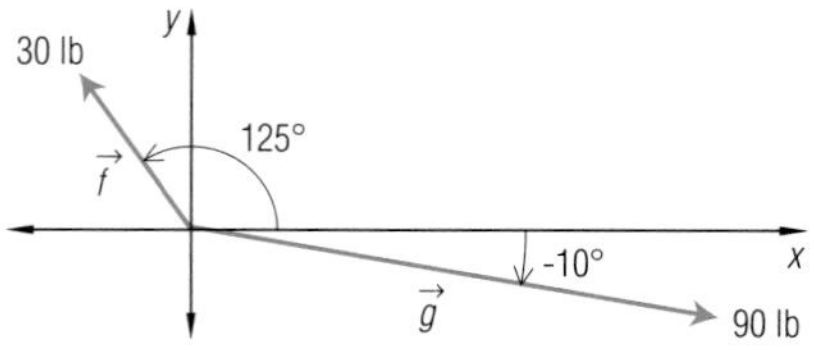

5. A motorboat is crossing a river at an angle of 60° to the downstream bank at a speed of 20 mph in the water. The river is flowing at 6 mph. Describe the resultant velocity (speed and direction) of the boat.

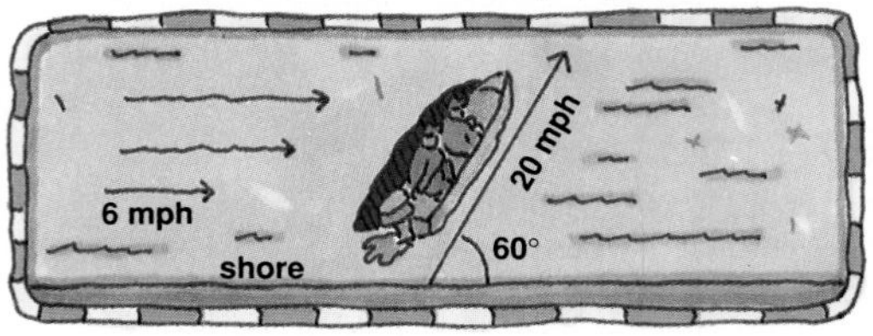

See margin.

6. Two tugboats push a barge with the forces shown at the right. Describe the resultant force.
≈ 2157 lb in the direction of ≈ 52.5° N of E

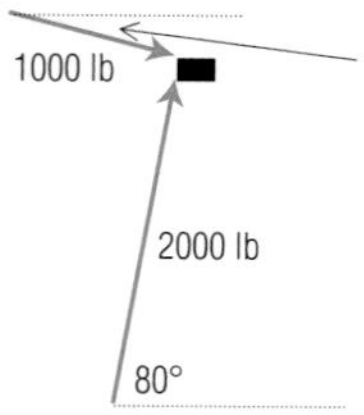

7. A ship sails for 10 hours at a speed of 14 knots (nautical miles per hour) at a heading of 72° north of east. It then turns to a heading of 57° north of east and travels for 2 hours at 11 knots. Find its position north and east of its starting point. (Assume the Earth is flat.)
See margin.

8. If $\vec{w} = [6, 20°]$, express $-\vec{w}$ in
a. polar form **b.** component form.
a) [-6, 20°] or [6, 200°]; b) (6 cos 200°, 6 sin 200°) ≈ (-5.64, -2.05)

In 9 and 10, use the arrows shown below to sketch the indicated vectors.

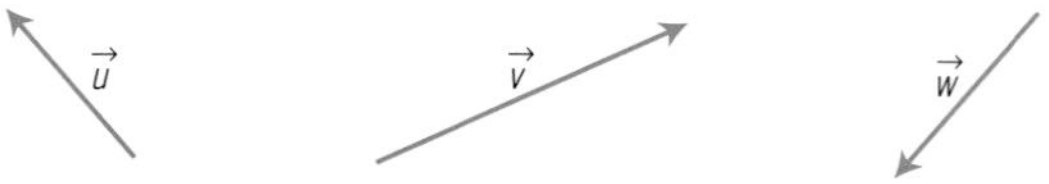

9. **a.** $\vec{u} + \vec{v}$ **b.** $\vec{v} + \vec{u}$ **c.** $\vec{v} + \vec{v}$ See margin.

10. **a.** $\vec{v} + \vec{w}$ **b.** $-\vec{w}$ **c.** $\vec{v} - \vec{w}$ See margin.

Applying the Mathematics

11. Show that this statement is false: If $\vec{u} = [r, \alpha]$ and $\vec{v} = [s, \beta]$, then $\vec{u} + \vec{v} = [r + s, \alpha + \beta]$. Sample counterexample: If $\vec{u} = [1, 45°]$ and $\vec{v} = [1, 45°]$, then $\vec{u} + \vec{v} = [2, 45°] \neq [2, 90°]$.

12. Use the polar representation of $-\vec{v}$ and trigonometric identities to prove that if $\vec{v} = (v_1, v_2)$ then $-\vec{v} = (-v_1, -v_2)$. See margin.

13. **a.** Prove: *For any vector $\vec{v}$, $\vec{v} - \vec{v}$ is the zero vector.* See margin.
b. Give a geometric interpretation for the result of part **a**.
Sample: The arrow for $\vec{v} - \vec{v}$ is a point.

14. If $\vec{s} = (-6, 5)$ and $\vec{t} = (-3, 1)$, find the components of $\vec{v}$ so that
a. $\vec{s} + \vec{v} = \vec{t}$. (3, -4) **b.** $\vec{v} - \vec{s} = \vec{t}$. (-9, 6)

15. A boat leaves point A on the south bank of a river and heads at a 70° angle with an engine speed of 15 knots.

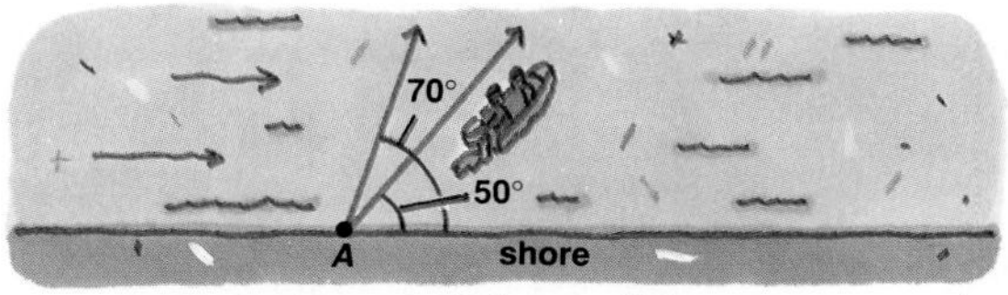

However, the eastward force of the current carries the boat along so it actually travels at a 50° angle with the shore.
a. How fast is the current? The current is about 6.7 knots.
b. How fast does the boat actually travel?
The boat travels about 18.4 knots.

Review

16. If $\vec{v} = (v_1, v_2)$ and k is a real number, show that the length of the vector (kv_1, kv_2) is $|k| \bullet |\vec{v}|$. *(Lesson 12-1)* See margin.

17. Four women and four men are to be seated at a table as shown at the right so that no two women and no two men are beside each other. In how many ways can this be done? *(Lessons 10-3, 10-2)* 144

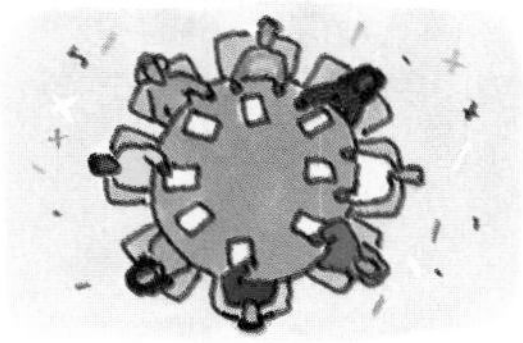

18. Suppose the voltage in an AC circuit is 120V and the impedance is $6 - 5i$ ohms.
a. Find the current. $\approx 11.8 + 9.84i$ amps
b. Interpret the answer to part **a** in terms of the graph of the current shown by an oscilloscope. (Hint: See Question 13 of Lesson 8-3.) *(Lessons 8-3, 8-1)* a sine curve with amplitude $\approx$ 15.4 and phase shift $\approx$ -39.8°

19. Simplify $\dfrac{4}{x^2 + 2x - 8} - \dfrac{3}{x^2 + 5x + 4}$. *(Lesson 5-1)* $\dfrac{x + 10}{(x + 4)(x - 2)(x + 1)}$

20. Approximate the smallest positive solution to $\tan x = (x - 1)^2$ to within 0.1. *(Lesson 3-5)* 0.4

Exploration

21. Describe some of the forces involved in the motions of the following objects.
a. an apple falling from a tree
b. a plane doing acrobatics
c. a pitcher throwing a baseball
d. a sled coasting down a hill

samples: a) gravity; b) gravity, engine thrust, lift due to wing design; c) gravity, torque, initial velocity of ball; d) gravity, friction, initial push

7. $\approx$55.2 nautical miles east and $\approx$151.6 nautical miles north of its starting point.

9., 10., 12., 13.a., 16. See Additional Answers in the back of this book.

NAME ____________

LESSON MASTER 12–2
QUESTIONS ON **SPUR** OBJECTIVES

■ **SKILLS** *Objective B (See pages 756–758 for objectives.)*
In 1–4, let $\vec{u} = (0, -5)$, $\vec{v} = (2, 3)$, and $\vec{w} = (-1, -6)$. Find the sum or difference.

1. $\vec{u} + \vec{v}$ (2, -2)
2. $-\vec{w}$ (1, 6)
3. $\vec{v} - \vec{w}$ (3, 9)
4. $\vec{v} - \vec{u} + \vec{w}$ (1, 2)

In 5 and 6, let $\vec{s} = [3, 15°]$ and $\vec{t} = [2, 160°]$. Compute and express the answer in its polar representation.

5. $\vec{s} + \vec{t}$ [1.8, 55°]
6. $-\vec{s}$ [3, 195°]

■ **PROPERTIES** *Objective E*
7. If $\vec{u} + \vec{v}$ is the zero vector, prove that $\vec{v}$ is the opposite of $\vec{u}$.
Let $\vec{u} = (u_1, u_2)$ and $\vec{v} = (v_1, v_2)$. Then $\vec{u} + \vec{v} = (u_1 + v_1, u_2 + v_2) = (0, 0)$. Thus, $u_1 + v_1 = 0$ and $u_2 + v_2 = 0$. So $u_1 = -v_1$ and $u_2 = -v_2$. Therefore, $\vec{u} = (u_1, u_2) = (-v_1, -v_2)$, the opposite of $\vec{v}$.

8. Show that $\vec{v} = (2 \cos 215°, 2 \sin 215°)$ is the opposite of $\vec{u} = (2 \cos 35°, 2 \sin 35°)$.
$\vec{v} = (2 \cos 215°, 2 \sin 215°) = (-1.6, -1.1)$;
$\vec{u} = (2 \cos 35°, 2 \sin 35°) = (1.6, 1.1)$;
Thus, $\vec{u}$ and $\vec{v}$ are opposites.

Continued 129

NAME ____________
Lesson MASTER 12-2 (page 2)

■ **USES** *Objective H*
9. Two people push a disabled car with the forces shown at the right.

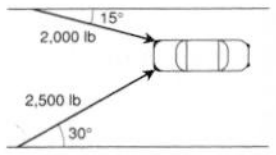

a. Give the resultant force in polar form.
[4,162, 10°]

b. Interpret your answer to part **a**.
The total force applied is 4,162 lb in the direction 10° north of east.

c. How much more forward force (to the right) is provided by the person exerting 2,500 lb of force than the person exerting 2,000 lb of force? 233 lb

10. Relative to the water, a boat is moving with a speed of 20 mph in the direction 15° west of north. Due to the current which is moving in the direction 60° south of east, the boat is actually heading north relative to the land.

a. Find the speed of the current. 10.35 mph
b. Find the speed of the boat moving north. 10.35 mph

■ **REPRESENTATIONS** *Objective J*
11. The vectors $\vec{v}$ and $\vec{w}$ are shown at the right. Sketch the following.

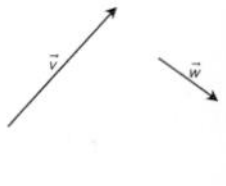

a. $\vec{v} + \vec{w}$

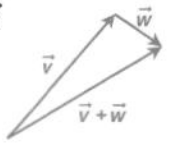

b. $-\vec{w}$

c. $\vec{w} - \vec{v}$

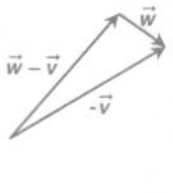

130

LESSON 12-3

RESOURCES

- ■ Lesson Master 12-3
- ■ Quiz for Lessons 12-1 through 12-3
- Computer Master 21

OBJECTIVES

B Find sums, opposites, and scalar products of two-dimensional vectors.
E Prove or disprove statements about vector operations.
F Identify parallel vectors.
I Represent two-dimensional vectors in their component or polar representation, or as directed segments.
J Represent addition, subtraction, and scalar multiplication of two-dimensional vectors graphically.
L Represent lines in a plane using vector or parametric equations.

TEACHING NOTES

The fact that a vector $\vec{v} = (v_1, v_2)$ and its scalar multiple $k\vec{v} = (kv_1, kv_2)$ have the same direction if $k > 0$ can be seen by examining the similar right triangles associated with each vector.

$$\tan \theta = \frac{v_2}{v_1} = \frac{kv_2}{kv_1}$$

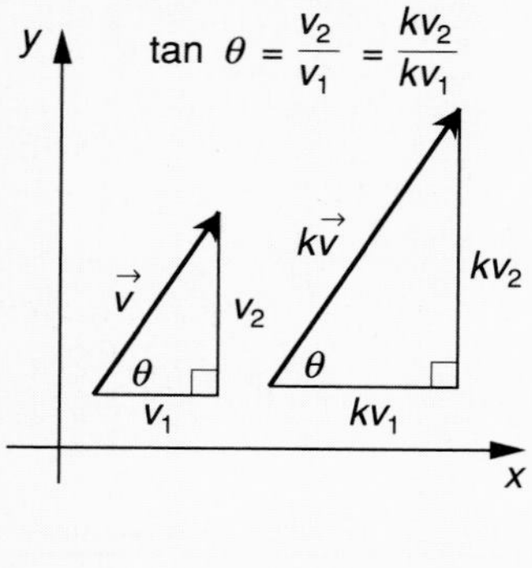

LESSON

12-3

Parallel Vectors and Equations of Lines

Two vectors with the same or opposite directions are called **parallel.** That is, the polar representations of parallel vectors differ by a multiple of 180° or π radians.

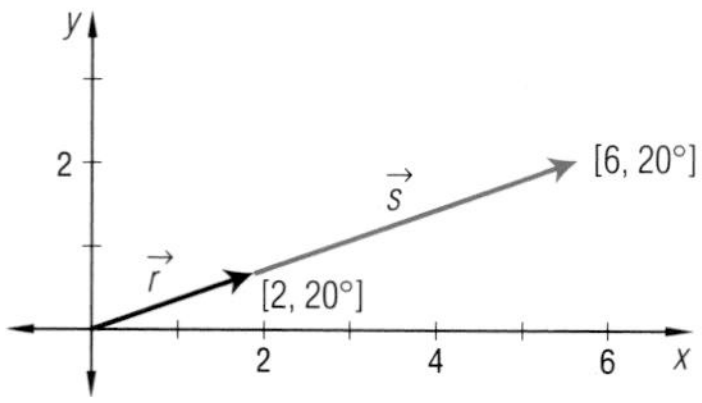

Consider two such vectors, $\vec{r} = [2, 20°]$ and $\vec{s} = [6, 20°]$. When drawn in standard position, it is clear that they lie on the same ray with initial point at the origin. Clearly the length of $\vec{s}$ is three times that of $\vec{r}$. The component representations of the two vectors can be found by converting their polar representations to rectangular:

$$\vec{r} = (2 \cos 20°, 2 \sin 20°) \text{ and } \vec{s} = (6 \cos 20°, 6 \sin 20°).$$

Thus the x- and y-components of $\vec{s}$ are three times those of $\vec{r}$. $\vec{s}$ is called a *scalar multiple* of $\vec{r}$ because both components of $\vec{r}$ are *scaled*, or multiplied, by the same factor. That scale factor is called a **scalar** to emphasize that the multiplier is not a vector.

Definition

A vector $\vec{w} = (w_1, w_2)$ is a **scalar multiple** of a vector $\vec{u} = (u_1, u_2)$, if and only if ∃ a real number k such that $(w_1, w_2) = (ku_1, ku_2)$. We write $\boldsymbol{\vec{w} = k\vec{u}}$.

For $\vec{r}$ and $\vec{s}$ above, the scale factor $k = 3$, and $\vec{s} = 3\vec{r}$. Also, $\vec{r} = \frac{1}{3}\vec{s}$, so that $\vec{r}$ is a scalar multiple of $\vec{s}$ with $k = \frac{1}{3}$. Notice that the definition above allows a scalar to be negative.

Example 1 The vector $\vec{u}$ is pictured at the right. Sketch the vector $\vec{w} = -2\vec{u}$ in standard position.

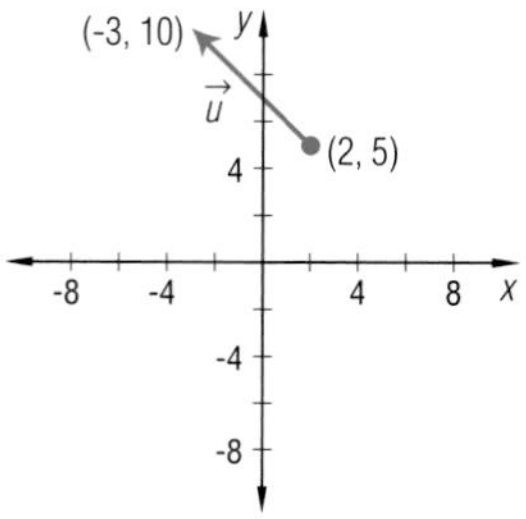

712

Solution Subtracting the coordinates of the endpoint and initial point of the arrow for $\vec{u}$ gives

$\vec{u} = (-3 - 2, 10 - 5) = (-5, 5).$

Therefore

$\vec{w} = -2\vec{u}$

$= -2(-5, 5)$

$= (10, -10).$

$\vec{w}$ is sketched at the right.

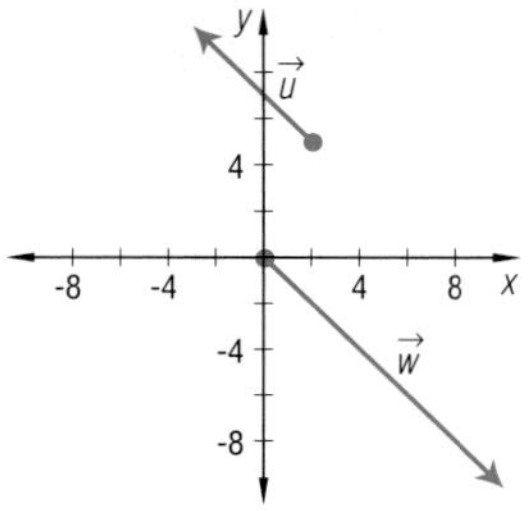

Note that in Example 1, $\vec{u}$ and $\vec{w}$ have opposite directions, and that $\vec{w}$ is twice as long as $\vec{u}$. In fact, Question 16 of Lesson 12-2 shows that multiplying a vector by a scalar multiplies its length by the absolute value of the scalar. If $\vec{v}$ is a vector and k is a real number, then $|k\vec{v}| = |k||\vec{v}|$.

Notice also that in the graph of Example 1, $\vec{u}$ and $\vec{w}$ lie on parallel lines. They are parallel vectors. In general, it is possible to prove the following.

Theorem

Nonzero vectors $\vec{u}$ and $\vec{v}$ are parallel if and only if one of the vectors is a nonzero scalar multiple of the other.

Example 2 Show that the vectors (2, 5) and (18, 45) are parallel.

Solution Write (18, 45) as a scalar multiple of (2, 5).

$(18, 45) = (9 \bullet 2, 9 \bullet 5)$

$= 9(2, 5)$

The diagram below shows several vectors that are parallel to $\vec{m} = (3, -1)$. Each one lies on a line with slope $-\frac{1}{3}$. If $a \neq 0$, a vector (a, b) will always lie on a line with slope $\frac{b}{a}$. If $a = 0$, the vector $(0, b)$ lies on a vertical line.

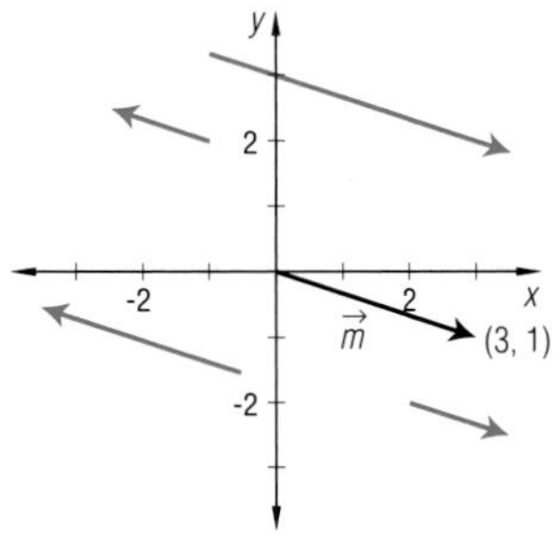

Making Connections In earlier UCSMP courses, the topics of scale factor multiplication and scale changes with negative magnitude were discussed. Students who have taken these courses should be accustomed to the idea that a negative scale factor reverses direction.

Some students may notice that the vector and parametric forms of the equation of a line look very similar to the point-slope form. Both are related to the definition of slope. If the line is not vertical, the quotient $\frac{v_2}{v_1}$ of the components (v_1, v_2) of the vector is its slope, as follows: From the definition of slope, $\frac{y - y_1}{x - x_1} = m$. Thus, $\frac{y - y_1}{x - x_1} = \frac{v_2}{v_1} = \frac{tv_2}{tv_1}$, from which there exists a t such that $y - y_1 = tv_2$ and $x - x_1 = tv_1$. Similarly, the point-slope form can be derived from the parametric form by solving each equation for t and substituting.

Alternate Approach
The parametric form of the equation of a line can be made more understandable to students by having them see the following example. A puppy weighs 8 pounds and is 15 inches long. Each month it grows $\frac{1}{2}$ inch and gains 3 pounds. After t months, its weight is $8 + 3t$ pounds and its length is $15 + \frac{1}{2}t$ inches. The pairs $(8 + 3t, 15 + \frac{1}{2}t)$ lie on a line. The parametric form examines the changes in the two quantities of weight and length individually.

ADDITIONAL EXAMPLES

1. If $\vec{v} = (3, -4)$ calculate $|5\vec{v}|$.
25

ADDITIONAL EXAMPLES

2. Show that the vectors (-1, 6) and (6, -1) are not parallel.

If there is a k such that $k(6, -1) = (-1, 6)$, then $6k = -1$ and $-1k = 6$, from which $k = \frac{-1}{6}$ and $k = -6$ simultaneously, which is impossible.

3. Find parametric equations for the line through (2, 8) parallel to the vector $\vec{v} = (-5, 4)$.

sample: $x = 2 - 5t$, $y = 8 + 4t$

Since vectors are related to slopes, vectors can be used to describe lines. In Euclidean geometry, Playfair's Parallel Postulate states: There is exactly one line parallel to a given line through a given point. Consider the line through the point $P = (-4, 5)$ parallel to the vector $\vec{v} = (7, 2)$.

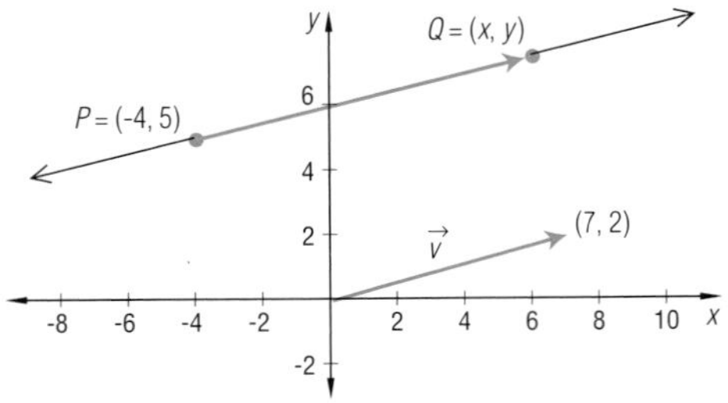

For any point $Q = (x, y)$ on the line, the vector $\overrightarrow{PQ}$ connecting $P = (-4, 5)$ to $Q = (x, y)$, satisfies the equation

$$\overrightarrow{PQ} = t\,\vec{v}$$

for some number t. Now $\overrightarrow{PQ}$ has x-component $x - (-4)$ and y-component $y - 5$. By substitution,

$$(x + 4,\ y - 5) = t(7, 2).$$

This argument can be generalized to any given point and vector.

Theorem

The line through $P = (x_0, y_0)$ parallel to the vector $\vec{v} = (v_1, v_2)$ is the set of points $Q = (x, y)$ such that $\overrightarrow{PQ} = t\,\vec{v}$, or, equivalently, $(x - x_0, y - y_0) = t(v_1, v_2)$ for some real number t.

$\overrightarrow{PQ} = t\,\vec{v}$ is a **vector equation for a line.** The variable t that appears in the vector equation is an independent variable called a **parameter**. In many physical situations this variable represents time.

The vector equation $(x - x_0, y - y_0) = t(v_1, v_2)$ can be written as two equations, one relating the x-components of the two vectors and one relating the y-components.

$$x - x_0 = tv_1$$

and

$$y - y_0 = tv_2.$$

When these are solved for x and y, the result is another form of an equation for a line, the **parametric form**.

Theorem

The line through (x_0, y_0) that is parallel to the vector $\vec{v} = (v_1, v_2)$ has parametric equations

$$\begin{cases} x = x_0 + tv_1 \\ y = y_0 + tv_2 \end{cases}$$

where t may be any real number.

714

For example, the line through (-4, 5) parallel to $\vec{v} = (7, 2)$, pictured on the previous page, has parametric equations

$$\begin{cases} x = -4 + 7t \\ y = 5 + 2t. \end{cases}$$

Think of a point moving at a constant speed along the line. The location of the point at time t is given by $(x, y) = (-4 + 7t, 5 + 2t)$. Each value of t corresponds to a different point on the line. Some specific values of t and their corresponding points are shown below.

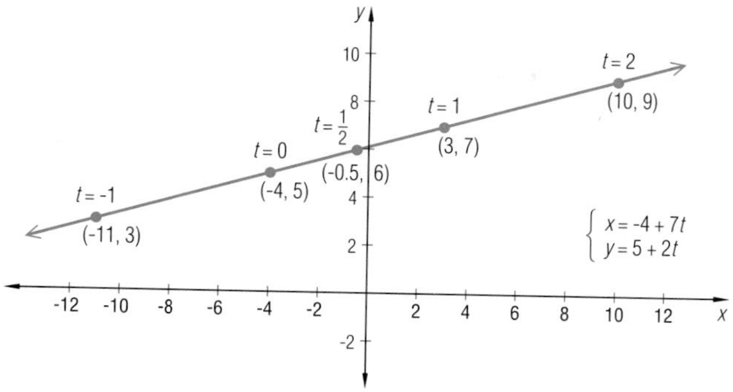

Parametric equations can exist for curves other than lines. Their advantage is that the position of a point is determined by the value of only one variable.

Questions

Covering the Reading

1. Suppose $\vec{s} = [5, 52°]$. Describe the vector with the same direction as $\vec{s}$ and magnitude $\frac{1}{4}$:
   a. in polar form $\left[\frac{5}{4}, 52°\right]$ **b.** in component form. See below.

2. Let $\vec{u}$ be the vector from (-3, 1) to (1, -5). Find the component form for the vector $\vec{v} = -3\vec{u}$. **(-12, 18)**

3. Give an example of a vector parallel to (-6, 14).
samples: (-12, 28), (6, -14); in general, k(-6, 14) where k is any real number.

4. Is the vector (8, -3) parallel to the vector (-48, 18)? Why or why not?
(-48, 18) = -6(8, -3), so the vectors are parallel.

5. A vector equation of a line is $(x + 3, y - 8) = t(2, 5)$.
   a. Give parametric equations for this line. **$x = -3 + 2t$; $y = 8 + 5t$**
   b. Graph the line and identify the points on it determined by $t = 0, 1, 2, -1$, and $-\frac{3}{4}$. **See margin.**

6. **a.** Sketch the line through $P = (-3, -1)$ that is parallel to $\vec{w} = (-4, 9)$.
   b. Write parametric equations for this line.
   a) See margin. b) Sample: $x = -3 - 4t$; $y = -1 + 9t$

1.b) $\left(\frac{5}{4}\cos 52°, \frac{5}{4}\sin 52°\right)$

NOTES ON QUESTIONS
Questions 5a, 6b, 8, and 9: Because there are many sets of parametric equations for a line, students could have different answers for these questions. The given answer is typically the one that would be found by directly applying the techniques of the lesson. Any point on the line, however, could be (x_0, y_0) in the parametric equations for it, and t could be replaced by any nonzero multiple kt.

ADDITIONAL ANSWERS
5.b.

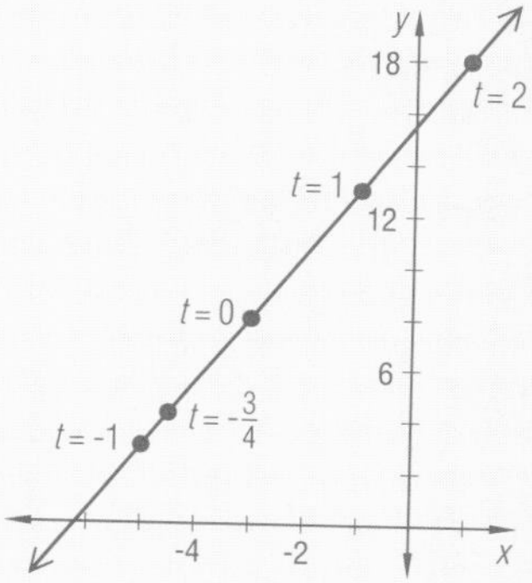

t	0	1	2	-1	$-\frac{3}{4}$
x	-3	-1	1	-5	-4.5
y	8	13	18	3	4.25

6.a.

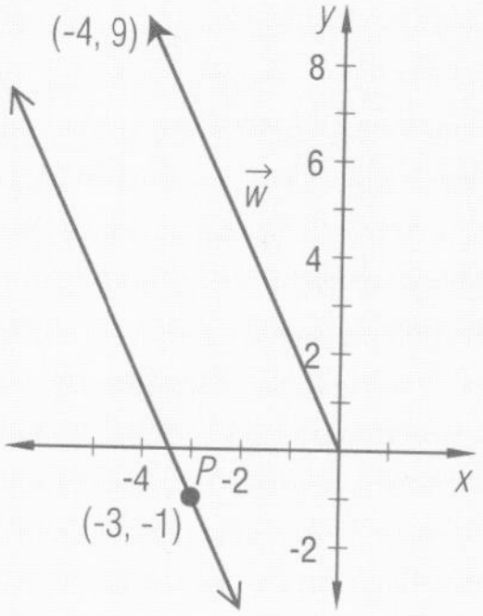

NOTES ON QUESTIONS
Question 7: Students will have different answers depending on which point they use as a starting point. The values of t will be opposites.

Question 12: Another type of generalization is that the point determined when $t = \frac{a}{b}$, $0 < \frac{a}{b} < 1$, is the point that is $\frac{a}{b}$ of the way from the points determined when $t = 0$ and $t = 1$. This generalization is found in **Question 21**.

ADDITIONAL ANSWERS
11.d. $a(b\vec{v}) = a(bv_1, bv_2)$
$= (abv_1, abv_2)$
$= ab(v_1, v_2)$
$= ab\,\vec{v}$

12.a.

t	0	$\frac{1}{2}$	1
x	x_0	$x_0 + \frac{v_1}{2}$	$x_0 + v_1$
y	y_0	$y_0 + \frac{v_2}{2}$	$y_0 + v_2$

$$\frac{(x_0, y_0) + (x_0 + v_1, y_0 + v_2)}{2} = \frac{(2x_0 + v_1, 2y_0 + v_2)}{2} = \left(x_0 + \frac{v_1}{2}, y_0 + \frac{v_2}{2}\right)$$

17.a. -0.267 words per sec; the subjects forget 8 words per 30-second period when the wait-time changes from 0 sec to 30 sec.
b. -.13 words per sec; the subjects forget 6 words per 30-second period when the wait-time changes from 30 sec to 60 sec.

Applying the Mathematics

7. Find parametric equations for the line through (-1, 5) and (4, 10).
Sample: $x = -1 - 5t$, $y = 5 - 5t$

8. Write a vector equation for the line that contains (-8, 5) and is parallel to the line with vector equation $(x - 4, y + 3) = t(7, 2)$.
Sample: $(x + 8, y - 5) = t(7, 2)$

9. A ship is currently 2 miles east and 8 miles north of port. It is moving in the direction parallel to the vector (3, 1). Find parametric equations for the path of this ship. Sample: $x = 2 + 3t$; $y = 8 + t$

10. Show that for any vector $\vec{v} = (v_1, v_2)$, $-1\vec{v} = -\vec{v}$. Thus $\vec{v}$ and $-1\vec{v}$ have opposite directions. $-1\vec{v} = -1(v_1, v_2) = (-v_1, -v_2) = -\vec{v}$

11. Let $\vec{v} = (3, -4)$.
a. Find $-5\vec{v}$. (-15, 20)
b. Multiply your answer to part **a** by the scalar 2. (-30, 40)
c. Find $-10\vec{v}$. (-30, 40)
d. Note that part **b** gave you $2(-5\vec{v})$, while part **c** gave you $(2 \cdot -5)\vec{v}$. Generalize parts **a-c** by proving that for any vector $\vec{v} = (v_1, v_2)$ and real numbers a and b, $a(\overrightarrow{bv}) = (ab)\vec{v}$. See margin.
e. *Multiple choice.* What name would be most appropriate for the result in part **d**?
i. associative property of scalar multiplication
ii. commutative property of scalar multiplication
iii. distributive property of scalar multiplication (i)

12. Let $\begin{cases} x = x_0 + tv_1 \\ y = y_0 + tv_2 \end{cases}$ be parametric equations for a line.
a. Prove: the point determined when $t = \frac{1}{2}$ is the midpoint of the points determined when $t = 0$ and $t = 1$. See margin.
b. Generalize the result of part **a**. (You do not have to prove your generalization.) Sample: The midpoint of the points determined by $t = a$ and $t = b$ is the point determined by $t = \frac{a+b}{2}$.

Review

13. Find the resultant force $\vec{h}$ for the forces $\vec{f}$ and $\vec{g}$ shown below. *(Lesson 12-2)* about 360.6 lb in the direction about 26.3°

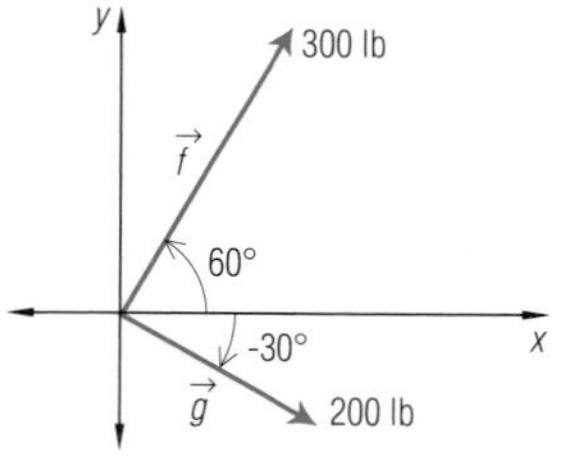

14. The instrument panel of an airplane indicates that its windspeed is 250 miles per hour and its compass heading is northwest. There is a steady 20-mph wind blowing from the north. Find the plane's actual speed and direction. *(Lesson 12-2)* about 236.3 mph in the direction about 138.4°

15. A vector $\vec{w} = (w_1, w_2)$ in standard position is perpendicular to $\vec{u} = (7, 1)$ and twice as long. Find two sets of values for (w_1, w_2). *(Lesson 12-1)* (-2, 14), (2, -14)

16. In the matrix of 4 cities shown at the right, a "1" indicates that a nonstop flight goes from the city on the left to the city above. A "0" indicates that no such flight exists. Find the number of different 2-stop (that is, 3 leg) routes through these cities from Los Angeles to Paris. *(Lesson 11-5)* 4

$$\begin{array}{r} \\ \text{Los Angeles} \\ \text{New York} \\ \text{Paris} \\ \text{London} \end{array} \begin{array}{c} \begin{array}{cccc} \text{LA} & \text{NY} & \text{Paris} & \text{London} \end{array} \\ \begin{bmatrix} 0 & 1 & 1 & 1 \\ 1 & 0 & 1 & 0 \\ 0 & 1 & 0 & 1 \\ 1 & 1 & 0 & 0 \end{bmatrix} \end{array}$$

17. A psychologist conducts a study in which subjects (the people being studied) are given a list of 40 words to memorize. Some of the subjects are then asked to recall as many words as they can. Others are told to wait 30 seconds, 1 minute, or 5 minutes, and then asked to recall the words.

As shown in the table below, on the average, the longer that the subjects are told to wait, the fewer words they remember.

Wait-time (sec)	Mean number of words remembered
0	25
30	17
60	13
300	10

a. Find and interpret the average rate of change in words remembered with respect to wait-time from 0 sec to 30 sec.

b. Repeat part a for wait-times from 30 sec to 60 sec. *(Lesson 9-1)* **a-b) See margin.**

18. Here is a "proof" that all positive integers are even. Where does the argument go wrong?

Let $$S(n): \quad n \text{ is even.}$$

Use the Strong Form of the Principle of Mathematical Induction and start with the inductive assumption that all of the statements $S(1)$, $S(2)$, ..., $S(k)$ are true. Show that this assumption implies that $S(k + 1)$ is true:

$$S(k+1): \quad k + 1 \text{ is even.}$$

The number k + 1 can be expressed as

$$k + 1 = k - 1 + 2.$$

Since $k - 1$ is less than k, $k - 1$ is even by the inductive assumption. That is, $k - 1$ can be expressed as $2m$ for some integer m. Substituting into the equation for $k + 1$, yields

$$k + 1 = 2m + 2,$$

or

$$= 2(m + 1)$$

by the distributive property. Thus, $k + 1$ is twice the integer, $m + 1$. Thus $k + 1$ is even, so $S(k + 1)$ is true. By the Principle of Mathematical Induction, $S(n)$ is true for all positive integers n. All positive integers are even! *(Lesson 7-7)*

The initial conditions do not hold, because s(1), s(3), etc., are not true.

NAME ____________

LESSON MASTER 12–3
QUESTIONS ON **SPUR** OBJECTIVES

■ SKILLS *Objective B (See pages 756–758 for objectives.)*

In 1–4, let $\vec{u} = (6, -2)$ and $\vec{v} = (-1, 5)$, and compute.

1. $\frac{1}{2}\vec{u}$ (3, -1)
2. $2\vec{v}$ (-2, 10)
3. $\frac{1}{2}\vec{u} + 2\vec{v}$ (1, 9)
4. $2\vec{u} - 3\vec{v}$ (15, -19)

■ PROPERTIES *Objective E*

5. Show that $\vec{v} = (3, 7)$ and $\vec{u} = (-18, -42)$ are parallel.
$\vec{u} = (-18, -42) = -6(3, 7) = -6\vec{v}$

6. Show that the line passing through (-1, 2) and (3, 4) is parallel to the vector $\vec{v} = (-2, -1)$.
$(3 - (-1), 4 - 2) = (4, 2) = -2(-2, -1) = -2\vec{v}$

■ PROPERTIES *Objective F*

In 7–10, determine whether or not $\vec{u}$ and $\vec{v}$ are parallel.

7. $\vec{u} = (2, -3)$, $\vec{v} = (4, 6)$ not parallel
8. $\vec{u} = \left(-1, -\frac{1}{2}\right)$, $\vec{v} = (10, 5)$ parallel
9. $\vec{u} = (7, 3)$, $\vec{v} = (7, 3)$ parallel
10. $\vec{u} = [3, 20°]$, $\vec{v} = [5, 200°]$ parallel

■ REPRESENTATIONS *Objective I*

11. Sketch $\vec{v} = (5, 1)$ and $-2\vec{v}$ on the same set of axes.

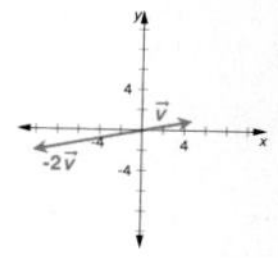

NAME ____________
Lesson MASTER 12–3 (page 2)

12. a. Give the endpoint of a vector $\vec{w}$ which has polar representation [2, 150°] and starts at the point (-1, 1).
$(-1 - \sqrt{3}, 2)$

b. Give the endpoint of the vector $3\vec{w}$ if it starts at (-1, 1).
$(-1 - 3\sqrt{3}, 4)$

c. Sketch both vectors.

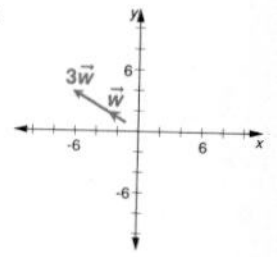

■ REPRESENTATIONS *Objective J*

13. Given the vectors $\vec{u}$ and $\vec{v}$ shown at the right. Sketch the following.

a. $2\vec{u}$

b. $-3\vec{v}$

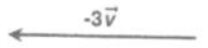

c. $2\vec{u} - 3\vec{v}$

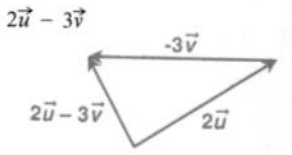

■ REPRESENTATIONS *Objective L*

14. a. Find parametric equations of the line through (-3, 5) that is parallel to the vector $\vec{v} = (3, 3)$.
$\begin{cases} x = -3 + 3t \\ y = 5 + 3t \end{cases}$

b. Graph the line.

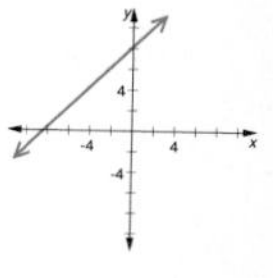

15. Write a vector equation for the line through $P = (-5, 2)$ that is parallel to $\vec{v} = 2, -1)$.
$(x + 5, y - 2) = t(2, -1)$

NOTES ON QUESTIONS
Question 20: This question prepares students for the discussion of perpendicular vectors in the next lesson.

Question 22: This question helps students understand the value of parametric equations.

FOLLOW-UP

MORE PRACTICE
For more questions on SPUR Objectives, use *Lesson Master 12-3*, shown on page 717.

EXTENSION
In Chapter 8, students learned that certain curves are described easily with polar equations. It is also the case that certain curves are described easily with parametric equations, as shown in **Question 22**. Ask students to find the parametric equations for the cycloid, the trochoid, and the helix, and to graph these figures. The last of these requires three dimensions. (cycloid: $x = a(t - \sin t)$, $y = a(t - \cos t)$, $a \in R$; trochoid: $x = at - b \sin t$, $y = a - b \cos t$, $a, b \in R$; helix: $x = a \sin t$, $y = a \cos t$, $z = bt$, $a, b \in R$)

EVALUATION
A quiz covering Lessons 12-1 through 12-3 is provided in the Teacher's Resource File.

ADDITIONAL ANSWERS
21. Since $x = x_0 + v_1 t$ and $y = y_0 + v_2 t$, when $t = 0$, $P = (x_0, y_0)$ is determined. When $t = 1$, $Q = (x_0 + v_1, y_0 + v_2)$ is determined. Create a number line with P at 0 and Q at 1. Then each value of t will determine the corresponding point on the number line.

22. See Additional Answers in the back of this book.

19. *Multiple choice.* If a is even and b is odd, then $ab + a + b$ is
(a) always even
(b) always odd
(c) sometimes even, sometimes odd. *(Lesson 1-7)* (b)

20. Given $A = (3, -4)$, $B = (5, 1)$, $C = (-1, 6)$, and $D = (8, -2)$. Of the lines $\overleftrightarrow{AB}$, $\overleftrightarrow{AC}$, and $\overleftrightarrow{AD}$, which two are perpendicular? *(Previous course)* $\overleftrightarrow{AC}$ and $\overleftrightarrow{AD}$

Exploration

21. In the parametric equations for a line, there is a 1-1 correspondence between values of t and points on the line. Using Questions 5 and 12 as a guide, how can you tell geometrically, given x_0, y_0, v_1, and v_2, what points will be determined by various values of t? That is, what point is given when $t = 0$, when $t = 1$, etc.? See margin.

22. Curves can be described by parametric equations. What curve is described by each pair of equations? How does the location of the point on the curve change as t increases?

a. $\begin{cases} x = \cos t \\ y = \sin t \end{cases}$ **b.** $\begin{cases} x = 3 \cos t \\ y = 5 \sin t \end{cases}$ a, b) See margin.

718

LESSON

12-4

The Dot Product and the Angle Between Two Vectors

Because of the difference between the force of gravity on the Earth and on the moon, a person's weight on the moon is only one-sixth of what it is on the earth. In the 1960s, when the United States decided to send astronauts to the moon, NASA engineers built an apparatus that simulated the effect of gravity on the moon, so astronauts could practice walking and jumping. The apparatus consisted of an inclined plane with a trolley that ran back and forth along the top of the plane. A cable from the trolley to a harness held the astronauts' bodies perpendicular to the plane and they were able to move back and forth along the plane. The idea was to fix the angle between the horizontal and the incline so that the force of the astronauts' feet against the plane was equal to one-sixth of their weight. Thus the force on their feet would feel like the force they would experience on the moon.

Example 1 Compute the angle of inclination θ for the lunar gravity simulator.

Partial Gravity Simulator, used to train astronauts to perform tasks on the lunar surface

Solution The weight of the astronaut is given by the length of a vector $\vec{w}$ pointing straight downward toward the center of the Earth. The force $\vec{f}$ exerted by the astronaut's feet against the inclined plane ($\overline{CA}$ in the diagram) is a vector perpendicular to the plane. The angle $\theta \cong \angle BCA$ (the angle of inclination of the plane) must be chosen so that $|\vec{f}| = \frac{1}{6}$ (astronaut's weight) $= \frac{1}{6}|\vec{w}|$. Since $\overline{BA} \parallel \overline{CD}$, $\angle BAC \cong \angle DCE$. This makes $\Delta ABC \sim \Delta CED$ and thus θ is equal to the angle between $\vec{f}$ and $\vec{w}$. It follows that

$$\cos \theta = \frac{DE}{CD} = \frac{|\vec{f}|}{|\vec{w}|}.$$

Since $|\vec{f}| = \frac{1}{6}|\vec{w}|$, $\cos \theta = \frac{1}{6}$. So the angle of inclination θ of the lunar simulator plane is given by $\theta = \cos^{-1}\left(\frac{1}{6}\right) \approx 80.4°$.

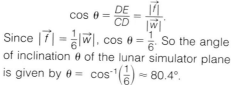

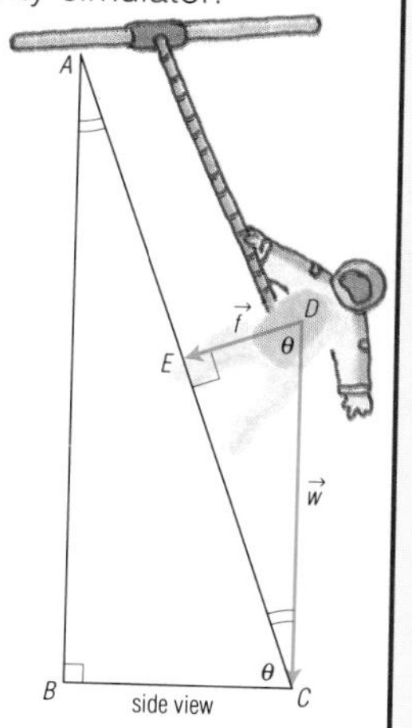

In Example 1, the angle between $\vec{f}$ and $\vec{w}$ could be determined from $|\vec{f}|$ and $|\vec{w}|$ because a right triangle is formed. This, of course, is not always the case. However, the angle between two nonzero vectors can always be determined from the components of the vectors. The process begins by first defining another operation on vectors.

LESSON 12-4

RESOURCES

- Lesson Master 12-4
- Teaching Aid 90 can be used with **Example 1**.

OBJECTIVES

B Find sums, opposites, scalar products, and dot products of two-dimensional vectors.
D Find the measure of the angle between two vectors.
E Prove or disprove statements about vector operations.
F Identify parallel and orthogonal vectors.
L Represent lines in a plane using vector or parametric equations.

TEACHING NOTES

Example 1 is designed to show that the angle between two vectors is a significant idea. In the lunar gravity simulator, finding this angle is part of the process of equalizing gravitational forces.

Summarize the remainder of the lesson by asking students for the three major properties of the dot product. Although these properties are found in the three theorems of the lesson, there is a great deal of reading in the lesson and students may not focus on them.

Alternate Approach Students may appreciate having a picture of the theorem on this page.
$\vec{w} \cdot \vec{w} = |\vec{w}|^2 = w_1^2 + w_2^2$

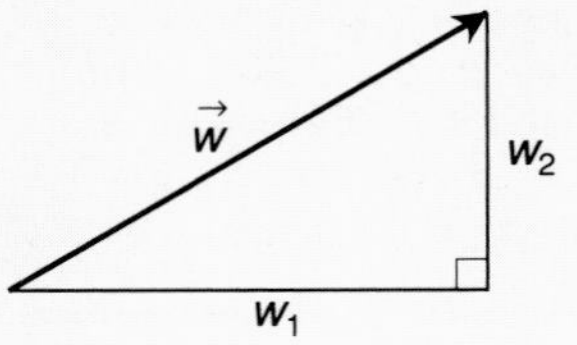

When dealing with vectors, the word *orthogonal* is more often used than the word *perpendicular*. It is related to orthodontia ("straight tooth") and orthodox ("having right opinions"). One nonobvious application of orthogonal vectors is given in Lesson 12-7. In deriving an equation for a plane, a vector orthogonal to the plane is used.

Technology The formula for cos θ in the Angle Between Vectors Theorem is used so often that it is worth discussing calculator keystrokes. Students should realize that $a \div (bc) = a \div b \div c$.

Example 4 asks for a geometric meaning of a negative dot product. **Question 6** asks for a geometric meaning of a positive dot product. The following is a more inclusive geometric interpretation. The product $\vec{u} \cdot \vec{v}$ is the product of two lengths: the length of $\vec{u}$ and the length of the projection of $\vec{v}$ onto u. If $\vec{u}$ and $\vec{v}$ are perpendicular, then the length of the projection of $\vec{v}$ onto $\vec{u}$ is zero, so the dot product is zero.

$\vec{u} \cdot \vec{v} = |u|\,|v| \cos \theta$

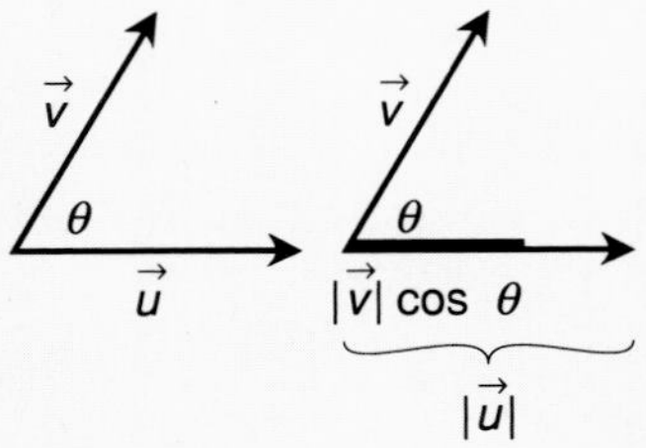

Definition

The **dot product** of $\vec{u} = (u_1, u_2)$ and $\vec{v} = (v_1, v_2)$, denoted by $\vec{u} \cdot \vec{v}$, is the real number $u_1v_1 + u_2v_2$.

Notice that the dot product of two vectors is a number, not a vector.

Example 2 If $\vec{u} = (2, 3)$ and $\vec{v} = (7, -1)$, calculate $\vec{u} \cdot \vec{v}$.

Solution $\vec{u} \cdot \vec{v} = 2 \cdot 7 + 3 \cdot -1 = 11$

The dot product of a vector with itself is related to its length.

Theorem

The dot product of a vector with itself equals the square of its length. In symbols, for all vectors $\vec{w}$, $\vec{w} \cdot \vec{w} = |\vec{w}|^2$.

Proof Let $\vec{w} = (w_1, w_2)$.
Then $\vec{w} \cdot \vec{w} = w_1w_1 + w_2w_2 = w_1^2 + w_2^2 = |\vec{w}|^2$.

With the dot product, the angle between two nonzero vectors $\vec{u}$ and $\vec{v}$ can easily be described. Consider the triangle ABC formed by the standard position arrows for $\vec{u}$ and $\vec{v}$, as shown at the right. The arrow extending from B to C represents the vector $\vec{u} - \vec{v}$ and is the third side of this triangle. If θ is the angle between $\vec{u}$ and $\vec{v}$, then according to the Law of Cosines,

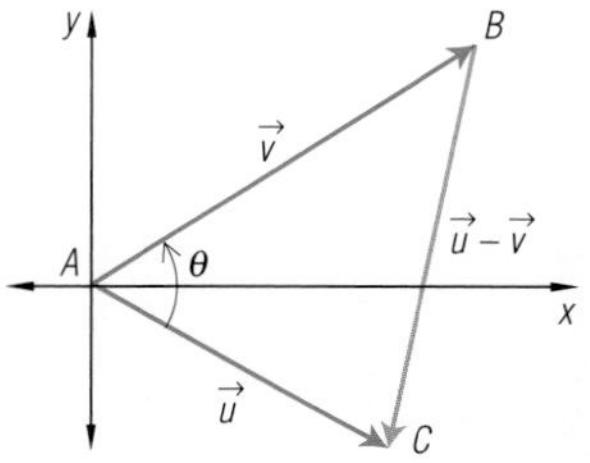

$$BC^2 = AC^2 + AB^2 - 2 \cdot AC \cdot AB \cdot \cos \theta.$$

In the language of vectors, this becomes

$$|\vec{u} - \vec{v}|^2 = |\vec{u}|^2 + |\vec{v}|^2 - 2|\vec{u}|\,|\vec{v}| \cos \theta.$$

Now rewrite the squares of the lengths of the vectors $|\vec{u} - \vec{v}|^2$, $|\vec{u}|^2$ and $|\vec{v}|^2$ as the sums of the squares of their components.

$$(u_1 - v_1)^2 + (u_2 - v_2)^2 = u_1^2 + u_2^2 + v_1^2 + v_2^2 - 2|\vec{u}|\,|\vec{v}| \cos \theta$$
$$u_1^2 - 2u_1v_1 + v_1^2 + u_2^2 - 2u_2v_2 + v_2^2 = u_1^2 + u_2^2 + v_1^2 + v_2^2 - 2|\vec{u}|\,|\vec{v}| \cos \theta$$
$$-2u_1v_1 - 2u_2v_2 = -2|\vec{u}|\,|\vec{v}| \cos \theta$$
$$u_1v_1 + u_2v_2 = |\vec{u}|\,|\vec{v}| \cos \theta$$

Now apply the definition of dot product.

$$\vec{u} \cdot \vec{v} = |\vec{u}|\,|\vec{v}| \cos \theta$$

Dividing both sides by $|\vec{u}|\,|\vec{v}|$ yields the following result.

720

16. Suppose that a pilot wants to fly directly east at a ground speed of 225 mph, but that there is a wind from the northwest at a speed of 40 mph. What air speed and compass heading should the pilot select? *(Lesson 12-2)* air speed $\approx$ 198.7 mph, compass heading $\approx$ 8.18° N of E

17. The following equation is called the Parallelogram Identity.

$$|\vec{u} + \vec{v}|^2 + |\vec{u} - \vec{v}|^2 = 2(|\vec{u}|^2 + |\vec{v}|^2)$$

a. Of the arrows numbered 1, 2, 3, or 4 in the diagram below, identify those representing the vectors $\vec{u} + \vec{v}$ and $\vec{u} - \vec{v}$. i) 3; ii) 4

b. Verify that the Parallelogram Identity is true for two vectors of your choosing. *(Lesson 12-2)* See below.

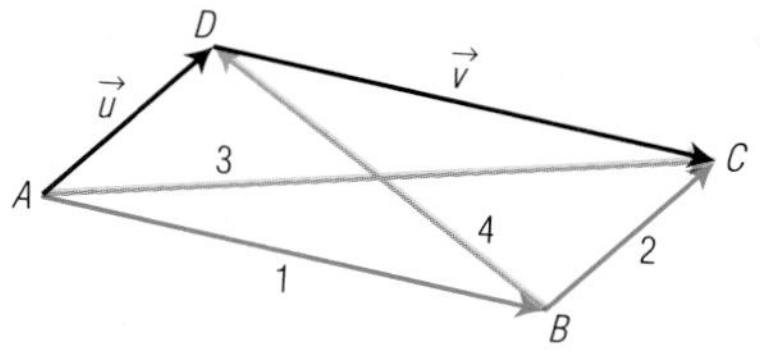

18. **a.** Find $\sum_{k=0}^{12} 4\left(\frac{6}{5}\right)^k$ $\approx$ 194

b. Does $\sum_{k=0}^{\infty} 4\left(\frac{6}{5}\right)k$ converge? If so, find its value. If not, tell why not. *(Lesson 7-6)* No, $r = \frac{6}{5} > 1$.

19. Prove that if the units digit of an even number is nonzero, then the units digit of its fourth power is 6. *(Lesson 4-3)* See below.

20. Solve: $\log_3(x + 2) = 2 - \log_3 x$. *(Lesson 3-2)* $\approx$ 2.16

Exploration

21. Let $\vec{v} = (-3, 4)$, $\vec{i} = (1, 0)$, and $\vec{j} = (0, 1)$.

a. Find $\vec{v} \cdot \vec{i}$. -3

b. Find $\vec{v} \cdot \vec{j}$. 4

c. Find $(\vec{v} \cdot \vec{i})\,\vec{i} + (\vec{v} \cdot \vec{j})\,\vec{j}$. (-3, 4)

d. Generalize the result of part **c** by letting $\vec{v} = (v_1, v_2)$.
$(\vec{v} \cdot \vec{i})\,\vec{i} + (\vec{v} \cdot \vec{j})\,\vec{j} = \vec{v}$

17. b) sample: $\vec{u} = (3, 4)$ and $\vec{v} = (12, 5)$
$|\vec{u} + \vec{v}|^2 + |\vec{u} - \vec{v}|^2 = 306 + 82 = 388$
$2(|\vec{u}|^2 + |\vec{v}|^2) = 2(25 + 169) = 2(194) = 388$

19. The units digit of the fourth power of a number is the fourth power of the units, and 2^4, 4^4, 6^4, and 8^4 each have units digit 6.

FOLLOW-UP

MORE PRACTICE

For more questions on SPUR Objectives, use *Lesson Master 12-4*, shown below.

NAME ______

LESSON **MASTER 12–4**
QUESTIONS ON **SPUR** OBJECTIVES

■ **SKILLS** *Objective B (See pages 756–758 for objectives.)*
In 1–3, let $\vec{u} = (5, 10)$, $\vec{v} = (-1, 3)$, and $\vec{w} = (4, 7)$, and compute.

1. $\vec{u} \cdot \vec{v}$ 25
2. $(\vec{u} \cdot \vec{v})\vec{w}$ (100, 175)
3. $(\vec{u} + \vec{v}) \cdot (\vec{v} - \vec{w})$ -72

■ **SKILLS** *Objective D*
In 4–6, find the measure of the angle between the two vectors.

4. (-1, 1) and (4, 2) 108°
5. (0, 8) and (-4, 0) 90°
6. [18, 63°] and [42, 172°] 109°

■ **PROPERTIES** *Objective E*

7. Show that $\vec{v} = (20, 18)$ and $\vec{u} = (-6, 10)$ are not orthogonal.
$\frac{\vec{u} \cdot \vec{v}}{|\vec{u}||\vec{v}|} = \frac{-120 + 180}{|\vec{u}||\vec{v}|} \neq 0$

8. If $\vec{u}$ and $\vec{v}$ are vectors in a plane, prove that $(\vec{u} + \vec{v}) \cdot (\vec{u} - \vec{v}) = \vec{u} \cdot \vec{u} - \vec{v} \cdot \vec{v}$.
Let $\vec{u} = (x, y)$ and $\vec{v} = (z, w)$. Then
$$\begin{aligned}(\vec{u} + \vec{v}) \cdot (\vec{u} - \vec{v}) &= (x + z, y + w) \cdot (x - z, y - w) \\ &= (x + z)(x - z) + (y + w)(y - w) \\ &= x^2 - z^2 + y^2 - w^2 \\ &= x^2 + y^2 - (z^2 + w^2) \\ &= (x, y) \cdot (x, y) - (z, w) \cdot (z, w) \\ &= \vec{u} \cdot \vec{u} - \vec{v} \cdot \vec{v}\end{aligned}$$

 Continued 133

NAME ______
Lesson MASTER 12–4 (page 2)

■ **PROPERTIES** *Objective F*
In 9–11, determine whether $\vec{u}$ and $\vec{v}$ are perpendicular, parallel, or neither.

9. $\vec{u} = (-2, 1)$, $\vec{v} = (10, -5)$ parallel
10. $\vec{u} = (3, -4)$, $\vec{v} = (6, 2)$ neither
11. $\vec{u} = (7, -3)$, $\vec{v} = (-6, -14)$ perpendicular

■ **REPRESENTATIONS** *Objective L*
In 12 and 13, let $\vec{v} = (4, -4)$. Write a vector equation for the line through $P = (-1, 8)$ that is

12. parallel to $\vec{v}$. $(x + 1, y - 8) = t(4, -4) = 0$
13. perpendicular to $\vec{v}$. $(x + 1, y - 8) \cdot (4, -4) = 0$

134

LESSON 12-5

RESOURCES
- Lesson Master 12-5
- Teaching Aid 91 provides a 3-dimensional coordinate system.
- Teaching Aid 92 can be used with the theorems of the lesson.

OBJECTIVE

M Represent lines, planes, and spheres in 3-space using parametric or coordinate equations.

TEACHING NOTES

Making Connections
The following connections can be made between 2-D and 3-D ideas.

Points from (x, y) to (x, y, z);

Distance from $\sqrt{(x_2 - x_1)^2 + (y_2 - y_1)^2}$ to $\sqrt{(x_2-x_1)^2+(y_2-y_1)^2+(z_2-z_1)^2}$;

Simple equations from the horizontal and vertical lines $x = a, y = b$ to the planes $x = a, y = b$, and $z = c$;

Circle from the circle $(x - a)^2 + (y - b)^2 = r^2$ to the sphere $(x - a)^2 + (y - b)^2 + (z - c)^2 = r^2$;

Diagonals from rectangle $\sqrt{a^2 + b^2}$ to box $\sqrt{a^2 + b^2 + c^2}$.

Graphing 3-dimensional figures on a 2-dimensional sheet of paper is quite difficult for many students. They will need a great deal of practice to master this skill.

LESSON

12-5

Three-Dimensional Coordinates

A communications satellite is usually launched so that it stays in a fixed position above Earth. Reception dishes on the ground can then be aimed directly at the satellite. Three numbers can determine the position of the satellite: latitude, longitude, and height above Earth. When three numbers are needed to determine the position of a point, the space in which the point lies is called **3-space** or **three-dimensional space.**

For many applications in 3-space, rectangular coordinates are easier to work with than coordinates which reference a sphere, like those above for the satellite. A corner of a room provides a model for how three numbers can locate a point in the room using rectangular coordinates. Suppose you want a light to be 3 feet from the side wall, 1 foot from the back wall, and 7 feet above the floor. If you were to consider the corner from which you measured as the origin of a coordinate system, you could write the position of the light as the ordered triple (3, 1, 7). Note that the intersections of the walls and floor are lines which are perpendicular to each other at the corner of the room.

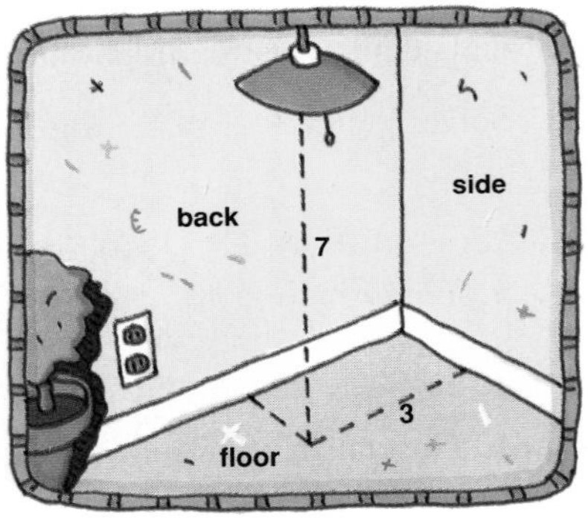

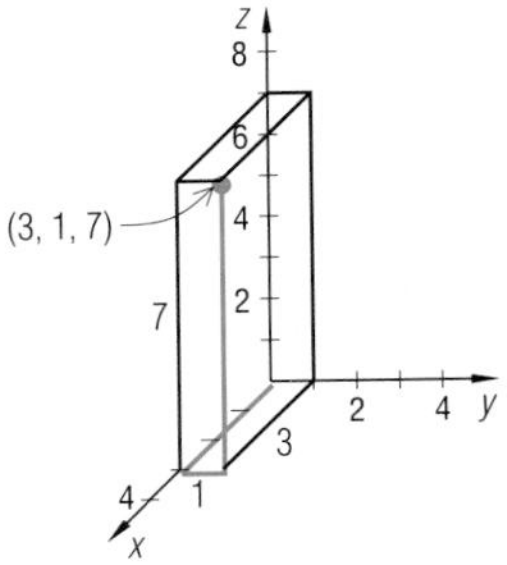

Customarily, two-dimensional coordinate axes are drawn as if the plane is vertical with respect to the ground. A typical three-dimensional coordinate system is drawn with the customary x- and y-axes in a horizontal plane and located so that the positive x-axis points toward the viewer.

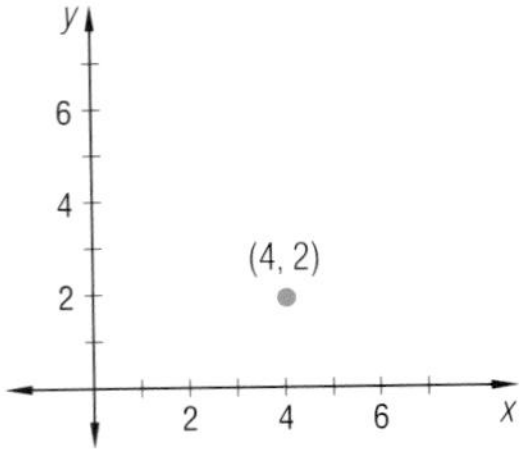

made horizontal and turned

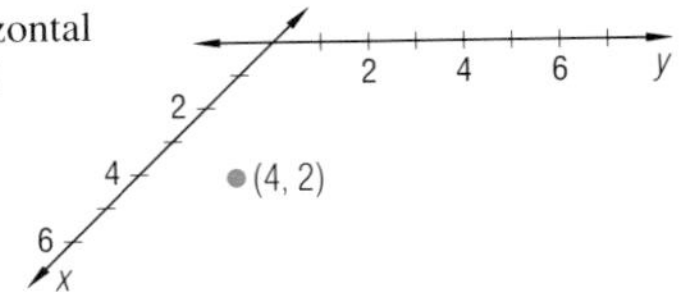

Then a z-axis is put into the picture as a vertical axis coming up from the origin. The location of a point is determined by three numbers, its x-, y-, and z-coordinates. The point (4, 2, 5), for instance, has x-coordinate 4, y-coordinate 2, and z-coordinate 5. To graph this point, think of graphing (4, 2) as you normally would in the xy-plane. Then go up 5 units from (4, 2) to graph (4, 2, 5). Many people like to think of a box, as shown here.

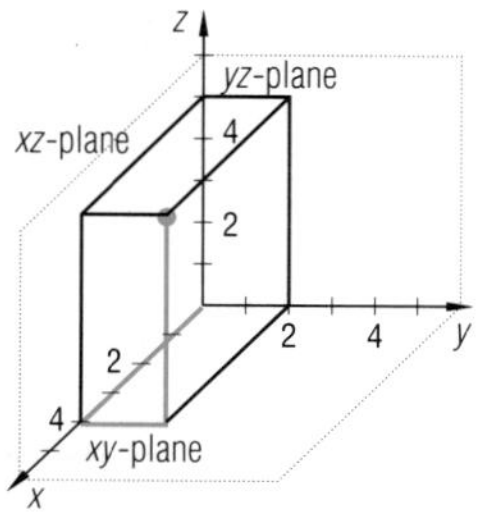

The edges of the box are parallel to the axes. The origin (0, 0, 0) is one corner of the box and the point (4, 2, 5) is the opposite corner. In the drawing, the back of the box is in the ***yz*-plane,** the left side of the box is in the ***xz*-plane,** and the bottom of the box is in the ***xy*-plane.** Notice that the xy-plane is the set of points for which the z-coordinate is 0; it has the equation $z = 0$. All horizontal and vertical planes have simple equations, just like their counterparts in 2-space, horizontal and vertical lines. Similarly, the axes can be characterized by equations.

Example 1 In a three-dimensional coordinate system,

a. Sketch the graph of all points (x, y, z) with $z = 5$.

b. Describe the y-axis with equations.

Solution

a. Points like (3, 2, 5), (2, -1, 5), (0, 0, 5), and $(x, y, 5)$ are all on the graph of $z = 5$. The graph is therefore a plane 5 units above the xy-plane. To sketch it, draw a horizontal line through (0, 0, 5). Then draw a line parallel to the x-axis intersecting the first line. Then complete the parallelogram so that it appears to "float" above the xy-plane.

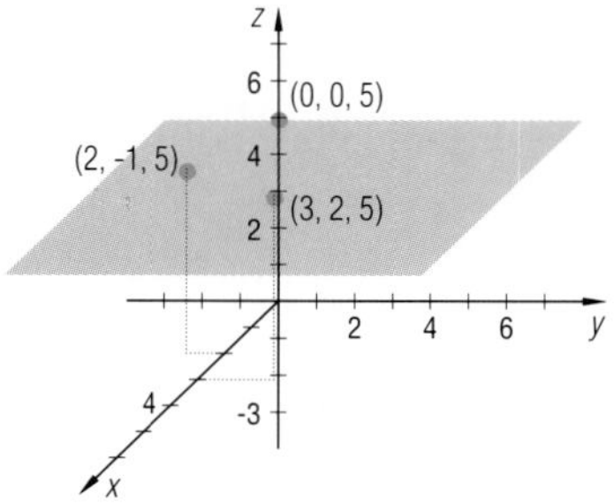

b. The y-axis contains points of the form $(0, a, 0)$, where a is any real number. So the y-axis is

$$\{(x, y, z): x = 0 \text{ and } z = 0\}.$$

In this way a horizontal plane or vertical plane parallel to the x- or y-axis can be described by a single linear equation in one variable; an axis can be described by a system of two linear equations.

Technology If you have computer software that graphs three-dimensional functions, some of the graphs in this lesson can be illustrated. The sphere $x^2 + y^2 + z^2 = r^2$ can be graphed using the functions $z = \sqrt{r^2 - x^2 - y^2}$ and $z = -\sqrt{r^2 - x^2 - y^2}$. Have your students solve the equation in **Example 3** for z so that it can be graphed on the computer.
$(z = \pm\sqrt{36 - (x - 1)^2 - y^2} - 5)$

ADDITIONAL EXAMPLES

1.a. Describe the graph of all points (x, y, z) with $y = -1$.
a plane 1 unit to the left of the xz-plane (the plane determined by the x-axis and z-axis)

b. Give equations for the x-axis.
$y = 0$ and $z = 0$; you can think of the x-axis as the intersection of the planes with these equations.

ADDITIONAL EXAMPLES

2. Sketch $A = (3, 4, 2)$ and $B = (-1, 5, 0)$, and find AB.

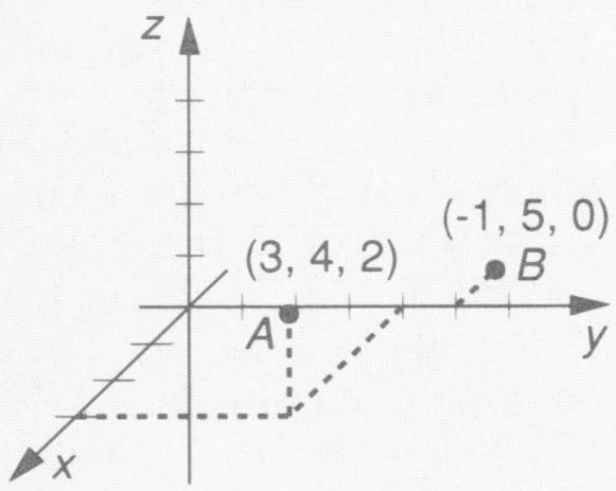

$AB = \sqrt{21}$

3. Find the center and radius of the sphere with equation $x^2 + y^2 + z^2 + 6x + 5y - 2z - 4 = 0$.

center (-3, -2.5, 1), radius 4.5

The longest diagonal of a box with dimensions a, b, and c, as shown below, has length $\sqrt{a^2 + b^2 + c^2}$.

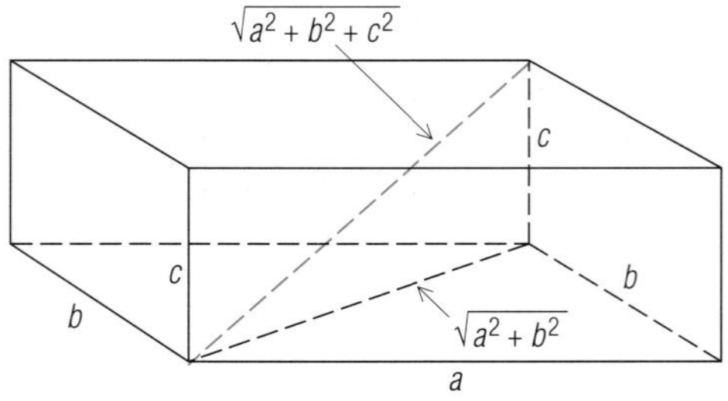

The formula in 3-space for the distance of a point from the origin is an immediate consequence of this fact.

Theorem

The distance of the point (x, y, z) from the origin is $\sqrt{x^2 + y^2 + z^2}$.

For instance, the light pictured on page 726 is

$$\sqrt{3^2 + 1^2 + 7^2} = \sqrt{59} \approx 7.7 \text{ feet}$$

from the corner of the walls and floor.

The formula for the distance between two points in space is a generalization of the preceding theorem.

Theorem (Distance in Space)

The distance between $P = (x_1, y_1, z_1)$ and $Q = (x_2, y_2, z_2)$ is given by

$$PQ = \sqrt{(x_2 - x_1)^2 + (y_2 - y_1)^2 + (z_2 - z_1)^2}$$

Proof The sides of the box below have lengths $|x_2 - x_1|$, $|y_2 - y_1|$, and $|z_2 - z_1|$. The proof is left for you to finish.

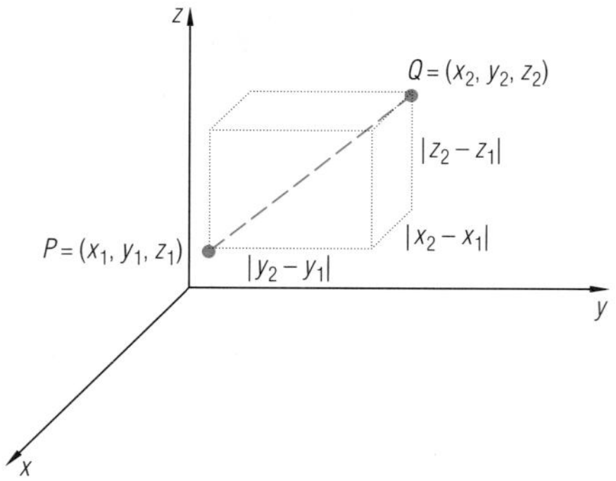

728

Example 2 Consider the points $A = (4, 3, 0)$ and $B = (1, -3, 5)$.
a. Sketch the points in a three-dimensional coordinate system.
b. Find AB.

Solution

a. Point A is in the xy-plane, so locate it in the xy-plane at (4, 3). Since the y-coordinate of B is negative, show the negative part of the y-axis in the sketch. Find (1, -3) in the xy-plane, then go up 5 units.

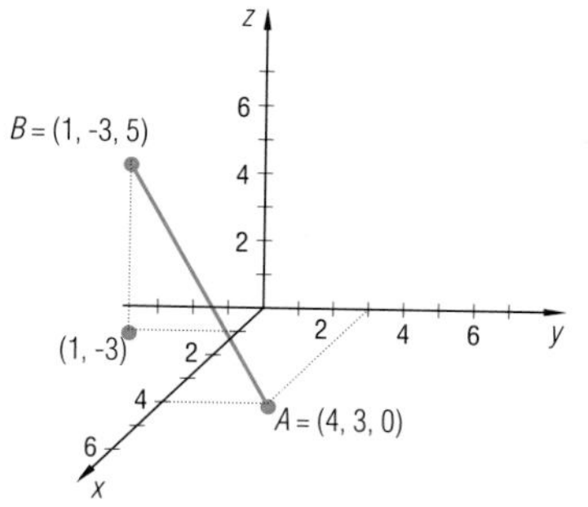

b. $AB = \sqrt{(4-1)^2 + (3 - -3)^2 + (0-5)^2}$
$= \sqrt{70}.$

Since a sphere is the set of all points in 3-space at a fixed distance from a center point, the distance formula leads to an equation for a sphere.

Theorem (Equation of a Sphere)

A point (x, y, z) is on the sphere with center (a, b, c) and radius r if and only if
$$r^2 = (x-a)^2 + (y-b)^2 + (z-c)^2.$$

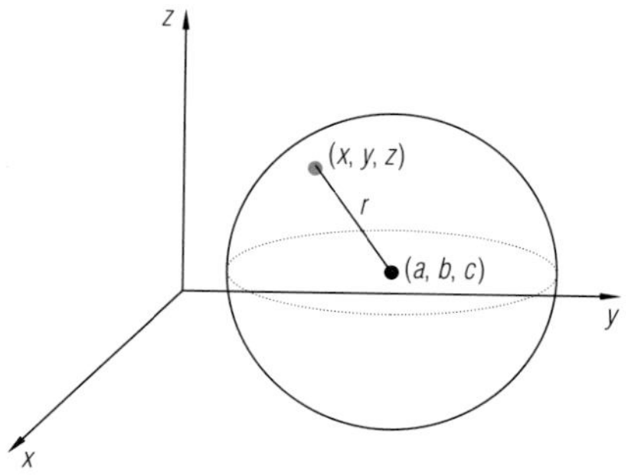

You can deduce the center and radius of a sphere from its equation.

NOTES ON QUESTIONS

Question 6: Students will probably find it difficult to make the drawing. Suggest that they draw lines parallel to the x- and z-axes as the outlines of the plane. Making drawings such as this one helps students to interpret the diagrams they see and makes use of fundamental ideas such as "parallel planes are everywhere equidistant."

Question 13: The rod is called thin so that students do not have to worry about its thickness. A thicker rod could not be as long.

Question 14: A diagram can be used to help explain the geometry involved. All points satisfying $(5, y, 2)$ lie in a line parallel to the y-axis. The question asks for the two points of intersection of this line and the sphere with radius $\sqrt{33}$ and center A.

Question 15: With the mathematics of this lesson, students can calculate how close airplanes come to each other in "near misses."

Question 21: The surface in this question is an example of a quadric surface. The equations and graphs of all such surfaces often are given in calculus textbooks. They can be graphed with computer software that graphs 3-D functions.

ADDITIONAL ANSWERS

2.

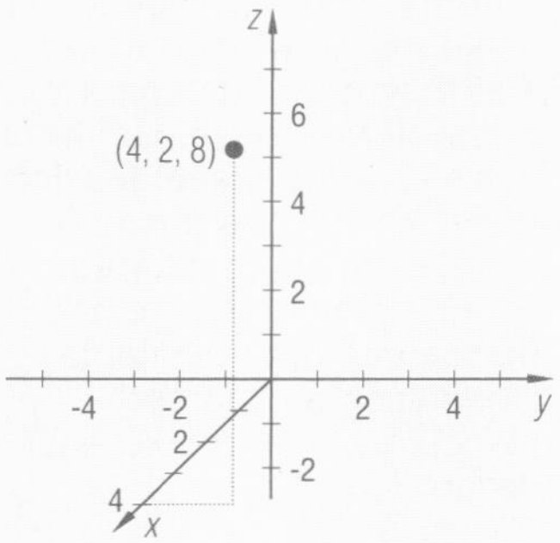

3.

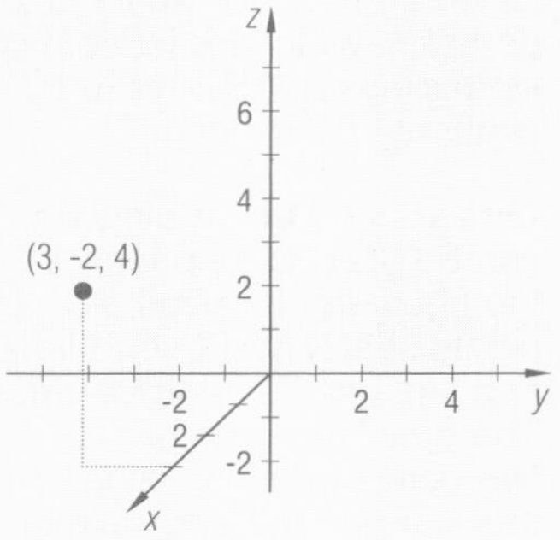

4.

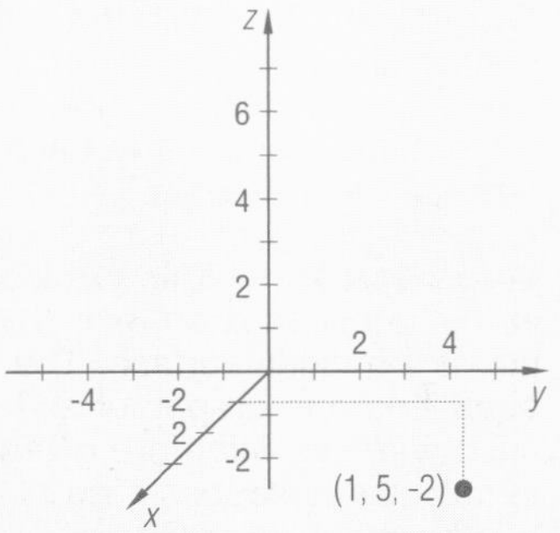

6.

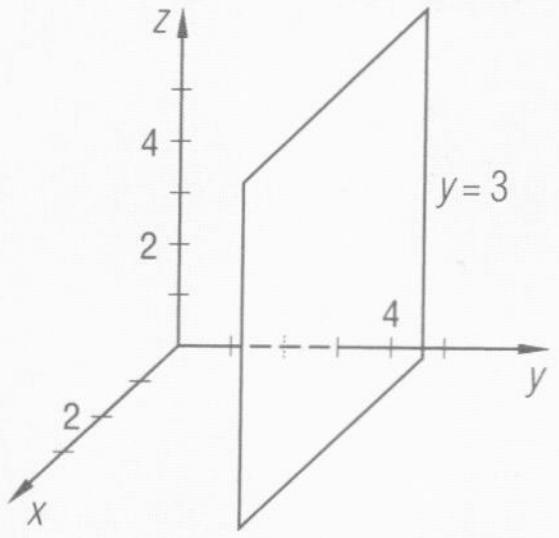

12., 18.a., 19., 20. See Additional Answers in the back of this book.

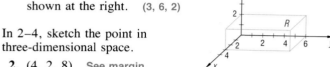

Example 3 Verify that the equation $x^2 + y^2 + z^2 - 2x + 10z = 10$ describes a sphere, and find its center and radius.

Solution Regroup and complete the square for the expressions in x and z.

$$\begin{aligned} x^2 - 2x \quad + y^2 + z^2 + 10z &= 10 \\ (x^2 - 2x + 1) + y^2 + (z^2 + 10z + 25) &= 10 + 1 + 25 \\ (x - 1)^2 \quad + y^2 + \quad (z + 5)^2 &= 36 \end{aligned}$$

This is an equation for the sphere with center (1, 0, -5) and radius 6.

Questions

Covering the Reading

1. Give the coordinates of point R shown at the right. **(3, 6, 2)**

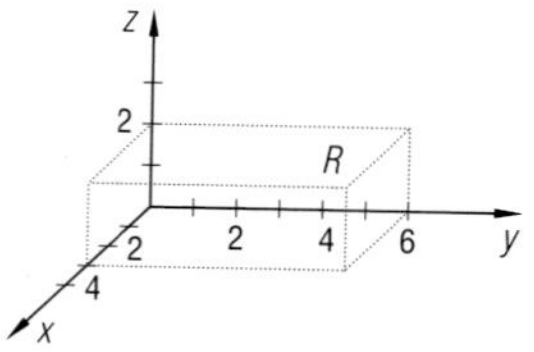

In 2–4, sketch the point in three-dimensional space.

2. (4, 2, 8) **See margin.**

3. (3, -2, 4) **See margin.**

4. (1, 5, -2) **See margin.**

5. *Multiple choice.* In 3-space, the equation $x = 0$ describes
(a) the x-axis (b) the z-axis
(c) the xy-plane (d) the yz-plane **(d)**

6. Sketch the plane $y = 3$. **See margin.**

7. Describe the z-axis with linear equations. **$x = 0$ and $y = 0$**

8. How far is (-1, 5, -8) from the origin? **$3\sqrt{10}$**

9. What is the distance between (1, 0, -1) and (4, -4, 11)? **13**

10. Write an equation for the sphere with radius 5 and center (-1, 2, 8). **$(x + 1)^2 + (y - 2)^2 + (z - 8)^2 = 25$**

11. Find the center and radius of the sphere with equation
$$x^2 + y^2 + z^2 - 2x + 6y + 8z = 10.$$
center: (1, -3, -4); radius: 6

Applying the Mathematics

12. Finish the proof of the Distance in Space Theorem. **See margin.**

13. A wooden crate is 60 cm long, 40 cm wide, and 32 cm high in its internal dimensions. What is the length of the longest thin steel rod that can fit inside the crate? **≈ 78 cm**

14. If $AB = \sqrt{33}$, $A = (1, 2, 3)$, and $B = (5, y, 2)$, find y. **$y = -2$ or $y = 6$**

15. At a particular point in time, one plane is 3 km south and 2 km west of another. One of the planes is flying at an altitude of 8500 meters altitude, the other at 6500 meters. How far are the planes from each other? ≈ 4.1 km

Review

16. *True or false?* If $\vec{u} = (-2, 3)$, $\vec{v} = (5, -1)$, and $\vec{w} = (-7, -2)$, then $(\vec{u} \cdot \vec{v})\vec{w} = (\vec{u} \cdot \vec{w})\vec{v}$. *(Lesson 12-4)* False

17. Suppose point P has coordinates (2, 1) and line ℓ has parametric equations

$$\begin{cases} x = -7 - 4t \\ y = 3 + 2t \end{cases}$$

a. Find parametric equations for the line through P and parallel to ℓ.
b. Find a vector with length 3 perpendicular to ℓ. See below.
c. Find parametric equations for the line through P and perpendicular to ℓ. *(Lessons 12-4, 12-3)* a) $x = 2 - 4t$; $y = 1 + 2t$; c) $x = 2 + t$; $y = 1 + 2t$

18. a. Sketch the polar graph of $r = 1 - 2 \sin \theta$. See margin.
b. Identify the shape of the curve that results. *(Lesson 8-4)* limaçon

19. Prove or disprove: For all real numbers x for which $\tan x$ is defined,

$$\tan^2 x - \sin^2 x = \tan^2 x \sin^2 x.$$

(Lessons 6-2, 6-1) See margin.

20. Prove that 3 is a factor of $n^3 + 2n$ for all positive integers n. *(Lesson 7-5)* See margin.

Exploration

21. Describe the shape of the three-dimensional figure whose equation is $\frac{x^2}{a^2} + \frac{y^2}{b^2} + \frac{z^2}{c^2} = 1$. an ellipsoid that intersects the x-axis at ($\pm a$, 0, 0), the y-axis at (0, $\pm b$, 0), and the z-axis at (0, 0, $\pm c$)

17. a) $\left(\frac{3\sqrt{5}}{5}, \frac{6\sqrt{5}}{5}\right)$

FOLLOW-UP

MORE PRACTICE
For more questions on SPUR Objectives, use *Lesson Master 12-5*, shown below.

EXTENSION
This is an appropriate time for students to learn about higher dimensions. Interested students might want to read two or more of the following books: *Flatland* by Edwin A. Abbott; *One Two Three ... Infinity* by George Gamow; *Geometry, Relativity and the Fourth Dimension* by Rudolf V. B. Rucker; *Beyond the Third Dimension: Geometry, Computer Graphics, and Higher Dimensions*, by Thomas F. Banchoff.

NAME ____________

LESSON **MASTER 12–5**
QUESTIONS ON **SPUR** OBJECTIVES

■ **REPRESENTATIONS** *Objective M* *(See pages 756–758 for objectives.)*

1. Give an equation that describes the *xy*-plane. $z = 0$

2. Write an equation for the sphere with radius 6 and center (6, 0, -1). $(x - 6)^2 + y^2 + (z + 1)^2 = 36$

3. Let M be the plane parallel to the *xz*-plane that is 2 units in the positive direction from the *xz*-plane. Give a system of two linear equations that describes the intersection of M and the *yz*-plane. $\begin{cases} x = 0 \\ y = 2 \end{cases}$

4. Find the center and radius of the sphere with equation $x^2 + y^2 + z^2 - 4x + 6z = 12$. (2, 0, -3), $r = 5$

5. a. Sketch $A(2, 8, 0)$ and $B(0, 1, 2,)$ in three-dimensional space.

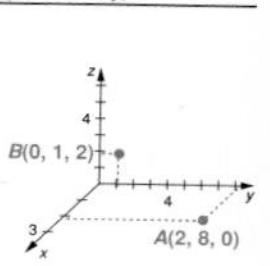

b. Write an equation for the sphere with center A and radius AB. $(x - 2)^2 + (y - 8)^2 + z^2 = 57$

6. Write a system of two linear equations that describes the line parallel to the *x*-axis passing through the point (9, -4, 7). $\begin{cases} y = -4 \\ z = 7 \end{cases}$

LESSON 12-6

RESOURCES
- Lesson Master 12-6
- Quiz for Lessons 12-4 through 12-6
- Teaching Aid 93 can be used when discussing the orbiter body axis system and **Example 2**.
- Teaching Aid 94 can be used with **Question 9**.

OBJECTIVES

C Find sums, lengths, scalar products, dot products, and cross products of vectors in 3-space.
D Find the measure of the angle between two vectors.
E Prove or disprove statements about vector operations.
F Identify parallel and orthogonal vectors.
K Geometrically interpret three-dimensional vectors and their operations.

TEACHING NOTES

Students may wonder why the origin of the orbiter body axis system is not placed at the nose of the shuttle. The reason seems to be that, by placing the origin in front of the nose, objects more often do not intersect the origin; if objects did intersect the origin, then any of the three coordinates could change as they are moved, and keeping track of the coordinates would be confusing.

Example 1 only requires that students extend their knowledge of vectors in 2 dimensions to 3-dimensional vectors.

LESSON

12-6 Vectors in 3-Space

One of the purposes of the space shuttle is to deliver communications satellites to their operating positions above the Earth. The shuttle also retrieves damaged or inoperable satellites and returns them to Earth for repair. A robot arm is used to move satellites to and from the cargo bay of the shuttle. The crew uses a variety of devices for controlling the arm from the safety of the cockpit. Their manipulation of levers and buttons is translated by microcomputers into movement of the arm, using a three-dimensional coordinate system.

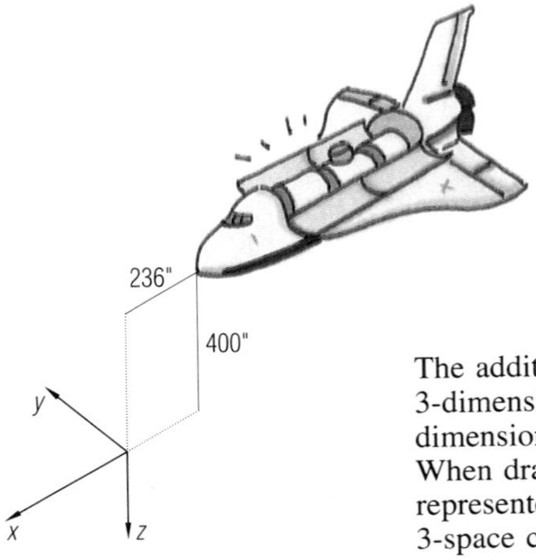

In the actual *orbiter body axis system* used by the shuttle, shown at the left, the origin is not positioned at the nose of the shuttle. Instead, it is placed ahead of and below the nose. Also, the positive z-axis points downward and the positive y-axis points to the right (for someone sitting in the cockpit). Measured in inches, the nose is at the point (-236, 0, -400). The left side of the cargo bay (again, as seen from the cockpit) has a clip to secure a payload. That clip is about 691 inches behind the nose, 98 inches left of center, and 14 inches higher than the nose. In this coordinate system, the clip is at the point with coordinates (-236, 0, -400) + (-691, -98, -14) = (-927, -98, -414).

The addition of ordered triples, component by component, is the 3-dimensional analog to the addition of vectors in a plane. Vectors in three dimensions, like vectors in two dimensions, have direction and magnitude. When drawn in standard position (beginning at the origin), they can be represented by the coordinates of their endpoints. Operations on vectors in 3-space can then be generalized from those on vectors in a plane.

Definitions

Let $\vec{u} = (u_1, u_2, u_3)$ and $\vec{v} = (v_1, v_2, v_3)$ be vectors in 3-space, and let k be a real number.

a. The **sum** of $\vec{u}$ and $\vec{v}$ is the vector
$$\vec{u} + \vec{v} = (u_1 + v_1, u_2 + v_2, u_3 + v_3).$$

b. The **scalar multiple** of $\vec{u}$ by the real number k is the vector
$$k\vec{u} = (ku_1, ku_2, ku_3).$$

c. The **dot product** of $\vec{u}$ and $\vec{v}$ is the number
$$\vec{u} \cdot \vec{v} = u_1v_1 + u_2v_2 + u_3v_3.$$

It follows from the Distance in Space Theorem that the length of $\vec{u} = (u_1, u_2, u_3)$ is given by $|\vec{u}| = \sqrt{u_1^2 + u_2^2 + u_3^2} = \sqrt{\vec{u} \cdot \vec{u}}$. That is, just as in 2-space, the norm of a 3-space vector is the square root of the dot product of a vector with itself. The opposite of a 3-space vector $\vec{u} = (u_1, u_2, u_3)$ is $-\vec{u} = (-u_1, -u_2, -u_3)$ and can be used for subtraction of 3-space vectors just as the opposite of a vector is used in the subtraction of vectors in 2-space.

Example 1 Suppose that $\vec{u} = (5, -1, 1)$ and $\vec{v} = (2, 0, 4)$. Find the following.

a. $\vec{u} - \vec{v}$ **b.** $\vec{u} + \frac{1}{2}\vec{v}$ **c.** $\vec{u} \cdot \vec{v}$ **d.** $|\vec{v}|$

Solution

a. $-\vec{v} = (-2, 0, -4)$, so $\vec{u} - \vec{v} = \vec{u} + -\vec{v}$

$$= (5, -1, 1) + (-2, 0, -4)$$
$$= (5 + -2, -1 + 0, 1 + -4)$$
$$= (3, -1, -3)$$

b. $\frac{1}{2}\vec{v} = \frac{1}{2}(2, 0, 4)$

$$= \left(\frac{1}{2} \cdot 2, \frac{1}{2} \cdot 0, \frac{1}{2} \cdot 4\right)$$
$$= (1, 0, 2)$$

$$\vec{u} + \frac{1}{2}\vec{v} = (5, -1, 1) + (1, 0, 2)$$
$$= (6, -1, 3)$$

c. $\vec{u} \cdot \vec{v} = 5 \cdot 2 + -1 \cdot 0 + 1 \cdot 4 = 14$

d. $|\vec{v}| = \sqrt{2^2 + 0^2 + 4^2} = \sqrt{20} = 2\sqrt{5}$

Notice that the answers to parts **a** and **b** of Example 1 are vectors, while the answers to parts **c** and **d** are real numbers. This is the same as it would be if $\vec{u}$ and $\vec{v}$ were two-dimensional vectors.

Using the definition of dot product for 3-dimensional vectors, theorems about the angle between two vectors can be proved for 3-dimensional vectors in precisely the same way they were deduced in Lesson 12-4 for two-dimensional vectors.

Theorem

Let $\vec{u}$ and $\vec{v}$ be any nonzero three-dimensional vectors and let θ be the measure of the angle between them. Then:

(1) $\cos \theta = \dfrac{\vec{u} \cdot \vec{v}}{|\vec{u}||\vec{v}|}$

(2) $\vec{u}$ and $\vec{v}$ are orthogonal $\Leftrightarrow \vec{u} \cdot \vec{v} = 0$.

Example 2 does the same as Example 1 for the angle between two vectors but in three dimensions. Students may have difficulty visualizing the angle. Remind them that two intersecting lines determine a plane, so the angle they are measuring is indeed an angle in a plane. In part **b** of the example, the fact that the distance is the norm of the vector makes sense because the components of the vector are exactly the quantities that are used in the distance formula.

The cross product yields a vector orthogonal to two others. This idea is applied in the next lesson. **Example 3c** shows that the cross product is a valuable concept by illustrating how difficult it is to work without it.

Alternate Approach
The answer to **Question 21** describes an algorithm for computing the cross product of $\vec{u}$ and $\vec{v}$. Students need to know that the determinant of the 2×2 matrix $\begin{bmatrix} a & b \\ c & d \end{bmatrix}$ is $ad - bc$. Suggest that students think of the components of $\vec{u}$ and $\vec{v}$ as entries in a 2×3 matrix. The components of $\vec{u} \times \vec{v}$ are the determinants of the arrays and are found as follows: (1) by deleting the first column; (2) by deleting the second column and taking the opposite of the resulting determinant; and (3) by deleting the third column. For instance, to find $\vec{u} \times \vec{v}$ in **Example 4**, think of the matrix $\begin{bmatrix} 3 & -1 & 2 \\ 4 & -5 & 6 \end{bmatrix}$.
Then $(3, -1, 2) \times (4, -5, 6) =$
$\left(\begin{bmatrix} -1 & 2 \\ -5 & 6 \end{bmatrix}, -\begin{bmatrix} 3 & 2 \\ 4 & 6 \end{bmatrix}, \begin{bmatrix} 3 & -1 \\ 4 & -5 \end{bmatrix}\right)$
$= (-6 - -10, -(18 - 8), -15 - (-4))$
$= (4, -10, -11)$

ADDITIONAL EXAMPLES

1. If $\vec{u} = (2, 3, 4)$ and $\vec{v} = (-5, -3, -1)$, find:

a. $\vec{u} + 2\vec{v}$

(-8, -3, 2)

b. $-\vec{v}$

(5, 3, 1)

c. $\vec{u} \cdot \vec{v}$

-23

d. $|\vec{u} - \vec{v}|$

$\sqrt{110}$

2. Suppose the robot arm in **Example 2** moves an object from (3.50, 12.50, 20) to (7.50, 16, -5.50).

a. At which point, start or end, is the arm more extended?

start

b. What is the distance between the starting and ending positions?

$\sqrt{678.5} \approx 26.05$ **cm**

c. Through what angle did the arm rotate?

$\approx 74.7°$

3. Let $\vec{u} = (4, -2, 7)$ and $\vec{v} = (-3, 1, 2)$.

a. Are $\vec{u}$ and $\vec{v}$ orthogonal?

Yes

b. Find z if $(6, 3, z)$ is orthogonal to v.

$z = 7.5$

c. Find all vectors orthogonal both to $\vec{u}$ and to $\vec{v}$.

any vector of the form $(11k, 29k, 2k)$

4. Find $\vec{u} \times \vec{v}$ using the vectors from Additional Example 3 and check your answer using the dot product.

$\vec{u} \times \vec{v} = (-11, -29, -2)$; **both** $(\vec{u} \times \vec{v}) \cdot \vec{u}$ **and** $(\vec{u} \times \vec{v}) \cdot \vec{v}$ **equal 0.**

Example 2 In an automobile factory, a robot arm is monitored with a coordinate system whose origin is at the base of the arm, and where distance is measured in centimeters. Suppose the arm moves an object from a starting position of (5.00, 25.00, 32.00) to an ending position of (92.00, -17.00, 10.00).

a. At which point, start or end, is the arm more extended?

b. What is the distance between the starting and ending positions?

c. Through what angle did the arm rotate?

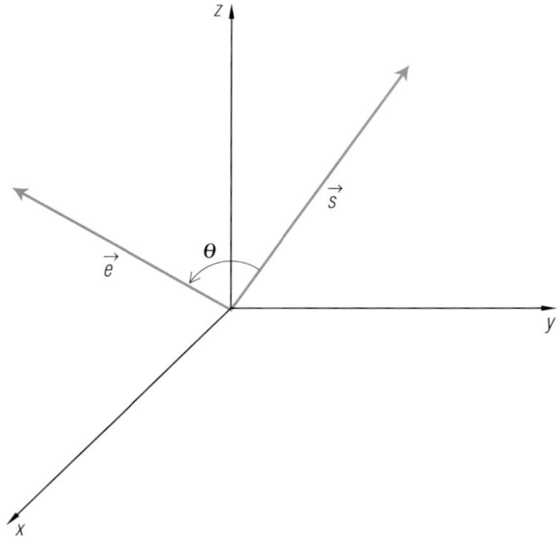

Solution Let the starting position be given by $\vec{s} = (5, 25, 32)$ and the ending position by $\vec{e} = (92, -17, 10)$. (Note: The diagram above is not drawn to scale.)

a. $|\vec{s}| = \sqrt{5^2 + 25^2 + 32^2} \approx 40.91$ cm

$|\vec{e}| = \sqrt{92^2 + (-17)^2 + 10^2} \approx 94.09$ cm

The arm is extended more in its final position.

b. The distance from start to end is the length of the difference vector.

$|\vec{e} - \vec{s}| = \sqrt{(92-5)^2 + (-17-25)^2 + (10-32)^2}$

$= \sqrt{87^2 + 42^2 + 22^2} \approx 99.08$ cm

The distance between start and end is almost one meter.

c. The angle is found using part (1) of the above theorem.

$\vec{s} \cdot \vec{e} = 5(92) + 25(-17) + 32(10)$

$= 355$

$$\cos\theta = \frac{\vec{s} \cdot \vec{e}}{|\vec{s}||\vec{e}|} \approx \frac{355}{(40.91)(94.09)} \approx .0922$$

$\therefore \quad \theta \approx \cos^{-1}(.0922) \approx 84.7°$

The second part of the theorem on page 733 provides a test for orthogonality of three-dimensional vectors.

Example 3 Consider the vectors $\vec{u} = (3, -1, 2)$ and $\vec{v} = (4, -5, 6)$.
a. Are they orthogonal?
b. Find p so that $(3, p, -2)$ is orthogonal to $\vec{u}$.
c. Find a vector that is orthogonal to both $\vec{u}$ and $\vec{v}$.

Solution

a. $\vec{u} \bullet \vec{v} = 3 \bullet 4 + -1 \bullet -5 + 2 \bullet 6 = 29 \neq 0$, so they are not orthogonal.

b. $\vec{u} \bullet (3, p, -2) = 3 \bullet 3 + -1 \bullet p + 2 \bullet -2 = 5 - p$. This must be zero for the vectors to be orthogonal, so $p = 5$.

$$5 - p = 0 \Leftrightarrow p = 5$$

c. Let an orthogonal vector be $\vec{w} = (w_1, w_2, w_3)$. Then
$\vec{u}$ and $\vec{w}$ are orthogonal $\Leftrightarrow \vec{u} \bullet \vec{w} = 0 \Leftrightarrow 3w_1 - w_2 + 2w_3 = 0$
$\vec{v}$ and $\vec{w}$ are orthogonal $\Leftrightarrow \vec{v} \bullet \vec{w} = 0 \Leftrightarrow 4w_1 - 5w_2 + 6w_3 = 0$

This system of two equations with three variables has an infinite number of solutions. But think of it as a system of two linear equations in the variables w_2 and w_3. Solve for w_2 and w_3 in terms of w_1.

$$\begin{cases} -w_2 = -3w_1 - 2w_3 \\ 6w_3 = -4w_1 + 5w_2 \end{cases} \Rightarrow \begin{cases} w_2 = 3w_1 + 2w_3 \\ w_3 = -\frac{2}{3}w_1 + \frac{5}{6}w_2 \end{cases}$$

Substituting the expression for w_3 in the first equation and the expression for w_2 in the second equation, then solving yields:

$$w_2 = -\frac{5}{2}w_1$$

$$w_3 = -\frac{11}{4}w_1$$

So all vectors orthogonal to $\vec{u}$ and $\vec{v}$ have the form $(w_1, -\frac{5}{2}w_1, -\frac{11}{4}w_1)$. Letting $w_1 = 1$, one such vector is $(1, -\frac{5}{2}, -\frac{11}{4})$. Letting $w_1 = -4$, another vector is $(-4, 10, 11)$. All the orthogonal vectors are parallel to each other.

A gyroscope

The vectors found in part **c** of Example 3 are perpendicular to the plane determined by $\vec{u}$ and $\vec{v}$. Perpendiculars to planes are used in many applications. For example, a rotating object, such as a rotating disk or wheel (like that found in a gyroscope), produces a force called *torque* which is represented by a vector perpendicular to the plane of rotation. In computer graphics, vectors perpendicular to surfaces are used in the computations of shadings which represent light intensities and reflections. The vector operation called the *cross product* computes orthogonal vectors directly and so is useful in these and many other applications.

Definition

The **cross product** of two 3-space vectors $\vec{u} = (u_1, u_2, u_3)$ and $\vec{v} = (v_1, v_2, v_3)$ is

$$\vec{u} \times \vec{v} = (u_2v_3 - u_3v_2, u_3v_1 - u_1v_3, u_1v_2 - u_2v_1)$$

NOTES ON QUESTIONS
Questions 2f and 3. After doing Question 3, students should go back and check their answer to Question 2f by finding the dot products of the result with the two given vectors. The dot products should equal 0.

Making Connections for Question 9: This question leads nicely into the next lesson by working with two vectors that determine a plane and a third vector perpendicular to it (and them). Each part of the question leads into the next part. Students may not realize how to use the result from **d** to answer **e**. Suggest that they think of cos θ as the fraction of the maximum brightness at P. Then they need to find that fraction of 255, the maximum brightness in the computer's scale. Thus, they should find 255 cos θ.

Question 11: This question asks for a proof of the check used in **Questions 2f and 3** and a generalization of **Example 4**.

FOLLOW-UP

MORE PRACTICE
For more questions on SPUR Objectives, use *Lesson Master 12-6*, shown on page 737.

EXTENSION
There are two properties of cross products you might wish to explore and prove. If two vectors are parallel, then what can be said about their cross product? (It is the zero vector.) Is the operation of cross product commutative? (No, the cross product $\vec{v} \times \vec{w}$ is the opposite of $\vec{w} \times \vec{v}$.)

PROJECTS
The projects for Chpater 12 are described on pages 751–752. **Project 3** is related to the content of this lesson.

EVALUATION
A quiz covering Lessons 12-4 through 12-6 is provided in the Teacher's Resource File.

Alternative Assessment
Prove the following for 3-dimensional vectors.
(1) $a\vec{u} \times b\vec{v} = (ab)\, \vec{u} \times \vec{v}$. (Let $\vec{u} = (u_1, u_2, u_3)$ and $\vec{v} = (v_1, v_2, v_3)$. Then $a\vec{u} \times b\vec{v} = (au_1, au_2, au_3) \times (bv_1, bv_2, bv_3) = (au_2bv_3 - au_3bv_2, au_3bv_1 - au_1bv_3, au_1bv_2 - au_2bv_1) = (ab)(u_2v_3 - u_3v_2, u_3v_1 - u_1v_3, u_1v_2 - u_2v_1) = (ab)\, \vec{u} \times \vec{v}$.)

(2) $\vec{u} \cdot (\vec{v} + \vec{w}) = \vec{u} \cdot \vec{v} + \vec{u} \cdot \vec{w}$ (Let $\vec{u} = (u_1, u_2, u_3)$, $\vec{v} = (v_1, v_2, v_3)$, and $\vec{w} = (w_1, w_2, w_3)$. Then $\vec{u} \cdot (\vec{v} + \vec{w}) = (u_1, u_2, u_3) \cdot (v_1 + w_1, v_2 + w_2, v_3 + w_3) = u_1v_1 + u_1w_1 + u_2v_2 + u_2w_2 + u_3v_3 + u_3w_3 = (u_1v_1 + u_2v_2 + u_3v_3) + (u_1w_1 + u_2w_2 + u_3w_3) = \vec{u} \cdot \vec{v} + \vec{u} \cdot \vec{w}$.)

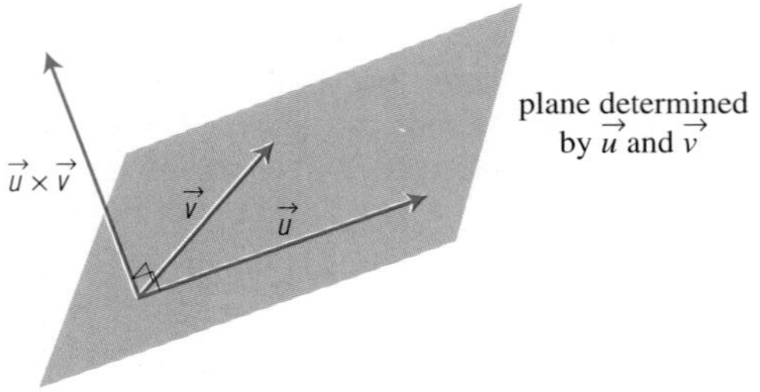

Example 4 Find $\vec{u} \times \vec{v}$ for the vectors $\vec{u} = (3, -1, 2)$ and $\vec{v} = (4, -5, 6)$ of Example 3, and verify that $\vec{u} \times \vec{v}$ is perpendicular to both $\vec{u}$ and $\vec{v}$.

Solution

$$\begin{aligned}\vec{u} \times \vec{v} &= (3, -1, 2) \times (4, -5, 6)\\ &= (-1 \bullet 6 - 2 \bullet -5,\ 2 \bullet 4 - 3 \bullet 6,\ 3 \bullet -5 - -1 \bullet 4)\\ &= (4, -10, -11)\end{aligned}$$

The dot product may be used to verify that the vectors are orthogonal.
$(3, -1, 2) \bullet (4, -10, -11) = 3 \bullet 4 + (-1)(-10) + 2(-11) = 12 + 10 - 22 = 0$, so $\vec{u} \times \vec{v}$ is perpendicular to $\vec{u}$.
$(4, -5, 6) \bullet (4, -10, -11) = 16 + 50 - 66 = 0$, which verifies that $\vec{v}$ is also orthogonal to the cross product.

Note that the cross-product vector is a scalar multiple of the vector $\left(1, -\frac{5}{2}, -\frac{11}{4}\right)$ found to be perpendicular to $\vec{u}$ and $\vec{v}$ in Example 3**c**. So it is perpendicular to them. This is true in general. (See Question 10.)

Questions

Covering the Reading

1. The right-hand clip in the space shuttle cargo bay is 691 inches behind the nose, 14 inches higher, and 98 inches right of center. Give the coordinates of this clip using the orbiter body axis system. **(-927, 98, −414)**

2. Let $\vec{u} = (5, -1, -1)$ and $\vec{v} = (-2, 0, 7)$. Compute.
a. $\vec{u} + \vec{v}$ **(3, -1, 6)**
b. $\vec{v} - \vec{u}$ **(-7, 1, 8)**
c. $|\vec{u}|$ **$3\sqrt{3}$**
d. $\vec{u} \bullet \vec{v}$ **-17**
e. $2\vec{u} + 3\vec{v}$ **(4, -2, 19)**
f. $\vec{u} \times \vec{v}$ **(-7, -33, -2)**

3. When $u = (3, 4, 5)$ and $v = (-2, 5, 9)$, verify that $\vec{u} \times \vec{v}$ is orthogonal to $\vec{u}$. **$\vec{u} \times \vec{v} = (11, -37, 23)$ $\vec{u} \bullet (\vec{u} \times \vec{v}) = 3 \bullet 11 + 4 \bullet -37 + 5 \bullet 23 = 0$, so $\vec{u}$ is orthogonal to $\vec{u} \times \vec{v}$.**

4. The length of a vector is the __?__ of its dot product with itself. **square root**

5. Let $\vec{u} = (1, 3, 2)$ and $\vec{v} = (5, 0, -1)$. Find the measure of the angle between $\vec{u}$ and $\vec{v}$. **≈ 81°**

6. Suppose the robot arm in Example 2 had moved an object from (10.00, 50.00, 64.00) to (23.00, -4.25, 2.50).
   a. How far was the finishing point from the starting point? $\approx$ 83 cm
   b. Through what angle did the arm rotate? $\approx 84.7°$

7. Tell why $\vec{r} = (1, 0, 1)$ is orthogonal to $\vec{s} = (\sqrt{2}, 10, -\sqrt{2})$. $\vec{r} \cdot \vec{s} = 0$

8. a. Find z so that $\vec{u} = (3, -1, z)$ is orthogonal to $\vec{v} = (2, 5, -2)$. $z = \frac{1}{2}$
   b. Find a vector orthogonal to both $\vec{u}$ and $\vec{v}$. sample: $\left(-\frac{1}{2}, 7, 17\right)$

Applying the Mathematics

9. Three-dimensional vectors are used in computer graphics. Suppose a *matte surface* (one which reflects and scatters light equally in all directions) is illuminated by a light source. Then the brightness of a point P on that surface depends on the angle at which the surface faces the light. In fact, if $\vec{s}$ is the vector from P to the light source, and $\vec{n}$ is a vector perpendicular to the surface at P, then the brightness of P is proportional to the cosine of the angle θ between $\vec{n}$ and $\vec{s}$. Thus, when the light hits the surface head on, so that $\theta = 0°$, the point has maximum brightness because $\theta = 1$.

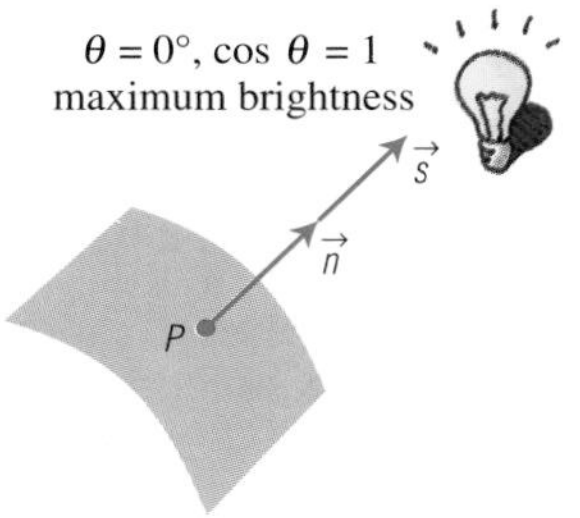

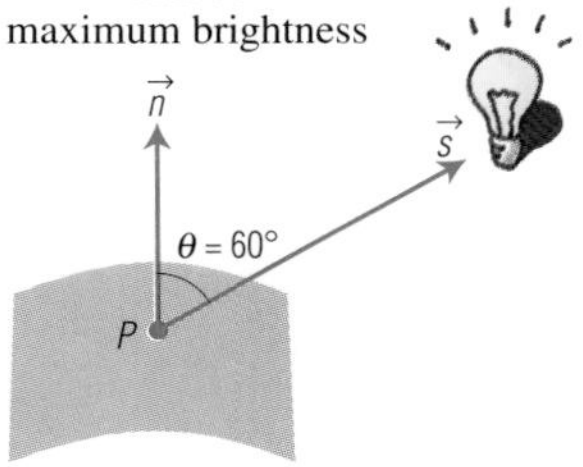

When the surface is tilted so that $\theta = 60°$, the brightness is only half of the maximum because $\cos \theta = \frac{1}{2}$.

Suppose $P = (-3, 5, 1)$, and the surface is a plane which contains P as well as the points $Q = (-4, 5, 2)$ and $R = (-3, 4, 4)$.

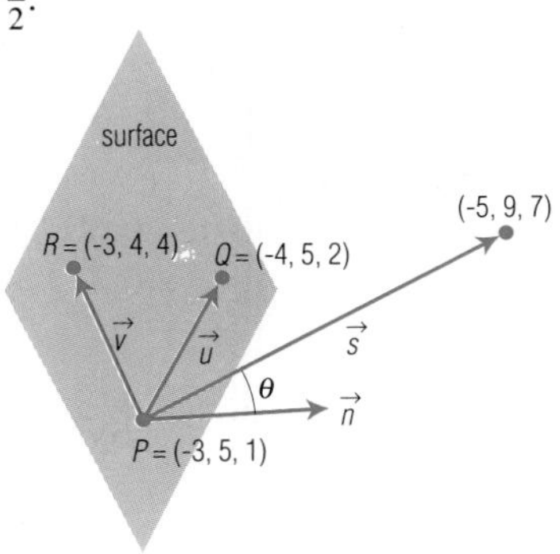

   a. Give the components of the vectors $\vec{u} = \overrightarrow{PQ}$ and $\vec{v} = \overrightarrow{PR}$ which determine the plane (see the figure at the right).
   b. Give the components of $\vec{n}$, a vector perpendicular to the surface. sample: (1, 3, 1)
   c. Give the components of $\vec{s}$, the vector from P to the light source at (-5, 9, 7). (-2, 4, 6)
   d. Find $\cos \theta$, where θ is the angle between $\vec{n}$ and $\vec{s}$. ≈ 0.645
   e. Brightness is represented in many computers by an integer from 0 (pitch black) to 255 (maximum brightness). What integer would be used to represent the brightness of the surface here? 164

   a) $\vec{u} = (-1, 0, 1)$, $\vec{v} = (0, -1, 3)$

NAME __________

LESSON **MASTER 12–6**
QUESTIONS ON **SPUR** OBJECTIVES

■ SKILLS *Objective C (See pages 756–758 for objectives.)*
In 1–5, let $\vec{u} = (-1, 2, 5)$ and $\vec{v} = (3, -4, -6)$, and compute.

1. $\vec{u} + \vec{v}$ (2, -2, -1)
2. $7\vec{u} - 2\vec{v}$ (-13, 22, 47)
3. $|\vec{v}|$ $\sqrt{61}$
4. $\vec{u} \cdot \vec{v}$ -41
5. Find a vector orthogonal to both $\vec{u}$ and $\vec{v}$. (-8, -9, 2) or any nonzero scalar multiple thereof

■ SKILLS *Objective D*
In 6–8, find the measure of the angle between the two vectors.

6. $\vec{u} = (2, 0, 3)$, $\vec{v} = (9, -5, -6)$ 90°
7. $\vec{u} = (-7, 1, -4)$, $\vec{v} = (14, -2, 8)$ 180°
8. $\vec{u} = (3, -1, -1)$, $\vec{v} = (4, 2, 5)$ 77°

■ PROPERTIES *Objective E*

9. Prove that if $\vec{u}$, $\vec{v}$, and $\vec{w}$ are vectors in three-dimensional space, then $\vec{u} \cdot (\vec{v} + \vec{w}) = \vec{u} \cdot \vec{v} + \vec{u} \cdot \vec{w}$.

Let $\vec{u} = (u_1, u_2, u_3)$, $\vec{v} = (v_1, v_2, v_3)$, and $\vec{w} = (w_1, w_2, w_3)$.
Then $\vec{u} \cdot (\vec{v} + \vec{w}) = (u_1, u_2, u_3) \cdot (v_1 + w_1, v_2 + w_2, v_3 + w_3)$
$= (u_1(v_1 + w_1), u_2(v_2 + w_2), u_3(v_3 + w_3))$
$= (u_1v_1 + u_1w_1, u_2v_2 + u_2w_2, u_3v_3 + u_3w_3)$
$= (u_1v_1, u_2v_2, u_3v_3) + (u_1w_1, u_2w_2, u_3w_3)$
$= \vec{u} \cdot \vec{v} + \vec{u} \cdot \vec{w}$

136 *Continued*

NAME __________
Lesson MASTER 12–6 (page 2)

■ PROPERTIES *Objective F*
In 10 and 11, determine whether $\vec{u}$ and $\vec{v}$ are perpendicular, parallel, or neither.

10. $\vec{u} = (4, 1, 2)$ and $\vec{v} = (-1, 2, 1)$ perpendicular
11. $\vec{u} = (-1, 3, -5)$ and $\vec{v} = (2, -6, 10)$ parallel
12. Let $\vec{v} = (x, 4, 3)$ and $\vec{u} = (x - 4, 4, -4)$. If $\vec{u}$ and $\vec{v}$ are orthogonal, find x. $x = 2$

■ REPRESENTATIONS *Objective K*
In 13 and 14, let $\vec{u} = (-1, 5, 3)$ and $\vec{v} = (2, 7, 0)$.

13. Sketch the vectors $\vec{u}$ and $\vec{v}$.
14. Find $\vec{u} - \vec{v}$ and sketch the vector $\vec{u} - \vec{v}$ in standard position on the same graph as vectors $\vec{u}$ and $\vec{v}$. (-3, -2, 3)

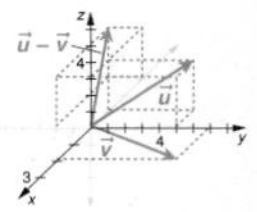

 137

ADDITIONAL ANSWERS

10.a. $\vec{u} \cdot \vec{v} = u_1v_1 + u_2v_2 + u_3v_3 = v_1u_1 + v_2u_2 + v_3u_3 = \vec{v} \cdot \vec{u}$

11. See Additional Answers in the back of this book.

12. $(a\vec{u}) \cdot (b\vec{v}) = (au_1, au_2, au_3) \cdot (bv_1, bv_2, bv_3) = abu_1v_1 + abu_2v_2 + abu_3v_3 = ab(u_1v_1 + u_2v_2 + u_3v_3) = ab(\vec{u} \cdot \vec{v})$

13. $|k\vec{w}| = \sqrt{(kw_1)^2 + (kw_2)^2 + (kw_3)^2} = |k|\sqrt{w_1^2 + w_2^2 + w_3^2} = |k||\vec{w}|$

14.a. The angle between $a\vec{u}$ and bv is $\cos^{-1}\left(\frac{a\vec{u} \cdot b\vec{v}}{|a\vec{u}||b\vec{v}|}\right) = \cos^{-1}\left(\frac{ab(\vec{u} \cdot \vec{v})}{a|\vec{u}|b|\vec{v}|}\right) = \cos^{-1}\left(\frac{\vec{u} \cdot \vec{v}}{|\vec{u}||\vec{v}|}\right)$, which is the angle between $\vec{u}$ and $\vec{v}$.

20. $\vec{u} \times \vec{v} = (10, -8, 0)$ whose length is $2\sqrt{41}$. The parallelogram with $\vec{u}$ and $\vec{v}$ as sides has base $|\vec{u}| = 2\sqrt{41}$ and height $= 1$, so its area is $2\sqrt{41}$.

10. a. Prove that $\vec{u} \bullet \vec{v} = \vec{v} \bullet \vec{u}$ for three-dimensional vectors $\vec{u}$ and $\vec{v}$.
b. What property does this illustrate?
a) See margin. b) commutative property of the dot product

11. Prove that $\vec{u} \times \vec{v}$ is orthogonal to $\vec{u}$, and $\vec{u} \times \vec{v}$ is orthogonal to $\vec{v}$, for all three-dimensional vectors $\vec{u}$ and $\vec{v}$. See margin.

12. Prove that $(a\vec{u}) \bullet (b\vec{v}) = (ab)(\vec{u} \bullet \vec{v})$ for all three-dimensional vectors $\vec{u}$ and $\vec{v}$ and real numbers a and b. See margin.

13. Suppose k is a real number and $\vec{w}$ is a three-dimensional vector. Show that $|k\vec{w}| = |k||\vec{w}|$. See margin.

14. a. Prove that the angle between $a\vec{u}$ and $b\vec{v}$ is the same as the angle between $\vec{u}$ and $\vec{v}$, for any positive real numbers a and b.
b. Give a geometric interpretation for part **a**. a) See margin.
b) Scalar multiplication does not change the measure of the angle between two vectors.

Review

15. Find the center and radius of the sphere whose equation is $x^2 - 4x + y^2 + 3y + z^2 = -2$. *(Lesson 12-5)* center: $(2, -\frac{3}{2}, 0)$; radius: $\frac{1}{2}\sqrt{17}$

16. Find all vectors in a plane that are orthogonal to $\vec{u} = (-4, 2)$ and that have length 5. *(Lesson 12-4)* $(\sqrt{5}, 2\sqrt{5})$ and $(-\sqrt{5}, -2\sqrt{5})$

17. Let ℓ be the line through the points (-3, 5) and (1, -2) in the plane.
a. Find a vector equation for ℓ. Sample: $(x - 1, y + 2) = t(-4, 7)$
b. Find parametric equations for ℓ. *(Lesson 12-3)*
Sample: $x = 1 - 4t$; $y = -2 + 7t$

18. Write the base 10 representation of 11111111_2. (This is the largest integer that can be stored by a computer in a single *byte*.) *(Lesson 4-7)* 255

19. Suppose f is the function defined by $f(x) = \lfloor 2x \rfloor$ on the domain $0 \le x \le 1$.
a. Find the range of f. {0, 1, 2}
b. Is f discrete? *(Lessons 2-2, 2-1)* No

Exploration

20. Let $\vec{u} = (8, 10, 0)$ and $\vec{v} = (4, 5, 1)$. Show that the length of $\vec{u} \times \vec{v}$ is the area of the parallelogram having $\vec{u}$ and $\vec{v}$ as sides.
See margin.

21. Find out how the definition of cross product relates to determinants of 2×2 matrices.
$\vec{u} \times \vec{v} = \left(\det\begin{bmatrix} u_2 & u_3 \\ v_2 & v_3 \end{bmatrix}, -\det\begin{bmatrix} u_1 & u_3 \\ v_1 & v_3 \end{bmatrix}, \det\begin{bmatrix} u_1 & u_2 \\ v_1 & v_2 \end{bmatrix}\right)$

738

LESSON

12-7

Lines and Planes in 3-Space

From the last lesson, you may realize that one advantage of using vectors is that they enable three-dimensional figures to be dealt with just like their two-dimensional counterparts. Addition, scalar multiplication, the dot product, and orthogonality are developed in the same way regardless of dimension. This is also true for vector equations of lines. (You may wish to compare what you read in this lesson with Lesson 12-3.)

Suppose that $P = (x_0, y_0, z_0)$ is a fixed point in three-space and that $\vec{v} = (v_1, v_2, v_3)$ is a fixed nonzero vector. Then there is one and only one line l in space that passes through P and is parallel to $\vec{v}$.

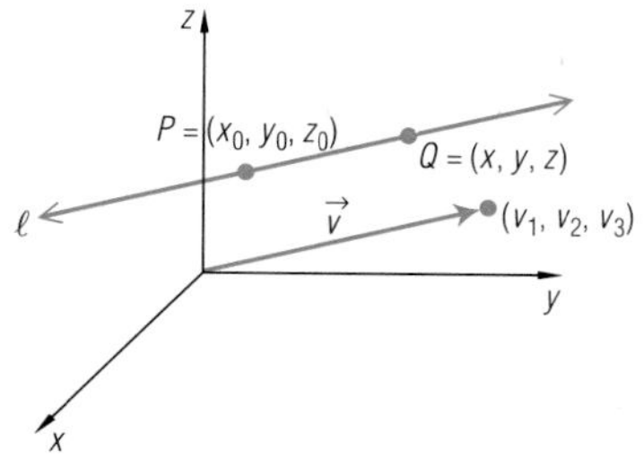

A point $Q = (x, y, z)$ will lie on the line ℓ if and only if the vector $\overrightarrow{PQ}$ joining P to Q is parallel to the vector $\vec{v}$. However, recall that two nonzero vectors are parallel to one another if and only if one is a nonzero scalar multiple of the other. Therefore, the point $Q = (x, y, z)$ is on the line ℓ if and only if there is a real number t such that

$$\overrightarrow{PQ} = t\vec{v}.$$

This is a vector equation for ℓ.

The vector $\overrightarrow{PQ}$ has the component representation

$$\overrightarrow{PQ} = (x - x_0, y - y_0, z - z_0).$$

Substituting the components for $\overrightarrow{PQ}$ and $\vec{v}$, the vector equation for ℓ becomes

$$(x - x_0, y - y_0, z - z_0) = t(v_1, v_2, v_3).$$

Equating the components on the two sides of the equation yields parametric equations for the line ℓ in 3-space.

Theorem

The line ℓ through $P = (x_0, y_0, z_0)$ parallel to $\vec{v} = (v_1, v_2, v_3)$ is the set of points (x, y, z) such that there is a real number t with

a. $\overrightarrow{PQ} = t\vec{v}$ (Vector equation for ℓ) or

b. $\begin{cases} x = x_0 + tv_1 \\ y = y_0 + tv_2 \\ z = z_0 + tv_3. \end{cases}$ (Parametric equations for ℓ)

LESSON 12-7

RESOURCES

■ Lesson Master 12-7

OBJECTIVE

M Represent lines, planes, and spheres in 3-space using parametric, vector, or coordinate equations.

TEACHING NOTES

Making Connections
Whereas Lesson 12-6 contained topics for which the results concerning three-dimensional vectors were virtually identical to those in two-dimensions, the equations of lines and planes given here are analogous to their two-dimensional counterparts but are not identical. In 2-space, $\{(x, y): ax + by = c\}$ is a line, but in 3-space, $\{(x, y, z): ax + by + cz = d\}$ is a plane. In two dimensions, the slope of a line helps to characterize it, but in 3-D, a plane contains infinitely many vectors with infinitely many different directions. To describe the plane, we use a vector that is not even in the plane. There is a more direct connection between the parametric and vector equations of lines in 2 and 3 dimensions.

Go through the proof of the Equation of the Plane Theorem so that students see why the coefficients a, b, and c in the euqation for a plane are also the components of a vector perpendicular to the plane.

Following **Example 3**, the plane M is graphed by finding its points of intersection with the axes. The lines connecting these points lie entirely in the xy-, xz-, and yz-planes and are sometimes called *traces* of the plane M.

Example 1 Let ℓ be the line passing through the two points (-1, 2, 3) and (4, -1, 5). Describe ℓ with **a.** a vector equation, **b.** parametric equations.

Solution

a. The vector $\vec{v}$ that joins (-1, 2, 3) to (4, -1, 5) has the component representation

$$\begin{aligned}\vec{v} &= (4 - (-1), -1 - 2, 5 - 3)\\ &= (5, -3, 2).\end{aligned}$$

If P is the point (-1, 2, 3) and $Q = (x, y, z)$ is any point on ℓ, then

$$\overrightarrow{PQ} = (x + 1, y - 2, z - 3).$$

Since ℓ is parallel to $\vec{v}$, a vector equation of ℓ is

$$(x + 1, y - 2, z - 3) = t(5, -3, 2).$$

b. Equating components in the vector equation yields $x + 1 = 5t$, $y - 2 = -3t$, and $z - 3 = 2t$. Thus, parametric equations for ℓ are

$$\begin{cases} x = -1 + 5t \\ y = 2 - 3t \\ z = 3 + 2t \end{cases}.$$

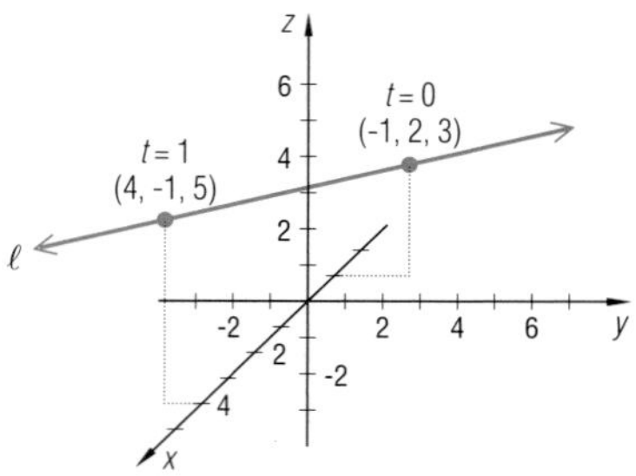

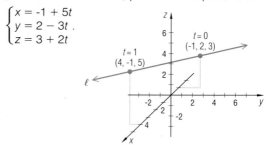

As in 2-space, each value of t in the parametric equations for a line determines a point on the line. In Example 1, $t = 0$ yields the point (-1, 2, 3), and $t = 1$ yields (4, -1, 5). Values of t between 0 and 1 yield points between (-1, 2, 3) and (4, -1, 5).

Recall from your work in geometry that a plane M is perpendicular to a line ℓ at a point P if and only if every line in M through P is perpendicular to ℓ (Imagine a pencil on its tip perpendicular to a table. Any line on the table through the tip is perpendicular to the pencil.)

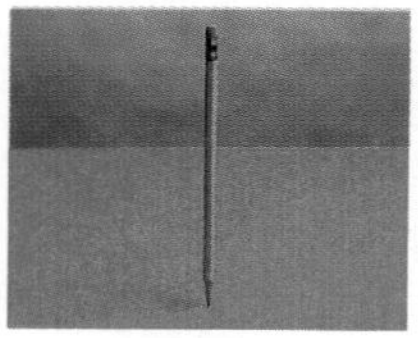

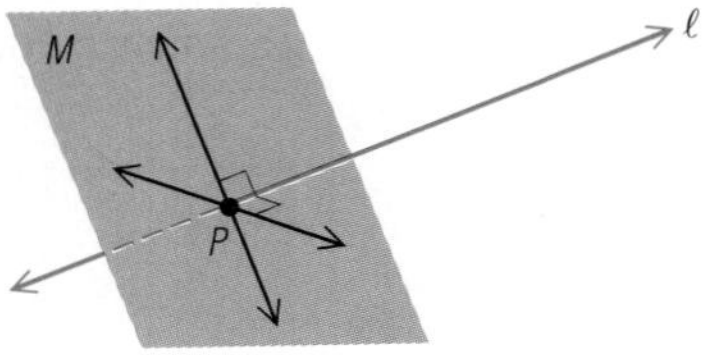

In Lesson 12-6, it was shown that if $\vec{u}$ and $\vec{v}$ are nonzero three-dimensional vectors, then $\vec{u} \cdot \vec{v} = 0$ if and only if $\vec{u}$ and $\vec{v}$ are orthogonal. This fact is very helpful for working with planes in space.

Example 2 Find vector and rectangular-coordinate equations for the plane M that passes through the point $P = (2, 3, 4)$ and is perpendicular to the vector $\vec{w} = (1, -1, 4)$.

Solution Suppose that $Q = (x, y, z)$ is a point on M, as shown below. Then

$$\overrightarrow{PQ} = (x - 2, y - 3, z - 4).$$

Now $\overrightarrow{PQ}$ lies in M if and only if $\vec{w}$ is perpendicular to $\overrightarrow{PQ}$. But $\vec{w} \perp \overrightarrow{PQ} \Leftrightarrow \vec{w} \cdot \overrightarrow{PQ} = 0$. Therefore, a vector equation of the plane M is

$$\vec{w} \cdot \overrightarrow{PQ} = 0$$
$$(1, -1, 4) \cdot (x - 2, y - 3, z - 4) = 0.$$

A rectangular coordinate equation for M can be obtained by using the definition of dot product.

$$(x - 2) - (y - 3) + 4(z - 4) = 0$$
$$x - y + 4z = 15$$

Thus, the plane M is the set of all points (x, y, z) in space that satisfy the equation $x - y + 4z = 15$.

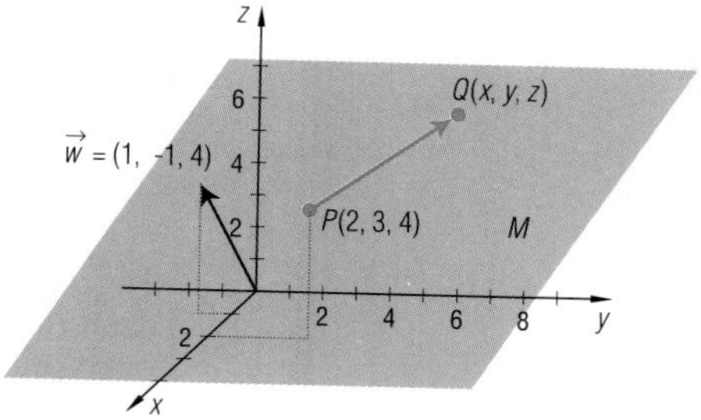

Notice that the coefficients of x, y, and z in Example 2 are the components of $\vec{w}$. This wonderfully simple pattern is true in general.

Equation of a Plane Theorem

The set of points $\{(x, y, z): ax + by + cz = d\}$, where at least one of the coefficients a, b, or c is nonzero, is a plane perpendicular to the vector $\vec{v} = (a, b, c)$.

Proof Let M be the plane containing a fixed point $P = (x_0, y_0, z_0)$ and perpendicular to a fixed nonzero vector $\vec{v} = (a, b, c)$. Let $Q = (x, y, z)$ be any point. Since $\vec{v}$ is nonzero, Q lies on M if and only if $\vec{v} \cdot \overrightarrow{PQ} = 0$.

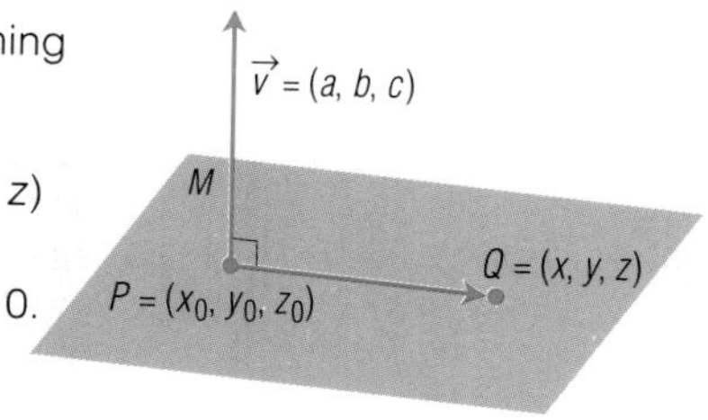

The last equation can be expressed in vector form as

$$(a, b, c) \cdot (x - x_0, y - y_0, z - z_0) = 0.$$

Calculating the dot product gives

$$a(x - x_0) + b(y - y_0) + c(z - z_0) = 0$$
$$\therefore ax + by + cz = ax_0 + by_0 + cz_0.$$

The right side is a constant. Call this d. Thus (x, y, z) lies on the plane M if and only if

$$ax + by + cz = d,$$

where $d = ax_0 + by_0 + cz_0$.

ADDITIONAL EXAMPLES

1. For the line containing the points (6, 0, -4) and (-5, 1, 2), find

a. a vector equation;
sample: $(x - 6, y, z + 4) = t(-11, 1, 6)$

b. a parametric equation.
sample: $x = 6 - 11t$, $y = t$, $z = -4 + 6t$

2. Find an equation for the plane that contains $P = (4, 2, 4)$ and is perpendicular to the vector $\vec{v} = (1, 4, 3)$.
$x + 4y + 3z = 24$

ADDITIONAL EXAMPLES

3.a. Find parametric equations for the line that contains (-5, 4, 2) and is perpendicular to the plane with equation $-2x + 3y + 9z = 7$.
sample: $x = -5 - 2t$, $y = 4 + 3t$, $z = 2 + 9t$

b. Is (-5, 4, 2) in that plane?
No

NOTES ON QUESTIONS

Questions 1-3: Emphasize that the vector that appears in the equation of a line is parallel to the line, but that in the equation of a plane, the vector is perpendicular to the plane (as in **Questions 6-9**).

Question 4: The plane in this question has the same orientation as the one graphed in the lesson. You may want to have your students graph a variety of planes. Planes with positive coefficients are the easiest to graph.

Question 8: Point out to students that a method is being described. This method is applied in **Question 9**.

Question 9: Encourage students to check their answer by substituting the given points into the equation they obtain.

Question 11: One way to show that M_1 and M_2 are not parallel is to find a point in common. An easier way (usually) is to realize that vectors perpendicular to parallel planes must be scalar multiples of each other. Thus, the coefficients a_i, b_i, and c_i must be proportional. The latter method is done in part **a** and the former method is asked for in part **b**.

Question 12: All possible ordered triples are the points in the plane whose coordinates are all nonnegative integers.

Example 3 **a.** Verify that the point $P = (1, 2, 4)$ lies on the plane M defined by the equation

$$3x - y + 2z = 9.$$

b. Find parametric equations of the line ℓ through P that is perpendicular to M.

Solution **a.** To show that the point (1, 2, 4) lies on the plane M, it is sufficient to check that the coordinates of P satisfy its equation. Since $3(1) - (2) + 2(4) = 9$, the point (1, 2, 4) is on the plane M.
b. According to the Equation of a Plane Theorem, the coefficients in the equation of M give the components of the vector $\vec{v} = (3, -1, 2)$ that is perpendicular to M. Any line ℓ through P perpendicular to M must be parallel to $\vec{v}$. Thus, if (x, y, z) is on ℓ, $(x - 1, y - 2, z - 4) = t(3, -1, 2)$ for any real number t, and so parametric equations of ℓ are

$$\begin{cases} x = 1 + 3t \\ y = 2 - t \\ z = 4 + 2t \end{cases}$$

To graph the plane M of Example 3, recall that three noncollinear points determine a plane. Three points on M that are easy to find are its points of intersection with the axes. To find its point of intersection with the z-axis, note that a point (x, y, z) is on the z-axis if and only if $x = 0$ and $y = 0$. Suppose the point is also on M. Then $3 \cdot 0 - 0 + 2z = 9$, so $z = 4.5$. Consequently (0, 0, 4.5) is on M. In a similar manner, (0, -9, 0) is the point where M intersects the y-axis, and (3, 0, 0) is the point where M intersects the x-axis. The numbers 3, -9, and 4.5 are the x-, y-, and z-intercepts of plane M, respectively.

Below is a sketch of M, along with the perpendicular vector $\vec{v}$. The three intercepts have been connected and the resulting region has been shaded to produce a look of flatness.

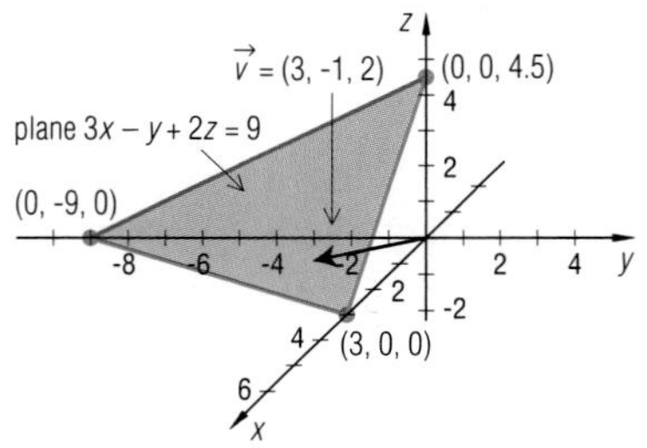

Questions

Covering the Reading

1. Consider the line containing (1, 5, -2) parallel to the vector (-3, 0, 4).
a. Find a vector equation for the line. $(x - 1, y - 5, z + 2) = t(-3, 0, 4)$
b. Find parametric equations for the line. $x = 1 - 3t$; $y = 5$; $z = -2 + 4t$
c. Find a point other than (1, 5, -2) that lies on the line. sample: (-2, 5, 2)

2. Let ℓ be the line with parametric equations

$$\begin{cases} x = -2 + 3t \\ y = 3 - 2t \\ z = 2 - 3t. \end{cases}$$

a. Is the point (10, -5, -8) on the line ℓ? Explain how you arrived at your answer. No, when $x = 10$, $t = 4$, but then $z \neq -8$.
b. Find a vector $\vec{v}$ that is parallel to ℓ. (3, -2, -3)
c. Find parametric equations for the line ℓ' that is parallel to ℓ and contains the point (1, 2, 3). Sample: $x = 1 + 3t$; $y = 2 - 2t$; $z = 3 - 3t$

3. Find an equation for the plane that is perpendicular to $\vec{u} = (2, 1, -2)$ and contains the point (0, 5, 3). $2x + y - 2z = -1$

In 4–7, let N be the plane defined by the equation $2x - y + 3z = 4$.

4. Sketch N. See margin.

5. Show that the vector $\vec{w} = (-4, 2, -6)$ is perpendicular to N. See margin.

6. Find parametric equations for the line m that is perpendicular to N and that passes through the point (2, 3, -1). Sample: $x = 2 + 2t$; $y = 3 - t$; $z = -1 + 3t$

7. Find an equation for the plane K that is parallel to N and that passes through the point (2, 3, -1). $2x - y + 3z = -2$

Applying the Mathematics

8. If P_1, P_2, and P_3 are three distinct noncollinear points in space, explain why a point $Q = (x, y, z)$ lies on the plane M through P_1, P_2, and P_3 if and only if

$$\overrightarrow{P_1Q} \cdot (\overrightarrow{P_1P_2} \times \overrightarrow{P_1P_3}) = 0.$$

(This is a vector equation of the plane through P_1, P_2, P_3.) See margin.

9. Find an equation of the plane M through the points $P_1 = (1, 0, -1)$, $P_2 = (2, 3, -2)$, and $P_3 = (1, 2, 3)$:
a. using the result of Question 8; $7x - 2y + z = 6$
b. by solving the system of equations which results from substituting the coordinates of P_1, P_2, and P_3 into $ax + by + cz = d$. (Note that substitution yields three linear equations with four variables. Therefore, you can solve for a, b, and c in terms of d. If $d \neq 0$, then any value of d can be used to find specific values of a, b, and c.) $7x - 2y + z = 6$

ADDITIONAL ANSWERS

4.

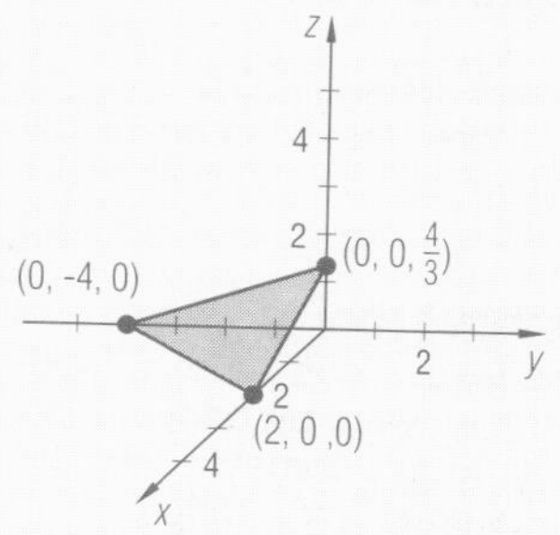

5. $\vec{w} = (-4, 2, -6) = -2(2, -1, 3)$; $\vec{w}$ is parallel to (2, -1, 3), since it is a scalar multiple of it. (2, -1, 3) is normal to the plane N, and hence $\vec{w}$ is normal to N.

8. Since $\overrightarrow{P_1P_2}$ and $\overrightarrow{P_1P_3}$ lie in M, $\overrightarrow{P_1P_2} \times \overrightarrow{P_1P_3}$ gives the vector perpendicular to M. Thus, Q is on M (so that $\overrightarrow{P_1Q}$ is on M) if and only if $\overrightarrow{P_1Q}$ is perpendicular to $\overrightarrow{P_1P_2} \times \overrightarrow{P_1P_3}$.

FOLLOW-UP

MORE PRACTICE
For more questions on SPUR Objectives, use *Lesson Master 12-7*, shown on page 745.

EXTENSION
This lesson can be extended in many ways by bringing in ideas from geometry. Recall with students the different ways of determining a plane: 3 noncollinear points, a line and a point not on it, two parallel lines, and two intersecting lines. Given the equations describing one of these situations, students can derive the equation of the plane. In three dimensions, lines can either be intersecting, parallel, or skew. Given the equation of a line and a point on the line, have students find the equation of a parallel line, an intersecting line, and a skew line.

PROJECTS
The projects for Chapter 12 are described on page 751–752. **Project 2** is related to the content of this lesson.

ADDITIONAL ANSWERS

10. M_1 is perpendicular to (a_1, b_1, c_1), and M_2 is perpendicular to (a_2, b_2, c_2), so ℓ is perpendicular to both (a_1, b_1, c_1) and (a_2, b_2, c_2). Since $\vec{w}$ is also perpendicular to both (a_1, b_1, c_1) and (a_2, b_2, c_2), $\vec{w}$ is parallel to ℓ.

14. $\vec{u} \cdot (\vec{v} + \vec{w}) = (u_1, u_2, u_3) \cdot (v_1 + w_1, v_2 + w_2, v_3 + w_3) = (u_1v_1 + u_1w_1) + (u_2v_2 + u_2w_2) + (u_3v_3 + u_3w_3) = (u_1v_1 + u_2v_2 + u_3v_3) + (u_1w_1 + u_2w_2 + u_3w_3) = \vec{u} \cdot \vec{v} + \vec{v} \cdot \vec{w}$

16.a.

	W	I
W	.96	.04
I	.5	.5

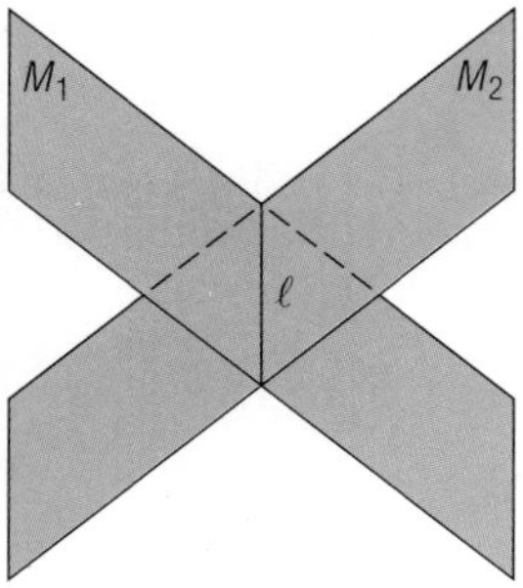

10. Suppose that the planes M_1 and M_2 are not parallel and that equations for M_1 and M_2 are

M_1: $a_1x + b_1y + c_1z = d_1$
M_2: $a_2x + b_2y + c_2z = d_2$.

Explain geometrically why the vector

$\vec{w} = (a_1, b_1, c_1) \times (a_2, b_2, c_2)$

is parallel to l, the line of intersection of M_1 and M_2. See margin.

11. Suppose that the planes M_1 and M_2 are defined by the equations

M_1: $3x - 2y + z = 4$
M_2: $x + 2y - z = 3$.

a. Show that M_1 and M_2 are not parallel planes.
b. Find a point $P = (x_0, y_0, z_0)$ on ℓ, the line of intersection of M_1 and M_2. sample: $\left(\frac{7}{4}, 1, \frac{3}{4}\right)$
c. Use the result of Question 10 to determine parametric equations for ℓ. a) (3, -2, 1) and (1, 2, -1), the vectors perpendicular to the planes M_1 and M_2 are not parallel. c) See below.

12. A person has n nickels, d dimes, and q quarters whose total value is \$10.00. What is an equation for the plane containing all possible ordered triples (n, d, q)? $5n + 10d + 25q = 1000$

Review

13. The angle between $\vec{u} = (1, 2, 0)$ and $\vec{v} = (1, 2, p)$ has measure $\frac{\pi}{4}$. Find p. *(Lesson 12-6)* $p = \pm\sqrt{5}$

14. Let $\vec{u} = (u_1, u_2, u_3)$, $\vec{v} = (v_1, v_2, v_3)$, and $\vec{w} = (w_1, w_2, w_3)$. Prove that $\vec{u} \cdot (\vec{v} + \vec{w}) = \vec{u} \cdot \vec{v} + \vec{u} \cdot \vec{w}$. *(Lesson 12-6)* See margin.

15. *Multiple choice.* In 3-space, $\{(x, y, z)\colon x = 0 \text{ and } y = 0\}$ is the
(a) yz-plane (b) xy-plane
(c) x-axis (d) y-axis
(e) z-axis. *(Lesson 12-5)* (e)

16. Suppose that 4% of all students who are well one day are ill the next day, and that 50% of all students who are ill one day are well the next day.
a. Set up a stochastic matrix to model the situation. See margin.
b. In the long run, what proportion of students will be ill on any given day? *(Lesson 11-6)* ≈ 7.4%

17. **a.** Use summation notation to write the sum

$$\left(\frac{1}{n}\right)^2 \cdot \frac{1}{n} + \left(\frac{2}{n}\right)^2 \cdot \frac{1}{n} + \left(\frac{3}{n}\right)^2 \cdot \frac{1}{n} + \ldots + \left(\frac{n}{n}\right)^2 \cdot \frac{1}{n}. \quad \sum_{i=1}^{n} \left(\frac{i}{n}\right)^2 \cdot \left(\frac{1}{n}\right)$$

b. Simplify the sum in part **a**. *(Lesson 7-3)* See below.

18. Solve the equation $\frac{x+3}{3x} - \frac{x+6}{3x+2} = 0$. *(Lesson 5-7)* $x = \frac{6}{7}$

11. c) Sample: $x = \frac{7}{4}$; $y = 1 + 4t$; $z = \frac{3}{4} + 8t$
17. b) $\frac{(n+1)(2n+1)}{6n^2}$

19. a. Sketch the region in the coordinate plane bounded by the x-axis, the line $y = \frac{1}{2}x$, and the line $x = 4$. **See margin.**

b. What geometric figure is generated by rotating this region about the x-axis? **a cone**

c. Find the volume of the figure in part **b.** $\frac{16\pi}{3}$ **cubic units**

(Previous course)

Exploration

20. a. Describe the plane in 3-space whose equation is $5z = 5$.

b. What happens to the plane with equation $ax + 5z = 5$ as a varies from 0 to 5? (Hint: What happens to its perpendicular vectors?)

a) a plane parallel to and 1 unit above the xy-plane; b) As a increases, vectors perpendicular to the plane rotate from the z-axis toward the x-axis, so the plane tilts more steeply.

19.a.

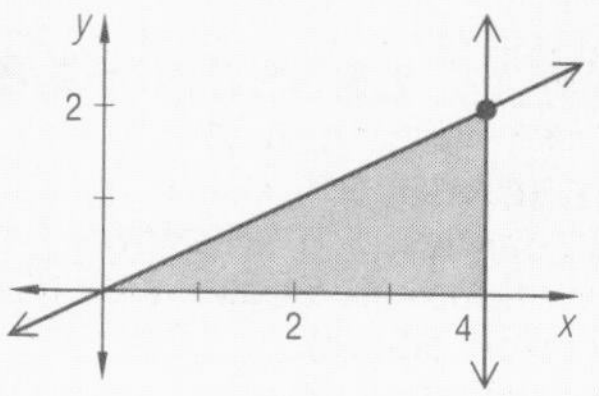

NAME

LESSON **MASTER 12–7**
QUESTIONS ON **SPUR** OBJECTIVES

■ **REPRESENTATIONS** *Objective M* (See pages 756–758 for objectives.)

1. a. Find the intercepts of the plane defined by the equation $x + 4y + 2z = 8$.

$x = 8, y = 2, z = 4$

b. Sketch the plane.

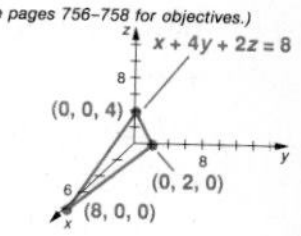

2. Find a vector equation for the line through (2, -1, 3) that is perpendicular to the plane given by $x - 4y + 2z = 8$.

$(x - 2, y + 1, z - 3) = t(1, -4, 2)$

3. Let l be a line passing through the two points (7, 12, -8) and (-4, 6, -9). Describe l with

a. a vector equation. $(x + 4, y - 6, z + 9) = t(11, 6, 1)$

b. parametric equations. $\begin{cases} x = -4 + 11t \\ y = 6 + 6t \\ z = -9 + t \end{cases}$

4. Find a vector perpendicular to the plane defined by the equation $2x + 5y - 3z = 30$. **(2, 5, -3)**

5. Find parametric equations for the line in 3-space through the point (6, -3, 5) that is parallel to the vector (-4, 12, -7). $\begin{cases} x = 6 - 4t \\ y = -3 + 12t \\ z = 5 - 7t \end{cases}$

6. Find an equation for the plane that is perpendicular to $\vec{u}$ = (-3, -2, 14) and that contains the point (7, 0, -1).

$-3(x - 7) - 2y + 14(z + 1) = 0$

138 *Continued*

NAME

Lesson MASTER 12-7 (page 2)

7. Let N be the plane defined by the equation $4x - 3y - 6z = 12$.

a. Sketch N.

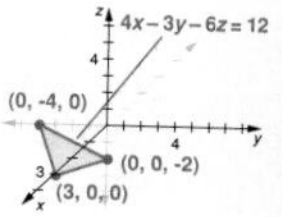

b. Show that (3, 2, -1) is a point on the plane.

4(3) − 3(2) − 6(-1) = 12 − 6 + 6 = 12; since the point satisfies the equation, it is on the plane.

c. Show that the vector $\vec{u}$ = (-2, 1.5, 3) is perpendicular to N.

The vector $\vec{w}$ = (4, -3, -6) is perpendicular to N. Since $\vec{u} = (-2, 1.5, 3) = -\frac{1}{2}(4, -3, -6) = -\frac{1}{2}\vec{w}$, $\vec{u}$ is parallel to $\vec{w}$ and thus perpendicular to N.

d. Find an equation for the plane K that is parallel to N and passes through the point (2, 3, -1).

$4(x - 2) - 3(y - 3) - 6(z + 1) = 0$

 139

LESSON 12-8

RESOURCES

- Teaching Aid 95 shows various intersections of three planes.

TEACHING NOTES

Making Connections
We end this chapter by returning to more familiar topics. Solving systems of two equations in three unknowns to obtain a line parallels the 2-dimensional situation of solving systems of two equations in two unknowns to obtain a point. Students also could compare the cases in 2 and 3 dimensions that result in no solution, infinitely many solutions, or a unique solution.

Regarding the solving of a 3×3 system of linear equations, it is not necessary to strive for mastery. The analysis is done most efficiently in a linear algebra course using row reduction of matrices. Students should simply be aware of the possible types of solutions. A list is given in the Perspectives for this lesson.

Students may think that a linear equation in three variables is the equation of a line. Point out that an equation is linear if each of its variable terms has degree 1, so that
$3x - 1 = 7$,
$2x + y = 12$,
$4x - 6y + z + w = 10$,
and so on are all linear equations.

LESSON

12-8 The Geometry of Systems of Linear Equations in Three Variables

In Lesson 12-7, you saw that a linear equation

$$ax + by + cz = d$$

in which at least one of the coefficients a, b, and c is nonzero, is the equation of a plane M in 3-space, and that the vector

$$\vec{v} = (a, b, c)$$

is perpendicular to M. Consequently, the solution set to a system of k linear equations in three unknowns

$$\begin{cases} a_1x + b_1y + c_1z = d_1 \\ a_2x + b_2y + c_2z = d_2 \\ \;\cdot \quad\; \cdot \quad\; \cdot \quad\; \cdot \\ \;\cdot \quad\; \cdot \quad\; \cdot \quad\; \cdot \\ a_kx + b_ky + c_kz = d_k \end{cases}$$

has a geometric interpretation; it is the set of points common to the k planes defined by the k equations in the system.

When there are two planes, that is, when $k = 2$, then there are three possibilities for the location of the planes: they intersect in a line, or they are parallel and not coincident, or they coincide.

For instance, consider the following system of two equations in three unknowns.

$$\begin{cases} 2x + 5y - z = 6 & (1) \\ x - 8y + 4z = 7 & (2) \end{cases}$$

Its solution is the intersection of the two planes. Notice that plane (1) is perpendicular to the vector (2, 5, -1) and plane (2) is perpendicular to the vector (1, -8, 4). Since these vectors are not parallel, the planes are neither parallel nor do they coincide. So the planes intersect in a line. To describe this line, solve for x and z in terms of y.

$$\begin{aligned} 2x - z &= 6 - 5y \quad (3) \\ x + 4z &= 7 + 8y \quad (4) \end{aligned}$$

To solve for x, eliminate z. Multiply equation (3) by 4 and add the result to equation (4).

$$\begin{aligned} 8x - 4z &= 24 - 20y \qquad 4 \bullet (3) \\ x + 4z &= 7 + 8y \qquad (4) \\ 9x &= 31 - 12y \end{aligned}$$

So, $$x = \frac{31 - 12y}{9}.$$

Now repeat the process to solve for z. This time eliminate x. Multiply equation (4) by -2 and add the result to equation (3).

$$\begin{aligned} -2x - 8z &= -14 - 16y \qquad -2 \bullet (4) \\ 2x - z &= 6 - 5y \qquad (3) \\ -9z &= -8 - 21y \end{aligned}$$

So, $$z = \frac{8 + 21y}{9}.$$

Thus $$(x, y, z) = \left(\frac{31 - 12y}{9}, y, \frac{8 + 21y}{9}\right).$$

Because all three components depend on y, y is a parameter in this situation. If we replace y with t on the right side, the result is parametric equations for the line as given below with t as parameter.

$$\begin{cases} x = \dfrac{31 - 12t}{9} \\ y = t \\ z = \dfrac{8 + 21t}{9} \end{cases}$$

To find a point on the line, pick a value of t. For example, $t = 0$ gives the point $\left(\frac{31}{9}, 0, \frac{8}{9}\right)$, while $t = 2$ gives $\left(\frac{7}{9}, 2, \frac{50}{9}\right)$. These two points satisfy the equations for both planes, as do all points generated by the above parametric equations.

If for two planes,

$$a_1x + b_1y + c_1z = d_1$$

and

$$a_2x + b_2y + c_2z = d_2$$

there is a constant k such that

$$k(a_1, b_1, c_1) = (a_2, b_2, c_2),$$

then the vectors perpendicular to these planes are parallel. Thus the planes are parallel or coincident. If $kd_1 = d_2$, then the equations are equivalent and so the planes coincide. If $kd_1 \neq d_2$, then the planes are parallel and not coincident so there is no point in common.

Now consider a system (*) of 3 linear equations in 3 unknowns.

$$(*) \quad \begin{cases} a_1x + b_1y + c_1z = d_1 \\ a_2x + b_2y + c_2z = d_2 \\ a_3x + b_3y + c_3z = d_3 \end{cases}$$

If the solution set to (*) consists of a single point, then no two of the three planes M_1, M_2, and M_3 defined by these equations respectively, are parallel. This can be determined by examining the vectors perpendicular to the planes, (a_1, b_1, c_1), (a_2, b_2, c_2), and (a_3, b_3, c_3).

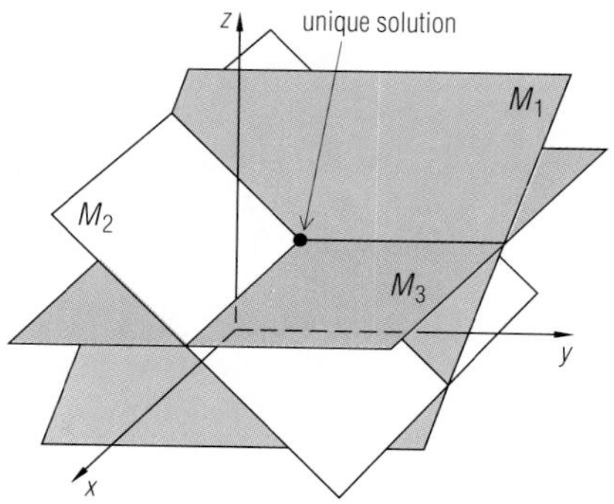

If the system (*) does not have a unique solution, then it either has infinitely many solutions or no solutions. The following three configurations show the geometric situations possible when a system has no solution.

Three parallel planes. All pairs of equations in the system have no solution. For example:

$$\begin{cases} x + \ \ y + 3z = 1 \\ x + \ \ y + 3z = 2 \\ 2x + 2y + 6z = 7 \end{cases}$$

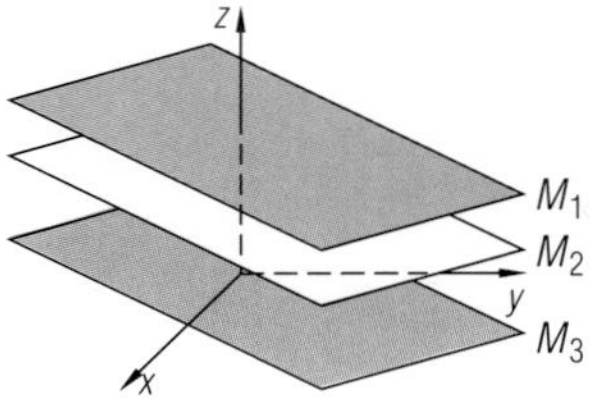

Two parallel planes with the third plane intersecting the parallel pair. One pair of equations in the system has no solution while each of the other two pairs has infinitely many solutions.

For example: $\begin{cases} x + y + 3z = 1 \\ x + y + 3z = 2 \\ 2x - y + \ \ z = 8 \end{cases}$

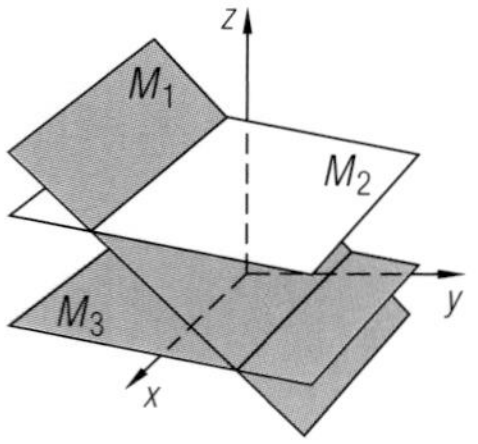

Pairs of planes intersect along parallel lines. Each pair of equations in the system has infinitely many solutions.

For example: $\begin{cases} x + 2y + \ \ z = 4 \\ 2x - \ \ y + \ \ z = 3 \\ 3x + \ \ y + 2z = -2 \end{cases}$

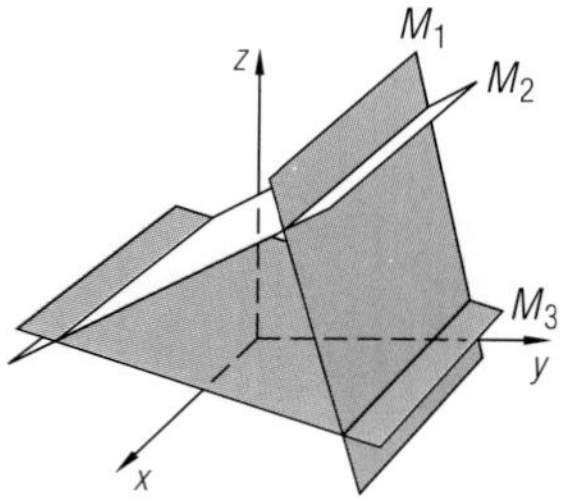

You are asked to explore the situation where the system (*) has infinitely many solutions in Question 6.

Questions

Covering the Reading

1. **a.** Solve the equations for planes (1) and (2) on page 746 for x and y in terms of z.
   b. Show that the line of intersection is the same line as found on page 747.
   See margin.
2. **a.** Sketch the planes corresponding to the two equations in the system $\begin{cases} x + y + z = 1 \\ \quad\ \ y - z = 0 \end{cases}$.
   b. Give parametric equations for the line of intersection of the planes.
   See margin.
3. Consider the plane M with equation $3x - 2y + 4z = 1$.
   a. Write an equation for a plane N that is parallel to (but not coincident with) M. sample: $3x - 2y + 4z = 2$
   b. Solve the system consisting of the equations for M and N to show that there is no solution. See margin.

NOTES ON QUESTIONS

Question 2: If students have difficulty making the drawing, suggest that they make individual sketches of the two planes, and then combine them into one drawing.

Questions 4 and 5: Suggest students determine whether the planes are parallel, coincident, or intersecting before trying to solve the system. Again, they should make use of the fact that if planes are parallel or coincident, then their coefficients will be proportional.

Question 12: The parabola goes through the points on the graph of the sequence of triangular numbers.

Question 13: In geometrical terms, this problem asks for the point of intersection of three planes.

ADDITIONAL ANSWERS

1.a. $x = \frac{-12z + 83}{21}$, $y = \frac{9z - 8}{21}$

b. When $z = \frac{8}{9}$, the point $\left(\frac{31}{9}, 0, \frac{8}{9}\right)$ is generated; when $z = \frac{50}{9}$, the point $\left(\frac{7}{9}, 2, \frac{50}{9}\right)$ is derived. These points are the same as those on the intersection line shown on page 747. Since two points determine a line, this is the same line.

2.a.

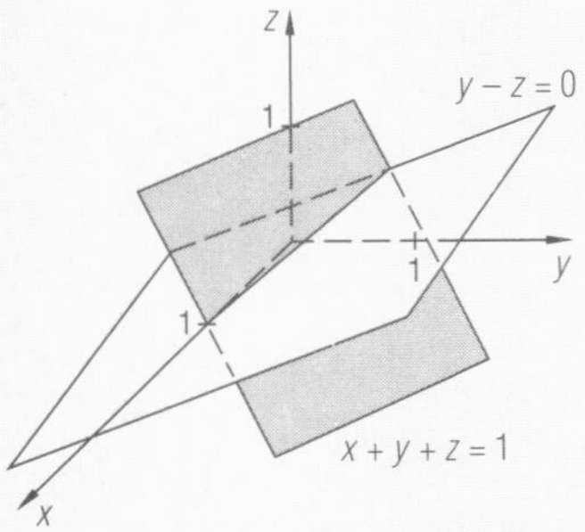

b. sample: $\begin{cases} x = t \\ y = \frac{1}{2} - \frac{1}{2}t \\ z = \frac{1}{2} - \frac{1}{2}t \end{cases}$

748

In 4–5, describe all solutions to the system. See margin.

4. $\begin{cases} 5x - 3y + z = 12 \\ x + y - 4z = 1 \end{cases}$ **5.** $\begin{cases} -x + 3y - 2z = -6 \\ 2x - 6y + 4z = 12 \end{cases}$

6. When the system (*) in this lesson has infinitely many solutions, draw all possible configurations of the three planes M_1, M_2, and M_3. See margin.

7. Consider the following system of equations given in the lesson:

$$\begin{cases} x + y + 3z = 1 \\ x + y + 3z = 2 \\ 2x - y + z = 8. \end{cases}$$

a. Which pair of equations corresponds to parallel planes? How do you know?

b. How can you tell that the other equation corresponds to a plane that intersects the parallel planes? a,b) See margin.

8. Consider the following system given in the lesson:

$$\begin{cases} x + 2y + z = 4 \\ 2x - y + z = 3 \\ 3x + y + 2z = -2. \end{cases}$$

a. Show that the system has no solution. See margin.

b. Choose one pair of the equations, and show that their graphs intersect in a line. The first two equations intersect at the line $x = \frac{10 - 3t}{5}$, $y = \frac{5 - t}{5}$, $z = t$.

In 9–11, describe the graph of all solutions to the system.

9. $\begin{cases} a + b + c = 1 \\ 4a + 2b + c = 3 \\ 9a + 3b + c = 6 \end{cases}$ the point $\left(\frac{1}{2}, \frac{1}{2}, 0\right)$

10. $\begin{cases} 5u - 3v = w \\ 10u - 2w = 6v \\ u + v = 9 \end{cases}$ the line $\begin{cases} u = t \\ v = 9 - t \\ w = -27 + 8t \end{cases}$

11. $\begin{cases} x + y - z = 2 \\ x + 3y - z = 16 \\ -x + y + z = 12 \end{cases}$ the line $\begin{cases} x = 0 + t = t \\ y = 7 + 0t = 7 \\ z = 5 + t \end{cases}$

12. The points (1, 1), (2, 3), and (3, 6) are on the graph of $y = ax^2 + bx + c$. Find a, b, and c. $a = b = \frac{1}{2}$, $c = 0$

13. To get from the O'Neill farm to the city, you must travel d miles over a dirt road, g miles over a gravel road, and h miles over a highway. When Mr. O'Neill averages 20 mph on dirt, 15 mph on gravel, and 60 mph on the highway, it takes him 68 minutes to get to the city. Mrs. O'Neill averages 5 mph slower than Mr. O'Neill on dirt and gravel, but averages 60 mph on the highway, and it takes her 82 minutes. The total distance is 42 miles. How many of these miles are over each surface?
d = 10 mi, g = 2 mi, h = 30 mi

3.b.
$3x - 2y + 4z = 2$ **plane** N
$3x - 2y + 4z = 1$ **plane** M
$0 = 1$
Therefore, the system has no solution.

4. the line $\begin{cases} x = \frac{49 + 11t}{21} \\ y = t \\ z = \frac{7 + 8t}{21} \end{cases}$

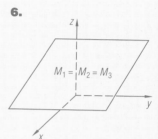

5. the plane $-x + 3y - 2z = -6$

6.

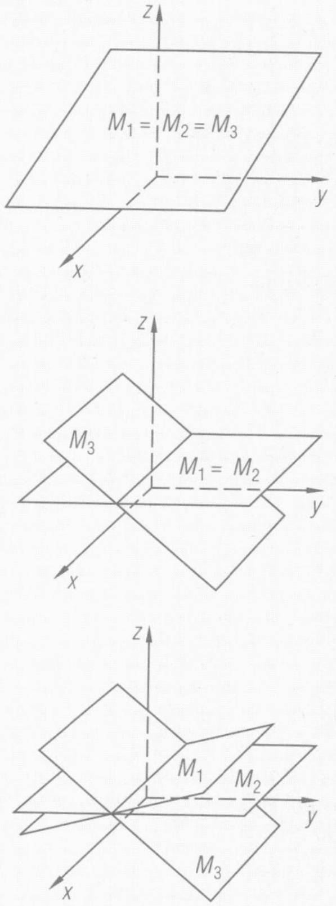

7.a. the first and second equations, because the vector (1, 1, 3) is perpendicular to both
b. (2, -1, 1) is perpendicular to the third plane, so that plane is not parallel to the first two, and must then intersect both of them.

8.a. See the margin on page 750.

FOLLOW-UP

EXTENSION

As an individual project, some students may wish to consult a book on linear algebra to find the answers to the following questions. How are vectors used to describe hyperplanes? What are the possible intersections of 4-dimensional hyperplanes?

ADDITIONAL ANSWERS

8.a. Subtracting the second equation from the first, we get $-x + 3y = 1$. Subtracting the third equation from twice the second equation, we get $x - 3y = 8$, or $x = 3y - 8$. Substituting for x, $-(3y - 8) + 3y = 1$, which simplifies to $8 = 1$. Hence, there is no solution to this system.

Review

14. Describe the graph of the set of points (x, y, z) such that $y + z = 10$. *(Lesson 12-7)* a plane parallel to the x-axis containing (0, 0, 10) and (0, 10, 0)

15. Find parametric equations for the line ℓ passing through (-7, 2, 0) and perpendicular to the plane with equation $3x - 4y - z = -1$. *(Lesson 12-7)* See below.

16. a. Find an equation for the plane containing the point (-7, 2, 0) which is perpendicular to line ℓ given in Question 15.

b. How is this plane related to the plane in Question 15? *(Lesson 12-7)* a) $3x - 4y - z = -29$ b) The planes are parallel.

17. Explain why three nonzero three-dimensional vectors $\vec{u_1}$, $\vec{u_2}$, and $\vec{u_3}$ are coplanar if and only if $\vec{u_1} \cdot (\vec{u_2} \times \vec{u_3}) = 0$. *(Lesson 12-7)* See below.

18. Suppose g is a polynomial function such that
$g'(x) < 0$ when $-4 < x < -1$ or $x > 4$
and $g'(x) > 0$ when $x < -4$ or $-1 < x < 4$.
What is the smallest possible degree of g? *(Lessons 9-5, 4-6)* 4

19. Is the following argument valid? Explain.
If school is not closed, then I go to school.
I don't go to school.
∴ *School is closed.* *(Lessons 1-8, 1-6)*
Yes, it is an example of modus tollens.

Exploration

20. a. Find an equation for the plane containing the lines ℓ_1 and ℓ_2 with the following parametric equations.

$$\ell_1: \begin{cases} x = 1 + 4t \\ y = 2 - 3t \\ z = 3 + t \end{cases} \qquad \ell_2: \begin{cases} x = 1 - 2t \\ y = 2 + t \\ z = 3 + 4t \end{cases}$$

b. Generalize part **a** to find an equation for the plane containing the lines with the following parametric equations.

$$\ell_1: \begin{cases} x = x_0 + u_1t \\ y = y_0 + u_2t \\ z = z_0 + u_3t \end{cases} \qquad \ell_2: \begin{cases} x = x_0 + v_1t \\ y = y_0 + v_2t \\ z = z_0 + v_3t \end{cases}$$

a) $13x + 18y + 2z = 55$; b) $(\vec{u} \times \vec{v}) \cdot (x - x_0, y - y_0, z - z_0) = 0$ or $(u_2v_3 - u_3v_2)x + (u_3v_1 - u_1v_3)y + (u_1v_2 - u_2v_1)z = (u_2v_3 - u_3v_2)x_0 + (u_3v_1 - u_1v_3)y_0 + (u_1v_2 - u_2v_1)z_0$

15. $\begin{cases} x = -7 + 3t \\ y = 2 - 4t \\ z = -t \end{cases}$

17. $\vec{u_2} \times \vec{u_3}$ is perpendicular to both $\vec{u_2}$ and $\vec{u_3}$. $\vec{u_1} \cdot (\vec{u_2} \times \vec{u_3}) = 0$ if and only if $\vec{u_1}$ is also perpendicular to $\vec{u_2} \times \vec{u_3}$. Thus $\vec{u_1}$, $\vec{u_2}$, and $\vec{u_3}$, being all perpendicular to the same vector, are coplanar.

Projects

1. Simulate your route from home to school on a computer by doing the following.
   a. Draw a map of your route from home to school. (Change any curved paths to segments). Choose a measurement system (miles, feet, kilometers, or meters) which seems appropriate. Use your home as the origin of a coordinate system with North being the positive part of the y-axis and East being the positive part of the x-axis.
   b. Write each segment of your trip as a vector in rectangular form. Verify that the sum of the vectors gives the location of your school with respect to your home.
   c. Repeat part **b**, but this time write each segment of your path as a vector in polar form. Use this form to write instructions in English giving direction and distance for each segment of your path to school.
2. In the water sport of windsurfing, the windsurfer usually sets the boom at shoulder height and holds the boom perpendicular to his or her arms. There are three main forces, P, W, and R, which must balance in order for the windsurfer not to fall. Each force can be represented by a vector. P is the pull in the arms (a force perpendicular to the boom), W is the weight of the windsurfer (a force downward), and R is the resultant reaction between the board and the windsurfer's feet. Let ϕ be the angle from the horizontal of his or her arms, and let θ be the angle between the force on the board and the horizontal. ϕ can be accurately measured from photographs or observations, but θ cannot because the windsurfer might bend his or her legs or put most pressure on the balls of his or her feet. (However, θ can be conjectured from other angles.) If the windsurfer does not fall off the board, then the following relationships must hold.
   $R \cos \theta - P \cos \phi = 0$
   $R \sin \theta + P \sin \phi = W$
   a. Use vectors to explain why the equations must hold.
   b. Solve the equations for P. You will get an expression involving the weight of the surfer and trigonometric functions of θ and ϕ.
   c. Assume that a value of P that equals more than half the weight of the surfer is hard to sustain over any length of time. Compute P for the following situations and determine which would be hard.
      i. Weight of surfer = 140 lb, $\phi = 45°$, $\theta = 50°$.
      ii. Your weight in pounds, $\phi = 45°$, $\theta = 50°$.
      iii. Your weight in pounds, $\phi = 30°$, $\theta = 70°$.
   d. Select values for ϕ and θ which you might believe to be reasonable in a very strong wind, and compute P for these values.

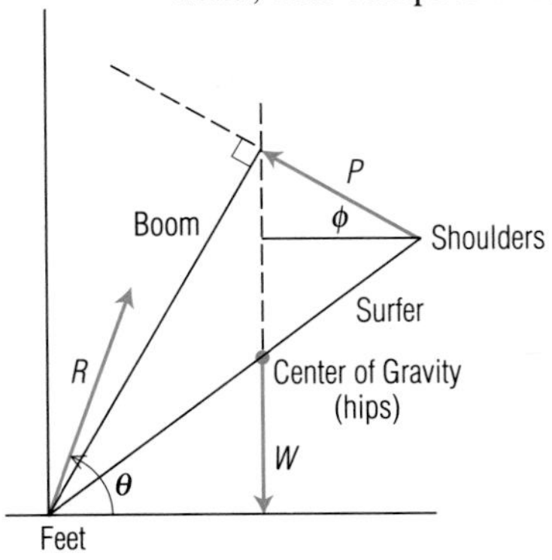

PROJECTS

The projects provide an opportunity for extended work on topics related to the content of the chapter. (See pages 70 and T37 for more information on the purposes of projects.) Some or all of these projects may be suitable for **small group work**. Solutions for the projects can be found at the end of the Additional Answers section in the back of this book.

If it is impossible to obtain the data for part **c** of **Project 4**, you could make up a data set, or ask the students to make up one of their own. To solve part **c**, students may need to expand some of the ideas from **Project 3**.

3. This chapter discussed vectors in two and three dimensions. Develop formulas for the following in four dimensions:

sum of vectors,
distance between points,
length of a vector,
dot product,
scalar product,
angle between vectors,
determination of perpendicular vectors, and
parametric equation of a line.

(Picturing these is difficult. But surprisingly, four-dimensional vectors simplify computational procedures in computer graphics.)

4. Given a vector, its corresponding *adjusted vector* is found by subtracting from each of its coordinates the average of those coordinates. For example, the vector (3, 2, 4) has the adjusted vector (3 − 3, 2 − 3, 4 − 3) = (0, -1, 1). Given two vectors, their *correlation coefficient* is the cosine of the angle between their adjusted vectors. Do two of the following.

a. Look up the formula for correlation coefficient in a statistics book. Relate the formula given in the book to the cosine of the angle between adjusted vectors. Report the formula and an example from the statistics book that shows how data is used with it. Note that some books might not give the formula directly. Also, many books use capital letters (such as X_i) for data coordinates, bars over the letters to indicate means (such as $\overline{X}$), and small letters to represent data adjusted by the mean (such as $x_i = X_i - \overline{X}$).

b. Write a computer program which inputs two vectors and calculates their correlation coefficient.

c. Ask your counselor for SAT Quantitative and Verbal scores (or other pairs of scores) on seven to ten randomly selected people. (Of course, you should not ask for the names of the students.) Let the coordinates of one vector be the quantitative scores, and the coordinates of a second vector be the verbal scores. Use the formula from part **a** or the computer program in part **b** to calculate their correlation coefficient. Compute the angle between the adjusted vectors. If the angle is under 30°, then the scores are considered to be "highly correlated" and are measuring nearly the same thing. If the angle is over 60°, then the scores are measuring relatively independent characteristics. Your result, of course, is highly dependent on the people chosen for your sample.

Summary

A vector is a quantity with magnitude and direction. Vectors have applications in physics and engineering, for instance, in their use to represent velocity and force. Two- and three-dimensional vectors were discussed in this chapter, although the concept of vector can be extended to any number of dimensions.

In two dimensions, vectors can be represented as ordered pairs of real numbers (a, b), polar coordinates $[r, \theta]$, or arrows. These representations are related by the equations $r = \sqrt{a^2 + b^2}$ and $(a, b) = (r \cos \theta, r \sin \theta)$, where r is the length (or norm or magnitude) of the vector, and θ is its direction.

Sums, opposites, and scalar multiples of vectors can be found, respectively, by adding corresponding components, taking the opposites of components, and multiplying each component by the same real number.

Vector operations have graphical representations. To add vectors in a coordinate plane, place them in standard position. They then form adjacent sides of a parallelogram. The sum is the diagonal of the parallelogram with one endpoint at the origin. The opposite of a vector is the collinear vector of the same length drawn in the opposite direction. If $\vec{v}$ is a vector and $k > 0$, the scalar multiple $k\vec{v}$ has the same direction as $\vec{v}$ but is k times as long.

Vector operations also have physical interpretations. If vectors represent forces, their sum represents the resultant force. Their opposites represent forces of equal magnitude in the opposite direction. Scalar multiples represent forces k times as intense in the same direction.

Two vectors are parallel if and only if one is a scalar multiple of the other. As a result, the vector equation of the line through a point P parallel to a vector $\vec{v}$ is $\overrightarrow{PQ} = k\vec{v}$, where Q is a variable point on the line. Equating components yields parametric equations for the line.

The dot product of two vectors $\vec{u}$ and $\vec{v}$, denoted by $\vec{u} \cdot \vec{v}$, is the sum of the products of corresponding coordinates. The Law of Cosines implies that the angle θ between $\vec{u}$ and $\vec{v}$ can be found from $|\vec{u}|\ |\vec{v}| \cos \theta = \vec{u} \cdot \vec{v}$. Two vectors are orthogonal if and only if their dot product is zero.

The formula for the distance between two points in 3-space leads to the equation of a sphere: $(x - a)^2 + (y - b)^2 + (z - c)^2 = r^2$, where (a, b, c) is the center and r is the radius.

The concepts of addition, scalar multiplication, opposites, length, and the angle between two vectors can be generalized to three dimensions. The criteria for parallel and orthogonal vectors is unchanged from two dimensions, and vector and parametric equations for a line through a point parallel to a vector are essentially the same as in two dimensions. The geometric idea of finding a perpendicular to the plane determined by two nonzero, noncollinear vectors gives rise to the operation of cross product. The fact that the dot product of two orthogonal vectors is 0 leads to the result that a plane perpendicular to the vector $\vec{v} = (a, b, c)$ has an equation of the form $ax + by + cz = d$.

The solution to a system of linear equations in three variables is the intersection of the planes corresponding to those equations. A unique solution exists if and only if the planes intersect at a single point. No solution exists if two of the planes are parallel or if each pair of planes intersects at a different parallel line. An infinite number of solutions exists if the intersection of all the planes is a straight line or if the planes are identical.

SUMMARY

The Summary gives an overview of the entire chapter and provides an opportunity for students to consider the material as a whole. Thus, the Summary can be used to help students relate the various concepts presented in the chapter.

VOCABULARY

Terms, symbols, and properties are listed by lesson to provide a checklist of concepts a student must know. (See page 70 for more information.)

ADDITIONAL ANSWERS

7. (-10, -24, -4)

9.

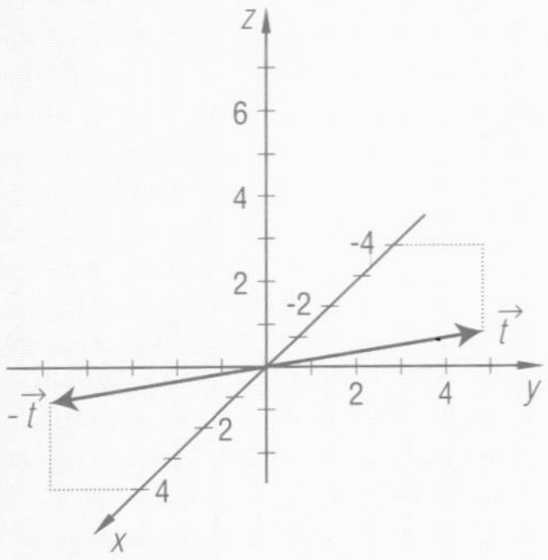

10.a.

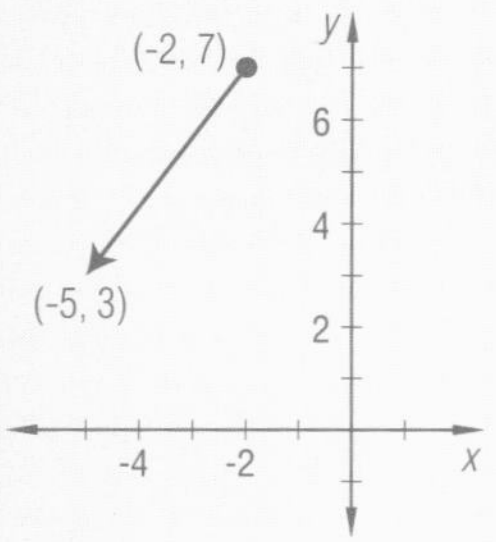

b. (-3, -4)
c. $\approx$ **[5, 233°]**

Vocabulary

For the starred (*) terms you should be able to give a definition of the term.
For the other terms you should be able to give a general description and a specific example of each.

Lesson 12-1
vector
initial point, endpoint
plane vector, two-dimensional vector
polar representation of a vector
magnitude, direction of a vector
standard position
component representation
x- and y-components
horizontal and vertical components
*norm of a vector, $|\vec{u}|$
zero vector
unit vector

Lesson 12-2
*sum of two vectors, $\vec{u} + \vec{v}$
force
resultant force
*opposite of a vector, $-\vec{v}$
difference of vectors, $\vec{u} - \vec{v}$

Lesson 12-3
*parallel vectors
scalar
*scalar multiple, $k\vec{v}$
vector equation of a line
parameter
parametric equations for a line

Lesson 12-4
*dot product, $\vec{u} \cdot \vec{v}$
Angle Between Vectors Theorem
orthogonal vectors

Lesson 12-5
3-space, three-dimensional space
xy-plane, xz-plane, yz-plane
Distance in Space Theorem
Equation of a Sphere Theorem

Lesson 12-6
sum, scalar multiples, dot product of three-dimensional vectors
cross product, $\vec{u} \times \vec{v}$

Lesson 12-7
vector equation of a line in 3-space
parametric equations for a line in 3-space
Equation of a Plane Theorem

Progress Self-Test

See margin for answers not shown below.

Take this test as you would in class. Then check the test yourself using the solutions at the back of the book.

In 1–5, let $\vec{u} = (3, -4)$ and $\vec{v} = (-5, 2)$, and compute.

1. $|\vec{v}|$ $\sqrt{29}$ **2.** $\vec{u} - \vec{v}$ (8, -6)

3. $-2\vec{u} + 5\vec{v}$ (-31, 18) **4.** $\vec{u} \cdot \vec{v}$ -23

5. the angle between $\vec{u}$ and $\vec{v}$ $\approx 149°$

In 6–9, let $\vec{s} = (-2, 0, 5)$ and $\vec{t} = (-4, 2, -2)$. Find the component representation of the indicated vector.

6. $\vec{s} + 3\vec{t}$ (-14, 6, -1)

7. a vector orthogonal to both $\vec{s}$ and $\vec{t}$

8. a vector parallel to $\vec{s}$ sample: (-4, 0, 10)

9. Sketch the vectors $\vec{t}$ and $-\vec{t}$ in 3-space. How do their graphs compare?

10. **a.** Sketch the vector $\vec{v}$ from the point (-2, 7) to the point (-5, 3).

b. Find the component representation of $\vec{v}$.

c. Find the polar representation of $\vec{v}$.

11. Prove that if $\vec{v}$ and $\vec{w}$ are two-dimensional vectors and k is a real number, then $k(\vec{v} - \vec{w}) = k\vec{v} - k\vec{w}$.

12. Find z so that (2, -4, 3) is orthogonal to (3, -2, z). $\frac{-14}{3}$

13. According to its instrument panel, an airplane is traveling 25° west of north at an airspeed of 270 mph. Suppose a 15 mph wind is coming from the south.

a. Find the airplane's actual speed and direction.

b. Interpret the x- and y-components of the vector representing the airplane's actual velocity in terms of its motion.

In 14 and 15, sketch the vector using the vectors drawn below.

14. $\vec{u} + \vec{v}$ **15.** $\vec{u} - \vec{v}$

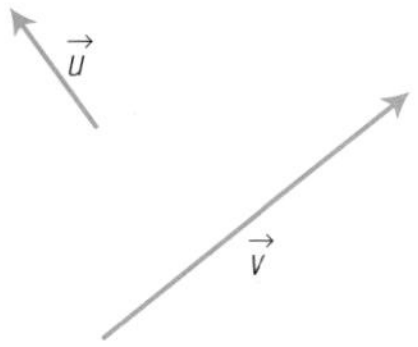

16. Write parametric equations for the line through the point $P = (-3, 2)$ parallel to the vector $\vec{v} = (4, -3)$. $\begin{cases} x = -3 + 4t \\ y = 2 - 3t \end{cases}$

17. Let Q be the point (4, 0, -2) in 3-space and $\vec{u}$ be the vector (1, 2, -3). Write an equation for the plane through Q perpendicular to $\vec{u}$.

18. Write an equation for the sphere with center (2, 0, -3) and radius 4.
$(x - 2)^2 + y^2 + (z + 3)^2 = 16$

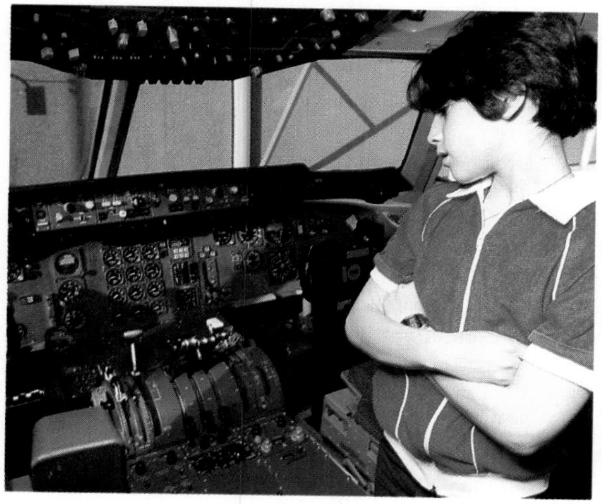

PROGRESS SELF-TEST

The Progress Self-Test provides the opportunity for feedback and correction; the Chapter Review provides additional opportunities for practice.

USING THE PROGRESS SELF-TEST

Assign the Progress Self-Test as a one-night assignment. (See pages 72 and T35 for more information.)

11. See Additional Answers in the back of this book.

13.a. speed $\approx$ 284 mph and direction $\approx$ 23.7° West of North.

b. The actual velocity of the plane is (-114, 260), which means the plane is traveling 114 mph west and 260 mph north.

14.

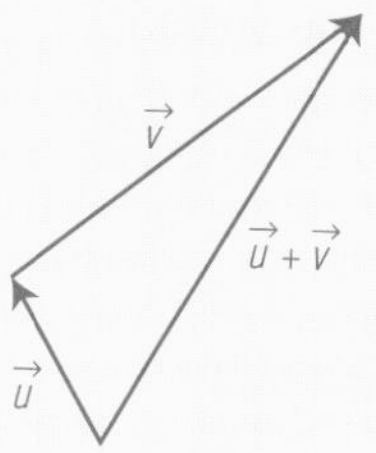

15.

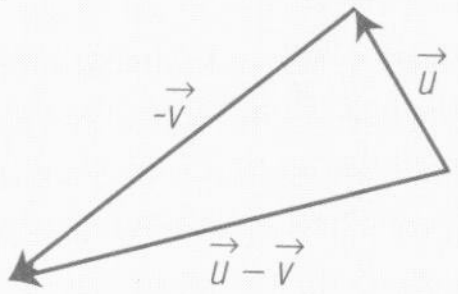

17. $x + 2y - 3z = 10$

RESOURCES
- Chapter 12 Test, Form A
- Chapter 12 Test, Form B
- Chapter 12 Test, Cumulative Form
- Quiz and Test Writer

CHAPTER REVIEW

The main objectives for the chapter are organized here into sections corresponding to the four main types of understanding this book promotes: Skills, Properties, Uses, and Representations. We call these the SPUR objectives.

USING THE CHAPTER REVIEW

Students should be able to answer questions like these with about 85% accuracy by the end of the chapter. (See pages 74 and T38–39 for more information.)

ADDITIONAL ANSWERS

2. $\approx 18.4°$ and $198.4°$

10. sample: $[4.14, 78.0°]$

11. sample: $[3, 285°]$

12. sample: $[3, 35°]$

13. sample: $[20, 35°]$

23. $\vec{v} \cdot (\vec{u} \times \vec{v}) = (-2, -3, 4) \cdot (11, 18, 19) = -22 - 54 + 76 = 0$

26. $\approx 146.8°$

27. 180°; they have opposite directions.

28. $\vec{u} \cdot (a\vec{v}) = (u_1, u_2) \cdot (av_1, av_2) = u_1av_1 + u_2av_2 = a(u_1v_1 + u_2v_2) = a(\vec{u} \cdot \vec{v})$

29. The vector from (x, y) to $(x + a, y + ma)$ is $(a, ma) = a(1, m)$. Therefore, by definition, the vectors are parallel.

Chapter Review

Questions on SPUR Objectives

SPUR stands for **S**kills, **P**roperties, **U**ses, and **R**epresentations.
The Chapter Review questions are grouped according to the SPUR Objectives for this chapter.
See margin for answers not shown below.

SKILLS deal with the procedures used to get answers.

Objective A: *Find the magnitude and direction of two-dimensional vectors.* (Lesson 12-1)

1. Find the magnitude and direction of the vector $(11, -3)$. magnitude: $\sqrt{130}$ direction: $\approx -15.3°$

2. Find all possible directions for a vector whose x-component is three times its y-component.

3. Let $A = (2, 3)$, $B = (4, t)$, and $\vec{s}$ be the vector from A to B. If $|\vec{s}| = \sqrt{40}$, give the two possible values of t. 9 and -3

4. Find the polar representation of the vector $(2, -5)$. $\approx [\sqrt{29}, 292°]$

Objective B: *Find sums, opposites, scalar products, and dot products of two-dimensional vectors.* (Lessons 12-2, 12-3, 12-4)

In 5–9, let $\vec{u} = (3, -2)$, $\vec{v} = (-5, 0)$, and $\vec{w} = (2, 6)$, and compute.

5. $\vec{u} + \vec{v}$ (-2, -2)
6. $-3\vec{u}$ (-9, 6)
7. $\vec{v} \cdot \vec{w}$ -10
8. $\vec{u} - \vec{w}$ (1, -8)
9. $4\vec{w} - 2\vec{u}$ (2, 28)

In 10–13, let $\vec{r} = [2, 35°]$, $\vec{s} = [3, 105°]$, and $\vec{t} = [5, 35°]$. Compute and express the answer in its polar representation.

10. $\vec{r} + \vec{s}$
11. $-\vec{s}$
12. $\vec{t} - \vec{r}$
13. $4\vec{t}$

14. If $\vec{v} = (3, -5)$ and $\vec{w} = (-2, 8)$, find the components of $\vec{u}$ so that $\vec{u} + \vec{v} = \vec{w}$. (-5, 13)

15. If $\vec{u}$ is the vector from (-2, 6) to (-5, 8), find the vector parallel to $\vec{u}$ and with half the magnitude. sample: $\left(-\frac{3}{2}, 1\right)$

16. Find the polar representation of a vector whose length is two-thirds that of $[12, 125°]$ and which has the opposite direction. $[8, 305°]$

Objective C: *Find sums, lengths, scalar products, dot products, and cross products of vectors in 3-space.* (Lesson 12-6)

In 17–22, let $\vec{u} = (-5, 2, 1)$ and $\vec{v} = (-2, -3, 4)$, and compute.

17. $\vec{u} + \vec{v}$ (-7, -1, 5)
18. $\vec{u} - \vec{v}$ (-3, 5, -3)
19. $2\vec{v} - 3\vec{u}$ (11, -12, 5)
20. $|\vec{u}|$ $\sqrt{30}$
21. $\vec{u} \cdot \vec{v}$ 8
22. $\vec{u} \times \vec{v}$ (11, 18, 19)

23. For the vectors $\vec{u}$ and $\vec{v}$ in Questions 17–22, verify that $\vec{u} \times \vec{v}$ is orthogonal to $\vec{v}$.

24. Find a vector orthogonal to both $\vec{s} = (3, 0, -1)$ and $\vec{t} = (1, -2, 4)$. sample: (-2, -13, -6)

Objective D: *Find the measure of the angle between two vectors.* (Lessons 12-4, 12-6)

25. Find the measure of the angle between the vectors $\vec{u} = (-5, 3)$ and $\vec{v} = (1, -6)$. $\approx 130.4°$

26. Find the measure of the angle between the vectors $\vec{s} = (2, 0, -4)$ and $\vec{t} = (-3, 1, 2)$.

27. Let $\vec{u} = (1, 2, -3)$ and $\vec{v} = -3\vec{u}$. Find the measure of the angle between $\vec{u}$ and $\vec{v}$. What does your answer mean?

PROPERTIES deal with the principles behind the mathematics.

Objective E: *Prove or disprove statements about vector operations.* (Lessons 12-2, 12-3, 12-4, 12-6)

28. If a is a real number and $\vec{u}$ and $\vec{v}$ are vectors in a plane, prove that $\vec{u} \cdot (a\vec{v}) = a(\vec{u} \cdot \vec{v})$.

29. Show that the vector $(1, m)$ is parallel to the vector from (x, y) to $(x + a, y + ma)$.

30. Prove that if $\vec{v}$ is a vector in 3-space and k and m are real numbers, then $(k + m)\vec{v} = k\vec{v} + m\vec{v}$.

31. When $\vec{u} = (1, 0, 1)$ and $\vec{v} = (0, 2, 3)$, does $\vec{u} \times \vec{v} = \vec{v} \times \vec{u}$? No

32. For any two vectors $\vec{u}$ and $\vec{v}$ in 3-space and any nonzero real number k, show that if $\vec{u}$ is orthogonal to $\vec{v}$, then $k\vec{u}$ is orthogonal to $\vec{v}$.

Objective F: *Identify parallel and orthogonal vectors. (Lessons 12-3, 12-4, 12-6)*

In 33–35, determine whether $\vec{u}$ and $\vec{v}$ are perpendicular, parallel, or neither.

33. $\vec{u} = (-7, 3)$ and $\vec{v} = (-2, 5)$ neither

34. $\vec{u} = (-6, 3)$ and $\vec{v} = (4, -2)$ parallel

35. $\vec{u} = (5, -3, 4)$ and $\vec{v} = (2, -2, -4)$

In 36 and 37, find x so that the vectors $\vec{s} = (x, 2, -4)$ and $\vec{t} = (15, -3, 6)$ are

36. orthogonal 2

37. parallel. -10

38. Find all vectors orthogonal to $\vec{u} = (2, -3)$ with magnitude 5.

USES deal with applications of mathematics in real situations.

Objective G: *Use vectors in a plane to decompose motion or force into x- and y-components. (Lesson 12-1)*

39. A ship's velocity is represented by [16, 25°], where the magnitude is measured in miles per hour and the direction is in degrees north of east.
   a. Sketch the vector for the velocity.
   b. Give the vector in component form.
   c. Interpret the components.

40. A kite is at the end of a 50-meter string. Its angle with the horizontal is 52°. Assume the string is straight.
   a. Write a polar representation for the kite's position. [50, 52°]
   b. Compute a component representation.
   c. Interpret the components.

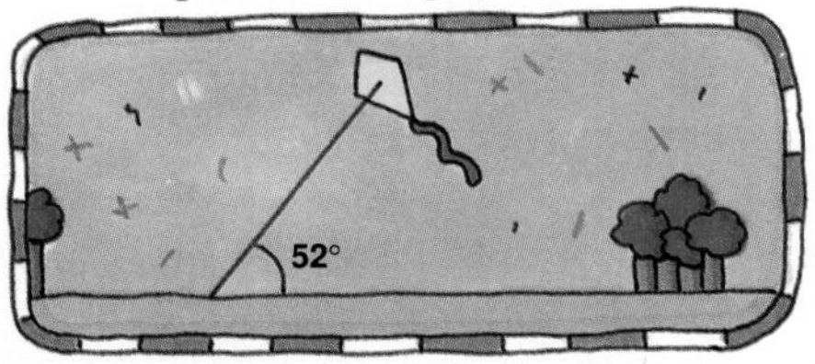

Objective H: *Use addition of vectors in a plane to solve problems involving forces or velocities. (Lesson 12-2)*

41. A plane travels at an airspeed of 400 km per hour with compass heading northwest. It encounters a 60-km/hr wind from the north.
   a. Graph a vector for the plane's airspeed and compass heading, and another for the velocity of the wind.
   b. Graph the resultant vector for the plane.
   c. Compute and interpret the polar representation of the plane's actual velocity.

42. George and Sarah push a go-cart containing Sam to the starting line of a race. George pushes with 20 lb of force, Sarah with 18 lb of force. Their angles with respect to a line behind the cart are shown.
   a. What is the magnitude and direction of the resultant force?
   b. Who contributes more to the forward force, George or Sarah? Sarah

43. A ferry leaves a pier for a dock one mile directly north across the river. The river current is 2 mph heading east. If the ferry boat captain maintains a speed of 10 mph in the direction 10° west of north, will the boat reach its destination? If so, how long will it take? No

44. An airplane pilot wishes to achieve an actual speed of 300 mph in the direction 25° south of east. If a 35 mph wind is blowing from the southwest, what airspeed and compass bearing should the pilot maintain on the instrument panel? ≈ 290 mph at 31.5° South of East

30. $(k + m)\vec{v} = (k + m)(v_1, v_2, v_3) = (kv_1 + mv_1, kv_2 + mv_2, kv_3 + mv_3) = (kv_1, kv_2, kv_3) + (mv_1, mv_2, mv_3) = k\vec{v} + m\vec{v}$

32. $(k\vec{u}) \cdot \vec{v}$
$= (ku_1, ku_2, ku_3) \cdot (v_1, v_2, v_3)$
$= ku_1v_1 + ku_2v_2 + ku_3v_3$
$= k(u_1v_1 + u_2v_2 + u_3v_3)$
$= k(\vec{u} \cdot \vec{v})$
Therefore, $k\vec{u}$ is orthogonal to $\vec{v}$ if $\vec{u}$ is orthogonal to $\vec{v}$.

35. perpendicular

38. $\left(\frac{15}{13}\sqrt{13}, \frac{10}{13}\sqrt{13}\right)$ **and** $\left(-\frac{15}{13}\sqrt{13}, -\frac{10}{13}\sqrt{13}\right)$

39.a.

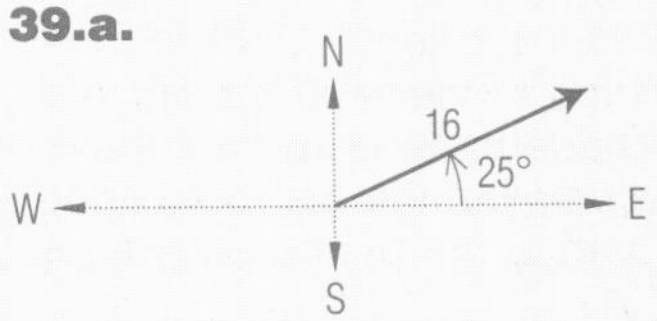

b. (14.5, 6.76)
c. The ship is going 14.5 mph towards the east and 6.76 mph towards the north.

40.b. (30.8, 39.4)
c. The kite is 39.4 m above a spot on the ground, which is 30.8 m away from the owner.

41.a., b.

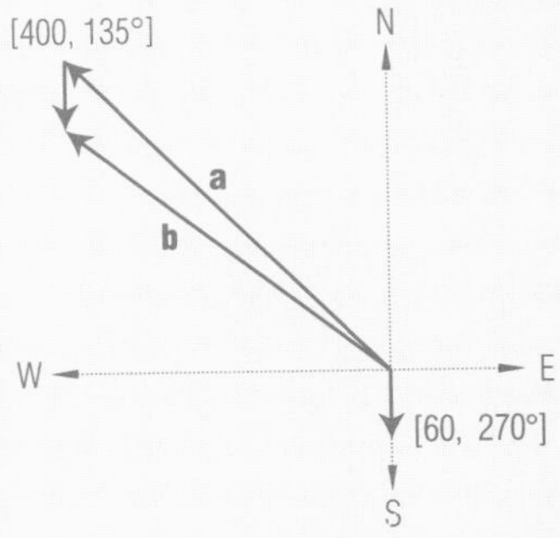

c. [360, 141.8°]; Relative to the ground, the plane's speed is 360 km/hr, and its heading is 38.2° North of West.

42.a. **See the margin on page 758.**

EVALUATION
Three forms of a Chapter Test are provided in the Teacher's Resource File. Chapter 12 Test, Forms A and B cover just Chpater 12. The third test is Chapter 12 Test, Cumulative Form. About 50% of this test covers Chapter 12, 25% covers Chapter 11, and 25% covers previous chapters. For information on grading, see *General Teaching Suggestions: Grading* on page T43 in the Teacher's Edition.

ASSIGNMENT RECOMMENDATION
We strongly recommend that you assign the reading and some questions from Lesson 13-1 for homework the evening of the test. It gives students work to do if they complete the test before the end of the period and keeps the class moving.

ADDITIONAL ANSWERS
42.a. 32.1 lb of force with direction 80.6° counterclockwise from the positive x-axis

45. (-2.5, 4.3)

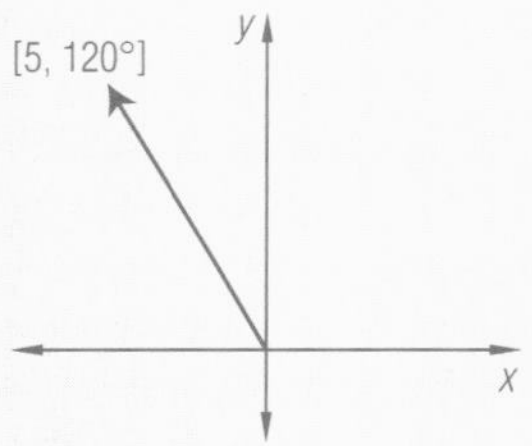

46.

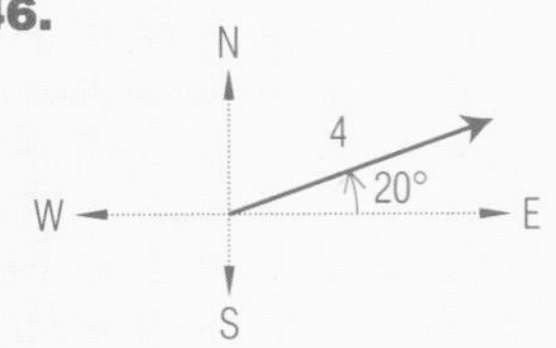

47.b. length ≈ 6.7; direction ≈ 26.6°

48.b., 49.-52., 53.a.,c., 54.a.ii., iii., b., c., 58., 60.-64., 66., 67. See Additional Answers in the back of this book.

REPRESENTATIONS deal with pictures, graphs, or objects that illustrate concepts.

Objective I: *Represent two-dimensional vectors in their component or polar representation, or as directed segments. (Lessons 12-1, 12-3)*

45. Find the component representation of [5, 120°] and sketch the vector.

46. Sketch the vector described by (4 cos 20°, 4 sin 20°).

47. a. Find the component representation of the vector pictured here. (6, 3)
b. Find its length and direction.

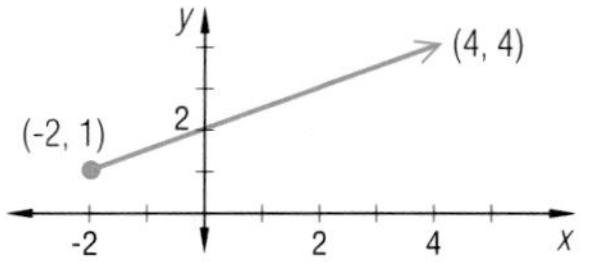

48. a. Give the endpoint of a vector which has polar representation [7, -200°] and starts at the point (2, 1). ≈ (-4.6, 3.4)
b. Sketch the vector.

Objective J: *Represent addition, subtraction, and scalar multiplication of two-dimensional vectors graphically. (Lessons 12-2, 12-3)*

49. Given the vectors sketched below, sketch the following:
a. $\vec{u} + \vec{v}$
b. $\vec{u} - \vec{v}$
c. $-\vec{u}$
d. $2\vec{v}$

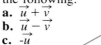

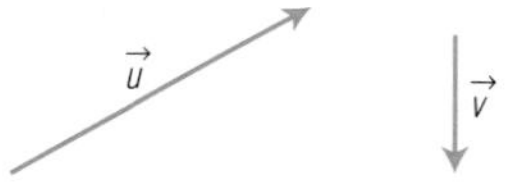

50. Sketch the vectors $\vec{u} = [5, 30°]$ and $\vec{v} = [3, -60°]$. Then sketch the results of the following operations.
a. $\vec{u} + \vec{v}$ **b.** $-\vec{v}$ **c.** $-\vec{u} - \vec{v}$

Objective K: *Geometrically interpret three-dimensional vectors and their operations. (Lesson 12-6)*

In 51 and 52, let $\vec{u} = (5, 1, 3)$ and $\vec{v} = (-3, -4, 1)$.

51. Sketch the vectors $\vec{u}$ and $\vec{v}$ and the angle θ between them.

52. Find the components of $\vec{v} - \vec{u}$. Interpret your answer in terms of the sketch you made.

53. Given $\vec{u} = (3, 2, 0)$ and $\vec{v} = (-1, -2, 0)$.
a. Sketch the plane determined by $\vec{u}$ and $\vec{v}$.
b. Compute $\vec{v} \times \vec{u}$. (0, 0, 4)
c. Add $\vec{v} \times \vec{u}$ to your sketch.

Objective L: *Represent lines in a plane using vector or parametric equations. (Lessons 12-3, 12-4)*

54. The vector equation of a line is $(x - 1, y + 8) = t(3, 1)$.
a. Find the points on the line corresponding to
i. $t = 0$ (1, -8) **ii.** $t = 1$ **iii.** $t = -3$.
b. Find a vector that is parallel to the line.
c. Sketch the line.

In 55 and 56, let $\vec{v} = (-5, 5)$. Write a vector equation for the line through $P = (1, 2)$ that is:

55. parallel to $\vec{v}$ $(x - 1, y - 2) = t(-5, 5)$

56. orthogonal to $\vec{v}$. sample: $(x - 1, y - 2) = t(1, 1)$

In 57 and 58, let $\vec{w}$ be the vector (-5, -3) and Q be the point (1, -2).

57. Find an equation of the line through Q perpendicular to $\vec{w}$. $5x + 3y + 1 = 0$

58. Find parametric equations for the line through Q parallel to $\vec{w}$.

59. Find parametric equations for the line through (5, 0) that is parallel to the line with vector equation $(x - 2, y + 1) = t(3, -4)$.
$\begin{cases} x = 5 + 3t \\ y = -4t \end{cases}$

Objective M: *Represent lines, planes, and spheres in 3-space using parametric, vector, or coordinate equations. (Lessons 12-5, 12-7)*

60. Find the center and radius of the sphere with equation $x^2 + y^2 - 4y + z^2 + 8z = -15$.

61. Find the equations of two planes parallel to and 3 units away from the xz-plane.

62. Find the intercepts and sketch the plane with equation $2x - 3y + z = 12$.

63. Give a system of two linear equations that describes the z-axis in 3-space.

64. Find parametric equations for the line in 3-space through the point (3, 0, -1) that is parallel to the vector (2, -4, 0).

65. Find an equation for the plane containing the point (3, 0, -1) that is perpendicular to the vector (2, -4, 0). $2x - 4y = 6$

66. Find two vectors perpendicular to the plane given by $-3x + y + 5z = 8$.

67. Find a vector equation for the line through (5, -1, 2) that is perpendicular to the plane given by $2x - 6y + z = 7$.

CHAPTER 13 THE INTEGRAL IN CALCULUS

OBJECTIVES

The objectives listed here are the same as in the Chapter 13 Review on pages 810–813 of the student text. The Progress Self-Test on page 809 and the tests in the Teacher's Resource File cover these objectives. For recommendations regarding the handling of this end-of-chapter material, see the notes in the margin on the corresponding pages of this Teacher's Edition.

OBJECTIVES FOR CHAPTER 13 (Organized into the SPUR Categories—Skills, Properties, Uses, and Representations)	Progress Self-Test Questions	Chapter Review Questions	Teacher's Resource File	
			Lesson Masters*	Chapter Test Forms A & B
SKILLS				
A Calculate Riemann sums of functions over specified intervals.	2	1–3	13-2	2
B Evaluate definite integrals.	3, 4, 5	4–11	13-3, 13-4, 13-5	5, 6
PROPERTIES				
C Apply properties of definite integrals.	9, 11	12–15	13-4	7, 8
USES				
D Find the distance traveled by a moving object, given its rate.	10	16–18	13-1, 13-2, 13-5	9
E Use the definite integral to solve application problems.	6	19–21	13-4, 13-6	11
F Estimate the distance under a velocity-time graph.	1	22–23	13-1	1
REPRESENTATIONS				
G Express areas in integral notation.	7	24–26	13-3	3
H Find areas bounded by curves.	8	27–29	13-4, 13-5	4
I Find volumes of solids.	12	30–31	13-6	10

*** The masters are numbered to match the lessons.**

OVERVIEW ■ CHAPTER 13

This chapter introduces students to a fundamental idea in calculus, namely, the integral. The approach taken is mathematically precise but informal. For many students, it is important for future success that they have some knowledge of the basic ideas of calculus before taking a full course in the subject.

Mathematically, the integral is approached through Riemann sums. In Lessons 13-1 and 13-2, Riemann sums are used to approximate the area under a curve and the distance traveled by a car whose speed at each instant is known. A computer program to evaluate Riemann sums is given in the second lesson. You might wish to supplement this work with calculus software that can illustrate and evaluate Riemann sums.

In Lesson 13-3, the definite integral is defined as the limit of a sequence of Riemann sums as the number of subintervals $n \to \infty$. Most of the examples in the lesson deal with functions that are nonnegative on an interval or nonpositive on an interval. Students should be able to estimate whether a given Riemann sum or definite integral will be positive or negative on an interval.

In some introductions to calculus, students are given a formula for $\int_0^a x^n\,dx$, and then the formula is used to solve problems. We do not take this approach in this chapter because students may then rely upon rote learning and not try to understand the basic concepts. Emphasis is placed upon properties that can be proved by using what students know about area or sums of infinite series. Several general properties relating integrals are given in Lesson 13-4, and then some formulas are given in Lesson 13-5, culminating in a formula for finding $\int_0^a (c_2x^2 + c_1x + c_0)\,dx$. Using this formula, the approximations of Lesson 13-1 can be compared with the exact values found by evaluating integrals.

In Lesson 13-6, a different application of integrals is given. Volumes of solids of revolution are approximated by slicing the solids into sections approximated by cylinders and then adding the volumes of those cylinders. One application yields an alternate derivation of the formula for the volume of a sphere, $V = \frac{4}{3}\pi r^3$.

The last lesson of the chapter introduces students to the Fundamental Theorem of Calculus, which is the theorem that relates derivatives to integrals.

PERSPECTIVES ■ CHAPTER 13

The Perspectives provide the rationale for the inclusion of topics or approaches, provide mathematical background, and make connections with other lessons and within UCSMP.

13-1

FROM THE DISCRETE TO THE CONTINUOUS

The integral has interpretations under each of the four SPUR categories. For each category, connections are made between discrete and continuous ideas. The lesson begins with a "Uses" example—the total distance of a trip if times and speeds for each of a finite number of segments (subintervals) of it are known. The "Representations" of this distance is the total area of a finite set of rectangles. The "Skills" is the calculation of the sum, which is of the form $\sum_{i=1}^{n} r_it_i$. The "Properties" of such sums and areas have been studied by students previously.

As a car accelerates, it is not at a constant speed over any of its subintervals. Still, the distance traversed can be approximated by dividing the trip into more and more subintervals and finding their sum. As the number of subintervals increases, the sum has more terms, the rectangles have smaller widths, and the total distance represented by the sum becomes a more accurate approximation to the actual total distance. The examples of this lesson involve two different formulas for acceleration from 0 to 60 mph in 10 seconds.

The limit is easy to understand in the context of the "Uses" (the total distance traveled when the speed is known at each instant) and the "Representations" (the total area under a curve). The symbolic definition is delayed until Lesson 13-3.

13-2

RIEMANN SUMS

If speed is a function of time, that is, if $r = f(t)$, then the sum $\left(\sum_{i=1}^{n} r_it_i\right)$ described above has the form $\sum_{i=1}^{n} f(z_i)(x_i - x_{i-1})$, where x_i and x_{i-1} are the endpoints of the subintervals, and each z_i is a particular value in the ith subinterval. This is a Riemann sum. In this lesson, Riemann sums are defined with subintervals of any lengths, but all applications are done with equal subintervals. If the lengths of the subintervals are a constant Δx, the sum becomes $\sum_{i=1}^{n} f(z_i)\Delta x$. In this lesson, Riemann sums are calculated using various choices of z_i from the subintervals. Usually, the z_i are either the left endpoints, the right endpoints, or the midpoints of the intervals.

13-3

THE DEFINITE INTEGRAL

The definite integral is the limit of Riemann sums as $n \to \infty$. The definite integral is a number, whereas the indefinite integral, not discussed in this book, is a function. An analogy is with the derivative at a point (a number) and the derivative of a function (a function).

With the development of the previous two lessons, the definite integral has the following interpretations: it is the area under a curve, and it yields the total distance. In this lesson, only the area application is considered.

13-4

PROPERTIES OF DEFINITE INTEGRALS

The area interpretation of integrals enables various properties of definite integrals to be deduced. In this lesson, there are four theorems that relate integrals to each other.
(1) The sum of the integrals of a function from a to b and from b to c equals the integral from a to c. This property enables an integral to be broken down into subintervals or built up from those subintervals.
(2) The integral of a function from a to b equals the integral from 0 to b less the integral from 0 to a. Thus, a general formula can be derived for an integral if a formula can be found for the special case over the interval from 0 to a.
(3) The integral of the sum of two functions is the sum of the integrals of the functions.
(4) The integral of a constant multiple of a function is that constant multiple of the integral. The last two properties are applied immediately in the next lesson to obtain the definite integral of any quadratic function.

13-5

THE AREA UNDER A PARABOLA

By adding $\int_0^a c_0\,dx$, $\int_0^a c_1 x\,dx$, and $\int_0^a c_2 x^2\,dx$, the definite integral of any quadratic function over the interval from 0 to a can be found. The first two have been calculated directly in previous lessons using area formulas for rectangles and triangles, but the integral of the quadratic requires more work. This lesson first deduces a formula for $\int_0^a x^2\,dx$. Then the sum and constant multiple properties of integrals are employed to obtain a formula for the definite integral of a general quadratic over this interval. In turn, this formula is applied to give an exact answer to a problem posed in Lesson 13-1, namely, the distance traveled by a car when its speed is described by a quadratic function. A bonus is that the area of a region bounded by a parabola can now be calculated.

13-6

DEFINITE INTEGRALS AND VOLUMES

Integrals can be used to find volumes of solids in much the same way they are used to find areas of regions. Students who have studied Cavalieri's Principle (in UCSMP *Geometry* or other texts) have seen figures cut into thin cross-sections as is done in this lesson.

The idea is to create a Riemann sum that approximates the volume and whose limit is the volume. In this lesson, the only figures considered are solids of revolution, in particular, solids formed by rotating the region under a curve about the x-axis. Then, if the curve is sufficiently smooth, the solid can be sliced into 3-dimensional regions that are approximated by cylinders. The volume cf the solid is approximately equal to the sum of the volumes of the regions.

The first example of the lesson creates and evaluates such a Riemann sum for a curve that is drawn and whose equation is not known. The second example shows how the method can be used to obtain the exact volume of the solid of revolution formed when a line is rotated about the x-axis. This requires taking the limit of the Riemann sum. The third example shows how the integral of a quadratic function can be applied to derive the volume formula for a sphere.

13-7

THE FUNDAMENTAL THEOREM OF CALCULUS

We have tried to end all UCSMP texts with a lesson that involves many of the concepts developed in the book and also indicates the direction of future work. This lesson follows that pattern: it summarizes this chapter and relates it to Chapter 9 by pointing out the nearly inverse relationship between derivatives and integrals. At the same time, it demonstrates the remarkable unity of mathematics and helps to prepare students for one of the major ideas in calculus.

DAILY PACING CHART 13

DAILY PACING CHART ■ CHAPTER 13

Every chapter of UCSMP *Precalculus and Discrete Mathematics* includes lessons, a Progress Self-Test, and a Chapter Review. For optimal student performance, the self-test and review should be covered. (See *General Teaching Suggestions: Mastery* on page T35 of this Teacher's Edition.) By following the pace of the Full Course given here, students can complete the entire text by the end of the year. Students following the pace of the Minimal Course spend more time when there are quizzes and on the Chapter Review and will generally not complete all of the chapters in this text.

When chapters are covered in full (the recommendation of the authors), then students in the Minimal Course can cover 11 chapters of the book. For more information on pacing, see *General Teaching Suggestions: Pace* on page T34 of this Teacher's Edition.

DAY	MINIMAL COURSE	FULL COURSE
1	13-1	13-1
2	13-2	13-2
3	13-3	13-3
4	13-4	13-4
5	Quiz (TRF); Start 13-5.	Quiz (TRF); 13-5
6	Finish 13-5.	13-6
7	13-6	13-7
8	13-7	Progress Self-Test
9	Progress Self-Test	Chapter Review
10	Chapter Review	Chapter Test (TRF)
11	Chapter Review	Comprehensive Test (TRF)
12	Chapter Test (TRF)	
13	Comprehensive Test (TRF)	

TESTING OPTIONS

TESTING OPTIONS

- Quiz for Lessons 13-1 through 13-4
- Chapter 13 Test, Form A
- Chapter 13 Test, Form B
- Chapter 13 Test, Cumulative Form
- Comprehensive Test, Chapters 1–13

A Quiz and Test Writer is available for generating additional questions, additional quizzes, or additional forms of the Chapter Test.

INDIVIDUAL DIFFERENCES

PROVIDING FOR INDIVIDUAL DIFFERENCES

The student text is written for, and tested with, average students. It also has been successfully used with better and more poorly prepared students.

The Lesson Notes often include Error Analysis and Alternate Approach features to help you with those students who need more help. Students of all abilities often learn from their peers and may benefit from small group work referenced as appropriate throughout the Notes. A blackline Lesson Master (in the Teacher's Resource File), keyed to the chapter objectives, is provided for each lesson to allow more practice. (However, since it is important to keep up with the daily pace, you are not expected to use all of these masters. Again, refer to the suggestions for pacing on page T34.) Extension activities are provided in the Lesson Notes for those students who have completed the particular lesson in a shorter amount of time than is expected, even in the Full Course.

The Integral in Calculus

CHAPTER 13

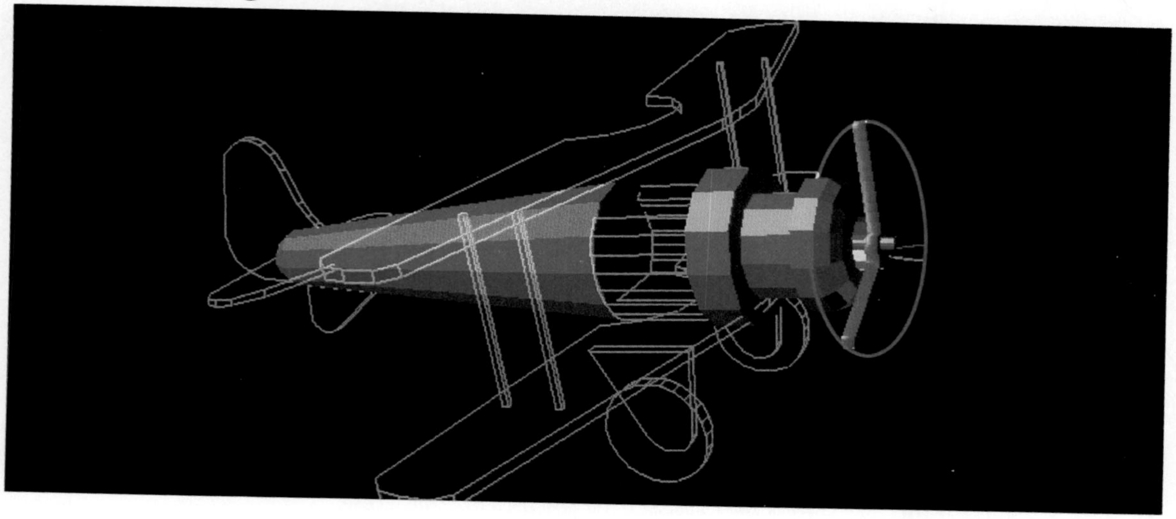

The study of calculus has two main branches: differential calculus and integral calculus. In Chapter 9, you were introduced to differential calculus, whose fundamental concept is the derivative. The fundamental concept in integral calculus is the integral. The symbol for integral is $\int$, an elongated "s," which stands for sum. The symbol is used in recognition of the fact that an integral is a certain kind of limit of a sum. You will learn its specific meaning in this chapter.

The process of finding the integral of a function is called *integration*. This chapter focuses on these four types of integration problems:

1. Given the (changing) velocity of an object at every instant, find how far the object travels in a given time period.
2. (a more general form of type 1) Given the derivative of a function, determine the original function.
3. Find the area between the graph of a function and the x-axis as illustrated in the diagram at the left below.

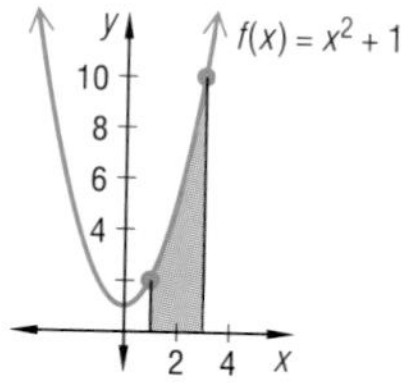

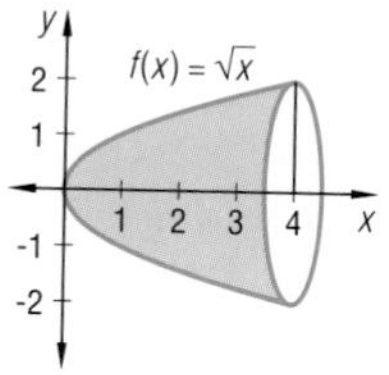

4. (an extension of type 3) Find the volume of a solid of revolution as illustrated in the diagram at the right above.

As with the derivative, the integral is a powerful tool. The connection of these two ideas, made by Isaac Newton and Gottfried Leibniz in the 17th century, paved the way for many of the advances in the physical sciences, engineering, and statistics which have taken place since then.

CHAPTER 13

We recommend 11 to 13 days for this chapter: 7 or 8 days for the lessons and quiz; 1 day for the Progress Self-Test; 1 or 2 days for the Chapter Review; 1 day for a Chapter Test; and 1 day for the Comprehensive Test. (See the Daily Pacing Chart on page 759D.)

USING PAGE 759

Most students have probably seen an integral sign. They may be glad to learn what it means and how it can be calculated. Point out that what is presented in this chapter will not be as detailed as the contents of a calculus course, but it will be an excellent foundation for that course.

LESSON 13-1

RESOURCES

- Lesson Master 13-1
- Teaching Aid 96 provides the graphs showing distance as the area under a rate curve.
- Teaching Aid 97 can be used with **Example 1**.
- Teaching Aid 98 can be used with **Questions 7 and 8**.
- Teaching Aid 99 can be used with **Question 16**.

OBJECTIVES

D Find the distance traveled by a moving object, given its rate.

F Estimate the distance under a velocity-time graph.

TEACHING NOTES

The key applied idea in this lesson is that distance traveled can be found by determining the area under a speed curve. If the speed function is linear (and it will be if the acceleration is constant), the area can be computed exactly using familiar formulas. If the velocity function is nonlinear, the area can be estimated by adding areas of rectangles.

Students need to understand the summation notation in this lesson, as it is used throughout the chapter. They should be able to relate the notation to the appropriate graphs.

LESSON

13-1

From the Discrete to the Continuous

When you calculate travel time for a trip, you take into account that you travel faster on some roads than on others.

If you travel for 3.5 hours at a constant speed of 50 miles per hour, the familiar formula $d = rt$ gives the total distance traveled, 175 miles. If your speed then changes to 30 miles per hour, and you travel for an additional 2 hours, then the total distance traveled is

$$50 \cdot 3.5 + 30 \cdot 2 = 235 \text{ miles.}$$

The formula for the total distance could be written as $r_1t_1 + r_2t_2$ or $\sum_{i=1}^{2} r_it_i$, a sum of two terms. Here t_1 and t_2 add up to the total time and r_1 and r_2 are rates for the two parts of the trip.

Of course, few things travel at constant speeds for any long period of time—not cars, nor people, nor objects tossed into the air. To get a better approximation of the real situation, the total time can be split into n time intervals $t_1, t_2, \ldots, t_n$. Then, if the rates in each interval of time are $r_1, r_2, \ldots, r_n$, the total distance traveled is

$$r_1t_1 + r_2t_2 + \ldots + r_nt_n, \text{ or } \sum_{n=1}^{n} r_it_i,$$

a sum of n terms.

But even this still gives only an approximation. The rate of a moving object usually changes continuously. For instance, as the speed changes from 50 to 30 mph, every number between 30 and 50 is a rate of the object at some instant in time. So there are (in theory) infinitely many terms of the form r_it_i to add, one term for each different rate. The situation is continuous; the set of terms to add is not a discrete set. This problem of the distance traveled by objects in continuous motion is one of the kinds of problems that led to the invention of calculus.

760

Graphically, when time is the x-coordinate and rate the y-coordinate, the distance traveled can be represented as the area under a rate curve. Here is a graph of the situation described in the first sentence of this lesson, where an object travels at 50 miles per hour for 3.5 hours.

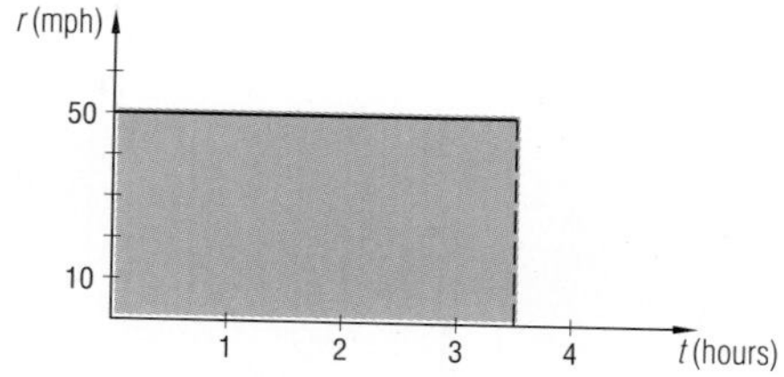

$$\begin{aligned}\text{Distance} &= \text{Area of rectangle}\\ &= 50\ \frac{\text{miles}}{\text{hr}} \cdot 3.5 \text{ hours}\\ &= 175 \text{ miles}\end{aligned}$$

Here is a graph of the situation in which part of a trip is at 50 mph, another part at 30 mph.

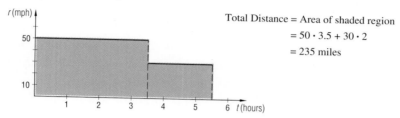

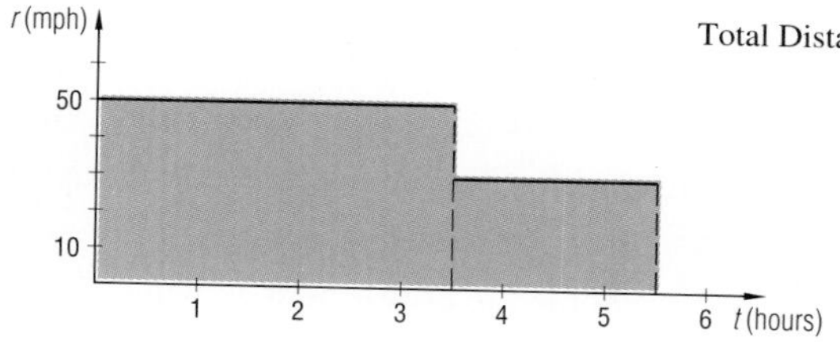

$$\begin{aligned}\text{Total Distance} &= \text{Area of shaded region}\\ &= 50 \cdot 3.5 + 30 \cdot 2\\ &= 235 \text{ miles}\end{aligned}$$

When there are 10 different constant rates $r_1, r_2, \ldots, r_{10}$, for times of lengths $t_1, t_2, \ldots, t_{10}$, here is a possible graph. The height of each rectangle is just the rate for the corresponding time interval. (Note that r_5 is zero; the object is not moving during that time interval.)

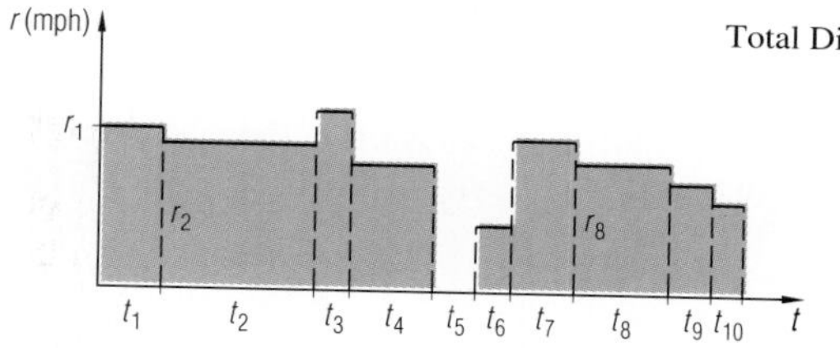

$$\begin{aligned}\text{Total Distance} &= \text{Area of shaded region}\\ &= \sum_{i=1}^{10} r_i t_i\end{aligned}$$

Now suppose $r = f(t)$ is the function giving the rate at which an object is moving at time t. For the functions represented by the three previous graphs, the range is a discrete set, containing 1, 2, and 10 elements, respectively. An actual situation is better modeled by a continuous function, as shown below. In this case, the range (the rates) consist of all real values from 0 to the maximum value of the function. Again, the area between the graph and the horizontal axis gives the total distance, as is explained below.

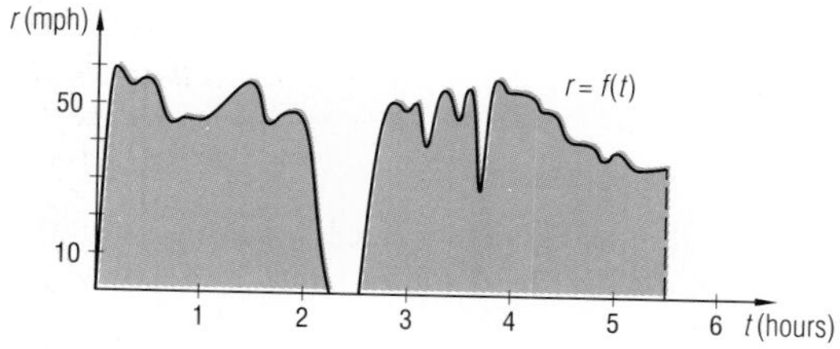

Point out that the speed functions f and g in **Examples 1 and 2** are not the only possible ways to accelerate from 0 to 60 mph $\left(0 \text{ to } 88\ \frac{\text{ft}}{\text{sec}}\right)$. It is likely that an actual acceleration does not have such a simple formula. Note that the answer to **Example 1** is greater than the answer to **Example 2**. Ask students why this is so. (The car in Example 1 is accelerating faster at first, so it should go farther.)

Students may wonder why $\frac{\text{ft}}{\text{sec}}$ is used as the unit rather than $\frac{\text{mi}}{\text{hr}}$. If the latter were used, then the area would be in miles and the answers would be fractions of a mile. It is easier to use feet.

To approximate this area, it is convenient to split the region under the graph of the function into rectangles whose horizontal dimensions (widths) are equal. For instance, with the function $r = f(t)$ above, the area under the curve from $t = 0$ to $t = 5.5$ can be estimated by using eleven rectangles with constant width $\frac{1}{2}$ and height equal to the value of f at the right side of the rectangle.

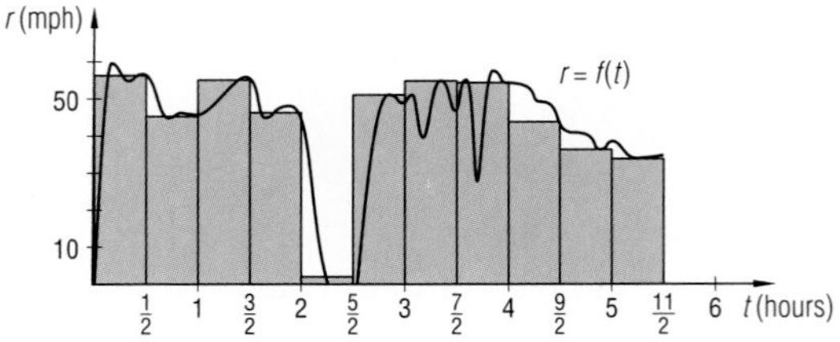

The estimate is $\frac{1}{2} \cdot f\left(\frac{1}{2}\right) + \frac{1}{2} \cdot f(1) + \frac{1}{2} \cdot f\left(\frac{3}{2}\right) + \ldots + \frac{1}{2} \cdot f\left(\frac{11}{2}\right) = \sum_{i=1}^{11} \frac{1}{2} f\left(\frac{i}{2}\right)$. This is like recording the speed every $\frac{1}{2}$ hour, assuming rate as an approximation for the average rate during the preceding $\frac{1}{2}$ hour, and calculating the total distance based only on this information. If the speed were recorded every $\frac{1}{10}$ hour, a better approximation of the distance traveled would be obtained. The width of each rectangle would be $\frac{1}{10}$, there would be 55 values in the 5.5 hours, and the estimate would be

$$\sum_{i=1}^{55} \frac{1}{10} f\left(\frac{i}{10}\right).$$

Now consider a situation in which there is a formula for the speed function. Imagine a car that accelerates from 0 to 60 mph (88 ft/sec) in ten seconds. If the acceleration is quicker at first (the driver presses hard on the pedal, then eases off), the speed might be given by the quadratic function $g(x) = -.88(x - 10)^2 + 88$. This results in the table and graph shown below.

x (seconds)	0	1	2	3	4	5	6	7	8	9	10
speed (ft/sec) after x seconds	0	16.72	31.68	44.88	56.32	66	73.92	80.08	84.48	87.12	88

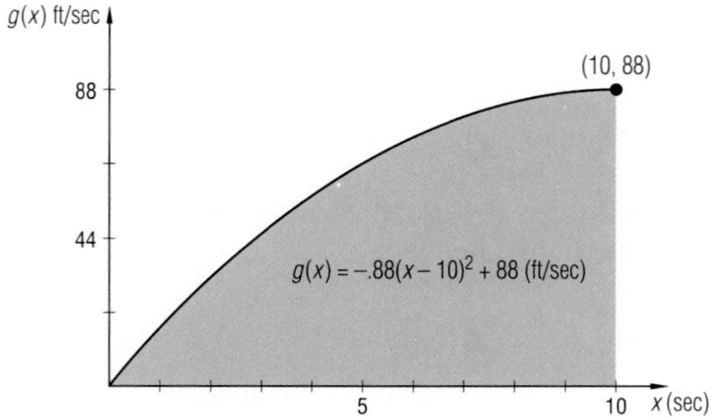

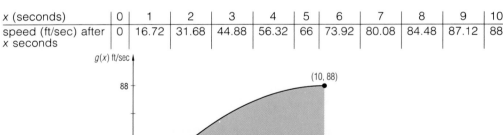

In this nonconstant acceleration situation, the speed curve is part of a parabola. The total distance traveled is given by the area between the parabola and the x-axis, but at this point you probably do not know any formula to find such a region. Still, the total distance can be estimated.

Example 1 A car accelerates from 0 to 88 $\frac{\text{ft}}{\text{sec}}$ with a speed of $g(x) = -.88(x - 10)^2 + 88$ $\frac{\text{ft}}{\text{sec}}$ after x seconds. Estimate the distance the car travels in 10 seconds.

Solution For our estimate, we choose to divide the interval $0 \le x \le 10$ into 5 subintervals.

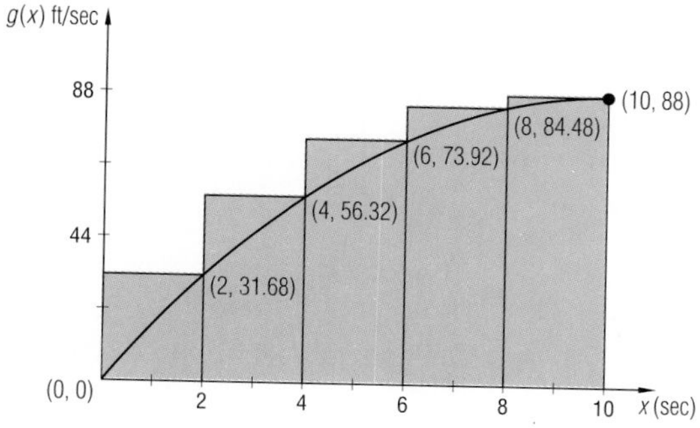

The distance the car travels is less than the sum of the areas of the 5 rectangles shown above. This area is

$$\sum_{i=1}^{5} 2 \cdot g(2i) = 2 \cdot 31.68 + 2 \cdot 56.32 + 2 \cdot 73.92 + 2 \cdot 84.48 + 2 \cdot 88$$
$$= 668.8 \text{ feet.}$$

The estimate in Example 1 could be improved by splitting the interval into more subintervals. We found a high estimate for the area; a low estimate can be obtained by using rectangles that lie completely under the curve.

The narrower the rectangles, the closer the sum of their areas is to the area between the graph of f and the x-axis. So the narrower the rectangles, the closer the sum of their areas is to the actual total distance traveled. In general, if f is a continuous function giving the rate at time t, then the total distance traveled equals the area between the graph of f and the x-axis over the time interval. In certain situations, this distance can be calculated exactly.

ADDITIONAL EXAMPLES

1. Suppose a car accelerates from 0 to 88 $\frac{\text{ft}}{\text{sec}}$ in 10 seconds such that its speed y after t seconds is given by $y = .088t^3$.

a. Estimate the total distance traveled by dividing the interval from 0 to 10 into 5 intervals and drawing rectangles with the values of y at the midpoints of the intervals as heights of the rectangles.
215.6 feet

b. Is this estimate too high or too low?
too low

2. A car accelerates from 0 to 88 $\frac{\text{ft}}{\text{sec}}$ in 11 seconds. If the acceleration is constant, how far will the car have traveled in that time?
484 feet

3. Recall from Chapter 9 that if the height (in feet) of a projectile after t seconds is given by $h(t) = 800t - 16t^2$, then its velocity is given by $h'(t) = v(t) = 800 - 32t$. Find the distance traveled between 8 and 20 seconds.
4224 feet

NOTES ON QUESTIONS
Question 3: Be generous in accepting estimates. The important idea is the method of estimating, not the accuracy of the estimate.

Question 4: This is the same situation as in **Example 2**. The car travels more in the last 5 seconds because its speed is greater.

Question 6: This is the situation of **Example 1**. Part **b** may be easier to interpret in the form $\sum_{i=1}^{10} 1 \cdot g(i)$, where the constant 1 indicates a time interval of 1 second. If you have appropriate software, you could use it to show part **c**.

Question 7: This question makes the point that you can estimate how far a vehicle has traveled by taking speedometer readings even if you do not have its odometer (distance) readings.

Example 2 A car accelerates from 0 to 60 mph $\left(88 \frac{\text{ft}}{\text{sec}}\right)$ in 10 seconds. If the acceleration is constant, how far will the car travel (in feet) in this time?

Solution Since the acceleration is constant, the following table gives the speed after each second.

x (seconds)	0	1	2	3	4	5	6	7	8	9	10
speed (ft/sec) after x seconds	0	8.8	17.6	26.4	35.2	44	52.8	61.6	70.4	79.2	88

So the speed is given by the equation $f(x) = 8.8x$. This speed function is graphed below.

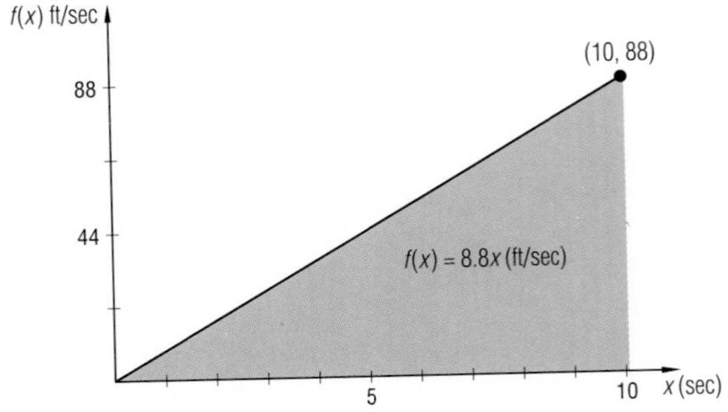

The total distance traveled by the car is given by the area between the graph of f and the x-axis. Using the formula for the area of a triangle ($A = \frac{1}{2}bh$), the distance is found to be $\frac{1}{2} \cdot 10 \text{ seconds} \cdot 88 \frac{\text{ft}}{\text{sec}}$ $= 440$ feet.

Questions

Covering the Reading

1. Of the 8 graphs of functions in this lesson, in how many is the range a discrete set? 3

2. For 20 minutes, you travel on an interstate highway at 65 mph. Then for 10 minutes, you travel at 55 mph, as you near a city. Then for 30 minutes, you only average 30 mph as you go through the city.
   a. What is the total distance traveled? ≈ 45.8 mi
   b. Graph the rate over time and indicate the total distance on the graph. See margin
   c. Generalize part **a** if you travel at a rate r_1 for a time of length t_1, r_2 for a time of length t_2, and r_3 for a time of length t_3. $D = r_1t_1 + r_2t_2 + r_3t_3 = \sum_{i=1}^{3} r_it_i$

3. Estimate $\sum_{i=1}^{11} \frac{1}{2} f\left(\frac{i}{2}\right)$ for the function f on page 762 by first estimating values of the function from its graph. See margin.

764

4. A car accelerates from 0 to 88 feet per second in 10 seconds. Assume the acceleration is constant. How far will the car travel
a. in the first second? 4.4 ft
b. in the first five seconds? 110 ft
c. in the total time of ten seconds? 440 ft

5. If a car accelerates from 0 to 28 meters per second in 8 seconds under constant acceleration, how far does it travel in those 8 seconds?
112 m

6. A car accelerates from 0 to 88 feet per second with a speed of
$g(x) = -.88(x - 10)^2 + 88 \frac{\text{ft}}{\text{sec}}$ after x seconds.
a. Will this car or the car of Question 4 travel farther? this car
b. Estimate the distance the car travels in the 10 seconds by calculating $\sum_{i=1}^{10} g(i)$. 629.2 ft
c. Trace the graph of g and draw the rectangles whose areas add to the estimate in part **b**. See margin.
d. Is the estimate in part **b** better or worse than the estimate found in Example 1? better
e. Estimate the distance the car travels in the 10 seconds by calculating $\sum_{i=1}^{20} \frac{1}{2} g\left(\frac{i}{2}\right)$. 608.3 ft

Applying the Mathematics

7. As a freight train travels eastward through North Dakota, an instrument in the locomotive records on a paper scroll the train's velocity at any time.

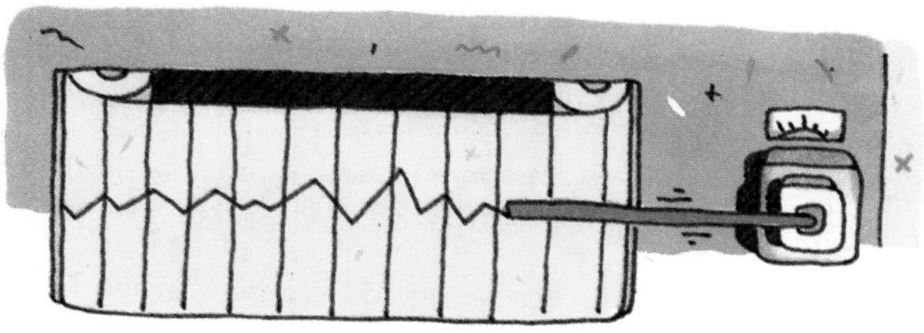

Speed-time recorder

The velocity-time record for the trip is pictured below.

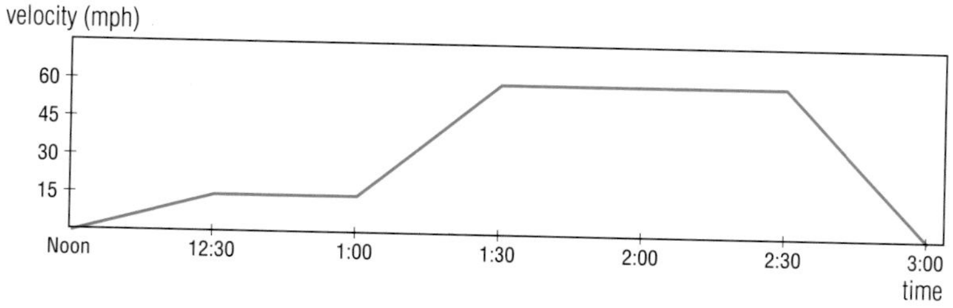

At 3:00, how far east of the starting point is the train? 105 mi

ADDITIONAL ANSWERS
2.b.

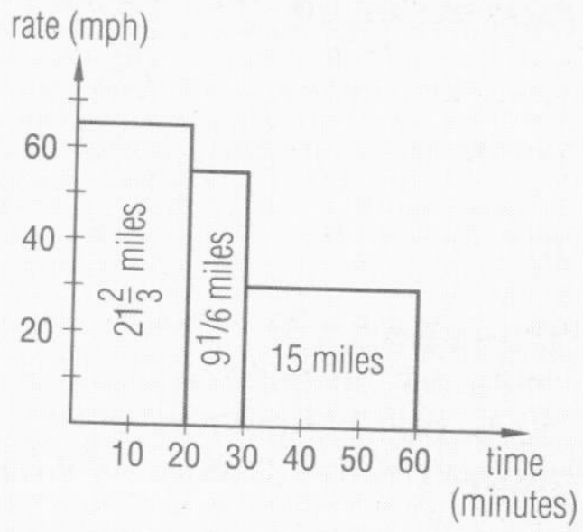

3. sample:

i	$\frac{i}{2}$	$f\left(\frac{i}{2}\right)$	$.5f\left(\frac{i}{2}\right)$
1	0.5	58	29
2	1	47	23.5
3	1.5	56	28
4	2	48	24
5	2.5	3	1.5
6	3	51	25.5
7	3.5	55	27.5
8	4	54	27
9	4.5	43	21.5
10	5	37	18.5
11	5.5	32	16

$\sum_{i=1}^{11} .5f\left(\frac{i}{2}\right) \approx 242 \text{ units}^2$

6.c.

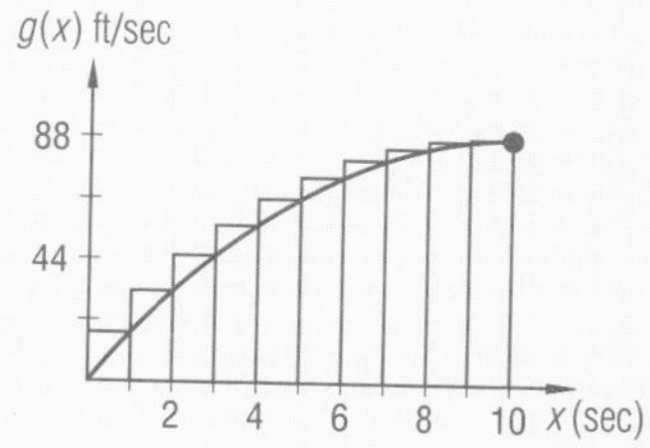

NOTES ON QUESTIONS
Question 9a: As with **Question 6b**, you may wish to write $\sum_{k=1}^{8} 1 \cdot f(k)$, where the constant 1 indicates a time interval of 1 minute.

Question 10: This formula will be used in Lesson 13-5, but it should be discussed in this lesson as well.

Question 15: The ideas in this question are important for Lesson 13-3.

ADDITIONAL ANSWERS
13.a., b. sample:

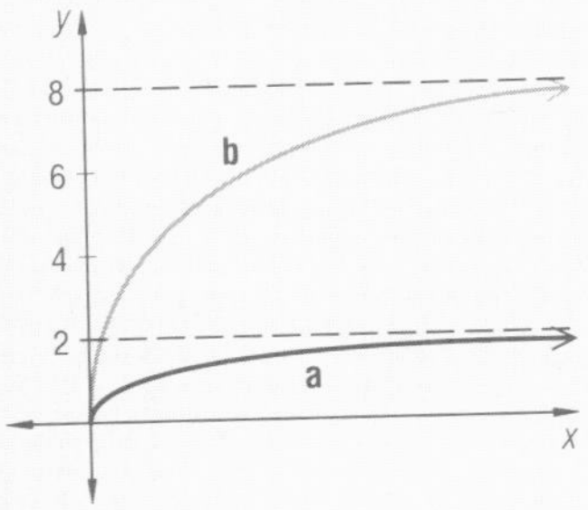

8. The graph below shows the velocity of a rocket moving vertically during an eight-minute interval.

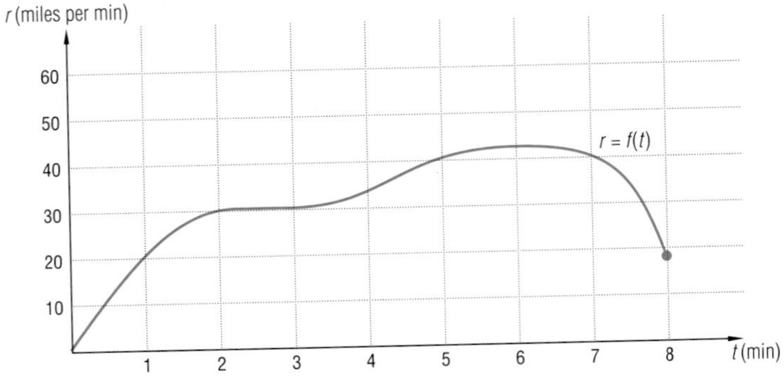

a. Estimate the distance traveled by evaluating $\sum_{k=1}^{8} f(k)$. **≈ 260 mi**

b. Estimate the distance traveled by evaluating $\sum_{k=1}^{4} 2 f(2k)$. **≈ 250 mi**

c. Which estimate, the one from part **a** or the one from part **b,** is likely to be more accurate, and why? **The estimate in part a should be more accurate because it has more rectangles, so it is closer to the actual graph.**

Review

9. Suppose that on a particular day during a flu epidemic, the probability that any given student has the flu is 20%. What is the probability that in a class of 25 students, at least 3 have the flu? *(Lesson 10-6)* **≈ .9**

10. Use the formula $\sum_{i=1}^{n} i^2 = \frac{n(n+1)(2n+1)}{6}$ to find $1^2 + 2^2 + 3^2 + \ldots + 25^2$. *(Lesson 7-4)* **5525**

11. Solve the equation $2 \sin x = \cos x$ over the interval $-2\pi \le x \le 2\pi$. *(Lesson 6-7)* **$x \approx -5.82$ or $x \approx -2.68$ or $x \approx 0.46$ or $x \approx 3.61$**

12. Describe the end behavior of the function f, where $f(x) = \frac{2x^2 + 3x - 1}{x - 2}$. Write equations for any horizontal, vertical, or oblique asymptotes. *(Lessons 5-5, 5-4)* **$\lim_{x \to +\infty} f(x) = +\infty$; $\lim_{x \to -\infty} f(x) = -\infty$; oblique asymptote: $y = 2x + 7$; vertical asymptote: $x = 2$**

13. a. Sketch the graph of a function with horizontal asymptote $y = 2$.
b. Sketch the result of transforming your graph by the scale change $S_{3,4}$.
c. Write the equation of the asymptote in your graph from part **b**.
d. Suppose a function has horizontal asymptote $y = c$. Write the equation of the asymptote when the graph is transformed by $S_{a,b}$.
(Lessons 3-9, 2-5) **a, b) See margin. c) $y = 8$; d) $y = bc$**

14. Explain why 60 miles per hour equals 88 feet per second. *(Previous course)* **$60 \frac{mi}{hr} = 60 \frac{mi}{hr} \cdot 5280 \frac{ft}{mi} \cdot \frac{1}{3600} \frac{hr}{sec} = 88 \frac{ft}{sec}$**

15. a. Suppose the interval from 3 to 7 is split into 5 equal parts. What are the endpoints of each of those parts? **3.0, 3.8, 4.6, 5.4, 6.2, 7.0**

b. Suppose the interval from 3 to 7 is split into n equal parts. What are the endpoints of the n parts? $3 + \frac{4k}{n}$, **where** $k = 0, 1, ..., n$

c. Suppose the interval from a to b is split into n equal parts. What is the length of each part? *(Previous course)* $\frac{b-a}{n}$

Exploration

16. The drawings below suggest two other solutions to Example 1.
a. Obtain these estimates. b. Which do you think is a better estimate, and why?

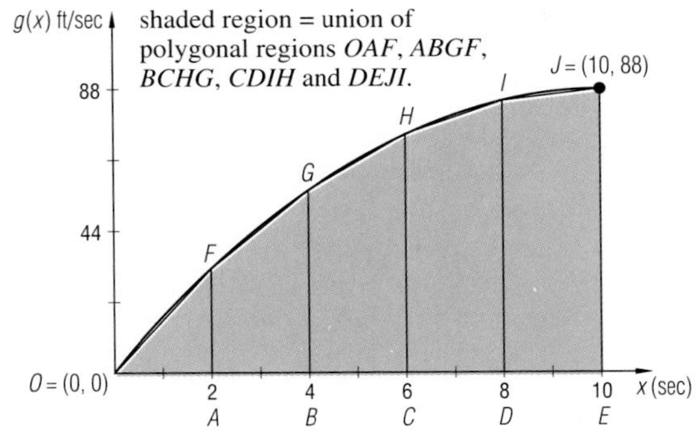

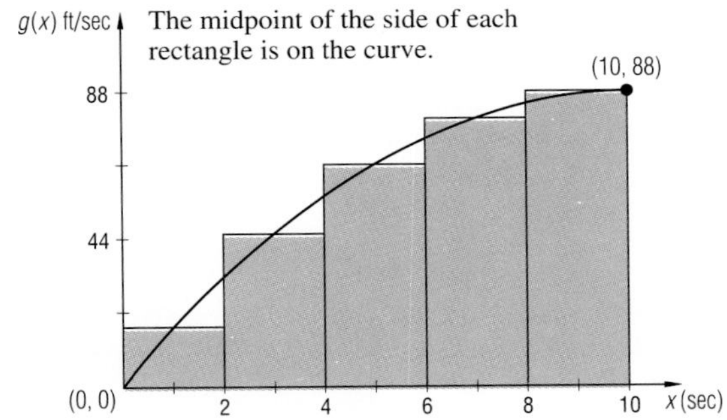

a) 580.8 ft and 589.6 ft; b) Answers may vary.

FOLLOW-UP

MORE PRACTICE
For more questions on SPUR Objectives, use *Lesson Master 13-1*, shown below.

PROJECTS
The projects for Chapter 13 are described on pages 806–807. **Project 5** is related to the content of this lesson.

NAME

LESSON **MASTER 13–1**
QUESTIONS ON **SPUR** OBJECTIVES

■**USES** *Objective D (See pages 810–813 for objectives.)*

1. What is the total distance traveled by a car which travels at a rate of 65 mph for 1.5 hours, 15 mph for 30 minutes, and 40 mph for 45 minutes? **135 mi**

2. Use summation notation to express the total distance traveled by an object whose rate-time graph is given at the right. $\sum_{i=1}^{5} r_i t_i$

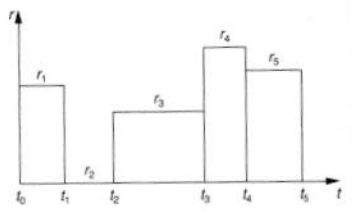

3. Suppose a space probe travels on a straight line with an initial speed of 100 m/sec and a constant acceleration of 9.8 m/sec². Then its velocity at time t seconds is given by $100 + 9.8t$. Find the distance it will have traveled in 10 seconds. **1,490 m**

■**USES** *Objective F*

In 4 and 5, each rate-time graph depicts a runner competing in a track event. From the graph, estimate the distance of the race.

4. **≈ 100 m** 5. **≈ 1,500 m**

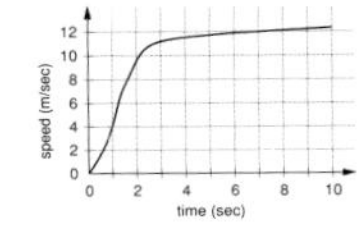

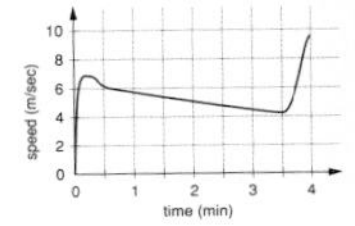

140 *Precalculus and Discrete Mathematics* © Scott, Foresman and Company

LESSON 13-2

RESOURCES

- Lesson Master 13-2
- Teaching Aid 100 displays the definition of Riemann sum.
- Teaching Aid 101 can be used with **Questions 5, 11, and 15**.

OBJECTIVES

A Calculate Riemann sums of functions over specified intervals.

D Find the distance traveled by a moving object, given its rate.

TEACHING NOTES

The goal of this lesson is to have students understand the definition and notation of Riemann sums. This is difficult to do without having a visual image of a Riemann sum as the sum of the areas of rectangles. The figure that follows illustrates such a Riemann sum $\left(\sum_{i=1}^{n}(x_i - x_{i-1})f(z_i)\right)$ with the following characteristics. The ith subinterval has left endpoint x_{i-1} and right endpoint x_i, and $\sum_{i=1}^{6}(x_i - x_{i-1})f(z_i)$ is the area of the shaded rectangles.

$n = 6$

$x_i - x_{i-1}$ is constant.

z_3 is the midpoint of the 3rd subinterval.

z_4 is the right endpoint of the 4th subinterval.

z_6 is the left endpoint of the 6th subinterval.

z_1, z_2, and z_5 are any values inside the subintervals.

LESSON

13-2 Riemann Sums

In the last lesson, you saw that the total distance traveled by an object moving with a speed $f(x)$ during a time interval $a \le x \le b$ equals the area of the region between the graph of f and the x-axis over the interval from a to b. This area (and this distance) can be estimated by adding areas of rectangles. If f is a continuous function, then by using more and more rectangles, the estimate can be made closer and closer to the area of the region.

For example, consider the function used in Lesson 13-1 to describe the speed of a car accelerating from 0 to 88 feet per second in 10 seconds at the particular nonconstant rate such that $g(x) = -.88(x - 10)^2 + 88$.

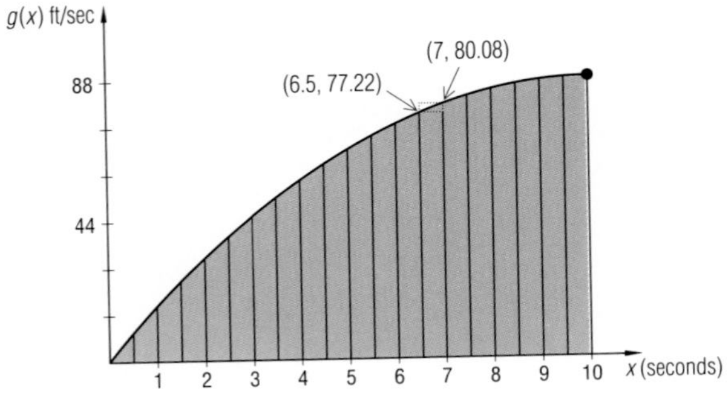

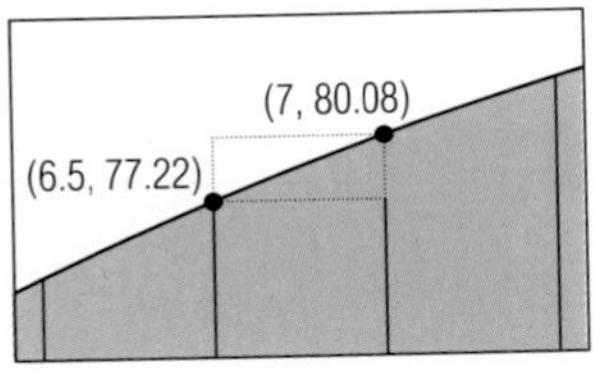

enlargement

The area of each rectangle is an estimate of the distance traveled in that time interval. For instance, to calculate the distance traveled between 6.5 and 7 seconds, note that the rectangle has width 0.5 second. If the height of the graph on the right side of the interval is used, the height of the rectangle is $g(7) = 80.08$ ft/sec. If the height of the graph on the left side of the interval is used, the height is $g(6.5) = 77.22$ ft/sec. Then the distance traveled between 6.5 and 7 seconds is estimated by

$$0.5 \text{ sec} \cdot 80.08 \frac{\text{feet}}{\text{sec}} = 40.04 \text{ feet} \qquad \text{if } g(7) \text{ is used,}$$

or by

$$0.5 \text{ sec} \cdot 77.22 \frac{\text{feet}}{\text{sec}} = 38.61 \text{ feet} \qquad \text{if } g(6.5) \text{ is used.}$$

768

Any other value for the height between $g(6.5)$ and $g(7)$ would give another estimate for the distance traveled. (Try $g(6.75)$ to see what estimate it gives.) An estimate of the total distance in the 10 seconds is found by adding the areas of the 20 rectangles. For each rectangle, the width is $\frac{1}{2}$ and the height is $g(z_i)$, where the z_i are values in the 20 intervals. The total area,

$$\sum_{i=1}^{20} \frac{1}{2} g(z_i),$$

is an example of a *Riemann sum,* named in honor of the German mathematician Georg Friedrich Bernhard Riemann (1826–66). In Question 6**e** of Lesson 13-1, you were asked to find the values of Riemann sums using the right endpoints of the intervals as the z_i.

The process described above can be generalized as follows:

Definition

Let f be a function defined over the interval from a to b. Suppose this interval is partitioned into n subintervals: the first from a to x_1, the second from x_1 to x_2, the third from x_2 to x_3, ..., the nth from x_{n-1} to b. (The lengths of these intervals are $x_1 - a$, $x_2 - x_1$, $x_3 - x_2$, ..., $b - x_{n-1}$). Let $z_1, z_2, z_3, \ldots, z_n$ be numbers in these intervals. Then $f(z_1)(x_1 - a) + f(z_2)(x_2 - x_1) + f(z_3)(x_3 - x_2) + \ldots + f(z_n)(b - x_{n-1})$ is a **Riemann sum of the function *f* over the interval from *a* to *b*.**

By letting $x_0 = a$ and $x_n = b$, the Riemann sum above can be written as

$$\sum_{i=1}^{n} f(z_i)(x_i - x_{i-1}).$$

The points $z_1, z_2, \ldots, z_n$ in the intervals are called the *intermediate points* for the Riemann sum. Thus, the value of a particular Riemann sum depends not only on the function f and the endpoints a and b of the interval, but also on the choice of the endpoints $x_0, x_1, \ldots, x_n$ of the subintervals and on the intermediate points $z_1, z_2, \ldots, z_n$ selected from these subintervals.

When you calculated Riemann sums in Lesson 13-1, the widths $x_i - x_{i-1}$ of the intervals were constant. This need not be the case, but it often is, because it eases computation. When the width is constant, it is often written as $\boldsymbol{\Delta x}$, the change in x. The Riemann sum is then

$$\sum_{i=1}^{n} f(z_i)\,\Delta x.$$

In the example graphed at the beginning of this lesson, the entire interval from 0 to 10 was partitioned into 20 subintervals, so Δx was $\frac{10}{20} = \frac{1}{2}$. In general, if the interval from a to b is split into n subintervals of equal width, then $\Delta x = \frac{b-a}{n}$, and a Riemann sum of g from a to b is

$$\sum_{i=1}^{n} g(z_i)\frac{b-a}{n}.$$

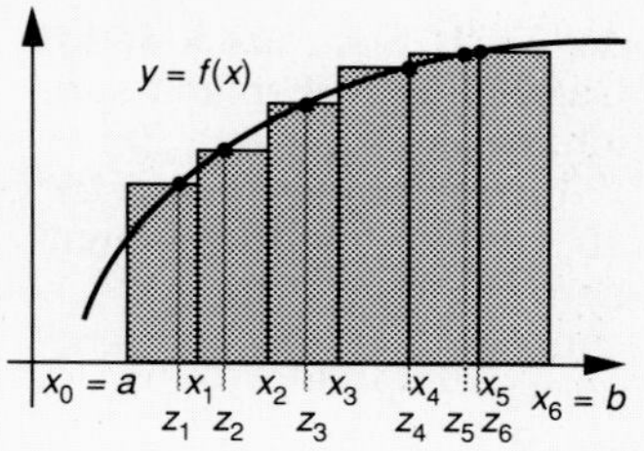

Emphasize that there are many Riemann sums for a given curve over a given closed interval, even if the subintervals are the same width, because there are many choices for z_i. However, if the function is continuous, the limit of those sums as $n \to \infty$ is unique.

For the program on page 770, the exact value of the limit is $586\frac{2}{3}$.

The discussion on page 771 shows that Riemann sums can be negative. Thus, they are a tool that can be employed to calculate area, but Riemann sums should not be equated with area. You might suggest that students think of them as "directed area."

The program in this lesson is used in the next lesson also. As you discuss the lessons, give students time to explore tables of Riemann sum values as $n \to \infty$ using the program.

ADDITIONAL EXAMPLE
Calculate the Riemann sum for the function f with $f(x) = x^3$ on the interval from 0 to 10, using the following numbers of subintervals and choosing z_i as the right endpoints.

a. 10
3025
b. 100
≈ 2550
c. 500
≈ 2510
d. 1000
≈ 2505
The limit is 2500.

Most calculus software and some automatic graphers have the capability to calculate Riemann sums and to show graphically the rectangles whose areas are being calculated. It is also relatively easy to write programs to calculate Riemann sums.

The BASIC computer program below computes Riemann sums for the function $g(x) = -.88(x - 10)^2 + 88$ using the right (upper) endpoints of the subintervals as the z_i.

```
10   DEF FNF(X) = -.88*(X – 10)*(X – 10) + 88
20   PRINT "INPUT LEFT ENDPOINT"
30   INPUT A
40   PRINT "INPUT RIGHT ENDPOINT"
50   INPUT B
60   PRINT "INPUT NUMBER 0F SUBINTERVALS"
70   INPUT N
80   REM CALCULATE DX, THE WIDTH OF EACH SUBINTERVAL
90   LET DX = (B – A)/N
100  LET SUM = 0
110  LET ZI = A
120  FOR I = 1 TO N
130    LET ZI = ZI + DX
140    LET ITHPRD = FNF(ZI)*DX
150    LET SUM = SUM + ITHPRD
160  NEXT I
170  PRINT "THE SUM IS"; SUM
180  END
```

We ran the above program several times for A = 0 and B = 10, and for increasingly larger values of N, the number of subintervals. Our output appears below:

N	Sum
10	629.2001
20	608.3
50	595.408
100	591.0519
500	587.5457
1000	587.1078
5000	586.753
10,000	586.7047

As n gets larger and larger, the Riemann sum gets closer and closer to the exact distance traveled by the car. It seems to be approaching a number slightly larger than 586. This is a good estimate of the distance traveled. In Lesson 13-5, the exact value for the distance is calculated.

Example Calculate Riemann sums for the cosine function $f(x) = \cos x$ on the interval from 0 to $\frac{\pi}{3}$, using various numbers of subintervals and choosing the z_i to be the right endpoints of the subintervals.

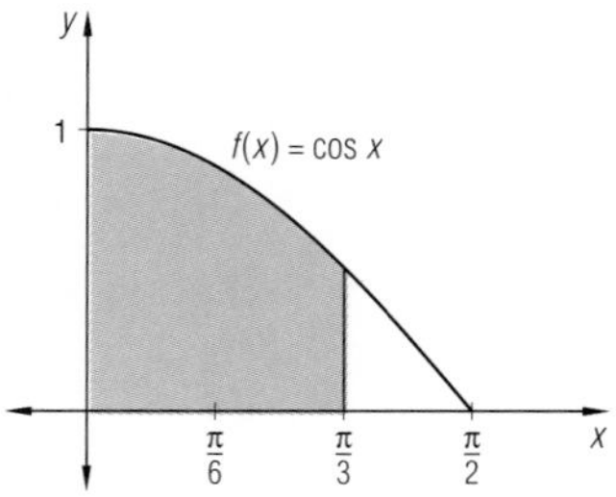

770

Solution This is tedious without a computer or programmable calculator. We used the above program with FNF(X) redefined to be COS X. Here, A = 0 and B = $\frac{\pi}{3} \approx 1.0471976$. We let N = 10, 50, 100, and 500 and obtained the following output, rounded to five decimal places.

N	Sum
10	.83905
50	.86076
100	.86340
500	.86550

That is, the area under the function $f(x) = \cos x$ between $x = 0$ and $x = \frac{\pi}{3}$ is approximately 0.866.

To use z_i = the midpoint of the ith subinterval, change line 110 in the above program to LET ZI = A − DX/2. You should try to run the program with this change for both of the functions defined in this lesson.

If a continuous real function has negative values, then the value of the Riemann sum may be negative.

For instance, the function $g(x) = -.88(x - 10)^2 + 88$ has negative values for $x > 20$, as shown in the graph at the right. To obtain a Riemann sum for the interval from 20 to 30, let A = 20 and B = 30 in the above program. For N = 10, the sum is -1306.8. This is negative because that part of the parabola lies below the x-axis. Its absolute value, 1306.8, is a good approximation to the area of the shaded region at the right.

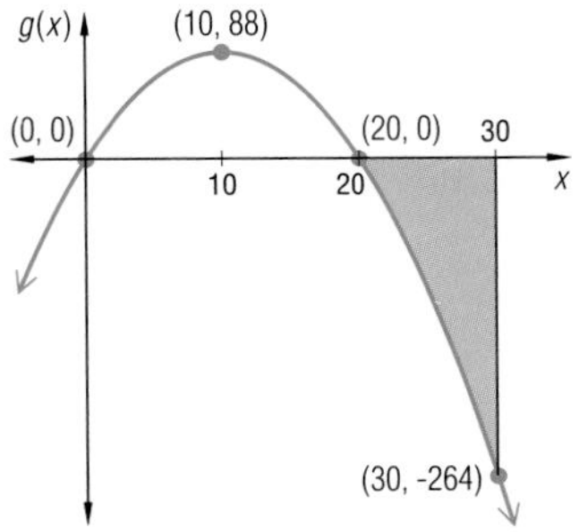

If a Riemann sum is calculated for this function on the interval from 10 to 30, then it will have some positive terms and some negative terms. We found, using the program, that the sum is about -657.535 when N = 50 and -593.711 when N = 500.

Questions

Covering the Reading

In 1–3, use the speed function $g(x) = -.88(x - 10)^2 + 88$ examined in this lesson.

1. Estimate the distance (in feet) traveled between 6.5 and 7 seconds using one rectangle with a height of $g(6.75)$. **39.3525 ft**

2. Give two estimates of the distance traveled between 3 and 3.5 seconds. **sample: using g(3): 22.44 ft; using g(3.5): 25.41 ft**

NOTES ON QUESTIONS
Technology for Questions 6-10: Encourage students to try both computer programs and available calculus software. The calculus software shows the rectangles being used in the calculations. However, the software may not allow values of n as large as 500 or 1000, as required in **Questions 9 and 10**.

ADDITIONAL ANSWERS
9.

N	Sum
10	0.89141
50	0.87123
100	0.86864
500	0.86655

10.

N	Sum
10	541.2
20	564.3
50	577.808
100	582.252
500	585.786
1000	586.227

12.a.

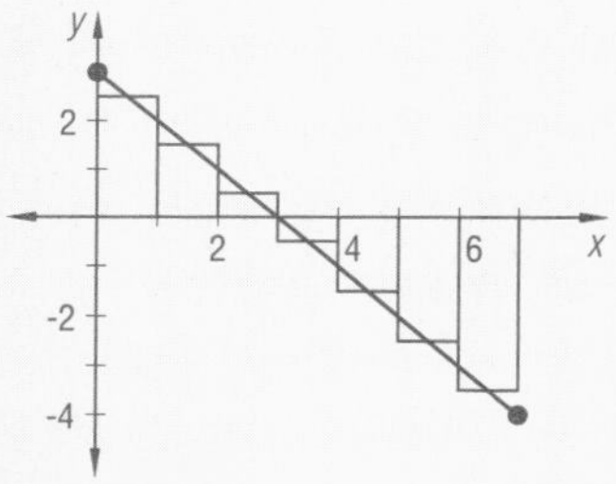

13.a.

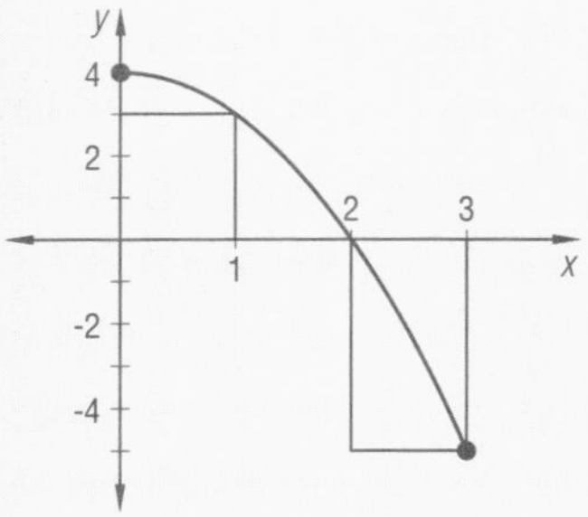

3. Find the Riemann sum of g over the interval from 0 to 10, partitioning the interval into 5 subintervals and using the function values at the midpoints of the subintervals for the heights of the rectangles. **589.6 ft**

4. If an interval with endpoints a and b is split into n subintervals of equal width, what is the width Δx of each subinterval? $\frac{b-a}{n}$

5. A function h is defined by the graph below. The interval $3 \le x \le 11$ has been divided into 4 subintervals of equal width Δx. A point z_i has been chosen in each subinterval. Evaluate $\sum_{i=1}^{4} h(z_i)\,\Delta x$. **4**

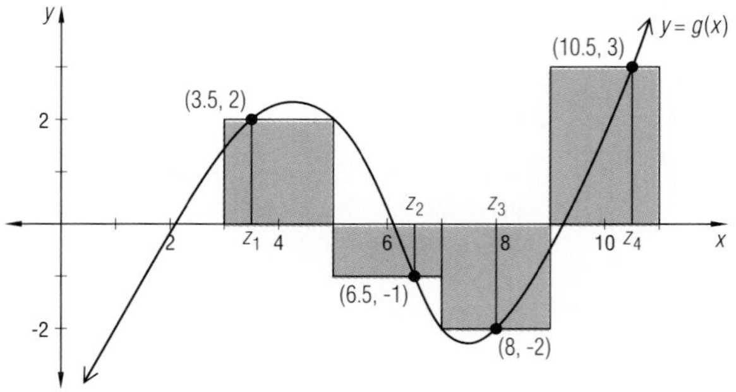

In 6–8, use the computer program of this lesson or other software that can calculate Riemann sums.

6. Repeat Question 3 using 200 subintervals. **588.8631**

7. a. Find a Riemann sum of $g(x) = -.88(x - 10)^2 + 88$ over the interval from -10 to 5, splitting the interval into 30 subintervals of equal length. **-908.0501 (using right endpoints)**

b. Why is the answer to part **a** a negative number? **The region between the graph of g and the x-axis has more area below than above the x-axis.**

8. Find a Riemann sum of $f(x) = \cos x$ over the interval from 0 to $\frac{\pi}{3}$, using 1000 equal-width subintervals. **≈ 0.8658 (using right endpoints)**

Applying the Mathematics

In 9 and 10, renumber line 130 of the computer program to be line 155. Then, the function values at the left endpoints of the subintervals will be used for the heights of the rectangles.

9. Refer to the Example. Let z_i = the left endpoint of the ith subinterval. Evaluate $\sum_{i=1}^{n} f(z_i)\,\Delta x$ for n = 10, 50, 100, and 500. **See margin.**

10. For $g(x) = -.88(x - 10)^2 + 88$ on the interval from 0 to 10, evaluate $\sum_{i=1}^{n} g(z_i)\,\Delta x$ for n = 10, 20, 50, 100, 500, and 1000 if z_i = the left endpoint of the ith subinterval. **See margin.**

772

11. A graph of $y = h(x)$ is given below.

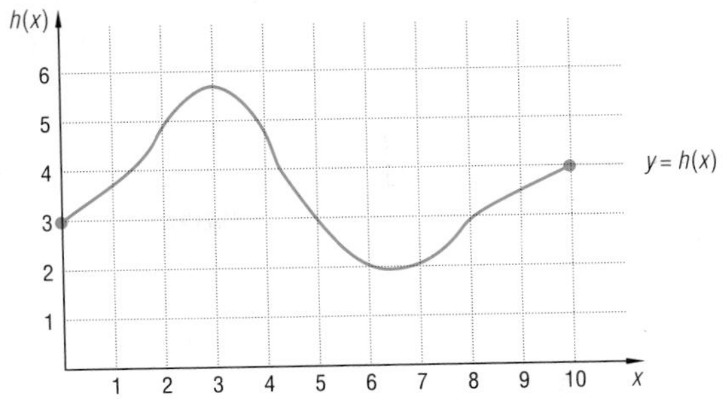

Partition the interval from 0 to 10 into 5 subintervals of equal length Δx. Evaluate $\sum_{i=1}^{5} h(z_i)\,\Delta x$, estimating each $h(z_i)$ to the nearest integer, when

a. z_i = the right endpoint of the ith subinterval **≈ 38**

b. z_i = the left endpoint of the ith subinterval **≈ 36**

c. z_i = the midpoint of the ith subinterval. **≈ 35.4**

In 12–14, **a.** sketch the function on the specified interval. **b.** Estimate whether the given Riemann sum will be positive, negative, or zero without actually computing it.

12. $f(x) = -x + 3$ on the interval $0 \le x \le 7$; $\Delta x = 1$; z_i = the midpoint of the ith subinterval; Riemann sum = $\sum_{i=1}^{7} f(z_i)\,\Delta x$ **a) See margin.** **b) negative**

13. $g(x) = -x^2 + 4$ on the interval $0 \le x \le 3$; $\Delta x = 1$; z_i = the right endpoint of the ith subinterval; Riemann sum = $\sum_{i=1}^{3} g(z_i)\,\Delta x$
a) See margin. b) negative

14. $h(x) = -x^2 + 4$ on the interval $0 \le x \le 3$; $\Delta x = 1$; z_i = the left endpoint of the ith subinterval; Riemann sum = $\sum_{i=1}^{3} h(z_i)\,\Delta x$
a) See margin. b) positive

15. A construction company is building an airport runway and needs to level the ground. The low spots are to be filled in using some of the dirt from the high spots, and the rest of the dirt is to be hauled away. This plan shows a cross-section of the region. Estimate the area that remains above the horizontal axis after the low spot below has been filled in. **≈ 10,500 ft²**

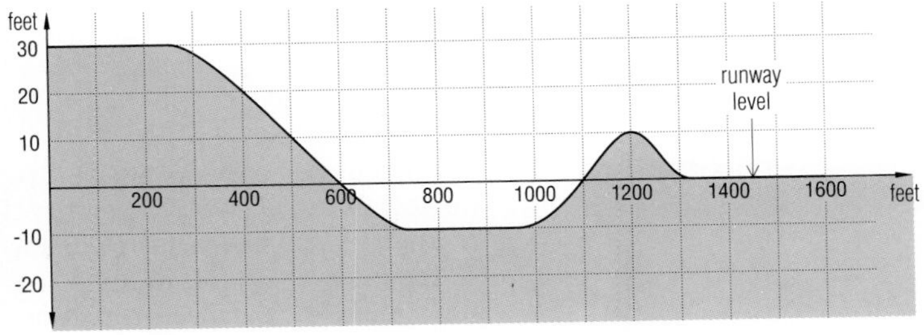

14.a.

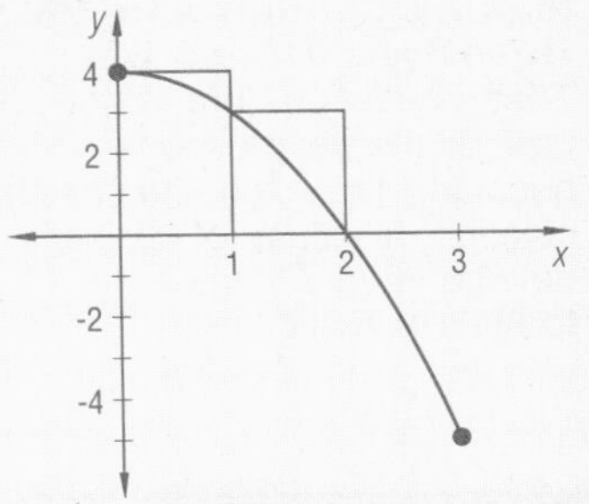

NAME

LESSON **MASTER 13–2**
QUESTIONS ON **SPUR** OBJECTIVES

■ SKILLS *Objective A (See pages 810–813 for objectives.)*

1. For the function $f(x) = 3x^3 - 1$, calculate the Riemann sum over the interval $0 \le x \le 2$ for $\Delta x = .25$ when

a. z_i = the left endpoint of the ith subinterval. **7.19**

b. z_i = the right endpoint of the ith subinterval. **13.19**

2. a. For the function $g(x) = 2 \sin x$, evaluate the Riemann sum $\sum_{i=1}^{n} g(z_i)\Delta x$ over the interval from 0 to $\frac{\pi}{2}$ with n = 4, 8, and 16. Let z_i be the right endpoint of the ith subinterval. **2.367 ($n = 4$); 2.190 ($n = 8$); 2.097 ($n = 16$)**

b. Which value of n provides an answer that is nearest the area under the graph of g? Why?
n = 16; By making the rectangles narrower, the amount of error is decreased.

c. To what value might you expect the Reimann sum to converge as n grows larger and larger? **2**

3. Use a computer or programmable calculator to evaluate the Reimann sum for the function $g(x) = x^2(\sin x - 2\cos x)$ over the interval from 0 to $\frac{\pi}{2}$ with n = 10, 50, 100, and 500. Choose z_i to be the right endpoint of the ith subinterval.
.1472 ($n = 10$); .2462 ($n = 50$); .2263 ($n = 100$); .2107 ($n = 500$)

NAME
Lesson MASTER 13–2 (page 2)

■ USES *Objective D*

4. The graph below indicates the velocity of a train during a 1 hour interval.

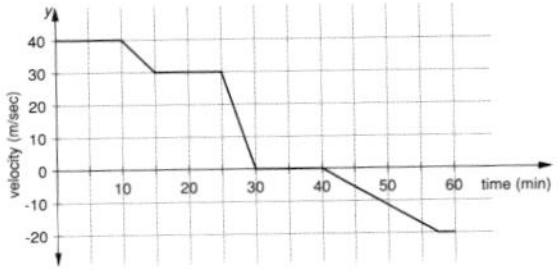

a. How far did the train travel in the first half hour? **57,000 m**

b. At the end of the hour, what is the train's distance from its position at the beginning of the interval? **43,500 m**

5. A cyclist accelerates from 36 ft/sec to 60 ft/sec during the last 5 seconds of the race. The cyclist's velocity t seconds after beginning to accelerate is given by $v(t) = .96t^2 + 36$. Estimate the distance the cyclist travels during these 5 seconds using a Riemann sum where $n = 5$ and

a. z_i = the right endpoint of the ith subinterval. **232.8 ft**

b. z_i = the left endpoint of the ith subinterval. **208.8 ft**

c. Which of the above answers is closer to the exact distance? Why? (Hint: Sketch the velocity-time graph.)
b; Since the velocity function is concave up, the lower Riemann sum provides a better estimate.

NOTES ON QUESTIONS
Making Connections for Questions 22 and 23:
These questions are important for the next lessons because they allow the calculation of values of certain definite integrals when no formula is available.

FOLLOW-UP

MORE PRACTICE
For more questions on SPUR Objectives, use *Lesson Master 13-2*, shown on page 773.

PROJECTS
The projects for Chapter 13 are described on pages 806–807. **Projects 1 and 6** are related to the content of this lesson.

ADDITIONAL ANSWERS
18.a. sample:

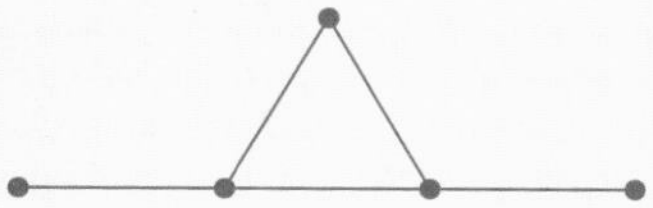

21.a.

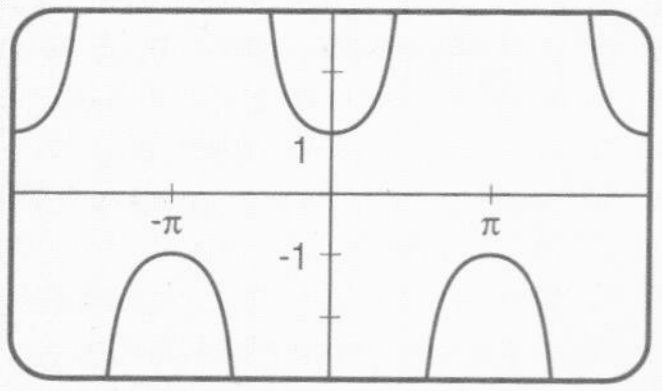

$y = (\tan\theta)(\sin\theta + \cot\theta\cos\theta) = \sec\theta$

It appears to be an identity.
b. See Additional Answers in the back of this book.

Review

16. The area between a velocity-time curve and the time axis represents what quantity? *(Lesson 13-1)* **distance**

17. A velocity-time graph for a freight train traveling along a straight stretch of track is pictured below. How far and in what direction is the train from its starting point at the end of the trip? *(Lesson 13-1)*

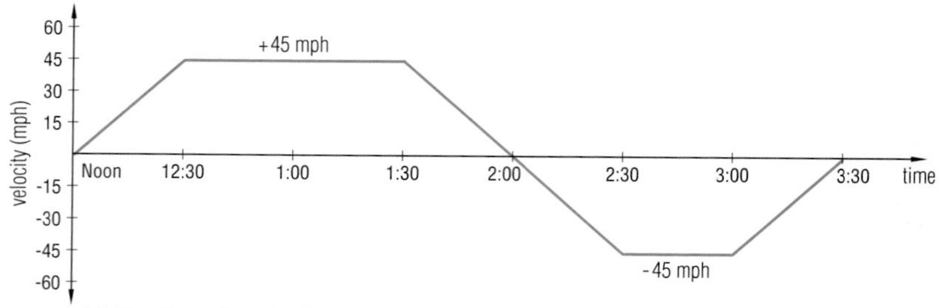

22.5 miles ahead of where it started

18. a. Draw a picture of a graph with exactly one circuit. **See margin.**
b. Identify the length of the circuit. **Answers will vary.**
c. Identify each edge which could be removed while keeping the graph connected. How many such edges are there? *(Lesson 11-4)* **c) any single edge of the circuit**

19. Find all zeros and the corresponding multiplicities for the polynomial function g with $g(x) = x^4 - 5x^3 + 6x^2 + 4x - 8$. *(Lessons 8-8, 4-6)*
x = -1 has multiplicity 1, and x = 2 has multiplicity 3.

20. Use the formula $\sum_{i=1}^{n} i^3 = \left[\frac{n(n+1)}{2}\right]^2$ to find $1^3 + 2^3 + \ldots + 25^3$. *(Lesson 7-3)* **105,625**

21. a. Use an automatic grapher to conjecture whether $(\tan\theta)(\sin\theta + \cot\theta\cos\theta) = \sec\theta$ is an identity. **See margin.**
b. If it appears to be an identity, prove it. If it is not, give a counterexample. *(Lessons 6-2, 6-1)* **See margin.**

In 22 and 23, find the area of the shaded region. *(Previous course)*

22.

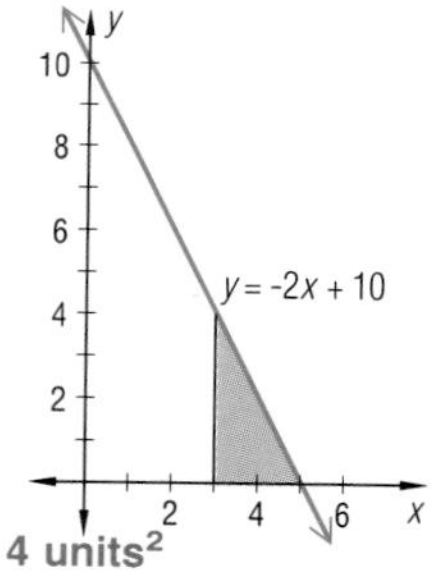

4 units^2

23.

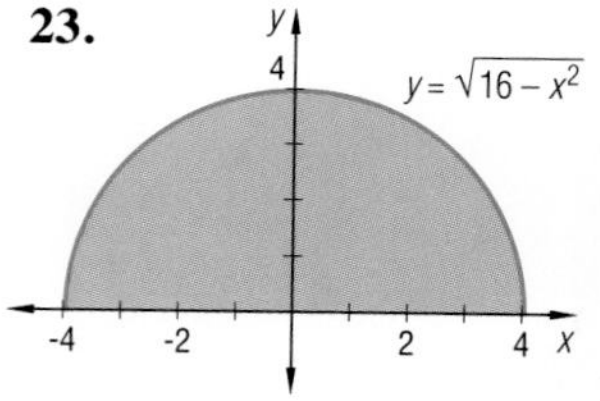

8π units^2

Exploration

24. Again consider $g(x) = -.88(x - 10)^2 + 88$, and use the program of this lesson to help you answer the following questions.
a. As $n \to \infty$, what seems to be the limit of Riemann sums of g over the interval from 0 to 30? **0**
b. Explain your answer to part **a.** **The region between the graph of *g* and the *x*-axis has just as much area above the *x*-axis as below it.**

LESSON

13-3

The Definite Integral

In the two previous lessons, you calculated Riemann sums of functions over intervals. The algorithm you used for subintervals of equal length can be described as follows.

Step 1: Identify the function f.
2: Identify a and b, the endpoints of the interval.
3: Identify n, the number of subintervals. Then $\Delta x = \frac{b-a}{n}$.
4: Pick values $z_1, z_2, z_3, \ldots, z_n$ in each subinterval.
5: Calculate the Riemann sum

$$\sum_{i=1}^{n} f(z_i)\frac{b-a}{n} = \sum_{i=1}^{n} f(z_i)\,\Delta x.$$

When $f(x) \geq 0$ for all x between a and b, the Riemann sum estimates the area between the graph of f and the x-axis. If f is a speed function, the Riemann sum estimates the total distance traveled. (It has other applications you have not yet studied.) When f is a nonconstant continuous function on $a \leq x \leq b$, it can be proved that f has a largest and a smallest value on each of the subintervals involved in the Riemann sum. If each $f(z_i)$ is the largest value of the function on the subinterval, then the estimate is too high and the sum is called an **upper Riemann sum.** If each $f(z_i)$ is the smallest value on the subinterval, then the estimate is too low and the sum is called a **lower Riemann sum.**

For the increasing function $f(x) = \sin x$ on the interval sketched below, taking the z_i to be the right endpoints of the subintervals yields an upper Riemann sum. Using the left endpoints yields a lower Riemann sum.

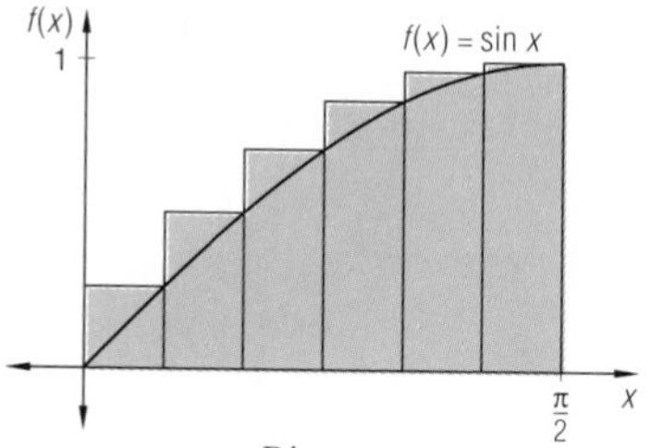

upper Riemann sum

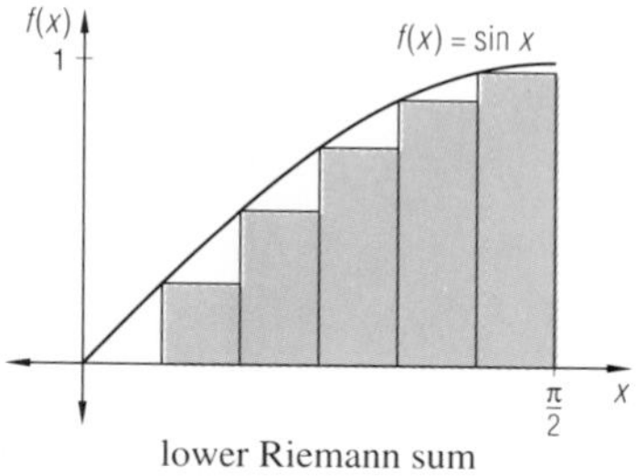

lower Riemann sum

We used the program from Lesson 13-2 to calculate these upper and lower Riemann sums from A = 0 to B = $\frac{\pi}{2}$ for N = 6, 10, 100, and 1000, and obtained the following results, rounded to five decimal places.

N	upper Riemann sum	lower Riemann sum
6	1.12518	.86338
10	1.07648	.91940
100	1.00783	.99213
1000	1.00079	.99922

LESSON 13-3

RESOURCES
- Lesson Master 13-3
- Teaching Aid 102 displays the upper and lower Riemann sums.
- Teaching Aid 103 can be used with **Examples 1-4**.
- Computer Master 22

OBJECTIVES

B Evaluate definite integrals.
G Express areas in integral notation.

TEACHING NOTES

Technology In this lesson, we look at Riemann sums as n approaches infinity and obtain definite integrals. Some calculus software draws the rectangles for Riemann sums for values of $n \leq 300$ and lets the user look at upper sums, lower sums, and the difference between upper and lower sums. Using this or similar software enables students to visualize what happens to Riemann sums as $n \to \infty$. In the table on this page, the upper Riemann sums appear to be decreasing toward a limiting value of 1, and the lower Riemann sums appear to be increasing toward a limiting value of 1 as $n \to \infty$. That is, in this example,

$$\lim_{n\to\infty} \sum_{i-1}^{n} \sin(z_i)\,\Delta x = \int_0^{\frac{\pi}{2}} \sin(x)\,dx = 1.$$

Point out to students that when they use a computer program to find upper and lower sums, they first will need to determine whether to use left or right endpoints. If a function is increasing on an interval, right endpoints give upper sums and left endpoints give lower sums. However, if a function is decreasing on an interval, the opposite is true.

Ask students if they can determine the purpose of each example in the lesson. (**Example 1**: to see if integral notation is understood; **Example 2**: to show that the integral of a linear function can be calculated exactly; **Example 3**: to show that the integral of a constant function can be calculated exactly; **Example 4**: to show that some other integrals can be calculated or estimated using area formulas from geometry.)

If you discuss the program from Lesson 13-2 with this lesson, give students time to work with both upper and lower Riemann sums so they can see the convergence.

Note that as n increases, the upper and lower Riemann sums get closer and closer to each other. This always happens when f is a continuous function. The choices of the z_i make less and less of a difference, so that the upper and lower Riemann sums approach the same limit. This limit is called the **definite integral of *f* from *a* to *b*,** written

$$\int_a^b f(x)\,dx$$

and read "the integral of f from a to b." That is, if f is a continuous function, then $\int_a^b f(x)\,dx = \lim_{n\to\infty} \sum_{i=1}^{n} f(z_i)\,\Delta x$ for any choices of the z_i in the subintervals.

The symbol "$\int$" is the integral sign. It is an elongated "s," which should remind you that the definite integral is a limit of sums. The a and b at the bottom and top of this symbol signify the endpoints of the interval. The symbol "dx" indicates that the independent variable is x.

While a Riemann sum $\sum_{i=1}^{n} f(z_i)\,\Delta x$ usually approximates a quantity such as distance or area, the corresponding definite integral $\int_a^b f(x)\,dx$ gives the *exact* value of this quantity.

Example 1 Express the area of the shaded region below with integral notation.

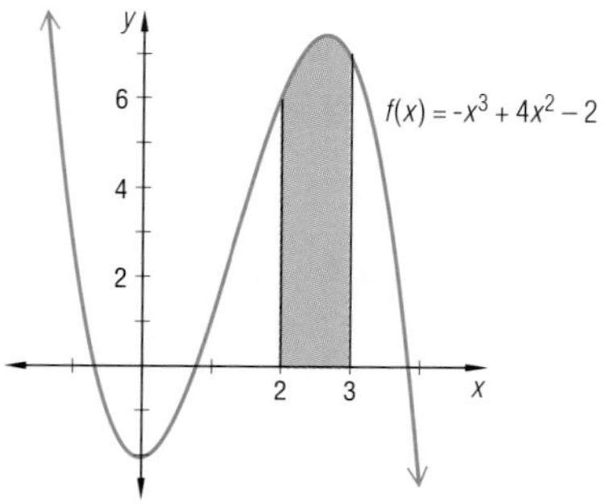

Solution The shaded area is the limit of Riemann sums of the form $\sum_{i=1}^{n} (-z_i^3 + 4z_i^2 - 2)\,\Delta x$, from $x = 2$ to $x = 3$. This is written $\int_2^3 (-x^3 + 4x^2 - 2)\,dx$.

If f is a constant or linear function, the exact value of $\int_a^b f(x)\,dx$ can be found using familiar area formulas.

Example 2 Find the exact value of $\int_2^8 \left(\frac{1}{2}x + 3\right) dx$.

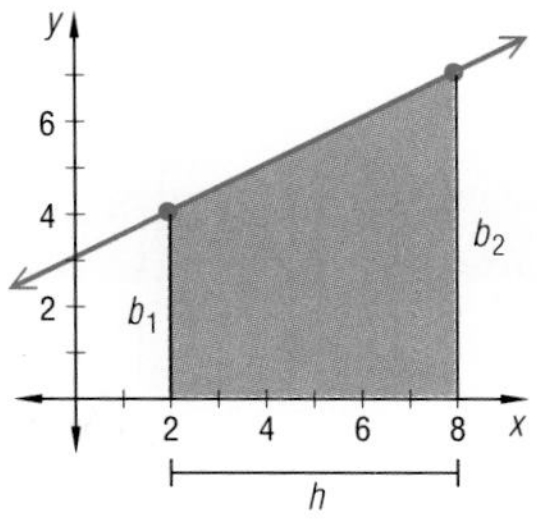

Solution Sketch $y = \frac{1}{2}x + 3$ on the interval from 2 to 8.

$\int_2^8 \left(\frac{1}{2}x + 3\right) dx$ = the area under $y = \frac{1}{2}x + 3$ on this interval. The shaded region is a trapezoid with

area $= \frac{1}{2}(b_1 + b_2) \cdot h = \frac{1}{2}(f(2) + f(8)) \cdot 6 = \frac{1}{2}(4 + 7) \cdot 6 = 33$.

Since a Riemann sum is negative when the function values are all negative, so is the value of the corresponding definite integral. As Example 3 illustrates, the definite integral represents the opposite of the area of the region, if that region is below the x-axis.

Example 3 Find the exact value of $\int_{-2}^{3} (-4)\, dx$.

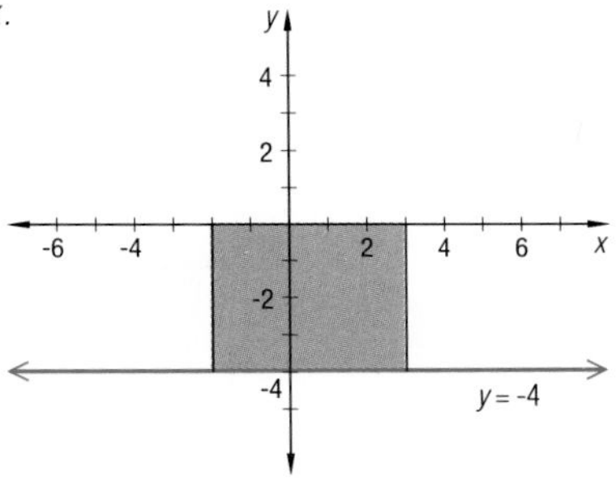

Solution Graph $y = -4$ on the interval from -2 to 3; it is a 5-by-4 rectangle. Since the region between $y = -4$ and the x-axis is below the x-axis, the value of the definite integral is negative.

$\int_{-2}^{3} (-4)\, dx = 5 \cdot (-4) = -20$

The formula for the area of a circle can help to find values of definite integrals of some other functions.

Example 4 Estimate $\int_2^3 \sqrt{9 - x^2}\, dx$.

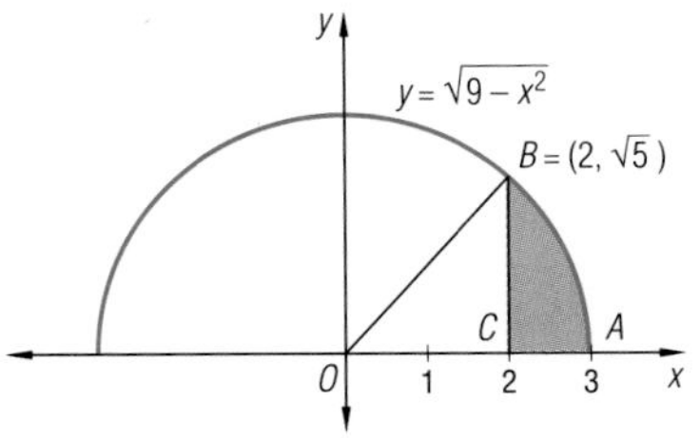

Solution The graph of $f(x) = \sqrt{9 - x^2}$ is a semicircle with radius 3. The given integral represents the area of the region bounded by the semicircle, the x-axis, and the lines $x = 2$ and $x = 3$. Call the region ABC. (It is shaded above.) Then $m\angle AOB = \tan^{-1} \frac{\sqrt{5}}{2}$, and so $m\angle AOB \approx 48.2°$. Thus the area of sector AOB is about $\frac{48.2}{360}$ of the area of the circle, or $\frac{48.2}{360} \cdot 9\pi$. The area of ΔOBC is $\frac{1}{2} \cdot 2 \cdot \sqrt{5} = \sqrt{5}$. Thus $\int_2^3 \sqrt{9 - x^2}\, dx \approx \frac{48.2}{360} \cdot 9\pi - \sqrt{5} \approx 1.55$.

ADDITIONAL EXAMPLES

1. Express the area between the graph of $y = 16x - x^2$ and the x-axis with integral notation.

$\int_0^{16} (16x - x^2)\, dx$

2. Find the exact value of $\int_4^{11} (3x + 1)\, dx$.

164.5

3. Find the exact value of $\int_{-5}^{5} 7\, dx$.

70

4. Estimate $\int_3^5 \sqrt{25 - x^2}\, dx$.

≈ 5.59

NOTES ON QUESTIONS
Questions 1 and 2:
Since the parabola is the subject of Lesson 13-5, these questions should be discussed in preparation for that lesson.

ADDITIONAL ANSWERS
2.a. lower sum: 0.328350; upper sum: 0.338350
b. lower sum: 0.332829; upper sum: 0.333828

11.a.

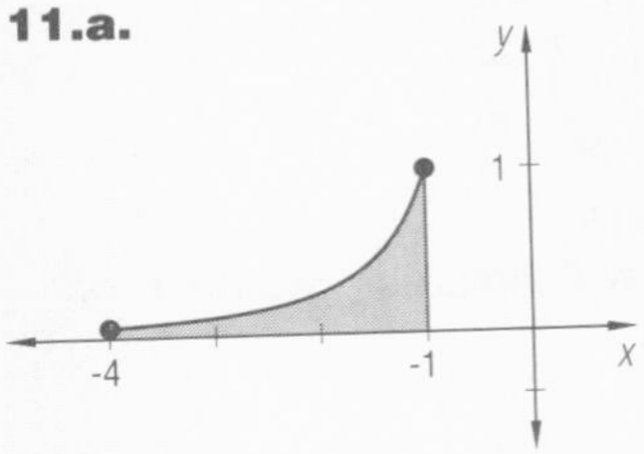

12.a.

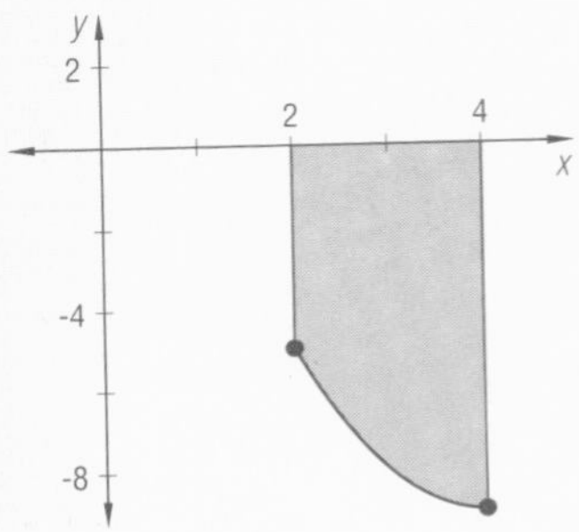

Questions

Covering the Reading

In 1 and 2, use the parabola $y = x^2$ graphed below. Consider Riemann sums on the interval from 0 to 1.

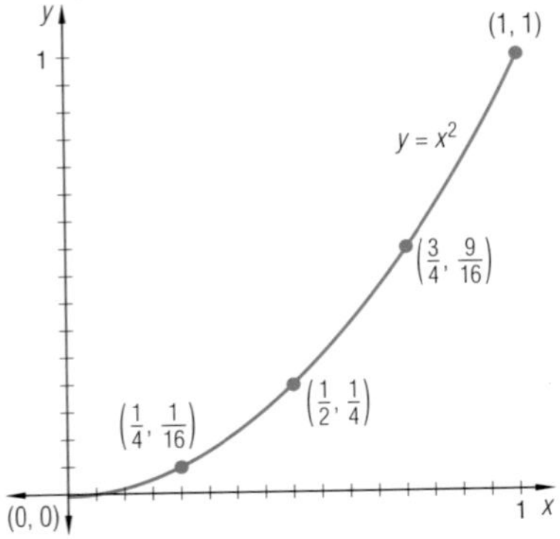

1. Give the values of the upper and lower Riemann sums with 4 equal-width subintervals. **upper sum: $\frac{15}{32}$; lower sum: $\frac{7}{32}$**

2. Use a computer to find the upper and lower Riemann sums with
a. 100 subintervals; **b.** 1000 subintervals. **See margin.**

3. a. For continuous functions f, _?_ and _?_ Riemann sums on an interval $a \le x \le b$ always approach the same limit.
b. What is that limit called? **the definite integral of *f* from *a* to *b***
a) upper; lower

In 4–6, express the area of the shaded region as a definite integral. (You do not have to find the value of the integral.)

4.

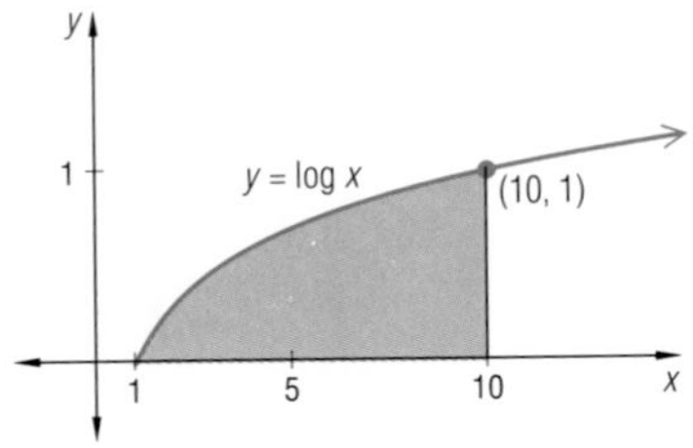

5.

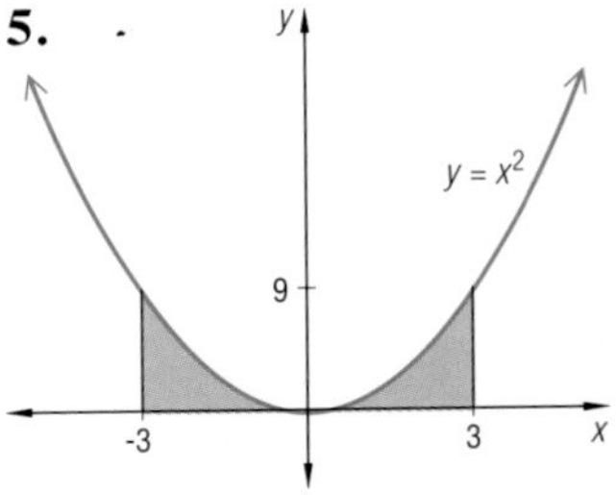

6.

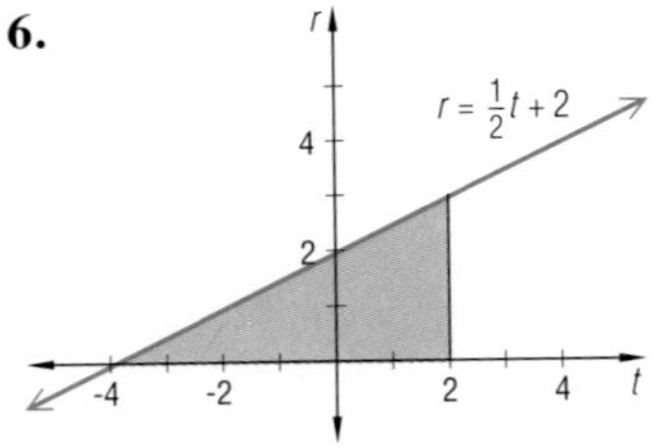

4. $\int_1^{10} (\log x)\, dx$

5. $\int_{-3}^{3} x^2\, dx$

6. $\int_{-4}^{2} \left(\frac{1}{2}t + 2\right) dt$

778

In 7–9, find the exact value of the definite integral.

7. $\int_1^4 (3x + 2)\, dx$ **28.5** **8.** $\int_{-100}^{100} 3\, dx$ **600**

9. $\int_{-3}^{3} \left(-\sqrt{9 - x^2}\right) dx$ **-4.5π**

10. Estimate $\int_1^5 \sqrt{25 - x^2}\, dx$ to the nearest hundredth. **14.67**

Applying the Mathematics

In 11–13, a definite integral of a function is given. **a.** Sketch the graph of the function over the indicated interval. **b.** Does the value of the integral appear to be positive or negative?

11. $\int_{-4}^{-1} \frac{1}{x^2}\, dx$ **a) See margin. b) positive**

12. $\int_2^4 (x^2 - 8x + 7)\, dx$ **a) See margin. b) negative**

13. $\int_0^7 (x^2 - 8x + 7)\, dx$ **a) See margin. b) negative**

14. Use area to find a formula for $\int_0^a mx\, dx$, where a and m are positive real numbers. $\frac{ma^2}{2}$

15. Use area to find a formula for $\int_a^b c\, dx$, where a, b, and c are positive and $a < b$. **c(b – a)**

Review

16. The graph shows the velocity of a boat traveling in a straight path on a river over a five-hour interval.

a. Use a Riemann sum with 5 subintervals and the grid to find the approximate distance covered. Let z_i = the right endpoint of the ith subinterval. **≈ 156 miles**

b. What does the area between the curve $r = f(t)$ and the t-axis represent? *(Lessons 13-2, 13-1)*

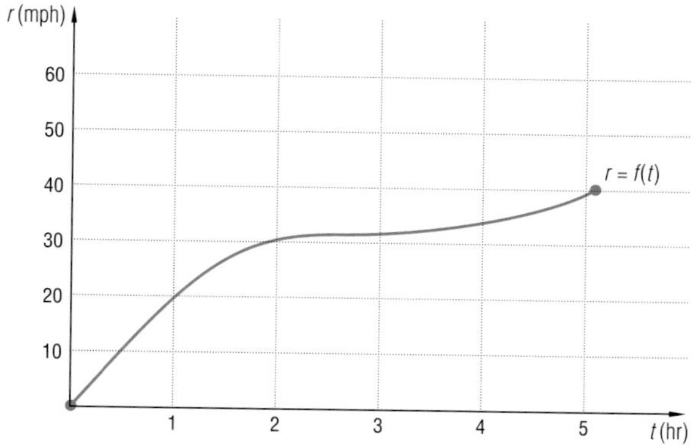

the distance from the starting to ending point for the 5-hour period

17. A machine offers 6 kinds of candy bars. In how many different ways can 4 candy bars be selected if

a. they must all be different kinds? **15**

b. more than one of the same kind can be chosen? **126**

(Lessons 10-7, 10-4)

13.a.

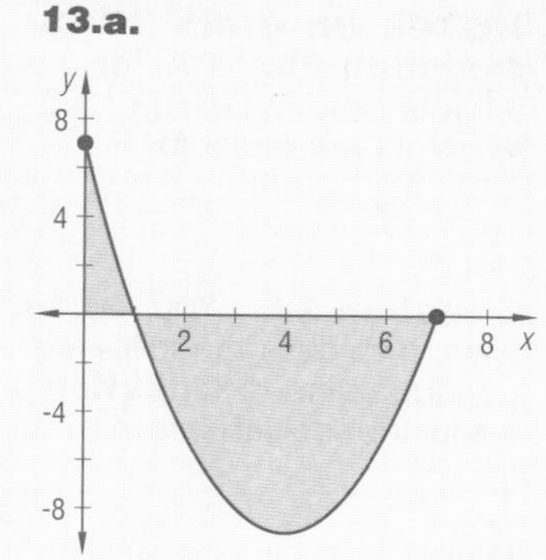

NAME ____________

LESSON **MASTER 13–3**
QUESTIONS ON **SPUR** OBJECTIVES

■ SKILLS *Objective B (See pages 810–813 for objectives.)*
In 1–4, find the exact value of the definite integral.

1. $\int_2^5 4\, dx$ **12**
2. $\int_{-1}^{3} x\, dx$ **4**
3. $\int_0^5 \sqrt{25 - x^2}\, dx$ $\frac{25\pi}{4}$
4. $\int_0^2 (2x + 2)\, dx$ **8**

5. Estimate the value of $\int_3^5 \sqrt{25 - x^2}\, dx$ to the nearest hundredth. **5.58**

■ REPRESENTATIONS *Objective G*
In 6–9, express the shaded area using integral notation.

6.

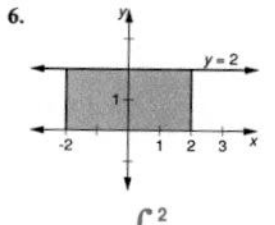

$\int_{-2}^{2} 2\, dx$

7.

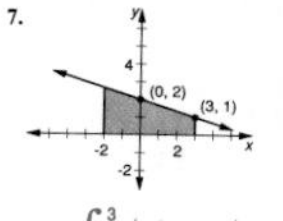

$\int_{-2}^{3} \left(-\frac{1}{3}x + 2\right) dx$

8.

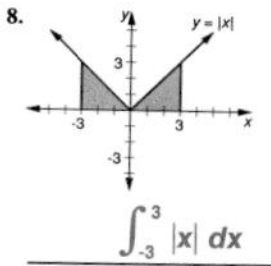

$\int_{-3}^{3} |x|\, dx$

9.

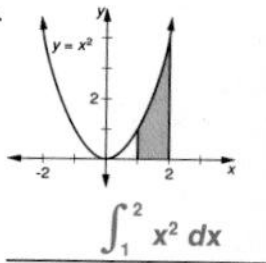

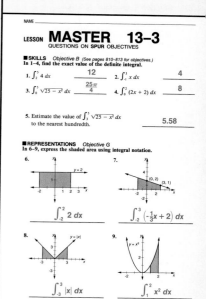

$\int_1^2 x^2\, dx$

NOTES ON QUESTIONS
Question 20: This formula is used in Lesson 13-5 to obtain a formula for $\int_0^a x^2\,dx$.

Question 22: You might wish to discuss the following generalization of **Question 14** for the special case $m = 1$.
$$\int_a^b x\,dx = \frac{b^2}{2} - \frac{a^2}{2}.$$
Make certain that students realize that a and/or b could be negative. For **Question 15**, a generalization is:
$$\int_a^b c\,dx = c(b - a).$$

FOLLOW-UP

MORE PRACTICE
For more questions on SPUR Objectives, use *Lesson Master 13-3*, shown on page 779.

PROJECTS
The projects for Chapter 13 are described on pages 806–807. **Projects 1, 2, and 6** are related to the content of this lesson.

ADDITIONAL ANSWERS
19.

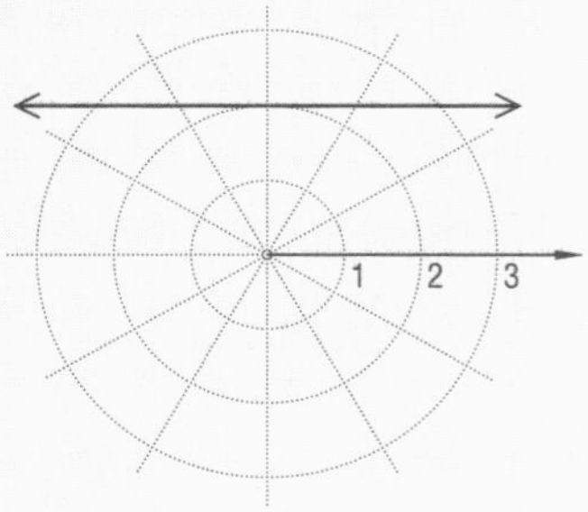

18. If a bank pays 6% interest, compounded continuously, then the amount A in an account t years after P dollars has been deposited is
$$A = Pe^{.06t}.$$
Find the average rate of change of the amount in the account from $t = 5$ years to $t = 10$ years if the following amount is deposited. *(Lesson 9-1)*

a. \$100 **\$9.45/yr** **b.** \$200 **\$18.89/yr** **c.** \$$P$ **.09445 P/yr**

19. Sketch the graph of the polar equation $r \sin \theta = 2$ and identify the figure you get. *(Lesson 8-4)* **See margin. The graph is a line.**

20. Given that $\sum_{i=1}^{n} i^2 = \frac{n(n+1)(2n+1)}{6}$, find $100^2 + 101^2 + 102^2 + \ldots + 200^2$. *(Lesson 7-4)*
2,358,350

21. A machine makes nails of length $1\frac{1}{2}$ inches. If the relative error can be no more than .5%, what nail lengths are acceptable? *(Lesson 3-8)*
$1.4925 \le \ell \le 1.5075$

Exploration

22. In Questions 14 and 15, you used ideas of area to explain why a particular property of definite integrals is true. Based on your knowledge of graphs and transformations, conjecture one or more additional properties of definite integrals. Then use an area argument to explain why the property is true.

sample: $k\int_0^a f(x)\,dx = \int_0^{\frac{a}{k}} f(x)\,dx$.
Under a vertical scale change of magnitude k, the area is multiplied by k.

LESSON

13-4

Properties of Definite Integrals

Because definite integrals can be interpreted as areas, properties of area can be used to deduce properties of these integrals. Four properties of definite integrals are developed in this lesson. The first two involve combining definite integrals of the same function over different intervals.

The sum of the areas under a curve over the adjacent intervals $a \leq x \leq b$ and $b \leq x \leq c$ equals the area under the curve above the union of the intervals $a \leq x \leq c$. If the curve is the graph of a continuous function f, then the following theorem, stated in the notation of definite integrals, holds.

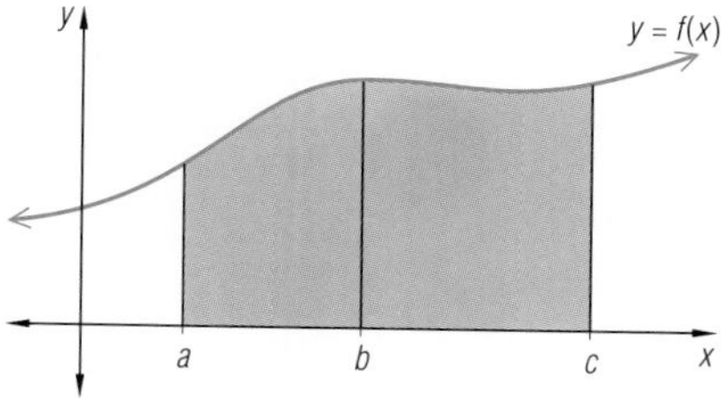

Theorem

If f is a continuous function on the interval $a \leq x \leq c$, and $a < b < c$, then $\int_a^b f(x)\,dx + \int_b^c f(x)\,dx = \int_a^c f(x)\,dx$.

From this first theorem, a second theorem can be deduced. Substitute 0 for a, a for b, and b for c in the equation.

$$\int_0^a f(x)\,dx + \int_a^b f(x)\,dx = \int_0^b f(x)\,dx$$

Now subtract the left integral from both sides.

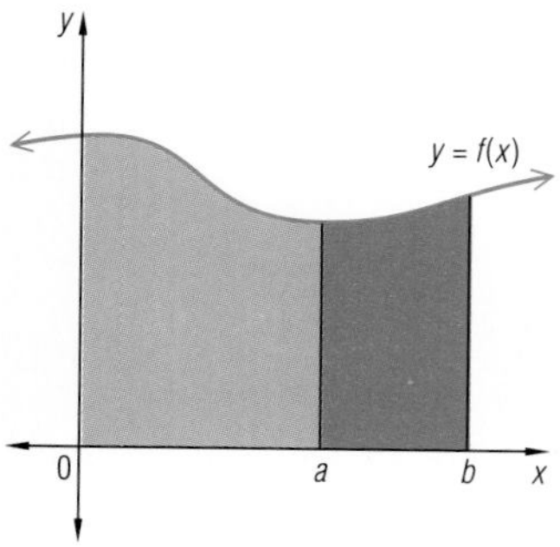

Theorem

If f is a continuous function on the interval $0 \leq x \leq b$, and $0 < a < b$, then $\int_a^b f(x)\,dx = \int_0^b f(x)\,dx - \int_0^a f(x)\,dx$.

The area interpretation of this theorem is as follows: If the area under the curve over the interval $0 \leq x \leq a$ (in blue at the left) is subtracted from the area over the interval $0 \leq x \leq b$, then the result is the area under the curve over the interval from $a \leq x \leq b$ (in green). The importance of this theorem is that if a formula is known for the area between the graph of a curve and the x-axis over the interval $0 \leq x \leq b$ for any b, then the formula can be utilized to find the area over any interval in the positive part of the x-axis.

LESSON 13-4

RESOURCES

- Lesson Master 13-4
- Quiz for Lessons 13-1 through 13-4
- Teaching Aid 104 can be used with **Questions 12 and 14**.

OBJECTIVES

B Evaluate definite integrals.
C Apply properties of definite intervals.
E Use the definite integal to solve application problems.
H Find areas bounded by curves.

TEACHING NOTES

The theorems on this page have simple area interpretations. One measure of students' understanding of integral notation is their ability to give the correct interpretations. The interpretation of the first theorem precedes it and the interpretation of the second theorem follows it.

Making Connections Point out to students that the situation of **Example 2** is quite realistic. Rate of flow is proportional to water pressure and can be known even if the total amount is not known. Point out the similarity between the following:

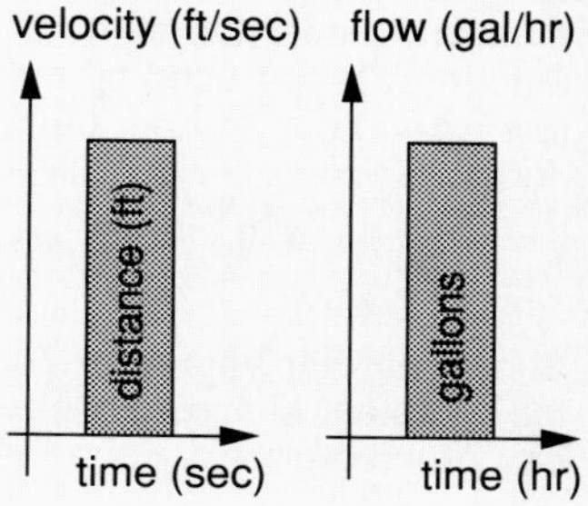

Extending this idea to other examples can show some of the many applications of integration to real-world problems; for example force = mass × acceleration; work = force × distance; and so on.

ADDITIONAL EXAMPLES

1. Verify the first theorem of the lesson when $f(x) = 2x + 6$, $a = -3$, $b = 9$, and $c = 20$.

$\int_{-3}^{9} (2x + 6)\,dx = 144;$
$\int_{9}^{20} (2x + 6)\,dx = 385;$
$\int_{-3}^{20} (2x + 6)\,dx = 529$, **so the sum of the first two integrals equals the third.**

2. Use the data from **Example 2** of the lesson.

a. Estimate how much more water comes from the pipe with rate of flow $f(t)$ than from the pipe with rate of flow $g(t)$.
Dividing into 12 subintervals and using the right endpoints, an estimate is 1550 gallons.

b. Use integral notation to represent the exact difference.
$\int_{0}^{24} [f(t) - g(t)]\,dt$, **or** $\int_{0}^{24} f(t)dt - \int_{0}^{24} g(t)\,dt$

Show both answers to exhibit the property that the integral of the difference of two functions is equal to the difference of the integrals of the individual functions.

3. If, due to drought, the rate of flow in the pipes in **Example 2** is reduced to 75% of what it was, use integral notation to describe the total amount of water used by the first pipe in the 24-hour period.
$\frac{3}{4}\int_{0}^{24} f(t)\,dt$, **or** $\int_{0}^{24} \frac{3}{4}f(t)\,dt$

Show both answers to exhibit the Constant Multiple Property of Integrals.

Example 1 Verify the second theorem on page 781 in the case that $a = 2$, $b = 7$, and $f(x) = 3x + 6$.

Solution Sketch the function f over the interval $0 \le x \le 7$. Then the three integrals in the theorem can be evaluated by finding areas of trapezoids.

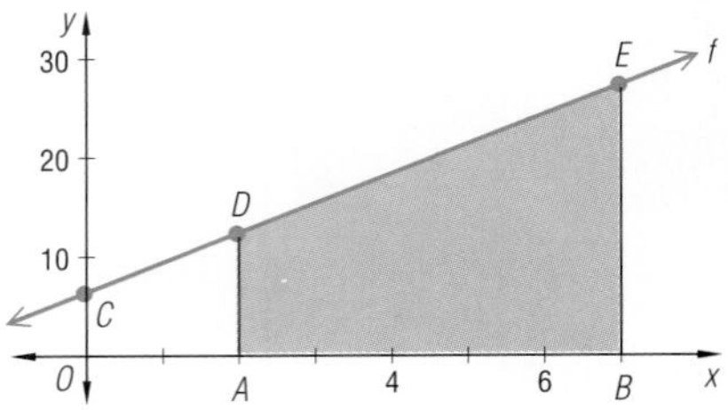

$$\int_a^b f(x)\,dx = \int_2^7 (3x + 6)\,dx = \text{area of } ABED = \frac{1}{2}(f(2) + f(7)) \cdot 5 = \frac{1}{2}(12 + 27) \cdot 5 = 97.5$$

$$\int_0^b f(x)\,dx = \int_0^7 (3x + 6)\,dx = \text{area of } OBEC = \frac{1}{2}(6 + 27) \cdot 7 = 115.5$$

$$\int_0^a f(x)\,dx = \int_0^2 (3x + 6)\,dx = \text{area of } OADC = \frac{1}{2}(6 + 12) \cdot 2 = 18$$

Then $\int_0^b f(x)\,dx - \int_0^a f(x)\,dx = 115.5 - 18 = 97.5 = \int_a^b f(x)\,dx$.

The third and fourth properties concern integrals of different functions over the same interval. Consider the following situation which involves the definite integral of the sum of two functions.

Example 2 Suppose a field in the Great Plains is irrigated by two pipes. Due to changes in water pressure, the rate of flow in these pipes varies; the rate is recorded by devices attached to the pipes.

a. Given the rate-of-flow charts for these two pipes for a day, estimate the total amount of water used to irrigate the field for that 24-hour period.

b. If $f(t)$ and $g(t)$ represent the rates of flow (in hundred gallons per hour) in the two pipes at time t, use integral notation to represent the exact amount of water used.

t	$f(t)$	$g(t)$	$(f + g)(t)$
0	5.0	2.0	7.0
2	4.5	2.5	7.0
4	4.0	3.0	7.0
6	4.0	3.5	7.5
8	4.5	3.5	8.0
10	5.0	4.0	9.0
12	5.5	3.5	9.0
14	5.0	2.5	7.5
16	4.5	2.5	7.0
18	4.5	3.0	7.5
20	4.5	3.5	8.0
22	4.0	3.5	7.5
24	4.0	3.5	7.5

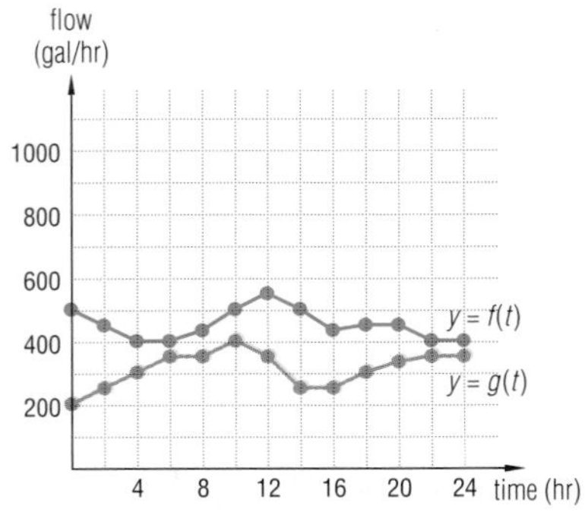

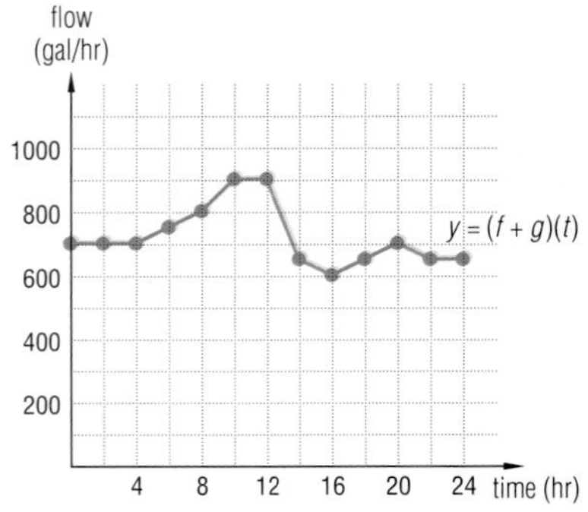

Solution **a.** The rate-of-flow charts give some of the values of the functions f and g. The sum $f + g$ gives the gallons of water per hour flowing through both pipes together. The number of gallons used in a short time period can be approximated by multiplying the estimated rate of flow for that time period by the length of the time period: $\frac{\text{gal}}{\text{hr}} \cdot \text{hr} = \text{gal}$. The total amount of water is then approximated by the sum of these products. If Δt is the length of each time period, then the approximation for the total water is the Riemann sum

$$[f(z_1) + g(z_1)]\,\Delta t + [f(z_2) + g(z_2)]\,\Delta t + \ldots + [f(z_n) + g(z_n)]\,\Delta t$$
$$= \sum_{i=1}^{n} [f(z_i) + g(z_i)]\,\Delta t.$$

If $n = 12$ and the right endpoints of the intervals are used as the z_i (so $z_1 = 2$, $z_2 = 4$, ..., $z_{12} = 24$), then $\Delta t = 2$ and the numbers $f(z_i) + g(z_i)$ are given in the right column of the table on page 782. Thus, the approximation is $7 \cdot 2 + 7 \cdot 2 + 7.5 \cdot 2 + 8 \cdot 2 + \ldots + 7.5 \cdot 2$ $= 185$ hundred gal $= 18{,}500$ gal.

b. The exact amount of water used is the limit of the Riemann sums $\sum_{i=1}^{n} [f(z_i) + g(z_i)]\,\Delta t$ as n gets arbitrarily large. Thus, it is given by the definite integral

$$\int_0^{24} (f(t) + g(t))\,dt = \lim_{n\to\infty} \sum_{i=1}^{n} (f(z_i) + g(z_i))\,\Delta t.$$

This integral can be computed in a different way, by adding the integrals of the individual functions. This can be seen as follows.

For each i, $[f(z_i) + g(z_i)]\,\Delta t = f(z_i)\,\Delta t + g(z_i)\,\Delta t$.
So

$$\sum_{i=1}^{n} [f(z_i) + g(z_i)]\,\Delta t = \sum_{i=1}^{n} f(z_i)\,\Delta t + \sum_{i=1}^{n} g(z_i)\,\Delta t.$$

Consequently, for the limit,

$$\int_0^{24} [f(t) + g(t)]\,dt = \int_0^{24} f(t)\,dt + \int_0^{24} g(t)\,dt.$$

Thus the amount of water used to irrigate the field is also the sum of the two integrals based on the separate rate-of-flow functions for the individual pipes.

The property in Example 2**b** is an instance of a general property of definite integrals whose proof follows the ideas of the example.

Theorem (Sum Property of Integrals)

If f and g are continuous functions on the interval from a to b, then

$$\int_a^b (f(x) + g(x))\,dx = \int_a^b f(x)\,dx + \int_a^b g(x)\,dx.$$

NOTES ON QUESTIONS

Question 1: After students work through the details of this question, the first theorem of the lesson becomes rather obvious.

Question 4: You could ask students to estimate the total amount of water used from, say, 4 AM to 10 AM by using a Riemann sum with $\Delta t = 2$ hours and z_i = right endpoint of ith subinterval. Then $z_1 = 6$, $z_2 = 8$, and $z_3 = 10$, so $f(z_1) = 400$, $f(z_2) \approx 450$, $f(z_3) = 500$, $g(z_1) = g(z_2) = 350$, and $g(z_3) = 400$. Thus, $(f + g)(z_1) = 750$, $(f + g)(z_2) = 800$, and $(f + g)(z_3) = 900$. From this, $\sum_{i=1}^{3} f(z_i)\Delta t + \sum_{i=1}^{3} g(z_i)\Delta t =$ $(400 + 450 + 500) \cdot 2 +$ $(350 + 350 + 400) \cdot 2 =$ $(750 + 800 + 900) \cdot 2 =$ $\sum_{i=1}^{3} [f(z_i) + g(z_i)]\Delta t$

Question 10: A special case is given in Additional Example 2 of this lesson.

Question 11: It is not easy to find functions f and g and a value of a for which equality holds.

Error Analysis for Question 12c: Students may write the integral as $\int_{-2}^{3} [g(x) - f(x)]\,dx$ in this type of problem, which would give a "negative area." This can happen especially if they do not reason carefully from a clear sketch of the two functions. Stress that the height of each rectangle in the Riemann sum is found by subtracting the y-value of g from that of f, since g is below f throughout the region in question.

Question 19: This exploration is helpful for the next lesson and could be used as a lead-in to that lesson.

ADDITIONAL ANSWERS

1. $\int_{-2}^{3} (8 - 2x)\,dx + \int_{3}^{4} (8 - 2x)\,dx$
$= 35 + 1 = 36$
$= \int_{-2}^{4} (8 - 2x)\,dx$

4.b. the amount of water that flowed through the pipes during the 24-hour period for each pipe and for both pipes together

5.

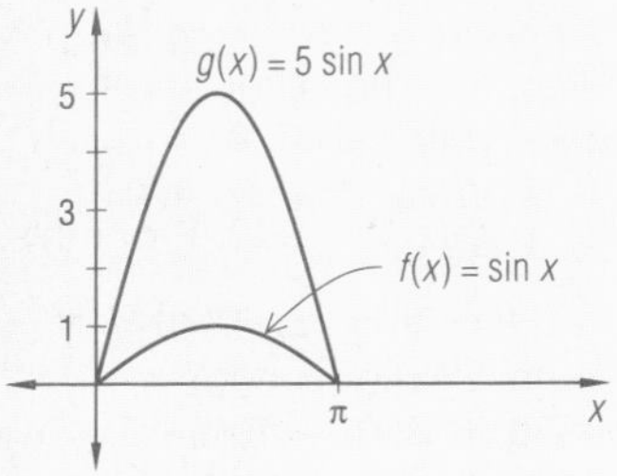

For each x_i, $g(x_i) = 5f(x_i)$, so $\int_0^{\pi} g(x)\,dx$, the area between g(x) and the x-axis, will be $\int_0^{\pi} 5f(x)\,dx = 5\int_0^{\pi} f(x)\,dx$.

7. $2\int_a^b \sin x\,dx$

8. $3\int_3^4 \log x\,dx$

9.c. answer from part b:
$3 \cdot \frac{4^2}{2} + 7(4) = 52$;
area of trapezoid:
$4 \cdot \frac{7 + 19}{2} = 52$

10.b. The area between the graphs of *f* and *g* (from *a* to *b*) is the area between *f* and the x-axis minus the area between *g* and the x-axis.

Example 3 Suppose the field of Example 2 is irrigated by three computer-controlled pipes, all with the same rate of flow. If $y = f(t)$ represents the rate of flow of each pipe, use integral notation to describe the total water used by the three pipes in a period of 24 hours.

Solution $3f(t)$ represents the rate of flow of the 3 pipes working simultaneously. Thus, the total amount of water used is $\int_0^{24} 3f(t)\,dt$. However, the total used by 3 pipes is 3 times the total used by 1 pipe, or $3\int_0^{24} f(t)\,dt$. Thus,

$$\int_0^{24} 3f(t)\,dt = 3\int_0^{24} f(t)\,dt.$$

The property in Example 3 is an instance of a fourth general property of definite integrals.

Theorem (Constant Multiple Property of Integrals)

If f is a continuous function on the interval from a to b, and c is a real number, then

$$\int_a^b c\,f(x)\,dx = c\int_a^b f(x)\,dx.$$

Proof The graph of the function g, where $g(x) = c \cdot f(x)$, is a scale change image of f. For every Riemann sum of f on the interval from a to b, there is a corresponding Riemann sum of g with the same subintervals and same z_i. The value of each term of the sum for g will be c times the value of the corresponding term of the sum for f. Consequently, the limit of the Riemann sums (the definite integral) for g will be c times the limit of the Riemann sums (the definite integral) for f.

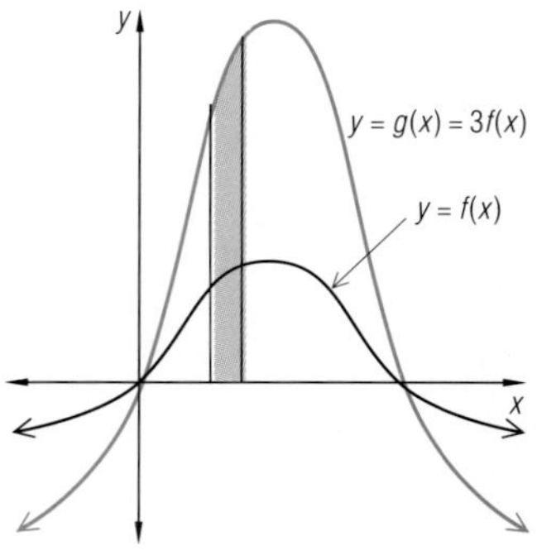

A picture of *f* and *g* when $c = 3$

784

Questions

Covering the Reading

1. Verify the first theorem of this lesson for the case $a = -2$, $b = 3$, $c = 4$, and $f(x) = 8 - 2x$.
See margin.

2. Suppose g is a continuous function such that $\int_{-3}^{-1} g(x)\,dx = 2$, $\int_{-3}^{0} g(x)\,dx = 6$, and $\int_{0}^{2} g(x)\,dx = 5$. Find:

a. $\int_{-3}^{2} g(x)\,dx$ **11** **b.** $\int_{-1}^{0} g(x)\,dx$ **4** **c.** $\int_{-1}^{2} g(x)\,dx$ **9**

In 3 and 4, refer to Example 2.

3. Give the units of t, $f(t)$, Δt, and $f(t)\Delta t$.
t hours, $f(t)$ gal/hr, Δt hours, $f(t)\ \Delta t$ gallons

4. a. Use the left endpoints of the 12 two-hour subintervals from 0 to 24 to calculate a Riemann sum for the given function.
i. f **11,000 gal** **ii.** g **7400 gal** **iii.** $f + g$ **18,400 gal**
b. What do your answers to part **a** represent?
c. How are they related to the answer found in Example 2**a**?
b) See margin. c) They are nearly equal.

5. Graph $f(x) = \sin x$ and $g(x) = 5 \sin x$ on the interval from 0 to π and use these graphs to explain why $\int_{0}^{\pi} g(x)\,dx = 5\int_{0}^{\pi} f(x)\,dx$. **See margin.**

Applying the Mathematics

In 6–8, use properties of integrals to write the expression as one integral.

6. $\int_{0}^{14} x^2\,dx - \int_{0}^{6} x^2\,dx$ **$\int_{6}^{14} x^2\,dx$** **7.** $3\int_{a}^{b} \sin x\,dx - \int_{a}^{b} \sin x\,dx$ **See margin.**

8. $\int_{3}^{4} \log x\,dx + \int_{3}^{4} (\log x^2)\,dx$ (Hint: Use properties of logarithms.) **See margin.**

9. a. Use area formulas to evaluate the integrals $\int_{a}^{b} 7\,dx$ and $\int_{0}^{a} x\,dx$.

b. Use your answers to part **a** and properties of integrals to evaluate $\int_{0}^{a} (3x + 7)\,dx$. **$3 \cdot \frac{a^2}{2} + 7a$**

c. Verify your answer to part **b** for $a = 4$ by finding the area of a trapezoid. **a) $7(b - a)$ and $\frac{a^2}{2}$; c) See margin.**

10. a. Fill in the blank with a single integral.
$\int_{a}^{b} f(x)\,dx - \int_{a}^{b} g(x)\,dx = \underline{\ ?\ }$ **$\int_{a}^{b} [f(x) - g(x)]\,dx$**

b. Interpret your answer to part **a** in terms of area. **See margin.**

11. Find functions f and g and a real number a for which
$\int_{0}^{a} f(x)\,dx \cdot \int_{0}^{a} g(x)\,dx \neq \int_{0}^{a} (f \cdot g)(x)\,dx$. **See margin.**

11. Sample: Let $f(x) = x$, $g(x) = x$, and $a = 1$. Then $\int_0^1 x\,dx \cdot \int_0^1 x\,dx = \frac{1^2}{2} \cdot \frac{1^2}{2} = \frac{1}{4}$, but $\int_0^1 x^2\,dx = \frac{1^3}{3} = \frac{1}{3}$.

NAME ____________

LESSON **MASTER 13–4**
QUESTIONS ON **SPUR** OBJECTIVES

■ **SKILLS** *Objective B (See pages 810–813 for objectives.)*
In 1–4, evaluate the definite integrals.

1. $\int_3^7 3(x - 2)\,dx$ — **36**

2. $\int_2^5 (2x + 2)\,dx + 2\int_5^{10} (x + 1)\,dx$ — **112**

3. $\int_a^b 4\,dx + \int_b^c 4\,dx$ — **$4(c - a)$**

4. $\int_a^b 5x\,dx + \int_a^b 4x\,dx$ — **$\frac{9}{2}(b^2 - a^2)$**

■ **PROPERTIES** *Objective C*
In 5–8, use properties of integrals to write the expression as a single integral.

5. $\int_0^5 (x^2 + 2)\,dx + 3\int_0^5 x\,dx$ — **$\int_0^5 (x^2 + 3x + 2)\,dx$**

6. $\int_0^3 (x + 4)\,dx + \int_3^5 (x + 4)\,dx$ — **$\int_0^5 (x + 4)\,dx$**

7. the expression given in Question 2 — **$2\int_2^{10} (x + 1)\,dx$**

8. the expression given in Question 4 — **$9\int_a^b x\,dx$**

144 *Continued* *Precalculus and Discrete Mathematics* © Scott, Foresman and Company

NAME ____________
Lesson MASTER 13–4 (page 2)

■ **USES** *Objective E*
9. A candy manufacturer has two machines which produce a chocolate mixture at different rates. Let $f(t)$ and $g(t)$ represent the production rates (in gallons per hour) of the two machines at time t (in hours).

a. Use integral notation in two different ways to write the total output of chocolate mixture during their first 12 hours of operation.
$\int_0^{12} f(t)\,dt + \int_0^{12} g(t)\,dt$, $\int_0^{12} [f(t) + g(t)]\,dt$

b. Approximate the value of your answer to part **a** by evaluating the appropriate Riemann sum, first using the left endpoints as the x_i's and then using the right endpoints as the x_i's.
63.9, 70.7

t	$f(t)$	$g(t)$
0	3.50	2.00
2	3.00	2.15
4	2.75	3.00
6	3.15	2.75
8	3.05	2.85
10	3.50	3.00
12	3.40	2.75

■ **REPRESENTATIONS** *Objective H*
In 10 and 11, express the area of each shaded region using integral notation and find its value.

10.
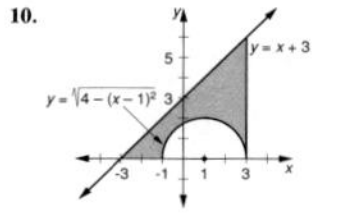

$\int_{-3}^{3} (x + 3)\,dx - \int_{-1}^{-3} \sqrt{4 - (x - 1)^2}\,dx = 18 - 2\pi$, or ≈ 11.72 unit^2

11.
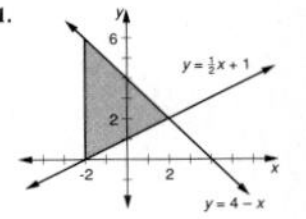

$\int_{-2}^{2} (4 - x)\,dx - \int_{-2}^{2} \left(\frac{1}{2}x + 1\right)\,dx = \int_{-2}^{2} \left[(4 - x) - \left(\frac{1}{2}x + 1\right)\right]\,dx = 12$ units^2

Precalculus and Discrete Mathematics © Scott, Foresman and Company **145**

FOLLOW-UP

MORE PRACTICE

For more questions on SPUR Objectives, use *Lesson Master 13-4*, shown on page 785.

EXTENSION

The Sum and Constant Multiple Properties of Integrals can be combined into the single statement: "The integral is a *linear operator*." In symbols, $\int [af(x) + bg(x)]\,dx = a\int f(x)\,dx + b\int g(x)\,dx$.

(1) If $\int_2^5 f(x)\,dx = \frac{13}{2}$ and $\int_2^5 g(x)\,dx = -7$, what is $\int_2^5 [2f(x) + 4g(x)]\,dx$? (-15)

(2) The definite integral is a linear operator because summation is. Complete the following two statements.

(a) $\sum_{i=1}^{n} kf(i) = \underline{\ ?\ }$ $\left(k\sum_{i=1}^{n} f(i)\right)$

(b) $\sum_{i=1}^{n} [s(i) + t(i)] = \underline{\ ?\ }$ $\left(\sum_{i=1}^{n} s(i) + \sum_{i=1}^{n} t(i)\right)$

(3) Find a counterexample to show that $\sqrt{}$ is not a linear operator. (Sample: $\sqrt{9 + 16} = \sqrt{25} = 5$, but $\sqrt{9} + \sqrt{16} = 3 + 4 = 7 \neq 5$.)

EVALUATION

A quiz covering Lesson 13-1 through 13-4 is provided in the Teacher's Resource File.

ADDITIONAL ANSWERS

12.c. $\int_{-3}^{2} (f(x) - g(x))\,dx$

15. sample:
a. (24, 30, 26)
b. $12x + 15y + 13z = -10$
c. $\begin{cases} x = 1 + 12t \\ y = 2 + 15t \\ z = -4 + 13t \end{cases}$

16. See Additional Answers in the back of this book.

12. Consider the region enclosed by the parabola $f(x) = -x^2 + 4$ and the line $g(x) = x - 2$.

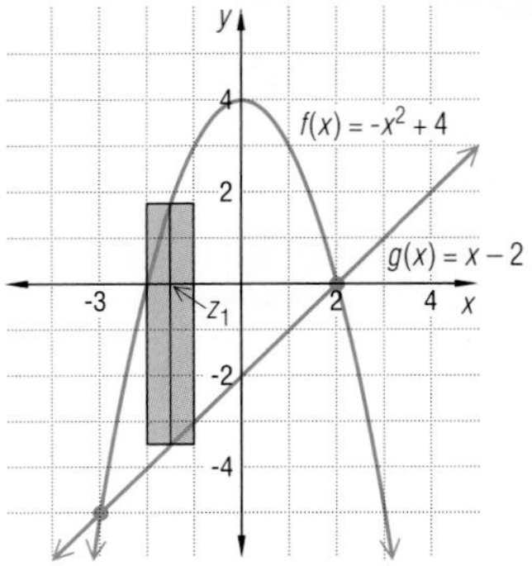

a. If the shaded rectangle at the right has width Δx, write its area in terms of $f(z_i)$, $g(z_i)$, and Δx.
b. Write a Riemann sum that approximates the area of the region enclosed by the graphs of f and g.
c. Write an integral that gives the exact area of the region.

a) $(f(z_i) - g(z_i))\,\Delta x$ b) $\sum_{i=1}^{n} (f(z_i) - g(z_i))\,\Delta x$ c) See margin.

Review

13. Find the value of the integral to the nearest hundredth. *(Lesson 13-3)*
a. $\int_{-5}^{5} \sqrt{25 - x^2}\,dx$ 39.27 **b.** $\int_{4}^{5} \sqrt{25 - x^2}\,dx$ 2.04

14. After sighting its prey, a hawk descends for 7 seconds with a speed described by the graph at the right. Find the distance traveled by the hawk in this time. *(Lesson 13-1)* 175 ft

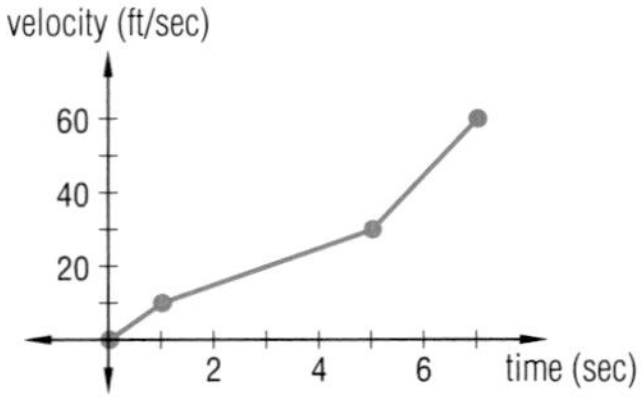

15. Let $P = (1, 2, -4)$, $Q = (-3, 0, 2)$, and $R = (2, -4, 2)$ be points in 3-space.
a. Find a vector orthogonal to vectors $\overrightarrow{PQ}$ and $\overrightarrow{PR}$.
b. Find an equation of the plane M containing P, Q, and R.
c. Find parametric equations of the line through P orthogonal to M.
(Lessons 12-7, 12-6)
See margin.

16. A survey of students at a college shows that 45% are science majors; the rest are humanities majors. Of the science majors, 40% plan to attend graduate school, compared to 30% of the humanities majors. Draw a graph which represents this situation, and find what percent of those who plan to attend graduate school are science majors. *(Lesson 11-1)*
See margin. ≈ 52%

In 17 and 18, use basic geometry formulas to find the volume of the figure. *(Previous course)*

17.

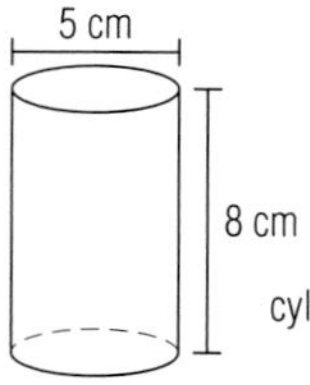

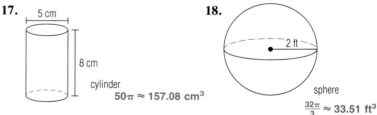

cylinder
$50\pi \approx 157.08\ \text{cm}^3$

18.

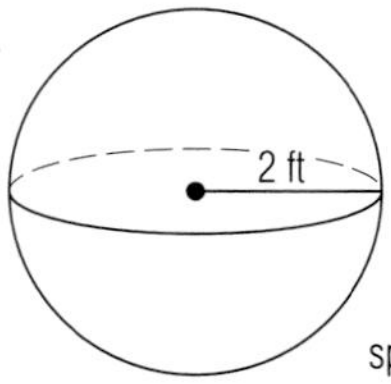

sphere
$\frac{32\pi}{3} \approx 33.51\ \text{ft}^3$

Exploration

19. Develop a formula for $\int_0^a (mx + c)\,dx$.

(Hint: One way is to use the answers to Questions 14 and 15 of Lesson 13-3.) $\frac{ma^2}{2} + ca$

786

LESSON

13-5

The Area Under a Parabola

Archimedes, who lived from about 287 to 212 B.C., is thought by many to have been the greatest mathematician of antiquity. He is certainly one of the brilliant minds of all time. He was known during his lifetime for his inventions, including the Archimedean screw (still used today in irrigation) and the catapult, and he made basic discoveries in physics. But he considered himself to be a mathematician, and he was the first person to find a formula for the area of a figure bounded by part of a parabola. It is said that he cut parabolic regions out of wood and weighed them to conjecture a formula, and then set about to deduce the formula. His methods were quite similar to the use of Riemann sums, calculating better and better approximations to the actual area, but he had none of today's notation.

Knowledge about parent functions can usually be extended to obtain knowledge about their offspring. So it is with integrals of functions. In a calculus course, you will learn how to obtain the integrals of all the parent functions you have studied in this course. Here, we examine the parent function $y = x^2$, and begin by asking for the area between $x = 0$ and $x = 1$. That is, we find the definite integral

$$\int_0^1 x^2\,dx.$$

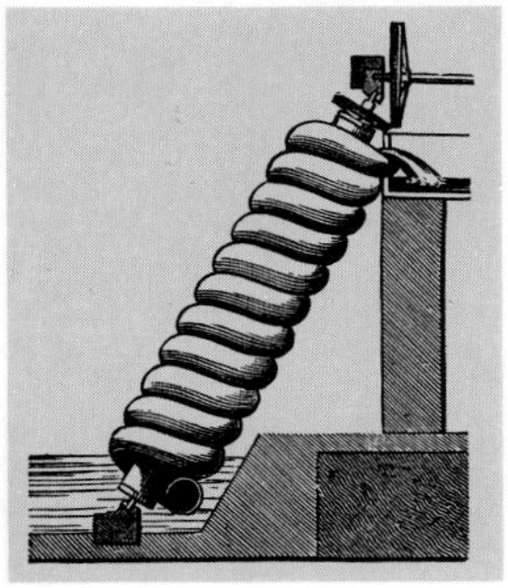

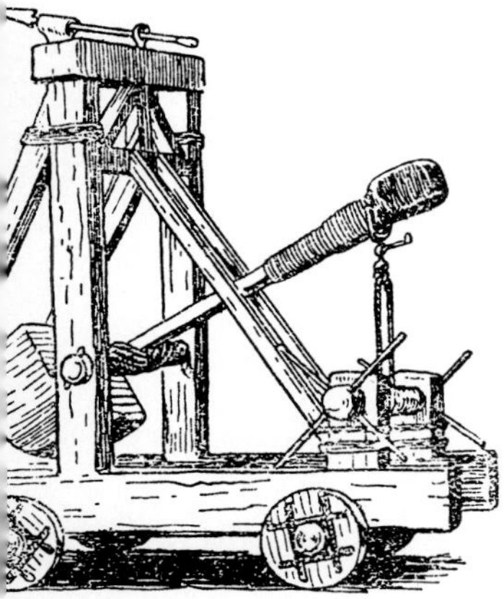

Archimedean screw and catapult

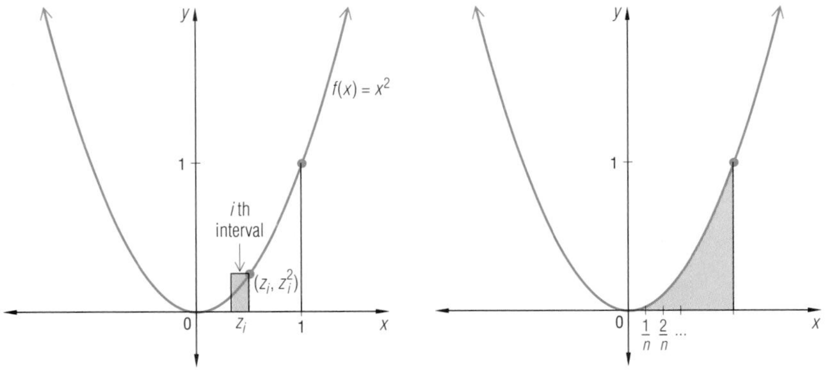

Since the integral is defined as the limit of Riemann sums, we begin by calculating those sums. Let

$$\sum_{i=1}^{n} f(z_i)\,\Delta x$$

be the Riemann sum obtained by dividing the interval $0 \le x \le 1$ into n equal subintervals, and let z_i be the right-hand endpoint of the ith subinterval. Since $\Delta x = \frac{1-0}{n} = \frac{1}{n}$, the n values (one for each subinterval) at which the function f will be evaluated are $z_1 = \frac{1}{n}$, $z_2 = \frac{2}{n}$, $z_3 = \frac{3}{n}$, ... , $z_n = 1 = \frac{n}{n}$.

LESSON 13-5

RESOURCES

- Lesson Master 13-5
- Teaching Aid 105 can be used with **Questions 11 and 12**.
- Computer Master 23

OBJECTIVES

B Evaluate definite integrals.
D Find the distance traveled by a moving object, given its rate.
H Find areas bounded by curves.

TEACHING NOTES

Evaluating $\int_0^1 x^2\,dx$ as a limit of Riemann sums is difficult for some students. Part of the difficulty is reconciling the facts that the sum goes from 1 to n, while the integral goes from 0 to 1. Remind students that n is the number of terms in the Riemann sum, that is, the number of rectangles whose areas are being added. The idea is to write $\sum_{i=1}^{n} f(z_i)\,\Delta x$ in terms of n, and possibly some constants. Then, it is possible to take the limit as n approaches infinity. To write the sum in terms of n, divide the interval from 0 to 1 into n subintervals—this is where the endpoints of the integral are used. Pick the z_i in the subintervals and calculate the values of the $f(z_i)$. Try to pick the z_i so that there is a simple pattern for the $f(z_i)$. On this page, the right endpoints of the subintervals are chosen. Not surprisingly, given that $f(x) = x^2$, it is necessary to find a sum of squares. In an earlier chapter (and in earlier reviews in this chapter), mathematical induction was used to deduce a formula for the required sum. Thus, the Riemann sum is expressed in terms of n. Finally, take the limit of the Riemann sum as $n \to \infty$; it turns out to be $\frac{1}{3}$.

Alternate Approach
The following is an alternate way to evaluate the limit on this page.

$$\lim_{n\to\infty} \frac{(n+1)(2n+1)}{6n^2}$$
$$= \lim_{n\to\infty} \frac{2n^2 + 3n + 1}{6n^2}$$
$$= \lim_{n=1} \frac{2 + \frac{3}{n} + \frac{1}{n^2}}{6} = \frac{2}{6} = \frac{1}{3}$$

You may want to have students use the method illustrated in this lesson to show that $\int_0^1 x\,dx = \frac{1}{2}$. If right endpoints are used for the n subintervals in the Riemann sum $\sum_{i=1}^{n} f(z_i)\Delta x$, then we obtain $\sum_{i=1}^{n} \frac{1}{n}\left(\frac{i}{n}\right)$. This is the sum of the integers from 1 to n divided by n^2, and using $\sum_{i=1}^{n} i = \frac{n(n+1)}{2}$, we see that the sum is $\frac{n(n+1)}{2n^2}$. Taking the limit as $n \to \infty$ yields the desired value. More generally, you can show that $\int_0^a x\,dx = \frac{a^2}{2}$. This follows the procedure for $\int_0^1 x\,dx = \frac{1}{2}$ but with a constant multiple of a^2 in each term of the sum, just as a^3 is found in the derivation of the theorem on page 789.

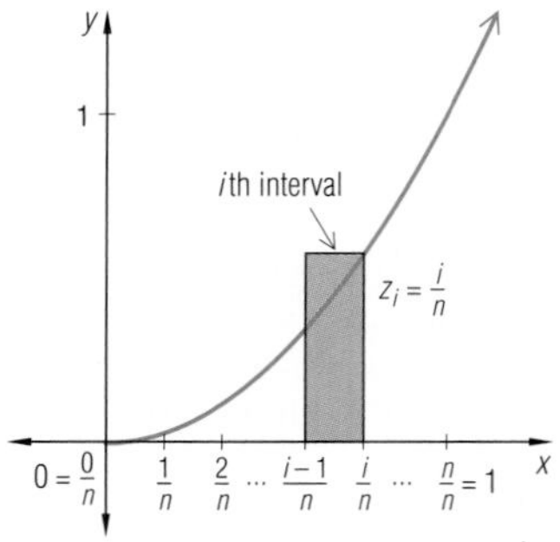

Therefore, for all i, $z_i = \frac{i}{n}$ and so $f(z_i) = \left(\frac{i}{n}\right)^2 = \frac{i^2}{n^2}$. So the Riemann sum formula is

$$\sum_{i=1}^{n} f(z_i)\,\Delta x = \sum_{i=1}^{n} \left(\frac{i^2}{n^2} \cdot \frac{1}{n}\right) = \sum_{i=1}^{n} \frac{i^2}{n^3} = \frac{1}{n^3} \sum_{i=1}^{n} i^2.$$

(Since n is a nonzero constant with respect to the sum, so is $\frac{1}{n^3}$, and so $\frac{1}{n^3}$ can be factored out of the sum.) In Lesson 7-4, it was proved that

$$\sum_{i=1}^{n} i^2 = \frac{n(n+1)(2n+1)}{6}.$$

Therefore,

$$\frac{1}{n^3} \sum_{i=1}^{n} i^2 = \frac{1}{n^3} \cdot \frac{n(n+1)(2n+1)}{6}$$
$$= \frac{n(n+1)(2n+1)}{6n^3} = \frac{(n+1)(2n+1)}{6n^2}.$$

Although this result gives the Riemann sum, it is the limit of this sum that gives the value of the definite integral. To find the limit of this sum as $n \to \infty$, we rewrite the right most expression:

$$= \frac{1}{6} \cdot \frac{n+1}{n} \cdot \frac{2n+1}{n} = \frac{1}{6}\left(1 + \frac{1}{n}\right)\left(2 + \frac{1}{n}\right).$$

Now take the limit of the above Riemann sum using this new expression.

$$\lim_{n\to\infty} \sum_{i=1}^{n} f(z_i)\,\Delta x = \lim_{n\to\infty} \frac{1}{n^3} \sum_{i=1}^{n} i^2$$
$$= \lim_{n\to\infty} \frac{1}{6}\left(1 + \frac{1}{n}\right)\left(2 + \frac{1}{n}\right)$$
$$= \frac{1}{6} \cdot (1 + 0)(2 + 0) \qquad \text{since } \lim_{n\to\infty} \frac{1}{n} = 0$$
$$= \frac{1}{3}$$

This limit is the definite integral of f, where $f(x) = x^2$, over the interval from 0 to 1. Consequently,

$$\int_0^1 x^2\,dx = \frac{1}{3}$$

and so the area under this parabola from 0 to 1 is $\frac{1}{3}$.

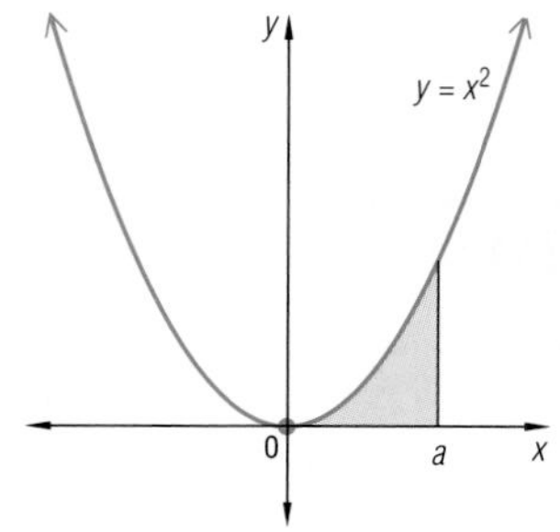

Using the program in Lesson 13-2 to approximate this integral with Riemann sums confirms this result. With 100 subintervals, the estimate is .33835, and with 1000 subintervals, it is .33383.

When $a > 0$, the definite integral $\int_0^a x^2\,dx$ can be evaluated using the same technique. This time, $\Delta x = \frac{a - 0}{n} = \frac{a}{n}$.

Again, let z_i = the right endpoint of the ith subinterval:
$z_1 = \frac{a}{n}$, $z_2 = \frac{2 \bullet a}{n}$, $z_3 = \frac{3 \bullet a}{n}$, ... , $z_n = \frac{n \bullet a}{n} = a$, and so for all i, $z_i = \frac{i \bullet a}{n}$.

$$\text{Thus } \sum_{i=1}^{n} f(z_i)\,\Delta x = \sum_{i=1}^{n} \left(\frac{i \bullet a}{n}\right)^2 \bullet \frac{a}{n} = \sum_{i=1}^{n} \frac{i^2 \bullet a^3}{n^3} = \frac{a^3}{n^3} \sum_{i=1}^{n} i^2$$
$$= a^3\left[\frac{1}{n^3} \sum_{i=1}^{n} i^2\right].$$

The limit of this expression is the value of the definite integral. Now use the result obtained above.

$$\lim_{n \to \infty} \sum_{i=1}^{n} f(z_i)\,\Delta x = \lim_{n \to \infty} a^3\left[\frac{1}{n^3} \sum_{i=1}^{n} i^2\right] = a^3 \lim_{n \to \infty}\left[\frac{1}{n^3} \sum_{i=1}^{n} i^2\right] = a^3 \bullet \frac{1}{3} = \frac{a^3}{3}$$

Thus $\int_0^a x^2\,dx = \frac{a^3}{3}$.

We have just proved the following theorem.

Theorem

When $a > 0$, $\int_0^a x^2\,dx = \frac{a^3}{3}$.

For instance, $\int_0^5 x^2\,dx = \frac{125}{3}$ and $\int_0^{15} x^2\,dx = \frac{15^3}{3} = 1125$.

To obtain a formula for $\int_a^b x^2\,dx$ when $0 < a < b$, apply the second theorem from Lesson 13-4.

$$\int_a^b x^2\,dx = \int_0^b x^2\,dx - \int_0^a x^2\,dx$$
$$= \frac{b^3}{3} - \frac{a^3}{3} = \frac{b^3 - a^3}{3}$$

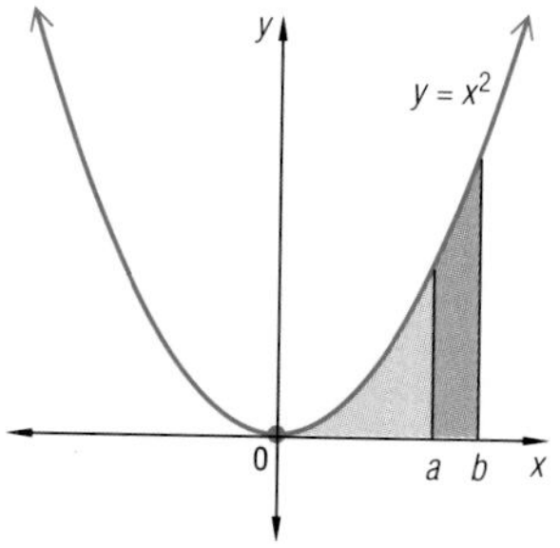

One more result is needed to find the definite integral for all offspring of $f(x) = x^2$. Note that the integral $\int_0^a x\,dx$ is the area of the triangle bounded by the x-axis, the line $y = x$, and the line $x = a$. This area is $\frac{1}{2}\,a \bullet a$, or $\frac{a^2}{2}$. Combined with the theorems of this and the previous lesson, the integral of the general quadratic function $y = c_2x^2 + c_1x + c_0$ can be calculated.

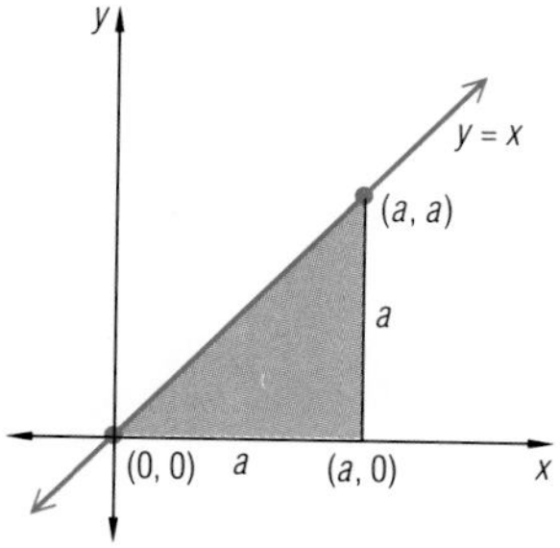

Making Connections
The limits of the integral in the theorem on this page are from 0 to a, but after the theorem, a formula is deduced for the area under the parabola from a to b. In **Question 16**, students are asked to find the corresponding formula $\int_a^b x\,dx = \frac{b^2}{2} - \frac{a^2}{2}$.
In Question 15 of Lesson 13-3, students found the formula $\int_a^b c\,dx = cb - ca$.
With all of the above, the Integral of a Quadratic Function Theorem on page 790 can be generalized:
If $0 < a < b$, then
$\int_a^b (c_2x^2 + c_1x + c_0)\,dx = c_2\left(\frac{b^3}{3} - \frac{a^3}{3}\right) + c_1\left(\frac{b^2}{2} - \frac{a^2}{2}\right) + c_0(b - a)$. You may wish to give this formula to students, but at this point it may be easier for many students simply to work with the integral from 0 to a.

ADDITIONAL EXAMPLES

1. Suppose that a car accelerates from 0 to 88 $\frac{\text{ft}}{\text{sec}}$ in 8 seconds such that its speed $f(x)$ after x seconds is given by $f(x) = -1.375(x - 8)^2 + 88$. Find the total distance traveled.
≈ 469.3 ft

2. Find a formula for the area between the line $y = ax$ and the parabola $y = x^2$, when $a > 0$.
$\frac{a^3}{6}$

Theorem (Integral of a Quadratic Function)

If $a > 0$, $\int_0^a (c_2x^2 + c_1x + c_0)\, dx = c_2\frac{a^3}{3} + c_1\frac{a^2}{2} + c_0a$.

Proof By the Sum and Constant Multiple properties of integrals,
$$\int_0^a (c_2x^2 + c_1x + c_0)\, dx = c_2\int_0^a x^2\, dx + c_1\int_0^a x\, dx + \int_0^a c_0\, dx.$$

The first and second terms on the right side are evaluated previously in this lesson. The third term is the area of the rectangle bounded by the line $y = c_0$, the x-axis, the y-axis, and the line $x = a$. So the area is c_0a. Thus
$$\int_0^a (c_2x^2 + c_1x + c_0)\, dx = c_2\frac{a^3}{3} + c_1\frac{a^2}{2} + c_0a.$$

In the proof, we used the fact that $\int_0^a c_0\, dx = c_0a$. More generally, the definite integral of the constant function $f(x) = c$ over the interval from a to b is $\int_a^b c\, dx = cb - ca$.

In Lessons 13-1 and 13-2 there was a question about the distance traveled by an object whose acceleration was not constant but was quicker at first. Because this particular rate function was quadratic, the question can now be answered exactly.

Example 1 Suppose a car accelerates from 0 to 88 ft/sec in 10 seconds and its speed in this time interval is given by $g(x) = -.88(x - 10)^2 + 88$. What is the total distance (in feet) traveled?

Solution
The distance is given by $\int_0^{10} g(x)\, dx$. Convert the formula for $g(x)$ into standard form.
$$\begin{aligned} g(x) &= -.88(x - 10)^2 + 88 \\ &= -.88(x^2 - 20x + 100) + 88 \\ &= -.88x^2 + 17.6x \end{aligned}$$
Now use the above theorem with $a = 10$ and $c_2 = -.88$, $c_1 = 17.6$, and $c_0 = 0$.
$$\begin{aligned} \int_0^{10} (-.88x^2 + 17.6x)\, dx &= -.88 \cdot \frac{10^3}{3} + 17.6 \cdot \frac{10^2}{2} + 0 \cdot 10 \\ &= \frac{-880}{3} + 880 \\ &= 586\tfrac{2}{3} \end{aligned}$$
This result agrees with the calculations done in Lesson 13-2.

Example 2 Find the area of the shaded region at the right.

Solution Subtract the area between the parabola and the x-axis from the area of the rectangle with vertices (-2, 4), (-2, 0), (2, 0), and (2, 4).

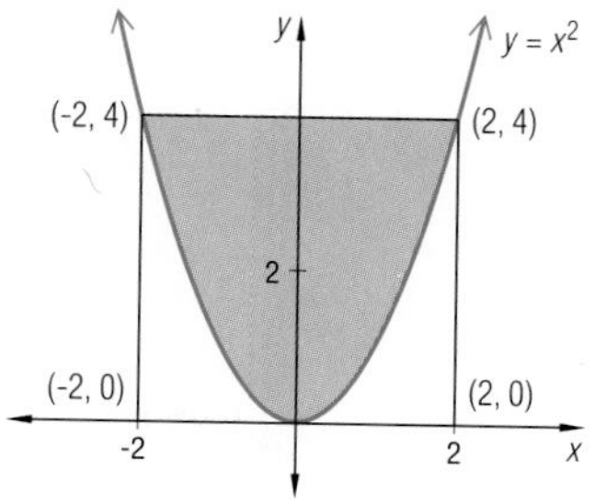

$$\text{Area of shaded region} = \text{Area of rectangle} - \int_{-2}^{2} x^2\,dx$$

$$= \text{Area of rectangle} - 2\int_{0}^{2} x^2\,dx$$

$$= 16 - 2\left(\frac{2^3}{3}\right)$$

$$= 16 - \frac{16}{3}$$

$$= \frac{32}{3} \text{ square units}$$

Questions

Covering the Reading

1. Consider the problem of finding the area between the parabola $y = x^2$ and the x-axis, from $x = 0$ to $x = 1$.

a. Suppose the interval is divided into 25 subintervals. If the z_i are the right-hand endpoints of the subintervals, what are the values for the z_i? $\frac{i}{25}$, with $i = 1, 2, 3, \ldots, 25$

b. Give a Riemann sum approximation to this area, letting the z_i be the right endpoints of each interval. (Use the formula for $\sum_{i=1}^{n} i^2$ to help calculate the value.) .3536 units

In 2 and 3, find the value of the integral.

2. $\int_0^3 x^2\,dx$ 9

3. $\int_{-6}^{8} x^2\,dx$ $242\frac{2}{3}$

4. Find the area of the region bounded by the parabola $y = x^2$ and the line $y = 100$. (Hint: Sketch the region and use the model of Example 2.) See margin. $\frac{4000}{3}$ units^2

5. Consider the following sum: $\int_2^5 x^2\,dx + \int_5^6 x^2\,dx$.

a. Rewrite the sum as a single integral. $\int_2^6 x^2\,dx$

b. Evaluate the result in part **a**. $\frac{208}{3}$

NOTES ON QUESTIONS

Question 5: Have students check the answer by evaluating the two integrals separately and then adding the results.

Question 11: It is interesting to ask which areas can be found exactly and which can only be estimated, given the formulas known to students. The area under the parabola (in white) can be found exactly using the theorem of this lesson. The area of the quarter-circle also can be found exactly. So it is possible to get an exact value for the total shaded area. But to get an exact value for either part requires knowing a formula for $\int_0^a \sqrt{1 - x^2}$, which is beyond the scope of this course.

Question 16: The answer given is in a form found when area has been used in the derivation of the formula. The equivalent $\frac{b^2}{2} - \frac{a^2}{2}$ is more familiar to those who know the formula for the integral of any polynomial.

Question 27: Since the pattern is surprisingly simple, students might wonder why they did not learn such a formula in earlier courses. The reason is that the introduction of formulas without derivations tends to encourage rote learning without understanding.

ADDITIONAL ANSWERS

4.

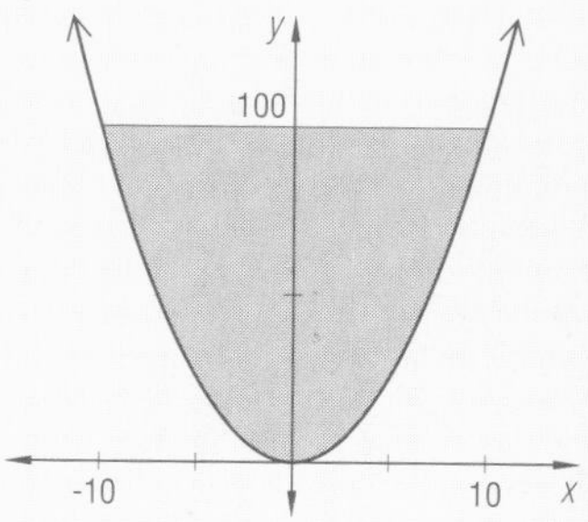

ADDITIONAL ANSWERS

14.a. $\int_0^7 t^2\,dt$

16.

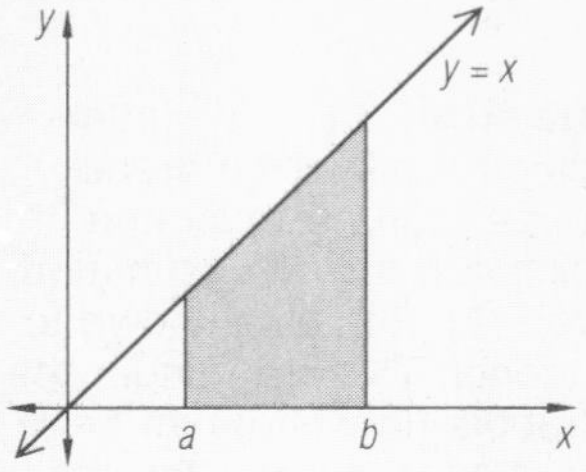

17.

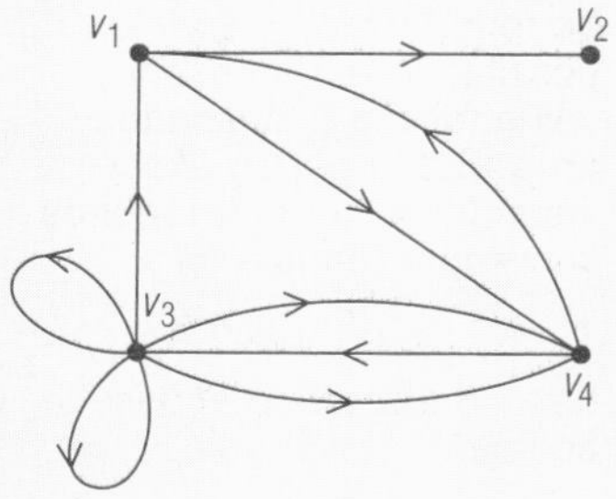

19.c.

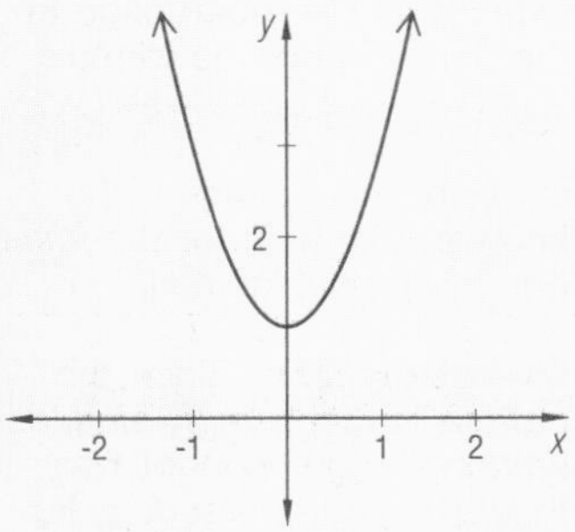

20.b.

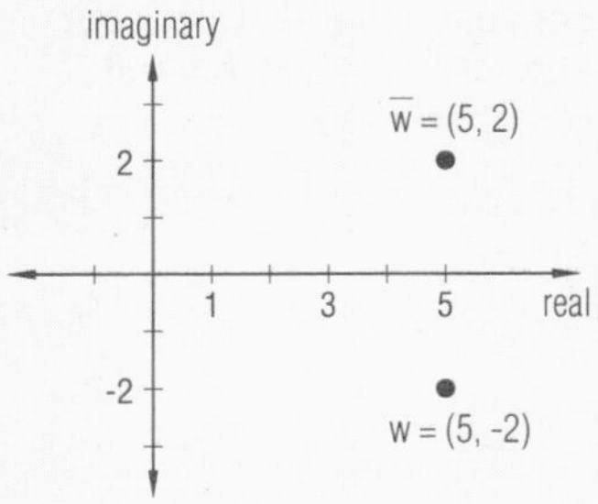

In 6–10, evaluate the integral.

6. $\int_0^5 3\,dx$ **15** **7.** $\int_{-2}^7 dx$ **9** **8.** $\int_0^7 x\,dx$ $\frac{49}{2}$

9. $\int_0^5 (3x^2 - 2x + 5)\,dx$ **125** **10.** $\int_0^1 ax^2\,dx$ $\frac{a}{3}$

Applying the Mathematics

11. Consider the shaded portion of the graph at the right.

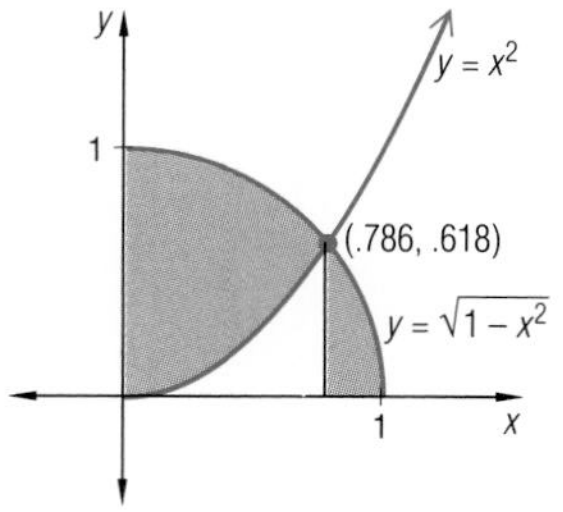

a. Evaluate $\int_0^1 \sqrt{1 - x^2}\,dx$.

(Hint: Use basic geometry formulas.) $\frac{\pi}{4}$

b. Evaluate $\int_0^{.786} x^2\,dx$. **≈ 0.162**

c. Use the results from parts **a** and **b** to estimate the area of the shaded portion to the nearest hundredth.
≈ 0.624 units²

12. A cylindrical trough has a base in the shape of a parabola, 1 meter high and 2 meters across. If the trough is 6 meters long, what is its volume? (Hint: You need to use the formula for the volume of a cylindric solid.) **8 m³**

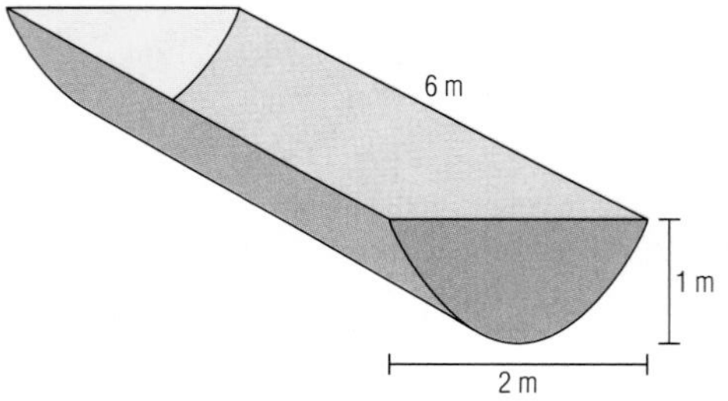

13. For what value of k is $\int_3^k x^2\,dx = 567$? **12**

14. The velocity (in feet per second) of an object at time t seconds is given by the function $f(t) = t^2$.

a. What integral gives the distance traveled by the object from $t = 0$ to $t = 7$ seconds? **See margin.**

b. Find the distance in part **a.** $\frac{7^3}{3} = 114.\overline{3}$ **ft**

Review

15. Find the exact value of $\int_5^9 (4x + 1)\,dx$. *(Lesson 13-3)* **116**

16. Find a formula for $\int_a^b x\,dx$. (Hint: Draw a picture.) *(Lesson 13-3)*

See margin. $\frac{(b + a)(b - a)}{2}$

792

17. Draw a directed graph whose adjacency matrix is $\begin{bmatrix} 0 & 1 & 0 & 1 \\ 0 & 0 & 0 & 0 \\ 1 & 0 & 2 & 2 \\ 1 & 0 & 1 & 0 \end{bmatrix}$. *(Lesson 11-2)* **See margin.**

18. A salad bar contains 20 items in addition to lettuce. How many different salads can be created if 12 items, not counting the lettuce, are selected? *(Lesson 10-4)* **125,970**

19. a. Use the definition of derivative to compute the derivative of f when $f(x) = 2x^2 + 1$. **$f'(x) = 4x$**

b. Use your answer to part **a** to find the values of f at which a relative maximum or minimum could occur. **$x = 0$**

c. Check your answer to part **b** by graphing f. **See margin.**

(Lessons 9-5, 9-3)

20. Given the complex number $w = 5 - 2i$.

a. Give $\overline{w}$. **$5 + 2i$**

b. Graph w and $\overline{w}$ in the complex plane. *(Lesson 8-1)* **See margin.**

21. a. Show by direct computation that

$$\sum_{i=1}^{7}(i^2 - 5i + 2) = \sum_{i=1}^{7} i^2 - \sum_{i=1}^{7} 5i + \sum_{i=1}^{7} 2.$$ **See margin.**

b. Give the general property. *(Lesson 8-3)* **See margin.**

In 22–24, evaluate without a calculator. *(Lessons 6-6, 2-7, 2-6)*

22. $e^{3 \ln x}$ $(x > 0)$ **x^3** **23.** $\log \sqrt{1000}$ **1.5** **24.** $\sin\left(\sin^{-1}\frac{\sqrt{2}}{2}\right)$ **$\frac{\sqrt{2}}{2}$**

Exploration

25. Find out what the Archimedean screw actually does. **See margin.**

26. Explain how Archimedes could use the weight of a parabolic region cut out of wood to arrive at a conjecture concerning the area bounded by a parabola. **See margin.**

27. Given: $\int_0^a dx = a$

$$\int_0^a x\,dx = \frac{a^2}{2}$$

$$\int_0^a x^2\,dx = \frac{a^3}{3}.$$

a. Make a conjecture about $\int_0^a x^3\,dx$. **sample: $\frac{a^4}{4}$**

b. Test your conjecture for various values of a, using the BASIC program of Lesson 13-2 to approximate the integral. Use various values of n. **Answers may vary.**

FOLLOW-UP

MORE PRACTICE

For more questions on SPUR Objectives, use *Lesson Master 13-5*, shown below.

PROJECTS

The projects for Chapter 13 are described on pages 806–807. **Projects 1 and 5** are related to the content of this lesson.

21.a. $\sum_{i=1}^{7}(i^2 - 5i + 2) =$ $(-2) + (-4) + (-4) + (-2) + 2 + 8 + 16 = 14;$ $\sum_{i=1}^{7} i^2 = 140,$ $\sum_{i=1}^{7} 5i = 140,$ $\sum_{i=1}^{7} 2 = 14,$ so $\sum_{i=1}^{7} i^2 - \sum_{i=1}^{7} 5i + \sum_{i=1}^{7} 2 = 14.$

b. sample: $\sum_{i=1}^{n}(ai^2 + bi + c) = \sum_{i=1}^{n} ai^2 + \sum_{i=1}^{n} bi + \sum_{i=1}^{n} c$

25., 26. See Additional Answers in the back of this book.

NAME ____________

LESSON **MASTER 13–5**
QUESTIONS ON **SPUR** OBJECTIVES

■ **SKILLS** *Objective B (See pages 810–813 for objectives.)*
In 1–3, evaluate the integral.

1. $\int_0^{15} x^2\,dx - \int_{10}^{15} x^2\,dx$ **$\frac{1{,}000}{3}$**

2. $\int_{-3}^{5} (x^2 + 4)\,dx$ **$\frac{248}{3}$**

3. $\int_2^{10} (x^2 + 5x + 2)\,dx$ **$\frac{1{,}760}{3}$**

■ **USES** *Objective D*

4. Suppose a car accelerates from 0 to 100 ft/sec in 5 seconds so that its velocity in ft/sec after t seconds is given by $v(t) = .25(t - 5)^2 + 100$. What is the total distance traveled in the 5-second interval? **510.42 ft**

■ **REPRESENTATIVES** *Objective H*
In 5 and 6, express the area of the shaded region using integral notation and find its value.

5.
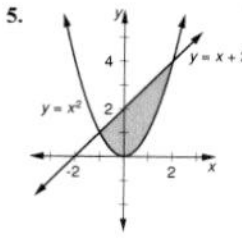

a. $\int_{-1}^{2} [(x - 2) - x^2]\,dx$

b. $\frac{9}{2}$

6.
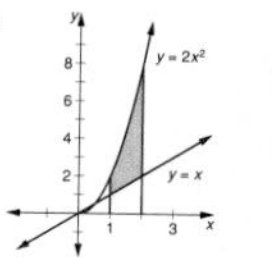

a. $\int_1^2 (2x^2 - x)\,dx$

b. $\frac{19}{6}$

LESSON 13-6

RESOURCES
- ■ Lesson Master 13-6
- ■ Teaching Aid 106 provides drawings for use with this lesson.

OBJECTIVES

E Use the definite integral to solve application problems.
I Find volumes of solids.

TEACHING NOTES

Alternate Approach
Physical models can help students to visualize solids of revolution. Make a stack of pennies, nickels, dimes, and quarters (in random order) and hold them between two fingers horizontally to illustrate how a solid can be made of thin cylinders perpendicular to a horizontal axis. This model will resemble the chair leg in **Example 1** if the coins are strategically placed. It is helpful to remove one coin and hold it flat so that students can see the proper relationships: Δx = the width, or height of the cylinders and $f(z_i)$ = the radius of the cylinders; thus, the volume $= \pi r^2 h = \pi(f(z_i))^2 \Delta x$.

LESSON
13-6

Definite Integrals and Volumes

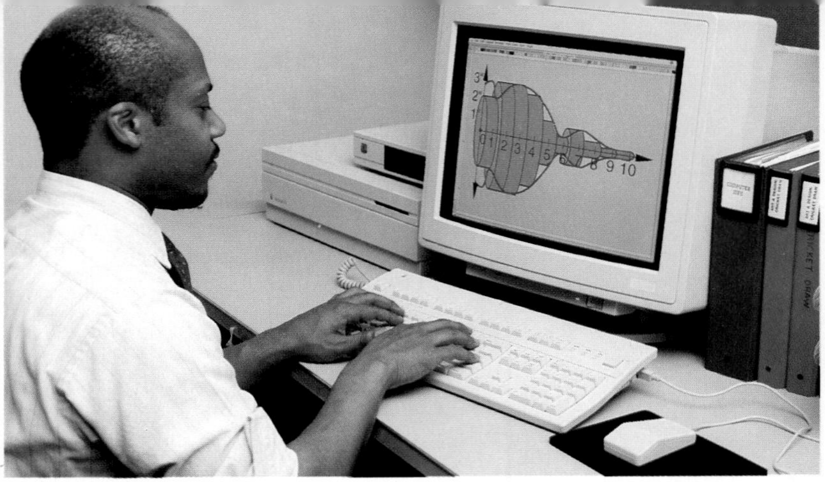

You have seen that definite integrals can be used to compute areas and total distances. In fact, definite integrals can be used to compute many quantities, as long as those quantities are limits of Riemann sums of appropriate functions on appropriate intervals. This lesson illustrates how to use Riemann sums to approximate volumes of some solids.

Example 1 This decorative chair leg is going to be molded out of plastic. Use the figures below to approximate the volume of the chair leg. Dimensions are expressed in inches.

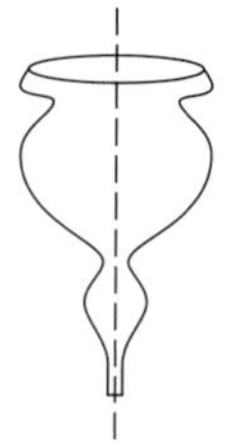

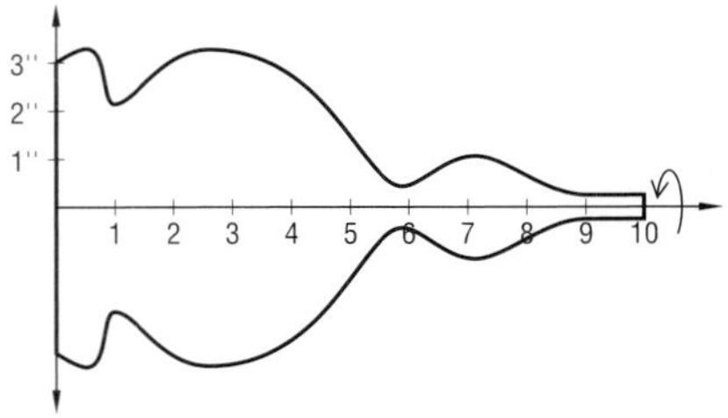

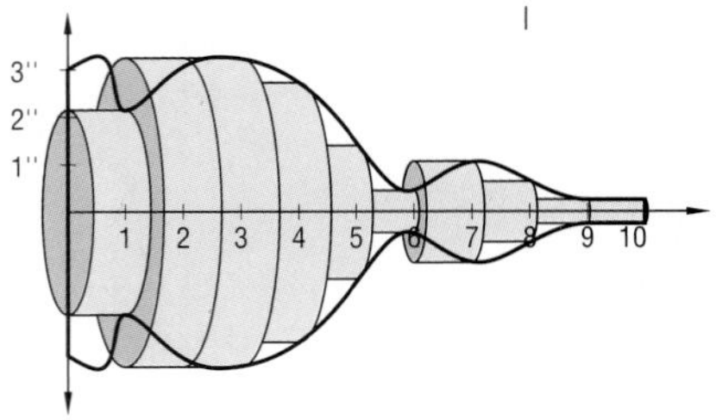

Solution In Figure 1, the axis of the chair leg has been made to coincide with the x-axis in a 2-dimensional graph. Imagine generating the chair leg by rotating Figure 1 about the x-axis. Figure 2 shows a three-dimensional approximation to the chair leg. Now, think of slicing the three-dimensional figure into sections, each of which can be approximated by a cylinder. In Figure 2 there are 10 sections, and the radius r_i of the cylinder on the ith section is given by the height (to the nearest half inch) of the graph at the right endpoint of the ith subinterval. The volume ΔV_i of the cylinder on the ith subinterval is given by $\Delta V_i = \pi r_i^2 h$.

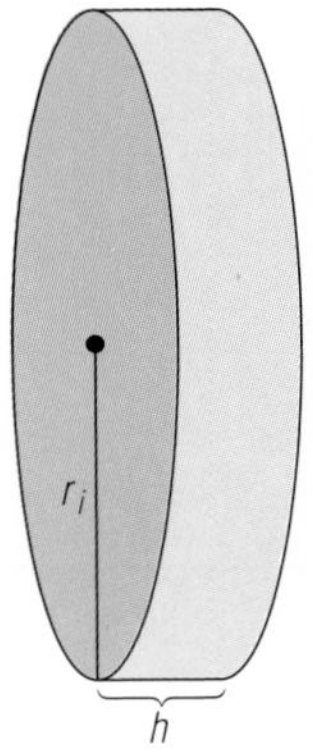

794

The height h of each cylinder is the width of the subinterval, which is 1 inch.

i	1	2	3	4	5	6	7	8	9	10
r_i	2	3	3	2.5	1.5	.5	1	.5	.5	.5
ΔV_i	4π	9π	9π	6.25π	2.25π	$.25\pi$	1π	$.25\pi$	$.25\pi$	$.25\pi$

So the total volume of the chair leg is approximately

$$\sum_{i=1}^{10} \Delta V_i = 32.5\pi \approx 102.1 \text{ cubic inches.}$$

For all problems of this type, you should draw a sketch and analyze a representative ith subinterval, as was done in Example 1.

Example 2 Consider the region bounded by the line $f(x) = x + 3$, the x- and y-axes, and the line $x = 7$. Find the volume of the solid generated by revolving this region about the x-axis.

Solution Slice the solid into n sections of equal width Δx. If $(z_i, 0)$ is a point contained in one of the sections, then the volume of the section is approximately that of a cylinder with width Δx and radius $f(z_i)$, or $z_i + 3$. Thus the volume of that section is $\pi(z_i + 3)^2 \Delta x$. The total volume of the solid can be approximated by the sum of the volumes of these cylinders,

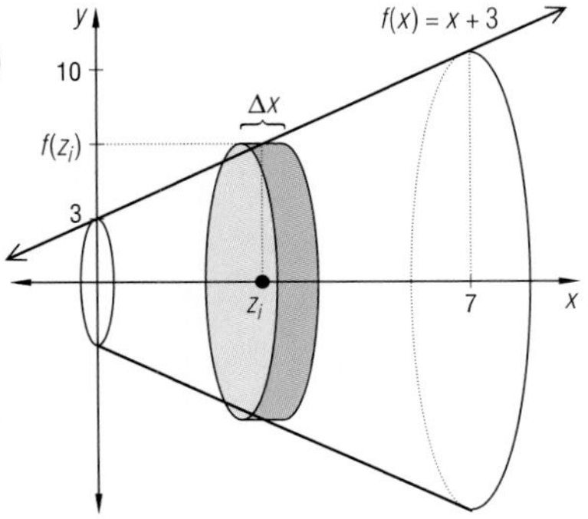

$$\sum_{i=1}^{n} \pi(z_i + 3)^2 \Delta x.$$

Notice that this is a Riemann sum for the function $f(x) = \pi(x + 3)^2$ on the interval $0 \le x \le 7$. As $n \to \infty$, that is, as more and more smaller slices are used, the sum becomes closer to the actual volume of the solid. Thus

$$\text{Volume} = \lim_{n \to \infty} \sum_{i=1}^{n} \pi(z_i + 3)^2 \Delta x.$$

This is the limit of a Riemann sum for the function on the interval. Thus, the volume is an integral:

$$\text{Volume} = \int_0^7 \pi(x + 3)^2\, dx = \pi\left[\int_0^7 (x^2 + 6x + 9)\, dx\right].$$

This can be evaluated by using the Integral of a Quadratic Function Theorem derived in Lesson 13-5, with $c_2 = 1$, $c_1 = 6$, and $c_0 = 9$:

$$\text{Volume} = \pi\left[\frac{7^3}{3} + 6 \cdot \frac{7^2}{2} + 9 \cdot 7\right] = \frac{973}{3}\pi.$$

The solid of **Example 2** is a truncated cone. Its volume can be found by using standard Euclidean geometry formulas. In this example, the "width" of the cylinder is the cylinder's height for the purpose of calculating its volume. Also, the point $(z_i, 0)$ is on the axis of the cylinder, which is why $f(z_i)$ is the cylinder's radius.

In the proof of the theorem for the volume of a sphere, you might point out that it also would be possible to double the volume of half a sphere: $V = 2\left[\pi \int_0^r (r^2 - x^2)\, dx\right]$. Rather than going through the details of the proof of the theorem for the volume of sphere, you might wish to give only the general idea. The sphere is sliced into sections. Because the heights of these sections involve square roots of quadratic expressions, the volumes of the sections involve quadratic expressions. The limit of the sum then becomes an integral of a quadratic function and the formula of the preceding lesson can be applied. This formula indicated that the integral of a quadratic function is a cubic function. Thus, there is yet another reason why the formula for the volume of a sphere involves a cubic.

You might wish to mention the kinds of solids of revolution for which volumes can now be found. For any solid, if the solid can be placed as in **Example 1**, then its volume can be estimated. The methods of **Example 2** can be generalized to find the volume of any solid formed by revolving a line about the x-axis. The methods of the theorem for finding the volume of a sphere can be generalized to find the volume of any ellipsoid.

ADDITIONAL EXAMPLES

1. Suppose the region between the parabola $y = x^2$ and the x-axis, and between the y-axis and the line $x = 10$, is revolved about the x-axis.

a. Split the region into five cylinders of equal width and use right endpoints of the subintervals to give an estimate for the volume of the generated solid.
$31{,}328\pi$

b. Repeat part **a** using 10 subintervals.
$25{,}333\pi$

2. Find the volume generated when the area bounded by $y = \frac{1}{2}x$, the x-axis, and $x = 6$ is revolved about the x-axis.
18π

Definite integrals can help to obtain formulas for the volumes of some solids. Here is a calculus proof of a theorem you may have first seen years ago.

Theorem

The volume of a sphere of radius r is $\frac{4}{3}\pi r^3$.

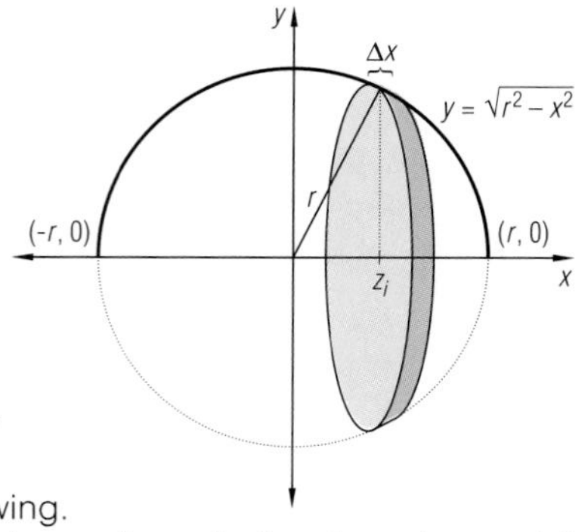

Proof This sphere can be obtained by revolving the semicircle $y = \sqrt{r^2 - x^2}$ about the x-axis.

Slice the sphere into sections of equal width Δx. If a section contains the point $(z_i, 0)$, then the volume of the section is approximately that of a cylinder with width Δx and radius $\sqrt{r^2 - z_i^2}$, as shown in the drawing. Then from the formula for the volume of a cylinder, the volume of the section is approximately

$$\pi(\sqrt{(r^2 - z_i^2)})^2 \, \Delta x = \pi(r^2 - z_i^2) \, \Delta x.$$

The sum of the volumes of these sections is

$$\sum_{i=1}^{n} \pi(r^2 - z_i^2) \, \Delta x.$$

The limit of this sum as $n \to \infty$ is the volume of the sphere. But this sum is a Riemann sum of the function $y = \pi(r^2 - x^2)$ over the interval $-r \le x \le r$. Thus its limit is an integral from $-r$ to r.

$$\text{Volume} = \lim_{n \to \infty} \sum_{i=1}^{n} \left(\pi(r^2 - z_i^2) \, \Delta x\right) = \int_{-r}^{r} \pi(r^2 - x^2) \, dx$$

Using the Constant Multiple and Sum properties of integrals,

$$\text{Volume} = \pi\left[\int_{-r}^{r} r^2 \, dx - \int_{-r}^{r} x^2 \, dx\right].$$

Now, because r is a constant, the first integral is the integral of a constant function and can be evaluated by using $\int_a^b c \, dx = c(b - a)$.

Also, by symmetry, $\int_{-r}^{r} x^2 \, dx = 2\int_0^r x^2 \, dx$. Thus,

$$\begin{aligned}\text{Volume} &= \pi\left[r^2(r - (-r)) - 2\left(\frac{r^3}{3}\right)\right]\\ &= \pi\left(r^2(2r) - \frac{2r^3}{3}\right)\\ &= \pi\left(\frac{4}{3} r^3\right),\end{aligned}$$

which is equivalent to the desired formula.

796

Questions

Covering the Reading

1. Refer to Example 1. Suppose r_i is given by the height (to the nearest half inch) of the graph at the left endpoint of the ith subinterval. Evaluate $\sum_{i=1}^{10} \Delta V_i$ in this case. **$41.25\pi \approx 129.59$ in.^{3}**

2. To approximate the volume of the jug below, the axis of the jug has been made to coincide with the x-axis as on the graph. Then the jug can be generated by rotating this figure about the x-axis.

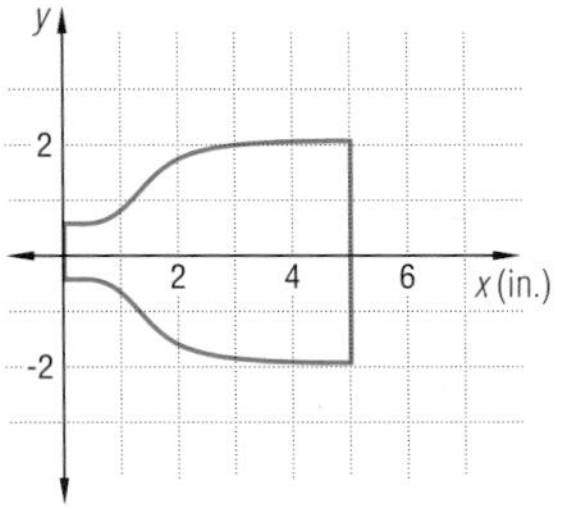

a. If sections are formed as in Example 1 of this lesson, what figure does each section approximate? **cylinder**

b. Using five subintervals, estimate the volume of the jug. (Take r_i to be the height, to the nearest fourth of an inch, of the graph at the right endpoint of the ith subinterval.) **$15.625\pi \approx 49.1$ in.^{3}**

3. Consider the region bounded by the line $f(x) = x + 5$, the x- and y-axes, and the line $x = 4$. Find the volume of the solid generated by revolving this region about the x-axis. **$\frac{604\pi}{3} \approx 632.5$ units^3**

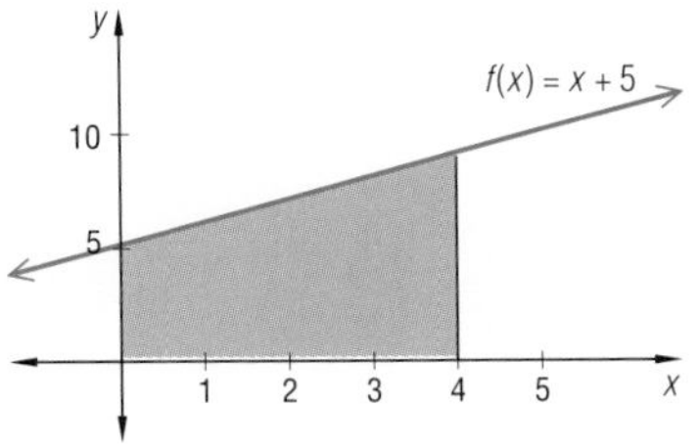

In 4–6, refer to the proof of the theorem on the volume of a sphere.

4. Explain why the cross-section of the sphere containing the point $(z_i, 0)$ has radius equal to $\sqrt{r^2 - z_i^2}$. **See margin.**

5. Explain why the following statement is true:

$\int_{-r}^{r} r^2\, dx = r^2 \int_{-r}^{r} dx.$ **See margin.**

6. Why are the two integrals in the proof evaluated from $-r$ to r?
$(-r, 0)$ and $(r, 0)$ are the left and right endpoints of the semicircle.

NOTES ON QUESTIONS

Question 3: This question is similar to **Example 2**. This can be checked by noting that the solid is a truncated cone. The original cone has height 9 and radius 9; the top 5 units of its height are cut off from it.

Question 5: Each side has value $2r^3$. Point out that in general the "dx," which seems to have little to do with the calculation of values of definite integrals, is critical, since it identifies the variable with respect to the integral being considered. If is left side were $\int_{-r}^{r} r^2\, dr$, then the integral would be that of a quadratic and would equal $\frac{2r^3}{3}$, while $r^2 \int_{-r}^{r} dx = 2r^3$.

ADDITIONAL ANSWERS

4. The radius of the cross-section is the height of the graph where $x = z_i$ since $f(x) = \sqrt{r^2 - x^2}$, so $f(z_i) = \sqrt{r^2 - z_i^2}$.

5. $\int_a^b kf(x)\, dx = k \int_a^b f(x)\, dx$ for k is constant. But r is a constant, and so r^2 is a constant. So $\int_{-r}^{r} r^2\, dx = r^2 \int_{-r}^{r} 1\, dx$ or $r^2 \int_{-r}^{r} dx$.

ADDITIONAL ANSWERS

7.a.

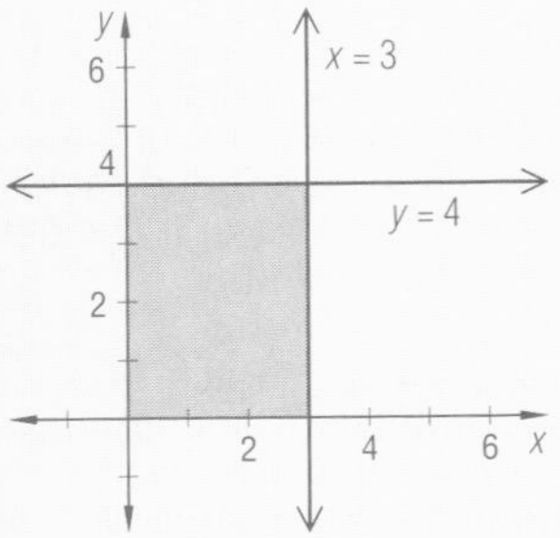

d. $\int_0^3 \pi r_i^2\, dx = 16\pi \int_0^3 dx =$ 48π units^3

8.a.

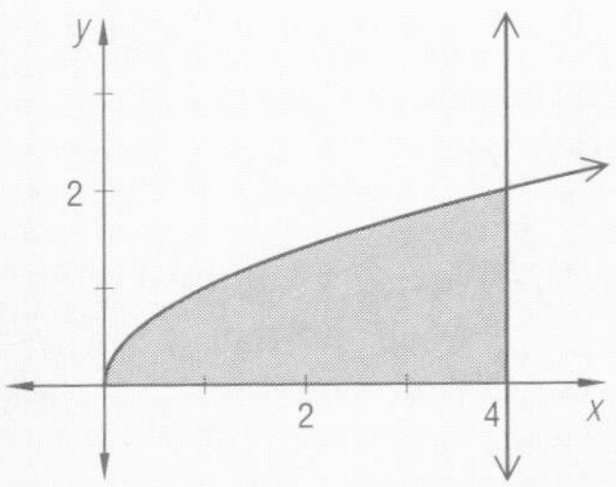

9.a.

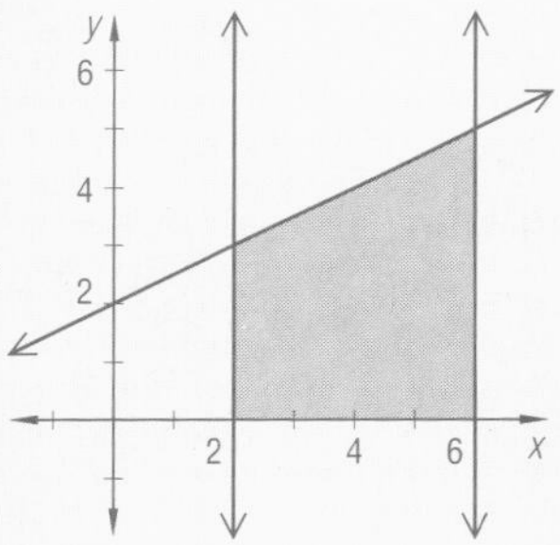

Applying the Mathematics

7. Consider the region bounded by the x- and y-axes and the lines $y = 4$ and $x = 3$.

a. Sketch a graph of this region. See margin.

b. Suppose the region is revolved about the x-axis. What type of solid is formed? cylinder

c. Let r_i be the radius of the cylindrical section on the ith subinterval. If r_i is the height of the graph at the right endpoint of the ith subinterval, find r_i. 4 units

d. Find the volume of the solid in part **b** by setting up and evaluating an integral. See margin.

e. Check your answer in part **b** by using geometry. $\pi r^2 h = \pi(4^2)(3) = 48\pi$ units^3

8. **a.** Sketch a graph of the region bounded by $f(x) = \sqrt{x}$, $x = 4$, and the x-axis. See margin.

b. Calculate the volume of the solid generated when this region revolves about the x-axis. 8π units^3

9. **a.** Sketch a graph of the region bounded by $f(x) = \frac{1}{2}x + 2$, $x = 2$, $x = 6$, and the x-axis. See margin.

b. Calculate the volume of the solid generated when this region revolves about the x-axis. $\frac{196\pi}{3} \approx 205.25$ units^3

Review

10. At the right is the graph of $y = 2x^2 - 16x + 33$.

a. Write an integral that describes the area of the shaded region.

b. Evaluate your integral in part **a**.
(Lessons 13-5, 13-4) 9 units^2

a) $\int_2^5 (2x^2 - 16x + 33)\, dx$

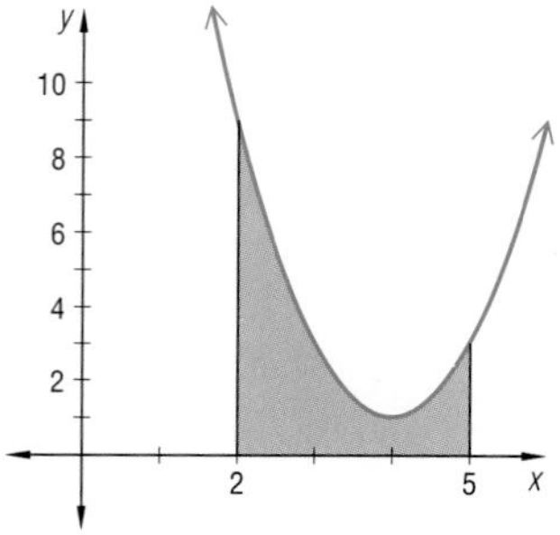

In 11–13, evaluate the definite integral. *(Lessons 13-5, 13-4, 13-3)*

11. $\int_0^7 5x\, dx$ $\frac{245}{2}$ **12.** $\int_2^3 (4x + 1)\, dx$ 11 **13.** $\int_0^2 \sqrt{4 - x^2}\, dx$ π

14. Write as a single integral:
$\int_3^{11} \log x^3\, dx + \int_3^{11} \log x^4\, dx$. *(Lesson 13-4)* $\int_3^{11} \log x^7\, dx$ or $7\int_3^{11} \log x\, dx$

15. The school bookstore carries only one size of three-ring binders but carries it in three colors—black, blue, and red. A student plans to purchase 5 binders. In how many different ways may the student do this? *(Lesson 10-7)* 21

16. Let f be the function $f(x) = ax^2 + bx$. Calculate the average rate of change in f from x to $x + \Delta x$. *(Lesson 9-1)* $2ax + b + a\Delta x$

798

17. Consider the graph of a polynomial function p shown below.

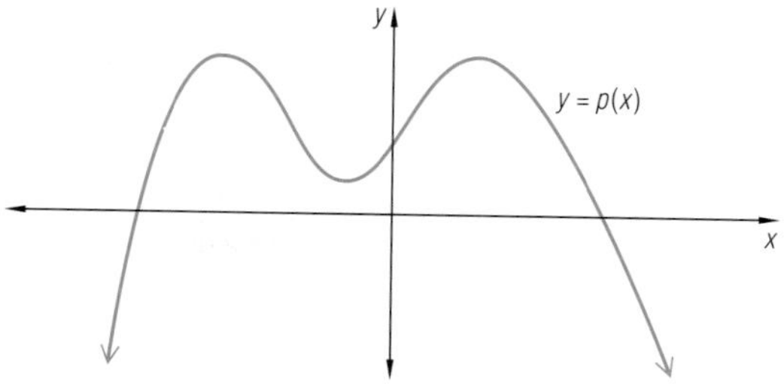

a. What does $\lim_{x \to \infty} p(x)$ appear to be? $-\infty$

b. What does $\lim_{x \to -\infty} p(x)$ appear to be? $-\infty$

c. What is the smallest possible degree of $p(x)$? *(Lessons 8-8, 8-9)* 4

18. Prove the following identity:

$\forall$ real numbers x for which $\cos x \neq 0$,

$\dfrac{\cos 3x}{\cos x} = 4\cos^2 x - 3$. *(Lessons 6-3, 6-2)* **See margin.**

19. Prove or disprove the following statement:

For all real numbers a, b, and c, if a is a factor of b and a is a factor of c, then a is a factor of bc.

(Lesson 4-1) **a is a factor of $c \Leftrightarrow c = am$ for some integer m. So $bc = b(am) = a(bm)$, so a is a factor of bc. (a is a factor of b is not needed.)**

Exploration

20. Graphs of $y = 2$ and $y = \sqrt{x}$ appear below. Determine the volume of the solid generated when the shaded region is revolved about the x-axis. (Hint: Think of the solid as a cylinder out of which a bullet-shaped region has been cut.) **8π units^3**

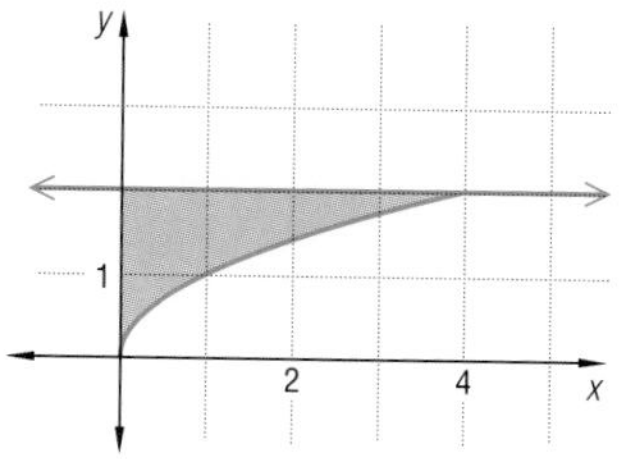

FOLLOW-UP

MORE PRACTICE

For more questions on SPUR Objectives, use *Lesson Master 13-6*, shown below.

18. See Additional Answers in the back of this book.

NAME ______

LESSON MASTER 13–6

QUESTIONS ON **SPUR** OBJECTIVES

■ **USES** *Objective E* *(See pages 810–813 for objectives.)*

1. How much water is required to completely fill five glasses, each of which is formed by rotating the line $f(x) = \frac{1}{2}x + 1$ from $x = 0$ to $x = 10$ around the x-axis? All coordinates are in centimeters. Give your answer in liters. **≈ 2.26 liters**

2. The parabolic cross-section of a trough is 2 feet wide and 2 feet high. If the trough is 9 feet long, find its volume. **24 ft^3**

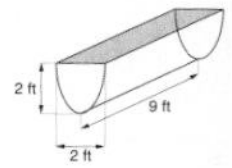

■ **REPRESENTATIONS** *Objective I*

In 3 and 4, a. sketch a graph of the region described and b. calculate the volume of the solid generated when the region is revolved about the x-axis.

3. the region bounded by the x-axis, the y-axis, and the line $y = -\frac{5}{4}x + 5$

a.

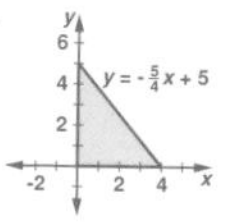

b. $\dfrac{100\pi}{3}$

4. the region bounded by the lines $y = 4 - \frac{x}{2}$, $x = 2$, $x = 4$, and the x-axis

a.

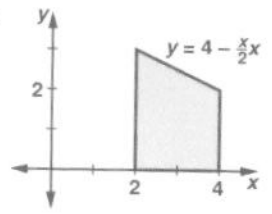

b. $\dfrac{38\pi}{3}$

LESSON 13-7

RESOURCES
- Teaching Aid 107 can be used with **Questions 15 and 16**.

TEACHING NOTES

The fact that many functions have the same derivative can be illustrated by asking students to name several functions whose derivative is $2x$. The goal is for students to understand that functions that differ by a constant have the same derivative. In order for students to see this, have them trace and then sketch tangents to the two functions in the graph at the top of page 801. Since corresponding tangents at each value of x are parallel, their slopes are equal and thus the derivatives at each value of x are equal.

Another illustration of part **1** of the Fundamental Theorem uses projectile motion. The position function $h(t) = 800t - 16t^2$ has as its derivative the velocity function $h'(t) = v(t) = 800 - 32t$. The distance traveled from $t = 8$ to $t = 20$ is given by the area under the velocity curve from 8 to 20, $\int_8^{20} v(t)\,dt$, which is $\int_8^{20} h'(t)\,dt$. But the distance traveled also is given by $h(20) - h(8)$. Therefore, $\int_8^{20} h'(t)\,dt = h(20) - h(8)$.

LESSON

13-7

The Fundamental Theorem of Calculus

Archimedes and Johannes Kepler (the person who discovered that planets travel around the sun in elliptical orbits) used ideas of integration, and rates of change were studied by Galileo and others. However, Isaac Newton and Gottfried Leibniz are usually mentioned as the founders or discoverers of calculus. Newton and Leibniz independently discovered that derivatives and integrals of functions are related in a simple way, and thus they created a unified branch of mathematics. Newton seems to have discovered this idea, which today we call the Fundamental Theorem of Calculus, in the 1660s, and Leibniz in the 1670s, but Leibniz published his work in 1684 and 1686, while Newton waited until 1687. The two men engaged in bitter quarrels about who was first; today they are both generally credited with the discovery of calculus.

Sir Isaac Newton

Gottfried von Liebniz

You have seen a number of examples of the relationship between derivatives and integrals.

In Chapter 9, you saw that the derivative of a position or distance function is a rate function. In this chapter, you saw that the integral of a rate function gives total distance.

In Chapter 9, you also saw that if f is the function defined by $f(x) = ax^2 + bx$, then its derivative function f' is defined by $f'(x) = 2ax + b$. Now examine $\int_0^v (2ax + b)\,dx$. From the Sum and Constant Multiple properties,

$$\begin{aligned}\int_0^v (2ax + b)\,dx &= 2a\int_0^v x\,dx + \int_0^v b\,dx\\ &= 2a\left(\frac{v^2}{2}\right) + bv\\ &= av^2 + bv.\end{aligned}$$

So the integral of the derivative function $f'(x) = 2ax + b$ gives the original function, when v is replaced by x.

In Lesson 13-2, you calculated a Riemann sum for $f(x) = \cos x$ on the interval from 0 to $\frac{\pi}{3}$. With $n = 500$, the value of that sum was approximately 0.8660, which is $\sin\frac{\pi}{3}$. In general,

$$\int_0^a \cos x\,dx = \sin a.$$

Do you recall from Chapter 9 that when $f(x) = \sin x$, $f'(x) = \cos x$? Thus the derivative of the sine function is the cosine function and, in a way (to be detailed below), the integral of the cosine function is the sine.

The fact that many functions have the same derivative prevents derivatives and integrals from being related in a 1-1 manner. Specifically, adding a constant c to each value of a function merely translates the function up c units and does not affect its derivative at any point.

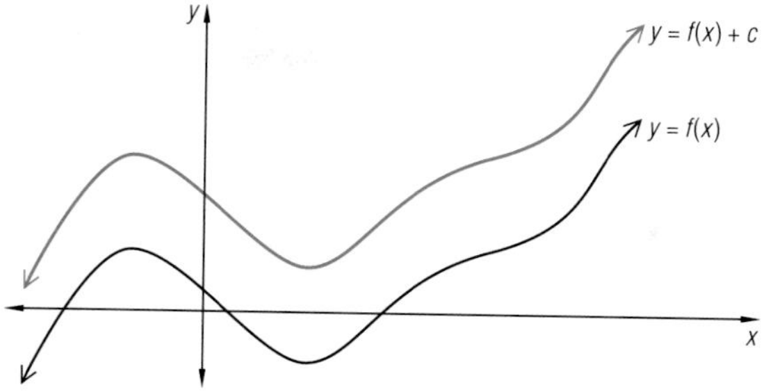

Still, the relationships between derivatives and integrals are most elegant.

Fundamental Theorem of Calculus

Let f be a continuous function on the interval from a to b.

1. If g is a function whose derivative is f, then $$\int_a^b f(x)\,dx = g(b) - g(a).$$
2. If $g(x) = \int_a^x f(t)\,dt$ for all x from a to b, then $g'(x) = f(x)$ for all such x.

The second part of the Fundamental Theorem of Calculus can be interpreted as saying that the derivative of an integral function is the original function. For instance, in Lesson 13-5 it was found that $\int_a^b x^2\,dx = \frac{b^3}{3} - \frac{a^3}{3}$. Let $x = t$ and $b = x$ and let $g(x) = \int_a^x t^2\,dt = \frac{x^3}{3} - \frac{a^3}{3}$. Then by Part 2 of the Fundamental Theorem $g'(x) = x^2$.

To see why the second part of the theorem is true in general when f is a continuous function that is positive on the interval $a \le x \le b$, define a function g as follows: $g(x) = \int_a^x f(t)\,dt$ for all x on $a \le x \le b$. Then $g(x)$ is the area between the graph of f and the horizontal axis from a to x.

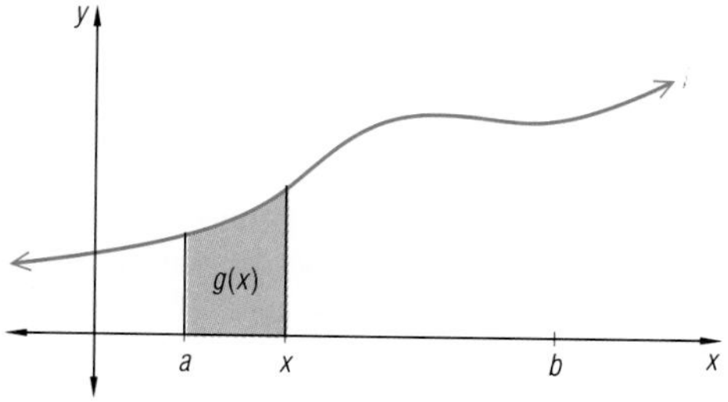

The use of many examples can help students to understand the Fundamental Theorem. For instance, you can prove (using the definition of derivative) that if $f(x) = \frac{1}{3}x^3$, then $f'(x) = x^2$. Now evaluate $\int_a^b x^2\,dx$ using the Fundamental Theorem. It is $f(b) - f(a)$, or $\frac{b^3}{3} - \frac{a^3}{3}$. This evaluation is quite simple and elegant compared with the derivation in Lesson 11-3, and gives many students an appreciation of the Fundamental Theorem.

Remember that $g'(x) = \lim_{\triangle x \to 0} \frac{g(x + \triangle x) - g(x)}{\triangle x}$. To evaluate the difference quotient when $\triangle x > 0$, notice that $g(x + \triangle x)$ is the area between the graph of f and the horizontal axis from a to $x + \triangle x$.

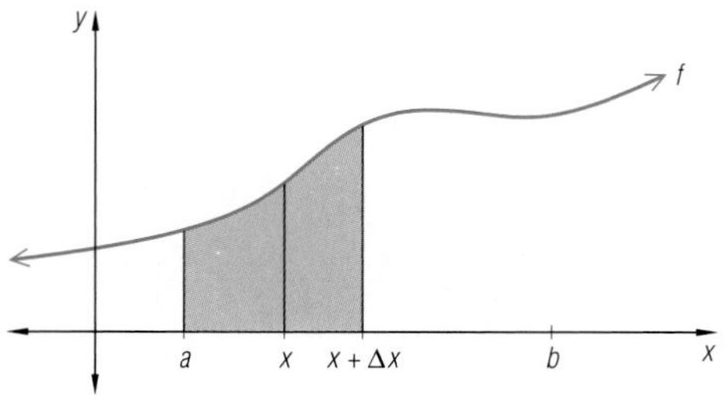

So $g(x + \triangle x) - g(x)$ is the area of the region between the graph of f and the horizontal axis from x to $x + \triangle x$.

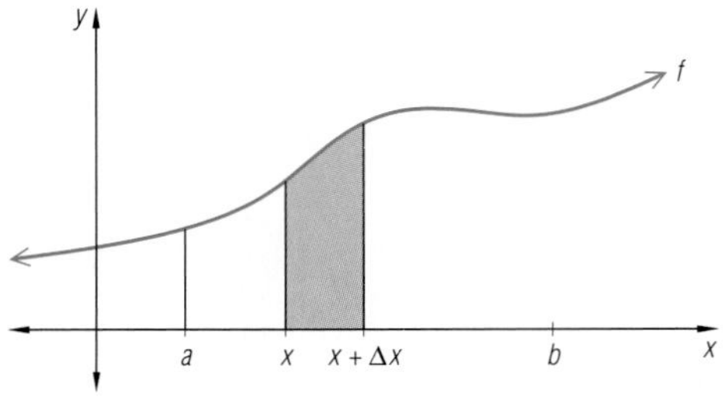

If $\triangle x$ is small, then the region is nearly a rectangle, and if we divide this area by $\triangle x$, we are dividing by the width of the rectangle. The quotient is approximately the height of the rectangle, which, as $\triangle x$ gets smaller and smaller, and since f is continuous, approaches the value $f(x)$. That is, $\lim_{\triangle x \to 0} \frac{g(x + \triangle x) - g(x)}{\triangle x} = f(x)$, or, in other symbols, $g'(x) = f(x)$.

In the case of the first part of the Fundamental Theorem of Calculus, g is called an **antiderivative** of f. This part indicates that the definite integral is found by evaluating an antiderivative. For instance, since the derivative of the sine is the cosine, if $f(x) = \cos x$, then one of its antiderivatives is $g(x) = \sin x$. Because of this,

$$\int_a^b \cos x \, dx = \sin b - \sin a.$$

Then $$\int_0^{\pi/3} \cos x \, dx = \sin \frac{\pi}{3} - \sin 0 = \frac{\sqrt{3}}{2},$$

which agrees with the estimate found in the Example in Lesson 13-2.

Together, the two parts of the Fundamental Theorem of Calculus imply that if you have found the derivative of a function then you can evaluate integrals of the derivative, and if you have the integral of a function, then you know the derivative of that integral.

NOTES ON QUESTIONS
Question 3: Students are expected to use the result that if $f(x) = \sin x$, then $f'(x) = \cos x$.

Thus, integration and differentiation are almost inverse operations on functions. Perhaps this is not so surprising, for in many cases integrals can be interpreted as areas and volumes, which are basic models for multiplication. Derivatives are rates, and rate is a basic model for division. And, just as multiplication and division undo each other, so do integration and differentiation.

It is perhaps fitting that we have ended this book by discussing a very important theorem, for throughout this book you have studied some of the most important theorems in all of mathematics. Three of them even are called ''fundamental''! If you take additional courses in mathematics, and we certainly hope you do, you will encounter the ideas in this course again and again, for discrete mathematics and the mathematics needed for calculus are found in virtually all fields of mathematics.

Why are mathematical ideas so interrelated? Perhaps because they originate from ideas in an inherently ordered world. No one knows for sure.

We hope that these and other ideas you have studied in this book have interested you, and have stimulated your curiosity to learn more. And, of course, we hope you have enjoyed and learned much from this course.

Questions

Covering the Reading

1. What two mathematicians are credited with discovering calculus? Isaac Newton, Gottfried Leibniz
2. Why are derivatives and integrals not related in a 1-1 manner? Many different functions have the same derivative.
3. Use the first part of the Fundamental Theorem of Calculus to evaluate

$$\int_0^{\pi/2} \cos x \, dx. \quad 1$$

4. You know that if $g(x) = x^2$, then $g'(x) = 2x$. So $g(x) = x^2$ is called a(n) _?_ of $f(x) = 2x$. antiderivative

5. If $g(x) = \int_a^x \ln t \, dt$, then $g'(x) = $ _?_. $\ln x$

Applying the Mathematics

6. *Multiple choice.* If $h(x) = \int_a^x (t^2 - 5t + 4)\, dt$, then $h'(x) =$
   (a) $t^2 - 5t + 4$
   (b) $\left(\frac{x^3}{3} - \frac{5x^2}{2} + 4x\right) - \left(\frac{a^3}{3} - \frac{5a^2}{2} + 4a\right)$
   (c) $x^2 - 5x + 4$
   (d) $(x^2 - 5x + 4) - (a^2 - 5a + 4)$. (c)

NOTES ON QUESTIONS
Making Connections for Question 8: The derivative of the sine is the cosine; the derivative of the cosine is the opposite of the sine. Thus, the sequence of nth derivatives of the sine is: sine, cosine, -sine, -cosine, sine, ... , repeating with a period of 4. It also means that the sine and cosine functions each satisfy the second order differential equation $f'' = f$. This makes the sine and cosine functions very important in the study of differential equations.

Question 17: By falling in different ways, a person can maximize or minimize the air resistance. It is similar to seeing how far down in the water you go on a dive; if you "belly-flop," you do not go down as far as if you dive in the form of a bullet. The first diver was unconscious and tumbling through the air; the second diver was able to be more efficient and catch up with the first.

Question 18: You might wish to inform students that there are tables of integrals and derivatives. There are also computer programs that can perform differentiation and integration symbolically.

7. If $f(x) = \frac{1}{x}$, then one of its antiderivatives is $g(x) = \ln x$. Use this result to evaluate $\int_5^{20} \frac{1}{x}\, dx$. **$\ln 20 - \ln 5 \approx 1.386$**

8. If $f(x) = \sin x$, then one of its antiderivatives is $g(x) = -\cos x$.

a. Use this relationship to evaluate $\int_0^{\pi/3} \sin x\, dx$. **0.5**

b. Use a computer to approximate the integral using a Riemann sum with 100 subintervals. How well does your result agree with the answer in part **a**? **0.5043; the result agrees very well; the relative error is only about 0.8%.**

9. What are the two other "fundamental theorems" discussed in this book? **See margin.**

Review

10. a. Sketch the region bounded by the line $y = x$, the x-axis, and the line $x = 4$. **See margin.**

b. Write an integral that finds the volume of the solid formed when the region in part **a** is rotated about the x-axis. *(Lesson 13-6)* **$\int_0^4 \pi x^2\, dx$**

11. A cylindrical trough has a base in the shape of a parabola, 4 feet high and 6 feet across. If the trough is 7 feet long, what is its volume? *(Lesson 13-5)* **112 ft^3**

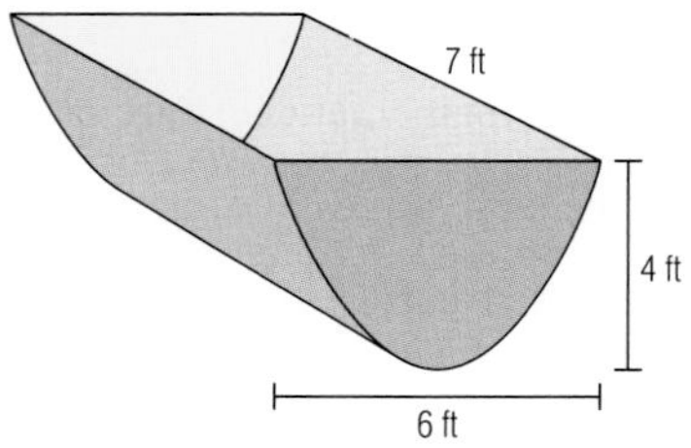

In 12 and 13, evaluate the definite integral. *(Lessons 13-5, 13-4, 13-3)*

12. $\int_{-3}^{2} (5x - 7)\, dx$ **-47.5**

13. $\int_1^2 (3x^2 + 4x)\, dx$ **13**

14. Express the shaded area below with integral notation. *(Lesson 13-3)* **$\int_0^{\frac{5\pi}{4}} (1 + \sin 2x)\, dx$**

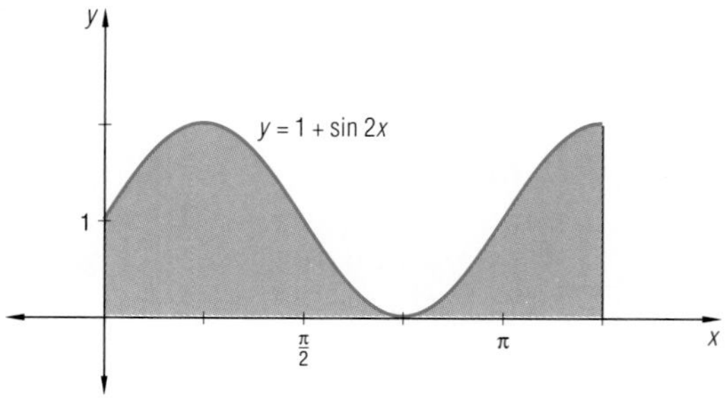

804

In 15 and 16, find an Euler circuit, if one exists. *(Lesson 11-4)*

15.

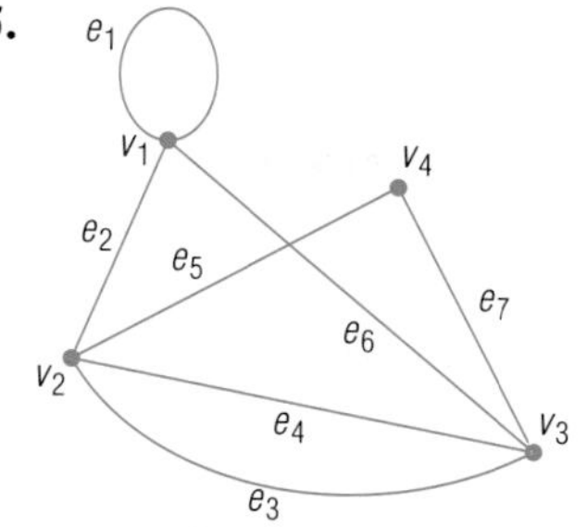

sample: $e_1e_2e_3e_4e_5e_7e_6$

16.

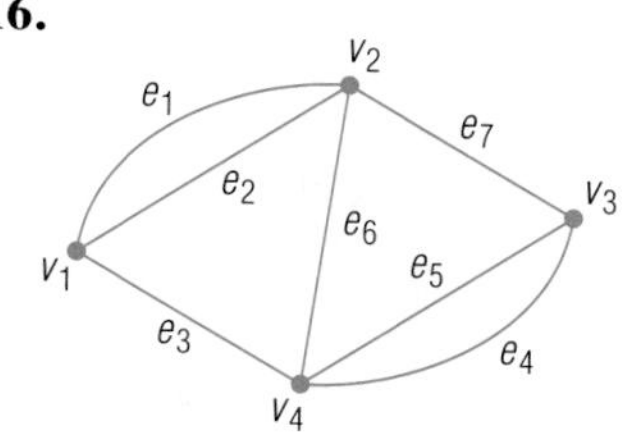

No Euler circuit exists.

17. A TV commercial aired in 1990 tells the story of a skydiving situation in which one individual saved another's life. When the first friend jumped, he collided with another person, was knocked unconscious, and started plummeting to Earth. His friend jumped and directed his fall toward the unconscious diver. When he caught the unconscious diver and pulled the cord, the two were 10 seconds from hitting the ground. If the jump occurred at a height of 10,000 feet, the height (in feet) at any time t during freefall is $h(t) = -16t^2 + 10{,}000$.

a. How long would it take someone to hit the ground if no parachute is pulled? 25 sec

b. How high above the ground did the individual catch the unconscious diver? *(Lesson 3-5)* 6400 ft

Exploration

18. Given the following relationships:

if $f(x) = \tan x$ then $f'(x) = \sec^2 x$;
if $f(x) = \ln x$ then $f'(x) = \frac{1}{x}$;
if $f(x) = e^x$ then $f'(x) = e^x$.

Make up one or more area problems to be evaluated using the Fundamental Theorem of Calculus. Check your results by using a computer to calculate Riemann sums with larger and larger numbers of subintervals. Answers may vary.

FOLLOW-UP

EXTENSION

A discussion of why mathematical ideas are so interrelated could be quite interesting. There may be a wide range of opinions, but there is no "right" answer. It really is the case that no one knows for sure why mathematical ideas are so interconnected.

PROJECTS

The projects for Chapter 13 are described on pages 806–807. **Projects 3 and 4** are related to the content of this lesson.

ADDITIONAL ANSWERS

9. Fundamental Theorem of Arithmetic: Suppose that n is an integer and that $n > 1$. Then either n is a prime number or n has a prime factorization which is unique except for the order of the factors. Fundamental Theorem of Algebra: Every polynomial of degree $n \ge 1$ with real or complex coefficients has at least one complex zero.

10.a.

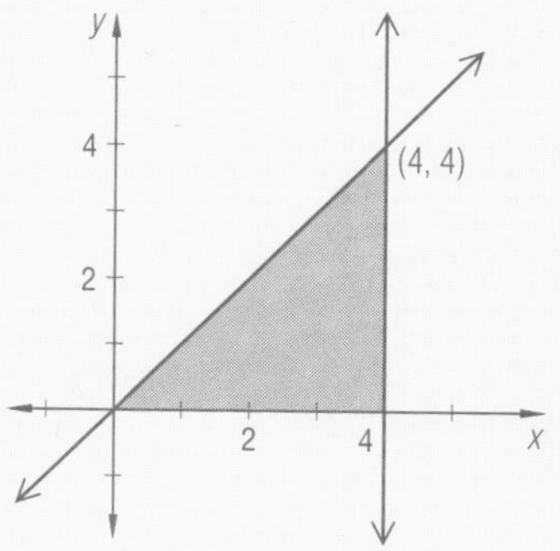

PROJECTS

The projects provide an opportunity for extended work on topics related to the content of the chapter. (See pages 70 and T37 for more information on the purposes of projects.) Some or all of these projects may be suitable for **small group work**. Solutions for the projects can be found at the end of the Additional Answers section in the back of this book.

Projects

1. The BASIC program below approximates the area under the curve $y = x^2$ from 0 to 10 using Riemann sums with subintervals of equal width. The height of each rectangle is determined by the value of the function at the *midpoint* of the interval. The program below uses 100 rectangles for an interval from 0 to 10.

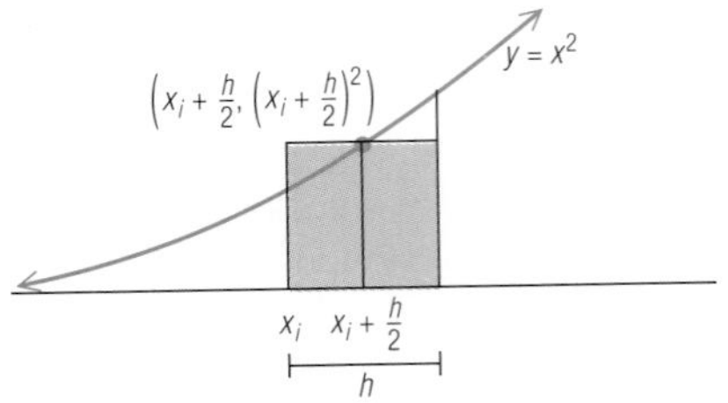


```
100 REM LEFT AND RIGHT ENDPOINTS
110 LET A = 0
120 LET B = 10
130 REM NUMBER OF SUBINTERVALS
140 LET N = 100
150 REM SUBINTERVAL SIZE
160 LET DX = (B – A)/N
170 REM INITIALIZE SUM
180 LET SUM = 0
190 REM LOOP THROUGH SUBINTERVALS
200 FOR I = 1 TO N
210   REM COMPUTE MIDPOINT OF
      SUBINTERVAL
220   M = A + (I – 0.5)*DX
230   REM COMPUTE THE VALUE OF THE
      FUNCTION
240   V = M^2
250   REM INCREMENT THE SUM
260   SUM = SUM + V*DX
270 NEXT I
280 PRINT "SUM = ",SUM
290 END
```

a. Run the program and make a table of values for right endpoint values $b = 1, 2, 3, 4, \ldots, 10$. Compare the values with the corresponding values of $\frac{b^3}{3}$.

b. Replace line 240 with V = M^3 to evaluate the area under the curve $y = x^3$. Make a table as in **a.** Make a conjecture as to the formula for the area under the curve from 0 to b.

2. Use the program in Project 1.

a. Investigate the area under $y = \sin x$ from 0 to b where $b = \frac{\pi}{6}, \frac{\pi}{4}, \frac{\pi}{3}, \frac{\pi}{2}, \frac{2\pi}{3}, \frac{3\pi}{4}, \frac{5\pi}{6}$, and π. Match your results to values of another trigonometric function and make a conjecture about the area under the curve $y = \sin x$ from 0 to b.

b. Investigate the area under $y = e^x$. Make a table for the area as the right endpoint increases by one-half unit intervals from 0.5 to 4.0 (the left endpoint is $x = 0$). Graph the curve $y = e^x$ and make a conjecture about a formula for the area under $y = e^x$ from 0 to b.

3. Johannes Kepler (1571–1630) made significant contributions to the development of integral calculus. Write a short paper on his life and his work. Include his studies of wine barrels and the orbits of planets. Sources you might use include encyclopedias, calculus books, and books in your school or community library.

4. Emilie du Châtelet (1706–1749) translated the work of Newton into French and thus helped calculus to be known among the mathematicians of the European continent. Find out more about her life.

Emilie du Châtelet

806

5. Archimedes used physical models to make conjectures about areas under curves. You can do an experiment similar to his to find the area under the parabola $y = x^2$ from 0 to b. You will need an accurate scale such as those found in chemistry or physics laboratories. Take a piece of cardboard and draw a grid. If you have an 8.5×11 inch cardboard, mark it in centimeter squares. You will get 21 squares along the width and 27 along the length. Mark centimeters on the width as -5, -4, ..., 4, 5, using the lower edge of the cardboard as your x-axis. Graph $y = x^2$ using the height of the cardboard as your vertical dimension. You should be able to get the points (-5, 25) and (5, 25) on your graphs. Draw the parabola very carefully, because you will be cutting it out later. Trim the cardboard so that you have a rectangle which is 10 units along the base and 25 units high. This has area 250 cm^2.

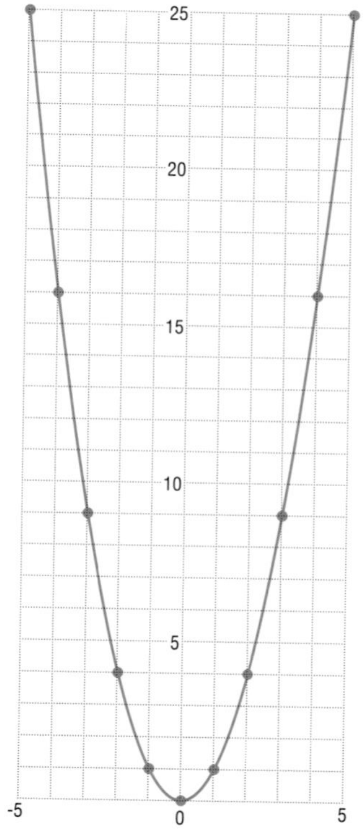

Weigh this piece of cardboard. You will get W grams. Record the weight, because you will need it to convert back to area later. (W grams corresponds to 250 cm^2.) Cut along the parabola from (-5, 25) through (0, 0) to (5, 25). Weigh the single, bullet-shaped piece. Suppose the weight is p grams. Then the area of the piece is $\frac{p}{W} \cdot 250$ cm^2. The area of the entire region *under* the curve is $250 - \frac{p}{W} \cdot 250$, and the area under the curve from 0 to 5 is one-half of that. You should compare your answer from this computation to $\frac{5^3}{3}$. One value is not sufficient, however. Carefully cut across the parabola from (-4, 16) to (4, 16). You can now weigh the parabola and deduce the area under the curve from 0 to 4. Repeat this process for the x-intervals 0 to 3 and 0 to 2. If you have a piece of the parabola left that seems to be big enough, use the x-interval 0 to 1. Compare each value to $\frac{b^3}{3}$ by plotting your data points, then graphing $y = \frac{x^3}{3}$.

6. Riemann sums are often used to calculate areas of irregular shapes. Below, the area of Hawaii can be estimated by adding up the areas above and below the x-axis. With a finer grid, the estimate becomes better. Find the area of your state, province, or country using Riemann sums. (Note: You may have to add and subtract several areas.)

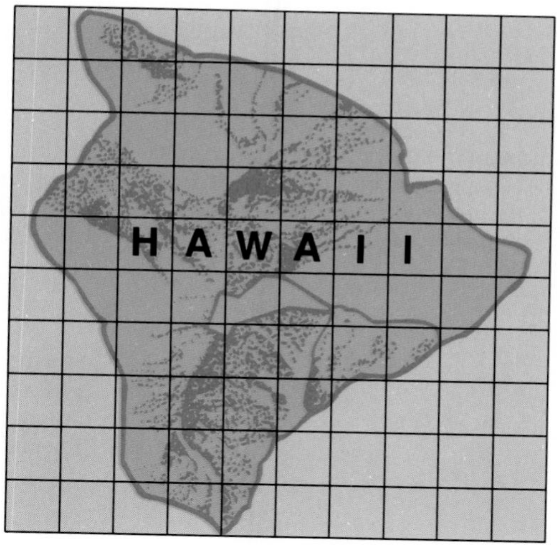

Summary

This chapter has focused on calculus ideas related to integration, the second of the two main ideas of calculus studied in this text.

Four types of problems were studied in this chapter. The first involves finding the distance traveled by an object in a given time period when the velocity at each instant is known. If the velocity during time period t_i is the constant r_i and there are n time periods, then the total distance is just $\sum_{i=1}^{n} r_i t_i$. If the velocity varies continuously, the distance is just the total area under a velocity-time graph.

A second type of problem involves finding the area of the region between the graph of a function and the x-axis over a particular interval. This area can be expressed as a definite integral, defined to be the limit of Riemann sums. A Riemann sum for the area is obtained by partitioning the interval into subintervals. For each subinterval, a rectangle is formed whose height is the value of the function at some point in the subinterval. The Riemann sum is then the sum of the areas of these n rectangles. As the number of rectangles increases without bound, the sums give a better and better approximation of the exact area, which is the definite integral.

An extension of the area problem involves finding the volume of the solid obtained when the region below a curve is rotated about the x-axis. The volume is partitioned into n nearly cylindrical sections, each of whose volume can be approximated. As the number of sections increases without bound, the sum of their approximate volumes gives a better and better approximation of the exact volume, which is expressed as a definite integral.

A more general version of the distance, area, and volume problems involves finding the original function when its derivative is known. The Fundamental Theorem of Calculus describes the relationship between integration and differentiation.

Vocabulary

For the starred (*) terms you should be able to give a definition of the term.
For the other terms you should be able to give a general description and a specific example of each.

Lesson 13-2
*Riemann sum of a function f over the interval from a to b
intermediate points for a Riemann sum, z_i
Δx

Lesson 13-3
upper Riemann sum
lower Riemann sum
*definite integral of f from a to b, $\int_a^b f(x)\,dx$

Lesson 13-4
Sum Property of Integrals Theorem
Constant Multiple Property of Integrals Theorem

Lesson 13-5
Integral of a Quadratic Function Theorem

Lesson 13-7
Fundamental Theorem of Calculus
antiderivative

SUMMARY

The Summary gives an overview of the entire chapter and provides an opportunity for students to consider the material as a whole. Thus, the Summary can be used to help students relate the various concepts presented in the chapter.

VOCABULARY

Terms, symbols, and properties are listed by lesson to provide a checklist of concepts a student must know. (See page 71 for more information.)

ADDITIONAL ANSWERS

3.a.

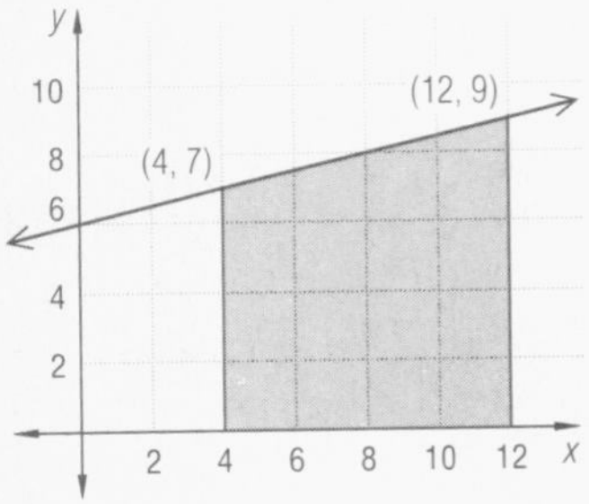

4.a.

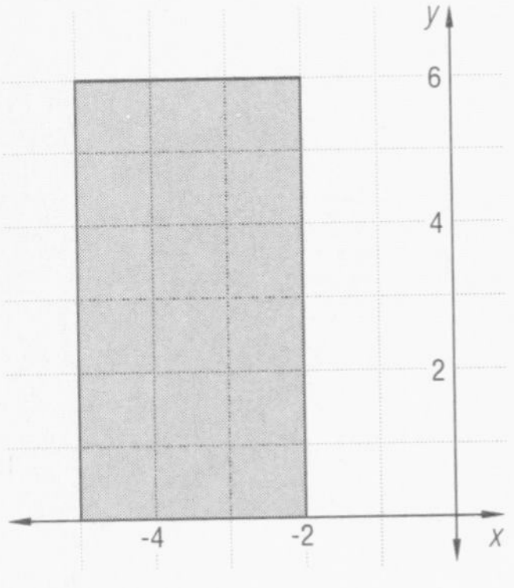

6. 49π ft^3

Progress Self-Test

See margin for answers not shown below.

Take this test as you would take a test in class. Then check the test yourself using the answers at the back of the book.

1. The velocity-time record for a train is given below. How far did the train travel from 8:00 A.M. to 1:00 P.M.? **175 miles**

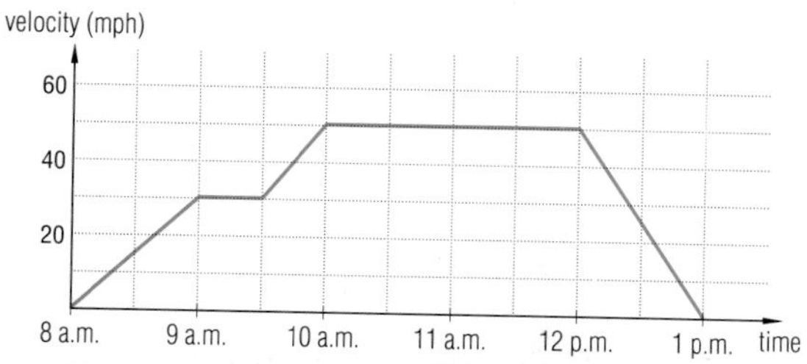

2. Below is a graph of $y = t(x)$. Partition the interval from 0 to 4 into 4 subintervals of equal length Δx. Let z_i = the right endpoint of the ith subinterval. Evaluate $\sum_{i=1}^{4} t(z_i)\,\Delta x$. **7.5 units²**

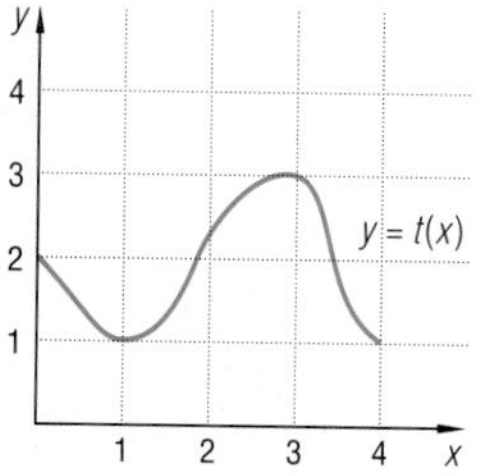

In 3 and 4, **a.** sketch the region whose area is indicated by the interval, and **b.** find the exact value of the integral.

3. $\int_{4}^{12}\left(\frac{1}{4}x + 6\right)\,dx$ **b) 64** **4.** $\int_{-5}^{-2} 6\,dx$ **b) 18**

5. Evaluate $\int_{-3}^{3}\sqrt{9 - x^2}\,dx$. **4.5π ≈ 14.137**

6. Use the grid and left endpoints of the rectangles to find a Riemann sum to estimate the volume of the nose cone at the right.

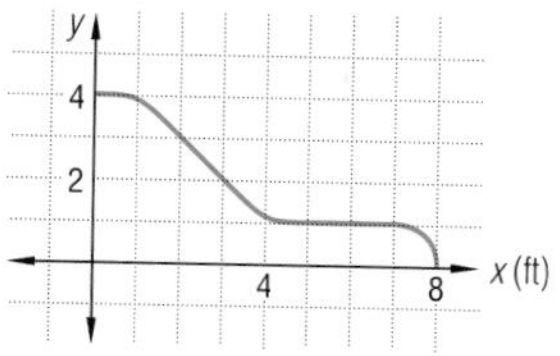

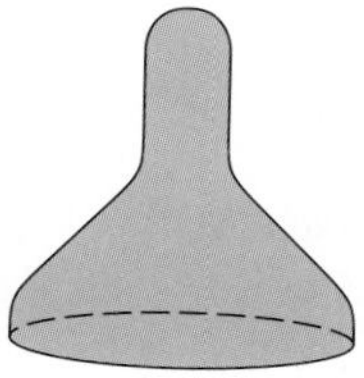

7. Sketch the region described by $\int_{-5}^{-3}\frac{1}{x}\,dx$. Does the value of the integral appear to be positive or negative? **negative**

8. Find the area of the shaded region below.

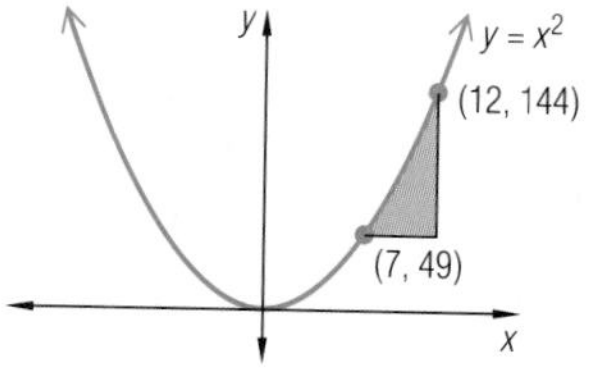

9. Consider the following theorem:

$$\int_a^b cf(x)\,dx = c\int_a^b f(x)\,dx.$$

Make a sketch and explain why the theorem is true when $0 < a < b$, $0 < c$, and $f(x) \geq 0$ on the interval from a to b.

10. Over a 15-second period, a car accelerates in such a way that its velocity (in feet per second) at time t is given by $f(t) = -.4(t - 15)^2 + 90$.

a. What integral gives the distance traveled by the car in the 15 seconds?

b. Find the distance in part **a.** **900 ft**

11. Rewrite as a single integral:

$$\int_0^{\frac{\pi}{2}} \cos x\,dx + \int_0^{\frac{\pi}{2}} 4\cos x\,dx.$$

12. a. Sketch the region bounded by the x-axis, the y-axis, and the line $y = 1 - x$.

b. Find the volume of the solid obtained when this region is revolved about the x-axis. **$\frac{\pi}{3}$ units³**

PROGRESS SELF-TEST

The Progress Self-Test provides the opportunity for feedback and correction; the Chapter Review provides additional opportunities for practice.

USING THE PROGRESS SELF-TEST

Assign the Progress Self-Test as a one-night assignment. (See pages 72 and T35 for more information.)

Teaching Aid 108 can be used with Progress Self-Test **Questions 1 and 2**.

7.

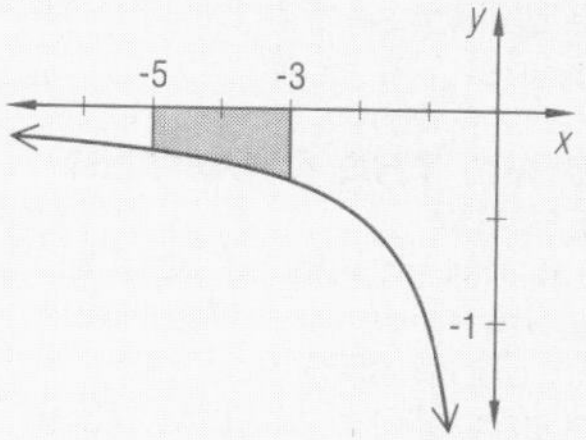

8. $\frac{650}{3}$ **units²**

9. See Additional Answers in the back of this book.

10.a. $\int_0^{15}(-.4(t - 15)^2 + 90)\,dt$

11. $5\int_0^{\frac{\pi}{2}}\cos x\,dx$ $\left(\text{or } \int_0^{\frac{\pi}{2}} 5\cos x\,dx\right)$

12.a.

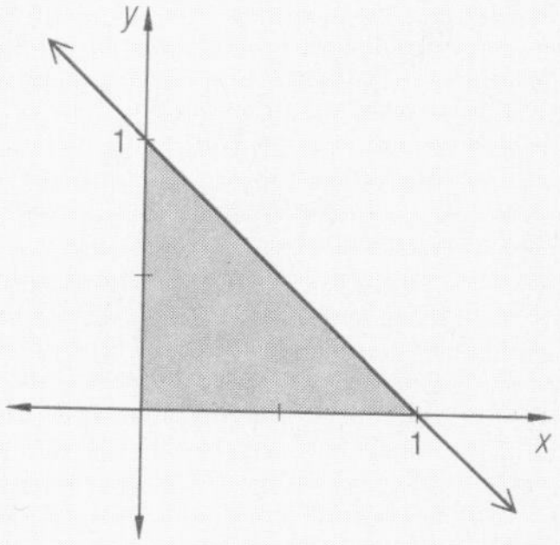

RESOURCES
- Chapter 13 Test, Form A
- Chapter 13 Test, Form B
- Chapter 13 Test, Cumulative Form
- Comprehensive Test, Chapters 1–13
- Quiz and Test Writer

CHAPTER REVIEW

The main objectives for the chapter are organized here into sections corresponding to the four main types of understanding this book promotes: Skills, Properties, Uses, and Representations. We call these the SPUR objectives.

USING THE CHAPTER REVIEW
Students should be able to answer questions like these with about 85% accuracy by the end of the chapter. (See pages 74 and T38–39 for more information.)

In **Question 20** of the Chapter Review, the rates of water flow given in the table are in gallons/hour.

ADDITIONAL ANSWERS

3.a. $\sum_{i=1}^{100} \frac{\pi}{600} \sin\left(\frac{\pi}{3} + i\frac{\pi}{600}\right)$

7.a.

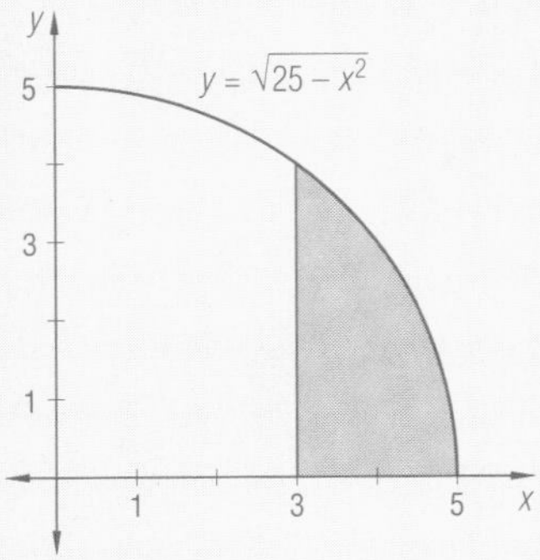

Chapter Review

Questions on SPUR Objectives

SPUR stands for **S**kills, **P**roperties, **U**ses, and **R**epresentations.
The Chapter Review questions are grouped according to the SPUR Objectives for this chapter.
See margin for answers not shown below.

SKILLS deal with the procedures used to get answers.

Objective A: *Calculate Riemann sums of functions over specified intervals. (Lesson 13-2)*

In 1 and 2, use the function $f(x) = x^3$ over the interval from 0 to 10.

1. Let z_i = the right endpoint of the ith subinterval.
Evaluate $\sum_{i=1}^{5} f(z_i)\,\Delta x$. **3600**

2. Let z_i = the left endpoint of the ith subinterval.
Evaluate $\sum_{i=1}^{5} f(z_i)\,\Delta x$. **1600**

3. Let $g(x) = \sin x$ over the interval from $\frac{\pi}{3}$ to $\frac{\pi}{2}$.
Let z_i = the right endpoint of the ith subinterval.
  a. Write the appropriate Riemann sum for 100 subintervals.
  b. Use a computer to evaluate that Riemann sum. **0.500**

Objective B: *Evaluate definite integrals. (Lessons 13-3, 13-4, 13-5)*

In 4–6, find the exact value of the definite integral.

4. $\int_3^7 (x + 3)\,dx$ **32** **5.** $\int_{-3}^{-1} -x\,dx$ **4**

6. $\int_0^1 \sqrt{1 - x^2}\,dx$ $\frac{\pi}{4}$

7. a. Sketch a picture of the region described by $\int_3^5 \sqrt{25 - x^2}\,dx$.
  b. Evaluate the integral. **≈ 5.59**

In 8–11, evaluate the integral.

8. $\int_0^6 x^2\,dx$ **72** **9.** $\int_4^9 x^2\,dx$ **$221.\overline{6}$**

10. $\int_1^3 x^2\,dx + \int_3^4 x^2\,dx$ **21**

11. $\int_0^6 (4x^2 - 3x + 1)\,dx$ **240**

PROPERTIES deal with the principles behind the mathematics.

Objective C: *Apply properties of definite integrals. (Lesson 13-4)*

In 12 and 13, rewrite each expression as a single integral.

12. $\int_4^7 2^x\,dx + \int_7^{10} 2^x\,dx$ $\int_4^{10} 2^x\,dx$

13. $\int_1^3 \cos^2 x\,dx + \int_1^3 \sin^2 x\,dx$ $\int_1^3 1\,dx$

14. *True* or *false?*
$\int_0^a (f \cdot g)(x)\,dx = \int_0^a f(x)\,dx \cdot \int_0^a g(x)\,dx$ **False**

15. Use area to explain why the following property is true. If f is continuous and $f(x) \ge 0$ on $a \le x \le b$, and $a < c < b$, then
$\int_a^c f(x)\,dx + \int_c^b f(x)\,dx = \int_a^b f(x)\,dx.$

810

USES deal with applications of mathematics in real situations.

Objective D: *Find the distance traveled by a moving object, given its rate.* *(Lessons 13-1, 13-2, 13-5)*

16. A car decelerates from 30 $\frac{\text{meters}}{\text{second}}$ to 18 $\frac{\text{meters}}{\text{second}}$ in 10 seconds. Assume the deceleration is constant.

a. How far will the car travel in the first second? **29.4 m**

b. How far will the car travel in six seconds?

c. How far will the car travel in the ten seconds? **b) 158.4 m; c) 240 m**

17. Another car decelerates from 88 ft/sec to 44 ft/sec in 5 seconds with the velocity at x seconds given by $h(x) = 1.76(x - 5)^2 + 44$. Subdivide the interval into five subintervals. Let z_i be the left endpoint of the ith subinterval. Use these values to estimate the distance traveled by the car in the five seconds. **316.8 ft**

18. The velocity (in ft/sec) of an object at t seconds is $3t^2 - 2t + 1$. Find the distance traveled by the object from $t = 7$ to $t = 12$ seconds. **1295 ft**

Objective E: *Use the definite integral to solve application problems.* *(Lessons 13-4, 13-6)*

19. A cylindrical trough has a base in the shape of a parabola, 3 meters high and 4 meters across. If the trough is 12 meters long, what is its volume?

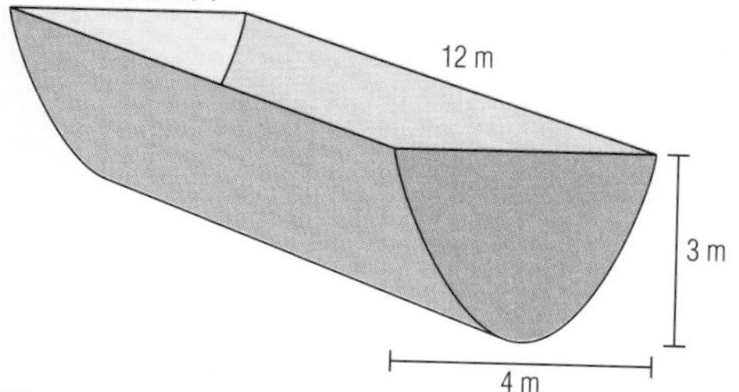

96 m³

20. When a hard freeze is expected, strawberry farmers spray their fields with water so that a thin layer of ice covers the fruit, because the ice insulates the fruit from temperatures below 32° F. The table at the top of the next column gives the rate of water flow (in gallons) through two sprayers for a 12-hour period.

a. Write an integral that describes the total amount of water used to spray the berries during this 12-hour period.

b. Approximate your answer to part **a** by evaluating the appropriate Reimann sum

i. using the left endpoints as the z_i;

ii. using the right endpoints as the z_i.

t	$f(t)$	$g(t)$
0	3.5	2.0
2	3.0	2.5
4	2.75	3.0
6	3.25	2.75
8	3.0	3.5
10	3.5	2.0
12	2.5	1.5

i) 69.5 gal
ii) 66.5 gal

21. A lamp base has a shape as below. To find its volume, the outer curve of the lamp base is drawn on a two-dimensional graph as shown below. The shape is obtained by revolving the curve about the x-axis. Let the solid be divided into six sections approximated by cylinders, each with width 3″. Using r_i as the height of the graph at the right endpoint of the ith subinterval, approximate the volume of the lamp. **465.75π ≈ 1463 in.³**

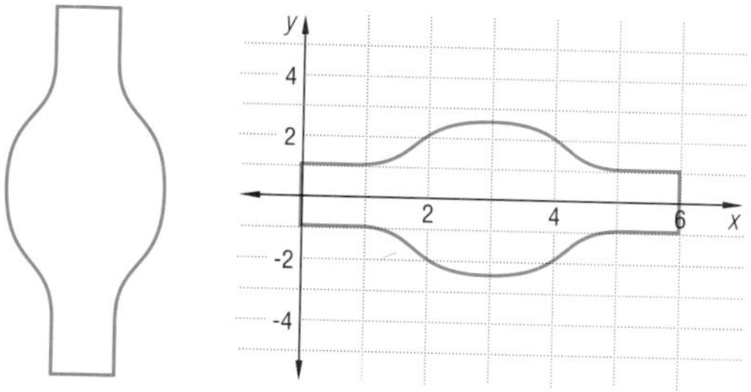

15.

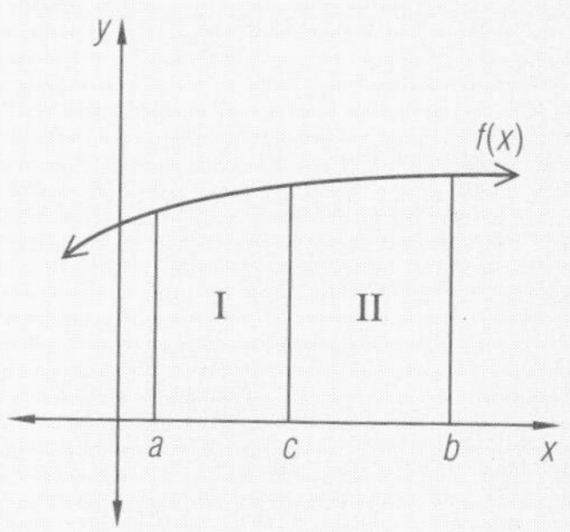

Area I is $\int_a^c f(x)\,dx$, Area II is $\int_c^b f(x)\,dx$, and the union of the regions represented by Areas I and II is $\int_a^b f(x)\,dx$. Since Area I + Area II equals the area of the union of the regions represented by Areas I and II, then $\int_a^c f(x)\,dx + \int_c^b f(x)\,dx = \int_a^b f(x)\,dx$.

20.a. $\int_0^{12} (f(t) + g(t))\,dt$

ADDITIONAL ANSWERS

26.a. The integral from -2 to 1 x^3 dx

29.a.

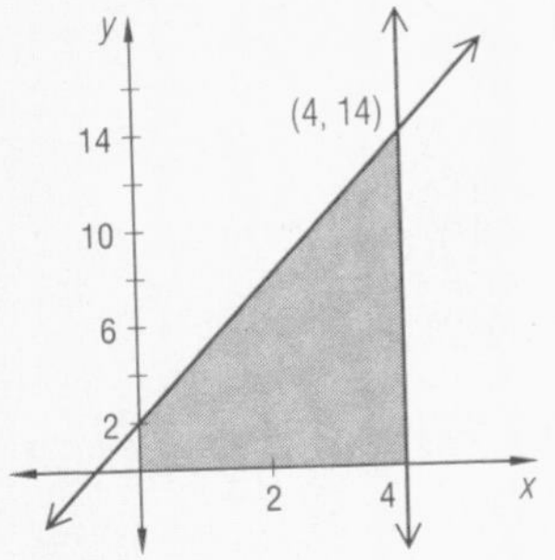

30.a.

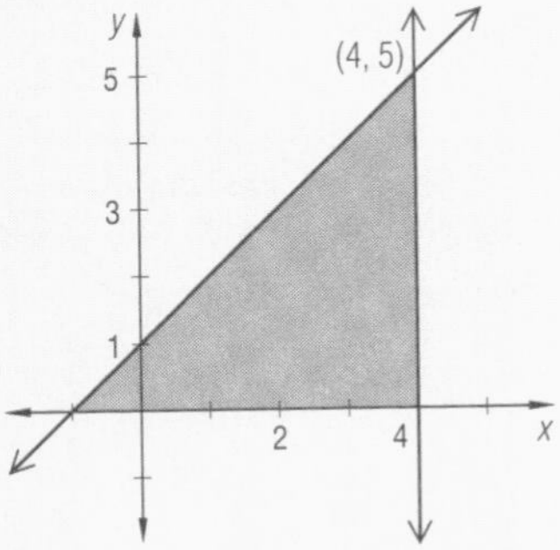

31.a.

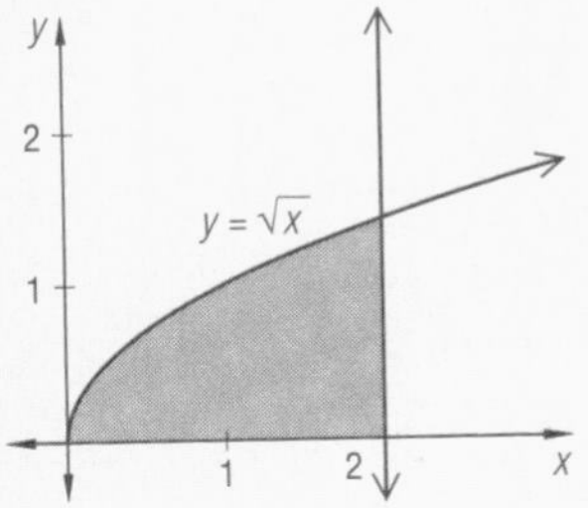

■ **Objective F:** *Estimate the distance under a velocity-time graph. (Lesson 13-1)*

22. A stunt driver is being filmed for a chase scene in a movie. The driver travels at a constant rate of speed as the cameras begin to roll. Two seconds into the scene the driver applies the brakes, going into a skid. Five seconds into the scene the vehicle comes to a stop. Use the velocity-time graph below to answer the following questions.

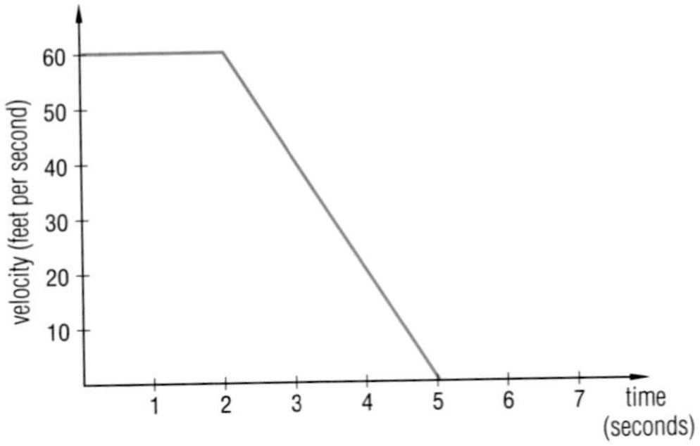

a. What was the driver's velocity before the brakes were applied? **60 ft/sec**

b. How far did the driver travel before applying the brakes? **120 ft**

c. How far did the driver travel after applying the brakes? **90 ft**

d. What is the total distance traveled by the driver while the cameras were rolling? **210 ft**

23. A velocity-time graph for a skydiver is shown below. In this case the downward direction is taken to be positive. How high was the airplane if the skydiver lands exactly 200 seconds after jumping? **4635 ft**

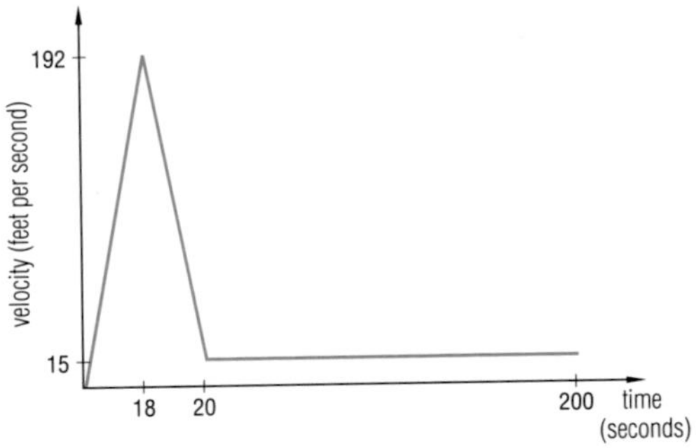

REPRESENTATIONS deal with pictures, graphs, or objects that illustrate concepts.

■ **Objective G:** *Express areas in integral notation. (Lesson 13-3)*

In 24 and 25, express the area of the shaded region with integral notation.

24.

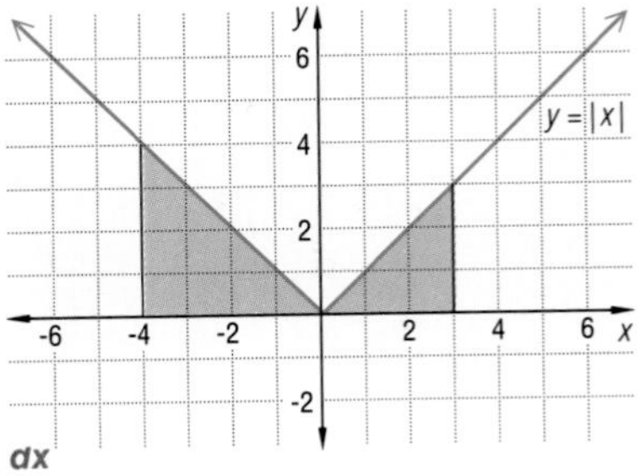

$\int_{-4}^{3} |x|\, dx$

25.

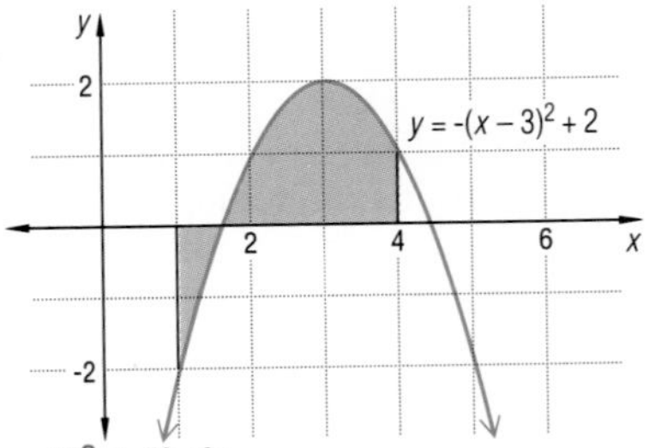

$\int_{1}^{4} (-(x-3)^2 + 2)\, dx$

812

26. Refer to the graph of g below.
  a. Express the area of the shaded region with integral notation.
  b. Does the value of the integral appear to be positive or negative? **negative**

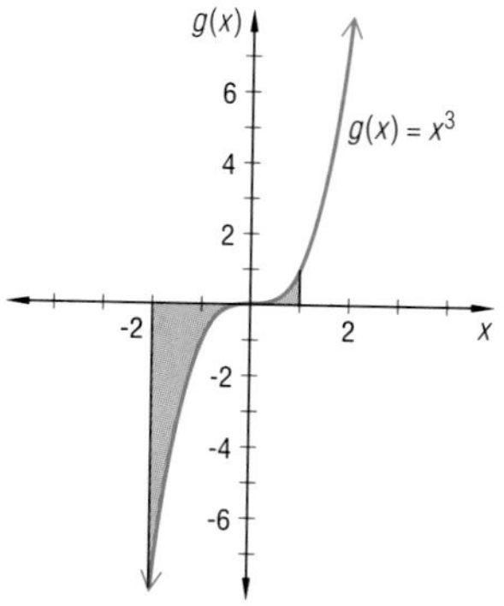

a) $\int_{-2}^{1} x^3\,dx$

■ **Objective H:** *Find areas bounded by curves.* *(Lessons 13-4, 13-5)*

27. Consider the shaded region below.
  a. Approximate its area using a Riemann sum with 5 equal subintervals and with z_i = the right endpoint of the ith subinterval.
  b. Find the exact area using an integral.

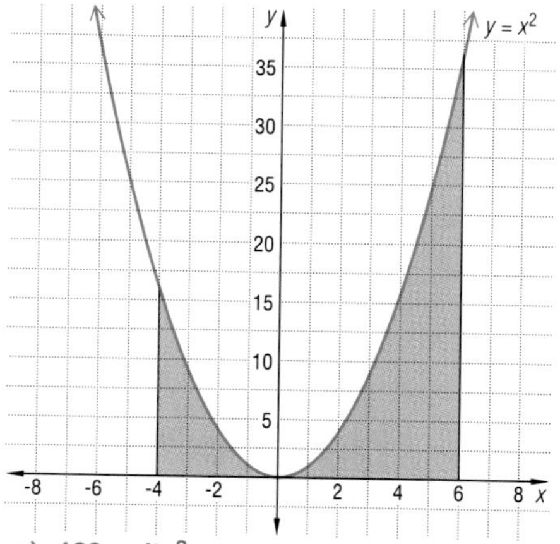

a) 120 units²
b) 93.$\overline{3}$ units²

28. Find the area of the shaded region below.

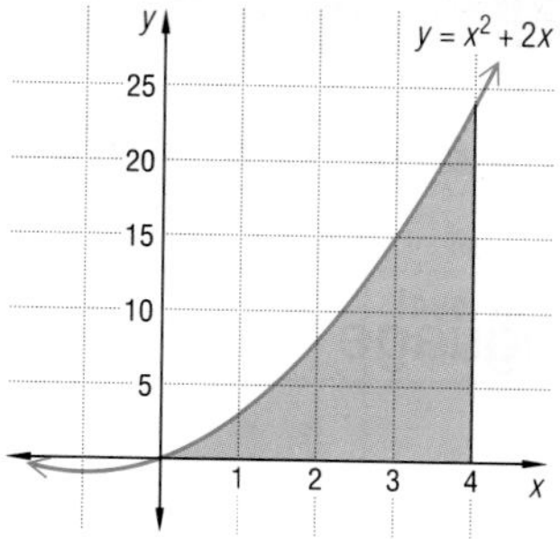

37.$\overline{3}$ units²

29. Consider the region bounded by the x-axis, the y-axis, the line $y = 3x + 2$, and the line $x = 4$.
  a. Sketch the region.
  b. Set up and evaluate an integral that describes the area of this region.
  $\int_{0}^{4} (3x + 2)\,dx = 32$

■ **Objective I:** *Find volumes of solids.* *(Lesson 13-6)*

30. a. Sketch a graph of the region bounded by the x-axis, the line $x = 4$, and the line $y = x + 1$.
  b. Find the volume of the solid obtained when the region of part **a** is revolved about the x-axis. $\frac{125\pi}{3} \approx 130.9$ **units³**

31. a. Sketch a graph of the region bounded by $y = \sqrt{x}$, the line $x = 2$, and the x-axis.
  b. Calculate the volume of the solid generated when this region is revolved about the x-axis. 2π **units³**

EVALUATION
Three forms of a Chapter Test are provided in the Teacher's Resource File. Chapter 13 Test, Forms A and B cover just Chapter 13. The third test is Chapter 13 Test, Cumulative Form. About 50% of this test covers Chapter 13, 25% covers Chapter 12, and 25% covers previous chapters. A fourth test, Comprehensive Test, Chapters 1–13, that is multiple choice in format, is also provided. For information on grading, see *General Teaching Suggestions: Grading* on page T43 in the Teacher's Edition.

Type of Function	Parent Function f	Graph of f	Inverse Function f^{-1}	Graph of f^{-1}
exponential natural base	$f(x) = e^x$	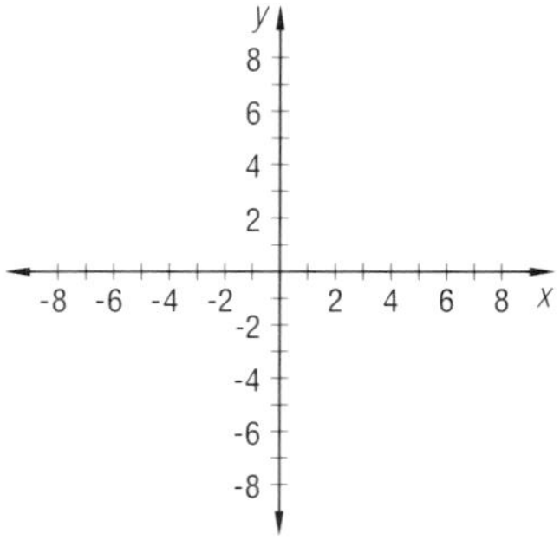	$f^{-1}(x) = \ln x$	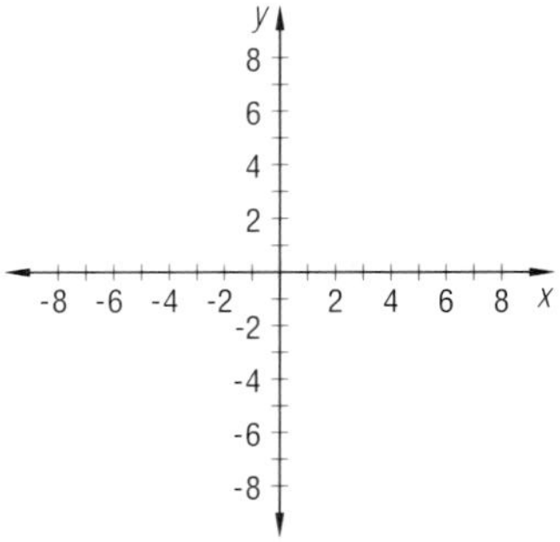
circular—sine	$f(x) = \sin x$	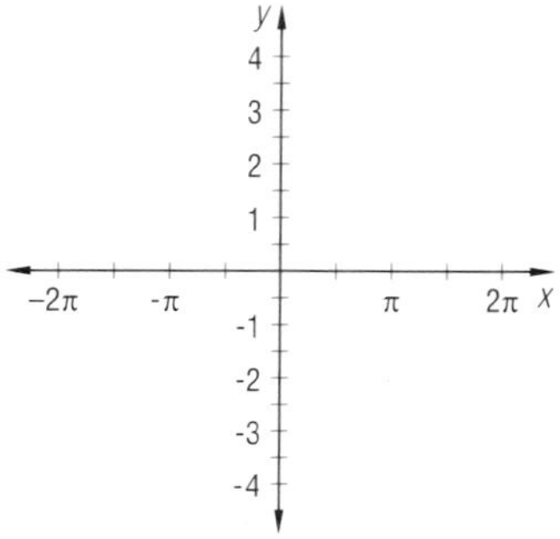	$f^{-1}(x) = \sin^{-1} x$ $-\frac{\pi}{2} \leq y \leq \frac{\pi}{2}$	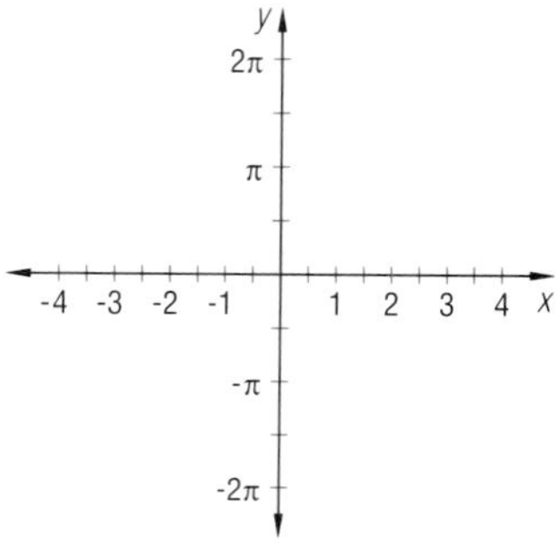
circular—cosine	$f(x) = \cos x$	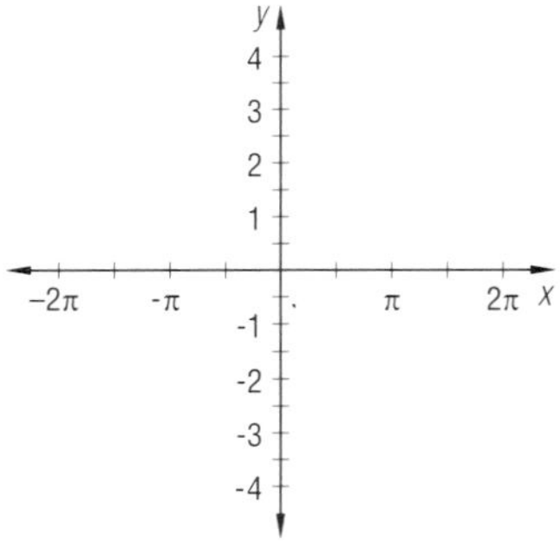	$f^{-1}(x) = \cos^{-1} x$ $0 \leq y \leq \pi$	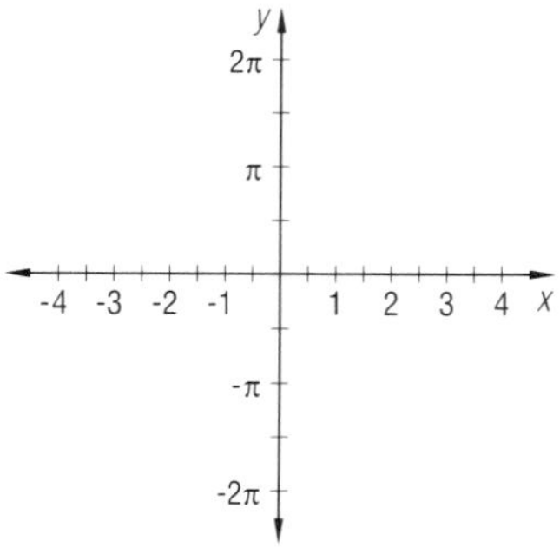
circular—tangent	$f(x) = \tan x$	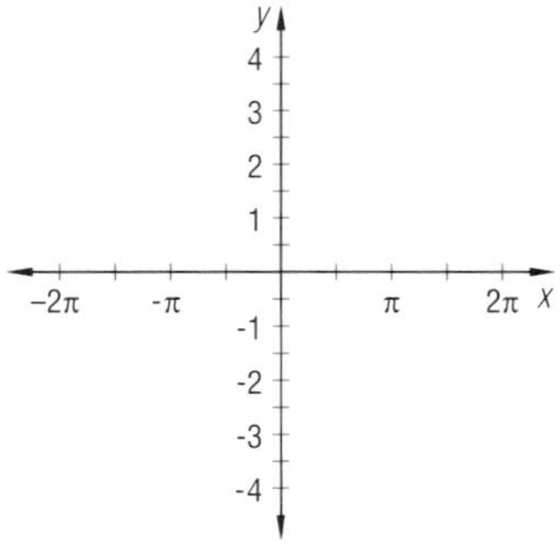	$f^{-1}(x) = \tan^{-1} x$ $-\frac{\pi}{2} \leq y \leq \frac{\pi}{2}$	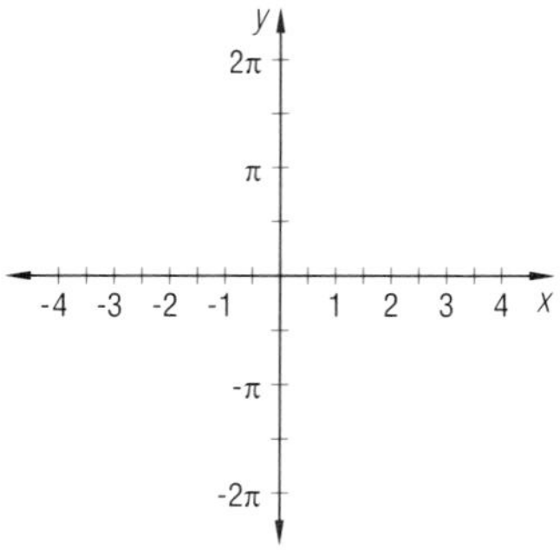

Parent Function *f*	Reciprocal Function *g*	Graph of Reciprocal Function *g*
$f(x) = \sin x$	$g(x) = \csc x$	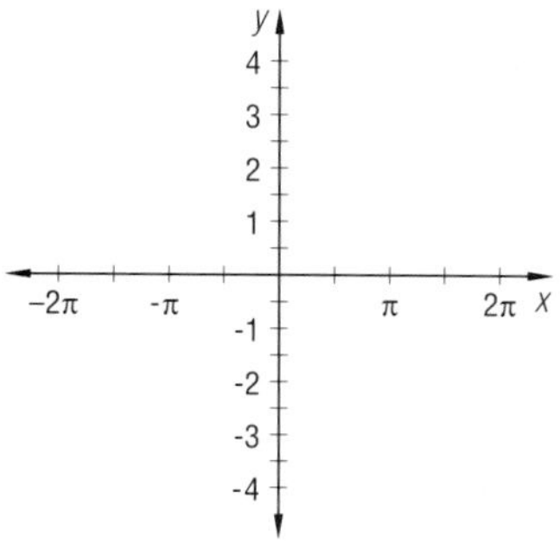
$f(x) = \cos x$	$g(x) = \sec x$	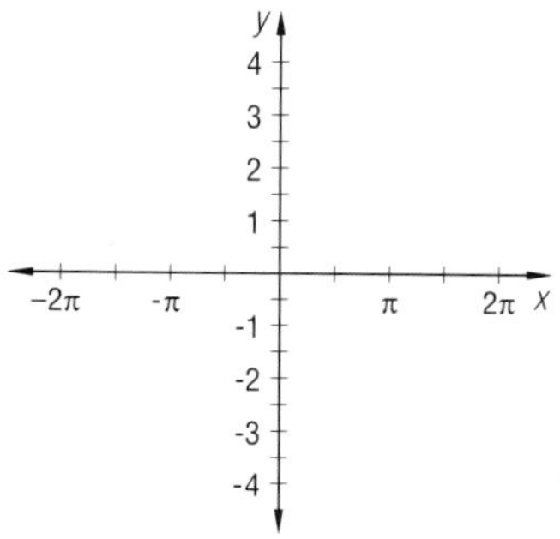
$f(x) = \tan x$	$g(x) = \cot x$	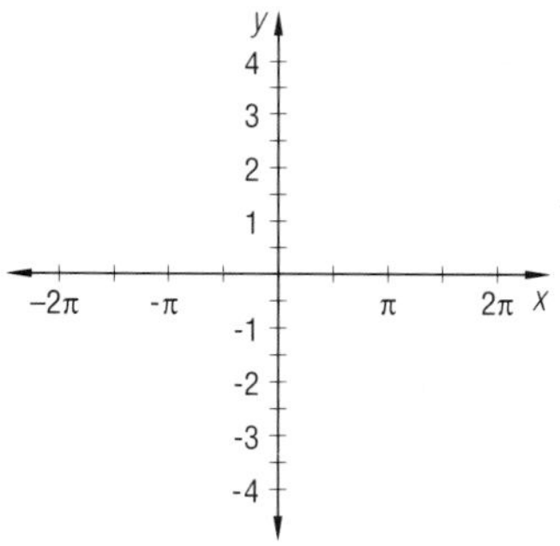

APPENDIX B

The BASIC Language

In the computer programming language BASIC (Beginner's All Purpose Symbolic Instruction Code), the arithmetic symbols are + (addition), − (subtraction), * (multiplication), / (division), and ^ (powering). In some versions of BASIC, ↑ is used for powering.

Variables are represented by strings of letters and digits. The first character of a variable name must be a letter, but later characters may be digits. Examples of variable names are X, X2, SDEV, and WEIGHT. Variable names are generally chosen to describe the variable they represent. In some versions of BASIC, only the first two characters of a variable are read by the computer. (So ARC and AREA would both be read as AR, for example.) Check to see if this is true for your computer.

The computer evaluates expressions according to the usual order of operations. Parentheses ( ) may be used for clarity or to change the usual order of operations. The comparison symbols, =, >, <, are also used in the standard way, but BASIC uses <= instead of ≤, >= instead of ≥, and <> instead of ≠.

Each time you enter information on the screen, whether it is a command, an input to a program, or a new program line, finish by pressing the key marked RETURN or ENTER.

Commands

The BASIC commands used in this course and examples of their uses are given below.

REM This command allows remarks to be inserted in a program. These may describe what the program does and how it works. REM statements are especially important for complex programs, programs that others will use and programs you may not use for long periods.

10 REM PYTHAGOREAN THEOREM	The statement has no effect when the program is run. It appears when LIST is used.

LET A value is assigned to a given variable. Some versions of BASIC allow you to omit the word LET in the assignment statement.

LET X = 5	The number 5 is stored in a memory location called X.
LET NEXT = NEXT + 2	The value in the memory location called NEXT is increased by 2 and then restored in the location called NEXT.

INPUT The computer prompts the user to enter a value and stores that value to the variable named. Using a message helps the user know what to type.

INPUT X	The computer prompts you with a question mark and then stores the value you type in memory location X.
INPUT "HOW OLD"; AGE	The computer prints HOW OLD? and stores your response in the memory location AGE.

Command	Description
GOTO	The computer goes to whatever line of the program is indicated. GOTO statements should be avoided wherever possible, as they interrupt smooth program flow and make programs hard to interpret.
	`GOTO 70` — The computer goes to line 70 and executes that command.
DEF FN	With this command, the computer allows the user to define a function for later use. Whenever the function is referred to in the program, it is written as FN, followed by the name of the function and an argument in parentheses.

	`10 DEF FN C(X) = 5 * (X-32)/9`
	`20 INPUT "FAHRENHEIT";F`
	`30 PRINT F,FN C(F)`
	Line 10 defines a function C to convert temperatures in degrees Fahrenheit to degrees Celsius, according to the rule $C(X) = \frac{5(X-32)}{9}$. The function can have any variable or number for its argument.

IF... THEN	The computer performs the consequent (the THEN part) only if the antecedent (the IF part) is true. When the antecedent is false, the computer *ignores* the consequent and goes directly to the next line of the program.

	`10 IF X>6 THEN END`
	`20 PRINT X`
	If the X value is less than or equal to 6, the computer ignores END, goes to the next line, and prints the value stored in X. If the X value is greater than 6, the computer stops and the value stored in X is not printed.

FOR ... NEXT STEP	The FOR command assigns a beginning and ending value to a variable using the word TO. The first time through the loop, the variable has the beginning value in the FOR command. When the computer reaches the line reading NEXT, the value of the variable is increased by the amount indicated by STEP. If STEP is not written, the computer increases the variable by 1 each time through the loop. The commands between FOR and NEXT are then repeated. When the incremented value of the variable is larger than the ending value in the FOR command, the computer leaves the loop and executes the next line of the program.

	`10 FOR N = 3 TO 6 STEP 2`
	`20 PRINT N`
	`30 NEXT N`
	`40 END`
	The computer assigns 3 to N and then prints the value of N. On reaching NEXT, the computer increases N by 2 (the STEP amount), and the second time through the loop prints 5. The next N would be 7 which is too large. So, the computer executes the command after NEXT, ending the program.

PRINT	The computer prints on the screen what follows the PRINT command. If what follows is a constant or variable the computer prints the value of that constant or variable. If what follows is in quotes, the computer prints exactly what is in quotes. Commas or semicolons are used to space the output.
	`PRINT COST` — The computer prints the number stored in memory location COST.
	`PRINT "X-VALUES"` — The computer prints the phrase X-VALUES.
	`PRINT X,Y,Z` — The computer prints the numbers stored in the memory locations X, Y and Z in that order with a tab space between each.
	`PRINT "MEAN IS ";XBAR` — The computer prints the phrase MEAN IS, followed immediately by the number in memory location XBAR.
END	The computer stops running the program. A program should not contain more than one END statement.

Functions

The following built-in functions are available in most versions of BASIC. These can be used directly, in a PRINT statement, or within a program.

ABS The absolute value of the number or expression that follows is calculated.

LET X = ABS(- 8) — The computer calculates $|-8|$ and stores the value 8 in the memory location X.

ATN The arctangent or inverse tangent of the number or expression that follows is calculated. The result is always in radians.

LET P = ATN(1) — The computer calculates $\tan^{-1} 1 = \frac{\pi}{4}$ and stores the value 0.785398 in the memory location P.

EXP The number e raised to the power of the number or expression that follows is calculated.

LET K = EXP(2) — The computer calculates e^2 and stores the value 7.389056 in the memory location K.

LOG The natural logarithm (that is the logarithm to the base e) of the number or expression that follows is calculated.

LET J = LOG(6) — The computer calculates ln 6 and stores the value 1.791759 in the memory location J.

INT The greatest integer less than or equal to the number or expression that follows is calculated.

LET W = INT(- 7.3) — The computer calculates $\lfloor -7.3 \rfloor$ and stores the value - 8 in the memory location W.

RND(1) A random number greater than 0 and less than 1 is generated. The argument of the RND function is always 1.

LET D = 2 * RND(1) — The computer generates a random number between 0 and 1 and stores twice the value in the memory location D.

SIN
COS
TAN The sine, cosine or tangent of the number or expression that follows is calculated. The argument of these circular functions is always in radians. To convert degrees to radians, multiply by (3.14159/180).

LET R = 5*SIN(0.7) — The computer finds 5 sin (0.7) and stores the value 3.221088 in the memory location R.

SQR The square root of the number or expression that follows is calculated.

LET C = SQR(A * A + B * B) — The computer calculates $\sqrt{A^2 + B^2}$ using the values stored in A and B and stores the result in C.

SGN The signum function of the number or expression that follows is calculated. The function has value 1 for positive numbers, - 1 for negative numbers and 0 for zero.

PRINT SGN(- 6) — The computer prints - 1.

Programs

In many versions of BASIC, every line in a BASIC program must begin with a line number. It is common to start numbering at 10 and count by tens, so that intermediate lines can be added later if necessary. The computer reads and executes a BASIC program in the order of the line numbers. It will not go back to a previous line unless told to do so.

To run a new program, first type NEW to erase the previous one; then type in the lines of the program. Finally, type RUN.

To run a program already saved on a disk you must know the exact name of the program, including any spaces. To run a program called TABLE SOLVE, type RUN TABLE SOLVE, then press the RETURN or ENTER key.

If you type LIST, the program currently in the computer memory will be listed on the screen. A program line can be changed by re-typing the line, including the line number.

In most versions of BASIC, you can stop a program completely while it is running by typing Control-C (CTRL C). This is done by pressing the C key while holding down the CTRL key. A program may be stopped temporarily (for example, to see part of the output) by typing CTRL S and resumed by typing CTRL Q. Some versions of BASIC use different commands for these purposes.

10 PRINT "A DIVIDING SEQUENCE"	The computer prints A DIVIDING SEQUENCE.
20 INPUT X	You must give a value to store in the location X. Suppose you use 16. (Press the RETURN or ENTER key after entering 16.)
30 LET Y = 2	2 is stored in location Y.
40 FOR Z = -5 TO 2	Z is given the value -5. Each time through the loop, the value of Z will be increased by 1.
50 IF Z = 0 THEN GO TO 70	When Z = 0 the computer goes directly to line 70. If Z ≠ 0 the computer executes line 60.
60 PRINT (X * Y)/Z	On the first pass through the loop, the computer prints -6.4 because (16 · 2)/(-5) is -6.4.
70 NEXT Z	The value in Z is increased by 1 and the computer goes back to line 50.
80 END	After going through the FOR ... NEXT loop with Z = 2, the computer stops.
-6.4 -8 -10.666667 -16 -32 32 16	The output of this program is shown.

Content/Pages T19-T159

Professional Sourcebook For UCSMP

1 Overview of UCSMP

The Reasons for UCSMP

Recommendations for Change

The mathematics curriculum has undergone changes in every country of the world throughout this century, as a result of an increasing number of students staying in school longer, a greater number of technically competent workers and citizens being needed, and because of major advances in mathematics itself. In the last generation, these developments have been accelerated due to the widespread appearance of computers with their unprecedented abilities to handle and display information.

In the last 100 years, periodically there have been national groups examining the curriculum in light of these changes in society. (A study of these reports can be found in *A History of Mathematics Education in the United States and Canada,* the 30th Yearbook of the National Council of Teachers of Mathematics, 1970.) The most recent era of reports can be said to have begun in the years 1975–1980, with the publication of reports by various national mathematics organizations calling attention to serious problems in the education of our youth.

Beginning in 1980, these reports were joined by governmental and private reports on the state of American education with broad recommendations for school practice. Two of these are notable for their specific remarks about mathematics education.

1983: National Commission on Excellence in Education,
A Nation At Risk

''The teaching of mathematics in high school should equip graduates to: (a) understand geometric and algebraic concepts; (b) understand elementary probability and statistics; (c) apply mathematics in everyday situations; and (d) estimate, approximate, measure, and test the accuracy of their calculations. In addition to the traditional sequence of studies available for college-bound students, new, equally demanding mathematics curricula need to be developed for those who do not plan to continue their formal education immediately.'' (p. 25)

1983: College Board (Project Equality),
Academic Preparation for College:
What Students Need to Know and Be Able to Do

All students (college-bound or not) should have:
''The ability to apply mathematical techniques in the solution of real-life problems and to recognize when to apply those techniques.
Familiarity with the language, notation, and deductive nature of mathematics and the ability to express quantitative ideas with precision.
The ability to use computers and calculators.
Familiarity with the basic concepts of statistics and statistical reasoning.
Knowledge in considerable depth and detail of algebra, geometry, and functions.'' (p. 20)

The specific remarks about school mathematics in these documents for the most part mirror what appeared in the earlier reports. Thus, **given what seemed to be a broad consensus on the problems and desirable changes in pre-college mathematics instruction, it was decided at the outset of UCSMP, that UCSMP would not attempt to form its own set of recommendations, but undertake the task of translating the existing recommendations into the reality of classrooms and schools.**

At the secondary (7–12) level, these reports respond to two generally perceived problems pursuant to mathematics education.

GENERAL PROBLEM 1: Students do not learn enough mathematics by the time they leave school.

Specifically:

(A) Many students lack the mathematics background necessary to succeed in college, on the job, or in daily affairs.

(B) Even those students who possess mathematical skills are not introduced to enough applications of the mathematics they know.

(C) Students do not get enough experience with problems and questions that require some thought before answering.

(D) Many students terminate their study of mathematics too soon, not realizing the importance mathematics has in later schooling and in the marketplace.

(E) Students do not read mathematics books and, as a result, do not learn to become independent learners capable of acquiring mathematics outside of school when the need arises.

These situations lead us to want to **upgrade students' achievement.**

GENERAL PROBLEM 2: The school mathematics curriculum has not kept up with changes in mathematics and the ways in which mathematics is used.

Specifically:

(A) Current mathematics curricula have not taken into account today's calculator and computer technology.

The UCSMP grades 7–12 can be considered to be the first mathematical sciences curriculum.

(B) Students who do succeed in secondary school mathematics are prepared for calculus, but are not equipped for the other mathematics they will encounter in college.

(C) Statistical ideas are found everywhere, from newspapers to research studies, but are not found in most secondary school mathematics curricula.

(D) The emergence of computer science has increased the importance of a background in discrete mathematics.

(E) Mathematics is now applied to areas outside the realm of the physical sciences, as much as within the field itself, but these applications are rarely taught and even more rarely tested.

(F) Estimation and approximation techniques are important in all of mathematics, from arithmetic on.

These existing situations lead us to a desire to **update the mathematics curriculum.**

Since the inception of UCSMP, reports from national groups of mathematics educators have reiterated the above problems, and research has confirmed their existence. Three reports are of special significance to UCSMP.

Universities have for many years had to recognize that mathematics encompasses far more than algebra, geometry, and analysis. The term **mathematical sciences** is an umbrella designation which includes traditional mathematics as well as a number of other disciplines. The largest of these other disciplines today are statistics, computer science, and applied mathematics. In 1983, the Conference Board of the Mathematical Sciences produced a report, *The Mathematical Sciences Curriculum: What Is Still Fundamental and What Is Not*. THE UCSMP GRADES 7–12 CAN BE CONSIDERED TO BE THE FIRST MATHEMATICAL SCIENCES CURRICULUM.

The Second International Mathematics Study (SIMS) was conducted in 1981–82 and involved 23 populations in 21 countries. At the eighth-grade level, virtually all students attend school in all those countries. At the 12th-grade level, the population tested consisted of those who are in the normal college preparatory courses, which, in the United States, include precalculus and calculus classes.

At the eighth-grade level, our students scored at or below the international average on all five subtests: arithmetic, measurement, algebra, geometry, and statistics. We are far below the top: Japan looked at the test and decided it was too easy for their 8th-graders, and so gave it at 7th grade. Still, the median Japanese 7th-grader performed at the 95th percentile of United States 8th-graders. These kinds of results have been confirmed in other studies, comparing students at lower-grade levels.

At the twelfth-grade level, about 13% of our population is enrolled in precalculus or calculus; the mean among developed countries is about 16%. Thus, the United States no longer keeps more students in mathematics than other developed countries, yet our advanced placement students do not perform well when compared to their peers in other countries. SIMS found:

1987: Second International Mathematics Study (SIMS), *The Underachieving Curriculum*

In the U.S., the achievement of the calculus classes, the nation's **best** mathematics students, was at or near the average achievement of the advanced secondary school mathematics students in other countries. (In most countries, **all** advanced mathematics students take calculus. In the U.S., only about one-fifth do.) The achievement of the U.S. precalculus students (the majority of twelfth grade college-preparatory students) was substantially below the international average. In some cases the U.S. ranked with the lower one-fourth of all countries in the study, and was the lowest of the advanced industrialized countries. (*The Underachieving Curriculum, p. vii.*)

The situation is, of course, even worse for those who do not take precalculus mathematics in high school. Such students either have performed poorly in their last mathematics course, a situation which has caused them not to go on in mathematics, or they were performing poorly in junior high school and had to take remedial mathematics as 9th-graders. If these students go to college, they invariably take remedial mathematics, which is taught at a faster pace than in high school, and the failure rates in such courses often exceed 40%. If they do not go to college but join the job market, they lack the mathematics needed to understand today's technology. It is no understatement to say that UCSMP has received its funding from business and industry because those who leave schooling to join the work force are woefully weak in the mathematics they will need.

SIMS recommended steps to renew school mathematics in the United States. **The UCSMP secondary curriculum implements the curriculum recommendations of the Second International Mathematics Study.**

In 1986, the National Council of Teachers of Mathematics began an ambitious effort to detail the curriculum it would like to see in schools. The "NCTM Standards," as they have come to be called, involve both content and methodology. The *Standards* document is divided into four sections, K–4, 5–8, 9–12, and Evaluation. Space limits our discussion here to just a few quotes from the 5–8 and 9–12 standards.

1989: National Council of Teachers of Mathematics, *Curriculum and Evaluation Standards for School Mathematics*

"The 5–8 curriculum should include the following features:

- Problem situations that establish the need for new ideas and motivate students should serve as the context for mathematics in grades 5–8. Although a specific idea might be forgotten, the context in which it is learned can be remembered and the idea can be re-created. In developing the problem situations, teachers should emphasize the application to real-world problems as well as to other settings relevant to middle school students.
- Communication with and about mathematics and mathematical reasoning should permeate the 5–8 curriculum.
- A broad range of topics should be taught, including number concepts, computation, estimation, functions, algebra, statistics, probability, geometry, and measurement. Although each of these areas is valid mathematics in its own right, they should be taught together as an integrated whole, not as isolated topics; the connections between them should be a prominent feature of the curriculum.

It is no understatement to say that UCSMP has received its funding from business and industry because those who leave schooling to join the work force are woefully weak in the mathematics they will need.

- Technology, including calculators, computers, and videos, should be used when appropriate. These devices and formats free students from tedious computations and allow them to concentrate on problem solving and other important content. They also give them new means to explore content. As paper-and-pencil computation becomes less important, the skills and understanding required to make proficient use of calculators and computers become more important.'' (pp. 66–67)

''The standards for grades 9–12 are based on the following assumptions:

- Students entering grade 9 will have experienced mathematics in the context of the broad, rich curriculum outlined in the K–8 standards.
- The level of [paper and pencil] computational proficiency suggested in the K–8 standards will be expected of all students; however, no student will be denied access to the study of mathematics in grades 9–12 because of a lack of computational facility.
- Although arithmetic computation will not be a direct object of study in grades 9–12, conceptual and procedural understandings of number, numeration, and operations, and the ability to make estimations and approximations and to judge the reasonableness of results will be strengthened in the context of applications and problem solving, including those situations dealing with issues of scientific computation.
- Scientific calculators with graphing capabilities will be available to all students at all times.
- A computer will be available at all times in every classroom for demonstration purposes, and all students will have access to computers for individual and group work.
- At least three years of mathematical study will be required of all secondary school students.
- These three years of mathematical study will revolve around a core curriculum differentiated by the depth and breadth of the treatment of topics and by the nature of applications.
- Four years of mathematical study will be required of all college-intending students.
- These four years of mathematical study will revolve around a broadened curriculum that includes extensions of the core topics and for which calculus is no longer viewed as *the* capstone experience.
- All students will study appropriate mathematics during their senior year.'' (pp. 124–125)

THE UCSMP SECONDARY CURRICULUM IS THE FIRST FULL MATHEMATICS CURRICULUM THAT IS CONSISTENT WITH THE RECOMMENDATIONS OF THE NCTM *STANDARDS*.

In 1989, the Mathematical Science Education Board (MSEB), a committee of the National Research Council that coordinates efforts for improvement of mathematics education in the United States, came out with the report *Everybody Counts*, emphasizing the need for the mathematics curriculum to be appropriate for as many students as possible. This thrust reflects the UCSMP position that as many students as possible be accommodated with the curriculum taken by those who go to college.

The entire movement of the past 15 years is summarized in a second MSEB report, *Reshaping School Mathematics: A Philosophy and Framework for Curriculum*, which appeared in 1991. Six changes are identified there as affecting the context of mathematics eduction:

- Changes in the need for mathematics.
- Changes in mathematics and how it is used.
- Changes in the role of technology.
- Changes in American society.
- Changes in understanding of how students learn.
- Changes in international competitiveness.

As the reader can plainly see, in the development of the UCSMP secondary curriculum we have attempted to respond to each of these changes.

Accomplishing the Goals

We at UCSMP believe that the goals of the various reform groups since 1975 can be accomplished, but not without a substantial reworking of the curriculum. It is not enough simply to insert applications, a bit of statistics, and take students a few times a year to a computer. Currently the greatest amount of time in arithmetic is spent on calculation,

The UCSMP secondary curriculum is the first full mathematics curriculum that is consistent with the recommendations of the NCTM *Standards*.

in algebra on manipulating polynomials and rational expressions, in geometry on proof, in advanced algebra and later courses on functions. These topics—the core of the curriculum—are the most affected by technology.

It is also not enough to raise graduation requirements, although that is the simplest action to take. Increases in requirements characteristically lead to one of two situations. If the courses are kept the same, the result is typically a greater number of failures and even a greater number of dropouts. If the courses are eased, the result is lower performance for many students as they are brought through a weakened curriculum.

The fundamental problem, as SIMS noted, is the curriculum, and the fundamental problem in the curriculum is **time**. There is not enough time in the current 4-year algebra-geometry-algebra-precalculus curriculum to prepare students for calculus, and the recommendations are asking students to learn even more content.

Fortunately, there is time to be had, because the existing curriculum wastes time. It underestimates what students know when they enter the classroom and needlessly reviews what students have already learned. This needless review has been documented by Jim Flanders, a UCSMP staff member (''How Much of the Content in Mathematics Textbooks is New?'' *Arithmetic Teacher,* September, 1987). Examining textbooks of the early 1980s, Flanders reports that at grade 2 there is little new. In grades 3–5, about half the pages have something new on them. But over half the pages in grades 6–8 are totally review.

And then in the 9th grade the axe falls. Flanders found that almost 90% of the pages of first-year algebra texts have content new to the student. The student, having sat for years in mathematics classes where little was new, is overwhelmed. Some people interpret the overwhelming as the student ''not being ready'' for algebra, but we interpret it as the student being swamped by the pace. When you have been in a classroom in which at most only 1 of 3 days is devoted to anything new, you are not ready for a new idea every day.

This amount of review in grades K–8, coupled with the magnitude of review in previous years, effectively decelerates students at least 1–2 years compared to students in other countries. It explains why almost all industrialized countries of the world, except the U.S. and Canada (and some French-speaking countries who do geometry before algebra), can begin concentrated study of algebra in the 7th or 8th grade.

Thus we believe that algebra should be taught one year earlier to most students than is currently the case.

However, we do not believe students should take calculus one year earlier than they do presently. It seems that most students who take four years of college preparatory mathematics successfully in high schools do not begin college with calculus. As an example, consider the data reported by Bert Waits and Frank Demana in the *Mathematics Teacher* (January, 1988). Of students entering Ohio State University with exactly four years of college preparatory high-school mathematics, only 8% placed into calculus on the Ohio State mathematics placement test. The majority placed into precalculus, with 31% requiring one semester and 42% requiring two semesters of work. The remaining 19% placed into remedial courses below precalculus.

Those students who take algebra in the 8th grade and are successful in calculus at the 12th grade are given quite a bit more than the normal four years of college preparatory mathematics in their ''honors'' or ''advanced'' courses. It is not stretching the point too much to say that they take five years of mathematics crammed into four years.

Thus, even with the current curriculum, four years are not enough to take a typical student from algebra to calculus. Given that the latest recommendations ask for students to learn more mathematics, WE BELIEVE FIVE YEARS OF COLLEGE PREPARATORY MATHEMATICS *BEGINNING WITH ALGEBRA* ARE NECESSARY TO PROVIDE THE TIME FOR STUDENTS TO LEARN THE MATHEMATICS THEY NEED FOR COLLEGE IN THE 1990s. The UCSMP secondary curriculum is designed with that in mind.

. . . we believe five years of college preparatory mathematics *beginning with algebra* are necessary to provide the time for students to learn the mathematics they need for college in the 1990s.

The UCSMP Secondary Curriculum

The UCSMP curriculum for grades 7–12 consists of these six courses:

Transition Mathematics

Algebra

Geometry

Advanced Algebra

Functions, Statistics, and Trigonometry

Precalculus and Discrete Mathematics

EACH COURSE IS MEANT TO STAND ALONE. Each course has also been tested alone. HOWEVER, TO TAKE BEST ADVANTAGE OF THESE MATERIALS, AND TO HAVE THEM APPROPRIATE FOR THE GREATEST NUMBER OF STUDENTS, IT IS PREFERABLE TO USE THEM IN SEQUENCE.

Each course is meant to stand alone. . . . However, to take best advantage of these materials, and to have them appropriate for the greatest number of students, it is preferable to use them in sequence.

Content Features

Transition Mathematics: This text weaves three themes—applied arithmetic, pre-algebra and pre-geometry—by focusing on arithmetic operations in mathematics and the real world. Variables are used as pattern generalizers, abbreviations in formulas, and unknowns in problems, and are represented on the number line and graphed in the coordinate plane. Basic arithmetic and algebraic skills are connected to corresponding geometry topics.

Algebra: This text has a scope far wider than most other algebra texts. It uses statistics and geometry as settings for work with linear expressions and sentences. Probability provides a context for algebraic fractions, functions, and set ideas. There is much work with graphing. Applications motivate all topics, and include exponential growth and compound interest.

Geometry: This text integrates coordinates and transformations throughout, and gives strong attention to measurement formulas and three-dimensional figures in the first two-thirds of the book. Work with proof-writing follows a carefully sequenced development of the logical and conceptual precursors to proof.

Advanced Algebra: This course emphasizes facility with algebraic expressions and forms, especially linear and quadratic forms, powers and roots, and functions based on these concepts. Students study logarithmic, trigonometric, polynomial, and other special functions both for their abstract properties and as tools for modeling real-world situations. A geometry course or its equivalent is a prerequisite, for geometric ideas are utilized throughout.

Functions, Statistics, and Trigonometry (FST): *FST* integrates statistical and algebraic concepts, and previews calculus in work with functions and intuitive notions of limits. Computers or sophisticated calculators are assumed available for student use in plotting functions, analyzing data, and simulating experiments. Enough trigonometry is available to constitute a standard precalculus course in trigonometry and circular functions.

Section 1: Overview of UCSMP

Precalculus and Discrete Mathematics (PDM): *PDM* integrates the background students must have to be successful in calculus (advanced work with functions and trigonometry, an introduction to limits and other calculus ideas), with the discrete mathematics (number systems, combinatorics, recursion, graphs) helpful in computer science. Mathematical thinking, including specific attention to formal logic and proof, is a theme throughout. It is assumed that computers or graphing calculators are available to students throughout this course.

General Features

Wider Scope: Geometry and discrete mathematics are present in all courses. Substantial amounts of statistics are integrated into the study of algebra and functions. The history of concepts and recent developments in mathematics and its applications are included as part of the lessons themselves.

Reality Orientation: Each mathematical idea is studied in detail for its applications to the understanding of real-world situations, or the solving of problems like those found in the real world. The reality orientation extends also to the approaches allowed the student in working out problems. Students are expected to use scientific calculators. Calculators are assumed throughout the series (and should be allowed on tests), because virtually all individuals who use mathematics today use calculators.

Problem Solving: Like skills, problem solving must be practiced. When practiced, problem solving becomes far less difficult. All lessons contain a variety of questions so that students do not blindly copy one question to do the next. Explorations are a feature of the first four years, and Projects are offered in the last two years. Some problem-solving techniques are so important that at times they (rather than the problems) are the focus of instruction.

Enhancing Performance: Each book's format is designed to maximize the acquisition of both skills and concepts, with lessons meant to take one day to cover. Within each lesson there is review of material from previous lessons from that chapter or from previous chapters. This gives the student more time to learn the material. The lessons themselves are sequenced into carefully constructed chapters. Progress Self-Test and Chapter Review questions, keyed to objectives in all the dimensions of understanding, are then used to solidify performance of skills and concepts from the chapter, so that they may be applied later with confidence. (See pages T34–T35 for more detail.)

Reading: Reading is emphasized throughout. Students can read; they must learn to read mathematics in order to become able to use mathematics outside of school. Every lesson has reading and contains questions covering that reading. In the last two courses, each chapter ends with a special reading lesson. (See pages T36–T37 for more detail.)

Understanding: Four dimensions of understanding are emphasized: skill in carrying out various algorithms; developing and using mathematical properties and relationships; applying mathematics in realistic situations; and representing or picturing mathematical concepts. We call this the SPUR approach: **S**kills, **P**roperties, **U**ses, **R**epresentations. On occasion, a fifth dimension of understanding, the historical dimension, is discussed. (See pages T38–T39 for more detail.)

Technology: Scientific calculators are recommended because they use an order of operations closer to that found in algebra and have numerous keys that are helpful in understanding concepts at this level. Work with computers is carefully sequenced within each year and between the years, with gradual gain in sophistication until *FST,* where computers are an essential element. In all courses, integrated computer exercises show how the computer can be used as a helpful tool in doing mathematics. Students are expected to run and modify programs, but are not taught programming. (See pages T40–T42 for more detail.)

Target Populations

We believe that all high-school graduates should take courses through *Advanced Algebra,* that all students planning to go to college should take courses through *Functions, Statistics, and Trigonometry*, and that students planning majors in technical areas should take all six UCSMP courses.

The fundamental principle in placing students into the first of these courses is that entry should not be based on age, but on mathematical knowledge. Our studies indicate that about 10% of students nationally are ready for *Transition Mathematics* at 6th grade, about another 40% at 7th grade, another 20% at 8th grade, and another 10–15% at 9th grade. We caution that these percentages are national, not local percentages, and the variability in our nation is enormous. We have tested the materials in school districts where few students are at grade level, where *Transition Mathematics* is appropriate for no more than the upper half of 8th-graders. We have tested also in school districts where as many as 90% of the students have successfully used *Transition Mathematics* in 7th grade.

However, the percentages are not automatic. Students who do not reach 7th-grade competence until the 9th-grade level often do not possess the study habits necessary for successful completion of these courses. At the 9th-grade level, *Transition Mathematics* has been substituted successfully either for a traditional pre-algebra course or for the first year of an algebra course spread out over two years. It does not work as a substitute for a general mathematics course in which there is no expectation that students will take algebra the following year.

On page T27 is a description of this curriculum and the populations for which it is intended. The percentiles are national percentiles on a 7th-grade standardized mathematics test using 7th-grade norms, and apply to students entering the program with *Transition Mathematics*. Some school districts have felt that students should also be reading at least at a 7th-grade reading level as well. See page T28 for advice when starting with a later course.

Top 10%: The top 10% nationally reach the 7th-grade level of competence a year early. They are ready for *Transition Mathematics* in 6th grade and take it then. They proceed through the entire curriculum by 11th grade and can take calculus in 12th grade. We recommend that these students be expected to do the Extensions suggested in this Teacher's Edition. Teachers may also wish to enrich courses for these students further with problems from mathematics contests.

50th–90th percentile: These students should be expected to take mathematics at least through the 11th grade, by which time they will have the mathematics needed for all college majors except those in the hard sciences and engineering. For that they need 12th-grade mathematics.

30th–70th percentile: These students begin *Transition Mathematics* one year later, in 8th grade. The college-bound student in this curriculum is more likely to take four years of mathematics because the last course is hands-on with computers and provides the kind of mathematics needed for any major.

15th–50th percentile: Students who do not reach the 7th-grade level in mathematics until 9th grade or later should not be tracked into courses that put them further behind. Rather, they should be put into this curriculum and counseled on study skills. The logic is simple: mathematics is too important to be ignored. If one is behind in one's mathematical knowledge, the need is to work more at it, not less.

Even if a student begins with *Transition Mathematics* at 9th grade, that student can finish *Advanced Algebra* by the time of graduation from high school. That would be enough mathematics to enable the student to get into most colleges.

UCSMP Target Populations in Grades 6–12

Each course is meant to stand alone. However, to take best advantage of these materials, and have them appropriate for the greatest number of students, it is preferable to use them in sequence. Although it is suggested that students begin with *Transition Mathematics,* students may enter the UCSMP curriculum at any point. Below is a brief description of the UCSMP curriculum and the populations for which it is intended. All percents should be considered using national, not local, comparisons.

The top 10% of students are ready for *Transition Mathematics* at 6th grade. These students can proceed through the entire curriculum by 11th grade and take calculus in the 12th grade.

Students who reach 7th-grade level in mathematics by 7th grade **(in the 50th–90th percentile** on a standardized test) should take *Transition Mathematics* in 7th grade.

Students who do not reach the 7th-grade level in mathematics until the 8th grade **(in the 30th–70th percentile)** begin *Transition Mathematics* in 8th grade.

Students who don't reach the 7th-grade level in mathematics until the 9th grade **(in the 15th–50th percentile)** begin *Transition Mathematics* in the 9th grade.

Grade	Top 10%	50th–90th percentile	30th–70th percentile	15th–50th percentile
6	Transition Mathematics			
7	Algebra	Transition Mathematics		
8	Geometry	Algebra	Transition Mathematics	
9	Advanced Algebra	Geometry	Algebra	Transition Mathematics
10	Functions, Statistics, and Trigonometry	Advanced Algebra	Geometry	Algebra
11	Precalculus and Discrete Mathematics	Functions, Statistics, and Trigonometry	Advanced Algebra	Geometry
12	Calculus (Not part of UCSMP)	Precalculus and Discrete Mathematics	Functions, Statistics, and Trigonometry	Advanced Algebra

Starting in the Middle of the Series

From the beginning, every UCSMP course has been designed so that it could be used independently of other UCSMP courses. Accordingly, about half of the testing of UCSMP courses after *Transition Mathematics* has been with students who have not had any previous UCSMP courses. We have verified that any of the UCSMP courses can be taken successfully following the typical prerequisite courses in the standard curriculum.

ALGEBRA

No additional prerequisites other than those needed for success in any algebra course are needed for success in UCSMP *Algebra*. Students who have studied *Transition Mathematics* tend to cover more of UCSMP *Algebra* than other students because they tend to know more algebra, because they are accustomed to the style of the book, and because they have been introduced to more of the applications of algebra.

UCSMP *Algebra* prepares students for any standard geometry course.

GEOMETRY

No additional prerequisites other than those needed for success in any geometry course are needed for success in UCSMP Geometry. UCSMP *Geometry* can be used with faster, average, and slower students who have these prerequisites. Prior study of *Transition Mathematics* and UCSMP *Algebra* insures this background, but this content is also found in virtually all existing middle school or junior high school texts.

Classes of students who have studied UCSMP *Algebra* tend to cover more UCSMP *Geometry* than other classes because they know more geometry and are better at the algebra used in geometry.

Students who have studied UCSMP *Geometry* are ready for any second-year algebra text.

ADVANCED ALGEBRA

***UCSMP Advanced Algebra* should not be taken before a geometry course but can be used following any standard geometry text.** Students who have studied UCSMP *Advanced Algebra* are prepared for courses commonly found at the senior level, including trigonometry or precalculus courses.

FUNCTIONS, STATISTICS, AND TRIGONOMETRY

FST assumes that students have completed a second-year algebra course. **No additional prerequisites other than those found in any second-year algebra text are needed for success in *FST*.**

PRECALCULUS AND DISCRETE MATHEMATICS

PDM can be taken successfully by students who have had *FST*, by students who have had typical senior level courses that include study of trigonometry and functions, and by top students who have successfully completed full advanced algebra and trigonometry courses.

PDM provides the background necessary for any typical calculus course, either at the high school or college level, including advanced placement calculus courses.

Section 1: Overview of UCSMP

Development Cycle for UCSMP Texts

The development of each text has been in four stages. First, the overall goals for each course are created by UCSMP in consultation with a national advisory board of distinguished professors, and through discussion with classroom teachers, school administrators, and district and state mathematics supervisors.

The Advisory Board for the Secondary Component at the time of this planning consisted of Arthur F. Coxford, Jr., University of Michigan; David Duncan, University of Northern Iowa; James Fey, University of Maryland; Glenda Lappan, Michigan State University; Anthony Ralston, State University of New York at Buffalo; and James Schultz, Ohio State University.

As part of this stage, UCSMP devoted an annual School Conference, whose participants were mathematics supervisors and teachers, to discuss major issues in a particular area of the curriculum. Past conferences have centered on the following issues:

1984 Changing the Curriculum in Grades 7 and 8

1985 Changing Standards in School Algebra

1986 Functions, Computers, and Statistics in Secondary Mathematics

1987 Pre-College Mathematics

1988 Mathematics Teacher Education for Grades 7–12

At the second stage, UCSMP selects authors who write first drafts of the courses. Half of all UCSMP authors currently teach mathematics in secondary schools, and all authors and editors for the first five courses have secondary school teaching experience. The textbook authors or their surrogates initially teach the first drafts of Secondary Component texts, so that revision may benefit from first-hand classroom experience.

After revision by the authors or editors, materials enter the third stage in the text development. Classes of teachers not connected with the project use the books, and independent evaluators closely study student achievement, attitudes, and issues related to implementation. For the first three years in the series, this stage involved a formative evaluation in six to ten schools, and all teachers who used the materials periodically met at the university to provide feedback to UCSMP staff for a second revision. For the last three years in the series, this stage has involved a second pilot.

The fourth stage consists of a wider comparative evaluation. For the first three books, this evaluation has involved approximately 40 classrooms and thousands of students per book in schools all over the country. For the last three books in the series, this stage has involved a careful formative evaluation. As a result of these studies, the books have been revised for commercial publication by Scott, Foresman and Company, into the edition you are now reading. (See pages T45–T49 for a summary of this research.)

2 Precalculus and Discrete Mathematics

Problems *Precalculus and Discrete Mathematics* is Trying to Address

UCSMP has received its funding because (1) there are major problems with school mathematics and (2) the solutions require a significant departure from current practice. *Precalculus and Discrete Mathematics* (PDM) is meant to follow UCSMP *Functions, Statistics, and Trigonometry* (FST) or an equivalent study of two years of algebra, one year of geometry, a half year of trigonometry, and an in-depth study of functions. It is quite different from most precalculus or high school analysis courses. The differences are due to its attempt to respond to six serious problems in mathematics education that cannot be treated by small differences in content or approach. Here are the problems that this book addresses and the UCSMP response to each.

PROBLEM 1: The traditional sequence of four years of high school mathematics does not prepare most students for a successful study of calculus.

Many high school teachers think that most of their successful 12th-grade students take calculus as college freshmen. The evidence is quite to the contrary. Most students need to take one or two courses in order to qualify for college calculus. Furthermore, international comparisons show that the mean performance of U.S. precalculus students on tests of algebra, geometry, and functions ranks among the lowest quarter of industrialized nations (SIMS, 1987; see page T21).

As a result, many students who do take calculus fail that course. Since calculus is a prerequisite for many technical fields, failure in calculus effectively limits a students' career options. This problem is detailed in the report *Calculus for the 21st Century: A Pump Not a Filter.*

The UCSMP response to this problem is to provide an extra year of mathematics for students. We offer a more challenging, less repetitive program (*Transition Mathematics*) for average students in grade 7 and recommend starting the study of *Algebra* in grade 8. This leaves the average student an additional year to prepare for college mathematics, and so UCSMP has created two courses to replace the traditional year of precalculus mathematics. In the first of these courses, *Functions, Statistics, and Trigonometry,* students study the various meanings of function and they study polynomial, exponential, logarithmic, and trigonometric functions in depth. They also extend the work of previous courses with quadratic relations, and the concepts of inverse and composition of functions. In this book, students continue the study of functions, trigonometry, sequences, limits, and continuity and study other precalculus topics, and also are introduced to the derivative and the integral. The graphical representation of functions is reviewed and extended. Specific attention is given to the formal language of calculus. And the mechanics of algebraic manipulation is integrated throughout.

PROBLEM 2: Even when students are prepared for calculus, they are often not prepared for the other mathematics they will encounter in college. In particular, the traditional curriculum ignores fundamental ideas of discrete mathematics.

The term ''discrete'' is usually used to contrast with the word ''continuous.'' The real line is considered continuous; the set of integers on that line is considered to be discrete. A ''discrete set'' is a set that can be put in one-to-one correspondence with the set of integers or a subset thereof. ''Discrete mathematics'' refers to that mathematics that deals with the integers or other discrete sets (sequences, mathematical induction, the steps in algorithms or proofs, or the subsets of a finite set). Discrete sets are critical to computer programming and computer science.

In 1965, computer science was unknown as a college major. By the early 1980s, there were many times more computer science majors than mathematics majors. In 1965, a business major needed no more than second-year algebra to succeed. But by the 1980s, such majors needed to take statistics, computer science, and often even some linear algebra, in addition to calculus. Today even nonphysical science majors require statistics and computers.

The emergence of computer science as a field of study has increased the importance of a background in topics from discrete mathematics that are not usually taught in high schools. Throughout the UCSMP secondary curriculum, topics from discrete mathematics are found. Some of these topics, such as the counting principles found in *Transition Mathematics* and *Algebra*, are traditional content. Other topics, such as the networks studied in UCSMP *Geometry*, are seldom found in contemporary curricula.

In *PDM*, students extend their study of probability, counting, sequences, and series. They also study properties of iteration and recursion. Furthermore, the logic and axiomatic nature of mathematics is studied to provide justifications for the techniques students have used for quite some time. Students are expected to follow and write simple direct proofs, proof by contradiction, and mathematical induction. They are also expected to recognize fallacious reasoning.

PROBLEM 3: Many students do not realize the importance mathematics has in later schooling, on the job, or in daily affairs.

At this point in their education, students are making career choices. Not all students are aware of the mathematics needed by college majors. In fact, they are surprised that mathematics has any applications in these fields.

Fundamentally, calculus and computer science are required for most students not because they are beautiful subjects but because they have many important applications.

Throughout UCSMP courses, pure and applied mathematics are taught together. We believe it necessary for students to see real-world applications of what they learn, to use these applications to motivate, develop, reinforce, and expand their knowledge. The evidence is strong that this approach gives them deeper appreciation for the value of mathematics and enhances their performance.

PROBLEM 4: Despite the nearly universal availability of calculators and widespread availability of computers, most contemporary texts do not take full advantage of the existence of this technology.

This problem has not been with us for a long time. Four-function hand-held calculators first appeared in 1971. The first personal computer appeared in 1977. The first easy-to-use calculator with the capability to graph functions appeared in 1985. Only in recent years has there been effective, easy-to-use computer software for graphing and for statistics. Many books do not accommodate contemporary technology because that technology was not available when the book was written.

From its inception, UCSMP has had a policy of using the latest in technology. Scott, Foresman has developed software to ensure that appropriate computer technology is available with each of the UCSMP texts.

Scientific calculators and computers allow students to work with realistic numbers, to practice estimation skills, to avoid tedious or repetitive calculations, and to solve problems that would not be solvable otherwise. For these reasons all UCSMP secondary courses refer to calculators and computers in the lessons, and assume that students have access to scientific calculators for every lesson and on all tests.

Access to an automatic grapher (a graphics calculator or function graphing software) is required for this course. With the aid of an automatic grapher, students can accurately approximate relative extrema and zeros of virtually any function. Students are able to graph many more functions than they could by hand, and build their understanding of the properties and uses of functions. In particular, the automatic grapher helps students see the parallel effects of transformations on equations and graphs of functions and to represent the solving of equations and inequalities, and to prove identities.

This course is not geared to a particular kind of technology. The technology is used to promote students' ability to visualize functions, to explore relations between equations and their graphs, to simulate experiments, to generate and/or analyze data, and to develop the concept of limit.

See pages T40–T42 for more technology information.

PROBLEM 5: Students do not get enough experience with problems which require some thought.

In general, the evidence is that students have a great deal of difficulty with complicated numbers, different wordings, or new contexts. In all UCSMP courses, questions are given with all sorts of numbers, a variety of wordings, and the many different contexts which arise naturally from applications. Exploration questions confront students with open-ended problems, problems requiring more sophisticated generalization, or with problems requiring outside references.

In preparation for the more sophisticated analysis and synthesis expected in college, the end of each chapter contains a set of Projects. These activities require more elaborate data collection or more extensive analyses than the questions in the regular lessons and help students develop reasoning and communication skills. (See p. T37 for more detail.)

PROBLEM 6: Students do not understand how, when, or why mathematicians do proofs.

Rigorous deductive thinking is how mathematicians determine whether statements are true relative to a set of postulates; it is the way in which new mathematics is established. Without such thinking, mathematics would only be a collection of assorted principles whose connection with each other would not have been established and whose truth with respect to each other is subject to question.

Despite the importance of proof in mathematics, most students are expected to do proofs only in connection with their study of geometry. They associate proofs, therefore, with geometry and not with mathematics as a whole, and they are not very good at doing them.

In *PDM*, we attempt to rectify that situation by examining proofs in a manner that takes advantage of the students' greater mathematical knowledge and prior experience. Still, we proceed carefully over a long period of time so that students have time to learn from their prior work.

PROBLEM 7: Students do not read mathematics textbooks.

Students tell us that they do not read because (1) the text is uninteresting, and (2) they do not need to read—their teacher explains everything for them. But students *must* learn to read for future success in mathematics. Our response to (1) is to include informative and interesting reading in every lesson. The text develops concepts and principles, provides examples and non-examples, and shows relations between ideas. Many questions require reading to set the context of a problem. Also, the last lesson of every chapter is a lesson that simulates the reading a student might do in non-mathematics courses. These special reading lessons cover a wide range of applications, discussions, and themes.

Our response to (2) above is to encourage teachers not to explain everything *in advance* to students. The teacher has the freedom to teach in a variety of ways. It is not necessary for the teacher to explain the text every day. The teacher can concentrate on helping students with difficult symbols or vocabulary, and on developing further examples and explanations specifically tailored to his or her class.

It is critical that students learn to read mathematics so that they are adequately prepared to read any college text.

More detailed information about reading and the reading lessons can be found on pages T36–T37.

Goals of *PRECALCULUS AND DISCRETE MATHEMATICS*

Obviously, among our goals is the removal of all the problems detailed above, while studying the content areas named in the title of the text. But we have been guided by other goals as well. We want to convey to the student, through historical references and references to recent mathematical work that mathematics was developed by people and continues to develop. Through applications, WE WANT STUDENTS TO VIEW THEIR STUDY OF MATHEMATICS AS WORTHWHILE, AS FULL OF INTERESTING AND ENTERTAINING INFORMATION, AS RELATED TO ALMOST EVERY ENDEAVOR. Through explorations and projects, we want students to look for and recognize mathematics in places they haven't before, to use the library, to search through newspapers or almanacs, to get excited by knowledge.

Guiding principles for including a topic in the curriculum of this book were:

(1) it is important for daily living, career development, or future study of mathematics;

(2) it contributes to a balanced preparation for college mathematics courses in calculus or discrete mathematics;

(3) it is at a suitable level of difficulty for students with the appropriate prerequisites (see below) and is accessible to them with current technology;

(4) it can be done by all students who have successfully completed UCSMP *FST* or the equivalent.

In this course, visual information derived from the use of an automatic grapher is employed to obtain conjectures about functions, equations, and inequalities which are then verified through algebraic means. Technology also allows students to explore calculus concepts. The book reviews polynomial, exponential, logarithmic, and trigonometric functions, and gives strong attention to rational functions. With technology students are able to build their understanding of the properties and uses of functions in ways that were inaccessible a generation ago.

The discrete mathematics in this book plays a number of roles. Some of it, such as the work with algorithms, combinatorics, and graphs and circuits, is studied primarily for its own sake. Some topics such as recursion and integers are important not only for their own sake but for the insights they provide into other mathematics at this level. We often study continuous functions by examining their values at a set of discrete points. The transition from the discrete to the continuous is a natural way to approach ideas from calculus.

Mathematical thinking is a theme throughout this course, and as such, we expect students to gain competence with algebraic proofs. The subject matter of logic and proof encompasses all the content of the book. The logic is connected with everyday thinking and the ways computers operate. Logic provides a higher-level look at the solving of equations and inequalities, and provides the basis for the direct proofs, indirect proofs, and proofs by induction that students are expected to handle in this course.

Who Should Take *Precalculus and Discrete Mathematics*?

All students who plan to major in technical fields or who will take college calculus next should take this course.

Students who have no UCSMP background should be strong in algebra, functions, and trigonometry. Students should be adept at solving linear and quadratic equations and linear systems. They need to know how to find an equation for a line given the coordinates of two points on that line. They should know how to use a calculator to evaluate polynomial, exponential, logarithmic, and trigonometric functions. They should be able to make and read coordinate graphs of linear, quadratic, logarithmic, and exponential functions and the power functions $f(x) = x^n$, for n a positive integer. Students should have studied the trigonometric functions from two standpoints—right triangle ratios and circular functions. They should be able to solve triangles and graph the circular functions. Some familiarity with the BASIC computer language and transformations is assumed.

If students do not have all the prerequisite skills, you may need to spend more time on certain chapters to assure success. We have made comments in the Teacher's Edition for topics where teachers had to slow down because their students had little or no previous background. For example, if your students have not studied functions in some depth, Chapter 2 will take longer than the recommended pace. If they have not studied polynomial functions in depth, Chapter 4 will take more time. If students are not familiar enough with trigonometric functions, Chapter 6 will take longer. Chapter 8's pace assumes that students have seen various counting problems, including permutations and combinations.

We do not recommend that *PDM* follow a strong course in second-year algebra and trigonometry unless the students are exceptional. For almost all students, *Functions, Statistics, and Trigonometry* is the appropriate course to follow a second course in algebra—even if it contains some trigonometry.

Managing technology

We have tried to make the use of technology in the classroom as easy as possible. The first lesson involving technology is Lesson 2-1. It gives you and your students at least two weeks to get settled into a class routine and to obtain the technology. While computers and calculators are not required every day, the mathematics they are used for is not predictable. In some chapters they are not very helpful, but in others, students can use them every day and on homework assignments. In general, the use of technology is embededed in the lesson, both in text and questions.

If you have never done much work with computers, do not be alarmed. This is a mathematics course, not a programming course. Computer language is explained in the text whenever needed.

There are two management suggestions regarding computer use which we cannot overemphasize. (1) Orient yourself by running programs in advance of class. This will better prepare you to anticipate student problems with hardware or software. The various versions of BASIC and each computer have slightly different characteristics, and our generic programs may need to be modified slightly for your system. Check with a computer expert in your school (perhaps a student) on how this might be done. (2) When you use the computer to demonstrate statistical analysis, or if students are all going to be analyzing the same data set during class, enter the data into a file on a disk before class begins. We recommend finding student assistants who can help you with both these tasks.

In addition to the references in the text to BASIC and graphing software, Computer Masters are provided in the Teacher's Resource File. These blackline masters are keyed to specific lessons in each of Chapters 1–13. They are meant for use by a class in a laboratory setting, or by individuals for extra credit work. It is not necessary to do these masters to be successful in *Precalculus and Discrete Mathematics*. Since each one stands alone, you may do as many or as few as you think appropriate for your class.

Evaluating Learning

Grading

No problem seems more difficult than the question of grading. If a teacher has students who perform so well that they all deserve *A*s and the teacher gives them *A*s as a result, the teacher will probably not be given plaudits for being successful but will be accused of being too easy. This suggests that the grading scale ought to be based on a fixed level of performance, which is what we recommend. We recommend this because the performance that gives an *A* in one school or with one teacher may only rate a *C* in another, and we think it unfair.

Seldom in this book are there ten similar questions in a row. To teach students to be flexible, the wording of questions is varied, and principles are applied in many contexts. The problems are varied because that is the way problems come in later courses and in life outside of school. Learning to solve problems in a variety of contexts or to discern relationships between properties is more difficult than learning to perform a routine skill. Thus, a natural question that arises is "How should I grade students in UCSMP courses?"

First, we believe in a multifaceted approach to evaluation. Such areas as class participation and preparation, homework, presentation of Projects, along with performance on written tests and quizzes, should be taken into account when grades are determined. The Follow-Up section of the lesson notes in the Teacher's Edition often provides recommendations for Alternative Assessment.

Second we believe a student should be able to do each set of objectives at about the 85% level. An 85% score on a test deserves no less than a high *B*, and probably an *A*. In the past, our tests have often led us to the following grading scale: 85–100 = *A*, 72–84 = *B*, 60–71 = *C*, 50–59 = *D*, 0–49 = *F*. Such a scale may alarm some teachers, but STUDENTS IN UCSMP COURSES GENERALLY LEARN MORE MATHEMATICS OVERALL THAN STUDENTS IN COMPARISON CLASSES. We believe that the above grading policy rewards students fairly for work well done.

Some teachers are accustomed to the grading scale 90–100 = *A*, 80–90 = *B*, 70–79 = *C*, 60–69 = *D*, and 0–59 = *F*. In fact, in some districts, this is the required grading scale. If this is your situation, you may want to subtract points from, say, 110.

One January a teacher of *Transition Mathematics* remarked, "I've never had a class that learned so much, but my grades are lower." Later she said, "I have students who are failing. But I can't switch them to another class [using another book at the same level] because they know too much." Her problem is not unusual: percentages correct on tests of higher order thinking are generally lower than scores on tests of routine skills. We often make a basketball analogy. In many courses all of the shots students ever have are lay-ups (exercises) and an occasional free throw (easy problems). They shoot these over and over again, from the same spot ("Do the odd-numbered exercises from 1–49."). In *Precalculus and Discrete Mathematics* almost every question is a different shot (a problem), some close in, some from middle distance, a few from half-court. To expect percentages of correct shots to be the same is unrealistic.

In basketball, there are times when a person wants to practice one specific shot to make it automatic, and so it is at times with the content of *Precalulus and Discrete Mathematics*. We suggest focussing in on a few topics for quizzes. The Teacher's Resource File contains quizzes and tests for each chapter. If students perform well on quizzes and tests, it has a positive effect on interest and motivation. You should endeavor to use grading as a vehicle for breeding success as well as for evaluating students.

Students in UCSMP courses generally learn more mathematics overall than students in comparison classes.

In the light of all this, we strongly urge you to let students know what they need to know in order to get good grades. All research on the subject indicates that telling students what they are supposed to learn increases the amount of material covered and tends to increase performance. Also, let students know that you expect *all of them* to perform well. Do not arbitrarily consign some of your students to low grades. Let them know that it is possible for all of them to get *A*s if they learn the material.

Some teachers have found that because of the way the Review questions maintain and improve performance, cumulative tests at the end of each marking period give students an opportunity to do well. For your convenience, the Teacher's Resource File has Chapter Tests in Cumulative Form (in addition to the regular Chapter Tests) for each chapter, beginning with Chapter 2. These Cumulative Tests allow students a chance to show what they have learned about topics studied in a previous chapter, along with the topics of the current chapter. Beginning with Chapter 3, the questions in the Cumulative Tests are about 50% from the current chapter, 25% from the preceding chapter, and 25% from all other earlier chapters.

Standardized Tests

At our first conference with school teachers and administrators in the early days of UCSMP, we were told in the strongest terms, "Be bold, but remember that you will be judged by old standardized tests."

Precalculus and Discrete Mathematics should be viewed as a new text, not as an experimental text.

There are many types of standardized tests in high school mathematics: standardized tests for a particular course; general test batteries in which there may be a mathematics subtest; local school or state assessments; and college entrance examinations. There is strong evidence that performance on any of these can be enhanced by practice on questions like those which will appear. Since these tests often have bearing on student placement you should, when possible, inform students of the kinds of questions they are likely to encounter.

Based on our testing and what teachers and schools have reported, here is what we conclude about these various kinds of tests. Students in *Precalculus and Discrete Mathematics* will perform as well as (if not better than) students from other classes on traditional standardized tests. Yet, since UCSMP students outperform other students on applications, graphing, and statistics, it is unfair to UCSMP students if they are judged only by a traditional standardized test.

From our studies, we believe that students studying UCSMP courses are likely to do better on the College Board SAT-M than other students. The SAT-M is a problem-solving test, and far more questions on the SAT-M cover content likely to be unfamiliar to other students than to UCSMP students. The ACT exams seem at least as favorable to our students as to students in other books. Furthermore, if UCSMP suggestions are followed and more students take *Geometry* in the 9th grade, then there is no question that students in the UCSMP curriculum will far outscore comparable students who are a course behind. Preliminary evidence with PSAT scores indicates that 10th graders who have had all UCSMP courses through *Advanced Algebra* score as well as non-UCSMP juniors of equivalent ability in second-year algebra.

Finally, a note about curriculum projects. Many people treat books from curriculum projects as "experimental." Few commercial textbooks, however, have been developed on the basis of large scale testing. Few have been produced as part of a coherent 7–12 curriculum design. This text has gone through extensive prepublication analysis and criticism. It has been revised on the basis of comments and evaluations of the many teachers who have been in our evaluation studies. It is the intention of UCSMP to provide materials which are successful as well as teachable. *PRECALCULUS AND DISCRETE MATHEMATICS* SHOULD BE VIEWED AS A NEW TEXT, NOT AS AN EXPERIMENTAL TEXT.

4 Research and Development of *Precalculus and Discrete Mathematics*

Planning, Selection of Authors

From its inception, UCSMP believed that students who finished the curriculum should be prepared for *all* the mathematics they would encounter in college, including statistics, discrete mathematics, linear algebra, and applied mathematics—in addition to a thorough grounding for calculus. In the early days of UCSMP, the last two courses in the secondary curriculum were entitled simply "Pre-College Mathematics 1" and "Pre-College Mathematics 2." After some discussions with the advisory board to the project, it was decided that, because of their importance to all college-bound students regardless of major, functions and statistics should be the two major themes of the first course. (See the corresponding material in the course *Functions*, *Statistics*, *and Trigonometry* for details on the development of that course.)

The second of these courses was planned to be for college-bound students who would major in technical subjects, including those who might major in mathematics, engineering, computer science, or physical sciences, or other areas that required a strong background in mathematics. The precalculus theme was seen as a natural continuation of the functions theme of the previous course, providing a two-year background for calculus where one year was the norm. A background in discrete mathematics was seen as appropriate not only for those who wished to major in computer science, but also for any other technical subjects. Two major themes for the second course, precalculus and discrete mathematics, were thus determined. Linear algebra, through work with transformations and matrices, was considered to have been covered rather well in previous courses, but it was felt that vectors should be discussed in this course. Applied mathematics is a theme of all UCSMP courses, and this theme was to be continued here. Finally, it was considered critical that this book provide a smooth transition to the discourse of college mathematics, and thus also include work both with mathematical reasoning and proof, and with the manipulative algebra required in some calculus courses.

It was natural, consequently, that both high school and college mathematics faculty were sought as members of the writing team. The two college members of the first writing team had important qualifications for such a task. Anthony Peressini, Professor of Mathematics at the University of Illinois at Urbana-Champaign, had been in charge of the university-wide placement examinations in mathematics and thus was acutely aware of the needs and possible shortcomings of entering college students with regard to calculus and a precalculus preparation. Susanna Epp, Associate Professor of Mathematics at DePaul University, had been involved in a conference on reform of the collegiate curriculum and was in the process of completing a textbook on discrete mathematics. The two high school teachers, Kathleen Hollowell who taught at Newton North H.S., in Newton, MA, and Jack Sorteberg of Burnsville H.S., Burnsville, MN, had both taught many advanced high school courses, including advanced placement calculus, and were familiar with the use of technology in the classroom.

First Pilot Study

Writing began in the summer of 1987 and continued during the 1987–88 school year. The manuscript was edited at UCSMP by Denisse Thompson and Dora Aksoy, each of whom had experience with teaching this kind of content. Overall direction for UCSMP was given by Zalman Usiskin. Two of the *PDM* authors and an author of another UCSMP book used the materials. The materials, entitled *Pre-College Mathematics,* were loose-leaf and sent to schools one chapter at a time. No formal testing was done at this stage, but the teachers commented on each lesson and twice were brought to the University of Chicago for full-day meetings.

Several themes emerged from the first draft and the first pilot. One was that discrete mathematics provided a wonderful setting for work with proof and allowed a theme of mathematical thinking to be sustained through the course. The second was that discussion of the precalculus ideas was quite affected by the existence of technology which gave the student the ability to graph functions, and though computers might not be required in such a course, graphics calculators should be recommended. The third was that the manipulative algebra in this first pilot (work with the kinds of complex rational expressions one sometimes sees in calculus) was hard for the students, even though these tended to be stronger students who had taken two years of algebra from texts and teachers who devoted a great deal of time to such manipulation. It became clear that, even at this level, manipulative algebra could not be successfully taught for its own sake; for maximum performance, it had to be in context.

Second Pilot Study

A major revision was made of the materials during the summer and fall of 1988. Greg McRill and Jeff Birky joined the editorial team during this year. In addition to the considerations given to the difficulty of lessons, to the clarity of the expositions, and to the value of the questions throughout the book, the second pilot reflected the continuing increase in the use of graphing technology. Access to an automatic grapher was at first strongly recommended, but then became required. In this study, three schools in three states used the materials, none of them a school of an author. The materials were again loose-leaf, sent to schools one chapter at a time. The current title, *Precalculus and Discrete Mathematics*, reflecting the two major themes of the course, was used for the first time.

As in all other UCSMP evaluations, the teachers were asked to keep detailed records on each lesson. These were sent back to the project as the year progressed. Again, all teachers were invited to visit the University of Chicago during the year for full-day meetings to discuss all aspects of teaching and learning from the materials. There were three of these meetings.

Continuing a pattern found in earlier UCSMP courses, the use of technology seemed to make it possible for the materials to work well with a wide variety of students, from slow students at this level to the best-prepared. Teachers enjoyed the variety of content that the two themes engendered and particularly liked the work with proof. Some commented that, for the first time, their students were feeling successful with proof. Length of the manuscript, however, was seen as a problem.

Formative Evaluation and Test Results

In the summer of 1989, the materials were again revised into what was termed the Field Trial Edition. Because of the need to have more materials ready by the beginning of the school year, two experienced UCSMP authors, John McConnell and Susan Brown, were asked to join the writing team. Wade Ellis, Jr., of West Valley College in Saratoga, California, an expert on the use of computers to teach mathematics, and an experienced teacher of students at the community college level, joined the team because of the increasing need to incorporate technology. The materials were printed in three spiral-bound parts with students and teachers receiving the course one part at a time.

Many studies of implementation of materials look only at students. From previous UCSMP studies, we have learned that teachers utilize materials with such diversity that, for many questions, the teacher or the school is the appropriate level of study. Nine schools in eight states participated in the formative evaluation of this course. These schools responded to a call for participants and were selected merely because they were the first to inform the project that they planned to use the materials in class-size quantities. Three schools were private, two were magnet public schools in large cities, two were public schools in mid-sized cities, and two were in suburban areas. At five of these schools, *PDM* was intended as a fifth-year course (with algebra considered as the first year) for students unprepared or unwilling to take calculus. At one site, calculus was not offered, and *PDM* replaced a fifth-year honors precalculus course. At three sites, *PDM* was a fourth-year course taken mostly by sophomore and junior students after an enriched second-year algebra/trigonometry course.

One teacher participated from each school. Four of the nine teachers had previously taught from UCSMP texts, but only one from a previous version of *PDM*. One was a first-year teacher. No special training was given the teachers. During the school year, all nine teachers met in Chicago in November and May to discuss how things were going in their classrooms and how they felt about the materials. Their classes were observed and the teachers and some students were interviewed by Denisse Thompson, who directed the evaluation.

A total of 180 students participated in the study, 54% male and 46% female. Most (70%) were seniors, 27% were juniors, and 3% were sophomores. The racial/ethnic background of the students was 73% white, 13% black, 9% Asian or Pacific Islanders, 4% Hispanic or Latino, and 1% not known. All but one of these students indicated definite plans to go to college. Almost all (96%) had used a scientific calculator before, and 15% had used a graphing calculator. A large percentage (69%) of these students had computers at home. Of the 178 students who completed a form detailing their mathematics backgrounds, 29 (16%) had one previous UCSMP course—27 had FST; 2 had *Advanced Algebra*—and 14 (8%) had both these courses.

No comparison classes were involved in this study, for it was impossible to give a test whose content was fair to both UCSMP and comparison classes.

During the first week of school, one of two forms of a 29-item pretest was given to all students. The pretest was designed to measure prerequisite skills and also to provide a base to see if student performance improved. Different items on functions and algebraic skills were on the forms, and common to both forms were items on logic or proof, discrete mathematics, trigonometry, and calculus. The range of school means was wide, from 10.1 to 20.1. The three schools in which *PDM* was a fourth-year course ranked 2nd, 8th, and 9th among the nine schools on the pretest. In the lower scoring schools on the pretest, the students typically did not have the opportunity to master some of the function and trigonometry topics that are expected to have been mastered by entering *PDM* students.

All of the classes utilized technology for graphing, finding zeros, and comparing functions. In seven of the nine schools, graphics calculators were used for these purposes; in six schools computers were used for these purposes. (Four schools used both.) In six of the seven schools with graphing calculators, these calculators were distributed to students. Specifically, of 146 students who completed a survey at the end of the year, 122 reported using graphics calculators and 78 reported using computers. In general, teachers and students who had graphics calculators used them most days in the week, compared to an average of once a week for those who had computers.

The posttest consisted of two parts, a 32–question multiple-choice part that was the same for all students, and an open-ended part that had two forms each taken by half the students. On the multiple-choice part, 17 items were repeated from one or both forms of the pretest. These items covered a variety of precalculus and discrete mathematics topics. The mean percent correct on these items was 34% on the pretest and 65% on the posttest; the gains on items ranged from 5% to 58%. In general, the gains were about equal on discrete mathematics and precalculus topics. It is clear that students learned quite a bit in this course.

Each open-ended test contained five items, including one proof of a trigonometric identity, one other proof, two questions about functions that required graphing technology, and a question applying the mathematics. The questions were not easy, perhaps comparable to an advanced placement test in difficulty. The students were quite successful on the trigonometric identities, with 83% success on an easy identity and 71% success on a more difficult identity. Success was lower on the other proofs, with 33% of the students successful on a divisibility proof and 12% of students successful on a mathematical induction proof, that in hindsight, was too difficult for a timed test.

Almost all of the students (95%) used automatic graphing technology on this test. Half of the students were able to graph complicated polynomial functions and could use the graph to determine zeros or relative extrema. Half could graph other rather complicated functions, but only a third could find zeros of such a function, and only an eighth could do this and give the end behavior of a given rational function. A max-min application of a type found in calculus classes was answered correctly by 58% of the students, and 51% answered correctly or made significant headway on a combinatorics application.

Teacher attitudes towards the book were compared with some of the attitudes of teachers in a study by Iris Weiss of the Research Triangle Institute for the 1985–86 National Survey of Science and Mathematics Education. (The Weiss study deals with mathematics texts in all the grades 10–12, not merely texts at this level.)

Percent of teachers giving favorable ratings about textbooks:

Item	UCSMP $n = 9$	Weiss $n = 517$
develops problem-solving skills	89	68
good suggestions for computers	89	31
good suggestions for calculators*	89	27
good suggestions for activities/ assignments	78	55
explains concepts clearly	78	73
appropriate reading level	67	87
interesting to students	33	43
high quality supplements	33	33

(*The Weiss item was modified for this study to read "good suggestions for automatic graphers".)

The teachers were asked to compare *PDM* in general to other texts at this level. Seven of nine responded, as follows:

much better	slightly better	about the same	slightly worse	much worse
4	1	2	0	0

Continuing Research and Development

The study of *Precalculus and Discrete Mathematics* convinced us that the approach we have taken is fundamentally sound. And, as was desired and is to be expected, the study suggested ways to improve the course. This Scott, Foresman edition differs from the edition used in the formative evaluation in many ways. Many lessons and a few chapters have been reworked to simplify the exposition and clarify connections between ideas. One chapter was deleted—some of its content was integrated with other chapters—in order to shorten the overall length of the text.

The present version of *Precalculus and Discrete Mathematics* also includes the appearance of four colors, of attractive pictures, an expanded Teacher's Edition, and many more supplements for teachers and students. These supplements include Lesson Masters for teachers who wish more questions on a particular lesson, Teaching Aids to help in explanations, and quizzes and tests. Because of these changes, we expect students to perform as well, if not better, with the Scott, Foresman version than with our previous versions.

Each November since 1985, we have held a conference at the University of Chicago at which users and prospective users of secondary component UCSMP materials can meet with each other and with authors. Beginning in August 1989, we have conducted a free full day in-service for any teachers or supervisors who will be using the materials. We encourage new users to attend one of these meetings.

We desire to know of any studies school districts conduct using these materials. UCSMP will be happy to assist school districts by supplying a copy of the noncommercial tests we have used in our evaluations (except for certain open-ended items) and other information as needed.

Both Scott, Foresman and UCSMP welcome comments on these books. Please address comments either to Mathematics Product Manager, Scott, Foresman, 1900 East Lake Avenue, Glenview, IL 60025, or to Zalman Usiskin, University of Chicago, 5835 S. Kimbark Avenue, Chicago, IL 60637.

5 Bibliography

REFERENCES for Sections 1–4 of Professional Sourcebook:

Board of Mathematical Sciences and the Mathematical Sciences Education Board. ***Calculus for a New Century: A Pump Not a Filter.*** MAA Notes, No. 8. Washington, D.C.: Mathematical Association of America, 1988.

College Board. ***Academic Preparation for College: What Students Need To Know and Be Able To Do.*** New York: College Board, 1983.

Flanders, James. **"How Much of the Content in Mathematics Textbooks Is New?"** *Arithmetic Teacher,* September 1987, pp. 18–23.

Jones, Philip, and Coxford, Arthur F. ***A History of Mathematics Education in the United States and Canada.*** 30th Yearbook of the National Council of Teachers of Mathematics. Reston, VA: NCTM, 1970.

Mathematical Sciences Education Board. ***Reshaping School Mathematics: A Philosophy and Framework for Curriculum.*** Washington, D.C.: National Academy Press, 1990.

McKnight, Curtis, et al. ***The Underachieving Curriculum: Assessing U.S. School Mathematics from an International Perspective.*** Champaign, IL: Stipes Publishing Company, 1987.

National Commission on Excellence in Education. ***A Nation at Risk: The Imperative for Educational Reform.*** Washington, D.C.: U.S. Department of Education, 1983.

National Council of Teachers of Mathematics. ***Curriculum and Evaluation Standards for School Mathematics.*** Reston, VA: NCTM, 1989.

National Council of Teachers of Mathematics. ***Professional Standards for Teaching Mathematics.*** Reston, VA: NCTM, 1991.

National Research Council. ***Everybody Counts.*** Washington, D.C.: National Academy Press, 1989.

Waits, Bert, and Demana, Franklin. **"Is Three Years Enough?"** *Mathematics Teacher,* January 1988, pp. 11–15.

Garfunkel, Solomon, and Steen, Lynn A., editors. ***For All Practical Purposes: An Introduction to Contemporary Mathematics.*** New York: W.H. Freeman, 1988.

Hanson, Viggo P., and Zweng, Marilyn J., editors. ***Computers in Mathematics Education.*** 1984 Yearbook of the National Council of Teachers of Mathematics. Reston, VA: NCTM, 1984.

Hoffman, Mark, editor. ***The World Almanac and Book of Facts.*** New York: World Almanac, yearly.

Johnson, Otto, executive ed. ***The Information Please Almanac.*** Boston: Houghton Mifflin Company, yearly.

Joint Committee of the Mathematical Association of America and the National Council of Teachers of Mathematics. ***A Sourcebook of Applications of School Mathematics.*** Reston, VA: NCTM, 1980.

Kastner, Bernice. ***Applications of Secondary School Mathematics.*** Reston, Va.: National Council of Teachers of Mathematics, 1978.

Kastner, Bernice. ***Space Mathematics.*** Washington, D.C.: U.S. Government Printing Office, 1985.

Sharron, Sidney, and Reys, Robert E., editors. ***Applications in School Mathematics.*** 1979 Yearbook of the National Council of Teachers of Mathematics. Reston, Va.: NCTM, 1979.

USA Today. Arlington, VA: Gannett & Co., Inc., daily.

U.S. Bureau of the Census. ***Statistical Abstract of the United States: 1990.*** 110th ed. Washington, D.C., 1990.

Additional References on Discrete Mathematics

Bogart, Kenneth B. ***Discrete Mathematics.*** Boston: D.C. Heath, 1988.

Dossey, John. ***Discrete Mathematics.*** Glenview, IL: Scott, Foresman, 1987.

Epp, Susanna S. ***Discrete Mathematics with Applications.*** Belmont, CA: Wadsworth, 1990.

Grimaldi, Ralph P. ***Discrete and Combinatorial Mathematics.*** Menlo Park, CA: Addison-Wesley, 1985.

Hirschfelder, R., and Hirschfelder, J. ***Introduction to Discrete Mathematics.*** Brooks-Cole, 1991.

Johnsonbaugh, Richard. ***Discrete Mathematics.*** Second edition. New York: Macmillan, 1990.

Lefton, Phyllis. **"Number Theory and Public-Key Cryptography"** *Mathematics Teacher,* January 1991, pp. 54–62.

Liu, C.L. ***Elements of Discrete Mathematics.*** Second Edition. New York: McGraw Hill, 1985.

Maurer, Stephen B., and Ralston, Anthony. ***Discrete Algorithmic Mathematics*** Menlo Park, CA: Addison-Wesley, 1991

Roman, Steven. ***An Introduction to Discrete Mathematics.*** Second edition. Harcourt Brace Jovanovich, 1989.

Rosen, Kenneth. ***Discrete Mathematics and Its Applications.*** Second Edition. New York: McGraw, 1988.

Ross, Kenneth A., and Wright, Charles R.B. ***Discrete Mathematics.*** Second edition. Englewood Cliffs, NJ: Prentice-Hall, 1988.

SOURCES for Data or Additional Problems

Consumer Reports. Yonkers, NY: Consumers Union of U.S. Inc., monthly.

The Diagram Group. ***Comparisons.*** New York. St. Martin's Press, 1980.

Eves, Howard. ***An Introduction to the History of Mathematics.*** 5th ed. Philadelphia. Saunders College Publishing, 1983.

UCSMP SOFTWARE

GraphExplorer. Scott, Foresman and Company. 1900 East Lake Avenue, Glenview, IL 60025. (Macintosh, IBM)

StatExplorer. Scott, Foresman and Company. 1900 East Lake Avenue, Glenview, IL 60025. (Macintosh, IBM)

UCSMP Algebra/Advanced Algebra Graphing Software. Scott, Foresman and Company, 1900 East Lake Avenue, Glenview, IL 60025. (Apple II, IBM)

OTHER SOFTWARE

ANUGraph. The Australian Association of Mathematics Teachers, 18 Blue Hills Road, O'Halloran Hill, South Australia 5158. (Macintosh)

Calculus. Broderbund Software, Inc. 17 Paul Drive, San Rafael, CA 94903. (Macintosh)

Calculus T/L. Brooks/Cole Publishing. 511 Forest Lodge Road, Pacific Grove, CA 93950. (Macintosh Plus)

Calculus Toolkit. Addison-Wesley Publishing Company. c/o Consumer Software Division, Route 128, Reading, MA 01867. (IBM, Apple II, Macintosh)

Calculus. TrueBASIC, Inc., 39 South Main Street, Hanover, NH 03755. (IBM, Macintosh)

Scarborough, Steve. ***Grapher.*** 7433 Densmore Avenue, Van Nuys, CA 91406. (Macintosh, IBM)

Schwartz, Judah. ***The Function Analyzer.*** Sunburst Communications, Pleasantville, NY 10570. (IBM)

Waits, Bert, and Demana, Frank. ***Master Grapher.*** Addison-Wesley Publishing Company. (Apple II, IBM, Macintosh)

Full-size versions of these pages appear in the Student Edition.

Full-size versions of the Progress Self-Test charts are found on pages T118–T123 in this Teacher's Edition.

LESSON 1-1 (pp. 4–11)
1. statement **3.** not a statement **5. a.** $4^3 - 1 = (4 - 1)(4^2 + 4 + 1)$; True **b.** $(\sqrt{a})^3 - 1 = (\sqrt{a} - 1)(a + \sqrt{a} + 1)$ **c.** Yes **7.** universal statement **9.** for all (∀) and there exists (∃) **11.** Sample: $x = 1$. **13. a.** an even integer; x is prime **b.** even integer; prime **c.** even integer; prime **15. a.** Yes **b.** No **c.** for 0, 1, and all integers greater than or equal to 5 **17.** Sample: *There exists a student in my math class who owns a car.*
19. False, for example $\log(\frac{1}{10}) = -1$ and $-1 < 0$. **21.** Let $x = 0$; $\forall\, y$, $0 \bullet y = 0 \neq 1$. **23.** circles; x is not a parabola **25. a.** 3 **b.** $c^2 - c + 1$ **c.** $a^2 - 2ab + b^2 - a + b + 1$

LESSON 1-2 (pp. 12–18)
3. False **5.** ∃ a fraction which is not a rational number. **7.** p **9.** ∀, not $p(x, y)$ **11. a.** people x; a person y; x loves y **b.** ∃ a person x such that ∀ people y, x does not love y. **13. a.** ∃ a man who is not mortal. **b.** the given statement **15. a.** *∃ n in S such that $n \geq 11$.* **b.** the negation, 11 is in S and $11 = 11$ **17. a.** *∃ a real number x such that ∀ real numbers y, tan $x \neq y$.* **b.** The negation is true. For $x = \frac{\pi}{2}$, $\tan \frac{\pi}{2}$ is undefined so that ∀ real numbers y, $\tan \frac{\pi}{2} \neq y$.
19. The flaw is in going from the fourth to the fifth line. Since $x - y = 0$, you cannot divide both sides of the equation by $x - y$. **21. a.** True, every student in the sample is on at least one sports team. **b.** True, no student from the sample is in the Spanish club. **c.** True, every student in the sample is in the math club. **d.** False, each academic club has at least one member in the sample. **e.** False, for example, Dave is not in a foreign language club. **23.** Sample: $\sqrt{(-1)^2} = 1 \neq -1$ **25. a.** 4 **b.** -3 **c.** 0

LESSON 1-3 (pp. 19–24)
1. L is greater than 12 or L equals 12. **3.** False **5. a. See below.** **b.** $p \equiv p$ or (p and q) **7.** True **9. See next column.** **11.** $\sim(3 < x \leq 4) \equiv \sim(3 < x \text{ and } x \leq 4) \equiv \sim(3 < x) \text{ or } \sim(x \leq 4) \equiv 3 \geq x \text{ or } x > 4$ **13.** sample: $x > 5$ and $x \leq 11$ **15.** inclusive or **17. a. See next column.** **b. See next column.** **19. a.** *∀ positive real numbers x, $\log_{10} x \neq 0$.* **b.** the statement **21.** $\sin(\frac{7\pi}{12}) = \sin(\frac{\pi}{3} + \frac{\pi}{4}) = \sin\frac{\pi}{3}\cos\frac{\pi}{4} + \cos\frac{\pi}{3}\sin\frac{\pi}{4} = \frac{\sqrt{3}}{2} \bullet \frac{\sqrt{2}}{2} + \frac{1}{2} \bullet \frac{\sqrt{2}}{2} = \frac{\sqrt{6} + \sqrt{2}}{4}$
23. See next column.

5. a.

p	q	p and q	p or (p and q)
T	T	T	T
T	F	F	T
F	T	F	F
F	F	F	F

same truth values (p and p or (p and q))

9.

p	q	p or q	not (p or q)	not p	not q	(not p) and (not q)
T	T	T	F	F	F	F
T	F	T	F	F	T	F
F	T	T	F	T	F	F
F	F	F	T	T	T	T

same truth values (not (p or q) and (not p) and (not q))

17. a.

p	q	p xor q
T	T	F
T	F	T
F	T	T
F	F	F

b.

p xor q	p or q	p and q	not (p and q)	(p or q) and (not(p and q))
F	T	T	F	F
T	T	F	T	T
T	T	F	T	T
F	F	F	T	F

same truth values (p xor q and (p or q) and (not(p and q)))

23.

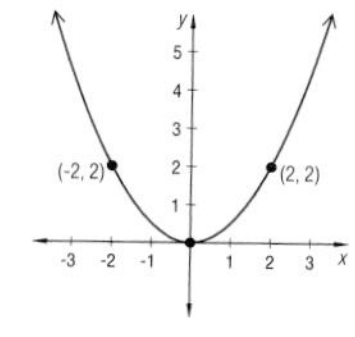

Lesson 1-4 (pp. 25–31)
3. (p or (not q)) and r **5.** ((not p) or q) and (not ((not q) and r)) **7. a.** 11 cents **b.** 7 cents **c.** the network of Question 6 **9. See below.** **11.** *There is a symphony orchestra with a full-time banjo player.* **13.** False; sample counterexample: let n be 5.

9.

p	q	not q	p and (not q)
T	T	F	F
T	F	T	T
F	T	F	F
F	F	T	F

LESSON 1-5 (pp. 32–40)
1. a. antecedent, hypothesis: $x > 1$; conclusion, consequent: $2x^2 + 3x^3 > 1$ **b.** True **3.** False **5.** True **7. a.** *If $m = 0$, then the graph of $y = mx + b$ is not an oblique line.* **b.** True **9.** converse: *If it will rain tomorrow, then it will rain today.*; inverse: *If it does not rain today, then it will not rain tomorrow.* **11.** $\log_2 32 = 5$ **13. a.** Yes **b.** No **c.** Yes **d.** No **15.** *If one can, then one does.* **17. See below.** **19.** *If a satellite is in orbit, then it is at a height of at least 200 miles above the earth.* **21.** *If one is elected to the honor society, then one must have a GPA of at least 3.5.* **23.** *∀ lines L, if L is vertical, then the slope of L is undefined.* **25. a.** 1 **b.** 1 **c.** 0

17.

p	q	$p \Rightarrow q$	$q \Rightarrow p$
T	T	T	T
T	F	F	T
F	T	T	F
F	F	T	T

equivalent ($p \Rightarrow q$ and $q \Rightarrow p$)

LESSON 1-6 (pp. 41–48)
1. a. *For all integers n, if n is divisible by 3, then its square is divisible by 9. 10 is divisible by 3.* **b.** *10^2 is divisible by 9.* **c.** integers n, if $p(n)$, then $q(n)$; $p(c)$, for a particular c; $\therefore q(c)$. **d.** No **e.** Yes **3.** (c) **5.** valid; Law of Transitivity **7.** Laws of Indirect Reasoning and Transitivity **9. a.** $p \Rightarrow q$; $\sim q$; $\therefore \sim p$. **b. See below.** **11. a.** If p, then q. If q, then r. ∴ if p, then r **b.** Yes, from the Law of Transitivity. **13. a.** Law of Indirect Reasoning **b.** Yes **15.** -3 and -1 are not positive real numbers, so the universal statement does not apply. **17.** (c) **19. a.** *∃ a real number y, such that $y^2 + 3 < 3$.* **b.** the statement **21.** Yes **23. a.** center: (3,-5); radius: 7 **b. See next column.**

9. b. Prove using a truth table. Must show that $((p \Rightarrow q) \text{ and } \sim q) \Rightarrow \sim p$ is always true.

p	q	r	$p \Rightarrow q$	$-q$	$((p \Rightarrow q)$ and $-q)$	$-p$	$((p \Rightarrow q)$ and $-q) \Rightarrow -p$
T	T	T	T	F	F	F	T
T	T	F	T	F	F	F	T
T	F	T	F	T	F	F	T
T	F	F	F	T	F	F	T
F	T	T	T	F	F	T	T
F	T	F	T	F	F	T	T
F	F	T	T	T	T	T	T
F	F	F	T	T	T	T	T

23. b.

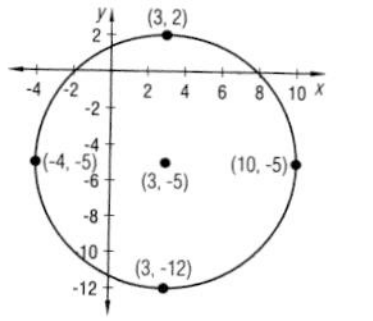

LESSON 1-7 (pp. 49–55)
1. even; $270 = 2(135)$ **3.** odd; $-59 = 2(-30) + 1$ **5.** Since b is even, let $b = 2m$. $t = a \bullet 2m = 2 \bullet (am)$. **7.** $6r + 4s^2 + 3 = 2(3r + 2s^2 + 1) + 1$; $3r + 2s^2 + 1$ is an integer by closure properties. Hence $6r + 4s^2 + 3$ is odd by definition.
9. a. Addition Property of Equality **b.** Distributive Property **11. a.** Sample: Suppose c and d are any even integers. **b.** $c - d$ is an even integer. **13.** counterexample: Let $r = 4$ and $s = 5$. $r \bullet s = 4 \bullet 5 = 20$ is an even integer. But s is not an even integer. **15.** Suppose m and n are any odd integers. There exists integers r and s such that $m = 2r + 1$ and $n = 2s + 1$ according to the definition of an odd integer. Then $m \bullet n = (2r + 1)(2s + 1) = 4rs + 2r + 2s + 1 = 2(2rs + r + s) + 1$. Because $(2rs + r + s)$ is an integer, by closure properties, $m \bullet n$ is an odd integer by definition.
17. a. Let p: *Devin is a boy.* Let q: *Devin plays baseball.* Let r: *Devin is a pitcher.* $p \Rightarrow q$; $q \Rightarrow r$; $\sim r$; $\therefore \sim p$ **b.** The argument correctly uses the Law of Indirect Reasoning and the Law of Transitivity. **19.** (b) **21. a.** $f(-2) = 3(-2)^2 - 5(-2) = 12 + 10 = 22$ **b.** $f(y + 2) = 3(y + 2)^2 - 5(y + 2) = 3y^2 + 12y + 12 - 5y - 10 = 3y^2 + 7y + 2$ **c.** $f(m + n) = 3(m + n)^2 - 5(m + n) = 3m^2 + 6mn + 3n^2 - 5m - 5n$
23. *∀ integers n, if n is even, then $(-1)^n = 1$.*
25. 3,628,800

LESSON 1-8 (pp. 56–62)
1. True **3.** False **5. a.** improper induction **b.** invalid **7. a.** Law of Indirect Reasoning **b.** valid **9. a.** improper induction **b.** invalid **11. a.** Let p: *not home*; and q: *answering machine is on.* $p \Rightarrow q$; q; $\therefore p$ **b.** invalid; converse error **13. a.** Let $q(p)$: *p is President of the U.S.* Let $r(p)$: *p is at least 35 years old.* Let Q be Queen Elizabeth. ∀ p, $q(p) \Rightarrow r(p)$; $r(Q)$; $\therefore q(Q)$ **b.** Yes, No **c.** invalid; converse error **15.** Let $p(x)$: *x is a real number.* Let $q(x)$: *$x^2 \geq 0$.* Let $r(x)$: *x is an imaginary number.* Let c: $x = 2i$. ∀ x, $p(x) \Rightarrow q(x)$; ∀ x, $r(x) \Rightarrow \sim q(x)$; $r(c)$; $\therefore \sim p(c)$; valid. ($r(c) \Rightarrow \sim q(c)$ by the Law of Detachment, and $\sim q(c) \Rightarrow \sim p(c)$ by the Law of Indirect Reasoning.) **17. a. See below.**
b. inverse error

17. a.

p	q	$p \Rightarrow q$	$-p$	$(p \Rightarrow q)$ and $-p$	$-q$	$((p \Rightarrow q)$ and $-p) \Rightarrow -q$
T	T	T	F	F	F	T
T	F	F	F	F	T	T
F	T	T	T	T	F	F
F	F	T	T	T	T	T

19. Suppose that m is any even integer and n is any odd integer. According to the definitions of even and odd, there exists integers r and s, such that $m = 2r$ and $n = 2s + 1$. Then $m \bullet n = 2r(2s + 1) = 4rs + 2r = 2(2rs + r)$. By closure properties, $(2rs + r)$ is an integer, and it follows by definition that $m \bullet n$ is even. **21.** Sample: Let p: $2 < 1$. Let q: $3 < 2$. Both statements are false. (p and q) is false, but ($p \Rightarrow q$) is true. **23.** *Vanna White is the hostess, and the show is not Wheel of Fortune.* **25. a.** ∀ real numbers x and a, $x^2 - a^2 = (x - a)(x + a)$ **b.i.** Let $a = 4$. $x^2 - 16 = (x - 4)(x + 4)$ **ii.** Let $x = 3y^2$ and $a = z$. $9y^4 - z^2 = (3y^2 - z)(3y^2 + z)$ **c.** Let $x = 48$ and $a = 52$. $48^2 - 52^2 = (48 - 52)(48 + 52) = -4 \bullet 100 = -400$

LESSON 1-9 (pp. 63–69)
1. Inductive reasoning makes generalizations based on the evidence of many examples. Deductive reasoning uses valid forms of argument, and takes accepted definitions or known theorems as premises. **3. a.** No operation starts when operation buttons are pushed. **b.** No, the POWER Button may be off, or the TIMER Button may be on. **5.** Sample: A car makes a squeaking noise under the hood when the A/C is turned on. The mechanic uses diagnostic reasoning to conclude that there is a problem with the A/C drive belt.
7. $\frac{1}{2^n}$ **9. a.** probabilistic reasoning **b.** No, they could test more bananas. **11.** invalid; inverse error **13.** Inverse: *If the temperature inside a refrigerator is not above 40° F, then the cooling system is not activated.* Converse: *If the cooling system is activated, then the temperature inside a refrigerator is above 40° F.* Contrapositive: *If the cooling system is not activated, then the temperature inside a refrigerator is not above 40° F.* **15.** *∀ real numbers x and y, $xy \neq 0$.*
17. Suppose that m is any odd integer. According to the definition of odd, there exists an integer r such that $m = 2r + 1$. Then $m^2 = (2r + 1)^2 = 4r^2 + 4r + 1 = 2(2r^2 + 2r) + 1$. $(2r^2 + 2r)$ is an integer by closure properties, so m^2 is an odd integer by definition. **19. a.** -84 **b.** $-4h^2 - 19h - 21$ **c.** $-4t^2 - 11t - 6$

CHAPTER 1 PROGRESS SELF-TEST (pp. 72–73)
1. (d) **2.** (c) **3.** True **4.** (d) and (e) **5.** True **6.** *∃ a real number y, such that $0 + y \neq y$.* **7.** $\sqrt{(-7c)^2} = |-7c|$ (which could be simplified to $|-7|\,|c| = 7\,|c|$) **8.** *x is greater than or equal to -8 and less than 12.* **9.** *The bald eagle is not the national bird or "The Star-Spangled Banner" is not the national anthem.* **10.** IF (C <> 2) AND (C <> 3) THEN 10 **11.** *If a person can be admitted to an R-rated movie, then they are at least 17 years old.* **12.** if s < 4 and |s| ≥ 4 (this happens when s ≤ -4) **13.** (c) **14.** (b) **15.** *If two lines cut by a transversal have corresponding angles with the same measure, then they are parallel. If two lines are parallel, then when cut by a transversal, corresponding angles have the same measure.* **16.** (b); the argument form is *modus tollens*, so the argument is valid. **17.** (e); this is an example of the inverse error, and is therefore invalid.
18. valid; for example, use the Law of Detachment followed by the Law of Indirect Reasoning **19.** Suppose that n is any odd integer. Thus there exists an integer r such that $n = 2r + 1$. Then $n^2 - 1 = (2r + 1)^2 - 1 = 4r^2 + 4r + 1 - 1 = 4r^2 + 4r = 2(2r^2 + 2r)$. Since $(2r^2 + 2r)$ is an integer, $n^2 - 1$ is an even integer by definition. **20.** (p or q) and $\sim q$

21.

p	q	p OR q	NOT q	(p OR q) AND (NOT q)
1	1	1	0	0
1	0	1	1	1
0	1	1	0	0
0	0	0	1	0

22.

p	q	p or q	not (p or q)
T	T	T	F
T	F	T	F
F	T	T	F
F	F	F	T

23. False, counterexample: let $x = 4$ and $y = 4$. $\sqrt{4 + 4} = \sqrt{8}$ and $\sqrt{4} + \sqrt{4} = 4$, but $\sqrt{8} \neq 4$, so $\sqrt{4 + 4} \neq \sqrt{4} + \sqrt{4}$.

The chart below keys the **Progress Self-Test** questions to the objectives in the **Chapter 1 Review** on pages 74–78. This will enable you to locate those **Chapter 1 Review** questions that correspond to questions you missed on the **Progress Self-Test.** The lesson where the material is covered is also indicated in the chart.

Question	**1, 2**	**3**	**4**	**5**	**6**	**7**	**8**
Objective	A	I	B	D	C	F	B
Lesson	1-1	1-1	1-1	1-1	1-2	1-1	1-3
Question	**9**	**10**	**11**	**12**	**13, 14**	**15**	**16**
Objective	C	K	B	E	A	B	G
Lesson	1-3	1-3	1-5	1-5	1-5	1-5	1-6
Question	**17**	**18**	**19**	**20**	**21**	**22**	**23**
Objective	G	J	H	L	L	M	D
Lesson	1-8	1-6	1-7	1-4	1-4	1-3	1-1

CHAPTER 1 REVIEW (pp. 74–78)
1. universal **3.** existential **5.** *∀ countries x, x has never landed people on Mars.* **7.** *∀ composite numbers x, ∃ a positive integer y, such that y ≠ x, y ≠ 1, but y is a factor of x.* **9.** (b) and (c) **11.** *If you pass a state's bar exam, then you may practice law in that state. If you practice law in a state, then you must pass that state's bar exam.* **13.** *If log x > 0, then x > 1.* **15.** *There is a British bobby that carries a gun.* **17.** *A person wants to travel from the U.S. to Europe and the person does not have to fly and the person does not have to travel by ship.* **19.** True **21.** True **23.** False, counterexample: rhombus **25. a.** *∃ a real number x, such that* $\sin^2 x + \cos^2 x \neq 1$. **b.** the statement **27.** No **29.** False **31.** False **33.** $[3x + 4]^3 = 27x^3 + 108x^2 + 144x + 64$ **35.** IV; invalid **37.** III; valid **39. a.** Yes **b.** No **41.** True **43.** All even numbers are real numbers. Law of Transitivity **45.** even; $a = 2k$ for some integer k, so $3a \bullet b = 3(2k)b = 2(3kb)$ **47.** using the same value k for expressing m and n **49.** Counterexample: Let $r = 3$ and $s = 1$. $r \bullet s = 3 \bullet 1 = 3 = 4k + 1$; $\therefore k = \frac{1}{2}$, but $\frac{1}{2}$ is not an integer. **51.** False, no camper participates in all the arts and crafts activities. **53.** False, Oscar does not participate in an arts and crafts activity. **55.** True, Kenji participates in entomology, but not hiking. **57.** IV; invalid **59.** II; valid **61. a.** A: True; B: True **b.** A: False; B: False **c.** A: True; B: False

63.

p	q	r	G (output)	NOT G	(NOT G) AND r
1	1	1	0	1	1
1	0	1	0	1	1
0	1	1	1	0	0
0	0	1	1	0	0
1	1	0	0	1	0
1	0	0	0	1	0
0	1	0	1	0	0
0	0	0	1	0	0

65. a. not (p and not q) **b.** (not p) or q **c. See below.** **d.** $\sim(p \text{ and } \sim q) \equiv (\sim p) \text{ or } \sim(\sim q) \equiv (\sim p) \text{ or } q$ **67. See below.** **69. See below.**

65. c.

p	q	NOT p	NOT q	p AND (NOT q)	NOT (p AND (NOT q))	(NOT p) OR q
1	0	0	1	1	0	0
1	1	0	0	0	1	1
0	0	1	1	0	1	1
0	1	1	0	0	1	1

same truth values

67.

p	q	$p \Rightarrow q$
T	F	F
T	T	T
F	F	T
F	T	T

69.

p	q	r	q or r	p and (q or r)
T	T	T	T	T
T	T	F	T	T
T	F	T	T	T
T	F	F	F	F
F	T	T	T	F
F	T	F	T	F
F	F	T	T	F
F	F	F	F	F

LESSON 2-1 (pp. 80–86)
1. table, graph, rule, arrow diagram **3.** 62.5 hours **5.** $\{z: z \neq 2, z \neq -2\}$ **7.** $f(x_1) < 0$ **9. a.** (iii) **b.** (i) $y = \frac{1}{3}\sin(x)$ **11.** Yes, this set can be put into a 1-1 correspondence with a subset of the set of integers. **13. a.(i)** Answers may vary. **(ii) See below.** **(iii) See next column.** **b.** 2 **15. a.,b. See next column.** **c.** The resulting graph shows over 3 full periods of each function.

13. a.ii.

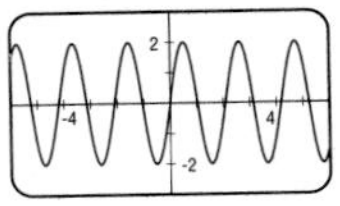

13. a.iii.

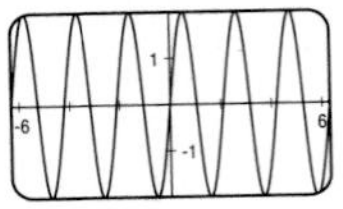

15. a.,b.

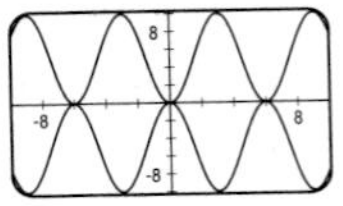

17. a., b. See below. **c.** The resulting graph looks like a segment. **19.** the set of real numbers **21. a.** \$8.25 **b.** not possible **c.** \$11.25 **d.** \$1.50 **e.** No, each element in C corresponds to more than one element in T. **f.** Yes, each element in T corresponds to exactly one element in C. **g.i)** time **ii)** cost **iii)** $\{t: 0 < t \leq 24\}$ **23.** $\frac{2}{3}$ **25.** 4, 3.2, 2.56, 2.048, 1.6384

17. a., b.

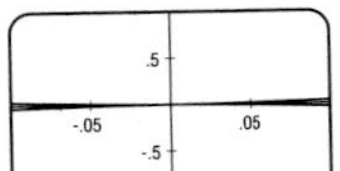

LESSON 2-2 (pp. 87–92)
1. a. the set of positive even integers **b.** the set of positive real numbers **c.** the set of real numbers **3.** $g(z) \leq M$ **5.** 0 **7. a.** $S(r) = 2\pi r^2 + \frac{2000}{r}$ **b.** $r \approx 5.4$ cm; $h \approx 10.9$ cm **9.** minimum value: -9.5 **See below.** **11. a.** $l = \frac{2000}{w}$ **b.** $P(w) = 2w + \frac{4000}{w}$ **c.** $P(40) = 180$; $P(45) = 178.9$; $P(50) = 180$ **d.** (44.7,178.9) **e.** 178.9 m **13. a.** domain: $\{x: -5 \leq x \leq 6\}$; range: $\{y: -3 \leq y \leq 3\}$ **b.** No, the domain cannot be put into a 1-1 correspondence with a subset of the set of integers. **c.** minimum: -3; maximum: 3 **d.** $x = 0$ and $x = 3$ **e.** $-5 \leq x < -4$ and $-1 < x < 4$ **15. a.** $y \geq -1$ **b.** $y > 1$ **c.** {real numbers y: $y > 1$} **17. a.** *n is an integer, and f(n) is not an odd integer.* **b.** the original conditional **c.** Suppose n is any integer. Then $f(n) = 2n + 3 = 2n + 2 + 1 = 2(n + 1) + 1$. By the closure property of addition, $(n + 1)$ is an integer; and $f(n)$ is odd, according to the definition of odd.

9.

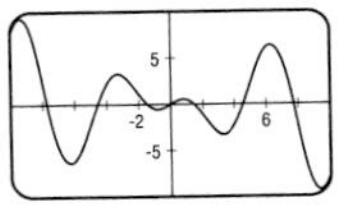

LESSON 2-3 (pp. 93–99)
1. a. $x = 1975$ **b.** $1979 \leq x \leq 1983$ **c.** 78.90 **d.** 70.50 **3.** y is any number from -2 to 2. **5. a.** $0 \leq x \leq 3$ **b.** $x \leq 0$ **c.** $x \geq 3$ **d.** relative maximum value: 3; relative minimum values: 0 and 3 **7.** relative maximum **9.** increasing: $0 \leq x \leq 2.0$ and $4.9 \leq x \leq 8$; decreasing: $2 \leq x \leq 4.9$ and $8 \leq x \leq 10$. **11. a.** $1954 \leq x \leq 1957$, $1958 \leq x \leq 1970$, $1975 \leq x \leq 1980$, $1986 \leq x \leq 1988$ **b.** $1972 \leq x \leq 1975$, $1980 \leq x \leq 1983$, $1984 \leq x \leq 1986$ **c.** 41.65, 62.07, 62.42, 64.76, 65.81 **d.** 38.81, 61.29, 59.86, 61.19, 64.25 **13. a.** $t \geq 2.55$ sec **b.** At $t = 2.55$ sec, the object reaches maximum height and starts to descend. **15.** 1,000,000,000,000,000 or 1×10^{15} **17. a. See next column.** **b.** $\{y: -\frac{1}{12} \leq y \leq 52\}$ **19.** sample: (X > = Y) OR (Y > = Z) **21.** $(ab)^5$

17. a.

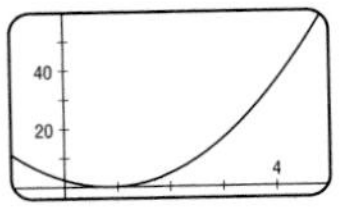

LESSON 2-4 (pp. 100–107)
1. the set of all positive (or nonnegative) integers **3. a.** none of these **b.** starting with $n = 2$: 2, $\frac{3}{2}$, $\frac{4}{3}$, $\frac{5}{4}$, $\frac{6}{5}$ **c.** decreasing **d.** 1 **5. a.** alternating **b.** -2, 1, $-\frac{2}{3}$, $\frac{1}{2}$, $-\frac{2}{5}$ **c.** neither **d.** 0 **7. a. See below.** **b.** the 101st term **9. a.** recursive **b. See below.** **c.** A **11.** sample: $s_n = \frac{2n - 1}{n}$ **13. a.** domain: {integers n: $n > 0$}; rule: $f(n) = \frac{1}{n^2}$ **b.** 0 **15. a.** $\{x: x \geq -1\}$ **b.** increasing: $0.23 \leq x \leq 5$; decreasing: $-1 \leq x \leq 0.23$ **c.** relative minimum: ≈ -1.06 **17. a. See below.** **b.** The period decreases. **c.** They do not change. **19.** Suppose that n is any odd integer. According to the definition of odd, there exists an integer k such that $n = 2k + 1$. Then $n^2 = (2k + 1)^2 = 4k^2 + 4k + 1 = 2(2k^2 + 2k) + 1$. By closure properties, $(2k^2 + 2k)$ is an integer and n^2 is odd, by definition of odd. **21.** 23°, 67°, 90°

7. a.

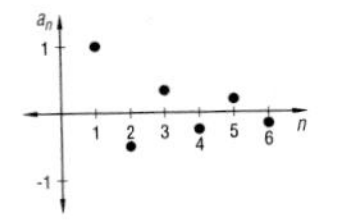

9. b.

17. a.

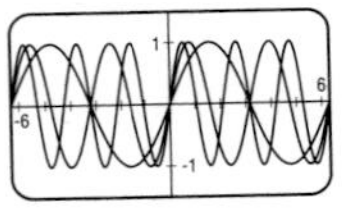

LESSON 2-5 (pp. 108–113)
1. Yes, the domain is a discrete set, the set of positive integers.

3. a.

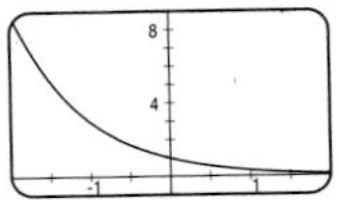

3. b. As the values of x increase without bound, the values of 3^{-x} become closer and closer to 0. **c.** ∞ **d.** $y = 0$ **5. a.** The limit does not exist. **b.** even; $\forall x \cos(-x) = \cos(x)$ **7.** neither **9. See below.** as $x \to \infty$, $y \to \infty$; as $x \to -\infty$, $y \to \infty$ **11. a. See below.** **b.** 5 **c.** $y = 5$ **d.** $x > 10$ **13. a. See below.** **b.** $y = 2.71828$ or $y = 2.71829$ or $y = e$ **c. See below.** **15.** geometric; $-\infty$ **17.** $\{y: y \leq -\frac{2}{3}\}$ **19.** 2 **21.** $81^{-1/4}$

9.

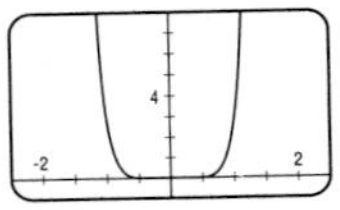

11. a.

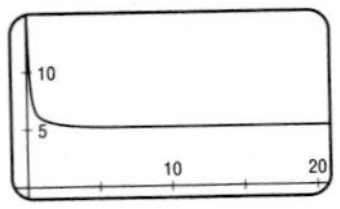

13. a.

x	10	100	1000
$(1 + \frac{1}{x})^x$	2.59374	2.70481	2.71692

x	10,000	100,000	1,000,000
$(1 + \frac{1}{x})^x$	2.71815	2.71827	2.71828

13. c.

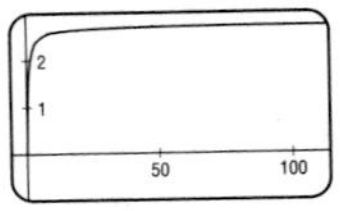

LESSON 2-6 (pp. 114–121)
1. (b) **3.** $y = 3^x$, $\forall$ positive integers x **5.** \$46,966.66 **7.** Usual domain: the set of real numbers; range: the set of positive real numbers; maxima or minima: none; increasing or decreasing: decreasing over its entire domain; end behavior: $\lim_{x \to \infty} b^x = 0$; $\lim_{x \to -\infty} b^x = \infty$; graph: **See below.**; model: decay; special properties: values of the function are related by the laws of exponents.

7.

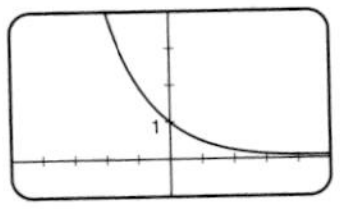

9. a. See below. **b.** True **11.** about 0.1 mg **13. a. See below.** **b.** increasing: $-4 \leq x \leq 0$; decreasing: $0 \leq x \leq 4$ **c.** relative maximum: $\frac{1}{\sqrt{2\pi}}$ **d.** as $x \to \infty$, $y \to 0$; as $x \to \infty$, $y \to 0$ **e.** $\{y: 0 < y \leq 0.4\}$ **15.** 3 **17.** $\frac{x^3}{y^2}$ **19.** $1 + 4x^2$ **21. a.** $\frac{1}{2}$ **b.** $\frac{\sqrt{3}}{2}$ **c.** $\frac{\sqrt{2}}{2}$

9. a.

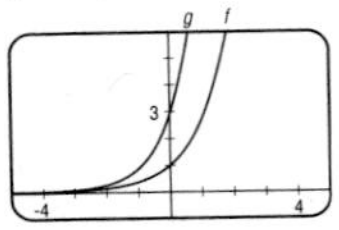

13. a.

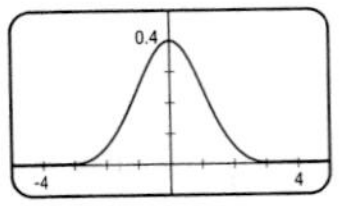

LESSON 2-7 (pp. 122–128)
1. 4 **3.** -1 **5.** $\frac{2}{3}$; $\frac{4}{9}$; 2 **7.** Usual domain: set of positive real numbers; range: set of real numbers; maxima or minima: none; increasing or decreasing: increasing over entire domain; end behavior: $\lim_{x \to \infty} \log_b x = \infty$; graph: **See below.**; model: sound intensity and loudness, logarithmic scales; special properties: related to exponential functions, Change of Base Theorem. **9.** $x = 25$ **11.** $\frac{1}{4}\log_b n + \frac{3}{4}\log_b w$ **13.** ≈ 4.77 decibels **15.** Loudness in decibels is .10 $\log(I \bullet 10^{12})$. If sound has a negative loudness, then $I \bullet 10^{12} < 1$ so, $I < 10^{-12}$. But since the weakest intensity that can still be heard is $I = 10^{-12}$ watts/m², the sound cannot be heard. **17. a.** ≈ 1.845 **b.** ≈ 0.542 **c.** $\log_a b = \frac{1}{\log_b a}$ **19. a.** $r \approx -1.21 \times 10^{-4}$ **b.** $\approx 13{,}000$ years old **21.** True **23.** True **25. a.** $h_n = \frac{n + 1}{2n}$ **b.** $h_{10} = .55$; $h_{100} = .505$; $h_{1000} = .5005$ **c.** 0.5 **27. a.** x^2 is equal to 4 but x is not equal to 2. **b.** $x = -2$

7.

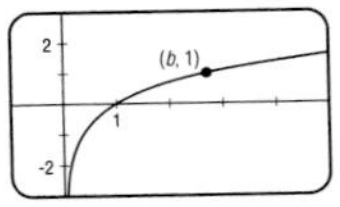

LESSON 2-8: (pp. 129–135)
1. a. $\frac{40}{41}$ **b.** $\frac{9}{41}$ **c.** $\frac{40}{41}$ **d.** $\frac{9}{41}$ **3.** $\cos x = \frac{1}{2}$; $\sin x = \frac{\sqrt{3}}{2}$ **5.** $\cos x = \frac{1}{2}$; $\sin x = -\frac{\sqrt{3}}{2}$ **7.** $\cos x = 1$; $\sin x = 0$

9. Consider a right triangle whose acute angle θ is at (0, 0) with one side along the x-axis and the other side intersecting the unit circle at (x, y). By the right triangle definition, $\cos\theta = \frac{\text{adjacent side}}{\text{hypotenuse}} = \frac{x}{1} = x$, which corresponds to the unit circle definition. **11.** when $0 \leq x \leq \frac{\pi}{2}$ and $\frac{3\pi}{2} \leq x \leq 2\pi$ **13.** $h(0) = 0$ ft; $h(0.25) = 3.5$ ft; $h(0.5) = 7.0$ ft; $h(0.75) = 3.5$ ft; $h(1) = 0$ ft **15. a. See below.** **b.** $\frac{4}{5}$ **17.** $-\frac{4}{5}$ **19.** ≈ 276 m **21. a.** $p(x) = .9^x$ **b.** $\approx 67\%$ **c.** ≈ 8.0 m **23.** $x = \sqrt{6}$ **25.** $\frac{q}{p}$

15. a.

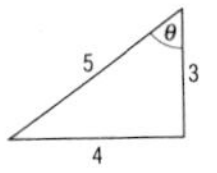

LESSON 2-9 (pp. 136–141)
1. $\aleph_0$ **3.** c **5.** Consider a countably infinite hotel in which all the rooms are filled. Thus, there are $\aleph_0$ people in the hotel. Suppose 100 new people want to check in. Move the person currently in room 1 to room 101, the person in room 2 to room 102, and in general move the person in room n to room $n + 100$. Then, put the new guests in rooms 1–100. Therefore, $\aleph_0 + 100 = \aleph_0$. **7.** Consider the set of positive odd integers and the set of positive even integers. Both sets have a cardinality of $\aleph_0$. Combining the two sets into one set forms the set of positive integers which also has a cardinality of $\aleph_0$. Therefore, $\aleph_0 + \aleph_0 = \aleph_0$. **9.** Take any two line segments, such as $\overline{AB}$ and $\overline{CD}$ below. Extend a line through A and C, then another through B and D. Since $\overline{AB}$ and $\overline{CD}$ have different lengths, these lines intersect at some point P. To establish a 1–1 correspondence, pair every point E on $\overline{AB}$ with the intersection of $\overleftrightarrow{PE}$ and $\overline{CD}$. **See below.** **11.** Time has cardinality c, and $c + 10{,}000 = c$. **13.** 10.85 yards **15.** False **17.** False **19.** Let the first term of the arithmetic sequence, a_1, be any odd integer. There exists an integer r such that $a_1 = 2r + 1$ by definition of odd. By definition of even, there exists an integer s such that the constant difference, d, is $2s$. The explicit form of an arithmetic sequence is $a_n = a_1 + (n - 1)d$. By substitution, $a_n = (2r + 1) + (n - 1)2s$. Then, $a_n = 2(r + sn - s) + 1$. By closure properties, $r + sn - s$ is an integer; hence, every term a_n of the sequence is odd by definition. **21. a.** Yes, the domain is a discrete set, a set of 88 integers. **b.** 5 keys below "middle A"

9.

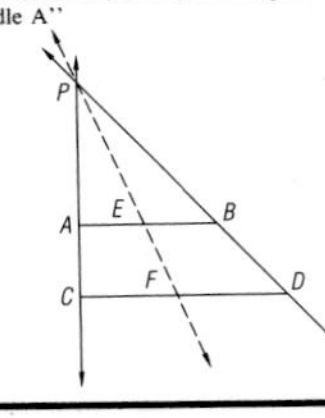

23. sample:

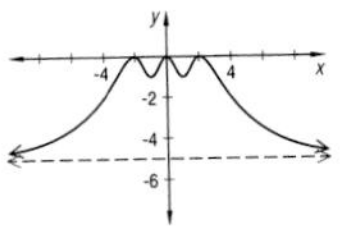

CHAPTER 2 Progress Self-Test (p. 145)
1. a. $z + 5 > 0 \Leftrightarrow z > -5$, so the domain is $\{z: z > -5\}$. **b.** w **c.** z **2.** $f(x) = 2x^2 + 4x + 3 = 2(x^2 + 2x) + 3 = 2(x^2 + 2x + 1) + 3 - 2 = 2(x + 1)^2 + 1$. The vertex of the graph of this parabola is (-1, 1), so the minimum value is 1 and the range is $\{y: y \geq 1\}$. **3. a.** increasing: $x \leq -1$ and $x \geq 1$; decreasing: $-1 \leq x \leq 1$ **b.** relative maximum: 1 occurs at (-1, 1); relative minimum: -1 occurs at (1, -1) **c.** odd **d.** ∞ **e.** $x \leq -2$ and $0 \leq x \leq 2$ **4.** $\pi \leq x \leq 2\pi$ **5. a. See below.** **b.** increasing **c.** 0 **6. a.** $l = \frac{300}{w}$ **b.** $f(w) = 2w + \frac{300}{w}$ **c. See below.** **d.** f has a relative minimum at $w \approx 12.25$, so the width is 12.25 ft, and the length is $\frac{300}{12.25} \approx 24.50$ ft. **7. a.** $\log_5 \frac{a}{b^2} = \log_5 a - \log_5 b^2 = \log_5 a - 2\log_5 b$ **b.** $\log 5\, x = -2 \Leftrightarrow x = 5^{-2} = \frac{1}{5^2} = \frac{1}{25}$ **c.** all real numbers

5. a.

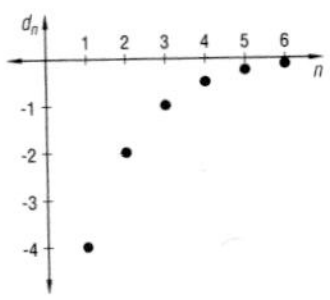

6. c.

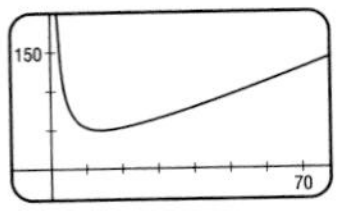

8. Usual domain: set of real numbers; Range: $-2 \leq y \leq 2$; Relative maximum: 2 occurs at $x = \frac{3\pi}{2} + 2\pi n$, $\forall$ integers n; Relative minimum: -2 occurs at $x = \frac{\pi}{2} + 2\pi n$, $\forall$ integers n; Increasing: $\frac{\pi}{2} + 2n\pi \leq x \leq \frac{3\pi}{2} + 2n\pi$, $\forall$ integers n; Decreasing: $\frac{\pi}{2} + 2n\pi \leq x \leq \frac{\pi}{2} + 2n\pi$, $\forall$ integers n; End behavior: the limit does not exist because f oscillates between -2 and 2 for arbitrarily large x and arbitrarily small x. Graph: **See next page.** Model: sound waves, phenomena based on rotations; Special Properties: odd function, periodic function with period 2π

8. 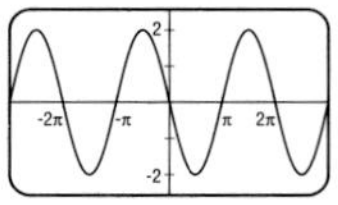

9. Find t such that $A(t) = 0.18$. $A(t) = 2e^{-.000124t} = .18 \Leftrightarrow e^{-.000124t} = .09 \Leftrightarrow -.000124t = \ln .09 \Leftrightarrow -.000124t \approx -2.41$. So $t \approx 19{,}000$ years old.

The chart below keys the **Progress Self-Test** questions to the objectives in the **Chapter 2 Review** on pages 146–150. This will enable you to locate those **Chapter 2 Review** questions that correspond to questions you missed on the **Progress Self-Test.** The lesson where the material is covered is also indicated in the chart.

Question	**1**	**2**	**3**	**3d**	**4**	**5**
Objective	C	C	G	D	A	G
Lesson	2-1	2-2	2-5	2-5	2-8	2-4
Question	**5c**	**6**	**7**	**8**	**9**	
Objective	D	F	B	G	E	
Lesson	2-4	2-2	2-7	2-8	2-6	

CHAPTER 2 REVIEW (pp. 146–150)
1. a. increasing: $1900 \le x \le 1930$, $1950 \le x \le 1970$; decreasing: $1930 \le x \le 1950$, $1970 \le x \le 1987$ **b.** No, E is not increasing over the entire interval, because E is decreasing over the interval $1930 \le x \le 1940$. **3. a.** 0, 1, 0, -1, 0, 1, 0, -1, 0 **b.** $3 \le n \le 5$ **c.** $1 \le n \le 3$ **d.** $n = 1, n = 5, n = 9$ **e.** $n = 3, n = 7, n = 11$ **5. a.** arithmetic **b.** decreasing **7. a.** harmonic **b.** decreasing **9.** $2^x = 8$, $x = 3$ **11.** $b^2 = 9$, $b = 3$ **13.** $3^{-2} = \frac{2z}{6}$, $z = \frac{5}{18}$ **15.** $\frac{1}{2}\log_{10} N + \log_{10} M - \frac{3}{2}\log_{10} P$ **17.** Yes, each element in R corresponds to exactly one element in S. **19.** the set of real numbers **21.** $\{z: z \ge 7\}$ **23.** minimum value: 1; maximum value: none; range: the set of positive odd integers **25. a.** $\{y: -17 \le y \le -\frac{19}{4}\}$ **b.** $\{y: y \le -\frac{19}{4}\}$ **27.** As $x \to \infty$, $f(x) \to \infty$; as $x \to -\infty$, $f(x) \to 0$. **29. a.i)** $n \ge 53$ **ii)** $n \ge 503$ **b.** $-\infty$ **31.** 0 **33. a.** $\lim_{y\to\infty} f(y) = -4$; $\lim_{y\to-\infty} f(y) = -4$ **b.** $y = -4$ **35.** -10 **37. a.** Yes **b.** $2^n(.1)$ **c.** 12.8 mm **d.** 42 **39. a.** ≈ 11.6 **b.** 1.0×10^{-7} moles/liter **41.** 3 planes **43. a.** $p - 5$ **b.** $f(p) = 150p - 10p^2 - 500$ **c.** price: \$7.50; profit: \$62.50 **45. a.** increasing: $-4 \le x \le -2$, $0 \le x \le 3$; decreasing: $x \le -4$, $-2 \le x \le 0$, $x \ge 3$ **b.** relative maxima: 4 occurs at $x = -2$, and .5 occurs at $x = 3$; relative minima: -3 occurs at $x = -4$, and -4.5 occurs at $x = 0$ **c.** neither **47. a.** increasing: $x \le -4$, $-3 \le x \le -2$, $2 \le x \le 3$, $x \ge 4$; decreasing: $-4 \le x \le -3$, $-1 \le x \le 1$, $3 \le x \le 4$ **b.** relative maxima: 3 occurs at $x = -4$, -1.5 occurs at $x = 3$, 2 for all x such that $-2 \le x \le -1$, and -2 for all x such that $1 \le x \le 2$; relative minima: 1.5 occurs at $x = -3$, -3 occurs at $x = 4$, 2 for all x such that $-2 \le x \le -1$, and -2 for all x such that $1 \le x \le 2$ **c.** odd **49. a.** $x = -6, x = -3, x = -1, x = 2, x = 4$ **b.** $-6 < x < -3$, $-1 < x < 2$, $x > 4$ **c.** $x < -6$, $-3 < x < -1$, $2 < x < 4$ **d.** $x = -7, x = -2.5, x = -1.5$ **e.** $x < -7$, $-2.5 < x < -1.5$

51. sample: 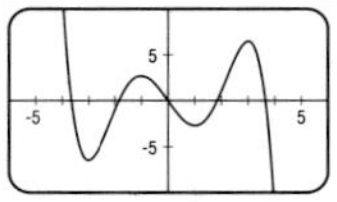

53. a. 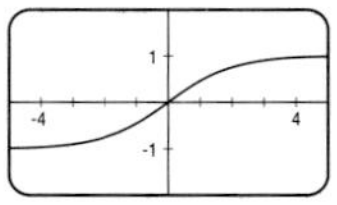

b. increasing: for all x; decreasing: nowhere **c.** relative maxima: none; relative minima: none **d.** $\lim_{x\to\infty} f(x) = 1$; $\lim_{x\to\infty} f(x) = -1$ **e.** $\{y: -1 < y < 1\}$ **f.** odd **55.** Usual domain: set of real numbers; range: $1 \le y \le 5$; maximum: 5 ; minimum: 1 ; increasing: when $2n\pi \le x \le (2n + 1)\pi$ $\forall$ integers n; decreasing: when $(2n - 1)\pi \le x \le 2n\pi$ $\forall$ integers n; end behavior: the limit does not exist ; graph: **See below.** ; model: sound waves, phenomena based on rotations; special properties: periodic, even function.

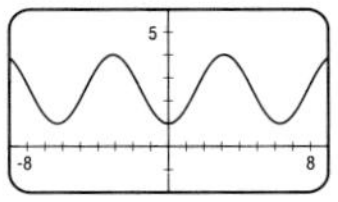

LESSON 3-1 (pp. 152–159)
1. $\forall$ real numbers x, $2x + 5 = 3x - 2$ (given) $\Leftrightarrow 5 = x - 2$ (addition property of equality) $\Leftrightarrow 7 = x$ (addition property of equality) $\therefore \forall$ real numbers x, $2x + 5 = 3x - 2 \Leftrightarrow x = 7$, by the transitive property. **3. a.** Yes **b.** No **c.** Yes **d.** No **5.** The function $f: x \to x^3$ is a 1–1 function. **7. a.** $f: x \to x^2$ **b.** $\sqrt{2x^2 + 3x + 1} = x - 1 \Rightarrow 2x^2 + 3x + 1 = (x - 1)^2 \Leftrightarrow 2x^2 + 3x + 1 = x^2 - 2x + 1 \Leftrightarrow x^2 + 5x = 0 \Leftrightarrow x(x + 5) = 0 \Leftrightarrow x = 0$ or $x = -5$ **c.** Neither of these is an actual solution. They arose because the first step is not reversible, thus the solution sets of the original and the final equations are different. There is no solution to this equation. **9.** $y = -2$ **11.** 528 miles **13.** no real solution **15.** $x = 0$ or $x = 2$

17. a. 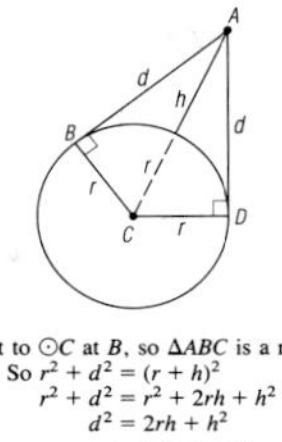

$\overline{AB}$ is tangent to $\odot C$ at B, so ΔABC is a right triangle.

$$\text{So } r^2 + d^2 = (r + h)^2$$
$$r^2 + d^2 = r^2 + 2rh + h^2$$
$$d^2 = 2rh + h^2$$
$$d = \sqrt{2rh + h^2}$$
$$d = \sqrt{7920h + h^2}$$

b. 168.92 miles **19. a.** The first step is nonreversible. **b.** $(2z + 1)^2 = z^2 \Leftrightarrow 4z^2 + 4z + 1 = z^2 \Leftrightarrow 3z^2 + 4z + 1 = 0 \Leftrightarrow (3z + 1)(z + 1) = 0 \Leftrightarrow 3z = -1$ or $z = -1 \Leftrightarrow z = -\frac{1}{3}$ or $z = -1$ **21.** no real solution **23.** The theorem was not proven for the general case, but only for two specific examples. This is the error of improper induction. **25. a.** 22 **b.** 485

LESSON 3-2 (pp. 160–165)
1. a. $-x^4 + 3$ **b.** -2398 **c.** $\frac{1}{2}(-2x + 3)^4$ **d.** 7320.5 **e.** No
3. $f \circ g(x) = 6x^3 + 1$; domain: the set of real numbers
5. a. the set of nonnegative real numbers

5. b. 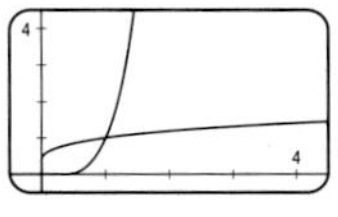

7. $f \circ g(x) = f(g(x)) = k(\frac{x}{k}) = x$; $g \circ f(x) = g(f(x)) = \frac{(kx)}{k} = x$; $\therefore f \circ g = g \circ f = I$ **9.** $x = \sqrt{7}$ or $x = -\sqrt{7}$ **11.** $x = \sqrt{2}$

13. k is a 1–1 function.

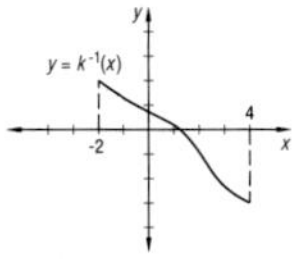

15. An increasing function is a 1–1 function, and all 1–1 functions have inverses. **17.** (b) **19. a.** Yes, $f(x)$ is increasing over its entire domain, the positive reals. **b.** $x = 1$ **21. a.** $\{x: x \ne 4\}$ **b.** x-intercept: $(\frac{7}{2}, 0)$; y-intercept: $(0, \frac{7}{4})$ **c.** $\lim_{x\to\infty} f(x) = 2$; $\lim_{x\to-\infty} f(x) = 2$ **23.** Suppose n is any odd integer. By definition there exists an integer k such that $n = 2k + 1$. Then $n^2 + 1 = (2k + 1)^2 + 1 = 4k^2 + 4k + 2 = 2(2k^2 + 2k + 1)$. Since $(2k^2 + 2k + 1)$ is an integer by closure properties, $n^2 + 1$ is even by definition.

LESSON 3-3 (pp. 166–171)
1. a. $C(700) = \$102{,}500$; $S(700) = \$169$; $R(700) = \$118{,}300$; $P(700) = \$15{,}800$ **b.** sample: 625 machines; $C(625) = \$93{,}125$; $R(625) = \$109{,}375$; $P(625) = \$16{,}250$ **3.** The domain of f is the set of real numbers. Let $h(x) = \cos x$, then, since $f = I \cdot h$, whenever $h(x) = \cos x = 1$, $f(x) = I(x) = x$, so the range of f is the set of real numbers. f crossses the x-axis whenever $I(x) = 0$ or $h(x) = \cos x = 0$. f is an odd function with relative maxima and minima that get larger as x is farther from the origin. **5. a. See below. b.** Sample: $f \cdot g$ is the line $y = 1$. $f - g$ is an odd, increasing function.

5. a. 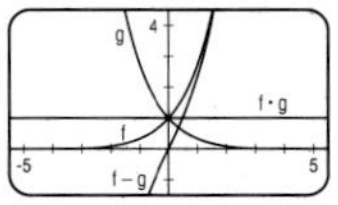

7. 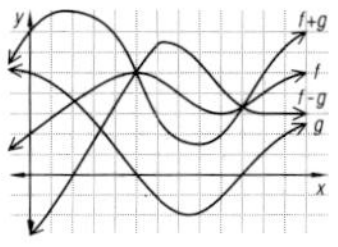

9. a. positive **b.** negative **c.** zero **d.** positive
11. Sample: the function $h(x) = x^4$ is an even function that can be expressed as the product of the two odd functions, $f(x) = x^3$ and $g(x) = x$.

13. a. $s(t) = \sin(.5t) - \frac{1}{60}t$ **b.** linear **c. See below.**
15. $x = 2$ **17.** $x \approx 3.616$ **19.** sample: **See below.**
21. $(x^2 + 4)(x^2 + 2)$

13. c. 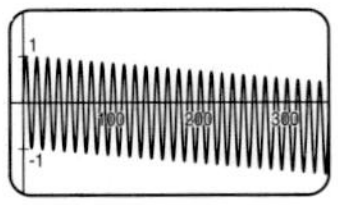

19. 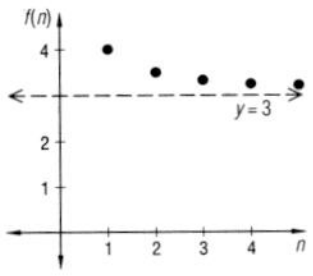

LESSON 3-4 (pp. 172–176)
1. a. 8^x **b.** $x = -\frac{1}{3}$ **3.** A zero of a function $h(x)$ is a number c such that $h(c) = 0$. **5.** $h(x) = 3x - 2^x - 1$ **7.** $t = 1$ and $t = -1$ **9.** $x = 2$ and $x = -(\frac{4}{3})^{1/3} \approx -1.1$ **11.** 99 °C **13.** $t = \pm\sqrt{\frac{1}{2}}$ **15.** $n = 1$ or $n = -\frac{59}{9}$ **17.** $\frac{2\pi}{3}, \frac{4\pi}{3}$

19. a. Yes **b.** $x = \frac{1}{2}$ or $x = -1$ **21. a. See below.**
b. The zeros of f and the zeros of g will also be zeros of $f \cdot g$. So the x-intercepts of $f \cdot g$ will be the x-intercepts of f and the x-intercepts of g. **23. a.** $0 \le t \le .625$ **b.** At $t = .625$ seconds the object reaches its maximum height, changes direction, and comes back down. **25. a.** $\frac{\pi}{6}, \frac{5\pi}{6}$ **b.** $\frac{\pi}{6}, \frac{5\pi}{6}, \frac{7\pi}{6}, \frac{11\pi}{6}$

21. a. 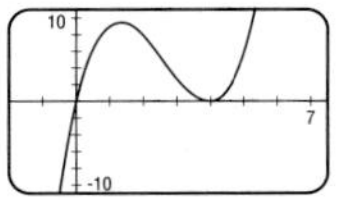

LESSON 3-5 (pp. 177–183)
1. a. $x \approx 174$ and $x \approx 1076$ **b.** They can be found where the graphs of R and C intersect and where the graph of P intersects the x-axis. **3. a.** Yes **b.** No **5. a.** -1 and 0 (or 0 and 1) **b.** 2 **c.** -2 and -1 (or 2 and 3) **7. a.** $x = .75$ **b.** $h(.625) = .37034 > 0$. Since $h(.75) < 0$ and $h(.5) > 0$, the solution is between .625 and .75 and is therefore closer to .75 than to .5. **9. a.** sample: $1 \le x \le 2$ **b.** sample: $1.5 \le x \le 1.6$ **11. a.** Yes **b.** No, g is not defined at $x = 0$. **13.** after approximately 1.9 minutes **15.** $n = 400$ **17. a.** *$\exists$ x such that 2 is a factor of x and 6 is not a factor of x.* **b.** sample: $x = 4$ **c.** True

19. 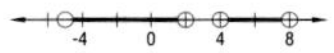

LESSON 3-6 (pp. 184–189)
1. (b) **3.** $y \le -4$ **5.** $0 \le t < 6.3$ **7.** $x > 5$ or $x < \frac{2}{3}$ **9.** $2 < x < 6$ **11.** Yes, if and only if the number is between -1 and 0. **13.** $0 < x < 1{,}000{,}000$ **15.** 10,000 to 13,000 years old **17.** $h(x)$ is continuous on the interval $2 \le x \le 3$. By the Intermediate Value Theorem, for every real number y_0 between $h(2)$ and $h(3)$ there exists at least one real number x_0 such that $f(x_0) = y_0$. Since $h(2) \approx -.693$ and $h(3) \approx 3.901$, then 0 is between $h(2)$ and $h(3)$, hence the function has a zero between 2 and 3. **19. a.** No **b.** Yes **c.** $x = 3.0016$
21. (b)

LESSON 3-7 (pp. 190–195)
1. a. $x < -2$ or $x > -\frac{5}{3}$ **b.** $-2 < x < -\frac{5}{3}$ **3. a.** $0 < x < 3$ or $x > 4$ **b.** $0 < x < 3$ or $x > 4$ **5.** $c < -4$ or $c > 1$ **7.** $g(x)$ is discontinuous at $x = 0$, so the theorem does not apply.
9. $\{x: -2 < x < -1$ or $x > 2\}$ **11.** $x < -1.9$ or $0 < x < 1.9$

13. a. 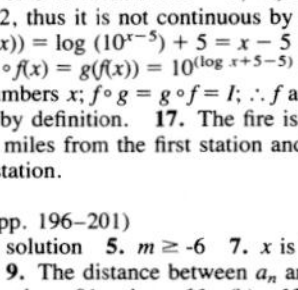

b. No, the graph of C is broken at $x = -2, -1, 0$, and 1 in the interval $-2 \le x \le 2$, thus it is not continuous by definition. **15.** $f \circ g(x) = f(g(x)) = \log(10^{x-5}) + 5 = x - 5 + 5 = x$, $\forall$ real numbers x; $g \circ f(x) = g(f(x)) = 10^{(\log x + 5 - 5)} = 10^{\log x} = x$, $\forall$ positive real numbers x; $f \circ g = g \circ f = I$; $\therefore f$ and g are inverse functions by definition. **17.** The fire is approximately 34 miles from the first station and 29 miles from the second station.

LESSON 3-8 (pp. 196–201)
1. $v = -3$ **3.** no solution **5.** $m \ge -6$ **7.** x is 7 units away from the origin. **9.** The distance between a_n and 6 on a number line is less than .01 units. **11.** (b) **13.** rejected lengths L satisfy $|L - 15.7| > 0.5$ **15.** $x > \frac{4}{5}$
17. $|x - 57| < 2$ **19. a.** $-12 \le x \le -6$ or $-4 \le x \le 2$

b. 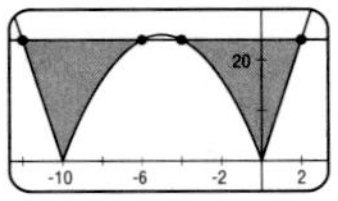

21. It was used in the proof of the theorem: $|x| > a$ if and only if $x < -a$ or $x > a$. **23. a.** Yes **b.** $x < -2$ or $x > 3$ **c. See below. 25. a. See below. b.** 3

23. c. 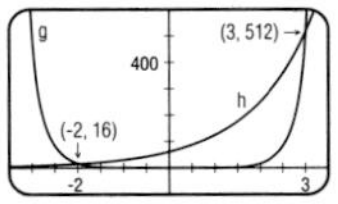

25. a. 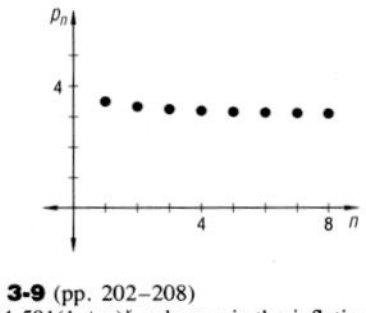

LESSON 3-9 (pp. 202–208)
1. a. $C = 4{,}581(1 + r)^x$, where r is the inflation rate. **b.** $C = 4{,}581(1 + r)^{x-1989}$, where r is the inflation rate. **c.** \$7,052 **d.** \$12,544 **3.** The graph of the ellipse is the graph of the unit circle stretched by a factor of 3 in the x-direction and by a factor of 4 in the y-direction. **5. a.** $y = \frac{1}{x - 3} + 2$ **b.** $y = \frac{5}{3x}$ **7. a.** 2 **b.** 2π **c.** π **d.** $y = 2\sin(x - \pi)$ **9. a.** $x = \frac{1}{8}$ **b.** $x = -\frac{7}{8}$ **c.** $x = -\frac{1}{2}$ **11.** Apply $S_{a,b}$, where $a = \frac{1}{\sqrt{2}}$ and $b = \frac{1}{\sqrt{2\pi}}$. **13.** $c = -\frac{\pi}{4}$ **15.** $f(x) = 370 + 1.30x$ **17.** $-\frac{\sqrt{7}}{4}$ **19. a.** $x = -\frac{33}{2}$ or $x = \frac{17}{2}$ **b.** $x < -\frac{33}{2}$ or $x > \frac{17}{2}$ **21. a.** $x < -5$ or $0 < x < 5$ **b.** The solutions to part **a** are the points of the graph which lie below the x-axis.

LESSON 3-10 (pp. 209–213)
1. $f(x) = \frac{9}{2}(\text{Fract } \frac{x}{3})^2$ **3. a.** h subtracts the greatest integer less than or equal to x from x. **b.** $h(x) = |x - \lceil x \rceil|$ **5. a.** $f(x) = \sqrt{16^2 - (x + 7)^2} + 9$ or $\sqrt{-x^2 - 14x + 207} + 9$ **b.** $g(x) = -\sqrt{16^2 - (x + 7)^2} + 9$ or $-\sqrt{-x^2 - 14x + 207} + 9$ **c.** Yes **7. See below. 9.** $f \circ g(x) = x^2 + 15x + 56$; $f \cdot g(x) = x^3 + 8x^2 + 7x$ **11.** $x = \ln 6 \approx 1.792$ **13.** $t = 12.5 \ln \frac{A}{1000}$

7. 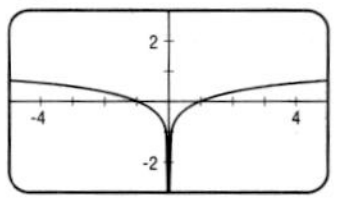

CHAPTER 3 PROGRESS SELF-TEST (p. 216)
1. $\sqrt{8x + 12} = x - 1 \Rightarrow 8x + 12 = x^2 - 2x + 1 \Leftrightarrow x^2 - 10x - 11 = 0 \Leftrightarrow (x - 11)(x + 1) = 0 \Leftrightarrow x = 11$ or $x = -1$. -1 does not check, and so the only solution is $x = 11$, which checks. **2.** $2\log x = \log(6x - 8) \Leftrightarrow \log x^2 = \log(6x - 8) \Leftrightarrow x^2 = 6x - 8 \Leftrightarrow x^2 - 6x + 8 = 0 \Leftrightarrow (x - 2)(x - 4) = 0 \Leftrightarrow x = 2$ or $x = 4$ **3.** $|4z - 3| \le 7 \Leftrightarrow -7 \le 4z - 3 \le 7 \Leftrightarrow -4 \le 4z \le 10 \Leftrightarrow -1 \le z \le \frac{5}{2}$ **4.** (c) **5. See below. 6.** $(a \cdot b)(x) = (5x + 4)(3x^2 + 9) = 15x^3 + 12x^2 + 45x + 36$ **7.** $f \circ g(x) = f(g(x)) = f(\sqrt{x}) = (\sqrt{x})^2 = x$; domain: $\{x: x \ge 0\}$
8. a. height of a person in inches **b.** the number of quarts of water in the body of a person who is x inches tall
9. $f(-1) = -18$, $f(0) = -7$, $f(1) = -10$, $f(2) = -15$, $f(3) = -10$, $f(4) = 17$, $f(5) = 78$, since $f(3) < 0 < f(4)$, (c) is the answer. **10.** Let $\sqrt[3]{x} = y$. $y^2 + 2y = 3 \Leftrightarrow y^2 + 2y - 3 = 0 \Leftrightarrow (y + 3)(y - 1) = 0 \Leftrightarrow y = -3$ or $y = 1$. So $\sqrt[3]{x} = -3$ or $\sqrt[3]{x} = 1 \Leftrightarrow x = -27$ or $x = 1$. **11.** True **12.** $x^2 - 8 > x + 22 \Leftrightarrow x^2 - x - 30 > 0 \Leftrightarrow (x - 6)(x + 5) > 0$. Let $g(x) = (x - 6)(x + 5)$. Using the Test Point Method, there are three intervals to check, $x < -5$, $-5 < x < 6$, and $x > 6$, since the zeros of $g(x)$ are -5 and 6. $g(-6) = 12 > 0$, so for $x < -5$, $g(x) > 0$. $g(0) = -30$, so for $-5 < x < 6$, $g(x) < 0$. $g(7) = 12$, so for $x > 6$, $g(x) > 0$. Therefore, the solution is $x < -5$ or $x > 6$. **13.** $2x(x - 1)(x + 3)^2 = 0 \Leftrightarrow x = 0$ or $x = 1$ or $x = -3$
14. No, the function is not unbroken; it jumps from above the x-axis to below it.
15. $\frac{1}{3}L_0 = L_0\sqrt{1 - (\frac{v}{c})^2} \Leftrightarrow \frac{1}{3} = \sqrt{1 - (\frac{v}{c})^2} \Rightarrow \frac{1}{9} = 1 - (\frac{v}{c})^2 \Leftrightarrow \frac{8}{9} = (\frac{v}{c})^2 \Leftrightarrow \frac{2\sqrt{2}}{3} = \frac{v}{c} \Leftrightarrow v = \frac{2\sqrt{2}}{3}c$

Check: $L_0\sqrt{1 - \left(\frac{\frac{2\sqrt{2}}{3}c}{c}\right)^2} = L_0\sqrt{1 - \frac{8}{9}} = \frac{1}{3}L_0.$

16. a. amplitude: 4; period: π; phase shift: $\frac{\pi}{4}$ **b.** $y = 4\sin(2(x - \frac{\pi}{4}))$ **17.** Using an automatic grapher, the zeros are found to be $x \approx 0.6$ and $x \approx 1.5$. **18.** $A_0e^{-.00012t} < .3A_0 \Leftrightarrow e^{-.00012t} < .3$ (since $A_0 > 0$) $\Leftrightarrow -.00012t < \ln .3$ (since $f(x) = e^x$ is increasing) $\Leftrightarrow t > \frac{\ln .3}{-.00012} \Leftrightarrow t > 10033$. $\therefore$ it is more than 10,000 years old.

5. 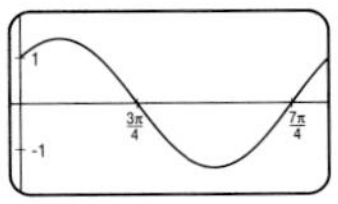

The chart below keys the **Progress Self-Test** questions to the objectives in the **Chapter 3 Review** on pages 217–220. This will enable you to locate those **Chapter 3 Review** questions that correspond to questions you missed on the **Progress Self-Test.** The lesson where the material is covered is also indicated in the chart.

Question	**1-2**	**3**	**4**	**5**	**6-7**	**8**	**9**	**10**	**11**
Objective	A	E	F	M	B	K	D	C	G
Lesson	3-1	3-8	3-1	3-2	3-2	3-3	3-5	3-4	3-2
Question	**12**	**13**	**14**	**15**	**16a**	**16b**	**17**	**18**	
Objective	E	C	H	J	I	N	O	L	
Lesson	3-7	3-4	3-5	3-1	3-9	3-9	3-5	3-6	

CHAPTER 3 REVIEW (pp. 217–220)
1. $x = 5$ **3.** $t = 1$ **5.** $x = -1$ or 2 **7.** $v = \frac{2}{9}$ **9.** $z = 10$ **11.** $x = \frac{3}{2}$ or -2 **13.** $t = 1$ or $t = \frac{5}{7}$ **15.** $f \cdot g = 1 - \frac{6}{x}$, domain: $\{x: x \neq 0\}$; $\frac{f}{g} = \frac{1}{x^2 - 6x}$, domain: $\{x: x \neq 0$ and $x \neq 6\}$;

$f \circ g = \circ\frac{1}{x - 6}$, domain: $\{x: x \neq 6\}$; $g \circ f = \frac{1}{x} - 6$, domain: $\{x: x \neq 0\}$ **17.** $f \circ g(x) = \sqrt{x^3}$ **19.** $x \approx .631$ **21.** -5, -4, 0 **23.** $x = 26$ or $\frac{6}{5}$ **25. a.** 2 **b.** $-1 < x < 0$ and $0 < x < 1$ **27.** $3 < x < 3.25$ **29.** $-2 < x < 0$ or $x > 1$ **31.** $z < -1$ or $z > \frac{1}{3}$ **33.** $t > -1$ **35.** (b), (c) **37. a.** No, $g(2) = g(-2) = 16$, but $2 \neq -2$. **b.** No, it is not a reversible operation since it is not a 1 − 1 function. **39.** True **41.** $h \circ m(x) = \log(10^{x-7}) + 7 = x$; $m \circ h(x) = 10^{\log x + 7 - 7} = x$; ∴ since $h \circ m = m \circ h = I$, h and m are inverse functions. **43. a.** -1 **b.** 1 **c.** No, $g(x)$ is undefined at $x = -2$, thus it is not continuous. Therefore, the Intermediate Value Theorem cannot be applied in the interval $-3 \leq x \leq -1$. **45.** h has a zero between a and b. **47.** 2π **49. a.** 6 **b.** $\frac{2\pi}{3}$ **c.** $-\frac{\pi}{3}$

d.

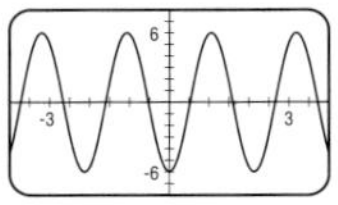

51. $v = \frac{\sqrt{3}}{2}c$ **53.** about 8 years **55. a.i.** $C(x) = 10000 + 55x$ **ii.** $S(x) = 150 - .06x$ **iii.** $R(x) = 150x - .06x^2$ **iv.** $P(x) = 95x - 10000 - .06x^2$

b.

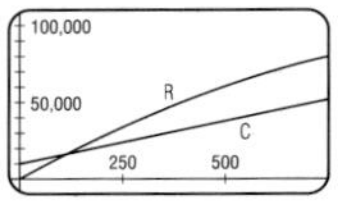

c.

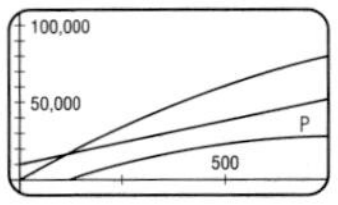

57. $0 < t < 3.56$

59.

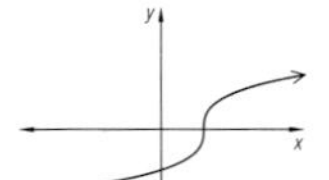

61. a.

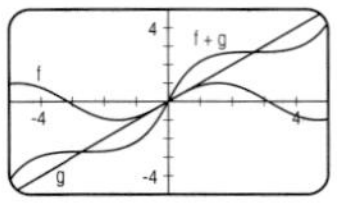

61. b.

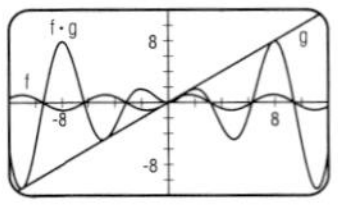

61. c. Since sin x varies from -1 to 1, $x \cdot \sin x$ will always be less than or equal to x, but greater than or equal to $-x$. That is why the graph of $f \cdot g$ oscillates between the lines $y = x$ and $y = -x$. Similarly, $f + g$ oscillates between the line $y = x + 1$ and $y = x - 1$.

63.

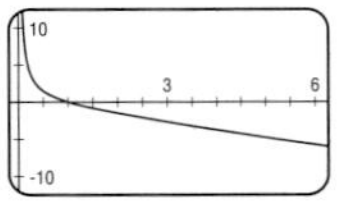

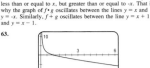

65. a. $\frac{x^2}{9} + 9y^2 = 1$ **b.** $\frac{(x-2)^2}{9} + 9(y+5)^2 = 1$

67. $y = \frac{1}{2}\cos(2x + \frac{\pi}{2})$ **69. a.** $g(x) = 3 \sin 8x$ **b. See below. 71.** sample: $.975 \leq x \leq 1.025$

69. b.

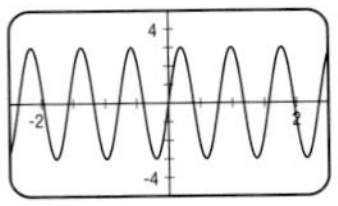

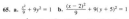

LESSON 4-1 (pp. 222–228)
1. True, there exists an integer, 12, such that $11 \cdot 12 = 132$. **3.** True, there exists an integer, 0, such that $2 \cdot 0 = 0$. **5.** True, $n^2 - 17n + 66 = (n - 6)(n - 11)$. **7. a.** 12 **b.** 10 **c.** 10 **9.** Then $b + c = (a \cdot q) + (a \cdot r) = a(q + r)$. Therefore, since $q + r$ is an integer by closure properties, a is a factor of $b + c$, by definition. **11. a.** Counterexample: let $a = 1$ and $b = -1$. Then a is divisible by b and b is divisible by a, but $a \neq b$. **b.** ∃ integers a and b such that a is divisible by b and b is divisible by a and $a \neq b$. **13.** 4 **15.** Suppose n is any integer. Then $n + (n + 1) + (n + 2) = 3n + 3 = 3(n + 1)$. Since $n + 1$ is an integer by closure properties, the sum of any three consecutive integers is divisible by 3. **17. a.** $(8x - 1)(4x + 3)(x + 5)$ **b.** $32x^3 + 180x^2 + 97x - 15$ **19. a.** domain: the set of real numbers; range: $\{y: y \geq 3\}$ **b.** $x \geq 2$ **c.** neither **d.** $T_{2,3}$ **21.** *b is not divisible by a, and c is not divisible by a.*

LESSON 4-2 (pp. 229–235)
1. a. Ms. Smith can make 43 copies, and she will have 1¢ left over. **b.** $n = 130$, $q = 43$, $r = 1$, $d = 3$ **c.** $130 = 43 \cdot 3 + 1$ **3. a.** 12.4 **b.** $q = 12$, $r = 2$

LESSON 4-2 (cont.)
5. a. 27.97368... **b.** $q = 27$, $r = 37$ **7. See below.** **9. a.** integer **b.** real number **c.** integer **d.** neither, not a division problem **e.** integer **11.** Let n, $n + 1$, and $n + 2$ be any 3 consecutive integers. Then by the results of Question 10, ∃ some integer q such that

$n = 3q$,	$n = 3q + 1$, or	$n = 3q + 2$
So, $n + 1 = 3q + 1$,	$n + 1 = 3q + 2$, or	$n + 1 = 3q + 3$
and $n + 2 = 3q + 2$,	$n + 2 = 3q + 3$, or	$n + 2 = 3q + 4$.

Note that in each of the three columns listed above, exactly one of the integers is divisible by 3. **13.** $q = -43$, $r = 5$ **15.** Yes, $(2m)! = 1 \cdot 2 \cdot 3 \cdot \ldots \cdot m \cdot (m + 1) \cdot \ldots \cdot 2m = m![(m + 1) \cdot \ldots \cdot 2m]$. Therefore, by definition $(2m)!$ is divisible by $m!$. **17.** True, $27x^2 + 9x + 12 = 3(9x^2 + 3x + 4)$ **19.** Let a and b be any integers such that a is divisible by b. Then, by definition, $a = q \cdot b$ for some integer q. Then $a^2 = (q \cdot b)^2 = q^2 \cdot b^2$. Since q^2 is an integer by closure properties, b^2 is a factor of a^2, and thus a^2 is divisible by b^2. **21.** $z^8 + z^7 + z^6 + z^5 + z^4 + z^3 + z^2 + z + 1$ **23.** 94.3% **25. a.** The vessel that spotted the boat at an angle of 42° is closer. **b.** ≈ 0.4 mi

7. $n = q \cdot 4 + r$

n	-10	-9	-8	-7	-6	-5	-4	-3	-2	-1	0	1	2	3	4	5	6	7	8	9	10	11	12	13	14	15
q	-3	-3	-2	-2	-2	-2	-1	-1	-1	-1	0	0	0	0	1	1	1	1	2	2	2	2	3	3	3	3
r	2	3	0	1	2	3	0	1	2	3	0	1	2	3	0	1	2	3	0	1	2	3	0	1	2	3

LESSON 4-3 (pp. 236–241)
1. *RI* **3.** *RI* **5.** Friday **7.** 2 **9.** 16 **11.** 5 hours after 9:00 P.M. is 2:00 A.M. **13. a.** 357 **b.** 624 **15.** 04 **17.** Suppose a, b, c, and d are any integers and m is a positive integer such that $a \equiv b(\text{mod } m)$ and $c \equiv d(\text{mod } m)$. According to the Congruence Theorem, m is a factor of $a - b$ and a factor of $c - d$. Thus there exist integers k_1 and k_2 such that $a - b = k_1m$ and $c - d = k_2m$. Using the Subtraction Property of Equality, $(a - b) - (c - d) = k_1m - k_2m = (a - c) - (b - d) = (k_1 - k_2)m$. Because $k_1 - k_2$ is an integer by closure properties, m is a factor of $(a - c) - (b - d)$. So $a - c \equiv (b - d)(\text{mod } m)$ by the Congruence Theorem. **19.** Let $n = 2k + 1$ be an odd integer. Then $n^2 = (2k + 1)^2 = 4k^2 + 4k + 1 = 4k(k + 1) + 1$. Either k or $k + 1$ is an even integer having a factor of 2, thus $4k(k + 1) = 8m$, for some integer m. Therefore, $n^2 = 8m + 1$, and $n^2 \equiv 1(\text{mod} 8)$. **21.** True **23. a.** $3x^2$ **b.** $2y^4$ **c.** $10z^2$ **25. a.** $x \leq -3$ or $x \geq \frac{1}{2}$ **b.** $x \leq -3$ or $x \geq \frac{1}{2}$ **27.** valid; Law of Detachment

LESSON 4-4 (pp. 242–247)
1. integer division and polynomial division; rational expression division and rational number division **3.** $q(x) = 3x + 8$; $r(x) = 20$ **5.** Let $f(x) = x^5 - x^3 - 2x$. By long division, $f(x) = (x^3 + 2x)(x^2 - 3) + 4x$. So, dividing by $x^2 - 3$, $\frac{f(x)}{x^2 - 3} = h(x) = x^3 + 2x + \frac{4x}{x^2 - 3}$. **7.** $f(x) = 1x + 7$ **9.** $5x^3 - 4x^2 - 10x - 4 = (5x^2 + 6x + 2)(x - 2)$ **11.** $q(x) = x^2 - xy + y^2$; $r(x) = 0$ **13. a.** 12 **b.** 4 **15.** Suppose m is any integer. When m is divided by 4, the four possible remainders are 0, 1, 2, and 3. Then, by the Quotient-Remainder Theorem, there exists an integer k such that $m = 4k$, or $m = 4k + 1$, or $m = 4k + 2$, or $m = 4k + 3$. **17.** invalid, converse error **19. a.** $x = \frac{-1 \pm \sqrt{5}}{2}$ **b.** no real numbers

LESSON 4-5 (pp. 248–253)
1. a. $p(x) = ((2x - 4)x + 1)x - 2$ **b.** 0 **c.** 2 [x] 7.54 [−] 4 [=] [x] 7.54 [+] 1 [=] [x] 7.54 [−] 2 [=]

3. The $x \cdot x \cdot x \cdot x \cdot x$ calculation may result in greater accuracy, since the powering key is evaluated by using the exponential and logarithmic functions. **5.** $q(x) = 3x^2 + 2x + 3$; $r(x) = 1$ **7.a.** $q(x) = 2x^3 + 3x^2 - 4x + 5$; $r(x) = 0$ **b.** $p(x) = (2x^3 + 3x^2 - 4x + 5)(x - 3) + 0$ **c. i.** True **ii.** True **iii.** True **iv.** False

9.
```
1| 1 0 0 0 0 -1
     1 1 1 1  1
  ---------------
   1 1 1 1 1 [0]
```
Since the remainder is 0, $x - 1$ is a factor of $x^5 - 1$. **11.** $q(x) = 2x^3 - x^2 - \frac{1}{2}x + \frac{7}{4}$; $r(x) = -\frac{3}{4}$ **13. a. See below. b.** 4

13. a.

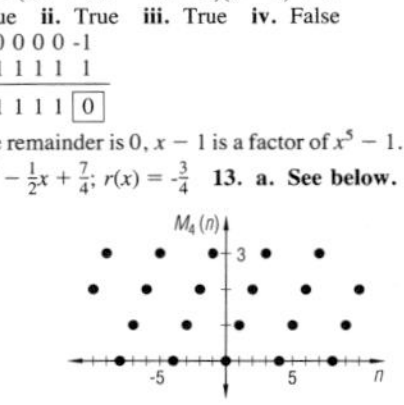

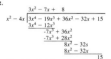

15. a. $C(n) = \begin{cases} 0.75n & n \leq 1500 \\ 225 + 0.60n & n > 1500 \end{cases}$

b. $R(n) = 1.25n$ **c.** $P(n) = \begin{cases} 0.50n & n \leq 1500 \\ 0.65n - 225 & n > 1500 \end{cases}$

d. 4962 **17.** period: π; amplitude: 4; phase shift: $\frac{\pi}{4}$; **See below.**

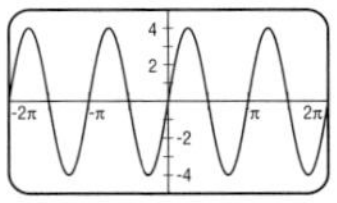

LESSON 4-6 (pp. 254–260)
1. a. 4 **b.** -4 **3.** 2355 **5. a.** No **b.** Yes **7. a.** $p(-1) = (-1)^4 - 5(-1)^2 - 10(-1) - 6 = 0$; $p(3) = (3)^4 - 5(3)^2 - 10(3) - 6 = 0$ **b.** $x = -1 + i$ and $x = -1 - i$ **9.** (b) and (e); There are horizontal lines the graphs pass through more than 4 times, but the polynomial is only of degree 4. **11.** $-\frac{110}{27}$ **13. a.** Given $p(x) = x^n - d^n$. The Factor Theorem tells us that $x - d$ is a factor of $p(x)$ if and only if $p(d) = 0$. $p(d) = d^n - d^n = 0$, so $x - d$ is a factor of $x^n - d^n$. If $x = c$, then $c - d$ is a factor of $c^n - d^n$. **b.** For all $n > 1$, $4^n - 1 = 4^n - 1^n$ which has $(4 - 1)$, or 3, as a factor by part a. Since $4^n - 1 > 3$ for $n > 1$ and 3 is a factor of $4^n - 1$, then $4^n - 1$ is not prime. **15.** Consider $p(x) = p_1(x) - p_2(x)$. The degree of this polynomial is at most n. But $p(x)$ has more than n zeros. However, a polynomial of degree n has at most n zeros, so $p(x)$ must be the zero polynomial. So $p(x) = 0$, and $p_1(x) - p_2(x) = 0$. Hence $p_1(x) = p_2(x)$ for all x. **17.** $q(x) = x^3 + 4x^2 - 12x + 34$; $r(x) = -101$ **19.** $(x + 3y)(x - y)^3(3x^2 + 2xy + 7y^2)$ **21.** $-2 < x < -1$

23. a. f: $\{x: x \geq -2\}$; g: $\{x: x \neq 0\}$ **b.** $g \circ f(x) = \frac{1}{x + 2}$ **c.** $\{x: x > -2\}$ **d.** f is a 1–1 function, but g is not.

LESSON 4-7 (pp. 261–267)
1. 6,000,200,300
3.

original problem	sum of digits	mod 9
2947	22	4
× 6551	17	× 8
19295797	49	32

The sum of the digits of the product is 49. This should be equivalent to 32(mod 9). But $49 \not\equiv 32(\text{mod } 9)$, so the multiplication is incorrect. **5.** 2 is not a digit in base 2. **7.** 50 **9. a.** 7 **b.** 1001000_2 **c.** $1 \cdot 2^6 + 0 \cdot 2^5 + 0 \cdot 2^4 + 1 \cdot 2^3 + 0 \cdot 2^2 + 0 \cdot 2^1 + 0 \cdot 2^0 = 2^6 + 2^3 = 64 + 8 = 72$ **11. a.** 4 **b.** 100 **13.** 100110_2 **15.** (e) **17.** 41 **19.** A number is divisible by 3 if and only if it is congruent to 0(mod 3). In base 10, the number has the value: $a_n10^n + a_{n-1}10^{n-1} + \ldots + a_0$. By the addition property of congruence, $a_n \cdot 10^n + a_{n-1} \cdot 10^{n-1} + \ldots + a_0 \equiv a_n + a_{n-1} + \ldots + a_0(\text{mod } 3)$. Either the value of the number and the sum of its digits are congruent to 0(mod 3) or neither is. **21. a.** $x(x(x(x + 2) - 5)) + 1$ **b.** -21.7999 **23.** $q(x) = \frac{1}{3}x^2 + \frac{7}{9}x + \frac{34}{27}$; $r(x) = \frac{88}{27}x + 1$ **25.** quotient: 662; remainder: 66 **27.** phase shift: $\frac{\pi}{8}$; period: $\frac{\pi}{4}$; amplitude: 7 **29.** Suppose m is any even integer and n is any odd integer. By definition there exist integers r and s such that $m = 2r$ and $n = 2s + 1$. Then $m \cdot n = 2r(2s + 1) = 4rs + 2r = 2(2rs + r)$. Since $(2rs + r)$ is an integer by closure properties, $m \cdot n$ is even by definition. **31.** False, let $x = -1$. Then $|x| = |-1| = 1 \neq -1$.

LESSON 4-8 (pp. 268–274)
1. There are no unicorns. **3.** 2, 3, 5, 7, 11, 13, 17, 19, 23, 29, 31, 37, 41, 43, 47 **5.** True **7.** $2^5 \cdot 3 \cdot 5$ **9.** $y^{98}(3y + 1)(3y - 1)$ **11.** $p(x) = x(x + 1)(x^2 + 1)$ **13.** Every even number is divisible by 2, thus it always has a factor of 2 and cannot be prime. **15.** Assume that there is a smallest positive real number, S. Consider $\frac{1}{2}S$. Since the real numbers are closed under multiplication, $\frac{1}{2}S$ is a real number. $\frac{1}{2}S < S$ and positive, so the assumption is invalid, and therefore there is no smallest positive real number. **17.** $6(y^2 + 1)^2(y - 1)^2(y + 1)^2$ **19.** The Fundamental Theorem of Arithmetic states that if a number is not prime it has a unique prime factorization. 1,000,000,000 has a prime factorization of $2^9 \cdot 5^9$, therefore 11 is not a prime factor. **21.** (c) **23.** Let p be a factor of a and p be a factor of b. Thus there exist integers m and n such that $a = m \cdot p$ and $b = n \cdot p$, by definition of factor. Then $a - b = m \cdot p - n \cdot p = (m - n)p$. Since $m - n$ is an integer by closure properties, p is a factor of $a - b$. **25. a.** sine function **b.** period: $\frac{1}{60}$ seconds; amplitude: 15 **c.** $c(t) = 15\sin(120\pi t)$ **d.** 0 amperes **27.** $x = -1$, $x = 2$, and $x = 3$ **29. a.** $h = \frac{544}{s^2}$ **b.** $A(s) = s^2 + \frac{2176}{s}$ **c. See below.;** $s \approx 10.29$ inches ; $h \approx 5.14$ inches

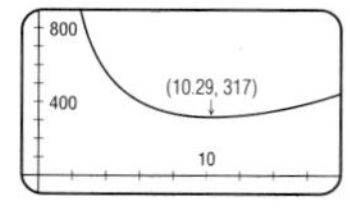

LESSON 4-9 (pp. 275–278)
1. ≈ 0.024 **3. a. See below. b.** Since $f_3(x)$ approaches infinity near $x = -1$, there may not be a polynomial function which can approximate f_3 near $x = -1$. **5.** $2 \cdot 3^2 \cdot 5^3 \cdot 79$ **7. a.** $6n = 3(2n)$, so 3 is a factor and $6n$ is not prime. $6n + 2 = 2(3n + 1)$, so 2 is a factor and $6n + 2$ is not prime. $6n + 3 = 3(2n + 1)$, so 3 is a factor and $6n + 3$ is not prime. $6n + 4 = 2(3n + 2)$, so 2 is a factor and $6n + 4$ is not prime. **b.** $6n + 5 = 6n + 6 - 1 = 6(n + 1) - 1 = 6k - 1$, for some integer k. **c.** $41 = (6 \cdot 7) - 1$; $43 = (6 \cdot 7) + 1$; $47 = (6 \cdot 8) - 1$; $53 = (6 \cdot 9) - 1$; $59 = (6 \cdot 10) - 1$ **9.** 100010_2 **11. a.** period: .002; amplitude: 100 **b.** $y = 100\sin(1000\pi t)$ **13.** $p(x) \Rightarrow q(x)$; $q(11)$; ∴ $p(11)$. invalid, converse error **15. a.** $x \leq -1$, $x \geq 1$ **b.** No, $f(0) = f(2)$, but $0 \neq 2$ **c.** No, f is not 1–1, therefore it cannot have an inverse function.

3. a.

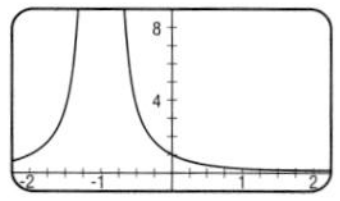

CHAPTER 4 PROGRESS SELF-TEST (p. 282)
1. a. $\lfloor\frac{145230}{8}\rfloor = 18{,}153$ 8-packs **b.** $145230 - 18153 \cdot 8 = 6$ bottles unpackaged **c.** $145230 = 18153(8) + 6$; that is, in the formula $n = q \cdot d + r$ where $0 \leq r < d$, $n = 145230$, $d = 8$, $q = 18153$, and $r = 6$.

2.
$$\begin{array}{r} 3x^2 - 7x + 8 \\ x^2 - 4x \overline{)3x^4 - 19x^3 + 36x^2 - 32x + 15} \\ \underline{3x^4 - 12x^3} \\ -7x^3 + 36x^2 \\ \underline{-7x^3 + 28x^2} \\ 8x^2 - 32s \\ \underline{8x^2 - 32x} \\ 15 \end{array}$$

∴ $q(x) = 3x^2 - 7x + 8$ and $r(x) = 15$. **3.** 46 **4.** False, since the remainder is nonzero.

5.
```
1.5 | 7   -2     11      12
         10.5   12.75   35.625
    -----------------------------
      7   8.5   23.75  [47.625]
```

∴ $p(1.5) = 47.625$ **6.** Let n, p, and m be integers such that n is a factor of p and n is a factor of m. By definition, there exist integers r and s such that $p = r \cdot n$ and $m = s \cdot n$. Then, $p \cdot m = r \cdot n \cdot s \cdot n = (r \cdot s \cdot n)n$. Since $r \cdot s \cdot n$ is an integer by closure properties, n is a factor of $p \cdot m$ by definition. **7.** $2^{10} = 1024 \equiv 1024$ (mod 10000); $2^{20} = 1048576 \equiv 8576$ (mod 10000); $2^{40} = 2^{20} \cdot 2^{20} \equiv 8576 \cdot 8576$ (mod 10000) $\equiv 73547776$ (mod 10000) $\equiv 7776$ (mod 10000); $2^{50} = 2^{40} \cdot 2^{10} \equiv 1024 \cdot 7776$ (mod 10000) $\equiv 7962624$ (mod 10000) $\equiv 2624$ (mod 10000). So the last four digits of 2^{50} are 2624. **8.** $\lfloor\frac{151}{11}\rfloor = 13$, $151 - 13 \cdot 11 = 8$, so $151 \equiv 8$ (mod 11). Hence, the smallest positive integer solution is $y = 8$. **9.** n is a factor of $(a - 5)$. **10.** $819 = 9 \cdot 91 = (3 \cdot 3) \cdot (7 \cdot 13) = 3^2 \cdot 7 \cdot 13$ **11.** False, counterexample: let $m = 25$. Then m is not divisible by any prime number less than $\sqrt{m} = \sqrt{25} = 5$, but 25 is not prime. **12.** $7x^4 - 3x^3 - 4x^2 = x^2(7x^2 - 3x - 4) = x^2(7x + 4)(x - 1)$ **13.** $33 = 32 + 1 = 2^5 + 2^0$, so $33 = 100001_2$. **14.** $101_2 + 110_2 = 1011_2$ **15.** $312_5 = 3 \cdot 5^2 = 1 \cdot 5^1 + 2 = 82$ **16.** Counterexample: let $a = 5$ and $b = 8$. Then b is a factor of $(5)^2 - 1 = 24$, but b is not a factor of $5 - 1 = 4$ or $5 + 1 = 6$. **17.** 0 **18.** Assume there is a largest even integer, N. $N = 2k$ for some integer k. Consider $N + 2$. Since the integers are closed under addition, $N + 2$ is an integer. Further, since $N + 2 = (2k) + 2 = 2(k + 1)$, $N + 2$ is even. $N + 2 > N$, so N is not the largest even integer. This contradicts the initial assumption. Therefore, the original statement is true, there does not exist a largest even integer.

The chart below keys the **Progress Self-Test** questions to the objectives in the **Chapter 4 Review** on pages 283–286. This will enable you to locate those **Chapter 4 Review** questions that correspond to questions you missed on the **Progress Self-Test.** The lesson where the material is covered is also indicated in the chart.

Question	**1a,b**	**1c**	**2**	**3**	**4**	**5**	**6**	**7**
Objective	K	A	C	C	H	B	F	L
Lesson	4-2	4-2	4-4	4-5	4-6	4-5	4-1	4-3
Question	**8**	**9**	**10-11**	**12**	**13-15**	**16**	**17**	**18**
Objective	D	G	J	E	M	F	H	I
Lesson	4-3	4-3	4-8	4-8	4-7	4-1	4-6	4-8

CHAPTER 4 REVIEW (pp. 283–286)

1. $q = 5, r = 6$ **3.** $q = 83, r = 63$ **5.** 66 **7.** $a = -5$ **9.** -2.875 **11.** True **13.** $q(x) = 3x^2 - 7x + 2$; $r(x) = 8$ **15.** $q(x) = x^3 + 3x^2$; $r(x) = 5$ **17.** $q(x) = x^3 - 8x^2 + 11x - 15$; $r(x) = 24$ **19.** $q(x) = x^4 + \frac{1}{2}x^3 - \frac{5}{4}x^2 + \frac{13}{8}x - \frac{29}{16}$; $r(x) = -\frac{67}{32}$ **21.** 12 **23.** 18 **25.** $5(x + y)(x - y)$ **27.** $x(3x - 5)(x + 2)$ **29.** $2(3v^2 + 5)^2$ **31.** $(2x + 1)(2x^2 - 4x - 9)$ **33.** $7(x - 3)(x + \frac{3 - \sqrt{37}}{14})(x - \frac{3 - \sqrt{37}}{14})$ **35.** True, $18 \cdot 5 = 90$. **37.** True, a and b may be represented by $2k$ and $2m$, respectively, since they are both even integers. Thus $2a + 2b = 2(2k) + 2(2m) = 4(k + m)$, so 4 is a factor by definition. **39.** True **41.** Let a, b, c, and d be integers such that $a = b - c$ and a and c are divisible by d. Then there exist integers r and s such that $a = d \cdot r$ and $c = d \cdot s$, by definition. Then $b = a + c = d \cdot r + d \cdot s = d(r + s)$. Since $(r + s)$ is an integer by closure properties, b is divisible by d. **43.** Let $2k + 1$ and $2k + 3$ represent any two consecutive odd integers. Then $(2k + 1)(2k + 3) = 4k^2 + 8k + 3$. Adding one gives $4k^2 + 8k + 4 = 4(k^2 + 2k + 1)$. $k^2 + 2k + 1$ is an integer for all integers k, so 4 is a factor.

45. $215 = 2(107) + 1$, so 215 is odd. **47.** $8s^2 + 4s + 3 = 2(4s^2 + 2s + 1) + 1$. Let $k = 4s^2 + 2s + 1$, then $8s^2 + 4s + 3 = 2k + 1$. So $8s^2 + 4s + 3$ is odd. **49.** The same integer k should not be used for both m and n. **51.** $x \equiv 0 \pmod 5$ **53. a.** $c - d \equiv -2 \pmod{11}$ **b.** $c \cdot d \equiv 35 \pmod{11}$ **55.** $0 \le$ degree of $r(x) <$ degree of $d(x)$ or $r(x) = 0$

57.

$$\begin{array}{r|l} & x^2 + 2x - 7 \\ \hline x^4 - 3 & x^6 + 2x^5 - 7x^4 - 3x^2 - 6x + 21 \\ & x^6 \qquad\qquad\quad - 3x^2 \\ \hline & 2x^5 - 7x^4 \qquad - 6x + 21 \\ & 2x^5 \qquad\qquad\quad - 6x \\ \hline & - 7x^4 \qquad\quad + 21 \\ & - 7x^4 \qquad\quad + 21 \\ \hline & 0 \end{array}$$

Therefore, $x^6 + 2x^5 - 7x^4 - 3x^2 - 6x + 21 = (x^4 - 3)(x^2 + 2x - 7)$, and $x^4 - 3$ is a factor. **59.** $x - 7$ **61.** 7 **63.** p is a prime, p is a factor of n^2, and p is not a factor of n. **65.** 2311 **67.** True **69.** $19 \cdot 29$ **71.** (c) **73.** 1001 **75.** 22 laps and 49.5 meters **77.** 5 months with 73 days each, 73 months with 5 days each, 365 months with 1 day, or 1 month with 365 days **79.** 143 **81.** 5 **83.** 10100101_2 **85.** 42 **87.** $11111_2 + 11111_2 = 111110_2$; $31 + 31 = 62$

LESSON 5-1 (pp. 288–294)

1. False, counterexample: $\frac{2}{3}$ is a rational number, but it is not an integer. **3.** True **5. a.** $\frac{400}{N}$ **b.** $\frac{800}{N^2 + 2N} = \frac{800}{N(N + 2)}$ **c.** $\frac{400}{5} = 80$. If each of the 5 people does 80 envelopes, then $5 \cdot 80 = 400$ are done, so part **a** is correct. For part **b**, $\frac{800}{35} \approx 22.9$ per person, and $80 - 22.9 = 57.1$. $7 \cdot 57.1 = 399.7$. So the formula works. (A few people will check 58 to make up the slack represented by the decimals.) **7. a.** The student eliminated identical terms, not identical factors. **b.** $x - 5$ **9. a.** $\frac{2}{y + 4}$ **b.** restrictions: $y \ne -4$, $y \ne 6$, and $y \ne -3$ **11. a.** $\frac{4 - a}{x(x - a)} + \frac{x - 4}{x^2 - ax} = \frac{4 - a}{x(x - a)} + \frac{x - 4}{x(x - a)} = \frac{4 - a + x - 4}{x(x - a)} = \frac{x - a}{x(x - a)} = \frac{1}{x}$ **b.** $\{x: x \ne 0$ and $x \ne a\}$ **13.** $\frac{3y}{2y + 2}$, $y \ne -1$, $y \ne 0$ **15.** a, $a \ne 1$ **17.** 12 and 35 do not have any common factors.

19.

$$\begin{array}{rl} \text{Let } N = & 0.\overline{012345679} \\ 10^9N = & 10^9(.012345679\overline{012345679}) \end{array}$$

Subtract N from 10^9N and divide by the coefficient of N:

$$\begin{array}{rl} 10^9N = & 12{,}345{,}679.\overline{012345679} \\ - \quad N = & \overline{.012345679} \\ \hline 999{,}999{,}999N = & 12{,}345{,}679 \\ N = & \frac{12{,}345{,}679}{999{,}999{,}999} \end{array}$$

21. $x = \ln 2$ or $x = \ln 3$ **23. a.** If n is not an even integer, then n^2 is not an even integer. **b.** Suppose n is any odd integer. By definition there exists an integer r such that $n = 2r + 1$. Then $n^2 = (2r + 1)^2 = 4r^2 + 4r + 1 = 2(2r^2 + 2r) + 1$. Since $2r^2 + 2r$ is an integer by closure properties, n^2 is an odd integer by definition. **c.** Yes **25.** 98

LESSON 5-2 (pp. 295–301)

1. True **3.** True **5. a.** $a^2 = 3b^2$ **b.** a^2 and a have a factor of 3. **c.** $b^2 = 3k^2$ **d.** b^2 and b have a factor of 3. **e.** Both a and b have a common factor of 3. However, from the beginning of the proof, $\frac{a}{b}$ is assumed in lowest terms, thus having no common factors. There is a contradiction, so the negated statement is false, and the original statement ($\sqrt{3}$ is irrational) must be true. **7. a.** irrational, the decimal expansion neither terminates nor repeats. **b.** rational, the decimal expansion repeats. **c.** rational, the decimal expansion terminates. **9.** Assume the negation of the original statement is true. Thus there is a rational number r and an irrational number i whose difference is a rational number d. Then, $r - i = d$, and so $r - d = i$. However, by the closure properties of rational numbers i is rational because the difference of two rational numbers is a rational number. Thus there is a contradiction, so the assumption must be false, and the original statement is therefore true. **11. a.** $5 - \sqrt{3}$ **b.** $\frac{25 - 5\sqrt{3}}{11}$ **c.** $\frac{10}{5 + \sqrt{3}} = 1.4854315$, $\frac{25 - 5\sqrt{3}}{11} = 1.4854315$ **13.** $\frac{2\sqrt{6} + 4\sqrt{3}}{-3}$ **15.** False, counterexample: let $a = \sqrt{2}$ and $b = \sqrt{8}$. Then $a \cdot b = \sqrt{2} \cdot \sqrt{8} = \sqrt{16} = 4$, but 4 is a rational number. **17.** reciprocal of $\sqrt{7} + \sqrt{6} = \frac{1}{\sqrt{7} + \sqrt{6}} = \frac{1}{\sqrt{7} + \sqrt{6}} \cdot \frac{\sqrt{7} - \sqrt{6}}{\sqrt{7} - \sqrt{6}} = \frac{\sqrt{7} - \sqrt{6}}{7 - 6} = \frac{\sqrt{7} - \sqrt{6}}{1} = \sqrt{7} - \sqrt{6}$ **19.** Suppose r and s are rational numbers. By definition there exist integers a, b, c, and d with $b \ne 0$ and $d \ne 0$ such that $r = \frac{a}{b}$ and $s = \frac{c}{d}$. Then $r \cdot s = \frac{a}{b} \cdot \frac{c}{d} = \frac{ac}{bd}$. Since $b \ne 0$ and $d \ne 0$, $bd \ne 0$. Also, ac and bd are integers by closure properties. Thus, since $r \cdot s$ has an integer numerator and denominator, $r \cdot s$ is a rational number by definition.

21. a. $\frac{3(x^2 + x - 4)}{x(x - 4)}$ **b.** restrictions: $x \ne 0$, $x \ne 4$ **c.** Let $x = 2$ in the original and in the simplified expressions. $\frac{3}{2} + \frac{3(2)}{2 - 4} = -\frac{3}{2}$, and $\frac{3((2)^2 + 2 - 4)}{2(2 - 4)} = -\frac{6}{4}$, which checks. **23.** $y < -1$ or $y > \frac{5}{2}$ **25.** (b)

LESSON 5-3 (pp. 302–306)

1. True **3.** See below. **5. a.** $T_{-6,0}$ **b.** $x = -6$ **c.** $\lim_{x \to -6^+} h(x) = +\infty$ and $\lim_{x \to -6^-} h(x) = -\infty$ **7. a.** $\lim_{t \to 0^+} g(t) = +\infty$ and $\lim_{t \to 0^-} g(t) = +\infty$ **b.** $\lim_{t \to +\infty} g(t) = 0$ and $\lim_{t \to -\infty} g(t) = 0$ **c.** See below. **9.** The sun's brightness seen from Earth is 2.25 times the brightness seen from Mars. **11.** The statement is false. Counterexample: π and $\pi + 3$ are both irrational, yet $\pi - (\pi + 3) = -3$ is rational. **13. a.** The decimal expansion of x is nonterminating and the decimal expansion of x is nonrepeating. **b.** If the decimal expansion of x is nonterminating and nonrepeating, then x is not a rational number. **c.i)** rational **ii)** irrational **15. a.** $\frac{a}{2a + 5}$ **b.** $a \ne \frac{1}{3}$, $a \ne -2$, $a \ne -\frac{5}{2}$ **17.** Suppose a, b, and c are any integers such that a is a factor of b and a is a factor of $b + c$. By definition there exists integers r and s such that $b = a \cdot r$ and $b + c = a \cdot s$. Then $c = (b + c) - b = a \cdot s - a \cdot r = a(s - r)$. Since $s - r$ is an integer by closure properties, a is a factor of c by definition. **19. a.** $\frac{f}{g} = t + 1$ **b.** domain: $\{t: t \ne 1\}$

3.

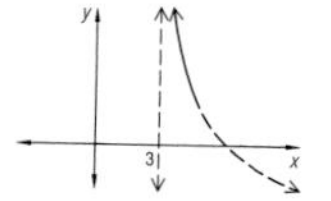

7. c.

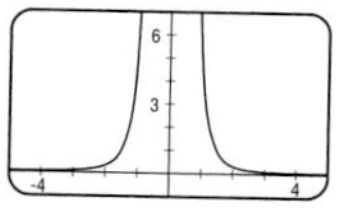

LESSON 5-4 (pp. 307–313)

1. \$350 **3.** Yes, domain: $\{x: x \ne -1\}$ **5.** Yes, domain: the set of real numbers **7. a.** removable **b.** See next column. **9. a.** $\{u: u \ne -\frac{5}{3}\}$ **b.** essential **c.** $u = -\frac{5}{3}$ **d.** See next column. **11. a.** the set of real numbers **b.** none **c.** none **d.** See next column. **13.** $g(x)$ is a function with a removable discontinuity at $x = 1$. Since $\lim_{x \to 1} g(x) = \lim_{x \to 1} (x + 5) = 6$, redefining $g(1) = 6$ makes $g(x)$ continuous for all x. **15. a.** domain: $\{x: x \ne 0\}$; range: $\{y: y > 0\}$ **b.** $\lim_{x \to 0^+} h(x) = +\infty$; $\lim_{x \to 0^-} h(x) = +\infty$ **c.** $\lim_{x \to +\infty} h(x) = 0$; $\lim_{x \to -\infty} h(x) = 0$ **d.** $\forall x$, $h(-x) = \frac{5}{(-x)^4} = \frac{x^4}{5} = h(x)$. **e.** See next column. **17.** $1 - x\sqrt{3}$ **19.** $\frac{35}{4(a - b)}$

7. b.

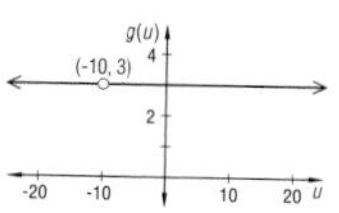

9. d.

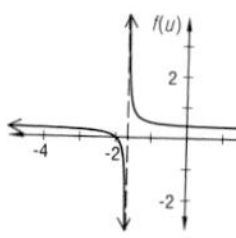

11. d.

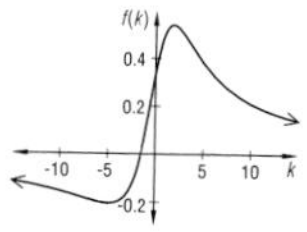

15. e.

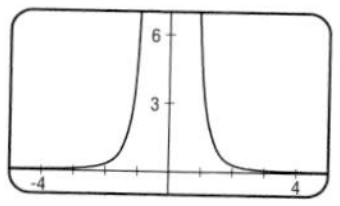

21. a. Let $d(x) = \cos x - 0.2x^2$. Since $d(1) > 0$, $d(1.5) < 0$, and $d(x)$ is continuous, the Intermediate Value Theorem insures that there exists a zero of $d(x)$ in the interval from 1 to 1.5. Where $d(x) = 0$, $\cos x - 0.2x^2 = 0$, so $\cos x = 0.2x^2$, and hence there is a solution to the equation $\cos x = 0.2x^2$ in the interval from 1 to 1.5. **b.** $1.25 \le x \le 1.26$ **23. a.** *The duplicating machine works, and it is not Sunday.* **b.** *The country is not in Southeast Asia, or it is in Philadelphia.* **25.** $q(x) = x^3 + 3x^2 + 2x - 2$; $r(x) = 2x - 8$

LESSON 5-5 (pp. 314–320)

1. $4 - \frac{11}{x} + \frac{6}{x^2} - \frac{2}{x^3}$ **3. a.** $y = \frac{1}{3}$ **b.** like the function $f(x) = \frac{1}{3}$; $\lim_{x \to +\infty} r(x) = \frac{1}{3}$; $\lim_{x \to -\infty} r(x) = \frac{1}{3}$ **5.** like the function $f(x) = x^4$; $\lim_{x \to +\infty} g(x) = +\infty$; $\lim_{x \to -\infty} g(x) = +\infty$

7.

$$\begin{array}{r|l} & 2v^2 + \frac{5}{2}v + \frac{21}{4} \\ \hline 2v - 1 & 4v^3 + 3v^2 + 8v - 2 \\ & 4v^3 - 2v^2 \\ \hline & \quad 5v^2 + 8v - 2 \\ & \quad 5v^2 - \frac{5}{2}v \\ \hline & \qquad \frac{21}{2}v - 2 \\ & \qquad \frac{21}{2}v - \frac{21}{4} \\ \hline & \qquad\qquad \frac{13}{4} \end{array}$$

9.

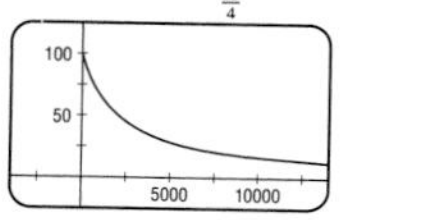

11. a.

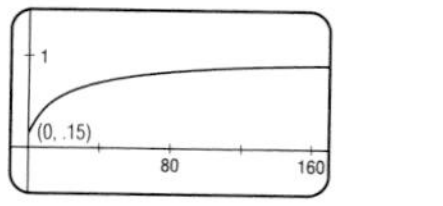

b. As the number of total points increases, Viola's grade approaches 100%. **13. a.** $z = 3$ **b.** $t(z) = \frac{z^2 - z - 6}{z^2 + z - 12} = \frac{(z - 3)(z + 2)}{(z - 3)(z + 4)} = \frac{z + 2}{z + 4}$; $\therefore t(z)$ is always undefined at $z = -4$. **15.** $\frac{-4}{x(x + h)}$, $x \ne 0$, $h \ne 0$, and $x \ne -h$ **17.** $(31 - 7y)^3 (8t + 5y)^3 (-8t^2 + 3t - 7y - 13ty - 5y^2)$ **19. a.** $\frac{\sqrt{2}}{2}$ **b.** $-\frac{1}{2}$ **c.** -1 **d.** $-\frac{\sqrt{3}}{2}$

LESSON 5-6 (pp. 321–325)

1. $\tan x = \frac{4}{3}$; $\cot x = \frac{3}{4}$; $\sec x = \frac{5}{3}$; $\csc x = \frac{5}{4}$ **3.** sample: $(\frac{\pi}{4}, \sqrt{2})$, $(\frac{\pi}{2}, 1)$, $(\frac{3\pi}{4}, \sqrt{2})$, $(\frac{3\pi}{2}, -1)$ **5.** Sample: $(0, 0)$, $(\frac{\pi}{6}, \frac{\sqrt{3}}{3})$, $(\frac{\pi}{4}, 1)$, $(\frac{\pi}{3}, \sqrt{3})$ **7. a.** sin, cos, sec, and csc have a period of 2π. cot and tan have a period of π. **b.** sin, csc, tan, and cot are odd functions. cos and sec are even functions. **c.** none **d.** sin and cos

e.

	$\frac{\pi}{6}$	$\frac{\pi}{4}$	$\frac{\pi}{3}$
sin	$\frac{1}{2}$	$\frac{\sqrt{2}}{2}$	$\frac{\sqrt{3}}{2}$
cos	$\frac{\sqrt{3}}{2}$	$\frac{\sqrt{2}}{2}$	$\frac{1}{2}$
tan	$\frac{\sqrt{3}}{3}$	1	$\sqrt{3}$
cot	$\sqrt{3}$	1	$\frac{\sqrt{3}}{3}$
sec	$\frac{2\sqrt{3}}{3}$	$\sqrt{2}$	2
csc	2	$\sqrt{2}$	$\frac{2\sqrt{3}}{3}$

9. $\tan(-\frac{\pi}{4}) = -1$; $\cot(-\frac{\pi}{4}) = -1$; $\sec(-\frac{\pi}{4}) = \sqrt{2}$; $\csc(-\frac{\pi}{4}) = -\sqrt{2}$ **11. a.** The area of the triangle ABO is $\frac{1}{2}(AB)h$. So, the area of the regular n-gon is the sum of the areas of n congruent triangles, $n\frac{1}{2}(AB)h$. $m\angle AOB = \frac{2\pi}{n}$. The altitude h, splits the $\angle AOB$ into two smaller angles measuring $\frac{\pi}{n}$. $\tan(\frac{\pi}{n}) = \frac{\frac{1}{2}AB}{h}$. So $AB = 2h \tan \frac{\pi}{n}$. Hence, the area of the n-gon is $n\frac{1}{2}(2h \tan \frac{\pi}{n})\, h = nh^2 \tan \frac{\pi}{n}$. **b.** Let the radius, r, be 6 and $n = 4$. Then $AB = 6\sqrt{2}$, and $h = 3\sqrt{2}$ by the Pythagorean Theorem. Using the formula from part a, the area is $4(3\sqrt{2})^2 \tan \frac{\pi}{4} = 72$. The area of the square is $(AB)^2 = (6\sqrt{2})^2 = 72$ and so the formula checks. **13.** $f: z \to \frac{11z}{6}$ **15. a.** See below. **b.** $\lim_{x \to 2^+} f(x) = +\infty$; $\lim_{x \to 2^-} f(x) = -\infty$ **c.** $\lim_{x \to +\infty} f(x) = 5$; $\lim_{x \to -\infty} f(x) = 5$ **17. a.** 1 **b.** 10 **19.** $x = \ln 3$ or $x = 0$

15. a.

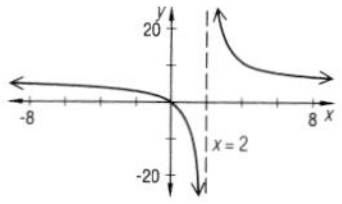

LESSON 5-7 (pp. 326–331)

1. a. $4(x + 5)(x - 5)$ **b.** $x = 5$ and $x = -5$ **3. a.** $\{\sqrt{8}, \sqrt{8} + 2, \sqrt{8} + 4\}$ and $\{-\sqrt{8}, -\sqrt{8} + 2, -\sqrt{8} + 4\}$ **b.** $\{2.83, 4.83, 6.83\}$ and $\{-2.83, -0.83, 1.17\}$ **c.** No **5.** $x = 0$ **7.** $t = -4$ **9. a.** ≈ 171 mph **b.** ≈ 342 miles **11.** $BC = \sqrt{2} + 1$; $AB = \sqrt{2} + 2$ **13.** $\frac{3(r^3 - 16r)}{3r^2 - 2r - 8}$ **15.** $\tan \frac{5\pi}{6} = -\frac{\sqrt{3}}{3}$; $\cot \frac{5\pi}{6} = -\sqrt{3}$; $\sec \frac{5\pi}{6} = -\frac{2\sqrt{3}}{3}$; $\csc \frac{5\pi}{6} = 2$ **17. a.** $x = \frac{7}{2}$ **b.** essential discontinuity at $x = \frac{7}{2}$ **c.** $x = \frac{7}{2}$ **d.** $y = \frac{1}{2}$ **e.** x-intercept: $(0, -\frac{3}{7})$; y-intercept: $(-3, 0)$ **f.** See below. **g.** $g(x) = \frac{x^2 - 2x - 15}{2x^2 - 17x + 35}$ **19. a.** amplitude: 4; period: $\frac{2\pi}{3}$; phase shift: $\frac{\pi}{3}$ **b.** $y = 4 \sin(3x - \pi)$ **21.** ≈ 7.0 %

17. f.

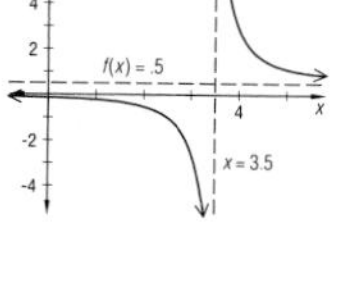

LESSON 5-8 (pp. 332–337)

1. sample: See below. **3.** $(\frac{1}{3}, \frac{1}{3})$, $(\frac{1}{3}, \frac{1}{4})$, $(\frac{1}{3}, \frac{1}{5})$, $(\frac{1}{3}, \frac{1}{6})$, $(\frac{1}{3}, \frac{1}{7})$, $(\frac{1}{3}, \frac{1}{8})$, $(\frac{1}{4}, \frac{1}{3})$, $(\frac{1}{4}, \frac{1}{4})$, $(\frac{1}{5}, \frac{1}{3})$, $(\frac{1}{6}, \frac{1}{3})$, $(\frac{1}{7}, \frac{1}{3})$, $(\frac{1}{8}, \frac{1}{3})$ **5. a.** $a_n = \frac{(n - 2)180°}{n}$ **b.** 60°, 90°, 108°, 120°, $128\frac{4}{7}$° **c.** 180° **7. a.** regular tetrahedron **b.** regular octahedron **c.** regular icosahedron **d.** cube **e.** regular dodecahedron **9.** Let n be an integer greater than 2. Suppose $(n - 2)$ is a factor of $2n$. Then $2n = (n - 2)k$ for some integer k. Then $\frac{2n}{n - 2} = k$, since $n \ne 2$. By taking the reciprocal, $\frac{n - 2}{2n} = \frac{1}{k}$. So $\frac{1}{2} - \frac{1}{n} = \frac{1}{k}$, or $\frac{1}{2} = \frac{1}{k} + \frac{1}{n}$. This is equivalent to the equation in Problem 2 which is also equivalent to Problem 1. **11. a.** $\{x: x \ne 5$ and $x \ne -5\}$ **b.** near $x = 5$: $\lim_{x \to 5^-} h(x) = -\infty$; $\lim_{x \to 5^+} h(x) = +\infty$; near $x = -5$: $\lim_{x \to -5^-} h(x) = -\infty$; $\lim_{x \to -5^+} h(x) = +\infty$ **c.** $\lim_{x \to -\infty} h(x) = 0$; $\lim_{x \to +\infty} h(x) = 0$ **d.** x-intercept: $(-2, 0)$; y-intercept: $(0, -\frac{2}{25})$ **e.** See below. **13.** $\tan \frac{5\pi}{4} = 1$; $\cot \frac{5\pi}{4} = 1$; $\sec \frac{5\pi}{4} = -\sqrt{2}$; $\csc \frac{5\pi}{4} = -\sqrt{2}$ **15. a.** $\frac{5x^2 + 3x - 2}{7x^2 + 8x + 1} \cdot \frac{7x^2 - 27x - 4}{5x^2 + 8x - 4} = \frac{(5x - 2)(x + 1)}{(7x + 1)(x + 1)} \cdot \frac{(7x + 1)(x - 4)}{(5x - 2)(x + 2)} = \frac{x - 4}{x + 2}$ **b.** $x \ne -\frac{1}{7}$, $x \ne -1$, $x \ne \frac{2}{5}$, and $x \ne -2$ **17.** $2^2 \cdot 1097$

1.

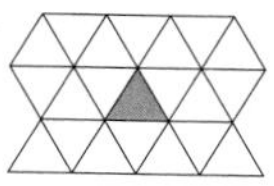

11. e.

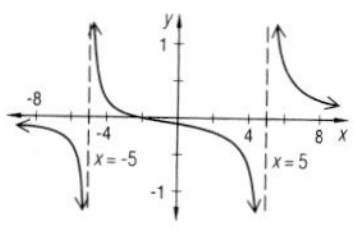

CHAPTER 5 PROGRESS SELF-TEST (p. 340)

1. a. $\frac{6x}{(x + 3)(x + 1)} + \frac{2x}{(x + 2)(x + 1)} = \frac{6x(x + 2) + 2x(x + 3)}{(x + 1)(x + 2)(x + 3)} = \frac{6x^2 + 12x + 2x^2 + 6x}{(x + 1)(x + 2)(x + 3)} = \frac{8x^2 + 18x}{(x + 1)(x + 2)(x + 3)} = \frac{2x(4x + 9)}{(x + 1)(x + 2)(x + 3)}$ **b.** $x \ne -3$, $x \ne -2$, and $x \ne -1$

2. a. $\frac{2t^2 - t - 1}{3t^2 - 2t - 5} \cdot \frac{3t^2 + 7t - 20}{t^2 + 3t - 4} = \frac{(2t + 1)(t - 1)}{(3t - 5)(t + 1)} \cdot \frac{(3t - 5)(t + 4)}{(t - 1)(t + 4)} = \frac{2t + 1}{t + 1}$

b. $t \ne \frac{5}{3}$, $t \ne -1$, $t \ne 1$, and $t \ne -4$ **3.** $f = \frac{1}{\frac{1}{p} + \frac{1}{q}} \cdot \frac{pq}{pq} = \frac{pq}{\frac{pq}{p} + \frac{pq}{q}} = \frac{pq}{p + q}$ **4. a.** R (standard rational form) **b.** I (because 24 is not a perfect square) **c.** R ($\sqrt{49} = 7$) **d.** R (repeating decimal) **e.** I (nonterminating, nonrepeating decimal) **5.** Assume the negation is true, that $\sqrt{11}$ is rational. By definition there exists integers a and b, with $b \ne 0$, such that $\sqrt{11} = \frac{a}{b}$, where $\frac{a}{b}$ is in lowest terms. Then $11 = \frac{a^2}{b^2} \Rightarrow a^2 = 11b^2$. Thus a^2 has a factor of 11. And a has a factor of 11, because if a is an integer and a^2 is divisible by a prime, then a is divisible by that prime. Therefore, let $a = 11k$ for some integer k. Then $11b^2 = (11k)^2 \Rightarrow b^2 = 11k^2$. So b^2 and b have a factor of 11 by similar argument. Thus a and b have a common factor of 11. This is a contradiction since $\frac{a}{b}$ is in lowest terms. Hence the assumption must be false, and so $\sqrt{11}$ is irrational. **6.** $\frac{8}{\sqrt{10} - \sqrt{5}} = \frac{8(\sqrt{10} + \sqrt{5})}{(\sqrt{10} - \sqrt{5})(\sqrt{10} + \sqrt{5})} = \frac{8(\sqrt{10} + \sqrt{5})}{10 - 5} = \frac{8(\sqrt{10} + \sqrt{5})}{5}$ **7.** As x approaches 2 from the left, $f(x)$ approaches positive infinity. **8. a.** $\lim_{x \to 2^+} g(x) = +\infty$ **b.** $\lim_{x \to 2^-} g(x) = -\infty$ **c.** $\lim_{x \to +\infty} g(x) = 3$; $\lim_{x \to -\infty} g(x) = 3$ **d.** See below. **9.** essential discontinuity

10.

$$\begin{aligned} \left(\frac{2}{y - 1} + \frac{3y}{y + 4}\right)(y - 1)(y + 4) &= 3(y - 1)(y + 4) \\ 2(y + 4) + 3y(y - 1) &= 3(y2 + 3y - 4) \\ 2y + 8 + 3y^2 - 3y &= 3y^2 + 9y - 12 \\ 20 &= 10y \\ 2 &= y \end{aligned}$$

Check: for $y = 2$ does $\frac{2}{2 - 1} + \frac{3(2)}{2 + 4} = 3$? Yes

11. a. $\tan \frac{2\pi}{3} = \frac{\sin \frac{2\pi}{3}}{\cos \frac{2\pi}{3}} = \frac{\frac{\sqrt{3}}{2}}{-\frac{1}{2}} = -\sqrt{3}$ **b.** $\csc \frac{\pi}{4} = \frac{1}{\sin \frac{\pi}{4}} = \frac{\frac{1}{1}}{\frac{1}{\sqrt{2}}} = \sqrt{2}$ **12.** See below. **13.** $\tan \alpha = \frac{7}{24}$ **14. a.** $-\infty$ **b.** $-\infty$ **c.** 1 **d.** 1

8. d.

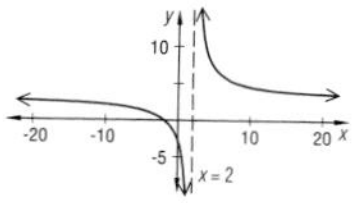

12.

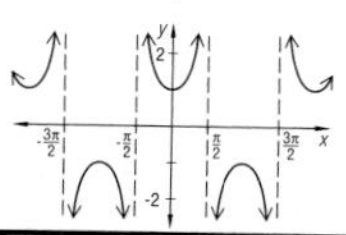

The chart below keys the **Progress Self-Test** questions to the objectives in the **Chapter 5 Review** on pages 341–344. This will enable you to locate those **Chapter 5 Review** questions that correspond to questions you missed on the **Progress Self-Test.** The lesson where the material is covered is also indicated in the chart.

Question	1-2	3	4	5	6	7	8
Objective	A	J	B	G	C	D	H
Lesson	5-1	5-1	5-2	5-2	5-2	5-5	5-3
Question	9	10	11	12-13	14	15	16
Objective	H	I	F	E	K	M	L
Lesson	5-5	5-4	5-7	5-6	5-6	5-7	5-3

CHAPTER 5 REVIEW (pp. 341–344)

1. a. -1 **b.** $x \neq \frac{2}{3}$ **3. a.** $\frac{3}{(z+1)(z+4)}$ **b.** $z \neq -1, z \neq 2, z \neq 3$, and $z \neq -4$ **5. a.** $\frac{-2t^2+20t+13}{(t-5)(t+4)}$ **b.** $t \neq 5$ and $t \neq -4$ **7.** $\frac{3(x-5)}{(x-6)}$ **9.** $a-3$, $a \neq -3$ and $a \neq 0$

11. a. $\frac{z^2-z-2}{z^2-4z-5} \cdot \frac{z-5}{z^2+z-6} = \frac{(z+1)(z-2)}{(z+1)(z-5)} \cdot \frac{z-5}{(z-2)(z+3)} = \frac{1}{z+3}$ **b.** $z \neq -1, z \neq 5, z \neq 2$, and $z \neq -3$ **13.** $\frac{\frac{1}{x}+\frac{2}{x^2}}{1-\frac{4}{x^2}} = \frac{x+2}{x^2-4} = \frac{x+2}{(x+2)(x-2)} = \frac{1}{x-2}$; $x \neq 0, x \neq -2$, and $x \neq 2$ **15.** rational, because -7 is an integer. **17.** rational, since 0 is an integer. **19.** rational, equals $\frac{26}{3}$. **21.** $\frac{15+5\sqrt{5}}{4}$ **23.** $12-4\sqrt{6}$ **25.** $\frac{7}{15+5\sqrt{2}}$ **27.** $\tan x = -\frac{3}{4}$; $\sec x = \frac{5}{4}$; $\cot x = -\frac{4}{3}$; $\csc x = -\frac{5}{3}$ **29.** -1 **31.** -1 **33.** 0 **35.** $x = 1$ **37.** $t = -\frac{3}{4}$ and $t = 1$ **39.** $y = \frac{1}{2}$ and $y = \frac{1}{4}$ **41.** $\frac{245}{99}$ **43.** Assume the negation is true, that $\sqrt{13}$ is rational. By definition there exists integers a and b, with $b \neq 0$, such that $\sqrt{13} = \frac{a}{b}$, where $\frac{a}{b}$ is in lowest terms. Then $13 = \frac{a^2}{b^2} \Rightarrow a^2 = 13b^2$. Thus a^2 has a factor of 13. And a has a factor of 13, because if a is an integer and a^2 is divisible by a prime, then a is divisible by that prime. Therefore, let $a = 13k$ for some integer k. Then $13b^2 = (13k)^2 \Rightarrow b^2 = 13k^2$. So b^2 and b have a factor of 13 by similar argument. Thus a and b have a common factor of 13. This is a contradiction since $\frac{a}{b}$ is in lowest terms. Hence the assumption must be false, and so $\sqrt{13}$ is irrational.
45. True, if $\frac{a}{b}$ and $\frac{c}{d}$ are two rational numbers, where $b \neq 0$ and $d \neq 0$, then $\frac{a}{b} \cdot \frac{c}{d} = \frac{ac}{bd}$. ac and bd are integers and $bd \neq 0$ since $b \neq 0$ and $d \neq 0$. Hence $\frac{ac}{bd}$ is rational.

47. Assume the negation is true. Thus the difference of a rational number p and an irrational number q is a rational number r. Then $p - q = r$. So $p - r = q$. However, by the closure property of the rational numbers, the differce between two rational numbers is another rational number. Hence there is a contradiction, and so the assumption is false which proves the original statement. **49.** The limit of $f(x)$ as x decreases without bound is 4.
51. a. $\lim_{x \to 6^+} h(x) = +\infty$ **b.** $\lim_{x \to 6^-} h(x) = -\infty$ **c.** $\lim_{x \to +\infty} h(x) = 2$; $\lim_{x \to -\infty} h(x) = 2$ **53.** $y = \frac{3}{4}x + \frac{1}{8}$ **55.** $h(y) = \frac{7}{3}y^2$ **57.** $q(z) = \frac{1}{4z}$ **59.** sample: $\frac{x^2-16}{x-4}$ **61.** True **63.** $v = f\lambda$ **65.** 61.6 mph **67. a.** $x = 3$ **b.** $x = 3$: essential **c.** none **d.** x-intercept: $-\sqrt[3]{6} \approx -1.8$; y-intercept: -2 **e.** $\lim_{x \to +\infty} f(x) = \infty$; $\lim_{x \to -\infty} f(x) = \infty$ **f.** See below. **69.** See below.

67. f

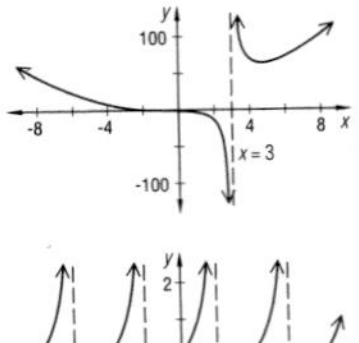

69.

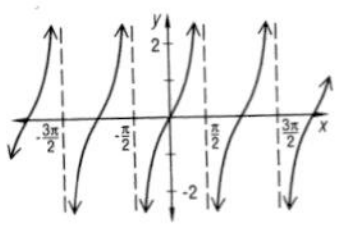

71. a. sample: **See below. b.** $x = 5$ and $y = -2$

71. a.

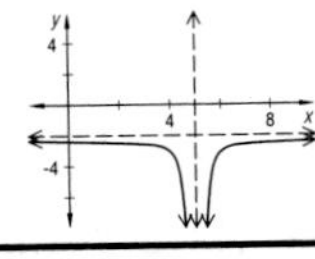

73. See below. $x = -1$ and $y = 0$ **75.** $\frac{13}{12}$ **77.** $\frac{12}{5}$

73.

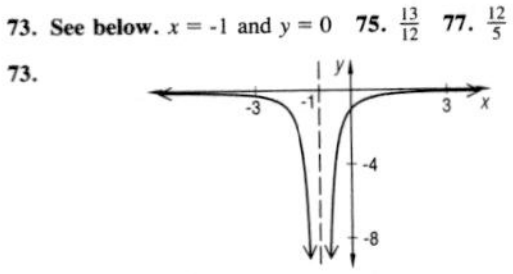

LESSON 6-1 (pp. 346–351)
1. an equation which is true for all values of the variable for which both sides are defined **3.** all real numbers **5.** $2 \sin x \sin y = \cos(x-y) - \cos(x+y)$ **7.** $\sin^2 x + \cos^2 x = 1$ **9.** $\tan x \csc x = \sec x$ **11. a. See below. b.** identity, domain: $x \neq \frac{(2n+1)\pi}{2}$, $\forall$ integers n
13. a. It is not an identity. **b.** Sample: let $x = 0.2$. Then $\sin(0.2\pi) \approx 0.588$, but $4 \cdot 0.2 \cdot (1-0.2) = 0.64$. **15. a. See below. b. See below. c. See below. d.** Yes

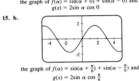

11. a.

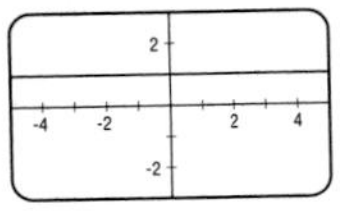

15. a.

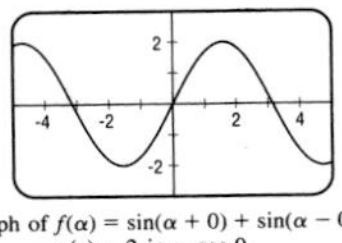

the graph of $f(\alpha) = \sin(\alpha + 0) + \sin(\alpha - 0)$ and $g(x) = 2\sin \alpha \cos 0$

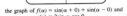

15. b.

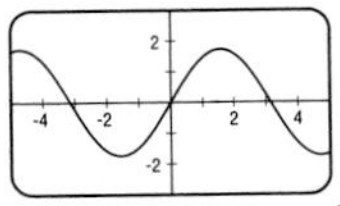

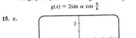

the graph of $f(\alpha) = \sin(\alpha + \frac{\pi}{6}) + \sin(\alpha - \frac{\pi}{6})$ and $g(x) = 2\sin \alpha \cos \frac{\pi}{6}$

15. c.

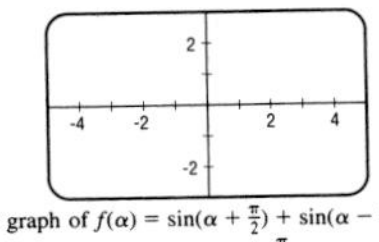

the graph of $f(\alpha) = \sin(\alpha + \frac{\pi}{2}) + \sin(\alpha - \frac{\pi}{2})$ and $g(x) = 2\sin \alpha \cos \frac{\pi}{2}$

17. $-\frac{\sqrt{3}}{2}$ **19.** $\frac{1}{2}, \pm 2$ **21.** $(f+g)(x) = x^2 + 2x + 1, x \neq 0$; $(f-g)(x) = 2x + 1 - x^2 - \frac{2}{x}, x \neq 0$; $(f \cdot g)(x) = 2x^3 + x^2 x + 2 + \frac{1}{x} - \frac{1}{x^2}, x \neq 0$; $\left(\frac{f}{g}\right)(x) = \frac{2x-1}{x^2-x+1}, x \neq 0, -1$
23. $OC = OA$ and $OD = OB$ since they are radii of the same respective circle. So by SAS, $\Delta COD \cong \Delta AOB$. Therefore, $AB = CD$ since corresponding parts in congruent figures are congruent. **25.** angles in Quadrant II: $\frac{\pi}{2} < \theta < \pi$; angles in Quadrant III: $\pi < \theta < \frac{3\pi}{2}$; angles in Quadrant IV: $\frac{3\pi}{2} < \theta < 2\pi$

LESSON 6-2 (pp. 352–356)
1. a. Left side $= \cot^2 x + 1$

$= \frac{\cos^2 x}{\sin^2 x} + \frac{\sin^2 x}{\sin^2 x}$	Definition of cot
$= \frac{\cos^2 x + \sin^2 x}{\sin^2 x}$	Addition of fractions
$= \frac{1}{\sin^2 x}$	Pythagorean Identity
$= \csc^2 x$	Definition of csc
= Right side.	

$\therefore \cot^2 x + 1 = \csc^2 x$ for all real numbers x for which both sides are defined. **b.** $x \neq n\pi$, $\forall$ integers n
3. Using Technique 3: $\sin^2 x + \cos^2 x = 1$

$\Rightarrow \frac{\sin^2 x}{\sin^2 x} + \frac{\cos^2 x}{\sin^2 x} = \frac{1}{\sin^2 x}$	Mult. property of equations provided $\sin x \neq 0$
$\Rightarrow 1 + \cot^2 x = \csc^2 x$	Definition of cot and csc

$\therefore \cot^2 x + 1 = \csc^2 x$

5.

	$\cos x \tan x$	$\sin x$
Def. of tan	$\cos x \cdot \frac{\sin x}{\cos x}$	
Simplification	$\sin x$	

$\therefore \cos x \tan x = \sin x$, domain: $x \neq \frac{2n+1}{2}\pi$, $\forall$ integers n

7.

	$\csc^2 x \sin x$	$\frac{\sec^2 x - \tan^2 x}{\sin x}$	
Def. of csc	$\frac{1}{\sin^2 x} \cdot \sin x$	$\frac{\frac{1}{\cos^2 x} - \frac{\sin^2 x}{\cos^2 x}}{\sin x}$	Def. of sec and tan
Simplify	$\frac{1}{\sin x}$	$\frac{1 - \sin^2 x}{\sin x \cos^2 x}$	Simplify
		$\frac{\cos^2 x}{\sin x \cos^2 x}$	Pythagorean Identity
		$\frac{1}{\sin x}$	Simplify

$\therefore \csc^2 x \sin x = \frac{\sec^2 x - \tan^2 x}{\sin x}$; domain: $x \neq \frac{n\pi}{2}$, $\forall$ integers n

9. an identity:
Left side $= \sin^2 x (\cot^2 x + 1)$

$= \sin^2 x \left(\frac{\cos^2 x}{\sin^2 x} + 1\right)$	Definition of cot
$= \frac{\sin^2 x \cos^2 x}{\sin^2 x} + \sin^2 x$	Distributive property
$= \cos^2 x + \sin^2 x$	Simplifying
$= 1$	Pythagorean identity
= Right side	

11.

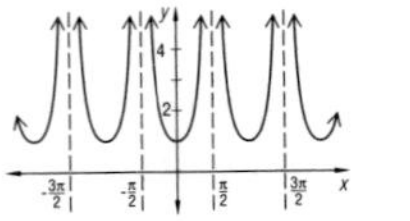

13. a. $\sin x, \tan x, \csc x, \cot x$ **b.** $\cos x, \sec x$
15. $\frac{9(x-3)}{x(1-3x)}, x \neq 0, x \neq \frac{1}{3}$ **17. a.** $\{n: n = 3 + 7q$ for some integer $q\}$ **b.** The set in part **a** is the set of all integers with remainder of 3 when divided by 7. **19. a.** $a = \cos \theta$; $b = \sin \theta$ **b.** $d = \sqrt{(a-1)^2 + b^2}$ or $\sqrt{2 - 2 \cdot \cos \theta}$ **21.** valid, Law of Indirect Reasoning

LESSON 6-3 (pp. 357–361)
1. OP, OQ, OR, OS are all radii, so they are equal. $m\angle POR = \alpha + \beta = m\angle QOS$. So $\Delta ROP \cong \Delta QOS$ (Side-Angle-Side congruence). Hence, $PR = QS$.
3. $\frac{\sqrt{6}+\sqrt{2}}{4}$ **5.** $\cos(\frac{3\pi}{2} + x) = \cos \frac{3\pi}{2} \cos x - \sin \frac{3\pi}{2} \sin x = 0 - (-1) \cdot \sin x = \sin x$ **7.** $\cos(x-y) - \cos(x+y) = \cos x \cos y + \sin x \sin y - (\cos x \cos y - \sin x \sin y) = 2 \sin x \sin y$ **9.** $\sin(\frac{\pi}{2} - x) = \cos x$ **11.** $\frac{\sqrt{2}-\sqrt{6}}{4}$
13. $\frac{-\sqrt{34}}{3}$ **15.** $\frac{1}{1+\sin x} + \frac{1}{1-\sin x} = \frac{1 - \sin x + 1 + \sin x}{(1+\sin x)(1-\sin x)} = \frac{2}{1-\sin^2 x} = \frac{2}{\cos^2 x} = 2\sec^2 x$, $x \not\equiv \frac{\pi}{2} (\text{mod } \pi)$
17. a. $h(x) = 4 + \frac{11}{x^2+3}$ **b.** $\lim_{x \to +\infty} h(x) = 4$ and $\lim_{x \to -\infty} h(x) = 4$ **c.** No, h is defined for all real numbers. **d.** Yes; **See below.**

17. d.

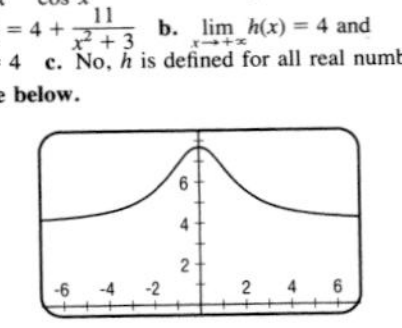

LESSON 6-4 (pp. 362–366)
1. a. See next column. $f(\alpha) = \sin \alpha + \sin \frac{\pi}{4}$; $g(\alpha) = \sin(\alpha + \frac{\pi}{4})$ **b.** not an identity **3.** $\sin(\alpha + (-\beta)) = \sin \alpha \cos(-\beta) + \cos \alpha \sin(-\beta) = \sin \alpha \cos \beta - \cos \alpha \sin \beta$
5. Step 1: Definition of tangent; Step 2: Identities for $\sin(\alpha + \beta)$ and $\cos(\alpha + \beta)$; Step 4: Simplifying factors of 1, and definition of tangent

7. $\sqrt{3} - 2$ **9. a.** $\tan(x + \pi) = \frac{\tan x + \tan \pi}{1 - \tan x \tan \pi} = \frac{\tan x + 0}{1 - (\tan x) \cdot 0} = \tan x$ **b.** The period is no larger than π.
11. $\tan(\alpha - \beta) = \tan(\alpha + (-\beta)) = \frac{(\tan \alpha + \tan(-\beta))}{1 - \tan \alpha \tan(-\beta)} = \frac{\tan \alpha - \tan \beta}{1 + \tan \alpha \tan \beta}$ **13.** $\cos(x + \frac{\pi}{2}) = \cos x \cos \frac{\pi}{2} - \sin x \sin \frac{\pi}{2} = (\cos x) \cdot 0 - (\sin x) \cdot 1 = -\sin x$
15. Counterexample: $\sin \frac{3\pi}{2} = -1$, but $\sqrt{1 - \cos^2 \frac{3\pi}{2}} = 1$.
17. a. rational **b.** irrational **c.** rational
19. $27.0875 \text{ cm} \leq x_m \leq 27.9125 \text{ cm}$ **21.** 18 ft

1. a.

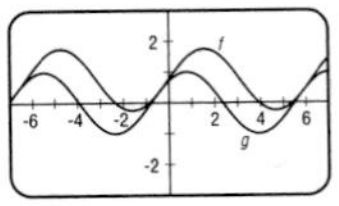

LESSON 6-5 (pp. 367–372)
1. $\cos 2x = \cos^2 x - \sin^2 x = (1 - \sin^2 x) - \sin^2 x = 1 - 2\sin^2 x$
3. a. $\cos(2 \cdot \frac{\pi}{6}) = \cos^2(\frac{\pi}{6}) - \sin^2(\frac{\pi}{6}) = \frac{3}{4} - \frac{1}{4} = \frac{1}{2}$; $\sin(2 \cdot \frac{\pi}{6}) = 2\sin(\frac{\pi}{6})\cos(\frac{\pi}{6}) = 2 \cdot \frac{1}{2} \cdot \frac{\sqrt{3}}{2} = \frac{\sqrt{3}}{2}$ **b.** $\cos(2 \cdot \frac{\pi}{6}) = \cos \frac{\pi}{3} = \frac{1}{2}$; $\sin(2 \cdot \frac{\pi}{6}) = \sin \frac{\pi}{3} = \frac{\sqrt{3}}{2}$ **5. a.** $-\frac{7}{8}$ **b.** $-\frac{\sqrt{15}}{8}$ **7.** $-\frac{\sqrt{26}}{26}$
9. Both f and g are the function $x \to \cos 2x$.

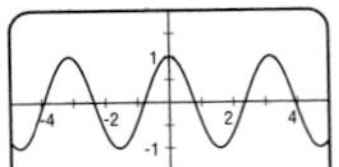

11. Left side

$= \sin 3x$	
$= \sin(2x + x)$	
$= \sin 2x \cos x + \cos 2x \sin x$	Sine of a sum identity
$= (2 \sin x \cos x) \cos x + (\cos^2 x - \sin^2 x) \sin x$	Double angle identities
$= 2 \sin x \cos^2 x + \cos^2 x \sin x - \sin^3 x$	Multiplication
$= 3 \sin x \cos^2 x - \sin^3 x$	Addition
$= 3 \sin x (1 - \sin^2 x) - \sin^3 x$	Pythagorean Identity
$= 3 \sin x - 3 \sin^3 x - \sin^3 x$	Multiplication
$= 3 \sin x - 4 \sin^3 x$	Addition
= Right side	

13. a. $-\frac{3}{5}$ **b.** $\frac{2\sqrt{2}}{3}$ **c.** $\frac{3+8\sqrt{2}}{15}$ **d.** $\frac{4-6\sqrt{2}}{15}$ **e.** Does $\sin^2(x+y) + \cos^2(x+y) = 1$? Does $\left(\frac{3+8\sqrt{2}}{15}\right)^2 + \left(\frac{4-6\sqrt{2}}{15}\right)^2 = 1$? Does $\frac{137+48\sqrt{2}}{225} + \frac{88-48\sqrt{2}}{225} = 1$? Does $\frac{225}{225} = 1$? Yes **15. a.** $-\frac{3}{5}$ **b.** $\frac{4}{5}$ **c.** $-\frac{4}{3}$ **d.** ≈ 2.214 **17.** $t = \pm 3, \pm 2$ **19.** invalid, converse error

LESSON 6-6 (pp. 373–378)
1. a. $x = \sin y$ and $\frac{-\pi}{2} \leq y \leq \frac{\pi}{2}$ **b.** $x = \tan y$ and $-\frac{\pi}{2} < y < \frac{\pi}{2}$
3. a.

$y = \sin x$	$(-\frac{\pi}{2}, -1)$	$(-\frac{\pi}{3}, -\frac{\sqrt{3}}{2})$	$(-\frac{\pi}{4}, \frac{\sqrt{2}}{2})$	$(-\frac{\pi}{6}, -\frac{1}{2})$
$y = \sin^{-1} x$	$(-1, -\frac{\pi}{2})$	$(-\frac{\sqrt{3}}{2}, -\frac{\pi}{3})$	$(-\frac{\sqrt{2}}{2}, -\frac{\pi}{4})$	$(-\frac{1}{2}, -\frac{\pi}{6})$

$y = \sin x$	(0, 0)	$(\frac{\pi}{6}, \frac{1}{2})$	$(\frac{\pi}{4}, \frac{\sqrt{2}}{2})$	$(\frac{\pi}{3}, \frac{\sqrt{3}}{2})$	$(\frac{\pi}{2}, 1)$
$y = \sin^{-1} x$	(0,0)	$(\frac{1}{2}, \frac{\pi}{6})$	$(\frac{\sqrt{2}}{2}, \frac{\pi}{4})$	$(\frac{\sqrt{3}}{2}, \frac{\pi}{3})$	$(1, \frac{\pi}{2})$

3. b.

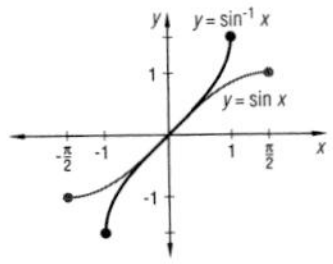

5. a. $-\frac{\pi}{4}$ **b.** -45° **7. a.** $-\frac{\pi}{3}$ **b.** -60° **9.** -1.120
11. a. the sine of the number whose cosine is $\frac{3}{5}$ **b.** $\frac{4}{5}$

13.

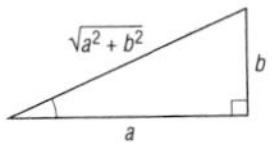

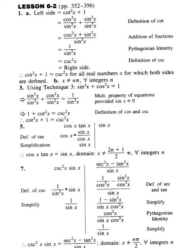

$\frac{b}{\sqrt{a^2+b^2}}$ **15.** $-\frac{\sqrt{2}}{2}$ **17.** .8 **19.** $\theta = \tan^{-1}(\frac{h}{10})$, where h = altitude (in miles) **21.** $\sin 2x = -\frac{4\sqrt{5}}{9}$, $\cos 2x = \frac{1}{9}$
23. $\sin(\alpha + \frac{\pi}{3}) + \sin(\alpha - \frac{\pi}{3}) = \sin \alpha \cos \frac{\pi}{3} + \sin \frac{\pi}{3} \cos \alpha + \sin \alpha \cos \frac{\pi}{3} - \sin \frac{\pi}{3} \cos \alpha = 2 \cdot \sin \alpha \cdot \frac{1}{2} = \sin \alpha$ **25.** sample: 15, 26, 37, 48, 59, 70, 81, 92, 103, 114

LESSON 6-7 (pp. 379–385)
1. $\frac{\pi}{2} + 2\pi n$, n an integer or $x \equiv \frac{\pi}{2} \text{ mod } 2\pi$ **3. a.** 0.841 and 5.442 **b.** $\pm 0.841 + 2n\pi$, n an integer **5. a.** $\frac{7\pi}{6}, \frac{11\pi}{6}$ **b.** $x = \frac{7\pi}{6} + 2n\pi$ or $x = \frac{11\pi}{6} + 2n\pi$, where n is any integer, or equivalently $x \equiv \frac{7\pi}{6} \text{ mod } 2\pi$ or $x \equiv \frac{11\pi}{6} \text{ mod } 2\pi$.
7. $0 \leq x \leq .848$ or $2.29 \leq x \leq 2\pi$ **9.** $\frac{\pi}{3} \leq x \leq \frac{5\pi}{3}$
11. $x = \frac{\pi}{3} + 2\pi n$ or $x = \frac{5\pi}{3} + 2\pi n$, n an integer
13. a. $\approx 42.22°$ **b.** 41.40° **15. a.** $-\frac{\pi}{4}$ **b.** $\frac{\pi}{3}$ **c.** $\frac{2\pi}{3}$

17. a. for all real numbers $x \neq \frac{n\pi}{2}$, where n is any integer
b.

	$\frac{\tan x}{\sin x} + \frac{1}{\cos x}$	$2\sec x$	
Def. of tan	$\frac{\sin x}{\cos x \sin x} + \frac{1}{\cos x}$	$\frac{2}{\cos x}$	Def. of sec
Simplify	$\frac{1}{\cos x} + \frac{1}{\cos x}$		
Adding	$\frac{2}{\cos x}$		

$\therefore \frac{\tan x}{\sin x} + \frac{1}{\cos x} = 2\sec x$
19. 29 **21. See below. a.** 2π **b.** Values of the function repeat every interval of length 2π, that is $f(x + 2\pi) = f(x)$, $\forall$ real numbers x. **23. a.** *If a parallelogram has one right angle, then it is a rectangle.* **b.** True

21.

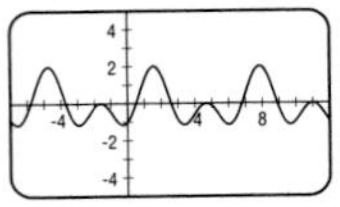

LESSON 6-8 (pp. 386–392)
1. a. See below. b. ≈ -3.26

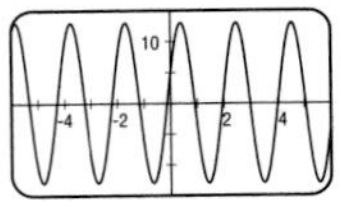

3. False **5.** (d) **7.** The weight oscillates with a maximum amplitude of 4 which decreases by a factor of $e^{-t/3}$ after t seconds. The weight is at the equilibrium position at $t = \frac{\pi}{4}$ seconds. **9.** $-\frac{5\pi}{6} \leq x \leq -\frac{\pi}{6}$ or $\frac{\pi}{6} \leq x \leq \frac{5\pi}{6}$, but $x \neq \frac{\pi}{2}$ and $x \neq -\frac{\pi}{2}$ **11.** $\theta = \tan^{-1}(\frac{d}{20})$ **13.** $\frac{\sqrt{3}-1}{1+\sqrt{3}}$
15. Left side

$= \sin(x+y) + \sin(x-y)$	
$= \sin x \cos y + \cos x \sin y + \sin x \cos y - \cos x \sin y$	Formulas for $\sin(x+y)$ and $\sin(x-y)$
$= 2\sin x \cos y$	Addition
= Right side	

$\therefore \sin(x+y) + \sin(x-y) = 2\sin x \cos y$, $\forall$ real numbers x and y. **17. a. i.** $-\cos x$ **ii.** $-\cos x$ **iii.** $\cos x$ **b.** The x-coordinates of A and D are $\cos x$, while those of B and C are $-\cos x$. But $D = (\cos(2\pi - x), \sin(2\pi - x))$, $B = (\cos(\pi - x), \sin(\pi - x))$, $C = (\cos(\pi + x), \sin(\pi + x))$.

19. identity: Left side
$= \sin^2x(\sec^2x + \csc^2x)$
$= \sin^2x\left(\frac{1}{\cos^2x} + \frac{1}{\sin^2x}\right)$ — Def. of sec and csc
$= \frac{\sin^2x}{\cos^2x} + \frac{\sin^2x}{\sin^2x}$ — Multiplication
$= \tan^2x + 1$ — Def. of tan and simplification
$= \sec^2x$ — Trigonometric Identity
= Right side.
$\therefore \sin^2x(\sec^2x + \csc^2x) = \sec^2x$, $x \neq \frac{\pi n}{2}$, n an integer.

CHAPTER 6 PROGRESS SELF-TEST (p. 395)

1. Since $\sin\alpha < 0$, $\sin\alpha = -\sqrt{1 - (\frac{x}{3})^2} = -\frac{1}{3}\sqrt{9 - x^2}$.
2. (b) 3. $\sin(\cos^{-1}\frac{1}{2}) = \sin\frac{\pi}{3} = \frac{\sqrt{3}}{2}$ 4. **See below.**
$\cos(\tan^{-1}\frac{2}{3}) = \frac{3}{\sqrt{13}}$

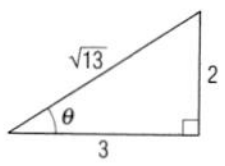

5. $\cos\frac{7\pi}{12} = \cos(\frac{\pi}{3} + \frac{\pi}{4}) = \cos\frac{\pi}{3} \cdot \cos\frac{\pi}{4} - \sin\frac{\pi}{3} \cdot \sin\frac{\pi}{4} = \frac{1}{2} \cdot \frac{\sqrt{2}}{2} - \frac{\sqrt{3}}{2} \cdot \frac{\sqrt{2}}{2} = \frac{\sqrt{2} - \sqrt{6}}{4}$ 6. $\sin\frac{\pi}{12} = \sin(\frac{\pi}{3} - \frac{\pi}{4}) = \sin\frac{\pi}{3} \cdot \cos\frac{\pi}{4} - \sin\frac{\pi}{4} \cdot \cos\frac{\pi}{3} = \frac{\sqrt{3}}{2} \cdot \frac{\sqrt{2}}{2} - \frac{\sqrt{2}}{2} \cdot \frac{1}{2} = \frac{\sqrt{6} - \sqrt{2}}{4}$

7. Not an identity, for example: $\tan(-\frac{\pi}{6}) = \frac{-1}{\sqrt{3}} \neq \sqrt{3} = \tan(-\frac{\pi}{6} + \frac{\pi}{2})$ **See below.**

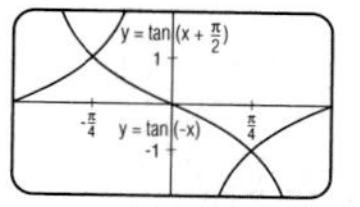

8. domain: $x \neq (2n + 1)\frac{\pi}{2}$, n an integer; $\cos x + \tan x \sin x = \cos x + \frac{\sin^2x}{\cos x} = \frac{\cos^2x + \sin^2x}{\cos x} = \frac{1}{\cos x} = \sec x$ 9. domain: $\alpha \neq (2n + 1)\frac{\pi}{2}$, $\beta \neq (2n + 1)\frac{\pi}{2}$, n an integer; $\tan\alpha + \tan\beta = \frac{\sin\alpha}{\cos\alpha} + \frac{\sin\beta}{\cos\beta} = \frac{\sin\alpha\cos\beta + \sin\beta\cos\alpha}{\cos\alpha\cos\beta} = \frac{\sin(\alpha + \beta)}{\cos\alpha\cos\beta}$
10. a. Use chunking for $\sin x$: $2\sin^2x - \sin x - 1 = 0$; $(2\sin x + 1)(\sin x - 1) = 0$; $\sin x = -\frac{1}{2}$ or 1; $x = \frac{7\pi}{6}, \frac{11\pi}{6}$, or $\frac{\pi}{2}$
b. $x = \frac{7\pi}{6} + 2\pi n, \frac{11\pi}{6} + 2\pi n, \frac{\pi}{2} + 2\pi n$; n an integer
11. $\cos 2x = \sin x$; $1 - 2\sin^2x = \sin x$; $2\sin^2x + \sin x - 1 = 0$; $\sin x = -1$ or $\sin x = \frac{1}{2}$; For $\sin x = -1$, there is no solution in the given interval. For $\sin x = \frac{1}{2}$, $x = \frac{\pi}{6}$. 12. In the given interval, $\cos 2x = \sin x$ when $x = -\frac{1}{2}, -\frac{7\pi}{6}$, or $-\frac{11\pi}{6}$. So, from the graph it can be seen that $\sin x \leq \cos 2x$ in then intervals: $0 \geq x \geq -\frac{7\pi}{6}$ or $-\frac{11\pi}{6} \geq x \geq -2\pi$ 13. $1.0 \sin 20° = 1.33 \sin\theta$; $0.2572 = \frac{\sin 20°}{1.33} = \sin\theta$; $14.9° \approx \theta$ 14. $\tan\theta = \frac{150}{d}$, so $\theta = \tan^{-1}(\frac{150}{d})$.

The chart below keys the **Progress Self-Test** questions to the objectives in the **Chapter 6 Review** on pages 396–398. This will enable you to locate those **Chapter 6 Review** questions that correspond to questions you missed on the **Progress Self-Test.** The lesson where the material is covered is also indicated in the chart.

Question	**1**	**2**	**3-4**	**5**	**6**	**7**
Objective	A	A	B	A	A	G
Lesson	6-2	6-3	6-6	6-3	6-4	6-1
Question	**8**	**9**	**10-11**	**12**	**13**	**14**
Objective	D	D	C	H	F	E
Lesson	6-2	6-4	6-7	6-7	6-7	6-6

CHAPTER 6 CHAPTER REVIEW (pp. 396–398)

1. $-\frac{\sqrt{55}}{8}$ 3. $-\frac{8}{3}$ 5. $\frac{\sqrt{2 + \sqrt{2}}}{2}$ 7. (b) 9. $\frac{2\sqrt{42} + 2}{15}$
11. $\frac{\sqrt{6}}{3}$ 13. $-\frac{\pi}{4}$ 15. $\frac{\sqrt{2}}{2}$ 17. $\frac{\sqrt{21}}{5}$ **See next column.**

17.

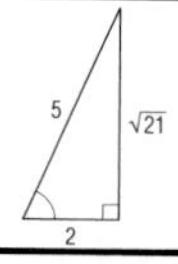

19. $x = \frac{3\pi}{4}, \frac{7\pi}{4}$ 21. $x = \frac{-\pi}{2} + 2\pi n, \frac{\pi}{4} + \pi n$, n an integer
23. $0 \leq x \leq 0.644$ or $5.640 \leq x \leq 2\pi$, approximately
25. $\frac{\pi}{6} < x < \frac{5\pi}{6}$ 27. $\sin(y - x)$ 29. $-\sin^2 x$
31. Left side
$= \cos\left(\frac{3\pi}{2} + x\right)$
$= \cos\frac{3\pi}{2}\cos x - \sin\frac{3\pi}{2}\sin x$ — Identity for the cosine of a sum
$= 0 - (-\sin x)$ — Evaluating trig. functions
$= \sin x$ — Multiplication
= Right side
$\therefore \cos\left(\frac{3\pi}{2} + x\right) = \sin x$; domain: all real numbers
33. Left side
$= \sin\left(\frac{\pi}{2} + x\right)$
$= \sin\frac{\pi}{2} \cdot \cos x + \cos\frac{\pi}{2} \cdot \sin x$ — Formula for sine of a sum
$= 1 \cdot \cos x + 0 \cdot \sin x$ — Evaluating trig. functions
$= \cos x$ — Multiplication
= Right side
$\therefore \sin\left(\frac{\pi}{2} + x\right) = \cos x$; domain: all real numbers
35. Left side
$= \cos(\alpha - \beta) - \cos(\alpha + \beta)$
$= \cos\alpha\cos\beta + \sin\alpha\sin\beta - \cos\alpha\cos\beta + \sin\alpha\sin\beta$ — Sum and difference identities
$= 2\sin\alpha\sin\beta$ — Addition
= Right side
$\therefore \cos(\alpha - \beta) - \cos(\alpha + \beta) = 2\sin\alpha\sin\beta$; domain: all real numbers

37. Left side
$= \cos 4x = \cos^2 2x - \sin^2 2x$ — Identity for cos 2x
$= (\cos^2x - \sin^2x)^2 - (2\sin x\cos x)^2$ — Identities for cos 2x and sin 2x
$= \cos^4x - 2\sin^2x\cos^2x + \sin^4x - 4\sin^2x\cos^2x$ — Multiplication
$= \cos^4x - 6\sin^2x\cos^2x + \sin^4x$ — Addition
= Right side
$\therefore \cos 4x = \cos^4x - 6\sin^2x\cos^2x + \sin^4x$; domain: all real number 39. $\theta = \sin^{-1}\left(\frac{h - 3}{200}\right)$, where h = height of the kite above the ground in feet. 41. $\theta = \tan^{-1}(\frac{x}{100})$
43. $\approx 34.8°$ or $55.2°$ 45. An identity.

	$1 + \cot^2x$	$\csc^2x$	
Def. of cot	$1 + \frac{\cos^2x}{\sin^2x}$	$\frac{1}{\sin^2x}$	Def. of csc
Addition	$\frac{\sin^2x + \cos^2x}{\sin^2x}$		
Pythagorean Identity	$\frac{1}{\sin^2x}$		

$\therefore 1 + \cot^2x = \csc^2x$, $x \neq n\pi$, n an integer
47. An identity. Left side
$= \tan(\pi + \gamma)$
$= \frac{\sin(\pi + \gamma)}{\cos(\pi + \gamma)}$ — Definition of tan
$= \frac{\sin\pi\cos\gamma + \sin\gamma\cos\pi}{\cos\pi\cos\gamma - \sin\pi\sin\gamma}$ — Sum identities for sin and cos
$= \frac{-\sin\gamma}{-\cos\gamma}$ — Evaluating specific trig. functions
$= \tan\gamma$ — Definition of tan
= Right side
49. You can check various different cases by holding α constant and graphing the resulting functions in β. For example, when $\alpha = \frac{\pi}{3}$ you could graph $y = \sin\left(\frac{\pi}{3} + \beta\right)$ and $y = \sin\frac{\pi}{3}\cos\beta - \cos\frac{\pi}{3}\sin\beta$. 51.a. $x \approx 1.1$ or $x \approx 3.6$
b. $0 \leq x < 1.1$, $3.6 < x \leq 2\pi$ 53. $0 \leq x \leq 0.675$

LESSON 7-1 (pp. 400–406)
1. $T_7 = 127$; $T_8 = 255$ 3. samples: 2, 8, 26, 80, 242, … and 5, 17, 53, 161, 485, … 5. a. 3, 12, 48, 192, 768
b. $a_1 = 3$, $a_{k+1} = 4a_k$, ∀ integers $k \geq 1$ c. $a_n = 3(4)^{n-1}$, ∀ integers $n \geq 1$ 7. 1, 1, 2, 3, 5, 8, 13, 21, 34, 55, 89, 144, 233 9. 5, 2, 11, 35, 116, 383 11. The initial conditions are different. $x_1 = 1$, $x_2 = 1$ for the first sequence, and $x_1 = 1$, $x_2 = 0$ for the second. 13. explicit; −1, 1, −1, 1, −1, 1
15. a. 1, 2, 3, 4, 5 b. 1, 2, 3, 4, 29 c. The first four terms of each sequence are identical, but the fifth terms are different. One must be cautious when generalizing because many different sequence formulas may generate the same first few terms. 17. a. **See next column.** b. $C_n = 20n + 5$, ∀ integers $n \geq 1$ c. $C_1 = 25$, $C_{k+1} = C_k + 20$, ∀ integers $k \geq 1$
19. **See next column.** $\tan x = \frac{1}{\cot x}$. One way to obtain the graph of $\cot x$ from the graph of $\tan x$ is to reflect $\tan x$ with respect to the y-axis, then apply the transformation $T_{-\pi/2,0}$.

17. a.

Weight ≤ (oz)	Rate (cents)
1	25
2	45
3	65
4	85
5	105
6	125
7	145
8	165

19.

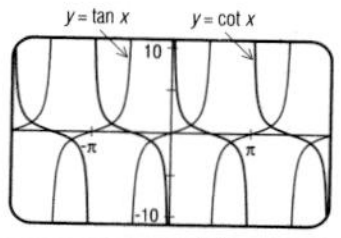

21. $x \approx 1.585$ 23. b. Let 1 be the top (smallest) disk, 2 the middle disk, and 3 the bottom (largest) disk. Let a be the original needle, and b and c the other 2 needles. Move 1 to b, 2 to c, 1 to c, 3 to b, 1 to a, 2 to b, and 1 to b (7 steps).
c. Let 4 now be the bottom (largest) disk. Move 1 to b, 2 to c, 1 to c, 3 to b, 1 to a, 2 to b, 1 to b, 4 to c, 1 to c, 2 to a, 1 to a, 3 to c, 1 to b, 2 to c, and 1 to c (15 steps).

LESSON 7-2 (pp. 407–412)
1. $S_1 = 11$, $S_{k+1} = S_k + 4$, ∀ integers $k \geq 1$ 3. ≈ 585 billion years 5. a. 3, 7, 11, 15, 19, 23 b. $t_n = 4n - 1$
c. $t_1 = 4(1) - 1 = 3$, so the initial condition is met. $t_{n+1} = 4(n + 1) - 1 = 4n + 4 - 1 = t_n + 4$, so the recursive relationship is satisfied. Therefore, the explicit formula is correct. 7. $1, \frac{1}{2}, \frac{1}{3}, \frac{1}{4}, \frac{1}{5}, \frac{1}{6}$; $a_n = \frac{1}{n}$, ∀ integers $n \geq 1$
9. a. $a_1 = 2$; $a_{k+1} = 3a_{k+2}$, ∀ integers $k \geq 1$
b. $a_n = 3^n - 1$, integers $n \geq 1$, conjectured from 2, 8, 26, 80, … . c. $a_1 = 3^1 - 1 = 2$, so the initial condition is met. $a_{n+1} = 3^{n+1} - 1 = 3 \cdot 3^n - 1 = 3 \cdot 3^n - 3 + 3 - 1 = 3(3^n - 1) + 2 = 3a_n + 2$, so the recurrence relationship is met. Therefore the explicit formula is correct. 11. a. explicit
b. $x, \frac{x^2}{2}, \frac{x^3}{6}, \frac{x^4}{24}, \frac{x^5}{120}$ 13. a. $\lim_{x \to +\infty} f(x) = 2$; $\lim_{x \to -\infty} f(x) = 2$
b. There are essential discontinuities at $x = 3$ and $x = -3$.
c. $x = 3$, $x = -3$, $y = 2$ 15. $\frac{n + 1}{n + 2}$, for $n \neq -1$
17. a. If there exists an x such that $p(x)$ is true and $q(x)$ is false. b. False

LESSON 7-3 (pp. 413–417)
1. a. $-2 + 2 + 8 + 16$ b. 24 3. True, the only difference is that different letters are used for the indices. 5. $\sum_{j=1}^{k} \frac{1}{j + 1}$
7. a. $\frac{37}{15} \approx 2.47$ b. $\frac{n + 1}{2n - 1} + \frac{n + 1}{2n} + \frac{n + 1}{2n + 1}$ 9. a. $\sum_{i=1}^{1} a_i = 1$, $\sum_{i=1}^{2} a_i = 5$, $\sum_{i=1}^{3} a_i = 14$, $\sum_{i=1}^{4} a_i = 30$ b. $\sum_{i=1}^{k+1} i^2 = 1^2 + 2^2 + \ldots + k^2 + (k + 1)^2 = (1^2 + 2^2 + \ldots + k^2) + (k + 1)^2 = \left(\sum_{i=1}^{k} i^2\right) + (k + 1)^2$ 11. $\sum_{i=1}^{k+1} i(i - 1) = \left(\sum_{i=1}^{k} i(i - 1)\right) + (k + 1)k$
13. It does. 15. Sample: let $a_n = 1$ ∀ n. Then $\sum_{n=1}^{4} (a_n^2) = 1^2 + 1^2 + 1^2 + 1^2 = 4$, and $\left(\sum_{n=1}^{4} a_n\right)^2 = (1 + 1 + 1 + 1)^2 = 16$. 17. $\frac{1}{2}, \frac{1}{3}, \frac{1}{4}, \frac{1}{5}, \frac{1}{6}, \frac{1}{7}$; $a_n = \frac{1}{n + 1}$, ∀ integers $n \geq 1$ 19. Left side $= \sin^2x - \sin^2y = (1 - \cos^2x) - (1 - \cos^2y) = 1 - \cos^2x - 1 + \cos^2y = \cos^2y - \cos^2x$ = Right side. Therefore, $\sin^2x - \sin^2y = \cos^2y - \cos^2x$ ∀ real numbers x and y. 21. $g \circ f(k) = \frac{(k + 1)(k + 2)}{2}$

LESSON 7-4 (pp. 418–425)
1. a. $\frac{1(1 + 1)}{2} = 1$ b. $1 + 2 + 3 + \ldots + k = \frac{k(k + 1)}{2}$ c. $1 + 2 + 3 + \ldots + k + k + 1 = \frac{(k + 1)(k + 2)}{2}$ d. $1 + 2 + 3 + \ldots + k + k + 1$ e. Use inductive assumption.

f. $\frac{k(k + 1) + 2(k + 1)}{2}$ g. $\frac{(k + 1)(k + 2)}{2}$ h. the Principle of Mathematical Induction 3. a. $\sum_{i=1}^{1} 2i = 2(1) = 2$ and $1(1 + 1) = 2$, so $S(1)$ is true. b. $S(k)$: $\sum_{i=1}^{k} 2i = k(k + 1)$; $S(k + 1)$: $\sum_{i=1}^{k+1} 2i = (k + 1)(k + 2)$ c. $\sum_{i=1}^{k+1} 2i = \sum_{i=1}^{k} 2i + 2(k + 1) = k(k + 1) + 2(k + 1) = (k + 1)(k + 2)$, so $S(k + 1)$ is true. d. $S(n)$ is true for all integers $n \geq 1$. 5. a. $S(1)$: $5 < 4$; $S(3)$: $13 < 16$; $S(5)$: $29 < 36$ b. $S(1)$ is false. $S(3)$ and $S(5)$ are true. c. $S(k + 1)$: $(k + 1)^2 + 4 < (k + 2)^2$
7. a. $S(1)$: $1 = \frac{1(2)(3)}{6}$, so $S(1)$ is true. b. Assume that $S(k)$ is true for a particular but arbitrarily chosen integer $k \geq 1$. $S(k)$: $1^2 + 2^2 + \ldots + k^2 = \frac{k(k + 1)(2k + 1)}{6}$. Show $S(k + 1)$: $1^2 + 2^2 + \ldots + k^2 + (k + 1)^2 = \frac{(k + 1)(k + 2)(2(k + 1) + 1)}{6}$ is true. $1^2 + 2^2 + \ldots + k^2 + (k + 1)^2 = \frac{k(k + 1)(2k + 1)}{6} + (k + 1)^2 = \frac{k(k + 1)(2k + 1) + 6(k + 1)^2}{6} = \frac{(k + 1)(k(2k + 1) + 6(k + 1))}{6} = \frac{(k + 1)(2k^2 + 7k + 6)}{6} = \frac{(k + 1)(k + 2)(2k + 3)}{6} = \frac{(k + 1)(k + 2)(2(k + 1) + 1)}{6}$
c. Since $S(1)$ is true, and $S(k) \Rightarrow S(k + 1)$, by mathematical induction, $S(n)$ is true ∀ integers $n \geq 1$.
9. $S(n)$: $a_n = 2n^2 - 1$, ∀ integers $n \geq 1$. (1) $a_1 = 1$ from the recursive definition. $a_1 = 2 \cdot 1^2 - 1 = 1$ from the explicit definition. Hence $S(1)$ is true. (2) Assume $S(k)$: $a_k = 2k^2 - 1$ for some integer $k \geq 1$. Show $S(k + 1)$: $a_{k+1} = 2(k + 1)^2 - 1$ is true. $a_{k+1} = a_k + 4k + 2 = 2k^2 - 1 + 4k + 2 = 2(k^2 + 2k + 1) - 1 = 2(k + 1)^2 - 1$. Therefore, $S(k + 1)$ is true if $S(k)$ is true, and (1) and (2) prove by the Principle of Mathematical Induction that the explicit formula does describe the sequence. 11. $\sum_{i=1}^{n} \frac{i}{n}$ 13. a. Since $x - y$ is a factor of $x^4 - y^4$, and $x^4 - y^4$ is a factor of $x^5 - xy^4 = x(x^4 - y^4)$, then by the Transitive Property of Factors, $x - y$ is a factor of $x^5 - xy^4$ b. $xy^4 - y^5 = y^4(x - y)$
c. Since $x - y$ is a factor of $x^5 - xy^4$ by part **a**, and $x - y$ is a factor of $xy^4 - y^5$ by part **b**, then $x - y$ is a factor of $(x^5 - xy^4) + (xy^4 - y^5)$ by the Factor of a Sum Theorem.
15. a., b., c **See below.** d.(i) $\frac{\pi}{4}$ (ii) $\frac{\pi}{3}$ (iii) $\frac{5\pi}{6}$
15. a., b., c.

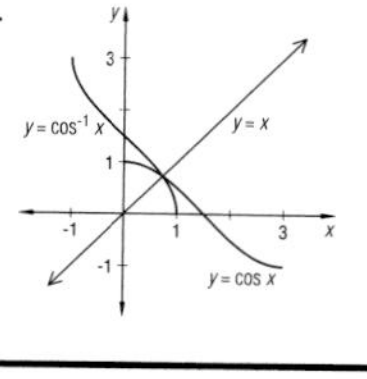

LESSON 7-5 (pp. 426–430)
1. $5^3 - 4 \cdot 3 - 1 = 112 = 16 \cdot 7$, so 16 is a factor.
3. a. $S(1)$: 2 is a factor of 2; since $2 \cdot 1 = 2$, $S(1)$ is true. $S(13)$: 2 is a factor of 158; since $2 \cdot 79 = 158$, $S(13)$ is true. $S(20)$: 2 is a factor of 382; since $2 \cdot 191 = 382$, $S(20)$ is true.
b. $S(k)$: 2 is a factor of $k^2 - k + 2$. $S(k + 1)$: 2 is a factor of $(k + 1)^2 - (k + 1) + 2$. c. $k^2 + k + 2$ d. $k^2 + k + 2 = (k^2 - k + 2) + 2k$; 2 is a factor of $k^2 - k + 2$ and $2k$, so 2 is a factor of $(k + 1)^2 - (k + 1) + 2$ by the Factor of a Sum Theorem. e. 2 is a factor of $n^2 - n + 2$ ∀ integers $n \geq 1$
5. Since $9 - 3 = 6$, substituting $x = 9$ and $y = 3$ into the result of Example 3 yields $9 - 3 = 6$ is a factor of $9^n - 3^n$ ∀ $n \geq 1$. 7. $S(1)$ is true since $3^3 + 14 \cdot 3 + 3 = 24 \cdot 3$. Assume $S(k)$: 3 is a factor of $k^3 + 14k + 3$. Show $S(k + 1)$: 3 is a factor of $(k + 1)^3 + 14(k + 1) + 3$ is true. $(k + 1)^3 + 14(k + 1) + 3 = k^3 + 3k^2 + 17k + 18 = (k^3 + 14k + 3) + 3k^2 + 3k + 15$. Since 3 is a factor of $k^3 + 14k + 3$ and of $3(k^2 + k + 5)$, $S(k + 1)$ is true. Hence, 3 is a factor of $n^3 + 14n + 3$ ∀ $n \geq 1$, by the Principle of Mathematical Induction.
9. a. $S(1)$ is false. $S(2)$ and $S(3)$ are true. b. All we can conclude is that $S(2)$ and $S(3)$ are true. c. $12^n - 8^n = 4^n(3^n - 2^n)$. 8 is a factor of 4^n if $n \geq 2$. Hence, $S(n)$ holds if $n \geq 2$. 11. $S(1)$ is true because $\sum_{i=1}^{1} i^3 = 1^3 = 1$ and $\left[\frac{1(1 + 1)}{2}\right]^2 = 1^2 = 1$. Assume $S(k)$: $\sum_{i=1}^{k} i^3 = \left[\frac{k(k + 1)}{2}\right]^2$ is true for some integer $k \geq 1$. Show that $S(k + 1)$: $\sum_{i=1}^{k+1} i^3 = \left[\frac{(k + 1)(k + 2)}{2}\right]^2$ is true. Now $\sum_{i=1}^{k+1} i^3 = \sum_{i=1}^{k} i^3 + (k + 1)^3 = \left[\frac{k(k + 1)}{2}\right]^2 + (k + 1)^3 = \frac{k^2(k + 1)^2}{4} + \frac{4(k + 1)^3}{4} = \frac{(k + 1)^2[k^2 + 4(k + 1)]}{4} = \frac{[(k + 1)(k + 2)]^2}{4}$. Since $S(1)$ is true and $S(k) \Rightarrow S(k + 1)$, by the Principle of Mathematical Induction, $S(n)$ is true for all integers $n \geq 1$. 13. True
15. Left side
$= \frac{a^4 - b^4}{a - b}$
$= \frac{(a^2 - b^2)(a^2 + b^2)}{a - b}$ — Factoring
$= \frac{(a - b)(a + b)(a^2 + b^2)}{a - b}$ — Factoring
$= (a + b)(a^2 + b^2)$ — Division
$= a^3 + a^2b + ab^2 + b^3$ — Multiplication
= Right side
Therefore, $\frac{a^4 - b^4}{a - b} = a^3 + a^2b + ab^2 + b^3$ when $a \neq b$, by the Transitive Property. 17. $f(t) = 30 \cdot 3^t$

LESSON 7-6 (pp. 431–437)
1. ≈ 1016.6 3. $S_3 = \frac{11}{6}$, $S_4 = \frac{25}{12}$, $S_5 = \frac{137}{60}$
5. a. ≈ 2.00000 b. 6 7. No, it diverges since the ratio $r = \frac{10}{9} > 1$ and $\lim_{k \to \infty} (\frac{10}{9})^k = +\infty \neq 0$.

9. a. $a\left(\frac{1 - r^n}{1 - r}\right) = a \cdot \frac{-1}{-1} \cdot \frac{1 - r^n}{1 - r} = a\left(\frac{r^n - 1}{r - 1}\right)$ b. $a = 1$, $r = 2$: Left side $= 1 \cdot \left(\frac{1 - 2^n}{-1}\right) = 2^n - 1$; Right side $= 1 \cdot \left(\frac{2^n - 1}{1}\right) = 2^n - 1$. $a = 1$, $r = \frac{1}{2}$: Left side $= 1 \cdot \left(\frac{1 - (\frac{1}{2})^n}{\frac{1}{2}}\right) = 2 - 2(\frac{1}{2})^n$; Right side $= 1 \cdot \left(\frac{(\frac{1}{2})^n - 1}{-\frac{1}{2}}\right) = 2 - 2(\frac{1}{2})^n$.
c. the right d. the left 11. a. $a_1 = 3$, $a_k = \frac{1}{2}a_{k-1}$, for all integers $2 \leq k \leq 25$ b. $a_n = 3(\frac{1}{2})^{n-1}$ c. $\sum_{j=1}^{25} 3(\frac{1}{2})^{j-1}$
d. ≈ 5.99999982 e. 6 f. The result confirms **d** and **e**.
13. a. $S_1 = a$ and $S_{n+1} = S_n + ar^n$, ∀ integers $n \geq 1$.
b. $S_1 = \frac{a(1 - r^1)}{1} = a$; so the initial condition is met. $S_{n+1} = \frac{a(1 - r^{n+1})}{1 - r} = a + ar + ar^2 + \ldots + ar^n = (a + ar + \ldots + ar^{n-1}) + ar^n = \sum_{i=1}^{n} ar^{i-1} + ar^n = S_n + ar^n$ 15. a. $S(3)$: $\sum_{i=1}^{3} i(i + 1) = \frac{3 \cdot 4 \cdot 5}{3}$.
Does $1 \cdot 2 + 2 \cdot 3 + 3 \cdot 4 = 4 \cdot 5$? Does $2 + 6 + 12 = 20$? Does $20 = 20$? Yes. Hence $S(3)$ is true. b. $S(1)$: $1 \cdot 2 = \frac{1 \cdot 2 \cdot 3}{3}$ which is true. Assume $S(k)$: $\sum_{i=1}^{k} i(i + 1) = \frac{k(k + 1)(k + 2)}{3}$ is true for some integer $k \geq 1$. Show that $S(k + 1) = \sum_{i=1}^{k+1} i(i + 1) = \frac{(k + 1)(k + 2)(k + 3)}{3}$ is true. Then $\sum_{i=1}^{k+1} i(i + 1) = \sum_{i=1}^{k} i(i + 1) + (k + 1)(k + 2) = \frac{k(k + 1)(k + 2)}{3} + (k + 1)(k + 2) = (k + 1)(k + 2)(\frac{k}{3} + 1) = \frac{(k + 1)(k + 2)(k + 3)}{3}$. Hence, $S(n)$ holds ∀ $n \geq 1$.
17. a. **See below.** The graph represents the transformation $T_{0.5\pi,0}$ applied to $y = \tan\theta$.
b. $-\pi < \theta \leq -\frac{\pi}{4}$ or $0 < \theta \leq \frac{3\pi}{4}$
19. a. 9.867 b. ≈ 3.141 c. π
17. a.

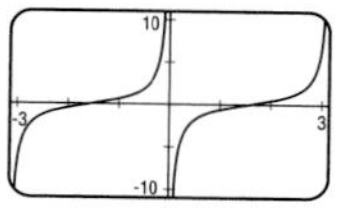

LESSON 7-7 (pp. 438–444)
1. In the Strong Form of Mathematical Induction, the assumption that each of $S(1), S(2), \ldots, S(k)$ are true is used to show that $S(k + 1)$ is true. In the original form, only the assumption that $S(k)$ is true is used to prove $S(k + 1)$ is true.
3. The 7-piece and 13-piece blocks needed 6 and 12 steps respectively by the inductive assumption. That is 18 steps; joining them is the 19th step.

5. a. 5, 15, 20, 35 b. $a_1 = 5$ and $a_2 = 15$ are multiples of 5. Assume $a_1, a_2, \ldots, a_k$ are all multiples of 5. Show that a_{k+1} is a multiple of 5. Now $a_{k+1} = a_k + a_{k-1}$. 5 is a factor of a_k and a_{k-1}, so it is a factor of their sum, a_{k+1}. Hence, by the Strong Form of Mathematical Induction, a_n is a multiple of 5 $\forall\, n \geq 1$. 7. a. $S(2)$: $L_2 = F_3 + F_1$. $3 = 2 + 1$ so $S(2)$ is true. $S(3)$: $L_3 = F_4 + F_2$. $4 = 3 + 1$ so $S(3)$ is true. b. $\forall$ integers j such that $2 \leq j \leq k$, $L_j = F_{j+1} + F_{j-1}$ c. $S(k+1)$: $L_{k+1} = F_{k+2} + F_k$ d. $L_{k+1} = L_k + L_{k-1} = (F_{k+1} + F_{k-1}) + (F_k + F_{k-2})$ e. $L_{k+1} = (F_{k-1} + F_{k-2}) + (F_{k+1} + F_k) = F_{k+2} + F_k$ 9. Let $S(n)$: $n^3 + 3n^2 + 2n$ is divisible by 3. Then $S(1)$: $1^3 + 3 \bullet 1^2 + 2 \bullet 1$ is divisible by 3. $1 + 3 + 2 = 6$ which is divisible by 3, so $S(1)$ is true. Assume $S(k)$: $k^3 + 3k^2 + 2k$ is divisible by 3 is true for some integer $k \geq 1$. Show $S(k+1)$: $(k+1)^3 + 3(k+1)^2 + 2(k+1)$ is divisible by 3 is true. $(k+1)^3 + 3(k+1)^2 + 2(k+1) = k^3 + 3k^2 + 3k + 1 + 3k^2 + 6k + 3 + 2k + 2 = k^3 + 6k^2 + 11k + 6 = (k^3 + 3k^2 + 2k) + 3(k^2 + 3k + 2)$. $k^3 + 3k^2 + 2k$ is divisible by 3 by the inductive assumption, and $3(k^2 + 3k + 2)$ is divisible by 3; thus, their sum is divisible by 3. Therefore by mathematical induction, $n^3 + 3n^2 + 2n$ is divisible by 3 $\forall\, n \geq 1$.

11. a. **See below.** b. $\tan x$

c. Left side

$= \csc x \sec x - \cot x$	
$= \frac{1}{\sin x} \bullet \frac{1}{\cos x} - \frac{\cos x}{\sin x}$	Def. of csc, sec, and cot
$= \frac{1}{\sin x \cos x} - \frac{\cos^2 x}{\sin x \cos x}$	Multiplication
$= \frac{1 - \cos^2 x}{\sin x \cos x}$	Subtraction of fractions
$= \frac{\sin^2 x}{\sin x \cos x}$	Pythagorean Identity
$= \frac{\sin x}{\cos x}$	Simplifying fractions
$= \tan x$	Definition of tangent
= Right side	

Therefore, $\csc x \sec x - \cot x = \tan x$ $\forall\, x$ for which both sides are defined. 13. $x = -5, 2$

11. a.

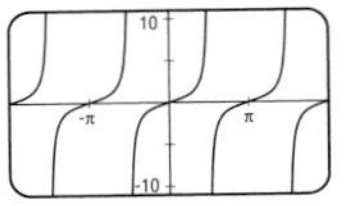

LESSON 7-8 (pp. 445–452)

1. a. 2 and 5 b. 2 and 5 are not exchanged 3. 1, 6, 4, 9 5. initial order: 5, -7, 1.5, -1, 13, 6; after first pass: -7, 1.5, -1, 5, 6, 13; after second pass: -7, -1, 1.5, 5, 6, 13 7. If no interchanges are necessary, adjacent numbers are in order. Hence, by the Transitive Property, the entire list is in order. 9. sample: 6, 5, 4, 3, 2, 1 11. Quicksort 13. a_1 and a_2 are even. Assume $a_1, a_2, \ldots, a_{k-1}, a_k$ are all even. Show that a_{k+1} is even. $a_{k+1} = c \bullet a_k + a_{k-1} = c\,(2p) + 2r = 2(cp + r)$ for some integers p and r. Since $cp + r$ is an integer by closure properties, a_{k+1} is even. Therefore a_n is even for all integers $n \geq 1$ by the Strong Form of Mathematical Induction.

15. Prove $S(n)$: $\sum_{i=1}^{n} \frac{1}{i(i+1)} = \frac{n}{n+1}$ for all integers $n \geq 1$. $S(1)$: $\sum_{i=1}^{1} \frac{1}{i(i+1)} = \frac{1}{1+1}$ is true. Assume $S(k)$: $\sum_{i=1}^{k} \frac{1}{i(i+1)} = \frac{k}{k+1}$ is true. Show that $S(k+1)$: $\sum_{i=1}^{k+1} \frac{1}{i(i+1)} = \frac{k+1}{k+2}$ is true. $\sum_{i=1}^{k+1} \frac{1}{i(i+1)} = \sum_{i=1}^{k} \frac{1}{i(i+1)} + \frac{1}{(k+1)(k+2)} = \frac{k}{k+1} + \frac{1}{k+1)(k+2)} = \frac{k(k+2)}{(k+1)(k+2)} + \frac{1}{(k+1)(k+2)} = \frac{(k+1)^2}{(k+1)(k+2)} = \frac{k+1}{k+2}$. So by mathematical induction, $S(n)$ is true $\forall$ integers $n \geq 1$. 17. a. $R = \frac{R_1 R_2}{R_1 + R_2}$ b. $\frac{70}{17} \approx 4.12$ ohms 19. The argument is valid. The three premises are: (1) $p \Rightarrow q$; (2) $q \Rightarrow r$; (3) $\sim r$. By the Transitive Property and statements (1) and (2), $p \Rightarrow r$. Then by modus tollens and (3), $\sim p$. This is the conclusion, and so the argument is valid.

LESSON 7-9 (pp. 453–457)

1. $E(n) = n$ 3. a. repeated multiplication: 9; sum of powers of 2: 4 b. repeated multiplication: 9999; sum of powers of 2: 14 5. $E(10) = 499{,}500$ 7. See below. 9. $S(n)$: $5^n - 4n - 1$ is divisible by 16. $S(1)$: $5^1 - 4 \bullet 1 - 1$ is divisible by 16. $S(1)$ is true since 0 is divisible by 16. Assume $S(k)$: $5^k - 4k - 1$ is divisible by 16 is true for some positive integer k. Show that 16 is a factor of $5^{k+1} - 4(k+1) - 1$. $5^{k+1} - 4(k+1) - 1 = 5(5^k - 4k - 1 + 4k + 1) - 4(k+1) - 1 = 5(5^k - 4k - 1) + 20k + 5 - 4(k+1) - 1 = 5(5^k - 4k - 1) + 16k$. 16 is a factor of $5^k - 4k - 1$ (by the inductive assumption) and a factor of $16k$. Hence, by the Factor of a Sum Theorem, 16 is also a factor of $5(5^k - 4k - 1) + 16k$. Therefore, by mathematical induction, $5^n - 4n - 1$ is divisible by 16 for all positive integers n. 11. $\sum_{i=0}^{4} (a_i + cb_i) = a_0 + cb_0 + a_1 + cb_1 + \ldots + a_4 + cb_4 = a_0 + a_1 + \ldots + a_4 + cb_0 + cb_1 + \ldots + cb_4 = (a_0 + a_1 + \ldots + a_4) + c(b_0 + b_1 + \ldots + b_4) = \sum_{i=0}^{4} a_i + c\sum_{i=0}^{4} b_i$ 13. a. not (p and (q or r)) b. sample: $p = 0$, $q = 1$, $r = 0$ 15. a. 2 b. 3 c. Each of the n digits of the second number are multiplied by up to n digits of the first number. Hence, there are at most n^2 multiplications. d. $2n$ column additions e. $n^2 + 2n$; for the problem, $n = 2$, so the efficiency should be 8. All 4 multiplication steps listed above must be carried out, as well as 4 column additions. $4 + 4 = 8$, so the algorithm checks.

7.

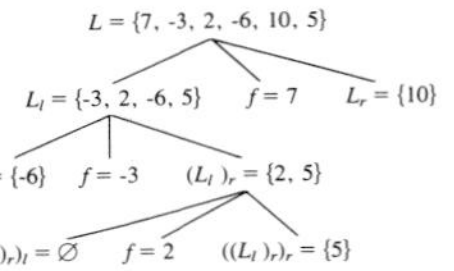

The sorted list is -6, -3, 2, 5, 7, 10.

CHAPTER 7 PROGRESS SELF-TEST (p. 461)

1. a. 1, 2, 4, 8, 16 b. $a_n = 2^{n-1}$ 2. $r_1 = 5$, $r_k = 3r_{k-1}$ for $k \geq 2$ 3. (c) 4. $\sum_{i=1}^{k+1} i^2 = \left(\sum_{i=1}^{k} i^2\right) + (k+1)^2$ 5. Let $S(n)$: $c_n = 3 \bullet 2^n - 3$ for integers $n \geq 1$. $S(1)$: $c_1 = 3 \bullet 2^1 - 3$. This matches the recursive definition since $c_1 = 3$, so $S(1)$ is true. Assume $S(k)$: $c_k = 3 \bullet 2^k - 3$ is true for an arbitrary integer k. Show that $S(k+1)$: $c_{k+1} = 3 \bullet 2^{k+1} - 3$ is true. From the recursive definition, $c_{k+1} = 2c_k + 3 = 2(3 \bullet 2^k) - 3 = 3 \bullet 2^{k+1} - 3$. Therefore, $S(k+1)$ is true. Hence by mathematical induction, $S(n)$ is true for all $n \geq 1$, and so the explicit formula $c_n = 3 \bullet 2^n - 3$ yields the correct definition of the sequence. 6. a. $2\left(\frac{1 - \left(\frac{1}{3}\right)^{11}}{1 - \frac{1}{3}}\right) \approx 2.9999831$ b. $\frac{2}{1 - \frac{1}{3}} = 3$ 7. a. recursive b. $b_1 = 1$, $b_{j+1} = 2b_j + 5$, for $j \geq 1$ 8. Let $S(n)$: $\sum_{i=1}^{n} i^2 = \frac{n(n+1)(2n+1)}{6}$.
(1) $\sum_{i=1}^{1} i^2 = 1^2 = 1$. $\frac{1 \bullet 2 \bullet 3}{6} = 1$ so $S(1)$ is true.
(2) Assume that $S(k)$ is true for an arbitrary integer $k \geq 1$ where $S(k)$: $\sum_{i=1}^{k} i^2 = \frac{k(k+1)(2k+1)}{6}$. Now use the assumption that $S(k)$ is true to prove $S(k+1)$ is true:

$\sum_{i=1}^{k+1} i^2 = \left(\sum_{i=1}^{k} i^2\right) + (k+1)^2 = \frac{k(k+1)(2k+1)}{6} + (k+1)^2 = \frac{k(k+1)(2k+1)}{6} + \frac{6(k+1)^2}{6} = \left(\frac{k+1}{6}\right)(2k^2 + k + 6k + 6) = \frac{(k+1)(k+2)(2k+3)}{6} = \frac{(k+1)((k+1)+1)(2(k+1)+1}{6}$.

Hence, $S(n)$ is true for all integers $n \geq 1$, by the Principle of Mathematical Induction. 9. Let $S(n)$: 4 is a factor of $2n^2 + 2n + 8$. (1) $2 \bullet 1^2 + 2 \bullet 1 + 8 = 12$ and 4 is a factor of 12, so $S(1)$ is true. (2) Assume $S(k)$ is true. That is, 4 is a factor of $2k^2 + 2k + 8$ for some arbitrary integer $k \geq 1$. Show $S(k+1)$ is true. $2(k+1)^2 + 2(k+1) + 8 = 2k^2 + 4k + 2 + 2k + 2 + 8 = (2k^2 + 2k + 8) + 4k + 4 = (2k^2 + 2k + 8) + 4(k+1)$. 4 is a factor of $2k^2 + 2k + 8$ by the inductive assumption and 4 is clearly a factor of $4(k+1)$. Hence, 4 is a factor of $2(k+1)^2 + 2(k+1) + 8$ by the Factor of a Sum Theorem. Therefore, $S(n)$ is true for all positive integers n by mathematical induction. So 4 is a factor of $2n^2 + 2n + 8$ for all positive integers n. 10. Assume $S(1), S(2), \ldots, S(k)$ are true for some integer $k \geq 1$

11.

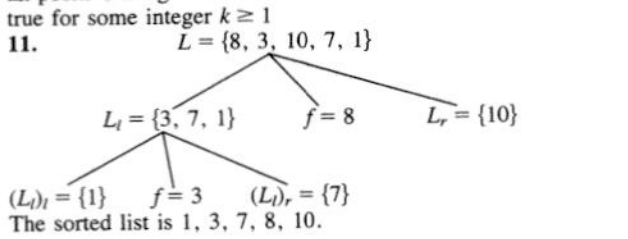

The sorted list is 1, 3, 7, 8, 10.

The chart below keys the **Progress Self-Test** questions to the objectives in the **Chapter 7 Review** on pages 462–464. This will enable you to locate those **Chapter 7 Review** questions that correspond to questions you missed on the **Progress Self-Test.** The lesson where the material is covered is also indicated in the chart.

Question	1a	1b	2	3	4	5	6	7
Objective	A	B	H	C	D	F	E	J
Lesson	7-1	7-2	7-1	7-3	7-3	7-4	7-6	7-2
Question	8	9	10	11				
Objective	G	G	G	I				
Lesson	7-4	7-5	7-7	7-8				

CHAPTER 7 CHAPTER REVIEW (pp. 462–464)

1. 7, 13, 31, 85, 247 3. 0, 2, 2, 4, 4 5. 3, 5, -2, -24, -40 7. $-3c$ 9. a. $1, \frac{1}{2}, \frac{1}{6}, \frac{1}{24}, \frac{1}{120}$ b. $a_n = \frac{1}{n!}$ for integers $n \geq 1$ 11. $I_n = n(n+1)$ 13. a. $(n-3) + (n-2) + \ldots + (n+n)$ b. 9 15. $\frac{1}{7}$ 17. $\sum_{i=-n}^{-1} \frac{1}{i}$ 19. $\sum_{j=1}^{k+1} (j-1)(2j+1) = \left[\sum_{j=1}^{k} (j-1)(2j+1)\right] + (k)(2k+3)$ 21. a. $3 + 9 + 27 + 81 = 120$ and $\frac{3}{2}(3^4 - 1) = \frac{3}{2}(81 - 1) = 120$, so the formula works for $n = 4$. b. $120 + 243 = 363$ c. $\frac{3}{2}(3^5 - 1) = \frac{3}{2}(243 - 1) = \frac{3}{2} \bullet 242 = 363$ which agrees with part b. 23. a. $\frac{4t}{3}\left(1 - \left(\frac{1}{4}\right)^{n+1}\right)$ b. ≈ 2.6667 25. a. $S_n =$

$10\left(1 - \left(\frac{4}{5}\right)^n\right)$ b. ≈ 7.9028 c. 10 27. a. 0, 2, 6, 12, 20 b. Let $S(n)$: $b_n = n(n-1)$ for all integers $n \geq 1$. (1) $S(1)$: $b_1 = 1(1-1)$. $b_1 = 0$ so $S(1)$ is true. (2) Assume $S(k)$: $b_k = k(k-1)$ is true for some arbitrary integer $k \geq 1$. Show that $S(k+1)$: $b_{k+1} = (k+1)((k+1)-1)$ is true. From the recursive definition, $b_{k+1} = b_k + 2k = k(k-1) + 2k = k^2 + k = (k+1)k = (k+1)((k+1)-1)$. Hence by mathematical induction, $S(n)$ is true $\forall$ integers $n \geq 1$, and the explicit formula is correct. 29. (1) $S(1)$: $\sum_{i=1}^{1} 3i(i+2) = \frac{1 \bullet 2 \bullet 9}{2}$. $3 \bullet 1(1+2) = 9$ and $\frac{1 \bullet 2 \bullet 9}{2} = 9$, so $S(1)$ is true. (2) Assume that $S(k)$: $\sum_{i=1}^{k} 3i(i+2) = \frac{k(k+1)(2k+7)}{2}$ is true for an arbitrary integer $k \geq 1$. Show $S(k+1)$: $\sum_{i=1}^{k+1} 3i(i+2) = \frac{(k+1)(k+2)(2(k+1)+7)}{2}$ is true. $\sum_{i=1}^{k+1} 3i(i+2) = \left(\sum_{i=1}^{k} 3i(i+2)\right) + 3(k+1)(k+3) = \frac{k(k+1)(2k+7)}{2} + (k+1)(3k+9) = \frac{k(k+1)(2k+7)}{2} + \frac{2(k+1)(3k+9)}{2} = \left(\frac{k+1}{2}\right)(2k^2 + 7k + 6k + 18) = \left(\frac{k+1}{2}\right)(k+2)(2k+9) = \frac{(k+1)(k+2)(2(k+1)+7)}{2}$. Hence by mathematical induction, $S(n)$ is true $\forall$ integers $n \geq 1$.

31. (1) $S(2)$: 3 is a factor of $2 \bullet 2^3 - 5 \bullet 2$. $16 - 10 = 6$, so $S(2)$ is true. (2) Assume that $S(k)$: 3 is a factor of $2k^3 - 5k$ is true for some integer $k \geq 2$. Prove $S(k+1)$: 3 is a factor of $2(k+1)^3 - 5(k+1)$ is true. $2(k+1)^3 - 5(k+1) = 2k^3 + 6k^2 + 6k + 2 - 5k - 5 = (2k^3 - 5k) + 3(2k^2 + 2k - 1)$. Since 3 is a factor of $2k^3 - 5k$ and $3(2k^2 + 2k - 1)$, 3 is a factor of their sum by the Factor of a Sum Theorem. Hence by mathematical induction, $S(n)$ is true $\forall$ integers $n \geq 2$. 33. a. 0, 4, 4, 16, 28 b. Let $S(n)$: 4 is a factor of b_n.. (1) $S(1)$: 4 is a factor of b_1. $b_1 = 0$ so $S(1)$ is true. $S(2)$: 4 is a factor of b_2. $b_2 = 4$ so $S(2)$ is true. (2) Assume $S(1), S(2), \ldots$, and $S(k)$ are true for some integer $k \geq 1$. So 4 is a factor of $b_1, b_2, \ldots, b_k$. Show $S(k+1)$: 4 is a factor of b_{k+1} is true. Since b_{k-1} and b_k have 4 as a factor, there exists integers p and q such that $b_{k-1} = 4p$ and $b_k = 4q$. Substituting into the recurrence relation, $b_{k+1} = 4q + 3(4p) = 4q + 12p = 4(q + 3p)$. $q + 3p$ is an integer by closure properties, so 4 is a factor of b_{k+1}. Hence by the Strong Form of Mathematical Induction, $S(n)$ is true for all integers $n \geq 1$.

35. a. $\begin{cases} A_1 = 80{,}000 \\ A_{k+1} = 1.01A_k - 900 \end{cases}$ for all integers $k \geq 1$ b. \$ 79,489.90 37. a. initial order: 1, 3, 5, 2, 4; 1st pass: 1, 3, 2, 4, 5; 2nd pass: 1, 2, 3, 4, 5 b. 5, 4, 3, 2, 1 39. a. $a_1 = 1$, $a_{k+1} = \frac{1}{2}a_k$ for $k \geq 1$ b. $a_n = \left(\frac{1}{2}\right)^{n-1}$ c. $\sum_{i=1}^{20} \left(\frac{1}{2}\right)^{i-1}$ d. $S_{20} = 2\left(1 - \left(\frac{1}{2}\right)^{20}\right)$

LESSON 8-1 (pp. 466–472)

1. a. Leonhard Euler; 18th b. Cardano; 16th c. Wessel; 18th 3. a. -21 b. -4 5. $10 + 15i$ 7. $3 - 4i$ 9. a. $8 + i$ b. $8 - i$ c. $2 + i$ d. $3 - 2i$ e. $8 - i$; they are equal. 11. **See below.** 13. a. $5 - 6i$ b. **See below.** c. parallelogram; sample: slopes of opposite sides are equal.

11.

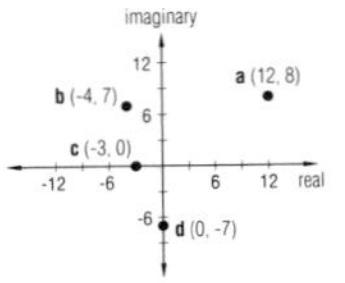

13. b.

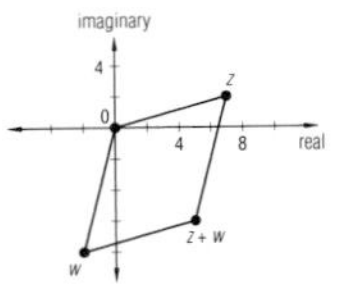

15. a. $\frac{5}{2} + 3i$ ohms b. $5i$ ohms 17. $(1 + i)^2 - 2(1 + i) + 2 = 1 + i + i + i^2 - 2 - 2i + 2 = 1 + i + i - 1 - 2 - 2i + 2 = 0$. So $1 + i$ is a solution of $z^2 - 2z + 2 = 0$. 19. a. $\frac{5}{29} + \frac{2}{29}i$ b. $z \bullet \frac{1}{z} = (5 - 2i)\left(\frac{5}{29} + \frac{2}{29}i\right) = \frac{25}{29} + \frac{10}{29}i - \frac{10}{29}i - \frac{4}{29}i^2 = \frac{25}{29} + \frac{4}{29} = 1$ 21. Suppose there is a smallest integer, s. Then $s - 1 < s$, and $s - 1$ is an integer. This contradicts the assumption that s is the smallest integer. So the assumption is false, and there is no smallest integer. 23. a. > 5 b. < 3

LESSON 8-2 (pp. 473–479)

1.

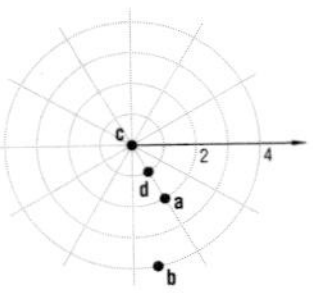

3. a. By case **b** of the Polar Representation Theorem in this lesson, $\left[4, \frac{\pi}{3}\right] = \left[4, \frac{\pi}{3} + 2(-1)\pi\right] = \left[4, -\frac{5\pi}{3}\right]$. Then by case **b** again, $\left[4, -\frac{5\pi}{3} + 2k\pi\right]$ for any integer k is a coordinate representation of $\left[4, \frac{\pi}{3}\right]$.

b. By case **c** of the Polar Representation Theorem in this lesson, $\left[4, \frac{\pi}{3}\right] = \left[-4, \frac{\pi}{3} + \pi\right] = \left[-4, \frac{4\pi}{3}\right]$. So by case **b**, $\left[-4, \frac{4\pi}{3} + 2k\pi\right]$ for any integer k is a coordinate representation of $\left[4, \frac{\pi}{3}\right]$. 5. Given any point $P = [r, \theta]$. First plot P and $Q = [1, \theta] = (\cos\theta, \sin\theta)$. **See below.** Because $[r, \theta]$ is r times as far from the origin as Q, its rectangular coordinates are $(r\cos\theta, r\sin\theta)$. Thus the rectangular coordinates of P are given by $x = r\cos\theta$ and $y = r\sin\theta$. 7. $\approx (1.7, -1.5)$ 9. sample: $\approx [-\sqrt{13}, 56.3°]$ 11. $-3 + 4i$ 13. $\left[\frac{10\sqrt{3}}{3}, \frac{\pi}{6}\right]$, $\left(5, \frac{5\sqrt{3}}{3}\right)$ 15. $Q_1 = \left[-4, \frac{\pi}{6}\right]$, $Q_2 = \left[-3, \frac{\pi}{6}\right]$, $Q_3 = \left[-2, \frac{\pi}{6}\right]$, $Q_4 = \left[-1, \frac{\pi}{6}\right]$, $Q_5 = \left[0, \frac{\pi}{6}\right]$, $Q_6 = \left[1, \frac{\pi}{6}\right]$, $Q_7 = \left[2, \frac{\pi}{6}\right]$, $Q_8 = \left[3, \frac{\pi}{6}\right]$, $Q_9 = \left[4, \frac{\pi}{6}\right]$; $\theta = \frac{\pi}{6}$ 17. $-\frac{5}{2}$ amps 19. Yes, 10 21. Let the vertices in clockwise order be $A(-1, 0)$, $B(-5, -5)$, $C(-11, -5)$, and $D(-7, 0)$. The slope of $\overline{AD} = 0$; the slope of $\overline{BC} = 0$. The slope of $\overline{AB} = \frac{-5 - 0}{-5 - (-1)} = \frac{5}{4}$. The slope of $\overline{DC} = \frac{-5 - 0}{-11 - (-7)} = \frac{5}{4}$. Since $ABCD$ is composed of two pairs of parallel lines, it is a parallelogram.

5.

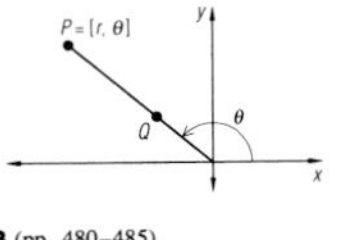

LESSON 8-3 (pp. 480–485)

1. $z = 7 + 3i = (7, 3) = Z$; $o = 0 + 0i = (0, 0) = O$; $w = 4 - 9i = (4, -9) = W$; $z + w = 11 - 6i = (11, -6) = P$. The slope of $\overline{OZ} = \frac{3 - 0}{7 - 0} = \frac{3}{7}$; the slope of $\overline{WP} = \frac{-6 - (-9)}{11 - 4} = \frac{3}{7}$; and so $\overline{OZ} \parallel \overline{WP}$. The slope of $\overline{ZP} = \frac{-6 - 3}{11 - 7} = -\frac{9}{4}$; the slope of $\overline{OW} = \frac{-9 - 0}{4 - 0} = -\frac{9}{4}$; and so $\overline{ZP} \parallel \overline{OW}$. Therefore, the figure is a parallelogram. 3. a. $\left[3\sqrt{2}, \frac{3\pi}{4}\right]$ b. $3\sqrt{2}\left(\cos\frac{3\pi}{4} + i\sin\frac{3\pi}{4}\right)$ 5. a. $\approx [5, 127°]$ b. $\approx 5(\cos 127° + i\sin 127°)$ 7. $[6, 210°]$ 9. $-11 - 10i$ 11. modulus: 3; argument: 270° 13. a. $|w| = \sqrt{5}$, $-\theta \approx -63.4°$

13. b.

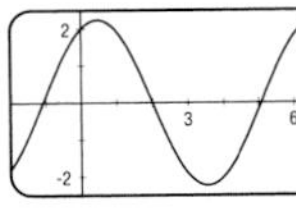

15. a. $E' = 5 + i$, $F' = 1 + 4i$, $G' = 3 + 6i$. b. $EF = 5$, $EG = \sqrt{29}$, and $FG = 2\sqrt{2}$, while $E'F' = 5$, $E'G' = \sqrt{29}$, and $F'G' = 2\sqrt{2}$. So $\triangle EFG \cong \triangle E'F'G'$ by the SSS Congruence Theorem. c. $T_{1,1}$

17. a. **See below.** b. $x^2 + y^2 = 16$

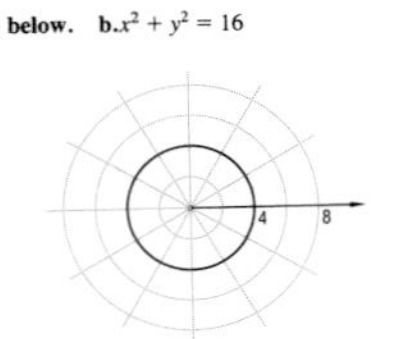

19. Let $z = a + bi$, where a and b are real numbers. Then $\bar{z} = a - bi$. $z + \bar{z} = (a + bi) + (a - bi) = 2a$. Since a is real, $2a$ is real. So for all complex numbers, the sum of the number and its complex conjugate is a real number. 21. a. Yes b. domain: the set of real numbers; range: the set of integers 23. Geometric Subtraction Theorem: Let $z = a + bi$ and $w = c + di$ be two complex numbers that are not collinear with the origin. Then the point representing $z - w$ is the fourth vertex of a parallelogram with consecutive vertices $z = a + bi$, 0, and $-w = -c - di$.

LESSON 8-4 (pp. 486–492)

1. θ	0	$\frac{\pi}{6}$	$\frac{\pi}{4}$	$\frac{\pi}{3}$	$\frac{\pi}{2}$	π	$\frac{3\pi}{2}$	2π
r	0	$\frac{3}{2}$	$\frac{3\sqrt{2}}{2}$	$\frac{3\sqrt{3}}{2}$	3	0	-3	0

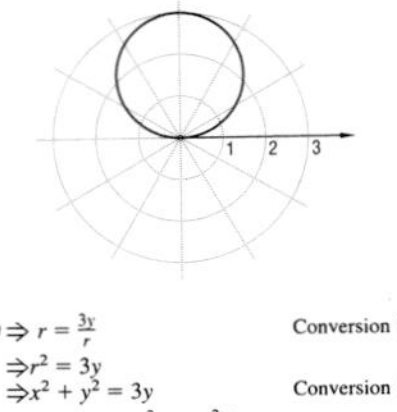

3. $r = 3\sin\theta \Rightarrow r = \frac{3y}{r}$	Conversion formula
$\Rightarrow r^2 = 3y$	
$\Rightarrow x^2 + y^2 = 3y$	Conversion formula
$\Rightarrow (x)^2 + \left(y - \frac{3}{2}\right)^2 = \left(\frac{3}{2}\right)^2$	

This verifies that the curve in Question 1 is a circle and has center $\left(0, \frac{3}{2}\right)$ and radius $\frac{3}{2}$. 5. **See next page.** 7. not periodic 9. a. **See next page.** b. **See next page.** c. i. Around $\theta = \pi$, r has negative values, causing a loop. ii. $r > 0$ for all θ.

5.

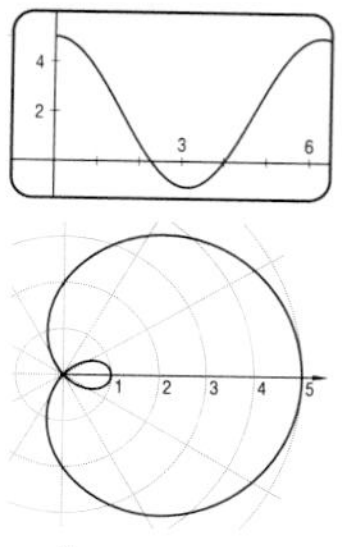

9. a.

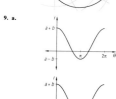

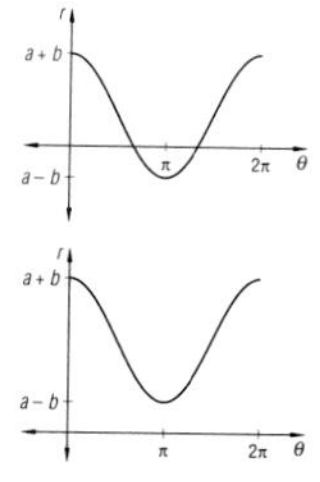

9. b.

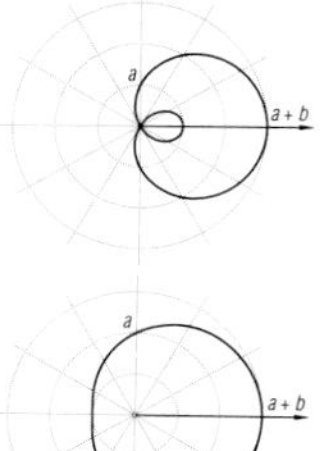

11. a. The polar graph of $r = k \sin\theta$ is the graph of $r = \sin\theta$ taken through a scale change of k. **b.** It is a circle with radius $\frac{k}{2}$ and center $\left(0, \frac{k}{2}\right)$. **13. a.** $Z' = \left(-\frac{3}{2}, -9\right)$, $W' = \left(-\frac{9}{2}, -5\right)$ **b.** $T_{-7/2,-8}$ **c.** 5 **15. a.** $\frac{1}{2} - \frac{\sqrt{3}}{2}i$ **b.** They are the same. They are both complex conjugates of z. **c.** $z = [1, 60°]$, $w = [1, -60°]$; their arguments are opposites.

17.

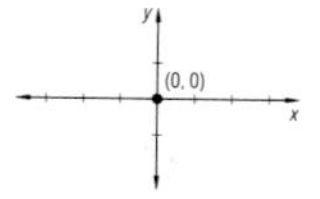

LESSON 8-5 (pp. 493–497)
1. **See below.** **3.** **See below.** **5.** sample: $[1, 0]$, $\left[2^{\pi/6}, \frac{\pi}{6}\right]$, $\left[2^{\pi/4}, \frac{\pi}{4}\right]$, $\left[2^{\pi/2}, \frac{\pi}{2}\right]$, $[2^{\pi}, \pi]$ **7.** $r = \frac{\theta}{\pi} + 2$

1.

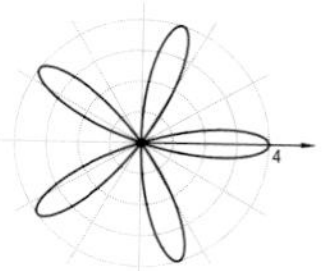

3.

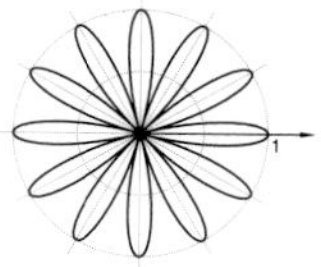

9. a. b. $\theta = \frac{\pi}{6}$, $\theta = \frac{\pi}{2}$, $\theta = \frac{5\pi}{6}$

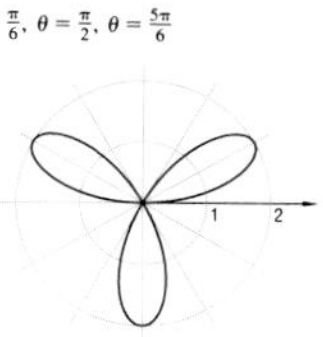

11. a. The second graph is a phase shift copy of the first one by $\frac{\pi}{4}$ to the right. **b.** The second graph is a rotated image of the first by $\frac{\pi}{4}$ counterclockwise. **c.** **See below.**
13. a. $\left[\frac{1}{2}, \frac{4\pi}{3}\right]$ **b.** $-\frac{1}{4} - \frac{\sqrt{3}}{4}i$ **c.** $u' = \frac{1}{2} + \frac{\sqrt{3}}{2}i$, $v' = -\frac{3}{2} - \frac{3\sqrt{3}}{2}i$, $w' = -\frac{3}{2} + \sqrt{3} + \left(-1 - \frac{3\sqrt{3}}{2}\right)i$, $z' = \frac{1}{2} + \sqrt{3} + \left(\frac{\sqrt{3}}{2} - 1\right)i$ **d.** **See below.** **15.** $0 < x < 0.92$

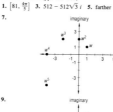

11. c.

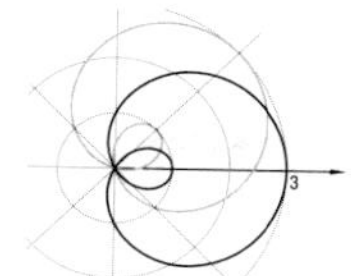

13. d.

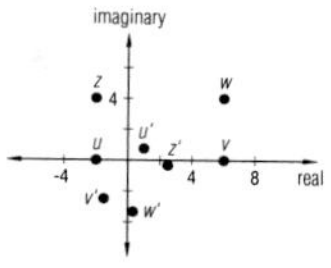

LESSON 8-6 (pp. 498–503)
1. $\left[81, \frac{4\pi}{5}\right]$ **3.** $512 - 512\sqrt{3}\,i$ **5.** farther

7.

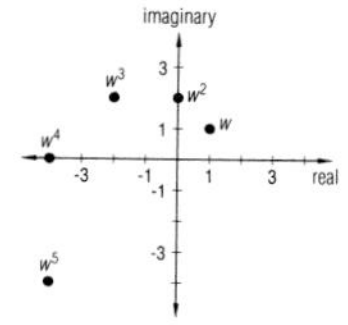

9.

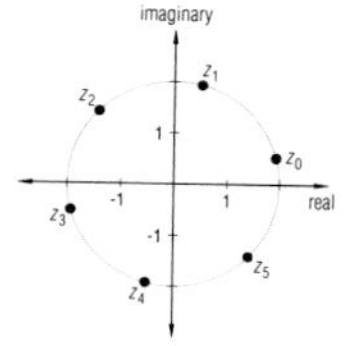

11. -625 **13.** Sample: Using DeMoivre's Theorem, for $z = [r, \theta]$ with $r \geq 0$, $z^n = [r^n, n\theta]$, and $|z| = r$. So $|z^n| = |r^n| = r^n$, and $|z|^n = r^n$. Sample: Using mathematical induction, let $S(n)$: $|z^n| = |z|^n$. $S(1)$: $|z^1| = |z| = |z|^1$, so $S(1)$ is true. Assume $S(k)$: $|z^k| = |z|^k$. Then $|z^{k+1}| = |z^k \bullet z| = |z^k||z| = |z|^k|z|^1 = |z|^{k+1}$. So $S(k) \Rightarrow S(k+1)$, and $S(n)$ is true for all integers n. **15. a.** **See below.** **b.** rose curve **17.** Let $z = a + bi$. Then $\bar{z} = a - bi$, and $z - \bar{z} = a + bi - (a - bi) = a + bi - a + bi = 0 + 2bi$ which is an imaginary number. **19.** (b) **21. a.** $\sqrt[3]{13}$ **b.** $-\sqrt[3]{13}$ **c.** $\sqrt[4]{13}$, $-\sqrt[4]{13}$ **d.** no real solution

15. a.

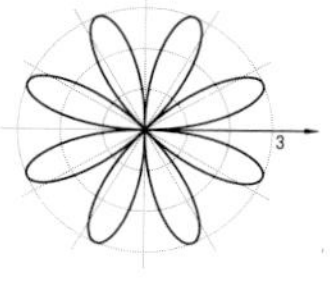

LESSON 8-7 (pp. 504–509)
1. For $k = 4$, $\left[3, \frac{\pi}{6} + \frac{2\pi k}{3}\right] = \left[3, \frac{17\pi}{6}\right] = \left[3, \frac{5\pi}{6}\right]$, which is z_1. **3.** $z_0 = 3[\cos(63°) + i\sin(63°)]$; $z_1 = 3[\cos(135°) + i\sin(135°)]$; $z_2 = 3[\cos(207°) + i\sin(207°)]$; $z_3 = 3[\cos(279°) + i\sin(279°)]$; $z_4 = 3[\cos(351°) + i\sin(351°)]$ **See below.**
5. a. square **b.** **See below.** **7. a.** 1 **b.** 8 **9.** -729; -9, $9(\cos 300° + i \sin 300°)$

3.

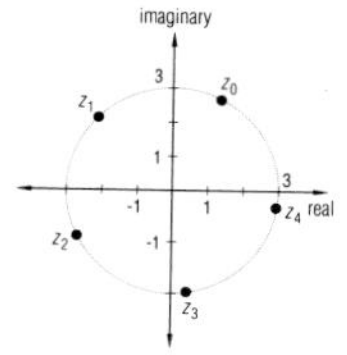

5. b.

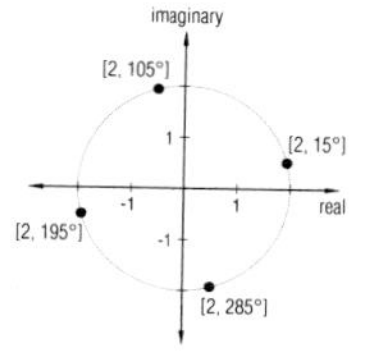

11. a.

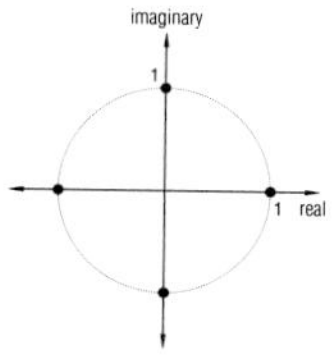

b.

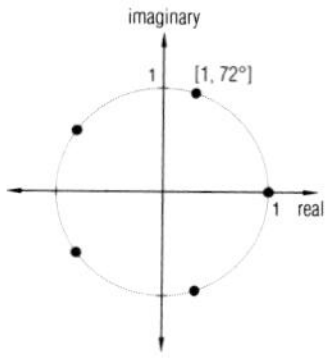

c. They are the vertices of a regular n-gon, with center at the origin, and a vertex at $(1, 0)$. **13. a.** $z = \pm 3, \pm 3i$ **See below.** **b.** $\frac{3\sqrt{2}}{2} \pm \frac{3\sqrt{2}}{2}i$, $-\frac{3\sqrt{2}}{2} \pm \frac{3\sqrt{2}}{2}i$ **See below.**

13. a.

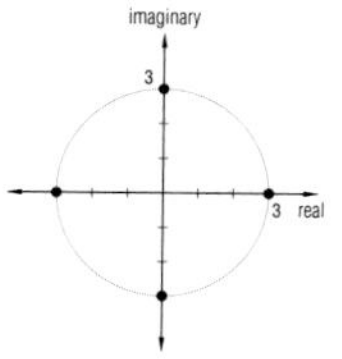

13. b.

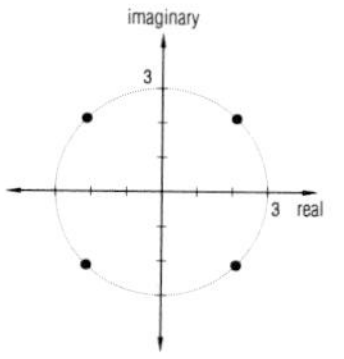

15. The 3 cube roots of 8 are 2, $-1 + \sqrt{3}\,i$, $-1 - \sqrt{3}\,i$; their sum is $2 + (-1 + \sqrt{3}\,i) + (-1 - \sqrt{3}\,i) = 0$. **17.** **See below.** The graph of $r = 3\theta$ is a spiral of Archimedes, and that of $r = 3^{\theta}$ is a logarithmic spiral. **19. a.** $a_n = 2^n + 1$ **b.** $S(n)$: $a_n = 2^n + 1$. $S(1)$: $a_1 = 2^1 + 1 = 3$, so $S(1)$ is true. Assume $S(k)$: $a_k = 2^k + 1$. Then $a_{k+1} = 2a_k - 1 = 2(2^k + 1) - 1 = 2^{k+1} + 2 - 1 = 2^{k+1} + 1$. So $S(k) \Rightarrow S(k+1)$. Therefore by mathematical induction, $S(n)$ is true for all positive integers n. **21. a.** $x - 1$ **b.** -4 **c.** 1 **d.** 1 and -4, by the Transitive Property of Factors

17.

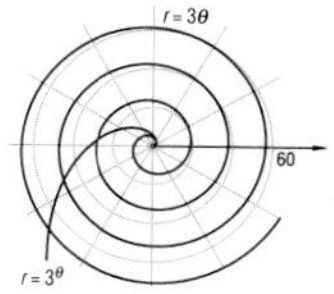

LESSON 8-8 (pp. 510–514)
1. a. $\lim_{x\to-\infty} p(x) = -\infty$, and $\lim_{x\to\infty} p(x) = \infty$ **b.** Since $p(x) < 0$ for a small enough real number x, $p(x) > 0$ for a large enough real number x, and p is continuous, the Intermediate Value Theorem ensures there exists a real number c such that $p(c) = 0$. **3.** zeros: 0 (with muliplicity 4), -1, 5 **5.** zeros: i, $-\frac{i}{2}$, both with multiplicity 1 **7.** sample: $p(x) = (x - 6)(x - 4i)x$
9. a. 5 **b.** $p(x) = (x - 1 - i)(x - 1 + i)(x + 1)(x - 3)^2$
11. It has three more zeros. There can be 3 more simple zeros, or one zero with multiplicity 3, or one simple zero and one zero with multiplicity two. **13. a.** 256 **b.** **See below.**
15. $(2 + i)^2 - 4(2 + i) + 5 = 4 + 4i - 1 - 8 - 4i + 5 = 0$; $(2 - i)^2 - 4(2 - i) + 5 = 4 - 4i - 1 - 8 + 4i + 5 = 0$
17. a. $f(x) = 3x^4$ **b.** $p(x) = 3x^4\left(1 - \frac{2}{3x} + \frac{1}{3x^3}\right)$
c. $\lim_{x\to-\infty} p(x) = \infty$, $\lim_{x\to\infty} p(x) = \infty$ **19.** any one of: $-5 < x < -4$; $-2 < x < -1$; or $3 < x < 4$

13. b.

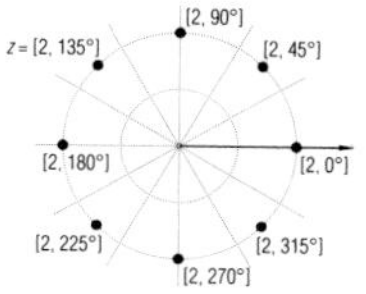

LESSON 8-9 (pp. 515–521)
1. a. $4 + 7i$ **b.** 0 **3. a.** $p(i) = i^2 + 4i^2 + 5 = -1 - 4 + 5 = 0$; $p(-5i) = 25i^2 - 20i^2 + 5 = -25 + 20 + 5 = 0$ **b.** No, the coefficients of p are not all real numbers. **5.** $\frac{1 - i\sqrt{3}}{2}$, $4 - 5i$ **7.** sample: $p(x) = x^3 - x^2 - 7x + 15$

9. sample:

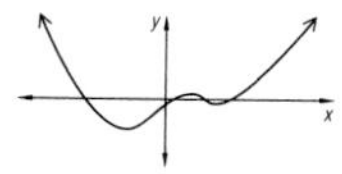

11. 1, -1, i, $-i$ **13.** sample: $p(x) = 12x^3 - 8x^2 + 3x - 2$
15. zeros: $\pm\sqrt{3}\,i$ (each with multiplicity 2), 1

17.

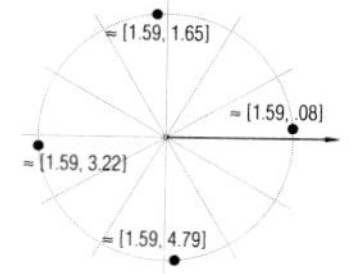

19. a. 3 **b.** $2x^2 - 4x + 2$ **21. a.** for $c = -5$, the zeros are $\pm\sqrt{5}$, $\pm i$; for $c = 0$, the zeros are 0, ± 2; for $c = 3$, the zeros are $\pm\sqrt{3}$, ± 1. As c slides from -5 to 0, its two real zeros move closer to the origin, and its two imaginary zeros converge at the origin. As c then slides from 0 to 3, the polynomial has 4 real zeros. **b.** For $c = 4$, the zeros are $\pm\sqrt{2}$; for $c = \frac{25}{4}$, the zeros are $\frac{3 \pm i}{2}$. As c slides from 3 to 4, its positive zeros converge to $\sqrt{2}$, and its negative zeros converge to $-\sqrt{2}$. As c slides from 4 to $\frac{25}{4}$, its four complex zeros converge to two complex zeros, $\frac{3 \pm i}{2}$.

LESSON 8-10 (pp. 522–528)
1. To 4 digits, the sequence is .5000, .7071, .8409, .9170, … . The sequence approaches 1. **3.** 7, 49, 2401, 5764801
5. $[1, 1]$, $[1, 2]$, $[1, 4]$, $[1, 8]$, $[1, 16]$, $[1, 32]$ **See below.**

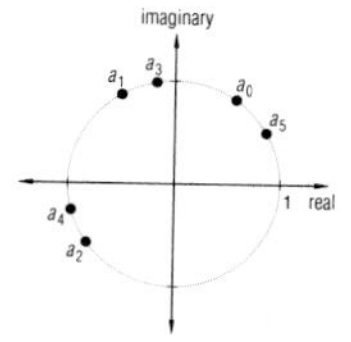

7. 1, 0 **9.** For all $x \neq 1$, let $a_0 = x$. Then $a_1 = \frac{1}{1 - x}$. And if $x \neq 0$, $a_2 = \frac{1}{1 - \frac{1}{1 - x}} = \frac{1 - x}{1 - x - 1} = \frac{1 - x}{-x} = \frac{x - 1}{x}$, and $a_3 = \frac{1}{1 - \frac{x - 1}{x}} = \frac{x}{x - (x - 1)} = \frac{x}{1} = x$. So $a_3 = a_0$, hence the period is 3.

11. sample: $x^4 - 4x^3 + 9x^2 - 10x$ **13.** $z_0 = \sqrt[4]{7}\left(\cos\frac{\pi}{5} + i\sin\frac{\pi}{5}\right)$; $z_1 = \sqrt[4]{7}\left(\cos\frac{7\pi}{10} + i\sin\frac{7\pi}{10}\right)$; $z_2 = \sqrt[4]{7}\left(\cos\frac{6\pi}{5} + i\sin\frac{6\pi}{5}\right)$; $z_3 = \sqrt[4]{7}\left(\cos\frac{17\pi}{10} + i\sin\frac{17\pi}{10}\right)$ **See below.** **15. a.** **See below.** **b.** $\theta = 0°$ **17.** $\frac{1152}{\pi^2} \approx 116.7$ ft

13.

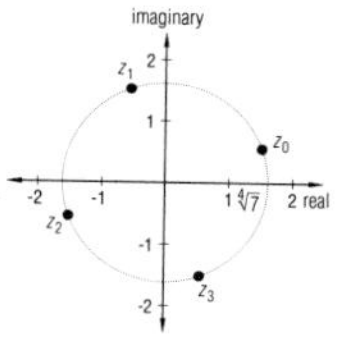

15. a.

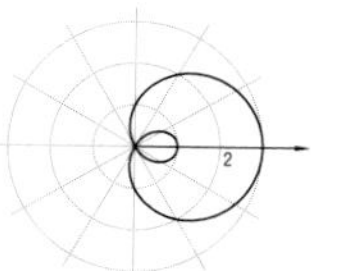

CHAPTER 8 PROGRESS SELF-TEST (p. 533)
1. a. $(8 - 5i) - (-2 + 3i) = 10 - 8i$ **b.** $(8 - 5i)(-2 + 3i) = -16 + 10i - 15i^2 + 24i = -1 + 34i$ **c.** $\frac{8 - 5i}{-2 + 3i} = \frac{(8 - 5i)(-2 - 3i)}{(-2 + 3i)(-2 - 3i)} = \frac{-16 + 10i + 15i^2 - 24i}{4 - 9i^2} = \frac{-31 - 14i}{13} = -\frac{31}{13} - \frac{14}{13}i$ **d.** $-2 - 3i$ **2.** $\left(8\cos\frac{5\pi}{6}, 8\sin\frac{5\pi}{6}\right) = \left(8 \bullet \frac{-\sqrt{3}}{2}, 8 \bullet \frac{1}{2}\right) = (-4\sqrt{3}, 4)$
3. a. $(4\sqrt{3}, -4)$ **b.** $\sqrt{(4\sqrt{3})^2 + (-4)^2} = \sqrt{48 + 16} = \sqrt{64} = 8$ **c.** $\tan^{-1}\left(-\frac{4}{4\sqrt{3}}\right) = \tan^{-1}\left(\frac{1}{\sqrt{3}}\right) = -\frac{\pi}{6}$, so $\theta = \frac{11\pi}{6}$. **d.** $\left[8, \frac{11\pi}{6}\right]$ **4.** $6 = \frac{-3 - 8i}{z}$, so $z = \frac{-3 - 8i}{6} = -\frac{1}{2} - \frac{4}{3}i$ ohms

5.

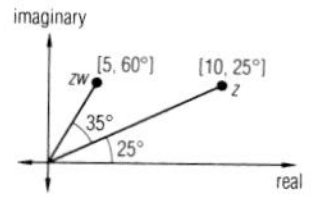

6. a. Using DeMoivre's Theorem, $z^1 = 2(\cos\frac{\pi}{3} + i\sin\frac{\pi}{3})$.

8. $p(x) = (x^2+9)(x^2-9)(x^2+6x+9) = (x^2+9)(x-3)$

43. a. $z^n = [r^n, n\theta]$; $z^m = [r^m, m\theta]$; $z^{n+m} = [r^{n+m}, (n+m)\theta]$

63. See below. 8-leaved rose 65. See below. $r = 4^\theta$ is a

CHAPTER 10 REVIEW (pp. 636–638)
1. Unordered symbols; repetition is allowed. 3. Ordered symbols; repetition is not allowed. 5. Ordered symbols; repetition is allowed. 7. 2002 9. 60,480 11. $x^8 + 8x^7y + 28x^6y^2 + 56x^5y^3 + 70x^4y^4 + 56x^3y^5 + 28x^2y^6 + 8xy^7 + y^8$
13. -10240 15. $C(n,r) = \frac{n!}{r!(n-r)!} = \frac{n!}{(n-(n-r))!(n-r)!} = C(n,n-r)$ 17. For each of the $C(n,r)$ combinations of n objects taken r at a time, there are $r!$ arrangements. So $C(n,r) \cdot r! = P(n,r)$, or $P(n,r) = r!\, C(n,r)$
19. $\binom{20}{4}$ 21. 5040 23. $\approx 1.2165 \times 10^{17}$ 25. 362,880
27. 120 29. a. 1140 b. 8000 31. 210 33. 26 35. 66
37. a. ≈ 0.016 b. ≈ 0.181 c. ≈ 0.632

39. 14 outcomes **See below.**

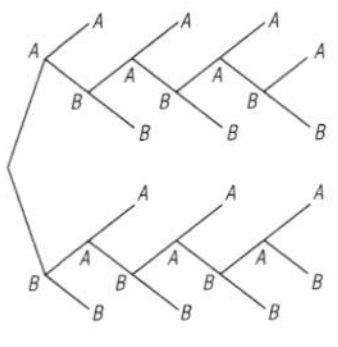

LESSON 11-1 (pp. 640–647)
1. a. **See below.** b. sample: $A,a,B,b,A,e,D,g,C,c,A,d,C$; No 3. edge, vertex 5. sample: **See below.** 7. 320 students 9. $\approx .43$ 11. Euler's 13. sample: **See below.**
15. a. 39 b. 38 17. a. 3×4 b. 8 c. 11 19. 7

1. a.

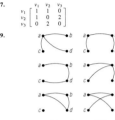

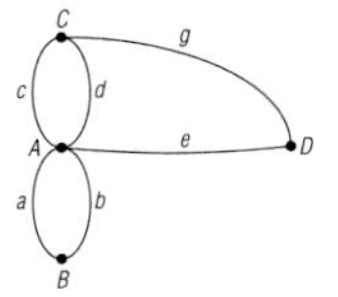

5.

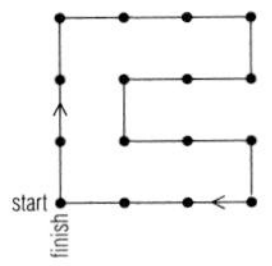

13.

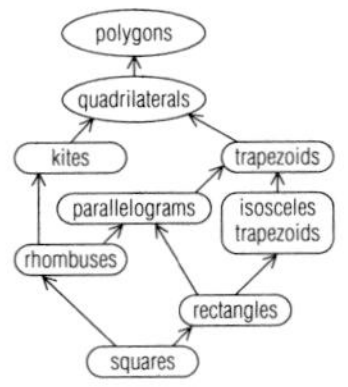

LESSON 11-2 (pp. 648–655)
1. a. Yes b. 4 edges, 4 vertices 3. a. Yes b. No c. v_4 d. e_3 and e_4, e_1 and e_2 e. e_3, e_4 5. **See below.**
7. **See below.** 9. **See below.** 11. **See below.**

5.
$$\begin{array}{c|cccc} & v_1 & v_2 & v_3 & v_4 \\ \hline v_1 & 0 & 1 & 0 & 0 \\ v_2 & 0 & 0 & 1 & 1 \\ v_3 & 0 & 0 & 1 & 1 \\ v_4 & 1 & 1 & 0 & 0 \end{array}$$

7.
$$\begin{array}{c|ccc} & v_1 & v_2 & v_3 \\ \hline v_1 & 1 & 1 & 0 \\ v_2 & 1 & 0 & 2 \\ v_3 & 0 & 2 & 0 \end{array}$$

9.

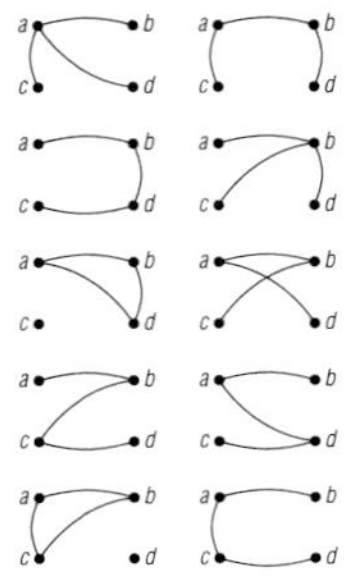

11.

edge	endpoint
e_1	$\{v_1, v_2\}$
e_2	$\{v_2, v_4\}$
e_3	$\{v_2, v_4\}$
e_4	$\{v_3\}$

13. a. ∃ a graph G such that G does not have any loops and G is not simple. b. True, G could have parallel edges and no loops. 15. The tester prefers A to C, and prefers C to P. But in a direct comparison of A and P, the tester prefers P.
17. a. **See below.** b. $\approx .396$ c. $\approx .474$ 19. quotient: $6x^3 - 42x^2 + 287x - 2006$; remainder: 14043

17. a.

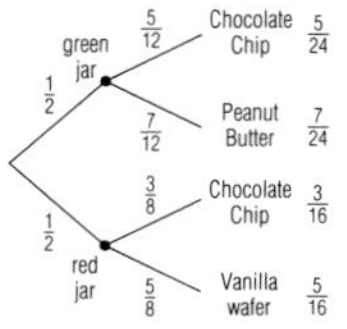

Lesson 11-3 (pp. 656–661)
1. a. $\deg(v_1) = 2$; $\deg(v_2) = 3$; $\deg(v_3) = 0$; $\deg(v_4) = 3$
b. 8 3. a. 2 b. v_1 c. 1 d. v_2 e. 2 f. v_1
g. contributes 1 to the degree of v_2 and 1 to the degree of v_1.
h. contributes 2 to the degree of v_3. 5. **See below.** 7. 42
9. Assume that each of the 9 people could shake hands with exactly three others. Represent each person as the vertex of a graph, and draw an edge joining each pair of people who shake hands. To say that a person shakes hands with three other people is equivalent to saying that the degree of the vertex representing that person is 3. The graph would then have an odd number of vertices of odd degree. This contradicts Corollary 2 of the Total Degree of a Graph Theorem. Thus the given situation is impossible. 11. 2 13. impossible; it can't have an odd number of odd vertices.
15. **See below.** 17. a. $n - 1$ b. $n(n-1)$ c. $\frac{n(n-1)}{2}$
19. a. $\frac{(n+1)n}{2}$ b. It is the same problem but with one more person.

5.

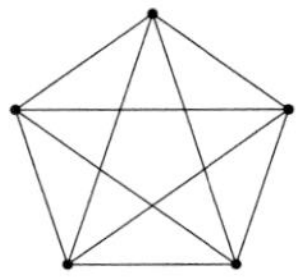

15.

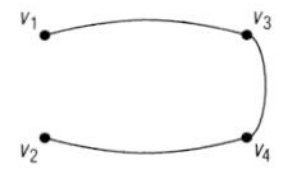

21. The set of even vertices contributes an even number to the total degrees of the graph. Since that total degree is even, the set of odd vertices must also contribute an even number to the total degree. An odd number of odd vertices would contribute an odd number, so the number of odd vertices must be even.
23. **See below.** 10 hours 25. a. $x < -3$, $1 < x < 3$
b. $-3 < x < 1$, $x > 3$ c. $-1 < x < 2$

23.

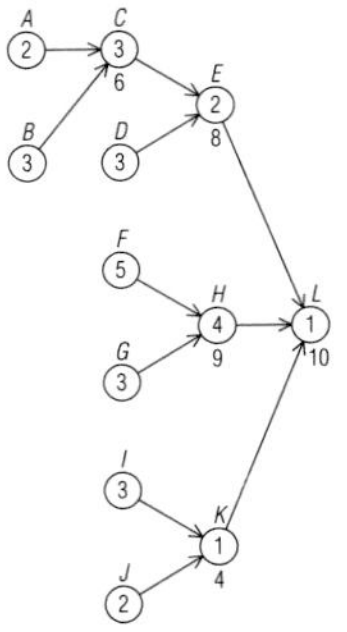

LESSON 11-4 (pp. 662–669)
1. a. No b. No c. No 3. a. If at least one vertex of a graph has an odd degree, then the graph does not have an Euler circuit. b. The presence of an odd vertex is sufficient to show that a graph cannot have an Euler circuit.
5. $e_1e_2e_9e_8e_7e_6e_5$ 7. a. $e_1, e_2, e_3, e_4, e_5, e_6$ b. 2
9. Yes; if the graph is not connected, there is no way a circuit could contain every vertex. 11. Yes; think of replacing each bridge by two bridges. Such a walk exists by the sufficient condition for an Euler Circuit Theorem, since every vertex will have an even degree. 13. Yes, if the walk repeats an edge, then there is a circuit. Remove edges from the graph until there is no circuit. Then connect v to w. 15. sample: **See below.** 17. a. **See next page.** b. 2.48% c. $\approx 19.76\%$ d. $\approx 80.24\%$ 19. $\approx -.204$

15.

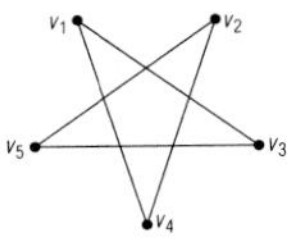

17. a.

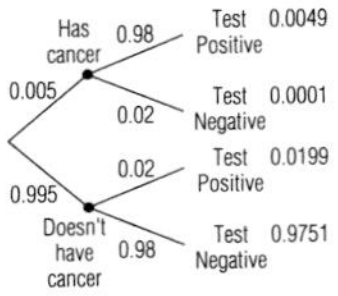

LESSON 11-5 (pp. 670–674)
1. False 3. False; the matrix at the right represents a directed graph. 5. a. 2 b. 5 7. $e_4e_1e_4$; $e_4e_2e_5$; $e_4e_3e_5$; $e_5e_2e_4$; $e_5e_3e_4$
9. $A^3 = \begin{bmatrix} 1 & 2 & 1 \\ 2 & 0 & 1 \\ 1 & 1 & 0 \end{bmatrix}\begin{bmatrix} 1 & 2 & 1 \\ 2 & 0 & 1 \\ 1 & 1 & 0 \end{bmatrix}\begin{bmatrix} 1 & 2 & 1 \\ 2 & 0 & 1 \\ 1 & 1 & 0 \end{bmatrix} = \begin{bmatrix} 6 & 3 & 3 \\ 3 & 5 & 2 \\ 3 & 2 & 2 \end{bmatrix}\begin{bmatrix} 1 & 2 & 1 \\ 2 & 0 & 1 \\ 1 & 1 & 0 \end{bmatrix} = \begin{bmatrix} 15 & 15 & 9 \\ 15 & 8 & 8 \\ 9 & 8 & 5 \end{bmatrix}$
11. If the main diagonal is all zeros, there are no loops, and if all other entries are zero or one, there are no parallel edges, so the graph is simple. 13. 4 15. No, add edge i. **See below.** 17. **See below.** 19. **See below.** 21. Left network $\equiv \sim(p \text{ and } (q \text{ or } \sim r)) \equiv \sim p \text{ or } \sim(q \text{ or } \sim r) \equiv \sim p \text{ or } (\sim q \text{ and } r) \equiv$ Right network. Therefore, the left network is equivalent to the right network.

15.

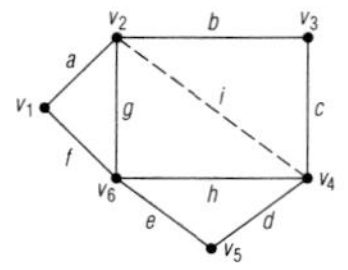

17.

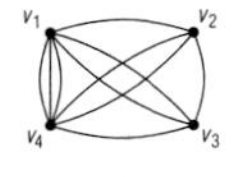

19.

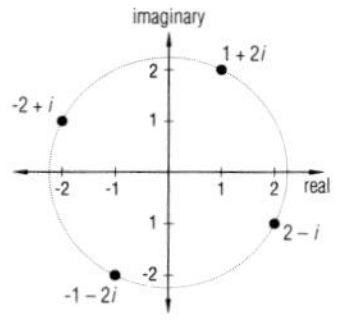

LESSON 11-6 (pp. 675–681)
1. a. 15% b. 75% 3. Yes 5. 0.70 7. $T^{20} = \begin{bmatrix} \frac{3}{7} & \frac{4}{7} \\ \frac{3}{7} & \frac{4}{7} \end{bmatrix}$
In 20 generations, $\frac{3}{7}$ of the seeds will produce pale flowers, and $\frac{4}{7}$ brilliant flowers, no matter what seeds you start with. 9. a. 15% b. 27.4% c. 32.1% tall, 32.7% medium, 35.2% short 11. a. Yes
b. $T^2 = \begin{bmatrix} 0.45 & 0.55 \\ 0.44 & 0.56 \end{bmatrix}$; $T^4 = \begin{bmatrix} 0.4445 & 0.5555 \\ 0.4444 & 0.5556 \end{bmatrix}$; $T^8 \approx \begin{bmatrix} 0.4444 & 0.5556 \\ 0.4444 & 0.5556 \end{bmatrix}$; $T^{16} \approx \begin{bmatrix} 0.4444 & 0.5556 \\ 0.4444 & 0.5556 \end{bmatrix}$
c. $a = \frac{4}{9}$ and $b = \frac{5}{9}$ d. Over the long term the proportion of occurences stabilize to a and b. 13. 10
15. No; a graph cannot have an odd number of odd vertices.
17. no solution 19. Let n, $n+1$, $n+2$, and $n+3$ represent the four consecutive integers. By the Quotient-Remainder Theorem, $n = 4q + r$ where q is an integer and $r =$ 0, 1, 2, or 3. If $r = 0$, then n is divisible by 4. If $r = 1$, $n + 3 = (4q+1) + 3 = 4(q+1)$ is divisible by 4. If $r = 2$, $n + 2 = (4q+2) + 2 = 4(q+1)$ is divisible by 4. If $r = 3$, $n + 1 = (4q+3) + 1 = 4(q+1)$ is divisible by 4. So, exactly one of every four consecutive integers is divisible by 4.

LESSON 11-7 (pp. 682–688)
1. $V - E + F = 9 - 16 + 9 = 2$

3. a.

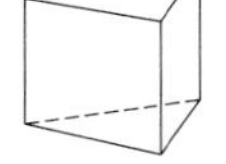

3. b.

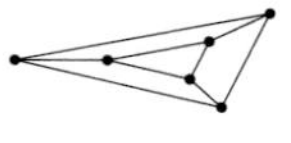

c. $V - E + F = 6 - 9 + 5 = 2$ 5. $V - E + F = 5 - 8 + 5 = 2$ 7. a. The vertex of degree 1 and its adjacent edge was removed. b. Both V and E were reduced by 1, so $V - E$ stayed the same. Since F did not change, $V - E + F$ did not change. 9. True, by the contrapositive of the second theorem of this lesson: Let G be a graph with at least one edge. If G has no vertex of degree 1, then G has a cycle.
11. a. sample: **See below.** b. Since the graph in **a** has no crossings and 6 edges, $V - E + F = 2$ holds true. Remove a vertex of degree 1 and its adjacent edge does not change the value of $V - E$. This was done in part **a**, and F did not change, so $V - E + F = 2$ holds for the original graph.

11. a.

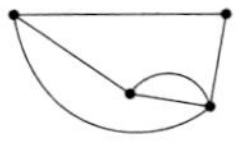

13. a. 9 b. 5 15. a. **See below.**
b. $\begin{array}{c|cc} & r & u \\ \hline r & 0.95 & 0.05 \\ u & 0.02 & 0.98 \end{array}$
c. urban $\approx 71\%$; rural $\approx 29\%$
d. $0.95a + 0.02b = a$
$$\begin{aligned} -5a + 2b &= 0 \\ 5a + 5b &= 5 \\ \hline 7b &= 5 \\ b = \tfrac{5}{7} &\approx 0.714 \\ a = \tfrac{2}{7} &\approx 0.286 \end{aligned}$$
17. a. No, there are two vertices with odd degrees.
b. sample: **See below.** 19. $\approx 20\%$

15. a.

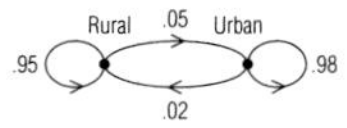

17. b.

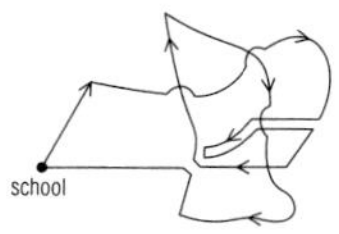

CHAPTER 11 PROGRESS SELF-TEST
(pp. 692–693)
1. v_3 2. e_5 and e_6 3. Remove e_7 and e_8 and either e_5 or e_6. 4. 5 5. $2 + 3 + 2 + 2 + 5 + 2 = 16$ 6. a. Yes b. Yes c. No 7. all except e_3 8. No, there are vertices of odd degree. 9. **See below.** 10. **See below.** 11. 2, Chicago-Detroit-Pittsburgh and Chicago-New York-Pittsburgh

9.

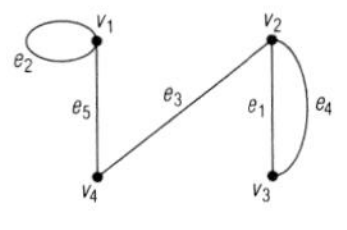

10.

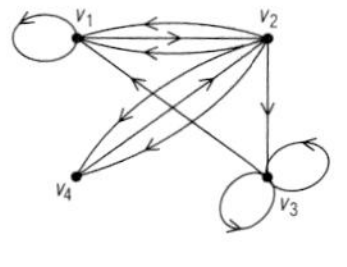

12. a.
$$\begin{array}{c|cccc} & v_1 & v_2 & v_3 & v_4 \\ \hline v_1 & 0 & 1 & 0 & 1 \\ v_2 & 1 & 0 & 1 & 1 \\ v_3 & 0 & 1 & 0 & 1 \\ v_4 & 1 & 1 & 1 & 0 \end{array}$$
b. $\begin{bmatrix} 0 & 1 & 0 & 1 \\ 1 & 0 & 1 & 1 \\ 0 & 1 & 0 & 1 \\ 1 & 1 & 1 & 0 \end{bmatrix}\begin{bmatrix} 0 & 1 & 0 & 1 \\ 1 & 0 & 1 & 1 \\ 0 & 1 & 0 & 1 \\ 1 & 1 & 1 & 0 \end{bmatrix} = \begin{bmatrix} 2 & 1 & 2 & 1 \\ 1 & 3 & 1 & 2 \\ 2 & 1 & 2 & 1 \\ 1 & 2 & 1 & 3 \end{bmatrix}$, so from the second row, the numbers of 1-stop routes to Chicago, Pittsburg, and New York are 1, 1, and 2, respectively.
13. No, this would be an Euler circuit, which is impossible since there are vertices with odd degree. 14. No, the total degree of a graph must be even.

15. a.

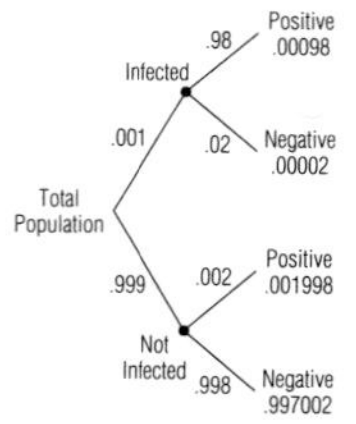

b. has AIDS and tests positive: .00098; has AIDS but tests negative: .00002; doesn't have AIDS but tests positive: .001998; doesn't have AIDS and tests negative: .997002;
c. $\frac{.001998}{.00098 + .001998} \approx 67\%$ 16. A graph cannot have an odd number of odd vertices. 17. a. **See below.**
b. $\begin{array}{c|cc} & Q & NQ \\ \hline Q & 0.0 & 1.0 \\ NQ & 0.5 & 0.5 \end{array}$
c. $\begin{bmatrix} 0 & 1 \\ .5 & .5 \end{bmatrix}^8 \approx \begin{bmatrix} .336 & .664 \\ .332 & .668 \end{bmatrix}$, so the probability of a pop quiz is $\approx \frac{1}{3}$.

17. a.

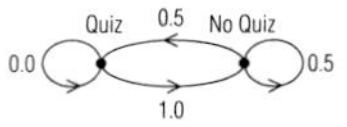

Bisection Method A method for finding zeros of a continuous function by splitting into successive subintervals an interval which has the property that its endpoints produce function values with opposite signs. A zero will be between two such endpoints. At each step, the subinterval with this same property is split in half. By continuing to subdivide, the zero can be estimated to any degree of accuracy desired. (181)

branch point A node of a tree corresponding to a step in which several choices or results are possible. (592)

Bubblesort A particular iterative sorting algorithm whereby terms are compared with neighboring terms and "bubble up" to the beginning of the list. (447)

***C*(*n, r*) Calculation Theorem** $C(n, r) = \frac{n!}{r!(n-r)!}$. (605)

***C*(*n, r*), $\binom{n}{r}$, $_nC_r$** The number of combinations of n elements taken r at a time. (604)

cardioid The polar graph of any equation of the form $r = a \pm a \sin \theta$ or $r = a \pm a \cos \theta$. (492)

Change of Base Theorem Let a and b be positive real numbers both unequal to 1; then for all $x > 0$, $\log_b x = \log_b a \bullet \log_a x$. (125)

chaos A property of a system in which small differences in the initial conditions for the system can lead to radically different results. (525)

chunking Treating an expression as if it were a single object. (172)

circuit A path in a graph that starts and ends at the same vertex. (662)

Circuits and Connectedness Theorem If a connected graph contains a circuit and an edge is removed from the circuit, then the resulting graph is also connected. (666)

circuits connected in series Circuits connected in such a way that current flows through one circuit and then through the other. (470)

closed interval See *interval*.

Cofunction identity $\cos\left(\frac{\pi}{2} - x\right) = \sin x$ or $\sin\left(\frac{\pi}{2} - x\right) = \cos x$. (359)

combination of *n* elements taken *r* at a time A subset of r elements from a set S with n elements. (604)

combinatorics The science of counting. (585)

common logarithm, log A logarithm with base 10. (124)

complement of a subset *A* of a set *S* The elements of S that are not in A. (594)

complete graph A graph in which every pair of vertices is joined by exactly one edge. (656)

complex conjugates The complex numbers $a + bi$ and $a - bi$, where a and b are real numbers. (468)

complex fraction A fraction whose numerator or denominator contains a fraction. (291)

Complex *n*th Roots Theorem (Polar Form): The n nth roots of $[r, \theta]$ are

$$\left[\sqrt[n]{r}, \frac{\theta}{n} + k \bullet \frac{2\pi}{n}\right], \text{ where } k = 0, 1, 2, \ldots, n-1.$$

(Trigonometric Form): The n nth roots of $r(\cos \theta + i \sin \theta)$ are

$$\sqrt[n]{r}\left(\cos\left(\frac{\theta}{n} + k \bullet \frac{2\pi}{n}\right) + i \sin\left(\frac{\theta}{n} + k \bullet \frac{2\pi}{n}\right)\right),$$

where $k = 0, 1, 2, \ldots, n - 1$. (505)

complex number A number that can be written in the form $a + bi$ where a and b are real numbers and $i^2 = -1$. The **real part** of $a + bi$ is a and the **imaginary part** is b. (467)

complex plane A coordinate plane for representing complex numbers. (469)

component representation For a vector, the ordered pair (u_1, u_2), the rectangular coordinates of the point at the tip of the standard position arrow of the vector. (702)

composite, $f \circ g$ The function with the rule $(f \circ g)(x) = f(g(x))$, whose domain is the set of all x in the domain of g for which $g(x)$ is in the domain of f. (160)

composition of functions The operation of first applying one function, then another. Denoted by the symbol $\circ$. (160)

concluding the inverse Using the inverse error form of argument to arrive at a conclusion. (66)

conclusion q, in the conditional $p \Rightarrow q$. Also called *consequent*. (32)

conclusion of an argument See *argument*.

conditional statement, $p \Rightarrow q$ A statement of the form *If p, then q*. Also read *p implies q*. $p \Rightarrow q$ is true except when p is true and q is false. (32)

congruence class For a given modulus, the set of numbers congruent to each other. (236)

Congruence Theorem $\forall$ integers a and b and positive integers m, $a \equiv b \pmod{m}$ if and only if m is a factor of $a - b$. (237)

congruent modulo *m*, $a \equiv b \pmod{m}$ a is congruent to b modulo m, denoted $a \equiv b \pmod{m}$, if and only if a and b have the same integer remainder when they are divided by m (the **modulus** or **mod**). (236)

conjecture A statement believed to be true but not proved. (49)

GLOSSARY

Conjugate Zeros Theorem Let $p(x)$ be a polynomial with real coefficients. If $z = a + bi$ is a zero of $p(x)$, then its complex conjugate $\bar{z} = a - bi$ is also a zero of $p(x)$. (515)

connected graph A graph G such that $\forall$ vertices v and w in G, $\exists$ a walk from v to w. (665)

connected vertices Two vertices of a graph for which there is a walk from one to the other. (665)

consequent q, in the conditional $p \Rightarrow q$. Also called *conclusion*.) (32)

constant difference See *arithmetic sequence*.

Constant Multiple Property of Integrals If f is a continuous function on the interval from a to b, and c is a real number, then $\int_a^b c\, f(x)\, dx = c \int_a^b f(x)\, dx$. (784)

constant ratio See *geometric sequence*.

Continuous Change Model If a quantity grows or decays continuously at an annual rate r, the amount $A(t)$ after t years is given by $A(t) = Be^{rt}$ where $B = A(0)$. (117)

continuous function on an interval A function f on an interval I such that every value in I is in the domain of f and the graph of f on I is an unbroken curve. (178)

contrapositive of $p \Rightarrow q$ $(\sim q) \Rightarrow (\sim p)$. (35)

Contrapositive Theorem A conditional and its contrapositive are logically equivalent. (36)

Convergence of Powers Theorem Let T be an $n \times n$ stochastic matrix with no zero entries. Then $\lim_{k \to \infty} T^k$ is a stochastic matrix with n identical rows. (677)

convergent series An infinite series whose sequence of partial sums has a finite limit. (436)

converse error An invalid argument of either of the following forms: (57)

Simple form	Universal form
If p, then q.	*For all x, if p(x) then q(x).*
q	*q(c) for a particular c.*
$\therefore p$	$\therefore p(c)$.

converse of $p \Rightarrow q$ $q \Rightarrow p$. (36)

cosecant (csc) of a real number x $\frac{1}{\sin x}$, $\forall x$ such that $\sin x \neq 0$. (321)

cosine (cos) of a real number *x* The first coordinate of the image of (1,0) under a rotation of magnitude x about the origin. (130)

cosine of an acute angle in a right triangle, cos θ The ratio $\frac{\text{side adjacent to } \theta}{\text{hypotenuse}}$. (129)

cotangent (cot) of a real number *x* $\frac{\cos x}{\sin x}$, $\forall x$ such that $\sin x \neq 0$. (321)

counterexample Given a universal statement $\forall$ *x in S, p(x)*, a value of x in S for which $p(x)$ is false. (6)

cross-product of three-dimensional vectors $\vec{u} = (u_1, u_2, u_3)$ and $\vec{v} = (v_1, v_2\ v_3)$ The vector $\vec{u} \times \vec{v} = (u_2v_3 - u_3v_2, u_3v_1 - u_1v_3, u_1v_2 - u_2v_1)$. (735)

crossing A place in a picture of a graph in which two edges seem to intersect not at their endpoints. (650)

current The rate of flow of electric charge through a circuit; measured in amps. (470)

deceleration Negative acceleration. (562)

decreasing function on an interval *S* A function f such that $\forall$ x_1 and x_2 in S, if $x_1 < x_2$ then $f(x_1) > f(x_2)$. (94)

deductive reasoning Using valid argument forms to arrive at conclusions. (63)

default window The viewing window which is set into the automatic grapher until changed by the user. (81)

definite integral of *f* from *a* to *b* The limit of upper and lower Riemann sums as the number of subintervals increases; denoted by $\int_a^b f(x)\, dx$. (776)

Degree of a Product Theorem Let $p(x)$ be a nonzero polynomial of degree m and $q(x)$ be a nonzero polynomial of degree n. The degree of $p(x) \cdot q(x)$ is $m + n$. (224)

Degree of a Sum Theorem Let $p(x)$ be a nonzero polynomial of degree m and $q(x)$ be a nonzero polynomial of degree n. The degree of $p(x) + q(x)$ is less than or equal to the larger of m and n, or $p(x) + q(x)$ is the zero polynomial. (224)

degree of a vertex, deg (*v*) The number of edges that have the vertex as an endpoint, with each edge that is a loop counted twice. (657)

DeMoivre's Theorem (Polar Form) For all positive integers n, if $z = [r, \theta]$, then $z^n = [r^n, n\theta]$. (Trigonometric form) For all positive integers n, if $z = r(\cos\theta + i \sin\theta)$, then $z^n = r^n(\cos n\theta + i \sin n\theta)$. (498)

De Morgan's Laws For all statements p and q,
1. $\sim(p \text{ and } q) \equiv (\sim p) \text{ or } (\sim q)$
2. $\sim(p \text{ or } q) \equiv (\sim p) \text{ and } (\sim q)$. (21)

dense One set is dense in another if there is an element of the first set as close as one wishes to every element of the second set. (525)

dependent variable The variable representing range values of a function. (80)

derivative function The function f' whose value at each point x is the value of the derivative $f'(x)$ of the function f at that point. Also called *first derivative*. (555)

derivative of a function at x Denoted by $f'(x)$ and given by $f'(x) = \lim_{\Delta x \to 0} \frac{f(x + \Delta x) - f(x)}{\Delta x}$, provided this limit exists and is finite. (548)

Derivative of a Quadratic Function Theorem If $f(x) = ax^2 + bx + c$, where a, b, and c are real numbers and $a \neq 0$, then $f'(x) = 2ax + b$ for all real numbers x. (557)

diagnostic reasoning Using the converse error form of an argument to arrive at a conclusion. (64)

difference of two functions For two functions f and g with domain S, the function $f - g$ defined $\forall x$ in S by $(f - g)(x) = f(x) - g(x)$. (167)

difference of vectors, $\vec{u} - \vec{v}$ If $\vec{u} = (u_1, u_2)$ and $\vec{v} = (v_1, v_2)$ then $\vec{u} - \vec{v}$, is the vector $\vec{u} + (-\vec{v})$. (709)

difference quotient $\frac{f(x_1 + \Delta x) - f(x_1)}{\Delta x}$ (541)

differential calculus The study of rates of change in continuous functions. (537)

differential equation An equation involving functions and their derivatives. (574)

directed graph, digraph A graph in which each edge has a direction. (643)

Discontinuity Theorem Given a rational function f with $f(x) = \frac{p(x)}{q(x)}$, where $p(x)$ and $q(x)$ are polynomials over the reals, if $\exists$ a real number c such that $q(c) = 0$ but $p(c) \neq 0$, then f has an essential discontinuity at $x = c$ and the line $x = c$ is a vertical asymptote to its graph. (310)

discrete dynamical system A set D together with a function f from D into itself that is repeatedly applied. (523)

discrete function A function whose domain is a discrete set. (83)

discrete set A set that can be put into 1–1 correspondence with a subset of the set of integers. (83)

Distance in Space Theorem The distance between $P = (x_1, y_1, z_1)$ and $Q = (x_2, y_2, z_2)$ is given by $PQ = \sqrt{(x_2 - x_1)^2 + (y_2 - y_1)^2 + (z_2 - z_1)^2}$. (728)

divergent An infinite series whose sequence of partial sums does not have a finite limit. (436)

divisor See *factor*.

domain See *function*.

domain of an identity The set of all values of the variables for which both sides in an identity are defined. (289)

dot product of three-dimensional vectors $\vec{u} = (u_1, u_2, u_3)$ and $\vec{v} = (v_1, v_2\ v_3)$ The number $\vec{u} \cdot \vec{v} = u_1v_1 + u_2v_2 + u_3v_3$. (732)

dot product of $\vec{u} = (u_1, u_2)$ and $\vec{v} = (v_1, v_2)$, The real number $\vec{u} \cdot \vec{v} = u_1v_1 + u_2v_2$. (720)

e The irrational number equal to $\lim_{x \to \infty} (1 + \frac{1}{x})^x$, which is approximately 2.71828... . (117)

edge An arc on a graph. (640)

edge-endpoint function See *graph*.

efficiency of an algorithm The maximum number $E(n)$ of significant operations necessary for the algorithm to solve a given problem of size n. (453)

elementary functions The basic functions from which more complicated functions may be defined, including polynomial, exponential, trigonometric, and logarithmic functions. (79)

end behavior A description of what happens to the values $f(x)$ of a function f as $x \to \infty$ and as $x \to -\infty$. (108)

endpoint See *graph*.

equality of complex numbers $a + bi = c + di$ if and only if $a = c$ and $b = d$. (467)

Equation of a Plane Theorem The set of points $\{(x, y, z): ax + by + cz = d\}$, where at least one of the coefficients a, b, or c is nonzero, is a plane perpendicular to the vector $\vec{v} = (a, b, c)$. (741)

Equation of a Sphere Theorem The sphere with center (a, b, c) and radius r has equation $r^2 = (x - a)^2 + (y - b)^2 + (z - c)^2$. (729)

equilibrium position A position of a weight hanging from a spring where the downward pull of gravity on the weight is exactly counterbalanced by the upward pull on the weight by the stretched spring. (386)

equivalent graphs Two graphs for which there is a 1–1 correspondence between their vertices and edges under which the edge-endpoint function of one corresponds to the edge-endpoint function of the other. (641)

essential discontinuity A discontinuity that cannot be removed by insertion of a single point. (309)

essential features of a counting problem Whether order makes a difference in the objects to be counted, and whether or not the objects can be repeated. (586)

Euler Circuit Theorem If a graph has an Euler circuit, then every vertex of the graph has even degree. (664)

Euler circuit Of a graph, a circuit that contains every edge and every vertex of the graph. (662)

Euler's formula In a connected graph with no crossings, if V is the number of vertices, E is the number of edges, and F is the number of faces, then $V - E + F = 2$. (685)

GLOSSARY

even function A function such that $\forall\, x$ in its domain, $f(-x) = f(x)$. (111)

even integer An integer n such that $n = 2k$ for some integer k. (49)

exclusive *or* One or the other but not both. (20)

existential statement A statement of the form *There exists x in S such that p(x).* (7)

expanded form of a sum The form in which $\sum_{i=m}^{n} a_i$ is rewritten as $a_m + a_{m+1} + \ldots + a_n$. (413)

explicit formula Of a sequence, a formula for the nth term of in terms of n. (100)

exponential function with base *b* The function defined by $f\colon x \rightarrow b^x,\ \forall\, x$. (114)

faces of a planar graph The regions of the plane into which the edges of the graph divide the plane. (683)

factor Given integers n and d with $d \neq 0$, d is a factor (or **divisor**) of n if and only if there is an integer q such that $n = q \cdot d$. (222)

Factor of a Product Theorem For all integers m, n, and p, if m is a factor of n, then m is a factor of $n \cdot p$. (227)

Factor of a Sum Theorem For all integers a, b, and c, if a is a factor of b and a is a factor of c, then a is a factor of $b + c$. (225)

Factor Search Theorem If an integer n has no prime factors between 1 and $\sqrt{n}$ inclusive, then n is prime. (270)

Fibonacci numbers The integers 1, 2, 3, 5, 8, ... that are terms of the Fibonacci sequence. (403)

Fibonacci sequence The infinite sequence 1, 1, 2, 3, 5, 8, 13, 21, ..., in which each term after the second is the sum of the two previous terms. (101)

field of complex numbers The complex numbers with the operations of addition and multiplication. (468)

field properties Addition and multiplication are closed, commutative, and associative. There is an identity for addition and a different identity for multiplication. Every number has an additive inverse and every number but the additive identity has a multiplicative inverse. Multiplication is distributive over addition. (467)

finite sequence A function whose domain is the set of integers from an integer m to an integer n. (100)

finite series The indicated sum of finitely many consecutive terms of a sequence. (431)

first derivative The derivative function or a value of that function. (564)

fixed point For a function f in a dynamical system, a value of x for which $f(x) = x$. (527)

floor function, $\lfloor\ \rfloor$ The function f defined on the real numbers such that $f(x)$ is the greatest integer less than or equal to x. Also called *greatest integer function*. (178)

force An influence that affects the motion of an object. (707)

Frac, Fract A function that maps x onto $x - \lfloor x \rfloor$. (210)

function A correspondence from a set A (the **domain**) to a set B in which each element in A corresponds to exactly one element of B. (80)

Function Inequality Theorem Suppose that f is a continuous real function. If f has zeros a and b and no zeros between them, then either $f(x) > 0$ for all x between a and b or $f(x) < 0$ for all x between a and b. (191)

functionally equivalent networks Two networks that produce the same output for each combination of input signals. (28)

fundamental region See *tessellation*.

Fundamental Theorem of Algebra If $p(x)$ is any polynomial of degree $n \geq 1$ with complex coefficients, then $p(x)$ has at least one complex zero. (510)

Fundamental Theorem of Arithmetic Suppose that n is an integer and that $n > 1$. Then either n is a prime number or n has a prime factorization which is unique except for the order of the factors. (270)

Fundamental Theorem of Calculus Let f be a continuous function on the interval from a to b.

1. If g is a function whose derivative is f, then $\int_a^b f(x)\,dx = g(b) - g(a)$.
2. If $g(x) = \int_a^x f(t)\,dt$, then $g'(x) = f(x)$. (801)

Geometric Addition Theorem Let z and w be two complex numbers that are not on the same line through the origin in the complex plane. Then the point representing $z + w$ is the fourth vertex of a parallelogram with consecutive vertices z, 0, and w. (481)

Geometric Multiplication Theorem Let z and w be complex numbers. If $z = [r, \theta]$ and $w = [s, \phi]$, then $zw = [rs, \theta + \phi]$. That is, multiplying a complex number z by w applies to z the composite of a size change of magnitude s and a rotation of ϕ about the origin. (483)

Geometric *n*th Roots Theorem When graphed in the complex plane, the nth roots of any complex number z are the vertices of a regular n-gon whose center is at (0, 0). (506)

geometric sequence A sequence g in which $\forall n > 1$, $g_n = rg_{n-1}$, where r is a constant. That is, each term beyond the first is r (the **constant ratio**) times the preceding term. (101)

geometric series The indicated sum of consecutive terms of a geometric sequence. (431)

graph A finite set of vertices, a finite set of edges, and a function (the **edge-endpoint function**) that maps each edge to a set of either one or two vertices (the **endpoints** of the edge). (648)

graph of a function For a function f, the set of all points (x, y) such that x is in the domain of f and $y = f(x)$. (81)

graph theory The study of graphs and networks and their properties. (639)

Graph Transformation Theorems

Translation For an equation relating real variables x and y, the following two processes yield the same graph for any specific values of h and k:
(1) replacing x by $x - h$ and y by $y - k$ in the equation and graphing the result;
(2) applying the translation $(x, y) \rightarrow (x + h, y + k)$ to the graph of the original relation.

Scale Change For an equation relating real variables x and y, the following two processes yield the same graph for any specific nonzero values of a and b:
(1) replacing x by $\frac{x}{a}$ and y by $\frac{y}{b}$ in the equation and graphing the result;
(2) applying the scale change $(x, y) \rightarrow (ax, by)$ to the graph of the original relation. (203)

greatest integer function See *floor function*.

half-adder A network that takes two binary digits as input and produces the two digits of their sum as output. (264)

half-open interval See *interval*.

handshake problem Suppose n people are at a party. If each person shakes hands with every other person, how many handshakes are required? (656)

harmonic sequence A sequence whose terms are reciprocals of terms of an arithmetic sequence. (101)

height of a graph The value of $f(x)$ at x; the directed distance from $(x, 0)$ to $(x, f(x))$. (81)

hexadecimal Base 16. (264)

horizontal asymptote A horizontal line to which the graph of a function gets closer and closer as $x \rightarrow \infty$ or as $x \rightarrow -\infty$. (109)

horizontal component of a vector See *x-component of a vector*.

hypothesis p, in the conditional $p \Rightarrow q$. Also called *antecedent*. (32)

Identities for cos 2*x* For all real numbers x, $\cos 2x = \cos^2 x - \sin^2 x = 2\cos^2 x - 1 = 1 - 2\sin^2 x$. (367)

identity An equation that is true for all values of the variables for which both sides are defined. (289)

Identity for cos($\alpha + \beta$) For all real numbers α and β, $\cos(\alpha + \beta) = \cos\alpha\cos\beta - \sin\alpha\sin\beta$. (357)

Identity for cos($\alpha - \beta$) For all real numbers α and β, $\cos(\alpha - \beta) = \cos\alpha\cos\beta + \sin\alpha\sin\beta$. (359)

Identity for sin 2*x* For all real numbers x, $\sin 2x = 2\sin x\cos x$. (368)

Identity for sin($\alpha + \beta$) For all real numbers α and β, $\sin(\alpha + \beta) = \sin\alpha\cos\beta + \cos\alpha\sin\beta$. (362)

Identity for sin($\alpha - \beta$) For all real numbers α and β, $\sin(\alpha - \beta) = \sin\alpha\cos\beta - \cos\alpha\sin\beta$. (362)

Identity for tan($\alpha + \beta$) For all real numbers α and β such that $\tan\alpha$, $\tan\beta$, and $\tan(\alpha + \beta)$ are defined, $\tan(\alpha + \beta) = \dfrac{\tan\alpha + \tan\beta}{1 - \tan\alpha\tan\beta}$. (363)

Identity for tan($\alpha - \beta$) For all real numbers α and β such that $\tan\alpha$, $\tan\beta$, and $\tan(\alpha - \beta)$ are defined, $\tan(\alpha - \beta) = \dfrac{\tan\alpha - \tan\beta}{1 + \tan\alpha\tan\beta}$. (363)

identity function A function defined by the rule $f(x) = x$. (163)

identity transformation A transformation which maps each point onto itself. (204)

if and only if "*p if and only if q*" is denoted by $p \Leftrightarrow q$ and is equivalent to (*if p then q*) and (*if q then p*) or, symbolically, $p \Rightarrow q$ *and* $q \Rightarrow p$. (37)

imaginary axis The vertical axis in the complex plane. (469)

imaginary number A number of the form bi, where b is a real number and i is the imaginary unit. (466)

imaginary part of a complex number See *complex number*.

imaginary unit, *i* The complex number $i = \sqrt{-1}$. (466)

impedance The opposition to the flow of current caused by components called resistors, coils, and capacitors; measured in ohms. (470)

implies See *conditional statement*.

improper induction Concluding a universal statement from instances of it. (58)

inclusive *or* One or the other or both. (20)

increasing function on an interval *S* A function f such that $\forall x_1$ and x_2 in S, if $x_1 < x_2$ then $f(x_1) < f(x_2)$. (94)

independent variable A variable representing domain values of a function. (80)

index A subscript in summation notation. (413)

GLOSSARY

inductive assumption In mathematical induction, the assumption that $S(k)$ is true for an integer $k \geq 1$. (419)

inductive reasoning Generalization based on the evidence of examples; using improper induction to arrive at a conclusion. (63)

inductive step See *Principle of Mathematical Induction*.

infinite sequence A function whose domain is the set of all nonnegative integers greater than or equal to a fixed integer. (100)

infinite series The indicated sum of the terms of an infinite sequence. (431)

infinite sum, $\sum_{k=1}^{\infty} a_k$ The limit of the partial sums of a sequence as $n \to \infty$, provided this limit exists and is finite. (434)

Infinitude of Primes Theorem There are infinitely many prime numbers. (269)

initial conditions See *recursive definition for a sequence*.

input-output table A table indicating what the output will be from a gate or network for any possible combination of input signals. (25)

instantaneous rate of change of *f* at *x* The derivative of f at x. (549)

instantaneous velocity at time *x* The limit as $\Delta x \to 0$ of the average velocity of an object between times x and Δx, provided this limit exists and is finite. (546)

Integral of a Quadractic Function Theorem If $a \geq 0$, $\int_0^a (c_2x^2 + c_1x + c_0)\,dx = c_2\frac{a^3}{3} + c_1\frac{a^2}{2} + c_0a$. (790)

Intermediate Value Theorem Suppose that f is a continuous real function on an interval I, and that a and b are in I. Then for every real number y_0 between $f(a)$ and $f(b)$, there is at least one real number x_0 between a and b such that $f(x_0) = y_0$. (179)

interval The set of numbers between two given numbers, including neither (an **open interval**) or one (a **half-open interval**) or both (a **closed interval**) of those two numbers. (94)

invalid argument An argument or argument form for which there exist instances in which the premises are true and the conclusion is false. (56)

inverse cosine function, $\cos^{-1}$ The function that maps x onto the number y whose cosine is x, for $0 \leq y \leq \pi$. (374)

inverse error An invalid argument of either of the following forms: (58)

Simple form	Universal form
If p, then q.	*For all x, if p(x) then q(x).*
not p	*not p(c), for a particular c.*
∴ not q	*∴ not q(c).*

inverse functions Two functions f and g such that $f \circ g(x) = x$ and $g \circ f(x) = x$ for all x in the domains of both f and g. (162)

inverse of $p \Rightarrow q$ $\sim p \Rightarrow \sim q$. (36)

inverse sine function, $\sin^{-1}$ The function that maps x onto the number y whose sine is x, for $-\frac{\pi}{2} \leq y \leq \frac{\pi}{2}$. (374)

inverse tangent function, $\tan^{-1}$ The function that maps x onto the number y whose tangent is x, for $-\frac{\pi}{2} < y < \frac{\pi}{2}$. (374)

irrational conjugates The real numbers $c + \sqrt{d}$ and $c - \sqrt{d}$, where c and d are rational numbers and $\sqrt{d}$ is not a rational number. (298)

irrational number A real number that cannot be written as a ratio of two integers. (295)

isolated vertex A vertex of a graph that is not the endpoint of any edge. (649)

iterate To repeat over and over. (447)

iterative algorithm An algorithm in which the same steps are repeated again and again. (447)

Königsberg bridge problem In the city of Königsberg (now Kaliningrad), two branches of the Pregol'a River come together. In the 1700s, parts of Königsberg were on the banks of the river, another part was on a large island in the middle, and a final part was between the two branches of the river. Seven bridges connected these four parts of the city. Is it possible for a person to walk around the city crossing each bridge exactly once, starting and ending at the same point? (639)

Law of Cosines For all triangles ABC with sides a, b, and c: $c^2 = a^2 + b^2 - 2ab \cos C$. (132)

Law of Detachment, *modus ponens* A valid argument of either of the following forms: (42)

Simple form	Universal form
If p then q.	*∀ x, if p(x) then q(x).*
p	*p(c), for a particular c.*
∴ q.	*∴ q(c)*

Law of Indirect Reasoning, *modus tollens* A valid argument of either of the following forms: (44)

Simple form	Universal form
If p then q.	$\forall$ *x, if p(x) then q(x).*
not q	*not q(c) for a particular c.*
$\therefore$ *not p.*	$\therefore$ *not p(c)*

Law of Sines For all triangles *ABC* with sides *a*, *b*, and *c*: $\frac{\sin A}{a} = \frac{\sin B}{b} = \frac{\sin C}{c}$. (132)

Law of Substitution If a universal statement is true for a given set, then it is true for any particular element of that set. (5)

Law of Transitivity A valid argument of either of the following forms: (43)

Simple form	Universal form
If p then q.	$\forall$ *x, if p(x), then q(x).*
If q then r.	$\forall$ *x, if q(x), then r(x).*
$\therefore$ *If p then r.*	$\therefore$ $\forall$ *x, if p(x), then r(x).*

Laws of Exponents

$b^x \bullet b^y = b^{x+y}$	(Product of Powers)
$\frac{b^x}{b^y} = b^{x-y}$	(Quotient of Powers)
$(b^x)^y = b^{xy}$	(Power of a Power) (115)

leaves The ends of the branches in a tree. (592)

length of a walk The number of edges in the walk. (670)

limaçon The polar graph of an equation of the form $r = a + b \cos \theta$ or $r = a + b \sin \theta$, where a and b are nonzero real numbers. (488)

limit of a sequence, $\lim_{n \to \infty} a_n$ A number *L* such that, for any number *d* picked, all terms beyond a certain point in the sequence are within *d* of *L*. (103)

logarithm function with base *b* The function defined by the rule $x \rightarrow \log_b x$ for all positive real numbers *x*. (122)

Logarithm of a Power Theorem $\log_b(u^s) = s \log_b u$. (123)

Logarithm of a Product Theorem $\log_b(u \bullet v) = \log_b u + \log_b v$. (123)

Logarithm of a Quotient Theorem $\log_b\left(\frac{u}{v}\right) = \log_b u - \log_b v$. (123)

logarithm of *x* base *b*, $\log_b x$ The power to which *b* must be raised to equal *x*; that is, the number *y* such that $b^y = x$. (122)

logarithmic scale A scale in which the units are spaced so that the ratio between successive units is the same. (126)

logarithmic spiral The polar graph of $r = ab^{\theta}$, where $a > 0$ and $b > 1$. (495)

logical expression A formula in which variables representing statements are combined in an unambiguous way with *and*, *or*, *not*, or *if-then*. (20)

logically equivalent expressions, $\equiv$ Two logical expressions with the same truth values for all substitutions of statements for their statement variables. (20)

loop An edge of a graph whose endpoints are the same point. (648)

lower Riemann sum The Riemann sum of a function *f* where each $f(z_i)$ is the smallest value on the subinterval. (775)

Markov chain Of a situation that can exist in only a finite number of states, when the probabilities of proceeding from one state to the next depend only on the first state. (677)

mathematical induction See *Principle of Mathematical Induction*.

maximum value of a function *f* with domain *A* A number *m* such that $\exists$ *x* in *A* with $f(x) = m$ and $\forall$ *x* in *A*, $m \geq f(x)$. (88)

Method of Direct Proof of a Universal Statement Express the statement to be proved in the form $\forall$ *x in S, if p(x) then q(x)*. Start the proof by supposing that *x* is any element of *S* for which the antecedent *p(x)* is true. Use the definitions of the terms that occur in *p(x)* and *q(x)* and other known properties, to make a chain of deductions ending with *q(x)*. (51)

minimum value of a function *f* with domain *A* A number *m* such that $\exists$ *x* in *A* with $f(x) = m$ and $\forall$ *x* in *A*, $m \leq f(x)$. (88)

modulus See *absolute value of a complex number*.

modulus, modulo, mod See *congruent modulo m*.

modus ponens See *Law of Detachment*.

modus tollens See *Law of Indirect Reasoning*.

multinomial coefficient A coefficient in the series expansion of $(a_1 + a_2 + \ldots + a_k)^n$ (627)

multiple Given integers *n* and *d* with $d \neq 0$, *n* is a multiple of *d* if and only if *d* is a factor of *n*. (222)

Multiplication Counting Principle Suppose that strings result from a procedure which consists of *k* successive steps and that:

the 1st step can be done in n_1 ways,
the 2nd step can be done in n_2 ways,
.
.
.
and the *k*th step can be done in n_k ways.

Then the number of possible strings is $n_1 \bullet n_2 \bullet \ldots \bullet n_k$. (593)

GLOSSARY

multiplication of complex numbers Let a, b, c, and d be real numbers and let $z = a + bi$ and $w = c + di$. Then $zw = (ac - bd) + (ad + bc)i$. (467)

Multiplication Property of Congruence See *Properties of Congruence*.

multiplicity of a zero For a zero c of a polynomial $p(x)$ of degree at least 1, the largest positive integer m such that $(x - c)^m$ is a factor of $p(x)$. (511)

natural logarithm, ln A logarithm with base e. (124)

necessary condition "p is a necessary condition for q" means $q \Rightarrow p$. (39)

negation The statement, denoted *not p*, that, if true, exactly expresses what it would mean for p to be false. (12)

negation of $\forall$ *x in S, if p(x) then q(x)* $\exists$ *x in S such that p(x) and not q(x)*. (35)

negation of $p \Rightarrow q$ *p and (not q)*. (34)

nested form A polynomial $a_nx^n + a_{n-1}x^{n-1} + a_{n-2}x^{n-2} + \ldots + a_1x + a_0$ in the form $((\ldots((a_nx + a_{n-1})x + a_{n-2})x + \ldots)x + a_1)x + a_0$. (223)

network of logic gates NOT, AND, and OR gates connected in such a way that the output signals from some of the gates become input signals for other gates. (27)

node A vertex in a tree; corresponds to a step in which several choices or results are possible. (592)

nonreversible operation A reasoning step in solving an equation or inequality whose converse is not true for some values of the variables for which the expressions in the equation are defined. (153)

norm of a vector The length of the vector; denoted by $|\vec{u}|$. (703)

not p See *negation*.

***n*th root** Of a complex number w, a number z such that $z^n = w$. (504)

$O(d)$ The orbit with initial point d. (523)

octal Base 8. (264)

odd function A function such that $\forall$ x in its domain, $f(-x) = -f(x)$. (111)

odd integer An integer n such that $n = 2k + 1$ for some integer k. (49)

one-to-one function, 1–1 A function g such that $g(x) = g(y) \Rightarrow x = y$. (154)

only if "*p only if q*" is equivalent to *if p then q*. (39)

open interval See *interval*.

opposite of a vector, $-\vec{v}$ The vector with the same magnitude and direction opposite that of the given vector. (708)

optimization problem A problem in which the value of one variable is sought to obtain the optimal, or most desirable, value of another. (568)

orbit with initial point *d* For a set D and a function $f: D \to D$ constituting a discrete dynamical system, the sequence a_0, a_1, a_2, defined by

$$\begin{cases} a_0 = d \\ a_{k+1} = f(a_k) \text{ for all integers } k \geq 0. \end{cases} (523)$$

orthogonal vectors Vectors whose directions are perpendicular. (722)

$P(n, r)$ Calculation Theorem $P(n, r) = \dfrac{n!}{(n-r)!}$. (600)

$P(n, r)$, $_nP_r$ The number of permutations of n elements taken r at a time. (600)

parallel edges Two edges of a graph with both endpoints in common. (648)

parallel vectors Two vectors with the same or opposite directions. (712)

parameter An independent variable on which other variables depend. (714)

parametric equations for a line Equations for a line in which each coordinate is expressed in terms of the same variable. (714)

partial sum The sum of the first n terms of a sequence. (433)

Pascal's Triangle A triangular array of binomial coefficients (equivalently, combinations) in which the rth element in row n is the sum of the $(r - 1)$st and rth elements in row $n - 1$. (610)

	Row
1	0
1 1	1
1 2 1	2
1 3 3 1	3
1 4 6 4 1	4
1 5 10 10 5 1	5
. . .	

path A walk from one vertex to another in which no edge is repeated. (662)

period of a function The smallest positive number p such that $f(x + p) = f(x)$ for all x in the domain of f. (131)

periodic function A real function f with the property that there is a positive number p such that $f(x + p) = f(x)$ for all x in the domain of f. (131)

permutation A string of all of the symbols $a_1, a_2, \ldots, a_n$ without repetition. (598)

permutation of the *n* elements of a set *S* taken *r* at a time A string of r elements from S without repetition. (600)

Permutation Theorem There are $n!$ permutations of n different elements. (599)

phase shift The least positive or the greatest negative horizontal translation that maps $y = \cos x$ or $y = \sin x$ onto a translation image. (205)

Phase-Amplitude Theorem Let a, b, and c be real numbers with $c \neq 0$ and $a^2 + b^2 \neq 0$. Let $\cos \theta = \frac{a}{\sqrt{a^2 + b^2}}$, $\sin \theta = \frac{b}{\sqrt{a^2 + b^2}}$, and $0 < \theta < 2\pi$. Then, for all t, $a \cos ct + b \sin ct = \sqrt{a^2 + b^2} \cos(ct - \theta)$, a cosine function with period $\frac{2\pi}{c}$, amplitude $\sqrt{a^2 + b^2}$, and phase shift $\frac{\theta}{c}$. (387)

plane vector See *two-dimensional vector.*

polar axis A ray, usually horizontal, through the pole of a polar coordinate system, from which rotations are measured. (473)

polar coordinate system A system in which a point is identified by a pair of numbers $[r, \theta]$ where $|r|$ is the distance of the point from a fixed point (the **pole**), and θ is a magnitude of rotation from the polar axis. (473)

polar coordinates, $[r, \theta]$ Description of a point in a polar coordinate system. (473)

polar equations Equations in polar coordinates. (486)

polar form of a complex number The representation of the number in polar coordinates. (482)

polar grid A background of circles and rays emanating from the pole, of use in sketching graphs in a polar coordinate system. (474)

polar representation of a vector The representation of a two-dimensional vector with positive magnitude r and direction θ by the polar coordinates $[r, \theta]$. (700)

Polar-Rectangular Conversion Theorem If $[r, \theta]$ is a polar coordinate representation of a point P, then the rectangular coordinates (x, y) of P are given by $x = r \cos \theta$ and $y = r \sin \theta$. (476)

pole See *polar coordinate system.*

possibility tree A diagram used to display the possible outcomes of an experiment. (592)

power function A function with an equation of the form $y = ax^n$. (110)

Power of a Power See *Laws of Exponents.*

premises See *argument.*

prime An integer $n > 1$ whose only positive integer factors are 1 and n. (268)

prime factorization A representation of a number as a product of primes. (270)

Principle of Mathematical Induction Suppose that for each positive integer n, $S(n)$ is a sentence in n. If
(1) $S(1)$ is true (the **basis step**), and
(2) for all integers $k \geq 1$, the assumption that $S(k)$ is true emplies that $S(k + 1)$ is true (the **inductive step**),
then $S(n)$ is true for all positive integers n. (418)

Principle of Strong Mathematical Induction Suppose that for each positive integer n, $S(n)$ is a sentence in n. If
(1) $S(1)$ is true, and
(2) for all integers $k \geq 1$, the assumption that $S(1)$, $S(2)$, ..., $S(k - 1)$, $S(k)$ are all true implies that $S(k + 1)$ is also true, then $S(n)$ is true for all positive integers n. (438)

probabilistic reasoning Using evidence from probabilities to arrive at a conclusion. (66)

probability tree A digraph in which each vertex represents an event, and the edge leading from vertex A to vertex B is labeled with the probability that B occurs if A occurs. (643)

Product of Powers See *Laws of Exponents.*

product of two functions For two functions f and g with domain S, the function $f \cdot g$ defined $\forall x$ in S by $(f \cdot g)(x) = f(x) \cdot g(x)$. (167)

proof by contradiction A proof in which, if s is to be proven, one reasons from *not s* until a contradiction is deduced; from this it is concluded that *not s* is false, which means that s is true. (268)

Properties of Congruence (Modular Arithmetic) Let a, b, c, and d be any integers and let m be a positive integer. If $a \equiv b \pmod{m}$ and $c \equiv d \pmod{m}$, then

$a + c \equiv b + d \pmod{m}$	(Addition Property of Congruence),
$a - c \equiv b - d \pmod{m}$	(Subtraction Property of Congruence), and
$ac \equiv bd \pmod{m}$	(Multiplication Property of Congruence). (238)

Quicksort A particular recursive sorting algorithm whereby terms are divided into three sets and those sets are then sorted. (447)

quotient The answer to a division problem. For integers n and d with $d \neq 0$, an integer q such that $n = q \cdot d$. (222)

quotient See *Quotient-Remainder Theorem (for Integers)* and *Quotient-Remainder Theorem (for Polynomials).*

Quotient of Powers See *Laws of Exponents.*

GLOSSARY

quotient of two functions For two functions f and g with domain S, the function $\frac{f}{g}$ defined $\forall x$ in S by $\left(\frac{f}{g}\right)(x) = \frac{f(x)}{g(x)}$ provided $g(x) \neq 0$. (167)

Quotient-Remainder Theorem (for Integers) If n is an integer and d is a positive integer, then there exist unique integers q (the **quotient**) and r (the **remainder**) such that $n = q \cdot d + r$ and $0 \leq r < d$. (230)

Quotient-Remainder Theorem (for Polynomials) If $p(x)$ is a polynomial and $d(x)$ is a nonzero polynomial, then there exist unique polynomials $q(x)$ (the **quotient**) and $r(x)$ (the **remainder**) such that $p(x) = q(x) \cdot d(x) + r(x)$ and either $0 \leq$ degree of $r(x) <$ degree of $d(x)$ or $r(x) = 0$. (242)

range The set of possible values of the dependent variable of a function. Symbolically, for a function $f: A \rightarrow B$, the set of all elements y in B such that $\exists\ x$ in A with $f(x) = y$. (87)

rational equation An equation of the form $f(x) = g(x)$ where $f(x)$ and $g(x)$ are rational expressions. (326)

rational expression The indicated quotient of two polynomials. (289)

rational function A function f such that for all values of x in the domain of f, $f(x) = \frac{p(x)}{q(x)}$, where $p(x)$ and $q(x)$ are polynomials. (308)

rational number A real number such that there exist integers a and b $(b \neq 0)$ with $r = \frac{a}{b}$. (288)

real axis The horizontal axis in the complex plane. (469)

real function A function whose independent and dependent variables have only real number values. (80)

real part of a complex number See *complex number*.

recurrence relation See *recursive definition for a sequence*.

recursive algorithm An algorithm that refers back to a smaller version of itself. (447)

recursive definition for a sequence A definition of a sequence consisting of a specification of one or more initial terms of the sequence (the **initial conditions**) and an equation that relates each of the other terms of the sequence to one or more of the previous terms (a **recurrence relation**). (401)

recursive formula A formula for a sequence in which the first term or first few terms are given, and the nth term is expressed in terms of the preceding term(s). (100)

regular polygon A convex polygon whose sides are all the same length and whose angles are all the same measure. (332)

regular polyhedron (plural *polyhedra*) A convex polyhedron whose faces are all congruent regular polygons. (334)

relative maximum value for a function f with domain A A number m such that $\exists\ x$ in A with $f(x) = m$ and f is increasing on an interval immediately to the left of x and decreasing on an interval immediately to the right of x. (95)

relative minimum value for a function f with domain A A number m such that $\exists\ x$ in A with $f(x) = m$ and f is decreasing on an interval immediately to the left of x and increasing on an interval immediately to the right of x. (95)

remainder See *Quotient-Remainder Theorem (for Integers)* and *Quotient-Remainder Theorem (for Polynomials)*.

removable discontinuity at x For a function f, the existence of a hole in the graph of f at x such that it is possible to redefine f at x in a way that removes that hole. (309)

restrictions The set of values that the variable(s) in an expression or function cannot have. (290)

resultant force The combined effect of two or more forces. (707)

reversible step A reasoning step in solving an equation or inequality whose converse is true for all values of the variables for which the expressions in the equation are defined. (153)

Riemann sum of a function f over the interval from a to b The sum $f(z_1)(x_1 - a) + f(z_2)(x_2 - x_1) + f(z_3)(x_3 - x_2) + \ldots + f(z_n)(b - x_{n-1})$, where f is a function defined over the interval from a to b and the interval is partitioned into n subintervals: the first from a to x_1, the second from x_1 to x_2, the third from x_2 to x_3, ..., the nth from x_{n-1} to b and each z_i is a value in the ith subinterval. Letting $x_0 = a$ and $x_n = b$, this Riemann sum can be written as $\sum_{i=1}^{n} f(z_i)(x_i - x_{i-1})$. (769)

rose curve The polar graphs of equations of the form $r = a\cos(n\theta)$, $a > 0$, n a positive integer or $r = a\sin(n\theta)$, $a > 0$, n a positive integer. (494)

scalar A real number (used in conjunction with vectors, matrices, and transformations). (714)

scalar multiple of a three-dimensional vector $\vec{u} = (u_1, u_2, u_3)$ and a scalar k The vector $k\vec{u} = (ku_1, ku_2, ku_3)$. (732)

scalar multiple, $k \cdot \vec{v}$ See *scalar multiplication*.

scalar multiplication The operation of multiplying a vector $\vec{v} = (v_1, v_2)$ by a real number k (the **scalar**) resulting in a **scalar multiple** $k \cdot \vec{v} = (kv_1, kv_2)$ of the original vector. (712)

Scale Change Theorem See *Graph Transformation Theorems*.

secant (sec) of a real number x $\frac{1}{\cos x}$ $\forall x$ such that $\cos x \neq 0$. (321)

secant line for the graph of a function A line passing through two distinct points on the graph of a continuous function. (539)

secant line to a circle A line that intersects the circle at two distinct points. (539)

second derivative The derivative function of a derivative function. (564)

sequence A function whose domain is the set of integers greater than or equal to a fixed integer. (100)

series The indicated sum of consecutive terms of a sequence. (431)

set An unordered list of symbols with no repetitions. (587)

significant operations The number of major operations needed in a problem. (453)

simple graph A graph with no loops and no parallel edges. (650)

simple zero Of a polynomial, a zero that has multiplicity 1. (511)

sine (sin) of a real number x The second coordinate of the image of (1, 0) under a rotation of magnitude x about the origin. (130)

sine of an acute angle in a right triangle, sin θ The ratio $\frac{\text{side opposite } \theta}{\text{hypotenuse}}$. (129)

sine wave An image of the graph of the sine or cosine function under a composite of scale changes or translations. (204)

size of a problem The number of operations needed to do a problem, or an estimate of that number. (453)

sorting algorithm An algorithm whose purpose is to arrange or sort a given list of items in some desired order. (445)

spiral of Archimedes The polar graph of $r = a\theta + b$, where a is positive and b is nonnegative. (495)

standard position An arrow for a vector whose initial point is at the origin or pole of the coordinate system. (701)

standard prime factorization The prime factorization of an integer $n > 1$ in which all like factors are combined using exponents, and the prime factors are arranged in increasing order of magnitude. (271)

statement A sentence that is either true or false and not both. (4)

stochastic matrix A matrix in which each element is nonnegative, and the entries in each row add to 1. (676)

string An ordered list of symbols. (587)

Subtraction Property of Congruence See *Properties of Congruence*.

sufficient condition "p is a sufficient condition for q" means $p \Rightarrow q$. (39)

Sufficient Condition for an Euler Circuit Theorem If a graph G is connected and every vertex of G has even degree, then G has an Euler circuit. (665)

Sum of Binomial Coefficients Theorem $\forall$ integers $n \geq 0$, $\binom{n}{0} + \binom{n}{1} + \binom{n}{2} + \ldots + \binom{n}{k} + \ldots + \binom{n}{n} = 2^n$. (615)

Sum of the First n Integers Theorem For each integer $n \geq 1$, $\sum_{i=1}^{n} i = 1 + 2 + \ldots + n = \frac{n(n+1)}{2}$. (423)

Sum of the First n Powers If $r \neq 1$, then $1 + r + r^2 + \ldots + r^{n-1} = \frac{1 - r^n}{1 - r}$ $\forall$ integers $n \geq 1$. (432)

sum of three-dimensional vectors $\vec{u} = (u_1, u_2, u_3)$ and $\vec{v} = (v_1, v_2, v_3)$ The vector $\vec{u} + \vec{v} = (u_1 + v_1, u_2 + v_2, u_3 + v_3)$. (732)

sum of two functions For two functions f and g with domain S, the function $f + g$ defined $\forall x$ in S by $(f + g)(x) = f(x) + g(x)$. (167)

sum of two vectors, $\vec{u} + \vec{v}$ If $\vec{u} = (u_1, u_2)$ and $\vec{v} = (v_1, v_2)$ then $\vec{u} + \vec{v}$ is the vector $(u_1 + v_1, u_2 + v_2)$. (707)

Sum Property of Integrals If f and g are continuous functions on the interval from a to b, then $\int_a^b (f(x) + g(x))\,dx = \int_b^a f(x)\,dx + \int_b^a g(x)\,dx$ (784)

summation notation, $\sum_{i=m}^{n}$ Suppose m and n are integers with $m < n$. Then $\sum_{i=m}^{n} a_i = a_m + a_{m+1} + \ldots + a_n$. (413)

symmetric matrix A matrix whose element in row i, column j equals its element in row j, column i $\forall$ i, j. (670)

synthetic division The use of synthetic substitution for finding the quotient and remainder when a polynomial $p(x)$ is divided by $x - c$. (251)

synthetic substitution An algorithm for evaluating a polynomial that combines aspects of long division and the Remainder Theorem. (249)

tangent (tan) of a real number x $\frac{\sin x}{\cos x}$ $\forall x$ such that $\cos x \neq 0$. (321)

tangent line to the graph of a function at the point $(x, f(x))$ A line that intersects the graph of a function at $(x, f(x))$ and whose slope equals $\lim_{\Delta x \to 0} \frac{f(x + \Delta x) - f(x)}{\Delta x}$. (547)

term of a sequence A value of a sequence. (100)

tessellation A covering of the plane with congruent copies of the same region (the **fundamental region**), with no holes and no overlaps. (332)

Test Point Method for Solving Inequalities A method for solving inequalities in which the real line is split into intervals by the zeros of an appropriate function, a value is chosen in each of these intervals, and the interval is part of the solution to the inequality if and only if the value satisfies the inequality. (192)

three-dimensional space, 3-space A space in which three numbers are needed to determine the position of a point. (726)

total degree of a graph The sum of the degrees of all the vertices of the graph. (657)

Total Degree of a Graph Theorem The total degree of any graph equals twice the number of edges in the graph. (657)

Tower of Hanoi problem According to legend, at the time of Creation, 64 golden disks were placed on one of the three golden needles in a temple in Hanoi. No two of the disks were the same size, and they were placed on the needle in such a way that no larger disk was on top of a smaller disk. The Creator ordained that the monks of the temple were to move all 64 disks one by one to one of the other needles, never placing a larger disk on top of a smaller disk. When all the disks are stacked on the other needle, the world will end, the faithful will be rewarded, and the unfaithful will be punished. If the monks work very rapidly, moving one disk every second, how long will it be until the end of the world? (400)

Tower of Hanoi sequence The sequence defined by

$$\begin{cases} T_1 = 1 \\ T_{k+1} = 2T_k + 1 \text{ for each integer } k \geq 1. \end{cases}$$ (401).

trace key A key or function on an automatic grapher that activates a cursor that indicates a point on the graph along with its approximate coordinates, and can move, pixel by pixel, along the screen. (87)

transition probability A probability that one event will be followed by another. (676)

Transitive Property of Factors For all integers a, b, and c, if a is a factor of b, and b is a factor of c, then a is a factor of c. (224)

Translation Theorem See *Graph Transformation Theorems*.

translation-symmetric figure A set of points that coincides with its own image under a translation. (131)

trial A probabilistic situation that is repeated in an experiment. (616)

trigonometric form of a complex number The form $r(\cos \theta + i \sin \theta)$ of the complex number $[r, \theta]$. (482)

trigonometric functions The sine, cosine, tangent, cotangent, secant, and cosecant functions and their offspring. (132)

truth table A table that gives the truth values for a logical expression for all possible truth values of the statements in that expression. (12)

two-dimensional vector A vector that can be characterized by two numbers. (700)

unit circle The circle with center (0, 0) and radius 1. (130)

unit fraction A fraction of the form $\frac{1}{n}$, where n is a positive integer. (332)

unit vector A vector whose length is 1. (704)

universal statement A statement asserting that a certain property holds for all elements in some set. A statement of the form *For all x in S, p(x)*. (5)

upper Riemann sum The Riemann sum of a function f where each $f(z_i)$ is the largest value of the function on the subinterval. (775)

valid argument An argument with the property that no matter what statements are substituted in the premises, the truth value of the form is true. If the premises are true, then the conclusion is true. (42)

valid conclusion The conclusion of a valid argument. (43)

Value of a Finite Geometric Series If a is any real number and r is any real number other than 1, then for all integers $n \geq 1$,

$$a + ar + ar^2 + \ldots + ar^{n-1} = a\left(\frac{1 - r^n}{1 - r}\right).$$ (432)

Value of an Infinite Geometric Series If a is any real number and r is a real number with $0 < |r| < 1$, then

$$\sum_{k=0}^{\infty} ar^k = \frac{a}{1 - r}.$$ (434)

vector A quantity that can be characterized by its direction and its magnitude. (700)

vector equation for a line through P parallel to $\vec{v}$ The set of all points Q satisfying the equation $\overrightarrow{PQ} = t\vec{v}$ for some real number t. (714)

vertex A point on a graph. Also called *node*. (640)

vertical asymptote A vertical line $x = a$ to which the graph of a function approaches as x approaches a either from the right or from the left. (303)

vertical component of a vector See *y-component of a vector*.

voltage The electrical potential between two points in an electrical circuit; measured in volts. (470)

walk An alternating sequence of adjacent vertices and edges from one vertex of a graph to another. (662)

window of an automatic grapher The part of the coordinate grid that is displayed on the screen of an automatic grapher. (81)

x-component of a vector The first component of the rectangular coordinates of a vector in standard position. Also called the *horizontal component*. (702)

xy-plane The set of points in 3-space for which the z-coordinate is 0; it has the equation $z = 0$. (727)

xz-plane The set of points in 3-space for which the y-coordinate is 0; it has the equation $y = 0$. (727)

y-component of a vector The second component of the rectangular coordinates of a vector in standard position. Also called the *vertical component*. (702)

yz-plane The set of points in 3-space for which the x-coordinate is 0; it has the equation $x = 0$. (727)

Zeno's Paradox A paradox dealing with the impossibility of adding up an infinite number of quantities to achieve a finite sum. (413)

zero of a function For a function f, a number c such that $f(c) = 0$. (173)

zero vector The vector with same initial and endpoint. (703)

zoom A feature that enables a window of an automatic grapher to picture a smaller part of the coordinate grid. (83)

SYMBOLS

Arithmetic and Algebra

$\approx$	is approximately equal to
$\pm$	positive or negative
e	the irrational number 2.71828...
π	pi
∞	infinity
$\aleph_0$	the cardinality of a countably infinite set
c	the cardinality of an uncountable set
$!$	factorial
$\lvert x \rvert$	absolute value of x
$\sqrt[n]{x}$	nth root of x
$a + bi$	complex number
(a, b)	rectangular form of a complex number
$[r, \theta]$	polar form of a complex number
$r(\cos\theta + i\sin\theta)$	trigonometric form of a complex number
$\overline{z}$	complex conjugate of a complex number
$\lvert z \rvert$	modulus of a complex number
i	imaginary unit, $\sqrt{-1}$

Logic

$\Rightarrow$	if-then (conditional)
$\Leftrightarrow$	if and only if (biconditional)
$\forall$	for all
$\exists$	there exists
$\sim$	negation
$\equiv$	logically equivalent
$\therefore$	therefore

Coordinates and Vectors

(x, y)	ordered pair
(x, y, z)	ordered triple
$[r, \theta]$	polar coordinate
$\overrightarrow{AB}$ or $\vec{v}$ or v	vector
(v_1, v_2)	component representation of vector
$[r, \theta]$	polar representation of vector
$k\vec{v}$	scalar k times vector
$\lvert\vec{v}\rvert$	length of vector
$\vec{0}$	zero vector
$\vec{u} \cdot \vec{v}$	dot product of vectors
$\vec{u} \times \vec{v}$	cross product of vectors

Trigonometry

$\sin x$	sine of angle x
$\cos x$	cosine of angle x
$\tan x$	tangent of angle x
$\csc x$	cosecant of angle x
$\sec x$	secant of angle x
$\cot x$	cotangent of angle x
$\sin^{-1} x$	inverse sine of x
$\cos^{-1} x$	inverse cosine of x
$\tan^{-1} x$	inverse tangent of x

Geometry

$\overleftrightarrow{AB}$	line through A and B
$\overline{AB}$	segment with endpoints A and B
AB	distance from A to B
$\angle ABC$	angle ABC
$m\angle ABC$	measure of angle ABC
$\triangle ABC$	triangle with vertices A, B, and C
$ABCD$	polygon with vertices A, B, C, and D
$/\!/$	is parallel to
$\cong$	is congruent to
$T_{h,k}$	translation
$S_{a,b}$	scale change

Functions and Sequences

$\lim_{n\to\infty} a_n$	limit of sequence a
a_n	nth term of sequence a
Σ	summation notation
$\log x$	common logarithm of x
$\log_b x$	base b logarithm of x
$\ln x$	natural logarithm
$\lfloor x \rfloor$	floor function of x
$\lceil x \rceil$	ceiling function of x
$\text{Fract}(x)$	fractional part of x
f^{-1}	inverse function of f
$f \circ g$	composite of functions f and g
$x \to \infty$	x approaches infinity
$x \to a^-$	x approaches a from the left
$x \to a^+$	x approaches a from the right
$\lim_{x\to a} f(x)$	limit of function f as x approaches a
f'	first derivative of f
f''	second derivative of f
Δ	delta x, change in x
$\int_a^b f(x)\,dx$	definite integral of f from a to b

Combinatorics and Graphs

$P(n, r)$	permutations of n elements taken r at a time
${}_nP_r$	permutations of n elements taken r at a time
$C(n, r)$	combinations of n elements taken r at a time
${}_nC_r$	combinations of n elements taken r at a time
$\binom{n}{r}$	binomial coefficient
$\deg(v)$	degree of vertex v
e_i	the ith edge of a graph
v_i	the ith vertex of a graph
K_n	complete graph with n vertices
mod	modulo
$\equiv$	modular congruence
Rn	modulo class n

LESSON 1-3 (pages 19–24)

9.

p	*q*	*p or q*	*not (p or q)*	*not p*	*not q*	*(not p) and (not q)*
T	T	T	F	F	F	F
T	F	T	F	F	T	F
F	T	T	F	T	F	F
F	F	F	T	T	T	T

(*not (p or q)* and *(not p) and (not q)*: same truth values)

17.b.

p xor q	*p or q*	*p and q*	*not(p and q)*	*(p or q) and (not(p and q))*
F	T	T	F	F
T	T	F	T	T
T	T	F	T	T
F	F	F	T	F

(*p xor q* and *(p or q) and (not(p and q))*: same truth values)

25.a. Sample: The waiter gives you a choice of coffee, tea, or milk. He then comes back to tell you that he has run out of all three. Therefore, you can't have coffee and you can't have tea and you can't have milk.

LESSON 1-4 (pages 25–31)

2.a.

p	*q*	*p OR q*	*NOT (p OR q)*
1	1	1	0
1	0	1	0
0	1	1	0
0	0	0	1

b.

p	*q*	*NOT p*	*NOT q*	*(NOT p) AND (NOT q)*
1	1	0	0	0
1	0	0	1	0
0	1	1	0	0
0	0	1	1	1

c. Given the same inputs, both networks have the same output. Hence, they are functionally equivalent.

6. See answer at the right.

6. The network in this question corresponds to the logical expression *q or ((not p) and (not r))*. The truth table for the logical expressions corresponding to the networks in Question 5 and 6 is shown below.

p	*q*	*r*	*not p*	*(not p) or q*	*not q*	*(not q) and r*	*not((not q) and r)*	output for Question 5	*not r*	*(not p) and (not r)*	output for Question 6
1	1	1	0	1	0	0	1	1	0	0	1
1	1	0	0	1	0	0	1	1	1	0	1
1	0	1	0	0	1	1	0	0	0	0	0
1	0	0	0	0	1	0	1	0	1	0	0
0	1	1	1	1	0	0	1	1	0	0	1
0	1	0	1	1	0	0	1	1	1	1	1
0	0	1	1	1	1	1	0	0	0	0	0
0	0	0	1	1	1	0	1	1	1	1	1

(output for Question 5 and output for Question 6: same truth values)

Hence, for the same input, the networks of Questions 5 and 6 have the same output. This shows they are functionally equivalent.

9.

p	*q*	*~q*	*(p and ~q)*
T	T	F	F
T	F	T	T
F	T	F	F
F	F	T	F

15. Sample: While working at Bell Laboratories (1941–1957), he developed a mathematical theory of communication known as "information theory."

16. Sample: For circuits in series, consider a string of lights. If one light fails, none of the lights will work. Each light must work for the string of lights to work. This is analogous to the AND gate. For circuits in parallel, consider the lights in a house. A light in one room may work regardless of whether any other lights in the house work or not. The house is completely dark only when all the lights are off. This is analogous to the OR gate.

LESSON 1-5 (pages 32–40)

18.

p	*q*	*q ⇒ p*	*~p*	*~q*	*~p ⇒ ~q*
T	T	T	F	F	T
T	F	T	F	T	T
F	T	F	T	F	F
F	F	T	T	T	T

(*q ⇒ p* and *~p ⇒ ~q*: equivalent)

LESSON 1-6 (pages 41–48)

8.

p	q	r	$p \Rightarrow q$	$q \Rightarrow r$	$(p \Rightarrow q)$ *and* $(q \Rightarrow r)$	$p \Rightarrow r$	$((p \Rightarrow q)$ *and* $(q \Rightarrow r)) \Rightarrow (p \Rightarrow r)$
T	T	T	T	T	T	T	T
T	T	F	T	F	F	F	T
T	F	T	F	T	F	T	T
T	F	F	F	T	F	F	T
F	T	T	T	T	T	T	T
F	T	F	T	F	F	T	T
F	F	T	T	T	T	T	T
F	F	F	T	T	T	T	T

9.b. Prove using a truth table. Must show $((p \Rightarrow q)$ *and* $\sim q) \Rightarrow \sim p$ is always true.

p	q	r	$p \Rightarrow q$	$\sim q$	$((p \Rightarrow q)$ *and* $\sim q)$	$\sim p$	$((p \Rightarrow q)$ *and* $\sim q) \Rightarrow \sim p$
T	T	T	T	F	F	F	T
T	T	F	T	F	F	F	T
T	F	T	F	T	F	F	T
T	F	F	F	T	F	F	T
F	T	T	T	F	F	T	T
F	T	F	T	F	F	T	T
F	F	T	T	T	T	T	T
F	F	F	T	T	T	T	T

LESSON 1-7 (pages 49–55)

22. False; *∃ an astronaut t, such that t is not a member of the military.* (For example, some astronauts who are mission specialists are civilians.)

LESSON 1-8 (pages 56–62)

17.a.

p	q	$p \Rightarrow q$	$\sim p$	$(p \Rightarrow q)$ *and* $\sim p$	$\sim q$	$((p \Rightarrow q)$ *and* $\sim p) \Rightarrow \sim q$
T	T	T	F	F	F	T
T	F	F	F	F	T	T
F	T	T	T	T	F	F
F	F	T	T	T	T	T

19. Suppose that m is any even integer and n is any odd integer. According to the definitions of even and odd, there exists integers r and s, such that $m = 2r$ and $n = 2s + 1$. Then $m \cdot n = 2r(2s + 1) = 4rs + 2r = 2(2rs + r)$. By closure properties, $(2rs + r)$ is an integer, and it follows by definition that $m \cdot n$ is even.

20.

$p \Rightarrow q$	(4)
$q \Rightarrow r$	(2)
$r \Rightarrow s$	contrapositive of (5)
$s \Rightarrow t$	contrapositive of (3)
$\therefore p \Rightarrow t$	Law of Transitivity
p	(1)
$\therefore t$	Law of Detachment

21. Sample: Let p: $2 < 1$. Let q: $3 < 2$. Both statements are false. (p and q) is false, but $(p \Rightarrow q)$ is true.

22. *∀ integers a and b, if* $\frac{a}{b} = \sqrt{2}$, *then* $\frac{a^2}{b^2} = 2$.

26.a.i. valid

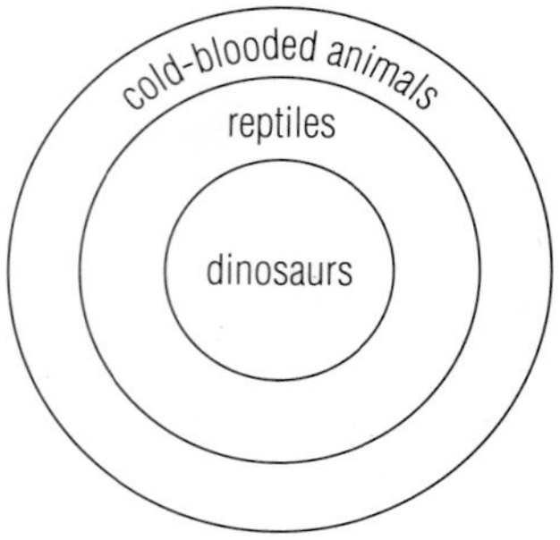

ii. invalid

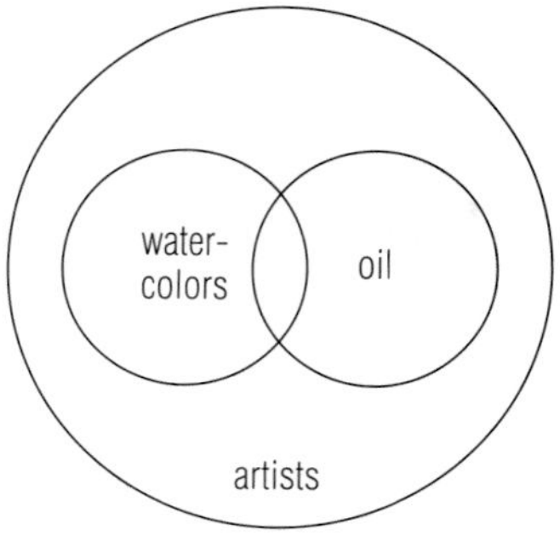

iii. valid

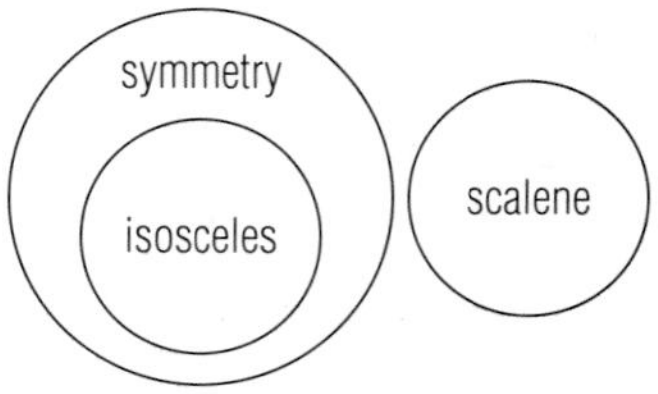

CHAPTER 1 PROGRESS SELF-TEST
(pages 72–73)

21.

p	q	p OR q	NOT q	(p OR q) AND (NOT q)
1	1	1	0	0
1	0	1	1	1
0	1	1	0	0
0	0	0	1	0

CHAPTER 1 REVIEW (pages 74–78)

63.

p	*q*	*r*	*G* (output)	*NOT G*	*(NOT G) AND r*
1	1	1	0	1	1
1	0	1	0	1	1
0	1	1	1	0	0
0	0	1	1	0	0
1	1	0	0	1	0
1	0	0	0	1	0
0	1	0	1	0	0
0	0	0	1	0	0

65.c.

p	*q*	*NOT p*	*NOT q*	*p AND (NOT q)*	*NOT(p AND (NOT q))*	*NOT p OR q*
1	0	0	1	1	0	0
1	1	0	0	0	1	1
0	0	1	1	0	1	1
0	1	1	0	0	1	1

same truth values

69.

p	*q*	*r*	*q or r*	*p and (q or r)*
T	T	T	T	T
T	T	F	T	T
T	F	T	T	T
T	F	F	F	F
F	T	T	T	F
F	T	F	T	F
F	F	T	T	F
F	F	F	F	F

70.

p	*q*	*not p*	*not q*	*(not p) or (not q)*	*p and q*	*not (p and q)*
T	F	F	T	T	F	T
T	T	F	F	F	T	F
F	F	T	T	T	F	T
F	T	T	F	T	F	T

same truth values

LESSON 2-1 (pages 79–86)

14.a.

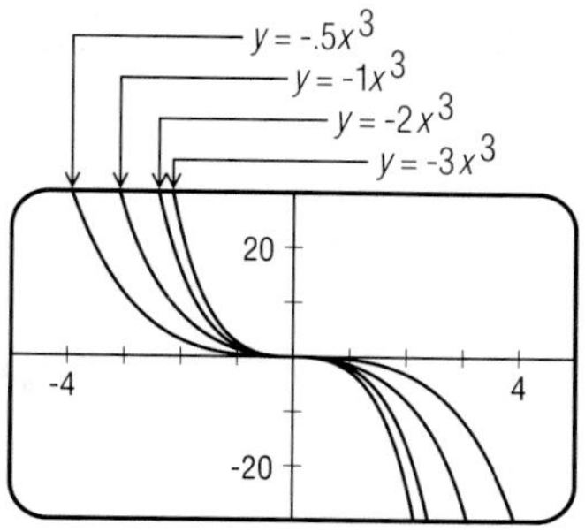

15.a., b.

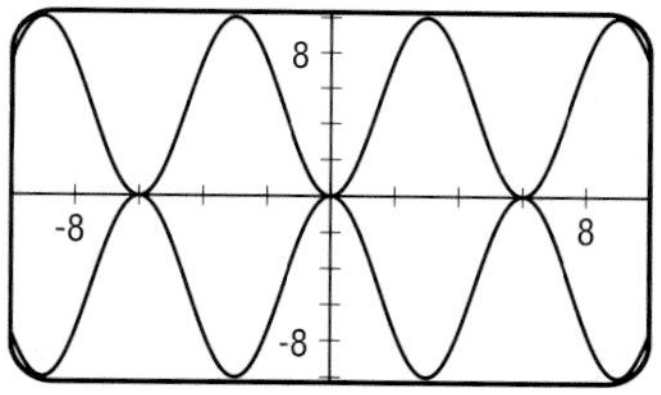

c. The resulting graph shows over three full periods of each function.

16.a., b.

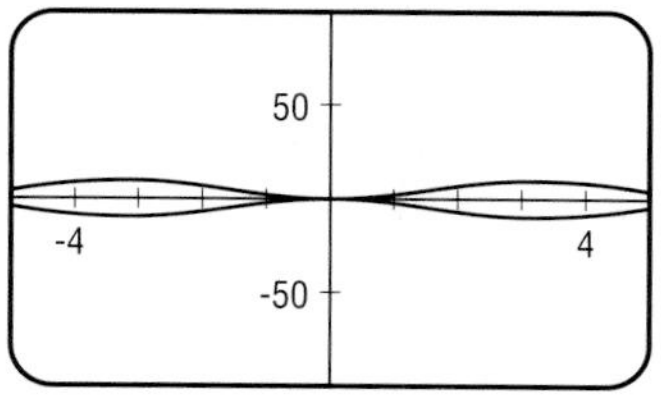

c. The resulting graph looks flatter and stretched horizontally; less than two periods are displayed.

17.a, b.

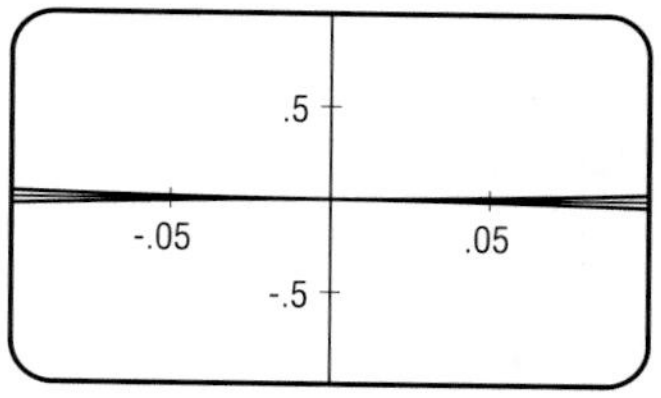

c. The resulting graph looks like a segment.

LESSON 2-2 (pages 87–92)

9. minimum value: -9.5

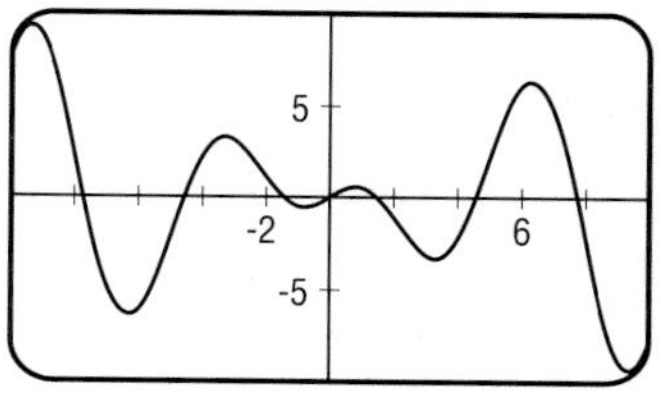

12.a. domain:
$\{-1, -\frac{1}{2}, 0, \frac{1}{2}, 1, \frac{3}{2}, 2\}$;
range: $\{-\frac{1}{2}, 0, \frac{1}{2}, 1, \frac{3}{2}, 2\}$

b. Yes, the domain is a discrete set.

c. minimum: $-\frac{1}{2}$; maximum: 2

13.a. domain: $\{x: -5 \le x \le 6\}$;
range: $\{y: -3 \le y \le 3\}$

b. No, the domain cannot be put into a 1–1 correspondence with a subset of the set of integers.

c. minimum: -3; maximum: 3

d. $x = 0$ and $x = 3$

e. $-5 \le x < -4$ and $-1 < x < 4$

CHAPTER 2 REVIEW (pages 146–150)

54.b. usual domain: set of real numbers;
range: set of nonpositive real numbers;
maximum: 0;
increasing: $x \le 0$; decreasing: $x \ge 0$;
end behavior: $\lim_{x \to \infty} f(x) = -\infty$, $\lim_{x \to -\infty} f(x) = -\infty$
graph: See next page.

54.b. graph:

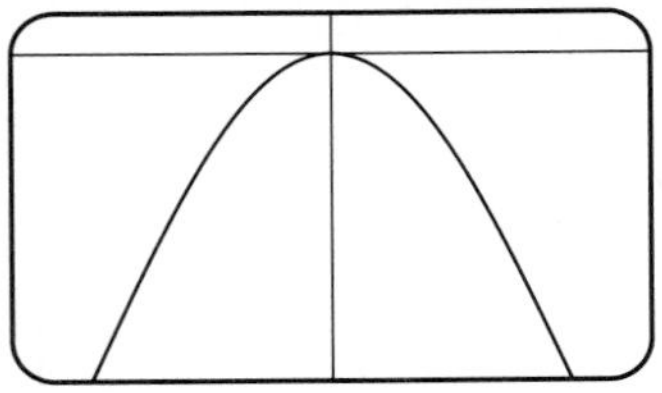

model: optics, projectile motion;
special properties: even function

55. usual domain: set of real numbers;
range: $1 \le y \le 5$;
increasing: when $2n\pi \le x \le (2n + 1)\pi$ $\forall$ integers n; decreasing: when $(2n - 1)\pi \le x \le 2n\pi$ $\forall$ integers n;
minimum: 1; maximum: 5;
end behavior: The limit does not exist.
graph:

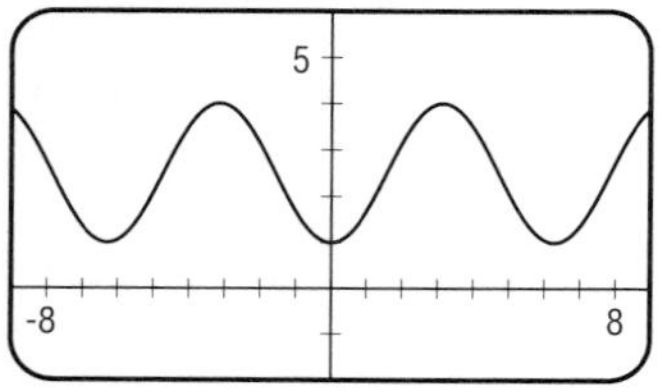

model: sound waves, phenomena based on rotations;
special properties: periodic, even function

LESSON 3-1 (pages 151–159)

1. $\forall$ real numbers x,

$2x + 5 = 3x - 2$	Given
$\Leftrightarrow = x - 2$	Addition Prop. of Equality
$\Leftrightarrow 7 = x$	Addition Prop. of Equality

$\therefore$ $\forall$ real numbers x, $2x + 5 = 3x - 2 \Leftrightarrow x = 7$ by the transitive property.

7.b. $\sqrt{2x^2 + 3x + 1} = x - 1$
$\Rightarrow 2x^2 + 3x + 1 = (x - 1)^2$
$\Leftrightarrow 2x^2 + 3x + 1 = x^2 - 2x + 1$
$\Leftrightarrow x^2 + 5x = 0$
$\Leftrightarrow x(x + 5) = 0$
$\Leftrightarrow x = 0$ or $x = -5$

18. It is sufficient to prove the contrapositive: if $u \neq v$, then $h(u) \neq h(v)$. If $u \neq v$, then either (1) $u < v$, or (2) $u > v$. (1) If $u < v$, then since h is a decreasing function, $h(u) > h(v)$. (2) If $u > v$, then since h is a decreasing function, $h(u) < h(v)$. Thus, in all cases when $u \neq v$, $h(u) \neq h(v)$. Since the contrapositive is true, if h is a decreasing function throughout its domain, then $h(u) = h(v) \Rightarrow u = v$.

22. Suppose m and n are any even integers. Thus, there exist integers r and s such that $m = 2r$ and $n = 2s$ by definition of even. Then $m \cdot n = 2r \cdot 2s = 4rs$. Since rs is an integer by closure properties, $m \cdot n = 4k$, for some integer k.

23. The theorem was not proven for the general case but only for two specific examples. This is the error of improper induction.

LESSON 3-2 (pages 160–165)

24.b.

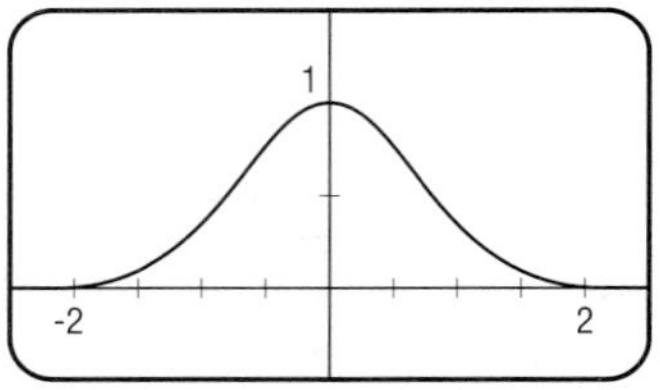

c. f_1 insures $\lim_{x \to \infty} f(x) = 0$. f_2 makes the graph quickly approach this limit. f_3 makes the function symmetric about the y-axis.

LESSON 3-3 (pages 166–171)

8.a.

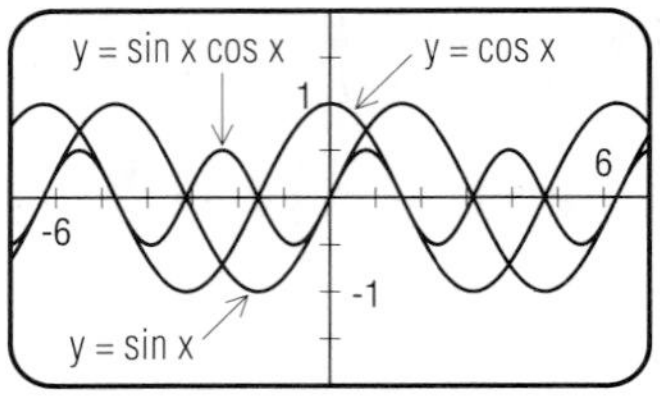

11. Sample: The function $h(x) = x^4$ is an even function that can be expressed as the product of two odd functions, $f(x) = x^3$ and $g(x) = x$.

12. Counterexample: Let $f(x) = 2x + 1$ and $g(x) = 3x - 4$, both of which are increasing. Then $f \cdot g = (2x + 1)(3x - 4) = 6x^2 - 5x - 4$, which is only increasing on the interval $x > \frac{5}{12}$.

13.c.

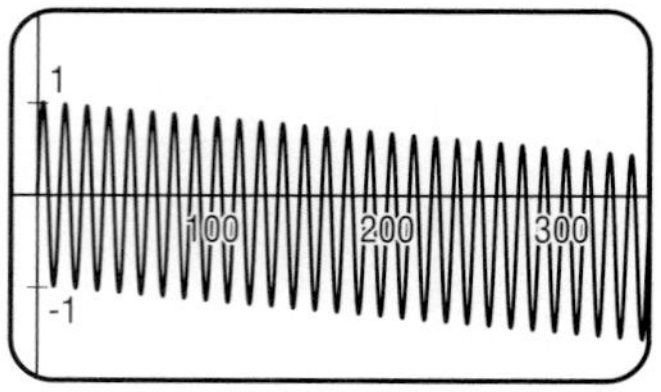

19. sample:

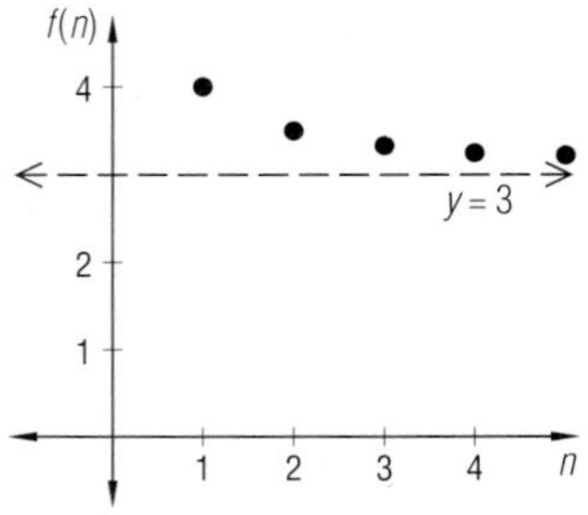

LESSON 3-6 (pages 184–189)

24.a. There is a discontinuity at $x = 0$.

b. Yes

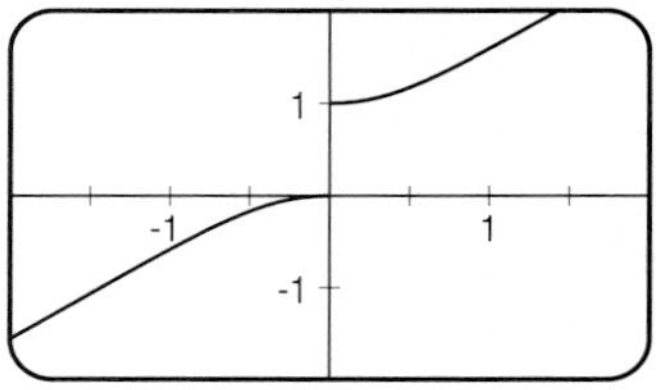

c. The values on the graph jump from 0 to 1, and so nowhere on $-2 \le x \le 2$ does $f(x) = \frac{1}{2}$.

d. Solving,

$\frac{1}{2} = \frac{1}{1 - e^{(-1/x)}}$	Set $f(x) = \frac{1}{2}$.
$\Rightarrow 2 = 1 - e^{(-1/x)}$	Take the reciprocal of both sides.
$\Rightarrow -1 = e^{(-1/x)}$	Add -1 to both sides, then multiply by -1.

$e^{(-1/x)} > 0$ $\forall$ x, so there is no real solution to $f(x) = \frac{1}{2}$.

CHAPTER 3 REVIEW (pages 217–220)

61.a.

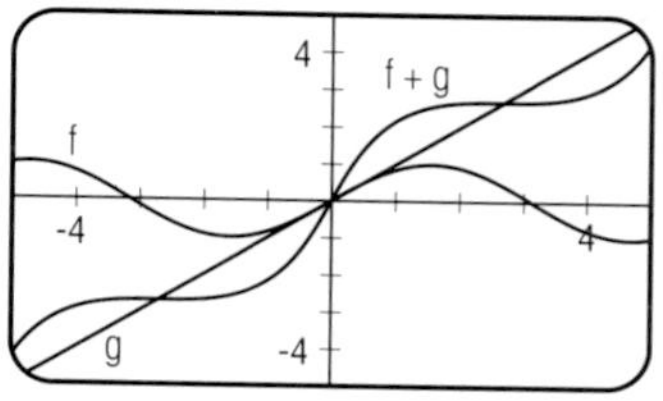

b.

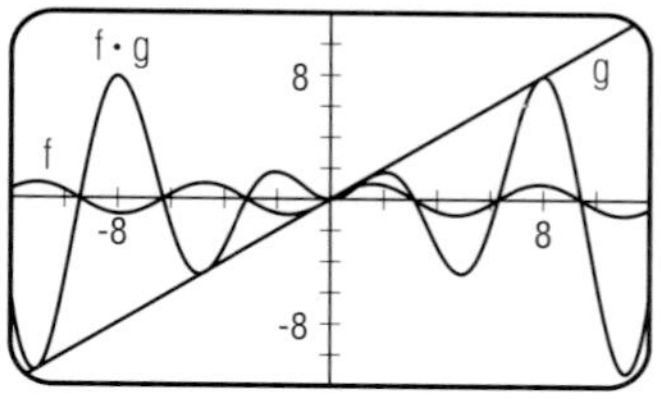

c. Since sin x varies from -1 to 1, $x \cdot \sin x$ will always be less than or equal to x but greater than or equal to $-x$. That is why the graph $f \cdot g$ oscillates between the lines $y = x$ and $y = -x$. Similarly, $f + g$ oscillates between the lines $y = x + 1$ and $y = x - 1$.

62.

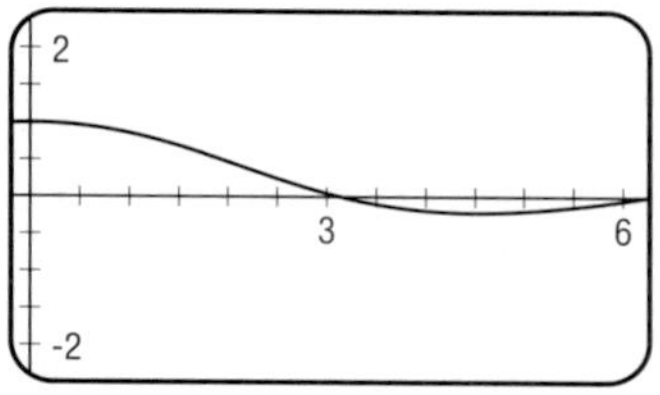

63.

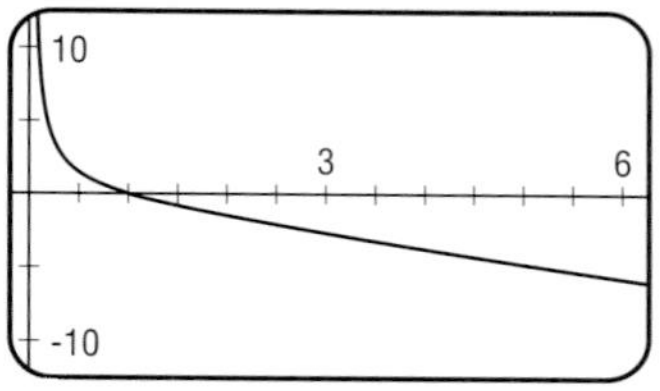

64.

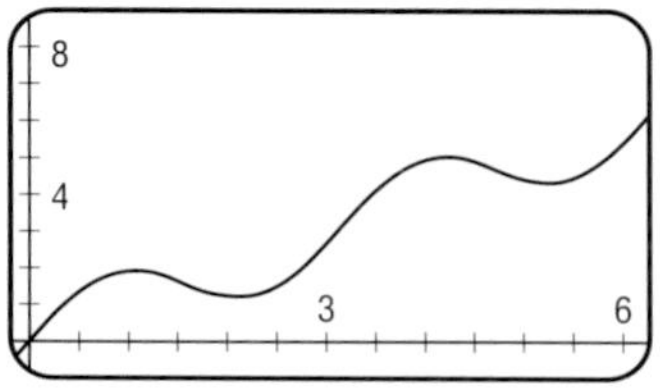

65.b. $\frac{(x - 2)^2}{9} + 9(y + 5)^2 = 1$

68. $xy = 1$; the scale changes on both axes canceled each other out.

69.a. $g(x) = 3 \sin 8x$

b.

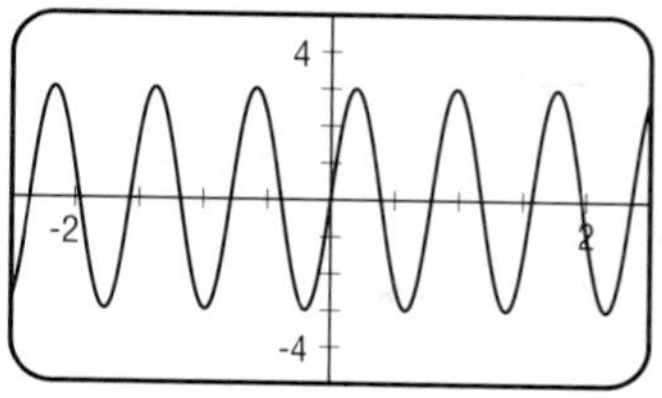

LESSON 4-2 (pages 229–235)

7.

$n = q \cdot 4 + r$

n	-10	-9	-8	-7	-6	-5	-4	-3	-2	-1	0	1	2	3	4	5	6	7	8	9	10	11	12	13	14	15
q	-3	-3	-2	-2	-2	-2	-1	-1	-1	-1	0	0	0	0	1	1	1	1	2	2	2	2	3	3	3	3
r	2	3	0	1	2	3	0	1	2	3	0	1	2	3	0	1	2	3	0	1	2	3	0	1	2	3

11. Let n, $n + 1$, and $n + 2$ be any 3 consecutive integers. Then, by the results of Question 10, ∃ some integer q such that

$n = 3q$, $\quad n = 3q + 1$, or $\quad n = 3q + 2$.

So, $n + 1 = 3q + 1$, $n + 1 = 3q + 2$, or $n + 1 = 3q + 3$;
and $n + 2 = 3q + 2$, $n + 2 = 3q + 3$, or $n + 2 = 3q + 4$.

Note that in each of the three columns listed above, exactly one of the integers is divisible by 3.

LESSON 4-3 (pages 236–241)

29. The calculator's error is due to the cumulative effects of rounding or truncating at intermediate steps where the result was larger than the calculator's storage capacity.

LESSON 4-4 (pages 242–247)

21. sample: First long division:

$$\frac{6x^3 - 9x^2 + 8x + 1}{2x + 1} = 3x^2 - 6x + 7 + \frac{-6}{2x + 1}.$$

Let $x = 5$.

$p(5) = 6(5)^3 - 9(5)^2 + 8(5) + 1 = 566$

$d(5) = 2 \cdot 5 + 1 = 11$

566 divided by 11 yields a quotient of 51 and a remainder of 5. But $q(5) = 52$ and $r(5) = -6$.

Second long division: $\frac{6x^5 - x^4 + x + 1}{2x^2 + x} =$

$$3x^3 - 2x^2 + x - \frac{1}{2} + \frac{\frac{3}{2}x + 1}{2x^2 + x}.$$

Let $x = 3$.

$p(3) = 6(3)^5 - (3)^4 + 3 + 1 = 1381$

$d(3) = 2(3)^2 + 3 = 21$

1381 divided by 21 yields a quotient of 65 and a remainder of 16. But $q(3) = 3(3)^3 - 2(3)^2 + 3 - \frac{1}{2} = 65.5$ and $r(3) = \frac{3}{2}(3) + 1 = 5.5$. The numerical quotients and remainders do not necessarily agree with the values one would get using polynomial substitution. However, the Quotient-Remainder Theorem does hold. In the first long division, $q(5) \cdot d(5) + r(5) = 52 \cdot 11 + -6 = 566 = p(5)$. In the second, $q(3) \cdot d(3) + r(s) = 65.5 \cdot 21 + 5.5 = 1381 = p(3)$.

LESSON 4-5 (pages 248–253)

15.a. $C(n) = \begin{cases} 0.75n & n \le 1500 \\ 225 + 0.60n & n > 1500 \end{cases}$

c. $P(n) = \begin{cases} 0.50n & n \le 1500 \\ 0.65n - 225 & n > 1500 \end{cases}$

19.

```
10   DIM A(7), B(6)
20   PRINT "THIS PROGRAM DIVIDES
     P(X) BY X – C"
30   PRINT "WHERE P(X) IS OF THE
     FORM"
40   PRINT "A7*X^7 + A6*X^6 +
     A5*X^5 + A4*X^4 + A3*X^3 +
     A2*X^2 + A1*X + A0"
50   FOR I = 7 to 0 STEP -1
60   PRINT "ENTER A"; I," ";
70   INPUT A(I)
80   NEXT I
90   PRINT "ENTER C";
100  INPUT C
110  REM COEFFICIENTS OF
     QUOTIENT WILL BE B6, B5, ..., B1, B0
120  LET B(6) = A(7): REM BRING DOWN
     FIRST COEFFICIENT
```

```
130 FOR I = 5 to 0 STEP -1
140 LET B(I) = B(I + 1)*C + A(I + 1)
150 NEXT I
160 LET R = B(0)*C + A(0)
170 PRINT "THE QUOTIENT IS Q(X) ="
180 FOR I = 6 to 0 STEP -1
190 IF I = 0 THEN PRINT B(O): GOTO 230
200 PRINT B(I); "*X^"; I;
210 PRINT "+";
220 NEXT I
230 PRINT "THE REMAINDER IS R(X)
    ="; R
240 END
```

LESSON 4-8 (pages 268–274)

23. Let p be a factor of a and p be a factor of b. Thus, there exist integers m and n such that $a = m \cdot p$ and $b = n \cdot p$ by definition of factor. Then $a - b = m \cdot p - n \cdot p = (m - n)p$. Since $m - n$ is an integer by closure properties, p is a factor of $a - b$.

24.a. original problem:

$$\begin{array}{r} 123456789 \\ \times \quad\quad\quad 9 \\ \hline 1111111111 \end{array}$$

sum of digits: 45 mod 9: 0

$$\begin{array}{rr} 45 & 0 \\ 9 & \times 0 \\ \hline 10 & 0 \end{array}$$

In order for the multiplication to be correct, 1111111111 must be congruent to 0(mod 9). The sum of the digits is 10. $10 \not\equiv 0 \pmod 9$, so the multiplication is incorrect.

31. The ultimate polyalphabetic scheme was invented by the Germans in World War II. One of their devices, called the *Enigma*, was captured by the Allies early in the war. Brilliant computational work led by the English mathematician Alan Turing in the ULTRA project enabled the Allies to monitor almost all critical German communications, even though the rules for coding would not show repetitions in messages shorter than 456,976 letters, and could be started in 1,305,093,289,500 unique ways. The theory and devices produced by the American and English code-breakers set the stage for the post-war development of digital computers.

CHAPTER 4 REVIEW (pages 283–286)

57.

$$\begin{array}{r} x^2 + 2x - 7 \\ x^4 - 3 \overline{) x^6 + 2x^5 - 7x^4 - 3x^2 - 6x + 21} \\ \underline{x^6 \quad\quad\quad\quad - 3x^2 \quad\quad\quad\quad} \\ 2x^5 - 7x^4 \quad\quad - 6x + 21 \\ \underline{2x^5 \quad\quad\quad\quad\quad - 6x \quad\quad} \\ - 7x^4 \quad\quad\quad\quad + 21 \\ \underline{- 7x^4 \quad\quad\quad\quad + 21} \\ 0 \end{array}$$

Therefore, $x^6 + 2x^5 - 7x^4 - 3x^2 - 6x + 21 = (x^4 - 3)(x^2 + 2x - 7)$, and $x^4 - 3$ is a factor.

76.c. $n = 387500$, $q = 9$, $d = 40000$, $r = 27500$; $387500 = 40000(9) + 27500$

77. 5 months with 73 days each, 73 months with 5 days each, 365 months with 1 day, or 1 month with 365 days

LESSON 5-1 (pages 288–294)

4. Suppose r and s are any two rational numbers. By definition, there exist integers a, b, c, and d, where $b \neq 0$ and $d \neq 0$ such that $r = \frac{a}{b}$ and $s = \frac{c}{d}$. So,

$r - s = \frac{a}{b} - \frac{c}{d}$

$= \frac{a}{b} \cdot \frac{d}{d} - \frac{c}{d} \cdot \frac{b}{b}$ Multiplication Prop. of 1

$= \frac{ad}{bd} - \frac{cb}{bd}$ Multiplication of fractions

$= \frac{ad - cb}{bd}$ Addition of fractions

Since by closure properties the product of two integers and the difference between two integers are integers, $ad - cb$ and bd are integers. Also, $bd \neq 0$ since $b \neq 0$ and $d \neq 0$. Therefore, since $\frac{ad - cb}{bd}$ is a ratio of integers, $r - s$ is a rational number by definition.

5.c. $\frac{400}{5} = 80$; If each of the 5 people does 80 envelopes, then $5 \cdot 80 = 400$ are done, so part **a** is correct. For part **b**, $\frac{800}{35} \approx 22.9$ per person, and $80 - 22.9 = 57.1$. $7 \cdot 57.1 = 399.7$. So, the formula works. (A few people will check 58 to make up the slack represented by the decimals.)

9.a. $\frac{2}{y + 4}$

b. restrictions: $y \neq -4$, $y \neq 6$, and $y \neq -3$

11.a. $\frac{4 - a}{x(x - a)} + \frac{x - 4}{x^2 - ax}$

$= \frac{4 - a}{x(x - a)} + \frac{x - 4}{x(x - a)}$

$= \frac{4 - a + x - 4}{x(x - a)}$

$= \frac{x - a}{x(x - a)}$

$= \frac{1}{x}$

16.b. Yes, a, b, c, d, and f are integers, so by the closure properties of integers, the numerator and the denominator are also integers. Thus, the expression is a rational number by definition.

18. Let $N = 0.\overline{148}$, then $1000N = 1000(.\overline{148148})$. Subtract N from $1000N$ and divide by the coefficient of N:

$$\begin{array}{rcl} 1000N & = & 148.\overline{148} \\ - \quad N & = & 0.\overline{148} \\ \hline 999N & = & 148 \\ N & = & \frac{148}{999} = 0.\overline{148} \end{array}$$

19. Let $N = \overline{.012345679}$. $10^9N = 10^9(\overline{.012345679012345679})$. Subtract N from 10^9N and divide by the coefficient of N:

$$\begin{array}{rcl} 10^9N & = & 12{,}345{,}679.\overline{012345679} \\ - \quad N & = & \overline{.012345679} \\ \hline 999{,}999{,}999N & = & 12345679 \\ N & = & \frac{12{,}345{,}679}{999{,}999{,}999} \end{array}$$

27.

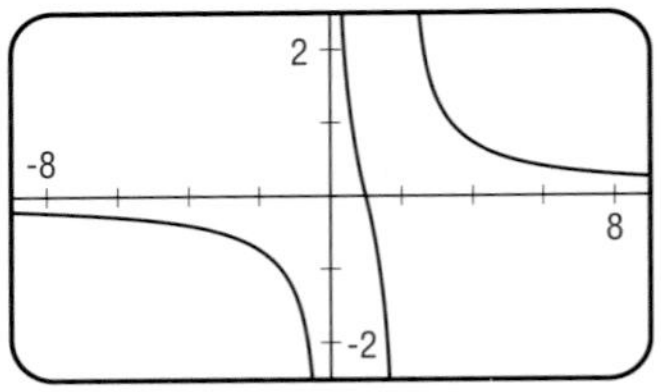

f and g have congruent graphs. They have vertical asymptotes $x = 0$ and $x = 2$.

LESSON 5-2 (pages 295–301)

17. reciprocal of $\sqrt{7} + \sqrt{6} = \dfrac{1}{\sqrt{7} + \sqrt{6}} =$

$\dfrac{1}{\sqrt{7} + \sqrt{6}} \cdot \dfrac{\sqrt{7} - \sqrt{6}}{\sqrt{7} - \sqrt{6}} = \dfrac{\sqrt{7} - \sqrt{6}}{7 - 6} =$

$\dfrac{\sqrt{7} - \sqrt{6}}{1} = \sqrt{7} - \sqrt{6}$

18.a. The Quotient-Remainder Theorem states that the remainder r for this division is an integer in the range $0 \le r < 54$. Therefore, after 54 steps at least one remainder must repeat since there are only 54 unique remainders.
b. The number of steps taken before repeat of the remainder is equal to the number of digits in the repeating number sequence in the decimal expansion. Each time the remainder repeats, the repeated digit sequence is begun again.
c. The Quotient-Remainder Theorem requires the remainder to be an integer r in the range $0 \le r < d$ for any long division. Hence, after d steps there must be a zero or a repeated remainder. If there is a zero remainder, then the decimal expansion terminates. If there is a repeated remainder, then the decimal expansion is repeating.

19. Suppose r and s are rational numbers. By definition, there exist integers a, b, c, and d, with $b \neq 0$ and $d \neq 0$ such that $r = \frac{a}{b}$ and $s = \frac{c}{d}$. Then $r \cdot s = \frac{a}{b} \cdot \frac{c}{d} = \frac{ac}{bd}$. Since $b \neq 0$ and $d \neq 0$, $bd \neq 0$. Also, ac and bd are integers by closure properties. Thus, since $r \cdot s$ has an integer numerator and denominator, $r \cdot s$ is a rational number by definition.

LESSON 5-3 (pages 302–306)
3.

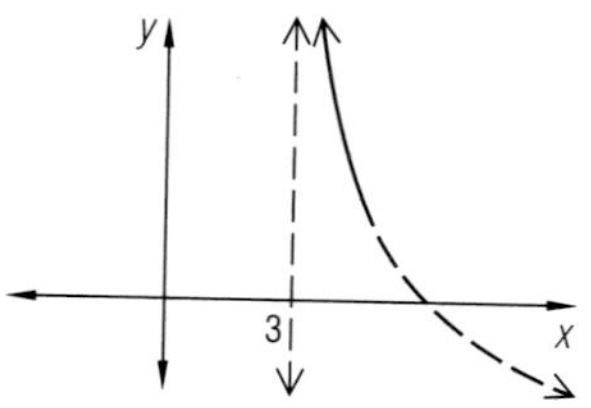

4.a. The limit of $f(x)$ as x approaches 0 from the left is negative infinity.
sample:

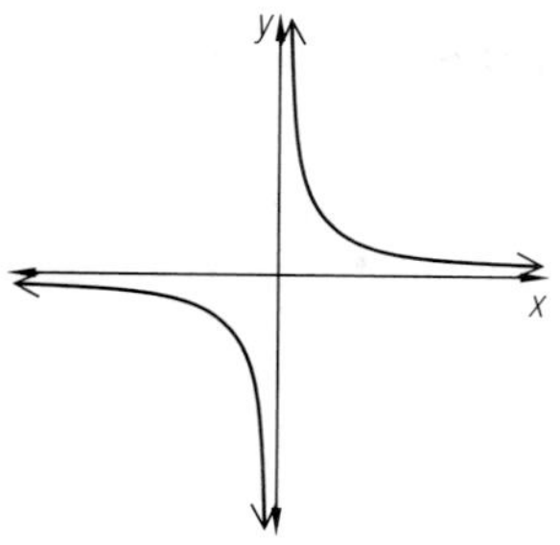

b. The limit of $f(x)$ as x approaches 0 from the left is positive infinity.
sample:

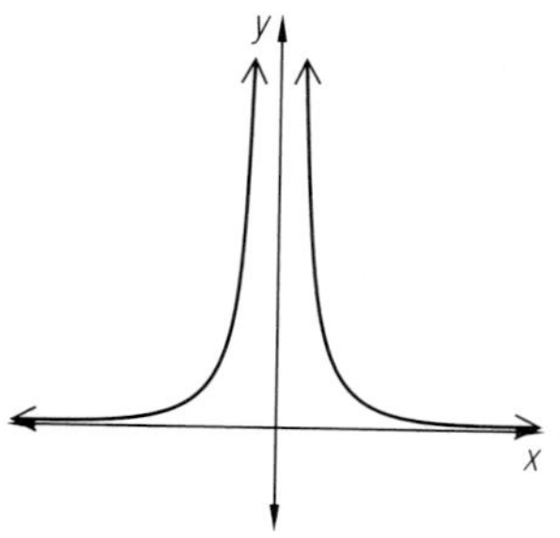

7.c.

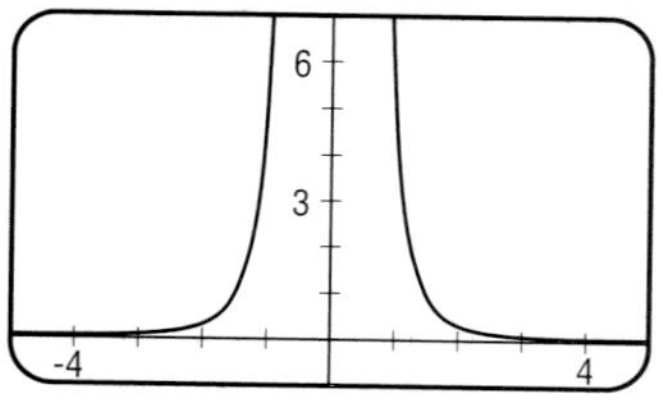

8.d.
$\lim\limits_{x \to 0^+} f(x) = +\infty$ and $\lim\limits_{x \to 0^-} f(x) = -\infty$
f.

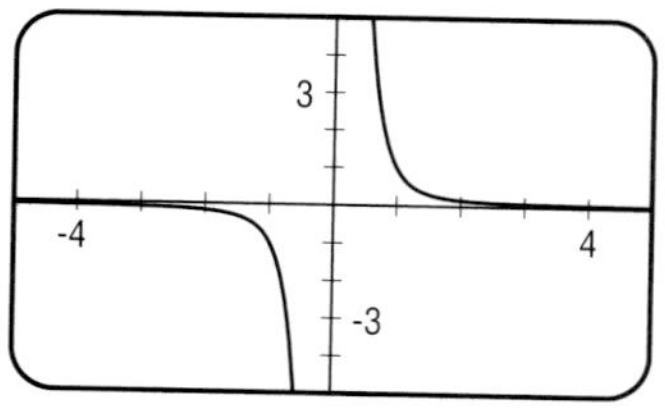

10.a.

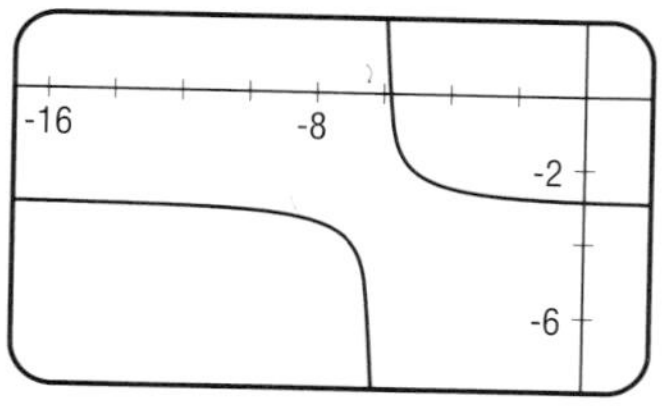

11. The statement is false. Counterexample: π and $\pi + 3$ are both irrational, yet $\pi - (\pi + 3) = -3$ is rational.

13.a. *The decimal expansion of x is nonterminating and the decimal expansion of x is nonrepeating.*
b. *If the decimal expansion of x is nonterminating and nonrepeating, then x is not a rational number.*
c.i. rational; **ii.** irrational

17. Suppose a, b, and c are any integers such that a is a factor of b and a is a factor of $b + c$. By definition, there exist integers r and s such that $b = a \cdot r$ and $b + c = a \cdot s$. Then, $c = (b + c) - b = a \cdot s - a \cdot r = a(s - r)$. Since $s - r$ is an integer by closure properties, a is a factor of c by definition.

20.b. sample:

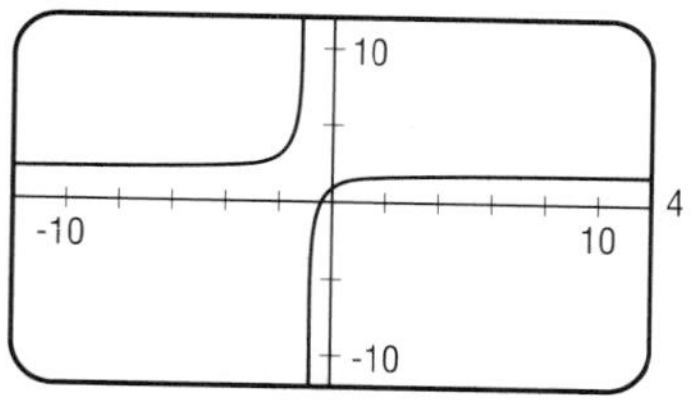

The above graph does not have the same problem, but graphs may vary.

LESSON 5-4 (pages 307–313)
10.a. $\{x: x \neq 1 \text{ and } x \neq -1\}$
b. essential
c. $x = 1$ and $x = -1$
d. See graph on next page.

34. Left side $= \dfrac{1}{1 + \cos\alpha} + \dfrac{1}{1 - \cos\alpha}$

$= \dfrac{2}{1 - \cos^2\alpha}$ Adding fractions

$= \dfrac{2}{\sin^2\alpha}$ Pythagorean Identity

$= 2\csc^2\alpha$ Def. of csc

= Right side

$\therefore \dfrac{1}{1 + \cos\alpha} + \dfrac{1}{1 - \cos\alpha} = 2\csc^2\alpha$

domain: $\alpha \neq n\pi$, n an integer

35. Left side $= \cos(\alpha - \beta) - \cos(\alpha + \beta)$

$= \cos\alpha\cos\beta + \sin\alpha\sin\beta - \cos\alpha\cos\beta + \sin\alpha\sin\beta$ Sum and difference identities

$= 2\sin\alpha\sin\beta$ Addition

= Right side

$\therefore \cos(\alpha - \beta) - \cos(\alpha + \beta) = 2\sin\alpha\sin\beta$

domain: all real numbers

36.

		$\sec x + \cot x \csc x$	$\sec x \csc^2 x$	
Trigonometric definitions	=	$\dfrac{1}{\cos x} + \dfrac{\cos x}{\sin^2 x}$	$= \dfrac{1}{\cos x \sin^2 x}$	Trigonometric definitions
Addition of fractions	=	$\dfrac{\sin^2 x + \cos^2 x}{\cos x \sin^2 x}$		
Pythagorean Identity	=	$\dfrac{1}{\cos x \sin^2 x}$		

$\therefore \sec x + \cot x \csc x = \sec x \csc^2 x$

domain: $x \neq \dfrac{n\pi}{2}$, n an integer

37. Left side $= \cos 4x$

$= \cos^2 2x - \sin^2 2x$ Identity for $\cos 2x$

$= (\cos^2 x - \sin^2 x)^2 - (2\sin x\cos x)^2$ Identities for $\cos 2x$ and $\sin 2x$

$= \cos^4 x - 2\sin^2 x\cos^2 x + \sin^4 x - 4\sin^2 x\cos^2 x$ Multiplication

$= \cos^4 x - 6\sin^2 x\cos^2 x + \sin^4 x$ Addition

= Right side

$\therefore \cos 4x = \cos^4 x - 6\sin^2 x\cos^2 x + \sin^4 x$

domain: all real numbers

38. Right side $= \dfrac{1 - \cos 2x}{1 + \cos 2x}$

$= \dfrac{1 - (1 - 2\sin^2 x)}{1 + (2\cos^2 x - 1)}$ Identities for $\cos 2x$

$= \dfrac{2\sin^2 x}{2\cos^2 x}$ Simplification

$= \tan^2 x$ Simplification and def. of tan

= Left side

$\therefore \dfrac{1 - \cos 2x}{1 + \cos 2x} = \tan^2 x$

domain: $x \neq \frac{\pi}{2} + n\pi$, n an integer

45. an identity;

		$1 + \cot^2 x$	$\csc^2 x$	
Def. of cot	=	$1 + \dfrac{\cos^2 x}{\sin^2 x}$	$= \dfrac{1}{\sin^2 x}$	Def. of csc
Addition	=	$\dfrac{\sin^2 x + \cos^2 x}{\sin^2 x}$		
Pythagorean Identity	=	$\dfrac{1}{\sin^2 x}$		

$\therefore 1 + \cot^2 x = \csc^2 x$, $x \neq n\pi$, n an integer

47. an identity;

Left side $= \tan(\pi + \gamma)$

$= \dfrac{\sin(\pi + \gamma)}{\cos(\pi + \gamma)}$ — Def. of tan

$= \dfrac{\sin \pi \cos \gamma + \sin \gamma \cos \pi}{\cos \pi \cos \gamma - \sin \pi \sin \gamma}$ — Sum identities for sine and cosine

$= \dfrac{-\sin \gamma}{-\cos \gamma}$ — Evaluating specific values of trigonometric functions

$= \tan \gamma$ — Def. of tan

$=$ Right side

49. You can check various different cases by holding α constant and graphing the resulting functions in β. For example, when $\alpha = \frac{\pi}{3}$, you could graph $y = \sin\left(\frac{\pi}{3} + \beta\right)$ and $y = \sin \frac{\pi}{3} \cos \beta - \cos \frac{\pi}{3} \sin \beta$.

LESSON 7-1 (pages 400–406)

19.

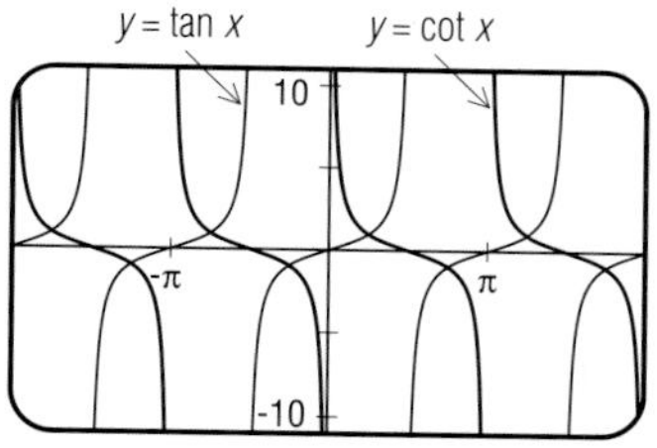

Sample: $\tan x = \dfrac{1}{\cot x}$. One way to obtain the graph of $\cot x$ from the graph of $\tan x$ is to reflect $\tan x$ with respect to the y-axis, then apply the transformation $T_{-\pi/2,\,0}$.

23.a., b. Let 1 be the top (smallest) disk, 2 the middle disk, and 3 the bottom (largest) disk. Let *a* be the original needle, and *b* and *c* the other 2 needles. Move 1 to *b*, 2 to *c*, 1 to *c*, 3 to *b*, 1 to *a*, 2 to *b*, and 1 to *b* (7 steps).

c. Let 4 now be the bottom (largest) disk. Move 1 to *b*, 2 to *c*, 1 to *c*, 3 to *b*, 1 to *a*, 2 to *b*, 1 to *b*, 4 to *c*, 1 to *c*, 2 to *a*, 1 to *a*, 3 to *c*, 1 to *b*, 2 to *c*, and 1 to *c* (15 steps).

LESSON 7-2 (pages 407–412)

6.a. sample:

```
10 FOR N=1 TO 50
20   TERM=SIN(3.1415*N/2)
30   PRINT TERM
40 NEXT N
50 END
```

8.b. 1, 2, 1, 2, 1, 2

$a_n = \begin{cases} 1 & \forall \text{ odd integers } n > 0 \\ 2 & \forall \text{ even integers } n > 1 \end{cases}$

c. 1, $c - 1$, 1, $c - 1$, 1, $c - 1$

$a_n = \begin{cases} 1 & \forall \text{ odd integers } n > 0 \\ c - 1 & \forall \text{ even integers } n > 1 \end{cases}$

9.c. $a_1 = 3^1 - 1 = 2$, so the initial condition is met. $a_{n+1} = 3^{n+1} - 1 = 3 \cdot 3^n - 1 = 3 \cdot 3^n - 3 + 3 - 1 = 3(3^n - 1) + 2 = 3a_n + 2$, so the recurrence relationship is met. Therefore, the explicit formula is correct.

13.a. $\lim\limits_{x \to +\infty} f(x) = 2$; $\lim\limits_{x \to -\infty} f(x) = 2$

14. $\dfrac{(k + 1)^2(k + 2)^2}{4}$

LESSON 7-3 (pages 413–417)

10.a. $p(1)$: $\sum_{i=1}^{1} (2i - 1) = 1^2$. Does $2 \cdot 1 - 1 = 1$? Yes, so $p(1)$ is true.

b. $\sum_{i=1}^{k} (2i - 1) = k^2$

c. $\sum_{i=1}^{k+1} (2i - 1) = (k + 1)^2$

d. $\sum_{i=1}^{k+1} (2i - 1) = \left(\sum_{i=1}^{k} (2i - 1)\right) + [2(k + 1) - 1] = k^2 + [2(k + 1) - 1]$

11. $\sum_{i=1}^{k+1} i(i - 1) = \left(\sum_{i=1}^{k} i(i - 1)\right) + (k + 1)k$

14. $\sum_{k=1}^{5} \left(\frac{1}{k} - \frac{1}{k + 1}\right) = \left(1 - \frac{1}{2}\right) + \left(\frac{1}{2} - \frac{1}{3}\right) + \left(\frac{1}{3} - \frac{1}{4}\right) + \left(\frac{1}{4} - \frac{1}{5}\right) + \left(\frac{1}{5} - \frac{1}{6}\right) = 1 + \left(-\frac{1}{2} + \frac{1}{2}\right) + \left(-\frac{1}{3} + \frac{1}{3}\right) + \left(-\frac{1}{4} + \frac{1}{4}\right) + \left(-\frac{1}{5} + \frac{1}{5}\right) - \frac{1}{6} = 1 + 0 + 0 + 0 + 0 - \frac{1}{6} = 1 - \frac{1}{6}$

15. Sample: Let $a_n = 1 \ \forall n$. Then $\sum_{n=1}^{4} (a_n{}^2) = 1^2 + 1^2 + 1^2 + 1^2 = 4$, and $\left(\sum_{n=1}^{4} a_n\right)^2 = (1 + 1 + 1 + 1)^2 = 16$.

16. $\dfrac{1}{1000} \sum_{i=-5}^{5} i^2$

19. Left side $= \sin^2 x - \sin^2 y$

$= (1 - \cos^2 x) - (1 - \cos^2 y)$

$= 1 - \cos^2 x - 1 + \cos^2 y$

$= \cos^2 y - \cos^2 x$

$=$ Right side

$\therefore \forall$ real numbers x and y, $\sin^2 x - \sin^2 y = \cos^2 y - \cos^2 x$

LESSON 7-4 (pages 418–425)

2.b. Assume that $S(k)$ is true for a particular but arbitrarily chosen integer $k \geq 1$, where $S(k)$: $\sum_{i=1}^{k} (2i - 1) = k^2$.

3.a. $\sum_{i=1}^{1} 2i = 2(1) = 2$ and $1(1 + 1) = 2$, so $S(1)$ is true.

b. $S(k)$: $\sum_{i=1}^{k} 2i = k(k + 1)$;

$S(k + 1)$: $\sum_{i=1}^{k+1} 2i = (k + 1)(k + 2)$

c. $\sum_{i=1}^{k+1} 2i = \sum_{i=1}^{k} 2i + 2(k + 1) = k(k + 1) + 2(k + 1) = (k + 1)(k + 2)$, so $S(k + 1)$ is true.

4.a. $S(1)$: 3 is a factor of 3; $S(3)$: 3 is a factor of 33; $S(5)$: 3 is a factor of 135.

b. All are true.

c. $S(k + 1)$: 3 is a factor of $(k + 1)^3 + 2(k + 1)$.

5.a. $S(1)$: $5 < 4$; $S(3)$: $13 < 16$; $S(5)$: $29 < 36$

b. $S(1)$ is false. $S(3)$ and $S(5)$ are true.

c. $S(k + 1)$: $(k + 1)^2 + 4 < (k + 2)^2$

6.a. $S(1)$: $1 = \frac{1(2)}{2}$;

$S(3)$: $1 + 4 + 7 = \frac{3(8)}{2}$;

$S(5)$: $1 + 4 + 7 + 10 + 13 = \frac{5(14)}{2}$

b. All are true.

c. $S(k + 1)$:
$\sum_{i=1}^{k+1} (3i - 2) = \frac{(k + 1)(3k + 2)}{2}$

7. b. Assume that $S(k)$ is true for a particular but arbitrarily chosen integer $k \geq 1$. $S(k)$:
$1^2 + 2^2 + \ldots + k^2 = \frac{k(k + 1)(2k + 1)}{6}$.
Show $S(k + 1)$: $1^2 + 2^2 + \ldots + k^2 + (k + 1)^2 = \frac{(k + 1)(k + 2)(2(k + 1) + 1)}{6}$
is true.
$1^2 + 2^2 + \ldots + k^2 + (k + 1)^2 =$
$\frac{k(k + 1)(2k + 1)}{6} + (k + 1)^2 =$
$\frac{k(k + 1)(2k + 1) + 6(k + 1)^2}{6} =$
$\frac{(k + 1)(k(2k + 1) + 6(k + 1))}{6} =$
$\frac{(k + 1)(2k^2 + 7k + 6)}{6} =$
$\frac{(k + 1)(k + 2)(2k + 3)}{6} =$
$\frac{(k + 1)(k + 2)(2(k + 1) + 1)}{6}$

c. Since $S(1)$ is true and $S(k) \Rightarrow S(k + 1)$ by mathematical induction, $S(n)$ is true $\forall$ integers $n \geq 1$.

8.c. Let $S(n)$: $a_n = \frac{3^n - 1}{2}$ $\forall$ integers $n \geq 1$.

(1) $a_1 = \frac{3^1 - 1}{2} = 1$. This agrees with the recursive formula, hence $S(1)$ is true.

(2) Assume $S(k)$: $a_k = \frac{3^k - 1}{2}$ is true for some integer $k \geq 1$.

Show $S(k + 1)$: $a_{k+1} = \frac{3^{k+1} - 1}{2}$ is true.

$a_{k+1} = 3a_k + 1 = 3\left(\frac{3^k - 1}{2}\right) + 1 =$

$\frac{3^{k+1} - 3 + 2}{2} = \frac{3^{k+1} - 1}{2}$.

Therefore, for all integers $k \geq 1$, if $S(k)$ is true, then $S(k + 1)$ is true. Thus, from (1) and (2) above, using the Principle of Mathematical Induction, $S(n)$ is true for all integers $n \geq 1$. Hence, the explicit formula describes the same sequence as the recursive formula.

9. $S(n)$: $a_n = 2n^2 - 1$ $\forall$ integers $n \geq 1$.
(1) $a_1 = 1$ from the recursive definition; $a_1 = 2 \cdot 1^2 - 1 = 1$ from the explicit definition. Hence, $S(1)$ is true.
(2) Assume $S(k)$: $a_k = 2k^2 - 1$ for some integer $k \geq 1$. Show $S(k + 1)$: $a_{k+1} = 2(k + 1)^2 - 1$ is true.
$a_{k+1} = a_k + 4k + 2 =$
$2k^2 - 1 + 4k + 2 =$
$2(k^2 + 2k + 1) - 1 = 2(k + 1)^2 - 1$.
Therefore, $S(k + 1)$ is true if $S(k)$ is true, and (1) and (2) prove by the Principle of Mathematical Induction that the explicit formula does describe the sequence.

10. (1) $\sum_{i=1}^{1} (3i - 2) = 1$.

$\frac{1 \cdot (3 - 1)}{2} = 1$. Hence, $S(1)$ is true.

(2) Assume $S(k)$: $\sum_{i=1}^{k} (3i - 2) =$

$\frac{k(3k - 1)}{2}$ is true for some integer $k \geq 1$.

Show $S(k + 1)$: $\sum_{i=1}^{k+1} (3i - 2) =$

$\frac{(k + 1)(3(k + 1) - 1)}{2}$ is true.

Now $\sum_{i=1}^{k+1} (3i - 2) = \sum_{i=1}^{k} (3i - 2) +$

$3(k + 1) - 2 = \frac{k(3k - 1)}{2} + 3k + 1 =$

$\frac{3k^2 - k + 6k + 2}{2} = \frac{3k^2 + 5k + 2}{2} =$

$\frac{(k + 1)(3k + 2)}{2} = \frac{(k + 1)(3(k + 1) - 1)}{2}$,

so the inductive step is true. Thus, (1) and (2) prove by the Principle of Mathematical Induction that $S(n)$ is true for all integers $n \geq 1$.

13.a. Since $x - y$ is a factor of $x^4 - y^4$, and $x^4 - y^4$ is a factor of $x^5 - xy^4 = x(x^4 - y^4)$, then by the Transitive Property of Factors, $x - y$ is a factor of $x^5 - xy^4$.
c. Since $x - y$ is a factor of $x^5 - xy^4$ by part **a**, and $x - y$ is a factor of $xy^4 - y^5$ by part **b**, then $x - y$ is a factor of $(x^5 - xy^4) + (xy^4 - y^5)$ by the Factor of a Sum Theorem.

15.a.-c.

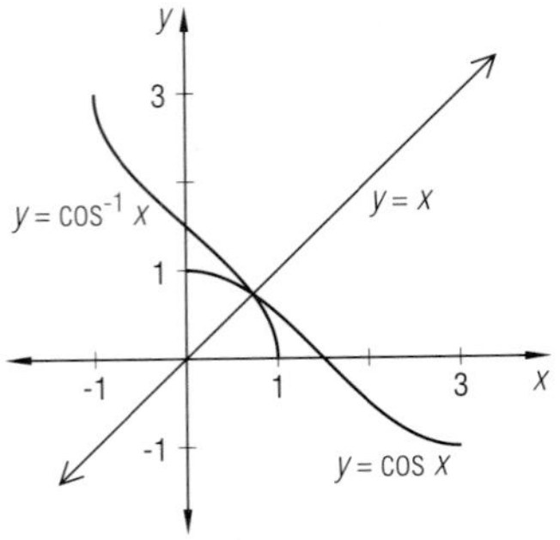

LESSON 7-5 (pages 426–430)

2. $(n + 1)^2 + (n + 1) = n^2 + 3n + 2 = (n^2 + n) + 2(n + 1)$. It is clear that 2 is a factor of $2(n + 1)$. Since 2 is also a factor of $n^2 + n$, then 2 will be a factor of their sum, $(n^2 + n) + 2(n + 1)$, by the Factor of a Sum Theorem.

3.a. $S(1)$: 2 is a factor of 2; since $2 \cdot 1 = 2$, $S(1)$ is true.
$S(13)$: 2 is a factor of 158; since $2 \cdot 79 = 158$, $S(13)$ is true.
$S(20)$: 2 is a factor of 382; since $2 \cdot 191 = 382$, $S(20)$ is true.
b. $S(k)$: 2 is a factor of $k^2 - k + 2$.
$S(k + 1)$: 2 is a factor of $(k + 1)^2 - (k + 1) + 2$.
d. $k^2 + k + 2 = (k^2 - k + 2) + 2k$; 2 is a factor of $k^2 - k + 2$ and $2k$, so 2 is a factor of $(k + 1)^2 - (k + 1) + 2$ by the Factor of a Sum Theorem.

4.d. $6^{k+1} - 1 = 6 \cdot 6^k - 1 = 6 \cdot 6^k - 6 + 5 = 6(6^k - 1) + 5$; 5 is a factor of $6^k - 1$ and of 5, so 5 is a factor of their sum. Thus, $S(k + 1)$ is true.
e. Hence, $S(n)$: 5 is a factor of $6^n - 1$ $\forall$ $n \geq 1$ is true by the Principle of Mathematical Induction.

5. Since $9 - 3 = 6$, substituting $x = 9$ and $y = 3$ into the result of Example 3 yields $9 - 3 = 6$ is a factor of $9^n - 3^n$ $\forall$ $n \geq 1$.

6. Since $13 - 1 = 12$, substituting $x = 13$ and $y = 1$ into the result of Example 3 yields $13 - 1 = 12$ is a factor of $13^n - 1^n$ $\forall$ $n \geq 1$.

7. $S(1)$ is true since $1^3 + 14 \cdot 1 + 3 = 27 = 3 \cdot 9$. Assume $S(k)$: 3 is a factor of $k^3 + 14k + 3$. Show $S(k + 1)$: 3 is a factor of $(k + 1)^3 + 14(k + 1) + 3$ is true. $(k + 1)^3 + 14(k + 1) + 3 = k^3 + 3k^2 + 17k + 18 = (k^3 + 14k + 3) + 3k^2 + 3k + 15$. Since 3 is a factor of $k^3 + 14k + 3$ and of $3(k^2 + k + 5)$, $S(k + 1)$ is true. Hence, 3 is a factor of $n^3 + 14n + 3$ $\forall$ $n \geq 1$ by the Principle of Mathematical Induction.

8. (1) 6 is a factor of $1^3 + 11 = 12 = 2 \cdot 6$, hence $S(1)$ is true.
(2) Assume $S(k)$: 6 is a factor of $k^3 + 11k$ is true for some positive integer k. Show $S(k + 1)$: 6 is a factor of $(k + 1)^3 + 11(k + 1)$ is true. Now $(k + 1)^3 + 11(k + 1) = k^3 + 3k^2 + 3k + 1 + 11k + 11 = (k^3 + 11k) + 3k(k + 1) + 12$. Because 6 is a factor of $k^3 + 11k$ by the inductive assumption, a factor of $3k(k + 1)$ by the theorem, and a factor of 12, it follows that 6 is a factor of $(k + 1)^3 + 11(k + 1)$. Since $S(1)$ is true and $S(k) \Rightarrow S(k + 1)$, by the Principle of Mathematical Induction, 6 is a factor of $n^3 + 11n$ $\forall$ $n \geq 1$.

9.b. All we can conclude is that $S(2)$ and $S(3)$ are true.
c. $12^n - 8^n = 4^n(3^n - 2^n)$. 8 is a factor of 4^n if $n \geq 2$. Hence, $S(n)$ holds if $n \geq 2$.

11. $S(1)$ is true because

$$\sum_{i=1}^{1} i^3 = 1^3 = 1 \text{ and } \left[\frac{1(1 + 1)}{2}\right]^2 = 1^2 = 1.$$

Assume $S(k)$: $\sum_{i=1}^{k} i^3 = \left[\frac{k(k + 1)}{2}\right]^2$ is true for some integer $k \geq 1$. Show that $S(k + 1)$: $\sum_{i=1}^{k+1} i^3 = \left[\frac{(k + 1)(k + 2)}{2}\right]^2$ is true. Now

$$\begin{aligned}\sum_{i=1}^{k+1} i^3 &= \sum_{i=1}^{k} i^3 + (k + 1)^3\\ &= \left[\frac{k(k + 1)}{2}\right]^2 + (k + 1)^3\\ &= \frac{k^2 (k + 1)^2}{4} + \frac{4(k + 1)^3}{4}\\ &= \frac{(k + 1)^2[k^2 + 4(k + 1)]}{4}\\ &= \frac{[(k + 1)(k + 2)]^2}{4}.\end{aligned}$$

Since $S(1)$ is true and $S(k) \Rightarrow S(k + 1)$, by the Principle of Mathematical Induction, $S(n)$ is true for all integers $n \geq 1$.

14.

$$\begin{aligned}\text{Right side} &= \cos(2x) & &\\ &= \cos^2 x - \sin^2 x & &\text{Formula for } \cos(2x)\\ &= (\cos^2 x - \sin^2 x) \cdot 1 & &\text{Multiplication by 1}\\ &= (\cos^2 x - \sin^2 x)(\cos^2 x + \sin^2 x) & &\text{Pythagorean Identity}\\ &= \cos^4 x - \sin^4 x & &\text{Multiplication}\\ &= \text{Left side} & &\end{aligned}$$

Therefore, $\cos^4 x - \sin^4 x = \cos(2x)$ by the Transitive Property.

15.

$$\begin{aligned}\text{Left side} &= \frac{a^4 - b^4}{a - b} & &\\ &= \frac{(a^2 - b^2)(a^2 + b^2)}{a - b} & &\text{Factoring}\\ &= \frac{(a - b)(a + b)(a^2 + b^2)}{a - b} & &\text{Factoring}\\ &= (a + b)(a^2 + b^2) & &\text{Division}\\ &= a^3 + a^2b + ab^2 + b^3 & &\text{Multiplication}\\ &= \text{Right side} & &\end{aligned}$$

Therefore, $\frac{a^4 - b^4}{a - b} = a^3 + a^2b + ab^2 + b^3$ when $a \neq b$ by the Transitive Property.

16.a.

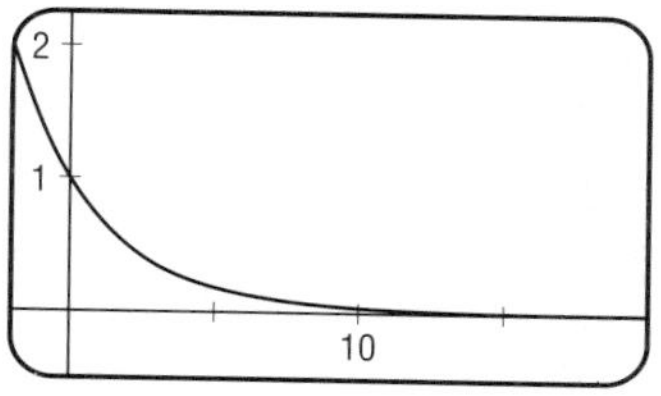

LESSON 7-6 (pages 431–437)

9.b. $a = 1, r = 2$:

Left side $= 1 \cdot \left(\frac{1 - 2^n}{-1}\right) = 2^n - 1$,

Right side $= 1 \cdot \left(\frac{2^n - 1}{1}\right) = 2^n - 1$;

$a = 1, r = \frac{1}{2}$:

Left side $= 1 \cdot \left(\frac{1 - \left(\frac{1}{2}\right)^n}{\frac{1}{2}}\right) = 2 - 2\left(\frac{1}{2}\right)^n$,

Right side $= 1 \cdot \left(\frac{\left(\frac{1}{2}\right)^n - 1}{-\frac{1}{2}}\right) = 2 - 2\left(\frac{1}{2}\right)^n$

11.a. $a_1 = 3$, $a_k = \frac{1}{2}a_{k-1}$, for all integers $2 \leq k \leq 25$.
c. $\sum_{j=1}^{25} 3\left(\frac{1}{2}\right)^{j-1}$

12. For all integers $k \geq 1$, $S_{k+1} = \sum_{i=1}^{k+1} a_i = \left(\sum_{i=1}^{k} a_i\right) + a_{k+1} = S_k + a_{k+1}$.

13.a. $S_1 = a$ and $S_{n+1} = S_n + ar^n$, $\forall$ integers $n \geq 1$.
b. $S_1 = \frac{a(1 - r^1)}{1 - r} = a$; so the initial condition is met. $S_{n+1} = \frac{a(1 - r^{n+1})}{1 - r} = a + ar + ar^2 + \ldots + ar^n = (a + ar + \ldots + ar^{n-1}) + ar^n = \sum_{i=1}^{n} ar^{i-1} + ar^n = S_n + ar^n$

14. Let $S(n)$: 3 is a factor of $n^3 + 14n$. $S(1)$: 3 is a factor of $1^3 + 14 \cdot 1$ is true. Assume $S(k)$: 3 is a factor of $k^3 + 14k$ is true for some integer $k \geq 1$. Show that $S(k + 1)$: 3 is a factor of $(k + 1)^3 + 14(k + 1)$ is true. Now $(k + 1)^3 + 14(k + 1) = k^3 + 3k^2 + 3k + 1 + 14k + 14 = (k^3 + 14k) + 3k^2 + 3k + 15 = (k^3 + 14k) + 3(k^2 + k + 5)$. 3 is a factor of $k^3 + 14k$ by the inductive assumption, and a factor of $3(k^2 + k + 5)$; so by mathematical induction, 3 is a factor of $n^3 + 14n$ $\forall$ $n \geq 1$.

15.a. $S(3)$: $\sum_{i=1}^{3} i(i + 1) = \frac{3 \cdot 4 \cdot 5}{3}$. Does $1 \cdot 2 + 2 \cdot 3 + 3 \cdot 4 = 4 \cdot 5$? Does $2 + 6 + 12 = 20$? Does $20 = 20$? Yes. Hence, $S(3)$ is true.
b. See answer on the next page.

15.b. $S(1)$: $1 \cdot 2 = \frac{1 \cdot 2 \cdot 3}{3}$, which is true. Assume $S(k)$: $\sum_{i=1}^{k} i(i+1) = \frac{k(k+1)(k+2)}{3}$ is true for some integer $k \geq 1$. Show that $S(k+1) = \sum_{i=1}^{k+1} i(i+1) = \frac{(k+1)(k+2)(k+3)}{3}$ is true. Then $\sum_{i=1}^{k+1} i(i+1) = \sum_{i=1}^{k} i(i+1) + (k+1)(k+2) = \frac{k(k+1)(k+2)}{3} + (k+1)(k+2) = (k+1)(k+2)\left(\frac{k}{3} + 1\right) = \frac{(k+1)(k+2)(k+3)}{3}$. Hence, $S(n)$ holds $\forall\ n \geq 1$.

17.a.

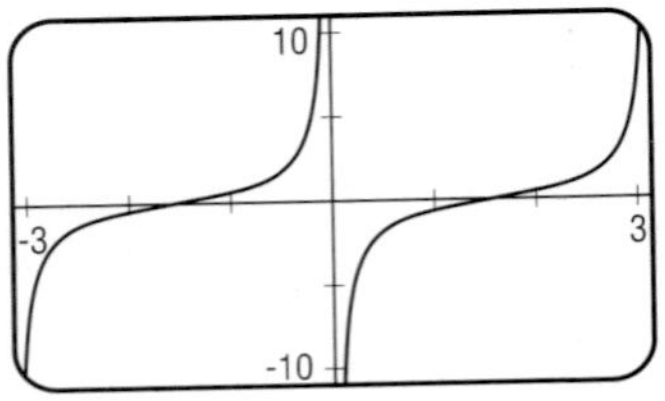

The graph represents the transformation $T_{0.5\pi,0}$ applied to $y = \tan\theta$.

20.b.

```
10  SUM = 0
20  FOR TERM = 1 TO 100
30    LET A = (-1)^(TERM + 1)
      *4/(2*TERM - 1)
40    SUM = SUM + A
50    PRINT SUM
60  NEXT TERM
70  END
```

LESSON 7-7 (pages 438–444)

9. Let $S(n)$: $n^3 + 3n^2 + 2n$ is divisible by 3. Then $S(1)$: $1^3 + 3 \cdot 1^2 + 2 \cdot 1$ is divisible by 3. $1 + 3 + 2 = 6$, which is divisible by 3, so $S(1)$ is true. Assume $S(k)$: $k^3 + 3k^2 + 2k$ is divisible by 3 is true for some integer $k \geq 1$. Show $S(k+1)$: $(k+1)^3 + 3(k+1)^2 + 2(k+1)$ is divisible by 3 is true. $(k+1)^3 + 3(k+1)^2 + 2(k+1) = k^3 + 3k^2 + 3k + 1 + 3k^2 + 6k + 3 + 2k + 2 = k^3 + 6k^2 + 11k + 6 = (k^3 + 3k^2 + 2k) + 3(k^2 + 3k + 2)$. $k^3 + 3k^2 + 2k$ is divisible by 3 by the inductive assumption, and $3(k^2 + 3k + 2)$ is divisible by 3; thus, their sum is divisible by 3. Therefore, by mathematical induction, $n^3 + 3n^2 + 2n$ is divisible by 3 $\forall\ n \geq 1$.

10. $S(n)$: $a_n = 4(3)^{n-1}$. $S(1)$: $a_1 = 4(3)^0$. $a_1 = 4$, so $S(1)$ is true. Assume $S(k)$: $a_k = 4(3)^{k-1}$ is true. Show $S(k+1)$: $a_{k+1} = 4(3)^k$ is true. $a_{k+1} = 3a_k = 3(4 \cdot 3^{k-1}) = 4 \cdot 3^k$. By mathematical induction, $S(n)$ is true for all $n \geq 1$. Hence, $a_n = 4(3)^{n-1}$ is an explicit formula for the sequence.

11.a.

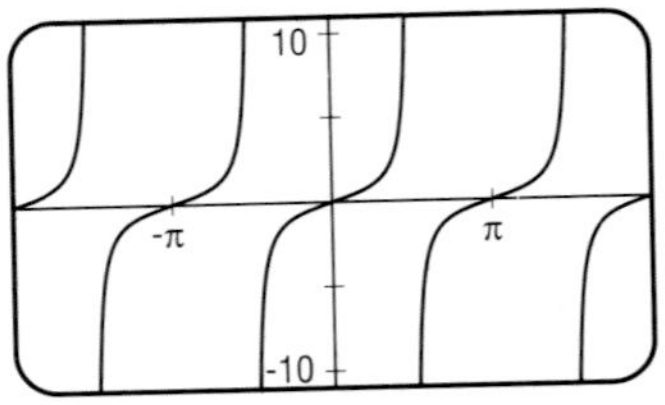

c. Left side $= \csc x \sec x - \cot x$

$= \frac{1}{\sin x} \cdot \frac{1}{\cos x} - \frac{\cos x}{\sin x}$	Definition of cosecant, secant, and cotangent
$= \frac{1}{\sin x \cos x} - \frac{\cos^2 x}{\sin x \cos x}$	Multiplication
$= \frac{1 - \cos^2 x}{\sin x \cos x}$	Subtraction of fractions
$= \frac{\sin^2 x}{\sin x \cos x}$	Pythagorean Identity
$= \frac{\sin x}{\cos x}$	Simplifying fractions
$= \tan x$	Definition of tangent
$=$ Right side	

Therefore, $\csc x \sec x - \cot x \tan x\ \forall\ x$ for which both sides are defined.

LESSON 7-8 (pages 445–452)

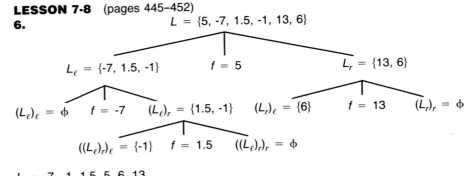

6.

$L = \{5, -7, 1.5, -1, 13, 6\}$

$L_\ell = \{-7, 1.5, -1\}$ $\quad f = 5 \quad$ $L_r = \{13, 6\}$

$(L_\ell)_\ell = \phi$ $\quad f = -7 \quad$ $(L_\ell)_r = \{1.5, -1\}$ $\quad (L_r)_\ell = \{6\}$ $\quad f = 13 \quad$ $(L_r)_r = \phi$

$((L_\ell)_r)_\ell = \{-1\}$ $\quad f = 1.5 \quad$ $((L_\ell)_r)_r = \phi$

$L = -7, -1, 1.5, 5, 6, 13$

LESSON 7-9 (pages 453–457)

7.

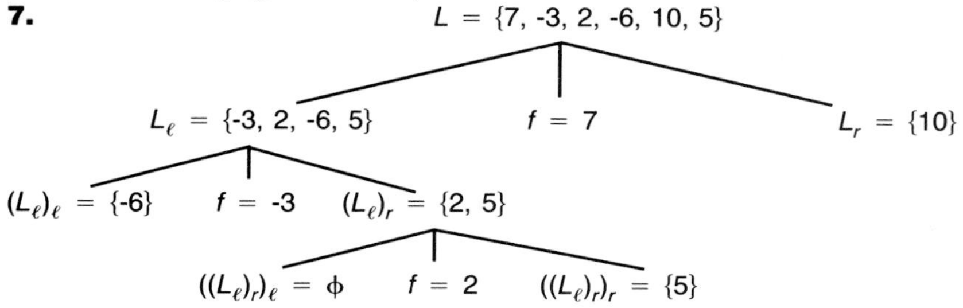

The sorted list is -6, -3, 2, 5, 7, 10.

16. For Merge sort, $E(n) = n \log_2 n$. For Selection sort, $E(n) = 2n$. Approximate $E(n)$:

	$n = 10$	$n = 100$	$n = 10{,}000$	$n = 1{,}000{,}000$
Merge sort	33	664	133,000	2×10^7
Selection sort	20	200	20,000	2×10^6
Bubblesort	45	4950	5×10^7	5×10^{11}
Quicksort	25	444	113,000	1.8×10^7

CHAPTER 7 PROGRESS SELF-TEST

(pages 461)

9. Let $S(n)$: 4 is a factor of $2n^2 + 2n + 8$.
(1) $2 \cdot 1^2 + 2 \cdot 1 + 8 = 12$ and 4 is a factor of 12, so $S(1)$ is true.
(2) Assume $S(k)$ is true. That is, 4 is a factor of $2k^2 + 2k + 8$ for some arbitrary integer $k \geq 1$. Show $S(k + 1)$ is true.
$2(k + 1)^2 + 2(k + 1) + 8 = 2k^2 + 4k + 2 + 2k + 2 + 8 = (2k^2 + 2k + 8) + 4k + 4 = (2k^2 + 2k + 8) + 4(k + 1)$. 4 is a factor of $2k^2 + 2k + 8$ by the inductive assumption, and 4 is clearly a factor of $4(k + 1)$. Hence, 4 is a factor of $2(k + 1)^2 + 2(k + 1) + 8$ by the Factor of a Sum Theorem. Therefore, $S(n)$ is true for all positive integers n by mathematical induction. So 4 is a factor of $2n^2 + 2n + 8$ for all positive integers n.

11.

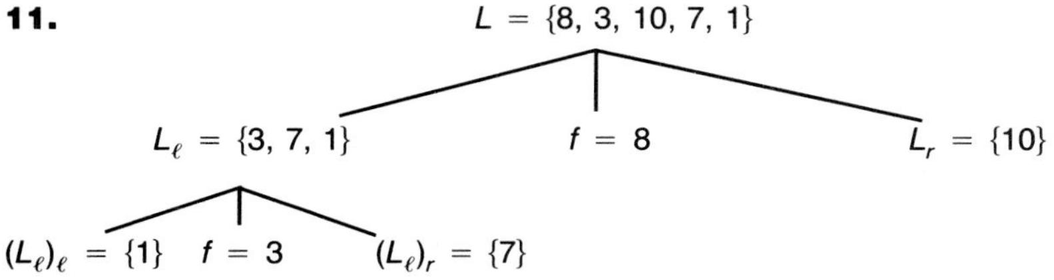

The sorted list is 1, 3, 7, 8, 10.

CHAPTER 7 REVIEW (pages 462–464)

27.a. 0, 2, 6, 12, 20

b. Let $S(n)$: $b_n = n(n - 1)$ for all integers $n \geq 1$.
(1) $S(1)$: $b_1 = 1(1 - 1)$. $b_1 = 0$, so $S(1)$ is true.
(2) Assume $S(k)$: $b_k = k(k - 1)$ is true for some arbitrary integer $k \geq 1$. Show that $S(k + 1)$: $b_{k+1} = (k + 1)((k + 1) - 1)$ is true. From the recursive definition,
$b_{k+1} = b_k + 2k$
$= k(k - 1) + 2k$
$= k^2 + k$
$= (k + 1)k$
$= (k + 1)((k + 1) - 1)$.
Hence, by mathematical induction, $S(n)$ is true $\forall$ integers $n \geq 1$, and the explicit formula is correct.

28. Let $S(n)$: $3 + 7 + 11 + \ldots + (4n - 1) = n(2n + 1)$.
(1) $S(1)$: $3 = 1(2 \cdot 1 + 1)$. $S(1)$ is true.
(2) Assume $S(k)$: $3 + 7 + 11 + \ldots + (4k - 1) = k(2k + 1)$ is true for some arbitrary integer $k \geq 1$. Show that $S(k + 1)$: $3 + 7 + 11 + \ldots + (4(k + 1) - 1) = (k + 1)(2(k + 1) + 1)$ is true. From the inductive assumption.
$3 + 7 + 11 + \ldots + (4k - 1) + (4(k + 1) - 1)$
$= k(2k + 1) + (4(k + 1) - 1)$
$= 2k^2 + k + 4k + 3$
$= 2k^2 + 5k + 3$
$= (k + 1)(2k + 3)$
$= (k + 1)(2(k + 1) + 1)$
Hence, by mathemmatical induction, $S(n)$ is true for all integers $n \geq 1$.

29. (1) $S(1)$: $\sum_{i=1}^{1} 3i(i + 2) = \frac{1 \cdot 2 \cdot 9}{2}$.

$3 \cdot 1(1 + 2) = 9$ and $\frac{1 \cdot 2 \cdot 9}{2} = 9$, so $S(1)$ is true.

(2) Assume that $S(k)$: $\sum_{i=1}^{k} 3i(i + 2) = \frac{k(k+1)(2k+7)}{2}$ is true for an arbitrary integer $k \geq 1$. Show $S(k + 1)$:

$\sum_{i=1}^{k+1} 3i(i + 2) = \frac{(k+1)(k+2)(2(k+1)+7)}{2}$ is true.

$\sum_{i=1}^{k+1} 3i(i + 2) = \left(\sum_{i=1}^{k} 3i(i + 2)\right) +$

$3(k + 1)(k + 3) = \frac{k(k+1)(2k+7)}{2} +$

$(k + 1)(3k + 9)$

$= \frac{k(k+1)(2k+7)}{2} + \frac{2(k+1)(3k+9)}{2}$

$= \left(\frac{k+1}{2}\right)(2k^2 + 7k + 6k + 18)$

$= \left(\frac{k+1}{2}\right)(k + 2)(2k + 9)$

$= \frac{(k+1)(k+2)(2(k+1)+7)}{2}$

Hence, by mathematical induction, $S(n)$ is true $\forall$ integers $n \geq 1$.

30. (1) $S(1)$: 3 is a factor of $1^3 + 14(1)$. $1^3 + 14(1) = 15$, so $S(1)$ is true.

(2) Assume $S(k)$: 3 is a factor of $k^3 + 14k$ is true for some integer $k \geq 1$. Show that $S(k + 1)$: 3 is a factor of $(k + 1)^3 + 14(k + 1)$ is true. Expanding,

$(k + 1)^3 + 14(k + 1)$

$= (k^3 + 3k^2 + 3k + 1) + (14k + 14)$

$= (k^3 + 14k) + (3k^2 + 3k + 15)$

$= (k^3 + 14k) + 3(k^2 + k + 5)$.

By the inductive assumption, 3 is a factor of $k^3 + 14k$, and 3 is a factor of $3(k^2 + k + 5)$. Therefore, 3 is a factor of their sum by the Factor of a Sum Theorem. Hence, by mathematical induction, $S(n)$ is true $\forall$ integers $n \geq 1$.

31. (1) $S(2)$: 3 is a factor of $2 \cdot 2^3 - 5 \cdot 2$. $16 - 10 = 6$, so $S(2)$ is true.

(2) Assume that $S(k)$: 3 is a factor of $2k^3 - 5k$ is true for some integer $k \geq 2$. Prove $S(k + 1)$: 3 is a factor of $2(k + 1)^3 - 5(k + 1)$ is true.

$2(k + 1)^3 - 5(k + 1)$

$= 2k^3 + 6k^2 + 6k + 2 - 5k - 5$

$= (2k^3 - 5k) + 3(2k^2 + 2k - 1)$

Since 3 is a factor of $2k^3 - 5k$ and $3(2k^2 + 2k - 1)$, 3 is a factor of their sum by the Factor of a Sum Theorem. Hence, by mathematical induction, $S(n)$ is true $\forall$ integers $n \geq 2$.

32. Let $S(n)$: t_n is an odd integer. $S(1)$: t_1 is an odd integer. $t_1 = 3$, so $S(1)$ is true. $S(2)$: t_2 is an odd integer. $t_2 = 9$, so $S(2)$ is true. Assume $S(1)$, $S(2)$, ... , and $S(k)$ are true for some integer $k \geq 1$. So $t_1, t_2, \ldots, t_k$ are odd integers. Show $S(k + 1)$: t_{k+1} is an odd integer. Since t_{k-1} and t_k are odd integers, there exist integers p and q such that $t_{k-1} = 2p + 1$ and $t_k = 2q + 1$. Hence, $t_{k+1} = (2q + 1) + 2(2p + 1) = 2p + 1 + 4q + 2 = 2(q + 2p + 1) + 1$ where $q + 2p + 1$ is an integer by closure properties, so t_{k+1} is an odd integer. Hence, by the Strong Form of Mathematical Induction, $S(n)$ is true for all integers $n \geq 1$.

33.b. Let $S(n)$: 4 is a factor of b_n.

(1) $S(1)$: 4 is a factor of b_1. $b_1 = 0$, so $S(1)$ is true. $S(2)$: 4 is a factor of b_2. $b_2 = 4$, so $S(2)$ is true.

(2) Assume $S(1)$, $S(2)$, ... , and $S(k)$ are true for some integer $k \geq 1$. So 4 is a factor of $b_1, b_2, \ldots, b_k$. Show $S(k + 1)$: 4 is a factor of b_{k+1} is true. Since b_{k-1} and b_k have 4 as a factor, there exist integers p and q such that $b_{k-1} = 4p$ and $b_k = 4q$. Substituting into the recurrence relation, $b_{k+1} = 4q + 3(4p) = 4q + 12p = 4(q + 3p)$.

$q + 3p$ is an integer by closure properties, so 4 is a factor of b_{k+1}. Hence, by the Strong Form of Mathematical Induction, $S(n)$ is true for all integers $n \geq 1$.

38. $L = \{21, 1, 8, 13, 1, 5\}$

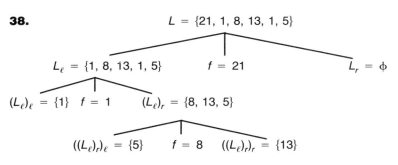

$L_\ell = \{1, 8, 13, 1, 5\}$ $f = 21$ $L_r = \phi$

$(L_\ell)_\ell = \{1\}$ $f = 1$ $(L_\ell)_r = \{8, 13, 5\}$

$((L_\ell)_r)_\ell = \{5\}$ $f = 8$ $((L_\ell)_r)_r = \{13\}$

So the sorted list is 1, 1, 5, 8, 13, 21.

LESSON 8-1 (pages 466–472)

11.

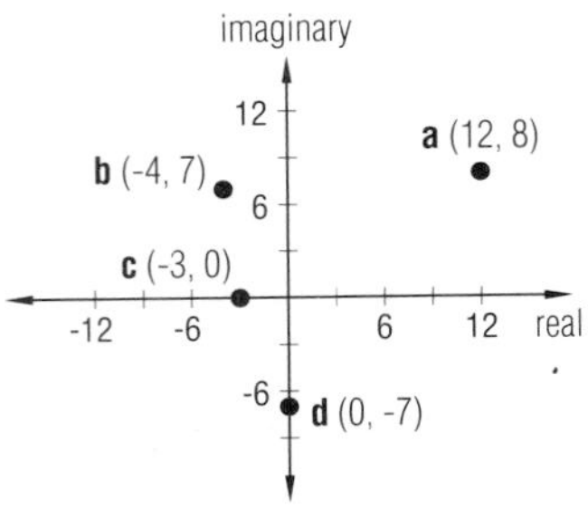

13.b.

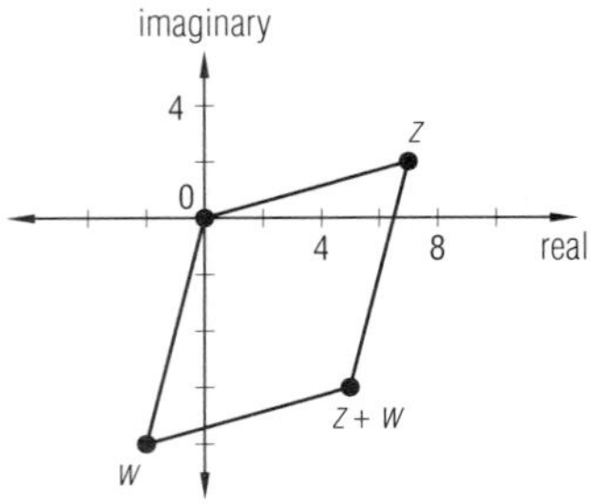

21. Suppose there is a smallest integer, s. Then $s - 1 < s$, and $s - 1$ is an integer. This contradicts the assumption that s is the smallest integer. So the assumption is false, and there is no smallest integer.

24.b.

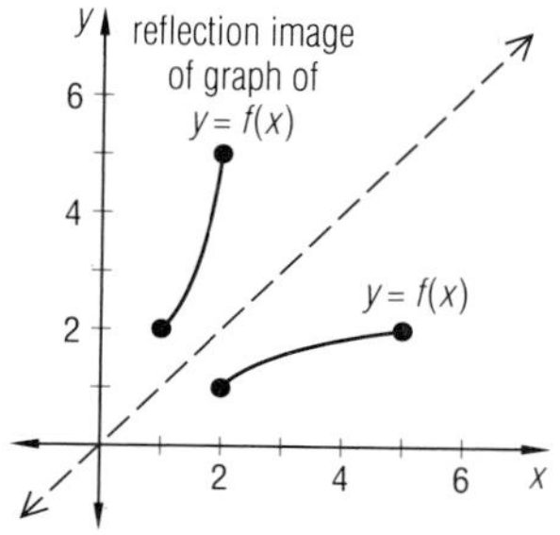

c. domain: $\{x: 1 \le x \le 2\}$;
range: $\{y: 2 \le y \le 5\}$

25.a. $z_1 = -\frac{3}{4} + \frac{\sqrt{31}}{4}i$, $z_2 = -\frac{3}{4} - \frac{\sqrt{31}}{4}i$

c. $z_1 + z_2 = -\frac{3}{2}$, $z_1 \cdot z_2 = \frac{5}{2}$;
The sum is the opposite of the x-coefficient divided by the x^2-coefficient; the product is the constant divided by the x^2-coefficient.

d. For $ax^2 + bx + c = 0$, the roots are

$$z_1 = \frac{-b + \sqrt{b^2 - 4ac}}{2a} \text{ and}$$

$$z_2 = \frac{-b - \sqrt{b^2 - 4ac}}{2a}.$$

So $z_1 + z_2 = \frac{-b + \sqrt{b^2 - 4ac}}{2a} + \frac{-b - \sqrt{b^2 - 4ac}}{2a} = \frac{-2b}{2a} = -\frac{b}{a}$.

$z_1 \cdot z_2 = \frac{-b + \sqrt{b^2 - 4ac}}{2a} \cdot \frac{-b - \sqrt{b^2 - 4ac}}{2a} = \frac{b^2 - (b^2 - 4ac)}{4a^2} = \frac{4ac}{4a^2} = \frac{c}{a}$.

LESSON 8-2 (pages 473-479)

1.a.-d.

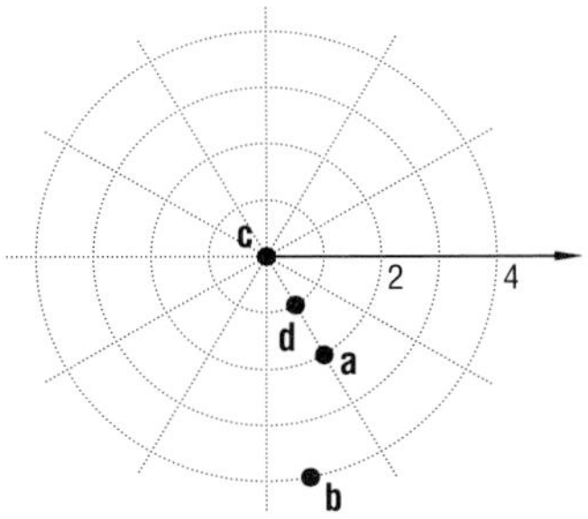

3.a. By case **b** of the polar representation theorem in this lesson, $[4, \frac{\pi}{3}] = [4, \frac{\pi}{3} + 2(-1)\pi] = [4, -\frac{5\pi}{3}]$. Then by case **b** again, $[4, -\frac{5\pi}{3} + 2k\pi]$ for any integer k is a coordinate representation of $[4, \frac{\pi}{3}]$.

b. By case **c** of the polar representation theorem in this lesson, $[4, \frac{\pi}{3}] = [-4, \frac{\pi}{3} + \pi] = [-4, \frac{4\pi}{3}]$. So by case **b**, $[-4, \frac{4\pi}{3} + 2k\pi]$ for any integer k is a coordinate representation of $[4, \frac{\pi}{3}]$.

18.b. $z \cdot w = (6 - 3i)(2 + 4i) = 24 + 18i$. So $\overline{z \cdot w} = 24 - 18i$. $\overline{z} \cdot \overline{w} = (6 + 3i)(2 - 4i) = 24 - 18i$.
$\therefore \overline{z \cdot w} = \overline{z} \cdot \overline{w}$.

c. $\frac{z}{w} = \frac{(6 - 3i)}{(2 + 4i)} = 0 - \frac{3}{2}i$, so $\overline{\left(\frac{z}{w}\right)} = 0 + \frac{3}{2}i$. $\frac{\overline{z}}{\overline{w}} = \frac{6 + 3i}{2 - 4i} = 0 + \frac{3}{2}i$.
$\therefore \overline{\left(\frac{z}{w}\right)} = \frac{\overline{z}}{\overline{w}}$.

LESSON 8-3 (pages 480-485)

14.b. $AB = \sqrt{5}$, $BC = 2$, and $AC = \sqrt{17}$, while $A'B' = 5$, $B'C' = \sqrt{20}$, and $A'C' = \sqrt{85}$.
Then $\frac{A'C'}{AC} = \frac{\sqrt{85}}{\sqrt{17}} = \sqrt{5}$,
$\frac{A'B'}{AB} = \frac{5}{\sqrt{5}} = \sqrt{5}$,
and $\frac{B'C'}{BC} = \frac{\sqrt{20}}{2} = \sqrt{5}$. So the triangles are similar by the SSS Similarity Theorem.

17.a.

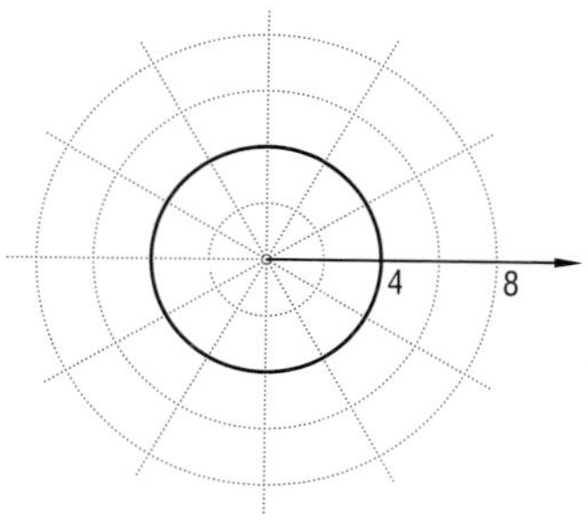

23. Geometric Subtraction Theorem: Let $z = a + bi$ and $w = c + di$ be two complex numbers that are not collinear with the origin. Then the point representing $z - w$ is the fourth vertex of a parallelogram with consecutive vertices $z = a + bi$, 0, and $-w = -c - di$.

24. Geometric Division Theorem: Let z and w be complex numbers. If $z = [r, \theta]$ and $w = [s, \phi]$, then $\frac{z}{w} = [\frac{r}{s}, \theta - \phi]$ $(s \neq 0)$. That is, dividing a complex number z by w applies to z a size change of magnitude $\frac{1}{s}$ and a rotation of $-\phi$ about the origin.

LESSON 8-4 (pages 486-492)

1.

θ	0	$\frac{\pi}{6}$	$\frac{\pi}{4}$	$\frac{\pi}{3}$	$\frac{\pi}{2}$	π	$\frac{3\pi}{2}$	2π
r	0	$\frac{3}{2}$	$\frac{3\sqrt{2}}{2}$	$\frac{3\sqrt{3}}{2}$	3	0	-3	0

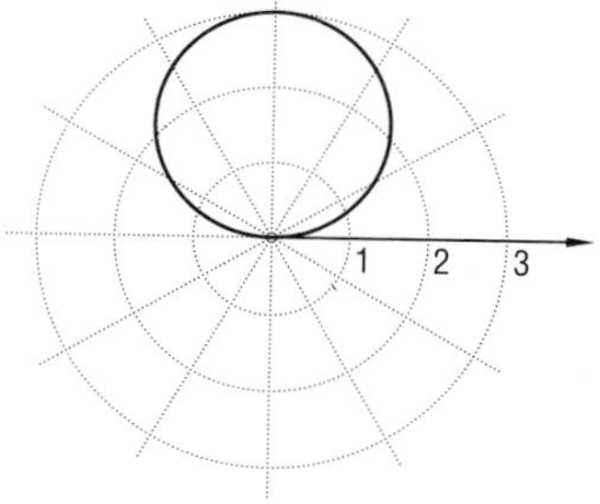

2.

θ	0	$\frac{\pi}{6}$	$\frac{\pi}{4}$	$\frac{\pi}{3}$	$\frac{\pi}{2}$	$\frac{3\pi}{4}$	π	$\frac{3\pi}{2}$	2π
r	2	0	$-\sqrt{2}$	-2	0	$\sqrt{2}$	-2	0	2

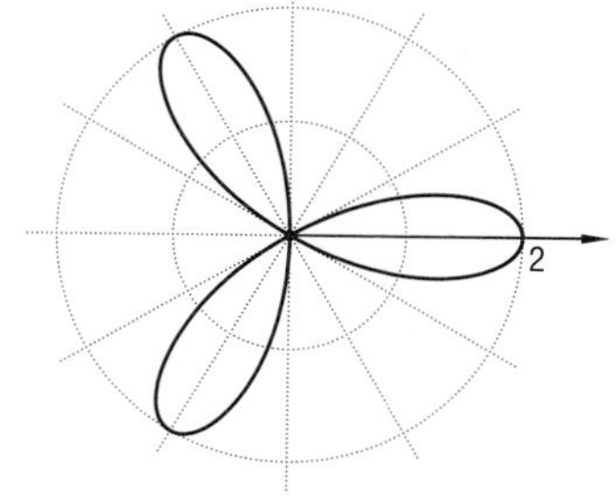

3. $r = 3 \sin \theta$

$\Rightarrow r = \dfrac{3y}{r}$ Conversion formula

$\Rightarrow r^2 = 3y$

$\Rightarrow x^2 + y^2 = 3y$ Conversion formula

$\Rightarrow (x)^2 + \left(y - \frac{3}{2}\right)^2 = \left(\frac{3}{2}\right)^2$

This verifies that the curve in Question 1 is a circle and has center $\left(0, \frac{3}{2}\right)$ and radius $\frac{3}{2}$.

4.

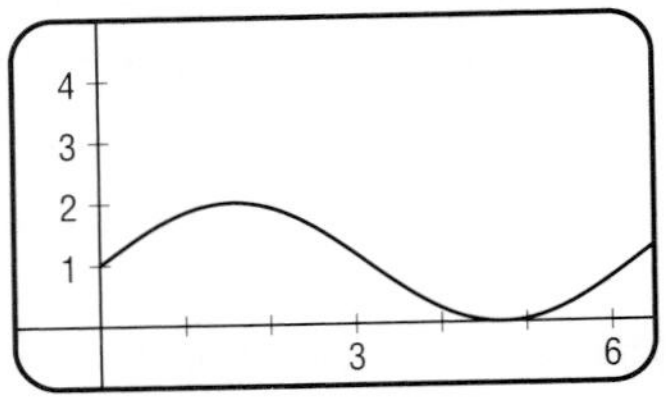

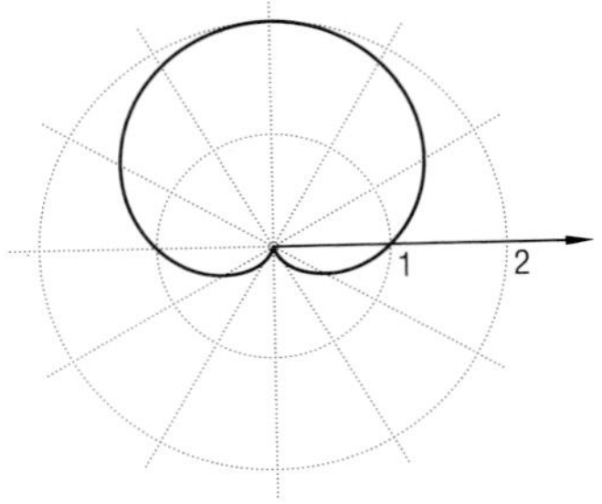

5.

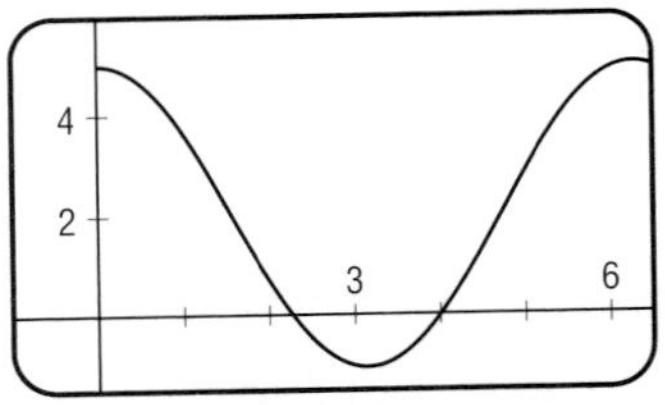

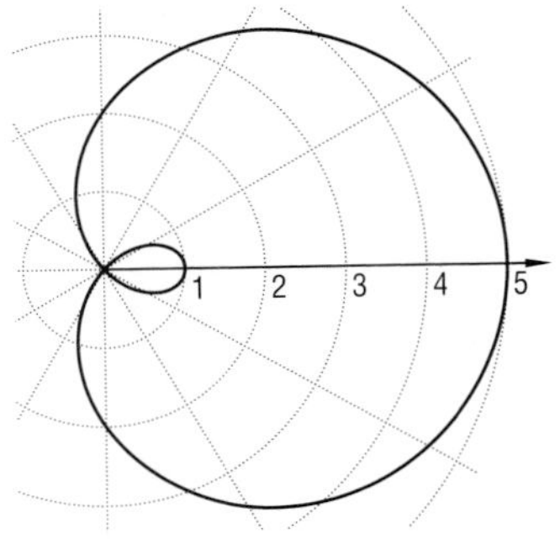

6.b.

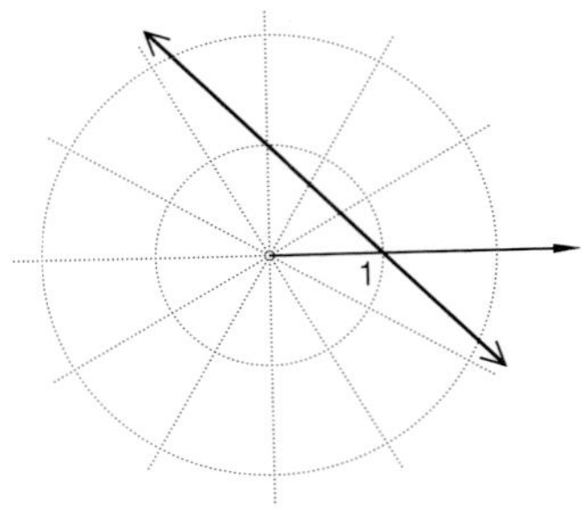

9.a. $r = a + b \cos$, where $a < b$

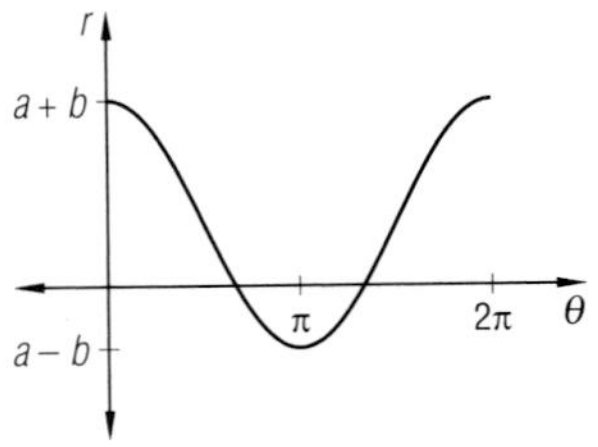

$r = a + b \cos$, where $a > b$

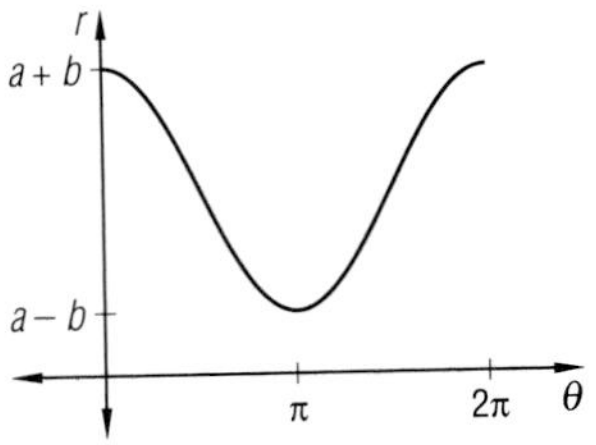

b. $r = a + b \cos$, where $a < b$

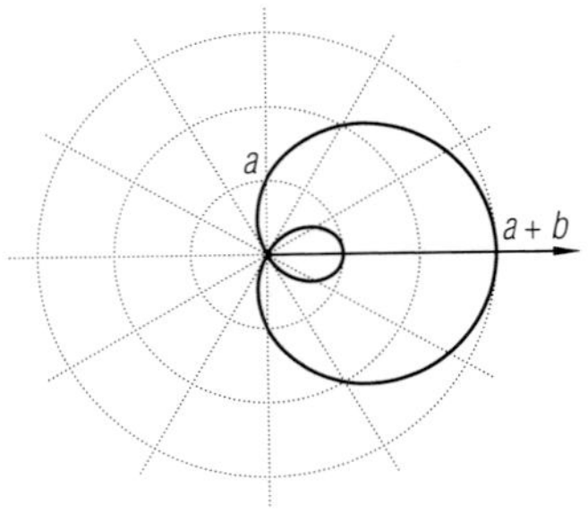

$r = a + b \cos$, where $a > b$

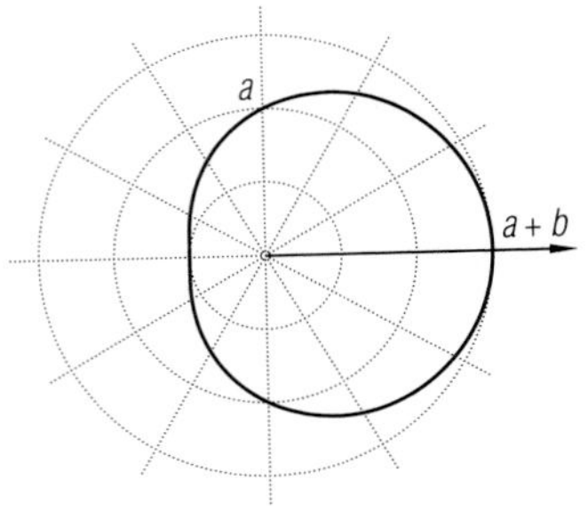

c.i. Around $\theta = \pi$, r has negative values, causing a loop.

10.

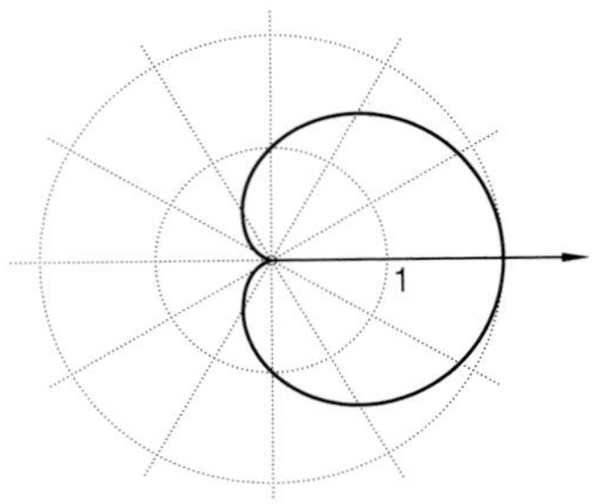

Cardio- is a prefix meaning "heart." Cardioid curves resemble hearts.

11.a. The polar graph of $r = k \sin \theta$ is the graph of $r = \sin \theta$ taken through a scale change of k.

12.

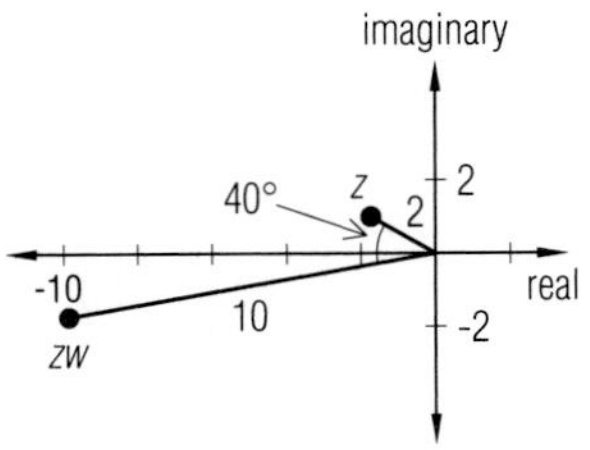

14.b.

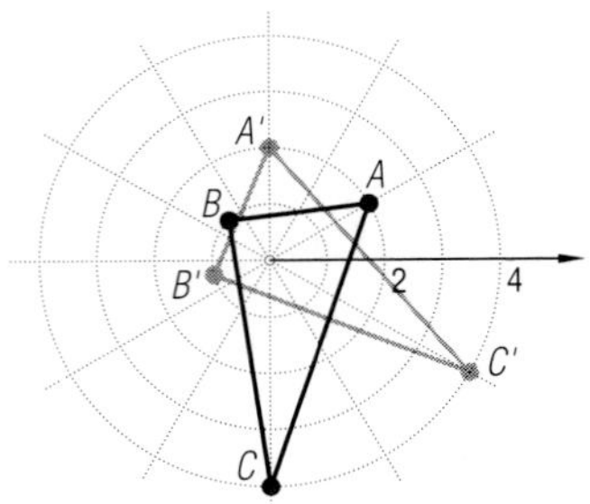

15.b. They are the same. They are both complex conjugates of z.

17.

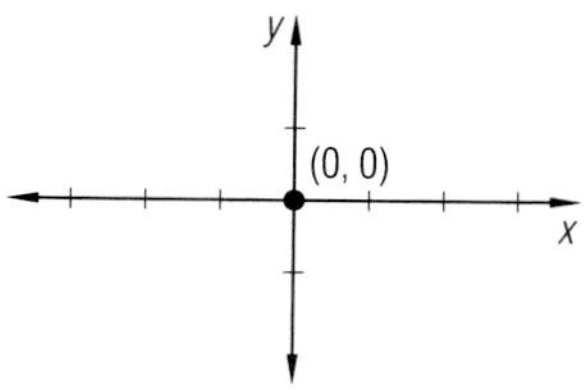

LESSON 8-5 (pages 493–497)

1.

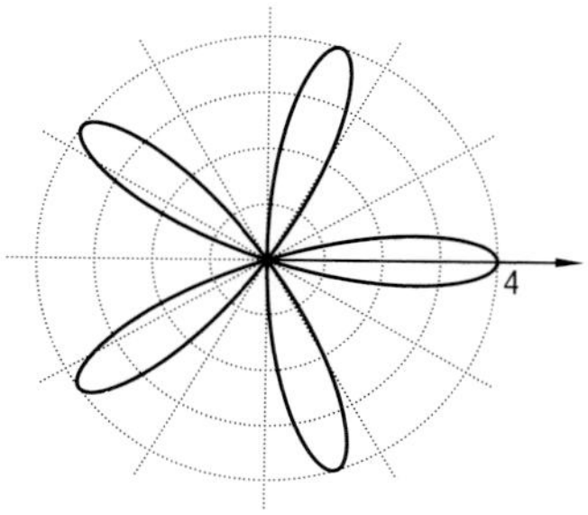

2.

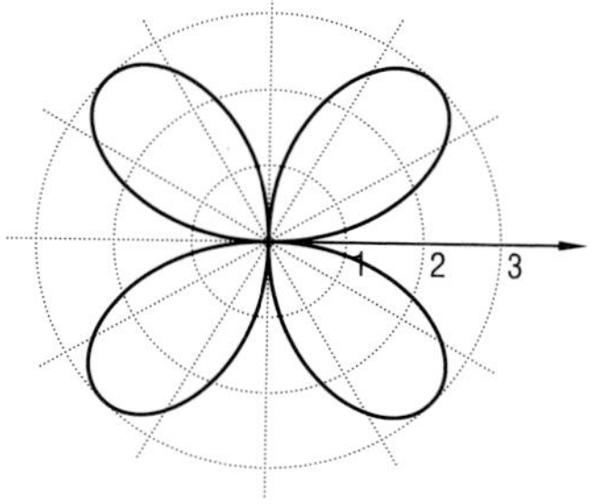

3.

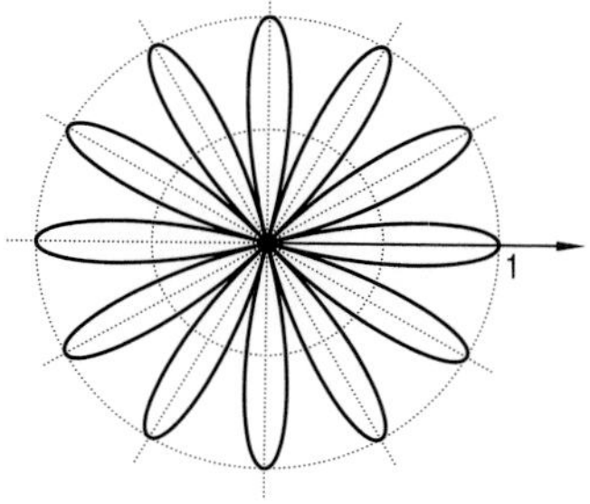

8.b.

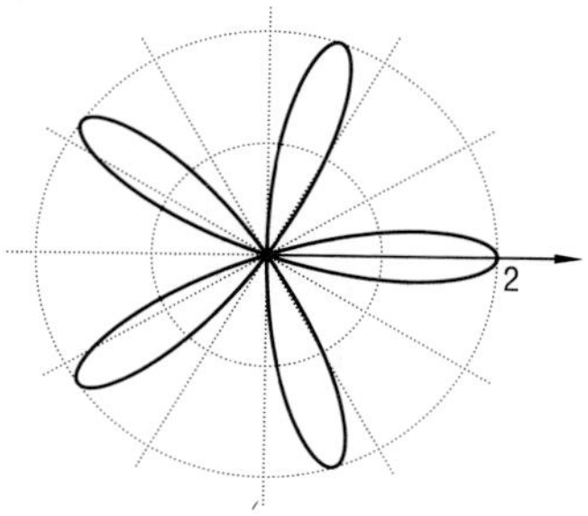

9.a.

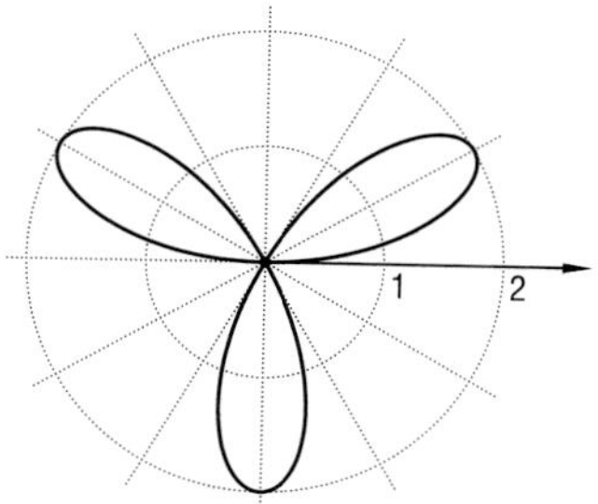

b. $\theta = \frac{\pi}{6}, \theta = \frac{\pi}{2}, \theta = \frac{5\pi}{6}$

10.a.

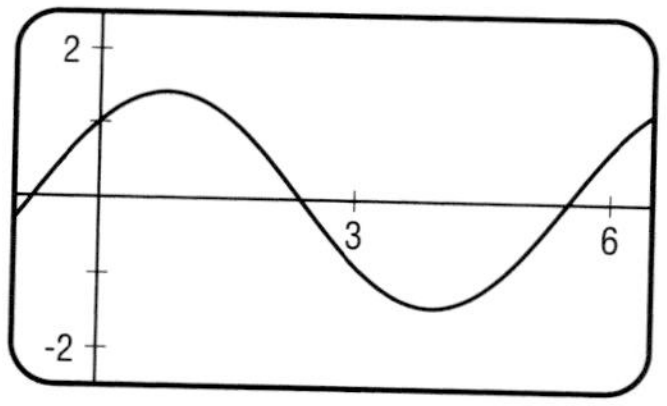

$r = \cos\theta + \sin\theta$
$r = \sqrt{2}\cos(\theta - \frac{\pi}{4})$

b. $r = \sqrt{2}\cos\left(\theta - \frac{\pi}{4}\right)$
$= \sqrt{2}\left(\cos\theta\cos\frac{\pi}{4} + \sin\theta\sin\frac{\pi}{4}\right)$
$= \sqrt{2}\left(\cos\theta \cdot \frac{\sqrt{2}}{2} + \sin\theta \cdot \frac{\sqrt{2}}{2}\right)$
$= \cos\theta + \sin\theta$

c.

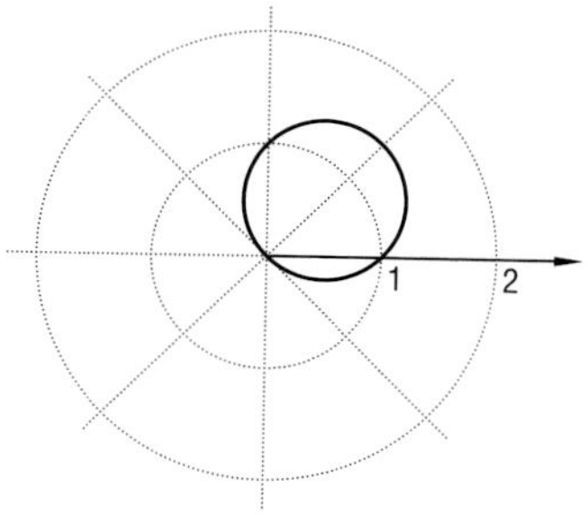

$r = \cos\theta + \sin\theta \Rightarrow r = \frac{x}{r} + \frac{y}{r}$
$\Rightarrow r = \frac{x + y}{r} \Rightarrow r^2 = x + y$
$\Rightarrow x^2 + y^2 = x + y$
$\Rightarrow x^2 - x + y^2 - y = 0$
$\Rightarrow x^2 - x + \frac{1}{4} + y^2 - y + \frac{1}{4} = \frac{1}{2}$
$\Rightarrow \left(x - \frac{1}{2}\right)^2 + \left(y - \frac{1}{2}\right)^2 = \left(\frac{1}{\sqrt{2}}\right)^2$
Hence, the graph is a circle.

11.a. The second graph is a phase shift copy of the first one by $\frac{\pi}{4}$ to the right.

c.

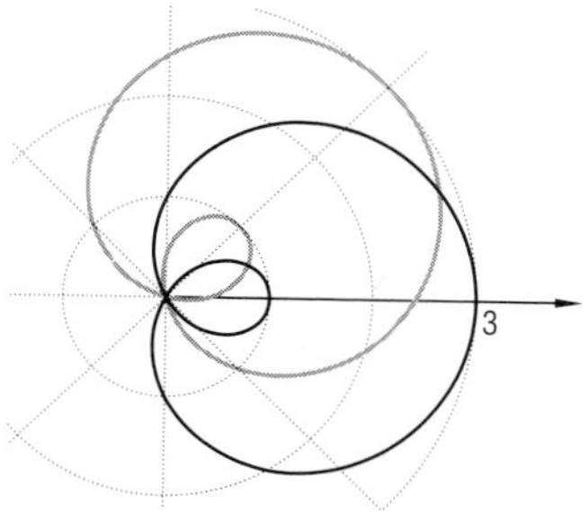

12.a.

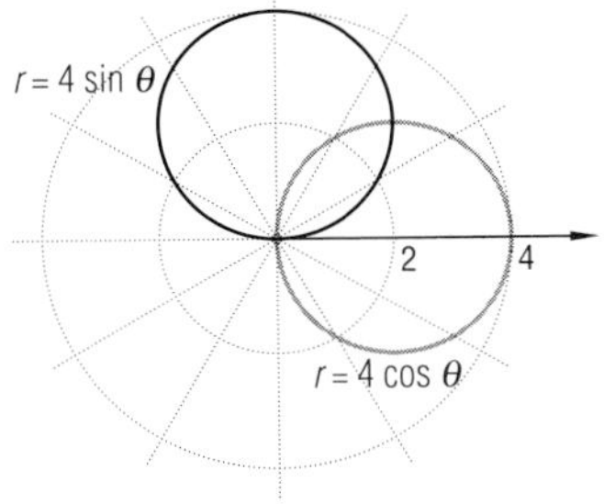

13.a. $\left[\frac{1}{2}, \frac{4\pi}{3}\right]$

c. $u' = \frac{1}{2} + \frac{\sqrt{3}}{2}i$,
$v' = -\frac{3}{2} - \frac{3\sqrt{3}}{2}i$,
$w' = -\frac{3}{2} + \sqrt{3} + \left(-1 - \frac{3\sqrt{3}}{2}\right)i$,
$z' = \frac{1}{2} + \sqrt{3} + \left(\frac{\sqrt{3}}{2} - 1\right)i$

d.

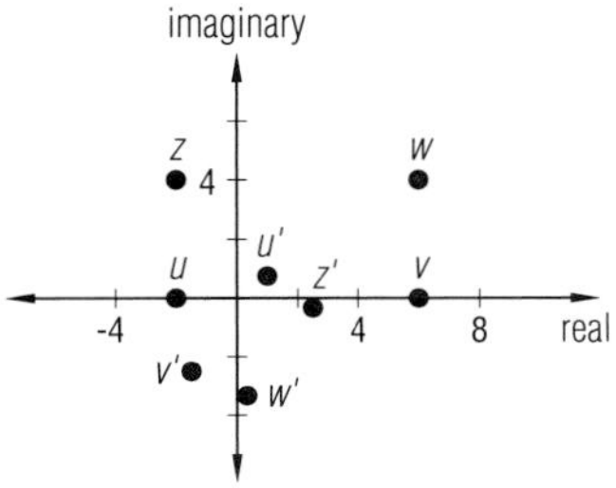

LESSON 8-6 (pages 498–503)

9.

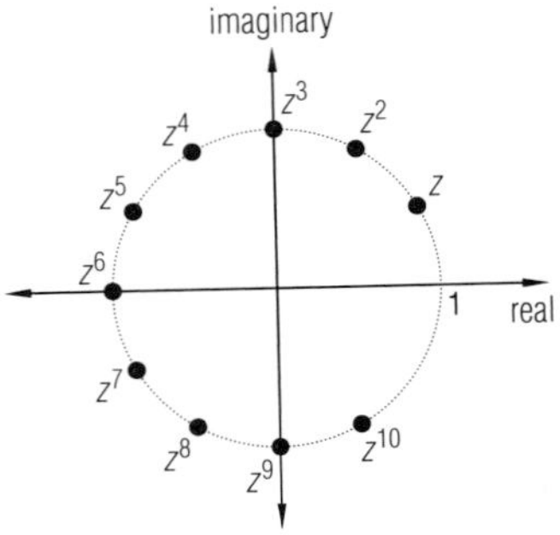

10.a. $-128 - 128\sqrt{3}\,i$

b. $\left[4, \frac{\pi}{3}\right]^4 = \left[256, \frac{4\pi}{3}\right] = -128 - 128\sqrt{3}\,i$

12.a. $z^7 = \left[2^7, \frac{14\pi}{3}\right]$; $z^{13} = \left[2^{13}, \frac{26\pi}{3}\right]$; $z^{19} = \left[2^{19}, \frac{38\pi}{3}\right]$; $z^{25} = \left[2^{25}, \frac{50\pi}{3}\right]$

b. $\frac{14\pi}{3} = \frac{2\pi}{3} + 2(2)\pi$; $\frac{26\pi}{3} = \frac{2\pi}{3} + 2(4)\pi$; $\frac{38\pi}{3} = \frac{2\pi}{3} + 2(6)\pi$; $\frac{50\pi}{3} = \frac{2\pi}{3} + 2(8)\pi$

13. Sample: Using DeMoivre's Theorem, for $z = [r, \theta]$ with $r \geq 0$, $z^n = [r^n, n\theta]$, and $|z| = r$. So $|z^n| = |r^n| = r^n$, and $|z|^n = r^n$. Sample: Using mathematical induction, let $S(n)$: $|z^n| = |z|^n$. $S(1)$: $|z^1| = |z| = |z|^1$, so $S(1)$ is true. Assume $S(k)$: $|z^k| = |z|^k$. Then $|z^{k+1}| = |z^k \cdot z| = |z^k|\,|z| = |z|^k|z|^1 = |z|^{k+1}$. So $S(k) \Rightarrow S(k+1)$, and $S(n)$ is true for all integers n.

15.a.

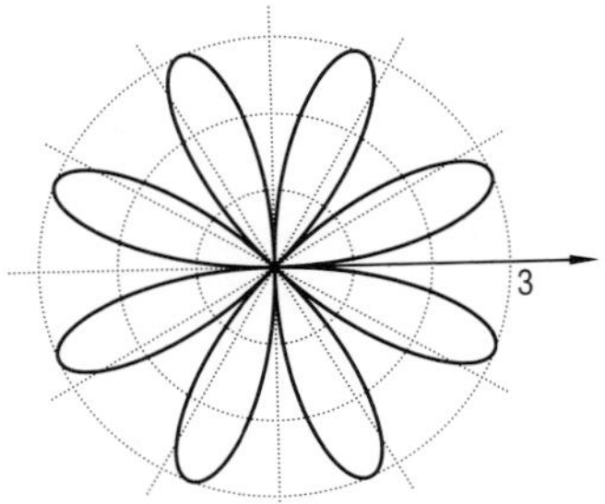

16.a.

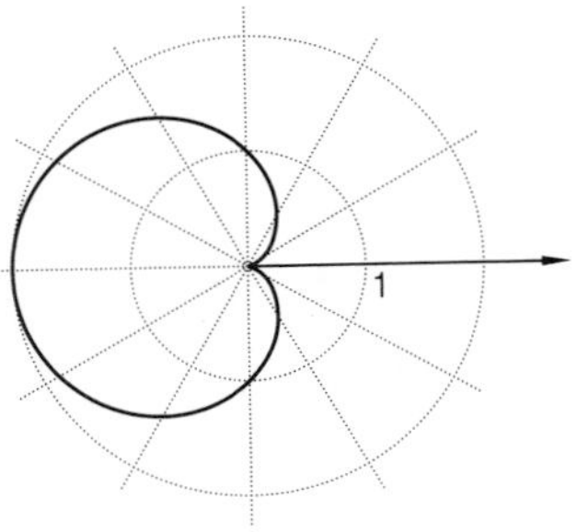

18.

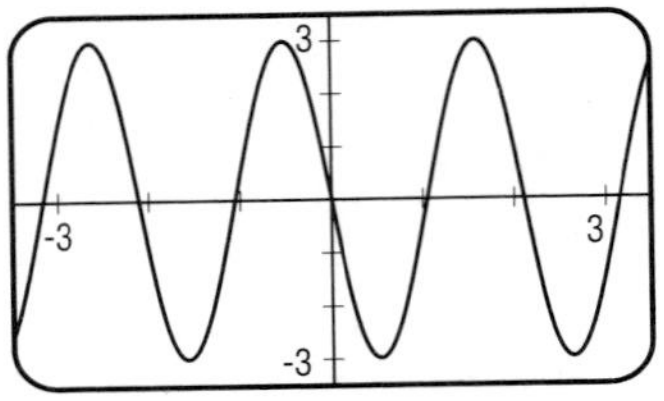

LESSON 8-7 (pages 504–509)

5.b.

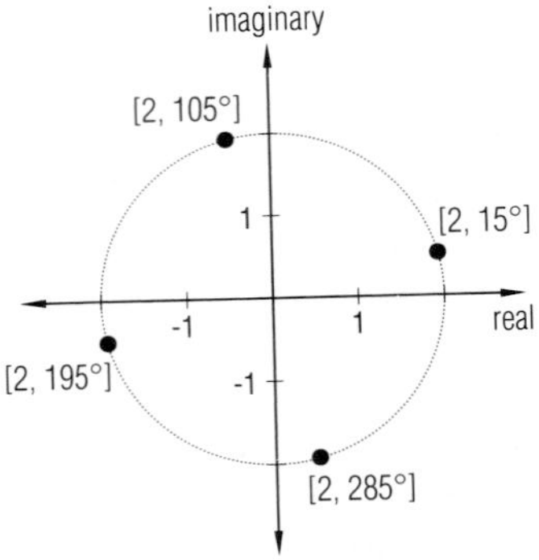

11.a.

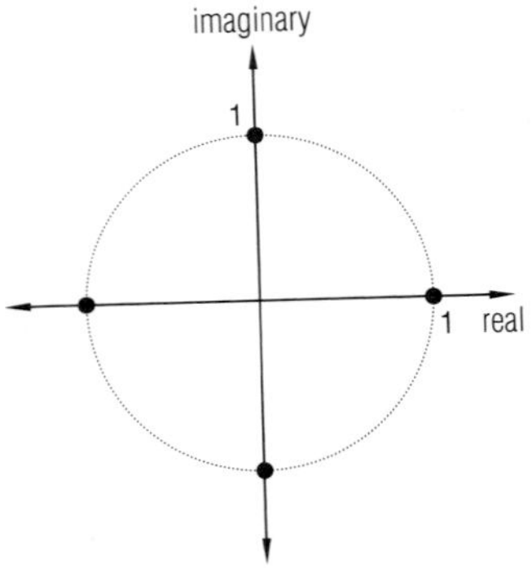

b.

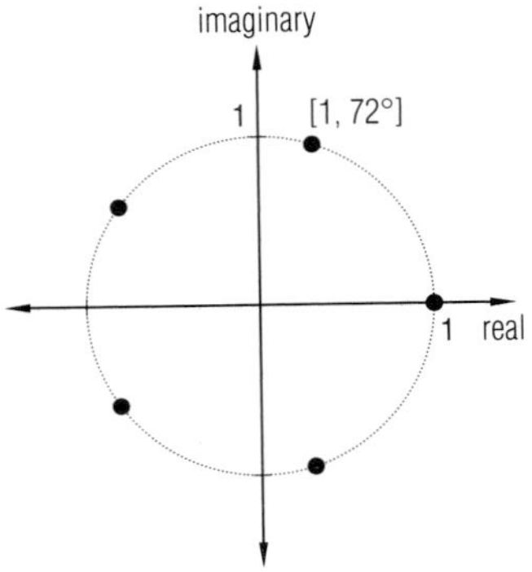

12.a. $2, -1 + \sqrt{3}i, -1 - \sqrt{3}i$

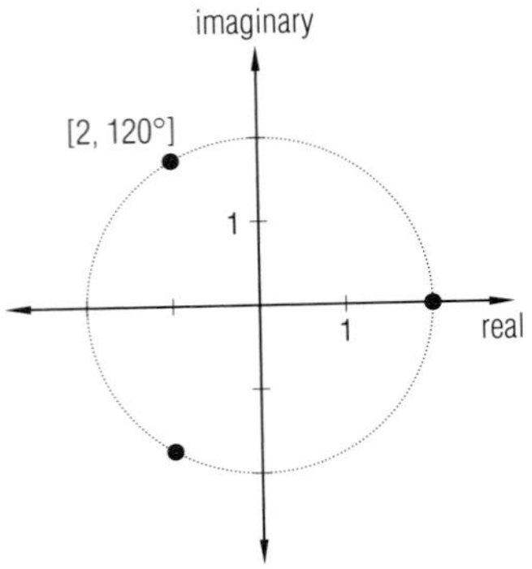

b. $-2, 1 + \sqrt{3}i, 1 - \sqrt{3}i$

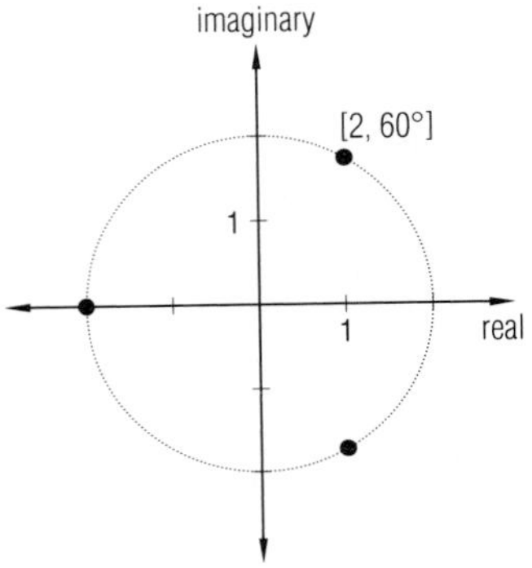

13.a. $z = \pm 3, \pm 3i$

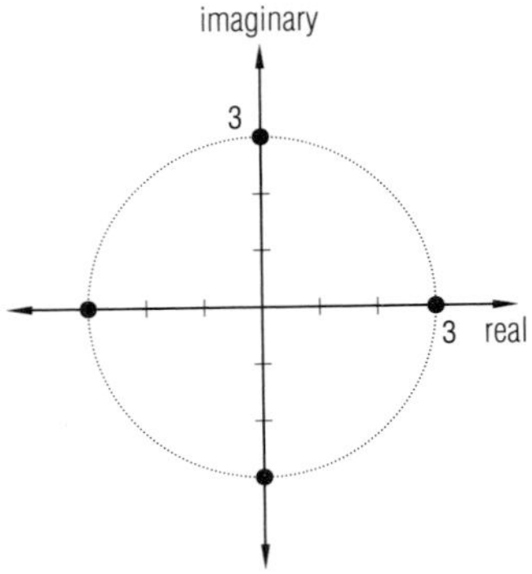

b. $\frac{3\sqrt{2}}{2} \pm \frac{3\sqrt{2}}{2}i$, $-\frac{3\sqrt{2}}{2} \pm \frac{3\sqrt{2}}{2}i$

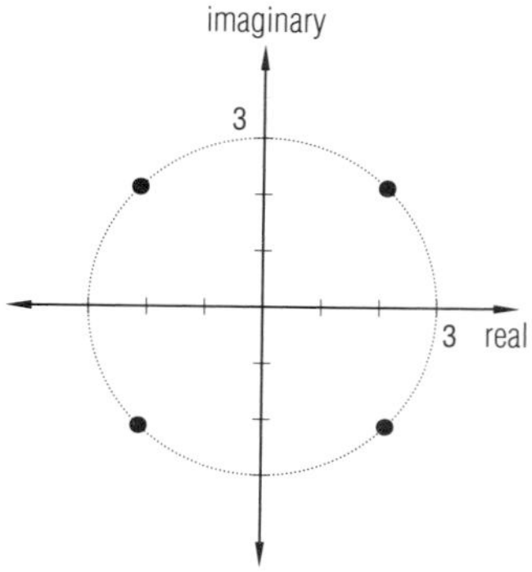

16.a. [.7, 80°], [$(.7)^2$, 160°], [$(.7)^3$, 240°], [$(.7)^4$, 320°], [$(.7)^5$, 400°], [$(.7)^6$, 480°], [$(.7)^7$, 560°], [$(.7)^8$, 640°]

b.

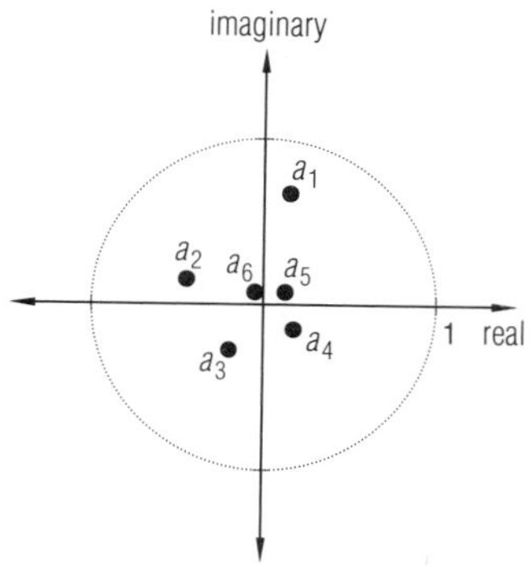

17.

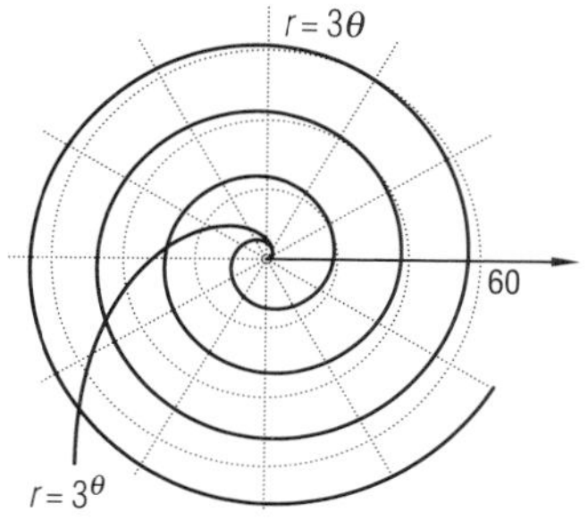

The graph of $r = 3\theta$ is a spiral of Archimedes, and that of $r = 3^\theta$ is a logarithmic spiral.

18. Let $z = [r, \theta]$. Then $z^n = [r^n, n\theta]$ and $(\overline{z^n}) = [r^n, -n\theta]$. Also $\overline{z} = [r, -\theta]$, so $(\overline{z})^n = [r^n, n(-\theta)] = [r^n, -n\theta]$. Thus, $(\overline{z^n}) = (\overline{z})^n$.

22. sample:

```
5   REM ENTER A COMPLEX NUMBER,
    A + BI, AND DESIRED ROOT, N
10  PRINT "WHAT IS THE REAL
    COMPONENT, A, OF THE COMPLEX
    NUMBER";
20  INPUT A
30  PRINT "WHAT IS THE IMAGINARY
    COMPONENT, B";
40  INPUT B
50  PRINT "WHICH ROOT DO YOU
    WANT";
60  INPUT N
70  LET PI = 3.14159265359
80  IF A=0 and B=0 THEN PRINT "0 IS
    THE ONLY ROOT.":GOTO 190
85  REM CALCULATE THE ARGUMENT,
    D, IN RADIANS OF A + BI
90  IF A=0 AND B<0 THEN LET D = -PI/2
100 IF A=0 AND B>0 THEN LET D = PI/2
110 IF A>0 THEN LET D = ATN(B/A)
120 IF A<0 THEN LET D = ATN(B/A) + PI
125 REM CALCULATE THE ABSOLUTE
    VALUE OF A + BI
130 LET L = SQR(A*A + B*B)
135 REM OUTPUT THE N NTH ROOTS
    OF A + BI
140 PRINT "THE ABSOLUTE VALUE OF
    EACH ROOT IS"; L^(1/N)
150 PRINT "THE ARGUMENTS ARE"
160 FOR I = 0 TO (N-1)
170   PRINT (D/N) + I*(2*PI/N)
180 NEXT I
190 END
```

LESSON 8-8 (pages 510–514)

13.b.

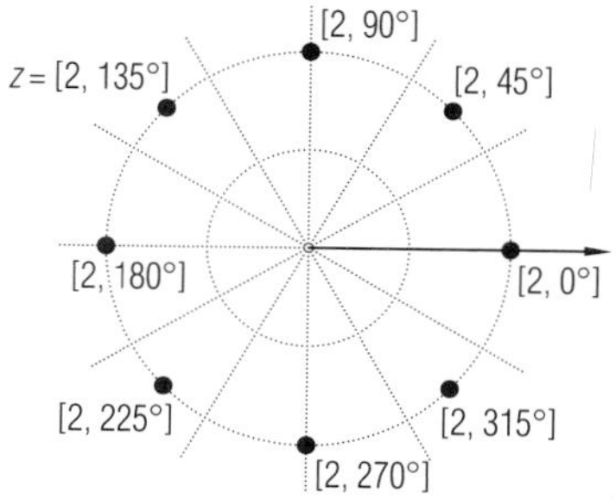

LESSON 8-9 (pages 515–521)

8. sample:

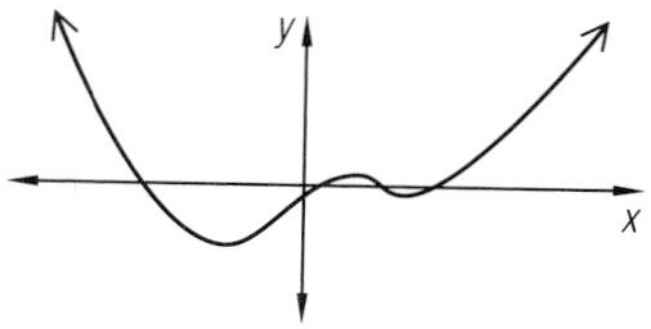

9. sample:

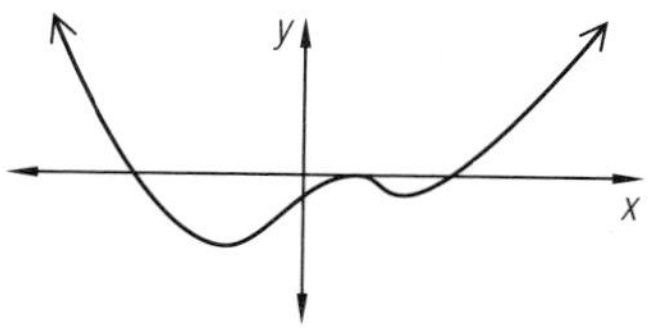

10.a. sample:

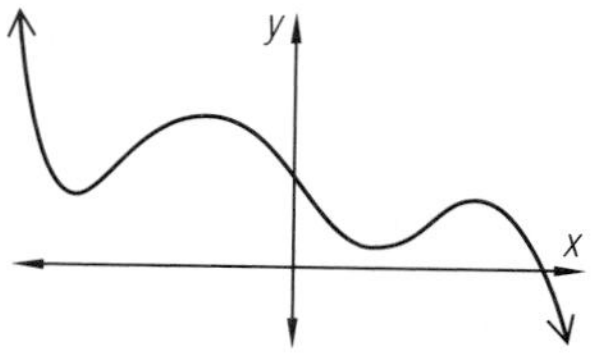

There are four nonreal zeros: either two conjugate pairs (each zero of multiplicity 1), or one conjugate pair (each zero of multiplicity 2).

b. sample:

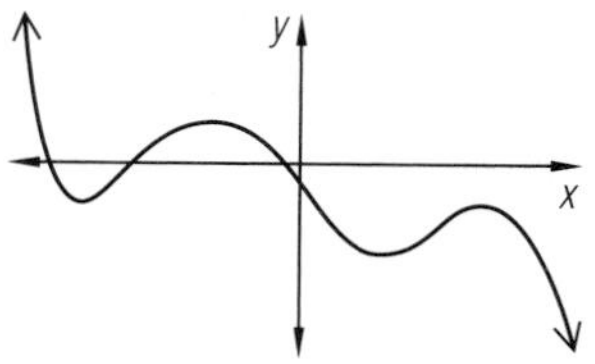

There is one pair of nonreal conjugate zeros, each of multiplicity 1.

c. sample:

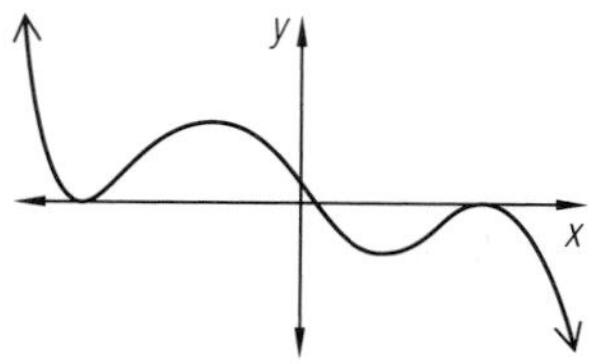

There are no nonreal zeros.

d. sample:

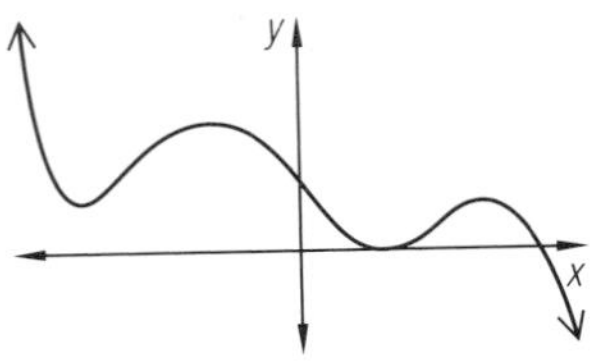

There is one pair of nonreal conjugate zeros, each of multiplicity 1.

12.b. If $p(x)$ does not have real coefficients, then $1 + 3i$ is not necessarily a zero of $p(x)$, and $p(x)$ may have degree 2.

17.

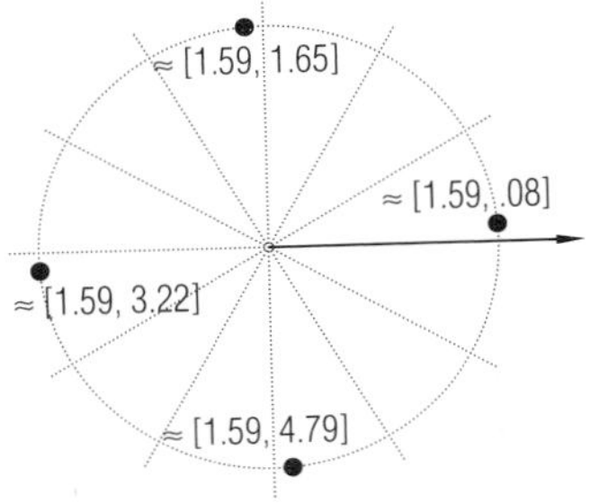

18.a., b.

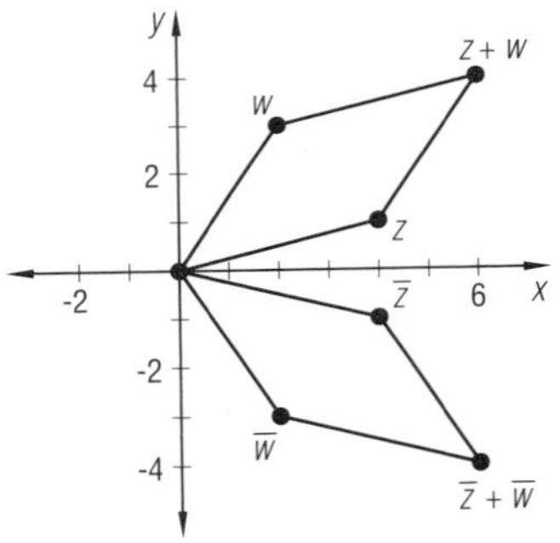

21.a. For $c = -5$, the zeros are $\pm\sqrt{5}$, $\pm i$; for $c = 0$, the zeros are 0, ± 2; for $c = 3$, the zeros are $\pm\sqrt{3}$, ± 1. As c slides from -5 to 0, its two real zeros move closer to the origin, and its two imaginary zeros converge at the origin. As c then slides from 0 to 3, the polynomial has 4 real zeros.

b. For $c = 4$, the zeros are $\pm\sqrt{2}$; for $c = \frac{25}{4}$, the zeros are $\frac{3 \pm i}{2}$.
As c slides from 3 to 4, its positive zeros converge to $\sqrt{2}$, and its negative zeros converge to $-\sqrt{2}$. As c slides from 4 to $\frac{25}{4}$, its four complex zeros converge to two complex zeros, $\frac{3 \pm i}{2}$.

22.

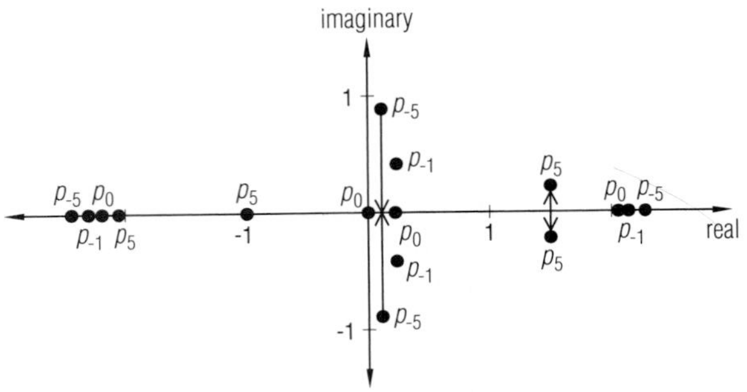

For $c = -5$, the zeros are approximately -2.50, 2.34, and $.08 \pm .93i$. As c increases, the real zeros move slowly toward the origin and the nonreal zeros move toward each other until they merge into approximately .11 (a zero of multiplicity 2) when $c \approx -.05$; at that time, the other zeros are -2.33 and 2.13. As c increases, the zero at .11 splits into two real zeros which move apart along the real axis, while the other real zeros continue moving toward the origin. When $c \approx 4.69$, the two largest zeros merge into 1.53, then split into complex conjugates.

c	approximate zeros
-5	-2.50, $.08 - .93i$, $.08 + .93i$, 2.34
-1	-2.36, $.10 - .39i$, $.10 + .39i$, 2.17
-.05	-2.33, .11, .11, 2.13
0	-2.32, 0.00, .20, 2.12
4.5	-2.11, -.93, 1.37, 1.67
4.69	-2.10, -.96, 1.53, 1.53
5	-2.08, -1.00, $1.54 - .19i$, $1.54 + .19i$

LESSON 8-10 (pages 522–528)

14.b. $(zw)^5 = [6, 110°]^5 = [6^5, 5 \cdot 110°] = [7776, 550°]$. $z^5 \cdot w^5 = [243, 200°] \cdot [32, 350°] = [(243)(32), 200° + 350°] = [7776, 550°]$. So $(zw)^5 = z^5 \cdot w^5$.

15.a.

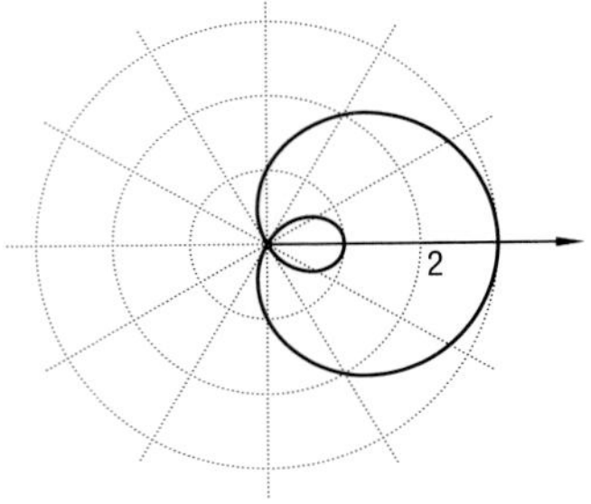

18. If m is odd, then $m = 2k + 1$ for some integer k.
$m^2 + m - 3 = (2k + 1)^2 + (2k + 1) - 3$
$= (4k^2 + 4k + 1) + (2k + 1) - 3$
$= (4k^2 + 6k - 2) + 1$
$= 2(2k^2 + 3k - 1) + 1$
Since $(2k^2 + 3k - 1)$ is an integer, $m^2 + m - 3$ is an odd integer by definition.
$\therefore$ If m is any odd integer, then $m^2 + m - 3$ is an odd integer.

19. $f(z)$ is constructed by squaring the absolute value of the complex number z and doubling its argument to obtain z^2. The point is then translated by adding c to give $f(z)$.

CHAPTER 8 REVIEW (pages 534–536)

54.a.–e.

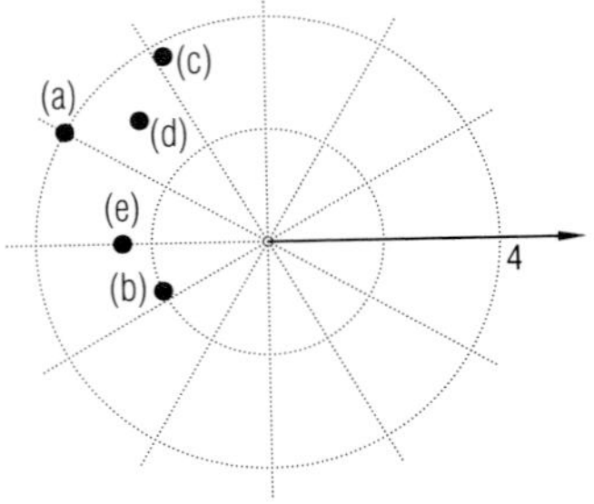

56.b. $D = B + C = 7 - 5i$

slope of $\overline{CA}$ is $\frac{1 - 0}{5 - 0} = \frac{1}{5}$;

slope of $\overline{CD} = \frac{-5 - 1}{7 - 5} = -3$;

slope of $\overline{DB} = \frac{-5 - (-6)}{7 - 2} = \frac{1}{5}$;

slope of $\overline{AB} = \frac{-6 - 0}{2 - 0} = -3$

Since the slopes of opposite sides are equal, *CABD* is a parallelogram.

57.

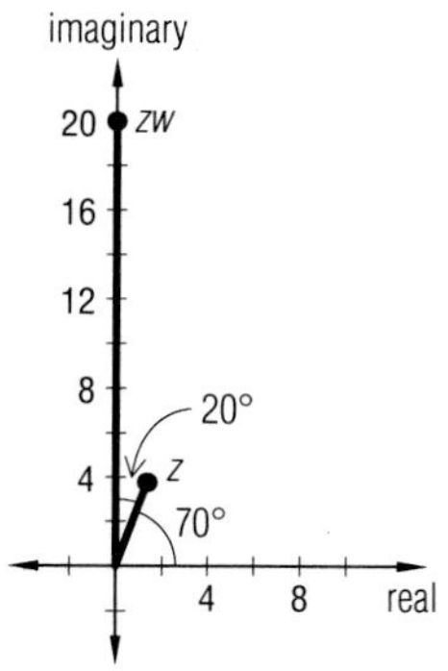

58.a., b.

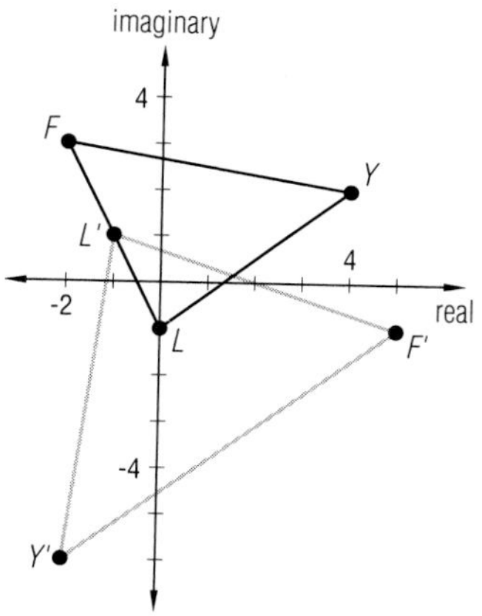

c. $FY = \sqrt{37}$, $LY = 5$, $FL = \sqrt{20}$, $F'Y' = \sqrt{74}$, $L'Y' = \sqrt{50}$, $F'L' = \sqrt{40}$. So $\frac{F'Y'}{FY} = \frac{\sqrt{74}}{\sqrt{37}} = \sqrt{2}$, $\frac{L'Y'}{LY} = \frac{\sqrt{50}}{5} = \sqrt{2}$, and $\frac{F'L'}{FL} = \frac{\sqrt{40}}{\sqrt{20}} = \sqrt{2}$. The ratio of similitude is $\sqrt{2}$.

In polar form, $z = \left[\sqrt{2}, \frac{5\pi}{4}\right]$, and so all distances are multiplied by $\sqrt{2}$.

d. The argument of F' is $\approx 348.7°$, which is 225° greater than the argument of F, which is $\approx 123.7°$.
$(-2 + 3i) \cdot (-1 - i) = [\sqrt{13}, 123.7°] \cdot [\sqrt{2}, 225°] = [\sqrt{26}, 348.7°]$ or applying a size change of $\sqrt{2}$ and a rotation of 225° to F to obtain F'.

59. $z + w = 3 - i$; vertices: $A(0, 0)$, $B(2, -5)$, $C(3, -1)$, $D(1, 4)$;

slope of $\overline{AB}$ is $\frac{-5 - 0}{2 - 0} = -\frac{5}{2}$,

slope of $\overline{AD} = \frac{4 - 0}{1 - 0} = 4$,

slope of $\overline{DC} = \frac{4 - (-1)}{1 - 3} = -\frac{5}{2}$,

slope of $\overline{BC} = \frac{-5 - (-1)}{2 - 3} = 4$

Since the slopes of opposite sides are equal, *ABCD* is a parallelogram.

60.

θ	0°	30°	45°	60°	90°	120°	135°	180°	240°	270°
r	6	≈5.2	≈4.2	3	0	-3	≈ -4.2	-6	-3	0

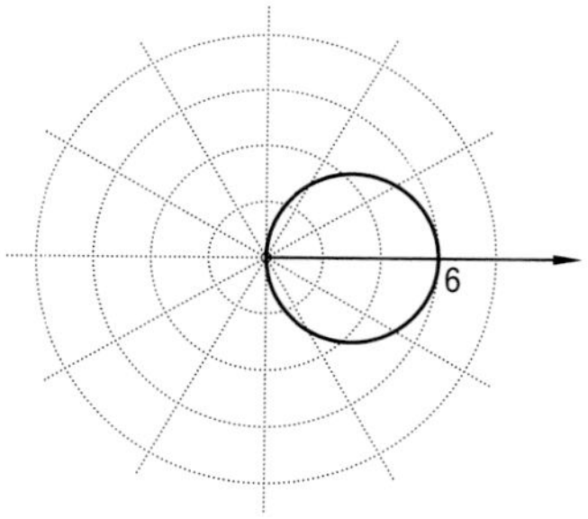

61.

θ	2	$\frac{\pi}{6}$	$\frac{\pi}{2}$	π	$\frac{3\pi}{2}$	2π
r	1	≈ 3.8	≈ 1.3	≈ .64	≈ .42	≈ .32

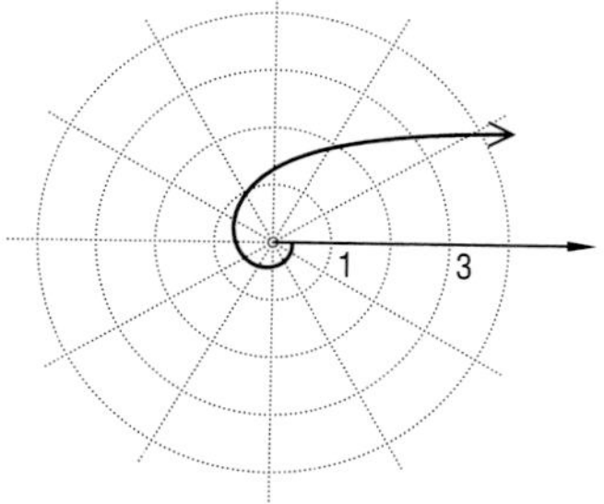

62.

θ	0°	30°	60°	120°	180°	330°
r	5	≈ 5.8	10	-10	-5	≈ 5.8

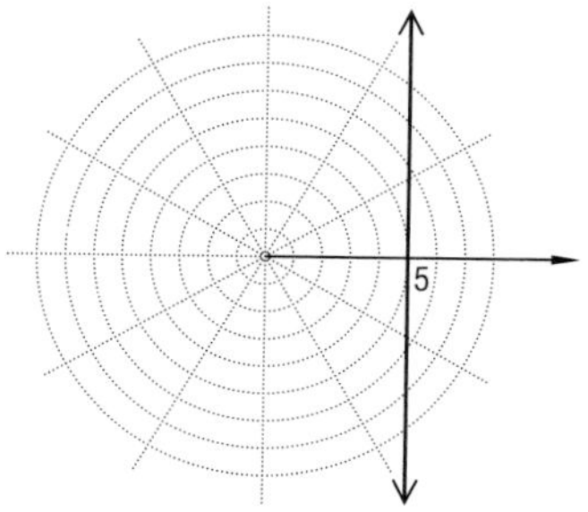

63.

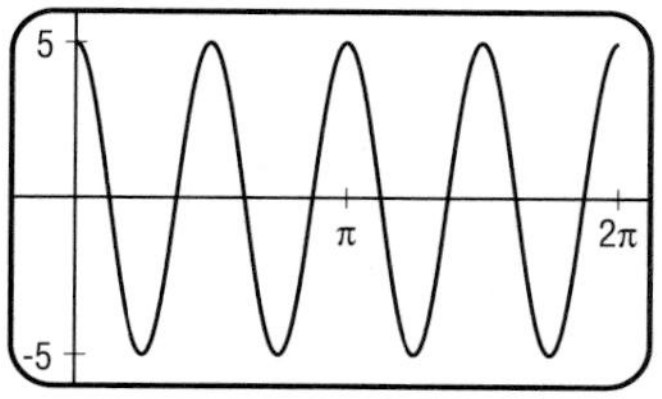

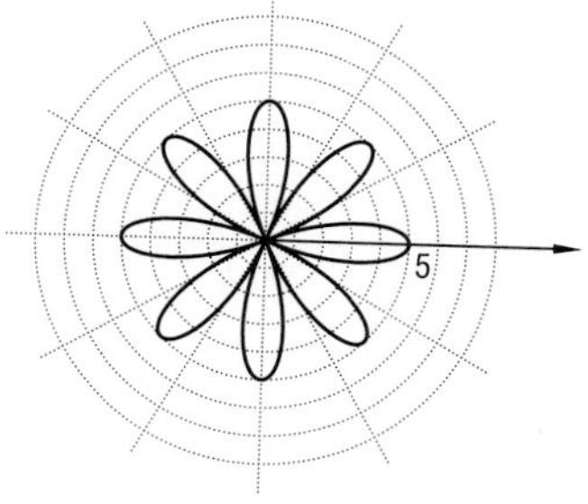

8-leaved rose

64.

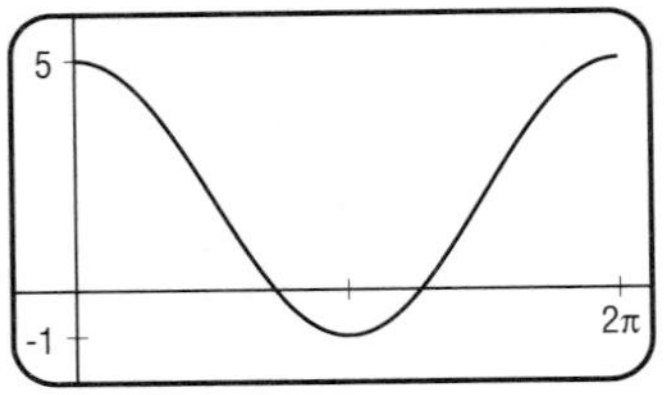

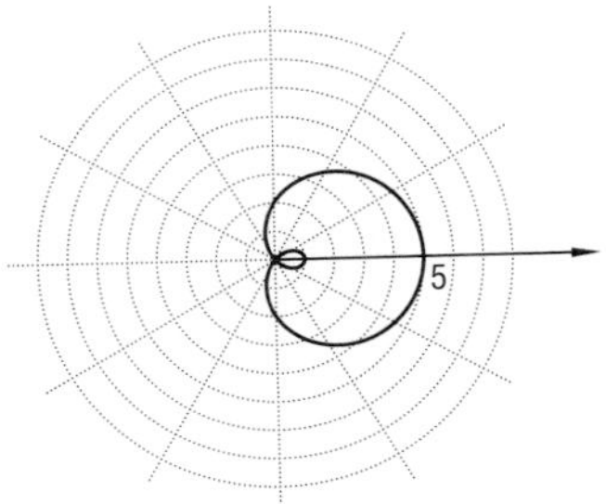

limaçon

65.

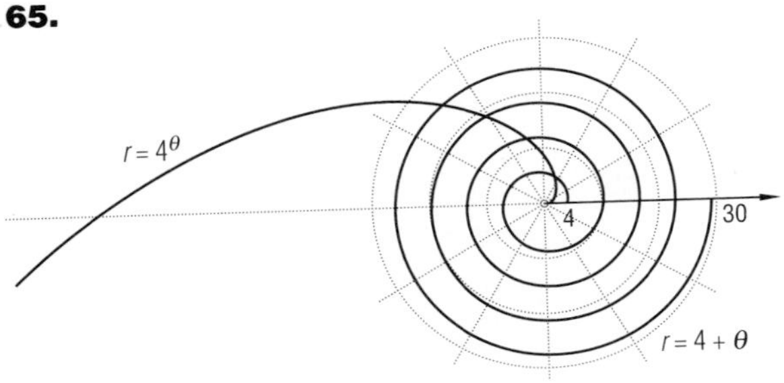

$r = 4^{\theta}$ is a logarithmic spiral, and
$r = 4 + \theta$ is a spiral of Archimedes.

66.a.

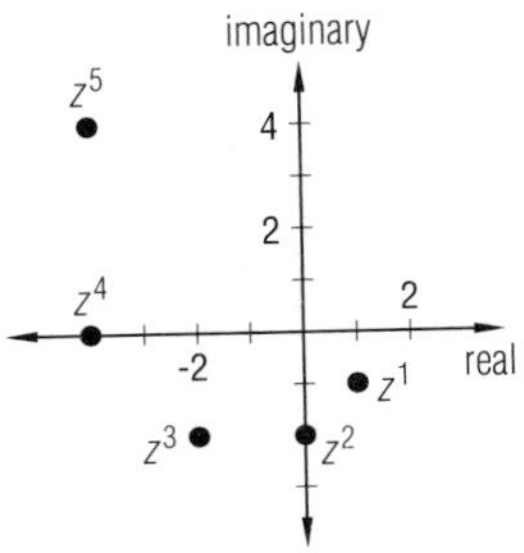

67.a.

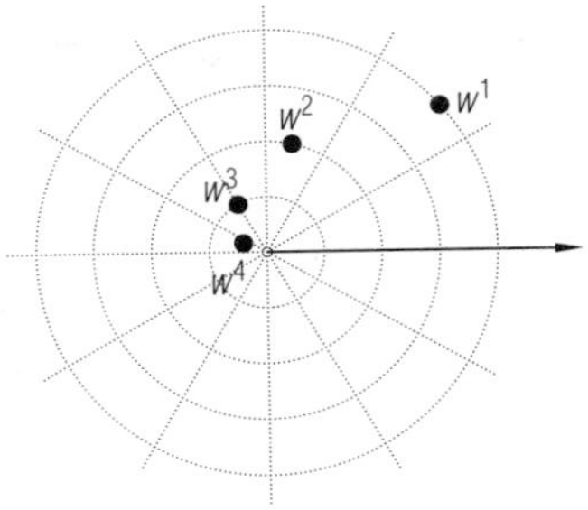

68.b.

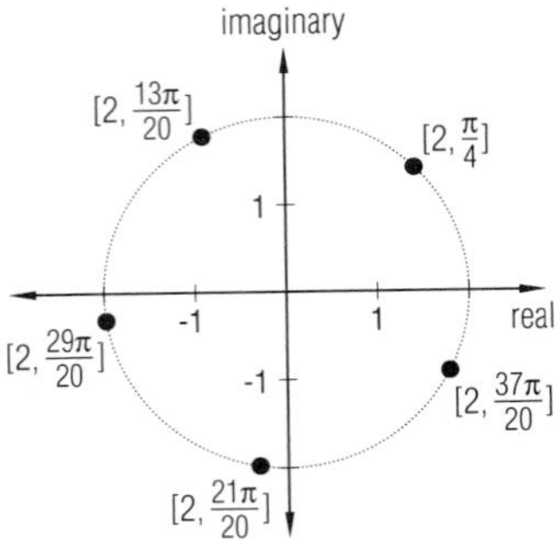

69.

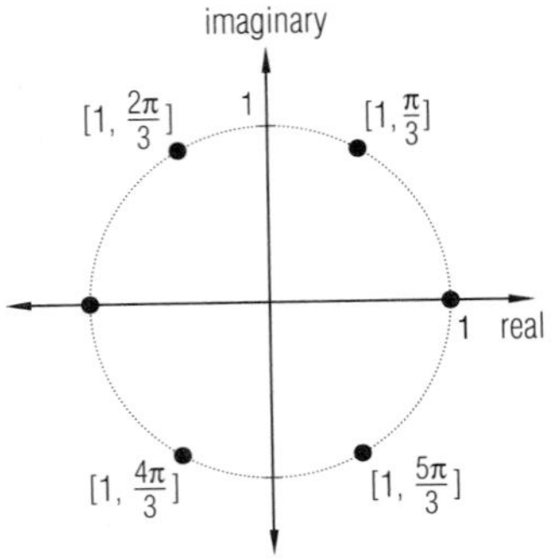

LESSON 9-2 (pages 546–553)

9. $V'(2)$

$= \lim_{\Delta x \to 0} \frac{V(2 + \Delta x) - V(2)}{\Delta x}$

$= \lim_{\Delta x \to 0} \frac{\frac{4}{3}\pi(2 + \Delta x)^3 - \frac{4}{3}\pi(2)^3}{\Delta x}$

$= \lim_{\Delta x \to 0} \frac{\frac{4}{3}\pi(8 + 12\Delta x + 6\Delta x^2 + \Delta x^3) - \frac{4}{3}\pi(8)}{\Delta x}$

$= \lim_{\Delta x \to 0} \frac{\frac{4}{3}\pi(12\Delta x + 6\Delta x^2 + \Delta x^3)}{\Delta x}$

$= \lim_{\Delta x \to 0} \frac{4}{3}\pi(12 + 6\Delta x + \Delta x^2)$

$= 16\pi$

23.a. inches/second
b.i. when the bar is being slowly lifted
ii. during rapid lifting of the bar
iii. when the bar is being lowered
iv. when the bar is being held steady, or while it is on the floor

24.a. feet/second
b.i. when the jogger is moving slowly
ii. when the jogger is running at top speed
iii. never
iv. when the jogger stops or runs in place

LESSON 9-3 (pages 554–561)

5.a. $g'(x)$

$= \lim_{\Delta x \to 0} \frac{5(x + \Delta x)^2 + 2(x + \Delta x) - (5x^2 + 2x)}{\Delta x}$

$= \lim_{\Delta x \to 0} \frac{10x\Delta x + \Delta x^2 + 2\Delta x}{\Delta x}$

$= \lim_{\Delta x \to 0} (10x + \Delta x + 2)$

$= 10x + 2$

b. $g(x) = 5x^2 + 2x + 0$, so $g'(x) = 10x + 2$.

12. $V'(r) = \lim_{\Delta r \to 0} \frac{V(r + \Delta r) - V(r)}{\Delta r}$

$= \lim_{\Delta r \to 0} \frac{\frac{4}{3}\pi(r + \Delta r)^3 - \frac{4}{3}\pi r^3}{\Delta r}$

$= \lim_{\Delta r \to 0} \frac{\frac{4}{3}\pi(r^3 + 3r^2\Delta r + 3r\Delta r^2 + \Delta r^3) - \frac{4}{3}\pi r^3}{\Delta r}$

$= \lim_{\Delta r \to 0} \frac{\frac{4}{3}\pi(3r^2\Delta r + 3r\Delta r^2 + \Delta r^3)}{\Delta r}$

$= \lim_{\Delta r \to 0} \frac{4}{3}\pi(3r^2 + 3r\Delta r + \Delta r^2)$

$= 4\pi r^2$

LESSON 9-4 (pages 562–566)

10. graph of f

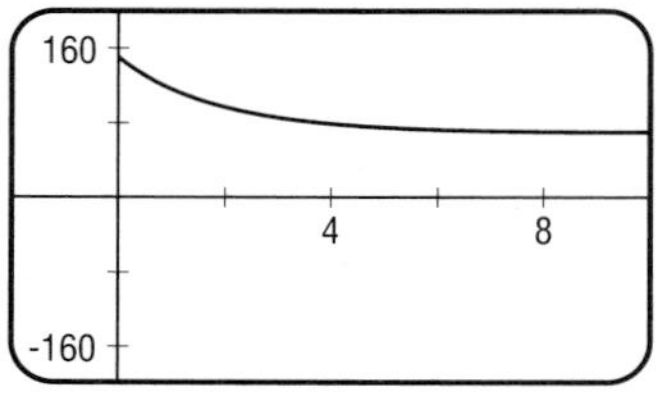

graph of f'

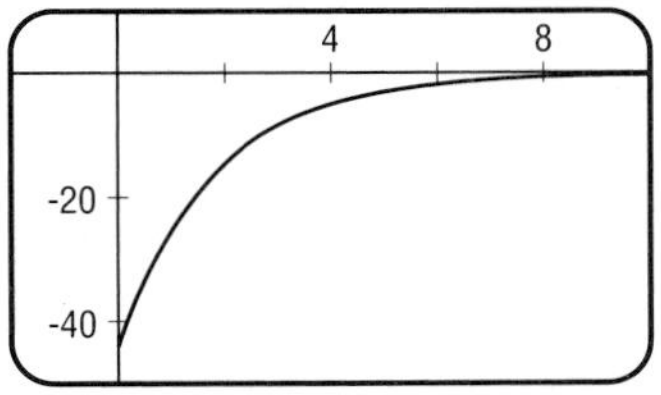

graph of f''

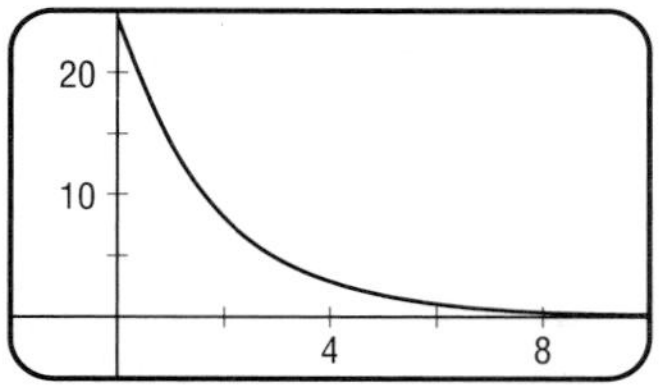

12.

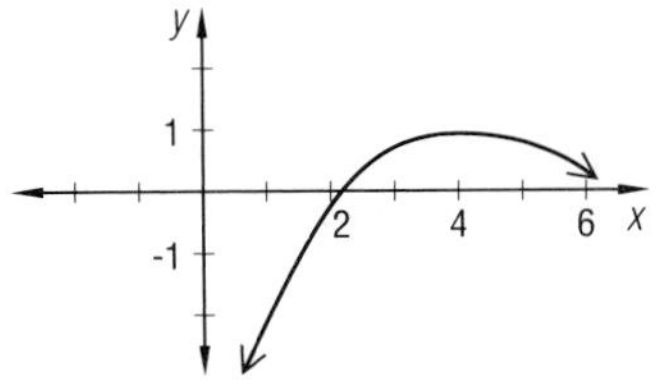

15.

$$\cos 2x = \cos^2 x - \sin^2 x$$
$$\cos 2x = (1 - \sin^2 x) - \sin^2 x \quad \text{Pyth. Id.}$$
$$\cos 2x = 1 - 2\sin^2 x \quad \text{Addition}$$
$$2\sin^2 x = 1 - \cos 2x$$
$$\sin^2 x = \frac{1 - \cos 2x}{2}$$

LESSON 9-5 (pages 567–571)

3.a.

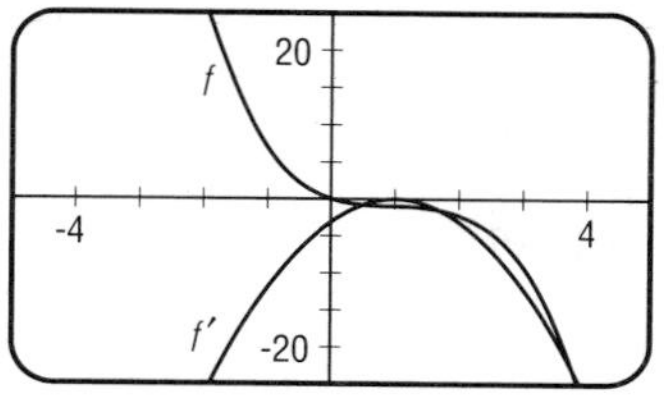

b. all values; $f(x)$ is everywhere decreasing.
d. No; as the graph shows, $f(x)$ merely "flattens out" at $x = 1$. Since it is everywhere decreasing, $f(x)$ has no relative maxima or minima.

8.d. $\left(-1, \frac{20}{3}\right)$, $(3, -4)$
e.

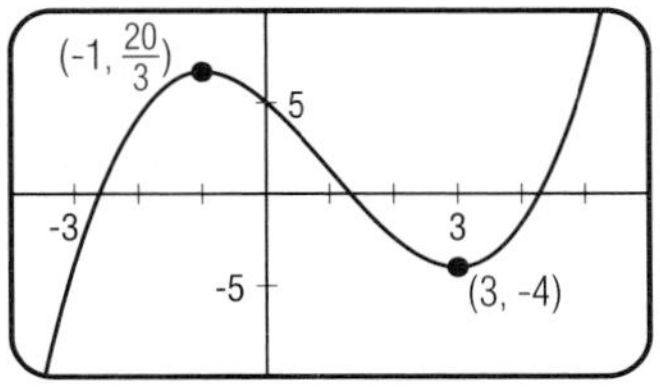

9. No; for example, consider the function $y = x^2$ graphed below.

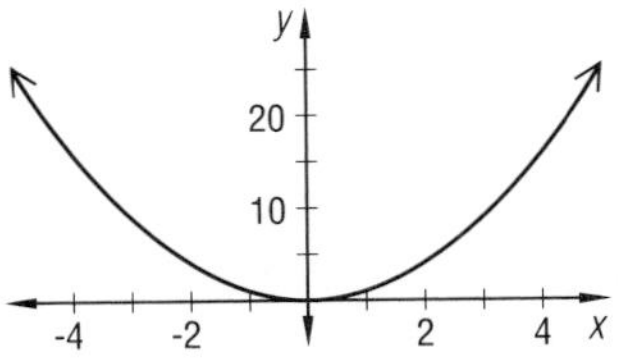

As x goes from -4 to -3 to -2 to -1 to 0, $f'(x)$ goes from -8 to -6 to -4 to -2 to 0. Those slopes are increasing, but the function is decreasing.

15.a.

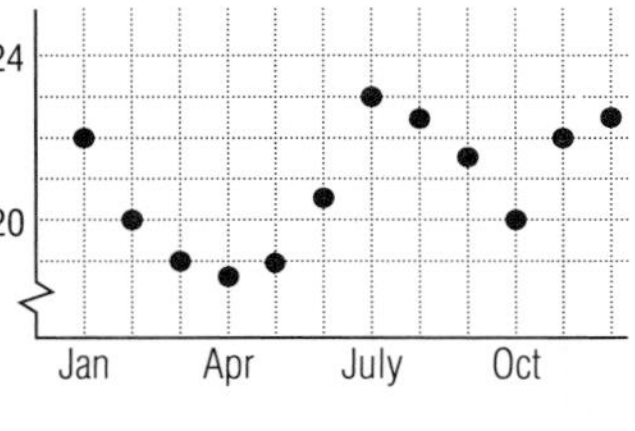

17.b.

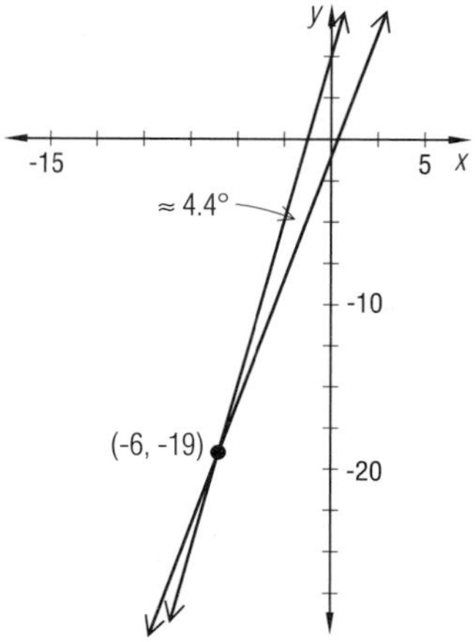

LESSON 10-1 (pages 586–590)

15.

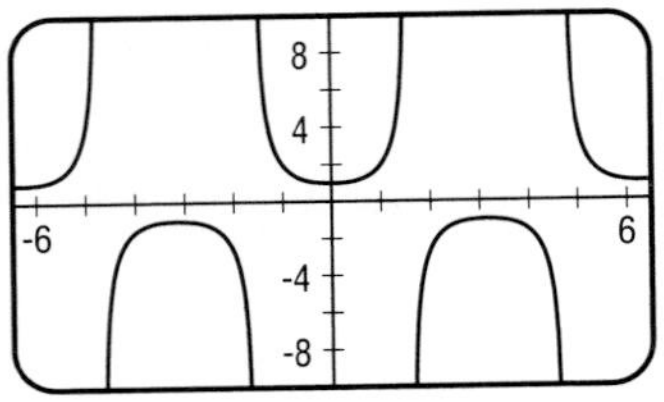

It is an identity.

Left side $= (\tan\theta)(\sin\theta + \cot\theta\cos\theta)$

$= \dfrac{\sin\theta}{\cos\theta}\left(\sin\theta + \dfrac{\cos\theta\cos\theta}{\sin\theta}\right)$ Definition of tangent and cotangent

$= \dfrac{\sin\theta}{\cos\theta}\left(\dfrac{\sin^2\theta + \cos^2\theta}{\sin\theta}\right)$ Common denominator

$= \dfrac{\sin\theta}{\cos\theta}\left(\dfrac{1}{\sin\theta}\right)$ Pythagorean Identity

$= \dfrac{1}{\cos\theta}$ Multiplication

$= \sec\theta$ Definition of secant

$=$ Right side

14. sample:

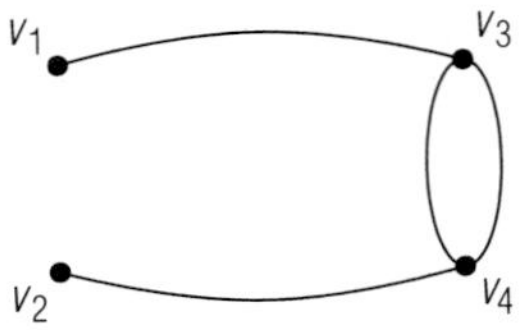

15.

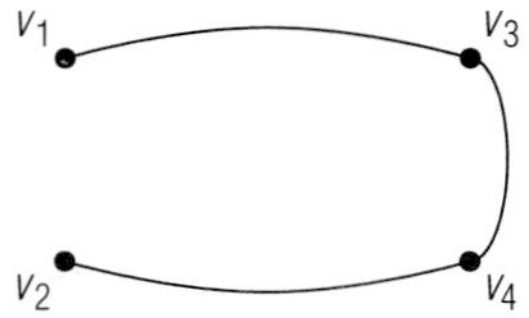

16. Impossible; one of the degree 3 vertices goes to each of the other vertices. But the other degree 3 vertex cannot connect to itself (the graph is simple), to the first degree 3 vertex (no parallel edges), or to the other two vertices (they already have one edge).

19.a. $\dfrac{(n+1)n}{2}$

b. It is the same problem but with one more person.

21. The set of even vertices contributes an even number to the total degrees of the graph. Since that total degree is even, the set of odd vertices must also contribute an even number to the total degree. An odd number of odd vertices would contribute an odd number, so the number of odd vertices must be even.

22. $\begin{bmatrix} 0 & 1 & 2 \\ 1 & 0 & 9 \\ 2 & 7 & 0 \end{bmatrix}$

23. 10 hours

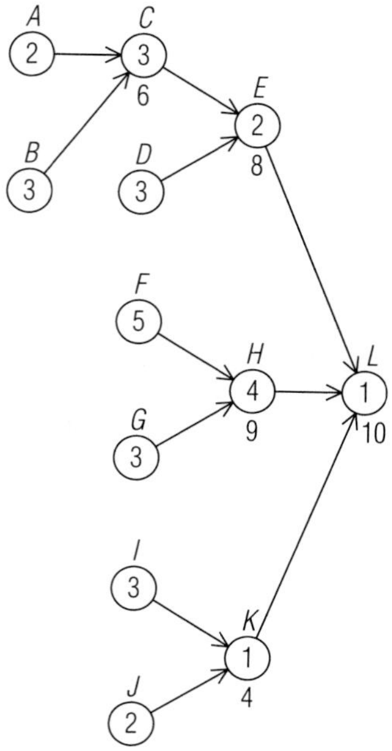

27.a. *If G is a graph with m edges and n vertices and* $m > \dfrac{n(n-1)}{2}$, *then G is not a simple graph.*

b. Proof of statement: Given a simple graph G with n vertices and m edges, let $k =$ maximum number of edges possible (that is, $m \le k$). We know from the lesson that $k = \dfrac{n(n-1)}{2}$, so $m \le \dfrac{n(n-1)}{2}$.

LESSON 11-1 (pages 640–647)

6.a.

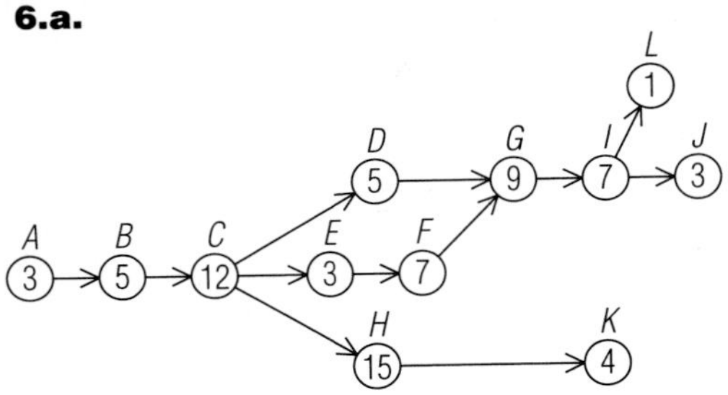

LESSON 11-3 (pages 656–661)

9. Assume that each of the nine people could shake hands with exactly three others. Represent each person as the vertex of a graph, and draw an edge joining each pair of people who shake hands. To say that a person shakes hands with three other people is equivalent to saying that the degree of the vertex representing that person is 3. The graph would then have an odd number of vertices of odd degree. This contradicts Corollary 2 of the Total Degree of a Graph Theorem. Thus, the given situation is impossible.

LESSON 11-5 (pages 670–674)

9. $A^3 =$

$$\begin{bmatrix} 1 & 2 & 1 \\ 2 & 0 & 1 \\ 1 & 1 & 0 \end{bmatrix}\begin{bmatrix} 1 & 2 & 1 \\ 2 & 0 & 1 \\ 1 & 1 & 0 \end{bmatrix}\begin{bmatrix} 1 & 2 & 1 \\ 2 & 0 & 1 \\ 1 & 1 & 0 \end{bmatrix} =$$

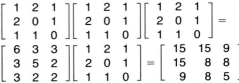

$$\begin{bmatrix} 6 & 3 & 3 \\ 3 & 5 & 2 \\ 3 & 2 & 2 \end{bmatrix}\begin{bmatrix} 1 & 2 & 1 \\ 2 & 0 & 1 \\ 1 & 1 & 0 \end{bmatrix} = \begin{bmatrix} 15 & 15 & 9 \\ 15 & 8 & 8 \\ 9 & 8 & 5 \end{bmatrix}$$

21.

Left network $\equiv \sim (p \text{ and } (q \text{ or } \sim r))$
$\equiv \sim p \text{ or } \sim (q \text{ or } \sim r)$
$\equiv \sim p \text{ or } (\sim q \text{ and } r)$
$\equiv$ Right network

Therefore, the left network is equivalent to the right network.

23.a. $A^2 = \begin{bmatrix} 0 & 0 & 1 \\ 1 & 0 & 0 \\ 0 & 1 & 0 \end{bmatrix}$, $A^3 = \begin{bmatrix} 1 & 0 & 0 \\ 0 & 1 & 0 \\ 0 & 0 & 1 \end{bmatrix}$, $A^4 = \begin{bmatrix} 0 & 1 & 0 \\ 0 & 0 & 1 \\ 1 & 0 & 0 \end{bmatrix}$

$A^4 = A^1$, $A^5 = A^2$, $A^6 = A^3$. The pattern is $A^n = A^{n(\text{mod}3)}$.

b. The paths between vertices are circular.

c.

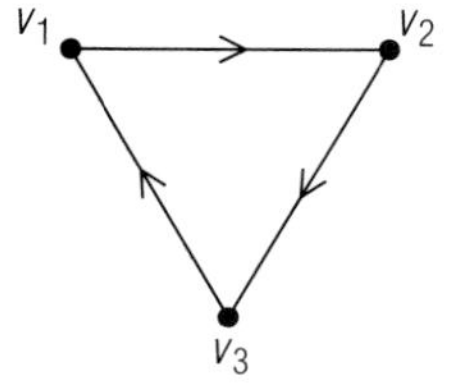

LESSON 11-6 (pages 675–681)

10. Let $A = \begin{bmatrix} a_1 & a_2 \\ a_3 & a_4 \end{bmatrix}$, and $B = \begin{bmatrix} b_1 & b_2 \\ b_3 & b_4 \end{bmatrix}$. If A and B are stochastic, then each row sums to 1 and each entry is nonnegative. $AB = \begin{bmatrix} a_1b_1 + a_2b_3 & a_1b_2 + a_2b_4 \\ a_3b_1 + a_4b_3 & a_3b_2 + a_4b_4 \end{bmatrix}$. Row 1 sums to $a_1b_1 + a_2b_3 + a_1b_2 + a_2b_4 = a_1b_1 + a_1b_2 + a_2b_3 + a_2b_4 = a_1(b_1 + b_2) + a_2(b_3 + b_4) = a_1 + a_2 = 1$, since $b_1 + b_2 = b_3 + b_4 = a_1 + a_2 = 1$. Row 2 sums to $a_3b_1 + a_4b_3 + a_3b_2 + a_4b_4 = a_3b_1 + a_3b_2 + a_4b_3 + a_4b_4 = a_3(b_1 + b_2) + a_4(b_3 + b_4) = a_3 + a_4 = 1$, since $b_1 + b_2 = b_3 + b_4 = a_3 + a_4 = 1$. Each entry of AB is nonnegative since it is the sum of two terms, each of which is the product of two nonnegative numbers. Hence, the product of two 2×2 stochastic matrices is stochastic.

LESSON 11-7 (pages 682–688)

11.a. sample:

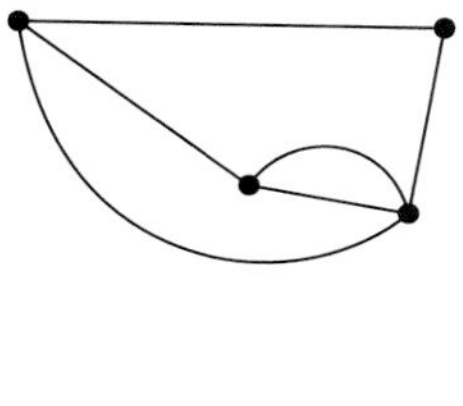

b. Since the graph in part **a** has no crossings and 6 edges, $V - E + F = 2$ holds true. Remove a vertex of degree 1 and its adjacent edge does not change the value of $V - E$. This was done in part **a**, and F did not change, so $V - E + F = 2$ holds for the original graph.

12.b. It is not connected and contains crossings.

14.a. 6 vertices, 9 edges

d. In this graph, a face cannot have 1 edge, since this would mean a line connects a house or utility to itself. A face cannot have 2 edges, since this would mean a house and a utility have two lines connecting them. Finally, a face cannot have 3 edges, since the restrictions prevent two houses or two utilities to be connected to each other. So a face must have at least 4 edges.

e. Since each edge borders exactly two faces, if we sum for every face the number of edges bordering it, we get $2E$. Since there are at *least* 4 edges bordering each face, there must be at *most* $\frac{2E}{4}$ faces. Hence, $F \le \frac{2E}{4}$. Multiplying, $4F \le 2E$, or $2F \le E$.

f. By part **a**, $E = 9$. If there are no crossings, by part **c**, $F = 5$. This contradicts part **e**, since $2(5) \le 9$ does not hold true. Therefore, it is impossible to connect three houses and three utilities without lines crossing.

18. No; for example, let v_1, v_2, v_3 have degree 3, and v_4 have degree 1. Then $\{v_1, v_2\}$, $\{v_1, v_3\}$, and $\{v_1, v_4\}$ are the three edges from v_1. $\{v_2, v_1\}$ and $\{v_2, v_3\}$ are 2 edges from v_2. Now there must be another edge from v_2. But that edge cannot connect v_2 to v_4, since v_4 must remain with degree 1. It cannot connect to v_1 or v_3, since a simple graph cannot contain parallel edges. And it cannot connect to itself, since a simple graph does not have any loops. So, there is no such graph.

20. To prove $S(n)$: $\sum_{i=1}^{n} i(i + 3) = \frac{n(n + 1)(n + 5)}{3}$.

$S(1)$: $\sum_{i=1}^{1} i(i + 3) = 1(4) = 4$, and $\frac{1(2)(6)}{3} = 4$, so $S(1)$ is true. Assume $S(k)$: $\sum_{i=1}^{k} i(i + 3) = \frac{k(k + 1)(k + 5)}{3}$. Then $\sum_{i=1}^{k+1} i(i + 3) = \sum_{i=1}^{k} i(i + 3) + (k + 1)((k + 1) + 3)$

$= \frac{k(k + 1)(k + 5)}{3} + (k + 1)(k + 4)$

$= \frac{k(k + 1)(k + 5)}{3} + \frac{3(k + 1)(k + 4)}{3}$

$= \frac{(k + 1)[k(k + 5) + 3(k + 4)]}{3}$

$= \frac{(k + 1)(k^2 + 5k + 3k + 12)}{3}$

$= \frac{(k + 1)(k^2 + 8k + 12)}{3}$

$= \frac{(k + 1)(k + 2)(k + 6)}{3}$

$= \frac{(k + 1)((k + 1) + 1)(k + 1 + 5)}{3}$

So, $S(k) \Rightarrow S(k + 1)$, and by mathematical induction $S(n)$ is true for all integers n.

21.a. sample:

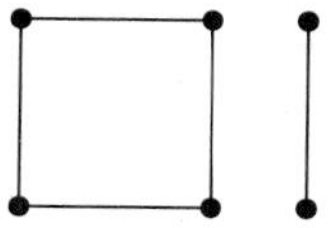

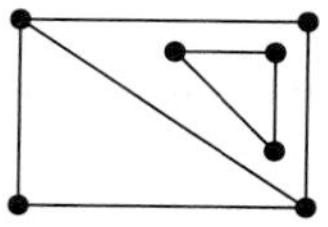

b. 3

c. sample:

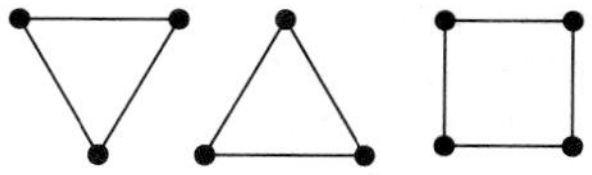

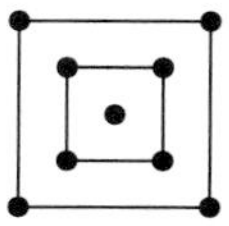 ; 4

19. Left side $= \tan^2 x - \sin^2 x$

$$= \frac{\sin^2 x}{\cos^2 x} - \sin^2 x$$

$$= \frac{\sin^2 x - \sin^2 x \cos^2 x}{\cos^2 x}$$

$$= \frac{\sin^2 x\,(1 - \cos^2 x)}{\cos^2 x}$$

$$= \frac{\sin^2 x \sin^2 x}{\cos^2 x}$$

$= \tan^2 x \sin^2 x$

$=$ Right side

20. For $n = 1$, 3 is a factor of $n^3 + 2n = 3$. Assume 3 is a factor of $n^3 + 2n$ for $n = k$. For $n = k + 1$, $n^3 + 2n = (k + 1)^3 + 2(k + 1) = k^3 + 3k^2 + 5k + 3 = (k^3 + 2k) + 3(k^2 + k + 1)$. As both terms are divisible by 3, so is the sum, $n^3 + 2n$. Hence, by mathematical induction, 3 is a factor of $n^3 + 2n$, for all positive integers $n \geq 1$.

LESSON 12-6 (pages 732–738)

11. $(\vec{u} \times \vec{v}) \cdot \vec{u} =$
$(u_2v_3 - u_3v_2, u_3v_1 - u_1v_3, u_1v_2 - u_2v_1) \cdot \vec{u} =$
$u_1u_2v_3 - u_1u_3v_2 + u_2u_3v_1 - u_1u_2v_3 +$
$u_1u_3v_2 - u_2u_3v_1 = 0$;
$(\vec{u} \times \vec{v}) \cdot \vec{v} =$
$(u_2v_3 - u_3v_2, u_3v_1 - u_1v_3, u_1v_2 - u_2v_1) \cdot \vec{v} =$
$v_1u_2v_3 - v_1u_3v_2 + v_2u_3v_1 - v_2u_1v_3 +$
$v_3u_1v_2 - v_3u_2v_1 = 0$

CHAPTER 12 PROGRESS SELF-TEST
(page 755)

11. Left side $= k(\vec{v} - \vec{w})$

$= k(v_1 - w_1, v_2 - w_2)$

$= (kv_1 - kw_1, kv_2 - kw_2)$

$= k(v_1, v_2) - k(w_1, w_2)$

$= k\vec{v} - k\vec{w}$

$=$ Right side

CHAPTER 12 REVIEW
(pages 756–758)

48.b.

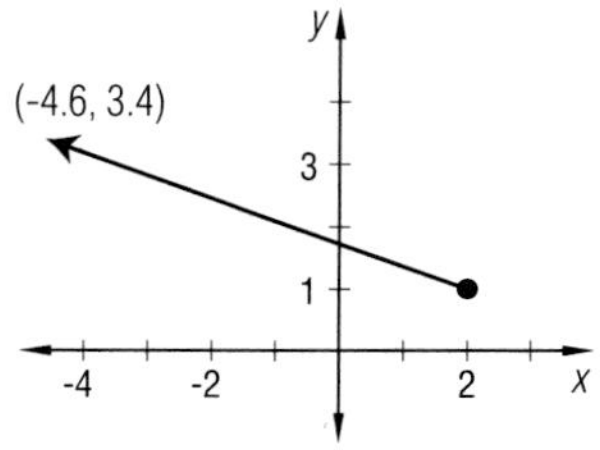

49.a.

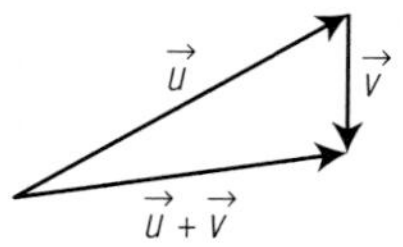

b.

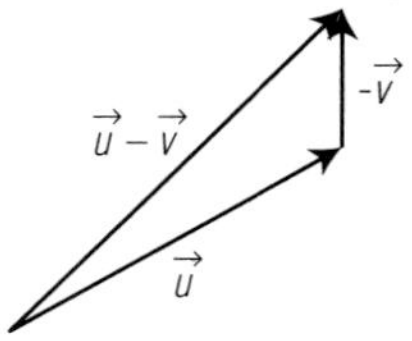

c.

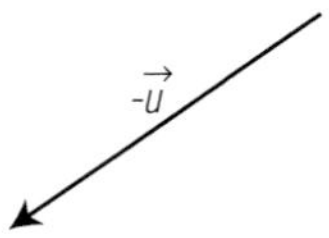

d.

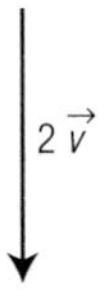

50.a.-c.

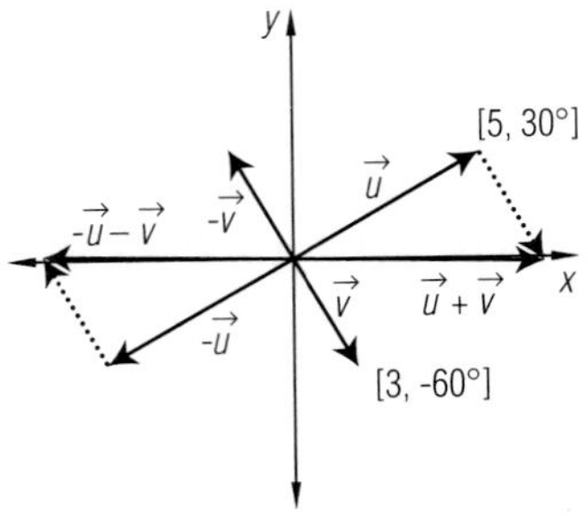

51.

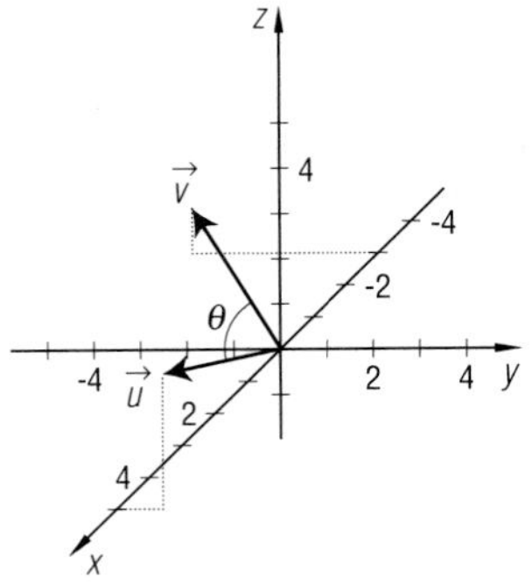

52. (-8, -5, -2); This vector can be pictured by an arrow starting at the endpoint of $\vec{u}$ and ending at the endpoint of $\vec{v}$, providing $\vec{u}$ and $\vec{v}$ have the same initial points; or putting the vector in standard position, it is the diagonal of the figure having vertices (-3, -4, 1), (-5, -1, -3), (0, 0, 0), and (-8, -5, -2).

53.a., c.

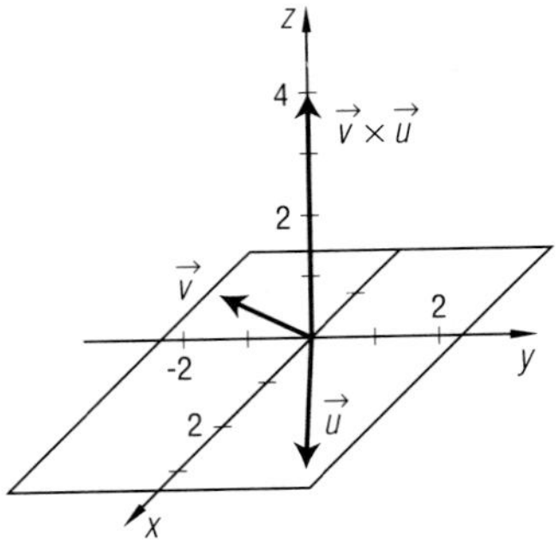

54.a.ii. (4, -7); **iii.** (-8, -11)
b. sample: (3, 1)
c.

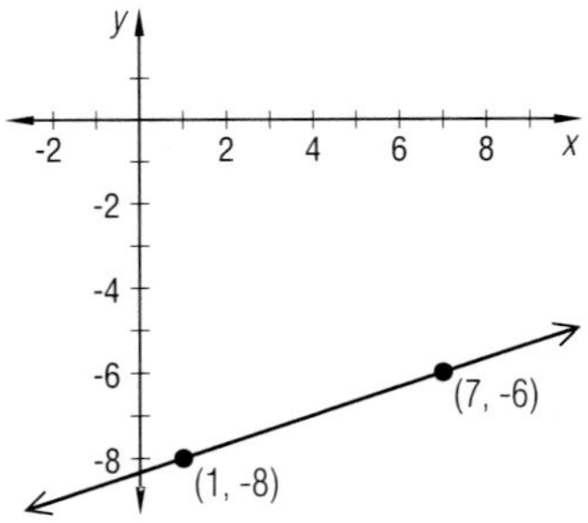

58. $\begin{cases} x = 1 - 5t \\ y = -2 - 3t \end{cases}$

60. center: (0, 2, -4); radius: $\sqrt{5}$

61. $y = 3$ and $y = -3$

62. (6, 0, 0), (0, -4, 0), (0, 0, 12)

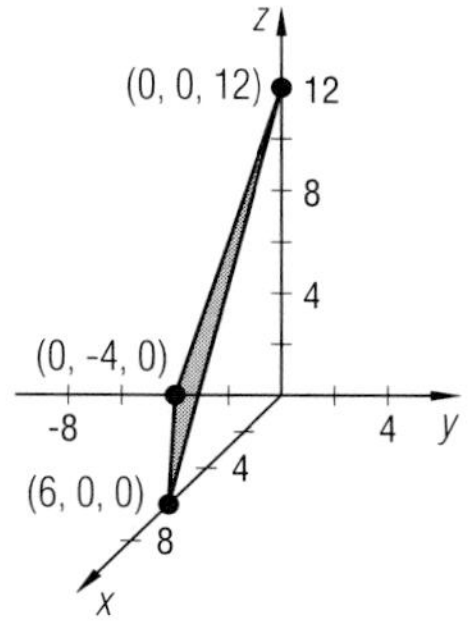

63. $x = 0$ and $y = 0$

64. $\begin{cases} x = 3 + 2t \\ y = -4t \\ z = -1 \end{cases}$

66. sample: (-3, 1, 5) and (3, -1, -5)

67. $(x - 5, y + 1, z - 2) = t(2, -6, 1)$

LESSON 13-2 (pages 768–774)

21.b.

$(\tan\theta)(\sin\theta + \cot\theta\cos\theta)$	$\sec\theta$	
$\left(\frac{\sin\theta}{\cos\theta}\right)\left(\sin\theta + \frac{\cos\theta}{\sin\theta}\cos\theta\right)$	$\frac{1}{\cos\theta}$	Definition of tan, cot, and sec
$\left(\frac{\sin\theta}{\cos\theta}\right)\left(\frac{\sin^2\theta}{\sin\theta} + \frac{\cos^2\theta}{\sin\theta}\right)$		Common denominators
$\left(\frac{\sin\theta}{\cos\theta}\right)\left(\frac{1}{\sin\theta}\right)$		Pythagorean Identity
$\frac{1}{\cos\theta}$		Simplifying

$\therefore (\tan\theta)(\sin\theta + \cot\theta\cos\theta) = \sec\theta$

LESSON 13-4 (pages 787–793)

16.

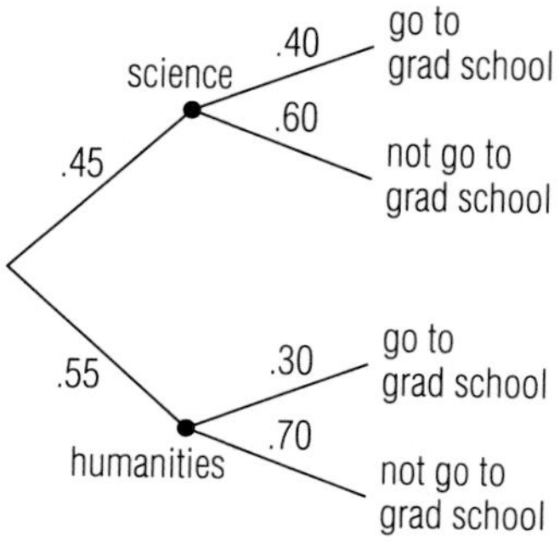

LESSON 13-5 (pages 787–793)

25. An Archimedean screw consists of a spiral passage within an inclined cylinder. It is used for raising water to a certain height. This is achieved by rotating the cylinder.

26. Using wood with a reasonably consistent density, Archimedes could have weighed a rectangular piece of wood, and measured its area. Then he could have cut out a parabolic region and weighed it. The weight of the parabolic region as compared to the rectangular block would be proportional to the areas of these wood blocks.

LESSON 13-6 (pages 794–799)

18. Left side $= \frac{\cos 3x}{\cos x}$

$= \frac{\cos(2x + x)}{\cos x}$

$= \frac{\cos 2x \cos x - \sin 2x \sin x}{\cos x}$

$= \frac{(2\cos^2 x - 1)\cos x - (2\sin x \cos x)\sin x}{\cos x}$ Formulas for $\cos(\alpha + \beta)$ and $\sin(\alpha + \beta)$

$= \frac{2\cos^3 x - \cos x - 2\sin^2 x \cos x}{\cos x}$

$= 2\cos^2 x - 1 - 2\sin^2 x$ $\forall$ real numbers for which $\cos x \neq 0$

$= 2\cos^2 x - 1 - 2(1 - \cos^2 x)$ Since $\sin^2 x = 1 - \cos^2 x$

$= 2\cos^2 x - 1 - 2 + 2\cos^2 x$

$= 4\cos^2 x - 3$

= Right side

CHAPTER 13 PROGRESS SELF-TEST (page 809)

9.

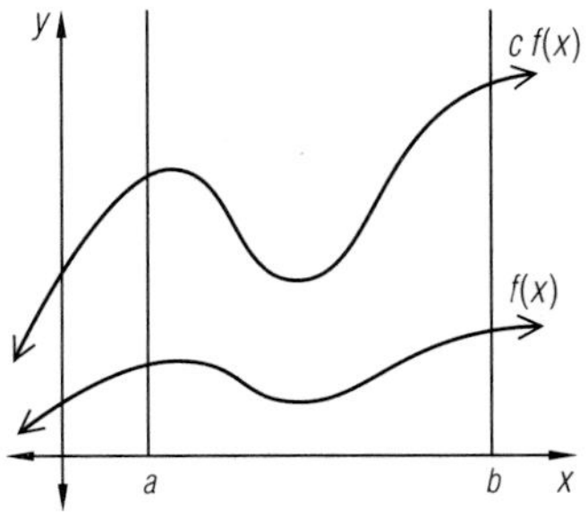

For every Riemann sum of f on the interval from a to b, there is a corresponding Riemann sum of $c \cdot f$ with the same subintervals and same z_i. The value of each term of the sum of $c \cdot f$ will be c times the value of the corresponding term of the sum for f.

$$\lim_{n\to\infty} \sum_{i=1}^{n} c \cdot f(z_i)\,\Delta x = c \cdot \lim_{n\to\infty} \sum_{i=1}^{n} f(z_i)\,\Delta x,$$

so $\int_a^b c \cdot f(x)\,dx = c \int_a^b f(x)\,dx$.

The chart below keys the **Progress Self-Test** questions to the objectives in the **Chapter 1 Review** on pages 74–78. This will enable you to locate those **Chapter 1 Review** questions that correspond to questions you missed on the **Progress Self-Test.** The lesson where the material is covered is also indicated in the chart.

Question	**1, 2**	**3**	**4**	**5**	**6**	**7**	**8**
Objective	A	I	B	D	C	F	B
Lesson	1-1	1-1	1-1	1-1	1-2	1-1	1-3
Question	**9**	**10**	**11**	**12**	**13, 14**	**15**	**16**
Objective	C	K	B	E	A	B	G
Lesson	1-3	1-3	1-5	1-5	1-5	1-5	1-6
Question	**17**	**18**	**19**	**20**	**21**	**22**	**23**
Objective	G	J	H	L	L	M	D
Lesson	1-8	1-6	1-7	1-4	1-4	1-3	1-1

The chart below keys the **Progress Self-Test** questions to the objectives in the **Chapter 2 Review** on pages 146–150. This will enable you to locate those **Chapter 2 Review** questions that correspond to questions you missed on the **Progress Self-Test.** The lesson where the material is covered is also indicated in the chart.

Question	**1**	**2**	**3**	**3d**	**4**	**5**
Objective	C	C	G	D	A	G
Lesson	2-1	2-2	2-5	2-5	2-8	2-4
Question	**5c**	**6**	**7**	**8**	**9**	
Objective	D	F	B	G	E	
Lesson	2-4	2-2	2-7	2-8	2-6	

The chart below keys the **Progress Self-Test** questions to the objectives in the **Chapter 3 Review** on pages 217–220. This will enable you to locate those **Chapter 3 Review** questions that correspond to questions you missed on the **Progress Self-Test.** The lesson where the material is covered is also indicated in the chart.

Question	**1-2**	**3**	**4**	**5**	**6-7**	**8**		**10**	**11**
Objective	A	E	F	M	B	K	D	C	G
Lesson	3-1	3-8	3-1	3-2	3-2	3-3	3-5	3-4	3-2
Question	**12**	**13**	**14**	**15**	**16a**	**16b**	**17**	**18**	
Objective	E	C	H	J	I	N	O	L	
Lesson	3-7	3-4	3-5	3-1	3-9	3-9	3-5	3-6	

The chart below keys the **Progress Self-Test** questions to the objectives in the **Chapter 4 Review** on pages 283–286. This will enable you to locate those **Chapter 4 Review** questions that correspond to questions you missed on the **Progress Self-Test.** The lesson where the material is covered is also indicated in the chart.

Question	**1a,b**	**1c**	**2**	**3**	**4**	**5**	**6**	**7**
Objective	K	A	C	C	H	B	F	L
Lesson	4-2	4-2	4-4	4-5	4-6	4-5	4-1	4-3
Question	**8**	**9**	**10-11**	**12**	**13-15**	**16**	**17**	**18**
Objective	D	G	J	E	M	F	H	I
Lesson	4-3	4-3	4-8	4-8	4-7	4-1	4-6	4-8

The chart below keys the **Progress Self-Test** questions to the objectives in the **Chapter 5 Review** on pages 341–344. This will enable you to locate those **Chapter 5 Review** questions that correspond to questions you missed on the **Progress Self-Test.** The lesson where the material is covered is also indicated in the chart.

Question	**1-2**	**3**	**4**	**5**	**6**	**7**	**8**
Objective	A	J	B	G	C	D	H
Lesson	5-1	5-1	5-2	5-2	5-2	5-5	5-3
Question	**9**	**10**	**11**	**12-13**	**14**	**15**	**16**
Objective	H	I	F	E	K	M	L
Lesson	5-5	5-4	5-7	5-6	5-6	5-7	5-3

The chart below keys the **Progress Self-Test** questions to the objectives in the **Chapter 6 Review** on pages 396–398. This will enable you to locate those **Chapter 6 Review** questions that correspond to questions you missed on the **Progress Self-Test.** The lesson where the material is covered is also indicated in the chart.

Question	**1**	**2**	**3-4**	**5**	**6**	**7**
Objective	A	A	B	A	A	G
Lesson	6-2	6-3	6-6	6-3	6-4	6-1
Question	**8**	**9**	**10-11**	**12**	**13**	**14**
Objective	D	D	C	H	F	E
Lesson	6-2	6-4	6-7	6-7	6-7	6-6

The chart below keys the **Progress Self-Test** questions to the objectives in the **Chapter 7 Review** on pages 462–464. This will enable you to locate those **Chapter 7 Review** questions that correspond to questions you missed on the **Progress Self-Test.** The lesson where the material is covered is also indicated in the chart.

Question	**1a**	**1b**	**2**	**3**	**4**	**5**	**6**	**7**
Objective	A	B	H	C	D	F	E	J
Lesson	7-1	7-2	7-1	7-3	7-3	7-4	7-6	7-2
Question	**8**	**9**	**10**	**11**				
Objective	G	G	G	I				
Lesson	7-4	7-5	7-7	7-8				

The chart below keys the **Progress Self-Test** questions to the objectives in the **Chapter 8 Review** on pages 534–536. This will enable you to locate those **Chapter 8 Review** questions that correspond to questions you missed on the **Progress Self-Test.** The lesson where the material is covered is also indicated in the chart.

Question	**1**	**2**	**3**	**4**	**5**	**6**
Objective	A	C	B	H	I	K
Lesson	8-1	8-2	8-3	8-1	8-3	8-6
Question	**7**	**8**	**9**	**10**	**11**	**12**
Objective	D	E	F	G	J	J
Lesson	8-7	8-8	8-1	8-9	8-4	8-5

The chart below keys the **Progress Self-Test** questions to the objectives in the **Chapter 9 Review** on pages 581–584. This will enable you to locate those **Chapter 9 Review** questions that correspond to questions you missed on the **Progress Self-Test.** The lesson where the material is covered is also indicated in the chart.

Question	**1**	**2**	**3**	**4**	**5**
Objective	G	D	A	H	B
Lesson	9-1	9-1	9-1	9-3	9-2
Question	**6**	**7**	**8ab**	**8cd**	**9**
Objective	I	I	E	F	C
Lesson	9-5	9-5	9-4	9-5	9-5

The chart below keys the **Progress Self-Test** questions to the objectives in the **Chapter 10 Review** on pages 636–638. This will enable you to locate those **Chapter 10 Review** questions that correspond to questions you missed on the **Progress Self-Test.** The lesson where the material is covered is also indicated in the chart.

Question	**1a**	**1b**	**2a**	**2b**	**3**	**4–5**	**6**	**7**
Objective	A	G	A	G	I	F	B	B
Lesson	10-1	10-7	10-1	10-4	10-2	10-2	10-3	10-4
Question	**8–9**	**10**	**11**	**12**	**13**	**14**	**15**	**16**
Objective	F	G	C	H	E	G	D	H
Lesson	10-2	10-4	10-5	10-6	10-6	10-7	10-3	10-6

The chart below keys the **Progress Self-Test** questions to the objectives in the **Chapter 11 Review** on pages 694–698. This will enable you to locate those **Chapter 11 Review** questions that correspond to questions you missed on the **Progress Self-Test.** The lesson where the material is covered is also indicated in the chart.

Question	**1-3**	**4-5**	**6-7**	**8**	**9**	**10**	**11**
Objective	B	B	B	D	A	I	B
Lesson	11-2	11-3	11-4	11-4	11-2	11-2	11-4
Question	**12**	**13**	**14**	**15**	**16**	**17**	
Objective	J	G	C	E	F	H	
Lesson	11-5	11-4	11-3	11-1	11-3	11-6	

The chart below keys the **Progress Self-Test** questions to the objectives in the **Chapter 12 Review** on pages 756–758. This will enable you to locate those **Chapter 12 Review** questions that correspond to questions you missed on the **Progress Self-Test.** The lesson where the material is covered is also indicated in the chart.

Question	**1**	**2**	**3**	**4**	**5**	**6-8**	**7**	**8**	**9**	**10**
Objective	A	B	B	B	D	C	K	I		
Lesson	12-1	12-2	12-3	12-4	12-4	12-6	12-6	12-3		
Question	**11**	**12**	**13a**	**13b**	**14-15**	**16**	**17**	**18**		
Objective	E	F	H	G	J	L	M	M		
Lesson	12-3	12-4	12-2	12-1	12-2	12-3	12-7	12-5		

The chart below keys the ***Progress Self-Test*** questions to the objectives in the **Chapter 13 Review** on pages 810–813. This will enable you to locate those **Chapter 13 Review** questions that correspond to questions you missed on the **Progress Self-Test.** The lesson where the material is covered is also indicated in the chart.

Question	**1**	**2**	**3-4**	**5**	**6**	**7**
Objective	F	A	B	B	E	G
Lesson	13-1	13-2	13-3	13-5	13-6	13-3

Question	**8**	**9**	**10**	**11**	**12**
Objective	H	C	D	C	I
Lesson	13-4	13-4	13-5	13-4	13-6

PROJECT SOLUTIONS

CHAPTER 1

1.a. For a new product which has a unique chemical structure, the FDA requires a long testing period before it is allowed on the market. It is released only after the FDA feels the major side effects and any possible dangers have been identified and weighed against the benefits of the product's use. New products composed of substances already approved by the FDA, like a new "all natural" breakfast cereal, have a much easier time getting FDA approval.

b. In an industry producing highly complex goods which have a high probability of defects, every item may be run through a diagnostic program which has been designed to find defects. For example, manufacturers of VLSI semiconductor chips used in calculators, computers, and many other electronic devices first examine each chip with x-rays to spot defects and then test the chip electronically. Items which are mass produced, like food, nails, or sheets of paper, use another method of quality control. Statistically significant samples are selected from each batch to examine for defects. If the defect count in the sample is below an acceptable level set by the particular manufacturer, the entire batch passes.

2. Answers will vary.

3. Charles Dodgson developed methods of expressing classical logic in terms of symbols. He dealt particularly with syllogisms and sorites. He also developed a diagrammatic method of drawing conclusions from propositions. Dodgson's favorite problems were thought provoking puzzles. Sample problems students may use come from the book *Symbolic Logic and the Game of Logic* by Lewis Carroll, which also contains all the information needed to use Carroll's methods.

4. sample:

BASIC	logical variables:	integers
	logical operators:	NOT, AND, OR
	logical statements:	IF-THEN
		IF-THEN-ELSE
Pascal	logical constants:	TRUE, FALSE
	logical variables:	boolean
	logical operators:	NOT, AND, OR
	logical statements:	IF-THEN
		IF-THEN-ELSE
		CASE
FORTRAN	logical constants:	.TRUE., .FALSE.
	logical variables:	logical
	logical operators:	.AND., .OR., .NOT.
	logical statements:	IF (logical or arithmetic) statement

Note that in some versions of programming languages syntax and semantics may vary.

Here is a sample of possible syntax and semantic descriptions of the conditional statement for three programming languages:

In BASIC, the syntax is

IF/condition/THEN/statement/.

The condition may be constructed from variables, logical operators, relational operators (i.e., $<$, $>$, etc.), and equality operators (i.e., $=$ or $<>$). The statement is any valid BASIC statement and is often a GOTO statement to change the flow of the program. To execute an IF-THEN statement, the condition is evaluated and if it is true the statement following THEN is executed. Otherwise, the condition is false, and the program advances to the next line in the program without executing the statement following THEN. Some BASIC languages have been enhanced with the IF-THEN-ELSE construct. Its syntax is

IF/condition/THEN/statement/ELSE/statement/.

The semantics are the same as described for the IF-THEN statement with the exception that when the condition is false, the program executes the statement following the ELSE before advancing.

In Pascal, the syntax is

IF/condition/THEN/then-statement/ELSE/else-statement/.

In Pascal, the /then-statement/ is executed if the condition is true and the /else-statement/ is executed if the condition is false. One notable feature of Pascal that makes it powerful is that each statement may be a complex statement. A complex statement is a series of statements surrounded by the key words "BEGIN" and "END."

In FORTRAN, the syntax is

IF (/logical expression/) /statement/

or IF (/arithmetic expression/) label1, label2, label3.

In the first case, the logical expression is evaluated and if it is .TRUE. the statement is executed, otherwise the program continues to the next line in sequence. The arithmetic IF evaluates an expression, and the program jumps to label1 if the result is less than 0, label2 if the result equals 0, and label3 if the result is greater than 0.

5. In the following postulates and theorems, the letters A, B, and C symbolize any subsets of any universe I. The most important interpretations are binary Boolean Algebra where I consists of a single element. If a set contains the element, it can be represented by a 1; if it does not, the ϕ will be represented by a 0. $\oplus$ represents logical addition, and $\otimes$ represents logical multiplication.

Addition		
$\oplus$	0	1
0	0	1
1	1	1

Multiplication		
$\otimes$	0	1
0	0	0
1	0	1

Complementation	
x	x'
0	1
1	0

Postulates

Commutative Laws	$A \oplus B = B \oplus A$
	$A \otimes B = B \otimes A$
Associative Laws	$A \oplus (B \oplus C) = (A \oplus B) \oplus C$
	$A \otimes (B \otimes C) = (A \otimes B) \otimes C$
Distributive Laws	$A \otimes (B \oplus C) = (A \otimes B) \oplus (A \otimes C)$
	$A \oplus (B \otimes C) = (A \oplus B) \otimes (A \oplus C)$
Idempotent Laws	$A \oplus A = A$
	$A \otimes A = A$
Complementation Laws	$A \oplus A' = 1$
	$A \otimes A' = 0$
DeMorgan's Laws	$(A \oplus B)' = A' \otimes B'$
	$(A \otimes B)' = A' \oplus B'$
Double Complementation	$(A')' = A$
Laws involving ϕ and I	$A \oplus 1 = 1$
	$A \oplus 0 = 0$
	$A \otimes 0 = A$
	$A \otimes 1 = A$
	$1' = 0$
	$0' = 1$
Laws of Absorption	$A \otimes (A \oplus B) = A$
	$A \oplus (A \otimes B) = A$

The following are the building blocks of logic networks.

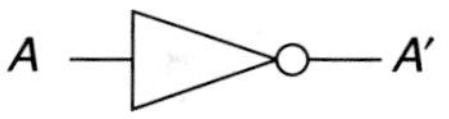

(NOT)

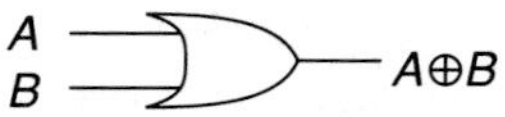

(OR)

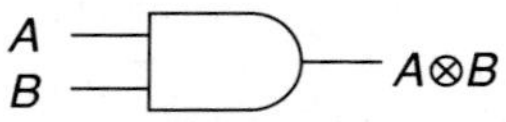

(AND)

All of the above theorems can be constructed by combining the gates above. For example:

Law of Absorption

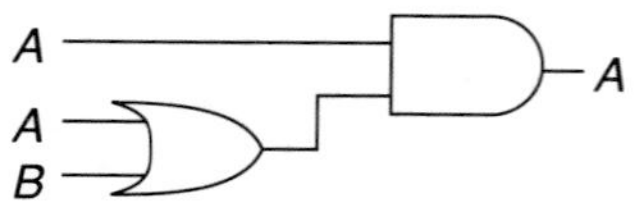

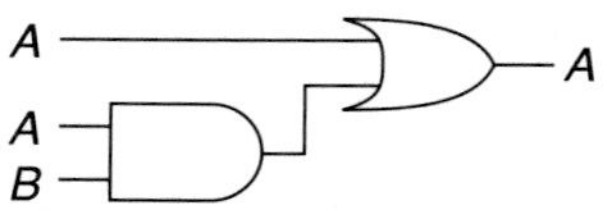

6. The original version of this logical predicament is that of the village barber's boasting that of course he does not shave those people who shave themselves but does shave all those who do not shave themselves. Who shaves the barber? If he does not shave himself, then by his boast he must shave himself. If he does shave himself, then he cannot shave himself.

Contradictions in set theory led mathematicians to question the validity of deductive mathematics. Russell and Whitehead attempted to derive mathematics from self-evident universal logical truths (logicism). Hilbert's formalism hoped to derive mathematics from a set of basic assumptions. Kronecker's intuitionism avoided all use of numbers which could not be constructed from natural numbers—it was free from contradiction because it was limited to constructive methods.

7. From the example found in Lesson 1-9:

```
10 PRINT:PRINT "Sylvania Troubleshooting Guide"
20 PRINT
30 PRINT "1) No power"
40 PRINT "2) Video cassette cannot be inserted"
50 PRINT "3) No operation starts when operation buttons are pushed"
60 PRINT "4) TV programs cannot be recorded"
70 PRINT "5) Timer recording cannot be performed"
80 PRINT "6) Other"
90 PRINT "7) Quit program": PRINT
100 INPUT "Enter the number of the VCR's problem:";PROBLEM
110 PRINT
120 IF PROBLEM < 1 GOTO 10
130 IF PROBLEM > 1 GOTO 180
140 PRINT "Check if power plug is completely connected to a live AC outlet."
150 PRINT "Make sure that the power button is ON and that the timer button"
160 PRINT "is OFF."
170 GOTO 10
180 IF PROBLEM > 2 GOTO 240
190 PRINT "Check that the power button is ON and that the timer button"
200 PRINT "is OFF. Insert the cassette with the window side up and the"
210 PRINT "erasure tab facing you. If the cassette-IN indicator goes on"
220 PRINT "or flashes, there is a cassette in the unit."
230 GOTO 10
240 IF PROBLEM > 3 GOTO 290
250 PRINT "Check that the power button is ON and that the timer button"
260 PRINT "is OFF. Check the DEW indicator. If it is displayed, refer to"
270 PRINT "page 9 of your owner's manual."
280 GOTO 10
290 IF PROBLEM > 4 GOTO 350
300 PRINT "Check the connections of the VCR external antenna and your TV."
310 PRINT "Make sure that the receiving channel of the VCR is properly tuned."
320 PRINT "Make sure that the erasure prevention tab of the cassette tape"
330 PRINT "is still in tact."
340 GOTO 10
350 IF PROBLEM > 5 GOTO 390
360 PRINT "Check the timer setting for the Timer Recording."
370 PRINT "Make sure that the timer button is ON."
380 GOTO 10
390 IF PROBLEM > 6 GOTO 420
400 PRINT "Please seek service at an authorized Sylvania service center."
410 GOTO 10
420 IF PROBLEM > 7 GOTO 10
430 END
```

1.a.

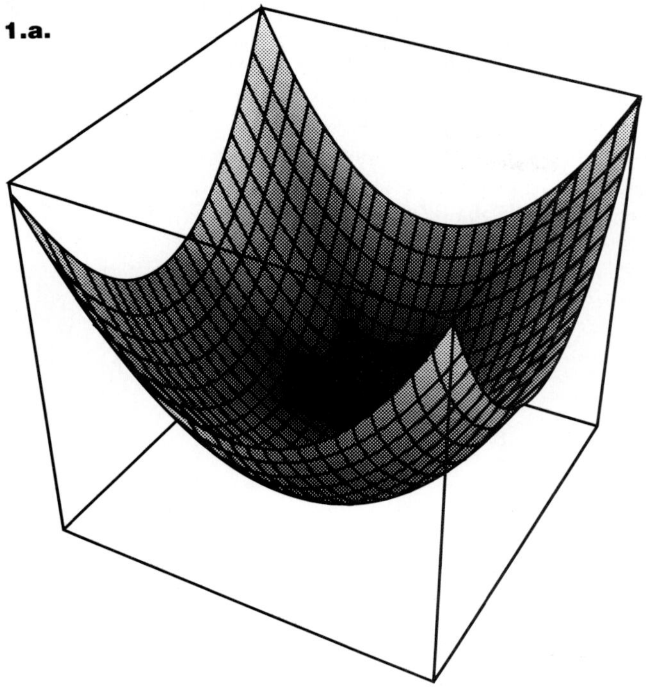

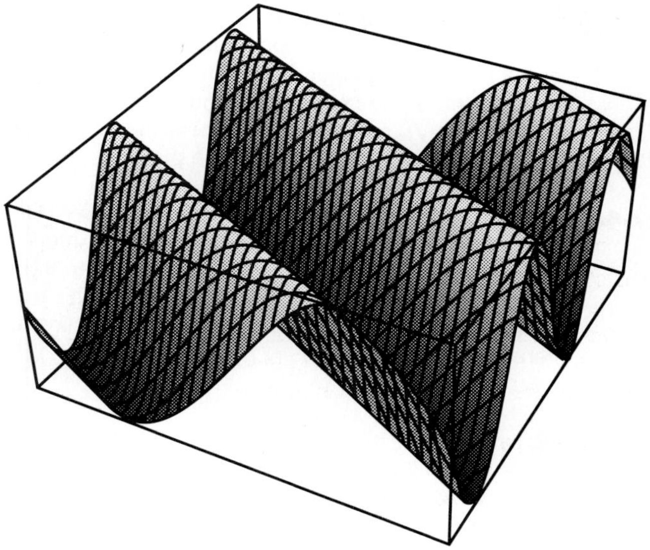

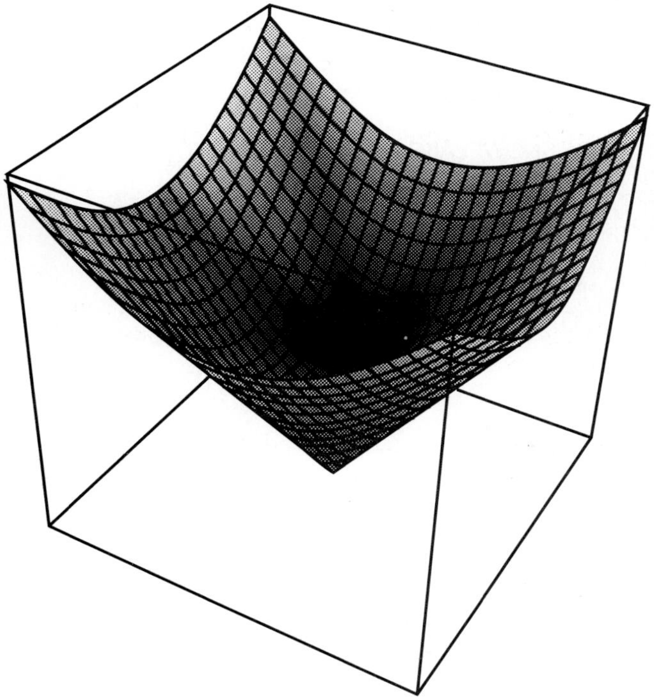

b. Sample (for $z = \sin x \cos y$):

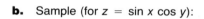

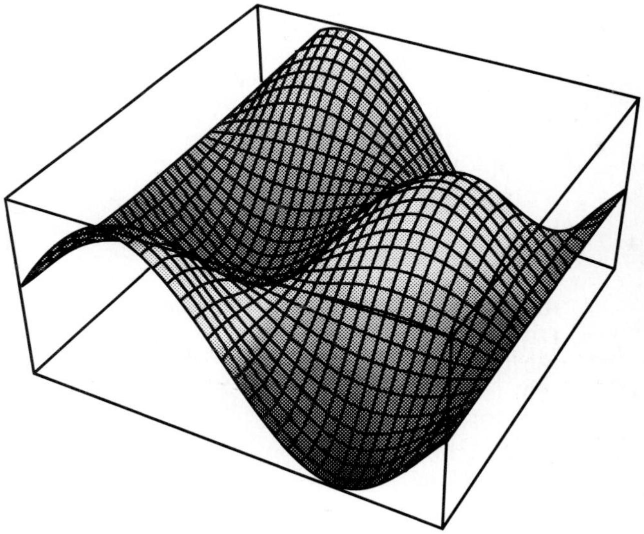

c. relative maximum: a point such that the function does not have a greater value near the point.
relative minimum: a point such that the function does not have a lesser value near the point.
saddle point: at the saddle point the tangent plane to the surface $z = f(x, y)$ is horizontal, but near the point the surface is partly above and partly below the plane.

2.a. From the continuous change model, $A(t) = Be^{rt}$. Rhonda begins with $7,053. Let $r = .086$ be the annual interest rate, t be the number of years, n be the amount taken from the account at the beginning of each year, and $A(t)$ be the amount available at the beginning of the tth year on which to accrue interest.

$A(1) = 7053 - n$
$A(2) = (7053 - n)\, e^{.086(1)} - n = 7686.4040 - 2.0898063n$
$A(3) = (7686.4040 - 2.0898063n)\, e^{.086(1)} - n = 8376.6917 - 3.2774841n$
$A(4) = (8376.6917 - 3.2774841n)\, e^{.086(1)} - n = 9128.971 - 4.5718229n$

Now $A(4)$ must be zero, so $9128.971 - 4.5718229n = 0$, and $n = \dfrac{9128.971}{4.5718229} \approx \$1,996.79$.

b.

```
10 REM retirement program
20 INPUT "d:",d
30 INPUT "r:",r
40 INPUT "y:",y
50 nm = 12*y : mr = (r/100)/12
60 k = EXP(mr)
70 a = d
80 b = -1
90 FOR m = 2 TO nm
100 a = a*k
110 b = b*k - 1
120 NEXT m
130 w = -a/b
140 w = (INT (w*100)/100)
150 PRINT "monthly withdrawal:",w:PRINT
160 PRINT "month", "beginning balance"
170 p = d
180 FOR m = 1 TO nm
190   p = (INT(p*100)/100)
200   PRINT m, p
210   p = (p - w)*k
220 NEXT m
230 PRINT m,0
240 END
```

Here is output from a sample run with $d = \$10,000$, $r = 5.5\%$, and $y = 1$ year.
monthly withdrawal: 854.50

month	beginning balance
1	10000
2	9187.51
3	8371.29
4	7551.32
5	6727.58
6	5900.06
7	5068.73
8	4233.58
9	3394.60
10	2551.76
11	1705.05
12	854.45
13	0

3. Students' reports may include some of the following main points:

1. Cantor's definitions of infinity, transfinite numbers, transfinite arithmetic
2. cardinal numbers
3. cardinality of the set of whole numbers, $\aleph_0$
4. cardinality of the set of real numbers, c
5. ordinal numbers
6. hierarchy of ordinals and cardinals yielding $\aleph_1, \aleph_2, \aleph_3 \ldots$
7. The set of subsets of a set S has a higher cardinal number than S.
8. sum of two cardinal numbers
9. product of two cardinal numbers
10. powers of cardinal numbers

4.a.

```
10  REM SNOWFLAKE SEQUENCE CALCULATION
20  DIM A(100)
30  A(1) = 1
40  FOR N = 2 TO 100
50    A(N) = A(N-1) + 3*(4^(N-2))*((1/9)^(N-1))
60    PRINT "A(";N;") = ";A(N)
70  NEXT N
80  END
```

b.

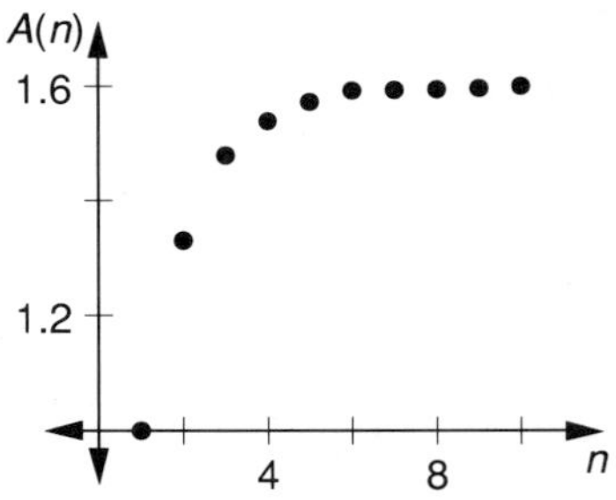

c. As $n \to +\infty$, $A \to 1.6$, which means the size of the snowflake approaches a finite limit.

d.
$$A_1 = 1$$
$$A_2 = 1 + 3 \cdot 4^0 \cdot \left(\frac{1}{9}\right)^1$$
$$A_3 = 1 + 3 \cdot 4^0 \cdot \left(\frac{1}{9}\right)^1 + 3 \cdot 4^1 \cdot \left(\frac{1}{9}\right)^2$$
$$\vdots$$
$$A_n = 1 + 3 \cdot 4^0 \cdot \left(\frac{1}{9}\right)^1 + 3 \cdot 4^1 \cdot \left(\frac{1}{9}\right)^2 + \ldots + 3 \cdot 4^{n-2} \cdot \left(\frac{1}{9}\right)^{n-1}$$
$$A_n = 1 + 3\left(4^0 \cdot \left(\frac{1}{9}\right)^1 + 4^1 \cdot \left(\frac{1}{9}\right)^2 + \ldots + 4^{n-2} \cdot \left(\frac{1}{9}\right)^{n-1}\right)$$
$$A_n = 1 + 3\left(\frac{1}{9}\right)\left(\left(\frac{4}{9}\right)^0 + \left(\frac{4}{9}\right)^1 + \ldots + \left(\frac{4}{9}\right)^{n-2}\right)$$
$$A_n = 1 + \left(\frac{1}{3}\right) \sum_{i=1}^{n-1} \left(\frac{4}{9}\right)^{i-1}$$
$$A_n = 1 + \frac{1}{3} \cdot \frac{1 - \left(\frac{4}{9}\right)^{n-1}}{1 - \left(\frac{4}{9}\right)}$$
$$A_n = 1 + \frac{1}{3} \cdot \frac{1 - \left(\frac{4}{9}\right)^{n-1}}{\frac{5}{9}}$$
$$A_n = 1 + \frac{1}{3} \cdot \frac{9}{5} \cdot \left(1 - \left(\frac{4}{9}\right)^{n-1}\right)$$
$$A_n = 1 + \frac{3}{5} \cdot \left(1 - \left(\frac{4}{9}\right)^{n-1}\right)$$
$$A_n = 1 + \frac{3}{5} - \frac{3}{5} \cdot \left(\frac{4}{9}\right)^{n-1}$$

And so the explicit formula is $A_n = \frac{8}{5} - \frac{3}{5} \cdot \left(\frac{4}{9}\right)^{n-1}$.

The $\lim\limits_{n \to \infty} A_n = \lim\limits_{n \to \infty} \frac{8}{5} - \frac{3}{5} \cdot \left(\frac{4}{9}\right)^{n-1} = \lim\limits_{n \to \infty} \frac{8}{5} = \frac{8}{5} = 1.6$ and verifies the observation in part **c** that the limit of the areas of the snowflakes approaches 1.6.

e. Answers may vary.

5. Sample (for acidity):
The pH scale is used to measure the acidity or alkalinity of a solution. pH is an abbreviation for the potential of hydrogen. Acids have a higher concentration of H^+ ions than OH^- ions. Bases have a higher concentration of OH^- ions than H^+ ions. A solution where the concentration of H^+ and OH^- ions are equal is a neutral solution, and it is assigned the pH value of 7.0. Lower pH values indicate increasingly acidic solutions, while higher pH values indicate increasingly basic solutions.

The Weber-Fechner Law states that the physical sensation produced by a stimulus is proportional to the logarithm of the stimulus. The pH scale is logarithmic, so a solution with a pH of 6.0 is 10 times more acidic than a solution with a pH of 7.0. On an ionic level, there are 10 times more H^+ ions than OH^- ions in the solution.

The pH scale is useful because it provides a measure of the ratio of the ionic concentrations which is independent of the volume of the solution. An absolute measure of the number of H^+ and OH^- ions could be used to find the difference in the number of each ion and compared to the volume of liquid to determine how acidic or basic the solution is, but this is probably more complicated than it sounds. Also, the testing process would use the entire solution which would not be very practical.

6. Figurate numbers refer to the number of dots in certain geometric configurations.

Triangular numbers:

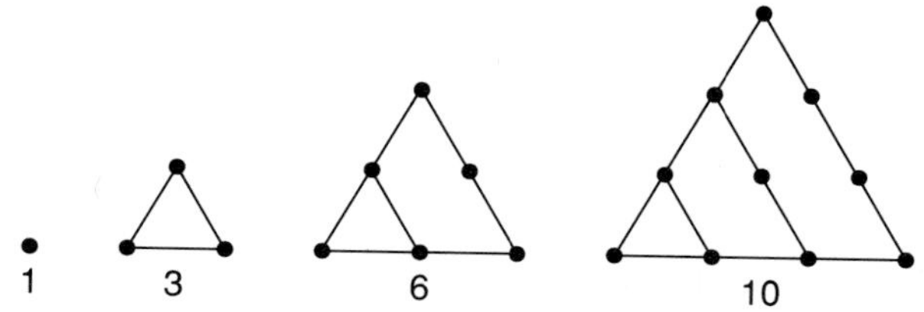

Explicit formula: $T_n = \frac{n(n+1)}{2}$
Recursive formula: $T_n = T_{n-1} + n$

Square numbers:

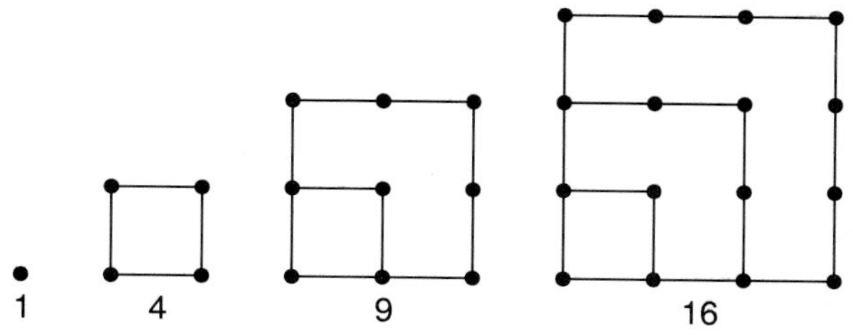

Explicit formula: $S_n = n^2$
Recursive formula: $S_n = S_{n-1} + \frac{2n - 1}{2}$

Pentagonal numbers:

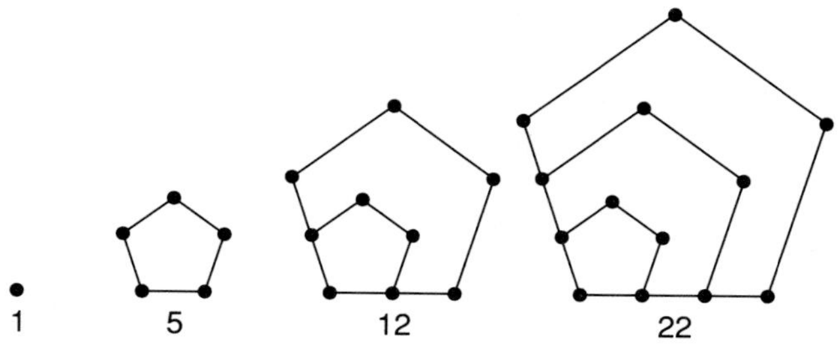

Explicit formula: $P_n = \dfrac{n(3n - 1)}{2}$

Recursive formula: $P_n = P_{n-1} + (3n - 2)$

Two relationships between these sequences are $S_n = T_n + T_{n-1}$ and $P_n = n + 3T_{n-1}$.

CHAPTER 3

1. Answers will vary from region to region.

2.a.

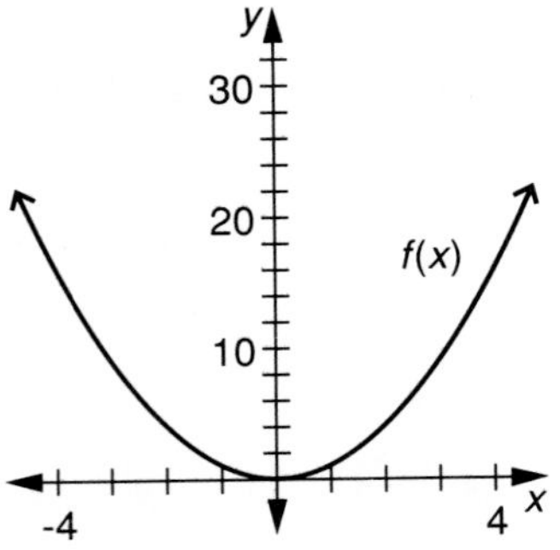

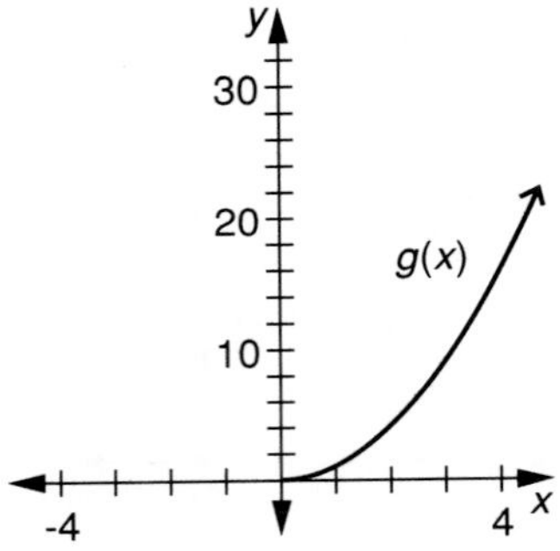

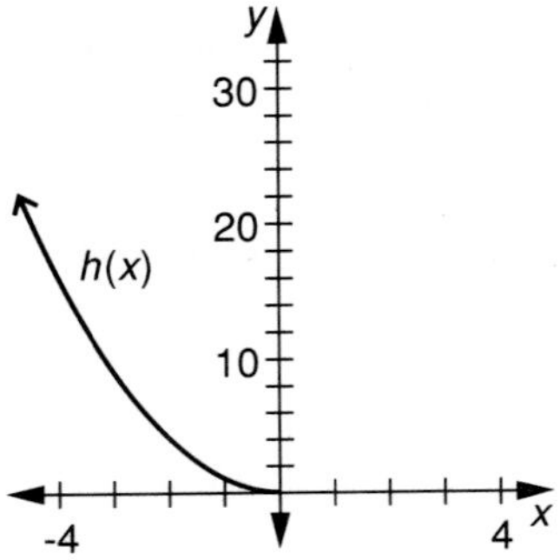

The terms added with absolute values in the denominator result in undefined values of the function for either all nonpositive or all nonnegative real numbers. This serves as a way to restrict the domain of a function.

b. $x > 5$ for $f = x$, $f + g$ yields a domain $x > 5$.

c. $f(x) = \dfrac{0}{|x - a| + (x - a)} + \dfrac{0}{|x - b| - (x - b)}$

d. For $x_1 < x_2$, $f(x) = \dfrac{0}{|x - x_1| + (x - x_1)} + \dfrac{0}{|x - x_2| + (x - x_2)}$.

$g(x)$: $(y - y_1) = \dfrac{y_2 - y_1}{x_2 - x_1}(x - x_1)$, so $g(x) = y = \dfrac{y_2 - y_1}{x_2 - x_1}(x - x_1) + y_1$.

e. head(x): $x^2 + (y - 10)^2 = 1$

arms (x) $= -|x| + 7$

legs(x) $= -|x|$

$R(x) = \dfrac{0}{|x + 2| + (x + 2)} + \dfrac{0}{|x - 2| - (x - 2)}$

With the body formed by the y-axis, the head by head(x), the arms by arms(x) + $R(x)$, and the legs by legs(x) + $R(x)$.

3.a. Line 110 evaluates *FN F(A)* and *FN F(B)*. If they have different signs, then there is a zero between *A* and *B*. The code multiplies *FN F(A)* and *FN F(B)*; a negative product indicates the existence of a zero in the interval (*A, B*).

b. From the program, the zero is ≈ 0.3320313 which differs from the actual zero by 0.39%.

c. From the program, the zero is ≈ 0.3867188 which differs from the actual zero by 0.98%.

d. Answers may vary.

4. Sample (for A, B, and C):

$RG0(x) = \dfrac{0}{|x| + x}$

$RL2(x) = \dfrac{0}{|x - 2| - (x - 2)}$

$RG2(x) = \dfrac{0}{|x - 2| + (x - 2)}$

$RL4(x) = \dfrac{0}{|x - 4| - (x - 4)}$

$RG1(x) = \dfrac{0}{|x - 1| + (x - 1)}$

$RL3(x) = \dfrac{0}{|x - 3| - (x - 3)}$

$A1(x) = 2.5x$

$A2(x) = -2.5x + 10$

$A3(x) = 2.5$

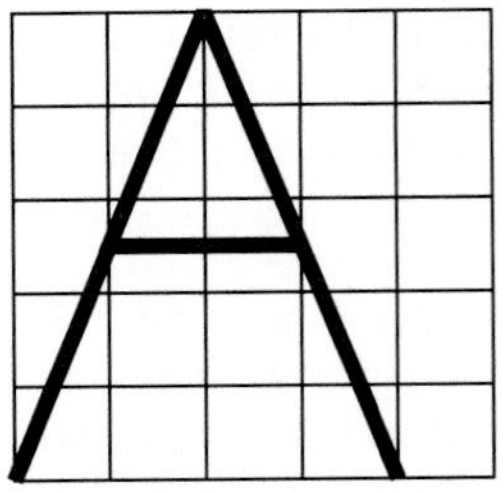

The letter A: $f1 = A1 + RG0 + RL2$,
$f2 = A2 + RG2 + RL4$, and
$f3 = A3 + RG1 + RL3$.

$B1(x) = 5$
$B2(x) = 0$
$B3(x) = \sqrt{1.5^2 - (x - 2)^2} + 3.5$
$B4(x) = -\sqrt{1.5^2 - (x - 2)^2} + 3.5$
$B5(x) = \sqrt{1.5^2 - (x - 2)^2} + 1.5$
$B6(x) = -\sqrt{1.5^2 - (x - 2)^2} + 1.5$

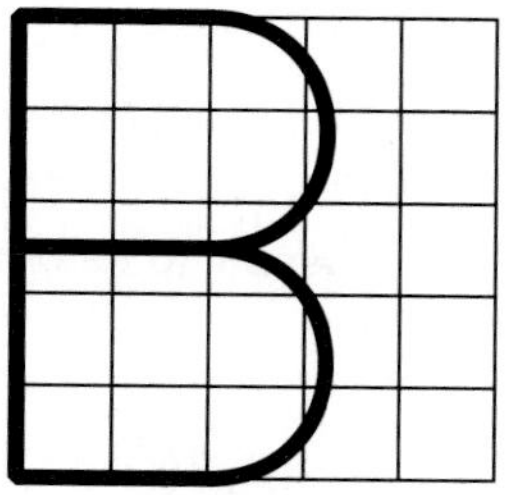

The letter B: $f1 = B1 + RG0 + RL0$,
$f2 = B2 + RG0 + RL0$,
$f3 = B3 + RG0 + RL0$,
$f4 = B3 + RG2$,
$f5 = B4 + RG2$,
$f6 = B5 + RG2$, and
$f7 = B6 + RG2$.

$C1(x) = \sqrt{2.5^2 - (x - 2.5)^2} + 2.5$

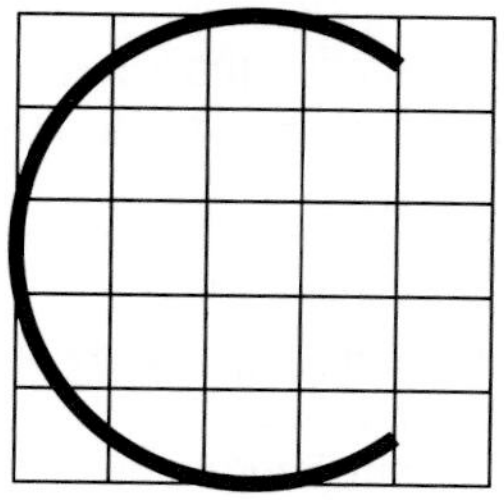

The letter C: $f1 = C1 + RL4$.

CHAPTER 4

1. Students' reports may include some of the following main points:
1. devised method of least squares
2. constructed a regular polygon of 17 sides
3. proved that every positive integer is the sum of three triangular numbers
4. proved the Fundamental Theorem of Algebra (4 different ways)
5. invented the heliotrope
6. co-invented the electromagnetic telegraph
7. anticipated non-Euclidean geometry by papers on surface theory
8. contributed also to astronomy, geodesy, and electricity

2.ai. $k = \text{GCF}\{a, d\} = \text{GCF}\{5, 12\} = 1$
$\frac{b}{k} = \frac{8}{1} = 8 \therefore$ There is 1 solution to $5x \equiv 8 \pmod{12}$ in the set $\{0, 1, \ldots, 11\}$: $x = 4$.

ii. $k = \text{GCF}\{3, 9\} = 3$
$\frac{b}{k} = \frac{12}{3} = 4 \therefore$ There are 3 solutions to $3x \equiv 12 \pmod{9}$ in the set $\{0, 1, \ldots, 8\}$: $x = 1$, $x = 4$, and $x = 7$.

iii. $k = \text{GCF}\{4, 6\} = 2$
$\frac{b}{k} = \frac{5}{2} = 2.5 \therefore$ There are no solutions to $4x \equiv 5 \pmod{6}$ in the set $\{0, 1, \ldots, 5\}$.

iv. $k = \text{GCF}\{6, 12\} = 6$
$\frac{b}{k} = \frac{24}{6} = 4 \therefore$ There are 6 solutions to $6x \equiv 24 \pmod{12}$ in the set $\{0, 1, \ldots, 11\}$: $x = 0$, $x = 2$, $x = 4$, $x = 6$, $x = 8$, and $x = 10$.

If d is prime and $d \neq a$, then there will be exactly one solution for every equation of the form $ax \equiv b \pmod{d}$, $a \not\equiv 0$.

b. Any value d will have at least 1 solution if
$\frac{-b + \sqrt{b^2 - 4ac}}{2a}$ or $\frac{-b - \sqrt{b^2 - 4ac}}{2a}$
is an integer. If they are both integers then there are 2 solutions (although they could be the same).

3. Answers will vary.

4. Sample (using method in part **a**):

```
10 REM factoring quadratics over the integers
20 a = 1:nf = 0:PRINT "a       b"
30 FOR b = -9 TO 9
40 FOR c = -9 TO 9
50 d = b^2 - 4*a*c
60 IF (d < 0) THEN f = 0:GOTO 100
70 x1 = (-b + SQR(d))/(2*a)
80 x2 = (-b - SQR(d))/(2*a)
90 IF (x1 = INT(x1)) AND (x2 = INT(x2)) THEN f = 1 ELSE
     f = 0
100 PRINT b;c;
110 IF (f=1) THEN PRINT "factorable (x - ";x1;")
     (x - ";x2;")":nf = nf + 1
120 IF (f= 0) THEN PRINT
130 NEXT c
140 NEXT b
150 PRINT
160 PRINT nf; "out of 361 were factorable."
170 END
```

It turns out that 66 expressions out of 361, or ≈18% are factorable.

5. Sample (using a calculator):
Must approximate all real zeros of
$p(x) = 5x^3 + 4x^2 - 551x + 110$.

a.

-2	5	4	-551	110
		-10	12	1078
	5	-6	-539	1188

So $p(-2) = +1188$.

2	5	4	-551	110
		10	28	-1046
	5	14	-523	-936

So $p(2) = -936$.

b.

i	t_i	remainder
3	0	110
4	1	-432
5	.5	-163.875
6	.25	-27.421875
7	.125	41.19726563
8	.1875	20.78295898
9	.21875	-10.2875061
10	.203125	-1.714931488
11	.1953125	2.572653294
12	.19921875	.4287542701
13	.201171875	-.64311537892
14	.2001953125	-.10718723293
15	.19970703125	.16078185069
16	.199951171875	.02679689168
17	.2000732421875	-.04019527494
18	.20001220703125	-.00669921771
19	.19998168945313	.01004883047
20	.19999694824219	.00167480475
21	.20000457763672	-.00251220688
22	.20000076293946	-.00041870117

.2	5	4	-551	110
	1	1	1	-110
	5	5	-550	0

Hence, one root is $x = .2$.

c. $5x^2 - 5x - 550 = 0$ is the reduced polynomial.
Using the quadratic formula, $x =$
$\frac{-5 \pm \sqrt{25 - 4 \cdot 5 \cdot (-550)}}{2 \cdot 5} = \frac{-5 \pm 105}{10} = 10, -11.$
So the remaining roots are $x = 10$ and $x = -11$.

CHAPTER 5

1. Students' reports may include some of the following events:

240 BC Archimedes developed the classical method of computing π. He found it to be between $\frac{223}{71}$ and $\frac{22}{7}$.

150 AD Ptolemy estimated π as $\frac{377}{120} \approx 3.1416$.

480 Tsu Ch'ung-chih estimated π as $\frac{355}{113} \approx 3.1415929$ (correct to six places).

1429 Al-Kashi computed π to 16 decimal places.

1610 Ludolph van Ceulen computed π to 35 decimal places.

1699 Abraham Sharp found 71 correct decimal places.

1706 John Machin found 100 correct decimal places.

1841 William Rutherford computed π to 208 places (152 of which were correct).

1844 Zacharias Dase found π correct to 200 decimal places.

1853 Rutherford found π to 400 correct decimal places.

1873 William Shanks computed π to 707 places (528 correct).

1892 A writer announced a long lost secret that leads to 3.2 as the exact value of π.

1892 Ferdinand Lindenman gives the first proof π is a transcendental number.

1897 Indiana State Legislature almost passed a bill stating, "It has been found that a circular area is to the square on a line equal to the quadrant of the circumference, as the area of an equilateral rectangle is the square on one side."

1931 Another book purported to demonstrate $\pi = 3\frac{13}{81}$.

1948 D. F. Ferguson and J. W. Wrench Jr. published π correct to 808 places.

1949 The ENIAC computer found π to 2,037 decimal places.

1959 Francois Genuys found π to 16,167 decimal places with an IBM 704.

1961 Wrench and Daniel Shanks found π to 100,265 places on an IBM 7090.

1966 M. Jean Guilloud approximated π to 250,000 places on a Stretch computer.

1967 M. Jean Guilloud approximated π to 500,000 places on a CDC 6600.
1974 M. Jean Guilloud approximated π to 1,000,000 places on a CDC 7600.

2. $y = \dfrac{ax + b}{cx + d}$ may be rewritten as $y = \dfrac{a}{c} + \dfrac{b - \frac{ad}{c}}{cx + d}$, which is equivalent to $y - \dfrac{a}{c} = \dfrac{\frac{b}{c} - \frac{ad}{c^2}}{x + \frac{d}{c}}$. This form makes the scaling and translation evident. First scale by $S_{1, b/c - ad/c^2}$, then translate by $T_{-d/c, a/c}$.
Sample test:
Let $g(x) = \dfrac{1x + 2}{1x + 0} = \dfrac{x + 2}{x}$, so $a = 1$, $b = 2$, $c = 1$, and $d = 0$. Hence $f(x)$ should be transformed to $g(x)$ by $S_{1,2}$ then $T_{0,1}$. $f(x) = \dfrac{1}{x}$. Applying $S_{1,2}$ yields $\dfrac{y}{2} = \dfrac{1}{x}$ which is $y = \dfrac{2}{x}$. Then applying $T_{0,1}$ yields $y - 1 = \dfrac{2}{x - 0}$ which is $y = 1 + \dfrac{2}{x} = \dfrac{x}{x} + \dfrac{2}{x} = \dfrac{x + 2}{x} = g(x)$, and so the tranformations check.

3.ai. using a calculator, ≈ 1.8478
ii. using a calculator, ≈ 1.9616
iii. Continuing the pattern from parts **i** and **ii** yields the values 1.9904, 1.9976, ... , so the expression equals 2.

b. Using the approach of part **a**, the sequence generated is 2.9068, 2.9844, 2.9974, ... , so the expression equals 3.

c. sample:
$a = 8$, $x \approx 3.37$
$a = 9$, $x \approx 3.54$
$a = 10$, $x \approx 3.7$
$a = 20$, $x = 5$

d.

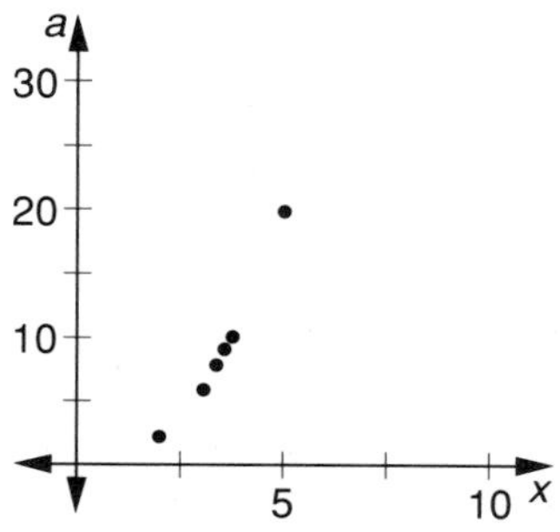

The graph of a and x values found appears to be a parabola. Hence the conjecture: $a = x^2 - x$.

e. $x = \sqrt{a + x}$ is given. Squaring both sides yields $x^2 = a + x$. And so by addition, $a = x^2 - x$, and the conjecture is proven.

4.a. $p(x) = a_3x^3 + a_2x^2 + a_1x + a_0$

i. $p\left(\frac{m}{k}\right) = a_3\left(\frac{m}{k}\right)^3 + a_2\left(\frac{m}{k}\right)^2 + a_1\frac{m}{k} + a_0 = 0$
$a_3m^3 + a_2m^2k + a_1mk^2 + a_0k^3 = 0$
$a_3m^3 + a_2m^2k + a_1mk^2 = -a_0k^3$
$m[a_3m^2 + a_2mk + a_1k^2] = -a_0k^3$

ii. Since m is a factor on the left side, m is a factor on the right side. But m and k have no common factor, so m is a factor of a_0 rather than k^3.

iii. $p\left(\frac{m}{k}\right) = a_3\left(\frac{m}{k}\right)^3 + a_2\left(\frac{m}{k}\right)^2 + a_1\left(\frac{m}{k}\right) + a_0 = 0$
$a_3m^3 + a_2m^2k + a_1mk^2 + a_0k^3 = 0$
$a_3m^3 = -a_2m^2k - a_1mk^2 - a_0k^3$
$a_3m^3 = -k(a_2m^2 + a_1mk + a_0k^2)$

iv. Since k is a factor on the right side, k is a factor on the left side. But m and k have no common factors, so k is a factor of a_3 rather than m^3.

b.i. $p(x) = a_nx^n + a_{n-1}x^{n-1} + \ldots + a_1x + a_0;\ a_n \neq 0$
Replace x by the root $\frac{m}{k}$.
$p\left(\frac{m}{k}\right) = a_n\left(\frac{m}{k}\right)^n + a_{n-1}\left(\frac{m}{k}\right)^{n-1} + \ldots + a_1\left(\frac{m}{k}\right) + a_0 = 0$
Multiply both sides by k^n.
$a_nm^n + a_{n-1}m^{n-1}k + \ldots + a_1mk^{n-1} + a_0k^n = 0$
Factor m from the first n terms and subtract a_0k^n from both sides.
$m(a_nm^{n-1} + a_{n-1}m^{n-2}k + \ldots + a_1k^{n-1}) = -a_0k^n$
Factor k from the last n terms and subtract a_nm^n from both sides.
$-k(a_{n-1}m^{n-1} + \ldots + a_1mk^{n-2} + a_0k^{n-1}) = a_nm^n$

By the same arguments in **ii** and **iv**, substituting k^n for k^3 and m^n for m^3, m is a factor of a_0 and k is a factor of a_n.

ii. $\pm\frac{2}{3}, \pm\frac{1}{2}, \pm\frac{1}{3}$

iii. $\frac{2}{3}$ and $\frac{1}{2}$ are rational zeros. The other roots are $+i\sqrt{6}$ and $-i\sqrt{6}$.

5. Sample:

The gas mileage for three popular cars was obtained:

Car	City	MPG Highway	Average
X	53	58	55.5
Y	24	34	29.0
Z	16	25	20.5

PROJECT SOLUTIONS

The price of gasoline was assumed to be a low \$1.00 per gallon.

Car	Miles per Year	Yearly Cost of Fuel	If you drove … Car X	Car Y	Car Z
X	10,000	\$ 180		lose 160	lose 308
	20,000	360		lose 330	lose 616
	30,000	540		lose 495	lose 924
Y	10,000	345	save 160		lose 143
	20,000	690	save 330		lose 286
	30,000	1,035	save 495		lose 429
Z	10,000	488	save 308	save 143	
	20,000	976	save 616	save 286	
	30,000	1,464	save 924	save 429	

CHAPTER 6

1.a.

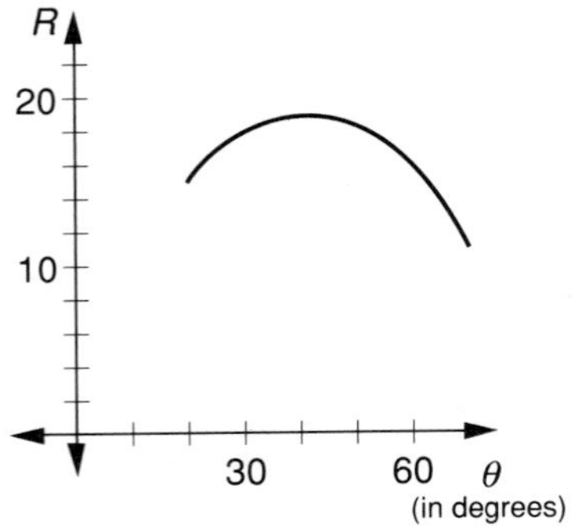

maximum distance: ≈ 19.14 m

b.

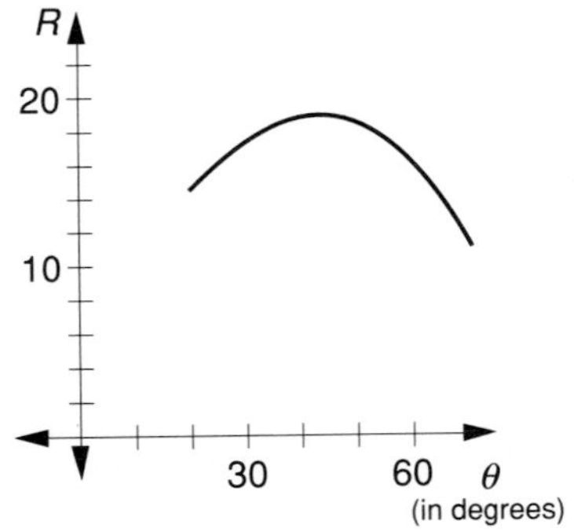

maximum distance: ≈ 18.86 m

c.i. Answers will vary from school to school.

ii. Answers may vary.

2.a. Sample (for Chicago):
Chicago's latitude is $41° 52' 28'' \approx 41.874°$.
$g = 9.78049\,(1 + 0.005288 \sin^2(41.874°) - 0.000006 \sin^2(2 \cdot 41.874°)) \approx 9.803$

b. $g = 9.78049\,(1 + 0.005288 \sin^2\theta - 0.000006\,(2 \sin\theta \cos\theta)^2)$
$= 9.78049\,(1 + 0.005288 \sin^2\theta - 0.000024 \sin^2\theta\,(1 - \sin^2\theta))$
$= 9.78049\,(1 + 0.005288 \sin^2\theta - 0.000024 \sin^2\theta + 0.000024 \sin^4\theta)$
$= 9.78049\,(1 + 0.005264 \sin^2\theta + 0.000024 \sin^4\theta)$

c. $9.8 = 9.78049\,(1 + 0.005264 \sin^2\theta + 0.000024 \sin^4\theta)$
$9.8 = 9.78049 + 0.051484 \sin^2\theta + 0.000234732 \sin^4\theta$
$0.000234732 \sin^4\theta + 0.051484 \sin^2\theta - 0.01951 = 0$

$$\sin^2\theta \approx \frac{-0.051484 \pm \sqrt{0.051484^2 - 4(0.000234732)(-0.01951)}}{2(0.000234732)}$$

$$\sin^2\theta \approx \frac{-0.051484 \pm 0.0516616}{0.000469464}$$

$\sin^2\theta \approx 0.378304$ or -219.709
$\sin^2\theta$ must be positive, so $\sin\theta \approx \sqrt{0.378304} \approx 0.615064$.
θ is then $\sin^{-1}(0.615064) \approx 37.96°$.
The answer found for Review Question 42a was 38.0°; they are very close.

d. To find g at any altitude, $g = \dfrac{4 \times 10^4}{(6.38 \times 10^6 + h)^2}$ m/sec^2, where h is the distance above sea level or $-h$ is the distance below sea level in meters. (Note: sea level is the average radius of the Earth.)

3.ai. $\text{pcos}^2 x - \text{psin}^2 x = (2^{x-1} + 2^{-x-1})^2 - (2^{x-1} - 2^{-x-1})^2 = 2^{2x-2} + 2^{-1} + 2^{-2x-2} - 2^{2x-2} + 2^{-1} - 2^{-2x-2} = 2(2^{-1}) = 1$

ii.
$$\frac{\text{psin}\,x}{\text{pcos}\,x} = \frac{2^{x-1} - 2^{-x-1}}{2^{x-1} + 2^{-x-1}} = \frac{\frac{2^{x-1}}{2^{-x-1}} - \frac{2^{-x-1}}{2^{-x-1}}}{\frac{2^{x-1}}{2^{-x-1}} + \frac{2^{-x-1}}{2^{-x-1}}} = \frac{2^{2x} - 1}{2^{2x} + 1} = \text{ptan}\,x$$

iii.
$$\frac{1}{\text{pcos}\,x} = \frac{1}{2^{x-1} + 2^{-x-1}} = \frac{\frac{1}{2^{-x-1}}}{\frac{2^{x-1}}{2^{-x-1}} + \frac{2^{-x-1}}{2^{-x-1}}} = \frac{2^{x+1}}{2^{2x} + 1} = \text{psec}\,x$$

b. $\text{pcot}\,x = \dfrac{\text{pcos}\,x}{\text{psin}\,x} = \dfrac{2^{x-1} + 2^{-x-1}}{2^{x-1} - 2^{-x-1}} = \dfrac{2^{2x} + 1}{2^{2x} - 1}$, and

$\text{pcsc}\,x = \dfrac{1}{\text{psin}\,x} = \dfrac{1}{2^{x-1} - 2^{-x-1}} = \dfrac{2^{x+1}}{2^{2x} - 1}$.

Sample identity: $\text{pcot}^2 x - 1 = \text{pcsc}^2 x$.

$$\text{pcot}^2 x - 1 = \left(\frac{2^{2x} + 1}{2^{2x} - 1}\right)^2 - 1 =$$

$$\frac{2^{4x} + 2^{2x+1} + 1 - 2^{4x} + 2^{2x+1} - 1}{(2^{2x} - 1)^2} =$$

$$\frac{2^{2(x+1)}}{(2^{2x} - 1)^2} = \left(\frac{2^{x+1}}{2^{2x} - 1}\right)^2 = \text{pcsc}^2 x$$

ci. psin x pcos y + pcos x psin y

$= (2^{x-1} - 2^{-x-1})(2^{y-1} + 2^{-y-1}) + (2^{x-1} + 2^{-x-1}) \cdot (2^{y-1} - 2^{-y-1})$

$= 2^{x+y-2} + 2^{x-y-2} - 2^{-x+y-2} - 2^{-x-y-2} + 2^{x+y-2} - 2^{x-y-2} + 2^{-x+y-2} - 2^{-x-y-2}$

$= 2^{x+y-2} - 2^{-x-y-2} + 2^{x+y-2} - 2^{-x-y-2}$

$= 2^{(x+y)-1} - 2^{-(x+y)-1}$

$=$ psin $(x + y)$

ii. pcos x pcos y + psin x psin y

$= (2^{x-1} + 2^{-x-1})(2^{y-1} + 2^{-y-1}) + (2^{x-1} - 2^{-x-1}) \cdot (2^{y-1} - 2^{-y-1})$

$= 2^{x+y-2} + 2^{x-y-2} + 2^{-x+y-2} + 2^{-x-y-2} + 2^{x+y-2} - 2^{x-y-2} - 2^{-x+y-2} + 2^{-x-y-2}$

$= 2^{x+y-2} + 2^{-x-y-2} + 2^{x+y-2} + 2^{-x-y-2}$

$= 2^{(x+y)-1} + 2^{-(x+y)-1}$

$=$ pcos$(x + y)$

iii. $\dfrac{\text{ptan } x + \text{ptan } y}{1 + \text{ptan } x \text{ ptan } y}$

$$= \frac{\frac{2^{2x}-1}{2^{2x}+1} + \frac{2^{2y}-1}{2^{2y}+1}}{1 + \frac{2^{2x}-1}{2^{2x}+1} \cdot \frac{2^{2y}-1}{2^{2y}+1}}$$

$$= \frac{\frac{(2^{2x}-1)(2^{2y}+1) + (2^{2y}-1)(2^{2x}+1)}{(2^{2x}+1)(2^{2y}+1)}}{\frac{(2^{2x}+1)(2^{2y}+1) + (2^{2x}-1)(2^{2y}-1)}{(2^{2x}+1)(2^{2y}+1)}}$$

$$= \frac{(2^{2x}-1)(2^{2y}+1) + (2^{2y}-1)(2^{2x}+1)}{(2^{2x}+1)(2^{2y}+1) + (2^{2x}-1)(2^{2y}-1)}$$

$$= \frac{2^{2x+2y} - 2^{2y} + 2^{2x} - 1 + 2^{2x+2y} - 2^{2x} + 2^{2y} - 1}{2^{2x+2y} + 2^{2x} + 2^{2y} + 1 + 2^{2x+2y} - 2^{2x} - 2^{2y} + 1}$$

$$= \frac{2^{2x+2y} - 1 + 2^{2x+2y} - 1}{2^{2x+2y} + 1 + 2^{2x+2y} + 1}$$

$$= \frac{2(2^{2(x+y)} - 1)}{2(2^{2(x+y)} + 1)}$$

$$= \frac{2^{2(x+y)} - 1}{2^{2(x+y)} + 1}$$

$=$ ptan $(x + y)$

d. samples:

psin x pcos y − pcos x psin y

$= (2^{x-1} - 2^{-x-1})(2^{y-1} + 2^{-y-1}) - (2^{x-1} + 2^{-x-1}) - (2^{y-1} - 2^{-y-1})$

$= 2^{x+y-2} + 2^{x-y-2} - 2^{-x+y-2} - 2^{-x-y-2} - 2^{x+y-2} + 2^{x-y-2} - 2^{-x+y-2} + 2^{-x-y-2}$

$= 2^{x-y-2} - 2^{-x+y-2} + 2^{x-y-2} - 2^{-x+y-2}$

$= 2^{(x-y)-1} - 2^{-(x-y)-1}$

$=$ psin$(x - y)$;

pcos x pcos y − psin x psin y

$= (2^{x-1} + 2^{-x-1})(2^{y-1} + 2^{-y-1}) - (2^{x-1} - 2^{-x-1}) \cdot (2^{y-1} - 2^{-y-1})$

$= 2^{x+y-2} + 2^{x-y-2} + 2^{-x+y-2} + 2^{-x-y-2} - 2^{x+y-2} + 2^{x-y-2} + 2^{-x+y-2} - 2^{-x-y-2}$

$= 2^{x-y2} + 2^{-x+y-2} + 2^{x-y-2} + 2^{-x+y-2}$

$= 2^{(x-y)-1} + 2^{-(x-y)-1}$

$=$ pcos$(x - y)$;

$\dfrac{\text{ptan } x - \text{ptan } y}{1 - \text{ptan } x \text{ ptan } y}$

$$= \frac{\frac{2^{2x}-1}{2^{2x}+1} - \frac{2^{2y}-1}{2^{2y}+1}}{1 - \frac{2^{2x}-1}{2^{2x}+1} \cdot \frac{2^{2y}-1}{2^{2y}+1}}$$

$$= \frac{\frac{(2^{2x}-1)(2^{2y}+1) - (2^{2y}-1)(2^{2x}+1)}{(2^{2x}+1)(2^{2y}+1)}}{\frac{(2^{2x}+1)(2^{2y}+1) - (2^{2x}-1)(2^{2y}-1)}{(2^{2x}+1)(2^{2y}+1)}}$$

$$= \frac{(2^{2x}-1)(2^{2y}+1) - (2^{2y}-1)(2^{2x}+1)}{(2^{2x}+1)(2^{2y}+1) - (2^{2x}-1)(2^{2y}-1)}$$

$$= \frac{2^{2x+2y} - 2^{2y} + 2^{2x} - 1 - 2^{2x+2y} + 2^{2x} - 2^{2y} + 1}{2^{2x+2y} + 2^{2x} + 2^{2y} + 1 - 2^{2x+2y} + 2^{2x} + 2^{2y} - 1}$$

$$= \frac{-2^{2y} + 2^{2x} + 2^{2x} - 2^{2y}}{2^{2x} + 2^{2y} + 2^{2x} + 2^{2y}}$$

$$= \frac{2^{2x+1} - 2^{2y+1}}{2^{2x+1} + 2^{2y+1}}$$

$$= \frac{\frac{2^{2x+1}}{2^{2y+1}} - 1}{\frac{2^{2x+1}}{2^{2y+1}} + 1}$$

$$= \frac{2^{2(x-y)} - 1}{2^{2(x-y)} + 1}$$

$=$ ptan $(x - y)$

4. samples:

$$\sin 15° = \sin(45° - 30°) = \sin 45° \cos 30° - \cos 45° \sin 30° = \frac{\sqrt{2}}{2} \cdot \frac{\sqrt{3}}{2} - \frac{\sqrt{2}}{2} \cdot \frac{1}{2} = \frac{\sqrt{6} - \sqrt{2}}{4}$$

$$\sin 75° = \sin(45° + 30°) = \sin 45° \cos 30° + \cos 45° \sin 30° = \frac{\sqrt{2}}{2} \cdot \frac{\sqrt{3}}{2} + \frac{\sqrt{2}}{2} \cdot \frac{1}{2} = \frac{\sqrt{6} + \sqrt{2}}{4}$$

5. For any acute or obtuse $\triangle ABC$,
$\tan A + \tan B + \tan C$

$$= \frac{\sin A}{\cos A} + \frac{\sin B}{\cos B} + \frac{\sin C}{\cos C}$$

$$= \frac{\sin A \cos B \cos C + \sin B \cos A \cos C + \sin C \cos A \cos B}{\cos A \cos B \cos C}$$

$$= \frac{\cos C\ (\sin A \cos B + \sin B \cos A) + \sin C \cos A \cos B}{\cos A \cos B \cos C}$$

$$= \frac{\cos C \sin(A + B) + \sin C \cos A \cos B}{\cos A \cos B \cos C}$$

$$= \frac{\cos C \sin(180° - C) + \sin C \cos A \cos B}{\cos A \cos B \cos C}$$

$$= \frac{\cos C \sin C + \sin C \cos A \cos B}{\cos A \cos B \cos C}$$

$$= \frac{\sin C\ (\cos C + \cos A \cos B)}{\cos A \cos B \cos C}$$

$$= \frac{\sin C\ (\cos(180° - (A + B)) + \cos A \cos B)}{\cos A \cos B \cos C}$$

$$= \frac{\sin C(\cos 180° \cos(A + B) + \cos A \cos B)}{\cos A \cos B \cos C}$$

$$= \frac{\sin C\ (-\cos A \cos B + \sin A \sin B + \cos A \cos B)}{\cos A \cos B \cos C}$$

$$= \frac{\sin C\ (\sin A \sin B)}{\cos A \cos B \cos C}$$

$$= \frac{\sin A \sin B \sin C}{\cos A \cos B \cos C}$$

$= \tan A \tan B \tan C.$

CHAPTER 7

1.a. The program asks for an initial value, then based on a recursive definition, outputs terms of a sequence. If a term is 1, the sequence terminates; otherwise, the next term is computed using the following rule:

$$a_{k+1} = \begin{cases} 3a_k + 1 & \text{when } a_k \text{ is odd} \\ \frac{a_k}{2} & \text{when } a_k \text{ is even} \end{cases}$$

b. sample:

N	Output
3	10, 5, 16, 8, 4, 2, 1
4	2, 1
5	16, 8, 4, 2, 1
6	3, 10, 5, 16, 8, 4, 2, 1
7	22, 11, 34, 17, 52, 26, 13, 40, 20, 10, 5, 16, 8, 4, 2, 1

2.ai. $3 \cdot 5,\ 5 \cdot 8,\ 8 \cdot 13$

ii. $S(n)$: $\sum_{i=1}^{n} F_i^2 = F_n \cdot F_{n+1}$

iii. sample (for $n = 7$):
$1^2 + 1^2 + 2^2 + 3^2 + 5^2 + 8^2 + 13^2 = 273 = 13 \cdot 21$

iv. (1) $1^2 = 1 = 1 \cdot 1 = F_1 \cdot F_2$, so $S(1)$ is true.
(2) Assume $S(k)$ is true for some integer k. Must show $S(k + 1)$ is true.
$F_1^2 + F_2^2 + F_3^2 + \ldots + F_k^2 + F_{k+1}^2$
$= F_k \cdot F_{k+1} + F_{k+1}^2$ Since $S(k)$ is true
$= F_{k+1}(F_k + F_{k+1})$ Distributive Property
$= F_{k+1} \cdot F_{k+2}$ Definition of Fibonacci sequence
∴ By the Principle of Mathematical Induction, $S(n)$ is true for all natural numbers n.

bi. 7, 12, 20

ii. $S_n = \sum_{i=1}^{n+1} F_i$. $S(n)$: $S_n = S_{n-1} + S_{n-2} + 1$

iii. sample (for $n = 6$):
$S_6 = 1 + 1 + 2 + 3 + 5 + 8 + 13 = 33 = 20 + 12 + 1 = S_5 + S_4 + 1$

iv. (1) $S_2 + S_1 + 1 = 4 + 2 + 1 = 7 = S_3$, so $S(3)$ is true.
(2) Assume $S(k)$ is true for some positive integer $k \geq 3$.
Must show $S(k + 1)$ is true.

$S_{k+1} = \sum_{i=1}^{k+2} F_i$	Def. of S
$= F_{k+2} + \sum_{i=1}^{k+1} F_i$	Rewrite summation.
$= F_{k+2} + S_k$	Def. of S
$= F_{k+2} + S_{k-1} + S_{k-2} + 1$	By inductive assumption
$= F_{k+1} + F_k + S_{k-1} + S_{k-2} + 1$	Def. of Fibonacci seq.
$= (F_{k+1} + S_{k-1}) + (F_k + S_{k-2}) + 1$	Commutativity
$= \left(F_{k+1} + \sum_{i=1}^{k} F_i\right) + \left(F_k + \sum_{i=1}^{k-1} F_i\right) + 1$	Def. of S
$= \left(\sum_{i=1}^{k+1} F_i\right) + \left(\sum_{i=1}^{k} F_i\right) + 1$	Rewrite summations.
$= S_k + S_{k-1} + 1$	Def. of S

∴ By the Principle of Mathematical Induction, $S(n)$ is true for all natural numbers $n \geq 3$.

ci. 5, 13, 34, 89

ii. $S_n = \sum_{i=1}^{n} (F_i^2 + F_{i+1}^2)$. $S(n)$: $S_n = 3S_{n-1} - S_{n-2}$

iii. sample (for $n = 6$):
$S_6 = 8^2 + 13^2 = 233 = 267 - 34 = 3(89) - 34 = 3S_5 + S_4$

iv. (1) $3S_2 + S_1 = 3(5) - 2 = 15 - 2 = 13 = 4 + 9 = 2^2 + 3^2 = S_3$, so $S(3)$ is true.
(2) Assume $S(k)$ is true for some positive integer $k \geq 3$.
Must show $S(k + 1)$ is true.
$S_{k+1} = \sum_{i=1}^{k+1} (F_i^2 + F_{i+1}^2)$
$= F_{k+1}^2 + F_{k+2}^2 + \sum_{i=1}^{k} (F_i^2 + F_{i+1}^2)$

$= F_{k+1}^2 + F_{k+2}^2 + S_k$
$= F_{k+1}^2 + F_{k+2}^2 + 3S_{k-1} - S_{k-2}$
$= F_{k+1}^2 + (F_k + F_{k+1})^2 + 3S_{k-1} - S_{k-2}$
$= F_{k+1}^2 + F_k^2 + 2F_kF_{k+1} + F_{k+1}^2 + 3S_{k-1} - S_{k-2}$
$= 2F_{k+1}^2 + F_k^2 + 2F_kF_{k+1} + 3S_{k-1} - S_{k-2}$
$= 2F_{k+1}^2 + F_k^2 + 2F_k(F_k + F_{k-1}) + 3S_{k-1} - S_{k-2}$
$= 2F_{k+1}^2 + F_k^2 + 2F_k^2 + 2F_kF_{k-1} + 3S_{k-1} - S_{k-2}$
$= 2F_{k+1}^2 + 2F_k^2 + F_k^2 + 2F_kF_{k-1} + 3S_{k-1} - S_{k-2}$
$= 2F_{k+1}^2 + 2F_k^2 + F_k^2 + 2F_kF_{k-1} + F_{k-1}^2 - F_{k-1}^2 + 3S_{k-1} - S_{k-2}$
$= 2F_{k+1}^2 + 2F_k^2 + (F_k + F_{k-1})^2 - F_{k-1}^2 + 3S_{k-1} - S_{k-2}$
$= 2F_{k+1}^2 + 2F_k^2 + F_{k+1}^2 - F_{k-1}^2 + 3S_{k-1} - S_{k-2}$
$= 3F_{k+1}^2 + 2F_k^2 - F_{k-1}^2 + 3S_{k-1} - S_{k-2}$
$= 3F_{k+1}^2 + 2F_k^2 + F_k^2 - F_k^2 - F_{k-1}^2 + 3S_{k-1} - S_{k-2}$
$= 3F_{k+1}^2 + 3F_k^2 - F_k^2 - F_{k-1}^2 + 3S_{k-1} - S_{k-2}$
$= 3(F_{k+1}^2 + F_k^2) - (F_k^2 + F_{k-1}^2) + 3S_{k-1} - S_{k-2}$
$= 3(F_{k+1}^2 + F_k^2) + 3S_{k-1} - (F_k^2 + F_{k-1}^2) - S_{k-2}$
$= 3(F_{k+1}^2 + F_k^2) + 3\sum_{i=1}^{k-1}(F_i^2 + F_{i+1}^2) - (F_k^2 + F_{k-1}^2) - \sum_{i=1}^{k-2}(F_i^2 + F_{i+1}^2)$
$= 3\sum_{i=1}^{k}(F_i^2 + F_{i+1}^2) - \sum_{i=1}^{k-1}(F_i^2 + F_{i+1}^2)$
$= 3S_k - S_{k-1}$

∴ By the Principle of Mathematical Induction, $S(n)$ is true for all natural numbers $n \geq 3$.

d. Sequence defined by $S_k = \sum_{i=1}^{k} (-1)^{i+1}F_{i+1}$

$$1 = 1$$
$$1 - 2 = -1$$
$$1 - 2 + 3 = 2$$
$$1 - 2 + 3 - 5 = -3$$
$$1 - 2 + 3 - 5 + 8 = 5$$
$$1 - 2 + 3 - 5 + 8 - 13 = -8$$

Claim: $S(n)$: $S_n = (-1)^{n+1}F_n$
(1) $S_1 = 1 = 1 \cdot 1 = (-1)^2F_2$, so $S(1)$ is true.
(2) Assume $S(k)$ is true for some integer k. Must show $S(k + 1)$ is true.

$S_{k+1} = \sum_{i=1}^{k+1} (-1)^{i+1}F_{i+1}$	Def. of S_{k+1}
$= S_k + (-1)^{k+2}F_{k+2}$	Simplify
$= (-1)^{k+1}F_k + (-1)^{k+2}F_{k+2}$	Since $S(k)$ is true
$= (-1)^{k+1}F_k + (-1)^{k+2}(F_k + F_{k+1})$	Def. of Fibonacci seq.
$= (-1)^{k+1}F_k - (-1)^{k+1}F_k - (-1)^{k+1}F_{k+1}$	Distributive Property
$= -(-1)^{k+1}F_{k+1}$	Subtraction
$= (-1)^{k+2}F_{k+1}$	

∴ By the principle of Mathematical Induction, $S(n)$ is true for all natural numbers.

3.a. Let $p(i)$ = the perimeter of the ith square.
$p(1) = 4s$
$p(2) = 4\frac{s}{2}\sqrt{2} = 2\sqrt{2}\,s$
$p(3) = 4\frac{\sqrt{2}\,s}{4}\sqrt{2} = 2s$
$p(4) = 4\frac{s}{4}\sqrt{2} = \sqrt{2}\,s$
$p(5) = 4\frac{\sqrt{2}\,s}{8}\sqrt{2} = s$
$p(6) = 4\frac{s}{8}\sqrt{2} = \frac{\sqrt{2}}{2}s$

It can be seen that $p(i)$ is a geometric sequence with $p(k + 1) = \frac{\sqrt{2}}{2}p(k)$.

So with $p(1) = 4s$ and $r = \frac{\sqrt{2}}{2}$, the sum $\sum_{i=1}^{\infty} p(i)$ is

$\frac{4s}{1 - \frac{\sqrt{2}}{2}} = \frac{4s}{\frac{2-\sqrt{2}}{2}} = \frac{8s}{2 - \sqrt{2}} = \frac{8s(2 + \sqrt{2})}{2} = 4s(2 + \sqrt{2}) = (8 + 4\sqrt{2})s.$

b. Let $A(i)$ = the area of the ith triangle.
$A(1) = \frac{1}{2}\left(\frac{s}{2}\right)^2 = \frac{s^2}{8}$
$A(2) = \frac{1}{2}\left(\frac{\sqrt{2}\,s}{4}\right)^2 = \frac{s^2}{16}$
$A(3) = \frac{1}{2}\left(\frac{s}{4}\right)^2 = \frac{s^2}{32}$
$A(4) = \frac{1}{2}\left(\frac{\sqrt{2}\,s}{8}\right)^2 = \frac{s^2}{64}$
$A(5) = \frac{1}{2}\left(\frac{s}{8}\right)^2 = \frac{s^2}{128}$

It can be seen that $A(i)$ is a geometric sequence with $A(k + 1) = \frac{1}{2}A(k)$.

So with $A(1) = \frac{s^2}{8}$ and $r = \frac{1}{2}$, the sum $\sum_{i=1}^{\infty} A(i)$ is

$\frac{\frac{s^2}{8}}{1 - \frac{1}{2}} = \frac{\frac{s^2}{8}}{\frac{1}{2}} = \frac{s^2}{4}.$

c. Answers may vary.

4.a. $M = \begin{bmatrix} 1 & 1 \\ 1 & 0 \end{bmatrix}$

$M^2 = \begin{bmatrix} 2 & 1 \\ 1 & 1 \end{bmatrix}$

$M^3 = \begin{bmatrix} 3 & 2 \\ 2 & 1 \end{bmatrix}$

$M^4 = \begin{bmatrix} 5 & 3 \\ 3 & 2 \end{bmatrix}$

$M^5 = \begin{bmatrix} 8 & 5 \\ 5 & 3 \end{bmatrix}$

$M^6 = \begin{bmatrix} 13 & 8 \\ 8 & 5 \end{bmatrix}$

b. Conjecture: $S(n)$: $M^n = \begin{bmatrix} F_{n+1} & F_n \\ F_n & F_{n-1} \end{bmatrix}$ for $n \geq 2$.

PROJECT SOLUTIONS

c. (1) $M^2 = \begin{bmatrix} 2 & 1 \\ 1 & 1 \end{bmatrix} = \begin{bmatrix} F_3 & F_2 \\ F_2 & F_1 \end{bmatrix}$, so $S(1)$ is true.

(2) Assume $S(k)$ is true for some integer k. Must show $S(k+1)$ is true.

$M^{k+1} = M^k \cdot M$ — Matrix multiplication

$= \begin{bmatrix} F_{k+1} & F_k \\ F_k & F_{k-1} \end{bmatrix}\begin{bmatrix} 1 & 1 \\ 1 & 0 \end{bmatrix}$ — Since $S(k)$ is true

$= \begin{bmatrix} F_{k+1} + F_k & F_{k+1} \\ F_k + F_{k-1} & F_k \end{bmatrix}$ — Matrix multiplication

$= \begin{bmatrix} F_{k+2} & F_{k+1} \\ F_{k+1} & F_k \end{bmatrix}$ — Def. of Fibonacci sequence

∴ By the Principle of Mathematical Induction, $S(n)$ is true for all natural numbers n.

d. $N = \begin{bmatrix} 2 & 1 \\ 1 & 0 \end{bmatrix}$

$N^2 = \begin{bmatrix} 5 & 2 \\ 2 & 1 \end{bmatrix}$

$N^3 = \begin{bmatrix} 12 & 5 \\ 5 & 2 \end{bmatrix}$

$N^4 = \begin{bmatrix} 29 & 12 \\ 12 & 5 \end{bmatrix}$

So the sequence is 2, 5, 12, 29, 70, 169, ...

e. $G_n = 2G_{n-1} + G_{n-2}$ for $n \geq 2$.

f. The sequence generated by $\begin{bmatrix} a & 1 \\ 1 & 0 \end{bmatrix}$ may be recursively defined $G_n = aG_{n-1} + G_{n-2}$ for $n \geq 2$.

5.a. $A_1 = 5000 + 50 - 166 = \$4,884$
$A_2 = 4884 + 48.84 - 166 = \$4,766.84$
$A_3 = 4766.84 + 47.6684 - 166 = \$4,648.51$

b. $A_n = A_{n-1} + 0.1A_{n-1} - 166 = 1.01A_{n-1} - 166$

c. $A_3 = 1.01A_2 - 166$
$= 1.01(1.01A_1 - 166) - 166$
$= 1.01(1.01(1.01A_0 - 166) - 166) - 166$
$= 1.01(1.01(1.01 \cdot 5000 - 166) - 166) - 166$
$= 1.01(1.01^2 \cdot 5000 - 1.01 \cdot 166) - 166) - 166$
$= 1.01^3 \cdot 5000 - 1.01^2 \cdot 166 - 1.01 \cdot 166 - 166$

d. Factoring out 166 from the formula in part **c**,
$A_3 = 5000 \cdot 1.01^3 - 166(1.01^2 + 1.01 + 1)$
$= 5000 \cdot 1.01^3 - 166(1.01^2 + 1.01^1 + 1.01^0)$
$= 5000 \cdot 1.01^3 - 166(1.01^0 + 1.01^1 + 1.01^2)$.

e. Conjecture: $A_n = 5000 \cdot 1.01^n - 166(1.01^0 + 1.01^1 + 1.01^2 + \ldots + 1.01^{n-1})$

f. $A_n = 5000 \cdot 1.01^n - 166\left(\frac{1 - 1.01^n}{1 - 1.01}\right)$
$= 5000 \cdot 1.01^n - 166\left(\frac{1 - 1.01^n}{-.01}\right)$
$= 5000 \cdot 1.01^n + 16600(1 - 1.01n)$

g. $A_3 = 5000 \cdot 1.01^3 + 16600(1 - 1.01^3) = 4648.5084 \approx \$4,648.51$

h. Following the same steps,
$A_n = B \cdot (1 + r)^n + \frac{166}{r}(1 - (1 + r)^n)$.

i. $0 = B \cdot (1 + r)^n + \frac{P}{r}(1 - (1 + r)^n)$

$\frac{P}{r}(1 - (1 + r)^n) = -B \cdot (1 + r)^n$

$P = \frac{-rB \cdot (1 + r)^n}{1 - (1 + r)^n}$

j. Sample:
The annual interest rate is 10%, and I would need to borrow \$8000 to buy the car I want. The monthly rate, r, is approximately 0.8333%.

$P = \frac{-rB \cdot (1 + r)^n}{1 - (1 + r)^n} = \frac{-.008333 \cdot 8000 \cdot (1.008333)^{36}}{1 - 1.008333^{36}} =$
$\frac{-70.6664 \cdot 1.372457}{1 - 1.372457} = \frac{-96.986592}{-0.372457} = \$260.40$/month.

After 3 years, I would have paid $36 \cdot (260.40) = \$9,374.40$, which is \$1,374.40 more than I borrowed.

CHAPTER 8

1.a. Addition holds:
$\begin{bmatrix} a & -b \\ b & a \end{bmatrix} + \begin{bmatrix} c & -d \\ d & c \end{bmatrix} = \begin{bmatrix} a + c & -(b + d) \\ b + d & a + c \end{bmatrix}$
$(a + bi) + (c + di) = (a + c) + (b + d)i$

Multiplication holds:
$\begin{bmatrix} a & -b \\ b & a \end{bmatrix} \cdot \begin{bmatrix} c & -d \\ d & c \end{bmatrix} = \begin{bmatrix} (ac - bd) & (-ad - bc) \\ (bc + ad) & (-bd + ac) \end{bmatrix}$
$= \begin{bmatrix} (ac - bd) & -(ad + bc) \\ (ad + bc) & (ac - bd) \end{bmatrix}$
$(a + bi) \cdot (c + di) = ac - bd + adi + bci$
$= (ac - bd) + (ad + bc)i$

b. $\det\left(\begin{bmatrix} a & -b \\ b & a \end{bmatrix}\right) = a^2 + b^2$, and the modulus of $a + bi$ is $\sqrt{a^2 + b^2}$. Hence, the modulus of $a + bi$ is the square root of the determinant of its matrix.

c. The additive inverse of $\begin{bmatrix} a & -b \\ b & a \end{bmatrix}$ is $\begin{bmatrix} -a & b \\ -b & -a \end{bmatrix}$, which is the matrix for $-a - bi$, the additive inverse of $a + bi$.

The multiplicative inverse of $\begin{bmatrix} a & -b \\ b & a \end{bmatrix}$ is

$\begin{bmatrix} \frac{a}{(a^2 + b^2)} & \frac{b}{(a^2 + b^2)} \\ \frac{-b}{(a^2 + b^2)} & \frac{a}{(a^2 + b^2)} \end{bmatrix}$, which is the matrix for $\frac{a - bi}{a^2 + b^2}$, and $(a + bi)\frac{a - bi}{a^2 + b^2} = \frac{a^2 - b^2 i^2}{a^2 + b^2} = \frac{a^2 + b^2}{a^2 + b^2} =$

1. Hence, the multiplicative inverse of a complex number is the multiplicative inverse of its matrix.

d. The multiplicative identity matrix is $\begin{bmatrix} 1 & 0 \\ 0 & 1 \end{bmatrix}$, because $\begin{bmatrix} 1 & 0 \\ 0 & 1 \end{bmatrix}\begin{bmatrix} a & -b \\ b & a \end{bmatrix} = \begin{bmatrix} a & -b \\ b & a \end{bmatrix}$. The additive identity matrix is $\begin{bmatrix} 0 & 0 \\ 0 & 0 \end{bmatrix}$, because $\begin{bmatrix} 0 & 0 \\ 0 & 0 \end{bmatrix} + \begin{bmatrix} a & -b \\ b & a \end{bmatrix} = \begin{bmatrix} a & -b \\ b & a \end{bmatrix}$.

e. The argument of $a + bi$ is $\tan^{-1} \frac{b}{a}$ or $\sin^{-1} \frac{b}{\sqrt{a^2 + b^2}}$ or $\cos^{-1} \frac{a}{\sqrt{a^2 + b^2}}$. For the matrix with determinant d, the argument is $\sin^{-1} \frac{b}{\sqrt{d}}$ or $\cos^{-1} \frac{a}{\sqrt{d}}$.

2. Sample (for part **a**, electricity):
Students' reports may include the following main points:
1. use of letter $j = \sqrt{-1}$ instead of i in expressing complex numbers
2. use of complex numbers to represent current, voltage, and power
3. use of Euler's identity to deal with certain time functions

Sample (for part **c**, differential equations):
Students' reports may include the following main points:
1. use of Euler's formula to determine roots of characteristic equation
2. the occurrence of roots of differential equations is often as complex conjugates

3. The general equation for all conic sections in polar coordinates is $p = \frac{(eq)}{1 + e\cos\theta}$ where e is the eccentricity: $e = 1$ (parabola), $e < 1$ (ellipse), and $e > 1$ (hyperbola). The focus is at the pole, and the directrix perpendicular to the polar axis is at a distance q from the pole.
For any curve which is the locus of a point which moves so that the ratio of its distance from a fixed point to its distance from a fixed line is constant, the ratio is the *eccentricity* of the curve, the fixed point is the *focus*, and the fixed line is the *directrix*.

4. Sample:

(1) Cylindrical Coordinates
A point in space may be described in a similar manner to rectangular coordinates. Keep the third, z, coordinate, but replace the first two (x and y) coordinates with their polar equivalents. Hence a cylindrical coordinate is (r, θ, z), as shown at the right above.

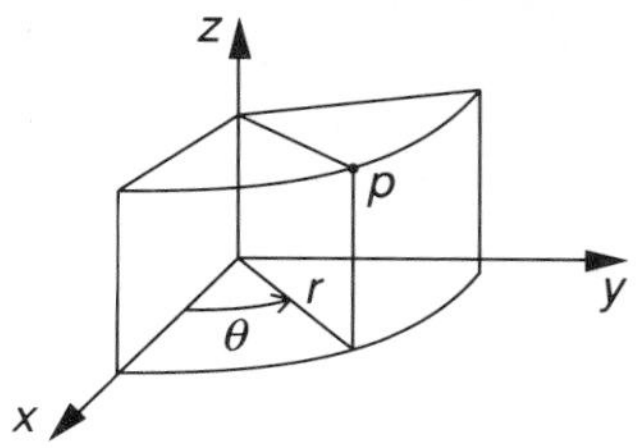

When r is fixed and z and θ vary, a cylinder is formed. When z is fixed and r and θ vary, a plane parallel to the x-y plane is formed. When θ is fixed and z and r vary, a plane containing the z-axis is formed. The intersection of these surfaces defines the point $P(r, \theta, z)$. To convert from cylindrical to rectangular coordinates:
$x = r \cos \theta$
$y = r \sin \theta$
$z = z$

(2) Spherical Coordinates
The position of a point P in space may be assigned coordinates by its radius vector, r (the distance from P to a fixed origin) and two angles–θ(colatitude angle) which is the angle made by a vector from the origin to point P with the polar or y-axis, and ϕ (longitude) which is the angle between the θ plane and a fixed plane through the polar axis (the initial meridian plane).

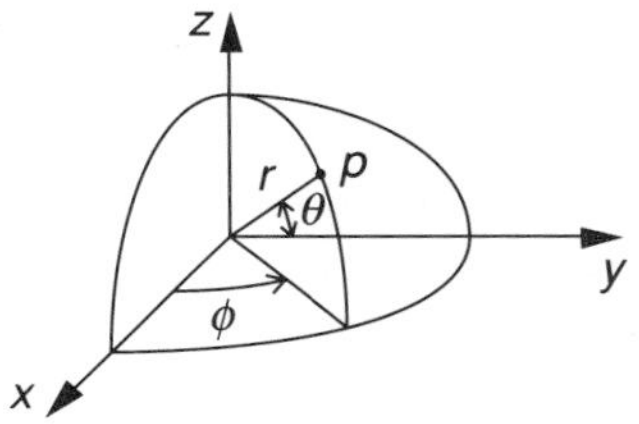

5. Answers will vary. One good source of information on spirals in nature is *Patterns in Nature* by Peter S. Stevens.

6. Quaternion numbers are numbers of the form $x = x_0 + x_1 i + x_2 j + x_3 k$, where x_0, x_1, x_2, and x_3 are real numbers. The following relationships hold among the quaternionic units: $i^2 = j^2 = k^2 = -1$, and $ij = jk = ki = -ji = -kj = ik = -1$. Addition is defined as $x + y = (y_0 + x_0) + (y_1 + x_1)i + (y_2 + x_2)j + (y_3 + x_3)k$. Scalar multiplication is defined as $cx = cx_0 + cx_1 i + cx_2 j + cx_3 k$. Multiplication is defined as $xy = (x_0 y_0 - x_1 y_1 - x_3 y_1 - x_1 y_2 + x_3 y_2 + x_1 y_3 - x_2 y_3 - x_3 y_3) + (x_1 y_0 + x_0 y_1)i + (x_2 y_0 + x_0 y_2)j + (x_3 y_0 + x_0 y_3)k$.

7. Students' reports may contain some of the following main topics:

1. chaotic phenomena in weather (Edward Lorenz)
2. the great red spot of Jupiter
3. chaotic phenomena in wildlife populations (Robert May)
4. chaotic phenomena and earthquakes
5. chaotic phenomena and clouds
6. chaotic phenomena in physiology (blood vessels, bronchial tubes)
7. chaotic phenomena in astronomy (Michael Henon)
8. the work of Benoit Mandelbrot
9. the work of Michael Burnsley

8.a. sample:

```
 10 REM ORBITS OF DYNAMICAL SYSTEMS
 20 REM A is the number of terms
 30 REM The window displayed will have corners
    (-MAX, -MAX) and (MAX, MAX)
 40 A = 100:MAX = 2
 50 DIM ORE(A), OLM(A)
 60 INPUT "ENTER C:";CR, CI
 70 INPUT "ENTER D";DR, DI
 80 ZR = DR:ZI = DI
 90 REM Draw axes then plot points
100 LINE (0,128) – (512,128) : LINE (256,0) – (256,256)
110 FOR N = 1 TO A
120 ZR = ZR*ZR – ZI*ZI + CR
130 ZI = 2*ZR*ZI + CI
140 ORE(N) = ZR: OLM(N) = ZI:PR=
    (128/MAX)*ZR:PI=(128/MAX)*ZI
150 IF (PR < -256) OR (PR > 256) OR (PI <-128) OR
    (PI>128) THEN 170
160 PSET (256+PR,128–PI)
170 NEXT N
180 REM Output list of values in orbit
190 INPUT "HIT <RETURN>";A$
200 FOR N = 1 TO A
210 PRINT ORE(N);"+";OLM(N);"i"
220 NEXT N
230 END
```

b. sample (for c = -.39054 − .58679i, d = 0):

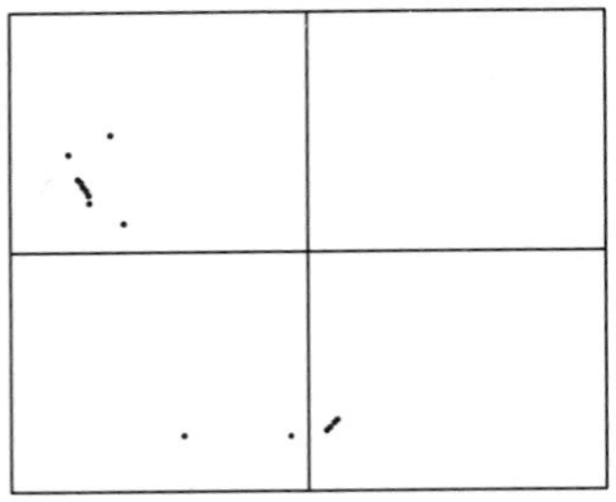

sample (for c = -.11 + .67i, d = 0):

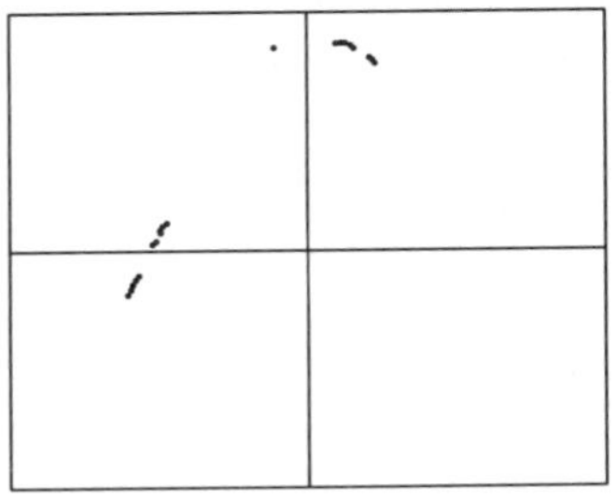

c. sample (for c = -.39054 − .58679i):

sample (for c = -.11 + .67i):

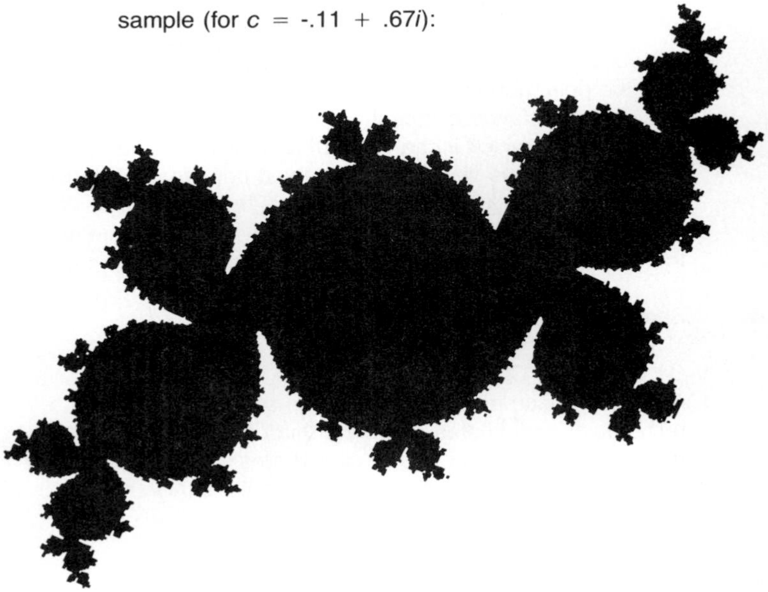

d. The Mandelbrot Set is the set of all complex numbers which when iterated by the rule $z \rightarrow z^2 + c$ do not diverge, where c is the complex number being tested.

sample program:

```
 10 REM MANDELZOOM
 20 INPUT "ENTER THE REAL AND IMAGINARY PARTS
    OF C:",ACORNER,BCORNER
 30 INPUT "ENTER THE SIZE OF THE VIEWING
    WINDOW:",SIZE
 40 CLS
 50 P = 200
 60 GAP = SIZE/P
 70 FOR J = 1 TO P
 80 FOR K = 1 TO P
 90   AC = ACORNER + J*GAP
100   BC = BCORNER + K*GAP
110   AZ = 0: BZ = 0
120   FOR COUNT = 1 TO 200
130     OLDAZ = AZ
140     AZ = AZ*AZ - BZ*BZ + AC
150     BZ = 2*OLDAZ*BZ + BC
160     MAGZ = SQR(AZ*AZ + BZ*BZ)
170     IF MAGZ > 2 THEN 200
180   NEXT COUNT
190   PSET (J, P-K)
200 NEXT K
210 NEXT J
220 END
```

sample (for $c = -2 - 1.25i$):

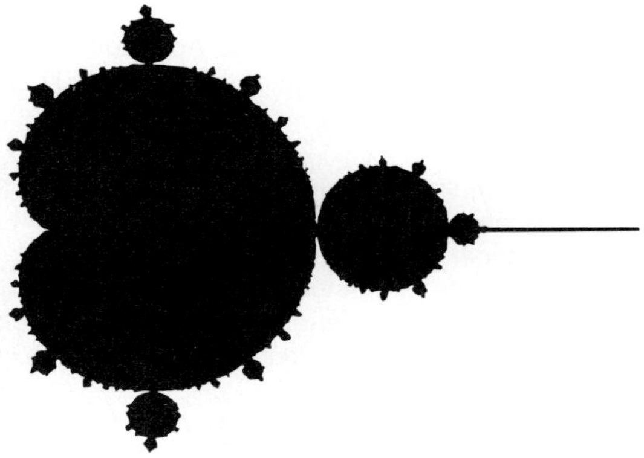

CHAPTER 9

1.ai. sample:

$y_1(x) = x^2 + 4$
$y_2(x) = x^2 - x + 2$
$y_3(x) = x^3$
$y_4(x) = \sin x$
$y_5(x) = \frac{1}{2}x - 5$

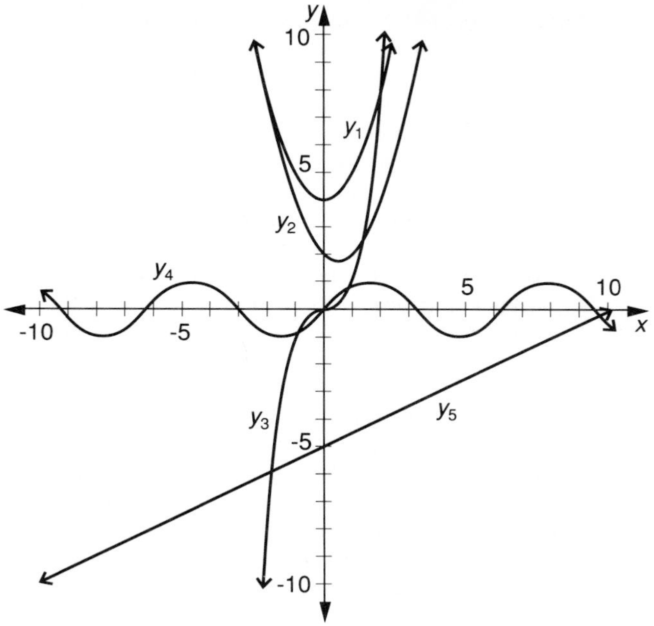

ii. relative maxima: for y_4: 1 which occurs at $x = \frac{\pi}{2} + 2\pi n$ for n an integer

relative minima: for y_1: 4 which occurs at $x = 0$

for y_2: $\frac{7}{4}$ which occurs at $x = \frac{1}{2}$

for y_4: -1 which occurs at $x = -\frac{\pi}{2} + 2\pi n$ for n an integer

iii. $y_4'(x) = \cos x$, and $\cos\left(\frac{\pi}{2} + 2n\pi\right) = 0$ for n an integer.

$y_1'(x) = 2x$, and $2 \cdot 0 = 0$.

$y_2'(x) = 2x - 1$, and $(2)\frac{1}{2} - 1 = 1 - 1 = 0$.

$y_4'(x) = \cos x$, and $\cos\left(-\frac{\pi}{2} + 2n\pi\right) = 0$ for n an integer.

iv. The graph looks more and more like a horizontal line.

v. The derivative at a point is the slope of the tangent line to the function at that point. Since the derivatives were zero at each minimum and maximum point as shown in part **iii**, we would expect the tangent line at each maximum or minimum point to have a slope of zero (i.e., be a horizontal line). Zooming in on a section of the graph causes its image to appear more and more like the tangent at that point.

b. The graph of f remains V-shaped when zooming in on its minimum point (2, 3), while the graphs in part **a.iv.** got closer and closer to a horizontal line.

c. If when zooming in on a portion of a graph, the graph does not approach a line, then the derivative does not exist at that point. If however, it does approach a line, the derivative is equal to the slope of that line; if such a line is horizontal, then the derivative of the function is zero at that point.

2. Students' reports may include some of the following main points:

Newton

1. devised corpuscular theory of light
2. verified laws of gravitation
3. justified Kepler's laws of planetary motion
4. wrote *Method of Fluxions* which contained differential calculus, differential equations which allowed him to find maxima, minima, tangents to curves, points of inflection, convexity and concavity of curves, and Newton's method for approximating the values of real roots of an algebraic or transcendental numerical equation
5. wrote *Arithmetica Universalis* which contained many important results about the theory of equations: imaginary roots of real polynomials must occur in conjugate pairs; rules for finding an upper bound for roots of a real polynomial; an extension of Descartes' rule of signs to give limits on the number of imaginary roots of a real polynomial; and formulas expressing the sum of *n*th powers of the roots of a polynomial in terms of the coefficients of the polynomial

Liebniz

1. developed *characteristica generalis*, a univeral language, using mathematical and symbolic logic created from formal rules that would obviate the need for thinking (e.g., logical addition, multiplication, negation, null class, and class inclusion)
2. independently from Newton created calculus and introduced the $\int$ integral notation
3. introduced the theory of determinants
4. generalized the binomial theorem to the multinomial theorem

3.a. sample:

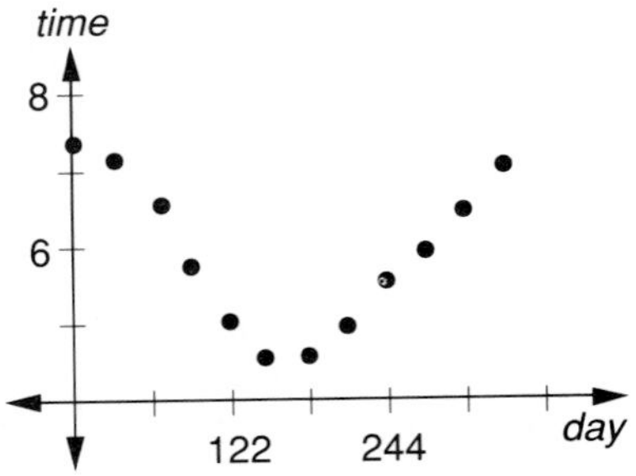

The graph appears to be a cosine curve.

$y = 6 + 1.5 \cos\left(\frac{2\pi}{365}x\right)$, where y is the time of sunrise and x is the number of days.

b. Let $f(x) = 6 + 1.5 \cos\left(\frac{2\pi}{365}x\right)$.

Month	First Day	Last Day	Average Rate of Change $\frac{f(\text{last day}) - f(\text{first day})}{\text{last day} - \text{first day}}$
Jan	0	31	-0.0068
Feb	31	59	-0.0179
Mar	59	90	-0.0245
Apr	90	120	-0.0247
May	120	151	-0.0213
Jun	151	181	-0.0043
Jul	181	212	0.0061
Aug	212	243	0.0177
Sep	243	273	0.0247
Oct	273	304	0.0248
Nov	304	334	0.018
Dec	334	365	0.0067

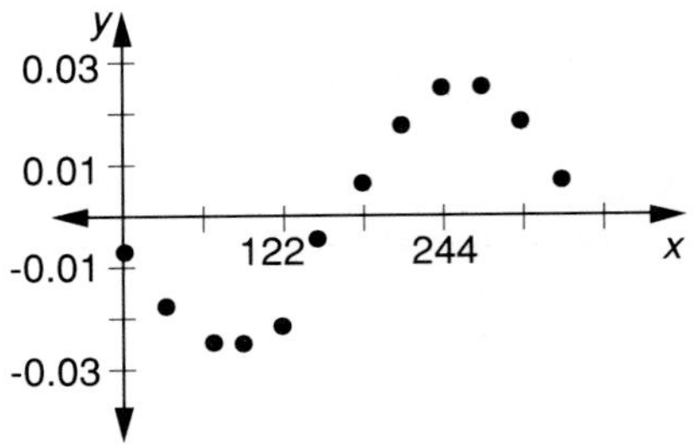

A formula that fits these data is $y = -0.025 \sin\left(\frac{2\pi}{365}x\right)$.

c. The fastest change is March to April and September to October. The slowest change is June to July and December to January.

d. The time of sunrise is later when the length of day is shorter, and the time of sunrise is earlier when the length of day is longer.
When the rate of change in time of sunrises is the slowest: (1) June to July, the length of day is longest; and (2) December to January, the length of day is shortest.

4.a. Let $x = x_2$ and $y = 0$ in the equation for the tangent line, $y - f(x_1) = f'(x_1)(x - x_1)$. So $0 - f(x_1) = f'(x_1)(x_2 - x_1)$,

$$x_2 - x_1 = \frac{-f(x_1)}{f'(x_1)},$$

$$x_2 = x_1 - \frac{f(x_1)}{f'(x_1)}.$$

b.

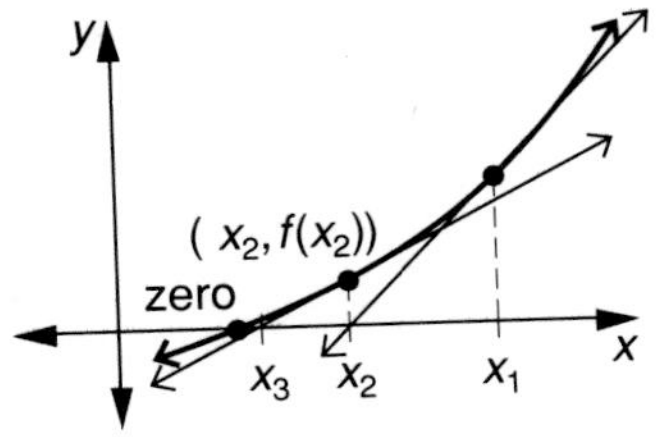

The equation of the tangent line is $y - f(x_2) = f'(x_2)(x - x_2)$.

c. Similarly to part **a**, x_3 is found to be $x_2 - \frac{f(x_2)}{f'(x_2)}$.

d. $x_{n+1} = x_n - \frac{f(x_n)}{f'(x_n)}$

e. $f(x) = 2x^3 + 3x^2 + 1$, and $f'(x) = 6x^2 + 6x$.
Arbitrarily, let $x_1 = -2$.

n	$f(x_n)$	$f'(x_n)$	$x_{n+1} = x_n - \frac{f(x_n)}{f'(x_n)}$
1	-3	12	-1.75
2	-0.53125	7.875	≈ -1.68
3	-0.016064	6.8544	≈ -1.68

The Intermediate Value Theorem allows us to use the Bisection Method (found in Chapter 3) to estimate the roots.

For example, it may be observed from a graph, that a zero lies between $x = -2$ and $x = -1$.

step	interval to check	value of endpoints	midpoint	value at midpoint
1	(-2, -1)	(-3, +2)	-1.5	+1
2	(-2, -1.5)	(-3, +1)	-1.75	-0.53
3	(-1.75, -1.5)	(-0.53, +1)	-1.625	+0.339
4	(-1.75, -1.625)	(-0.53, +0.339)	-1.6875	-0.68
5	(-1.6875, -1.625)	(-0.68, +0.339)	-1.6563	+0.143
6	(-1.6875, -1.6563)	(-0.68, +0.143)	-1.6719	+0.039
7	(-1.6875, -1.6719)	(-0.68, +0.039)	-1.6797	-0.014
8	(-1.6797, -1.6719)	(-0.014, +0.039)	-1.6758	+0.013
9	(-1.6797, -1.6758)	(-0.014, +0.013)		

So, to two decimal places, the zero is at $x = -1.68$. It took the Bisection Method 8 steps to find the zero to two decimal places, and it took Newton's Method only 3.

CHAPTER 10

1. Answers will vary from state to state.

2. Answers will vary from school to school.

3. Sample:
There are 11 partitions of the number 6:

6
1 + 5
2 + 4
3 + 3
1 + 1 + 4
1 + 2 + 3
2 + 2 + 2
1 + 1 + 1 + 3
1 + 1 + 2 + 2
1 + 1 + 1 + 1 + 1 + 1

There are 15 partitions of the number 7:

7
1 + 6
2 + 5
3 + 4
1 + 1 + 5
1 + 2 + 4
1 + 3 + 3
2 + 2 + 3
1 + 1 + 1 + 4
1 + 1 + 2 + 3
1 + 2 + 2 + 2
1 + 1 + 1 + 1 + 3
1 + 1 + 1 + 2 + 2
1 + 1 + 1 + 1 + 1 + 2
1 + 1 + 1 + 1 + 1 + 1 + 1

The following program is an example of a method which generates partitions of n in increasing length, in dictionary order for each fixed length.

```
 10 REM PARTITIONING INTEGERS PROGRAM
 20 DIM P(100)
 30 INPUT "ENTER A POSITIVE INTEGER TO
    PARTITION:";N
 40 PRINT
 50 NUMP = 0
100 L = 1
110 P(1) = N
120 P(0) = -1
130 IF (L>N) THEN 280
140   FOR J=1 TO L
150     PRINT P(J);" ";
160   NEXT J
170   PRINT
180   NUMP = NUMP + 1
190   I = L - 1
200   IF ((P(L)-P(I)) > = 2) THEN GOTO 230
210     I = I - 1
220     GOTO 200
230   IF (I <> 0) THEN FOR J = L TO I STEP -1:P(J) =
      P(I) + 1:NEXT J
240   IF (I = 0) THEN FOR J = 1 to L:P(J) = 1:NEXT J:
      L = L + 1
250   SUM = 0:FOR J = 1 TO L - 1: SUM = SUM +
      P(J):NEXT J
```

```
260   P(L) = N − SUM
270 GOTO 130
280 PRINT:PRINT "THERE ARE";NUMP;"PARTITIONS."
290 END
```

Students may also attempt a method which generates/counts partitions of an integer in dictionary order regardless of length.

4.a. $(x + y + z)^3$
$= (x + y + z)(x + y + z)^2$
$= (x + y + z)(x^2 + 2xz + z^2 + 2yz + y^2 + 2xy)$
$= x^3 + 3x^2z + 3xz^2 + 6xyz + 3xy^2 + 3x^2y + 3yz^2 + 3y^2z + y^3 + z^3$

So the fourth layer is

$$\begin{array}{ccccccc} & & & 1z^3 & & & \\ & & 3xz^2 & & 3yz^2 & & \\ & 3x^2z & & 6xyz & & 3y^2z & \\ 1x^3 & & 3x^2y & & 3xy^2 & & 1y^3. \end{array}$$

b. Calculated similarly are the fifth and sixth layers shown below.

fifth layer:

$$\begin{array}{ccccccccc} & & & & 1z^4 & & & & \\ & & & 4xz^3 & & 4yz^3 & & & \\ & & 6x^2z^2 & & 12xyz^2 & & 6y^2z^2 & & \\ & 4x^3z & & 12x^2yz & & 12xy^2z & & 4y^3z & \\ 1x^4 & & 4x^3y & & 6x^2y^2 & & 4xy^3 & & 1y^4; \end{array}$$

sixth layer:

$$\begin{array}{ccccccccccc} & & & & & 1z^5 & & & & & \\ & & & & 5xz^4 & & 5yz^4 & & & & \\ & & & 10x^2z^3 & & 20xyz^3 & & 10y^2z^3 & & & \\ & & 10x^3z^2 & & 30x^2yz^2 & & 30xy^2z^2 & & 10y^3z^2 & & \\ & 5x^4z & & 20x^3yz & & 30x^2y^2z & & 20xy^3z & & 5y^4z & \\ 1x^5 & & 5x^4y & & 10x^3y^2 & & 10x^2y^3 & & 5xy^4 & & 1y^5 \end{array}$$

c. As positioned in the text, powers of z increase from the bottom to the top of each layer. Powers of x increase from right to left, and powers of y increase from left to right. A coefficient in the center of a layer is equal to the sum of the three coefficients in the previous layer which form a triangle immediately above it.

d. The students' models might be constructed out of cardboard, light wood, or plastic. The triangular layers would be of increasing sizes and could be connected with string for a mobile style or solid pillars for a free standing version.

e. Pascal's triangle appears on each of the 3 side faces of the tetrahedron but not on the base. One represents $(x + y)^n$, another $(x + z)^n$, and the third $(y + z)^n$.

5. Sample (for poker): The following odds may be found in an almanac.

Hand	Odds
Nothing	1 to 1
One pair	1.37 to 1
Two pairs	20 to 1
Three of a kind	46 to 1
Straight	254 to 1
Flush	508 to 1
Full house	693 to 1
Four of a kind	4,164 to 1
Straight flush	72,192 to 1
Royal flush	649,739 to 1

The odds against being dealt a particular hand is simply $\dfrac{\text{total number of hands}}{\text{number possible of particular hand}} - 1$ to 1. Combinatorics may be used to compute the number possible of each poker hand as discussed below.

Royal flush: There are four possible suits, so 4 royal flushes are possible.

Straight flush: The first card may be an A, 2, 3, 4, … , 9, and there are 4 suits, so there are $4 \cdot 9 = 36$ possible other straight flushes.

Four of a kind: There are 13 different ranks possible. For the fifth card there are $52 - 4 = 48$ that could be dealt. So there are $13 \cdot 48 = 624$ possible hands which contain four of a kind.

Full house: There are 13 ranks possible for the triple and $\binom{4}{3}$ ways the suits could appear on these three cards. This leaves 12 ranks from which the pair may be dealt. There are $\binom{4}{2}$ ways the suits could appear on the pair. So there are

$13 \cdot \binom{4}{3} \cdot 12 \cdot \binom{4}{2} = 13 \cdot 4 \cdot 12 \cdot 6 = 3744$ possible

full houses.

Flush: There are $\binom{13}{5}$ ways that 5 cards from one suit could be dealt, and there are 4 suits. Subtracting the number of royal flushes and other straight flushes yields $4 \cdot \binom{13}{5} - 4 - 36 = 4 \cdot 1287 - 40 = 5108$ possible flushes.

Straight: The lowest ranked card in a straight may be A, 2, 3, … , 10. There is only one way the ranks of the four remaining cards may appear. But, the cards need not be of the same suit, so there are 4^5 ways the suits may be uniquely dealt. Subtracting the number of royal flushes and straight flushes yields $10 \cdot 4^5 - 4 - 36 = 10{,}240 - 40 = 10{,}200$ possible straights.

Three of a kind: There are 13 ranks which may be dealt for the triple and $\binom{4}{3}$ ways the suits may be dealt for these three cards. The remaining two cards dealt must not be a pair and must not match the rank of the first three cards; this yields $\binom{12}{2}$ possible ways the ranks may appear on the last two cards. Since any suit is possible for these last two cards, there are 4^2 ways the suits may appear on them. So there are $13 \cdot \binom{4}{3} \cdot \binom{12}{2} \cdot 4^2 = 54{,}912$ possible three of a kind hands.

Two pairs: There are $\binom{13}{2}$ different ways the ranks may be dealt for two pairs. And there are $\binom{4}{2}$ different ways that the suits may appear for each pair of cards. The fifth card in the hand must have unique rank and may have any suit which means there are $11 \cdot 4$ ways it may be dealt. So there are $\binom{13}{2} \cdot \binom{4}{2} \cdot \binom{4}{2} \cdot 11 \cdot 4 = 78 \cdot 6 \cdot 6 \cdot 11 \cdot 4 = 123{,}552$ possible hands that contain two pairs.

One pair: There are 13 ranks possible for the pair and $\binom{4}{2}$ ways that suits may appear on these two cards. The three remaining cards must each have a different rank to avoid having a full house, four of a kind, three of a kind, or two pairs. There are $\binom{12}{3}$ ways that this may occur. The last three cards may be of any suit, hence there are 4^3 ways that the suits may appear. So there are $13 \cdot \binom{4}{2} \cdot \binom{12}{3} \cdot 4^3 = 13 \cdot 6 \cdot 220 \cdot 64 = 1{,}098{,}240$ possible one pair hands.

Nothing: The total number of hands that can be dealt is $\binom{52}{5} = \frac{52!}{47!\,5!} = \frac{52 \cdot 51 \cdot 50 \cdot 49 \cdot 48}{5!} = 2{,}598{,}960$. A nothing hand is simply none of the hands described above and may be calculated by subtracting these from the total number of hands. So there are $2{,}598{,}960 - 4 - 36 - 624 - 3744 - 5108 - 10{,}200 - 54{,}912 - 123{,}552 - 1{,}098{,}240 = 1{,}302{,}540$ possible nothing hands.

6.a. $10 \cdot 12 = 120$ bulbs, so there should be $120(.01) = 1.2$, or approximately 1 defective bulb. Sample: None appeared in the first simulation.

b. $\binom{12}{2}(.99)^{10}(.01)^2 \approx 0.00597$ is the probability of exactly 2 defective bulbs in a carton, by Example 2 of Lesson 10-6. $10 \cdot 10 \cdot (0.00597) = 0.597$. So about 1 carton should be expected to have exactly 2 defective bulbs after 10 runs of the program. Sample: Running the program 10 times, there was one carton that appeared with 2 defective bulbs as expected.

c. Change line 40 to:

```
40 IF RND(1) < .95 THEN PRINT "G";
                    ELSE PRINT "D";
```

The probability of a carton with exactly two defective bulbs becomes $\binom{12}{2}(.95)^{10}(.01)^2 \approx 0.0983$. So $10 \cdot 10 \cdot (0.0983) = 9.83 \approx 10$ such cartons should appear after running the program 10 times. Sample: In 10 trial runs, 5 cartons appeared that had 2 defective bulbs.

d. Sample: In the first run, there were 5 defective bulbs out of 120. So the quotient is $\frac{5}{120} \approx 0.0417$. This is close to the probability of any one bulb being defective (0.05). The results after several runs are summarized below.

Number of Runs	Quotient
2	.0375
3	.0389
4	.0354
5	.0417
6	.0444
7	.0452

The quotient gets closer to the probability that any one bulb is defective after more and more runs. This is true because $\left(\text{probability any one bulb is defective}\right) \cdot \left(\text{total number of bulbs}\right) \approx \left(\text{total number of defective bulbs}\right)$.

So in the long run, $\left(\text{probability any one bulb is defective}\right) = \dfrac{\left(\text{total number of defective bulbs}\right)}{\left(\text{total number of bulbs}\right)}$.

e. Sample: After 24 runs of the program, the quotient calculated was ≈ 0.0167. Assuming my partner put a whole number probability for defective bulbs, I guessed p to be .02 and was correct.

CHAPTER 11

1. Sample:
Ringel's conjecture can be verified on some smaller graphs, for example:

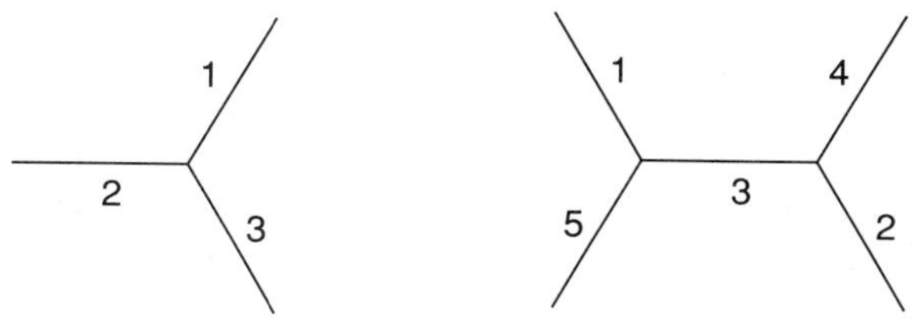

PROJECT SOLUTIONS

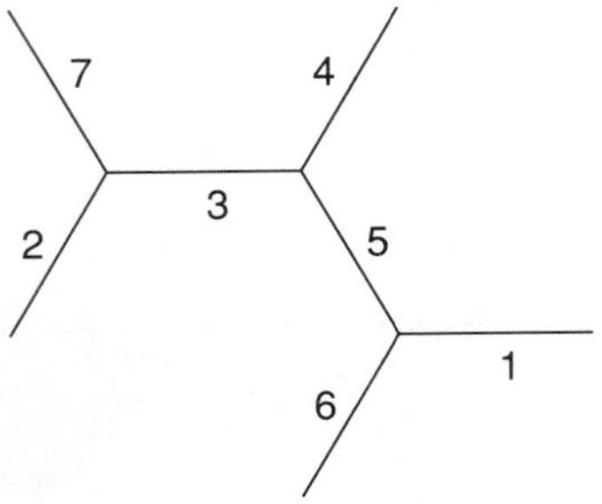

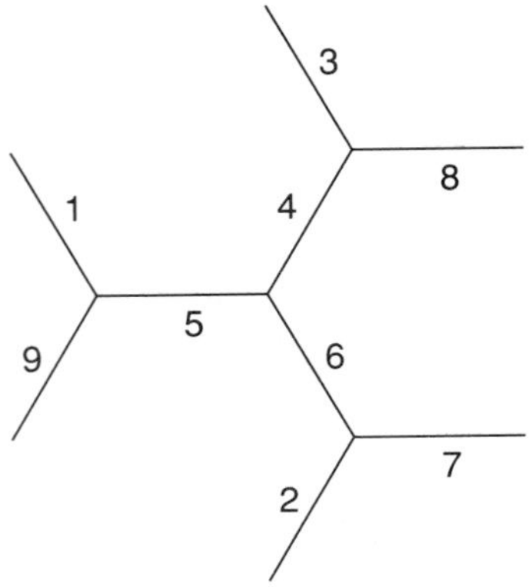

One method of numbering the edges of the graph is to note that for a graph with n edges, each vertex of degree 3 appears to have edges whose numbers sum to $\frac{3}{2}(n + 1)$.

2. Dynamic programming is a method for finding the shortest path between two nodes of a graph. It is based on the *dynamic-programming principle* which states that, when searching for the shortest path from a node S to a node G, all but the shortest path from S to any intermediate node I can be ignored. Here is an algorithm for this method. It builds a queue which holds partial paths. Notice that at each stage of the search, redundant paths are discarded, leaving only the shortest one.

1. Put the path of length zero from the starting node to nowhere in the queue.
2. If the queue is empty or the goal has been reached then go to step 4, otherwise, do the following:
   a. Remove the first path from the queue.
   b. Form new paths by extending the path you just removed by one step.
   c. Add these new paths to the queue.
   d. Sort the queue, putting the shortest paths in front.
   e. If two or more paths reach a common node, delete all but those paths except the shortest one.
3. Return to step 2.
4. If the goal has been reached, then the procedure was successful; otherwise it failed.

3.a. A good route may be found by moving from city to city by choosing the next city to be the nearest city from among all possible neighboring cities. Through some trial and error, the following route *appears* to be the shortest: New York–Washington–Pittsburgh–Charleston–Cincinnati–Columbus–Detroit–Buffalo–New York, which covers a total distance of 2030 miles.

b. A salesperson starting in Seattle or Portland must travel through Spokane to get to Boise and then must return through Spokane. In the same way, a salesperson starting in Missoula or Boise must travel through Spokane to get to Portland and then must return through Spokane. Similarly, a salesperson starting in Spokane will "visit" Spokane three times over such a trip. So the described trip is impossible.

c. Students' reports may include the following main points:
1. discussion of Hamiltonian circuits
2. formulation and physical interpretation of the Traveling Salesman Problem
3. mathematical difficulty in its solution
4. methods used to solve it (e.g., nearest-neighbor method)

4. Students' reports may contain some of the following applications:
1. description matching
2. goal reduction
3. exploring alternatives
4. problem solving
5. theorem proving
6. understanding images
7. learning

5.a. sample:

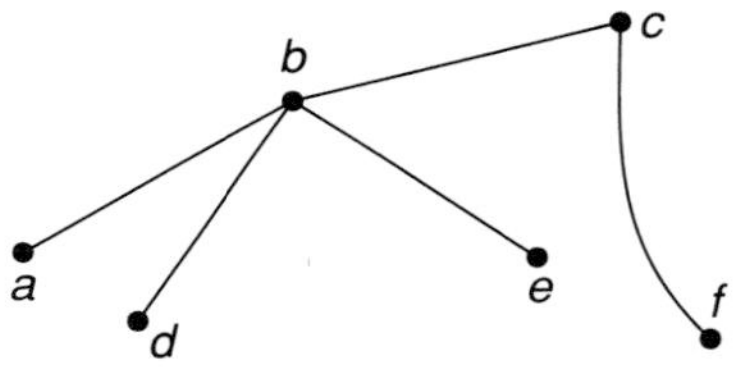

b. Sample: Examine each edge of the graph one at a time and keep those edges which do not form a circuit with the edges already chosen.

sample graph:

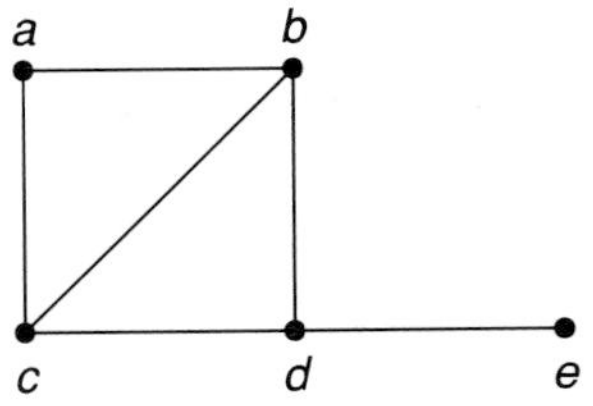

edge	decision
$\{a, b\}$	keep
$\{a, c\}$	keep
$\{b, c\}$	discard
$\{b, d\}$	keep
$\{c, d\}$	discard
$\{d, e\}$	keep

Spanning tree:

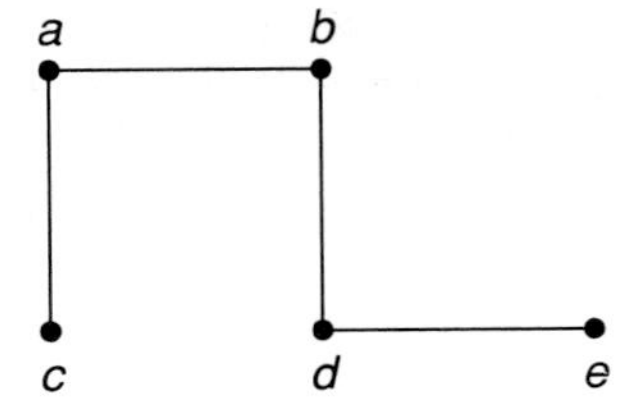

c.

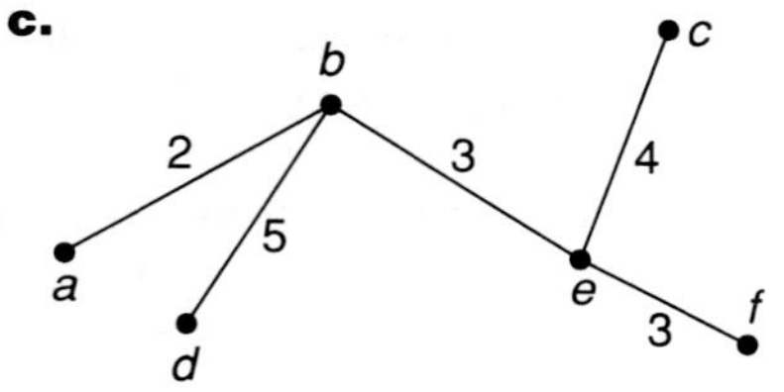

d. Sample: Use the method described in part **b**, but examine the edges in order of increasing label numbers.

sample graph:

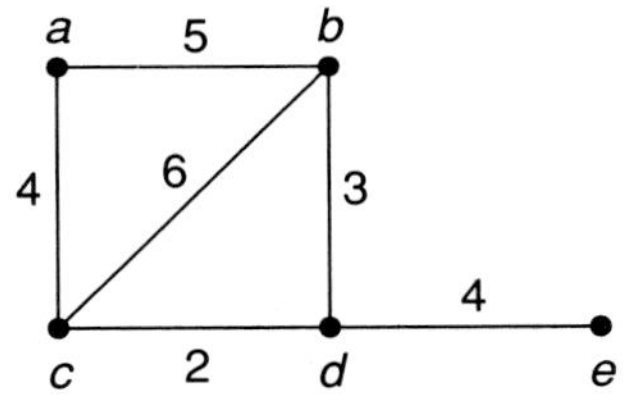

label	edge	decision
2	{*c*, *d*}	keep
3	{*b*, *d*}	keep
4	{*a*, *c*}	keep
4	{*d*, *e*}	keep
5	{*a*, *b*}	discard
6	{*b*, *c*}	discard

Minimal spanning tree:

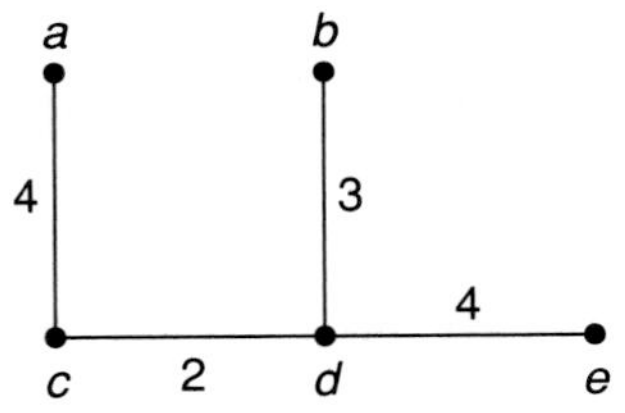

e. One real-life application could involve considering the vertices of a graph to be cities. The labeled numbers could be considered the cost of setting up and maintaining communication or transport links between the cities. Determining a minimal spanning tree would correspond to setting up a communication or transport network with the minimum cost.

CHAPTER 12

1.a. sample:

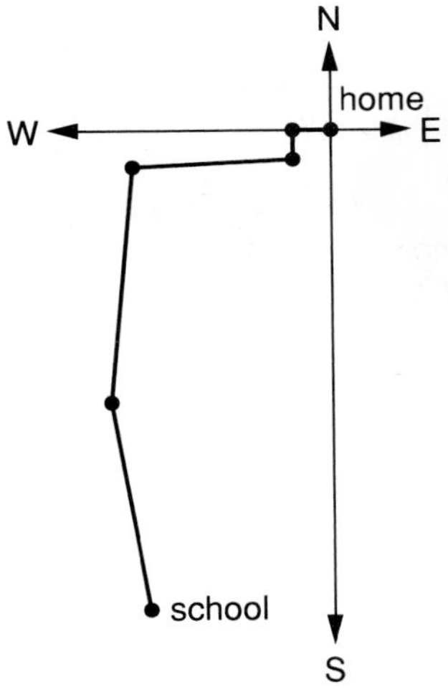

b. Sample: Vector representation of path to school: (-0.5, 0), (0, -0.4), (-2.1, -0.1), (-0.3, -3.2), (0.7, -2.8). The location of the school is then (-0.5 − 2.1 − 0.3 + 0.7, -0.4 − 0.1 − 3.2 − 2.8) = (-2.2, -6.5) which checks.

c. Sample: In polar form, the path becomes: [0.5, 180°], [0.4, 270°], [2.10, 183°], [3.21, 265°], [2.89, 104°]. The location of the school is then [6.86, 251°] which checks. These vectors can be translated into directions in English as follows: From home go 0.5 km west. Then go 0.4 km south. Next go 2.1 km 3° south of west. Next go 3.21 km 5° west of south. Finally, go 2.89 km 14° east of south to arrive at school.

2.a. Draw the vectors *P, R,* and *W* with a common initial point. These vectors must balance (sum to zero) in order for the windsurfer to stay on the board.

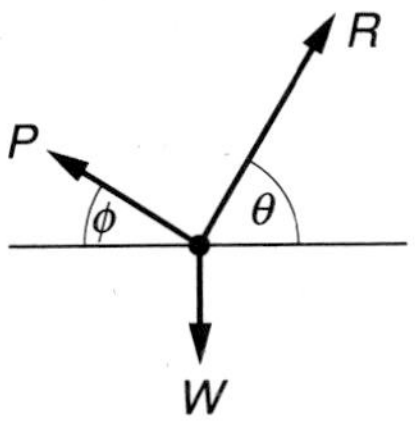

For the horizontal component to balance,

$P\cos(180° - \phi) + R\cos\theta + W\cos 270° = 0$

$\Leftrightarrow P(\cos 180° \cos(-\phi) - \sin 180° \sin(-\phi)) + R\cos\theta + W \cdot 0 = 0$

$\Leftrightarrow P(-1 \cdot \cos\phi - 0 \cdot -\sin\phi) + R\cos\theta = 0$

$\Leftrightarrow P \cdot -\cos\phi - P \cdot 0 + R\cos\theta = 0$

$\Leftrightarrow R\cos\theta - P\cos\phi = 0.$

3. Students' papers may include the following main points:
1. Kepler's humble beginnings
2. his education at the University of Tübingen
3. his association with Tycho Brahe
4. Kepler's laws of planetary motion
5. his study of wine barrels by using solids of revolution

4. Students' reports may include the following main points:
1. Emilie's interest in mathematics, physics, and philosophy
2. her marriage and many affairs
3. her association with Voltaire
4. her translation of Newton's *Principia Mathematica*
5. her own published works

5.

b	Weight of cardboard (grams)	Area under $y = x^2$ from $x = 0$ to $x = b$ calculated from experiment
2	0.5	2.3
3	2.0	4.3
4	3.5	24.2
5	7.25	42.6

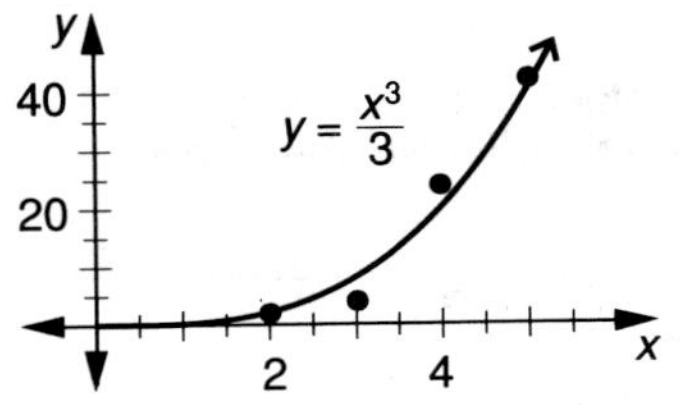

The values computed experimentally come close to the actual values as seen from the graph.

6. sample (for Texas):

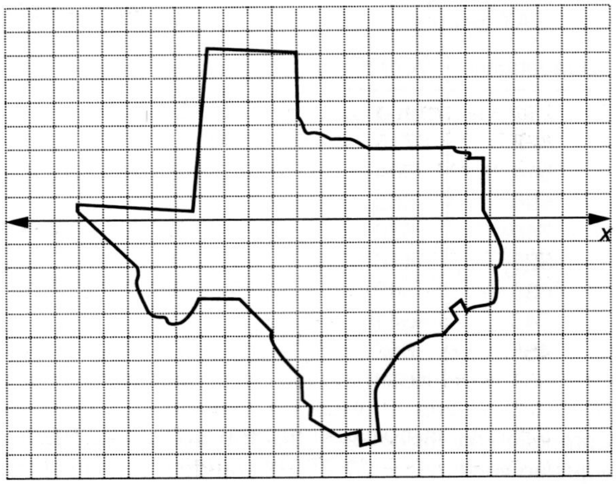

Using left endpoints, the area of the state above the x-axis is $.7 + .6 + .5 + .4 + .3 + 4.3 + 7.2 + 7.2 + 7.1 + 7.0 + 3.6 + 3 + 3 + 3 + 3 + 2.6 = 56.8$ units^2. Also using left endpoints, the area of the state below the x-axis is $.8 + 1.8 + 4.2 + 4.5 + 3.4 + 3.4 + 4 + 5.5 + 6.6 + 8.9 + 9.2 + 9.5 + 6 + 5.2 + 4.8 + 4 + 3.6 = 85.4$ units^2.

1 unit $\approx$ 44 mi, so the total area of Texas is approximately

$(56.8 \text{ units}^2 + 85.4 \text{ units}^2) \cdot \frac{44^2 \text{ sq mi}}{\text{unit}^2} = 275299.2 \approx$ 275,000 square miles. (Note: Actual area found in a dictionary is 267,000 square miles.)

INDEX

Page references in **bold face type** indicate the location of the definition of the term.

INDEX

INDEX

INDEX

INDEX

ACKNOWLEDGMENTS

For permission to reproduce indicated information on the following pages, acknowledgment is made to:

32 (a) Quote from *Peggy Sue*, words and music by Jerry Allison, Norman Petty and Buddy Holly. Copyright © 1957 MPL COMMUNICATIONS, INC. and WREN MUSIC CO. Renewed 1985 MPL COMMUNICATIONS, INC. and WREN MUSIC CO. International Copyright Secured. All Rights Reserved. (b) Excerpt from *If I Loved You* by Richard Rodgers and Oscar Hammerstein II. Copyright 1945 by Williamson Music Company. Copyright Renewed. Used by permission. ALL RIGHTS RESERVED. (c) Quote from *If I Had a Hammer* (The Hammer Song) Words and Music by Lee Hays and Pete Seeger. TRO © copyright 1958 (renewed) and 1962 (renewed) Ludlow Music, Inc., New York, NY. Used by permission. (d) From *Man in the Mirror* words and music by Glen Ballard and Siedah Garrett. Copyright © 1987 by MCA MUSIC PUBLISHING, A Division of MCA Inc., AERO-STATION CORPORATION and YELLOWBRICK ROAD MUSIC. Rights of AEROSTATION CORPORATION administered by MCA MUSIC PUBLISHING, A Division of MCA Inc., New York, NY 10010. All Rights on behalf of YELLOW BRICK ROAD MUSIC, for the world excluding Japan, Brazil and Hong Kong, administered by WB MUSIC CORP. All Rights Reserved. Used by permission.

Extant Materials Unless otherwise acknowledged, all photos are the property of Scott, Foresman. Page abbreviations are as follows: (T)top, (C)center, (B)bottom, (L)left, (R)right, (INS)inset.

Cover: Original robotic hand photo by Nubar Alexanian/Woodfin Camp & Associates, Inc. Enhanced image © Scott, Foresman

3 Milt & Joan Mann/Cameramann International, Ltd. **23** Courtesy U.S. Postal Service **29** The Granger Collection, New York **37** Tony Duffy/ALLSPORT USA **39** NASA **41** Hulton/UPI/Bettmann **49** Jean-Claude Figenwald **54** NASA **60** M. P. Kahl/DRK Photo **63** Ronny Jacques/Photo Researchers **67** Milt & Joan Mann/Cameramann International, Ltd. **70** Sidney Harris **72** Zoological Society of San Diego **75** Milt & Joan Mann/Cameramann International, Ltd. **79** Thomas Hovland from Grant Heilman Photography **87** Courtesy U.S. Can Co., Elgin, Il. Photo: Milt & Joan Mann/Cameramann International, Ltd. **91** Joe Munroe/Photo Researchers **93** Milt & Joan Mann/Cameramann International, Ltd. **99** Joseph Nettis/Stock Boston **101** The Granger Collection, New York **108** Jared/Focus On Sports **120** E. R. Degginger/Bruce Coleman Inc. **122** The Granger Collection, New York **122INS** Culver Pictures **134** Lynn Stone/Bruce Coleman Inc. **136** Don Dixon Spacescapes **149T** Mark Antman/The Image Works **149B** Courtesy Little Caesar's Pizza, Des Plaines, Il. Photo: Milt & Joan Mann/Cameramann International, Ltd. **151** Tony Freeman/Photo Edit **166** Courtesy Speed Queen Company, Photo: Milt & Joan Mann/Cameramann International, Ltd. **170** Milt & Joan Mann/Cameramann International, Ltd. **175** Focus On Sports **176** Milt & Joan Mann/Cameramann International, Ltd. **177** Courtesy Speed Queen Company, Photo: Milt & Joan Mann/Cameramann International, Ltd. **202** Jim Wright/The University of Chicago **221** Tom Gorman Photo **224** Steve Hill Photo **233** Steve Kaufman/Peter Arnold, Inc. **236** The Granger Collection, New York **260** Universal Press Syndicate. Reprinted with permission. All Rights Reserved. **263** Milt & Joan Mann/Cameramann International, Ltd. **276** The Granger Collection, New York **282** Milt & Joan Mann/Cameramann International, Ltd. **287** NASA **297ALL** Courtesy Gregory and David Chudnovsky **318** NASA **328** Milt & Joan Mann/Cameramann International, Ltd. **331** Bob Daemmrich **332** © 1991 M. C. Escher Heirs/Cordon Art-Baarn-Holland. **360** The Granger Collection, New York **366** Milt & Joan Mann/Cameramann International, Ltd. **368** David Madison **379** Sidney Harris **392** AP/Wide World **397** Bruce M. Wellman/Stock Boston **417** Tony Freeman/Photo Edit **418** Milt & Joan Mann/Cameramann International, Ltd. **430** David Young-Wolff/Photo Edit **433** The Granger Collection, New York **437** MUGSHOTS, Gabe Palmer/The Stock Market **453** Sidney Harris **458** Sidney Harris **465** Courtesy Wolfram Research, Inc. **466** The Granger Collection, New York **467** The Granger Collection, New York **486L** Dr. E. R. Degginger **486R** Tim Rock/ANIMALS ANIMALS **493** Dr. E. R. Degginger **496** Dr. E. R. Degginger **498T** Phil Degginger/National Optical Astronomy Observatories **498B** The Granger Collection, New York **515** Battelle/Pacific Northwest Labs **526** Courtesy Edward N. Lorenz **529** Dr. E. R. Degginger **530** Courtesy of Dr. Chudnovsky **537** Milt & Joan Mann/Cameramann International, Ltd. **538L** Brian S. Sytmyk/Masterfile **538R** Bill Brooks/Masterfile **546** NASA **552** Rafael Macia/Photo Researchers **572** Rhoda Sidney/Photo Edit **575** Charles Palek/ANIMALS ANIMALS **579T** Grant Heilman Photography **579B** Milt & Joan Mann/Cameramann International, Ltd. **583** Milt & Joan Mann/Cameramann International, Ltd. **585** Jose Carrillo **587** Bob Amft **593** Courtesy Secretary of State Office, Illinois **609** Tony Freeman/Photo Edit **614** Vicki Silbert/Photo Edit **620** The Granger Collection, New York **632** Tony Freeman/Photo Edit **636** Richard Hutchings/InfoEdit **642** David Young-Wolff/Photo Edit **656** Mark Antman/The Image Works **661** Milt & Joan Mann/Cameramann International, Ltd. **674** The Image Works **678** Courtesy Wayside Gardens **693** Reprinted by permission of United Feature Syndicate, Inc. **699** David Muench **700** Bob Amft **705** Milt & Joan Mann/Cameramann International, Ltd. **707** David Madison **712** Tony Freeman/Photo Edit **719** NASA **723** NASA **726** NASA **732** NASA **734** Milt & Joan Mann/Cameramann International, Ltd. **751** J. Nicholson/ALLSPORT USA **755** Milt & Joan Mann/Cameramann International, Ltd. **760** Milt & Joan Mann/Cameramann International, Ltd. **763** Milt & Joan Mann/Cameramann International, Ltd. **773** Brent Jones **787ALL** The Bettmann Archive **800ALL** The Granger Collection, New York **806** The Granger Collection, New York **809** Milt & Joan Mann/Cameramann International, Ltd. **811** Milt & Joan Mann/Cameramann International, Ltd. **812** Brent Jones